A GLOBAL COMPENDIUM
of WEEDS

A GLOBAL COMPENDIUM
of WEEDS

R.P. Randall

R.G. and F.J. Richardson
Melbourne

Published by and available from

R.G. and F.J. Richardson, PO Box 42, Meredith, Victoria 3333, Australia

Tel/fax +61 (0)3 5286 1533, Email richardson@weedinfo.com.au

Website www.weedinfo.com.au

National Library of Australia Cataloguing-in-Publication entry

Randall, R.P. (Roderick Peter), 1960–

A global compendium of weeds.

Bibliography.

Includes index.

ISBN 0 9587439 8 3.

1. Weeds. 2. Weeds – Identification. I. Title.

632.5

Supported by

Cooperative Research Centre for Australian Weed Management

Department of Agriculture, Western Australia

Missouri Botanical Garden Press

United States Geological Survey

Designed and typeset by

R.G. and F.J. Richardson, Meredith, Victoria, Australia

Printed by

Shannon Books, Melbourne, Victoria, Australia

Images by

R.G. Richardson   *Lavandula stoechas* L. and *Phalaris aquatica* L.

R.C.H. Shepherd   *Papaver somniferum* L.

CONTENTS

'There must have been plenty of them about,
growing up quietly and inoffensively, with
nobody taking particular notice of them…
And so the one in our garden continued its
growth peacefully, as did thousands like it
in neglected spots all over the world…
It was some little time later that the first
one picked up its roots and walked.'

John Wyndham, *The Day of the Triffids*

FOREWORD

Weeds are a major problem in natural and agricultural ecosystems throughout the world. They pose one of the greatest threats to biodiversity and weed control is often a major cost of production in both developing and developed countries. Many of today's weed problems have resulted from earlier, well-intentioned introductions of plants to regions remote from their native ranges. Human-mediated translocations of plants have been undertaken for a variety of purposes and will no doubt continue to occur as a component of global trade. One of the objectives of the developing discipline of weed risk assessment is to identify the potential costs associated with proposed plant introductions. Where necessary, it should be possible to identify and utilise alternative species, thereby reducing consequent costs without foregoing the benefits that can be gained through the introduction process (Panetta *et al.* 2001).

A considerable amount of research effort has been devoted to gaining a better understanding of the biological and ecological characteristics that promote weediness. However, little has emerged that might help to identify species that have a high risk of becoming weeds, beyond the simple generalisation that those that have become weeds when introduced in one part of the world are likely to cause problems if introduced elsewhere. This generalisation is a core component of the weed risk assessment systems that are currently in place in a number of countries. Its importance also suggests that databases that capture the weed histories of a wide range of plants will be invaluable tools to those who must make decisions regarding the importation of plants or the management of recent incursions.

To date, the most comprehensive coverage of the world's weed flora was produced by Holm *et al.* (1979) and listed 6400 species. In the present volume, Rod Randall has compiled a list comprising nearly three times that number. The database that he used to construct this document contains 992,000 taxa records from over 700 data sources, including the complete floras of Australia, the USA and North America and numerous naturalised floras from other countries. Although the book will undoubtedly be useful worldwide, it is primarily intended as a resource for developing countries to use in undertaking weed risk assessments. Thus, it represents a significant step in meeting the challenge of assisting developing countries to develop a coordinated approach to the problems posed by invasive species.

F.D. Panetta
Brisbane, January 2002

Holm, L., Pancho, J.V., Herberger, J.P. and Plucknett, D.L. (1979). *A Geographical Atlas of World Weeds*. John Wiley and Sons, New York.

Panetta, F.D., Mackey, A.P., Virtue, J.G. and Groves, R.H. (2001). Weed Risk Assessment: Core Issues and Future Directions. In *Weed Risk Assessment*, eds R.H. Groves, F.D. Panetta and J.G. Virtue, pp. 231-240. CSIRO Publishing, Melbourne.

INTRODUCTION

The collection of data for this project began back in 1996 with a species list from Martin Hanf's *The Arable Weeds of Europe, with their Seedlings and Seeds*. As the book was only on short term loan, the list of the taxa it covered allowed for quicker referencing in any future weed assessments. As with many collecting-type situations, this method of documenting weed lists quickly became cumbersome, with flora lists, nursery stock lists (in fact any list of plant names referenced to a source) being collected and poured into a database. This database forms the backbone of the Department of Agriculture of Western Australia's weed potential assessment process. It allows rapid determinations of weedy histories for any species under assessment and aids in determining a plant's origin, where it has naturalised, if it is used in the horticultural trade and numerous other useful bits of data.

The information presented in this compendium is specifically designed to give a weed risk assessor, or anyone interested in the weed potential of a plant, a condensed report of the status of a species with, most importantly, further avenues for finding more information through the extensive reference listing. The single most important indicator of a species' weed potential, over all other attributes, is a documented weedy history and this is indicated in the status line of each record.

This data is a distillation of many years work, sourcing as many significant weed references as possible. While there are some journal articles included, the main thrust has been to capture sources with large numbers of weed species documented. A search of various journal abstracts for the last 10 years will find over 30,000 journal articles that contain the key word 'weed'. Indexing and extracting a species list from this dataset would take a major effort and still end up with an agriculturally biased list that is unlikely to cover more than a fraction of the weeds covered here. This compendium has 20,672 entries comprising 18,146 discrete weedy taxa plus 2526 alternate name records, a significant increase on Holm *et al.*'s *A Geographical Atlas of World Weeds* (1979), the previous benchmark in weed compendia. The comprehensive index contains 15,078 alternate scientific names and 27,108 common names in numerous languages. There are 288 referenced data sources covering 11 continental regions and roughly 47 countries, not counting the many countries covered in *A Geographical Atlas of World Weeds*.

This compendium represents a huge increase in the number of documented weed species globally but by no means covers the entire weed flora of the world. While people continue to move plants around the world with little regard for the consequences of their actions, new weeds will continue to appear.

While the internet and computers can sometimes make life very easy, a book requires no power to read, apart from literacy, is more portable and eventually may reach a greater potential audience than any other format would. While the internet may be with us for a long time, its content will change and there are no guarantees that specific data will be available forever. Some of the web sources cited in this compendium have already disappeared, highlighting the ephemeral nature of data on the internet and the value of hardcopy publications.

From a very personal viewpoint, if this compendium helps prevent the establishment of one new weed anywhere, it will have been worth all the effort.

Rod Randall
Perth, February 2002

UNDERSTANDING THE RECORD INFORMATION

Each record comprises:

Genus, Species, Author

The taxonomy of weeds is very fragmented and many texts use old names. Rather than update these old names they have been left as published by the source author and linkages between related names are indicated by the use of the equals symbol (=), (see) and (NoR). While all effort has been made to determine all the appropriate linkages between related records, it is possible some may have been overlooked or, indeed, never determined. Furthermore, some accepted names are often controversial and readers may not agree with those selected in this text. With the large number of sources used to determine synonymy it was common to encounter differences of opinion as to the correct name.

Family

Alternate Names or = (see this record)

All known alternate names from the main dataset are listed. Wherever a synonym has (see) after it, this name should also be consulted in the text. The equals symbol (=) is used to indicate which name has priority in this text and if (NoR) is listed after this name, it is not cited elsewhere in the text. Any records cited that are synonymous with this name follow in square brackets.

Common Names

All common names from the database are presented including French, German, Finnish, Italian, Japanese (phonetic), Portuguese and Spanish.

Status

A group of relatively standard descriptors were used to form the status field of the dataset.

◆ **Weed**   Most common term used. These plants are nearly always economic weeds (i.e., pests of agriculture, horticulture, turf, nurseries etc.). The source details can usually give some idea of the type of weed, but are not conclusive in all cases.

◆ **Sleeper Weed**   Species that have been identified as present and posing a future threat. Often the source of these references has already been proved correct by other publications acknowledging the impacts of the species.

◆ **Quarantine Weed**   Species prohibited entry under a country's quarantine regulations.

◆ **Noxious Weed**   Species subject to legal restrictions (i.e., control, eradication, containment, etc.). For some countries this term also encompasses quarantine species (i.e., US Federal Noxious Weeds).

◆ **Naturalised**   Species has self-sustaining and spreading populations with no human assistance, but does not necessarily impact upon the environment. A species' capacity to naturalise in foreign environments, however, is a good indicator of its weed potential.

◆ **Native Weed**   Species that are native to the country in which they are considered weedy. Sometimes it is difficult to determine if the species has spread outside its native range within its country of origin or is weedy within its native range, as sources are often state or regionally based.

◆ **Introduced**   Species that have been released (planted) that may or may not have become naturalised. Introduced taxa obviously include many species deemed desirable by humans for one purpose or another, and many weeds enter countries via this pathway. Forestry, agriculture and horticulture are traditionally the biggest advocates of species introduction programs. Lately, overseas aid agencies have become involved in regeneration and rehabilitation projects. Rather than promoting non-invasive native species, such agencies often introduce exotic species with little consideration for weed potential.

◆ **Garden Escape**   Garden species known to have escaped, either directly by seed or other propagules moving out of the garden, or indirectly by establishing from dumped garden waste. Other garden escapes originate from abandoned gardens, graveyards and commercial waste disposal sites, to name just a few.

◆ **Environmental Weed**   Species that invade native ecosystems. Many of these can be easily determined from the source references. Whilst in the past, most attention has been focused on weeds of agriculture, this dataset provides information on over 2000 environmental weed species.

◆ **Cultivation Escape**   Species may have escaped from gardens, cultivation or both; source not specific but includes some crop and pasture species.

◆ **Casual Alien**   These species appear with no direct (apparent) human assistance, survive, possibly set seed, but do not persist. They then may appear again some seasons later, but do not develop persistent populations.

Source Codes

References cited (see Tables 1 and 2).

Life Form

This section was added to give a more complete picture of each species. More detailed life form categorisations have been developed, but this system was designed to provide a quick overview of each species and is available for over half of all entries. Categories are listed overleaf.

Arid/Aqua

This section is a basic measure of a plant's ability to survive, by avoidance or adaptation, in either environmental extreme. Some species can survive either extreme while others can establish only during a period of good conditions. This listing, like many of the following, is not complete and is only an indication of a capacity to grow

Life Form Category	Modifiers	Examples	
H = Herbaceous species	w = Aquatics	pG	= perennial grass
G = Grasses. In the context of this document this includes the families Poaceae, Cyperaceae, Juncaceae, Restionaceae	a = Annuals	a/pC	= annual to perennial climber
	b = Biennials		
	p = Perennials	aH	= annual herb
C = Climbers. Encompasses all kinds of climbers, including climbing shrubs, not just vines and creepers		a/bH	= annual/biennial herb
		wpG	= aquatic perennial grass
S = Shrubs. To describe perennial or short lived small, woody plants			
		waH	= aquatic annual herb
T = Tree. Either perennial or short lived		S/T	= shrub to tree

under the two extremes of moisture availability. Some species noted as arid may be annuals that only appear in good conditions, set seed and then die off. Perennials may only establish under good conditions, then survive extreme droughts by accessing ground water, use of water loss avoidance mechanisms, specialised physiology or water storage mechanisms. The use of aqua is to flag a species' capacity to survive under very wet conditions. Any species that are considered aquatic in the life form section will not have an aqua notation.

Cultivated or Promoted

Plants that are cultivated or promoted as species worthy of commercial cultivation. In many cases these are garden plants, but the category also includes crops. Cultivated plants, with their generally long association with humans, are often the first species to invade.

Herbal

Plants that are used for ceremonial, medicinal or culinary purposes, often only by traditional users. This should not be considered the same as 'cultivated', as often these species will be wild harvested and used only under extreme conditions of drought or famine (i.e., some of these plants are not eaten by preference but according to availability).

Toxic

Where a species is documented as toxic. Degree and parts considered toxic are often difficult to determine, as sources rarely indicate this. Degree of toxicity and the varying susceptibility of different species from human to livestock to wildlife should all be considered.

Origin

This can be difficult to determine, as often sources will quote conflicting origins. Where this occurs an origin is cited only if there is a clear majority among sources. Conversely, if there is only one source indicating an origin, its validity may be difficult to determine. Many thousands of origins were not cited because of the difficulty in validating some sources.

An example from the text

Opuntia ficus-indica (L.) Mill.
Cactaceae
Cactus ficus-indica L., *Opuntia castillae* Griffiths, *Opuntia compressa* (Salisb.) J.F.Macbr. (see), *Opuntia incarnadilla* Griffiths, *Opuntia megacantha* Salm-Dyck (see), *Opuntia occidentalis* Engelm., *Opuntia vulgaris* Mill. (see)
♦ Indian fig, tuna cactus, sweet prickly pear, mission prickly pear, prickly pear, spineless cactus, Boereturksvy, grootdoringturksvy, spiny pest pear
♦ Weed, Quarantine Weed, Noxious Weed, Naturalised, Introduced, Garden Escape, Environmental Weed, Cultivation Escape
♦ 10, 34, 51, 63, 72, 76, 86, 87, 88, 95, 98, 101, 121, 151, 152, 158, 198, 203, 228, 261, 269, 272, 278, 279, 283, 287
♦ S, arid, cultivated, herbal, toxic.
Origin: Central America.

NOMENCLATURE AND ERRORS

Every attempt has been made to retain the original species names as supplied by the authors in each source. At times, where there have been spelling errors, some records have been changed and more than likely, but hopefully not too frequently, errors may have been made at this end of the data daisy chain. Some spelling errors or variations may not have been detected in the screening process when adding new data sources. Moreover some taxa may have more than one record but the sheer number of names to check and recheck means there are bound to be some errors of this type. For these and any other errors the author takes full responsibility and apologises in advance. It would be appreciated if readers finding errors could advise the author.

Author's email address:
rprandall@agric.wa.gov.au

CASCADING SOURCES

Knowledgeable readers will quickly recognise that several sources regularly appear together. This should in no way be taken as a reliable measure of the weediness of a species, it should rather lend greater weight to the original source, which has prompted this cascade of reporting (i.e., an original reference being continually cited by other authors). This phenomenon is certainly not new, nor restricted to the study of weeds, but could be misleading. Two references which are copiously quoted are No. 218 (Darrow, R.A., Erickson, L.C., Holstrum, J.T. Jnr., Miller, J.F., Scudder, W.F. and Williams, J.L. Jnr. (1966). Report of the Terminology Committee, Standardized Names of Weeds. *WSSA* (14) 346-386. Weed Science Society of America, USA). A large list of weed names also included in No. 87 (Holm, L.G., Pancho, J.V., Herberger, J.P. and Plucknett, D.L. (1979). *A Geographical Atlas of World Weeds*, John Wiley and Sons, New York, USA), another heavily quoted weed reference. Hence these two source codes appear throughout this text as well as No. 88 (FAO, Global Pest Plant Information Service, now EcoPort), which also cites No. 87 (Holm *et al.* 1979).

NATURALISED AND INTRODUCED – DEFINITIONS

The use of both 'Naturalised' and 'Introduced' as status terms may confuse some people and in many cases are freely interchanged by authors. A great deal of effort was taken to determine whether or not a species was actually considered naturalised– 'has self sustaining and spreading populations with no human assistance' or introduced– 'plant populations are self sustaining but have not yet spread beyond their original point of introduction'. However, in many cases this could not be determined. Accordingly, where both terms are used for the same species, then both are presented. In most cases it is more than likely the author means 'naturalised' but the source has not made this sufficiently clear. Hopefully in the future the terminology associated with the dispersal and naturalisation of plants will become more standardised along the lines of Richardson *et al.*'s (2000) paper Naturalization of alien plants: concepts and definitions.

INTERPRETING THE DATA

The best measure of the weediness of a species that can be drawn from this compendium is the number of sources, combined with the range of weed types that have been allocated. For example, a species may have twelve sources attributed (taking into account any cascaded sources), but is only ever considered a 'weed' in the status section.

Another species may have six sources, but be considered a noxious, quarantine and environmental weed, plus a garden escape and casual alien. All other considerations being considered equal, the second species should pose a greater potential threat because of its demonstrated plasticity.

DATA SOURCES

For most published books and articles the conventional citation is usually sufficient, but where it would assist the reader any other appropriate information has been included. There are several personal communication sources, these will provide as much information and background as necessary to validate the credentials of the source. Also provided is a geographically sorted list of the source codes to assist people in compiling regional weed lists from the data sources.

The database (Randall 2002) that was used to construct this document contains 992,000 taxa records from over 700 data sources, including the complete floras of Australia, the USA and North America and numerous naturalised floras from various other countries and states. These range in size from a handful of records from personal communications, up to the 85,000 record dataset from the Flora of North America. This database was the main source of the common names, alternate names, toxicity data, origins, herbal and cultivation data and the environmental extremes data.

SEARCHING THE INTERNET

Searching the internet can yield a whole plethora of information, but how you search the internet can make a huge difference to the data you find and where you find it.

There are numerous search engines available on the internet and any of them will do a pretty good job of finding data. What you ask for, though, determines just how useful the results you get back are. Google is a simple interface and its rapid response to queries makes life on a slow connection bearable. Google can be found at: www.google.com/ (Note: any address that starts with www you do not need the <http://> prefix.)

Google has an advanced search option on this page and this allows you to better tailor your search to get appropriate data faster.

The direct address for this page is:
www.google.com/advanced_search

For any plant search, use the scientific name in full and enter the name into the 'with the exact phrase' option box. This means that a search for *Emex australis* will find only those pages that have those two words together and in that order. *Emex* and *australis* in the same web page or either word alone would not be returned as a search result. This means that the number of useful results in any search will be far higher.

It is important to remember that these types of searches do not detect online databases, only web pages. Databases have to be queried individually and so these sites can sometimes elude web surfers. Correct spelling of the plant name is critical, of course, and it helps to have a few good reference books at hand to make sure you're using the correct spelling. Caution should be exercised when using some of the online language translators. Google's translator operating in German to English mode translated the phrase 'irano-turan' referring to a region of Iran and Turkey as 'Irish Republican Army NOT uranium'.

ACKNOWLEDGMENTS

Over the six years this publication has been compiled a great many people have had input, either directly or indirectly. For those whom I can remember as providing any direct input, assistance or comments my heartfelt thanks and gratitude. Equally so, my sincere apologies for anyone not mentioned who deserves to be.

Dave Albrecht, Northern Territory, Australia

Margaret Allan, Western Australia, Australia

Channa Bambaradeniya, Sri Lanka

Pierre Binggeli, Ireland

Kate Blood, Victoria, Australia

Chris Buddenhagen, Galapagos

David Cooke, South Australia, Australia

Jack Craw, New Zealand

Steve Csurhes, Queensland , Australia

Patricia Dalton, United States of America

Jon Dodd, Western Australia, Australia

Richard Groves, Canberra, Australia

Lesley Henderson, South Africa

John Hosking, New South Wales, Australia

Penny Hussey, Western Australia, Australia

Greg Keighery, Western Australia, Australia

Yong Woong Kwong, South Korea

Sandy Lloyd, Western Australia, Australia

Mark Lonsdale, Canberra, Australia

Sarah Lowe, New Zealand

Andrew Mitchell, Northern Territory, Australia

Hirohiko Morita, Japan

Michael Mulvaney, Canberra, Australia

Dane Panetta, Queensland, Australia

Geoff Sainty, New South Wales, Australia

Anisur Rahman, New Zealand

Fiona Richardson, Victoria, Australia

Rob Richardson, Victoria, Australia

Ethel San Román, Argentina

Phil Short, Northern Territory, Australia

Charles Stirton, Wales

Mark Stuart, Western Australia, Australia

Philip Thomas, Maui, United States of America

Craig Walton, Queensland, Australia

Ricky Ward, Thailand

Barbara Waterhouse, Queensland, Australia

Randy Westbrooks, United States of America

Peter Williams, New Zealand

Colin Wilson, Northern Territory, Australia

Aaron Wilton, New Zealand

The production of this publication would not have been possible without the financial assistance of the Cooperative Research Centre for Australian Weed Management, Western Australian Department of Agriculture, Missouri Botanical Garden Press, United States Geological Survey and North Carolina Botanical Garden.

OTHER SIGNIFICANT REFERENCES USED IN THE
DEVELOPMENT OF THIS BOOK

The year followed by a hyphen (–) indicates a web site database that is in continual development and can be expected to change over time. Bold acronyms are the more commonly used descriptor of the database reference.

Agricultural Research Service, USDA (2002–). Germplasm Resources Information Network (**GRIN**). taxonomic data provide the structure and nomenclature for the accessions of the National Plant Germplasm System. URL: http://www.ars-grin.gov/npgs/tax/index.html

Australia, New Zealand Food Standards Code – Draft (2000–). Prohibited and Restricted Plants and Fungi. Standard 1.4.4 Prohibited and Restricted Plants and Fungi. Purpose. This standard regulates plants and fungi. It lists the species of plants and fungi that must not be added to food or offered for sale as food. It also lists the species of plants and fungi that may not be used in food except as a flavouring. URL: http://www.anzfa.gov.au/foodstandardscodecontents/standard14/standard144.cfm

Barnard, S.M. (1996). *Reptile Keeper's Handbook*. Krieger Publishing Company. From URL: www.bmts.com/~csz/MingsExoticZoo/Mingstoxicplants.html

Centre for Plant Biodiversity Research (2002–). Australian Plant Name Index (**APNI**). URL: http://www.anbg.gov.au/cpbr/databases/apni.html (based on A.D. Chapman's *Australian Flora and Fauna Series* Nos 12–15. Australian Plant Name Index. Australian Biological Resources Study, Canberra).

Farr, E. and Zijlstra, G. (eds) (2002–). The Index Nominum Genericorum (**ING**), a collaborative project of the International Association for Plant Taxonomy (IAPT) and the Smithsonian Institution initiated in 1954 as a compilation of generic names published for all organisms covered by the International Code of Botanical Nomenclature. URL: http://rathbun.si.edu/botany/ing/

Hnatiuk, R.J. (1990). Census of Australian Vascular Plants. *Australian Flora and Fauna Series* No. 11. Bureau of Flora and Fauna, Canberra, ACT.

Holm, L., Pancho, J.V., Herberger, J.P. and Plucknett, D.L. (1979). *A Geographical Atlas of World Weeds*. John Wiley and Sons, New York.

Hunt, D. (1992). *CITES Cactaceae Checklist*. Royal Botanic Gardens Kew and the International Organization for Succulent Plant Study, UK.

Huxley, A. (1999). *Dictionary of Gardening*, Vols 1–4. The New Royal Horticulture Society. Macmillan Reference Ltd., London, UK.

Mabberley, D.J. (1998). *The Plant-Book. A Portable Dictionary of the Higher Plants*, 2nd edition. Cambridge University Press, UK.

Missouri Botanical Garden's (2002–). **W3TROPICOS**. Provides access to the Missouri Botanical Garden's VAST (VAScular Tropicos) nomenclatural database and associated authority files. URL: http://mobot.mobot.org/W3T/Search/vast.html

Randall, R.P. (2002–). Plant Database. Department of Agriculture, Western Australia. This database holds 992,000 records from 700+ global plant related sources (not available online).

Richardson, D.M., Pysek, P., Rejmanek, M., Barbour, M.G., Panetta, F.D. and West, C.J. (2000). Naturalization of alien plants: concepts and definitions. *Diversity and Distributions* 6, 93-107.

Royal Botanic Garden Edinburgh (2002–). Flora Europaea. Data provided has been extracted from the digital version of the Flora Europaea. URL: http://www.rbge.org.uk/forms/fe.html

Royal Botanic Gardens Kew, Harvard, University Herbaria and Australian National Herbarium (2002–). International Plant Names Index (**IPNI**). The data in the IPNI comes from three sources: the Index Kewensis, the Gray Card Index and the Australian Plant Names Index. URL: http://www.ipni.org/

USDA, NRCS (2002–). The **PLANTS** Database, Version 3.1 (URL: http://plants.usda.gov). National Plant Data Center, Baton Rouge, LA 70874-4490, USA.

Table 1. Source codes with references.

Source No.	Reference	Region	Country
1	Noxious Weeds of Whitman County, Washington State. www.wa.gov/agr/weedboard/weed_list/weed_listhome.html	North America	United States, Washington State
2	Boose, A.B. and Holt, J.S. (1999). Environmental effects on asexual reproduction in *Arundo donax*. *Weed Research* 39 (2): 117-27.		
3	Actual and Prospective Weeds. Pacific Islands Ecosystems at Risk Project. www.hear.org/pier/	Pacific	Pacific Islands
4	Invasive Plants Of Canada, Melinda Thompson, Canadian Botanical Conservation Network, August 1997. www.rbg.ca/cbcn/en/invasives/invade1.html	North America	Canada
5	Miscanthus: A Review of European Experience with a Novel Energy Crop. www.esd.ornl.gov/bfdp/reports/miscanthus/toc.html	Europe	
6	Waterhouse, D.F. (1997). *The Major Invertebrate Pests and Weeds of Agriculture and Plantation Forestry in the Southern and Western Pacific*. ACIAR, Canberra.	Pacific	Pacific Islands
7	Keighery, G. (1996). *A Checklist of the Naturalised Vascular Plants of Western Australia*. With annotations for weedy WA and eastern states species. CALM, Western Australia.	Australasia	Australia, Western Australia
8	Foster, S. and Duke, J.A. (1990). *A Field Guide to Medicinal Plants. Eastern and Central North America*. Houghton Mifflin Co., New York, USA.	North America	eastern and central North America
9	Gibson, N., Keighery, B., Keighery, G., Burbidge, A. and Lyons, M. (1994). *A floristic survey of the Southern Swan Coastal Plain*. Unpublished report for the Australian Heritage Commission prepared by the Department of Conservation and Land Management and the Conservation Council of Western Australia (Inc.).	Australasia	Australia, Western Australia
10	Vermeulen, J.B., Dreyer, M., Grobler, H. and van Zyl, K. (1996). *A Guide to the Use of Herbicides*, 15th Edition. National Department of Agriculture, Republic of South Africa.	Africa	South Africa
11	Mitchell, A.A. pers. comm. (1995–2001). Botanist for the Northern Australian Quarantine Service.	Australasia	northern Australia and immediate northern neighbours
12	Weeds in Rubber Plantations (2002). Thailand Department of Agriculture. www.doa.go.th/botany/rubber.html	Asia	Thailand
13	Backer, C.A. (1973). *Atlas of 220 Weeds of Sugarcane Fields in Java*. (A reprint of an original Dutch publication from the early 1930s.)	Asia	Java
14	Acuna, G.J. (1974). *Plantas Indeseables en Los Cultivos Cubanos*. Academia de Ciencias, Insitituto de Investigaciones de Cuba, Havana.	Central America	Cuba
15	Adventive Plants Collected in Department of Conservation's (DOC) Wanganui Conservancy, New Zealand (1988–98). Data supplied by DOC, New Zealand.	Australasia	New Zealand, North Island
16	Alberta Research Council, Biological Control of Weeds using Insects and Mites. Target Weeds. www.arc.ab.ca/crop/weed/BiocontrolMain.html	North America	Canada, Alberta

Source No.	Reference	Region	Country
17	Virginia Native Plant Society (1998). *Invasive Alien Plant Species of Virginia.* Virginia Native Plant Society (PO Box 844, Annandale, VA 22003) and the VA Division of Natural Heritage. www.state.va.us/~vaher.html	North America	United States, Virginia
18	Aliens List Group. An initial list of invasive weeds nominated by members after a request by Sarah Lowe. This list was further refined and a later list presented (see ref 132) (Invasive Species Specialist Group, ISSG). www.issg.org/	Global	
19	Aliens List Group. Species brought to the attention of the group by list members or in response to queries. These responses were used, in part, to help compile the ISSG 100 Worst Invaders Booklet. www.issg.org/booklet.pdf	Global	
20	Sainty, G.R., Hosking, J. and Jacobs, S.W.L. (1998). *Alps Invaders, Weeds of the Australian High Country.* Australian Alps Liaison Committee, Sainty and Associates, Darlinghurst, NSW.	Australasia	Australia, Australian alps
21	Chris, R. and McLendon, T. (1998). An Assessment of Exotic Plant Species of Rocky Mountain National Park Rutledge. Department of Rangeland Ecosystem Science, Colorado State University, 97 pp. Northern Prairie Wildlife Research Center. http://www.npwrc.usgs.gov/resource/othrdata/explant/explant.htm	North America	United States, Rocky Mountains
22	Binggeli, P., Hall, J.B. and Healey, J.R. (199x). An Overview of Invasive Woody Plants in the Tropics, http://members.tripod.co.uk/WoodyPlantEcology/invasive/index.html	Pantropics	
23	Andersen, R.N. (1968). *Germination and Establishment of Weeds for Experimental Purposes.* Weed Science Society of America Handbook. WSSA, Illinois.	Global	
24	Annual/Biennial Seed Germination Database. Mentions weeds in passing but does contain a lot of useful information on seed germination requirements and expected germination levels over time. www.anet.com/~manytimes/index.htm	Global	
25	AQIS Draft List of Quarantine Plants. A list of weeds complied in an AQIS paper from the early 1990s.	Australasia	Australia
26	Arizona State Designated Exotic Plant Species. http://ag.arizona.edu/OALS/agnic/weeds/home.html	North America	United States, Arizona
27	B & T Seeds Catalogue (toxic species noted). www.b-and-t-world-seeds.com/letters.htm	Global	
28	Randall, R.P. (1999). *Banana Weeds.* Report to Hortguard, Department of Agriculture, Western Australia. A compilation of weeds found in banana crops.	Global	
29	Waterhouse, B. pers. comm. (1997–2001). Botanist for the Northern Australian Quarantine Service.	Australasia	Australia, northern
30	Behrendt, S. and Hanf, M. (1979). *Grass Weeds in World Agriculture.* BASF Aktiengesellschaft, Ludwigshaten am Rhein, Germany.	Global	
31	Vacant		

Source No.	Reference	Region	Country
32	Boggan, J., Funk, V., Kelloff, C., Hoff, M., Cremers, G. and Feuillet, C. (1997). *Biological Diversity of the Guianas (BDG) Guyana, Surinam, French Guiana. The Checklist of the Plants of the Guianas*, 2nd Edition. Produced as a cooperative project between the Biological Diversity of the Guianas Program (Smithsonian Institution, Washington, DC, USA) and the ORSTOM Herbarium (Cayenne, French Guiana) under the auspices of the Centre for the Study of Biological Diversity (University of Guyana, Georgetown, Guyana). www.nmnh.si.edu/biodiversity/checklst.htm	South America	Guyana, Surinam and French Guiana
33	See below		

Acacia nilotica	Alikodra, H.S. (1987). The Exotic Plantation of *Acacia nilotica* and its Problems in the Savanna Ecosystem of the Baluran National Park. *Dosen Fakultas Kehutanan*, IPB, Indonesia.	South East Asia	Indonesia
Asystasia intrusa	Mohamad, R.B. (1990). Weeds and Weed Management in Malaysia Agriculture. *BIOTROP Special Publication* No. 38, 41-52. A symposium on weed management held in Bogor, Indonesia, 7–9 June 1990, 37 ref.	South East Asia	Malaysia
Borreria latifolia	Lam, C.H., Lim, J.K. and Jantan, B. (1993). Comparative Studies of a Paraquat Mixture and Glyphosate and/or its Mixtures on Weed Succession in Plantation Crops. *Planter* 69: 812, 525-535.		
Cynanchum komarovii	Hatziminaoglou, J. (1996). Ningxias Steppe Vegetation: Floristic Composition and Utilization by Sheep and Goats. The optimal exploitation of marginal Mediterranean areas by extensive ruminant production systems. Proceedings of an international symposium organized by HSAP and EAAP and sponsored by EU(DGVI), FAO and CIHEAM, Thessaloniki, Greece, 18–20 June 1994. 1996, 201-5; EAAP Publication No. 83, 5.		
Eryngium horridum	Ferri, M.V.W., Eltz, F.L.F. and Kruse, N.D. (1998). Desiccation of native pastures for direct sowing of soyabeans. *Ciencia-Rural* 1998, 28: 2, 235-40; 12 ref.		
Mikania cordata	Gangwar, B., Singh, D. and Dharam, S. (1987). *Mikania cordata* (Burm.f.) serious weed of South Andaman. *Journal of the Andaman Science Association* 3: 2, 135-7; 2 ref.		
Peganum nigellastrum	Hatziminaoglou, J. (1996). Ningxias Steppe Vegetation: Floristic Composition and Utilization by Sheep and Goats. The optimal exploitation of marginal Mediterranean areas by extensive ruminant production systems. Proceedings of an international symposium organized by HSAP and EAAP and sponsored by EU(DGVI), FAO and CIHEAM, Thessaloniki, Greece, 18–20 June, 1994. 1996, 201-5; EAAP Publication No. 83, 5.		
Saccharum spontaneum	Sison, C.M. and Mendoza. S.P. Jr. (1993). Control of Wild Sugarcane in Pineapple on the Del Monte Philippines, Inc. plantation. *Acta Horticulturae* No. 334, 337-9.	South East Asia	Philippines
Vernonia nudiflora	Ferri, M.V.W., Eltz, F.L.F. and Kruse, N.D. (1998). Desiccation of Native Pastures for Direct Sowing of Soyabeans. *Ciencia Rural* 1998, 28: 2, 235-40.		
Vernonia polyanthes	Ferri, M.V.W., Eltz, F.L.F. and Kruse, N.D. (1998). Desiccation of Native Pastures for Direct Sowing of Soyabeans. *Ciencia Rural* 1998, 28: 2, 235-40.		
34	CALFLORA Database Information about the 8375 currently recognized vascular plants in California, including scientific and common names, synonymy, distribution from literature sources, legal status, wetland codes, habitat info, and more. http://elib.cs.berkeley.edu/calflora/	North America	United States, California

Source No.	Reference	Region	Country
35	California Noxious Weed Control Projects Inventory (CNWCPI). www.caleppc.org/symposia/97symposium/schoenig.html	North America	United States, California
36	Canada Noxious Weed List (1986). Weed Seeds Order, 1986. Weed Seeds. The seeds of the species of plants set out in the schedule are deemed to be weed seeds for the purpose of establishing grades under the Seeds Act. Schedule (s.2). www.aosaseed	North America	Canada
37	Binggeli, P. (1999). Case Histories of Highly Invasive Woody Species in the Tropics. http://members.tripod.co.uk/WoodyPlantEcology/invasive/	Pantropics	
38	Brako, L. and Zarucchi, J.L. (1993). Catalogue of the Flowering Plants and Gymnosperms of Peru, Vol. 45. *Monographs in Systematic Botany from the Missouri Botanical Garden.* W3TROPICOS – Peru Checklist (2001–). Search facilities provided on the Missouri Botanical Garden's Peru Checklist. http://mobot.mobot.org/W3T/Search/peru.html	South America	Peru
39	Poisonous Plant Database (1998). Center for Food Safety and Applied Nutrition, FDA, Office of Plant and Dairy Foods and Beverages, USA. http://vm.cfsan.fda.gov/~djw/plantnam.html	Global	
40	French, C.N. and Murphy, R.J. (1994). Check-list of the Flowering Plants and Ferns of Cornwall and the Isles of Scilly, University of Exeter. www.ex.ac.uk/~cnfrench/ics/cbru/checklist/a1menu.htm	Europe	United Kingdom, Cornwall and Scilly Isles
41	Crook, C.S. (1998). Checklist of Conifers in the British Isles. Preston, Lancashire, UK. http://ourworld.compuserve.com/homepages/cameronscrook/	Europe	United Kingdom
42	Kurtto, A. and Lahti, T. (1987). Checklist of the Vascular Plants of Finland (based on). Suomen putkilokasvien luettelo. *Pamphl. Bot. Mus. Univ. Helsinki* 11, I-VI + 1-163. The list includes all the vascular plants found in Finland up to 1987, excluding those only in cultivation (i.e., taxa found during the past few years—mainly casual plants or plants escaping from cultivation—are missing, and erroneous records of the 1987 list have not been corrected).	Europe	Finland
43	Arias, J., Martin, M.E., and Gimenez, M.J. (1983). Chemical Control of a New Weed in North Western Argentina. *Tithonia tubaeformis* (Jacq.)-Cass. *Malezas* 11: 5, 177-81.	South America	Argentina
44	Häfliger, E. and Brun-Hool, J. (1968–). *Ciba-Geigy Weed Tables.* CIBA-GEIGY Ltd., Agrochemicals Division, Basel, Switzerland.	Global	
45	City of Portland, Oregon. Nuisance Plant List: www.planning.ci.portland.or.us/lib_plantlist.html	North America	United States, Oregon
46	Dupe, B. (1999). Collection of suggested control methods for *Crataegus monogyna* in environmental areas. Enviroweeds Email List.		
47	Collier County Natural Resources Department Exotic Plant Management Collier County Land Development Code: Section 2.4. http://co.collier.fl.us/natresources/exotics/default.htm	North America	United States, Florida
48	Colorado State Designated Exotic Plant Species. www.usgs.nau.edu/swemp/Info_pages/states/colorado/codespp.html (now '404 not found')	North America	United States, Colorado

Source No.	Reference	Region	Country
49	Thornton, B.J. and Durrell, L.W. (1933). *Colorado Weeds*. Bulletin 403, Colorado Agricultural College, Colorado Experimental Station, Fort Collins.	North America	United States, Colorado
50	Wild, H. (1955). *Common Rhodesian Weeds*. Government Printer, Salisbury, Rhodesia.	Africa	Kenya (Rhodesia)
51	Henderson, M. and Anderson, J.G. (1966). *Common Weeds of South Africa*. Memoirs of the Botanical Surveys of South Africa No. 37, Department of Agricultural Technical Services Republic of South Africa.	Africa	South Africa
52	Mulligan, G.A. (1987). *Common Weeds of Canada*. McClelland and Stewart, the Department of Agriculture and the Publishing Centre, Supply and Services Canada.	North America	Canada
53	Terry, P.J. and Michieka, R.W. (1987). *Common Weeds of East Africa*. Food and Agriculture Organization of the United Nations, Rome.	Africa	East Africa
54	Groves, R.H. and Hosking, J.R. (1997). *Recent Incursions of Weeds to Australia 1971–1995*. Technical Series No. 3, CRC for Weed Management Systems, Australia.	Australasia	Australia
55	Wilson, B.J., Hawton, D. and Duff, A.A. (1995). *Crop Weeds of Northern Australia*. Department of Primary Industries, Queensland.	Australasia	Australia, northern
56	Rich, T.C.G. (1991). *Crucifers of Great Britain and Ireland*. BSBI Handbook No. 6 Botanical Society of the British Isles, London.	Europe	United Kingdom
57	Lazarides, M., Cowley, K. and Hohnen, P. (1997). *CSIRO Handbook of Australian Weeds*. CSIRO Publications.	Australasia	Australia
58	Vacant		
59	Dafni, Amots and Heller, David (1990). Invasions of adventive plants in Israel. Section 8 in *Biological Invasions in Europe and the Mediterranean Basin*, eds F. Di Castri, A.J. Hansen and M. Debussche, Kluwer Academic Publishers, Dordrecht.	Middle East	Israel
60	Vacant		
61	McLaren, D. pers. comm. (2000). Comments on recent horse poisonings attributed to *Phalaris coerulescens*. Natural Resources and Environment, Victoria.	Australasia	Australia
62	Declared Plants of Western Australia (Noxious Weeds). Plants declared under the Agriculture and Related Resources Protection Act, 1976. One of the regulatory duties of the Department of Agriculture of Western Australia.	Australasia	Australia, Western Australia
63	Declared Weeds and Invader Plants of South Africa (2000). www.nda.agric.za/docs/Act43/Act43.html	Africa	South Africa
64	Delisted from US List Noxious Weeds of the United States. USDA, NRCS 1999. The PLANTS database. http://plants.usda.gov/plants. National Plant Data Center, Baton Rouge, LA 70874-4490, USA.	North America	United States
65	Garcia, A.E., Chaila, S., Vega, Y.M. de la. (1991). Distribution of *Tithonia tubaeformis* (Jack.) Cass. in Tucuman and the Crops Affected. Proceedings of the 12th Argentine Meeting on Weeds and their Control. Facultad de Agronomia y Zootecnica, Universidad Nacional de Tucuman, San Miguel de Tucuman 400, Argentina.	South America	Argentina

Source No.	Reference	Region	Country
66	Diwakar, P.G. and Ansari, A.A. (1995). Weed Flora of Buldhana District of Maharashtra. *J. Econ. Tax. Bot.* Vol. 19 No. 3.	Central Asia	India
67	Patterson, D. (1999). Weed Specialist North Carolina Department of Agriculture and Consumer Services Plant Protection Section. www.ncagr.com/plantind/plant/weed/weedprog.htm	Global	
68	Dr. Ian Heap. Herbicide Resistant Plants. www.weedscience.org/	Global	
69	Khedr, A.H.A. and Hegazy, A.K. (1998). Ecology of the Rampant Weed *Nymphaea lotus* (L. Willdenow) in Natural and Ricefield Habitats of the Nile Delta, Egypt. *Hydrobiologia* 386, 119-29.	Africa	Egypt
70	Williams, G.H. (1982). *Elseviers's Dictionary of Weeds of Western Europe.* Elsevier Scientific Publishing Company, Amsterdam, Holland.	Europe	western
71	Baumer, M. (1983). *EMASAR PHASE II. Notes on Trees and Shrubs in Arid and Semi-Arid Regions.* FAO and UNEP, Rome.	Global	
72	Carr, G.W., Yugovic, J.V. and Robinson, K.E. (1992). *Environmental Weed Invasions in Victoria.* Department of Conservation and Environment, Melbourne.	Australasia	Australia, Victoria
73	Nagle, J. (1995). Environmental Weeds of North Coast NSW. Greening Australia. www.nor.com.au/environment/greenwork/enweed.htm	Australasia	Australia, New South Wales
74	Enviroweeds. See below		
Acacia melanoxylon	Luckhurst, G. (2000). Request for help to control *Acacia melanoxylon* in Sintra region of Portugal. Enviroweeds Email List.	Europe	Portugal
Monadenia bracteata	Blood, K. (2000). Email Press Release on the incursion of *Monadenia bracteata* in Victoria. Enviroweeds Email List.	Australasia	Australia, Victoria
75	Lomer, F. (2000). Ephemeral Introductions of Vascular Plants around Vancouver, British Columbia (Part 1). From a posting to Botanical Electronic News (BEN).	North America	Canada, Vancouver
76	Western Australian Quarantine Species List. Plant species that have been assessed and are not approved for entry to Western Australia on the basis of their weed potential. Weed Science Group, Department of Agriculture WA. This list is updated several times a year. www.agric.wa.gov.au/progserv/plants/weeds/Weedsci.htm	Global	
77	Miller, J.H. (1999). Exotic Invasive Plants in Southeastern Forests. Southern Research Station, USDA Forest Service Auburn University, AL. www.srs.fs.fed.us/pubs/rpc/1999-03/rpc_99mar_27.htm	North America	United States
78	Exotic Pest Plants of Ecological Concern in California. www.caleppc.org/info/plantlist.html	North America	United States, California
79	Toney, C.J., Rice, P.M and Forcella, F. (1998). Exotic Plant Records in the Northwest United States 1950–1996: an Ecological Assessment. www.mrsars.usda.gov/morris/recntpub/98reprnt/jctpmrff.htm	North America	United States, north west
80	Campbell, F.T. (1999). 'Worst' Invasive Plant Species in the Conterminous United States. American Lands Alliance. www.americanlands.org/forestweb/invasive.htm	North America	United States

Source No.	Reference	Region	Country
81	FAO, Grassland Index. www.fao.org/ag/AGP/AGPC/doc/ GBASE/Default.htm	Global	
82	Flora of North America. www.bonap.org/	North America	
83	Florida Exotic Pest Plant Council. Online weed forum, Downy Rose Myrtle in Sarasota County. www.fleppc.org/	North America	United States, Florida
84	Florida Weeds. A part of the Florida Agricultural Information Retrieval System (FAIRS). http://edis.ifas.ufl.edu/scripts/ MENU_FW_SCI	North America	United States, Florida
85	G.R.I.N. TAXON Database. GRIN frequently mentions weed references in its species report outputs. www.ars-grin.gov/npgs/tax/index.html	Global	
86	Garden Thugs and Naturalised Plants Weed List. This list was complied as the base list to derive the environmental weeds of Australia that are still grown or on sale in Australian nurseries. The top 100 species were published in Roush, R., Groves, R.H., Blood, K., Randall, R.P., Walton, C., Thorp, J. and Csurhes, S. (1999). *Garden Plants Under the Spotlight. An Australian Strategy for Invasive Garden Plants.* (Draft Released for Public Comment) Cooperative Research Centre for Weed Management Systems and Nursery Industry Association of Australia. A more up to date version of the environmental weeds of Australia sourced from gardens was published as: Randall, R.P. (2001). Garden Thugs, a National List of Invasive and Potentially Invasive Garden Plants. *Plant Protection Quarterly* 16(4), 138-71.	Australasia	Australia
87	Holm, L.G., Pancho, J.V., Herberger, J.P. and Plucknett, D.L. (1979). *A Geographical Atlas of World Weeds.* John Wiley and Sons, New York, USA.	Global	
88	Ecocrop (Ecoport, Global Pest Plant Information Service) an FAO database. http://pppis.fao.org/EC_Content.htm	Global	
89	Sedivec, K.S., Tober, D.A. and Berdahl, J.D. (2001, revised). Grass Varieties for North Dakota. Sustainable Agriculture Research and Education Program, Utah State University. www.ext.nodak.edu/extpubs/plantsci/hay/r794w.htm	North America	United States, North Dakota
90	Häfliger, E. and Scholz, H. (1980). *Grassweeds 1.* Ciba-Geigy Ltd., Basel, Switzerland.	Global	
91	Häfliger, E. and Scholz, H. (1981). *Grassweeds 2.* Ciba-Geigy Ltd., Basel, Switzerland.	Global	
92	Vacant		
93	Short, P. (1998). Guide to the Weeds of the Northern Territory. Darwin Herbarium (internal paper) with additions by C. Wilson, NT Parks and Wildlife.	Australasia	Australia, Northern Territory
94	Hanf, M. (1983). *The Arable Weeds of Europe, with their Seedlings and Seeds.* BASF Aktiengesellschaft, D-6700 Ludwigshafen, Germany.	Europe	
95	Henderson, L. (1995). *Plant Invaders of Southern Africa.* Plant Protection Research Institute Handbook No. 5. Agriculture Research Council.	Africa	South Africa
96	Holisticopia (1997). *Herbage. A guide to herbs.* An online database of herb information. www.herbweb.com/	Global	
97	Whitten, G. (1999). *Herbal Harvest.* 2nd Edition. Bloomings Books, Hawthorn, Victoria.	Australasia	Australia

Source No.	Reference	Region	Country
98	Hnatiuk, R.J. (1990). *Census of Australian Vascular Plants*. Australian Flora and Fauna Series No. 11. Bureau of Flora and Fauna, Canberra, ACT.	Australasia	Australia
99	Rozefelds, A.C.F., Cave, L., Morris, D.I., Buchanan, A.M. (1999). The weed invasion in Tasmania since 1970. *Aust. J. Bot.* 47, 23-48.	Australasia	Australia
100	International Legume Database and Information Service. www.ildis.org/LegumeWeb	Global	
101	Introduced (Naturalised) Species to the United States. USDA, NRCS (1999). The PLANTS database http://plants.usda.gov/plants. National Plant Data Center, Baton Rouge, LA 70874-4490, USA.	North America	United States
102	Invasive Exotic Pest Plants in Tennessee. Report from the Tennessee Exotic Pest Plant Council. www.exoticpestplantcouncil.org/states/TN/TNIList.html	North America	United States, Tennessee
103	Invasive Exotic Plants of Canada. An initiative of Parks Canada, Canadian Heritage CWS, BCO, EMAN, Environment Canada, GeoAccess Division, Natural Resources Canada. http://infoweb.magi.com/~ehaber/ipcan.html	North America	Canada
104	Invasive Plants of Natural Habitats in Canada, An Integrated Review of Wetland and Upland Species and Legislation Government their Control. Environment Canada and the Canadian Wildlife Service. www.cws-scf.ec.gc.ca/habitat/inv/index_e.html	North America	Canada
105	Westbrooks, R.G. (1998). *Invasive Plants, Changing the Landscape of America: Fact Book*. Federal Interagency Committee for the Management of Noxious and Exotic Weeds, Washington, DC.	North America	United States
106	Invasive Species in China. China Species Information System, http://monkey.ioz.ac.cn/bwg-cciced/english/cesis/invasive.htm	Asia	China
107	Space, J. (2002–). Invasive Species Present on Pohnpei, Federated States of Micronesia. A product of the Pacific Island Ecosystems at Risk project (PIER). www.hear.org/pier/pohnpei.htm	Pacific	Micronesia
108	Fern, K. (1992–97). *Plants for a Future*. A resource centre for edible and other useful plants. The Field, Penpol, Lostwithiel, Cornwall, PL22 0NG, England. www.comp.leeds.ac.uk/pfaf/index.html	Global	
109	King's American Dispensatory by Harvey Wickes Felter, M.D., and John Uri Lloyd, Phr.M., Ph.D. www.ibiblio.org/herbmed/eclectic/kings/euphorbia-hype.html	Global	
110	Allred, K.W. and Lee, R.D. (1999). Knapweeds, Starthistles, and Basketflowers of New Mexico. *The New Mexico Botanist*, Issue No. 10. http://web.nmsu.edu/~kallred/herbweb/newpage26.htm. Other thistle information can be found on www.largocanyon.org/s	North America	United States, New Mexico
111	Soufi, Z. (1988). Les Principales Mauvaises Herbes des Vergers dans la Region Maritime de Syrie. *Weed Research*, Vol. 2.	Middle East	Syria
112	List of Florida's Most Invasive Species – Florida Exotic Pest Plant Council. www.fleppc.org/99list.htm	North America	United States, Florida
113	Vacant		

Source No.	Reference	Region	Country
114	Malik, R.K. and Tsedev, D. (1996). *Major Weeds of Mongolia.* FAO, Rome.	Central Asia	Mongolia
115	Robson, T.O., Americanos, P.J. and Abu-Irmaileh, B.E. (1991). *Major Weeds of the Near East.* Paper 104. FAO Plant Production and Protection, Rome.	Middle East	
116	Santa Catalina Island Conservancy (1997). Management Plan for the Control and Eradication of Wildland Weeds Ecological Restoration Department. URL: www.catalinaconservancy.org/	North America	United States, California
117	Hitchcock, A.S. (1935). *Manual of the Grasses of the United States,* Vol. 2, Second Edition, revised by Agnes Chase Dover Publications, Inc., New York.	North America	United States
118	Anon (1989). Manuale per il Riconoscimento delle Principali erbe Infestanti. Società Italiana per lo Studio della Lotta alle Malerbe, S.I.L.M. Italy. http://users.unimi.it/weed/sez01/libri/libri.html	Europe	Italy
119	Medit Plants. This email list group is a gardening enthusiast group who grow mostly Mediterranean plants and frequently discuss weedy plants. See below		
Heracleum mantegazzianum	Dufresne, R.F. (1999). Comments on the various and extreme reactions to humans from contact with *Heracleum mantegazzianum.* Medit Plants Email List.		
Ononis spinosa	Osorio, R.X. (1999). Comments on the escape and naturalisation of *Ononis spinosa* in the Chicago area. Medit Plants Email List.	North America	United States
120	Clarke, I. pers. comm. (1999). Comments on the naturalisation of *Cistus monspeliensis.* Melbourne Herbarium, Victoria.	Australasia	Australia, Victoria
121	Wells, M.J., Balsinhas, A.A., Joffe, H., Englebrecht, V.M., Harding, G. and Stirton, C.H. (1986). A Catalogue of Problem Plants in Southern Africa. *Memoirs of the Botanical Surveys of South Africa* No. 53. Botanical Research Institute, Pretoria, South Africa.	Africa	South Africa
122	Plants to avoid in Miami-Dade County. Web Site www.co.miami-dade.fl.us/derm/environment/land/badplants/scientific.htm	North America	United States, Florida
123	Hartwig, N.L., Kuhns, L.J., McCormick, L.H. and Neal, J.C. (1995). Controlling Mile-a-Minute Weed with Pre- and Post-emergence Herbicides. Proceedings of the 49th Annual Meeting of the Northeastern Weed Science Society, Boston, Massachusetts, USA, 2.	North America	United States, Pennsylvania
124	Vacant		
125	Monksilver Nursery Catalogue, Cottenham, Cambridge. www.monksilver.com/	Europe	United Kingdom
126	Häfliger, E., Kühn, U., Hämet-Ahti, L., Cook, C.D.K., Faden, R. and Speta, F. (1982). *Monocot Weeds 3.* Ciba-Geigy Ltd., Basel, Switzerland.	Global	
127	Monterey Bay Nursery Home Page. http://montereybaynsy.com/A.htm	North America	United States, California
128	Mulvaney, M. pers. comm. (1999). Comments on the weediness of *Pistacia chinensis* in Canberra. Weed Consultant/Ecologist, Canberra, ACT.	Australasia	Australia, Canberra

Source No.	Reference	Region	Country
129	Native Plant Conservation Initiative, Alien Plant Working Group. www.nps.gov/plants/alien/fact/eufo1.htm	North America	United States
130	New Mexico State Exotic Plant Candidate Species List. www.usgs.nau.edu/swemp/Info_pages/states/new_mexico/nmdespp.html (now '404 not found')	North America	United States, New Mexico
131	Harris, G. (1998). Invasive New Zealand Weeds. *CalEPPC Newsletters*, Fall 1998 (newsletter is incorrectly dated Winter 1998) Vol. 6 No. 4 (539 kb). www.caleppc.org/publications/newsletters/vol6no4.pdf	Australasia	New Zealand
132	Various (1999). *Nominated Dominating Weed Species*. Weed species nominated as most invasive by members of the Aliens Email List. There are species on this list not on the previous weed list (see ref 18), opinions do change.	Global	
133	Mehrhoff, L.J. (1999). Non-Native Invasive Plant Species Occurring in Connecticut, Revised Edition. George Safford Torrey Herbarium. Connecticut Invasive Plant Working Group. PLUS The Non-Native Invasive and Potentially Invasive Vascular Plants in Connecticut www.eeb.uconn.edu/research/invasives/ind_spec.html	North America	United States, Connecticut
134	Green, P. (1994). Norfolk Island Species List. In *Flora of Australia* Vol. 49, AGPS. Australian Biological Resources Study.	Australasia	Australia, Norfolk Island
135	Mitchell, A.A. and Waterhouse, B.M. (1998). *Northern Australian Quarantine Service, Weeds Target List*. 2nd Edition. Miscellaneous Pub. No. 6/98. AQIS, Canberra. NAQS species included in a targeted surveillance operation that includes Australia's nearby northern neighbours and small offshore islands. The NAQS scheme not only targets weeds but a range of pests and diseases.	Australasia	Australia, northern
136	Taylor, R.J. (1990). *Northwest Weeds, The Ugly and Beautiful Villains of Fields, Gardens and Roadsides*. Mountain Press Publishing Company, Missoula, Montana.	North America	United States
137	Noxious Invasive Vegetation of the Willamette Valley. Native Plant Society of Oregon. http://csf.colorado.edu/northwest/nwnatives/dec97/msg00594.html	North America	United States, Oregon
138	Noxious Weeds and Non-Native Plants (1999). Colorado Weed Management Association. www.cwma.org	North America	United States, Colorado
139	Noxious Weeds in Washington State, USA. www.wa.gov/agr/weedboard/weed_list/overview.html	North America	United States, Washington State
140	Federal Noxious Weeds of the United States. USDA, NRCS (1999). The PLANTS database http://plants.usda.gov/plants. National Plant Data Center, Baton Rouge, LA 70874-4490, USA.	North America	United States
141	Noxious Weeds of Utah (1999). www.blm.gov:80/utah/resources/weeds/ weed2.htm#officially	North America	United States, Utah
142	Randall, J.M. and Marinelli, J. (eds) (1996). *Invasive Plants, Weeds of the Global Garden*. Brooklyn Botanic Garden Publications, USA	North America	United States
143	Overview of Noxious Freshwater Weeds in Washington, Washington State Department of Ecology – Aquatic Plants and Lakes. www.wa.gov/ecology/wq/plants/weeds/exotic.html	North America	United States, Washington State

Source No.	Reference	Region	Country
144	See below		
Azolla filiculoides	Hill, M.P. and Cilliers, C.J. (1999). *Azolla filiculoides* Lamarck (Pteridophyta: Azollaceae), Its Status in South Africa and Control. *Hydrobiologia* 415, 203-6.	Africa	South Africa
Hirschfeldia incana	Darmency, H. and Fleury, A. (2000). Mating System in *Hirschfeldia incana* and Hybridization to Oilseed Rape. *Weed Research* 40(2), 231-8.		
145	Space, J. (2002–). Pacific Island Ecosystems at Risk (PIER). www.hear.org/pier	Pacific	
146	Pacific Northwest Exotic Pest Plant Council. URL: www.wnps.org/eppclet.html	North America	United States
147	Parsons, W.T. and Cuthbertson, E.G. (1992). *Noxious Weeds of Australia*. Inkata Press, Melbourne and Sydney.	Australasia	Australia
148	Vacant		
149	McKinnell, F. pers. comm. (2000). Forester with Department of Conservation and Land Management, Western Australia.	Australasia	Australia
150	Marigo, G. and Pautou, G. (1998). Phenology, Growth and Ecophysiological Characteristics of *Fallopia sachalinensis*. *Journal of Vegetation Science* 9(3), 379-86.		
151	Plant Invaders of Parks and Natural Areas, US National Parks. www.nps.gov/plants/nfwf/nfwf97.htm (site suspended, Jan 2002).	North America	United States
152	Cronk, Q.C.B and Fuller, J.L. (1995). *Plant Invaders. The Threat to Natural Ecosystems*. Chapman and Hall, UK.	Global	
153	Plantas Invasoras Mas Frecuentes En Las Pasturas De La Zona De Pucallpa. (Frequently Invasive Plants of Pastures in the Pucallpa Region of Peru). www.idrc.ca/library/document/099396/index_s.html	South America	Peru
154	*Plants Toxic To Cockatiels*. Was www.cockatiels.org/badplants.html – now '404 not found'.	Global	
155	Csurhes, S. and Edwards, R. (1998). *Potential Environmental Weeds in Australia, Candidate Species for Preventative Control*. Biodiversity Group, Environment Australia, Canberra, ACT.	Australasia	Australia
156	Rice, P.M., Toney, C. and Sacco, B. (1997). Potential Exotic Plant Species Invading the Blackfoot Drainage, Montana. http://invader.dbs.umt.edu/blackfoot/	North America	United States, Montana
157	Garcia, J.G., MacBryde, B., Molina, A.R. and Herrera-MacBryde, O. (1975). *Prevalent Weeds of Central America*. International Plant Protection Centre, El Salvador.	Central America	
158	Bromilow, C. (1995). *Problem Plants of South Africa*. Briza Publications, Arcadia, South Africa.	Africa	South Africa
159	Project to identify the riparian and aquatic weeds of Montana. The Montana State University Herbarium. http://gemini.oscs.montana.edu/~mlavin/herb/ripweed.htm	North America	United States, Montana
160	Duke, J.A. (1983). Handbook of Energy Crops, unpublished, Purdue University. www.hort.purdue.edu/newcrop/duke_energy/Euphorbia_lathyris.html		
161	RAPID. Common Weeds and Poisonous Plants of North America. University of Idaho, College of Agriculture. http://sdg.ag.uidaho.edu/rapid/	North America	United States

Source No.	Reference	Region	Country
162	Scott, L. (1988). Restricted, Noxious and Nuisance Weeds of Alberta. Weed Invasion. In *Assessment of species diversity in the Montane Cordillera Ecozone, Burlington*, eds I.M. Smith and G.G.E. Scudder, in http://eqb-dqe.cciw.ca/eman/reports/publications/99_m	North America	Canada, Alberta
163	Randall, R.P. (2002–). *Plants Database*. A Database of Plant Information Focusing on Invasive Plant Species. Department of Agriculture, Western Australia. Currently (Feb 2002) the database contains 989,800 plant records with over 96,000 weed related records.		
164	OVID Citations. OVID automatically delivers to subscribers any new journal articles that meet a predetermined set of keywords (below).		
Apocynum cannabinum	Webster, T.M. and Cardina, J. (1999). *Apocynum cannabinum* Seed Germination and Vegetative Shoot Emergence. *Weed Science* 47(5), 524-8.		
Pueraria thunbergiana	Susko, D.J. and Mueller, J.P. (1999). Influence of environmental factors on germination and emergence of *Pueraria lobata*. *Weed Science* 47(5), 585-8.		
165	Roy, B., Popay, I., Champion, P., James, T., and Rahman, A. (1998). *An Illustrated Guide to the Common Weeds of New Zealand*. New Zealand Plant Protection Society.	Australasia	New Zealand
166	Royal Botanic Gardens, Kew (1999). Survey of Economic Plants for Arid and Semi-Arid Lands (SEPASAL) database. Published on the Internet. www.rbgkew.org.uk/ceb/sepasal/internet/	Global	
167	Swinney, D. (1999). *San Dimas Experimental Forest Flora Nomenclature Revision*. From the publication *The SDEF: 50 Years of Research* by Dunn *et al.*, *c.* 1988; GTR PSW-104; 49 pp. according to the Jepson Manual; Higher Plants of Calif., *c.* 1993. www.rfl.psw.fs.fed.us/prefire/sdefhtml/sdefflora.html	North America	Mexico
168	Sharp Bros Seed Co. www.sharpseed.com/	North America	United States
169	Snowy River Shire Council Noxious Weeds. www.snowyriver.nsw.gov.au/s9s_weed.htm	Australasia	Australia, New South Wales
170	Soerjani, M., Kostermans, A.J.G.H. and Tjitrosoepomo, G. (eds) (1987). *Weeds of Rice in Indonesia*. Balai Pustaka, Jakata.	South East Asia	Indonesia
171	South Australian 'Proclaimed Plants' (Noxious Weeds). Check for latest list at www.weeds.org.au/index.html	Australasia	Australia, South Australia
172	South Cone Plant Protection Committee (COSAVE). Quarantine plants (weed). www.cosave.org.py/	South America	
173	Vacant		
174	Stubbendieck, J., Friisoe, G.Y. and Bolick M.R. (1994). *Weeds of Nebraska and the Great Plains*. Nebraska Department of Agriculture, Bureau of Plant Industry. Lincoln, Nebraska, 589 pp. See USDA Plants Database for listing of species. http://plants.us	North America	United States, Nebraska
175	Suzuki, K., Hirose, K., Kawase, K and Okitsu, B. (1988). Studies on Weed Control in a Citrus Orchard. III. Control of Broad Leaf Perennial Weeds. *Bulletin of the Fruit Tree Research Station*, Shizuoka, Japan, No. 15, 21-34.	Asia	Japan
176	Introduced Species of Tasmania (1999). Data supplied by the Resource Management and Conservation Department of Primary Industry, Water and Environment, Hobart, Tasmania.	Australasia	Australia, Tasmania
177	Groves, R.H., Shepherd, R.C.H. and Richardson, R.G. (eds) (1995). *The Biology of Australian Weeds*, Vol. 1. R.G. and F.J. Richardson, Melbourne.	Australasia	Australia

Source No.	Reference	Region	Country
178	Panetta, F.D., Groves, R.H. and Shepherd, R.C.H. (eds) (1998). *The Biology of Australian Weeds*, Vol. 2. R.G. and F.J. Richardson, Melbourne.	Australasia	Australia
179	The Exotic Plants of Southern Florida. Exotic Specifics. The Institute for Regional Conservation, George D. Gann and Keith A. Bradley, irc@regionalconservation.org, 22601 SW 152 Ave. Miami, Florida 33170. www.regionalconservation.or g/sfe3/sfehome.h	North America	United States, Florida
180	Anon (1992). *The Grower's Weed Identification Handbook*. Publication 4030. University of California, Division of Agriculture and Natural Resources, California.	North America	United States, California
181	Reid, V.A. (1998). *The Impact of Weeds on Threatened Plants*. Internal Report No. 164. Department of Conservation New Zealand, Science and Research, New Zealand.	Australasia	New Zealand
182	The Nature Conservancy Weed Alert! Wildland Invasive Species Program. *Invasives on the Web* protecting the native biodiversity of our wild lands from harmful invaders. http://tncweeds.ucdavis.edu/alert/archive.html	North America	United States
183	Mabberley, D.J. (1987). *The Plant-Book. A Portable Dictionary of the Higher Plants*. Cambridge University Press, UK.	Global	
184	The Royal Veterinary College: Virtual Gardens, Plants Generally Recognised as Putting the Environment at Risk. www.rvc.ac.uk/general/virt_gar/plants/heracleu.htm	Global	
185	Boulos, L. and Nabil el-Hadidi, M. (1994). *The Weed Flora of Egypt*. The American University in Cairo Press, Cairo, Egypt.	Africa	Egypt
186	Holm, L.G., Plucknett, D.L., Pancho, J.V. and Herberger, J.P. (1977). *The World's Worst Weeds. Distribution and Biology*. University Press of Hawaii, Honolulu, Hawaii.	Global	
187	Vacant		
188	Vacant		
189	Guide to Poisonous and Toxic Plants. US Army Center for Health Promotion and Preventive Medicine. http://chppm-www.apgea.army.mil/ento/plant.htm (same data is also available at http://toxicplants.com).	Global	
190	Palgrave, K.C. (1996). *Trees of Southern Africa*. Struik Publishers, Cape Town, South Africa.	Africa	Southern Africa
191	Randall, R.P., Mitchell, A.A. and Waterhouse, B.M. (1999). *Tropical Weeds*. Internal Report to the Manager of Plant Industry Protection, Department of Agriculture, Western Australia.	South East Asia	Tropics
192	Missouri Botanical Garden's (2002–). W3TROPICOS. Provides access to the Missouri Botanical Garden's VAST (VAScular Tropicos) nomenclatural database and associated authority files. http://mobot.mobot.org/W3T/Search/vast.html	Middle East	Iran, Iraq
193	University of Florida, Center for Aquatic and Invasive Plants. http://aquat1.ifas.ufl.edu/	North America	United States, Florida
194	Toxicity of Common Houseplants. University of Nebraska Cooperative Extension. http://lancaster.unl.edu/factsheets/031.htm	North America	United States, Nebraska

Source No.	Reference	Region	Country
195	USDA Forest Service Eastern Region. www.fs.fed.us/r9/weed/index.htm	North America	United States, eastern
196	USDA, NRCS (2002–). The PLANTS Database, Version 3.1 (http://plants.usda.gov). National Plant Data Center, Baton Rouge, LA 70874-4490, USA.	North America	United States
197	USGS Nonindigenous Aquatic Species Database. http://nas.er.usgs.gov/nas	North America	United States
198	The Exotic Flora of Victoria (1998). Data supplied by the Department of Natural Resources and Environment, Victoria.	Australasia	Australia, Victoria
199	Villaseñor Ríos, J.L. and Espinosa García, F.J. (1998). *Catálogo De Malezas De México*. Ediciones Científicas Universitarias, Mexico.	Central America	Mexico
200	Sainty, G.R. and Jacobs, S.W.L. (1994). *Waterplants in Australia, A Field Guide*. 3rd Edition. Sainty and Associates, Darlinghurst, Australia.	Australasia	Australia
201	Noosa Council Environmental Services. Issues Affecting Noosa's Natural Environment. http://www.noosa.qld.gov.au/environment/IssuesEnv1.htm	Australasia	Australia, Queensland
202	Ouedraogo, O., Neumann, U., Raynal Roques, A., Salle G., Tuquet, C. and Dembele, B. (1999). New Insights Concerning the Ecology and the Biology of *Rhamphicarpa fistulosa* (Scrophulariaceae). *Weed Research Oxford* 39: 2, 159-69.	Africa	tropical Africa
203	Weed Science List, Naturalised and Non Naturalised Weeds in Australia (1995). Identified by AQIS and AgWA. A preliminary attempt to catalogue weeds in or threatening Australia.	Global	
204	Tetangco, M.H. (ed) (1981). *Weeds and Weed Control in Asia*. Series No. 20, Food and Fertilizer Technology Center, Taipei.	Asia	
205	Albrecht, D. (1998). Weeds of Alice Springs. Northern Territory Herbarium, unpublished document.	Australasia	Australia, Northern Territory
206	Fournet, J. (1993). Phytoecological Characteristics of Weed Populations in Sugar Cane and Banana Plantations in Basse Terre (Guadeloupe). *Weed Research Oxford* 33: 5, 383-95.	Central America	Guadelope
207	Haragan, P.D. (1991). *Weeds of Kentucky and the Adjacent States, A Field Guide*. The University Press of Kentucky.	North America	United States, Kentucky
208	Taylor, R.L. (1980). *Weeds of Ponds and Streams in New Zealand*. R.L. Taylor, New Zealand.	Australasia	New Zealand
209	Matchacheep, S. (1995). *Weeds of Thailand*. Thai Publication. Author Asst. Prof. Dr. Surachai Matchacheep of the Ratchamongkul Technology Institute, Thailand (list source: Ricky Ward, Enviroweeds subscriber).	Asia	Thailand
210	Weeds of the North Central States. North Central Regional Research Publication No. 50. Cooperative Extension Service. Agricultural Experiment Station, University of Illinois. www.ag.uiuc.edu/~vista/abstracts/aWEEDS.html	North America	United States
211	Uva, R.H., Neal, J.C. and DiTomaso, J.M. (1997). *Weeds of the Northeast*. Cornell University Press, USA.	North America	United States
212	Whitson, T.D. (ed), Burrill, L.C., Dewey, S.A., Cudney, D.W., Nelson, B.E., Richard, D.L. and Parker, R.P. (1996). *Weeds of the West*. The Western Society of Weed Science, Newark.	North America	United States

Source No.	Reference	Region	Country
213	Anon. (1998). Weed of Rangelands in Saratoga, Wyoming. An article in the *Weed Science Society of America Proceedings* Vol. 51. (The highly competitive brush species *Chrysothamnus viscidflorus* (Douglas rabbitbrush) and *Ericameria nauseosa* (gray rabbitbrush)).	North America	United States
214	Weir, J.R. (1927). The Problem of *Dichrostachys nutans*, a Weed Tree, in Cuba with Remarks on its Pathology. *Phytopathology* 17, 137-46.	Central America	Cuba
215	Burnie, D. (1995). *Wild Flowers of the Mediterranean*. Dorling Kindersley, London.	Europe	Mediterranean
216	Hossain, M.K. and Pasha, M.K. (2001). Alien exotics in Bangladesh which have a detrimental impact on the Ecosystem. *ALIENS* No. 13, pp. 12-13. ISSG, New Zealand.	Central Asia	Bangladesh
217	Holm, L.G., Doll, J., Holm, E., Pancho, J.V., and Herberger, J.P. (1997). *World Weeds. Natural Histories and Distribution*. John Wiley and Sons, New York, USA.	Global	
218	Darrow, R.A., Erickson, L.C., Holstrum, J.T. Jnr., Miller, J.F., Scudder, W.F. and Williams, J.L. Jnr. (1966). Report of the Terminology Committee, Standardized Names of Weeds. *WSSA* (14), 346-86. Weed Science Society of America, USA.	North America	United States
219	Anon. (19xx). *Weed Handbook*. Series 1-55, Wyoming Weed and Pest Council, Douglas, Wyoming. A small handbook of 55 cards describing weeds found in the Wyoming region.	North America	United States, Wyoming
220	ICON Database (Nov 2001). This quarantine database holds all the conditions of entry for all approved and non-approved plants animals and products normally imported into Australia. www.aqis.gov.au/icon/asp/ex_querycontent.asp	Global	
221	'Weeds' of Egypt. The Arasi company is a supplier of essential oils extracted from cultivated plants and 'wild' plants (or weeds). This definition is not that which is normally applied to weeds, but it was decided to include this dataset as one which contains many plants which are known weeds elsewhere but obviously have some utility purposes in this situation. http://arasi.freeservers.com/ Arasi Lawrence Company TM.	Africa	Egypt
222	Hoffman, R. and Kearns, K. (eds) (1997). Wisconsin Manual of Control Recommendations for Ecologically Invasive Plants. Wisconsin Dept. Natural Resources, Madison, Wisconsin, 102 pp. See species list on the USDA Plants Database. http://plants.usda.gov/pl	North America	United States, Wisconsin
223	Mienis, H.K. (2002). *Hibiscus asper*. In response to a query on PacificPestNet. Dept. Evolution, Systematics and Ecology, Hebrew University of Jerusalem, Israel.	Africa	West Africa
224	Non-Native Invasive and Potentially Invasive Vascular Plants in Connecticut. List and criteria developed by the George Safford Torrey Herbarium at the University of Connecticut in conjunction with the State Geological and Natural History Survey of Connecticut and the Connecticut Invasive Plant Working Group. www.eeb.uconn.edu/research/invasives/ind_spec.html	North America	United States, Connecticut
225	Owen, S.J. (1996). *Weeds of Concern on Conservation Lands in New Zealand. Ecological Weeds on Conservation Land in New Zealand: a Database*, 118 p. Department of Conservation, Wellington.	Australasia	New Zealand

Source No.	Reference	Region	Country
226	Alien Invasive Plant Species of Jamaica. www.jamaicachm.org.jm/aliens_i_pl.htm	Central America	Jamaica
227	Summary of non-native plant species that have become established in the Cockpit Country region of Jamaica. www.cockpitcountry.com/Non-native%20plants.html	Central America	Jamaica
228	Royal Botanic Gardens, Kew (1999). Survey of Economic Plants for Arid and Semi-Arid Lands (SEPASAL) database. www.rbgkew.org.uk/ceb/sepasal/internet/	Global	Arid and Semi-Arid Lands
229	State Noxious Weeds of the USA. USDA, NRCS (1999). The PLANTS database. http://plants.usda.gov/plants. National Plant Data Center, Baton Rouge, LA 70874-4490, USA.	North America	United States
230	Dahl, C. (1997). *Flora List for Pohnpei.* College of Micronesia-FSM Botany 250. www.geocities.com/TheTropics/Cabana/4705/Botany.html	Pacific	Pohnpei
231	Brossard, C.C., Randall, J.M. and Hoshovsky, M.C. (2000). *Invasive Plants of California's Wildlands.* University of California Press, USA.	North America	United States, California
232	Australian Capital Territory Land (Planning and Environment) Act 1991 Declaration of Pest Plants. Declaration No. 1 OF 1999 under Subsection 254 (1) of the Land (Planning and Environment) Act 1991. www.publishing.act.gov.au/legsales/inst/inst99	Australasia	Australia, ACT
233	Hawaii's Most Invasive Horticultural Plants. This list of the worst invasive horticultural plants in Hawaii was provided by the Hawaii State Alien Species Coordinator (Department of Land and Natural Resources, Division of Forestry and Wildlife). www.state.hi.us/dlnr/dofaw/hortweeds/	North America	United States, Hawaii
234	Tye, A. (1997). Charles Darwin Research Station, Isla Santa Cruz. In response to an email on Aliens.	Pacific	Galapagos Islands, Isla Santa Cruz
235	Edited from Miyake and Kore (1937–1938) Source Formosan Weed Seed Morphology. Investigation on the Seed Morphology of Taiwan Weed Species. Formosa Agricultural Review. http://seed.agron.ntu.edu.tw/ENG/tech/Eweed.htm	Asia	Taiwan
236	Agro Mercado – via Rural – Arventis (2001). Weeds of Cultivation in Argentina. *Malezas en Cultivos de Argentina.* www.viarural.com.ar/viarural.com.ar/agricultura/malezas-nombrelatino.htm	South America	Argentina
237	AGRO (2001). Management Malezas Incluidas (Management of Weeds). Argentinian Software Development Company. www.agromanagement.com/malezas_incluidas.htm	South America	Argentina
238	Weeds in the Highlands of Northern Thailand (2001). Botany and Weed Science Division of the Thai Department of Agriculture. www.disc.doa.go.th/botany/dicot1.html	Asia	Thailand, northern
239	Major Weeds of Thailand (2001). Botany and Weed Science Division of the Thai Department of Agriculture. www.disc.doa.go.th/botany/dicot1.html	Asia	Thailand
240	Tamado, T. and Milberg, P. (2000). Weed Flora in Arable Fields of Eastern Ethiopia with Emphasis on the Occurrence of *Parthenium hysterophorus. Weed Research* 40, 507-21.	Africa	Ethiopia
241	Marticorena, C. and Quezada, A. (1985) Catálogo de la flora vascular de Chile. *GAYANA, BOTANICA* Vol. 42 No. 1-2. Universidad de Concepcion-Chile.	South America	Chile

Source No.	Reference	Region	Country
242	Braun, M., Burstaller, H., Hamdoun, A.M. and Walter, H. (1991). *Common Weeds of Central Sudan.* Verlag Joseph Margraf, Scientific Books, Germany.	Africa	Sudan
243	HortGuard and GrainGuard Threat Lists (Baseline Data). For more information on Hortguard and Grainguard and other asociated programs. www.agric.wa.gov.au/programs/app/Industry/index.htm	Global	
244	Pariyar, D. Country Pasture/Forage Resource Profiles, NEPAL. FAO. www.fao.org/WAICENT/FAOINFO/AGRICULT/AGP/AGPC/doc/Counprof/Nepal.htm	Asia	Nepal
245	AgroInformacoes Plantas Daninhas – Brasil (Problematic Plants of Agriculture in Brazil). www.agronet.eng.br/informa/planta_daninha.html	South America	Brazil
246	Buddenhagen, C. and Newfield, M. pers. comm. (2001). A list of potential and actual environmental weeds for New Zealand. Department of Conservation, New Zealand.	Australasia	New Zealand
247	Russell, A.B., Hardin, J.W., Grand, L. and Fraser, A. (1997). *Poisonous Plants of North Carolina.* Department of Horticultural Science, Botany, Plant Pathology and Family and Consumer Sciences, North Carolina State University. www.ces.ncsu.edu/depts/hort/consumer/poison/poison.htm	North America	United States, North Carolina
248	Marwat, Q. and Hussain, F. (1988). Ecological Assessment of Apple and Apricot Weeds in Hanna – Urak Valley Quetta. *Pakistan J. Agric. Res.* Vol. 9, No. 2.	Central Asia	Pakistan
249	Murphy T.R. and Johnson B.J. Weeds of Southern Turfgrasses. The College of Agricultural and Environmental Sciences, Griffin, Georgia. www.griffin.peachnet.edu/cssci/TURF/turf.htm	North America	United States, southern
250	Sá, G., Vasconcelos, T. and Filipe, N. (1989). Weed Flora of Some Orchards in Portugal. Influence of Ecological Factors. Proceedings of the 4th European Weed Research Society Symposium on Weed Problems in the Mediterranean Climates, Valencia, Spain.	Europe	Portugal
251	Kate Blood pers. comm. (2001). Garden thug additions (see comments next reference).	Australasia	Australia
252	Garden Thug Additions. Species added to the list of known environmental weeds still grown or sold in Australia. The full list is published in Randall, R.P. (2001). Garden Thugs, a National List of Invasive and Potentially Invasive Garden Plants. *Plant Protection Quarterly* 16(4), 138-71.	Australasia	Australia
253	Lonchamp, J.P. (2000). Unité de Malherbologie and Agronomie Weed Science and Agronomy INRA-Dijon. This encyclopedic database on plant protection catalogues the main weeds of western Europe (580), describes the species at two stages: mature plants and seedlings, and provides information on their taxonomy, their distribution and their ecology. www.inra.fr/Dijon/malherbo/hyppa/hyppa-a/noms_sca.htm	Europe	western
254	Croft, J. (1982). Ferns and Man in New Guinea. Paper presented to Papua New Guinea Botany Society. (Ferns and their Allies Used By or Affecting Man in Papuasia). www.anbg.gov.au/projects/fern/ferns-man-ng.html	South East Asia	New Guinea

Source No.	Reference	Region	Country
255	Lorenzi, H. (2000). *Plantas daninhas do Brasil. Terrestres, Aquaticas, Parasitas e Toxicas*, 3rd Edition. Instituto Plantarum De Estudos Da Flora Ltda.	South America	Brazil
256	Qiang, S. and Cao, X. (2000). Survey and Analysis of Exotic Weeds in China. *Journal of Plant Resources* 9(4), 34-47.	Asia	China
257	Tye, A. (2001). Invasive Plant Problems and Requirements for Weed Risk Assessment in the Galapagos Islands. In *Weed Risk Assessment*, eds R.H. Groves, F.D. Panetta and J.G. Virtue. CSIRO Publishing, Melbourne.	Pacific	Galapagos Islands
258	Puerto Rico: Summary of Plant Protection Regulations Updated September, 1999. Puerto Rico Department of Agriculture Plant Quarantine Services, Noxious Weeds. www.aphis.usda.gov/npb/FandSQS/prsq.html	Central America	Puerto Rico
259	Swarbrick, J.T. and Hart, R. (2000). Environmental Weeds of Christmas Island (Indian Ocean) and their Management. *Plant Protection Quarterly* 16(2), 54-7.	Australasia	Australia, Christmas Island
260	Queensland Tree Selector Index QDNRPI. www.forests.qld. gov.au/qts/treetext.html	Australasia	Australia, Queensland
261	Liogier, H.A. (2000). *Flora of Puerto Rico and Adjacent Islands. A Systematic Synopsis*, 2nd Edition. Universidad de Puerto Rico.	Central America	Puerto Rico
262	Koo, S.K., Chin, Y.W., Kwon, Y.W. and Cung, H.A. (2000). *Common Weeds in Vietnam*. Agriculture Publishing House, Vietnam.	Asia	Vietnam
263	Morita, H. (1997). *Handbook of Arable Weeds in Japan*. Kumiai Chemical Company.	Asia	Japan
264	Heil, K.D. (2000). *Four Corners Invasive and Poisonous Plant Field Guide*. San Juan College, Bureau of Land Management.	North America	United States, south east
265	Oleskevich, C., Shamoun, S.F. and Punja, Z.K. (1996). The Biology of Canadian Weeds. 105. *Rubus strigosus* Michx., *Rubus parviflorus* Nutt. and *Rubus spectabilis* Pursh. *Can. J. Pl. Sci.* 76, 187-201.	North America	Canada
266	Anderson, W.P. (1999). *Perennial Weeds*. Iowa State University Press, Iowa, USA.	North America	United States
267	Sheley, R.L. and Petroff, J.K. (eds) (1999). *Biology and Management of Noxious Rangeland Weeds*. Oregon State University Press.	North America	United States, rangelands
268	Bambaradeniya, C. (2000). List of Alien Invasive Plant Species from Sri Lanka. Email to Aliens Listserver group from Dr. Channa Bambaradeniya, Head Biodiversity Unit, IUCN – The World Conservation Union, Sri Lanka.	Central Asia	Sri Lanka
269	Auld, B.A. and Medd, R.W. (1992). *Weeds. An Illustrated Botanical Guide to the Weeds of Australia*. Inkata Press, Melbourne.	Australasia	Australia
270	Molina, A.R. (1998). *Malezas Presentes en Cultivos de Verano*, Vol. 1. www.molinaanibal.com.ar/molinaanibal/manuales.htm	South America	South America
271	Molina, A.R. (1999). *Malezas Presentes en la Zona (Templada), Subtropical y Tropical de, América del Sur*, Parte 1. www.molinaanibal.com.ar/molinaanibal/manuales.htm	South America	South America

Source No.	Reference	Region	Country
272	Williams, G. and Hunyadi, K. (1987). *Dictionary of Weeds of Eastern Europe: Their Common Names and Importance in Latin, Albanian, Bulgarian, Czech, German, English, Greek, Hungarian, Polish, Romanian, Russian, Serbo-Croat and Slovak.* Elsevier, Amsterdam.	Europe	eastern
273	Chiang, M.Y. and Shi, L.M. (2000). *Lawn Weeds in Taiwan.* Council of Agriculture Executive, Yuen, Taiwan.	Asia	Taiwan
274	Horng, H.C. and Leu, L.S. (1980). *Weeds of Cultivated Land in Taiwan.* Weed Science Society of the Republic of China.	Asia	Taiwan
275	Zhirong, W., Mingyuan, X., Dehui, M., Shunzu, S., Xianfeng, W., Chunbo, Y., Dianjing, Z., Weizhuo, F., Enhui, M. and Jixian, C. (1990). *Farmland Weeds in China. A Collection of Coloured Illustrative Plates.* Agricultural Publishing House, China.	Asia	China
276	Henty, E.E. and Pritchard, G.S. (1973). Weeds of New Guinea and their Control. Botany Bulletin No. 7. Department of Forests, Division of Botany, Lae, PNG.	South East Asia	New Guinea
277	A summary of a study of the status and impacts of invading exotic plants in South Africa. Summary of information on prominent weed species in South Africa after Dean *et al.* 1986 and Henderson 1995 (see ref 95). http://fred.csir.co.za/plants/global/continen/afric	Africa	South Africa
278	Henderson, M., Fourie, D.M.C., Wells, M.J. and Henderson, L. (1987). *Declared Weeds and Alien Invader Plants in South Africa.* Department of Agriculture and Water Supply, Pretoria, South Africa.	Africa	South Africa
279	Henderson, L. and Musil, K.J. (1987). *Plant Invaders of the Transvaal.* Department of Agriculture and Water Supply, Pretoria, South Africa.	Africa	South Africa
280	Landcare Research New Zealand (2001). Plant Names Database (includes the naturalised flora of New Zealand). nzflora.landcare.cri.nz/plantnames	Australasia	New Zealand
281	Principales Malezas De Centro America Y El Caribe Controladas Con Roundup. MAX. Per Label Registration in Peru and Ecuador. (Data supplied by Chris Buddenhagen, Galapagos Islands, 2001.)	Central and South America	Peru, Ecuador
282	Marine Invasives in Hawai'i. Functional forms and invasive species. URL: www.botany.hawaii.edu/Invasive/default.htm	Pacific	Hawaii
283	Henderson, L. (2001). *Alien Weeds and Invasive Plants.* Plant Protection Research Institute and Agricultural Research Council, South Africa.	Africa	South Africa
284	Randall Stocker, Director, Center for Aquatic and Invasive Plants University of Florida (Enviroweeds posting, 2001).	North America	United States
285	Colin Wilson, Weed Management Officer, Parks and Wildlife Commission of the NT (Enviroweeds posting, 2001).	Australasia	Australia
286	The Research Institute for Bioresources, Okayama University, Laboratory of Wild Plant Science (2001). Weeds of Japan. www.rib.okayama-u.ac.jp/wild/zassou/z_table.htm	Asia	Japan
287	The Research Institute for Bioresources, Okayama University, Laboratory of Wild Plant Science (2001). Naturalized plants from foreign country into Japan. www.rib.okayama-u.ac.jp/wild/kika/kika_table.htm	Asia	Japan

Source No.	Reference	Region	Country
288	James Asa Strong pers. comm. (2001). Portaferry Marine Station, The Queen's University of Belfast, Northern Ireland.	Europe	United Kingdom
289	Blood, K. (2001). *Environmental Weeds: A Field Guide for SE Australia*. C.H. Jerram and Associates.	Australasia	Australia
290	Blood, K. (2001). Environmental Weeds in south eastern Australia. Species collated but not used in her book.	Australasia	Australia
291	Cavers, P.B. (ed.) (2000). *The Biology of Canadian Weeds*. IV. Contributions 84-102. Agricultural Institute of Canada.	North America	Canada
292	Cavers, P.B. (ed.) (1995). *The Biology of Canadian Weeds*. III. Contributions 62-83. Agricultural Institute of Canada.	North America	Canada
293	Mulligan, G.A. (ed.) (1984). *The Biology of Canadian Weeds*. II. Contributions 33-61. Agriculture Canada.	North America	Canada
294	Mulligan, G.A. (ed.) (1979). *The Biology of Canadian Weeds*. I. Contributions 1-32. Agriculture Canada.	North America	Canada
295	Marzocca, A. (1994). *Guia Descriptiva De Malezas Del Cono Sur*. Instituto Nacional De Tecnologia Agropecuaria.	South America	South America
296	Muyt, A. (2001). *Bush Invaders of South-East Australia. A Guide to the Identification and Control of Environmental Weeds Found in South-East Australia*. R.G. and F.J. Richardson, Melbourne.	Australasia	Australia
297	Zhang, Z.P. and Hirota, S. (eds) (2000). *Chinese Colored Weed Illustrated Book*. Institute for the Control of Agrochemicals, Ministry of Agriculture, P.R. China and the Japan Association for Advancement of Phyto-Regulators.	Asia	China
298	Espie, P.R. (2001). *Hieracium in New Zealand: Ecology and Management*. AgResearch Ltd., Mosgiel, New Zealand.	Australasia	New Zealand
299	Royer, F. and Dickinson, R. (1999). *Weeds of Canada and the Northern United States*. The University of Alberta Press.	North America	Canada and northern USA
300	Marticorena, C. (2000?). *Naturalised Plants of Chile*. University of Santiago Herbarium, Chile.	South America	Chile

Table 2. Reference codes for each region represented.

Region	Country	Codes
Africa	East Africa	53
Africa	Egypt	69, 185, 221
Africa	Ethiopia	240
Africa	Kenya (Rhodesia)	50
Africa	South Africa	10, 51, 63, 95, 121, 144, 158, 277, 278, 279, 283
Africa	Southern Africa	190
Africa	Sudan	242
Africa	tropical and west Africa	202, 223
Asia	China	106, 256, 275, 297
Asia	Japan	175, 263, 286, 287
Asia	Nepal	244
Asia	Taiwan	235, 273, 274
Asia	Thailand	12, 209, 238, 239
Asia	Vietnam	262
Asia	Java	13
Asia		204
Australasia	Australia	25, 54, 57, 61, 86, 97, 98, 99, 147, 149, 155, 177, 178, 200, 232, 251, 252, 269, 285, 289, 290, 296
Australasia	Australia, Australian alps	20
Australasia	Australia, Canberra	128
Australasia	Australia, Christmas Island	259
Australasia	Australia, New South Wales	73, 169
Australasia	Australia, Norfolk Island	134
Australasia	Australia, northern	29, 55, 135
Australasia	Australia, Northern Territory	93, 205
Australasia	Australia, Queensland	260
Australasia	Australia, South Australia	171
Australasia	Australia, Tasmania	176
Australasia	Australia, Victoria	72, 74, 120, 198
Australasia	Australia, Western Australia	7, 9, 62
Australasia	New Zealand	15, 131, 165, 181, 208, 225, 246, 280, 298
Australasia	northern Australia and immediate northern neighbours	11
Central America	Cuba	14, 214
Central America	Guadeloupe	206
Central America	Jamaica	226, 227
Central America	Mexico	199
Central America	Puerto Rico	258, 261
Central America		157
Central and South America	Peru, Ecuador	281 (see 38, 153)

Region	Country	Codes
Central Asia	India	66
Central Asia	Mongolia	114
Central Asia	Pakistan	248
Central Asia	Sri Lanka	268
Europe	eastern	272
Europe	Finland	42
Europe	Italy	118
Europe	Mediterranean	215
Europe	Portugal	74, 250
Europe	United Kingdom	41, 56, 125, 288
Europe	United Kingdom, Cornwall and Scilly Isles	40
Europe	western	70, 253
Europe		5, 94
Global	Arid and semi-arid lands	228
Global		18, 19, 23, 24, 27, 28, 30, 39, 44, 67, 68, 71, 76, 81, 85, 87, 88, 90, 91, 96, 100, 108, 109, 126, 132, 152, 154, 166, 183, 184, 186, 189, 203, 217, 220, 243
Middle East	Iran, Iraq	192
Middle East	Israel	59
Middle East	Syria	111
Middle East		115
North America	Canada	4, 36, 52, 103, 104, 265, 291, 292, 293, 294
North America	Canada and northern USA	299
North America	Canada, Alberta	16, 162
North America	Canada, Vancouver	75
North America	eastern and central North America	8
North America	Mexico	167
North America	United States	64, 77, 77, 80, 101, 105, 117, 119, 129, 136, 140, 142, 146, 151, 161, 168, 182, 196, 197, 210, 211, 212, 213, 218, 229, 266, 284
North America	United States, Arizona	26
North America	United States, California	34, 35, 78, 116, 127, 180, 231
North America	United States, Colorado	48, 49, 138
North America	United States, Connecticut	133, 224
North America	United States, eastern	195
North America	United States, Florida	47, 83, 84, 112, 122, 179, 193
North America	United States, Kentucky	207
North America	United States, Montana	156, 159
North America	United States, Nebraska	174, 194
North America	United States, New Mexico	110, 130
North America	United States, North Carolina	247

Region	Country	Codes
North America	United States, North Dakota	89
North America	United States, north west	79
North America	United States, Oregon	45, 137
North America	United States, Pennsylvania	123
North America	United States, rangelands	267
North America	United States, Rocky Mountains	21
North America	United States, south east	264
North America	United States, southern	249
North America	United States, Tennessee	102
North America	United States, Utah	141
North America	United States, Virginia	17
North America	United States, Washington State	1, 139, 143
North America	United States, Wisconsin	222
North America	United States, Wyoming	219
North America		82
Pacific	Galapagos Islands, Isla Santa Cruz	234, 257
Pacific	Hawaii	233, 282
Pacific	Micronesia	107
Pacific	Pohnpei	230
Pacific		3, 6, 145
Pantropics		22, 37
South America	Argentina	43, 65, 236, 237
South America	Brazil	245, 255
South America	Chile	241, 300
South America	Guyana, Surinam and French Guiana	32
South America	Peru	38, 153 (see 281)
South America	South America	172, 270, 271, 295
South East Asia	Indonesia	33, 170
South East Asia	Malaysia	33
South East Asia	New Guinea	254, 276
South East Asia	Philippines	33
South East Asia	Tropics	191

A GLOBAL COMPENDIUM
of WEEDS

A

Abelia × grandiflora (Rovelli ex André) Rehder
Linnaeaceae/Caprifoliaceae
= *Abelia chinensis* R.Br. × *Abelia uniflora* R.Br.
♦ glossy abelia, largeflower abelia
♦ Naturalised, Casual Alien
♦ 101, 198, 280
♦ cultivated, herbal. Origin: horticultural hybrid.

˙Abelmoschus esculentus (L.) Moench
Malvaceae
Hibiscus esculentus L. (see)
♦ lady's finger, okra, gombo, gumbo
♦ Weed, Naturalised, Introduced, Casual Alien
♦ 101, 121, 161, 179, 228, 242, 261
♦ aH, arid, cultivated, herbal. Origin: Eurasia.

Abelmoschus ficulneus Wight & Arn.
Malvaceae
♦ native rosella
♦ Native Weed
♦ 55
♦ cultivated. Origin: Australia, Malaysia, India, Pakistan, Madagascar.

Abelmoschus manihot (L.) Medik.
Malvaceae
Hibiscus manihot L.
♦ sweet hibiscus, aibika, bele, pele, ailan kapis
♦ Weed, Naturalised
♦ 98, 203, 257
♦ pH, cultivated, herbal. Origin: south-east Asia.

Abelmoschus manihot (L.) Medik. ssp. manihot
Malvaceae
♦ Naturalised
♦ 86, 98
♦ Origin: south-east Asia.

Abelmoschus manihot (L.) Medik. ssp. *tetraphyllus* (Roxb. ex Hornem.) Borss.Waalk.
Malvaceae
♦ Naturalised
♦ 86, 98
♦ Origin: south-east Asia.

Abelmoschus mindanaensis Warb. ex Perkins
Malvaceae
♦ Weed
♦ 87, 88

Abelmoschus moschatus Medik.
Malvaceae
Hibiscus abelmoschus L. (see)
♦ musk okra, musk mallow, algalia, almizcle vegetal, ambretta semi
♦ Weed, Naturalised, Introduced, Cultivation Escape
♦ 38, 88, 101, 261, 262, 286
♦ a/pH, cultivated, herbal. Origin: China, Malaysia, Indonesia, India.

Abelmoschus moschatus Medik. ssp. moschatus
Malvaceae
Hibiscus moschatus Salisb.
♦ Naturalised
♦ 86
♦ cultivated. Origin: Australia.

Abelmoschus moschatus Medik. var. haenkeanus (Presl) Kurz
Malvaceae
♦ musk okra
♦ Weed
♦ 23

Abies Mill. spp.
Pinaceae/Abietaceae
♦ fir
♦ Weed, Naturalised
♦ 39, 54, 86, 88
♦ T, toxic.

Abies balsamea (L.) Mill.
Pinaceae/Abietaceae
♦ balsam fir, palsamipihta
♦ Weed, Cultivation Escape
♦ 39, 42, 87, 88, 218
♦ T, cultivated, herbal, toxic.

Abies concolor (Gord. & Glend.) Lindl. ex Hildebr.
Pinaceae/Abietaceae
Picea concolor Gordon
♦ white fir
♦ Weed
♦ 87, 88, 218
♦ T, cultivated, herbal. Origin: south-western North America.

Abies fraseri (Pursh) Poir.
Pinaceae/Abietaceae
♦ Fraser fir, Fraser balsam fir, she balsam
♦ Weed
♦ 87, 88, 218
♦ T, cultivated, herbal. Origin: south-eastern North America.

Abies grandis (Douglas ex D.Don) Lindl.
Pinaceae/Abietaceae
Abies excelsior Franco
♦ grand fir, giant fir
♦ Weed, Naturalised
♦ 15, 87, 88, 218, 280
♦ T, cultivated, herbal. Origin: western North America.

Abies homolepis Sieb. & Zucc.
Pinaceae/Abietaceae
♦ nikko fir
♦ Naturalised
♦ 101
♦ T, cultivated, herbal. Origin: east Asia, Japan.

Abies lasiocarpa (Hook.) Nutt.
Pinaceae/Abietaceae
♦ subalpine fir
♦ Weed
♦ 87, 88, 218
♦ T, cultivated, herbal. Origin: western North America.

Abies nordmanniana (Steven) Spach
Pinaceae/Abietaceae
♦ Caucasian fir, Nordman fir
♦ Naturalised
♦ 280
♦ T, cultivated, herbal.

Abies pinsapo Boiss.
Pinaceae/Abietaceae
♦ Spanish fir
♦ Weed
♦ 80
♦ T, cultivated, herbal. Origin: south-west Europe.

Abies sibirica Ledeb.
Pinaceae/Abietaceae
♦ Siberian fir, siperianpihta
♦ Cultivation Escape
♦ 42
♦ T, cultivated, herbal. Origin: Russia, China.

Abildgaardia ovata (Burm.f.) Kral
Cyperaceae
Abildgaardia monostachya (L.) Vahl, *Fimbristylis monostachya* (L.) Hassk. (see)
♦ flatspike sedge
♦ Native Weed
♦ 121
♦ pG. Origin: southern Africa.

Abrus Adans. spp.
Fabaceae/Papilionaceae
♦ Quarantine Weed
♦ 220

Abrus fruticulosus Wall. ex Wight & Arn.
Fabaceae/Papilionaceae
♦ Introduced
♦ 32
♦ cultivated. Origin: India.

Abrus precatorius L.
Fabaceae/Papilionaceae
♦ rosary pea, jequirity seeds, crab's eyes, love nut, prayer beads, coral bead vine, jequirity bean, prayer bean, matamoso, love bean, lucky bean, jequirity, kabeko kaikes
♦ Weed, Quarantine Weed, Naturalised, Introduced, Environmental Weed
♦ 38, 39, 76, 80, 84, 86, 87, 88, 151, 154, 161, 179, 189, 203, 209, 218, 220, 228, 247, 287
♦ pC, arid, cultivated, herbal, toxic. Origin: Madagascar.

Abrus precatorius L. ssp. africanus Verdc.
Fabaceae/Papilionaceae
♦ coral bead plant, crab's eyes, jequirity bean, love bean, lucky bean, minnie minnies, prayer bean, rosary pea
♦ Native Weed
♦ 121
♦ pC, toxic. Origin: Africa.

Abutilon asiaticum (L.) Sweet
Malvaceae
♦ Weed
♦ 87, 88

Abutilon auritum (Wall. ex Link) Sweet
Malvaceae
♦ Asian Indian mallow
♦ Naturalised
♦ 101
♦ cultivated. Origin: Asia.

Abutilon avicennae Gaertn. *nom. illeg.*
Malvaceae
= *Abutilon theophrasti* Medik.
♦ Weed
♦ 88
♦ herbal. Origin: Australasia.

Abutilon bidentatum A.Rich.
Malvaceae
♦ Weed
♦ 221

Abutilon darwinii Hook.f. × *pictum*
(Gillies ex Hook. & Arn.) Walp.
Malvaceae
♦ Chinese lantern
♦ Weed, Sleeper Weed, Naturalised,
Environmental Weed
♦ 15, 225, 246, 280

Abutilon fruticosum Guill. & Perr.
Malvaceae
♦ Texas Indian mallow
♦ Weed
♦ 87, 221

Abutilon glaucum G.Don
Malvaceae
Abutilon pannosum (Forsk.f.) Schlecht.
(see)
♦ Weed
♦ 87, 88

Abutilon grandifolium (Willd.) Sweet
Malvaceae
♦ hairy Indian mallow
♦ Weed, Naturalised
♦ 86, 98, 101, 134, 203, 280, 300
♦ cultivated, herbal. Origin: South
America.

Abutilon graveolens (Roxb. ex Hornem.)
Wight & Arn.
Malvaceae
= *Abutilon hirtum* (Lam.) Sweet
♦ Weed
♦ 87, 88
♦ herbal.

Abutilon guineense (Schumach.) E.G.Bak.
& Exell
Malvaceae
♦ Weed
♦ 87, 88

Abutilon hemsleyanum Rose
Malvaceae
♦ Weed
♦ 157
♦ pH.

Abutilon hirtum (Lam.) Sweet
Malvaceae
Abutilon graveolens (Roxb. ex Hornem.)
Wight & Arn. (see)

♦ Florida Keys Indian mallow, buenos
días
♦ Weed, Introduced
♦ 179, 228, 261
♦ arid. Origin: Old World Tropics.

Abutilon indicum (L.) Sweet
Malvaceae
Abutilon mauritianum (Jacq.) Medik.
(see), *Sida indica* L.
♦ monkey bush, country mallow,
khrop chak krawaan, buenas tardes,
Indian abutilon
♦ Weed, Introduced
♦ 14, 87, 88, 209, 228, 239, 243, 261, 262,
286, 297
♦ a/pS, arid, cultivated, herbal.
Origin: Old World Tropics.

Abutilon indicum (L.) Sweet ssp.
guineense Borss.
Malvaceae
♦ Weed
♦ 286

Abutilon mauritianum (Jacq.) Medik.
Malvaceae
= *Abutilon indicum* (L.) Sweet
♦ Weed
♦ 88

Abutilon megapotamicum A.St.-Hil. &
Naudin
Malvaceae
♦ trailing abutilon
♦ Naturalised
♦ 280
♦ S, cultivated, herbal.

Abutilon megapotamicum (Spreng.)
A.St.-Hil. & Naudin × *pictum* (Hook. &
Arn.) Walp.
Malvaceae
♦ Casual Alien
♦ 280

Abutilon molle Sweet
Malvaceae
Sida mollis Ortega *nom. illeg.*
♦ hairy abutilon
♦ Weed
♦ 87, 88, 218
♦ herbal.

Abutilon oxycarpum F.Muell.
Malvaceae
♦ lantern bush
♦ Weed
♦ 87, 88
♦ cultivated. Origin: Australia.

Abutilon pannosum (G.Foster) Schltdl.
Malvaceae
= *Abutilon glaucum* G.Don
♦ Weed
♦ 88, 221, 242
♦ arid.

Abutilon pauciflorum Sweet
Malvaceae
= *Abutilon hulseanum* (Torr. & Gray)
Torr. ex Gray (NoR)
♦ country mallow
♦ Weed
♦ 161
♦ cultivated.

Abutilon sonneratianum (Cav.) Sweet
Malvaceae
♦ wild hibiscus
♦ Weed, Native Weed
♦ 88, 121, 158
♦ pS, cultivated. Origin: southern
Africa.

Abutilon striatum Dicks. ex Lindl.
Malvaceae
♦ redvein Indian mallow
♦ Weed, Naturalised
♦ 87, 88, 287
♦ herbal. Origin: South America.

Abutilon theophrasti Medik.
Malvaceae
Abutilon abutilon (L.) Rusby, *Abutilon
avicennae* Gaertn. *nom. illeg.* (see), *Sida
abutilon* L.
♦ swamp Chinese lantern, butterprint
velvetleaf, velvetleaf, Indian
mallow, chingma lantern, pie maker,
buttonweed, butterprint, velvet weed,
butterweed, Indian hemp, cottonweed,
wild cotton, China jute, flower of an
hour
♦ Weed, Noxious Weed, Naturalised,
Introduced, Environmental Weed,
Casual Alien, Cultivation Escape
♦ 1, 7, 23, 24, 34, 36, 42, 52, 68, 80, 86,
87, 88, 93, 94, 98, 101, 118, 146, 151, 156,
161, 174, 179, 180, 195, 198, 203, 205,
207, 210, 211, 212, 218, 229, 235, 243,
252, 253, 263, 272, 275, 280, 286, 287,
291, 297, 299
♦ aH, cultivated, herbal. Origin:
southern Asia, India.

Abutilon trisulcatum (Jacq.) Urb.
Malvaceae
♦ anglestem Indian mallow
♦ Weed
♦ 87, 88
♦ herbal.

Acacia Mill. spp.
Fabaceae/Mimosaceae
♦ acacia, mimosa, guajillo, cat's claw
♦ Weed, Environmental Weed
♦ 3, 18, 88, 116, 155
♦ herbal, toxic. Origin: dry, tropical
and subtropical regions of Australia,
Africa and South America.

Acacia abyssinica Hochst. ex Benth.
Fabaceae/Mimosaceae
Acacia xiphocarpa Hochst. ex Benth.
♦ Weed, Quarantine Weed, Introduced
♦ 76, 88, 203, 220, 228
♦ arid. Origin: east Africa.

Acacia acatalensis Benth.
Fabaceae/Mimosaceae
♦ Weed, Quarantine Weed
♦ 76, 88, 220
♦ Origin: Mexico.

Acacia acuminata Benth.
Fabaceae/Mimosaceae
♦ raspberry jam wood
♦ Naturalised, Native Weed,
Introduced, Garden Escape,
Cultivation Escape
♦ 7, 228

♦ arid, cultivated, herbal. Origin: Australia.

Acacia adansonii Guill. & Perr.
Fabaceae/Mimosaceae
= *Acacia nilotica* (L.) Willd. ex Del. ssp. *adstringens* (Schum. & Thonn.) Roberty
♦ Weed, Quarantine Weed
♦ 76, 88, 203

Acacia adenocalyx Brenan & Exell
Fabaceae/Mimosaceae
♦ Weed, Quarantine Weed
♦ 76, 88, 203, 220

Acacia adhaerens Benth.
Fabaceae/Mimosaceae
♦ Weed, Quarantine Weed
♦ 76, 88, 203, 220

Acacia alata R.Br.
Fabaceae/Mimosaceae
♦ winged wattle
♦ Weed, Quarantine Weed, Naturalised
♦ 86, 88
♦ arid, cultivated.

Acacia albicorticata Burkart
Fabaceae/Mimosaceae
♦ Weed
♦ 87, 88

Acacia albida Delile
Fabaceae/Mimosaceae
= *Faidherbia albida* (Delile) A.Chev. (NoR)
♦ apple ring acacia
♦ Weed, Quarantine Weed
♦ 76, 88, 203, 220, 221
♦ cultivated, herbal.

Acacia amazonica Benth.
Fabaceae/Mimosaceae
♦ Weed, Quarantine Weed
♦ 76, 88, 203, 220

Acacia ampliceps Maslin
Fabaceae/Mimosaceae
♦ salt wattle, acacia
♦ Introduced
♦ 228
♦ arid, cultivated.

Acacia amythethophylla Steud. ex A.Rich.
Fabaceae/Mimosaceae
Acacia buchananii Harms, *Acacia dalzielii* Craib, *Acacia macrothyrsa* Harms (see), *Acacia prorsispinula* Stapf
♦ Weed, Quarantine Weed
♦ 76, 88, 203, 220
♦ arid, cultivated. Origin: east Africa.

Acacia ancistroclada Brenan
Fabaceae/Mimosaceae
♦ Weed, Quarantine Weed
♦ 76, 88, 203, 220

Acacia andamanica I.Nielsen
Fabaceae/Mimosaceae
♦ Weed, Quarantine Weed
♦ 76, 88, 203, 220

Acacia andongensis Welw. ex Hiern
Fabaceae/Mimosaceae
♦ Weed, Quarantine Weed
♦ 76, 88, 203, 220

Acacia aneura F.Muell. ex Benth.
Fabaceae/Mimosaceae
Racosperma aneurum (Benth.) Pedley
♦ mulga
♦ Naturalised, Introduced
♦ 101, 228
♦ T, arid, cultivated, herbal. Origin: Australia.

Acacia angustissima (Mill.) Kuntze
Fabaceae/Mimosaceae
Acacia angustissima (Mill.) Kuntze ssp. *typica* Wiggins, *Acacia angustissima* (Mill.) Kuntze var. *cuspidata* (Schlecht.) L.Benson *p.p.*, *Acacia cuspidata* Schlecht. *p.p.*, *Acacia elegans* M.Martens & Galeotti, *Acacia glabrata* Schldl., *Acacia hirta* Nutt., *Acacia hirta* Nutt. ssp. *lemmonii* (Rose) Wiggins, *Acacia lemmonii* Rose, *Acacia texensis* Torr. & Gray, *Acaciella angustissima* Britton & Rose, *Acaciella breviracemosa* Britton & Rose, *Acaciella hirta* (Nutt.) Britton & Rose, *Acaciella salvadorensis* Britton & Rose, *Acaciella shrevei* Britton & Rose, *Mimosa angustissima* Mill., *Mimosa augustissima* Mill.
♦ prairie acacia, guajillo, palo de pulque, timbe, xaux
♦ Weed, Naturalised, Introduced
♦ 86, 157, 228
♦ S/T, arid, cultivated, herbal. Origin: America.

Acacia ankokib Chiov.
Fabaceae/Mimosaceae
♦ Weed, Quarantine Weed
♦ 76, 88, 203, 220
♦ arid.

Acacia antunesii Harms
Fabaceae/Mimosaceae
♦ Weed, Quarantine Weed
♦ 76, 88, 203, 220

Acacia arabica (Lam.) Willd.
Fabaceae/Mimosaceae
= *Acacia nilotica* (L.) Delile
♦ wattle
♦ Weed, Quarantine Weed
♦ 39, 76, 87, 88, 203, 220, 221
♦ herbal, toxic.

Acacia arenaria Schinz
Fabaceae/Mimosaceae
♦ sand acacia
♦ Weed, Quarantine Weed
♦ 76, 88, 203, 220
♦ cultivated.

Acacia armata R.Br.
Fabaceae/Mimosaceae
= *Acacia paradoxa* DC.
♦ kangaroo thorn, prickly acacia, acacia hedge
♦ Weed, Naturalised
♦ 87, 88, 121, 241, 300
♦ S/T, cultivated, herbal. Origin: Australia.

Acacia aroma Gill. ex Hook. & Arn.
Fabaceae/Mimosaceae
= *Acacia macracantha* Humb. & Bonpl. ex Willd.
♦ tusca

♦ Weed, Naturalised
♦ 87, 88, 236, 241, 295, 300
♦ herbal. Origin: South America.

Acacia articulata Ducke
Fabaceae/Mimosaceae
♦ Weed, Quarantine Weed
♦ 76, 88, 203, 220

Acacia asak (Forssk.) Willd.
Fabaceae/Mimosaceae
Acacia glaucophylla A.Rich., *Acacia triacantha* A.Rich
♦ Weed, Quarantine Weed
♦ 76, 88, 203, 220
♦ arid.

Acacia ataxacantha DC.
Fabaceae/Mimosaceae
♦ flame thorn, flame acacia
♦ Weed, Quarantine Weed, Native Weed
♦ 76, 88, 121, 203, 220
♦ S/T, arid, cultivated, herbal. Origin: southern Africa.

Acacia atramentaria Benth.
Fabaceae/Mimosaceae
Prosopis astringens (Gill.) Speg., *Acacia farnesiana* (L.) Willd. var. *atramentaria* (Benth.) Kuntze, *Acacia astringens* Gill. ex Hook. & Arn.
♦ garabato negro
♦ Weed, Quarantine Weed
♦ 76, 87, 88, 203, 220, 295
♦ Origin: Argentina.

Acacia auriculiformis A.Cunn. ex Benth.
Fabaceae/Mimosaceae
Racosperma auriculiforme (Benth.) Pedley
♦ earleaf acacia, Papuan wattle, auri, earpod wattle
♦ Weed, Noxious Weed, Naturalised, Introduced, Garden Escape, Environmental Weed
♦ 3, 22, 80, 88, 101, 107, 112, 122, 142, 151, 179, 191, 216, 228, 230
♦ T, arid, cultivated, herbal. Origin: northern Australia, Papua New Guinea, eastern Indonesia.

Acacia bahiensis Benth.
Fabaceae/Mimosaceae
♦ Weed, Quarantine Weed
♦ 76, 88, 203, 220

Acacia baileyana F.Muell.
Fabaceae/Mimosaceae
Racosperma baileyanum (F.Muell.) Pedley (see)
♦ Cootamundra wattle, Bailey's wattle, Bailey se wattel
♦ Weed, Noxious Weed, Naturalised, Native Weed, Introduced, Garden Escape, Environmental Weed, Cultivation Escape
♦ 7, 63, 72, 86, 88, 95, 101, 116, 121, 198, 228, 260, 277, 279, 280, 283, 289, 296
♦ S/T, arid, cultivated, herbal. Origin: south-east Australia.

Acacia baileyana × decurrens F.Muell.
Fabaceae/Mimosaceae
♦ Naturalised
♦ 86

3

Acacia baileyana × *leucoclada* F.Muell.
Fabaceae/Mimosaceae
♦ Naturalised
♦ 86

Acacia bavazzanoi Pic.Serm.
Fabaceae/Mimosaceae
♦ Weed, Quarantine Weed
♦ 76, 88, 203, 220

Acacia berlandieri Benth.
Fabaceae/Mimosaceae
Acacia emoryana Benth. (see), *Acacia tephroloba* A.Gray
♦ guajillo
♦ Weed, Quarantine Weed
♦ 39, 76, 87, 88, 161, 203, 218, 220
♦ arid, herbal, toxic. Origin: Mexico.

Acacia binervia (Wendl.) J.F.Macbr.
Fabaceae/Mimosaceae
Acacia glaucescens Willd. (see), *Mimosa binervia* Wendl.
♦ coast myall
♦ Native Weed, Introduced
♦ 228, 269
♦ arid, cultivated. Origin: Australia.

Acacia bivenosa DC.
Fabaceae/Mimosaceae
Acacia bivenosa DC. var. *borealis* Hochr., *Acacia elliptica* A.Cunn. ex Benth., *Acacia xanthina* Benth.
♦ two nerved wattle
♦ Introduced
♦ 228
♦ arid, cultivated.

Acacia blakelyi Maiden
Fabaceae/Mimosaceae
♦ Naturalised, Native Weed
♦ 7, 86
♦ cultivated. Origin: Australia.

Acacia boliviana Rusby
Fabaceae/Mimosaceae
♦ Weed, Sleeper Weed, Naturalised, Environmental Weed
♦ 3, 86, 155, 191
♦ S/T. Origin: South America.

Acacia bonariensis Gill.
Fabaceae/Mimosaceae
♦ arranha gato
♦ Weed, Quarantine Weed
♦ 76, 87, 88, 203, 220, 237, 255, 295
♦ T. Origin: South America.

Acacia borleae Burtt Davy
Fabaceae/Mimosaceae
♦ Weed, Quarantine Weed
♦ 76, 88, 203, 220
♦ cultivated.

Acacia brevispica Harms
Fabaceae/Mimosaceae
♦ Weed, Quarantine Weed
♦ 76, 87, 88, 203, 220
♦ arid, cultivated.

Acacia bricchettiana Chiov.
Fabaceae/Mimosaceae
♦ Weed, Quarantine Weed
♦ 76, 88, 203

Acacia browniana H.L.Wendl.
Fabaceae/Mimosaceae

♦ Weed, Quarantine Weed, Naturalised
♦ 86, 88
♦ cultivated.

Acacia bullockii Brenan
Fabaceae/Mimosaceae
♦ Weed, Quarantine Weed
♦ 76, 88, 203, 220

Acacia burkei Benth.
Fabaceae/Mimosaceae
♦ black monkey thorn
♦ Weed, Quarantine Weed, Native Weed
♦ 10, 76, 88, 121, 203, 220
♦ T, cultivated. Origin: southern Africa.

Acacia burttii E.G.Baker
Fabaceae/Mimosaceae
♦ Weed, Quarantine Weed
♦ 76, 88, 203, 220

Acacia bussei Harms ex Sjostedt
Fabaceae/Mimosaceae
♦ Weed, Quarantine Weed
♦ 76, 88, 203, 220
♦ arid.

Acacia caesia (L.) Willd.
Fabaceae/Mimosaceae
Acacia intsia (L.) Willd. (see), *Acacia intsia* var. *caesia* (L.) Wright & Arn. ex Baker, *Mimosa caesia* L.
♦ Weed, Quarantine Weed
♦ 76, 88, 203, 220
♦ cultivated, herbal. Origin: Asia.

Acacia caffra (Thunb.) Willd.
Fabaceae/Mimosaceae
♦ cat thorn, common hookthorn, kaffir thorn, whitethorn, hookthorn, haakdoring
♦ Weed, Quarantine Weed, Native Weed
♦ 39, 63, 76, 88, 121, 203, 220
♦ S/T, arid, cultivated, herbal, toxic. Origin: southern Africa.

Acacia callicoma Meissn.
Fabaceae/Mimosaceae
♦ Weed, Quarantine Weed
♦ 76, 88, 203, 220

Acacia cambagei R.T.Baker
Fabaceae/Mimosaceae
Racosperma cambagei (R.T.Baker) Pedley
♦ gidgee, stinking wattle
♦ Introduced
♦ 228
♦ arid, cultivated, herbal. Origin: Australia.

Acacia campylacantha Hochst. ex A.Rich.
Fabaceae/Mimosaceae
= *Acacia catechu* (L.f.) Willd.
♦ Weed
♦ 203

Acacia caraniana Chiov.
Fabaceae/Mimosaceae
♦ Weed, Quarantine Weed
♦ 76, 88, 203, 220

Acacia cardiophylla Cunn. ex Benth.
Fabaceae/Mimosaceae

♦ wyalong wattle
♦ Naturalised
♦ 198
♦ cultivated, herbal. Origin: Australia.

Acacia catechu (L.f.) Willd.
Fabaceae/Mimosaceae
Acacia campylacantha Hochst. ex A.Rich. (see), *Acacia polyacantha* Willd. (see), *Mimosa catechu* L.f.
♦ cutch tree, black cutch, khair, catechu
♦ Weed, Noxious Weed, Quarantine Weed, Introduced
♦ 22, 39, 54, 76, 87, 88, 93, 147, 191, 203, 220, 228
♦ T, arid, cultivated, herbal, toxic. Origin: Indo Malaysia.

Acacia catechu (L.f.) Willd. var. *sundra* (Roxb.) Kurz
Fabaceae/Mimosaceae
♦ cutch tree
♦ Noxious Weed, Naturalised
♦ 86
♦ Origin: Sri Lanka, Burma, India.

Acacia caven (Molina) Molina
Fabaceae/Mimosaceae
Acacia cavenia (Molina) Hook. & Arn. (see), *Acacia farnesiana* (L.) Willd. var. *cavenia* Hook. & Arn.
♦ caven, churco, churque, espino, espino maulino, quiringa
♦ Weed, Quarantine Weed, Introduced
♦ 76, 88, 203, 220, 228, 237, 295
♦ arid, cultivated. Origin: South America.

Acacia cavenia (Molina) Hook. & Arn
Fabaceae/Mimosaceae
= *Acacia caven* (Molina) Molina
♦ Weed, Quarantine Weed
♦ 88, 203
♦ promoted, herbal.

Acacia chariessa Milne-Redh.
Fabaceae/Mimosaceae
♦ Weed, Quarantine Weed
♦ 76, 88, 203, 220

Acacia cheilanthifolia Chiov.
Fabaceae/Mimosaceae
♦ Weed, Quarantine Weed
♦ 76, 88, 203, 220

Acacia chiapensis Saff.
Fabaceae/Mimosaceae
♦ Weed, Quarantine Weed
♦ 76, 88, 203, 220

Acacia ciliolata Brenan & Exell
Fabaceae/Mimosaceae
♦ Weed, Quarantine Weed
♦ 76, 88, 203, 220

Acacia cochliacantha Humb. & Bonpl. ex Willd.
Fabaceae/Mimosaceae
♦ Weed, Quarantine Weed
♦ 76, 88, 203, 220
♦ herbal.

Acacia concinna (Willd.) DC.
Fabaceae/Mimosaceae
= *Acacia sinuata* (Lour.) Merr.

♦ soap pod, chikakai, piquant sappan, sappan, soap
♦ Weed, Quarantine Weed
♦ 39, 87, 88, 203, 220
♦ cultivated, herbal, toxic. Origin: Asia.

***Acacia condyloclada* Chiov.**
Fabaceae/Mimosaceae
♦ Weed, Quarantine Weed
♦ 76, 88, 203, 220

***Acacia confusa* Merr.**
Fabaceae/Mimosaceae
Acacia richii auct. non A.Gray,
Racosperma confusum (Merr.) Pedley
♦ small Philippine acacia, Formosa koa, Formosa acacia, sosigi
♦ Weed, Naturalised, Introduced, Cultivation Escape
♦ 3, 22, 80, 101, 107, 191, 228, 230, 233, 287
♦ S/T, arid, cultivated, herbal. Origin: northern Philippines.

***Acacia constricta* Benth.**
Fabaceae/Mimosaceae
Acaciopsis constricta Britt. & Rose,
Acaciopsis constricta var. *paucispina* Moldenke
♦ whitethorn, common whitethorn, desert acacia
♦ Weed, Quarantine Weed
♦ 76, 88, 161, 203, 218, 220
♦ arid, cultivated, herbal, toxic.

***Acacia constricta* Benth. var. *vernicosa* (Standl.) L.D.Benson**
Fabaceae/Mimosaceae
= *Acacia neovernicosa* Isely
♦ whitethorn, common whitethorn, desert acacia
♦ Weed
♦ 218
♦ toxic.

***Acacia cornigera* (L.) Willd.**
Fabaceae/Mimosaceae
Acacia furcella Saff., *Acacia hernandezi* Saff., *Acacia spadicigera* Schldl. & Cham. (see), *Mimosa cornigera* L.
♦ bullhorn wattle
♦ Weed, Quarantine Weed, Naturalised, Introduced
♦ 76, 88, 101, 203, 220, 228
♦ arid, cultivated, herbal.

***Acacia coulteri* Benth.**
Fabaceae/Mimosaceae
Senegalia coulteri (Benth.) Britton & Rose
♦ Weed, Quarantine Weed
♦ 76, 88, 203, 220
♦ arid.

***Acacia crassifolia* A.Gray**
Fabaceae/Mimosaceae
♦ Weed, Quarantine Weed
♦ 76, 88, 203, 220

***Acacia cultriformis* A.Cunn. ex G.Don**
Fabaceae/Mimosaceae
♦ knifeleaf wattle
♦ Naturalised
♦ 86
♦ S, cultivated. Origin: Australia.

***Acacia curvifructa* Burkart**
Fabaceae/Mimosaceae
♦ Weed, Quarantine Weed
♦ 76, 88, 203, 220

***Acacia cyanophylla* Lindl.**
Fabaceae/Mimosaceae
= *Acacia saligna* (Labill.) H.L.Wendl.
♦ Port Jackson acacia, blue leaved acacia
♦ Weed, Introduced
♦ 23, 51, 87, 88, 228
♦ arid, cultivated, herbal. Origin: Australia.

***Acacia cyclops* A.Cunn. ex G.Don**
Fabaceae/Mimosaceae
♦ redeye, rooikrans acacia, western coastal wattle, redwreath acacia, cyclops acacia
♦ Weed, Noxious Weed, Naturalised, Introduced, Garden Escape, Environmental Weed, Cultivation Escape
♦ 10, 22, 23, 51, 63, 86, 87, 88, 95, 101, 121, 152, 158, 198, 228, 277, 278, 283
♦ pS, arid, cultivated, toxic. Origin: Western Australia.

***Acacia davyi* N.E.Br.**
Fabaceae/Mimosaceae
♦ corky thorn, paperbark thorn
♦ Weed, Quarantine Weed, Native Weed
♦ 76, 88, 121, 203, 220
♦ S/T, cultivated. Origin: southern Africa.

***Acacia dealbata* Link**
Fabaceae/Mimosaceae
Acacia decurrens Willd. var. *dealbata* (Link) F.Muell., *Racosperma dealbatum* (Link) Pedley (see)
♦ silver wattle, black wattle, Tasmania mimosa, blue wattle
♦ Weed, Noxious Weed, Naturalised, Native Weed, Introduced, Garden Escape, Environmental Weed, Cultivation Escape
♦ 7, 10, 15, 22, 51, 63, 86, 87, 88, 95, 101, 121, 134, 152, 158, 225, 228, 241, 246, 277, 278, 279, 280, 283, 287, 300
♦ S/T, arid, cultivated, herbal. Origin: Australia.

***Acacia deanei* (R.Baker) Welch, Coombs & McGlynn**
Fabaceae/Mimosaceae
Racosperma deanei (R.T.Baker) Pedley
♦ Deane's acacia
♦ Weed, Naturalised, Introduced
♦ 86, 88, 228
♦ arid, cultivated. Origin: Australia.

***Acacia decora* Rchb.**
Fabaceae/Mimosaceae
Racosperma decorum (Rchb.) Pedley
♦ western golden wattle
♦ Introduced
♦ 228
♦ arid, cultivated, herbal. Origin: Australia.

***Acacia decurrens* (J.C.Wendl.) Willd.**
Fabaceae/Mimosaceae

Acacia mollissima Willd., *Acacia doratoxylon* Cunn. var. *doratoxylon*, *Mimosa decurrens* J.C.Wendl., *Racosperma decurrens* (Willd.) Pedley (see)
♦ green wattle, early black wattle, black wattle
♦ Weed, Noxious Weed, Naturalised, Native Weed, Introduced, Garden Escape, Environmental Weed, Cultivation Escape
♦ 7, 10, 15, 22, 34, 63, 72, 86, 87, 88, 95, 101, 116, 121, 158, 176, 198, 228, 260, 277, 279, 280, 283, 290
♦ T, arid, cultivated, herbal. Origin: Australia.

***Acacia dodonaeifolia* (Pers.) Balb.**
Fabaceae/Mimosaceae
♦ sticky hop wattle, hop leaved wattle
♦ Naturalised
♦ 86, 198
♦ cultivated. Origin: Australia.

***Acacia dolichocephala* Saff.**
Fabaceae/Mimosaceae
= *Acacia sphaerocephala* Schltdl. & Cham.
♦ Weed, Quarantine Weed
♦ 88, 203, 220

***Acacia dolichostachya* S.F.Blake**
Fabaceae/Mimosaceae
Senegalia dolichostachya (S.F.Blake) Britton & Rose
♦ Weed, Quarantine Weed
♦ 76, 88, 203, 220

***Acacia drepanolobium* (Harms) Sjostedt.**
Fabaceae/Mimosaceae
Acacia formicarum Harms, *Acacia lathouwersii* Staner
♦ whistling thorn, black acacia
♦ Weed, Quarantine Weed
♦ 76, 87, 88, 203, 220
♦ arid, cultivated. Origin: Africa.

***Acacia drewiana* W.Fitzg.**
Fabaceae/Mimosaceae
♦ Weed, Naturalised
♦ 86, 88
♦ cultivated.

***Acacia dudgeoni* Craib. ex Holl.**
Fabaceae/Mimosaceae
Acacia samoryana A.Chev., *Acacia senegal* (L.) Willd. var. *samoryana* (A.Chev.) Roberty
♦ Weed, Quarantine Weed
♦ 76, 88, 203, 220
♦ arid, cultivated.

***Acacia dunnii* (Maiden) Turrill**
Fabaceae/Mimosaceae
♦ elephant's ear wattle
♦ Weed
♦ 93
♦ cultivated, herbal. Origin: Australia.

***Acacia eburnea* (L.f.) Willd.**
Fabaceae/Mimosaceae
♦ Weed, Quarantine Weed
♦ 76, 88, 203, 220
♦ Origin: India, Middle East, Pakistan.

Acacia edgeworthii T.Anderson
Fabaceae/Mimosaceae
Acacia erythraea Chiov., *Acacia humifusa*
Chiov., *Acacia pseudosocotrana* Chiov.,
Acacia socotrana Balf.f., *Acacia sultani*
Chiov.
♦ Weed, Quarantine Weed
♦ 76, 88, 203, 220
♦ arid.

Acacia ehrenbergiana Hayne
Fabaceae/Mimosaceae
Acacia ehrenbergii Nees, *Acacia flava*
(Forssk.) Schweinf., *Acacia flava* Forssk.
var. *ehrenbergiana* (Hayne) Roberty,
Acacia seyal sensu A.Chev., *Mimosa flava*
Forssk.
♦ Weed, Quarantine Weed
♦ 76, 88, 203, 220, 221
♦ T, arid, cultivated.

Acacia elata A.Cunn. ex Benth.
Fabaceae/Mimosaceae
Acacia terminalis non auct. pl.,
Racosperma elatum (Benth.) Pedley (see)
♦ cedar wattle, mountain cedar wattle,
peppertree wattle, peperboomwattel
♦ Weed, Noxious Weed, Naturalised,
Native Weed, Garden Escape,
Environmental Weed, Cultivation
Escape
♦ 7, 63, 72, 86, 88, 95, 101, 198, 280, 283,
289
♦ T, cultivated, herbal. Origin:
Australia.

Acacia elatior Brenan
Fabaceae/Mimosaceae
♦ atat, bura, burkuke, burra,
hemnialiliet, esanyanait, munga,
muuga, ollerai, saetch, sesiai
♦ Weed, Quarantine Weed
♦ 76, 88, 203, 220
♦ arid.

Acacia emoryana Benth.
Fabaceae/Mimosaceae
= *Acacia berlandieri* Benth.
♦ Weed, Quarantine Weed
♦ 76, 88, 203, 220
♦ toxic.

Acacia eriocarpa Brenan
Fabaceae/Mimosaceae
♦ Weed, Quarantine Weed
♦ 76, 88, 203, 220

Acacia erioloba E.Mey.
Fabaceae/Mimosaceae
Acacia giraffae sensu auct. mult. non
Willd.
♦ camel thorn, giraffe thorn, mimosa,
Transvaal camelthorn
♦ Weed, Quarantine Weed, Native
Weed, Introduced
♦ 10, 76, 88, 121, 203, 220, 228
♦ T, arid, cultivated, herbal, toxic.
Origin: southern Africa.

Acacia eriopoda Maiden & Blakely
Fabaceae/Mimosaceae
♦ Broome pindan wattle
♦ Introduced
♦ 228
♦ arid, cultivated.

Acacia erubescens Welw. ex Oliv.
Fabaceae/Mimosaceae
Acacia dulcis Marloth ex Engl., *Acacia
kwebensis* N.E.Br.
♦ blue thorn, blouhaak
♦ Weed,
Quarantine Weed, Native Weed
♦ 10, 63, 76, 87, 88, 121, 203, 220
♦ S/T, arid, cultivated. Origin:
southern Africa.

Acacia erythrocalyx J.P.M.Brenan
Fabaceae/Mimosaceae
♦ Weed, Quarantine Weed
♦ 76, 88, 203, 220
♦ cultivated.

Acacia erythrophloea Brenan
Fabaceae/Mimosaceae
♦ Weed, Quarantine Weed
♦ 76, 88, 203, 220

Acacia etbaica Schweinf.
Fabaceae/Mimosaceae
♦ Weed, Quarantine Weed
♦ 76, 88, 203, 220, 221
♦ arid.

Acacia etilis Speg.
Fabaceae/Mimosaceae
♦ Weed, Quarantine Weed
♦ 76, 88, 203, 220

Acacia excelsa Benth.
Fabaceae/Mimosaceae
Racosperma excelsum (Benth.) Pedley
♦ ironwood, ironwood wattle
♦ Weed, Quarantine Weed, Introduced
♦ 88, 228
♦ arid, cultivated. Origin: Australia.

Acacia exuvialis Verd.
Fabaceae/Mimosaceae
♦ flakythorn, lowveld thorn, scaly
acacia
♦ Weed, Quarantine Weed, Native
Weed
♦ 10, 76, 88, 121, 203, 220
♦ S/T. Origin: southern Africa.

Acacia falcata Willd.
Fabaceae/Mimosaceae
Racosperma falcatum (Willd.) Mart.
♦ Introduced
♦ 39, 228
♦ arid, cultivated, toxic. Origin:
Australia.

Acacia farnesiana (L.) Willd.
Fabaceae/Mimosaceae
Acacia acicularis Willd., *Acacia smallii*
Isely (see), *Mimosa farnesiana* L.,
Pithecellobium minutum M.E.Jones,
Vachellia farnesiana Wight & Arn.
♦ huisache, mimosa bush, Ellington
curse, perfumed wattle, cassie flower,
sponge flower, sweet acacia, arapiraca,
corona christi, aroma, klu, popinac,
kandaroma, cassie, vaivai vaka vatona,
vaivai vakavotona, ban baburi, oki, te
kaibakoa, debena, kolu
♦ Weed, Quarantine Weed,
Naturalised, Native Weed, Introduced,
Garden Escape, Environmental Weed
♦ 3, 6, 7, 14, 22, 80, 86, 87, 88, 98, 121,
151, 166, 203, 218, 220, 228, 258, 269

♦ S/T, arid, cultivated, herbal. Origin:
obscure, possibly south-east Asia.

Acacia fasciculifera Benth.
Fabaceae/Mimosaceae
♦ rosewood, scrub ironbark
♦ Introduced
♦ 228
♦ arid, cultivated. Origin: Australia.

Acacia fischeri Harms
Fabaceae/Mimosaceae
♦ Weed, Quarantine Weed
♦ 76, 88, 203, 220

Acacia fleckii Schinz
Fabaceae/Mimosaceae
Acacia caffra (Thunb.) Willd. var.
tomentosa sensu Bak.f., *Acacia cinerea*
Schinz
♦ blade thorn, plate thorn, bladdoring,
geelhaak
♦ Weed, Quarantine Weed, Native
Weed
♦ 10, 63, 76, 88, 121, 203, 220
♦ S/T, arid, cultivated. Origin:
southern Africa.

Acacia floribunda (Vent.) Willd.
Fabaceae/Mimosaceae
Racosperma floribundum (Vent.) Pedley
(see)
♦ catkin wattle, white sallow wattle,
sally wattle
♦ Weed, Naturalised, Native Weed,
Garden Escape, Environmental Weed
♦ 7, 72, 86, 88, 280
♦ S/T, cultivated. Origin: Australia.

Acacia furcatispina Burkart
Fabaceae/Mimosaceae
♦ Weed, Quarantine Weed, Introduced
♦ 76, 88, 203, 220, 228
♦ arid. Origin: South America.

Acacia galpinii Burtt Davy
Fabaceae/Mimosaceae
Acacia senegal sensu O.B.Mill.
♦ monkey thorn
♦ Weed, Quarantine Weed, Native
Weed, Introduced
♦ 76, 88, 121, 203, 220, 228
♦ T, arid, cultivated. Origin: southern
Africa.

Acacia gaumeri S.F.Blake
Fabaceae/Mimosaceae
♦ Weed, Quarantine Weed
♦ 76, 88, 203, 220

Acacia georginae F.M.Bailey
Fabaceae/Mimosaceae
♦ Georgina gidgee, Georgina gidyea
♦ Weed
♦ 39, 87, 88
♦ arid, cultivated, herbal, toxic. Origin:
Australia.

Acacia gerrardii Benth.
Fabaceae/Mimosaceae
♦ red thorn
♦ Weed, Quarantine Weed
♦ 10, 76, 87, 88, 203, 220
♦ arid, cultivated. Origin: Africa.

Acacia gerrardii Benth. var. *gerrardii*
Fabaceae/Mimosaceae

- red thorn
- Native Weed
- 121
- T. Origin: southern Africa.

Acacia giraffae Willd.
Fabaceae/Mimosaceae
= *Acacia haematoxylon* Willd. × *Acacia
erioloba* E.Mey.
- camel thorn
- Weed, Quarantine Weed,
Naturalised
- 39, 76, 86, 88, 203
- cultivated, herbal, toxic. Origin:
southern Africa.

Acacia glandulifera S.Watson
Fabaceae/Mimosaceae
- Weed, Quarantine Weed
- 76, 88, 203, 220

Acacia glaucescens Willd.
Fabaceae/Mimosaceae
= *Acacia binervia* (Wendl.) J.F.Macbr.
- Weed, Native Weed
- 87, 88, 269
- cultivated. Origin: Australia.

Acacia globulifera Saff.
Fabaceae/Mimosaceae
- Weed, Quarantine Weed
- 76, 88, 203, 220

Acacia glomerosa Benth.
Fabaceae/Mimosaceae
Senegalia glomerosa (Benth.) Britton
& Rose, *Senegalia langlassei* Britton &
Rose
- white tamarind
- Weed, Quarantine Weed
- 76, 87, 88, 203, 220
- T. Origin: Central America.

Acacia goetzei Harms
Fabaceae/Mimosaceae
- Weed, Quarantine Weed
- 76, 88, 203
- Origin: Africa.

Acacia gourmaensis A.Chev.
Fabaceae/Mimosaceae
- Weed, Quarantine Weed
- 76, 88, 203, 220
- arid, cultivated.

Acacia grandicornuta Gerstner
Fabaceae/Mimosaceae
- horned thorn
- Weed, Quarantine Weed
- 76, 88, 203, 220
- arid, cultivated.

Acacia grandistipula Benth.
Fabaceae/Mimosaceae
- Weed, Quarantine Weed
- 76, 88, 203, 220

Acacia greggii A.Gray
Fabaceae/Mimosaceae
Acacia durandiana Buckley, *Senegalia
greggia* (A.Gray) Britton & Rose
- cat's claw acacia
- Weed, Quarantine Weed, Introduced
- 76, 87, 88, 154, 161, 203, 218, 220, 228
- S, arid, cultivated, herbal, toxic.

Acacia gummifera Willd.
Fabaceae/Mimosaceae

- acacia
- Weed, Quarantine Weed
- 76, 88, 203, 220
- arid, herbal. Origin: Africa.

Acacia haematoxylon Willd.
Fabaceae/Mimosaceae
- grey camelthorn
- Weed, Quarantine Weed, Native
Weed
- 10, 76, 88, 121, 203, 220
- S/T, arid, cultivated. Origin:
southern Africa.

Acacia hamulosa Benth.
Fabaceae/Mimosaceae
- Weed, Quarantine Weed
- 76, 88, 203, 220

Acacia harmandiana Gagnep.
Fabaceae/Mimosaceae
- Weed, Quarantine Weed
- 76, 88, 203, 220

Acacia harpophylla Benth.
Fabaceae/Mimosaceae
Racosperma harpolhyllum (F.Muell. ex
Benth.) Pedley
- brigalow
- Weed
- 87, 88
- arid, cultivated, herbal. Origin:
Australia.

Acacia hebeclada DC.
Fabaceae/Mimosaceae
Acacia stolonifera Burch., *Acacia
stolonifera* Burch. var. *chobiensis* Mill.
- Weed, Quarantine Weed
- 76, 88, 203, 220
- arid. Origin: Africa.

Acacia hebeclada DC. ssp. hebeclada
Fabaceae/Mimosaceae
- mousebush, candle thorn,
trassiedoring, trassiebos, muisdoring,
candle acacia
- Weed, Native Weed
- 10, 63, 121
- S/T. Origin: southern Africa.

Acacia hebecladoides Harms
Fabaceae/Mimosaceae
- Weed, Quarantine Weed
- 76, 87, 88, 203, 220

Acacia hecatophylla Steud. ex A.Rich.
Fabaceae/Mimosaceae
- Weed, Quarantine Weed
- 76, 88, 203, 220

Acacia hereroensis Engl.
Fabaceae/Mimosaceae
Acacia mellei Verd.
- mountain thorn
- Weed, Quarantine Weed, Native
Weed
- 10, 76, 88, 121, 203, 220
- S/T, arid. Origin: southern Africa.

Acacia heteracantha Burch.
Fabaceae/Mimosaceae
= *Acacia tortilis* (Forssk.) Hayne ssp.
heteracantha (Burch.) Brenan
- Weed, Quarantine Weed
- 76, 87, 88, 203

Acacia hindsii Benth.
Fabaceae/Mimosaceae
- Weed
- 87, 88
- herbal.

Acacia hockii De Wild.
Fabaceae/Mimosaceae
Acacia chariensis A.Chev., *Acacia
oerfota* Brenan, *Acacia seyal* Delile var.
multijuga Baker f., *Acacia stenocarpa*
sensu auct.
- umugenge
- Weed, Quarantine Weed
- 76, 87, 88, 203, 220
- arid, cultivated.

Acacia holosericea A.Cunn. ex G.Don
Fabaceae/Mimosaceae
Acacia mangium Willd. var. *holosericea*
(G.Don) C.White, *Acacia neurocarpa*
A.Cunn. ex Hook.
- candelabra wattle, wah roon, woolly
wattle
- Introduced
- 228
- arid, cultivated.

Acacia homalophylla A.Cunn. ex Benth.
Fabaceae/Mimosaceae
Acacia omalophylla Cunn. ex Benth. in
error, *Racosperma omalophyllum* (Cunn.
ex Benth.) Pedley
- yarran
- Introduced
- 228
- arid, cultivated, herbal.

Acacia hooperiana Zipp. ex Miq.
Fabaceae/Mimosaceae
- Weed, Quarantine Weed
- 76, 88, 203

Acacia horrida (L.) Willd.
Fabaceae/Mimosaceae
Acacia latronum (L.f.) Willd. (see)
- Weed, Quarantine Weed,
Naturalised
- 76, 86, 88, 203, 241, 300
- arid, cultivated. Origin: Africa,
Middle East, India.

Acacia implexa Benth.
Fabaceae/Mimosaceae
Racosperma implexum (Benth.) Pedley
- screw pod wattle, lightwood,
hickory wattle
- Weed, Noxious Weed, Introduced,
Cultivation Escape
- 63, 121, 228, 283
- T, arid, cultivated, herbal. Origin:
Australia.

Acacia inopinata Prain
Fabaceae/Mimosaceae
- Weed, Quarantine Weed
- 76, 88, 203, 220

Acacia insolita E.Pritz.
Fabaceae/Mimosaceae
- Weed, Quarantine Weed,
Naturalised
- 86, 88
- cultivated.

Acacia intsia (L.) Willd.
Fabaceae/Mimosaceae
= *Acacia caesia* (L.) Willd.
♦ Weed, Quarantine Weed
♦ 87, 88, 203, 220
♦ Origin: Asia.

Acacia iteaphylla F.Muell. ex Benth.
Fabaceae/Mimosaceae
♦ Flinder's Range wattle
♦ Weed, Naturalised, Native Weed,
Garden Escape, Environmental Weed
♦ 7, 72, 86, 88, 198
♦ S, arid, cultivated. Origin: Australia.

Acacia ixiophylla Benth.
Fabaceae/Mimosaceae
♦ Weed
♦ 87, 88
♦ cultivated.

Acacia jacquemontii Benth.
Fabaceae/Mimosaceae
♦ Weed, Quarantine Weed
♦ 76, 88, 203, 220
♦ herbal.

Acacia karroo Hayne
Fabaceae/Mimosaceae
Acacia dekindtiana A.Chev., *Acacia
hirtella* E.Mey., *Acacia horrida sensu
auct. mult. non* (L.) Willd., *Acacia
inconflagrabilis* Gerstner, *Acacia natalitia*
E.Mey.
♦ karroo thorn, cape gum, gum
arabic tree, mimosa thorn, sour thorn,
whitethorn, umbrella thorn, sweet
thorn, soetdoring, pendoring, doorn
boom
♦ Weed, Quarantine Weed,
Naturalised, Native Weed, Introduced
♦ 7, 10, 54, 63, 76, 86, 87, 88, 98, 121,
203, 215, 220, 228
♦ S/T, arid, cultivated. Origin:
southern Africa.

**Acacia kelloggiana A.M.Carter &
V.E.Rudd**
Fabaceae/Mimosaceae
♦ Weed, Quarantine Weed
♦ 76, 88, 203, 220

Acacia kirkii Oliv.
Fabaceae/Mimosaceae
♦ seyal
♦ Weed, Quarantine Weed
♦ 76, 88, 203, 220
♦ cultivated. Origin: Africa.

Acacia kirkii Oliv. × seyal Delile
Fabaceae/Mimosaceae
♦ Quarantine Weed
♦ 220

Acacia klugii Standl. ex J.F.Macbr.
Fabaceae/Mimosaceae
♦ Weed, Quarantine Weed
♦ 76, 88, 203, 220

Acacia kraussiana Meisn. ex Benth.
Fabaceae/Mimosaceae
♦ coast climbing thorn
♦ Weed, Quarantine Weed, Native
Weed
♦ 76, 88, 121, 203, 220
♦ pC. Origin: southern Africa.

Acacia lacerans Benth.
Fabaceae/Mimosaceae
♦ Weed, Quarantine Weed
♦ 76, 88, 203, 220

Acacia laeta R.Br. ex Benth.
Fabaceae/Mimosaceae
Acacia senegal (L.) Willd. var. *laeta* (R.Br.
ex Benth.) Roberty, *Acacia trintigniani*
A.Chev.
♦ Weed, Quarantine Weed, Introduced
♦ 76, 88, 203, 220, 221, 228
♦ S/T, arid, cultivated. Origin: Africa.

**Acacia laevigata Humb. & Bonpl. ex
Willd.**
Fabaceae/Mimosaceae
= *Prosopsis laevigata* (Humb. & Bonpl.
ex Willd.) M.C.Johnst. (NoR)
♦ Weed, Quarantine Weed
♦ 203

Acacia lahai Steud. & Hochst. ex Benth.
Fabaceae/Mimosaceae
♦ Weed, Quarantine Weed
♦ 76, 87, 88, 203, 220
♦ Origin: Africa.

Acacia langsdorffii Benth.
Fabaceae/Mimosaceae
♦ Weed, Quarantine Weed
♦ 76, 88, 203, 220

Acacia laricina Meissn.
Fabaceae/Mimosaceae
♦ Weed, Quarantine Weed,
Naturalised
♦ 86, 88
♦ cultivated.

Acacia lasiocalyx C.R.P.Andrews
Fabaceae/Mimosaceae
♦ wilyurwur
♦ Naturalised, Native Weed
♦ 7, 86
♦ cultivated. Origin: Australia.

Acacia lasiopetala Oliv.
Fabaceae/Mimosaceae
♦ Weed, Quarantine Weed
♦ 76, 88, 203, 220

Acacia latistipulata Harms
Fabaceae/Mimosaceae
♦ Weed, Quarantine Weed
♦ 76, 88, 203, 220

Acacia latronum (L.f.) Willd.
Fabaceae/Mimosaceae
= *Acacia horrida* (L.) Willd.
♦ Weed, Quarantine Weed
♦ 76, 88, 203, 220

Acacia lenticularis Buch.-Ham. ex Benth.
Fabaceae/Mimosaceae
♦ Weed, Quarantine Weed
♦ 76, 88, 203, 220
♦ Origin: Asia.

Acacia leptospermoides Benth.
Fabaceae/Mimosaceae
♦ Weed, Quarantine Weed,
Naturalised
♦ 86, 88
♦ cultivated.

Acacia leucoclada Tindale
Fabaceae/Mimosaceae

♦ Weed, Naturalised
♦ 86, 88
♦ cultivated. Origin: Australia.

Acacia leucophloea (Roxb.) Willd.
Fabaceae/Mimosaceae
Acacia arcuata Decne., *Acacia leucophaea*
Willd., *Acacia melanochaetes* Zoll.,
Delaportea ferox Gagnep., *Delaportea
microphylla* Gagnep., *Mimosa
leucophloea* Roxb.
♦ arjuna, aronja, arunja, bilijali,
keru, orabjia, raung, reonj, rewar,
safed kikar, shira, tellatuma, urajio,
velvayalam
♦ Weed, Quarantine Weed, Introduced
♦ 76, 88, 203, 220, 228
♦ arid, cultivated. Origin: Asia.

Acacia ligulata A.Cunn. ex Benth.
Fabaceae/Mimosaceae
Acacia bivenosa DC. ssp. *wayi* (Maiden)
Pedley, *Racosperma ligulatum* (Cunn. ex
Benth.) Pedley
♦ dune wattle, small cooba, umbrella
bush, umbrella wattle
♦ Introduced
♦ 228
♦ arid, cultivated.

Acacia linarioides Benth.
Fabaceae/Mimosaceae
♦ Introduced
♦ 228
♦ arid. Origin: Australia.

Acacia longifolia (Andrews) Willd.
Fabaceae/Mimosaceae
Racosperma longifolium (Andrews)
C.Mart. (see)
♦ Sydney golden wattle, long leaved
wattle, langblaarwattel, sallow wattle,
Port Jackson acacia
♦ Weed, Noxious Weed, Naturalised,
Native Weed, Introduced, Garden
Escape, Environmental Weed,
Cultivation Escape
♦ 3, 7, 10, 34, 39, 51, 63, 72, 80, 87, 88,
95, 101, 116, 121, 132, 152, 158, 228, 246,
277, 278, 279, 280, 283, 289
♦ S/T, arid, cultivated, herbal, toxic.
Origin: Australia.

**Acacia longifolia (Andrews) Willd. var.
longifolia**
Fabaceae/Mimosaceae
♦ Sydney golden wattle
♦ Naturalised, Native Weed, Garden
Escape, Environmental Weed
♦ 7, 86, 296
♦ cultivated, toxic. Origin: Australia.

**Acacia longifolia (Andrews) Willd. var.
sophorae (Labill.) Benth.**
Fabaceae/Mimosaceae
♦ sallow wattle
♦ Naturalised, Garden Escape,
Environmental Weed
♦ 7, 86
♦ cultivated, toxic. Origin: Australia.

Acacia luederitzii Engl.
Fabaceae/Mimosaceae
Acacia gillettiae Burtt Davy, *Acacia
goeringii* Schinz, *Acacia retinens* Sim,

Acacia uncinata sensu auct.
- ◆ Weed, Quarantine Weed
- ◆ 76, 88, 203, 220
- ◆ arid. Origin: Africa.

Acacia luederitzii **Engl. var.** *luederitzii*
Fabaceae/Mimosaceae
- ◆ golden wattle, long leaved wattle, Port Jackson acacia, sallow wattle, Sydney golden wattle, bastard umbrella thorn
- ◆ Weed, Quarantine Weed, Native Weed
- ◆ 10, 121, 220
- ◆ S/T. Origin: southern Africa.

Acacia luederitzii **Engl. var.** *retinens* **(Sim) Ross & Brenan**
Fabaceae/Mimosaceae
- ◆ belly thorn, swollen spined acacia
- ◆ Weed, Quarantine Weed, Native Weed
- ◆ 10, 121, 220
- ◆ S/T. Origin: southern Africa.

Acacia lujaei **Wildem. & Th. Dur.**
Fabaceae/Mimosaceae
- ◆ Weed, Quarantine Weed
- ◆ 76, 88, 203, 220

Acacia macalusoi **Mattei**
Fabaceae/Mimosaceae
- ◆ Weed, Quarantine Weed
- ◆ 76, 88, 203, 220

Acacia macilenta **Rose**
Fabaceae/Mimosaceae
- ◆ Weed, Quarantine Weed
- ◆ 76, 88, 203, 220

Acacia macracantha **Humb. & Bonpl. ex Willd.**
Fabaceae/Mimosaceae
Acacia aroma Gillies ex Hook. & Arn. (see), *Acacia flexuosa* Willd., *Acacia lutea* (Mill.) Britton, *Acacia macracantha* Vogel, *Acacia macracantha* Willd. var. *glabrens* Eggers, *Acacia macracanthoides* Bertol. (see), *Acacia obtusa* Willd., *Acacia pellecantha* Meyen ex J.Vogel, *Acacia subinermis* DC., *Mimosa lutea* Mill., *Poponax lutea* (Mill.) Britton & Rose, *Poponax macracantha* (Humb. & Bonpl.) Killip, *Poponax macracanthoides* (DC.) Britton & Rose
- ◆ steel acacia, porknut
- ◆ Weed, Quarantine Weed, Naturalised, Introduced
- ◆ 32, 76, 87, 88, 203, 220, 228, 241, 300
- ◆ arid, cultivated, herbal. Origin: South America.

Acacia macracanthoides **Bertol.**
Fabaceae/Mimosaceae
= *Acacia macracantha* Humb. & Bonpl. ex Willd.
- ◆ Weed, Quarantine Weed
- ◆ 76, 88, 203, 220

Acacia macrostachya **DC.**
Fabaceae/Mimosaceae
Acacia ataxacantha sensu P.Sousa
- ◆ Weed, Quarantine Weed
- ◆ 76, 88, 203, 220
- ◆ arid, cultivated.

Acacia macrothyrsa **Harms**
Fabaceae/Mimosaceae
= *Acacia amythethophylla* Steud. ex A.Rich.
- ◆ large leaved acacia
- ◆ Weed, Quarantine Weed
- ◆ 76, 88, 203, 220
- ◆ cultivated.

Acacia mangium **Willd.**
Fabaceae/Mimosaceae
- ◆ mangium, tuhkehn pwelmwahu, silkleaf acacia
- ◆ Weed, Introduced, Environmental Weed
- ◆ 3, 87, 88, 93, 107, 152, 191, 230
- ◆ T, cultivated. Origin: northern Australia, Papua New Guinea, eastern Indonesia.

Acacia manubensis **J.H.Ross**
Fabaceae/Mimosaceae
- ◆ Weed, Quarantine Weed
- ◆ 76, 88, 203, 220

Acacia martii **Benth.**
Fabaceae/Mimosaceae
- ◆ Weed, Quarantine Weed
- ◆ 76, 88, 203, 220

Acacia mauroceana **DC.**
Fabaceae/Mimosaceae
- ◆ Weed, Quarantine Weed
- ◆ 76, 88, 203, 220

Acacia mbuluensis **Brenan**
Fabaceae/Mimosaceae
- ◆ Weed, Quarantine Weed
- ◆ 76, 88, 203, 220

Acacia mearnsii **De Wild.**
Fabaceae/Mimosaceae
Acacia mollissima sensu auct., *Racosperma mearnsii* (De Wild.) Pedley (see)
- ◆ black wattle, swartwattel
- ◆ Weed, Noxious Weed, Noxious Weed, Naturalised, Introduced, Environmental Weed, Cultivation Escape
- ◆ 3, 10, 15, 22, 51, 63, 80, 87, 88, 95, 101, 121, 151, 152, 158, 228, 229, 234, 246, 277, 278, 279, 280, 283
- ◆ S/T, arid, cultivated, herbal. Origin: Australasia.

Acacia megaladena **Desv.**
Fabaceae/Mimosaceae
- ◆ Weed, Quarantine Weed
- ◆ 76, 88, 203, 220

Acacia melanoxylon **R.Br.**
Fabaceae/Mimosaceae
Racosperma melanoxylon (R.Br.) Mart. (see)
- ◆ Australian blackwood, blackwood, blackwood acacia, Australiese swarthout, Tasmanian blackwood
- ◆ Weed, Noxious Weed, Naturalised, Native Weed, Introduced, Garden Escape, Environmental Weed, Cultivation Escape
- ◆ 3, 7, 10, 15, 18, 34, 40, 51, 63, 74, 80, 86, 87, 88, 95, 101, 116, 121, 151, 152, 158, 228, 241, 246, 277, 278, 279, 280, 283, 290, 300

- ◆ T, arid, cultivated, herbal. Origin: Australia.

Acacia mellifera **(Vahl) Benth.**
Fabaceae/Mimosaceae
- ◆ Weed, Quarantine Weed
- ◆ 76, 87, 88, 203, 220, 221
- ◆ arid, cultivated. Origin: Africa, southern Arabia.

Acacia mellifera **(Vahl) Benth. ssp.** *detinens* **(Burch.) Brenan**
Fabaceae/Mimosaceae
Acacia detinens Burch.
- ◆ black thorn, noebush, blouhaak, hakiesdoring, hookthorn, monga, mongana, monkana, mukona, omusaona, swart haak, swarthaak, wynruit
- ◆ Weed, Native Weed
- ◆ 10, 63, 121
- ◆ S/T. Origin: southern Africa.

Acacia microbotrya **Benth.**
Fabaceae/Mimosaceae
- ◆ gum wattle
- ◆ Naturalised, Native Weed
- ◆ 7, 86
- ◆ arid, cultivated, herbal. Origin: Australia.

Acacia miersii **Benth.**
Fabaceae/Mimosaceae
- ◆ Weed, Quarantine Weed
- ◆ 76, 88, 203, 220

Acacia mikanii **Benth.**
Fabaceae/Mimosaceae
- ◆ Weed, Quarantine Weed
- ◆ 76, 88, 203, 220

Acacia millefolia **S.Wats.**
Fabaceae/Mimosaceae
- ◆ milfoil wattle
- ◆ Weed, Quarantine Weed
- ◆ 76, 88, 203, 220

Acacia modesta **Wall.**
Fabaceae/Mimosaceae
- ◆ Weed, Quarantine Weed, Introduced
- ◆ 76, 88, 203, 220, 228
- ◆ arid, herbal. Origin: Middle East, India.

Acacia monacantha **Willd.**
Fabaceae/Mimosaceae
- ◆ Weed, Quarantine Weed
- ◆ 76, 88, 203, 220

Acacia montigena **Brenan & Exell**
Fabaceae/Mimosaceae
- ◆ luoyeoye
- ◆ Weed, Quarantine Weed
- ◆ 76, 88, 203, 220

Acacia montis-usti **Merxm. & A.Schreib.**
Fabaceae/Mimosaceae
- ◆ red peeling bark
- ◆ Weed, Quarantine Weed
- ◆ 76, 88, 203, 220
- ◆ arid, cultivated.

Acacia mountfordiae **Specht**
Fabaceae/Mimosaceae
- ◆ Weed
- ◆ 93, 191
- ◆ cultivated. Origin: Australia.

***Acacia mucronata* H.L.Wendl.**
Fabaceae/Mimosaceae
♦ variable sallow wattle
♦ Weed, Naturalised
♦ 86, 88
♦ T, cultivated. Origin: Australia.

***Acacia nebrownii* Burtt Davy**
Fabaceae/Mimosaceae
Acacia glandulifera Schinz, *Acacia rogersii* Burtt Davy, *Acacia walteri* Suess.
♦ waterthorn
♦ Weed, Quarantine Weed, Native Weed
♦ 10, 76, 88, 121, 203, 220
♦ S/T, arid. Origin: southern Africa.

***Acacia negrii* Pic.Serm.**
Fabaceae/Mimosaceae
♦ Weed, Quarantine Weed
♦ 76, 88, 203, 220

***Acacia neovernicosa* Isely**
Fabaceae/Mimosaceae
Acacia constricta var. *vernicosa* (Standl.) L.D.Benson (see), *Acacia vernicosa* Standl.
♦ whitethorn, stickyleaf, Chihuahua whitethorn, viscid acacia
♦ Quarantine Weed
♦ 161, 220
♦ herbal, toxic. Origin: Central America.

***Acacia neriifolia* A.Cunn. ex Benth.**
Fabaceae/Mimosaceae
Racosperma neriifolium (A.Cunn ex Benth.) Pedley
♦ white wattle, silver wattle
♦ Introduced
♦ 228
♦ arid, cultivated. Origin: Australia.

***Acacia nigrescens* Oliv.**
Fabaceae/Mimosaceae
Acacia mellifera sensu Henckel, *Acacia pallens* Rolfe, *Acacia passargei* Harms, *Acacia schliebenii* Harms
♦ knob thorn, knoppiesdoring
♦ Weed, Quarantine Weed, Native Weed, Introduced
♦ 10, 63, 76, 87, 88, 121, 203, 220, 228
♦ T, arid, cultivated. Origin: southern Africa.

***Acacia nigripilosa* Maiden**
Fabaceae/Mimosaceae
♦ Weed, Quarantine Weed, Naturalised
♦ 86, 88, 203, 220
♦ cultivated.

***Acacia nilotica* (L.) Del.**
Fabaceae/Mimosaceae
Acacia arabica (Lam.) Willd. (see), *Mimosa nilotica* (L.) Del. (see)
♦ prickly acacia, gum arabic tree, algaroba, tiare, babul, black thorn
♦ Weed, Quarantine Weed, Noxious Weed, Naturalised, Environmental Weed, Cultivation Escape
♦ 3, 6, 10, 11, 18, 22, 33, 37, 62, 76, 88, 93, 98, 101, 147, 191, 203, 220, 221, 257, 261
♦ S/T, cultivated, herbal, toxic. Origin:

arid and semi arid regions of Africa and west Asia.

***Acacia nilotica* (L.) Willd. ex Del. ssp. *adstringens* (Schum. & Thonn.) Roberty**
Fabaceae/Mimosaceae
Acacia adansonii Guill. & Perr. (see), *Acacia adstringens* (Schum. & Thonn.) Berhaut, *Acacia scorpioides* (L.) W.Wight var. *adstringens* (Schum. & Thonn.) A.Chev.
♦ Introduced
♦ 228

***Acacia nilotica* (L.) Willd. ex Del. ssp. *indica* (Benth.) Brenan**
Fabaceae/Mimosaceae
Acacia arabica sensu Brenan
♦ deshi babul, prickly acacia, black thorn, gum arabic tree, scented pod acacia
♦ Weed, Naturalised, Introduced, Environmental Weed
♦ 86, 152, 178
♦ cultivated. Origin: Arabian peninsula, Pakistan, India, Burma.

***Acacia nilotica* (L.) Willd. ex Del. ssp. *kraussiana* (Benth.) Brenan**
Fabaceae/Mimosaceae
Acacia arabica (Lam.) Willd. var. *kraussiana* Benth., *Acacia benthamiana* Rochebr., *Acacia benthamii* Rochebr., *Acacia nilotica* ssp. *subalata null sensu* auct., *Acacia nilotica* (L.) Willd. ex Del. var. *kraussiana* (Benth.) A.F.Hill, *Acacia subalata sensu* auct.
♦ black thorn, Egyptian mimosa, gum acacia, redheart acacia, red heart thorn, lekkerruikpeul, snuifpeul, stinkpeul, scented thorn, redheart
♦ Weed, Native Weed
♦ 63, 121
♦ T, toxic. Origin: Africa.

***Acacia notabilis* F.Muell.**
Fabaceae/Mimosaceae
♦ Flinder's wattle
♦ Introduced
♦ 228
♦ arid, cultivated. Origin: Australia.

***Acacia nubica* Benth.**
Fabaceae/Mimosaceae
= *Acacia perfota* (Forssk.) Schweinf. (NoR)
♦ Weed, Quarantine Weed
♦ 76, 88, 203, 220, 221
♦ arid, cultivated, herbal.

***Acacia nubica* Benth. × *paolii* Chiov.**
Fabaceae/Mimosaceae
♦ Weed, Quarantine Weed
♦ 76, 88, 220

***Acacia ogadensis* Chiov.**
Fabaceae/Mimosaceae
Albizia ogadensis (Chiov.) Chiov. (see)
♦ Weed, Quarantine Weed
♦ 76, 88, 203, 220

***Acacia oliveri* Vatke**
Fabaceae/Mimosaceae
♦ Weed, Quarantine Weed
♦ 76, 88, 203, 220

***Acacia orfota* Schweinf.**
Fabaceae/Mimosaceae
♦ Weed, Quarantine Weed
♦ 76, 88, 203, 220

***Acacia ornithophora* Sweet**
Fabaceae/Mimosaceae
♦ Weed, Quarantine Weed
♦ 76, 88, 203

***Acacia paniculata* Willd.**
Fabaceae/Mimosaceae
= *Acacia tenuifolia* (L.) Willd.
♦ Weed, Quarantine Weed
♦ 76, 87, 88, 203, 220

***Acacia paolii* Chiov.**
Fabaceae/Mimosaceae
♦ Weed, Quarantine Weed
♦ 76, 88, 203, 220

***Acacia paradoxa* DC.**
Fabaceae/Mimosaceae
Acacia armata R.Br. (see), *Racosperma paradoxum* (DC.) Mart. (see)
♦ prickly acacia, acacia hedge, hedge acacia, hedge wattle, kangaroo acacia, kangaroo thorn, paradox acacia
♦ Weed, Noxious Weed, Naturalised, Garden Escape, Environmental Weed, Cultivation Escape
♦ 9, 35, 63, 86, 86, 88, 101, 147, 161, 176, 203, 229, 246, 280, 283, 290
♦ S, cultivated, herbal. Origin: Australia.

***Acacia parramattensis* Tindale**
Fabaceae/Mimosaceae
Racosperma parramattense (Tindale) Pedley (see)
♦ New South Wales wattle, Parramatta green wattle, Sydney green wattle
♦ Weed, Naturalised
♦ 15, 101, 134, 280
♦ cultivated. Origin: Australia.

***Acacia pedicellata* Benth.**
Fabaceae/Mimosaceae
♦ Weed, Quarantine Weed
♦ 76, 88, 203, 220

***Acacia pendula* G.Don**
Fabaceae/Mimosaceae
Racosperma pendulum (G.Don) Pedley
♦ boree, weeping myall, myall, weeping acacia
♦ Introduced
♦ 228
♦ arid, cultivated, herbal. Origin: Australia.

***Acacia pennata* (L.) Willd.**
Fabaceae/Mimosaceae
Acacia pendata (L.) Willd., *Mimosa pennata* L.
♦ bala
♦ Weed, Quarantine Weed
♦ 39, 76, 87, 88, 203, 220
♦ arid, cultivated, herbal, toxic. Origin: Africa, Asia.

***Acacia pennatula* (Schltdl. & Cham.) Benth.**
Fabaceae/Mimosaceae
Inga pennatula Schldl. & Cham., *Pithecellobium minutissimum* M.E.Jones, *Poponax pennatula* Britton & Rose

♦ carbon blanco
♦ Weed, Quarantine Weed
♦ 76, 88, 203, 220
♦ arid, cultivated, herbal. Origin: Americas.

Acacia penninervis DC.
Fabaceae/Mimosaceae
Racosperma penninerve (Sieber ex DC.) Pedley
♦ hickory wattle, mountain hickory
♦ Weed, Introduced
♦ 39, 88, 228
♦ arid, cultivated, toxic. Origin: Australia.

Acacia pentagona (Schumach. & Thonn.) Hook.f.
Fabaceae/Mimosaceae
♦ Weed, Quarantine Weed
♦ 76, 88, 203, 220
♦ herbal. Origin: Africa.

Acacia permixta Burtt Davy
Fabaceae/Mimosaceae
♦ mimosa thorn, thorn tree, slender thorn
♦ Weed, Quarantine Weed, Native Weed
♦ 76, 88, 121, 203, 220
♦ S/T. Origin: southern Africa.

Acacia persiciflora Pax
Fabaceae/Mimosaceae
♦ Weed, Quarantine Weed
♦ 76, 88, 203, 220

Acacia peuce F.Muell.
Fabaceae/Mimosaceae
♦ waddywood, casuarina
♦ Introduced
♦ 228
♦ arid, cultivated. Origin: Australia.

Acacia piauhiensis Benth.
Fabaceae/Mimosaceae
♦ Weed, Quarantine Weed
♦ 76, 88, 203, 220
♦ arid.

Acacia pilispina Pic.Serm.
Fabaceae/Mimosaceae
♦ Weed, Quarantine Weed
♦ 76, 88, 203, 220

Acacia planifrons Wight & Arn.
Fabaceae/Mimosaceae
♦ Weed, Quarantine Weed
♦ 76, 88, 203, 220
♦ arid.

Acacia plumosa Lowe
Fabaceae/Mimosaceae
♦ arranha gato
♦ Weed, Quarantine Weed
♦ 76, 88, 203, 220, 255
♦ T. Origin: Brazil.

Acacia pluriglandulosa B.Verdc.
Fabaceae/Mimosaceae
♦ Weed, Quarantine Weed
♦ 76, 88, 203, 220

Acacia podalyriifolia A.Cunn. ex G.Don
Fabaceae/Mimosaceae
Racosperma podalyriifolium (G.Don) Pedley (see)
♦ Queensland silver wattle, pearl

acacia, Mount Morgan wattle, vaalmimosa
♦ Weed, Noxious Weed, Naturalised, Native Weed, Introduced, Garden Escape, Environmental Weed, Cultivation Escape, Casual Alien
♦ 7, 63, 72, 86, 88, 95, 101, 121, 198, 228, 277, 279, 280, 283
♦ S/T, arid, cultivated, herbal. Origin: Australia.

Acacia polyacantha Willd.
Fabaceae/Mimosaceae
= *Acacia catechu* (L.f.) Willd.
♦ catechu tree, falcon's claw acacia
♦ Weed, Quarantine Weed, Naturalised
♦ 76, 88, 101, 203, 220, 261
♦ cultivated, herbal. Origin: tropical Asia and Africa.

Acacia polyacantha Willd. ssp. polyacantha
Fabaceae/Mimosaceae
Acacia catechu (L.f.) Willd. ssp. *suma* (Roxb.) Roberty, *Acacia suma* Broun & Massey (see)
♦ Introduced
♦ 228

Acacia polyphylla DC.
Fabaceae/Mimosaceae
♦ Weed, Quarantine Weed
♦ 76, 87, 88, 203, 220
♦ T, herbal.

Acacia pravissima F.Muell.
Fabaceae/Mimosaceae
♦ Ovens wattle, alpine wattle
♦ Weed, Naturalised, Native Weed, Garden Escape, Environmental Weed
♦ 72, 86, 88
♦ S/T, cultivated, herbal. Origin: Australia.

Acacia prominens Cunn. ex G.Don
Fabaceae/Mimosaceae
♦ Gosford wattle, golden rain wattle
♦ Weed, Naturalised, Native Weed, Garden Escape, Environmental Weed
♦ 72, 86, 88, 198
♦ S/T, cultivated, herbal. Origin: Australia.

Acacia pruinocarpa Tindale
Fabaceae/Mimosaceae
♦ black gidgee
♦ Introduced
♦ 228
♦ arid, cultivated.

Acacia pseudo-arabica Blume ex Miq.
Fabaceae/Mimosaceae
♦ Weed, Quarantine Weed
♦ 76, 88, 203, 220

Acacia pseudofistula Harms
Fabaceae/Mimosaceae
♦ Weed, Quarantine Weed
♦ 76, 88, 203, 220

Acacia pseudointsia Miq.
Fabaceae/Mimosaceae
♦ Weed, Quarantine Weed
♦ 76, 88, 203, 220
♦ herbal.

Acacia pulchella R.Br.
Fabaceae/Mimosaceae
♦ western prickly Moses, prickly Moses, prickly bastard
♦ Quarantine Weed, Naturalised, Introduced
♦ 39, 86, 228
♦ arid, cultivated, toxic.

Acacia pycnantha Benth.
Fabaceae/Mimosaceae
Acacia petiolaris Lehm.
♦ golden wattle, Australian golden wattle, blackwood, gouewattel, broad leaved wattle
♦ Weed, Noxious Weed, Naturalised, Native Weed, Introduced, Garden Escape, Environmental Weed, Cultivation Escape
♦ 7, 9, 10, 34, 51, 63, 86, 87, 88, 95, 101, 121, 158, 176, 228, 277, 283
♦ S/T, arid, cultivated, herbal. Origin: Australia.

Acacia quintanilhae Torre
Fabaceae/Mimosaceae
♦ Weed, Quarantine Weed
♦ 76, 88, 203, 220

Acacia raddiana Savi
Fabaceae/Mimosaceae
= *Acacia tortilis* ssp. *raddiana* (Savi) Brenan
♦ Weed, Quarantine Weed
♦ 220, 221
♦ arid, cultivated.

Acacia recurva Benth.
Fabaceae/Mimosaceae
♦ Weed, Quarantine Weed
♦ 76, 88, 203, 220

Acacia redacta J.H.Ross
Fabaceae/Mimosaceae
♦ Weed, Quarantine Weed
♦ 76, 88, 203, 220

Acacia redolens Maslin
Fabaceae/Mimosaceae
♦ bank cat's claw, vanilla wattle
♦ Naturalised, Introduced
♦ 101, 228
♦ arid, cultivated.

Acacia reficiens Wawra & Peyr.
Fabaceae/Mimosaceae
♦ Weed, Quarantine Weed
♦ 76, 88, 203, 220
♦ cultivated.

Acacia reficiens Wawra & Peyr. ssp. reficiens
Fabaceae/Mimosaceae
Acacia cf. *uncinata sensu* Torre
♦ red thorn, vals haak en steek, geelhaak, false umbrella thorn
♦ Weed, Quarantine Weed
♦ 10, 63, 228

Acacia rehmanniana Schinz
Fabaceae/Mimosaceae
♦ acacia
♦ Weed, Quarantine Weed
♦ 76, 87, 88, 203, 220
♦ cultivated. Origin: Africa.

Acacia retinodes Schlecht.
Fabaceae/Mimosaceae
♦ ever blooming acacia, water wattle, wirilda, wirilda wattle
♦ Weed, Naturalised, Native Weed, Introduced, Garden Escape, Environmental Weed, Cultivation Escape
♦ 34, 72, 80, 86, 87, 88, 101, 179, 228
♦ T, arid, cultivated, herbal. Origin: Australia.

Acacia rigidula Benth.
Fabaceae/Mimosaceae
Acaciopsis rigidula (Benth.) Britton & Rose
♦ blackbrush acacia
♦ Weed, Quarantine Weed
♦ 76, 87, 88, 203, 218, 220
♦ arid, herbal. Origin: Americas.

Acacia riparia (Kunth) Britton & Rose ex Britton & Killip
Fabaceae/Mimosaceae
♦ Weed, Quarantine Weed
♦ 76, 88, 203, 220

Acacia robusta Burch.
Fabaceae/Mimosaceae
♦ Weed, Quarantine Weed
♦ 76, 87, 88, 203, 220
♦ cultivated. Origin: Africa.

Acacia robusta Burch. ssp. *robusta*
Fabaceae/Mimosaceae
♦ enkeldoring, brosdoring, splendid thorn, false umbrella thorn, splendid acacia
♦ Weed, Native Weed
♦ 63, 121
♦ T. Origin: southern Africa.

Acacia robynsiana Merxm. & A.Schreib.
Fabaceae/Mimosaceae
♦ Weed, Quarantine Weed
♦ 76, 88, 203, 220

Acacia roemeriana Scheele
Fabaceae/Mimosaceae
Acacia malacophylla A.Gray, *Senegalia roemeriana* (Scheele) Britton & Rose
♦ roundflower cat's claw
♦ Weed, Quarantine Weed
♦ 76, 88, 203, 220
♦ arid, herbal.

Acacia rovumae Oliv.
Fabaceae/Mimosaceae
Acacia morondavensis Drake, *Acacia chrysothrix* Taub.
♦ Weed, Quarantine Weed
♦ 76, 88, 203, 220
♦ Origin: Madagascar.

Acacia rugata (Lam.) Buch.-Ham. ex Voigt
Fabaceae/Mimosaceae
= *Acacia sinuata* (Lour.) Merr.
♦ Weed, Quarantine Weed
♦ 76, 88, 203, 220

Acacia salicina Lindl.
Fabaceae/Mimosaceae
Acacia varians Benth., *Racosperma salignum* (Labill.) Pedley
♦ cooba, native willow, doolan, Broughton willow, willow wattle

♦ Introduced
♦ 39, 228
♦ arid, cultivated, toxic. Origin: Australia.

Acacia saligna (Labill.) H.L.Wendl.
Fabaceae/Mimosaceae
Acacia cyanophylla Lindl. (see), *Mimosa saligna* Labill., *Racosperma saligna* (Labill.) Pedley
♦ Port Jackson willow, golden wreath wattle, blue leaved wattle, orange wattle
♦ Weed, Noxious Weed, Naturalised, Native Weed, Introduced, Garden Escape, Environmental Weed, Cultivation Escape
♦ 3, 10, 22, 63, 72, 86, 88, 95, 101, 121, 152, 158, 198, 228, 277, 278, 283, 289, 296
♦ S/T, arid, cultivated, herbal. Origin: south-west Australia.

Acacia sarcophylla Chiov.
Fabaceae/Mimosaceae
♦ Weed, Quarantine Weed
♦ 76, 88, 203, 220

Acacia schaffneri (S.Watson) F.J.Herm.
Fabaceae/Mimosaceae
♦ Schaffner's wattle
♦ Quarantine Weed
♦ 220

Acacia schinoides Benth.
Fabaceae/Mimosaceae
♦ frosty wattle
♦ Naturalised
♦ 86, 198
♦ cultivated. Origin: Australia.

Acacia schlechteri Harms
Fabaceae/Mimosaceae
♦ Weed, Quarantine Weed
♦ 76, 88, 203, 220

Acacia schottii Torr.
Fabaceae/Mimosaceae
♦ Schott's wattle
♦ Quarantine Weed
♦ 220
♦ herbal.

Acacia schweinfurthii Brenan & Exell
Fabaceae/Mimosaceae
♦ Weed, Quarantine Weed
♦ 76, 88, 203, 220
♦ cultivated. Origin: Africa.

Acacia schweinfurthii Brenan & Exell var. *schweinfurthii*
Fabaceae/Mimosaceae
♦ river climbing thorn
♦ Native Weed
♦ 121
♦ C S/T. Origin: southern Africa.

Acacia sclerophylla Lindl.
Fabaceae/Mimosaceae
♦ hardleaf wattle
♦ Weed, Naturalised
♦ 86, 88
♦ cultivated. Origin: Australia.

Acacia sclerosperma F.Muell.
Fabaceae/Mimosaceae
♦ limestone wattle, silver bark wattle

♦ Weed, Quarantine Weed, Naturalised, Introduced
♦ 86, 88, 220, 228
♦ arid, cultivated.

Acacia scorpioides (L.) W.Wight
Fabaceae/Mimosaceae
= *Acacia nilotica* (L.) Willd. ex Del. ssp. *nilotica* (NoR)
♦ Weed, Quarantine Weed
♦ 76, 88, 203, 220

Acacia senegal (L.) Willd.
Fabaceae/Mimosaceae
Acacia circummargiata Chiov., *Acacia cufodontii* Chiov., *Acacia glaucophlla sensu* Brenan, *Acacia kinionge sensu* Brenan, *Acacia oxyosprion* Chiov., *Acacia somalensis sensu* Brenan, *Acacia spinosa* Marloth & Engl., *Acacia thomasii sensu* Brenan, *Acacia verek* Guill. & Perr., *Acacia volkii* Suess.
♦ gum tree, gum arabic tree, threehook thorn
♦ Weed, Quarantine Weed, Naturalised
♦ 10, 76, 86, 88, 203, 220
♦ arid, cultivated, herbal. Origin: Africa, Arabia, Middle East, West India.

Acacia senegal (L.) Willd. var. *leiorhachis* Brenan
Fabaceae/Mimosaceae
♦ slender threehook thorn
♦ Native Weed
♦ 121
♦ S/T. Origin: southern Africa.

Acacia senegal (L.) Willd. var. *rostrata* Brenan
Fabaceae/Mimosaceae
♦ slender threehook thorn, threehook thorn, three thorned acacia
♦ Weed, Native Weed
♦ 63, 121
♦ S/T. Origin: southern Africa.

Acacia serra Benth.
Fabaceae/Mimosaceae
♦ Weed, Quarantine Weed
♦ 76, 88, 203, 220

Acacia seyal Delile
Fabaceae/Mimosaceae
Acacia boboensis Aubrev., *Acacia fistula* Schweinf., *Acacia stenocarpa* A.Rich.
♦ gum arabic
♦ Weed, Quarantine Weed, Introduced
♦ 76, 87, 88, 203, 220, 221, 228
♦ arid, cultivated, herbal. Origin: Africa.

Acacia seyal Delile var. *fistula* (Schweinf.) Oliv.
Fabaceae/Mimosaceae
♦ Quarantine Weed
♦ 220
♦ arid.

Acacia seyal Delile var. *seyal*
Fabaceae/Mimosaceae
♦ Quarantine Weed
♦ 220
♦ arid.

Acacia sieberiana DC.
Fabaceae/Mimosaceae
♦ paperbark thorn
♦ Weed, Sleeper Weed, Quarantine Weed, Naturalised
♦ 76, 86, 88, 203, 220
♦ cultivated. Origin: Africa.

Acacia sieberiana DC. var. *sieberiana*
Fabaceae/Mimosaceae
Acacia blommaertii De Wild., *Acacia nefasia sensu* Lebrun, *Acacia purpurascens* Vatke, *Acacia sieberana* DC. fo. *eusiberana* Roberty, *Acacia sieberana* DC. var. *sing* (Guill. & Perr.) Roberty, *Acacia sing* Guill. & Perr., *Acacia verrugera* Schweinf., *Acacia verrugera* Schweinf. var. *africana* Defl., *Acacia verrugera* Schweinf. var. *subinermis* A.Chev.
♦ mgunga, mgumga duu
♦ Quarantine Weed
♦ 220
♦ arid.

Acacia sieberiana DC. var. *villosa* A.Chev.
Fabaceae/Mimosaceae
♦ Quarantine Weed
♦ 220

Acacia sieberiana DC. var. *woodii* (Burtt Davy) Keay & Brenan
Fabaceae/Mimosaceae
Acacia amboensis Schinz, *Acacia lasiopetala sensu* Bartt Davy, *Acacia nefasia* (Hochst. ex A.Rich.) Schweinf. var. *vermoesenii* Keay & Brenan, *Inga nefasia* Hochst. ex A.Rich.
♦ paperbark thorn, achara, cherin, chiak, eyesura, leldet, mgunga kuu, ol asiti, ol debesi, seep, flat topped thorn, Natal camelthorn, whitethorn, Wood's acacia
♦ Quarantine Weed, Native Weed
♦ 121, 220
♦ T, arid, cultivated, toxic. Origin: southern Africa.

Acacia sinuata (Lour.) Merr.
Fabaceae/Mimosaceae
Acacia concinna (Willd.) DC. (see), *Acacia rugata* (Lam.) Buch.-Ham. ex Voigt (see), *Mimosa concinna* Willd., *Mimosa rugata* Lam., *Mimosa sinuata* Lour., *Senegalia rugata* (Lam.) Britton & Rose
♦ Naturalised
♦ 86, 287
♦ Origin: China.

Acacia smallii Isely
Fabaceae/Mimosaceae
= *Acacia farnesiana* (L.) Willd.
♦ Weed, Quarantine Weed
♦ 88, 203, 220
♦ herbal.

Acacia somalensis Vatke
Fabaceae/Mimosaceae
♦ Weed, Quarantine Weed
♦ 76, 88, 203, 220

Acacia sophorae (Labill.) R.Br.
Fabaceae/Mimosaceae

Acacia longifolia (Andrews) Willd. var. *sophorae* (Labill.) F.Muell., *Mimosa sophorae* Labill., *Racosperma sophorae* (Labill.) Mart. (see)
♦ coast wattle, coastal wattle
♦ Weed, Naturalised, Native Weed, Garden Escape, Environmental Weed
♦ 7, 15, 72, 86, 88, 246, 280
♦ S, arid, cultivated. Origin: Australia.

Acacia spadicigera Schltdl. & Cham.
Fabaceae/Mimosaceae
= *Acacia cornigera* (L.) Willd.
♦ Weed, Quarantine Weed
♦ 76, 88, 203

Acacia sphaerocephala Schlecht. & Cham.
Fabaceae/Mimosaceae
Acacia dolichocephala Saff. (see), *Acacia veracruzensis* Schenck
♦ bee wattle
♦ Weed, Quarantine Weed, Naturalised
♦ 76, 88, 101, 179, 203, 220
♦ Origin: North America.

Acacia spirocarpa Hochst. ex A.Rich.
Fabaceae/Mimosaceae
♦ Weed, Quarantine Weed
♦ 76, 88, 203, 220

Acacia stricta (Andrews) Willd.
Fabaceae/Mimosaceae
Racosperma strictum (Andrews) Mart. (see)
♦ hop wattle
♦ Casual Alien
♦ 280
♦ cultivated.

Acacia stuhlmannii Taub.
Fabaceae/Mimosaceae
♦ Weed, Quarantine Weed
♦ 76, 87, 88, 203, 220
♦ arid.

Acacia subalata Vatke
Fabaceae/Mimosaceae
♦ Weed
♦ 87, 88

Acacia sulcata R.Br.
Fabaceae/Mimosaceae
♦ Weed, Quarantine Weed, Naturalised
♦ 86, 88
♦ cultivated.

Acacia suma Broun & Massey
Fabaceae/Mimosaceae
= *Acacia polyacantha* Willd. ssp. *campylacantha* (Hochst. ex A.Rich.) Brenan
♦ Weed, Quarantine Weed
♦ 76, 88, 203, 220

Acacia sundra (Roxb.) DC.
Fabaceae/Mimosaceae
♦ Weed, Quarantine Weed
♦ 76, 88, 203, 220

Acacia sutherlandii (F.Muell.) F.Muell.
Fabaceae/Mimosaceae
Acacia melaleucoides Bailey, *Albizia sutherlandii* F.Muell.

♦ Introduced
♦ 228
♦ arid, cultivated. Origin: Australia.

Acacia swazica Burtt Davy
Fabaceae/Mimosaceae
♦ Weed, Quarantine Weed
♦ 76, 88, 203, 220
♦ cultivated. Origin: southern Africa.

Acacia tanganyikensis Brenan
Fabaceae/Mimosaceae
♦ mgunga
♦ Weed, Quarantine Weed
♦ 76, 88, 203, 220
♦ cultivated.

Acacia taylori Brenan & Exell
Fabaceae/Mimosaceae
♦ Weed, Quarantine Weed
♦ 76, 88, 203, 220

Acacia tenuifolia (L.) Willd.
Fabaceae/Mimosaceae
Acacia paniculata Willd. (see)
♦ Weed, Quarantine Weed
♦ 14, 76, 88, 203

Acacia tenuispina Verd.
Fabaceae/Mimosaceae
♦ fyndoring
♦ Weed, Quarantine Weed, Native Weed
♦ 63, 76, 88, 121, 203, 220
♦ pS. Origin: southern Africa.

Acacia tephrodermis J.P.M.Brenan
Fabaceae/Mimosaceae
♦ Weed, Quarantine Weed
♦ 76, 88, 203, 220

Acacia terminalis (Salisb.) J.F.Macbr.
Fabaceae/Mimosaceae
♦ elata wattle, sunshine wattle, cedar wattle, mountain hickory, peppermint tree wattle
♦ Weed
♦ 88, 121, 158
♦ T, arid, cultivated, herbal. Origin: Australia.

Acacia thomasii Harms
Fabaceae/Mimosaceae
♦ Weed, Quarantine Weed
♦ 76, 88, 203, 220

Acacia tomentosa Willd.
Fabaceae/Mimosaceae
Acacia chrysocoma Miq, *Acacia tomentosa* Willd. var. *chrysocoma* (Miq) Backer
♦ Weed, Quarantine Weed
♦ 76, 88, 203, 220
♦ arid.

Acacia torrei Brenan
Fabaceae/Mimosaceae
♦ Weed, Quarantine Weed
♦ 76, 88, 203, 220

Acacia tortilis (Forssk.) Hayne
Fabaceae/Mimosaceae
Mimosa tortilis Forssk.
♦ umbrella thorn
♦ Weed, Quarantine Weed, Introduced
♦ 76, 88, 203, 220, 221, 228
♦ arid, cultivated. Origin: North Africa and Near East.

**Acacia tortilis (Forssk.) Hayne ssp.
heteracantha (Burch.) Brenan**
Fabaceae/Mimosaceae
Acacia heteracantha Burch. (see), *Acacia
litakuensis* Burch., *Acacia maras* Engl.,
Acacia spirocarpoides Engl.
♦ curly pod acacia, umbrella thorn,
haak en steek
♦ Weed, Native Weed
♦ 10, 63, 121
♦ S/T. Origin: southern Africa.

**Acacia tortilis (Forssk.) Hayne ssp.
raddiana (Savi) Brenan**
Fabaceae/Mimosaceae
Acacia fasciculata Guill. & Perr.,
Acacia raddiana Savi (see), *Acacia
tortilis* (Forssk.) Hayne var. *pubescens*
Aylmer, *Acacia tortilis* (Forssk.) Hayne
var. *pubescens* A.Chev., *Acacia tortilis*
(Forssk.) Hayne var. *lenticellosa* Chiov.,
Acacia tortilis (Forssk.) Hayne fo.
raddiana (Savi) Roberty
♦ mu gaa, dadach, faux gommier,
kure, mgunga, samr, siyal, umbrella
thorn
♦ Introduced
♦ 228

Acacia tortuosa (L.) Willd.
Fabaceae/Mimosaceae
♦ poponax
♦ Weed, Quarantine Weed
♦ 76, 87, 88, 203, 220
♦ herbal.

Acacia trachycarpa E.Pritz.
Fabaceae/Mimosaceae
♦ sweet scented miniritchie
♦ Introduced
♦ 228
♦ arid, cultivated.

Acacia tumida Benth.
Fabaceae/Mimosaceae
♦ pindan wattle
♦ Introduced
♦ 228
♦ arid, cultivated.

Acacia turnbulliana Brenan
Fabaceae/Mimosaceae
♦ Weed, Quarantine Weed
♦ 76, 88, 203, 220

Acacia unicinella Desf.
Fabaceae/Mimosaceae
♦ Quarantine Weed
♦ 220

Acacia velutina Benth.
Fabaceae/Mimosaceae
♦ Weed, Quarantine Weed
♦ 76, 88, 203, 220

Acacia venosa Hochst. ex Benth.
Fabaceae/Mimosaceae
♦ Weed, Quarantine Weed
♦ 76, 88, 203

Acacia verticillata (L'Hér.) Willd.
Fabaceae/Mimosaceae
Racosperma verticillatum (L'Hér.) Mart.
(see)
♦ prickly Moses
♦ Weed, Naturalised, Introduced

♦ 15, 39, 88, 101, 228, 280
♦ S/T, cultivated, herbal, toxic. Origin:
Australia.

Acacia vestita Ker Gawl.
Fabaceae/Mimosaceae
♦ hairy wattle
♦ Naturalised
♦ 86
♦ cultivated. Origin: Australia.

Acacia victoriae Benth.
Fabaceae/Mimosaceae
♦ bardi bush, bramble wattle, elegant
wattle, gundabluey, prickly acacia,
prickly wattle
♦ Weed, Introduced
♦ 228
♦ arid, cultivated, herbal. Origin:
Australia.

Acacia villosa (Sw.) Willd.
Fabaceae/Mimosaceae
♦ Quarantine Weed
♦ 220
♦ herbal.

Acacia visco Lorentz ex Griseb.
Fabaceae/Mimosaceae
♦ Quarantine Weed, Naturalised,
Introduced
♦ 38, 220, 241, 300
♦ T, cultivated. Origin: South America.

Acacia walwalensis Gilliland
Fabaceae/Mimosaceae
♦ Weed, Quarantine Weed
♦ 76, 88, 203, 220

Acacia welwitschii Oliv.
Fabaceae/Mimosaceae
♦ Weed, Quarantine Weed
♦ 76, 88, 203, 220

**Acacia welwitschii Oliv. ssp. delagoensis
(Harms) J.Ross & Brenan**
Fabaceae/Mimosaceae
♦ delagoa thorn
♦ Native Weed
♦ 121
♦ T. Origin: southern Africa.

Acacia wightii Wight & Arn.
Fabaceae/Mimosaceae
♦ Weed, Quarantine Weed
♦ 76, 88, 203, 220

Acacia willardiana Rose
Fabaceae/Mimosaceae
♦ Quarantine Weed
♦ 220

Acacia woodii Burtt Davy
Fabaceae/Mimosaceae
♦ camel thorn, paperbark
♦ Weed, Quarantine Weed
♦ 76, 88, 203

Acacia wrightii Benth. ex A.Gray
Fabaceae/Mimosaceae
Acacia greggii A.Gray var. *wrightii*
(Benth.) Isely, *Senegalia wrightii* (Benth.)
Britton & Rose
♦ Weed, Quarantine Weed
♦ 76, 88, 203, 220
♦ arid. Origin: Americas.

Acacia xanthophloea Benth.
Fabaceae/Mimosaceae

♦ fever tree
♦ Weed, Quarantine Weed, Introduced
♦ 76, 88, 203, 220, 228
♦ arid, cultivated. Origin: southern
Africa.

Acacia zanzibarica (S.Moore) Taub.
Fabaceae/Mimosaceae
Pithecellobium zanzibaricum S.Moore
♦ Weed, Quarantine Weed
♦ 76, 87, 88, 203,
220

Acacia zizyphispina Chiov.
Fabaceae/Mimosaceae
♦ Weed, Quarantine Weed
♦ 76, 88, 203, 220

Acaena agnipila Gand.
Rosaceae
Acaena ovina A.Cunn. (see)
♦ Australian sheep's burr, sheep's burr
♦ Weed, Naturalised, Native Weed,
Garden Escape, Environmental Weed
♦ 7, 86, 165, 181, 246, 269, 280
♦ pH, cultivated. Origin: Australia,
New Zealand.

**Acaena agnipila Gand. var. aequispina
Orchard**
Rosaceae
♦ sheep's burr
♦ Weed, Naturalised
♦ 15, 280

**Acaena agnipila Gand. var. protenta
Orchard**
Rosaceae
♦ Naturalised
♦ 280

**Acaena agnipila Gand. var. tenuispica
(Bitter) Orchard**
Rosaceae
♦ Naturalised
♦ 280

**Acaena anserinifolia (J.R. & G.Forst.)
Druce**
Rosaceae
Acaena montana Hook.f., *Acaena
sanguisorbae* L.f. (see) [*Acaena novae-
zelandiae* Kirk has been misapplied to]
♦ biddy biddy
♦ Weed, Noxious Weed, Introduced
♦ 35, 87, 88, 161, 228
♦ arid, cultivated. Origin: Australia,
New Zealand.

Acaena argentea Ruiz & Pav.
Rosaceae
♦ Weed
♦ 87, 88
♦ cultivated.

Acaena arvensis Poepp. & Endl.
Rosaceae
♦ Weed
♦ 87, 88

Acaena caesiglauca (Bitter) Bergmans
Rosaceae
♦ piripiri
♦ Weed, Quarantine Weed
♦ 76, 88, 220
♦ cultivated.

Acaena ciliata Forssk.
Rosaceae
♦ Weed
♦ 87, 88

Acaena echinata Nees
Rosaceae
♦ Naturalised, Casual Alien
♦ 7, 86, 280
♦ cultivated. Origin: Australia.

Acaena fallax Müll.Arg.
Rosaceae
♦ Weed
♦ 87, 88

Acaena fissistipula Bitter
Rosaceae
♦ Weed, Quarantine Weed
♦ 76, 88, 220
♦ cultivated.

Acaena hispida Burm.f.
Rosaceae
♦ Weed
♦ 87, 88

Acaena indica L.
Rosaceae
♦ Weed
♦ 87, 88

Acaena macrostachya Jacq.
Rosaceae
♦ Weed
♦ 87, 88

Acaena magellanica (Lam.) Vahl
Rosaceae
♦ acaena magallánica
♦ Weed, Quarantine Weed
♦ 76, 88, 220, 237, 295
♦ cultivated. Origin: South America.

Acaena neomexicana Müll.Arg.
Rosaceae
♦ Weed
♦ 87, 88

Acaena novae-zelandiae Kirk
Rosaceae
Acaena novae-zelandica Kirk
♦ bidy bidy, pirri pirri burr, New
Zealand acaena, biddy biddy, red
bidibidi
♦ Weed, Noxious Weed, Naturalised,
Native Weed, Environmental Weed
♦ 7, 15, 35, 40, 86, 88, 101, 146, 165, 181,
229, 229, 269
♦ S, cultivated. Origin: Australia, New
Zealand.

Acaena ostryaefolia Riddell.
Rosaceae
♦ Weed
♦ 87, 88

Acaena ovalifolia Ruiz & Pav.
Rosaceae
♦ abrojo
♦ Weed, Quarantine Weed
♦ 76, 87, 88, 220, 237, 295
♦ cultivated.

Acaena ovina A.Cunn.
Rosaceae
= *Acaena agnipila* Gand.
♦ sheep's burr, Australian sheep's burr

♦ Weed, Quarantine Weed, Native
Weed
♦ 87, 88, 269
♦ pH, cultivated. Origin: eastern
Australia, New Zealand.

Acaena pallida (T.Kirk) J.W.Dawson
Rosaceae
♦ pale biddy biddy
♦ Weed, Noxious Weed
♦ 35, 88, 229

Acaena pinnatifida Ruiz & Pav.
Rosaceae
♦ sheepbur
♦ Weed, Quarantine Weed
♦ 76, 87, 88, 220, 237, 295
♦ cultivated, herbal.

Acaena poiretti Spreng.
Rosaceae
♦ Weed
♦ 87, 88

Acaena rhomboidea Raf.
Rosaceae
♦ Weed
♦ 87, 88

Acaena saccaticupula Bitter
Rosaceae
♦ Weed, Quarantine Weed
♦ 76, 88, 220
♦ cultivated.

Acaena sanguisorbae L.f.
Rosaceae
= *Acaena anserinifolia* (J.R. & G.Forst.)
Druce
♦ piripiri
♦ Weed
♦ 87, 88, 218
♦ herbal.

Acaena schiedeana Schlecht.
Rosaceae
♦ Weed
♦ 87, 88

Acaena segetalis Müll.Arg.
Rosaceae
♦ Weed
♦ 87, 88

Acaena setosa A.Rich.
Rosaceae
♦ Weed
♦ 87, 88

Acaena splendens Hook. & Arn.
Rosaceae
♦ cadillo de la sierra
♦ Weed
♦ 87, 88, 237, 295
♦ herbal.

Acaena virginica L.
Rosaceae
♦ Weed
♦ 87, 88

Acaena wilkesiana Müll.Arg.
Rosaceae
♦ Weed
♦ 87, 88

Acalypha L. spp.
Euphorbiaceae
♦ copperleaf, three seeded mercury

♦ Quarantine Weed
♦ 220

Acalypha alopecuroides Jacq.
Euphorbiaceae
♦ algodoncito, foxtail copperleaf
♦ Weed, Quarantine Weed
♦ 76, 87, 88, 153, 203, 220
♦ herbal.

Acalypha amentacea Roxb.
Euphorbiaceae
♦ match me if you can, copperleaf
♦ Weed, Naturalised
♦ 101, 161
♦ cultivated, toxic.

**Acalypha amentacea Roxb. ssp.
wilkesiana (Müll.Arg.) Fosberg**
Euphorbiaceae
Acalypha wilkesiana Müll.Arg. (see)
♦ Wilkes' acalypha, copperleaf
♦ Weed, Naturalised, Introduced
♦ 101, 179, 230
♦ toxic.

Acalypha arvensis Poepp. & Endl.
Euphorbiaceae
♦ acalypha
♦ Weed, Naturalised
♦ 28, 101, 206, 243

Acalypha australis L.
Euphorbiaceae
♦ Australian acalypha
♦ Weed, Naturalised
♦ 86, 88, 98, 101, 203, 204, 235, 263, 273,
275, 286, 297
♦ aH, cultivated, herbal. Origin: Asia.

Acalypha boehmerioides Miq.
Euphorbiaceae
♦ Weed
♦ 276
♦ aH.

Acalypha brachystachya Hornem.
Euphorbiaceae
♦ lobed bract threeseed mercury
♦ Weed
♦ 297

Acalypha caperonioides Baill.
Euphorbiaceae
♦ false nettle bush
♦ Native Weed
♦ 121
♦ a/pH. Origin: southern Africa.

**Acalypha cardiophylla Merr. var.
ponapensis (Kaneh. & Hatus.)
F.R.Fosberg**
Euphorbiaceae
Alcalypha ponapensis Kaneh. & Hatus.
♦ Introduced
♦ 230
♦ S.

Acalypha ciliata Fosk.
Euphorbiaceae
♦ Weed, Quarantine Weed
♦ 76, 88, 203, 220

Acalypha communis Müll.Arg.
Euphorbiaceae
♦ Weed
♦ 255
♦ aH. Origin: Americas.

Acalypha crenata **A.Rich.**
Euphorbiaceae
♦ Weed
♦ 240
♦ aH.

Acalypha ecklonii **Baill.**
Euphorbiaceae
♦ acalypha
♦ Weed, Native Weed
♦ 88, 121, 158
♦ a/pH. Origin: south-eastern Africa.

Acalypha fallax **Müll.Arg.**
Euphorbiaceae
♦ Weed, Quarantine Weed
♦ 76, 88, 203, 220

Acalypha gracilens **A.Gray**
Euphorbiaceae
♦ slender threeseed mercury
♦ Naturalised
♦ 287
♦ herbal.

Acalypha guatemalensis **Pax & Hoffm.**
Euphorbiaceae
♦ Weed
♦ 157
♦ pH.

Acalypha havanensis **Müll.Arg.**
Euphorbiaceae
♦ Weed
♦ 14

Acalypha hispida **Burm.f.**
Euphorbiaceae
♦ bristly copperleaf, chenille plant, berica, rabo de gato, red hot cat's tail
♦ Weed, Naturalised, Introduced, Casual Alien
♦ 101, 161, 194, 230, 247, 261
♦ S, cultivated, herbal, toxic.

Acalypha indica **L.**
Euphorbiaceae
♦ Indian copperleaf, haang krarok daeng
♦ Weed, Naturalised
♦ 32, 39, 88, 101, 239, 242, 287
♦ aH, arid, cultivated, herbal, toxic. Origin: Malaysia, Kenya, Indonesia, India, South Africa, Philippines, Madagascar, Australia.

Acalypha infesta **Poepp. & Endl.**
Euphorbiaceae
♦ Weed
♦ 199

Acalypha lanceolata **Willd.**
Euphorbiaceae
♦ ogo mûmû
♦ Naturalised
♦ 32
♦ herbal.

Acalypha neomexicana **Müll.Arg.**
Euphorbiaceae
♦ New Mexico copperleaf
♦ Weed
♦ 88, 161, 218
♦ herbal.

Acalypha ostryifolia **Riddell**
Euphorbiaceae
♦ hophornbeam copperleaf, pineland threeseed mercury
♦ Weed
♦ 88, 161, 207, 218, 261
♦ herbal. Origin: West Indies.

Acalypha petiolaris **Hochst.**
Euphorbiaceae
Acalypha senensis Klotzsch
♦ Native Weed
♦ 121
♦ pH. Origin: southern Africa.

Acalypha poiretii **Spreng.**
Euphorbiaceae
♦ Poiret's copperleaf
♦ Weed
♦ 261
♦ Origin: Old World Tropics.

Acalypha rhomboidea **Raf.**
Euphorbiaceae
♦ rhombic copperleaf, Virginia threeseed mercury
♦ Weed
♦ 88, 161, 218
♦ herbal.

Acalypha segetalis **Müll.Arg.**
Euphorbiaceae
♦ threeseed mercury
♦ Weed, Quarantine Weed, Native Weed
♦ 50, 76, 88, 121, 203, 220
♦ a/bH. Origin: Africa.

Acalypha setosa **A.Rich.**
Euphorbiaceae
♦ Cuban copperleaf
♦ Weed
♦ 261

Acalypha virginica **L.**
Euphorbiaceae
♦ Virginia copperleaf, three sided mercury, wax balls, copperleaf, mercury weed
♦ Weed
♦ 39, 88, 161, 210, 211, 218, 243, 253
♦ aH, herbal, toxic.

Acalypha wilkesiana **Müll.Arg.**
Euphorbiaceae
= *Acalypha amentacea* ssp. *wilkesiana* (Müll.Arg.) Fosberg
♦ Jacob's coat
♦ Naturalised, Introduced
♦ 134, 228, 261
♦ arid, cultivated, herbal. Origin: Oceania.

Acanthocereus **(Engelm. ex Berg.) Britt. & Rose spp.**
Cactaceae
Dendrocereus Britt. & Rose spp. (see), *Monvillea* Britt. & Rose spp. (see), *Pseudocanthocereus* Ritt. spp. (see)
♦ triangle cactus
♦ Weed, Quarantine Weed
♦ 76, 88, 203, 220

Acanthocereus pentagonus **(L.) Britton & Rose**
Cactaceae
= *Acanthocereus tetragonus* (L.) Humm. (NoR)
♦ Weed, Naturalised

♦ 86, 87, 88, 98, 203
♦ cultivated.

Acanthopanax lasiogyne **Harms**
Araliaceae
♦ Quarantine Weed
♦ 220

Acanthopanax sciadophylloides **Franch. & Sav.**
Araliaceae
♦ koshiabura
♦ Quarantine Weed
♦ 220
♦ cultivated.

Acanthophora spicifera **(Vahl) Børgesen**
Rhodomelaceae
♦ soft spineweed, red algae, spiny seaweed
♦ Weed
♦ 197
♦ algae.

Acanthosicyos horridus **Welw. ex Hook.f.**
Cucurbitaceae
Acanthosicyos horrida Welw.
♦ nara, botterpitte, naras
♦ Introduced
♦ 228
♦ arid.

Acanthosicyos naudinianus **(Sond.) C.Jeffrey**
Cucurbitaceae
Citrullus naudinianus (Sond.) Hook.f., *Colocynthis naudinianus* (Sond.) Kuntze, *Cucumis naudinianus* Sond.
♦ cha, gemsbok komkommer, wild melon, herero cucumber, ruputui, lungwatanga, thonga, sirakarana, chirakaraka, mokapana
♦ Weed, Quarantine Weed
♦ 76, 88, 220
♦ pH, cultivated. Origin: southern Africa.

Acanthospermum australe **(Loefl.) Kuntze**
Asteraceae
Acanthospermum brasilum Schrank (see), *Melampodium australe* Loefl.
♦ Paraguay starbur, creeping starbur, Australian starbur, Brazilian starbur, eight seeded prostrate starbur, Paraguay burr, prostrate starbur, sheepbur, amor de negro, carrapicho rasteiro
♦ Weed, Naturalised
♦ 50, 51, 86, 87, 88, 98, 101, 121, 158, 179, 203, 218, 237, 245, 255, 270, 275, 295, 297
♦ pH, cultivated, herbal. Origin: tropical America.

Acanthospermum brasilum **Schrank**
Asteraceae
= *Acanthospermum australe* (Loefl.) Kuntze
♦ Brazilian starbur, five seeded starbur
♦ Weed
♦ 51, 87, 88, 158
♦ Origin: South America.

Acanthospermum glabratum **(DC.) Willd**
Asteraceae
♦ creeping starbur, five seeded

prostrate starbur, prostrate starbur, sheepbur
♦ Weed
♦ 121
♦ aH. Origin: South America.

Acanthospermum hispidum **DC.**
Asteraceae
♦ bristly starbur, upright starbur, siBama yauli, goat's head, Texas cockspur, kati, torito, starburr
♦ Weed, Noxious Weed, Quarantine Weed, Naturalised, Environmental Weed
♦ 7, 39, 50, 51, 53, 55, 66, 76, 84, 86, 87, 88, 93, 98, 121, 147, 157, 158, 161, 170, 179, 203, 205, 216, 217, 218, 236, 237, 240, 245, 255, 261, 270, 287, 295
♦ aH, arid, cultivated, herbal, toxic. Origin: tropical America.

Acanthospermum humile **(Sw.) DC.**
Asteraceae
♦ low starbur
♦ Naturalised
♦ 101
♦ herbal.

Acanthospermum microcarpum **B.L.Rob.**
Asteraceae
♦ Naturalised
♦ 257

Acanthospermum xanthioides **DC.**
Asteraceae
♦ southern starbur
♦ Naturalised
♦ 101

Acanthostyles buniifolius **(Hook. & Arn.) R.M.King & H.Rob.**
Asteraceae
Eupatorium bunifolium H.B.K.
♦ Weed, Quarantine Weed
♦ 76, 87, 88, 203, 220
♦ Origin: South America.

Acanthus balcanicus **Heywood & Richards.**
Acanthaceae
♦ long leaved bear's breech
♦ Weed
♦ 272
♦ cultivated, herbal.

Acanthus caudatus **Lindau**
Acanthaceae
♦ Weed
♦ 87, 88

Acanthus ebracteatus **Vahl**
Acanthaceae
♦ acanthus
♦ Weed
♦ 87, 88, 209, 262
♦ S, herbal.

Acanthus ilicifolius **L.**
Acanthaceae
♦ Weed
♦ 13, 87, 88
♦ S, cultivated, herbal. Origin: Australia.

Acanthus mollis **L.**
Acanthaceae
♦ bear's breeches, artist's acanthus

♦ Weed, Naturalised, Garden Escape, Environmental Weed
♦ 15, 40, 70, 72, 86, 87, 88, 165, 176, 198, 280, 290
♦ pH, cultivated, herbal. Origin: south-west Europe.

Acanthus montanus **T.Anders.**
Acanthaceae
♦ mbaka
♦ Weed, Quarantine Weed
♦ 76, 87, 88, 203, 220
♦ cultivated.

Acanthus pubescens **(Thoms. ex Oliv.) Engl.**
Acanthaceae
♦ prickly acanthus, lurodu
♦ Weed, Quarantine Weed
♦ 76, 88, 121, 203, 220
♦ pS, cultivated. Origin: Eurasia.

Acanthus spinosus **L.**
Acanthaceae
♦ spiny bear's breech, spine acanthus, bear's breech
♦ Weed
♦ 272
♦ cultivated, herbal.

Acanthus syriacus **Boiss.**
Acanthaceae
♦ Syrian acanthus
♦ Weed
♦ 87, 88
♦ Origin: Middle East.

Acer barbatum **Michx.**
Aceraceae
♦ Florida maple, southern sugar maple
♦ Weed
♦ 87, 88, 218
♦ cultivated.

Acer buergerianum **Miq.**
Aceraceae
♦ trident maple, kolmihammas-vaahtera
♦ Naturalised
♦ 287
♦ cultivated, herbal.

Acer campestre **L.**
Aceraceae
♦ hedge maple, field maple, erable champêtre
♦ Weed, Naturalised
♦ 80, 101
♦ T, cultivated, herbal. Origin: Europe.

Acer circinatum **Pursh**
Aceraceae
♦ vine maple
♦ Weed
♦ 87, 88, 218
♦ S/T, cultivated, herbal. Origin: North America.

Acer ginnala **Maxim.**
Aceraceae
Acer tataricum L. ssp. *ginnala* (Maxim.) Wesm.
♦ Amur maple, Mongolian vaahtera
♦ Weed, Naturalised, Introduced, Garden Escape, Environmental Weed, Cultivation Escape

♦ 42, 80, 88, 101, 133, 142, 151, 195, 222, 224
♦ T, cultivated, herbal. Origin: east Asia, China, Japan, Manchuria.

Acer macrophyllum **Pursh**
Aceraceae
♦ bigleaf maple, Oregon maple
♦ Weed
♦ 87, 88, 218
♦ T, cultivated, herbal. Origin: North America.

Acer negundo **L.**
Aceraceae
♦ box elder, ashleaf maple, box elder maple, Manitoba maple
♦ Weed, Naturalised, Introduced, Garden Escape, Environmental Weed, Cultivation Escape
♦ 15, 42, 54, 72, 80, 86, 87, 88, 98, 104, 133, 155, 198, 203, 211, 218, 228, 243, 280, 289
♦ T, arid, cultivated, herbal. Origin: North America.

Acer negundo **L. var.** *negundo*
Aceraceae
Negundo aceroides (L.) Moench
♦ box elder
♦ Weed, Naturalised
♦ 132, 280

Acer palmatum **Thunb.**
Aceraceae
♦ Japanese maple, box elder maple, ash leaved maple
♦ Weed, Naturalised, Casual Alien
♦ 15, 54, 86, 88, 101, 195, 198, 280
♦ T, cultivated, herbal. Origin: east Asia.

Acer pensylvanicum **L.**
Aceraceae
♦ striped maple, moosewood
♦ Weed
♦ 87, 88, 218
♦ T, cultivated, herbal. Origin: North America.

Acer platanoides **L.**
Aceraceae
♦ Norway maple
♦ Weed, Noxious Weed, Naturalised, Introduced, Garden Escape, Environmental Weed
♦ 4, 45, 80, 87, 88, 101, 104, 133, 137, 142, 151, 195, 211, 218, 222, 224, 243
♦ T, cultivated, herbal. Origin: Eurasia.

Acer pseudoplatanus **L.**
Aceraceae
Acer pseudo-platanus L.
♦ sycamore, great maple, planetree maple, erable sycamore
♦ Weed, Naturalised, Garden Escape, Environmental Weed, Cultivation Escape
♦ 15, 20, 40, 42, 70, 72, 80, 86, 88, 98, 101, 133, 151, 152, 155, 165, 198, 224, 225, 246, 280, 289, 296
♦ T, cultivated, herbal. Origin: south and central Europe and the Caucasus.

Acer rubrum L.
Aceraceae
♦ red maple
♦ Weed
♦ 39, 87, 88, 154, 161, 211, 218, 243
♦ T, cultivated, herbal, toxic. Origin:
North America.

Acer saccharinum L.
Aceraceae
Acer dasycarpum Ehrh., *Argentacer*
saccharinum (L.) Small
♦ silver maple
♦ Weed
♦ 88, 218
♦ T, cultivated, herbal. Origin: North
America.

Acer saccharum Marsh
Aceraceae
Acer saccharinum Wangenh., *Acer*
saccharinum non L.
♦ sugar maple
♦ Weed
♦ 87, 88, 218
♦ T, cultivated, herbal. Origin: North
America.

Acer spicatum Lam.
Aceraceae
♦ mountain maple
♦ Weed
♦ 87, 88, 218
♦ T, cultivated, herbal. Origin: North
America.

Acer tataricum L.
Aceraceae
♦ tataarivaahtera, Tatarian maple,
javor tatársky
♦ Naturalised, Cultivation Escape
♦ 42, 101
♦ T, cultivated, herbal. Origin: Eurasia.

Aceras anthropophorum L.
Orchidaceae
♦ man orchid
♦ Weed
♦ 70
♦ herbal.

Acetosa sagittata (Thunb.)
L.A.S.Johnson & B.G.Briggs
Polygonaceae
Rumex sagittatus Thunb. (see)
♦ rambling dock, climbing sorrel,
turkey rhubarb
♦ Weed, Noxious Weed, Naturalised,
Garden Escape, Environmental Weed
♦ 86, 98, 176, 198, 203, 269, 289, 296
♦ pC, cultivated. Origin: South Africa.

Acetosa vesicaria (L.) Á.Löve
Polygonaceae
Rumex vesicarius L. (see)
♦ bladder dock, wild hops, rosy dock,
ruby dock
♦ Weed, Naturalised, Garden Escape,
Environmental Weed
♦ 86, 93, 98, 198, 203, 205, 251, 269, 290
♦ aH, cultivated, toxic. Origin:
Mediterranean.

Acetosella vulgaris (Koch) Fourr.
Polygonaceae
= *Rumex acetosella* L. ssp. *acetosella*

(NoR)
♦ sheep sorrel
♦ Weed, Naturalised, Environmental
Weed
♦ 20, 86, 87, 88, 176, 198, 289
♦ pH, cultivated, herbal. Origin:
Eurasia.

Achatocarpus nigricans Triana
Achatocarpaceae/Phytolaccaceae
♦ Introduced
♦ 228
♦ arid, herbal.

Achatocarpus praecox Griseb.
Achatocarpaceae/Phytolaccaceae
♦ Introduced
♦ 228
♦ S/T, arid.

Achetaria guianensis Pennell
Scrophulariaceae
♦ Weed
♦ 87, 88

Achillea L. spp.
Asteraceae
♦ yarrow
♦ Weed, Naturalised
♦ 198, 272
♦ herbal.

Achillea ageratum L.
Asteraceae
Achillea decolorans Schrad., *Achillea*
serrata hort.
♦ sweet yarrow
♦ Weed, Naturalised
♦ 70, 101
♦ pH, cultivated, herbal. Origin:
Europe.

Achillea alpina L.
Asteraceae
Achillea sibirica Ledeb.
♦ nokogirisou
♦ Weed
♦ 275, 286, 297

Achillea biebersteinii Afan.
Asteraceae
Achillea micrantha Willd. *nom. illeg.*
(see)
♦ Weed
♦ 272

Achillea borealis Bong.
Asteraceae
= *Achillea millefolium* L. var. *borealis*
(Bong.) Farw. (NoR)
♦ Weed
♦ 87, 88
♦ herbal.

Achillea collina Becker ex Rchb.
Asteraceae
Achillea millefolium L. ssp. *collina*
(Becker) Weiss
♦ mountain yarrow
♦ Weed
♦ 272
♦ cultivated, herbal.

Achillea conyzoides L.
Asteraceae
♦ Naturalised
♦ 241

Achillea distans Waldst. & Kit. ex Willd.
Asteraceae
♦ tansyleaf milfoil, Alps yarrow,
ahokärsämö
♦ Weed, Sleeper Weed, Naturalised,
Garden Escape, Environmental Weed,
Casual Alien
♦ 42, 86, 98, 101, 155, 176, 203, 252, 290
♦ cultivated. Origin: Europe.

Achillea distans Waldst. & Kit. ex Willd.
ssp. *tanacetifolia* Janch.
Asteraceae
♦ tansyleaf milfoil
♦ Naturalised
♦ 86
♦ Origin: Europe.

Achillea filipendulina Lam.
Asteraceae
♦ fernleaf yarrow, cloth of gold,
angervokärsämö
♦ Naturalised, Casual Alien
♦ 42, 101, 280, 287
♦ cultivated, herbal. Origin: west Asia.

Achillea fragrantissima (Forssk.) Sch.Bip.
Asteraceae
♦ Weed
♦ 221
♦ arid, herbal.

Achillea lanulosa Nutt.
Asteraceae
= *Achillea millefolium* L. var. *occidentalis*
DC. (NoR)
♦ western yarrow
♦ Weed
♦ 23, 87, 88, 161, 212, 218
♦ cultivated, herbal, toxic.

Achillea ligustica All.
Asteraceae
♦ ligurian yarrow, yarrow
♦ Naturalised
♦ 101
♦ cultivated, herbal.

Achillea micrantha Willd. *nom. illeg.*
Asteraceae
= *Achillea biebersteinii* Afan.
♦ keltakärsämö
♦ Weed, Casual Alien
♦ 42, 87, 88

Achillea millefolium L.
Asteraceae
♦ common yarrow, yarrow, milfoil,
thousand leaf, bloodwort, sanguinary
♦ Weed, Noxious Weed, Naturalised,
Garden Escape, Environmental Weed
♦ 8, 15, 20, 23, 39, 49, 52, 70, 72, 86, 87,
88, 94, 97, 98, 101, 121, 136, 158, 161,
165, 174, 176, 180, 198, 203, 207, 210,
211, 218, 237, 241, 243, 247, 269, 272,
280, 286, 287, 289, 293, 295, 296, 297,
299, 300
♦ pH, arid, cultivated, herbal, toxic.
Origin: Eurasia.

Achillea millefolium L. var. *millefolium*
Asteraceae
♦ common yarrow
♦ Weed, Naturalised
♦ 80, 101

Achillea nobilis L.
Asteraceae
♦ noble yarrow, creamy yarrow
♦ Weed, Naturalised, Casual Alien
♦ 42, 80, 87, 88, 101, 272, 280
♦ cultivated, herbal.

Achillea ptarmica L.
Asteraceae
♦ sneezewort, adder's tongue, wildfire, sneezeweed
♦ Weed, Naturalised, Environmental Weed
♦ 39, 80, 86, 87, 88, 101, 155, 203, 272, 280, 300
♦ pH, cultivated, herbal, toxic. Origin: Europe.

Achillea salicifolia Besser
Asteraceae
♦ isokärsämö
♦ Naturalised
♦ 42

Achillea santolina L.
Asteraceae
♦ Weed
♦ 87, 88, 221, 243, 248
♦ pH, arid, promoted, herbal. Origin: east Asia.

Achillea stricta Schleich. ex Koch
Asteraceae
Achillea tanacetifolia var. *stricta* Koch, *Achillea taurica* M.Bieb.
♦ Naturalised
♦ 287

Achillea tomentosa L.
Asteraceae
♦ woolly yarrow, nukkakärsämö
♦ Weed, Naturalised, Casual Alien
♦ 42, 86, 98, 203
♦ cultivated, herbal. Origin: Europe.

Achlaena piptostachya Griseb.
Poaceae
♦ Weed
♦ 14
♦ G.

Achnatherum brachychaetum (Godr.) Barkworth
Poaceae
Nassella brachychaeta (Godr.) Barkworth (see), *Stipa brachychaeta* Godr. (see)
♦ punagrass, espartillo, short bristled needlegrass
♦ Weed, Quarantine Weed, Noxious Weed, Naturalised, Environmental Weed
♦ 35, 76, 86, 88, 101, 229, 290
♦ pG. Origin: South America.

Achnatherum calamagrostis (L.) P.Beauv.
Poaceae
Agrostis calamagrostis L., *Stipa calamagrostis* (L.) Wahlenb. (see)
♦ silver spike grass, speargrass, cannella argentea
♦ Environmental Weed
♦ 290
♦ G, cultivated, herbal.

Achnatherum caudatum (Trin.) Barkworth comb. nov. ined.
Poaceae

Stipa caudata Trin. (see)
♦ Chilean ricegrass, espartillo, broad kernel espartillo
♦ Quarantine Weed, Noxious weed, Naturalised, Environmental Weed
♦ 76, 86, 101, 176, 198, 232, 280, 290
♦ G. Origin: South America.

Achnatherum clandestinum (Hack.) Barkworth
Poaceae
Stipa clandestina Hack. (see)
♦ Mexican ricegrass
♦ Naturalised
♦ 101
♦ G.

Achnatherum splendens (Trin.) Nevski
Poaceae
Stipa splendens Trin., *Lasiagrostis splendens* Kunth
♦ Weed, Quarantine Weed
♦ 76, 275
♦ pG, arid, toxic. Origin: temperate Asia.

Achras zapota L. nom. illeg.
Sapotaceae
= *Manilkara zapota* (L.) van Royen.
♦ sapodilla
♦ Weed
♦ 22
♦ T, cultivated, herbal.

Achyrachaena mollis Schauer
Asteraceae
♦ blow wives
♦ Weed
♦ 161
♦ aH, cultivated, herbal.

Achyranthes aspera L.
Amaranthaceae
Achyranthes annua Dinter, *Achyranthes argentea* Lam., *Achyranthes aspera* L. var. *indica* L., *Achyranthes aspera* L. fo. *excelsa* Cavaco, *Achyranthes aspera* L. fo. *rubella* Suess., *Achyranthes aspera* L. fo. *robustiformis* Suess., *Achyranthes aspera* L. var. *nigro-olivacea* Suess., *Achyranthes aspera* L. var. *obtusifolia* (Lam.) Suess., *Achyranthes aspera* L. var. *argentea* (Lam.) C.B.Clarke, *Achyranthes bidentata sensu* Baker & Clarke, *Achyranthes fruticosa* Desf. var. *pubescens* Moq., *Achyranthes indica* (L.) Mill. (see), *Achyranthes obovata* Peter, *Achyranthes obtusifolia* Lam., *Achyranthes robusta* C.H.Wright, *Achyranthes sicula* (L.) All. (see), *Achyranthes virgata* Poir., *Centrostachys aspera* (L.) Standl., *Centrostachys indica* (L.) Standl.
♦ chafflower, dombo, tamatama, aerofai, lautafifi, talamoa fisi, devil's horsewhip, piripiri, sono ivi, rough chaff flower, lau tamatama, kongo lokosi
♦ Weed, Quarantine Weed, Noxious Weed, Naturalised, Introduced, Garden Escape, Environmental Weed
♦ 3, 6, 12, 14, 32, 50, 51, 66, 86, 87, 88, 101, 121, 157, 158, 209, 217, 220, 221, 228, 243, 257, 275, 276, 280, 283, 297

♦ pH, arid, cultivated, herbal. Origin: obscure.

Achyranthes aspera L. var. aspera
Amaranthaceae
♦ devil's horsewhip
♦ Weed, Naturalised
♦ 101, 179, 261

Achyranthes aspera L. var. pubescens (Moq.) Towns.
Amaranthaceae
♦ devil's horsewhip
♦ Weed, Naturalised
♦ 101, 179

Achyranthes bidentata Bl.
Amaranthaceae
♦ oxknee, nui xi, pig's knee
♦ Weed
♦ 39, 275, 297
♦ pH, arid, cultivated, herbal, toxic.

Achyranthes bidentata Bl. var. tomentosa Hara
Amaranthaceae
♦ Weed
♦ 286

Achyranthes fauriei Lev. & Van.
Amaranthaceae
♦ hinatainokozuchi
♦ Weed
♦ 87, 88
♦ herbal.

Achyranthes indica (L.) Mill.
Amaranthaceae
= *Achyranthes aspera* L.
♦ devil's horsewhip
♦ Weed
♦ 80, 88
♦ herbal.

Achyranthes japonica (Miq.) Nakai
Amaranthaceae
♦ Japanese chaff flower
♦ Weed, Naturalised
♦ 87, 88, 101, 204
♦ herbal. Origin: east Asia.

Achyranthes japonica (Miq.) Nakai var. hachijoensis Honda
Amaranthaceae
♦ Japanese chaff flower
♦ Naturalised
♦ 101

Achyranthes longifolia Makino
Amaranthaceae
♦ Weed
♦ 87, 88
♦ cultivated.

Achyranthes sicula (L.) All.
Amaranthaceae
= *Achyranthes aspera* L.
♦ Weed, Quarantine Weed
♦ 76, 88, 121, 203
♦ a/bH. Origin: North America.

Achyrocline satureioides (Lam.) DC.
Asteraceae
Achyrocline candicans (Kunth) DC.
♦ alecrim de parede
♦ Weed
♦ 87, 88, 245, 255
♦ Origin: Brazil.

Achyrocline vargasiana **DC.**
Asteraceae
♦ Weed
♦ 87, 88

Achyropappus anthemoides **H.B.K.**
Asteraceae
♦ Weed
♦ 199

Acicarpha spathulata **R.Br.**
Calyceraceae
Acanthosperma littorale Vell.
♦ espinho de roseta
♦ Native Weed
♦ 255
♦ Origin: Brazil.

Acicarpha tribuloides **Juss.**
Calyceraceae
♦ madam gorgon
♦ Weed, Naturalised, Casual Alien
♦ 87, 88, 101, 237, 271, 280, 295

Acinos arvensis **(Lam.) Dandy**
Lamiaceae
Acinos thymoides Moench, *Calamintha acinos* (L.) Clairv. (see)
♦ basil thyme, mother of thyme
♦ Weed, Naturalised, Environmental Weed
♦ 4, 80, 88, 94, 101, 104, 253, 272, 280
♦ a/pH, cultivated, herbal. Origin: Eurasia.

Acinos rotundifolius **Pers.**
Lamiaceae
♦ Weed
♦ 272
♦ a/pH, promoted. Origin: Eurasia.

Aciotis caulialata **(Ruiz & Pav.) Triana**
Melastomataceae
♦ Weed
♦ 87, 88

Aciotis dichotoma **Cogn.**
Melastomataceae
♦ Weed
♦ 87, 88

Aciphylla **J.R. & G.Forst. spp.**
Apiaceae
♦ Spaniard
♦ Weed
♦ 165
♦ cultivated.

Acisanthera quadrata **Juss.**
Melastomataceae
♦ dustseed
♦ Weed
♦ 14

Acmella decumbens **(Sm.) R.K.Jansen**
Asteraceae
♦ creeping spotflower
♦ Naturalised
♦ 101

Acmella pilosa **R.K.Jansen**
Asteraceae
♦ hairy spotflower
♦ Weed, Naturalised
♦ 101, 179

Acmella pusilla **(Hook. & Arn.) R.K.Jansen**
Asteraceae
♦ dwarf spotflower
♦ Naturalised, Cultivation Escape
♦ 101, 261
♦ cultivated. Origin: Brazil, Argentina.

Acmena smithii **(Poir.) Merr. & Perry**
Myrtaceae
♦ lilly pilly, monkey apple
♦ Quarantine Weed, Naturalised, Environmental Weed
♦ 225, 246, 280
♦ cultivated, herbal. Origin: Australia.

Acnida altissima **(Riddell) Moq. ex Standl.**
Amaranthaceae
= *Amaranthus tuberculatus* (Moq.) Sauer.
♦ tall waterhemp
♦ Weed, Naturalised
♦ 87, 88, 218, 287

Acokanthera oblongifolia **(Hochst.) Codd**
Apocynaceae
Carissa acokanthera Pinchon
♦ bushman's poison bush, dune poison bush, wintersweet
♦ Weed, Native Weed
♦ 24, 24, 39, 121, 161
♦ S/T, cultivated, herbal, toxic. Origin: southern Africa.

Acokanthera oppositifolia **(Lam.) Codd**
Apocynaceae
= *Acokanthera venenata auct. non* G.Don (NoR)
♦ bushman's poison bush, common poison bush
♦ Weed, Native Weed
♦ 39, 121, 161
♦ S/T, cultivated, toxic. Origin: southern Africa.

Aconitum barbatum **Pers. var.** *puberulum* **Ledeb.**
Ranunculaceae
♦ puberulent monkshood
♦ Weed
♦ 297

Aconitum × cammarum **L.**
Ranunculaceae
= *Aconitum variegatum* L. × *Aconitum napellus* L. [most probable combination]
♦ hybrid monkshood
♦ Naturalised
♦ 40
♦ herbal, toxic.

Aconitum carmichaelii **Debeaux**
Ranunculaceae
Aconitum carmichaeli var. *wilsonii* Debeaux, *Aconitum fischeri* Reichb.
♦ Carmichael's monkshood, Japanese wolfsbane, autumn fig, azure monkshood, leopard's bane, monkshood, Debeaux
♦ Weed, Quarantine Weed
♦ 76, 80, 88, 220
♦ cultivated, herbal, toxic. Origin: Asia.

Aconitum columbianum **Nutt.**
Ranunculaceae
♦ monkshood, Columbian monkshood, aconite
♦ Weed
♦ 161
♦ pH, cultivated, herbal, toxic.

Aconitum ferox **Wall. ex Ser.**
Ranunculaceae
♦ Indian aconite
♦ Weed, Quarantine Weed
♦ 39, 76, 88, 220
♦ pH, promoted, herbal, toxic. Origin: east Asia.

Aconitum gymnandrum **Maxim.**
Ranunculaceae
♦ nakedstem monkshood
♦ Weed
♦ 297

Aconitum jeholense **Nakai & Kitag. var.** *angustius* **(W.T.Wang) Y.Z.Zhao**
Ranunculaceae
♦ north China monkshood
♦ Weed
♦ 297

Aconitum kusnezoffii **Reichb.**
Ranunculaceae
♦ Kusnezoff monkshood
♦ Weed
♦ 297
♦ herbal.

Aconitum napellus **L.**
Ranunculaceae
♦ Venus's chariot, garden monkshood, aitoukonhattu, aconite, monkshood
♦ Weed, Naturalised, Cultivation Escape, Casual Alien
♦ 23, 39, 42, 80, 87, 101, 161, 272, 280
♦ pH, cultivated, herbal, toxic. Origin: Europe.

Aconitum reclinatum **Gray**
Ranunculaceae
♦ trailing wolfsbane, trailing white monkshood
♦ Weed
♦ 161
♦ herbal, toxic.

Aconitum rotundifolium **Kar. & Kir.**
Ranunculaceae
♦ Quarantine Weed
♦ 220
♦ pH, promoted.

Aconitum septentrionale **Koelle**
Ranunculaceae
♦ lehtoukonhattu
♦ Weed
♦ 70
♦ pH, cultivated, herbal. Origin: northern Europe.

Aconitum soongaricum **Stapf**
Ranunculaceae
♦ Quarantine Weed
♦ 220

Aconitum uncinatum **L.**
Ranunculaceae
♦ wild monkshood, southern blue monkshood

- Weed
- 39, 161
- pH, promoted, herbal, toxic.

Aconitum variegatum L.
Ranunculaceae
- tarhaukonhattu
- Weed, Cultivation Escape
- 42, 272
- cultivated, herbal. Origin: Europe.

Aconitum volubile Pall. ex Koelle
Ranunculaceae
- monkshood
- Quarantine Weed
- 220
- pC, cultivated, herbal.

Acorus calamus L.
Acoraceae
- sweet flag, calmus, cinnamon sedge, gladdon, kalmoes, kalmus, myrtle grass, myrtle sedge, sweet cane, sweet myrtle, sweet root, sweet rush, sweet sedge
- Weed, Naturalised, Environmental Weed
- 39, 42, 70, 87, 88, 121, 152, 218, 272, 275, 286, 297
- pH, aqua, cultivated, herbal, toxic. Origin: Eurasia.

Acourtia microcephala DC.
Asteraceae
- sacapellote
- Quarantine Weed
- 220
- pH, cultivated.

Acrachne racemosa (Roem. & J.A.Schult.) Ohwi
Poaceae
Acrachne verticillata (Roxb.) Lindl. ex Chiov. (see), *Eleusine racemosa* Heyne ex Roem. & Schult., *Eleusine verticillata* Roxb.
- goosegrass
- Naturalised
- 101, 243
- G, arid.

Acrachne verticillata (Roxb.) Lindl. ex Chiov.
Poaceae
= *Acrachne racemosa* (Roem. & J.A.Schult.) Ohwi
- Weed
- 87, 88
- G.

Acrocephalus indicus (Burm.f.) Kuntze
Lamiaceae
- Weed
- 13

Acroceras macrum Stapf
Poaceae
- Nile grass
- Naturalised
- 32
- G, cultivated. Origin: Africa.

Acroceras zizanioides (H.B.K.) Dandy
Poaceae
Panicum zizanioides H.B.K.
- Weed, Quarantine Weed, Naturalised

- 32, 76, 87, 88, 203, 220
- G. Origin: South America.

Acrocomia aculeata (Jacq.) Lodd. ex Mart.
Arecaceae
Acrocomia lasiospatha Mart.
- macaw palm, grugru palm
- Weed
- 179
- cultivated. Origin: South America.

Acrocomia totai C.Mart.
Arecaceae
- grugru palm
- Naturalised
- 101
- cultivated, herbal.

Acroptilon repens (L.) DC.
Asteraceae
Acroptilon angustifolium Cass., *Acroptilon australe* Iljin, *Acroptilon obtusifolium* Cass., *Acroptilon picris* (Pall.) C.A.Mey., *Acroptilon serratum* Cass., *Acroptilon subdentatum* Cass., *Centaurea picris* Pall. ex Willd., *Centaurea repens* L. (see), *Serratula picris* (Pall.) M.Bieb.
- creeping knapweed, hardheads, hardhead thistle, Russian knapweed, Turkestan thistle, arokaunokki, blue weed
- Weed, Quarantine Weed, Noxious Weed, Naturalised, Introduced, Casual Alien
- 1, 7, 26, 34, 35, 42, 48, 49, 55, 62, 76, 86, 88, 93, 98, 101, 110, 116, 130, 138, 147, 161, 171, 174, 180, 198, 203, 229, 237, 243, 248, 264, 269, 272, 280, 293
- pH, herbal, toxic. Origin: Eurasia.

Acroptilon repens (L.) DC. ssp. *australe* (Iljin) Rech.f.
Asteraceae
- Casual Alien
- 280

Acrostichum aureum L.
Pteridaceae
- leather fern, golden leatherfern
- Weed
- 14, 87, 88, 209
- cultivated, herbal. Origin: Australia.

Acrotome inflata Benth.
Lamiaceae
- tumbleweed
- Native Weed
- 121
- aH, toxic. Origin: southern Africa.

Actaea alba (L.) Mill.
Ranunculaceae
- white baneberry, white cohosh
- Weed, Quarantine Weed
- 39, 76, 88, 220
- pH, cultivated, herbal, toxic. Origin: North America.

Actaea pachypoda Ell.
Ranunculaceae
Actaea alba auct. non (L.) P.Mill.
- white baneberry, doll's eyes
- Weed

- 154, 161
- cultivated, herbal, toxic.

Actaea rubra (Aiton) Willd.
Ranunculaceae
Actaea arguta Nutt.
- baneberry, red baneberry, lännenkonnanmarja
- Weed, Quarantine Weed, Cultivation Escape
- 8, 39, 42, 76, 87, 88, 154, 161, 218, 220
- pH, cultivated, herbal, toxic. Origin: North America.

Actaea spicata L.
Ranunculaceae
- baneberry, herb Christopher, mustakonnanmarja
- Weed, Quarantine Weed
- 39, 76, 88, 220, 272
- pH, cultivated, herbal, toxic. Origin: Europe.

Actinea odorata Kuntze
Asteraceae
- Weed
- 23, 88

Actinidia arguta (Sieb. & Zucc.) Planch. ex Miq.
Actinidiaceae
Actinidia giraldii Diels, *Actinidia megalocarpa* Nakai., *Trochostigma arguta* Sieb. & Zucc.
- tara vine, bower actinidia
- Weed, Naturalised
- 101, 195
- pC, cultivated, herbal. Origin: east Asia.

Actinidia chinensis Planch.
Actinidiaceae
- kiwi fruit
- Naturalised
- 287
- pC, cultivated, herbal. Origin: China.

Actinidia deliciosa (A.Chev.) C.F.Liang & A.R.Ferguson
Actinidiaceae
Actinidia chinensis hort. non Planch.
- kiwi fruit
- Weed, Naturalised
- 15, 280
- pC, cultivated, herbal. Origin: east Asia.

Actinocyclus normanii Hust. fo. *subsalsa* (Juhl.-Dannf.) Hust.
Hemidiscaceae
- Weed
- 197
- diatom.

Actinostemma lobatum (Maxim.) Maxim
Cucurbitaceae
Mitrosicyos lobatus Maxim.
- Weed
- 87, 88, 286

Actinostemma tenerum Griff.
Cucurbitaceae
- lobed actinostemma
- Weed
- 297

Actinostrobus pyramidalis Miq.
Cupressaceae
♦ swamp cypress
♦ Introduced
♦ 228
♦ arid, cultivated, herbal.

Actinotus helianthi Labill.
Apiaceae
♦ flannel flower
♦ Weed, Naturalised, Native Weed, Garden Escape, Environmental Weed
♦ 72, 86, 88, 198, 252
♦ pH, cultivated, herbal. Origin: Australia.

Acuan acuminata (Benth.) Kuntze
Fabaceae/Mimosaceae
= *Desmanthus acuminatus* Benth.
♦ Quarantine Weed
♦ 220

Acuan arsenei Britton & Rose
Fabaceae/Mimosaceae
= *Desmanthus arsenei* Standl.
♦ Quarantine Weed
♦ 220

Acuan bicornutum Britton & Rose
Fabaceae/Mimosaceae
= *Desmanthus bicornutus* S.Watson
♦ Quarantine Weed
♦ 220

Acuan fallax Small
Fabaceae/Mimosaceae
= *Desmanthus fallax* K.Schum.
♦ Quarantine Weed
♦ 220

Acuan fruticosum Standl.
Fabaceae/Mimosaceae
= *Desmanthus fruticosus* Rose
♦ Quarantine Weed
♦ 220

Acuan glandulosa A.A.Heller
Fabaceae/Mimosaceae
Mimosa glandulosa Michx. (see)
♦ Quarantine Weed
♦ 220

Acuan guadeloupense Britton & Rose
Fabaceae/Mimosaceae
♦ Quarantine Weed
♦ 220
♦ Origin: Guadeloupe Island.

Acuan illinoense (Michx.) Kuntze
Fabaceae/Mimosaceae
= *Desmanthus illinoensis* (Michx.) MacMill. ex B.L.Rob. & Fernald
♦ Quarantine Weed
♦ 220

Acuan insulare Britton & Rose
Fabaceae/Mimosaceae
= *Desmanthus insularis* (Britton & Rose) León
♦ Quarantine Weed
♦ 220
♦ Origin: Bermuda, Puerto Rico, Cuba, Jamaica.

Acuan interior Britton & Rose
Fabaceae/Mimosaceae
= *Desmanthus interior* (Britton & Rose) Bullock

♦ Quarantine Weed
♦ 220

Acuan latum Britton & Rose
Fabaceae/Mimosaceae
= *Desmanthus latus* Standl. (NoR)
♦ Quarantine Weed
♦ 220

Acuan leptoloba (Torr. & A.Gray) Kuntze
Fabaceae/Mimosaceae
= *Desmanthus leptolobus* Torr. & A.Gray
♦ Quarantine Weed
♦ 220

Acuan obtusa A.Heller
Fabaceae/Mimosaceae
= *Desmanthus obtusus* S.Watson
♦ Quarantine Weed
♦ 220

Acuan painterii Britton & Rose
Fabaceae/Mimosaceae
= *Desmanthus painterii* Standl.
♦ Quarantine Weed
♦ 220

Acuan palmeri Britton & Rose
Fabaceae/Mimosaceae
= *Desmanthus palmeri* (Britton & Rose) Wiggins ex Turner
♦ Quarantine Weed
♦ 220

Acuan pringlei Britton & Rose
Fabaceae/Mimosaceae
= *Desmanthus pringlei* (Britton & Rose) F.J.Herm.
♦ Quarantine Weed
♦ 220

Acuan reticulata (Benth.) Kuntze
Fabaceae/Mimosaceae
= *Desmanthus reticulatus* Benth.
♦ Quarantine Weed
♦ 220

Acuan sublatum Britton & Rose
Fabaceae/Mimosaceae
= *Desmanthus bicornutus* S.Watson
♦ Quarantine Weed
♦ 220

Acuan velutinum (Scheele) Kuntze
Fabaceae/Mimosaceae
= *Desmanthus velutinus* Scheele
♦ Quarantine Weed
♦ 220

Acuan virgatum (L.) Medik.
Fabaceae/Mimosaceae
= *Desmanthus virgatus* (L.) Willd.
♦ Quarantine Weed
♦ 220

Acuania acuminata Kuntze
Fabaceae/Mimosaceae
= *Desmanthus acuminatus* Benth.
♦ Quarantine Weed
♦ 220

Acuania illinoense Kuntze
Fabaceae/Mimosaceae
= *Desmanthus illinoensis* (Michx.) MacMill. ex B.L.Rob. & Fern.
♦ Quarantine Weed
♦ 220

Acuania reticulata Kuntze
Fabaceae/Mimosaceae
= *Desmanthus reticulatus* Benth.
♦ Quarantine Weed
♦ 220

Adansonia digitata L.
Bombacaceae
Adansonia somalensis Chiov., *Adansonia sphaerocarpa* A.Chev., *Adansonia sulcata* A.Chev.
♦ baobab, sour gourd, monkey bread, boabab
♦ Naturalised, Introduced
♦ 39, 166, 228, 261
♦ arid, cultivated, herbal, toxic. Origin: Africa, Madagascar.

Adenanthera macrocarpa (Benth.) Bren.
Fabaceae/Mimosaceae
Piptadenia cebil (Gr.) Gr.
♦ cebil
♦ Weed
♦ 295

Adenanthera pavonina L.
Fabaceae/Mimosaceae
♦ coral bean tree, red sandalwood tree, red bead tree, lopa, pomea, bead tree, false wiliwili, kaikes, colales, culalis, metekam, telengtúngd, mwetkwem, lera, ndamu, vaivai ni vavalangi, Barbados pride
♦ Weed, Noxious Weed, Naturalised, Introduced, Environmental Weed, Cultivation Escape
♦ 3, 22, 39, 80, 86, 88, 101, 107, 112, 122, 179, 227, 228, 230, 259, 261
♦ T, arid, cultivated, herbal, toxic. Origin: south-east Asia, India, Sri Lanka.

Adenia digitata (Harv.) Engl.
Passifloraceae
Adenia angustisecta Burtt Davy, *Adenia buchananii* Harms, *Adenia multiflora* Pott, *Adenia senensis* (Klotzsch) Engl., *Adenia stenophylla* Harms, *Clemanthus senensis* Klotzsch, *Modecca digitata* Harv., *Modecca senensis* (Klotzsch) Mast.
♦ wild granadilla
♦ Weed, Native Weed
♦ 39, 121
♦ aC, arid, cultivated, toxic. Origin: southern Africa.

Adenia gracilis Harms
Passifloraceae
♦ Weed
♦ 87, 88

Adenium obesum (Forssk.) Roem. & Schult.
Apocynaceae
♦ desert rose
♦ Weed
♦ 39, 161
♦ arid, cultivated, herbal, toxic.

Adenocalymna alliaceum Miers
Bignoniaceae
♦ Weed, Quarantine Weed
♦ 76, 87, 88, 203, 220
♦ herbal.

Adenocalymna bracteatum DC.
Bignoniaceae
♦ Weed
♦ 87, 88

Adenophora lilifolia (L.) Ledeb.
Campanulaceae
♦ lilyleaf ladybell
♦ Weed
♦ 272
♦ pH, promoted. Origin: central Europe to Siberia.

Adenophora polyantha Nakai
Campanulaceae
♦ manyflower ladybell
♦ Weed
♦ 297

Adenophora potaninii Korsh.
Campanulaceae
♦ ladybells
♦ Weed
♦ 275
♦ pH, cultivated.

Adenophora stricta Miq.
Campanulaceae
♦ Naturalised
♦ 287
♦ pH, promoted, herbal.

Adenophora triphylla (Thunb.) A.DC. var. *japonica* (Regel) H.Hara
Campanulaceae
♦ Weed
♦ 286
♦ cultivated.

Adenophora wawreana A.Zahlbr
Campanulaceae
♦ wawre ladybell
♦ Weed
♦ 297

Adenoropium gossypifolium (L.) Pohl.
Euphorbiaceae
= *Jatropha gossypiifolia* L.
♦ Weed
♦ 3, 191

Adenostemma lavenia (L.) Kuntze
Asteraceae
♦ pepepepe, common medicine plant, club wort
♦ Weed
♦ 13, 87, 88, 276, 286
♦ pH, herbal.

Adenostemma platyphyllum Cass.
Asteraceae
♦ Naturalised
♦ 257
♦ aH.

Adenostoma fasciculatum Hook. & Arn.
Rosaceae
♦ common chamise, chamise, greasewood chamise, chamiso
♦ Weed
♦ 87, 88, 218
♦ S/T, cultivated, herbal. Origin: North America.

Adenostoma sparsifolium Torr.
Rosaceae
♦ redshank chamise, redshank
♦ Weed
♦ 87, 88, 218
♦ T, cultivated, herbal.

Adenostyles alliariae (Gouan) Kern.
Asteraceae
♦ Weed
♦ 70
♦ cultivated.

Adenostyles alpina (L.) Bluff & Fingerh.
Asteraceae
♦ alpine plantain
♦ Weed, Quarantine Weed
♦ 220, 272
♦ cultivated, herbal.

Adesmia aueri Burkart
Fabaceae/Papilionaceae
♦ Introduced
♦ 228
♦ arid.

Adesmia horridiuscula Burkart
Fabaceae/Papilionaceae
♦ Introduced
♦ 228
♦ arid.

Adesmia incana Vogel
Fabaceae/Papilionaceae
Adesmia angulata Hook.f., *Adesmia grisea* Hook.f.
♦ Introduced
♦ 228
♦ arid.

Adesmia muricata (Jacq.) DC.
Fabaceae/Papilionaceae
♦ Quarantine Weed
♦ 220
♦ arid. Origin: South America.

Adesmia spinosissima Meyen ex J.Vogel
Fabaceae/Papilionaceae
Adesmia rupicola Wedd., *Patagonium alcicornutum* Rusby
♦ Introduced
♦ 228
♦ arid.

Adiantum anceps Maxon & Morton
Pteridaceae/Adiantaceae
♦ large leaved maidenhair, double edge maidenhair
♦ Weed
♦ 179
♦ cultivated. Origin: South America.

Adiantum capillus-veneris L.
Pteridaceae/Adiantaceae
♦ maidenhair, Venus's hair, maidenhair fern
♦ Weed, Naturalised, Introduced
♦ 38, 87, 88, 185, 221, 241, 300
♦ pH, arid, cultivated, herbal. Origin: Europe.

Adiantum caudatum L.
Pteridaceae/Adiantaceae
♦ tailed maidenhair, trailing maidenhair fern, walking fern
♦ Weed, Naturalised
♦ 101, 179
♦ cultivated, herbal.

Adiantum cristatum L.
Pteridaceae/Adiantaceae
= *Adiantum pyramidale* (L.) Willd. (NoR)
♦ Weed
♦ 87, 88

Adiantum cuneatum Langsd. & Fisch.
Pteridaceae/Adiantaceae
= *Adiantum raddianum* C.Presl
♦ Weed
♦ 87, 88
♦ cultivated.

Adiantum hirsutum Bory
Pteridaceae/Adiantaceae
♦ hairy maidenhair
♦ Introduced
♦ 261
♦ Origin: Macarene Islands.

Adiantum hispidulum Sw.
Pteridaceae/Adiantaceae
♦ rough maidenhair, Australian maidenhair, finger adiantum
♦ Naturalised
♦ 101
♦ cultivated, herbal. Origin: Malaysia, India, New Zealand, Australia.

Adiantum raddianum Presl
Pteridaceae/Adiantaceae
Adiantum cuneatum Langsd. & Fisch. (see)
♦ delta maidenhair fern, fine maidenhair fern
♦ Weed, Naturalised, Introduced
♦ 15, 101, 121, 261
♦ pH, cultivated, herbal. Origin: South America.

Adiantum tenerum Sw.
Pteridaceae/Adiantaceae
♦ brittle maidenhair fern, fan maidenhair
♦ Weed
♦ 87, 88
♦ cultivated, herbal.

Adiantum tenerum Sw. var. *farleyense* (Moore) André
Pteridaceae/Adiantaceae
♦ Introduced
♦ 261
♦ Origin: Barbados.

Adiantum trapeziforme L.
Pteridaceae/Adiantaceae
♦ diamond maidenhair
♦ Weed, Naturalised
♦ 39, 101, 179
♦ cultivated, herbal, toxic.

Adiantum villosum L.
Pteridaceae/Adiantaceae
♦ woolly maidenhair
♦ Weed
♦ 206, 243

Adonis L. spp.
Ranunculaceae
♦ pheasant's eye, adonis
♦ Weed, Naturalised
♦ 198, 247, 272
♦ herbal, toxic.

Adonis aestivalis L.
Ranunculaceae
Adonis miniatus Jacq., *Adonis phoeniceus*
Frit.
♦ summer pheasant's eye, summer
adonis, red pheasant's eye, adonis
♦ Weed, Naturalised, Casual Alien
♦ 39, 42, 70, 87, 88, 94, 101, 161, 243,
248, 253, 272
♦ aH, cultivated, herbal, toxic. Origin:
Mediterranean.

Adonis amurensis Regel & Radde
Ranunculaceae
♦ Amur adonis
♦ Weed
♦ 39, 161
♦ pH, cultivated, herbal, toxic. Origin:
east Asia, Siberia to Japan.

Adonis annua L.
Ranunculaceae
Adonis autumnalis L. (see)
♦ pheasant's eye, adonis, blood drops
♦ Weed, Naturalised, Cultivation
Escape, Casual Alien
♦ 39, 40, 42, 70, 87, 88, 94, 101, 161, 218,
272
♦ aH, cultivated, herbal, toxic. Origin:
Europe.

Adonis autumnalis L. nom. illeg.
Ranunculaceae
= *Adonis annua* L.
♦ false hellebore
♦ Weed
♦ 87, 88
♦ herbal.

Adonis baetica Coss.
Ranunculaceae
♦ Weed
♦ 87, 88

Adonis dentata Delile
Ranunculaceae
♦ Weed
♦ 87, 88, 221

Adonis flammea Jacq.
Ranunculaceae
♦ burning pheasant's eye, flame
adonis
♦ Weed
♦ 70, 87, 88, 94, 272
♦ cultivated. Origin: Europe.

Adonis microcarpa DC.
Ranunculaceae
Adonis annua L. ssp. *carinata* Vierth.,
Adonis dentata auct. *eur. non* Delile
♦ small fruited pheasant's eye,
pheasant's eye, red chamomile,
yelllow pheasant's eye
♦ Weed, Quarantine Weed, Noxious
Weed, Naturalised, Garden Escape
♦ 7, 70, 86, 87, 88, 94, 98, 147, 171, 176,
198, 203, 220, 243
♦ cultivated, toxic. Origin:
Mediterranean.

Adonis vernalis L.
Ranunculaceae
♦ spring pheasant's eye, spring adonis,
false hellebore

♦ Weed, Naturalised
♦ 39, 87, 88, 101, 161, 272
♦ pH, cultivated, herbal, toxic. Origin:
central and southern Europe.

Adoxa moschatellina L.
Adoxaceae
♦ moschatel, tesmayrtti, renpukusou,
townhall clock, muskroot
♦ Weed
♦ 272
♦ cultivated, herbal.

Aegialophila Boiss. & Heldr. spp.
Asteraceae
♦ Weed
♦ 221

Aegilops L. spp.
Poaceae
♦ goatgrass
♦ Weed, Quarantine Weed
♦ 76, 88, 203, 220, 272
♦ G.

Aegilops bicornis (Forssk.) Jaub. & Spach
Poaceae
♦ goatgrass
♦ Weed
♦ 221
♦ G.

Aegilops biuncialis Vis.
Poaceae
Triticum biunciale (Vis.) K.Richt. *nom.
illeg.*, *Aegilops ovata* L. var. *biuncialis*
(Vis) Fiori & Paol., *Aegilops geniculata*
ssp. *biuncialis* (Vis.) Zange., *Aegilops
lorentii* Hochst., *Aegilops connata* Steud.,
Aegilops intermedia Steud., *Aegilops
macrochaeta* Shuttlew. & E.Huet ex
Duval-Jouve, *Aegilops triaristata* Willd.
ssp. *trispiculata* Hack. ex Trab, *Aegilops
biaristata* Lojac, *Aegilops biuncialis* Vis.
var. *velutina* Zhuk., *Aegilops biuncialis*
Vis. var. *archipelagica* Eig, *Aegilops
macrochaeta* Shuttlew. & E.Huet ex
Duval-Jouve. ssp. *pontica* Degen.
♦ Weed
♦ 111, 243
♦ G. Origin: southern Europe.

Aegilops crassa Boiss.
Poaceae
Triticum crassum (Boiss.) Aitch. &
Hemsl., *Gastropyrum crassum* (Boiss.)
Á.Löve, *Aegilops platyathera* Jaub. &
Spach
♦ Persian goatgrass
♦ Naturalised
♦ 101, 243
♦ G. Origin: Middle East.

Aegilops cylindrica Host.
Poaceae
Aegilops tauschii auct. *non* Coss.,
Cylindropyrum cylindricum (Host)
Á.Löve, *Triticum cylindricum* (Host)
Ces., *Triticum caudatum* (L.) Godr. &
Gren., *Aegilops caudata* L. ssp. *cylindrica*
(Host) Hegi, *Aegilops caudata* L. var.
cyclindrica (Host) Fiori
♦ jointed goatgrass, jointgrass,
goatgrass
♦ Weed, Quarantine Weed, Noxious

Weed, Naturalised, Introduced, Casual
Alien
♦ 1, 23, 26, 34, 35, 40, 42, 62, 76, 80, 87,
88, 91, 101, 105, 138, 146, 156, 161, 174,
180, 203, 210, 212, 218, 219, 229, 243,
264, 272, 287
♦ aG, herbal. Origin: southern Europe.

Aegilops geniculata Roth
Poaceae
Aegilops altera Cam. ex Roth., *Aegilops
brachyathera* Pomel., *Aegilops divaricata*
Jord. & Fourr., *Aegilops echinus* Godr.,
Aegilops erigens Jord. & Fourr., *Aegilops
erratica* Jord. & Fourr., *Aegilops fonsii*
Sennen., *Aegilops narbonensis* Lobel
ex Honck. *nom. inval.*, *Aegilops ovata*
auct., *Aegilops ovata* L. fo. *nudiglumis*
Nábelek., *Aegilops ovata* ssp. *atlantica*
Eig, *Aegilops ovata* ssp. *gibberosa* Zhuk.,
Aegilops ovata ssp. *globulosa* Zhuk.,
Aegilops ovata ssp. *planiuscula* Zhuk.,
Aegilops ovata ssp. *umbonata* Zhuk.,
Aegilops ovata var. *africana* Eig, *Aegilops
ovata* var. *eventricosa* Eig, *Aegilops ovata*
var. *hirsuta* Eig, *Aegilops ovata* var.
lanuginosa Zhuk., *Aegilops ovata* var.
puberulla Zhuk., *Aegilops ovata* var.
vernicosa Zhuk., *Aegilops parvula* Jord.
& Fourr., *Aegilops peregrina* Tabern.
ex Honck. *nom. inval.*, *Aegilops procera*
Jord. & Fourr., *Aegilops pubiglumis* Jord.
& Fourr., *Aegilops sicula* Jord. & Fourr.,
Aegilops triaristata Willd. fo. *submutica*
Batt. & Trab., *Aegilops veterum* J.Bauhin
ex Honck. *nom. inval.*, *Festuca italica* J.Gerard ex
Honck. *nom. inval.*, *Phleum aegylops*
Scop., *Triticum ovatum* auct., *Triticum
sylvestre* Cesalpino ex Bubani, *Triticum
vagans* (Jord. & Fourr.) Greuter.
♦ ovate goatgrass
♦ Weed, Noxious Weed, Naturalised,
Introduced
♦ 70, 91, 101, 161, 222, 229, 272
♦ G, cultivated, herbal. Origin:
southern Europe.

Aegilops kotschyi Boiss.
Poaceae
Aegilops triuncialis L. var. *kotschyi*
(Boiss.) Boiss., *Triticum triunciale*
(L.) Raspail ssp. *kotschyi* (Boiss.)
Asch. & Graebn., *Aegilemma kotschyi*
(Boiss.) Á.Löve, *Aegilops triuncialis*
ssp. *kotschyi* (Boiss.) Zhuk., *Aegilops
geniculata* Fig. & De Not. *nom. illeg.*,
Aegilops glabriglumis Gand., *Triticum
ovatum* Rasp. var. *bispiculatum* auct.
non Kuntze, *Aegilops triuncialis* L. var.
leptostachya Bornm., *Triticum kotschyi*
(Boiss.) Bowden
♦ Weed
♦ 23, 88, 221
♦ G. Origin: Middle East.

Aegilops ovata L.
Poaceae
= *Aegilops neglecta* Req. ex Bertol.
(NoR)
♦ ovate goatgrass
♦ Weed, Noxious Weed
♦ 34, 35, 87, 88, 253
♦ aG, cultivated, herbal.

Aegilops squarrosa L.
Poaceae
= *Aegilops tauschii* Coss.
♦ Weed
♦ 256, 275
♦ G, promoted.

Aegilops tauschii Coss.
Poaceae
Aegilops squarrosa L. (see), *Aegilops squarrosa* L. ssp. *salinum* Zhuk., *Aegilops squarrosa* ssp. *strangulata* Eig, *Aegilops squarrosa* var. *anathera* Eig, *Aegilops squarrosa* var. *meyeri* Griseb., *Triticum tauschii* (Coss.) Schmalh. (see), *Patropyrum tauschii* (Coss.) Á.Löve, *Aegilops squarrosa* L. var. *pubescens* Regel., *Aegilops squarrosa* L. var. *albescens* Pop., *Aegilops squarrosa* L. var. *brunnea* Pop., *Aegilops squarrosa* L. var. *ferruginea* Pop.
♦ Tausch's goatgrass
♦ Weed, Noxious Weed
♦ 87, 88, 229, 243, 272
♦ G, cultivated. Origin: Middle East.

Aegilops triuncialis L.
Poaceae
Aegilops elongata Lam. *nom. illeg.*, *Aegilops triaristata* Req. ex Bertol. *nom. illeg.*, *Aegilops squarrosa* auct. *non* L., *Triticum triuncile* (L.) Raspail, *Aegilopodes triuncialis* (L.) Á.Löve
♦ barb goatgrass
♦ Weed, Noxious Weed, Naturalised
♦ 23, 34, 35, 78, 80, 87, 88, 91, 101, 116, 161, 218, 229, 253, 272
♦ aG, cultivated. Origin: southern Europe.

Aegilops ventricosa Tausch
Poaceae
Aegilops fragilis Parl., *Triticum ventricosum* (Tausch) Ces., Pass. & Gibelli., *Gastropyrum ventricosum* (Tausch) Á.Löve, *Aegilops squarrosa* L. var. *comosa* Coss., *Aegilops squarrosa* L. var. *truncata* Coss, *Aegilops subulata* Pomel.
♦ barbed goatgrass
♦ Weed, Noxious Weed
♦ 87, 88, 229
♦ G. Origin: southern Europe.

Aeginetia L. spp.
Orobanchaceae
♦ aeginetia
♦ Weed
♦ 67, 88
♦ H parasitic.

Aeginetia indica L.
Orobanchaceae
♦ aeginetia, ye gu, nanbangiseru
♦ Weed, Noxious Weed
♦ 87, 88, 140, 229, 286
♦ herbal. Origin: Asia.

Aegiphila martinicensis Jacq.
Lamiaceae/Verbenaceae
♦ Caribbean spiritweed
♦ Weed
♦ 28, 206, 243
♦ herbal.

Aegiphila sellowiana Cham.
Lamiaceae/Verbenaceae
♦ minura
♦ Native Weed
♦ 255
♦ T. Origin: Brazil.

Aegle marmelos (L.) Corr. Serr.
Rutaceae
♦ bael fruit, Indian bael, bael
♦ Naturalised, Introduced
♦ 86, 228, 230
♦ arid, cultivated, herbal. Origin: Asia.

Aegopodium podagraria L.
Apiaceae
♦ goutweed, bishop's weed, ground elder, herb Gerard, ashweed
♦ Weed, Naturalised, Garden Escape, Environmental Weed
♦ 4, 15, 40, 44, 70, 80, 86, 87, 88, 94, 98, 101, 104, 133, 151, 155, 156, 161, 165, 176, 195, 203, 224, 243, 253, 272, 280, 287
♦ pH, aqua, cultivated, herbal. Origin: Eurasia.

Aegopogon cenchroides Humb. & Bonpl.
Poaceae
♦ relaxgrass
♦ Naturalised
♦ 101
♦ G.

Aegopogon tenellus (DC.) Trin.
Poaceae
♦ fragilegrass, relaxgrass
♦ Weed
♦ 157, 199
♦ a/pG.

Aeluropus lagopoides (L.) Trin. ex Thwaites
Poaceae
Aeluropus laevis Trin., *Aeluropus mucronatus* (Forrsk.) Boiss., *Aeluropus repens* (Desf.) Parl., *Aeluropus villosus* Hook.f., *Calotheca niliaca* Spreng., *Calotheca repens* (Desf.) Spreng., *Dactylis brevifolia* Koenig ex Willd., *Dactylis lagopoides* L., *Dactylis repens* Desf., *Festuca mucronata* Forssk., *Koeleria brevifolia* (Koenig ex Willd.) Spreng., *Koeleria lagopoides* (L.) Panzer ex Spreng., *Poa repens* (Desf.) M.Bieb.
♦ Weed
♦ 221
♦ G, arid.

Aeluropus littoralis (Willd.) Parl.
Poaceae
♦ Weed
♦ 87, 88, 221, 272, 275
♦ pG, arid. Origin: Eurasia.

Aeluropus sinensis (Debeaux) Tzvel.
Poaceae
♦ Chinese aeluropus
♦ Weed
♦ 297
♦ G.

Aeluropus villosus Trin.
Poaceae
♦ Weed

♦ 87, 88
♦ G.

Aeonium arboreum (L.) Webb & Berthel.
Crassulaceae
♦ golden aeonium, tree aenium
♦ Weed, Naturalised, Garden Escape, Environmental Weed, Casual Alien
♦ 7, 40, 54, 86, 88, 101, 198
♦ cultivated, herbal. Origin: Canary Islands.

Aeonium arboreum (L.) Webb & Berthel. cv. 'Atropurpureum'
Crassulaceae
♦ golden aeonium, tree aenium
♦ Naturalised
♦ 280

Aeonium castello-paivae Bolle
Crassulaceae
♦ Naturalised
♦ 86
♦ cultivated. Origin: Canary Islands.

Aeonium cuneatum Webb & Berthel.
Crassulaceae
♦ aeonium
♦ Naturalised
♦ 40
♦ cultivated.

Aeonium × floribundum A.Berger
Crassulaceae
♦ Casual Alien
♦ 280

Aeonium haworthii Salm-Dyck ex Webb & Berthel.
Crassulaceae
♦ pinwheel aeonium, Haworth's aeonium
♦ Weed, Naturalised
♦ 15, 54, 86, 88, 101, 198, 280
♦ S, cultivated. Origin: Teneriffe.

Aeonium haworthii Salm-Dyck ex Webb & Berthel. cv. 'Major'
Crassulaceae
♦ pinwheel aeonium
♦ Weed
♦ 15
♦ S, cultivated. Origin: horticultural.

Aeonium haworthii hybrids Webb & Berthel.
Crassulaceae
♦ Haworth's aeonium, pinwheel aeonium
♦ Naturalised
♦ 280

Aeonium leucoblepharum Webb ex A.Rich.
Crassulaceae
Sempervivum chrysanthum Hochst., *Sempervivum leucoblepharum* (Webb ex A.Rich.) Hutch. & E.A.Bruce
♦ Introduced
♦ 228
♦ arid, cultivated.

Aeonium undulatum Webb & Berthel.
Crassulaceae
♦ Naturalised
♦ 280
♦ cultivated.

**Aeonium urbicum (C.A.Sm.) Webb &
Berthel.**
Crassulaceae
♦ Naturalised
♦ 280
♦ cultivated.

Aeonium × velutinum Praeger
Crassulaceae
♦ Naturalised
♦ 280

Aerva javanica (Burm.f.) Juss. ex Schult.
Amaranthaceae
Aerva tomentosa Lam.
♦ kapok bush, snowbush, Java aerva
♦ Weed, Naturalised, Environmental
Weed
♦ 7, 86, 87, 88, 93, 98, 203, 205, 221
♦ S, herbal. Origin: Africa,
Madagascar, Asia.

Aerva lanata (L.) Juss. ex Schult.
Amaranthaceae
Achyranthes lanata L., *Achyranthes
villosa* Forssk., *Aerva lanata* (L.) Juss. ex
Schult var. *citrina* Suess., *Aerva lanata*
(L.) Juss. ex Schult. var. *intermedia*
Suess., *Aerva lanata* (L.) Juss. ex
Schult. var. *leucuroides* Suess., *Aerva
mozambicensis* Gand., *Aerva sansibarica*
Suess., *Illecebrum lanatum* (L.) L.
♦ aerva
♦ Weed, Naturalised, Introduced
♦ 66, 86, 87, 88, 221, 228
♦ arid, cultivated, herbal. Origin:
Africa, India, south-east Asia,
Australia, Madagascar.

Aeschynomene afraspera J.Léonard
Fabaceae/Papilionaceae
Aeschynomene aspera sensu auct.
♦ Weed
♦ 88

Aeschynomene americana L.
Fabaceae/Papilionaceae
Aeschynomene americana L. var.
glandulosa (Poir.) Rudd (see)
♦ vergonzosa blanca, American
jointvetch, shyleaf, jointvetch, sano
khon
♦ Weed, Naturalised
♦ 14, 87, 88, 93, 98, 153, 203, 209, 239,
255, 262, 276, 287
♦ S, cultivated. Origin: tropical
America.

**Aeschynomene americana L. var.
americana**
Fabaceae/Papilionaceae
♦ Naturalised
♦ 86
♦ Origin: Americas.

**Aeschynomene americana L. var.
glandulosa (Poir.) Rudd**
Fabaceae/Papilionaceae
= *Aeschynomene americana* L.
♦ Naturalised
♦ 86
♦ Origin: Americas.

Aeschynomene aspera L.
Fabaceae/Papilionaceae

= *Aeschynomene indica* L.
♦ sola pith plant, sano khaang khok
♦ Weed, Quarantine Weed,
Naturalised
♦ 76, 86, 87, 88, 98, 191, 203, 204, 239,
262
♦ aH. Origin: south-east Asia.

Aeschynomene brasiliana (Poir.) DC.
Fabaceae/Papilionaceae
♦ Naturalised
♦ 86
♦ Origin: Central and South America.

Aeschynomene brevifolia L. ex Poir.
Fabaceae/Papilionaceae
♦ Naturalised
♦ 86
♦ Origin: Madagascar.

Aeschynomene denticulata Rudd
Fabaceae/Papilionaceae
♦ angiquinho, paquinha
♦ Weed
♦ 236, 255
♦ aH. Origin: South America.

**Aeschynomene elaphroxylon (Guill. &
Perr.) Taub.**
Fabaceae/Papilionaceae
Herminiera elaphroxylon Guill. & Perr.,
Smithia grandidieri Baill.
♦ ambatch, balsawood, pith tree
♦ Introduced
♦ 228
♦ arid. Origin: Madagascar.

Aeschynomene evenia C.Wright
Fabaceae/Papilionaceae
♦ shrubby jointvetch
♦ Naturalised
♦ 101
♦ Origin: South America.

Aeschynomene histrix Poir.
Fabaceae/Papilionaceae
♦ porcupine jointvetch
♦ Weed, Naturalised
♦ 101, 157, 255
♦ pH. Origin: tropical America.

**Aeschynomene histrix Poir. var. incana
(Vogel) Benth.**
Fabaceae/Papilionaceae
♦ porcupine jointvetch
♦ Naturalised
♦ 101
♦ Origin: South America.

Aeschynomene indica L.
Fabaceae/Papilionaceae
Aeschynomene aspera L. (see)
♦ Indian jointvetch, kat sola, knuckle
bean bush, jointvetch, sano haag kai,
budda pea, sensitive vetch
♦ Weed, Noxious Weed, Native Weed,
Naturalised, Introduced
♦ 86, 87, 88, 121, 161, 170, 204, 217, 218,
228, 229, 230, 235, 239, 262, 263, 275,
276, 286, 297
♦ a/bH, arid, cultivated, herbal, toxic.
Origin: pantropical.

Aeschynomene marginata Benth.
Fabaceae/Papilionaceae

♦ Weed
♦ 87, 88

Aeschynomene micranthos (Poir.) DC.
Fabaceae/Papilionaceae
♦ Naturalised
♦ 86
♦ Origin: Madagascar.

Aeschynomene paniculata Willd. ex Vog.
Fabaceae/Papilionaceae
♦ pannicle jointvetch, sensitiva mansa
♦ Weed, Naturalised
♦ 101, 255
♦ pH. Origin: Brazil.

Aeschynomene portoricensis Urb.
Fabaceae/Papilionaceae
= *Aeschynomene gracilis* Vogel (NoR)
♦ Weed
♦ 87, 88

Aeschynomene rudis Benth.
Fabaceae/Papilionaceae
♦ rough jointvetch, zigzag jointvetch,
pinheirinho, angiquinho
♦ Weed, Quarantine Weed, Noxious
Weed
♦ 35, 76, 88, 203, 229, 245, 255, 270
♦ a/pH. Origin: South America.

Aeschynomene sensitiva Sw.
Fabaceae/Papilionaceae
Aeschynomene belvesii DC.,
Aeschynomene fistulosa Bello,
Aeschynomene honesta Nees &
C.Mart., *Aeschynomene macropoda* var.
belvisii DC., *Aeschynomene sensitiva*
fo. *paucifoliolata* Chodat & Hassl.,
Aeschynomene sensitiva P.Beauv.,
Aeschynomene sulcata Kunth, *Cassia
paramariboensis* Miq.
♦ sensitive jointvetch
♦ Weed
♦ 14, 87, 88
♦ herbal. Origin: South America.

Aeschynomene villosa Poir.
Fabaceae/Papilionaceae
♦ hairy jointvetch
♦ Weed, Naturalised
♦ 86, 93, 98, 203
♦ cultivated. Origin: Central and
South America.

**Aeschynomene virginica (L.) Britton,
Sterns & Pogg.**
Fabaceae/Papilionaceae
♦ northern jointvetch, Virginia
jointvetch
♦ Weed, Naturalised
♦ 88, 161, 218, 287
♦ herbal.

Aesculus arguta Buckl.
Hippocastanaceae
= *Aesculus glabra* Willd. var. *arguta*
(Buckl.) B.L.Robins. (NoR)
♦ Texas buckeye
♦ Weed
♦ 88, 218

Aesculus californica (Spach) Nutt.
Hippocastanaceae
♦ California buckeye shrub, California
buckeye, California horse chestnut

♦ Weed
♦ 39, 88, 161, 218, 247
♦ T, cultivated, herbal, toxic. Origin:
North America.

Aesculus flava Aiton
Hippocastanaceae
Aesculus octandra Marshall (see)
♦ sweet buckeye
♦ Weed
♦ 39, 161, 247
♦ T, cultivated, herbal, toxic.

Aesculus glabra Willd.
Hippocastanaceae
♦ Ohio buckeye
♦ Weed
♦ 8, 39, 88, 161, 218, 247
♦ T, cultivated, herbal, toxic. Origin:
North America.

Aesculus hippocastanum L.
Hippocastanaceae
Hippocastanum vulgare Gaertn.
♦ horse chestnut, conkers, Marronnier
d'Inde
♦ Weed, Naturalised, Cultivation
Escape
♦ 4, 8, 15, 39, 40, 42, 80, 86, 88, 97, 101,
161, 247, 252, 280
♦ T, cultivated, herbal, toxic. Origin:
Europe.

Aesculus indica Colebr. ex Wall.
Hippocastanaceae
♦ Indian horse chestnut, horse
chestnut
♦ Casual Alien
♦ 280
♦ T, cultivated, herbal.

Aesculus octandra Marshall
Hippocastanaceae
= *Aesculus flava* Aiton
♦ yellow buckeye, sweet buckeye
♦ Weed
♦ 39, 88, 218
♦ cultivated, herbal, toxic.

Aesculus parviflora Walter
Hippocastanaceae
Aesculus macrostachya Michx., *Pavia
macrostachya* DC., *Pavia macrostachys*
Loisel.
♦ bottlebrush buckeye, horse chestnut
♦ Weed
♦ 161, 247
♦ S, cultivated, herbal, toxic.

Aesculus pavia L.
Hippocastanaceae
♦ red buckeye
♦ Weed
♦ 39, 88, 161, 218, 247
♦ S, cultivated, herbal, toxic. Origin:
North America.

Aesculus sylvatica Bartr.
Hippocastanaceae
♦ painted buckeye
♦ Weed
♦ 88, 161, 218, 247
♦ cultivated, herbal, toxic.

Aetheorhiza bulbosa (L.) Cass.
Asteraceae

= *Crepis bullosa* Tausch
♦ Weed
♦ 70, 221

Aethusa cynapium L.
Apiaceae
♦ fool's parsley, cicuta aglina
♦ Weed, Naturalised
♦ 39, 44, 70, 87, 88, 94, 101, 118, 161,
218, 243, 253, 272
♦ aH, cultivated, herbal, toxic. Origin:
Europe.

Agalinis fasciculata (Ell.) Raf.
Scrophulariaceae
♦ beach false foxglove
♦ Weed
♦ 87, 88
♦ herbal.

Agapanthus L'Hér. spp.
Liliaceae/Agapanthaceae/Alliaceae/
Amaryllidaceae
♦ agapanthus, ladybells
♦ Naturalised, Garden Escape,
Environmental Weed
♦ 86, 198
♦ aH, cultivated.

Agapanthus africanus (L.) Hoffmanns.
Liliaceae/Agapanthaceae/Alliaceae/
Amaryllidaceae
Agapanthus umbellatus p.p.
♦ African lily, lily of the Nile
♦ Weed
♦ 39, 161
♦ pH, cultivated, herbal, toxic.

Agapanthus orientalis F.M.Leight.
Liliaceae/Agapanthaceae/Alliaceae/
Amaryllidaceae
♦ agapanthus
♦ Weed
♦ 15, 247
♦ cultivated, toxic.

Agapanthus praecox Willd.
Liliaceae/Agapanthaceae/Alliaceae/
Amaryllidaceae
♦ orientalis lily, blue African, African
lily
♦ Weed, Naturalised, Garden Escape,
Environmental Weed
♦ 7, 72, 88, 98, 161, 203, 225, 246, 280,
289
♦ pH, cultivated, herbal, toxic. Origin:
South Africa.

**Agapanthus praecox Willd. ssp. orientalis
(F.M.Leight.) F.M.Leight.**
Liliaceae/Agapanthaceae/Alliaceae/
Amaryllidaceae
♦ agapanthus, African lily
♦ Weed, Naturalised, Environmental
Weed
♦ 40, 54, 86, 198, 280, 296
♦ cultivated. Origin: South Africa.

Agapanthus praecox Willd. ssp. praecox
Liliaceae/Agapanthaceae/Alliaceae/
Amaryllidaceae
♦ African lily
♦ Weed, Naturalised, Environmental
Weed

♦ 54, 86
♦ cultivated. Origin: South Africa.

**Agardhiella subulata (C.Agardh) Kraft &
M.J.Wynne**
Solieriaceae
Rhabdonia tenera (J.Agardh) J.Agardh,
Agardhiella tenera (J.Agardh) F.Schmitz
♦ Introduced
♦ 288
♦ algae.

Agastache nepetoides (L.) Kuntze
Lamiaceae
♦ catnip, giant hyssop, yellow giant
hyssop
♦ Weed
♦ 161
♦ cultivated, herbal.

Agastache urticifolia (Benth.) Kuntze
Lamiaceae
Lophanthus urticifolius Benth.
♦ horsemint, giant hyssop, nettleleaf
giant hyssop
♦ Weed
♦ 23, 88
♦ pH, cultivated, herbal. Origin: North
America.

Agathis robusta (F.Muell.) Bailey
Araucariaceae
♦ Queensland kauri, kauri
♦ Cultivation Escape
♦ 261
♦ cultivated, herbal. Origin: Australia.

**Agathisanthemum bojeri Klotzsch var.
bojeri**
Rubiaceae
♦ Native Weed
♦ 121
♦ pH. Origin: southern Africa.

Agathosma crenulata (L.) Pillans
Rutaceae
♦ buchu, oval buchu
♦ Naturalised
♦ 86
♦ cultivated, herbal. Origin: South
Africa.

Agathosma ovata (Thunb.) Pillans
Rutaceae
♦ false buchu
♦ Native Weed
♦ 121
♦ pS, cultivated, herbal. Origin:
southern Africa.

Agave L. spp.
Agavaceae
♦ agave
♦ Naturalised, Garden Escape,
Environmental Weed
♦ 86, 198, 201, 247, 279
♦ cultivated, herbal, toxic.

Agave aktites Gentry
Agavaceae
♦ Introduced
♦ 228
♦ arid.

Agave americana L.
Agavaceae
♦ American agave, century plant, maguey, American aloe
♦ Weed, Sleeper Weed, Naturalised, Introduced, Garden Escape, Environmental Weed, Cultivation Escape
♦ 6, 7, 15, 39, 72, 73, 88, 95, 98, 121, 161, 179, 198, 203, 215, 225, 246, 269, 279, 280, 283, 287, 289
♦ pS, cultivated, herbal, toxic. Origin: North America.

Agave americana L. cv. 'Marginata'
Agavaceae
♦ century plant
♦ Naturalised, Environmental Weed
♦ 86
♦ pS, cultivated, herbal, toxic. Origin: horticultural.

Agave americana L. var. americana
Agavaceae
♦ century plant
♦ Naturalised, Environmental Weed
♦ 86, 198
♦ cultivated. Origin: Mexico.

Agave americana L. var. expansa (Jacobi) Gentry
Agavaceae
♦ century plant
♦ Naturalised, Introduced, Garden Escape, Environmental Weed
♦ 86, 228
♦ arid, cultivated. Origin: southern North America.

Agave americana L. var. picta (Salm.) Terrac.
Agavaceae
♦ variegated century plant
♦ Naturalised, Environmental Weed
♦ 86, 198
♦ cultivated. Origin: Mexico.

Agave angustifolia Haw.
Agavaceae
Agave aboriginum Trel., *Agave kirchneriana* A.Berger, *Agave owenii* I.M.Johnst., *Agave pacifica* Trel., *Agave yaquiana* Trel., *Agave zapupe* Trel.
♦ century plant
♦ Introduced
♦ 228
♦ arid, cultivated, herbal.

Agave angustifolia Haw. cv. 'Marginata'
Agavaceae
♦ Weed
♦ 179
♦ cultivated, herbal. Origin: horticultural.

Agave applanata Koch ex Jacobi
Agavaceae
♦ maguey de ixtle, maguey de la casa, socolume
♦ Introduced
♦ 228
♦ arid.

Agave attenuata Salm-Dyck
Agavaceae

Agave cernua Berger, *Agave glaucescens* Hook., *Agave pruinosa* Lem. ex Jacobi
♦ Mexican century plant
♦ Weed, Naturalised, Introduced
♦ 86, 93, 161, 191, 228
♦ arid, cultivated, toxic. Origin: Mexico.

Agave bracteosa S.Wats. ex Engelm.
Agavaceae
♦ Introduced
♦ 228
♦ arid, cultivated.

Agave cantala Roxb.
Agavaceae
Agave cantalabrum Tod., *Agave rumphii* Jacobi
♦ cantala
♦ Introduced
♦ 228
♦ arid.

Agave celsii Hook. var. celsii
Agavaceae
Agave botterii Baker, *Agave boucheri* Jacobi, *Agave haseloffii* Jacobi, *Agave micrantha* Salm-Dyck, *Agave mitis* H.Monac. ex Salm-Dyck
♦ Introduced
♦ 228
♦ arid.

Agave colimana Gentry
Agavaceae
Agave angustissima Engelm. var. *ortgiesiana* Trel., *Agave schidigera* Lem. var. *ortgiesiana* Trel.
♦ Introduced
♦ 228
♦ arid, cultivated.

Agave colorata Gentry
Agavaceae
♦ Introduced
♦ 228
♦ arid, cultivated.

Agave decipiens Baker
Agavaceae
♦ false sisal
♦ Naturalised
♦ 101

Agave desmettiana Jacobi
Agavaceae
♦ dwarf century plant
♦ Weed, Naturalised
♦ 101, 179

Agave difformis Berger
Agavaceae
Agave haynaldii Tod.
♦ Introduced
♦ 228
♦ arid.

Agave durangensis Gentry
Agavaceae
♦ Introduced
♦ 228
♦ arid.

Agave filifera Salm-Dyck
Agavaceae
Agave filamentosa Salm-Dyck, *Agave filamentosa null* var. *filamentosa* Baker

♦ Introduced
♦ 228
♦ arid, cultivated, herbal.

Agave flexispina Trel. ex Standl.
Agavaceae
♦ Introduced
♦ 228
♦ arid, cultivated.

Agave fortiflora Gentry
Agavaceae
♦ Introduced
♦ 228
♦ arid.

Agave fourcroydes Lem.
Agavaceae
Agave sullivani Trel.
♦ henequen
♦ Introduced
♦ 228
♦ arid, cultivated.

Agave funkiana K.Koch & C.D.Bouche
Agavaceae
♦ Introduced
♦ 228
♦ arid, cultivated.

Agave ghiesbreghtii K.Koch
Agavaceae
Agave huehueteca Standl., *Agave purpusorum* A.Berger, *Agave roezliana* Baker
♦ Introduced
♦ 228
♦ arid.

Agave gigantensis Gentry
Agavaceae
♦ Introduced
♦ 228
♦ arid.

Agave guiengola Gentry
Agavaceae
♦ Introduced
♦ 228
♦ arid.

Agave jaiboli Gentry
Agavaceae
♦ Introduced
♦ 228
♦ arid.

Agave kerchovei Lem.
Agavaceae
Agave convallis Trel., *Agave dissimulans* Trel., *Agave expatriata* Rose, *Agave inopinabilis* Trel., *Agave noli-tangere* Berger
♦ Introduced
♦ 228
♦ arid.

Agave lecheguilla Torr.
Agavaceae
♦ lechuguilla
♦ Weed
♦ 39, 87, 88, 161, 218
♦ cultivated, herbal, toxic.

Agave lophantha Schiede
Agavaceae
Agave heteracantha Zucc., *Agave*

mezortillo hort., *Agave univittata* Haw.
♦ thorncrest century plant
♦ Introduced
♦ 228
♦ arid, cultivated, herbal.

Agave lurida Aiton
Agavaceae
Agave vera-cruz Mill., *Agave vernae*
A.Berger, *Agave vernae-crucis* Haw.
♦ Introduced
♦ 228
♦ arid.

Agave macroacantha Zucc.
Agavaceae
Agave besseriana Van Houtte, *Agave*
flavescens Salm-Dyck, *Agave flavescens*
Salm-Dyck var. *macroacantha* Jacobi,
Agave pugioniformis Zucc.
♦ Introduced
♦ 228
♦ arid, cultivated.

Agave ocahui Gentry var. *longifolia*
Gentry
Agavaceae
♦ Introduced
♦ 228
♦ arid.

Agave ocahui Gentry var. *ocahui*
Agavaceae
♦ Introduced
♦ 228
♦ arid.

Agave parryi Engelm. var. *huachucensis*
(Baker) Little
Agavaceae
Agave applanata K.Koch var.
huachuecensis (Baker) Mulford, *Agave*
huachuecensis Baker
♦ Introduced
♦ 228
♦ arid.

Agave pedunculifera Trel. ex Standl.
Agavaceae
♦ Introduced
♦ 228
♦ arid.

Agave pendula Schnittsp.
Agavaceae
Agave aldina Koch, *Agave coespitosa*
Tod., *Agave sartorii* Koch
♦ Introduced
♦ 228
♦ arid.

Agave polyacantha Haw.
Agavaceae
Agave chlorocantha Salm, *Agave*
densiflora Hook., *Agave engelmannii*
Trel., *Agave uncinata* Jacobi
♦ Introduced
♦ 228
♦ arid.

Agave salmiana Otto ex Salm-Dyck ssp.
***crassispina* (Trel.) Gentry**
Agavaceae
Agave crassispina Trel.
♦ Introduced
♦ 228

Agave schottii Engelm.
Agavaceae
♦ Schott agave, Schott's century plant
♦ Weed, Environmental Weed
♦ 88, 151, 218
♦ cultivated, herbal.

Agave shrevei Gentry ssp. *shrevei*
Agavaceae
Agave shrevei Gentry
♦ Introduced
♦ 228

Agave sisalana Perrine
Agavaceae
♦ sisal hemp, sisal, garingboom, hemp
plant
♦ Weed, Noxious Weed, Naturalised,
Introduced, Garden Escape,
Environmental Weed, Cultivation
Escape
♦ 7, 10, 19, 54, 63, 80, 86, 87, 88, 93, 95,
98, 101, 112, 121, 179, 191, 228, 261, 279,
283
♦ pS, arid, cultivated, herbal, toxic.
Origin: Central America.

Agave tequilana A.Weber
Agavaceae
Agave palmaris Trel., *Agave pedrosana*
Trel., *Agave pes-mulae* Trel., *Agave*
pseudotequilana Trel., *Agave subtilis* Trel.
♦ tequila agave
♦ Introduced
♦ 228
♦ arid, cultivated, herbal.

Agave vivipara L.
Agavaceae
♦ Naturalised
♦ 98
♦ cultivated, herbal.

Agave vivipara L. var. *vivapara*
Agavaceae
♦ Naturalised, Cultivation Escape
♦ 86, 252
♦ cultivated. Origin: obscure.

Agave weberi Cels ex Poisson
Agavaceae
Agave franceschiana Trel.
♦ Weber's century plant
♦ Naturalised, Introduced
♦ 101, 228
♦ arid, herbal.

Agave wercklei Weber ex Berger
Agavaceae
♦ Introduced
♦ 228
♦ arid.

Agdestis clematidea Moç. & Sessé ex DC.
Phytolaccaceae/Agdestidaceae
♦ rock root
♦ Weed, Naturalised, Cultivation
Escape
♦ 101, 179, 261
♦ cultivated. Origin: Mexico and
Central America.

Ageratina adenophora (Spreng.) King
& Rob.
Asteraceae

Eupatorium adenophorum Spreng. (see),
Eupatorium glandulosum H.B.K. (see),
Eupatorium pasadense Parish
♦ crofton weed, cat weed, hemp
agrimony, Mexican devil, sticky
agrimony, sticky eupatorium, sticky
snakeroot, maui pamakani
♦ Weed, Quarantine Weed, Noxious
Weed, Naturalised, Garden Escape,
Environmental Weed, Cultivation
Escape
♦ 18, 22, 34, 35, 78, 80, 86, 87, 88, 95,
98, 101, 116, 140, 147, 152, 165, 181, 191,
203, 220, 225, 229, 231, 246, 269, 279,
280, 283, 290
♦ pH, cultivated, herbal, toxic. Origin:
Mexico.

Ageratina altissima (L.) R.M.King &
H.Rob.
Asteraceae
Ageratum altissimum L., *Eupatorium*
rugosum Houtt. (see), *Eupatorium*
ageratoides L.f.
♦ white snakeroot, richweed,
snakeroot
♦ Weed, Quarantine Weed
♦ 39, 76, 82, 87, 88
♦ cultivated, herbal, toxic.

Ageratina aromatica (L.) Spach
Asteraceae
Eupatorium aromaticum L.
♦ lesser snakeroot
♦ Quarantine Weed
♦ 220
♦ pH, cultivated, herbal, toxic.

Ageratina riparia (Regel) R.M.King &
H.Rob
Asteraceae
Eupatorium riparium Regel (see),
Eupatorium cannabinum L. (often
misapplied)
♦ mistflower, cat's paw, creeping
crofton weed, river eupatorium, small
crofton weed, whiteweed
♦ Weed, Quarantine Weed, Noxious
Weed, Naturalised, Garden Escape,
Environmental Weed
♦ 18, 39, 76, 86, 87, 88, 93, 98, 134, 147,
152, 181, 191, 203, 225, 229, 243, 246,
269, 280, 283, 290
♦ cultivated, herbal, toxic. Origin:
Central America.

Ageratina vernalis (Vatke & Kurtz)
R.M.King & H.Rob.
Asteraceae
Ageratina subcoriacea R.M.King &
H.Rob., *Ageratina subinclusa* (Klatt)
R.M.King & H.Rob., *Ageratina*
subpenninervia (Sch.Bip. ex Klatt)
R.M.King & H.Rob., *Eupatorium*
chiapense B.L.Rob., *Eupatorium*
grandiflorum André, *Eupatorium*
monticola L.O.Williams, *Eupatorium*
vernale Vatke & Kurtz
♦ Quarantine Weed
♦ 220

Ageratina wrightii (Gray) King & H.E.Rob.
Asteraceae
- boneset, Wright's snakeroot
- Weed
- 161
- herbal, toxic.

Ageratum L. spp.
Asteraceae
- flossflower, whiteweed
- Weed
- 88

Ageratum conyzoides L.
Asteraceae
Ageratum hirsutum Lam.
- tropic ageratum, goatweed, billygoat weed, tropical whiteweed, bluetop, Mother Brinkly, winter weed, whiteweed, flossflower, munyavi, invading ageratum, saapraegn saapkaa, yaa saap raeng, botoncillo, mejorana, mentrasto
- Weed, Quarantine Weed, Noxious Weed, Naturalised, Introduced, Environmental Weed
- 6, 12, 13, 23, 24, 32, 50, 53, 55, 66, 76, 87, 88, 93, 95, 98, 121, 134, 157, 158, 170, 179, 185, 186, 203, 204, 206, 216, 218, 221, 228, 230, 235, 238, 239, 243, 245, 255, 256, 257, 262, 263, 269, 273, 274, 275, 276, 281, 283, 286, 287, 295, 297
- aH, arid, cultivated, herbal, toxic. Origin: tropical America.

Ageratum conyzoides L. var. *latifolium* (Cav.) M.F.Johnson
Asteraceae
- Naturalised
- 257

Ageratum houstonianum Mill.
Asteraceae
Ageratum mexicanum Sims
- Mexican ageratum, invading ageratum, goatweed, tropic ageratum, ageratum, Todd's curse, garden ageratum, bluemink, flossflower
- Weed, Noxious Weed, Naturalised, Garden Escape, Environmental Weed
- 32, 55, 87, 88, 95, 98, 101, 121, 158, 179, 186, 203, 235, 246, 256, 269, 273, 274, 276, 280, 283, 286, 287
- aH, cultivated, herbal, toxic. Origin: tropical South and Central America.

Aglaia ponapensis Kaneh.
Meliaceae
- marasau
- Introduced
- 230
- T.

Aglaonema Schott spp.
Araceae
- Chinese evergreen, evergreens, aglaonema
- Weed
- 161
- toxic.

Aglaonema commutatum Schott
Araceae
- Philippine evergreen, Chinese evergreen
- Naturalised, Cultivation Escape
- 101, 261
- cultivated, herbal. Origin: Philippines.

Aglaonema commutatum Schott var. *maculatum* (Hook.f.) Nicolson
Araceae
- Philippine evergreen
- Weed, Naturalised
- 101, 179

Aglaonema pictum (Roxb.) Kunth
Araceae
- cara de caballo
- Naturalised, Cultivation Escape
- 101, 261
- cultivated, herbal. Origin: East Indies.

Agonis flexuosa (Willd.) Sweet
Myrtaceae
Metrosideros flexuosa Willd.
- Western Australian willow myrtle, willow myrtle, peppermint willow myrtle
- Weed, Naturalised, Native Weed, Garden Escape, Environmental Weed
- 7, 72, 86, 88, 198
- T, cultivated, herbal. Origin: Australia.

Agonis juniperina Schauer
Myrtaceae
- juniper myrtle
- Casual Alien
- 280
- cultivated, herbal. Origin: Australia.

Agoseris grandiflora (Nutt.) Greene
Asteraceae
- mountain dandelion, bigflower agoseris, grand mountain dandelion
- Weed
- 23, 88
- pH, cultivated, herbal.

Agrimonia eupatoria L.
Rosaceae
- agrimony, church steeples, common agrimony
- Weed, Naturalised
- 39, 70, 86, 87, 88, 94, 98, 101, 203, 272
- pH, cultivated, herbal, toxic. Origin: Europe.

Agrimonia gryposepala Wallr.
Rosaceae
- agrimony, tall hairy agrimony
- Weed
- 88, 218
- pH, cultivated, herbal.

Agrimonia japonica (Miq.) Koidz.
Rosaceae
- Weed
- 286

Agrimonia odorata (Gouan) Mill.
Rosaceae
- agrimony
- Weed
- 51, 87, 88, 121
- pH, herbal. Origin: Eurasia.

Agrimonia pilosa Ledeb.
Rosaceae
Agrimonia dahurica Ser.
- shaggy speedwell, Chinese agrimony
- Weed, Quarantine Weed
- 87, 88, 220, 297
- pH, cultivated, herbal. Origin: east Europe to east Asia.

Agrimonia procera Wallr.
Rosaceae
Agrimonia odorata auct. non Mill.
- agrimony, scented agrimony, fragrant agrimony
- Weed
- 88, 158
- cultivated. Origin: Europe.

Agrimonia striata Michx.
Rosaceae
- roadside agrimony
- Weed
- 88, 218
- cultivated, herbal.

Agriophyllum arenarium M.Bieb. ex C.Mey.
Chenopodiaceae
= *Agriophyllum squarrosum* (L.) Moq.
- Weed
- 275
- aH.

Agriophyllum squarrosum (L.) Moq.
Chenopodiaceae
Agriophyllum arenarium M.Bieb. ex C.Mey. (see), *Agriophyllum pungens* Link, *Corispermum pungens* Vahl, *Corispermum squarrosum* Pall.
- squarrose agriophyllum
- Weed
- 297
- arid.

× *Agropogon littoralis* (Sm.) C.E.Hubb.
Poaceae
= *Agrostis stolonifera* L. (see) × *Polypogon monspeliensis* (L.) Desf. (see) [*Polypogon littoralis* Sm.]
- coast agropogon
- Naturalised
- 86, 98, 101, 176, 198, 280, 300
- G. Origin: south-west Europe, England.

Agropyron caninum (L.) Beauv.
Poaceae
= *Elymus caninus* (L.) L.
- bearded couchgrass, fibrous wheatgrass, agropyre des chiens
- Weed
- 30, 88, 91
- pG, cultivated, herbal.

Agropyron cristatum (L.) Gaertn.
Poaceae
Agropyron cristatum (L.) P.Beauv., *Agropyron deweyi* Á.Löve, *Bromus cristatus* L., *Triticum cristatum* (L.) Schreb.
- crested wheatgrass, fairway wheatgrass
- Weed, Naturalised, Introduced
- 21, 80, 91, 228, 228, 272, 297, 300

♦ pG, arid, cultivated, herbal. Origin: Eurasia.

Agropyron cristatum (L.) Gaertn. ssp. pectinatum (Bieb.) Tzvelev
Poaceae
Agropyron pectinatum (Bieb.) Beauv.,
Agropyron pectiniforme Roem. & Schult.
(see)
♦ crested wheatgrass
♦ Naturalised
♦ 101
♦ G.

Agropyron desertorum (Fisch. ex Link) J.A.Schult.
Poaceae
Triticum desertorum Fisch. ex Link
♦ desert wheatgrass
♦ Weed, Naturalised, Introduced
♦ 34, 101, 228
♦ pG, arid, herbal.

Agropyron distichum (Thunb.) P.Beauv.
Poaceae
= *Elytrigia disticha* (Thunb.) Prokudin ex Á.Löve (NoR)
♦ Naturalised
♦ 7
♦ G.

Agropyron elongatum (Host) Beauv.
Poaceae
= *Elytrigia elongata* (Host) Nev.
♦ tall wheatgrass
♦ Weed
♦ 80
♦ G, herbal.

Agropyron fragile (Roth) Candargy
Poaceae
♦ Siberian wheatgrass
♦ Naturalised
♦ 101
♦ pG. Origin: Asia.

Agropyron intermedium (Host) Beauv.
Poaceae
= *Elytrigia intermedia* ssp. *intermedia* (Host) Nevski (NoR) [see *Thinopyrum intermedium* (Host) Barkworth & D.R.Dewey, *Elymus hispidus* ssp. *barbulatus* (Opiz) Melderis, *Elymus hispidus* (Opiz) Meld.]
♦ chiendent glauque
♦ Weed
♦ 30, 80, 88
♦ G, herbal.

Agropyron kamoji Ohwi
Poaceae
= *Elymus tsukushiensis* Honda var. *transiens* (Hack.) Osada (NoR) [see *Roegneria kamoji* (Ohwi) Ohwi ex Keng, *Agropyron tsukushiense* Honda var. *transiens* (Honda) Ohwi]
♦ Weed
♦ 87, 88
♦ G.

Agropyron littorale (Host) Dur.
Poaceae
= *Elytrigia pungens* (Pers.) Tutin
♦ Weed
♦ 203
♦ G.

Agropyron pectiniforme Roem. & Schult.
Poaceae
= *Agropyron cristatum* (L.) Gaertn. ssp. *pectinatum* (Bieb.) Tzvelev
♦ crested wheatgrass
♦ Weed, Environmental Weed
♦ 4, 88, 104
♦ G. Origin: Europe.

Agropyron ramosum (Trin.) K.Richt.
Poaceae
= *Leymus ramosus* (Trin.) Tzvelev
♦ Weed
♦ 87, 88
♦ G.

Agropyron repens (L.) Beauv.
Poaceae
= *Elytrigia repens* (L.) Nev.
♦ couchgrass, twitch, quackgrass, falsa gramigna, agropiro invasor
♦ Weed, Naturalised, Noxious Weed, Introduced, Environmental Weed
♦ 7, 8, 21, 23, 30, 35, 36, 44, 49, 52, 80, 87, 88, 91, 114, 118, 121, 136, 141, 146, 159, 162, 180, 186, 203, 207, 210, 218, 219, 236, 237, 243, 263, 269, 294, 295, 299
♦ pG, cultivated, herbal. Origin: Eurasia.

Agropyron semicostatum (Nees ex Steud.) Boiss.
Poaceae
= *Elymus semicostatum* (Nees ex Steud.) Á.Löve
♦ drooping wheatgrass
♦ Weed, Quarantine Weed
♦ 76, 87, 88, 203, 220
♦ G.

Agropyron smithii Rydb.
Poaceae
= *Pascopyrum smithii* (Rydb.) Á.Löve
♦ western wheatgrass
♦ Weed
♦ 87, 88, 218
♦ G, cultivated, herbal.

Agropyron triticeum Gaertn.
Poaceae
= *Eremopyrum triticeum* (Gaertn.) Nevski
♦ Weed
♦ 80
♦ G, herbal.

Agropyron tsukushiense (Honda) Ohwi
Poaceae
= *Elymus tsukushiensis* Honda var. *tsukushiensis* (NoR)
♦ Weed, Quarantine Weed
♦ 76, 87, 88, 203, 220
♦ G.

Agropyron tsukushiense Honda var. transiens (Honda) Ohwi
Poaceae
= *Elymus tsukushiensis* Honda var. *transiens* (Hack.) Osada (NoR) [see *Agropyron kamoji* Ohwi, *Roegneria kamoji* (Ohwi) Ohwi ex Keng]
♦ kamojigusa
♦ Weed
♦ 263

♦ pG.

Agrostemma brachyloba (Fenzl) Hammer
Caryophyllaceae
♦ narrow corncockle
♦ Naturalised
♦ 101

Agrostemma githago L.
Caryophyllaceae
Githago segetum auct., *Lychnis githago* (L.) Scop. (see)
♦ corncockle, common corncockle, purple cockle, corn rose, corn campion, crown of the field, corn mullien, old maid's pink, bastard nigella, cockle, corn pink, mullen pink, purple cockle, rose campion, wild savager
♦ Weed, Naturalised, Casual Alien
♦ 7, 8, 23, 24, 34, 39, 42, 44, 51, 70, 80, 86, 87, 88, 98, 101, 121, 136, 161, 176, 179, 203, 210, 211, 212, 218, 237, 241, 243, 247, 253, 269, 272, 280, 287, 295, 300
♦ a/pH, cultivated, herbal, toxic. Origin: eastern Mediterranean.

Agrostemma linicola Terech.
Caryophyllaceae
♦ Weed
♦ 87, 88
♦ cultivated.

Agrostis L. spp.
Poaceae
♦ bentgrass
♦ Weed
♦ 80, 272
♦ G.

Agrostis alba auct. Amer. non L.
Poaceae
= *Agrostis gigantea* Roth [auct. Eur. = *Agrostis stolonifera* L.]
♦ redtop
♦ Weed, Naturalised
♦ 87, 88, 136, 204, 218, 236, 241, 286, 287
♦ G, herbal.

Agrostis alba L. fo. aristigera Fern.
Poaceae
♦ Naturalised
♦ 287
♦ G.

Agrostis arvensis L.
Poaceae
♦ Naturalised
♦ 241
♦ G.

Agrostis avenacea J.F.Gmel.
Poaceae
♦ bentgrass, blown grass, Pacific bentgrass
♦ Weed, Naturalised, Introduced
♦ 34, 101, 121, 228, 287
♦ pG, arid, cultivated, herbal. Origin: Australia, Chile, New Zealand.

Agrostis avenacea J.Gmel. var. avenacea
Poaceae
♦ blown grass
♦ Native Weed
♦ 269
♦ aG. Origin: Australia.

Agrostis canina L.
Poaceae
♦ brown bent, velvet bentgrass, brown bentgrass
♦ Weed, Naturalised
♦ 70, 80, 87, 88, 91, 241, 272, 287
♦ pG, cultivated, herbal. Origin: Europe.

Agrostis capillaris L.
Poaceae
Agrostis tenuis Sibth. (see), *Agrostis vulgaris* With.
♦ colonial bentgrass, common bent, browntop bent
♦ Weed, Naturalised, Environmental Weed, Cultivation Escape
♦ 7, 15, 20, 34, 70, 72, 80, 86, 88, 98, 101, 146, 151, 176, 181, 198, 203, 225, 241, 243, 246, 252, 269, 272, 280, 290, 296, 300
♦ pG, cultivated, herbal. Origin: Eurasia.

Agrostis capillaris L. var. *aristata* (Parn.) Druce
Poaceae
♦ browntop bent
♦ Naturalised, Environmental Weed
♦ 86, 198
♦ G. Origin: Eurasia.

Agrostis capillaris L. var. *capillaris*
Poaceae
♦ browntop bent
♦ Naturalised, Environmental Weed
♦ 86, 198
♦ G. Origin: Eurasia.

Agrostis castellana Boiss. & Reut.
Poaceae
♦ bentgrass, highland bent
♦ Weed, Naturalised
♦ 87, 88, 250, 280, 300
♦ G, cultivated, herbal. Origin: Europe.

Agrostis clavata Trin.
Poaceae
♦ clavate bentgrass, hoikkarölli, yamanukabo
♦ Weed
♦ 87, 88, 204, 286
♦ G.

Agrostis clavata Trin. ssp. *matsumurae* Tateoka
Poaceae
♦ Weed
♦ 286
♦ G.

Agrostis delicatula Steud. ex Lechl.
Poaceae
= *Aciachne pulvinata* Benth. (NoR)
♦ Weed
♦ 70
♦ G.

Agrostis exarata Trin.
Poaceae
♦ spike bentgrass
♦ Weed
♦ 87, 88
♦ pG, herbal.

Agrostis filifolia Link
Poaceae
♦ Weed
♦ 87, 88
♦ G.

Agrostis geniculata L.
Poaceae
♦ Weed, Naturalised
♦ 98, 203
♦ G.

Agrostis gigantea Roth
Poaceae
Agrostis alba auct. non L. (see), *Agrostis nigra* With. (see), *Agrostis stolonifera* L. ssp. *gigantea* (Roth) Maire & Weiller
♦ black bent, redtop, giant mountain dandelion, fiorin, common bentgrass
♦ Weed, Naturalised, Introduced, Environmental Weed, Cultivation Escape
♦ 7, 30, 34, 38, 44, 70, 80, 86, 87, 88, 91, 101, 151, 176, 198, 243, 252, 253, 272, 280, 290, 295, 300
♦ pG, cultivated, herbal. Origin: Eurasia.

Agrostis hiemalis (Walter) Britton, Sterns & Pogg.
Poaceae
♦ Weed, Naturalised
♦ 98, 203
♦ G, cultivated.

Agrostis inaequiglumis Griseb.
Poaceae
♦ Weed
♦ 244
♦ G.

Agrostis interrupta L.
Poaceae
= *Apera interrupta* (L.) Beauv.
♦ Weed
♦ 80
♦ G, herbal.

Agrostis lachnantha Nees var. *lachnantha*
Poaceae
Agrostis huttoniae (Hack.) C.E.Hubb.
♦ South African bentgrass
♦ Native Weed
♦ 121
♦ a/pG. Origin: southern Africa.

Agrostis montevidensis Spreng. ex Nees
Poaceae
♦ foggrass
♦ Weed
♦ 121, 295
♦ aG. Origin: South America.

Agrostis nigra With.
Poaceae
= *Agrostis gigantea* Roth
♦ Naturalised
♦ 287
♦ G.

Agrostis palustris Huds.
Poaceae
= *Agrostis stolonifera* L.
♦ marsh bentgrass, creeping bentgrass
♦ Weed
♦ 161

♦ G, aqua, herbal.

Agrostis pilosula Trin.
Poaceae
♦ Weed
♦ 244
♦ G.

Agrostis pourretti Willd.
Poaceae
♦ kalpearölli
♦ Weed, Casual Alien
♦ 42, 70, 253
♦ G.

Agrostis salmantica (Lag.) Kunth
Poaceae
♦ Weed
♦ 87, 88
♦ G.

Agrostis scabra Willd.
Poaceae
♦ rough bentgrass, rikkarölli, ticklegrass, ezonukabo
♦ Weed, Naturalised
♦ 42, 88, 218, 286, 300
♦ pG, herbal.

Agrostis semiverticillata (Forssk.) Christ.
Poaceae
= *Polypogon viridis* (Gouan) Breistr.
♦ Weed
♦ 87, 88, 199
♦ G, herbal.

Agrostis spica-venti L.
Poaceae
= *Apera spica-venti* (L.) Beauv.
♦ loose silky bent, silky apera, windgrass
♦ Weed, Quarantine Weed
♦ 44, 76, 80, 87, 88, 203, 220, 243
♦ G.

Agrostis stolonifera L.
Poaceae
Agrostis palustris Huds. (see)
♦ creeping bent, redtop, creeping bentgrass, seaside bentgrass
♦ Weed, Naturalised, Native Weed, Introduced, Garden Escape, Environmental Weed, Cultivation Escape
♦ 7, 15, 21, 30, 34, 38, 44, 68, 70, 72, 80, 86, 87, 88, 91, 98, 146, 159, 161, 174, 176, 181, 198, 203, 212, 243, 253, 272, 280, 286, 287, 290
♦ pG, aqua, cultivated, herbal. Origin: Eurasia, North America.

Agrostis stolonifera L. var. *palustris* (Huds.) Farw.
Poaceae
♦ Naturalised
♦ 300
♦ G.

Agrostis stolonifera L. var. *stolonifera*
Poaceae
♦ creeping bentgrass
♦ Naturalised
♦ 241, 300
♦ G.

Agrostis tandilensis (Kuntze) Parodi
Poaceae

♦ Kennedy's bentgrass
♦ Naturalised
♦ 101
♦ aG.

Agrostis tenuis Sibth.
Poaceae
= *Agrostis capillaris* L.
♦ common bentgrass, browntop, sand couch, Rhode Island bentgrass, colonial bentgrass
♦ Weed, Naturalised
♦ 30, 44, 80, 87, 88, 91, 161, 243, 287
♦ pG, aqua, cultivated, herbal.

Agrostis verticillata Vill.
Poaceae
= *Polypogon viridis* (Gouan) Breistr.
♦ Naturalised
♦ 287
♦ G.

Agrostis vinealis Schreb.
Poaceae
♦ brown bentgrass, brown bent, jäykkärölli
♦ Naturalised
♦ 300
♦ G.

Agrostis viridis Gouan
Poaceae
= *Polypogon viridis* (Gouan) Breistr.
♦ water bent, green bentgrass
♦ Weed
♦ 34, 93, 191, 205
♦ pG, cultivated.

Ailanthus altissima (Mill.) Swingle
Simaroubaceae
Ailanthus glandulosa Desf., *Toxicodendron altissimum* Mill.
♦ tree of heaven, copal tree, varnish tree, hemelboom, Chinese sumac, stinking cedar
♦ Weed, Naturalised, Noxious Weed, Introduced, Garden Escape, Environmental Weed, Cultivation Escape
♦ 3, 4, 7, 8, 15, 17, 34, 35, 39, 45, 72, 78, 80, 86, 87, 88, 95, 98, 101, 102, 116, 121, 129, 129, 132, 133, 137, 142, 146, 147, 151, 152, 161, 179, 195, 198, 203, 211, 218, 222, 224, 228, 231, 269, 272, 280, 283, 286, 287, 289, 296, 300
♦ T, arid, cultivated, herbal, toxic. Origin: temperate and subtropical China.

Ainsworthia trachycarpa Boiss.
Apiaceae
♦ Weed
♦ 111, 243
♦ cultivated.

Aira cappillaris Host
Poaceae
♦ Weed
♦ 87, 88
♦ G.

Aira caryophyllea L.
Poaceae
♦ silver hairgrass, ovsienka mnohokvetá, English hairgrass, silvery hairgrass

♦ Weed, Naturalised, Environmental Weed
♦ 7, 9, 15, 72, 86, 87, 88, 98, 101, 121, 133, 161, 176, 195, 198, 203, 218, 224, 241, 280, 286, 287, 290, 300
♦ aG, cultivated, herbal. Origin: Eurasia.

Aira caryophyllea L. ssp. caryophyllea
Poaceae
♦ silver hairgrass
♦ Naturalised
♦ 280
♦ G.

Aira caryophyllea L. ssp. multiculmis (Dumort.) Bonnier & Layens
Poaceae
♦ Naturalised
♦ 280
♦ G.

Aira cupaniana Guss.
Poaceae
♦ small hairgrass, hairgrass
♦ Weed, Naturalised, Environmental Weed, Casual Alien
♦ 7, 9, 72, 86, 88, 98, 134, 198, 203, 280
♦ aG. Origin: southern Europe, North Africa.

Aira elegans Willd. ex Gaudin
Poaceae
= *Aira elegantissima* Schur
♦ elegant hairgrass, annual silver hairgrass, hairgrass
♦ Naturalised, Environmental Weed
♦ 86, 101, 176, 198, 280, 286, 287
♦ G, herbal.

Aira elegans Willd. ex Gaudin ssp. ambigua (Arcang.) Holub
Poaceae
♦ Naturalised
♦ 287
♦ G.

Aira elegantissima Schur
Poaceae
Aira capillaris Host, *Aira elegans* Willd. ex Kunth (see)
♦ hairgrass, hapsilauha, lace hairgrass, elegant European hairgrass, silver hairgrass
♦ Weed, Naturalised, Environmental Weed, Casual Alien
♦ 7, 42, 72, 88, 98, 203, 241, 272, 280, 300
♦ aG, cultivated, herbal. Origin: Eurasia.

Aira multiculmis Dumort.
Poaceae
Aira caryophyllea L. var. *multiculmis* Dumort.
♦ Weed
♦ 253
♦ G.

Aira praecox L.
Poaceae
♦ early hairgrass, kääpiölauha, yellow hairgrass
♦ Weed, Naturalised, Environmental Weed
♦ 7, 9, 23, 42, 72, 86, 88, 98, 101, 176,

198, 203, 241, 280, 300
♦ aG, cultivated, herbal. Origin: Eurasia.

Aira provincialis Duval-Jouve
Poaceae
♦ Weed, Naturalised
♦ 86, 98, 203
♦ G. Origin: southern Europe.

Aizoon canariense L.
Aizoaceae
♦ Weed
♦ 39, 88, 221
♦ cultivated, herbal, toxic.

Aizoon glinoides L.f.
Aizoaceae
♦ aizoon
♦ Weed
♦ 88, 158
♦ Origin: South Africa.

Ajuga australia R.Br.
Lamiaceae
♦ Weed
♦ 87, 88

Ajuga bracteosa Wall. ex Benth.
Lamiaceae
♦ Weed
♦ 87, 88, 243
♦ pH, promoted, herbal. Origin: east Asia.

Ajuga chamaepitys (L.) Schreb.
Lamiaceae
Teucrium chamaepitys L.
♦ ground pine, yellow bugle, ajuga pianta
♦ Weed, Naturalised
♦ 23, 44, 70, 87, 88, 94, 101, 243, 253, 272
♦ a/pH, cultivated, herbal. Origin: Europe.

Ajuga chia Schreb.
Lamiaceae
Ajuga chamaepitys ssp. *chia* (Schreb.) Arcang., *Ajuga chamaepitys* ssp. *chia* (Schreb.) Murb., *Chamaepitys chia* (Schreb.) Holob
♦ Weed
♦ 87, 88
♦ herbal.

Ajuga decumbens Thunb.
Lamiaceae
♦ kiransou
♦ Weed
♦ 87, 88, 286
♦ pH, promoted, herbal. Origin: east Asia.

Ajuga genevensis L.
Lamiaceae
♦ hammasakankaali, blue bugle, standing bugle
♦ Weed, Naturalised
♦ 42, 80, 101, 272
♦ cultivated, herbal.

Ajuga iva (L.) Schreb.
Lamiaceae
♦ Weed, Naturalised
♦ 70, 86, 98, 203, 221
♦ herbal. Origin: Mediterranean.

Ajuga pyramidalis **L.**
Lamiaceae
♦ pyramidal bugle, erect bugle, kartioakankaali
♦ Weed
♦ 70, 272
♦ cultivated, herbal.

Ajuga remota **Benth.**
Lamiaceae
♦ Weed
♦ 87, 88

Ajuga reptans **L.**
Lamiaceae
♦ bugle, bugleweed, common bugle, creeping bugleweed, carpet bugle, rönsyakankaali, blue bugle
♦ Weed, Naturalised, Garden Escape, Environmental Weed, Cultivation Escape
♦ 4, 42, 44, 54, 70, 80, 86, 87, 88, 94, 101, 121, 133, 176, 195, 272, 280, 286, 287
♦ pH, cultivated, herbal. Origin: Eurasia.

Akebia × *pentaphylla* **(Makino) Makino**
Lardizabalaceae
= *Akebia quinata* (Houtt.) Decne. × *Akebia trifoliata* (Thunb.) Koidz.
♦ akebia
♦ Naturalised
♦ 101
♦ pC, promoted. Origin: east Asia.

Akebia quinata **(Houtt.) Decne.**
Lardizabalaceae
♦ fiveleaf akebia, chocolate vine, akebia vine, akebia
♦ Weed, Quarantine Weed, Naturalised, Environmental Weed
♦ 15, 80, 88, 101, 129, 133, 151, 195, 246, 280, 286
♦ pC, arid, cultivated, herbal. Origin: central China, Korea, Japan.

Akebia trifoliata **(Thunb.) Koidz.**
Lardizabalaceae
Akebia lobata Decne., *Clematis trifoliata* Thunb.
♦ akebia, mitsubaakebi
♦ Weed
♦ 286
♦ pC, cultivated, herbal. Origin: China, Japan, Korea.

Albizia adinocephala **(Donn.Sm.) Britt. & Rose**
Fabaceae/Mimosaceae
♦ cream albizia
♦ Naturalised
♦ 101, 261
♦ Origin: Central America.

Albizia amara **(Roxb.) Boivin**
Fabaceae/Mimosaceae
♦ Weed
♦ 87, 88
♦ cultivated.

Albizia carbonaria **Britt.**
Fabaceae/Mimosaceae
Pithecellobium carbonarium (Britt.) Niez. & Nevl. (see)
♦ naked albizia
♦ Naturalised

♦ 101
♦ T.

Albizia chinensis **(Osb.) Merr.**
Fabaceae/Mimosaceae
♦ Chinese albizia
♦ Weed, Naturalised
♦ 3, 22, 87, 88, 101, 191
♦ T, cultivated, herbal.

Albizia coriaria **Oliv.**
Fabaceae/Mimosaceae
♦ Introduced
♦ 228
♦ arid.

Albizia falcataria **(L.) Fosb.**
Fabaceae/Mimosaceae
= *Paraserianthes falcataria* (L.) I.Nielson
♦ Weed
♦ 3, 22, 80, 87, 88, 191
♦ T.

Albizia harveyi **Fourn.**
Fabaceae/Mimosaceae
Albizia pospischilii Harms
♦ common false thorn, sickle leaved albizia
♦ Weed, Native Weed
♦ 10, 121
♦ T, arid, cultivated. Origin: southern Africa.

Albizia julibrissin **Durazz.**
Fabaceae/Mimosaceae
Acacia mollis Wall., *Sericandra julibrissin* (Durazz.) Raf.
♦ silk tree, silktree albizzia, Persian silk tree, pink mimosa
♦ Weed, Naturalised, Environmental Weed, Casual Alien
♦ 77, 80, 88, 101, 102, 112, 129, 151, 179, 218, 280
♦ S/T, cultivated, herbal. Origin: Iran to Japan.

Albizia lebbeck **(L.) Benth.**
Fabaceae/Mimosaceae
Albizia lebbek sensu auct., *Mimosa lebbeck* L. (see)
♦ lebbeck tree, Indian siris, siris tree, Indian albizia, East Indian walnut, bois noir, kokko, trongkon mames, woman's tongue tree
♦ Weed, Noxious Weed, Naturalised, Introduced, Garden Escape, Environmental Weed, Cultivation Escape
♦ 3, 22, 39, 47, 80, 86, 87, 88, 95, 98, 101, 107, 112, 121, 122, 151, 152, 179, 203, 228, 230, 260, 261, 283
♦ T, arid, cultivated, herbal, toxic. Origin: Old World Tropics.

Albizia lebbekoides **(DC.) Benth.**
Fabaceae/Mimosaceae
Acacia lebbekoides DC.
♦ Indian albizia
♦ Naturalised
♦ 101

Albizia lophantha **(Willd.) Benth.**
Fabaceae/Mimosaceae
= *Paraserianthes lophantha* (Willd.) I.Nielsen
♦ albizia, plume albizia, Australian

albizia, cape wattle, plume albizia, silk tree, sirus, stinkbean, evergreen silk tree
♦ Weed, Naturalised, Environmental Weed, Cultivation Escape
♦ 3, 34, 40, 78, 87, 88, 116, 121, 152, 241
♦ S/T, cultivated, herbal. Origin: Australia.

Albizia odorotissima **(L.f.) Benth.**
Fabaceae/Mimosaceae
♦ Introduced
♦ 228
♦ arid.

Albizia ogadensis **(Chiov.) Chiov.**
Fabaceae/Mimosaceae
= *Acacia ogadensis* Chiov.
♦ Quarantine Weed
♦ 76

Albizia polyantha **(A.Spreng.) G.P.Lewis**
Fabaceae/Mimosaceae
Acacia multiflora Spreng., *Acacia polyantha* A.Spreng., *Arthrosamanea multiflora* (Kunth) Kleinhoonte, *Arthrosamanea polyantha* (A.Spreng.) Burkart, *Cathormion polyanthum* (A.Spreng.) Burkart, *Pithecellobium multiflorum* (Kunth) Benth.
♦ Introduced
♦ 228
♦ arid.

Albizia procera **(Roxb.) Benth.**
Fabaceae/Mimosaceae
Acacia procera Willd., *Mimosa procera* Roxb.
♦ tall albizia, safe siris, bastard lebbeck
♦ Weed, Quarantine Weed, Noxious Weed, Naturalised, Introduced, Environmental Weed, Cultivation Escape
♦ 39, 101, 121, 152, 179, 228, 258, 261, 283
♦ T, arid, cultivated, herbal, toxic. Origin: north-east Africa.

Albizia saman **(Jacq.) F.Muell.**
Fabaceae/Mimosaceae
= *Samanea saman* (Jacq.) Merr.
♦ rain tree
♦ Weed, Introduced
♦ 3, 191, 228
♦ S/T, arid, cultivated.

Albizia saponaria **(Lour.) Blume ex Miq.**
Fabaceae/Mimosaceae
♦ whiteflower albizia
♦ Naturalised
♦ 101
♦ herbal.

Albizia tanganyicensis **Bak.f.**
Fabaceae/Mimosaceae
♦ Weed, Quarantine Weed
♦ 39, 76, 88, 203
♦ cultivated, toxic.

Albizia tanganyicensis **Bak.f. ssp. tanganyicensis**
Fabaceae/Mimosaceae
♦ fever tree, paperbark, paperbark false thorn, sneezewort
♦ Weed, Quarantine Weed, Native Weed

♦ 121, 220
♦ T, toxic. Origin: southern Africa.

Albizia zygia (DC.) J.F.Macbr.
Fabaceae/Mimosaceae
♦ ebamba, albizia
♦ Weed
♦ 88
♦ arid.

Albuca canadensis (L.) F.M.Leight.
Liliaceae/Hyacinthaceae
♦ sentry in box
♦ Weed, Naturalised, Environmental Weed
♦ 7, 86, 98, 203
♦ pH, cultivated. Origin: southern Africa.

Alcea ficifolia L.
Malvaceae
considered a 'doubtful taxon' by most
♦ Antwerp hollyhock, perennial hollyhock
♦ Naturalised, Casual Alien
♦ 101, 280
♦ bH, cultivated, herbal.

Alcea pallida (Waldst. & Kit. ex Willd.) Waldst. & Kit.
Malvaceae
♦ hollyhock
♦ Weed
♦ 88, 94, 272

Alcea rosea L.
Malvaceae
♦ hollyhock, tarhasalkoruusu
♦ Weed, Naturalised, Garden Escape, Cultivation Escape, Casual Alien
♦ 15, 34, 40, 42, 86, 98, 101, 161, 252, 261, 272, 280
♦ pH, cultivated, herbal, toxic. Origin: possibly China.

Alcea striata (DC.) Alef.
Malvaceae
♦ Weed
♦ 221

Alchemilla abyssinica Fresen.
Rosaceae
♦ Weed
♦ 240
♦ aH.

Alchemilla arvensis (L.) Scop.
Rosaceae
= *Aphanes arvensis* L.
♦ parsley piert
♦ Weed
♦ 23, 44, 87, 88, 161, 243, 249
♦ cultivated, herbal.

Alchemilla auriculata Juz.
Rosaceae
♦ korvakepoimulehti
♦ Casual Alien
♦ 42

Alchemilla cymatophylla Juz.
Rosaceae
♦ suppilopoimulehti
♦ Casual Alien
♦ 42

Alchemilla glabricaulis H.Lindb.
Rosaceae

♦ kaljuvarsipoimulehti
♦ Casual Alien
♦ 42

Alchemilla gracilipes Engl.
Rosaceae
♦ Weed, Quarantine Weed
♦ 76, 87, 88, 203, 220

Alchemilla gracilis Opiz
Rosaceae
♦ silkkipoimulehti
♦ Naturalised
♦ 280

Alchemilla heptagona Juz.
Rosaceae
♦ seitsenkulmapoimulehti
♦ Casual Alien
♦ 42

Alchemilla leiophylla Juz.
Rosaceae
♦ vjatkanpoimulehti
♦ Casual Alien
♦ 42

Alchemilla microcarpa Boiss. & Reut.
Rosaceae
= *Aphanes microcarpa* (Boiss. & Reut.) Rothm.
♦ slender parsley piert
♦ Weed
♦ 161
♦ herbal.

Alchemilla mollis (Buser) Rothm.
Rosaceae
♦ lady's mantle
♦ Naturalised
♦ 40
♦ cultivated, herbal.

Alchemilla monticola Opiz
Rosaceae
Alchemilla vulgaris ssp. *pastoralis* auct. (Buser) Murb., *Alchemilla vulgaris* ssp. *palmata* auct. (Gilib.) Gams, *Alchemilla pascualis* S.E.Fröhner, *Alchemilla pastoralis* Buser
♦ hairy lady's mantle, laidunpoimulehti
♦ Naturalised
♦ 101
♦ cultivated, herbal.

Alchemilla pectinata Kunth
Rosaceae
= *Alchemilla orbiculata* Ruiz & Pav. (NoR)
♦ Weed
♦ 199

Alchemilla polemochora Fröhner
Rosaceae
♦ sotapoimulehti
♦ Casual Alien
♦ 42

Alchemilla sarmatica Juz.
Rosaceae
♦ vastakarvapoimulehti
♦ Naturalised
♦ 42

Alchemilla semilunaris Alechin
Rosaceae
♦ puolikuupoimulehti

♦ Casual Alien
♦ 42

Alchemilla subcrenata Buser
Rosaceae
♦ broadtooth lady's mantle, hakamaapoimulehti
♦ Weed, Naturalised
♦ 70, 101

Alchemilla vulgaris auct.
Rosaceae
= *Alchemilla xanthochlora* Rothm.
♦ lady's mantle, lion's foot, alchemilla pianta
♦ Weed
♦ 80
♦ pH, cultivated, herbal. Origin: Europe.

Alchemilla xanthochlora Rothm.
Rosaceae
Alchemilla sylvestris auct., *Alchemilla pratensis* auct. *vix* Opiz, *Alchemilla vulgaris* auct. (see), *Alchemilla vulgaris* ssp. *pratensis* auct. *sensu* Gams
♦ lady's mantle, intermediate lady's mantle, dewcup, pyökkipoimulehti
♦ Weed, Naturalised
♦ 40, 42, 70, 86, 98, 203, 272
♦ cultivated, herbal.

Alchornea cordifolia (Schu. & Thon.) Müll.Arg.
Euphorbiaceae
♦ lungusu, bonji
♦ Weed, Quarantine Weed
♦ 76, 87, 88, 203, 220
♦ herbal.

Alchornea latifolia Sw.
Euphorbiaceae
♦ achiotillo
♦ Weed
♦ 87, 88
♦ T, herbal.

Alchornea laxiflora (Benth.) Pax & K.Hoffm.
Euphorbiaceae
♦ Weed
♦ 88

Alectra Thunb. spp.
Scrophulariaceae
♦ alectra
♦ Weed
♦ 67, 88
♦ H parasitic.

Alectra fluminensis (Vell.) Stearn
Scrophulariaceae
♦ yerba de hierro
♦ Noxious Weed
♦ 140, 229

Alectra kirkii Hemsl.
Scrophulariaceae
♦ Kirk's tobacco witchweed
♦ Weed
♦ 50, 87, 88

Alectra orobanchoides Benth.
Scrophulariaceae
♦ Native Weed
♦ 121
♦ H parasitic. Origin: southern Africa.

Alectra senegalensis Benth.
Scrophulariaceae
♦ Weed
♦ 87, 88

Alectra vogelii Benth.
Scrophulariaceae
♦ cowpea witchweed, Vogel alectra, yellow witchweed
♦ Weed, Native Weed
♦ 50, 51, 87, 88, 121
♦ H, parasitic. Origin: southern Africa.

Alectryon coriaceus (Benth.) Radlk.
Sapindaceae
♦ smooth rambutan
♦ Weed
♦ 179
♦ cultivated. Origin: Australia.

Alectryon tomentosus (F.Muell.) Radlk.
Sapindaceae
♦ woolly rambutan
♦ Naturalised, Garden Escape, Environmental Weed
♦ 86
♦ cultivated. Origin: Australia.

Aleurites cordata (Thunb.) R.Br.
Euphorbiaceae
= *Vernicia cordata* (Thunb.) Airy Shaw (NoR)
♦ Japan wood oil tree, Japan wood oil, tung oil
♦ Naturalised
♦ 287
♦ T, cultivated, herbal, toxic. Origin: China.

Aleurites fordii Hemsl.
Euphorbiaceae
= *Vernicia fordii* (Hemsl.) Airy Shaw
♦ tung oil tree, China wood oil tree, kalonut, tung nut
♦ Weed, Naturalised, Environmental Weed, Casual Alien
♦ 22, 39, 80, 88, 112, 121, 151, 247, 280, 287
♦ T, cultivated, herbal, toxic. Origin: Eurasia.

Aleurites moluccana (L.) Willd.
Euphorbiaceae
Aleurites triloba J.R. & G.Forst (see), *Croton moluccanus* L. (see)
♦ Indian walnut, candlenut tree, candlenut, bancoulier
♦ Weed, Naturalised, Introduced, Environmental Weed, Casual Alien
♦ 3, 22, 39, 86, 101, 107, 161, 228, 259, 261, 280
♦ T, arid, cultivated, herbal, toxic. Origin: Malaysia, western Polynesia.

Aleurites montana (Lour.) P.Wilson
Euphorbiaceae
♦ mu oil tree, tung
♦ Naturalised
♦ 39, 101
♦ cultivated, toxic.

Aleurites triloba J.R. & G.Forst.
Euphorbiaceae
= *Aleurites moluccana* (L.) Willd.
♦ candlenut
♦ Weed

♦ 3, 191
♦ herbal.

Alhagi camelorum Fisch.
Fabaceae/Papilionaceae
= *Alhagi maurorum* Medik.
♦ camel thorn, Caspian manna, kameeldoringos
♦ Weed, Noxious Weed
♦ 10, 51, 80, 87, 88, 121, 218, 278
♦ pS, herbal. Origin: central Asia.

Alhagi graecorum Boiss.
Fabaceae/Papilionaceae
♦ camel thorn, mannatree, Persian mannaplant
♦ Weed
♦ 88, 185
♦ Origin: North Africa, Middle East, south-east Europe.

Alhagi maurorum Medik.
Fabaceae/Papilionaceae
Alhagi camelorum Fisch. (see), *Alhagi mannifera* Desv., *Alhagi persarum* Boiss. & Buhse (see), *Alhagi pseudalhagi* Desv. *nom. nud.* (see), *Hedysarum alhagi* L., *Hedysarum pseud-alhagi* M.Bieb.
♦ kameeldoringbos, camelthorn bush
♦ Weed, Quarantine Weed, Noxious Weed, Naturalised, Introduced
♦ 1, 7, 26, 35, 62, 63, 76, 86, 87, 88, 95, 98, 101, 130, 147, 158, 191, 198, 203, 220, 221, 228, 229, 243, 264, 283
♦ pH, arid, cultivated, herbal. Origin: Eurasia.

Alhagi persarum Boiss. & Buhse
Fabaceae/Papilionaceae
= *Alhagi maurorum* Medik.
♦ Weed
♦ 87, 88

Alhagi pseudalhagi (M.Bieb.) Desv.
Fabaceae/Papilionaceae
= *Alhagi maurorum* Medik.
♦ camel thorn
♦ Weed, Noxious Weed, Quarantine Weed
♦ 34, 35, 78, 80, 80, 87, 88, 116, 146, 161, 212, 220, 231, 272, 275
♦ S, promoted, herbal. Origin: west Asia.

Aliella helichrysoides (Ball) M.Qaiser & H.W.Lack
Asteraceae
♦ Quarantine Weed
♦ 220

Alisma L. spp.
Alismataceae
♦ waterplantain
♦ Weed
♦ 181
♦ H.

Alisma canaliculatum A.Br. & Bouche
Alismataceae
♦ heraomodaka
♦ Weed
♦ 87, 88, 204, 263, 286
♦ wpH, cultivated. Origin: southern China, Japan.

Alisma flava L.
Alismataceae

= *Limnocharis flava* (L.) Buch.
♦ Quarantine Weed
♦ 220
♦ wpH.

Alisma gramineum K.C.Gmel.
Alismataceae
♦ narrowleaf waterplantain, ribbon leaved waterplantain
♦ Weed
♦ 87, 88, 218, 221, 272, 275, 297
♦ wpH, cultivated, herbal.

Alisma lanceolata With.
Alismataceae
= *Alisma lanceolatum* With.
♦ waterplantain
♦ Weed, Naturalised, Environmental Weed
♦ 7, 15, 72, 86, 88, 198, 253
♦ wpH. Origin: Eurasia.

Alisma lanceolatum With.
Alismataceae
Alisma lanceolata With. (see)
♦ lanceleaf waterplantain, narrow leaved waterplantain, waterplantain
♦ Weed, Quarantine Weed, Naturalised, Introduced, Environmental Weed
♦ 34, 70, 76, 87, 88, 98, 101, 155, 200, 203, 272, 280, 300
♦ wpH, cultivated, herbal. Origin: Mediterranean, west Asia.

Alisma orientale (Sam.) Juzep
Alismataceae
= *Alisma plantago-aquatica* L.
♦ waterplantain
♦ Weed
♦ 88, 275, 297
♦ wpH, herbal.

Alisma plantago-aquatica L.
Alismataceae
Alisma orientale (Sam.) Juzep (see)
♦ waterplantain, common waterplantain, mad dog weed, mud plantain, water alisma, American waterplantain
♦ Weed, Quarantine Weed, Naturalised, Native Weed
♦ 39, 68, 70, 76, 86, 87, 88, 98, 101, 121, 126, 159, 161, 165, 176, 180, 181, 203, 208, 220, 253, 269, 272, 280, 300
♦ wpH, cultivated, herbal, toxic. Origin: Europe, west Asia, north and central Africa, Australia.

Alisma plantago-aquatica L. var. orientale Samuels.
Alismataceae
♦ Weed
♦ 263, 286
♦ wpH.

Alisma triviale Pursh
Alismataceae
♦ common waterplantain, northern waterplantain
♦ Weed
♦ 23, 87, 88, 218
♦ wH, herbal.

Alkanna orientalis (L.) Boiss.
Boraginaceae
♦ Weed
♦ 87, 88, 221
♦ cultivated.

Alkanna tinctoria (L.) Tausch
Boraginaceae
♦ dyer's bugloss, dyer's alkanet, alkanna, alkanet
♦ Weed
♦ 221, 272
♦ pH, arid, cultivated, herbal.

Allagoptera arenaria (Gomes) Kuntze
Arecaceae
Allagoptera littorale (Mart.) Kuntze, *Allagoptera pumila* Nees, *Cocos arenaria* Gomes, *Diplothemium arenarium* (Gomes) Vasc. & Franco, *Diplothemium littorale* Mart., *Diplothemium maritimum* Mart.
♦ buri de praia, coqueiro do praia, coqueiro guriry, guriri, imburg, puranan
♦ Introduced
♦ 228
♦ arid, cultivated.

Allamanda blanchetii A.DC.
Apocynaceae
♦ purple allamanda
♦ Naturalised, Introduced
♦ 101, 261
♦ cultivated, herbal. Origin: Brazil.

Allamanda cathartica L.
Apocynaceae
Allamanda hendersonii Bull ex Dombrain (see)
♦ yellow trumpet vine, allamanda, golden allamanda, golden cup, lani ali'I, pua tanofo
♦ Weed, Sleeper Weed, Naturalised, Introduced, Garden Escape, Environmental Weed, Cultivation Escape
♦ 3, 39, 54, 86, 88, 98, 101, 107, 155, 161, 179, 203, 230, 247, 261
♦ pC/S, cultivated, herbal, toxic. Origin: north-eastern South America.

Allamanda hendersonii Bull ex Dombrain
Apocynaceae
= *Allamanda cathartica* L.
♦ pahtoh
♦ Weed, Introduced
♦ 3, 191, 230
♦ S, cultivated.

Allamanda laevis Markgraf
Apocynaceae
♦ Quarantine Weed
♦ 220
♦ cultivated.

Allamanda neriifolia Hook.
Apocynaceae
♦ cautiva
♦ Introduced
♦ 39, 261
♦ cultivated, herbal, toxic. Origin: Brazil.

Allamanda oenotherifolia Pohl.
Apocynaceae
Allamanda oenotheraefolia Pohl.
♦ Introduced
♦ 230
♦ cultivated.

Allamanda schottii Pohl
Apocynaceae
♦ bush allamanda
♦ Naturalised
♦ 101
♦ cultivated.

Allenrolfea vaginata (Griseb.) Kuntze
Chenopodiaceae
♦ Introduced
♦ 228
♦ arid.

Alliaria officinalis Andrz. ex M.Bieb. nom. illeg.
Brassicaceae
= *Alliaira petiolata* (Bieb.) Cavara & Grande
♦ garlic mustard
♦ Weed
♦ 45, 88, 132, 218
♦ herbal.

Alliaria petiolata (Bieb.) Cavara & Grande
Brassicaceae
Alliaria officinalis Andrz. ex Bieb. *nom. illeg.* (see), *Sisymbrium alliaria* Scop.
♦ garlic mustard, hedge garlic, sauce alone, Jack by the hedge, cesnaáka lekárska, poor man's mustard, Jack in the bush, garlic root, garlicwort, mustard root
♦ Weed, Naturalised, Introduced, Environmental Weed
♦ 4, 17, 70, 80, 87, 88, 101, 102, 103, 104, 129, 129, 133, 151, 182, 195, 222, 224, 243, 272, 280, 293
♦ bH, cultivated, herbal, toxic. Origin: Europe.

Allionia incarnata L.
Nyctaginaceae
♦ trailing four o'clock, trailing allionia, trailing windmills
♦ Weed
♦ 88, 161, 218
♦ pH, arid, cultivated, herbal.

Allium acuminatum Hook.
Liliaceae/Alliaceae
♦ tapertip onion, Hooker's onion
♦ Weed
♦ 88, 218
♦ pH, cultivated, herbal. Origin: western North America.

Allium ampeloprasum L.
Liliaceae/Alliaceae
Allium myrianthum Boiss. (see)
♦ wild leek, broadleaf wild leek, elephant garlic
♦ Weed, Naturalised, Introduced, Garden Escape, Environmental Weed, Casual Alien
♦ 7, 70, 86, 87, 88, 98, 101, 176, 198, 203, 221, 228, 272, 280

♦ pH, arid, cultivated, herbal. Origin: southern Europe to west Asia.

Allium ampeloprasum L. var. ampeloprasum
Liliaceae/Alliaceae
♦ broadleaf wild leek
♦ Naturalised
♦ 101

Allium ampeloprasum L. var. atroviolaceum (Boiss.) Regel
Liliaceae/Alliaceae
♦ broadleaf wild leek
♦ Naturalised
♦ 101

Allium amplectens Torr.
Liliaceae/Alliaceae
♦ narrowleaf onion
♦ Weed
♦ 161
♦ pH, cultivated, herbal.

Allium angulosum L.
Liliaceae/Alliaceae
Allium acutangulum Schrad., *Allium inodorum* Willd, *Allium laxum* Don
♦ särmälaukka, mouse garlic
♦ Weed, Casual Alien
♦ 42, 87, 88, 272
♦ pH, cultivated, herbal. Origin: Eurasia.

Allium aschersonianum Barbey
Liliaceae/Alliaceae
♦ Weed
♦ 221
♦ arid.

Allium atropurpureum Waldst. & Kit.
Liliaceae/Alliaceae
♦ Weed
♦ 272
♦ pH, cultivated. Origin: Eurasia.

Allium atroviolaceum Boiss.
Liliaceae/Alliaceae
♦ sotet hagyma
♦ Weed
♦ 272
♦ cultivated.

Allium caeruleum Pall.
Liliaceae/Alliaceae
♦ blue allium
♦ Weed
♦ 272
♦ cultivated, herbal.

Allium canadense L.
Liliaceae/Alliaceae
Allium mutabile Michx. (see)
♦ wild onion, meadow leek, meadow garlic
♦ Weed, Native Weed
♦ 23, 39, 88, 161, 174, 210, 218, 266
♦ pH, cultivated, herbal, toxic. Origin: North America.

Allium carinatum L.
Liliaceae/Alliaceae
♦ keeled garlic
♦ Weed, Casual Alien
♦ 40, 70
♦ pH, cultivated, herbal. Origin: Europe.

Allium cepa L.
Liliaceae/Alliaceae
- ruokasipuli, garden onion, shallot, cultivated onion, ngengi
- Weed, Naturalised, Introduced, Cultivation Escape, Casual Alien
- 39, 42, 87, 88, 98, 101, 161, 203, 230, 280
- pH, cultivated, herbal, toxic. Origin: western Asia.

Allium cepa L. var. cepa
Liliaceae/Alliaceae
- garden onion
- Naturalised
- 101

Allium cepa L. var. viviparum M.C.Metz
Liliaceae/Alliaceae
- garden onion
- Naturalised
- 101
- cultivated, herbal.

Allium cernuum Roth.
Liliaceae/Alliaceae
- wild onion, lady's leek, nodding onion
- Weed
- 161
- pH, cultivated, herbal, toxic.

Allium decipiens Fisch. ex Schult. & Schult.f.
Liliaceae/Alliaceae
- Weed
- 272

Allium desertorum Forssk.
Liliaceae/Alliaceae
- Weed
- 221

Allium fistulosum L.
Liliaceae/Alliaceae
- Welsh onion
- Naturalised
- 101
- pH, cultivated, herbal. Origin: obscure.

Allium flavum L.
Liliaceae/Alliaceae
- small yellow onion
- Weed
- 87, 88
- a/pH, cultivated, herbal.

Allium geyeri S.Watson
Liliaceae/Alliaceae
- Geyer's onion, meadow onion
- Weed
- 161
- pH, cultivated, herbal. Origin: western North America.

Allium giganteum Regel
Liliaceae/Alliaceae
- Weed
- 80
- cultivated, herbal.

Allium grayi Regel
Liliaceae/Alliaceae
Allium nipponicum Franch. & Sav.
- Weed
- 286

- pH, promoted, herbal. Origin: Japan.

Allium macrostemon Bunge
Liliaceae/Alliaceae
- nobiru
- Weed
- 87, 88, 275, 297
- pH, promoted. Origin: east Asia.

Allium modestum Boiss.
Liliaceae/Alliaceae
- Weed
- 221

Allium moly L.
Liliaceae/Alliaceae
- keltasipuli, lily leek, golden garlic
- Casual Alien
- 39, 42
- pH, cultivated, herbal, toxic. Origin: Mediterranean.

Allium mongolicum Regel
Liliaceae/Alliaceae
- Mongolica onion
- Weed
- 297

Allium mutabile Michx.
Liliaceae/Alliaceae
= *Allium canadense* L.
- purple onion, wild onion
- Weed
- 88, 218
- pH, promoted, herbal. Origin: south-eastern North America.

Allium myrianthum Boiss.
Liliaceae/Alliaceae
= *Allium ampeloprasum* L.
- Weed
- 221

Allium neapolitanum Cirillo
Liliaceae/Alliaceae
Nothoscordum inodorum (Ait.) Nichols. (see)
- white garlic, flowering onion, false garlic, Naples onion, daffodil garlic
- Weed, Noxious Weed, Naturalised, Garden Escape, Environmental Weed
- 34, 40, 72, 86, 88, 98, 101, 116, 126, 161, 176, 198, 203, 229, 272, 280
- pH, cultivated, herbal. Origin: Mediterranean.

Allium nigrum L.
Liliaceae/Alliaceae
Allium multibulbosum Jacq.
- black onion, black garlic
- Weed, Naturalised
- 44, 70, 87, 88, 101, 126
- cultivated, herbal.

Allium oleraceum L.
Liliaceae/Alliaceae
- field garlic
- Weed, Naturalised
- 44, 70, 87, 88, 98, 101, 126, 198, 203, 253, 272
- pH, cultivated, herbal. Origin: Europe.

Allium oreophilum C.A.Mey.
Liliaceae/Alliaceae
- lehtolaukka, dwarf rose leek
- Cultivation Escape

- 42
- cultivated, herbal.

Allium orientale Boiss.
Liliaceae/Alliaceae
- Weed, Naturalised
- 98, 203
- pH, promoted. Origin: North Africa to west Asia.

Allium paniculatum L.
Liliaceae/Alliaceae
- Mediterranean onion, panicled onion
- Weed, Noxious Weed, Naturalised
- 35, 88, 98, 101, 161, 203, 229, 253
- cultivated.

Allium paniculatum L. ssp. paniculatum
Liliaceae/Alliaceae
- leek
- Naturalised, Cultivation Escape
- 86, 198, 252
- cultivated. Origin: Eurasia.

Allium polyanthum Schult. & Schult.f
Liliaceae/Alliaceae
- Weed
- 253

Allium porrum L.
Liliaceae/Alliaceae
- garden leek
- Weed, Naturalised, Casual Alien
- 39, 98, 101, 203, 280
- cultivated, herbal, toxic.

Allium ramosum L.
Liliaceae/Alliaceae
- branchy onion, Chinese chives, doftlk, tuoksulaukka
- Weed
- 297
- pH, cultivated, herbal.

Allium roseum L.
Liliaceae/Alliaceae
- rusosipuli, rosy garlic
- Weed, Naturalised, Cultivation Escape, Casual Alien
- 40, 42, 86, 87, 88, 98, 203, 252, 253, 280
- pH, cultivated, herbal. Origin: Mediterranean.

Allium roseum ssp. bulbiferum (DC.) E.F.Warb.
Liliaceae/Alliaceae
- Naturalised
- 280

Allium rotundum L.
Liliaceae/Alliaceae
= *Allium scorodoprasum* ssp. *rotundum* (L.) Stearn
- Weed
- 30, 44, 87, 88, 126, 243
- cultivated, herbal.

Allium rubellum Bieb.
Liliaceae/Alliaceae
Allium albanum Grossh.
- Weed
- 87, 88
- pH, promoted. Origin: Eurasia.

Allium sativum L.
Liliaceae/Alliaceae

- garlic
- Weed, Naturalised, Garden Escape
- 39, 80, 87, 88, 101, 161
- pH, cultivated, herbal, toxic. Origin: Asia.

Allium saxatile Bieb.
Liliaceae/Alliaceae
- stone leek
- Weed
- 272

Allium schoenoprasum L.
Liliaceae/Alliaceae
- chives
- Weed, Naturalised
- 39, 70, 101, 161
- pH, cultivated, herbal, toxic. Origin: Europe, north Asia.

Allium schoenoprasum L. var. schoenoprasum Mansf.
Liliaceae/Alliaceae
- wild chives, ezonegi
- Naturalised
- 101

Allium scorodoprasum L.
Liliaceae/Alliaceae
- sand leek, käärmeenlaukka, rocambole
- Weed, Naturalised
- 98, 101, 198, 203
- pH, cultivated, herbal. Origin: Europe.

Allium scorodoprasum L. ssp. rotundum (L.) Stearn
Liliaceae/Alliaceae
Allium rotundum L. (see), *Porrum rotundum* Riechb.
- purple flowered garlic, round headed sand leek
- Weed, Naturalised, Garden Escape
- 86, 252, 272
- cultivated.

Allium scorodoprasum L. ssp. scorodoprasum
Liliaceae/Alliaceae
- sand leek
- Weed, Naturalised, Garden Escape
- 86, 252, 272
- cultivated.

Allium scorodoprasum L. ssp. waldsteinii (Don) Stearn
Liliaceae/Alliaceae
Allium waldsteinii Don
- Weed
- 272

Allium sibiricum L.
Liliaceae/Alliaceae
= *Allium schoenoprasum* L. var. *sibiricum* (L.) Hartm. (NoR)
- Weed
- 70

Allium sinaiticum Boiss.
Liliaceae/Alliaceae
- Weed
- 221

Allium sphaerocephalon L.
Liliaceae/Alliaceae

Allium sphaerocephalum Land., *Allium veronense* Poll.
- round headed leek, ballhead onion
- Weed
- 30, 44, 70, 88, 126, 221, 253, 272
- pH, cultivated, herbal. Origin: Europe.

Allium subhirsutum L.
Liliaceae/Alliaceae
- hairy garlic
- Weed, Casual Alien
- 40, 87, 88
- pH, cultivated. Origin: Mediterranean.

Allium textile A.Nelson & J.F.Macbr.
Liliaceae/Alliaceae
- prairie onion, textile onion
- Weed
- 161, 212
- pH, cultivated, herbal. Origin: North America.

Allium tricoccum Aiton
Liliaceae/Alliaceae
- wild leek, ramp
- Weed
- 39, 88, 161, 218
- pH, cultivated, herbal, toxic. Origin: eastern North America.

Allium triquetrum L.
Liliaceae/Alliaceae
- three cornered leek, angled onion, flowering onion, three corner garlic, triangular stalked garlic
- Weed, Noxious Weed, Naturalised, Garden Escape, Environmental Weed
- 7, 15, 34, 40, 72, 80, 86, 87, 88, 98, 101, 116, 147, 165, 171, 176, 198, 203, 225, 246, 269, 280, 289, 296
- pH, cultivated, herbal. Origin: western Mediterranean.

Allium tuberosum Rottler ex Spreng.
Liliaceae/Alliaceae
- Chinese chives, garlic chives
- Naturalised
- 101
- pH, cultivated, herbal. Origin: obscure, possibly east Asia.

Allium ursinum L.
Liliaceae/Alliaceae
- wood garlic, ramsons
- Weed, Quarantine Weed, Naturalised
- 23, 39, 70, 76, 86, 88, 272
- pH, cultivated, herbal, toxic. Origin: Europe.

Allium validum S.Watson
Liliaceae/Alliaceae
- swamp onion, Pacific onion
- Weed
- 39, 161
- pH, cultivated, herbal, toxic.

Allium vineale L.
Liliaceae/Alliaceae
- crow garlic, wild garlic, field garlic, wild onion, stag's garlic, scallions, ail des vignes
- Weed, Quarantine Weed, Noxious

Weed, Naturalised, Garden Escape, Environmental Weed
- 7, 23, 30, 35, 39, 44, 70, 80, 86, 87, 88, 98, 101, 102, 121, 126, 133, 147, 151, 161, 171, 176, 195, 198, 203, 207, 210, 211, 212, 218, 220, 224, 229, 243, 249, 253, 266, 269, 272, 280, 300
- pH, cultivated, herbal, toxic. Origin: Europe, North Africa, Asia Minor.

Allium vineale L. ssp. compactum (Thuill.) Coss. & Germ.
Liliaceae/Alliaceae
- wild garlic, compact onion
- Noxious Weed, Naturalised
- 101, 229

Allium vineale L. ssp. vineale
Liliaceae/Alliaceae
- wild garlic
- Noxious Weed, Naturalised
- 101, 229

Allmania nodiflora (L.) R.Br. ex Wight
Amaranthaceae
Celosia nodiflora L., *Allmania albida* R.Br.
- vallikeerai, kumuttikeerai, thoyyan keerai
- Weed, Quarantine Weed
- 76, 87, 88, 170, 191, 203, 220
- herbal.

Allocasuarina littoralis (Salisb.) L.A.S.Johnson
Casuarinaceae
- Casual Alien
- 280
- cultivated.

Allocasuarina torulosa (Aiton) L.A.S.Johnson
Casuarinaceae
- forest oak
- Naturalised
- 198
- cultivated, herbal. Origin: Australia.

Alloplectus schultzei Mansf.
Gesneriaceae
- Quarantine Weed
- 220

Alloteropsis cimicina (L.) Stapf
Poaceae
Axonopus cimicinus (L.) P.Beauv., *Milium cimicinum* L.
- summergrass
- Weed, Naturalised, Introduced
- 66, 87, 88, 101, 228
- G, arid. Origin: Africa, Asia, Australia.

Alnus Mill. spp.
Betulaceae
- alder
- Weed, Quarantine Weed
- 161, 258

Alnus cordata (Loisel.) Duby
Betulaceae
- Italian alder
- Naturalised, Garden Escape
- 86, 98, 176, 251
- T, cultivated, herbal. Origin: southern Europe.

Alnus crispa (Aiton) Pursh
Betulaceae
= *Alnus viridis* (Vill.) Lam. & DC. ssp.
crispa (Ait.) Turrill (NoR)
♦ American green alder
♦ Weed
♦ 88, 218
♦ cultivated, herbal.

Alnus glutinosa (L.) Gaertn.
Betulaceae
Alnus rotundifolia Mill., *Betula glutinosa*
(L.) Lam.
♦ black alder, European alder, alder,
common alder, sticky alder
♦ Weed, Naturalised, Introduced,
Garden Escape, Environmental Weed
♦ 4, 15, 20, 54, 70, 80, 86, 88, 101, 102,
104, 133, 151, 155, 165, 195, 222, 225,
241, 246, 280, 290, 300
♦ T, cultivated, herbal. Origin: Eurasia.

Alnus incana (L.) Moench
Betulaceae
♦ grey alder, white alder
♦ Weed
♦ 70
♦ T, cultivated, herbal. Origin: Europe.

Alnus nepalensis D.Don
Betulaceae
♦ Nepal alder
♦ Naturalised
♦ 101
♦ T, cultivated. Origin: China,
Himalayas.

Alnus pendula Matsum.
Betulaceae
♦ himeyashabushi
♦ Weed
♦ 286
♦ cultivated. Origin: Japan.

Alnus × pubescens Tausch
Betulaceae
= *Alnus glutinosa* (L.) Gaertn. × *Alnus
incana* Medik.
♦ hybrid alder
♦ Cultivation Escape
♦ 40
♦ cultivated, herbal.

Alnus rhombifolia Nutt.
Betulaceae
♦ white alder
♦ Weed
♦ 88, 218
♦ T, cultivated, herbal. Origin: North
America.

Alnus rubra Bong.
Betulaceae
Alnus oregona Nutt.
♦ red alder, smooth alder, Oregon
alder
♦ Weed, Sleeper Weed, Naturalised,
Garden Escape, Environmental Weed
♦ 86, 88, 155, 218
♦ S/T, cultivated, herbal. Origin:
coastal north-western North America.

Alnus rugosa (Du Roi) Spreng.
Betulaceae
= *Alnus incana* (L.) Moench ssp. *rugosa*
(Du Roi) R.T.Clausen (NoR)

♦ speckled alder
♦ Weed
♦ 88, 218
♦ T, cultivated, herbal. Origin: North
America.

Alnus serrulata (Aiton) Willd.
Betulaceae
Betula serrulata Aiton, *Alnus
noveboracensis* Britton, *Alnus rubra*
Desf. ex Spach, *Alnus rugosa* (Du Roi)
Spreng. var. *serrulata* (Ait.) Winkler.
♦ hazel alder, smooth alder, tag alder
♦ Weed
♦ 88, 218
♦ S, cultivated, herbal. Origin: eastern
North America.

Alnus sinuata (Reg.) Rydb.
Betulaceae
= *Alnus viridis* (Vill.) Lam. & DC. ssp.
sinuata (Regel) Á. & D.Löve (NoR)
♦ sitka alder
♦ Weed
♦ 88, 218
♦ S, promoted, herbal. Origin: western
North America.

Alnus viridis (Chaix) DC.
Betulaceae
♦ sitka alder, green alder
♦ Naturalised
♦ 280
♦ cultivated, herbal.

**Alocasia brisbanensis (F.M.Bailey)
Domin.**
Araceae
Alocasia macrorrhiza auct. Aust. *non* (L.)
G.Don. (see)
♦ elephant's ear, cunjevoi, spoon lily
♦ Weed, Quarantine Weed,
Naturalised, Environmental Weed
♦ 86, 165, 225, 246, 280
♦ cultivated. Origin: Australia.

Alocasia indica (Lour.) Koch
Araceae
= *Alocasia macrorrhizos* (L.) G.Don
♦ Weed
♦ 39, 87, 88
♦ aqua, cultivated, herbal, toxic.

Alocasia macrorrhiza (L.) Schott
Araceae
= *Alocasia macrorrhizos* (L.) G.Don
[in Australia = *Alocasia brisbanensis*
(F.M.Bailey) Domin]
♦ elephant's ear, giant taro
♦ Weed, Naturalised, Introduced,
Garden Escape, Environmental Weed
♦ 7, 15, 39, 87, 88, 152, 230, 261, 262
♦ aqua, cultivated, herbal, toxic.
Origin: tropical Asia.

Alocasia macrorrhizos (L.) G.Don
Araceae
Alocasia macrorrhiza (L.) Schott (see),
Alocasia indica (Lour.) Koch (see)
♦ giant taro
♦ Weed, Naturalised
♦ 101, 179
♦ cultivated, herbal. Origin: tropical
Asia.

Alocasia plumbea K.Koch ex Van Houtte
Araceae
♦ metallic taro
♦ Naturalised, Introduced, Garden
Escape
♦ 38, 101, 261
♦ cultivated, herbal. Origin: Malaysia.

Aloe L. spp.
Aloeaceae
♦ aloe
♦ Weed, Naturalised
♦ 161, 198, 247
♦ herbal, toxic.

Aloe aculeata Pole-Evans
Aloeaceae
♦ Native Weed
♦ 121
♦ pH, cultivated. Origin: southern
Africa.

Aloe arborescens Mill.
Aloeaceae
♦ candelabra aloe, torch aloe
♦ Weed, Naturalised, Garden Escape
♦ 39, 86, 98, 198, 203, 251
♦ cultivated, herbal, toxic. Origin:
southern Africa.

Aloe barbadensis Mill.
Aloeaceae
= *Aloe vera* L.
♦ aloe vera
♦ Weed
♦ 87, 88
♦ cultivated, herbal.

Aloe cameronii Hemsl.
Aloeaceae
♦ aloe
♦ Weed, Naturalised, Garden Escape,
Environmental Weed, Cultivation
Escape
♦ 54, 86, 88, 155
♦ cultivated. Origin: Africa.

Aloe castanea Schönl.
Aloeaceae
♦ Native Weed
♦ 121
♦ S, cultivated. Origin: southern
Africa.

Aloe davyana Schönl. var. davyana
Aloeaceae
= *Aloe greatheadii* (Schönl.) Schönl. var.
davyana Glen & Hardy (NoR)
♦ Native Weed
♦ 121
♦ pH. Origin: southern Africa.

Aloe ferox Mill.
Aloeaceae
♦ bitter aloe, cape aloe
♦ Weed, Sleeper Weed, Quarantine
Weed, Naturalised, Native Weed,
Garden Escape, Environmental Weed
♦ 39, 76, 86, 88, 121, 155, 203
♦ S, arid, cultivated, herbal, toxic.
Origin: southern Africa.

Aloe latifolia (Haw.) Haw.
Aloeaceae
= *Aloe maculata* All.
♦ Weed, Naturalised

♦ 98, 203
♦ herbal.

Aloe lutescens Groenewald
Aloeaceae
♦ Native Weed
♦ 121
♦ pH, cultivated. Origin: southern Africa.

Aloe maculata All.
Aloeaceae
Aloe saponaria (Aiton) Haw. (see), *Aloe latifolia* (Haw.) Haw. (see), *Aloe perfoliata* var. *saponaria* Ait.
♦ broadleaf aloe, common soap aloe, soap aloe
♦ Naturalised, Garden Escape, Environmental Weed
♦ 86, 98, 251, 289
♦ cultivated. Origin: southern Africa.

Aloe marlothii Berger var. *marlothii*
Aloeaceae
♦ mountain aloe
♦ Native Weed
♦ 121
♦ pS. Origin: southern Africa.

Aloe mutans Reynolds
Aloeaceae
♦ Native Weed
♦ 121
♦ pH, cultivated. Origin: southern Africa.

Aloe parvibracteata Schönl.
Aloeaceae
♦ Weed, Naturalised, Native Weed
♦ 86, 98, 121, 203
♦ pH, arid, cultivated. Origin: southern Africa.

Aloe saponaria (Aiton) Haw.
Aloeaceae
= *Aloe maculata* All.
♦ soap aloe
♦ Weed, Naturalised, Garden Escape, Environmental Weed
♦ 39, 54, 72, 88, 98, 198, 203, 280
♦ S, cultivated, herbal, toxic. Origin: South Africa.

Aloe saponaria (Aiton) Haw. var. *ficksburgensis* Reynolds
Aloeaceae
♦ Naturalised, Environmental Weed
♦ 86

Aloe saponaria (Aiton) Haw. var. *saponaria*
Aloeaceae
♦ soap aloe
♦ Naturalised, Environmental Weed
♦ 86

Aloe vera (L.) Burm.f.
Aloeaceae
Aloe barbadensis Mill. (see), *Aloe perfoliata* var. *vera* L., *Aloe vulgaris* Lam. nom. illeg., *Aloe vera* Tourn. ex L.
♦ Barbados aloe, aloe vera, star cactus, sabila, sempervivum, sinkle bible, zabila, zavila, West Indian aloes
♦ Weed, Naturalised, Introduced, Cultivation Escape

♦ 39, 101, 179, 189, 228, 247, 261
♦ pH, arid, cultivated, herbal, toxic. Origin: Mediterranean.

Aloe wickensii Pole Evans
Aloeaceae
♦ Native Weed
♦ 121
♦ pH, cultivated. Origin: obscure, possibly Africa.

Alopecurus L. spp.
Poaceae
♦ foxtail grass, foxtail
♦ Weed
♦ 243
♦ G, herbal.

Alopecurus aequalis Sobol.
Poaceae
Alopecurus aristulatus Michx., *Alopecurus fulvus* J.E.Sm. (see)
♦ short awned foxtail, orange foxtail, marsh foxtail, rantapuntarpää
♦ Weed, Naturalised
♦ 23, 86, 87, 88, 91, 98, 198, 203, 204, 217, 243, 275, 280, 286, 297, 300
♦ wpG, promoted, herbal. Origin: Eurasia, North America.

Alopecurus aequalis Sobol. var. *aequalis*
Poaceae
♦ shortawn foxtail
♦ Naturalised
♦ 287
♦ G.

Alopecurus aequalis Sobol. var. *amurensis* (Kom.) Ohwi
Poaceae
Alopecurus amurensis Kom. (see)
♦ shortawn foxtail
♦ Weed
♦ 235, 243, 263, 274
♦ pG.

Alopecurus agrestis L.
Poaceae
= *Alopecurus myosuroides* Huds.
♦ Weed
♦ 87, 88
♦ G.

Alopecurus amurensis Kom.
Poaceae
= *Alopecurus aequalis* Sobol. var. *amurensis* (Kom.) Ohwi
♦ Weed
♦ 87, 88
♦ G.

Alopecurus anthoxantoides Boiss.
Poaceae
♦ Weed
♦ 87, 88
♦ G.

Alopecurus arundinaceus Poir.
Poaceae
Alopecurus armenus (K.Koch) Grossh., *Alopecurus arundinaceus* ssp. *armenus* (K.Koch) Tzvelev, *Alopecurus brachystachyus* M.Bieb., *Alopecurus candicans* Salzm. ex Steud., *Alopecurus castellanus* Boiss. & Reut., *Alopecurus exaltatus* Less., *Alopecurus lasiostachyus*

Link, *Alopecurus muticus* Karav. & Kir., *Alopecurus nigricans* Hornem., *Alopecurus nigricans* var. *ventricosus* (Pers.) Reichb., *Alopecurus pratensis* ssp. *arundinaceus* (Poir.) Douin ex Bonn., *Alopecurus pratensis* ssp. *ventricosus* (Pers.) Thell., *Alopecurus pratensis* var. *armenus* K.Koch, *Alopecurus pratensis* var. *arundinaceus* (Poir.) Kuntze, *Alopecurus pratensis* var. *exalatus* (Less.) Griseb., *Alopecurus pratensis* var. *nigricans* (Hornem.) Wahlenb., *Alopecurus pratensis* var. *ventricosus* (Pers.) Coss. & Durand, *Alopecurus repens* M.Bieb., *Alopecurus ruthenicus* Weinm., *Alopecurus salvatoris* Losc. ex Willk., *Alopecurus sibiricus* hort. ex Roem. & Schult., *Alopecurus ventricosus* Pers., *Alopecurus ventricosus* var. *exserens* (Griseb.) Asch. & Graebn.
♦ creeping meadow foxtail, garrison creeping foxtail
♦ Weed, Naturalised, Introduced
♦ 80, 87, 88, 101, 168
♦ pG, cultivated. Origin: Eurasia.

Alopecurus carolinianus Walter
Poaceae
♦ Carolina foxtail
♦ Weed
♦ 34, 161
♦ aG, herbal.

Alopecurus creticus Trin.
Poaceae
♦ Cretan meadow foxtail
♦ Naturalised
♦ 101
♦ G.

Alopecurus fulvus J.E.Sm.
Poaceae
= *Alopecurus aequalis* Sobol.
♦ Weed
♦ 87, 88
♦ G.

Alopecurus geniculatus L.
Poaceae
Alopecurus australis Nees
♦ water foxtail, kneed foxtail, marsh foxtail
♦ Weed, Naturalised, Environmental Weed
♦ 7, 15, 70, 72, 86, 87, 88, 91, 98, 101, 176, 198, 203, 218, 243, 246, 269, 272, 280
♦ pG, aqua, cultivated, herbal. Origin: northern hemisphere.

Alopecurus geniculatus L. var. *geniculatus*
Poaceae
♦ water foxtail
♦ Naturalised
♦ 241, 300
♦ G.

Alopecurus japonicus Steud.
Poaceae
♦ Japanese foxtail
♦ Weed
♦ 68, 87, 88, 243, 275, 286, 297
♦ G.

Alopecurus myosuroides Huds.
Poaceae
Alopecurus agrestis L. (see)
♦ slender foxtail, blackgrass, mousetail grass
♦ Weed, Noxious Weed, Naturalised, Environmental Weed, Casual Alien
♦ 1, 7, 30, 34, 42, 44, 68, 70, 80, 86, 87, 88, 91, 98, 101, 111, 118, 146, 161, 176, 203, 212, 217, 229, 243, 253, 263, 272, 280, 286, 300
♦ aG, cultivated, herbal. Origin: Mediterranean.

Alopecurus pratensis L.
Poaceae
♦ meadow foxtail, meadow foxtail grass
♦ Weed, Naturalised, Introduced, Environmental Weed
♦ 7, 15, 21, 70, 72, 80, 86, 87, 88, 98, 101, 176, 198, 203, 218, 243, 272, 280, 286, 287, 300
♦ pG, aqua, cultivated, herbal. Origin: Eurasia.

Alopecurus rendlei Eig
Poaceae
Alopecurus utriculatis (L.) Sol., *Tozzettia pratensis* Savi, *Tozzettia utriculata* Savi
♦ Rendle's meadow foxtail
♦ Weed, Naturalised
♦ 91, 101
♦ G.

Alopecurus utriculatus Sol.
Poaceae
♦ rakkopuntarpää
♦ Weed, Casual Alien
♦ 42, 272
♦ G, herbal.

Aloysia gratissima (Gillies & Hook.) Tronc.
Verbenaceae
♦ Texas whitebrush
♦ Weed
♦ 161
♦ herbal, toxic.

Aloysia looseri Mold.
Verbenaceae
♦ Naturalised
♦ 241

Aloysia lycioides Cham.
Verbenaceae
= *Aloysia gratissima* (Gill. & Hook.) Tronc. var. *gratissima* (NoR)
♦ whitebrush
♦ Weed
♦ 39, 87, 88, 218
♦ herbal, toxic.

Aloysia triphylla (L'Hér.) Britt.
Verbenaceae
Aloysia citrodora Palau, *Lippia citrodora* (Lam.) Kunth, *Lippia triphylla* (L'Hér.) Kuntze, *Verbena triphylla* L'Hér.
♦ lemon beebrush, lemon verbena, lippia lippia
♦ Naturalised, Casual Alien
♦ 86, 101, 241, 261
♦ S, cultivated, herbal. Origin: South America.

Alpinia allughas (Retz.) Roscoe
Zingiberaceae
♦ Environmental Weed
♦ 226

Alpinia carolinensis Koidz.
Zingiberaceae
♦ iuiu
♦ Introduced
♦ 230
♦ S.

Alpinia mutica Roxb.
Zingiberaceae
♦ small shell ginger
♦ Naturalised
♦ 101
♦ cultivated.

Alpinia purpurata (Vieill.) K.Schum.
Zingiberaceae
Alpinia purpurea (Vieill.) K.Schum.
♦ red ginger, pink ginger, double red flame ginger
♦ Naturalised, Introduced, Garden Escape
♦ 101, 230, 261
♦ H, cultivated, herbal. Origin: Pacific Islands.

Alpinia zerumbet (Pers.) Burtt & R.M.Sm.
Zingiberaceae
♦ shellplant, variegated shell ginger, shell ginger
♦ Weed, Naturalised, Cultivation Escape
♦ 101, 261, 283
♦ H, cultivated. Origin: China, Japan, India.

Alsophila latebrosa Wall. ex Presl
Cyatheaceae
= *Cyathea latebrosa* (Wall. ex Hook.) Copel. (NoR)
♦ Weed
♦ 87, 88

Alstonia constricta F.Muell.
Apocynaceae
♦ alstonia
♦ Weed
♦ 87, 88
♦ cultivated, herbal, toxic. Origin: Australia.

Alstonia macrophylla Wall. ex G.Don
Apocynaceae
♦ devil tree
♦ Weed, Naturalised, Environmental Weed
♦ 80, 88, 101, 112, 179
♦ herbal.

Alstroemeria aurantiaca D.Don ex Sweet
Liliaceae/Alstroemeriaceae
♦ Peruvian lily
♦ Weed
♦ 15
♦ cultivated.

Alstroemeria aurea Graham
Liliaceae/Alstroemeriaceae
Alstroemeria aurantiaca D.Don
♦ alstroemeria, yellow alstroemeria
♦ Weed, Naturalised, Garden Escape, Environmental Weed

♦ 20, 72, 86, 88, 98, 198, 203, 251, 280, 289, 296
♦ pH, cultivated, herbal. Origin: South America.

Alstroemeria haemantha Ruiz & Pav.
Liliaceae/Alstroemeriaceae
Alstroemeria simsii Spreng.
♦ purplespot parrotlily, Peruvian lily
♦ Naturalised
♦ 101
♦ pH, cultivated, herbal. Origin: South America.

Alstroemeria psittacina Lehm.
Liliaceae/Alstroemeriaceae
♦ decorative Peruvian lily
♦ Naturalised, Garden Escape
♦ 101, 269
♦ cultivated, herbal. Origin: South America.

Alstroemeria pulchella L.f.
Liliaceae/Alstroemeriaceae
♦ parrot alstroemeria
♦ Weed, Naturalised, Garden Escape, Environmental Weed
♦ 7, 86, 98, 101, 134, 179, 198, 203, 251, 280
♦ cultivated, herbal.

Alternanthera Forssk. spp.
Amaranthaceae
♦ joyweed
♦ Quarantine Weed
♦ 220, 258

Alternanthera achyrantha (L.) R.Br. ex Sweet
Amaranthaceae
= *Alternanthera pungens* Kunth
♦ Weed
♦ 87, 88

Alternanthera amabilis Lam.
Amaranthaceae
♦ Weed
♦ 87, 88

Alternanthera angustifolia R.Br.
Amaranthaceae
♦ Weed, Naturalised
♦ 98, 203

Alternanthera bettzickiana (Regel) Nicholson
Amaranthaceae
Alternanthera amoena (Lem.) Voss, *Alternanthera bettzichiana* (Regel) Voss, *Alternanthera ficoidea* var. *amoena* (Lem.) L.B.Sm. & Downs, *Alternanthera versicolor* (Lem.) Regel, *Telanthera bettzickiana* Regel
♦ calico plant, papageienblatt, nature of the world, pajarito, sanguinaria
♦ Weed, Naturalised, Garden Escape
♦ 54, 86, 88, 101, 261, 262
♦ pH, aqua, cultivated. Origin: Brazil.

Alternanthera brasiliana (L.) Kuntze
Amaranthaceae
Alternanthera dentata (Moench) Stuchlik ex R.E.Fr. (see)
♦ Brazilian joyweed
♦ Weed, Naturalised, Garden Escape
♦ 101, 179, 255, 261
♦ cultivated. Origin: Brazil.

Alternanthera caracasana Kunth
Amaranthaceae
♦ washerwoman
♦ Weed, Cultivation Escape
♦ 34, 261
♦ pH, arid, cultivated, herbal.

Alternanthera dentata (Moench) Stuchlik ex R.E.Fr.
Amaranthaceae
= *Alternanthera brasiliana* (L.) Kuntze
♦ Weed
♦ 191
♦ cultivated.

Alternanthera denticulata R.Br.
Amaranthaceae
♦ lesser joyweed
♦ Weed, Native Weed
♦ 87, 88, 269
♦ pH, arid, cultivated. Origin: Australia.

Alternanthera echinata Sm.
Amaranthaceae
= *Alternanthera pungens* Kunth
♦ Weed
♦ 87, 88

Alternanthera ficoidea (L.) R.Br. ex Roem. & Schult.
Amaranthaceae
Alternanthera paronychioides A.St.Hil., *Alternanthera tenella* Colla (see), *Bucholzia ficoidea* (L.) Mart., *Gomphrena ficoidea* L., *Telanthera ficoidea* (Lam.) Moq.
♦ calico plant, copperleaf, Joseph's coat, apaga fogo
♦ Weed, Naturalised, Environmental Weed
♦ 86, 87, 88, 216, 245
♦ aH, cultivated. Origin: Central America, northern South America.

Alternanthera ficoidea R.Br. ex Roem. & Schult. var. bettzickiana Backer
Amaranthaceae
♦ Naturalised
♦ 287

Alternanthera flavescens Kunth
Amaranthaceae
♦ marcela, yellow joyweed
♦ Weed
♦ 161, 249

Alternanthera frutescens R.Br. ex Spreng.
Amaranthaceae
♦ Weed
♦ 87, 88

Alternanthera halimifolia Standl. ex Pittier
Amaranthaceae
= *Alternanthera ficoidea* R.Br. ex Roem. & Schult. var. *halimifolia* Kuntze (NoR)
♦ Weed
♦ 87, 88
♦ arid.

Alternanthera lanceolata (Benth.) Schinz
Amaranthaceae
= *Alternanthera mexicana* (Schltdl.) Hieron. (NoR) [see *Alternanthera lehmannii* Hieron.]
♦ Environmental Weed
♦ 257

Alternanthera lehmannii Hieron.
Amaranthaceae
= *Alternanthera mexicana* (Schltdl.) Hieron. (NoR) [see *Alternanthera lanceolata* (Benth.) Schinz]
♦ Naturalised
♦ 39, 257
♦ toxic.

Alternanthera nana R.Br.
Amaranthaceae
♦ hairy joyweed
♦ Weed, Native Weed
♦ 87, 88, 269
♦ aH, cultivated. Origin: Australia.

Alternanthera nodiflora R.Br.
Amaranthaceae
♦ common joyweed
♦ Weed, Native Weed
♦ 87, 88, 235, 269, 273, 274, 286
♦ aH, cultivated. Origin: Australia.

Alternanthera paronichyoides St.-Hil.
Amaranthaceae
Alternanthera paronychoides St.-Hil.
♦ smooth joyweed, whitlowwort, yerba del pollo
♦ Weed, Naturalised
♦ 101, 161, 179, 237, 243, 249, 261, 295
♦ pH. Origin: Old World Tropics.

Alternanthera paronichyoides St.-Hil. var. amazonica Huber
Amaranthaceae
♦ smooth joyweed
♦ Naturalised
♦ 101

Alternanthera philoxeroides (Mart.) Griseb.
Amaranthaceae
Bucholzia philoxeroides Mart., *Telanthera philoxeroides* (Mart.) Moq.
♦ alligator weed, pigweed, alligator grass, phak pet nam
♦ Weed, Quarantine Weed, Noxious Weed, Naturalised, Garden Escape, Environmental Weed
♦ 3, 17, 23, 26, 35, 62, 76, 80, 86, 87, 88, 93, 98, 101, 102, 106, 112, 147, 151, 152, 155, 161, 165, 171, 177, 179, 181, 191, 197, 198, 200, 203, 208, 209, 217, 218, 225, 229, 232, 236, 237, 239, 246, 251, 255, 258, 261, 262, 263, 268, 269, 270, 273, 274, 275, 280, 287, 289, 295, 296, 297
♦ wpH, cultivated, herbal. Origin: South America.

Alternanthera polygonoides (L.) R.Br.
Amaranthaceae
♦ Weed
♦ 14, 87, 88
♦ herbal.

Alternanthera pungens Kunth
Amaranthaceae
Achyranthes achyrantha R.Br., *Achyranthes leiantha* (Seub.) Standl., *Achyranthes repens* L., *Alternanthera achyrantha* (L.) Sweet (see), *Alternanthera echinata* Sm. (see), *Alternanthera repens* (L.) Link *non* J.F.Gmel., *Illecebrum achyrantha* L.

♦ khakiweed, creeping chaffweed, spingflower alternanthera
♦ Weed, Quarantine Weed, Noxious Weed, Naturalised, Introduced, Casual Alien
♦ 7, 51, 66, 76, 86, 88, 93, 98, 106, 121, 147, 158, 161, 171, 198, 203, 205, 212, 228, 237, 240, 243, 249, 255, 269, 276, 280, 295, 297
♦ a/pH, arid, toxic. Origin: South America.

Alternanthera repens (L.) Kuntze
Amaranthaceae
♦ khakiweed
♦ Weed, Naturalised
♦ 50, 88, 199, 218, 221, 287
♦ cultivated, herbal.

Alternanthera sessilis (L.) R.Br. ex DC.
Amaranthaceae
Achyranthes sessilis (L.) Desf. *extend.*, *Alternanthera achyranthoides* Forsk., *Alternanthera glabra* Moq., *Alternanthera repens* Gmel., *Gomphrena sessilis* L., *Illecebrum sessile* (L.) L.
♦ sessile flowered globe amaranth, sessile joyweed, phak pet thai
♦ Weed, Quarantine Weed, Noxious Weed, Naturalised
♦ 6, 67, 86, 87, 88, 134, 140, 157, 161, 170, 179, 185, 204, 206, 209, 217, 221, 229, 235, 239, 243, 257, 258, 262, 273, 274, 275, 276, 280, 286, 287, 297
♦ pH, aqua, cultivated, herbal. Origin: obscure.

Alternanthera tenella Colla
Amaranthaceae
= *Alternanthera ficoidea* (L.) R.Br. ex Roem. & Schult.
♦ sanguinaria
♦ Weed
♦ 179, 255
♦ Origin: Brazil.

Alternanthera tenella Colla var. bettzickiana (Regel) Veldk.
Amaranthaceae
♦ Introduced
♦ 228
♦ arid.

Althaea acaulis Cav.
Malvaceae
♦ Weed
♦ 87, 88

Althaea cannabina L.
Malvaceae
Althaea kotschyi Boiss., *Althaea narbonensis* Pourr. ex Cav.
♦ palmleaf marshmallow
♦ Weed
♦ 88, 94, 272
♦ pH, cultivated, herbal. Origin: Europe.

Althaea hirsuta L.
Malvaceae
♦ rough marshmallow, hairy marshmallow, hispid marshmallow, rough mallow
♦ Weed, Naturalised
♦ 88, 94, 101, 253, 272
♦ herbal.

Althaea longiflora Boiss. & Reut.
Malvaceae
♦ Weed
♦ 87, 88

Althaea ludwigii L.
Malvaceae
♦ Weed
♦ 87, 88

Althaea officinalis L.
Malvaceae
♦ common marshmallow, sweet weed, althaea, guimauve
♦ Weed, Naturalised, Cultivation Escape
♦ 42, 70, 87, 88, 101, 272
♦ pH, cultivated, herbal. Origin: Eurasia.

Althaea sulphurea Boiss. & Hohen.
Malvaceae
♦ Weed
♦ 87, 88

Alvaradoa amorphoides Liebm.
Picramniaceae/Simaroubaceae
♦ Mexican alvaradoa
♦ Weed
♦ 14
♦ herbal.

Alysicarpus bupleurifolius (L.) A.DC.
Fabaceae/Papilionaceae
♦ Weed
♦ 262
♦ cultivated. Origin: Asia.

Alysicarpus longifolius Wight. & Arn.
Fabaceae/Papilionaceae
♦ Weed
♦ 87, 88
♦ Origin: Australia, India, Pakistan.

Alysicarpus monilifer (L.) DC.
Fabaceae/Papilionaceae
♦ Weed
♦ 87, 88
♦ cultivated.

Alysicarpus nummularifolius (L.) DC.
Fabaceae/Papilionaceae
♦ Weed
♦ 87, 88

Alysicarpus ovalifolius (Schumach.) J.Léonard
Fabaceae/Papilionaceae
Alysicarpus vaginalis sensu auct. mult. non (L.) DC., *Hedysarum ovalifolium* Schum. & Thonn.
♦ alyce clover
♦ Weed, Naturalised, Introduced
♦ 87, 88, 101, 179, 228
♦ arid, cultivated. Origin: Asia, Madagascar.

Alysicarpus procumbens (Roxb.) Schindl.
Fabaceae/Papilionaceae
♦ Weed
♦ 66

Alysicarpus quartinianus A.Rich
Fabaceae/Papilionaceae
♦ Weed
♦ 240
♦ aH.

Alysicarpus rugosus (Willd.) DC.
Fabaceae/Papilionaceae
♦ red moneywort, rough chainpea
♦ Weed, Naturalised, Introduced
♦ 87, 88, 101, 179, 228
♦ arid, cultivated, herbal. Origin: Africa, Asia, Australia, Madagascar.

Alysicarpus vaginalis (L.) DC.
Fabaceae/Papilionaceae
Hedysarum vaginale L.
♦ white moneywort, alyce clover, oneleaf clover
♦ Weed, Naturalised, Introduced
♦ 7, 14, 32, 86, 87, 88, 93, 101, 161, 179, 204, 218, 228, 235, 249, 255, 261, 262, 273, 275, 286, 297
♦ pH, arid, cultivated, herbal. Origin: India.

Alyssum L. spp.
Brassicaceae
♦ madwort, alyssum
♦ Weed, Naturalised
♦ 198, 272

Alyssum alyssoides (L.) L.
Brassicaceae
♦ yellow alyssum, pale madwort, small alison, pale alyssum, hoary alyssum
♦ Weed, Naturalised, Introduced, Environmental Weed, Casual Alien, Cultivation Escape
♦ 21, 34, 42, 56, 80, 86, 88, 94, 101, 161, 212, 237, 241, 252, 253, 272, 280, 287, 295, 300
♦ aH, cultivated, herbal. Origin: Eurasia.

Alyssum blepharocarpum Dudley & Hub-Mor.
Brassicaceae
♦ Weed
♦ 243

Alyssum campestre (L.) L.
Brassicaceae
♦ Weed, Casual Alien
♦ 40, 87, 88

Alyssum dasycarpum Stephan ex Willd.
Brassicaceae
♦ Weed
♦ 243

Alyssum desertorum Stapf
Brassicaceae
Alyssum minimum auct. non Willd.
♦ desert madwort, pikkukilpiruoho
♦ Weed, Naturalised, Casual Alien
♦ 42, 80, 87, 88, 101, 248, 272
♦ aH, herbal.

Alyssum desertorum Stapf var. desertorum
Brassicaceae
♦ desert madwort
♦ Naturalised
♦ 101

Alyssum desertorum Stapf var. himalayensis Dudley
Brassicaceae
♦ desert madwort
♦ Naturalised

♦ 101

Alyssum hirsutum M.Bieb.
Brassicaceae
♦ huopakilpiruoho
♦ Weed, Casual Alien
♦ 42, 272

Alyssum incanum L.
Brassicaceae
= *Berteroa incana* (L.) DC.
♦ Weed
♦ 87, 88

Alyssum linifolium Stephan ex Willd.
Brassicaceae
♦ flaxleaf alyssum
♦ Weed, Naturalised
♦ 7, 86, 98, 198, 203
♦ cultivated. Origin: Eurasia.

Alyssum macropodium Boiss & Bal.
Brassicaceae
♦ Weed
♦ 243

Alyssum maritimum (L.) Lam.
Brassicaceae
= *Lobularia maritima* (L.) Desv.
♦ seaside lobularia, sweet alyssum
♦ Weed
♦ 87, 88
♦ cultivated, herbal.

Alyssum minimum Willd.
Brassicaceae
♦ Weed
♦ 87, 88

Alyssum minus (L.) Rothm.
Brassicaceae
♦ alyssum
♦ Weed, Naturalised
♦ 101, 272
♦ herbal.

Alyssum minus (L.) Rothm. var. micranthum (C.A.Mey.) Dudley
Brassicaceae
♦ alyssum
♦ Naturalised
♦ 101

Alyssum minus (L.) Rothm. var. strigosum (Banks & Sol.) Zohary
Brassicaceae
Alyssum strigosum Banks & Sol. (see)
♦ alyssum
♦ Naturalised
♦ 101

Alyssum minutum Schltdl. ex DC.
Brassicaceae
♦ Weed
♦ 272

Alyssum montanum L.
Brassicaceae
♦ vuorikilpiruoho
♦ Casual Alien
♦ 42
♦ cultivated.

Alyssum murale Waldst. & Kit.
Brassicaceae
♦ hopeakilpiruoho, yellowtuft, silver alyssum
♦ Weed, Naturalised, Casual Alien

♦ 42, 101, 272
♦ cultivated, herbal.

Alyssum saxatile L.
Brassicaceae
= *Aurinia saxatilis* (L.) Desv.
♦ golden Alison
♦ Weed, Naturalised, Casual Alien
♦ 40, 80, 280
♦ cultivated, herbal.

Alyssum simplex Rudolphi
Brassicaceae
♦ kenttäkilpiruoho
♦ Weed, Casual Alien
♦ 42, 253

Alyssum strigosum Banks & Sol.
Brassicaceae
= *Alyssum minus* (L.) Rothm. var.
strigosum (Banks & Sol.) Zohary
♦ hairy madwort, alyssum
♦ Weed
♦ 272
♦ aH.

Alyssum szovitsianum Fisch. & C.A.Mey.
Brassicaceae
♦ Szowits's madwort
♦ Weed, Naturalised
♦ 101, 248

Amaranthus L. spp.
Amaranthaceae
♦ grain amaranth, amaranthus,
pigweed, love lies bleeding, yuyo
colorado, bledos
♦ Weed
♦ 39, 106, 236, 243, 281
♦ H, herbal, toxic.

Amaranthus albus L.
Amaranthaceae
♦ tumble pigweed, white pigweed,
prostrate pigweed, stiff tumbleweed
♦ Weed, Naturalised, Native Weed,
Introduced, Environmental Weed,
Garden Escape, Casual Alien
♦ 7, 34, 40, 42, 44, 46, 70, 86, 87, 88, 94,
98, 116, 161, 174, 176, 180, 198, 203, 210,
211, 212, 218, 228, 237, 243, 253, 269,
272, 275, 280, 287, 295, 297, 300
♦ aH, arid, cultivated, herbal. Origin:
Americas.

Amaranthus angustifolius Lam.
Amaranthaceae
= *Amaranthus graecizans* L.
♦ Weed
♦ 87, 88

Amaranthus arenicola I.M.Johnst.
Amaranthaceae
♦ sandhill amaranth
♦ Naturalised
♦ 287
♦ aH, herbal.

Amaranthus ascendens Loisel.
Amaranthaceae
= *Amaranthus blitum* L.
♦ Weed
♦ 88, 221

Amaranthus australis (A.Gray) J.D.Sauer
Amaranthaceae
♦ southern amaranth

♦ Weed
♦ 32, 87, 88
♦ herbal.

Amaranthus blitoides S.Watson
Amaranthaceae
Amaranthus graecizans auct.
♦ prostrate pigweed, prostrate
amaranth, mat amaranth
♦ Weed, Quarantine Weed, Noxious
Weed, Naturalised, Introduced, Casual
Alien
♦ 34, 42, 44, 49, 68, 70, 76, 87, 88, 94,
101, 115, 121, 161, 174, 180, 210, 211,
212, 218, 229, 243, 253, 272, 287, 297
♦ aH, promoted, herbal. Origin: North
America.

Amaranthus blitum L.
Amaranthaceae
Amaranthus ascendens Loisel. (see),
Amaranthus lividus L. emend. Thell.
(see), *Amaranthus lividus* L. ssp.
polygonoides (Moq.) Probst (see),
Amaranthus oleraceus L. (see)
♦ purple amaranth
♦ Weed, Naturalised, Casual Alien
♦ 32, 40, 42, 101, 134, 179, 237, 249, 253
♦ herbal. Origin: Eurasia.

**Amaranthus blitum L. ssp. emarginatus
(Moq. ex Uline & W.L.Bray) Carretero,
Muñoz Garm. & Pedrol**
Amaranthaceae
♦ Naturalised
♦ 300

Amaranthus bouchonii Thell.
Amaranthaceae
= *Amaranthus powellii* S.Wats.
♦ Weed
♦ 44, 68, 88, 250, 253

Amaranthus capensis Thell. ssp. capensis
Amaranthaceae
♦ Native Weed
♦ 121
♦ aH, toxic. Origin: southern Africa.

**Amaranthus capensis Thell. ssp.
uncinatus (Thell.) Brenan**
Amaranthaceae
♦ Native Weed
♦ 121
♦ aH, toxic. Origin: southern Africa.

Amaranthus caudatus L.
Amaranthaceae
Amaranthus edulis Speg., *Amaranthus
leucospermus* S.Watson, *Amaranthus
manteguzzianus* Pass., *Amaranthus
retroflexus* L. ssp. *quitensis* (Kunth)
O.Bolos & Vigo, *Amaranthus sanguineus*
L.
♦ love lies bleeding, Inca wheat,
cifogot, dimesitu, red hot cat's tail,
yeteffre, foxtail, jataco, tasselflower,
velvet flower, amarante caudée,
amarante queue de renard, discipline
des religieux, queue de renard,
Gartenfuchsschwanz, Inkaweizen,
sennin koku, moncos de Peru, coimi,
cuipa, achita, bledo francés, kiwicha,
trigo del Inca

♦ Weed, Naturalised, Introduced,
Garden Escape, Environmental Weed,
Cultivation Escape
♦ 7, 32, 34, 42, 70, 86, 87, 88, 98, 101,
198, 203, 221, 228, 252, 256, 257, 261,
280, 287, 297
♦ aH, arid, cultivated, herbal. Origin:
obscure.

Amaranthus chlorostachys Willd.
Amaranthaceae
= *Amaranthus hybridus* L.
♦ slim amaranth
♦ Weed, Casual Alien
♦ 40, 87, 88, 221
♦ aH, cultivated, herbal.

Amaranthus crassipes Schlecht.
Amaranthaceae
♦ spreading amaranth
♦ Weed, Naturalised
♦ 87, 88, 179, 287
♦ herbal.

**Amaranthus crispus (Lesp. & Thev.)
N.Terracc.**
Amaranthaceae
♦ crispleaf amaranth
♦ Weed, Naturalised, Casual Alien
♦ 42, 87, 88, 94, 101, 237, 295
♦ herbal.

Amaranthus cruentus L.
Amaranthaceae
Amaranthus hybridus L. ssp. *cruentus*
(L.) Thell. (see), *Amaranthus paniculatus*
L. (see)
♦ red shank, smooth pigweed, red
amaranth, prince's feather
♦ Weed,
Naturalised, Garden Escape, Casual
Alien
♦ 7, 39, 42, 68, 70, 86, 87, 88, 93, 98, 101,
198, 203, 221, 241, 253, 261, 272, 280,
287
♦ aH, cultivated, herbal, toxic. Origin:
obscure.

Amaranthus deflexus L.
Amaranthaceae
♦ prostrate amaranth, low amaranth,
pigweed, perennial pigweed, largefruit
amaranth, spreading amaranth
♦ Weed, Naturalised, Casual Alien
♦ 15, 34, 40, 42, 51, 70, 86, 87, 88, 94, 98,
101, 121, 158, 161, 165, 176, 198, 203,
237, 241, 245, 253, 255, 269, 272, 280,
286, 287, 295, 300
♦ a/pH, cultivated, herbal, toxic.
Origin: South America.

**Amaranthus dinteri Schinz ssp.
brevipetiolatus Brenan**
Amaranthaceae
♦ Native Weed
♦ 121
♦ aH. Origin: southern Africa.

Amaranthus dinteri Schinz ssp. dinteri
Amaranthaceae
♦ Native Weed
♦ 121
♦ aH, toxic. Origin: southern Africa.

Amaranthus dubius Mart. ex Thell.
Amaranthaceae
Amaranthus tristis L. (see)
♦ amaranth, te uekeueke, spleen amaranth, doodo, hondi, onvoko, red green
♦ Weed, Quarantine Weed, Naturalised, Introduced, Cultivation Escape
♦ 3, 11, 76, 86, 87, 88, 121, 135, 179, 203, 206, 220, 228, 243, 252, 257
♦ aH, arid, cultivated, herbal, toxic. Origin: tropical America.

Amaranthus fimbriatus (Torr.) Benth. ex S.Wats.
Amaranthaceae
♦ fringed pigweed, fringed amaranth
♦ Weed
♦ 161, 243
♦ aH, arid, herbal.

Amaranthus galii Sennen & Gonzalo ex Priszter
Amaranthaceae
♦ Naturalised
♦ 287

Amaranthus gangeticus L.
Amaranthaceae
= *Amaranthus tricolor* L.
♦ spleen amaranth, elephant's head amaranth, amaranth greens
♦ Weed
♦ 39, 87, 88, 218
♦ aH, cultivated, herbal, toxic.

Amaranthus giganticus L.
Amaranthaceae
♦ Weed
♦ 87, 88

Amaranthus gracilis Desf. *nom. nud.*
Amaranthaceae
= *Amaranthus viridis* L.
♦ pigweed, largefruit amaranth, phak khom
♦ Weed, Naturalised, Environmental Weed
♦ 34, 88, 170, 191, 238, 243, 257, 287
♦ aH, herbal.

Amaranthus graecizans L.
Amaranthaceae
Amaranthus angustifolia Lam., *Amaranthus angustifolia* Lam. var. *silvester* Thell., *Amaranthus angustifolia* Lam. var. *silvestris* (Vill.) Thell., *Amaranthus angustifolius* Lam. (see), *Amaranthus angustifolius* Lam. var. *graecizans* (L.) Thell., *Amaranthus angustifolius* Lam. ssp. *aschersonianus* Thell., *Amaranthus angustifolius* Lam. var. *silvestris* (Vill.) Asch., *Amaranthus angustifolius* Lam. var. *silvestris* (Vill.) Thell., *Amaranthus angustifolius* Lam. ssp. *silvestris* (Vill.) Henkels, *Amaranthus angustifolius* Lam. ssp. *polygonoides* (Moq.) Maire & Weiller, *Amaranthus angustifolus* Lam. ssp. *graecizans* (L.) Probst., *Amaranthus aschersonianus* (Thell.) Chiov., *Amaranthus blitum auct. non* L., *Amaranthus blitum* L. var. *polygonoides*

Moq., *Amaranthus blitum* L. var. *silvestris* (Vill.) Moq., *Amaranthus blitum sensu* Baker & Clarke var. *polygonoides* Moq., *Amaranthus oleraceus sensu* Baker & Clarke, *Amaranthus paolii* Chiov., *Amaranthus parvulus* Peter, *Amaranthus polygamus sensu* Baker & Clarke, *Amaranthus silvestris* Vill., *Amaranthus thellungianus* Nevski, *Amaranthus thunbergii* Moq. var. *grandifolius* Suess.
♦ prostrate pigweed, spreading pigweed, white pigweed, short tepalled pigweed
♦ Weed, Noxious Weed, Naturalised, Introduced, Casual Alien
♦ 23, 40, 42, 44, 53, 70, 87, 88, 94, 98, 111, 115, 176, 185, 203, 218, 221, 228, 240, 242, 243, 253, 272, 280, 299
♦ aH, arid, cultivated, herbal, toxic.

Amaranthus graecizans L. ssp. *sylvestris* (Vill.) Brenan
Amaranthaceae
♦ amaranth
♦ Naturalised
♦ 86, 198

Amaranthus hybridus L.
Amaranthaceae
Amaranthus chlorostachys Willd. (see), *Amaranthus frumentaceus* Buch.-Ham. ex Roxb., *Amaranthus incurvatus* Timeroy ex Gren. & Godr., *Amaranthus patulus* Bertol. (see)
♦ smooth pigweed, rough pigweed, slim amaranth, redshank, prince's feather, green amaranth, bwamanga, imBuya, imBuya yamabize, slender pigweed
♦ Weed, Naturalised
♦ 7, 8, 23, 34, 39, 44, 50, 51, 53, 55, 68, 70, 86, 87, 88, 93, 94, 98, 111, 115, 157, 158, 161, 179, 180, 185, 186, 195, 198, 203, 207, 210, 211, 218, 236, 240, 241, 243, 245, 249, 253, 262, 269, 272, 280, 287, 293, 299, 300
♦ aH, arid, cultivated, herbal, toxic. Origin: tropical America.

Amaranthus hybridus L. ssp. *cruentus* (L.) Thell.
Amaranthaceae
= *Amaranthus cruentus* L.
♦ cape pigweed, common pigweed, pigweed, prince's feather, redshank
♦ Weed, Introduced
♦ 121, 228
♦ aH, toxic. Origin: Americas.

Amaranthus hybridus L. ssp. *hybridus*
Amaranthaceae
♦ cape pigweed, cock's comb, common pigweed, hell's curse, pigweed, smooth pigweed, wild beet
♦ Weed, Introduced
♦ 121, 228
♦ aH, toxic. Origin: Americas.

Amaranthus hybridus L. var. *erythrostachys* Moq.
Amaranthaceae
= *Amaranthus hypochondriacus* L.
♦ Naturalised

♦ 280
♦ herbal.

Amaranthus hybridus L. var. *hybridus*
Amaranthaceae
♦ Naturalised
♦ 280

Amaranthus hybridus L. var. *paniculatus* (L.) Thell.
Amaranthaceae
♦ caruru roxo, caruru, bredo
♦ Weed
♦ 255
♦ Origin: Tropical America.

Amaranthus hybridus L. var. *patulus* (Betol.) Thell.
Amaranthaceae
♦ caruru branco, caruru bravo, caruru
♦ Weed
♦ 255
♦ Origin: Tropical America.

Amaranthus hypochondriacus L.
Amaranthaceae
Amaranthus hybridus var. *erythrostachys* Moq. (see), *Amaranthus hybridus* var. *hypochondriacus* (L.) B.L.Rob.
♦ prince's feather, chua, ramdana, huantli, alegría
♦ Weed, Naturalised, Casual Alien
♦ 39, 42, 87, 88, 101
♦ aH, promoted, herbal, toxic. Origin: obscure, possibly Mexico.

Amaranthus inamoenus Willd.
Amaranthaceae
♦ Weed
♦ 23, 88

Amaranthus interruptus R.Br.
Amaranthaceae
♦ native amaranth
♦ Weed
♦ 6, 55, 88
♦ cultivated. Origin: Australia.

Amaranthus lineatus R.Br.
Amaranthaceae
♦ Australian amaranth
♦ Naturalised
♦ 101

Amaranthus lividus L.
Amaranthaceae
= *Amaranthus blitum* L.
♦ livid amaranth, purple amaranth, boa, chaulai, epinard, horsetooth amaranth, norpa, vleeta
♦ Weed, Naturalised, Introduced
♦ 7, 15, 44, 68, 70, 86, 87, 88, 94, 98, 161, 185, 203, 204, 218, 228, 236, 243, 255, 257, 263, 270, 272, 274, 275, 276, 280, 286, 287, 295, 297
♦ aH, arid, cultivated, herbal. Origin: obscure.

Amaranthus lividus L. ssp. *polygonoides* (Moq.) Probst
Amaranthaceae
= *Amaranthus blitum* L.
♦ livid amaranth
♦ Weed
♦ 121
♦ a/pH. Origin: South America.

Amaranthus macrocarpus Benth.
Amaranthaceae
♦ dwarf amaranth
♦ Weed, Quarantine Weed, Native Weed
♦ 39, 55, 76, 87, 88, 203, 269
♦ aH, arid, cultivated, toxic. Origin: Australia.

Amaranthus mangostanus L.
Amaranthaceae
= *Amaranthus tricolor* L.
♦ hijau salad amaranth
♦ Weed, Naturalised
♦ 87, 88, 204, 286, 287
♦ aH, promoted, herbal.

Amaranthus mangostanus L. fo. ruber Mak.
Amaranthaceae
= *Amaranthus tricolor* L.
♦ Naturalised
♦ 287

Amaranthus mangostanus L. fo. versicolor Mak.
Amaranthaceae
= *Amaranthus tricolor* L.
♦ Naturalised
♦ 287

Amaranthus mitchellii Benth.
Amaranthaceae
♦ boggabri
♦ Weed, Naturalised
♦ 39, 55, 86, 87, 88, 243
♦ S, arid, cultivated, toxic. Origin: Australia.

Amaranthus muricatus (Gillies ex Moq.) Gillies ex Hicken
Amaranthaceae
♦ African amaranth, roughfruit amaranth
♦ Weed, Naturalised
♦ 86, 87, 88, 94, 98, 101, 121, 198, 203, 237
♦ pH, herbal. Origin: South America.

Amaranthus oleraceus L.
Amaranthaceae
= *Amaranthus blitum* L.
♦ amaranth
♦ Weed
♦ 87, 88
♦ herbal.

Amaranthus palmeri S.Wats.
Amaranthaceae
♦ Palmer amaranth, careless weed
♦ Weed, Naturalised, Introduced
♦ 28, 34, 39, 68, 87, 88, 161, 199, 212, 218, 228, 243, 286, 287
♦ aH, arid, promoted, herbal, toxic.

Amaranthus paniculatus L.
Amaranthaceae
= *Amaranthus cruentus* L.
♦ Weed, Naturalised
♦ 87, 88, 98, 203
♦ aH, promoted, herbal. Origin: obscure, possibly tropical America.

Amaranthus patulus Bertol.
Amaranthaceae
= *Amaranthus hybridus* L.

♦ spleen amaranth
♦ Weed, Naturalised
♦ 44, 87, 88, 204, 263, 286, 287
♦ aH.

Amaranthus polygamus L.
Amaranthaceae
= *Amaranthus tricolor* L.
♦ Weed
♦ 87, 88
♦ aH, promoted, herbal.

Amaranthus polygonoides L.
Amaranthaceae
♦ tropical amaranth
♦ Weed
♦ 179
♦ herbal.

Amaranthus powellii S.Watson
Amaranthaceae
Amaranthus bouchonii Thell. (see)
♦ Powell's amaranth, redroot, smooth pigweed, rough pigweed, amaranth pigweed, green amaranth, careless weed
♦ Weed, Naturalised
♦ 7, 15, 34, 55, 68, 86, 87, 88, 98, 161, 176, 198, 203, 211, 218, 243, 269, 280, 287, 293
♦ aH, promoted, herbal. Origin: North or Central America.

Amaranthus praetermissus Brenan
Amaranthaceae
♦ Weed
♦ 121
♦ aH, toxic.

Amaranthus quitensis Kunth
Amaranthaceae
♦ South American amaranth, ataco, yuyo Colorado
♦ Weed, Naturalised, Introduced
♦ 86, 87, 88, 98, 203, 228, 236, 237, 257, 269, 270, 295
♦ aH, arid, promoted, herbal. Origin: South America.

Amaranthus retroflexus L.
Amaranthaceae
Amaranthus delilei Loret, *Amaranthus retroflexus* L. var. *salicifolius* I.M.Johnst.
♦ redroot pigweed, rough pigweed, green amaranth, redroot, redroot amaranth, amaranto comune, reflexed amaranth
♦ Weed, Noxious Weed, Naturalised, Introduced, Environmental Weed, Casual Alien
♦ 8, 21, 23, 34, 39, 40, 42, 44, 49, 52, 68, 70, 86, 87, 88, 94, 98, 101, 114, 115, 118, 136, 138, 159, 161, 162, 165, 174, 179, 180, 195, 198, 203, 210, 211, 212, 217, 218, 228, 229, 236, 243, 245, 253, 255, 256, 269, 272, 275, 280, 286, 287, 293, 297, 299, 300
♦ aH, arid, cultivated, herbal, toxic. Origin: tropical America.

Amaranthus retroflexus L. var. delilei
Amaranthaceae
♦ green amaranth, green pigweed, pigweed, redroot, redroot pigweed, reflexed amaranth, rough pigweed

♦ Weed
♦ 121, 297
♦ aH. Origin: North America.

Amaranthus retroflexus L. var. retroflexus
Amaranthaceae
♦ prince's feather, redroot, redroot pigweed, reflexed amaranth, rough pigweed
♦ Weed
♦ 121
♦ aH, toxic. Origin: Americas.

Amaranthus roxburghianus Kung.
Amaranthaceae
♦ Weed
♦ 275
♦ aH.

Amaranthus rudis Sauer
Amaranthaceae
♦ common waterhemp, tall amaranth
♦ Weed, Naturalised, Native Weed, Casual Alien
♦ 42, 68, 88, 161, 174, 287
♦ herbal. Origin: North America.

Amaranthus schinzianus Thell.
Amaranthaceae
♦ Native Weed
♦ 121
♦ aH, toxic. Origin: southern Africa.

Amaranthus spinosus L.
Amaranthaceae
♦ spiny amaranth, spiny pigweed, bledo macho, spring pigweed, phak khom nam, needle burr
♦ Weed, Quarantine Weed, Naturalised, Introduced, Garden Escape, Environmental Weed, Casual Alien
♦ 6, 14, 23, 32, 34, 39, 42, 50, 51, 76, 86, 87, 88, 93, 98, 121, 152, 157, 158, 161, 170, 176, 179, 186, 191, 203, 207, 209, 210, 218, 228, 235, 237, 238, 239, 243, 245, 255, 256, 257, 261, 262, 269, 271, 273, 274, 275, 276, 280, 286, 287, 295, 297
♦ aH, arid, cultivated, herbal, toxic. Origin: tropical America.

Amaranthus standleyanus L.R.Parodi ex Covas.
Amaranthaceae
Amaranthus vulgatissimus Speg.
♦ Argentiinanrevonhäntä
♦ Weed, Casual Alien
♦ 42, 87, 88, 121, 237
♦ aH, promoted, toxic. Origin: South America.

Amaranthus thunbergii Moq.
Amaranthaceae
♦ cape pigweed, pigweed, poor man's spinach, red pigweed, small pigweed, ape pigweed, ekuaka, ekwakwa, elopa, hanekam, kalkoenslurp, red devil, rooiduiwel, Thunberg's amaranthus, mohwa
♦ Weed, Quarantine Weed, Naturalised, Native Weed, Introduced
♦ 50, 51, 76, 87, 88, 101, 121, 158, 203, 220, 228
♦ aH, arid, promoted, toxic. Origin: Africa.

Amaranthus tricolor L.
Amaranthaceae
Amaranthus gangeticus L. (see),
Amaranthus inamornus Willd.,
Amaranthus mangostanus L. (see),
Amaranthus mangostanus L. fo. *ruber*
Mak. (see), *Amaranthus mangostanus*
L. fo. *versicolor* Mak. (see), *Amaranthus
melancholius* L., *Amaranthus polygamus*
L. (see), *Amaranthus tricolor* L. var.
tristis (L.) Thell., *Amaranthus tricolor*
L. ssp. *tristis* (L.) Allen, *Amaranthus
tricolor* L. ssp. *mangostanus* (L.) Aellen
♦ Joseph's coat, Chinese amaranth,
tampala, ala de penco, dwarf Chinese,
quelite morado, spinach, Chinese
spinach, amarante comestible,
amarante tricolore, gemüseamarant,
math, aupa, hageito, hiyu, amaranto,
moco de pavo
♦ Weed, Naturalised, Introduced
♦ 7, 66, 86, 87, 88, 93, 101, 228, 256
♦ aH, arid, cultivated, herbal. Origin:
obscure.

Amaranthus tristis L.
Amaranthaceae
= *Amaranthus dubius* Mart. ex Thell.
♦ Weed
♦ 87, 88
♦ herbal.

Amaranthus tuberculatus (Moq.) Sauer.
Amaranthaceae
Acnida altissima (Riddell) Moq. ex
Standl. (see)
♦ tall waterhemp, roughfruit amaranth
♦ Weed
♦ 87, 88, 161, 210
♦ herbal.

Amaranthus viridis L.
Amaranthaceae
Amaranthus gracilis Desf. *nom.
nud.* (see), *Amaranthus gracilis*
Poir., *Amaranthus viridis* Hook.f.,
Chenopodium caudatum Jacq., *Albertia
caudata* (Jacq.) Boiss, *Euxolus viridus*
Moq.
♦ slender amaranth, green amaranth,
Prince of Wales feather, green
pigweed, pigweed, pak khom, chulai,
calalu, caruru de mancha, caruru
verde, brendo verdaddeiro, citaco
♦ Weed, Naturalised, Introduced,
Garden Escape, Environmental Weed
♦ 6, 7, 23, 32, 39, 55, 86, 87, 88, 93, 98,
121, 157, 158, 161, 179, 198, 203, 204,
205, 209, 217, 218, 221, 228, 235, 237,
239, 242, 243, 245, 249, 255, 256, 261,
263, 269, 270, 273, 274, 275, 276, 280,
286, 287, 295, 297, 300
♦ pH, arid, cultivated, herbal, toxic.
Origin: obscure, possibly tropical
America.

Amarella acuta (Michx.) Raf.
Gentianaceae
= *Gentianella amarella* (L.) Boerner ssp.
acuta (Michx.) J.M.Gillett (NoR)
♦ Quarantine Weed
♦ 220

Amaryllis belladonna L.
Liliaceae/Amaryllidaceae
♦ belladonna lily, kapamaryllis
amaryllis, Jersey lily
♦ Weed, Naturalised, Garden Escape,
Environmental Weed
♦ 7, 39, 40, 70, 86, 98, 101, 161, 198, 203,
280, 300
♦ cultivated, herbal, toxic. Origin:
South Africa.

Amasonia campestrus (Aubl.) Moldenke
Lamiaceae/Verbenaceae
♦ Weed
♦ 87, 88

Amberboa lippii (L.) DC.
Asteraceae
♦ Weed
♦ 221

Amberboa moschata (L.) DC.
Asteraceae
Centaurea moschata L.
♦ sweet sultan
♦ Naturalised
♦ 101
♦ aH, herbal.

**Amblygonocarpus andogensis (Welw. ex
Oliv.) Exell & Torre**
Fabaceae/Mimosaceae
Amblygonocarpus obtusangulus (Welw.
ex Oliv.) Harms, *Tetrapleura andongensis*
Welw. ex Oliv., *Tetrapleura obtusangula*
Welw. ex Oliv.
♦ Quarantine Weed
♦ 220
♦ herbal. Origin: Africa.

Ambrosia L. spp.
Asteraceae
Franseria Cav. spp.
♦ ragweed, perennial ragweed
♦ Weed, Quarantine Weed, Noxious
Weed, Naturalised
♦ 39, 86, 88, 106, 171, 203, 220
♦ cultivated, toxic.

Ambrosia acanthicarpa Hook.
Asteraceae
Franseria acanthicarpa (Hook.) Cov.
(see)
♦ annual bursage, annual burrweed,
burrweed, franseria povertyweed,
franseria ragweed, flatspine burr
ragweed
♦ Weed
♦ 39, 161, 167, 180, 212
♦ aH, arid, herbal, toxic. Origin: North
America.

Ambrosia arborescens Mill.
Asteraceae
♦ Naturalised
♦ 300
♦ Origin: South America.

Ambrosia artemisiifolia L.
Asteraceae
Ambrosia media Rydb., *Ambrosia
monophylla* (Walt.) Rydb.
♦ common ragweed, ragweed, wild
tansy, hogweed, Roman wormwood,
wild tansy, bitterweed, mayweed,
hayfever weed, black weed, Roman

wormweed, annual ragweed
♦ Weed, Noxious Weed, Naturalised,
Native Weed, Environmental Weed,
Casual Alien
♦ 7, 8, 23, 34, 36, 40, 42, 44, 49, 52, 68,
70, 86, 87, 88, 93, 94, 98, 104, 121, 146,
147, 161, 174, 203, 207, 210, 211, 212,
218, 229, 241, 243, 249, 253, 256, 257,
269, 272, 275, 280, 290, 294, 297, 300
♦ aH, arid, cultivated, herbal. Origin:
North America.

**Ambrosia artemisiifolia L. var.
artemisiifolia**
Asteraceae
♦ common ragweed, ragweed, annual
ragweed
♦ Noxious Weed
♦ 229

**Ambrosia artemisiifolia L. var. elatior
(L.) Descourt.**
Asteraceae
Ambrosia elatior L. (see)
♦ common ragweed, ragweed, annual
ragweed
♦ Weed, Naturalised, Noxious Weed
♦ 229, 263, 286, 287
♦ aH.

**Ambrosia artemisiifolia L. var.
paniculata (Michx.) Blank.**
Asteraceae
Ambrosia paniculata Michx. (see)
♦ common ragweed, ragweed, annual
ragweed
♦ Noxious Weed
♦ 229

Ambrosia bidentata Michx.
Asteraceae
♦ lanceleaf ragweed
♦ Weed
♦ 87, 88, 161, 210, 218
♦ herbal.

Ambrosia bipinnatifida (Nutt.) E.Greene
Asteraceae
♦ beach burr
♦ Weed
♦ 161
♦ pH.

Ambrosia chamissonis (Less.) Greene
Asteraceae
♦ Chamisso's burrweed, silver burr
ragweed, beach burr
♦ Weed
♦ 161
♦ pH, cultivated, herbal.

Ambrosia confertiflora DC.
Asteraceae
Franseria confertiflora (DC.) Rydb.
(see), *Franseria strigulosa* Rydb. (see),
Gaertneria tenuifolia Harv. & Gray
♦ burr ragweed, slimleaf bursage,
altamisa de playa, weakleaf burr
ragweed
♦ Weed, Noxious Weed, Quarantine
Weed, Naturalised
♦ 76, 86, 88, 98, 147, 161, 199, 203, 243,
261, 269
♦ pH, herbal. Origin: southern USA,
Mexico.

Ambrosia cumanensis Kunth
Asteraceae
♦ Weed
♦ 87, 88
♦ herbal.

Ambrosia discolor Nutt.
Asteraceae
= *Ambrosia tomentosa* Nutt.
♦ silverleaf povertyweed, white leaved franseria, creeping ragweed, bursage, skeletonleaf, skeletonleaf bursage
♦ Weed
♦ 39, 49
♦ toxic. Origin: North America.

Ambrosia dumosa (Gray) Payne
Asteraceae
Franseria dumosa A.Gray
♦ burroweed, white bursage, white burrobush
♦ Weed
♦ 161
♦ S, arid, cultivated, herbal.

Ambrosia elatior L.
Asteraceae
= *Ambrosia artemisiifolia* L. var. *elatior* (L.) Descourt.
♦ artemisia de terra
♦ Weed
♦ 237, 255, 295
♦ herbal. Origin: Americas.

Ambrosia grayi (A.Nels.) Shinners
Asteraceae
Franseria tomentosa A.Gray (see)
♦ woollyleaf burr ragweed, woollyleaf bursage, burr ragweed, woollyleaf franseria, lagoonweed, woollyleaf povertyweed
♦ Weed,
Quarantine Weed, Native Weed
♦ 76, 88, 161, 174, 210, 212
♦ herbal. Origin: North America.

Ambrosia helenae Rouleau
Asteraceae
♦ Naturalised
♦ 287

Ambrosia maritima L.
Asteraceae
♦ sea ambrosia
♦ Weed, Casual Alien
♦ 40, 87, 88, 185, 221
♦ promoted, herbal.

Ambrosia paniculata Michx.
Asteraceae
= *Ambrosia artemisiifolia* L. var. *paniculata* (Michx.) Blank.
♦ Weed
♦ 87, 88
♦ herbal.

Ambrosia paniculata Michx. var. peruviana (Willd.) O.E.Schulz
Asteraceae
♦ Weed
♦ 14

Ambrosia peruviana Willd.
Asteraceae
♦ Peruvian ragweed

♦ Weed, Naturalised
♦ 87, 88, 101, 300
♦ herbal.

Ambrosia polystachya DC.
Asteraceae
Ambrosia maritima Vell.
♦ cravorana
♦ Weed
♦ 255

Ambrosia psilostachya DC.
Asteraceae
Ambrosia coronopifolia Torr. & Gray, *Ambrosia psilostachya* DC. var. *coronopifolia* (Torr. & Gray) Farw. (see)
♦ western ragweed, perennial ragweed, cuman ragweed
♦ Weed, Noxious Weed, Naturalised, Native Weed, Garden Escape
♦ 7, 34, 49, 86, 87, 88, 93, 98, 121, 147, 161, 174, 180, 198, 203, 210, 218, 243, 269, 287, 294
♦ pH, arid, cultivated, herbal. Origin: western North America.

Ambrosia psilostachya DC. var. coronopifolia (Torr. & Gray) Farw.
Asteraceae
= *Ambrosia psilostachya* DC.
♦ perennial ragweed
♦ Weed
♦ 218

Ambrosia tarapacana Phil.
Asteraceae
♦ Weed
♦ 87, 88

Ambrosia tenuifolia Spreng.
Asteraceae
Franseria tenuifolia Harv. & Gray (see)
♦ lacy ragweed, slimleaf burr ragweed, cravorana, ragweed, hogweed, altamisa, ajenjo de campo
♦ Weed, Quarantine Weed, Noxious Weed, Naturalised, Casual Alien
♦ 76, 86, 87, 88, 98, 147, 198, 203, 236, 237, 241, 243, 255, 269, 271, 280, 295
♦ cultivated. Origin: Americas.

Ambrosia tomentosa Nutt.
Asteraceae
Ambrosia discolor Nutt. (see), *Franseria discolor* Nutt. (see), *Gaertneria discolor* (Nutt.) Kuntze
♦ skeletonleaf bursage, woolly leaved povertyweed, woolly franseria, burr ragweed, silverleaf, skeletonleaf burr ragweed
♦ Weed, Noxious Weed, Environmental Weed
♦ 49, 161, 210, 212, 219, 229, 243
♦ herbal. Origin: North America.

Ambrosia trifida L.
Asteraceae
♦ giant ragweed, great ragweed, horseweed, tall ambrosia, big bitterweed, kinghead
♦ Weed, Quarantine Weed, Noxious Weed, Naturalised, Native Weed, Introduced, Casual Alien

♦ 23, 34, 35, 36, 40, 42, 49, 52, 68, 76, 87, 88, 161, 174, 203, 210, 211, 212, 218, 228, 229, 256, 263, 272, 275, 286, 287, 293, 297, 299
♦ aH, arid, cultivated, herbal. Origin: North America.

Ambrosia trifida L. fo. integrifolia Fern.
Asteraceae
♦ Weed, Naturalised
♦ 286, 287

Ambrosia trifida L. var. texana Scheele
Asteraceae
♦ Texan great ragweed, blood ragweed, giant ragweed
♦ Noxious Weed, Naturalised
♦ 229, 287

Ambrosia trifida L. var. trifida
Asteraceae
♦ giant ragweed, great ragweed
♦ Noxious Weed
♦ 229

Amburana cearensis (Allemão) A.C.Sm.
Fabaceae/Papilionaceae
Amburana claudii Schwacke & Taub., *Torresea cearensis* Allemão
♦ Introduced
♦ 228
♦ T, arid, herbal.

Amelanchier alnifolia Nutt.
Rosaceae
♦ Saskatoon serviceberry, marjatuomipihlaja
♦ Weed, Cultivation Escape
♦ 42, 87, 88, 218
♦ S, cultivated, herbal.

Amelanchier arborea (Michx.f.) Fern.
Rosaceae
♦ downy serviceberry, common serviceberry
♦ Weed
♦ 87, 88, 218
♦ T, cultivated, herbal.

Amelanchier canadensis (L.) Med.
Rosaceae
♦ thicket serviceberry, juneberry, shadblow serviceberry, Canadian serviceberry
♦ Weed
♦ 87, 88, 218
♦ S, cultivated, herbal.

Amelanchier florida Lindl.
Rosaceae
= *Amelanchier alnifolia* (Nutt.) Nutt. var. *semiintegrifolia* (Hook.) C.Hitchc. (NoR)
♦ Pacific serviceberry
♦ Weed
♦ 87, 88, 218
♦ herbal.

Amelanchier laevis Wieg.
Rosaceae
♦ Allegheny serviceberry, sirotuomipihlaja
♦ Weed, Cultivation Escape
♦ 42, 87, 88, 218
♦ S, cultivated, herbal.

Amelanchier lamarckii F.G.Schroed.
Rosaceae
= *Amelanchier laevis* Wieg. ×
Amelanchier arborea (Michx.f.) Fern.
[most probable combination]
♦ apple serviceberry, rusotuomipihlaja
♦ Cultivation Escape, Casual Alien
♦ 42, 280
♦ S, cultivated, herbal. Origin: obscure,
possibly North America.

Amelanchier sanguinea (Pursh) DC.
Rosaceae
Pyrus sanguinea Pursh.
♦ roundleaf serviceberry, Utah
serviceberry
♦ Weed
♦ 87, 88, 218
♦ S, cultivated, herbal.

Amelanchier spicata (Lam.) Koch
Rosaceae
Amelanchier ovalis auct. non Medik.
Borkh., *Crataegus spicata* Lam.
♦ isotuomipihlaja
♦ Cultivation Escape
♦ 42
♦ S, cultivated, herbal.

Amellus anisatus Cass.
Asteraceae
♦ Quarantine Weed
♦ 220

Amellus asteroides Druce
Asteraceae
♦ Quarantine Weed
♦ 220

Amellus fruticosus Linn. ex Jacks.
Asteraceae
♦ Quarantine Weed
♦ 220

Amellus lychnitis L.
Asteraceae
Amellus pallidus Salisb. (see)
♦ Quarantine Weed
♦ 220

Amellus pallidus Salisb.
Asteraceae
= *Amellus lychnitis* L.
♦ Quarantine Weed
♦ 220

Amellus tenuifolius Burm.f.
Asteraceae
♦ Quarantine Weed
♦ 220

Amethystea caerulea L.
Lamiaceae
♦ rurihakka
♦ Weed, Quarantine Weed
♦ 76, 87, 88, 220, 275, 297
♦ aH.

**Amianthium muscaetoxicum (Walter)
A.Gray**
Liliaceae/Melanthiaceae
♦ staggergrass
♦ Weed
♦ 39, 161
♦ herbal, toxic.

Ammannia arenaria Kunth
Lythraceae

♦ Weed
♦ 275
♦ aH.

Ammannia auriculata Willd.
Lythraceae
Ammannia auriculata Willd. var. *arenaria*
(H.B.K.) Koehne (see)
♦ redstem, eared redstem
♦ Weed, Naturalised
♦ 68, 86, 87, 88, 93, 218, 286, 287
♦ herbal. Origin: Americas, Africa,
Asia.

**Ammannia auriculata Willd. var.
arenaria (H.B.K.) Koehne**
Lythraceae
= *Ammannia auriculata* Willd.
♦ redstems, ceibalillo
♦ Weed
♦ 236, 237, 295

Ammannia baccifera L.
Lythraceae
♦ monarch redstem, yaa raknaa, mafai
nok khum
♦ Weed, Naturalised
♦ 39, 66, 87, 88, 101, 170, 204, 238, 239,
272, 274, 275
♦ aH, herbal, toxic.

Ammannia coccinea Rothb.
Lythraceae
♦ purple ammania, valley redstem, red
ammania, long leaved ammania, red
stem, redberry
♦ Weed, Quarantine Weed,
Naturalised, Introduced
♦ 14, 38, 70, 76, 87, 88, 161, 180, 203,
218, 220, 243, 253, 263, 286, 287, 300
♦ aH, arid, herbal. Origin: Europe.

Ammannia latifolia L.
Lythraceae
♦ pink redstem
♦ Weed, Naturalised
♦ 87, 88, 300
♦ herbal.

Ammannia microcarpa DC.
Lythraceae
♦ Weed
♦ 88, 170, 191

Ammannia multiflora Roxb.
Lythraceae
♦ Weed
♦ 87, 88, 263, 274, 286, 297
♦ aH. Origin: Australia.

Ammannia octandra L.f.
Lythraceae
♦ Weed
♦ 87, 88, 170, 191

Ammannia pentandra Roxb.
Lythraceae
= *Rotala rosea* (Poir.) Cook
♦ Weed
♦ 87, 88

Ammannia peploides Spreng.
Lythraceae
= *Rotala indica* (Willd.) Koehne
♦ Weed, Quarantine Weed
♦ 76, 87, 88, 203, 220

Ammannia prieuriana Guill. & Perr.
Lythraceae
♦ Weed, Quarantine Weed
♦ 76, 87, 88, 203, 220

**Ammannia rotundifolia Buch.-Ham. ex
Roxb.**
Lythraceae
= *Rotala rotundifolia* Koehne
♦ Weed
♦ 87, 88

Ammannia senegalensis Lam.
Lythraceae
♦ red ammannia
♦ Weed
♦ 87, 88
♦ aqua, cultivated, herbal.

Ammannia verticillata (Ard.) Lam.
Lythraceae
♦ ammania a fiori sessili
♦ Weed
♦ 272
♦ herbal.

Ammi majus L.
Apiaceae
♦ greater ammi, bullwort, bishop's
weed, large bullwort, common
bishop's weed, lace flower, Queen
Anne's lace
♦ Weed, Naturalised, Introduced,
Casual Alien
♦ 7, 34, 39, 40, 42, 70, 86, 87, 88, 94, 98,
101, 115, 118, 158, 161, 176, 179, 185,
198, 199, 203, 218, 221, 228, 236, 237,
243, 253, 269, 271, 272, 280, 287, 295,
300
♦ a/bH, arid, cultivated, herbal, toxic.
Origin: Mediterranean.

Ammi majus L. var. glaucifolius L.
Apiaceae
♦ bishop's weed, lace flower, Queen
Anne's lace
♦ Weed
♦ 121
♦ a/pH. Origin: Eurasia.

Ammi visnaga (L.) Lam.
Apiaceae
Daucus visnaga L.
♦ toothpick ammi, toothpick bullwort,
toothpick weed, visnaga, honeyplant
♦ Weed, Naturalised, Introduced,
Casual Alien
♦ 34, 38, 39, 42, 70, 86, 87, 88, 94, 98,
101, 161, 185, 203, 218, 228, 236, 237,
241, 253, 271, 287, 295, 300
♦ a/bH, arid, cultivated, herbal, toxic.
Origin: Eurasia.

Ammobium alatum R.Br.
Asteraceae
♦ sandflower, winged everlasting
♦ Naturalised, Casual Alien
♦ 86, 176, 280
♦ cultivated, herbal. Origin: Australia.

Ammophila Host spp.
Poaceae
♦ beachgrass, marram, sand reed
♦ Weed
♦ 18, 88
♦ G.

Ammophila arenaria (L.) Link
  Poaceae
  ♦ European beachgrass, marram grass,
  rantakaura
  ♦ Weed, Quarantine Weed, Noxious
  Weed, Naturalised, Introduced,
  Garden Escape, Environmental Weed
  ♦ 7, 15, 35, 70, 72, 76, 78, 80, 86, 87, 88,
  98, 101, 116, 134, 137, 146, 151, 152, 176,
  181, 198, 203, 218, 221, 225, 228, 231,
  241, 246, 280, 289, 296, 300
  ♦ pG, arid, cultivated, herbal. Origin:
  western Europe.

Ammophila breviligulata Fern.
  Poaceae
  ♦ American beachgrass, beachgrass
  ♦ Weed, Naturalised
  ♦ 87, 88, 218, 287
  ♦ pG, promoted, herbal.

Ammoselinum popei Torr. & A.Gray
  Apiaceae
  ♦ kääpiöputki, plains sandparsley
  ♦ Casual Alien
  ♦ 42
  ♦ herbal.

**Amomyrtus luma (Molina) Legrand &
Kausel**
  Myrtaceae
  Myrtus luma Molina
  ♦ Chilean myrtle, luma, palo madrona,
  cauchao
  ♦ Naturalised
  ♦ 40
  ♦ cultivated.

Amorpha canescens Pursh.
  Fabaceae/Papilionaceae
  ♦ leadplant
  ♦ Weed
  ♦ 161
  ♦ S, cultivated, herbal.

Amorpha fruticosa L.
  Fabaceae/Papilionaceae
  Amorpha occidentalis Abrams
  ♦ false indigo, indigobush, bastard
  indigo, amorpha, desert indigobush
  ♦ Weed, Quarantine Weed, Noxious
  Weed, Naturalised, Introduced
  ♦ 1, 76, 80, 88, 133, 143, 146, 161, 195,
  224, 228, 229, 272, 286, 287
  ♦ S, aqua, cultivated, herbal.

**Amorphophallus campanulatus (Roxb.)
Bl. ex Decne.**
  Araceae
  = *Amorphophallus paeoniifolius* (Dennst.)
  Nicolson (NoR)
  ♦ dragon arum
  ♦ Weed
  ♦ 87, 88
  ♦ herbal.

Ampelamus albidus (Nutt.) Britt.
  Asclepiadaceae/Apocynaceae
  = *Cynanchum laeve* (Michx.) Pers (NoR)
  ♦ honeyvine milkweed, sandvine,
  milkweed
  ♦ Weed
  ♦ 87, 88, 161, 195, 207, 210, 218
  ♦ herbal. Origin: North America.

**Ampelodesmos mauritanica (Poir.) T.Dur.
& Schinz**
  Poaceae
  = *Ampelodesmos tenax* (Vahl) Link
  (NoR)
  ♦ Mauritanian grass
  ♦ Naturalised
  ♦ 101
  ♦ pG, promoted.

Ampelopsis aconitifolia Bunge
  Vitaceae
  ♦ monkshood vine
  ♦ Naturalised
  ♦ 101
  ♦ cultivated.

**Ampelopsis aconitifolia Bunge var.
glabra Diels**
  Vitaceae
  ♦ Weed
  ♦ 275

Ampelopsis arborea (L.) Koehne
  Vitaceae
  ♦ peppervine
  ♦ Weed, Cultivation Escape
  ♦ 80, 87, 88, 161, 218, 261
  ♦ pC, cultivated, herbal.

**Ampelopsis brevipedunculata (Maxim.)
Trautv.**
  Vitaceae
  Ampelopsis heterophylla (Thunb.) Sieb. &
  Zucc., *Cissus brevipedunculata* Maxim.
  ♦ Chinaberry, heartleaf ampelopsis,
  porcelain berry, Amur peppervine,
  wild grape creeper, porcelain
  ampelopsis
  ♦ Weed, Naturalised, Introduced,
  Garden Escape, Environmental Weed
  ♦ 17, 80, 88, 101, 133, 142, 151, 195, 222,
  224
  ♦ pC, cultivated, herbal. Origin: east
  Asia, China, Japan, Korea, eastern
  Russia.

Ampelopsis cordata Michx.
  Vitaceae
  ♦ raccoon grape, heartleaf peppervine
  ♦ Weed
  ♦ 80
  ♦ herbal.

**Ampelopsis glandulosa (Wall.) Momiy.
var. heterophylla (Thunb.) Momiy.**
  Vitaceae
  ♦ Weed
  ♦ 286

**Ampelopsis glandulosa (Wall.) Momiy.
var. heterophylla (Thunb.) Momiy. fo.
citrulloides (Lebas) Momiy.**
  Vitaceae
  ♦ Weed
  ♦ 286

Ampelopsis humulifolia Bunge
  Vitaceae
  ♦ hopleaf ampelopsis
  ♦ Weed
  ♦ 297

Amphibromus neesii Steud.
  Poaceae
  ♦ swamp wallaby grass

  ♦ Weed
  ♦ 87, 88
  ♦ G, arid, cultivated. Origin: Australia.

**Amphibromus scabrivalvis (Trin.)
Swallen**
  Poaceae
  ♦ swamp wallaby grass
  ♦ Naturalised
  ♦ 101
  ♦ G.

Amphicarpaea bracteata (L.) Fernald
  Fabaceae/Papilionaceae
  Amphicarpaea monoica (L.) Ell., *Falcata
  comosa* (L.) Kuntze
  ♦ sianpapu, hog peanut, American
  hog peanut
  ♦ Casual Alien
  ♦ 42
  ♦ pC, cultivated, herbal. Origin:
  eastern North America.

Amphicarpaea edgeworthii Benth.
  Fabaceae/Papilionaceae
  ♦ Weed
  ♦ 87, 88
  ♦ pH, promoted.

**Amphicarpaea edgeworthii Benth. var.
japonica Oliv.**
  Fabaceae/Papilionaceae
  ♦ yabumame
  ♦ Weed
  ♦ 286

Amphilophis glabra (Roxb.) Stapf
  Poaceae
  ♦ Weed
  ♦ 87, 88
  ♦ G.

Amphilophium oxylophium Donn.Sm.
  Bignoniaceae
  ♦ Weed
  ♦ 87, 88

Amphilophium paniculatum (L.) Kunth
  Bignoniaceae
  Bignonia paniculata L.
  ♦ liana de cuello, cipo de agua
  ♦ Weed
  ♦ 255
  ♦ Origin: South America.

Amsinckia Lehm. spp.
  Boraginaceae
  ♦ yellow burrweed, amsinckia,
  fiddleneck
  ♦ Weed, Quarantine Weed, Noxious
  Weed, Naturalised
  ♦ 62, 76, 86, 88, 147, 171, 203, 220, 243
  ♦ aH, toxic.

Amsinckia angustifolia Lehm.
  Boraginaceae
  = *Amsinckia hispida* (Ruiz & Pav.)
  I.M.Johnst.
  ♦ Weed
  ♦ 87, 88

Amsinckia barbata Greene
  Boraginaceae
  = *Amsinckia lycopsoides* Lehm.
  ♦ Naturalised
  ♦ 287

Amsinckia calycina (Moris) Chater
Boraginaceae
♦ fiddleneck, yellow gromwell, hairy fiddleneck, ryhmykeltalemmikki, yellow burrweed
♦ Weed, Naturalised, Environmental Weed, Casual Alien
♦ 7, 42, 72, 86, 88, 93, 98, 121, 165, 176, 198, 203, 243, 269, 280
♦ aH, cultivated, toxic. Origin: Americas.

Amsinckia douglasiana A.DC.
Boraginaceae
♦ Douglas fiddleneck
♦ Weed
♦ 87, 88, 161, 218
♦ aH, herbal.

Amsinckia hispida (Ruiz & Pav.) I.M.Johnst.
Boraginaceae
Amsinckia angustifolia Lehm. (see)
♦ cuello de violín
♦ Weed, Introduced
♦ 38, 87, 88, 295

Amsinckia intermedia Fisch. & Mey.
Boraginaceae
♦ coast fiddleneck, common fiddleneck, fiddleneck, finger weed, yellow burnweed, yellow burrweed, yellow forget me not, yellow tarweed, kurttukeltalemmikki
♦ Weed, Naturalised, Casual Alien
♦ 23, 39, 42, 70, 86, 87, 88, 98, 161, 180, 198, 203, 212, 218, 243, 269
♦ aH, arid, cultivated, herbal, toxic. Origin: Americas.

Amsinckia lycopsoides Lehm.
Boraginaceae
Amsinckia barbata Greene (see)
♦ tarweed fiddleneck, bugloss fiddleneck, scarce fiddleneck, yellow burrweed
♦ Weed, Naturalised, Environmental Weed
♦ 7, 39, 72, 86, 87, 88, 98, 161, 198, 203, 218, 269, 287
♦ aH, promoted, herbal, toxic. Origin: western North America.

Amsinckia menziesii (Lehm.) A.Nels. & J.F.Macbr.
Boraginaceae
♦ fiddleneck, Menzies' fiddleneck
♦ Weed, Naturalised
♦ 54, 86, 88, 121, 158, 198
♦ aH, herbal. Origin: North America.

Amsinckia menziesii (Lehm.) Nelson & J.F.Macbr. var. *intermedia* (Fisch. & C.A.Mey.) Ganders
Boraginaceae
♦ coast fiddleneck, common fiddleneck, fiddleneck
♦ Weed
♦ 243
♦ aH.

Amsinckia tessellata Gray
Boraginaceae
Amsinckia macra Suksd.
♦ tessellate fiddleneck, western

fiddleneck, bristly fiddleneck, checker fiddleneck, devil's lettuce
♦ Weed, Naturalised
♦ 80, 87, 88, 161, 218, 287
♦ aH, arid, herbal.

Anabasis aphylla L.
Chenopodiaceae
♦ Weed
♦ 39, 87, 88
♦ arid, toxic.

Anabasis articulata (Forssk.) Moq.
Chenopodiaceae
♦ Weed
♦ 221
♦ arid.

Anabasis setifera Moq.
Chenopodiaceae
Seidlitzia lanigera Post
♦ Weed
♦ 221
♦ arid.

Anacamptis pyramidalis (L.) Rich
Orchidaceae
Orchis pyramidalis L.
♦ pyramidal orchid
♦ Weed
♦ 70, 272
♦ pH, promoted, herbal.

Anacardium occidentale L.
Anacardiaceae
♦ cashew nut, apu initia, kesiu, cashew, cashew apple, cashew nut tree, korocho nut, hijlibadam, kaju
♦ Weed, Naturalised, Introduced, Garden Escape, Environmental Weed
♦ 3, 22, 39, 54, 86, 87, 88, 93, 101, 121, 155, 161, 228
♦ T, arid, cultivated, herbal, toxic. Origin: South America.

Anacharis canadensis (Michx.) Planch.
Hydrocharitaceae
= *Elodea canadensis* Michx.
♦ Cultivation Escape
♦ 261
♦ wpH, cultivated. Origin: North America.

Anacharis densa (Planch.) Vict.
Hydrocharitaceae
= *Egeria densa* Planch.
♦ elodea, tomillo de agua
♦ Cultivation Escape
♦ 261
♦ wpH, cultivated. Origin: South America.

Anacyclus L. spp.
Asteraceae
♦ Weed
♦ 272

Anacyclus alexandrinus Willd.
Asteraceae
♦ Weed
♦ 221

Anacyclus clavatus (Desf.) Pers.
Asteraceae
Anacyclus tomentosus (All.) DC. (see)
♦ valkoraimikki, whitebuttons
♦ Weed, Quarantine Weed,

Naturalised, Casual Alien
♦ 40, 42, 70, 76, 87, 88, 94, 101, 243, 253
♦ cultivated.

Anacyclus maroccanus (Ball) Ball
Asteraceae
Anacyclus valentinus L. (see)
♦ Quarantine Weed
♦ 76

Anacyclus radiatus Loisel.
Asteraceae
♦ yellow anacyclus, keltaraimikki
♦ Weed, Quarantine Weed, Naturalised, Casual Alien
♦ 42, 76, 86, 87, 88, 94, 98, 203, 253
♦ cultivated. Origin: southern Europe.

Anacyclus tomentosus (All.) DC.
Asteraceae
= *Anacyclus clavatus* (Desf.) Pers.
♦ anacyclus
♦ Weed
♦ 243
♦ aH.

Anacyclus valentinus L.
Asteraceae
= *Anacyclus maroccanus* (Ball) Ball.
♦ kehräraimikki
♦ Weed, Casual Alien
♦ 42, 87, 88, 253
♦ cultivated, herbal.

Anadenanthera colubrina (Vell.Conc.) Brenan var. *cebil* (Griseb.) Altschul
Fabaceae/Mimosaceae
Acacia cebil Griseb., *Anadenanthera macrocarpa* (Benth.) Brenan, *Piptadenia hassleriana* Chodat, *Piptadenia hassleriana* Chodat var. *fruticosa* Chodat & Hassl., *Piptadenia macrocarpa* Benth., *Piptadenia macrocarpa* Benth. var. *plurifoliate* Hoehne, *Piptadenia macrocarpa* Benth. var. *genuina* Chod. & Hass., *Piptadenia macrocarpa* Benth. var. *vestita* Chod. & Hass., *Piptadenia macrocarpa* Benth. var. *cebil* (Griseb.) Chod. & Hass., *Piptadenia microphylla* Benth.
♦ angico jacaré, angico preto, angico vermelho
♦ Introduced
♦ 228
♦ S/T, arid.

Anadenanthera peregrina (L.) Speg.
Fabaceae/Mimosaceae
Acacia peregrina (L.) Willd., *Mimosa peregrina* L., *Niopa peregrina* (L.) Britton & Rose, *Piptadenia peregrina* (L.) Benth. (see)
♦ cohoba yope, niopo
♦ Naturalised, Introduced
♦ 32, 39, 228
♦ arid, herbal, toxic.

Anadendrum montanum (Blume) Schott
Araceae
♦ Weed
♦ 87, 88
♦ herbal.

Anagallis arvensis L.
Primulaceae

Anagallis phoenicea (Gouan) Scop., *Anagallis pulchella* Salisbury, *Anagallis parviflora* Hoffm. & Link (see)
♦ scarlet pimpernel, blue pimpernel, pimpernel
♦ Weed, Naturalised, Introduced, Garden Escape, Environmental Weed
♦ 7, 9, 23, 24, 34, 38, 39, 44, 51, 55, 66, 70, 72, 86, 87, 88, 94, 98, 101, 111, 115, 115, 118, 121, 134, 136, 157, 158, 161, 167, 176, 179, 180, 185, 186, 198, 203, 207, 211, 218, 221, 235, 237, 240, 241, 243, 250, 253, 255, 261, 263, 269, 271, 272, 280, 286, 287, 289, 295, 300
♦ aH, arid, cultivated, herbal, toxic. Origin: Mediterranean.

Anagallis arvensis L. fo. *caerulea* (L.) **Ludi**
Primulaceae
♦ Weed, Naturalised
♦ 286, 287

Anagallis arvensis L. ssp. *arvensis*
Primulaceae
♦ scarlet pimpernel
♦ Weed, Naturalised
♦ 101, 165, 243, 280
♦ cultivated, toxic. Origin: Eurasia.

Anagallis arvensis L. ssp. *arvensis* var. *arvensis*
Primulaceae
♦ scarlet pimpernel
♦ Weed, Naturalised
♦ 15, 280

Anagallis arvensis L. ssp. *arvensis* var. *coerulea* (L.) **Gouan**
Primulaceae
♦ blue pimpernel
♦ Naturalised
♦ 280

Anagallis arvensis L. ssp. *foemina* (Mill.) **Schinz & Thell.**
Primulaceae
= *Anagallis foemina* Mill.
♦ poor man's weatherglass
♦ Naturalised
♦ 101

Anagallis arvensis L. ssp. *parviflora* (Hoffmanns. & Link) **Arcang.**
Primulaceae
♦ Naturalised
♦ 280

Anagallis arvensis L. var. *caerulea* (Schreb.) **Gren. & Godr.**
Primulaceae
= *Anagallis foemina* Mill.
♦ scarlet pimpernel, no me olvides, amurajes, blue pimpernel
♦ Weed, Naturalised
♦ 176, 236, 237, 271

Anagallis caerulea **Schreb.**
Primulaceae
= *Anagallis foemina* Mill.
♦ pimpernel
♦ Weed
♦ 23, 87, 88, 243

Anagallis foemina **Mill.**
Primulaceae

Anagallis caerulea Schreb. (see), *Anagallis coerulea* Nath., *Anagallis arvensis* L. ssp. *caerulea* (Schreb.) Gren. & Godr. (see), *Anagallis arvensis* L. ssp. *foemina* (Mill.) Schinz & Thell. (see)
♦ blue pimpernel, drchniãka belasá, poor man's weatherglass
♦ Weed
♦ 44, 70, 87, 88, 94, 115, 243, 253, 272
♦ aH, cultivated, herbal.

Anagallis minima (L.) **Krause**
Primulaceae
Centunculus minimus L. (see)
♦ chaffweed, pikkupunka
♦ Weed, Naturalised, Environmental Weed
♦ 7, 44, 72, 86, 88, 94, 98, 198, 203, 272, 280
♦ aH, herbal. Origin: Mediterranean.

Anagallis monelli **L.**
Primulaceae
♦ flaxleaf pimpernel
♦ Naturalised
♦ 101
♦ cultivated.

Anagallis parviflora **Hoffm. & Link**
Primulaceae
= *Anagallis arvensis* L.
♦ Naturalised
♦ 86
♦ Origin: obscure.

Anagallis pumila **Sw.**
Primulaceae
♦ Florida pimpernel
♦ Weed, Naturalised
♦ 86, 98, 157, 203
♦ aH.

Anagyris foetida **L.**
Fabaceae/Papilionaceae
♦ purging trefoil
♦ Weed, Naturalised
♦ 39, 54, 86, 88, 221, 272
♦ cultivated, herbal, toxic. Origin: Eurasia.

Ananas comosus (L.) **Merr.**
Bromeliaceae
♦ pineapple, pweinper, abacaxi, gravata, caraguatá
♦ Weed, Naturalised, Introduced, Cultivation Escape
♦ 39, 86, 101, 230, 247, 255, 261
♦ pH, cultivated, herbal, toxic. Origin: South America.

Ananas microstachys **Lindm.**
Bromeliaceae
♦ Weed, Quarantine Weed
♦ 76, 87, 88, 203, 220

Anaphalis adnata **Wall. ex DC.**
Asteraceae
♦ naat khao
♦ Weed
♦ 238
♦ herbal.

Anaphalis margaritacea (L.) **Benth. & Hook.**
Asteraceae
Antennaria margaritacea (L.) R.Br. ex

DC., *Gnaphalium margaritaceum* L.
♦ pearly everlasting, helminukkajäkkärä, everlasting, naat doi, large flowered everlasting, silver button
♦ Weed, Quarantine Weed, Naturalised, Cultivation Escape
♦ 23, 40, 42, 76, 87, 88, 136, 161, 209, 218, 238
♦ pH, cultivated, herbal. Origin: North America, Eurasia.

Anarrhinum pubescens **Fresen.**
Scrophulariaceae
♦ Weed
♦ 221

Anastatica hierochuntica **L.**
Brassicaceae
♦ rose of Jericho
♦ Weed
♦ 221
♦ cultivated, herbal.

Anchusa **L. spp.**
Boraginaceae
♦ bugloss
♦ Weed
♦ 243, 272
♦ herbal.

Anchusa aegyptiaca **DC.**
Boraginaceae
Anchusa flava Forssk.
♦ Weed
♦ 88, 221
♦ herbal.

Anchusa arvensis (L.) **Bieb.**
Boraginaceae
Lycopsis arvensis L. (see)
♦ bugloss, small bugloss, annual bugloss, lesser bugloss, alkanet
♦ Weed, Noxious Weed, Naturalised
♦ 1, 70, 80, 86, 88, 94, 98, 101, 139, 146, 156, 161, 176, 198, 203, 229, 241, 243, 253, 272, 300
♦ aH, cultivated, herbal. Origin: Europe.

Anchusa azurea **Mill.**
Boraginaceae
Anchusa italica Retz. (see)
♦ Italian bugloss, large blue alkanet, anchusa, garden anchusa, hemhem, khail
♦ Weed, Quarantine Weed, Naturalised, Garden Escape
♦ 40, 70, 76, 80, 86, 88, 94, 101, 115, 121, 220, 243, 272
♦ pH, arid, cultivated, herbal. Origin: Eurasia.

Anchusa barrelieri (All.) **Vitman**
Boraginaceae
♦ tarharasti, Barrelier's bugloss
♦ Weed, Naturalised, Cultivation Escape
♦ 42, 101, 272, 280
♦ cultivated, herbal.

Anchusa calcarea **Boiss.**
Boraginaceae
♦ Weed
♦ 70

Anchusa capensis **Thunb.**
Boraginaceae
♦ cape forget me not, cape bugloss, bugloss
♦ Weed, Naturalised
♦ 7, 86, 98, 101, 198, 203
♦ a/bH, cultivated, herbal. Origin: southern Africa.

Anchusa hybrida **Ten.**
Boraginaceae
♦ Weed
♦ 87, 88

Anchusa italica **Retz.**
Boraginaceae
= *Anchusa azurea* Mill.
♦ Italianrasti', smohla talianska
♦ Weed, Quarantine Weed, Casual Alien
♦ 42, 76, 87, 88, 203, 253
♦ cultivated, herbal.

Anchusa ochroleuca **M.Bieb.**
Boraginaceae
♦ yellow alkanet
♦ Weed, Naturalised
♦ 40, 272

Anchusa officinalis **L.**
Boraginaceae
Anchusa procera Besser & Link
♦ common bugloss, alkanet
♦ Weed, Noxious Weed, Naturalised, Introduced
♦ 1, 34, 39, 40, 70, 80, 87, 88, 94, 98, 101, 139, 146, 156, 161, 203, 218, 229, 243, 272
♦ pH, arid, cultivated, herbal, toxic.

Anchusa undulata **L.**
Boraginaceae
♦ undulate alkanet
♦ Weed
♦ 70
♦ b/pH, cultivated.

Ancrumia cuspidata **Harv. ex Baker**
Liliaceae/Alliaceae
♦ Quarantine Weed
♦ 220
♦ arid.

Andira humilis **Mart. ex Benth.**
Fabaceae/Papilionaceae
♦ Weed, Quarantine Weed
♦ 76, 87, 88, 203, 220

Andira inermis **(Wright) Kunth ex DC.**
Fabaceae/Papilionaceae
♦ cabbagebark tree, cabbage tree
♦ Weed
♦ 39, 87, 88
♦ T, herbal, toxic.

Andrachne aspera **Spreng.**
Euphorbiaceae
♦ Weed
♦ 221
♦ herbal.

Andrachne telephioides **L.**
Euphorbiaceae
♦ rakaa
♦ Weed
♦ 87, 88, 221, 272

Andrographis echioides **(L.) Nees**
Acanthaceae
♦ false waterwillow
♦ Weed, Naturalised
♦ 87, 88, 101

Andrographis paniculata **(Burm.f.) Wall. ex Nees**
Acanthaceae
♦ Weed, Naturalised
♦ 13, 86, 87, 88, 93, 98, 191, 203
♦ bH, herbal. Origin: India, Sri Lanka.

Andropogon barbinodis **Lag.**
Poaceae
= *Bothriochloa barbinodis* (Lag.) Herter
♦ cane bluestem
♦ Weed
♦ 87, 88, 199, 218
♦ G, herbal.

Andropogon bicornis **L.**
Poaceae
Anatherum bicorne (L.) P.Beauv., *Saccharum bicorne* (L.) Griseb., *Sorghum bicorne* (L.) Kuntze
♦ West Indian foxtail grass, cola de caballo, barbas de Indio
♦ Weed, Quarantine Weed, Noxious Weed
♦ 14, 76, 87, 88, 90, 153, 157, 161, 203, 218, 220, 229, 255
♦ pG. Origin: Americas.

Andropogon brevifolius **Sw.**
Poaceae
= *Schizachyrium brevifolium* (Sw.) Nees ex Büse
♦ Weed
♦ 263
♦ aG.

Andropogon caricosus **L.**
Poaceae
= *Dichanthium caricosum* (L.) A.Camus
♦ Weed
♦ 14
♦ G.

Andropogon chinensis **(Nees) Merr.**
Poaceae
Andropogon ascinodis C.B.Clarke
♦ Weed
♦ 238
♦ pG.

Andropogon condensatus **Kunth**
Poaceae
= *Schizachyrium condensatum* (Kunth) Nees
♦ bush beardgrass, little bluestem
♦ Weed, Quarantine Weed
♦ 76, 87, 88, 203, 220
♦ G.

Andropogon distachyos **L.**
Poaceae
♦ gamba grass
♦ Weed, Naturalised, Environmental Weed
♦ 7, 86, 98, 203
♦ G. Origin: Africa.

Andropogon eucomus **Nees**
Poaceae
♦ old man's beard, silverthread grass,

small silver andropogon, snowflake grass
♦ Native Weed
♦ 121
♦ pG. Origin: southern Africa.

Andropogon gayanus **Kunth**
Poaceae
♦ gamba grass, bluestem, Rhodesian andropogon, Rhodesian bluegrass
♦ Weed, Naturalised, Introduced, Environmental Weed
♦ 3, 7, 11, 86, 93, 155, 191, 228
♦ pG, arid, cultivated, herbal. Origin: tropical Africa.

Andropogon gayanus **Kunth var. bisquamulatus (Hochst.) Hack.**
Poaceae
♦ Introduced
♦ 228
♦ G, arid.

Andropogon gayanus **Kunth var.** *gayanus*
Poaceae
Andropogon gayanus Kunth var. *genuinus* Hack.
♦ Weed
♦ 88
♦ G, arid.

Andropogon gayanus **Kunth var. tridentatus Hack.**
Poaceae
♦ Introduced
♦ 228
♦ G, arid.

Andropogon gerardii **Vitman**
Poaceae
Andropogon furcatus Muhlenb., *Andropogon provincialis* Lam.
♦ big bluestem, turkey foot
♦ Weed, Introduced
♦ 88, 90, 218, 228
♦ pG, arid, cultivated, herbal.

Andropogon glomeratus **(Walter) Britton, Sterns & Pogg.**
Poaceae
= *Andropogon marcrourus* Michx. (NoR)
♦ bushy bluestem, bush beardgrass
♦ Weed, Introduced, Environmental Weed
♦ 3, 80, 87, 88, 90, 152, 161, 191, 218, 228, 249
♦ pG, aqua, herbal. Origin: North America.

Andropogon huillensis **Rendle**
Poaceae
♦ large silver andropogon, old man's beard
♦ Native Weed
♦ 121
♦ pG. Origin: southern Africa.

Andropogon kelleri **Hack.**
Poaceae
Andropogon cyrtocladus Stapf, *Schizachyrium kelleri* (Hack.) Stapf
♦ Introduced
♦ 228
♦ G, arid.

Andropogon lateralis **Nees**
Poaceae

♦ Weed, Quarantine Weed
♦ 76, 87, 88, 90, 203, 220
♦ G.

Andropogon leucostachyus Kunth
Poaceae
♦ matojillo bluestem
♦ Weed
♦ 87, 88, 255
♦ pG. Origin: Americas.

Andropogon pertusus (L.) Willd.
Poaceae
= *Bothriochloa pertusa* (L.) A.Camus
♦ Weed
♦ 14, 88
♦ G, herbal.

Andropogon pseudapricus Stapf
Poaceae
♦ Introduced
♦ 228
♦ G, arid.

Andropogon pumilus Roxb.
Poaceae
Andropogon demissus Steud.,
Andropogon pachyarthrus Hack.
♦ Introduced
♦ 228
♦ G, arid.

Andropogon saccharoides Sw.
Poaceae
= *Bothriochloa saccharoides* (Sw.) Rydb.
♦ silver beardgrass
♦ Weed
♦ 87, 88, 218
♦ G, herbal.

**Andropogon saccharoides Sw. var.
laguroides (DC.) Hack.**
Poaceae
= *Bothriochloa laguroides* (DC.) Herter
♦ Weed
♦ 199
♦ G.

Andropogon scoparius Michx.
Poaceae
= *Schizachyrium scoparium* (Michx.)
Nash
♦ little bluestem
♦ Weed
♦ 88, 218
♦ G, herbal.

Andropogon selloanus (Hack.) Hack.
Poaceae
♦ Weed
♦ 90
♦ G.

Andropogon tectorum Schum. & Thonn.
Poaceae
Andropogon spectabilis Schum.,
Andropogon tenuiculmis Reznik
♦ Weed
♦ 87, 88
♦ G, arid.

Andropogon ternarius Michx.
Poaceae
Andropogon argyraeus Schult.
♦ splitbeard bluestem, silver
beardgrass
♦ Weed

♦ 87, 88, 90, 218
♦ pG, herbal.

Andropogon virginicus L.
Poaceae
Andropogon dissitiflorus Michx.
♦ broomsedge, whisky grass, yellow
bluestem, broomsedge bluestem, sedge
grass, beardgrass
♦ Weed, Sleeper Weed, Quarantine
Weed, Noxious Weed, Naturalised,
Garden Escape, Environmental Weed
♦ 3, 76, 80, 86, 87, 88, 90, 98, 147, 152,
161, 191, 198, 203, 207, 210, 211, 218,
225, 229, 243, 246, 249, 269, 280, 286,
287, 290
♦ pG, cultivated, herbal. Origin: North
America.

**Andropogon virginicus L. var. decipiens
C.Campbell**
Poaceae
♦ broomsedge, broomsedge bluestem
♦ Noxious Weed
♦ 229
♦ G.

Andropogon virginicus L. var. virginicus
Poaceae
♦ broomsedge, broomsedge bluestem
♦ Weed, Noxious Weed
♦ 34, 229
♦ pG.

Androsace diffusa Small
Primulaceae
= *Androsace septentrionalis* L. ssp.
subulifera (Gray) G.T.Robbins (NoR)
♦ Weed
♦ 23, 88

Androsace elongata L.
Primulaceae
♦ California rockjasmine
♦ Weed
♦ 88, 94, 272
♦ cultivated, herbal.

Androsace filiformis Retz.
Primulaceae
♦ filiform rockjasmine, idännukki
♦ Weed, Casual Alien
♦ 42, 87, 88, 297
♦ aH, herbal.

Androsace lactiflora Pall.
Primulaceae
♦ Weed
♦ 275
♦ bH, cultivated.

Androsace maxima L.
Primulaceae
Aretia maxima Bubani
♦ greater rockjasmine
♦ Weed, Naturalised
♦ 87, 88, 94, 101, 272
♦ cultivated.

Androsace septentrionalis L.
Primulaceae
♦ pygmyflower rockjasmine, northern
rockjasmine, ketonukki
♦ Weed, Noxious Weed
♦ 23, 87, 88, 94, 218, 272, 299
♦ cultivated, herbal.

Androsace umbellata (Lour.) Merr.
Primulaceae
♦ Weed
♦ 87, 88, 297

Andryala integrifolia L.
Asteraceae
♦ keiriö
♦ Weed, Casual Alien
♦ 42, 70, 250, 253
♦ cultivated.

Andryala laxiflora DC.
Asteraceae
♦ Weed
♦ 70

Aneilema aequinoctiale (P.Beauv.) Kunth
Commelinaceae
Aneilema tacazzeanum Hochst. ex
A.Rich. (see)
♦ djendjeko
♦ Weed
♦ 88
♦ cultivated, herbal.

Aneilema beniniense (P.Beauv.) Kunth
Commelinaceae
♦ gangulu
♦ Weed
♦ 87, 88
♦ pH, herbal.

Aneilema japonica (Thunb.) Kunth
Commelinaceae
♦ Weed, Quarantine Weed
♦ 76, 87, 88, 191, 203, 204, 220

Aneilema keisak Hassk.
Commelinaceae
= *Murdannia keisak* (Hassk.) Hand.-
Mazz.
♦ Weed, Quarantine Weed
♦ 87, 88, 203, 220
♦ herbal.

Aneilema nudiflorum (L.) R.Br.
Commelinaceae
= *Murdannia nudiflora* (L.) Brenan
♦ Weed
♦ 262
♦ pH.

Aneilema sinicum Ker Gawl.
Commelinaceae
= *Commelina sinica* (Ker Gawl.) Roem.
& Schult. (NoR)
♦ Weed
♦ 87, 88

Aneilema spiratum (L.) Sweet
Commelinaceae
= *Murdannia spirata* (L.) Brückn.
♦ Weed
♦ 87, 88

**Aneilema tacazzeanum Hochst. ex
A.Rich.**
Commelinaceae
= *Aneilema aequinoctiale* (P.Beauv.)
Kunth
♦ Weed
♦ 221

Aneilema umbrosum (Vahl) Kunth
Commelinaceae
♦ Weed
♦ 87, 88

Aneilema vitiense **Seem.**
Commelinaceae
♦ Weed
♦ 87, 88

Anemone **L. spp.**
Ranunculaceae
♦ windflower, pasqueflower, thimbleweed, anemone
♦ Weed
♦ 189, 247, 272
♦ herbal, toxic.

Anemone apennina **L.**
Ranunculaceae
♦ blue anemone
♦ Naturalised
♦ 39, 40
♦ cultivated, toxic.

Anemone blanda **Schott & Kotschy**
Ranunculaceae
♦ Greek thimbleweed, poppy flowered anemone
♦ Weed, Naturalised
♦ 101, 161
♦ cultivated, herbal, toxic.

Anemone canadensis **L.**
Ranunculaceae
♦ kanadanvuokko, meadow anemone, Canadian anemone
♦ Weed, Cultivation Escape
♦ 23, 42, 88, 161
♦ pH, cultivated, herbal, toxic.

Anemone cernua **Thunb.**
Ranunculaceae
♦ nodding anemone
♦ Weed
♦ 87, 88
♦ herbal.

Anemone coronaria **L.**
Ranunculaceae
♦ crown anemone
♦ Weed, Casual Alien
♦ 39, 87, 88, 221, 280
♦ cultivated, herbal, toxic.

Anemone cylindrica **Gray**
Ranunculaceae
♦ candle anemone, thimbleweed
♦ Weed
♦ 23, 88
♦ pH, cultivated, herbal.

Anemone dichotoma **L.**
Ranunculaceae
♦ dichotomous anemone, futamataichige
♦ Weed
♦ 39, 297
♦ toxic.

Anemone flaccida **F.Schmidt**
Ranunculaceae
♦ nirinsou
♦ Weed
♦ 286
♦ pH, promoted. Origin: north and west China, Japan.

Anemone hupehensis (**hort. ex Lem.**) **Lem. ex Boynton**
Ranunculaceae
♦ Hupeh anemone, Japanese thimbleweed

♦ Weed, Naturalised, Environmental Weed
♦ 80, 101, 152
♦ cultivated, herbal. Origin: China.

Anemone hupehensis (**hort. ex Lem.**) **Lem. ex Boynton var.** *japonica* (**Thunb.**) **Bowles**
Ranunculaceae
♦ Japanese thimbleweed
♦ Naturalised
♦ 101, 287

Anemone × *hybrida* **Paxton**
Ranunculaceae
= *Anemone hupehensis* Lemoine × *Anemone vitifolia* Buch-Ham ex DC. [most probable combination]
♦ Japanese anemone
♦ Weed, Naturalised, Casual Alien
♦ 15, 40, 280
♦ cultivated, herbal.

Anemone multifida **Poir.**
Ranunculaceae
♦ Pacific anemone, flikanemon Amerikanvuokko
♦ Weed
♦ 23, 88
♦ pH, cultivated, herbal.

Anemone narcissiflora **L.**
Ranunculaceae
♦ narcissus flowered anemone, veternica narcisokvetá
♦ Weed
♦ 161
♦ pH, cultivated, herbal, toxic.

Anemone nemorosa **L.**
Ranunculaceae
♦ European thimbleweed, wood anemone, valkovuokko, veternica hájna
♦ Weed, Naturalised
♦ 39, 87, 88, 101, 272, 280
♦ pH, cultivated, herbal, toxic.

Anemone pulsatilla **L.**
Ranunculaceae
= *Pulsatilla vulgaris* Mill.
♦ pasqueflower, pulsatilla, anemone, wind flower
♦ Weed
♦ 39, 87, 88
♦ cultivated, herbal, toxic.

Anemone sylvestris **L.**
Ranunculaceae
♦ arovuokko, veternica lesná
♦ Cultivation Escape
♦ 39, 42
♦ cultivated, herbal, toxic.

Anemone tuberosa **Rydb.**
Ranunculaceae
♦ tuber anemone
♦ Weed
♦ 161
♦ pH, cultivated, herbal, toxic.

Anemone virginiana **L.**
Ranunculaceae
♦ thimbleweed, tall thimbleweed
♦ Weed
♦ 8, 23, 88

♦ pH, cultivated, herbal, toxic.

Anemopsis californica (**Nutt.**) **Hook. & Arn.**
Saururaceae
♦ yerb mansa
♦ Weed
♦ 87, 88, 161, 180, 218
♦ wpH, cultivated, herbal. Origin: south-western North America, northern Mexico.

Anethum graveolens **L.**
Apiaceae
Peucedanum graveolens Bail. (*non* Watson)
♦ dill, aneto frutti
♦ Weed, Naturalised, Introduced, Cultivation Escape, Casual Alien
♦ 34, 38, 40, 42, 70, 86, 87, 88, 94, 98, 101, 121, 203, 228, 241, 252, 261, 262, 287
♦ aH, arid, cultivated, herbal. Origin: Eurasia.

Angadenia berterii (**A.DC.**) **Miers**
Apocynaceae
♦ Weed
♦ 14

Angelica archangelica **L.**
Apiaceae
♦ angelica
♦ Weed
♦ 39, 87, 88, 272
♦ bH, cultivated, herbal, toxic. Origin: Syria.

Angelica atropurpurea **L.**
Apiaceae
♦ purplestem angelica, American angelica
♦ Weed
♦ 87, 88, 161, 218
♦ pH, cultivated, herbal.

Angelica pachycarpa **Lange**
Apiaceae
♦ Weed, Naturalised
♦ 15, 280
♦ cultivated.

Angelica sylvestris **L.**
Apiaceae
♦ angelica, woodland angelica, wild angelica, angelica nostrana
♦ Weed, Environmental Weed
♦ 4, 70, 80, 87, 88, 104, 272
♦ bH, cultivated, herbal.

Angelonia angustifolia **Benth.**
Scrophulariaceae
♦ narrowleaf angelon
♦ Weed, Naturalised, Introduced
♦ 101, 179, 230
♦ herbal.

Angelonia salicariifolia **Humb. & Bonpl.**
Scrophulariaceae
Angelonia salicariaefolia Humb. & Bonpl.
♦ willowleaf angelon, grannie's bonnets, angelón
♦ Weed, Naturalised, Cultivation Escape
♦ 86, 87, 88, 261
♦ cultivated, herbal. Origin: Brazil.

Angiopteris evecta (J.R.Forst.) Hoffm.
Dryopteridaceae
♦ mule's foot, Madagascar tree fern,
oriental vessel fern, giant fern
♦ Weed, Naturalised, Cultivation
Escape
♦ 3, 39, 101, 191, 233
♦ H, cultivated, herbal, toxic. Origin:
Malaysia, Australia, Madagascar.

Angophora costata (Gaertn.) Britten
Myrtaceae
♦ apple jack, smooth angophora, red
gum
♦ Weed, Naturalised, Native Weed,
Garden Escape, Environmental Weed,
Cultivation Escape, Casual Alien
♦ 7, 72, 86, 88, 280
♦ T, cultivated, herbal. Origin:
Australia.

Anisantha madritensis (L.) Nev.
Poaceae
= *Bromus madritensis* L.
♦ compact brome
♦ Naturalised
♦ 40
♦ G.

Anisantha rigida (Roth) Hyl.
Poaceae
= *Bromus rigidus* Roth
♦ ripgut brome, bromegrass
♦ Naturalised
♦ 40
♦ G.

Anisantha sterilis (L.) Nev.
Poaceae
= *Bromus sterilis* L.
♦ barren brome
♦ Weed
♦ 243
♦ G.

Aniseia martinicensis (Jacq.) Choisy
Convolvulaceae
♦ Weed
♦ 13, 179
♦ C.

Anisocereus Backeb. spp.
Cactaceae
= *Pachycereus* (Berg.) Britt. & Rose spp.
♦ Weed, Quarantine Weed
♦ 88, 203, 220

Anisomeles indica (L.) Kuntze
Lamiaceae
♦ Weed
♦ 13, 87, 88, 276
♦ herbal.

Anisomeles ovata R.Br.
Lamiaceae
♦ Weed
♦ 87, 88

Anisomeles salviifolia R.Br.
Lamiaceae
♦ Weed
♦ 276
♦ cultivated.

Anisostachya tenella (Nees) Lindau
Acanthaceae
♦ Weed, Naturalised

♦ 98, 203

Annona cherimola Mill.
Annonaceae
♦ chirimoya, cherimoya
♦ Naturalised, Introduced,
Environmental Weed
♦ 101, 230, 257, 261, 280
♦ T, cultivated, herbal, toxic.

Annona coriacea Mart.
Annonaceae
♦ araticum, cabeca de negro, marolo
♦ Weed
♦ 87, 88, 255
♦ Origin: Brazil.

Annona glabra L.
Annonaceae
♦ pond apple, alligator apple,
bullock's heart, cherimoyer, uto ni
mbulumakau, kaitambo, custard apple
♦ Weed, Quarantine Weed,
Naturalised, Introduced,
Environmental Weed
♦ 3, 76, 86, 87, 88, 98, 107, 155, 191, 230,
268
♦ T, cultivated, herbal. Origin: tropical
Americas, coastal West Africa.

Annona muricata L.
Annonaceae
♦ soursop, prickly custard apple,
laguaná, sei, truka shai, jojaab, sasaf,
sausab
♦ Weed, Introduced, Environmental
Weed
♦ 39, 87, 88, 107, 230, 257
♦ S/T, cultivated, herbal, toxic. Origin:
tropical America.

Annona purpurea Moç. & Sessé ex Dunal
Annonaceae
♦ annona
♦ Weed
♦ 87, 88
♦ cultivated, herbal.

Annona reticulata L.
Annonaceae
♦ custard apple, bullock's heart
♦ Weed, Naturalised, Introduced
♦ 39, 86, 87, 88, 93, 98, 191, 203, 228
♦ arid, cultivated, herbal, toxic. Origin:
Central America.

Annona senegalensis Pers.
Annonaceae
♦ custard apple, wild custard apple,
lufila mtopetope
♦ Weed, Quarantine Weed, Native
Weed
♦ 76, 87, 88, 121, 203, 220
♦ S/T, herbal. Origin: southern Africa,
Madagascar.

Annona squamosa L.
Annonaceae
♦ sugar apple, sweetsop, custard apple
♦ Weed, Naturalised, Introduced,
Environmental Weed
♦ 22, 39, 86, 87, 88, 98, 107, 151, 179,
191, 228, 230
♦ S/T, arid, cultivated, herbal, toxic.
Origin: tropical America.

Anoda acerifolia Cav.
Malvaceae
= *Anoda cristata* (L.) Schltdl.
♦ Weed, Naturalised
♦ 87, 88, 257
♦ aH.

Anoda cristata (L.) Schltdl.
Malvaceae
Anoda acerifolia auct. *non* (Zucc. ex
Roem. & Schult.) DC. (see), *Anoda
arizonica* A.Gray, *Anoda brachynatha*
Rchb., *Anoda hastata* Cav. (see), *Anoda
dilleniana* Cav., *Anoda lavatevioides*
Medik., *Anoda populifolia* Phil, *Anoda
triangularis* (Willd.) DC., *Anoda triloba*
Cav., *Canavillea hastata* (Cav.) Medik.,
Sida centrata Spreng., *Sida cristata* L.,
Sida deltoidea Hornem., *Sida dilleniana*
(Cav.) Willd., *Sida hastata* (Cav.) Willd.,
Sida mexicana Scop., *Sida quinqueagulata*
D.Pietr., *Sida triangularis* Willd.
♦ anoda weed, spurred anoda,
sinianoda
♦ Weed, Naturalised, Casual Alien
♦ 7, 24, 28, 34, 42, 55, 86, 87, 88, 98, 157,
161, 199, 203, 207, 212, 218, 236, 237,
243, 255, 270, 295, 300
♦ a/pH, arid, cultivated, herbal.
Origin: Americas.

Anoda hastata Cav.
Malvaceae
= *Anoda cristata* (L.) Schltdl.
♦ Weed, Naturalised
♦ 87, 88, 241, 287

Anoda pentaschista Gray
Malvaceae
♦ field anoda
♦ Weed
♦ 243
♦ aH.

**Anogeissus leiocarpus (DC.) Guill. &
Perr.**
Combretaceae
Anogeissus schimperi Hochst. ex Hutch
& Dalziel
♦ Introduced
♦ 228
♦ arid, herbal.

Anomatheca laxa (Thunb.) Goldblatt
Iridaceae
= *Freesia laxa* (Thunb.) Goldblatt &
J.C.Manning
♦ anomatheca
♦ Weed, Naturalised, Cultivation
Escape
♦ 86, 98, 134, 198, 203, 252
♦ cultivated. Origin: southern Africa.

Anomianthus dulcis (Dunal) J.Sinclair
Annonaceae
♦ Quarantine Weed
♦ 220

Anotis spermacoce L.
Rubiaceae
♦ Weed, Quarantine Weed
♦ 76, 87, 88, 203, 220

Anredera baselloides (H.B.K) Baill.
Basellaceae
Boussingaultia basselloides Kunth (see)
♦ bridal wreath, cascade creeper, lamb's tail, Madeira vine
♦ Weed, Garden Escape
♦ 88, 95, 121, 255, 261
♦ pC, cultivated. Origin: South America.

Anredera cordifolia (Ten.) Steenis
Basellaceae
♦ Madeira vine, heartleaf madeiravine, mignonette vine, lamb's tail, madeiraranka
♦ Weed, Quarantine Weed, Noxious Weed, Naturalised, Garden Escape, Environmental Weed, Cultivation Escape
♦ 3, 7, 15, 22, 34, 39, 63, 72, 73, 86, 87, 88, 98, 101, 134, 152, 155, 158, 165, 176, 181, 198, 201, 203, 225, 229, 246, 269, 280, 283, 289, 296
♦ pC, cultivated, herbal, toxic. Origin: subtropical South America.

Anredera vesicaria (Lam.) Gaertn.
Basellaceae
♦ Texas madeiravine
♦ Weed, Garden Escape, Cultivation Escape
♦ 32, 179, 261
♦ cultivated, herbal.

Antennaria dioica (L.) Gaertn.
Asteraceae
♦ cat's foot, mountain everlasting, stoloniferous pussytoes
♦ Weed
♦ 272
♦ pH, cultivated, herbal.

Antennaria neglecta Greene
Asteraceae
♦ field pussytoes, pussytoes
♦ Weed
♦ 87, 88, 218
♦ herbal.

Antennaria neodioica Greene
Asteraceae
= *Antennaria howellii* Greene ssp. *neodioica* (Greene) R.J.Bayer (NoR)
♦ common pussytoes
♦ Weed
♦ 23, 87, 88, 218
♦ herbal.

Antennaria parvifolia Nutt.
Asteraceae
♦ smallleaf pussytoes
♦ Weed
♦ 23, 88
♦ cultivated, herbal.

Antennaria plantaginea R.Br.
Asteraceae
= *Antennaria plantaginifolia* (L.) Hook
♦ plantainleaf pussytoes
♦ Weed
♦ 87, 88

Antennaria plantaginifolia (L.) Hook.
Asteraceae
Antennaria caroliniana Rydb., *Antennaria decipiens* Greene, *Antennaria*

denikeana Boivin, *Antennaria nemoralis* Greene, *Antennaria petiolata* Fern., *Antennaria pinetorum* Greene, *Antennaria plantaginea* R.Br. (see), *Gnaphalium plantaginifolium* L.
♦ plantainleaf pussytoes, woman's tobacco, white plantain
♦ Weed
♦ 161, 210, 218
♦ cultivated, herbal.

Antennaria rosea Greene
Asteraceae
♦ rosy pussytoes
♦ Weed
♦ 23, 88
♦ pH, cultivated, herbal.

Antenoron filiforme (Thunb.) Roberty & Vautier
Polygonaceae
♦ Weed, Naturalised
♦ 280, 286

Anthemis L. spp.
Asteraceae
♦ chamomile
♦ Weed
♦ 243, 272
♦ aH, herbal.

Anthemis altissima L.
Asteraceae
Anthemis cota L.
♦ isosauramo, tall chamomile
♦ Weed, Naturalised, Casual Alien
♦ 39, 42, 88, 94, 101, 253, 272
♦ herbal, toxic.

Anthemis arvensis L.
Asteraceae
Anthemis agrestis Wallr.
♦ corn chamomile, field chamomile, wild chamomile, peltosauramo
♦ Weed, Quarantine Weed, Noxious Weed, Naturalised
♦ 23, 24, 39, 44, 70, 76, 86, 87, 88, 94, 98, 101, 121, 136, 158, 161, 176, 198, 203, 218, 219, 229, 241, 243, 253, 269, 272, 280, 286, 287, 300
♦ aH, cultivated, herbal, toxic. Origin: Eurasia.

Anthemis aurea Brot.
Asteraceae
♦ keltavuonansilmä
♦ Casual Alien
♦ 42

Anthemis austriaca Jacq.
Asteraceae
Cota austriaca Schultz
♦ Austrian chamomile, itävallansauramo
♦ Weed, Naturalised, Casual Alien
♦ 39, 42, 44, 70, 87, 88, 94, 101, 253, 272
♦ cultivated, toxic.

Anthemis chia L.
Asteraceae
♦ Weed
♦ 39, 87, 88
♦ toxic.

Anthemis cotula L.
Asteraceae

Maruta cotula DC., *Chamaemelum cotula* (L.) All.
♦ stinking mayweed, dillweed, dog's camomile, dog daisy, dogfennel, mather, mayweed, stinking chamomile, manzanilla cimarrona
♦ Weed, Noxious Weed, Naturalised, Introduced, Garden Escape
♦ 7, 15, 23, 24, 34, 36, 39, 42, 44, 49, 51, 52, 68, 70, 86, 87, 88, 94, 98, 101, 116, 121, 136, 147, 158, 161, 165, 174, 176, 180, 198, 203, 204, 207, 210, 211, 212, 218, 228, 229, 236, 237, 241, 243, 253, 263, 269, 271, 272, 280, 286, 287, 295, 300
♦ aH, arid, cultivated, herbal, toxic. Origin: Eurasia.

Anthemis fumariifolia Boiss.
Asteraceae
♦ berberisauramo
♦ Casual Alien
♦ 42

Anthemis fuscata Brot.
Asteraceae
= *Chamaemelum fuscatum* (Brot.) Vasc.
♦ Weed
♦ 87, 88

Anthemis graveolens Boiss.
Asteraceae
♦ Weed
♦ 87, 88

Anthemis hyalina DC.
Asteraceae
♦ kalvosauramo
♦ Weed, Casual Alien
♦ 42, 87, 88

Anthemis melampodina Delile
Asteraceae
♦ Weed
♦ 221
♦ Origin: Israel, Sinai.

Anthemis melanolepis Boiss.
Asteraceae
♦ Weed
♦ 87, 88, 243

Anthemis mixta L.
Asteraceae
= *Chamaemelum mixtum* (L.) All.
♦ Weed
♦ 87, 88

Anthemis montana L.
Asteraceae
♦ Weed
♦ 39, 87, 88
♦ cultivated, toxic.

Anthemis nobilis L.
Asteraceae
= *Chamaemelum nobile* (L.) All.
♦ oil chamomile, Roman chamomile
♦ Weed, Naturalised
♦ 39, 87, 88, 287
♦ cultivated, herbal, toxic.

Anthemis odontostephana Boiss.
Asteraceae
♦ Weed
♦ 243

Anthemis palaestina Reut.
Asteraceae
♦ Palestine chamomile
♦ Weed
♦ 88, 111, 115, 243

Anthemis pseudocotula Boiss.
Asteraceae
♦ Weed
♦ 87, 88, 185, 221

Anthemis punctata Vahl
Asteraceae
♦ Casual Alien
♦ 280
♦ cultivated.

Anthemis punctata ssp. cupaniana (Tod. ex Nyman) R.R.Fern.
Asteraceae
Anthemis cupaniana Tod. ex Nyman
♦ Italian mayweed
♦ Naturalised, Casual Alien
♦ 40, 280

Anthemis retusa Delile
Asteraceae
= *Anthemis cairica* Vis. (NoR)
♦ Weed
♦ 221

Anthemis ruthenica Bieb.
Asteraceae
♦ ruman rusínsky, kaakonsauramo
♦ Weed, Casual Alien
♦ 40, 42, 87, 88, 94, 253, 272
♦ cultivated.

Anthemis secundiramea Biv.
Asteraceae
♦ prostrate chamomile
♦ Naturalised
♦ 101

Anthemis segetalis Ten.
Asteraceae
Anthemis brachycentros Gay ex Ko.
♦ Weed
♦ 88, 94

Anthemis tinctoria L.
Asteraceae
♦ yellow chamomile, golden chamomile, keltasauramo, golden marguerite, cota des teinturiers, dyer's chamomile
♦ Weed, Naturalised, Cultivation Escape, Casual Alien
♦ 23, 34, 40, 42, 70, 86, 87, 88, 94, 98, 101, 176, 203, 218, 243, 252, 272, 280, 287
♦ pH, cultivated, herbal. Origin: Mediterranean.

Anthephora ampullacea Stapf & G.E.Hubb.
Poaceae
♦ Weed
♦ 88
♦ G.

Anthephora hermaphrodita (L.) Kuntze
Poaceae
= *Cenchrus pilosus* Kunth
♦ oldfield grass, falsa caminadora
♦ Weed, Naturalised

♦ 88, 157, 257, 281
♦ G, arid.

Anthephora pubescens Nees
Poaceae
Anthephora abyssinica A.Rich., *Anthephora cenchroides* Schum., *Anthephora hochstetteri* Nees ex Hochst., *Anthephora kotschyi* Hochst.
♦ bottle brush grass, cat's tail grass, wool grass
♦ Weed, Quarantine Weed, Native Weed
♦ 76, 88, 121, 220
♦ aG, arid, cultivated. Origin: southern Africa.

Anthericum galpinii Bak. var. galpinii
Liliaceae/Anthericaceae
♦ Native Weed
♦ 121
♦ pH. Origin: southern Africa.

Anthericum liliago L.
Liliaceae/Anthericaceae
Ornithogalum gramineum Lam., *Phalangium liliago* Schreb.
♦ St. Bernard's lily
♦ Weed
♦ 272
♦ cultivated, herbal.

Anthericum ramosum L.
Liliaceae/Anthericaceae
Antheriscus ramosus Asch. & Graebn., *Ornithogalum ramosum* Lam., *Pessularia ramosa* Salisb., *Phalangium ramosum* Poir.
♦ liten sandlilja, haarahietalilja
♦ Weed
♦ 272
♦ cultivated, herbal.

Anthocephalus chinensis (Lam.) A.Rich. ex Walp.
Rubiaceae/Naucleaceae
♦ kadam
♦ Weed, Naturalised
♦ 86, 98, 101, 203
♦ cultivated, herbal.

Anthoceros L. emend. Prosk. spp.
Anthoceratoaceae
♦ hornwort
♦ Weed
♦ 243
♦ moss.

Anthonotha macrophylla (P.Beauv.) J.Léonard
Fabaceae/Caesalpiniaceae
♦ lomuma mwasi
♦ Weed
♦ 88

Anthospermum aethiopicum L. var. aethiopicum
Rubiaceae
♦ Native Weed
♦ 121
♦ pS. Origin: southern Africa.

Anthoxanthum amarum Brot.
Poaceae
♦ Weed
♦ 70
♦ G.

Anthoxanthum aristatum Boiss.
Poaceae
Anthoxanthum puelii Lecoq & Lamotte (see)
♦ annual vernalgrass, vihnesimake, vernalgrass
♦ Weed, Naturalised, Environmental Weed, Cultivation Escape, Casual Alien
♦ 34, 42, 54, 70, 86, 87, 88, 91, 101, 198, 243, 252, 253, 272, 280, 287, 290
♦ aG, cultivated, herbal. Origin: Europe.

Anthoxanthum odoratum L.
Poaceae
♦ sweet vernalgrass, scented vernalgrass, flouve odorante, tuoksusimake
♦ Weed, Naturalised, Introduced, Garden Escape, Environmental Weed, Cultivation Escape
♦ 3, 7, 9, 15, 18, 20, 34, 38, 39, 70, 72, 80, 86, 87, 88, 91, 98, 101, 121, 134, 146, 151, 152, 155, 161, 176, 181, 191, 195, 198, 203, 204, 218, 241, 249, 269, 280, 286, 287, 289, 296, 300
♦ pG, cultivated, herbal, toxic. Origin: Eurasia.

Anthoxanthum puelii Lecoq & Lamotte
Poaceae
= *Anthoxanthum aristatum* Boiss.
♦ annual vernalgrass
♦ Weed
♦ 30, 88, 243
♦ pG, herbal. Origin: Europe.

Anthriscus Pers. spp.
Apiaceae
♦ chervil
♦ Weed
♦ 272
♦ herbal.

Anthriscus caucalis Bieb.
Apiaceae
Anthriscus scandicina (Web.) Mansf. (see)
♦ burr chervil, piikkikirveli, beaked parsley
♦ Weed, Naturalised, Casual Alien
♦ 34, 42, 86, 98, 101, 161, 165, 176, 212, 241, 253, 272, 280, 286, 287, 300
♦ aC, cultivated, herbal. Origin: Eurasia.

Anthriscus cerefolium (L.) Hoffm.
Apiaceae
♦ garden chervil, maustekirveli, salad chervil, chervil
♦ Weed, Naturalised, Casual Alien
♦ 39, 40, 42, 101, 121, 272
♦ aH, cultivated, herbal, toxic. Origin: Eurasia.

Anthriscus neglecta Boiss. & Reut. ex Lange
Apiaceae
♦ Weed, Naturalised
♦ 87, 88, 287
♦ herbal.

Anthriscus nitida (Wahlenb.) Garcke
Apiaceae
Chaerophyllum nitidum Wahlenb.,
Cerefolium nitidum Celak.
♦ glossy cow parsley
♦ Weed
♦ 272
♦ cultivated.

Anthriscus scandicina (Web.) Mansf.
Apiaceae
= *Anthriscus caucalis* Bieb.
♦ Weed
♦ 87, 88, 121
♦ aH. Origin: Eurasia.

Anthriscus sylvestris (L.) Hoffm.
Apiaceae
♦ woodland beakchervil, cow parsley,
wild chervil, keck, koiranputki
♦ Weed, Quarantine Weed, Noxious
Weed, Naturalised
♦ 1, 39, 70, 76, 79, 87, 88, 101, 121, 146,
161, 195, 218, 229, 240, 253, 272, 300
♦ aH, cultivated, herbal, toxic. Origin:
Eurasia.

Anthriscus vulgaris Bernh.
Apiaceae
♦ Weed
♦ 39, 87, 88
♦ toxic.

Anthurium Schott spp.
Araceae
♦ tailflower
♦ Weed
♦ 161, 247
♦ toxic.

Anthurium acaule (Hook.) Schott
Araceae
♦ Weed
♦ 87, 88

Anthyllis L. spp.
Fabaceae/Papilionaceae
♦ kidneyvetch
♦ Weed
♦ 272

Anthyllis barba-jovis L.
Fabaceae/Papilionaceae
♦ Naturalised, Introduced
♦ 86, 228
♦ arid, cultivated. Origin: Europe.

Anthyllis lotoides L.
Fabaceae/Papilionaceae
♦ Weed
♦ 250

Anthyllis montana L.
Fabaceae/Papilionaceae
Vulneraria montana Scop.
♦ Quarantine Weed
♦ 220
♦ cultivated.

Anthyllis vulneraria L.
Fabaceae/Papilionaceae
♦ kidneyvetch, common kidneyvetch
♦ Weed, Naturalised, Casual Alien
♦ 39, 70, 98, 101, 272, 280, 287
♦ a/pH, cultivated, herbal, toxic.

Anticharis glandulosa Asch.
Scrophulariaceae

♦ Weed
♦ 221

Antidaphne wrightii (Griseb.) Kuijt
Eremolepidaceae
♦ Wright's catkin mistletoe
♦ Naturalised
♦ 101

Antidesma ghaesembilla Gaertn.
Euphorbiaceae/Stilaginaceae
Antidesma frutescens Jack
♦ black currant tree
♦ Weed
♦ 88
♦ cultivated.

Antigonon guatemalense Meisn.
Polygonaceae
♦ bellisima grande
♦ Naturalised, Cultivation Escape
♦ 101, 261
♦ cultivated. Origin: Central America.

Antigonon leptopus Hook. & Arn.
Polygonaceae
♦ coral vine, Mexican rose, mountain
rose, Mexican creeper, queen's jewels
♦ Weed, Naturalised, Introduced,
Garden Escape, Environmental Weed,
Cultivation Escape
♦ 3, 6, 7, 11, 80, 86, 87, 88, 93, 98, 101,
107, 112, 152, 155, 179, 203, 228, 230,
233, 257, 259, 261, 283
♦ pC, arid, cultivated, herbal. Origin:
Mexico, Central America.

Antirrhinum majus L.
Scrophulariaceae
♦ garden snapdragon, snapdragon,
leijonankita
♦ Weed, Naturalised, Cultivation
Escape
♦ 15, 34, 40, 42, 70, 87, 88, 101, 243, 272,
280
♦ a/pH, cultivated, herbal.

Antirrhinum orontium L.
Scrophulariaceae
= *Misopates orontium* (L.) Raf.
♦ weasel's snout, wild antirrhinum,
linearleaf snapdragon, lesser
snapdragon, small snapdragon, calf's
snout
♦ Weed, Naturalised, Cultivation
Escape
♦ 15, 23, 34, 44, 87, 88, 121, 185, 212,
221, 240, 243, 280, 287
♦ aH, cultivated, herbal. Origin:
Eurasia.

Antirrhinum pedatum Desf.
Scrophulariaceae
♦ simpiö
♦ Casual Alien
♦ 42

**Antithamnion pectinatum (Montague)
Brauner**
Ceramiaceae
♦ red alga
♦ Weed
♦ 197
♦ algae.

**Antithamnionella spirographidis
(Schiffn.) Woll.**

Ceramiaceae
Antithamnion spirographidis Schiffn.,
Antithamnion tenuissimum Gardner
♦ Weed
♦ 282, 288
♦ algae.

**Antithamnionella ternifolia (J.D.Hook. &
Harv.) Lyle**
Ceramiaceae
Antithamnion ternifolium (J.D.Hook. &
Harv.) De Toni, *Antithamnion sarniensis*
(Lyle) Feldm.-Maz.
♦ Weed
♦ 288
♦ algae.

Apargia autumnale (L.) Hoffm.
Asteraceae
♦ Weed
♦ 87, 88

Apera interrupta (L.) Beauv.
Poaceae
Agrostis interrupta L. (see)
♦ dense silkybent, interrupted apera,
interrupted windgrass
♦ Weed, Quarantine Weed,
Naturalised
♦ 30, 76, 80, 88, 101, 146, 161, 241, 272,
287, 300
♦ aG, cultivated, herbal.

Apera spica-venti (L.) Beauv.
Poaceae
Agrostis spica-venti L. (see)
♦ loose silky bent, silky bentgrass,
windgrass, silky apera
♦ Weed, Quarantine Weed,
Naturalised
♦ 23, 30, 68, 70, 76, 87, 88, 91, 101, 118,
203, 220, 243, 253, 272, 287, 292
♦ aG, cultivated, herbal.

Aphanes arvensis L.
Rosaceae
Alchemilla arvensis (L.) Scop. (see),
Alchemilla occidentalis Nutt.
♦ parsley piert, peltopoimulehti, field
parsley piert
♦ Weed, Naturalised, Casual Alien
♦ 7, 15, 42, 70, 86, 88, 94, 98, 101, 176,
198, 203, 241, 243, 253, 269, 272, 280,
287, 300
♦ aH, cultivated, herbal. Origin:
Europe.

Aphanes australiana (Rothm.) Rothm.
Rosaceae
♦ Naturalised
♦ 280

Aphanes inexspectata Lippert
Rosaceae
♦ parsley piert, slender parsley piert
♦ Weed, Naturalised
♦ 15, 165, 176, 253, 280
♦ Origin: Eurasia.

**Aphanes microcarpa (Boiss. & Reut.)
Rothm.**
Rosaceae
Alchemilla microcarpa Boiss. & Reut.
(see)
♦ slender parsley piert, small piert
♦ Weed, Naturalised, Casual Alien

♦ 42, 70, 86, 88, 94, 98, 101, 198, 203, 243
♦ Origin: Europe.

Aphanopetalum resinosum Endl.
Cunoniaceae
♦ gum vine
♦ Weed, Casual Alien
♦ 15, 39, 280
♦ cultivated, toxic. Origin: Australia.

Aphanostephus ramosissimus DC.
Asteraceae
♦ plains dozedaisy
♦ Weed
♦ 199
♦ herbal.

Apios americana Medik.
Fabaceae/Papilionaceae
Apios tuberosa Moench, *Glycine apios* L.
♦ American potatobean, groundnut
♦ Weed, Naturalised, Introduced
♦ 87, 88, 218, 228, 287
♦ pC, arid, cultivated, herbal. Origin: North America.

Apium ammi (Jacq.) Urb.
Apiaceae
= *Cyclospermum leptophyllum* (Pers.) Sprague ex Britt. & Wilson
♦ Weed
♦ 87, 88
♦ aH.

Apium australe Thou.
Apiaceae
♦ Weed
♦ 87, 88
♦ pH, cultivated.

Apium graveolens L.
Apiaceae
♦ celery, garden celery, wild celery, selleri
♦ Weed, Naturalised, Introduced, Garden Escape, Environmental Weed, Cultivation Escape
♦ 7, 34, 38, 39, 42, 72, 86, 88, 93, 98, 101, 116, 121, 134, 161, 179, 198, 199, 203, 221, 228, 241, 280, 287, 300
♦ a/bH, aqua, cultivated, herbal, toxic. Origin: Eurasia.

Apium graveolens L. var. dulce (Mill.) DC.
Apiaceae
♦ wild celery, bladselleri lehtiselleri, celery
♦ Naturalised
♦ 101
♦ bH, cultivated, herbal.

Apium leptophyllum (Pers.) F.Muell.
Apiaceae
= *Cyclospermum leptophyllum* (Pers.) Sprague ex Britt. & Wilson
♦ wild celery, Amerikanselleri, slender celery, marsh parsley, wild parsley
♦ Weed, Naturalised, Introduced, Casual Alien
♦ 39, 42, 50, 51, 87, 88, 161, 180, 185, 199, 218, 228, 236, 243, 255, 256, 257, 263, 269, 271, 286, 287, 295
♦ aH, arid, herbal, toxic. Origin: Eurasia.

Apium nodiflorum (L.) Lag.
Apiaceae
♦ water celery, fool's watercress, European marshwort, koiranselleri, procumbent marshwort
♦ Weed, Naturalised, Environmental Weed, Casual Alien
♦ 15, 39, 42, 70, 87, 88, 101, 165, 246, 280, 300
♦ pH, herbal, toxic. Origin: Eurasia.

Apium prostratum Labill.
Apiaceae
♦ prostrate marshwort
♦ Naturalised
♦ 101
♦ pH, cultivated. Origin: Australia, South America, New Zealand.

Apium repens (Jacq.) Lag.
Apiaceae
♦ creeping marshwort
♦ Naturalised
♦ 101

Aplopappus laricifolius A.Gray
Asteraceae
♦ false damiana
♦ Quarantine Weed
♦ 220
♦ herbal.

Aplopappus macronema A.Gray
Asteraceae
♦ Quarantine Weed
♦ 220

Apluda mutica L.
Poaceae
Apluda aristata L., *Apluda mutica* L. var. *aristata* (L.) Pilg.
♦ Mauritian grass, apluda
♦ Weed, Naturalised, Introduced
♦ 66, 87, 88, 101, 228
♦ G, arid. Origin: Australia.

Apluda varia Hack.
Poaceae
♦ Weed
♦ 87, 88
♦ G.

Apocynum androsaemifolium L.
Apocynaceae
♦ spreading dogbane, dogbane
♦ Weed, Noxious Weed
♦ 8, 39, 87, 88, 136, 161, 162, 212, 218, 243, 299
♦ pH, cultivated, herbal, toxic. Origin: northern North America.

Apocynum cannabinum L.
Apocynaceae
Apocynum sibiricum Jacq. (see), *Apocynum suksdorfii* Greene
♦ hemp dogbane, Indian hemp, common dogbane, dogbane, Canadian hemp
♦ Weed, Noxious Weed, Native Weed
♦ 8, 23, 39, 87, 88, 161, 164, 174, 207, 210, 212, 218, 229, 247, 266, 299
♦ pH, arid, cultivated, herbal, toxic. Origin: north-eastern North America.

Apocynum sibiricum Jacq.
Apocynaceae

= *Apocynum cannabinum* L.
♦ prairie dogbane
♦ Weed
♦ 87, 88, 161, 218
♦ herbal, toxic.

Apocynum venetum L.
Apocynaceae
Trachomitum venetum (L.) Woodson
♦ Weed
♦ 39, 275, 297
♦ pH, arid, promoted, herbal, toxic.

Aponogeton L. spp.
Aponogetonaceae
♦ aponogeton
♦ Quarantine Weed
♦ 220
♦ aqua, cultivated, herbal.

Aponogeton abyssinicus Hochst. ex A.Rich.
Aponogetonaceae
♦ Weed, Quarantine Weed
♦ 76, 88, 203, 220
♦ aqua, cultivated.

Aponogeton desertorum Zeyh. ex Spreng.f.
Aponogetonaceae
♦ dog with two tails
♦ Native Weed
♦ 121
♦ wpH, cultivated. Origin: southern Africa.

Aponogeton distachyos L.f.
Aponogetonaceae
♦ cape water hawthorn, cape pondweed, cape pond lily, dog with two tails
♦ Weed, Quarantine Weed, Naturalised, Native Weed, Garden Escape, Environmental Weed
♦ 15, 72, 76, 86, 87, 88, 98, 101, 121, 165, 176, 198, 200, 203, 208, 220, 241, 246, 280, 300
♦ wH, cultivated. Origin: southern Africa.

Aponogeton junceus ssp. junceus Lehm. ex Schlechtd.
Aponogetonaceae
♦ ram's horn
♦ Native Weed
♦ 121
♦ wpH. Origin: southern Africa.

Aponogeton monostachyon L.f.
Aponogetonaceae
♦ Weed, Quarantine Weed
♦ 76, 87, 88, 203, 220
♦ herbal.

Aponogeton natans Engl. & K.Krause
Aponogetonaceae
♦ drifting swordplant
♦ Weed, Quarantine Weed
♦ 76, 88, 220
♦ aqua, cultivated.

Aponogeton rigidifolius H.W.E.Bruggen
Aponogetonaceae
♦ Weed, Quarantine Weed
♦ 76, 88, 220
♦ aqua, cultivated.

× *Aporoselenicereus* **Fearn spp.**
Cactaceae
= *Disocactus* Lindl. spp. × *Heliocereus*
(Berg.) Britt. & Rose spp.
♦ Weed, Quarantine Weed
♦ 76, 88, 203, 220

Aptenia cordifolia **(L.f.) Schwantes**
Aizoaceae/Mesembryanthemaceae
♦ heartleaf iceplant, red apple, baby
sun rose
♦ Weed, Noxious Weed, Naturalised,
Garden Escape, Environmental Weed,
Cultivation Escape
♦ 7, 15, 34, 35, 40, 72, 78, 80, 86, 88, 98,
101, 116, 176, 179, 198, 203, 231, 246,
280
♦ pH, cultivated. Origin: South Africa.

Apuleia leiocarpa **(Vogel) J.F.Macbr.**
Fabaceae/Caesalpiniaceae
Apuleia praecox C.Mart.
♦ Introduced
♦ 228
♦ T, arid, cultivated.

Aquilegia adoxoides **(DC.) Ohwi**
Ranunculaceae
♦ Weed
♦ 286

Aquilegia viridiflora **Pall.**
Ranunculaceae
♦ greenflower columbine
♦ Weed
♦ 297

Aquilegia vulgaris **L.**
Ranunculaceae
♦ columbine, European columbine,
lehtoakileija
♦ Weed, Naturalised, Garden Escape,
Environmental Weed, Cultivation
Escape
♦ 39, 42, 54, 72, 86, 88, 101, 198, 243,
272, 280, 290, 300
♦ pH, cultivated, herbal, toxic. Origin:
Europe.

Aquilegia yabeana **Kitag.**
Ranunculaceae
♦ yabe columbine
♦ Weed
♦ 297
♦ cultivated.

Arabidopsis kneuckeri **O.E.Schulz**
Brassicaceae
= *Crucihimalaya kneuckeri* (Bornm.) Al-
Shehbaz, O'Kane & R.A.Price (NoR)
♦ Weed
♦ 221

Arabidopsis pumila **(Stephan) N.Busch**
Brassicaceae
♦ Weed
♦ 221, 272

Arabidopsis thaliana **(L.) Heynh.**
Brassicaceae
Sisymbrium thalianum (L.) J.Gay &
Monnard, *Stenophragma thalianum* (L.)
Celak., *Arabis thaliana* L.
♦ mouse ear cress, wall cress, thale
cress
♦ Weed, Naturalised, Cultivation
Escape

♦ 15, 23, 34, 44, 70, 86, 87, 88, 94, 98,
101, 161, 176, 203, 207, 218, 243, 252,
253, 272, 280, 286
♦ aH, cultivated, herbal. Origin:
Eurasia, east Africa.

Arabis **L. spp.**
Brassicaceae
♦ rockcress
♦ Weed
♦ 272
♦ herbal.

Arabis caucasica **Willd.**
Brassicaceae
Arabis albida Steven
♦ garden arabis, kaukasianpitkäpalko,
rockcress, gray rockcress
♦ Naturalised, Cultivation Escape
♦ 40, 42, 101
♦ pH, cultivated, herbal. Origin:
Mediterranean.

Arabis glabra **(L.) Bernh.**
Brassicaceae
= *Turritis glabra* L.
♦ tower mustard, glabrous tower
cress, tower rockcress, pölkkyruoho
♦ Weed
♦ 23, 24, 88, 272, 286
♦ pH, cultivated, herbal. Origin:
Eurasia.

Arabis hirsuta **(L.) Scop.**
Brassicaceae
♦ hairy rockcress
♦ Weed, Naturalised
♦ 23, 88, 272
♦ pH, cultivated, herbal.

Arabis hirsuta **(L.) Scop. var.** *hirsuta*
Brassicaceae
♦ hairy rockcress
♦ Naturalised
♦ 101

Arabis holboellii **Hornem.**
Brassicaceae
♦ Holboell's rockcress
♦ Weed
♦ 23, 88
♦ pH, cultivated, herbal.

Arabis pendula **L.**
Brassicaceae
♦ ezohatazao
♦ Weed
♦ 272, 275, 297
♦ bH, promoted.

Arabis procurrens **Waldst. & Kit.**
Brassicaceae
♦ running rockcress
♦ Naturalised
♦ 101
♦ cultivated, herbal.

Arabis recta **Vill.**
Brassicaceae
Arabis auriculata DC.
♦ annual rockcress
♦ Weed
♦ 272

Arabis stricta **Huds.**
Brassicaceae
♦ Bristol rockcress

♦ Weed
♦ 23, 88

Arabis turrita **L.**
Brassicaceae
♦ tower rockcress, korkeapitkäpalko,
tower cress, arábka ovisnutá
♦ Naturalised, Cultivation Escape,
Casual Alien
♦ 40, 42, 56
♦ cultivated, herbal.

Arabis verna **(L.) R.Br.**
Brassicaceae
♦ spring rockcress
♦ Weed
♦ 272

Arachis hagenbeckii **Harms**
Fabaceae/Papilionaceae
♦ Hagenbeck's peanut
♦ Naturalised
♦ 101

Arachis hypogaea **L.**
Fabaceae/Papilionaceae
♦ maapähkinä, peanut, mani,
groundnut, goober, monkey nut
♦ Weed, Naturalised, Casual Alien
♦ 39, 40, 42, 86, 87, 88, 98, 101, 179, 203,
261, 270
♦ aH, cultivated, herbal, toxic. Origin:
Brazil, Argentina.

Arachis prostrata **Benth.**
Fabaceae/Papilionaceae
♦ grass nut
♦ Weed, Naturalised, Cultivation
Escape
♦ 101, 179, 261
♦ cultivated. Origin: Brazil.

Arachis pusila **Benth.**
Fabaceae/Papilionaceae
♦ Introduced
♦ 228
♦ arid.

Arachniodes simplicior **(Makino) Ohwi**
Dryopteridaceae
♦ simpler hollyfern
♦ Naturalised
♦ 101
♦ cultivated.

Aragallus abbreviatus **Greene**
Fabaceae/Papilionaceae
= *Astragalus lambertii* var. *abbreviatus*
(Greene) Shinners (NoR)
♦ Quarantine Weed
♦ 220

Aralia chinensis **L.**
Araliaceae
= *Aralia stipulata* Franch. (NoR)
♦ puistoaralia, Chinese angelica tree,
angelica tree
♦ Naturalised, Cultivation Escape
♦ 42
♦ S, cultivated, herbal.

Aralia elata **(Miq.) Seem.**
Araliaceae
Dimorphanthus elatus Miq.
♦ Japanese angelica tree, Chinese
aralia, angelica tree, taranoki
♦ Weed, Naturalised

♦ 101, 133, 195, 286
♦ T, cultivated, herbal.

Aralia spinosa L.
Araliaceae
♦ devil's walkingstick, Hercules's club, angelica tree, taggaralia, prickly elder
♦ Weed
♦ 8, 39, 87, 88, 133, 161, 195, 218, 247
♦ S/T, cultivated, herbal, toxic.

Araucaria angustifolia (Bertol.) Kuntze
Araucariaceae
Columbea angustifolia Bertol.
♦ parana pine
♦ Weed, Introduced
♦ 22, 228, 261
♦ T, arid, cultivated, herbal. Origin: Brazil.

Araucaria bidwillii Hook.
Araucariaceae
♦ bungabunga, bunya pine, bunya bunya
♦ Introduced, Casual Alien
♦ 261, 280
♦ T, cultivated, herbal. Origin: Australia.

Araucaria columnaris (Forst.f.) Hook.
Araucariaceae
♦ New Caledonia pine
♦ Introduced
♦ 230
♦ T, cultivated.

Araucaria heterophylla (Salisb.) Franco
Araucariaceae
♦ pino de Norfolk
♦ Weed, Naturalised, Introduced, Environmental Weed
♦ 19, 80, 88, 151, 203, 230, 261, 280
♦ T, cultivated, herbal. Origin: Norfolk Islands.

Araujia hortorum Fourn.
Asclepiadaceae/Apocynaceae
= *Araujia sericifera* Brot.
♦ moth plant, white bladder flower, mothvine
♦ Weed, Naturalised, Environmental Weed
♦ 72, 73, 87, 88, 98, 155, 203, 269
♦ C, cultivated. Origin: South America.

Araujia sericifera Brot.
Asclepiadaceae/Apocynaceae
Araujia hortorum Fourn. (see)
♦ moth catcher, white bladder flower, moth plant, moth vine, motvanger, ladder flower, cruel plant, glehold plant, milkweed, stranglehold plant
♦ Weed, Quarantine Weed, Noxious Weed, Naturalised, Garden Escape, Environmental Weed, Cultivation Escape
♦ 3, 7, 15, 35, 39, 51, 63, 73, 80, 86, 87, 88, 95, 101, 121, 158, 161, 165, 180, 198, 225, 229, 243, 246, 279, 280, 283, 289, 296
♦ pC, cultivated, herbal, toxic. Origin: South America.

Arbutus menziesii Pursh
Ericaceae
♦ Pacific madrone
♦ Weed
♦ 87, 88, 218
♦ T, cultivated, herbal.

Arbutus unedo L.
Ericaceae
♦ strawberry tree
♦ Weed, Naturalised, Garden Escape, Environmental Weed
♦ 40, 72, 86, 88, 98, 155, 198, 203, 280, 289
♦ T, cultivated, herbal. Origin: southern Europe.

Arceuthobium abietinum Engelm. ex Munz
Viscaceae
♦ fir dwarf mistletoe
♦ Weed
♦ 88
♦ pH parasitic.

Arceuthobium americanum Nutt. ex Engelm.
Viscaceae
♦ American dwarf mistletoe, lodgepole dwarf mistletoe
♦ Weed
♦ 88
♦ pH parasitic, herbal.

Arceuthobium campylopodum Engelm.
Viscaceae
♦ western dwarf mistletoe
♦ Weed
♦ 88
♦ pH parasitic, herbal.

Arceuthobium douglasii Engelm.
Viscaceae
♦ Douglas fir dwarf mistletoe
♦ Weed
♦ 88
♦ pH parasitic, herbal.

Arceuthobium laricis (Piper) St.John
Viscaceae
♦ larch dwarf mistletoe
♦ Weed
♦ 88
♦ H parasitic.

Arceuthobium minutissimum Hook.f.
Viscaceae
♦ Indian dwarf mistletoe
♦ Weed
♦ 88
♦ H parasitic.

Arceuthobium occidentale Engelm.
Viscaceae
♦ digger pine dwarf mistletoe
♦ Weed
♦ 88
♦ pH parasitic, cultivated.

Arceuthobium oxycedri (DC.) M.Bieb.
Viscaceae
♦ Weed
♦ 87, 88
♦ H parasitic.

Arceuthobium pusillum Peck
Viscaceae
♦ eastern dwarf mistletoe
♦ Weed
♦ 88

♦ H parasitic, herbal.

Arceuthobium tsugense (Rosendahl) G.N.Jones
Viscaceae
♦ hemlock dwarf mistletoe
♦ Weed
♦ 88
♦ pH parasitic.

Arceuthobium vaginatum (Willd.) J.Presl
Viscaceae
♦ pineland dwarf mistletoe
♦ Weed
♦ 88
♦ H parasitic, herbal.

Archontophoenix alexandrae (F.Muell.) H.Wendl. & Drude
Arecaceae
♦ Alexandra palm
♦ Naturalised
♦ 101
♦ cultivated, herbal. Origin: Australia.

Archontophoenix cunninghamiana (H.Wendl.) H.Wendl. & Drude
Arecaceae
♦ bungalow palm, piccabeen palm, kuningatarpalmu
♦ Environmental Weed
♦ 246
♦ cultivated, herbal. Origin: Australia.

Arctium lappa L.
Asteraceae
Arctium chaorum Klokov, *Arctium edule* (Siebold ex Miq.) Nakai, *Arctium majus* Bernh. (see), *Lappa edulis* Siebold ex Miq., *Lappa glabra* Lam., *Lappa major* Gaertn., *Lappa officinalis* All.
♦ greater burdock
♦ Weed, Naturalised, Introduced
♦ 8, 23, 34, 39, 70, 80, 86, 87, 88, 94, 98, 101, 108, 161, 198, 203, 210, 218, 228, 243, 272, 280, 286, 287, 293, 297
♦ pH, arid, cultivated, herbal, toxic. Origin: Eurasia.

Arctium majus Bernh.
Asteraceae
= *Arctium lappa* L.
♦ Weed
♦ 87, 88

Arctium minus (Hill) Bernh.
Asteraceae
Lappa minor Hill
♦ common burdock, lesser burdock, burdock, pikkutakiainen
♦ Weed, Noxious Weed, Naturalised, Introduced, Environmental Weed
♦ 8, 23, 24, 39, 49, 52, 70, 80, 86, 87, 88, 94, 98, 101, 136, 138, 146, 151, 159, 161, 165, 174, 176, 195, 203, 207, 210, 211, 212, 218, 219, 222, 229, 237, 241, 243, 246, 255, 272, 280, 293, 295, 299, 300
♦ pH, cultivated, herbal, toxic. Origin: Eurasia.

Arctium tomentosum Mill.
Asteraceae
♦ cotton burdock, woolly burdock
♦ Weed, Noxious Weed, Naturalised
♦ 70, 88, 94, 101, 161, 218, 272, 299
♦ cultivated, herbal, toxic.

Arctium vulgare (Hill) Evans
Asteraceae
♦ woodland burdock
♦ Naturalised
♦ 101
♦ herbal.

Arctopus echinatus L.
Apiaceae
♦ poxthorn
♦ Native Weed
♦ 121
♦ pH, herbal. Origin: obscure.

Arctostaphylos canescens Eastw.
Ericaceae
♦ hoary manzanita
♦ Weed
♦ 87, 88, 218
♦ S, herbal.

Arctostaphylos columbiana Piper
Ericaceae
♦ hairy manzanita
♦ Weed
♦ 87, 88, 218
♦ S/T, cultivated, herbal.

Arctostaphylos glandulosa Eastw.
Ericaceae
♦ Eastwood's manzanita
♦ Weed
♦ 87, 88, 218
♦ S, cultivated, herbal.

Arctostaphylos glauca Lindl.
Ericaceae
♦ bigberry manzanita
♦ Weed
♦ 87, 88, 218
♦ S, cultivated, herbal.

Arctostaphylos hispidula Howell
Ericaceae
♦ Howell manzanita, Gasquet manzanita
♦ Weed
♦ 87, 88, 218
♦ S.

Arctostaphylos manzanita Parry
Ericaceae
♦ big manzanita, whiteleaf manzanita, common manzanita
♦ Weed
♦ 87, 88, 218
♦ S, cultivated, herbal.

Arctostaphylos nevadensis A.Gray
Ericaceae
♦ pinemat manzanita
♦ Weed
♦ 87, 88, 218
♦ S, cultivated, herbal.

Arctostaphylos parryana Lemmon
Ericaceae
♦ pine manzanita, Parry manzanita
♦ Weed
♦ 87, 88
♦ S, promoted.

Arctostaphylos parryana Lemm. var. pinetorum (Rollins) Wiesl. & Shreib.
Ericaceae
= *Arctostaphylos patula* Greene
♦ pine manzanita

♦ Weed
♦ 218

Arctostaphylos patula Greene
Ericaceae
Arctostaphylos parryana Lemm. var. *pinetorum* (Rollins) Wiesl. & Shreib. (see)
♦ greenleaf manzanita
♦ Weed
♦ 87, 88, 218
♦ S, promoted, herbal.

Arctostaphylos pungens Kunth
Ericaceae
♦ Mexican manzanita, pointleaf manzanita
♦ Weed
♦ 87, 88, 218
♦ S, cultivated, herbal.

Arctostaphylos uva-ursi (L.) Spreng.
Ericaceae
♦ bearberry, kinnikinick, uva ursi
♦ Weed
♦ 8, 39, 87, 88, 218, 272
♦ S, cultivated, herbal, toxic.

Arctostaphylos viscida Parry
Ericaceae
♦ whiteleaf manzanita, sticky whiteleaf manzanita
♦ Weed
♦ 87, 88, 218
♦ S, cultivated, herbal.

Arctotheca calendula (L.) Levyns
Asteraceae
Cryptostemma calendulaceum (L.) R.Br.
♦ capeweed, cape daisy, cape marigold, marigold, silver spreader
♦ Weed, Sleeper Weed, Noxious Weed, Naturalised, Native Weed, Introduced, Garden Escape, Environmental Weed
♦ 7, 9, 15, 35, 39, 51, 68, 70, 72, 78, 80, 86, 87, 88, 93, 98, 101, 116, 121, 134, 151, 158, 165, 176, 198, 203, 205, 225, 228, 229, 231, 241, 243, 246, 253, 269, 280, 287, 289, 296, 300
♦ a/bH, arid, cultivated, herbal, toxic. Origin: southern Africa.

Arctotheca populifolia (Berg.) Norl.
Asteraceae
♦ coast capeweed, beach daisy
♦ Weed, Naturalised, Native Weed, Environmental Weed
♦ 7, 72, 86, 88, 98, 121, 176, 198, 203, 289
♦ pH. Origin: southern Africa.

Arctotheca prostrata (Salisb.) Britten
Asteraceae
♦ creeping bear's ears, creeping capeweed
♦ Weed, Naturalised, Environmental Weed
♦ 72, 86, 88, 98, 198, 203
♦ pH. Origin: South Africa.

Arctotheca repens Wendl.
Asteraceae
Arctotheca grandiflora Schrad.
♦ Weed
♦ 87, 88

Arctotis acaulis L.
Asteraceae
♦ African daisy
♦ Naturalised
♦ 287
♦ cultivated. Origin: southern Africa.

Arctotis arctotoides (L.f.) O.Hoffm.
Asteraceae
Venidium decurrens Less.
♦ Native Weed
♦ 121
♦ pH. Origin: southern Africa.

Arctotis leiocarpa Harv.
Asteraceae
♦ karoo daisy
♦ Native Weed
♦ 121
♦ aH, toxic. Origin: southern Africa.

Arctotis leptorhiza DC.
Asteraceae
♦ Native Weed
♦ 121
♦ H. Origin: southern Africa.

Arctotis stoechadifolia Berg.
Asteraceae
♦ African daisy, white arctotis, arctotis
♦ Weed, Naturalised, Environmental Weed
♦ 7, 15, 34, 86, 101, 176, 198, 280, 296
♦ pH, cultivated, herbal. Origin: South Africa.

Arctotis venusta Norl.
Asteraceae
♦ Free State daisy, white arctotis, blue eyed African daisy, silverga hopeasilm
♦ Weed, Naturalised, Native Weed, Garden Escape, Environmental Weed
♦ 51, 72, 86, 87, 88, 98, 121, 158, 203
♦ pH, cultivated, herbal, toxic. Origin: southern Africa.

Ardisia Sw. spp.
Myrsinaceae
♦ marlberry, coralberry
♦ Weed
♦ 18, 88

Ardisia crenata Sims
Myrsinaceae
Ardisia bicolor E.Walker, *Ardisia crenata* var. *bicolor* (E.Walker) C.Y.Wu & C.Chen, *Ardisia crispa* var. *taquetii* H.Lév., *Ardisia konishii* Hayata, *Ardisia kusukusensis* Hayata, *Ardisia labordei* H.Lév., *Ardisia lentiginosa* Ker Gawl., *Ardisia linangensis* C.M.Hu, *Ardisia miaoliensis* Lu, *Bladhia crenata* (Sims) H.Hara, *Bladhia crispa* var. *taquetii* (H.Lév.) Nakai, *Bladhia lentiginosa* var. *lanceolata* Masam.
♦ coralberry, coral ardisia, hen's eyes, hilo holly
♦ Weed, Noxious Weed, Naturalised, Garden Escape, Environmental Weed, Cultivation Escape
♦ 3, 22, 54, 73, 86, 88, 101, 112, 151, 152, 155, 179, 233, 283
♦ S, cultivated, herbal. Origin: northeast India to Japan.

Ardisia crenulata Vent.
Myrsinaceae
= *Parathesis crenulata* (Vent.) Hook.f.
(NoR)
♦ coral ardisia
♦ Environmental weed
♦ 80, 201
♦ S, Cultivated.

Ardisia crispa (Thunb.) A.DC.
Myrsinaceae
Ardisia dielsii H.Lév., *Ardisia henryi*
Hemsl., *Ardisia henryi* var. *dielsii*
(H.Lév.) E.Walker, *Ardisia hortorum*
Maxim. ex Regel, *Ardisia hortorum*
var. *brachysepala* Hand.-Mazz., *Ardisia*
multicaulis Z.Y.Zhu, *Ardisia penduliflora*
Mez, *Ardisia simplicicaulis* Hayata,
Bladhia crispa Thunb., *Bladhia crispa*
var. *dielsii* (H.Lév.) Nakai, *Tinus crispa*
(Thunb.) Kuntze, *Tinus henryi* (Hemsl.)
Kuntze
♦ coral ardisia
♦ Weed, Sleeper Weed, Naturalised,
Garden Escape, Environmental Weed
♦ 3, 22, 86, 98, 155, 191, 203
♦ S, cultivated, herbal. Origin:
southern China, Japan, India.

Ardisia dentata (A.DC.) Mez.
Myrsinaceae
♦ Weed
♦ 14

Ardisia elliptica Thunb.
Myrsinaceae
Ardisia humilis Vahl (see), *Ardisia*
solanacea Roxb. (see), *Ardisia*
squamulosa C.Presl (see)
♦ shoebutton ardisia, shoebutton
♦ Weed, Noxious Weed, Naturalised,
Garden Escape, Environmental Weed,
Cultivation Escape
♦ 3, 80, 88, 101, 112, 122, 132, 142, 151,
152, 179, 191, 229, 233
♦ S/T, cultivated, herbal. Origin:
south-east Asia.

Ardisia humilis Vahl
Myrsinaceae
= *Ardisia elliptica* Thunb.
♦ shoebutton ardisia
♦ Weed, Noxious Weed,
Environmental Weed
♦ 3, 22, 80, 87, 88, 122, 152, 155
♦ S/T, cultivated. Origin: northern
India.

Ardisia solanacea Roxb.
Myrsinaceae
= *Ardisia elliptica* Thunb.
♦ shoebutton ardisia
♦ Weed, Naturalised, Environmental
Weed, Cultivation Escape
♦ 3, 86, 88, 101, 151, 191, 261
♦ cultivated, herbal. Origin: Asia.

Ardisia squamulosa C.Presl
Myrsinaceae
= *Ardisia elliptica* Thunb.
♦ Weed
♦ 3, 191

Areca catechu L.
Arecaceae
♦ betel nut, areca nut, betel palm
♦ Introduced
♦ 39, 39, 230
♦ T, cultivated, herbal, toxic.

Arecastrum romanzoffianum (Cham.)
Becc.
Arecaceae
= *Syagrus romanzoffiana* (Cham.)
Glassman
♦ Cocos palm, queen palm
♦ Weed, Environmental Weed
♦ 73, 88, 201
♦ cultivated, herbal.

Arenaria L. spp.
Caryophyllaceae
♦ sandwort
♦ Weed
♦ 272

Arenaria balearica L.
Caryophyllaceae
♦ mossy sandwort, Balearic pearlwort
♦ Naturalised
♦ 40
♦ cultivated.

Arenaria biflora L.
Caryophyllaceae
♦ alppiarho'
♦ Cultivation Escape
♦ 42
♦ cultivated.

Arenaria ciliata L.
Caryophyllaceae
Alsinella ciliata S.F.Gray, *Alsine ciliata*
(L.) Crantz
♦ hairy sandwort, fringed sandwort,
tunturiarho
♦ Weed
♦ 272

Arenaria fendleri A.Gray
Caryophyllaceae
♦ Fendler's sandwort
♦ Introduced
♦ 228
♦ arid, herbal.

Arenaria graveolens Schreb.
Caryophyllaceae
♦ Weed
♦ 221

Arenaria groenlandica (Retz.) Spreng.
Caryophyllaceae
= *Minuartia groenlandica* (Retz.) Ostenf.
(NoR)
♦ Weed
♦ 23, 88
♦ herbal.

Arenaria lanuginosa (Michx.) Rohrb.
Caryophyllaceae
♦ spreading sandwort
♦ Weed
♦ 157
♦ a/pH, herbal.

Arenaria leptoclados (Rchb.) Guss.
Caryophyllaceae
♦ lesser thyme leaved sandwort,
slender sandwort
♦ Naturalised
♦ 86, 176, 198

♦ Origin: Eurasia.

Arenaria serpyllifolia L.
Caryophyllaceae
♦ thymeleaf sandwort, sandwort,
mäkiarho
♦ Weed, Naturalised, Garden Escape,
Environmental Weed
♦ 7, 9, 15, 23, 39, 44, 68, 70, 86, 87, 88,
94, 98, 101, 161, 165, 176, 195, 198, 203,
218, 241, 243, 253, 263, 272, 275, 280,
286, 297, 300
♦ aH, cultivated, herbal, toxic. Origin:
Eurasia.

Arenaria serpyllifolia L. ssp. serpyllifolia
Caryophyllaceae
♦ thymeleaf sandwort
♦ Weed
♦ 34
♦ aH. Origin: Eurasia.

Arenga pinnata (Wurmb) Merr.
Arecaceae
♦ sugar palm, kaong
♦ Weed, Introduced
♦ 19, 230
♦ T, cultivated, herbal.

Argania spinosa (L.) Skeels
Sapotaceae
Argania sideroxylon Roem. & Schult.,
Sideroxylon spinosum L.
♦ argam
♦ Introduced
♦ 228
♦ arid, cultivated.

Argemone albiflora Hornem.
Papaveraceae
♦ bluestem prickle poppy, prickly
poppy
♦ Weed
♦ 8, 161
♦ aH, cultivated, herbal, toxic.

Argemone glauca (Nutt. ex Prain) Pope
Papaveraceae
♦ smooth prickle poppy
♦ Weed
♦ 87, 88, 218
♦ cultivated, herbal.

Argemone intermedia L.
Papaveraceae
♦ bluestem prickle poppy
♦ Weed
♦ 87, 88, 210, 218
♦ herbal, toxic.

Argemone mexicana L.
Papaveraceae
♦ Mexican prickle poppy, Mexican
poppy, devil's fig, golden thistle
of Peru, Mexican thistle, prickly
poppy, white thistle, yellow poppy,
umJelemani, piikkiunikko
♦ Weed, Quarantine Weed, Noxious
Weed, Naturalised, Introduced,
Environmental Weed, Casual Alien
♦ 6, 14, 23, 39, 42, 50, 53, 55, 62, 66, 76,
87, 88, 98, 121, 147, 157, 158, 161, 185,
186, 203, 218, 220, 228, 237, 242, 243,
247, 255, 257, 261, 269, 283
♦ aH, arid, cultivated, herbal, toxic.
Origin: South America.

Argemone munita **Dur. & Hilg.**
Papaveraceae
♦ chicalote, flatbud prickly poppy
♦ Weed
♦ 161
♦ a/pH, cultivated, herbal, toxic.

Argemone ochroleuca **Sweet**
Papaveraceae
♦ Mexican poppy, devil's fig, golden
thistle of Peru, Mexican prickle poppy,
Mexican thistle, prickly poppy, white
thistle, yellow poppy
♦ Weed, Quarantine Weed, Noxious
Weed, Naturalised, Garden Escape
♦ 7, 39, 55, 62, 76, 87, 88, 98, 147, 158,
165, 176, 203, 220, 240, 243, 280
♦ aH, cultivated, herbal, toxic. Origin:
Central America.

Argemone ochroleuca **Sweet ssp.**
ochroleuca
Papaveraceae
♦ Mexican poppy, witblom bloudissel,
white flowered Mexican poppy
♦ Weed, Noxious Weed, Naturalised
♦ 63, 86, 93, 198, 269, 283
♦ aH, cultivated, toxic. Origin:
Americas.

Argemone platyceras **Link & Otto**
Papaveraceae
♦ crested prickle poppy, rough prickly
poppy
♦ Weed
♦ 87, 88, 218, 243
♦ aH, cultivated, herbal.

Argemone polyanthemos **(Fedde)**
G.B.Ownbey
Papaveraceae
♦ annual prickle poppy, crested
prickly poppy
♦ Weed, Native Weed
♦ 161, 174, 212, 264
♦ a/bH, promoted, herbal. Origin:
North America.

Argemone squarrosa **Greene**
Papaveraceae
♦ hedgehog prickly poppy
♦ Weed
♦ 161
♦ herbal.

Argemone subfusiformis **G.B.Ownbey**
Papaveraceae
♦ Mexican poppy, devil's fig, golden
thistle of Peru, Mexican prickle poppy,
Mexican thistle, prickly poppy, white
thistle, yellow poppy
♦ Weed, Quarantine Weed, Noxious
Weed, Naturalised, Garden Escape
♦ 39, 51, 86, 87, 88, 98, 121, 134, 147,
203, 220, 237, 243, 295
♦ aH, arid, cultivated, herbal, toxic.
Origin: Mexico.

Argemone subfusiformis **G.B.Ownbey**
ssp. *subfusiformis*
Papaveraceae
♦ Weed
♦ 269

Argentina anserina **(L.) Rydb.**
Rosaceae
Argentina argentea (L.) Rydb., *Potentilla
anserina* L. (see)
♦ silver weed, goose tansy, wild tansy,
silverweed cinquefoil
♦ Weed
♦ 49
♦ herbal. Origin: North America.

Argusia sibirica **(L.) Dandy**
Boraginaceae
♦ Siberian sea rosemary
♦ Weed, Naturalised
♦ 101, 272

Argyranthemum foeniculaceum **(Willd.)**
Sch.Bip.
Asteraceae
Argyranthemum foeniculum (Willd.)
Sch.Bip. (see), *Chrysanthemum
anethifolium non* (Willd.) Steud.,
Chrysanthemum foeniculaceum (Willd.)
Desf., *Pyrethrum anethifolium* Willd.
♦ ripsumarkette, argyranthemum
♦ Naturalised
♦ 7, 86
♦ pH, cultivated, herbal. Origin:
Canary Islands.

Argyranthemum foeniculum **(Willd.)**
Sch.Bip.
Asteraceae
= *Argyranthemum foeniculaceum* (Willd.)
Sch.Bip.
♦ dill daisy
♦ Naturalised
♦ 101

Argyranthemum frutescens **(L.) Sch.Bip.**
Asteraceae
Chrysanthemum frutescens L. (see)
♦ marguerite, buskmargerit, marketta
♦ Weed, Naturalised, Environmental
Weed
♦ 7, 9, 15, 86, 98, 134, 203, 280
♦ cultivated, herbal.

Argyreia mollis **(Burm.f.) Choisy.**
Convolvulaceae
♦ Weed
♦ 13
♦ C, herbal.

Argyreia nervosa **(Burm.f.) Bojer**
Convolvulaceae
Convolvulus speciosus L.f., *Convolvulus
nervosus* Burm.f.
♦ elephant creeper, woolly
morningglory
♦ Weed, Sleeper Weed, Quarantine
Weed, Noxious Weed, Naturalised,
Introduced
♦ 76, 86, 88, 101, 179, 220, 229, 261, 287
♦ cultivated, herbal, toxic. Origin:
India.

Argyrolobium abyssinicum **Jaub. & Spach**
Fabaceae/Papilionaceae
♦ Weed
♦ 221

Argythamnia neomexicana **Müll.Arg.**
Euphorbiaceae
Ditaxis neomexicana (Müll.Arg.) Heller
(see)

♦ New Mexico silverbush
♦ Weed
♦ 28, 243
♦ herbal.

Arisaema dracontium **(L.) Schott**
Araceae
♦ green dragon, dragon arum, green
arum
♦ Weed
♦ 39, 161, 247
♦ pH, promoted, herbal, toxic.

Arisaema erubescens **(Wall.) Schott**
Araceae
♦ reddish Jack in the pulpit
♦ Weed
♦ 297
♦ cultivated.

Arisaema flavum **(Forssk.) Schott**
Araceae
Arisaema abbreviatum Schott, *Arum
flavum* Forssk., *Dochafa flava* (Forssk.)
Schott
♦ Weed
♦ 88
♦ pH, cultivated.

Arisaema triphyllum **(L.) Schott**
Araceae
Arisaema atrorubens (Aiton) Blume,
Arisaema stewardsonii Britton, *Arum
atrorubens* Aiton, *Arum triphyllum* L.
♦ Jack in the pulpit, bog onion, brown
dragon, cuckoo plant, Indian Jack in
the pulpit, memory root, pepperturnip,
petit precheur, priests pentle, small
Jack in the pulpit, starchwort, three
leaved Indian turnip, wake robin
♦ Weed
♦ 23, 39, 88, 154, 161, 189, 247
♦ pH, cultivated, herbal, toxic.

Arisarum vulgare **O.Targ.Tozz.**
Araceae
♦ friar's cowl
♦ Weed, Quarantine Weed, Garden
Escape, Naturalised
♦ 39, 54, 70, 76, 88, 98, 203, 221, 243,
251, 253, 272
♦ pH, cultivated,
herbal, toxic. Origin: Mediterranean.

Aristea africana **(L.) Hoffmanns.**
Iridaceae
♦ Native Weed
♦ 121
♦ pH, cultivated. Origin: southern
Africa.

Aristea ecklonii **Baker**
Iridaceae
♦ blue stars, blue cornlily, aristea
♦ Weed, Naturalised, Garden Escape,
Environmental Weed
♦ 40, 54, 72, 86, 88, 155, 165, 198, 225,
246, 280
♦ pH, cultivated. Origin: west and
southern Africa.

Aristea ensifolia **Muir**
Iridaceae
♦ Naturalised
♦ 198
♦ cultivated.

Aristea gerrardii Weim.
Iridaceae
♦ Gerrard's aristea
♦ Naturalised
♦ 101
♦ cultivated.

Aristida adscensionis L.
Poaceae
Aristida bromoides Kunth, *Aristida fasciculata* Torr., *Aristida submucronata* Schumach.
♦ sixweeks threeawn, sixweeks needlegrass, annual bristlegrass
♦ Weed, Naturalised, Introduced
♦ 87, 88, 91, 199, 221, 228, 241, 242, 243, 297
♦ a/pG, arid, cultivated, herbal.

Aristida adscensionis L. ssp. guineensis (Trin. & Rupr.) Henrard
Poaceae
♦ annual bristlegrass, annual three awned grass, threeawn
♦ Native Weed
♦ 121
♦ aG. Origin: southern Africa.

Aristida arenaria Gaudich.
Poaceae
= *Aristida contorta* F.Muell.
♦ Weed
♦ 39, 87, 88
♦ G, toxic.

Aristida balansae Henr.
Poaceae
♦ Weed
♦ 87, 88
♦ G.

Aristida benthamii Henrard
Poaceae
♦ Naturalised, Environmental Weed
♦ 86, 176
♦ G, arid. Origin: Australia.

Aristida bipartita (Nees) Trin. & Rupr.
Poaceae
♦ rolling three awned grass
♦ Native Weed
♦ 121
♦ pG. Origin: southern Africa.

Aristida coerulescens Desf.
Poaceae
♦ Weed
♦ 221
♦ G.

Aristida congesta Roem. & Schult.
Poaceae
♦ Weed, Quarantine Weed
♦ 76, 87, 88, 203, 220
♦ G.

Aristida congesta Roem. & Schult. ssp. barbicollis (Trin. & Rupr.) de Winter
Poaceae
♦ buffalograss, piercing grass, spreading threeawn, stickgrass, tassel bristlegrass, tassel threeawn, white stick grass
♦ Weed, Quarantine Weed, Native Weed
♦ 88, 121, 158, 220
♦ pG. Origin: southern Africa.

Aristida congesta Roem. & Schult. ssp. congesta
Poaceae
♦ tassel threeawn
♦ Weed, Quarantine Weed, Native Weed
♦ 51, 88, 121, 158, 220
♦ pG. Origin: southern Africa.

Aristida contorta F.Muell.
Poaceae
Aristida arenaria Gaudich. (see)
♦ Native Weed
♦ 269
♦ G, arid, cultivated. Origin: Australia.

Aristida dichotoma Michx.
Poaceae
♦ churchmouse threeawn
♦ Weed
♦ 39, 87, 88, 161, 218
♦ G, herbal, toxic.

Aristida divaricata Humb. & Bonpl. ex Willd.
Poaceae
♦ poverty threeawn
♦ Weed
♦ 199
♦ pG, herbal.

Aristida erecta Hitchc.
Poaceae
Aristida curtifolia Hitchc.
♦ Weed
♦ 14
♦ G.

Aristida funiculata Trin. & Rupr.
Poaceae
Aristida macranthera A.Rich.
♦ Weed
♦ 221
♦ G, arid.

Aristida hordeacea Kunth
Poaceae
Aristida steudeliana Trin. & Rupr.
♦ Weed
♦ 87, 88
♦ G, arid.

Aristida junciformis Trin. & Rupr.
Poaceae
Aristida contractinodis Stent & Rattray, *Aristida dewildemanii* Henrard, *Aristida macilenta* Henrard, *Aristida pardyi* Stent & Rattray, *Aristida schliebenii* Henrard, *Aristida textilis* Mez, *Aristida welwitschii* Rendle, *Aristida welwitschii* var. *minor* Rendle
♦ ngongoni threeawn
♦ Weed, Quarantine Weed
♦ 76, 88, 158, 203, 220
♦ G, arid.

Aristida junciformis Trin. & Rupr. ssp. junciformis
Poaceae
♦ bristlegrass, gongoni grass, gongoni threeawn, ngongoni bristlegrass, wiregrass
♦ Weed, Quarantine Weed, Native Weed
♦ 121, 220
♦ pG. Origin: southern Africa.

Aristida latifolia Domin
Poaceae
♦ feathertop wiregrass
♦ Weed
♦ 87, 88
♦ G, arid, cultivated.

Aristida laxa Cav.
Poaceae
♦ Weed
♦ 199
♦ G.

Aristida leptopoda Benth.
Poaceae
♦ Weed
♦ 87, 88
♦ G. Origin: Australia.

Aristida longespica Poir.
Poaceae
♦ slimspike threeawn
♦ Weed, Naturalised
♦ 161, 287
♦ G, herbal.

Aristida longiseta Steud.
Poaceae
= *Aristida purpurea* Nutt. var. *longiseta* (Steud.) Vasey (NoR)
♦ red threeawn
♦ Weed
♦ 87, 88, 91, 218, 255
♦ pG, herbal. Origin: South America.

Aristida meccana Hochst. ex Trin. & Rupr.
Poaceae
♦ Weed
♦ 221
♦ G.

Aristida oligantha Michx.
Poaceae
♦ prairie threeawn, oldfield threeawn
♦ Weed, Native Weed
♦ 39, 87, 88, 91, 161, 174, 210, 218
♦ aG, herbal, toxic. Origin: North America.

Aristida pallens Cav.
Poaceae
♦ pasto amargo
♦ Weed
♦ 87, 88, 295
♦ G.

Aristida purpurascens Poir.
Poaceae
♦ arrowfeather threeawn
♦ Weed
♦ 87, 88, 218
♦ G, herbal.

Aristida purpurea Nutt.
Poaceae
♦ purple threeawn
♦ Native Weed
♦ 161, 174
♦ pG, herbal. Origin: North America.

Aristida ramosa R.Br.
Poaceae
♦ Naturalised, Native Weed, Introduced
♦ 228, 269, 280
♦ G, arid, cultivated. Origin: Australia.

Aristida schiedeana **Trin. & Rupr.**
Poaceae
♦ single threeawn
♦ Weed
♦ 199
♦ G, herbal.

Aristida sciurus **Stapf**
Poaceae
♦ bristlegrass, tall three awned grass
♦ Native Weed
♦ 121
♦ pG.

Aristida setacea **Retz.**
Poaceae
Aristida depressa Trin.
♦ Weed
♦ 87, 88
♦ G.

Aristida setifolia **Kunth**
Poaceae
♦ Weed
♦ 87, 88
♦ G.

Aristida suringarii **Henr.**
Poaceae
♦ St. Eustatius threeawn
♦ Naturalised
♦ 101
♦ G.

Aristida swartziana **Steud.**
Poaceae
♦ Swartz's threeawn
♦ Naturalised
♦ 101
♦ G.

Aristida ternipes **Cav.**
Poaceae
♦ spidergrass
♦ Weed
♦ 14
♦ G, herbal.

Aristida vagans **Cav.**
Poaceae
♦ Naturalised
♦ 280
♦ G.

Aristolochia arcuata **Mast.**
Aristolochiaceae
♦ jarrinha, cipo mil homens
♦ Weed
♦ 255
♦ Origin: Brazil.

Aristolochia baetica **L.**
Aristolochiaceae
♦ Weed
♦ 70
♦ cultivated, herbal.

Aristolochia bilobata **L.**
Aristolochiaceae
♦ twolobe Dutchman's pipe
♦ Naturalised
♦ 101

Aristolochia bracteata **Retz.**
Aristolochiaceae
Aristolochia bracteolata Lam. (see)
♦ birthwort
♦ Weed, Quarantine Weed

♦ 76, 87, 88, 203, 220
♦ herbal.

Aristolochia bracteolata **Lam.**
Aristolochiaceae
= *Aristolochia bracteata* Retz.
♦ Weed
♦ 88, 242
♦ arid.

Aristolochia brasiliensis **Mart. & Zucc.**
Aristolochiaceae
♦ birthwort
♦ Weed
♦ 39, 87, 88
♦ cultivated, herbal, toxic.

Aristolochia clematitis **L.**
Aristolochiaceae
♦ birthwort
♦ Weed, Quarantine Weed,
Naturalised
♦ 39, 44, 70, 76, 86, 87, 88, 94, 101, 203,
220, 243, 253, 272
♦ pH, cultivated, herbal, toxic.

Aristolochia contorta **Bunge**
Aristolochiaceae
♦ northern Dutchman's pipe
♦ Weed
♦ 297
♦ pH, promoted, herbal. Origin:
China, Japan, Korea, Manchuria.

Aristolochia durior **Hill.**
Aristolochiaceae
Aristolochia macrophylla Lam.,
Aristolochia sipho L'Hérit.
♦ Dutchman's pipe
♦ Environmental weed
♦ 201
♦ Cultivated.

Aristolochia elegans **Mast.**
Aristolochiaceae
Aristolochia littoralis non Parodi
♦ Dutchman's pipe, calico flower
♦ Weed, Quarantine Weed,
Naturalised, Garden Escape,
Environmental Weed, Cultivation
Escape
♦ 3, 39, 73, 76, 86, 87, 88, 98, 101, 121,
155, 191, 201, 203, 261, 279, 283
♦ pC, cultivated, herbal, toxic. Origin:
South America.

Aristolochia erecta **L.**
Aristolochiaceae
♦ swan flower
♦ Naturalised
♦ 101

Aristolochia floribunda **Lem.**
Aristolochiaceae
♦ Naturalised
♦ 86

Aristolochia galeata **Mart. & Zucc.**
Aristolochiaceae
♦ jarrinha, cipo mil homens
♦ Weed
♦ 255
♦ Origin: Brazil.

Aristolochia grandiflora **Sw.**
Aristolochiaceae
♦ pelican flower

♦ Weed, Naturalised
♦ 39, 87, 88, 101, 179
♦ cultivated, herbal, toxic.

Aristolochia labiata **Willd.**
Aristolochiaceae
♦ mottled Dutchman's pipe
♦ Naturalised
♦ 101
♦ cultivated.

Aristolochia lingua **Malme**
Aristolochiaceae
♦ Weed
♦ 87, 88

Aristolochia littoralis **Parodi**
Aristolochiaceae
♦ calico flower, elegant Dutchman's
pipe
♦ Weed, Environmental Weed
♦ 22, 80, 88, 112, 179
♦ pC, cultivated.

Aristolochia longa **L.**
Aristolochiaceae
♦ birthwort
♦ Weed
♦ 39, 70, 88, 94
♦ cultivated, herbal, toxic.

Aristolochia macrophylla **Lam.**
Aristolochiaceae
♦ Virginia snakeroot, pipevine
♦ Environmental weed
♦ 201
♦ Cultivated, herbal.

Aristolochia maurorum **L.**
Aristolochiaceae
♦ Weed, Quarantine Weed
♦ 76, 87, 88, 203, 220, 243
♦ herbal.

Aristolochia maxima **Jacq.**
Aristolochiaceae
♦ Florida Dutchman's pipe
♦ Weed, Naturalised
♦ 101, 179
♦ herbal.

Aristolochia odoratissima **L.**
Aristolochiaceae
♦ fragrant Dutchman's pipe
♦ Naturalised
♦ 39, 101
♦ cultivated, herbal, toxic. Origin:
Central and South America.

Aristolochia peltata **L.**
Aristolochiaceae
♦ peltate Dutchman's pipe
♦ Naturalised
♦ 101

Aristolochia pistolochia **L.**
Aristolochiaceae
♦ Spanish birthwort, pistolochia
♦ Weed
♦ 39, 70
♦ herbal, toxic.

Aristolochia ringens **Vahl**
Aristolochiaceae
♦ gaping Dutchman's pipe
♦ Weed, Naturalised, Cultivation
Escape
♦ 86, 101, 191, 252, 261

♦ cultivated, herbal. Origin: South
America.

Aristolochia rotunda L.
Aristolochiaceae
♦ snakeroot
♦ Weed
♦ 39, 87, 88, 94
♦ pH, cultivated, herbal, toxic. Origin:
Mediterranean.

Aristolochia trilobata L.
Aristolochiaceae
♦ bejuco de santiago
♦ Naturalised
♦ 101
♦ cultivated, herbal.

Arjona tuberosa Cav.
Santalaceae
♦ Weed
♦ 87, 88

**Arjona tuberosa Cav. var. tandilensis
(Kuntze) Dawson**
Santalaceae
♦ macachín del trigo
♦ Weed
♦ 237, 295

Armatocereus Bakeb. spp.
Cactaceae
♦ Weed, Quarantine Weed
♦ 76, 88, 203, 220
♦ cultivated.

**Armeria alliacea (Cav.) Hoffmanns. &
Link**
Plumbaginaceae
♦ Jersey thrift
♦ Casual Alien
♦ 280
♦ cultivated, herbal.

Armeria maritima (Mill.) Willd.
Plumbaginaceae
♦ thrift, thrift seapink
♦ Casual Alien
♦ 280
♦ pH, cultivated, herbal.

Armeria rumelica Boiss.
Plumbaginaceae
♦ Weed
♦ 272
♦ cultivated.

Armeria welwitschii Boiss.
Plumbaginaceae
♦ Weed
♦ 70
♦ cultivated.

**Armoracia rusticana P.Gaertn., B.Mey.
& Scherb.**
Brassicaceae
Armoracia lapathifolia Gilib., *Cochlearia
armoracia* L., *Nasturtium armoriaca* (L.)
Fr., *Rorippa armoracia* Hitchc.
♦ horse radish, red cole, piparjuuri
♦ Weed, Naturalised, Cultivation
Escape
♦ 8, 34, 39, 40, 42, 70, 88, 94, 101, 243,
247, 272, 280, 287
♦ pH, cultivated, herbal, toxic. Origin:
Europe.

**Arnebia hispidissima (Sieber & Lehm.)
DC.**
Boraginaceae
Arnebia asperrima (Del.) Hutch. &
Dalziel, *Lithospermum hispidissimum*
Sieber & Lehm.
♦ Weed
♦ 221
♦ arid.

Arnebia linearifolia DC.
Boraginaceae
♦ Weed
♦ 221

Arnebia tetrastigma Forssk.
Boraginaceae
♦ Weed
♦ 221
♦ Origin: Egypt.

Arnica amplexicaulis Nutt.
Asteraceae
♦ clasping arnica
♦ Quarantine Weed
♦ 220
♦ pH, herbal.

Arnica angustifolia Vahl
Asteraceae
♦ narrowleaf arnica
♦ Naturalised
♦ 241, 300

Arnica cordifolia Hook.
Asteraceae
♦ heartleaf arnica
♦ Weed
♦ 161
♦ pH, promoted, herbal, toxic.

Arnica fulgens Pursh
Asteraceae
♦ foothill arnica
♦ Weed
♦ 23, 88
♦ pH, promoted, herbal.

Arnica montana L.
Asteraceae
♦ etelänarnikki, leopard's bane,
mountain arnica, European arnica
♦ Weed, Cultivation Escape
♦ 23, 39, 42, 70, 88, 161, 247, 272
♦ pH, cultivated, herbal, toxic.

Arnoseris minima (L.) Schweig. & Koerte
Asteraceae
♦ lamb's succory, swine's succory,
kultakaunokit
♦ Weed, Naturalised, Casual Alien
♦ 42, 70, 88, 94, 101, 243, 253, 272
♦ herbal.

Aronia × prunifolia (Marshall) Rehder
Rosaceae
= *Aronia arbutifolia* (L.) Pers. × *Aronia
melanocarpa* (Michx.) Elliot
♦ chokecherry, hybrid chokeberry,
purpurne apfelbeere
♦ Environmental Weed
♦ 152
♦ cultivated, herbal. Origin: hybrid.

**Arrabidaea brachypoda (DC.) Bur. &
K.Schum.**
Bignoniaceae

Bignonia brachypoda A.DC., *Arrabidaea
platyphylla* (Cham.) Bureau &
K.Schum., *Bignonia platyphylla* Cham.
♦ cipo una
♦ Weed
♦ 255
♦ Origin: South America.

Arrabidaea florida DC.
Bignoniaceae
Arrabidaea cardenasii Rusby, *Arrabidaea
divaricata* Bureau & K.Schum.,
Arrabidaea panamensis Sprague
♦ cipo neve
♦ Weed
♦ 255
♦ Origin: South America.

Arracacia xanthorrhiza E.N.Bancr.
Apiaceae
Arracacia esculenta DC.
♦ arracacha
♦ Naturalised, Introduced
♦ 101, 261
♦ pH, promoted, herbal. Origin:
Colombia.

Arrhenatherum album (Vahl) Clay.
Poaceae
Arrhenatherum erianthum Boiss. & Reut.
(see)
♦ Weed
♦ 70
♦ G.

Arrhenatherum elatius (L.) J. & C.Presl
Poaceae
Arrhenatherum americanum P.Beauv.,
Avena avenaceum (Scop.) P.Beauv.,
Avena alata Salisb., *Avena elatior* L.,
Avenastrum elatius (L.) Jess., *Holcus
avenaceus* Scop., *Hordeum avenaceum*
Wigg. ex P.Beauv.
♦ tall oatgrass, false oatgrass, French
oatgrass
♦ Weed, Quarantine Weed,
Naturalised, Introduced,
Environmental Weed, Cultivation
Escape
♦ 7, 15, 23, 34, 72, 80, 86, 87, 88, 91, 98,
101, 118, 121, 146, 151, 155, 161, 176,
181, 195, 203, 218, 220, 225, 228, 243,
246, 252, 272, 280, 286, 287, 290
♦ pG, arid, cultivated, herbal. Origin:
Eurasia.

**Arrhenatherum elatius (L.) P.Beauv. ex J.
& C.Presl ssp. bulbosum (Willd.) Schübl.
& G.Martens**
Poaceae
♦ onion couch, tall oatgrass, false
oatgrass
♦ Weed, Naturalised
♦ 30, 70, 243, 253, 280, 300
♦ pG, herbal.

**Arrhenatherum elatius (L.) J. & C.Presl
ssp. elatius**
Poaceae
♦ false oatgrass
♦ Weed, Naturalised
♦ 70, 241, 280, 300
♦ G, cultivated.

Arrhenatherum elatius (L.) Beauv. ex J.& C.Presl var. *biaristatum* (Peterm.) Peterm.
Poaceae
= *Arrhenatherum elatius* (L.) J. & C.Presl var. *elatius*
♦ tall oatgrass
♦ Naturalised
♦ 287
♦ G.

Arrhenatherum elatius (L.) J. & C.Presl var. *bulbosum* (Willd.) Spenn.
Poaceae
♦ tall oatgrass, tuber oatgrass, false oatgrass
♦ Weed, Naturalised, Environmental Weed
♦ 86, 101, 176, 198, 212, 269, 287
♦ pG. Origin: Eurasia.

Arrhenatherum elatius (L.) J. & C.Presl var. *elatius*
Poaceae
Arrhenatherum elatius (L.) Beauv. ex J.& C.Presl var. *biaristatum* (Peterm.) Peterm. (see)
♦ tall oatgrass, false oatgrass
♦ Naturalised, Environmental Weed
♦ 86, 101
♦ G. Origin: Eurasia.

Arrhenatherum erianthum Boiss. & Reut.
Poaceae
= *Arrhenatherum album* (Vahl) Clay.
♦ Weed
♦ 87, 88
♦ G.

Artabotrys hexapetalus (L.f.) Bhandari
Annonaceae
♦ climbing ilang ilang, climbing ylang ylang lanalana
♦ Naturalised, Cultivation Escape
♦ 101, 233
♦ cultivated.

Artanema longifolium (L.) Merr.
Scrophulariaceae
Artanema sesamoides Benth., *Sesamum javanicum* Burm.f.
♦ Weed
♦ 13, 88, 170, 191

Artemisia L. spp.
Asteraceae
♦ woolflower, sagebrush
♦ Weed
♦ 39, 243, 272
♦ herbal, toxic.

Artemisia abrotanum L.
Asteraceae
♦ southernwood, old man southernwood, aaprottimaruna
♦ Weed, Naturalised, Cultivation Escape, Casual Alien
♦ 39, 40, 42, 101, 272
♦ S, cultivated, herbal, toxic. Origin: obscure, possibly southern Europe.

Artemisia absinthium L.
Asteraceae
♦ wormwood, absinth sagewort, absinth wormwood, assenzio romano, palina pravá

♦ Weed, Noxious Weed, Naturalised, Introduced, Garden Escape, Environmental Weed, Cultivation Escape
♦ 1, 4, 39, 42, 52, 54, 70, 80, 86, 87, 88, 98, 101, 104, 136, 139, 146, 151, 161, 203, 210, 218, 219, 228, 229, 241, 243, 247, 272, 280, 287, 292, 299, 300
♦ pH, arid, cultivated, herbal, toxic. Origin: Eurasia.

Artemisia abyssinica Sch.Bip.
Asteraceae
♦ Weed
♦ 88
♦ arid.

Artemisia afra Jacq. ex Willd.
Asteraceae
♦ wild wormwood, wormwood
♦ Native Weed
♦ 121
♦ pH, arid, cultivated, herbal. Origin: southern Africa.

Artemisia anethoides Mattf.
Asteraceae
♦ Weed
♦ 275
♦ aH.

Artemisia annua L.
Asteraceae
♦ sweet sagewort, annual wormwood, sweet wormwood, sweet Annie, rikkamaruna, palina roāná
♦ Weed, Naturalised, Casual Alien
♦ 42, 80, 87, 88, 94, 101, 161, 207, 210, 218, 237, 272, 275, 286, 287, 297
♦ aH, cultivated, herbal. Origin: Eurasia.

Artemisia apiacea Hance
Asteraceae
Artemisia thunbergiana Maxim.
♦ Chinese wormwood, kawaraninjin
♦ Weed
♦ 87, 88, 275, 286
♦ bH, promoted, herbal. Origin: east Asia.

Artemisia arborescens L.
Asteraceae
♦ hedge artemisia, silver wormwood, tree wormwood
♦ Weed, Naturalised, Environmental Weed
♦ 7, 15, 72, 86, 88, 98, 203, 280
♦ S, cultivated, herbal. Origin: southern Europe.

Artemisia arbuscula Nutt.
Asteraceae
♦ low sagebrush, little sagebrush
♦ Weed
♦ 87, 88, 218
♦ S, herbal.

Artemisia argyi H.Lév & Vanihot
Asteraceae
♦ Weed
♦ 275, 297
♦ pH, arid, promoted, herbal.

Artemisia austriaca Jacq.
Asteraceae
♦ tonavanmaruna, palina rakúska

♦ Weed, Casual Alien
♦ 42, 87, 88, 272

Artemisia biennis Willd.
Asteraceae
♦ biennial wormwood, sahamaruna
♦ Weed, Naturalised, Casual Alien
♦ 34, 40, 42, 87, 88, 101, 161, 174, 212, 218
♦ a/pH, promoted, herbal. Origin: north-west North America.

Artemisia biennis Willd. var. *biennis*
Asteraceae
♦ biennial wormwood
♦ Naturalised
♦ 101

Artemisia californica Less.
Asteraceae
♦ California sagebrush, coast sagebrush
♦ Weed
♦ 87, 88, 218
♦ S, arid, cultivated, herbal.

Artemisia campestris L.
Asteraceae
♦ common sagewort, field caudata sagewort, field wormwood, Breckland wormwood, field southernwood, field sagewort, ketomaruna
♦ Weed
♦ 70, 161, 212, 272
♦ b/pH, cultivated, herbal.

Artemisia cana Pursh
Asteraceae
♦ silver sagebrush
♦ Weed
♦ 39, 87, 88, 161, 212, 218
♦ herbal, toxic.

Artemisia canariensis (Besser) Less.
Asteraceae
♦ Naturalised
♦ 86
♦ cultivated.

Artemisia capillaris Thunb.
Asteraceae
♦ Weed, Introduced
♦ 87, 88, 228, 286
♦ S, arid, cultivated, herbal. Origin: China, Japan, Korea, Manchuria.

Artemisia douglasiana Bess.
Asteraceae
♦ California mugwort, western mugwort, Douglas's mugwort, wormwood, Douglas's sagewort
♦ Weed, Introduced
♦ 136, 161, 180, 228, 243
♦ pH, arid, promoted, herbal. Origin: Europe.

Artemisia dracunculus L.
Asteraceae
♦ tarragon, green sagewort, false tarragon, silky wormwood, rakuuna, French tarragon, wormwood
♦ Weed, Naturalised, Native Weed, Cultivation Escape, Casual Alien
♦ 40, 42, 87, 88, 161, 174, 280
♦ pH, cultivated, herbal. Origin: Eurasia.

Artemisia dubia **Wall. ex DC.**
Asteraceae
Artemisia vulgaris L. var. *indica* Maxim.
♦ phak hia, mugwort
♦ Weed
♦ 238
♦ pH.

Artemisia eriopoda **Bunge**
Asteraceae
♦ woollystalk wormwood
♦ Weed
♦ 297

Artemisia filifolia **Torr.**
Asteraceae
♦ sand sagebrush, sandhill sage, silvery wormwood
♦ Weed, Native Weed
♦ 39, 87, 88, 161, 174, 212, 218
♦ S, arid, promoted, herbal, toxic. Origin: North America.

Artemisia frigida **Willd.**
Asteraceae
♦ fringed sagebrush, prairie sagewort, fringed sagewort
♦ Weed, Native Weed
♦ 87, 88, 161, 174, 212, 218, 297
♦ pH, arid, cultivated, herbal. Origin: North America.

Artemisia gmelinii **Webb ex Stechm.**
Asteraceae
♦ Gmelin's wormwood, Russian wormwood
♦ Weed, Naturalised
♦ 101, 297
♦ a/bH, arid, promoted. Origin: Himalayas.

Artemisia herba-alba **Asso.**
Asteraceae
Artemisia aethiopica L., *Artemisia aragonensis* Lam., *Artemisia billardieriana* Besser, *Artemisia lippii* Besser, *Artemisia oliveriana* Gay ex Besser, *Artemisia ontina* Dufour, *Artemisia sieberi* Besser
♦ Barbary santonica, Barbary wormseed, Moroccan wormwood, armoise blanche, bo'aihran, chih, desert wormwood, ifessi, izri, shih, tirkha
♦ Weed
♦ 87, 88
♦ arid, herbal.

Artemisia incompta **Nutt.**
Asteraceae
= *Artemisia ludoviciana* Nutt. ssp. *incompta* (Nutt.) Cronquist (NoR)
♦ Weed
♦ 23, 88

Artemisia inculta **Delile**
Asteraceae
= *Seriphidium incultum* (Delile) Y.R.Ling (NoR)
♦ Weed
♦ 221

Artemisia indica **Willd.**
Asteraceae
= *Artemisia vulgaris* L.
♦ Weed

♦ 286
♦ a/pH, promoted. Origin: China, Japan, India.

Artemisia japonica **Thunb.**
Asteraceae
Artemisia mandschurica (Kom.) Kom., *Artemisia subintegra* Kitam., *Chrysanthemum japonicum* Thunb.
♦ otoko yomogi
♦ Weed
♦ 87, 88, 263, 286
♦ pH, promoted, herbal. Origin: east Asia.

Artemisia judaica **L.**
Asteraceae
♦ wormwood
♦ Weed
♦ 221
♦ arid, herbal.

Artemisia klotzschiana **Besser**
Asteraceae
♦ Weed
♦ 199

Artemisia lavandulaefolia **Miq.**
Asteraceae
= *Artemisia vulgaris* L.
♦ lavandulaefolia wormwood
♦ Weed
♦ 275, 297
♦ pH.

Artemisia ludoviciana **Nutt.**
Asteraceae
♦ Louisiana wormwood, cudweed sagewort, white sagebrush, mugwort wormwood, prairie sage, gray sagewort, white sage
♦ Weed, Naturalised, Native Weed
♦ 8, 54, 86, 87, 88, 161, 174, 218, 243
♦ pH, cultivated, herbal. Origin: North America.

Artemisia ludoviciana **Nutt. ssp.** *mexicana* **(Spreng.) Keck**
Asteraceae
Artemisia mexicana Willd. ex Spreng. (see)
♦ white sagebrush
♦ Weed
♦ 199

Artemisia maritima **L.**
Asteraceae
♦ sea wormwood, merimaruna, levant wormseed
♦ Weed, Casual Alien
♦ 39, 42, 70, 243, 248
♦ S, arid, cultivated, herbal, toxic.

Artemisia mexicana **Willd. ex Spreng.**
Asteraceae
= *Artemisia ludoviciana* Nutt. ssp. *mexicana* (Spreng.) Keck
♦ Weed
♦ 39, 87, 88
♦ pH, promoted, herbal, toxic.

Artemisia mongolica **Fisch.**
Asteraceae
♦ Mongolian wormwood
♦ Weed
♦ 297

Artemisia monosperma **Delile**
Asteraceae
♦ Weed
♦ 221
♦ arid.

Artemisia montana **(Nak.) Pamp.**
Asteraceae
♦ mountain mugwort, ooyomogi
♦ Weed
♦ 88, 204, 286
♦ pH, promoted. Origin: east Asia.

Artemisia nova **A.Nels.**
Asteraceae
♦ black sagebrush
♦ Weed
♦ 87, 88, 218
♦ S, arid, promoted, herbal.

Artemisia ordosica **Krasch.**
Asteraceae
♦ Weed
♦ 275, 297
♦ arid.

Artemisia palustris **L.**
Asteraceae
♦ swampy wormwood
♦ Weed
♦ 297

Artemisia pontica **L.**
Asteraceae
♦ Roman wormwood, palina pontická, assenzio gentile
♦ Naturalised
♦ 39, 101
♦ cultivated, herbal, toxic.

Artemisia princeps **Pamp.**
Asteraceae
♦ ragweed, yomogi
♦ Weed, Quarantine Weed
♦ 76, 87, 88, 203, 204, 220, 243, 263, 286
♦ pS, promoted, herbal. Origin: Japan.

Artemisia princeps **Pamp. var.** *orientalis* **(Pamp.) Hara.**
Asteraceae
♦ Asiatic wormwood
♦ Weed
♦ 274

Artemisia rubripes **Nakai**
Asteraceae
♦ yabuyomogi
♦ Weed
♦ 286
♦ pH, promoted. Origin: China, southern Japan.

Artemisia scoparia **Waldst. & Kit.**
Asteraceae
♦ redstem wormwood, siromaruna, palina metlovitá
♦ Weed, Naturalised, Introduced, Casual Alien
♦ 40, 42, 86, 101, 228, 272, 275, 297
♦ aH, arid, promoted. Origin: Eurasia.

Artemisia selengensis **Turcz. ex Bess.**
Asteraceae
= *Artemisia vulgaris* L. var. *selengensis* (Turcz. ex Besser) Maxim. (NoR)
♦ Naturalised
♦ 287

Artemisia sieversiana **Ehrh. ex Wild.**
Asteraceae
♦ sagebrush, palina sieversova, idänmaruna
♦ Weed, Naturalised, Casual Alien
♦ 42, 87, 88, 114, 272, 275, 287, 297
♦ aH, promoted, herbal.

Artemisia sphaerocephala **Krasch.**
Asteraceae
♦ roundhead wormwood
♦ Weed
♦ 297
♦ arid.

Artemisia spinescens **D.Eaton**
Asteraceae
= *Picrothamnus desertorum* Nutt. (NoR)
♦ spring sage, bud sage, bud sagebrush
♦ Weed
♦ 39, 161
♦ S, arid, herbal, toxic.

Artemisia stelleriana **Besser**
Asteraceae
♦ beach wormwood, old woman, hoary mugwort
♦ Weed
♦ 179
♦ pH, cultivated, herbal.

Artemisia suksdorfii **Piper**
Asteraceae
♦ coastal wormwood
♦ Weed
♦ 23, 88
♦ pH, herbal.

Artemisia sylvatica **Maxim.**
Asteraceae
♦ Weed
♦ 275
♦ pH, promoted.

Artemisia tridentata **Nutt.**
Asteraceae
♦ big sagebrush, sagebrush, purple sage
♦ Weed
♦ 23, 39, 87, 88, 161, 212, 218, 243
♦ pS, arid, promoted, herbal, toxic.

Artemisia tripartita **Rydb.**
Asteraceae
Seriphidium tripartitum (Rydb.) W.A.Weber.
♦ threetip sagebrush
♦ Weed
♦ 87, 88, 218
♦ S, promoted, herbal.

Artemisia umbelliformis **Lam.**
Asteraceae
Artemisia mutelliana Vill. *non* S.G.Gmel. *nom. illeg.*, *Artemisia laxa* (Lam.) Fritsch
♦ alps wormwood, alpine wormwood
♦ Naturalised
♦ 101
♦ pH, arid, promoted, herbal.

Artemisia verlotiorum **Lamotte**
Asteraceae
Artemisia verlotorum Lamotte (see)
♦ mugwort, Chinese wormwood, verlot's mugwort, palina verlotovcov

♦ Weed, Naturalised, Environmental Weed
♦ 15, 44, 72, 86, 88, 98, 165, 198, 203, 246, 253, 280
♦ pH, cultivated, herbal. Origin: south-west China.

Artemisia verlotorum **Lamotte**
Asteraceae
= *Artemisia verlotiorum* Lamotte
♦ mugwort, yuyo San Vicente
♦ Weed
♦ 236, 237, 255, 269, 295
♦ pH, herbal. Origin: Eurasia.

Artemisia vulgaris **L.**
Asteraceae
Artemisia indica Willd. (see), *Artemisia lavandulaefolia* Miq. (see)
♦ mugwort, common wormwood, felon herb, chrysanthemum weed, motherwort, green ginger, mugweed, sailor's tobacco, pujo
♦ Weed, Naturalised, Environmental Weed
♦ 8, 13, 23, 39, 44, 70, 80, 87, 88, 94, 101, 102, 121, 136, 151, 161, 211, 217, 218, 236, 243, 249, 253, 272
♦ pH, cultivated, herbal, toxic. Origin: Eurasia.

Artemisia vulgaris **L. var. *vulgaris***
Asteraceae
♦ common wormwood
♦ Naturalised
♦ 101

Arthraxon castratus **(Griffiths) Naray. ex Bor**
Poaceae
♦ castrate carpgrass
♦ Naturalised
♦ 101
♦ G. Origin: Australia.

Arthraxon hispidus **(Thunb.) Makino**
Poaceae
Arthraxon hispidus (Thunb.) Makino var. *cryptatherus* (Hack.) Honda (see)
♦ hairy jointgrass, small carpgrass, Greek grass, joint headed arthraxon
♦ Weed, Naturalised
♦ 19, 80, 87, 88, 101, 102, 133, 161, 195, 224, 263, 275, 286, 297
♦ aG, aqua, herbal. Origin: Asia, Australia.

Arthraxon hispidus **(Thunb.) Makino var. *cryptatherus* (Hack.) Honda**
Poaceae
= *Arthraxon hispidus* (Thunb.) Makino
♦ Weed
♦ 275
♦ aG, herbal.

Arthraxon lanceolatus **(Roxb.) Hochst.**
Poaceae
♦ Weed, Naturalised
♦ 287, 297
♦ G.

Arthrocnemum glaucum **Delile**
Chenopodiaceae
Salicornia macrostachya Moric.
♦ Weed

♦ 221
♦ arid.

Arthrostemma ciliatum **Pav. ex D.Don**
Melastomataceae
♦ pink fringe
♦ Naturalised, Cultivation Escape
♦ 101, 233
♦ cultivated.

Arthrostemma fragile **Lindl.**
Melastomataceae
Arthrostema ciliatum Ruiz & Pav. ex Cogn.
♦ Weed
♦ 157
♦ a/pH.

Arthrostylidium capillifolium **Griseb.**
Poaceae
= *Arthrostylidium farctum* (Aubl.) Soderstr. & Lourteig (NoR)
♦ old man's beard
♦ Weed
♦ 14
♦ G.

Artocarpus altilis **(Park.) Fosb.**
Moraceae
Artocarpus incisa (Thunb.) L.f.
♦ breadfruit, mahi
♦ Weed, Naturalised, Introduced, Casual Alien
♦ 38, 87, 88, 101, 230, 261
♦ T, cultivated, herbal. Origin: Pacific Islands.

Artocarpus altilis **(Parkinson) Fosb. × *marianensis* Trécul**
Moraceae
♦ mahi
♦ Introduced
♦ 230
♦ T.

Artocarpus heterophyllus **Lam.**
Moraceae
♦ jackfruit
♦ Naturalised, Introduced
♦ 38, 101, 230, 261
♦ T, cultivated, herbal. Origin: India.

Artocarpus lakoocha **Roxb.**
Moraceae
♦ Introduced
♦ 228
♦ arid, herbal.

Arum discoridis **Sibth. & Sm.**
Araceae
♦ Weed
♦ 111, 243

Arum italicum **Mill.**
Araceae
Arum numidicum Schott
♦ Italian cuckoo pint, Italian arum, Italian lords and ladies, Italian lily
♦ Weed, Quarantine Weed, Naturalised, Introduced, Garden Escape, Environmental Weed
♦ 7, 15, 34, 39, 70, 72, 76, 80, 86, 87, 88, 98, 101, 127, 161, 165, 198, 203, 225, 246, 247, 253, 272, 280
♦ pH, cultivated, herbal, toxic. Origin: Eurasia.

Arum maculatum L.
Araceae
Arum vulgare Lam., *Arum pyrenaeum* Dufour, *Arum trapezuntinum* Schotter ex Engl., *Arum zelebori* Schott.
♦ lords and ladies, cuckoo pint, Jack in the pulpit, cuckoo plant
♦ Weed
♦ 23, 39, 87, 88, 161, 247, 272
♦ pH, cultivated, herbal, toxic.

Arum neglectum (Towns.) Ridl.
Araceae
♦ Weed
♦ 23, 88

Aruncus dioicus (Walter) Fern.
Rosaceae
Aruncus silvestris L.
♦ goat's beard, töyhtöangervo, bride's feathers
♦ Weed, Naturalised, Cultivation Escape
♦ 42, 70, 86, 98, 101, 195, 203, 252, 272
♦ pH, cultivated, herbal. Origin: Europe, Asia, North America.

Aruncus dioicus (Walter) Fern. var. vulgaris (Maxim.) Hara
Rosaceae
♦ bride's feathers
♦ Naturalised
♦ 101

Arundina graminifolia (D.Don) Hochr.
Orchidaceae
♦ bamboo orchid
♦ Weed, Naturalised, Cultivation Escape
♦ 19, 101, 261
♦ cultivated, herbal.

Arundinaria gigantea (Walter) Muhl.
Poaceae
♦ giant cane, cane reed
♦ Weed
♦ 8, 80
♦ G, promoted, herbal, toxic.

Arundinaria pusilla A.Chev. & A.Camus
Poaceae
♦ Weed
♦ 209
♦ G.

Arundinaria pygmaea (Miq.) Asch. & Graebn.
Poaceae
♦ Weed
♦ 88, 204
♦ G, cultivated.

Arundinaria simonii (Carr.) A.& C. Rivière
Poaceae
♦ Simon bamboo
♦ Naturalised
♦ 101
♦ G, cultivated.

Arundinaria tecta (Walter) Muhl.
Poaceae
= *Arundinaria gigantea* (Walter) Muhl. ssp. *tecta* (Walter) McClure (NoR)
♦ switch cane
♦ Weed

♦ 87, 88, 218
♦ G, herbal.

Arundinella bengalensis (Spreng) Druce
Poaceae
♦ Weed, Quarantine Weed
♦ 76, 87, 88, 203, 220
♦ G.

Arundinella hirta (Thunb.) Tanaka
Poaceae
♦ todashiba
♦ Weed
♦ 87, 88, 204, 286
♦ G. Origin: China, Japan.

Arundinella hookeri Munro. ex Keng
Poaceae
♦ Weed
♦ 244
♦ G.

Arundinella leptochloa (Nees) Hook.f.
Poaceae
♦ Weed, Quarantine Weed
♦ 76, 87, 88, 203, 220
♦ G.

Arundinella setosa Trin.
Poaceae
♦ Weed
♦ 87, 88
♦ G, arid. Origin: Australia.

Arundinella sinensis Rendle
Poaceae
♦ Weed
♦ 87, 88
♦ G.

Arundo donax L.
Poaceae
Arundo bifaria Retz., *Arundo glauca* Bubani, *Arundo latifolia* Salisb., *Arundo sativa* Lam., *Cynodon donax* (L.) Raspail., *Donax donax* (L.) Asch & Graebn., *Scolochloa arundinacea* (P.Beauv.) Mert. & Koch., *Scolochloa donax* (L.) Gaudin
♦ giant reed, arundo giant reed, gasau ni vavalagi, bamboo reed, Spanish reed, grand roseau, canne de Provence, Spaanse riet, giant Danube reed, bamboo reed, false bamboo, elephant grass, wild cane, fiso papâlagi, canna
♦ Weed, Noxious Weed, Naturalised, Introduced, Garden Escape, Environmental Weed, Cultivation Escape
♦ 2, 3, 15, 17, 18, 32, 35, 38, 63, 72, 78, 80, 87, 88, 91, 95, 98, 101, 102, 116, 121, 129, 134, 142, 151, 158, 161, 179, 80, 203, 218, 221, 225, 228, 231, 236, 238, 241, 246, 261, 269, 272, 279, 280, 283, 286, 289, 295, 296, 300
♦ pG, aqua, cultivated, herbal. Origin: Eurasia.

Arundo donax L. var. donax
Poaceae
♦ giant reed, false bamboo
♦ Weed, Naturalised, Introduced
♦ 7, 93, 98, 205, 230
♦ pG. Origin: Asia.

Arundo donax L. var. *versicolor* (Mill.) Stokes
Poaceae
♦ giant reed
♦ Naturalised
♦ 7, 98, 287
♦ G. Origin: Asia.

Arundo formosana Hack.
Poaceae
♦ Weed
♦ 87, 88
♦ G.

Arundo madagascariensis Kunth
Poaceae
♦ Weed
♦ 87, 88
♦ G.

Asaemia axillaris (Thunb.) Harv. ex B.D.Jacks.
Asteraceae
♦ Weed, Native Weed
♦ 39, 121
♦ S, toxic. Origin: southern Africa.

Asarina procumbens Mill.
Scrophulariaceae
Antirrhinum asarina L.
♦ trailing snapdragon, chickabiddy
♦ Naturalised
♦ 40
♦ a/pH, cultivated.

Asarum canadense L.
Aristolochiaceae
♦ wild ginger, snakeroot, Canadian wild ginger
♦ Weed
♦ 39, 161
♦ pH, cultivated, herbal, toxic.

Asarum europaeum L.
Aristolochiaceae
♦ European ginger, asarabacca, taponlehti
♦ Weed
♦ 39, 272
♦ pH, cultivated, herbal, toxic.

Asclepias L. spp.
Asclepiadaceae/Apocynaceae
♦ milkweed
♦ Weed
♦ 23, 88, 154, 243, 247
♦ herbal, toxic.

Asclepias asperula (Decne.) Woods.
Asclepiadaceae/Apocynaceae
♦ spider milkweed
♦ Weed
♦ 39, 161
♦ pH, cultivated, herbal, toxic.

Asclepias burchellii Schltr.
Asclepiadaceae/Apocynaceae
♦ wild cotton, milkweed
♦ Native Weed
♦ 121
♦ pH. Origin: southern Africa.

Asclepias campestris Decne.
Asclepiadaceae/Apocynaceae
♦ quiebra arado blanco
♦ Weed
♦ 87, 88, 237, 295

Asclepias cordifolia (Benth.) Jeps.
Asclepiadaceae/Apocynaceae
♦ purple milkweed, heartleaf
milkweed
♦ Weed
♦ 161
♦ pH, herbal, toxic.

Asclepias curassavica L.
Asclepiadaceae/Apocynaceae
♦ red head cottonbush, bloodflower,
redhead, butterfly weed, bloodflower
milkweed, scarlet milkweed, bandera
Española, bloodflower milkweed
♦ Weed, Naturalised, Introduced,
Garden Escape, Environmental Weed,
Cultivation Escape, Casual Alien
♦ 7, 14, 34, 39, 86, 87, 88, 93, 98, 101,
157, 161, 179, 198, 203, 217, 218, 230,
241, 255, 257, 269, 276, 280, 287, 295,
300
♦ pH, cultivated, herbal, toxic. Origin:
tropical America.

Asclepias curassavica L. fo. *flava* Tawada
Asclepiadaceae/Apocynaceae
♦ Naturalised
♦ 287

Asclepias engelmanniana Woods.
Asclepiadaceae/Apocynaceae
Acerates auriculata Engelm., *Asclepias
auriculata* Holzinger *non* Kunth
♦ Engelmann's milkweed
♦ Weed
♦ 161
♦ arid, herbal, toxic.

Asclepias eriocarpa Benth.
Asclepiadaceae/Apocynaceae
Asclepias fremontii Torr. ex Gray,
Asclepias kotolo Eastw.
♦ woollypod milkweed
♦ Weed
♦ 39, 87, 88, 161, 218
♦ pH, cultivated, herbal, toxic.

Asclepias fascicularis Decne.
Asclepiadaceae/Apocynaceae
♦ Mexican whorled milkweed,
narrowleaf milkweed
♦ Weed
♦ 87, 88, 161, 180, 212, 218
♦ pH, cultivated, herbal, toxic. Origin:
western America.

Asclepias fruticosa L.
Asclepiadaceae/Apocynaceae
= *Gomphocarpus fruticosus* (L.) R.Br. ex
W.T.Ait.
♦ cotton milkbush, fire sticks,
milkweed, narrow leaved cottonbush,
shrubby milkweed, swan plant, wild
cotton
♦ Weed, Native Weed, Introduced
♦ 38, 39, 51, 88, 121, 161
♦ pS, cultivated, herbal, toxic. Origin:
southern Africa.

Asclepias galioides auct. Amer.
Asclepiadaceae/Apocynaceae
= *Asclepias subverticillata* (Gray) Vail
♦ poison milkweed, whorled
milkweed, bedstraw milkweed
♦ Weed

♦ 49, 88, 218
♦ pH, promoted, herbal, toxic. Origin:
North America.

Asclepias incarnata L.
Asclepiadaceae/Apocynaceae
♦ swamp milkweed, rose milkweed
♦ Weed, Quarantine Weed, Native
Weed
♦ 8, 23, 39, 76, 87, 88, 108, 161, 174, 210,
212, 218
♦ pH, arid/aqua, cultivated, herbal,
toxic. Origin: North America.

Asclepias labriformis Jones
Asclepiadaceae/Apocynaceae
♦ labriform milkweed, Jones
milkweed, Utah milkweed
♦ Weed
♦ 39, 87, 88, 161, 212, 218
♦ herbal, toxic. Origin: North America.

Asclepias latifolia (Torr.) Raf.
Asclepiadaceae/Apocynaceae
♦ broadleaf milkweed
♦ Weed
♦ 39, 87, 88, 161, 218
♦ pH, arid, promoted, herbal, toxic.

Asclepias mellodora A.St.-Hil.
Asclepiadaceae/Apocynaceae
= *Asclepias nervosa* Decne.
♦ yerba de la víbora
♦ Weed
♦ 237, 295
♦ Origin: Argentina, Brazil, Uruguay,
Venezuela.

Asclepias mexicana Cav.
Asclepiadaceae/Apocynaceae
♦ Mexican milkweed
♦ Weed
♦ 39, 87, 88, 218
♦ pH, cultivated, toxic.

Asclepias nervosa Decne.
Asclepiadaceae/Apocynaceae
Asclepias mellodora A.St.-Hil. (see)
♦ Weed
♦ 87, 88

Asclepias nivea L.
Asclepiadaceae/Apocynaceae
♦ Caribbean milkweed
♦ Weed
♦ 14, 87, 88
♦ herbal.

Asclepias notha W.D.Stevens
Asclepiadaceae/Apocynaceae
♦ Weed
♦ 199

**Asclepias oenotheroides Cham. &
Schlecht.**
Asclepiadaceae/Apocynaceae
♦ zizotes milkweed
♦ Weed
♦ 157
♦ pH, herbal.

Asclepias ovalifolia Decne.
Asclepiadaceae/Apocynaceae
♦ dwarf milkweed, ovalleaf milkweed
♦ Weed
♦ 87, 88, 218
♦ pH, promoted, herbal.

Asclepias physocarpa (E.Mey.) Schlecht.
Asclepiadaceae/Apocynaceae
= *Gomphocarpus physocarpus* E.Mey.
♦ bindweed, milkweed, wild cotton,
balloonplant, balloon cottonbush
♦ Weed, Naturalised, Native Weed,
Environmental Weed
♦ 39, 87, 88, 101, 121, 158, 161, 246
♦ H, cultivated, herbal, toxic. Origin:
southern Africa.

Asclepias pumila (A.Gray) Vail
Asclepiadaceae/Apocynaceae
♦ plains milkweed, low milkweed
♦ Weed
♦ 39, 161
♦ pH, promoted, herbal, toxic.

Asclepias purpurascens L.
Asclepiadaceae/Apocynaceae
♦ purple milkweed
♦ Weed
♦ 161
♦ pH, promoted, herbal, toxic.

Asclepias rostrata N.E.Br.
Asclepiadaceae/Apocynaceae
♦ Weed, Quarantine Weed
♦ 76, 88, 220

Asclepias rotundifolia Mill.
Asclepiadaceae/Apocynaceae
= *Gomphocarpus cancellatus* (Burm.f.)
Bruyns
♦ Weed, Naturalised
♦ 98, 203

Asclepias speciosa Torr.
Asclepiadaceae/Apocynaceae
Asclepias douglasii Hook.
♦ showy milkweed
♦ Weed, Noxious Weed
♦ 23, 39, 49, 87, 88, 136, 161, 212, 218,
299
♦ pH, arid, cultivated, herbal, toxic.
Origin: western North America.

Asclepias stenophylla Gray
Asclepiadaceae/Apocynaceae
♦ narrow leaved milkweed, slimleaf
milkweed
♦ Weed
♦ 161
♦ herbal, toxic.

Asclepias subverticillata (Gray) Vail
Asclepiadaceae/Apocynaceae
Asclepias galioides auct. Amer. (see)
♦ western whorled milkweed,
horsetail milkweed
♦ Weed, Native Weed
♦ 39, 87, 88, 161, 212, 218, 264
♦ pH, arid, herbal, toxic. Origin:
western United States, Mexico.

Asclepias sullivantii Engelm. ex A.Gray
Asclepiadaceae/Apocynaceae
♦ smooth milkweed
♦ Weed
♦ 161
♦ toxic.

Asclepias syriaca L.
Asclepiadaceae/Apocynaceae
Asclepias cornutii Decne.
♦ common milkweed, wild cotton,

Virginia silk, silk weed, cottonweed
♦ Weed, Quarantine Weed, Noxious Weed, Naturalised, Native Weed, Introduced
♦ 8, 23, 24, 39, 52, 76, 86, 87, 88, 161, 174, 203, 207, 210, 211, 218, 220, 228, 229, 243, 247, 266, 272, 294
♦ pH, arid, cultivated, herbal, toxic. Origin: North America.

Asclepias tuberosa L.
Asclepiadaceae/Apocynaceae
♦ butterfly milkweed, pleurisy root
♦ Weed
♦ 8, 23, 39, 87, 88, 161, 218, 247
♦ pH, arid, cultivated, herbal, toxic.

Asclepias verticillata L.
Asclepiadaceae/Apocynaceae
Asclepias galioides H.B.K.
♦ eastern whorled milkweed
♦ Weed, Native Weed
♦ 39, 87, 88, 161, 174, 210, 218
♦ herbal, toxic. Origin: North America.

Asclepias viridiflora Raf.
Asclepiadaceae/Apocynaceae
♦ green milkweed, green comet milkweed
♦ Weed
♦ 161
♦ pH, arid, promoted, herbal, toxic.

Asclepias viridis Walter
Asclepiadaceae/Apocynaceae
♦ Ozark milkweed, green antelopehorn
♦ Weed
♦ 161
♦ herbal, toxic.

Asimina triloba (L.) Dunal
Annonaceae
♦ pawpaw, papaw
♦ Weed
♦ 8, 39, 161, 247
♦ S, cultivated, herbal, toxic.

Aspalathus L. spp.
Fabaceae/Papilionaceae
♦ prickly pea bushes, South African gorse
♦ Native Weed
♦ 121
♦ pS. Origin: southern Africa.

Aspalthium bituminosum (L.) Kuntze
Fabaceae/Papilionaceae
= *Bituminaria bituminosa* (L.) Stirt.
♦ Weed, Quarantine Weed
♦ 76, 88, 220
♦ cultivated.

Asparagopsis armata Harv.
Liliaceae/Asparagaceae
Falkenbergia rufolanosa (Harv.) Schmitz
♦ harpoon weed, algae
♦ Weed
♦ 282, 288
♦ algae.

Asparagus L. spp.
Liliaceae/Asparagaceae
♦ asparagus
♦ Weed, Naturalised
♦ 198, 272
♦ herbal.

Asparagus acutifolius L.
Liliaceae/Asparagaceae
♦ lesser asparagus
♦ Weed
♦ 70
♦ pH, arid, cultivated, herbal.

Asparagus aethiopicus L.
Liliaceae/Asparagaceae
= *Asparagus densiflorus* (Kunth) Jessop
♦ Naturalised, Environmental Weed
♦ 134, 296
♦ cultivated.

Asparagus africanus Lam.
Liliaceae/Asparagaceae
Asparagus asiaticus auct., *Protasparagus africanus* (Lam.) Oberm. (see)
♦ asparagus fern, lukungwisa, climbing asparagus
♦ Weed, Naturalised, Garden Escape, Environmental Weed
♦ 54, 86, 88, 155, 290
♦ cultivated, herbal. Origin: South Africa.

Asparagus aphyllus L.
Liliaceae/Asparagaceae
♦ greater asparagus
♦ Weed
♦ 70, 221
♦ pH, cultivated.

Asparagus asparagoides (L.) W.Wight
Liliaceae/Asparagaceae
Myrsiphyllum asparagoides (L.) Willd. (see), *Asparagus medeoloides* (L.f.) Thunb. (see), *Dracaena medeoloides* L.f., *Medeola asparagoides* L.
♦ bridal creeper, bridal veil creeper, baby smilax, African asparagus fern, smilax
♦ Weed, Quarantine Weed, Noxious Weed, Naturalised, Garden Escape, Environmental Weed
♦ 9, 15, 72, 76, 86, 101, 165, 198, 225, 246, 269, 280, 289, 296
♦ pC, cultivated, herbal. Origin: South Africa.

Asparagus davuricus Fisch. & Link
Liliaceae/Asparagaceae
♦ Dahurian asparagus
♦ Weed
♦ 297

Asparagus declinatus L.
Liliaceae/Asparagaceae
Myrsiphyllum declinatum (L.) Oberm. (see), *Asparagus crispus* Lam.
♦ pale berry asparagus fern, asparagus fern, bridal vale
♦ Naturalised, Garden Escape, Environmental Weed
♦ 86, 198, 251, 289
♦ cultivated. Origin: South Africa.

Asparagus densiflorus (Kunth) Jessop
Liliaceae/Asparagaceae
Asparagopsis densiflora Kunth, *Asparagus aethiopicus* L. (see), *Asparagus myriocladus* Baker, *Asparagus sarmentosus* L., *Asparagus sprengeri* Regel, *Protasparagus aethiopicus* (L.) Oberm. (see), *Protasparagus densiflorus* (Kunth) Oberm. (see)

♦ emerald feather, asparagus fern, sprengeri fern, bushy asparagus, protasparagus, Sprenger's asparagus fern
♦ Weed, Sleeper Weed, Noxious Weed, Naturalised, Garden Escape, Environmental Weed, Cultivation Escape
♦ 3, 80, 86, 88, 101, 112, 151, 155, 161, 179, 198, 225, 233, 246, 247, 269, 289
♦ pC, cultivated, toxic. Origin: South Africa.

Asparagus filicinus Buch.-Ham. ex D.Don
Liliaceae/Asparagaceae
♦ fern asparagus
♦ Naturalised
♦ 198
♦ pH, cultivated, herbal.

Asparagus lucidus Lindl.
Liliaceae/Asparagaceae
♦ Chinese asparagus, shiny asparagus
♦ Weed
♦ 87, 88
♦ pH, promoted, herbal.

Asparagus medeoloides (L.f.) Thunb.
Liliaceae/Asparagaceae
= *Asparagus asparagoides* (L.) W.Wight
♦ Naturalised
♦ 86

Asparagus officinalis L.
Liliaceae/Asparagaceae
♦ asparagus, asparágus lekársky
♦ Weed, Naturalised, Garden Escape, Environmental Weed
♦ 7, 15, 39, 42, 72, 80, 86, 87, 88, 98, 101, 116, 161, 176, 179, 180, 195, 198, 203, 243, 247, 253, 269, 272, 280, 290
♦ pH, cultivated, herbal, toxic. Origin: Eurasia.

Asparagus officinalis L. ssp. *officinalis*
Liliaceae/Asparagaceae
♦ garden asparagus
♦ Weed, Cultivation Escape, Casual Alien
♦ 34, 40
♦ pH, cultivated.

Asparagus plumosus Baker
Liliaceae/Asparagaceae
Asparagus comorensis hort., *Protasparagus plumosus* (Baker) Oberm. (see)
♦ climbing asparagus fern
♦ Weed, Noxious Weed, Naturalised, Garden Escape, Environmental Weed, Cultivation Escape
♦ 86, 134, 198, 233, 262, 290, 296
♦ pC, cultivated, herbal.

Asparagus scandens Thunb.
Liliaceae/Asparagaceae
Myrsiphyllum scandens (Thunb.) Oberm. (see)
♦ asparagus fern, climbing asparagus
♦ Weed, Quarantine Weed, Naturalised, Garden Escape, Environmental Weed
♦ 15, 86, 165, 198, 225, 246, 280, 289
♦ pH, cultivated. Origin: South Africa.

Asparagus setaceus (Kunth) Jessop
Liliaceae/Asparagaceae
Asparagopsis setacea Kunth, *Asparagus graminifolius* L., *Protasparagus setaceus* (Kunth) Oberm. (see)
♦ common asparagus fern, asparagus fern, abeto, ala de pájaro, helecho plumosa
♦ Weed, Naturalised, Garden Escape
♦ 15, 86, 101, 161, 179, 261, 280
♦ pC, cultivated, herbal, toxic. Origin: southern Africa.

Asparagus stipularis Forssk.
Liliaceae/Asparagaceae
♦ Weed
♦ 221
♦ pH, cultivated, herbal.

Asparagus tenuifolius Lam.
Liliaceae/Asparagaceae
♦ narrow leaved asparagus
♦ Weed
♦ 70
♦ pH, cultivated, herbal.

Asparagus virgatus Baker
Liliaceae/Asparagaceae
Protasparagus virgatus (Bak.) Oberm. (see)
♦ asparagus fern
♦ Naturalised, Cultivation Escape
♦ 86, 252
♦ cultivated. Origin: South Africa.

Asperugo procumbens L.
Boraginaceae
♦ madwort, catchweed, German madwort
♦ Weed, Naturalised, Introduced, Casual Alien
♦ 34, 39, 40, 70, 87, 88, 94, 101, 159, 161, 212, 221, 272, 287, 300
♦ aH, arid, cultivated, herbal, toxic.

Asperula L. spp.
Rubiaceae
♦ woodruff, witloof
♦ Weed
♦ 39, 272
♦ herbal, toxic.

Asperula arvensis L.
Rubiaceae
Asperula ciliata Moench
♦ blue woodruff, peltomaratti
♦ Weed, Naturalised, Casual Alien
♦ 40, 42, 70, 86, 87, 88, 94, 98, 101, 115, 203, 243, 253, 272
♦ aH, cultivated, herbal. Origin: Europe.

Asperula cynanchica L.
Rubiaceae
♦ squinancywort, marinka psia
♦ Weed
♦ 39, 70, 272
♦ pH, cultivated, herbal, toxic.

Asperula humifusa (Bieb.) Bess.
Rubiaceae
= *Galium humifusum* Bieb.
♦ Weed
♦ 87, 88

Asperula odorata L.
Rubiaceae

= *Galium odoratum* (L.) Scop.
♦ sweet woodruff, kurumabasou
♦ Weed
♦ 4, 39, 88
♦ cultivated, herbal, toxic.

Asperula orientalis Boiss. & Hohen.
Rubiaceae
Asperula azurea Jaub. & Spach
♦ tarhamaratti, oriental asperula
♦ Naturalised, Casual Alien
♦ 42, 101
♦ aH, cultivated, herbal.

Asperula tinctoria L.
Rubiaceae
= *Galium tinctorium* (L.) Scop.
♦ dyer's woodruff, värimaratti
♦ Weed
♦ 272
♦ pH, cultivated, herbal.

Asphodelus aestivus Brot.
Liliaceae/Asphodelaceae
Asphodelus microcarpus Viv. (see)
♦ tall asphodel, common asphodel
♦ Weed
♦ 126, 272
♦ pH, arid, cultivated.

Asphodelus albus Willd.
Liliaceae/Asphodelaceae
Asphodelus macrocarpus Parl.
♦ rimmed lichen, common asphodel
♦ Weed
♦ 87, 88
♦ pH, cultivated, herbal, toxic.

Asphodelus fistulosus L.
Liliaceae/Asphodelaceae
Asphodelus tenuifolius Cav. (see)
♦ onion weed, asphodel, hollow stemmed asphodel, wild onion
♦ Weed, Noxious Weed, Naturalised, Introduced, Garden Escape, Environmental Weed, Casual Alien
♦ 7, 34, 40, 67, 72, 86, 87, 88, 93, 98, 101, 126, 140, 147, 165, 176, 198, 199, 203, 221, 229, 269, 280, 290, 296, 300
♦ pH, cultivated, herbal. Origin: Mediterranean.

Asphodelus microcarpus Viv.
Liliaceae/Asphodelaceae
= *Asphodelus aestivus* Brot.
♦ Weed
♦ 87, 88, 221
♦ herbal.

Asphodelus ramosus L.
Liliaceae/Asphodelaceae
Asphodelus lusitanicus Henriq.
♦ branching asphodel
♦ Weed, Quarantine Weed, Naturalised
♦ 70, 76, 86, 88, 203, 220
♦ cultivated, herbal.

Asphodelus tenuifolius Cav.
Liliaceae/Asphodelaceae
= *Asphodelus fistulosus* L.
♦ asphodelus
♦ Weed, Quarantine Weed
♦ 76, 87, 88, 203, 217, 220, 243
♦ herbal.

Aspidistra elatior Blume
Liliaceae/Convallariaceae
♦ cast iron plant, aspidistra
♦ Naturalised
♦ 287
♦ pH, cultivated, herbal. Origin: China.

Aspidosperma polyneuron Müll.Arg.
Apocynaceae
Aspidosperma peroba Allemão
♦ Introduced
♦ 228
♦ T, arid.

Aspidosperma quebracho-blanco Schlecht.
Apocynaceae
♦ quebracho blanco, quebracho
♦ Weed
♦ 39, 295
♦ cultivated, herbal, toxic.

Aspilia africana (Pers.) Adams
Asteraceae
♦ Weed, Quarantine Weed
♦ 76, 87, 88, 203, 220
♦ herbal.

Aspilia bupthalmiflora Griseb.
Asteraceae
♦ Weed, Quarantine Weed
♦ 76, 87, 88, 203, 220

Aspilia helianthoides (Schumach. & Thonn.) Oliv. & Hiern
Asteraceae
Coronocarpus helianthoides Schumach. & Thonn.
♦ Weed
♦ 87, 88

Aspilia montevidensis (Spreng.) Kuntze
Asteraceae
Verbesina montevidensis Spreng.
♦ mal me quer
♦ Weed
♦ 255
♦ Origin: Brazil.

Asplenium adiantum-nigrum L.
Aspleniaceae
♦ black spleenwort, doradille noire
♦ Weed
♦ 70
♦ pH, cultivated, herbal.

Asplenium flabellulatum Kunze
Aspleniaceae
♦ Latin American spleenwort
♦ Naturalised
♦ 101

Asplenium pellucidum Lam. var. ponapensis Hosk.
Aspleniaceae
♦ Introduced
♦ 230
♦ H.

Asplenium trichomanes L.
Aspleniaceae
♦ maidenhair spleenwort, zanokcica skalna
♦ Weed
♦ 87, 88, 272
♦ pH, cultivated, herbal.

Aster ageratoides **Turcz.**
Asteraceae
= *Aster trinervius* Desf. ssp. *ageratoides*
(Turcz.) Grierson (NoR) [see *Aster*
trinervius Desf.]
♦ Weed
♦ 88, 191, 204, 297
♦ Origin: China, Japan, Korea, eastern
Russia.

Aster ageratoides **Turcz. ssp.** *ovatus*
Kitam.
Asteraceae
♦ Weed
♦ 286

Aster amellus **L.**
Asteraceae
♦ summer Michaelmas daisy
♦ Weed, Naturalised, Casual Alien
♦ 86, 272, 280
♦ pH, arid, cultivated, herbal. Origin:
Eurasia.

Aster cordifolius **Michx.**
Asteraceae
= *Symphyotrichum cordifolium* (L.)
Nesom (NoR)
♦ hertta asteri
♦ Casual Alien
♦ 42
♦ pH, cultivated, herbal.

Aster dumosus **L.**
Asteraceae
= *Symphyotrichum dumosum* (L.) Nesom
var. *dumosum* (NoR)
♦ bushy aster, oktoberaster
reunusasteri
♦ Weed
♦ 161, 249
♦ cultivated, herbal.

Aster ericoides **L.**
Asteraceae
= *Symphyotrichum ericoides* (L.) Nesom
var. *ericoides* (NoR)
♦ heath aster
♦ Weed
♦ 39, 87, 88, 161, 210, 218
♦ cultivated, herbal, toxic.

Aster exilis **Ell., nomen dubium**
Asteraceae
= *Symphyotrichum divaricatum* (Nutt.)
Nesom (NoR)
♦ slender aster
♦ Weed
♦ 14, 87, 88, 161, 218
♦ herbal.

Aster falcatus **Lindl.**
Asteraceae
= *Symphyotrichum falcatum* (Lindl.)
Nesom var. *falcatum* (NoR)
♦ white prairie aster
♦ Weed
♦ 161
♦ herbal.

Aster grisebachii **Britt.**
Asteraceae
♦ Weed
♦ 14

Aster lanceolatus **Willd.**
Asteraceae
= *Symphyotrichum lanceolatum* (Willd.)
Nesom ssp. *lanceolatum* var. *lanceolatum*
(NoR) [see *Aster simplex* Willd.]
♦ narrow leaved Michaelmas daisy,
säiläasteri, astra kopijovitolistá
♦ Weed, Naturalised, Casual Alien
♦ 15, 40, 42, 272, 280
♦ cultivated, herbal.

Aster linariifolius **L.**
Asteraceae
= *Ionactis linariifolius* (L.) Greene (NoR)
♦ savoryleaf aster
♦ Weed
♦ 87, 88, 218
♦ herbal.

Aster novae-angliae **L.**
Asteraceae
= *Symphyotrichum novae-angliae* (L.)
Nesom (NoR)
♦ New England aster, astra
novoanglická, luktaster tuoksuasteri
♦ Weed, Naturalised
♦ 23, 87, 88, 218, 280, 287
♦ cultivated, herbal.

Aster novi-belgii **L.**
Asteraceae
♦ New York aster, Michaelmas
daisy, New Belgium aster, confused
Michaelmas daisy, syysasteri
♦ Weed, Naturalised, Cultivation
Escape
♦ 40, 42, 86, 87, 88, 98, 132, 198, 203,
218, 252, 269, 272, 287
♦ pH, cultivated, herbal. Origin:
eastern North America.

Aster novi-belgii **L. hybrids**
Asteraceae
♦ New York aster, New Belgium
aster, Michaelmas daisy, common
Michaelmas daisy
♦ Naturalised
♦ 280

Aster occidentalis **(Nutt.) Torr. & Gray**
Asteraceae
= *Symphyotrichum spathulatum* (Lindl.)
Nesom var. *spathulatum* (NoR)
♦ western aster
♦ Weed
♦ 39, 87, 88, 136, 218
♦ pH, cultivated, herbal, toxic. Origin:
western North America.

Aster paniculatus **Lam.**
Asteraceae
♦ vihta asteri
♦ Casual Alien
♦ 42

Aster parryi **A.Gray**
Asteraceae
= *Xylorhiza glabriuscula* Nutt. var.
glabriuscula (NoR)
♦ woody aster
♦ Weed
♦ 87, 88, 218

Aster pilosus **Willd.**
Asteraceae

= *Symphyotrichum pilosum* (Willd.)
Nesom var. *pilosum* (NoR)
♦ white heath aster, awl aster, subulate
bracted aster
♦ Weed, Naturalised
♦ 87, 88, 161, 207, 210, 211, 218, 243,
286, 287
♦ pH, herbal. Origin: North America.

Aster puniceus **L.**
Asteraceae
= *Symphyotrichum puniceum* (L.) Á. &
D.Löve var. *puniceum* (NoR)
♦ purplestem aster
♦ Weed
♦ 87, 88, 218
♦ cultivated, herbal.

Aster × *salignus* **Willd.**
Asteraceae
= *Aster lanceolatus* Willd. × *Aster novi-
belgii* L.
♦ common Michaelmas daisy,
pajuasteri
♦ Naturalised, Cultivation Escape
♦ 40, 42
♦ cultivated, herbal.

Aster scaber **Thunb.**
Asteraceae
♦ aster, shirayamagiku
♦ Weed
♦ 87, 88, 204, 286
♦ pH, promoted. Origin: east Asia.

Aster sedifolius **L.**
Asteraceae
♦ pikkuasteri
♦ Cultivation Escape
♦ 42
♦ cultivated, herbal.

Aster sibiricus **L.**
Asteraceae
= *Eurybia sibirica* (L.) Nesom (NoR)
♦ siperianasteri, sibirisk aster
♦ Cultivation Escape
♦ 42
♦ cultivated, herbal.

Aster simplex **Willd.**
Asteraceae
= *Symphyotrichum lanceolatum* (Willd.)
Nesom ssp. *lanceolatum* var. *lanceolatum*
(NoR) [see *Aster lanceolatus* Willd.]
♦ white field aster
♦ Weed
♦ 87, 88, 218
♦ herbal.

Aster spectabilis **Aiton**
Asteraceae
= *Eurybia spectabilis* (Ait.) Nesom
(NoR)
♦ showy aster
♦ Weed
♦ 87, 88, 218
♦ herbal.

Aster spinosus **Benth.**
Asteraceae
= *Chloracantha spinosa* (Benth.) Nesom
♦ spiny aster
♦ Weed
♦ 87, 88, 161, 218
♦ arid, herbal.

Aster squamatus (Spreng.) Hieron.
Asteraceae
= *Symphyotrichum squamatum* (Spreng.)
Nesom (NoR) [see *Aster subulatus*
Michx. var. *sandwicensis* (Gray)
A.G.Jones]
♦ swamp aster
♦ Weed
♦ 70, 87, 88, 121, 185, 237, 253, 255, 295
♦ aH, herbal, toxic. Origin: Americas.

Aster subulatus Michx.
Asteraceae
Symphyotrichum subulatum (Michx.)
Nesom
♦ slender aster, aster, slim aster, aster
weed, bushy starwort, wild aster
♦ Weed, Naturalised, Garden Escape,
Environmental Weed
♦ 7, 9, 15, 51, 72, 86, 87, 88, 93, 98, 134,
152, 165, 176, 180, 198, 203, 205, 243,
261, 269, 280, 286, 287, 297
♦ a/bH, cultivated, herbal. Origin:
North America.

Aster subulatus Michx. var. *elongatus*
Bosserdet
Asteraceae
= *Symphyotrichum bahamense* (Britt.)
Nesom (NoR)
♦ Bahaman aster
♦ Weed, Naturalised
♦ 286, 287

Aster subulatus Michx. var. *sandwicensis*
(Gray) A.G.Jones
Asteraceae
= *Symphyotrichum squamatum* (Spreng.)
Nesom (NoR) [see *Aster squamatus*
(Spreng.) Hieron.]
♦ annual saltmarsh aster
♦ Weed, Naturalised
♦ 286, 287

Aster tataricus L.f.
Asteraceae
♦ Tatarian aster
♦ Naturalised
♦ 101
♦ pH, promoted, herbal.

Aster trinervius Desf.
Asteraceae
= *Aster trinervius* ssp. *ageratoides*
(Turcz.) Grierson (NoR) [see *Aster*
ageratoides Turcz.]
♦ Weed
♦ 87, 88
♦ herbal.

Aster tripolium L.
Asteraceae
= *Tripolium pannonicum* (Jacq.)
Dobrocz.
♦ sea aster, meriasteri, uragiku
♦ Weed
♦ 23, 87, 88
♦ a/pH, cultivated, herbal. Origin:
Europe.

Aster × versicolor Willd.
Asteraceae
= *Aster laevis* L. × *Aster novi-belgii* L.
♦ late Michaelmas daisy
♦ Naturalised

♦ 40

Aster yomena (Kitamura) Honda
Asteraceae
♦ Weed
♦ 87, 88, 191, 204
♦ pH, promoted.

Asteriscium chilense Cham. & Schlecht.
Apiaceae
♦ Weed
♦ 87, 88
♦ arid.

Asteriscus Mill. spp.
Asteraceae
♦ Weed
♦ 272

Asteriscus aquaticus (L.) Less.
Asteraceae
♦ Weed
♦ 70

Asteriscus graveolens (Forssk.) Less.
Asteraceae
Nauplius graveolens (Forssk.) Wiklund
♦ Weed
♦ 221
♦ arid.

Asteriscus maritimus (L.) Less.
Asteraceae
Odontospernum maritimum (L.) Sch.Bip.
♦ Weed
♦ 70
♦ cultivated.

Asteriscus spinosus Sch.Bip.
Asteraceae
= *Pallenis spinosa* (L.) Cass.
♦ Weed, Naturalised
♦ 98, 203
♦ cultivated.

Asterolinon linum-stellatum (L.) Duby
Primulaceae
♦ asterolinon
♦ Weed, Naturalised
♦ 86, 88, 94, 98, 198, 203
♦ cultivated. Origin: Mediterranean,
Middle East.

Asteromoea indica (L.) Bl.
Asteraceae
Aster indicus L.
♦ Weed
♦ 87, 88

Asteropterus leyseroides (Desf.) Rothm.
Asteraceae
♦ Weed
♦ 221

Asthenatherum forskalii (Vahl) Nevski
Poaceae
= *Centropodia forskalii* (Vahl) Cope
♦ Weed
♦ 221
♦ G. Origin: North Africa, Middle
East.

Astilbe japonica (Morr. & Decne.) Gray
Saxifragaceae
♦ florist's spiraea
♦ Naturalised
♦ 101
♦ cultivated, herbal.

Astoma seselifolium DC.
Apiaceae
Astomaea seselifolium DC.
♦ Weed
♦ 88

Astragalus L. spp.
Fabaceae/Papilionaceae
♦ milkvetch, locoweed
♦ Weed
♦ 154, 221, 272
♦ toxic.

Astragalus accidens S.Wats.
Fabaceae/Papilionaceae
♦ Rogue River milkvetch, hendersonii
milkvetch
♦ Weed
♦ 161
♦ toxic.

Astragalus adsurgens Pall.
Fabaceae/Papilionaceae
♦ standing milkvetch
♦ Weed
♦ 275, 297
♦ pH, promoted, herbal.

Astragalus afghanus Boiss.
Fabaceae/Papilionaceae
♦ Weed
♦ 243

Astragalus agnicidus Barneby
Fabaceae/Papilionaceae
♦ lambkill milkvetch, Humboldt
County milkvetch
♦ Weed
♦ 161
♦ pH, cultivated, toxic.

Astragalus agrestis Douglas ex G.Don
Fabaceae/Papilionaceae
♦ field milkvetch, purple milkvetch
♦ Weed
♦ 161
♦ pH, herbal, toxic.

Astragalus aleppicus Boiss.
Fabaceae/Papilionaceae
♦ Weed
♦ 87, 88

Astragalus allochrous A.Gray
Fabaceae/Papilionaceae
♦ halfmoon loco, halfmoon vetch,
halfmoon milkvetch
♦ Weed
♦ 87, 88, 161, 218
♦ aH, herbal, toxic.

Astragalus andersonii A.Gray
Fabaceae/Papilionaceae
♦ Anderson's milkvetch
♦ Weed
♦ 161
♦ pH, toxic.

Astragalus annularis Forssk.
Fabaceae/Papilionaceae
♦ Weed
♦ 221
♦ Origin: North Africa.

Astragalus arenarius L.
Fabaceae/Papilionaceae
♦ hietakurjenherne
♦ Casual Alien
♦ 42

Astragalus asterias Steven ex Ledeb.
Fabaceae/Papilionaceae
♦ Weed
♦ 221
♦ Origin: North Africa.

Astragalus asymmetricus Sheld.
Fabaceae/Papilionaceae
♦ San Joaquin locoweed, horse milk,
loco, locoweed, woolly loco, woolly
leaved loco
♦ Weed
♦ 161, 180
♦ pH, toxic.

Astragalus austriacus Jacq.
Fabaceae/Papilionaceae
♦ Austrian milkvetch
♦ Weed
♦ 272

Astragalus balansae Boiss.
Fabaceae/Papilionaceae
♦ Weed
♦ 87, 88

Astragalus bergii Hieron.
Fabaceae/Papilionaceae
♦ Weed
♦ 87, 88

Astragalus bicristatus A.Gray
Fabaceae/Papilionaceae
♦ two crested milkvetch, two keeled
milkvetch
♦ Weed
♦ 161
♦ pH, toxic.

Astragalus bisulcatus (Hook.) A.Gray
Fabaceae/Papilionaceae
♦ two grooved milkvetch
♦ Weed, Noxious Weed
♦ 36, 39, 87, 88, 161, 212, 218
♦ herbal, toxic.

**Astragalus bisulcatus (Hook.) Gray var.
haydenianus (Gray) Barneby**
Fabaceae/Papilionaceae
*Astragalus haydenianus Gray ex Brand.,
Tragacantha haydenianthus (Gray)
Kuntze, Diholcos haydenianus (Gray)
Rydb.*
♦ two grooved milkvetch, Hayden's
milkvetch
♦ Native Weed
♦ 264
♦ pH, toxic. Origin: USA.

Astragalus boeticus L.
Fabaceae/Papilionaceae
♦ Swedish coffee
♦ Weed, Casual Alien
♦ 40, 221, 243
♦ aH, arid, promoted.

Astragalus bolanderi A.Gray
Fabaceae/Papilionaceae
♦ Bolander's milkvetch
♦ Weed
♦ 161
♦ pH, herbal, toxic.

Astragalus bombycinus Boiss.
Fabaceae/Papilionaceae
♦ Weed
♦ 221

**Astragalus californicus (A.Gray)
E.Greene**
Fabaceae/Papilionaceae
♦ Klamath milkvetch, Klamath Basin
milkvetch
♦ Weed
♦ 161
♦ pH, toxic.

Astragalus calycosus Torr. ex S.Wats.
Fabaceae/Papilionaceae
♦ Torrey's milkvetch
♦ Weed
♦ 34, 82, 161
♦ cultivated, herbal, toxic.

Astragalus canadensis L.
Fabaceae/Papilionaceae
♦ Canadian milkvetch, short toothed
brevidens milkvetch
♦ Weed
♦ 161
♦ pH, cultivated, herbal, toxic. Origin:
central and eastern North America.

Astragalus chinensis L.f.
Fabaceae/Papilionaceae
♦ Chinese milkvetch, sha yuan
♦ Weed, Naturalised
♦ 101, 297
♦ pH, promoted.

Astragalus cicer L.
Fabaceae/Papilionaceae
*Astragalus mucronatus DC., Astragalus
microphyllus L., Astragalus cicer L. var.
microphyllus (L.) Asch. & Graebn.*
♦ pulleakurjenherne, cicer milkvetch,
chickpea milkvetch, wild lentil
♦ Weed, Naturalised, Casual Alien
♦ 42, 87, 88, 94, 101, 272
♦ cultivated, herbal.

Astragalus clevelandii E.Greene
Fabaceae/Papilionaceae
♦ Cleveland's milkvetch
♦ Weed
♦ 161
♦ pH, toxic.

**Astragalus collinus (Hook.) Douglas ex
G.Don**
Fabaceae/Papilionaceae
♦ hillside milkvetch
♦ Weed
♦ 87, 88

Astragalus complanatus R.Br.
Fabaceae/Papilionaceae
♦ Weed
♦ 275, 297
♦ pH, promoted, herbal.

Astragalus contortuplicatus L.
Fabaceae/Papilionaceae
♦ Hungarian milkvetch
♦ Naturalised
♦ 101

Astragalus convallarius Greene
Fabaceae/Papilionaceae
♦ lesser rushy milkvetch
♦ Weed
♦ 39, 161
♦ herbal, toxic.

Astragalus corrugatus Bertol.
Fabaceae/Papilionaceae

♦ Weed
♦ 243

**Astragalus curvicarpus (A.Heller)
J.F.Macbr.**
Fabaceae/Papilionaceae
♦ sickle milkvetch, curvepod
milkvetch
♦ Weed
♦ 161
♦ herbal, toxic.

Astragalus dahuricus (Pall.) DC.
Fabaceae/Papilionaceae
♦ milkvetch
♦ Weed
♦ 88, 114

Astragalus decumbens (Nutt.) A.Gray
Fabaceae/Papilionaceae
= *Astragalus miser* Dougl. var.
decumbens (Nutt. ex Torr. & Gray)
Cronquist (NoR)
♦ timber milkvetch
♦ Noxious Weed
♦ 36
♦ herbal.

Astragalus deinacanthus Boiss.
Fabaceae/Papilionaceae
♦ Weed
♦ 87, 88

Astragalus diphysus A.Gray
Fabaceae/Papilionaceae
♦ blue loco
♦ Weed
♦ 39, 87, 88, 161, 218
♦ pH, promoted, toxic.

Astragalus distinens G.Macloskie
Fabaceae/Papilionaceae
Astragalus bergii auct. non Hieron.
♦ garbancillo
♦ Weed
♦ 237, 295

Astragalus douglasii (Torr. & Gray) Gray
Fabaceae/Papilionaceae
♦ Douglas loco, Jacumba milkvetch
♦ Weed
♦ 161
♦ pH, cultivated, toxic.

Astragalus earlei Greene ex Rydb.
Fabaceae/Papilionaceae
♦ bigbend loco
♦ Weed
♦ 39, 87, 88, 161, 218
♦ toxic.

Astragalus echinops Aucher ex Boiss.
Fabaceae/Papilionaceae
♦ Weed
♦ 87, 88

Astragalus echinus DC.
Fabaceae/Papilionaceae
♦ Weed
♦ 221
♦ pH, arid, promoted.

Astragalus emoryanus (Rydb.) Cory
Fabaceae/Papilionaceae
♦ Emory's loco, Emory's milkvetch
♦ Weed
♦ 39, 161
♦ herbal, toxic.

Astragalus eremophilus Boiss.
Fabaceae/Papilionaceae
♦ Weed
♦ 221

Astragalus falcatus Lam.
Fabaceae/Papilionaceae
♦ Russian milkvetch, Russian sickle, milkvetch
♦ Naturalised
♦ 101
♦ cultivated, herbal, toxic. Origin: temperate Asia.

Astragalus fatmensis Choiv.
Fabaceae/Papilionaceae
Asparagus arabicus Ehrenb. ex Bunge
♦ Weed
♦ 88

Astragalus filipes Torr. ex A.Gray
Fabaceae/Papilionaceae
♦ basalt milkvetch, threadstalk milkvetch
♦ Weed
♦ 161
♦ pH, herbal, toxic.

Astragalus flavus Nutt. ex Torr. & Gray
Fabaceae/Papilionaceae
Cnemidophacos flavus (Nutt.) Rydb., *Tragacantha flaviflora* Kuntze, *Astragalus confertifolius* var. *flaviflorus* (Kuntze) Jones
♦ yellow milkvetch
♦ Native Weed
♦ 264
♦ pH, herbal, toxic. Origin: USA.

Astragalus fresenii Decne.
Fabaceae/Papilionaceae
♦ Weed
♦ 221

Astragalus garbancillo Cav.
Fabaceae/Papilionaceae
Astragalus unifultus L'Hér.
♦ garbancillo
♦ Weed, Quarantine Weed
♦ 39, 76, 87, 88, 203, 220, 295
♦ pH, promoted, herbal, toxic.

Astragalus gibbsii Kellogg
Fabaceae/Papilionaceae
♦ Gibbs' milkvetch
♦ Weed
♦ 161
♦ pH, herbal, toxic.

Astragalus glycyphyllos L.
Fabaceae/Papilionaceae
♦ imeläkurjenherne, wild liquorice, licorice milkvetch, common milkvetch
♦ Weed, Naturalised, Casual Alien
♦ 39, 40, 42, 70, 101, 272
♦ pH, cultivated, herbal, toxic.

Astragalus gyzensis Delile
Fabaceae/Papilionaceae
♦ Weed
♦ 221

Astragalus hamosus L.
Fabaceae/Papilionaceae
♦ milkvetch
♦ Weed, Naturalised
♦ 39, 86, 87, 88, 98, 101, 203, 221

♦ aH, arid, promoted, herbal, toxic. Origin: Eurasia.

Astragalus hornii A.Gray
Fabaceae/Papilionaceae
♦ sheep loco, Horn's milkvetch
♦ Weed
♦ 39, 161
♦ toxic.

Astragalus intermedius Boiss.
Fabaceae/Papilionaceae
♦ Quarantine Weed
♦ 220

Astragalus kahiricus DC.
Fabaceae/Papilionaceae
♦ Weed
♦ 221
♦ arid.

Astragalus kirkukensis Eig.
Fabaceae/Papilionaceae
= *Astragalus suberosus* Banks & Sol. ssp. *ancyleus* (Boiss.) V.Matthews (NoR)
♦ Introduced
♦ 100
♦ Origin: Middle East.

Astragalus lambertii (Pursh) Spreng.
Fabaceae/Papilionaceae
= *Oxytropis lambertii* Pursh
♦ Quarantine Weed
♦ 220

Astragalus layneae E.Greene
Fabaceae/Papilionaceae
♦ Layne milkvetch, widow's milkvetch
♦ Weed
♦ 161
♦ pH, herbal, toxic.

Astragalus leansanicus Ulbr.
Fabaceae/Papilionaceae
♦ Weed
♦ 275
♦ pH.

Astragalus lentiginosus Douglas ex Hook.
Fabaceae/Papilionaceae
♦ spotted loco, freckled milkvetch, specklepod milkvetch
♦ Weed
♦ 39, 87, 88, 161, 218
♦ a/pH, herbal, toxic.

Astragalus lentiginosus Douglas ex Hook. var. diphysus (Gray) M.E.Jones
Fabaceae/Papilionaceae
♦ freckled milkvetch, specklepod milkvetch
♦ Native Weed
♦ 264
♦ pH, toxic. Origin: USA.

Astragalus lusitanicus Lam.
Fabaceae/Papilionaceae
♦ Iberian milkvetch
♦ Weed
♦ 70
♦ cultivated.

Astragalus macrocarpus DC.
Fabaceae/Papilionaceae
♦ Weed
♦ 221

Astragalus massiliensis (Mill.) Lam.
Fabaceae/Papilionaceae
Astragalus tragacantha L. *p.p.*
♦ Weed
♦ 70
♦ S, promoted.

Astragalus melilotoides Pall.
Fabaceae/Papilionaceae
♦ sweetclover like milkvetch
♦ Weed
♦ 297

Astragalus michauxii (Kuntze) F.J.Herm.
Fabaceae/Papilionaceae
♦ Michaux's loco, sandhills milkvetch
♦ Weed
♦ 39, 161
♦ toxic.

Astragalus microcephalus Willd.
Fabaceae/Papilionaceae
♦ Introduced
♦ 166, 228
♦ S, arid, promoted. Origin: west Asia.

Astragalus miser Douglas ex Hook.
Fabaceae/Papilionaceae
♦ Yellowstone hylophilus milkvetch, timber milkvetch
♦ Weed
♦ 39, 87, 88, 161
♦ herbal, toxic.

Astragalus missouriensis Nutt.
Fabaceae/Papilionaceae
♦ Missouri milkvetch
♦ Native Weed
♦ 161, 174
♦ cultivated, herbal. Origin: North America.

Astragalus mollissimus Torr.
Fabaceae/Papilionaceae
♦ woolly loco, halfmoon coryi loco, purple locoweed, woolly locoweed, woolly milkvetch
♦ Weed, Native Weed
♦ 39, 87, 88, 161, 174, 218
♦ herbal, toxic. Origin: North America.

Astragalus nothoxys A.Gray
Fabaceae/Papilionaceae
♦ sheep loco, sheep milkvetch
♦ Weed
♦ 39, 161
♦ herbal, toxic.

Astragalus nuttallianus DC.
Fabaceae/Papilionaceae
♦ Nuttall milkvetch, small flowered milkvetch
♦ Weed
♦ 161
♦ aH, herbal.

Astragalus nuttallii (Torr. & Gray) J.T.Howell
Fabaceae/Papilionaceae
♦ gray loco, Nuttall's milkvetch
♦ Weed
♦ 161
♦ pH, herbal, toxic.

Astragalus onobrychis L.
Fabaceae/Papilionaceae
Astragalus chlorocarpus Griseb.,

Astragalus rochelianus Heuff.
- ♦ sainforth milkvetch
- ♦ Weed
- ♦ 272
- ♦ herbal.

Astragalus oocarpus A.Gray
Fabaceae/Papilionaceae
- ♦ smooth loco, Descanso milkvetch
- ♦ Weed
- ♦ 39, 161
- ♦ pH, toxic.

Astragalus pachypus E.Greene
Fabaceae/Papilionaceae
- ♦ bush milkvetch, thickpod milkvetch
- ♦ Weed
- ♦ 161
- ♦ pH, herbal, toxic.

Astragalus pectinatus (Hook.) Douglas ex G.Don
Fabaceae/Papilionaceae
- ♦ narrowleaf milkvetch
- ♦ Weed
- ♦ 39, 87, 88, 161, 218
- ♦ herbal, toxic.

Astragalus pectinatus (Gray) M.E.Jones var. *serotinus* (Hook.) Douglas
Fabaceae/Papilionaceae
- ♦ narrow leaved milkvetch
- ♦ Noxious Weed
- ♦ 36

Astragalus peregrinus Vahl
Fabaceae/Papilionaceae
- ♦ Weed
- ♦ 221

Astragalus pomonensis M.E.Jones
Fabaceae/Papilionaceae
- ♦ Pomona milkvetch
- ♦ Weed
- ♦ 34
- ♦ pH, herbal.

Astragalus prolixus Sieber ex Bunge
Fabaceae/Papilionaceae
- ♦ Naturalised
- ♦ 7, 86

Astragalus pubentissimus Torr. & A.Gray
Fabaceae/Papilionaceae
- ♦ Green River milkvetch
- ♦ Weed
- ♦ 39, 161
- ♦ toxic.

Astragalus racemosus Pursh
Fabaceae/Papilionaceae
- ♦ creamy poison vetch, cream milkvetch
- ♦ Weed
- ♦ 82, 161
- ♦ herbal, toxic.

Astragalus scaberrimus Bunge
Fabaceae/Papilionaceae
- ♦ scabrousleaf milkvetch
- ♦ Weed
- ♦ 297
- ♦ herbal.

Astragalus schimperi Boiss.
Fabaceae/Papilionaceae
- ♦ Weed
- ♦ 221

Astragalus serenoi (Kuntze) Sheldon
Fabaceae/Papilionaceae
- ♦ naked milkvetch
- ♦ Weed
- ♦ 161
- ♦ herbal, toxic.

Astragalus sesameus L.
Fabaceae/Papilionaceae
- ♦ Weed, Naturalised
- ♦ 86, 98, 203
- ♦ Origin: Eurasia.

Astragalus sieberi DC.
- ♦ Weed
- ♦ 221

Astragalus sinicus L.
Fabaceae/Papilionaceae
- ♦ Weed, Naturalised
- ♦ 235, 286, 287
- ♦ bH, promoted, herbal.

Astragalus squarrosus Bunge
Fabaceae/Papilionaceae
- ♦ Weed
- ♦ 243

Astragalus tetrapterus A.Gray
Fabaceae/Papilionaceae
- ♦ fourwing milkvetch
- ♦ Weed
- ♦ 39, 161
- ♦ toxic.

Astragalus thurberi A.Gray
Fabaceae/Papilionaceae
- ♦ Thurber's milkvetch, Thurber's loco
- ♦ Weed
- ♦ 39, 161
- ♦ herbal, toxic.

Astragalus tibetanus Benth. ex Bunge
Fabaceae/Papilionaceae
- ♦ Tibet milkvetch
- ♦ Naturalised
- ♦ 101

Astragalus tomentosus Sessé & Moç.
Fabaceae/Papilionaceae
= *Astragalus fruticosus* Forssk. (NoR)
- ♦ Weed
- ♦ 221
- ♦ Origin: Egypt, Israel, Jordan, Sinai.

Astragalus tribuloides Delile
Fabaceae/Papilionaceae
- ♦ Weed
- ♦ 87, 88, 221
- ♦ herbal.

Astragalus tweedyi Canby
Fabaceae/Papilionaceae
- ♦ Tweedy's milkvetch
- ♦ Weed
- ♦ 161

Astragalus umbracticus E.Sheld.
Fabaceae/Papilionaceae
- ♦ sylvan milkvetch
- ♦ Weed
- ♦ 161
- ♦ toxic.

Astragalus vogelii (Webb) Bornm.
Fabaceae/Papilionaceae
- ♦ Weed
- ♦ 221

Astragalus voloratum L.
Fabaceae/Papilionaceae
- ♦ Weed
- ♦ 87, 88

Astragalus webberi Gray ex Brewer & S.Wats.
Fabaceae/Papilionaceae
- ♦ Webber's milkvetch
- ♦ Weed
- ♦ 161
- ♦ pH, toxic.

Astragalus whitneyi A.Gray
Fabaceae/Papilionaceae
- ♦ balloon milkvetch, balloonpod milkvetch
- ♦ Weed
- ♦ 161
- ♦ pH, herbal, toxic.

Astragalus wootonii Sheldon
Fabaceae/Papilionaceae
= *Astragalus allochrous* var. *playanus* Gray (Isely) (NoR)
- ♦ Wooton loco, milkvetch
- ♦ Weed
- ♦ 39, 87, 88, 161, 218
- ♦ aH, herbal, toxic.

Astrantia major L.
Apiaceae
Astrantia biebersteinii Trautv., *Astrantia carinthiaca* D.H.Hoppe, *Astrantia carniolica* Jacq., *Astrantia trifida* Hoffm.
- ♦ astrantia, great masterwort, melancholy gentleman, isotähtiputki
- ♦ Weed, Naturalised, Cultivation Escape
- ♦ 39, 40, 42, 272
- ♦ pH, cultivated, herbal, toxic.

Astrebla elymoides F.Muell. ex F.M.Bailey
Poaceae
Astrebla pectinata F.Muell. var. *elymoides* Bailey
- ♦ hoop Mitchell grass
- ♦ Introduced
- ♦ 228
- ♦ G, arid, cultivated.

Astrebla lappacea (Lindl.) Domin
Poaceae
Astrebla triticoides (Lindl.) F.Muell. ex Benth., *Danthonia lappacea* Lindl.
- ♦ curly Mitchell grass
- ♦ Introduced
- ♦ 228
- ♦ G, arid, cultivated.

Astrebla pectinata (Lindl.) F.Muell. ex Benth.
Poaceae
Danthonia pectinata Lindl.
- ♦ barley Mitchell grass
- ♦ Introduced
- ♦ 228
- ♦ G, arid, cultivated.

Astrephia chaerophylloides (Sm.) DC.
Valerianaceae
- ♦ Naturalised
- ♦ 257
- ♦ arid.

Astripomoea hyoscyamoides (Vatke) Verde
Convolvulaceae
Convolvulus hyoscyamoides Vatke
♦ Weed
♦ 87, 88

Astrolepis cochisensis (Goodd.) D.M.Benham & Windham
Pteridaceae/Adiantaceae
♦ wavy cloakfern, Cochise scaly cloakfern, scaly cloakfern
♦ Weed
♦ 161
♦ pH, toxic.

Astronidium ponapense (Kaneh.) Markg.
Melastomataceae
♦ Introduced
♦ 230
♦ T.

Astronium urundeuva (Allemão) Engl.
Anacardiaceae
♦ aroeira, urinde´va, urunde´va, aroeira do sertão, aroeira legítima
♦ Introduced
♦ 228
♦ arid, herbal.

Asystasia gangetica (L.) T.Anderson
Acanthaceae
Asystasia coromandeliana Nees
♦ Ganges primrose, Chinese violet, Philippine violet, coromande, katikamonga
♦ Weed, Quarantine Weed, Naturalised, Environmental Weed, Cultivation Escape
♦ 3, 13, 32, 80, 86, 87, 88, 101, 112, 179, 220, 243, 261
♦ pH, cultivated, herbal. Origin: Africa.

Asystasia intrusa (Forssk.) Blume
Acanthaceae
= *Asystasia gangetica* (L.) T.Anderson ssp. *micrantha* (Nees) Ensermu (NoR)
♦ Weed, Quarantine Weed
♦ 33, 88, 203, 220, 243
♦ cultivated, herbal.

Asystasia nemorum Nees
Acanthaceae
Isochoriste javanica Miq.
♦ Weed
♦ 13
♦ herbal.

Asystasia schimperi T.Anderson
Acanthaceae
♦ acwer
♦ Weed, Quarantine Weed, Native Weed
♦ 53, 76, 87, 88, 121, 203, 220, 240
♦ aH. Origin: southern Africa.

Asystasia welwitschii S.Moore
Acanthaceae
♦ Weed
♦ 87, 88

Atalaya hemiglauca F.Muell. ex Benth.
Sapindaceae
♦ whitewood
♦ Weed

♦ 39, 87, 88
♦ cultivated, herbal, toxic.

Atamisquea emarginata Miers ex Hook. & Arn.
Capparaceae
♦ vomitbush
♦ Introduced
♦ 228
♦ arid, herbal.

Athanasia crithmifolia L.
Asteraceae
♦ Native Weed
♦ 121
♦ S. Origin: southern Africa.

Athanasia trifurcata (L.) L.
Asteraceae
♦ Weed, Native Weed
♦ 39, 121
♦ pS, herbal, toxic. Origin: southern Africa.

Athrixia phylicoides DC.
Asteraceae
♦ bushman's tea, bush tea, kaffir tea
♦ Native Weed
♦ 121
♦ S, arid. Origin: southern Africa.

Athroisma laciniatum DC.
Asteraceae
Athroisma viscida Zoll. & Mor.
♦ Weed
♦ 13, 88, 170, 191

Athroisma stuhlmannii (O.Hoffm.) Mattf.
Asteraceae
♦ Weed
♦ 87, 88

Athyrium distentifolium Tausch ex Opiz
Dryopteridaceae/Woodsiaceae
Athyrium alpestre (Hoppe) Rylands
♦ alpine lady fern, papradka alpínska
♦ Weed
♦ 70
♦ cultivated, herbal.

Athyrium fallaciosum Milde
Dryopteridaceae/Woodsiaceae
♦ fallacious lady fern
♦ Weed
♦ 297

Athyrium filix-femina (L.) Roth
Dryopteridaceae/Woodsiaceae
♦ lady fern
♦ Weed
♦ 70, 272
♦ pH, cultivated, herbal.

Athyrium thelypteroides (Michx.) Desv.
Dryopteridaceae/Woodsiaceae
♦ silvery spleenwort
♦ Weed
♦ 87, 88, 218

Athysanus pusillus (Hook.) E.Greene
Brassicaceae
♦ athysanus, common sandweed
♦ Weed
♦ 161
♦ aH, herbal.

Atractylis cancellata L.
Asteraceae
♦ distaff thistle

♦ Weed
♦ 70
♦ cultivated.

Atractylis carduus (Forssk.) C.Christ.
Asteraceae
♦ Weed
♦ 221

Atractylis gummifera L.
Asteraceae
♦ pine thistle
♦ Weed
♦ 39, 70, 87, 88
♦ pH, arid, promoted, herbal, toxic.

Atractylis preauxiana Sch.Bip.
Asteraceae
♦ Quarantine Weed
♦ 220

Atractylodes japonica Koidz. ex Kitam.
Asteraceae
♦ atractylodis rhizoma
♦ Quarantine Weed
♦ 220
♦ pH, promoted.

Atractylodes lancea (Thunb.) DC.
Asteraceae
♦ swordlike atractylodes
♦ Weed
♦ 297
♦ pH, promoted, herbal.

Atractylodes macrocephala Koidz.
Asteraceae
Atractylis lancea var. *chinensis* (Bunge) Kitam., *Atractylis macrocephala* (Koidz.) Hand.-Mazz., *Atractylis macrocephala* var. *hunanensis* Ling
♦ bai zhu
♦ Quarantine Weed
♦ 220
♦ pH, promoted, herbal.

Atraphaxis frutescens (L.) Eversm.
Polygonaceae
♦ Introduced
♦ 228
♦ arid, cultivated.

Atraphaxis spinosa L.
Polygonaceae
♦ Weed
♦ 221
♦ S, promoted.

Atriplex L. spp.
Chenopodiaceae
♦ saltbush, spreading orach, atriplex, orache
♦ Weed
♦ 236, 243, 272
♦ H, herbal.

Atriplex argentea Nutt.
Chenopodiaceae
♦ silverscale saltbush, silvery orach
♦ Weed
♦ 87, 88, 218
♦ aH, promoted, herbal.

Atriplex canescens (Pursh) Nutt.
Chenopodiaceae
♦ fourwing saltbush, grey sagebrush
♦ Weed, Introduced
♦ 39, 87, 88, 218, 228, 243, 248

♦ S, arid, cultivated, herbal, toxic.
Origin: North America.

Atriplex centralasiatica Iljin
Chenopodiaceae
♦ Weed
♦ 275, 297
♦ aH.

Atriplex chenopodioides Batt.
Chenopodiaceae
♦ Weed
♦ 87, 88

**Atriplex confertifolia (Torr. & Frem.)
S.Wats.**
Chenopodiaceae
Obione confertifolia Torr. & Frem.
♦ shadscale, shadscale saltbush, spiny
saltbush
♦ Weed
♦ 87, 88, 218
♦ S, arid, promoted, herbal.

Atriplex coriacea Forssk.
Chenopodiaceae
♦ Weed
♦ 221
♦ arid.

Atriplex crassifolia C.A.Mey.
Chenopodiaceae
♦ Weed
♦ 87, 88
♦ herbal.

Atriplex dimorphostegia Kar. & Kir.
Chenopodiaceae
Atriplex bracteosum Trautv., *Atriplex
transcaspica* Bornm. & Sint.
♦ Weed
♦ 221
♦ aH, arid, promoted.

Atriplex eardleyae Aellen
Chenopodiaceae
♦ small saltbush
♦ Naturalised
♦ 86, 101
♦ Origin: Australia.

Atriplex elegans (Moq.) Dietr.
Chenopodiaceae
♦ wheelscale saltbush, fasciculata
saltbush, salton fasciculata saltbush
♦ Weed
♦ 28, 87, 88, 161, 199, 218, 243
♦ aH, arid, promoted, herbal.

Atriplex farinosa Forssk.
Chenopodiaceae
♦ Weed
♦ 221
♦ arid.

Atriplex fera (L.) Bunge
Chenopodiaceae
♦ Weed
♦ 275
♦ aH.

Atriplex halimus L.
Chenopodiaceae
♦ shrubby orache, sea orach, tree
purslane, atriplex, saltbush
♦ Weed, Naturalised, Introduced
♦ 15, 39, 40, 70, 221, 228, 280

♦ S, arid, cultivated, toxic. Origin:
Mediterranean.

Atriplex hastata L.
Chenopodiaceae
♦ spear leaved orache, hastate orache,
halberd leaved orach, iron root
♦ Weed, Naturalised
♦ 44, 70, 87, 88, 94, 272, 275, 286, 287,
295, 297
♦ aH, cultivated, herbal.

Atriplex heterosperma Bunge
Chenopodiaceae
= *Atriplex micrantha* Ledeb.
♦ weedy orache, twoscale saltbush
♦ Weed
♦ 88, 156
♦ aH, cultivated, herbal.

Atriplex holocarpa F.Muell.
Chenopodiaceae
♦ pop saltbush
♦ Naturalised
♦ 86, 101
♦ arid, cultivated. Origin: Australia.

Atriplex hortensis L.
Chenopodiaceae
♦ garden orache, orach, mountain
spinach, loboda záhradná
♦ Weed, Noxious Weed, Naturalised,
Introduced, Garden Escape,
Cultivation Escape
♦ 7, 23, 42, 70, 86, 87, 88, 94, 98, 101,
161, 176, 198, 203, 218, 228, 252, 272,
280, 299, 300
♦ aH, arid/aqua, cultivated, herbal.
Origin: horticultural.

Atriplex laciniata L.
Chenopodiaceae
= *Atriplex tatarica* L.
♦ frosted orache
♦ Weed, Naturalised
♦ 39, 87, 88, 101
♦ toxic.

Atriplex laevis C.A.Mey.
Chenopodiaceae
♦ silomaltsa
♦ Casual Alien
♦ 42

Atriplex lampa (Moq.) Gillies ex Small
Chenopodiaceae
♦ South American saltbush
♦ Naturalised
♦ 101

Atriplex lasiantha Boiss.
Chenopodiaceae
♦ hilsemaltsa
♦ Casual Alien
♦ 42

Atriplex lentiformis (Torr.) S.Wats.
Chenopodiaceae
Obione lentiformis Torr.
♦ quail brush, big saltbush
♦ Weed, Introduced
♦ 22, 228
♦ S, arid/aqua, promoted, herbal.

Atriplex leptocarpa F.Muell.
Chenopodiaceae
♦ Weed, Naturalised

♦ 7, 86, 98, 203
♦ arid, cultivated. Origin: Australia.

Atriplex leucoclada Boiss.
Chenopodiaceae
Obione leucoclada (Boiss.) Ulbn.
♦ Weed
♦ 221
♦ arid.

Atriplex lindleyi Moq.
Chenopodiaceae
♦ Lindley's saltbush
♦ Weed, Naturalised, Cultivation
Escape
♦ 34, 88, 101
♦ pH, arid, cultivated. Origin:
Australia.

**Atriplex lindleyi Moq. ssp. *inflata*
(Muell.) P.G.Wilson**
Chenopodiaceae
Blackiella inflata (Muell.) Aellen (see)
♦ blasiesoutbos, spongefruit saltbush
♦ Weed, Noxious Weed
♦ 63, 88, 95, 158, 283
♦ Origin: Australia.

Atriplex maximowicziana Makino
Chenopodiaceae
♦ Maximowicz's saltbush
♦ Naturalised
♦ 101
♦ aH, promoted.

Atriplex micrantha Ledeb.
Chenopodiaceae
Atriplex heterosperma Bunge (see)
♦ twoscale saltbush
♦ Naturalised
♦ 101

Atriplex montevidensis Spreng.
Chenopodiaceae
Atriplex grisebachii Kurtz ex Gand.
♦ cachiyuyo
♦ Weed
♦ 237, 295

Atriplex muelleri Benth.
Chenopodiaceae
♦ cape saltbush, vaalbrak saltbush,
Mueller's saltbush
♦ Weed, Naturalised, Introduced
♦ 39, 55, 101, 121, 228
♦ pH, arid, cultivated, herbal, toxic.
Origin: Australia.

Atriplex nitens Schkuhr
Chenopodiaceae
possible ancestor to *Atriplex hortensis*
L.
♦ hoary orache, kiiltomaltsa, shining
orache
♦ Weed, Naturalised, Casual Alien
♦ 42, 44, 101, 272
♦ cultivated, herbal.

Atriplex nummularia Lindl.
Chenopodiaceae
♦ bluegreen saltbush, old man
saltbush
♦ Weed, Naturalised, Introduced
♦ 88, 95, 101, 228, 241, 300
♦ S, arid, cultivated, herbal. Origin:
Australia.

Atriplex nummularia Lindl. ssp.
nummularia
Chenopodiaceae
♦ oumansoutbos, old man saltbush
♦ Weed, Noxious Weed, Cultivation
Escape
♦ 63, 283
♦ cultivated. Origin: central and south-
east Australia.

Atriplex oblongifolia Waldst. & Kit.
Chenopodiaceae
♦ oblongleaf orache, loboda
podlhovastolistá
♦ Weed, Naturalised
♦ 44, 101, 272

Atriplex patens (Litv.) Iljin
Chenopodiaceae
Atriplex subcordata Kitag.
♦ patent saltbush
♦ Weed
♦ 297

Atriplex patula (L.) Gray
Chenopodiaceae
♦ halberdleaf orach, kylämaltsa,
spreading orach, common oroche,
spear leaved orache, spearscale,
fathen, halberdleaf hastata orach, spear
orache, orach
♦ Weed, Naturalised, Native Weed
♦ 23, 39, 44, 68, 70, 86, 87, 88, 94, 98,
118, 121, 136, 161, 176, 198, 203, 210,
218, 241, 243, 253, 272, 280, 292, 300
♦ waH, cultivated, herbal, toxic.
Origin: southern Africa.

Atriplex patula L. var. *hastata* auct. non
(L.) Gray
Chenopodiaceae
= *Atriplex prostrata* Boucher ex DC.
♦ halberdleaf orach, atriplex, fathen,
saltbush
♦ Weed
♦ 180, 218

Atriplex polycarpa (Torr.) S.Watson
Chenopodiaceae
♦ allscale, cattle saltbush, cattle
spinach, cow spinach
♦ Introduced
♦ 228
♦ S, arid, herbal.

Atriplex prostrata Boucher ex DC.
Chenopodiaceae
Atriplex patula L. var. *hastata* auct. non
(L.) Gray (see), *Atriplex triangularis*
Willd. (see)
♦ hastate orache, orache, spear leaved
orache, triangle orache, isomaltsa,
loboda rozprestretá
♦ Weed, Naturalised, Garden Escape,
Environmental Weed
♦ 7, 9, 15, 72, 86, 88, 165, 176, 198, 237,
253, 280, 292, 300
♦ aH, cultivated. Origin: Eurasia,
North Africa.

Atriplex pseudocampanulata Aellen
Chenopodiaceae
♦ Naturalised
♦ 86
♦ cultivated. Origin: Australia.

Atriplex pumilio R.Br.
Chenopodiaceae
♦ Naturalised
♦ 86
♦ cultivated. Origin: Australia.

Atriplex rhagodioides F.Muell.
Chenopodiaceae
♦ saltbush
♦ Introduced
♦ 228
♦ arid, cultivated. Origin: Australia.

Atriplex rosea L.
Chenopodiaceae
♦ red orach, tumbling saltweed,
hopeamaltsa
♦ Weed, Naturalised, Introduced,
Casual Alien
♦ 23, 39, 42, 87, 88, 101, 161, 218, 228,
237, 241, 272, 280, 292, 295, 300
♦ aH, arid, cultivated, herbal, toxic.

Atriplex semibaccata R.Br.
Chenopodiaceae
♦ Australian saltbush, berry saltbush,
creeping saltbush, wild lucerne,
saltbush
♦ Weed, Noxious Weed, Naturalised,
Introduced, Environmental Weed
♦ 35, 39, 78, 80, 87, 88, 101, 116, 121,
134, 161, 180, 218, 228, 231, 237, 241,
295, 300
♦ pH, arid, cultivated, herbal, toxic.
Origin: Australia.

Atriplex serenana A.Nels.
Chenopodiaceae
♦ bracted saltbush, bract scale,
saltbush
♦ Weed
♦ 161, 180
♦ aH, promoted, herbal.

Atriplex sibirica L.
Chenopodiaceae
♦ Siberian saltbush
♦ Weed,
Naturalised
♦ 101, 297

Atriplex sphaeromorpha Iljin
Chenopodiaceae
♦ Weed
♦ 272

Atriplex spinibractea R.Anderson
Chenopodiaceae
♦ spinyfruit saltbush
♦ Naturalised
♦ 86
♦ cultivated. Origin: Australia.

Atriplex suberecta I.Verd.
Chenopodiaceae
♦ peregrine saltbush
♦ Weed, Naturalised
♦ 101, 199, 241, 300
♦ aH, herbal. Origin: Australia.

Atriplex tatarica L.
Chenopodiaceae
Atriplex laciniata L. (see)
♦ Tatarian orache, tataarimaltsa,
loboda tatárska
♦ Weed, Naturalised, Casual Alien

♦ 39, 42, 44, 87, 88, 94, 101, 272, 275,
300
♦ aH, arid, cultivated, toxic.

Atriplex thunbergiifolia (Boiss. & Noe)
Boiss.
Chenopodiaceae
♦ susannamaltsa
♦ Casual Alien
♦ 42

Atriplex triangularis Willd.
Chenopodiaceae
= *Atriplex prostrata* Boucher ex DC.
♦ spearscale
♦ Naturalised
♦ 241
♦ aH, herbal.

Atriplex undulata (Moq.) D.Dietr.
Chenopodiaceae
Obione undulata Moq.
♦ wavyleaf saltbush
♦ Introduced
♦ 228
♦ arid.

Atriplex vesicaria Heward ex Benth.
Chenopodiaceae
♦ aboriginal saltbush, bladder saltbush
♦ Naturalised, Introduced
♦ 39, 101, 228
♦ arid, cultivated, herbal, toxic. Origin:
Australia.

Atriplex vestita (Thunb.) Aell.
Chenopodiaceae
Atriplex capensis Moq.
♦ cape saltbush, vaalbrak saltbush
♦ Native Weed
♦ 121
♦ pS. Origin: southern Africa.

Atriplex wrightii S.Wats.
Chenopodiaceae
♦ Wright's saltbush
♦ Weed
♦ 87, 88, 161, 218
♦ herbal.

Atropa acuminata Royle
Solanaceae
♦ Indian belladonna
♦ Weed
♦ 23, 88
♦ pH, promoted, herbal.

Atropa bella-donna L.
Solanaceae
Atropa lethalis Salisb., *Belladonna
baccifera* Lam., *Belladonna trichotoma*
Scop.
♦ deadly nightshade, belladonna,
dwale
♦ Weed, Naturalised, Cultivation
Escape
♦ 23, 39, 42, 70, 87, 88, 101, 154, 161,
272, 280
♦ pH, cultivated, herbal, toxic.

Attalea exigua Drude
Arecaceae
♦ Weed
♦ 87, 88

Attalea geraensis Barb.Rodr.
Arecaceae

♦ coquinho
♦ Weed
♦ 255
♦ Origin: Brazil.

Attalea phalerata Mart. ex Spreng.
Arecaceae
♦ bacuri, acuri
♦ Weed
♦ 87, 88, 255
♦ Origin: Brazil.

Attalea speciosa C.Mart.
Arecaceae
= *Orbignya barbosiana* Burret (NoR)
♦ Weed
♦ 255
♦ cultivated, herbal. Origin: South
America.

Atylosia scarabaeoides (L.) Benth.
Fabaceae/Papilionaceae
Cantharospermum scarabaeoides Baill.
♦ Weed, Environmental Weed
♦ 87, 88, 216
♦ cultivated, herbal.

Aubrieta deltoidea (L.) DC.
Brassicaceae
♦ aubretia, lilac bush
♦ Naturalised, Cultivation Escape
♦ 40, 56, 101
♦ pH, cultivated, herbal. Origin:
Europe.

Aucuba japonica Thunb.
Cornaceae/Aucubaceae
♦ spotted laurel, Japanese aucuba,
aukuba japonská
♦ Naturalised
♦ 39, 101, 247
♦ S, cultivated, herbal, toxic.

Aurinia petraea (Ard.) Schur
Brassicaceae
♦ goldentuft
♦ Naturalised
♦ 101

Aurinia saxatilis (L.) Desv.
Brassicaceae
Alyssum saxatile L. (see)
♦ golden Alison, basket of gold
♦ Naturalised, Garden Escape,
Cultivation Escape
♦ 56, 101
♦ pH, cultivated, herbal.

Australina acuminata Wedd.
Urticaceae
♦ Weed, Quarantine Weed
♦ 76, 87, 88, 203, 220

**Australopyrum pectinatum (Labill.)
Á.Löve**
Poaceae
♦ Naturalised
♦ 280
♦ G.

**Australopyrum retrofractum (Vickery)
Á.Löve**
Poaceae
♦ Naturalised
♦ 280
♦ G.

Austrocylindropuntia Bakeb. spp.
Cactaceae
= *Opuntia* Mill. spp.
♦ Weed, Quarantine Weed
♦ 76, 88, 203

**Austrocylindropuntia subulata (Engelm.)
Backeb.**
Cactaceae
= *Opuntia subulata* (Muehlenpf.)
Engelm.
♦ cholla
♦ Introduced
♦ 228
♦ arid, cultivated.

**Austroeupatorium inulaefolium (H.B.K.)
R.M.King & H.Rob.**
Asteraceae
Eupatorium inulaefolium H.B.K. (see)
♦ austroeupatorium
♦ Weed, Quarantine Weed, Noxious
Weed, Naturalised
♦ 3, 76, 86, 87, 88, 135, 191, 203, 220,
243
♦ S/T.

**Austrostipa bigeniculata (Hughes)
S.W.L.Jacobs & J.Everett**
Poaceae
♦ Naturalised
♦ 280
♦ G, cultivated.

**Austrostipa blackii (C.E.Hubb.)
S.W.L.Jacobs & J.Everett**
Poaceae
♦ Naturalised
♦ 280
♦ G, cultivated.

**Austrostipa flavescens (Labill.)
S.W.L.Jacobs & J.Everett**
Poaceae
♦ Naturalised
♦ 280
♦ G, cultivated.

**Austrostipa nitida (Summerh. &
C.E.Hubb.) S.W.L.Jacobs & J.Everett**
Poaceae
♦ Naturalised
♦ 280
♦ G, cultivated.

**Austrostipa nodosa (S.T.Blake)
S.W.L.Jacobs & J.Everett**
Poaceae
♦ Naturalised
♦ 280
♦ G, cultivated.

**Austrostipa rudis (Spreng.) S.W.L.Jacobs
& J.Everett**
Poaceae
♦ Naturalised
♦ 280
♦ G, cultivated.

**Austrostipa rudis (Spreng.) S.W.L.Jacobs
& J.Everett ssp. *rudis***
Poaceae
♦ Naturalised
♦ 280
♦ G.

**Austrostipa scabra (Lindl.) S.W.L.Jacobs
& J.Everett**
Poaceae
♦ Naturalised
♦ 280
♦ G, cultivated.

**Austrostipa scabra (Lindl.) S.W.L.Jacobs
& J.Everett ssp. *falcata* (Hughes)
S.W.L.Jacobs & J.Everett**
Poaceae
♦ Naturalised
♦ 280
♦ G.

**Austrostipa scabra (Lindl.) S.W.L.Jacobs
& J.Everett ssp. *scabra***
Poaceae
♦ Naturalised
♦ 280
♦ G.

**Austrostipa stuposa (Hughes)
S.W.L.Jacobs & J.Everett**
Poaceae
♦ Naturalised
♦ 280
♦ G, cultivated.

**Austrostipa verticillata (Nees ex
Spreng.) S.W.L.Jacobs & J.Everett**
Poaceae
♦ Naturalised
♦ 280
♦ G, cultivated.

Avellinia michelii (Savi) Parl.
Poaceae
♦ avellinia, Rivieranheinä
♦ Weed, Naturalised, Environmental
Weed, Casual Alien
♦ 9, 42, 72, 86, 88, 198
♦ aG, cultivated. Origin:
Mediterranean.

Avena L. spp.
Poaceae
♦ oat
♦ Weed, Naturalised, Environmental
Weed
♦ 198, 243, 290, 296
♦ G.

Avena abyssinica Hochst.
Poaceae
♦ Abyssinian oat
♦ Weed, Naturalised
♦ 7, 98, 203
♦ G, cultivated. Origin: North Africa.

Avena alba Vahl
Poaceae
♦ Weed
♦ 87, 88
♦ G.

Avena barbata Pott. ex Link
Poaceae
Avena fatua L. var. *barbata* (Pott ex
Link) Fiori & Paol., *Avena sterilis* L. ssp.
barbata (Pott ex Link) Gillet & Magne,
Avena strigosa Schreb. ssp. *barbata* (Pott
ex Link) Thell., *Avena sativa* L. ssp.
fatua var. *barbata* (Pott ex Link) Fiori,
Avena alba var. *barbata* (Pott ex Link)
Maire & Weiller

♦ slender oat, bearded oat, barbed oat, slender wild oat, partakaura
♦ Weed, Naturalised, Introduced, Environmental Weed, Casual Alien
♦ 7, 9, 15, 18, 23, 34, 38, 42, 70, 72, 86, 87, 88, 91, 98, 101, 121, 158, 161, 167, 176, 198, 203, 218, 221, 237, 241, 243, 253, 272, 280, 287, 295, 300
♦ a/pG, arid, cultivated, herbal.
Origin: Eurasia.

Avena barbata **Pott ex Link ssp.** *barbata*
Poaceae
♦ Weed
♦ 250
♦ G.

Avena barbata **Pott ex Link ssp.** *wiestsii* **(Steud.) Tsvelev**
Poaceae
♦ Weed
♦ 185, 243
♦ G.

Avena byzantina **C.Koch**
Poaceae
♦ red oats, Algerian oats
♦ Weed, Naturalised, Casual Alien
♦ 7, 87, 88, 98, 121, 158, 203, 237, 280
♦ aG, promoted. Origin: Eurasia.

Avena cultiformis **Malz.**
Poaceae
♦ Weed
♦ 87, 88
♦ G.

Avena fatua **L.**
Poaceae
Avena fatua L. var. *glabrata* Peterm. (see)
♦ wild oat, common wild oat, wheat oats, oatgrass, flax grass, spring wild oat, avena selvatica, hukkakaura
♦ Weed, Noxious Weed, Naturalised, Introduced, Environmental Weed
♦ 7, 9, 15, 23, 30, 34, 36, 38, 39, 40, 42, 44, 49, 51, 52, 53, 55, 68, 70, 72, 86, 87, 88, 91, 93, 98, 101, 114, 118, 121, 136, 158, 161, 162, 176, 180, 185, 186, 198, 203, 204, 205, 210, 211, 212, 218, 219, 229, 236, 237, 240, 241, 243, 248, 253, 263, 269, 271, 272, 275, 280, 286, 287, 294, 295, 297, 299, 300
♦ aG, arid, cultivated, herbal, toxic. Origin: Eurasia.

Avena fatua **L. var.** *glabrata* **Peterm.**
Poaceae
= *Avena fatua* L.
♦ Naturalised
♦ 287
♦ G.

Avena fatua **L. var.** *sativa* **(L.) Hausskn.**
Poaceae
= *Avena sativa* L.
♦ Weed
♦ 179
♦ G.

Avena fatua **L. × *sativa* L.**
Poaceae
♦ wild oat
♦ Naturalised
♦ 98
♦ G.

Avena hybrida **Peterm. ex Rchb.** *p.p.*
Poaceae
♦ Weed
♦ 243
♦ G.

Avena ludoviciana **Dur.**
Poaceae
= *Avena sterilis* L. ssp. *ludoviciana* (Dur.) Nyman
♦ winter wild oat, wilder rothafer, avena cimarrona, oat
♦ Weed, Naturalised
♦ 30, 44, 87, 88, 236, 243, 287, 295
♦ G, promoted. Origin: Mediterranean.

Avena occidentalis **Durieu**
Poaceae
♦ western oat
♦ Naturalised
♦ 101
♦ G.

Avena sativa **L.**
Poaceae
Avena fatua L. var. *sativa* (L.) Hausskn. (see)
♦ wild oat, common oat, oat, sativa oat, ovos siaty
♦ Weed, Naturalised, Environmental Weed, Cultivation Escape
♦ 7, 15, 34, 39, 40, 42, 80, 86, 87, 88, 98, 101, 121, 134, 167, 176, 198, 199, 203, 237, 241, 243, 245, 252, 261, 272, 280, 287, 295, 300
♦ aG, cultivated, herbal, toxic. Origin: Eurasia.

Avena sterilis **L.**
Poaceae
♦ animated oat, sterile oat, rikkakaura, red wild oat, tall wild oat, winter wild oat
♦ Weed, Noxious Weed, Naturalised, Environmental Weed, Casual Alien
♦ 7, 40, 42, 67, 68, 72, 87, 88, 91, 98, 101, 111, 115, 118, 121, 140, 158, 161, 176, 185, 198, 203, 229, 236, 237, 241, 243, 280, 295, 300
♦ aG, arid, cultivated, herbal. Origin: Eurasia.

Avena sterilis **L. ssp.** *ludoviciana* **(Dur.) Nyman**
Poaceae
Avena trichophylla C.Koch, *Avena ludoviciana* Dur. (see)
♦ sterile oat, wild oat, winter wild oat
♦ Weed, Naturalised, Environmental Weed
♦ 55, 68, 70, 86, 88, 176, 198, 243, 253, 272, 280
♦ G. Origin: Eurasia.

Avena sterilis **L. ssp.** *sterilis*
Poaceae
♦ sterile oat
♦ Weed, Naturalised, Environmental Weed
♦ 70, 86, 198, 253, 272, 280
♦ G. Origin: Europe.

Avena sterilis **L. var.** *ludoviciana* **(Dur.) Husn.**
Poaceae

♦ ludo wild oat
♦ Naturalised
♦ 98
♦ G.

Avena strigosa **Schreb.**
Poaceae
♦ bristle oat, lopsided oats, sand oat, small oat, black oat, ukonkaura
♦ Weed, Naturalised, Casual Alien
♦ 7, 15, 30, 40, 42, 44, 70, 86, 87, 88, 91, 98, 101, 176, 198, 203, 241, 243, 245, 272, 280, 300
♦ aG, cultivated. Origin: Europe.

Avenula pratensis **(L.) Dumort.**
Poaceae
Arrhenatherum pratense (L.) Samp., *Avena pratensis* L., *Helictotrichon pratense* (L.) Pilg.
♦ meadow oatgrass
♦ Weed
♦ 272
♦ G, cultivated, herbal.

Avenula pubescens **(Huds.) Dumort.**
Poaceae
= *Helictotrichon pubescens* (Huds.) Bess. ex Pilg.
♦ downy oatgrass, mäkikaura, hairy oat
♦ Weed
♦ 272
♦ G, cultivated.

Averrhoa bilimbi **L.**
Oxalidaceae/Averrhoaceae
♦ bilimbi
♦ Introduced
♦ 230
♦ T, cultivated, herbal.

Averrhoa carambola **L.**
Oxalidaceae/Averrhoaceae
♦ carambola, ansu
♦ Introduced, Cultivation Escape
♦ 230, 261
♦ S/T, cultivated, herbal.

Avicennia germinans **(L.) L.**
Verbenaceae/Avicenniaceae
Avicennia africana P.Beauv., *Avicennia nitida* Jacq. (see), *Avicennia officinalis* L., *Bontia germanicus* L., *Hilairanthus nitidus* (Jacq.) Tiegh., *Hilairanthus tomentosus* (Jacq.) Tiegh., *Hilairanthus tomentosa* Jacq.
♦ black mangrove
♦ Weed
♦ 14
♦ T, herbal.

Avicennia marina **(Forssk.) Vierh.**
Verbenaceae/Avicenniaceae
♦ gray mangrove
♦ Weed, Naturalised
♦ 101, 221
♦ cultivated. Origin: Australia.

Avicennia marina **(Forssk.) Vierh. var.** *resinifera* **(G.Forst.) Bakh.**
Verbenaceae/Avicenniaceae
♦ gray mangrove, red mangrove, resinous avicennia
♦ Naturalised

♦ 101
♦ T.

Avicennia nitida Jacq.
Verbenaceae/Avicenniaceae
= *Avicennia germinans* (L.) L.
♦ black mangrove
♦ Weed
♦ 87, 88, 218
♦ herbal.

Avrainvillea amadelpha (Mont.) A. & E.S.Gepp
Udoteaceae
♦ algae
♦ Weed
♦ 282
♦ algae.

Axonopus affinis A.Chase
Poaceae
= *Axonopus fissifolius* (Raddi) Kuhlm.
♦ common carpetgrass, Swazi grass, narrow leaved carpetgrass
♦ Weed, Naturalised
♦ 3, 7, 15, 34, 87, 88, 90, 93, 98, 121, 161, 198, 203, 218, 249, 269, 280, 287
♦ pG, cultivated, herbal. Origin: Americas.

Axonopus compressus (Sw.) Beauv.
Poaceae
Agrostis compressa (Sw.) Poir., *Anastrophus compressus* (Sw.) Schlecht., *Milium compressum* Sw., *Panicum platycaulon* (Poir.) Kuntze, *Paspalum compressum* Raf., *Paspalum platycaulon* Poir.
♦ blanket grass, carpetgrass, savannah grass, broad leaved carpetgrass
♦ Weed, Naturalised, Introduced, Environmental Weed
♦ 30, 86, 87, 88, 90, 93, 98, 153, 170, 186, 203, 204, 230, 237, 243, 255, 256, 257, 263, 273, 286, 287, 295, 297
♦ pG, cultivated, herbal. Origin: tropical America.

Axonopus fissifolius (Raddi) Kuhlm.
Poaceae
Axonopus affinis A.Chase (see)
♦ narrow leaved carpetgrass, common carpetgrass
♦ Weed, Naturalised, Environmental Weed
♦ 3, 86, 134, 191, 280
♦ G, herbal. Origin: South America.

Axonopus scoparius (Flugge) Kuhlm.
Poaceae
♦ carpetgrass
♦ Weed
♦ 87, 88
♦ G, cultivated.

Axyris amaranthoides L.
Chenopodiaceae
♦ Russian pigweed, hörtsö
♦ Weed, Noxious Weed, Naturalised, Casual Alien
♦ 23, 40, 42, 52, 87, 88, 101, 161, 218, 272, 275, 287, 297, 299
♦ aH, cultivated.

Azadirachta indica A.Juss.
Meliaceae

Antelaea azadirachta (L.) Adelb., *Melia azadirachta* L., *Melia indica* (A.Juss.) Brand
♦ neem, Indian lilac, margosa tree, nim tree
♦ Weed, Naturalised, Introduced, Garden Escape, Environmental Weed
♦ 3, 22, 39, 86, 87, 88, 93, 155, 228, 261
♦ T, arid, cultivated, herbal, toxic. Origin: Bangladesh, India, Burma, Sri Lanka, northern Myanmar.

Azanza garckeana (F.Hoffm.) Exell & Hillc.
Malvaceae
= *Thespesia garckeana* F.Hoffm.
♦ mtobo
♦ Introduced
♦ 228
♦ arid, herbal.

Azara dentata Ruiz & Pav.
Flacourtiaceae
♦ Quarantine Weed
♦ 220
♦ cultivated, herbal. Origin: Peru, Chile.

Azara microphylla Hook.f.
Flacourtiaceae
♦ Naturalised
♦ 280
♦ S, cultivated, herbal.

Azima tetracantha Lam.
Salvadoraceae
♦ beehanger, bee stinger bush, needle bush, stinkbush, four thorns
♦ Native Weed
♦ 121
♦ pS, promoted, herbal. Origin: southern Africa.

Azolla rubra R.Br.
Azollaceae
♦ azolla
♦ Weed
♦ 87, 88, 208
♦ wpH, herbal. Origin: New Zealand.

Azolla africana Desv.
Azollaceae
= *Azolla pinnata* R.Br. ssp. *africana* (Desv.) R.M.K.Saunders & K.Fowler (NoR) [see *Azolla pinnata* var. *africana* (Desv.) Baker.]
♦ Weed
♦ 88
♦ wpH.

Azolla caroliniana Willd.
Azollaceae
Salvinia azolla Raddi
♦ Atlantic azolla, fairy moss, water velvet, Carolina mosquito fern, samaabaia aquatica, azol, almiscarvegetal, ambar vegetal, tapete d agua, murue rendado, musgo d'agua
♦ Weed, Quarantine Weed, Naturalised
♦ 76, 87, 88, 203, 218, 220, 237, 255, 287
♦ wpH, cultivated, herbal. Origin: South America.

Azolla filiculoides Lam.
Azollaceae

♦ Pacific azolla, mosquito fern, red water fern, red azolla, water fern, rooiwatervaring, azola papraìovitá
♦ Weed, Quarantine Weed, Noxious Weed, Naturalised, Garden Escape, Environmental Weed
♦ 40, 63, 87, 88, 95, 121, 144, 158, 167, 197, 218, 220, 237, 283, 295
♦ wpH, cultivated, herbal. Origin: South America.

Azolla filiculoides Lam. var. rubra (R.Br.) Strasb.
Azollaceae
♦ Pacific azzola
♦ Native Weed
♦ 269
♦ wpH. Origin: Australia.

Azolla imbricata (Roxb. ex Griff.) Nakai
Azollaceae
♦ azolla, Pacific azolla
♦ Weed
♦ 87, 88, 204, 263, 275, 286, 297
♦ wpH.

Azolla japonica Fr. & Sav. ex Nakai
Azollaceae
♦ Weed
♦ 87, 88, 204, 263, 286
♦ wpH.

Azolla nilotica Decne. ex Mett.
Azollaceae
♦ Weed
♦ 87, 88
♦ wpH.

Azolla pinnata R.Br.
Azollaceae
♦ ferny azolla, mosquito fern, water velvet, feathered mosquito fern, azolla, water fern, nae daeng
♦ Weed, Sleeper Weed, Noxious Weed, Native Weed, Environmental Weed
♦ 67, 87, 88, 140, 170, 191, 193, 209, 217, 225, 229, 239, 246, 262, 269, 274
♦ wpH, cultivated. Origin: tropical Africa, Asia, Australia.

Azolla pinnata R.Br. var. africana (Desv.) Baker.
Azollaceae
= *Azolla pinnata* ssp. *africana* (Desv.) R.M.K.Saunders & K.Fowler (NoR) [see *Azolla africana* Desv.]
♦ Native Weed
♦ 121
♦ wpH. Origin: southern Africa.

Azolla Lam. spp.
Azollaceae
♦ mosquito fern, ferny azolla
♦ Quarantine Weed, Cultivation Escape
♦ 220, 233
♦ wpH, cultivated, herbal.

Azukia angularis (Willd.) Ohwi
Fabaceae/Papilionaceae
Dolichos angularis Willd., *Phaseolus angularis* W.F.Wight
♦ Weed
♦ 87, 88
♦ herbal.

B

Babiana adpressa G.J.Lewis
Iridaceae
♦ Quarantine Weed
♦ 220

Babiana ambigua (Roem. & Schult.) G.Lewis
Iridaceae
♦ Quarantine Weed
♦ 220
♦ cultivated.

Babiana angustifolia Sweet
Iridaceae
Babiana disticha Ker Gawl. (see
– misapplied), *Babiana pulchra* (Salisb.)
G.J.Lewis, *Babiana stricta* (Ait.) Ker
Gawl. (see – misapplied)
♦ baboon flower, babiana
♦ Naturalised, Garden Escape,
Environmental Weed
♦ 86
♦ cultivated. Origin: South Africa.

Babiana curviscapa G.Lewis
Iridaceae
♦ Quarantine Weed
♦ 220
♦ cultivated.

Babiana disticha Ker Gawl.
Iridaceae
[misapplied in Western Australia =
Babiana angustifolia Sweet]
♦ baboon flower, babiana
♦ Weed, Naturalised
♦ 7, 9, 98, 203
♦ cultivated. Origin: South Africa.

Babiana flabellifolia Harv. ex Klatt
Iridaceae
♦ Quarantine Weed
♦ 220

Babiana framesii L.Bolus
Iridaceae
♦ Quarantine Weed
♦ 220

Babiana gawleri N.E.Br.
Iridaceae
♦ Quarantine Weed
♦ 220

Babiana klaverensis G.J.Lewis
Iridaceae
♦ Quarantine Weed
♦ 220

Babiana longibracteata G.J.Lewis
Iridaceae
♦ Quarantine Weed

♦ 220

Babiana mucronata (Jacq.) Ker Gawl.
Iridaceae
♦ Quarantine Weed
♦ 220
♦ cultivated.

Babiana nana (Andrews) Spreng.
Iridaceae
♦ Naturalised, Cultivation Escape
♦ 86, 86, 252
♦ cultivated. Origin: South Africa.

Babiana occidentalis Baker
Iridaceae
♦ Quarantine Weed
♦ 220

Babiana odorata L.Bolus
Iridaceae
♦ Quarantine Weed
♦ 220
♦ cultivated.

Babiana ringens (L.) Ker Gawl.
Iridaceae
♦ Quarantine Weed
♦ 220

Babiana sambucina (Jacq.) Ker Gawl.
Iridaceae
♦ Quarantine Weed
♦ 220
♦ cultivated.

Babiana scabrifolia Brehmer ex Klatt
Iridaceae
♦ Quarantine Weed
♦ 220
♦ cultivated.

Babiana spathacea (L.) Ker Gawl.
Iridaceae
♦ Quarantine Weed
♦ 220
♦ cultivated.

Babiana stellata Eckl.
Iridaceae
♦ Quarantine Weed
♦ 220

Babiana stricta (Aiton) Ker Gawl.
Iridaceae
[misapplied in Western Australia =
Babiana angustifolia Sweet]
♦ baboon flower, babiana
♦ Weed, Naturalised, Native Weed,
Environmental Weed
♦ 7, 15, 72, 88, 98, 121, 198, 203, 280
♦ pH, cultivated, herbal. Origin: South
Africa.

Babiana thunbergii Ker Gawl.
Iridaceae
♦ Quarantine Weed
♦ 220

Babiana truncata G.J.Lewis
Iridaceae
♦ Quarantine Weed
♦ 220

Babiana tubulosa (Burm.f.) Ker Gawl.
Iridaceae
♦ Weed, Naturalised, Cultivation
Escape
♦ 7, 54, 86, 88, 252

♦ cultivated.

Babiana undulato-venosa Klatt
Iridaceae
♦ Quarantine Weed
♦ 220

Babiana vanzyliae L.Bolus
Iridaceae
♦ Quarantine Weed
♦ 220
♦ cultivated.

Baccharis articulata (Lam.) Pers.
Asteraceae
Conyza articulata Lam., *Molina articulata*
(Lam.) Less.
♦ carquejinha
♦ Weed
♦ 87, 88, 255
♦ herbal. Origin: Brazil.

Baccharis coridifolia DC.
Asteraceae
Baccharis cordifolia DC., *Eupatorium
montevidense* Spreng.
♦ romerillo, toxic groundsel, mio mio
♦ Weed, Quarantine Weed, Noxious
Weed, Naturalised
♦ 39, 76, 86, 87, 88, 203, 220, 237, 255,
295
♦ toxic. Origin: South America.

Baccharis douglasii DC.
Asteraceae
♦ saltmarsh baccharis, marsh baccharis
♦ Weed, Quarantine Weed
♦ 76, 88, 203, 220
♦ pH, arid/aqua, herbal.

Baccharis dracunculifolia DC.
Asteraceae
Baccharis leptospermoides DC.
♦ cilca, vassourinha
♦ Weed
♦ 87, 88, 255
♦ Origin: South America.

Baccharis floribunda Kunth
Asteraceae
= *Baccharis latifolia* (Ruiz & Pav.) Pers.
(NoR)
♦ sachahuaca
♦ Weed
♦ 153
♦ herbal.

Baccharis gillesii A.Gray
Asteraceae
♦ Weed
♦ 87, 88, 237

Baccharis glomeruliflora Pers.
Asteraceae
♦ groundsel tree, silverling
♦ Weed
♦ 161
♦ toxic.

Baccharis glutinosa Pers.
Asteraceae
= *Baccharis salicifolia* (Ruiz & Pav.) Pers.
♦ seepwillow baccharis, seepwillow
♦ Weed
♦ 87, 88, 161, 218
♦ arid, herbal.

Baccharis halimifolia L.
Asteraceae
♦ tree groundsel, groundsel bush, eastern baccharis, groundsel, groundsel baccharis, groundsel tree
♦ Weed, Quarantine Weed, Noxious Weed, Naturalised, Introduced, Garden Escape, Environmental Weed
♦ , 7, 14, 39, 73, 76, 86, 87, 88, 93, 98, 147, 152, 161, 203, 218, 228, 246, 269, 280, 290
♦ S, arid, cultivated, herbal, toxic. Origin: Central and North America.

Baccharis lanceolata Kunth
Asteraceae
= *Baccharis salicifolia* (Ruiz & Pav.) Pers.
♦ Weed
♦ 87, 88

Baccharis medullosa DC.
Asteraceae
♦ Weed
♦ 295

Baccharis microphylla H.B.K.
Asteraceae
♦ Weed
♦ 87, 88
♦ herbal.

Baccharis neglecta Britton
Asteraceae
♦ Rooseveltweed
♦ Weed, Quarantine Weed
♦ 76, 88, 203, 220
♦ herbal.

Baccharis notosergila Griseb.
Asteraceae
♦ Weed
♦ 87, 88
♦ herbal.

Baccharis phyteumoides (Less.) DC.
Asteraceae
♦ baccharis fiteumoides
♦ Weed
♦ 237, 295

Baccharis pilularis DC.
Asteraceae
Baccharis pilularis DC. ssp. *consanguinea* (DC.) C.B.Wolf (see)
♦ coyote brush, dwarf chaparral broom, dwarf baccharis
♦ Weed, Quarantine Weed, Naturalised
♦ 76, 86, 87, 88, 203, 218, 220
♦ S, arid, cultivated, herbal. Origin: western North America.

Baccharis pilularis DC. ssp. consanguinea (DC.) C.B.Wolf
Asteraceae
= *Baccharis pilularis* DC.
♦ Quarantine Weed
♦ 220

Baccharis pingraea DC.
Asteraceae
♦ chiquilla
♦ Weed
♦ 87, 88, 237, 295
♦ Origin: South America.

Baccharis pteronioides DC.
Asteraceae
Baccharis ramulosa (DC.) A.Gray (see)
♦ yerba de pasmo, chill weed
♦ Weed
♦ 39, 161
♦ arid, herbal, toxic.

Baccharis ramulosa (DC.) A.Gray
Asteraceae
= *Baccharis pteronioides* DC.
♦ yerba de pasmo
♦ Weed
♦ 87, 88, 218
♦ herbal.

Baccharis salicifolia (Ruiz & Pav.) Pers.
Asteraceae
Baccharis alamani DC., *Baccharis coerulescens* DC., *Baccharis farinosa* Spreng., *Baccharis glutinosa* Pers. (see), *Baccharis lanceolata* Kunth (see), *Baccharis longifolia* DC., *Baccharis viminea* DC., *Molina viscosa* Ruiz & Pav.
♦ mule's fat, guatamate, huatamate
♦ Weed, Quarantine Weed, Introduced
♦ 76, 88, 203, 220, 228, 295
♦ S, arid, herbal.

Baccharis salicina Torr. & Gray
Asteraceae
♦ willow baccharis, Great Plains falsewillow
♦ Weed
♦ 87, 88, 161, 218
♦ herbal.

Baccharis sarothroides A.Gray
Asteraceae
♦ desert broom
♦ Weed
♦ 87, 88, 218
♦ S, arid, cultivated, herbal.

Baccharis spicata (Lam.) Baill.
Asteraceae
♦ baccharis
♦ Weed
♦ 295

Baccharis trimera (Less.) DC.
Asteraceae
Molina trimera Less., *Baccharis genistelloides* (Lam.) Pers. var. *trimera* (Less.) Baker
♦ carqueja
♦ Weed
♦ 87, 88, 237, 255, 295
♦ herbal. Origin: Brazil.

Baccharis trinervis (Lam.) Pers.
Asteraceae
Baccaris rhexioides Kunth
♦ assapeixe fino
♦ Weed
♦ 255
♦ cultivated, herbal. Origin: tropical America.

Baccharis ulicina Hook. & Arn.
Asteraceae
♦ Weed
♦ 87, 88

Bacopa Aubl. spp.
Scrophulariaceae

♦ water hyssop
♦ Quarantine Weed
♦ 258

Bacopa amplexicaulis (Pursh) Wettst.
Scrophulariaceae
= *Bacopa caroliniana* (Walter) B.L.Rob.
♦ Weed, Naturalised
♦ 98, 203
♦ H, aqua, cultivated.

Bacopa aquatica Aubl.
Scrophulariaceae
♦ Weed
♦ 87, 88
♦ wH, cultivated.

Bacopa calycina (Benth.) Engl. ex De Wild.
Scrophulariaceae
Herpestis calycina Benth.
♦ Weed
♦ 87, 88

Bacopa caroliniana (Walter) B.L.Rob.
Scrophulariaceae
Bacopa amplexicaulis (Pursh) Wettst. (see), *Hydrotrida caroliniana* Small, *Herpestis amplexicaulis* Pursh
♦ Carolina water hyssop, blue water hyssop, giant red bacopa, bacopa amplexicaulis, blue water hyssop, lemon bacopa
♦ Weed, Quarantine Weed, Naturalised
♦ 54, 76, 86, 87, 88, 218
♦ H, aqua, cultivated, herbal. Origin: south-east North America.

Bacopa crenata (Beauv.) Hepper
Scrophulariaceae
♦ Weed, Quarantine Weed
♦ 76, 87, 88, 203, 220
♦ H, aqua, cultivated.

Bacopa cuneifolia (Michx.) Wettst.
Scrophulariaceae
Moniera cuneifolia Michx.
♦ Weed
♦ 87, 88

Bacopa dianthera (Sw.) Descole & Borsini
Scrophulariaceae
= *Mecardonia procumbens* (Mill.) Small
♦ Weed
♦ 87, 88

Bacopa egensis (Poepp.) Pennell
Scrophulariaceae
♦ Brazilian water hyssop
♦ Weed, Naturalised
♦ 101, 197
♦ H, aqua.

Bacopa eisenii (Kell.) Pennell
Scrophulariaceae
♦ eisen water hyssop, Gila River water hyssop
♦ Weed
♦ 87, 88, 161, 180, 218
♦ pH.

Bacopa erecta Hutch. & Dalz.
Scrophulariaceae
♦ Weed, Quarantine Weed
♦ 76, 87, 88, 203, 220

Bacopa floribunda (R.Br.) Wettst.
Scrophulariaceae
Herpestis floribunda R.Br., *Mella*
floribunda (R.Br.) Pennell
♦ Weed
♦ 88, 170
♦ H, aqua, cultivated.

Bacopa monnieri (L.) Pennell
Scrophulariaceae
Bacopa monnieri (L.) Wettst., *Herpestis*
monnieria (L.) Kunth, *Herpestis*
spathulata Blume
♦ dwarf bacopa, herb of grace, brahmi
herb
♦ Weed, Naturalised
♦ 7, 13, 14, 87, 88, 170, 262, 287
♦ pH, aqua, cultivated, herbal.

Bacopa procumbens (Mill.) Greenm.
Scrophulariaceae
= *Mecardonia procumbens* (Mill.) Small
♦ water hyssop
♦ Weed, Naturalised, Introduced
♦ 86, 87, 88, 170, 191, 230
♦ H, herbal. Origin: Central and South
America.

Bacopa rotundifolia (Michx.) Wettst.
Scrophulariaceae
♦ disc water hyssop, bound bacopa,
roundleaf water hyssop
♦ Weed, Naturalised
♦ 87, 88, 161, 170, 191, 263, 286, 287
♦ wa/pH, cultivated, herbal.

Bacopa sessiflora Pulle
Scrophulariaceae
♦ Weed
♦ 87, 88

Baeckea virgata (J.R. & G.Forst.)
Andrews
Myrtaceae
Babingtonia virgata (J.R. & G.Forst)
F.Muell.
♦ tall baeckea
♦ Weed, Naturalised, Native Weed,
Garden Escape, Environmental Weed
♦ 72, 86, 88
♦ S, cultivated. Origin: Australia.

Baeometra uniflora (Jacq.) G.Lewis
Liliaceae/Colchicaceae
♦ baeometra
♦ Weed, Naturalised, Environmental
Weed
♦ 7, 39, 86, 98, 198, 203
♦ cultivated, toxic. Origin: South
Africa.

Baeria chrysostoma Fisch. & C.A.Mey.
Asteraceae
= *Lasthenia californica* DC. ex Lindl.
(NoR)
♦ Weed
♦ 23, 88
♦ herbal.

Bahia oppositifolia (Nutt.) Gray
Asteraceae
= *Picradeniopsis oppositifolia* (Nutt.)
Rydb. ex Britt.
♦ bahia
♦ Weed

♦ 39, 87, 88
♦ herbal, toxic.

Bahia schaffneri S.Wats.
Asteraceae
♦ Schaffner's bahia
♦ Naturalised
♦ 101

Baileya multiradiata Harv. & Gray
Asteraceae
Baileya pleniradiata Harv. & Gray (see)
♦ desert baileya, wild marigold, desert
marigold
♦ Weed, Quarantine Weed
♦ 39, 87, 88, 161, 218, 220, 243
♦ cultivated, herbal, toxic.

Baileya pauciradiata Harv. & Gray ex
Gray
Asteraceae
♦ Colorado desert marigold, laxflower
♦ Weed
♦ 161
♦ aH, herbal, toxic.

Baileya pleniradiata Harv. & Gray
Asteraceae
= *Baileya multiradiata* Harv. & Gray
♦ woolly marigold, woolly desert
marigold
♦ Weed
♦ 161
♦ a/pH, cultivated, herbal, toxic.

Balanites aegyptiaca (L.) Del.
Balanitaceae/Zygophyllaceae
Ximenia aegyptiaca L.
♦ hingota, soapberry tree, thorn tree,
desert date
♦ Weed, Introduced
♦ 39, 221, 228
♦ arid, herbal, toxic.

Baldellia ranunculoides (L.) Parl.
Alismataceae
Alisma ranunculoides L., *Echinodorus*
ranunculoides (L.) Engelm.
♦ lesser waterplantain
♦ Weed, Naturalised
♦ 101, 253
♦ wH, cultivated, herbal.

Balfourodendron riedelianum (Engl.)
Engl.
Rutaceae
♦ Introduced
♦ 228
♦ arid, cultivated.

Ballota africana (L.) Benth.
Lamiaceae
♦ cat herb
♦ Native Weed
♦ 121
♦ pS, cultivated. Origin: southern
Africa.

Ballota damascena Boiss.
Lamiaceae
♦ Weed
♦ 221

Ballota kaiseri V.Tackh.
Lamiaceae
♦ Weed
♦ 221

Ballota nigra L.
Lamiaceae
Marrubium nigrum Crantz
♦ black horehound, porro, stinking
horehound, balota äierna
♦ Weed, Naturalised
♦ 39, 42, 70, 87, 88,
98, 101, 203, 272, 280
♦ pH, cultivated, herbal, toxic. Origin:
Europe.

Ballota nigra L. ssp. foetida (Lam.)
Hayek
Lamiaceae
Ballota borealis Schweigg.
♦ black horehound
♦ Weed, Naturalised
♦ 86, 198, 269
♦ pH, cultivated. Origin: western
Europe.

Ballota nigra L. var. alba (L.) Sm.
Lamiaceae
♦ black horehound
♦ Naturalised
♦ 101

Ballota nigra L. var. foetida (Hayek) Vis.
Lamiaceae
Ballota nigra L. ssp. *foetida* (Lam.)
Hayek (see)
♦ black horehound
♦ Naturalised
♦ 101

Ballota nigra L. var. nigra
Lamiaceae
♦ black horehound
♦ Naturalised
♦ 101

Ballota saxatilis Sieber ex Benth.
Lamiaceae
♦ Weed
♦ 221

Ballota undulata (Sieber ex Fresen.)
Benth.
Lamiaceae
♦ Weed
♦ 221

Baloghia inophylla (G.Forst.) P.S.Green
Euphorbiaceae
♦ Naturalised
♦ 280
♦ cultivated.

Balsamita major Desf.
Asteraceae
♦ costmary
♦ Naturalised
♦ 101
♦ herbal.

Balsamorhiza sagittata (Pursh) Nutt.
Asteraceae
♦ arrow leaved balsamroot,
balsamroot, Oregon sunflower
♦ Weed
♦ 161
♦ pH, promoted, herbal, toxic.

Baltimora recta L.
Asteraceae
♦ beautyhead, flor amarilla

♦ Weed, Quarantine Weed,
Naturalised
♦ 76, 87, 88, 101, 157, 203, 220, 281
♦ aH.

Bambusa Schreb. spp.
Poaceae
♦ bamboo, bambou, pehri en sapahn,
bambuu, bambu, piao, piao palaoan,
clumping bamboo
♦ Weed, Naturalised, Environmental
Weed
♦ 3, 22, 45, 80, 86, 88, 107, 151, 191, 198,
201
♦ pG, herbal. Origin: Asia.

Bambusa arundinacea (Retz.) Willd.
Poaceae
♦ thorny bamboo, bans, bamboo
♦ Weed, Naturalised
♦ 98, 179, 203
♦ G, arid, cultivated, herbal.

Bambusa atra Lindl.
Poaceae
♦ bamboos, bambou, pehri en sapahn,
bambuu, bambu, piao, piao palaoan
♦ Weed
♦ 107
♦ G.

Bambusa balcooa Roxb.
Poaceae
♦ bamboo, cape bamboo, bambou,
pehri en sapahn, bambuu, bambu,
piao, piao palaoan
♦ Weed, Naturalised, Environmental
Weed
♦ 86, 107, 121, 279
♦ T, cultivated. Origin: Eurasia.

Bambusa blumeana Schult.f.
Poaceae
♦ bamboos, bambou, pehri en sapahn,
bambuu, bambu, piao, piao palaoan
♦ Weed
♦ 107
♦ pG, cultivated.

**Bambusa glaucescens (Willd.) Sieb. ex
Munro**
Poaceae
♦ golden goddess bamboo, hedge
bamboo
♦ Naturalised, Garden Escape
♦ 101, 261, 280
♦ G, cultivated, herbal. Origin: China.

Bambusa guadua Kunth
Poaceae
Guadua angustifolia Kunth (see)
♦ Environmental Weed
♦ 257
♦ G.

**Bambusa multiplex (Lour.) Raeusch. ex
J.A. & J.H.Schult.**
Poaceae
Arundo multiplex Lour., *Bambos nana*
Roxb. var. *alphonso-karri*, *Bambusa
alphonse-karri* Mitford, *Bambusa nana*
Roxb.
♦ bamboos, bambou, pehri en sapahn,
bambuu, bambu, piao, piao palaoan,

hedge bamboo, Chinese dwarf
bamboo, oriental hedge
♦ Weed, Naturalised, Introduced
♦ 101, 107, 179, 230, 280, 287
♦ pG, cultivated. Origin: east Asia.

Bambusa oldhamii Munro
Poaceae
♦ ryoku chiku
♦ Naturalised
♦ 280
♦ G, cultivated.

Bambusa polymorpha Munro
Poaceae
♦ bamboos, bambou, pehri en sapahn,
bambuu, bambu, piao, piao palaoan,
polymorph bamboo
♦ Weed
♦ 107
♦ pG, cultivated.

Bambusa tulda Roxb.
Poaceae
♦ bamboo, bambou, pehri en sapahn,
bambuu, bambu, piao, piao palaoan
♦ Weed
♦ 107
♦ pG, cultivated, herbal.

Bambusa vulgaris Schrad. ex Wendl.
Poaceae
♦ bamboo grass, bamboo, common
bamboo, pehri, golden bamboo, grand
bambou
♦ Weed, Naturalised, Introduced,
Environmental Weed, Cultivation
Escape
♦ 38, 86, 87, 88, 101, 107, 179, 226, 227,
230, 261
♦ pG, cultivated, herbal. Origin:
tropical Asia.

Bangia atropurpurea (Roth.) C.Agardh
Bangiaceae
♦ red alga
♦ Weed
♦ 197
♦ algae.

Banisteriopsis oxyclada (A.Juss.) B.Gates
Malpighiaceae
♦ Weed
♦ 255
♦ pH. Origin: Brazil.

Banksia caleyi R.Br.
Proteaceae
♦ Naturalised, Native Weed
♦ 7, 86
♦ cultivated. Origin: Australia.

Banksia integrifolia L.f.
Proteaceae
♦ coastal banksia, banksia
♦ Weed, Sleeper Weed, Naturalised,
Environmental Weed
♦ 15, 225, 246, 280
♦ T, cultivated, herbal. Origin:
Australia.

Baptisia alba (L.) Vent.
Fabaceae/Papilionaceae
♦ macrophylla wild indigo, white wild
indigo

♦ Weed
♦ 161
♦ promoted, herbal, toxic.

Baptisia tinctoria (L.) R.Br.
Fabaceae/Papilionaceae
Baptisia gibbesii Small, *Baptisia tinctoria*
(L.) R.Br. ex Ait.f. var. *crebra* Fernald,
Baptisia tinctoria (L.) R.Br. ex Ait.f. var.
projecta Fernald
♦ wild indigo, horsefly weed, yellow
wild indigo, American indigo, false
indigo, indigo broom, yellow broom,
yellow indigo
♦ Weed, Environmental Weed
♦ 39, 87, 88, 108, 161, 216, 218
♦ pH, cultivated, herbal, toxic. Origin:
Canada, USA.

Barbarea Aiton f. spp.
Brassicaceae
♦ wintercress, mustard, common
wintercress, yellow rocket, field
mustard
♦ Weed, Noxious Weed,
Environmental Weed
♦ 36, 80, 88, 151

**Barbarea arcuata (Opiz ex J.& C.Presl)
Rchb.**
Brassicaceae
= *Barbarea vulgaris* R.Br.
♦ barborka oblúkovitá
♦ Weed
♦ 87, 88

Barbarea intermedia Boreau
Brassicaceae
Barbarea augustana Boiss.
♦ medium flowered wintercress,
wintercress, yellow rocket
♦ Weed, Naturalised, Cultivation
Escape
♦ 15, 40, 70, 86, 88, 94, 98, 165, 176, 198,
203, 252, 280
♦ bH, cultivated. Origin: Europe,
north and east Africa.

Barbarea orthoceras Ledeb.
Brassicaceae
Barbarea americana Rydb., *Barbarea
stricta non* Andrz.
♦ wintercress, northern wintercress,
American yellow rocket
♦ Weed
♦ 23, 88, 159, 161, 212
♦ pH, promoted, herbal.

Barbarea praecox (Sm.) R.Br.
Brassicaceae
= *Barbarea verna* (Mill.) Asch.
♦ Weed
♦ 87, 88
♦ herbal.

Barbarea stricta Andrz.
Brassicaceae
♦ upright wintercress, small flowered
wintercress, rantakanankaali, barborka
tuhá
♦ Naturalised
♦ 280
♦ aqua, cultivated.

Barbarea verna (Mill.) Asch.
Brassicaceae
Barbarea praecox (Sm.) R.Br. (see)
♦ early wintercress, American
wintercress, scurvy grass, early yellow
rocket, land cress, early flowering
wintercress, upland cress, cress
♦ Weed, Naturalised, Cultivation
Escape
♦ 23, 34, 40, 56, 70, 86, 87, 88, 94, 98,
101, 121, 161, 176, 198, 203, 218, 241,
252, 253, 280, 287, 300
♦ pH, cultivated, herbal. Origin:
Eurasia.

Barbarea vulgaris R.Br.
Brassicaceae
Barbarea arcuata (Opiz ex J.& C.Presl)
Reichen. (see), *Campe barbarea* (L.)
W.Wight ex Piper, *Campe vulgaris*
Dulac, *Erysimum barbarea* L.
♦ yellow rocket, wintercress, common
wintercress, bitter wintercress, garden
yellow rocket, peltokanankaali, St.
Barbara's cress, bittercress, rocket
cress, barborka obyáajná
♦ Weed, Noxious Weed, Naturalised,
Introduced, Environmental Weed
♦ 21, 23, 24, 34, 39, 42, 52, 70, 87, 88,
94, 101, 161, 174, 207, 210, 211, 218, 229,
253, 272, 280, 286, 287, 291
♦ pH, cultivated, herbal, toxic. Origin:
Europe.

Barbula convoluta Hedw.
Pottiaceae
♦ convoluted barbula moss, lesser
bird's claw beard moss
♦ Naturalised
♦ 280
♦ moss.

Barbula unguiculata Hedw.
Pottiaceae
♦ fúzatka nechtovitá
♦ Naturalised
♦ 198, 280
♦ moss.

Barclaya kunstleri (King) Ridl.
Nymphaeaceae/Barclayaceae
Hydrostemma kunstleri (Ridl.) B.C.Stone
(see)
♦ Quarantine Weed
♦ 220
♦ wH, cultivated.

Barclaya motleyi Hook.f.
Nymphaeaceae/Barclayaceae
Hydrostemma motleyi (Hook.f.) Mabb.
(see)
♦ Quarantine Weed
♦ 220
♦ wH, cultivated.

Barleria acanthoides Vahl
Acanthaceae
♦ Weed
♦ 221
♦ arid, cultivated.

Barleria cristata L.
Acanthaceae
♦ barleria, crested Philippine violet,
Philippine violet

♦ Weed, Sleeper Weed, Naturalised,
Introduced, Garden Escape,
Environmental Weed, Cultivation
Escape
♦ 13, 86, 101, 155, 179, 230, 261
♦ H, cultivated, herbal. Origin: India.

Barleria elegans S.Moore
Acanthaceae
♦ Native Weed
♦ 121
♦ pH. Origin: southern Africa.

Barleria eranthemoides R.Br.
Acanthaceae
♦ Weed
♦ 240
♦ S, cultivated.

Barleria hochstetteri Nees
Acanthaceae
♦ Weed
♦ 221

Barleria lugardii C.B.Clarke
Acanthaceae
♦ Weed
♦ 121
♦ S.

Barleria lupulina Lindl.
Acanthaceae
♦ hophead Philippine violet, barleria
♦ Weed, Sleeper Weed, Naturalised,
Environmental Weed
♦ 32, 86, 93, 101, 155, 179, 191
♦ cultivated, herbal. Origin: India.

Barleria mysorensis Heyne
Acanthaceae
♦ Weed
♦ 87, 88

Barleria obtusa Nees
Acanthaceae
♦ barleria, bush violet, south coast
bush violet
♦ Weed
♦ 121
♦ S.

Barleria prionitis L.
Acanthaceae
♦ barleria, porcupine flower
♦ Weed, Noxious Weed, Naturalised,
Environmental Weed, Cultivation
Escape
♦ 3, 13, 54, 86, 87, 88, 93, 98, 101, 155,
191, 203, 261
♦ S, arid, cultivated, herbal. Origin:
tropical Asia and Africa, Madagascar.

Barleria rotundifolia Oberm.
Acanthaceae
♦ Weed
♦ 121
♦ a/bS.

Barleria strigosa Willd.
Acanthaceae
♦ Weed, Naturalised
♦ 29, 86, 191
♦ cultivated, herbal. Origin: India.

**Barnardiella spiralis (N.E.Br.)
P.Goldblatt**
Iridaceae
Gynandrisis spiralis (Baker) R.C.Foster,

Homeria herrei L.Bolus
♦ Weed, Quarantine Weed
♦ 76, 88, 220

Barringtonia asiatica (L.) Kurz
Lecythidaceae/Barringtoniaceae
♦ futu, wih
♦ Naturalised, Environmental Weed,
Cultivation Escape
♦ 39, 101, 259, 261
♦ T, cultivated, herbal, toxic. Origin:
tropical Asia, Madagascar.

**Bartlettina sordida (Less.) R.King &
H.Rob.**
Asteraceae
= *Eupatorium sordidum* Less. (NoR)
♦ bartlettina
♦ Sleeper Weed, Quarantine Weed,
Naturalised, Environmental Weed
♦ 225, 246, 280
♦ cultivated.

Bartsia atifolia (L.) Sibth. & Sm.
Scrophulariaceae
♦ Weed
♦ 87, 88

Bartsia odontites (L.) Huds.
Scrophulariaceae
= *Odontites verna* (Bell.) Dumort.
♦ red bartsia
♦ Weed
♦ 87, 88
♦ herbal.

Bartsia trixago L.
Scrophulariaceae
= *Bellardia trixago* (L.) All.
♦ Naturalised, Introduced
♦ 38, 86, 300
♦ Origin: Mediterranean.

Basella alba L.
Basellaceae
Basella cordifolia Lam., *Basella rubra* L.
(see)
♦ Ceylon spinach, Indian spinach,
Malabar nightshade, nderema,
masingu
♦ Weed, Sleeper Weed, Naturalised,
Environmental Weed, Cultivation
Escape
♦ 32, 54, 86, 88, 101, 155, 261, 287
♦ pH, cultivated, herbal. Origin:
tropical Africa and Asia.

Basella rubra L.
Basellaceae
= *Basella alba* L.
♦ red vinespinach
♦ Weed, Naturalised
♦ 87, 88, 262, 274, 287
♦ cultivated, herbal.

Basilicum polystachyon (L.) Moench
Lamiaceae
Moschosma polystachyon (L.) Benth.
(see), *Ocimum polystachyon* L., *Ocimum
tashiroi* Hayata, *Ocimum tenuiflorum*
Burm.f., *Plectranthus parviflorus* R.Br.
(see)
♦ musk-basil, basilic musqué
♦ Weed
♦ 13, 88, 170

♦ bH. Origin: tropical Africa, Asia and Madagascar.

Bassia bicornis (Lindl.) F.Muell.
Chenopodiaceae
♦ Weed
♦ 87, 88

Bassia birchii (F.Muell.) F.Muell.
Chenopodiaceae
= *Sclerolaena birchii* (F.Muell.) Domin
♦ Weed
♦ 87, 88

Bassia dasyphylla (Fisch. & C.A.Mey.) Kuntze
Chenopodiaceae
♦ divaricate bassia
♦ Weed
♦ 297

Bassia eriophora (Schrad.) Asch.
Chenopodiaceae
♦ Weed
♦ 221

Bassia hirsuta (L.) Asch.
Chenopodiaceae
♦ hairy smotherweed
♦ Naturalised
♦ 101
♦ herbal.

Bassia hyssopifolia (Pall.) Ktze.
Chenopodiaceae
Echinopsilon hyssopifolium (Pall.) Moq., *Kochia hyssopifolia* Pall., *Salsola hyssopifolia* Pall.
♦ fivehook bassia, bassia, hyssop leaved echinopsilon, smother weed, fivehorn smotherweed
♦ Weed, Noxious Weed, Naturalised, Introduced, Environmental Weed
♦ 34, 35, 78, 80, 86, 87, 88, 98, 101, 116, 136, 146, 151, 161, 180, 198, 203, 218, 228, 231, 237, 241, 243, 272, 295, 300
♦ aH, arid, herbal, toxic. Origin: Eurasia.

Bassia muricata (L.) Asch.
Chenopodiaceae
Echinopsilon muricata (Schrad.) Moq., *Kochia muricata* Schrad.
♦ Weed
♦ 221
♦ arid.

Bassia quinquecuspis (F.Muell.) F.Muell.
Chenopodiaceae
= *Sclerolaena muricata* (Moq.) Domin
♦ Weed
♦ 39, 87, 88
♦ toxic.

Bassia scoparia (L.) Scott
Chenopodiaceae
Chenopodium scoparia L., *Kochia childsii* hort. ex anon., *Kochia scoparia* (L.) Schrad. (see), *Kochia scoparia* fo. *trichophylla* (hort. ex Voss) Schinz & Thell., *Kochia scoparia* (L.) Schrad. var. *culta* (see), *Kochia trichophylla* hort. ex Voss
♦ kochia, mock cypress, Mexican firebush, summer cypress, burningbush, Mexican fireweed, besenkraut, hokigi, mirabela, mirabel

♦ Weed, Quarantine Weed, Noxious Weed, Naturalised
♦ 39, 62, 76, 85, 86, 88, 98, 176, 203, 220, 237, 243
♦ aH, cultivated, toxic. Origin: Eurasia.

Bassia tetracuspis C.T.White
Chenopodiaceae
♦ Weed
♦ 87, 88

Bastardia viscosa (L.) Kunth
Malvaceae
♦ viscid mallow
♦ Weed
♦ 32
♦ herbal.

Batis maritima L.
Bataceae
♦ turtleweed, saltwort
♦ Weed, Introduced
♦ 14, 87, 88, 228
♦ S, arid/aqua, herbal.

Batrachium bungei (Steud.) L.Liou
Ranunculaceae
♦ bunge batrachium
♦ Weed
♦ 297

Batrachium pekinense L.Liou
Ranunculaceae
♦ Beijing batrachium
♦ Weed
♦ 297

Batrachium trichophyllum (Chaix) Bosch
Ranunculaceae
= *Ranunculus trichophyllus* Chaix var. *trichophyllus*
♦ water fennel
♦ Weed, Naturalised
♦ 86, 98, 198, 203
♦ herbal. Origin: Eurasia, North America.

Bauhinia aculeata L.
Fabaceae/Caesalpiniaceae
♦ Weed
♦ 179
♦ arid.

Bauhinia acuminata L.
Fabaceae/Caesalpiniaceae
♦ Weed, Introduced
♦ 93, 191, 230
♦ T, cultivated, herbal.

Bauhinia candicans Benth.
Fabaceae/Caesalpiniaceae
= *Bauhinia forficata* Link ssp. *pruinosa* (Vogel) Fortunato & Wunderlin (NoR) [see *Bauhinia forficata* Link]
♦ pata de vaca
♦ Weed
♦ 295
♦ cultivated.

Bauhinia cunninghamii (Benth.) Benth.
Fabaceae/Caesalpiniaceae
Lysiphyllum cunninghamii (Benth.) de Wit
♦ Introduced
♦ 228
♦ arid. Origin: Australia.

Bauhinia × blakeana S.T.Dunn
Fabaceae/Caesalpiniaceae
= *Bauhinia variegata* L. × *Bauhinia purpurea* L. [most probable combination]
♦ Blake's bauhinia
♦ Naturalised
♦ 101

Bauhinia cuyabensis Steud.
Fabaceae/Caesalpiniaceae
♦ Weed
♦ 87, 88

Bauhinia divaricata L.
Fabaceae/Caesalpiniaceae
Bauhinia adansonia Guill. & Perr., *Bauhinia amblyophylla* Harms, *Bauhinia americana* Laun., *Bauhinia aurita* Aiton, *Bauhinia caribaea* Jennings, *Bauhinia confusa* Rose, *Bauhinia furcata* Desv., *Bauhinia goldmanii* Rose, *Bauhinia lamarkiana* DC., *Bauhinia latifolia* Cav., *Bauhinia mexicana* Vog., *Bauhinia penincularis* Brandeg, *Bauhinia racemifera* Desv., *Bauhinia retusa* Poir., *Bauhinia schlechtendaliana* Mart., *Bauhinia spathacea* DC., *Bauhinia versicolor* Bertol.
♦ Introduced
♦ 228
♦ arid, herbal.

Bauhinia forficata Link
Fabaceae/Caesalpiniaceae
♦ pata de vaca
♦ Weed
♦ 255
♦ cultivated. Origin: South America.

Bauhinia galpinii N.E.Br.
Fabaceae/Caesalpiniaceae
Bauhinia punctata Bolle nom. illeg. (see)
♦ lowveld bauhinia, pride of De Kaap, red bauhinia, African plume, nasturtium bauhinia
♦ Weed, Naturalised, Native Weed
♦ 101, 121
♦ pS, cultivated, herbal. Origin: southern Africa.

Bauhinia integrifolia Roxb.
Fabaceae/Caesalpiniaceae
Bauhinia brachyxypha Bak., *Bauhinia cumingiana* Benth., *Bauhinia flammifera* Ridl., *Bauhinia holosericea* Ridl., *Bauhinia pierrei* Gagnep., *Bauhinia polyantha* Miq., *Phanera integrifolia* (Roxb.) Benth.
♦ Introduced
♦ 228
♦ arid.

Bauhinia monandra Kurz.
Fabaceae/Caesalpiniaceae
♦ orchid tree, St. Thomas tree, flamboyant, flores mariposa, mariposa, pine fua loloa, pink butterfly tree, Napoleon's plume, pilampwoia, Jerusalem date, vae povi
♦ Weed, Naturalised, Introduced, Cultivation Escape
♦ 3, 22, 86, 101, 107, 230, 261
♦ S/T, cultivated, herbal. Origin: southern Asia.

Bauhinia multinervia (Kunth) DC.
Fabaceae/Caesalpiniaceae
Bauhinia megalandra Griseb.
- petite flamboyant bauhinia
- Naturalised, Casual Alien
- 101, 261

Bauhinia pauletia Pers.
Fabaceae/Caesalpiniaceae
- railroadfence
- Naturalised
- 101

Bauhinia petersiana Bolle ssp. *macrantha* (Oliv.) Brummitt & J.H.Ross
Fabaceae/Caesalpiniaceae
Bauhinia petersiana Bolle ssp. *serpae* (Ficalho & Hiern) Brummitt & J.H.Ross, *Bauhinia macrantha* Oliv.
- camel's foot, coffee neat's foot, wild coffee bean
- Native Weed
- 121
- S/T. Origin: southern Africa.

Bauhinia pottsii G.Don var. *subsessilis* (Craib) De. Wit.
Fabaceae/Caesalpiniaceae
- Weed
- 12

Bauhinia punctata Bolle *nom. illeg.*
Fabaceae/Caesalpiniaceae
= *Bauhinia galpinii* N.E.Br.
- Introduced
- 261
- cultivated, herbal. Origin: tropical Asia.

Bauhinia purpurea L.
Fabaceae/Caesalpiniaceae
- butterfly tree, orchid tree, camel's foot tree, khairwal, koiral, koliar, palo de orqujdeas, pie de cabra, poor man's orchid, purple bauhinia, purple orchid tree
- Weed, Noxious Weed, Naturalised, Introduced, Garden Escape, Cultivation Escape
- 3, 101, 122, 179, 228, 261, 283
- arid, cultivated, herbal. Origin: south-east Asia.

Bauhinia racemosa Lam.
Fabaceae/Caesalpiniaceae
- endro
- Introduced
- 228
- arid, herbal.

Bauhinia rufescens Lam.
Fabaceae/Caesalpiniaceae
Adenolobus rufescens (Lam.) Schmitz, *Bauhinia andansoniana* Guill. & Perr., *Bauhinia parviflora* Vahl
- Introduced
- 228
- arid, cultivated, herbal. Origin: Africa.

Bauhinia tomentosa L.
Fabaceae/Caesalpiniaceae
Bauhinia tomentosa L. var. *glabrata* Hook.f., *Bauhinia volkensii* Taub., *Bauhinia wituensis* Harms, *Pauletia tomentosa* (L.) Schmitz

- St. Thomas tree, yellow bauhinia
- Naturalised, Introduced, Cultivation Escape
- 101, 228, 261
- arid, cultivated, herbal. Origin: India, China.

Bauhinia vahlii Wight & Arn.
Fabaceae/Caesalpiniaceae
- malu creeper
- Introduced
- 228
- arid, herbal.

Bauhinia variegata L.
Fabaceae/Caesalpiniaceae
- orchid tree, butterfly tree, mountain ebony
- Weed, Noxious Weed, Naturalised, Introduced, Garden Escape, Environmental Weed, Cultivation Escape
- 3, 39, 80, 88, 101, 112, 122, 151, 179, 191, 228, 260, 261, 279, 283
- arid, cultivated, herbal, toxic. Origin: India, China.

Bauhinia yunnanensis Franch.
Fabaceae/Caesalpiniaceae
- Yunnan bauhinia
- Weed, Naturalised
- 101, 179
- cultivated.

Becium obovatum (E.Mey. ex Benth.) N.E.Br. var. *galpinii* (Guerke) N.E.Br.
Lamiaceae
Ocimum galpini Guerke
- cat's whiskers
- Native Weed
- 121
- pH. Origin: southern Africa.

Becium obovatum (E.Mey. ex Benth.) N.E.Br. var. *obovatum*
Lamiaceae
- cat's whiskers
- Native Weed
- 121
- pH. Origin: southern Africa.

Beckeropsis uniseta (Nees) K.Schum.
Poaceae
- beckeropsis, Duncan grass, Natal grass, silky grass, kantentwa
- Native Weed
- 121
- pG. Origin: southern Africa.

Beckmannia eruciformis (L.) Host
Poaceae
- idäntörökki
- Weed, Casual Alien
- 42, 272
- pG, aqua, cultivated, herbal.

Beckmannia syzigachne (Steud.) Fern.
Poaceae
- American sloughgrass, sloughgrass, Amerikantörökki, kazunokogusa
- Weed, Casual Alien
- 42, 68, 87, 88, 159, 243, 263, 272, 275, 286, 297
- aG, cultivated, herbal. Origin: China, Japan.

Begonia coccinea Hook.
Begoniaceae
- scarlet begonia
- Naturalised, Cultivation Escape
- 101, 261
- cultivated. Origin: South America.

Begonia convolvulacea A.DC.
Begoniaceae
- morningglory begonia
- Naturalised, Cultivation Escape
- 101, 261
- cultivated. Origin: Brazil.

Begonia corallina Carr.
Begoniaceae
- Casual Alien
- 280
- cultivated, herbal.

Begonia cucullata Willd.
Begoniaceae
Begonia cucullata Willd. var. *hookeri* (A.DC.) L.B.Sm. & Schub. (see)
- clubed begonia
- Weed, Naturalised, Environmental Weed
- 101, 179, 255
- cultivated, herbal. Origin: Brazil.

Begonia cucullata Willd. var. *hookeri* (A.DC.) L.B.Sm. & Schub.
Begoniaceae
= *Begonia cucullata* Willd.
- Cultivation Escape
- 261
- cultivated. Origin: Brazil.

Begonia decandra Pav. ex A.DC.
Begoniaceae
- native begonia
- Weed
- 87, 88

Begonia evansiana Andrews
Begoniaceae
- Naturalised
- 287
- herbal.

Begonia foliosa Kunth
Begoniaceae
- fuchsia begonia
- Naturalised
- 101
- cultivated, herbal.

Begonia foliosa Kunth var. *miniata* (Planch.) L.B.Sm. & Schub.
Begoniaceae
- fuchsia begonia
- Naturalised
- 101
- herbal.

Begonia heracleifolia Cham. & Schlecht.
Begoniaceae
- starleaf begonia, liuskabegonia
- Naturalised, Introduced, Cultivation Escape
- 38, 101, 261
- cultivated, herbal. Origin: Mexico, Central America.

Begonia hirtella Link
Begoniaceae
- Brazilian begonia

♦ Weed, Naturalised, Cultivation
Escape
♦ 28, 101, 179, 206, 243, 261
♦ cultivated. Origin: tropical South
America.

Begonia nelumbiifolia **Cham. & Schlecht.**
Begoniaceae
♦ lilypad begonia
♦ Naturalised, Cultivation Escape
♦ 101, 261
♦ cultivated. Origin: Mexico,
Colombia.

Begonia reniformis **Dry.**
Begoniaceae
♦ grapeleaf begonia
♦ Naturalised
♦ 101
♦ cultivated.

Begonia semperflorens **Link & Otto cv.**
hybrids
Begoniaceae
♦ Naturalised
♦ 247, 280
♦ cultivated, herbal, toxic. Origin:
horticultural.

Belamcanda chinensis **(L.) DC.**
Iridaceae
Belamcanda punctata Moench,
Gemmingia chinensis Kuntze, *Ixia
chinensis* L, *Pardanthus chinensis* (L.)
Ker Gawl.
♦ blackberry lily, leopard lily, shenan,
maravilla
♦ Naturalised, Garden Escape
♦ 101, 247, 261
♦ pH, cultivated, herbal, toxic. Origin:
China, Japan, India.

Bellardia trixago **(L.) All.**
Scrophulariaceae
Bartsia trixago L. (see)
♦ bellardia, Mediterranean lineseed,
bartsia
♦ Weed, Noxious Weed, Naturalised,
Introduced, Environmental Weed
♦ 7, 9, 34, 35, 72, 78, 80, 86, 87, 88, 94,
98, 101, 116, 151, 161, 176, 198, 203, 231,
241, 287
♦ aH, cultivated.

Bellevalia flexuosa **Boiss.**
Liliaceae/Hyacinthaceae
♦ Weed
♦ 221

Bellevalia glauca **(Lind) Kunth**
Liliaceae/Hyacinthaceae
= *Hyacinthus ciliatus* Cyrill. (NoR)
♦ Weed
♦ 243

Bellevalia macrobotrys **Boiss.**
Liliaceae/Hyacinthaceae
♦ Weed
♦ 221

Bellevalia sessiliflora **(Viv.) Kunth**
Liliaceae/Hyacinthaceae
♦ Weed
♦ 221

Bellis **L. spp.**
Asteraceae
♦ common daisy, daisy, English daisy

♦ Weed, Naturalised
♦ 198, 272
♦ herbal.

Bellis annua **L.**
Asteraceae
♦ annual daisy
♦ Weed
♦ 88, 94
♦ bH.

Bellis perennis **L.**
Asteraceae
♦ English daisy, lawndaisy, European
daisy
♦ Weed, Naturalised, Garden Escape,
Environmental Weed, Cultivation
Escape
♦ 7, 15, 23, 34, 42, 86, 87, 88, 94, 98, 101,
116, 136, 161, 165, 176, 198, 203, 211,
212, 218, 241, 272, 280, 290, 300
♦ pH, arid. cultivated, herbal. Origin:
Eurasia.

Benincasa hispida **(Thunb.) Cogn.**
Cucurbitaceae
Benincasa cerifera Savi
♦ fagufagu, wax gourd, Chinese
winter melon, doan gwa, talvimeloni
♦ Naturalised, Introduced
♦ 86, 101, 230
♦ aH, cultivated, herbal. Origin: south-
east Asia.

Berberis **L. spp.**
Berberidaceae
♦ barberry, berberis, happomarjat
♦ Quarantine Weed, Naturalised
♦ 39, 198, 220
♦ herbal, toxic.

Berberis aggregata **C.K.Schneid.**
Berberidaceae
♦ clustered barberry
♦ Naturalised
♦ 40
♦ S, cultivated.

Berberis aquifolium **Pursh**
Berberidaceae
= *Mahonia aquifolium* (Pursh) Nutt.
♦ mountain grape, holly leaved
barberry
♦ Weed
♦ 39, 87, 88
♦ S, herbal, toxic.

Berberis aristata **DC.**
Berberidaceae
Berberis floribunda hort. [confused with
in cult.]
♦ chitra, Nepal barberry
♦ Weed, Naturalised, Cultivation
Escape
♦ 39, 86, 98, 203, 252
♦ S, cultivated, herbal, toxic. Origin:
India, Nepal.

Berberis canadensis **Mill.**
Berberidaceae
Berberis angulizans hort. ex Massias
♦ American barberry, Allegheny
barberry
♦ Weed
♦ 87, 88, 218
♦ S/T, promoted, herbal.

Berberis cretica **L.**
Berberidaceae
♦ Weed
♦ 272

Berberis darwinii **Hook.**
Berberidaceae
♦ Darwin's barberry, berberis,
barberry
♦ Weed, Quarantine Weed,
Naturalised, Garden Escape,
Environmental Weed
♦ 15, 40, 72, 86, 88,
98, 101, 152, 155, 176, 181, 198, 203, 225,
246, 280, 289, 296
♦ S, cultivated, herbal. Origin: South
America.

Berberis fendleri **A.Gray**
Berberidaceae
♦ Colorado barberry
♦ Weed
♦ 87, 88, 218
♦ S, promoted, herbal.

Berberis glaucocarpa **Stapf**
Berberidaceae
♦ barberry, great barberry
♦ Weed, Naturalised, Environmental
Weed
♦ 15, 165, 181, 225, 246, 280
♦ T. Origin: western Himalayas.

Berberis haematocarpa **Woot.**
Berberidaceae
= *Mahonia haematocarpa* (Woot.) Fedde
♦ red barberry
♦ Weed
♦ 87, 88
♦ S, cultivated, herbal.

Berberis japonica **R.Br.**
Berberidaceae
♦ Japanese barberry
♦ Weed
♦ 80

Berberis julianiae **Schneid.**
Berberidaceae
♦ Julian's berberis
♦ Naturalised
♦ 101
♦ herbal.

Berberis ruscifolia **Lam.**
Berberidaceae
♦ quebrachillo
♦ Weed
♦ 295
♦ S, promoted.

Berberis soulieana **C.K.Schneid.**
Berberidaceae
♦ Naturalised
♦ 280
♦ S, cultivated.

Berberis thunbergii **DC.**
Berberidaceae
♦ Japanese barberry,
Japaninhappomarja, berberis
♦ Weed, Naturalised, Introduced,
Garden Escape, Environmental Weed,
Cultivation Escape
♦ 4, 23, 42, 80, 87, 88, 101, 102, 132, 133,
142, 151, 161, 195, 218, 222, 224
♦ S, cultivated, herbal. Origin: Japan.

***Berberis trifoliolata* Moric.**
Berberidaceae
= *Mahonia trifoliolata* (Moric.) Fedde
♦ Weed
♦ 87, 88
♦ herbal.

***Berberis vulgaris* L.**
Berberidaceae
Berberis acutifolia Prantl
♦ common barberry, European barberry, epine vinette, vinetteier, epine vinette commune, ruostehappomarja
♦ Weed, Noxious Weed, Naturalised, Introduced, Cultivation Escape
♦ 23, 39, 40, 42, 49, 54, 80, 86, 87, 88, 101, 133, 161, 195, 218, 222, 224, 243, 252, 269, 272, 280, 299
♦ S/T, cultivated, herbal, toxic. Origin: Eurasia.

***Berberis wilsonae* Hemsl.**
Berberidaceae
♦ Naturalised
♦ 280
♦ cultivated.

***Berchemia scandens* (Hill) K.Koch**
Rhamnaceae
♦ Alabama supplejack, supplejack, rattan vine
♦ Weed
♦ 87, 88, 218, 247
♦ cultivated, herbal, toxic.

***Bergenia cordifolia* (Haw.) Sternb.**
Saxifragaceae
♦ herttavuorenkilpi
♦ Cultivation Escape
♦ 42
♦ pH, cultivated, herbal.

***Bergenia crassifolia* (L.) Fritsch**
Saxifragaceae
Bergenia bifolia (Haw.) A.Braun, *Saxifraga crassifolia* L.
♦ elephant's ears, soikkovuorenkilpi
♦ Naturalised, Cultivation Escape
♦ 40, 42, 198
♦ pH, cultivated, herbal.

***Bergerocactus* Britt. & Rose spp.**
Cactaceae
♦ snake cactus
♦ Weed, Quarantine Weed
♦ 76, 88, 203, 220

***Bergia ammannioides* Roxb.**
Elatinaceae
Bergia oryzetorum Fenzl, *Elatine ammannioides* (Roth) Wright & Arn.
♦ Weed
♦ 87, 88, 170, 204

***Bergia capensis* L.**
Elatinaceae
Bergia aquatica Roxb., *Bergia repens* Blume., *Bergia verticillata* Willd., *Elatine verticillata* Willd.
♦ Weed, Quarantine Weed, Introduced
♦ 38, 76, 87, 88, 170, 191, 203, 220
♦ H, aqua, cultivated.

***Bergia texana* (Hook.) Seub. ex Walp.**
Elatinaceae
♦ bergia, Texas bergia

♦ Weed
♦ 161
♦ aH.

***Berkheya annectens* Harv.**
Asteraceae
♦ thistle thorn
♦ Native Weed
♦ 121
♦ pH. Origin: southern Africa.

***Berkheya erysithales* (DC.) Roessler**
Asteraceae
♦ Weed
♦ 88, 158
♦ Origin: South Africa.

***Berkheya heterophylla* O.Hoffm.**
Asteraceae
♦ prickly gousblom
♦ Naturalised
♦ 101

***Berkheya heterophylla* O.Hoffm. var. *heterophylla* (Thunb.) O.Hoffm.**
Asteraceae
♦ Native Weed
♦ 121
♦ pH. Origin: southern Africa.

***Berkheya macrocephala* J.M.Wood**
Asteraceae
♦ Weed, Quarantine Weed
♦ 88, 158, 220
♦ cultivated. Origin: South Africa.

***Berkheya multijuga* (DC.) Roessler**
Asteraceae
Crocodilodes multijugum Kuntze (see), *Stobaea multijuga* DC. (see)
♦ Quarantine Weed
♦ 220

***Berkheya pinnatifida* (Thunb.) Thell.**
Asteraceae
♦ Native Weed
♦ 121
♦ pH. Origin: southern Africa.

***Berkheya purpurea* Benth. & Hook.f. ex Mast.**
Asteraceae
Crocodilodes purpureum Kuntze (see), *Stobaea purpurea* DC. (see)
♦ Quarantine Weed
♦ 220

***Berkheya radula* (Harv.) de Willd.**
Asteraceae
♦ Native Weed
♦ 121
♦ pH. Origin: southern Africa.

***Berkheya rigida* (Thunb.) Bol. & Woll.**
Asteraceae
Stobaea rigida Thunb.
♦ African thistle, Augusta thistle, berkheya thistle, Hamelin thistle
♦ Weed, Quarantine Weed, Noxious Weed, Naturalised, Native Weed, Environmental Weed
♦ 7, 62, 72, 76, 86, 87, 88, 98, 121, 147, 158, 176, 198, 203
♦ pH. Origin: southern Africa.

***Berkheya setifera* DC.**
Asteraceae
♦ Quarantine Weed

♦ 220
♦ arid.

***Berkheya speciosa* O.Hoffm.**
Asteraceae
Stobaea speciosa DC. (see)
♦ Quarantine Weed
♦ 220

***Berkheyopsis linearifolia* (Bolus) Burtt Davy**
Asteraceae
♦ Quarantine Weed
♦ 220

***Berteroa incana* (L.) DC.**
Brassicaceae
Alyssum incanum L. (see)
♦ hoary alyssum, hoary false alyssum, hoary false madwort
♦ Weed, Noxious Weed, Naturalised, Introduced, Environmental Weed, Casual Alien
♦ 4, 21, 23, 40, 42, 56, 70, 80, 87, 88, 94, 101, 104, 161, 174, 210, 218, 229, 272
♦ b/pH, cultivated, herbal. Origin: Eurasia.

***Berteroa mutabilis* (Vent.) DC.**
Brassicaceae
♦ roadside false madwort
♦ Weed, Naturalised
♦ 101, 272

***Bertholletia excelsa* Humb. & Bonpl.**
Lecythidaceae
♦ Brazil nuts, castanheira, castanheira do Brasil
♦ Introduced
♦ 230
♦ T, herbal.

***Bertya rosmarinifolia* Planch.**
Euphorbiaceae
♦ Weed, Naturalised
♦ 98, 203
♦ cultivated.

***Berula erecta* (Huds.) Coville**
Apiaceae
Apium berula Caruel, *Berula angustifolia* Mert. & W.D.J.Koch, *Siella erecta* (Huds.) M.Pimen., *Sium angustifolium* L. *nom. illeg.* (see), *Sium erectum* Huds. (see)
♦ cutleaf water parsnip, water parsnip, narrow leaved water parsnip, lesser water parsnip
♦ Weed, Naturalised
♦ 7, 80, 86, 98, 161, 203, 272
♦ wpH, cultivated, herbal, toxic. Origin: northern hemisphere.

***Beta procumbens* L.**
Chenopodiaceae
♦ cultivated beet
♦ Naturalised
♦ 101

***Beta vulgaris* L.**
Chenopodiaceae
Beta cicla L., *Beta maritima* L., *Beta vulgaris* L. ssp. *maritima* (L.) Arcang. (see)
♦ sea beet, silver beet, wild beet, beet, juurikas
♦ Weed, Naturalised, Environmental

Weed, Cultivation Escape
- 15, 34, 39, 42, 72, 87, 88, 94, 98, 101, 115, 116, 161, 176, 181, 185, 199, 203, 221, 243, 253, 272, 280
- aH, cultivated, herbal, toxic. Origin: Europe.

Beta vulgaris L. ssp. maritima (L.) Arcang.
Chenopodiaceae
= *Beta vulgaris* L.
- wild beet, silver beet, sea beet
- Weed, Naturalised, Environmental Weed
- 70, 86, 176, 198, 300
- cultivated. Origin: North Africa, Eurasia.

Beta vulgaris L. ssp. vulgaris
Chenopodiaceae
- root beet
- Naturalised
- 40, 300
- cultivated. Origin: horticultural.

Betonica officinalis L.
Lamiaceae
= *Stachys officinalis* (L.) Trevis.
- betony
- Quarantine Weed
- 220
- herbal.

Betula alba L. nom. utique rej. prop.
Betulaceae
= *Betula pubescens* Ehrh.
- silver birch
- Weed
- 87, 88
- cultivated, herbal.

Betula alleghaniensis Britt.
Betulaceae
- yellow birch
- Weed
- 87, 88, 218
- T, cultivated, herbal.

Betula lenta L.
Betulaceae
Betula carpinifolia Ehrh.
- sweet birch, black birch, cherry birch
- Weed
- 8, 39, 80, 87, 88, 218
- T, cultivated, herbal, toxic.

Betula nigra L.
Betulaceae
- river birch, red birch, black birch
- Weed, Naturalised, Garden Escape, Environmental Weed
- 54, 86, 87, 88, 218
- T, cultivated, herbal. Origin: eastern North America.

Betula occidentalis Hook.
Betulaceae
Betula fontinalis Sarg.
- water birch
- Weed
- 87, 88, 218
- S, cultivated, herbal.

Betula papyrifera Marsh.
Betulaceae
Betula alba L. var. *papyrifera* (Marsh.) Spach, *Betula papyracea* Aiton.

- paper birch
- Weed
- 87, 88, 218
- T, cultivated, herbal.

Betula pendula Roth
Betulaceae
- silver birch, European white birch, weeping birch, warty birch
- Weed, Naturalised, Environmental Weed
- 4, 40, 70, 80, 86, 88, 101, 104, 181, 280
- T, cultivated, herbal. Origin: Eurasia.

Betula platyphylla Sukaczev
Betulaceae
Betula verrucosa (Sukaczev) Lindl. ex Jansen var. *platyphylla*
- Asian white birch
- Naturalised
- 101
- T, cultivated, herbal.

Betula platyphylla Sukaczev var. platyphylla
Betulaceae
- Asian white birch
- Naturalised
- 101

Betula platyphylla Sukaczev var. szechuanica Schneid.
Betulaceae
- Szechuan white birch
- Naturalised
- 101

Betula populifolia Marsh.
Betulaceae
- gray birch, grey birch
- Weed
- 87, 88, 218
- T, cultivated, herbal.

Betula pubescens Ehrh.
Betulaceae
Betula alba L. nom. utique rej. prop. (see), *Betula glauca* Wend., *Betula tomentosa* Reither & Abel
- downy birch, birch, white birch, breza biela
- Weed, Naturalised
- 54, 70, 86, 88, 101, 272
- T, cultivated, herbal. Origin: Europe.

Betula pubescens Ehrh. ssp. pubescens
Betulaceae
- downy birch
- Naturalised
- 40

Bewsia biflora (Hack.) Gooss.
Poaceae
Diplachne biflora Hack.
- Native Weed
- 121
- pG. Origin: southern Africa.

Biddulphia laevis Ehrenb.
Biddulphiaceae
- Weed
- 197
- diatom.

Bidens alba (L.) DC.
Asteraceae
- beggar's ticks, romerillo
- Weed

- 3, 6, 84, 88, 191, 243, 249, 255
- herbal.

Bidens alba (L.) DC. var. radiata (Sch.-Dip.) Ballard
Asteraceae
Bidens pilosa L. var. *radiata* Sch.Bip. (see)
- margarita, romerillo, bidens, aceitilla
- Weed
- 261, 281

Bidens aristosa (Michx.) Britt.
Asteraceae
Bidens polylepis S.F.Blake (see)
- bearded beggar's ticks
- Weed, Naturalised
- 161, 207, 287
- herbal. Origin: North America.

Bidens aurea (Aiton) Sherff
Asteraceae
- Arizona beggar's ticks
- Weed, Quarantine Weed, Naturalised
- 70, 76, 87, 88, 220, 253, 287, 300
- pH, cultivated.

Bidens beckii Torr. ex Spreng.
Asteraceae
= *Megalodonta beckii* (Torr. ex Spreng.) Greene var. *beckii* (NoR)
- beggar's ticks
- Weed
- 159
- aqua, cultivated.

Bidens bipinnata L.
Asteraceae
Bidens pilosa L. var. *bipinnata* (L.) Hook.f., *Kerneria bipinnata* Gren. & Godr.
- Spanish needles, cobbler's pegs, beggar's ticks, black fellows, blackjack, burr marigold, pitchfork, Spanish blackjack, sweet hearts, soapbush needles, beggar's lice
- Weed, Naturalised, Native Weed, Environmental Weed
- 7, 23, 24, 49, 51, 86, 87, 88, 93, 94, 98, 121, 158, 161, 174, 203, 205, 210, 218, 243, 269, 272, 274, 275, 286, 287, 297
- aH, arid, cultivated, herbal. Origin: Eurasia.

Bidens biternata (Lour.) Merr. & Sherff ex Sherff
Asteraceae
- yellow flowered blackjack, blackjack, five leaved blackjack, kon cham, beggar's ticks
- Weed
- 13, 50, 51, 66, 87, 88, 121, 238, 263, 286
- aH, promoted. Origin: eastern Asia.

Bidens cernua L.
Asteraceae
Bidens cernuus L.
- nodding beggar's ticks, burr marigold, sticktight, nodding burr marigold, nuokkurusokki
- Weed, Native Weed
- 44, 87, 88, 136, 159, 161, 174, 212, 218, 243, 253, 272, 297
- H, herbal. Origin: Europe.

Bidens chrysanthemoides Michx.
Asteraceae
♦ Weed
♦ 87, 88

Bidens comosa (Gray) Wieg.
Asteraceae
= *Bidens tripartita* L.
♦ leafybract beggar's ticks, beggar's ticks
♦ Weed
♦ 87, 88, 159, 218
♦ herbal.

Bidens connata Muhl. ex Willd.
Asteraceae
Bidens decipiens Warnst. (see), *Bidens tripartita* L. var. *fallax* Warnst.
♦ purplestem beggar's ticks
♦ Weed
♦ 87, 88, 218
♦ cultivated, herbal.

Bidens cynapiifolia Kunth
Asteraceae
♦ West Indian beggar's ticks
♦ Weed, Environmental Weed
♦ 14, 87, 88, 257
♦ herbal.

Bidens decipiens Warnst.
Asteraceae
= *Bidens connata* Muhl. ex Willd.
♦ Weed
♦ 44

Bidens formosa (Bonato) Sch.Bip.
Asteraceae
= *Cosmos bipinnatus* Cav.
♦ cosmos
♦ Weed
♦ 121, 158
♦ aH, toxic. Origin: Central America, West Indies.

Bidens frondosa L.
Asteraceae
♦ devil's beggar's ticks, beggar's ticks, sticktights, devil's boot jack, burr marigold, pitchfork weed, tickseed sunflower, rayless marigold, beggar's lice
♦ Weed, Naturalised, Native Weed, Casual Alien
♦ 15, 23, 34, 42, 49, 70, 87, 88, 159, 161, 165, 174, 210, 211, 218, 253, 263, 280, 286, 287, 297
♦ aH, promoted, herbal, toxic. Origin: North America.

Bidens gardneri Baker
Asteraceae
♦ ridge beggar's ticks
♦ Weed, Naturalised
♦ 101, 255
♦ Origin: Brazil.

Bidens humilis Kunth
Asteraceae
= *Bidens triplinervia* Kunth var. *triplinervia* (NoR)
♦ Weed
♦ 87, 88

Bidens laevis (L.) Britton, Sterns & Pogg.
Asteraceae

♦ smooth beggar's ticks
♦ Weed, Naturalised
♦ 241, 287, 295, 300
♦ pH, aqua, herbal.

Bidens megapotamica Spreng.
Asteraceae
= *Thelesperma megapotamicum* (Spreng.) Kuntze
♦ Weed
♦ 87, 88

Bidens mitis (Michx.) Sherff
Asteraceae
♦ smallfruit beggar's ticks
♦ Quarantine Weed
♦ 258
♦ herbal.

Bidens odorata Cav.
Asteraceae
♦ Weed
♦ 243

Bidens pachyloma (Oliv. & Hiern) Cuf.
Asteraceae
♦ Weed
♦ 240
♦ aH.

Bidens parviflora Willd.
Asteraceae
♦ Weed, Naturalised
♦ 275, 287, 297
♦ aH, promoted, herbal.

Bidens pilosa L.
Asteraceae
Bidens chinensis Willd., *Bidens leucantha* Willd., *Bidens pilosa* L. var. *minor* (Blume) Sherff (see)
♦ hairy beggar's ticks, cobbler's pegs, beggar's ticks, pitchforks, sticktights, Spanish needle, fisi'uli, kofe tonga, tae puaka, matua kamate, blackjack, burr marigold, cadillo de huerta, chipaca, maswquia, mozote, papunga, puen nok sai, nyamaradza, amor seco, coambi, mozote aceitilla, picão preto, kalasa
♦ Weed, Naturalised, Introduced, Garden Escape, Environmental Weed, Casual Alien
♦ 3, 6, 7, 14, 23, 42, 50, 51, 53, 55, 68, 80, 87, 88, 93, 94, 98, 101, 107, 116, 121, 134, 157, 158, 161, 167, 186, 198, 203, 204, 205, 206, 228, 236, 237, 238, 240, 243, 245, 249, 255, 257, 261, 262, 269, 270, 273, 275, 276, 280, 281, 286, 287, 290, 295, 297
♦ aH, arid, cultivated, herbal. Origin: tropical America.

Bidens pilosa L. fo. decumbens Sherff
Asteraceae
♦ Naturalised
♦ 287

Bidens pilosa L. var. bisetosa S.Ohtani & S.Suzuki
Asteraceae
♦ Naturalised
♦ 287

Bidens pilosa L. var. minor (Blume) Sherff
Asteraceae

= *Bidens pilosa* L.
♦ blackjack, Spanish needles, beggar's ticks, cobbler's pegs
♦ Weed, Naturalised
♦ 12, 13, 170, 273, 274, 286, 287

Bidens pilosa L. var. pilosa
Asteraceae
♦ common beggar's ticks
♦ Weed, Naturalised, Introduced
♦ 34, 86, 230, 241, 300
♦ aH.

Bidens pilosa L. var. radiata Sch.Bip.
Asteraceae
= *Bidens alba* (L.) DC. var. *radiata* (Sch.Bip.) Ballard ex T.E.Melchert
♦ hairy beggar's ticks
♦ Weed, Naturalised
♦ 218, 263, 273, 287
♦ aH.

Bidens polylepis S.F.Blake
Asteraceae
= *Bidens aristosa* (Michx.) Britt.
♦ long bracted beggar's ticks, kultarusokki, coreopsis beggar's ticks
♦ Weed, Naturalised, Environmental Weed, Casual Alien
♦ 42, 80, 87, 88, 151, 218, 287
♦ herbal.

Bidens pseudocosmos Sherff
Asteraceae
♦ Naturalised
♦ 300

Bidens radiata Thuill.
Asteraceae
Bidens fastigiatus Michal., *Bidens platycephalus* Oerst.
♦ säderusokki, burr marigold
♦ Weed
♦ 87, 88, 272

Bidens radiata Thuill. var. pinnatifida (Turcz.) Kitam.
Asteraceae
♦ ezonotaukogi
♦ Weed
♦ 286

Bidens reptans (L.) G.Don
Asteraceae
♦ manzanilla trepador
♦ Weed
♦ 87, 88

Bidens riparia Kunth
Asteraceae
♦ Weed
♦ 87, 88

Bidens schimperi Sch.Bip. ex Walp.
Asteraceae
♦ Weed
♦ 87, 88

Bidens squarrosa Kunth
Asteraceae
♦ Weed
♦ 157
♦ pC.

Bidens steppia (Steetz) Sherff
Asteraceae
Bidens insecta (S.Moore) Sherff
♦ lisanda

♦ Weed, Native Weed
♦ 87, 88, 121
♦ aH. Origin: southern Africa.

Bidens subalternans DC.
Asteraceae
Bidens megapotamica O.E.Schulz (*non* Spreng.), *Bidens platensis* Manganaro, *Bidens quadrangularis* DC.
♦ hairy beggar's ticks, amor seco, cobbler's pegs, picao do campo, fura capa
♦ Weed, Naturalised
♦ 86, 98, 203, 236, 237, 253, 255, 269, 270, 295, 300
♦ cultivated. Origin: Americas.

Bidens subalternans DC. var. *simulans* Scherff.
Asteraceae
♦ Weed
♦ 270

Bidens subalternans DC. var. *subalternans*
Asteraceae
♦ Weed
♦ 270

Bidens sulphurea (Cav.) Sch.Bip.
Asteraceae
Cosmus sulphureus Cav.
♦ cosmo amarelo
♦ Weed
♦ 255
♦ Origin: Mexico.

Bidens tripartita L.
Asteraceae
Bidens comosa (Gray) Wieg. (see), *Bidens nodiflora* L., *Bidens taquetii* Lév. & Van., *Bidens tripartitus* L., *Bidens shimadai* Hay.
♦ triffid burr marigold, tripartite burr marigold, burr beggar's ticks, dreiteiliger zweizahn, tummarusokki, threelobe beggar's ticks
♦ Weed, Naturalised, Introduced
♦ 23, 34, 44, 68, 70, 86, 87, 88, 94, 98, 198, 203, 243, 253, 263, 269, 272, 275, 280, 286, 297
♦ aH, cultivated, herbal. Origin: Americas.

Bidens vulgata E.Greene
Asteraceae
♦ big devil's beggar's ticks, tall beggar's ticks
♦ Weed, Introduced
♦ 23, 34, 87, 88, 161, 218
♦ aH, herbal.

Bifora Hoffm. spp.
Apiaceae
♦ carrot weed, bishop, bifora
♦ Weed
♦ 243
♦ H.

Bifora radians M.Bieb.
Apiaceae
Anidrum radians Calest., *Bifora radicans* Hinterh., *Biforis radians* Spreng., *Coriandrum radians* E.H.L.Krause, *Selinum radians* E.H.L.Krause
♦ wild bishop, poilikki

♦ Weed, Naturalised, Casual Alien
♦ 42, 44, 70, 87, 88, 94, 101, 243, 253, 272
♦ cultivated, herbal.

Bifora testiculata (L.) Roth
Apiaceae
Anidrum testiculatum Kuntze, *Bifora testicularis* Bubani, *Corion testiculatum* Hoffmanns. & Link
♦ bifora, carrot weed, bird's eye
♦ Weed, Quarantine Weed, Noxious Weed, Naturalised
♦ 70, 76, 86, 87, 88, 94, 98, 101, 171, 203, 272, 287
♦ cultivated, herbal. Origin: Eurasia.

Bigelowia cooperi A.Gray
Asteraceae
= *Ericameria cooperi* (A.Gray) H.M.Hall (NoR)
♦ Quarantine Weed
♦ 220

Bigelowia macronema (A.Gray) M.E.Jones
Asteraceae
Haplopappus macronema A.Gray (see)
♦ Quarantine Weed
♦ 220

Bigelowia nelsonii Fernald
Asteraceae
= *Ericameria nelsonii* S.F.Blake
♦ Quarantine Weed
♦ 220

Bignonia capreolata L.
Bignoniaceae
♦ crossvine
♦ Weed, Quarantine Weed, Naturalised
♦ 39, 76, 86, 87, 88, 218, 220
♦ cultivated, herbal, toxic.

Bignonia exoleta Vell.
Bignoniaceae
= *Macfadyena unguis-cati* (L.) A.H.Gentry
♦ Weed
♦ 87, 88

Bignonia unguis-cati L.
Bignoniaceae
= *Macfadyena unguis-cati* (L.) A.H.Gentry
♦ Weed
♦ 87, 88
♦ herbal.

Bilderdykia convolvulus (L.) Dumort
Polygonaceae
= *Fallopia convolvulus* (L.) Á.Löve
♦ bear bind, black bindweed, black knotweed, climbing knotweed, corn bindweed, ivy bindweed, knot bindweed, wild buckwheat
♦ Weed, Naturalised
♦ 70, 121, 241, 243, 272, 287
♦ aC, herbal. Origin: Eurasia.

Bilderdykia dumetorum (L.) Dumort.
Polygonaceae
= *Fallopia dumetorum* (L.) Holub (NoR)
[see *Polygonum dumetorum* L.]
♦ copse bindweed
♦ Weed

♦ 70, 272

Billbergia pyramidalis (Sims) Lindl.
Bromeliaceae
♦ foolproofplant, pyramid billbergia, kartiopapinkaura
♦ Weed, Naturalised, Cultivation Escape
♦ 101, 179, 261
♦ cultivated, herbal. Origin: Brazil.

Binghamia Britton & Rose spp.
Cactaceae
♦ Weed, Quarantine Weed
♦ 76, 88, 203, 220

Biophytum reinwardtii (Zucc.) Klotz
Oxalidaceae
♦ Weed
♦ 87, 88
♦ herbal.

Biophytum sensitivum (L.) DC.
Oxalidaceae
Biophytum petersianum Klotz.
♦ sikerpud
♦ Weed
♦ 66, 87, 88
♦ aH, cultivated, herbal.

Bischofia javanica Blume
Euphorbiaceae/Bischofiaceae
♦ bischofia, Javanese bishopwood, bishop weed, Java wood, o'a, toog, koka, tongotongo, koka ndamu, tongo, tongatonga, tea
♦ Weed, Noxious Weed, Naturalised, Garden Escape, Environmental Weed
♦ 3, 22, 80, 88, 101, 112, 122, 142, 151, 179, 191
♦ T, cultivated, herbal. Origin: southern Asia, Polynesia, Malaysia.

Biscutella auriculata L.
Brassicaceae
Iondraba auriculata (L.) Webb & Berthel. (see)
♦ Weed
♦ 88, 94
♦ cultivated, herbal. Origin: North Africa, southern Europe.

Biscutella didyma L.
Brassicaceae
Biscutella columnae (Ten.) Halacs.
♦ Weed, Quarantine Weed
♦ 76, 87, 88, 203, 220
♦ cultivated.

Biscutella laevigata L.
Brassicaceae
Crucifera biscutella E.H.L.Krause
♦ Weed
♦ 87, 88, 94
♦ cultivated, herbal.

Biserrula pelecinus L.
Fabaceae/Papilionaceae
♦ biserrula
♦ Weed
♦ 70
♦ cultivated.

Bistella digyna (Retz.) Bullock
Vahliaceae/Saxifragaceae
= *Vahlia digyna* (Retz.) Kuntze
♦ Weed
♦ 221

Bistorta milletii **H.Lév.**
Polygonaceae
= *Polygonum milletii* (H.Lév.) H.Lév.
♦ Quarantine Weed
♦ 220

Bistorta taipaishanensis **(H.W.Kung)**
Yonekura & H.Ohashi
Polygonaceae
♦ Quarantine Weed
♦ 220

Bituminaria bituminosa **(L.) Stirt.**
Fabaceae/Papilionaceae
Aspalthium bituminosum (L.) Kuntze
(see), *Psoralea bituminosa* L. (see)
♦ Arabian pea
♦ Weed, Quarantine Weed,
Naturalised, Introduced
♦ 76, 88, 101, 220, 228
♦ arid.

Bixa orellana **L.**
Bixaceae
♦ annatto, lipstick tree, lubenga,
eyango
♦ Weed, Naturalised, Introduced,
Environmental Weed, Cultivation
Escape
♦ 22, 86, 98, 101, 203, 228, 230, 257, 261
♦ T, arid, cultivated, herbal. Origin:
continental tropical America.

Blackiella inflata **(F.Muell.) Aellen**
Chenopodiaceae
= *Atriplex lindleyi* Moq. ssp. *inflata*
(Muell.) P.G.Wilson
♦ little saltbush, spongefruit saltbush
♦ Weed, Introduced
♦ 121, 228
♦ S, arid. Origin: Australia.

Blackstonia perfoliata **(L.) Huds.**
Gentianaceae
Chlora perfoliata L., *Gentiana perfoliata*
L., *Seguiera perfoliata* Kuntze
♦ yellow wort
♦ Weed, Naturalised
♦ 86, 88, 94, 98, 198, 203, 272, 280
♦ aH, cultivated, herbal. Origin:
Mediterranean.

Blackstonia serotina **G.Beck**
Gentianaceae
= *Blackstonia perfoliata* (L.) Huds. ssp.
serotina (W.D.J.Koch ex Rchb.) Vollm.
(NoR)
♦ Weed
♦ 87, 88

Blainvillea acmella **(L.) Philipson**
Asteraceae
♦ Weed
♦ 66
♦ Origin: Africa, China, India.

Blainvillea biaristata **DC.**
Asteraceae
Olygogyne megapotamica DC.,
Calyptocarpus megapotamica Sch.Bip.
♦ picao, erva palha
♦ Weed
♦ 88, 255
♦ Origin: South America.

Blainvillea gayana **Cass.**
Asteraceae
Wedelia gossweileri S.Moore
♦ Introduced
♦ 29
♦ aH. Origin:
Africa.

Blainvillea rhomboidea **Cass.**
Asteraceae
♦ erva palha
♦ Weed
♦ 87, 88, 255
♦ Origin: Brazil.

Blechnum capense **(L.) Schlecht.**
Blechnaceae
♦ Weed
♦ 87, 88
♦ cultivated.

Blechnum orientale **L.**
Blechnaceae
♦ Weed
♦ 87, 88
♦ H, cultivated, herbal. Origin: Asia.

Blechnum spicant **(L.) Roth**
Blechnaceae
Lomaria spicant (L.) Desv.
♦ hard fern, kampasaniainen, deer
fern
♦ Weed
♦ 70, 272
♦ pH, cultivated, herbal.

Blechnum tabulare **(Thunb.) Kuhn**
Blechnaceae
Pteris tabularis Thunb.
♦ Native Weed
♦ 121
♦ pH, cultivated. Origin: southern
Africa.

Blechum brownei **Juss.**
Acanthaceae
= *Blechum pyramidatum* (Lam.) Urb.
♦ Weed
♦ 191
♦ herbal.

Blechum brownei **Juss. fo. *puberulum***
Leonard
Acanthaceae
♦ Introduced
♦ 230
♦ Origin: Mexico, Central America,
West Indies & northern South America.

Blechum pyramidatum **(Lam.) Urb**
Acanthaceae
Blechum brownei Juss. (see)
♦ Browne's blechum
♦ Weed, Quarantine Weed
♦ 6, 14, 76, 87, 88, 157, 179, 203, 206,
220, 243
♦ pH, arid, herbal.

Blepharis ciliaris **(L.) Burtt**
Acanthaceae
♦ Weed
♦ 221, 240
♦ aH.

Blepharis exigua **(Zoll.) Valeton ex**
Backer
Acanthaceae

♦ Weed
♦ 13

Blepharis javanica **Brem.**
Acanthaceae
♦ Weed
♦ 13

Blepharis maderaspatensis **(L.) Roth**
Acanthaceae
♦ Weed, Quarantine Weed
♦ 76, 87, 88, 203, 220
♦ pH.

Blepharis molluginifolia **Pers.**
Acanthaceae
♦ Weed
♦ 87, 88

Blepharis subvolubilis **C.B.Clarke var.**
subvolubilis
Acanthaceae
♦ Native Weed
♦ 121
♦ S. Origin: southern Africa.

Bletilla striata **(Thunb.) Rchb.f.**
Orchidaceae
Bletia hyacinthina R.Br.
♦ urn orchid, hyacinth orchid
♦ Naturalised, Casual Alien
♦ 101, 280
♦ pH, cultivated, herbal.

Blighia sapida **König**
Sapindaceae
♦ akee, seso vegetal
♦ Weed, Naturalised, Introduced
♦ 22, 39, 87, 88, 101, 161, 261
♦ T, herbal, toxic. Origin: west tropical
Africa.

Blumea alata **(D.Don) DC.**
Asteraceae
Laggera alata (D.Don) Sch.Bip. ex Oliv.
♦ ihoko
♦ Native Weed
♦ 121
♦ pS. Origin: southern Africa.

Blumea aurita **DC.**
Asteraceae
= *Laggera aurita* Sch.Bip. ex C.B.Clarke
♦ Weed
♦ 88

Blumea balsamifera **(L.) DC.**
Asteraceae
♦ nagi camphor
♦ Weed
♦ 13, 87, 88
♦ herbal.

Blumea eriantha **DC.**
Asteraceae
♦ Weed
♦ 66

Blumea gariepina **DC.**
Asteraceae
♦ wolbos
♦ Weed, Native Weed
♦ 88, 121, 158
♦ pS, arid. Origin: southern Africa.

Blumea hieracifolia **(Spreng.) DC.**
Asteraceae
♦ Weed
♦ 235

Blumea lacera (Burm.f.) DC.
Asteraceae
Conyza lacera Burm.f., *Conyza javanica*
Blume
♦ blumea, batard
♦ Weed
♦ 13, 87, 88, 170, 276
♦ pH, arid, herbal.

Blumea laciniata (Roxb.) DC.
Asteraceae
♦ cutleaf false oxtongue
♦ Weed, Naturalised
♦ 87, 88, 101

Blumea lanceolaria (Roxb.) Druce
Asteraceae
♦ Weed
♦ 88, 191, 204

Blumea mollis (D.Don) Merr.
Asteraceae
Blumea wrightiana DC.
♦ Weed
♦ 13

Blumea napifolia DC.
Asteraceae
♦ naat noi
♦ Weed
♦ 209, 238
♦ aH.

Blumea riparia (Bl.) DC.
Asteraceae
Blumea chinensis (L.) DC.
♦ Weed
♦ 13

Blumea saxatilis Zoll. & Mor.
Asteraceae
♦ Weed
♦ 13

Blumea sessiliflora Decne.
Asteraceae
♦ sessileleaf false oxtongue
♦ Weed, Naturalised
♦ 13, 101

Blumea sinuata (Lour.) Merr.
Asteraceae
♦ Weed
♦ 87, 88

Blumea tenella DC.
Asteraceae
Blumea humifusa Boerl.
♦ Weed
♦ 13, 88, 170

Blumea viscosa (Mill.) Badillo
Asteraceae
♦ clammy false oxtongue
♦ Naturalised
♦ 101

Blumea wightiana DC.
Asteraceae
♦ Weed
♦ 87, 88

Blysmus compressus (L.) Panz.
Cyperaceae
Scirpus caricinus Schrad., *Schoenus compressus* L.
♦ flatsedge, litteäkaisla, sedgelike clubrush, broad blysmus

♦ Weed
♦ 272
♦ G, cultivated, herbal.

Blyxa aubertii Rich.
Hydrocharitaceae
Blyxa ecaudata Hayata, *Blyxa octandra* (Roxb.) Planch. ex Thwaites, *Blyxa roxburghii* Rich., *Diplosiphon orycetorum* Decne., *Valisneria octandra* Roxb.
♦ roundfruit blyxa, ohlot en pil
♦ Weed, Naturalised
♦ 87, 88, 101, 126, 191, 197, 286
♦ wa/pH, cultivated. Origin: Japan to Madagascar.

Blyxa aubertii Rich. var. *echinosperma* (Clarke) Cook & Luond.
Hydrocharitaceae
♦ Weed
♦ 170
♦ wH.

Blyxa ceratosperma Maxim. ex Asch. & Gurk.
Hydrocharitaceae
= *Blyxa echinosperma* (Clarke) Hook.f.
♦ Weed
♦ 87, 88, 286
♦ wH.

Blyxa echinosperma (Clarke) Hook.f.
Hydrocharitaceae
Blyxa bicaudata Nakai, *Blyxa ceratosperma* Maxim. ex Asch. & Gurk. (see), *Blyxa shimadai* Hay. (see), *Hydrotrophus echinospermus* C.B.Clarke
♦ Weed
♦ 87
♦ wH, cultivated.

Blyxa japonica (Miq.) Asch. & Guerke
Hydrocharitaceae
Blyxa laevissima Hayata
♦ bamboo plant
♦ Weed, Quarantine Weed
♦ 76, 87, 88, 126, 203, 220, 286, 297
♦ wa/pH, cultivated.

Blyxa shimadai Hay.
Hydrocharitaceae
= *Blyxa echinosperma* (Clarke) Hook.f.
♦ Weed
♦ 87, 88, 274
♦ wH.

Bocconia frutescens L.
Papaveraceae
Bocconia pearcei Hutch.
♦ plume poppy, bocconia, parrotweed
♦ Weed, Quarantine Weed, Noxious Weed
♦ 3, 22, 76, 80, 145, 191, 229
♦ S/T, cultivated, herbal. Origin: Central and South America.

Boehmeria japonica Miq. var. *longispica* Yahara
Urticaceae
♦ Weed
♦ 286

Boehmeria macrophylla D.Don
Urticaceae
♦ false nettle
♦ Weed, Environmental Weed
♦ 3, 152, 191

♦ S, cultivated. Origin: Himalayas, west China.

Boehmeria nivea (L.) Gaudich.
Urticaceae
Urtica nivea L., *Urtica candicans* Burn.f., *Boehmeria tenacissima* Gaudin
♦ Chinese grass, ramie
♦ Weed, Naturalised, Introduced
♦ 86, 87, 88, 101, 134, 179, 255, 261, 287
♦ S, cultivated, herbal. Origin: tropical Asia.

Boehmeria nivea Gaud. ssp. *nipononivea* Kitam.
Urticaceae
♦ Weed
♦ 286

Boehmeria nivea Gaud. var. *candicans* Wedd.
Urticaceae
♦ Naturalised
♦ 287

Boehmeria platyphylla D.Don
Urticaceae
Boehmeria spicata (Thunb.) Thunb. (see)
♦ Weed
♦ 87, 88
♦ S, cultivated. Origin: Australia.

Boehmeria spicata (Thunb.) Thunb.
Urticaceae
= *Boehmeria platyphylla* D.Don
♦ koakaso
♦ Weed
♦ 286
♦ S, cultivated. Origin: China, Japan.

Boehmeria sylvestrii W.T.Wang
Urticaceae
♦ Weed
♦ 286

Boerhavia coccinea Mill.
Nyctaginaceae
= *Boerhavia diffusa* L.
♦ boerhavia, hogweed, hog feed, mata pavo, patagon, scarlet spiderling
♦ Weed, Naturalised, Native Weed
♦ 3, 7, 14, 86, 88, 218, 243, 257, 271
♦ pH, herbal. Origin: Australia.

Boerhavia cordobensis Kuntze
Nyctaginaceae
♦ Weed
♦ 121
♦ pH. Origin: South America.

Boerhavia coulteri (Hook.f.) S.Wats.
Nyctaginaceae
♦ Coulter's spiderling
♦ Weed
♦ 88, 161, 218
♦ aH, herbal.

Boerhavia diffusa L.
Nyctaginaceae
Boerhavia adscendens Willd., *Boerhavia caribaea* Jacq., *Boerhavia coccinea* Mill. (see), *Boerhavia paniculata* Rich. (see), *Boerhavia repens* L. (see)
♦ tar vine, spreading hogweed, sow weed, creeping spiderling, red spiderling, ufi, spiderling, phak khom hin

♦ Weed, Naturalised, Introduced
♦ 12, 50, 66, 87, 88, 157, 158, 161, 217, 221, 228, 239, 249, 255, 261, 262, 276, 287, 295
♦ pH, arid, promoted, herbal. Origin: tropical South America.

Boerhavia diffusa L. var. *diffusa*
Nyctaginaceae
♦ erect boerhavia, spiderling
♦ Weed
♦ 121
♦ pH. Origin: South America.

Boerhavia dominii Meikle & Hewson
Nyctaginaceae
♦ tarvine
♦ Weed
♦ 55
♦ Origin: Australia.

Boerhavia erecta L.
Nyctaginaceae
♦ erect spiderling, spindlepod, erect boerhavia, upright spiderling
♦ Weed, Quarantine Weed, Naturalised
♦ 11, 12, 39, 51, 76, 87, 88, 121, 135, 157, 158, 170, 191, 199, 203, 204, 217, 218, 220, 240, 242, 243, 261, 276, 287
♦ aH, arid, herbal, toxic. Origin: tropical America.

Boerhavia paniculata Rich.
Nyctaginaceae
= *Boerhavia diffusa* L.
♦ Weed
♦ 87, 88
♦ herbal.

Boerhavia repanda Willd.
Nyctaginaceae
♦ Weed
♦ 87, 88
♦ herbal.

Boerhavia repens L.
Nyctaginaceae
= *Boerhavia diffusa* L.
♦ creeping spiderling, anena
♦ Weed, Native Weed
♦ 39, 88, 121, 218, 242
♦ H, arid, herbal, toxic. Origin: obscure.

Boerhavia scandens L.
Nyctaginaceae
Commicarpus scandens (L.) Standl. (see)
♦ climbing wartclub
♦ Weed
♦ 87, 88

Boerhavia schomburgkiana Oliv.
Nyctaginaceae
♦ Naturalised
♦ 7, 86
♦ Origin: Australia.

Boerhavia spicata Choisy
Nyctaginaceae
♦ creeping spiderling
♦ Weed
♦ 243
♦ aH, herbal.

Boerhavia vulvarifolia Poir.
Nyctaginaceae

♦ Weed
♦ 88

Boissiera pumilio (Trin.) Hack.
Poaceae
Pappophorum pumilio Trin.
♦ Weed
♦ 221
♦ G.

Bolboschoenus caldwellii (V.Cook) Sojak
Cyperaceae
♦ Weed
♦ 200
♦ pG, aqua, cultivated. Origin: Australia.

Bolboschoenus fluviatilis (Torr.) Sojak
Cyperaceae
= *Schoenoplectus fluviatilis* (Torr.) Strong (NoR) [see *Scirpus fluviatilis* (Torr.) Gray]
♦ marsh clubrush
♦ Weed
♦ 200
♦ pG, aqua, cultivated. Origin: Australia.

Bolboschoenus maritimus (L.) Palla
Cyperaceae
Bolboschoenus compactus (Hoffm.) Drob., *Bolboschoenus macrostachys* (Willd.) Grossh., *Bolboschoenus maritimus* ssp. *paludosus* (A.Nelson) Á. & D.Löve, *Bolboschoenus paludosus* (A.Nelson) Soó, *Schoenoplectus maritimus* (L.) Lye, *Scirpus maritimus* L. (see), *Scirpus maritimus* var. *paludosus* (A.Nelson) Kük. (see), *Scirpus paludosus* A.Nels. (see)
♦ sea clubrush
♦ Weed
♦ 88, 253
♦ G, herbal.

Boldoa purpurascens Cav. ex Lag.
Nyctaginaceae
♦ Weed
♦ 157
♦ pH.

Bolivicereus Cárdenas spp.
Cactaceae
= *Cleistocactus* Lem. spp.
♦ Weed, Quarantine Weed
♦ 76, 88, 203, 220

Boltonia asteroides (L.) L'Hér.
Asteraceae
♦ white doll's daisy
♦ Naturalised
♦ 287
♦ cultivated, herbal.

Bolusanthus speciosus (Bolus) Harms
Fabaceae/Papilionaceae
Lonchocarpus speciosus Bolus
♦ Introduced
♦ 228
♦ arid, cultivated.

Bomarea edulis (Tussac) Herb.
Liliaceae/Alstroemeriaceae
Bomarea hirtella (H.B.K.) Herb
♦ Quarantine Weed
♦ 220
♦ pC, cultivated.

Bomarea hirtella (Kunth) Herb.
Liliaceae/Alstroemeriaceae
♦ Quarantine Weed
♦ 220

Bomarea multiflora (L.) Mirb.
Liliaceae/Alstroemeriaceae
♦ Weed, Quarantine Weed, Naturalised, Environmental Weed, Casual Alien
♦ 15, 246, 280
♦ pC, cultivated.

Bombax malabaricum DC.
Bombacaceae
= *Bombax ceiba* L. (NoR)
♦ red silk cottontree
♦ Introduced
♦ 228
♦ arid, cultivated, herbal.

Bombycilaena erecta (L.) Smoljan.
Asteraceae
♦ micropus
♦ Weed
♦ 88, 94, 272
♦ cultivated.

Bongardia chrysogonum (L.) Boiss.
Berberidaceae/Podophyllaceae
= *Bongardia rauwolfia* C.A.Mey (NoR)
♦ Weed, Quarantine Weed
♦ 76, 87, 88, 203, 220
♦ pH, arid, promoted.

Bonnaya multiflora Bonati
Scrophulariaceae
♦ Weed
♦ 87, 88

Bonnaya oppositifolia Spreng.
Scrophulariaceae
♦ Weed
♦ 87, 88

Bonnaya veronicaefolia Spreng.
Scrophulariaceae
♦ Weed
♦ 39, 87, 88
♦ toxic.

Bonnemaisonia hamifera Har.
Bonnemaisoniaceae
Asparagopsis hamifera (Har.) Okamura
♦ Weed
♦ 288
♦ algae.

Bontia daphnoides L.
Myoporaceae
♦ Barbados olive, white alling
♦ Weed
♦ 179
♦ cultivated, herbal.

Boophane disticha (L.f.) Herb.
Liliaceae/Amaryllidaceae
♦ bushman poison bulb, candelabra flower, cape poison bulb, century plant, fan leaved boophane, kaffir onion, poison bulb, red posy, sore eye flower
♦ Weed, Native Weed
♦ 39, 121
♦ pH, herbal, toxic. Origin: southern Africa.

Boopis anthemoides Juss.
Calyceraceae
Boopis anthemoides var. *rigidula* (Miers)
Griseb., *Boopis anthemoides* var.
subscandens Speg., *Boopis rigidula* Miers
- boopis
- Weed
- 295

Borago officinalis L.
Boraginaceae
- borage, common borage, borák
lekársky, purasruoho
- Weed, Naturalised, Garden Escape,
Environmental Weed, Cultivation
Escape
- 7, 15, 34, 39, 40, 42, 70, 86, 87, 88, 94,
98, 101, 165, 176, 198, 203, 241, 253, 272,
280, 287, 300
- aH, cultivated, herbal, toxic. Origin:
southern Europe.

Boreava orientalis Jaub. & Spach
Brassicaceae
- Weed
- 87, 88, 243

Borreria alata (Aubl.) DC.
Rubiaceae
= *Spermacoce alata* Aubl.
- broadleaf buttonweed
- Weed, Quarantine Weed,
Naturalised
- 12, 13, 67, 87, 88, 170, 191, 217, 245,
258, 286, 287

Borreria articularis (L.f.) F.N.Williams
Rubiaceae
Borreria hispida (L.) Schum.
- Weed, Naturalised
- 13, 87, 88, 262, 286, 287
- aH, herbal.

Borreria bartlingiana DC.
Rubiaceae
- Weed
- 87, 88

Borreria brachystema (Bth.) Val.
Rubiaceae
- Weed
- 13

Borreria centranthoides Cham. & Schl.
Rubiaceae
- Weed
- 87, 88

Borreria distans (H.B.K.) Cham. &
Schltdl.
Rubiaceae
- Weed
- 13, 88, 170, 191

Borreria laevis (Lam.) Griseb.
Rubiaceae
= *Spermacoce tenuoir* L. (NoR)
- borreria negra, buttonplant,
yaa khamen lek, woodland
false buttonweed, hierba buena,
chiquizacillo
- Weed, Quarantine Weed,
Naturalised
- 12, 13, 76, 87, 88, 153, 170, 191, 199,
203, 209, 220, 238, 239, 276, 281, 286,
287

- pH, herbal.

Borreria latifolia (Aubl.) Schum.
Rubiaceae
Spermacoce caerulescens Aubl.,
Spermacoce cephalophora Rusby,
Spermacoce latifolia Aubl. (see)
- broadleaf buttonweed, kradum bai
yai, hierba buena, chiquizacillo
- Weed
- 28, 33, 209, 238, 239, 243, 262, 281
- aH, herbal.

Borreria natalensis (Hochst.) K.Schum.
ex S.Moore
Rubiaceae
Diodia natalensis (Hochst.) Garcia
- buttonweed
- Native Weed
- 121
- pH. Origin: southern Africa.

Borreria ocymoides (Burm.f.) DC.
Rubiaceae
- borreria blanca, hierba buena,
chiquizacillo
- Weed, Quarantine Weed
- 13, 14, 76, 87, 88, 153, 157, 203, 220,
281
- aH, herbal.

Borreria poaya DC.
Rubiaceae
- Weed, Quarantine Weed
- 76, 87, 88, 203, 220
- herbal.

Borreria princeae K.Schum.
Rubiaceae
- Weed, Quarantine Weed
- 76, 87, 88, 203, 220

Borreria radiata DC.
Rubiaceae
- Weed
- 87, 88

Borreria ramisparsa DC.
Rubiaceae
- Weed
- 87, 88

Borreria repens DC.
Rubiaceae
- Weed
- 13, 87, 88, 170, 191, 276
- pH.

Borreria ruelliae Schum.
Rubiaceae
= *Borreria scabra* (Schum. & Thonn.)
K.Schum.
- Weed
- 87, 88

Borreria scabra (Schum. & Thonn.)
Schum.
Rubiaceae
Borreria ruelliae K.Schum. (see)
- buttonweed
- Weed, Native Weed
- 50, 51, 88, 121
- aH. Origin: southern Africa.

Borreria setidens (Miq.) Bold.
Rubiaceae
- Weed
- 13

Borreria spinosa (L.) Cham. & Schlecht.
Rubiaceae
- Weed
- 237

Borreria stachydea (DC.) Hutch. & Dalz.
Rubiaceae
- Weed
- 87, 88

Borreria stricta (L.f.) G.F.W.Mey.
Rubiaceae
- Weed
- 87, 88, 297

Borreria verticillata (L.) G.F.W.Mey.
Rubiaceae
= *Spermacoce verticillata* L.
- botoncito blanco
- Weed, Quarantine Weed
- 76, 87, 88, 203, 220, 295
- herbal.

Borzicactella Ritter spp.
Cactaceae
= *Cleistocactus* Lem. spp.
- Weed, Quarantine Weed
- 76, 88, 203

Borzicactus Riccobono spp.
Cactaceae
= *Cleistocactus* Lem. spp.
- Weed, Quarantine Weed
- 76, 88, 203

Boscia albitrunca (Burch.) Gilg & Ben.
Capparaceae
- caper bush, shepherd's tree, white
stem
- Native Weed
- 121
- T, cultivated. Origin: southern
Africa.

Boswellia frereana Birdw.
Burseraceae
Boswellia hildebrandtii sensu Chiov.
- African elemi, elemi frankincense,
gekar, inaidi, luban, luban matti, luban
meiti, lufod, maidi, meydi, uban,
yagar, yagcar, yegaar, yigaar
- Introduced
- 228
- arid.

Bothriochloa Kuntze spp.
Poaceae
- Caucasian bluestem, Eurasian
bluestem, beardgrass
- Weed
- 80, 195
- G.

Bothriochloa ambigua S.T.Blake
Poaceae
- Weed
- 87, 88
- G, arid.

Bothriochloa barbinodis (Lag.) Herter
Poaceae
Andropogon barbinodis Lag. (see),
Andropogon saccharoides Sw. var.
barbinodis Hack.
- cane bluestem, cane beardgrass
- Introduced
- 228
- pG, arid, herbal.

Bothriochloa biloba S.T.Blake
Poaceae
♦ Naturalised
♦ 86
♦ G, cultivated. Origin: Australia.

Bothriochloa bladhii (Retz.) S.T.Blake
Poaceae
Amphilophis intermedia (R.Br.)
Stapf, *Andropogon intermedius*
R.Br., *Bothriochloa intermedia* (R.Br.)
A.Camus (see), *Bothriochloa intermedia*
var. *punstata* (R.Br.) A.Camus (see),
Bothriochloa glabra (Roxb.) A.Camus
♦ Caucasian bluestem, Australian
beardgrass
♦ Weed, Naturalised, Native Weed,
Introduced
♦ 32, 98, 101, 121, 228, 280
♦ pG, arid, cultivated, herbal. Origin:
southern Africa.

Bothriochloa bladhii (Retz.) S.T.Blake
ssp. *glabra* (Roxb.) B.K.Simon
Poaceae
Andropogon glaber Roxb.
♦ Naturalised
♦ 86
♦ G.

Bothriochloa decipiens (Hack.) C.E.Hubb.
Poaceae
Andropogon pertusus (L.) Willd. var.
decipiens Hack.
♦ crown beardgrass
♦ Weed
♦ 87, 88
♦ G, arid, cultivated. Origin: Australia.

Bothriochloa insculpta (Hochst.)
A.Camus
Poaceae
Andropogon insculpta (A.Rich.) Stapf,
Andropogon insculptus A.Rich.,
Andropogon pertusus (L.) Willd. var.
capensis Hack., *Andropogon pertusus* var.
insculptus (A.Rich.) Hack., *Bothriochloa*
insulpta (A.Rich.) A.Camus var. *hirta*
(Chiov.) Cuf., *Dichanthium insulpta*
(A.Rich.) W.Clayton
♦ pinhole grass, tassel grass, turf grass,
sweet pitted grass, creeping bluegrass
♦ Weed, Naturalised, Native Weed,
Introduced
♦ 86, 88, 121, 158, 228
♦ pG, arid, cultivated. Origin:
southern Africa.

Bothriochloa intermedia (R.Br.) A.Camus
Poaceae
= *Bothriochloa bladhii* (Retz.) S.T.Blake
♦ Weed
♦ 87, 88
♦ G.

Bothriochloa intermedia (R.Br.) A.Camus
var. *punstata* Keng
Poaceae
= *Bothriochloa bladhii* (Retz.) S.T.Blake
♦ Naturalised
♦ 287
♦ G.

Bothriochloa ischaemum (L.) Keng
Poaceae

Amphilophis ischaemum (L.) Nash,
Andropogon ischaemum L., *Andropogon*
ischaemum var. *genuinus* Hack.,
Andropogon ischaemum var. *radicans*
(Lehm.) Hack., *Andropogon ischaemum*
var. *songaricus* Rupr. ex Fisch. &
Meyen, *Andropogon taiwanensis* Ohwi,
Bothriochloa ischaemum (L.) Henrard,
Bothriochloa ischaemum (L.) Keng var.
songarica (Rupr. ex Fisch. & C.A.Mey.)
Celarier & Harlan (see), *Dichanthium*
ischaemum (L.) Rob. (see), *Sorgum*
ischaemum (L.) Kuntze
♦ yellow bluestem, Turkestan
bluestem, King Ranch bluestem
♦ Weed, Naturalised, Introduced
♦ 80, 87, 88, 90, 101, 228, 241, 275, 287,
297
♦ pG, arid, cultivated, herbal.

Bothriochloa ischaemum (L.) Keng var.
ischaemum
Poaceae
♦ yellow bluestem
♦ Naturalised
♦ 101
♦ G.

Bothriochloa ischaemum (L.) Keng var.
songarica (Rupr. ex Fisch. & C.A.Mey.)
Celarier & Harlan
Poaceae
= *Bothriochloa ischaemum* (L.) Keng
♦ yellow bluestem
♦ Weed, Naturalised
♦ 101, 179
♦ pG.

Bothriochloa laguroides (DC.) Herter
Poaceae
Andropogon saccharoides Sw. var.
laguroides (DC.) Hack. (see)
♦ silver beardgrass
♦ Weed, Naturalised
♦ 241, 295, 300
♦ G, herbal.

Bothriochloa (DC.) Herter laguroides ssp.
laguroides
Poaceae
♦ silver beardgrass
♦ Naturalised
♦ 101
♦ G.

Bothriochloa macra (Steud.) S.T.Blake
Poaceae
♦ Naturalised, Introduced,
Environmental Weed
♦ 86, 134, 176, 228, 280
♦ G, arid, cultivated. Origin: Australia.

Bothriochloa pertusa (L.) A.Camus
Poaceae
Amphilophis pertusa (L.) Nash ex Stapf,
Andropogon pertusus (L.) Willd (see),
Dichanthium pertusum (L.) W.Clayton
(see)
♦ Indian bluegrass, pitted beardgrass
♦ Weed, Naturalised, Introduced
♦ 7, 86, 87, 88, 93, 101, 179, 228, 261
♦ G, arid, cultivated. Origin: Old
World Tropics.

Bothriochloa saccharoides (Sw.) Rydb.
Poaceae
Adropogon saccharoides Sw. (see)
♦ silver beardgrass, silver bluegrass
♦ Weed, Naturalised, Introduced
♦ 80, 90, 228, 241, 295, 300
♦ G, arid, herbal.

Bothriospermum chinense Bunge
Boraginaceae
♦ Weed
♦ 275, 297
♦ aH.

Bothriospermum kusnezowii Bunge
Boraginaceae
♦ kusnezow bothriospermum
♦ Weed
♦ 297

Bothriospermum tenellum (Hornem.)
Fisch. & Mey.
Boraginaceae
♦ leaf between flower, hanaibana,
small bothriospermum
♦ Weed, Quarantine Weed,
Naturalised
♦ 76, 87, 88, 101, 203, 204, 220, 235, 263,
274, 275, 286, 297
♦ aH.

Bothriospermum zeylanicum (J.Jacq.)
Druce
Boraginaceae
♦ hanaibana
♦ Weed
♦ 273

Bouchea agristis Schauer & Mart.
Verbenaceae
♦ Naturalised
♦ 287

Bouchea ehrenbergii Cham.
Verbenaceae
♦ Weed
♦ 87, 88

Bouchea prismatica (L.) Kuntze
Verbenaceae
Verbena prismatica L.
♦ prism bouchea
♦ Weed
♦ 14
♦ herbal.

Bougainvillea Comm. ex Juss. spp.
Nyctaginaceae
♦ bougainvillea, ihmekynnkset
♦ Environmental weed
♦ 201
♦ pC, Cultivated.

Bougainvillea glabra Choisy
Nyctaginaceae
♦ bougainvillea, paperflower
♦ Weed, Naturalised, Introduced,
Garden Escape
♦ 80, 86, 101, 230, 261, 280
♦ S, cultivated, herbal. Origin: Brazil.

Bougainvillea glabra Choisy cv.
'Magnifica'
Nyctaginaceae
♦ trinitaria, paperflower, bougainvillea
♦ Naturalised
♦ 280

- ♦ S, cultivated, herbal. Origin: horticultral.

Bougainvillea spectabilis Willd.
Nyctaginaceae
- ♦ bougainvillea, great bougainvillea
- ♦ Garden Escape, Environmental Weed
- ♦ 257, 261
- ♦ pC/S, cultivated, herbal. Origin: Brazil.

Bourreria acimoides (Burm.) DC.
Boraginaceae/Ehretiaceae
- ♦ Weed, Quarantine Weed
- ♦ 76, 87, 88, 203, 220

Bourreria andreuxii Hemsl.
Boraginaceae/Ehretiaceae
- ♦ Weed
- ♦ 87, 88

Bourreria microphylla Griseb.
Boraginaceae/Ehretiaceae
- ♦ Weed
- ♦ 14

Boussingaultia baselloides Kunth
Basellaceae
= *Anredera baselloides* (H.B.K) Baill.
- ♦ Weed
- ♦ 87, 88
- ♦ herbal.

Boussingaultia cordifolia Ten.
Basellaceae
- ♦ Weed, Naturalised
- ♦ 87, 88, 287
- ♦ herbal.

Boussingaultia gracilis Miers
Basellaceae
- ♦ Weed
- ♦ 87, 88
- ♦ herbal.

Bouteloua americana (L.) Scribn.
Poaceae
- ♦ American grama
- ♦ Weed
- ♦ 32
- ♦ G.

Bouteloua aristidoides (Kunth) Griseb.
Poaceae
- ♦ sixweeks needle grama, needle grama, sixweeks grama
- ♦ Weed, Introduced
- ♦ 87, 88, 91, 161, 218, 228, 243
- ♦ aG, arid, herbal.

Bouteloua barbata Lag.
Poaceae
= *Chondrosum barbatum* (Lag.) Clayton
- ♦ sixweeks grama
- ♦ Weed
- ♦ 87, 88, 161, 218
- ♦ aG, herbal.

Bouteloua curtipendula (Michx.) Torr.
Poaceae
Chloris curtipendula Michx.
- ♦ gama grass, sideoats grama
- ♦ Naturalised, Introduced
- ♦ 228, 287
- ♦ pG, arid, cultivated, herbal.

Bouteloua disticha (Kunth) Benth.
Poaceae

- ♦ Naturalised
- ♦ 257
- ♦ G, arid.

Bouteloua gracilis (Willd. ex Kunth) Lag. ex Griffiths
Poaceae
- ♦ blue grama
- ♦ Weed, Introduced
- ♦ 91, 228
- ♦ pG, arid, cultivated, herbal.

Bouteloua pilosa Benth. ex S.Wats.
Poaceae
- ♦ Weed
- ♦ 87, 88
- ♦ G.

Bouteloua simplex Lag.
Poaceae
Chondrosum simplex (Lag.) Kunth
- ♦ matted grama, mat gama
- ♦ Weed, Introduced
- ♦ 199, 228
- ♦ G, arid, herbal.

Bowiea volubilis Harv. ex Hook.f.
Liliaceae/Hyacinthaceae
- ♦ climbing onion, Zulu potato, sea onion, mukelo
- ♦ Weed, Native Weed
- ♦ 39, 121, 161, 247
- ♦ pC, arid, cultivated, herbal, toxic. Origin: southern Africa.

Bowlesia incana Ruiz & Pav.
Apiaceae
Bowlesia tenera Spreng. (see)
- ♦ hairy bowlesia
- ♦ Weed
- ♦ 87, 88, 161, 218, 236, 237, 255, 271, 295
- ♦ aH, arid, herbal. Origin: South America.

Bowlesia tenera Spreng.
Apiaceae
= *Bowlesia incana* Ruiz & Pav.
- ♦ Weed
- ♦ 87, 88

Bowlesia tropaeolifolia Gillies & Hook.
Apiaceae
- ♦ Naturalised
- ♦ 280

Brachiaria advena Vickery
Poaceae
= *Urochloa advena* (Vick.) R.Webster
- ♦ Weed, Naturalised, Native Weed
- ♦ 86, 121
- ♦ aG. Origin: southern Africa.

Brachiaria annulatum (Forssk.) Stapf
Poaceae
- ♦ ringed signalgrass
- ♦ Naturalised
- ♦ 101
- ♦ G.

Brachiaria arizonica (Scribn. & Merr.) S.T.Blake
Poaceae
= *Urochloa arizonica* (Scribn. & Merr.) Morrone & Zuloaga (NoR) [see *Panicum arizonicum* Scribn. & Merr.]
- ♦ Arizona panicum
- ♦ Weed

- ♦ 161
- ♦ G, arid, herbal.

Brachiaria arrecta (Hack. ex Dur. & Schinz) Strent
Poaceae
= *Urochloa arrecta* (Hack. ex T.Durand & Schinz) Morrone & Zuloaga
- ♦ tanner grass
- ♦ Weed, Native Weed, Introduced
- ♦ 121, 228, 245
- ♦ pG, arid. Origin: southern Africa.

Brachiaria brizantha (Hochst. ex A.Rich.) Stapf
Poaceae
= *Urochloa brizantha* (Horst. ex Rich.) Webster
- ♦ bread grass, bread signalgrass, broad leaved false paspalum, common signalgrass, large seeded milletgrass, large seeded panicgrass, upright brachiaria, upright false paspalum, palisade grass, signalgrass, mbute
- ♦ Weed, Naturalised, Native Weed, Introduced
- ♦ 39, 86, 87, 88, 98, 121, 203, 228, 245, 255
- ♦ pG, arid, cultivated, herbal, toxic. Origin: southern Africa.

Brachiaria ciliatissima (Buckl.) Chase
Poaceae
= *Urochloa ciliatissima* (Buckl.) R.Webster (NoR)
- ♦ fringed signalgrass
- ♦ Weed
- ♦ 87, 88, 218
- ♦ G.

Brachiaria decumbens Stapf
Poaceae
= *Urochloa decumbens* (Stapf) R.Webster
- ♦ signalgrass, brachiaria
- ♦ Weed, Naturalised, Environmental Weed
- ♦ 39, 86, 88, 226, 227, 245, 255
- ♦ pG, arid, cultivated, toxic. Origin: southern Africa.

Brachiaria deflexa (Schum.) C.E.Hubb. ex Robyns
Poaceae
= *Urochloa deflexa* (Schum.) H.Scholz (NoR) [see *Brachiaria deflexa* (Schum.) C.E.Hubb. ex Robyns]
- ♦ deflexed brachiaria
- ♦ Weed, Quarantine Weed
- ♦ 76, 87, 88, 90, 158, 203, 220, 242
- ♦ aG, arid. Origin: South Africa.

Brachiaria distachya (L.) Stapf
Poaceae
= *Urochloa distachya* (L.) T.Q.Nguyen (NoR)
- ♦ armgrass millet, green summergrass
- ♦ Weed, Introduced
- ♦ 12, 87, 88, 90, 228
- ♦ G, arid.

Brachiaria distichophylla (Trin.) Stapf
Poaceae
= *Urochloa villosa* (Lam.) T.Q.Nguyen
- ♦ Weed
- ♦ 88
- ♦ G.

Brachiaria eruciformis **(J.E.Sm.) Griseb.**
Poaceae
Brachiaria iaschne (Roth) Stapf, *Panicum eruciforme* J.E.Sm., *Panicum isachne* Roth
♦ sweet summer grass, sweet signalgrass
♦ Weed, Naturalised, Native Weed, Introduced
♦ 51, 55, 86, 87, 88, 90, 98, 101, 121, 158, 170, 185, 191, 203, 228, 242, 243, 256, 272, 287
♦ aG, aqua. Origin: Africa, Mediterranean, India.

Brachiaria extensa **Chase**
Poaceae
= *Urochloa platyphylla* (Nash) R.D.Webster (NoR) [see *Brachiaria platyphylla* (Griseb.) Nash]
♦ pasto brachiaria
♦ Weed, Naturalised
♦ 14, 236, 287, 295
♦ G.

Brachiaria fasciculata **(Sw.) Parodi**
Poaceae
= *Urochloa fasciculata* (Sw.) Webster
♦ browntop millet, bamboo grass
♦ Weed, Naturalised
♦ 28, 88, 90, 98, 203, 206, 243
♦ G, arid.

Brachiaria fasciculata **(Sw.) Parodi var. reticulata** **(Torr.) Vickery**
Poaceae
= *Urochloa fasciculata* (Sw.) Webster
♦ browntop millet, bamboo grass, browntop signalgrass
♦ Naturalised
♦ 86
♦ G. Origin: Central America.

Brachiaria gilesii **(Benth.) Chase**
Poaceae
Panicum gilesii Benth., *Urochloa gilesii* ssp. *gilesii* Hughes
♦ Introduced
♦ 39, 228
♦ G, arid, toxic.

Brachiaria humidicola **(Rendle) Schweick.**
Poaceae
= *Urochloa humidicola* (Rendle) Morrone & Zuloaga (NoR)
♦ creeping false paspalum, creeping signalgrass, koronivia grass
♦ Weed, Naturalised, Native Weed, Introduced
♦ 54, 86, 88, 121, 228, 255
♦ pG, arid, cultivated. Origin: Africa.

Brachiaria jubata **(Fig. & De Not.) Stapf**
Poaceae
Brachiaria bamaensis Vanderyst, *Brachiaria brevis* Stapf, *Brachiaria fulva* Stapf, *Brachiaria soluta* Stapf, *Panicum jubatum* Fig. & De Not., *Urochloa jubata* (Fig. & De Not.) Sosef
♦ Weed
♦ 88
♦ G, arid.

Brachiaria lata **(Schumach.) C.E.Hubb.**
Poaceae
= *Urochloa lata* (Schum.) C.E.Hubb. (NoR)
♦ Weed, Introduced
♦ 87, 88, 90, 228
♦ G, arid.

Brachiaria marlothii **(Hack.) Stent**
Poaceae
Panicum marlothii Hack.
♦ small signalgrass
♦ Native Weed
♦ 121
♦ a/pG. Origin: southern Africa.

Brachiaria meziana **Hitch.**
Poaceae
= *Urochloa meziana* (Hitch.) Morrone & Zuloaga (NoR)
♦ Weed
♦ 199
♦ G.

Brachiaria milliformis **(Presl) A.Chase**
Poaceae
Panicum miliiforme J.Presl
♦ thurston grass
♦ Weed
♦ 3, 87, 88, 191
♦ G.

Brachiaria mutica **(Forssk.) Stapf**
Poaceae
= *Urochloa mutica* (Forssk.) Nguyen
♦ para grass, buffalograss, California grass, Mauritius grass, puakatau, Scotch grass, panicum grass, yaa khon
♦ Weed, Sleeper Weed, Naturalised, Environmental Weed
♦ 3, 6, 30, 32, 39, 80, 86, 87, 88, 90, 98, 112, 152, 157, 170, 186, 197, 203, 204, 221, 225, 239, 243, 246, 255, 256, 257, 262, 274, 280, 286, 287
♦ pG, aqua, cultivated, herbal, toxic. Origin: Africa.

Brachiaria nigropedata **(Fical. & Hiern) Stapf**
Poaceae
Panicum nigropedatum Munro ex Ficalho & Hiern
♦ black footed brachiaria, spotted false paspalum, spotted signalgrass, sweetgrass
♦ Native Weed
♦ 121
♦ pG, arid. Origin: southern Africa.

Brachiaria paspaloides **(J.Presl ex C.Presl) C.E.Hubb.**
Poaceae
Panicum ambiguum Trin. (see), *Urochloa paspaloides* J.Presl
♦ common brachiaria, thurston grass, common signalgrass
♦ Weed, Quarantine Weed
♦ 76, 87, 88, 90, 135, 170, 191, 203, 220, 243, 286
♦ G, aqua, herbal.

Brachiaria piligera **(F.Muell. ex Benth.) D.K.Hughes**
Poaceae

= *Urochloa piligera* (F.Muell. ex Benth.) Webster
♦ wattle signalgrass
♦ Weed
♦ 87, 88
♦ G, arid.

Brachiaria plantaginea **(Link) A.S.Hitchc.**
Poaceae
= *Urochloa plantaginea* (Link) R.D.Webster
♦ Alexander grass, marmalade grass, plaintain signalgrass
♦ Weed, Quarantine Weed
♦ 30, 68, 76, 87, 88, 90, 161, 203, 220, 236, 237, 243, 245, 249, 255, 295
♦ aG. Origin: Africa.

Brachiaria platyphylla **(Griseb.) Nash**
Poaceae
= *Urochloa platyphylla* (Nash) R.D.Webster (NoR) [see *Brachiaria extensa* Chase]
♦ broadleaf signalgrass
♦ Weed, Naturalised
♦ 23, 30, 87, 88, 90, 161, 207, 218, 237, 243, 249, 263, 286, 287
♦ aG. Origin: North America.

Brachiaria purpurascens **(Raddi) Henr.**
Poaceae
= *Urochloa mutica* (Forssk.) Nguyen
♦ Weed, Quarantine Weed
♦ 3, 172, 191
♦ G.

Brachiaria ramosa **(L.) Stapf**
Poaceae
= *Urochloa ramosa* (L.) R.D.Webster
♦ browntop millet
♦ Weed, Naturalised, Introduced
♦ 66, 86, 87, 88, 90, 161, 228, 243
♦ G, aqua, herbal. Origin: tropical Africa.

Brachiaria regularis **(Nees) Stapf**
Poaceae
Panicum regulare Nees
♦ Weed
♦ 221
♦ G, herbal.

Brachiaria reptans **(L.) Gard. & C.E.Hubb.**
Poaceae
= *Urochloa reptans* (L.) Stapf
♦ running grass, cent per cent, fine armgrass, yaa ton tit
♦ Weed, Naturalised
♦ 6, 12, 87, 88, 90, 170, 221, 239, 262, 276, 286, 287
♦ a/pG, arid.

Brachiaria ruziziensis **R.Germ. & Evrard**
Poaceae
= *Urochloa brizantha* (Hochst. ex Rich.) Webster
♦ ruzi grass
♦ Weed
♦ 255
♦ pG, cultivated. Origin: tropical Africa.

Brachiaria serrata **(Thunb.) Stapf**
Poaceae

Holcus serratus Thunb., *Panicum serratum* (Thunb.) Spreng.
- velvetgrass, velvet signalgrass
- Native Weed
- 121
- pG, arid. Origin: southern Africa.

Brachiaria subquadripara (Trin.) Hitchc.
Poaceae
= *Urochloa subquadripara* (Trin.) R.D.Webster
- tropical signalgrass, small flowered Alexander grass, green summergrass
- Weed, Quarantine Weed, Naturalised, Introduced
- 3, 6, 55, 76, 86, 87, 88, 161, 203, 228, 249, 255, 273, 274, 286
- pG, arid, cultivated, herbal. Origin: tropical Africa and Asia.

Brachiaria texana (Buckley) Blake
Poaceae
= *Urochloa texana* (Buckley) R.D.Webster
- Texas millet, Colorado grass, conchograss, Texas panicum
- Weed, Naturalised
- 86, 90
- G, arid. Origin: Central America.

Brachiaria windersii C.E.Hubb.
Poaceae
- velvet leaved summer grass
- Weed
- 55
- G. Origin: tropical and subtropical Australia.

Brachiaria xantholeuca (Schinz) Stapf
Poaceae
= *Urochloa xantholeuca* (Schniz) H.Scholz (NoR)
- Native Weed
- 121
- aG, arid. Origin: southern Africa.

Brachyachne convergens (F.Muell.) Stapf
Poaceae
Cynodon convergens F.Muell.
- Weed, Naturalised, Introduced
- 39, 86, 87, 88, 228
- G, arid, toxic. Origin: Australia.

Brachyactis ciliata (Ledeb.) Ledeb.
Asteraceae
= *Symphyotrichum ciliatum* (Ledeb.) Nesom (NoR) [see *Brachyactis ciliata* (Ledeb.) Ledeb. ssp. *angusta* (Lindl.) A.G.Jones]
- alkali rayless aster
- Weed
- 275
- aH.

Brachyactis (Ledeb.) Ledeb. *ciliata* ssp. *angusta* (Lindl.) A.G.Jones
Asteraceae
= *Symphyotrichum ciliatum* (Ledeb.) Nesom (NoR) [see *Brachyactis ciliata* (Ledeb.) Ledeb.]
- alkali rayless aster
- Native Weed
- 49
- Origin: North America.

Brachyapium involucratum (Maire) Maire
Apiaceae
- Weed
- 87, 88

Brachycereus Britton & Rose spp.
Cactaceae
- Weed, Quarantine Weed
- 76, 88, 203, 220

Brachychiton australis (Schott & Endl.) A.Terracc.
Sterculiaceae
Brachychiton platanoides R.Br., *Brachychiton trichosiphon* (Benth.) J.W.Audas, *Sterculia trichosiphon* Benth., *Trichosiphon australe* Schott & Endl.
- Introduced
- 228
- arid, cultivated. Origin: Australia.

Brachychiton discolor F.Muell.
Sterculiaceae
- hat tree, pink flame tree, pink sterculia, Queensland lacebark, scrub bottletree, white kurrajong
- Weed
- 121
- arid, cultivated, herbal. Origin: Australia.

Brachychiton populneum (Schott & Endl.) R.Br.
Sterculiaceae
= *Brachychiton populneus* (Schott & Endl.) R.Br.
- whiteflower kurrajong, brachychiton
- Weed, Naturalised
- 39, 101, 161
- arid, toxic.

Brachychiton populneus (Schott & Endl.) R.Br.
Sterculiaceae
Brachychiton populneum (Schott & Endl.) R.Br. (see), *Poecilodermis populnea* Schott, *Sterculia diversifolia* G.Don
- kurrajong, bottletree, brachychiton
- Naturalised, Native Weed, Introduced, Garden Escape, Environmental Weed
- 7, 86, 228
- T, arid, cultivated, herbal. Origin: Australia.

Brachychiton rupestre (Lindl.) K.Schum.
Sterculiaceae
Delabechea rupestris Lindl.
- Introduced
- 39, 228
- arid, herbal, toxic.

Brachyelytrum erectum (Schreb. ex Spreng.) Beauv.
Poaceae
- short huskgrass, bearded shorthusk
- Weed
- 161
- G, herbal.

Brachypodium Beauv. spp.
Poaceae

- brachypodium, falsebrome
- Weed, Naturalised
- 198, 272
- G.

Brachypodium distachyon (L.) Beauv.
Poaceae
Trachynia distachya (L.) Link. (see)
- aroluste, purple falsebrome, falsebrome, stiff brome
- Weed, Noxious Weed, Naturalised, Introduced, Environmental Weed, Casual Alien
- 7, 34, 35, 38, 40, 42, 68, 72, 78, 86, 88, 98, 101, 116, 176, 198, 203, 253, 280, 287, 300
- a/pG, cultivated, herbal. Origin: Eurasia.

Brachypodium phoenicoides (L.) Roem. & J.A.Schult.
Poaceae
- thinleaf falsebrome
- Weed, Naturalised
- 101, 253
- G.

Brachypodium pinnatum (L.) Beauv.
Poaceae
- chalk falsebrome, tor grass, heath falsebrome
- Weed, Quarantine Weed, Naturalised
- 30, 70, 87, 88, 101, 220, 272, 280
- pG, cultivated, herbal. Origin: Europe, North Africa, Middle East.

Brachypodium sylvaticum (Huds.) P.Beauv.
Poaceae
Brevipodium silvaticum Á. & D.Löve, *Festuca silvatica* Huds. (*non* Vill.), *Triticum silvaticum* Moench (*non* Salisbury), *Bromus silvaticus* Pollich
- falsebrome, slender falsebrome, wood falsebrome
- Weed, Naturalised
- 70, 80, 88, 101, 132, 146, 272, 280
- pG, cultivated, herbal.

Brachyscome iberidifolia Benth.
Asteraceae
- Casual Alien
- 280
- cultivated.

Brachyscome perpusilla (Steetz) J.M.Black
Asteraceae
- Casual Alien
- 280

Brachyscome perpusilla (Steetz) J.M.Black var. *tenella* (Turcz.) G.L.R.Davis
Asteraceae
- Casual Alien
- 280

Brachystegia spiciformis Benth.
Fabaceae/Caesalpiniaceae
- mtulu mwombo
- Weed, Quarantine Weed
- 76, 87, 88, 203, 220
- cultivated, herbal.

***Brachythecium albicans* (Hedw.) Bruch & Schimp.**
　Brachytheciaceae
　♦ brachythecium moss, whitish feather moss
　♦ Naturalised
　♦ 198, 280
　♦ moss.

***Brachythecium campestre* (Müll.Hal.) Bruch & Schimp.**
　Brachytheciaceae
　♦ brachythecium moss
　♦ Naturalised
　♦ 280
　♦ moss.

***Bracteantha bracteata* (Vent.) Anderb. & Haegi**
　Asteraceae
　Helichrysum bracteatum (Vent.) Andrews (see)
　♦ strawflower
　♦ Naturalised
　♦ 7
　♦ aH, cultivated.

***Brahea aculeata* (Brandegee) H.E.Moore**
　Arecaceae
　Erythea aculeata Brandegee, *Glaucothea aculeata* (Brandegee) I.M.Johnst.
　♦ Sinaloa hesper palm
　♦ Introduced
　♦ 228
　♦ arid, cultivated.

***Brahea armata* S.Watson**
　Arecaceae
　Brahea glauca hort., *Brahea roezlii* Linden, *Erythea armata* S.Watson, *Erythea roezlii* (Linden) Becc., *Glaucothea armata* (S.Watson) O.F.Cook
　♦ Mexican blue fan palm, Mexican blue palm, big blue hesper palm, blue palm, fan palm, great blue hesper palm, palma blanca, palma negra, rock palm
　♦ Introduced
　♦ 228
　♦ T, arid, cultivated.

***Brahea edulis* H.Wendl. ex S.Watson**
　Arecaceae
　Erythea edulis (H.Wendl.) S.Watson
　♦ Guadeloupe palm, fan palm
　♦ Introduced
　♦ 228
　♦ T, arid, cultivated. Origin: south-western North America.

***Brahea elegans* (Franceschi) H.E.Moore**
　Arecaceae
　Erythea elegans Franceschi, *Glaucothea elegans* (Franceschi) I.M.Johnst.
　♦ Franceschi palm
　♦ Introduced
　♦ 228
　♦ arid.

***Brasenia schreberi* Gmel.**
　Cabombaceae
　Brasenia peltata Pursh
　♦ watershield, watertarget
　♦ Weed, Quarantine Weed
　♦ 80, 87, 88, 161, 218, 258

　♦ wpH, cultivated, herbal. Origin: tropical regions.

***Brasiliopuntia* (K.M.Schum.) A.Berger spp.**
　Cactaceae
　♦ Weed, Quarantine Weed
　♦ 76, 88, 203

***Brassaia actinophylla* Endl.**
　Araliaceae
　= *Schefflera actinophylla* (Endl.) Harms
　♦ Queensland umbrella tree
　♦ Weed, Introduced
　♦ 3, 87, 88, 191, 261
　♦ cultivated, herbal. Origin: Australia.

***Brassica* L. spp.**
　Brassicaceae
　♦ mustard, wild mustard, turnip, kale, kohlrabi
　♦ Weed, Naturalised
　♦ 243
　♦ H, herbal, toxic.

***Brassica adpressa* (Moench) Boiss.**
　Brassicaceae
　= *Hirschfeldia incana* (L.) Lagr.-Foss.
　♦ Weed
　♦ 87, 88

***Brassica alba* (L.) Boiss.**
　Brassicaceae
　♦ white mustard, senape gialia
　♦ Weed
　♦ 39, 80, 87, 88
　♦ arid, herbal, toxic.

***Brassica arabica* Fiori**
　Brassicaceae
　♦ Weed
　♦ 88

***Brassica armoracioides* Czern. ex Turcz.**
　Brassicaceae
　= *Brassica elongata* Ehrh. ssp. *integrifolia* (Boiss.) Breistr. (NoR)
　♦ Weed
　♦ 87, 88

***Brassica arvensis* (L.) Rabenh.**
　Brassicaceae
　= *Sinapis arvensis* L.
　♦ annual wild mustard
　♦ Weed
　♦ 39, 80
　♦ herbal, toxic.

***Brassica barrelieri* (L.) Janka**
　Brassicaceae
　♦ Weed, Naturalised
　♦ 7, 98, 203, 250
　♦ cultivated.

***Brassica barrelieri* (L.) Janka ssp. oxyrrhina* (Coss.) Regel**
　Brassicaceae
　Brassica oxyrrhina (Coss.) Coss. (see)
　♦ Naturalised
　♦ 86
　♦ Origin: western Mediterranean.

***Brassica campestris* L.**
　Brassicaceae
　= *Brassica rapa* L. ssp. *campestris* (L.) Clapham

　♦ wild turnip, bird rape, birdsrape mustard
　♦ Weed, Naturalised
　♦ 23, 52, 68, 87, 88, 136, 217, 218, 236, 243, 245, 248, 257, 270, 286
　♦ aH, arid, cultivated, herbal.

***Brassica carinata* A.Braun**
　Brassicaceae
　Brassica integrifolia (West) Rupr. var. *carinata* (A.Braun) O.Schulz, *Melanosinapis abyssinica* hort. ex Regel, *Sinapis abyssinica* A.Braun
　♦ Abyssinian mustard, Ethiopian rape, Abyssinian cabbage
　♦ Introduced
　♦ 228
　♦ aH, arid, promoted.

***Brassica chinensis* L.**
　Brassicaceae
　♦ Chinese cabbage, pak choi
　♦ Naturalised, Cultivation Escape
　♦ 86, 252
　♦ cultivated, herbal.

***Brassica deflexa* Boiss.**
　Brassicaceae
　♦ persiankaali
　♦ Casual Alien
　♦ 42

***Brassica deserti* Danin & Hedge**
　Brassicaceae
　♦ Weed
　♦ 88

***Brassica elongata* Ehrh.**
　Brassicaceae
　♦ elongated mustard, kapusta predælená
　♦ Weed, Naturalised, Cultivation Escape, Casual Alien
　♦ 42, 86, 88, 94, 98, 101, 203, 252, 269, 272
　♦ b/pH, cultivated. Origin: Eurasia.

***Brassica fruticulosa* Cirillo**
　Brassicaceae
　♦ twiggy turnip, Mediterranean cabbage
　♦ Weed, Naturalised, Garden Escape, Environmental Weed
　♦ 7, 15, 72, 86, 88, 98, 101, 198, 203, 269, 280
　♦ b/pH, cultivated. Origin: Mediterranean.

***Brassica fruticulosa* Cirillo ssp. mauritanica* (Coss.) Maire**
　Brassicaceae
　♦ Naturalised
　♦ 280

***Brassica hirta* Moench**
　Brassicaceae
　= *Sinapis alba* L.
　♦ white mustard, sarshapa, yellow mustard
　♦ Weed
　♦ 39, 80, 87, 88, 161, 218, 243
　♦ aH, cultivated, herbal, toxic.

***Brassica incana* Ten.**
　Brassicaceae
　♦ shortpod mustard

♦ Weed
♦ 87, 88, 218

Brassica integrifolia (DC.) O.E.Schultz var. *timoriana*
Brassicaceae
♦ Introduced
♦ 230
♦ H.

Brassica juncea (L.) Czern.
Brassicaceae
Brassica japonica Thunb., *Sinapis juncea* L. (see)
♦ Indian mustard, brown mustard, leaf mustard, sareptansinappi, mustard greens, sarepta mustard, kapusta sitinová, curled mustard, swollenstem mustard, root mustard, green in the snow, large Chinese mustard
♦ Weed, Naturalised, Introduced, Casual Alien
♦ 7, 23, 32, 34, 38, 39, 40, 42, 49, 56, 70, 80, 86, 87, 88, 93, 94, 98, 101, 121, 134, 161, 174, 176, 179, 198, 203, 210, 218, 228, 237, 243, 261, 269, 272, 280, 286, 287, 295
♦ aH, arid, cultivated, herbal, toxic. Origin: Eurasia, Asia, east Africa.

Brassica kaber (DC.) Wheeler
Brassicaceae
= *Sinapis arvensis* L.
♦ wild mustard, charlock mustard, kaber mustard, California rape, charlock
♦ Weed, Noxious Weed
♦ 23, 39, 80, 87, 88, 114, 121, 161, 199, 210, 211, 212, 243, 299
♦ aH, herbal, toxic. Origin: Eurasia.

Brassica kaber (DC.) Wheeler var. *pinnatifida* (Stokes) Wheeler
Brassicaceae
= *Sinapis arvensis* L.
♦ wild mustard, charlock
♦ Weed
♦ 217, 218, 243
♦ Origin: Eurasia.

Brassica napus L.
Brassicaceae
Sinapis napus Brot., *Brassica sativa* Clavaud ssp. *napus* (L.) Bonnier & Layens, *Raphanus napus* Crantz, *Crucifera napus* E.H.L.Krause, *Brassica campestris* Hook. ssp. *campestris*, *Brassica campestris* Hook. ssp. *napus*, *Rapa napus* Mill., *Napus campestris* Schimp. & Spenn., *Brassica napa* St.-Lag., *Brassica napella* auct. *fl. jap. non* Chaix *ap.* Vill.
♦ swede, kale, rape, canola, turnip
♦ Weed, Naturalised, Garden Escape, Environmental Weed, Cultivation Escape
♦ 7, 34, 39, 40, 42, 80, 86, 87, 88, 94, 98, 101, 134, 161, 176, 198, 203, 236, 241, 243, 280, 286, 287, 300
♦ aH, cultivated, herbal, toxic. Origin: obscure.

Brassica napus L. ssp. *napus*
Brassicaceae

♦ rape, turnip
♦ Weed
♦ 243
♦ H.

Brassica nigra (L.) Koch
Brassicaceae
Sinapis nigra L.
♦ black mustard, shortpod mustard, worlock, Trieste mustard, brown mustard, wild black mustard, mustasinappi, senape nera, kapusta äierna
♦ Weed, Noxious Weed, Naturalised, Introduced, Environmental Weed, Casual Alien
♦ 7, 23, 24, 35, 39, 42, 56, 70, 78, 80, 86, 87, 88, 94, 98, 101, 116, 121, 136, 151, 161, 176, 180, 185, 198, 199, 203, 210, 212, 218, 221, 236, 237, 243, 253, 269, 272, 280, 287, 295, 300
♦ aH, arid, cultivated, herbal, toxic. Origin: Eurasia.

Brassica oleracea L.
Brassicaceae
Brassica oleracea L. var. *capitata* L. (see), *Crucifera brassica* E.H.L.Krause, *Napus oleracea* Schimp. & Spenn., *Raphanus brassica-officinalis* Crantz
♦ cabbage, cauliflower, Brussels sprout, broccoli, kohlrabi, kale
♦ Weed, Naturalised, Cultivation Escape
♦ 7, 39, 42, 86, 88, 93, 94, 98, 101, 198, 203, 205, 241, 252, 261, 280, 300
♦ b/p, cultivated, herbal, toxic. Origin: Europe.

Brassica oleracea L. ssp. *oleracea* Mansf.
Brassicaceae
♦ wild cabbage, sea cabbage, cabbage
♦ Cultivation Escape
♦ 56
♦ cultivated.

Brassica oleracea L. var. *capitata*
Brassicaceae
= *Brassica oleracea* L.
♦ red cabbage
♦ Naturalised
♦ 40
♦ herbal.

Brassica oleracea L. var. *oleracea*
Brassicaceae
♦ cabbage
♦ Introduced
♦ 230
♦ H.

Brassica oxyrrhina (Coss.) Coss.
Brassicaceae
= *Brassica barrelieri* (L.) Janka ssp. *oxyrrhina* (Coss.) Regel
♦ smoothstem turnip
♦ Naturalised
♦ 280

Brassica pekinensis (Lour.) Rupr.
Brassicaceae
= *Brassica rapa* L. var. *amplexicaulis* Tanaka & Ono
♦ Chinese cabbage
♦ Introduced

♦ 230
♦ H, cultivated, herbal.

Brassica rapa L.
Brassicaceae
Sinapis rapa Brot., *Crucifera rapa* E.H.L.Krause
♦ birdsrape mustard, field mustard, wild mustard, wild turnip, wild rutabaga, common mustard, rutabaga, turnip, wild kale, wild turnip, yellow mustard
♦ Weed, Naturalised, Garden Escape
♦ 7, 15, 34, 39, 40, 80, 87, 88, 94, 98, 101, 121, 157, 161, 167, 176, 179, 180, 198, 203, 212, 218, 237, 241, 243, 253, 255, 280, 295, 300
♦ aH, cultivated, herbal, toxic. Origin: Eurasia.

Brassica rapa L. ssp. *campestris* (L.) Clapham
Brassicaceae
Brassica campestris L. (see), *Brassica campestris* L. ssp. *campestri*, *Brassica rapa* L. var. *campestris* (L.) Clapham, *Brassica rapa* ssp. *silvestris* (Lam.) Janch. (see), *Brassica rapa* var. *silvestris* (Lam.) Briggs
♦ Introduced
♦ 38

Brassica rapa L. ssp. *chinensis* (L.) Hanelt
Brassicaceae
♦ Casual Alien
♦ 280
♦ herbal.

Brassica rapa L. ssp. *oleifolia* (DC.) Meteg.
Brassicaceae
♦ Naturalised
♦ 86

Brassica rapa L. ssp. *sylvestris* (Lam.) Janch.
Brassicaceae
= *Brassica rapa* ssp. *campestris* (L.) Clapham
♦ wild turnip, navew, bargeman's cabbage
♦ Weed, Naturalised, Introduced
♦ 56, 70, 165, 269, 272, 280
♦ a/bH. Origin: Europe.

Brassica rapa L. var. *amplexicaulis* Tanaka & Ono
Brassicaceae
Brassica pekinensis (Lour.) Rupr. (see)
♦ field mustard
♦ Naturalised
♦ 101

Brassica rapa L. var. *rapa*
Brassicaceae
♦ field mustard, Chinese mustard
♦ Naturalised, Environmental Weed
♦ 86, 101
♦ herbal.

Brassica rapa L. var. *sylvestris* (L.) Briggs
Brassicaceae
♦ Naturalised, Cultivation Escape
♦ 86, 252
♦ cultivated.

Brassica rugosa (Roxb.) L.H.Bailey
Brassicaceae
♦ Weed
♦ 87, 88
♦ herbal.

Brassica sinapistrum Boiss.
Brassicaceae
= *Sinapis arvensis* L.
♦ Weed
♦ 87, 88
♦ herbal.

Brassica tournefortii Gouan
Brassicaceae
♦ wild turnip, Mediterranean mustard, Mediterranean turnip, Asian mustard, Moroccan mustard, African mustard, prickly turnip, Sahara mustard, välimerenkaali, turnip weed
♦ Weed, Noxious Weed, Naturalised, Introduced, Garden Escape, Environmental Weed, Casual Alien
♦ 7, 9, 15, 34, 35, 42, 55, 68, 72, 78, 80, 86, 87, 88, 93, 98, 101, 116, 121, 151, 161, 176, 185, 198, 203, 205, 221, 231, 243, 269, 280
♦ aH, cultivated, herbal. Origin: Eurasia.

Brayulinea densa (H.& B.) Small
Amaranthaceae
= *Guilleminea densa* (Humb. & Bonpl. ex Willd.) Moq.
♦ small matweed, carrot weed
♦ Weed, Native Weed
♦ 51, 87, 88, 121, 218
♦ herbal, toxic. Origin: tropical America.

Breea setosa (Bieb.) Kitam.
Asteraceae
= *Cirsium arvense* (L.) Scop.
♦ ezonokitsuneazami
♦ Weed, Naturalised
♦ 87, 88, 286, 287

Breonadia salicina (Vahl) Hepper & Wood
Rubiaceae/Naucleaceae
♦ Quarantine Weed
♦ 220

Breynia disticha J.R. & G.Forst.
Capparaceae
Breynia nivosa (Bull) Small
♦ snowbush, hiutalepensas, carnaval, nevado
♦ Weed, Naturalised, Introduced
♦ 80, 101, 179, 261
♦ cultivated, herbal. Origin: Pacific islands.

Brickellia diffusa (Vahl) A.Gray
Asteraceae
♦ Naturalised
♦ 257

Bridgesia incisifolia M.Bertero ex Cambess.
Sapindaceae
♦ Introduced
♦ 228
♦ arid.

Brillantaisia lamium Benth.
Acanthaceae

Leucorhaphis lamium Nees
♦ Weed, Sleeper Weed, Quarantine Weed, Naturalised, Environmental Weed
♦ 3, 76, 86, 87, 88, 155, 191, 203, 220, 243
♦ cultivated. Origin: West Africa.

Brillantaisia nitens Lindau
Acanthaceae
♦ lutolotolo kumonga
♦ Weed, Quarantine Weed
♦ 76, 87, 88, 203, 220

Brittonia hort. spp.
Cactaceae
= *Ferocactus* Britt. & Rose spp. (NoR)
♦ Quarantine Weed
♦ 220

Briza humilis M.Bieb.
Poaceae
Briza spicata Sm.
♦ quaking grass
♦ Introduced
♦ 228
♦ G, arid.

Briza maxima L.
Poaceae
Briza major C.Presl
♦ large quaking grass, blowfly grass, broncho grass, lady's heart grass, large fairy bells, quaking grass, shaky grass, fairy bells, quivering grass, great quaking grass, big quaking grass
♦ Weed, Naturalised, Introduced, Environmental Weed, Cultivation Escape
♦ 7, 9, 15, 34, 40, 42, 70, 72, 86, 87, 88, 91, 93, 98, 101, 121, 134, 158, 161, 176, 198, 203, 205, 241, 243, 250, 252, 269, 272, 280, 286, 287, 289, 296, 300
♦ aG, cultivated, herbal. Origin: Eurasia.

Briza media L.
Poaceae
Briza clusii Schult., *Briza tremula* Koeler
♦ perennial quaking grass, intermediate quaking grass, tottergrass, shivery grass, quaking grass, doddering dillies, common quaking grass, kraslica prostredná
♦ Weed, Naturalised
♦ 15, 23, 70, 87, 88, 91, 101, 218, 272, 280, 287
♦ pG, cultivated, herbal.

Briza minor L.
Poaceae
Briza gracilis hort., *Briza minima* hort. ex Nichols.
♦ little quaking grass, pikkuräpelö, little fairy bells, quaking grass, small quaking grass, shivery grass, fairy bells, lesser quaking grass
♦ Weed, Naturalised, Introduced, Garden Escape, Environmental Weed, Casual Alien
♦ 7, 9, 15, 34, 40, 42, 70, 72, 86, 87, 88, 91, 98, 101, 121, 134, 158, 161, 176, 198, 203, 218, 228, 241, 243, 250, 251, 253, 269, 272, 280, 286, 287, 290, 295, 300

♦ aG, arid, cultivated, herbal. Origin: Eurasia.

Briza rufa (J.Presl) Steud.
Poaceae
= *Poidium rufum* (J.Presl) Matthei (NoR)
♦ Naturalised
♦ 280
♦ G.

Briza subaristata Lam.
Poaceae
= *Chascolytrum subaristatum* (Lam.) Desv.
♦ Weed, Naturalised
♦ 86, 98, 203
♦ G. Origin: South America.

Bromelia antiacantha Bertol.
Bromeliaceae
♦ gravata
♦ Weed
♦ 255

Bromelia fastuosa Lindl.
Bromeliaceae
= *Bromelia pinguin* L.
♦ Weed
♦ 87, 88

Bromelia hieronymi Mez
Bromeliaceae
♦ Introduced
♦ 228
♦ arid.

Bromelia laciniosa C.Mart. ex Schult.f.
Bromeliaceae
♦ Introduced
♦ 228
♦ arid, herbal.

Bromelia pinguin L.
Bromeliaceae
Bromelia fastuosa Lindl. (see)
♦ pinguin
♦ Weed
♦ 14, 39, 87, 88
♦ cultivated, herbal, toxic.

Bromelia serra Griseb.
Bromeliaceae
♦ Introduced
♦ 228
♦ arid, cultivated.

Bromus L. spp.
Poaceae
♦ bromegrass, brome, rescue grass, smooth bromegrass
♦ Weed, Environmental Weed
♦ 18, 243, 272, 296
♦ G, herbal, toxic.

Bromus adoensis Hochst. ex Steud.
Poaceae
♦ Weed
♦ 87, 88, 221
♦ G.

Bromus alopecuros Poir.
Poaceae
♦ weedy brome
♦ Weed, Naturalised
♦ 7, 86, 87, 88, 98, 101, 176, 203
♦ G. Origin: Mediterranean.

Bromus arenarius Labill.
  Poaceae
  ♦ Australian chess, Australian brome
  ♦ Weed, Naturalised, Introduced
  ♦ 86, 87, 88, 98, 101, 161, 167, 203, 218, 228, 280
  ♦ aG, arid, cultivated, herbal. Origin: Australia.

Bromus arizonicus (Shear) Stebbins
  Poaceae
  ♦ Arizona brome
  ♦ Weed
  ♦ 161, 180, 243
  ♦ aG, herbal.

Bromus arvensis L.
  Poaceae
  Bromus altissimus Gilib. (*non* Weber), *Bromus versicolor* Pollich, *Forasaccus arvensis* Bubani var. *arvensis* Huds.
  ♦ field brome, field chess, corn bromegrass, pyörtänökattara, soft brome
  ♦ Weed, Naturalised, Casual Alien
  ♦ 30, 42, 44, 70, 80, 87, 88, 91, 101, 161, 218, 243, 253, 272, 287
  ♦ aG, cultivated, herbal.

Bromus berterianus Colla
  Poaceae
  Trisetobromus hirtus (Trin.) Nev. (see)
  ♦ Chilean chess
  ♦ Weed, Naturalised
  ♦ 101, 161
  ♦ G, arid.

Bromus brevis Nees & Steud.
  Poaceae
  = *Bromus catharticus* Vahl
  ♦ cebadilla pampeana
  ♦ Weed, Naturalised, Introduced
  ♦ 86, 87, 88, 98, 176, 203, 228, 237, 280, 295
  ♦ G, arid. Origin: South America.

Bromus brizaeformis Fisch. & Mey.
  Poaceae
  = *Bromus briziformis* Fisch. & Mey.
  ♦ rattlegrass, rattlesnake chess, rattlesnake grass
  ♦ Weed, Naturalised
  ♦ 23, 80, 87, 88, 136, 146, 218, 287
  ♦ G, herbal.

Bromus briziformis Fisch. & Mey.
  Poaceae
  Bromus brizaeformis Fisch. & Mey. (see)
  ♦ rattlesnake brome, rattlesnake chess
  ♦ Weed, Naturalised, Introduced
  ♦ 34, 101, 161
  ♦ aG, cultivated.

Bromus canadensis Michx.
  Poaceae
  = *Bromus ciliatus* L. var. *ciliatus* (NoR)
  ♦ Canadian brome
  ♦ Introduced
  ♦ 228
  ♦ G, arid.

Bromus cappadocicus Boiss. & Balansa
  Poaceae
  ♦ Introduced
  ♦ 228
  ♦ G, arid.

Bromus carinatus Hook. & Arn.
  Poaceae
  Ceratochloa carinata (Hook. & Arn.) Tutin
  ♦ California brome, mountain brome
  ♦ Weed, Naturalised, Introduced
  ♦ 161, 212, 228, 286, 287
  ♦ aG, arid, promoted, herbal.

Bromus catharticus J.Vahl
  Poaceae
  Bromus brevis Nees & Steud. (see), *Bromus haenkeanus* (Presl) Kunth, *Bromus mathewsii* Steud., *Bromus schraderi* Kunth, *Bromus strictus* Brongn. *nom. illeg.*, *Bromus unioloides* Kunth (see), *Bromus willdenowii* Kunth (see), *Ceratochloa cathartica* (Vahl) Herter (see), *Ceratochloa unioloides* (Willd.) Beauv., *Festuca unioloides* Willd.
  ♦ prairie grass, rescue grass, bromegrass, Schrader's bromegrass
  ♦ Weed, Naturalised, Introduced, Environmental Weed, Casual Alien
  ♦ 7, 34, 55, 72, 80, 86, 87, 88, 91, 93, 98, 101, 102, 134, 158, 161, 198, 203, 205, 212, 228, 243, 245, 253, 255, 269, 280, 286, 287, 289, 295, 297
  ♦ a/pG, arid, cultivated, herbal. Origin: Americas.

Bromus cebadilla Steud.
  Poaceae
  ♦ Chilean brome
  ♦ Weed, Naturalised
  ♦ 86, 98, 176, 198, 203
  ♦ G. Origin: Chile.

Bromus coloratus Steud.
  Poaceae
  ♦ Weed, Naturalised
  ♦ 7, 98, 203
  ♦ G.

Bromus commutatus Schrad.
  Poaceae
  Bromus pratensis Ehrh. (*non* Spreng.), *Serrafalcus commutatus* (Schrad.) Babington
  ♦ meadow brome, hairy chess
  ♦ Weed, Quarantine Weed, Naturalised, Casual Alien
  ♦ 23, 30, 39, 42, 70, 76, 80, 88, 91, 101, 102, 121, 161, 218, 220, 237, 243, 253, 272, 280, 287, 295
  ♦ aG, cultivated, herbal, toxic. Origin: Eurasia.

Bromus danthoniae (Desf.) Trin.
  Poaceae
  ♦ oat brome
  ♦ Weed, Naturalised
  ♦ 87, 88, 287
  ♦ G.

Bromus diandrus Roth
  Poaceae
  Anisantha diandra (Roth) Tutin, *Anisantha gussonii* (Parl.) Nevski, *Bromus gussonii* Parl., *Bromus rigens* Druce (*non* L.)
  ♦ bromegrass, great brome, ripgut brome, ripgut, wild oat, broncho grass
  ♦ Weed, Naturalised, Introduced,

Environmental Weed
  ♦ 7, 9, 15, 34, 51, 70, 72, 86, 87, 88, 91, 93, 98, 101, 121, 134, 158, 161, 176, 177, 180, 198, 199, 203, 205, 228, 243, 250, 253, 269, 280, 300
  ♦ aG, arid/aqua, cultivated, herbal. Origin: Eurasia.

Bromus erectus Huds.
  Poaceae
  Bromopsis erecta (Huds.) Fourr., *Festuca erecta* Wallr., *Schenodorus erectus* Fr.
  ♦ upright brome, pystykattara, erect brome, meadow brome
  ♦ Weed, Naturalised, Casual Alien
  ♦ 30, 42, 44, 80, 88, 91, 101, 272, 280, 300
  ♦ pG, cultivated, herbal.

Bromus fasciculatus C.Presl
  Poaceae
  ♦ Weed
  ♦ 221
  ♦ G.

Bromus fonkii Phil
  Poaceae
  = *Bromus lithobius* Trin.
  ♦ Naturalised, Environmental Weed
  ♦ 86
  ♦ G.

Bromus grossus Desf. ex DC.
  Poaceae
  Bromus secalinus L. ssp. *grossus* (Desf.) Binz
  ♦ whiskered brome
  ♦ Naturalised
  ♦ 101
  ♦ G.

Bromus hordeaceus L.
  Poaceae
  ♦ bromegrass, soft brome, soft chess, forasacco pelso, lop grass, mäkikattara
  ♦ Weed, Naturalised, Environmental Weed
  ♦ 7, 9, 15, 34, 70, 72, 80, 88, 91, 98, 101, 102, 118, 134, 167, 176, 203, 241, 243, 250, 253, 269, 280, 300
  ♦ aG, cultivated, herbal.

Bromus hordeaceus L. ssp. hordeaceus
  Poaceae
  ♦ soft brome
  ♦ Weed, Naturalised, Environmental Weed
  ♦ 86, 101, 198, 272
  ♦ G. Origin: Eurasia.

Bromus hordeaceus L. ssp. molliformis (Lloyd) Maire & Weiller
  Poaceae
  Bromus molliformis Lloyd (see)
  ♦ soft brome, soft chess
  ♦ Naturalised, Introduced
  ♦ 101, 228
  ♦ aG. Origin: southern Europe.

Bromus hordeaceus L. ssp. pseudothominii P.Sm.
  Poaceae
  ♦ soft brome
  ♦ Naturalised
  ♦ 101
  ♦ G.

Bromus hordeaceus L. ssp. thominei (Hardham ex Nyman) Braun-Blanq.
Poaceae
♦ soft brome
♦ Naturalised
♦ 101
♦ G.

Bromus inermis Leyss.
Poaceae
Forasaccus inermis Lunell, *Schenodorus inermis* P.Beauv., *Zerna inermis* (Leyss.) Lindm.
♦ smooth brome, Hungarian chess, Hungarian bromegrass, idänkattara, awnless brome
♦ Weed, Noxious Weed, Naturalised, Environmental Weed
♦ 4, 21, 38, 42, 70, 80, 86, 87, 88, 91, 98, 101, 102, 104, 146, 151, 159, 161, 174, 195, 203, 218, 222, 228, 272, 275, 280, 287, 297, 299
♦ pG, arid, cultivated, herbal. Origin: Eurasia, North America.

Bromus inermis Leyss. ssp. inermis
Poaceae
♦ smooth brome
♦ Weed, Naturalised
♦ 34, 101
♦ pG.

Bromus inermis Leyss. ssp. inermis var. divaricatus Rohlena
Poaceae
♦ smooth brome
♦ Naturalised
♦ 101
♦ G.

Bromus inermis Leyss. ssp. inermis var. inermis
Poaceae
♦ smooth brome
♦ Naturalised
♦ 101
♦ G.

Bromus japonicus Thunb. ex Murr.
Poaceae
Serrafalcus japonicus (Thunb.) Wilmott
♦ Japanese bromegrass, Japanese chess, wild oats, meadow chess, Japaninkattara, Thunberg's brome, Japanese cheat
♦ Weed, Naturalised, Introduced, Environmental Weed, Casual Alien
♦ 7, 21, 23, 30, 34, 40, 42, 80, 86, 87, 88, 91, 98, 101, 102, 121, 146, 158, 161, 174, 176, 198, 203, 207, 210, 212, 218, 221, 243, 264, 272, 275, 280, 286, 297
♦ aG, cultivated, herbal. Origin: Eurasia.

Bromus japonicus Thunb. ex Murr. var. vestitus (Schrad.) Halácsy
Poaceae
♦ Casual Alien
♦ 280
♦ G.

Bromus lanceolatus Roth
Poaceae
Bromus macrostachys Desf. (see)
♦ lanceolate brome, purple awned brome, Mediterranean brome, hiirenkattara
♦ Weed, Naturalised, Introduced, Environmental Weed, Casual Alien
♦ 7, 42, 70, 72, 86, 88, 91, 98, 101, 198, 203, 228, 253, 272, 300
♦ aG, arid, cultivated, herbal. Origin: Eurasia.

Bromus lanceolatus Roth. ssp. macrostachys (Desf.) Maire
Poaceae
♦ Weed
♦ 111, 243
♦ G.

Bromus lanceolatus Roth. var. lanceolatus
Poaceae
♦ Weed
♦ 88
♦ G.

Bromus lepidus Holmb.
Poaceae
Bromus britannicus I.A.Williams
♦ slender soft brome, slender brome, sirokattara
♦ Naturalised, Casual Alien
♦ 40, 42, 101
♦ G. Origin: Europe.

Bromus lithobius Trin.
Poaceae
Bromus andinus Phil., *Bromus chilensis* Trin., *Bromus collinus* Phil., *Bromus fonkii* Phil. (see), *Bromus scaber* Phil., *Bromus unioloides* Kunth var. *pubescens* Hack. ex Stuck.
♦ Chilean brome
♦ Weed, Naturalised
♦ 15, 86, 176, 198, 280
♦ G. Origin: Central and South America.

Bromus macrostachys Desf.
Poaceae
= *Bromus lanceolatus* Roth
♦ Mediterranean brome
♦ Naturalised
♦ 287
♦ G, cultivated, herbal.

Bromus madritensis L.
Poaceae
Anisantha madritensis (L.) Nev. (see)
♦ Madrid brome, Spanish brome, compact brome, compact chess
♦ Weed, Noxious Weed, Naturalised, Introduced, Environmental Weed, Casual Alien
♦ 34, 35, 42, 70, 72, 87, 88, 93, 98, 101, 111, 161, 176, 198, 203, 228, 241, 243, 250, 253, 269, 280, 287, 300
♦ aG, arid, cultivated, herbal.

Bromus madritensis L. ssp. madritensis
Poaceae
♦ compact brome
♦ Weed
♦ 34
♦ aG.

Bromus madritensis L. ssp. rubens (L.) Husn.
Poaceae
= *Bromus rubens* L.
♦ red brome, foxtail chess, compact brome, Spanish brome
♦ Weed, Environmental Weed
♦ 34, 78, 116, 231
♦ aG.

Bromus madritensis L. var. ciliatus Guss.
Poaceae
♦ compact brome
♦ Naturalised
♦ 176
♦ G.

Bromus marginatus Nees ex Steud.
Poaceae
♦ mountain brome
♦ Introduced
♦ 228
♦ pG, arid, promoted, herbal.

Bromus molliformis Lloyd
Poaceae
= *Bromus hordeaceus* L. ssp. *molliformis* (Lloyd) Maire & Weiller
♦ soft brome
♦ Weed, Naturalised
♦ 86, 121, 269, 287
♦ aG. Origin: Eurasia.

Bromus mollis L.
Poaceae
Forasaccus mollis Bubani
♦ soft chess, soft brome, soft cheat, lop grass
♦ Weed, Naturalised
♦ 23, 30, 39, 44, 80, 87, 88, 161, 180, 212, 218, 237, 243, 287, 295
♦ G, herbal, toxic. Origin: Europe.

Bromus pacificus Shear
Poaceae
♦ Pacific brome
♦ Naturalised
♦ 287
♦ G, cultivated, herbal.

Bromus pectinatus Thunb.
Poaceae
Bromus vestitus Schrad., *Bromus gedrosianus* Penzés, *Bromus rechingeri* Melderis
♦ bromegrass
♦ Weed
♦ 53, 88
♦ aG.

Bromus × pseudothominii P.M.Sm.
Poaceae
♦ Naturalised
♦ 86, 98
♦ G.

Bromus racemosus L.
Poaceae
Serrafalcus racemosus (L.) Parl.
♦ upright brome, myllykattara, bald brome, smooth brome
♦ Weed, Naturalised, Casual Alien
♦ 30, 42, 70, 87, 88, 98, 101, 161, 203, 218, 241, 272, 280, 287, 295, 300
♦ G, herbal.

Bromus racemosus L. ssp. commutatus (Schrad.) Maire & Weiller
Poaceae

♦ meadow brome
♦ Naturalised
♦ 198
♦ G.

Bromus racemosus L. × commutatus Schrad.
Poaceae
♦ hybrid brome
♦ Cultivation Escape
♦ 40
♦ G, cultivated.

Bromus ramosus Huds.
Poaceae
Zerna ramosa (Huds.) Lindm., *Schenodorus asper* Trin.
♦ hairy brome, woodland brome, varjokattara
♦ Weed, Naturalised, Casual Alien
♦ 42, 70, 101, 272
♦ pG, cultivated, herbal.

Bromus rigidus Roth
Poaceae
Anisantha rigida (Roth) Hyl. (see), *Bromus diandrus* Roth ssp. *rigidus* (Roth) Laínz
♦ ripgut brome, great brome, broncho grass, villakattara
♦ Weed, Naturalised, Casual Alien
♦ 7, 23, 30, 42, 70, 80, 86, 87, 88, 91, 98, 101, 177, 212, 218, 237, 241, 243, 286, 287, 295
♦ aG, promoted, herbal. Origin: Europe, North Africa.

Bromus rubens L.
Poaceae
Bromus madritensis L. ssp. *rubens* (L.) Husn. (see)
♦ red brome, foxtail chess, tupsukattara, foxtail brome, tufted brome, foxtail bromegrass
♦ Weed, Naturalised, Introduced, Environmental Weed, Casual Alien
♦ 7, 23, 30, 42, 72, 80, 86, 87, 88, 91, 98, 101, 151, 161, 180, 198, 203, 218, 221, 228, 243, 253, 264, 269, 280, 287
♦ aG, arid, cultivated, herbal. Origin: southern Europe, Middle East.

Bromus scoparius L.
Poaceae
♦ kimppukattara, broom brome
♦ Weed, Naturalised, Casual Alien
♦ 42, 101, 134, 221, 300
♦ G.

Bromus scoparius L. var. scoparius
Poaceae
♦ Naturalised
♦ 241
♦ G.

Bromus secalinus L.
Poaceae
Avena secalinus Salisb., *Forasaccus secalinus* Bubani, *Serrafalcus secalinus* (L.) Babington
♦ bromegrass, rye brome, cheat grass, cheat, chess, rye brome, ruiskattara
♦ Weed, Naturalised, Casual Alien
♦ 23, 24, 30, 34, 39, 40, 44, 70, 80, 86, 87, 88, 91, 98, 101, 102, 161, 203, 210, 212, 218, 237, 241, 243, 253, 272, 280, 286,
287, 295, 300
♦ aG, cultivated, herbal, toxic. Origin: Europe.

Bromus setifolius J.Presl var. brevifolius Nees
Poaceae
Bromus macranthus Meyen
♦ Introduced
♦ 228
♦ G, arid.

Bromus sinaicus (Hack.) Täckh.
Poaceae
Bromus patulus Mert. & Koch ssp. *sinaicus* Hack.
♦ Weed
♦ 221
♦ G.

Bromus sitchensis Trin.
Poaceae
♦ sitka brome, Alaska brome, mountain brome
♦ Naturalised
♦ 280, 287
♦ pG, herbal.

Bromus squarrosus L.
Poaceae
Serrafalcus squarrosus (L.) Babington
♦ pörrökattara, corn brome
♦ Weed, Naturalised, Casual Alien
♦ 30, 40, 42, 80, 87, 88, 98, 195, 203, 253, 272, 300
♦ pG, cultivated, herbal.

Bromus stamineus Desv.
Poaceae
♦ Harlan brome, roadside brome
♦ Weed, Naturalised
♦ 34, 98, 101, 203, 280
♦ a/pG, cultivated, herbal.

Bromus sterilis L.
Poaceae
Anisantha sterilis (L.) Nev. (see)
♦ barren bromegrass, sterile brome, barren brome, poverty brome, ripgut brome, hietakattara
♦ Weed, Naturalised, Casual Alien
♦ 15, 30, 34, 42, 44, 70, 80, 86, 87, 88, 91, 98, 101, 102, 111, 115, 118, 146, 176, 181, 198, 203, 218, 241, 243, 253, 272, 280, 287, 295, 300
♦ aG, cultivated, herbal. Origin: Eurasia.

Bromus tectorum L.
Poaceae
Anisantha tectorum (L.) Nev., *Bromus tectorum* L. var. *glabratus* Spenn. (see), *Bromus tectorum* L. var. *hirsutus* Regel (see)
♦ cheat grass, downy brome, thatch bromegrass, drooping bromegrass
♦ Weed, Quarantine Weed, Noxious Weed, Naturalised, Introduced, Environmental Weed, Casual Alien
♦ 18, 21, 23, 30, 34, 35, 39, 42, 44, 49, 52, 68, 70, 76, 78, 80, 86, 87, 88, 91, 98, 101, 102, 105, 115, 116, 133, 136, 138, 146, 151, 152, 161, 162, 174, 195, 198, 203, 210, 211, 212, 218, 221, 224, 225, 229, 231, 241, 243, 246, 250, 253, 263, 264,
267, 272, 280, 286, 287, 292, 295, 299, 300
♦ aG, cultivated, herbal, toxic. Origin: Eurasia.

Bromus tectorum L. var. glabratus Spenn.
Poaceae
= *Bromus tectorum* L.
♦ Naturalised
♦ 287
♦ G, herbal.

Bromus tectorum L. var. hirsutus Regel
Poaceae
= *Bromus tectorum* L.
♦ Weed
♦ 88
♦ G.

Bromus trinianus Schult. nom. illeg.
Poaceae
= *Bromus tomentosus* Trin. (NoR)
♦ Introduced
♦ 228
♦ G, arid.

Bromus unioloides Kunth
Poaceae
= *Bromus catharticus* Vahl
♦ broncho grass, orchard grass, prairie grass, rescue grass, priebes prairie grass
♦ Weed
♦ 39, 121, 236, 237, 263
♦ a/pG, herbal, toxic. Origin: South America.

Bromus valdivianus Phil.
Poaceae
♦ Weed, Naturalised
♦ 15, 280
♦ G.

Bromus willdenowii Kunth
Poaceae
= *Bromus catharticus* Vahl
♦ rescue grass, litteäkattara, prairie grass, prairie brome
♦ Weed, Naturalised, Casual Alien
♦ 15, 42, 51, 70, 86, 87, 88, 176, 180, 218, 280
♦ G, cultivated, herbal. Origin: South America.

Brosimum alicastrum Sw.
Moraceae
♦ breadnut, Jamaican breadnut, snakewood
♦ Weed, Naturalised
♦ 101, 179
♦ cultivated, herbal.

Broussonetia papyrifera (L.) Vent.
Fabaceae/Papilionaceae
Morus papyrifera L., *Papyrius papyriferus* (L.) Kuntze
♦ paper mulberry, tapa cloth tree, mûrier à papier, papiermaulbeerbaum, aka, kodzu, amoreira do papel, moral de la China, morera de papel, papelero
♦ Weed, Naturalised, Introduced, Garden Escape, Environmental Weed
♦ 3, 6, 22, 80, 87, 88, 101, 102, 112, 142, 151, 152, 179, 218, 228, 230, 286
♦ T, arid, cultivated, herbal. Origin: Japan.

Browallia americana L.
Solanaceae
Browallia cordata D.Don., *Browallia
demissa* L., *Browallia dombreyana*
Dammer, *Browallia elata* L., *Browallia
melanotricha* Brandegee, *Browallia
nervosa* Miers, *Browallia peduncularis*
Benth., *Browallia viscosa* Kunth,
Nierembergia petunioides Dunal
♦ Jamaican forget me not, Andean
forget me not, catalina, teresita
♦ Weed, Naturalised, Cultivation
Escape
♦ 86, 87, 88, 101, 157, 179, 257, 261
♦ aH, arid, cultivated, herbal. Origin:
tropical America.

Browallia viscida H.B.K.
Solanaceae
♦ Naturalised
♦ 287

Brownea crawfordii S.Watson
Fabaceae/Caesalpiniaceae
♦ Quarantine Weed
♦ 220

Brugmansia Pers. spp.
Solanaceae
♦ brugmansia, angel's trumpet
♦ Quarantine Weed
♦ 220, 247
♦ toxic.

Brugmansia × candida Pers.
Solanaceae
Datura arborea L. (see), *Datura × candida*
(Pers.) Saff. (see) [= *Brugmansia aurea*
Lagerh. × *Brugmansia versicolor* Lagerh.
most likely combination]
♦ angel's trumpet tree, belladonna,
campana, cornucopia, floripondio,
angel's trumpet, white angel's trumpet
♦ Weed, Naturalised, Environmental
Weed, Cultivation Escape
♦ 39, 86, 101, 161, 165, 189, 257, 261,
280
♦ S/T, cultivated, herbal, toxic. Origin:
South America.

**Brugmansia sanguinea (Ruiz & Pav.)
D.Don**
Solanaceae
♦ red angel's trumpet, red floripontio
♦ Weed, Naturalised
♦ 39, 161, 280
♦ cultivated, herbal, toxic.

**Brugmansia suaveolens (Humb. & Bonpl.
ex Willd.) Bercht. & J.Presl**
Solanaceae
Datura suaveolens Humb. & Bonpl. ex
Willd. (see)
♦ angel's tears, angel's trumpet,
trombeteira, campana blanca
♦ Weed, Naturalised, Cultivation
Escape
♦ 39, 101, 134, 161, 255, 261, 280
♦ S, cultivated, herbal, toxic. Origin:
South and Central America.

Bruguiera gymnorhiza (L.) Savigny
Rhizophoraceae
Bruguiera capensis Blume, *Bruguiera
conjugata* Merr., *Bruguiera eriopetata*

Wight & Arn., *Bruguiera rhedii* Tul.,
Bruguiera rumphii Blume, *Bruguiera
wightii* Blume, *Bruguiera zippelii* Blume,
Rhizophora gymnorhiza L., *Rhizophora
palun* DC., *Rhizophora tinctoria* Blanco
♦ oriental mangrove, crimson
flowered mangrove, Burmese
mangrove
♦ Weed, Naturalised
♦ 22, 101
♦ T, cultivated, herbal. Origin: Africa,
Asia, Australia.

Brunfelsia americana L.
Solanaceae
♦ aguacero, trompeta de ángel,
American brunfelsia
♦ Cultivation Escape
♦ 261
♦ cultivated.

Brunfelsia nitida Benth.
Solanaceae
♦ lady of the night
♦ Naturalised, Cultivation Escape
♦ 101, 261
♦ cultivated. Origin: Cuba.

**Brunfelsia pauciflora (Cham. & Schltdl.)
Benth.**
Solanaceae
♦ Casual Alien
♦ 280
♦ cultivated, herbal.

Brunnera macrophylla (Bieb.) I.M.Johnst.
Boraginaceae
Myosotis macrophylla Bieb.
♦ largeleaf brunnera, Siberian bugloss,
largeleaf brunnera, great forget me not
♦ Naturalised, Casual Alien
♦ 40, 101, 280
♦ cultivated, herbal.

Brunnichia cirrhosa Gaertn.
Polygonaceae
= *Brunnichia ovata* (Walt.) Shinners
♦ redvine
♦ Weed
♦ 87, 88, 218
♦ herbal.

Brunnichia ovata (Walter) Shinners
Polygonaceae
Brunnichia cirrhosa Gaertn. (see)
♦ redvine, American buckwheat vine
♦ Weed
♦ 161
♦ herbal.

Brya ebenus (L.) DC.
Fabaceae/Papilionaceae
♦ ebony coccuswood, grenadilla,
granadilla, Jamaican ebony, West
Indian ebony
♦ Weed, Naturalised
♦ 14, 101

Bryonia alba L.
Cucurbitaceae
Bryonia nigra Dumort.
♦ white bryony, mustakoiranköynnös,
posed biely, European white bryony
♦ Weed, Noxious Weed, Naturalised,
Cultivation Escape
♦ 1, 39, 42, 70, 79, 80, 88, 101, 146, 156,

161, 229, 272
♦ pC, cultivated, herbal, toxic.

Bryonia cretica L.
Cucurbitaceae
♦ common bryony, Cretan bryony,
punakoiranköynnös, white bryony,
white dioica bryonia
♦ Weed, Naturalised
♦ 70, 88, 101, 161, 221, 272, 280
♦ herbal, toxic.

Bryonia cretica L. ssp. dioica (Jacq.) Tutin
Cucurbitaceae
Bryonia dioica Jacq.
♦ Cretan bryony, white bryony
♦ Weed, Naturalised
♦ 15, 101, 280
♦ herbal.

Bryonia dioica Jacq.
Cucurbitaceae
Bryonia dioeca Jacq., *Bryonia digyna*
Pomel
♦ white bryony, red bryony, Cretan
bryony, bryony
♦ Weed, Sleeper Weed, Quarantine
Weed, Environmental Weed,
Cultivation Escape, Casual Alien
♦ 34, 39, 40, 80, 225, 246, 253
♦ pC, cultivated, herbal, toxic.

Bryonia grandis L.
Cucurbitaceae
= *Coccinia grandis* (L.) Voigt
♦ Weed
♦ 3

Bryonopsis laciniosa (L.) Naud.
Cucurbitaceae
♦ Weed
♦ 87, 88
♦ promoted, herbal.

Bryophyllum Salisb. spp.
Crassulaceae
♦ mother of millions
♦ Environmental weed
♦ 201
♦ Cultivated.

**Bryophyllum beauverdii (Raym.-Hamet)
A.Berger**
Crassulaceae
Bryophyllum juelii (Raym.-Hamet
& H.Perrier) A.Berger, *Kalanchoe
beauverdii* Raym.-Hamet (see),
Kalanchoe juelii Raym.-Hamet &
H.Perrier
♦ Weed, Naturalised, Environmental
Weed
♦ 54, 86, 88
♦ Origin: Madagascar.

**Bryophyllum daigremontianum (Raym.-
Hamet & H.Perrier) A.Berger**
Crassulaceae
Kalanchoe daigremontiana Raym.-Hamet
& H.Perrier (see)
♦ mother of millions
♦ Noxious Weed, Naturalised, Casual
Alien
♦ 86, 86, 98, 203, 280
♦ pH, cultivated, toxic. Origin:
Madagascar.

Bryophyllum daigremontianum
**(Raym.-Hamet & H.Perrier) A.Berger.
× *delagoense* (Eckl. & Zeyh.) Schinz cv.
'Hougtonii'**
　Crassulaceae
　= *Bryophyllum daigremontianum*
　(Raym.-Hamet & H.Perrier) A.Berger ×
　delagoense (Eckl. & Zeyh.) Schinz
　♦ Naturalised
　♦ 98, 280
　♦ cultivated. Origin: horticultural.

Bryophyllum delagoense **(Eckl. & Zeyh.)
Schinz**
　Crassulaceae
　Bryophyllum tubiflorum Harv. nom.
　illeg. (see), *Bryophyllum verticillatum*
　(Scott-Elliot) Berger (see), *Kalanchoe
　delagoensis* Eckl. & Zeyh. (see),
　Kalanchoe tubiflora (Haw.) Raym.-
　Hamet (see), *Kalanchoe verticillata*
　Scott-Elliot (see)
　♦ mother of millions, chandelier plant
　♦ Weed, Quarantine Weed, Noxious
　Weed, Naturalised, Garden Escape,
　Environmental Weed
　♦ 73, 86, 88, 121, 220, 251, 269, 280, 283,
　290, 296
　♦ pH, cultivated, toxic. Origin:
　Madagascar.

Bryophyllum fedtschenkoi **(Hamet &
Perrier) Cheng**
　Crassulaceae
　Kalanchoe fedtschenkoi Hamet & Perrier
　(see)
　♦ Weed, Naturalised, Garden Escape
　♦ 54, 86, 88, 261
　♦ cultivated. Origin: Madagascar.

Bryophyllum pinnatum **(Lam.) Oken**
　Crassulaceae
　Bryophyllum calycinum Salisb.,
　Cotyledon pinnata Lam., *Crassula
　pinnata* L.f., *Kalanchoe pinnata* Pers.
　(see), *Sedum madagascaricum* Clus.
　♦ live plant, live leaf, bruja, life plant,
　yerba de bruja
　♦ Weed, Noxious Weed, Naturalised,
　Garden Escape, Environmental Weed
　♦ 14, 86, 88, 98, 157, 203, 251, 261, 280,
　286, 287, 290
　♦ cultivated, herbal. Origin:
　Madagascar.

Bryophyllum proliferum **Bowie ex Hook.**
　Crassulaceae
　Kalanchoe prolifera Raym.-Hamet
　♦ Weed, Naturalised, Garden Escape
　♦ 86, 98, 121, 203, 251
　♦ pH, cultivated. Origin: Indian Ocean
　Islands.

Bryophyllum tubiflorum **Harv. nom. illeg.**
　Crassulaceae
　= *Bryophyllum delagoense* (Eckl. &
　Zeyh.) Schinz
　♦ mother of millions
　♦ Weed, Quarantine Weed,
　Naturalised, Environmental Weed,
　Cultivation Escape
　♦ 73, 88, 98, 198, 203, 220, 261
　♦ cultivated, herbal. Origin: South
　Africa, Madagascar.

Bryophyllum verticillatum **(Scott-Elliot)
Berger**
　Crassulaceae
　= *Bryophyllum delagoense* (Eckl. &
　Zeyh.) Schinz
　♦ Quarantine Weed
　♦ 220

Bryopsis **J.V.F.Lamour. spp.**
　Bryopsidaceae
　♦ green alga
　♦ Weed
　♦ 197
　♦ algae.

Bryum argenteum **L.**
　Bryaceae
　♦ silver thread, silvery bryum,
　silvergreen bryum moss
　♦ Weed
　♦ 211, 286
　♦ moss.

Buchloë dactyloides **(Nutt.) Engelm.**
　Poaceae
　Sesleria dactyloides Nutt.
　♦ buffalograss
　♦ Weed, Naturalised, Introduced
　♦ 87, 88, 161, 218, 228, 280
　♦ G, arid, cultivated, herbal.

Buchnera longifolia **Kunth**
　Scrophulariaceae
　♦ elongated bluehearts
　♦ Weed
　♦ 14

Buchnera obliqua **Benth.**
　Scrophulariaceae
　♦ Weed
　♦ 199

Buchnera pusilla **Kunth**
　Scrophulariaceae
　♦ pygmy bluehearts
　♦ Naturalised
　♦ 101

Buchnera ternifolia **H.B.K.**
　Scrophulariaceae
　♦ Weed
　♦ 87, 88

Buchnera urticifolia **R.Br.**
　Scrophulariaceae
　♦ Weed
　♦ 87, 88
　♦ Origin: Australia.

Bucida buceras **L.**
　Combretaceae
　♦ black olive, gregorywood
　♦ Weed
　♦ 179
　♦ cultivated.

Buddleja alternifolia **Maxim.**
　Loganiaceae/Buddlejaceae
　♦ fountain butterfly bush
　♦ Weed, Noxious Weed, Naturalised
　♦ 80, 88, 101, 137
　♦ cultivated, herbal.

Buddleja americana **L.**
　Loganiaceae/Buddlejaceae
　♦ Weed
　♦ 14
　♦ S/T, herbal.

Buddleja asiatica **Lour.**
　Loganiaceae/Buddlejaceae
　♦ dog's tail, Raachaawadee paa, white
　butterfly bush
　♦ Weed, Naturalised, Cultivation
　Escape
　♦ 3, 22, 39, 54, 86, 88, 101, 191, 238, 252
　♦ S, cultivated, herbal, toxic. Origin:
　China, south-east Asia to Pakistan.

Buddleja australis **Vell.**
　Loganiaceae/Buddlejaceae
　♦ dog's tail
　♦ Naturalised, Cultivation Escape
　♦ 86, 252
　♦ cultivated. Origin: South America.

Buddleja brasiliensis **Jacq. ex Spreng.**
　Loganiaceae/Buddlejaceae
　♦ barbasco
　♦ Weed
　♦ 39, 255
　♦ pH, cultivated, herbal, toxic. Origin:
　Brazil.

Buddleja curviflora **Hook. & Arn.**
　Loganiaceae/Buddlejaceae
　♦ Naturalised
　♦ 287
　♦ herbal, toxic.

Buddleja davidii **Franch.**
　Loganiaceae/Buddlejaceae
　Buddleja variabilis Hemsl.
　♦ orange eye butterfly bush, summer
　lilac, buddleia, purple buddleia,
　butterfly bush
　♦ Weed, Quarantine Weed, Noxious
　Weed, Naturalised, Garden Escape,
　Environmental Weed, Cultivation
　Escape
　♦ 3, 15, 34, 40, 72, 80, 86, 88, 101, 137,
　142, 146, 151, 155, 165, 176, 181, 198,
　225, 233, 246, 261, 280, 287, 289, 296
　♦ S/T, aqua, cultivated, herbal. Origin:
　west and central China.

Buddleja dysophylla **(Benth.) Radlk.**
　Loganiaceae/Buddlejaceae
　Chilianthus dysophyllus (Benth.) A.DC.
　♦ buddleia
　♦ Weed, Naturalised, Cultivation
　Escape
　♦ 7, 15, 54, 86, 88, 198, 252, 280
　♦ cultivated. Origin: South Africa.

Buddleja globosa **Hope**
　Loganiaceae/Buddlejaceae
　♦ orange ball tree
　♦ Weed, Naturalised
　♦ 15, 39, 280
　♦ S, cultivated, herbal, toxic.

Buddleja indica **Lam.**
　Loganiaceae/Buddlejaceae
　Buddleja diversifolia Vahl, *Buddleja
　loniceroides* (Moldenke) Moldenke,
　Buddleja nepalensis Colla, *Buddleja
　rondeletiaeflora* Benth., *Nicodemia
　diversifolia* (Vahl) Ten., *Nicodemia
　diversifolia* var. *lucida* Scott-Elliot,
　Nicodemia grandifolia Scott-Elliot,
　Nicodemia hermanniana Cordem.,
　Nicodemia isleana Cordem., *Nicodemia
　rondeletiaeflora* (Benth.) Benth.,

Nicodemia rufescens Soler.
- Weed
- 179
- cultivated, herbal.

Buddleja lindleyana Fortune ex Lindl.
Loganiaceae/Buddlejaceae
- Lindley's butterfly bush, butterfly bush
- Weed, Naturalised
- 39, 80, 86, 101, 179
- cultivated, herbal, toxic. Origin: Asia.

Buddleja madagascariensis Lam.
Loganiaceae/Buddlejaceae
Nicodemia madagascariensis (Lam.) R.Parker (see)
- butterfly bush, smokebush, buddleia, buddleja bush, smokebush
- Weed, Naturalised, Garden Escape, Environmental Weed, Cultivation Escape
- 3, 7, 15, 39, 73, 86, 88, 98, 101, 179, 201, 203, 233, 246, 261, 280
- S, cultivated, herbal, toxic. Origin: Madagascar.

Buddleja officinalis Maxim.
Loganiaceae/Buddlejaceae
- pole butterfly bush
- Naturalised
- 101
- S, cultivated, herbal.

Buddleja saligna Willd.
Loganiaceae/Buddlejaceae
- squarestem butterfly bush
- Weed, Naturalised, Cultivation Escape
- 34, 101
- S, cultivated, herbal.

Buddleja salviifolia (L.) Lam.
Loganiaceae/Buddlejaceae
- butterfly bush, mountain sage, sagewood, wild sage
- Weed, Native Weed
- 121, 280
- S/T, cultivated, herbal. Origin: southern Africa.

Buddleja × weyeriana Weyer
Loganiaceae/Buddlejaceae
= *Buddleja davidii* Franch. × *Buddleja globosa* Hope
- Naturalised
- 40
- cultivated, herbal. Origin: horticultural hybrid.

Bufonia multiceps Decne.
Caryophyllaceae
- Weed
- 221

Bufonia tenuifolia L.
Caryophyllaceae
- Weed
- 272

Buglossoides arvense (L.) I.M.Johnst.
Boraginaceae
= *Buglossoides arvensis* (L.) I.M.Johnst.
- corn gromwell
- Weed, Naturalised

- 98, 102, 203, 269
- aH. Origin: Eurasia.

Buglossoides arvensis (L.) I.M.Johnst.
Boraginaceae
Buglossoides arvense (L.) I.M.Johnst. (see), *Lithospermum arvense* L. (see)
- corn gromwell, field gromwell, puccoon
- Weed, Naturalised, Introduced
- 7, 39, 70, 80, 86, 88, 93, 94, 101, 174, 176, 198, 243, 253, 272
- H, herbal, toxic. Origin: Eurasia.

Buglossoides purpurocaerulea (L.) Johnst.
Boraginaceae
Lithospermum purpuro-caeruleum L., *Rhytispermum purpureo-caeruleum* Link
- erba perla azzurra
- Weed
- 272
- herbal.

Bulbine bulbosa (R.Br.) Haw.
Liliaceae/Asphodelaceae
- native leek, bulbine lily
- Native Weed
- 39, 269
- pH, cultivated, toxic. Origin: Australia.

Bulbine frutescens (L.) Willd.
Liliaceae/Asphodelaceae
- copaiva, snake flower
- Native Weed
- 121
- pH. Origin: southern Africa.

Bulbine narcissifolia Salm-Dyck
Liliaceae/Asphodelaceae
- snake flower
- Native Weed
- 121
- pH, herbal. Origin: southern Africa.

Bulbine semibarbata (R.Br.) Haw.
Liliaceae/Asphodelaceae
Bulbinopsis semibarbata (R.Br.) Borzí (see)
- Casual Alien
- 39, 280
- cultivated, toxic.

Bulbinella floribunda (Aiton) T.Durand & Schinz
Liliaceae/Asphodelaceae
- Weed, Naturalised
- 7, 98, 203
- cultivated.

Bulbinella robusta Kunth
Liliaceae/Asphodelaceae
- Weed, Naturalised
- 98, 203

Bulbinopsis semibarbata (R.Br.) Borzí
Liliaceae/Asphodelaceae
= *Bulbine semibarbata* (R.Br.) Haw.
- Casual Alien
- 280

Bulbocodium vernum L.
Liliaceae/Colchicaceae
Colchicum bulbocodium Ker Gawl.
- spring meadow saffron
- Weed
- 161, 247

- cultivated, herbal, toxic.

Bulbophyllum micronesiacum Schltr.
Orchidaceae
- Introduced
- 230
- H.

Bulbophyllum ponapense Schltr.
Orchidaceae
- Introduced
- 230
- H.

Bulbostylis barbata (Rottb.) C.B.Clarke
Cyperaceae
Fimbristylis barbata (Rottb.) Benth, *Stenophyllus barbatus* (Rottb.) Cooke
- dense bulbostylis, watergrass
- Weed, Naturalised
- 12, 87, 88, 101, 126, 170, 179, 286, 297
- aG, herbal.

Bulbostylis burchellii (Fical. & Hiern) C.B.Clarke
Cyperaceae
- Native Weed
- 121
- pG. Origin: southern Africa.

Bulbostylis capillaris (L.) C.B.Clarke
Cyperaceae
Scirpus capillaris L.
- alecrim da praia, threadleaf beakseed, densetuft hairsedge
- Weed
- 255, 295
- aG, herbal. Origin: South America.

Bulbostylis contexta (Nees) Bodard
Cyperaceae
Bulbostylis collina (Kunth) C.B.Clarke, *Bulbostylis burkei* C.B.Clarke, *Bulbostylis kirkii* C.B.Clarke
- Native Weed
- 121
- pG. Origin: southern Africa.

Bulbostylis densa (Wall.) Hand.-Mazz.
Cyperaceae
Bulbostylis capillaris (*non* L.) Boeck. var. *trifida* C.B.Clarke
- bulbostylis barbu
- Weed, Native Weed
- 121, 126
- aG, aqua. Origin: southern Africa.

Bulbostylis filamentosa C.B.Clarke
Cyperaceae
- Weed
- 87, 88
- G.

Bulbostylis funckii (Steud.) Clarke
Cyperaceae
- Funck's hairsedge
- Weed
- 199
- G.

Bulbostylis hirta (Thunb.) Svens.
Cyperaceae
- rough hairsedge
- Naturalised
- 101
- G. Origin: Madagascar.

Bulbostylis hispidula (Vahl) R.Haines
Cyperaceae
= *Fimbristylis hispidula* (Vahl) Kunth
♦ Weed
♦ 88
♦ G.

Bulbostylis humilis (Kunth) C.B.Clarke
Cyperaceae
♦ shy sedge
♦ Native Weed
♦ 121
♦ aG. Origin: southern Africa.

Bulbostylis puberula (Poir.) C.B.Clarke
Cyperaceae
Isolepis puberula (Poir.) Kunth, *Scirpus puberulus* Michx., *Stenophyllus puberulus* Killip
♦ Weed, Quarantine Weed
♦ 76, 87, 88, 170, 191, 203, 220
♦ G.

Bulbostylis striatella C.B.Clarke
Cyperaceae
♦ Weed, Sleeper Weed, Naturalised, Environmental Weed
♦ 54, 86, 88, 155
♦ a/pG. Origin: South Africa.

Bulnesia retamo Griseb.
Zygophyllaceae
♦ Introduced
♦ 228
♦ arid.

Bumelia lanuginosa (Michx.) Pers.
Sapotaceae
= *Sideroxylon lanuginosum* Michx. ssp. *lanuginosum* (NoR)
♦ gum bumelia, chittamwood
♦ Weed
♦ 87, 88, 218
♦ T, cultivated, herbal.

Bumelia lycioides (L.) Gaertn.
Sapotaceae
= *Sideroxylon lycioides* L. (NoR)
♦ buckthorn bumelia, shittamwood
♦ Weed
♦ 87, 88, 218
♦ T, cultivated, herbal.

Bunchosia armeniaca (Cav.) DC.
Malpighiaceae
♦ Introduced
♦ 228
♦ T, arid.

Bunchosia glandulifera (Jacq.) Kunth
Malpighiaceae
♦ cafe falso
♦ Naturalised
♦ 86
♦ T, cultivated, herbal. Origin: Central and South America.

Bunias erucago L.
Brassicaceae
Myagrum erucago L.
♦ crested bunias, corn rocket, southern warty cabbage, etelänukonpalko
♦ Weed, Naturalised, Casual Alien
♦ 42, 70, 87, 88, 94, 101, 253, 272
♦ a/bH, cultivated, herbal.

Bunias orientalis L.
Brassicaceae
Laelia orientalis Desv.
♦ warty cabbage, Turkish warty cabbage
♦ Weed, Quarantine Weed, Naturalised, Introduced
♦ 42, 56, 70, 87, 88, 94, 101, 220, 253, 272
♦ pH, cultivated, herbal.

Bunium bulbocastanum L.
Apiaceae
Aegopodium bulbocastanum Michot, *Apium bulbocastanum* Caruel, *Carvi bulbocastanum* Bubani
♦ great pignut, tuberous carraway, maakastanja
♦ Weed, Quarantine Weed, Naturalised, Casual Alien
♦ 42, 70, 76, 88, 94, 253, 272, 287
♦ pH, cultivated, herbal.

Buphthalmum salicifolium L.
Asteraceae
Buphthalmum grandiflorum L.
♦ häränkukka
♦ Casual Alien
♦ 42
♦ cultivated, herbal.

Buphthalmum speciosum (Baumg.) Schreb.
Asteraceae
♦ Weed, Quarantine Weed
♦ 76, 88, 220
♦ herbal.

Bupleurum L. spp.
Apiaceae
♦ bupleurum
♦ Weed, Naturalised
♦ 198, 272
♦ cultivated, herbal.

Bupleurum chinense DC.
Apiaceae
= *Bupleurum falcatum* L.
♦ Chinese thoroughwax, bei chai hu, chai hu
♦ Weed, Quarantine Weed
♦ 220, 275
♦ pH, herbal. Origin: temperate Asia.

Bupleurum croceum Fenzl
Apiaceae
♦ kultajänönputki
♦ Casual Alien
♦ 42

Bupleurum falcatum L.
Apiaceae
Bupleurum chinense DC. (see)
♦ bupleurum, chai hu, sickle leaved hare's ear, thorow wax
♦ Weed, Quarantine Weed
♦ 70, 76, 87, 88, 220, 221
♦ pH, cultivated, herbal. Origin: Eurasia.

Bupleurum fontanesii Guss. ex Caruel
Apiaceae
= *Bupleurum odontites* L.
♦ äimäjänönputki
♦ Casual Alien
♦ 42

Bupleurum gerardi Jacq.
Apiaceae
♦ Weed
♦ 87, 88
♦ herbal.

Bupleurum lancifolium Hornem.
Apiaceae
Bupleurum protractum Hoffm. & Link (see)
♦ suippujänönputki, false thorow wax, lanceleaf thorow wax
♦ Weed, Naturalised, Casual Alien
♦ 7, 34, 42, 70, 86, 88, 94, 98, 101, 203, 221, 243, 287
♦ aH. Origin: central to south-west Europe.

Bupleurum longifolium L.
Apiaceae
Diaphyllum longifolium Hoffm.
♦ Weed
♦ 272
♦ cultivated, herbal.

Bupleurum odontites L.
Apiaceae
Bupleurum fontanesii Guss. ex Caruel (see)
♦ narrowleaf thorow wax
♦ Naturalised
♦ 101, 287

Bupleurum protractum Hoffm. & Link
Apiaceae
= *Bupleurum lancifolium* Hornem.
♦ Weed
♦ 87, 88

Bupleurum rotundifolium L.
Apiaceae
Tenorea rotundifolia Bubani
♦ hound's ear, hare's ear, thorow wax, pyöröjänönputki
♦ Weed, Naturalised, Casual Alien
♦ 23, 42, 70, 80, 86, 87, 88, 94, 98, 101, 102, 203, 243, 253, 272, 287
♦ aH, cultivated, herbal. Origin: Eurasia.

Bupleurum semicompositum L.
Apiaceae
♦ hare's ear
♦ Weed, Naturalised, Environmental Weed
♦ 7, 86, 87, 88, 98, 198, 203
♦ Origin: Mediterranean.

Bupleurum smithii Wolff
Apiaceae
♦ black thorowax
♦ Weed
♦ 297

Bupleurum subovatum Spreng.
Apiaceae
♦ false thorow wax, narrow thorow wax
♦ Naturalised
♦ 280

Bupleurum tenuissimum L.
Apiaceae
♦ smallest hare's ear, slender hare's ear
♦ Weed, Naturalised
♦ 87, 88, 272, 280
♦ herbal.

Burkea africana Hook.
Fabaceae/Caesalpiniaceae
♦ wild seringa, red seringa, Rhodesian
ash, Rhodesian seringa, seringa,
seringa tree, white seringa
♦ Weed, Native
Weed
♦ 10, 121
♦ T, arid, cultivated, toxic. Origin:
southern Africa.

Bursaria spinosa Cav.
Pittosporaceae
♦ sweet bursaria, Christmas bush,
sweet box
♦ Weed, Native Weed
♦ 87, 88, 269
♦ S/T, cultivated, herbal. Origin:
Australia.

Bursera inaguensis Britt.
Burseraceae
♦ Weed
♦ 14

Butea monosperma (Lam.) Taub.
Fabaceae/Papilionaceae
Butea frondosa J.König ex Roxb.,
Erythrina monosperma Lam.
♦ flame of the forest, Bengal kino
♦ Introduced
♦ 39, 228
♦ arid, promoted, herbal, toxic.

Butia capitata (Mart.) Becc.
Arecaceae
Butia bonnetii (Linden ex Chabaud)
Becc., *Butia capitata* var. *deliciosa*
Prosch., *Butia capitata* var. *elegantissima*
(Chabaud) Becc., *Butia capitata* var.
erythrospatha (Chabaud) Becc., *Butia
capitata* var. *lilaceiflora* (Chabaud)
Becc., *Butia capitata* var. *nehrlingiana*
(L.H.Bailey) L.H.Bailey, *Butia capitata*
var. *odorata* (Barb.Rodr.) Becc., *Butia
capitata* var. *pulposa* (Barb.Rodr.) Becc.,
Butia capitata var. *pygmaea* Prosch.,
Butia capitata var. *subglobosa* Becc.,
Butia capitata var. *virescens* Becc.,
Butia leiospatha (Barb.Rodr.) Becc.,
Butia nehrlingiana L.H.Bailey, *Cocos
bonnetii* Linden ex Chabaud, *Cocos
capitata* Mart., *Cocos campestris* Mart.,
Cocos odorata Barb.Rodr., *Cocos pulposa*
Barb.Rodr., *Syagrus capitata* (Mart.)
Glassman
♦ Brazil butia palm, butia palm, jelly
palm, pindo palm, yatay palm, wine
palm
♦ Introduced
♦ 228
♦ T, arid, cultivated, herbal.

Butomus umbellatus L.
Butomaceae
Butomus caesalpini Neck., *Butomus
floridus* Gaertn.
♦ flowering rush, sarjarimpi, okrasa
okolíkatá
♦ Weed, Quarantine Weed,
Naturalised, Introduced,
Environmental Weed
♦ 4, 23, 70, 79, 80, 87, 88, 101, 103, 104,
133, 156, 195, 197, 222, 224, 246, 253,

263, 272, 275, 297
♦ wpH, cultivated, herbal, toxic.

Buxus sempervirens L.
Buxaceae
♦ common box, box, boxwood, bosso
♦ Weed,
Naturalised
♦ 39, 101, 161, 247, 280
♦ S/T, cultivated, herbal, toxic.

Byrsonima crassifolia (L.) Kunth
Malpighiaceae
Byrsonima cotinifolia Kunth, *Byrsonima
karwinskiana* A.Juss., *Byrsonima
oaxacana* A.Juss., *Malpighia crassifolia* L.
♦ maricao cimun
♦ Introduced
♦ 14, 39, 228
♦ arid, cultivated, herbal, toxic.

Byrsonima intermedia A.Juss
Malpighiaceae
♦ murici do campo
♦ Weed
♦ 255
♦ pH. Origin: Brazil.

Byrsonima orbignyana A.Juss.
Malpighiaceae
♦ canjiqueira
♦ Weed
♦ 255
♦ S. Origin: South America.

Byrsonima wrightiana Urb. & Niedz
Malpighiaceae
Byrsonima pinetorum Griseb.
♦ Weed
♦ 14

C

Cabomba spp. Aubl.
Cabombaceae
♦ cabomba, fanwort
♦ Quarantine Weed, Noxious Weed,
Naturalised
♦ 86, 161, 198, 220, 258
♦ wH, cultivated, herbal.

Cabomba aquatica Aubl.
Cabombaceae
♦ yellow cabomba
♦ Quarantine Weed
♦ 220
♦ wH, cultivated.

Cabomba australis Speg.
Cabombaceae
♦ Weed, Quarantine Weed
♦ 76, 88, 220, 237
♦ wpH, cultivated.

Cabomba caroliniana A.Gray
Cabombaceae
♦ fanwort, cabomba, Carolina
watershield, fishgrass, Washington
grass, watershield, green cabomba
♦ Weed, Quarantine Weed, Noxious
Weed, Naturalised, Garden Escape,
Environmental Weed
♦ 1, 54, 62, 72, 76, 80, 86, 87, 88, 93, 98,
103, 133, 139, 146, 147, 178, 191, 195,
197, 198, 200, 203, 218, 220, 224, 229,
237, 246, 255, 258, 286, 287, 290, 295
♦ wpH, cultivated, herbal. Origin:
North America.

**Cabomba caroliniana Gray var.
caroliniana**
Cabombaceae
♦ fanwort, Carolina fanwort
♦ Noxious Weed
♦ 229
♦ wpH.

**Cabomba caroliniana Gray var.
pulcherrima Harper**
Cabombaceae
♦ fanwort, Carolina fanwort
♦ Noxious Weed
♦ 229
♦ wpH.

Cabomba furcata J.A. & J.H.Schult.
Cabombaceae
Cabomba piauhyensis Gardner (see)
♦ forked fanwort
♦ Naturalised
♦ 101
♦ wH.

Cabomba haynesii Wiersema
Cabombaceae

- fishgrass
- Weed
- 179
- wH.

Cabomba piauhyensis Gardner
Cabombaceae
= *Cabomba furcata* J.A. & J.H.Schult.
- red cabomba
- Quarantine Weed
- 220
- wH, cultivated.

Cacabus prostratus (L'Hér.) Bernh.
Solanaceae
= *Exodeconus prostratus* (L'Hér.) Raf.
(NoR)
- Weed
- 87, 88

Cacalia hastata L.
Asteraceae
- keihäsnauhus, cacalia, kakalia
- Cultivation Escape
- 42
- cultivated, herbal.

Cacalia tangutica (Maxim.) Hand.-Mazz.
Asteraceae
- liuskanauhus
- Cultivation Escape
- 42
- cultivated, herbal.

Cacalia tuberosa Nutt.
Asteraceae
= *Arnoglossum plantagineum* Raf. (NoR)
- Indian plantain
- Weed
- 87, 88, 218
- herbal.

Cachrys eriantha DC.
Apiaceae
- Weed
- 87, 88

Cachrys trifida Mill.
Apiaceae
- Weed
- 70

Cadaba farinosa Forssk.
Capparaceae
- cadaba
- Weed
- 221
- arid, herbal.

Cadaba glandulosa Forssk.
Capparaceae
- Weed
- 221
- arid.

Cadaba rotundifolia Forssk.
Capparaceae
- Weed
- 221
- herbal.

Caesalpinia L. spp.
Fabaceae/Caesalpiniaceae
- bird of paradise bush, nicker
- Weed
- 116, 247
- herbal, toxic.

Caesalpinia bonduc (L.) Roxb.
Fabaceae/Caesalpiniaceae
Caesalpinia bonducella (L.) Fleming.
nom. illeg., *Caesalpinia crista* auct. Amer.,
Guilandina bonduc L. (see), *Guilandina bonducella* L.
- yellow nickers, gray nickers, wait a bit, pacap, togodulik, talamoa, ash coloured nicker, bonduc, Brazilian redwood, fever nut, nicker nut, pernambuco redwood, nicker seed
- Weed, Naturalised, Native Weed, Introduced, Environmental Weed
- 3, 14, 39, 87, 88, 107, 121, 161, 191, 218, 228, 230, 241, 257
- S/T, arid, cultivated, herbal, toxic. Origin: tropical Africa and Asia.

Caesalpinia bracteosa Tul.
Fabaceae/Caesalpiniaceae
- Introduced
- 228
- arid.

Caesalpinia coriaria (Jacq.) Willd.
Fabaceae/Caesalpiniaceae
- divi divi
- Weed, Introduced
- 39, 87, 88, 228
- arid, herbal, toxic. Origin: Central America.

Caesalpinia crista L.
Fabaceae/Caesalpiniaceae
Caesalpinia kwangtungensis Merr.,
Caesalpinia nuga (L.) W.T.Aiton,
Caesalpinia szechuenensis Craib,
Guilandina bonduc L. var. *minus* DC.,
Guilandina crista (L.) Small, *Guilandina nuga* L., *Guilandina semina* Lour.,
Ticanto nuga (L.) Medik. (see)
- bonduc nut, nicker nut, nicker bean
- Weed
- 3, 87, 88, 179, 191
- cultivated, herbal. Origin: Asia.

Caesalpinia decapetala (Roth) Alston
Fabaceae/Caesalpiniaceae
Biancaea decapetela (Roth) O.Deg.,
Biancaea scandens Tod., *Biancaea sepiaria* (Roxb.) Tod., *Caesalpinia decapetala* var. *japonica* (Siebold & Zucc.) H.Ohashi, *Caesalpinia decapetala* var. *japonica* (Siebold & Zucc.) Isely, *Caesalpinia japonica* Siebold & Zucc., *Caesalpinia sepiaria* Roxb. (see), *Caesalpinia sepiaria* var. *japonica* (Siebold & Zucc.) Gagnep., *Caesalpinia sepiaria* var. *japonica* (Siebold & Zucc.) Makino, *Reichardia decapetala* Roth
- Mauritius thorn, mysore thorn, thorny poinciana, shoofly, wait a while, whoa back, chithari, cat's claw, tiger stopper, uña de gato, yuen shin, liane sappan, cassie, sappan
- Weed, Quarantine Weed, Noxious Weed, Naturalised, Garden Escape, Environmental Weed, Cultivation Escape
- 3, 10, 22, 39, 50, 51, 63, 73, 86, 87, 88, 95, 98, 101, 121, 134, 152, 158, 191, 203, 225, 246, 269, 277, 278, 279, 280, 283
- pC, cultivated, toxic. Origin: India, Sri Lanka.

Caesalpinia ferrea Mart. ex Tul.
Fabaceae/Caesalpiniaceae
- leopard tree
- Introduced
- 228
- arid, cultivated, herbal.

Caesalpinia gilliesii (Wall. ex Hook.) Dietr.
Fabaceae/Caesalpiniaceae
Poinciana gilliesii Wall. ex Hook. (see)
- bird of paradise, cat's claw, mysore thorn, bird of paradise shrub, paradise poinciana
- Weed, Naturalised, Introduced, Environmental Weed
- 7, 39, 80, 86, 87, 88, 93, 98, 101, 151, 161, 203, 218, 228, 241, 261, 279, 295, 300
- S/T, arid, cultivated, herbal, toxic. Origin: South America.

Caesalpinia major (Medik.) Dandy & Exell
Fabaceae/Caesalpiniaceae
- nickers, pakao, Hawai'i pearls
- Weed, Naturalised
- 3, 134, 191
- cultivated, herbal. Origin: Madagascar.

Caesalpinia mexicana Gray
Fabaceae/Caesalpiniaceae
- Mexican holdback, retamilia, Mexican poinsettia
- Weed, Naturalised
- 101, 161
- cultivated, herbal, toxic.

Caesalpinia mollis (Kunth) Spreng.
Fabaceae/Caesalpiniaceae
Brasilettia mollis (Kunth) Britton & Killip, *Caesalpinia acutiflora* I.M.Johnst., *Coulteria mollis* Kunth
- Introduced
- 228
- arid.

Caesalpinia pulcherrima (L.) Sw.
Fabaceae/Caesalpiniaceae
Poinciana pulcherrima L.
- flowerfence, sehmwida, pride of Barbados, dwarf poinciana, lau pa
- Weed, Naturalised, Introduced, Environmental Weed, Cultivation Escape
- 22, 39, 88, 101, 151, 161, 179, 228, 230, 261
- S/T, arid, cultivated, herbal, toxic. Origin: South America.

Caesalpinia sepiaria Roxb.
Fabaceae/Caesalpiniaceae
= *Caesalpinia decapetala* (Roth) Alston
- Weed, Quarantine Weed
- 3, 76, 80, 88, 191, 203
- herbal.

Caesalpinia spinosa (Molina) Kuntze
Fabaceae/Caesalpiniaceae
Caesalpinia pectinata Cav., *Caesalpinia tinctoria* (Kunth) Benth.
- spiny holdback
- Naturalised, Introduced
- 101, 228
- S, arid, cultivated.

Cajanus cajan (L.) Millsp.
Fabaceae/Papilionaceae
Cajanus indicus Spreng.
♦ cajan, catjan pea, Congo pea, dhal, dhal bean, noeye pea, pigeon pea, red gram, kabalama mbazi
♦ Weed, Naturalised, Introduced, Casual Alien
♦ 80, 86, 87, 88, 98, 101, 121, 179, 203, 228, 230, 261
♦ pS, arid, cultivated, herbal. Origin: Old World Tropics.

Cajanus scarabaeoides (L.) Thouars
Fabaceae/Papilionaceae
♦ cajanus
♦ Naturalised
♦ 101
♦ Origin: Madagascar.

Cakile edentula (Bigelow) Hook.
Brassicaceae
♦ sea rocket, American sea rocket
♦ Weed, Naturalised, Garden Escape, Environmental Weed
♦ 7, 72, 86, 88, 98, 176, 198, 203, 269, 280, 287
♦ pH, cultivated, herbal. Origin: North America.

Cakile maritima Scop.
Brassicaceae
Cakile edentula non Hook., *Bunias cakile* L.
♦ sea rocket, European sea rocket
♦ Weed, Naturalised, Garden Escape, Environmental Weed
♦ 7, 72, 87, 88, 98, 101, 116, 165, 176, 203, 237, 269, 280, 289
♦ pH, cultivated, herbal. Origin: Europe.

Cakile maritima Scop. ssp. *maritima*
Brassicaceae
♦ sea rocket
♦ Naturalised, Environmental Weed
♦ 86, 198, 280
♦ cultivated.

Caladium arboreum H.B.K.
Araceae
♦ Weed
♦ 87, 88

Caladium bicolor (Aiton) Vent.
Araceae
Cyrtospadix bicolor Britton & Wils., *Caladium × hortulanum* Birdsey
♦ elephant's ear, heart of Jesus, caladium, angel wings, caladio, mother in law plant, paleta de pintor
♦ Weed, Naturalised, Introduced, Garden Escape
♦ 39, 87, 88, 101, 189, 230, 261, 262
♦ H, aqua, cultivated, herbal, toxic. Origin: South America.

Calamagrostis Adans. spp.
Poaceae
♦ reedgrass
♦ Weed
♦ 272
♦ G, herbal.

Calamagrostis arundinacea (L.) Roth.
Poaceae

♦ metsäkastikka, nogariyasu
♦ Weed
♦ 87, 88
♦ G, cultivated, herbal.

Calamagrostis arundinacea (L.) Roth var. brachytricha (Steud.) Hack.
Poaceae
♦ Weed
♦ 286
♦ G.

Calamagrostis canadensis (Michx.) P.Beauv.
Poaceae
Arundo canadensis Michx.
♦ bluejoint reedgrass, Canada reedgrass, reedgrass, bluejoint
♦ Weed
♦ 87, 88, 218
♦ pG, herbal.

Calamagrostis epigeios (L.) Roth.
Poaceae
= *Calamagrostis epigejos* (L.) Roth
♦ chee reedgrass, yamaawa
♦ Weed, Naturalised
♦ 87, 88, 101, 286
♦ G, cultivated, herbal.

Calamagrostis epigeios (L.) Roth var. capensis (Stapf) N.N.Tsvelev
Poaceae
♦ reedgrass, saltpan grass
♦ Native Weed
♦ 121
♦ pG. Origin: southern Africa.

Calamagrostis epigeios (L.) Roth var. epigeios
Poaceae
♦ chee reedgrass
♦ Naturalised
♦ 101
♦ G.

Calamagrostis epigeios (L.) Roth var. georgica (K.Koch) Ledeb.
Poaceae
= *Calamagrostis epigeios* (L.) Roth ssp. *glomerata* (Boiss. & Buhse) Tzvelev (NoR)
♦ chee reedgrass
♦ Naturalised
♦ 101
♦ G.

Calamagrostis epigejos (L.) Roth
Poaceae
Arundo epigejos L., *Calamagrostis epigeios* (L.) Roth. (see), *Calamagrostis macrolepis* Litv., *Calamagrostis georgica* K.Koch, *Calamagrostis glomerata* Boiss. & Buhse, *Calamagrostis koibalensis* Reverd., *Calamagrostis lenkoranensis* Steud.
♦ wood smallreed, chee reedgrass, bushgrass, hietakastikka, reedgrass
♦ Weed, Naturalised, Introduced, Garden Escape, Environmental Weed, Casual Alien
♦ 15, 54, 70, 86, 98, 176, 203, 228, 272, 280, 300
♦ G, arid,
cultivated, herbal. Origin: Eurasia, Africa.

Calamagrostis fulva (Griseb.) Kuntze
Poaceae
Deyeuxia fulva (Griseb.) L.Parodi
♦ Introduced
♦ 228
♦ G, arid.

Calamagrostis hakonensis Franch. & Sav.
Poaceae
♦ himenogariyasu
♦ Weed
♦ 88, 204, 286
♦ G.

Calamagrostis montevidensis Nees
Poaceae
= *Deyeuxia viridiflavescens* (Poir.) Steud. var. *montevidensis* (Nees) Kämpf
♦ Weed
♦ 87, 88
♦ G.

Calamagrostis neglecta (Ehrh.) C.F.Gaertn., B.Mey. & Scherb.
Poaceae
= *Calamagrostis stricta* (Timm) Koel. ssp. *stricta* (NoR)
♦ Weed
♦ 87, 88
♦ G, herbal.

Calamagrostis pseudophragmites (Haller f.) Koeler
Poaceae
Arundo litorea Schrad., *Calamagrostis litorea* P.Beauv.
♦ cannella spondicola
♦ Weed
♦ 272, 275, 297
♦ pG, cultivated, herbal.

Calamagrostis purpurea (Trin.) Trin.
Poaceae
♦ Scandinavian small reed
♦ Quarantine Weed
♦ 220
♦ G, cultivated. Origin: Europe.

Calamagrostis varia (Schrad.) Host
Poaceae
Arundo varia Schrad., *Deyeuxia varia* Kunth
♦ variegated smallreed
♦ Weed
♦ 272
♦ G, cultivated, herbal.

Calamintha Mill. spp.
Lamiaceae
♦ rmische minze
♦ Weed
♦ 272
♦ herbal.

Calamintha acinos (L.) Clairv.
Lamiaceae
= *Acinos arvensis* (Lam.) Dandy
♦ Weed
♦ 87, 88
♦ herbal.

Calamintha grandiflora (L.) Moench
Lamiaceae
Melissa grandiflora L., *Satureja grandiflora* (L.) Scheele (see)
♦ large flowered calamint, showy calamint

♦ Weed
♦ 272
♦ pH, cultivated, herbal.

Calamintha nepeta (L.) Savi
Lamiaceae
Satureia nepeta Scheele
♦ lesser calamint, nepeta calamint, nepetella, lesser calamint, nipitella
♦ Weed, Naturalised
♦ 86, 98, 101, 161, 203, 280
♦ pH, cultivated, herbal. Origin: Europe.

Calamintha nepeta (L.) Savi ssp. glandulosa (Riquien) P.W.Ball
Lamiaceae
♦ lesser calamint
♦ Naturalised
♦ 101

Calamintha nepeta (L.) Savi ssp. nepeta
Lamiaceae
♦ lesser calamint
♦ Naturalised
♦ 101

Calamintha sylvatica Bromf.
Lamiaceae
Calamintha montana Lam., *Calamintha montana* Lam. ssp. *montana*, *Calamintha rouyana* Rouy, *Satureja vulgaris* Rouy *p.p. non* (L.) Fritsch
♦ woodland calamint, calamint
♦ Weed, Naturalised
♦ 101, 272
♦ pH, cultivated, herbal.

Calamintha sylvatica Bromf. ssp. ascendens (Jord.) P.W.Ball
Lamiaceae
♦ woodland calamint
♦ Naturalised
♦ 101

Calamus L. spp.
Arecaceae
♦ calamus, rattans
♦ Weed, Introduced
♦ 3, 191, 230
♦ T.

Calandrinia caulescens Kunth
Portulacaceae
= *Calandrinia ciliata* (Ruiz & Pav.) DC.
♦ Weed
♦ 88

Calandrinia caulescens (Ruiz & Pav.) DC. var. menziesii (Hook.) Macbr.
Portulacaceae
♦ redmaids rock purslane
♦ Weed
♦ 218

Calandrinia ciliata (Ruiz & Pav.) DC.
Portulacaceae
Calandrinia arizonica Rydb., *Calandrinia caulescens* Kunth (see), *Calandrinia ciliata* (Ruiz & Pav.) DC. var. *menziesii* (Hook.) J.F.Macbr. (see), *Calandrinia micrantha* Schltdl. (see), *Calandrinia phacosperma* DC., *Talinum caulescens* (Kunth) Spreng., *Talinum ciliatum* Ruiz & Pav.

♦ desert menziesii rock purslane, fringed redmaids, redmaids, pil, saapah, sapx
♦ Weed, Introduced
♦ 161, 228, 243
♦ aH, arid, cultivated, herbal.

Calandrinia ciliata (Ruiz & Pav.) DC. var. menziesii (Hook.) J.F.Macbr.
Portulacaceae
= *Calandrinia ciliata* (Ruiz & Pav.) DC.
♦ redmaids desert rock purslane
♦ Weed
♦ 180, 212
♦ Origin: North America.

Calandrinia compressa DC.
Portulacaceae
♦ Naturalised
♦ 280
♦ aH.

Calandrinia menziesii (Hook.) Torr. & A.Gray
Portulacaceae
♦ purple purslane, Curnow's curse, etelänsailio
♦ Weed, Naturalised, Casual Alien
♦ 7, 42, 86, 98, 165, 176, 198, 203, 280
♦ cultivated. Origin: north-west North America.

Calandrinia micrantha Schltdl.
Portulacaceae
= *Calandrinia ciliata* (Ruiz & Pav.) DC.
♦ Weed
♦ 157, 199

Calathea crotalifera S.Wats.
Marantaceae
♦ rattlesnake plant
♦ Naturalised
♦ 101
♦ cultivated.

Calathea zebrina (Sims) Lindl.
Marantaceae
♦ galatea, zebra, zebraplant
♦ Cultivation Escape
♦ 261
♦ cultivated, herbal. Origin: Brazil.

Calceolaria mexicana Benth.
Scrophulariaceae
♦ slipper flower
♦ Weed
♦ 157
♦ aH.

Calceolaria tripartita Ruiz & Pav.
Scrophulariaceae
♦ splitleaf slipperwort
♦ Weed, Naturalised
♦ 86, 98, 134, 203, 280, 300
♦ arid, cultivated. Origin: Central America.

Calea jamaicensis L.
Asteraceae
♦ Weed
♦ 87, 88

Calea zacatechichi Schltdl.
Asteraceae
♦ Weed, Quarantine Weed
♦ 39, 76, 88, 220
♦ promoted, herbal, toxic.

Calendula L. spp.
Asteraceae
♦ marigold
♦ Weed, Naturalised
♦ 116, 198
♦ herbal.

Calendula aegyptiaca Pers.
Asteraceae
♦ Weed
♦ 87, 88

Calendula algeriensis Boiss. & Reut.
Asteraceae
♦ Weed
♦ 87, 88

Calendula arvensis L.
Asteraceae
Calendula aegyptiaca Desf., *Calendula persica* C.A.Mey, *Calendula gracilis* DC., *Calendula micrantha* Tin. & Guss.
♦ field marigold, peltokehäkukka
♦ Weed, Naturalised, Introduced, Garden Escape, Casual Alien
♦ 34, 40, 42, 44, 70, 86, 87, 88, 94, 98, 101, 115, 176, 185, 198, 203, 241, 243, 250, 253, 269, 272, 280, 287, 300
♦ aH, cultivated, herbal. Origin: Europe.

Calendula officinalis L.
Asteraceae
Caltha officinalis Moench *non* Scop.
♦ pot marigold, calendula, marigold, English marigold, African marigold, garden marigold
♦ Weed, Naturalised, Introduced, Garden Escape, Cultivation Escape
♦ 15, 24, 34, 40, 42, 86, 98, 101, 176, 198, 203, 215, 228, 241, 252, 256, 269, 272, 280, 287, 300
♦ aH, arid, cultivated, herbal. Origin: obscure, possibly southern Europe.

Calendula palaestina Boiss.
Asteraceae
♦ Palestine marigold
♦ Weed, Naturalised
♦ 86, 98, 203, 269
♦ Origin: Middle East.

Calendula tripterocarpa Rupr.
Asteraceae
♦ Naturalised
♦ 300
♦ Origin: southern Europe.

Calepina corvini (All.) Desv.
Brassicaceae
= *Calepina irregularis* (Asso) Thell.
♦ Weed
♦ 87, 88

Calepina irregularis (Asso) Thell.
Brassicaceae
Calepina corvini (All.) Desv. (see), *Myagrum irregulare* Asso
♦ white ball mustard, calepina
♦ Weed, Naturalised, Introduced, Garden Escape, Environmental Weed, Cultivation Escape
♦ 54, 70, 86, 88, 94, 101, 155, 228, 243, 253, 272
♦ aH, arid, cultivated, herbal. Origin: Europe.

Calhounia nelsonae **A.Nelson**
Asteraceae
- Quarantine Weed
- 220

Calibrachoa parviflora **(Juss.) D'Arcy**
Solanaceae
Petunia parviflora Juss. (see)
- seaside petunia
- Naturalised
- 86
- herbal. Origin: tropical America.

Calicotome spinosa **(L.) Link**
Fabaceae/Papilionaceae
Cytisus spinosus Desc., *Calycotome spinosa* Link (see)
- spiny broom, thorny broom
- Weed, Noxious Weed, Naturalised, Garden Escape, Environmental Weed
- 15, 86, 86, 198, 246, 280, 289
- S, cultivated. Origin: western Mediterranean.

Calla palustris **L.**
Araceae
Calla aethiopica Gaertn.
- water arum, bog arum, wild calla, vehka
- Weed
- 23, 39, 88, 161, 247
- pH, aqua, cultivated, herbal, toxic.

Calliandra **Benth. spp.**
Fabaceae/Mimosaceae
- stickpea, powder puff
- Naturalised, Garden Escape, Environmental Weed, Cultivation Escape
- 86, 155
- cultivated.

Calliandra anomala **(Kunth) J.F.Macbr.**
Fabaceae/Mimosaceae
Acacia callistemon Schldl., *Calliandra grandiflora* (L'Hér.) Benth., *Calliandra kunthii* Benth., *Inga anomala* Kunth
- cabello de angel
- Introduced
- 228
- arid, promoted, herbal.

Calliandra calothyrsus **Meisn.**
Fabaceae/Mimosaceae
Anneslia calothyrus (Meisn.) J.D.Sm., *Anneslia confusa* (Sprague & Riley) Britton & Rose, *Calliandra confusa* Sprague & Riley
- Weed, Introduced
- 32, 155, 228
- arid.

Calliandra haematocephala **Hassk.**
Fabaceae/Mimosaceae
- red powder puff, powder puff bush, stickpea
- Weed, Introduced, Environmental Weed
- 3, 107, 155, 191, 230
- S, cultivated.

Calliandra surinamensis **Benth.**
Fabaceae/Mimosaceae
- Surinamese stickpea
- Sleeper Weed, Naturalised, Cultivation Escape

- 86, 101, 261
- cultivated. Origin: northern South America.

Callicarpa americana **L.**
Lamiaceae/Verbenaceae
- American beautyberry, French mulberry, American beautybush
- Weed
- 14, 39, 87, 88, 218
- S, cultivated, herbal, toxic.

Callicarpa candicans **(Burm.f.) Hochr. var.** *ponapensis* **Fosb.**
Lamiaceae/Verbenaceae
- Introduced
- 230
- S.

Callicarpa dichotoma **(Lour.) K.Koch**
Lamiaceae/Verbenaceae
- purple beautyberry
- Weed, Naturalised
- 80, 101
- cultivated, herbal.

Callicarpa japonica **Thunb.**
Lamiaceae/Verbenaceae
Callicarpa murasaki Siebold
- Japanese callicarpa, beautyberry, Japanese beautyberry
- Naturalised
- 101
- S, cultivated.

Callichilia subsessilis **Stapf**
Apocynaceae
Tabernaemontana subsessilis Benth.
- Quarantine Weed
- 220

Calligonum polygonoides **L.**
Polygonaceae
- phog
- Introduced
- 228
- arid, herbal.

Callilepis laureola **DC.**
Asteraceae
- oxeye daisy
- Native Weed
- 121
- pH, toxic. Origin: southern Africa.

Callirhoe alcaeoides **(Michx.) Gray**
Malvaceae
- plains poppymallow, light poppymallow
- Weed
- 161
- herbal.

Callirhoe involucrata **(Torr. & A.Gray) A.Gray**
Malvaceae
Callirhoe lineariloba (Torr. & A.Gray) A.Gray, *Malva involucrata* Torr. & A.Gray
- low poppymallow, purple poppymallow, poppymallow
- Native Weed
- 161, 174
- pH, promoted, herbal. Origin: North America.

Callisia fragrans **(Lindl.) Woods.**
Commelinaceae

- fragrant inchplant, basketplant, inchplant, spironema, cohitre morado
- Weed, Sleeper Weed, Naturalised, Garden Escape, Environmental Weed
- 3, 80, 86, 88, 98, 101, 112, 155, 179, 191, 201, 203, 261
- cultivated, herbal. Origin: Mexico.

Callisia repens **(Jacq.) L.**
Commelinaceae
- creeping inchplant
- Weed, Naturalised
- 86, 157, 179
- cultivated, herbal. Origin: Central and South America.

Callistachys lanceolata **Vent.**
Fabaceae/Papilionaceae
Oxylobium lanceolatum (Vent.) Druce (see)
- greenbush
- Naturalised, Environmental Weed
- 86, 198, 280
- cultivated. Origin: Australia.

Callistemon **R.Br. spp.**
Myrtaceae
- Australian bottlebrush, bottlebrush
- Weed, Naturalised
- 121
- S/T. Origin: Australia.

Callistemon citrinus **(Curtis) Stapf**
Myrtaceae
- crimson bottlebrush
- Naturalised, Introduced
- 101, 261
- cultivated, herbal. Origin: Australia.

Callistemon rigidus **R.Br.**
Myrtaceae
- bottlebrush, stiff leaved bottlebrush
- Weed, Naturalised, Native Weed, Garden Escape, Environmental Weed
- 72, 86, 88, 198, 280
- S, cultivated, herbal. Origin: Australia.

Callistemon rugulosus **(Link) DC.**
Myrtaceae
- bottlebrush
- Weed, Naturalised, Native Weed, Environmental Weed
- 72, 86, 88
- S, aqua, cultivated. Origin: Australia.

Callistemon salignus **(Sm.) Sweet**
Myrtaceae
- bottlebrush
- Naturalised
- 198
- cultivated, herbal. Origin: Australia.

Callistephus chinensis **(L.) Nees**
Asteraceae
Aster chinensis L., *Callistemma hortense* Nees
- kiinanasteri, China aster, aster
- Weed, Naturalised, Cultivation Escape
- 42, 101, 280, 297
- aH, cultivated, herbal.

Callithamnion byssoides **Arn. ex Harv.**
Ceramiaceae

= *Aglaothamnion furcellariae* (Agardh)
Feldm.-Maz. (NoR)
♦ red alga
♦ Weed
♦ 197
♦ algae.

Callitriche deflexa **A.Braun ex Hegelm.**
Callitrichaceae
= *Callitriche terrestris* Raf. (NoR)
♦ Naturalised
♦ 257
♦ herbal.

Callitriche fallax **Petrov.**
Callitrichaceae
♦ Weed, Quarantine Weed
♦ 76, 87, 88, 203, 220

Callitriche hamulata **Kütz.**
Callitrichaceae
Callitriche autumnalis Kütz., *Callitriche
intermedia* Hoffm. *auct.*
♦ thread water starwort, water
starwort, starwort
♦ Weed, Naturalised, Environmental
Weed
♦ 7, 72, 86, 88, 198, 280
♦ wa/pH, cultivated, herbal. Origin:
Europe, North America.

Callitriche heterophylla **Pursh emend.
Darby**
Callitrichaceae
♦ larger water starwort, two headed
water starwort
♦ Naturalised
♦ 280
♦ pH, herbal.

Callitriche japonica **Engelm. ex Hegelm.**
Callitrichaceae
♦ awagoke
♦ Weed
♦ 87, 88, 286

Callitriche palustris **L.**
Callitrichaceae
Callitriche verna L. (see)
♦ vernal water starwort,
pikkuvesitähti
♦ Weed
♦ 87, 88, 272
♦ pH, aqua, promoted, herbal. Origin:
Australia.

Callitriche platycarpa **Kütz.**
Callitrichaceae
= *Callitriche stagnalis* Scop.
♦ common water starwort, various
leaved water starwort, long styled
water starwort, pallevesitähti
♦ Weed
♦ 87, 88

Callitriche stagnalis **Scop.**
Callitrichaceae
Callitriche platycarpa Kütz. (see)
♦ water starwort, pond water
starwort, common water starwort,
starwort, mud water starwort
♦ Weed, Quarantine Weed,
Naturalised, Introduced,
Environmental Weed, Casual Alien
♦ 7, 9, 42, 70, 72, 86, 87, 88, 98, 101, 133,

155, 159, 165, 176, 195, 197, 198, 200,
203, 208, 220, 224, 246, 280, 287
♦ waH, cultivated, herbal. Origin:
Europe, North Africa.

Callitriche verna **L.**
Callitrichaceae
= *Callitriche palustris* L.
♦ water starwort, vernal water
starwort, mizuhakobe
♦ Weed
♦ 87, 88, 218, 263, 274, 286
♦ wa/pH, cultivated, herbal.

Callitris columellaris **F.Muell.**
Cupressaceae
♦ coastal pine
♦ Naturalised, Native Weed
♦ 7, 86
♦ T, arid, cultivated. Origin: Australia.

Callitris endlicheri **(Parl.) F.M.Bailey**
Cupressaceae
♦ black cypress pine, red cypress pine
♦ Weed, Naturalised, Native Weed,
Garden Escape, Environmental Weed
♦ 72, 86, 88
♦ T, cultivated, herbal. Origin:
Australia.

Callitris glaucophylla **J.Thomps. &
L.A.S.Johnson**
Cupressaceae
Callitris hugelii auct. non (Carr.) Franco
(see)
♦ white cypress pine
♦ Weed, Naturalised, Native Weed
♦ 7, 86, 101, 179
♦ cultivated, herbal. Origin: Australia.

Callitris hugelii **auct. non (Carr.) Franco**
Cupressaceae
= *Callitris glaucophylla* J.Thomps. &
L.Johnson
♦ blue cypress pine
♦ Weed, Environmental Weed
♦ 88, 151
♦ cultivated.

Callitris oblonga **A.Rich. & Rich.**
Cupressaceae
♦ Casual Alien
♦ 280
♦ cultivated.

Callitris rhomboidea **R.Br. ex Rich.**
Cupressaceae
♦ oyster bay pine
♦ Weed, Naturalised, Native Weed,
Garden Escape, Environmental Weed
♦ 72, 86, 88, 280
♦ T, cultivated, herbal. Origin:
Australia.

Callitris robusta **(Parl.) F.M.Bailey**
Cupressaceae
Callitris gracilis R.Baker
♦ Naturalised, Native Weed
♦ 7, 86
♦ cultivated.

Callitris verrucosa **(Endl.) F.Muell.**
Cupressaceae
♦ Naturalised,
Native Weed
♦ 7, 86

♦ cultivated. Origin: Australia.

Calluna vulgaris **(L.) Hull**
Ericaceae
Calluna erica DC., *Erica vulgaris* L.,
Calluna sagittaefolia S.F.Gray, *Erica
vulgaris* Thal
♦ heather, ling, Scotch heather,
kanerva
♦ Weed, Quarantine Weed,
Naturalised, Garden Escape,
Environmental Weed, Cultivation
Escape
♦ 15, 80, 86, 87, 88, 101, 152, 155, 165,
176, 181, 225, 246, 272, 280, 289
♦ S, cultivated, herbal. Origin: Europe,
Asia Minor, North Africa.

Calocarpum mammosum **(L.) Pierre**
Sapotaceae
Achradelpha mammosa Cook, *Achras
mammosa* L., *Lucuma mammosa*
Gaertn.f., *Sideroxylon sapota* Jacq.,
Vitellaria mammosa Radlk.
♦ Introduced
♦ 228
♦ arid.

Calochortus weedii **Alph. Wood var.
intermedius Ownbey**
Liliaceae/Calochortaceae
♦ Weed's mariposa lily
♦ Weed
♦ 34
♦ pH, cultivated.

Calochortus weedii **Alph. Wood var.
vestus Purdy**
Liliaceae/Calochortaceae
♦ Weed's mariposa lily
♦ Weed
♦ 34
♦ pH.

Calochortus weedii **Alph. Wood var.
weedii**
Liliaceae/Calochortaceae
♦ Weed's mariposa lily
♦ Weed
♦ 34
♦ pH, cultivated.

Calonyction aculeatum **(L.) House**
Convolvulaceae
= *Ipomoea alba* L.
♦ giant moonflower
♦ Naturalised
♦ 287
♦ cultivated, herbal.

Calonyction muricatum **(L.) G.Don**
Convolvulaceae
♦ small moonflower
♦ Weed
♦ 88, 218
♦ herbal.

Calophyllum antillanum **Britt.**
Clusiaceae
Calophyllum calaba L. (see)
♦ beauty leaf, Santa Maria, mast
wood, Alexandrian laurel, Antilles
calophyllum
♦ Weed, Environmental Weed
♦ 88, 112, 151, 179

Calophyllum calaba L.
Clusiaceae
= *Calophyllum antillanum* Britt.
♦ Santa Maria, mast wood, Alexandrian laurel, Brazil beauty leaf
♦ Weed, Cultivation Escape
♦ 3, 39, 80, 122, 191
♦ cultivated, herbal, toxic.

Calophyllum inophyllum L.
Clusiaceae
♦ mast wood, Alexandrian laurel, India oil nut, fetau
♦ Weed, Naturalised, Introduced, Environmental Weed
♦ 39, 80, 101, 161, 179, 261
♦ cultivated, herbal, toxic. Origin: Madagascar, Africa, Asia, Australia.

Calopogonium caeruleum (Benth.) Suav.
Fabaceae/Papilionaceae
♦ calopogonium, jicama
♦ Weed
♦ 3, 14, 191
♦ cultivated.

Calopogonium mucunoides Desv.
Fabaceae/Papilionaceae
♦ calopo, akankan guakag, frisofilla
♦ Weed, Naturalised, Introduced, Environmental Weed
♦ 3, 86, 87, 88, 93, 98, 107, 155, 157, 191, 203, 206, 230, 243
♦ pH, cultivated, herbal. Origin: tropical America.

Calopogonium muscioides L.
Fabaceae/Papilionaceae
♦ calopogonio
♦ Weed
♦ 153

Calothamnus graniticus Hawkeswood
Myrtaceae
♦ Naturalised, Native Weed, Garden Escape, Cultivation Escape
♦ 7
♦ cultivated.

Calothamnus quadrifidus R.Br.
Myrtaceae
♦ Naturalised, Native Weed
♦ 7, 86
♦ cultivated. Origin: Australia.

Calothamnus validus S.Moore
Myrtaceae
♦ Naturalised, Native Weed
♦ 7, 86
♦ cultivated. Origin: Australia.

Calotis cuneata (Benth.) G.Davis
Asteraceae
♦ daisy burr
♦ Native Weed
♦ 269
♦ cultivated. Origin: Australia.

Calotis cuneifolia R.Br.
Asteraceae
♦ wedgeleaf, bogan flea
♦ Naturalised, Introduced
♦ 101, 228, 287
♦ arid, cultivated, herbal.

Calotis lappulacea Benth.
Asteraceae
♦ daisy burr
♦ Weed, Naturalised, Native Weed, Introduced, Environmental Weed
♦ 87, 88, 228, 246, 269, 280
♦ arid, cultivated. Origin: Australia.

Calotropis gigantea (L.) Dryand. ex W.T.Aiton
Asclepiadaceae/Apocynaceae
Asclepias gigantea L.
♦ giant milkweed, calotrope, bowstring hemp, crownplant, madar, mudar, mercure végétal, mudarpflanze, lechoso
♦ Weed, Naturalised, Introduced, Garden Escape, Environmental Weed, Cultivation Escape
♦ 3, 11, 39, 66, 85, 86, 87, 88, 93, 98, 101, 155, 161, 203, 209, 228
♦ arid, cultivated, herbal, toxic. Origin: tropical Africa, India, China, south-east Asia.

Calotropis procera (Aiton) Dryand. ex W.T.Aiton
Asclepiadaceae/Apocynaceae
Asclepias procera Aiton
♦ calotropis, rubber bush, apple of Sodom, Indian milkweed, king's crown kapok, rubber tree, roostertree, Dead Sea apple, poumpoumssé
♦ Weed, Quarantine Weed, Noxious Weed, Naturalised, Introduced, Garden Escape, Environmental Weed
♦ 3, 7, 39, 62, 66, 76, 86, 87, 88, 93, 98, 101, 147, 152, 161, 191, 203, 205, 220, 228, 242, 255, 258
♦ S/T, arid, cultivated, herbal, toxic. Origin: tropical Africa and Asia.

Caltha leptosepala DC.
Ranunculaceae
♦ mountain marsh marigold, western marsh marigold, marsh marigold, white marsh marigold
♦ Weed
♦ 39, 161, 189
♦ pH, aqua, cultivated, herbal, toxic.

Caltha palustris L.
Ranunculaceae
♦ marshmarigold, kingcup, mollyblobs, rentukka, cowslip
♦ Weed
♦ 8, 39, 87, 88, 161, 189, 218, 247, 272, 297
♦ pH, aqua, cultivated, herbal, toxic.

Calycanthus floridus L.
Calycanthaceae
♦ Carolina allspice, eastern sweetshrub
♦ Weed
♦ 8, 39, 161, 247
♦ S, cultivated, herbal, toxic.

Calycanthus occidentalis Hook. & Arn.
Calycanthaceae
Butneria occidentalis (Hook. & Arn.) Greene
♦ western sweetshrub, Californian allspice, western spicebush

♦ Weed
♦ 39, 161
♦ S, cultivated, herbal, toxic.

Calycera balsamitifolia A.Rich.
Calyceraceae
♦ calycera
♦ Naturalised
♦ 101

Calycera leucanthema (Poepp. ex Less.) Kuntze
Calyceraceae
♦ Weed
♦ 87, 88

Calycocarpum lyonii (Pursh) Gray
Menispermaceae
♦ cupseed
♦ Weed
♦ 87, 88, 218
♦ herbal.

Calycophyllum candidissimum (Vahl) DC.
Rubiaceae
♦ degame
♦ Naturalised, Introduced
♦ 101, 261
♦ herbal.

Calycotome spinosa (L.) Link
Fabaceae/Papilionaceae
= *Calicotome spinosa* (L.) Link
♦ spiny broom
♦ Weed, Noxious Weed, Quarantine Weed, Environmental Weed
♦ 72, 76, 88, 147, 203
♦ S, cultivated. Origin: Mediterranean.

Calymmanthium F.Ritter spp.
Cactaceae
♦ Weed, Quarantine Weed
♦ 76, 88, 203, 220

Calymmodon ponapensis Copel.
Grammitidaceae
♦ Introduced
♦ 230
♦ H.

Calyptocarpus vialis Less.
Asteraceae
Blainvillea tampicana Hemsl., *Synedrellopsis grisebachii* auct. non Hieron. & Kuntze (see)
♦ straggler daisy, sprawling horseweed
♦ Weed, Naturalised
♦ 14, 86, 87, 88, 98, 101, 157, 161, 179, 203, 249
♦ pH, herbal. Origin: Central America.

Calyptocarpus wendlandii Sch.Bip.
Asteraceae
♦ cachito
♦ Weed
♦ 281

Calystegia dahurica (Herb.) Choisy
Convolvulaceae
♦ dahuria glorybind
♦ Weed
♦ 297

Calystegia fraterniflora Burmon
Convolvulaceae
= *Calystegia silvatica* (Kit.) Griseb. ssp.

fraterniflora (Mack. & Bush) Brummitt (NoR)
♦ Naturalised
♦ 287

Calystegia hederacea Wall.
Convolvulaceae
♦ Japanese false bindweed, kohirugao
♦ Weed, Naturalised
♦ 87, 88, 101, 204, 263, 275, 286, 297
♦ pC, cultivated. Origin: east Asia.

Calystegia japonica Choisy
Convolvulaceae
Calystegia subvolubilis non Don.
♦ Japanese bindweed, hirugao
♦ Weed, Quarantine Weed
♦ 76, 87, 88, 203, 204, 220, 263, 286
♦ pC, promoted, herbal. Origin: east Asia.

Calystegia occidentalis (A.Gray) Brummitt
Convolvulaceae
♦ western morningglory, chaparral false bindweed
♦ Weed
♦ 161
♦ pH, herbal.

Calystegia pellita (Ledeb.) G.Don
Convolvulaceae
Convolvulus pellitus Ledeb., *Convolvulus japonicus* Thunb. (see)
♦ calystegia, hairy false bindweed
♦ Weed, Noxious Weed, Naturalised
♦ 101, 229, 275, 297
♦ pH.

Calystegia pulchra Brummitt & Heywood
Convolvulaceae
♦ hairy bindweed, kehtokarhunköynnös, povoja pekná
♦ Naturalised, Cultivation Escape
♦ 40, 42
♦ cultivated.

Calystegia sepium (L.) R.Br.
Convolvulaceae
Convolvulus sepium L. (see)
♦ hedge false bindweed, bindweed, large bindweed, bellbine, hedge bindweed, larger bindweed, lady's nightcap, bell bind, Rutland beauty, wild morningglory, devil's vine, great bindweed, hedgebell, bear bind, old man's night cap, devil's guts
♦ Weed, Naturalised
♦ 15, 38, 70, 87, 88, 94, 101, 161, 165, 174, 207, 211, 212, 253, 272, 275, 280, 287, 297
♦ pH, arid/aqua, cultivated, herbal. Origin: eastern USA.

Calystegia sepium (L.) R.Br. ssp. *americana* (Sims) Brummitt
Convolvulaceae
Convolvulus americanus (Sims) Greene
♦ hedge bindweed, wild morningglory, large flowered morningglory, great bindweed, bracted bindweed, hedge false bindweed
♦ Weed, Naturalised
♦ 49, 300

Calystegia sepium (L.) R.Br. ssp. *angulata* Brummitt
Convolvulaceae
Convolvulus repens L.
♦ trailing bindweed, hedge false bindweed
♦ Weed
♦ 49

Calystegia sepium (L.) R.Br. ssp. *sepium*
Convolvulaceae
Convolvulus nashii House, *Calystegia sepium* R.Br. var. *communis* Hara (see)
♦ hedge false bindweed
♦ Noxious Weed, Naturalised
♦ 101, 229, 300

Calystegia sepium R.Br. var. *communis* Hara
Convolvulaceae
= *Calystegia sepium* (L.) R.Br. ssp. *sepium* (NoR)
♦ Weed
♦ 286

Calystegia sepium (L.) R.Br. var. *sepium*
Convolvulaceae
♦ Naturalised
♦ 241

Calystegia silvatica (Kit.) Griseb.
Convolvulaceae
Convolvulus silvaticus Kit., *Convolvulus silvester* Wald. & Kit.
♦ shortstalk false bindweed, greater bindweed, great bindweed, large bindweed
♦ Weed, Naturalised, Garden Escape, Environmental Weed
♦ 15, 40, 72, 86, 88, 98, 101, 165, 176, 198, 203, 280
♦ C, cultivated. Origin: southern Europe.

Calystegia silvatica (Kit.) Griseb. ssp. *silvatica*
Convolvulaceae
♦ shortstalk false bindweed
♦ Naturalised
♦ 101

Calystegia soldanella (L.) R.Br.
Convolvulaceae
♦ sea bindweed, seashore false bindweed, beach morningglory, shore bindweed
♦ Weed
♦ 70
♦ pH, cultivated, herbal. Origin: Eurasia.

Camelina Crantz spp.
Brassicaceae
♦ false flax
♦ Weed, Noxious Weed, Naturalised
♦ 36, 198

Camelina alyssum (Mill.) Thell.
Brassicaceae
Camelina dentata (Willd.) Pers. (see), *Camelina foetida* (Schkuhr) Fr., *Camelina linicola* Schimp. & Spenn. (see), *Camelina macrocarpa* Wierzb. ex Rchb. (see), *Camelina pinnatifida* Hornem., *Camelina sativa* auct. ital., *Camelina*

sativa (L.) Crantz var. *sublinicola* N.W.Zinger, *Myagrum alyssum* Mill.
♦ pellavatankio
♦ Weed, Naturalised
♦ 86, 87, 88, 94, 98, 176, 203, 237, 243, 272, 280, 287, 295
♦ cultivated. Origin: Europe.

Camelina alyssum (Mill.) Thell. ssp. *integerrima* (Celak) Smejkal
Brassicaceae
♦ Introduced
♦ 56

Camelina dentata (Willd.) Pers.
Brassicaceae
= *Camelina sativa* (L.) Crantz ssp. *alyssum* (Mill.) E.Schmid
♦ flatseed false flax
♦ Weed
♦ 23, 87, 88, 218

Camelina glabrata (DC.) Fritsch
Brassicaceae
= *Camelina sativa* (L.) Crantz
♦ Weed
♦ 87, 88

Camelina hispida Boiss.
Brassicaceae
♦ turkintankio
♦ Casual Alien
♦ 42

Camelina linicola Schimp. & Spenn.
Brassicaceae
= *Camelina alyssum* (Mill.) Thell.
♦ Weed
♦ 87, 88

Camelina macrocarpa Wierzb. ex Rchb.
Brassicaceae
= *Camelina alyssum* (Mill.) Thell.
♦ liinatankio
♦ Weed
♦ 88, 94, 272

Camelina microcarpa Andrz. ex DC.
Brassicaceae
Myagrum sativum L. var. *silvestre* Murray
♦ smallseed false flax, littlepod false flax, false flax, hairy gold of pleasure, small fruited false flax, littleseed false flax
♦ Weed, Noxious Weed, Naturalised, Introduced, Environmental Weed, Casual Alien
♦ 21, 23, 34, 42, 56, 70, 87, 88, 94, 101, 161, 174, 210, 212, 218, 237, 243, 253, 272, 287, 299
♦ aH, cultivated, herbal. Origin: Europe.

Camelina parodii Ibar. & La Porte
Brassicaceae
= *Camelina sativa* (L.) Crantz ssp. *sativa* Mansf.
♦ flat seeded false flax
♦ Weed
♦ 87, 88, 161

Camelina pilosa (DC.) Zinger
Brassicaceae
♦ Weed
♦ 87, 88

Camelina rumelica Vel.
Brassicaceae
♦ graceful false flax
♦ Weed
♦ 87, 88, 94, 243, 272

Camelina sativa (L.) Crantz
Brassicaceae
Alyssum sativum Scop., *Camelina glabrata* (DC.) Fritsch (see), *Myagrum sativum* L. *p.p.*
♦ largeseed false flax, gold of pleasure, ruistankio, false flax
♦ Weed, Naturalised, Introduced, Casual Alien
♦ 7, 23, 42, 56, 70, 86, 87, 88, 94, 98, 101, 161, 176, 198, 199, 203, 210, 218, 236, 237, 272, 280, 287, 295
♦ aH, cultivated, herbal. Origin: Eurasia.

Camelina sativa (L.) Crantz ssp. alyssum (Mill.) E.Schmid
Brassicaceae
♦ gold of pleasure
♦ Naturalised
♦ 101
♦ cultivated.

Camelina sativa (L.) Crantz ssp. sativa Mansf.
Brassicaceae
Camelina parodii Ibar. & La Porte (see), *Cochlearia sativa* Cav.
♦ gold of pleasure
♦ Naturalised
♦ 101

Camellia japonica L.
Theaceae
Thea japonica (L.) Baill.
♦ camellia, Japanese camellia, tea
♦ Naturalised, Casual Alien
♦ 39, 101, 280
♦ S, cultivated, herbal, toxic.

Camellia sasanqua Thunb.
Theaceae
Thea sasanqua (Thunb.) Nois. ex Cels
♦ Sasanqua camellia, camellia
♦ Naturalised
♦ 39, 101
♦ S, cultivated, herbal, toxic.

Camellia sinensis (L.) Kuntze
Theaceae
Camellia bohea Lindl., *Camellia thea* Link, *Camellia theifera* Griff., *Thea sinensis* L. (see)
♦ tea, tebuske, teepensas, tee
♦ Sleeper Weed, Naturalised, Garden Escape, Environmental Weed
♦ 86, 98, 101, 155
♦ S/T, cultivated, herbal. Origin: China to India.

Campanula L. spp.
Campanulaceae
♦ bellflower, Canterbury bells
♦ Weed
♦ 272
♦ herbal.

Campanula alliariifolia Willd.
Campanulaceae
♦ Cornish bellflower

♦ Naturalised
♦ 40, 287
♦ pH, cultivated, herbal. Origin: eastern Europe to west Asia.

Campanula americana L.
Campanulaceae
♦ tall bellflower, American bellflower
♦ Weed
♦ 161
♦ a/bH, cultivated, herbal.

Campanula aparinoides Pursh
Campanulaceae
♦ sammakonkello, marsh bellflower
♦ Naturalised
♦ 42
♦ herbal.

Campanula bononiensis L.
Campanulaceae
♦ European bellflower
♦ Weed, Naturalised
♦ 101, 272
♦ cultivated, herbal.

Campanula carpatica Jacq.
Campanulaceae
♦ karpaattienkello, tussock bellflower
♦ Naturalised, Cultivation Escape
♦ 42, 101
♦ pH, cultivated, herbal.

Campanula erinus L.
Campanulaceae
♦ hiirenkello
♦ Casual Alien
♦ 42

Campanula floridana S.Wats. ex Gray
Campanulaceae
♦ Florida bellflower
♦ Weed
♦ 161, 249
♦ aqua.

Campanula glomerata L.
Campanulaceae
♦ clustered bellflower, Dane's blood
♦ Weed, Naturalised
♦ 87, 88, 101, 272
♦ pH, cultivated, herbal.

Campanula lanata Friv.
Campanulaceae
♦ woolly bellflower
♦ Naturalised
♦ 101
♦ cultivated.

Campanula latifolia L.
Campanulaceae
♦ ukonkello, greater bellflower, giant bellflower
♦ Naturalised
♦ 39, 42, 101
♦ pH, cultivated, herbal, toxic.

Campanula lingulata Waldst. & Kit.
Campanulaceae
♦ bellflower
♦ Weed
♦ 272

Campanula lusitanica Loefl.
Campanulaceae
♦ rastaankello
♦ Casual Alien

♦ 42
♦ cultivated.

Campanula macrostachya Waldst. & Kit.
Campanulaceae
♦ Weed
♦ 272
♦ cultivated.

Campanula medium L.
Campanulaceae
♦ Canterbury bells
♦ Weed, Naturalised
♦ 15, 40, 101, 280
♦ bH, cultivated, herbal.

Campanula patula L.
Campanulaceae
♦ spreading bellflower, harakankello
♦ Weed, Naturalised
♦ 70, 87, 88, 101, 272
♦ bH, cultivated, herbal.

Campanula persicifolia L.
Campanulaceae
♦ peachleaf bellflower, harebell, kurjenkello
♦ Naturalised
♦ 101
♦ pH, cultivated, herbal.

Campanula portenschlagiana Schult.
Campanulaceae
Campanula muralis Portenschl
♦ Adria bellflower
♦ Naturalised, Casual Alien
♦ 40, 280
♦ pH, cultivated, herbal.

Campanula poscharskyana Degen
Campanulaceae
♦ trailing bellflower
♦ Naturalised
♦ 40, 280
♦ pH, cultivated, herbal.

Campanula punctata Lam.
Campanulaceae
♦ spotted bellflower, Chinese rampion
♦ Weed, Naturalised
♦ 101, 286, 297
♦ pH, cultivated, herbal.

Campanula ramosissima Sm.
Campanulaceae
Campanula loreyi Pollich
♦ ketunkello
♦ Casual Alien
♦ 42

Campanula rapunculoides L.
Campanulaceae
♦ creeping bellflower, rover bellflower, creeping campanula, rampion bellflower
♦ Weed, Noxious Weed, Naturalised, Cultivation Escape
♦ 4, 23, 40, 44, 70, 86, 87, 88, 94, 98, 101, 138, 161, 162, 176, 195, 210, 212, 218, 243, 272, 280, 299
♦ pH, cultivated, herbal. Origin: Eurasia.

Campanula rapunculus L.
Campanulaceae
Rapunculus verus Fourr.
♦ rampion, kauriinkello, rampion

bellflower, European bellflower, rampion harebell
- Weed, Naturalised, Casual Alien
- 42, 44, 70, 272, 280
- bH, cultivated, herbal.

Campanula rotundifolia L.
Campanulaceae
- harbell, bluebell
- Weed, Naturalised
- 23, 70, 87, 88, 272, 280
- pH, cultivated, herbal.

Campanula scutellata L.
Campanulaceae
- Weed
- 272

Campanula sibirica L.
Campanulaceae
- Siberian bellflower
- Weed
- 272
- cultivated.

Campanula strigosa Banks & Sol.
Campanulaceae
- bellflower
- Weed
- 87, 88
- aH.

Campanula trachelium L.
Campanulaceae
- bats in the belfry, nettle leaved bellflower, varsankello
- Naturalised, Casual Alien
- 40, 101
- cultivated, herbal.

Campelia zanonia (L.) H.B.K.
Commelinaceae
= *Tradescantia zanonia* (L.) Sw. (NoR)
- Weed
- 157
- pH, cultivated, herbal.

Campsis radicans (L.) Seem.
Bignoniaceae
- trumpetcreeper, cow itch, trumpet vine
- Weed, Quarantine Weed, Naturalised, Garden Escape
- 39, 54, 76, 86, 87, 88, 161, 207, 210, 211, 218, 243, 247, 286, 287
- pC, cultivated, herbal, toxic. Origin: eastern North America.

Campsis × tagliabuana (Vis.) Rehder
Bignoniaceae
= *Campsis grandiflora* (Thunb.) Schum.
× *Campsis radicans* (L.) Seem.
- trumpet creeper
- Naturalised, Garden Escape
- 280
- cultivated, herbal. Origin: horticultural hybrid.

Camptandra latifolia Ridl.
Zingiberaceae
- Quarantine Weed
- 220

Camptorrhiza strumosa (Bak.) Oberm.
Liliaceae/Colchicaceae
- Native Weed
- 121
- pH. Origin: southern Africa.

Camptosema grandiflorum Benth.
Fabaceae/Papilionaceae
- Weed
- 87, 88

Campuloclinium macrocephalum (Less.) DC.
Asteraceae
Eupatorium macrocephalum Less. (see)
- pom pom weed, pom pom bossie
- Weed, Quarantine Weed, Noxious Weed
- 63, 88, 95, 220, 283
- Origin: South America.

Campyloneurum phyllitidis (L.) Presl
Polypodiaceae
Polypodium phyllitidis L.
- long strapfern
- Weed
- 87, 88
- cultivated, herbal.

Campylopus introflexus (Hedw.) Brid.
Dicranaceae
= *Campylopus pilifer* Brid. (NoR)
- campylopus moss
- Environmental Weed
- 152
- moss. Origin: South America, Africa, Australia, Pacific Islands.

Cananga odorata (Lam.) Hook.f. & Thomson
Annonaceae
- ylang ylang, chiráng, irang, seir en wai, makasoi, moso'oi
- Weed, Naturalised, Introduced
- 3, 101, 107, 191, 230, 261
- cultivated, herbal. Origin: southern Asia.

Canarium commune L.
Burseraceae
= *Canarium indicum* L. (NoR)
- Java almond
- Introduced
- 39, 230
- herbal, toxic.

Canarium ovatum Engl.
Burseraceae
Canarium pachyphyllum Perkins, *Canarium melioides* Elmer.
- pili nut, pili, anangi, basiad, liputi, pilaui, pili pilauai
- Introduced
- 230
- T, cultivated, herbal. Origin: Philippines.

Canavalia africana Dunn
Fabaceae/Papilionaceae
Canavalia ferruginea Piper, *Canavalia gladiata* Robyns, *Canavalia polystachya* Schweinf. *p.p.*, *Canavalia virosa* Sauer, *Canavalia virosa* (Roxb.) Wight & Arn. *p.p.*, *Dolichos virosus* Roxb. *p.p.*
- karuthamma
- Introduced
- 228
- arid.

Canavalia brasiliensis Mart. ex Benth.
Fabaceae/Papilionaceae
Canavalia anomala Piper, *Canavalia*

campylocarpa Piper, *Canavalia caribaea* Urb., *Canavalia fendleri* Piper, *Canavalia leptophylla* Piper, *Canavalia mexicana* Piper, *Canavalia panamensis* Piper, *Canavalia paraguayensis* Piper, *Canavalia prolifera* Piper ex Ricker
- Brazilian jackbean
- Weed, Introduced
- 179, 228
- arid.

Canavalia cathartica Thou.
Fabaceae/Papilionaceae
Canavalia bouquete Montrouzier, *Canavalia ensiformis* (L.) DC. var. *turgida* (Graham ex A.Gray) Baker, *Canavalia glandifolia* A.Gray ex Streets, *Canavalia microcarpa* (DC.) Piper (see), *Canavalia turgida* Graham ex A.Gray, *Canavalia virosa* (Roxb.) Wight & Arn., *Dolichos virosus* Roxb., *Lablab microcarpus* DC.
- maunaloa
- Weed, Naturalised, Introduced
- 87, 88, 101, 230
- herbal. Origin: Asia, Australia, East Africa.

Canavalia dictyota Piper
Fabaceae/Papilionaceae
- Naturalised
- 257

Canavalia ensiformis (L.) DC.
Fabaceae/Papilionaceae
- wonder bean, jackbean, horse bean, sword bean, boyengo, haba de burro
- Weed, Naturalised, Cultivation Escape, Casual Alien
- 32, 39, 86, 87, 88, 98, 101, 179, 203, 252, 261
- cultivated, herbal, toxic.

Canavalia gladiata (Jacq.) DC.
Fabaceae/Papilionaceae
- sword jackbean, sword bean
- Naturalised
- 101
- cultivated, herbal. Origin: Australia.

Canavalia gladiata (Jacq.) DC. var. *gladiata*
Fabaceae/Papilionaceae
- Weed, Cultivation Escape
- 32
- cultivated.

Canavalia maritima (Aubl.) Thouars.
Fabaceae/Papilionaceae
= *Canavalia rosea* (Sw.) DC.
- wild jackbean, baybean
- Weed
- 14, 87, 88
- herbal.

Canavalia microcarpa (DC.) Piper
Fabaceae/Papilionaceae
= *Canavalia cathartica* Thou.
- Weed
- 209

Canavalia plagiosperma Piper
Fabaceae/Papilionaceae
- oblique seeded jackbean
- Weed, Cultivation Escape
- 32
- cultivated.

Canavalia rosea (Sw.) DC.
Fabaceae/Papilionaceae
Canavalia apiculata Piper, *Canavalia arenicola* Piper, *Canavalia baueriana* Endl., *Canavalia emarginata* (Jacq.) G.Don, *Canavalia maritima* (Aubl.) Thouars (see), *Canavalia miniata* (Kunth) DC., *Canavalia moneta* Welw., *Canavalia obcordata* Voigt, *Canavalia obtusifolia* (Lam.) DC., *Canavalia obtusifolia* (Lam.) DC. var. *insularis* Ridl., *Canavalia obtusifolia* (Lam.) DC. var. *emarginata* (Jacq.) DC., *Canavalia podocarpa* Dunn, *Clitoria rotundifolia* (Vahl) Sessé & Moç., *Dolichos emarginatus* Jacq., *Dolichos littoralis* Vell., *Dolichos maritimus* Aubl., *Dolichos miniatus* Kunth, *Dolichos obcordatus* Roxb., *Dolichos obovatus* Schum. & Thonn., *Dolichos obtusifolius* Lam., *Dolichos roseus* Sw., *Dolichos rotundifolius* Vahl
♦ fuefue fai va'a, fue vili, baybean
♦ Weed
♦ 6, 88
♦ C, arid, cultivated, herbal.

Canavalia sericea A.Gray
Fabaceae/Papilionaceae
♦ silky jackbean
♦ Weed, Naturalised
♦ 87, 88, 101
♦ Origin: Australia.

Canella alba Murr.
Canellaceae
= *Canella winterana* Gaertn. (NoR)
♦ white cinnamon
♦ Weed
♦ 39, 87, 88
♦ herbal, toxic.

Canna coccinea Mill.
Cannaceae
= *Canna indica* L.
♦ achira roja
♦ Weed, Naturalised
♦ 87, 88, 286, 287, 295
♦ cultivated.

Canna edulis Ker Gawl.
Cannaceae
= *Canna indica* L.
♦ achira, Indian shot
♦ Weed
♦ 87, 88
♦ pH, cultivated, herbal.

Canna flaccida Salisb.
Cannaceae
♦ bandanna of the Everglades
♦ Quarantine Weed
♦ 258
♦ cultivated, herbal.

Canna × generalis L.H.Bailey
Cannaceae
Canna indica L. (see), *Canna × orchiodes* L.H.Bailey (see), *Canna orchiodes* L.Bailey (see) and others; a complex group of hybrids
♦ canna lily, canna
♦ Naturalised, Garden Escape
♦ 86, 98, 101, 283, 287

♦ aH, cultivated, herbal. Origin: horticultural hybrid.

Canna glauca L.
Cannaceae
♦ maraca amarilla
♦ Weed
♦ 87, 88
♦ pH, aqua, cultivated, herbal.

Canna hybrida hort. ex Back.
Cannaceae
♦ Weed
♦ 87, 88

Canna indica L.
Cannaceae
Canna coccinea Mill. (see), *Canna edulis* Ker Gawl. (see), *Canna lutea* Mill. (see)
♦ Indian canna, canna, Indian shot, wild canna, canna lily, mongos halum tano, fanamanu, apeellap, oruuru
♦ Weed, Sleeper Weed, Noxious Weed, Naturalised, Introduced, Garden Escape, Environmental Weed
♦ 3, 73, 86, 87, 88, 98, 101, 107, 121, 134, 179, 201, 218, 225, 230, 246, 262, 280, 283, 286, 287
♦ pH, cultivated, herbal. Origin: South America.

Canna indica L. var. flava Roxb.
Cannaceae
♦ Weed, Naturalised
♦ 286, 287

Canna lutea Mill.
Cannaceae
= *Canna indica* L.
♦ yellow canna
♦ Environmental Weed
♦ 257
♦ cultivated, herbal.

Canna neglecta Steud.
Cannaceae
♦ broadleaf canna
♦ Naturalised
♦ 101

Canna orchiodes L.Bailey
Cannaceae
= *Canna × orchiodes* L.H.Bailey
♦ Naturalised
♦ 98

Canna × orchiodes L.H.Bailey
Cannaceae
= *Canna × generalis* L.H.Bailey [a hybrid complex no longer distinct, see *Canna orchiodes* L.Bailey]
♦ orchid canna, canna
♦ Naturalised
♦ 86
♦ cultivated. Origin: horticultural hybrid.

Cannabis L. spp.
Cannabaceae
♦ hemp, marijuana
♦ Weed, Quarantine Weed, Naturalised, Cultivation Escape
♦ 76, 86, 88, 198
♦ cultivated, toxic.

Cannabis ruderalis Jan.
Cannabaceae

♦ Weed
♦ 23, 87, 88, 272

Cannabis sativa L.
Cannabaceae
♦ Indian hemp, marijuana, dacha, grass, pot, redroot, Russian hemp, pakalolo, dagga, dagga canopy, fragrant weed, gallow grass, grass, hemp, native hemp, soft hemp, hashish, Mary Jane, bangui, canapa, konopa siata, mbanje, marryjoanna, nie dagga nie
♦ Weed, Quarantine Weed, Noxious Weed, Naturalised, Introduced, Garden Escape, Casual Alien, Cultivation Escape
♦ 23, 34, 39, 40, 42, 50, 51, 63, 76, 80, 86, 87, 88, 94, 98, 101, 121, 147, 158, 161, 169, 174, 179, 189, 198, 203, 210, 212, 218, 219, 228, 229, 243, 247, 256, 261, 269, 272, 278, 280, 287, 297
♦ aH, arid, cultivated, herbal, toxic. Origin: west Asia, Iran to India.

Cannabis sativa L. (Lam.) ssp. indica E.Small & Cronquist
Cannabaceae
♦ marijuana, hemp
♦ Noxious Weed, Naturalised
♦ 101, 229

Cannabis sativa L. ssp. sativa
Cannabaceae
♦ marijuana, hemp
♦ Noxious Weed, Naturalised
♦ 101, 229

Cannabis sativa L. ssp. sativa var. sativa
Cannabaceae
♦ marijuana, hemp
♦ Noxious Weed, Naturalised
♦ 101, 229

Cannabis sativa L. ssp. sativa var. spontanea Vavilov
Cannabaceae
♦ marijuana, hemp
♦ Noxious Weed, Naturalised
♦ 101, 229

Canotia holacantha Torr.
Celastraceae/Canotiaceae
♦ canotia, crucifixion thorn
♦ Weed
♦ 87, 88, 218
♦ herbal.

Canscora decussata Schult.
Gentianaceae
♦ Weed, Quarantine Weed
♦ 76, 87, 88, 203, 220
♦ herbal. Origin: Madagascar.

Canscora diffusa (Vahl) R.Br. ex Roem. & Schult.
Gentianaceae
♦ Weed
♦ 66
♦ cultivated, herbal.

Canthium barbatum (Forst.) Seem. var. korrorense (Val.) Fosb.
Rubiaceae
♦ Introduced
♦ 230
♦ S.

Canthium hispidum **Benth.**
Rubiaceae
♦ luntafwanengwa
♦ Weed, Quarantine Weed
♦ 76, 87, 88, 203, 220

Caperonia castaneifolia **(L.) St.-Hil.**
Euphorbiaceae
Caperonia castanaefolia (L.) St.-Hil.
♦ chestnutleaf false croton, Mexican
weed
♦ Weed, Naturalised
♦ 88, 101

Caperonia palustris **(L.) St.Hail.**
Euphorbiaceae
♦ Texas weed, sacatrapo
♦ Weed
♦ 14, 87, 88, 161, 179, 243, 271
♦ aH, herbal.

Caperonia serrata **Presl**
Euphorbiaceae
♦ Weed
♦ 87, 88

Capillipedium filiculme **(Hook.f.) Stapf**
Poaceae
Andropogon filiculmis Hook.f.
♦ Weed
♦ 87, 88
♦ G.

Capillipedium parviflorum **(R.Br.) Stapf**
Poaceae
Anatherum parviflorum (R.Br.)
Spreng., *Andropogon alternans*
J.Presl, *Andropogon capilliflorus*
Steud., *Andropogon micranthus*
Kunth, *Andropogon parviflorus*
Roxb., *Andropogon parvispica* Steud.,
Andropogon quartinianus A.Rich.,
Andropogon serratus Miq., *Andropogon*
violascens (Trin.) Nees ex Steud.,
Chrysopogon parviflorus (R.Br.) Benth.,
Chrysopogon parvispicus (Steud.)
W.Watson, *Chrysopogon violascens* Trin.,
Holcus caerulescens Gaudich., *Holcus*
parviflorus R.Br., *Rhaphis caerulescens*
(Gaudich.) Desv., *Rhaphis microstachya*
Nees ex Steud., *Sorghum parviflorum*
P.Beauv., *Sorghum quartinianum*
(A.Rich.) Asch.
♦ scented goldenbeard, scented top
♦ Weed
♦ 87, 88, 90, 238, 286, 297
♦ pG, cultivated. Origin: Africa, Asia,
Australia.

Capillipedium parviflorum **(R.Br.) Stapf**
var. *spicigera* **(Benth.) Roberty**
Poaceae
♦ Naturalised
♦ 287
♦ G.

Capillipedium spicigerum **S.T.Blake**
Poaceae
♦ scented top grass
♦ Weed,
Naturalised
♦ 86, 93, 191
♦ G, cultivated. Origin: northern
Australia.

Capnophyllum peregrinum **(L.) Lange**
Apiaceae
♦ Weed
♦ 87, 88

Capparis **L. spp.**
Capparaceae
♦ caper, pulu, caper bush, kapar
♦ Weed
♦ 272
♦ herbal.

Capparis aphylla **Roth**
Capparaceae
♦ Weed
♦ 87, 88

Capparis erythrocarpos **Isert**
Capparaceae
♦ Weed
♦ 87, 88
♦ herbal.

Capparis frutescens **L.**
Capparaceae
♦ Weed
♦ 87, 88
♦ herbal.

Capparis incana **Kunth**
Capparaceae
♦ hoary caper
♦ Naturalised
♦ 101
♦ herbal.

Capparis micrantha **A.Rich.**
Capparaceae
♦ Weed
♦ 209
♦ herbal.

Capparis retusa **Griseb.**
Capparaceae
♦ Introduced
♦ 228
♦ arid.

Capparis sandwichiana **DC.**
Capparaceae
♦ smooth caperbush, native caper
♦ Weed
♦ 87, 88, 218
♦ herbal.

Capparis sepiaria **L. var.** *citrifolia* **(Lam.)**
Toelken
Capparaceae
♦ cape capers, wild caper bush
♦ Native Weed
♦ 121
♦ pS. Origin: southern Africa.

Capparis spinosa **L.**
Capparaceae
♦ caper, caper bush
♦ Weed
♦ 87, 88
♦ S, cultivated, herbal.

Capparis tomentosa **Lam.**
Capparaceae
♦ Weed
♦ 87, 88
♦ promoted, herbal.

Capparis tweediana **Eichler**
Capparaceae

♦ Introduced
♦ 228
♦ arid, herbal.

Capraria biflora **L.**
Scrophulariaceae
♦ goatweed
♦ Weed
♦ 14, 39, 87, 88, 261
♦ arid, herbal, toxic.

Capsella bursa-pastoris **(L.) Medik.**
Brassicaceae
Bursa pastoris Weber ex Wigg., *Capsella*
rubella Reut. (see), *Iberis bursa-pastoris*
Crantz, *Lepidium bursa-pastoris* (L.)
Willd., *Thlaspi bursa-pastoris* L.
♦ shepherd's purse, lady's purse,
pepperplant, St. James weed,
shepherd's pouch, mother's heart, case
weed, pick weed
♦ Weed, Noxious Weed, Naturalised,
Introduced, Garden Escape,
Environmental Weed
♦ 7, 15, 21, 23, 34, 38, 39, 44, 51, 52, 53,
55, 68, 70, 86, 87, 88, 93, 94, 97, 101, 114,
115, 118, 121, 134, 136, 157, 158, 161,
162, 165, 174, 176, 179, 180, 185, 186,
195, 198, 203, 204, 205, 207, 210, 211,
212, 218, 235, 236, 237, 241, 243, 248,
249, 253, 263, 269, 271, 272, 273, 274,
275, 280, 286, 295, 297, 299, 300
♦ aH, arid, cultivated, herbal, toxic.
Origin: Eurasia.

Capsella rubella **Reut.**
Brassicaceae
= *Capsella bursa-pastoris* (L.) Medik.
♦ pink shepherd's purse,
punertavalutukka
♦ Weed, Naturalised, Casual Alien
♦ 40, 42, 70, 87, 88, 94, 250, 253, 272,
287
♦ cultivated, herbal.

Capsicum annuum **L.**
Solanaceae
Capsicum frutescens L. (see)
♦ green pepper, chilli pepper,
capsicum, balake, sweet pepper,
polo, cayenne pepper, paprika,
ruokapaprika
♦ Weed, Naturalised, Introduced,
Casual Alien
♦ 11, 39, 42, 87, 88, 98, 101, 161, 203,
230, 247, 280
♦ a/pH, arid, cultivated, herbal, toxic.
Origin: obscure.

Capsicum annuum **L. var.** *annuum*
Solanaceae
♦ cayenne pepper
♦ Weed, Naturalised, Cultivation
Escape
♦ 32, 101, 261
♦ cultivated.

Capsicum annuum **L. var.** *glabriusculum*
(Dunal) Heiser & Pickersgill
Solanaceae
♦ cayenne pepper
♦ Naturalised
♦ 86
♦ cultivated. Origin: tropical America.

Capsicum baccatum **L.**
Solanaceae
♦ habanero chile
♦ Weed
♦ 87, 88, 255
♦ pH, promoted, herbal. Origin:
Americas.

Capsicum frutescens **L.**
Solanaceae
= *Capsicum annuum* L.
♦ African chilli, African pepper, bird
pepper, capsicum, cayenne pepper,
chilli bean, green chilli, guinea pepper,
Natal chilli, tabasco pepper, red
pepper, sele
♦ Weed, Naturalised, Introduced,
Cultivation Escape
♦ 13, 39, 86, 98, 121, 179, 203, 230, 257,
261, 287
♦ pH/S, cultivated, herbal, toxic.
Origin: tropical America.

Capsicum frutescens **L. var.** *baccatum*
(L.) Irish
Solanaceae
♦ Weed
♦ 14

Capsicum pendulum **Willd.**
Solanaceae
= *Capsicum baccatum* (Willd.) Eshbaugh
var. *pendulum* (Willd.) Eshbaugh (NoR)
♦ aji Colorado chile
♦ Naturalised
♦ 257
♦ herbal.

Caragana arborescens **Lam.**
Fabaceae/Papilionaceae
♦ siperianhernepensas, Siberian
peashrub, pea tree
♦ Weed, Naturalised, Environmental
Weed, Cultivation Escape
♦ 4, 42, 80, 87, 88, 101, 104, 195, 218
♦ S/T, cultivated, herbal.

Caragana aurantiaca **Koehne**
Fabaceae/Papilionaceae
♦ dwarf peashrub
♦ Naturalised
♦ 101
♦ cultivated, herbal.

Caragana brevispina **Royle ex Benth.**
Fabaceae/Papilionaceae
♦ Weed, Quarantine Weed
♦ 76, 88, 220
♦ S, cultivated.

Caragana chamlagu **Lam.**
Fabaceae/Papilionaceae
♦ Naturalised
♦ 287
♦ herbal.

Caragana frutex **(L.) K.Koch**
Fabaceae/Papilionaceae
♦ Euroopanhernepensas, Russian
peashrub
♦ Naturalised, Cultivation Escape
♦ 42, 101
♦ cultivated, herbal.

Cardamine **spp. L.**
Brassicaceae

♦ bittercress
♦ Weed
♦ 272
♦ herbal.

Cardamine africana **L.**
Brassicaceae
♦ Weed, Introduced
♦ 38, 276

Cardamine amara **L.**
Brassicaceae
♦ large bittercress
♦ Weed
♦ 39, 272
♦ pH, promoted, herbal, toxic.

Cardamine bonariensis **Pers.**
Brassicaceae
Cardamine flaccida Cham. & Schltdl.
♦ agriao bravo, agraozinho
♦ Weed
♦ 255
♦ aH, promoted. Origin: Europe.

Cardamine corymbosa **Hook.f.**
Brassicaceae
♦ New Zealand bittercress
♦ Naturalised, Environmental Weed
♦ 86, 198
♦ Origin: New Zealand.

Cardamine debilis **D.Don**
Brassicaceae
♦ roadside bittercress
♦ Weed, Naturalised
♦ 87, 88, 101
♦ pH, promoted.

Cardamine flexuosa **With.**
Brassicaceae
Cardamine drymeia Schur, *Cardamine
hirsuta* L. ssp. *flexuosa* Forbes & Hemsl.
♦ greater bittercress, wavy bittercress,
woodland bittercress, wood bittercress,
common bittercress
♦ Weed, Naturalised, Environmental
Weed
♦ 15, 86, 87, 88, 98, 101, 176, 198, 203,
204, 263, 273, 275, 280, 286, 297
♦ aH, aqua, cultivated, herbal. Origin:
Eurasia.

Cardamine glacialis **(G.Forst.) DC.**
Brassicaceae
♦ berro cimarrón
♦ Weed
♦ 295
♦ pH, promoted.

Cardamine hirsuta **L.**
Brassicaceae
♦ hairy bittercress, common
bittercress, hoary bittercress, popping
cress, Pennsylvania bittercress
♦ Weed, Naturalised, Introduced,
Garden Escape, Environmental Weed
♦ 7, 9, 15, 44, 70, 72, 86, 87, 88, 93, 94,
98, 101, 134, 161, 165, 176, 181, 198, 199,
203, 205, 207, 211, 218, 228, 238, 243,
249, 253, 272, 275, 276, 286, 287, 290,
297, 300
♦ aH, arid, cultivated, herbal. Origin:
northern hemisphere.

Cardamine impatiens **L.**
Brassicaceae

Ghinia impatiens Bubani
♦ narrowleaf bittercress, bushy
rockcress
♦ Weed, Naturalised
♦ 101, 121, 133, 195, 224, 272
♦ aH, cultivated, herbal. Origin:
Eurasia.

Cardamine lyrata **Bunge**
Brassicaceae
♦ Chinese ivy, Japanese cress
♦ Weed
♦ 87, 88, 286
♦ pH, aqua, cultivated.

Cardamine occidentalis **(S.Wats. ex
B.L.Robins.) T.J.Howell**
Brassicaceae
♦ bittercress, big western bittercress
♦ Weed
♦ 159
♦ pH, herbal.

Cardamine oligosperma **Nutt.**
Brassicaceae
♦ little bittercress, few seeded
bittercress, little western bittercress,
peppercress, Idaho bittercress, spring
cress
♦ Weed
♦ 136, 161, 180, 243
♦ a/pH, promoted, herbal.

Cardamine parviflora **L.**
Brassicaceae
♦ small flowered bittercress, sand
bittercress, rantalitukka
♦ Weed, Naturalised
♦ 87, 88, 161, 274, 286, 287
♦ herbal.

Cardamine paucijuga **Turcz.**
Brassicaceae
♦ Weed, Naturalised
♦ 9, 98, 203

Cardamine pensylvanica **Muhl. ex Willd.**
Brassicaceae
♦ Pennsylvania bittercress
♦ Weed
♦ 87, 88, 161, 218
♦ pH, herbal.

Cardamine pratensis **L.**
Brassicaceae
Cardamine nemorosa Lej.
♦ cuckoo bittercress, lady's smock,
cuckoo flower, may flower
♦ Weed, Naturalised
♦ 15, 39, 70, 87, 88, 195, 218, 272, 280
♦ pH, cultivated, herbal, toxic.

Cardamine raphanifolia **Pourr.**
Brassicaceae
Cardamine latifolia non Lej.
♦ greater cuckoo flower
♦ Naturalised, Cultivation Escape
♦ 40, 56
♦ pH, cultivated.

Cardamine resedifolia **L.**
Brassicaceae
Arabis bellidioides Lam, *Cardamine
heterophylla* Host
♦ billeri pennato
♦ Weed

♦ 272
♦ herbal.

***Cardamine sarmentosa* Forst.f.**
Brassicaceae
= *Rorippa sarmentosa* (G.Forst. ex DC.)
J.F.Macbr.
♦ Weed
♦ 87, 88
♦ herbal.

***Cardamine scutata* Thunb.**
Brassicaceae
= *Cardamine regeliana* Miq.
♦ oobatanetsukegana
♦ Weed
♦ 87, 88, 286
♦ a/bH, promoted. Origin: east Asia.

***Cardamine trifolia* L.**
Brassicaceae
♦ Naturalised, Garden Escape
♦ 56
♦ pH, cultivated, herbal. Origin:
central and southern Europe.

***Cardaminopsis arenosa* (L.) Hayek**
Brassicaceae
Arabis arenosa (L.) Scop. *p.p.* (*non auct.*
helv.), *Cardaminopsis arcuata* (Hayek)
Dutilly
♦ sand rockcress
♦ Weed, Naturalised
♦ 42, 70, 88, 94, 243, 272
♦ cultivated.

***Cardaminopsis halleri* (L.) Hayek**
Brassicaceae
Arabis halleri L., *Arabis tenella* Host
♦ etelänpitkäpalko
♦ Casual Alien
♦ 42

***Cardanthera difformis* (L.f.) Druce**
Acanthaceae
♦ Weed, Quarantine Weed
♦ 76, 87, 88, 203, 220

***Cardanthera uliginosa* Buch.-Ham.**
Acanthaceae
♦ Weed
♦ 87, 88

***Cardaria* spp. Desv.**
Brassicaceae
♦ hoary cress, cardaria, whitetop
♦ Weed, Noxious Weed, Naturalised,
Environmental Weed
♦ 36, 80, 88, 151, 162, 198

***Cardaria chalepensis* (L.) Hand-Mazz.**
Brassicaceae
= *Cardaria draba* (L.) Desv. ssp.
chalepensis (L.) Schulz
♦ lens podded hairy cress, whitetop,
hoary cress, lens podded whitetop
♦ Weed, Noxious Weed, Naturalised
♦ 26, 35, 78, 80, 87, 88, 101, 116, 161,
212, 229, 231, 237, 266, 294, 299
♦ pH, toxic.

***Cardaria draba* (L.) Desv.**
Brassicaceae
Lepidium draba L. (see)
♦ hoary cress, whiteweed,
pepperweed whitetop, whitetop,
hoary pepperwort, thanet cress,
lepidium, perennial peppergrass, heart

podded hoary cress, cardaria, hoary
cardaria, whitlow pepperwort
♦ Weed, Quarantine Weed, Noxious
Weed, Naturalised, Introduced,
Garden Escape, Environmental Weed
♦ 1, 7, 23, 26, 34, 35, 39, 42, 51, 62, 63,
70, 76, 78, 80, 86, 87, 88, 94, 98, 101, 115,
116, 121, 130, 136, 138, 146, 147, 151,
155, 161, 165, 171, 174, 176, 180, 198,
203, 210, 212, 217, 218, 219, 228, 229,
231, 237, 241, 243, 253, 264, 266, 269,
272, 280, 287, 294, 295, 300
♦ pH, arid, cultivated, herbal, toxic.
Origin: Eurasia.

***Cardaria draba* (L.) Desv. ssp. *chalepensis*
(L.) Schulz**
Brassicaceae
Cardaria chalepensis (L.) Hand.-Mazz.
(see), *Hymenophysa fenestrata* Boiss.,
Hymenophysa persica Gilli, *Lepidium
chalepense* L. (see), *Lepidium draba*
L. ssp. *chalepense* (L.) Schulz (see),
Lepidium propinquum Fisch. & C.Mey.,
Lepidium repens (Schrenk) Boiss. (see),
Physolepidium repens Schrenk
♦ Introduced
♦ 228

***Cardaria pubescens* (C.A.Mey.) Jarm.**
Brassicaceae
Cardaria pubescens var. *elongata*
(C.A.Mey.) Rollins (see), *Hymenophysa
pubescens* C.A.Mey. (see)
♦ hairy whitetop, whitetop, lens
peppergrass, whiteweed, hoary cress
♦ Weed, Noxious Weed, Naturalised
♦ 1, 23, 26, 34, 35, 49, 80, 87, 88, 101,
146, 161, 212, 229, 237, 266, 267, 294
♦ pH, herbal. Origin: Russia, Iraq,
Afghanistan.

***Cardaria pubescens* (C.A.Mey.) Jarm. var.
elongata (C.A.Mey.) Rollins**
Brassicaceae
= *Cardaria pubescens* (C.A.Mey.) Jarm.
♦ hairy whitetop
♦ Weed
♦ 218

***Cardiocrinum giganteum* (Wall.) Makino**
Liliaceae
♦ giant lily
♦ Casual Alien
♦ 280
♦ pH, cultivated, herbal.

***Cardiospermum* L. spp.**
Sapindaceae
♦ balloon vine
♦ Sleeper Weed, Quarantine Weed,
Environmental Weed
♦ 225, 246
♦ pC.

***Cardiospermum corindum* L.**
Sapindaceae
♦ farolito, faux persil
♦ Weed
♦ 261
♦ arid.

***Cardiospermum grandiflorum* Sw.**
Sapindaceae
♦ balloon vine, heartseed,
blaasklimop, showy balloonvine, large

balloon creeper
♦ Weed, Noxious Weed, Naturalised,
Garden Escape, Environmental Weed
♦ 63, 73, 86, 87, 88, 95, 98, 101, 121, 158,
201, 203, 269, 283, 290, 296
♦ C, cultivated, herbal. Origin: tropical
America.

***Cardiospermum halicacabum* L.**
Sapindaceae
Cardiospermum microcarpum Kunth
♦ balloon vine, heart pea,
wintercherry, bombilla, bombilla
menor, small balloon creeper
♦ Weed, Naturalised, Introduced,
Garden Escape
♦ 3, 6, 7, 22, 39, 80, 87, 88, 95, 98, 102,
121, 157, 161, 179, 203, 207, 209, 218,
228, 255, 261, 262, 269, 283, 286, 287,
295, 297
♦ C, arid, cultivated, herbal, toxic.
Origin: tropical Asia, Africa, America.

***Cardiospermum halicacabum* L. var.
*halicacabum***
Sapindaceae
♦ love in a puff, balloon vine
♦ Naturalised, Environmental Weed
♦ 86
♦ cultivated. Origin: South America,
Caribbean, Africa, Pakistan.

***Cardiospermum halicacabum* L. var.
microcarpum (Kunth) Blume**
Sapindaceae
= *Cardiospermum microcarpum* Kunth
(NoR)
♦ Naturalised, Environmental Weed
♦ 86, 286, 287
♦ Origin: Africa.

***Cardopatium corymbosum* Pers.**
Asteraceae
♦ Weed
♦ 39, 87, 88
♦ herbal, toxic.

***Carduncellus caeruleus* (L.) Presl**
Asteraceae
♦ blue safflower
♦ Weed
♦ 70

***Carduus acanthoides* L.**
Asteraceae
Carduus axillaris Gaudin, *Carduus
polyanthus* Schreb.
♦ curled thistle, plumeless thistle,
welted thistle, acanthus thistle, giant
plumeless thistle, spiny plumeless
thistle
♦ Weed, Quarantine Weed, Noxious
Weed, Naturalised, Introduced,
Environmental Weed, Casual Alien
♦ 1, 23, 26, 35, 42, 52, 54, 67, 70, 76, 78,
80, 86, 87, 88, 94, 101, 116, 138, 146, 151,
156, 161, 174, 195, 203, 210, 212, 218,
219, 220, 222, 229, 236, 237, 243, 246,
271, 272, 280, 291, 295
♦ pH, cultivated, herbal. Origin:
Eurasia.

***Carduus argentatus* L.**
Asteraceae
♦ Weed
♦ 87, 88

Carduus arvensis (L.) E.Robson
Asteraceae
= *Cirsium arvense* (L.) Scop.
♦ creeping plume thistle
♦ Weed
♦ 87, 88
♦ herbal.

Carduus candicans Waldst. & Kit.
Asteraceae
♦ Weed
♦ 272

Carduus corymbosus Ten.
Asteraceae
♦ Weed
♦ 88, 94

Carduus crispus L.
Asteraceae
♦ curled thistle, welted thistle, curly welted thistle, curly plumeless thistle
♦ Weed, Quarantine Weed, Noxious Weed, Naturalised
♦ 23, 70, 76, 80, 86, 87, 88, 101, 161, 218, 220, 229, 243, 272, 275, 286, 287, 297
♦ bH, cultivated, herbal.

Carduus crispus L. fo. albus Makino
Asteraceae
♦ Naturalised
♦ 287

Carduus hamulosus Ehrh.
Asteraceae
♦ Weed
♦ 87, 88, 272

Carduus kikuyorum R.E.Fr.
Asteraceae
♦ Weed
♦ 87, 88

Carduus lanatus Roxb. ex Willd.
Asteraceae
♦ Weed
♦ 87, 88

Carduus macrocephalus Desf.
Asteraceae
= *Carduus nutans* L. ssp. *macrocephalus* (Desf.) Nyman
♦ musk thistle
♦ Weed, Naturalised
♦ 87, 88, 98, 203, 267
♦ bH. Origin: Eurasia.

Carduus meonanthus Hoffmanns. & Link
Asteraceae
♦ Weed
♦ 87, 88

Carduus myriacanthus Salzm.
Asteraceae
♦ Weed
♦ 87, 88

Carduus nutans L.
Asteraceae
Carduus thoermeri J.A.Weinm. (see)
♦ nodding thistle, musk thistle, nodding plumeless thistle, chardon penche, musk thistle, plumeless thistle
♦ Weed, Quarantine Weed, Noxious Weed, Naturalised, Introduced, Garden Escape, Environmental Weed
♦ 1, 4, 15, 20, 21, 23, 34, 35, 36, 42, 48, 52, 62, 67, 68, 70, 76, 80, 87, 88, 94, 98, 101, 102, 104, 129, 130, 133, 136, 138, 146, 147, 151, 152, 161, 162, 165, 169, 174, 176, 195, 198, 203, 207, 210, 211, 212, 217, 218, 219, 220, 222, 225, 229, 236, 241, 243, 246, 249, 264, 267, 271, 272, 280, 290, 291, 299
♦ a/bH, cultivated, herbal. Origin: Eurasia.

Carduus nutans L. ssp. leiophyllus (Petrovic) Stoj. & Stef.
Asteraceae
♦ musk thistle, nodding plumeless thistle, nodding thistle
♦ Noxious Weed, Naturalised
♦ 101, 229

Carduus nutans L. ssp. macrocephalus (Desf.) Nyman
Asteraceae
Carduus macrocephalus Desf. (see)
♦ musk thistle, nodding plumeless thistle, nodding thistle
♦ Noxious Weed, Naturalised
♦ 101, 229

Carduus nutans L. ssp. macrolepis (Peterm.) Kazmi
Asteraceae
♦ musk thistle, nodding plumeless thistle, nodding thistle
♦ Noxious Weed, Naturalised
♦ 101, 229
♦ herbal.

Carduus nutans L. ssp. nutans
Asteraceae
Carduus thoermeri Weinm. ssp. *armenus* (Boiss.) Kazmi
♦ musk thistle, nodding plumeless thistle, nodding thistle
♦ Weed, Noxious Weed, Naturalised
♦ 86, 101, 177, 229, 269
♦ a/bH, cultivated. Origin: Eurasia, Africa.

Carduus × orthocephalus Wallr.
Asteraceae
= *Carduus acanthoides* L. × *Carduus nutans* L.
♦ plumeless thistle
♦ Weed, Naturalised
♦ 101, 291

Carduus personata (L.) Jacq.
Asteraceae
Arctium personata L., *Cirsium arctioides* Vill., *Cirsium lappaceum* Lam.
♦ great marsh thistle
♦ Weed
♦ 272
♦ herbal.

Carduus platypus Lange.
Asteraceae
♦ Weed
♦ 88, 94

Carduus pteracanthus Dur.
Asteraceae
♦ Weed
♦ 87, 88

Carduus pycnocephalus L.
Asteraceae

Carduus albidus Lam.
♦ Italian thistle, shore thistle, slender thistle, slender flowered thistle, slender winged thistle, compact headed thistle, Italian plumeless thistle, Plymouth thistle, rabbit thistle
♦ Weed, Noxious Weed, Naturalised, Environmental Weed, Casual Alien
♦ 1, 7, 9, 15, 34, 35, 39, 40, 51, 68, 70, 72, 78, 80, 86, 87, 88, 94, 98, 101, 116, 139, 146, 147, 151, 161, 176, 180, 198, 203, 212, 217, 218, 229, 231, 236, 237, 241, 243, 253, 269, 272, 280, 287, 295, 300
♦ aH, herbal, toxic. Origin: Europe.

Carduus tenuiflorus Curt.
Asteraceae
Carduus crispus Gouan (*non* L.)
♦ slenderflower thistle, shore thistle, sheep thistle, winged thistle, winged slender thistle, Italian thistle, winged plumeless thistle
♦ Weed, Quarantine Weed, Noxious Weed, Naturalised, Introduced, Environmental Weed, Casual Alien
♦ 1, 7, 34, 35, 42, 70, 72, 76, 80, 86, 87, 88, 94, 98, 101, 121, 134, 139, 146, 147, 151, 161, 165, 167, 171, 176, 198, 203, 218, 229, 237, 253, 269, 280, 287, 295
♦ aH, herbal. Origin: Eurasia.

Carduus theodori R.E.Fr.
Asteraceae
♦ Weed
♦ 87, 88

Carduus thoermeri J.A.Weinm.
Asteraceae
= *Carduus nutans* L.
♦ idänkarhiainen
♦ Weed, Quarantine Weed, Naturalised, Casual Alien
♦ 42, 76, 86, 98, 203, 237, 267, 295, 300
♦ bH. Origin: Eurasia.

Carduus uncinatus Bieb.
Asteraceae
♦ Weed
♦ 87, 88, 272

Carex L. spp.
Cyperaceae
♦ sedge
♦ Weed, Quarantine Weed
♦ 23, 70, 88, 181, 203, 220, 243, 272
♦ G, herbal.

Carex acuta L.
Cyperaceae
Carex gracilis Curt.
♦ slender spiked sedge, slender tufted sedge
♦ Weed
♦ 87, 88
♦ G, cultivated, herbal.

Carex acutiformis Ehrh.
Cyperaceae
♦ lesser pond sedge, swamp sedge
♦ Naturalised
♦ 101
♦ pG, cultivated, herbal.

Carex albata Boott ex Franch.
Cyperaceae

= *Carex nubigena* D.Don ssp. *albata*
(Boott ex Franch.) T.Koyama
♦ Weed
♦ 286
♦ G.

Carex albo-nigra Mack.
Cyperaceae
♦ Weed
♦ 23, 88
♦ G.

Carex albula Allan
Cyperaceae
Carex comans Bergg. (see)
♦ New Zealand hair sedge
♦ Noxious Weed, Naturalised, Garden
Escape
♦ 86, 176
♦ pG, cultivated. Origin: New
Zealand.

Carex aphanolepis Fr. & Sav.
Cyperaceae
♦ enashihigokusa
♦ Weed
♦ 286
♦ G.

Carex appressa R.Br.
Cyperaceae
♦ tussock sedge, tall sedge
♦ Native Weed
♦ 269
♦ pG, arid, cultivated. Origin:
Australia.

Carex aquatilis Wahlenb.
Cyperaceae
♦ water sedge, mountain water sedge
♦ Weed
♦ 87, 88, 161, 218
♦ pG, herbal.

Carex arenaria L.
Cyperaceae
Vignea arenaria (L.) Rchb.
♦ sand sedge
♦ Weed, Naturalised
♦ 54, 86, 88
♦ pG, cultivated, herbal. Origin:
Europe.

Carex atherodes Spreng.
Cyperaceae
♦ sugargrass sedge, wheat sedge,
slough sedge
♦ Weed
♦ 87, 88, 218
♦ pG, herbal.

Carex baccans Nees
Cyperaceae
♦ Weed
♦ 87, 88
♦ G.

Carex bichenoviana Boott
Cyperaceae
♦ Casual Alien
♦ 280
♦ G, cultivated.

Carex bonariensis Desf. ex Poir.
Cyperaceae
♦ carex
♦ Weed

♦ 87, 88, 237, 295
♦ G.

Carex breviculmis R.Br.
Cyperaceae
♦ aosuge
♦ Weed
♦ 286
♦ G.

Carex brizoides L.
Cyperaceae
Vignea brizoides Rchb.
♦ alpine grass
♦ Weed
♦ 126
♦ pG, promoted.

Carex brongniartii Kunth
Cyperaceae
♦ Weed
♦ 87, 88
♦ G.

Carex brownii Tuck.
Cyperaceae
♦ awabosuge
♦ Naturalised
♦ 280
♦ G.

Carex brunnea Thunb.
Cyperaceae
♦ greater brown sedge
♦ Naturalised
♦ 101
♦ G, herbal. Origin: Asia, Australia,
Madagascar.

Carex buchananii Bergg.
Cyperaceae
♦ New Zealand sedge
♦ Weed, Quarantine Weed, Noxious
Weed
♦ 76, 88, 147, 203
♦ G, cultivated.

Carex buxbaumii Wahlenb.
Cyperaceae
♦ club sedge, Buxbaum's sedge
♦ Naturalised
♦ 176
♦ pG, promoted, herbal. Origin:
Australia.

**Carex buxbaumii Wahlenb. ssp.
buxbaumii**
Cyperaceae
♦ New Zealand sedge
♦ Naturalised
♦ 198
♦ G.

Carex canescens L.
Cyperaceae
Vignea canescens Rchb.
♦ silvery sedge
♦ Weed
♦ 87, 88
♦ pG, cultivated, herbal. Origin:
Australia.

Carex caryophyllea Latourr.
Cyperaceae
♦ vernal sedge, spring sedge
♦ Naturalised

♦ 101
♦ G, herbal.

Carex cephalophora Muhl. ex Willd.
Cyperaceae
♦ ovalleaf sedge
♦ Weed
♦ 34, 161
♦ pG, herbal.

Carex cherokeensis Schwein.
Cyperaceae
♦ wolftail sedge, Cherokee sedge
♦ Weed
♦ 161
♦ G, cultivated, herbal.

Carex comans Bergg.
Cyperaceae
= *Carex albula* Allan.
♦ New Zealand sedge
♦ Weed, Quarantine Weed, Noxious
Weed, Naturalised
♦ 86, 88, 147, 203
♦ pG, cultivated.

Carex comosa Boott
Cyperaceae
♦ bristly sedge, bristle sedge, longhair
sedge
♦ Weed
♦ 161, 180
♦ pG, aqua, herbal.

Carex demissa Hornem.
Cyperaceae
♦ low sedge, common yellow sedge,
yellow sedge
♦ Weed, Naturalised
♦ 15, 86, 176, 280
♦ G, cultivated. Origin: Europe.

Carex dietrichiae Boeck.
Cyperaceae
♦ Weed
♦ 87, 88
♦ G.

Carex dimorpholepis Steud.
Cyperaceae
♦ azenarukosuge
♦ Weed
♦ 286
♦ G.

Carex dispalata Boott
Cyperaceae
♦ kasa suge, sedge
♦ Weed
♦ 286
♦ pG, promoted. Origin: Japan.

Carex distans L.
Cyperaceae
♦ distant sedge
♦ Naturalised
♦ 101
♦ G, cultivated, herbal.

Carex distenta Kunze
Cyperaceae
♦ widow's sedge
♦ Naturalised
♦ 101
♦ G.

Carex disticha **Huds.**
Cyperaceae
Carex intermedia Gooden., *Vignea intermedia* Rchb.
♦ creeping brown sedge, brown sedge
♦ Weed, Naturalised
♦ 86, 87, 88, 198
♦ pG, promoted. Origin: Europe.

Carex divisa **Huds.**
Cyperaceae
Carex hybrida Lam. (*non* Schkuhr ex Kunth.), *Vignea divisa* Rchb.
♦ separated sedge, divided sedge, New Zealand sedge, salt meadow sedge
♦ Weed, Naturalised, Environmental Weed
♦ 7, 86, 87, 88, 98, 101, 176, 198, 203, 280
♦ G, herbal. Origin: Europe.

Carex divulsa **Stokes**
Cyperaceae
♦ grassland sedge, grey sedge
♦ Weed, Naturalised
♦ 15, 87, 88, 101, 176, 237, 280, 295
♦ G, cultivated, herbal. Origin: Mediterranean.

Carex divulsa **Stokes ssp.** *divulsa*
Cyperaceae
♦ sedge, grey sedge
♦ Naturalised, Environmental Weed
♦ 86, 198
♦ G. Origin: Europe.

Carex duriuscula **C.A.Mey. ssp.** *rigescens* **(Fr.) S.Y.Liang & Y.C.Tang**
Cyperaceae
Carex rigescens (Fr.) Krecz. (see)
♦ rigescent sedge
♦ Weed
♦ 297
♦ G.

Carex ebenea **Rydb.**
Cyperaceae
♦ ebony sedge
♦ Weed
♦ 23, 88
♦ G, herbal.

Carex eurycarpa **Holm.**
Cyperaceae
= *Carex angustata* Boott
♦ widefruit sedge
♦ Weed
♦ 87, 88, 218
♦ G, herbal.

Carex extensa **Gooden.**
Cyperaceae
♦ longbract sedge, long bracted sedge
♦ Naturalised
♦ 101
♦ G, cultivated, herbal.

Carex filicina **Boeck. ex C.B.Clarke**
Cyperaceae
♦ Weed
♦ 88
♦ G.

Carex flacca **Schreb.**
Cyperaceae
Carex glauca Scop. (see)

♦ carnation sedge, sedge, heath sedge, carnation grass, glaucus sedge
♦ Weed, Sleeper Weed, Naturalised, Environmental Weed
♦ 15, 23, 86, 88, 101, 176, 198, 225, 246, 280
♦ G, cultivated, herbal. Origin: Europe.

Carex flagellifera **Col.**
Cyperaceae
♦ New Zealand sedge
♦ Weed, Quarantine Weed, Noxious Weed, Naturalised, Garden Escape
♦ 76, 86, 88, 147, 176, 203, 251
♦ G, cultivated. Origin: New Zealand.

Carex flava **L.**
Cyperaceae
♦ yellow sedge, large yellow sedge
♦ Weed
♦ 159
♦ G, cultivated, herbal.

Carex gibba **Wahlenb.**
Cyperaceae
♦ Weed
♦ 286
♦ G.

Carex glauca **Scop.**
Cyperaceae
= *Carex flacca* Schreb.
♦ Europe blue sedge
♦ Weed
♦ 87, 88
♦ G, cultivated.

Carex graeffeana **Boeck.**
Cyperaceae
♦ Weed
♦ 87, 88
♦ G.

Carex heterostachya **Bunge**
Cyperaceae
♦ Fuller's sedge
♦ Weed
♦ 275
♦ pG.

Carex hirta **L.**
Cyperaceae
Carex villosa Stokes
♦ hairy sedge, hammer sedge
♦ Weed, Naturalised
♦ 101, 126, 253, 280
♦ pG, cultivated, herbal.

Carex hispida **Willd. ex Schkuhr**
Cyperaceae
♦ Naturalised
♦ 300
♦ G.

Carex hudsonii **A.Benn.**
Cyperaceae
= *Carex elata* All. ssp. *elata* (NoR)
♦ Weed
♦ 87, 88
♦ G.

Carex inversa **R.Br.**
Cyperaceae
♦ knob sedge
♦ Naturalised, Native Weed
♦ 101, 269
♦ pG, cultivated. Origin: Australia.

Carex iynx **Nelmes**
Cyperaceae
♦ Naturalised
♦ 280
♦ G, cultivated.

Carex kobomugi **Ohwi**
Cyperaceae
♦ Asiatic sand sedge, Japanese sedge
♦ Weed, Naturalised, Environmental Weed
♦ 17, 80, 88, 101, 133, 151, 195
♦ pG, promoted, herbal.

Carex lacustris **Willd.**
Cyperaceae
♦ ripgut sedge, hairy sedge
♦ Weed
♦ 87, 88, 218
♦ G, herbal.

Carex lanceolata **Boott.**
Cyperaceae
♦ kasa suge, sedge
♦ Weed
♦ 88, 204, 286
♦ G.

Carex lanuginosa **Michx.**
Cyperaceae
♦ woolly sedge, bull sedge
♦ Weed
♦ 161
♦ pG, herbal.

Carex lasiocarpa **Ehrh.**
Cyperaceae
♦ bull sedge, slender leaved sedge, slender sedge, downy fruited sedge, woollyfruit sedge
♦ Weed
♦ 87, 88, 126
♦ pG, aqua, herbal.

Carex lasiocarpa **Ehrh. var.** *latifolia* **(Bock.) Gleason**
Cyperaceae
= *Carex pellita* Muhl. ex Willd. (NoR)
♦ bull sedge
♦ Weed
♦ 218
♦ G.

Carex leavenworthii **Dewey**
Cyperaceae
♦ Leavenworth's sedge
♦ Weed
♦ 34
♦ pG, herbal.

Carex leiorhyncha **C.A.Mey.**
Cyperaceae
♦ sharpbeak sedge
♦ Weed
♦ 297
♦ G.

Carex lepidocarpa **Tausch**
Cyperaceae
♦ long stalked yellow sedge, yellow sedge
♦ Weed, Naturalised, Environmental Weed
♦ 54, 86, 88, 176
♦ G. Origin: Europe.

Carex leporina **L.**
Cyperaceae

Vignea leporina Rchb.
- leporina sedge
- Weed
- 87, 88, 126
- pG, cultivated, herbal.

Carex leucochlora Bunge
Cyperaceae
- whitegreen sedge
- Weed
- 297
- G.

Carex longebrachiata Boeck.
Cyperaceae
- Australian sedge
- Naturalised, Environmental Weed
- 225, 246, 280
- G, cultivated. Origin: Australia.

Carex longii Mack.
Cyperaceae
- Long's sedge
- Naturalised
- 280
- G, herbal.

Carex lupulina Muhl. ex Willd.
Cyperaceae
- hop sedge
- Weed
- 87, 88, 161, 210, 218
- pG, cultivated, herbal.

Carex lurida Wahlenb.
Cyperaceae
- shallow sedge
- Naturalised
- 280
- G, cultivated, herbal.

Carex macrostachys Bertol.
Cyperaceae
- largespike sedge
- Naturalised
- 101
- G, herbal.

Carex maximowiczii Miq.
Cyperaceae
= *Carex picta* Steud.
- Weed
- 286
- G.

Carex muricata L.
Cyperaceae
Carex stellulata auct. non Gooden., *Carex pairaei* F.W.Schultz
- rough sedge, prickly sedge
- Weed, Quarantine Weed, Naturalised
- 15, 76, 88, 101, 280
- G, cultivated, herbal. Origin: Ecuador.

Carex myosurus Nees
Cyperaceae
Carex longebrachiata Steud.
- Weed
- 87, 88
- G.

Carex nebrascensis Dewey
Cyperaceae
Carex nebraskensis Dewey (see)
- Nebraska sedge

- Weed
- 159
- pG, herbal.

Carex nebraskensis Dewey
Cyperaceae
= *Carex nebrascensis* Dewey
- Nebraska sedge
- Weed
- 87, 88, 161, 218
- pG, herbal.

Carex nervata Franch. & Sav.
Cyperaceae
- shiba suge, sedge
- Weed
- 88, 204, 286
- G.

Carex neurocarpa Maxim.
Cyperaceae
- wingfruit sedge
- Weed
- 297
- G.

Carex nigra (L.) Reichard
Cyperaceae
- common sedge, smooth black sedge
- Weed
- 87, 88
- G, cultivated, herbal.

Carex nubigena D.Don ex Tilloch & Taylor
Cyperaceae
- Weed
- 87, 88
- pG, promoted.

Carex otrubae Podp.
Cyperaceae
- false foxsedge
- Naturalised
- 280
- G, cultivated, herbal.

Carex ovalis Gooden.
Cyperaceae
- oval sedge, eggbract sedge, sedge
- Weed, Naturalised
- 15, 86, 176, 181, 280
- pG, cultivated. Origin: Europe.

Carex pallescens L.
Cyperaceae
- pale sedge
- Naturalised
- 280
- G, cultivated, herbal.

Carex panicea L.
Cyperaceae
- carnation sedge, carnat tonggrass, grasslike sedge
- Weed, Naturalised
- 101, 126
- pG, aqua, herbal.

Carex paniculata L.
Cyperaceae
Vignea paniculata Rchb.
- greater tussock sedge, panicled sedge
- Weed, Naturalised, Casual Alien
- 87, 88, 126, 280
- pG, aqua, cultivated, herbal.

Carex pendula Huds.
Cyperaceae
- hanging sedge, pendulous sedge, drooping sedge
- Naturalised
- 101, 280
- G, aqua, cultivated, herbal. Origin: Africa, Eurasia.

Carex picta Steud.
Cyperaceae
Carex maximowiczii Miq. (see)
- Boott's sedge
- Weed
- 87, 88
- G, herbal.

Carex pilulifera L.
Cyperaceae
- pill headed sedge, pill sedge
- Weed, Naturalised, Environmental Weed
- 54, 86, 88, 176
- G, herbal. Origin: Europe.

Carex pumila Thunb.
Cyperaceae
- dwarf sedge, sedge, kouboushiba
- Weed, Naturalised
- 87, 88, 101
- G, cultivated.

Carex punctata Gaudin
Cyperaceae
- sedge, dotted sedge
- Weed, Naturalised
- 54, 86, 88, 198, 280
- G, cultivated, herbal. Origin: Europe.

Carex remota L.
Cyperaceae
- distant flowered sedge, remote sedge
- Weed
- 87, 88
- G, cultivated, herbal.

Carex retroflexa Willd. var. texensis (Torr.) Fern.
Cyperaceae
= *Carex texensis* (Torr.) Bailey
- Texas sedge
- Introduced
- 34
- pG.

Carex rigescens (Fr.) Krecz.
Cyperaceae
= *Carex duriuscula* C.A.Mey. ssp. *rigescens* (Fr.) S.Y.Liang & Y.C.Tang
- Weed
- 275
- pG.

Carex riparia Curt.
Cyperaceae
- great pond sedge, greater pond sedge
- Weed, Naturalised, Casual Alien
- 87, 88, 280
- pG, aqua, cultivated, herbal.

Carex rostrata Stokes ex With.
Cyperaceae
- beaked sedge
- Weed
- 159
- pG, promoted, herbal.

Carex scoparia Schkuhr ex Willd.
Cyperaceae
♦ broomsedge, pointed broomsedge
♦ Weed, Naturalised, Environmental
Weed
♦ 15, 54, 86, 88, 176, 280
♦ pG, herbal. Origin: North America.

Carex senta Boott
Cyperaceae
♦ rough sedge, swamp carex
♦ Weed
♦ 87, 88, 218
♦ pG, herbal.

Carex spicata Huds.
Cyperaceae
♦ prickly sedge, spiked sedge
♦ Naturalised
♦ 101, 280
♦ G.

Carex spicato-paniculata C.B.Clarke
Cyperaceae
♦ sedge
♦ Native Weed
♦ 121
♦ pG. Origin: southern Africa.

Carex sylvatica Huds.
Cyperaceae
♦ European woodland sedge, wood
sedge
♦ Naturalised
♦ 101, 280
♦ G, cultivated, herbal.

Carex testacea Sol. ex Boott
Cyperaceae
♦ New Zealand sedge
♦ Weed, Quarantine Weed, Noxious
Weed, Naturalised, Garden Escape
♦ 54, 76, 86, 88, 147, 176, 203, 251
♦ G, cultivated. Origin: New Zealand.

Carex thunbergii Steud.
Cyperaceae
♦ Maui sedge, azesuge
♦ Weed
♦ 286
♦ G, herbal.

Carex vulpina L.
Cyperaceae
Vignea vulpina Rchb.
♦ foxsedge, true foxsedge, great sedge
♦ Weed
♦ 39, 87, 88
♦ G, cultivated, herbal, toxic.

Carex vulpinoidea Michx.
Cyperaceae
♦ foxsedge
♦ Weed, Naturalised
♦ 15, 280, 287
♦ pG, aqua, cultivated, herbal.

Carex zuluensis C.B.Clarke
Cyperaceae
♦ sedge
♦ Native Weed
♦ 121
♦ pG. Origin: southern Africa.

Careya arborea Roxb.
Lecythidaceae/Barringtoniaceae
♦ patana oak

♦ Weed
♦ 87, 88
♦ herbal.

Carica papaya L.
Caricaceae
♦ papaya, pawpaw
♦ Weed, Naturalised, Introduced,
Environmental Weed, Cultivation
Escape
♦ 22, 39, 80, 86, 88, 98, 101, 151, 203,
230, 257, 261, 279
♦ T, cultivated, herbal, toxic. Origin:
continental tropical America.

Carica pubescens Lenné & K.Koch
Caricaceae
♦ mountain papaya
♦ Weed, Naturalised
♦ 22, 280
♦ S, cultivated.

Carissa bispinosa (L.) Desf. ex Brenan
Apocynaceae
♦ num num, red num num, small
amatungula, y thorned carissa
♦ Native Weed
♦ 121
♦ S/T, cultivated. Origin: southern
Africa.

Carissa edulis Vahl
Apocynaceae
♦ Egyptian carissa
♦ Weed
♦ 87, 88
♦ cultivated, herbal. Origin:
Madagascar.

Carissa haematocarpa (Eckl.) A.DC.
Apocynaceae
♦ karroo num num, large num num
♦ Native Weed
♦ 121
♦ pS. Origin: southern Africa.

Carissa macrocarpa (Eckl.) A.DC.
Apocynaceae
♦ amatungulu, large num num, Natal
plum
♦ Weed, Naturalised, Native Weed,
Introduced
♦ 80, 101, 121, 179, 261
♦ pS, cultivated. Origin: southern
Africa.

Carissa tetramera (Sacleux) Stapf
Apocynaceae
♦ Native Weed
♦ 121
♦ pS. Origin: southern Africa.

Carlina L. spp.
Asteraceae
♦ carline thistle
♦ Weed
♦ 272
♦ herbal.

Carlina acanthifolia All.
Asteraceae
Carlina utzka Hacquet
♦ acanthus leaved thistle
♦ Weed
♦ 272
♦ b/pH, cultivated, herbal. Origin:
Mediterranean to Balkans.

Carlina acaulis L.
Asteraceae
♦ dwarf carline thistle, silver thistle,
caline thistle, stemless carline thistle
♦ Weed
♦ 39, 272
♦ b/pH, promoted, herbal, toxic.

Carlina corymbosa L.
Asteraceae
♦ Weed
♦ 253
♦ b/pH.

Carlina involucrata Poir.
Asteraceae
♦ Weed, Quarantine Weed
♦ 76, 87, 88, 203, 220

Carlina lanata L.
Asteraceae
♦ Weed
♦ 87, 88
♦ herbal.

Carlina racemosa L.
Asteraceae
♦ carline thistle
♦ Weed
♦ 70, 87, 88, 253

Carlina vulgaris L.
Asteraceae
♦ carline thistle
♦ Weed, Naturalised
♦ 101, 272
♦ bH, cultivated, herbal.

Carlina vulgaris L. ssp. longifolia Nyman
Asteraceae
♦ carline thistle
♦ Naturalised
♦ 101

Carludovica palmata Ruiz & Pav.
Cyclanthaceae
♦ Panama hat plant
♦ Naturalised, Garden Escape
♦ 101, 261
♦ H, cultivated, herbal. Origin: tropical
America.

Carmichaelia arborea (G.Forst.) Druce
Fabaceae/Papilionaceae
♦ Quarantine Weed
♦ 220

Carmichaelia monroi Hook.f.
Fabaceae/Papilionaceae
♦ Quarantine Weed
♦ 220
♦ cultivated. Origin: New Zealand.

Carmichaelia ovata G.Simpson
Fabaceae/Papilionaceae
♦ Quarantine Weed
♦ 220
♦ Origin: New Zealand.

Carmona retusa (Vahl) Masam.
Boraginaceae
Carmona microphylla (Lam.) G.Don,
Ehretia buxifolia Roxb., *Ehretia
microphylla* Lam. (see)
♦ scorpion bush, Fukien tea,
Philippine tea
♦ Naturalised, Cultivation Escape
♦ 101, 233
♦ cultivated.

Carpanthea pomeridiana (L.) N.E.Br.
Aizoaceae/Mesembryanthemaceae
♦ Weed, Naturalised
♦ 98, 203
♦ cultivated, herbal. Origin: South Africa.

Carpesium abrotanoides L.
Asteraceae
Carpesium thunbergianum Sieb. & Zucc.
♦ yabutabako
♦ Weed
♦ 275, 286, 297
♦ pH, promoted, herbal.

Carpesium cernuum L.
Asteraceae
♦ koyabutabako
♦ Weed, Naturalised
♦ 86, 98, 203, 286, 297
♦ cultivated, herbal. Origin: Eurasia.

Carpesium divaricatum Sieb. & Zucc.
Asteraceae
♦ Weed
♦ 87, 88
♦ pH, promoted, herbal.

Carpha glomerata (Thunb.) Nees
Cyperaceae
♦ sedge
♦ Native Weed
♦ 121
♦ pG. Origin: southern Africa.

Carpinus americana Michx.
Betulaceae/Corylaceae/Carpinaceae
= *Carpinus caroliniana* Walt. ssp. *caroliniana* (NoR)
♦ American hornbeam
♦ Weed
♦ 87, 88

Carpinus betulus L.
Betulaceae/Corylaceae/Carpinaceae
♦ valkopyökki, hornbeam
♦ Weed, Naturalised, Cultivation Escape
♦ 42, 70, 101
♦ T, cultivated, herbal.

Carpinus caroliniana Walter
Betulaceae/Corylaceae/Carpinaceae
♦ American hornbeam
♦ Weed
♦ 218
♦ T, cultivated, herbal.

Carpobrotus N.E.Br. spp.
Aizoaceae/Mesembryanthemaceae
♦ carpobrotus, iceplant
♦ Weed, Environmental Weed
♦ 181, 257

Carpobrotus acinaciformis (L.) L.Bolus
Aizoaceae/Mesembryanthemaceae
♦ Sally my handsome, Hottentot fig
♦ Naturalised
♦ 40
♦ pH, cultivated.

Carpobrotus aequilaterus (Haw.) N.E.Br.
Aizoaceae/Mesembryanthemaceae
Carpobrotus chilensis (Molina) N.E.Br. (see), *Mesembryanthemum aequilaterum* Haw., *Mesembryanthemum chilense* Molina

♦ angled pigface, iceplant, sea fig
♦ Weed, Naturalised, Environmental Weed
♦ 7, 15, 72, 88, 98, 151, 176, 198, 203, 280, 296
♦ S, arid, promoted, herbal. Origin: possibly South America.

Carpobrotus aequilaterus (Haw.) N.E.Br. × Disphyma australe (W.T.Aiton) N.E.Br.
Aizoaceae/Mesembryanthemaceae
♦ Naturalised
♦ 280

Carpobrotus chilensis (Molina) N.E.Br.
Aizoaceae/Mesembryanthemaceae
= *Carpobrotus aequilaterus* (Haw.) N.E.Br.
♦ angled pigface
♦ Weed, Naturalised, Garden Escape, Environmental Weed
♦ 78, 80, 86, 101, 116
♦ pH, cultivated.

Carpobrotus edulis (L.) L.Bolus
Aizoaceae/Mesembryanthemaceae
Mesembryanthemum edule L.
♦ Hottentot fig, iceplant, freeway iceplant, sour fig, common hottentot fig, sea fig
♦ Weed, Noxious Weed, Naturalised, Native Weed, Introduced, Garden Escape, Environmental Weed
♦ 3, 7, 9, 15, 35, 40, 70, 72, 78, 80, 86, 88, 98, 101, 116, 121, 151, 152, 158, 165, 176, 179, 181, 198, 203, 228, 231, 280
♦ pH, arid, cultivated. Origin: southern Africa.

Carpobrotus edulis (L.) L.Bolus × Disphyma australe (W.T.Aiton) N.E.Br.
Aizoaceae/Mesembryanthemaceae
♦ Naturalised
♦ 280

Carpobrotus edulis (L.) L.Bolus × virescens (Haw.) Schwantes
Aizoaceae/Mesembryanthemaceae
♦ Naturalised
♦ 86

Carpodinus acida Sabine
Apocynaceae
♦ Quarantine Weed
♦ 220
♦ Origin: tropical Africa.

Carpodinus barteri Stapf
Apocynaceae
= *Landolphia dulcis* (Sabine) Pichon var. *barteri* (Stapf) Pichon (NoR) [see *Landolphia dulcis* (Sabine) Pichon]
♦ Quarantine Weed
♦ 220
♦ Origin: tropical Africa.

Carpodinus baumannii Hutch. & Dalziel
Apocynaceae
♦ Quarantine Weed
♦ 220
♦ Origin: tropical Africa.

Carpodinus dulcis Sabine
Apocynaceae
= *Landolphia dulcis* (Sabine) Pichon
♦ Quarantine Weed
♦ 220

Carpodinus flava Pierre
Apocynaceae
♦ Quarantine Weed
♦ 220
♦ Origin: tropical Africa.

Carpodinus oocarpus Stapf
Apocynaceae
♦ Quarantine Weed
♦ 220

Carpodinus parviflorus Stapf
Apocynaceae
♦ Quarantine Weed
♦ 220

Carpodinus parvifolia Pierre
Apocynaceae
♦ Quarantine Weed
♦ 220

Carpodinus pauciflora K.Schum.
Apocynaceae
♦ Quarantine Weed
♦ 220

Carpodinus tenuifolia Pierre
Apocynaceae
= *Landolphia tenuifolia* (Pierre ex Stapf) Pichon
♦ Quarantine Weed
♦ 220

Carpophillus Neck. spp.
Cactaceae
♦ Quarantine Weed
♦ 220

Carpophyllum Grev. spp.
Sargassaceae
♦ Weed, Quarantine Weed
♦ 76, 88, 220
♦ algae.

× Carpophyma G.D.Rowley sp.
Aizoaceae
= *Carpobrotus* N.E.Br. sp. × *Disphyma* N.E.Br. sp.
♦ Naturalised
♦ 280

Carrichtera annua (L.) DC.
Brassicaceae
Vella annua L. (see)
♦ Ward's weed
♦ Weed, Naturalised, Introduced, Garden Escape, Environmental Weed
♦ 7, 72, 86, 88, 93, 98, 176, 198, 203, 205, 228, 243, 269, 280, 287
♦ aH, arid, cultivated. Origin: Mediterranean.

Carthamus L. spp.
Asteraceae
♦ distaff thistle, krokosz
♦ Weed, Naturalised, Environmental Weed
♦ 88, 151, 198
♦ herbal.

Carthamus baeticus (Boiss. & Reut.) Lara
Asteraceae
= *Carthamus lanatus* L. ssp. *baeticus* (Boiss. & Reut.) Nyman
♦ woolly distaff thistle, smooth distaff thistle
♦ Weed, Noxious Weed
♦ 34, 35, 80, 88
♦ aH.

Carthamus caeruleus L.
Asteraceae
♦ Weed
♦ 87, 88

Carthamus calvus (Boiss. & Reut.) Batt.
Asteraceae
♦ Weed
♦ 87, 88

Carthamus dentatus (Forssk.) Vahl
Asteraceae
♦ thistle
♦ Weed, Naturalised
♦ 86, 98, 198, 203, 269, 272
♦ Origin: south-west Europe.

Carthamus flavescens Willd.
Asteraceae
♦ Weed, Quarantine Weed
♦ 76, 87, 88, 203, 220

Carthamus glaucus M.Bieb.
Asteraceae
♦ glaucous star thistle
♦ Weed, Quarantine Weed,
Naturalised
♦ 76, 86, 87, 88, 98, 198, 203, 220, 269,
272
♦ Origin: Eurasia.

Carthamus lanatus L.
Asteraceae
Atracyclis lanata Scop., *Cardunculus
lanatus* Moris, *Centaurea lanata* DC.,
Heracantha lanata Link, *Kentrophyllum
lanatum* DC., *Kentrophyllum luteum*
Cass.
♦ saffron thistle, distaff thistle, false
star thistle, woolly safflower, woolly
star thistle, downy safflower
♦ Weed, Quarantine Weed, Noxious
Weed, Naturalised, Introduced,
Garden Escape, Environmental Weed,
Casual Alien
♦ 7, 35, 40, 42, 62, 70, 72, 76, 80, 86, 87,
88, 93, 94, 98, 101, 121, 146, 147, 161,
165, 176, 177, 198, 203, 205, 212, 228,
229, 236, 237, 241, 243, 246, 253, 269,
272, 280, 287, 295, 300
♦ aH, arid, cultivated, herbal. Origin:
Eurasia.

**Carthamus lanatus L. ssp. baeticus
(Boiss. & Reut.) Nyman**
Asteraceae
Carthamus baeticus (Boiss. & Reut.)
Lara (see)
♦ woolly distaff thistle, smooth distaff
thistle
♦ Noxious Weed, Naturalised
♦ 101, 229, 280

Carthamus lanatus L. ssp. lanatus
Asteraceae
♦ woolly distaff thistle
♦ Noxious Weed, Naturalised
♦ 101, 229

Carthamus leucocaulos Sm.
Asteraceae
♦ yellow distaff thistle, whitestem
distaff thistle, glaucous star thistle
♦ Weed, Quarantine Weed, Noxious
Weed, Naturalised
♦ 7, 35, 54, 62, 76, 80, 86, 88, 98, 101,

147, 161, 203, 220, 229
♦ aH. Origin: Greece, Crete.

Carthamus oxyacantha Bieb.
Asteraceae
♦ wild safflower, jeweled distaff
thistle, carthamus
♦ Weed, Quarantine Weed, Noxious
Weed, Introduced
♦ 67, 76, 87, 88, 140, 203, 220, 228, 229,
243
♦ arid, herbal.

Carthamus syriacus (Boiss.) Celak.
Asteraceae
♦ Weed
♦ 87, 88

**Carthamus tenuis (Boiss. & Blanche)
Bornm.**
Asteraceae
♦ Weed
♦ 87, 88, 111, 221, 243

Carthamus tinctorius L.
Asteraceae
Carduus tinctorius (L.) Falk *non* Scop.,
Carduus inermis Hegi
♦ safflower, värisaflori
♦ Weed, Naturalised, Introduced,
Garden Escape, Environmental Weed,
Casual Alien
♦ 7, 40, 42, 70, 86, 87, 88, 93, 94, 98, 101,
176, 198, 203, 205, 228, 241, 280, 287,
300
♦ aH, arid, cultivated, herbal. Origin:
obscure.

**Carthamus tinctorius L. var. spinosus
Kitamura**
Asteraceae
♦ Naturalised
♦ 287

Carum carvi L.
Apiaceae
Apium carvi Crantz, *Bunium carvi*
M.Bieb., *Pimpinella carvi* Jess., *Sium
carvi* Bernh.
♦ caraway, caraway seed, kumina,
common caraway
♦ Weed, Noxious Weed, Naturalised,
Cultivation Escape, Casual Alien
♦ 8, 21, 24, 40, 70, 87, 88, 101, 121, 159,
161, 212, 218, 229, 272, 280
♦ a/bH, cultivated, herbal, toxic.
Origin: Eurasia.

Carum gairdneri (Hook. & Arn.) Gray
Apiaceae
= *Perideridia gairdneri* (Hook. & Arn.)
Mathias ssp. *gairdneri* (NoR)
♦ Quarantine Weed
♦ 220

Carya aquatica (Michx.f.) Nutt.
Juglandaceae
♦ water hickory, swamp hickory
♦ Weed
♦ 87, 88, 218
♦ T, promoted, herbal.

Carya cordiformis (Wangenh.) C.Koch
Juglandaceae
♦ bitternut hickory, bitter nut
♦ Weed
♦ 87, 88, 218

♦ T, cultivated, herbal.

Carya floridana Sarg.
Juglandaceae
♦ scrub hickory
♦ Weed
♦ 87, 88, 218
♦ T, promoted.

Carya glabra (Mill.) Sweet
Juglandaceae
Carya porcina Nutt., *Juglans glabra* Mill.
♦ pignut hickory, redheart hickory,
pignut
♦ Weed
♦ 87, 88, 218
♦ T, promoted, herbal.

Carya illinoensis (Wangenh.) K.Koch
Juglandaceae
= *Carya illinoinensis* (Wagenh.) K.Koch
♦ pecan
♦ Weed
♦ 87, 88, 218
♦ T, promoted, herbal.

Carya illinoinensis (Wagenh.) K.Koch
Juglandaceae
Carya illinoensis (Wangenh.) K.Koch
(see), *Carya olivaeformis* Nutt., *Carya
pecan* (Marshall) Engl. & Graebn.,
Hicoria pecan (Marshall) Britton, *Juglans
pecan* Marshall, *Juglans illinoensis*
Wangenh.
♦ pecan
♦ Introduced
♦ 228
♦ arid, cultivated, herbal.

Carya laciniosa (Michx.f.) Loudon
Juglandaceae
Carya sulcata Nutt., *Juglans laciniosa*
F.Michx.
♦ shellbark hickory
♦ Weed
♦ 87, 88, 218
♦ T, cultivated, herbal.

Carya × lecontei Little
Juglandaceae
= *Carya aquatica* (Michx.f.) Nutt. ×
Carya illinoinensis (Wangenh.) C.Koch
♦ bitter pecan
♦ Weed
♦ 88, 218

Carya leiodermis Sarg.
Juglandaceae
= *Carya glabra* (Mill.) Sweet var. *hirsuta*
(Ashe) Ashe (NoR)
♦ swamp hickory
♦ Weed
♦ 87, 88, 218

Carya ovata (Mill.) C.Koch
Juglandaceae
Carya alba (L.) Nutt. *non* Koch., *Juglans
ovata* Mill.
♦ shagbark hickory
♦ Weed
♦ 87, 88, 218
♦ T, cultivated, herbal.

Carya pallida (Ashe) Engl. & Graebn.
Juglandaceae
♦ sand hickory
♦ Weed

◆ 87, 88, 218
◆ T, promoted, herbal.

Carya texana Buckl.
Juglandaceae
Carya arkansana Sarg., *Carya buckleyi* Durand., *Carya villosa* Schneid.
◆ black hickory
◆ Weed
◆ 87, 88, 218
◆ T, promoted, herbal.

Carya tomentosa (Lam. ex Poir.) Nutt.
Juglandaceae
= *Carya alba* (L.) Nutt. ex Ell.
◆ mockernut hickory
◆ Weed
◆ 87, 88, 218
◆ T, promoted, herbal.

Caryocar brasiliense Camb.
Caryocaraceae
◆ pequi
◆ Weed
◆ 87, 88

Caryopteris × clandonensis A.Simmonds ex anon.
Lamiaceae/Verbenaceae
= *Caryopteris incana* (Thunb. ex Houtt.) Miq. × *Caryopteris mongholica* Bunge
◆ Naturalised
◆ 101
◆ cultivated, herbal.

Caryota mitis Lour.
Arecaceae
Caryota furfuracea Mart., *Thuessinkia speciosa* Miq.
◆ fishtail palm, clustered fishtail palm, Burmese fishtail palm
◆ Weed, Sleeper Weed, Naturalised, Introduced, Garden Escape, Environmental Weed
◆ 3, 39, 86, 93, 101, 155, 179, 228, 230, 247
◆ T, arid, cultivated, herbal, toxic. Origin: India, south-east Asia.

Caryota urens L.
Arecaceae
◆ jaggery palm, single fishtail
◆ Naturalised, Introduced
◆ 39, 101, 230
◆ T, cultivated, herbal, toxic.

Cascabela peruviana (Pers.) Raf.
Apocynaceae
= *Thevetia peruviana* (Pers.) K.Schum.
◆ yellow oleander, thevetia peruviana
◆ Weed, Naturalised, Environmental Weed
◆ 3, 7, 39, 93, 98, 155, 161, 191, 203
◆ cultivated, herbal, toxic.

Cascabela thevetia (L.) Lippold
Apocynaceae
= *Thevetia peruviana* (Pers.) K.Schum.
◆ yellow oleander, lucky nut
◆ Naturalised, Garden Escape, Cultivation Escape
◆ 86
◆ T, cultivated.

Casearia aculeata Jacq.
Flacourtiaceae

◆ rabo de ranton
◆ Weed
◆ 14, 88

Casearia guianensis (Abul.) Urb.
Flacourtiaceae
◆ Guyanese wild coffee
◆ Weed
◆ 14, 88

Casearia hirsuta Sw.
Flacourtiaceae
◆ Weed
◆ 14, 88

Casearia nitida Jacq.
Flacourtiaceae
◆ smooth casearia
◆ Weed, Naturalised
◆ 101, 179

Casearia spinescens (Sw.) Griseb.
Flacourtiaceae
◆ Weed
◆ 14

Casearia sylvestris Sw.
Flacourtiaceae
◆ crackopen
◆ Weed
◆ 14
◆ T, herbal.

Casimiroa edulis La Llave & Lex.
Rutaceae
◆ white sapote, sapote
◆ Naturalised
◆ 39, 86
◆ S/T, cultivated, herbal, toxic. Origin: Central America.

Cassia absus L.
Fabaceae/Caesalpiniaceae
= *Chamaecrista absus* (L.) Irwin & Barneby
◆ pig's senna, tropical sensitive pea
◆ Weed, Native Weed
◆ 50, 87, 88, 121, 276
◆ aH, herbal, toxic. Origin: southern Africa.

Cassia aculeata Pohl.
Fabaceae/Caesalpiniaceae
◆ Weed
◆ 14

Cassia afrofistula Brenan
Fabaceae/Caesalpiniaceae
◆ Kenyan shower
◆ Naturalised
◆ 101

Cassia alata L.
Fabaceae/Caesalpiniaceae
= *Senna alata* (L.) Roxb.
◆ candle bush, ringworm shrub, yellowtop weed, candelabra bush, Roman candle tree, bakau plant, tirakahonuki, andadose, candlestick bush
◆ Weed, Quarantine Weed, Noxious Weed, Introduced, Environmental Weed
◆ 3, 39, 62, 87, 88, 147, 155, 203, 209, 220, 230, 276
◆ S, cultivated, herbal, toxic. Origin: South America.

Cassia auriculata L.
Fabaceae/Caesalpiniaceae
= *Senna auriculata* (L.) Roxb.
◆ avaram
◆ Weed, Quarantine Weed
◆ 22, 76, 87, 88, 203, 220
◆ S, cultivated, herbal.

Cassia barclayana Sweet
Fabaceae/Caesalpiniaceae
◆ pepperleaf senna
◆ Weed, Naturalised
◆ 39, 98, 203
◆ cultivated, toxic. Origin: Australia.

Cassia bauhinioides A.Gray
Fabaceae/Caesalpiniaceae
= *Senna bauhinioides* (Gray) Irwin & Barneby
◆ twoleaf desert senna
◆ Weed
◆ 87, 88, 218
◆ herbal.

Cassia bicapsularis L.
Fabaceae/Caesalpiniaceae
= *Senna bicapsularis* (L.) Roxb.
◆ Christmas bush
◆ Weed, Environmental Weed
◆ 87, 88, 121, 257, 290
◆ S/T, cultivated, herbal. Origin: South America.

Cassia biflora L.
Fabaceae/Caesalpiniaceae
ambiguous name could apply equally well to *Senna pallida* (Vahl) Irwin & Barneby
◆ Weed
◆ 157
◆ S/T, herbal.

Cassia brewsterii F.Muell.
Fabaceae/Caesalpiniaceae
◆ Weed
◆ 87, 88

Cassia capensis Thunb. var. *flavescens* Thunb.
Fabaceae/Caesalpiniaceae
◆ Native Weed
◆ 121
◆ pH. Origin: southern Africa.

Cassia circinnata Benth.
Fabaceae/Caesalpiniaceae
◆ Introduced
◆ 228
◆ arid. Origin: Australia.

Cassia coluteoides Collad.
Fabaceae/Caesalpiniaceae
= *Senna pendula* (Humb. & Bonpl. ex Willd.) Irwin & Barneby var. *glabrata* (Willd.) Irwin & Barneby
◆ climbing cassia, Christmas cassia, Christmas senna
◆ Weed, Environmental Weed
◆ 73, 80, 88, 112, 201
◆ cultivated.

Cassia comosa (E.Mey.) Vogel var. *capricornia* Steyaert
Fabaceae/Caesalpiniaceae
◆ Native Weed
◆ 121
◆ pH. Origin: southern Africa.

Cassia corymbosa Lam.
Fabaceae/Caesalpiniaceae
= *Senna corymbosa* (Lam.) Irwin &
Barneby
♦ autumn cassia, buttercup tree,
golden senna, scrambled eggs, flowery
senna
♦ Weed
♦ 80, 121
♦ cultivated, herbal. Origin: South
America.

Cassia didymobotrya Fresen.
Fabaceae/Caesalpiniaceae
= *Senna didymobotrya* (Fresen.) Irwin
& Barneby
♦ peanut butter cassia, wild senna
♦ Weed, Naturalised, Environmental
Weed
♦ 39, 98, 121, 203, 279
♦ S/T, cultivated, herbal, toxic.

Cassia diffusa DC.
Fabaceae/Caesalpiniaceae
♦ Weed
♦ 14

Cassia diphylla L.
Fabaceae/Caesalpiniaceae
= *Chamaecrista diphylla* (L.) Greene
♦ Weed
♦ 14
♦ herbal.

Cassia emarginata L.
Fabaceae/Caesalpiniaceae
= *Senna bicapsularis* (L.) Roxb.
♦ Introduced
♦ 228
♦ arid, cultivated, herbal.

Cassia eremophila A.Cunn. ex Vog.
Fabaceae/Caesalpiniaceae
♦ desert cassia
♦ Weed
♦ 87, 88
♦ cultivated, herbal.

Cassia fasciculata Michx.
Fabaceae/Caesalpiniaceae
= *Chamaecrista fasciculata* (Michx.)
Greene
♦ partridge pea
♦ Weed
♦ 23, 84, 87, 88, 161, 207, 218, 247
♦ herbal, toxic. Origin: North America.

Cassia fistula L.
Fabaceae/Caesalpiniaceae
♦ golden shower, golden rain tree,
Indian laburnum, purging cassia
♦ Weed, Naturalised, Introduced,
Garden Escape, Environmental Weed,
Cultivation Escape
♦ 7, 32, 39, 86, 87,
88, 93, 98, 101, 161, 179, 203, 228, 230,
261
♦ T, arid, cultivated, herbal, toxic.
Origin: south-east Asia.

Cassia floribunda Cav.
Fabaceae/Caesalpiniaceae
= *Senna × floribunda* (Cav.) Irwin &
Barneby
♦ smooth cassia, arsenic bush, pod
bush

♦ Weed, Environmental Weed
♦ 39, 73, 87, 88, 121, 279
♦ pS, cultivated, herbal, toxic. Origin:
South America.

Cassia grandis L.f.
Fabaceae/Caesalpiniaceae
Bactyrilobium molle (Vahl) Schrad.,
Cassia brasiliana sensu Collad., *Cassia
brasiliana* Collad. var. *tomentosa sensu*
Miq., *Cassia brasiliensis* Buc'hoz, *Cassia
grandis sensu* Benth., *Cassia mollis* Vahl,
Cassia pachycarpa de Wit, *Cathartocarpus
erubescens* Ham., *Cathartocarpus grandis
sensu* G.Don
♦ pink shower
♦ Naturalised, Introduced
♦ 86, 228
♦ T, arid, cultivated, herbal. Origin:
Central and South America.

Cassia hirsuta L.
Fabaceae/Caesalpiniaceae
= *Senna hirsuta* (L.) Irwin & Barneby
var. *hirsuta*
♦ Weed, Naturalised
♦ 14, 39, 87, 88, 121, 257
♦ pS, toxic. Origin: South America.

Cassia hispidula Vahl
Fabaceae/Caesalpiniaceae
= *Chamaecrista hispidula* (Vahl)
H.S.Irwin & Barneby
♦ Weed
♦ 14

Cassia hookeriana Gill. ex Hook & Arn.
Fabaceae/Caesalpiniaceae
= *Senna birostris* (Vogel) H.S.Irwin
& Barneby var. *hookeriana* (Hook.)
H.S.Irwin & Barneby
♦ tanque
♦ Weed
♦ 295
♦ herbal.

Cassia italica (Mill.) Spreng.
Fabaceae/Caesalpiniaceae
♦ Weed
♦ 39, 88, 221
♦ herbal, toxic.

**Cassia italica (Mill.) Spreng. ssp.
arachoides (Burch.) Brenan**
Fabaceae/Caesalpiniaceae
♦ Eland's senna, wild senna
♦ Native Weed
♦ 121
♦ pH, toxic. Origin: southern Africa.

Cassia javanica L.
Fabaceae/Caesalpiniaceae
♦ apple blossom, Java shower, apple
blossom cassia
♦ Naturalised,
Cultivation Escape
♦ 39, 101, 261
♦ cultivated, herbal, toxic. Origin:
tropical Asia.

**Cassia javanica L. var. indochinensis
Gagnep.**
Fabaceae/Caesalpiniaceae
♦ apple blossom
♦ Weed, Naturalised
♦ 101, 179

Cassia lechenaultiana DC.
Fabaceae/Caesalpiniaceae
= *Chamaecrista lechenaultiana* (DC.)
Deg. (NoR) [see *Cassia mimosoides* L.
ssp. *lechenaultiana* Ohashi]
♦ Weed, Introduced
♦ 87, 88, 230, 262
♦ aH.

Cassia leiophylla Vogel
Fabaceae/Caesalpiniaceae
= *Senna leiophylla* (Vogel) H.S.Irwin &
Barneby (NoR)
♦ Weed
♦ 87, 88

Cassia leptocarpa Benth.
Fabaceae/Caesalpiniaceae
= *Senna hirsuta* (L.) Irwin & Barneby
var. *leptocarpa* (Benth.) Irwin &
Barneby (NoR)
♦ slimpod senna
♦ Weed
♦ 87, 88, 218, 255
♦ pH. Origin: tropical America.

Cassia ligustrina L.
Fabaceae/Caesalpiniaceae
= *Senna ligustrina* (L.) Irwin & Barneby
♦ Weed
♦ 87, 88

Cassia marilandica L.
Fabaceae/Caesalpiniaceae
= *Senna marilandica* (L.) Link (NoR)
♦ wild senna, American senna
♦ Weed
♦ 23, 39, 87, 88, 218
♦ herbal, toxic.

Cassia mimosoides L.
Fabaceae/Caesalpiniaceae
= *Chamaecrista mimosoides* (L.) Greene
♦ fishbone cassia
♦ Weed, Native Weed, Introduced
♦ 50, 87, 88, 121, 230, 240, 256
♦ aH, cultivated, herbal. Origin:
southern Africa.

**Cassia mimosoides L. ssp. lechenaultiana
Ohashi**
Fabaceae/Caesalpiniaceae
= *Chamaecrista lechenaultiana* (DC.)
Deg. (NoR) [see *Cassia lechenaultiana*
DC.]
♦ Naturalised
♦ 287

**Cassia mimosoides L. ssp. nomame
Honda**
Fabaceae/Caesalpiniaceae
= *Chamaecrista nomame* (Siebold)
H.Ohashi (NoR) [see *Cassia nomame*
(Sieb.) Honda]
♦ Weed
♦ 286

Cassia newtonii Mend. & Torre
Fabaceae/Caesalpiniaceae
♦ Weed
♦ 87, 88

Cassia nictitans L.
Fabaceae/Caesalpiniaceae
= *Chamaecrista nictitans* (L.) Moench
ssp. *nictitans* var. *nictitans*

♦ tuntokassia, sensitive partridge pea
♦ Weed, Casual Alien
♦ 42, 87, 88, 161, 218
♦ herbal.

Cassia nomame (Sieb.) Honda
Fabaceae/Caesalpiniaceae
= *Chamaecrista nomame* (Siebold)
H.Ohashi (NoR) [see *Cassia mimosoides*
L. ssp. *nomame* Honda]
♦ nomame senna, kawaraketsumei
♦ Weed
♦ 297
♦ herbal.

Cassia obtusa Roxb.
Fabaceae/Caesalpiniaceae
♦ Weed
♦ 87, 88
♦ herbal.

Cassia obtusifolia L.
Fabaceae/Caesalpiniaceae
= *Senna obtusifolia* (L.) Irwin & Barneby
♦ Afrikankassia, sicklepod, Java bean, habucha
♦ Weed, Quarantine Weed, Noxious Weed, Naturalised, Introduced, Casual Alien
♦ 14, 23, 42, 76, 80, 84, 88, 107, 147, 157, 161, 191, 201, 203, 218, 220, 230, 247, 286, 287
♦ pH, herbal, toxic.

Cassia occidentalis L.
Fabaceae/Caesalpiniaceae
= *Senna occidentalis* (L.) Link
♦ coffee senna, stinking weed, wild senna, wild coffee, negro coffee
♦ Weed, Quarantine Weed, Noxious Weed, Naturalised, Introduced, Environmental Weed
♦ 23, 39, 76, 84, 87, 88, 121, 147, 157, 161, 203, 216, 217, 218, 220, 230, 242, 247, 256, 262, 276, 286, 287, 295, 297
♦ H, arid, cultivated, herbal, toxic. Origin: tropical and subtropical America.

Cassia patellaria DC. ex Collad.
Fabaceae/Caesalpiniaceae
= *Chamaecrista nictitans* Moench ssp. *pattelaria* (Collad.) Irwin & Barneby
♦ Weed
♦ 88

Cassia petersiana Bolle
Fabaceae/Caesalpiniaceae
♦ dwarf cassia, eared cassia, monkey pod
♦ Native Weed
♦ 121
♦ S/T. Origin: southern Africa.

Cassia phyllodinea R.Br.
Fabaceae/Caesalpiniaceae
♦ Introduced
♦ 228
♦ arid, cultivated.

Cassia pleurocarpa F.Muell.
Fabaceae/Caesalpiniaceae
♦ Weed
♦ 87, 88
♦ cultivated. Origin: Australia.

Cassia quarrei (Ghesq.) Steyaert
Fabaceae/Caesalpiniaceae
♦ Native Weed
♦ 121
♦ pH. Origin: southern Africa.

Cassia quinquefolia L.
Fabaceae/Caesalpiniaceae
♦ Introduced
♦ 230
♦ S.

Cassia robiniaefolia Benth.
Fabaceae/Caesalpiniaceae
♦ Weed
♦ 14

Cassia rotundifolia Pers.
Fabaceae/Caesalpiniaceae
= *Chamaecrista rotundifolia* (Pers.) Greene
♦ Weed
♦ 14, 39
♦ cultivated, toxic.

Cassia roxburghii DC.
Fabaceae/Caesalpiniaceae
Cassia marginata Roxb. *non* Willd.
♦ Roxburgh's cassia, Ceylon senna
♦ Introduced
♦ 228
♦ arid, cultivated.

Cassia senna L.
Fabaceae/Caesalpiniaceae
= *Senna alexandrina* Mill.
♦ Weed
♦ 87, 88, 221, 242
♦ arid, herbal.

Cassia siamea Lam.
Fabaceae/Caesalpiniaceae
= *Senna siamea* (Lam.) H.S.Irwin & Barneby
♦ cassod tree, msonobali
♦ Weed, Naturalised, Garden Escape, Environmental Weed
♦ 3, 22, 86, 155, 230
♦ T, cultivated, herbal.

Cassia sophera L.
Fabaceae/Caesalpiniaceae
= *Senna sophera* (L.) Roxb. (NoR)
♦ Weed, Naturalised
♦ 87, 88, 121, 287
♦ pS, herbal.

Cassia sophera L. var. purpurea Roxb.
Fabaceae/Caesalpiniaceae
Cassia purpurea Roxb.
♦ Introduced
♦ 228
♦ arid.

Cassia spectabilis DC.
Fabaceae/Caesalpiniaceae
= *Senna spectabilis* (DC.) Irwin & Barneby var. *spectabilis*
♦ Weed
♦ 14, 22
♦ T, cultivated, herbal.

Cassia suffructicosa Koenig
Fabaceae/Caesalpiniaceae
♦ Weed
♦ 22
♦ T.

Cassia surattensis Burm.f.
Fabaceae/Caesalpiniaceae
= *Senna surattensis* (Burm.f.) Irwin & Barneby
♦ Weed, Naturalised
♦ 87, 88, 98, 203
♦ cultivated, herbal.

Cassia tomentosa L.f.
Fabaceae/Caesalpiniaceae
= *Senna multiglandulosa* (Jacq.) Irwin & Barneby
♦ wild senna, shower tree, senna, downy senna
♦ Weed
♦ 80, 87, 88, 121
♦ S/T, cultivated, herbal. Origin: South America.

Cassia tora L.
Fabaceae/Caesalpiniaceae
= *Senna tora* (L.) Roxb.
♦ Java bean, foetid cassia, sicklepod, peanut weed, wild pistache, vao pinati, kaumoce, sirppikassia, sickle senna
♦ Weed, Quarantine Weed, Naturalised, Introduced, Environmental Weed, Casual Alien
♦ 3, 39, 42, 66, 87, 88, 153, 209, 217, 218, 220, 230, 245, 256, 257, 262, 263, 270, 276, 287, 295, 297
♦ aH, herbal, toxic.

Cassia uniflora Mill.
Fabaceae/Caesalpiniaceae
= *Senna uniflora* (Mill.) Irwin & Barneby
♦ Weed
♦ 14, 157
♦ aH.

Cassinia aculeata (Labill.) R.Br.
Asteraceae
♦ dolly bush, common cassinia, dogwood
♦ Native Weed, Naturalised
♦ 269, 280
♦ S, cultivated. Origin: Australia.

Cassinia arcuata R.Br.
Asteraceae
♦ Chinese scrub, sifton bush, Australian tauhinu, biddy bush, Chinese shrub, drooping cassinia, sifting bush, tear shrub
♦ Weed, Quarantine Weed, Noxious Weed, Naturalised, Native Weed
♦ 76, 86, 87, 88, 147, 178, 203, 269
♦ S, arid, cultivated, herbal. Origin: southern Australia.

Cassinia fulvida Hook.f.
Asteraceae
♦ Weed
♦ 87, 88
♦ S, cultivated.

Cassinia laevis R.Br.
Asteraceae
♦ coughbush, curry bush
♦ Weed, Native Weed
♦ 87, 88, 269
♦ S, cultivated, herbal. Origin: Australia.

Cassinia leptophylla (Forst.f.) R.Br.
Asteraceae
♦ tauhinu
♦ Weed, Quarantine Weed
♦ 76, 87, 88, 165, 181, 203, 220
♦ cultivated.

Cassinia quinquefaria R.Br.
Asteraceae
♦ sifting bush, sifton bush, biddy bush
♦ Native Weed
♦ 269
♦ S, cultivated. Origin: Australia.

Cassinopsis ilicifolia (Hochst.) Kuntze
Icacinaceae
♦ Quarantine Weed
♦ 220
♦ cultivated, herbal.

Cassytha filiformis L.
Lauraceae
♦ false dodder, dodder, agasi, mayagas, mai'agas, kohtokot shau, techellela chull, tainoka, devil's gut, mbulabwimo, kodokodshau, dodder laurel
♦ Weed, Quarantine Weed
♦ 3, 6, 12, 14, 87, 88, 121, 203, 220, 276, 297
♦ pH parasitic, herbal, toxic. Origin: obscure.

Cassytha pubescens R.Br.
Lauraceae
♦ Naturalised
♦ 280

Castalis spectabilis (Schltr.) T.Norl.
Asteraceae
Dimorphotheca spectabilis Schltr.
♦ bietou, blue bietou, Transvaal bietou, Transvaal castalis
♦ Native Weed
♦ 121
♦ pH, toxic. Origin: southern Africa.

Castalis tragus (Aiton) Norl.
Asteraceae
♦ cape marigold
♦ Naturalised
♦ 101

Castanea alnifolia Nutt.
Fagaceae
= *Castanea pumila* (L.) Mill. var. *pumila*
♦ trailing chinquapin, bush chinkapin
♦ Weed
♦ 87, 88, 218
♦ S, promoted, herbal.

Castanea crenata Sieb. & Zucc.
Fagaceae
Castanea japonica Blume
♦ Japanese chestnut, kuri
♦ Naturalised
♦ 101
♦ T, promoted.
Origin: east Asia.

Castanea dentata (Marsh) Borkh.
Fagaceae
♦ American chestnut, American sweet chestnut
♦ Weed
♦ 87, 88, 218
♦ T, cultivated, herbal.

Castanea mollissima Blume
Fagaceae
Castanea bungeana Blume, *Castanea duclouxii* Dode, *Castanea hupehensis* Dode
♦ Chinese chestnut
♦ Naturalised
♦ 101
♦ T, promoted, herbal.

Castanea pumila (L.) Mill.
Fagaceae
Fagus pumila L.
♦ Allegheny chinquapin, chinquapin, Virginia chestnut
♦ Weed
♦ 87, 88, 218
♦ S, promoted, herbal.

Castanea sativa Mill.
Fagaceae
Castanea vesca Gaertn., *Castanea vulgaris* Lam., *Fagus castanea* L., *Fagus-Castanea castanea* Weston
♦ sweet chestnut, European chestnut, Spanish chestnut
♦ Naturalised, Casual Alien
♦ 40, 101, 280
♦ T, cultivated, herbal.

Castanopsis chrysophylla (Douglas ex Hook.) A.DC.
Fagaceae
= *Chrysolepis chrysophylla* (Douglas ex Hook.) Hjelmq.
var. *chrysophylla*
♦ golden chinquapin, golden chestnut
♦ Weed
♦ 87, 88, 218
♦ cultivated, herbal.

Castanopsis sempervirens (Kellogg) Dudley ex Merriam
Fagaceae
= *Chrysolepis sempervirens* (Kellogg) Hjelmq.
♦ California chinquapin, bush chinquapin
♦ Weed
♦ 87, 88, 218
♦ herbal.

Castanospermum australe A.Cunn. ex Mudie
Fabaceae/Papilionaceae
♦ Moreton Bay chestnut, bean tree, Australian black bean, puupapu
♦ Weed, Naturalised, Native Weed
♦ 39, 134, 161, 269
♦ T, arid, cultivated, herbal, toxic. Origin: Australia.

Castela coccinea Griseb.
Simaroubaceae
♦ Introduced
♦ 228
♦ arid.

Castellanosia caineana Cárdenas
Cactaceae
Browningia caineana (Cárdenas) D.R.Hunt
♦ Weed, Quarantine Weed
♦ 76, 88, 203, 220

Castilla elastica Sessé
Moraceae
♦ Panama rubber tree, castilla rubber tree, Mexican rubber tree, puluvao
♦ Weed, Sleeper Weed, Naturalised, Environmental Weed
♦ 3, 19, 86, 101, 155, 191, 259, 261
♦ T, herbal. Origin: southern Mexico to northern South America.

Castilleja Mutis ex L.f. spp.
Scrophulariaceae
♦ paintbrush, Indian paintbrush
♦ Weed, Naturalised
♦ 23, 39, 88, 198
♦ herbal, toxic.

Castilleja arvensis Schlecht. & Cham.
Scrophulariaceae
♦ field Indian paintbrush
♦ Weed, Naturalised
♦ 101, 157, 243
♦ aH, arid, herbal.

Castilleja coccinea (L.) Spreng.
Scrophulariaceae
♦ Indian paintbrush, scarlet Indian paintbrush, painted cup
♦ Weed
♦ 8, 87, 88, 218
♦ a/bH, herbal, toxic.

Castilleja exserta (A.A.Heller) Chuang & Heckard
Scrophulariaceae
♦ owl's clover, exserted Indian paintbrush, purple owl's clover
♦ Weed
♦ 161
♦ aH, cultivated.

Castilleja exserta (Heller) Chuang & Heckard ssp. exserta
Scrophulariaceae
Orthocarpus purpurascens Benth. (see)
♦ exserted Indian paintbrush, purple owl's clover
♦ Naturalised
♦ 86, 198
♦ aH. Origin: North America.

Castilleja haydenii (Gray) Cockerell
Scrophulariaceae
♦ Hayden's Indian paintbrush
♦ Quarantine Weed
♦ 220

Castilleja pallida (L.) Spreng.
Scrophulariaceae
♦ vaaleakastilja
♦ Quarantine Weed, Cultivation Escape
♦ 42, 220
♦ cultivated, herbal.

Castilleja sessiliflora Pursh
Scrophulariaceae
♦ downy paintbrush, downy painted cup
♦ Weed
♦ 87, 88, 218
♦ herbal.

Castilleja tenuifolia Benth.
Scrophulariaceae
♦ Weed
♦ 199

Casuarina L. spp.
Casuarinaceae
♦ ironwood Australian pine, sheoak, beefwood, toa
♦ Weed, Cultivation Escape
♦ 80, 116, 233
♦ T, cultivated, herbal.

Casuarina cristata F.Muell. ex Miq.
Casuarinaceae
= *Casuarina lepidophloia* F.Muell.
♦ scalybark casuarina, belah
♦ Weed, Introduced, Environmental Weed
♦ 88, 151, 228, 261
♦ T, arid, cultivated, herbal. Origin: Australia.

Casuarina cunninghamiana Miq.
Casuarinaceae
♦ Australian pine, small cone ironwood, beefwood, melanga, elephant's ear, river sheoak, kasuarisboom
♦ Weed, Noxious Weed, Naturalised, Introduced, Garden Escape, Environmental Weed
♦ 3, 7, 63, 80, 88, 95, 101, 112, 179, 228, 229, 261, 280, 283
♦ T, arid/aqua, cultivated, herbal. Origin: Australia.

Casuarina cunninghamiana Miq. ssp. *cunninghamiana*
Casuarinaceae
♦ Naturalised
♦ 86
♦ cultivated. Origin: Australia.

Casuarina equisetifolia L.
Casuarinaceae
Casuarina litorea L. ex Fosberg & Sachet
♦ Australian pine, common ironwood, common ru, horsetail tree, beefwood, bloodwood, coastal beefwood, mile tree, sheoak, south sea ironwood, beach sheoak, toa, casuarina
♦ Weed, Noxious Weed, Naturalised, Garden Escape, Environmental Weed
♦ 7, 10, 18, 22, 37, 63, 80, 88, 95, 101, 107, 112, 121, 129, 142, 151, 152, 179, 226, 229, 257, 261, 283, 287
♦ T, arid/aqua, cultivated, herbal. Origin: Australia, Melanesia, Polynesia.

Casuarina equisetifolia L. ssp. *incana* (Benth.) L.A.S.Johnson
Casuarinaceae
♦ Naturalised
♦ 86
♦ cultivated. Origin: Australia.

Casuarina glauca Sieb. ex Spreng.
Casuarinaceae
♦ Australian pine, saltmarsh ironwood, suckering casuarina, swamp oak, Brazilian oak, gray sheoak
♦ Weed, Noxious Weed, Naturalised, Introduced, Garden Escape, Environmental Weed
♦ 3, 7, 22, 80, 86, 88, 101, 112, 134, 151, 179, 228, 229, 280
♦ T, arid/aqua, cultivated, herbal. Origin: Australia.

Casuarina lepidophloia F.Muell.
Casuarinaceae
Casuarina cristata F.Muell. ex Miq. (see)
♦ belah
♦ Noxious Weed, Naturalised
♦ 101, 229
♦ aqua, herbal.

Casuarina littoralis L. var. *littoralis*
Casuarinaceae
♦ Introduced
♦ 230
♦ T.

Catabrosa aquatica (L.) P.Beauv.
Poaceae
Colpodium aquaticum Trin., *Molinia aquatica* Wibel
♦ brook grass, water whorl grass, vesihilpi, gramignone di palude, catabrose aquatique, whorl grass
♦ Weed, Naturalised
♦ 159, 272, 300
♦ G, herbal.

Catalepis gracilis Stapf & Stent
Poaceae
♦ gause grass
♦ Native Weed
♦ 121
♦ pG. Origin: southern Africa.

Catalpa bignonioides Walter
Bignoniaceae
Catalpa syringaefolia Bunge, *Bignonia catalpa* L.
♦ southern catalpa, Indian bean, Indian bean tree, common catalpa
♦ Weed, Quarantine Weed, Naturalised
♦ 80, 87, 88, 218, 220, 280, 287
♦ T, cultivated, herbal. Origin: south-eastern North America.

Catalpa longissima (Jacq.) Dum.Cours.
Bignoniaceae
♦ Haitian catalpa, yoke wood
♦ Naturalised
♦ 101
♦ herbal.

Catalpa ovata G.Don.
Bignoniaceae
Catalpa kaempferi Siebold & Zucc.
♦ Chinese catalpa, Indian bean tree
♦ Weed, Naturalised
♦ 80, 101, 287
♦ T, cultivated, herbal. Origin: China.

Catalpa speciosa (Warder) Warder ex Engelm.
Bignoniaceae
♦ northern catalpa, western catalpa, cigar tree, Shawnee wood
♦ Weed
♦ 80, 87, 88, 218
♦ T, cultivated, herbal.

Catapodium marinum (L.) C.E.Hubb.
Poaceae
Desmazeria marina (L.) Druce, *Festuca marina* L.
♦ stiff sand grass, sea fern grass, darnel poa
♦ Naturalised
♦ 86, 176, 198

♦ G. Origin: Mediterranean.

Catapodium rigidum (L.) C.E.Hubb.
Poaceae
Desmazeria rigida (L.) Tutin (see), *Diplachne rigida* (L.) Munro ex Chapm., *Festuca rigida* (L.) Rasp., *Glyceria rigida* (L.) Sm., *Poa rigida* L., *Sclerochloa rigida* Link, *Scleropoa rigida* (L.) Griseb. (see), *Synaphe rigida* (L.) Dulac
♦ fern grass, hardgrass, hard meadowgrass, hard poa
♦ Weed, Naturalised
♦ 7, 9, 86, 91, 111, 121, 134, 176, 198, 243, 253, 280, 300
♦ aG, cultivated, herbal. Origin: Eurasia.

Catha edulis (Vahl) Forssk.
Celastraceae
Celastrus edulis Vahl
♦ khat, Arabian tea, miraa, Abyssinian tea, African tea, bushman's tea, cat, catha, chafta, chat, crafta, djimma, four of paradise, ikwa, ischott, iubulu, kaad, kafta, kat, la salade, liss, liruti, mairongi, m'mke, masbukinja, msuvuti, ol nerra, qat, quat, salahin, seri, Somali tea, tohai, tshut, tumayot, waifo, warfi, warfo
♦ Introduced
♦ 39, 228
♦ arid, cultivated, herbal, toxic.

Catharanthus pusillus G.Don
Apocynaceae
♦ Weed
♦ 87, 88

Catharanthus roseus (L.) G.Don
Apocynaceae
Vinca rosea L.
♦ Madagascar periwinkle, vinca, pink periwinkle
♦ Weed, Naturalised, Introduced, Garden Escape, Environmental Weed, Cultivation Escape
♦ 3, 7, 32, 39, 80, 86, 87, 88, 93, 98, 101, 121, 122, 151, 155, 161, 179, 203, 230, 247, 261, 280, 283, 286, 287
♦ pH/S, cultivated, herbal, toxic. Origin: Madagascar.

Catopheria spicata Benth.
Lamiaceae
♦ Introduced
♦ 38

Catophractes alexandri D.Don
Bignoniaceae
♦ trumpet thorn
♦ Weed, Native Weed
♦ 10, 121
♦ pS. Origin: southern Africa.

Catunaregam spinosa (Thunb.) Tirveng. ssp. *spinosa*
Rubiaceae
Xeromphis obovata (Hochst.) Keay
♦ thicket xeromphis, thorny bone apple
♦ Native Weed
♦ 121
♦ S/T. Origin: southern Africa.

Caucalis bifrons Coss. & Dur. ex Ball
Apiaceae
♦ Weed
♦ 87, 88

Caucalis daucoides L.
Apiaceae
= *Orlaya daucoides* (L.) Greuter
♦ pata de cabra
♦ Weed, Naturalised
♦ 39, 87, 88, 287, 295
♦ toxic.

Caucalis lappula (Weber) Grande
Apiaceae
= *Caucalis platycarpos* L.
♦ Weed
♦ 87, 88

Caucalis latifolia L.
Apiaceae
= *Turgenia latifolia* (L.) Hoffm.
♦ great burr parsley
♦ Weed
♦ 44, 87, 88, 243

Caucalis leptophylla L.
Apiaceae
= *Torilis leptophylla* (L.) Rchb.f.
♦ Weed
♦ 87, 88

Caucalis platycarpos L.
Apiaceae
Caucalis lappula (Weber) Grande (see),
Daucus platycarpos Scop. *non* Celak.
♦ small burr parsley, carrot burr
parsley, purho
♦ Weed, Naturalised, Casual Alien
♦ 42, 44, 70, 87, 88, 94, 101, 237, 243,
253, 272
♦ aH, cultivated, herbal.

**Caulanthus lasiophyllus (Hook. & Arn.)
Payson**
Brassicaceae
= *Guillenia lasiophylla* (Hook. & Arn.)
E.Greene (NoR)
♦ cutleaf thelypody
♦ Weed
♦ 161
♦ herbal.

Caulerpa taxifolia (Vahl) C.Agardh
Caulerpaceae
♦ marine macro algae, notched
caulerpa, feather caulerpa
♦ Weed, Quarantine Weed, Noxious
Weed
♦ 18, 76, 88, 140, 220, 282
♦ algae.

Caulophyllum thalictroides (L.) Michx.
Berberidaceae/Leonticaceae
♦ blue cohosh, papoose root
♦ Weed
♦ 39, 161, 247
♦ pH, promoted, herbal, toxic.

Cayaponia floribunda Cogn.
Cucurbitaceae
♦ taiuia, tajuja, melanciazinha
♦ Weed
♦ 255
♦ pH. Origin: tropical America.

Cayaponia microdonta Blake
Cucurbitaceae
♦ Weed
♦ 157
♦ aC.

Cayaponia podantha Cogn.
Cucurbitaceae
♦ taiuia, tajuja, melanciazinha
♦ Weed
♦ 255
♦ pH. Origin: tropical America.

Caylusea canescens (L.) Walp.
Resedaceae
♦ Weed
♦ 88

Caylusea hexagyna (Forssk.) M.Green
Resedaceae
♦ Weed
♦ 221

**Cayratia ibuensis (Hook.f.) Suess. &
Suess.**
Vitaceae
♦ Weed
♦ 221

**Cayratia japonica (Thunb. ex Murray)
Gagnep.**
Vitaceae
Vitis japonica Thunb. ex Murray,
Causonis japonica (Thunb. ex Murray)
Raf., *Cayratia japonica* var. *dentata*
(Makino) Honda, *Cayratia japonica* var.
taiwaniana Masam., *Cissus japonica*
(Thunb. ex Murray) Willd., *Cissus
japonica* var. *dentata* Makino, *Cissus
tenuifolia* F.Heyne ex Planch., *Columella
japonica* (Thunb. ex Murray) Merr.,
Vitis leucocarpa Hayata, *Vitis tenuifolia*
(F.Heyne ex Planch.) Laws in Hook.f.
♦ bushkiller, sorrelvine, Java plum,
yabugarashi
♦ Weed, Naturalised, Environmental
Weed
♦ 80, 87, 88, 101, 175, 182, 204, 263, 275,
286, 297
♦ pC, cultivated, herbal. Origin:
temperate to subtropical Asia,
Australia.

Cayratia trifolia (L.) Domin
Vitaceae
♦ threeleaf cayratia
♦ Weed, Naturalised
♦ 87, 88, 101
♦ cultivated.

Ceanothus americanus L.
Rhamnaceae
♦ New Jersey tea, red root
♦ Weed
♦ 39, 87, 88
♦ S, cultivated, herbal, toxic.

Ceanothus cordulatus Kellogg
Rhamnaceae
♦ mountain whitehorn, whitethorn
ceanothus
♦ Weed
♦ 87, 88, 218
♦ S, arid, herbal.

Ceanothus cuneatus (Hook.) Nutt.
Rhamnaceae
Rhamnus cuneatus Hook.
♦ wedgeleaf ceanothus, buckbrush
♦ Weed
♦ 87, 88, 218
♦ S, arid, cultivated, herbal.

Ceanothus cyaneus Eastw.
Rhamnaceae
♦ San Diego ceanothus, San Diego
buckbrush
♦ Weed
♦ 87, 88, 218
♦ S, cultivated.

Ceanothus integerrimus Hook. & Arn.
Rhamnaceae
♦ deerbrush ceanothus, deerbrush
♦ Weed
♦ 39, 87, 88, 218
♦ S, arid, cultivated, herbal, toxic.

Ceanothus lemmonii Parry
Rhamnaceae
♦ Lemmon's ceanothus
♦ Weed
♦ 87, 88, 218
♦ S, herbal.

Ceanothus leucodermis Greene
Rhamnaceae
♦ chaparral whitethorn
♦ Weed
♦ 87, 88, 218
♦ S, arid, cultivated, herbal.

Ceanothus megacarpus Nutt.
Rhamnaceae
♦ bigpod ceanothus
♦ Weed
♦ 87, 88, 218
♦ S, cultivated, herbal.

Ceanothus prostratus Benth.
Rhamnaceae
♦ squawcarpet ceanothus
♦ Weed
♦ 87, 88, 218
♦ S, cultivated, herbal.

Ceanothus sanguineus Pursh
Rhamnaceae
Ceanothus oreganus Nutt.
♦ redstem ceanothus, Oregon teatree
♦ Weed
♦ 87, 88, 218
♦ S, promoted, herbal.

Ceanothus sorediatus Hook. & Arn.
Rhamnaceae
♦ jimbrush ceanothus, jimbrush
ceanothus
♦ Weed
♦ 87, 88, 218
♦ cultivated.

Ceanothus spinosus Nutt.
Rhamnaceae
♦ spiny ceanothus, redheart,
greenbark ceanothus
♦ Weed
♦ 87, 88, 218
♦ S, arid, cultivated, herbal.

Ceanothus thyrsiflorus Esch.
Rhamnaceae

- blueblossom ceanothus, blueblossom, blue brush, Californian lilac
- Weed
- 39, 87, 88, 218
- S, cultivated, herbal, toxic.

Ceanothus velutinus **Douglas ex Hook.**
Rhamnaceae
- snowbrush ceanothus
- Weed
- 39, 87, 88, 218
- S, cultivated, herbal, toxic.

Ceanothus velutinus **Douglas ex Hook. var.** *laevigatus* **Torr. & Gray**
Rhamnaceae
= *Ceanothus velutinus* Douglas ex Hook. var. *hookeri* M.C.Johnst. (NoR)
- varnishleaf ceanothus
- Weed
- 218

Cecropia adenopus **Mart. ex Miq.**
Cecropiaceae
- ambay pumpwood
- Weed, Naturalised
- 87, 88, 101

Cecropia ferreyrae **Cuatr.**
Cecropiaceae
- Weed
- 87, 88

Cecropia juranyiana **Alad. Richt.**
Cecropiaceae
= *Cecropia sciadophylla* Mart. (NoR)
- Weed
- 87, 88

Cecropia lyratiloba **Miq.**
Cecropiaceae
- Weed
- 87, 88

Cecropia obtusifolia **Bertol.**
Cecropiaceae
- trumpet tree, guarumo
- Weed, Naturalised
- 3, 80, 101, 191
- herbal.

Cecropia pachystachya **Trécul.**
Cecropiaceae
Ambaiba adenopus (Mart. ex Miq) Kuntze
- ibaituga, embauba
- Weed
- 255
- Origin: Americas.

Cecropia palmata **Willd.**
Cecropiaceae
- Weed
- 179
- T, cultivated, herbal.

Cecropia peltata **L.**
Cecropiaceae
= *Cecropia schreberiana* Miq. (NoR)
- trumpet tree, pumpwood
- Weed
- 6, 22, 37, 80, 87, 88, 191
- T, herbal.

Cecropia tessmannii **Mildbr.**
Cecropiaceae
= *Cecropia membranacea* Trécul (NoR)

- Weed
- 87, 88

Cedrela odorata **L.**
Meliaceae
Cedrela glaziovii C.DC., *Cedrela mexicana* M.Roem.
- West Indian cedar, cigar box cedar, Mexican cedar, Spanish cedar, cedar, Barbados cedar, Central American cedar
- Weed, Sleeper Weed, Naturalised, Introduced, Garden Escape, Environmental Weed
- 3, 22, 86, 121, 132, 152, 155, 191, 228, 257
- T, arid, cultivated, herbal. Origin: tropical America.

Cedrela toona **Roxb. ex Rottl. & Willd.**
Meliaceae
= *Toona ciliata* M.J.Roem.
- Burma cedar, cedrela, Indian mahogany, red cedar, red toon, toon, toon tree, tun tree, toona
- Weed, Environmental Weed
- 121, 279
- T, herbal. Origin: South America.

Cedronella canariensis **(L.) Webb & Berthel.**
Lamiaceae
Cedronella triphylla Moench
- balm of Gilead, canary balm, herb of Gilead
- Weed, Naturalised
- 15, 86, 101, 121, 176, 198, 280
- pS, cultivated, herbal. Origin: Atlantic Ocean islands.

Cedrus deodara **(Roxb. ex D.Don) G.Don f.**
Pinaceae/Abietaceae
- deodar cedar, Himalaya cedar, Indian cedar
- Naturalised
- 101
- T, cultivated, herbal. Origin: Afghanistan to Nepal.

Ceiba aesculifolia **(Kunth) Britton & Baker**
Bombacaceae
- pochote
- Introduced
- 261
- Origin: southern Mexico and Central America.

Ceiba pentandra **(L.) Gaertn.**
Bombacaceae
- kapok, kapok tree, algodon de Manila, silk cottontree
- Weed, Introduced, Environmental Weed
- 3, 22, 107, 191, 230, 259
- T, cultivated, herbal.

Celastrus orbiculata **Thunb.**
Celastraceae
= *Celastrus orbiculatus* Thunb.
- oriental bittersweet, Asiatic bittersweet
- Naturalised, Garden Escape,

Cultivation Escape
- 101, 142
- cultivated, herbal.

Celastrus orbiculatus **Thunb.**
Celastraceae
Celastrus articulatus Thunb., *Celastrus orbiculata* Thunb. (see)
- oriental bittersweet, Asiatic bittersweet
- Weed, Quarantine Weed, Naturalised, Environmental Weed
- 4, 15, 17, 19, 80, 88, 102, 129, 133, 151, 161, 195, 211, 222, 224, 225, 246, 280, 286
- S/T, cultivated, herbal, toxic.

Celastrus scandens **L.**
Celastraceae
- common bittersweet, American bittersweet, climbing bittersweet
- Weed
- 8, 39, 161, 247
- pC, cultivated, herbal, toxic.

Celastrus stephanotiifolius **Makino**
Celastraceae
- Quarantine Weed
- 220

Celmisia sessiliflora **Hook.f.**
Asteraceae
- Quarantine Weed
- 220
- cultivated.

Celosia argentea **L.**
Amaranthaceae
Celosia coccinea L., *Celosia debilis* S.Moore, *Celosia margaritacea* L., *Celosia plumosa* hort. ex Burv.
- quail grass, roponmalek, kukonharja, cock's comb, silver cock's comb, Lagos spinach, chilmil, imarti, kukari, makhamal, sarai, sarwari, sheiba, ngon kai Thai
- Weed, Naturalised, Introduced, Casual Alien
- 7, 12, 23, 42, 66, 86, 87, 88, 93, 98, 170, 203, 209, 228, 230, 235, 239, 242, 243, 262, 275, 276, 286, 287, 297
- aH, arid, cultivated, herbal. Origin: pantropical.

Celosia argentea **L. fo.** *spontanea* **Backer**
Amaranthaceae
- ngornkai dong
- Weed
- 238

Celosia argentea **L. var.** *childsii* **hort.**
Amaranthaceae
- Naturalised
- 287

Celosia argentea **L. var.** *cristata* **(L.) Kuntze**
Amaranthaceae
Celosia cristata L. (see)
- cock's comb, cresta de gallo, moco de pavo
- Garden Escape
- 261
- cultivated, herbal.

Celosia cristata L.
Amaranthaceae
= *Celosia argentea* L. var. *cristata* (L.)
Kuntze)
♦ crested cock's comb, cock's comb
♦ Weed, Naturalised
♦ 86, 87, 88, 101, 203, 262
♦ aH, cultivated, herbal.

Celosia laxa Schum. & Thonn.
Amaranthaceae
= *Celosia trigyna* L.
♦ Weed
♦ 87, 88

Celosia polygonoides Retz.
Amaranthaceae
♦ Weed
♦ 87, 88

Celosia trigyna L.
Amaranthaceae
Celosia digyna Suess., *Celosia laxa*
Schum. & Thonn. (see), *Celosia*
loandensis sensu Suess. ex Brenan,
Celosia melanocarpa Poir., *Celosia*
semperflorens Baker, *Celosia trigyna*
L. var. *convexa* Suess., *Celosia triloba*
E.Mey ex Meissn.
♦ silver spinach, isiHlabe, woolflower,
bambit, belbella, el bueida, lifuluka,
mfungu
♦ Weed, Quarantine Weed,
Naturalised, Native Weed, Introduced
♦ 39, 50, 76, 87, 88, 101, 121, 179, 203,
220, 228
♦ H, arid, herbal, toxic. Origin:
southern Africa.

Celsia cretica L.f.
Scrophulariaceae
♦ Weed, Naturalised
♦ 98, 203

Celsia ramosissima Benth.
Scrophulariaceae
♦ moilikki
♦ Casual Alien
♦ 42

Celtis africana Burm.f.
Celtidaceae/Ulmaceae
♦ African elm, camdeboo, camdeboo
stinkwood, cannibal stinkwood, white
stinkwood, ngehe, kahefu
♦ Native Weed
♦ 121
♦ T, cultivated, herbal. Origin:
southern Africa.

Celtis australis L.
Celtidaceae/Ulmaceae
Celtis lutea Pers.
♦ European hackberry, nettle tree
♦ Weed, Quarantine Weed,
Naturalised, Garden Escape,
Environmental Weed, Cultivation
Escape
♦ 76, 86, 88, 98, 101,
203, 280, 283, 289
♦ T, cultivated, herbal. Origin: Africa.

Celtis laevigata Willd.
Celtidaceae/Ulmaceae
Celtis integrifolia Lam.
♦ sugarberry

♦ Weed
♦ 87, 88, 218
♦ T, cultivated, herbal.

Celtis occidentalis L.
Celtidaceae/Ulmaceae
♦ western hackberry, hackberry,
sugarberry, Chinese elm, common
hackberry
♦ Weed, Quarantine Weed,
Naturalised, Garden Escape,
Environmental Weed, Cultivation
Escape
♦ 73, 76, 86, 87, 88, 98, 155, 203, 218,
283
♦ T, cultivated, herbal. Origin: North
America.

Celtis pallida Torr.
Celtidaceae/Ulmaceae
♦ granjeno, spiny hackberry
♦ Weed
♦ 87, 88, 218
♦ T, cultivated, herbal.

Celtis reticulata Torr.
Celtidaceae/Ulmaceae
= *Celtis laevigata* Willd. var. *reticulata*
(Torr.) L.Benson (NoR)
♦ netleaf hackberry, paloblanco,
western hackberry
♦ Weed
♦ 87, 88, 218
♦ T, promoted, herbal.

Celtis sinensis Pers.
Celtidaceae/Ulmaceae
Celtis japonica Planch.
♦ Chinese elm, celtis, Chinese celtis
♦ Weed, Quarantine Weed,
Naturalised, Garden Escape,
Environmental Weed, Cultivation
Escape, Casual Alien
♦ 3, 73, 76, 86, 88, 98, 155, 201, 203, 280,
283
♦ T, cultivated, herbal. Origin: China,
Korea, Japan.

Celtis spinosa Spreng.
Celtidaceae/Ulmaceae
= *Celtis iguanaea* (Jacq.) Sarg. (NoR)
♦ Weed, Quarantine Weed
♦ 76, 87, 88, 203, 220
♦ cultivated.

Celtis tala Gill. ex Planch.
Celtidaceae/Ulmaceae
= *Celtis ehrenbergiana* (Klotzsch) Liebm.
(NoR)
♦ tala
♦ Weed
♦ 295

Cenchrus L. spp.
Poaceae
♦ sandbur, birdwood grasses, buffel
grass
♦ Weed, Quarantine Weed,
Naturalised
♦ 14, 76, 198, 203, 220
♦ G.

Cenchrus australis R.Br.
Poaceae
♦ Weed
♦ 87, 88

♦ G. Origin: Australia.

Cenchrus barbatus Schumach.
Poaceae
= *Cenchrus biflorus* Roxb.
♦ Weed
♦ 221
♦ G.

Cenchrus biflorus Roxb.
Poaceae
Cenchrus barbatus Schum. (see)
♦ Indian sandbur, burrgrass, gallon's
curse
♦ Weed, Naturalised, Native Weed,
Introduced, Environmental Weed
♦ 7, 86, 87, 88, 90, 93, 98, 101, 121, 185,
203, 228, 242
♦ aG, arid, cultivated, herbal. Origin:
Africa.

Cenchrus brevisetus E.Fourn.
Poaceae
= *Cenchrus echinatus* L.
♦ Weed
♦ 3
♦ G.

Cenchrus brownii Roem. & Schult.
Poaceae
Cenchrus viridis Spreng. (see)
♦ burrgrass, fine bristled burrgrass,
fine burrgrass, slimbristle sandbur,
sandbur
♦ Weed, Quarantine Weed,
Naturalised, Introduced,
Environmental Weed
♦ 3, 51, 76, 86, 87, 88, 90, 93, 98, 107,
121, 157, 158, 203, 230, 243, 286, 287
♦ aG, cultivated, herbal. Origin:
tropical America.

Cenchrus calyculatus Cav.
Poaceae
♦ hillside burrgrass
♦ Weed, Quarantine Weed
♦ 76, 87, 88
♦ G.

Cenchrus capitatus L.
Poaceae
Echinaria capitata (L.) Desf. (see),
Panicastrella capitata (L.) Moench
♦ Weed
♦ 243
♦ G.

Cenchrus ciliaris L.
Poaceae
Pennisetum cenchroides A.Rich.,
Pennisetum ciliare (L.) Link (see),
Pennisetum incomptum Nees ex Steud.,
Cenchrus glaucus C.R.Mudaliar &
Sundararaj
♦ buffel grass, African foxtail, anjan
grass, dhaman, dhaman grass, koluk
katai, zacate buffel
♦ Weed, Naturalised, Native Weed,
Introduced, Environmental Weed
♦ 3, 7, 38, 39, 80, 86, 87, 88, 93, 98, 121,
152, 158, 179, 203, 205, 221, 228, 255
♦ pG, arid, cultivated, herbal, toxic.
Origin: south-west Asia, Africa.

Cenchrus echinatus L.
Poaceae
Cenchrus brevisetus E.Fourn. (see),

Cenchrus cavanillesii Tausch, *Cenchrus crinitus* Mez, *Cenchrus hexaflorus* Blanco, *Cenchrus hillebrandianus* Hitchc., *Cenchrus insularis* Scribn., *Cenchrus lechleri* Steud. ex Lechl., *Cenchrus macrocarpus* Ledeb. ex Steud., *Cenchrus pungens* Kunth (see), *Cenchrus quinquevalvis* Buch.-Ham. ex Wall. (see)
♦ Mossman River grass, burrgrass, southern sandbur, sandbur, herbe a cateaux, motie vihilago, Galland's curse, hedgehog grass, mozote, cadillo
♦ Weed, Noxious Weed, Naturalised, Introduced, Environmental Weed
♦ 3, 6, 7, 26, 30, 34, 35, 55, 84, 86, 87, 88, 90, 93, 98, 107, 147, 157, 161, 186, 199, 203, 205, 209, 218, 229, 230, 236, 237, 239, 241, 243, 245, 249, 255, 257, 263, 269, 272, 274, 276, 281, 286, 287, 290, 295, 297, 300
♦ aG, arid/aqua, cultivated, herbal. Origin: tropical America.

Cenchrus gracillimus **Nash**
Poaceae
♦ slender sandbur, burrgrass
♦ Weed
♦ 88
♦ G.

Cenchrus incertus **M.A.Curtis**
Poaceae
Cenchrus albertsonii Runyon, *Cenchrus humilis* Hitchc., *Cenchrus muricatus* Phil., *Cenchrus parviceps* Shinners, *Cenchrus pauciflorus* Benth. (see), *Cenchrus roseus* E.Fourn., *Cenchrus spinifex* Cav. (see), *Cenchrus strictus* Chapm., *Nastus strictus* (Walter) Lunell
♦ coastal sandbur, mat sandbur, field sandbur, innocent weed, spiny burrgrass, sandbur grass, American burrgrass, gentle Annie, hedgehog grass, longspine sandbur, sandbur
♦ Weed, Quarantine Weed, Noxious Weed, Naturalised, Introduced, Casual Alien
♦ 7, 26, 34, 35, 42, 49, 51, 76, 84, 86, 87, 88, 90, 98, 121, 147, 158, 161, 167, 171, 198, 199, 203, 228, 241, 249, 269, 287, 300
♦ aG, arid, cultivated, herbal. Origin: tropical America.

Cenchrus longispinus **(Hack.) Fern.**
Poaceae
Cenchrus carolinianus Walter, *Cenchrus echinatus* L. fo. *longispina* Hack., *Cenchrus pauciflorus* Benth. var. *longispinus* (Hack.) Jansen & Wacht., *Cenchrus tribuloides* L. (see), *Nastus carolinianus* (Walter) Lunell
♦ spiny burrgrass, innocent weed, American burrgrass, burrgrass, coast sandbur, field burr, field sandbur, gentle Annie, hedgehog grass, longspine sandbur, sandbur
♦ Weed, Quarantine Weed, Noxious Weed, Naturalised, Native Weed
♦ 1, 7, 35, 76, 80, 86, 87, 88, 93, 98, 136, 138, 146, 147, 161, 171, 174, 180, 198, 203, 205, 210, 211, 212, 218, 229, 243, 264, 269

♦ aG, cultivated, herbal. Origin: USA.

Cenchrus macrocephalus **(Döll) Scribn.**
Poaceae
Cenchrus tribuloides L. var. *macrocephalus* Döll
♦ Weed
♦ 87, 88
♦ G.

Cenchrus myosuroides **Kunth**
Poaceae
Cenchrus alopecuroides J.Presl, *Cenchrus ekmanianus* Hitchc., *Cenchrus elliotii* Kunth, *Cenchrus scabridus* Arechav., *Cenchrus setoides* Buckley, *Panicum cenchroides* Elliott, *Pennisetum myosuroides* (Kunth) Spreng., *Pennisetum pungens* Nutt., *Setaria elliottiana* Roem. & Schult.
♦ big sandbur
♦ Weed, Naturalised
♦ 87, 88, 90, 237, 241, 295, 300
♦ pG, arid, herbal.

Cenchrus pauciflorus **Benth.**
Poaceae
= *Cenchrus incertus* M.A.Curtis
♦ field sandbur, spiny burrgrass
♦ Weed
♦ 30, 87, 88, 90, 218, 236, 237, 295
♦ G, herbal.

Cenchrus pennisetiformis **Hochst. & Steud. ex Steud.**
Poaceae
Cenchrus aequiglumis Chiov., *Cenchrus echinoides* Wight ex Steud., *Cenchrus lappaceus* Tausch, *Pennisetum cenchroides* A.Rich var. *echinoides* (Hochst. & Steud. ex Steud.) Hook.f.
♦ Weed, Naturalised, Introduced, Environmental Weed
♦ 86, 221, 228
♦ G, arid, cultivated. Origin: east Africa, Arabia, India.

Cenchrus pilosus **Kunth**
Poaceae
Anthephora hermaphrodita (L.) Kuntze (see), *Cenchrus pallidus* E.Fourn., *Tripsacum hermaphroditum* L.
♦ Weed
♦ 87, 88
♦ G.

Cenchrus pungens **Kunth**
Poaceae
= *Cenchrus echinatus* L.
♦ Weed
♦ 3
♦ G.

Cenchrus quinquevalvis **Buch.-Ham. ex Wall.**
Poaceae
= *Cenchrus echinatus* L.
♦ Weed
♦ 3
♦ G.

Cenchrus setiger **Vahl**
Poaceae
= *Cenchrus setigerus* M.Vahl
♦ birdwood grass
♦ Weed, Naturalised, Environmental Weed

♦ 86, 98, 203
♦ G, cultivated. Origin: tropical Africa, Arabia, India.

Cenchrus setigerus **Vahl**
Poaceae
Cenchrus biflorus Hook.f., *Cenchrus bulbifer* Hochst. ex Boiss., *Cenchrus ciliaris* L. var. *setigerus* (Vahl) Maire & Weiler, *Cenchrus montanus* Nees ex Royle, *Cenchrus setiger* M.Vahl (see), *Cenchrus schimperi* Hochst. & Steud. ex Steud., *Cenchrus triflorus* Roxb. ex Aitch., *Cenchrus tripsacoides* R.Br., *Cenchrus uniflorus* Ehrenb. ex Boiss., *Pennisetum vahlii* Kunth
♦ cow sandbur, birdwood grass, dhaman, anjan
♦ Weed, Naturalised, Introduced
♦ 7, 93, 98, 205, 221, 228
♦ pG, arid, cultivated, herbal.

Cenchrus setosus **Sw.**
Poaceae
= *Pennisetum setosum* (Sw.) L.Rich
♦ Weed
♦ 3
♦ G.

Cenchrus spinifex **Cav.**
Poaceae
= *Cenchrus incertus* M.A.Curtis
♦ southern sandbur, field sandbur, coastal sandbur
♦ Noxious Weed
♦ 229
♦ G.

Cenchrus tribuloides **L.**
Poaceae
♦ dune sandbur, longspine sandbur, sanddune sandbur
♦ Weed, Naturalised
♦ 23, 87, 88, 90, 287
♦ G, arid, cultivated, herbal.

Cenchrus viridis **Spreng.**
Poaceae
= *Cenchrus brownii* Roem. & Schult.
♦ Weed
♦ 3
♦ G.

Cenia turbinata **(L.) Pers.**
Asteraceae
= *Cotula turbinata* L.
♦ daisy, goose daisy, mayweed
♦ Weed, Native Weed
♦ 88, 121
♦ H, cultivated. Origin: southern Africa.

Cenolophium denudatum **(Hornem.) Tutin**
Apiaceae
♦ tuholatva
♦ Casual Alien
♦ 42

Centaurea **L. spp.**
Asteraceae
♦ knapweed, star thistle, cornflower, sweet sultan
♦ Weed, Naturalised, Environmental Weed
♦ 18, 88, 151, 195, 198, 221, 237, 272
♦ herbal.

Centaurea acaulis L.
Asteraceae
♦ Weed, Quarantine Weed
♦ 76, 87, 88, 203, 220
♦ pH, promoted, herbal.

Centaurea aegyptiaca L.
Asteraceae
♦ Weed
♦ 221

Centaurea alpestris Hegetschw.
Asteraceae
Centaurea scabiosa L. ssp. *alpestris*
(Hegetschw.) Nyman, *Centaurea alpina*
auct. non L.
♦ alppikaunokki
♦ Casual Alien
♦ 42

Centaurea americana Nutt.
Asteraceae
♦ basketflower, American star thistle,
American basketflower, American
knapweed, thornless thistle
♦ Weed
♦ 87, 88, 161
♦ aH, herbal, toxic.

Centaurea araneosa Boiss.
Asteraceae
♦ Weed
♦ 221

Centaurea aspera L.
Asteraceae
♦ rough star thistle
♦ Weed, Naturalised
♦ 86, 87, 88, 98, 101, 203, 253
♦ cultivated, herbal. Origin: south-
west Europe.

Centaurea biebersteinii DC.
Asteraceae
Acosta maculosa auct. non Holub,
Centaurea maculosa auct. non Lam.
♦ knapweed, spotted knapweed
♦ Weed, Noxious Weed, Naturalised,
Introduced
♦ 1, 26, 48, 80, 88, 94, 101, 105, 130, 139,
174, 222, 229
♦ aH. Origin: Eurasia.

Centaurea bovina Velen.
Asteraceae
♦ pasture knapweed
♦ Naturalised
♦ 101

Centaurea calcitrapa L.
Asteraceae
♦ purple star thistle, red star thistle,
star thistle, caltrop, maize thorn
♦ Weed, Noxious Weed, Naturalised,
Garden Escape, Environmental Weed,
Casual Alien
♦ 1, 7, 26, 34, 35, 39, 40, 42, 70, 72, 78,
80, 86, 87, 88, 94, 98, 101, 115, 116, 121,
130, 147, 161, 165, 176, 185, 198, 203,
218, 221, 229, 231, 236, 237, 241, 243,
269, 272, 280, 287, 295, 300
♦ a/pH, cultivated, herbal, toxic.
Origin: Mediterranean, west Asia.

Centaurea calcitrapoides L.
Asteraceae

♦ smallhead star thistle
♦ Naturalised
♦ 101

Centaurea cineraria L.
Asteraceae
♦ dusty miller
♦ Weed, Naturalised
♦ 86, 98, 101, 203, 280
♦ cultivated. Origin: Italy.

Centaurea crupinoides Desf.
Asteraceae
= *Volutaria crupinoides* (Desf.) Maire
(NoR)
♦ Weed
♦ 221

Centaurea cyanoides Berger
Asteraceae
♦ Weed, Quarantine Weed,
Naturalised
♦ 7, 76, 88, 98, 203

Centaurea cyanus L.
Asteraceae
Cyanus arvensis Moench
♦ batchelor's buttons, bluebottle,
cornflower, blue centaurea, garden
cornflower, hurtsickle
♦ Weed, Naturalised, Garden Escape,
Environmental Weed, Cultivation
Escape
♦ 23, 34, 44, 51, 70, 80, 86, 87, 88, 94, 98,
101, 102, 121, 136, 142, 151, 158, 161,
167, 179, 180, 203, 212, 218, 236, 237,
243, 252, 253, 272, 280, 287, 295
♦ aH, cultivated, herbal. Origin:
Eurasia.

Centaurea dealbata Willd.
Asteraceae
♦ whitewash cornflower
♦ Weed, Naturalised
♦ 80, 101
♦ cultivated, herbal.

Centaurea debeauxii Gren. & Godr.
Asteraceae
♦ meadow knapweed
♦ Weed, Naturalised
♦ 101, 161

**Centaurea debeauxii Gren. & Godr. ssp.
thuillieri Dostál**
Asteraceae
Centaurea pratensis Thuill. *non* Salisb.
(see)
♦ meadow knapweed
♦ Noxious Weed, Naturalised
♦ 101, 229

Centaurea depressa Bieb.
Asteraceae
Centaurea pygmaea Hoffm.
♦ kääpiökaunokki
♦ Weed, Quarantine Weed,
Naturalised, Casual Alien
♦ 42, 76, 87, 88, 101, 203, 220, 243, 272
♦ a/pH, promoted, herbal.

Centaurea diffusa Lam.
Asteraceae
♦ diffuse knapweed, white knapweed,
musk thistle, plumeless thistle,
harakankaunokki
♦ Weed, Quarantine Weed, Noxious

Weed, Naturalised, Introduced,
Environmental Weed, Casual Alien
♦ 1, 21, 23, 26, 35, 36, 40, 42, 48, 52, 76,
80, 87, 88, 101, 130, 136, 138, 139, 141,
146, 151, 161, 162, 203, 212, 218, 219,
220, 229, 264, 267, 272, 294, 295
♦ pH, herbal, toxic. Origin: eastern
Mediterranean, west Asia.

Centaurea diluta Aiton
Asteraceae
♦ North African knapweed, lesser star
thistle
♦ Weed, Naturalised, Introduced,
Casual Alien
♦ 34, 40, 87, 88, 101
♦ aH, herbal.

Centaurea dubia Suter
Asteraceae
= *Centaurea transalpina* Schleich. ex DC.
♦ short fringed knapweed, Wocheiner
knapweed, Vochin knapweed
♦ Weed
♦ 80

Centaurea eriophora L.
Asteraceae
♦ wild sandheath
♦ Weed, Naturalised
♦ 54, 86, 88, 98, 101, 203, 243
♦ Origin: Spain, Portugal.

Centaurea eryngioides Lam.
Asteraceae
♦ Weed
♦ 221

Centaurea glomerata Vahl
Asteraceae
♦ Weed
♦ 221

Centaurea iberica Trev. ex Spreng.
Asteraceae
♦ Iberian knapweed, Iberian star
thistle, pale star thistle
♦ Weed, Quarantine Weed, Noxious
Weed, Introduced
♦ 26, 34, 35, 76, 80, 87, 88, 101, 115, 161,
203, 212, 218, 220, 229, 237, 272
♦ pH, promoted. Origin: Eurasia.

Centaurea jacea L.
Asteraceae
Jacea pratensis Lam.
♦ brown knapweed, brownray
knapweed
♦ Weed, Quarantine Weed, Noxious
Weed, Naturalised, Casual Alien
♦ 1, 23, 40, 70, 76, 80, 86, 86, 87, 88, 98,
101, 146, 146, 161, 176, 203, 218, 220,
229, 272, 280, 287, 295, 300
♦ pH, cultivated,
herbal. Origin: Eurasia.

Centaurea jacea L. × nigra L.
Asteraceae
♦ meadow knapweed
♦ Weed
♦ 1

Centaurea lippii L.
Asteraceae
♦ Weed
♦ 87, 88

Centaurea macrocephala **Puschk. ex Willd.**
Asteraceae
♦ bighead knapweed, keltakaunokki, golden thistle
♦ Weed, Noxious Weed, Naturalised, Cultivation Escape
♦ 1, 42, 80, 88, 101, 139, 146, 161, 229
♦ cultivated, herbal.

Centaurea maculosa **Lam.**
Asteraceae
♦ spotted knapweed, star thistle
♦ Weed, Quarantine Weed, Noxious weed, Naturalised, Introduced, Environmental Weed, Cultivation Escape
♦ 4, 17, 21, 34, 35, 36, 52, 76, 80, 86, 87, 88, 94, 102, 104, 133, 136, 141, 146, 151, 161, 162, 195, 203, 207, 210, 211, 212, 218, 219, 220, 224, 232, 252, 264, 267, 272, 280, 294, 299
♦ pH, cultivated, herbal, toxic. Origin: central Europe, east to central Russia.

Centaurea maroccana **Ball**
Asteraceae
♦ Weed
♦ 87, 88

Centaurea melitensis **L.**
Asteraceae
♦ Maltese star thistle, Napa star thistle, tocalote, Maltese centaury, Maltese cockspur, saucy Jack, Malta thistle
♦ Weed, Noxious Weed, Naturalised, Introduced, Garden Escape, Environmental Weed, Casual Alien
♦ 7, 9, 34, 35, 39, 40, 42, 51, 55, 70, 72, 78, 80, 86, 87, 88, 94, 98, 101, 116, 121, 130, 134, 158, 161, 176, 198, 203, 218, 228, 231, 236, 237, 241, 253, 264, 269, 280, 287, 295, 300
♦ aH, arid, cultivated, herbal, toxic. Origin: Mediterranean.

Centaurea montana **L.**
Asteraceae
Cyanus montanus (L.) Mill.
♦ mountain cornflower, perennial cornflower
♦ Weed, Naturalised, Cultivation Escape, Casual Alien
♦ 40, 42, 79, 80, 87, 88, 101, 136, 280
♦ pH, cultivated, herbal.

Centaurea muricata **L.**
Asteraceae
= *Cyanopsis muricata* (L.) Dostál
♦ Weed
♦ 87, 88

Centaurea napifolia **L.**
Asteraceae
♦ Weed
♦ 87, 88

Centaurea nigra **L.**
Asteraceae
♦ black knapweed, common knapweed, knapweed, lesser knapweed, Spanish buttons, mustakaunokki, hardheads
♦ Weed, Quarantine Weed, Noxious Weed, Naturalised, Introduced, Casual

Alien
♦ 1, 34, 42, 70, 76, 80, 86, 87, 88, 98, 101, 146, 147, 161, 165, 176, 198, 203, 218, 229, 241, 280, 287, 300
♦ pH, cultivated, herbal. Origin: Europe.

Centaurea nigrescens **Willd.**
Asteraceae
Centaurea vochinensis Bernh. ex Rchb. (see)
♦ Wocheiner knapweed, Vochin knapweed, short fringed knapweed, Tyrol knapweed
♦ Weed, Noxious Weed, Naturalised
♦ 1, 17, 80, 88, 98, 101, 161, 203, 229
♦ cultivated, herbal.

Centaurea nigrescens **Willd. ssp. nigrescens**
Asteraceae
♦ Naturalised
♦ 86
♦ Origin: Europe.

Centaurea ornata **Willd.**
Asteraceae
♦ Weed
♦ 70

Centaurea pallescens **Del.**
Asteraceae
♦ pale star thistle
♦ Weed
♦ 87, 88, 115, 221
♦ arid.

Centaurea paniculata **L.**
Asteraceae
♦ Jersey knapweed, panicled knapweed
♦ Weed, Naturalised
♦ 86, 98, 101, 198, 203
♦ Origin: Europe.

Centaurea phrygia **L.**
Asteraceae
Jacea phrygia (L.) Soják
♦ wig knapweed, nurmikaunokki
♦ Weed, Naturalised
♦ 87, 88, 101, 272
♦ herbal.

Centaurea phyllocephala **Boiss.**
Asteraceae
♦ Weed
♦ 87, 88

Centaurea pieris **Pall.**
Asteraceae
♦ Weed
♦ 87, 88

Centaurea × pouzinii **DC.**
Asteraceae
= *Centaurea aspera* L. × *Centaurea calcitrapa* L.
♦ Naturalised
♦ 101
♦ a/pH.

Centaurea praecox **Oliv. & Hiern**
Asteraceae
♦ Weed
♦ 87, 88

Centaurea pratensis **Thuill. *non* Salisb.**
Asteraceae

= *Centaurea debeauxii* Gren. & Godr. ssp. *thuillieri* Dostál
♦ meadow knapweed
♦ Weed, Noxious Weed
♦ 80, 88, 139, 146, 212, 219
♦ herbal.

Centaurea pulchella **Ledeb.**
Asteraceae
♦ Weed
♦ 221

Centaurea pullata **L.**
Asteraceae
♦ Weed
♦ 70

Centaurea repens **L.**
Asteraceae
= *Acroptilon repens* (L.) DC.
♦ Russian knapweed, Russian centaurea
♦ Weed, Noxious Weed, Naturalised
♦ 23, 35, 36, 39, 51, 80, 87, 88, 98, 121, 136, 141, 146, 158, 162, 195, 203, 210, 212, 218, 219, 243, 267
♦ pH, herbal, toxic. Origin: Eurasia.

Centaurea rhenana **Boreau**
Asteraceae
Centaurea stoebe L., *Centaurea paniculata* auct. non L.
♦ reininkaunokki
♦ Weed, Casual Alien
♦ 42, 88, 94, 272
♦ cultivated.

Centaurea rothrockii **Greenm.**
Asteraceae
♦ Rothrock's knapweed, Rothrock's basketflower
♦ Naturalised
♦ 101
♦ a/bH.

Centaurea salmantica **L.**
Asteraceae
= *Mantisalca salmantica* (L.) Briq. & Cav.
♦ Weed, Naturalised, Casual Alien
♦ 40, 51, 87, 88, 287

Centaurea scabiosa **L.**
Asteraceae
♦ greater knapweed, ketokaunokki
♦ Weed, Quarantine Weed, Naturalised
♦ 23, 70, 76, 86, 87, 88, 94, 101, 243, 272, 280
♦ pH, cultivated, herbal. Origin: Eurasia.

Centaurea sinaica **DC.**
Asteraceae
♦ yemrar
♦ Weed
♦ 88, 221
♦ arid.

Centaurea solstitialis **L.**
Asteraceae
Leucantha solstitialis (L.) Á. & D.Löve
♦ yellow star thistle, St. Barnaby's thistle, knapweed, yellow centaurea, golden star thistle, yellow cockspur
♦ Weed, Noxious Weed, Naturalised, Environmental Weed, Casual Alien

♦ 1, 7, 23, 26, 34, 35, 36, 39, 40, 42, 44,
51, 55, 68, 78, 80, 86, 87, 88, 94, 98, 101,
105, 116, 121, 130, 136, 138, 141, 146,
147, 151, 158, 161, 162, 176, 180, 198,
203, 212, 218, 219, 229, 231, 236, 237,
241, 243, 253, 264, 267, 269, 272, 280,
287, 295, 300
♦ aH, arid, cultivated, herbal, toxic.
Origin: Eurasia.

Centaurea squarrosa Willd. nom. illeg.
Asteraceae
= *Centaurea virgata* Lam. ssp. *squarrosa*
(Boiss.) Gugler
♦ squarrose knapweed
♦ Weed, Noxious Weed
♦ 34, 35, 87, 88, 141, 218
♦ pH.

Centaurea stenolepis A.Kern.
Asteraceae
♦ Weed
♦ 272

Centaurea sulphurea Willd.
Asteraceae
♦ Sicilian thistle, Sicilian star thistle,
sulphur knapweed
♦ Weed, Noxious Weed, Naturalised
♦ 26, 35, 87, 88, 101, 229
♦ aH.

Centaurea trichocephala Bieb. ex Willd.
Asteraceae
♦ feather head knapweed
♦ Weed, Noxious Weed, Naturalised
♦ 80, 101, 161, 229

Centaurea triumfettii All.
Asteraceae
Cyanus triumfettii (All.) Dostal,
Centaurea virgata Lam. (see), *Centaurea*
virgata Lam. var. *squarrosa* (Willd.)
Boiss. (see), *Centaurea variegata* Lam.
(see)
♦ squarrose knapweed
♦ Weed, Noxious Weed, Naturalised
♦ 26, 87, 88, 101, 105, 229
♦ cultivated, herbal.

Centaurea tweedieii Hook. & Arn.
Asteraceae
♦ Weed
♦ 87, 88, 237

Centaurea uniflora Turra
Asteraceae
♦ singleflower knapweed
♦ Naturalised
♦ 101
♦ promoted, herbal.

Centaurea uniflora Turra ssp. nervosa
(Willd.) Bonnier & Layens
Asteraceae
♦ singleflower knapweed
♦ Naturalised
♦ 101
♦ cultivated.

Centaurea variegata Lam.
Asteraceae
= *Centaurea triumfettii* All. ssp.
triumfettii (NoR)
♦ Weed
♦ 87, 88

Centaurea verutum L.
Asteraceae
♦ Weed, Casual Alien
♦ 40, 87, 88

Centaurea virgata Lam.
Asteraceae
= *Centaurea triumfettii* All.
♦ squarrose knapweed
♦ Weed
♦ 79, 80, 88, 161, 219

Centaurea virgata Lam. var. squarrosa
(Willd.) Boiss.
Asteraceae
= *Centaurea triumfettii* All.
♦ squarrose knapweed
♦ Weed
♦ 146, 212, 264, 267
♦ pH. Origin: Eurasia.

Centaurea vochinensis Bernh. ex Rchb.
Asteraceae
= *Centaurea nigrescens* Willd.
♦ Vochin knapweed
♦ Weed
♦ 87, 88, 218

Centaurea vulgaris Simonk.
Asteraceae
♦ Weed
♦ 88, 146

Centaurium Hill spp.
Gentianaceae
♦ centaury
♦ Naturalised, Environmental Weed
♦ 198, 289

Centaurium beyrichii (Torr. & Gray ex
Torr.) B.L.Robins.
Gentianaceae
♦ rock centaury, quinine weed
♦ Weed
♦ 39, 161
♦ herbal, toxic.

Centaurium calycosum (Buckl.) Fern.
Gentianaceae
♦ Buckley centaury, Arizona centaury
♦ Weed
♦ 39, 161
♦ aH, herbal, toxic.

Centaurium erythraea Rafn
Gentianaceae
Centaurium umbellatum auct. non Gilib.
(see), *Centaurodes centaurium* Kuntze,
Erythraea centaurium Pers. (see)
♦ common centaury, European
centaury, century, lesser centaury
♦ Weed, Naturalised, Introduced,
Garden Escape, Environmental Weed
♦ 7, 9, 15, 34, 38, 72, 86, 88, 94, 97, 98,
101, 165, 176, 198, 203, 269, 272, 280,
287
♦ a/bH, cultivated, herbal. Origin:
Europe.

Centaurium floribundum (Benth.)
B.L.Robins.
Gentianaceae
= *Centaurium muehlenbergii* (Griseb.)
W.Wight ex Piper (NoR)
♦ June centaury
♦ Weed, Naturalised

♦ 161, 287
♦ herbal, toxic.

Centaurium littorale (Turner) Gilmour
Gentianaceae
♦ seaside centaury, sea centaury,
erythrée littorale, isorantasappi
♦ Naturalised
♦ 300

Centaurium maritimum (L.) Fritsch
Gentianaceae
♦ sea centaury
♦ Weed, Naturalised
♦ 7, 86, 98, 198, 203
♦ Origin: western Europe,
Mediterranean.

Centaurium minus Moench
Gentianaceae
♦ centuria
♦ Weed
♦ 87, 88
♦ herbal.

Centaurium pulchellum (Sw.) Druce
Gentianaceae
Erythraea pulchella Fr., *Gentiana*
pulchella Sw.
♦ lesser centaury, branched centaury
♦ Weed, Naturalised, Introduced
♦ 38, 86, 88, 94, 101, 198, 243, 272, 287,
300
♦ cultivated, herbal. Origin: Eurasia.

Centaurium pulchellum (Sw.) Druce var.
altaicum Kitag. & Hara
Gentianaceae
♦ Altai mountain centaurium
♦ Weed
♦ 297

Centaurium quitense (Kunth) B.L.Rob.
Gentianaceae
♦ Britton's centaury
♦ Weed
♦ 157
♦ aH, arid.

Centaurium spicatum (L.) Fritsch ex
E.Jansen
Gentianaceae
♦ spiked centaury, centaury
♦ Weed, Naturalised, Garden Escape,
Environmental Weed
♦ 72, 86, 88, 98, 101, 198, 203, 221
♦ a/b, cultivated, herbal. Origin:
Eurasia, Australia.

Centaurium tenuiflorum (Hoffmanns. &
Link) Fritsch ex Janch.
Gentianaceae
♦ branched centaury, centaury, slender
centaury
♦ Weed,
Naturalised, Garden Escape,
Environmental Weed
♦ 7, 72, 86, 88, 98, 134, 176, 198, 203,
269, 280
♦ aH, cultivated. Origin: western
Europe, Mediterranean.

Centaurium umbellatum auct. non Gilib.
Gentianaceae
= *Centaurium erythraea* Raf.
♦ centuria zwyczajna
♦ Weed

♦ 243
♦ herbal.

Centella asiatica (L.) Urb.
Apiaceae
Hydrocotyle asiatica L.
♦ Indian pennywort, Asiatic pennywort, marsh pennywort, marsh pepperwort, pennywort, spadeleaf, gotu kola, togo
♦ Weed, Naturalised
♦ 39, 87, 88, 101, 121, 134, 161, 170, 191, 209, 218, 235, 249, 255, 262, 273, 275, 286, 297
♦ pH, aqua, cultivated, herbal, toxic. Origin: unknown pantropical.

Centella erecta (L.f.) Fern.
Apiaceae
♦ erect centella
♦ Weed
♦ 14
♦ herbal.

Centella triflora Nannf.
Apiaceae
♦ Weed
♦ 87, 88

Centipeda cunninghamii (DC.) A.Br. & Asch.
Asteraceae
♦ sneezeweed
♦ Weed, Naturalised
♦ 15, 165, 181, 280
♦ arid, cultivated. Origin: Australia.

Centipeda minima (L.) A.Br. & Asch.
Asteraceae
Artemisia minima L., *Centipeda minuta* (Less.) C.B.Clarke, *Centipeda orbicularis* Lour. (see)
♦ spreading sneezeweed, tokinsou
♦ Weed, Naturalised, Introduced
♦ 13, 87, 88, 101, 170, 204, 228, 263, 273, 274, 275, 286, 297
♦ aH, arid, cultivated, herbal.

Centipeda orbicularis Lour.
Asteraceae
= *Centipeda minima* (L.) A.Br. & Asch.
♦ Weed
♦ 39, 87, 88
♦ herbal, toxic.

Centotheca lappacea (L.) Desv.
Poaceae
Centosteca latifolia (Obs.) Trin.
♦ sefa
♦ Weed
♦ 12, 87, 88
♦ G, herbal. Origin: Asia, Australia.

Centranthera cochinchinensis (Lour.) Merr.
Scrophulariaceae
Centranthera hispida R.Br.
♦ Weed
♦ 13
♦ aH.

Centranthera muticum (H.B.K.) Less.
Scrophulariaceae
♦ Weed
♦ 87, 88

Centranthus DC. spp.
Valerianaceae

♦ valerian, centranth, ostrogowiec
♦ Weed, Naturalised
♦ 198, 272
♦ herbal.

Centranthus macrosiphon Boiss.
Valerianaceae
♦ Naturalised
♦ 7, 9, 86, 98
♦ aH, cultivated. Origin: Spain.

Centranthus ruber (L.) DC.
Valerianaceae
Kentranthus ruber DC.
♦ red valerian, spur valerian, Jupiter's beard
♦ Weed, Naturalised, Garden Escape, Environmental Weed, Cultivation Escape
♦ 7, 15, 34, 40, 70, 78, 80, 87, 88, 98, 101, 116, 146, 165, 176, 203, 280, 290
♦ a/pH, cultivated, herbal. Origin: southwest Europe, Mediterranean.

Centranthus ruber (L.) DC. ssp. ruber
Valerianaceae
♦ red valerian, Jupiter's beard
♦ Naturalised, Environmental Weed
♦ 86, 198
♦ cultivated. Origin: Eurasia.

Centratherum intermedium (Link) Less.
Asteraceae
= *Centratherum punctatum* Cass.
♦ Weed
♦ 262
♦ pH.

Centratherum punctatum Cass.
Asteraceae
Amperephis aristata Kunth, *Amperephis intermedia* Link, *Amperephis mutica* Kunth, *Amperephis pilosa* Cass., *Amperephis pulchella* Cass., *Amphibecis violacea* (Schrank) Schrank, *Baccharoides brachylepis* (Sch.Bip.) Kuntze, *Baccharoides holtonii* (Baker) Kuntze, *Baccharoides muticum* (Kunth) Kuntze, *Baccharoides punctatum* (Cass.) Kuntze, *Baccharoides violaceum* (Schrank) Kuntze, *Centratherum aristatum* Cass., *Centratherum brachylepis* Sch.Bip., *Centratherum brevispinum* Cass., *Centratherum camporum* (Hassl.) Malme, *Centratherum camporum* (Hassl.) Malme var. *longipes* (Hassl.) Malme, *Centratherum holotoni* Baker, *Centratherum intermedium* (Link) Less. (see), *Centratherum longispinum* Cass., *Centratherum muticum* (Kunth) Less., *Centratherum pulchellum* (Cass.) Steud., *Centratherum punctatum* Cass. fo. *brachyphyllum* Hassl., *Centratherum punctatum* fo. *foliosum* (Chod.) Hassl., *Centratherum punctatum* ssp. *camporum* Hassl., *Centratherum punctatum* var. *foliosa* Chod., *Centratherum punctatum* var. *longipes* Hassl., *Centratherum punctatum* var. *parviflorum* Baker, *Centratherum punctatum* var. *viscosissimum* Hassl., *Centratherum violaceum* (Schrank) Gleason, *Crantzia ovata* Vell., *Spixia violacea* Schrank

♦ larkdaisy
♦ Weed, Naturalised, Cultivation Escape
♦ 101, 179, 255, 261
♦ cultivated. Origin: tropical America.

Centratherum punctatum Cass. ssp. punctatum
Asteraceae
♦ Naturalised
♦ 86
♦ Origin: Central and South America, West Indies.

Centrolepis strigosa (R.Br.) Roem. & Schult.
Centrolepidaceae
♦ Naturalised
♦ 280
♦ arid, cultivated.

Centropogon surinamensis (L.) C.Presl
Campanulaceae/Lobeliaceae
= *Centropogon cornutus* (L.) Druce (NoR)
♦ Weed
♦ 87, 88

Centrosema brazilianum Benth.
Fabaceae/Papilionaceae
♦ Weed, Naturalised, Introduced
♦ 98, 203, 228

Centrosema pascuorum Mart. ex Benth.
Fabaceae/Papilionaceae
♦ centurion, centro, pascuorum
♦ Weed, Naturalised, Introduced
♦ 86, 93, 228
♦ arid, cultivated. Origin: Central and tropical South America.

Centrosema plumieri (Turp. & Pers.) Benth.
Fabaceae/Papilionaceae
♦ feefee
♦ Weed, Naturalised
♦ 14, 39, 86, 87, 88, 93, 157, 191
♦ pC, cultivated, herbal, toxic. Origin: Central and tropical South America, Caribbean.

Centrosema pubescens Benth.
Fabaceae/Papilionaceae
Bradburya pubescens (Benth.) Kuntze, *Centrosema virginianum* (L.) Benth. (see)
♦ centro, flor de conchitas
♦ Weed, Naturalised, Introduced, Environmental Weed
♦ 14, 86, 87, 88, 93, 98, 101, 203, 206, 228, 243
♦ arid, cultivated. Origin: Central and tropical South America.

Centrosema sagittatum (Willd.) Brandeg. ex Riley
Fabaceae/Papilionaceae
♦ arrowleaf butterfly pea
♦ Naturalised
♦ 101

Centrosema schiedeanum (Schltdl.) R.J.Williams & R.J.Clem.
Fabaceae/Papilionaceae
Clitoria schiedeana Schltdl.
♦ Introduced
♦ 228
♦ arid. Origin: Central America.

Centrosema schottii **(Millsp.) K.Schum.**
Fabaceae/Papilionaceae
Bradburya schottii (Schum.) Millsp.,
Centrosema haitiense Urb. & Ekman,
Centrosema kermesii Burkart, *Centrosema macranthum* Hoehne
♦ Introduced
♦ 228
♦ arid.

Centrosema virginianum **(L.) Benth.**
Fabaceae/Papilionaceae
= *Centrosema pubescens* Benth.
♦ spurred butterfly pea
♦ Weed, Introduced
♦ 87, 88, 228, 261
♦ arid, cultivated, herbal.

Centunculus minimus **L.**
Primulaceae
= *Anagallis minima* (L.) Krause
♦ chaffweed
♦ Weed, Introduced
♦ 38, 243
♦ aH, arid, herbal.

Cephalandra indica **(Wight & Arn.) Naudin**
Cucurbitaceae
= *Coccinia grandis* (L.) Voigt
♦ Weed
♦ 221
♦ herbal.

Cephalanthus occidentalis **L.**
Rubiaceae
♦ common buttonbush, buttonbush
♦ Weed, Quarantine Weed
♦ 8, 14, 39, 87, 88, 161, 218, 258
♦ S, cultivated, herbal, toxic.

Cephalaria alpina **(L.) Roem. & Schult.**
Dipsacaceae
♦ yellow cephalaria
♦ Weed
♦ 87, 88
♦ cultivated, herbal.

Cephalaria aristata **C.Koch**
Dipsacaceae
♦ otakirahvinkukka
♦ Casual Alien
♦ 42

Cephalaria gigantea **(Ledeb.) Bobrov**
Dipsacaceae
♦ isokirahvinkukka, giant scabious, yellow scabious, Tatarian cephalaria
♦ Naturalised, Cultivation Escape, Casual Alien
♦ 40, 42, 101
♦ cultivated, herbal.

Cephalaria joppensis **(Rchb.) Coult.**
Dipsacaceae
♦ Weed
♦ 111, 243

Cephalaria laevigata **(Waldst. & Kit.) Schrad.**
Dipsacaceae
♦ Weed
♦ 272

Cephalaria leucantha **(L.) Roem. & Schul.**
Dipsacaceae
♦ Weed

♦ 70
♦ cultivated, herbal.

Cephalaria syriaca **(L.) Scrad. ex Roem. & J.A. Schult.**
Dipsacaceae
♦ Syrian cephalaria, Syrian scabious
♦ Weed, Quarantine Weed, Naturalised
♦ 76, 87, 88, 94, 101, 115, 203, 220, 221, 243, 272
♦ aH, promoted.

Cephalaria transsylvanica **(L.) Roem. & Schult.**
Dipsacaceae
♦ kaakonkirahvinkukka
♦ Weed, Cultivation Escape
♦ 42, 272
♦ cultivated, herbal.

Cephalocereus **Pfeiff. spp.**
Cactaceae
♦ Weed, Quarantine Weed, Naturalised
♦ 86, 88, 220
♦ cultivated.

Cephalocereus dybowskii **(Gosselin) Britton & Rose**
Cactaceae
Cereus dybowskii Gosselin
♦ Quarantine Weed
♦ 220
♦ cultivated.

Cephalocereus fluminensis **(Miq.) Britt. & Rose**
Cactaceae
Cereus fluminensis Miq.
♦ Quarantine Weed
♦ 220

Cephalocereus russelianus **Rose**
Cactaceae
♦ Quarantine Weed
♦ 220

Cephalocereus smithianus **Britt. & Rose**
Cactaceae
Cereus smithianus (Britt. & Rose) Werderm., *Monvillea smithiana* (Britt. & Rose) Backeb., *Praecereus euchlorus* (Weber) N.P.Taylor ssp. *smithianus* (Britt. & Rose) N.P.Taylor, *Praecereus smithianus* (Britt. & Rose) Buxb.
♦ Quarantine Weed
♦ 220

Cephalocleistocactus **Ritter spp.**
Cactaceae
♦ Weed, Quarantine Weed
♦ 76, 88, 203

Cephalonoplos segetum **(Bunge) Kitam.**
Asteraceae
♦ Weed
♦ 88, 204, 275
♦ b/pH.

Cephalonoplos setosum **(Willd.) Kitam.**
Asteraceae
♦ Weed
♦ 275
♦ pH.

Cephalophyllum **(A.H.Haw.) N.E.Br. spp.**
Aizoaceae/Mesembryanthemaceae

♦ iceplant
♦ Weed
♦ 116

Cephalostigma perrottetii **A.DC.**
Campanulaceae
Wahlenbergia perrottetii (A.DC.) Thulin
♦ Weed
♦ 87, 88

Cerastium **L. spp.**
Caryophyllaceae
♦ mouse ear chickweed, chickweed
♦ Weed, Noxious Weed, Naturalised
♦ 36, 181, 198, 243, 272
♦ herbal.

Cerastium arvense **L.**
Caryophyllaceae
Cetunculus arvensis Scop., *Leucodonium arvense* (L.) Opiz
♦ field chickweed, field mouse ear, ketohärkki
♦ Weed, Quarantine Weed, Noxious Weed, Naturalised
♦ 42, 70, 76, 86, 87, 88, 94, 136, 161, 162, 203, 218, 220, 241, 243, 272, 280, 299, 300
♦ a/pH, cultivated, herbal.

Cerastium balearicum **F.Herm.**
Caryophyllaceae
♦ Balearic mouse ear chickweed
♦ Naturalised, Environmental Weed
♦ 86, 198
♦ Origin: Mediterranean.

Cerastium biebersteinii **DC.**
Caryophyllaceae
♦ nukkahärkki, boreal chickweed
♦ Naturalised, Cultivation Escape
♦ 42, 101
♦ cultivated, herbal.

Cerastium brachypetalum **Desp. ex Pers.**
Caryophyllaceae
Stellaria brachypetala (Pers.) Jess.
♦ grey mouse ear, gray chickweed, lyhytterähärkki
♦ Weed, Quarantine Weed, Naturalised, Casual Alien
♦ 42, 76, 88, 94, 101, 272
♦ cultivated, herbal.

Cerastium brachypetalum **Desp. ex Pers. ssp.** *brachypetalum*
Caryophyllaceae
♦ gray chickweed
♦ Naturalised
♦ 101

Cerastium brachypetalum **Desp. ex Pers. ssp.** *tauricum* **(Spreng.) Murb.**
Caryophyllaceae
♦ gray chickweed
♦ Naturalised
♦ 101

Cerastium caespitosum **Gilib.**
Caryophyllaceae
= *Cerastium fontanum* Baumg.
♦ Weed
♦ 243, 275
♦ bH.

Cerastium caespitosum **Gilib. var.** *ianthes* **Hara**

Caryophyllaceae
- Weed
- 263
- aH.

Cerastium capense Sond.
Caryophyllaceae
- cape cerastium
- Weed, Native Weed
- 51, 87, 88, 121, 158, 243
- aH, herbal. Origin: southern Africa.

Cerastium cerastioides (L.) Britt.
Caryophyllaceae
Arenaria trigynum (Vill.) Shinners,
Dichodon cerastioides (L.) Reich.,
Provencheria cerastioides (L.) Boivin,
Stellaria cerastioides L.
- starwort mouse ear
- Weed
- 243
- herbal.

Cerastium comatum Desv.
Caryophyllaceae
- levantine mouse ear chickweed
- Naturalised
- 86, 198
- Origin: Mediterranean.

Cerastium dahuricum Fisch. ex Spreng.
Caryophyllaceae
- idänhärkki
- Casual Alien
- 42

Cerastium dichotomum L.
Caryophyllaceae
- forked chickweed
- Weed, Naturalised, Introduced
- 34, 88, 101
- aH.

Cerastium diffusum Pers.
Caryophyllaceae
- sea mouse ear chickweed,
fourstamen chickweed, dark green
mouse ear chickweed, dark green
chickweed, sea mouse ear
- Weed, Naturalised
- 7, 86, 98, 101, 198, 203
- Origin: western Europe.

Cerastium dubium (Bastard) Guepin
Caryophyllaceae
Cerastium anomalum Waldst. & Kit.,
Stellaria viscida Bieb.
- doubtful chickweed
- Weed, Naturalised
- 80, 101, 272

Cerastium fontanum Baumg.
Caryophyllaceae
Cerastium caespitosum Gilib. (see)
- mouse ear chickweed, common
mouse ear
- Weed, Naturalised, Environmental
Weed
- 15, 70, 72, 86, 88, 94, 98, 101, 165, 176,
195, 203, 207, 243, 272, 280
- pH, promoted, herbal. Origin:
Eurasia.

**Cerastium fontanum Baumg. ssp. *vulgare*
(Hartm.) Greuter & Burdet**
Caryophyllaceae

Cerastium triviale Link. (see)
- common mouse ear chickweed, big
chickweed, mouse ear chickweed
- Weed, Naturalised, Introduced
- 34, 101, 134, 174, 176, 198, 253, 280,
300
- pH.

Cerastium glomeratum Thuill.
Caryophyllaceae
Stellaria glomerata (Thuill.) Jess.
- mouse ear chickweed, sticky
chickweed, sticky mouse ear
chickweed, clustered mouse ear
- Weed, Naturalised, Garden Escape,
Environmental Weed, Casual Alien
- 7, 9, 15, 23, 34, 42, 44, 70, 72, 86, 87,
88, 93, 94, 98, 101, 134, 157, 161, 165,
176, 179, 203, 204, 205, 237, 243, 249,
250, 253, 263, 269, 272, 275, 280, 286,
287, 295, 297, 300
- aH, arid, cultivated, herbal. Origin:
Mediterranean, west Asia.

**Cerastium glomeratum Thuill. fo.
apetalum Hegi**
Caryophyllaceae
- Naturalised
- 287

Cerastium glutinosum Fr.
Caryophyllaceae
- Weed
- 87, 88
- herbal.

Cerastium gracile Dufour
Caryophyllaceae
- slender chickweed
- Naturalised
- 101

Cerastium grandiflorum Waldst. & Kit.
Caryophyllaceae
- showy chickweed
- Weed
- 80
- cultivated, herbal.

Cerastium holosteoides Fr.
Caryophyllaceae
Cerastium vulgare Hartm. (see)
- common mouse ear
- Weed, Naturalised
- 87, 88, 241, 243
- aH, promoted, herbal.

**Cerastium holosteoides Fr. var.
angustifolium (Franch.) Mizushima**
Caryophyllaceae
- Weed
- 286

Cerastium ianthes Wall.
Caryophyllaceae
- Weed
- 235

Cerastium indicum Wight & Arn.
Caryophyllaceae
- Weed, Quarantine Weed
- 76, 87, 88, 203, 220

Cerastium nutans Raf.
Caryophyllaceae
- nodding chickweed

- Weed
- 161
- aqua, herbal.

Cerastium perfoliatum L.
Caryophyllaceae
- Weed
- 87, 88, 243, 272

Cerastium pumilum Curtis
Caryophyllaceae
- Curtis's mouse ear chickweed,
dwarf chickweed, dwarf mouse ear,
European chickweed
- Weed, Naturalised
- 70, 86, 101, 198, 272
- herbal. Origin: Mediterranean, west
Asia.

**Cerastium schizopetalum Maxim. var.
ibukiense Ohwi**
Caryophyllaceae
- Naturalised
- 287

Cerastium semidecandrum L.
Caryophyllaceae
Cetunculus semidecandrus (L.) Scop.,
Myosotis semidecandra (L.) Moench,
Stellaria semidecandra (L.) Link
- little mouse ear, fivestamen
chickweed, little mouse ear chickweed,
scarious chickweed, mäkihärkki
- Weed, Naturalised, Environmental
Weed
- 7, 70, 86, 98, 101, 176, 203, 253, 272,
280
- aH, promoted, herbal. Origin:
Mediterranean, west Asia.

Cerastium siculum Guss.
Caryophyllaceae
- dry chickweed
- Weed, Naturalised
- 23, 80, 88, 101

Cerastium strigosum Fr.
Caryophyllaceae
- Weed
- 23, 88

Cerastium tomentosum L.
Caryophyllaceae
Stellaria tomentosa (L.) Link
- dusty miller, snow in summer,
hopeahärkki
- Weed, Naturalised, Cultivation
Escape
- 40, 42, 80, 101, 280
- pH, cultivated, herbal.

Cerastium triviale Link.
Caryophyllaceae
= *Cerastium fontanum* Baumg. ssp.
vulgare (Hartm.) Greuter & Burdet
- common mouse ear chickweed
- Weed
- 44, 87, 88

Cerastium viscosum L.
Caryophyllaceae
- sticky chickweed, annual mouse ear
chickweed
- Weed
- 80, 87, 88, 136, 180, 218, 243
- aH, arid, promoted, herbal.

Cerastium vulcanicum **Schltdl.**
Caryophyllaceae
♦ Weed
♦ 199

Cerastium vulgare **Hartm.**
Caryophyllaceae
♦ Naturalised, Environmental Weed
♦ 86
♦ Origin: Northern Hemisphere.

Cerastium vulgatum **L.**
Caryophyllaceae
♦ mouse ear chickweed, larger mouse
ear chickweed
♦ Weed, Introduced, Environmental
Weed
♦ 21, 23, 49, 52, 80, 87, 88, 136, 161, 191,
204, 210, 211, 218, 243
♦ pH, herbal.

Ceratocarpus arenarius **L.**
Chenopodiaceae
♦ Weed
♦ 87, 88, 243, 272, 275
♦ aH.

Ceratocephala falcatus **(L.) Pers.**
Ranunculaceae
Ceratocephala glaberrima Klok.,
Ranunculus falcatus L. (see)
♦ Weed
♦ 88, 94, 243, 248
♦ herbal.

Ceratocephala testiculata **(Crantz) Roth.**
Ranunculaceae
Ceratocephala falcata auct. non (L.)
Pers., *Ceratocephala orthoceras* DC.,
Ranunculus testiculatus Crantz (see)
♦ curveseed butterwort, testiculate
buttercup, little burr, burr buttercup,
sarvileinikki
♦ Naturalised, Introduced
♦ 42, 101, 174, 272
♦ aH.

Ceratochloa cathartica **(Vahl) Herter**
Poaceae
= *Bromus catharticus* Vahl
♦ rescue brome
♦ Naturalised
♦ 40
♦ G.

Ceratoides arborescens **(Losinsk.) Tsien & C.G.Ma**
Chenopodiaceae
♦ arboresecent ceratoides, winterfat
♦ Weed
♦ 297

Ceratoides latens **(J.F.Gmel.) Rev. & Holmgren**
Chenopodiaceae
Krascheninnikovia ceratoides (L.)
Gueldenst. (see)
♦ Pamirian winterfat
♦ Weed
♦ 275

Ceratonia siliqua **L.**
Fabaceae/Caesalpiniaceae
♦ St. John's bread, carob tree, carob,
carruba
♦ Weed, Naturalised, Introduced,

Garden Escape, Environmental Weed
♦ 86, 98, 101, 182, 203, 228
♦ T, arid, cultivated, herbal. Origin:
Mediterranean, west Asia.

Ceratophyllum **L. spp.**
Ceratophyllaceae
♦ hornwort
♦ Weed, Quarantine Weed,
Naturalised
♦ 76, 86, 88, 220, 258, 272
♦ cultivated, herbal.

Ceratophyllum demersum **L.**
Ceratophyllaceae
Dichotophyllum demersum (L.) Moench
♦ common coontail, hortwort,
common hornweed, rigid hornwort,
coontail, saaraai haang maa
♦ Weed, Sleeper Weed, Quarantine
Weed, Noxious Weed, Naturalised,
Native Weed, Environmental Weed
♦ 15, 23, 86, 87, 88, 121, 147, 159, 161,
165, 186, 191, 203, 208, 209, 218, 221,
225, 239, 246, 255, 258, 269, 272, 274,
275, 280, 286, 297, 299
♦ wH, cultivated, herbal. Origin:
obscure.

Ceratophyllum demersum **L. var.
oxyacanthum Schum.**
Ceratophyllaceae
♦ cola de mono
♦ Weed
♦ 295

Ceratophyllum echinatum **Gray**
Ceratophyllaceae
♦ prickly coontail, spineless hornwort
♦ Weed
♦ 87, 88, 218
♦ aqua, cultivated, herbal.

Ceratophyllum muricatum **Cham.**
Ceratophyllaceae
♦ prickly hornwort
♦ Weed
♦ 272
♦ herbal.

Ceratophyllum submersum **L.**
Ceratophyllaceae
♦ soft hornwort, tropical hornwort
♦ Weed, Quarantine Weed
♦ 87, 88, 272
♦ aqua, cultivated, herbal.

Ceratopteris pteridoides **(Hook.) Hieron**
Pteridaceae/Parkeriaceae
♦ floating fern, floating antlerfern
♦ Weed
♦ 87, 88, 218, 262
♦ wa/bH.

Ceratopteris thalictroides **(L.) Brongn.**
Pteridaceae/Parkeriaceae
Acrostichum thalictroides L., *Ceratopteris
cornuta* Le Prieur, *Ceratopteris siliquosa*
(L.) Copel.
♦ water fern, water sprite, Indian fern,
Sumatra fern, broad leaved Indian
fern, floating staghorn fern, pod fern,
swamp fern, phak kuutnam
♦ Weed, Naturalised, Native Weed
♦ 7, 86, 87, 88, 170, 197, 239, 262, 286,
297

♦ wa/bH, cultivated, herbal. Origin:
pantropical and temperate regions.

Ceratostigma plumbaginoides **Bunge**
Plumbaginaceae
♦ blue leadwood, plumbago
♦ Naturalised
♦ 101
♦ pH, cultivated, herbal.

Ceratostylis **cf. Kaneh. sp.**
Orchidaceae
♦ Introduced
♦ 230
♦ H.

Ceratotheca integribracteata **Engl.**
Pedaliaceae
♦ Weed
♦ 87, 88

Ceratotheca sesamoides **Endl.**
Pedaliaceae
♦ false sesame
♦ Weed
♦ 87, 88
♦ herbal.

Ceratotheca triloba **(Bernh.) E.Mey. ex
Hook.f.**
Pedaliaceae
♦ African wild foxglove, Rhodesian
foxglove, wild foxglove
♦ Weed, Quarantine Weed,
Naturalised, Native Weed
♦ 76, 88, 101, 121, 158, 179
♦ a/bH, cultivated. Origin: southern
Africa.

Cercidiphyllum japonicum **Sieb. & Zucc.
ex J.Hoffm. & H.Schult.**
Cercidiphyllaceae
♦ katsura tree, katsura
♦ Naturalised
♦ 101
♦ T, cultivated, herbal. Origin: east
Asia.

Cercidium australe **I.M.Johnst.**
Fabaceae/Caesalpiniaceae
♦ brea
♦ Weed
♦ 295

Cercidium floridum **Benth. ex Gray**
Fabaceae/Caesalpiniaceae
= *Parkinsonia florida* (Benth. ex Gray)
S.Wats. (NoR)
♦ blue paloverde
♦ Weed, Introduced
♦ 87, 88, 218, 228
♦ arid, cultivated, herbal.

Cercidium macrum **I.M.Johnst.**
Fabaceae/Caesalpiniaceae
= *Parkinsonia texana* (Gray) S.Wats. var.
macra (I.M.Johnst.) Isely (NoR)
♦ border paloverde
♦ Weed
♦ 87, 88, 218

Cercidium microphyllum **(Torr.) Rose &
Johnst.**
Fabaceae/Caesalpiniaceae
= *Parkinsonia microphylla* Torr. (NoR)
♦ paloverde, smallleaf paloverde,
yellow paloverde

♦ Weed, Introduced
♦ 87, 88, 218, 228
♦ S, arid, herbal.

Cercidium praecox (Ruiz & Pav.) Harms
Fabaceae/Caesalpiniaceae
Parkinsonia praecox Ruiz & Pav.
♦ Weed, Introduced
♦ 228, 237
♦ S/T, arid, cultivated, herbal.

Cercidium praecox (Ruiz) Harms ssp. *glaucum* (Cav.) Burkart & Carter
Fabaceae/Caesalpiniaceae
Pomaria glauca Cav.
♦ Introduced
♦ 228

Cercis canadensis L.
Fabaceae/Caesalpiniaceae
♦ eastern redbud, redbud, North America redbud
♦ Weed
♦ 39, 87, 88, 218
♦ T, cultivated, herbal, toxic.

Cercis canadensis L. var. *texensis* (Wats.) M.Hopkins
Fabaceae/Caesalpiniaceae
Cercis occidentalis Torr. (see), *Cercis reniformis* Engelm., *Cercis texensis* S.Watson
♦ Texas redbud
♦ Weed
♦ 218
♦ arid.

Cercis occidentalis Torr.
Fabaceae/Caesalpiniaceae
= *Cercis canadensis* L. var. *texensis* (Wats.) M.Hopkins
♦ western redbud, California redbud
♦ Weed
♦ 87, 88, 218
♦ S, cultivated, herbal.

Cercis siliquastrum L.
Fabaceae/Caesalpiniaceae
Cercis siliquosa St.-Lag., *Siliquastrum orbiculatum* Moench
♦ Judas tree
♦ Naturalised, Introduced, Casual Alien
♦ 86, 228, 280
♦ T, arid, cultivated, herbal. Origin: southern Europe to east Asia.

Cercocarpus betuloides Nutt.
Rosaceae
♦ birchleaf mountain mahogany, mountain mahogany
♦ Weed
♦ 87, 88, 218
♦ S, arid, cultivated, herbal.

Cercocarpus ledifolius Nutt.
Rosaceae
♦ curlleaf mountain mahogany, desert mountain mahogany, curlleaf mountain mahogany
♦ Weed
♦ 87, 88, 218
♦ S, cultivated, herbal.

Cercocarpus montanus Raf.
Rosaceae
♦ mountain mahogany, alderleaf, hairy mountain mahogany, alderleaf mountain mahogany
♦ Weed
♦ 39, 87, 88, 161, 218
♦ S, cultivated, herbal, toxic.

Cereus Mill. spp.
Cactaceae
Cirinosum Neck. spp. (see), *Myrtillocereus* Fric & Kreuz. spp. (see), *Piptanthocereus* (Berg.) Riccob. spp. (see), *Subpilocereus* Backeb. spp. (see)
♦ Weed, Quarantine Weed, Naturalised
♦ 76, 86, 88, 203, 220
♦ cultivated, herbal.

Cereus dayami Speg.
Cactaceae
Piptanthocereus dayamii (Speg.) Ritter
♦ Introduced
♦ 228
♦ arid.

Cereus hexagonus (L.) Mill.
Cactaceae
♦ lady of the night cactus, cacto columnar
♦ Naturalised, Introduced
♦ 101, 261
♦ cultivated, herbal. Origin: Trinidad, Tobago, Venezuela.

Cereus hildmannianus K.Schum.
Cactaceae
Cereus alacriportanus Pfeiff., *Cereus neonesioticus* (Ritter) P.J.Braun, *Cereus peruvianus* auct. non (L.) Mill., *Cereus uruguayanus* Kiesling, *Cereus xanthocarpus* K.Schum., *Piptanthocereus uruguayanus* F.Ritter, *Piptanthocereus xanthoca* F.Ritter
♦ spint tree cactus, queen of the night, Peruvian apple cactus, koubu, eden, hedge cactus
♦ Weed, Quarantine Weed, Noxious Weed, Naturalised, Introduced
♦ 76, 88, 101, 158, 228, 229
♦ arid, cultivated. Origin: South America.

Cereus jamacaru DC.
Cactaceae
Cereus goiasensis (Ritter) P.J.Braun, *Cereus peruvianus* (L.) Mill. misappl., *Piptanthocereus goiasensis* F.Ritter, *Piptanthocereus jamacaru* (DC.) Riccob.
♦ queen of the night, nagblom
♦ Weed, Noxious Weed, Introduced, Cultivation Escape
♦ 10, 63, 88, 95, 228, 283
♦ arid, cultivated, herbal. Origin: South America.

Cereus peruvianus (L.) Mill.
Cactaceae
♦ apple cactus, Peruvian apple, Peruvian apple cactus, queen of the night, koubu, eden
♦ Weed, Noxious Weed, Environmental Weed

♦ 88, 121, 278, 279
♦ S/T, cultivated, herbal. Origin: South America.

Cereus undatus Haw.
Cactaceae
= *Hylocereus undatus* (Haw.) Britt. & Rose
♦ night blooming cereus, night scented glorybower
♦ Weed
♦ 80, 80, 88, 112
♦ herbal.

Cerinthe L. spp.
Boraginaceae
♦ honeywort
♦ Weed
♦ 272

Cerinthe major L.
Boraginaceae
Cerinthe aspera Roth
♦ honeywort
♦ Weed, Naturalised
♦ 70, 87, 88, 94, 272, 280
♦ a/bH, cultivated, herbal.

Cerinthe minor L.
Boraginaceae
Cerinthe acuta Moench
♦ pikkuvahakukka
♦ Weed, Casual Alien
♦ 42, 87, 88, 94, 243, 272
♦ aH, cultivated, herbal.

Ceropteris tartarea (Cav.) Link
Hemionitidaceae
♦ Weed
♦ 87, 88

Ceruana pratensis Forssk.
Asteraceae
♦ Weed
♦ 87, 88, 185, 221

Cestrum aurantiacum Lindl.
Solanaceae
♦ yellow cestrum, orange cestrum, orange jessamine, orange flowering jessamine, oranjesestrum
♦ Weed, Noxious Weed, Naturalised, Garden Escape, Environmental Weed
♦ 39, 63, 86, 87, 88, 95, 98, 101, 121, 161, 165, 203, 225, 246, 268, 269, 278, 279, 280, 283
♦ S/T, cultivated, herbal, toxic. Origin: South America.

Cestrum auriculatum L'Hér.
Solanaceae
♦ Environmental Weed
♦ 39, 257
♦ arid, toxic.

Cestrum corymbosum Schltdl.
Solanaceae
♦ coerana amarela
♦ Weed
♦ 255
♦ S.

Cestrum × cultum Francey
Solanaceae
♦ Casual Alien
♦ 280
♦ cultivated.

Cestrum diurnum L.
Solanaceae
♦ day jasmine, day blooming jessamine, Chinese inkberry, dama de dia, galan de dia
♦ Weed, Noxious Weed, Naturalised, Environmental Weed, Cultivation Escape
♦ 3, 14, 39, 80, 87, 88, 101, 112, 122, 151, 161, 179, 189, 191, 216, 218, 233, 261
♦ cultivated, toxic.

Cestrum diurnum L. var. diurnum
Solanaceae
♦ day jessamine
♦ Naturalised
♦ 101

Cestrum elegans (Brongn. ex Neumann) Schltdl.
Solanaceae
♦ red cestrum, jessamine, cestrum
♦ Weed, Noxious Weed, Naturalised, Garden Escape, Environmental Weed
♦ 15, 72, 73, 86, 88, 155, 161, 198, 225, 246, 280, 283
♦ S, cultivated, toxic. Origin: South America.

Cestrum fasciculatum (Schlecht.) Miers
Solanaceae
Cestrum newellii hort.
♦ early flowering jessamine, red cestrum, early jessamine
♦ Weed, Naturalised, Introduced
♦ 34, 101, 161, 280
♦ S, cultivated, toxic.

Cestrum fasciculatum (Schltdl.) Miers cv. 'Newellii'
Solanaceae
♦ early flowering jessamine, red cestrum, early jessamine
♦ Naturalised
♦ 280

Cestrum intermedium Sendtn.
Solanaceae
♦ peloteira preta
♦ Weed
♦ 255
♦ S. Origin: South America.

Cestrum laevigatum Schlechtd.
Solanaceae
Cestrum axillare Vell., *Cestrum foetidissimum* Dunal, *Cestrum multiflorum* Schott. ex Sendtn., *Cestrum pendulinum* hort. Monsp. ex Dunal, *Cestrum undulatum* Schltdl. var. *otites* Dunal
♦ inkberry, inkberry bush, poison berry, inkbessie
♦ Weed, Noxious Weed, Introduced, Garden Escape
♦ 10, 39, 63, 88, 95, 121, 228, 278, 283
♦ S/T, arid, cultivated, herbal, toxic. Origin: Central America.

Cestrum nocturnum L.
Solanaceae
♦ lady of the night, night flowering cestrum, queen of the night, dama de noche, iki he po, kara
♦ Weed, Sleeper Weed, Naturalised,

Introduced, Garden Escape, Environmental Weed, Cultivation Escape
♦ 3, 39, 54, 80, 86, 88, 101, 107, 152, 161, 179, 189, 191, 225, 230, 233, 246, 261, 280
♦ S, cultivated, herbal, toxic. Origin: West Indies, tropical America.

Cestrum parqui L'Hér.
Solanaceae
♦ green cestrum, willow jasmine, Chilean cestrum, green poison berry, willow leaved jessamine, Chilean flowering jessamine, Chilean jessamine
♦ Weed, Sleeper Weed, Quarantine Weed, Noxious Weed, Naturalised, Garden Escape, Environmental Weed
♦ 3, 39, 63, 76, 80, 86, 87, 88, 98, 101, 147, 155, 161, 191, 198, 203, 218, 225, 236, 237, 246, 269, 278, 280, 283, 290, 295, 296
♦ cultivated, herbal, toxic. Origin: Central America.

Ceterach officinarum Willd.
Aspleniaceae
= *Asplenium ceterach* L. (NoR)
♦ rustyback, rustyback fern
♦ Weed
♦ 39, 272
♦ cultivated, herbal, toxic.

Chaenactis douglasii Hook. & Arn.
Asteraceae
♦ Douglas's dusty maiden, morning brides
♦ Weed
♦ 23, 88
♦ a/pH, promoted, herbal.

Chaenomeles japonica (Thunb.) Lindl. ex Spach
Rosaceae
Chaenomeles maulei Schneid., *Cydonia japonica* Pers.
♦ Maule's quince, dwarf quince
♦ Weed, Naturalised
♦ 80, 101
♦ S, cultivated, herbal.

Chaenomeles speciosa (Sweet) Nak.
Rosaceae
Chaenomeles laganaria Koidzumi, *Cydonia speciosa* Sweet, *Pyrus japonica non* Thunb.
♦ Japanese quince, flowering quince
♦ Weed, Naturalised, Garden Escape
♦ 15, 101, 280
♦ S, cultivated, herbal.

Chaenorhinum minus (L.) Lange
Scrophulariaceae
Antirrhinum minor L., *Chaenorrhinum viscidum* Simonk., *Linaria minor* (L.) Desf. (see)
♦ dwarf snapdragon, small toadflax, kissankita, small snapdragon
♦ Weed, Noxious Weed, Naturalised, Casual Alien
♦ 1, 42, 70, 79, 80, 87, 88, 94, 101, 146, 156, 161, 218, 229, 253, 272, 280, 287, 299
♦ cultivated, herbal.

Chaenorhinum origanifolia (L.) Fourr.
Scrophulariaceae
♦ Casual Alien
♦ 280

Chaerophyllum L. spp.
Apiaceae
♦ chervil
♦ Weed
♦ 272

Chaerophyllum aromaticum L.
Apiaceae
♦ tuoksukirveli
♦ Weed, Casual Alien
♦ 42, 272
♦ cultivated, herbal.

Chaerophyllum aureum L.
Apiaceae
Myrrhis aurea All., *Selinum aureum* E.H.L.Krause
♦ kultakirveli
♦ Casual Alien
♦ 42
♦ cultivated, herbal.

Chaerophyllum bulbosum L.
Apiaceae
Scandix bulbosa Roth
♦ mukulakirveli, turnip root chervil
♦ Weed, Cultivation Escape
♦ 42, 70, 87, 88, 243, 272
♦ bH, cultivated, herbal.

Chaerophyllum hirsutum L.
Apiaceae
Scandix hirsuta Scop.
♦ karvakirveli
♦ Casual Alien
♦ 42
♦ cultivated, herbal.

Chaerophyllum prescottii DC.
Apiaceae
♦ idänkirveli
♦ Naturalised
♦ 42

Chaerophyllum tainturieri Hook.
Apiaceae
♦ hairyfruit chervil
♦ Weed, Casual Alien
♦ 42, 161
♦ herbal.

Chaerophyllum temulentum L.
Apiaceae
= *Chaerophyllum temulum* L.
♦ rough chervil
♦ Weed
♦ 39, 70, 272
♦ toxic.

Chaerophyllum temulum L.
Apiaceae
Chaerophyllum temulentum L. (see)
♦ myrkkykirveli, rough chervil
♦ Naturalised, Casual Alien
♦ 39, 42, 101, 280
♦ cultivated, herbal, toxic. Origin: North Africa, central Europe, Turkey.

Chaerophyllum villarsii Koch
Apiaceae
Myrrhis villarsii Bertoloni, *Selinum villarsii* E.H.L.Krause

♦ itävallankirveli
♦ Casual Alien
♦ 42
♦ cultivated.

Chaetacanthus burchellii Nees
Acanthaceae
♦ Native Weed
♦ 121
♦ S. Origin: southern Africa.

Chaetium bromoides (Presl) Benth.
Poaceae
♦ Weed
♦ 199
♦ G.

Chaetoceros hohnii Graebn. & Wujek
Chaetocerotaceae
♦ Weed
♦ 197
♦ diatom.

Chaetolepis cubensis (A.Rich) Triana
Melastomataceae
♦ Weed
♦ 14

Chaetopogon fasciculatus (Link) Hayek
Poaceae
♦ partaheinä
♦ Casual Alien
♦ 42
♦ G.

Chaffeyopuntia A.V.Fric ex K.Kreuz. spp. not validly published
Cactaceae
= *Opuntia* Mill. spp.
♦ Weed, Quarantine Weed
♦ 76, 88, 220

Chaiturus marrubiastrum (L.) Rchb.
Lamiaceae
Leonurus marrubiastrum L. (see)
♦ lion's tail
♦ Naturalised
♦ 101

Chamaebatia foliolosa Benth.
Rosaceae
♦ bearmat, bearclover, Sierran mountain misery, mountain misery
♦ Weed
♦ 87, 88, 218
♦ S, cultivated, herbal.

Chamaecrista absus (L.) Irwin & Barneby
Fabaceae/Caesalpiniaceae
Cassia absus L. (see)
♦ tropical sensitive pea
♦ Naturalised
♦ 101
♦ arid.

Chamaecrista absus (L.) Irwin & Barneby var. meonandra (Irwin & Barneby) Irwin
Fabaceae/Caesalpiniaceae
♦ tropical sensitive pea
♦ Naturalised
♦ 101

Chamaecrista aeschynomene (DC.) Greene
Fabaceae/Caesalpiniaceae
♦ Weed
♦ 87, 88

Chamaecrista desvauxii (Collad.) Killip
Fabaceae/Caesalpiniaceae
Cassia desvauxii Collad.
♦ mata pasto
♦ Weed
♦ 255
♦ pH. Origin: Brazil.

Chamaecrista flexuosa (L.) Greene
Fabaceae/Caesalpiniaceae
Cassia flexuosa L.
♦ peninha, Texas sensitive pea
♦ Weed
♦ 255
♦ pH. Origin: South America.

Chamaecrista mimosoides (L.) Greene
Fabaceae/Caesalpiniaceae
Cassia mimosoides L. (see)
♦ chamaecrista
♦ Introduced
♦ 228
♦ arid.

Chamaecrista nictitans (L.) Moench ssp. *nictitans* var. *diffusa* (DC.) Irwin & Barneby
Fabaceae/Caesalpiniaceae
♦ partridge pea
♦ Weed
♦ 261

Chamaecrista nictitans (L.) Moench ssp. *pattelaria* (Collad.) Irwin & Barneby
Fabaceae/Caesalpiniaceae
Cassia patellaria DC. ex Collad. (see)
♦ peninha
♦ Weed
♦ 255
♦ pH. Origin: Brazil.

Chamaecrista nictitans (L.) Moench var. *glabrata* (Vogel) Irwin & Barneby
Fabaceae/Caesalpiniaceae
♦ Naturalised
♦ 32

Chamaecrista nictitans (L.) Moench var. *mensalis* Greenm.) Irwin & Barneby
Fabaceae/Caesalpiniaceae
Cassia leptadenia Greenm., *Chamaecrista leptadenia* Cockerell
♦ Introduced
♦ 228
♦ arid.

Chamaecrista nigricans (Vahl) Greene
Fabaceae/Caesalpiniaceae
Cassia nigricans Vahl
♦ Naturalised
♦ 86
♦ Origin: Africa, Middle East.

Chamaecrista pilosa (L.) Greene
Fabaceae/Caesalpiniaceae
♦ hairy sensitive pea
♦ Weed, Naturalised
♦ 101, 179

Chamaecrista rotundifolia (Pers.) Greene
Fabaceae/Caesalpiniaceae
Cassia rotundifolia Pers. (see)
♦ roundleaf sensitive pea, pasto rasteiro
♦ Weed, Naturalised
♦ 32, 86, 134, 179, 255
♦ pH, arid. Origin: Brazil.

Chamaecrista rotundifolia (Pers.) Greene var. *rotundifolia*
Fabaceae/Caesalpiniaceae
♦ Weed
♦ 93, 191

Chamaecrista serpens (L.) Greene
Fabaceae/Caesalpiniaceae
♦ slender sensitive pea
♦ Weed
♦ 179

Chamaecyparis funebris (Endl.) Franco
Cupressaceae
Cupressus funebris Endl. (see)
♦ Chinese weeping chamaecyparis, Chinese weeping cypress, mourning cypress
♦ Introduced
♦ 261
♦ cultivated. Origin: China.

Chamaecyparis lawsoniana (A.Murr.) Parl.
Cupressaceae
♦ Port Orford cedar, Lawson's cypress
♦ Weed, Naturalised
♦ 15, 87, 88, 98, 218, 280
♦ T, cultivated, herbal.

Chamaecyparis pisifera (Siebold & Zucc.) Endl.
Cupressaceae
♦ sawara cypress
♦ Weed
♦ 80
♦ T, cultivated, herbal.

Chamaecyparis thyoides (L.) Britton, Sterns & Pogg.
Cupressaceae
Chamaecyparis sphaeroidea Spach, *Cupressus thyoides* L., *Thuja sphaeroidea* Spreng.
♦ white cypress, Atlantic white cedar
♦ Naturalised
♦ 98
♦ T, cultivated, herbal.

Chamaecytisus austriacus (L.) Link
Fabaceae/Papilionaceae
♦ Austrian clustered broom
♦ Weed
♦ 272

Chamaecytisus heuffelii (Wierzb.) Rothm.
Fabaceae/Papilionaceae
♦ Weed
♦ 272

Chamaecytisus hirsutus (L.) Link
Fabaceae/Papilionaceae
Cytisus hirsutus L.
♦ keltavihma
♦ Weed, Cultivation Escape
♦ 42, 272
♦ cultivated, herbal.

Chamaecytisus palmensis (Christ) Bisby & Nicholls
Fabaceae/Papilionaceae
Chamaecystis palmensis (Christ) Bisby & Nicholls (see), *Cytisus palmensis* (Christ) Hutch. (see), *Cytisus proliferus* L.f. var. *palmensis* Christ
♦ tree lucerne, tagasaste
♦ Weed, Naturalised, Introduced,

Environmental Weed
- 7, 15, 101, 165, 176, 198, 203, 228, 280, 289
- arid, cultivated. Origin: Canary Islands.

Chamaecytisus prolifera (L.f.) Link
Fabaceae/Papilionaceae
= *Chamaecytisus prolifer* (L.f.) Link (NoR) [see *Chamaecytisus proliferus* (L.) Link and *Cytisus proliferus* L.f.]
- escabon
- Naturalised
- 101

Chamaecytisus proliferus (L.f.) Link
Fabaceae/Papilionaceae
= *Chamaecytisus prolifer* (L.f.) Link (NoR) [see *Chamaecytisus prolifera* (L.f.) Link and *Cytisus proliferus* L.f.]
- tree lucerne
- Weed, Environmental Weed, Cultivation Escape
- 34, 39, 296
- S, cultivated, toxic.

Chamaedaphne calyculata (L.) Moench
Ericaceae
Andromeda calyculata L.
- leatherleaf, vaivero, yachitsutsuji
- Weed
- 87, 88, 218
- pS, cultivated, herbal.

Chamaedorea seifrizii Burret
Arecaceae
- Seifriz's chamaedorea
- Weed, Naturalised
- 101, 179
- cultivated.

Chamaelobivia Y.Ito spp.
Cactaceae
- Weed, Quarantine Weed
- 76, 88, 220

Chamaemelum fuscatum (Brot.) Vasc.
Asteraceae
Anthemis fuscata Brot. (see)
- dusky dogfennel, tummasuomusauramo
- Weed, Naturalised, Introduced, Casual Alien
- 34, 42, 70, 88, 94, 101, 250, 253
- aH.

Chamaemelum mixtum (L.) All.
Asteraceae
Anthemis mixta L. (see), *Ormenis mixta* (L.) Dum. (see)
- weedy dogfennel, kampasauramo
- Weed, Quarantine Weed, Naturalised, Casual Alien
- 42, 70, 76, 87, 88, 94, 101, 203, 220, 237, 243, 250, 253, 295, 300
- aH, cultivated.

Chamaemelum nobile (L.) All.
Asteraceae
Anthemis nobilis L. (see), *Matricaria nobilis* Baill., *Ormenis nobilis* (L.) Gay
- common chamomile, chamomile, Roman chamomile, English chamomile, garden chamomile, lawn chamomile
- Weed, Naturalised, Garden Escape,

Casual Alien
- 15, 42, 70, 86, 88, 94, 98, 101, 176, 198, 203, 280
- pH, cultivated, herbal. Origin: western Europe.

Chamaenerion angustifolium (L.) Scop.
Onagraceae
= *Epilobium angustifolium* L.
- fireweed
- Weed
- 253
- herbal.

Chamaeraphis brunoniana (Hook.f.) A.Camus
Poaceae
- Weed
- 87, 88
- G.

Chamaeraphis gracilis Hack.
Poaceae
- Weed
- 87, 88
- G.

Chamaeraphis spinescens Poir.
Poaceae
- Weed
- 87, 88
- G.

Chamaerops humilis L.
Arecaceae
- European fan palm, doum, dwarf palm, palmetto, windmill palm, dwarf fan palm
- Introduced
- 228
- T, arid, cultivated, herbal.

Chamaesaracha coronopus (Dunal) Gray
Solanaceae
Saracha coronopus A.Gray, *Solanum coronopus* Dunal
- small groundcherry, greenleaf five eyes, dwarf ground chervil
- Weed
- 87, 88, 161, 218
- pH, arid, herbal.

Chamaespartium sagittale (L.) P.Gibbs
Fabaceae/Papilionaceae
Genistella sagittalis (L.) Gams
- nuolike
- Weed, Cultivation Escape
- 42, 272
- cultivated.

Chamaesyce albomarginata (Torr. & Gray) Small
Euphorbiaceae
Euphorbia albomarginata Torr. & Gray (see)
- whitemargin spurge, whitemargin sandmat
- Weed
- 243
- a/pH.

Chamaesyce australis (Boiss.) D.C.Hassall
Euphorbiaceae
Euphorbia australis Boiss.
- Naturalised, Native Weed

- 86
- cultivated. Origin: Australia.

Chamaesyce buxifolia (Lam.) Small
Euphorbiaceae
= *Chamaesyce mesembrianthemifolia* (Jacq.) Dugand (NoR)
- Weed
- 14

Chamaesyce cordifolia (Ell.) Small
Euphorbiaceae
- roundleaf spurge, heartleaf sandmat
- Weed
- 161, 249
- herbal.

Chamaesyce cumulicola Small
Euphorbiaceae
- sanddune spurge, coastal dune sandmat
- Weed
- 161, 249
- herbal.

Chamaesyce dioeca (H.B.K.) Millsp.
Euphorbiaceae
- Weed
- 157
- pH.

Chamaesyce drummondii (Boiss.) D.C.Hassall
Euphorbiaceae
Euphorbia drummondii Boiss. (see)
- caustic weed
- Naturalised, Native Weed
- 86, 269
- Origin: Australia.

Chamaesyce hirta (L.) Millsp.
Euphorbiaceae
Euphorbia hirta L. (see), *Euphorbia pilulifera* L. (see)
- garden spurge, asthma plant, red milkweed, pillpod sandmat
- Weed, Naturalised, Environmental Weed
- 6, 14, 39, 86, 88, 157, 158, 243, 249, 255, 257, 261, 273
- aH, arid, herbal, toxic. Origin: tropical America.

Chamaesyce humistrata (Engelm.) Small
Euphorbiaceae
Euphorbia humistrata Engelm. (see)
- prostrate spurge, spreading sandmat
- Weed
- 243
- aH.

Chamaesyce hypericifolia (L.) Millsp.
Euphorbiaceae
Chamaesyce glomerifera Millsp., *Euphorbia glomerifera* (Millsp.) Wheeler (see), *Euphorbia hypericifolia* L. (see)
- graceful sandmat
- Weed
- 14, 88, 157, 261
- pH, herbal.

Chamaesyce hyssopifolia (L.) Small
Euphorbiaceae
Chamaesyce brasiliensis (Lam.) Small, *Euphorbia brasiliensis* Lam. (see), *Euphorbia hyssopifolia* L. (see)
- hyssopleaf sandmat

- Weed, Naturalised
- 14, 86, 88, 157, 243, 249, 255, 261
- aH, herbal. Origin: Americas.

Chamaesyce inaequilatera (Sond.) Sojak
Euphorbiaceae
Euphorbia inaequilatera Sond. (see)
- Weed
- 88

Chamaesyce lasiocarpa (Klotzsch) Arthur
Euphorbiaceae
Euphorbia lasiocarpa Klotzsch (see)
- roadside sandmat
- Weed, Naturalised
- 179, 257
- arid.

Chamaesyce maculata (L.) Small
Euphorbiaceae
Chamaesyce supina (Raf.) Moldenke (see), *Euphorbia maculata* L. (see), *Euphorbia supina* Raf. (see)
- milk purslane, prostrate spurge, prostrate spotted spurge, spurge, milk spurge, spotted spurge, spotted pursley, spotted sandmat
- Weed, Native Weed
- 34, 80, 174, 180, 243, 249
- aH, herbal. Origin: North America.

Chamaesyce micromera (Boiss. ex Engelm.) Woot. & Standl.
Euphorbiaceae
Euphorbia micromera Boiss. ex Engelm. (see)
- Sonoran sandmat, littleleaf spurge
- Weed
- 243
- aH.

Chamaesyce nutans (Lag.) Small
Euphorbiaceae
Euphorbia nutans Lag. (see), *Euphorbia preslii* Guss. (see)
- eyebane, nodding spurge
- Weed, Naturalised
- 34, 86, 207
- aH, toxic. Origin: eastern North America.

Chamaesyce ophthalmica (Pers.) Burch
Euphorbiaceae
Euphorbia hirta L. var. *procumbens* DC. N.E.Br. (see)
- Florida hammock sandmat
- Naturalised
- 257
- arid. Origin: Central America.

Chamaesyce prostrata (Aiton) Small
Euphorbiaceae
Euphorbia chamaesyce L. (see), *Euphorbia prostrata* Ait. (see)
- hairy creeping milkweed, prostrate sandmat
- Weed, Naturalised
- 14, 34, 86, 88, 157, 158, 243, 255, 269
- aH, herbal. Origin: tropical America.

Chamaesyce serpens (Kunth) Small
Euphorbiaceae
Euphorbia serpens H.B.K. (see)
- matted sandmat
- Weed
- 243

- aH, arid, herbal.

Chamaesyce serpyllifolia (Pers.) Small
Euphorbiaceae
- thyme leaved spurge, thymeleaf sandmat
- Weed
- 161
- aH, herbal, toxic.

Chamaesyce serrula (Engelm.) Woot. & Standl.
Euphorbiaceae
Euphorbia serrula Engelm. (see)
- sawtooth spurge, sawtooth sandmat
- Weed
- 161

Chamaesyce supina (Raf.) Moldenke
Euphorbiaceae
= *Chamaesyce maculata* (L.) Small
- Naturalised
- 86

Chamaesyce thymifolia (L.) Millsp.
Euphorbiaceae
Euphorbia thymifolia L. (see)
- red caustic creeper, thyme leaved spurge, gulf sandmat
- Weed
- 273
- herbal.

Chamelaucium uncinatum Schauer
Myrtaceae
- Geraldton wax
- Naturalised, Native Weed
- 7, 86, 280
- cultivated, herbal. Origin: Australia.

Chamerion dodonaei (Vill.) Holub
Onagraceae
Epilobium dodonaei Vill. (see)
- Quarantine Weed
- 220

Chamissoa altissima (Jacq.) Kunth
Amaranthaceae
- false chaff flower
- Weed
- 87, 88
- pC, arid, herbal.

Chamomilla S.F.Gray spp.
Asteraceae
= *Matricaria* L. spp.
- mayweed
- Weed
- 243
- H.

Chamomilla aurea (L.) J.Gay ex Coss. & Kral.
Asteraceae
= *Matricaria aurea* (L.) Sch.Bip.
- Weed
- 88, 94

Chamomilla recutita (L.) Rausch.
Asteraceae
= *Matricaria recutita* L.
- scented mayweed, German chamomile, chamomile, wild chamomile, rumianek, camomilla
- Weed, Naturalised
- 70, 88, 94, 241, 243, 272
- cultivated, herbal.

Chamomilla suaveolens (Pursh) Rydb.
Asteraceae
= *Matricaria matricarioides* (Less.) Porter
- pineapple weed, disc mayweed
- Weed, Naturalised
- 7, 34, 68, 70, 88, 94, 167, 241, 243, 272
- aH, cultivated, herbal.

Chaptalia integerrima (Vell.) Burkart
Asteraceae
Chaptalia integrifolia (Cass.) Baker.
- lingua de vaca
- Weed
- 255
- Origin: Americas.

Chaptalia nutans (L.) Polak
Asteraceae
- heal and draw
- Weed
- 28, 87, 88, 199, 206, 243, 255, 295
- herbal. Origin: Americas.

Chara L. spp.
Characeae
- chara, stonewort, musk grass
- Weed, Quarantine Weed
- 88, 218, 220, 258
- aqua.

Chara braunii S.G.Gmel.
Characeae
- stonewort, musk grass
- Weed
- 88, 204, 286, 297
- aqua.

Chara globularis Thuill.
Characeae
Chara fragilis Desv., *Chara kraussiana* J.Gr. & Steph.
- Weed, Quarantine Weed
- 76, 87, 88, 203

Chara vulgaris L.
Characeae
- Weed
- 87, 88

Chara zeylanica Willd.
Characeae
- chara, stonewort, saarai fai
- Weed, Quarantine Weed
- 76, 87, 88, 191, 203, 204, 239

Charieis heterophylla Cass.
Asteraceae
Charieis neesii Cass.
- Weed
- 23, 88
- cultivated.

Chasmanthe aethiopica (L.) N.E.Br.
Iridaceae
- African cornflag, chasmanthe
- Weed, Environmental Weed
- 87, 88, 155
- cultivated, herbal.

Chasmanthe bicolor (Gasp. ex Ten) N.E.Br.
Iridaceae
- chasmanthe
- Weed, Naturalised
- 40, 54, 86, 88, 280
- cultivated. Origin: South Africa.

Chasmanthe floribunda (Salisb.) N.E.Br.
Iridaceae
Antholyza floribunda Salisb.
♦ African cornflag, chasmanthe
♦ Weed, Naturalised, Garden Escape, Environmental Weed, Cultivation Escape
♦ 7, 34, 72, 86, 88, 98, 101, 176, 198, 203, 280, 289
♦ pH, cultivated. Origin: South Africa.

Cheilanthes austrotenuifolia Quirck & Chambers
Pteridaceae/Adiantaceae
Cheilanthes tenuifolia (Burm.f.) Sw. *auct.*
♦ rock fern
♦ Native Weed
♦ 269
♦ cultivated, toxic. Origin: Australia.

Cheilanthes catanensis (Cosent.) H.P.Fuchs
Pteridaceae/Adiantaceae
Acrostichum catanense Cosent.
♦ Weed
♦ 221

Cheilanthes distans (R.Br.) Mett.
Pteridaceae/Adiantaceae
♦ shaggy rock fern
♦ Native Weed
♦ 39, 269
♦ cultivated, toxic. Origin: Australia.

Cheilanthes lasiophylla Pic.Serm.
Pteridaceae/Adiantaceae
♦ Native Weed
♦ 269
♦ cultivated, toxic. Origin: Australia.

Cheilanthes marantae (L.) R.Br.
Pteridaceae/Adiantaceae
♦ hardy cloakfern
♦ Weed
♦ 272

Cheilanthes notholaenoides (Desv.) Maxon ex Weath.
Pteridaceae/Adiantaceae
♦ royal lipfern
♦ Naturalised
♦ 101

Cheilanthes sieberi Kunze
Pteridaceae/Adiantaceae
♦ Weed
♦ 39, 87, 88
♦ cultivated, toxic. Origin: Australia.

Cheilanthes Kunze *sieberi* ssp. *sieberi*
Pteridaceae/Adiantaceae
♦ Native Weed
♦ 269
♦ toxic. Origin:
Australia.

Cheilanthes viridis (Forssk.) Sw. var. *macrophylla* (Kuntze) Schelpe & N.C.Anthony
Pteridaceae/Adiantaceae
Pellaea viridis (Forssk.) Prantl var. *macrophylla* Sim
♦ green cliffbrake
♦ Native Weed
♦ 121
♦ pH. Origin: southern Africa.

Cheiranthus cheiri L.
Brassicaceae
= *Erysimum cheiri* (L.) Crantz
♦ wallflower
♦ Weed, Naturalised
♦ 39, 54, 86, 88, 198, 272, 280
♦ cultivated, herbal, toxic.

Chelidonium franchetianum Prain
Papaveraceae
= *Dicranostigma franchetianum* (Prain) Fedde
♦ Quarantine Weed
♦ 220

Chelidonium japonicum Thunb.
Papaveraceae
♦ Weed, Quarantine Weed
♦ 76, 88, 220
♦ aH, cultivated.

Chelidonium majus L.
Papaveraceae
♦ celandine, swallowwort, wartweed, greater celandine, common celandine, keltamo
♦ Weed, Naturalised, Introduced, Environmental Weed
♦ 4, 38, 39, 44, 45, 70, 80, 87, 88, 94, 97, 101, 104, 151, 161, 195, 218, 243, 247, 272, 275, 280, 297
♦ b/pH, cultivated, herbal, toxic.

Chelidonium majus L. var. *asiaticum* (Hara) Ohwi
Papaveraceae
♦ Weed
♦ 286

Chelidonium majus L. var. *laciniatum* (Mill.) Syme
Papaveraceae
♦ celandine
♦ Naturalised
♦ 101
♦ cultivated.

Chelidonium majus L. var. *majus*
Papaveraceae
♦ celandine
♦ Naturalised
♦ 101

Chelidonium majus L. var. *plenum* Wehrh.
Papaveraceae
♦ celandine
♦ Naturalised
♦ 101

Chenolea arabica Boiss.
Chenopodiaceae
♦ Weed
♦ 221
♦ arid.

Chenopodium spp. L.
Chenopodiaceae
♦ goosefoot, pigweed, lamb's quarters, wormseed
♦ Weed, Quarantine Weed
♦ 220, 243, 272
♦ H, herbal, toxic.

Chenopodium acerifolium Andrz.
Chenopodiaceae
= *Chenopodium berlandieri* Moq. var. *zschackii* (J.Murr) J.Murr ex Asch. (NoR)

♦ jokisavikka
♦ Casual Alien
♦ 42
♦ herbal.

Chenopodium acuminatum Willd.
Chenopodiaceae
♦ aasiansavikka
♦ Weed, Casual Alien
♦ 42, 87, 88, 275, 297
♦ aH, promoted.

Chenopodium album L.
Chenopodiaceae
Atriplex alba (L.) Crantz
♦ white goosefoot, common lamb's quarters, fathen, lamb's quarters, pigweed, baconweed, chou grass, frost bite, mealweed, pitseed goosefoot, white pigweed, wild spinach, netseed lamb's quarters, farinello comune
♦ Weed, Noxious Weed, Naturalised, Garden Escape, Environmental Weed
♦ 7, 15, 21, 23, 24, 34, 39, 44, 49, 50, 51, 52, 55, 68, 70, 80, 84, 86, 87, 88, 93, 94, 98, 101, 108, 114, 115, 118, 121, 134, 136, 151, 158, 161, 165, 167, 174, 176, 179, 180, 185, 186, 198, 203, 204, 205, 207, 210, 211, 218, 221, 229, 236, 237, 242, 243, 245, 248, 250, 253, 255, 263, 269, 270, 272, 273, 275, 280, 286, 294, 295, 297, 299, 300
♦ aH, arid, cultivated, herbal, toxic. Origin: Eurasia.

Chenopodium album L. fo. *colorans* Hiyama
Chenopodiaceae
♦ Naturalised
♦ 287

Chenopodium album L. ssp. *album* var. *polymorphum* Aellen
Chenopodiaceae
♦ Naturalised
♦ 241

Chenopodium album L. var. *album*
Chenopodiaceae
Chenopodium suecicum J.Murr. (see)
♦ common lamb's quarters, lamb's quarters
♦ Noxious Weed, Naturalised
♦ 101, 229

Chenopodium album L. var. *centrorubrum* Makino
Chenopodiaceae
♦ lamb's quarters
♦ Weed, Naturalised
♦ 263, 286, 287

Chenopodium album L. var. *microphyllum* Boenn.
Chenopodiaceae
♦ common lamb's quarters, lamb's quarters
♦ Noxious Weed, Naturalised
♦ 101, 229

Chenopodium album L. var. *missouriense* (Aellen) I.J.Bassett & C.W.Crompton
Chenopodiaceae
Chenopodium missouriense Aellen (see)
♦ common lamb's quarters, Missouri lamb's quarters

◆ Noxious Weed
◆ 229

Chenopodium album L. var. stevensii Aellen
Chenopodiaceae
◆ common lamb's quarters, Stevens's lamb's quarters
◆ Noxious Weed
◆ 229

Chenopodium album L. var. striatum (Krasan) Kartesz *comb. nov. ined.*
Chenopodiaceae
◆ common lamb's quarters, lateflowering goosefoot
◆ Noxious Weed
◆ 229

Chenopodium ambrosioides L.
Chenopodiaceae
Chenopodium ambrosioides L. var. *anthelminticum* (L.) A.Gray (see), *Chenopodium anthelminticum* L. (see)
◆ Mexican tea, Indian goosefoot, bitterweed, wormseed
◆ Weed, Naturalised, Environmental Weed, Casual Alien
◆ 7, 8, 9, 23, 34, 39, 40, 42, 50, 51, 70, 80, 84, 86, 87, 88, 98, 101, 106, 121, 134, 157, 158, 161, 165, 167, 179, 180, 185, 198, 203, 210, 217, 218, 221, 235, 237, 243, 245, 247, 255, 256, 257, 261, 263, 269, 270, 272, 274, 275, 280, 286, 287, 295, 297
◆ a/pH, arid, cultivated, herbal, toxic. Origin: tropical America.

Chenopodium ambrosioides L. var. ambrosioides
Chenopodiaceae
◆ Mexican tea
◆ Naturalised
◆ 101

Chenopodium ambrosioides L. var. anthelminticum (L.) A.Gray
Chenopodiaceae
= *Chenopodium ambrosioides* L.
◆ Weed, Naturalised
◆ 286, 287

Chenopodium ambrosioides L. var. obovatum Speg.
Chenopodiaceae
◆ Mexican tea
◆ Naturalised
◆ 101

Chenopodium anthelminticum L.
Chenopodiaceae
= *Chenopodium ambrosioides* L.
◆ American wormseed
◆ Weed
◆ 39, 87, 88
◆ herbal, toxic.

Chenopodium aristatum L.
Chenopodiaceae
Chenopodium minimum P.Y.Fu & Wang-Wei, *Chenopodium sinense* hort ex Moq., *Lecanocarpus aristatus* (L.) Zucc., *Teloxys aristata* (L.) Moq. (see)
◆ wormseed, sea foam flower
◆ Weed, Naturalised, Casual Alien
◆ 23, 40, 42, 87, 88, 101, 272, 275, 287,

297
◆ aH, herbal.

Chenopodium auricomum Lindl.
Chenopodiaceae
◆ Introduced
◆ 228
◆ aH, arid, cultivated.

Chenopodium berlandieri Moq.
Chenopodiaceae
◆ pitseed goosefoot, netseed lamb's quarters, teksasinsavikka
◆ Weed, Casual Alien
◆ 23, 40, 42, 87, 88, 161, 212
◆ aH, promoted, herbal.

Chenopodium bontei Aellen
Chenopodiaceae
= *Chenopodium carinatum* R.Br. × *Chenopodium cristatum* F.Muell.
◆ Weed, Quarantine Weed
◆ 51, 76, 87, 88, 203, 220

Chenopodium bonus-henricus L.
Chenopodiaceae
Anserina bonus-henricus (L.) Dumort., *Blitum bonus-henricus* (L.) C.A.Mey. ex Ledeb., *Orthosporum bonus-henricus* (L.) T.Nees
◆ perennial goosefoot, good King Henry, hyvänheikinsavikka
◆ Weed, Quarantine Weed, Naturalised, Casual Alien
◆ 23, 39, 40, 42, 44, 70, 87, 88, 101, 218, 220, 253, 272, 280
◆ pH, cultivated, herbal, toxic.

Chenopodium borbasii Murr
Chenopodiaceae
Chenopodium zobelii Ludw. & Aellen *nom. illleg.* (see)
◆ Weed
◆ 237

Chenopodium botrys L.
Chenopodiaceae
Ambrina botrys (L.) Moq., *Atriplex botrys* (L.) Crantz
◆ feather geranium, Jerusalem oak goosefoot
◆ Weed, Naturalised, Casual Alien
◆ 23, 34, 39, 42, 70, 80, 87, 88, 94, 101, 161, 180, 218, 221, 253, 272
◆ aH, arid, cultivated, herbal, toxic. Origin: Eurasia.

Chenopodium californicum (S.Watson) S.Watson
Chenopodiaceae
◆ soapplant, California goosefoot
◆ Weed
◆ 161
◆ pH, promoted, herbal.

Chenopodium capitatum (L.) Asch.
Chenopodiaceae
Blitum capitatum L.
◆ blite goosefoot, strawberry blite, strawberry pigweed, blite mulberry, mykerösavikka
◆ Weed, Naturalised, Casual Alien
◆ 34, 42, 49, 87, 88, 161, 218, 280
◆ aH, cultivated, herbal.

Chenopodium carinatum R.Br.
Chenopodiaceae

◆ creeping goosefoot, green goosefoot, keeled goosefoot
◆ Weed, Native Weed
◆ 39, 87, 88, 121, 158, 255, 269
◆ aH, cultivated, herbal, toxic. Origin: Australia.

Chenopodium carnosulum Moq.
Chenopodiaceae
◆ ridged goosefoot
◆ Naturalised
◆ 101

Chenopodium carnosulum Moq. var. patagonicum (Phil.) H.A.Wahl
Chenopodiaceae
◆ ridged goosefoot
◆ Naturalised
◆ 101
◆ aH.

Chenopodium cordobense Aellen
Chenopodiaceae
◆ quínoa de córdoba
◆ Weed
◆ 87, 88, 237, 295

Chenopodium cristatum (F.Muell.) F.Muell.
Chenopodiaceae
◆ crested goosefoot
◆ Weed, Naturalised
◆ 86, 87, 88, 101
◆ aH, arid, cultivated. Origin: Australia.

Chenopodium desiccatum A.Nelson
Chenopodiaceae
◆ aridland goosefoot, narrowleaf lamb's quarters
◆ Weed, Casual Alien
◆ 40, 161, 199
◆ aH, herbal.

Chenopodium desiccatum A.Nelson var. leptophylloides (Murr.) Wahl
Chenopodiaceae
◆ Weed
◆ 237

Chenopodium detestans Kirk
Chenopodiaceae
◆ Weed, Naturalised, Casual Alien
◆ 40, 86, 98, 203
◆ Origin: New Zealand.

Chenopodium erosum R.Br.
Chenopodiaceae
◆ Naturalised
◆ 280

Chenopodium fasciculosum Aellen
Chenopodiaceae
◆ Weed, Quarantine Weed
◆ 76, 87, 88, 203, 220
◆ arid.

Chenopodium ficifolium Sm.
Chenopodiaceae
Chenopodium blomianum Aell., *Chenopodium ficifolium* Sm. ssp. *blomianum* (Aell.) Aell. (see), *Chenopodium serotinum* auct. *non* L.
◆ fig leaved goosefoot, viikunanlehtisavikka
◆ Weed, Quarantine Weed, Naturalised, Casual Alien

♦ 23, 42, 44, 68, 70, 76, 87, 88, 94, 101,
185, 203, 220, 253, 263, 272, 274, 280,
286, 287, 300
♦ aH, cultivated. Origin: Europe,
Egypt, Turkey.

Chenopodium ficifolium **Sm. ssp.**
blomianum (Aellen) Aellen
Chenopodiaceae
= *Chenopodium ficifolium* Sm.
♦ goosefoot, fathen
♦ Weed
♦ 238

Chenopodium foetidum **Schrad.** *non* **Lam.**
Chenopodiaceae
= *Chenopodium schraderianum* Roem.
& Schult.
♦ Weed
♦ 87, 88, 275, 297
♦ aH, cultivated, herbal.

Chenopodium foliosum **(Moench) Asch.**
Chenopodiaceae
Morocarpus foliosus Moench
♦ leafy goosefoot, marjasavikka
♦ Weed, Naturalised, Casual Alien
♦ 34, 42, 80, 87, 88, 101
♦ aH, arid, cultivated, herbal.

Chenopodium giganteum **D.Don**
Chenopodiaceae
controversial name that may be
Chenopodium album L. ssp. *amaranticolor*
Coste & A.Reyn.
♦ jättisavikka, tree spinach, nettleleaf
goosefoot, lamb's quarters, magenta
lamb's quarters
♦ Weed, Naturalised, Garden Escape,
Cultivation Escape
♦ 42, 121, 275, 280
♦ aH, cultivated, herbal. Origin:
obscure, possibly the subcontinent.

Chenopodium gigantospermum **Aellen**
Chenopodiaceae
= *Chenopodium simplex* (Torr.) Raf.
♦ plataanisavikka, mapleleaf
goosefoot
♦ Weed, Noxious Weed, Casual Alien
♦ 23, 42, 80, 87, 88, 161, 299
♦ aH, promoted, herbal.

Chenopodium glaucum **L.**
Chenopodiaceae
♦ oak leaved goosefoot
♦ Weed, Naturalised, Introduced,
Environmental Weed
♦ 7, 21, 23, 39, 70, 80, 87, 88, 94, 98, 101,
161, 176, 199, 203, 218, 253, 272, 275,
286, 287, 297, 300
♦ aH, cultivated, herbal, toxic.

Chenopodium glaucum **L. ssp.** *ambiguum*
(R.Br.) Thell. & Aellen
Chenopodiaceae
♦ Naturalised
♦ 241

Chenopodium glaucum **Aellen ssp.**
glaucum
Chenopodiaceae
♦ Naturalised
♦ 86

Chenopodium graveolens **Willd.**
Chenopodiaceae
Chenopodium incisum Poir., *Teloxys*
graveolens (Willd.) W.A.Weber
♦ fetid goosefoot
♦ Weed
♦ 243
♦ aH, herbal.

Chenopodium hircinum **Schrad.**
Chenopodiaceae
♦ avian goosefoot, pukinsavikka
♦ Weed, Quarantine Weed,
Naturalised, Casual Alien
♦ 42, 76, 87, 88, 101, 203, 220, 236, 237,
241, 295, 300
♦ arid, herbal.

Chenopodium humile **Hook.**
Chenopodiaceae
♦ marshland goosefoot
♦ Naturalised
♦ 101

Chenopodium hybridum **L.**
Chenopodiaceae
♦ mapleleaf goosefoot, sowbane,
vaahterasavikka
♦ Weed, Quarantine Weed,
Naturalised
♦ 23, 39, 42, 44, 70, 76, 87, 88, 94, 210,
218, 243, 253, 272, 275, 287, 297
♦ aH, cultivated, herbal, toxic.

Chenopodium leptophyllum **Nutt.**
Chenopodiaceae
♦ slimleaf lamb's quarters, narrowleaf
goosefoot, thin leaved goosefoot
♦ Weed, Naturalised
♦ 87, 88, 161, 218, 243, 287
♦ aH, promoted, herbal.

Chenopodium macrospermum **Hook.f.**
Chenopodiaceae
♦ redstem goosefoot, largeseed
goosefoot
♦ Weed, Naturalised
♦ 7, 9, 54, 86, 87, 88, 98, 101, 198, 203,
237, 295
♦ herbal. Origin: South America.

Chenopodium macrospermum **Hook.f.**
var. *halophilum* **(Phil.) Standl.**
Chenopodiaceae
♦ saltloving goosefoot
♦ Naturalised
♦ 101
♦ aH.

Chenopodium melanocarpum **(J.Black)**
J.Black
Chenopodiaceae
♦ Naturalised
♦ 39, 86
♦ toxic. Origin: Australia.

Chenopodium missouriense **Aellen**
Chenopodiaceae
= *Chenopodium album* L. var.
missouriense (Aellen) I.J.Bassett &
C.W.Crompton
♦ missourinsavikka, Missouri lamb's
quarters
♦ Casual Alien
♦ 42
♦ aH, herbal.

Chenopodium multifidum **L.**
Chenopodiaceae
♦ stinking goosefoot, cutleaf
goosefoot, scented goosefoot
♦ Weed, Naturalised
♦ 34, 51, 86, 87, 88, 98, 101, 121, 161,
198, 203, 237, 241, 295, 300
♦ a/pH, cultivated, herbal. Origin:
South America.

Chenopodium murale **L.**
Chenopodiaceae
Chenopodium biforme Nees,
Chenopodium congestum Hook.f.
♦ nettleleaf goosefoot, green fathen,
sowbane, nettle leaved fathen,
swinebane, wall goosefoot, chuana
soap, goosefoot, lamb's quarters,
round leaved fathen, wheat bush,
green goosefoot, rauniosavikka
♦ Weed, Naturalised, Garden Escape,
Environmental Weed, Casual Alien
♦ 7, 15, 23, 34, 39, 40, 42, 44, 50, 51, 53,
55, 66, 70, 72, 80, 86, 87, 88, 93, 94, 98,
101, 115, 121, 134, 158, 161, 176, 179,
180, 185, 198, 203, 205, 212, 217, 218,
221, 236, 237, 241, 243, 253, 257, 261,
269, 272, 280, 287, 295, 300
♦ aH, arid, cultivated, herbal, toxic.
Origin: Eurasia.

Chenopodium nitrariaceum **(F.Muell.)**
F.Muell. ex Benth.
Chenopodiaceae
Chenopodium lycioides Gand.
♦ Naturalised, Introduced
♦ 86, 228
♦ arid, cultivated, herbal. Origin:
Australia.

Chenopodium oahuense **(Meyen) Aellen**
Chenopodiaceae
♦ alaweo
♦ Weed
♦ 87, 88
♦ herbal.

Chenopodium opulifolium **Schrad. ex**
Koch & Ziz
Chenopodiaceae
Chenopodium viride L. (see)
♦ grey goosefoot, round leaved
fathen, broad leaved goosefoot,
lamb's quarters, seaport goosefoot,
heisisavikka
♦ Weed, Naturalised, Introduced,
Casual Alien
♦ 34, 42, 44, 50, 70, 86, 87, 88, 94, 101,
228, 240, 253, 272, 280, 287
♦ aH, arid, cultivated, herbal. Origin:
Europe, North Africa.

Chenopodium paganum **Rchb.**
Chenopodiaceae
♦ pigweed goosefoot
♦ Weed
♦ 23, 87, 88, 218
♦ herbal.

Chenopodium pallidicaule **Aellen**
Chenopodiaceae
♦ kaniwa, canihua
♦ Quarantine Weed
♦ 220
♦ aH, arid, promoted, herbal.

Chenopodium paniculatum Hook.
Chenopodiaceae
= *Chenopodium petiolare* Kunth
♦ goosefoot
♦ Weed, Quarantine Weed
♦ 76, 87, 88, 203, 220, 243
♦ H, arid.

Chenopodium papulosum Moq.
Chenopodiaceae
♦ quínoa pampeana
♦ Weed
♦ 237, 295

Chenopodium polyspermum L.
Chenopodiaceae
♦ many seeded goosefoot, allseed, hentosavikka
♦ Weed, Naturalised
♦ 23, 39, 44, 68, 70, 87, 88, 94, 101, 218, 243, 253, 272
♦ aH, cultivated, herbal, toxic.

Chenopodium polyspermum L. var. acutifolium (Sm.) Gaud.
Chenopodiaceae
♦ manyseed goosefoot
♦ Naturalised
♦ 101

Chenopodium polyspermum L. var. obtusifolium Gaud.
Chenopodiaceae
♦ manyseed goosefoot
♦ Naturalised
♦ 101

Chenopodium pratericola Rydb.
Chenopodiaceae
= *Botrys pratericola* (Rydb.) Lunell (NoR)
♦ narrowleaf goosefoot, kapealehtisavikka, desert goosefoot
♦ Weed, Casual Alien
♦ 42, 87, 88, 218, 295
♦ aH, herbal.

Chenopodium preissmannii J.Murray
Chenopodiaceae
♦ Naturalised
♦ 287

Chenopodium probstii Aellen
Chenopodiaceae
♦ soijasavikka
♦ Casual Alien
♦ 42

Chenopodium procerum Hochst. ex Moq.
Chenopodiaceae
♦ Weed
♦ 87, 88
♦ herbal.

Chenopodium pumilio R.Br.
Chenopodiaceae
♦ clammy goosefoot, Tasmanian goosefoot, small crumbweed
♦ Weed, Naturalised, Native Weed, Introduced, Casual Alien
♦ 7, 15, 34, 39, 42, 80, 86, 101, 121, 161, 165, 167, 180, 228, 236, 237, 243, 269, 280, 286, 287, 295
♦ aH, arid, herbal, toxic. Origin: Australia.

Chenopodium quinoa Willd.
Chenopodiaceae
♦ canihua, chica, huauzontle, quinoa, quinua
♦ Weed, Introduced, Casual Alien
♦ 40, 228, 236, 295
♦ aH, arid, cultivated, herbal. Origin: Americas.

Chenopodium rubrum L.
Chenopodiaceae
Blitum rubrum C.A.Mey.
♦ red goosefoot, coast blite
♦ Weed
♦ 23, 39, 44, 70, 87, 88, 94, 218, 253, 272
♦ aH, cultivated, herbal, toxic.

Chenopodium schraderianum Roem. & Schult.
Chenopodiaceae
Chenopodium foetidum Schrad. *non* Lam. (see)
♦ Schrader's goosefoot, Schrader's spinach, Afrikansavikka
♦ Weed, Casual Alien
♦ 23, 42, 50, 51, 88, 121, 272
♦ aH, arid, promoted. Origin: Africa.

Chenopodium serotinum L.
Chenopodiaceae
♦ Weed
♦ 235, 275, 297
♦ aH, arid.

Chenopodium sosnowskyi Kapeller
Chenopodiaceae
♦ anatoliansavikka
♦ Casual Alien
♦ 42

Chenopodium strictum Roth
Chenopodiaceae
♦ late flowering goosefoot, intiansavikka
♦ Weed, Casual Alien
♦ 23, 42, 68, 87, 88, 161
♦ herbal.

Chenopodium strictum Roth var. glaucophyllum (Aellen) H.A.Wahl
Chenopodiaceae
♦ late flowering goosefoot
♦ Weed
♦ 68
♦ aH.

Chenopodium suecicum J.Murr.
Chenopodiaceae
= *Chenopodium album* L. var. *album*
♦ pohjanjauhosavikka
♦ Weed
♦ 88, 94, 272
♦ aH, promoted.

Chenopodium triangulare R.Br.
Chenopodiaceae
= *Einadia trigonos* (Roem. & Schult.) P.G.Wilson
♦ Weed
♦ 87, 88

Chenopodium urbicum L.
Chenopodiaceae
♦ city goosefoot, upright goosefoot, kyläsavikka
♦ Weed, Naturalised, Casual Alien

♦ 23, 40, 42, 70, 87, 88, 94, 101, 218, 241, 272, 275, 300
♦ aH, arid, cultivated, herbal.

Chenopodium urbicum L. ssp. sinicum Kung & G.L.Chu
Chenopodiaceae
♦ Weed
♦ 275
♦ aH.

Chenopodium virgatum Thunb.
Chenopodiaceae
♦ idänsavikka
♦ Weed, Casual Alien
♦ 42, 235
♦ aH, promoted.

Chenopodium viride L.
Chenopodiaceae
= *Chenopodium opulifolium* Schrad. ex Koch & Ziz
♦ Weed
♦ 44
♦ aH, promoted.

Chenopodium vulvaria L.
Chenopodiaceae
Chenopodium olidum Curtis
♦ stinking goosefoot, haisusavikka
♦ Weed, Naturalised, Introduced, Environmental Weed, Casual Alien
♦ 34, 39, 42, 70, 86, 87, 88, 94, 98, 101, 176, 198, 203, 221, 228, 243, 253, 272, 280, 300
♦ aH, arid, cultivated, herbal, toxic. Origin: obscure.

Chenopodium zahnii J.Murray
Chenopodiaceae
♦ Naturalised
♦ 287

Chenopodium zobelii Ludw. & Aellen nom. illleg.
Chenopodiaceae
= *Chenopodium borbasii* Murr
♦ quínoa blanca
♦ Weed
♦ 87, 88, 295

Chevreulia sarmentosa (Pers.) Blake
Asteraceae
♦ chevreulia
♦ Weed
♦ 237, 295

Chiliotrichum diffusum (Forst.f.) Kuntze
Asteraceae
Amellus diffusus Forst., *Aster magellanicus* Spreng., *Chiliotrichum amelloides* Cass., *Chiliotrichum amelloides* var. *lanceolatum* T.Nees, *Chiliotrichum diffusum* (Forst.f.) Kuntze fo. *media* Speg., *Chiliotrichum rosmarinifolium* Less., *Chiliotrichum virgatum* Phil.
♦ mata negra
♦ Weed
♦ 87, 88, 237, 295
♦ S, cultivated.

Chilopsis linearis (Cav.) Sweet
Bignoniaceae
♦ desert willow
♦ Introduced
♦ 228
♦ S, arid, cultivated, herbal.

Chimonobambusa quadrangularis
(Franceschi) Mak.
Poaceae
Arundinaria quadrangularis (Franceschi)
Makino, *Bambusa angulata* Munro,
Bambusa quadrangularis Franceschi
♦ square stemmed bamboo, square
bamboo
♦ Weed, Naturalised, Introduced,
Cultivation Escape
♦ 15, 40, 280
♦ G, cultivated.

Chiococca alba **(L.) Hitchc.**
Rubiaceae
♦ West Indian milkberry
♦ Weed
♦ 14, 39, 87, 88
♦ pC, herbal, toxic.

Chionachne hubbardiana **Henr.**
Poaceae
♦ hairy ribbon grass, river grass
♦ Weed
♦ 87, 88
♦ G, arid.

Chionanthus ramiflorus **Roxb.**
Oleaceae
♦ northern olive, native olive
♦ Environmental Weed
♦ 260
♦ cultivated. Origin: Australia.

Chionochloa beddiei **Zotov**
Poaceae
♦ Weed, Quarantine Weed
♦ 76, 88, 220
♦ G.

Chionochloa flavicans **Zotov**
Poaceae
Danthonia antarctica (G.Forst.) Spreng.
var. *elata* Hook.f.
♦ Weed, Quarantine Weed
♦ 76, 88, 220
♦ G.

Chionochloa rubra **Zotov**
Poaceae
Danthonia antarctica (G.Forst.) Spreng.
var. *minor* Hook.f.
♦ red tussock, snow grass
♦ Weed, Quarantine Weed
♦ 76, 88, 220
♦ G, cultivated.

Chionochloa spiralis **Zotov**
Poaceae
♦ Weed, Quarantine Weed
♦ 76, 88, 220
♦ G.

Chionodoxa luciliae **Boiss.**
Liliaceae/Hyacinthaceae
♦ kirjokevättähti, glory of the snow
♦ Naturalised, Cultivation Escape
♦ 42, 101
♦ cultivated, herbal.

Chloranthus spicatus **(Thunb.) Makino**
Chloranthaceae
♦ Naturalised
♦ 287

Chlorella **Beij. spp.**
Chlorellaceae

♦ algae, agua verde, nitela, limo
♦ Weed
♦ 255
♦ algae.

Chloris barbata **(L.) Sw.**
Poaceae
Chloris inflata Link (see), *Chloris
paraguayensis* Steud. (see)
♦ swollen fingergrass, plush grass, pea
cock plumegrass, airport grass, mau'u
lei, purpletop chloris, yaa rangnok
♦ Weed, Naturalised
♦ 3, 6, 7, 87, 88, 91, 107, 179, 209, 239,
255, 262, 286, 287
♦ a/pG, arid, herbal. Origin: South
America.

Chloris barbata **(L.) Sw. var. *formosana***
Honda
Poaceae
♦ Naturalised
♦ 287
♦ G.

Chloris canterai **Arechav. var. *canterai***
Poaceae
♦ Paraguayan windmill grass
♦ Naturalised
♦ 101
♦ G.

Chloris canterai **Arechav. var. *grandiflora***
(Roseng. & Izag.) D.E.Anderson
Poaceae
♦ Paraguayan windmill grass
♦ Naturalised
♦ 101
♦ G.

Chloris chloridea **(Presl) Hitchc.**
Poaceae
= *Enteropogon chlorideus* (J.Presl)
W.D.Clayton (NoR)
♦ chloris
♦ Weed, Quarantine Weed
♦ 76, 87, 88, 203, 220, 281
♦ G.

Chloris ciliata **Sw.**
Poaceae
♦ fringed windmill grass
♦ Weed, Naturalised
♦ 23, 86, 87, 88, 98, 203
♦ G. Origin: Central and South
America, Caribbean.

Chloris distichophylla **Lag.**
Poaceae
= *Eustachys distichophylla* (Lag.) Nees
♦ Weed
♦ 39, 87, 88, 255
♦ pG, toxic. Origin: South America.

Chloris divaricata **R.Br.**
Poaceae
♦ Australian stargrass, spreading
windmill grass
♦ Weed, Naturalised
♦ 87, 88, 101, 218, 286, 287
♦ G. Origin: Australia.

Chloris divaricata **R.Br. var. *divaricata***
Poaceae
♦ Introduced
♦ 228

♦ G, arid.

Chloris dolichostachya **Lag.**
Poaceae
♦ Naturalised
♦ 287
♦ G.

Chloris elata **Desv.**
Poaceae
♦ tall windmill grass
♦ Naturalised
♦ 101
♦ G.

Chloris gayana **Kunth**
Poaceae
♦ Rhodes grass, hunyanigrass,
Rhodesian bluegrass, Rhodes chloris
♦ Weed, Naturalised, Native Weed,
Introduced, Environmental Weed
♦ 7, 34, 38, 72, 86, 87, 88, 91, 93, 98, 101,
121, 134, 158, 179, 198, 199, 203, 204,
218, 228, 236, 241, 255, 280, 286, 287,
296
♦ pG, arid, cultivated, herbal. Origin:
Africa.

Chloris halophila **Parodi**
Poaceae
♦ Weed, Quarantine Weed
♦ 76, 87, 88, 203, 220
♦ G, arid.

Chloris inflata **Link**
Poaceae
= *Chloris barbata* Sw.
♦ swollen fingergrass, purpletop
chloris, purpletop Rhodes grass,
Mexican bluegrass, horquetilla morada
♦ Weed, Naturalised, Introduced,
Environmental Weed
♦ 3, 68, 86, 88, 93, 98, 203, 218, 230, 261
♦ G, herbal.

Chloris mollis **(Nees) Swallen**
Poaceae
= *Enteropogon mollis* (Nees) Clayton
(NoR)
♦ Naturalised
♦ 257
♦ G, arid.

Chloris paraguayensis **Steud.**
Poaceae
= *Chloris barbata* Sw.
♦ Weed
♦ 3, 191
♦ G.

Chloris pectinata **Benth.**
Poaceae
♦ comb windmill grass
♦ Naturalised, Introduced
♦ 101, 228, 287
♦ G, arid.

Chloris petraea **Sw.**
Poaceae
= *Eustachys petraea* (Sw.) Desv.
♦ rock fingergrass
♦ Weed
♦ 14, 161
♦ G, herbal.

Chloris pilosa **Schumach.**
Poaceae

Chloris breviseta Benth.
♦ Weed, Naturalised
♦ 86, 88, 91, 93
♦ aG. Origin: tropical Africa.

Chloris polydactyla **Sw. nom. illeg.**
Poaceae
= *Chloris dandyana* C.D.Adams
♦ pasto borla
♦ Weed
♦ 87, 88, 91, 255, 295
♦ pG. Origin: Americas.

Chloris prieurii **Kunth**
Poaceae
= *Enteropogon prieurii* (Kunth) Clayton
♦ Weed
♦ 88
♦ G.

Chloris pycnothrix **Trin.**
Poaceae
♦ spiderweb chloris, false stargrass, spinnerak chloris, orchard grass, radiate fingergrass, spiderweb grass
♦ Weed, Naturalised, Native Weed
♦ 51, 53, 88, 91, 121, 158, 238, 257
♦ aG. Origin: southern Africa.

Chloris radiata **(L.) Sw.**
Poaceae
♦ radiate fingergrass, plush grass, ilusión
♦ Weed, Naturalised
♦ 3, 87, 88, 107, 157, 179, 191, 218, 241, 255, 257, 281, 286, 287, 300
♦ a/pG, arid, cultivated. Origin: tropical Americas.

Chloris truncata **R.Br.**
Poaceae
Chloris elongata Poir., *Chloris megastachya* Schrad. ex Schult.
♦ windmill grass, Australian fingergrass, stargrass
♦ Weed, Naturalised, Introduced
♦ 39, 87, 88, 93, 101, 121, 205, 228, 280, 287
♦ pG, arid, cultivated, toxic. Origin: Australia.

Chloris ventricosa **R.Br.**
Poaceae
Chloris sclerantha Lindl.
♦ Australian windmill grass
♦ Naturalised, Introduced
♦ 39, 101, 228
♦ G, arid, cultivated, toxic. Origin: Australia.

Chloris verticillata **Nutt.**
Poaceae
♦ tumble windmill grass, windmill grass
♦ Weed, Native Weed, Cultivation Escape
♦ 34, 87, 88, 161, 174, 195, 218
♦ pG, cultivated, herbal. Origin: Americas.

Chloris virgata **Sw.**
Poaceae
Chloris alba J.S. & C.Presl, *Chloris elegans* Kunth, *Chloris meccana* Hochst. ex Steud. ex Schldl.
♦ feather fingergrass, feathertop

Rhodes grass, furry grass, bluegrass, haygrass, feathertop chloris, old landsgrass, sweetgrass, sweethay grass, white grass, windmill grass, showy chloris, silky chloris
♦ Weed, Naturalised, Native Weed, Introduced, Environmental Weed, Casual Alien
♦ 3, 7, 34, 42, 51, 55, 86, 87, 88, 91, 93, 98, 121, 158, 161, 180, 198, 203, 205, 218, 228, 241, 242, 257, 269, 275, 286, 287, 296, 297, 300
♦ a/pG, arid, cultivated, herbal. Origin: southern Africa.

Chlorocodon whitei **Hook.f.**
Asclepiadaceae/Apocynaceae
♦ Quarantine Weed
♦ 220

Chlorocyperus rotundus **Palla**
Cyperaceae
= *Cyperus rotundus* L.
♦ Weed
♦ 3, 191
♦ G.

Chlorogalum pomeridianum **(DC.) Kunth**
Liliaceae/Hyacinthaceae
♦ soapplant, wavyleaf soap plant, soaproot, amole lily
♦ Weed
♦ 87, 88, 218
♦ pH, cultivated, herbal.

Chloroleucon ebano **(Berland.) L.Rico**
Fabaceae/Mimosaceae
= *Ebenopsis ebano* (Berland.) Barneby & J.W.Grimes
♦ Introduced
♦ 228
♦ arid.

Chloroleucon mangense **(Jacq.) Britton & Rose**
Fabaceae/Mimosaceae
♦ Quarantine Weed
♦ 220

Chlorophytum borivilianum **Santapau & Fernández**
Liliaceae/Anthericaceae
♦ biskandri, safed musli
♦ Weed
♦ 19

Chlorophytum capense **(L.) Voss**
Liliaceae/Anthericaceae
♦ bracket plant, viherrnsylilja
♦ Naturalised
♦ 101
♦ herbal.

Chlorophytum comosum **(Thunb.) Jacques**
Liliaceae/Anthericaceae
Anthericum comosum Thunb.
♦ bracket plant, hen and chicks, ribbon plant, spider ivy, spiderplant, walking anthericum, spiderplant, airplane plant
♦ Weed, Naturalised, Native Weed, Garden Escape, Environmental Weed
♦ 15, 73, 86, 88, 98, 121, 198, 203, 290, 296

♦ pH, aqua, cultivated, herbal. Origin: southern Africa.

Choisya ternata **Kunth**
Rutaceae
♦ Mexican orange flower, Mexican orange blossom
♦ Naturalised
♦ 39, 86
♦ S, cultivated, herbal, toxic. Origin: southern North America.

Chondrilla graminea **Bieb.**
Asteraceae
♦ Weed
♦ 87, 88

Chondrilla juncea **L.**
Asteraceae
♦ rush skeletonweed, skeleton weed, hogbite, chondrilla, gum succory, naked weed
♦ Weed, Quarantine Weed, Noxious Weed, Naturalised, Environmental Weed
♦ 1, 7, 26, 34, 35, 62, 70, 72, 76, 80, 86, 87, 88, 94, 98, 101, 108, 136, 139, 146, 147, 151, 161, 171, 172, 176, 177, 198, 203, 212, 217, 218, 219, 220, 229, 237, 243, 246, 250, 253, 267, 269, 272, 280
♦ pH, arid, cultivated, herbal, toxic. Origin: Eurasia.

Chondropetalum tectorum **Pillans**
Restionaceae
Restio tectorum L.f.
♦ South African cape rush, cape rush, dakriet
♦ Quarantine Weed
♦ 72, 76
♦ G, arid, cultivated. Origin: South Africa.

Chondrosum barbatum **(Lag.) Clayton**
Poaceae
Bouteloua barbata Lag. (see), *Bouteloua micrantha* Scribn. & Merr.
♦ Introduced
♦ 228
♦ G, arid.

Chondrus crispus **(L.) Stackl.**
Gigartinaceae
♦ Irish moss, carrageen moss, carraigín, chondrus, carrahan
♦ Quarantine Weed
♦ 220
♦ algae, herbal.

Chonemorpha fragrans **(Moon) Alston**
Apocynaceae
♦ Naturalised
♦ 86
♦ cultivated. Origin: Asia.

Chorisia speciosa **A.St.-Hil.**
Bombacaceae
♦ floss silk tree, chorisia
♦ Introduced
♦ 228, 261
♦ T, arid, cultivated. Origin: Argentina, Brazil, Paraguay.

Chorispora syriaca **Boiss.**
Brassicaceae
♦ Weed
♦ 87, 88, 243

Chorispora tenella (Pall.) DC.
Brassicaceae
♦ blue mustard, tenella mustard, purple mustard, crossflower, chorispora, beanpodded mustard, irsokki
♦ Weed, Noxious Weed, Naturalised, Introduced, Casual Alien, Cultivation Escape
♦ 23, 34, 35, 42, 80, 86, 87, 88, 101, 136, 146, 161, 174, 198, 212, 218, 229, 243, 252, 272, 275, 287, 297
♦ aH, arid, cultivated, herbal. Origin: Eurasia.

Christella dentata (Forssk.) Brownsey & Jermy agg.
Thelypteridaceae
♦ Weed
♦ 15
♦ cultivated.

Christia verspertilionis (L.f.) Bakh.f.
Fabaceae/Papilionaceae
♦ East Indian island pea
♦ Naturalised
♦ 101

Chromolaena ivaefolia (L.) R.M.King & H.Rob.
Asteraceae
♦ Weed
♦ 179

Chromolaena odorata (L.) R.M.King & H.Rob.
Asteraceae
Eupatorium odoratum L. (see)
♦ Siam weed, paraffienbos, chromolaena, triffid weed, coacihuizpatli, Christmas bush, Armstrong's weed, eupatorium, kingweed, paraffinbush, paraffinweed, turpentine weed, bitter bush, herbe du Laos, kesengesil, masigsig, hagonoy, agonoi, huluhagonoi, mahsrihsrihk, Jack in the bush, Christmas rose, hemp agrimony
♦ Weed, Quarantine Weed, Noxious Weed, Introduced, Garden Escape, Environmental Weed
♦ 3, 6, 10, 11, 12, 18, 22, 37, 54, 54, 62, 63, 76, 87, 88, 93, 95, 107, 121, 135, 135, 147, 152, 155, 158, 186, 191, 203, 209, 220, 228, 229, 243, 277, 278, 279, 283
♦ pS, arid, cultivated, herbal, toxic. Origin: Americas.

Chroodactylon ramosum (Thwaites) Hansg.
Goniotrichaceae
♦ red algae
♦ Weed
♦ 197
♦ algae.

Chrozophora obliqua (Vahl) Spreng.
Euphorbiaceae
♦ Weed
♦ 88, 221

Chrozophora plicata (Vahl) A.Juss.
Euphorbiaceae
Croton plicatus Vahl
♦ Weed, Quarantine Weed

♦ 39, 76, 87, 88, 185, 203, 220, 221, 242
♦ arid, herbal, toxic.

Chrozophora rottleri (Giesel) A.Juss. ex Spreng.
Euphorbiaceae
♦ Weed
♦ 87, 88
♦ herbal.

Chrozophora tinctoria (L.) A.Juss.
Euphorbiaceae
♦ dyer's litmus, croton, giradol
♦ Weed, Quarantine Weed, Naturalised
♦ 7, 39, 70, 76, 86, 87, 88, 94, 98, 101, 115, 198, 203, 220, 221, 243, 272
♦ herbal, toxic. Origin: Mediterranean, Middle East.

Chrozophora verbascifolia (Willd.) A.Juss.
Euphorbiaceae
♦ Weed
♦ 87, 88
♦ herbal.

Chrysalidocarpus lutescens (Bory) H.Wendl.
Arecaceae
= *Dypsis lutescens* (H.Wendl.) Beentje & Dransf.
♦ gold cane palm
♦ Introduced, Garden Escape
♦ 230, 261
♦ T, cultivated, herbal. Origin: Madagascar.

Chrysanthellum americanum (L.) Vatke
Asteraceae
♦ Weed
♦ 14, 87, 88

Chrysanthemoides monilifera (L.) Norl.
Asteraceae
Osteospermum moniliferum L. (see)
♦ boneseed, bitou bush, brother berry, bush tickberry, South African star bush, jungle weed, jungle flower, saltbush, Higgin's curse, Mort's curse, bietou
♦ Weed, Quarantine Weed, Noxious Weed, Naturalised, Native Weed, Introduced, Garden Escape, Environmental Weed
♦ 7, 15, 62, 72, 73, 76, 88, 98, 121, 155, 165, 171, 178, 181, 198, 203, 220, 225, 228, 241, 246, 280, 300
♦ pS, arid, cultivated, herbal. Origin: South Africa.

Chrysanthemoides (L.) Norl. monilifera ssp. monilifera
Asteraceae
♦ boneseed
♦ Weed, Noxious Weed, Naturalised, Garden Escape, Environmental Weed
♦ 86, 147, 152, 155, 176, 198, 269, 289, 296
♦ S/T, cultivated. Origin: South Africa.

Chrysanthemoides monilifera (L.) Norl. ssp. rotundata (DC.) Norl.
Asteraceae
♦ bitou bush

♦ Weed, Noxious Weed, Naturalised, Garden Escape, Environmental Weed
♦ 72, 86, 88, 147, 152, 155, 198, 269, 289, 296
♦ S, cultivated. Origin: South Africa.

Chrysanthemum L. spp.
Asteraceae
♦ chrysanthemum, pyrethrum, daisy
♦ Weed
♦ 39, 272
♦ cultivated, herbal, toxic.

Chrysanthemum balsamita L.
Asteraceae
= *Tanacetum balsamita* L. ssp. *balsamitoides* (Sch.Bip.) Grierson (NoR)
♦ costmary
♦ Weed
♦ 39, 80
♦ cultivated, herbal, toxic.

Chrysanthemum carinatum Schousb.
Asteraceae
= *Glebionis carinata* (Schousb.) Tzvelev (NoR)
♦ tricolor daisy
♦ Weed, Naturalised, Cultivation Escape, Casual Alien
♦ 34, 101, 280
♦ aH, cultivated, herbal. Origin: obscure, possibly North Africa.

Chrysanthemum cinerariifolium (Trevir.) Vis.
Asteraceae
= *Tanacetum cinerariifolium* (Trevir.) Sch.Bip.
♦ chrysanthemum, Dalmatian insect flower, Dalmatian pyrethrum, pyrethrum
♦ Weed
♦ 23, 39, 88, 121
♦ pH, cultivated, herbal, toxic. Origin: Eurasia.

Chrysanthemum coccineum Willd.
Asteraceae
= *Tanacetum coccineum* (Willd.) Grierson
♦ pyrethrum daisy
♦ Naturalised
♦ 101
♦ cultivated, herbal.

Chrysanthemum coronarium L.
Asteraceae
♦ crown daisy, garland chrysanthemum, chop suey greens
♦ Weed, Naturalised, Introduced, Cultivation Escape, Casual Alien
♦ 7, 34, 40, 42, 51, 70, 86, 87, 88, 94, 98, 101, 111, 115, 121, 203, 221, 228, 241, 243, 252, 256, 272, 280, 297, 300
♦ aH, arid, cultivated, herbal. Origin: Eurasia.

Chrysanthemum corymbosum L.
Asteraceae
= *Tanacetum corymbosum* (L.) Sch.Bip.
♦ Weed
♦ 80
♦ cultivated, herbal.

Chrysanthemum frutescens L.
Asteraceae
= *Argyranthemum frutescens* (L.)

Sch.Bip.
♦ marguerite
♦ Naturalised
♦ 101
♦ cultivated, herbal.

Chrysanthemum indicum L.
Asteraceae
Dendranthema indicum (L.) Des Moul.
♦ Weed
♦ 87, 88
♦ cultivated, herbal.

Chrysanthemum leucanthemum L.
Asteraceae
= *Leucanthemum vulgare* Lam.
♦ oxeye daisy, white daisy, whiteweed, field daisy, marguerite, poorland flower, margarita, moon daisy
♦ Weed, Noxious Weed, Naturalised, Environmental Weed, Cultivation Escape
♦ 23, 36, 44, 52, 80, 87, 88, 102, 121, 136, 138, 146, 161, 162, 195, 210, 211, 212, 217, 218, 219, 264, 267, 286, 287, 290, 295, 299
♦ pH, cultivated, herbal. Origin: Eurasia.

Chrysanthemum leucanthemum L. var. pinnatifidum Lecoq & Lam.
Asteraceae
= *Leucanthemum vulgare* Lam.
♦ oxeye daisy, field oxeye daisy
♦ Weed
♦ 207, 218
♦ Origin: Europe
.

Chrysanthemum maximum Ramond
Asteraceae
= *Leucanthemum maximum* (Ramond) DC.
♦ jastrun wielki
♦ Weed, Cultivation Escape
♦ 80, 261
♦ cultivated, herbal. Origin: Europe.

Chrysanthemum myconis L.
Asteraceae
= *Coleostephus myconis* (L.) Cass.
♦ malo me quer, mal me quer amarelo
♦ Weed, Quarantine Weed
♦ 76, 87, 88, 203, 220, 255, 295
♦ Origin: Mediterranean.

Chrysanthemum paludosum Poir.
Asteraceae
= *Leucanthemum paludosum* (Poir.) Bonnet & Barratte
♦ whitebuttons
♦ Weed
♦ 15
♦ aH, cultivated, herbal.

Chrysanthemum parthenium (L.) Bernh.
Asteraceae
= *Tanacetum parthenium* (L.) Sch.Bip.
♦ feverfew, batchelor's buttons
♦ Weed
♦ 39, 80, 87, 88, 218
♦ cultivated, herbal, toxic.

Chrysanthemum segetum L.
Asteraceae

= *Glebionis segeta* (L.) Fourr. (NoR)
♦ corn chrysanthemum, corn marigold, keltapäivänkakkara, corn daisy
♦ Weed, Naturalised, Introduced, Casual Alien
♦ 15, 23, 24, 34, 40, 42, 44, 51, 70, 86, 87, 88, 94, 98, 101, 118, 121, 158, 161, 203, 243, 250, 253, 272, 280, 300
♦ aH, cultivated, herbal. Origin: Eurasia.

Chrysanthemum suaveolens Asch.
Asteraceae
♦ sweet feverfew
♦ Weed
♦ 87, 88
♦ herbal.

Chrysanthemum × superbum J.W.Ingram
Asteraceae
= *Leucanthemum lacustre* (Brot.) Samp. × *Leucanthemum maximum* (Ramond) DC. [*Leucanthemum × superbum* (Bergmans ex J.W.Ingram) Kent]
♦ shasta daisy
♦ Weed
♦ 80
♦ cultivated, herbal.

Chrysanthemum tanacetum Vis.
Asteraceae
= *Tanacetum vulgare* L.
♦ Weed
♦ 87, 88

Chrysanthemum vulgare (Lam.) Parsa
Asteraceae
= *Tanacetum vulgare* L.
♦ tansy
♦ Weed
♦ 87, 88
♦ cultivated, herbal.

Chrysobalanus icaco L.
Chrysobalanaceae
♦ coco plum, icaco, icaque, icaco coco plum
♦ Weed, Introduced
♦ 3, 22, 191, 228
♦ T, arid, cultivated, herbal.

Chrysocoma coma-aurea L.
Asteraceae
♦ shrub goldilocks
♦ Quarantine Weed, Naturalised
♦ 39, 40, 220
♦ cultivated, toxic. Origin: South Africa.

Chrysocoma tenuifolia Berg.
Asteraceae
♦ bitter bush
♦ Weed, Native Weed
♦ 39, 121
♦ S, herbal, toxic. Origin: southern Africa.

Chrysoma laricifolia Greene
Asteraceae
= *Aster laricifolius* Kuntze (NoR)
♦ Quarantine Weed
♦ 220

Chrysophyllum cainito L.
Sapotaceae

♦ starapple
♦ Weed, Introduced
♦ 179, 230
♦ cultivated, herbal.

Chrysophyllum oliviforme L.
Sapotaceae
♦ satin leaf, caimitillo
♦ Weed
♦ 3, 191
♦ herbal.

Chrysopogon aciculatus (Retz.) Trin.
Poaceae
Andropogon aciculatus Retz., *Rhaphis aciculatus* (Retz.) Honda
♦ Mackie's pest, seed grass, golden beardgrass, seedy grass, inifuk, matapekepeke, herbe plate, lovegrass, pilipiliula, golden false beardgrass, reh takai
♦ Weed, Noxious Weed, Naturalised
♦ 3, 12, 67, 87, 88, 90, 93, 101, 107, 140, 161, 191, 209, 229, 235, 262, 273, 274, 275, 286, 297
♦ pG, cultivated, herbal. Origin: Asia, Pacific.

Chrysopogon fulvus (Spreng.) Chiov.
Poaceae
Andropogon monticolor Roem. & Schult., *Chrysopogon montanus* Trin. ex Spreng. (see), *Pollinia fulva* Spreng., *Rhaphis montana* ined.
♦ red false beardgrass
♦ Naturalised, Introduced
♦ 101, 228
♦ G, arid.

Chrysopogon gryllus (L.) Trin.
Poaceae
Andropogon gryllus L.
♦ Introduced
♦ 228
♦ G, arid, cultivated, herbal.

Chrysopogon montanus Trin. ex Spreng.
Poaceae
= *Chrysopogon fulvus* (Spreng.) Chiov.
♦ Weed
♦ 87, 88
♦ G.

Chrysopogon orientalis (Desv.) A.Camus
Poaceae
♦ Weed
♦ 87, 88
♦ G.

Chrysopsis pilosa Nutt.
Asteraceae
♦ soft goldenaster
♦ Weed
♦ 161
♦ herbal.

Chrysothamnus graveolens (Nutt.) Greene
Asteraceae
= *Ericameria nauseosa* (Pall. ex Pursh) Nesom & Baird ssp. *nauseosa* var. *glabrata* (Gray) Nesom & Baird (NoR)
♦ greenplume rabbitbrush
♦ Weed
♦ 87, 88, 218
♦ S, promoted, herbal.

Chrysothamnus greenei (A.Gray) Greene
  Asteraceae
  ♦ Greene's rabbitbrush
  ♦ Weed
  ♦ 87, 88, 218
  ♦ S, arid, herbal.

Chrysothamnus nauseosus (Pall. ex Pursh) Britt.
  Asteraceae
  = *Ericameria nauseosa* (Pall. ex Pursh) Nesom & Baird ssp. *nauseosa* var. *nauseosa* (NoR)
  ♦ rubber rabbitbrush, gray rabbitbrush
  ♦ Weed
  ♦ 39, 87, 88, 161, 212, 218
  ♦ S, arid, promoted, herbal, toxic.

Chrysothamnus parryi (A.Gray) Greene
  Asteraceae
  = *Ericameria parryi* Nesom & Baird var. *parryi* (Gray) (NoR)
  ♦ Parry rabbitbrush
  ♦ Weed
  ♦ 87, 88, 218
  ♦ S, herbal.

Chrysothamnus pulchellus (A.Gray) Greene
  Asteraceae
  ♦ south-west rabbitbrush
  ♦ Weed
  ♦ 87, 88, 218
  ♦ arid, herbal.

Chrysothamnus viscidiflorus (Hook) Nutt.
  Asteraceae
  ♦ green rabbitbrush, Douglas rabbitbrush, stickyleaf rabbitbrush
  ♦ Weed
  ♦ 87, 88, 161, 212, 213, 218
  ♦ S, promoted, herbal, toxic.

Chrysothamnus viscidiflorus ssp. viscidiflorus (Hook.) Nutt. var. stenophyllus (Gray) Hall
  Asteraceae
  ♦ small rabbitbrush, yellow rabbitbrush
  ♦ Weed
  ♦ 218

Chrysothamnus viscidiflorus (Hook.) Nutt. var. lanceolatus (Nutt.) Greene
  Asteraceae
  = *Chrysothamnus viscidiflorus* (Hook.) Nutt. ssp. *lanceolatus* (Nutt.) Hall & Clem.
  ♦ lanceleaf rabbitbrush
  ♦ Weed
  ♦ 218

Chrysothemis pulchella (Donn ex Sims) Decne.
  Gesneriaceae
  ♦ aurinkokello
  ♦ Naturalised, Garden Escape
  ♦ 101, 261
  ♦ cultivated, herbal.

Chukrasia tabularis A.Juss.
  Meliaceae
  ♦ East Indian mahogany
  ♦ Environmental Weed
  ♦ 260

♦ T, herbal.

Chukrasia velutina M.Roem.
  Meliaceae
  ♦ Weed, Sleeper Weed, Naturalised, Environmental Weed
  ♦ 3, 86, 155, 191

Chusquea abietifolia Griseb.
  Poaceae
  ♦ climbing bamboo
  ♦ Weed
  ♦ 14
  ♦ G.

Chymococca empetroides Meissn.
  Thymelaeaceae
  = *Passerina ericoides* L.
  ♦ Quarantine Weed
  ♦ 220

Cicendia filiformis (L.) Delarbre
  Gentianaceae
  ♦ slender cicendia, yellow gentianella, yellow centaury
  ♦ Weed, Naturalised, Garden Escape, Environmental Weed
  ♦ 7, 9, 72, 86, 88, 98, 176, 198, 203
  ♦ aH, cultivated. Origin: Eurasia, North Africa.

Cicendia quadrangularis (Dombey ex Lambert) Griseb.
  Gentianaceae
  ♦ square cicendia, Oregon timwort
  ♦ Weed, Naturalised, Environmental Weed
  ♦ 7, 72, 86, 88, 98, 198, 203
  ♦ aH, herbal. Origin: western North America.

Cicer arietinum L.
  Fabaceae/Papilionaceae
  ♦ kahviherne, chickpea, Egyptian pea
  ♦ Naturalised, Casual Alien
  ♦ 42, 86, 98, 101, 261
  ♦ aH, cultivated, herbal. Origin: Mediterranean.

Cicerbita macrophylla (Willd.) Wallr.
  Asteraceae
  Lactuca macrophylla A.Gray, *Sonchus macrophyllus* Willd.
  ♦ blue sow thistle
  ♦ Weed
  ♦ 70
  ♦ cultivated, herbal.

Cichorium endivia L.
  Asteraceae
  Cichorium esculentum Salisb.
  ♦ chicory, endiivi, cultivated endive
  ♦ Weed, Naturalised, Casual Alien
  ♦ 42, 86, 87, 88, 94, 98, 101, 185, 203, 243
  ♦ aH, cultivated, herbal. Origin: Eurasia.

Cichorium endivia L. ssp. divaricatum (Schou.) Sell
  Asteraceae
  ♦ Weed
  ♦ 70

Cichorium intybus L.
  Asteraceae
  ♦ chicory, coffeeweed, bachelor's

button, blue daisy, blue dandelion, blue sailors, succory, bunk, witchgrass
  ♦ Weed, Noxious Weed, Naturalised, Introduced, Garden Escape, Environmental Weed, Casual Alien
  ♦ 7, 23, 24, 34, 36, 39, 40, 42, 44, 49, 51, 52, 55, 70, 80, 86, 87, 88, 94, 98, 101, 102, 121, 136, 138, 158, 161, 165, 174, 176, 180, 195, 198, 199, 203, 207, 210, 211, 212, 217, 218, 228, 229, 237, 241, 243, 249, 251, 253, 255, 256, 269, 272, 275, 280, 287, 290, 295, 300
  ♦ pH, arid, cultivated, herbal, toxic. Origin: Eurasia, North Africa.

Cichorium pumilum Jacq.
  Asteraceae
  ♦ Weed, Quarantine Weed
  ♦ 76, 87, 88, 203, 220, 221

Cichorium spinosum L.
  Asteraceae
  ♦ Weed
  ♦ 87, 88
  ♦ b/pH, promoted.

Ciclospermum leptophyllum (Pers.) Britton & E.Wilson
  Apiaceae
  = *Cyclospermum leptophyllum* (Pers.) Sprague ex Britt. & Wilson
  ♦ marsh parsley
  ♦ Weed, Naturalised, Environmental Weed
  ♦ 7, 34, 86, 88, 93, 121, 134, 158, 176, 198, 205, 280
  ♦ aH, herbal. Origin: Central America.

Cicuta bolanderi S.Watson
  Apiaceae
  = *Cicuta maculata* L. var. *bolanderi* (Wats.) Mulligan
  ♦ Boland's waterhemlock
  ♦ Weed
  ♦ 39, 161
  ♦ herbal, toxic.

Cicuta bulbifera L.
  Apiaceae
  ♦ bulblet bearing waterhemlock, bulbous waterhemlock
  ♦ Weed
  ♦ 39, 87, 88, 161
  ♦ herbal, toxic.

Cicuta douglasii (DC.) Coult. & Rose
  Apiaceae
  ♦ western waterhemlock, waterhemlock
  ♦ Weed, Quarantine Weed, Noxious Weed, Native Weed
  ♦ 36, 39, 76, 80, 87, 88, 161, 203, 212, 218, 220, 264, 293
  ♦ waH, herbal, toxic. Origin: North America.

Cicuta mackenzieana Raup
  Apiaceae
  = *Cicuta virosa* L.
  ♦ Weed, Quarantine Weed
  ♦ 76, 87, 88, 203, 220
  ♦ herbal.

Cicuta maculata L.
  Apiaceae

♦ spotted waterhemlock, waterhemlock, poison parsnip, common waterhemlock, spotted parsley, spotted cowbane
♦ Weed, Noxious Weed, Native Weed
♦ 8, 23, 39, 52, 87, 88, 161, 174, 189, 210, 218, 229, 247, 293, 299
♦ pH, herbal, toxic. Origin: North America.

Cicuta maculata L. var. *angustifolia* **Hook.**
Apiaceae
♦ waterhemlock, spotted waterhemlock
♦ Noxious Weed
♦ 229
♦ pH.

Cicuta maculata L. var. *bolanderi* **(Wats.) Mulligan**
Apiaceae
Cicuta bolanderi S.Watson (see)
♦ waterhemlock, spotted waterhemlock, saltmarsh waterhemlock
♦ Noxious Weed
♦ 229
♦ pH, aqua.

Cicuta maculata L. var. *maculata*
Apiaceae
♦ waterhemlock, spotted waterhemlock
♦ Noxious Weed
♦ 229

Cicuta virosa L.
Apiaceae
Cicuta mackenzieana Raup (see), *Sium cicuta* Weber
♦ cowbane, MacKenzie waterhemlock, myrkkykeiso
♦ Weed, Quarantine Weed, Naturalised
♦ 23, 39, 76, 86, 87, 88, 161, 203, 220, 272, 275, 286, 293, 297
♦ pH, cultivated, herbal, toxic.

Cinchona officinalis L.
Rubiaceae
♦ quinine
♦ Weed
♦ 22
♦ T, cultivated, herbal.

Cinchona pubescens **Vahl**
Rubiaceae
Cinchona succirubra Pav. ex Klotsch (see)
♦ quinine tree, quinine
♦ Weed, Naturalised, Environmental Weed
♦ 3, 22, 101, 132, 191, 257
♦ T, cultivated, herbal.

Cinchona succirubra **Pav. ex Klotsch**
Rubiaceae
= *Cinchona pubescens* Vahl
♦ red quinine tree, Peruvian bark
♦ Weed, Environmental Weed
♦ 3, 18, 22, 88, 132, 152, 191
♦ T, herbal. Origin: South America.

Cineraria abyssinica **Sch.Bip. ex A.Rich.**
Asteraceae
♦ Weed
♦ 88

Cineraria aspera **Thunb.**
Asteraceae
♦ Native Weed
♦ 121
♦ H, herbal. Origin: southern Africa.

Cineraria lobata **L'Hér.**
Asteraceae
♦ Native Weed
♦ 121
♦ pH. Origin: southern Africa.

Cineraria lyrata **DC.**
Asteraceae
♦ cineraria, African marigold, wild parsely
♦ Weed, Quarantine Weed, Noxious Weed, Native Weed, Naturalised
♦ 51, 54, 76, 86, 87, 88, 98, 121, 147, 203, 269
♦ aH, toxic. Origin: southern Africa.

Cinnamomum burmannii **(Nees & T.Nees) Nees ex Blume**
Lauraceae
♦ padang cassia, cinnamon tree
♦ Weed, Naturalised, Cultivation Escape
♦ 3, 101, 191, 233
♦ cultivated, herbal.

Cinnamomum camphora **(L.) J.Presl**
Lauraceae
Camphora camphora (L.) H.Karst., *Camphora officinarum* Nees, *Cinnamomum camphoroides* Hayata, *Cinnamomum nominale* (Hayata) Hayata, *Cinnamomum simondii* Lecomte, *Laurus camphora* L., *Persea camphora* (L.) Spreng.
♦ camphor tree, camphor laurel
♦ Weed, Noxious Weed, Naturalised, Introduced, Garden Escape, Environmental Weed
♦ 3, 7, 22, 73, 80, 86, 88, 95, 98, 101, 107, 112, 151, 152, 155, 161, 179, 201, 203, 230, 261, 269, 283, 290, 296
♦ T, cultivated, herbal, toxic. Origin: China, Taiwan, Japan.

Cinnamomum carolinense **Koidz.**
Lauraceae
♦ madeu
♦ Introduced
♦ 230
♦ T.

Cinnamomum sieboldii **Meisn.**
Lauraceae
♦ Naturalised
♦ 287
♦ herbal.

Cinnamomum verum **Presl**
Lauraceae
Cinnamomum zeylanicum Bl. (see)
♦ cinnamon tree, tigamoni, cinnamon, ochod ra ngebard, canela, cynamonowiec
♦ Weed, Naturalised, Introduced
♦ 3, 101, 107, 191, 230, 261
♦ T, cultivated, herbal, toxic. Origin: south-east Asia.

Cinnamomum zeylanicum **Breyn.**
Lauraceae
= *Cinnamomum verum* Presl
♦ cinnamon, cannella
♦ Weed, Environmental Weed
♦ 3, 22, 87, 88, 152
♦ T, cultivated, herbal. Origin: East Indies.

Cipadessa baccifera **(Roth) Miq.**
Meliaceae
♦ Quarantine Weed
♦ 220
♦ cultivated, herbal.

Cipadessa cinerascens **(Pellegr.) Hand.-Mazz.**
Meliaceae
Cipadessa baccifera (Roth) Miq. var. *sinensis* Rehder & E.H.Wilson, *Cipadessa fruticosa* Bl. var. *cinerascens* Pellegr., *Cipadessa sinensis* (Rehder & E.H.Wilson) Hand.-Mazz., *Rhus blinii* H.Lév.
♦ Quarantine Weed
♦ 220

Cipadessa fruticosa **Bl.**
Meliaceae
♦ Quarantine Weed
♦ 220

Cipocereus **F.Ritter spp.**
Cactaceae
♦ Weed, Quarantine Weed
♦ 76, 88, 203, 220

Circaea alpina L.
Onagraceae
♦ alpine enchanter's nightshade, small enchanter's nightshade, nightshade
♦ Weed
♦ 23, 88
♦ herbal.

Circaea lutetiana L.
Onagraceae
♦ enchanter's nightshade, broadleaf enchanter's nightshade
♦ Weed
♦ 39, 272
♦ cultivated, herbal, toxic.

Cirinosum **Neck. spp.**
Cactaceae
= *Cereus* Mill. spp.
♦ Quarantine Weed
♦ 220

Cirsium **Mill spp.**
Asteraceae
♦ thistle
♦ Weed, Naturalised, Environmental Weed
♦ 198, 225, 246, 272
♦ herbal.

Cirsium acarna **(L.) Moench**
Asteraceae
= *Picnomon acarna* (L.) Cass.
♦ soldier thistle
♦ Weed, Quarantine Weed, Noxious Weed, Naturalised
♦ 76, 86, 87, 88, 198, 203
♦ herbal.

Cirsium acaule Scop.
Asteraceae
Carduus acaulis L.
- dwarf thistle
- Weed
- 87, 88, 272
- cultivated, herbal.

Cirsium altissimum (L.) Spreng.
Asteraceae
- tall thistle, roadside thistle
- Weed, Noxious Weed, Native Weed
- 87, 88, 161, 174, 210, 218, 229
- bH, herbal. Origin: North America.

Cirsium arvense (L.) Scop.
Asteraceae
Breea setosa (Bieb.) Kitam. (see), *Breea setosum* (Bieb.) Kitam. (see), *Carduus arvensis* (L.) E.Robson (see), *Cirsium arvense* (L.) Scop. var. *argenteum* (Vest) Fiori (see), *Cirsium arvense* (L.) Scop. var. *integrifolium* (Wimm. & Grab.) (see), *Cirsium arvense* (L.) Scop. var. *mite* Wimm. & Grab. (see), *Cirsium arvense* (L.) Scop. var. *vestitum* Wimm. & Grab. (see), *Cirsium setosum* (Willd.) Bess. ex Bieb. (see), *Cirsium lanatum* (Roxb. ex Willd.) Spreng. (see), *Cirsium incanum* (S.G.Gmel.) Fisch. ex M.Bieb., *Cnicus arvensis* Hoff., *Serratula arvensis* L.
- Canada thistle, perennial thistle, Californian thistle, creeping thistle, field thistle, cursed thistle, small flowered thistle, perennial thistle, hard thistle, stoppione o cardo campestre
- Weed, Quarantine Weed, Noxious Weed, Naturalised, Introduced, Environmental Weed
- 1, 4, 7, 8, 15, 16, 21, 23, 26, 34, 35, 36, 39, 44, 45, 48, 49, 51, 52, 62, 67, 68, 76, 78, 80, 86, 87, 88, 94, 98, 101, 102, 103, 104, 114, 116, 118, 121, 129, 129, 130, 133, 136, 138, 139, 141, 146, 147, 151, 161, 162, 165, 169, 171, 172, 174, 176, 181, 186, 195, 198, 203, 204, 210, 211, 212, 218, 219, 220, 222, 224, 229, 231, 241, 243, 253, 263, 264, 266, 267, 269, 272, 280, 286, 287, 294, 295, 299, 300
- aH, promoted, herbal, toxic. Origin: Eurasia.

Cirsium arvense (L.) Scop. var. argenteum (Vest) Fiori
Asteraceae
= *Cirsium arvense* (L.) Scop.
- Naturalised
- 287

Cirsium arvense (L.) Scop. var. integrifolium Wimm. & Grab.
Asteraceae
= *Cirsium arvense* (L.) Scop.
- Canada thistle, field thistle, cursed thistle, small flowered thistle, perennial thistle, hard thistle
- Weed
- 49

Cirsium arvense (L.) Scop. var. mite Wimm. & Grab.
Asteraceae
= *Cirsium arvense* (L.) Scop.

Cirsium arvense var. (L.) Scop. vestitum Wimm. & Grab.
Asteraceae
= *Cirsium arvense* (L.) Scop.
- Canada thistle, field thistle, cursed thistle, small flowered thistle, perennial thistle, hard thistle
- Weed
- 49

Cirsium brachycephalum Jurat.
Asteraceae
- Weed
- 272

Cirsium brevistylum Cronquist
Asteraceae
- clustered thistle
- Naturalised
- 280
- pH, promoted, herbal.

Cirsium canescens Nutt.
Asteraceae
- Platte thistle, prairie thistle
- Weed, Native Weed
- 161, 174, 212
- herbal. Origin: North America.

Cirsium canum (L.) All.
Asteraceae
- Queen Anne's thistle
- Weed, Naturalised
- 101, 272

Cirsium centaureae (Rydb.) K.Schum.
Asteraceae
= *Cirsium remotifolium* (Hook.) DC. ssp. *oregonense* Petrak (NoR)
- Weed
- 23, 88
- herbal.

Cirsium costaricense (Polak.) Petrak.
Asteraceae
- Weed
- 157
- bH.

Cirsium discolor (Muhl. ex Willd.) Spreng.
Asteraceae
- field thistle, pasture thistle
- Weed
- 161, 207
- herbal. Origin: North America.

Cirsium durangense (Greenm.) G.B.Ownbey
Asteraceae
- Weed
- 199

Cirsium eriophorum Scop.
Asteraceae
Carduus eriophorus L., *Cnicus eripohorus* Roth
- woolly thistle
- Weed
- 70, 87, 88, 272
- bH, cultivated, herbal.

Cirsium erisithales (Jacq.) Scop.
Asteraceae
Cnicus erisithales L., *Cirsium*

ochroleucum DC.
- yellow melancholy thistle
- Weed
- 272
- cultivated, herbal.

Cirsium esculentum C.A.Mey.
Asteraceae
- Weed
- 275, 297
- pH.

Cirsium flodmanii (Rydb.) Arthur
Asteraceae
- Flodman's thistle
- Weed, Native Weed
- 49, 87, 88, 161, 174, 210, 218
- bH, herbal. Origin: North America.

Cirsium foliosum (Hook.) DC.
Asteraceae
- leafy thistle, elk thistle, Drummond's thistle, meadow thistle
- Weed
- 161, 212
- pH, promoted, herbal.

Cirsium helenioides (L.) Hill
Asteraceae
Cirsium heterophyllum (L.) Hill (see)
- common melancholy thistle, huopaohdake, melancholy thistle
- Weed
- 272
- cultivated, herbal.

Cirsium heterophyllum (L.) Hill
Asteraceae
= *Cirsium helenioides* (L.) Hill
- melancholy thistle
- Weed
- 23, 87, 88
- promoted, herbal.

Cirsium hillii (Canby) Fern.
Asteraceae
- Hill's thistle
- Weed
- 23, 88

Cirsium horridulum Michx.
Asteraceae
- yellow thistle
- Weed
- 23, 87, 88, 161, 218, 249
- herbal.

Cirsium japonicum DC.
Asteraceae
- noazami, Japanese thistle
- Weed, Quarantine Weed
- 76, 87, 88, 204, 220, 274, 286
- b/pH, cultivated, herbal. Origin: east Asia.

Cirsium lanatum (Roxb. ex Willd.) Spreng.
Asteraceae
= *Cirsium arvense* (L.) Scop.
- hairy thistle
- Weed
- 297

Cirsium lanceolatum (L.) Scop. non Hill
Asteraceae
= *Cirsium vulgare* (Savi) Ten.
- bull thistle, common thistle, spear

thistle, plume thistle
♦ Weed
♦ 49, 80, 88
♦ bH, herbal. Origin: Eurasia.

Cirsium leo Nakai & Kitag.
Asteraceae
♦ Weed
♦ 275
♦ pH.

Cirsium libanoticum DC.
Asteraceae
♦ Weed
♦ 87, 88

Cirsium maritimum Makino
Asteraceae
♦ Weed
♦ 286
♦ pH, promoted. Origin: Japan.

Cirsium mexicanum (DC.) Hemsl.
Asteraceae
♦ Mexican thistle
♦ Weed
♦ 14
♦ herbal.

Cirsium muticum Michx.
Asteraceae
♦ swamp thistle
♦ Weed
♦ 23, 88
♦ herbal.

Cirsium nipponicum (Maxim.) Mak.
Asteraceae
♦ Weed, Quarantine Weed
♦ 88, 204, 220
♦ pH, promoted.

Cirsium ochrocentrum Gray
Asteraceae
♦ yellowspine thistle
♦ Weed, Noxious Weed, Native Weed
♦ 35, 88, 161, 174, 212, 229
♦ pH, promoted, herbal. Origin: North America.

Cirsium oleraceum (L.) Scop.
Asteraceae
Carduus oleraceus Vill., *Cnicus oleraceus* L.
♦ cabbage thistle, keltaohdake
♦ Weed, Naturalised
♦ 23, 42, 70, 87, 88, 243, 272
♦ pH, cultivated, herbal.

Cirsium palustre (L.) Scop.
Asteraceae
Carduus palustris L., *Cnicus palustris* Willd.
♦ marsh thistle, European swamp thistle
♦ Weed, Naturalised
♦ 15, 44, 70, 87, 88, 101, 133, 165, 195, 272, 280
♦ bH, promoted, herbal. Origin: Eurasia, North Africa.

Cirsium pastoris J.T.Howell
Asteraceae
♦ snowy thistle
♦ Weed
♦ 161, 180
♦ herbal.

Cirsium pendulum Fisch.
Asteraceae
♦ takaazami
♦ Weed
♦ 275, 297
♦ pH, promoted, herbal.

Cirsium pitcheri (Torr. ex Eat.) Torr. & Gray
Asteraceae
♦ sanddune thistle
♦ Weed
♦ 23, 88
♦ herbal.

Cirsium pumilum Spreng.
Asteraceae
♦ pasture thistle
♦ Weed
♦ 87, 88, 218
♦ herbal.

Cirsium pyrenaicum (Jacq.) All.
Asteraceae
♦ Naturalised
♦ 101
♦ cultivated.

Cirsium rhaphilepis (Hemsl.) Petr.
Asteraceae
♦ Weed
♦ 199

Cirsium rivulare (Jacq.) All.
Asteraceae
Carduus rivularis Jacq.
♦ Weed
♦ 272
♦ cultivated, herbal.

Cirsium scabrum (Poir.) Bonnet & Barratte
Asteraceae
♦ rough thistle
♦ Naturalised
♦ 101

Cirsium segetum Bunge
Asteraceae
Breea segetum (Bunge) Kitam
♦ Weed
♦ 87, 88, 297
♦ pH, promoted.

Cirsium serrulatum Bieb.
Asteraceae
♦ Weed
♦ 87, 88

Cirsium setosum (Willd.) Bess. ex Bieb.
Asteraceae
= *Cirsium arvense* (L.) Scop.
♦ setose thistle
♦ Weed
♦ 297

Cirsium syriacum (L.) Gaertn.
Asteraceae
= *Notobasis syriaca* (L.) Cass.
♦ Weed
♦ 221

Cirsium tanakae (Franch. & Savat.) Matsum.
Asteraceae
♦ Weed
♦ 88, 204
♦ pH, promoted.

Cirsium undulatum (Nutt.) Spreng.
Asteraceae
♦ wavyleaf thistle, gray thistle
♦ Weed, Noxious Weed, Native Weed, Introduced
♦ 23, 34, 35, 87, 88, 161, 174, 212, 229
♦ pH, promoted, herbal. Origin: North America.

Cirsium undulatum (Nutt.) Spreng. var. tracyi (Rydb.) Welsh
Asteraceae
♦ wavyleaf thistle, Tracy's thistle
♦ Noxious Weed
♦ 229

Cirsium undulatum (Nutt.) Spreng. var. undulatum
Asteraceae
♦ wavyleaf thistle
♦ Noxious Weed
♦ 229

Cirsium vulgare (Savi) Ten.
Asteraceae
Carduus lanceolatus L., *Cirsium lanceolatum* (L.) Scop. *non* Hill (see), *Cnicus lanceolatus* Willd.
♦ bull thistle, spear thistle, Scotch thistle, black thistle, common thistle, common bull thistle, Fuller's thistle, swamp thistle, piikkiohdake, skotse dissel, speerdissel
♦ Weed, Noxious Weed, Naturalised, Introduced, Environmental Weed
♦ 1, 7, 9, 15, 20, 23, 24, 34, 35, 38, 44, 45, 51, 52, 55, 63, 70, 72, 78, 80, 86, 87, 88, 94, 95, 98, 101, 102, 116, 133, 139, 146, 147, 158, 161, 165, 171, 174, 176, 180, 195, 198, 203, 210, 211, 212, 217, 218, 219, 229, 231, 236, 237, 241, 243, 253, 255, 263, 267, 269, 271, 272, 278, 280, 283, 286, 287, 289, 295, 300
♦ b/pH, promoted, herbal. Origin: Eurasia.

Cissampelos glaberrima St.-Hil.
Menispermaceae
♦ Weed
♦ 87, 88
♦ herbal.

Cissampelos pareira L.
Menispermaceae
Cissampelos mucronata A.Rich.
♦ pareira brava
♦ Weed, Quarantine Weed, Naturalised, Introduced
♦ 39, 76, 86, 87, 88, 157, 203, 220, 228, 295
♦ pC, arid, cultivated, herbal, toxic.

Cissampelos pareira L. var. hirsuta (DC.) Forman
Menispermaceae
Cissampelos hirsuta DC., *Cissampelos pareira sensu* Hook.f. & Thomson
♦ Introduced
♦ 228
♦ arid.

Cissus cornifolia (Baker) Planch.
Vitaceae
♦ Weed
♦ 87, 88

Cissus incisa (Nutt. ex Torr. & A.Gray) Des Moul.
Vitaceae
- ivy treebine
- Weed
- 87, 88, 218
- herbal.

Cissus nodosa Blume
Vitaceae
- Javanese treebine, grape ivy
- Naturalised, Cultivation Escape
- 101, 233
- cultivated.

Cissus quadrangularis L.
Vitaceae
- Weed
- 19
- pC, cultivated, herbal.

Cissus repens Lam.
Vitaceae
- cissus
- Weed
- 262
- Origin: Asia, Australia.

Cissus rotundifolia (Forssk.) Vahl
Vitaceae
- Venezuelan treebine
- Naturalised, Introduced
- 86, 101, 228
- arid, cultivated.

Cissus sicyoides L.
Vitaceae
= *Cissus verticillata* (L.) Nicolson & C.E.Jarvis ssp. *verticillata*
- Weed
- 14, 87, 88
- pC, herbal.

Cissus striata Ruiz & Pav.
Vitaceae
- Casual Alien
- 280
- cultivated, herbal.

Cissus trifoliata (L.) L.
Vitaceae
- sorrelvine
- Weed
- 87, 88
- herbal.

Cissus verticillata (L.) Nicolson & C.E.Jarvis ssp. verticillata
Vitaceae
Cissus elliptica Schldl. & Cham., *Cissus sicyoides* L. (see)
- Introduced
- 228
- arid.

Cistanche phelypaea (L.) P.Cout.
Orobanchaceae
Lathraea phelypaea L.
- Weed
- 221

Cistanche tubulosa (Schenk) Wight
Orobanchaceae
- Weed
- 87, 88
- herbal.

Cistus creticus L.
Cistaceae
= *Cistus incanus* L. ssp. *creticus* (L.) Heywood
- Crete rockrose, rockrose
- Naturalised, Introduced
- 34, 86, 280
- S, cultivated.

Cistus crispus L.
Cistaceae
- Weed
- 70
- cultivated.

Cistus incanus L.
Cistaceae
Cistus villosus auct. non L.
- hairy rockrose, rockrose
- Weed, Naturalised
- 80, 101
- cultivated, herbal.

Cistus incanus L. ssp. corsicus (Loisel.) Heywood
Cistaceae
- hairy rockrose
- Naturalised
- 101

Cistus incanus L. ssp. creticus (L.) Heywood
Cistaceae
Cistus creticus L. (see)
- Cretan rockrose
- Naturalised
- 101
- Origin: Greece, Aegean region.

Cistus incanus L. ssp. incanus
Cistaceae
- hairy rockrose
- Naturalised
- 101

Cistus ladanifer L.
Cistaceae
Cistus ladaniferus L. (see)
- gum rockrose, gum cistus, sun rose, labdanum
- Weed, Noxious Weed, Naturalised
- 35, 70, 78, 88, 116, 280
- S, cultivated, herbal.

Cistus ladaniferus L.
Cistaceae
= *Cistus ladanifer* L.
- gum rockrose
- Naturalised
- 101
- herbal.

Cistus laurifolius L.
Cistaceae
- laurel leaved rockrose
- Naturalised
- 280
- cultivated, herbal.

Cistus monspeliensis L.
Cistaceae
- cistus, Montpellier cistus, narrow leaved cistus, rockrose
- Naturalised
- 86, 101, 120
- S, cultivated. Origin: southern Europe.

Cistus populifolius L.
Cistaceae
- poplar leaved cistus
- Weed
- 70
- cultivated.

Cistus psilosepalus Sweet
Cistaceae
- rockrose
- Weed, Naturalised, Environmental Weed
- 72, 86, 88, 98, 176, 203, 280
- S, cultivated. Origin: south-west Europe.

Cistus salvifolius L.
Cistaceae
= *Cistus salviifolius* L.
- sage leaved cistus, cistus
- Weed, Casual Alien
- 70, 280
- S, herbal.

Cistus salviifolius L.
Cistaceae
Cistus salvifolius L. (see), *Cistus salviaefolius* L.
- salvia cistus, sage leaved rockrose, rockrose
- Naturalised
- 101
- S, cultivated, herbal.

Citharexylum caudatum L.
Verbenaceae
- juniper berry, fiddlewood juniper berry
- Weed, Cultivation Escape
- 3, 22, 80, 191, 233
- T, cultivated.

Citharexylum fruticosum L.
Verbenaceae
= *Citharexylum spinosum* L.
- Florida fiddlewood
- Naturalised
- 86
- cultivated. Origin: tropical America.

Citharexylum quadrangulare Jacq.
Verbenaceae
= *Citharexylum spinosum* L.
- Weed
- 3, 191
- cultivated.

Citharexylum spinosum L.
Verbenaceae
Citharexylum fruticosum L. (see), *Citharexylum quadrangulare* Jacq. (see)
- fiddlewood, spiny fiddlewood
- Weed, Cultivation Escape
- 3, 87, 88, 191, 233
- S/T, cultivated.

Citrullus colocynthis (L.) Schrad.
Cucurbitaceae
Colocynthis vulgaris Schrad. (see), *Cucumis colocynthis* L.
- bitter paddy melon, colocynth, bitterapple, wild watermelon, Indian colocynth, bitter melon, cara, ekir, gare damer, gartoomba, ground gourd, handal, indravarooni, tagalate, tumba, turo, unun, wild watermelon

♦ Weed, Noxious Weed, Naturalised, Garden Escape, Environmental Weed
♦ 7, 23, 39, 72, 86, 87, 88, 93, 98, 101, 147, 198, 203, 205, 221, 269
♦ pH, arid, cultivated, herbal, toxic. Origin: Mediterranean, west Asia.

Citrullus colocynthis (L.) Schrad. var. *citroides* (L.Bailey) Mansf.
Cucurbitaceae
♦ citron, citron melon, preserving melon
♦ Weed
♦ 180, 243
♦ aH. Origin: Africa.

Citrullus colocynthis (L.) Schrad. var. *lanatus* (Thunb.) Matsum. & Nakai
Cucurbitaceae
♦ watermelon
♦ Weed, Cultivation Escape
♦ 34
♦ aH, cultivated.

Citrullus lanatus (Schrad.) Mansf.
Cucurbitaceae
Citrullus vulgaris Schrad., *Colocynthis citrullus* (L.) Kuntze, *Cucurbita citrullus* L., *Momordica lanata* Thunb.
♦ wild watermelon, bitterapple, colocynth, kaffir melon, watermelon, white watermelon, wild melon, paddy melon, pie melon, camel melon, volganpernaruoho, afglida melon, bastard melon
♦ Weed, Noxious Weed, Naturalised, Native Weed, Introduced, Environmental Weed, Casual Alien
♦ 7, 13, 32, 39, 40, 42, 55, 72, 84, 87, 88, 93, 98, 101, 121, 147, 158, 161, 179, 198, 203, 205, 228, 257, 261, 280
♦ aH, arid, cultivated, herbal, toxic. Origin: southern Africa.

Citrullus lanatus (Thunb.) Matsum. & Nakai var. *citroides* (Bailey) Mansf.
Cucurbitaceae
♦ watermelon
♦ Naturalised
♦ 101, 243

Citrullus lanatus (Thunb.) Matsum. & Nakai var. *lanatus*
Cucurbitaceae
♦ watermelon, bitter melon, Afghan melon, bastard melon, bitterapple, camel melon, mickey melon, pie melon
♦ Weed, Noxious Weed, Naturalised
♦ 86, 101, 269
♦ aC. Origin: southern Africa.

Citrus L. spp.
Rutaceae
♦ citrus tree
♦ Weed
♦ 279
♦ cultivated, herbal.

Citrus aurantiaca hort. ex Tanaka
Rutaceae
= *Citrus glaberrima* T.Tanaka (NoR)
♦ Weed
♦ 80

Citrus aurantiifolia (Christm.) Swingle
Rutaceae

Citrus acida Roxb., *Citrus hystrix* DC. ssp. *acida* (Roxb.) Engl., *Citrus lima* Lunan, *Citrus limetta* Risso var. *aromatica* Wester, *Citrus medica* L. var. *acida* (Roxb.) Hook.f., *Limonia aurantiifolia* Christm.
♦ lime, karer, orange sour, tipolo, key lime
♦ Weed, Naturalised, Introduced, Environmental Weed, Cultivation Escape
♦ 22, 39, 80, 88, 101, 151, 161, 179, 230, 257, 261
♦ T, cultivated, herbal, toxic. Origin: tropical Asia.

Citrus aurantium L.
Rutaceae
Citrus amara Link, *Citrus bigarradia* Loisel., *Citrus vulgaris* Risso
♦ sour orange, Seville orange, bigarade, lemon
♦ Weed, Naturalised, Introduced, Environmental Weed, Casual Alien
♦ 22, 39, 88, 101, 151, 161, 179, 230, 261
♦ T, cultivated, herbal, toxic. Origin: south-east Asia.

Citrus grandis (L.) Osbeck
Rutaceae
= *Citrus maxima* (Burm.f.) Merr.
♦ pomelo, pummelo
♦ Cultivation Escape
♦ 261
♦ cultivated, herbal. Origin: south-east Asia.

Citrus iriomotensis hort. ex Tanaka. nom. nud.
Rutaceae
♦ Naturalised
♦ 287

Citrus jambhiri Lush.
Rutaceae
♦ rough lemon
♦ Naturalised
♦ 134
♦ cultivated.

Citrus limetta Risso
Rutaceae
♦ sweet lime, bitter orange
♦ Naturalised, Environmental Weed, Cultivation Escape
♦ 101, 152, 257, 261
♦ T, cultivated, herbal. Origin: Eurasia, North Africa.

Citrus limon (L.) Burm.f.
Rutaceae
= *Citrus medica* L. × *Citrus aurantifolia* (Christm.) Swingle. [possible source]
♦ lemon, bush lemon, tipolo, limón de cabro
♦ Weed, Naturalised, Garden Escape, Environmental Weed, Cultivation Escape, Casual Alien
♦ 22, 80, 86, 101, 179, 257, 261, 280
♦ S, cultivated, herbal. Origin: obscure.

Citrus limonia Osbeck
Rutaceae
= *Citrus* × *limonia* Osbeck
♦ rough lemon

♦ Weed, Naturalised
♦ 73, 88, 98, 203
♦ cultivated, herbal.

Citrus × *limonia* Osbeck
Rutaceae
= *Citrus limon* (L.) Burm.f. × *Citrus reticulata* Blanco [most probable combination. see *Citrus limonia* Osbeck]
♦ lemandarin, mandarin, wild lemon, mandarin lime, Canton lemon, marmalade lime, Otaheite orange, Rangpur lime, red lemon, cravo lemon, citronnier de Canton, limettier Rangpur, Rangpur mandarinenlimette, Pinochio orange, Volkamer zitrone, sharbati, surkh nimboo, Sylhet lime, hime lemon
♦ Weed, Naturalised
♦ 86, 121, 179
♦ T. Origin: Eurasia.

Citrus maxima (Burm.f.) Merr.
Rutaceae
Citrus grandis (L.) Osbeck (see)
♦ shaddock, pomelo, pompelmus
♦ Naturalised
♦ 101
♦ cultivated, herbal.

Citrus medica L.
Rutaceae
♦ citron, suckatcitron
♦ Weed, Naturalised, Casual Alien
♦ 22, 39, 98, 101, 179, 203, 261
♦ S, cultivated, herbal, toxic. Origin: southern Asia.

Citrus mitis Blanco
Rutaceae
♦ karer tik, kalamondina
♦ Introduced
♦ 230
♦ T, herbal.

Citrus paradisi Macfad.
Rutaceae
= *Citrus* × *paradisi* Macfad. (*pro* sp.)
♦ karel lap, grapefruit
♦ Introduced
♦ 230
♦ T, cultivated, herbal.

Citrus × *paradisi* Macfad. (*pro* sp.)
Rutaceae
= *Citrus sinensis* (L.) Osbeck × *Citrus maxima* (Burm.f.) Merr. [most probable combination, see *Citrus paradisi* Macfad.]
♦ grapefruit, toronja, grapefrukt, greippi, pampelmuse
♦ Weed, Naturalised, Casual Alien
♦ 39, 101, 179, 261
♦ cultivated, herbal, toxic. Origin: produced in West Indies.

Citrus reticulata Blanco
Rutaceae
♦ tangerine, mandarin, king orange, mandarina
♦ Weed, Naturalised, Introduced
♦ 22, 101, 179, 261
♦ T, cultivated, herbal. Origin: east Asia.

Citrus rokugatsu hort. ex Yu.Tanaka
Rutaceae
♦ Naturalised
♦ 287

Citrus sinensis (L.) Osbeck
Rutaceae
Citrus aurantium L. var. *sinensis* L.,
Citrus macracantha Hassk.
♦ sweet orange, appelsiini, orange,
moli 'aina, blood orange, navel orange,
Valencia orange, navel, oranger,
oranger doux, sanguine, apfelsine,
apfelsinenbaum, orangenbaum,
arancio dolce, laranjeira, naranja
♦ Weed, Naturalised, Introduced,
Environmental Weed, Casual Alien
♦ 22, 39, 42, 80, 88, 101, 151, 179, 230,
261, 280
♦ T, cultivated, herbal, toxic. Origin:
China.

Cladium colocasia (L.) W.Wight
Cyperaceae
♦ malanga
♦ Naturalised
♦ 101
♦ G.

Cladium jamaicense Crantz
Cyperaceae
= *Cladium mariscus* (L.) Pohl ssp.
jamaicense (Crantz) K.Kenth.
♦ sawgrass
♦ Weed, Quarantine Weed
♦ 14, 87, 88, 218, 258
♦ G, herbal.

Cladium mariscoides (Muhl.) Torr
Cyperaceae
♦ smooth sawgrass
♦ Weed
♦ 87, 88, 218
♦ G, herbal.

Cladium mariscus (L.) Pohl
Cyperaceae
Cladium serratus Gilib., *Mariscus
cladium* (Sw.) Kuntze, *Mariscus serratus*
Gilib.
♦ smooth sawgrass, great fensedge,
twigrush
♦ Weed
♦ 23, 88, 126, 221, 272
♦ pG, aqua, cultivated, herbal. Origin:
Asia, Australia, Europe.

Cladium mariscus (L.) Pohl ssp.
jamaicense (Crantz) K.Kenth.
Cyperaceae
Cladium jamaicense Crantz (see)
♦ sawgrass, Jamaica swamp sawgrass
♦ Weed
♦ 121
♦ pG. Origin:
Central America.

Cladium procerum S.T.Blake
Cyperaceae
♦ Naturalised, Environmental Weed
♦ 86, 176
♦ G, cultivated. Origin: Australia.

Cladophora Kütz. spp.
Cladophoraceae
♦ cladophora

♦ Weed, Quarantine Weed
♦ 76, 88, 220
♦ algae.

Cladophora sericea (Huds.) Kütz.
Cladophoraceae
♦ algae
♦ Weed
♦ 282
♦ algae.

Cladoraphis cyperoides (Thunb.)
S.M.Phillips
Poaceae
♦ bristly lovegrass
♦ Naturalised
♦ 101
♦ G.

Cladrastis lutea (Michx.f.) K.Koch
Fabaceae/Papilionaceae
= *Cladrastis kentukea* (Dum.Cours.)
Rudd (NoR)
♦ Kentucky yellowwood, American
yellowwood, yellowwood
♦ Weed
♦ 88, 218
♦ T, cultivated, herbal.

Claoxylon carolinianum Pax & Hoff.
Euphorbiaceae
♦ koee
♦ Introduced
♦ 230
♦ S.

Clappertonia ficifolia (Willd.) Decne.
Tiliaceae
♦ bolo bolo
♦ Weed
♦ 88
♦ cultivated.

Clarkia amoena (Lehm.) A.Nelson &
J.F.Macbr.
Onagraceae
Godetia amoena (Lehm.) G.Don (see),
Oenothera amoena Lehm.
♦ farewell to spring
♦ Naturalised
♦ 280
♦ aH, cultivated, herbal.

Clarkia amoena (Lehm.) A.Nelson &
J.Macbr. ssp. *amoena*
Onagraceae
♦ farewell to spring
♦ Naturalised
♦ 280
♦ aH.

Clarkia pulchella Pursh
Onagraceae
♦ siroklarkia, pink fairies
♦ Cultivation Escape
♦ 42
♦ aH, cultivated, herbal.

Clarkia unguiculata Lindl.
Onagraceae
Clarkia elegans Douglas
♦ komeaklarkia, elegant clarkia,
elegant fairyfan, mountain garland
♦ Cultivation Escape, Casual Alien
♦ 42, 280
♦ aH, cultivated, herbal.

Clausena anisata (Willd.) Hook.f. ex
Benth.
Rutaceae
♦ horsewood, clausena
♦ Weed, Native Weed, Introduced
♦ 121, 228
♦ S/T, herbal. Origin: southern Africa.

Clausena excavata Burm.f.
Rutaceae
♦ clausena
♦ Weed, Environmental Weed
♦ 3, 191, 259
♦ herbal.

Claytonia perfoliata Donn. ex Willd.
Portulacaceae
Montia perfoliata (Donn) Howell (see)
♦ miner's lettuce, perhoskleitonia,
Indian lettuce, petota
♦ Weed, Quarantine Weed,
Naturalised, Introduced, Cultivation
Escape
♦ 40, 42, 76, 86, 98, 161, 198, 203, 228,
243, 252, 253, 280
♦ aH, arid, cultivated, herbal, toxic.
Origin: North to Central America.

Claytonia perfoliata Donn. ex Willd. ssp.
perfoliata
Portulacaceae
♦ miner's lettuce, Indian lettuce,
Spanish lettuce
♦ Weed
♦ 180
♦ aH, toxic. Origin: western USA.

Claytonia sibirica L.
Portulacaceae
♦ pink purslane, Siberian spring
beauty, candy flower
♦ Naturalised, Cultivation Escape
♦ 40, 42
♦ pH, cultivated, herbal.

Claytonia virginica L.
Portulacaceae
Claytonia grandiflora Sweet
♦ narrowleaf spring beauty, Virginia
spring beauty, spring beauty
♦ Weed, Garden Escape
♦ 161
♦ pH, promoted, herbal.

Cleistachne sorghoides Benth.
Poaceae
♦ false sorghum
♦ Native Weed
♦ 121
♦ aG. Origin: southern Africa.

Cleistocactus Lem. spp.
Cactaceae
Bolivicereus Cárdenas spp. (see),
Borzicactella Ritter spp. (see),
Borzicactus Riccobono spp. (see),
Demnosa A.V.Fric spp. (see),
Hildewintera Ritt. spp. (see),
Loxanthocereus Backeb. spp. (see),
Maritimocereus Ackers & Buining spp.
(see), *Seticereus* Backeb. spp. (see),
Seticleistocactus Backeb. spp. (see),
Winteria Ritt. spp. (see), *Winterocereus*
Backeb. spp. (see)
♦ silverpelare
♦ Weed, Quarantine Weed,

Naturalised
- 76, 86, 88, 203, 220
- cultivated, herbal.

Clematis L. spp.
Ranunculaceae
- clematis, leather flower
- Weed
- 154, 161, 247
- herbal, toxic.

Clematis aethusifolia Turcz.
Ranunculaceae
- longplume clematis
- Weed
- 297

Clematis apiifolia DC.
Ranunculaceae
- Weed
- 87, 88, 286
- pC, cultivated.

Clematis bonariensis Juss. ex DC.
Ranunculaceae
- Naturalised
- 300

Clematis brachiata Thunb.
Ranunculaceae
- bridal wreath, old man's beard, poobah's beard, traveller's joy
- Weed, Native Weed
- 39, 121
- pC, cultivated, herbal, toxic. Origin: southern Africa.

Clematis brevicaudata DC.
Ranunculaceae
- Weed
- 275, 297

Clematis denticulata Vell.
Ranunculaceae
Clematis hilarii Spreng., *Clematis montevidensis* Spreng. (see)
- cabello de angel
- Weed, Naturalised
- 241, 295
- Origin: Argentina, Bolivia, Brazil, Chile, Paraguay, Peru, Uruguay.

Clematis dioica L.
Ranunculaceae
- cabellos de angel
- Weed
- 87, 88, 157
- pC, herbal.

Clematis drummondii Torr. & Gray
Ranunculaceae
- Drummond's clematis
- Weed
- 87, 88
- herbal.

Clematis flammula L.
Ranunculaceae
- clematis, virgin's bower, fragrant virgin's bower, fragrant clematis
- Weed, Sleeper Weed, Quarantine Weed, Naturalised, Garden Escape, Environmental Weed
- 39, 40, 54, 86, 88, 155, 225, 246, 272, 280
- pC, cultivated, herbal, toxic.

Clematis florida Thunb.
Ranunculaceae

- Asian virgin's bower
- Naturalised
- 39, 101
- cultivated, herbal, toxic.

Clematis fusca Turcz.
Ranunculaceae
- Stanavoi clematis, kurobanahanshouzuru
- Weed
- 297
- Origin: east Asia.

Clematis hexapetala Pall.
Ranunculaceae
- sixpetal clematis
- Weed
- 297
- herbal.

Clematis integrifolia L.
Ranunculaceae
Clematis inclinata Scop., *Clematis nutans* Crantz
- solitary clematis
- Weed
- 39, 272
- cultivated, herbal, toxic.

Clematis intricata Bunge
Ranunculaceae
- intricate clematis
- Weed
- 297

Clematis × jackmanii T.Moore
Ranunculaceae
= *Clematis lanuginosa* Lindl. × *Clematis viticella* L.
- Jackman clematis
- Naturalised
- 101
- cultivated, herbal.

Clematis ligusticifolia Nutt.
Ranunculaceae
- western clematis, western white clematis, virgin's bower, yerba de chiva
- Weed
- 45, 87, 88, 136, 218, 243
- pC, cultivated, herbal.

Clematis maximowicziana Franch. & Sav.
Ranunculaceae
= *Clematis terniflora* DC.
- Weed, Naturalised
- 15, 87, 88, 280
- cultivated.

Clematis montana DC.
Ranunculaceae
- Himalayan clematis
- Naturalised
- 154, 280
- cultivated, herbal, toxic.

Clematis montevidensis Spreng
Ranunculaceae
= *Clematis denticulata* Vell.
- Weed, Naturalised
- 237, 300

Clematis orientalis L.
Ranunculaceae
- Chinese clematis, oriental virgin's bower
- Weed, Noxious Weed, Naturalised

- 39, 80, 101, 229, 275
- pC, cultivated, herbal, toxic.

Clematis oweniae Harv.
Ranunculaceae
- traveller's joy
- Native Weed
- 121
- pC. Origin: southern Africa.

Clematis recta L.
Ranunculaceae
Clematis erecta All.
- ground virgin's bower, pensaskärhö
- Weed, Naturalised, Cultivation Escape
- 39, 42, 101, 272
- pC, cultivated, herbal, toxic.

Clematis taiwaniana Hayata
Ranunculaceae
- Weed
- 87, 88

Clematis tangutica (Maxim.) Korsh.
Ranunculaceae
- orangepeel clematis, Chinese clematis, golden clematis
- Weed, Sleeper Weed, Naturalised, Casual Alien
- 40, 225, 280, 297
- cultivated, herbal.

Clematis terniflora DC.
Ranunculaceae
Clematis maximowicziana Franch. & Sav.
- yam leaved clematis, leatherleaf clematis, sweet autumn virgin's bower
- Weed, Naturalised
- 80, 88, 101, 102, 133, 195, 286
- pC, promoted, herbal.

Clematis tibetana Kuntze
Ranunculaceae
- Chinese clematis
- Environmental Weed, Casual Alien
- 246, 280
- pC, cultivated. Origin: Nepal, Tibet.

**Clematis tibetana Kuntze ssp. *vernayi*
(C.E.C.Fisch.) Grey-Wilson**
Ranunculaceae
- Naturalised
- 280

Clematis virginiana L.
Ranunculaceae
- Virginia clematis, virgin's bower, clematis, devil's darning needles
- Weed
- 8, 39, 87, 88, 211, 218
- cultivated, herbal, toxic.

Clematis vitalba L.
Ranunculaceae
- traveller's joy, old man's beard, evergreen clematis
- Weed, Quarantine Weed, Naturalised, Garden Escape, Environmental Weed
- 15, 18, 39, 45, 70, 72, 76, 80, 86, 88, 98, 101, 146, 152, 155, 165, 176, 181, 198, 225, 243, 246, 253, 272, 280, 289, 296
- pC, cultivated, herbal, toxic. Origin: Eurasia.

Clematis viticella L.
Ranunculaceae
Viticella deltoidea Moench
♦ Italian leather flower, viinikärhö
♦ Weed, Naturalised, Cultivation Escape
♦ 42, 101, 272
♦ cultivated, herbal. Origin: Mediterranean.

Clematopsis scabiosifolia (DC.) Hutch.
Ranunculaceae
Clematis scabiosaefolia DC.
♦ shock headed Peter, wild dog, mpisya
♦ Native Weed
♦ 121
♦ pH, cultivated. Origin: southern Africa.

Cleome aculeata L.
Capparaceae/Cleomaceae
♦ prickly spiderflower
♦ Weed, Naturalised
♦ 11, 55, 86, 87, 88, 93, 98, 203, 261
♦ Origin: tropical America.

Cleome affinis DC.
Capparaceae/Cleomaceae
Cleome fugax Schrad., *Cleome triphylla* Vell.
♦ mussambe, sojinha
♦ Weed
♦ 255
♦ Origin: Brazil.

Cleome africana Botsch.
Capparaceae/Cleomaceae
♦ Weed
♦ 88, 221

Cleome angustifolia A.Rich.
Capparaceae/Cleomaceae
♦ spiderflower, peultjiesbos
♦ Weed
♦ 88, 158
♦ aH, promoted, herbal. Origin: South Africa.

Cleome angustifolia A.Rich. ssp. petersiana (Klotzsch ex Sond.) Kers
Capparaceae/Cleomaceae
♦ Native Weed
♦ 121
♦ aH. Origin: southern Africa.

Cleome arabica L.
Capparaceae/Cleomaceae
♦ Weed
♦ 221

Cleome aspera Koen. ex DC.
Capparaceae/Cleomaceae
♦ Weed
♦ 87, 88

Cleome brachycarpa Vahl ex DC.
Capparaceae/Cleomaceae
♦ Weed
♦ 87, 88, 221
♦ herbal.

Cleome burmanni Wight & Arn.
Capparaceae/Cleomaceae
♦ Weed, Quarantine Weed
♦ 76, 87, 88, 203, 220

Cleome chelidonii L.f.
Capparaceae/Cleomaceae
Polanisia chelidonii DC.
♦ Weed
♦ 39, 87, 88, 170, 191
♦ herbal, toxic.

Cleome chrysantha Decne.
Capparaceae/Cleomaceae
♦ Weed
♦ 221

Cleome ciliata Schumach. & Thonn.
Capparaceae/Cleomaceae
= *Cleome rutidosperma* DC.
♦ Weed
♦ 88

Cleome diffusa Banks ex DC.
Capparaceae/Cleomaceae
♦ spreading spiderflower
♦ Weed, Naturalised
♦ 87, 88, 101

Cleome droserifolia (Forssk.) Delile
Capparaceae/Cleomaceae
♦ Weed
♦ 221

Cleome gynandra L.
Capparaceae/Cleomaceae
Gynandropsis gynandra (L.) Briq. (see), *Gynandropsis pentaphylla* (L.) DC., *Pedicellaria pentaphylla* Schrank
♦ spiderflower, spiderwisp, wild spiderflower, phak sian
♦ Weed, Quarantine Weed, Naturalised, Native Weed, Introduced
♦ 7, 14, 32, 38, 55, 76, 86, 87, 88, 93, 98, 101, 121, 158, 179, 203, 209, 217, 239, 242, 261
♦ aH, arid, cultivated, herbal. Origin: Africa.

Cleome hassleriana Chodat
Capparaceae/Cleomaceae
Cleome spinosa (hort.) nom. illeg.
♦ spiderplant, spiderflower, pink queen
♦ Weed, Naturalised, Cultivation Escape
♦ 55, 86, 98, 101, 179, 203, 255, 261, 280
♦ cultivated. Origin: tropical America.

Cleome hirta (Klotz.) Oliv.
Capparaceae/Cleomaceae
♦ Weed, Native Weed
♦ 87, 88, 121
♦ aH. Origin: southern Africa.

Cleome iberica DC.
Capparaceae/Cleomaceae
♦ Iberian spiderflower
♦ Naturalised
♦ 101

Cleome icosandra L.
Capparaceae/Cleomaceae
= *Cleome viscosa* L.
♦ Weed
♦ 87, 88, 286
♦ herbal.

Cleome integrifolia Torr. & Gray
Capparaceae/Cleomaceae
= *Cleome serrulata* Pursh
♦ Weed

♦ 87, 88

Cleome lutea Hook.
Capparaceae/Cleomaceae
♦ yellow spiderflower, yellow cleome
♦ Weed
♦ 87, 88, 161
♦ aH, promoted, herbal.

Cleome maculata (Sond.) Szyszyl.
Capparaceae/Cleomaceae
♦ Native Weed
♦ 121
♦ aH. Origin: southern Africa.

Cleome monophylla L.
Capparaceae/Cleomaceae
♦ spindlepod, msuShangishangi, single leaved cleome, spiderflower
♦ Weed, Naturalised, Native Weed
♦ 50, 51, 86, 87, 88, 98, 121, 158, 203, 240
♦ aH, promoted, herbal. Origin: southern Africa.

Cleome moritziana Klotz. ex Eichl.
Capparaceae/Cleomaceae
♦ Weed
♦ 87, 88

Cleome ornithopodioides L.
Capparaceae/Cleomaceae
♦ bird spiderflower
♦ Naturalised
♦ 101
♦ aH, promoted.

Cleome paradoxa R.Br.
Capparaceae/Cleomaceae
♦ Weed
♦ 221

Cleome rubella Burch.
Capparaceae/Cleomaceae
♦ cleome, pretty lady
♦ Weed, Native Weed
♦ 88, 121, 158
♦ aH. Origin: southern Africa.

Cleome rutidosperma DC.
Capparaceae/Cleomaceae
Cleome ciliata Schum. & Thonn. (see)
♦ spiderflower, fringed spiderflower, phak sian
♦ Weed, Quarantine Weed, Naturalised
♦ 12, 32, 76, 87, 88, 93, 101, 135, 170, 179, 191, 203, 206, 220, 239, 243, 262, 273
♦ aH. Origin: tropical Africa.

Cleome scaposa DC.
Capparaceae/Cleomaceae
= *Cleome papillosa* Steud. (NoR)
♦ Weed
♦ 88

Cleome serrata Jacq.
Capparaceae/Cleomaceae
♦ toothed spiderflower
♦ Weed, Naturalised
♦ 87, 88, 101, 199, 243, 261

Cleome serrulata Pursh
Capparaceae/Cleomaceae
Cleome integrifolia Torr. & Gray (see), *Peritoma integrifolium* Nutt.

♦ Rocky Mountain beeplant, pink cleome, stinking clover, stink weed, bee spiderflower, bee plant
♦ Weed, Native Weed
♦ 23, 49, 88, 161, 174, 210, 212, 218
♦ aH, cultivated, herbal, toxic. Origin: North America.

Cleome speciosa **Raf.**
Capparaceae/Cleomaceae
Gynandropsis speciosa (H.B.K.) DC.
♦ volantines preciosos
♦ Weed, Naturalised, Introduced, Casual Alien
♦ 101, 230, 261, 262
♦ H.

Cleome spinosa **Jacq.**
Capparaceae/Cleomaceae
♦ spiny spiderflower
♦ Weed, Naturalised
♦ 39, 87, 88, 218, 255, 261, 286, 287
♦ cultivated, herbal, toxic. Origin: tropical America.

Cleome stenophylla **Klotzsch ex Urb.**
Capparaceae/Cleomaceae
♦ tropical spiderflower
♦ Weed
♦ 87, 88

Cleome viscosa **L.**
Capparaceae/Cleomaceae
Cleome icosandra L. (see), *Polanisia icosandra* (L.) Wight & Arn., *Polanisia viscosa* DC.
♦ kuhyolung, tickweed, Asian spiderflower, wild caia, phak sian phee
♦ Weed, Quarantine Weed, Introduced, Environmental Weed
♦ 23, 55, 87, 88, 157, 170, 209, 221, 228, 230, 239, 257, 258, 261, 262, 276, 297
♦ aH, arid, cultivated, herbal.

Cleome welwitschii **Exell**
Capparaceae/Cleomaceae
♦ Weed
♦ 87, 88

Cleomella obtusifolia **Torr. & Frem.**
Capparaceae/Cleomaceae
♦ bushy stinkweed, Mojave cleomella, hairy stinkweed
♦ Weed
♦ 161
♦ aH, herbal.

Clerodendrum aculeatum **(L.) Schlecht.**
Lamiaceae/Verbenaceae
♦ haggarbush
♦ Weed
♦ 14
♦ cultivated.

Clerodendrum buchanani **Roxb. ex Wall.**
Lamiaceae/Verbenaceae
♦ pagoda flower lau'awa
♦ Weed, Cultivation Escape
♦ 87, 88, 233
♦ cultivated, herbal.

Clerodendrum buchanani **Roxb. ex Wall. var.** *fallax* **(Lindl.) Bakh.**
Lamiaceae/Verbenaceae
♦ Introduced
♦ 230

Clerodendrum bungei **Steud.**
Lamiaceae/Verbenaceae
♦ rose glorybower, tube flower, glory flower
♦ Weed, Naturalised, Garden Escape, Environmental Weed
♦ 80, 101, 179, 261, 287
♦ S, cultivated, herbal. Origin: China, Vietnam to north India.

Clerodendrum calamitosum **L.**
Lamiaceae/Verbenaceae
♦ Weed
♦ 13
♦ herbal.

Clerodendrum chinense **(Osb.) Mabb.**
Lamiaceae/Verbenaceae
Clerodendrum fragrans (*hort.* ex Vent.) Willd. (see), *Clerodendrum philippinum* Schauer (see)
♦ stickbush, fragrant clerodendrum, Honolulu rose, Spanish jasmine, glory bower
♦ Weed, Sleeper Weed, Naturalised, Environmental Weed
♦ 3, 6, 86, 88, 101, 107, 155, 179, 191
♦ S. Origin: southern China and northern Vietnam.

Clerodendrum colebrookianum **Walp.**
Lamiaceae/Verbenaceae
♦ Weed
♦ 275
♦ cultivated.

Clerodendrum fragrans **(*hort.* ex Vent.) Willd.**
Lamiaceae/Verbenaceae
= *Clerodendrum chinense* (Osb.) Mabb.
♦ fragrant clerodendrum, glory bower
♦ Weed, Cultivation Escape
♦ 3, 88, 191, 218, 233
♦ cultivated, herbal.

Clerodendrum glabrum **E.Mey.**
Lamiaceae/Verbenaceae
♦ Natal glorybower
♦ Weed, Naturalised
♦ 101, 179
♦ cultivated, herbal.

Clerodendrum heterophyllum **(Poir.) W.T.Aiton fo.** *baueri* **(Mold.) A.A.Munir**
Lamiaceae/Verbenaceae
♦ Naturalised
♦ 86

Clerodendrum indicum **(L.) Kuntze**
Lamiaceae/Verbenaceae
Clerodendrum siphonanthus R.Br.
♦ Turk's turban, tube flower
♦ Weed, Naturalised
♦ 80, 87, 88, 101, 179
♦ herbal.

Clerodendrum inerme **(L.) Gaertn.**
Lamiaceae/Verbenaceae
Clerodendrum buxifolium (Willd.) Spreng., *Clerodendrum neriifolium* (Roxb.) Schauer, *Volkameria buxifolia* Willd., *Volkameria inermis* L., *Volkameria nereifolia* Roxb.
♦ embrert, aloalo tai
♦ Weed, Naturalised
♦ 13, 101

♦ S, cultivated, herbal. Origin: Australia.

Clerodendrum inerme **(L.) Gaertn. var.** *oceanicum* **A.Gray**
Lamiaceae/Verbenaceae
♦ ilau
♦ Introduced
♦ 230
♦ S.

Clerodendrum infortunatum **Gaertn.**
Lamiaceae/Verbenaceae
♦ Weed
♦ 39, 87, 88
♦ herbal, toxic.

Clerodendrum japonicum **(Thunb.) Sweet**
Lamiaceae/Verbenaceae
♦ Japanese glorybower, glorybower
♦ Weed, Naturalised
♦ 3, 22, 80, 191, 287
♦ S, cultivated.

Clerodendrum kaempferi **(Jacq.) Sieb.**
Lamiaceae/Verbenaceae
♦ Kaempfer's glorybower
♦ Weed, Naturalised
♦ 101, 179

Clerodendrum lindleyi **Decne. ex Planch.**
Lamiaceae/Verbenaceae
♦ Lindley's clerodendrum
♦ Naturalised
♦ 101, 287

Clerodendrum lindleyi **Decne. var.** *paniculatum* **Moldenke**
Lamiaceae/Verbenaceae
♦ Lindley's clerodendrum
♦ Naturalised
♦ 101

Clerodendrum macrostegium **Schauer**
Lamiaceae/Verbenaceae
♦ velvetleaf glorybower
♦ Naturalised, Cultivation Escape
♦ 101, 233
♦ cultivated.

Clerodendrum navesianum **Vidal.**
Lamiaceae/Verbenaceae
Clerodendron navesianum Vidal.
♦ Quarantine Weed
♦ 220

Clerodendrum paniculatum **L.**
Lamiaceae/Verbenaceae
♦ pagoda plant, pagoda flower, butcherchár, butecherchar, butcherchár tukehn sousou
♦ Weed, Naturalised, Introduced
♦ 3, 86, 107, 191, 230
♦ S, cultivated, herbal. Origin: India, China, and Taiwan south to Malaysia.

Clerodendrum philippinum **Schauer**
Lamiaceae/Verbenaceae
= *Clerodendrum chinense* (Osb.) Mabb.
♦ Honolulu rose, losa Honolulu, pikake hohono
♦ Weed, Quarantine Weed, Naturalised, Cultivation Escape
♦ 3, 14, 76, 80, 86, 87, 88, 179, 191, 203, 261, 262
♦ arid, cultivated. Origin: Old World Tropics.

Clerodendrum phlomoidis L.f.
Lamiaceae/Verbenaceae
♦ Weed
♦ 87, 88
♦ herbal.

Clerodendrum quadriloculare (Blanco) Merr.
Lamiaceae/Verbenaceae
♦ bronze leaved clerodendrum, tuhkehn palau
♦ Weed, Quarantine Weed
♦ 3, 22, 76, 107, 191, 220
♦ S/T, cultivated.

Clerodendrum schweinfurthii Gürke var. bakeri (Gürke) Thomas
Lamiaceae/Verbenaceae
♦ Naturalised
♦ 86

Clerodendrum serratum (L.) Moon
Lamiaceae/Verbenaceae
♦ Weed
♦ 13
♦ cultivated, herbal.

Clerodendrum speciosissimum Van Geert ex Morr.
Lamiaceae/Verbenaceae
Clerodendrum fallax Lindl.
♦ Javanese glorybower, eldklerodendrum, tripa de coral
♦ Weed, Naturalised, Cultivation Escape
♦ 80, 101, 179, 261
♦ cultivated, herbal. Origin: Java.

Clerodendrum × speciosum Dombrain
Lamiaceae/Verbenaceae
= *Clerodendrum splendens* (Thunb.) G.Don × *Clerodendrum thomsoniae* Balf.
♦ Weed, Naturalised, Introduced
♦ 101, 179, 261
♦ cultivated. Origin: west tropical Africa.

Clerodendrum splendens (Thunb.) G.Don
Lamiaceae/Verbenaceae
♦ Weed
♦ 87, 88
♦ cultivated.

Clerodendrum thomsoniae Balf.f.
Lamiaceae/Verbenaceae
♦ bagflower, bleedingheart, bandera danesa
♦ Naturalised, Introduced
♦ 86, 101, 230, 261
♦ cultivated, herbal. Origin: tropical Africa.

Clerodendrum trichotomum Thunb.
Lamiaceae/Verbenaceae
♦ harlequin glorybower
♦ Weed, Naturalised
♦ 101, 280, 286
♦ T, cultivated, herbal.

Clerodendrum trichotomum Thunb. var. ferrugineum Nakai
Lamiaceae/Verbenaceae
♦ ferruginous clerodendrum
♦ Naturalised
♦ 101

Clerodendrum umbellatum Poir.
Lamiaceae/Verbenaceae
♦ umbel clerodendrum
♦ Naturalised
♦ 101
♦ herbal.

Clerodendrum uncinatum Schinz
Lamiaceae/Verbenaceae
♦ Native Weed
♦ 121
♦ aH. Origin: southern Africa.

Clerodendrum wallichii Merr.
Lamiaceae/Verbenaceae
Clerodendrum nutans Wall. ex D.Don nom. illeg.
♦ Wallich's glorybower, bandera danesa
♦ Naturalised, Cultivation Escape
♦ 101, 261
♦ cultivated. Origin: southern Asia.

Clethra arborea W.T.Aiton
Clethraceae
♦ Casual Alien
♦ 280
♦ cultivated.

Clethra mexicana DC.
Clethraceae
Clethra palmeri Britt.
♦ Quarantine Weed
♦ 258

Cleyera japonica Thunb.
Theaceae
♦ sakaki, Japanese cleyera
♦ Naturalised
♦ 101
♦ cultivated, herbal.

Clibadium surinamense L.
Asteraceae
♦ Weed
♦ 39, 87, 88
♦ herbal, toxic.

Clidemia dentata D.Don
Melastomataceae
♦ Weed
♦ 87, 88
♦ S.

Clidemia dependens D.Don
Melastomataceae
= *Clidemia capitellata* D.Don var. *dependens* (Pav. ex D.Don) J.F.Macbr. (NoR)
♦ Weed
♦ 87, 88

Clidemia hirta (L.) D.Don
Melastomataceae
Melastoma hirta L.
♦ Koster's curse, soap bush, kúi, mbulamakau, vuti
♦ Weed, Quarantine Weed, Noxious Weed, Naturalised, Environmental Weed
♦ 3, 6, 18, 22, 37, 76, 80, 86, 87, 88, 135, 151, 152, 153, 161, 191, 203, 206, 218, 220, 229, 243, 268, 287
♦ S, herbal. Origin: Central and South America.

Clidemia rubra (Aubl.) Mart.
Melastomataceae
♦ Weed
♦ 87, 88

Cliffortia ruscifolia L.
Rosaceae
♦ climber's friend, prickly bush
♦ Native Weed
♦ 121
♦ pS. Origin: southern Africa.

Cliftonia monophylla Sarg.
Cyrillaceae
Cliftonia ligustrina (Willd.) Spreng., *Ptelea monophylla* Lam.
♦ titi, ironwood, buckwheat tree
♦ Weed
♦ 87, 88, 218
♦ S, promoted.

Clinacanthus nutans (Burm.f.) Lindau
Acanthaceae
= *Clinacanthus burmanni* Nees (NoR)
♦ Weed
♦ 13
♦ herbal.

Clinelymus sibiricus (L.) Nevski
Poaceae
♦ Quarantine Weed
♦ 220
♦ G.

Clinopodium chinense (Benth.) Kuntze
Lamiaceae
Calamintha chinensis Benth.
♦ Chinese clinopodium
♦ Weed
♦ 87, 88, 297
♦ pH, promoted. Origin: China, Japan.

Clinopodium chinense (Benth.) Kuntze var. parviflorum (Kudo) Hara
Lamiaceae
♦ Weed
♦ 286

Clinopodium gracile (Benth.) Kuntze
Lamiaceae
♦ slender wild basil
♦ Weed, Naturalised
♦ 101, 235, 273, 275, 286, 297
♦ aH.

Clinopodium umbrosum (M.Bieb.) K.Koch
Lamiaceae
Calamintha umbrosa (M.Bieb.) Fisch. & C.A.Mey., *Melissa umbrosum* M.Bieb.
♦ Weed
♦ 235
♦ pH, promoted.

Clinopodium vulgare L.
Lamiaceae
Melissa vulgaris Trev., *Satureja vulgaris* (L.) Fritsch
♦ wild basil, cushion calamint
♦ Weed, Naturalised
♦ 15, 86, 165, 198, 272, 280
♦ pH, cultivated, herbal. Origin: Europe.

Clinostigma ponapensis (Becc.) Moore & Fosb.
Arecaceae

Exorrhiza ponapensis (Becc.) Burr.
- kotop
- Introduced
- 230, 230
- T.

Clistanthocereus Backeb. spp.
Cactaceae
- Weed, Quarantine Weed
- 76, 88, 203, 220

Clitoria fairchildiana R.A.Howard
Fabaceae/Papilionaceae
- Cultivation Escape
- 261
- cultivated. Origin: tropical America.

Clitoria laurifolia Poir.
Fabaceae/Papilionaceae
- laurelleaf pigeonwings
- Weed, Sleeper Weed, Naturalised, Environmental Weed
- 3, 14, 86, 155, 191
- Origin: tropical America.

Clitoria ternatea L.
Fabaceae/Papilionaceae
Ternatea vulgaris Kunth
- butterfly pea, blue pea, Asian pigeonwings, buikike, paokeke, capa de la reina, putitainubia, pepe, nawa, deleite, papito, conchitas
- Weed, Naturalised, Introduced, Garden Escape, Environmental Weed
- 3, 7, 39, 86, 87, 88, 93, 98, 101, 107, 179, 203, 221, 228, 230, 261, 276
- pC, arid, cultivated, herbal, toxic. Origin: Old World Tropics.

Clitoria ternatea L. var. *ternatea*
Fabaceae/Papilionaceae
- Weed, Cultivation Escape
- 32
- pC, cultivated.

Clusia rosea Jacq.
Clusiaceae
- monkey apple, Scotch attorney, klusia, autograph tree, signature tree, copey
- Weed, Cultivation Escape
- 3, 39, 191, 233, 268
- cultivated, herbal, toxic.

Clytostoma binatum (Thunb.) Sandwith
Bignoniaceae
- Weed, Quarantine Weed, Naturalised
- 76, 86, 88, 220
- cultivated.

Cnestis ferruginea DC.
Connaraceae
- elende, lihandjo
- Weed
- 88
- herbal.

Cnicus benedictus L.
Asteraceae
Carbenia benedicta (L.) Adans., *Carduus benedictus* Garsault, *Centaurea benedicta* L., *Hierapicra benedicta* Kuntze
- blessed thistle, bitter thistle, cursed thistle, holy thistle, karmedik, Our Lady's thistle, spotted thistle, St.

Benedicts's thistle
- Weed, Naturalised, Introduced
- 34, 51, 70, 86, 87, 88, 94, 98, 101, 121, 161, 203, 218, 228, 237, 241, 253, 272, 287, 295, 300
- aH, arid, cultivated, herbal. Origin: Eurasia.

Cnidium dubium (Schkuhr) Thell.
Apiaceae
- palleroputki
- Casual Alien
- 42

Cnidium formosanum K.Yabe
Apiaceae
- Weed
- 87, 88
- herbal.

Cnidium monnieri (L.) Cusson ex Juss.
Apiaceae
- Monnier's snowparsley
- Weed, Naturalised
- 101, 275, 297
- aH, promoted, herbal.

Cnidoscolus aconitifolius (Mill.) I.M.Johnst.
Euphorbiaceae
- tread softly
- Naturalised, Garden Escape
- 101, 261
- cultivated. Origin: Mexico.

Cnidoscolus albomaculatus (Pax) I.M.Johnst.
Euphorbiaceae
- ortiga brava
- Weed
- 237, 295

Cnidoscolus stimulosus (Michx.) Engelm. & Gray
Euphorbiaceae
- bullnettle, finger rot
- Weed
- 39, 87, 88, 161, 218, 247
- herbal, toxic.

Cnidoscolus texanus (Müll.Arg.) Small
Euphorbiaceae
- Texas bullnettle
- Weed
- 39, 87, 88, 161, 218
- herbal, toxic.

Cnidoscolus urens (L.) Arthur
Euphorbiaceae
Cnidoscolus adenophilus (Pax & K.Hoffm.) Pax & K.Hoffm., *Jatropha adenophila* Pax & K.Hoffm., *Jatropha urens* L. (see)
- cnidoscolus
- Weed
- 28, 39, 163, 243, 255
- S, herbal, toxic. Origin: tropical and subtropical America.

Cobaea lutea D.Don
Polemoniaceae/Cobaeaceae
- Weed
- 157
- pC.

Cobaea scandens Cav.
Polemoniaceae/Cobaeaceae

Rosenbergia scandens (Cav.) House
- cathedral bells
- Weed, Quarantine Weed, Naturalised, Environmental Weed
- 15, 39, 86, 98, 165, 203, 225, 246, 280
- a/pH, cultivated, herbal, toxic. Origin: tropical America.

Coccinia adoensis (Hochst. ex A.Rich.) Cogn.
Cucurbitaceae
- wild spinach
- Native Weed
- 121
- pC. Origin: Africa.

Coccinia cordifolia (L.) Cogn.
Cucurbitaceae
= *Mukia maderaspatana* (L.) M.J.Roem.
- ivy gourd
- Weed
- 3
- herbal.

Coccinia diversifolia Naud. ex Huber
Cucurbitaceae
- Weed
- 221

Coccinia grandis (L.) Voigt
Cucurbitaceae
Bryonia grandis L. (see), *Cephalandra indica* (Wight & Arn.) Naudin (see), *Coccinia indica* Wight & Arn. *nom. illeg.* (see), *Coccinia cordifolia auct. non* (L.) Cogn., *Coccinia moghadd* (J.F.Gmel.) Schweinf., *Coccinia palmatisecta* Kotschy, *Turia moghadd* J.F.Gmel.
- ivy gourd, scarlet fruited gourd, arakis, ekadala, mughad, roh, scarlet gourd, tindola, kundree, pepasan, pepino cimarrón, little gourd
- Weed, Quarantine Weed, Noxious Weed, Naturalised, Introduced, Environmental Weed, Cultivation Escape
- 3, 6, 7, 13, 18, 32, 62, 76, 80, 86, 88, 101, 107, 152, 155, 161, 179, 191, 203, 209, 212, 228, 229, 233, 243, 252
- pC, arid, cultivated. Origin: central Africa, India, Asia.

Coccinia indica Wight & Arn. *nom. illeg.*
Cucurbitaceae
= *Coccinia grandis* (L.) Voigt
- Weed
- 87, 88
- herbal.

Coccinia palmata (Sond.) Cogn.
Cucurbitaceae
- Native Weed
- 121
- pC. Origin: Africa.

Coccocypselum repens Sw.
Rubiaceae
Coccocypselum herbaceum Aubl.
- Weed
- 87, 88

Coccoloba acapulcensis Standl.
Polygonaceae
- Weed, Quarantine Weed
- 76, 87, 88, 203, 220

Coccoloba acuminata **Kunth**
Polygonaceae
♦ Weed
♦ 87, 88
♦ S/T.

Coccoloba polystachya **Wedd.**
Polygonaceae
Coccoloba mollis Casar.
♦ Weed
♦ 87, 88

Coccoloba schiedeana **Lindau**
Polygonaceae
♦ Weed, Quarantine Weed
♦ 76, 87, 88, 203, 220

Coccoloba uvifera **(L.) L.**
Polygonaceae
♦ seagrape, Jamaica kino
♦ Weed, Naturalised
♦ 22, 87, 88, 218, 287
♦ T, cultivated, herbal.

Cocculus carolinus **(L.) DC.**
Menispermaceae
♦ redberry moonseed, Carolina coralbead
♦ Weed
♦ 87, 88, 161, 218
♦ cultivated, herbal.

Cocculus orbiculatus **(L.) DC.**
Menispermaceae
♦ queen coralbead
♦ Weed
♦ 286, 297
♦ pC, cultivated. Origin: China, Japan.

Cocculus pendulus **(J.R. & G.Forst.) Diels**
Menispermaceae
♦ Weed
♦ 221
♦ herbal.

Cochlearia danica **L.**
Brassicaceae
♦ Danish scurvygrass, tanskankuirimo
♦ Weed
♦ 23, 88
♦ aH, cultivated, herbal.

Cochlearia officinalis **L.**
Brassicaceae
Crucifera cochlearia E.H.L.Krause
♦ ruijankuirimo, common scurvygrass, scurvy grass
♦ Naturalised
♦ 42
♦ b/pH, cultivated, herbal.

Cochlospermum insigne **St.-Hil.**
Bixaceae/Cochlospermaceae
♦ Weed
♦ 87, 88

Cochlospermum planchoni **Hook.f. ex Planch.**
Bixaceae/Cochlospermaceae
♦ Weed
♦ 88

Cochlospermum vitifolium **(Willd.) Willd. ex Spreng.**
Bixaceae/Cochlospermaceae
♦ silk cottontree, buttercup tree
♦ Naturalised, Cultivation Escape
♦ 101, 261

♦ S/T, cultivated, herbal.

Cocos **L. spp.**
Arecaceae
♦ coconut palm
♦ Quarantine Weed
♦ 220
♦ cultivated, herbal.

Cocos nucifera **L.**
Arecaceae
♦ coconut, coconut palm, nih, makapuno coconut, green Malay coconut, kokospalme, copra, nariyal, cocotier, khopar, coqueiro, cocotero
♦ Weed, Naturalised, Introduced, Environmental Weed, Casual Alien
♦ 22, 39, 40, 86, 88, 101, 151, 179, 230, 261
♦ T, cultivated, herbal, toxic. Origin: obscure.

Codariocalyx gyroides **(Roxb. ex Link) Hassk.**
Fabaceae/Papilionaceae
Desmodium gyroides (Roxb. ex Link) DC. (see)
♦ false tick trefoil
♦ Naturalised
♦ 101

Coddia rudis **E.Mey. ex Harv.**
Rubiaceae
Xeromphis rudis (E.Mey. ex Harv.) Codd, *Randida rudis* E.Mey.
♦ lesser xeromphis, small bone apple
♦ Native Weed
♦ 121
♦ pS. Origin: Africa.

Codiaeum variegatum **(L.) Juss.**
Euphorbiaceae
Croton pictus Lodd., *Croton variegatum* L.
♦ garden croton, variegated laurel, ave'ave
♦ Weed, Naturalised, Garden Escape
♦ 39, 86, 101, 161, 194, 247, 261
♦ cultivated, herbal, toxic. Origin: Malaysia, Pacific Islands.

Codiaeum variegatum **(L.) Bl. var. *pictum* (Lodd.) Müll.Arg.**
Euphorbiaceae
♦ kurodon
♦ Introduced
♦ 230
♦ S, cultivated, herbal.

Codium fragile **(Suringar) Har. ssp. *atlanticum* (A.D.Cotton) P.C.Silva**
Codiaceae
Codium atlanticum (A.D.Cotton) De Valéra, *Codium mucronatum* J.Agardh var. *atlanticum* Cotton, *Codium tomentosum* Stackhouse var. *atlanticum* (A.D.Cotton) L.Newton, *Codium fragile* fo. *atlanticum* (Cotton) Levring
♦ Weed
♦ 288
♦ algae.

Codium fragile **(Suringar) Har. ssp. *tomentosoides* (van Goor) P.C.Silva**
Codiaceae
Codium mucronatum J.Agardh var.

tomentosoides Goor
♦ green sea fingers, algae
♦ Weed
♦ 282, 288
♦ algae.

Codonopsis pilosula **(Fr.) Nannf.**
Campanulaceae
♦ tangshen, pilose Asia bell
♦ Weed
♦ 297
♦ pC, promoted, herbal. Origin: north-east Asia.

Coelachyrum brevifolium **Hochst. & Nees**
Poaceae
= *Eleusine brevifolia* (Hochst. & Nees) Steud. (NoR)
♦ Weed
♦ 221
♦ G.

Coelorachis glandulosa **(Trin.) Stapf**
Poaceae
♦ Weed
♦ 12, 87, 88
♦ G.

Coelorachis striata **A.Camus var. *pubescens* (Hack.) Bor.**
Poaceae
♦ hairy itchgrass
♦ Weed
♦ 297
♦ G.

Coffea **L. spp.**
Rubiaceae
♦ coffee
♦ Weed, Cultivation Escape
♦ 3, 39, 252
♦ cultivated, herbal, toxic.

Coffea arabica **L.**
Rubiaceae
♦ coffee, dwarf coffee, Arabian coffee, kove, kofe, koahpi
♦ Weed, Naturalised, Introduced, Garden Escape, Environmental Weed, Cultivation Escape, Casual Alien
♦ 22, 32, 39, 54, 73, 86, 88, 98, 101, 107, 134, 155, 203, 230, 257, 261, 280
♦ S/T, cultivated, herbal, toxic. Origin: Ethiopia, Kenya, Sudan.

Coffea dewevrei **Wildem. & T.Dur.**
Rubiaceae
= *Coffea liberica* Bull ex Hiern
♦ café excelsa, Liberian coffee
♦ Cultivation Escape
♦ 261
♦ cultivated. Origin: west Africa.

Coffea liberica **Bull ex Hiern**
Rubiaceae
Coffea dewevrei Wildem. & T.Dur. (see)
♦ Liberian coffee, café de Liberia
♦ Naturalised, Cultivation Escape
♦ 101, 261
♦ cultivated, herbal. Origin: Liberia.

Coffea robusta **L.Linden**
Rubiaceae
= *Coffea canephora* Pierre ex A.Froehner (NoR)
♦ koahpi

♦ Introduced
♦ 230
♦ herbal.

Cogniauxia trilobata **Cogn.**
Cucurbitaceae
♦ Weed
♦ 87, 88

Coincya cheiranthos **(Vill.) Greuter & Burdet**
Brassicaceae
= *Coincya monensis* (L.) Greuter & Burdet ssp. *recurvata* (All.) Leadlay
♦ nokkasinappi
♦ Casual Alien
♦ 42

Coincya monensis **(L.) Greuter & Burdet**
Brassicaceae
♦ star mustard, wallflower cabbage, tall wallflower cabbage, Isle of Man cabbage, coincya
♦ Naturalised, Environmental Weed
♦ 101, 182
♦ a/pH. Origin: western Europe.

Coincya monensis **(L.) Greuter & Burdet ssp.** *recurvata* **(All.) Leadlay**
Brassicaceae
Coincya cheiranthos (Vill.) Greuter & Burdet (see), *Rhynchosinapis cheiranthos* (Vill.) Dandy (see)
♦ star mustard, wallflower cabbage
♦ Naturalised, Casual Alien
♦ 40, 101

Coix aquatica **Roxb.**
Poaceae
♦ Job's tears
♦ Weed, Quarantine Weed
♦ 76, 87, 88, 135, 191, 203, 220
♦ G.

Coix gigantea **Koenig**
Poaceae
♦ Weed, Quarantine Weed, Naturalised
♦ 76, 86, 87, 88, 203, 220
♦ G. Origin: India, Sri Lanka to New Guinea.

Coix lacryma-jobi **L.**
Poaceae
Coix lachryma L., *Coix lacrima* L.
♦ Job's tears, adlay, adlay millet, sanasana
♦ Weed, Naturalised, Introduced, Casual Alien
♦ 7, 42, 66, 86, 87, 88, 90, 98, 101, 121, 203, 230, 255, 257, 280, 286, 287, 297
♦ aG, aqua, cultivated, herbal. Origin: tropical Asia.

Coix lacryma-jobi **L. var.** *maxima* **Makino**
Poaceae
♦ Naturalised
♦ 287
♦ G.

Cojoba arborea **(L.) Britton & Rose**
Fabaceae/Mimosaceae
Acacia arborea (L.) Willd., *Feuilleea filicifolia* (Lam.) Kuntze, *Mimosa arborea* L., *Mimosa filicifolia* Lam.,

Pithecellobium arboreum (L.) Urb., *Pithecellobium filicifolium* (Lam.) Benth., *Samanea arborea* (L.) Ricker
♦ Quarantine Weed
♦ 220
♦ T, arid.

Cola **Schott & Endl. spp.**
Sterculiaceae
♦ cola
♦ Quarantine Weed
♦ 220

Cola acuminata **(P.Beauv.) Schott & Endl.**
Sterculiaceae
♦ ligo, ngongoliya, abata cola, cola nut, nuez de cola
♦ Weed, Introduced
♦ 87, 88, 261
♦ cultivated, herbal. Origin: tropical Africa.

Cola nitida **(Vent.) Schott & Endl.**
Sterculiaceae
♦ ghanja kola, kola nut, cola nut, ligo
♦ Quarantine Weed
♦ 220
♦ herbal.

Colchicum alpinum **DC.**
Liliaceae/Colchicaceae
♦ alppimyrkkylilja
♦ Cultivation Escape
♦ 42
♦ cultivated, herbal.

Colchicum autumnale **L.**
Liliaceae/Colchicaceae
♦ syysmyrkkylilja, autumn crocus, meadow saffron, naked ladies, autumn grass, autumn crocus
♦ Weed, Naturalised, Cultivation Escape
♦ 23, 39, 42, 44, 70, 86, 87, 88, 101, 126, 161, 215, 252, 272, 280
♦ pH, cultivated, herbal, toxic. Origin: Europe.

Colchicum laetum **Stev.**
Liliaceae/Colchicaceae
Colchicum kotschyi Boiss.
♦ Weed
♦ 87, 88

Colchicum ritchii **R.Br.**
Liliaceae/Colchicaceae
♦ Weed
♦ 221
♦ herbal.

Colchicum speciosum **Steven**
Liliaceae/Colchicaceae
♦ fall crocus
♦ Weed
♦ 39, 161
♦ cultivated, herbal, toxic.

Coldenia procumbens **L.**
Boraginaceae/Ehretiaceae
♦ Weed
♦ 13, 87, 88
♦ aH, herbal. Origin: Madagascar.

Coleocephalocereus **Backeb. spp.**
Cactaceae
♦ Weed, Quarantine Weed
♦ 76, 88, 203, 220

Coleogyne ramosissima **Torr.**
Rosaceae
♦ blackbush
♦ Weed
♦ 87, 88, 218
♦ S, cultivated, herbal.

Coleonema album **(Thunb.) Bartl. & H.L.Wendl.**
Rutaceae
♦ diosma
♦ Weed, Naturalised
♦ 7, 86, 98, 203
♦ cultivated.

Coleonema pulchellum **I.Williams**
Rutaceae
♦ diosma, breath of heaven
♦ Weed, Naturalised, Environmental Weed
♦ 15, 72, 86, 88
♦ S, cultivated. Origin: South Africa.

Coleonema pulchrum **Hook.**
Rutaceae
♦ Weed, Naturalised, Casual Alien
♦ 54, 86, 88, 280
♦ cultivated.

Coleostephus multicaulis **(Desf.) Durieu**
Asteraceae
♦ Quarantine Weed
♦ 220
♦ herbal.

Coleostephus myconis **(L.) Cass.**
Asteraceae
Chrysanthemum myconis L. (see), *Pyrethrum myconis* (L.) Moench
♦ sitruunapäivänkakkara
♦ Weed, Naturalised, Casual Alien
♦ 42, 70, 88, 94, 237, 241, 250, 253, 300
♦ cultivated.

Coleus amboinicus **Lour.**
Lamiaceae
= *Plectranthus amboinicus* (Lour.) Spreng.
♦ militini, Jamaica thyme, broadleaf thyme
♦ Weed
♦ 262
♦ cultivated, herbal.

Coleus blumei **Benth.**
Lamiaceae
= *Coleus scutellarioides* (L.) Benth.
♦ pate
♦ Weed, Introduced
♦ 38, 87, 88
♦ cultivated, herbal.

Coleus pumilus **Blanco**
Lamiaceae
♦ dwarf coleus
♦ Weed, Naturalised
♦ 101, 179
♦ cultivated, herbal.

Coleus scutellarioides **(L.) Benth.**
Lamiaceae
Coleus blumei Benth. (see), *Plectranthus scutellarioides* (L.) R.Br. (see)
♦ common coleus, pate
♦ Naturalised
♦ 101
♦ cultivated, herbal. Origin: Australia.

Coleus spicatus **Benth.**
Lamiaceae
♦ Weed, Naturalised
♦ 98, 203
♦ cultivated.

Colletia paradoxa **(Spreng.) Escal**
Rhamnaceae
Colletia cruciata Gill. & Hook.
♦ curro
♦ Weed
♦ 237, 295
♦ S, cultivated, herbal.

Colletia spinosissima **J.F.Gmel.**
Rhamnaceae
Colletia ferox Gill. & Hook.
♦ quina quina
♦ Weed
♦ 237, 295

Collinsia heterophylla **Buist ex Graham**
Scrophulariaceae
♦ Chinese houses, verihanhikki, harlequin blue eyed Mary, purple Chinese houses
♦ Weed, Cultivation Escape
♦ 42, 161
♦ aH, cultivated, herbal.

Collinsia parviflora **Douglas ex Lindl.**
Scrophulariaceae
♦ maiden blue eyed Mary, smallflower blue eyed Mary
♦ Weed
♦ 87, 88
♦ aH, cultivated, herbal.

Collomia biflora **(Ruiz & Pav.) Brand**
Polemoniaceae
Phlox biflora Ruiz & Pav.
♦ Weed
♦ 87, 88
♦ aH, herbal.

Collomia cavanillesii **Hook. & Arn.**
Polemoniaceae
♦ Naturalised
♦ 280
♦ cultivated.

Collomia grandiflora **Lindl.**
Polemoniaceae
Gilia grandiflora A.Gray
♦ largeflower mountain trumpet, grand collomia, mountain collomia, orange mountain trumpet
♦ Weed, Naturalised
♦ 86, 88, 94, 98, 203
♦ aH, cultivated, herbal. Origin: North America.

Collomia linearis **Nutt.**
Polemoniaceae
♦ tahmikki, tiny trumpet, slenderleaf collomia, narrowleaf mountain trumpet
♦ Weed, Naturalised, Casual Alien
♦ 42, 86, 87, 88, 176
♦ aH, cultivated, herbal. Origin: Australia.

Colocasia esculenta **(L.) Schott**
Araceae
Colocasia antiquorum Schott, *Caladium esculenta* Vent.

♦ wild taro, dasheen, elephant's ear, coco yam, ikoma ole, sawa, green sedge, green taro
♦ Weed, Quarantine Weed, Naturalised, Introduced, Environmental Weed, Cultivation Escape
♦ 7, 39, 80, 87, 88, 98, 101, 112, 134, 179, 197, 209, 220, 230, 257, 258, 261, 262, 280
♦ H, aqua, cultivated, herbal, toxic. Origin: India.

Colocasia gigantea **Hook.**
Araceae
Caladium giganteum Blume, *Leucocasia gigantea* (Blume) Schott
♦ giant yam
♦ Weed
♦ 39, 87, 88
♦ cultivated, herbal, toxic.

Colocynthis vulgaris **Schrad.**
Cucurbitaceae
= *Citrullus colocynthis* (L.)
♦ bitter gourd, Sierra Leone gourd
♦ Weed
♦ 87, 88

Colophospermum mopane **(Benth.) Léonard**
Fabaceae/Caesalpiniaceae
Copaifera mopane J.Kirk ex Benth.
♦ balsam tree, black ironwood, butterfly tree, mapane, mopane, mopani, red Angola copal, Rhodesian ironwood, Rhodesian mahogany, turpentine tree, white ironwood
♦ Weed, Quarantine Weed, Native Weed, Introduced
♦ 10, 22, 63, 76, 87, 88, 121, 220, 228
♦ T, arid, cultivated. Origin: Africa.

Colpomenia peregrina **(Sauv.) Hamel**
Phaeophyceae/Punctariaceae
Colpomenia sinuosa (Mert. ex Roth) Derbès & Solier var. *peregrina* Sauv.
♦ oyster thief, algae
♦ Weed
♦ 282, 288
♦ algae.

Colquhounia coccinea **Wall. var. *mollis* (Schlecht.) Prain**
Lamiaceae
♦ Weed
♦ 297

Colubrina asiatica **(L.) Brongn.**
Rhamnaceae
♦ latherleaf, Asian nakedwood
♦ Weed, Noxious Weed, Environmental Weed
♦ 22, 39, 47, 80, 87, 88, 112, 122, 129, 129, 151, 179
♦ S, cultivated, herbal, toxic. Origin: Madagascar.

Colubrina texensis **(Torr. & Gray) Gray**
Rhamnaceae
♦ Texas colubrina, hog plum, Texan hog plum
♦ Weed
♦ 87, 88, 161, 218
♦ arid, herbal, toxic.

Colutea arborescens **L.**
Fabaceae/Papilionaceae
Colutea brevialata Lange
♦ bladder senna
♦ Weed, Naturalised
♦ 80, 101
♦ S, arid, cultivated, herbal.

Colutea istria **Mill.**
Fabaceae/Papilionaceae
Colutea halepica Lam., *Colutea pallida* Salisb., *Colutea pocockii* Aiton
♦ Weed, Introduced
♦ 221, 228
♦ S, arid, promoted.

Colutea media **Willd.**
Fabaceae/Papilionaceae
♦ Quarantine Weed
♦ 220

Comandra elegans **(Roch.) Rchb.**
Santalaceae
♦ Weed
♦ 272

Comarum palustre **L.**
Rosaceae
Potentilla comarum Nestl., *Potentilla palustris* (L.) Scop. (see)
♦ purple marshlocks
♦ Weed
♦ 87, 88
♦ cultivated.

Combretum apiculatum **Sond.**
Combretaceae
♦ red bushwillow, mlama
♦ Weed, Quarantine Weed
♦ 10, 76, 87, 88, 203, 220
♦ cultivated, herbal. Origin: southern Africa.

Combretum apiculatum **Sond. ssp. *apiculatum***
Combretaceae
♦ rooibos, red bushwillow
♦ Weed
♦ 63, 121
♦ Origin: southern Africa.

Combretum hereroense **Schinz**
Combretaceae
♦ Weed, Quarantine Weed
♦ 76, 87, 88, 203, 220
♦ cultivated.

Combretum hispidum **C.Lawson**
Combretaceae
♦ Weed
♦ 88

Combretum imberbe **Wawra**
Combretaceae
♦ bastard yellowwood, elephant's trunk, ivory tree, leadwood
♦ Weed, Native Weed
♦ 87, 88, 121
♦ T, promoted, herbal. Origin: Africa.

Combretum mechowianum **O.Hoffm.**
Combretaceae
♦ Weed
♦ 87, 88

Combretum micranthum **G.Don**
Combretaceae
Combretum altum Perr., *Combretum*

floribundum Engl. & Diels, *Combretum parviflorum* Rchb., *Combretum raimbaulti* Heckel
♦ kantalma, kinkeliba, kinkiliba, kwando, red withy
♦ Introduced
♦ 228
♦ arid. Origin: Africa.

Combretum paniculatum Vent.
Combretaceae
♦ likapo
♦ Weed, Naturalised
♦ 86, 87, 88
♦ cultivated. Origin: central Africa.

Combretum racemosum P.Beauv.
Combretaceae
♦ longelengele
♦ Weed
♦ 39, 88
♦ herbal, toxic.

Combretum zeyheri Sond.
Combretaceae
♦ large fruited bushwillow, large fruited combretum, Zeyher's bushwillow
♦ Weed, Native Weed
♦ 10, 88, 121
♦ S/T, promoted, herbal. Origin: southern Africa.

Cometes abyssinica R.Br.
Caryophyllaceae/Illecebraceae
♦ Weed
♦ 88, 221

Commelina spp. L.
Commelinaceae
♦ dayflower
♦ Weed, Quarantine Weed
♦ 88, 220, 221, 243
♦ herbal.

Commelina africana L.
Commelinaceae
Commelina subulata Roth. (see)
♦ yellow wandering Jew
♦ Weed, Quarantine Weed, Naturalised
♦ 54, 76, 86, 87, 88, 126, 158, 203, 220
♦ pH, herbal. Origin: southern Africa.

Commelina africana L. var. africana
Commelinaceae
♦ Native Weed
♦ 121
♦ pH. Origin: southern Africa.

Commelina agraria Kunth
Commelinaceae
= *Commelina diffusa* Burm.f.
♦ Weed
♦ 87, 88

Commelina auriculata Bl.
Commelinaceae
= *Commelina erecta* L.
♦ Weed
♦ 286

Commelina benghalensis L.
Commelinaceae
Commelina prostrata Regel
♦ wandering Jew, Mickey Mouse, dayflower, hairy wandering Jew,
tropical spiderwort, Benghal commelina, Benghal wandering Jew, blue wandering Jew, Benghal dayflower, jio, pak prarb
♦ Weed, Quarantine Weed, Noxious Weed, Naturalised, Native Weed, Environmental Weed
♦ 6, 32, 50, 51, 53, 55, 66, 67, 76, 86, 87, 88, 98, 101, 121, 126, 140, 158, 161, 170, 179, 186, 203, 209, 220, 229, 238, 240, 243, 245, 255, 262, 269, 270, 276, 286, 297
♦ a/pH, cultivated, herbal. Origin: Africa, India, south-east Asia.

Commelina boissieriana C.B.Clarke
Commelinaceae
♦ Weed
♦ 88, 221
♦ Origin: Africa.

Commelina caroliniana Walter
Commelinaceae
♦ Carolina dayflower
♦ Weed
♦ 179

Commelina coelestis Willd.
Commelinaceae
= *Commelina tuberosa* L.
♦ blue spiderwort, hierba de pollo
♦ Weed, Quarantine Weed
♦ 76, 87, 88, 220, 243, 281
♦ pH, cultivated, herbal.

Commelina communis L.
Commelinaceae
♦ dayflower, common dayflower, Asiatic dayflower, soljo
♦ Weed, Naturalised
♦ 8, 23, 80, 87, 88, 101, 126, 161, 195, 204, 207, 211, 218, 263, 275, 286, 297
♦ aH, cultivated, herbal. Origin: southern China, Japan, India.

Commelina communis L. var. communis
Commelinaceae
♦ Asiatic dayflower
♦ Naturalised
♦ 101

Commelina communis L. var. ludens (Miq.) C.B.Clarke
Commelinaceae
♦ Asiatic dayflower
♦ Naturalised
♦ 101

Commelina condensata C.B.Clarke
Commelinaceae
♦ ilungulungu
♦ Weed
♦ 87, 88

Commelina cyanea R.Br.
Commelinaceae
♦ scurvy weed
♦ Weed, Native Weed
♦ 87, 88, 269
♦ cultivated. Origin: Australia.

Commelina diffusa Burm.f.
Commelinaceae
Commelina agraria Kunth (see), *Commelina aquatica* J.K.Morton, *Commelina cayennensis* Rich., *Commelina*
cespitosa Roxb., *Commelina diffusa* var. *cordispatha* Rohweder, *Commelina gracilis* Ruiz & Pav., *Commelina longicaulis* Jacq. (see), *Commelina loureiroi* Kunth, *Commelina nudiflora* auct. non L., *Commelina ochreata* Schauer, *Commelina pacifica* Vahl, *Lechea chinensis* Lour.
♦ wandering Jew, dayflower, honohono, mau'utoga, spreading dayflower, watergrass, climbing dayflower, mau'u toga, phak plaap, tripa de pollo, hierba de pollo
♦ Weed, Quarantine Weed, Native Weed, Introduced
♦ 3, 6, 14, 68, 76, 87, 88, 121, 126, 161, 170, 186, 203, 206, 209, 211, 218, 220, 230, 239, 242, 243, 249, 255, 262, 263, 273, 274, 276, 281, 286
♦ a/pH, aqua, cultivated, herbal. Origin: southern Asia.

Commelina diffusa Burm.f. var. diffusa
Commelinaceae
♦ climbing dayflower
♦ Weed
♦ 179

Commelina elegans Kunth
Commelinaceae
= *Commelina erecta* L.
♦ Weed, Quarantine Weed
♦ 76, 87, 88, 203, 220
♦ cultivated, herbal.

Commelina erecta L.
Commelinaceae
Commelina auriculata Bl. (see), *Commelina elegans* Kunth (see), *Commelina gerrardii* C.B.Clarke (see), *Commelina virginica* auct. non L.
♦ whitemouth dayflower, santa luzia, tripa de pollo
♦ Weed
♦ 14, 28, 87, 88, 126, 157, 199, 237, 243, 255, 270, 281, 295
♦ pH, promoted, herbal. Origin: tropical America.

Commelina forskaolii Vahl
Commelinaceae
♦ rat's ear
♦ Weed, Naturalised
♦ 87, 88, 101, 126, 179
♦ herbal. Origin: Africa, Madagascar.

Commelina gambiae C.B.Clarke
Commelinaceae
♦ Weed
♦ 179

Commelina gerrardii C.B.Clarke
Commelinaceae
= *Commelina erecta* L.
♦ Weed, Quarantine Weed
♦ 76, 87, 88, 203, 220

Commelina hasskarlii C.B.Clarke
Commelinaceae
♦ Weed
♦ 66, 88

Commelina jacobi Fisch.
Commelinaceae
♦ Weed
♦ 87, 88

Commelina kotschyi Hassk.
Commelinaceae
♦ Weed, Quarantine Weed
♦ 76, 87, 88, 203, 220, 242
♦ arid.

Commelina lagosensis C.B.Clarke
Commelinaceae
♦ Weed
♦ 87, 88

Commelina latifolia Hochst. ex A.Rich.
Commelinaceae
♦ Weed, Quarantine Weed
♦ 76, 87, 88, 203, 220, 240
♦ pH.

Commelina livingstonii C.B.Clarke
Commelinaceae
♦ Native Weed
♦ 121
♦ pH. Origin: southern Africa.

Commelina longicaulis Jacq.
Commelinaceae
= *Commelina diffusa* Burm.f.
♦ Weed, Quarantine Weed
♦ 76, 87, 88, 203, 220
♦ arid.

Commelina maculata Edgew.
Commelinaceae
♦ Weed
♦ 88

Commelina nigritana Benth.
Commelinaceae
♦ African dayflower
♦ Naturalised
♦ 101
♦ pH.

Commelina nigritana Benth. var. gambiae (C.B.Clarke) Brenan
Commelinaceae
♦ Gambian dayflower
♦ Naturalised
♦ 101

Commelina nudiflora L.
Commelinaceae
= *Murdannia nudiflora* (L.) Brenan
♦ Weed
♦ 39, 88, 275
♦ aH, herbal, toxic.

Commelina palaeata Hassk.
Commelinaceae
♦ Weed
♦ 276
♦ pH.

Commelina salicifolia Roxb.
Commelinaceae
♦ Weed, Quarantine Weed
♦ 76, 87, 88, 203, 220
♦ herbal.

Commelina subulata Roth
Commelinaceae
= *Commelina africana* L.
♦ Weed, Quarantine Weed
♦ 76, 87, 88, 203, 220

Commelina tuberosa L.
Commelinaceae
Commelina alpestris Standl. &
Steyerm., *Commelina coelestis* Willd.

(see), *Commelina dianthifolia* Delile,
Commelina elliptica Kunth, *Commelina
graminifolia* Kunth
♦ commelina, blue spiderwort,
dayflower
♦ Weed, Quarantine Weed
♦ 39, 76, 199, 243
♦ pH, cultivated, herbal, toxic.

Commelina virginica L.
Commelinaceae
Commelina sulcata Hoffm.
♦ Virginia dayflower
♦ Weed
♦ 87, 88, 236, 245
♦ a/pH, promoted, herbal.

Commicarpus africanus (Lour.) Dandy
Nyctaginaceae
♦ Weed
♦ 221
♦ Origin: Egypt, Sudan.

Commicarpus boissieri (Heimerl) Cufod.
Nyctaginaceae
♦ Weed
♦ 88, 221

Commicarpus helenae (J.A.Schult.) Meikle
Nyctaginaceae
♦ Weed
♦ 221

Commicarpus pentandrus (Burch.) Heimerl
Nyctaginaceae
♦ Native Weed
♦ 121
♦ pH. Origin: southern Africa.

Commicarpus scandens (L.) Standl.
Nyctaginaceae
= *Boerhavia scandens* L.
♦ Weed
♦ 87, 88
♦ herbal.

Commiphora angolensis Engl.
Burseraceae
Commiphora kwebensis N.E.Br
♦ commiphora, poison grub, sand
commiphora, sand corkwood
♦ Native Weed
♦ 121
♦ S/T. Origin: southern Africa.

Commiphora gileadensis (L.) C.Chr.
Burseraceae
Amyris gileadensis L., *Amyris
opobalsamum* L., *Commiphora
opobalsamum* (L.) Engl. (see),
Commiphora opobalsamum (L.) Engl. var.
ehrenbergianum (Berg.) Engl.
♦ balm of Gilead, Mecca myrrh
♦ Introduced
♦ 228
♦ arid, cultivated. Origin: North
Africa, Middle East.

Commiphora madagascariensis Jacq.
Burseraceae
Commiphora agallocha Engl.,
Commiphora roxburghii (Arn.) Engl. var.
serratifolia Haines, *Amyris agallocha*
Roxb.

♦ Introduced
♦ 228
♦ arid.

Commiphora mollis Engl.
Burseraceae
♦ Weed
♦ 87, 88

Commiphora opobalsamum (L.) Engl.
Burseraceae
= *Commiphora gileadensis* (L.) C.Chr.
♦ Mecca myrrh, balsam of Gilead
♦ Weed
♦ 221
♦ herbal.

Commiphora pyracanthoides Engl.
Burseraceae
♦ gewone kanniedood, kurkbas,
corktree, common corkwood
♦ Weed, Native Weed
♦ 63, 121
♦ pS, arid, toxic. Origin: southern
Africa.

Comocladia dentata Jacq.
Anacardiaceae
♦ toothed maidenplum
♦ Weed
♦ 14
♦ herbal.

Comocladia glabra (Schult.) Spreng.
Anacardiaceae
♦ carrasco
♦ Weed
♦ 39, 87, 88
♦ herbal, toxic.

Comocladia platyphylla A.Rich
Anacardiaceae
♦ Weed
♦ 14

Comptonia peregrina (L.) Coult.
Myricaceae
♦ sweet fern
♦ Weed
♦ 87, 88, 218, 294
♦ S, promoted, herbal.

Conanthera minima Grau.
Liliaceae/Tecophilaeaceae
♦ Quarantine Weed
♦ 220

Conanthera trimaculata (D.Don) Meigen
Liliaceae/Tecophilaeaceae
♦ Quarantine Weed
♦ 220
♦ arid.

Condalia microphylla Cav.
Rhamnaceae
Condalia lineata A.Gray
♦ piquillín
♦ Weed
♦ 295

Condalia obovata Hook.
Rhamnaceae
= *Condalia hookeri* M.C.Johnst. var.
hookeri (NoR)
♦ bluewood condalia
♦ Weed
♦ 87, 88, 218

Condalia obtusifolia (Hook.) Weberb.
Rhamnaceae
= *Ziziphus obtusifolia* (Torr. & A.Gray)
A.Gray var. *obtusifolia* (NoR)
♦ lotebush condalia
♦ Weed
♦ 87, 88, 218
♦ arid, herbal.

Congea tomentosa Roxb.
Verbenaceae/Symphoremataceae
♦ lluvia de orquideas, shower of
orchids
♦ Naturalised, Introduced
♦ 101, 261
♦ cultivated. Origin: southern Asia.

Conicosia bijlii N.E.Br.
Aizoaceae/Mesembryanthemaceae
♦ Weed, Naturalised
♦ 98, 203

Conicosia pugioniformis (L.) N.E.Br.
Aizoaceae/Mesembryanthemaceae
Carpobrotus crassifolius auth.,
Herrea elongata (Haw.) L.Bolus,
Mesembryanthemum elongatum Haw
♦ narrow leaved iceplant, roundleaf
iceplant, false iceplant, conicosia
♦ Weed, Noxious Weed, Naturalised,
Environmental Weed
♦ 35, 78, 80, 88, 101, 116, 151, 231
♦ pH, cultivated.

Conium chaerophylloides (Thunb.) Sond.
Apiaceae
♦ Weed, Quarantine Weed
♦ 39, 76, 88, 220
♦ herbal, toxic.

Conium maculatum L.
Apiaceae
Coriandrum maculatum Roth, *Sium
conium* Vest
♦ poison hemlock, wild carrot, wild
parsnip, hemlock, bunk, California
fern, poison parsley, poison root,
snakeweed, spotted hemlock, spotted
parsley, winter fern, wode whistle.
♦ Weed, Quarantine Weed, Noxious
Weed, Naturalised, Introduced,
Garden Escape, Environmental Weed
♦ 1, 7, 8, 15, 20, 21, 34, 35, 36, 38, 39,
45, 70, 72, 76, 78, 80, 82, 86, 87, 88, 94,
98, 101, 102, 116, 136, 139, 146, 147, 151,
154, 159, 161, 165, 167, 174, 176, 180,
189, 195, 198, 203, 207, 210, 211, 212,
217, 218, 219, 228, 229, 231, 237, 241,
243, 246, 247, 253, 264, 267, 269, 271,
272, 280, 287, 289, 295, 300
♦ b/pH, aqua, cultivated, herbal,
toxic. Origin: Europe.

Conobea multifida (Michx.) Benth.
Scrophulariaceae
= *Leucospora multifida* (Michx.) Nutt.
(NoR)
♦ Weed
♦ 179

Conocarpus erectus L.
Combretaceae
♦ button mangrove, buttonwood,
buttonwood sea mulberry
♦ Cultivation Escape

♦ 233
♦ cultivated, herbal.

Conocarpus lancifolius Engl. & Diels
Combretaceae
♦ Introduced
♦ 228
♦ arid. Origin: Somalia.

Conopholis americana (L.) Wallr.f.
Orobanchaceae
♦ squawroot, American squawroot
♦ Weed
♦ 161
♦ herbal.

Conostegia xalapensis (Bonpl.) D.Don
Melastomataceae
♦ Weed
♦ 14
♦ cultivated.

**Conostomium natalense (Hochst.) Brem.
var. glabrum Brem.**
Rubiaceae
Oldenlandia natalensis (Hochst.) Kuntze
♦ Native Weed
♦ 121
♦ pH. Origin: southern Africa.

Conradina verticillata Jennison
Lamiaceae
♦ Cumberland rosemary, Cumberland
false rosemary
♦ Quarantine Weed
♦ 220
♦ S, promoted, herbal.

Conringia austriaca (Jacq.) Sweet
Brassicaceae
♦ konringia rakúska
♦ Weed
♦ 272

Conringia orientalis (L.) Dumort.
Brassicaceae
Brassica orientalis L., *Gorinkia orientalis*
(L.) Presl
♦ hare's ear cabbage, hare's ear
mustard, klinkweed, treacle mustard,
rabbit ears, vélar d'Orient
♦ Weed, Quarantine Weed,
Naturalised, Introduced, Casual Alien,
Cultivation Escape
♦ 23, 34, 40, 42, 49, 52, 76, 86, 87, 88, 94,
98, 101, 161, 198, 203, 210, 218, 228, 243,
252, 253, 272, 287
♦ aH, arid, cultivated, herbal, toxic.
Origin: Europe.

Conringia planisiliqua Fisch. & C.A.Mey.
Brassicaceae
♦ aasiansavuruoho
♦ Casual Alien
♦ 42

Consolea Lem. spp.
Cactaceae
= *Opuntia* Mill. spp.
♦ Weed, Quarantine Weed
♦ 76, 88, 203

Consolea macracantha (Griseb.) Berger
Cactaceae
= *Opuntia macracantha* Griseb. (NoR)
♦ Weed
♦ 14

Consolida ajacis (L.) Schur
Ranunculaceae
Consolida ambigua auct., *Delphinium
ajacis* L., *Delphinium ambiguum* auct.
♦ doubtful knight's spur, rocket
larkspur
♦ Weed, Naturalised
♦ 39, 87, 88, 101, 161
♦ cultivated, herbal, toxic. Origin:
southern Europe.

Consolida ambigua (L.) Ball & Heywood
Ranunculaceae
Delphinium ambiguum L.
♦ rocket larkspur, doubtful knight's
spur, tarhakukonkannus, larkspur
♦ Weed, Naturalised, Cultivation
Escape
♦ 7, 15, 34, 42, 70, 86, 88, 94, 98, 203,
252, 272, 280
♦ aH, cultivated, herbal.

Consolida orientalis (Gay) Schröd.
Ranunculaceae
Delphinium orientale J.Gay (see)
♦ oriental knight's spur,
idänkukokannus
♦ Weed, Naturalised, Casual Alien
♦ 42, 87, 88, 94, 101, 272
♦ cultivated, herbal.

Consolida pubescens (DC.) Soó
Ranunculaceae
♦ Naturalised
♦ 101

Consolida regalis S.F.Gray
Ranunculaceae
Delphinium consolida L. (see)
♦ forking larkspur, royal knight's spur
♦ Weed, Naturalised
♦ 39, 70, 88, 94, 101, 243, 253, 272
♦ a/bH, cultivated, herbal, toxic.

Consolida tenuissima (Sibth. & Sm.) Soó
Ranunculaceae
♦ Naturalised
♦ 101

Convallaria majalis L.
Liliaceae/Convallariaceae
Polygonatum majale All.
♦ lily of the valley, European lily of the
valley, kielo
♦ Weed, Naturalised, Introduced
♦ 4, 8, 23, 24, 39, 87, 88, 101, 161, 189,
194, 195, 218, 222, 247
♦ pH, cultivated, herbal, toxic.

Convallaria majuscula Greene
Liliaceae/Convallariaceae
♦ American lily of the valley
♦ Weed
♦ 161
♦ toxic.

Convolvulus L. spp.
Convolvulaceae
♦ bindweed, morningglory
♦ Weed, Noxious Weed
♦ 18, 80, 88, 141, 243, 272
♦ herbal.

Convolvulus aegyptius L.
Convolvulaceae
♦ hairy morningglory
♦ Weed
♦ 87, 88, 218

Convolvulus althaeoides L.
Convolvulaceae
♦ mallow leaved bindweed, mallow
bindweed
♦ Weed, Quarantine Weed, Noxious
Weed, Naturalised
♦ 34, 70, 76, 87, 88, 94, 101, 115, 185,
221, 229, 243
♦ pC, cultivated.

Convolvulus ammannii Desr.
Convolvulaceae
♦ ammann glorybind
♦ Weed
♦ 297

Convolvulus arvensis L.
Convolvulaceae
Convolvulus minor Gilib.
♦ field bindweed, bindweed, creeping
Jenny, morningglory, perennial
morningglory, small bindweed,
cornbine, wild morningglory, small
flowered morningglory, European
bindweed, cornbind, bear bind, green
vine, akkerwinde, klimop, laplove
♦ Weed, Quarantine Weed, Noxious
Weed, Naturalised, Introduced,
Garden Escape, Environmental Weed,
Cultivation Escape
♦ 1, 7, 15, 16, 16, 21, 23, 26, 34, 35, 36,
38, 44, 45, 49, 51, 52, 55, 62, 63, 68, 70,
76, 78, 80, 86, 87, 88, 94, 98, 101, 111,
115, 116, 118, 121, 136, 138, 139, 146,
147, 151, 158, 161, 162, 165, 171, 174,
176, 180, 185, 186, 195, 198, 199, 203,
207, 210, 211, 212, 216, 218, 219, 221,
222, 225, 229, 236, 237, 240, 241, 243,
246, 248, 250, 253, 264, 266, 269, 270,
272, 275, 280, 283, 286, 287, 293, 295,
297, 299, 300
♦ pH, arid, cultivated, herbal, toxic.
Origin: Eurasia.

Convolvulus betonicifolius Mill.
Convolvulaceae
♦ shaggy bindweed
♦ Weed
♦ 88, 115, 243, 272

Convolvulus cantabricus L.
Convolvulaceae
♦ Naturalised
♦ 287
♦ cultivated.

Convolvulus equitans Benth.
Convolvulaceae
♦ Texas bindweed
♦ Weed, Noxious
Weed
♦ 199, 229
♦ herbal.

Convolvulus erubescens Sims
Convolvulaceae
♦ Australian bindweed
♦ Weed, Naturalised
♦ 55, 87, 88, 101
♦ cultivated, herbal.

Convolvulus farinosus L.
Convolvulaceae
♦ wild bindweed
♦ Weed, Native Weed, Introduced

♦ 88, 121, 158, 228
♦ C, arid, cultivated. Origin: southern
Africa.

Convolvulus fatmensis Kunze
Convolvulaceae
♦ Weed
♦ 88

Convolvulus galaticus Rost. ex Choisy
Convolvulaceae
♦ Weed
♦ 87, 88, 243

Convolvulus glomeratus Choisy
Convolvulaceae
♦ Weed
♦ 221

Convolvulus hermanniae L'Hér.
Convolvulaceae
♦ Weed
♦ 295

Convolvulus hirsutus Bieb.
Convolvulaceae
♦ Weed
♦ 87, 88

Convolvulus humilis Jacq.
Convolvulaceae
♦ Weed
♦ 87, 88

Convolvulus hystrix Vahl
Convolvulaceae
♦ Weed
♦ 221

Convolvulus japonicus Thunb.
Convolvulaceae
= *Calystegia pellita* (Ledeb.) G.Don
♦ Weed
♦ 87, 88
♦ herbal.

Convolvulus lanatus Vahl
Convolvulaceae
♦ Weed
♦ 221

Convolvulus nodiflorus Desv.
Convolvulaceae
♦ aguinaldo blanco
♦ Noxious Weed
♦ 229

Convolvulus pilosellifolius Desr.
Convolvulaceae
♦ soft bindweed
♦ Weed, Noxious Weed, Naturalised
♦ 88, 229, 287

Convolvulus pluricaulis Choisy
Convolvulaceae
♦ Weed
♦ 87, 88
♦ cultivated.

Convolvulus prostratus Forssk.
Convolvulaceae
♦ Weed
♦ 221

Convolvulus sabatius Viv.
Convolvulaceae
♦ ampelvinda, riippakierto
♦ Naturalised
♦ 280
♦ cultivated, herbal.

Convolvulus sabatius Viv. ssp.
mauritanicus (Boiss.) Murb.
Convolvulaceae
♦ Naturalised
♦ 280

Convolvulus sagittatus Thunb. ssp.
sagittatus
Convolvulaceae
♦ wild bindweed
♦ Weed
♦ 88, 158
♦ Origin: South Africa.

Convolvulus sagittatus Thunb. ssp.
sagittatus var. phyllosepalus (Hallier f.)
A.Meeuse
Convolvulaceae
♦ Native Weed
♦ 121
♦ pH. Origin: southern Africa.

Convolvulus sagittatus Thunb. ssp.
ulosephalus (Hallier f.) Verdc.
Convolvulaceae
♦ wild bindweed
♦ Weed
♦ 158
♦ Origin: South Africa.

Convolvulus sagittatus Thunb. var.
ulosepalus (Hallier f.) Verdc.
Convolvulaceae
Convolvulus ulosepalus Hallier f. (see)
♦ wild bindweed
♦ Weed, Native Weed
♦ 88, 121
♦ pC. Origin: southern Africa.

Convolvulus scammonia L.
Convolvulaceae
♦ scammony, Syrian bindweed
♦ Noxious Weed
♦ 39, 229
♦ pH, arid, promoted, herbal, toxic.

Convolvulus sepium L.
Convolvulaceae
= *Calystegia sepium* (L.) R.Br.
♦ hedge bindweed, lady's nightcap,
bellbine, wild morningglory, larger
bindweed, wind morningglory,
devil's vine, great bindweed, bracted
bindweed, Rutland beauty
♦ Weed, Noxious Weed
♦ 23, 39, 44, 88, 136, 162, 210, 218, 299
♦ herbal, toxic.

Convolvulus thunbergii Roem. & Schult.
Convolvulaceae
♦ Native Weed
♦ 121
♦ pC. Origin: southern Africa.

Convolvulus tricolor L.
Convolvulaceae
♦ bindweed, dwarf morningglory,
kirjokierto
♦ Weed, Quarantine Weed,
Naturalised, Cultivation Escape
♦ 24, 42, 70, 76, 86, 87, 88, 94, 203, 220,
243, 252
♦ pC, cultivated, herbal.

Convolvulus ulosepalus Hallier f.
Convolvulaceae
= *Convolvulus sagittatus* Thunb. var.

ulosepalus (Hallier f.) Verdc.
- wild bindweed
- Weed
- 51, 87, 88
- herbal. Origin: South Africa.

Convolvulus wallichianus **Spreng.**
Convolvulaceae
- Wallich's bindweed
- Noxious Weed, Naturalised
- 101, 229

Conyza abyssinica **Sch.Bip. ex A.Rich.**
Asteraceae
- Weed
- 87, 88

Conyza aegyptiaca **(L.) Aiton**
Asteraceae
- Weed, Naturalised, Native Weed
- 86, 87, 88, 98, 121, 185, 203, 221, 276
- aH. Origin: tropical and subtropical Africa, Asia.

Conyza albida **Willd. ex Spreng.**
Asteraceae
Conyza sumatrensis (Retz.) Walker (see), *Erigeron sumatrensis* Retz. (see)
- tall fleabane, broad leaved fleabane, fleabane
- Weed, Naturalised, Environmental Weed
- 7, 9, 15, 72, 86, 86, 88, 158, 165, 176, 198, 269, 280
- b, herbal. Origin: subtropical South America.

Conyza ambigua **DC.**
Asteraceae
= *Conyza bonariensis* (L.) Cronquist
- Weed
- 87, 88

Conyza apurensis **Kunth**
Asteraceae
- manzanilla horseweed, botoncillo
- Weed
- 28, 157, 206, 243, 281
- aH.

Conyza bilbaoana **E.J.Remy**
Asteraceae
- asthma weed, fleabane
- Weed, Naturalised
- 15, 34, 86, 98, 198, 203, 269, 280
- aH. Origin: Americas.

Conyza bonariensis **(L.) Cronquist**
Asteraceae
Conyza ambigua DC. (see), *Conyza floribunda* Kunth (see), *Conyza linifolia* Willd. (see), *Erigeron bonariense* L., *Erigeron crispus* Pourr., *Erigeron linifolius* Willd. (see)
- fleabane, hairy fleabane, flax leaved fleabane, horseweed, asthma weed, Argentiinankoiransilmä
- Weed, Naturalised, Introduced, Environmental Weed, Casual Alien
- 7, 9, 15, 42, 53, 55, 68, 70, 86, 87, 88, 93, 94, 98, 101, 111, 115, 121, 134, 158, 161, 167, 176, 180, 185, 198, 199, 203, 205, 212, 228, 236, 237, 240, 243, 250, 253, 255, 256, 257, 263, 269, 271, 272, 273, 275, 280, 286, 287, 290, 295, 297, 300

- aH, arid, cultivated, herbal. Origin: tropical America.

Conyza bovei **DC.**
Asteraceae
= *Doellia bovei* (DC.) Anderb. (NoR)
- Weed
- 221

Conyza canadensis **(L.) Cronquist**
Asteraceae
Erigeron canadensis L. (see)
- horseweed fleabane, Canadian horseweed, Canada fleabane, butterweed, blood stanch, colt's tail, fireweed, hogweed, horseweed, mare's tail, pride weed, voaderia, buva
- Weed, Naturalised, Native Weed, Introduced, Garden Escape, Environmental Weed
- 14, 32, 34, 40, 42, 68, 70, 87, 88, 94, 98, 121, 136, 158, 159, 161, 167, 174, 176, 180, 198, 203, 207, 210, 211, 212, 217, 228, 243, 249, 250, 253, 255, 256, 257, 261, 272, 273, 275, 280, 297
- a/bH, arid, cultivated, herbal. Origin: North America.

Conyza canadensis **(L.) Cronquist var. canadensis**
Asteraceae
Leptilon canadense (L.) Britt.
- Canada fleabane, horseweed, mare's tail, blood stanch
- Weed, Naturalised, Environmental Weed
- 49, 55, 86, 98, 269
- aH, cultivated, herbal. Origin: North America.

Conyza canadensis **(L.) Cronquist var. pusilla (Nutt.) Cronquist**
Asteraceae
Conyza parva Cronquist (see), *Erigeron pusillus* Nutt. (see), *Leptilon pusillum* (Nutt.) Britton (see)
- Canadian fleabane
- Weed
- 55
- herbal.

Conyza chilensis **Spreng.**
Asteraceae
= *Conyza primulaefolia* (Lam.) Cuatrec. & Lourteig (NoR)
- Chilean fleabane
- Weed, Naturalised
- 86, 87, 88, 98, 121, 203, 295
- H. Origin: South America.

Conyza dioscoridis **Desf.**
Asteraceae
= *Pluchea dioscoridis* (L.) DC.
- Weed
- 87, 88, 221
- arid. Origin: Africa.

Conyza filaginoides **(DC.) Hieron.**
Asteraceae
= *Laennecia filaginoides* DC. (NoR)
- Weed
- 199
- herbal.

Conyza floribunda **Kunth**
Asteraceae

= *Conyza bonariensis* (L.) Cronquist
- horseweed, tall fleabane, asthma weed
- Weed, Naturalised, Introduced
- 34, 101, 121
- aH. Origin: South America.

Conyza japonica **(Thunb.) Less.**
Asteraceae
- Weed
- 87, 88, 204, 286

Conyza leucantha **(D.Don) Ludlow & P.H.Raven**
Asteraceae
- Naturalised
- 86
- Origin: Asia.

Conyza lineariloba **DC.**
Asteraceae
- Weed
- 87, 88

Conyza linifolia **Willd.**
Asteraceae
= *Conyza bonariensis* (L.) Cronquist
- horseweed
- Weed
- 68, 88, 221

Conyza parva **Cronquist**
Asteraceae
= *Conyza canadensis* (L.) Cronquist var. *pusilla* (Nutt.) Cronquist
- small fleabane
- Weed, Naturalised
- 7, 86, 98, 198, 203, 269, 280
- aH. Origin: Americas.

Conyza pinnata **(L.F.) Kuntze**
Asteraceae
- Native Weed
- 121
- pH, herbal. Origin: southern Africa.

Conyza podocephala **DC.**
Asteraceae
- conyza
- Weed, Native Weed
- 51, 87, 88, 121, 158
- pH, herbal. Origin: southern Africa.

Conyza primulifolia **(Lam.) Cuatrec. & Lourteig**
Asteraceae
- primroseleaf horseweed
- Weed, Naturalised
- 86, 237
- Origin: Central and South America.

Conyza pyrrhopappa **Sch.Bip. ex A.Rich.**
Asteraceae
- Weed
- 87, 88

Conyza ramosissima **Cronquist**
Asteraceae
Erigeron divaricatus Michx. (see)
- dwarf fleabane, dwarf horseweed
- Weed
- 161, 210
- herbal.

Conyza scabiosifolia **Remy**
Asteraceae
- rough conyza
- Naturalised
- 86, 198

Conyza scabrida DC.
Asteraceae
Conyza ivifolia (L.) Less.
♦ Native Weed
♦ 121
♦ pS. Origin: southern Africa.

Conyza Less. spp.
Asteraceae
♦ horseweed
♦ Weed, Naturalised, Environmental
Weed
♦ 106, 198, 237, 289

Conyza steudelii Sch.Bip. ex A.Rich.
Asteraceae
♦ Weed
♦ 87, 88

Conyza stricta Willd.
Asteraceae
Conyza schimperi Sch.Bip. ex A.Rich.
♦ Weed
♦ 87, 88

**Conyza stricta Willd. var. pinnatifida
(D.Don) Kitam.**
Asteraceae
Conyza chrysocoma (DC.) Vatke, *Conyza
pinnatifida* Roxb., *Conyza triloba* Decne.
(see), *Erigeron pinnatifidum* D.Don,
Erigeron trilobum (Decne.) Boiss.,
Nidorella chrysocoma DC., *Nidorella
triloba* (Decne.) DC.
♦ Weed
♦ 88

Conyza sumatrensis (Retz.) Walker
Asteraceae
= *Conyza albida* Willd. ex Spreng.
♦ tall fleabane, Sumatran fleabane, cho
lo
♦ Weed, Naturalised
♦ 55, 68, 88, 134, 238, 253, 263, 273, 286,
287
♦ aH.

Conyza triloba Decne.
Asteraceae
= *Conyza stricta* Willd. var. *pinnatifida*
(D.Don) Kitam.
♦ Weed
♦ 221

Conyza ulmifolia (Burm.f.) Kuntze
Asteraceae
♦ Native Weed
♦ 121
♦ pH. Origin: Africa.

Cooperia pedunculata Herb.
Liliaceae/Amaryllidaceae
Zephyranthes drummondii D.Don (see)
♦ giant rain lily, prairie lily
♦ Weed
♦ 161
♦ herbal, toxic.

Copaifera officinalis (Jacq.) L.
Fabaceae/Caesalpiniaceae
♦ copaiba
♦ Naturalised
♦ 39, 101
♦ herbal, toxic.

Copernicia australis Becc.
Arecaceae

= *Copernicia alba* Morong ex Morong &
Britton (NoR)
♦ Weed
♦ 87, 88

Coprosma repens A.Rich.
Rubiaceae
Coprosma baueri auct. non Endl.
♦ mirror bush, taupata, creeping
mirrorplant, lookingglass bush, New
Zealand mirror bush, tree bedstraw,
mirrorplant
♦ Weed, Naturalised, Garden Escape,
Environmental Weed
♦ 7, 40, 72, 78, 86, 88, 98, 101, 116, 131,
155, 176, 198, 203, 289, 296
♦ S, cultivated, herbal. Origin: New
Zealand.

Coprosma robusta Raoul
Rubiaceae
♦ karamu
♦ Weed, Naturalised, Garden Escape,
Environmental Weed
♦ 72, 86, 88, 176, 198, 251, 289
♦ S/T, cultivated. Origin: New
Zealand.

**Corallocarpus bainesii (Hook.f.)
A.Meeuse**
Cucurbitaceae
Corallocarpus bussei Gilg, *Corallocarpus
sphaerocarpus* Cogn.
♦ Native Weed
♦ 121
♦ pC, arid. Origin: southern Africa.

Corbichonia decumbens (Forssk.) Exell
Molluginaceae
♦ corbichonia
♦ Weed, Native Weed
♦ 50, 87, 88, 121
♦ S. Origin: Africa.

Corbicula Meunier spp.
Craspedophyceae/Acanthoecacea
Parvicorbicula Deflandre spp.
♦ microplankton
♦ Weed
♦ 18, 88
♦ microplankton.

Corchoropsis tomentosa Makino
Sterculiaceae
♦ Weed
♦ 87, 88, 286

Corchorus aestuans L.
Tiliaceae
Corchorus acutangulus Lam.
♦ jute
♦ Weed, Naturalised, Introduced
♦ 23, 87, 88, 179, 204, 228, 243, 276, 286,
287
♦ aH, arid, herbal. Origin: Central
America.

Corchorus angolensis Exell & Mendonca
Tiliaceae
♦ Weed
♦ 87, 88

Corchorus antichorus Raeusch.
Tiliaceae
♦ Weed
♦ 87, 88

Corchorus argutus H.B.K.
Tiliaceae
♦ Weed
♦ 87, 88

Corchorus asplenifolius Burch.
Tiliaceae
♦ Native Weed
♦ 121
♦ pH.

Corchorus capsularis L.
Tiliaceae
♦ cultivated jute, jute
♦ Weed, Introduced
♦ 39, 87, 88, 228, 262
♦ aH, arid, promoted, herbal, toxic.
Origin: southern China.

Corchorus depressus Stocks
Tiliaceae
♦ Weed
♦ 87, 88, 242
♦ arid, promoted, herbal.

Corchorus fascicularis Lam.
Tiliaceae
♦ Weed
♦ 87, 88, 242
♦ arid.

Corchorus hirsutus L.
Tiliaceae
♦ jackswitch
♦ Weed
♦ 87, 88

Corchorus hirtus L.
Tiliaceae
Corchorus orinocensis Kunth (see),
Corchorus hirtus L. var. *orinocensis*
(Kunth) K.Schum. (see)
♦ Orinoco jute
♦ Weed
♦ 87, 88

**Corchorus hirtus L. var. orinocensis
(Kunth) K.Schum.**
Tiliaceae
= *Corchorus hirtus* L.
♦ Weed
♦ 271

Corchorus olitorius L.
Tiliaceae
♦ Malta jute, Jew's mallow, tossa jute,
wild jute, ng'ombebanda, po krachao
♦ Weed, Naturalised, Native Weed,
Introduced, Environmental Weed
♦ 7, 23, 39, 55, 86, 87, 88, 93, 98, 101,
121, 185, 203, 217, 221, 228, 239, 242,
255, 262, 286, 287
♦ aH, arid, cultivated, herbal, toxic.
Origin: Old World tropics.

Corchorus orinocensis Kunth
Tiliaceae
= *Corchorus hirtus* L.
♦ red jute
♦ Weed
♦ 87, 88, 157
♦ aH.

Corchorus siliquosus L.
Tiliaceae
♦ slippery burr
♦ Weed, Introduced

- 14, 38, 87, 88
- herbal.

Corchorus tridens L.
Tiliaceae
- tridens corchorus, wild jute, nyenje
- Weed, Naturalised, Introduced
- 50, 87, 88, 101, 121, 228, 242
- aH, arid. Origin: Eurasia.

Corchorus trilocularis L.
Tiliaceae
Corchorus aestuans Forssk., *Corchorus asplenifolius* E.Mey. ex Harv. & Sond., *Corchorus serraefolius* DC., *Corchorus triflorus* Bojer
- native jute, wild jute, three locule corchorus
- Weed, Introduced
- 55, 87, 88, 93, 121, 158, 191, 221, 228, 240, 242
- aH, arid. Origin: Eurasia.

Cordeauxia edulis Hemsl.
Fabaceae/Caesalpiniaceae
- yeheb nut
- Introduced
- 228
- arid.

Cordia L. spp.
Boraginaceae/Ehretiaceae
- cordia
- Quarantine Weed
- 220

Cordia alliodora (Ruiz & Pav.) Oken
Boraginaceae/Ehretiaceae
Cerdana alliodora Ruiz & Pav., *Cordia gerascanthus* Jacq.
- laurel, clammy cherry, Spanish elm
- Weed, Introduced, Environmental Weed
- 3, 191, 228, 257
- T, arid, cultivated, herbal.

Cordia boissieri A.DC.
Boraginaceae/Ehretiaceae
- Texas wild olive, anacahuita
- Weed, Quarantine Weed
- 76, 87, 88, 203, 220
- herbal.

Cordia collococca L.
Boraginaceae/Ehretiaceae
Cordia glabra auct. non L. (see)
- red manjack
- Weed
- 3, 191
- T, herbal.

Cordia corymbosa (L.) G.Don
Boraginaceae/Ehretiaceae
- Weed
- 87, 88

Cordia curassavica Roem. & Schult.
Boraginaceae/Ehretiaceae
Cordia brevispicata M.Martens & Galeotti, *Cordia chacoensis* Chodat., *Cordia imparillis* Macbr., *Cordia macrostachya* (Jacq.) Roem. & Schult. (see), *Cordia palmeri* S.Watson, *Cordia socorrensis* Brandegee
- erva baleeira
- Weed, Quarantine Weed, Garden Escape, Environmental Weed

- 3, 76, 87, 88, 191, 203, 220, 243, 255, 259
- arid, cultivated, herbal. Origin: tropical Americas.

Cordia cylindristachya Roem. & Schult.
Boraginaceae/Ehretiaceae
- Weed, Quarantine Weed
- 76, 87, 88, 203, 220

Cordia dichotoma G.Forst.
Boraginaceae/Ehretiaceae
- fragrant manjack, clammy cherry, Indian cherry
- Weed, Naturalised, Introduced
- 13, 101, 228
- arid, cultivated, herbal. Origin: temperate and tropical Asia, Australasia, Pacific.

Cordia glabra auct. non L.
Boraginaceae/Ehretiaceae
= *Cordia collococca* L.
- broad leaved cordia, manjack
- Weed
- 3, 191

Cordia globosa (Jacq.) H.B.K.
Boraginaceae/Ehretiaceae
- curaciao bush
- Weed
- 14, 87, 88
- herbal.

Cordia macrostachya (Jacq.) Roem & Schlucht.
Boraginaceae/Ehretiaceae
= *Cordia curassavica* Roem. & Schult.
- Weed
- 22, 87, 88
- S.

Cordia myxa L.
Boraginaceae/Ehretiaceae
- Assyrian plum, sapistan plum, sapistan, sebesten plum, Sudan teak
- Weed, Naturalised
- 22, 101, 179
- S, arid, cultivated, herbal. Origin: central Asia.

Cordia obliqua Willd.
Boraginaceae/Ehretiaceae
- clammy cherry, lassora, seloo
- Naturalised, Garden Escape
- 101, 261
- cultivated, herbal. Origin: India.

Cordia oxyphylla DC.
Boraginaceae/Ehretiaceae
- Weed
- 87, 88

Cordia polycephala (Lam.) I.M.Johnst.
Boraginaceae/Ehretiaceae
- black sage
- Weed
- 87, 88, 255
- Origin: Brazil.

Cordia sebestena L.
Boraginaceae/Ehretiaceae
- geiger tree, largeleaf geigertree
- Weed, Naturalised, Garden Escape, Environmental Weed
- 22, 88, 101, 151, 261
- T, cultivated, herbal.

Cordia sinensis Lam.
Boraginaceae/Ehretiaceae
Cordia angustifolia Roxb., *Cordia gharaf* Ehrenb. ex Asch., *Cordia rebiculata* Roth., *Cordia rothii* Roem. & Schult., *Cordia subopposita* DC.
- gundi
- Weed
- 221
- arid. Origin: Africa, Asia, Madagascar.

Cordia subcordata Lam.
Boraginaceae/Ehretiaceae
- cordia, kou, kerosene wood, ikoik
- Weed, Quarantine Weed, Naturalised
- 6, 88, 101, 220
- T, cultivated, herbal.

Cordia trichotoma (Vell.) Arráb. ex Steud.
Boraginaceae/Ehretiaceae
Cordia excelsa A.DC., *Cordia frondosa* Schott, *Cordia hypoleuca* DC., *Cordia tomentosa* Cham.
- Introduced
- 228
- arid, cultivated.

Cordyline australis (G.Forst.) Endl.
Agavaceae/Asteliaceae
Dracaena australis Forst.f.
- New Zealand cabbage tree, cabbage tree, dracaena palm, ti kouka
- Weed, Noxious Weed, Naturalised, Garden Escape, Environmental Weed
- 35, 54, 72, 78, 86, 88, 116, 131, 198
- T, cultivated, herbal. Origin: New Zealand.

Cordyline fruticosa (L.) A.Chev.
Agavaceae/Asteliaceae
Cordyline terminalis (L.) Kunth
- dihng, tiplant, ti
- Weed, Naturalised, Introduced, Cultivation Escape, Casual Alien
- 22, 230, 261, 280
- T, cultivated, herbal. Origin: tropical Asia.

Corema album (L.) Don
Empetraceae/Ericaceae
Empetrum album Linn
- Portuguese crowberry
- Weed
- 70
- S, cultivated, herbal.

Coreopsis auriculata L.
Asteraceae
- korvakaunosilmä, lobed tickseed, tickseed
- Cultivation Escape
- 42
- pH, cultivated, herbal.

Coreopsis basalis (A.Dietr.) Blake
Asteraceae
- goldenmane tickseed, golden wave
- Naturalised
- 287
- aH, herbal.

Coreopsis grandiflora **Hogg ex Sweet**
Asteraceae
- large flowered coreopsis, largeflower tickseed, tickweed
- Weed, Naturalised
- 161, 203, 256, 287
- a/pH, cultivated, herbal.

Coreopsis lanceolata **L.**
Asteraceae
Coreopsis lanceolata L. var. *villosa* Michx. (see)
- tickseed, garden coreopsis, lanceleaf tickseed, tickseed coreopsis
- Weed, Naturalised, Native Weed, Garden Escape, Environmental Weed, Cultivation Escape
- 23, 34, 88, 98, 121, 155, 158, 161, 198, 203, 256, 269, 280, 283, 286, 287, 290, 296
- pH, cultivated, herbal. Origin: obscure.

Coreopsis lanceolata **L. var.** *angustifolia* **Torr. & Gray**
Asteraceae
- Naturalised
- 287

Coreopsis lanceolata **L. var.** *lanceolata*
Asteraceae
- Naturalised
- 287

Coreopsis lanceolata **L. var.** *villosa* **Michx.**
Asteraceae
= *Coreopsis lanceolata* L.
- Naturalised
- 287

Coreopsis tinctoria **Nutt.**
Asteraceae
- plains coreopsis, golden tickseed, garden tickseed, golden coreopsis, tiikerikaunosilmä
- Weed, Naturalised, Native Weed, Garden Escape, Cultivation Escape
- 34, 42, 49, 87, 88, 121, 161, 174, 179, 210, 218, 256, 280, 286, 287
- a/pH, cultivated, herbal. Origin: North America.

Coreopsis verticillata **L.**
Asteraceae
- whorled coreopsis, whorled tickseed
- Weed
- 23, 88, 161
- cultivated, herbal.

Coriandrum sativum **L.**
Apiaceae
Selinum coriandrum E.H.L.Krause
- coriander
- Weed, Naturalised, Introduced, Garden Escape, Casual Alien, Cultivation Escape
- 34, 39, 40, 42, 86, 87, 88, 94, 98, 101, 161, 179, 203, 221, 228, 252, 256, 261, 272, 280, 287
- aH, arid, cultivated, herbal, toxic. Origin: southern Europe.

Coriandrum tordylioides **Boiss.**
Apiaceae
- Weed

- 87, 88

Coriandrum vulgare **L.**
Apiaceae
- Chinese parsley, coriander
- Weed
- 121
- aH. Origin: Eurasia.

Coriaria **L. spp.**
Coriariaceae
- coriaria, tutu, puuhou, tuupaakihi
- Weed
- 161, 165
- toxic.

Coriaria nepalensis **Wall.**
Coriariaceae
- Quarantine Weed
- 220
- S, cultivated, herbal.

Corispermum candelabrum **Iljin**
Chenopodiaceae
- Weed
- 275, 297
- aH.

Corispermum declinatum **Stephan ex Steven**
Chenopodiaceae
- kaakonkurmio
- Weed, Casual Alien
- 42, 272, 275
- aH.

Corispermum hyssopifolium **L.**
Chenopodiaceae
- hyssopleaf tickseed, iisoppikurmio, common bugseed
- Weed, Naturalised, Casual Alien
- 23, 42, 87, 88, 218, 272, 287
- aH, cultivated, herbal.

Corispermum leptopterum **(Asch.) Iljin**
Chenopodiaceae
= *Corispermum pallasii* Steven
- rikkakurmio
- Casual Alien
- 42
- cultivated, herbal.

Corispermum marginale **Rydb.**
Chenopodiaceae
= *Corispermum americanum* (Nutt.) Nutt. var. *americanum* (NoR)
- Weed
- 23, 88

Corispermum marschallii **Steven**
Chenopodiaceae
- Weed
- 88, 94
- herbal.

Corispermum nitidum **Kit. ex J.A.Schult.**
Chenopodiaceae
- shiny bugseed
- Weed, Naturalised
- 88, 94, 101, 272
- herbal.

Corispermum pallasii **Steven**
Chenopodiaceae
Corispermum leptopterum (Asch.) Iljin (see)
- Siberian bugseed
- Naturalised
- 101

Corispermum patelliforme **Iljin**
Chenopodiaceae
- patelliform tickseed
- Weed
- 297

Corispermum villosum **Rydb.**
Chenopodiaceae
- hairy bugseed
- Weed
- 23, 88
- herbal.

Cornulaca ehrenbergii **Asch.**
Chenopodiaceae
- eed
- Weed
- 221

Cornulaca monacantha **Delile**
Chenopodiaceae
- Weed
- 221
- arid.

Cornus alba **L. p.p.**
Cornaceae
= *Cornus sericea* L. ssp. *sericea* (NoR)
- white dogwood, idänkanukka, Tatarian dogwood
- Weed, Cultivation Escape, Casual Alien
- 40, 42, 80
- T, cultivated, herbal.

Cornus canadensis **L.**
Cornaceae
Chamaepericlymenum canadense (L.) Asch. & Graebn.
- bunchberry dogwood, bunchberry, gozentachibana, creeping dogwood
- Weed
- 294
- pH, cultivated, herbal.

Cornus capitata **Wall.**
Cornaceae
Benthamia capitata (Wall.) Nakai, *Benthamidia capitata* (Wall.) H.Hara, *Cynoxylon capitata* (Wall.) Nakai, *Dendrobenthamia capitata* (Wall. ex Roxb.) Hutch.
- Himalayan strawberry tree, evergreen dogwood, Bentham's cornel
- Weed, Naturalised, Garden Escape, Environmental Weed
- 54, 72, 86, 88, 198, 289
- T, cultivated, herbal. Origin: China to Himalayas.

Cornus drummondii **C.A.Mey.**
Cornaceae
- roughleaf dogwood
- Weed
- 87, 88, 195, 218
- cultivated, herbal.

Cornus florida **L.**
Cornaceae
Benthamidia florida (L.) Spach, *Cynoxylon floridum* (L.) Britton & Shafer
- flowering dogwood, American boxwood
- Weed
- 87, 88, 218
- S, cultivated, herbal.

Cornus kousa Hance
Cornaceae
♦ kousa dogwood, Japanese dogwood
♦ Naturalised
♦ 101
♦ T, cultivated, herbal.

Cornus mas L.
Cornaceae
Cornus mascula Zorn, *Macrocarpium mas*
(L.) Nakai
♦ Cornelian cherry
♦ Weed, Naturalised
♦ 39, 80
♦ S, cultivated, herbal, toxic.

Cornus nuttallii Aud.
Cornaceae
♦ Pacific dogwood, mountain
dogwood
♦ Weed
♦ 87, 88, 218
♦ S, cultivated, herbal.

Cornus obliqua Raf.
Cornaceae
♦ silky dogwood
♦ Weed
♦ 195
♦ cultivated, herbal.

Cornus racemosa Lam.
Cornaceae
♦ gray dogwood
♦ Weed, Environmental Weed
♦ 88, 151, 195
♦ cultivated, herbal.

Cornus rugosa Lam.
Cornaceae
♦ roundleaf dogwood
♦ Weed
♦ 87, 88, 218
♦ S, cultivated, herbal.

Cornus sanguinea L.
Cornaceae
Swida sanguinea (L.) Opiz, *Thelycrania
sanguinea* (L.) Fourr.
♦ bloodtwig dogwood,
mustamarjakanukka, dogwood,
dogwood cornel, common dogwood,
skogskornell
♦ Naturalised, Cultivation Escape
♦ 39, 42, 101, 243
♦ S, cultivated, herbal, toxic.

Cornus sericea L.
Cornaceae
Cornus stolonifera Michx. (see)
♦ red osier dogwood
♦ Weed
♦ 195
♦ S, cultivated, herbal.

Cornus stolonifera Michx.
Cornaceae
= *Cornus sericea* L.
♦ red osier dogwood
♦ Weed
♦ 87, 88, 218
♦ cultivated, herbal.

Cornus torreyi S.Wats.
Cornaceae
♦ western dogwood
♦ Weed
♦ 87, 88, 218

Cornutia grandiflora Steud.
Lamiaceae/Verbenaceae
♦ pavilla
♦ Weed
♦ 179

Corokia buddleioides A.Cunn. ×
cotoneaster Raoul
Cornaceae/Escalloniaceae/
Corokiaceae
♦ Weed
♦ 15

Coronilla L. spp.
Fabaceae/Papilionaceae
♦ crownvetch
♦ Weed, Naturalised
♦ 198, 272
♦ toxic.

Coronilla cretica L.
Fabaceae/Papilionaceae
= *Securigera securidaca* (L.) Degen &
Dörfl.
♦ Cretan crownvetch, coronilla
♦ Naturalised
♦ 101

Coronilla globosa Lam.
Fabaceae/Papilionaceae
♦ white crownvetch
♦ Naturalised
♦ 101

Coronilla repanda (Poir.) Guss.
Fabaceae/Papilionaceae
♦ Weed
♦ 87, 88

Coronilla scorpioides (L.) Koch
Fabaceae/Papilionaceae
♦ annual scorpion vetch, yellow
crownvetch, keltanivelvirna
♦ Weed, Naturalised, Casual Alien
♦ 39, 40, 42, 87, 88, 94, 101, 111, 215,
221, 243, 253, 287
♦ aH, cultivated, herbal, toxic.

Coronilla securidaca L.
Fabaceae/Papilionaceae
Securigera securidaca (L.) Degen &
Dörfl. (see)
♦ goat pea
♦ Naturalised
♦ 101

Coronilla vaginalis Lam.
Fabaceae/Papilionaceae
Coronilla minima L. var. *vaginalis* Fiori
& Paol.
♦ small scorpion vetch
♦ Weed
♦ 272
♦ cultivated.

Coronilla valentina L.
Fabaceae/Papilionaceae
♦ scorpion vetch, Mediterranean
crownvetch
♦ Weed, Naturalised
♦ 70, 101
♦ cultivated, herbal.

Coronilla varia L.
Fabaceae/Papilionaceae
= *Securigera varia* (L.) Lassen
♦ crownvetch, trailing crownvetch,
purple crownvetch, rosenkronill

♦ Weed, Naturalised, Introduced,
Garden Escape, Environmental Weed
♦ 4, 17, 39, 42, 80, 86, 87, 88, 94, 98, 101,
102, 133, 142, 151, 156, 161, 176, 179,
195, 198, 203, 218, 222, 243, 252, 272,
280, 287
♦ pH, cultivated, herbal, toxic.

Coronopus Zinn spp.
Brassicaceae
♦ swinecress
♦ Weed, Naturalised
♦ 39, 198, 221, 243
♦ H, toxic.

Coronopus didymus (L.) J.E.Sm.
Brassicaceae
Lepidium didymum L., *Senebiera didyma*
Pers. (see)
♦ wart cress, bittercress, carrot
weed, hog's cress, land cress, lesser
swinecress, twin cress, wild carrot,
karvavariksenkrassi
♦ Weed, Naturalised, Introduced,
Environmental Weed, Casual Alien
♦ 7, 15, 32, 34, 40, 42, 55, 70, 86, 87, 88,
94, 98, 101, 121, 134, 158, 161, 165, 176,
179, 180, 185, 198, 203, 217, 218, 228,
236, 237, 241, 243, 249, 253, 255, 257,
263, 269, 271, 280, 286, 287, 295, 297
♦ a/bH, arid, cultivated, herbal.
Origin: South America.

Coronopus integrifolius (DC.) Spreng.
Brassicaceae
Coronopus englerianus Muschl.,
Senebiera integrifolia DC., *Senebiera
linoides* DC.
♦ Weed, Naturalised
♦ 86, 98, 203
♦ cultivated. Origin: Madagascar.

Coronopus niloticus (Del.) Spreng.
Brassicaceae
♦ Weed
♦ 87, 88, 185, 221

Coronopus procumbens Gilib.
Brassicaceae
= *Coronopus squamatus* (Forssk.) Asch.
♦ creeping watercress
♦ Weed
♦ 87, 88, 218
♦ herbal.

Coronopus squamatus (Forssk.) Asch.
Brassicaceae
Cochlearia coronopus L., *Coronopus
procumbens* Gilib. (see), *Coronopus
ruellii* All., *Coronopus verrucarius*
(Garsault) Muschl. & Thell., *Coronopus
verrucarius* ssp. *euverrucarius*
Muschl., *Lepidium squamatum* Forssk.,
Nasturtium verrucarium Garsault,
Senebiera coronopus (L.) Poir.
♦ greater swinecress, swinecress, wart
cress
♦ Weed, Noxious Weed, Naturalised,
Casual Alien
♦ 15, 26, 35, 42, 70, 86, 87, 88, 94, 98,
101, 165, 176, 185, 198, 203, 229, 253,
272, 280, 300
♦ a/pH, arid, cultivated, herbal.
Origin: Mediterranean, west Asia.

Corrigiola litoralis L.
Molluginaceae/Caryophyllaceae
Polygonifolia litoralis Kuntze
♦ strapwort, varsanpolvi
♦ Weed, Naturalised, Native Weed, Environmental Weed, Casual Alien
♦ 7, 9, 23, 42, 70, 86, 87, 88, 94, 98, 101, 121, 198, 240, 253
♦ H, cultivated. Origin: south-west Europe, north and east Africa, west Asia.

Corryocactus Britton & Rose spp.
Cactaceae
Erdisia Britt. & Rose spp. (see)
♦ Weed, Quarantine Weed
♦ 76, 88, 203, 220

Cortaderia Stapf spp.
Poaceae
♦ pampas grass
♦ Weed, Noxious Weed, Naturalised, Garden Escape, Environmental Weed
♦ 86, 142, 169, 181, 198, 296
♦ G, cultivated.

Cortaderia jubata (Lemoine) Stapf
Poaceae
♦ Andean pampas grass, jubatagrass, purple pampas grass, pampas grass, pink pampas grass, selloa pampas grass, Andes grass
♦ Weed, Quarantine Weed, Noxious Weed, Naturalised, Environmental Weed, Cultivation Escape
♦ 3, 15, 18, 35, 54, 63, 72, 76, 78, 80, 86, 88, 95, 101, 116, 147, 151, 171, 176, 181, 198, 203, 220, 225, 229, 231, 233, 246, 252, 280, 283, 289
♦ pG, cultivated, herbal. Origin: Argentina, Bolivia, Ecuador, Peru.

Cortaderia richardii (Endl.) Zotov
Poaceae
Arundo fulvida J.Buch.
♦ cortaderia, toe toe, pampas grass
♦ Weed, Quarantine Weed, Noxious Weed, Naturalised, Environmental Weed
♦ 76, 86, 88, 98, 147, 171, 176, 203, 220, 290
♦ G, cultivated. Origin: New Zealand.

Cortaderia selloana (Schult. & Schult.f.) Asch. & Graebn.
Poaceae
♦ pampas grass, common pampas grass, silver pampas grass, Uruguayan pampas grass, silvergrass
♦ Weed, Quarantine Weed, Noxious Weed, Garden Escape, Environmental Weed, Cultivation Escape
♦ 3, 7, 15, 18, 34, 35, 40, 45, 63, 72, 73, 76, 78, 80, 86, 88, 95, 98, 101, 116, 121, 147, 151, 152, 155, 176, 179, 181, 198, 201, 203, 220, 225, 231, 233, 246, 255, 269, 280, 283, 287, 289, 295
♦ pG, cultivated, herbal. Origin: South America.

Cortaderia splendens Connor
Poaceae
♦ toe toe

♦ Weed
♦ 181
♦ G.

Cortusa brotheri Pax ex Lipsky
Primulaceae
♦ Quarantine Weed
♦ 220

Corybas ponapensis (Hosok. & Fuk.) Hosok. & Fuk.
Orchidaceae
♦ Introduced
♦ 230
♦ H.

Corydalis Vent. spp.
Papaveraceae/Fumariaceae
♦ fumewort, fitweed
♦ Weed
♦ 272
♦ herbal, toxic.

Corydalis aurea Willd.
Papaveraceae/Fumariaceae
♦ golden corydalis, scrambled eggs
♦ Weed
♦ 8, 39, 87, 88, 161, 218
♦ a/pH, cultivated, herbal, toxic.

Corydalis bulbosa (L.) DC. comb. rej.
Papaveraceae/Fumariaceae
= *Corydalis solida* (L.) Clairv.
♦ purple corydalis
♦ Weed
♦ 272
♦ cultivated, herbal.

Corydalis bungeana Turcz.
Papaveraceae/Fumariaceae
♦ Weed
♦ 275, 297
♦ a/bH.

Corydalis capnoides (L.) Pers.
Papaveraceae/Fumariaceae
♦ kitkeräkirunkannus
♦ Casual Alien
♦ 42

Corydalis caseana Gray
Papaveraceae/Fumariaceae
♦ fitweed, Sierra fumewort
♦ Weed
♦ 161
♦ herbal, toxic.

Corydalis edulis Maxim.
Papaveraceae/Fumariaceae
♦ Weed
♦ 87, 88, 275
♦ aH, promoted.

Corydalis flavula (Raf.) DC.
Papaveraceae/Fumariaceae
♦ short spurred corydalis, yellow fumewort
♦ Weed
♦ 39, 161
♦ herbal, toxic.

Corydalis incisa (Thunb.) Pers.
Papaveraceae/Fumariaceae
♦ murasakikeman
♦ Weed
♦ 286
♦ bH, promoted, herbal. Origin: China, Japan.

Corydalis lutea (L.) DC.
Papaveraceae/Fumariaceae
= *Pseudofumaria lutea* (L.) Borkh.
♦ yellow fumitory, yellow corydalis, keltakiurunkannus
♦ Weed, Naturalised, Cultivation Escape
♦ 42, 98, 203, 280
♦ cultivated, herbal.

Corydalis nobilis (L.) Pers.
Papaveraceae/Fumariaceae
♦ jalokiurunkannus
♦ Cultivation Escape
♦ 42
♦ cultivated, herbal.

Corydalis pallida (Thunb.) Pers.
Papaveraceae/Fumariaceae
♦ Weed
♦ 275
♦ aH, promoted. Origin: China, Japan, Korea.

Corydalis sempervirens (L.) Pers.
Papaveraceae/Fumariaceae
Corydalis glauca Pursh
♦ Amerikankiurunkannus, rock harlequin, pink corydalis
♦ Weed, Cultivation Escape
♦ 42, 161
♦ a/bH, cultivated, herbal, toxic.

Corydalis solida (L.) Clairv.
Papaveraceae/Fumariaceae
Corydalis bulbosa (L.) DC. *comb. rej.* (see)
♦ spring fumewort, fumewort, bulbous corydalis, pystykiurunkannus, colombina solida, stor nunnert
♦ Weed, Naturalised
♦ 101, 272
♦ pH, cultivated, herbal.

Corylus americana Walter
Betulaceae/Corylaceae
♦ American hazel, American filbert
♦ Weed
♦ 87, 88, 218
♦ T, cultivated, herbal.

Corylus avellana L.
Betulaceae/Corylaceae
♦ hazel, common filbert, cob nut, European hazel, nocciolo
♦ Weed, Naturalised
♦ 70, 101
♦ T, cultivated, herbal.

Corylus cornuta Marsh.
Betulaceae/Corylaceae
Corylus rostrata Aiton
♦ beaked hazel
♦ Weed
♦ 87, 88, 218
♦ S, cultivated, herbal.

Corylus cornuta Marsh. var. californica (A.DC.) Sharp
Betulaceae/Corylaceae
♦ California hazelnut, hazelnut
♦ Weed
♦ 218
♦ S.

Corylus heterophylla Fisch. & Trautv.
Betulaceae/Corylaceae
♦ Siberian hazelnut, Japanese hazel,
Siberian filbert
♦ Naturalised
♦ 101
♦ T, promoted, herbal.

Corymbia ficifolia (F.Muell.) K.D.Hill &
L.A.S.Johnson
Myrtaceae
Eucalyptus ficifolia F.Muell. (see)
♦ flowering gum
♦ Naturalised
♦ 198
♦ cultivated.

Corymbia torelliana (F.Muell.) K.D.Hill
& L.A.S.Johnson
Myrtaceae
Eucalyptus torelliana F.Muell. (see)
♦ cadaghi gum
♦ Environmental weed
♦ 201
♦ Cultivated.

Corymbium africanum L.
Asteraceae
♦ Quarantine Weed
♦ 220
♦ Origin: South Africa.

Corymbium glabrum L.
Asteraceae
♦ Quarantine Weed
♦ 220
♦ Origin: South Africa.

Corynephorus canescens (L.) Beauv.
Poaceae
♦ hopeaheinä, gray clubawn grass,
silvergrass, grey hairgrass
♦ Weed, Quarantine Weed,
Naturalised, Casual Alien
♦ 42, 70, 101, 195, 220
♦ G, cultivated, herbal.

Corynephorus fasciculatus Boiss. & Reut.
Poaceae
♦ Weed
♦ 121
♦ aG. Origin: Eurasia.

Corynocarpus laevigatus J.R. & G.Forst.
Corynocarpaceae
♦ New Zealand laurel, karakara nut
♦ Weed, Naturalised, Environmental
Weed
♦ 3, 22, 39, 80, 88, 101, 131, 151, 191
♦ cultivated, herbal, toxic. Origin:
New Zealand.

Corynopuntia F.M.Knuth spp.
Cactaceae
= *Opuntia* Mill. spp.
♦ Weed, Quarantine Weed
♦ 76, 88, 203

Coryphantha (Engelm.) Lem. spp.
Cactaceae
♦ beehive cactus
♦ Weed, Quarantine Weed,
Naturalised
♦ 76, 86, 88, 220
♦ cultivated.

Cosmos bipinnatus Cav.
Asteraceae

Bidens formosa (Bonato) Sch.Bip. (see),
Coreopsis formosa Bonato
♦ cosmos, garden cosmos, common
cosmos
♦ Weed, Naturalised, Garden Escape,
Environmental Weed, Cultivation
Escape
♦ 15, 34, 42, 51, 72, 80, 86, 87, 88, 98,
101, 102, 179, 203, 243, 256, 269, 280,
287
♦ aH, cultivated, herbal. Origin:
Central America, West Indies.

Cosmos caudatus Kunth
Asteraceae
♦ wild cosmos
♦ Weed, Naturalised
♦ 13, 23, 32, 86, 87, 88, 98, 179, 203, 255,
269
♦ aH. Origin: tropical America.

Cosmos crithmifolius H.B.K.
Asteraceae
♦ Weed
♦ 199

Cosmos diversifolius Otto ex Knowles
& Westc.
Asteraceae
♦ Weed
♦ 199
♦ cultivated.

Cosmos parviflorus (Jacq.) Pers.
Asteraceae
♦ south-western cosmos
♦ Weed
♦ 199
♦ herbal.

Cosmos sulphureus Cav.
Asteraceae
♦ sulphur cosmos, cosmos, yellow
cosmos, kenikir
♦ Weed, Naturalised, Garden Escape,
Cultivation Escape, Casual Alien
♦ 34, 39, 80, 88, 101, 102, 179, 256, 261,
280, 287
♦ aH, cultivated, herbal, toxic. Origin:
Mexico, Central America.

Costus cylindricus Jacq.
Costaceae/Zingiberaceae
= *Costus spicatus* (Jacq.) Sw.
♦ Weed
♦ 87, 88
♦ herbal.

Costus sericeus Blume
Costaceae/Zingiberaceae
♦ Weed
♦ 3, 191
♦ herbal.

Costus speciosus (J.König) J.E.Sm.
Costaceae/Zingiberaceae
♦ crepe ginger, wild ginger, Malay
ginger, isebsab, spiral ginger, cane reed
♦ Weed, Cultivation Escape
♦ 3, 12, 107, 191, 209, 261
♦ H, cultivated, herbal. Origin: tropical
Asia.

Costus spicatus (Jacq.) Sw.
Costaceae/Zingiberaceae
Costus cylindricus Jacq. (see)

♦ spiked spiralflag
♦ Weed
♦ 179
♦ cultivated, herbal.

Cotinus coggygria Scop.
Anacardiaceae
Rhus cotinus L.
♦ European smoketree, smoketree
♦ Weed, Naturalised
♦ 39, 101, 161
♦ S, cultivated, herbal, toxic.

Cotoneaster Medik. spp.
Rosaceae
♦ cotoneaster, firethorn
♦ Weed, Noxious Weed, Naturalised,
Garden Escape, Environmental Weed
♦ 39, 80, 86, 88, 137, 142, 151, 181, 198,
279
♦ cultivated, herbal, toxic.

Cotoneaster acutifolius Turcz.
Rosaceae
♦ Peking cotoneaster
♦ Naturalised
♦ 101
♦ cultivated, herbal.

Cotoneaster adpressus Boiss.
Rosaceae
♦ creeping cotoneaster
♦ Naturalised
♦ 101
♦ cultivated, herbal.

Cotoneaster apiculatus Rehd. & Wilson
Rosaceae
♦ cranberry cotoneaster
♦ Naturalised
♦ 101
♦ cultivated, herbal.

Cotoneaster bullatus Boiss.
Rosaceae
♦ cotoneaster, hollyberry cotoneaster
♦ Weed, Naturalised, Casual Alien
♦ 19, 40, 80, 280
♦ cultivated, herbal.

Cotoneaster cf. *monopyrenus* Diels
Rosaceae
♦ Casual Alien
♦ 280

Cotoneaster conspicuus Marquand
Rosaceae
♦ Naturalised
♦ 280
♦ S, cultivated, herbal.

Cotoneaster dielsianus E.Pritz.
Rosaceae
♦ Diel's cotoneaster
♦ Naturalised
♦ 40
♦ cultivated, herbal.

Cotoneaster divaricatus Rehder &
E.H.Wilson
Rosaceae
♦ green cotoneaster, cotoneaster,
spreading cotoneaster
♦ Weed, Naturalised, Garden Escape,
Environmental Weed
♦ 54, 72, 86, 88, 101, 198, 290
♦ S, cultivated, herbal. Origin: China.

Cotoneaster franchetii Boiss.
Rosaceae
Cotoneaster mairei H.Lév.
♦ cotoneaster, Franch cotoneaster, orange cotoneaster, dwergmispel, silverleaf cotoneaster, rockspray cotoneaster
♦ Weed, Noxious Weed, Naturalised, Garden Escape, Environmental Weed, Cultivation Escape
♦ 15, 34, 40, 54, 63, 86, 88, 95, 101, 121, 165, 176, 198, 246, 280, 283, 289
♦ pS, cultivated, herbal, toxic. Origin: Asia.

Cotoneaster frigidus Wall. ex Lindl.
Rosaceae
♦ tree cotoneaster
♦ Naturalised
♦ 280
♦ S, cultivated. Origin: China, India, Nepal.

Cotoneaster glaucophyllus Franch.
Rosaceae
♦ bright bead cotoneaster, cotoneaster
♦ Weed, Naturalised, Garden Escape, Environmental Weed
♦ 7, 15, 20, 72, 88, 98, 165, 176, 203, 225, 246, 279, 280, 296
♦ S, cultivated. Origin: Asia.

Cotoneaster glaucophyllus Franch. fo. serotinus (Hutch.) Stapf
Rosaceae
♦ largeleaf cotoneaster, cotoneaster
♦ Naturalised, Environmental Weed
♦ 86, 198, 289
♦ S/T. Origin: China.

Cotoneaster hjelmqvistii K.E.Flinck & B.Hylmö
Rosaceae
♦ Hjelmqvist's cotoneaster
♦ Naturalised
♦ 40

Cotoneaster horizontalis Decne.
Rosaceae
♦ fishbone cotoneaster, prostrate cotoneaster, cotoneaster, rockspray cotoneaster
♦ Weed, Naturalised, Garden Escape, Environmental Weed
♦ 40, 54, 72, 86, 88, 155, 198, 246, 280, 289
♦ S, cultivated, herbal. Origin: China.

Cotoneaster hupehensis Rehd. & Wilson
Rosaceae
♦ Hupeh cotoneaster
♦ Naturalised
♦ 101

Cotoneaster integrifolius (Roxb.) Klotz
Rosaceae
♦ small leaved cotoneaster
♦ Naturalised
♦ 40
♦ cultivated.

Cotoneaster lacteus W.W.Sm.
Rosaceae
♦ milkflower cotoneaster, late cotoneaster

♦ Weed, Noxious Weed, Naturalised, Garden Escape, Environmental Weed, Casual Alien
♦ 15, 35, 40, 54, 78, 80, 86, 88, 101, 116, 246, 280, 290
♦ S, cultivated. Origin: China.

Cotoneaster lucidus Schlecht.
Rosaceae
♦ shiny cotoneaster, hedge cotoneaster, kiiltotuhkapensas
♦ Weed, Naturalised, Cultivation Escape
♦ 42, 80, 101
♦ cultivated, herbal.

Cotoneaster microphyllus Wall. ex Lindl.
Rosaceae
♦ smallleaf cotoneaster
♦ Weed, Naturalised, Garden Escape, Environmental Weed
♦ 15, 54, 86, 88, 98, 203, 246, 280, 290
♦ S, cultivated, herbal. Origin: China, Bhutan, India, Nepal.

Cotoneaster multiflora Bunge
Rosaceae
♦ cotoneaster
♦ Weed
♦ 80
♦ herbal.

Cotoneaster pannosus Franch.
Rosaceae
♦ show berry bushes, cotoneaster, silwerdwergmispel, silverleaf cotoneaster
♦ Weed, Noxious Weed, Naturalised, Garden Escape, Environmental Weed, Cultivation Escape, Casual Alien
♦ 7, 34, 35, 54, 63, 72, 78, 86, 88, 95, 98, 101, 116, 176, 198, 199, 203, 231, 233, 279, 280, 283, 289
♦ pS, cultivated, toxic. Origin: China.

Cotoneaster rehderi Pojark.
Rosaceae
♦ bullate cotoneaster
♦ Naturalised
♦ 40

Cotoneaster rotundifolius Wall. ex Lindl.
Rosaceae
♦ cotoneaster
♦ Weed, Naturalised, Garden Escape, Environmental Weed
♦ 86, 98, 203
♦ cultivated. Origin: China.

Cotoneaster salicifolius Franch.
Rosaceae
♦ willow leaved cotoneaster, cotoneaster
♦ Naturalised
♦ 40
♦ cultivated.

Cotoneaster simonsii Baker
Rosaceae
♦ Himalayan cotoneaster, Simon's cotoneaster, khasia berry
♦ Weed, Quarantine Weed, Naturalised, Garden Escape, Environmental Weed
♦ 15, 40, 86, 98, 101, 165, 198, 203, 225, 246, 280, 290

♦ S, cultivated, herbal. Origin: India, Burma, Bhutan, Nepal.

Cotoneaster tenuipes Rehder & E.H.Wilson
Rosaceae
♦ Weed
♦ 80

Cotoneaster tomentosus Lindl.
Rosaceae
♦ Weed
♦ 80
♦ cultivated, herbal.

Cotoneaster vulgaris Lindl.
Rosaceae
= *Cotoneaster integerrimus* Medik. (NoR)
♦ Weed
♦ 87, 88

Cotoneaster × watereri Exell
Rosaceae
= *Cotoneaster frigidus* Wall. ex Lindl. × *Cotoneaster salicifolius* Franch. × *Cotoneaster rugosus* E.Pritz. ex Diels
♦ Waterer's cotoneaster
♦ Naturalised
♦ 40
♦ S, cultivated.

Cotula abyssinica (Sch.Bip.) A.Rich.
Asteraceae
♦ Weed, Quarantine Weed
♦ 76, 87, 88, 203, 220

Cotula anthemoides L.
Asteraceae
♦ Weed, Quarantine Weed, Native Weed
♦ 76, 87, 88, 121, 185, 203, 220
♦ aH, arid, cultivated, herbal. Origin: Africa.

Cotula australis (Sieber ex Spreng.) Hook.f.
Asteraceae
Anacyclus australis Spreng.
♦ carrot weed, batchelor's button, common cotula, Australian waterbuttons, waterbuttons, southern brassbuttons, soldier's button
♦ Weed, Noxious Weed, Naturalised, Native Weed, Casual Alien
♦ 7, 34, 40, 55, 86, 87, 88, 93, 101, 147, 158, 161, 165, 167, 180, 199, 203, 205, 218, 236, 237, 241, 243, 269, 286, 287, 295, 300
♦ a/pH, arid, cultivated, herbal. Origin: Australia.

Cotula bipinnata Thunb.
Asteraceae
♦ ferny cotula, fern cotula
♦ Weed, Naturalised, Environmental Weed
♦ 7, 9, 72, 86, 88, 98, 198, 203
♦ aH, arid. Origin: South Africa.

Cotula cinerea Delile
Asteraceae
♦ Weed
♦ 221

Cotula coronopifolia L.
Asteraceae
♦ waterbuttons, common

brassbuttons, bachelor's button, brassbuttons, buttonweed
♦ Weed, Naturalised, Native Weed, Introduced, Garden Escape, Environmental Weed, Casual Alien
♦ 7, 38, 40, 72, 86, 87, 88, 94, 98, 101, 116, 121, 161, 165, 176, 197, 198, 203, 208, 241, 269, 287, 300
♦ a/pH, aqua, cultivated, herbal. Origin: South Africa [Australia, New Zealand?].

Cotula scariosa **Franch.**
Asteraceae
♦ Weed
♦ 87, 88
♦ cultivated.

Cotula tenella **E.Mey. ex DC.**
Asteraceae
♦ Native Weed
♦ 121
♦ aH. Origin: southern Africa.

Cotula turbinata **L.**
Asteraceae
Cenia turbinata (L.) Pers. (see)
♦ goose daisy, mayweed, brassbuttons
♦ Weed, Naturalised, Environmental Weed, Casual Alien
♦ 7, 9, 86, 88, 98, 158, 203, 280
♦ aH, cultivated, herbal. Origin: South Africa.

Cotula vulgaris **Levyns**
Asteraceae
♦ Casual Alien
♦ 280

Cotula vulgaris **Levyns var.** *australasica* **J.H.Willis**
Asteraceae
♦ Casual Alien
♦ 280

Cotyledon orbiculata **L.**
Crassulaceae
♦ pig's ears, cotyledon, pyrmehilehti
♦ Weed, Naturalised, Garden Escape, Environmental Weed
♦ 7, 15, 39, 72, 88, 98, 101, 161, 198, 203, 280
♦ S, cultivated, herbal, toxic. Origin: South Africa.

Cotyledon orbiculata **L. var.** *dactylopsis* **Toelken**
Crassulaceae
♦ pig's ears
♦ Weed, Native Weed
♦ 39, 121
♦ pS, toxic. Origin: southern Africa.

Cotyledon orbiculata **L. var.** *oblonga* **(Haw.) DC.**
Crassulaceae
♦ cotyledon, pig's ear
♦ Naturalised, Introduced, Environmental Weed
♦ 34, 39, 86
♦ pH, cultivated, toxic. Origin: southern Africa.

Cotyledon orbiculata **L. var.** *orbiculata*
Crassulaceae
♦ cotyledon, pig's ear
♦ Weed, Naturalised, Native Weed,

Environmental Weed
♦ 86, 121
♦ pS, cultivated, toxic. Origin: southern Africa.

Cotyledon wallichii **Harv.**
Crassulaceae
= *Tylecodon wallichii* (Harv.) Toelken
♦ nenta
♦ Weed
♦ 161
♦ cultivated, herbal, toxic.

Couepia uiti **(Mart. & Zucc.) Benth. ex Hook.f.**
Chrysobalanaceae
♦ pateiro
♦ Weed
♦ 255
♦ Origin: South America.

Couroupita guianensis **Aubl.**
Lecythidaceae
♦ cannonball tree, bala de cañón
♦ Naturalised, Introduced
♦ 101, 163, 261
♦ T, cultivated, herbal, toxic. Origin: tropical South America, West Indies.

Courtoisina cyperoides **(Roxb.) J.Soják**
Cyperaceae
Kyllinga cyperoides Roxb.
♦ Weed, Quarantine Weed
♦ 76, 87, 88, 203, 220
♦ G.

Cousinia minuta **Boiss.**
Asteraceae
♦ Weed
♦ 87, 88, 243

Coutoubea spicata **Aubl.**
Gentianaceae
♦ Weed
♦ 87, 88
♦ herbal.

Cowania mexicana **D.Don var.** *stansburiana* **(Torr.) Jeps.**
Rosaceae
= *Purshia stansburiana* (Torr.) Henr. (NoR)
♦ Introduced
♦ 228
♦ arid, cultivated.

Crabbea hirsuta **Harv.**
Acanthaceae
♦ Native Weed
♦ 121
♦ pH. Origin: South Africa.

Crambe abyssinica **Hochst. ex R.E.Fr.**
Brassicaceae
♦ colewort, crambe, Abyssinian kale, Abyssinian cabbage, krambe
♦ Weed, Naturalised
♦ 98, 203
♦ a/pH, cultivated.

Crambe filiformis **Jacq.**
Brassicaceae
♦ Naturalised
♦ 241
♦ toxic.

Crambe gigantea **(Ceballos & Ortuno) Bramwell**

Brassicaceae
♦ Quarantine Weed
♦ 220

Crambe hispanica **L.**
Brassicaceae
♦ hispanic crambe
♦ Weed, Introduced
♦ 121, 199, 228
♦ aH, arid, cultivated. Origin: Eurasia.

Crambe maritima **L.**
Brassicaceae
♦ sea kale
♦ Naturalised, Cultivation Escape
♦ 56, 101
♦ pH, cultivated, herbal.

Crambe orientalis **L.**
Brassicaceae
♦ Weed
♦ 87, 88
♦ pH, promoted.

Crambe sventenii **Pett. ex Bramwell & Sunding**
Brassicaceae
♦ Quarantine Weed
♦ 220
♦ Origin: south-west Europe.

Crambe tataria **Sebeök**
Brassicaceae
♦ Tartarian sea kale, katran tatársky
♦ Weed
♦ 272

Crantzia ambigua **(Urb.) Britton**
Apiaceae
= *Columnea ambigua* (Urb.) Morley (NoR)
♦ Weed
♦ 87, 88

Craspedia chrysantha **(Schltdl) Benth.**
Asteraceae
♦ golden billybuttons, yellow drumsticks
♦ Native Weed
♦ 39, 269
♦ pH, cultivated, toxic. Origin: Australia.

Crassocephalum crepidioides **(Benth.) S.Moore**
Asteraceae
Gynura crepidioides Benth. (see)
♦ thick head, redflower ragleaf, fireweed, pualele, fisi puna, phak phet maeo, hawksbeard velvetplant
♦ Weed, Quarantine Weed, Naturalised, Native Weed, Introduced, Environmental Weed
♦ 3, 6, 12, 55, 68, 76, 86, 87, 88, 98, 101, 107, 121, 134, 170, 179, 191, 203, 204, 209, 230, 238, 243, 261, 262, 263, 269, 273, 276, 286, 287, 290, 297
♦ aH, herbal. Origin: Madagascar, tropical Africa and Asia.

Crassocephalum rubens **(Juss. ex Jacq.) S.Moore**
Asteraceae
♦ fatheads, litiku
♦ Weed
♦ 50, 87, 88
♦ Origin: Madagascar.

Crassula L. spp.
Crassulaceae
♦ pygmyweed
♦ Weed
♦ 243
♦ H, herbal.

Crassula alata (Viv.) Berger
Crassulaceae
♦ Weed, Naturalised
♦ 7, 98, 203

Crassula alata (Viv.) A.Berger var. *alata*
Crassulaceae
♦ three part crassula
♦ Naturalised, Environmental Weed
♦ 86, 198
♦ Origin: Mediterranean.

Crassula biplanata Haw.
Crassulaceae
♦ Casual Alien
♦ 280

Crassula campestris (Eckl. & Zeyh.) Endl. ex Walp. ssp. *campestris*
Crassulaceae
♦ Native Weed
♦ 121
♦ aH. Origin: southern Africa.

Crassula ciliata L.
Crassulaceae
♦ Weed, Naturalised
♦ 54, 86, 88
♦ Origin: South Africa.

Crassula coccinea L.
Crassulaceae
♦ Naturalised
♦ 280
♦ cultivated, herbal.

Crassula connata (Ruiz & Pav.) Berg.
Crassulaceae
♦ sand pygmyweed
♦ Weed
♦ 161
♦ aH, arid.

Crassula connata (Ruiz & Pav.) Berger var. *connata*
Crassulaceae
♦ sand pygmyweed, tillaea, pygmyweed
♦ Weed
♦ 180

Crassula decumbens Thunb.
Crassulaceae
♦ Scilly pygmyweed, cape crassula
♦ Weed, Naturalised
♦ 9, 15, 40, 98, 203, 280

Crassula ericoides Haw.
Crassulaceae
♦ skilpadbossie
♦ Weed
♦ 54, 88
♦ cultivated.

Crassula ericoides Haw. ssp. *ericoides*
Crassulaceae
♦ skilpadbossie, reptile crassula
♦ Naturalised
♦ 86, 198

Crassula expansa Dryand. ssp. *expansa*
Crassulaceae

♦ Native Weed
♦ 121
♦ pH. Origin: southern Africa.

Crassula glomerata P.Bergius
Crassulaceae
♦ Weed, Naturalised, Environmental Weed
♦ 7, 9, 86, 98, 203
♦ Origin: South Africa.

Crassula helmsii (T.Kirk) Cockayne
Crassulaceae
♦ swamp stonecrop, New Zealand pygmyweed, swamp crassula, pygmyweed swamp stonecrop
♦ Weed, Noxious Weed, Naturalised, Environmental Weed
♦ 40, 67, 87, 88, 152, 200, 229
♦ wpH, cultivated. Origin: Australasia.

Crassula multicava Lem.
Crassulaceae
♦ Cape Province pygmyweed, fairy crassula, pitted crassula, crassula
♦ Weed, Sleeper Weed, Naturalised, Garden Escape, Environmental Weed
♦ 7, 15, 72, 88, 98, 101, 203, 225, 246, 280
♦ S, cultivated. Origin: South Africa.

Crassula multicava Lem. ssp. *multicava*
Crassulaceae
♦ shade crassula, prostrate cotoneaster, crassula
♦ Naturalised, Garden Escape, Environmental Weed
♦ 86, 155, 198, 251, 280, 289
♦ cultivated. Origin: South Africa.

Crassula muscosa L. var. *muscosa*
Crassulaceae
♦ clubmoss crassula
♦ Naturalised
♦ 86, 198
♦ Origin: southern Africa.

Crassula natans Thunb.
Crassulaceae
♦ crassula
♦ Weed, Naturalised, Environmental Weed
♦ 7, 9, 72, 88, 98, 176, 203
♦ aH, aqua, herbal.

Crassula natans Thunb. var. *minus* (Eckl. & Zeyh.) Rowley
Crassulaceae
♦ water crassula, crassula
♦ Naturalised, Environmental Weed
♦ 86, 198
♦ Origin: southern Africa.

Crassula oblanceolata Schönl. & Bak.f.
Crassulaceae
♦ Native Weed
♦ 121
♦ aH. Origin: southern Africa.

Crassula ovata (Mill.) Druce
Crassulaceae
Crassula argentea Thunb.
♦ jade plant, jade tree
♦ Weed, Naturalised
♦ 101, 161
♦ cultivated, herbal, toxic.

Crassula pellucida L.
Crassulaceae
♦ Casual Alien
♦ 280

Crassula pellucida L. ssp. *marginalis* (Dryand.) Toelken
Crassulaceae
♦ Casual Alien
♦ 280

Crassula sarmentosa Harv.
Crassulaceae
♦ Weed, Casual Alien
♦ 54, 88, 280
♦ cultivated.

Crassula sarmentosa Harv. var. *sarmentosa*
Crassulaceae
♦ Naturalised
♦ 86
♦ Origin: South Africa.

Crassula sieberiana (J.A.Schult.) Druce
Crassulaceae
♦ Siberian pygmyweed
♦ Naturalised
♦ 101
♦ cultivated. Origin: Australia.

Crassula spathulata Thunb.
Crassulaceae
♦ Weed, Naturalised
♦ 54, 86, 88, 280
♦ cultivated. Origin: South Africa.

Crassula strigosa L.
Crassulaceae
♦ Native Weed
♦ 121
♦ aH. Origin: southern Africa.

Crassula tetragona L.
Crassulaceae
♦ crassula
♦ Weed, Naturalised, Environmental Weed
♦ 7, 72, 88, 280
♦ cultivated, herbal. Origin: Australia.

Crassula tetragona L. ssp. *robusta* (Toelken) Toelken
Crassulaceae
♦ shrubby crassula, crassula
♦ Naturalised, Environmental Weed
♦ 86, 198, 280
♦ Origin: southern Africa.

Crassula thunbergiana Schult.
Crassulaceae
♦ crassula
♦ Weed, Naturalised, Environmental Weed
♦ 7, 9, 86, 88, 98, 158, 203, 243
♦ Origin: South Africa.

Crassula tillaea Lester-Garland
Crassulaceae
♦ sammalpaunikko, mossy stonecrop, mossy tillea, moss pygmyweed, Mediterranean pygmyweed
♦ Naturalised, Casual Alien
♦ 42, 101
♦ aH.

Crataegus L. spp.
Rosaceae

- hawthorn
- Noxious Weed, Naturalised, Garden Escape
- 39, 86, 161, 198
- S/T, cultivated, herbal, toxic.

Crataegus azarolus L.
Rosaceae
Crataegus aronia (L.) Bosc. ex DC.
- azarole
- Weed, Naturalised, Introduced, Cultivation Escape
- 86, 221, 228, 252
- T, arid, cultivated, herbal. Origin: southern Europe to west Asia.

Crataegus crenulata Roxb.
Rosaceae
- hawthorn
- Weed
- 279

Crataegus crus-galli L.
Rosaceae
- cockspur hawthorn
- Weed, Environmental Weed
- 87, 88, 155, 218, 292
- S, cultivated, herbal.

Crataegus douglasii Lindl.
Rosaceae
- black hawthorn
- Weed
- 87, 88, 218
- S, cultivated, herbal.

Crataegus grayana Eggl.
Rosaceae
= *Crataegus flabellata* (Spach) Kirchn. (NoR)
- aitaorapihlaja
- Cultivation Escape
- 42
- cultivated.

Crataegus laevigata (Poir.) DC.
Rosaceae
Crataegus oxyacanthoides Thuill.
- hawthorn, may, azzarola, smooth hawthorn
- Weed, Quarantine Weed, Noxious Weed, Naturalised, Garden Escape, Cultivation Escape
- 42, 76, 86, 88, 97, 101, 147, 203, 252, 269
- S, cultivated, herbal. Origin: Eurasia, North Africa.

Crataegus marshallii Eggl.
Rosaceae
- parsley hawthorn
- Weed
- 87, 88, 218
- herbal.

Crataegus monogyna Jacq.
Rosaceae
Mespilus monogyna All.
- English hawthorn, oneseed hawthorn, single seeded hawthorn, whitethorn, may, quickthorn
- Weed, Quarantine Weed, Noxious Weed, Naturalised, Garden Escape, Environmental Weed
- 15, 20, 35, 39, 46, 46, 70, 72, 76, 78, 80, 86, 88, 97, 98, 101, 116, 137, 142, 146, 147, 151, 152, 155, 165, 171, 176, 198,

203, 225, 231, 246, 269, 272, 280, 290, 296
- pS, cultivated, herbal, toxic. Origin: Europe.

Crataegus multiflora Bunge
Rosaceae
- inkberry hawthorn
- Naturalised
- 101

Crataegus oxyacantha L.
Rosaceae
- English hawthorn, biancospino
- Weed, Naturalised, Environmental Weed
- 87, 88, 98, 151, 155, 203
- cultivated, herbal. Origin: Europe, North Africa, western Asia.

Crataegus pubescens (Kunth) Steud.
Rosaceae
Crataegus stipulosa Steud., *Mespilus pubescens* Kunth
- manzanilla
- Introduced
- 228
- T, arid, cultivated, herbal.

Crataegus rivularis Nutt.
Rosaceae
- river hawthorn
- Weed
- 87, 88, 218
- T, promoted, herbal.

Crataegus saligna Greene
Rosaceae
- willow hawthorn
- Weed
- 87, 88, 218
- herbal.

Crataegus sanguinea Pall.
Rosaceae
- siperianorapihlaja
- Cultivation Escape
- 42
- T, cultivated, herbal.

Crataegus sinaica Boiss.
Rosaceae
Crataegus × sinaica Boiss. (see)
- hawthorn, may, azzarola, azarola thorn, neapolitan medlar
- Weed, Quarantine Weed, Noxious Weed, Naturalised, Garden Escape
- 76, 86, 88, 98, 147, 171, 203, 221
- cultivated. Origin: Mediterranean.

Crataegus × sinaica Boiss.
Rosaceae
= *Crataegus sinaica* Boiss.
- azzarola
- Naturalised
- 86, 198
- Origin: Mediterranean.

Crataegus submollis Sarg.
Rosaceae
- iso orapihlaja, Quebec hawthorn
- Cultivation Escape
- 42
- T, cultivated, herbal.

Crataegus succulenta Link
Rosaceae
- fleshy hawthorn

- Weed
- 87, 88, 218
- S, promoted, herbal.

Crateva adansonii DC.
Capparaceae
Crateva religiosa Forst.f.
- Weed, Introduced
- 93, 228
- arid. Origin: Madagascar.

Crateva tapia L.
Capparaceae
Cleome arborea Schrad., *Crateva acuminata* DC., *Crateva apetala* Urb., *Crateva benthamii* Eichler, *Crateva coriacea* Herzog, *Crateva gynandra* L., *Crateva radiatiflora* DC., *Crateva tapioides* DC.
- Introduced
- 228
- S/T, arid, herbal.

Crepidiastrum keiskeanum (Maxim.) Nakai
Asteraceae
- Quarantine Weed
- 220

Crepis L. spp.
Asteraceae
- hawksbeard, crepis
- Weed, Naturalised
- 272

Crepis aspera L.
Asteraceae
- crepis, hawksbeard
- Weed
- 243
- aH, arid. Origin: Eurasia.

Crepis biennis L.
Asteraceae
Hieracium bienne Karsch
- rough hawksbeard, piennarkeltto, greater hawksbeard
- Weed, Naturalised
- 23, 42, 70, 87, 88, 94, 98, 101, 203, 218, 272
- cultivated.

Crepis bullosa Tausch.
Asteraceae
- Weed, Quarantine Weed, Naturalised
- 76, 86, 87, 88, 203, 220

Crepis bursifolia L.
Asteraceae
- Italian hawksbeard
- Naturalised, Introduced
- 34, 101
- pH, promoted. Origin: Europe.

Crepis capillaris (L.) Schwägr.
Asteraceae
Crepis virens L. nom. illeg. (see)
- smooth hawksbeard, hoikkakeltto, hawksbeard
- Weed, Noxious Weed, Naturalised, Environmental Weed, Casual Alien
- 7, 15, 23, 24, 34, 42, 44, 52, 70, 72, 86, 87, 88, 94, 98, 101, 121, 136, 161, 165, 176, 181, 198, 203, 210, 218, 229, 241, 243, 250, 253, 269, 272, 280, 287, 300
- a/bH, cultivated, herbal. Origin: Eurasia.

Crepis capillaris (L.) Schwägr. var. glandulosa (L.) Wallr.
Asteraceae
- smooth hawksbeard
- Cultivation Escape
- 40
- cultivated.

Crepis dioscoridis L.
Asteraceae
= *Crepis setosa* Hallier f.
- Naturalised
- 86
- Origin: Greece, Aegean region.

Crepis foetida L.
Asteraceae
Barkhausia foetida F.W.Schmidt, *Picris foetida* Lam.
- stinking hawksbeard, kaakonkeltto
- Weed, Naturalised, Environmental Weed
- 7, 39, 70, 88, 94, 98, 101, 198, 203, 253, 272, 280
- a/bH, cultivated, herbal, toxic.

Crepis foetida L. ssp. foetida
Asteraceae
- Naturalised
- 280

Crepis fraasii Sch.Bip.
Asteraceae
- Weed
- 87, 88

Crepis japonica (L.) Benth
Asteraceae
= *Youngia japonica* (L.) DC.
- barba de falcao
- Weed
- 255
- herbal. Origin: China, Japan.

Crepis mollis (Jacq.) Asch.
Asteraceae
Hieracium molle Jacq.
- pehmytkeltto, northern hawksbeard, soft hawksbeard
- Weed, Casual Alien
- 42, 272
- cultivated.

Crepis neglecta L.
Asteraceae
- Weed, Quarantine Weed
- 76, 88
- herbal.

Crepis nicaeensis Balb. ex Pers.
Asteraceae
- ranskankeltto, Turkish hawksbeard, French hawksbeard
- Naturalised, Casual Alien
- 42, 101

Crepis occidentalis Nutt.
Asteraceae
- western hawksbeard, largeflower hawksbeard
- Weed
- 23, 87, 88, 218
- pH, herbal.

Crepis paludosa (L.) Moench
Asteraceae
Aracium paludosum Monn., *Geracium*

paludosum Reich., *Hieracium paludosum* L.
- marsh hawksbeard, suokeltto
- Weed
- 272
- herbal.

Crepis pannonica (Jacq.) K.Koch
Asteraceae
- pasture hawksbeard
- Weed, Naturalised
- 101, 272

Crepis parviflora Desf.
Asteraceae
- Weed
- 87, 88

Crepis pulchra L.
Asteraceae
Chondrilla pulchra Lam., *Lampsana pulchra* Vis., *Prenanthes hieracifolia* Willd.
- smallflower hawksbeard
- Weed, Naturalised
- 101, 253, 272, 300
- cultivated, herbal.

Crepis pusilla (Sommier) Merxm.
Asteraceae
- Weed, Naturalised
- 86, 98, 203
- Origin: Mediterranean.

Crepis rubra L.
Asteraceae
- red hawksbeard, hawksbeard
- Naturalised
- 101
- aH, cultivated, herbal.

Crepis rueppellii Sch.Bip.
Asteraceae
- Weed
- 88

Crepis sancta (L.) Babc.
Asteraceae
- lontimo
- Weed, Casual Alien
- 42, 88, 94, 253
- aH, promoted, herbal.

Crepis setosa Haller f.
Asteraceae
Barkhausia setosa (Haller f.) DC., *Crepis dioscoridis* L. (see)
- bristly hawksbeard, rough hawksbeard
- Weed, Naturalised, Casual Alien
- 34, 40, 42, 44, 70, 86, 87, 88, 94, 98, 101, 136, 161, 176, 198, 203, 212, 237, 241, 272, 280, 287, 295, 300
- aH, cultivated, herbal. Origin: Europe.

Crepis taraxacifolia Thuill.
Asteraceae
= *Crepis vesicaria* L. ssp. *taraxacifolia* (Thuill.) Thell. ex Schinz & R.Keller
- Weed
- 87, 88
- cultivated.

Crepis tectorum L.
Asteraceae
Hieracium tectorum Karsch

- narrowleaf hawksbeard, succory hawksbeard, ketokeltto
- Weed, Noxious Weed, Naturalised
- 23, 70, 87, 88, 94, 101, 114, 136, 161, 162, 218, 229, 243, 272, 286, 287, 293, 299
- cultivated, herbal.

Crepis vesicaria L.
Asteraceae
- beaked hawksbeard
- Weed, Naturalised
- 7, 70, 87, 88, 94, 98, 101, 253, 272, 280
- bH, promoted, herbal.

Crepis vesicaria L. ssp. haenseleri (Boiss. ex DC.) P.D.Sell
Asteraceae
= *Crepis vesicaria* L. ssp. *taraxacifolia* (Thuill.) Thell. ex Schinz & R.Keller
- bladder hawksbeard, voikukkakeltto
- Naturalised
- 86, 198, 300
- Origin: Europe.

Crepis vesicaria L. ssp. taraxacifolia (Thuill.) Thell. ex Schinz & R.Keller
Asteraceae
Crepis taraxacifolia Thuill. (see), *Crepis vesicaria* L. ssp. *haenseleri* (Boiss. ex DC.) P.D.Sell (see)
- Haenseler's hawksbeard, beaked hawksbeard
- Weed, Naturalised
- 15, 34, 40, 101, 241, 280
- a/pH. Origin: North Africa, Europe.

Crepis virens L. nom. illeg.
Asteraceae
= *Crepis capillaris* (L.) Wallr.
- Weed
- 87, 88

Crepis zacintha (L.) Babc.
Asteraceae
- striped hawksbeard
- Weed, Naturalised
- 88, 94, 101, 272
- cultivated.

Crescentia alata Kunth
Bignoniaceae
Parmentiera alata Miers
- morrito, Mexican calabash, gourd tree, tecomate
- Naturalised, Introduced
- 101, 228
- arid, herbal.

Crescentia cujete L.
Bignoniaceae
Crescentia acuminata Kunth, *Crescentia angustifolia* Willd. ex Seem., *Crescentia arborea* Raf., *Crescentia cujete* var. *puberula* Bureau & K.Schum., *Crescentia cuneifolia* Gardner, *Crescentia fasciculata* Miers, *Crescentia plectantha* Miers, *Crescentia spathulata* Miers
- calabash tree, common calabash tree, calebassier, kalebassenbaum, calabacero, crescencia, guacal, morro
- Weed, Naturalised, Introduced, Environmental Weed
- 39, 87, 88, 98, 151, 228

♦ T, arid, cultivated, herbal, toxic.

Cressa cretica L.
Convolvulaceae
♦ Weed
♦ 87, 88, 221
♦ herbal.

Cressa truxillensis Kunth
Convolvulaceae
Cressa truxillensis Kunth var. *vallicola* (Heller) Munz (see)
♦ alkali weed, spreading alkaliweed
♦ Weed
♦ 161
♦ pH, arid, herbal.

Cressa truxillensis Kunth var. *vallicola* (Heller) Munz
Convolvulaceae
= *Cressa truxillensis* Kunth
♦ alkali weed, alkali clover, cress
♦ Weed
♦ 180

Crinum L. spp.
Liliaceae/Amaryllidaceae
♦ swamplily, kiup
♦ Weed
♦ 161, 247
♦ toxic.

Crinum americanum L.
Liliaceae/Amaryllidaceae
♦ swamplily, seven sisters
♦ Quarantine Weed
♦ 258
♦ aqua, cultivated, herbal.

Crinum asiaticum L.
Liliaceae/Amaryllidaceae
Crinum pedunculatum R.Br.
♦ poison bulb, kiup, lau talotalo
♦ Weed, Naturalised, Introduced
♦ 39, 82, 101, 179, 230
♦ H, cultivated, herbal, toxic.

Crinum bulbispermum (Burm.f.) Milne-Redh. & Schweick.
Liliaceae/Amaryllidaceae
♦ hardy swamplily
♦ Naturalised
♦ 39, 101
♦ pH, cultivated, herbal, toxic.

Crinum defixum Ker Gawl.
Liliaceae/Amaryllidaceae
♦ Weed
♦ 87, 88
♦ herbal.

Crinum flaccidum Herb.
Liliaceae/Amaryllidaceae
♦ Darling lily, Murray lily
♦ Weed
♦ 269
♦ pH, cultivated. Origin: Australia.

Crinum powellii hort. ex Baker
Liliaceae/Amaryllidaceae
♦ Naturalised
♦ 7
♦ cultivated.

Crinum zeylanicum (L.) L.
Liliaceae/Amaryllidaceae
♦ Ceylon swamplily, lirio
♦ Naturalised, Cultivation Escape

♦ 39, 101, 261
♦ cultivated, herbal, toxic. Origin: tropical Asia.

Critesion glaucum (Steud.) Á.Löve
Poaceae
= *Hordeum murinum* L. ssp. *glaucum* (Steud.) Tzvelev
♦ Naturalised
♦ 280
♦ G.

Critesion hystrix (Roth) Á.Löve
Poaceae
= *Hordeum marinum* Huds. ssp. *gussoneanum* (Parl.) Thell.
♦ Mediterranean barleygrass
♦ Weed, Naturalised, Environmental Weed
♦ 72, 86, 88, 98, 198, 203, 280
♦ aG.

Critesion jubatum (L.) Nevski
Poaceae
= *Hordeum jubatum* L.
♦ Naturalised
♦ 241, 280
♦ G, herbal.

Critesion marinum (Huds.) Á.Löve
Poaceae
= *Hordeum marinum* Huds. ssp. *marinum*
♦ sea barleygrass
♦ Weed, Naturalised, Environmental Weed
♦ 72, 86, 88, 98, 198, 203, 241, 280
♦ aG, cultivated.

Critesion murinum (L.) Á.Löve
Poaceae
= *Hordeum murinum* L. ssp. *murinum*
♦ barleygrass
♦ Weed, Naturalised
♦ 98, 191, 198, 203, 280
♦ G, cultivated.

Critesion murinum (L.) Á.Löve ssp. glaucum (Steud.) W.A.Weber
Poaceae
= *Hordeum murinum* L. ssp. *glaucum* (Steud.) Tzvelev
♦ barleygrass, blue barleygrass
♦ Weed, Naturalised
♦ 93, 198, 205
♦ G.

Critesion murinum (L.) Á.Löve ssp. leporinum (Link) Á.Löve
Poaceae
= *Hordeum murinum* L. ssp. *leporinum* (Link) Arcang.
♦ northern barleygrass, wall barleygrass
♦ Weed, Naturalised
♦ 198, 205, 228, 280
♦ G.

Critesion murinum (L.) Á.Löve ssp. murinum
Poaceae
♦ barleygrass
♦ Weed, Naturalised
♦ 93, 241, 280
♦ G.

Critesion secalinum (Schreb.) Á.Löve
Poaceae
= *Hordeum secalinum* Schr.
♦ meadow barleygrass
♦ Weed, Naturalised
♦ 98, 198, 203, 280
♦ G, cultivated.

Crithmum maritimum L.
Apiaceae
Cachrys maritima (L.) Spreng.
♦ rock samphire, finocchio marino, samphire
♦ Introduced
♦ 39, 228
♦ pH, arid, cultivated, herbal, toxic.

Crockeria chrysantha Greene ex Gray
Asteraceae
= *Lasthenia chrysantha* (Greene ex Gray) Greene
♦ crockeria
♦ Weed
♦ 180

Crocodilodes multijugum Kuntze
Asteraceae
= *Berkheya multijuga* (DC.) Roessler
♦ Quarantine Weed
♦ 220

Crocodilodes purpureum Kuntze
Asteraceae
= *Berkheya purpurea* Benth. & Hook.f. ex Mast.
♦ Quarantine Weed
♦ 220

Crocosmia crocosmiiflora (Lem.) N.E.Br.
Iridaceae
= *Crocosmia* × *crocosmiiflora* (Lem. ex anon.) N.E.Br.
♦ montbretia
♦ Weed, Naturalised, Environmental Weed
♦ 73, 88, 98, 155, 203, 287
♦ cultivated.

Crocosmia × crocosmiiflora (Lem. ex anon.) N.E.Br.
Iridaceae
Crocosmia crocosmiiflora (Nicholson) N.E.Br. (see), *Tritonia crocosmaeflora* Lem. (see), *Tritonia crocosmiflora* Nichols (see), *Tritonia* × *crocosmiflora* (Lemoine) Nichols (see) [= *Crocosmia aurea* (Pappe ex Hook.) Planch. × *Crocosmia pottsii* (McNab ex Baker) N.E.Br. most probable combination]
♦ montbretia, crocosmia, garden montbretia
♦ Weed, Noxious Weed, Naturalised, Garden Escape, Environmental Weed
♦ 15, 40, 72, 79, 86, 88, 101, 165, 176, 181, 198, 225, 246, 255, 261, 269, 280, 289, 296, 300
♦ pH, cultivated, herbal. Origin: southern Africa, horticultural hybrid.

Crocosmia masoniorum (L.Bolus) N.E.Br.
Iridaceae
♦ giant montbretia
♦ Naturalised
♦ 40
♦ cultivated, herbal.

Crocosmia paniculata (Klatt) Goldblatt
Iridaceae
♦ Aunt Eliza
♦ Naturalised
♦ 280
♦ cultivated.

Crocus L. spp.
Iridaceae
♦ crocus
♦ Weed
♦ 272

Crocus angustifolius Weston
Iridaceae
Crocus vernus (L.) Hill (see)
♦ cloth of gold
♦ Naturalised
♦ 101
♦ cultivated.

Crocus biflorus Mill.
Iridaceae
Crocus circumscissus Haw., *Crocus pusillus* Ten.
♦ zafferano selvatico
♦ Weed
♦ 272
♦ cultivated, herbal.

Crocus flavus Weston
Iridaceae
♦ yellow crocus
♦ Naturalised, Casual Alien
♦ 40, 101, 280
♦ cultivated.

Crocus imperati Ten.
Iridaceae
♦ early crocus
♦ Naturalised
♦ 101
♦ cultivated, herbal.

Crocus sativus L.
Iridaceae
♦ saffron, zafferano
♦ Weed
♦ 39, 87, 88
♦ pH, cultivated, herbal, toxic.

Crocus serotinus Salisb.
Iridaceae
Crocus clusii J.Gay
♦ Weed
♦ 70
♦ pH, cultivated.

Crocus sieberi J.Gay
Iridaceae
♦ Sieber's crocus
♦ Naturalised
♦ 101
♦ cultivated, herbal.

Crocus tomasinianus Herb.
Iridaceae
♦ lilac crocus
♦ Weed
♦ 15
♦ cultivated, herbal.

Crocus vernus (L.) Hill
Iridaceae
= *Crocus angustifolius* Weston
♦ Dutch crocus, spring crocus, purple crocus

♦ Naturalised
♦ 40, 101
♦ cultivated, herbal.

Crotalaria aculeata De Wild.
Fabaceae/Papilionaceae
♦ Weed, Quarantine Weed
♦ 76, 87, 88, 203, 220

Crotalaria L. spp.
Fabaceae/Papilionaceae
♦ kaskabeles, cascanetas, rattlepod, rattlebox
♦ Weed
♦ 3, 23, 88, 107, 243, 247
♦ herbal, toxic.

Crotalaria aegyptiaca Benth.
Fabaceae/Papilionaceae
Crotalaria wissmannii Schwartz
♦ Weed
♦ 221
♦ arid.

Crotalaria agatiflora Schweinf.
Fabaceae/Papilionaceae
♦ canary bird bush, Queensland birdflower
♦ Weed, Naturalised, Garden Escape
♦ 98, 134, 161, 203, 280
♦ arid, cultivated, toxic.

Crotalaria agatiflora G.Schweinf. ssp. imperialis (Taub.) Polhill
Fabaceae/Papilionaceae
♦ bird flower, canarybird bush
♦ Weed
♦ 121, 279
♦ pS, toxic.

Crotalaria alata Buch.-Ham. ex D.Don
Fabaceae/Papilionaceae
♦ rattlebox
♦ Naturalised
♦ 39, 86
♦ herbal, toxic.

Crotalaria anagyroides Kunth
Fabaceae/Papilionaceae
= *Crotalaria micans* Link
♦ Weed, Naturalised
♦ 87, 88, 287
♦ Origin: Australia.

Crotalaria assamica Benth.
Fabaceae/Papilionaceae
♦ Indian rattlebox
♦ Naturalised
♦ 101, 287

Crotalaria berteriana DC.
Fabaceae/Papilionaceae
♦ Berteron's rattlebox
♦ Weed, Naturalised
♦ 87, 88, 101

Crotalaria bialata Schrank.
Fabaceae/Papilionaceae
♦ Naturalised
♦ 287

Crotalaria biflora L.
Fabaceae/Papilionaceae
♦ twoflower rattlebox
♦ Naturalised
♦ 101

Crotalaria bracteata Roxb.
Fabaceae/Papilionaceae

♦ Weed
♦ 87, 88

Crotalaria brevidens Benth.
Fabaceae/Papilionaceae
Crotalaria albertiana Bak.f., *Crotalaria brevidens* Benth. var. *intermedia* (Kotschy) Polhill (see), *Crotalaria intermedia* Kotschy (see), *Crotalaria intermedia* Kotschy var. *abyssinica* Engl., *Crotalaria intermedia* Kotschy var. *parviflora* (Bak.f.) Polhill, *Crotalaria intermedia* Kotschy var. *dorumaensis* (Wilczek) Polhill
♦ Ethiopian rattlebox
♦ Weed, Naturalised
♦ 101, 261
♦ arid. Origin: Old World Tropics.

Crotalaria brevidens Benth. var. brevidens
Fabaceae/Papilionaceae
♦ Ethiopian rattlebox
♦ Naturalised
♦ 101

Crotalaria brevidens Benth. var. intermedia (Kotschy) Polhill
Fabaceae/Papilionaceae
= *Crotalaria brevidens* Benth.
♦ Ethiopian rattlebox
♦ Naturalised
♦ 101

Crotalaria breviflora DC.
Fabaceae/Papilionaceae
♦ shortflower rattlebox
♦ Weed
♦ 87, 88

Crotalaria burkeana Benth.
Fabaceae/Papilionaceae
♦ rattlebush, sickness crotalaria, stiff sickness crotalaria
♦ Weed, Native Weed
♦ 39, 121
♦ pH, toxic. Origin: southern Africa.

Crotalaria chrysochlora Bak.f. ex Harms
Fabaceae/Papilionaceae
♦ Weed
♦ 87, 88

Crotalaria dissitiflora Benth.
Fabaceae/Papilionaceae
♦ grey rattlepod
♦ Weed
♦ 39, 55
♦ cultivated, toxic. Origin: Australia.

Crotalaria dissitiflora Benth. ssp. dissitiflora
Fabaceae/Papilionaceae
♦ grey rattlepod
♦ Weed
♦ 269
♦ S. Origin: Australia.

Crotalaria distans Benth.
Fabaceae/Papilionaceae
♦ Weed, Naturalised, Native Weed
♦ 86, 98, 121, 203
♦ aH, toxic. Origin: southern Africa.

Crotalaria dura J.M.Wood & M.S.Evans
Fabaceae/Papilionaceae
♦ wild lucerne
♦ Weed, Native Weed

♦ 39, 121
♦ pH, toxic. Origin: southern Africa.

Crotalaria filipes Benth.
Fabaceae/Papilionaceae
♦ Weed
♦ 66

Crotalaria flacata Vahl ex DC.
Fabaceae/Papilionaceae
♦ Weed
♦ 87, 88

Crotalaria fulva Roxb.
Fabaceae/Papilionaceae
♦ tawny crotalaria
♦ Weed
♦ 87, 88, 218

Crotalaria globifera E.Mey.
Fabaceae/Papilionaceae
♦ wild lucerne
♦ Weed, Native Weed
♦ 39, 121
♦ pH, toxic. Origin: southern Africa.

Crotalaria goreensis Guill. & Perr.
Fabaceae/Papilionaceae
♦ Gambia pea
♦ Weed, Naturalised, Introduced, Environmental Weed
♦ 7, 32, 39, 55, 86, 87, 88, 93, 98, 203, 228
♦ arid, cultivated, toxic. Origin: tropical Africa.

Crotalaria grahamiana Wight & Arn.
Fabaceae/Papilionaceae
♦ bushy rattlepod
♦ Weed, Quarantine Weed, Naturalised, Environmental Weed
♦ 76, 86, 98, 203
♦ Origin: tropical Asia.

Crotalaria incana L.
Fabaceae/Papilionaceae
♦ woolly rattlepod, chipilin, shakeshake
♦ Weed,
Naturalised
♦ 23, 39, 87, 88, 93, 98, 161, 179, 191, 203, 237, 240, 241, 255, 287, 295, 300
♦ aH, arid, cultivated, herbal, toxic. Origin: South America.

Crotalaria incana L. ssp. incana
Fabaceae/Papilionaceae
♦ Naturalised
♦ 86
♦ Origin: tropical America.

Crotalaria incana L. ssp. purpurascens (Lam.) Milne-Redh.
Fabaceae/Papilionaceae
♦ Naturalised
♦ 86

Crotalaria intermedia Kotschy
Fabaceae/Papilionaceae
= *Crotalaria brevidens* Benth.
♦ Weed
♦ 87, 88

Crotalaria juncea L.
Fabaceae/Papilionaceae
♦ deccan hemp, sunn hemp
♦ Weed, Naturalised, Cultivation Escape

♦ 7, 32, 39, 86, 87, 88, 98, 101, 121, 161, 203, 261, 287
♦ aH, cultivated, herbal, toxic. Origin: Eurasia.

Crotalaria laburnifolia L.
Fabaceae/Papilionaceae
♦ Weed, Naturalised
♦ 39, 86, 87, 88, 240
♦ cultivated, herbal, toxic. Origin: Africa.

Crotalaria laburnifolia L. ssp. australis (Bak.f.) Polhill
Fabaceae/Papilionaceae
Crotalaria australis Bak.f.
♦ rattlepod
♦ Native Weed
♦ 121
♦ pH, toxic. Origin: southern Africa.

Crotalaria lanata Bedd.
Fabaceae/Papilionaceae
♦ Weed, Naturalised
♦ 98, 203

Crotalaria lanceolata E.Mey.
Fabaceae/Papilionaceae
Crotalaria mossambicensis Klotzsch
♦ lanceleaf rattlebox
♦ Weed, Quarantine Weed, Naturalised, Native Weed
♦ 76, 87, 88, 98, 101, 121, 179, 203, 255
♦ aH, toxic. Origin: Africa.

Crotalaria lanceolata E.Mey. ssp. lanceolata
Fabaceae/Papilionaceae
♦ Naturalised
♦ 86
♦ Origin: Africa, Madagascar.

Crotalaria linifolia L.f.
Fabaceae/Papilionaceae
♦ Weed
♦ 39, 87, 88
♦ cultivated, toxic.

Crotalaria longirostrata Hook. & Arn.
Fabaceae/Papilionaceae
♦ longbeak crotalaria, longbeak rattlebox
♦ Weed, Quarantine Weed, Naturalised
♦ 76, 87, 88, 101, 203, 218, 220
♦ herbal.

Crotalaria lotoides Benth.
Fabaceae/Papilionaceae
♦ Native Weed
♦ 121
♦ pH. Origin: southern Africa.

Crotalaria lunata Bedd. ex Polhill
Fabaceae/Papilionaceae
♦ Naturalised, Introduced
♦ 86, 100
♦ pS. Origin: India.

Crotalaria macrocalyx Benth.
Fabaceae/Papilionaceae
♦ Weed
♦ 88
♦ arid, cultivated.

Crotalaria medicaginea Lam.
Fabaceae/Papilionaceae
♦ trefoil rattlepod, cloverleaf rattlepod

♦ Weed
♦ 23, 39, 66, 87, 88
♦ arid, cultivated, toxic.

Crotalaria micans Link
Fabaceae/Papilionaceae
Crotalaria anagyroides Kunth (see)
♦ Caracas rattlebox
♦ Weed, Naturalised
♦ 86, 101, 255
♦ aH. Origin: South America.

Crotalaria microphylla Vahl
Fabaceae/Papilionaceae
Crotalaria sennii Chiov.
♦ Weed
♦ 221

Crotalaria mucronata Desv.
Fabaceae/Papilionaceae
= *Crotalaria pallida* Ait. var. *obovata* (G.Don) Polhill
♦ Weed
♦ 39, 87, 88, 209, 262, 275, 297
♦ aH, herbal, toxic. Origin: Australia.

Crotalaria mysorensis Roth
Fabaceae/Papilionaceae
♦ rattlebox
♦ Weed
♦ 87, 88

Crotalaria natalitia Meisn.
Fabaceae/Papilionaceae
♦ Native Weed
♦ 121
♦ pS, arid. Origin: Africa.

Crotalaria ochroleuca G.Don
Fabaceae/Papilionaceae
♦ slenderleaf rattlebox
♦ Weed, Naturalised
♦ 98, 101, 179, 203
♦ cultivated.

Crotalaria pallida Aiton
Fabaceae/Papilionaceae
Crotalaria brownei Bert. in DC., *Crotalaria hookeri* Arn., *Crotalaria saltiana* Andr. (see)
♦ smooth rattlebox, mlyankoko, striped crotalaria, streaked rattlepod, xique xique, cascaveleira, krotalaria
♦ Weed, Quarantine Weed, Naturalised, Introduced
♦ 6, 32, 39, 76, 88, 93, 98, 101, 157, 161, 191, 203, 228, 230, 255, 287
♦ a/pH, arid, herbal, toxic. Origin: Africa.

Crotalaria pallida Aiton var. obovata (G.Don) Polhill
Fabaceae/Papilionaceae
Crotalaria falcata Vahl ex DC., *Crotalaria mucronata* Desv. (see), *Crotalaria striata* DC. (see)
♦ smooth rattlebox
♦ Weed, Naturalised
♦ 101, 179, 261
♦ Origin: Old World Tropics.

Crotalaria pallida Aiton var. pallida
Fabaceae/Papilionaceae
♦ smooth rattlebox
♦ Native Weed
♦ 121
♦ H, toxic. Origin: southern Africa.

Crotalaria podocarpa DC.
Fabaceae/Papilionaceae
♦ Native Weed
♦ 121
♦ H, arid, cultivated, toxic. Origin:
Africa.

Crotalaria polysperma Kotschy
Fabaceae/Papilionaceae
♦ Weed
♦ 87, 88

Crotalaria pumila Ortega
Fabaceae/Papilionaceae
Crotalaria lupulina Kunth
♦ low rattlebox
♦ Weed
♦ 14
♦ arid, herbal.

Crotalaria pycnostachya Benth.
Fabaceae/Papilionaceae
♦ Weed
♦ 87, 88

Crotalaria quinquefolia L.
Fabaceae/Papilionaceae
♦ Weed, Naturalised
♦ 32, 87, 88, 287
♦ Origin: Asia, Australia.

Crotalaria recta Steud. ex A.Rich.
Fabaceae/Papilionaceae
♦ Native Weed
♦ 121
♦ pH. Origin: southern Africa.

Crotalaria retusa L.
Fabaceae/Papilionaceae
Crotalaria tunguensis (Lima) Polhill
♦ wedgeleaf rattlepod, rattleweed,
rattlebox
♦ Weed, Naturalised
♦ 6, 14, 39, 86, 87, 88, 93, 157, 161, 179,
205, 261, 276
♦ a/pH, arid, cultivated, herbal, toxic.
Origin: Australia.

Crotalaria sagittalis L.
Fabaceae/Papilionaceae
♦ rattlebox, arrowhead rattlebox
♦ Weed
♦ 39, 87, 88, 161, 218
♦ arid, herbal, toxic.

Crotalaria saltiana Andrews
Fabaceae/Papilionaceae
= *Crotalaria pallida* Ait.
♦ African rattlebox
♦ Weed, Naturalised
♦ 87, 88, 101, 241, 242, 274
♦ arid.

Crotalaria semperflorens Vent.
Fabaceae/Papilionaceae
♦ Weed, Naturalised
♦ 86, 98, 203
♦ cultivated.

Crotalaria senegalensis (Pers.) DC.
Fabaceae/Papilionaceae
Crotalaria maxillaris sensu auct.,
Crotalaria shamvaensis sensu Torre
♦ Weed, Naturalised
♦ 86, 87, 88, 98, 203, 221, 242
♦ arid, cultivated. Origin: Africa.

Crotalaria sessiliflora L.
Fabaceae/Papilionaceae
♦ Weed
♦ 286
♦ herbal.

Crotalaria spartioides DC.
Fabaceae/Papilionaceae
♦ dune bush
♦ Weed, Native Weed
♦ 39, 121
♦ pS, toxic. Origin: South Africa.

Crotalaria spectabilis Roth
Fabaceae/Papilionaceae
♦ showy crotalaria
♦ Weed, Naturalised, Introduced
♦ 23, 38, 39, 86, 87, 88, 98, 101, 161, 179,
203, 218, 255, 261
♦ aH, cultivated, herbal, toxic. Origin:
India.

Crotalaria sphaerocarpa Perr. ex DC.
Fabaceae/Papilionaceae
♦ crotalaria, mealie crotalaria, wild
lucerne
♦ Weed, Native Weed
♦ 39, 88, 121, 158
♦ aH, toxic. Origin: southern Africa.

Crotalaria stipularia Desv.
Fabaceae/Papilionaceae
♦ cascabeiillo alado
♦ Weed, Introduced
♦ 87, 88, 228
♦ arid, herbal.

Crotalaria striata DC.
Fabaceae/Papilionaceae
= *Crotalaria pallida* Ait. var. *obovata*
(G.Don) Polhill
♦ striped crotalaria, showy crotalaria,
Hing men
♦ Weed
♦ 39, 88, 218, 239
♦ aH, herbal, toxic.

Crotalaria thebaica (Del.) DC.
Fabaceae/Papilionaceae
Spartium thebaicum Del.
♦ Weed
♦ 221
♦ arid, herbal.

Crotalaria uncinella Lam.
Fabaceae/Papilionaceae
♦ Naturalised
♦ 287

Crotalaria usaramoensis Baker f.
Fabaceae/Papilionaceae
= *Crotalaria zanzibarica* Benth.
♦ Weed
♦ 275

Crotalaria verrucosa L.
Fabaceae/Papilionaceae
♦ blue rattlesnake
♦ Weed
♦ 32, 39, 87, 88, 179
♦ cultivated, herbal, toxic.

Crotalaria virgulata Klotzsch
Fabaceae/Papilionaceae
♦ rattlebox
♦ Weed, Naturalised
♦ 98, 101, 203

**Crotalaria virgulata Klotzsch ssp.
grantiana (Harv.) Polhill**
Fabaceae/Papilionaceae
♦ Grant's rattlebox
♦ Weed, Naturalised
♦ 86, 101, 179
♦ Origin: Africa.

**Crotalaria virgulata Klotzsch ssp.
virgulata**
Fabaceae/Papilionaceae
♦ Naturalised
♦ 86
♦ Origin: Africa.

Crotalaria vitellina Ker Gawl.
Fabaceae/Papilionaceae
♦ rattlebox
♦ Weed
♦ 87, 88
♦ arid.

Crotalaria zanzibarica Benth.
Fabaceae/Papilionaceae
Crotalaria usaramoensis Baker (see)
♦ West Indian rattlebox
♦ Weed, Naturalised, Introduced
♦ 32, 38, 86, 87, 88, 98, 101, 179, 203,
228, 261, 287
♦ arid. Origin: tropical Africa.

Croton L. spp.
Euphorbiaceae
Julocroton Mart. spp. (see)
♦ croton
♦ Weed
♦ 39, 270
♦ toxic.

Croton argenteus L.
Euphorbiaceae
♦ silver July croton
♦ Naturalised
♦ 101

Croton bonplandianus Baill.
Euphorbiaceae
♦ Bonpland's croton
♦ Weed, Naturalised, Environmental
Weed
♦ 87, 88, 101, 216, 237
♦ cultivated, herbal.

Croton campestris St.-Hil.
Euphorbiaceae
♦ Weed
♦ 87, 88
♦ herbal.

Croton capitatus Michx.
Euphorbiaceae
♦ woolly croton, hogwort
♦ Weed, Naturalised
♦ 39, 86, 87, 88, 98, 161, 203, 210, 218,
247
♦ herbal, toxic. Origin: North America.

Croton ciliato-glanduliferus Ortega
Euphorbiaceae
Croton cilio-glanduliferus Ort., *Croton
ciliato-glanduliferum* Ort.
♦ Mexican croton
♦ Weed, Quarantine Weed
♦ 76, 87, 88, 203, 220
♦ cultivated, herbal, toxic.

Croton craspedotrichus Griseb.
Euphorbiaceae
♦ Weed
♦ 14

Croton dioicus Cav.
Euphorbiaceae
♦ grassland croton
♦ Weed
♦ 87, 88, 199
♦ herbal.

Croton glandulosus L.
Euphorbiaceae
♦ tropic croton, vente conmigo, gervão branco
♦ Weed, Naturalised
♦ 14, 86, 87, 88, 161, 210, 218, 245, 255
♦ aH, herbal. Origin: Americas.

Croton heterochrous Müll.Arg.
Euphorbiaceae
♦ Weed
♦ 157
♦ S.

Croton hirtus L'Hér.
Euphorbiaceae
Croton glandulosus L. ssp. *hirtus* (L'Hér.) Croizat, *Croton glandulosus* var. *hirtus* (L'Hér.) Müll.Arg.
♦ croton, croto
♦ Weed, Quarantine Weed
♦ 76, 87, 88, 135, 157, 170, 191, 203, 206, 220, 237, 243, 262, 276, 281, 295
♦ aH. Origin: tropical Americas.

Croton humilis L.
Euphorbiaceae
♦ pepperbush
♦ Weed
♦ 87, 88
♦ herbal.

Croton leptostachyus H.B.K.
Euphorbiaceae
♦ Weed
♦ 87, 88

Croton lindheimeri (Engelm. & Gray) Wood
Euphorbiaceae
= *Croton capitatus* Michx. var. *lindheimeri* (Engelm. & Gray) Müll.Arg. (NoR)
♦ Lindheimer croton
♦ Weed
♦ 88, 218

Croton lindheimerianus Scheele
Euphorbiaceae
♦ threeseed croton
♦ Weed
♦ 87, 88

Croton linearis Jacq.
Euphorbiaceae
♦ grannybush
♦ Weed
♦ 87, 88
♦ herbal.

Croton lobatus L.
Euphorbiaceae
♦ lobed croton, papayita, croton lobulado
♦ Weed, Quarantine Weed

♦ 14, 76, 87, 88, 203, 220, 237, 255, 261, 281, 295
♦ aH. Origin: tropical America.

Croton lucidus L.
Euphorbiaceae
♦ firebush
♦ Weed
♦ 14

Croton lundianus (Diedr.) Müll.Arg.
Euphorbiaceae
Podostachys lundianus Diedr.
♦ Weed
♦ 255
♦ aH. Origin: South America.

Croton menyhartii Pax
Euphorbiaceae
♦ rough leaved croton
♦ Native Weed
♦ 121
♦ pS. Origin: southern Africa.

Croton moluccanus L.
Euphorbiaceae
= *Aleurites moluccana* (L.) Willd.
♦ Weed
♦ 3, 191

Croton monanthogynus Michx.
Euphorbiaceae
♦ prairietea, prairietea croton
♦ Weed
♦ 87, 88, 161, 218
♦ herbal.

Croton ripense Kaneh. & Hatus.
Euphorbiaceae
♦ Introduced
♦ 230
♦ S.

Croton sparsiflorus Morong
Euphorbiaceae
♦ Weed
♦ 87, 88

Croton texensis (Klotzsch) Müll.Arg.
Euphorbiaceae
♦ Texas croton, doveweed, croton, goatweed, skunkweed
♦ Weed, Native Weed
♦ 39, 87, 88, 161, 174, 218, 243
♦ herbal, toxic. Origin: North America.

Croton tiglium L.
Euphorbiaceae
♦ purging croton, croton
♦ Weed, Introduced
♦ 39, 161, 230
♦ S, cultivated, herbal, toxic.

Croton tinctorius L.
Euphorbiaceae
♦ Weed, Quarantine Weed
♦ 76, 87, 88, 203, 220

Croton trinitatis Millsp.
Euphorbiaceae
♦ roadside croton, pichana
♦ Weed, Naturalised
♦ 87, 88, 101, 153

Croton verbascifolia Willd.
Euphorbiaceae
♦ Weed
♦ 87, 88

Croton wilsonii Griseb.
Euphorbiaceae
♦ Weed
♦ 87, 88

Crucianella angustifolia L.
Rubiaceae
♦ narrowleaf crucianella, crucianella
♦ Weed, Naturalised
♦ 40, 88, 94, 101
♦ aH, cultivated, herbal.

Crucianella ciliata Lam.
Rubiaceae
♦ Weed
♦ 221

Crucianella graeca Boiss.
Rubiaceae
♦ Weed
♦ 272

Crucianella macrostachya Boiss.
Rubiaceae
♦ Weed
♦ 111, 243

Crucianella membranacea Boiss.
Rubiaceae
♦ Weed
♦ 221

Cruciata articulata (L.) Ehrend.
Rubiaceae
♦ Weed
♦ 111, 243

Cruciata glabra (L.) Ehrend.
Rubiaceae
Valantia glabra L.
♦ kaljuristimatara
♦ Weed, Casual Alien
♦ 42, 272
♦ herbal.

Cruciata laevipes Opiz
Rubiaceae
Galium cruciata (L.) Scop.
♦ karvaristimatara, smooth bedstraw, crosswort, krížavka chlpatá
♦ Naturalised, Casual Alien
♦ 40, 42, 101, 243
♦ pH, cultivated, herbal.

Cruciata pedemontana (Bellardi) Ehrend.
Rubiaceae
Galium chloranthum Brot., *Galium pedemontanum* (Bellardi) All. (see)
♦ piedmont bedstraw, krížavka piemontská
♦ Weed, Naturalised
♦ 101, 161, 272

Crupina crupinastrum (Moris.) Vis.
Asteraceae
♦ Weed
♦ 87, 88

Crupina vulgaris (Pers.) Cass.
Asteraceae
Centaurea crupina L.
♦ common crupina, bearded creeper
♦ Weed, Noxious Weed, Naturalised, Environmental Weed
♦ 35, 36, 67, 79, 80, 86, 88, 94, 98, 101, 140, 146, 151, 161, 203, 212, 219, 229, 267, 272
♦ aH, cultivated, herbal. Origin: Mediterranean region.

Crusea parviflora Hook. & Arn.
Rubiaceae
♦ Weed
♦ 157
♦ pH.

Crypsis aculeata (L.) Aiton
Poaceae
♦ skrytka ostnatá
♦ Weed, Naturalised
♦ 87, 88, 272, 275, 287
♦ aG.

Crypsis alopecuroides (Piller & Mitterp.) Schrad.
Poaceae
Heleochloa alopecuroides (Piller & Mitterp.) Host ex Roem.
♦ foxtail pricklegrass
♦ Weed, Naturalised
♦ 101, 272
♦ aG.

Crypsis schoenoides (L.) Lam.
Poaceae
Heleochloa schoenoides (L.) Host ex Roem. (see), *Phleum schoenoides* L.
♦ swamp pricklegrass, swampgrass
♦ Weed, Naturalised
♦ 7, 34, 68, 86, 87, 88, 101, 272, 275
♦ aG, herbal. Origin: North Africa, Eurasia.

Crypsis vaginiflora (Forssk.) Opiz
Poaceae
♦ African pricklegrass
♦ Naturalised
♦ 101
♦ aG.

Cryptandra amara Sm.
Rhamnaceae
♦ bitter cryptandra
♦ Casual Alien
♦ 280
♦ cultivated.

Cryptantha albida (Kunth) I.M.Johnst.
Boraginaceae
♦ New Mexico cryptantha
♦ Weed
♦ 199
♦ herbal.

Cryptantha torreyana (Gray) Greene
Boraginaceae
♦ Torrey's cryptantha, Torrey's catseye
♦ Weed
♦ 23, 88
♦ aH, herbal.

Cryptocarya obovata R.Br.
Lauraceae
♦ Casual Alien
♦ 280
♦ cultivated.

Cryptocoryne beckettii Thwaites ex Trimen
Araceae
♦ Beckett's cryp, watertrumpet
♦ Environmental Weed
♦ 182, 284
♦ aqua, cultivated. Origin: Sri Lanka.

Cryptocoryne ciliata (Roxb.) Fisch. ex Schott
Araceae
♦ Weed
♦ 87, 88
♦ aqua, cultivated.

Cryptocoryne wendtii de Wit
Araceae
♦ Wendt's watertrumpet
♦ Naturalised
♦ 101
♦ aqua, cultivated.

Cryptolepis oblongifolia (Meisn.) Schltr.
Asclepiadaceae/Apocynaceae/Periplocaceae
♦ Native Weed
♦ 121
♦ pS. Origin: southern Africa.

Cryptomeria japonica (L.f.) D.Don
Taxodiaceae/Cupressaceae
♦ Japanese cedar, kryproméria japonská, sugi
♦ Weed, Naturalised
♦ 15, 101, 280
♦ T, cultivated, herbal. Origin: China, Japan.

Cryptostegia R.Br. spp.
Asclepiadaceae/Apocynaceae/Periplocaceae
♦ rubbervine, India rubbervine
♦ Cultivation Escape
♦ 233
♦ cultivated.

Cryptostegia grandiflora Roxb. ex R.Br.
Asclepiadaceae/Apocynaceae/Periplocaceae
Nerium grandiflorum Roxb.
♦ rubbervine, Palay rubbervine, India rubbervine, liane de gatope
♦ Weed, Quarantine Weed, Noxious Weed, Naturalised, Introduced, Garden Escape, Environmental Weed
♦ 3, 18, 22, 39, 62, 76, 80, 86, 87, 88, 93, 98, 101, 121, 147, 152, 161, 178, 179, 191, 203, 218, 228, 246, 261
♦ pC, arid, cultivated, herbal, toxic. Origin: Africa, Madagascar.

Cryptostegia madagascariensis Bojer ex Decne.
Asclepiadaceae/Apocynaceae/Periplocaceae
♦ rubbervine, Madagascar rubbervine
♦ Weed, Quarantine Weed, Noxious Weed, Naturalised, Garden Escape, Environmental Weed
♦ 11, 39, 62, 76, 80, 88, 101, 112, 155, 161, 179, 191, 203, 261
♦ cultivated, toxic. Origin: Madagascar.

Cryptostegia madagascariensis Bojer ex Decne. var. glaberrima (Hochr.) J.Marohasy & P.I.Forst.
Asclepiadaceae/Apocynaceae/Periplocaceae
♦ rubbervine
♦ Noxious Weed, Naturalised
♦ 86
♦ Origin: Madagascar.

Cryptostegia madagascariensis Bojer ex Decne. var. madagascariensis
Asclepiadaceae/Apocynaceae/Periplocaceae
♦ rubbervine
♦ Noxious Weed, Naturalised, Garden Escape
♦ 86, 252
♦ cultivated. Origin: Madagascar.

Cryptostemma calendula (L.) Druce
Asteraceae
♦ Weed
♦ 87, 88

Cryptotaenia canadensis (L.) DC.
Apiaceae
Cryptotaenia japonica Hassk. (see)
♦ Japanese parsley, honewort, Canadian honewort
♦ Weed, Introduced
♦ 38, 87, 88
♦ pH, aqua, cultivated, herbal.

Cryptotaenia japonica Hassk.
Apiaceae
= *Cryptotaenia canadensis* (L.) DC.
♦ Japanese parsley, mitsuba
♦ Weed, Quarantine Weed
♦ 76, 87, 88, 247, 286
♦ pH, cultivated, herbal, toxic.

Ctenolepis cerasiformis (Stocks) Naud.
Cucurbitaceae
♦ ctenolepis
♦ Naturalised
♦ 101

Ctenopsis pectinella (Delile) De Not.
Poaceae
Festuca pectinella Delile
♦ Weed
♦ 221
♦ G.

Cucubalus baccifer L.
Caryophyllaceae
♦ marjakohokki, berry catchfly, berry bearing catchfly
♦ Weed, Casual Alien
♦ 42, 272
♦ aH, cultivated, herbal.

Cucumis anguria L.
Cucurbitaceae
♦ burr gherkin, gherkin, West Indian gherkin, gooseberry gourd, burr cucumber, maxixe, pepino espinhoso, cohombro, pepinillo
♦ Weed, Quarantine Weed, Naturalised, Introduced
♦ 38, 55, 76, 87, 88, 98, 134, 161, 179, 203, 255, 261
♦ aH, cultivated, herbal. Origin: Africa.

Cucumis anguria L. var. anguria
Cucurbitaceae
Cucumis anguria L. ssp. *cubensis* Gand., *Cucumis anguria* L. ssp. *jamaicensis* Gand., *Cucumis angurioides* M.Roem., *Cucumis arada* L. ex Naudin & F.Muell., *Cucumis echinatus* Moench, *Cucumis erinaceus* Naudin ex Huber, *Cucumis longipes* Hook.f., *Cucumis macrocarpus* Wender., *Cucumis parviflorus* Salisb.,

Cucumis subhirsutus minor P.Browne
- West Indian gherkin, burr gherkin
- Naturalised, Introduced
- 86, 228, 243
- arid, cultivated. Origin: West Indies.

Cucumis anguria L. var. *longaculeatus* **J.H.Kirkbr.**
Cucurbitaceae
- West Indian gherkin
- Naturalised
- 101

Cucumis anguria L. var. *longipes* **A.Meeuse**
Cucurbitaceae
- West Indian gherkin
- Introduced
- 228
- arid.

Cucumis callosus (Rottler) Cogn.
Cucurbitaceae
- Weed
- 87, 88
- herbal.

Cucumis dipsaceus C.G.Ehrenb. ex Spach
Cucurbitaceae
- wild spiny cucumber, hedgehog gourd, ekaleruk, ibemba
- Weed, Naturalised, Introduced, Environmental Weed
- 38, 39, 86, 87, 88, 98, 101, 203, 218, 242, 257, 271
- aH, arid, cultivated, herbal, toxic. Origin: Africa.

Cucumis ficifollus A.Rich.
Cucurbitaceae
- Weed
- 240
- pH.

Cucumis hirsutus Sond.
Cucurbitaceae
- Weed
- 87, 88

Cucumis melo L.
Cucurbitaceae
Cucumis melo L. ssp. *agrestis* (Naudin) Pangalo (see), *Cucumis melo* L. var. *dudaim* (L.) Naud. (see)
- smellmelon, dudaim melon, cantaloupe, rock melon, melon, uhorka Ïltá
- Weed, Noxious Weed, Naturalised, Garden Escape, Cultivation Escape, Casual Alien
- 13, 26, 32, 35, 40, 42, 87, 88, 101, 161, 179, 203, 229, 261
- aC, cultivated, herbal. Origin: obscure, possibly Asia.

Cucumis melo L. ssp. *agrestis* (Naudin) **Pangalo**
Cucurbitaceae
= *Cucumis melo* L.
- rock melon, melon
- Naturalised
- 86
- Origin: Africa.

Cucumis melo L. var. *agrestis* Naud.
Cucurbitaceae

- Weed
- 242
- arid.

Cucumis melo L. var. *dudaim* (L.) Naud.
Cucurbitaceae
= *Cucumis melo* L.
- cantaloupe
- Weed
- 34, 243
- a/pH.

Cucumis metuliferus E.Mey. ex Naud.
Cucurbitaceae
Cucumis tinneanus Kotschy & Peyr.
- African horned cucumber, horned cucumber, jelly melon, kiwano
- Weed, Quarantine Weed, Naturalised
- 39, 76, 86, 87, 88, 98, 101, 203
- aC, arid, cultivated, herbal, toxic. Origin: tropical and southern Africa.

Cucumis myriocarpus Naudin
Cucurbitaceae
Cucumis africanus L.f. var. *acutilobus* Cogn., *Cucumis dissectifolius* Naudin, *Cucumis grossularia* hort., *Cucumis leptodermis* Schweick., *Cucumis merxmuelleri* Suess.
- paddy melon, prickly paddy melon, bitterapple, gooseberry cucumber, small thorny cucumber, small wild cucumber, small wild melon, striped wild cucumber, wild cucumber, gooseberry gourd, bitterappel, gifappel, isendelenja, mokapana, monyaku, thlare sa mpja, wilde komkommer
- Weed, Noxious Weed, Naturalised, Native Weed, Introduced, Garden Escape, Environmental Weed, Cultivation Escape
- 7, 15, 34, 35, 39, 51, 72, 87, 88, 93, 98, 101, 121, 147, 158, 161, 176, 180, 203, 205, 228, 229, 243, 269, 280
- aH, arid, cultivated, herbal, toxic. Origin: southern Africa.

Cucumis myriocarpus Naudin ssp. *leptodermis* (Schweick.) C.Jeffrey & **P.Halliday**
Cucurbitaceae
- paddy melon
- Naturalised
- 198

Cucumis pepo (L.) Dumort.
Cucurbitaceae
- Weed
- 203

Cucumis prophetarum L.
Cucurbitaceae
- mandera cucumber, ekaleruk, tegesrarit, wild cucumber
- Weed, Introduced
- 39, 221, 228, 242
- arid, herbal, toxic.

Cucumis prophetarum L. ssp. *zeyheri* (Sond.) C.Jeffrey
Cucurbitaceae
= *Cucumis zeyheri* Sond.
- wild cucumber

- Native Weed
- 121
- pH, toxic. Origin: South Africa.

Cucumis sativus L.
Cucurbitaceae
- garden cucumber, kurkku, cucumber
- Weed, Naturalised, Cultivation Escape, Casual Alien
- 42, 87, 88, 101, 179, 261
- aC, cultivated, herbal. Origin: East Indies.

Cucumis trigonus Roxb.
Cucurbitaceae
- Weed
- 39, 87, 88
- herbal, toxic.

Cucumis zeyheri Sond.
Cucurbitaceae
Cucumis africanus L.f. var. *zeyheri* (Sond.) Burtt Davy, *Cucumis diniae* L.W.D.van Raamsdonk & D.L.Visser, *Cucumis prophetarum* L. ssp. *zeyheri* (Sond.) C.Jeffrey
- tsinyagu, wild cucumber, wilde agurkie, wilde komkommer
- Naturalised
- 86
- arid, herbal. Origin: Africa.

Cucurbita andreana Naud.
Cucurbitaceae
= *Cucurbita maxima* Duch. ssp. *andreana* (Naud.) Filov
- zapallito amargo
- Weed
- 87, 88, 236, 270, 295

Cucurbita argyrosperma C.Huber
Cucurbitaceae
Cucurbita mixta Pangalo, *Cucurbita palmeri* L.H.Bailey, *Cucurbita sororia* L.H.Bailey
- Introduced
- 228
- arid.

Cucurbita digitata A.Gray
Cucurbitaceae
- fingerleaf gourd
- Weed
- 87, 88, 161, 218
- pH, arid, herbal.

Cucurbita ficifolia Bouché
Cucurbitaceae
Cucurbita melanosperma Gasp., *Pepo ficifolia* (Bouché) Britton
- figleaf gourd, chilacayote, blackseed squash, Malabar gourd, courge à feuilles de figuier, feigenblattkürbis, abóbora chila, chilacayote
- Naturalised
- 280
- pC, cultivated, herbal. Origin: obscure, possibly Central America.

Cucurbita foetidissima Kunth
Cucurbitaceae
Cucurbita perennis (E.James) A.Gray
- buffalo gourd, Missouri gourd, calabazilla loca, chili coyote, chilicote, fetid gourd, mock orange, wild pumkin, wild gourd

♦ Weed, Native Weed, Introduced
♦ 87, 88, 161, 174, 180, 199, 210, 218, 228, 243
♦ pC, arid, cultivated, herbal. Origin: North America.

Cucurbita maxima Lam.
Cucurbitaceae
♦ winter squash, lal kumra, lal bhopli, lal dudiya, metha kumra, parangikayi, pumpkin, red gourd, squash, zapallo, pwengkin, calabaza
♦ Naturalised, Introduced, Cultivation Escape
♦ 7, 39, 86, 98, 101, 228, 230, 261, 280
♦ aC, arid, cultivated, herbal, toxic. Origin: tropical, subtropical Americas.

Cucurbita maxima Duch. ssp. andreana (Naud.) Filov
Cucurbitaceae
Cucurbita andreana Naudin (see)
♦ Weed
♦ 237

Cucurbita moschata (Duch. ex Lam.) Duch. ex Poir.
Cucurbitaceae
♦ crookneck squash, squash, butternut squash, butternut pumkin
♦ Weed, Naturalised, Cultivation Escape
♦ 32, 101, 179, 261
♦ aC, cultivated, herbal. Origin: obscure.

Cucurbita pepo L.
Cucurbitaceae
♦ field pumpkin, kurpitsa, kiuri, marrow, squash, pumpkin, gourd, zucca, calabaza
♦ Weed, Naturalised, Introduced, Cultivation Escape, Casual Alien
♦ 39, 40, 42, 86, 87, 88, 98, 101, 230, 243, 252, 261, 280
♦ aC, cultivated, herbal, toxic. Origin: tropical Americas.

Cucurbita pepo L. var. pepo
Cucurbitaceae
♦ field pumpkin
♦ Naturalised
♦ 101

Cucurbita radicans Naud.
Cucurbitaceae
♦ Weed
♦ 199

Cucurbita texana (Scheele) A.Gray
Cucurbitaceae
= *Cucurbita pepo* L. var. *texana* (Scheele) D.Decker (NoR)
♦ Texas gourd
♦ Weed
♦ 161
♦ herbal.

Cucurbitella asperata (Gilles ex Hook.& Arn.) Walp.
Cucurbitaceae
Cucurbitella duriaei (Naudin) Cogn.
♦ Weed
♦ 237

Cudrania javanensis Trec.
Moraceae

♦ Weed
♦ 87, 88
♦ herbal.

Cudrania tricuspidata (Carr.) Bureau ex Lavallée
Moraceae
Cudrania triloba Hance, *Maclura tricuspidata* Carrière
♦ storehousebush, silkworm thorn
♦ Naturalised
♦ 101
♦ T, cultivated.

Culcasia liberica N.E.Br.
Araceae
♦ Quarantine Weed
♦ 220
♦ cultivated.

Cullen americana (L.) Rydb.
Fabaceae/Papilionaceae
Psoralea americana L. (see)
♦ American scurfpea
♦ Weed, Quarantine Weed, Naturalised
♦ 76, 88, 101, 220

Cullen corylifolia (L.) Medik.
Fabaceae/Papilionaceae
Psoralea corylifolia L. (see)
♦ Malaysian scurfpea
♦ Weed, Quarantine Weed, Naturalised
♦ 76, 88, 101, 220

Cuminum cyminum L.
Apiaceae
♦ cumin, spiskummin, roomankumina, cumino di Malta
♦ Naturalised
♦ 101
♦ aH, cultivated, herbal. Origin: Mediterranean.

Cunninghamia lanceolata (Lamb.) Hook.
Taxodiaceae/Cupressaceae
Cunninghamia sinensis R.Br., *Pinus lanceolata* Lamb.
♦ Chinese fir
♦ Naturalised, Casual Alien
♦ 101, 280
♦ T, cultivated, herbal.

Cupania macrophylla Mart.
Sapindaceae
= *Talisia macrophylla* (C.Mart.) Radlk. (NoR)
♦ Weed
♦ 14

Cupaniopsis anacardioides (A.Rich.) Radlk.
Sapindaceae
Cupania anacardioides A.Rich.
♦ carrot wood, carrot weed
♦ Weed, Noxious Weed, Naturalised, Introduced, Garden Escape, Environmental Weed
♦ 3, 80, 88, 101, 112, 122, 129, 142, 151, 179, 191, 228
♦ arid, cultivated. Origin: eastern Australia, Indonesia.

Cuphea aequipetala Cav.
Lythraceae
♦ Weed

♦ 199
♦ herbal.

Cuphea balsamona Cham. & Schlecht.
Lythraceae
= *Cuphea carthagenensis* (Jacq.) J.F.McBr.
♦ Weed
♦ 87, 88

Cuphea carthagenensis (Jacq.) J.F.McBr.
Lythraceae
Cuphea balsamona Cham. & Schlecht. (see)
♦ tarweed cuphea, tarweed, Colombian waxweed, cuphea, sete sangrias
♦ Weed, Naturalised, Environmental Weed
♦ 6, 28, 68, 86, 87, 88, 98, 101, 157, 179, 191, 203, 206, 218, 243, 255, 257, 287
♦ pH, herbal. Origin: South America.

Cuphea densiflora Koehne
Lythraceae
♦ Weed
♦ 87, 88

Cuphea glutinosa Cham. & Schlecht.
Lythraceae
Cuphea hyssopifolia Kunth var. *brachyphylla* Griseb., *Cuphea thymoides* Griseb.
♦ sticky waxweed, lavender lady, hardy Mexican feather
♦ Weed, Naturalised
♦ 87, 88, 101, 295
♦ herbal.

Cuphea hyssopifolia Kunth
Lythraceae
♦ false heather, sinitulitorvi, elfin herb
♦ Naturalised, Cultivation Escape
♦ 86, 101, 261, 280
♦ cultivated, herbal. Origin: Mexico, Central America.

Cuphea ignea A.DC.
Lythraceae
Cuphea platycentra Lem. *nom. illeg.*
♦ cigar flower
♦ Naturalised
♦ 101
♦ a/pH, cultivated, herbal.

Cuphea lanceolata W.T.Aiton
Lythraceae
♦ Casual Alien
♦ 280
♦ cultivated.

Cuphea parsonsia (L.) R.Br.
Lythraceae
♦ island waxweed
♦ Weed
♦ 14, 87, 88
♦ herbal.

Cuphea petiolata (L.) Koehne
Lythraceae
= *Cuphea viscosissima* Jacq. (NoR)
♦ clammy cuphea
♦ Weed
♦ 87, 88, 218
♦ herbal.

Cuphea pseudosilene Griseb.
Lythraceae

♦ Weed
♦ 14

***Cuphea racemosa* (L.f.) Spreng.**
Lythraceae
♦ sete sangrias
♦ Weed, Naturalised
♦ 87, 88, 255, 257
♦ pH, herbal. Origin: Brazil.

***Cuphea strigulosa* Kunth**
Lythraceae
♦ stiffhair waxweed
♦ Weed
♦ 261
♦ arid.

***Cuphea wrightii* A.Gray**
Lythraceae
♦ Wright's waxweed
♦ Weed, Quarantine Weed
♦ 76, 87, 88, 203, 220

**× *Cupressocyparis leylandii* (A.B.Jacks. &
Dallim.) Dallim.**
Cupressaceae
= *Chamaecyparis nootkatonsis* (Lam.)
Spach × *Cupressus macrocarpa* Hartw.
ex Gord.
♦ Leyland cypress
♦ Cultivation Escape
♦ 40
♦ cultivated, herbal.

***Cupressus arizonica* Greene**
Cupressaceae
♦ Arizona cypress
♦ Naturalised, Introduced
♦ 98, 261
♦ T, cultivated, herbal. Origin: North
America.

***Cupressus dupreziana* A.Camus**
Cupressaceae
♦ Introduced
♦ 228
♦ arid, cultivated.

***Cupressus funebris* Endl.**
Cupressaceae
= *Chamaecyparis funebris* (Endl.) Franco
♦ Chinese weeping cypress
♦ Naturalised
♦ 98
♦ T, cultivated.

***Cupressus glabra* Sudw.**
Cupressaceae
= *Cupressus arizonica* Greene ssp.
arizonica (NoR)
♦ smooth Arizona cypress
♦ Weed, Naturalised
♦ 54, 86, 88
♦ cultivated, herbal.

***Cupressus lusitanica* Mill.**
Cupressaceae
♦ Mexican cypress, Arizona cypress,
cedar of Goa
♦ Weed, Naturalised, Introduced,
Garden Escape, Environmental Weed
♦ 3, 72, 86, 88, 98, 134, 152, 155, 198,
228, 261, 279, 280
♦ T, arid, cultivated, herbal. Origin:
Mexico, Central America.

***Cupressus macrocarpa* Hartw. ex Gord.**
Cupressaceae

Cupressus lambertiana hort. ex Carr.
♦ Monterey cypress, ciprés de
Monterrey
♦ Weed, Naturalised, Introduced,
Garden Escape, Environmental Weed
♦ 15, 39, 54, 72, 86, 88, 98, 151, 155, 198,
261, 280
♦ T, cultivated, herbal, toxic. Origin:
California.

***Cupressus sempervirens* L.**
Cupressaceae
♦ Italian cypress, cipresso
♦ Weed, Naturalised, Introduced,
Garden Escape, Environmental Weed
♦ 72, 80, 86, 88, 98, 198, 261, 280
♦ T, cultivated, herbal. Origin:
Mediterranean.

***Cupressus sempervirens* L. var. *stricta* Ait.**
Cupressaceae
♦ Introduced
♦ 228
♦ arid.

***Cupressus torulosa* D.Don**
Cupressaceae
♦ Himalayan cypress
♦ Naturalised
♦ 98
♦ T, cultivated, herbal.

***Curatella americana* L.**
Dilleniaceae
♦ Weed
♦ 87, 88, 255
♦ T, herbal. Origin: Brazil.

***Curculigo capitulata* (Lour.) Kuntze**
Liliaceae/Hypoxidaceae
= *Molineria capitulata* (Lour.) Herb.
♦ palm grass, molineria
♦ Naturalised, Cultivation Escape
♦ 101, 261
♦ cultivated. Origin: Asia.

***Curculigo orchioides* Gaertn.**
Liliaceae/Hypoxidaceae
Hypoxis aurea Lour.
♦ Introduced
♦ 230
♦ H, herbal.

***Curcuma aromatica* Salisb.**
Zingiberaceae
♦ curcuma
♦ Naturalised
♦ 287
♦ herbal.

***Curcuma domestica* Val.**
Zingiberaceae
♦ turmeric, njano njano
♦ Weed
♦ 12
♦ cultivated, herbal.

***Curcuma longa* L.**
Zingiberaceae
♦ common turmeric
♦ Weed, Naturalised, Cultivation
Escape
♦ 86, 98, 101, 203, 261
♦ cultivated, herbal. Origin: India.

***Curcuma zedoaria* (Christm.) Roscoe**
Zingiberaceae

♦ zedoary
♦ Naturalised
♦ 101
♦ herbal.

***Cuscuta* L. spp.**
Cuscutaceae/Convolvulaceae
♦ dodder, love vine, strangleweed,
devil's gut, hellbind
♦ Weed, Quarantine Weed, Noxious
Weed
♦ 8, 26, 35, 36, 49, 52, 67, 76, 80, 86, 147,
161, 162, 169, 191, 210, 211, 220, 243,
253, 258, 262, 271, 272, 295, 299
♦ aC, parasitic, herbal.

***Cuscuta americana* L.**
Cuscutaceae/Convolvulaceae
♦ American dodder
♦ Weed, Quarantine Weed, Noxious
Weed
♦ 14, 39, 67, 76, 87, 88, 199, 203, 220,
229
♦ C, parasitic, herbal, toxic.

***Cuscuta applanata* Engelm.**
Cuscutaceae/Convolvulaceae
♦ Gila River dodder
♦ Weed, Noxious Weed
♦ 67, 88, 229
♦ C, parasitic.

***Cuscuta approximata* Bab.**
Cuscutaceae/Convolvulaceae
Cuscuta planiflora Ten. (see)
♦ clustered dodder, smoothseed alfalfa
dodder, alfalfa dodder
♦ Weed, Quarantine Weed, Noxious
Weed, Naturalised
♦ 1, 23, 67, 76, 80, 87, 88, 101, 146, 203,
220, 229, 272
♦ aC, parasitic.

***Cuscuta arvensis* Beyr. ex Hook.**
Cuscutaceae/Convolvulaceae
= *Cuscuta pentagona* Engelm. var.
pentagona
♦ Weed
♦ 87, 88
♦ C, parasitic, herbal.

***Cuscuta attenuata* Waterf.**
Cuscutaceae/Convolvulaceae
♦ tapertip dodder
♦ Weed, Noxious Weed
♦ 67, 229
♦ C, parasitic.

***Cuscuta australis* R.Br.**
Cuscutaceae/Convolvulaceae
♦ southern dodder, mamedaoshi
♦ Weed
♦ 13, 39, 70, 87, 88, 272, 275, 276, 286
♦ aC, parasitic, cultivated, toxic.

***Cuscuta babylonica* Aucher & Choisy**
Cuscutaceae/Convolvulaceae
♦ Weed
♦ 87, 88
♦ C, parasitic.

***Cuscuta boldinghii* Urb.**
Cuscutaceae/Convolvulaceae
♦ Boldingh's dodder
♦ Weed, Noxious Weed, Naturalised
♦ 67, 101, 229
♦ C, parasitic.

Cuscuta brachycalyx (Yunck.) Yunck.
Cuscutaceae/Convolvulaceae
♦ San Joaquin dodder
♦ Weed, Noxious Weed
♦ 67, 229
♦ C, parasitic, herbal.

Cuscuta brachycalyx (Yunck.) Yunck. var. apodanthera (Yunck.) Yunck.
Cuscutaceae/Convolvulaceae
♦ San Joaquin dodder
♦ Noxious Weed
♦ 229
♦ C, parasitic.

Cuscuta brachycalyx (Yunck.) Yunck. var. brachycalyx
Cuscutaceae/Convolvulaceae
♦ San Joaquin dodder
♦ Noxious Weed
♦ 229
♦ C, parasitic.

Cuscuta breviflora Vis.
Cuscutaceae/Convolvulaceae
♦ Weed
♦ 87, 88
♦ C, parasitic.

Cuscuta brevistyla A.Braun ex A.Rich.
Cuscutaceae/Convolvulaceae
♦ Weed
♦ 87, 88, 221
♦ C, parasitic.

Cuscuta burrellii Yunck.
Cuscutaceae/Convolvulaceae
♦ Weed
♦ 87, 88
♦ C, parasitic.

Cuscuta californica Hook. & Arn.
Cuscutaceae/Convolvulaceae
♦ California dodder, chaparral dodder
♦ Weed, Noxious Weed
♦ 67, 87, 88, 229
♦ aC, parasitic, herbal.

Cuscuta californica Hook. & Arn. var. apiculata Engelm.
Cuscutaceae/Convolvulaceae
♦ chaparral dodder
♦ Noxious Weed
♦ 229
♦ aC, parasitic.

Cuscuta californica Hook. & Arn. var. breviflora Engelm.
Cuscutaceae/Convolvulaceae
Cuscuta occidentalis Millsp. (see)
♦ California dodder
♦ Noxious Weed
♦ 229
♦ aC, parasitic.

Cuscuta californica Hook. & Arn. var. californica
Cuscutaceae/Convolvulaceae
♦ chaparral dodder, California dodder
♦ Noxious Weed
♦ 229
♦ aC, parasitic, herbal.

Cuscuta californica Hook. & Arn. var. papillosa Yunck.
Cuscutaceae/Convolvulaceae
♦ chaparral dodder

♦ Noxious Weed
♦ 229
♦ aC, parasitic.

Cuscuta campestris Yunck.
Cuscutaceae/Convolvulaceae
Grammica campestris (Yunck.) Hada & Chrtek
♦ field dodder, golden dodder, dodder, angel's hair, common dodder, strangle vine, gewone dodder
♦ Weed, Quarantine Weed, Noxious Weed, Naturalised, Environmental Weed, Casual Alien
♦ 3, 23, 42, 50, 51, 62, 63, 67, 68, 70, 72, 76, 86, 87, 88, 95, 98, 115, 121, 158, 171, 191, 198, 199, 203, 212, 217, 218, 272, 278, 280, 283
♦ a/pH parasitic, cultivated, herbal.
Origin: North America.

Cuscuta cassytoides Nees ex Engelm.
Cuscutaceae/Convolvulaceae
♦ dodder, African dodder
♦ Weed, Noxious Weed, Naturalised, Native Weed
♦ 13, 67, 101, 121, 229
♦ pC, parasitic. Origin: South Africa.

Cuscuta ceanothi Behr
Cuscutaceae/Convolvulaceae
Cuscuta subinclusa Dur. & Hilg. (see)
♦ canyon dodder
♦ Noxious Weed
♦ 67, 229
♦ C, parasitic, herbal.

Cuscuta cephalanthi Engelm.
Cuscutaceae/Convolvulaceae
♦ buttonbush dodder
♦ Weed, Noxious Weed
♦ 23, 67, 88, 229
♦ aC, parasitic, herbal.

Cuscuta chilensis Ker Gawl.
Cuscutaceae/Convolvulaceae
♦ Weed
♦ 87, 88
♦ C, parasitic, arid.

Cuscuta chinensis Lam.
Cuscutaceae/Convolvulaceae
♦ Chinese dodder, dodder, foi thong
♦ Weed, Quarantine Weed
♦ 87, 88, 209, 220, 239, 275, 297
♦ aC, parasitic, promoted, herbal.
Origin: Middle East, Asia.

Cuscuta compacta Juss. ex Choisy
Cuscutaceae/Convolvulaceae
♦ compact dodder
♦ Weed, Noxious Weed
♦ 67, 229
♦ C, parasitic, herbal.

Cuscuta compacta Juss. ex Choisy var. compacta
Cuscutaceae/Convolvulaceae
♦ compact dodder
♦ Noxious Weed
♦ 229
♦ C, parasitic.

Cuscuta compacta Juss. ex Choisy var. efimbriata Yunck.
Cuscutaceae/Convolvulaceae
♦ compact dodder

♦ Noxious Weed
♦ 229
♦ C, parasitic.

Cuscuta coryli Engelm.
Cuscutaceae/Convolvulaceae
♦ hazel dodder
♦ Weed, Noxious Weed
♦ 67, 87, 88, 218, 229
♦ C, parasitic, herbal.

Cuscuta corymbosa Ruiz & Pav. var. stylosa (Choisy) Engelm.
Cuscutaceae/Convolvulaceae
♦ Weed
♦ 199
♦ C, parasitic.

Cuscuta cupulata Engelm.
Cuscutaceae/Convolvulaceae
♦ Weed
♦ 87, 88
♦ C, parasitic.

Cuscuta cuspidata Engelm.
Cuscutaceae/Convoivulaceae
♦ cusp dodder
♦ Weed, Noxious Weed
♦ 67, 229
♦ C, parasitic, herbal.

Cuscuta decipiens Yunck.
Cuscutaceae/Convolvulaceae
♦ TransPecos dodder
♦ Weed, Noxious Weed
♦ 67, 229
♦ C, parasitic.

Cuscuta dentatasquamata Yunck.
Cuscutaceae/Convolvulaceae
♦ los pinitos dodder
♦ Weed, Noxious Weed
♦ 67, 229
♦ C, parasitic.

Cuscuta denticulata Engelm.
Cuscutaceae/Convolvulaceae
♦ desert dodder
♦ Weed, Noxious Weed
♦ 67, 229
♦ aC, parasitic, herbal.

Cuscuta denticulata Engelm. var. denticulata
Cuscutaceae/Convolvulaceae
♦ smalltooth dodder, desert dodder
♦ Noxious Weed
♦ 229
♦ C, parasitic.

Cuscuta denticulata Engelm. var. vetchii (Brandeg.) T.Beliz
Cuscutaceae/Convolvulaceae
Cuscuta nevadensis I.M.Johnst. (see), *Cuscuta vetchii* Brandegee (see)
♦ Vetch's dodder
♦ Noxious Weed
♦ 229
♦ C, parasitic.

Cuscuta epilinum Weihe
Cuscutaceae/Convolvulaceae
Epilinella cuscutoides Pfeiff.
♦ flax dodder, pellavanvieras
♦ Weed, Quarantine Weed, Noxious Weed, Naturalised
♦ 67, 70, 76, 87, 88, 101, 203, 218, 229, 272, 287

♦ C, parasitic, herbal.

Cuscuta epithymum (L.) Murr.
Cuscutaceae/Convolvulaceae
Cuscuta trifolii Bab. (see)
♦ dodder, European dodder, clover dodder, flax dodder, lesser dodder, apilanvier.as, common dodder
♦ Weed, Noxious Weed, Naturalised, Environmental Weed, Casual Alien
♦ 7, 9, 39, 42, 51, 67, 70, 80, 86, 87, 88, 98, 101, 121, 146, 165, 176, 198, 203, 217, 218, 229, 272, 280
♦ pH parasitic, promoted, herbal, toxic. Origin: Eurasia.

Cuscuta epithymum (L.) Murr. ssp. trifolii Hegi
Cuscutaceae/Convolvulaceae
♦ Naturalised
♦ 287
♦ C, parasitic.

Cuscuta erosa Yunck.
Cuscutaceae/Convolvulaceae
♦ Sonoran dodder
♦ Weed, Noxious Weed
♦ 67, 229
♦ C, parasitic.

Cuscuta europaea L.
Cuscutaceae/Convolvulaceae
♦ greater dodder
♦ Weed, Noxious Weed, Naturalised, Casual Alien
♦ 23, 39, 67, 70, 87, 88, 101, 229, 272, 275, 280, 297
♦ aC, parasitic, cultivated, herbal, toxic.

Cuscuta exaltata Engelm.
Cuscutaceae/Convolvulaceae
♦ tall dodder
♦ Noxious Weed
♦ 229
♦ C, parasitic.

Cuscuta fasciculata Yunck.
Cuscutaceae/Convolvulaceae
♦ clustered dodder
♦ Weed, Noxious Weed
♦ 67, 229
♦ C, parasitic.

Cuscuta gigantea Griff.
Cuscutaceae/Convolvulaceae
♦ Weed
♦ 87, 88
♦ C, parasitic.

Cuscuta glabior (Engelm.) Yunck.
Cuscutaceae/Convolvulaceae
♦ Weed
♦ 67
♦ C, parasitic.

Cuscuta glandulosa (Engelm.) Small
Cuscutaceae/Convolvulaceae
= *Cuscuta obtusiflora* Kunth var. *glandulosa* (Engelm.)
♦ Weed
♦ 87, 88
♦ C, parasitic.

Cuscuta globulosa Benth.
Cuscutaceae/Convolvulaceae
♦ West Indian dodder

♦ Weed, Noxious Weed
♦ 67, 229
♦ C, parasitic.

Cuscuta glomerata Choisy
Cuscutaceae/Convolvulaceae
♦ composite dodder, rope dodder
♦ Weed, Noxious Weed
♦ 67, 87, 88, 218, 229
♦ C, parasitic, herbal.

Cuscuta gronovii Willd. ex J.A.Schult.
Cuscutaceae/Convolvulaceae
♦ swamp dodder, scaldweed
♦ Weed, Noxious Weed
♦ 23, 67, 87, 88, 218, 229, 243, 272
♦ H parasitic, herbal.

Cuscuta gronovii Willd. ex J.A.Schult. var. *calyptrata* Engelm.
Cuscutaceae/Convolvulaceae
♦ scaldweed
♦ Noxious Weed
♦ 229
♦ C, parasitic.

Cuscuta gronovii Willd. ex J.A.Schult. var. *gronovii*
Cuscutaceae/Convolvulaceae
Cuscuta umbrosa Beyr. ex Hook. (see)
♦ love dodder, scaldweed
♦ Noxious Weed
♦ 229
♦ C, parasitic.

Cuscuta harperi Small
Cuscutaceae/Convolvulaceae
♦ Harper's dodder
♦ Weed, Noxious Weed
♦ 67, 229
♦ C, parasitic.

Cuscuta howelliana Rubtzov
Cuscutaceae/Convolvulaceae
♦ Boggs Lake dodder
♦ Weed, Noxious Weed
♦ 67, 229
♦ C, parasitic.

Cuscuta hyalina Roth
Cuscutaceae/Convolvulaceae
♦ Weed
♦ 87, 88, 242
♦ C, parasitic, arid.

Cuscuta indecora Choisy
Cuscutaceae/Convolvulaceae
♦ largeseed dodder, devil's hair, hairweed, love vine, strangleweed, bigseed alfalfa dodder
♦ Weed, Quarantine Weed, Noxious Weed
♦ 23, 67, 76, 87, 88, 171, 180, 203, 212, 218, 220, 229, 237, 243
♦ aC, parasitic, herbal.

Cuscuta indecora Choisy var. *bifida* Yunck.
Cuscutaceae/Convolvulaceae
♦ bigseed alfalfa dodder
♦ Noxious Weed
♦ 229
♦ C, parasitic.

Cuscuta indecora Choisy var. *indecora*
Cuscutaceae/Convolvulaceae
Cuscuta jepsonii Yunck. (see)
♦ bigseed alfalfa dodder

♦ Weed, Noxious Weed
♦ 167, 229
♦ aC, parasitic.

Cuscuta indecora Choisy var. *longisepala* Yunck.
Cuscutaceae/Convolvulaceae
♦ bigseed dodder
♦ Weed, Noxious Weed
♦ 229, 237
♦ C, parasitic.

Cuscuta indecora Choisy var. *neuropetala* (Engelm.) A.S.Hitchc.
Cuscutaceae/Convolvulaceae
♦ bigseed dodder
♦ Noxious Weed
♦ 229
♦ aC, parasitic.

Cuscuta japonica Choisy
Cuscutaceae/Convolvulaceae
Cuscuta systyla Maxim.
♦ Japanese dodder
♦ Weed, Noxious Weed, Naturalised
♦ 87, 88, 101, 140, 229, 263, 275, 286, 297
♦ aC, parasitic, promoted, herbal.

Cuscuta jepsonii Yunck.
Cuscutaceae/Convolvulaceae
= *Cuscuta indecora* Choisy var. *indecora*
♦ Weed
♦ 67
♦ aC, parasitic.

Cuscuta kilimanjari Oliv.
Cuscutaceae/Convolvulaceae
♦ dodder
♦ Native Weed
♦ 121
♦ pC, parasitic. Origin: southern Africa.

Cuscuta kotschyi Des Moul.
Cuscutaceae/Convolvulaceae
♦ Weed
♦ 87, 88
♦ C, parasitic.

Cuscuta lehmanniana Bunge
Cuscutaceae/Convolvulaceae
♦ Weed
♦ 87, 88
♦ C, parasitic.

Cuscuta leptantha Engelm.
Cuscutaceae/Convolvulaceae
♦ slender dodder
♦ Weed, Noxious Weed
♦ 67, 229
♦ C, parasitic.

Cuscuta lupuliformis Krock.
Cuscutaceae/Convolvulaceae
Monogynella lupuliformis (Krock.) Hada & Chrtek
♦ hop dodder
♦ Weed, Quarantine Weed
♦ 70, 76, 87, 88, 203, 220, 272
♦ C, parasitic.

Cuscuta maroccana Trabut
Cuscutaceae/Convolvulaceae
♦ Weed, Quarantine Weed
♦ 76, 87, 88, 203, 220
♦ C, parasitic.

Cuscuta megalocarpa Rydb.
Cuscutaceae/Convolvulaceae
Cuscuta curta Engelm.
♦ bigfruit dodder
♦ Noxious Weed
♦ 140, 229
♦ C, parasitic, promoted, herbal.

Cuscuta mitriformis Engelm.
Cuscutaceae/Convolvulaceae
♦ Cochise dodder
♦ Weed, Noxious Weed
♦ 67, 229
♦ C, parasitic.

Cuscuta monogyna Vahl
Cuscutaceae/Convolvulaceae
♦ eastern dodder
♦ Weed, Quarantine Weed
♦ 76, 87, 88, 115, 203, 220, 272
♦ C, parasitic.

Cuscuta nevadensis I.M.Johnst.
Cuscutaceae/Convolvulaceae
= *Cuscuta denticulata* Engelm. var.
vetchii (Brandeg.) T.Beliz
♦ Weed
♦ 67
♦ C, parasitic, herbal.

Cuscuta obtusiflora H.B.K.
Cuscutaceae/Convolvulaceae
♦ Peruvian dodder
♦ Noxious Weed
♦ 67, 87, 88, 229
♦ C, parasitic, herbal.

**Cuscuta obtusiflora Kunth var.
glandulosa Engelm.**
Cuscutaceae/Convolvulaceae
Cuscuta glandulosa (Engelm.) Small
(see)
♦ Peruvian dodder
♦ Noxious Weed
♦ 229
♦ C, parasitic.

Cuscuta occidentalis Millsp.
Cuscutaceae/Convolvulaceae
= *Cuscuta californica* Hook. & Arn. var.
breviflora Engelm.
♦ Weed
♦ 23, 67, 88
♦ C, parasitic.

Cuscuta odontolepis Engelm.
Cuscutaceae/Convolvulaceae
♦ Santa Rita Mountain dodder
♦ Weed, Noxious Weed
♦ 67, 229
♦ C, parasitic.

Cuscuta palaestina Boiss.
Cuscutaceae/Convolvulaceae
♦ Weed, Quarantine Weed
♦ 76, 87, 88, 203, 220
♦ C, parasitic.

Cuscuta pedicellata Ledeb.
Cuscutaceae/Convolvulaceae
♦ clover dodder
♦ Weed
♦ 87, 88, 185
♦ C, parasitic.

Cuscuta pellucida Butkov.
Cuscutaceae/Convolvulaceae

Cuscuta indica (Englem.) Petrov
♦ Weed
♦ 272
♦ C, parasitic.

Cuscuta pentagona Engelm.
Cuscutaceae/Convolvulaceae
♦ five angled dodder, dodder,
lespedeza dodder, field dodder
♦ Weed, Quarantine Weed, Noxious
Weed, Naturalised, Native Weed
♦ 23, 67, 76, 87, 88, 174, 203, 220, 229,
263, 272, 286, 287
♦ aC, parasitic, herbal. Origin: North
America.

**Cuscuta pentagona Engelm. var. calycina
Engelm.**
Cuscutaceae/Convolvulaceae
= *Cuscuta pentagona* Engelm. var.
pentagona
♦ Weed
♦ 237
♦ C, parasitic.

**Cuscuta pentagona Engelm. var. glabrior
(Engelm.) Gandhi, Thomas & Hatch**
Cuscutaceae/Convolvulaceae
♦ bushclover dodder
♦ Noxious Weed
♦ 229
♦ C, parasitic.

**Cuscuta pentagona Engelm. var.
pentagona**
Cuscutaceae/Convolvulaceae
Cuscuta arvensis Beyr. ex Hook. (see),
Cuscuta pentagona Engelm. var. *calycina*
(see)
♦ five angled dodder
♦ Noxious Weed
♦ 229
♦ C, parasitic.

**Cuscuta pentagona Engelm. var.
pubescens (Engelm.) Yunck.**
Cuscutaceae/Convolvulaceae
♦ bushclover dodder
♦ Noxious Weed
♦ 229
♦ C, parasitic.

Cuscuta planiflora Ten.
Cuscutaceae/Convolvulaceae
♦ dodder, smallseed dodder, red
dodder, smallseed alfalfa dodder
♦ Weed, Quarantine Weed, Noxious
Weed, Naturalised, Environmental
Weed, Casual Alien
♦ 54, 67, 72, 76, 80, 86, 87, 88, 98, 155,
171, 198, 203, 218, 272, 280
♦ H parasitic, herbal. Origin:
Mediterranean.

Cuscuta plattensis A.Nelson
Cuscutaceae/Convolvulaceae
♦ prairie dodder
♦ Weed, Noxious Weed
♦ 67, 229
♦ C, parasitic.

Cuscuta polygonorum Engelm.
Cuscutaceae/Convolvulaceae
♦ polygonum dodder, smartweed
dodder
♦ Weed, Noxious Weed

♦ 67, 87, 88, 218, 229
♦ C, parasitic, herbal.

Cuscuta potosina Schaffn.
Cuscutaceae/Convolvulaceae
♦ globe dodder
♦ Noxious Weed
♦ 140, 229
♦ C, parasitic.

**Cuscuta potosina Schaffn. var. globifera
Yunck.**
Cuscutaceae/Convolvulaceae
♦ globe dodder
♦ Noxious Weed
♦ 140, 229
♦ C, parasitic.

Cuscuta pulchella Engelm.
Cuscutaceae/Convolvulaceae
♦ Weed
♦ 87, 88
♦ C, parasitic.

Cuscuta racemosa C.Mart.
Cuscutaceae/Convolvulaceae
♦ branching dodder, Chilean dodder
♦ Weed, Naturalised
♦ 86, 87, 88, 98, 198, 203, 255
♦ C, parasitic, arid. Origin: South
America.

**Cuscuta racemosa C.Mart. var. chiliana
Engelm.**
Cuscutaceae/Convolvulaceae
♦ Chilean dodder
♦ Weed
♦ 218
♦ C, parasitic.

Cuscuta reflexa Roxb.
Cuscutaceae/Convolvulaceae
♦ giant dodder
♦ Weed, Quarantine Weed, Noxious
Weed
♦ 35, 76, 87, 88, 203, 220, 229, 297
♦ C, parasitic, promoted, herbal.

Cuscuta rhodesiana Yunck.
Cuscutaceae/Convolvulaceae
♦ Weed
♦ 87, 88
♦ C, parasitic.

Cuscuta rostrata Shuttlew. ex Engelm.
Cuscutaceae/Convolvulaceae
♦ beaked dodder
♦ Weed, Noxious Weed
♦ 67, 229
♦ C, parasitic, herbal.

Cuscuta runyonii Yunck.
Cuscutaceae/Convolvulaceae
♦ Runyon's dodder
♦ Weed, Noxious Weed
♦ 67, 229
♦ C, parasitic.

Cuscuta salina Engelm.
Cuscutaceae/Convolvulaceae
♦ saltmarsh dodder
♦ Weed, Noxious Weed
♦ 67, 229
♦ aC, parasitic, herbal.

**Cuscuta salina Engelm. var. major
Yunck.**
Cuscutaceae/Convolvulaceae

♦ goldenthread
♦ Noxious Weed
♦ 229
♦ aC, parasitic.

Cuscuta salina Engelm. var. *papillata* Yunck.
Cuscutaceae/Convolvulaceae
♦ goldenthread
♦ Noxious Weed
♦ 229
♦ aC, parasitic.

Cuscuta salina Engelm. var. *salina*
Cuscutaceae/Convolvulaceae
♦ saltmarsh dodder
♦ Noxious Weed
♦ 229
♦ aC, parasitic.

Cuscuta sandwichiana Choisy
Cuscutaceae/Convolvulaceae
♦ sandwich dodder, kauna'oa
♦ Weed, Quarantine Weed, Noxious Weed
♦ 67, 76, 87, 88, 203, 218, 220, 229
♦ C, parasitic, herbal.

Cuscuta scandens Brot.
Cuscutaceae/Convolvulaceae
♦ aasianvieras
♦ Casual Alien
♦ 42
♦ C, parasitic.

Cuscuta squamata Engelm.
Cuscutaceae/Convolvulaceae
♦ scaleflower dodder
♦ Weed, Noxious Weed
♦ 67, 229
♦ C, parasitic.

Cuscuta suaveolens Engelm.
Cuscutaceae/Convolvulaceae
♦ dodder, lucerne dodder, luserndodder, fringed dodder, Chilean dodder, Chile dodder
♦ Weed, Noxious Weed, Naturalised, Casual Alien
♦ 23, 40, 50, 51, 63, 67, 70, 86, 87, 88, 95, 98, 101, 121, 171, 176, 198, 203, 229, 237, 272, 280, 283
♦ aH parasitic. Origin: Americas.

Cuscuta subinclusa Dur. & Hilg.
Cuscutaceae/Convolvulaceae
= *Cuscuta ceanothi* Behr
♦ canyon dodder
♦ Weed
♦ 87, 88
♦ aC, parasitic.

Cuscuta suksdorfii Yunck.
Cuscutaceae/Convolvulaceae
♦ mountain dodder
♦ Weed, Noxious Weed
♦ 67, 229
♦ C, parasitic, herbal.

Cuscuta suksdorfii Yunck. var. *subpedicellata* Yunck.
Cuscutaceae/Convolvulaceae
♦ mountain dodder
♦ Noxious Weed
♦ 229
♦ C, parasitic.

Cuscuta suksdorfii Yunck. var. *suksdorfii*
Cuscutaceae/Convolvulaceae
♦ Suksdorf's dodder, mountain dodder
♦ Noxious Weed
♦ 229
♦ C, parasitic.

Cuscuta trifolii Bab.
Cuscutaceae/Convolvulaceae
= *Cuscuta epithymum* (L.) Murr.
♦ clover dodder
♦ Weed
♦ 87, 88
♦ C, parasitic, herbal.

Cuscuta triumvirati Lange
Cuscutaceae/Convolvulaceae
♦ Weed
♦ 87, 88
♦ C, parasitic.

Cuscuta tuberculata Brandeg.
Cuscutaceae/Convolvulaceae
♦ tubercle dodder
♦ Weed, Noxious Weed
♦ 67, 229
♦ C, parasitic.

Cuscuta umbellata H.B.K.
Cuscutaceae/Convolvulaceae
♦ umbrella dodder, flatglobe dodder
♦ Weed, Noxious Weed
♦ 67, 87, 88, 218, 229
♦ C, parasitic, promoted, herbal.

Cuscuta umbrosa Beyr. ex Hook.
Cuscutaceae/Convolvulaceae
= *Cuscuta gronovii* Willd. ex J.A.Schult. var. *gronovii*
♦ Weed
♦ 67, 87, 88
♦ C, parasitic, herbal.

Cuscuta vetchii Brandegee
Cuscutaceae/Convolvulaceae
= *Cuscuta denticulata* Engelm. var. *vetchii* (Brandeg.) T.Beliz
♦ Weed
♦ 67
♦ C, parasitic.

Cuscuta vivipara T.Beliz, sp. *nov. ined.*
Cuscutaceae/Convolvulaceae
♦ love vine
♦ Noxious Weed
♦ 229
♦ C, parasitic.

Cuscuta warneri Yunck.
Cuscutaceae/Convolvulaceae
♦ Warner's dodder
♦ Weed
♦ 67
♦ C, parasitic.

Cutandia memphitica (Spreng.) Benth.
Poaceae
Cutandia scleropoides Willk., *Dactylis memphitica* Spreng., *Scleropoa memphitica* (Spreng.) Parl.
♦ Memphis grass
♦ Weed, Naturalised
♦ 101, 221
♦ G, arid.

Cyamopsis senegalensis Guill. & Perr.
Fabaceae/Papilionaceae
♦ Native Weed
♦ 121
♦ aH. Origin: Africa.

Cyamopsis tetragonoloba (L.) Taub.
Fabaceae/Papilionaceae
Cyamopsis psoralioides DC., *Dolichos fabiformis* L'Hér., *Dolichos psoraloides* Lam., *Lupinus trifoliatus* Cav., *Psoralea tetragonoloba* L.
♦ guar, cluster bean, Calcutta lucerne, Siam bean, cyamopse à quatre ailes
♦ Weed
♦ 88
♦ cultivated, herbal.

Cyananthus spathulifolius Nannf.
Campanulaceae
♦ Quarantine Weed
♦ 220

Cyanella amboensis Schinz
Liliaceae/Tecophilaeaceae
♦ Quarantine Weed
♦ 220
♦ pH, promoted. Origin: Africa.

Cyanella hyacinthoides L.
Liliaceae/Tecophilaeaceae
♦ lady's hand
♦ Weed, Naturalised, Native Weed, Environmental Weed
♦ 7, 86, 98, 121, 203
♦ pH, cultivated. Origin: South Africa.

Cyanella lutea L.f.
Liliaceae/Tecophilaeaceae
Cyanella odoratissima Lindl.
♦ five fingers
♦ Native Weed
♦ 121
♦ pH, cultivated. Origin: southern Africa.

Cyanopsis muricata (L.) Dostál
Asteraceae
Centaurea muricata L. (see)
♦ Morocco knapweed
♦ Naturalised
♦ 101

Cyanotis axillaris (L.) D.Don
Commelinaceae
Commelina axillaris L.
♦ phak plaap naa
♦ Weed
♦ 12, 87, 88, 126, 170, 204, 239
♦ pH, aqua, cultivated, herbal.

Cyanotis barbata Roem. & Schult.
Commelinaceae
♦ Weed
♦ 12

Cyanotis cristata (L.) D.Don
Commelinaceae
♦ Weed
♦ 87, 88, 170, 191, 297
♦ herbal.

Cyanotis cucullata Kunth
Commelinaceae
♦ Weed
♦ 87, 88

Cyanotis lanata **Benth.**
Commelinaceae
♦ Weed
♦ 87, 88

Cyanotis papilionacea **Roem. & Schult.**
Commelinaceae
♦ Weed, Quarantine Weed
♦ 76, 87, 88, 203, 220

Cyanotis speciosa **(L.f.) Hassk.**
Commelinaceae
♦ doll's powder puff, Job's tears, wandering Jew
♦ Native Weed
♦ 121
♦ pH. Origin: southern Africa.

Cyanotis villosa **Schult.f.**
Commelinaceae
♦ Weed, Quarantine Weed
♦ 76, 87, 88, 203, 220

Cyanthillium cinereum **(L.) H.Rob.**
Asteraceae
Vernonia cinerea (L.) Less (see)
♦ little ironweed
♦ Naturalised
♦ 32, 101
♦ Origin: Australia.

Cyathea cooperi **(Hook. ex F.Muell.) Dom.**
Cyatheaceae
Sphaeropteris cooperi (F.Muell.) R.M.Tryon (see)
♦ Cooper's cyathea, tree fern, Australian tree fern
♦ Weed, Sleeper Weed, Quarantine Weed, Naturalised, Environmental Weed
♦ 3, 22, 80, 88, 101, 151, 203, 225, 246
♦ T, cultivated. Origin: Australia.

Cyathea nigricans **Mett.**
Cyatheaceae
♦ katar
♦ Introduced
♦ 230
♦ T.

Cyathea ponapeana **(Hosk.) Glassman**
Cyatheaceae
♦ Introduced
♦ 230
♦ T.

Cyathocline purpurea **(D.Don) Kuntze**
Asteraceae
♦ Weed
♦ 66

Cyathula achyranthoides **(Kunth) Moq.**
Amaranthaceae
Achyranthes geminata Thonn., *Cyathula geminata* (Thonn.) Moq., *Desmochaeta achyranthoides* Kunth, *Desmochaeta densiflora* Kunth
♦ Weed, Introduced
♦ 32, 38, 87, 88
♦ arid, herbal. Origin: Madagascar.

Cyathula cylindrica **Moq.**
Amaranthaceae
Cyathula albida Lopr., *Cyathula distorta* (Heirn) C.B.Clarke, *Cyathula mannii* Baker, *Cyathula schimperiana* Moq.,

Desmochaeta distorta Hiern, *Pupalia huillensis* Hiern
♦ Weed, Quarantine Weed
♦ 76, 87, 88, 203, 220
♦ arid, herbal.

Cyathula natalensis **Sond.**
Amaranthaceae
Cyathula spathulifolia Lopr.
♦ Native Weed
♦ 121
♦ pH. Origin: southern Africa.

Cyathula polycephala **Bak.**
Amaranthaceae
Cyathula cordifolia Chiov., *Cyathula echinulata* Hauman, *Cyathula schimperiana* Moq. var. *tomentosa* Suess.
♦ Weed
♦ 87, 88
♦ arid.

Cyathula prostrata **(L.) Bl.**
Amaranthaceae
Achyranthes prostrata L., *Cyathula pedicellata* C.B.Clarke, *Cyathula prostrata* (L.) Blume fo. *pedicellata* (C.B.Clarke) Hauman, *Desmochaeta prostrata* (L.) DC., *Pupalia prostrata* (L.) C.Mart.
♦ pasture weed, osonlikantikap, yaa phannguu daeng
♦ Weed, Quarantine Weed, Naturalised, Introduced
♦ 32, 38, 76, 87, 88, 101, 157, 203, 206, 209, 220, 238, 243, 255, 261, 276
♦ pH, arid, cultivated, herbal. Origin: tropical Asia and Africa.

Cyathula uncinulata **(Schrad.) Schinz**
Amaranthaceae
Achyranthes uncinulata Schrad., *Cyathula globulifera* Moq.
♦ globe cyathula
♦ Weed, Native Weed
♦ 51, 87, 88, 121, 158
♦ pH, arid. Origin: southern Africa.

Cycas **L. spp.**
Cycadaceae
♦ zamia, tree zamia, zamia palm, cycad, nut palm, dwarf sago, sago palm
♦ Weed, Noxious Weed, Naturalised
♦ 86, 88, 147
♦ cultivated, toxic.

Cycas circinalis **L.**
Cycadaceae
♦ queen sago, sago palm, cica
♦ Weed, Naturalised, Introduced
♦ 39, 101, 161, 179, 230, 261, 276
♦ S, cultivated, herbal, toxic.

Cycas revoluta **Thunb.**
Cycadaceae
♦ sago palm, Japanese sago palm, sago cycas, alcanfor, cica, plama de sagú
♦ Weed, Naturalised, Introduced
♦ 39, 101, 154, 161, 247, 261
♦ T, cultivated, herbal, toxic. Origin: tropical Asia.

Cyclamen latifolium **Sibth. & Sm.**
Primulaceae

Cyclamen persicum Mill. (see)
♦ Introduced
♦ 39, 228
♦ arid, toxic.

Cyclamen persicum **Mill.**
Primulaceae
= *Cyclamen latifolium* Sibth. & Sm.
♦ florist's cyclamen, cyclamen
♦ Weed
♦ 161, 194, 247
♦ aH, cultivated, herbal, toxic. Origin: North Africa, Middle East, Aegean region.

Cyclamen purpurascens **Mill.**
Primulaceae
♦ common cyclamen
♦ Weed
♦ 161
♦ cultivated, toxic.

Cyclanthera hystrix **Gill. ex Hook & Arn**
Cucurbitaceae
♦ Weed
♦ 237

Cyclea burmanni **Arn. ex Wight**
Menispermaceae
♦ Weed
♦ 87, 88

Cyclocarpa stellaris **Baker**
Fabaceae/Papilionaceae
♦ Naturalised
♦ 86
♦ Origin: tropical Africa, Malaysia.

Cycloloma atriplicifolium **(Spreng.) Coult.**
Chenopodiaceae
Chenopodium atriplicifolium (Spreng.) A.Ludw., *Cyclolepis platyphylla* (Michx.) Moq., *Cycloloma platyphylla* (Michx.) Moq. (see), *Kochia atriplicifolia* (Spreng.) Roth, *Kochia dentata* Willd., *Salsola atriplicifolia* Spreng., *Salsola platyphylla* Michx.
♦ winged pigweed, yuyo rodante, cicloloma comune
♦ Weed, Naturalised, Introduced
♦ 34, 86, 87, 88, 98, 133, 161, 195, 198, 203, 218, 228, 237, 295
♦ aH, arid, promoted, herbal. Origin: North America.

Cycloloma platyphylla **(Michx.) Moq.**
Chenopodiaceae
= *Cycloloma atriplicifolium* (Spreng.) Coult.
♦ Weed
♦ 87, 88

Cyclosorus afrus **(Christ) Ching**
Thelypteridaceae
♦ Weed
♦ 87, 88

Cyclosorus aridus **(Don) Ching**
Thelypteridaceae
Dryopteris arida Kuntze
♦ Weed
♦ 87, 88

Cyclosorus dentatus **(Forssk.) Ching**
Thelypteridaceae
= *Thelypteris dentata* (Forssk.) E.St.John

- lynyolo
- Weed
- 87, 88
- cultivated, herbal.

Cyclosorus gongylodes (Schkuhr) Link
Thelypteridaceae
= *Thelypteris interrupta* (Willd.) K.Iwats. (NoR)
- Weed
- 87, 88

Cyclosorus parasiticus (L.) Farw.
Thelypteridaceae
- Weed
- 87, 88

Cyclospermum leptophyllum (Pers.) Sprague ex Britt. & Wilson
Apiaceae
Apium ammi Urb. (see), *Apium leptophyllum* (Pers.) F.Muell. (see), *Ciclospermum leptophyllum* (Pers.) Britton & E.Wilson (see), *Pimpinella leptophylla* Pers.
- lawn celery, slender celery, wild celery, wild carrot, marsh parsley, perejil de pantano
- Weed, Naturalised
- 55, 101, 179, 237, 261
- Origin: tropical America.

Cyclotella atomus Hust.
Bacillariophyceae
- Weed
- 197
- diatom.

Cyclotella cryptica Reim. et al.
Bacillariophyceae
- Weed
- 197
- diatom.

Cyclotella pseudostelligera Hust.
Bacillariophyceae
- Weed
- 197
- diatom.

Cycnium adonense E.Mey. ex Benth.
Scrophulariaceae
- blotting paperflower, inkplant, mushroom flower
- Native Weed
- 121
- pH parasitic. Origin: southern Africa.

Cydonia oblonga Mill.
Rosaceae
- quince
- Weed, Naturalised, Garden Escape
- 39, 86, 98, 101, 198, 203, 280
- T, cultivated, herbal, toxic. Origin: Mediterranean, Middle East.

Cydonia sinensis Thouin
Rosaceae
= *Pseudocydonia sinensis* (Dum.Cours.) Schneid.
- Chinese quince
- Casual Alien
- 280
- cultivated, herbal.

Cylindropuntia (Engelm.) F.M.Knuth spp.
Cactaceae
= *Opuntia* Mill. spp.
- Weed, Quarantine Weed
- 76, 88, 203

Cylindropuntia imbricata (Haw.) Kunth
Cactaceae
= *Opuntia imbricata* (Harv.) DC.
- devil's rope, chain link cactus, devil's rope cactus
- Weed
- 269
- arid.

Cylindropuntia tunicata (Lehm.) Kunth
Cactaceae
= *Opuntia tunicata* (Lehm.) Link & Otto
- Weed
- 14
- cultivated.

Cymbalaria muralis Gaertn., Mey. & Scherb.
Scrophulariaceae
Antirrhinum cymbalaria L., *Cymbalaria cymbalaria* Wettst., *Linaria cymbalaria* (L.) Mill.
- Kenilworth ivy, Coliseum ivy, ivyleaf toadflax, mother of a thousand, pennywort, kilkkaruoho
- Weed, Naturalised, Garden Escape, Environmental Weed, Cultivation Escape
- 7, 9, 15, 34, 40, 42, 70, 86, 87, 88, 98, 101, 121, 165, 176, 198, 203, 218, 241, 280, 286, 287, 295, 300
- a/pC, cultivated, herbal. Origin: Eurasia.

Cymbaria mongolica Maxim.
Scrophulariaceae
- Weed
- 275, 297
- pH.

Cymbispatha comelinoides Roem. & Schult.
Commelinaceae
- Weed
- 199

Cymbocarpa refracta Miers
Burmanniaceae
- cat's milk
- Naturalised
- 101

Cymbonotus lawsonianus Gaudich.
Asteraceae
- bear's ear
- Weed, Native Weed
- 87, 88, 269
- pH, arid, cultivated. Origin: Australia.

Cymbonotus preissianus Steetz
Asteraceae
- austral bear's ear
- Native Weed
- 269
- pH, cultivated. Origin: Australia.

Cymbopogon afronardus Stapf
Poaceae
- Weed
- 87, 88
- G.

Cymbopogon caesius (Nees ex Hook. & Arn.) Stapf
Poaceae
Andropogon caesius Nees ex Hook. & Arn.
- Weed
- 87, 88
- G, arid.

Cymbopogon citratus (DC. ex Nees) Stapf
Poaceae
- lemon grass, lady's mantle, lion's foot, citronella
- Weed, Naturalised, Cultivation Escape
- 86, 87, 88, 98, 101, 179, 203, 241, 261, 300
- G, cultivated, herbal. Origin: India.

Cymbopogon coloratus (Nees) Stapf
Poaceae
Andropogon coloratus Nees
- Introduced
- 228
- G, arid.

Cymbopogon excavatus (Hochst.) Stapf ex Burtt Davy
Poaceae
- buchu grass, common turpentine grass, eau de Cologne grass, ginger grass, lemon grass, lemon scented grass, turpentine grass
- Native Weed
- 121
- pG. Origin: southern Africa.

Cymbopogon giganteus (Hochst.) Chiov.
Poaceae
Andropogon connatus (A.Rich.) Chiov. var. *benearmatus* Chiov., *Andropogon giganteus* Hochst. (*non* Ten.)
- Weed
- 87, 88
- G, arid.

Cymbopogon jwarancusa (Jones) Schult.
Poaceae
Andropogon himalayensis Gand., *Andropogon jwarancusa* Jones
- Iwarancusa grass
- Naturalised
- 101
- G, arid, herbal.

Cymbopogon marginatus (Steud.) Stapf ex Burtt Davy
Poaceae
- dobo grass, khuskhus, lemon grass, scented turpentine grass, tambookie grass
- Native Weed
- 121
- pG. Origin: southern Africa.

Cymbopogon martinii (Roxb.) J.F.Watson
Poaceae
Andropogon martinii Roxb., *Andropogon schoenanthus* L. var. *martinii* Hook.f.
- palmarosa, geranium grass, rosha grass
- Naturalised, Introduced
- 86, 228
- G, arid, promoted, herbal. Origin: India.

Cymbopogon nardus (L.) Rendle
Poaceae
♦ citronella grass
♦ Naturalised, Cultivation Escape
♦ 101, 261
♦ G, cultivated, herbal. Origin: India.

Cymbopogon nervatus (Hochst.) Chiov.
Poaceae
♦ Weed
♦ 242
♦ G, arid.

Cymbopogon plurinodis (Stapf) Stapf ex Burtt Davy
Poaceae
♦ bitter turpentine grass, bushveld turpentine grass
♦ Native Weed
♦ 121
♦ pG. Origin: southern Africa.

Cymbopogon proximus (Hochst.) Stapf
Poaceae
Andropogon proximus Hochst. ex A.Rich.
♦ Weed, Quarantine Weed
♦ 76, 87, 88, 203, 220, 221
♦ G, herbal. Origin: Africa.

Cymbopogon refractus (R.Br.) A.Camus
Poaceae
♦ barbwire grass
♦ Weed, Noxious Weed, Naturalised
♦ 87, 88, 101, 218, 229
♦ G, arid, cultivated, herbal. Origin: Australia.

Cymbopogon schoenanthus (L.) Spreng.
Poaceae
Andropogon schoenanthus L.
♦ camel grass
♦ Weed
♦ 87, 88, 221
♦ G, herbal.

Cymbopogon tortilis Hitchc. var. goeringii Mand.-Hazz.
Poaceae
♦ Weed
♦ 286
♦ G.

Cymbopogon validus (Stapf) ex Burtt Davy
Poaceae
♦ giant turpentine grass, tambookie grass, tambuti
♦ Weed, Native Weed
♦ 87, 88, 121
♦ pG. Origin: southern Africa.

Cymodocea ciliata (Setch.) Ehrenb. ex Asch.
Cymodoceaceae
Zostera marina L. fo. *latifolia* Setch.
♦ Weed
♦ 221

Cymodocea major Grande
Cymodoceaceae
♦ Weed
♦ 221
♦ herbal.

Cymophora accedens (S.F.Blake) B.L.Turner & Powell

Asteraceae
♦ Naturalised
♦ 101

Cymopterus watsonii (Coult. & Rose) Jones
Apiaceae
= *Cymopterus ibapensis* M.E.Jones (NoR)
♦ spring parsley
♦ Weed
♦ 39, 87, 88, 218
♦ toxic.

Cynanchum L. spp.
Asclepiadaceae/Apocynaceae
♦ dog strangling vine, swallowwort
♦ Environmental Weed
♦ 104

Cynanchum acutum L.
Asclepiadaceae/Apocynaceae
♦ stranglewort
♦ Weed
♦ 39, 87, 88, 185, 221, 243, 272
♦ pC, cultivated, herbal, toxic.

Cynanchum auriculatum Royle & Wight
Asclepiadaceae/Apocynaceae
♦ auriculate swallowwort
♦ Weed
♦ 297
♦ herbal.

Cynanchum caribaeum Alain
Asclepiadaceae/Apocynaceae
♦ Weed
♦ 14

Cynanchum caudatum (Miq.) Maxim.
Asclepiadaceae/Apocynaceae
♦ ikema
♦ Weed
♦ 286
♦ pC, promoted, herbal. Origin: China, Japan.

Cynanchum chinense R.Br.
Asclepiadaceae/Apocynaceae
♦ Chinese swallowwort
♦ Weed
♦ 275, 297
♦ pH.

Cynanchum compactum Choux
Asclepiadaceae/Apocynaceae
♦ Weed, Quarantine Weed
♦ 76, 88, 220
♦ cultivated.

Cynanchum ellipticum (Harv.) R.A.Dyer
Asclepiadaceae/Apocynaceae
Cynanchum capense Thunb.
♦ excelsior, monkey rope
♦ Weed, Native Weed
♦ 39, 121
♦ pC, toxic. Origin: South Africa.

Cynanchum komarovii Ijinsk.
Asclepiadaceae/Apocynaceae
♦ Weed
♦ 33, 275, 297
♦ pH, toxic.

Cynanchum louiseae Kartesz & Gandhi
Asclepiadaceae/Apocynaceae
Cynanchum nigrum (L.) Pers. *non* Cav. (see), *Vincetoxicum nigrum* (L.) Moench.

(see)
♦ Louis's swallowwort
♦ Naturalised
♦ 101

Cynanchum madagascariense K.Schum.
Asclepiadaceae/Apocynaceae
♦ Weed, Quarantine Weed
♦ 76, 88, 220

Cynanchum marnieranum Rauh
Asclepiadaceae/Apocynaceae
♦ Weed, Quarantine Weed
♦ 76, 88, 220
♦ arid, cultivated.

Cynanchum napiforme napiferum Choux
Asclepiadaceae/Apocynaceae
♦ Weed, Quarantine Weed
♦ 76, 88, 220

Cynanchum nigrum (L.) Pers., non Cav.
Asclepiadaceae/Apocynaceae
= *Cynanchum louiseae* Kartesz & Gandhi
♦ black swallowwort
♦ Weed, Quarantine Weed
♦ 4, 76, 87, 88, 132, 161, 211, 218, 220
♦ herbal.

Cynanchum perrieri Choux
Asclepiadaceae/Apocynaceae
♦ Weed, Quarantine Weed
♦ 76, 88, 220
♦ arid. Origin: Madagascar.

Cynanchum rossicum (Kleo.) Barb.
Asclepiadaceae/Apocynaceae
Vincetoxicum rossicum (Kleo.) Barb. (see)
♦ dog strangling vine, swallowwort, European swallowwort
♦ Weed, Naturalised, Environmental Weed
♦ 80, 80, 88, 101, 132, 151

Cynanchum sauvallei Alain
Asclepiadaceae/Apocynaceae
♦ Weed
♦ 14

Cynanchum savannarum Alain
Asclepiadaceae/Apocynaceae
♦ Weed
♦ 14

Cynanchum sibiricum Willd.
Asclepiadaceae/Apocynaceae
♦ Weed
♦ 275
♦ pH, cultivated, herbal. Origin: east Asia.

Cynanchum thesioides (Freyn) K.Schum.
Asclepiadaceae/Apocynaceae
♦ Weed
♦ 275, 297
♦ pH.

Cynanchum versicolor Bunge
Asclepiadaceae/Apocynaceae
♦ versicolorous swallowwort
♦ Weed
♦ 297

Cynanchum vincetoxicum (L.) Pers.
Asclepiadaceae/Apocynaceae
Vincetoxicum hirundinaria Medik. (see)

♦ white swallowwort
♦ Weed, Naturalised
♦ 87, 88, 101, 161, 218
♦ herbal, toxic.

Cynara L. spp.
Asteraceae
♦ cynara
♦ Weed, Naturalised
♦ 18, 198

Cynara algarbiensis Coss.
Asteraceae
♦ Weed
♦ 70

Cynara cardunculus L.
Asteraceae
♦ artichoke thistle, cardoon, wild
artichoke, Scotch thistle
♦ Weed, Quarantine Weed, Noxious
Weed, Naturalised, Introduced,
Garden Escape, Environmental Weed
♦ 7, 34, 35, 39, 62, 70, 72, 76, 78, 80, 86,
87, 88, 98, 101, 116, 142, 147, 151, 161,
171, 198, 203, 229, 231, 236, 237, 241,
251, 252, 269, 271, 280, 289, 295, 296,
300
♦ pH, arid, cultivated, herbal, toxic.
Origin: Mediterranean.

Cynara humilis L.
Asteraceae
♦ Weed
♦ 70, 87, 88
♦ cultivated, herbal.

Cynara scolymus L.
Asteraceae
♦ globe artichoke
♦ Weed, Naturalised, Cultivation
Escape
♦ 7, 15, 34, 86, 98, 101, 203, 252, 280
♦ pH, cultivated, herbal. Origin:
horticultural, possibly a form of
Cynara cardunculus L.

Cynara sibthorpiana Boiss. & Heldr.
Asteraceae
♦ Weed
♦ 221

Cynara syriaca Boiss.
Asteraceae
♦ Weed
♦ 87, 88

Cynara tournefortii Boiss. & Reut.
Asteraceae
♦ Weed
♦ 70

Cynodon L.C.Rich. spp.
Poaceae
♦ Bermuda grass
♦ Weed, Noxious Weed
♦ 35, 88, 243
♦ G.

Cynodon aethiopicus Clayton & Harlan
Poaceae
♦ cynodon couch, giant quickgrass,
giant stargrass, stargrass, kapamba,
Ethiopian dog's tooth grass, Bermuda
grass
♦ Weed, Noxious Weed, Naturalised,
Introduced

♦ 39, 55, 86, 98, 101, 121, 203, 228, 229
♦ pG, arid, cultivated, toxic. Origin:
tropical Africa.

Cynodon arcuatus Presl
Poaceae
♦ giant couchgrass
♦ Weed
♦ 276
♦ pG.

Cynodon dactylon (L.) Pers.
Poaceae
Capriola dactylon (L.) Kuntze, *Milium
dactylon* Moench, *Panicum dactylon* L.,
Paspalum umbellatum Lam.
♦ Bermuda grass, coarse kweek,
common couch, common quickgrass,
couchgrass, devil's grass, dog's tooth,
doob grass, dub grass, finegrass,
fingergrass, fingers, Florida grass,
Indian couch, quickgrass, running
grass, Scotch grass, stargrass, twitch
grass, white quickgrass, wiregrass,
Indian doab, grama Bermuda
♦ Weed, Sleeper Weed, Noxious Weed,
Naturalised, Introduced, Garden
Escape, Environmental Weed, Casual
Alien
♦ 3, 6, 14, 15, 30, 34, 38, 39, 42, 44, 50,
51, 55, 66, 70, 72, 80, 87, 88, 91, 93, 101,
107, 111, 115, 116, 118, 121, 134, 141,
151, 157, 158, 161, 167, 170, 176, 179,
180, 185, 186, 204, 205, 207, 210, 211,
212, 218, 219, 221, 225, 228, 229, 236,
237, 240, 241, 242, 243, 244, 245, 246,
248, 250, 253, 255, 257, 262, 263, 266,
269, 270, 272, 273, 274, 275, 276, 280,
281, 286, 295, 296, 297, 300
♦ pG, arid, cultivated, herbal, toxic.
Origin: tropical Africa.

Cynodon dactylon (L.) Pers. var. *dactylon*
Poaceae
♦ couch, Bermuda grass
♦ Naturalised, Introduced,
Environmental Weed
♦ 86, 198, 230
♦ G.

Cynodon hirsutus Stent
Poaceae
♦ dog grass, hairy couch, hairy
quickgrass, red quickgrass, Transvaal
quickgrass
♦ Weed, Native Weed
♦ 87, 88, 121, 237, 295
♦ pG, toxic. Origin: southern Africa.

Cynodon incompletus Nees
Poaceae
♦ dog grass, fine couch, fine quick,
hairy quickgrass, karoo quickgrass, red
quickgrass
♦ Weed, Naturalised, Native Weed
♦ 39, 98, 121, 203
♦ pG, toxic. Origin: South Africa.

Cynodon incompletus Nees var. *incompletus*
Poaceae
♦ Naturalised
♦ 86

♦ G, arid. Origin: South Africa.

Cynodon magennisii Hurcombe
Poaceae
♦ Bermuda grass, Magennis' dog's
tooth grass
♦ Noxious Weed, Naturalised
♦ 101, 229
♦ G.

Cynodon nlemfuensis Vanderyst
Poaceae
♦ East African couch, East African
stargrass, giant quickgrass, giant
stargrass, robust stargrass, stargrass,
Bermuda grass, African Bermudagrass
♦ Weed, Noxious Weed, Naturalised,
Cultivation Escape
♦ 39, 53, 88, 98, 101, 121, 158, 203, 209,
229, 238, 261
♦ pG, cultivated, toxic. Origin: tropical
East Africa.

Cynodon nlemfuensis Vanderyst var. *nlemfuensis*
Poaceae
♦ African Bermudagrass, Bermuda
grass
♦ Noxious Weed, Naturalised
♦ 86, 101, 229
♦ G. Origin: tropical Africa.

Cynodon nlemfuensis Vanderyst var. *robustus* W.D.Clayton & Harlan
Poaceae
♦ African Bermudagrass, Bermuda
grass
♦ Noxious Weed, Naturalised
♦ 86, 101, 229
♦ G. Origin: tropical Africa.

Cynodon plectostachyus (Schum.) Pilg.
Poaceae
♦ stargrass, feathery couch, giant
quickgrass, giant stargrass, Bermuda
grass, zacate estrella
♦ Weed, Noxious Weed, Naturalised,
Introduced
♦ 34, 39, 87, 88, 101, 121, 229, 255, 281,
287
♦ pG, cultivated, herbal, toxic. Origin:
Africa.

Cynodon radiatus Roth ex Roem. & Schult.
Poaceae
♦ Weed, Naturalised
♦ 86, 93, 191
♦ G. Origin: Madagascar.

Cynodon transvaalensis Burtt Davy
Poaceae
♦ soft couch, African dog's tooth grass,
Bermuda grass, African Bermudagrass
♦ Weed, Noxious Weed, Naturalised
♦ 86, 98, 101, 198, 203, 229
♦ G, cultivated. Origin: South Africa.

Cynoglossum L. spp.
Boraginaceae
♦ hound's tongue
♦ Weed
♦ 272
♦ herbal.

**Cynoglossum amabile Stapf &
J.R.Drumm.**
　　Boraginaceae
　　♦ Chinese hound's tongue, Chinese
　　forget me not, tiibetinkoirankieli
　　♦ Weed, Naturalised, Cultivation
　　Escape, Casual Alien
　　♦ 42, 101, 199, 261, 275, 280, 297
　　♦ aH, cultivated, herbal. Origin: west
　　China, northern India.

Cynoglossum australe R.Br.
　　Boraginaceae
　　♦ Naturalised
　　♦ 134
　　♦ cultivated.

Cynoglossum cheirifolium L.
　　Boraginaceae
　　♦ Weed
　　♦ 70
　　♦ bH, cultivated.

Cynoglossum clandestinum Desf.
　　Boraginaceae
　　♦ Weed
　　♦ 70, 88, 94

Cynoglossum coeruleum Hochst. ex DC.
　　Boraginaceae
　　♦ Weed, Quarantine Weed
　　♦ 76, 87, 88, 203, 220

Cynoglossum creticum Mill.
　　Boraginaceae
　　♦ blue hound's tongue
　　♦ Weed, Sleeper Weed, Naturalised,
　　Environmental Weed
　　♦ 54, 70, 86, 87, 88, 94, 155, 237, 241,
　　272, 295, 300
　　♦ bH, cultivated, herbal.

Cynoglossum divaricatum Steph.
　　Boraginaceae
　　♦ divaricate hound's tongue
　　♦ Weed
　　♦ 297

Cynoglossum enerve Turcz.
　　Boraginaceae
　　Cynoglossum hispidum Thunb. (see)
　　♦ hound's tongue, dog's tongue
　　♦ Weed, Native Weed
　　♦ 51, 87, 88, 121, 158
　　♦ pH, herbal. Origin: South Africa.

**Cynoglossum geometricum Bak. &
C.H.Wright**
　　Boraginaceae
　　♦ Weed
　　♦ 87, 88

**Cynoglossum glochidiatum Wall. ex
Benth.**
　　Boraginaceae
　　♦ intiankoirankieli, prickly hound's
　　tongue
　　♦ Naturalised, Cultivation Escape
　　♦ 42, 101, 243
　　♦ cultivated, herbal.

Cynoglossum hispidum Thunb.
　　Boraginaceae
　　= *Cynoglossum enerve* Turcz.
　　♦ Weed
　　♦ 88

Cynoglossum lanceolatum Forssk.
　　Boraginaceae
　　♦ hound's tongue
　　♦ Weed, Native Weed
　　♦ 23, 51, 87, 88, 121, 158
　　♦ pH, herbal. Origin: southern Africa.

Cynoglossum limense Willd.
　　Boraginaceae
　　♦ Naturalised
　　♦ 241, 300

Cynoglossum microglochin Benth.
　　Boraginaceae
　　♦ smallbristle hound's tongue
　　♦ Naturalised
　　♦ 101

Cynoglossum officinale L.
　　Boraginaceae
　　Cynoglossum hybridum Thuill.
　　♦ hound's tongue, common hound's
　　tongue, gypsyflower, purple hound's
　　tongue, beggar's lice, rohtokoirankielet
　　♦ Weed, Noxious Weed, Naturalised,
　　Introduced, Environmental Weed
　　♦ 1, 21, 23, 34, 39, 49, 70, 80, 87, 88, 94,
　　101, 136, 138, 146, 151, 159, 161, 162,
　　174, 180, 210, 212, 218, 219, 229, 264,
　　272, 291
　　♦ pH, cultivated, herbal, toxic. Origin:
　　Europe.

Cynoglossum pictum Sol.
　　Boraginaceae
　　♦ Weed
　　♦ 87, 88
　　♦ herbal.

Cynoglossum suaveolens R.Br.
　　Boraginaceae
　　♦ sweet hound's tongue
　　♦ Weed
　　♦ 269
　　♦ pH, arid, cultivated. Origin:
　　Australia.

Cynoglossum virginianum L.
　　Boraginaceae
　　♦ wild comfrey
　　♦ Weed
　　♦ 8, 23, 88
　　♦ herbal, toxic.

**Cynoglossum zeylanicum (Vahl) Thunb.
ex Lehm.**
　　Boraginaceae
　　♦ Ceylon hound's tongue
　　♦ Weed, Naturalised, Casual Alien
　　♦ 101, 179, 241, 261, 300
　　♦ herbal. Origin: India.

Cynomorium coccineum L.
　　Balanophoraceae/Cynomoriaceae
　　♦ Maltese mushroom
　　♦ Weed
　　♦ 88, 221
　　♦ H parasitic, herbal.

Cynosurus cristatus L.
　　Poaceae
　　Phleum cristatum Scop.
　　♦ crested dog's tail grass, crested dog's
　　tail
　　♦ Weed, Naturalised
　　♦ 7, 15, 70, 86, 87, 88, 91, 98, 101, 176,

198, 203, 218, 241, 272, 280, 287, 300
　　♦ pG, cultivated, herbal. Origin:
　　Eurasia.

Cynosurus echinatus L.
　　Poaceae
　　Phalona echinata Dumort.
　　♦ bristly dog's tail grass, rough dog's
　　tail, hedgehog dog's tail grass, annual
　　dog's tail, vihnesukapää
　　♦ Weed, Naturalised, Environmental
　　Weed, Casual Alien
　　♦ 7, 9, 15, 30, 34, 40, 42, 70, 72, 86, 87,
　　88, 91, 98, 101, 111, 121, 151, 161, 176,
　　198, 203, 241, 243, 250, 272, 280, 287,
　　300
　　♦ aG, cultivated, herbal. Origin:
　　Eurasia.

Cyperus L. spp.
　　Cyperaceae
　　♦ nutsedge, flatsedge
　　♦ Weed
　　♦ 221, 243
　　♦ G, herbal.

Cyperus acicularis Schrad. ex Nees
　　Cyperaceae
　　= *Cyperus odoratus* L.
　　♦ Weed
　　♦ 88
　　♦ G.

Cyperus aggregatus (Willd.) Endl.
　　Cyperaceae
　　Cyperus flavus (Vahl) Nees (see),
　　Mariscus flavus Vahl (see), *Cyperus
　　cayennensis* (Lam.) Britt. nom. illeg. (see)
　　♦ inflatedscale flatsedge
　　♦ Naturalised
　　♦ 86
　　♦ G. Origin: Mexico to Argentina.

Cyperus albostriatus Schrad.
　　Cyperaceae
　　♦ dwarf umbrella plant
　　♦ Weed, Naturalised
　　♦ 15, 86, 98, 134, 203, 280
　　♦ G, cultivated. Origin: South Africa.

Cyperus alopecuroides Rottb.
　　Cyperaceae
　　Juncellus alopecuroides (Rottb.)
　　C.B.Clarke (see)
　　♦ foxtail flatsedge
　　♦ Weed, Naturalised
　　♦ 80, 87, 88, 101, 179, 185, 221, 287
　　♦ G, cultivated. Origin: Madagascar.

Cyperus alternifolius L.
　　Cyperaceae
　　Cyperus flabelliformis Rottb. nom. illeg.,
　　Cyperus racemosus Poir.
　　♦ umbrella plant, umbrella flatsedge
　　♦ Weed, Naturalised, Garden Escape
　　♦ 23, 80, 87, 88, 112, 126, 194, 218, 261,
　　286, 287, 300
　　♦ pG, aqua, cultivated, herbal, toxic.
　　Origin: east and southern Africa,
　　Madagascar.

Cyperus alulatus Kern.
　　Cyperaceae
　　♦ Weed
　　♦ 66, 243
　　♦ G.

Cyperus amabilis Vahl
Cyperaceae
♦ foothill flatsedge
♦ Weed, Quarantine Weed
♦ 76, 87, 88, 126, 157, 203, 220
♦ G.

Cyperus amuricus Maxim.
Cyperaceae
Cyperus microiria Steud. (see)
♦ Asian flatsedge
♦ Weed, Naturalised
♦ 87, 88, 101, 275, 286, 297
♦ G.

Cyperus amuricus Maxim. var. *laxus*
Nakai
Cyperaceae
♦ Weed
♦ 204
♦ G.

Cyperus arenarius Retz.
Cyperaceae
♦ Weed, Naturalised
♦ 86, 98, 203
♦ G, arid. Origin: Asia.

Cyperus aristatus Rottb.
Cyperaceae
= *Cyperus squarrosus* L.
♦ awned cyperus
♦ Weed
♦ 87, 88, 180, 243
♦ pG, promoted, herbal.

Cyperus aromaticus (Ridl.) Mattf. & Kük.
Cyperaceae
Kyllinga aromatica Ridl.
♦ navua sedge
♦ Weed, Quarantine Weed, Noxious
Weed, Naturalised
♦ 54, 76, 86, 88, 147, 191, 203, 220
♦ G. Origin:
tropical Africa, Madagascar, Mauritius,
Seychelles.

Cyperus articulatus L.
Cyperaceae
♦ jointed flatsedge, chintul, aldure,
guinea rush, silolo
♦ Weed, Naturalised, Native Weed
♦ 14, 39, 87, 88, 121, 126, 185, 218, 221,
287, 300
♦ pG, aqua, herbal, toxic. Origin:
Africa.

Cyperus babakan Steud.
Cyperaceae
Cyperus babakensis C.B.Clarke
♦ Weed
♦ 87, 88, 170, 191
♦ G.

Cyperus bifax C.B.Clarke
Cyperaceae
♦ giant nutgrass
♦ Weed
♦ 276
♦ pG.

Cyperus blysmoides C.B.Clarke
Cyperaceae
Cyperus bulbosus Vahl var. *spicatus*
Boeck.
♦ watergrass

♦ Weed
♦ 53, 88
♦ pG.

Cyperus brevifolius (Rottb.) Endl. ex
Hassk.
Cyperaceae
= *Kyllinga brevifolia* Rottb.
♦ junquinho, capim de uma so cabeca,
Mullumbimby couch, globe kyllinga
♦ Weed, Naturalised, Native Weed,
Garden Escape, Environmental Weed
♦ 7, 15, 86, 87, 93, 98, 170, 198, 203, 204,
209, 217, 243, 249, 255, 262, 269, 276,
280, 286
♦ pG, cultivated, herbal. Origin:
pantropical.

Cyperus brevifolius (Rottb.) Endl. ex
Hassk. var. *leiolepis* (Franch. & Savigny)
T.Koyama
Cyperaceae
= *Kyllinga gracillima* Miq. (NoR) [see
Kyllinga brevifolioides (Delahoussaye &
Thieret) G.C.Tucker]
♦ Weed
♦ 263, 286
♦ pG.

Cyperus bulbosus Vahl
Cyperaceae
♦ galingale
♦ Weed
♦ 87, 88
♦ G, cultivated, herbal.

Cyperus capitatus Vand.
Cyperaceae
♦ Weed
♦ 221
♦ G.

Cyperus castaneus Willd.
Cyperaceae
♦ Weed
♦ 87, 88
♦ G. Origin: Australia.

Cyperus cayennensis (Lam.) Britt. nom.
illeg.
Cyperaceae
= *Cyperus aggregatus* (Willd.) Endl.
♦ cípero de cayena
♦ Weed
♦ 237, 295
♦ G.

Cyperus cephalotes Vahl
Cyperaceae
♦ Weed
♦ 88, 170, 191
♦ G. Origin: Australia.

Cyperus compactus Retz.
Cyperaceae
Cyperus dilutus Vahl, *Mariscus
compactus* (Retz.) Bold. (see), *Mariscus
dilutus* Nees, *Mariscus microcephalus* J.
& C.Presl
♦ yaa bai khom
♦ Weed
♦ 87, 88, 170, 191, 239, 262
♦ pG, aqua, herbal.

Cyperus compressus L.
Cyperaceae

♦ flatsedge, annual sedge, poorland
flatsedge
♦ Weed, Naturalised, Introduced,
Environmental Weed
♦ 7, 23, 28, 66, 86, 87, 88, 93, 98, 121,
126, 161, 170, 199, 203, 204, 230, 243,
249, 257, 262, 263, 273, 274, 286, 297
♦ aG, herbal. Origin: pantropical.

Cyperus concinnus R.Br.
Cyperaceae
♦ Naturalised
♦ 86
♦ G, cultivated. Origin: Australia.

Cyperus congestus Vahl
Cyperaceae
Mariscus congestus (Vahl) C.B.Clarke
(see)
♦ clustered flatsedge, dense flatsedge
♦ Weed, Naturalised, Environmental
Weed
♦ 7, 9, 15, 86, 98, 101, 176, 198, 203, 280
♦ G, cultivated. Origin: South Africa.

Cyperus conglomeratus Rottb.
Cyperaceae
♦ Weed
♦ 221
♦ G, arid.

Cyperus corymbosus Rottb.
Cyperaceae
♦ peri peri
♦ Weed
♦ 87, 88
♦ G, arid, cultivated, herbal.

Cyperus cuspidatus Kunth
Cyperaceae
♦ coastal plain flatsedge
♦ Weed
♦ 88, 126
♦ G, aqua. Origin: South America,
Africa, Asia.

Cyperus cyperinus (Retz.) J.Suringar
Cyperaceae
Mariscus cyperinus (Retz.) Vahl
♦ old world flatsedge, reh likarak
♦ Weed, Naturalised
♦ 87, 88, 101, 276
♦ G. Origin: Australia.

Cyperus cyperoides (L.) Kuntze
Cyperaceae
Kyllinga sumatrensis Retz., *Mariscus
alternifolius* Vahl (see), *Mariscus
sumatrensis* (Retz.) Raynal (see),
Mariscus umbellatus (Rottb.) Vahl (see),
Scirpus cyperoides L.
♦ Pacific island flatsedge, kok haang
ka rok
♦ Weed
♦ 87, 88, 170, 191, 238, 276, 286
♦ pG, herbal. Origin: Africa, Middle
East, Asia.

Cyperus denudatus L.f.
Cyperaceae
♦ Native Weed
♦ 121
♦ aG. Origin: southern Africa.

Cyperus denudatus L.f. var. sphaerospermum (Schrad.) Kük.
Cyperaceae
♦ Native Weed
♦ 121
♦ pG. Origin: southern Africa.

Cyperus diandrus Torr.
Cyperaceae
♦ low flatsedge, umbrella flatsedge
♦ Weed
♦ 87, 88, 218
♦ G, herbal.

Cyperus difformis L.
Cyperaceae
Cyperus complanatus Forssk., *Cyperus protractus* Link
♦ smallflower umbrella sedge, variable flatsedge, smallflower umbrella plant, kok khanaak
♦ Weed, Naturalised, Native Weed, Introduced
♦ 28, 34, 38, 55, 68, 70, 80, 87, 88, 101, 121, 126, 161, 170, 179, 180, 185, 186, 200, 204, 218, 221, 230, 239, 243, 248, 253, 255, 262, 263, 269, 272, 274, 275, 286, 297, 300
♦ wpG, cultivated, herbal. Origin: Old World tropics.

Cyperus diffusus Vahl
Cyperaceae
= *Cyperus laxus* Lam. (NoR)
♦ diffused flatsedge, kabongo
♦ Weed
♦ 87, 88
♦ G, cultivated, herbal. Origin: Australia.

Cyperus digitatus Roxb.
Cyperaceae
Cyperus giganteus Griseb., *Mariscus sieberianus* Nees var. *evolutior* C.B.Clarke
♦ finger flatsedge, digitate cyperus
♦ Weed
♦ 12, 87, 88, 126, 170, 191, 262
♦ a/pG, aqua. Origin: Australia.

Cyperus digitatus Roxb. var. obtusifructus Kük.
Cyperaceae
♦ Weed
♦ 271
♦ G.

Cyperus distans L.f.
Cyperaceae
Cyperus nutans Presl
♦ slender cyperus
♦ Weed, Native Weed
♦ 28, 87, 88, 121, 126, 170, 179, 206, 239, 243, 255, 262, 276
♦ wpG, promoted. Origin: obscure.

Cyperus dives Del.
Cyperaceae
Cyperus alopecuroides Rottb. var. *dives* Boeck., *Cyperus exaltatus* var. *dives* C.B.Clarke, *Cyperus fastigiatus* Forssk.
♦ Weed
♦ 221
♦ G.

Cyperus dubius Rottb.
Cyperaceae
♦ Weed, Naturalised
♦ 86, 98, 203
♦ G. Origin: Africa.

Cyperus duclouxii E.-G.Camus
Cyperaceae
♦ Weed
♦ 275
♦ pG.

Cyperus elatus L.
Cyperaceae
♦ Weed
♦ 88, 170, 191
♦ G.

Cyperus elegans L.
Cyperaceae
♦ royal flatsedge, galingale
♦ Weed
♦ 14
♦ G, herbal.

Cyperus entrerianus Boeck.
Cyperaceae
♦ woodrush flatsedge
♦ Weed
♦ 80, 88, 237
♦ G.

Cyperus eragrostis Lam.
Cyperaceae
♦ tall umbrella plant, umbrella sedge, umbrella grass, drain flatsedge, pale galingale, tall flatsedge
♦ Weed, Quarantine Weed, Noxious Weed, Naturalised, Environmental Weed, Casual Alien
♦ 7, 9, 15, 40, 70, 72, 76, 86, 87, 88, 98, 147, 176, 198, 200, 203, 218, 237, 253, 280, 286, 287, 295, 296
♦ pG, aqua, cultivated, herbal. Origin: North and South America.

Cyperus erythrorhizos Muhl.
Cyperaceae
♦ redroot flatsedge
♦ Weed
♦ 23, 87, 88, 161, 218
♦ aG, herbal.

Cyperus esculentus L.
Cyperaceae
Cyperus repens Ell., *Cyperus tuberosus* Pursh
♦ yellow nutsedge, yellow nutgrass
♦ Weed, Quarantine Weed, Noxious Weed, Naturalised
♦ 1, 15, 23, 28, 30, 34, 35, 50, 51, 53, 55, 70, 76, 84, 86, 87, 88, 98, 101, 115, 121, 126, 139, 157, 158, 161, 174, 179, 180, 186, 198, 199, 203, 207, 210, 211, 212, 218, 221, 229, 237, 240, 243, 245, 249, 253, 255, 263, 266, 270, 272, 280, 286, 287, 294, 295, 299, 300
♦ pG, arid, cultivated, herbal, toxic. Origin: North America.

Cyperus esculentus L. var. hermannii (Buckl.) Britt.
Cyperaceae
♦ yellow nutsedge, chufa flatsedge
♦ Noxious Weed
♦ 229

♦ G.

Cyperus esculentus L. var. leptostachyus Boeck.
Cyperaceae
♦ yellow nutsedge, chufa flatsedge
♦ Weed, Noxious Weed
♦ 229, 236
♦ G.

Cyperus esculentus L. var. macrostachyus Boeck.
Cyperaceae
♦ yellow nutsedge, chufa flatsedge
♦ Noxious Weed
♦ 229
♦ G.

Cyperus esculentus L. var. sativus Boeck.
Cyperaceae
♦ yellow nutsedge, chufa flatsedge
♦ Noxious Weed, Naturalised
♦ 101, 229
♦ G.

Cyperus exaltatus Retz.
Cyperaceae
♦ Weed
♦ 87, 88
♦ G, cultivated. Origin: Australia.

Cyperus fastigiatus Rottb.
Cyperaceae
♦ Native Weed
♦ 121
♦ pG, herbal. Origin: southern Africa.

Cyperus ferax L.C.Rich.
Cyperaceae
= *Cyperus odoratus* L.
♦ junquinho, capim de cheiro
♦ Weed
♦ 88, 255
♦ a/pG, arid, herbal. Origin: North and South America.

Cyperus ferruginescens Boeck.
Cyperaceae
= *Cyperus odoratus* L.
♦ Naturalised
♦ 287
♦ G, herbal.

Cyperus flabelliformis Rottb. nom. illeg.
Cyperaceae
= *Cyperus involucratus* Rottb.
♦ Weed
♦ 209
♦ G.

Cyperus flaccidus R.Br.
Cyperaceae
♦ Weed
♦ 263, 286
♦ G. Origin: Australia.

Cyperus flavescens L.
Cyperaceae
Pycreus flavescens (L.) Rchb. (see)
♦ yellow flatsedge
♦ Weed, Naturalised
♦ 23, 86, 88, 93, 98, 191, 203, 272
♦ G, cultivated, herbal.

Cyperus flavidus Retz.
Cyperaceae
= *Pycreus globosus* (All.) Rchb.
♦ globe sedge

♦ Weed
♦ 87, 88, 170, 191, 263, 272
♦ G. Origin: Australia.

Cyperus flavus (Vahl) Nees
Cyperaceae
= *Cyperus aggregatus* (Willd.) Endl.
♦ capim santo
♦ Weed, Naturalised
♦ 88, 98, 203, 245
♦ G.

Cyperus friburgensis Boeck.
Cyperaceae
♦ Weed
♦ 88
♦ G, aqua.

Cyperus fulvo-albescens Koy.
Cyperaceae
♦ Weed
♦ 262
♦ pG.

Cyperus fulvus R.Br.
Cyperaceae
♦ Weed
♦ 87, 88
♦ G. Origin: Australia.

Cyperus fuscus L.
Cyperaceae
Eucyperus fuscus Rikli
♦ brown flatsedge, brown cyperus
♦ Weed, Quarantine Weed,
Naturalised
♦ 76, 87, 88, 101, 203, 220, 221, 272, 275, 297
♦ aG, cultivated, herbal.

Cyperus giganteus Vahl
Cyperaceae
♦ giant flatsedge
♦ Weed
♦ 87, 88, 179, 255
♦ pG, aqua. Origin: North and South America.

Cyperus glaber L.
Cyperaceae
♦ zigolo glabro
♦ Weed
♦ 272
♦ G, herbal.

Cyperus globosus All.
Cyperaceae
= *Pycreus globosus* (All.) Rchb.
♦ Weed
♦ 87, 88, 286
♦ G.

Cyperus globulosus Aubl.
Cyperaceae
= *Cyperus luzulae* Rottb. ex Willd.
♦ globe sedge
♦ Weed
♦ 161, 249
♦ pG, herbal.

Cyperus glomeratus L.
Cyperaceae
Chlorocyperus glomeratus (L.) Palla.,
Cyperus cinnamomeus Retz.
♦ round headed nutsedge,
numagayatsuri
♦ Weed

♦ 272, 275, 286, 297
♦ G, herbal.

Cyperus gracilinux C.B.Clarke
Cyperaceae
♦ Weed
♦ 87, 88
♦ G.

Cyperus gracilis R.Br.
Cyperaceae
♦ slimjim flatsedge
♦ Weed, Naturalised
♦ 87, 88, 101, 134
♦ G, cultivated, herbal. Origin: Australia.

Cyperus gunnii Hook.f.
Cyperaceae
♦ Casual Alien
♦ 280
♦ G, cultivated.

Cyperus gymnocaulos Steud.
Cyperaceae
♦ spiny flatsedge
♦ Naturalised
♦ 86
♦ G, cultivated. Origin: Australia.

Cyperus hakonensis Franch. & Sav.
Cyperaceae
♦ Weed, Quarantine Weed
♦ 76, 88, 203, 220
♦ G.

Cyperus halpan L.
Cyperaceae
= *Cyperus haspan* L.
♦ Weed
♦ 80, 88, 126, 170, 191, 262
♦ pG, aqua.

Cyperus hamulosus M.Bieb.
Cyperaceae
♦ flatsedge, couchgrass
♦ Weed, Naturalised, Environmental Weed
♦ 7, 86, 93, 98, 198, 203, 205
♦ aG. Origin: Africa, Asia.

Cyperus haspan L.
Cyperaceae
Cyperus aphyllus Vahl, *Cyperus cayennensis* Link, *Cyperus efoliatus* Boeck., *Cyperus haspan* var. *americanus* Boeck., *Cyperus haspan* ssp. *juncoides* (Lam.) Kük., *Cyperus juncoides* Lam., *Cyperus tenuispicus* Steud., *Cyperus halpan* L., *Cyperus microcarpus* Boeck.
♦ sheathed cyperus, dwarf papyrus, haspan flatsedge, rice paddy weed, paddy field flatsedge
♦ Weed
♦ 12, 87, 217, 274, 275, 286, 297
♦ a/pG, aqua, cultivated, herbal. Origin: Australia.

Cyperus haspan L. ssp. juncoides (Lam.) Kük.
Cyperaceae
♦ Naturalised
♦ 300
♦ G.

Cyperus hermaphroditus (Jacq.) Standl.
Cyperaceae

♦ Weed, Quarantine Weed
♦ 76, 87, 88, 199, 203, 220
♦ G, arid, herbal.

Cyperus hokonensis Fr. & Sav.
Cyperaceae
♦ Weed
♦ 87, 88
♦ G.

Cyperus imbricatus Retz.
Cyperaceae
Cyperus radiatus Vahl, *Cyperus radicans* Vahl
♦ shingle flatsedge, kok
♦ Weed, Quarantine Weed
♦ 76, 87, 88, 126, 170, 191, 203, 204, 220, 239, 262, 275, 297
♦ pG, aqua.

Cyperus immensus C.B.Clarke
Cyperaceae
♦ Native Weed
♦ 121
♦ pG. Origin: southern Africa.

Cyperus involucratus Rottb.
Cyperaceae
= *Cyperus alternifolius* L. ssp. *alternifolius* Kük. (NoR)
♦ umbrella plant, umbrella sedge, haspan
♦ Weed, Naturalised, Introduced, Environmental Weed, Garden Escape, Cultivation Escape
♦ 7, 15, 32, 34, 38, 86, 93, 98, 101, 134, 179, 200, 203, 205, 252, 280
♦ wpG, cultivated. Origin: Africa.

Cyperus iria L.
Cyperaceae
♦ rice flatsedge, sedge, yellow sedge, umbrella sedge, grasshopper's cyperus, rice sedge, ricefield flatsedge, yaa rangkaa khaao
♦ Weed, Naturalised, Introduced
♦ 12, 14, 23, 32, 38, 55, 66, 86, 87, 88, 101, 126, 161, 170, 179, 186, 204, 218, 235, 236, 239, 243, 255, 262, 273, 274, 275, 286, 295, 297
♦ a/pG, aqua, herbal. Origin: pantropical Asia, Africa.

Cyperus javanicus Houtt.
Cyperaceae
♦ Javanese flatsedge
♦ Weed
♦ 87, 88
♦ G, herbal. Origin: Australia.

Cyperus kyllingia Endl.
Cyperaceae
= *Kyllinga nemoralis* (J.R. & G.Forst.) Dandy ex Hutch. & Dalziel
♦ white kyllinga, yaa tum huu
♦ Weed, Quarantine Weed, Casual Alien
♦ 3, 12, 76, 87, 88, 170, 218, 239, 276, 280
♦ pG. Origin: Australia.

Cyperus laetus J. & C.Presl
Cyperaceae
♦ Weed, Naturalised
♦ 87, 88, 287
♦ G. Origin: Australia.

Cyperus laevigatus L.
Cyperaceae
Cyperus distachyos All., *Juncellus laevigatus* (L.) C.B.Clarke (see)
♦ tawny sedge, smooth flatsedge
♦ Weed
♦ 87, 88, 185, 221
♦ pG, arid, cultivated, herbal.

Cyperus laevigatus L. var. laevigatus
Cyperaceae
♦ Naturalised
♦ 300
♦ G.

Cyperus lanceolatus Poir.
Cyperaceae
♦ epiphytic flatsedge
♦ Weed, Naturalised
♦ 179, 255, 300
♦ pG. Origin: Brazil.

Cyperus latifolius Poir.
Cyperaceae
♦ mushasha
♦ Native Weed
♦ 121
♦ pG. Origin: southern Africa.

Cyperus lentiginosus Millsp. & Chase
Cyperaceae
♦ Latin American flatsedge
♦ Weed
♦ 179
♦ G.

Cyperus ligularis L.
Cyperaceae
Cyperus glandulosus Rolfe, *Cyperus ligularis* L. (see), *Mariscus glandulosus* Bojer, *Mariscus rufus* H.B.K. (see)
♦ large cyperus, Alabama swamp flatsedge
♦ Weed
♦ 88, 161, 249
♦ pG, herbal.

Cyperus longus L.
Cyperaceae
Chlorocyperus longus (L.) Palla
♦ rough cyperus, sweet cyperus, sweet galingale, galingale, English galingale
♦ Weed, Naturalised, Quarantine Weed
♦ 76, 87, 88, 126, 185, 203, 220, 221, 253, 272, 280
♦ pG, aqua, cultivated, herbal.

Cyperus longus L. ssp. badius (Desf.) Murb.
Cyperaceae
♦ brown galingale
♦ Weed
♦ 70
♦ G.

Cyperus longus L. ssp. longus
Cyperaceae
♦ galingale
♦ Weed
♦ 70
♦ G.

Cyperus longus L. var. tenuiflorus (Rottb.) Kuk.
Cyperaceae

♦ sweet cyperus
♦ Native Weed
♦ 121
♦ pG. Origin: southern Africa.

Cyperus luzulae Rottb. ex Willd.
Cyperaceae
Cyperus globulosus Aubl. (see), *Scirpus luzulae* L.
♦ botoncito
♦ Weed, Quarantine Weed
♦ 76, 87, 88, 203, 220, 255, 295
♦ pG. Origin: North and South America.

Cyperus maculatus Boeck.
Cyperaceae
♦ Weed
♦ 221
♦ G, arid.

Cyperus malaccensis Lam.
Cyperaceae
Chlorocyperus malaccensis (Lam.) Palla, *Cyperus tegetiformis* auct.
♦ shichito matgrass
♦ Weed
♦ 87, 88, 126, 170, 191
♦ pG, aqua. Origin: Middle East, Asia, Australia.

Cyperus malaccensis Lamk. var. brevifolius Boeck.
Cyperaceae
♦ shichito matgrass
♦ Naturalised
♦ 287
♦ G.

Cyperus maranguensis K.Schum.
Cyperaceae
♦ Weed
♦ 87, 88
♦ G.

Cyperus margaritaceus Vahl
Cyperaceae
♦ Native Weed
♦ 121
♦ pG. Origin: southern Africa.

Cyperus maritimus Poir.
Cyperaceae
♦ Weed, Quarantine Weed
♦ 76, 87, 88, 203, 220
♦ G.

Cyperus melanospermus (Nees) J.V.Suringar
Cyperaceae
Kyllinga melanosperma Nees
♦ Weed
♦ 276
♦ G.

Cyperus metzii (Hochst. ex Steud.) Mattf. & Kük. ex Kük.
Cyperaceae
= *Kyllinga squamulata* Thonn. ex Vahl
♦ Naturalised
♦ 86
♦ G. Origin: Africa, west Asia.

Cyperus meyenianus Kunth
Cyperaceae

♦ Meyen's flatsedge
♦ Weed, Naturalised
♦ 88, 101, 255
♦ pG, herbal. Origin: North and South America.

Cyperus michelianus (L.) Delile
Cyperaceae
Dichostylis micheliana Nees, *Eleocharis micheliana* Reich., *Fimbrystylis micheliana* Reich., *Isolepis micheliana* Roem. & Schult., *Scirpus michelianus* L. (see)
♦ zigolo del micheli
♦ Weed
♦ 275, 297
♦ G, herbal.

Cyperus microiria Steud.
Cyperaceae
= *Cyperus amuricus* Maxim.
♦ kavatsurigusa
♦ Weed, Quarantine Weed
♦ 76, 87, 88, 203, 204, 220, 263, 286, 297
♦ G.

Cyperus monti L.
Cyperaceae
= *Cyperus serotinus* Rottb.
♦ Weed
♦ 87, 88
♦ G.

Cyperus mutisii (H.B.K.) Griseb.
Cyperaceae
Mariscus mutisii Kunth (see)
♦ Mutis' flatsedge
♦ Weed
♦ 157, 221
♦ pG.

Cyperus mutisii (H.B.K.) Griseb. var. mutisii
Cyperaceae
♦ Weed
♦ 199
♦ G.

Cyperus nipponicus Fr. & Sav.
Cyperaceae
Juncellus nipponicus (Fr. & Sav.) C.B.Clarke
♦ Weed
♦ 87, 88, 263, 275, 286, 297
♦ G.

Cyperus novae-hollandiae Boeck.
Cyperaceae
♦ Weed
♦ 87, 88
♦ G.

Cyperus obtusiflorus Vahl
Cyperaceae
♦ blunt flowered sedge
♦ Weed, Native Weed
♦ 23, 88, 121
♦ pG, cultivated. Origin: southern Africa.

Cyperus ochraceus Vahl
Cyperaceae
♦ pond flatsedge
♦ Weed
♦ 179, 243
♦ S, arid, herbal.

Cyperus odoratus L.

Cyperaceae

Cyperus acicularis Schrad. ex Nees (see), *Cyperus eggersii* Boeck., *Cyperus engelmannii* Steud., *Cyperus ferax* L.C.Rich. (see), *Cyperus ferruginescens* Boeck. (see), *Cyperus longispicatus* J.B.S.Norton, *Cyperus macrocephalus* Liebm., *Cyperus speciosus* Vahl, *Mariscus ferax* (L.C.Rich.) C.B.Clarke, *Mariscus huarmensis* Kunth, *Torulinium confertum* Desv. ex Ham., *Torulinium eggersii* (Boeck.) C.B.Clarke, *Torulinium ferax* (L.C.Rich.) Urb. (see), *Torulinium odoratum* (L.) Hooper (see)

♦ flatsedge, fragrant flatsedge, galingale, coyolillo
♦ Weed, Quarantine Weed, Naturalised
♦ 23, 87, 157, 161, 170, 191, 236, 243, 257, 258, 281, 286, 295, 300
♦ pG, arid/aqua, cultivated, herbal. Origin: pantropical.

Cyperus ohwii Kük.

Cyperaceae
♦ Naturalised
♦ 287
♦ G.

Cyperus orthostachyus Fr. & Sav.

Cyperaceae
= *Cyperus truncatus* C.A.Mey. ex Turcz.
♦ Weed
♦ 286
♦ G.

Cyperus ovularis (Michx.) Torr.

Cyperaceae
= *Cyperus echinatus* (L.) Wood (NoR)
♦ Weed, Naturalised
♦ 23, 88, 287
♦ G, herbal.

Cyperus oxylepis Nees ex Steud.

Cyperaceae
♦ sharpscale flatsedge
♦ Naturalised
♦ 287
♦ G, herbal.

Cyperus pangorei Rottb.

Cyperaceae
Cyperus tegetum Roxb. (see)
♦ Weed
♦ 87, 88
♦ G.

Cyperus papyrus L.

Cyperaceae
♦ paper plant, papyrus, Egyptian paper, Egyptian paper reed, ibongo
♦ Weed, Naturalised, Native Weed, Introduced
♦ 7, 86, 87, 88, 98, 101, 121, 179, 203, 209, 228
♦ pG, aqua, cultivated, herbal. Origin: Africa.

Cyperus phaeolepis Cherm.

Cyperaceae
♦ Madagascar flatsedge
♦ Naturalised
♦ 101
♦ G.

Cyperus pilosus Vahl

Cyperaceae
♦ fuzzy flatsedge, kok samliam lek
♦ Weed, Quarantine Weed, Naturalised
♦ 76, 80, 87, 88, 101, 170, 191, 209, 235, 239, 243, 262, 275, 286
♦ aG. Origin: Australia.

Cyperus platystylis R.Br.

Cyperaceae
♦ Weed
♦ 88, 170, 262
♦ pG, cultivated. Origin: Australia.

Cyperus polystachyos Rottb.

Cyperaceae
= *Pycreus polystachyos* (Rottb.) P.Beauv.
♦ Texas sedge, manyspike flatsedge
♦ Weed, Naturalised
♦ 7, 23, 86, 87, 88, 98, 161, 170, 203, 204, 249, 255, 263, 280, 286
♦ pG, arid, cultivated, herbal. Origin: Australia.

Cyperus procerus Rottb.

Cyperaceae
Cyperus ornatus R.Br.
♦ yaa ta krap
♦ Weed
♦ 87, 88, 170, 191, 204, 209, 239
♦ pG. Origin: Australia.

Cyperus prolifer Lam.

Cyperaceae
♦ dwarf papyrus, miniature flatsedge
♦ Weed, Naturalised, Environmental Weed
♦ 80, 86, 88, 98, 101, 112, 179, 197, 203, 287
♦ G, aqua, cultivated. Origin: south Africa.

Cyperus prolifer Lam. var. *isocladus* (Kunth) Kük.

Cyperaceae
Cyperus ioscladus Kunth
♦ Native Weed
♦ 121
♦ pG. Origin: southern Africa.

Cyperus prolixus H.B.K.

Cyperaceae
♦ mosquito flatsedge
♦ Weed, Naturalised
♦ 39, 88, 101
♦ G, toxic.

Cyperus pseudovegetus Steud.

Cyperaceae
Cyperus virens Michx. var. *arenicola* (Boeck.) Shinners
♦ marsh flatsedge, coyolillo
♦ Weed
♦ 281
♦ G, herbal.

Cyperus pulcherrimus Willd. ex Kunth

Cyperaceae
Cyperus silletensis Thw.
♦ elegant cyperus, kok lek
♦ Weed
♦ 87, 88, 170, 191, 204, 209, 239, 262
♦ pG.

Cyperus pumilus Nees.

Cyperaceae
Cyperus nitens Retz., *Pycreus nitens* (Retz.) Nees (see), *Pycreus pumilus* (L.) Nees
♦ low flatsedge
♦ Weed, Naturalised
♦ 87, 88, 101, 170, 179, 191, 262
♦ G. Origin: Australia.

Cyperus purpuro-variegatus Boeck.

Cyperaceae
= *Cyperus rotundus* L.
♦ Weed
♦ 3, 191
♦ G.

Cyperus pygmaeus Rottb.

Cyperaceae
Dichostylis pygmaea (Rottb.) Nees, *Juncellus pygmaeus* C.B.Clarke
♦ Weed
♦ 87, 88, 170
♦ G. Origin: Australia.

Cyperus radians Nees & Mey.

Cyperaceae
Mariscus radians (Nees & Meyen) Tang & F.T.Wang (see)
♦ Weed
♦ 87, 88
♦ G.

Cyperus reduncus Hochst. ex Boeck.

Cyperaceae
♦ Weed
♦ 87, 88
♦ G.

Cyperus reflexus Vahl

Cyperaceae
♦ flatsedge, bentawn flatsedge
♦ Weed, Naturalised
♦ 86, 98, 198, 203, 295
♦ G, herbal. Origin: North and South America.

Cyperus retrorsus Chapm.

Cyperaceae
♦ cylindric sedge, pine barren flatsedge
♦ Weed
♦ 161, 249
♦ pG, herbal.

Cyperus retzii Nees

Cyperaceae
♦ Weed
♦ 87, 88
♦ G.

Cyperus rigens Presl & C.Presl

Cyperaceae
♦ Weed, Naturalised
♦ 86, 98, 203
♦ G, arid. Origin: South America.

Cyperus rigidifolius Steud.

Cyperaceae
♦ highland nutsedge
♦ Weed, Quarantine Weed
♦ 53, 76, 87, 88, 203, 220
♦ G.

Cyperus rotundus L.
Cyperaceae
Chlorocyperus rotundus Palle (see),
Cyperus purpuro-variegatus Boeck. (see),
Cyperus stoloniferum pallidus Boeck.
(see), *Cyperus tetrastachyos* Desf. (see),
Cyperus tuberosus Rottb. (see)
♦ purple nutsedge, nutgrass,
nutsedge, cocograss, red nutsedge,
watergrass, red grass, yaa haeo muu,
coyolillo, coquito
♦ Weed, Noxious Weed, Naturalised,
Native Weed, Introduced,
Environmental Weed, Garden Escape
♦ 3, 6, 7, 12, 14, 15, 23, 30, 34, 35, 51, 55,
66, 70, 84, 86, 87, 88, 93, 98, 101, 105,
107, 116, 121, 126, 134, 146, 147, 157,
158, 161, 170, 171, 179, 180, 185, 186,
199, 203, 204, 209, 211, 212, 218, 221,
228, 229, 230, 235, 236, 237, 239, 240,
242, 243, 245, 246, 249, 253, 255, 257,
261, 262, 263, 266, 269, 270, 272, 273,
274, 275, 276, 280, 281, 286, 295, 297,
300
♦ pG, arid, cultivated, herbal. Origin:
Africa, Asia.

Cyperus rotundus L. ssp. rotundus
Cyperaceae
♦ purple nutsedge, nutgrass
♦ Weed
♦ 53, 115
♦ pG.

Cyperus sanguinolentus (Vahl) Nees
Cyperaceae
= *Pycreus snaguinolentus* (Vahl) Nees
♦ purpleglume flatsedge
♦ Weed, Naturalised
♦ 86, 87, 88, 101, 170, 191, 263, 280, 286
♦ G, cultivated.

Cyperus schimperianus Steud.
Cyperaceae
♦ Weed
♦ 221
♦ G.

Cyperus schweinfurthianus Boeck.
Cyperaceae
♦ Weed
♦ 87, 88
♦ G.

Cyperus seemannianus Boeck.
Cyperaceae
♦ Weed
♦ 87, 88
♦ G.

Cyperus serotinus Rottb.
Cyperaceae
Cyperus monti L. (see), *Duraljouvea
serotina* (Rottb.) Pall., *Juncellus serotinus*
(Rottb.) C.B.Clarke (see)
♦ tidalmarsh flatsedge
♦ Weed, Naturalised
♦ 88, 101, 204, 263, 272, 274, 286
♦ pG, herbal.

Cyperus seslerioides Kunth
Cyperaceae
♦ Texas flatsedge
♦ Weed

♦ 87, 88
♦ G, herbal.

Cyperus sesquiflorus (Torr.) Kük.
Cyperaceae
Kyllinga odorata Vahl (see), *Kyllinga
sesquiflora* Torr. (see)
♦ capim santo, manubre
♦ Weed, Naturalised
♦ 86, 87, 88, 93, 98, 191, 203, 245, 249,
255
♦ pG, herbal. Origin: pantropical.

Cyperus sexangularis Nees.
Cyperaceae
♦ Native Weed
♦ 121
♦ pG. Origin: southern Africa.

Cyperus spacelatus Rottb.
Cyperaceae
♦ Weed
♦ 87
♦ G.

Cyperus spectabilis Link
Cyperaceae
♦ spectacular flatsedge
♦ Weed
♦ 199
♦ G.

Cyperus sphacelatus Rottb.
Cyperaceae
Cyperus locuples C.B.Clarke
♦ roadside flatsedge
♦ Weed
♦ 28, 80, 88, 93, 126, 170, 179, 191, 206,
243
♦ pG, aqua. Origin: tropical Africa,
Americas.

Cyperus squarrosus L.
Cyperaceae
Chlorocyperus inflexus (Muhl.) Palla,
Cyperus aristatus Rottb. fo. *falciculosus*
(Liebm.) Kük., *Cyperus aristatus* Rottb.
fo. *inflexus* (Muhl.) C.B.Clarke, *Cyperus
aristatus* Rottb. (see), *Cyperus aristatus*
Rottb. var. *inflexus* (Muhl.) Boeck.
ex Kük., *Cyperus aristatus* Rottb. var.
runyonii O'Neill, *Cyperus falciculosus*
Liebm., *Cyperus inflexus* Muhl.,
Isolepis echinulata Kunth, *Mariscus
aristatus* auct., *Mariscus squarrosus* (L.)
C.B.Clarke (see)
♦ awned cyperus, bearded flatsedge
♦ Weed,
Naturalised
♦ 161, 300
♦ aG.

Cyperus stoloniferum pallidus Boeck.
Cyperaceae
= *Cyperus rotundus* L.
♦ Weed
♦ 3, 191
♦ G.

Cyperus strigosus L.
Cyperaceae
♦ false nutsedge, nutsedge, straw
coloured flatsedge
♦ Weed, Quarantine Weed,
Naturalised

♦ 23, 76, 87, 88, 161, 203, 218, 220, 249,
300
♦ pG, herbal.

Cyperus surinamensis Rottb.
Cyperaceae
♦ Surinam sedge, tropical flatsedge,
piri
♦ Weed
♦ 88, 161, 237, 249, 255, 295
♦ pG, herbal. Origin: tropical America.

Cyperus tegetiformis Roxb. ex Arn.
Cyperaceae
♦ Weed
♦ 87, 88
♦ pG, promoted.

Cyperus tegetum Roxb.
Cyperaceae
= *Cyperus pangorei* Rottb.
♦ galingale
♦ Weed, Quarantine Weed
♦ 76, 87, 88, 203, 220
♦ G, herbal.

Cyperus tenellus L.f.
Cyperaceae
♦ tiny flatsedge
♦ Weed, Naturalised, Environmental
Weed
♦ 7, 9, 15, 72, 86, 88, 98, 198, 203, 280
♦ aG, cultivated. Origin: southern
Africa.

Cyperus teneristolon Mattf. & Kük.
Cyperaceae
Cyperus transitorius Kük., *Kyllinga
pulchella* Kunth
♦ Weed
♦ 53, 88
♦ pG. Origin: eastern and southern
Africa.

Cyperus tenuiculmis Boeck.
Cyperaceae
♦ Weed, Naturalised
♦ 87, 88, 287
♦ G. Origin: Australia.

Cyperus tenuiflorus Rottb.
Cyperaceae
♦ scaly sedge
♦ Weed, Naturalised, Environmental
Weed
♦ 7, 86, 98, 203
♦ G. Origin: southern Africa.

Cyperus tenuis Sw.
Cyperaceae
♦ crevice flatsedge
♦ Weed
♦ 88, 157
♦ a/pG.

Cyperus tenuispica Steud.
Cyperaceae
♦ Weed
♦ 87, 88, 170, 191, 286
♦ G.

Cyperus tetrastachyos Desf.
Cyperaceae
= *Cyperus rotundus* L.
♦ Weed
♦ 3, 191
♦ G.

Cyperus textilis Thunb.
Cyperaceae
♦ matsedge, rushes
♦ Weed, Quarantine Weed, Native Weed
♦ 76, 121
♦ pG, cultivated. Origin: South Africa.

Cyperus trialatus (Boeck.) Kern
Cyperaceae
Cyperus bancanus Miq.
♦ Weed
♦ 12, 87, 88
♦ G.

Cyperus trinervis R.Br.
Cyperaceae
♦ Australian flatsedge
♦ Naturalised
♦ 101
♦ G. Origin: Australia.

Cyperus truncatus C.A.Mey. ex Turcz.
Cyperaceae
Cyperus orthostachyus Fr. & Sav. (see)
♦ Weed
♦ 87, 88
♦ G.

Cyperus tuberosus Rottb.
Cyperaceae
= *Cyperus rotundus* L.
♦ Weed, Naturalised
♦ 3, 86, 98, 191, 203
♦ G.

Cyperus uncinatus Poir.
Cyperaceae
♦ Weed
♦ 87, 88
♦ G.

Cyperus ustulatus A.Rich.
Cyperaceae
♦ giant umbrella sedge, toetoe upoko tangata
♦ Weed
♦ 208
♦ pG, aqua, cultivated.

Cyperus virens Michx.
Cyperaceae
Cyperus drummondii Torr. & Hook., *Cyperus robustus* Kunth
♦ green flatsedge
♦ Weed
♦ 23, 88
♦ G, herbal.

Cyperus vorsteri K.L.Wilson
Cyperaceae
♦ cyperus
♦ Weed, Naturalised, Garden Escape, Environmental Weed, Cultivation Escape
♦ 7, 54, 86, 88, 155
♦ pG, cultivated. Origin: South Africa.

Cyperus zollingeri Steud.
Cyperaceae
♦ Weed
♦ 87, 88
♦ G.

Cyphia bulbosa (L.) Bergam.
Campanulaceae/Lobeliaceae/ Cyphiaceae

Lobelia bulbosa L.
♦ Quarantine Weed
♦ 220

Cyphia elata Harv.
Campanulaceae/Lobeliaceae/ Cyphiaceae
♦ Quarantine Weed
♦ 220
♦ cultivated, herbal.

Cyphia linarioides Presl
Campanulaceae/Lobeliaceae/ Cyphiaceae
Cyphia campestris Presl
♦ Quarantine Weed
♦ 220

Cyphia longifolia N.E.Br.
Campanulaceae/Lobeliaceae/ Cyphiaceae
♦ Quarantine Weed
♦ 220

Cyphia phyteuma Willd.
Campanulaceae/Lobeliaceae/ Cyphiaceae
Lobelia phyteuma L.
♦ Quarantine Weed
♦ 220
♦ cultivated.

Cyphia subtubulata E.Wimm.
Campanulaceae/Lobeliaceae/ Cyphiaceae
♦ Quarantine Weed
♦ 220

Cyphia volubilis Willd.
Campanulaceae/Lobeliaceae/ Cyphiaceae
♦ Quarantine Weed
♦ 220
♦ cultivated.

Cypholophus moluccanus (Blume) Miq.
Urticaceae
♦ Hawai'i lopleaf
♦ Naturalised
♦ 101

Cyphomandra betacea (Cav.) Sendtn.
Solanaceae
Cyphomandra crassicaulis (Cav.) Sendtn. (see), *Cyphomandra crassifolia* (Ortega) J.F.Macbr., *Cyphomandra procera* Wawra, *Pionandra betacea* (Cav.) Miers, *Solanum betacea* Cav., *Solanum betaceum* Cav. (see), *Solanum crassifolium* Ortega, *Solanum insigne* Lowe
♦ tomato tree, tree tomato
♦ Weed, Naturalised
♦ 86, 121, 198, 280
♦ S/T, cultivated, herbal. Origin: South America.

Cyphomandra crassicaulis (Cav.) Sendtn.
Solanaceae
= *Cyphomandra betacea* (Cav.) Sendtn.
♦ tree tomato
♦ Weed
♦ 22
♦ T, cultivated.

Cypselea humifusa Turp.
Aizoaceae

♦ panal
♦ Weed
♦ 179
♦ aH.

Cyrtandra lanceolata Ridl.
Gesneriaceae
♦ Quarantine Weed
♦ 220

Cyrtandra pendula Bl.
Gesneriaceae
♦ Quarantine Weed
♦ 220
♦ herbal.

Cyrtandra rotundifolia Ridl.
Gesneriaceae
♦ Quarantine Weed
♦ 220

Cyrtanthus elatus (Jacq.) Traub
Liliaceae/Amaryllidaceae
♦ Scarborough lily
♦ Casual Alien
♦ 280
♦ cultivated.

Cyrtococcum accrescens (Trin.) Stapf
Poaceae
= *Cyrtococcum patens* (L.) A.Camus
♦ diffuse panicgrass
♦ Weed
♦ 12, 87, 88, 90
♦ pG, herbal.

Cyrtococcum deltoideum (Hack.) A.Camus
Poaceae
♦ Naturalised
♦ 86
♦ G. Origin: Madagascar.

Cyrtococcum oxyphyllum (Hochst. ex Steud.) Stapf
Poaceae
Cyrtococcum pilipes A.Camus, *Panicum hermaphroditum* Steud., *Panicum oxyphyllum* Hochst. ex Steud., *Panicum pilipes* Nees & Arn. ex Büse (see)
♦ shining panicgrass
♦ Weed
♦ 87, 88, 90
♦ pG. Origin: Asia, Australia.

Cyrtococcum patens (L.) A.Camus
Poaceae
Cyrtococcum accrescens (Trin.) Stapf (see), *Cyrtococcum radicans* (Retz.) Stapf, *Panicum accrescens* Trin., *Panicum carinatum* J.Presl, *Panicum obliquum* Roth ex Roem. & Schult., *Panicum patens* L., *Panicum radicans* Retz.
♦ bowgrass, reh maikol
♦ Weed, Quarantine Weed
♦ 76, 87, 88, 90, 203, 220
♦ pG.

Cyrtococcum trigonum (Retz.) A.Camus
Poaceae
Panicum trigonum Retz
♦ Weed
♦ 87, 88
♦ G.

Cyrtomium falcatum (L.f.) C.Presl
Dryopteridaceae
Polypodium falcatum L.f.
♦ hollyfern, Japanese netvein
hollyfern, Japanese hollyfern, house
hollyfern, oni yabusotetsu
♦ Weed, Naturalised, Cultivation
Escape
♦ 7, 34, 86, 98, 101, 198, 203, 252
♦ pH, cultivated, herbal. Origin:
China, Japan.

Cyrtomium fortunei J.Sm.
Dryopteridaceae
= *Polystichum fortunei* (J.Sm.) Nakai
(NoR)
♦ Asian netvein hollyfern
♦ Naturalised
♦ 101
♦ pH, cultivated.

Cyrtopodium glutiniferum Raddi
Orchidaceae
♦ Naturalised
♦ 101
♦ herbal.

Cyrtopodium paranaense Schltr.
Orchidaceae
♦ Weed
♦ 179

Cyrtosperma chamissonis (Schott) Merr.
Araceae
♦ mwahng, swamp taro
♦ Introduced
♦ 230
♦ pH, aqua.

Cyrtostachys lakka Becc.
Arecaceae
♦ sealing wax palm
♦ Introduced
♦ 230
♦ T, cultivated.

Cystopteris fragilis (L.) Bernh.
Dryopteridaceae/Woodsiaceae
Aspidium fragilis (L.) Sw.
♦ brittle bladder fern, bladder fern
♦ Weed
♦ 39, 87, 88
♦ pH, cultivated, herbal, toxic.

Cytisus albidus DC.
Fabaceae/Papilionaceae
Cytisus mollis (Cav.) Pau
♦ Introduced
♦ 228
♦ arid.

Cytisus × dallimorei Rolfe
Fabaceae/Papilionaceae
= *Cytisus multiflorus* (L'Hér.) Sweet
× *Cytisus scoparius* (L.) Link cv.
'Andreanus'
♦ Dallimore's Spanish broom
♦ Weed, Naturalised
♦ 88, 101

Cytisus decumbens (Durande) Spach
Fabaceae/Papilionaceae
♦ suikerovihma
♦ Cultivation Escape
♦ 42
♦ cultivated, herbal.

Cytisus glaber (L.f.) Rothm.
Fabaceae/Papilionaceae
♦ kultavihma
♦ Cultivation Escape
♦ 42
♦ cultivated.

**Cytisus leucanthus Waldst. & Kit. ex
Willd.**
Fabaceae/Papilionaceae
♦ Naturalised
♦ 287

Cytisus monspessulanus L.
Fabaceae/Papilionaceae
= *Genista monspessulana* (L.) L.Johnson
♦ Montpellier broom, French broom
♦ Weed, Noxious Weed, Naturalised,
Quarantine Weed, Garden Escape
♦ 35, 63, 76, 86, 87, 88, 95, 146, 161, 203,
220, 283
♦ cultivated, herbal, toxic. Origin:
Mediterranean.

Cytisus multiflorus (L'Hér.) Sweet
Fabaceae/Papilionaceae
♦ white Spanish broom, Spanish
broom
♦ Weed, Naturalised, Garden Escape,
Environmental Weed
♦ 39, 72, 80, 86, 88, 98, 101, 198, 203,
280, 289
♦ pS, cultivated, herbal, toxic. Origin:
Spain, Portugal.

Cytisus palmensis (Christ) Hutch.
Fabaceae/Papilionaceae
= *Chamaecytisus palmensis* (Christ)
Bisby & Nicholls
♦ tree lucerne, tagasaste
♦ Weed, Naturalised, Introduced,
Environmental Weed
♦ 72, 86, 88, 98, 155, 203, 228
♦ S/T, arid, cultivated. Origin: Canary
Islands.

**Cytisus procumbens (Waldst. & Kit.)
Spreng.**
Fabaceae/Papilionaceae
Genista procumbens Waldst. & Kit.
♦ Weed
♦ 272
♦ cultivated.

Cytisus proliferus L.f.
Fabaceae/Papilionaceae
= *Chamaecytisus prolifer* (L.f.) Link
(NoR) [see *Chamaecytisus prolifera* (L.f.)
Link and *Chamaecytisus proliferus* (L.)
Link]
♦ Weed, Naturalised, Introduced
♦ 39, 98, 203, 228
♦ arid, cultivated, toxic.

Cytisus scoparius (L.) Link
Fabaceae/Papilionaceae
Sarothamnus scoparius (L.) Wimm. ex
W.D.J.Koch
♦ Scotch broom, English broom,
broom, Spanish broom, jänönvihma,
monarch broom, Andreanus broom
♦ Weed, Quarantine Weed, Noxious
Weed, Naturalised, Introduced,
Garden Escape, Environmental Weed,
Cultivation Escape

♦ 1, 4, 15, 18, 20, 34, 35, 39, 42, 45, 70,
72, 76, 78, 80, 87, 88, 95, 97, 98, 101, 103,
104, 116, 121, 133, 136, 137, 139, 142,
146, 147, 151, 152, 155, 161, 163, 165,
169, 169, 171, 176, 181, 195, 198, 203,
212, 218, 219, 225, 229, 231, 232, 241,
246, 272, 280, 283, 286, 287, 300
♦ S/T, cultivated, herbal, toxic. Origin:
Eurasia.

Cytisus scoparius (L.) Link ssp. *scoparius*
Fabaceae/Papilionaceae
♦ English broom, Scotch broom,
broom, common broom, Scottish
broom, Spanish broom
♦ Weed, Naturalised, Garden Escape,
Environmental Weed
♦ 86, 178, 269, 289, 296
♦ S, cultivated. Origin: Europe.

**Cytisus scoparius (L.) Link var.
andreanus (Puiss.) Dippel**
Fabaceae/Papilionaceae
= *Cytisus scoparius* (L.) Link cv.
'Andreanus'
♦ Scotch broom
♦ Noxious Weed, Naturalised
♦ 101, 229

Cytisus scoparius (L.) Link var. *scoparius*
Fabaceae/Papilionaceae
♦ Scotch broom
♦ Noxious Weed, Naturalised
♦ 101, 229

Cytisus striatus (Hill) Rothm.
Fabaceae/Papilionaceae
♦ Portuguese broom, striated broom
♦ Weed, Noxious Weed, Naturalised,
Environmental Weed
♦ 35, 78, 80, 88, 101, 116, 146, 151, 229,
231, 241, 300
♦ S, toxic.

Cytisus villosus Pourr.
Fabaceae/Papilionaceae
Cytisus triflorus L'Hér.
♦ hairy broom
♦ Naturalised
♦ 101
♦ arid, cultivated.

D

Daboecia cantabrica (Huds.) K.Koch
Ericaceae
Daboecia polifolia D.Don
♦ St. Daboec's heath, Irish heath, heather, Connemara heath
♦ Naturalised
♦ 40, 280
♦ cultivated, herbal.

Dactylis glomerata L.
Poaceae
Dactylis hispanica Roth
♦ orchard grass, cock's foot, cat's grass, koiranheinä
♦ Weed, Naturalised, Introduced, Garden Escape, Environmental Weed, Cultivation Escape
♦ 3, 7, 15, 20, 21, 34, 38, 39, 70, 72, 80, 86, 87, 88, 98, 101, 111, 121, 134, 136, 146, 151, 152, 158, 159, 161, 174, 176, 181, 195, 198, 203, 211, 212, 218, 225, 241, 243, 246, 249, 253, 261, 263, 272, 280, 286, 287, 289, 296, 300
♦ pG, cultivated, herbal, toxic. Origin: Eurasia.

Dactylis glomerata L. ssp. *aschersoniana* (Graebn.) Thell.
Poaceae
♦ Ascherson's orchard grass
♦ Naturalised
♦ 101
♦ G.

Dactylis glomerata L. ssp. *glomerata*
Poaceae
♦ orchard grass
♦ Naturalised
♦ 101
♦ G.

Dactyloctenium aegyptium (L.) P.Beauv.
Poaceae
Chloris mucronata Michx., *Cynosurus aegypticus* L., *Dactyloctenium aegyptiacum* Willd., *Dactyloctenium mucronatum* (Michx.) Willd., *Eleusine aegyptia* (L.) Desf.
♦ buttongrass, coast buttongrass, muncho, Egyptian fingergrass, crow's foot grass, Durban crow's foot grass, beach wiregrass, yaa paak khwaai
♦ Weed, Naturalised, Introduced, Environmental Weed
♦ 6, 7, 12, 23, 30, 34, 53, 55, 66, 80, 86, 87, 88, 91, 93, 98, 101, 121, 157, 158, 161, 170, 179, 185, 186, 203, 205, 218, 228, 235, 239, 242, 243, 249, 255, 257, 261, 262, 273, 274, 286, 297
♦ a/pG, arid, cultivated, herbal, toxic. Origin: tropical Africa, Asia.

Dactyloctenium australe Steud.
Poaceae
♦ Durban grass, L.M. grass, Natal crow's foot
♦ Weed, Naturalised, Native Weed
♦ 7, 86, 88, 98, 121, 158, 203
♦ pG, cultivated. Origin: southern Africa.

Dactyloctenium germinatum Hack.
Poaceae
♦ Sudan crow's foot grass
♦ Naturalised
♦ 101
♦ G.

Dactyloctenium giganteum Fisch. & Schweick.
Poaceae
♦ giant crow's foot
♦ Weed, Naturalised, Native Weed
♦ 86, 87, 98, 121, 158, 203
♦ aG, cultivated. Origin: southern Africa.

Dactyloctenium radulans (R.Br.) Beauv.
Poaceae
♦ buttongrass
♦ Weed, Naturalised
♦ 39, 55, 101
♦ G, arid, cultivated, toxic.

Dactylorhiza maculata (L.) Soó
Orchidaceae
Orchis biermannii Ortm., *Orchis maculata* L., *Orchis solida* Moench
♦ heath spotted orchid, spotted orchid, moorland spotted orchid
♦ Weed
♦ 272
♦ pH, aqua, cultivated, herbal.

Dactylorhiza majalis (Rchb.) P.F.Hunt & Summerh.
Orchidaceae
♦ leveälehtikämmekkä, broad leaved marsh orchid, western marsh orchid, fan orchid, common marsh orchid
♦ Weed, Naturalised
♦ 42, 272
♦ herbal.

Dahlia coccinea Cav. × *pinnata* Cav.
Asteraceae
♦ Naturalised
♦ 280

Dahlia excelsa Benth.
Asteraceae
may be conspecific with *Dahlia imperialis* Roezl ex Ortgies
♦ tree dahlia
♦ Weed, Naturalised
♦ 15, 280
♦ cultivated.

Dahlia pinnata Cav.
Asteraceae
♦ pinnate dahlia
♦ Naturalised
♦ 101
♦ pH, cultivated.

Dalbergia armata E.Mey.
Fabaceae/Papilionaceae
♦ hluhluwe climber, monkey rope, thorny rope

♦ Native Weed
♦ 121
♦ S/T, cultivated. Origin: southern Africa.

Dalbergia bariensis Pierre
Fabaceae/Papilionaceae
♦ Quarantine Weed
♦ 220

Dalbergia frutescens (Vell.) Britton
Fabaceae/Papilionaceae
Dalbergia variabilis Vog., *Pterocarpus frutescens* Vell.
♦ jacaranda rosa
♦ Introduced
♦ 228
♦ arid.

Dalbergia horrida R.Grah.
Fabaceae/Papilionaceae
♦ Quarantine Weed
♦ 220

Dalbergia latifolia Roxb.
Fabaceae/Papilionaceae
♦ Indian rosewood, Malabar rosewood, sitsal, beete, shisham, Bombay blackwood, sonokeling, sonobrits
♦ Weed, Introduced
♦ 22, 228
♦ T, arid, cultivated, herbal.

Dalbergia melanoxylon Guill. & Perr.
Fabaceae/Papilionaceae
♦ African blackwood, atiyi, babanus, blackwood, Cape Damson ebony, Congowood, dalbergia, ebony, grenadille wood, moghano, Mozambique ebony, mpingo, mufunjo, pau preto, Senegal ebony, sibbe, Ungoro ebony, zebrawood
♦ Weed, Quarantine Weed, Native Weed, Introduced
♦ 121, 149, 228
♦ S/T, arid, cultivated. Origin: southern Africa.

Dalbergia sissoo Roxb. ex DC.
Fabaceae/Papilionaceae
Amerimnon sissoo (Roxb. ex DC.) Kuntze
♦ Indian dalbergia, sissoo, shisham, skuva, sissu, tali, dalbergia, Himalaya rain tree, Indian rosewood
♦ Weed, Quarantine Weed, Noxious Weed, Naturalised, Garden Escape, Environmental Weed, Cultivation Escape
♦ 3, 22, 76, 80, 86, 88, 93, 98, 101, 112, 122, 147, 155, 179, 203, 205, 220, 261
♦ T, arid, cultivated, herbal. Origin: India.

Dalea alopecuroides Willd.
Fabaceae/Papilionaceae
= *Dalea leporina* (Aiton) Bullock (NoR)
♦ Weed
♦ 87, 88
♦ arid.

Dalea annua (Mill.) Kuntze
Fabaceae/Papilionaceae
♦ Weed
♦ 157
♦ aH.

Dalea scandens (Mill.) R.T.Clausen
Fabaceae/Papilionaceae
♦ low prairie clover
♦ Weed
♦ 14

Dalechampia scandens L.
Euphorbiaceae
Dalechampia brevipes Mull.
♦ spurge creeper
♦ Weed
♦ 255
♦ pH, herbal, toxic. Origin: Brazil.

Damasonium alisma Mill.
Alismataceae
Damasonium stellatum Pers.,
Actinocarpus damasonium Sm.
♦ thrumwort, starfruit
♦ Weed
♦ 23, 87, 88
♦ wa/pH, cultivated.

Damasonium australe Salisb. nom. illeg.
Alismataceae
= *Damasonium minus* Buchen.
♦ Weed, Quarantine Weed
♦ 87, 88, 220
♦ wa/pH.

Damasonium minus (R.Br.) Buchen.
Alismataceae
Actinocarpus minor R.Br., *Damasonium
australe* Salisb. *nom. illeg.* (see)
♦ starfruit
♦ Weed, Quarantine Weed, Native
Weed
♦ 68, 88, 200, 269
♦ wa/pH, cultivated. Origin:
Australia.

**Daniellia oliveri (Rolfe) Hutch. &
Dalziel**
Fabaceae/Caesalpiniaceae
Paradaniellia oliveri Rolfe
♦ African copaiba balsam tree, Accra
copal, African copaiba balsam, Benin
gum copal, Ilorin balsam, West African
copal, West African gum copal, chahar,
denchi, ekhimi, galhanga, kaharlahi,
madié, popde, sambam, tschade,
ulugui, wood oil tree, yulundii
♦ Weed
♦ 88
♦ arid, herbal.

Danthonia decumbens (L.) DC.
Poaceae
Poa decumbens Scop., *Sieglingia
decumbens* (L.) Bernh. (see)
♦ common heathgrass, hina
♦ Weed, Naturalised
♦ 98, 101, 176, 203, 272
♦ G, cultivated, herbal.

Danthonia spicata (L.) Beauv.
Poaceae
♦ poverty oatgrass, poverty oats
♦ Weed
♦ 23, 87, 88, 218, 291
♦ G, cultivated, herbal.

Daphne cneorum L.
Thymelaeaceae
Thymelaea cneorum Scop.
♦ rose daphne, garland flower

♦ Weed
♦ 39, 161, 247, 272
♦ cultivated, herbal, toxic.

Daphne genkwa Siebold & Zucc.
Thymelaeaceae
Daphne fortunei Lindl.
♦ lilac daphne
♦ Weed
♦ 39, 161, 247
♦ S, cultivated, herbal, toxic.

Daphne laureola L.
Thymelaeaceae
♦ spurge laurel
♦ Weed, Noxious Weed, Naturalised
♦ 39, 80, 88, 101, 137, 146, 161, 280
♦ S, cultivated, herbal, toxic.

Daphne mezereum L.
Thymelaeaceae
♦ February daphne, paradise plant,
mezereon, bois gentil, bois joli, dwarf
bay, flax olive, lady laurel, spurge
laurel, spurge olive
♦ Weed, Naturalised
♦ 39, 80, 101, 161, 189, 247
♦ S, cultivated, herbal, toxic.

Daphne odora Thunb.
Thymelaeaceae
♦ winter daphne
♦ Weed
♦ 161, 247
♦ S, cultivated, herbal, toxic. Origin:
China, Taiwan.

Daphne oleoides Schreb.
Thymelaeaceae
♦ dafne spatolata
♦ Casual Alien
♦ 280
♦ S, promoted, herbal.

Daphne tangutica Maxim.
Thymelaeaceae
♦ Chinese garland flower
♦ Weed
♦ 161
♦ cultivated, herbal, toxic.

**Daphniphyllum oldhamii (Hemsl.)
Rosenthal**
Daphniphyllaceae
Daphniphyllum glaucescens Blume var.
oldhamii Hemsl.
♦ Quarantine Weed
♦ 220

Darlingtonia brachyloba (Willd.) DC.
Sarraceniaceae
Acacia brachyloba Willd.
♦ Quarantine Weed
♦ 220

Darlingtonia glandulosa (Michx.) DC.
Sarraceniaceae
♦ Quarantine Weed
♦ 220

Darlingtonia intermedia Torr.
Sarraceniaceae
= *Darlingtonia brachyloba* (Willd.) DC.
var. *intermedia* (Torr.) Torr. & A.Gray
(NoR)
♦ Quarantine Weed
♦ 220

Dasylirion serratifolium (Karw.) Zucc.
Agavaceae/Dracaenaceae/Nolinaceae
Yucca serratifolia Karw., *Roulinia
serratifolia* (Karw.) Brongn., *Dasylirion
laxiflorum* Baker
♦ sotol
♦ Introduced
♦ 228
♦ arid, cultivated.

Dasylirion texanum Scheele
Agavaceae/Dracaenaceae/Nolinaceae
♦ Texas sotol
♦ Weed
♦ 87, 88, 218
♦ cultivated, herbal.

Dasylirion wheeleri S.Wats.
Agavaceae/Dracaenaceae/Nolinaceae
♦ wheeler sotol, spoon flower,
common sotol, desert spoon
♦ Weed
♦ 87, 88, 218
♦ T, cultivated, herbal.

**Dasypyrum villosum (L.) Coss. & Durieu
ex P.Candargy**
Poaceae
♦ mosquito grass
♦ Weed, Naturalised
♦ 101, 243, 272
♦ G, cultivated.

Datisca glomerata (C.Presl) Baill.
Datiscaceae
♦ Durango root
♦ Weed
♦ 39, 161
♦ pH, cultivated, herbal, toxic.

Datura L. spp.
Solanaceae
♦ thornapple, jimsonweed, angel's
trumpet, tree bielu
♦ Weed, Quarantine Weed,
Naturalised
♦ 198, 203, 220
♦ herbal.

Datura arborea L.
Solanaceae
= *Brugmansia candida* Pers.
♦ angel's trumpet, tree datura
♦ Weed, Quarantine Weed, Noxious
Weed, Naturalised
♦ 39, 76, 86, 87, 88, 203
♦ cultivated, herbal, toxic.

Datura × candida (Pers.) Saff.
Solanaceae
= *Brugmansia candida* Pers.
♦ moonflower, angel's trumpet
♦ Weed, Naturalised
♦ 39, 86, 88, 98, 121, 203
♦ S, cultivated, herbal, toxic.

Datura cornigera Hook.
Solanaceae
= *Brugmansia cornigera* (Hook.) Lagerh.
(NoR)
♦ thornapple
♦ Weed, Quarantine Weed, Noxious
Weed, Naturalised
♦ 76, 86, 88, 203
♦ cultivated.

Datura discolor Bernh.
Solanaceae
♦ desert thornapple, small datura
♦ Weed
♦ 87, 88, 161, 212, 218, 243
♦ pS, cultivated, herbal, toxic.

Datura fastuosa L.
Solanaceae
= *Datura metel* L.
♦ Weed
♦ 221
♦ herbal.

Datura ferox L.
Solanaceae
= *Datura quercifolia* Kunth
♦ fierce thornapple, longspine thornapple, large thornapple, thornapple, white stinkweed, Chinese thornapple
♦ Weed, Quarantine Weed, Noxious Weed, Naturalised, Introduced
♦ 7, 34, 39, 50, 51, 55, 62, 63, 76, 86, 87, 88, 93, 95, 98, 121, 147, 158, 161, 176, 180, 198, 203, 205, 236, 237, 243, 269, 270, 278, 280, 283, 287, 295, 300
♦ aH, cultivated, herbal, toxic. Origin: Eurasia.

Datura inermis Jacq.
Solanaceae
♦ smoooth thornapple
♦ Naturalised
♦ 101
♦ herbal.

Datura innoxia Mill.
Solanaceae
= *Datura inoxia* Mill.
♦ harige stinkblaar, downy thornapple, barambal, black datura, enumu, semina daturae, stinkblar, thornapple, toloache, hairy thornapple, Indian apple, sacred datura, white thornapple
♦ Weed, Noxious Weed, Introduced, Environmental Weed
♦ 63, 121, 228, 242, 257, 283
♦ arid, cultivated, herbal, toxic.

Datura inoxia Mill.
Solanaceae
Datura innoxia Mill. (see), *Datura metel* auct. non L., *Datura meteloides* DC. ex Dunal (see)
♦ downy thornapple, recurved thornapple, prickly burr, angel's trumpet, sacred datura
♦ Weed, Quarantine Weed, Noxious Weed, Naturalised, Introduced, Environmental Weed
♦ 7, 38, 39, 51, 62, 76, 86, 87, 88, 93, 95, 98, 101, 147, 152, 161, 185, 198, 203, 205, 212, 243, 261, 269, 300
♦ aH, arid, cultivated, herbal, toxic. Origin: southern North America.

Datura leichhardtii F.Muell. ex Benth.
Solanaceae
♦ native thornapple, Leichhardt's thornapple, Mexican thornapple
♦ Weed, Quarantine Weed, Noxious Weed, Naturalised

♦ 55, 62, 76, 86, 87, 88, 147, 203, 269
♦ cultivated. Origin: obscure, possibly Mexico.

Datura metel L.
Solanaceae
Datura alba Rumph. ex Nees, *Datura cornucopaea* hort. ex W.W., *Datura fastuosa* L. (see), *Datura fastuosa* L. var. *alba* (Nees) C.B.Clarke, *Datura milhummatu* Dunal
♦ hindu datura, downy thornapple, hairy thornapple, horn of plenty, recurved thornapple, thornapple, white thornapple
♦ Weed, Quarantine Weed, Noxious Weed, Naturalised, Introduced, Garden Escape, Cultivation Escape
♦ 7, 13, 32, 39, 62, 76, 86, 87, 88, 98, 121, 147, 161, 203, 218, 228, 230, 247, 252, 256, 261, 287, 297
♦ aH, arid, cultivated, herbal, toxic. Origin: tropical Asia.

Datura meteloides DC. ex Dunal
Solanaceae
= *Datura inoxia* Mill.
♦ sacred datura, tolguacha, Indian apple, moon lily, thornapple, moonflower
♦ Weed, Noxious Weed, Naturalised
♦ 86, 87, 88, 98, 180, 203, 218, 286, 287
♦ cultivated, herbal, toxic.

Datura quercifolia Kunth
Solanaceae
Datura ferox L. (see)
♦ Chinese thornapple, oak leaved angel's trumpet
♦ Naturalised
♦ 101
♦ aH, cultivated, herbal.

Datura stramonium L.
Solanaceae
Datura tatula L. (see)
♦ jimsonweed, Jamestown weed, thornapple, common thornapple, mad apple, stinkwort, embaleki, astanargit, atafaris, boruti, boruto, chayotillo, chemogong, colenso weed, common stinkapple, devil's apple, duling'weki, ebune, estramonio, frizillo, hoja de tapa, hulluruoho, msiafu, olieboom, pula, sikran, silulu, somena, stramonium, tapa, tapate, taturah, tlapa, tlaquoal, toloache, vue luate loco, zambumba
♦ Weed, Quarantine Weed, Noxious Weed, Naturalised, Introduced, Garden Escape, Environmental Weed, Cultivation Escape, Casual Alien
♦ 7, 8, 14, 15, 23, 24, 32, 34, 36, 39, 40, 42, 44, 49, 50, 51, 53, 55, 62, 63, 68, 70, 72, 76, 84, 86, 87, 88, 94, 95, 98, 101, 121, 133, 134, 147, 154, 157, 158, 161, 165, 174, 176, 180, 185, 189, 195, 198, 199, 203, 207, 210, 211, 212, 215, 217, 218, 221, 224, 228, 229, 236, 237, 240, 241, 242, 243, 246, 247, 250, 253, 255, 257, 258, 261, 263, 269, 272, 275, 278, 280, 283, 286, 287, 292, 295, 297, 299, 300

♦ aH, arid, cultivated, herbal, toxic. Origin: obscure, possibly tropical America.

Datura stramonium L. var. *chalybaea* Koch.
Solanaceae
♦ thornapple, jimsonweed, purple thornapple
♦ Weed
♦ 88, 217
♦ aH.

Datura suaveolens Humb. & Bonpl. ex Willd.
Solanaceae
= *Brugmansia suaveolens* (Humb. & Bonpl. ex Willd.) Bercht. & J.Presl
♦ angel's trumpet tree
♦ Naturalised
♦ 39, 287
♦ herbal, toxic.

Datura tatula L.
Solanaceae
= *Datura stramonium* L.
♦ thornapple, purple stramonium
♦ Weed, Noxious Weed, Naturalised
♦ 50, 86
♦ herbal, toxic.

Datura wrightii Regel
Solanaceae
♦ hairy thornapple, sacred thornapple, toluaca
♦ Weed, Quarantine Weed, Noxious Weed, Naturalised, Native Weed
♦ 39, 62, 76, 86, 88, 93, 98, 147, 198, 203, 205, 264, 269
♦ a/pH, cultivated, herbal, toxic. Origin: North America.

Daubentonia punicea (Cav.) DC.
Fabaceae/Papilionaceae
= *Sesbania punicea* (Cav.) Benth.
♦ Weed
♦ 87, 88
♦ herbal.

Daubentonia texana Pierce
Fabaceae/Papilionaceae
= *Sesbania drummondii* (Rydb.) Cory
♦ coffeeweed
♦ Weed
♦ 87, 88, 218

Daucus aureus Desf.
Apiaceae
♦ Weed
♦ 87, 88, 94

Daucus carota L.
Apiaceae
♦ Queen Anne's lace, wild carrot
♦ Weed, Noxious Weed, Naturalised, Introduced, Garden Escape, Environmental Weed, Casual Alien
♦ 1, 7, 8, 15, 23, 24, 34, 38, 39, 44, 45, 52, 68, 70, 80, 86, 87, 88, 94, 98, 101, 102, 121, 136, 138, 139, 146, 151, 159, 161, 165, 171, 174, 176, 179, 180, 195, 198, 199, 203, 207, 210, 211, 212, 217, 218, 229, 241, 243, 247, 249, 253, 261, 272, 275, 280, 286, 287, 294, 297, 299, 300
♦ aH, cultivated, herbal, toxic. Origin: Eurasia.

Daucus carota L. fo. *epurpuratus* Farw.
Apiaceae
♦ Naturalised
♦ 287

Daucus carota L. fo. *roseus* Millsp.
Apiaceae
♦ Naturalised
♦ 287

Daucus carota L. ssp. *carota*
Apiaceae
♦ wild carrot
♦ Weed, Noxious Weed
♦ 36, 229
♦ cultivated.

Daucus carota L. ssp. *maritimus* (Lam.) Batt.
Apiaceae
♦ Weed
♦ 250

Daucus carota L. ssp. *maximus* (Desf.) Ball
Apiaceae
♦ wild carrot
♦ Weed
♦ 115, 243, 250

Daucus carota L. ssp. *sativus* (Hoffm.) Arcang.
Apiaceae
♦ wild carrot
♦ Noxious Weed
♦ 229

Daucus crinitus Desf.
Apiaceae
♦ Weed
♦ 88, 94

Daucus glochidiatus (Lab.) Fisch., Mey. & Avé-Lall.
Apiaceae
Daucus brachiatus Sieber ex DC.,
Scandix glochidiata Labill.
♦ Australian carrot, wild carrot
♦ Weed, Naturalised, Native Weed
♦ 55, 87, 88, 134, 269
♦ aH, arid, cultivated. Origin: Australia, New Zealand.

Daucus guttatus Sibth. & Sm.
Apiaceae
♦ Weed
♦ 87, 88

Daucus littoralis Sm.
Apiaceae
♦ Weed
♦ 221
♦ cultivated.

Daucus maximum Desf.
Apiaceae
♦ Weed
♦ 87, 88

Daucus montanus Humb. & Bonpl. ex Schult.
Apiaceae
♦ vuoriporkkana
♦ Weed, Naturalised, Casual Alien
♦ 42, 87, 88, 157, 241
♦ aH, arid.

Daucus montevidensis Link ex Spreng.
Apiaceae

♦ Weed
♦ 87, 88

Daucus muricatus (L.) L.
Apiaceae
♦ Weed
♦ 87, 88
♦ cultivated.

Daucus pusillus Michx.
Apiaceae
♦ American wild carrot, rattlesnake weed, wild carrot
♦ Weed, Naturalised, Introduced
♦ 23, 87, 88, 228, 237, 241, 295, 300
♦ aH, arid, promoted, herbal.

Daucus setulosus Guss. ex DC.
Apiaceae
♦ Weed
♦ 87, 88

Davallia bilabiata Hosok.
Davalliaceae
♦ Introduced
♦ 230
♦ H.

Deamia Britt. & Rose spp.
Cactaceae
= *Selenicereus* (Berger) Britt. & Rose spp.
♦ Weed, Quarantine Weed
♦ 76, 88, 203

Decodon aquaticum J.F.Gmel.
Lythraceae
♦ Quarantine Weed
♦ 220

Decodon verticillatus (L.) Ell.
Lythraceae
♦ swamp loosestrife, waterwillow
♦ Weed, Quarantine Weed
♦ 87, 88, 218, 220
♦ aqua, cultivated, herbal.

Deinostema violacea (Maxim.) T.Yamaz.
Scrophulariaceae
Gratiola violacea Maxim.
♦ sawatougarashi
♦ Weed
♦ 87, 88, 286

Delairea odorata Lem.
Asteraceae
Senecio scandens Buch.-Ham. ex D.Don (see), *Senecio mikanoides* Otto ex Walp. (see)
♦ cape ivy, German ivy, Italian ivy, African ivy, climbing groundsel
♦ Weed, Noxious Weed, Naturalised, Garden Escape, Environmental Weed, Cultivation Escape
♦ 3, 7, 40, 72, 73, 86, 88, 98, 101, 155, 176, 198, 203, 231, 233, 289, 296
♦ C, cultivated, toxic. Origin: South Africa.

Delarbrea paradoxa Vieill.
Araliaceae
♦ Naturalised
♦ 134

Delilia biflora (L.) Kuntze
Asteraceae
Delilia berteri Spreng., *Elvira biflora* (L.) DC. (see), *Elvira martyni* Cass., *Meratia*

sprengelii Cass., *Milleria biflora* L., *Myosotis palustris* (L.) L.
♦ Naturalised
♦ 257
♦ aH.

Delonix elata (L.) Gamble
Fabaceae/Caesalpiniaceae
Poinciana elata L.
♦ mseele
♦ Weed, Introduced
♦ 221, 228
♦ arid, cultivated, herbal.

Delonix regia (Bojer ex Hook) Raf.
Fabaceae/Caesalpiniaceae
Poinciana regia Bojer (see)
♦ flame tree, flamboyant, arbol del fuego, atbot, atbot det fuegu, nangiosákura, nangyo, pilampwoia weitahta, sakuranirow, sekoula, ohai', flame of the forest, flame tree, peacock flower, poinciana, royal poinciana
♦ Weed, Naturalised, Introduced, Garden Escape, Environmental Weed
♦ 3, 22, 54, 86, 88, 93, 98, 101, 107, 151, 155, 179, 203, 228, 230, 257, 259, 261
♦ T, arid, cultivated, herbal. Origin: Madagascar.

Delosperma N.E.Br. spp.
Aizoaceae/Mesembryanthemaceae
♦ iceplant
♦ Weed, Casual Alien
♦ 39, 116, 280
♦ cultivated, herbal, toxic. Origin: South Africa.

Delosperma ecklonis (Salm) Schwant. var. *latifolium* L.Bolus
Aizoaceae/Mesembryanthemaceae
♦ Native Weed
♦ 121
♦ pH. Origin: southern Africa.

Delosperma herbeum (N.E.Br.) N.E.Br.
Aizoaceae/Mesembryanthemaceae
♦ highveld white vygie
♦ Native Weed
♦ 121
♦ pH, cultivated. Origin: southern Africa.

Delosperma litorale (Kensit) L.Bolus
Aizoaceae/Mesembryanthemaceae
♦ seaside delosperma
♦ Naturalised
♦ 101
♦ S.

Delosperma lydenburgense L.Bolus
Aizoaceae/Mesembryanthemaceae
♦ Native Weed
♦ 121
♦ pH, cultivated. Origin: South Africa.

Delosperma pottsii (L.Bolus) L.Bolus
Aizoaceae/Mesembryanthemaceae
♦ Native Weed
♦ 121
♦ pH. Origin: southern Africa.

Delphinium L. spp.
Ranunculaceae
♦ delphinium, larkspur, staggerweed
♦ Weed, Casual Alien

♦ 15, 154, 247, 280
♦ cultivated, herbal, toxic.

Delphinium andersonii A.Gray
Ranunculaceae
♦ Anderson's larkspur
♦ Weed
♦ 39, 87, 88, 161
♦ pH, herbal, toxic.

Delphinium anthriscifolium Hance
Ranunculaceae
♦ Naturalised
♦ 287

Delphinium axilliflorum DC.
Ranunculaceae
♦ Weed
♦ 87, 88

Delphinium barbeyi (Huth) Huth
Ranunculaceae
♦ tall larkspur, subalpine larkspur
♦ Weed
♦ 23, 39, 87, 88, 161, 218
♦ cultivated, herbal, toxic.

Delphinium bicolor Nutt.
Ranunculaceae
♦ little larkspur, low larkspur,
Montana larkspur
♦ Weed, Noxious Weed
♦ 36, 39, 87, 88, 161, 218, 299
♦ cultivated, herbal, toxic.

Delphinium californicum Torr. & Gray
Ranunculaceae
♦ California larkspur, coast larkspur
♦ Weed
♦ 161
♦ pH, cultivated, herbal, toxic.

Delphinium carolinianum Walter
Ranunculaceae
♦ Carolina larkspur
♦ Weed
♦ 161
♦ herbal, toxic.

**Delphinium carolinianum Walter ssp.
virescens (Nutt.) Brooks**
Ranunculaceae
Delphinium virescens Nutt. (see)
♦ Carolina larkspur, plains larkspur,
prairie larkspur
♦ Native Weed
♦ 161, 174
♦ toxic. Origin: North America.

Delphinium cheilanthum Fisch. ex DC.
Ranunculaceae
♦ garland larkspur
♦ Weed
♦ 39, 161
♦ cultivated, toxic.

Delphinium consolida L.
Ranunculaceae
= *Consolida regalis* S.F.Gray
♦ larkspur
♦ Weed
♦ 23, 87, 88, 243
♦ aH, cultivated, herbal.

Delphinium cossonianum Batt.
Ranunculaceae
♦ Weed
♦ 87, 88

Delphinium × cultorum Voss
Ranunculaceae
= *Delphinium* × hybrids *hort.* (NoR)
♦ jaloritarinkannus
♦ Cultivation Escape
♦ 42
♦ cultivated.

Delphinium elatum L.
Ranunculaceae
Delphinium alpinum Wald. & Kit.,
Delphinium intermedium Ait.,
Delphinium tirolense Kern.
♦ isoritarinkannus, candle larkspur
♦ Weed, Naturalised, Cultivation
Escape
♦ 39, 42, 101, 161
♦ pH, cultivated, herbal, toxic.

Delphinium cf. elatum L.
Ranunculaceae
♦ candle larkspur
♦ Casual Alien
♦ 280

Delphinium exaltatum Aiton
Ranunculaceae
♦ tall larkspur
♦ Weed
♦ 39, 161
♦ cultivated, herbal, toxic.

Delphinium geyeri Greene
Ranunculaceae
♦ Geyer's larkspur
♦ Weed
♦ 39, 87, 88, 161, 212, 218
♦ toxic.

**Delphinium geyeri Greene var. wootonii
(Rydb.) K.C.Davis**
Ranunculaceae
= *Delphinium wootonii* Rydb.
♦ Wooton plains larkspur
♦ Weed
♦ 218

Delphinium glaucum S.Wats.
Ranunculaceae
♦ tall larkspur, Sierra larkspur
♦ Weed, Quarantine Weed, Noxious
Weed
♦ 36, 39, 76, 87, 88, 161, 203, 220
♦ pH, cultivated, herbal, toxic.

Delphinium grandiflorum L.
Ranunculaceae
Delphinium sinense Fisch. ex Link
♦ kiinanritarinkannus, Siberian
larkspur
♦ Weed, Naturalised, Cultivation
Escape
♦ 42, 101, 297
♦ a/pH, cultivated, herbal.

**Delphinium grandiflorum L. var.
glandulosum W.T.Wang**
Ranunculaceae
♦ Siberian larkspur
♦ Weed
♦ 275
♦ pH.

Delphinium halteratum Sibth. & Sm.
Ranunculaceae
♦ Weed
♦ 87, 88, 94

Delphinium hesperium A.Gray
Ranunculaceae
♦ foothill larkspur, western larkspur,
coast larkspur
♦ Weed
♦ 161
♦ pH, cultivated, herbal, toxic.

Delphinium megacarpum Nels. & MacBr.
Ranunculaceae
♦ sagebrush larkspur
♦ Weed
♦ 87, 88, 218

Delphinium menziesii DC.
Ranunculaceae
♦ Menzies' larkspur
♦ Weed
♦ 39, 161
♦ pH, cultivated, herbal, toxic.

Delphinium nelsonii Greene
Ranunculaceae
= *Delphinium nuttallianum* Pritz. ex
Walp.
♦ low larkspur
♦ Weed
♦ 87, 88, 218
♦ cultivated, herbal.

Delphinium nuttallianum Pritz. ex Walp.
Ranunculaceae
Delphinium nelsonii Greene (see)
♦ low larkspur, Nuttal's larkspur,
meadow larkspur, twolobe larkspur
♦ Weed, Native Weed
♦ 39, 161, 212, 264
♦ pH, cultivated, herbal, toxic. Origin:
North America.

Delphinium occidentale (Wats.) Wats.
Ranunculaceae
♦ duncecap larkspur
♦ Weed
♦ 88, 161, 212, 218
♦ cultivated, herbal, toxic.

Delphinium orientale J.Gay
Ranunculaceae
= *Consolida orientalis* (Gay) Schröd.
♦ Weed
♦ 87, 88
♦ cultivated, herbal.

Delphinium parryi A.Gray
Ranunculaceae
♦ Parry's larkspur, San Bernardino
larkspur
♦ Weed
♦ 39, 161
♦ pH, cultivated, herbal, toxic.

Delphinium patens Benth.
Ranunculaceae
♦ spreading larkspur, zigzag larkspur,
smooth larkspur
♦ Weed
♦ 161
♦ pH, cultivated, herbal, toxic.

Delphinium recurvatum E.Greene
Ranunculaceae
♦ Byron larkspur, alkali larkspur
♦ Weed
♦ 39, 161
♦ pH, toxic.

Delphinium scaposum **Greene**
Ranunculaceae
♦ barestem larkspur, tall mountain larkspur
♦ Weed
♦ 87, 88, 161, 218
♦ cultivated, herbal, toxic.

Delphinium scopulorum **A.Gray**
Ranunculaceae
♦ tall mountain larkspur, Rocky Mountain larkspur, pale larkspur
♦ Weed
♦ 39, 87, 88, 161, 218
♦ cultivated, herbal, toxic.

Delphinium staphisagria **L.**
Ranunculaceae
♦ licebane, stavesacre
♦ Weed
♦ 39, 70
♦ b/pH, cultivated, herbal, toxic.

Delphinium tricorne **Michx.**
Ranunculaceae
♦ dwarf larkspur
♦ Weed
♦ 39, 87, 88, 161, 218
♦ herbal, toxic.

Delphinium trolliifolium **A.Gray**
Ranunculaceae
♦ Columbian larkspur, cow poison
♦ Weed
♦ 39, 161
♦ pH, herbal, toxic.

Delphinium variegatum **Torr. & Gray**
Ranunculaceae
♦ royal larkspur
♦ Weed
♦ 161
♦ pH, toxic.

Delphinium virescens **Nutt.**
Ranunculaceae
= *Delphinium carolinianum* Walt. ssp. *virescens* (Nutt.) Brooks
♦ plains larkspur, prairie larkspur
♦ Weed
♦ 39, 87, 88, 218
♦ cultivated, herbal, toxic.

Delphinium wootonii **Rydb.**
Ranunculaceae
Delphinium geyeri Greene var. *wootonii* (Rydb.) K.C.Davis (see)
♦ Wooton plains larkspur, Organ Mountain larkspur
♦ Weed
♦ 161

Demnosa **A.V.Fric spp. not validly published**
Cactaceae
= *Cleistocactus* Lem. spp.
♦ Weed, Quarantine Weed
♦ 76, 88, 220

Dendranthema **(DC.) Des Moul. spp.**
Asteraceae
♦ Arctic daisy
♦ Casual Alien
♦ 280

Dendranthema boreale **(Makino) Ling**
Asteraceae

♦ Weed
♦ 275, 297
♦ pH.

Dendranthema × *grandiflorum* **(Ramat.) Kitam.**
Asteraceae
= *Chrysanthemum* × *morifolium* Ramat.
♦ florist's daisy, chrysanthemum
♦ Naturalised
♦ 101
♦ pH, cultivated. Origin: horticultural hybrid.

Dendranthema indicum **(L.) Desmoulins**
Asteraceae
♦ krysanteemi
♦ Cultivation Escape
♦ 42
♦ pH, cultivated. Origin: China, Japan, Korea, Taiwan.

Dendrobenthamia capitata **(Wall. ex Roxb.) Hutch.**
Cornaceae
= *Cornus capitata* Wall.
♦ strawberry dogwood
♦ Weed, Naturalised
♦ 15, 280

Dendrobium adamsii **Hawkes**
Orchidaceae
♦ Introduced
♦ 230
♦ H.

Dendrobium carolinense **Schltr.**
Orchidaceae
♦ rahngh
♦ Introduced
♦ 230
♦ H.

Dendrobium crumenatum **Sw.**
Orchidaceae
♦ white dove orchid
♦ Weed
♦ 19
♦ cultivated, herbal.

Dendrobium delicatulum **Kranzlin**
Orchidaceae
Dendrobium nanarauticolum Fuk.
♦ Introduced
♦ 230
♦ H, cultivated.

Dendrobium ponapense **Schltr.**
Orchidaceae
♦ Introduced
♦ 230
♦ H.

Dendrobium pseudo-kraemeri **Fuk.**
Orchidaceae
♦ Introduced
♦ 230
♦ H.

Dendrobium scopa **Lindl.**
Orchidaceae
Dendrobium amesianum Schltr.
♦ lily of the valley orchid
♦ Introduced
♦ 230
♦ H, cultivated.

Dendrobium violaceo-miniatum **Schltr.**
Orchidaceae

♦ Introduced
♦ 230
♦ H.

Dendrocalamus giganteus **Wall. ex Munro**
Poaceae
♦ bamboo, giant bamboo
♦ Weed
♦ 121
♦ pG, cultivated. Origin: Asia.

Dendrocereus **Britt. & Rose spp.**
Cactaceae
= *Acanthocereus* (Engelm. ex Berg.) Britt. & Rose spp.
♦ Weed, Quarantine Weed
♦ 76, 88

Dendrocnide kusaiana **(Kaneh.) Chew**
Urticaceae
Laportea kusiana Kaneh.
♦ Introduced
♦ 230
♦ C.

Dendropemon emarginatus **(Sw.) Steud.**
Loranthaceae
♦ Weed
♦ 14
♦ parasitic, herbal.

Dendrophthoe falcata **(L.f.) Ettingsh.**
Loranthaceae
♦ Weed
♦ 87, 88
♦ parasitic, cultivated, herbal. Origin: Australia.

Dendrophthora mancinella **(Wr.) Eichl.**
Viscaceae
♦ Weed
♦ 14
♦ H parasitic.

Dendrosenecio johnstonii **(Oliv.) B.Nord.**
Asteraceae
♦ Quarantine Weed
♦ 220

Dennstaedtia punctilobula **(Michx.) Moore**
Dennstaedtiaceae
♦ hay scented fern, eastern hay scented fern
♦ Weed
♦ 87, 88, 218, 294
♦ herbal.

Dentella repens **(L.) J.R. & G.Forst.**
Rubiaceae
Dentella matsudai Hayata, *Oldenlandia repens* L., *Dentella repens* var. *grandis* Pierre ex Pit.
♦ Weed
♦ 13, 87, 88, 170
♦ pH, herbal.

Deparia petersenii **(Kunze) M.Kato**
Dryopteridaceae/Woodsiaceae
Deparia petersonii (Kunze) M.Kato (see), *Diplazium japonicum* (Thunb.) Bedd. (see)
♦ Weed
♦ 80
♦ cultivated.

Deparia petersonii **(Kunze) M.Kato**
Dryopteridaceae/Woodsiaceae

= *Deparia petersenii* (Kunze) M.Kato
- ♦ Japanese false spleenwort
- ♦ Naturalised
- ♦ 101

Derris elliptica (Wall.) Benth.
Fabaceae/Papilionaceae
Deguelia elliptica (Roxb.) Taub.,
Galedupa elliptica Roxb., *Pongamia
elliptica* Wall.
- ♦ rotenone, peinuhp, derris, tubaroot
- ♦ Weed, Quarantine Weed,
Naturalised, Introduced
- ♦ 22, 39, 76, 86, 87, 88, 203, 220, 230
- ♦ pC, cultivated, herbal, toxic.

Derris indica (Lam.) Bennet
Fabaceae/Papilionaceae
= *Millettia pinnata* (L.) Panigrahi (NoR)
[see *Pongamia pinnata* (L.) Pierre]
- ♦ pongam, pongame oiltree, karum
tree
- ♦ Weed, Quarantine Weed,
Naturalised, Cultivation Escape
- ♦ 3, 76, 88, 101, 122, 161, 191, 203
- ♦ cultivated, toxic.

Deschampsia antartica Desv.
Poaceae
- ♦ Quarantine Weed
- ♦ 220
- ♦ G.

Deschampsia caespitosa (L.) Beauv.
Poaceae
= *Deschampsia cespitosa* (L.) Beauv.
- ♦ fescue leaved hairgrass, tufted
hairgrass, tussockgrass
- ♦ Weed, Naturalised
- ♦ 86, 87, 88, 91, 121
- ♦ pG, cultivated, herbal. Origin:
obscure.

Deschampsia cespitosa (L.) Beauv.
Poaceae
Deschampsia caespitosa (L.) Beauv. (see)
- ♦ tufted hairgrass
- ♦ Weed
- ♦ 70, 243, 272
- ♦ pG, cultivated, herbal. Origin:
Eurasia, Africa, Australia.

**Deschampsia cespitosa (L.) Beauv. var.
parviflora (Thuill.) Coss. & Germ.**
Poaceae
- ♦ small flowered ticklegrass
- ♦ Weed
- ♦ 195
- ♦ G.

**Deschampsia danthonioides (Trin.)
Munro ex Benth.**
Poaceae
- ♦ annual hairgrass
- ♦ Naturalised
- ♦ 287
- ♦ aG, arid, herbal.

**Deschampsia elongata (Hook.) Munro
ex Benth.**
Poaceae
- ♦ slender hairgrass
- ♦ Weed
- ♦ 87, 88, 161
- ♦ pG, herbal.

Deschampsia flexuosa (L.) Trin.
Poaceae
Aira flexuosa L., *Avenella flexuosa* (L.)
Parl., *Lerchenfeldia flexuosa* (L.) Schur
- ♦ crinkled hairgrass, wavy hairgrass,
common hairgrass, metsälauha
- ♦ Weed, Naturalised
- ♦ 70, 87, 88, 91, 121, 272, 280
- ♦ pG, cultivated, herbal. Origin:
obscure.

Descurainia argentina O.E.Schulz
Brassicaceae
- ♦ altamisa colorada
- ♦ Weed
- ♦ 87, 88, 237, 295

Descurainia longipedicellata O.E.Schulz
Brassicaceae
- ♦ slimstem tansymustard
- ♦ Weed
- ♦ 87, 88, 218

Descurainia pinnata (Wait.) Britt.
Brassicaceae
Sisymbrium canescens Walt., *Sophia
halictorum* Cockerell, *Sophia pinnata*
(Walt.) T.J.Howell
- ♦ tansymustard, sirolitutilli, green
tansymustard, pinnate tansymustard,
western tansymustard
- ♦ Weed, Noxious Weed, Naturalised,
Native Weed, Casual Alien
- ♦ 23, 39, 42, 87, 88, 136, 161, 162, 174,
210, 212, 218, 243, 249, 287, 300
- ♦ aH, arid, cultivated, herbal, toxic.
Origin: western North America.

**Descurainia pinnata (Walt.) Britt. var.
brachycarpa (Richards) Fern.**
Brassicaceae
= *Descurainia pinnata* (Walt.) Britt. ssp.
brachycarpa (Richards.) Detling (NoR)
- ♦ green tansymustard
- ♦ Weed
- ♦ 218

**Descurainia richardsonii (Sweet)
O.E.Schulz**
Brassicaceae
= *Descurainia incana* (Bernh. ex Fisch. &
C.A.Mey.) Dorn ssp. *incana* (NoR)
- ♦ Richardson tansymustard,
tansymustard, lännenlitutilli
- ♦ Weed, Casual Alien
- ♦ 42, 87, 88, 136, 161
- ♦ a/bH, arid, promoted, herbal, toxic.

Descurainia sophia (L.) Webb ex Prantl
Brassicaceae
Arabis sophia Bernh., *Descurea sophia*
Schur, *Sisymbrium sophia* L. (see),
Sophia vulgaris Fourn.
- ♦ flixweed, tansymustard, herb Sophia
- ♦ Weed, Noxious Weed, Naturalised,
Introduced, Casual Alien
- ♦ 23, 34, 35, 40, 52, 70, 78, 80, 86, 87, 88,
94, 98, 101, 136, 138, 146, 161, 162, 180,
203, 212, 218, 228, 229, 241, 243, 248,
253, 272, 275, 280, 287, 294, 297, 299,
300
- ♦ a/bH, arid, cultivated, herbal, toxic.
Origin: Eurasia.

Desmanthus acuminatus Benth.
Fabaceae/Mimosaceae
Acuan acuminata (Benth.) Kuntze (see),
Acuania acuminata Kuntze (see)
- ♦ Quarantine Weed
- ♦ 220

Desmanthus arsenei Standl.
Fabaceae/Mimosaceae
Acuan arsenei Britton & Rose (see)
- ♦ Quarantine Weed
- ♦ 220

Desmanthus balsensis J.L.Contr.
Fabaceae/Mimosaceae
- ♦ Quarantine Weed
- ♦ 220

Desmanthus bicornutus S.Wats.
Fabaceae/Mimosaceae
Acuan bicornutum Britton & Rose (see),
Acuan sublatum Britton & Rose (see)
- ♦ Quarantine Weed
- ♦ 220

Desmanthus brachylobus (Willd.) Benth.
Fabaceae/Mimosaceae
= *Desmanthus illinoensis* (Michx.)
MacMill. ex B.L.Rob. & Fern.
- ♦ Quarantine Weed
- ♦ 220

**Desmanthus covillei (Britt. & Rose)
Wiggins ex B.L.Turner**
Fabaceae/Mimosaceae
Acuan covillei Britton & Rose
- ♦ Coville's bundleflower
- ♦ Quarantine Weed
- ♦ 220
- ♦ arid.

**Desmanthus depressus Humb. & Bonpl.
ex Willd.**
Fabaceae/Mimosaceae
= *Desmanthus virgatus* (L.) Willd. var.
depressus (Humb. & Bonpl. ex Willd.)
B.L.Turner (NoR)
- ♦ Quarantine Weed
- ♦ 220
- ♦ arid.

Desmanthus falcatus Scheele
Fabaceae/Mimosaceae
- ♦ Quarantine Weed
- ♦ 220

Desmanthus fallax K.Schum.
Fabaceae/Mimosaceae
Acuan fallax Small (see)
- ♦ Quarantine Weed
- ♦ 220

Desmanthus fruticosus Rose
Fabaceae/Mimosaceae
Acuan fruticosum Standl. (see)
- ♦ Quarantine Weed
- ♦ 220
- ♦ arid, herbal.

**Desmanthus glandulosus (B.L.Turner)
Luckow**
Fabaceae/Mimosaceae
- ♦ glandular bundleflower
- ♦ Quarantine Weed
- ♦ 220
- ♦ herbal.

Desmanthus illinoensis (Michx.) MacMill. ex B.L.Rob. & Fern.
Fabaceae/Mimosaceae
Acuan illinoensis (Michx.) Kuntze
(see), *Acuania illinoense* Kuntze (see),
Desmanthus brachylobus (Willd.) Benth.
(see), *Mimosa illinoensis* Michx. (see)
♦ Illinois bundleflower, prairie
mimosa, prickleweed
♦ Weed, Quarantine Weed,
Naturalised
♦ 161, 220, 286, 287
♦ pH, cultivated, herbal.

Desmanthus insularis (Britton & Rose) León
Fabaceae/Mimosaceae
Acuan insulare Britton & Rose (see)
♦ Quarantine Weed
♦ 220

Desmanthus interior (Britton & Rose) Bullock
Fabaceae/Mimosaceae
Acuan interior Britton & Rose (see)
♦ Quarantine Weed
♦ 220

Desmanthus leptolobus Torr. & A.Gray
Fabaceae/Mimosaceae
Acuan leptoloba (Torr. & A.Gray)
Kuntze (see)
♦ slenderlobe bundleflower
♦ Quarantine Weed
♦ 220
♦ cultivated, herbal.

Desmanthus leptophyllus Kunth
Fabaceae/Mimosaceae
♦ slenderleaf bundleflower
♦ Weed, Quarantine Weed
♦ 179, 220

Desmanthus multiglandulosus Britton & Killip
Fabaceae/Mimosaceae
♦ Quarantine Weed
♦ 220

Desmanthus obtusus S.Watson
Fabaceae/Mimosaceae
Acuan obtusa A.Heller (see)
♦ bluntpod bundleflower
♦ Quarantine Weed
♦ 220

Desmanthus painteri Standl.
Fabaceae/Mimosaceae
Acuan painterii Britton & Rose (see)
♦ Quarantine Weed
♦ 220

Desmanthus palmeri (Britton & Rose) Wiggins ex Turner
Fabaceae/Mimosaceae
Acuan palmeri Britton & Rose (see)
♦ Quarantine Weed
♦ 220

Desmanthus paspalaceus (Lindm.) Burkart
Fabaceae/Mimosaceae
♦ Quarantine Weed
♦ 220

Desmanthus pedunculatus Buckl.
Fabaceae/Mimosaceae
♦ Quarantine Weed
♦ 220

Desmanthus pringlei (Britton & Rose) F.J.Herm.
Fabaceae/Mimosaceae
Acuan pringlei Britton & Rose (see)
♦ Quarantine Weed
♦ 220

Desmanthus pumilus (Schlecht.) J.F.Macbr.
Fabaceae/Mimosaceae
♦ Quarantine Weed
♦ 220

Desmanthus reticulatus Benth.
Fabaceae/*Mimosaceae*
Acuan reticulata (Benth.) Kuntze (see),
Acuania reticulata Kuntze (see)
♦ netleaf bundleflower
♦ Quarantine Weed
♦ 220

Desmanthus rhombifolius Buckl.
Fabaceae/Mimosaceae
♦ Quarantine Weed
♦ 220

Desmanthus rostruatus B.L.Turner
Fabaceae/Mimosaceae
♦ Quarantine Weed
♦ 220

Desmanthus sublatus (Britton & Rose) Wiggins ex B.L.Turner
Fabaceae/Mimosaceae
♦ Quarantine Weed
♦ 220

Desmanthus tatuhyensis Hoehne
Fabaceae/Mimosaceae
♦ Quarantine Weed
♦ 220

Desmanthus velutinus Scheele
Fabaceae/Mimosaceae
Acuan velutinum (Scheele) Kuntze (see)
♦ velvet bundleflower
♦ Quarantine Weed
♦ 220
♦ herbal.

Desmanthus virgatus (L.) Willd.
Fabaceae/Mimosaceae
Acuan virgatum (L.) Medik. (see),
Mimosa virgata L.
♦ ground tamarind, wild tantan, pena
da saracura, desmanthus
♦ Weed, Naturalised, Introduced
♦ 22, 86, 87, 88, 93, 98, 121, 157, 191,
203, 228, 286, 287, 295, 300
♦ pS, arid, cultivated, herbal. Origin:
South America.

Desmazeria rigida (L.) Tutin
Poaceae
= *Catapodium rigidum* (L.) C.E.Hubb.
♦ ferngrass, hard bluegrass, soulo,
rigid fescue
♦ Weed, Naturalised, Environmental
Weed, Casual Alien
♦ 15, 34, 42, 72, 86, 88, 98, 101, 203, 241,
272
♦ aG.

Desmodium Desv. spp.
Fabaceae/Papilionaceae
♦ tick clover, tick trefoil, pega pega
♦ Weed
♦ 281
♦ herbal.

Desmodium adscendens (Sw.) DC.
Fabaceae/Papilionaceae
Desmodium trifoliastrum Miq.
♦ tokiki, zarzabacoa galana, parecido
al pega pega, desmódio, carrapicho
♦ Weed
♦ 87, 88, 153, 157, 255
♦ pH, arid, herbal. Origin:
Madagascar.

Desmodium aparines (Link) DC.
Fabaceae/Papilionaceae
♦ Weed
♦ 199

Desmodium axillare (Sw.) DC.
Fabaceae/Papilionaceae
♦ zarzabacoa de monte, pega pega
grande
♦ Weed
♦ 87, 88, 153

Desmodium baccatum (Miq.) Schindl.
Fabaceae/Papilionaceae
♦ Weed
♦ 87, 88

Desmodium barbatum (L.) Benth. & Oerst.
Fabaceae/Papilionaceae
♦ zarzabacoa peluda, barbadinho
♦ Weed
♦ 14, 87, 88, 157, 255, 295
♦ pH, cultivated. Origin: Madagascar.

Desmodium cajanifolium (Kunth) DC.
Fabaceae/Papilionaceae
♦ tropical tick trefoil
♦ Weed, Naturalised
♦ 87, 88, 101
♦ herbal.

Desmodium canadense (L.) DC.
Fabaceae/Papilionaceae
♦ hoary tick clover, showy tick trefoil,
Canada tick trefoil
♦ Weed
♦ 87, 88, 218
♦ cultivated, herbal.

Desmodium canum Schinz & Thell. nom. illeg.
Fabaceae/Papilionaceae
= *Desmodium incanum* DC.
♦ kaimi clover, creeping beggarweed
♦ Weed, Naturalised
♦ 14, 87, 88, 161, 257, 287
♦ cultivated, herbal.

Desmodium cinerascens A.Gray
Fabaceae/Papilionaceae
Meibomia canbyi Schindl.
♦ spiked tick trefoil
♦ Introduced
♦ 228
♦ arid, herbal.

Desmodium cuspidatum (Muhl. ex Willd.) DC. ex Loudon
Fabaceae/Papilionaceae

Hedysarum cuspidatum Muhl. ex Willd.
♦ sticktights, beggar's ticks, largebract tick trefoil
♦ Weed, Quarantine Weed
♦ 76, 88
♦ herbal.

Desmodium dichotomum (Willd.) DC.
Fabaceae/Papilionaceae
♦ Weed
♦ 87, 88, 242
♦ arid.

Desmodium diffusum (Willd.) DC.
Fabaceae/Papilionaceae
= *Desmodium laxiflorum* DC. (NoR)
♦ Weed, Quarantine Weed
♦ 76, 87, 88, 203, 220

Desmodium discolor Vogel
Fabaceae/Papilionaceae
♦ carrapicho grande
♦ Weed
♦ 255
♦ pH, cultivated. Origin: tropical America.

Desmodium frutescens (Jacq.) Schindl.
Fabaceae/Papilionaceae
= *Desmodium incanum* DC.
♦ Weed
♦ 87, 88

Desmodium gangeticum (L.) DC.
Fabaceae/Papilionaceae
Desmodium gangeticum (L.) DC. var. *maculatum* (L.) Baker, *Desmodium natalitium* Sond., *Hedysarum lanceolatum* Schum. & Thonn.
♦ asumat, salparni
♦ Weed
♦ 87, 88, 275
♦ pH, arid, cultivated, herbal. Origin: Africa, Asia.

Desmodium glabrum (Mill.) DC.
Fabaceae/Papilionaceae
♦ zarzabacoa dulce
♦ Naturalised
♦ 257
♦ arid.

Desmodium gyroides (Roxb. ex Link) DC.
Fabaceae/Papilionaceae
= *Codariocalyx gyroides* (Roxb. ex Link) Hassk.
♦ Cultivation Escape
♦ 261
♦ cultivated, herbal. Origin: tropical Asia.

Desmodium heterocarpon (L.) DC.
Fabaceae/Papilionaceae
♦ Asian tick trefoil, variable desmodium
♦ Weed, Naturalised
♦ 87, 88, 101, 286
♦ herbal. Origin: Asia, Australia.

Desmodium heterocarpon (L.) DC. var. *strigosum* Meeuwen
Fabaceae/Papilionaceae
♦ Asian tick trefoil
♦ Naturalised
♦ 101

Desmodium heterophyllum (Willd.) DC.
Fabaceae/Papilionaceae
♦ variableleaf tick trefoil, hetero desmodium
♦ Weed, Naturalised
♦ 86, 87, 88, 101, 235
♦ cultivated, herbal. Origin: Asia.

Desmodium heterophyllum (Willd.) DC. var. *heterophyllum*
Fabaceae/Papilionaceae
♦ Introduced
♦ 230

Desmodium illinoense A.Gray
Fabaceae/Papilionaceae
♦ Illinois tick trefoil
♦ Weed, Naturalised
♦ 161, 287
♦ herbal.

Desmodium incanum DC.
Fabaceae/Papilionaceae
Desmodium canum Schinz & Thell. *nom. illeg.* (see), *Desmodium frutescens* (Jacq.) Schindl. (see), *Desmodium supinum* DC. (see)
♦ Spanish clover, zarzabacoa comun, amores do campo
♦ Weed, Naturalised, Introduced
♦ 6, 28, 86, 88, 134, 191, 206, 228, 243, 249, 255, 295
♦ pH, arid, herbal. Origin: tropical America.

Desmodium intortum (Mill.) Urb.
Fabaceae/Papilionaceae
Desmodium hjalmarsonii Standl., *Desmodium sonorae* A.Gray, *Desmodium trigonum* (Sw.) DC., *Hedysarum intortum* Mill., *Meibomia intorta* (Mill.) Blake
♦ greenleaf desmodium, beggar's lice
♦ Weed, Naturalised, Introduced, Environmental Weed
♦ 54, 86, 88, 98, 157, 203, 228, 287
♦ pC, arid, cultivated, herbal. Origin: Central America.

Desmodium lasiocarpum (P.Beauv.) DC.
Fabaceae/Papilionaceae
♦ Weed
♦ 87, 88
♦ cultivated.

Desmodium limense Hook.
Fabaceae/Papilionaceae
♦ Naturalised
♦ 257
♦ arid.

Desmodium paniculatum (L.) DC.
Fabaceae/Papilionaceae
♦ panicledleaf tick trefoil, panicled tick trefoil
♦ Weed, Naturalised
♦ 286, 287
♦ cultivated, herbal.

Desmodium penduliflorum Oudem.
Fabaceae/Papilionaceae
♦ Weed, Quarantine Weed
♦ 76, 88, 220

Desmodium podocarpum DC. ssp. *oxyphyllum* (DC.) H.Ohashi
Fabaceae/Papilionaceae
♦ nusubitohagi
♦ Weed
♦ 286

Desmodium polycarpum (Poir.) DC.
Fabaceae/Papilionaceae
♦ Weed
♦ 87, 88
♦ herbal.

Desmodium polygaloides Chod. & Hassl.
Fabaceae/Papilionaceae
♦ Weed
♦ 87, 88

Desmodium procumbens (Mill.) A.S.Hitchc.
Fabaceae/Papilionaceae
♦ western trailing tick trefoil
♦ Weed, Quarantine Weed
♦ 76, 87, 88, 203, 220

Desmodium procumbens (Mill.) A.S.Hitchc. var. *procumbens*
Fabaceae/Papilionaceae
♦ western trailing tick trefoil
♦ Introduced
♦ 228
♦ arid.

Desmodium psilocarpum A.Gray
Fabaceae/Papilionaceae
♦ Santa Cruz Island tick trefoil
♦ Introduced
♦ 228
♦ arid.

Desmodium purpureum (Mill.) Fawc. & Rendle *nom. illeg.*
Fabaceae/Papilionaceae
= *Desmodium tortuosum* (Sw.) DC.
♦ desmódio, pega pega
♦ Weed
♦ 245

Desmodium ramosissimum G.Don.
Fabaceae/Papilionaceae
♦ Weed
♦ 87, 88
♦ herbal. Origin: Madagascar.

Desmodium rigidum (Elliott) DC.
Fabaceae/Papilionaceae
= *Desmodium obtusum* (Muhl. ex Willd.) DC. (NoR)
♦ Naturalised
♦ 287
♦ herbal.

Desmodium salicifolium (Poir.) DC.
Fabaceae/Papilionaceae
♦ Weed
♦ 88
♦ pH. Origin: Madagascar.

Desmodium sandwicense E.Mey.
Fabaceae/Papilionaceae
♦ Hawai'i tick trefoil, Spanish clover
♦ Weed, Naturalised
♦ 87, 88, 101, 287
♦ cultivated, herbal.

Desmodium scorpiurus (Sw.) Desv.
Fabaceae/Papilionaceae
♦ scorpion tick trefoil
♦ Weed,
Naturalised
♦ 11, 14, 86, 87, 88, 93, 157, 179, 191,
287
♦ pH, arid, cultivated. Origin: tropical
Americas.

Desmodium sequax Wall.
Fabaceae/Papilionaceae
♦ Weed
♦ 87, 88

Desmodium sericophyllum Schlecht.
Fabaceae/Papilionaceae
♦ tick trefoil
♦ Weed
♦ 87, 88

Desmodium strigillosum Schindl.
Fabaceae/Papilionaceae
♦ Naturalised
♦ 86
♦ Origin: Burma, Cambodia, Laos,
Vietnam.

Desmodium styracifolium (Osbeck) Merr.
Fabaceae/Papilionaceae
Desmodium capitatum (Burm.f.) DC.
♦ Weed
♦ 87, 88
♦ pH, promoted, herbal.

Desmodium subsericeum Malme
Fabaceae/Papilionaceae
♦ Naturalised
♦ 241, 300

Desmodium supinum DC.
Fabaceae/Papilionaceae
= *Desmodium incanum* DC.
♦ Weed
♦ 87, 88

Desmodium thunbergii DC.
Fabaceae/Papilionaceae
♦ Weed, Quarantine Weed
♦ 76, 88, 220

Desmodium tortuosum (Sw.) DC.
Fabaceae/Papilionaceae
Desmodium purpureum (Mill.) Fawc. &
Rendle *nom. illeg.* (see)
♦ Florida beggarweed, pega pega,
Dixie tick trefoil, carrapicho
♦ Weed, Naturalised, Environmental
Weed
♦ 7, 23, 32, 86, 87, 88, 93, 98, 134, 153,
161, 203, 218, 245, 255
♦ aH, arid, cultivated, herbal. Origin:
tropical America.

Desmodium triflorum (L.) DC.
Fabaceae/Papilionaceae
♦ threeflower beggarweed, trebolillo,
creeping tick clover
♦ Weed, Naturalised, Introduced
♦ 14, 32, 86, 87, 88, 93, 153, 157, 161,
179, 218, 230, 249, 262, 273
♦ pH, cultivated, herbal. Origin:
pantropical.

Desmodium triquetrum (L.) DC.
Fabaceae/Papilionaceae
♦ Weed

♦ 12, 275
♦ herbal.

Desmodium uncinatum (Jacq.) DC.
Fabaceae/Papilionaceae
♦ Spanish tick clover, silverleaf
desmodium, ee nieu, pega pega
♦ Weed, Naturalised, Environmental
Weed
♦ 86, 87, 88, 98, 203, 218, 238, 255
♦ pH, cultivated, herbal. Origin:
tropical America.

Desmodium vargasianum B.G.Schub.
Fabaceae/Papilionaceae
♦ Weed
♦ 87, 88

Desmodium velutinum (Willd.) DC.
Fabaceae/Papilionaceae
♦ Weed
♦ 87, 88
♦ pH, arid, herbal. Origin: Africa, Asia,
Australia, Madagascar.

Desmostachya bipinnata (L.) Stapf
Poaceae
Briza bipinnata L., *Desmostachya
cynosuroides* (Retz.) Stapf ex Massey,
Coelachyrum longiglume Napper,
Cynosurus durus Forssk., *Eragrostis
bipinnata* (L.) Stapf, *Eragrostis
cynosuroides* (Retz.) Beauv., *Leptochloa
bipinnata* (L.) Hochst., *Poa cynosuroides*
Retz., *Pogonarthria bipinnata* (L.) Chiov.,
Stapfiola bipinnata (L.) Kuntze, *Uniola
bipinnata* L.
♦ halfa grass
♦ Weed, Quarantine Weed
♦ 76, 87, 88, 91, 185, 203, 220
♦ pG, arid, herbal.

Deuterocohnia chrysantha (Phil.) Mez
Bromeliaceae
Pitcairnia chrysantha Phil.
♦ Quarantine Weed
♦ 220
♦ arid.

Deutzia crenata Sieb. & Zucc.
Hydrangeaceae/Philadelphaceae
♦ crenate pride of Rochester
♦ Weed, Naturalised, Casual Alien
♦ 15, 101, 280, 286
♦ cultivated, herbal.

Deutzia parviflora Bunge
Hydrangeaceae/Philadelphaceae
♦ Mongolian pride of Rochester
♦ Naturalised
♦ 101

Deutzia scabra Thunb.
Hydrangeaceae/Philadelphaceae
Deutzia sieboldiana Maxim.
♦ fuzzy pride of Rochester
♦ Weed, Naturalised
♦ 80, 101
♦ S, cultivated, herbal. Origin: Japan.

**Deyeuxia viridiflavescens (Poir.) Steud.
var. montevidensis (Nees) Kämpf**
Poaceae
Calamagrostis montevidensis Nees (see),
Deyeuxia splendens Brongn.

♦ paja de plata
♦ Weed
♦ 237, 295
♦ G.

Dialium guineense Willd.
Fabaceae/Caesalpiniaceae
♦ velvet tamarind
♦ Weed
♦ 88
♦ arid, cultivated, herbal.

Diandrochloa glomerata (Walter) Burk.
Poaceae
= *Eragrostis japonica* (Thunb.) Trin.
♦ Weed
♦ 157
♦ G.

**Diandrochloa namaquensis (Nees) De
Winter**
Poaceae
= *Eragrostis japonica* (Thunb.) Trin.
♦ Native Weed
♦ 121
♦ aG. Origin: southern Africa.

Dianella caerulea Sims
Liliaceae/Phormiaceae
♦ cerulean flax lily, paroo lily, flax lily
♦ Naturalised
♦ 39, 101
♦ pH, cultivated, toxic. Origin:
Australia.

Dianella ensifolia (L.) DC.
Liliaceae/Phormiaceae
Dianella nemorosa Lam.
♦ common dianella
♦ Weed
♦ 12, 126, 179
♦ pH, cultivated, herbal.

Dianthera pectoralis (Jacq.) Murr.
Acanthaceae
= *Justicia pectoralis* Jacq.
♦ Weed
♦ 87, 88

Dianthus L. spp.
Caryophyllaceae
♦ carnation, clove pinks, rainbow
pinks, pink, sweet William
♦ Weed, Naturalised
♦ 198, 247, 272
♦ herbal, toxic.

Dianthus anatolicus Boiss.
Caryophyllaceae
♦ Weed
♦ 87, 88
♦ pH, cultivated, herbal.

Dianthus armeria L.
Caryophyllaceae
Caryophyllus armerius (L.) Moench
♦ Deptford pink, mykeröneilikka,
grass pink
♦ Weed, Naturalised, Introduced,
Environmental Weed, Cultivation
Escape, Casual Alien
♦ 21, 23, 42, 86, 87, 88, 98, 101, 165, 176,
195, 198, 203, 218, 272, 280, 286, 287,
300
♦ a/bH, cultivated, herbal. Origin:
Mediterranean.

Dianthus barbatus L.
Caryophyllaceae
Crantzaryophyllus barbatus Moench,
Tunica barbata (L.) Scop.
♦ sweet William, harjaneilikka
♦ Weed, Naturalised, Garden Escape,
Cultivation Escape
♦ 40, 42, 98, 101, 203, 280
♦ pH, cultivated, herbal. Origin:
southern Europe.

Dianthus barbatus L. ssp. barbatus
Caryophyllaceae
♦ sweet William
♦ Weed, Cultivation Escape
♦ 34
♦ pH, cultivated.

Dianthus carthusianorum L.
Caryophyllaceae
Caryophyllus carthusianorum Moench,
Tunica carthusianorum (L.) Scop.
♦ munkkineilikka, clusterhead,
clusterhead pink, Carthusian pink
♦ Weed, Naturalised, Casual Alien
♦ 42, 101, 272
♦ cultivated, herbal.

Dianthus caryophyllus L.
Caryophyllaceae
Dianthus coronarius Lam.
♦ tarhaneilikka, carnation, wild
carnation
♦ Naturalised, Cultivation Escape
♦ 39, 42, 101
♦ pH, cultivated, herbal, toxic.

Dianthus chinensis L.
Caryophyllaceae
♦ Chinese pink, rainbow pink
♦ Weed
♦ 39, 275, 297
♦ pH, cultivated, herbal, toxic.

Dianthus deltoides L.
Caryophyllaceae
Caryophyllus deltoides (L.) Moench
♦ maiden pink, ketoneilikka
♦ Weed, Naturalised
♦ 24, 101, 272, 280
♦ a/pH, cultivated, herbal.

Dianthus fragrans M.F.Adams
Caryophyllaceae
♦ fragrant pink
♦ Naturalised
♦ 101
♦ cultivated.

Dianthus plumarius L.
Caryophyllaceae
♦ feathered pink, wild pink, cottage
pink, pink
♦ Weed, Naturalised, Garden Escape,
Environmental Weed
♦ 39, 54, 86, 88, 101, 280
♦ pH, cultivated, herbal, toxic. Origin:
eastern Europe.

Dianthus prolifer L.
Caryophyllaceae
= *Petrorhagia prolifera* (L.) P.Ball &
Heywood
♦ Weed
♦ 87, 88

Dianthus sinaicus Boiss.
Caryophyllaceae
♦ Weed
♦ 221
♦ Origin: Israel, Sinai.

Dianthus superbus L.
Caryophyllaceae
Caryophyllus superbus (L.) Moench,
Dianthus wimmeri Wich.
♦ superb pink, fringed pink
♦ Weed
♦ 23, 88, 272, 297
♦ pH, cultivated, herbal.

**Dianthus superbus L. var. longicalycinus
(Maxim.) Will.**
Caryophyllaceae
♦ Weed
♦ 286
♦ cultivated.

Dianthus sylvestris Wulfen
Caryophyllaceae
Dianthus silvester Wulfen
♦ woodland pink, wood pink
♦ Naturalised
♦ 101
♦ cultivated, herbal.

Diarthron linifolium Turcz.
Thymelaeaceae
♦ Weed
♦ 275
♦ aH.

Diastatea micrantha (H.B.K.) McVaugh
Campanulaceae/Lobeliaceae
♦ Weed
♦ 157
♦ aH.

Diatoma ehrenbergii Kütz.
Bacillariophyceae
Diatoma grande W.Sm., *Diatoma vulgaris*
Bory var. *ehrenbergii* (Kütz) Grunow
♦ Weed
♦ 197
♦ diatom.

Dicentra canadensis (Goldie) Walp.
Papaveraceae/Fumariaceae
♦ squirrel corn, turkey corn
♦ Weed
♦ 39, 161
♦ pH, promoted, herbal, toxic.

**Dicentra chrysantha (Hook. & Arn.)
Walp.**
Papaveraceae/Fumariaceae
= *Ehrendorferia chrysantha* (Hook. &
Arn.) Rylander (NoR)
♦ golden eardrops
♦ Weed
♦ 161
♦ pH, cultivated, herbal, toxic.

Dicentra cucullaria (L.) Bernh.
Papaveraceae/Fumariaceae
♦ Dutchman's breeches
♦ Weed
♦ 8, 39, 161, 189
♦ pH, cultivated, herbal, toxic.

Dicentra eximia (Ker Gawl.) Torr.
Papaveraceae/Fumariaceae
♦ fringed bleedingheart, wild

bleedingheart, turkey corn,
bleedingheart
♦ Weed, Casual Alien
♦ 39, 40, 161
♦ cultivated, herbal, toxic.

Dicentra formosa (Haw.) Walp.
Papaveraceae/Fumariaceae
♦ western bleedingheart, Pacific
bleedingheart, bleedingheart
♦ Weed
♦ 39, 161
♦ pH, cultivated, herbal, toxic.

Dicentra ochroleuca Engelm.
Papaveraceae/Fumariaceae
= *Ehrendorferia ochroleuca* (Englem.)
Fukuhara (NoR)
♦ yellow bleedingheart, firehearts
♦ Weed
♦ 161
♦ pH, cultivated, herbal, toxic.

Dicentra pauciflora S.Watson
Papaveraceae/Fumariaceae
♦ small flowered bleedingheart,
shorthorn steershead
♦ Weed
♦ 161
♦ pH, herbal, toxic.

Dicentra spectabilis (L.) Lem.
Papaveraceae/Fumariaceae
= *Lamprocapnos spectabilis* (L.)
Fukuhara
♦ common bleedingheart,
bleedingheart
♦ Weed, Casual Alien
♦ 39, 40, 161, 189
♦ pH, cultivated, herbal, toxic.

Dicentra uniflora Kellogg
Papaveraceae/Fumariaceae
♦ steer's head, longhorn steer's head
♦ Weed
♦ 161
♦ pH, cultivated, herbal, toxic.

**Dicerocaryum eriocarpum (Decne.)
J.Abels.**
Pedaliaceae
= *Dicerocaryum zanguebarium* (Lour.)
Merr.
♦ Weed
♦ 88
♦ cultivated, herbal.

**Dicerocaryum zanguebarium (Lour.)
Merr.**
Pedaliaceae
Pretrea zanguebarica J.Gay, *Dicerocaryum
eriocarpum* (Decne.) J.Abels. (see)
♦ boot protector, muFeso, umGinga
ginga, devil's thorn
♦ Weed
♦ 50, 51, 87, 88, 158
♦ cultivated. Origin: South Africa.

**Dicerocaryum zanguebarium (Lour.)
Merr. ssp. zanguebarium**
Pedaliaceae
♦ boot protector plant, devil's thorn,
wild foxglove
♦ Native Weed
♦ 121
♦ pH. Origin: southern Africa.

Dichanthelium clandestinum **(L.) Gould**
Poaceae
= *Panicum clandestinum* L.
♦ deertongue, deertongue panicgrass
♦ Weed
♦ 87, 88
♦ G, herbal.

Dichanthium annulatum **(Forssk.) Stapf**
Poaceae
Andropogon annulatus Forssk.,
Dicanthium papillosum (Hochst.) Stapf
♦ Hindi grass, shedagrass, two flowered goldenbeard, Kleberg's bluestem, bluestem
♦ Weed, Naturalised
♦ 66, 86, 87, 88, 90, 93, 101, 185, 205, 221, 242, 286, 287
♦ pG, arid, cultivated. Origin: Africa to Malaysia.

Dichanthium annulatum **(Forssk.) Stapf var.** *annulatum*
Poaceae
Dichanthium nodosum Willemet
♦ Diaz bluestem
♦ Introduced
♦ 228
♦ G, arid.

Dichanthium annulatum **(Forssk.) Stapf var.** *paillosum* **(A.Rich.) De Wet & Harlan**
Poaceae
Dichanthium papillosum (Hochst.) Stapf
♦ Native Weed
♦ 121
♦ pG. Origin: southern Africa.

Dichanthium aristatum **(Poir.) C.E.Hubb.**
Poaceae
Andropogon aristatus Poir., *Andropogon caricosus* L. ssp. *mollicomus* (Kunth) Hack., *Andropogon caricosus* L. var. *mollicomus* (Kunth) Hack., *Andropogon mollicomus* Kunth, *Andropogon nodosus* (Willemet) Nash, *Diplasanthum lanosum* Desv., *Lepeocercis mollicoma* (Kunth) Nees
♦ angleton bluestem, angleton grass, bearded goldenbeard, bluestem, hierba angleton, puntero
♦ Weed, Naturalised, Introduced
♦ 7, 86, 87, 88, 90, 93, 98, 101, 121, 179, 203, 228, 261, 287
♦ pG, arid, cultivated. Origin: India.

Dichanthium bladhii **(Retz.) Clayton**
Poaceae
♦ reh nta
♦ Introduced
♦ 230
♦ G.

Dichanthium caricosum **(L.) A.Camus**
Poaceae
Andropogon caricosus L. (see)
♦ roadside bluestem, nadi bluegrass, angleton grass, antigua haygrass
♦ Weed, Naturalised, Introduced
♦ 86, 87, 88, 90, 98, 203, 228
♦ pG, arid, cultivated. Origin: India, south-east Asia.

Dichanthium ischaemum **(L.) Rob.**
Poaceae

= *Bothriochloa ischaemum* (L.) Keng
♦ beardgrass
♦ Weed
♦ 253, 272
♦ G.

Dichanthium pertusum **(L.) Clayton**
Poaceae
= *Bothriochloa pertusa* (L.) A.Camus
♦ Weed
♦ 66
♦ G.

Dichanthium saccharoides **(Sw.) Roberty**
Poaceae
♦ Weed
♦ 253
♦ G, cultivated.

Dichanthium sericeum **(R.Br.) A.Camus**
Poaceae
♦ Queensland bluegrass, silky bluestem
♦ Weed, Naturalised, Introduced, Environmental Weed
♦ 80, 86, 101, 176, 179, 228, 287
♦ G, arid, cultivated. Origin: Australia.

Dichanthium sericeum **(R.Br.) A.Camus ssp.** *sericeum*
Poaceae
Dichanthium affine (R.Br.) A.Camus
♦ Introduced
♦ 228
♦ G.

Dichanthium tenue **(R.Br.) A.Camus**
Poaceae
♦ slender bluestem
♦ Naturalised
♦ 101
♦ G. Origin: Australia.

Dichapetalum cymosum **(Hook.) Engl.**
Dichapetalaceae
Chailletia cymosa Hook.
♦ poison leaf, gifblaar, umKauzaan, ncusane
♦ Weed, Quarantine Weed, Native Weed
♦ 10, 39, 50, 63, 76, 87, 88, 121, 203, 220
♦ p, arid, herbal, toxic. Origin: southern Africa.

Dichapetalum guineense **(DC.) Keay**
Dichapetalaceae
♦ Weed
♦ 87, 88

Dichelachne crinata **(L.f.) Hook.f.**
Poaceae
♦ clovenfoot plumegrass
♦ Naturalised
♦ 101
♦ G.

Dichelachne micrantha **(Cav.) Domin**
Poaceae
♦ plumegrass
♦ Naturalised
♦ 101
♦ G, cultivated.

Dichelachne rara **(R.Br.) Vickery**
Poaceae
♦ Naturalised
♦ 280

♦ G.

Dichelachne sieberiana **Trin. & Rupr.**
Poaceae
♦ Naturalised
♦ 280
♦ G.

Dichelostemma congestum **(Sm.) Kunth**
Liliaceae/Alliaceae/Themidaceae
♦ wild hyacinth, forktooth ookow, ookow, congested snakelily, tiukukukka
♦ Weed
♦ 161
♦ pH, cultivated, herbal.

Dichelostemma pulchellum **(Salisb.) Heller**
Liliaceae/Alliaceae/Themidaceae
= *Dichelostemma capitatum* Alph. ssp. *capitatum* Wood (NoR)
♦ wild hyacinth, desert hyacinth
♦ Weed
♦ 180, 243
♦ pH, arid, cultivated, herbal.

Dichilus lebeckioides **DC.**
Fabaceae/Papilionaceae
♦ silver bullet
♦ Native Weed
♦ 121
♦ S. Origin: southern Africa.

Dichondra carolinensis **Michx.**
Convolvulaceae
Dichondra repens Forst. (Michx.) var. *carolinensis* Choisy (see)
♦ Carolina dichondra, Carolina pony's foot
♦ Weed, Naturalised
♦ 161, 249, 287
♦ herbal.

Dichondra micrantha **Urb.**
Convolvulaceae
♦ dichondra, kidneyweed, Mercury Bay weed, Asian pony's foot
♦ Weed, Naturalised
♦ 7, 15, 40, 86, 179, 198, 273, 280
♦ aqua, cultivated. Origin: South America.

Dichondra microcalyx **(Hallier f.) Fabris**
Convolvulaceae
Dichondra repens auct. non J.R. & G.Forst., *Dichondra repens* var. *microcalix* Hallier f.
♦ dinheiro em penca, dichondra
♦ Weed
♦ 237, 255, 295
♦ Origin: South America.

Dichondra repens **Forst.**
Convolvulaceae
♦ kidneyweed, dewdrop lawn, wonder lawn, Mercury Bay weed, dichondra
♦ Weed, Native Weed, Introduced
♦ 39, 50, 87, 88, 121, 161, 165, 261, 269, 286, 297
♦ pH, cultivated, herbal, toxic. Origin: Australia, New Zealand.

Dichondra repens **Forst. var.** *carolinensis* **(Michx.) Choisy**
Convolvulaceae

= *Dichondra carolinensis* Michx.
♦ dichondra
♦ Weed
♦ 218, 286

Dichorisandra thyrsiflora Mikan
Commelinaceae
♦ blue ginger
♦ Weed, Naturalised
♦ 86, 98, 203
♦ cultivated. Origin: Brazil.

Dichrocephala auriculata (Thunb.) Druce
Asteraceae
♦ Weed
♦ 275, 297
♦ aH.

Dichrocephala benthamii C.B.Clarke
Asteraceae
♦ Weed
♦ 275, 297
♦ aH.

Dichrocephala bicolor (Roth) Schlect.
Asteraceae
= *Dichrocephala latifolia* (L.f.) Kuntze
♦ Weed, Quarantine Weed
♦ 13, 76, 87, 88, 203, 220, 235, 276, 286
♦ aH.

Dichrocephala integrifolia (L.f.) Kuntze
Asteraceae
Dichrocephala bicolor (Roth) Schlect.
(see), *Dichrocephala latifolia* (Lam. ex
Poir.) DC., *Cotula latifolia* Pers., *Grangea
latifolia* Lam. ex Poir.
♦ phak chee doi
♦ Weed
♦ 88, 170, 191, 238
♦ Origin: Australia.

Dichromena Michx. spp.
Cyperaceae
= *Rhynchospora* Vahl spp.
♦ Quarantine Weed
♦ 258
♦ G.

Dichromena ciliata Vahl
Cyperaceae
= *Rhynchospora nervosa* (Vahl) Boeck.
ssp. *ciliata* T.Koyama (NoR)
♦ whitetop sedge, stargrass, starsedge,
cortadera estrellita, estrellita, coyolillo
♦ Weed
♦ 87, 88, 126, 153, 157, 281
♦ pG, aqua, herbal.

Dichrostachys cinerea (L.) Wight & Arn.
Fabaceae/Mimosaceae
Dichrostachys glomerata (Forssk.) Chiov.
(see), *Dichrostachys nutans* Benth. (see)
♦ aroma, marabou thorn, marabu, Sen
Domeng, Saint Domeng, acacia Saint
Domingue
♦ Weed, Quarantine Weed, Noxious
Weed, Naturalised, Introduced
♦ 3, 14, 76, 80, 86, 87, 88, 101, 179, 191,
203, 214, 228, 229
♦ S/T, arid, cultivated, herbal. Origin:
Africa.

**Dichrostachys cinerea (L.) Wight & Arn.
ssp. *africana* Brenan & Brumm.**
Fabaceae/Mimosaceae

♦ sekelbos, sickle bush
♦ Weed
♦ 10, 63

**Dichrostachys cinerea (L.) Wight & Arn.
ssp. *africana* Brenan & Brumm. var.
*africana***
Fabaceae/Mimosaceae
♦ Chinese lantern tree, Kalahari
Christmas tree, sickle bush
♦ Native Weed
♦ 121
♦ S/T. Origin: southern Africa.

**Dichrostachys cinerea (L.) Wight & Arn.
ssp. *africana* Brenan & Brumm. var.
pubescens Brenan & Brumm.**
Fabaceae/Mimosaceae
♦ Chinese lantern tree, sickle bush
♦ Native Weed
♦ 121
♦ S/T. Origin: southern Africa.

**Dichrostachys cinerea (L.) Wight &
Arn. ssp. *africana* Brenan & Brumm.
var. *setulosa* (Welw. ex Oliv.) Brenan &
Brumm.**
Fabaceae/Mimosaceae
♦ Kalahari Christmas tree, sickle bush
♦ Native Weed
♦ 121
♦ S/T. Origin: southern Africa.

**Dichrostachys cinerea (L.) Wight. & Arn.
ssp. *nyassana* (Taub.) Brenan**
Fabaceae/Mimosaceae
♦ Kalahari Christmas tree, large leaved
sickle bush
♦ Native Weed
♦ 121
♦ S/T. Origin: southern Africa.

Dichrostachys glomerata (Forssk.) Chiov.
Fabaceae/Mimosaceae
= *Dichrostachys cinerea* (L.) Wight &
Arn.
♦ Chinese lantern tree, umGagu,
muShashasha, chiZhuzhu
♦ Weed, Naturalised
♦ 32, 50, 87, 88

Dichrostachys nutans Benth.
Fabaceae/Mimosaceae
= *Dichrostachys cinerea* (L.) Wight &
Arn.
♦ marabu
♦ Weed
♦ 87, 88, 218

Dicksonia antarctica Labill.
Dicksoniaceae
♦ Tasmanian tree fern, soft tree fern
♦ Weed
♦ 121
♦ pH, cultivated, herbal. Origin:
Australia.

Dicliptera assurgens (L.) Juss.
Acanthaceae
= *Dicliptera sexangularis* (L.) Juss. (NoR)
♦ Weed
♦ 14

Dicliptera canescens Nees
Acanthaceae
♦ Weed

♦ 13

Dicliptera chinensis (L.) Juss.
Acanthaceae
♦ Chinese foldwing
♦ Weed, Naturalised
♦ 87, 88, 101, 297

Dicliptera peduncularis Nees
Acanthaceae
♦ Weed
♦ 199

Dicliptera resupinata (Vahl) Juss.
Acanthaceae
♦ Arizona foldwing
♦ Quarantine Weed
♦ 220
♦ herbal.

Dicliptera ungiculata Nees
Acanthaceae
♦ Weed
♦ 157
♦ pH.

Diclis ovata Benth.
Scrophulariaceae
♦ Weed
♦ 87, 88

Diclis reptans Benth.
Scrophulariaceae
♦ dwarf snapdragon, toadflax
♦ Native Weed
♦ 121
♦ H. Origin: southern Africa.

Dicoma anomala Sond. ssp. anomala
Asteraceae
♦ Native Weed
♦ 121
♦ pH. Origin: southern Africa.

**Dicoma anomala Sond. ssp. cirsoides
(Harv.) Willd.**
Asteraceae
♦ Native Weed
♦ 121
♦ pH. Origin: southern Africa.

Dicoma zeyheri Sond.
Asteraceae
♦ dicoma, doll's protea, toy sugarbush
♦ Native Weed
♦ 121
♦ pH, cultivated. Origin: southern
Africa.

Dicranella heteromalla (Hedw.) Schimp.
Dicranaceae
♦ dicranella moss, silky forklet moss
♦ Naturalised
♦ 280
♦ moss.

Dicranopteris Bernh. spp.
Gleicheniaceae
♦ fern
♦ Environmental Weed
♦ 227

Dicranopteris dichotoma (Thunb.) Bernh.
Gleicheniaceae
= *Dicranopteris linearis* (Burm.)
Underw.
♦ mang qi, ko shida
♦ Weed
♦ 297

Dicranopteris linearis **(Burm.) Underw.**
Gleicheniaceae
Dicranopteris dichotoma (Thunb.) Bernh.
(see), *Gleichenia linearis* (Burm.) Clarke
(see), *Polypodium dichotomum* Thunb.,
Polypodium lineare Burm.f.
♦ old world forkedfern
♦ Weed
♦ 12, 28, 243, 262, 286
♦ pH, cultivated, herbal. Origin:
Africa, Asia, Australia.

Dicranostigma franchetianum **(Prain)**
Fedde
Papaveraceae
Chelidonium franchetianum Prain (see)
♦ Quarantine Weed
♦ 220
♦ cultivated.

Dicranostigma leptopodum **(Maxim.)**
Fedde
Papaveraceae
♦ Weed
♦ 275, 297
♦ a/bH, cultivated.

Dictamnus albus **L.**
Rutaceae
Fraxinella dictamnus Moench
♦ mooseksenpalavapensas, gasplant,
burningbush
♦ Weed, Naturalised, Cultivation
Escape
♦ 39, 42, 101, 161, 272
♦ pH, cultivated, herbal, toxic.

Dictamnus dasycarpus **Turcz.**
Rutaceae
♦ densefruit pittany
♦ Weed
♦ 297

Dictyosphaeria cavernosa **(Forssk.)**
Børgesen
Valoniaceae
♦ algae
♦ Weed
♦ 282
♦ algae.

Dictyota **J.V.F.Lamour. spp.**
Dictyotaceae
♦ algae
♦ Weed
♦ 282
♦ algae.

Didelta carnosa **Ait.**
Asteraceae
♦ seegousblom
♦ Quarantine Weed
♦ 220
♦ cultivated.

Didelta spinosa **(L.f.) Aiton**
Asteraceae
Polymnia spinosa L.f.
♦ Quarantine Weed
♦ 220
♦ arid, cultivated.

Dieffenbachia **Schott spp.**
Araceae
♦ dumbcane, dieffenbachia
♦ Weed

♦ 154, 161, 247
♦ toxic.

Dieffenbachia maculata **(Lodd.) G.Don**
Araceae
= *Dieffenbachia seguine* (Jacq.) Schott
♦ spotted dieffenbachia, dumbcane,
yalu ni vavalagi
♦ Weed, Cultivation Escape
♦ 39, 107, 261
♦ cultivated, herbal, toxic. Origin:
tropical America.

Dieffenbachia seguine **(Jacq.) Schott**
Araceae
Arum seguine Jacq., *Dieffenbachia*
maculata (Lodd.) G.Don (see),
Dieffenbachia picta Schott
♦ dumbcane
♦ Weed, Introduced
♦ 39, 87, 88, 194, 230
♦ H, cultivated, herbal, toxic.

Dierama pendulum **(L.f.) Baker**
Iridaceae
♦ wandflower
♦ Weed, Naturalised
♦ 7, 86, 98, 203, 280
♦ cultivated. Origin: South Africa.

Dierama pulcherrimum **Baker**
Iridaceae
♦ angel's fishing rods, wandflower
♦ Weed, Naturalised
♦ 98, 203
♦ cultivated, herbal.

Diervilla lonicera **Mill.**
Diervillaceae/Caprifoliaceae
Diervilla canadensis Willd., *Diervilla*
humilis Pers.
♦ bush honeysuckle, northern bush
honeysuckle
♦ Weed
♦ 87, 88, 218
♦ S, promoted, herbal.

Diervilla sessilifolia **Buckl.**
Diervillaceae/Caprifoliaceae
♦ rusovuohenkuusama, southern bush
honeysuckle, bush honeysuckle
♦ Cultivation Escape
♦ 42
♦ cultivated, herbal.

Dietes bicolor **(Steud.) Sweet ex Klatt**
Iridaceae
♦ fortnight lily, wild yellow iris,
African iris, peacock flower
♦ Naturalised, Garden Escape,
Environmental Weed, Casual Alien
♦ 86, 280
♦ cultivated. Origin: southern Africa.

Dietes grandiflora **N.E.Br.**
Iridaceae
♦ wild iris
♦ Naturalised, Garden Escape
♦ 252
♦ pH, cultivated. Origin: South Africa.

Dietes iridioides **(L.) Sweet ex Klatt**
Iridaceae
Dietes vegeta (L.) N.E.Br., *Moraea*
catenulata Lindl., *Moraea iridioides* L.
♦ dietes, fortnight lily
♦ Naturalised, Cultivation Escape

♦ 86, 252
♦ cultivated. Origin: east and southern
Africa.

Dietes robinsoniana **(F.Muell.) Klatt**
Iridaceae
♦ Weed, Naturalised
♦ 54, 86, 88
♦ cultivated. Origin: Lord Howe Is.

Digera arvensis **Forssk.**
Amaranthaceae
= *Digera muricata* (L.) Mart. ssp.
muricata
♦ Weed, Quarantine Weed
♦ 76, 87, 88, 203, 220
♦ herbal.

Digera muricata **(L.) Mart.**
Amaranthaceae
Digera alternifolia (L.) Asch.
♦ Weed, Quarantine Weed
♦ 66, 76, 87, 88, 203, 220, 242, 243
♦ arid, herbal.

Digera muricata **(L.) Mart. ssp.** *muricata*
Amaranthaceae
Achyranthes alternifolia L., *Achyranthes*
muricata L., *Desmochaeta alternifolia* (L.)
DC., *Digera alternifolia* (L.) Asch. var.
ciliata (Lam.) Chiov., *Digera arvensis*
Forssk. (see)
♦ kohendro, kundra, lesua, lhasua,
lolaru, tandln
♦ Introduced
♦ 228
♦ arid.

Digitalis **L. spp.**
Scrophulariaceae
♦ foxglove
♦ Weed, Naturalised
♦ 198, 272
♦ herbal.

Digitalis grandiflora **Mill.**
Scrophulariaceae
Digitalis ochroleuca Lam., *Digitalis*
orientalis Lam. (see)
♦ keltasormustinkukka, large yellow
foxglove, yellow foxglove
♦ Weed, Naturalised, Cultivation
Escape
♦ 42, 101, 272
♦ pH, cultivated, herbal.

Digitalis lanata **Ehrh.**
Scrophulariaceae
♦ Grecian foxglove, woolly foxglove
♦ Weed, Naturalised
♦ 19, 23, 88, 101, 272
♦ b/pH, cultivated, herbal.

Digitalis lutea **L.**
Scrophulariaceae
♦ pikkusormustinkukka, small yellow
foxglove, straw foxglove
♦ Naturalised, Cultivation Escape,
Casual Alien
♦ 39, 40, 42, 101
♦ pH, cultivated, herbal, toxic.

Digitalis orientalis **Lam.**
Scrophulariaceae
= *Digitalis grandiflora* Mill.
♦ Weed
♦ 87, 88

♦ cultivated.

Digitalis purpurea L.
Scrophulariaceae
Digitalis purpurascens Lej.
♦ foxglove, purple foxglove, common foxglove
♦ Weed, Noxious Weed, Naturalised, Introduced, Garden Escape, Environmental Weed, Cultivation Escape
♦ 8, 15, 23, 24, 35, 38, 39, 42, 70, 72, 78, 80, 86, 87, 88, 98, 101, 136, 137, 142, 146, 151, 154, 155, 161, 165, 176, 181, 189, 198, 203, 212, 218, 231, 241, 247, 269, 272, 280, 287, 289, 295, 296, 300
♦ b/pH, cultivated, herbal, toxic. Origin: south-west and central Europe.

Digitalis purpurea L. var. *alba* hort.
Scrophulariaceae
♦ purple foxglove
♦ Naturalised
♦ 101

Digitalis purpurea L. var. *purpurea*
Scrophulariaceae
♦ purple foxglove
♦ Naturalised
♦ 101

Digitaria Haller spp.
Poaceae
♦ crabgrass, fingergrass
♦ Weed, Noxious Weed
♦ 23, 36, 243
♦ G.

Digitaria abyssinica (A.Rich.) Stapf
Poaceae
Digitaria scalarum (Schweinf.) Chiov. (see), *Panicum scalarum* Schweinf., *Syntherisma abyssinica* (Hochst.) Newbold
♦ African couchgrass, fingergrass, blue couch, Abyssinian fingergrass, blue couchgrass, couch fingergrass, Dunn's fingergrass, East African fingergrass
♦ Weed, Quarantine Weed, Noxious Weed, Naturalised, Native Weed
♦ 53, 76, 87, 88, 101, 121, 140, 229, 240, 243
♦ pG, arid, cultivated. Origin: Madagascar.

Digitaria adscendens (H.B.K.) Henr.
Poaceae
= *Digitaria ciliaris* (Retz.) Koch
♦ summer grass, bamboo grass, Henry's crabgrass, tarai, fingergrass, tropical crabgrass, yaa plong khaao nok
♦ Weed
♦ 30, 87, 88, 186, 204, 209, 239, 243, 273, 274, 275
♦ aG, arid, herbal.

Digitaria adscendens (H.B.K.) Henr. var. *fimbriata* Henr.
Poaceae
♦ Weed
♦ 286
♦ G.

Digitaria aequiglumis (Hack. & Arechav.) Parodi

Poaceae
♦ Weed, Naturalised
♦ 86, 98, 203, 241, 280, 300
♦ G. Origin: South America.

Digitaria ammophila (F.Muell.) Hughes
Poaceae
♦ Introduced
♦ 228
♦ G, arid, cultivated.

Digitaria bicornis (Lam.) Roem. & Schult.
Poaceae
Digitaria adscendens Henrard ssp. *chrysoblephara* (Fig. & De Not.) Henrard, *Digitaria barbata* Willd., *Digitaria barbulata* Desv., *Digitaria biformis* Willd. ssp. *desvauxii* Henrard, *Digitaria biformis* Willd. var. *chrysoblephara* (Fig. & De Not.) Beetle, *Digitaria biformis* Willd. (see), *Digitaria chrysoblephara* Fig. & De Not. (see), *Digitaria diversiflora* Swallen, *Digitaria marginata* Link var. *fimbriata* (Link) Stapf, *Digitaria queenslandica* Henrard, *Digitaria rottleri* Roem. & Schult., *Panicum adpressum* Willd., *Panicum barbatum* (Willd.) Kunth, *Panicum bicorne* (Lam.) Kunth, *Panicum biforme* (Willd.) Kunth, *Panicum ciliare* A.Rich., *Panicum sanguinale* L. var. *barcaldinense* Domin, *Panicum sanguinale* L. var. *biforme* (Willd.) T.Durand & Schinz, *Paspalum bicorne* Lam., *Syntherisma barbatum* (Willd.) Nash
♦ tropical crabgrass, hairy fingergrass, Asian crabgrass, fingergrass
♦ Weed, Introduced
♦ 28, 93, 161, 179, 205, 206, 230, 243, 249, 255
♦ aG, arid. Origin: Australia.

Digitaria biformis Willd.
Poaceae
= *Digitaria bicornis* (Lam.) Roem. & Schult.
♦ Weed
♦ 87, 88
♦ G.

Digitaria brownii (Roem. & Schult.) Hughes
Poaceae
Digitaria brownii (Roem. & Schult.) Hughes var. *monostachya* (Benth.) Hughes, *Panicum brownii* Roem. & Schult., *Panicum glareae* F.Muell., *Panicum laniflorum* Nees, *Panicum leucophaeum* Kunth var. *monostachyum* Benth., *Panicum villosum* R.Br., *Trichachne brownii* (Roem. & Schult.) Henrard
♦ Introduced
♦ 228
♦ G, arid, cultivated.

Digitaria californica (Benth.) Henrard
Poaceae
Panicum californica Benth., *Panicum lachnanthum* Torr., *Panicum saccharatum* Buckley, *Trichachne californica* (Benth.) Chase ex Hitchc., *Trichachne saccharatum* (Buckley) Nash
♦ Arizona cottontop
♦ Introduced
♦ 228
♦ G, arid, herbal.

Digitaria chrysoblephara Fig. & De Not.
Poaceae
= *Digitaria bicornis* (Lam.) Roem. & Schult.
♦ hairy crabgrass
♦ Weed
♦ 297
♦ G.

Digitaria ciliaris (Retz.) Koch
Poaceae
Asprella digitaria Lam., *Digitaria abortiva* Reeder, *Digitaria adscendens* (Kunth) Henrard (see), *Digitaria adscendens* ssp. *adscendens*, *Digitaria adscendens* ssp. *marginata* (Link) Henrard, *Digitaria adscendens* ssp. *nubica* (Stapf) Henrard, *Digitaria adscendens* var. *criniformis* Henrard, *Digitaria adscendens* var. *fimbriata* (Link) Cufod., *Digitaria adscendens* var. *pes-avis* (Büse) Henrard, *Digitaria adscendens* var. *typica* Henrard, *Digitaria brevifolia* Link, *Digitaria ciliaris* (Retz.) Koel. ssp. *chrysoblephara* (Fig. & De Not.) Henrard, *Digitaria ciliaris* ssp. *nubica* (Stapf) S.T.Blake, *Digitaria ciliaris* var. *chrysoblephara* (Fig. & De Not.) R.R.Stewart (see), *Digitaria ciliaris* var. *criniformis* (Link) R.R.Stewart, *Digitaria commutata* Schult., *Digitaria henryi* Rendle (see), *Digitaria marginata* Link var. *ciliaris* (Retz.) Hook.f. ex Ridl., *Digitaria marginata* var. *linkii* Stapf, *Digitaria marginata* var. *nubica* Stapf, *Digitaria pes-avis* Büse, *Digitaria sanguinalis* (L.) Scop. fo. *ciliaris* (Retz.) Hook.f. ex Haines, *Digitaria sanguinalis* ssp. *ciliaris* (Retz.) Arcang., *Digitaria sanguinalis* var. *ciliaris* (Retz.) Parl., *Digitaria sanguinalis* var. *fimbriata* (Link) Stapf ex Merr., *Digitaria sasakii* (Honda) Tuyama, *Digitaria tarapacana* Phil., *Milium ciliare* (Retz.) Moench, *Panicum brachyphyllum* Steud., *Panicum brevifolium* (Link) Kunth, *Panicum ciliare* Retz., *Panicum henryi* (Rendle) Makino & Nemoto, *Panicum marginellum* Schrad., *Panicum ornithopus* Trin. ex Spreng., *Panicum pes-avis* (Büse) Hook.f. ex Koord., *Panicum sanguinale* L. subvar. *marginatum* (Link) Döll, *Panicum sanguinale* var. *blepharanthum* Hack. ex T.Durand & Schinz, *Panicum sanguinale* var. *ciliare* (Retz.) St.-Amans, *Panicum sanguinale* var. *fimbriatum* (Link) Usteri, *Panicum sanguinale* var. *macrostachyum* Hack. ex T.Durand & Schinz, *Paspalum ciliare* (Retz.) DC., *Paspalum inaequale* Link, *Paspalum sanguinale* (L.) Lam. var. *ciliare* (Retz.) Hook.f., *Sanguinaria ciliaris* (Retz.) Bubani, *Spartina pubera* Hassk., *Syntherisma cilare* (Retz.) Schrad., *Syntherisma henryi* (Rendle) Newbold, *Syntherisma marginatum*

(Link) Nash, *Syntherisma sanguinalis*
(L.) Dulac ssp. *ciliaris* (Retz.) Masam.
& Yanagih., *Syntherisma sanguinalis*
var. *ciliaris* (Retz.) Honda, *Syntherisma
sasakii* Honda
♦ fingergrass, tropical crabgrass,
common summergrass, crabgrass,
summer grass, Henry's crabgrass,
violet crabgrass, large crabgrass,
saulangi, zacate cangrejo velludo,
riisiverihirssi
♦ Weed, Naturalised, Native Weed,
Introduced, Casual Alien
♦ 3, 6, 7, 12, 42, 55, 66, 86, 87, 88, 90, 93,
98, 107, 121, 134, 157, 161, 170, 176, 203,
205, 228, 243, 245, 249, 255, 262, 263,
269, 272, 275, 280, 286, 297, 300
♦ aG, aqua, cultivated. Origin:
tropical America, Asia, Africa.

**Digitaria ciliaris (Retz.) Koel. ssp.
chrysoblephara (Fig. & De Not.) Henrard**
Poaceae
= *Digitaria ciliaris* (Retz.) Koch
♦ Weed
♦ 242
♦ G, arid.

**Digitaria cognata (Schult.) Pilg. var.
cognata**
Poaceae
Leptoloma cognatum (Schult.) Chase
♦ fall witchgrass, Carolina crabgrass
♦ Introduced
♦ 228
♦ G, arid.

**Digitaria cruciata (Nees ex Steud.)
A.Camus**
Poaceae
♦ crabgrass
♦ Weed
♦ 243, 244
♦ aG.

Digitaria debilis (Desf.) Willd.
Poaceae
♦ fingergrass
♦ Weed, Quarantine Weed, Native
Weed
♦ 76, 87, 88, 121, 203, 220
♦ aG, arid. Origin: southern Africa.

Digitaria decumbens Stent
Poaceae
= *Digitaria eriantha* Steud.
♦ pangola grass, woolly fingergrass
♦ Weed, Introduced, Environmental
Weed, Cultivation Escape
♦ 39, 87, 88, 152, 155, 204, 228, 257, 261
♦ a/pG, arid, cultivated, herbal, toxic.
Origin: South Africa.

Digitaria didactyla Willd.
Poaceae
Digitaria caespitosa Boivin ex A.Camus,
Digitaria camusiana Henrard, *Digitaria
didactyla* var. *decalvata* Henrard,
Digitaria didactyla var. *penisula* (Ohwi)
Henrard, *Digitaria peninsulae* Ohwi,
Digitaria swazilandensis Stent (see),
Digitaria truncata Henrard ex A.Camus,
Panicum bicorne Sieber ex Steud.,
Panicum commutatum J.A.Schultes
var. *didactylum* (Willd.) Nees, *Panicum*

didactylum (Willd.) Kunth, *Panicum
gracile* Nees ex Spreng., *Panicum
sanguinale* L. var. *bicorne* Drake,
Panicum sanguinale var. *brevispicatum*
Maiden, *Panicum subtile* R.Br. ex Nees,
Paspalum sanguinale (L.) Lam. var.
didactylum (Willd.) A.Camus
♦ blue couchgrass, blue fingergrass,
bluegrass, Mauritius blue, Swazi grass,
Swaziland fingergrass, Queensland
blue couch, crabgrass
♦ Weed, Naturalised, Environmental
Weed
♦ 7, 39, 86, 87, 88, 98, 121, 203
♦ pG, cultivated, toxic. Origin:
Mauritius, Reunion.

Digitaria divaricatissima (R.Br.) Hughes
Poaceae
Digitaria macractinia (Benth.) Hughes,
Panicum divaricatissimum R.Br.
♦ Introduced
♦ 228
♦ G, arid, cultivated. Origin: Australia.

Digitaria diversinervis (Nees) Stapf
Poaceae
Digitaria albomarginata Stent, *Digitaria
diversinervis* var. *woodiana* Henr.
♦ Richmond fingergrass, Richmond
grass, Wynberg fingergrass
♦ Native Weed
♦ 121
♦ a/pG. Origin: southern Africa.

Digitaria eriantha Steud.
Poaceae
Digitaria decumbens Stent (see),
Digitaria eriantha Steud. ssp. *eriantha*,
Digitaria eriantha ssp. *pentzii* (Stent)
Kok (see), *Digitaria eriantha* ssp.
stolonifera (Stapf) Kok, *Digitaria
eriantha* var. *stolonifera* Stapf, *Digitaria
geniculata* Stent, *Digitaria glauca* Stent.
nom. illeg., *Digitaria pentzii* Stent (see),
Digitaria pentzii var. *minor* Stent,
Digitaria pentzii var. *stolonifera* (Stapf)
Henrard, *Digitaria polevansii* Stent,
Digitaria seriata Stapf, *Digitaria setivalva*
Stent, *Digitaria smutsii* Stent, *Digitaria
stentiana* Henrard, *Digitaria valida*
Stent, *Syntherisma eriantha* (Steud.)
Newbold
♦ pangola grass, common fingergrass,
fingergrass, pangola fingergrass, smuts
fingergrass, woolly fingergrass
♦ Weed, Naturalised, Native Weed
♦ 6, 88, 98, 101, 121, 203
♦ pG, arid. Origin: southern Africa.

Digitaria eriantha Steud. ssp. eranthia
Poaceae
= *Digitaria eriantha* Steud.
♦ Naturalised
♦ 86
♦ G.

**Digitaria eriantha Steud. ssp. pentzii
(Stent) Kok.**
Poaceae
= *Digitaria eriantha* Steud.
♦ pangola grass, woolly fingergrass
♦ Weed, Naturalised
♦ 3, 86

♦ G. Origin: southern Africa.

Digitaria fauriei Ohwi
Poaceae
♦ Weed
♦ 87, 88
♦ G.

Digitaria filiformis (L.) Koel.
Poaceae
= *Digitaria ischaemum* (Schr.) Muhl.
♦ slender crabgrass
♦ Weed
♦ 87, 88, 90, 218
♦ aG, herbal.

Digitaria fuscescens (J.Presl) Henr.
Poaceae
Digitaria pseudo-ischaemum Büse (see),
Panicum fuscescens J.Presl, *Panicum
pseudo-ischaemum* (Büse) Boerl.,
Paspalum fuscescens J.Presl, *Paspalum
micranthum* Desv., *Syntherisma
fuscescens* (J.Presl) Scribn.
♦ yellow crabgrass, common
crabgrass, crabgrass
♦ Weed, Quarantine Weed,
Naturalised, Introduced
♦ 32, 38, 76, 87, 88, 90, 101, 135, 170,
191, 203, 220, 243
♦ pG, herbal.

Digitaria hayatae Honda
Poaceae
♦ Weed
♦ 87, 88
♦ G.

Digitaria henryi Rendle
Poaceae
= *Digitaria ciliaris* (Retz.) Koch
♦ Henry's crabgrass
♦ Weed
♦ 88, 218, 274, 286
♦ G, herbal.

Digitaria heterantha (J.D.Hook.) Merr.
Poaceae
♦ Weed, Naturalised
♦ 98, 203
♦ G.

Digitaria horizontalis Willd.
Poaceae
Digitaria sanguinalis (L.) Scop. var.
horizontalis (Willd.) Rendle, *Digitaria
setosa* Desv., *Digitaria velutina* (Forssk.)
P.Beauv. (see), *Panicum hamiltonii*
Kunth, *Panicum horizontale* (Willd.)
G.Mey., *Panicum sanguinale* L. ssp.
horizontale (Willd.) Hack., *Panicum
sanguinale* var. *cognatum* Hack.
ex Schweinf., *Panicum sanguinale*
var. *horizontale* (Willd.) Schweinf.,
Panicum sanguinale var. *porranthum*
(Steud.) Franch., *Panicum stipatum*
J.Presl, *Paspalum oxyanthum* Steud.,
Syntherisma setosa (Desv.) Nash
♦ Jamaican crabgrass
♦ Weed, Naturalised, Quarantine
Weed, Introduced
♦ 76, 88, 90, 157, 203, 206, 220, 221, 228,
241, 243, 245, 255, 257
♦ aG, arid, herbal. Origin: Central and
South America.

Digitaria hubbardii Henrard
Poaceae
Digitaria neurachnoides Vickery
♦ Introduced
♦ 228
♦ G, arid. Origin: Australia.

Digitaria insularis (L.) Fedde
Poaceae
Acicarpa sacchariflora Raddi, *Andropogon fabricii* Ekman ex Henrard, *Andropogon insularis* L., *Digitaria insularis* (L.) Mez ex Ekman, *Digitaria leucophaea* (Kunth) Stapf, *Digitaria sacchariflora* (Nees) Henrard, *Leptocoryphium penicilligerum* Speg., *Milium hirsutum* P.Beauv., *Milium villosum* Sw., *Monachne unilateralis* Roem. & Schult., *Panicum duchaissingii* Steud., *Panicum falsum* Steud., *Panicum gavanianum* Steud. ex Lechl., *Panicum insulare* (L.) G.Mey., *Panicum insulare* var. *leucophaeum* (Kunth) Kuntze, *Panicum insulare* var. *typicum* Hack., *Panicum lanatum* Rottb., *Panicum leucophaeum* Kunth, *Panicum saccharoides* A.Rich., *Schoenus fabri* Rottb., *Scirpoides fabri* Rottb., *Syntherisma insularis* (L.) Millsp. & Chase, *Trichachne insularis* (L.) Nees (see), *Trichachne penicilligera* (Speg.) Parodi, *Trichachne sacchariflora* Nees, *Tricholaena insularis* (L.) Griseb., *Valota insularis* (L.) Chase, *Valota penicilligera* (Speg.) Chase ex Parodi
♦ cottongrass, sour grass, feathertop grass
♦ Weed, Quarantine Weed, Naturalised
♦ 3, 6, 86, 87, 88, 135, 157, 191, 237, 243, 245, 255, 271, 276, 295
♦ pG, herbal. Origin: tropical and subtropical South America.

Digitaria ischaemum (Schr.) Muhl.
Poaceae
Digitaria filiformis (L.) Koel. (see), *Panicum filiforme* Garcke, *Panicum lineare* Krock.
♦ smooth fingergrass, smooth crabgrass, smooth summergrass
♦ Weed, Quarantine Weed, Naturalised, Introduced, Casual Alien
♦ 15, 23, 30, 34, 40, 42, 44, 52, 68, 70, 76, 86, 87, 88, 90, 98, 101, 161, 174, 198, 203, 204, 210, 218, 235, 241, 243, 249, 253, 272, 275, 280, 297, 300
♦ aG, cultivated, herbal. Origin: Eurasia.

Digitaria leucites (Trin.) Henrard
Poaceae
♦ Weed
♦ 199
♦ G.

Digitaria longiflora (Retz) Pers.
Poaceae
Digitaria propinqua (R.Br.) P.Beauv., *Digitaria friesii* Pilg., *Paspalum brevifolium* Flüggé, *Paspalum longiflorum* Retz., *Syntherisma longiflora* (Retz.) Skeels
♦ lesser crabgrass, false couch

fingergrass, India crabgrass, wire crabgrass
♦ Weed, Naturalised, Native Weed, Introduced
♦ 28, 32, 38, 87, 88, 90, 121, 161, 170, 179, 203, 206, 217, 243, 249
♦ pG, herbal.
Origin: southern Africa.

Digitaria macractenia (Benth.) Hughes
Poaceae
♦ Weed
♦ 87, 88
♦ G.

Digitaria magna (Honda) Honda
Poaceae
♦ Weed
♦ 87, 88
♦ G.

Digitaria microbachne (Presl) Henr.
Poaceae
= *Digitaria setigera* Roth ex Roem. & Schult.
♦ Weed, Quarantine Weed
♦ 76, 87, 88, 203, 220
♦ G.

Digitaria milanjiana (Rendle) Stapf
Poaceae
Digitaria boivinii Henrard, *Digitaria bulbosa* Peter, *Digitaria endlichii* Mez, *Digitaria endlichii* Mez ssp. *meziana* Henrard, *Digitaria exasperata* Henrard, *Digitaria fusca* Chiov., *Digitaria gallaensis* Chiov., *Digitaria gracilenta* Henrard, *Digitaria kilimandscharica* Mez, *Digitaria milanjiana* (Rendle) Stapf var. *abscondita* Henrard, *Digitaria milanjiana* (Rendle) Stapf ssp. *eylesiana* Henrard, *Digitaria mombasana* C.E.Hubb., *Digitaria polevansii* Stent ssp. *peterana* (Peter) Henrard, *Digitaria stapfii* Henrard, *Digitaria swynnertonii* Rendle, *Panicum milanjianum* Rendle, *Panicum sanguinale* L. var. *scabriglume* Hack.
♦ makarikari fingergrass, milanje fingergrass, milanje grass, Madagascar crabgrass, fingergrass
♦ Weed, Naturalised, Native Weed
♦ 88, 98, 101, 121, 179, 203
♦ pG, arid, cultivated. Origin: southern Africa.

Digitaria monodactyla (Nees) Stapf
Poaceae
♦ one fingergrass
♦ Native Weed
♦ 121
♦ pG. Origin: southern Africa.

Digitaria natalensis Stent
Poaceae
Digitaria macroglossa Henr.
♦ Natal crabgrass
♦ Native Weed
♦ 121
♦ pG. Origin: southern Africa.

Digitaria nodosa Parl.
Poaceae
Digitaria commutata Schult. ssp. *nodosa* (Parl.) Maire

♦ Weed, Introduced
♦ 221, 228
♦ G, arid.

Digitaria nuda Schumach.
Poaceae
Agrostis digitata (Sw.) Poir., *Axonopus digitatus* (Sw.) P.Beauv., *Digitaria adscendens* (H.B.K.) Henrard var. *rachiseta* Henrard, *Digitaria borbonica* Desv., *Digitaria diamesum* (Steud.) Henrard, *Digitaria digitata* (Sw.) Urb., *Digitaria jamaicensis* Spreng., *Digitaria nuda* ssp. *schumacheriana* Henrard, *Milium digitatum* Sw., *Panicum diamesum* Steud., *Panicum digitatum* Asch. & Graebn., *Panicum sanguinale* L. var. *digitatum* (Sw.) Hack. ex Urb., *Paspalum digitatum* (Sw.) Kunth, *Syntherisma digitata* (Sw.) Hitchc.
♦ naked crabgrass
♦ Weed, Naturalised
♦ 32, 88, 101, 170, 191, 261
♦ G. Origin: tropical Africa.

Digitaria pentzii Stent
Poaceae
= *Digitaria eriantha* Steud.
♦ creeping fingergrass, woolly fingergrass
♦ Weed, Introduced
♦ 38, 87, 88, 179
♦ G, cultivated, herbal.

Digitaria pruriens (Fisch. ex Trin.) Büse
Poaceae
= *Digitaria setigera* Roth ex Roem. & Schult.
♦ slim crabgrass
♦ Weed
♦ 87, 88, 218, 276, 286
♦ aG, herbal.

Digitaria pseudo-ischaemum Büse
Poaceae
= *Digitaria fuscescens* (J.Presl) Henr.
♦ Weed
♦ 87, 88
♦ G.

Digitaria radicosa (Presl) Miq.
Poaceae
Digitaria chinensis (Nees) A.Camus var. *hirsuta* Ohwi, *Digitaria formosana* Rendle, *Digitaria formosana* var. *hirsuta* (Honda) Henrard, *Digitaria propinqua* Gaudich., *Digitaria sanguinalis* (L.) Scop. var. *multinervis* (Honda) Kitag., *Digitaria sanguinalis* var. *timorensis* (Kunth) Hayata, *Digitaria tenuispica* Rendle, *Digitaria timorensis* (Kunth) Balansa (see), *Digitaria timorensis* ssp. *kunthiana* Henrard, *Digitaria timorensis* var. *hirsuta* (Ohwi) Henrard, *Panicum formosanum* (Rendle) Makino & Nemoto, *Panicum formosanum* var. *hirsuta* (Honda) Makino & Nemoto, *Panicum radicosum* J.Presl, *Panicum radicosum* var. *procerior* J.Presl, *Panicum sanguinale* L. var. *multinerve* (Honda) Makino & Nemoto, *Panicum sanguinale* var. *timorense* (Kunth) Hack., *Panicum timorense* Kunth, *Spartina glabriuscula* Hassk., *Syntherisma formosana* (Rendle)

Honda, *Syntherisma formosana* ssp.
hirsuta (Honda) Masam. & Yanagih.,
Syntherisma formosana var. *hirsuta*
Honda, *Syntherisma multinervis*
(Honda) Honda, *Syntherisma*
sanguinalis (L.) Dulac var. *multinervis*
Honda, *Syntherisma tenuispica* (Rendle)
Keng
♦ trailing crabgrass, Chinese crabgrass
♦ Weed, Naturalised
♦ 90, 101, 274, 286
♦ aG. Origin: Asia.

Digitaria sanguinalis (L.) Scop.
Poaceae
Cynodon praecox (Walter) Roem. &
Schult., *Dactylon sanguinale* (L.) Vill.,
Digitaria fimbriata Link, *Digitaria*
marginata Link, *Digitaria marginata* var.
fimbriata (Link) Stapf, *Digitaria nealleyi*
Henrard, *Digitaria praecox* (Walter)
Willd., *Digitaria sanguinalis* ssp. *vulgaris*
(Schrad.) Henrard, *Digitaria sanguinalis*
var. *marginata* (Link) Fernald, *Digitaria*
vulgaris (Schrad.) Besser, *Panicum*
adscendens Kunth, *Panicum fimbriatum*
(Link) Kunth, *Panicum linkianum*
Kunth, *Panicum sanguinale* L., *Panicum*
sanguinale ssp. *marginatum* (Link)
Thell., *Panicum sanguinale* var. *vulgare*
(Schrad.) Döll, *Paspalum sanguinale* (L.)
Lam., *Syntherisma fimbriatum* (Link)
Nash, *Syntherisma praecox* Walter,
Syntherisma sanguinalis (L.) Dulac,
Syntherisma vulgaris Schrad.
♦ large crabgrass, hairy crabgrass,
fingergrass, twitch grass, summer
grass, crabgrass, crop grass, early
crabgrass, hairy fingergrass, landgrass,
manna, Polish millet, wild millet,
rikkaverihirssi, yaa teenkaa, capim
colchão ou milhã
♦ Weed, Noxious Weed, Naturalised,
Native Weed, Introduced,
Environmental Weed, Casual Alien
♦ 7, 14, 15, 23, 30, 34, 40, 42, 44, 49, 51,
68, 70, 86, 87, 88, 90, 98, 115, 118, 121,
136, 158, 161, 167, 174, 176, 180, 185,
186, 198, 203, 207, 210, 211, 212, 218,
221, 228, 236, 237, 239, 241, 243, 245,
249, 250, 253, 255, 269, 270, 272, 275,
280, 281, 295, 297, 299, 300
♦ aG, arid, cultivated, herbal. Origin:
Europe.

Digitaria scalarum (Schweinf.) Chiov.
Poaceae
= *Digitaria abyssinica* (A.Rich.) Stapf
♦ blue couch, couchgrass, couch,
African couchgrass, fingergrass
♦ Weed, Quarantine Weed
♦ 23, 30, 67, 87, 88, 90, 186, 203, 220,
243
♦ pG.

Digitaria sericea (Honda) Honda
Poaceae
♦ Weed, Quarantine Weed
♦ 76, 87, 88, 203, 220
♦ G.

Digitaria serotina (Walter) Michx.
Poaceae

♦ blanket crabgrass, dwarf crabgrass
♦ Weed
♦ 161, 249
♦ G.

Digitaria setigera Roth ex Roem. &
Schult.
Poaceae
Cynodon setigerus (Roth) A.Rich.
ex Hassk., *Digitaria consanguinea*
Gaudich., *Digitaria corymbosa*
(Roxb.) Merr., *Digitaria microbachne*
(Presl) Henr. (see), *Digitaria pruriens*
(Fisch. ex Trin.) Büse (see), *Panicum*
corymbosum Roxb., *Panicum fimbriatum*
(Link) Kunth var. *setigerum* (Roth)
E.Fourn., *Panicum pruriens* Fisch. ex
Trin., *Panicum setigerum* (Roth) Boerl.,
Paspalum sanguinale (L.) Lam. var.
pruriens, *Syntherisma consanguinea*
(Gaudich.) Skeels, *Syntherisma pruriens*
(Fisch.) Arthur
♦ itchy crabgrass, East Indian
crabgrass, hispid crabgrass
♦ Weed, Naturalised, Introduced
♦ 6, 32, 38, 88, 90, 101, 170, 179, 235,
243, 262, 274, 280, 287
♦ a/pG, herbal. Origin: tropical Asia,
Australia, Pacific islands.

Digitaria stricta Roth ex Roem. & Schult.
Poaceae
Agrostis pilosa Retz., *Digitaria*
denudata Link, *Digitaria puberula* Link,
Digitaria royleana (Nees ex Hook.f.)
Prain, *Panicum denudatum* (Link)
Kunth, *Panicum pseudosetaria* (Roth)
Steud., *Paspalum royleanum* Nees ex
Hook.f., *Setaria stricta* (Roth) Kunth,
Syntherisma royleana (Nees ex Steud.)
Newbold.
♦ Weed
♦ 243
♦ G.

Digitaria swazilandensis Stent
Poaceae
= *Digitaria didactyla* Willd.
♦ fingergrass
♦ Naturalised, Cultivation Escape
♦ 32, 98
♦ G.

Digitaria ternata (A.Rich) Stapf
Poaceae
Cynodon ternatus A.Rich., *Digitaria*
argyrostachya (Steud.) Fernald, *Digitaria*
argyrostachya var. *glabrescens* (Büse)
Henrard, *Digitaria argyrostachya*
var. *hirticulmis* Henrard, *Digitaria*
ropalotricha Büse, *Digitaria ropalotricha*
var. *glabrescens* Büse, *Digitaria ternata*
fo. *glabrispicula* Cufod., *Panicum*
argyrostachyum Steud., *Panicum*
phaeocarpum Nees var. *gracile* Nees,
Panicum rapalotrichum Büse ex Koord.,
Panicum ternatum (A.Rich.) Hochst. ex
Steud., *Paspalum ternatum* (A.Rich.)
Hook.f., *Syntherisma argyrostachya*
(Steud.) Hitchc. & Chase, *Syntherisma*
ternata (A.Rich.) Newbold
♦ crabgrass, blackseed fingergrass
♦ Weed, Naturalised, Native Weed,

Environmental Weed
♦ 54, 86, 87, 88, 90, 98, 121, 170, 176,
191, 203, 240
♦ aG. Origin: Asia, Africa.

Digitaria timorensis (Kunth) Balansa
Poaceae
= *Digitaria radicosa* (Presl) Miq.
♦ Weed, Quarantine Weed
♦ 76, 87, 88, 203,
220, 263
♦ aG, herbal.

Digitaria tricholaenoides Stapf
Poaceae
♦ purple fingergrass
♦ Native Weed
♦ 121
♦ pG. Origin: southern Africa.

Digitaria velutina (Forssk.) P.Beauv.
Poaceae
= *Digitaria horizontalis* Willd.
♦ velvet fingergrass, long plumed
fingergrass, annual couchgrass
♦ Weed, Quarantine Weed, Noxious
Weed, Naturalised, Native Weed
♦ 53, 67, 76, 87, 88, 98, 101, 121, 140,
203, 217, 221, 229
♦ aG. Origin: Africa, Asia.

Digitaria violascens Link
Poaceae
Digitaria bogoriensis Ohwi, *Digitaria*
chinensis (Nees) A.Camus, *Digitaria*
digitata Büse, *Digitaria filiculmis*
(Nees ex Miq.) Ohwi, *Digitaria fusca*
(J.Presl) Merr., *Digitaria ischaemum*
(Schreb.) Muhl. var. *asiatica* Ohwi,
Digitaria ischaemum var. *intersita* Ohwi,
Digitaria ischaemum var. *lasiophylla*
(Honda) Ohwi, *Digitaria ischaemum*
var. *violascens* (Link) Radford, *Digitaria*
pertenuis Büse, *Digitaria pertenuis*
var. *glabra* (Boerl.) Ohwi, *Digitaria*
pseudo-durva (Nees) Schltdl., *Digitaria*
recta Hughes, *Digitaria ropalotricha*
Büse var. *villosa* (Keng) Tuyama,
Digitaria thwaitesii (Hack.) Henrard
var. *tonkinensis* Henrard, *Digitaria*
violascens var. *intersita* (Ohwi) Ohwi,
Digitaria violascens var. *lasiophylla*
(Honda) Tuyama, *Digitaria violascens*
var. *villosa* Keng, *Digitaria zeyheri*
(Nees) Henrard, *Panicum digitatum*
(Büse) Koord., *Panicum pertenue* (Büse)
Boerl., *Panicum pertenue* var. *glabrum*
Boerl., *Panicum pseudo-durva* Nees var.
majus Nees, *Panicum pseudo-ischaemum*
(Büse) Boerl. var. *elongata* Boerl.,
Panicum steudelianum Domin, *Panicum*
violascens (Link) Kunth, *Paspalum*
chinense Nees, *Paspalum filiculme* Nees
ex Miq., *Paspalum fuscum* J.Presl,
Paspalum minutiflorum Steud., *Paspalum*
pertenue (Büse) Backer, *Syntherisma*
chinensis (Nees) Hitchc., *Syntherisma*
fusca (J.Presl) Scribn., *Syntherisma*
helleri Nash, *Syntherisma ischaemum*
(Schreb.) Nash var. *lasiophylla* Honda,
Syntherisma violascens (Link) Nash
♦ violet crabgrass, sau, smooth
crabgrass

♦ Weed, Naturalised, Introduced
♦ 3, 7, 86, 87, 88, 90, 93, 98, 101, 107, 170, 203, 218, 230, 243, 261, 263, 274, 280, 286
♦ a/pG, herbal. Origin: South America.

Digraphis arundinacea (L.) Trin.
Poaceae
= *Phalaris arundinacea* L.
♦ Weed
♦ 23
♦ G.

Diheteropogon filifolius (Nees) W.D.Clayton
Poaceae
Andropogon filifolius Steud.
♦ wire bluestem
♦ Native Weed
♦ 121
♦ pG. Origin: southern Africa.

Dillenia indica L.
Dilleniaceae
♦ chulta, elephant apple, Indian dillenia
♦ Naturalised, Introduced
♦ 101, 230, 261
♦ cultivated, herbal. Origin: India to Malaysia.

Dillenia suffruticosa (Griff.) Martelli
Dilleniaceae
♦ shrubby dillenia, shrubby simpoh
♦ Weed, Naturalised, Cultivation Escape
♦ 101, 233, 268
♦ cultivated.

Dimocarpus longan Lour.
Sapindaceae
Euphoria longan (Lour.) Steud., *Euphoria longana* Lam.
♦ longan
♦ Weed
♦ 179
♦ T, cultivated, herbal.

Dimorphandra mollis Benth.
Fabaceae/Caesalpiniaceae
♦ faveira
♦ Weed
♦ 255
♦ T. Origin: Brazil.

Dimorphotheca cuneata (Thunb.) Less.
Asteraceae
♦ bride's bouquet, bush dimorphotheca, karoo bietou, large bietou, weather prophet, white bietou, white gousblom
♦ Weed, Native Weed, Introduced
♦ 39, 121, 228
♦ pS, arid, cultivated, toxic. Origin: South Africa.

Dimorphotheca ecklonis DC.
Asteraceae
= *Osteospermum ecklonis* (DC.) Norl.
♦ dimorphotheca, African daisy
♦ Weed, Naturalised, Environmental Weed
♦ 39, 54, 72, 86, 88
♦ pH, cultivated, herbal, toxic. Origin: South Africa.

Dimorphotheca pluvialis (L.) Moench
Asteraceae
Calendula hybrida L., *Calendula pluvialis* L., *Dimorphotheca annua* Less, *Dimorphotheca calendulacea* Harv. var. *dubia* Phill., *Dimorphotheca hybrida* (L.) DC., *Dimorphotheca leptocarpa* DC.
♦ cape marigold, rain daisy, valkosääkukka, weather prophet
♦ Weed, Naturalised, Introduced, Garden Escape, Environmental Weed, Casual Alien
♦ 23, 39, 42, 72, 86, 88, 98, 203, 228
♦ pH, arid, cultivated, herbal, toxic. Origin: Africa.

Dimorphotheca sinuata DC.
Asteraceae
Dimorphotheca auranthiaca DC.
♦ Namaqualand daisy, glandular cape marigold, African daisy, bieto, keltasääkukka, cape marigold
♦ Weed, Naturalised, Native Weed, Introduced, Casual Alien
♦ 42, 86, 98, 101, 121, 203, 228
♦ aH, arid, cultivated, herbal, toxic. Origin: southern Africa.

Dimorphotheca zeyheri Sond.
Asteraceae
♦ bietou, white dimorphotheca
♦ Native Weed
♦ 121
♦ aH, arid, toxic. Origin: southern Africa.

Dinebra arabica Jacq.
Poaceae
♦ Weed, Naturalised
♦ 87, 88, 287
♦ G, herbal.

Dinebra retroflexa (Vahl) Panz.
Poaceae
Cynosurus retroflexus Vahl
♦ cat's tail grass, dinebra grass, viper grass, dinebra, häntäheinä
♦ Weed, Quarantine Weed, Naturalised, Native Weed, Cultivation Escape, Casual Alien
♦ 42, 55, 66, 76, 86, 87, 88, 91, 98, 101, 121, 185, 203, 221, 242, 243
♦ aG, arid, cultivated. Origin: tropical Africa, Iraq to India.

Dioclea reflexa Hook.f.
Fabaceae/Papilionaceae
= *Dioclea hexandra* (Ralph) Mabb. (NoR)
♦ Introduced
♦ 228
♦ arid, herbal. Origin: South America.

Dioclea wilsonii Standl.
Fabaceae/Papilionaceae
♦ Wilson's clusterpea
♦ Naturalised
♦ 101
♦ herbal.

Diodia alata Nees & C.Mart.
Rubiaceae
♦ erva de lagarto

♦ Weed
♦ 255
♦ pH. Origin: Brazil.

Diodia apiculata (Willd. ex Roem. & Schult.) K.Schum.
Rubiaceae
♦ stiff buttonweed
♦ Weed
♦ 14

Diodia dasycephala Cham. & Schlecht.
Rubiaceae
♦ carretilla
♦ Naturalised
♦ 101

Diodia lippioides Griseb.
Rubiaceae
♦ Weed
♦ 14

Diodia ocymifolia (Willd. ex Roem. & Schult.) Bremek.
Rubiaceae
Hemidiodia ocymifolia (Willd. ex Roem. & Schult.) K.Schum. (see), *Spermacoce ocymifolia* Willd. ex Roem. & Schult.
♦ slender buttonweed
♦ Weed
♦ 13, 28, 206, 243

Diodia radula Cham. & Schlect.
Rubiaceae
♦ diodia
♦ Naturalised
♦ 101, 257
♦ pH.

Diodia saponariifolia (Cham. & Schltdl) K.Shum.
Rubiaceae
♦ poaia do brejo
♦ Weed
♦ 255
♦ pH. Origin: Americas.

Diodia sarmentosa Sw.
Rubiaceae
♦ tropical buttonweed
♦ Weed, Quarantine Weed
♦ 13, 88, 135, 157, 191
♦ pH.

Diodia scandens Sw.
Rubiaceae
♦ bosambi, botaholu
♦ Weed
♦ 87, 88

Diodia serrulata (Beauv.) G.Taylor
Rubiaceae
♦ seaside buttonweed
♦ Naturalised
♦ 101

Diodia teres Walter
Rubiaceae
♦ poorjoe, nappimatara
♦ Weed, Naturalised, Casual Alien
♦ 42, 86, 87, 88, 98, 161, 203, 207, 210, 218, 249, 255, 286, 287
♦ aH, arid, herbal. Origin: North America.

Diodia virginiana L.
Rubiaceae
Diodia hirsuta Pursh, *Diodia tetragona*
Walt.
♦ Virginia buttonweed
♦ Weed, Naturalised
♦ 87, 88, 161, 211, 218, 249, 286, 287
♦ pH, herbal.

Dioscorea alata L.
Dioscoreaceae
♦ water yam, kehp, white yam,
winged yam
♦ Weed, Naturalised, Introduced,
Environmental Weed, Casual Alien
♦ 32, 39, 80, 86, 88, 93, 101, 112, 151,
179, 191, 230, 261, 262
♦ C, cultivated, herbal, toxic. Origin:
southern Asia.

Dioscorea altissima Lam.
Dioscoreaceae
♦ dunguey
♦ Naturalised, Casual Alien
♦ 101, 261
♦ Origin: Brazil.

Dioscorea batatas Decne.
Dioscoreaceae
= *Dioscorea oppositifolia* L.
♦ air potato, Chinese yam, cinnamon
yam, cinnamon vine
♦ Weed, Naturalised
♦ 17, 80, 88, 102, 195, 247, 287
♦ pH, cultivated, herbal, toxic.

Dioscorea bulbifera L.
Dioscoreaceae
Dioscorea anthropophagorum A.Chev.
ex Jum., *Dioscorea bulbifera* var.
anthropophagorum (A.Chev. ex Jum.)
Prain & Burkill ex Summerh., *Dioscorea
hoffa* Cordem., *Dioscorea hofika* Jum.
& H.Perrier, *Dioscorea latifolia* Benth.
(see), *Dioscorea longipetiolata* Baudon,
Dioscorea perrieri R.Knuth, *Dioscorea
sativa* Thunb., *Dioscorea violacea*
Baudon, *Helmia bulbifera* (L.) Kunth
♦ air potato, abobo, palai, air yam,
bojaka, lisaku
♦ Weed, Noxious Weed, Naturalised,
Introduced, Environmental Weed,
Casual Alien
♦ 32, 39, 47, 80, 86, 87, 88, 101, 112, 122,
151, 152, 179, 230, 261
♦ pC, cultivated, herbal, toxic. Origin:
southern Asia, Pacific and possibly
Africa.

Dioscorea cayennensis Lam.
Dioscoreaceae
♦ yellow yam, Goa potato, Tongo yam
♦ Weed, Naturalised, Casual Alien
♦ 32, 101, 261
♦ herbal. Origin: tropical Asia.

Dioscorea esculenta (Lour.) Burkill
Dioscoreaceae
♦ lesser yam, potato yam
♦ Naturalised, Casual Alien
♦ 101, 261
♦ herbal. Origin: tropical Asia.

**Dioscorea esculenta (Lour.) Burkill var.
esculenta**

Dioscoreaceae
♦ kehpalai
♦ Introduced
♦ 230
♦ C.

Dioscorea hispida Dennst.
Dioscoreaceae
♦ intoxicating yam, kehp en Hawaii
♦ Weed, Introduced
♦ 12, 39, 230
♦ C, herbal, toxic.

Dioscorea japonica Thunb.
Dioscoreaceae
♦ glutinous yam, Japanese yam,
Chinese yam
♦ Weed
♦ 88, 191, 204, 286
♦ pC, cultivated, herbal.

Dioscorea latifolia Benth.
Dioscoreaceae
= *Dioscorea bulbifera* L.
♦ akam
♦ Casual Alien
♦ 261
♦ Origin: west tropical Asia.

Dioscorea nipponica Makino
Dioscoreaceae
♦ Chuanlong yam, uchiwadokoro
♦ Weed
♦ 297
♦ herbal.

Dioscorea nummularia Lam.
Dioscoreaceae
♦ yam, kehpeneir
♦ Introduced
♦ 230
♦ C, cultivated, herbal.

Dioscorea oppositifolia L.
Dioscoreaceae
Dioscorea batatas Decne. (see)
♦ Chinese yam, cinnamon vine
♦ Weed, Naturalised
♦ 101, 161
♦ cultivated, herbal, toxic.

Dioscorea pentaphylla L.
Dioscoreaceae
♦ fiveleaf yam, kehp
♦ Weed, Naturalised, Introduced
♦ 87, 88, 101, 230
♦ C, cultivated, herbal, toxic.

Dioscorea rotundata Poir.
Dioscoreaceae
♦ white guinea yam, guinea yam
♦ Introduced
♦ 261
♦ Origin: tropical Africa.

Dioscorea sansibarensis Pax
Dioscoreaceae
♦ Zanzibar yam
♦ Weed, Naturalised
♦ 101, 179
♦ cultivated, herbal.

Dioscorea tokoro Makino ex Miyabe
Dioscoreaceae
♦ onidokoro
♦ Weed
♦ 286

♦ pC, promoted, herbal. Origin: Japan.

Dioscorea trifida L.f.
Dioscoreaceae
♦ Indian yam
♦ Naturalised, Casual Alien
♦ 101, 261
♦ Origin: South America.

Dioscorea villosa L.
Dioscoreaceae
♦ wild yam, jamssi
♦ Weed
♦ 23, 39, 88
♦ pC, cultivated, herbal, toxic.

**Diospyros crassinervis (Krug & Urb.)
Standl.**
Ebenaceae
♦ feather bed
♦ Weed
♦ 14
♦ herbal.

**Diospyros dichrophylla (Gand.) De
Winter**
Ebenaceae
♦ common starapple, monkey apple,
poison peach
♦ Weed, Native Weed
♦ 27, 121
♦ S/T, cultivated, toxic. Origin:
southern Africa.

Diospyros digyna Jacq.
Ebenaceae
Diospyros ebenaster Retz, *Diospyros
laurifolia* A.Rich., *Diospyros
membranacea* A.DC., *Diospyros nigra*
Perr., *Diospyros nigra* (Blanco) Blanco,
Diospyros obtusifolia Humb. & Bonpl.,
Diospyros obtusifolius Humb. & Bonpl.,
Diospyros sapota Roxb., *Diospyros
tiltzapotl* Sessé & Moç., *Sapota nigra*
Blanco, *Sapota nigra* J.F.Gmel.
♦ East Indian ebony, biaqui,
black sapote, ebano, guayabota,
mueque, tauch, tauch ya, tlilzapotl,
totcuitlutzapotl, zapote negro, zapote
prieto
♦ Weed, Introduced
♦ 179, 228
♦ arid, cultivated.

Diospyros discolor Willd.
Ebenaceae
♦ velvet apple
♦ Introduced
♦ 230
♦ cultivated.

Diospyros ebenum Koenig ex Retz.
Ebenaceae
♦ ebony
♦ Weed, Naturalised
♦ 39, 101, 179
♦ cultivated, herbal, toxic.

Diospyros kaki L.
Ebenaceae
♦ Japanese persimmon, persimmon
♦ Environmental weed
♦ 201
♦ T, Cultivated, herbal. Origin: China,
Japan.

Diospyros lotus L.
   Ebenaceae
   ♦ date plum
   ♦ Naturalised, Casual Alien
   ♦ 280, 287
   ♦ T, cultivated, herbal.

Diospyros lycioides Desf. ssp. *sericea*
(Bernh.) De Winter
   Ebenaceae
   Royena sericea Bernh. (see), *Royena*
   lycioides (Desf.) A.DC. ssp. *sericea*
   (Bernh.) De Winter
   ♦ bluebush, Natal bluebush, red
   starapple
   ♦ Native Weed
   ♦ 121
   ♦ S/T. Origin: southern Africa.

Diospyros maritima Blume
   Ebenaceae
   ♦ Weed, Naturalised
   ♦ 101, 179
   ♦ cultivated, herbal. Origin: Asia,
   Australia.

Diospyros mollis Griff.
   Ebenaceae
   ♦ maklua
   ♦ Introduced
   ♦ 228
   ♦ arid.

Diospyros scabrida (Harv. ex Hiern) De
Winter var. *cordata* (E.Mey. & A.DC.) De
Winter
   Ebenaceae
   ♦ hard leaved monkey plum
   ♦ Native Weed
   ♦ 121
   ♦ S/T, cultivated. Origin: southern
   Africa.

Diospyros simii (Kuntze) De Winter
   Ebenaceae
   ♦ climbing starapple, rub rub berry,
   starapple
   ♦ Native Weed
   ♦ 121
   ♦ S/T, cultivated. Origin: southern
   Africa.

Diospyros texana Scheele
   Ebenaceae
   ♦ Texas persimmon
   ♦ Weed
   ♦ 87, 88, 218
   ♦ cultivated, herbal.

Diospyros virginiana L.
   Ebenaceae
   ♦ persimmon, American persimmon,
   common persimmon
   ♦ Weed
   ♦ 8, 39, 87, 88, 218
   ♦ T, cultivated, herbal, toxic.

Dipcadi bakerianum Bol.
   Liliaceae/Hyacinthaceae
   ♦ Native Weed
   ♦ 121
   ♦ pH, toxic. Origin: southern Africa.

Dipcadi erythraeum Webb & Berthel.
   Liliaceae/Hyacinthaceae
   ♦ Weed

   ♦ 221
   ♦ herbal.

Dipcadi glaucum (Ker Gawl.) Bak.
   Liliaceae/Hyacinthaceae
   ♦ poison onion, wild onion
   ♦ Weed, Native Weed
   ♦ 39, 121
   ♦ pH, toxic. Origin: southern Africa.

Diphysa robinioides Benth. ex Benth. &
Oerst.
   Fabaceae/Papilionaceae
   Diphysa carthaginensis Benth. & Oerst.,
   Diphysa floribunda Peyr., *Diphysa*
   humilis Benth. & Oerst.
   ♦ cuachepil, cuauchepilli, guachipelí,
   guachipilín, huachipilín, macano,
   much, palo amarillo
   ♦ Introduced
   ♦ 228
   ♦ arid, herbal.

Diplachne fascicularis (Lam.) P.Beauv.
   Poaceae
   = *Leptochloa fusca* (L.) Kunth ssp.
   fascicularis (Lam.) N.Snow (NoR) [see
   Leptochloa fascicularis (Lam.) Gray]
   ♦ thread sprangletop, bearded
   sprangletop
   ♦ Weed, Naturalised
   ♦ 91, 287
   ♦ G, aqua, herbal.

Diplachne fusca (L.) P.Beauv. ex Roem.
& Schult.
   Poaceae
   = *Leptochloa malabarica* (L.) Veldkamp
   (NoR) [see *Diplachne malabarica* (L.)
   Merr. and *Leptochloa fusca* (L.) Kunth]
   ♦ brown beetlegrass, littoral
   sprangletop, beetlegrass
   ♦ Weed, Naturalised, Introduced,
   Casual Alien
   ♦ 87, 88, 91, 200, 221, 228, 280, 287, 297
   ♦ a/pG, aqua, cultivated.

Diplachne malabarica (L.) Merr.
   Poaceae
   = *Leptochloa malabarica* (L.) Veldkamp
   (NoR) [see *Diplachne fusca* (L.) P.Beauv.
   ex Roem. & Schult. and *Leptochloa fusca*
   (L.) Kunth]
   ♦ Weed
   ♦ 87, 88
   ♦ G.

Diplachne malayana C.B.Hubb.
   Poaceae
   ♦ Weed
   ♦ 12

Diplachne uninervia (Presl) Parodi
   Poaceae
   = *Leptochloa uninervia* (J.Presl) Hitchc.
   & Chase
   ♦ Mexican sprangletop, giant ryegrass,
   triguillo
   ♦ Weed, Quarantine Weed,
   Naturalised, Environmental Weed
   ♦ 54, 76, 88, 91, 155, 203, 220, 236, 286,
   287, 295
   ♦ G, aqua.

Diplanthera uninervis (Forssk.) Asch.
   Bignoniaceae

   Cymodocea australis Trimen, *Diplanthera*
   madagascariensis Steud., *Diplanthera*
   tridentata Steinh., *Halodule australis*
   Miq., *Halodule uninervis* (Forssk.) Asch.
   in Boiss., *Phucagrostis tridentata* Ehrenb.
   & Hempr., *Zostera tridentata* Ehrenb. &
   Hempr., *Zostera uninervis* Forssk.
   ♦ Weed
   ♦ 221

Diplazium esculentum (Retz.) Sw.
   Dryopteridaceae/Woodsiaceae
   ♦ vegetable fern
   ♦ Weed, Naturalised
   ♦ 12, 86, 87, 88, 101, 179
   ♦ cultivated, herbal. Origin: China,
   Japan, India, south-east Asia,
   Polynesia.

Diplazium japonicum (Thunb.) Bedd.
   Dryopteridaceae/Woodsiaceae
   = *Deparia petersenii* (Kunze) M.Kato
   ♦ Weed, Quarantine Weed
   ♦ 76, 87, 88, 203, 220

Diplazium ponapense (Copel.) Hosok.
   Dryopteridaceae/Woodsiaceae
   Athyrium ponapense Copel.
   ♦ Introduced
   ♦ 230
   ♦ H.

Diplazium sammatii (Kuhn) C.Chr.
   Dryopteridaceae/Woodsiaceae
   Asplenium crenato-serratum J.Bommer
   ex H.Christ, *Asplenium ottonis* Kuhn,
   Asplenium sammatii Kuhn, *Athyrium*
   sammatii (Kuhn) Tardieu, *Diplazium*
   bommeri H.Christ, *Diplazium zenkeri*
   Hieron.
   ♦ Weed
   ♦ 88

Diplocaulobium carolinense Hawkes
   Orchidaceae
   ♦ Introduced
   ♦ 230
   ♦ H.

Diplocaulobium flavicolle (Schltr.)
Hawkes
   Orchidaceae
   Dendrobium flavicolle Schltr.
   ♦ Introduced
   ♦ 230
   ♦ H.

Diplolaena dampieri Desf.
   Rutaceae
   ♦ southern diplolaena
   ♦ Naturalised, Native Weed
   ♦ 7
   ♦ cultivated. Origin: Australia.

Diplorhynchus angustifolia Stapf
   Apocynaceae
   ♦ Weed
   ♦ 87, 88

Diplotaxis DC. spp.
   Brassicaceae
   ♦ wallrocket
   ♦ Weed, Naturalised
   ♦ 198, 243, 272
   ♦ H.

Diplotaxis acris **(Forssk.) Boiss.**
Brassicaceae
Malcolmia arabica Velen.
♦ Weed
♦ 221
♦ arid.

Diplotaxis catholica **(L.) DC.**
Brassicaceae
♦ Weed
♦ 70, 87, 88, 94, 250, 253

Diplotaxis erucoides **(L.) DC.**
Brassicaceae
Crucifera erucoides E.H.L.Krause,
Euzomum erucoides Spach, *Sinapis
erucoides* Torner
♦ white rocket, white wallrocket,
white salad rocket
♦ Weed, Quarantine Weed,
Naturalised
♦ 39, 76, 87, 88, 94,
101, 203, 220, 243, 253
♦ aH, cultivated, herbal, toxic.

Diplotaxis harra **(Forssk.) Boiss.**
Brassicaceae
♦ Weed
♦ 221

Diplotaxis muralis **(L.) DC.**
Brassicaceae
Brassica muralis Boiss., *Caulis muralis*
E.H.L.Krause, *Crucifera diplotaxis*
E.H.L.Krause, *Sinapis muralis* R.Br.
♦ stinking wallrocket, stinking
wallrocket, stinkweed, wallmustard,
sandrocket, annual wallrocket,
stinking diplotaxis, pikkuhietasinappi
♦ Weed, Naturalised, Garden Escape,
Environmental Weed, Casual Alien
♦ 7, 15, 34, 40, 42, 51, 70, 86, 87, 88, 94,
98, 101, 121, 156, 161, 176, 179, 198, 203,
218, 237, 243, 253, 269, 272, 280, 295,
300
♦ a/bH, cultivated, herbal, toxic.
Origin: Eurasia.

Diplotaxis siifolia **Kuntze**
Brassicaceae
♦ wallrocket
♦ Naturalised
♦ 101

Diplotaxis tenuifolia **(L.) DC.**
Brassicaceae
Brassica tenuifolia Fr., *Caulis tenuifolius*
E.H.L.Krause, *Crucifera tenuifolia*
E.H.L.Krause, *Eruca tenuifolia* Moench,
Sinapis tenuifolia R.Br.
♦ slimleaf wallrocket, sand rocket,
sand mustard, Lincoln weed, large
sandrocket, perennial rocket, perennial
wallrocket, wallrocket, flor amarilla
♦ Weed, Quarantine Weed, Noxious
Weed, Naturalised, Garden Escape,
Environmental Weed, Casual Alien
♦ 7, 34, 39, 40, 42, 72, 76, 86, 87, 88, 94,
98, 101, 147, 161, 171, 172, 176, 198, 203,
218, 236, 237, 243, 253, 269, 271, 272,
280, 287, 295
♦ pH, cultivated, herbal, toxic. Origin:
Europe, Middle East.

Diplotaxis tenuisiliqua **Delile**
Brassicaceae

♦ Weed
♦ 87, 88

Diplotaxis viminea **(L.) DC.**
Brassicaceae
Brassica viminea Boiss., *Crucifera
viminea* E.H.L.Krause, *Sisymbrium
vimineum* L.
♦ Weed
♦ 88, 94, 272
♦ cultivated.

Diplotaxis virgata **(Cav.) DC.**
Brassicaceae
♦ mustard diplotaxis, sand mustard,
wild mustard
♦ Weed
♦ 87, 88, 94, 121
♦ aH. Origin: Eurasia.

Dipogon lignosus **(L.) Verdc.**
Fabaceae/Papilionaceae
Dolichos lignosus L. (see)
♦ mile a minute, dipogon, okie bean,
Australian pea
♦ Weed, Quarantine Weed,
Naturalised, Garden Escape,
Environmental Weed
♦ 3, 7, 15, 72, 86, 88, 98, 101, 176, 181,
198, 203, 225, 246, 280, 289, 296
♦ C, cultivated. Origin: South Africa.

Dipsacus **L. spp.**
Dipsacaceae
♦ teasel, Fuller's teasel
♦ Weed, Naturalised, Environmental
Weed
♦ 78, 80, 104, 116, 198, 272
♦ herbal.

Dipsacus asper **Wall. ex DC.**
Dipsacaceae
♦ Himalayan teasel
♦ Weed
♦ 24, 297
♦ bH, cultivated, herbal.

Dipsacus fullonum **L.**
Dipsacaceae
Dipsacus sylvestris Huds. (see)
♦ wild teasel, common teasel, card
teasel, Venus cup, card thistle,
gypsy combs, Fuller's teasel,
pikarikarttaohdake
♦ Weed, Quarantine Weed, Noxious
Weed, Naturalised, Introduced,
Garden Escape, Environmental Weed,
Cultivation Escape
♦ 4, 20, 24, 34, 42, 70, 72, 76, 80, 87, 88,
98, 101, 102, 121, 130, 161, 176, 180, 195,
203, 207, 211, 212, 220, 229, 237, 272,
295
♦ bH, cultivated, herbal. Origin:
southern and eastern Europe.

Dipsacus fullonum **L. ssp. *fullonum***
Dipsacaceae
♦ wild teasel, card thistle, Fuller's
teasel
♦ Weed, Noxious Weed, Naturalised,
Environmental Weed
♦ 86, 101, 147, 198, 229, 269
♦ pH, cultivated. Origin: southern
Europe.

Dipsacus fullonum **L. ssp. *sylvestris***
(Huds.) Clapham

Dipsacaceae
♦ Fuller's teasel, common teasel
♦ Noxious Weed, Naturalised,
Introduced
♦ 101, 222, 229

Dipsacus inermis **Wall.**
Dipsacaceae
♦ spineless teasel
♦ Quarantine Weed
♦ 220
♦ herbal.

Dipsacus japonicus **Miq.**
Dipsacaceae
♦ Japanese teasel
♦ Weed
♦ 297
♦ b/pH, promoted, herbal. Origin:
China, Japan.

Dipsacus laciniatus **L.**
Dipsacaceae
♦ cutleaf teasel, air potato, air yam
♦ Weed, Noxious Weed, Naturalised,
Introduced, Environmental Weed
♦ 70, 80, 87, 88, 101, 151, 151, 161, 195,
218, 222, 229, 272
♦ cultivated, herbal.

Dipsacus pilosus **L.**
Dipsacaceae
Cephalaria pilosa (L.) Gren., *Virga pilosa*
(L.) Hill
♦ small teasel
♦ Weed
♦ 24, 272
♦ bH, cultivated, herbal.

Dipsacus sativus **(L.) Honck.**
Dipsacaceae
♦ Fuller's teasel, Indian teasel
♦ Weed, Noxious Weed, Naturalised,
Cultivation Escape, Casual Alien
♦ 24, 34, 40, 98, 101, 199, 203, 229, 241,
272, 280, 295, 300
♦ bH, cultivated, herbal. Origin:
obscure.

Dipsacus strigosus **Roem. & Schult.**
Dipsacaceae
Virga strigosa (Willd. ex Roem. &
Schult) Holub
♦ sukakarttaohdake
♦ Cultivation Escape
♦ 42
♦ cultivated.

Dipsacus sylvestris **Huds.**
Dipsacaceae
= *Dipsacus fullonum* L.
♦ common teasel, wild teasel, teasel
♦ Weed, Naturalised
♦ 15, 80, 87, 88, 98, 136, 165, 195, 210,
218, 264, 280, 294
♦ b/pH, cultivated, herbal. Origin:
Eurasia.

Dipteracanthus prostratus **(Poir.) Nees**
Acanthaceae
= *Ruellia prostrata* Poir.
♦ Weed
♦ 13, 87, 88
♦ cultivated.

Dipteracanthus repens **(L.) Hassk.**
Acanthaceae

- Weed
- 12, 13

Dipterygium glaucum Decne.
Capparaceae/Brassicaceae
- Weed
- 88, 221

Diptychocarpus strictus (Fisch.) Trautv.
Brassicaceae
- Weed
- 272

Dirca palustris L.
Thymelaeaceae
- leatherwood, wicopy, eastern leatherwood, American mezereon
- Weed
- 8, 39, 87, 88, 161, 218, 247
- S, cultivated, herbal, toxic.

Disa bracteata Sw.
Orchidaceae
Monadenia bracteata (Sw.) T.Durand & Schinz (see)
- African weed orchid, South African orchid, mondenia
- Naturalised, Environmental Weed
- 289
- Origin: South Africa.

Discaria toumatou Raoul
Rhamnaceae
- matagouri, tumatakuru
- Weed
- 87, 88, 165
- cultivated.

Dischisma arenarium E.Mey.
Scrophulariaceae/Selaginaceae
- Weed, Naturalised, Environmental Weed
- 7, 9, 86, 98, 203
- Origin: South Africa.

Dischisma capitatum (Thunb.) Choisy
Scrophulariaceae/Selaginaceae
Hebenstretia capitata Thunb.
- Weed, Naturalised, Native Weed, Environmental Weed
- 7, 9, 86, 98, 121, 203
- aH. Origin: southern Africa.

Dischisma ciliatum (Berg.) Choisy ssp. ciliatum
Scrophulariaceae/Selaginaceae
- Native Weed
- 121
- S. Origin: South Africa.

Discocalyx ponapensis Mez
Myrsinaceae
- kartiel
- Introduced
- 230
- T.

Disocactus Lindl. spp.
Cactaceae
Bonifazia Standl. & Steyer. spp., *Chiapasia* Britt. & Rose spp., *Lobeira* Alex. spp., *Nopalxochia* Britt. & Rose spp., *Pseudonopalxochia* Backen. spp., *Wittia* Schumm. spp., *Wittiocactus* Rausch. spp.
- Weed, Quarantine Weed
- 203, 220

Disocactus Lindl. × Heliochia Rowley spp.
Cactaceae
= *Disocactus* Lindl. spp. × [*Heliocereus* (Berg.) Britt. & Rose spp. × *Disocactus* Lindl. spp.]
- Quarantine Weed
- 76, 220

Disocactus Lindl. × Helioselenius Rowley spp.
Cactaceae
= *Disocactus* Lindl. spp. × [*Heliocereus* (Berg.) Britt. & Rose spp. × *Selenicereus* (Berg.) Britt. & Rose spp.]
- Quarantine Weed
- 76, 220

Disocactus Lindl. × Heliphyllum Rowley spp.
Cactaceae
= *Disocactus* Lindl. spp. × [*Heliocereus* (Berg.) Britt. & Rose spp. × *Epiphyllum* Haw. spp.]
- Quarantine Weed
- 76, 220

Disphyma clavellatum (Haw.) Chinnock
Aizoaceae/Mesembryanthemaceae
- rounded noonflower
- Naturalised
- 280
- cultivated.

Disphyma crassifolium (L.) L.Bolus
Aizoaceae/Mesembryanthemaceae
- purple dewplant
- Naturalised, Introduced
- 40, 228
- arid, cultivated.

Dissotis debilis Triana
Melastomataceae
- Weed
- 87, 88

Dissotis erecta (Guill. & Perr.) Dandy
Melastomataceae
- Weed
- 88

Dissotis irvingiana Hook.f.
Melastomataceae
- Weed
- 87, 88

Dissotis rotundifolia (Sm.) Triana
Melastomataceae
- dissotis, pink lady, nhunga
- Weed, Naturalised, Garden Escape
- 3, 87, 88, 101, 107, 191, 261
- cultivated, herbal. Origin: tropical Africa.

Distichlis humilis Phil.
Poaceae
- Introduced
- 228
- G, arid.

Distichlis palmeri (Vasey) Fassett ex I.M.Johnst.
Poaceae
Uniola palmeri Vasey
- Introduced
- 228
- G, arid.

Distichlis scoparia (Kunth) Arechav.
Poaceae
- pasto salado
- Weed, Introduced
- 228, 295
- G, arid.

Distichlis spicata (L.) Greene
Poaceae
Distichlis distichophylla (Labill.) Fassett, *Distichlis stricta* (Torr.) Rydb. (see), *Distichlis maritima* Raf., *Distichlis thalassica* (Kunth) Desv. (see)
- seashore saltgrass, inland saltgrass, marsh spikegrass, desert saltgrass, saltgrass
- Weed, Quarantine Weed, Native Weed, Introduced, Casual Alien
- 76, 87, 88, 161, 174, 180, 212, 218, 228, 237, 280, 295
- pG, arid/aqua, cultivated, herbal. Origin: North America.

Distichlis stricta (Torr.) Rydb.
Poaceae
= *Distichlis spicata* (L.) Greene
- desert saltgrass
- Weed
- 87, 88, 218
- G.

Distichlis thalassica (Kunth) Desv.
Poaceae
= *Distichlis spicata* (L.) Greene
- Weed
- 87, 88
- G, arid.

Distictis buccinatoria (DC.) A.H.Gentry
Bignoniaceae
Phaedranthus buccinatorius Miers, *Bignonia buccinatoria* Mairet ex DC., *Phaedranthus lindleyanus* Miers
- red bignonia, blood trumpet, dynamic trumpet vine
- Naturalised
- 280
- cultivated.

Ditaxis neomexicana (Müll.Arg.) Heller
Euphorbiaceae
= *Argythamnia neomexicana* Müll.Arg.
- New Mexico silverbush
- Weed
- 199
- a/pH.

Dittrichia graveolens (L.) Greuter
Asteraceae
Inula graveolens (L.) Desf. (see)
- stinkwort, camphor inula, stinkweed
- Weed, Noxious Weed, Naturalised, Environmental Weed
- 7, 9, 72, 86, 88, 93, 94, 98, 101, 147, 176, 198, 203, 269, 272, 280
- aH. Origin: Mediterranean, Middle East.

Dittrichia viscosa (L.) Greuter
Asteraceae
Inula viscosa (L.) Ait. (see)
- false yellowhead, aromatic inula
- Weed, Naturalised, Garden Escape, Environmental Weed
- 7, 70, 86, 88, 94, 98, 203, 272
- pH, cultivated. Origin: Mediterranean.

Dodartia orientalis L.
Scrophulariaceae
♦ arokki
♦ Weed, Casual Alien
♦ 42, 87, 88, 272, 275

Dodonaea Mill. spp.
Sapindaceae
♦ Quarantine Weed
♦ 220

Dodonaea angustifolia L.f.
Sapindaceae
Dodonaea arborea Herter, *Dodonaea arizonica* Nelson, *Dodonaea bialata* Kunth, *Dodonaea lagurensis* M.E.Jones, *Dodonaea linearifolia* Linden & Turcz., *Dodonaea sandwicensis* Sherff, *Dodonaea stenoptera* Hillebr., *Dodonaea thunbergiana* Eckl. & Zeyh., *Dodonaea viscosa* Jacq. var. *angustifolia* (L.f.) Benth.
♦ cape sand olive, common sand olive, hopbush, sand olive, granadina, guaya billo, jarillo de loma
♦ Weed, Introduced, Native Weed
♦ 10, 121, 228
♦ S/T, arid, cultivated. Origin: southern Africa.

Dodonaea eriocarpa Sm.
Sapindaceae
= *Dodonaea viscosa* (L.) Jacq.
♦ Weed
♦ 87, 88
♦ herbal.

Dodonaea lobulata F.Muell.
Sapindaceae
♦ lobed leaf hop bush
♦ Introduced
♦ 228
♦ arid, cultivated.

Dodonaea viscosa Jacq.
Sapindaceae
Ptelea viscosa L., *Dodonaea burmanniana* DC., *Dodonaea dioica* Roxb., *Dodonaea eriocarpa* Sm. (see), *Dodonaea schiedeana* Schldl.
♦ Florida hopbush, togo vao, msangabale, akeake, hopbush
♦ Weed, Naturalised, Introduced
♦ 7, 39, 86, 87, 88, 198, 221, 228
♦ S, arid, cultivated, herbal, toxic. Origin: Australia.

Dolichos falciformis E.Mey.
Fabaceae/Papilionaceae
♦ Native Weed
♦ 121
♦ pH. Origin: southern Africa.

Dolichos formosus Hochst. ex A.Rich.
Fabaceae/Papilionaceae
♦ Weed
♦ 240
♦ pH.

Dolichos lablab L.
Fabaceae/Papilionaceae
= *Lablab purpureus* (L.) Sweet
♦ hyacinth bean, hyacinth bean vine
♦ Weed, Naturalised
♦ 87, 88, 247, 287
♦ cultivated, herbal, toxic.

Dolichos lignosus L.
Fabaceae/Papilionaceae
= *Dipogon lignosus* (L.) Verdc.
♦ Naturalised
♦ 241, 300
♦ cultivated.

Dolichos minimus L.
Fabaceae/Papilionaceae
= *Rhynchosia minima* (L.) DC.
♦ Weed
♦ 87, 88

Dombeya rotundifolia Planch.
Sterculiaceae
♦ wild pear, msubu
♦ Weed, Quarantine Weed
♦ 76, 87, 88, 203, 220
♦ cultivated.

Donax grandis Ridl.
Marantaceae
♦ bemban
♦ Weed
♦ 12

Dontostemon eglandulosus (DC.) Ledeb.
Brassicaceae
♦ eglandulose dontostemon
♦ Weed
♦ 297

Dopatrium junceum (Roxb.) Buch.-Ham.
Scrophulariaceae
Gratiola juncea Roxb. (see)
♦ horsefly's eye, dopatrium
♦ Weed, Naturalised
♦ 34, 87, 88, 101, 170, 191, 197, 204, 263, 274, 286, 297
♦ waH.

Doronicum L. spp.
Asteraceae
♦ leopard's bane, false leopard's bane
♦ Weed
♦ 39, 272
♦ herbal, toxic.

Doronicum pardalianches L.
Asteraceae
Doronicum cordatum Lam. (*non* Sch.Bip.)
♦ great leopard's bane, tarhavuohenjuuri, great false leopard's bane
♦ Naturalised, Cultivation Escape
♦ 39, 40, 42, 101
♦ pH, cultivated, herbal, toxic.

Doronicum plantagineum L.
Asteraceae
Doronicum hungaricum Rchb.f.
♦ plantain leaved leopard's bane, leopard's bane
♦ Naturalised, Casual Alien
♦ 40, 101, 280
♦ cultivated, herbal.

Dorotheanthus bellidiformis (Burm.f.) N.E.Br.
Aizoaceae/Mesembryanthemaceae
Mesembryanthemum criniflorum L.f.
♦ buckbay vygie, Livingstone daisy
♦ Weed, Naturalised, Native Weed, Casual Alien
♦ 7, 15, 86, 121, 280

♦ aH, cultivated, herbal. Origin: southern Africa.

Dorycnium hirsutum (L.) Ser.
Fabaceae/Papilionaceae
Bonjeania hirsuta (L.) Reichen., *Lotus hirsutus* L.
♦ hairy canary clover
♦ Weed, Casual Alien
♦ 272, 280
♦ cultivated, herbal.

Dorycnium pentaphyllum Scop.
Fabaceae/Papilionaceae
♦ socarrillo
♦ Weed
♦ 70, 87, 88, 272
♦ pH, cultivated, herbal.

Dovyalis caffra (J.D.Hook. & Harv.) J.D.Hook.
Flacourtiaceae
Aberia caffra Hook.f. & Harv.
♦ kei apple, umkokola
♦ Weed, Naturalised, Introduced
♦ 54, 86, 88, 98, 190, 190, 203, 228
♦ S/T, arid, cultivated, herbal. Origin: Africa.

Dovyalis hebecarpa (G.Gardn.) Warb.
Flacourtiaceae
♦ Ceylon gooseberry, tropical apricot, ketembila, quetembila
♦ Naturalised, Cultivation Escape
♦ 101, 261
♦ cultivated. Origin: India, Sri Lanka.

Draba atlantica Pomel
Brassicaceae
♦ Quarantine Weed
♦ 220

Draba brachycarpa Nutt. ex Torr. & Gray
Brassicaceae
♦ shortpod draba
♦ Weed
♦ 161
♦ herbal.

Draba hispanica Boiss.
Brassicaceae
♦ Quarantine Weed
♦ 220
♦ cultivated.

Draba muralis L.
Brassicaceae
Crucifera capselloides E.H.L.Krause, *Drabella muralis* Fourr.
♦ wall whitlowgrass, vallikynsimö
♦ Weed
♦ 23, 88, 272
♦ a/bH, promoted, herbal.

Draba nemorosa L.
Brassicaceae
Draba nemoralis Ehrh., *Draba lutea* Gilib.
♦ woodland draba, woodland whitlowgrass, inunazuna, keltakynsimö, yellow whitlowgrass, yellow draba
♦ Weed, Noxious Weed, Naturalised, Cultivation Escape
♦ 86, 87, 88, 98, 176, 203, 252, 275, 286, 297, 299
♦ aH, cultivated, herbal. Origin: Eurasia.

Draba nemorosa L. var. *hebecarpa* Ledeb.
  Brassicaceae
  ♦ Weed
  ♦ 263
  ♦ aH.

Draba nemorosa L. var. *hebecarpa* Ledeb.
fo. *leiocarpa* Kitag.
  Brassicaceae
  ♦ Naturalised
  ♦ 287

Draba sibirica (Pall.) Thell.
  Brassicaceae
  ♦ siperiankynsimö
  ♦ Cultivation
  Escape
  ♦ 42
  ♦ cultivated.

Draba verna L.
  Brassicaceae
  = *Erophila verna* (L.) Chev.
  ♦ whitlowwort, spring draba, spring
  whitlowgrass
  ♦ Weed, Naturalised
  ♦ 88, 101, 161, 207, 218, 300
  ♦ aH, herbal. Origin: Europe.

Dracaena fragrans (L.) Ker Gawl.
  Agavaceae/Dracaenaceae
  ♦ fragrant dracaena, corn plant,
  cocomacaco, drecina, Indian cane,
  Indian palm
  ♦ Naturalised, Garden Escape
  ♦ 101, 261
  ♦ cultivated, herbal. Origin: east
  tropical Africa.

Dracaena ombet Kotschy & Peyr.
  Agavaceae/Dracaenaceae
  = *Dracaena draco* L. (NoR)
  ♦ Weed
  ♦ 221
  ♦ cultivated.

Dracocephalum heterophyllum Benth.
  Lamiaceae
  ♦ diverseleaf dragonhead
  ♦ Weed
  ♦ 297
  ♦ pH, promoted. Origin: Himalayas.

Dracocephalum moldavica L.
  Lamiaceae
  ♦ dragonhead, tuoksuampiaisyrtti,
  Moldavian dragonhead, Moldavian
  balm
  ♦ Weed, Naturalised, Casual Alien
  ♦ 42, 86, 98, 101, 121, 203, 272, 275, 297
  ♦ aH, cultivated, herbal. Origin:
  Eurasia.

Dracocephalum nutans L.
  Lamiaceae
  ♦ nuokkuampiaisyrtti
  ♦ Casual Alien
  ♦ 42
  ♦ cultivated.

Dracocephalum parviflorum Nutt.
  Lamiaceae
  Moldavica parviflora (Nutt.) Britton.
  ♦ American dragonhead, dragonhead
  mint, Amerikanampiaisyrtti
  ♦ Weed, Noxious Weed, Casual Alien

  ♦ 23, 42, 49, 52, 87, 88, 161, 210, 218,
  237, 295, 299
  ♦ a/bH, cultivated, herbal.

Dracocephalum ruyschiana L.
  Lamiaceae
  ♦ isoampiaisyrtti, drakblomma
  ♦ Cultivation Escape
  ♦ 42
  ♦ cultivated, herbal.

Dracocephalum sibiricum (L.) L.
  Lamiaceae
  ♦ siperianampiaisyrtti, drakblomma,
  Siberian catnip
  ♦ Cultivation Escape
  ♦ 42
  ♦ cultivated, herbal.

Dracocephalum thymiflorum L.
  Lamiaceae
  ♦ thymeleaf dragonhead, thyme
  flowered dragonhead
  ♦ Weed, Naturalised
  ♦ 87, 88, 101, 161, 272
  ♦ herbal.

Dracocephalum triflorum L.
  Lamiaceae
  ♦ ketoampiaisyrtti
  ♦ Naturalised
  ♦ 42

Dracunculus vulgaris Schott.
  Araceae
  Arum dracunculus L.
  ♦ dragon arum, common dracunculus,
  stink lily, arum arrowroot
  ♦ Weed, Naturalised
  ♦ 15, 39, 40, 87, 88, 101, 272, 280
  ♦ cultivated, herbal, toxic.

Droguetia iners (Forssk.) Schweinf.
  Urticaceae
  ♦ Weed
  ♦ 240
  ♦ pH.

Drosanthemum (Haw.) Schwantes spp.
  Aizoaceae/Mesembryanthemaceae
  ♦ drosanthemum, dewflower, iceplant
  ♦ Weed, Naturalised
  ♦ 39, 116, 198
  ♦ cultivated, toxic.

Drosanthemum candens (Haw.)
Schwantes
  Aizoaceae/Mesembryanthemaceae
  ♦ rodondo creeper
  ♦ Weed, Naturalised, Environmental
  Weed
  ♦ 7, 72, 86, 88, 98, 198, 203
  ♦ S, cultivated. Origin: South Africa.

Drosanthemum floribundum (Haw.)
Schwant.
  Aizoaceae/Mesembryanthemaceae
  ♦ showy dewflower, rosea iceplant,
  pale dewplant
  ♦ Weed, Naturalised
  ♦ 15, 40, 101, 280
  ♦ pH, cultivated.

Drosanthemum hispidum (L.) Schwant.
  Aizoaceae/Mesembryanthemaceae
  ♦ hairy dewflower, rosea iceplant
  ♦ Naturalised

  ♦ 101
  ♦ cultivated.

Drosanthemum speciosum (Haw.)
Schwant.
  Aizoaceae/Mesembryanthemaceae
  ♦ royal dewflower, showy dewflower,
  dewflower
  ♦ Naturalised
  ♦ 101
  ♦ cultivated.

Drosera intermedia Hayne
  Droseraceae
  ♦ oblong leaved sundew, spoonleaf
  sundew, love nest sundew, spatulate
  leaved sundew, narrow leaved
  sundew, long leaved sundew
  ♦ Weed
  ♦ 39, 70
  ♦ cultivated, herbal, toxic.

Drosophyllum lusitanicum (L.) Link
  Droseraceae/Dioncophyllaceae
  ♦ Portuguese sundew
  ♦ Weed
  ♦ 39, 70
  ♦ cultivated, toxic.

Dryandra formosa R.Br.
  Proteaceae
  ♦ showy dryandra
  ♦ Naturalised
  ♦ 86
  ♦ cultivated. Origin: Australia.

Dryas drummondii Richardson ex Hook.
  Rosaceae
  ♦ Drummond's dryad, Drummond's
  mountain avens
  ♦ Quarantine Weed
  ♦ 220
  ♦ cultivated, herbal.

Drymaria arenarioides Humb. & Bonpl.
ex J.A.Schult.
  Caryophyllaceae
  ♦ alfombrilla, sandwort drymary,
  lightning weed
  ♦ Weed, Quarantine Weed, Noxious
  Weed
  ♦ 26, 39, 67, 76, 82, 87, 88, 140, 161, 229
  ♦ herbal, toxic.

Drymaria cordata (L.) Willd.
  Caryophyllaceae
  Drymaria diandra Bl. (see), *Holosteum
  cordatum* L.
  ♦ heartleaf drymary, tropical
  chickweed, chickweed, West Indian
  chickweed, whitesnow, yaa klet hoi,
  jaboticaa, ilovizna
  ♦ Weed
  ♦ 28, 55, 87, 88, 157, 161, 206, 209, 217,
  218, 238, 243, 249, 255, 274, 276, 281
  ♦ aH, aqua, cultivated, herbal. Origin:
  tropical America.

Drymaria diandra Bl.
  Caryophyllaceae
  = *Drymaria cordata* (L.) Willd.
  ♦ heartleaf drymary, tropical
  chickweed
  ♦ Weed
  ♦ 39, 273, 286
  ♦ toxic.

Drymaria hirsuta Bart. ex Presl
Caryophyllaceae
= *Drymaria villosa* Cham. & Schlecht.
♦ Weed
♦ 87, 88

Drymaria pachyphylla Woot. & Standl.
Caryophyllaceae
♦ inkweed drymary, thickleaf drymary
♦ Weed
♦ 39, 87, 88, 161, 218
♦ herbal, toxic.

Drymaria villosa Cham. & Schlecht.
Caryophyllaceae
Drymaria hirsuta Bart. ex Presl (see),
Drymaria palustris Cham. & Schltdl.,
Drymaria pauciflora Bartl.
♦ dew herb
♦ Weed
♦ 88, 170, 191, 199
♦ herbal.

Drymoglossum heterophyllum (L.) C.Chr.
Polypodiaceae
♦ Weed
♦ 87, 88
♦ herbal.

Dryopteris cristata (L.) Gray
Dryopteridaceae
Aspidium cristatum (L.) Sw., *Nephrodium cristatum* Michx.
♦ crested buckler fern, crested woodfern, korpialvejuuri
♦ Weed
♦ 272
♦ pH, cultivated, herbal.

Dryopteris filix-mas (L.) Schott
Dryopteridaceae
♦ male fern, kivikkoalvejuuri, papraì samäia
♦ Weed, Naturalised
♦ 8, 15, 39, 70, 87, 88, 161, 272, 300
♦ pH, cultivated, herbal, toxic.

Dryopteris spinulosa (Mull.) Kuntze
Dryopteridaceae
♦ prickly toothed shield fern
♦ Weed
♦ 87, 88
♦ herbal.

Dryopteris splendens (Hook.) Kuntze
Dryopteridaceae
Dryopteris sprengeli Kuntze
♦ Weed
♦ 87, 88

Dryopteris vivipara (Raddi) C.Chr.
Dryopteridaceae
Goniopteris prolifera Fee
♦ Weed
♦ 87, 88

Drypetes deplanchei (Brongn. & Griseb.) Merr.
Euphorbiaceae
♦ Casual Alien
♦ 280

Duboisia myoporoides R.Br.
Solanaceae
♦ corkwood tree
♦ Weed
♦ 39, 87, 88

♦ cultivated, herbal, toxic. Origin: Australia, New Zealand.

Duchesnea chrysantha (Zoll. & Moritzi) Miq.
Rosaceae
♦ mock strawberry, wild cape gooseberry
♦ Weed
♦ 88, 191, 204, 273, 286
♦ pH, promoted.

Duchesnea indica (Andrews) Focke
Rosaceae
Fragaria indica Andr. (see), *Potentilla indica* (Andrews) Th.Wolf
♦ Indian strawberry, mock strawberry, false strawberry, pajahoda indická
♦ Weed, Naturalised, Garden Escape, Environmental Weed, Cultivation Escape
♦ 15, 34, 40, 80, 86, 88, 98, 101, 121, 161, 165, 198, 203, 218, 241, 249, 252, 269, 275, 280, 286, 290, 295, 296, 297, 300
♦ pH, cultivated, herbal. Origin: India, east Asia.

Dugesia mexicana (A.Gray) A.Gray
Asteraceae
♦ Weed
♦ 199

Duggena hirsuta (Jacq.) Britt. ex Britt. & Wilson
Rubiaceae
= *Gonzalagunia hirsuta* (Jacq.) K.Schum. (NoR) [see *Gonzalagunia spicata* (Lam.) G.Maza]
♦ Weed
♦ 87, 88

Duguetia furfuracea (St.-Hil.) Benth. & Hook.
Annonaceae
♦ araticum miudo
♦ Weed
♦ 255
♦ Origin: Brazil.

Dunbaria villosa (Thunb.) Makino
Fabaceae/Papilionaceae
♦ Weed
♦ 286

Duranta erecta L.
Verbenaceae
Duranta repens L. (see)
♦ golden dewdrop duranta, pigeon berry, Brazilian skyflower, golden dewdrops, variegated dewdrop
♦ Weed, Naturalised, Cultivation Escape
♦ 39, 86, 161, 233, 252, 283
♦ S/T, arid, cultivated, toxic. Origin: Americas.

Duranta repens L.
Verbenaceae
= *Duranta erecta* L.
♦ duranta, forget me not bush, forget me not tree, golden dewdrop, golden tears, pigeon berry, skyflower
♦ Weed, Introduced, Environmental Weed
♦ 39, 73, 88, 106, 107, 121, 151, 201, 230, 279

♦ S/T, cultivated, herbal, toxic. Origin: tropical America.

Durio zibethinus Murray
Bombacaceae
♦ durian, civet fruit
♦ Introduced
♦ 230
♦ T, cultivated, herbal.

Dyckia brevifolia Bak.
Bromeliaceae
♦ sawblade, pineapple dyckia
♦ Naturalised
♦ 101
♦ arid, cultivated.

Dypsis lutescens (H.Wendl.) Beentje & Dransf.
Arecaceae
Chrysalidocarpus lutescens (Bory) H.Wendl. (see)
♦ yellow butterfly palm, rehazo
♦ Weed, Naturalised
♦ 101, 179
♦ cultivated. Origin: Madagascar.

Dyschoriste depressa Nees
Acanthaceae
♦ Weed
♦ 87, 88
♦ arid. Origin: India, Africa.

Dyschoriste quadrangularis (Oerst.) Kuntze
Acanthaceae
♦ Weed
♦ 157
♦ pH.

Dysophylla auricularia (L.) Bl.
Lamiaceae
= *Pogostemon auricularia* (L.) El Gazzar & L.Watson
♦ Weed
♦ 13
♦ herbal.

Dysphania plantaginella F.J.Muell.
Chenopodiaceae/Dysphaniaceae
♦ dysphania
♦ Naturalised
♦ 101

Dyssodia montana (Benth.) A.Gray
Asteraceae
♦ Weed
♦ 157

Dyssodia papposa (Vent.) A.Hitchc.
Asteraceae
♦ fetid marigold, false mayweed, false dogfennel, stink weed
♦ Weed, Native Weed
♦ 34, 49, 87, 88, 161, 174, 218
♦ aH, herbal, toxic. Origin: North America.

Dyssodia pinnata (Cav.) B.L.Rob.
Asteraceae
♦ Weed
♦ 199

Dyssodia tenuifolia (Cass.) Loes.
Asteraceae
♦ Weed
♦ 199

E

Ebenopsis ebano (Berland.) Barneby & J.W.Grimes
Fabaceae/Mimosaceae
Chloroleucon ebano (Berland.) L.Rico (see)
♦ Texas ebony
♦ Quarantine Weed
♦ 220

Ecballium ciliatum (Cogn.) A.Rich.
Cucurbitaceae
Elaterium ciliatum Cogn.
♦ Weed
♦ 87, 88

Ecballium elaterium (L.) A.Rich.
Cucurbitaceae
Elaterium cordifolium Moench,
Momordica elaterium L.
♦ squirting cucumber, spitting cucumber, springgurka, cocomero asinino, exploding cucumber
♦ Weed, Noxious Weed, Naturalised, Garden Escape
♦ 7, 39, 86, 87, 88, 98, 101, 147, 176, 198, 203, 272, 280
♦ pH, cultivated, herbal, toxic. Origin: Mediterranean.

Eccoilopus cotulifer (Thunb.) A.Camus
Poaceae
♦ Weed
♦ 286
♦ G.

Eccremocarpus scaber Ruiz & Pav.
Bignoniaceae
♦ Chilean glory flower
♦ Quarantine Weed, Naturalised, Environmental Weed, Casual Alien
♦ 40, 246, 280
♦ a/pH, cultivated, herbal.

Ecdysanthera rosea Hook. & Arn.
Apocynaceae
= *Urceola rosea* (Hook. & Arn.)
D.J.Middleton (NoR)
♦ Weed
♦ 87, 88

Echeandia gracilis Cruden
Liliaceae/Anthericaceae
♦ Weed
♦ 199

Echeandia mexicana Cruden
Liliaceae/Anthericaceae
♦ Weed
♦ 199

Echeandia reflexa (Cav.) Rose
Liliaceae/Anthericaceae
♦ reflexed craglily, amber lily
♦ Weed

♦ 87, 88

Echeveria gibbiflora DC.
Crassulaceae
♦ Weed
♦ 243
♦ cultivated, herbal.

Echeveria × imbricata Deleuil ex E.Morr.
Crassulaceae
♦ hen and chickens
♦ Casual Alien
♦ 280

Echeveria mucronata (Bak.) Schltdl.
Crassulaceae
♦ Weed
♦ 199
♦ cultivated.

Echeveria multicaulis Rose
Crassulaceae
♦ Casual Alien
♦ 280
♦ cultivated.

Echeveria secunda Booth
Crassulaceae
♦ Naturalised
♦ 280
♦ cultivated.

Echeveria setosa Rose & Purpus
Crassulaceae
♦ Naturalised
♦ 280
♦ cultivated.

Echinacea pallida (Nutt.) Nutt.
Asteraceae
♦ pale echinacea, pale purple coneflower, coneflower
♦ Weed
♦ 161
♦ pH, cultivated, herbal.

Echinacea paradoxa (J.B.S.Norton) Britt.
Asteraceae
♦ Bush's purple coneflower, yellow coneflower
♦ Weed, Quarantine Weed
♦ 76, 88, 220
♦ cultivated, herbal.

Echinaria capitata (L.) Desf.
Poaceae
= *Cenchrus capitatus* L.
♦ Weed
♦ 87, 88
♦ G, cultivated.

Echinocactus texensis Hopffer
Cactaceae
Homalocephala texensis (Hopffer) Britt. & Rose
♦ horse crippler
♦ Weed, Quarantine Weed
♦ 76, 88
♦ cultivated, herbal.

Echinochloa Beauv. spp.
Poaceae
♦ barnyardgrass, cockspur grass
♦ Weed, Naturalised
♦ 39, 198, 204
♦ G, toxic.

Echinochloa caudata Roshev.
Poaceae

♦ long awn barnyardgrass
♦ Weed
♦ 297
♦ G.

Echinochloa colona (L.) Link
Poaceae
Echinochloa colonum (L.) Link (see),
Panicum colonum L.
♦ jungle rice, awnless barnyardgrass, swampgrass, river grass, bird's rice, Kalahari watergrass, marsh grass, shama millet, arroz de monte
♦ Weed, Naturalised, Native Weed, Introduced, Environmental Weed, Casual Alien
♦ 6, 7, 34, 38, 40, 55, 68, 86, 87, 88, 93, 98, 101, 121, 158, 161, 179, 185, 198, 200, 203, 204, 205, 212, 228, 236, 237, 242, 243, 261, 262, 269, 273, 295, 300
♦ aG, aqua, cultivated, herbal. Origin: Africa, tropical Asia.

Echinochloa colonum (L.) Link
Poaceae
= *Echinochloa colona* (L.) Link
♦ jungle rice, shama millet, awnless barnyardgrass, bird's rice, swampgrass, yaa khaao nok, little barnyardgrass, kukonhirssi, liendripuerco, arrocillo
♦ Weed, Naturalised, Introduced, Environmental Weed, Casual Alien
♦ 12, 14, 23, 28, 30, 42, 88, 90, 115, 170, 180, 186, 204, 218, 221, 230, 239, 241, 243, 249, 253, 255, 257, 263, 270, 274, 275, 281, 286, 297
♦ aG, aqua, promoted, herbal. Origin: India.

Echinochloa crus-galli (L.) Beauv.
Poaceae
Echinochloa colonum (L.) auct.,
Echinochloa crus-galli fo. *vittata*
F.T.Hubb., *Echinochloa crus-galli* var.
aristata Gray, *Echinochloa crus-galli* var. *michauxii* House, *Echinochloa crus-galli* var. *muricata* Farw., *Echinochloa echinata* (Willd.) P.Beauv., *Echinochloa muricata* (P.Beauv.) Fernald, *Echinochloa pungens* (Poir.) Rydb. (see), *Echinochloa pungens* var. *coarctata* Fernald & Griscom, *Echinochloa subverticillata* Pilg., *Echinochloa zelayensis* (Kunth) Schult., *Milium crus-galli* (L.) Moench, *Oplismenus crus-galli* (L.) Dumort., *Oplismenus echinatus* (Willd.) Kunth, *Oplismenus muricatus* Kunth, *Oplismenus zelayensis* Kunth, *Orthopogon crus-galli* (L.) Spreng., *Orthopogon echinatus* (Willd.) Spreng., *Panicum crus-galli* L., *Panicum crus-galli* var. *aristatum* Pursh, *Panicum crus-galli* var. *echinatum* (Willd.) Döll, *Panicum crus-galli* var. *longisetum* Döll, *Panicum crus-galli* var. *vulgare* Döll, *Panicum crus-galli* var. *vulgare* Döll, *Panicum echinatum* Willd., *Panicum grossum* Salisb., *Panicum muricatum* Michx., *Panicum pungens* Poir., *Pennisetum crus-galli* (L.) Baumg., *Setaria muricata* P.Beauv.

♦ barnyardgrass, cockspur, Japanese
millet, cock's foot panicum, cockspur
grass, panicgrass, watergrass, chicken
panic, California watergrass, billion
dollar grass
♦ Weed, Noxious Weed, Naturalised,
Introduced, Environmental Weed,
Casual Alien
♦ 6, 7, 9, 12, 14, 15, 23, 30, 34, 39, 40, 42,
44, 49, 52, 55, 68, 70, 86, 87, 88, 90, 93,
98, 101, 118, 121, 134, 136, 151, 158, 159,
161, 167, 170, 174, 176, 179, 180, 185,
186, 195, 198, 200, 203, 204, 205, 207,
210, 211, 212, 218, 221, 228, 236, 237,
239, 241, 243, 245, 250, 253, 261, 262,
269, 270, 272, 275, 280, 286, 292, 295,
296, 297, 299
♦ aG, aqua, cultivated, herbal, toxic.
Origin: Eurasia.

Echinochloa crus-galli (L.) Beauv. var. *caudata* (Roshev.) Kitag.
Poaceae
♦ keinubie
♦ Weed
♦ 275, 286
♦ aG.

Echinochloa crus-galli (L.) Beauv. var. *crus-galli*
Poaceae
♦ large barnyardgrass
♦ Weed, Naturalised
♦ 255, 263, 300
♦ aG. Origin: Eurasia.

Echinochloa crus-galli (L.) Beauv. var. *formosensis* Ohwi
Poaceae
♦ Taiwan barnyardgrass
♦ Weed
♦ 235, 263, 274, 286
♦ aG.

Echinochloa crus-galli (L.) Beauv. var. *hispidula* (Retz.) Hack.
Poaceae
♦ tainubie
♦ Weed
♦ 275
♦ aG.

Echinochloa crus-galli (L.) Beauv. var. *mitis*
Poaceae
= *Echinochloa muricata* var. *microstachya* (Beauv.) Fern. Wieg.
♦ Weed, Naturalised
♦ 199, 255, 275, 300
♦ aG. Origin: Eurasia.

Echinochloa crus-galli (L.) Beauv. var. *oryzicola* (Vasinger) Ohwi
Poaceae
= *Echinochloa oryzoides* (Ard.) Fritsch
♦ barnyardgrass
♦ Weed
♦ 235, 274, 286
♦ G.

Echinochloa crus-galli (L.) Beauv. var. *praticola* Ohwi
Poaceae
♦ Weed
♦ 235, 286
♦ G.

Echinochloa crus-galli (L.) Beauv. var. *zelayensis* (Kunth) Hitchc.
Foaceae
= *Echinochloa crus-pavonis* (Kunth) J.A.Schult. var. *macera* (Wieg.) Gould (NoR)
♦ Weed, Naturalised
♦ 275, 300
♦ aG.

Echinochloa crus-pavonis (Kunth) Schult.
Poaceae
♦ gulf barnyardgrass, gulf cockspur
♦ Weed, Naturalised, Native Weed
♦ 15, 86, 87, 88, 90, 98, 101, 121, 203,
245, 253, 255, 270, 280, 295, 300
♦ aG, aqua, herbal. Origin: Eurasia.

Echinochloa crus-pavonis (Kunth) Schult. var. *crus-pavonis*
Poaceae
♦ gulf cockspur grass
♦ Naturalised
♦ 101
♦ G.

Echinochloa esculenta (A.Braun) H.Scholz
Poaceae
Echinochloa utilis Ohwi & Yabuno (see)
♦ barnyardgrass, cockspur grass, Japanese millet
♦ Weed, Naturalised, Introduced, Environmental Weed
♦ 86, 93, 101, 176, 198, 205, 228, 280
♦ aG, arid. Origin: obscure.

Echinochloa frumentacea Link
Poaceae
Echinochloa colonum (L.) Link var. *frumentacea* (Link) Ridl., *Hoplismenus frumentaceus* Kunth, *Oplismenus frumentaceus* Kunth
♦ Siberian millet, billion dollar grass, white millet, barnyardgrass
♦ Weed, Naturalised, Introduced, Casual Alien
♦ 7, 40, 86, 98, 101, 176, 198, 203, 228
♦ G, arid, cultivated, herbal. Origin: obscure.

Echinochloa glabrescens Munro ex Hook.
Poaceae
♦ barnyardgrass
♦ Weed, Quarantine Weed
♦ 76, 88, 135, 191, 203, 220
♦ G.

Echinochloa helodes (Hack.) Parodi
Poaceae
♦ pasto de laguna
♦ Weed
♦ 87, 88, 237, 255, 295
♦ pG. Origin: South America.

Echinochloa holubii (Stapf) Stapf
Poaceae
♦ Kalahari watergrass
♦ Weed, Quarantine Weed, Native Weed
♦ 76, 87, 88, 121, 203, 220
♦ pG. Origin: southern Africa.

Echinochloa macrocarpa Vas.
Poaceae

= *Echinochloa oryzoides* (Ard.) Fritsch
♦ Weed
♦ 87, 88
♦ G.

Echinochloa macrocorvi Nakai
Poaceae
♦ Weed
♦ 87, 88
♦ G.

Echinochloa microstachya (Wieg.) Rydb.
Poaceae
= *Echinochloa muricata* (Beauv.) Fern. var. *microstachya* Wieg.
♦ prickly barnyardgrass
♦ Weed, Naturalised, Casual Alien
♦ 7, 269, 280
♦ G, herbal.

Echinochloa muricata (Michx.) Fernald
Poaceae
♦ rough barnyardgrass
♦ Weed, Naturalised
♦ 98, 101, 203
♦ aG, cultivated, herbal.

Echinochloa muricata (Beauv.) Fern. var. *microstachya* Wieg.
Poaceae
Echinochloa crus-galli (L.) P.Beauv var. *mitis* (see), *Echinochloa microstachya* (Wieg.) Rydb. (see)
♦ barnyardgrass, rough barnyardgrass
♦ Naturalised
♦ 86, 198
♦ G. Origin: North America.

Echinochloa muricata (Beauv.) Fern. var. *muricata*
Poaceae
♦ rough barnyardgrass
♦ Naturalised
♦ 101
♦ G.

Echinochloa obtusiflora Stapf
Poaceae
♦ Weed
♦ 88
♦ G.

Echinochloa oryzicola (Vas.) Vas.
Poaceae
= *Echinochloa oryzoides* (Ard.) Fritsch
♦ rice barnyardgrass
♦ Weed, Naturalised
♦ 87, 88, 101, 204, 263, 297
♦ aG.

Echinochloa oryzoides (Ard.) Fritsch
Poaceae
Echinochloa coarctata Kossenko, *Echinochloa crus-galli* (L.) Beauv. var. *oryzicola* (Vasinger) Ohwi (see), *Echinochloa hostii* Link, *Echinochloa macrocarpa* Vas. (see), *Echinochloa oryzicola* (Vas.) Vas. (see), *Echinochloa phyllopogon* (Stapf) Stapf ex Koss. (see), *Panicum coarctatum* Steven ex Trin. nom. inval., *Panicum oryzicola* Vasinger, *Panicum oryzoides* Ard., *Panicum phyllopogon* Stapf (see)
♦ early watergrass, hairy barnyardgrass, rice cockspur
♦ Weed, Naturalised, Introduced,

Environmental Weed, Casual Alien
♦ 7, 38, 70, 86, 87, 88, 98, 101, 176, 203, 237, 253, 269, 272, 280, 295
♦ aG. Origin: southern Europe to east Asia.

Echinochloa phyllopogon (Stapf) Stapf ex Koss.
Poaceae
= *Echinochloa oryzoides* (Ard.) Fritsch
♦ late watergrass
♦ Weed, Quarantine Weed
♦ 68, 76, 87, 88, 203, 220, 253
♦ G.

Echinochloa picta (J.König) P.W.Michael
Poaceae
♦ variegated cockspur grass
♦ Naturalised
♦ 101
♦ G. Origin: Asia, Australia.

Echinochloa polystachya (Kunth) A.S.Hitchc.
Poaceae
Echinochloa spectabilis (Nees) Link, *Pseudechinolaena polystachya* (Kunth) Stapf (see), *Oplismenus polystachyus* Kunth
♦ aleman grass, creeping rivergrass, carib grass
♦ Weed, Quarantine Weed, Naturalised, Environmental Weed
♦ 3, 54, 76, 86, 88, 93, 155, 179, 191, 237, 241, 255, 295, 300
♦ pG, aqua, cultivated. Origin: tropical America.

Echinochloa praestans Michael
Poaceae
♦ Weed, Naturalised
♦ 98, 203
♦ G.

Echinochloa pungens (Poir.) Rydb.
Poaceae
= *Echinochloa crus-galli* (L.) Beauv.
♦ prickly barnyardgrass
♦ Weed
♦ 87, 88, 218
♦ G, herbal.

Echinochloa pyramidalis (Lam.) A.Hitch. & Chase
Poaceae
♦ antelope grass, Limpopo grass
♦ Weed, Naturalised, Native Weed, Introduced
♦ 7, 86, 87, 88, 90, 98, 121, 203, 228
♦ pG, aqua, cultivated, herbal. Origin: Madagascar.

Echinochloa stagnina (Retz.) Beauv.
Poaceae
Echinochloa scabra (Lam.) Roem. & Schult., *Panicum stagninum* Retz.
♦ barnyardgrass, hippo grass, long awned watergrass, umvuma grass
♦ Weed, Quarantine Weed, Naturalised, Native Weed
♦ 76, 87, 88, 101, 121, 135, 170, 191, 203, 220, 221
♦ pG, arid/aqua, herbal. Origin: Madagascar.

Echinochloa telmatophila Michael & Vick.
Poaceae
♦ swamp barnyardgrass
♦ Weed, Naturalised, Environmental Weed
♦ 7, 86, 200, 280
♦ G, aqua. Origin: Australia.

Echinochloa turneriana (Domin) J.M.Black
Poaceae
Panicum turnerianum Domin
♦ channel millet
♦ Introduced
♦ 228
♦ G, arid, cultivated. Origin: Australia.

Echinochloa utilis Ohwi & Yab.
Poaceae
= *Echinochloa esculenta* (A.Braun) H.Scholz
♦ Japanese millet
♦ Weed, Naturalised, Casual Alien
♦ 7, 39, 40, 98, 203
♦ G, cultivated, toxic.

Echinochloa walteri (Pursh) Heller
Poaceae
♦ coast cockspur grass, coast cockspur
♦ Weed, Naturalised
♦ 87, 88
♦ G, herbal.

Echinocystis araneosa Griseb.
Cucurbitaceae
♦ Weed
♦ 237

Echinocystis lobata (A.Mich.) Torr. & Gray
Cucurbitaceae
♦ wild cucumber, wild mock cucumber, wild balsam apple, balsam apple, piikkikurkku, creeping jenny, four seeded burr cucumber
♦ Weed, Noxious Weed, Quarantine Weed, Native Weed, Cultivation Escape
♦ 8, 23, 24, 42, 76, 87, 88, 161, 174, 203, 210, 218, 220, 272, 299
♦ aH, cultivated, herbal, toxic. Origin: North America.

Echinocystis milleflora (Naudin) Cogn.
Cucurbitaceae
= *Echinopepon milleflorus* Naudin
♦ Weed
♦ 243

Echinocystis oregana (Torr. & Gray) Cogn.
Cucurbitaceae
= *Marah oreganus* (Torr. & S.Wats) T.J.Howell
♦ western wild cucumber
♦ Weed
♦ 87, 88, 218

Echinodorus berteroi (Spreng.) Fassett
Alismataceae
♦ burhead, upright burhead, cellophane plant
♦ Weed
♦ 180

♦ va/pH, cultivated, herbal. Origin: Central and South America.

Echinodorus cordifolius (L.) Griseb.
Alismataceae
Echinodorus radicans (Nutt.) Engelm., *Echinodorus bathii* H.Mühlberg, *Echinodorus schlueteri* Rataj
♦ burrhead, radicans sword, creeping burhead, Texas mud baby, Honduras radicans, mini radicans
♦ Weed, Sleeper Weed, Quarantine Weed, Naturalised, Garden Escape, Environmental Weed
♦ 76, 86, 87, 88, 155, 161, 218
♦ wpH, cultivated, herbal. Origin: North America.

Echinodorus grandiflorus (Cham. & Schlecht.) Micheli
Alismataceae
Alisma floribundum Seub., *Alisma grandiflorum* Cham. & Schltdl., *Echinodorus argentinensis* Rataj, *Echinodorus floribundus* (Seub.) Seub., *Echinodorus grandiflorus* var. *aureus* Fassett, *Echinodorus grandiflorus* var. *floribundus* (Seub.) Micheli, *Echinodorus grandiflorus* var. *longibracteatus* Rataj, *Echinodorus grandiflorus* var. *ovatus* Micheli, *Echinodorus longiscapus* Arechav., *Echinodorus muricatus* Griseb., *Echinodorus sellowianus* Buchenau
♦ large flowered Amazon sword, cha de campanha, cha do brejo, cha mineiro, chapeu de couro, congonha do brejo, erva do brejo aguape
♦ Weed
♦ 237, 255, 295
♦ wpH, cultivated. Origin: Central and South America.

Echinolaena inflexa (Poir.) Chase
Poaceae
Cenchrus inflexus Poir., *Cenchrus marginalis* Rudge, *Panicum echinolaena* Nees
♦ Weed
♦ 255
♦ G. Origin: Brazil.

Echinopepon milleflorus Naudin
Cucurbitaceae
Echinocystis milleflora (Naudin) Cogn. (see)
♦ echinopepon
♦ Weed
♦ 243
♦ aH.

Echinophora sibthorpiana Guss.
Apiaceae
♦ Weed
♦ 87, 88, 243

Echinopogon ovatus (G.Forst.) P.Beauv.
Poaceae
♦ forest hedgehog grass
♦ Native Weed
♦ 39, 269
♦ pG, arid, cultivated, toxic. Origin: Australia.

Echinops **L. spp.**
Asteraceae
♦ globe thistle
♦ Weed
♦ 221, 272
♦ herbal.

Echinops bannaticus **Rochel ex Schrad.**
Asteraceae
♦ sinipallo ohdake, globe thistle
♦ Weed, Cultivation Escape
♦ 42, 272
♦ cultivated, herbal.

Echinops echinatus **Roxb.**
Asteraceae
♦ unt kantalo
♦ Weed
♦ 87, 88
♦ arid, herbal.

Echinops exaltatus **Schrad.**
Asteraceae
♦ isopallo ohdake, globe thistle, tall
globe thistle
♦ Naturalised, Cultivation Escape
♦ 42, 101
♦ cultivated, herbal.

Echinops gmelinii **Turcz.**
Asteraceae
♦ Weed
♦ 275, 297
♦ aH, cultivated.

Echinops hussoni **Boiss.**
Asteraceae
♦ Weed
♦ 221

Echinops latifolius **Tausch.**
Asteraceae
♦ broadleaf globe thistle
♦ Weed
♦ 297

Echinops microcephalus **Sibth. & Sm.**
Asteraceae
♦ Weed
♦ 272
♦ cultivated.

Echinops ritro **L.**
Asteraceae
♦ globe thistle, small globe thistle,
southern globe thistle
♦ Weed, Naturalised, Casual Alien
♦ 39, 87, 88, 101, 272, 280
♦ cultivated, herbal, toxic.

Echinops ritro **L. ssp. *ruthenicus* (Bieb.)
Nyman**
Asteraceae
♦ southern globe thistle, bright blue
globe thistle
♦ Naturalised
♦ 101
♦ herbal.

Echinops sphaerocephalus **L.**
Asteraceae
♦ globe thistle, great globe thistle,
valkopallo ohdake
♦ Weed, Naturalised, Introduced,
Cultivation Escape
♦ 34, 42, 80, 86, 87, 88, 98, 101, 161, 203,
252, 272

♦ pH, cultivated, herbal. Origin:
Eurasia.

Echinops spinosissimus **Turra**
Asteraceae
Echinops viscosus DC. *nom. illeg.* (see)
♦ Weed
♦ 221
♦ pH, promoted.

Echinops spinosus **L.**
Asteraceae
♦ akhshir
♦ Weed
♦ 87, 88
♦ arid.

Echinops strigosus **L.**
Asteraceae
♦ globe thistle
♦ Weed
♦ 70, 87, 88
♦ cultivated.

Echinops viscosus **DC. *nom. illeg.***
Asteraceae
= *Echinops spinosissimus* Turra
♦ Weed
♦ 87, 88
♦ herbal.

Echinopsilon divaricatum **Kar. & Kir.**
Chenopodiaceae
♦ Weed
♦ 275
♦ aH.

Echinopsis pasacana **(F.A.C.Weber)
Friedrich & Rowley**
Cactaceae
Cereus pasacana F.A.C.Weber, *Echinopsis
rivierei* (Backb.) Friedrich & Rowley,
Helianthocereus pasacana (A.Weber)
Backeb., *Leucostele rivierei* Backeb.,
Trichocereus pasacana (F.A.C.Weber)
Britton & Rose, *Trichocereus rivierei*
(Backeb.) Krainz
♦ Argentinian saguaro
♦ Introduced
♦ 228
♦ arid, cultivated, herbal.

Echinopsis schickendantzii **F.A.C.Weber**
Cactaceae
Echinopsis manguinii (Backb.) Friedrich
& Rowley, *Trichocereus schickendantzii*
(F.A.C.Weber) Britt. & Rose,
Trichocereus shaferi Britt. & Rose
♦ Introduced
♦ 228
♦ arid, cultivated.

Echinopsis spachiana **(Lem.) Friedr. &
Rowley**
Cactaceae
Cereus spachianus Lem., *Echinopsis
santiaguensis* (Speg.) Friedrich &
Rowley, *Trichocereus santiaguensis*
(Speg.) Backeb., *Trichocereus spachianus*
Riccob.
♦ torch cactus, orrelkaktus
♦ Weed, Noxious Weed, Cultivation
Escape
♦ 63, 88, 95, 283
♦ cultivated, herbal. Origin: South
America.

Echinopsis terscheckii **(Parm.) Friedrich
& Rowley**
Cactaceae
Cereus terscheckii Parm., *Echinopsis
werdermanniana* (Backeb.) Friedrich
& G.D.Rowley, *Trichocereus terscheckii*
(Parm.) Britton & Rose, *Trichocereus
werdermannianus* Backeb.
♦ Introduced
♦ 228
♦ arid, cultivated, herbal.

Echinospartum barnadesii **(Graells)
Rothm.**
Fabaceae/Papilionaceae
Genista barnadesii Graells (see)
♦ Weed, Quarantine Weed
♦ 76, 88, 220

Echinospermum lappula **(L.) Lehm.**
Boraginaceae
= *Lappula squarrosa* (Retz.) Dumort.
♦ Weed
♦ 87, 88
♦ herbal.

Echinospermum redowskii **Lehm.**
Boraginaceae
Lappula echinophora Kuntze
♦ Weed
♦ 243

Echiochilon fruticosum **Desf.**
Boraginaceae
♦ Weed
♦ 221
♦ arid.

Echites repens **Jacq.**
Apocynaceae
♦ Weed
♦ 87, 88

Echites umbellata **Jacq.**
Apocynaceae
♦ devil's potato, rubbervine
♦ Weed
♦ 87, 88
♦ cultivated.

Echium **L. spp.**
Boraginaceae
♦ viper's bugloss
♦ Weed, Naturalised
♦ 198, 221, 247, 272
♦ toxic.

Echium acutifolium **J.G.C.Lehm.**
Boraginaceae
♦ Quarantine Weed
♦ 220

Echium amoenum **Fisch. & Mey.**
Boraginaceae
♦ Weed
♦ 87, 88

Echium australe **Lam.**
Boraginaceae
= *Echium creticum* L.
♦ Weed
♦ 87, 88

Echium boissieri **Steud.**
Boraginaceae
♦ Quarantine Weed
♦ 220
♦ cultivated.

Echium brevirame Sprague & Hutch.
Boraginaceae
♦ Weed, Quarantine Weed
♦ 76, 88, 220
♦ cultivated.

Echium candicans L.f.
Boraginaceae
Echium fastuosum Aiton (see)
♦ pride of Madeira, tower of jewels
♦ Weed, Noxious Weed, Naturalised,
Garden Escape
♦ 86, 101, 116, 121, 251, 280
♦ pS, cultivated. Origin: Madeira
Islands.

Echium clavatum Willd. ex Lehm.
Boraginaceae
♦ Quarantine Weed
♦ 220

Echium coincyanum Lacaita
Boraginaceae
♦ smallstamen viper's bugloss
♦ Naturalised
♦ 101

Echium creticum L.
Boraginaceae
Echium australe Lam. (see)
♦ Cretan viper's bugloss
♦ Naturalised
♦ 101

Echium fastuosum Aiton
Boraginaceae
= *Echium candicans* L.f.
♦ pride of Madeira, beeshead
♦ Noxious Weed, Naturalised, Garden
Escape
♦ 86
♦ cultivated, herbal.

Echium glomeratum Poir.
Boraginaceae
♦ Weed
♦ 87, 88
♦ arid.

Echium horridum Batt.
Boraginaceae
♦ Weed
♦ 88

Echium italicum L.
Boraginaceae
Echium altissimum Jacq., *Echium
pyramidatum* DC.
♦ Italian bugloss, Italian viper's
bugloss, pale bugloss, hadinec
taliansky
♦ Weed, Quarantine Weed, Noxious
Weed, Naturalised, Garden Escape
♦ 76, 86, 87, 88, 98, 101, 203, 251, 272
♦ bH, cultivated, herbal. Origin:
Mediterranean.

Echium linearifolium C.Koch
Boraginaceae
♦ Quarantine Weed
♦ 220

Echium longifolium Delile
Boraginaceae
♦ Weed
♦ 221

Echium lusitanicum L.
Boraginaceae

♦ Weed
♦ 70
♦ cultivated.

Echium lycopsis L.
Boraginaceae
= *Echium plantagineum* L.
♦ Paterson's curse, purple echium,
salvation Jane
♦ Weed, Naturalised
♦ 39, 98, 121, 203
♦ bH, herbal, toxic. Origin: Eurasia.

Echium maculatum L.
Boraginaceae
♦ Quarantine Weed
♦ 220

Echium officinalis L.
Boraginaceae
♦ Naturalised
♦ 98

Echium papillosum Thunb.
Boraginaceae
= *Lobostemon swartzii* Buek (NoR)
♦ Quarantine Weed
♦ 220

Echium pininana Webb & Berthel.
Boraginaceae
♦ Canary echium, tree echium, pride
of Tenerife, giant bugloss
♦ Weed, Quarantine Weed,
Naturalised
♦ 15, 76, 88, 116, 220, 280
♦ bH, cultivated.

Echium plantagineum L.
Boraginaceae
Echium lycopsis L. (see)
♦ Paterson's curse, salvation Jane, blue
weed, Lady Campbell weed, purple
bugloss, purple echium, purple viper's
bugloss, Riverina bluebell, viper's
bugloss, ratamoneidonkieli
♦ Weed, Sleeper Weed, Quarantine
Weed, Noxious Weed, Naturalised,
Introduced, Garden Escape,
Environmental Weed, Casual Alien
♦ 7, 15, 20, 34, 42, 51, 62, 63, 68, 70, 72,
76, 86, 87, 88, 93, 94, 95, 98, 101, 134,
147, 158, 161, 165, 169, 171, 176, 177,
198, 203, 205, 220, 225, 236, 237, 241,
243, 246, 250, 251, 255, 269, 271, 272,
280, 283, 289, 295, 296, 300
♦ a/bH, arid, cultivated, herbal, toxic.
Origin: Eurasia.

Echium pomponium Boiss.
Boraginaceae
♦ Quarantine Weed
♦ 220

Echium pustulatum Sibth. & Sm.
Boraginaceae
♦ blue devil
♦ Naturalised
♦ 101

Echium rauwolfii Delile
Boraginaceae
♦ Weed
♦ 87, 88, 185, 221, 242
♦ arid.

Echium rosulatum Lange
Boraginaceae

♦ Weed
♦ 70
♦ cultivated.

Echium rubrum Jacq.
Boraginaceae
♦ Quarantine Weed
♦ 220

Echium russicum J.F.Gmel.
Boraginaceae
♦ Weed, Quarantine Weed
♦ 220, 272
♦ bH, cultivated.

Echium sericeum Vahl
Boraginaceae
♦ Weed
♦ 221

Echium simplex DC.
Boraginaceae
♦ pride of Tenerife
♦ Weed, Noxious Weed, Naturalised
♦ 86, 98, 203
♦ bH, cultivated. Origin: Canary
Islands.

Echium vulgare L.
Boraginaceae
Echium elegans Noe ex Nym.
♦ blue weed, viper's bugloss, blue
echium, blue thistle, blue devil,
common viper's bugloss, blou echium
♦ Weed, Quarantine Weed, Noxious
Weed, Naturalised, Garden Escape,
Environmental Weed
♦ 1, 8, 15, 20, 23, 39, 42, 51, 52, 63, 70,
72, 76, 80, 86, 87, 88, 94, 95, 98, 101, 102,
121, 136, 139, 146, 147, 156, 158, 161,
162, 165, 176, 181, 195, 198, 203, 218,
220, 225, 229, 241, 243, 246, 269, 272,
280, 283, 287, 289, 299, 300
♦ a/bH, cultivated, herbal, toxic.
Origin: Eurasia.

**Echium wildpretii H.Pearson ex Hook.f.
× pininiana Webb & Berthel.**
Boraginaceae
♦ Quarantine Weed
♦ 220

Eclipta alba (L.) Hassk.
Asteraceae
= *Eclipta prostrata* (L.) L.
♦ eclipta, yerba de tago, white heads,
false daisy, kameng, botoncillo, agriao
do brejo
♦ Weed, Naturalised
♦ 50, 88, 157, 185, 204, 218, 221, 239,
242, 255, 262, 281, 286, 287, 295
♦ aH, arid, herbal.

Eclipta erecta L.
Asteraceae
= *Eclipta prostrata* (L.) L.
♦ Weed
♦ 87, 88
♦ herbal.

Eclipta platyglossa Muell.
Asteraceae
♦ yellow twin heads
♦ Native Weed
♦ 269
♦ aH. Origin: Australia.

Eclipta prostrata (L.) L.
Asteraceae
Eclipta alba (L.) Hassk. (see), *Eclipta erecta* L. (see), *Verbesina alba* L., *Verbesina prostrata* L.
♦ white eclipta, false daisy, eclipta, white heads, yerba de tago
♦ Weed, Quarantine Weed, Naturalised, Native Weed
♦ 13, 14, 51, 66, 70, 76, 87, 88, 121, 158, 161, 170, 180, 186, 199, 204, 207, 209, 211, 235, 237, 241, 243, 249, 253, 261, 263, 269, 271, 273, 274, 275, 276, 297
♦ aH, arid/aqua, cultivated, herbal, toxic. Origin: Eurasia.

Eclipta thermalis (L.) Bunge
Asteraceae
♦ Weed
♦ 286

Edgeworthia chrysantha Lindl.
Thymelaeaceae
= *Edgeworthia papyrifera* Sieb. & Zucc.
♦ Naturalised
♦ 287
♦ S, cultivated. Origin: China.

Edgeworthia papyrifera Sieb. & Zucc.
Thymelaeaceae
Edgeworthia chrysantha Lindl. (see)
♦ oriental paperbush, paperbush
♦ Naturalised
♦ 101
♦ S, cultivated. Origin: China.

Edmondia sesamoides (L.) Hilliard
Asteraceae
Helichrysum sesamoides (L.) Willd., *Helichrysum sesamoides* var. *filiforme* (D.Don) Harv., *Helichrysum sesamoides* var. *heterophyllum* Harv., *Helichrysum sesamoides* var. *willdenowii* Harv., *Xeranthemum sesamoides* L.
♦ Quarantine Weed
♦ 220
♦ cultivated.

Egeria Planch. spp.
Hydrocharitaceae
♦ egeria
♦ Quarantine Weed, Naturalised
♦ 198, 220
♦ wH.

Egeria densa Planch.
Hydrocharitaceae
Anacharis densa (Planch.) Vict. (see), *Elodea densa* (Planch.) Casp. (see)
♦ Brazilian waterweed, leafy elodea, dense waterweed, egeria, ditch moss, water thyme, waterweed, giant elodea, Argentine acharis, Argentinian waterweed, vodomorec hust
♦ Weed, Quarantine Weed, Noxious Weed, Naturalised, Garden Escape, Environmental Weed
♦ 1, 15, 23, 35, 45, 62, 63, 72, 76, 78, 80, 86, 87, 88, 98, 101, 102, 116, 121, 133, 139, 147, 151, 152, 161, 165, 171, 176, 179, 191, 195, 197, 198, 200, 203, 204, 208, 218, 224, 225, 229, 231, 241, 246, 255, 258, 269, 278, 280, 283, 286, 287, 295, 300

♦ wpH, cultivated, herbal. Origin: South America.

Egeria naians Planch.
Hydrocharitaceae
♦ Weed
♦ 88
♦ wH.

Egletes prostrata (Sw.) Kuntze
Asteraceae
♦ prostrate tropic daisy
♦ Weed, Quarantine Weed
♦ 76, 87, 88, 203, 220

Egletes viscosa (L.) Less.
Asteraceae
♦ erect tropical daisy
♦ Weed
♦ 14
♦ herbal.

Ehretia microphylla Lam.
Boraginaceae/Ehretiaceae
= *Carmona retusa* (Vahl) Masam.
♦ Philippine tea
♦ Weed
♦ 179
♦ herbal. Origin: Australia.

Ehretia rigida (Thunb.) Druce
Boraginaceae/Ehretiaceae
Ehretia hottentottica Burch., *Capraria rigida* Thunb.
♦ cape lilac, Hottentot's lilac, kraaldog, puzzle bush, stamper wood
♦ Weed, Native Weed
♦ 10, 121
♦ S/T, arid, cultivated, herbal. Origin: southern Africa.

Ehrharta brevifolia Schrad.
Poaceae
♦ Weed, Naturalised, Native Weed, Environmental Weed
♦ 7, 86, 98, 121, 203
♦ aG. Origin: southern Africa.

Ehrharta calycina J.E.Sm.
Poaceae
♦ veld grass, common ehrharta, perennial veld grass
♦ Weed, Noxious Weed, Naturalised, Native Weed, Introduced, Environmental Weed
♦ 7, 9, 15, 35, 72, 78, 80, 86, 88, 98, 101, 116, 121, 152, 161, 176, 180, 198, 203, 228, 231, 280, 289, 296
♦ pG, arid, cultivated, herbal. Origin: South Africa.

Ehrharta dura Nees ex Trin
Poaceae
♦ Native Weed
♦ 121
♦ pG. Origin: southern Africa.

Ehrharta erecta Lam.
Poaceae
♦ panic veld grass, veld grass
♦ Weed, Noxious Weed, Naturalised, Introduced, Environmental Weed
♦ 7, 15, 34, 35, 72, 78, 80, 86, 88, 98, 101, 116, 176, 181, 198, 203, 225, 231, 246, 280, 289, 296
♦ pG, cultivated, herbal. Origin: Africa.

Ehrharta erecta Lam. ssp. *erecta*
Poaceae
♦ Lamarck's ehrharta, panic veld grass
♦ Native Weed
♦ 121
♦ pG. Origin: southern Africa.

Ehrharta erecta Lam. ssp. *natalensis* Stapf
Poaceae
♦ Native Weed
♦ 121
♦ pG. Origin: southern Africa.

Ehrharta longiflora J.E.Sm.
Poaceae
♦ annual veld grass, oat seed grass, long flowered veld grass
♦ Weed, Sleeper Weed, Noxious Weed, Naturalised, Native Weed, Environmental Weed
♦ 7, 9, 15, 35, 72, 78, 86, 88, 98, 101, 121, 158, 176, 198, 203, 225, 231, 243, 246, 280, 289, 296
♦ aG, cultivated. Origin: southern Africa.

Ehrharta pusilla Trin.
Poaceae
♦ pyp grass
♦ Weed, Naturalised, Environmental Weed
♦ 86, 98, 203
♦ G. Origin: South Africa.

Ehrharta stipoides Labill.
Poaceae
Microlaena stipoides (Labill.) R.Br. (see)
♦ meadow ricegrass, weeping grass
♦ Weed, Naturalised, Environmental Weed
♦ 3, 80, 101, 152, 191
♦ G, herbal. Origin: Australia.

Ehrharta villosa (L.f.) J.H.Schult.
Poaceae
♦ pyp grass, pipe grass
♦ Weed, Naturalised, Native Weed, Environmental Weed
♦ 7, 15, 98, 121, 176, 203, 225, 246, 280
♦ pG, cultivated. Origin: South Africa.

Ehrharta villosa (L.f.) J.H.Schult. var. *maxima* Stapf
Poaceae
♦ pyp grass
♦ Naturalised, Environmental Weed
♦ 86, 198
♦ G. Origin: South Africa.

Eichhornia Kunth spp.
Pontederiaceae
♦ water hyacinth
♦ Quarantine Weed, Naturalised
♦ 198, 220

Eichhornia azurea (Sw.) Kunth
Pontederiaceae
♦ anchored water hyacinth, blue water hyacinth
♦ Weed, Quarantine Weed, Noxious Weed, Naturalised, Environmental Weed
♦ 14, 26, 67, 76, 86, 87, 88, 101, 140, 161, 193, 203, 229, 237, 246, 255, 287, 295

♦ pH, aqua, cultivated, herbal. Origin: South America.

Eichhornia crassipes (Mart.) Solms
Pontederiaceae
Eichhornia speciosa Kunth (see),
Heteranthera formosa Miq., *Piaropus crassipes* (Mart.) Raf., *Piaropus mesomelas* Raf., *Pontederia azurea* Sw., *Pontederia crassipes* Mart., *Pontederia elongata* Balf.
♦ water hyacinth, floating water hyacinth, pickerelweed, Nile lily, water orchid, phak top chawaa, waterhiasint, jacinto de agua, lirio acuatico, jacinthe d'eau, bung el ralm, mbekambekairanga, ndambendambe ni nga, jai khumbe, bekabekairaga, dabedabe ne ga, jal khumbe, riri vai
♦ Weed, Quarantine Weed, Noxious Weed, Naturalised, Introduced, Garden Escape, Environmental Weed, Cultivation Escape
♦ 3, 6, 7, 10, 14, 18, 23, 26, 35, 50, 51, 62, 63, 70, 72, 76, 78, 80, 86, 87, 88, 93, 95, 98, 101, 106, 107, 112, 116, 121, 126, 134, 137, 142, 147, 151, 152, 158, 161, 169, 170, 171, 177, 179, 180, 181, 186, 191, 197, 198, 200, 203, 204, 208, 209, 216, 218, 220, 221, 226, 229, 230, 231, 232, 233, 237, 239, 241, 246, 255, 256, 258, 261, 262, 268, 269, 274, 275, 276, 278, 280, 283, 286, 287, 289, 295, 296, 297, 300
♦ wH, cultivated, herbal. Origin: north-east Brazil.

Eichhornia diversifolia (Vahl) Urb.
Pontederiaceae
♦ variableleaf water hyacinth
♦ Quarantine Weed, Noxious Weed
♦ 76, 229
♦ aqua, cultivated.

Eichhornia natans Solms
Pontederiaceae
♦ Weed, Quarantine Weed
♦ 76, 87, 88, 203
♦ aqua, cultivated.

Eichhornia paniculata (Spreng.) Solms
Pontederiaceae
♦ Brazilian water hyacinth
♦ Weed, Noxious Weed
♦ 32, 229, 255
♦ pH, aqua, cultivated. Origin: Brazil.

Eichhornia speciosa Kunth
Pontederiaceae
= *Eichhornia crassipes* (Mart.) Solms
♦ Quarantine Weed
♦ 220

Einadia nutans (R.Br.) A.J.Scott
Chenopodiaceae
♦ berry saltbush
♦ Weed, Naturalised
♦ 55, 280
♦ cultivated. Origin: Australia.

Einadia trigonos (Roem. & Schult.) P.G.Wilson
Chenopodiaceae
Chenopodium trigonon Roem. & Schult., *Chenopodium triangulare* R.Br. (see)

♦ fishweed
♦ Native Weed
♦ 269
♦ cultivated. Origin: Australia.

Einadia trigonos (Roem. & Schult.) P.G.Wilson ssp. stellulata (Benth.) Paul G.Wilson
Chenopodiaceae
♦ Naturalised
♦ 280

Ekebergia capensis Sparrm.
Meliaceae
Ekebergia meyeri C.Presl ex C.DC.
♦ cape ash
♦ Introduced
♦ 228
♦ arid, cultivated.

Elaeagnus angustifolia L.
Elaeagnaceae
Elaeagnus argentea Moench, *Elaeagnus hortensis* M.Bieb., *Elaeagnus moorcroftii* Wall.
♦ Russian olive, trebizond date, oleaster
♦ Weed, Noxious Weed, Naturalised, Introduced, Garden Escape, Environmental Weed, Cultivation Escape
♦ 4, 34, 35, 78, 80, 88, 101, 103, 116, 129, 133, 138, 142, 146, 151, 152, 156, 159, 161, 174, 195, 212, 222, 224, 228, 231, 264
♦ S/T, arid, cultivated, herbal. Origin: Eurasia.

Elaeagnus argentea Pursh
Elaeagnaceae
= *Elaeagnus commutata* Bernh. ex Rydb.
♦ oliwnik srebrzysty
♦ Weed
♦ 87, 88
♦ herbal.

Elaeagnus commutata Bernh. ex Rydb.
Elaeagnaceae
Elaeagnus argentea Pursh (see)
♦ silverberry, kilsepensas, hloäina striebristá
♦ Weed, Cultivation Escape
♦ 23, 42, 88, 218
♦ S, cultivated, herbal.

Elaeagnus × ebbingei Boom.
Elaeagnaceae
= *Elaeagnus macrophylla* Thunb. × *Elaeagnus pungens* Thunb.
♦ elaeagnus
♦ Cultivation Escape
♦ 40
♦ S, cultivated, herbal.

Elaeagnus multiflora Thunb.
Elaeagnaceae
Elaeagnus longipes A.Gray
♦ cherry silverberry, hloäina mnohokvetá
♦ Quarantine Weed, Naturalised
♦ 101, 258
♦ S, cuitivated, herbal.

Elaeagnus pungens Thunb.
Elaeagnaceae
♦ Russian olive, thorny olive, thorny

elaeagnus, elaeagnus, oleaster
♦ Weed, Naturalised, Environmental Weed
♦ 80, 88, 101, 102, 179
♦ S, cultivated, herbal.

Elaeagnus × reflexa Morr. & Decne.
Elaeagnaceae
= *Elaeagnus pungens* Thunb. × *Elaeagnus glabra* Thunb.
♦ elaeagnus
♦ Weed, Naturalised, Environmental Weed
♦ 15, 165, 225, 246, 280
♦ S, cultivated. Origin: horticultural hybrid.

Elaeagnus umbellata Thunb.
Elaeagnaceae
Elaeagnus crispa Thunb.
♦ autumn olive, oleaster
♦ Weed, Noxious Weed, Naturalised, Introduced, Garden Escape, Environmental Weed, Cultivation Escape
♦ 3, 4, 17, 22, 80, 88, 101, 102, 133, 142, 151, 161, 191, 195, 222, 224, 229, 233, 286
♦ S/T, cultivated, herbal.

Elaeagnus umbellata Thunb. var. parvifolia (Royle) Schneid.
Elaeagnaceae
♦ autumn olive
♦ Noxious Weed, Naturalised
♦ 101, 229

Elaeis guineensis Jacq.
Arecaceae
♦ African oil palm, apwiraiasi, oil palm, oil palm nut, mbile
♦ Weed, Naturalised, Introduced
♦ 3, 101, 107, 179, 191, 230
♦ T, cultivated, herbal. Origin: Madagascar.

Elaeis oleifera (Kunth) Cortés
Arecaceae
♦ American oil palm
♦ Naturalised
♦ 32
♦ cultivated, herbal.

Elaeocarpus kerstingianus Schltr.
Elaeocarpaceae
♦ Introduced
♦ 230
♦ T.

Elaeocarpus kusaiensis Kaneh.
Elaeocarpaceae
♦ Introduced
♦ 230
♦ T.

Elaeocarpus kusanoi Koidz.
Elaeocarpaceae
♦ opop maratte
♦ Introduced
♦ 230
♦ T.

Elaeocarpus serrata L.
Elaeocarpaceae
♦ Introduced
♦ 230
♦ T.

Elaphoglossum carolinense Hosok.
Lomariopsidaceae
♦ Introduced
♦ 230
♦ H.

Elatine alsinastrum L.
Elatinaceae
Elatine verticillata Lam.
♦ isovesirikko
♦ Weed
♦ 272

Elatine ambigua Wight
Elatinaceae
♦ Asian waterwort
♦ Weed, Naturalised
♦ 34, 101
♦ aH.

Elatine americana (Pursh) Arn.
Elatinaceae
♦ American waterwort
♦ Weed
♦ 23, 87, 88, 218
♦ herbal.

Elatine chilensis Gray
Elatinaceae
♦ Chilean waterwort
♦ Naturalised
♦ 101
♦ a/pH, aqua.

Elatine hexandra (Lapierre) DC.
Elatinaceae
Birolia paludosa Bellard., *Elatine paludosa* Seub.
♦ waterwort, six stamened waterwort
♦ Weed
♦ 272
♦ aqua, cultivated.

Elatine minima (Nutt.) Fisch. & Mey.
Elatinaceae
♦ small waterwort
♦ Weed
♦ 87, 88, 218
♦ herbal.

Elatine orientalis Makino
Elatinaceae
= *Elatine triandra* Schk.
♦ Weed, Quarantine Weed
♦ 76, 88

Elatine triandra Schk.
Elatinaceae
Elatine orientalis Makino (see)
♦ threestamen waterwort, kolmihedevesirikko
♦ Weed, Quarantine Weed
♦ 76, 87, 88, 170, 191, 203, 204, 220, 272, 274, 275, 297
♦ aqua, herbal.

Elatine triandra Schk. var. *pedicellata* Krylov.
Elatinaceae
♦ mizohakobe
♦ Weed
♦ 263, 286
♦ aH.

Elatostema flumineo-rupestre Hosok.
Urticaceae
♦ Introduced

♦ 230
♦ S.

Eleocharis R.Br. spp.
Cyperaceae
Heleocharis Lestib. spp. (see)
♦ spikerush
♦ Weed, Native Weed
♦ 88, 121, 220, 272
♦ aG.

Eleocharis acicularis (L.) Roem. & Schult.
Cyperaceae
Scirpus acicularis L.
♦ slender spikerush, needle spikerush, dwarf hairgrass
♦ Weed, Naturalised
♦ 87, 88, 126, 161, 170, 191, 204, 217, 218, 262, 272, 300
♦ wpG, cultivated, herbal.

Eleocharis acicularis Roem. & Schult. fo. *longiseta* (Svenson) T.Koyama
Cyperaceae
♦ spikerush
♦ Weed
♦ 274
♦ G.

Eleocharis acicularis Roem. & Schult. var. *longiseta* Svenson
Cyperaceae
♦ needle spikerush
♦ Weed
♦ 263, 286
♦ pG.

Eleocharis acuta R.Br.
Cyperaceae
♦ common spikerush
♦ Weed
♦ 87, 88
♦ pG, aqua, cultivated.

Eleocharis acutangula (Roxb.) Schult.
Cyperaceae
Eleocharis fistulosa Link., *Scirpus acutangulus* Roxb.
♦ acute spikerush
♦ Weed
♦ 87, 88, 126, 170, 191, 255, 262
♦ pG, aqua. Origin: South America.

Eleocharis afflata Steud.
Cyperaceae
= *Eleocharis pellucida* J. & C.Presl
♦ Weed, Quarantine Weed
♦ 76, 87, 88, 203, 220
♦ G.

Eleocharis atropurpurea (Retz.) Kunth
Cyperaceae
Heleocharis atropurpurea (Retz.) Kunth, *Scirpus atropurpureus* Retz.
♦ purple spikerush
♦ Weed
♦ 34, 87, 88, 126, 170
♦ aG, aqua, herbal.

Eleocharis attenuata (Fr. & Sav.) Palla
Cyperaceae
Eleocharis laeviseta Nakai
♦ Weed
♦ 87, 88
♦ G.

Eleocharis baldwinii (Torr.) Chapman
Cyperaceae

♦ Baldwin's spikerush
♦ Quarantine Weed
♦ 258
♦ G.

Eleocharis bonariensis Nees
Cyperaceae
♦ hairgrass
♦ Weed
♦ 236, 237, 295
♦ G.

Eleocharis calva Torr.
Cyperaceae
= *Eleocharis erythropoda* Steud. (NoR)
♦ Weed
♦ 23, 88
♦ G.

Eleocharis cellulosa Torr.
Cyperaceae
♦ gulfcoast spikerush, gulf spikerush
♦ Weed
♦ 87, 88, 218
♦ G, cultivated, herbal.

Eleocharis chaetaria Roem. & Schult.
Cyperaceae
= *Eleocharis retroflexa* (Poir.) Urb.
♦ Weed, Quarantine Weed
♦ 76, 87, 88, 203, 220
♦ G.

Eleocharis congesta D.Don
Cyperaceae
= *Eleocharis pellucida* J. & C.Presl
♦ spikerush
♦ Weed
♦ 87, 88, 170, 191, 263, 286
♦ a/pG, aqua.

Eleocharis dulcis (Burm.f.) Trin. ex Hensch.
Cyperaceae
Andropogon dulcis Burm.f., *Eleocharis equisetina* C.Presl (see), *Eleocharis indica* Druce, *Eleocharis plantaginea* (Retz.) Roem. & Schult., *Eleocharis plantaginoidea* W.F.Wight (see), *Eleocharis tuberosa* (Roxb.) Schult. (see), *Eleocharis tuberculosa* (Michx.) Roem. & Sch., *Heleocharis plantaginoidea* W.F.Wight., *Scirpus tuberosus* Roxb. *nom. illeg.* (see)
♦ waterchestnut, Chinese waterchestnut, ground chestnut, waternut, haeo song krathiam, châtaigne d'eau, Chinesische wassernuß, wasserkastanie, cabezas de negrito, nuez China
♦ Weed, Introduced
♦ 87, 88, 126, 170, 204, 209, 217, 230, 239, 262, 286
♦ pG, aqua, cultivated, herbal. Origin: Asia, Australasia.

Eleocharis dulcis Trin. var. *tuberosa* T.Koyama
Cyperaceae
♦ Naturalised
♦ 287
♦ G.

Eleocharis elegans (Kunth) Roem. & Schult.
Cyperaceae

- ♦ elegant spikerush
- ♦ Weed, Introduced
- ♦ 14, 88, 126, 228, 255
- ♦ pG, aqua. Origin: tropical America.

Eleocharis equisetina C.Presl
Cyperaceae
= *Eleocharis dulcis* (Burm.f.) Trin. ex Hensch.
- ♦ Weed
- ♦ 87, 88
- ♦ G, cultivated. Origin: Australia.

Eleocharis erecta Schumac.
Cyperaceae
- ♦ Weed, Quarantine Weed
- ♦ 76, 87, 88, 203, 220
- ♦ G.

Eleocharis filiculmis Kunth
Cyperaceae
Eleocharis sulcata (Roth) Nees, *Scirpus sulcatus* Roth
- ♦ Weed, Quarantine Weed
- ♦ 126
- ♦ pG, aqua.

Eleocharis geniculata (L.) Roem. & Schult.
Cyperaceae
Eleocharis capitata (L.) R.Br. var. *dispar* (E.J.Hill) Fern., *Eleocharis caribaea* (Rottb.) S.F.Blake, *Scirpus geniculatus* L.
- ♦ Canada spikesedge, spikerush
- ♦ Weed
- ♦ 6, 87, 88, 126, 170
- ♦ aG, aqua, herbal.

Eleocharis intersita Zinserl.
Cyperaceae
= *Eleocharis palustris* (L.) Roem. & Schult.
- ♦ intermediate spikesedge
- ♦ Weed
- ♦ 297
- ♦ G.

Eleocharis interstincta (Roem. & Schult.) R.Br.
Cyperaceae
Eleocharis articulata Kunth
- ♦ knotted spikerush
- ♦ Weed, Quarantine Weed
- ♦ 14, 76, 87, 88, 203, 220, 255
- ♦ pG, aqua, cultivated, herbal. Origin: tropical America.

Eleocharis kuroguwai Ohwi
Cyperaceae
- ♦ Weed, Quarantine Weed
- ♦ 76, 87, 88, 203, 204, 220, 263, 286
- ♦ pG.

Eleocharis mamillata Lind.f.
Cyperaceae
Heleocharis mamillata Lind.f.
- ♦ mutaluikka, numaharii
- ♦ Weed
- ♦ 87, 88
- ♦ G, herbal.

Eleocharis mamillata Lindb.f. var. cyclocarpa Kitag.
Cyperaceae
- ♦ Weed
- ♦ 286

- ♦ G.

Eleocharis minuta Boeck.
Cyperaceae
- ♦ variable spike sedge
- ♦ Weed, Naturalised
- ♦ 86, 98, 198, 203
- ♦ G, cultivated. Origin: Africa.

Eleocharis multicaulis Sm.
Cyperaceae
- ♦ many stemmed spikerush
- ♦ Weed
- ♦ 87, 88
- ♦ G.

Eleocharis mutata (L.) Roem. & Schult.
Cyperaceae
- ♦ scallion grass
- ♦ Weed, Quarantine Weed
- ♦ 76, 87, 88, 203, 220
- ♦ G.

Eleocharis nigrescens (Nees) Kunth
Cyperaceae
- ♦ black spikerush
- ♦ Weed
- ♦ 179
- ♦ G. Origin: Australia.

Eleocharis nodulosa Schult.
Cyperaceae
= *Eleocharis montana* (Kunth) Roem. & Schult. (NoR)
- ♦ Weed, Quarantine Weed
- ♦ 76, 87, 88, 203, 220
- ♦ G.

Eleocharis obtusa (Willd.) Schult.
Cyperaceae
- ♦ blunt spikerush, blunt spikesedge
- ♦ Weed
- ♦ 88, 161, 180, 218
- ♦ aG, aqua, cultivated, herbal.

Eleocharis ochreata (Nees) Steud.
Cyperaceae
Eleocharis flaccida Urb.
- ♦ Weed, Quarantine Weed
- ♦ 76, 87, 88, 203, 220
- ♦ G.

Eleocharis ochrostachys Steud.
Cyperaceae
Eleocharis subulata Boeck., *Scirpus ochrostachys* Kuntze
- ♦ spikerush
- ♦ Weed
- ♦ 88, 170, 191
- ♦ G. Origin: Australia.

Eleocharis ovata (Roth) Roem. & Schult.
Cyperaceae
Eleocharis annua House, *Eleocharis obtusa* (Willd.) J.A.Schult. var. *heuseri* Uechtr., *Eleocharis obtusa* var. *ovata* (Roth) Drapalik & Mohlenbr., *Eleocharis soloniensis* (Dubois) Hara, *Scirpus ovatus* Roth, *Scirpus soloniensis* (Dubois) Hara
- ♦ ovate spikerush, bahniăka vajcovitá
- ♦ Weed
- ♦ 87, 88, 286
- ♦ G, herbal.

Eleocharis pachycarpa Desv.
Cyperaceae

- ♦ black sand spikerush
- ♦ Weed, Naturalised
- ♦ 86, 98, 101, 203
- ♦ pG, aqua. Origin: Chile.

Eleocharis pallens S.T.Blake
Cyperaceae
- ♦ Naturalised
- ♦ 86
- ♦ G, arid, cultivated. Origin: Australia.

Eleocharis palustris (L.) Roem. & Schult.
Cyperaceae
Eleocharis crassa Fisch. & Mey. ex Zinserl., *Eleocharis eupalustris* H.Lindb., *Eleocharis filiculmis* Schur, *Eleocharis intersita* Zinserl. (see), *Eleocharis kasakstanica* Zinserl., *Eleocharis levinae* Zoz, *Eleocharis oxystachys* D.I.Sakalo, *Eleocharis palustris* ssp. *microcarpa* Walters, *Heleocharis palustris* (L.) R.Br. (see), *Scirpus palustris* L. (see)
- ♦ creeping spikerush, rantaluikka, common spikerush, wiregrass
- ♦ Weed, Quarantine Weed, Naturalised
- ♦ 23, 70, 76, 87, 88, 126, 161, 180, 203, 217, 218, 220, 237, 263, 272, 300
- ♦ pG, aqua, cultivated, herbal.

Eleocharis parodii Barros
Cyperaceae
- ♦ Weed, Naturalised
- ♦ 54, 86, 88
- ♦ G. Origin: Argentina.

Eleocharis parvula (Roem. & Schult.) Link
Cyperaceae
- ♦ dwarf spikerush, small spikerush, pikkuluikka
- ♦ Weed
- ♦ 87, 88, 218
- ♦ pG, aqua, herbal.

Eleocharis pellucida J & C.Presl
Cyperaceae
Eleocharis afflata Steud. (see), *Eleocharis congesta* D.Don (see), *Eleocharis japonica* Miq.
- ♦ Weed
- ♦ 126, 275
- ♦ a/pG, aqua.

Eleocharis philippinensis Svens.
Cyperaceae
- ♦ Weed
- ♦ 88, 170, 191
- ♦ G, aqua, cultivated. Origin: Australia.

Eleocharis plantaginoidea W.F.Wight
Cyperaceae
= *Eleocharis dulcis* (Burm.f.) Trin. ex Hensch.
- ♦ Weed, Quarantine Weed
- ♦ 76, 87, 88, 203, 220
- ♦ G.

Eleocharis quadrangulata (Michx.) Roem. & Schult.
Cyperaceae
- ♦ squarestem spikerush
- ♦ Weed, Introduced
- ♦ 34, 87, 88, 218
- ♦ pG, aqua, herbal.

Eleocharis retroflexa (Poir.) Urb.
Cyperaceae
Chaetocyperus niveus Liebm.,
Chaetocyperus rugulosus Nees,
Chaetocyperus viviparus Liebm.,
Eleocharis chaetaria Roem. & Schult.
(see), *Scirpus retroflexus* Poir.
♦ coastal plain spikerush
♦ Weed
♦ 87, 88, 170, 191
♦ G. Origin: Australia.

Eleocharis schaffneri Boeck.
Cyperaceae
♦ Schaffner's spikerush
♦ Naturalised
♦ 101
♦ G.

Eleocharis sellowiana Kunth
Cyperaceae
Elocharis galapagensis Sven.
♦ Weed
♦ 255
♦ pG, aqua. Origin: tropical America.

Eleocharis sphacelata R.Br.
Cyperaceae
♦ tall spikerush
♦ Weed
♦ 87, 88, 200, 269
♦ pG, aqua, cultivated. Origin:
Australia.

Eleocharis spiralis R.Br.
Cyperaceae
♦ Weed
♦ 239
♦ pG, aqua, cultivated.

Eleocharis subtilis Boeck.
Cyperaceae
♦ Weed
♦ 87, 88
♦ G.

Eleocharis tetraquetra Nees
Cyperaceae
Eleocharis wichurai Boeck.
♦ Weed
♦ 87, 88, 275
♦ pG. Origin: Australia.

Eleocharis tuberosa Schult.
Cyperaceae
= *Eleocharis dulcis* (Burm.f.) Trin. ex
Hensch.
♦ Chinese waterchestnut
♦ Weed, Quarantine Weed
♦ 76, 87, 88, 203, 220
♦ pG, aqua, cultivated.

Eleocharis uniglumis (Link) Schult.
Cyperaceae
Heleocharis uniglumis (Link) Schult.,
Scirpus uniglumis Link.
♦ meriluikka, onescale spikerush,
slender spikerush, one glimed
spikerush
♦ Weed
♦ 23, 88
♦ G, herbal.

**Eleocharis valleculosa Owhi fo. setosa
(Owhi) Kitag.**
Cyperaceae

♦ ribbedculm spikesedge
♦ Weed
♦ 275, 297
♦ G.

Eleocharis variegata Presl
Cyperaceae
♦ Weed
♦ 87, 88
♦ G.

Eleocharis wichurae Bocklr.
Cyperaceae
♦ Weed
♦ 286
♦ G.

Eleocharis wolfii (Gray) Gray ex Britt.
Cyperaceae
♦ Wolf's spikerush
♦ Weed
♦ 87, 88
♦ G, herbal.

**Eleocharis yokoscensis (Franch. & Sav.)
Tang & Wang**
Cyperaceae
♦ Weed
♦ 88, 275, 297
♦ pG.

Elephantella groenlandica (Retz.) Rydb.
Scrophulariaceae
= *Pedicularis groenlandica* Retz.
♦ Quarantine Weed
♦ 220

Elephantopus augustifolius Sw.
Asteraceae
♦ Weed
♦ 87, 88

Elephantopus mollis Kunth
Asteraceae
Elephantopus carolinianus Raeusch. var.
mollis (Kunth) Beurlin, *Elephantopus
hypomalacus* S.F.Blake, *Elephantopus
martii* Graham, *Elephantopus pilosus*
Philipson, *Elephantopus scaber* L.
var. *tomentosus* Sch.Bip. ex Bak.,
Elephantopus sericeus Graham,
Elephantopus serratus Blanco
♦ elephantopus, elephant's foot,
tobacco weed, papago vaca, papago
halomtano, papago' halom tano, lata
hina, tavako ni veikau, jangli tambaku,
tapua erepani, faux tabac, lau veveli
♦ Weed, Noxious Weed, Naturalised,
Introduced
♦ 3, 6, 54, 86, 88, 107, 191, 203, 229, 230,
235, 255, 274, 276, 287
♦ pH, herbal. Origin: tropical America.

Elephantopus scaber L.
Asteraceae
♦ elephant's foot
♦ Weed, Quarantine Weed,
Naturalised
♦ 3, 6, 12, 13, 76, 86, 87, 88, 191, 203,
218, 276, 287, 297
♦ pH, cultivated, herbal. Origin: Asia,
Australia.

Elephantopus spicatus Juss. ex Aubl.
Asteraceae
= *Pseudelephantopus spicatus* (Juss. ex

Aubl.) C.F.Bak.
♦ Weed
♦ 23, 88
♦ herbal. Origin: Australia.

Elephantopus tomentosus L.
Asteraceae
♦ tobacco weed, devil's grandmother
♦ Weed, Quarantine Weed
♦ 39, 76, 87, 88, 203, 297
♦ cultivated, herbal, toxic.

**Elephantorrhiza elephantina (Burch.)
Skeels**
Fabaceae/Mimosaceae
Acacia elephanthorhiza DC., *Acacia
elephantina* Burch., *Elephantorrhiza
burchellii* Benth.
♦ Eland's bean, Eland's wattle,
elephant's root
♦ Weed, Native Weed
♦ 39, 88, 121, 158
♦ S, arid, cultivated, herbal, toxic.
Origin: southern Africa.

Elettaria cardamomum (L.) Maton
Zingiberaceae
♦ cardamom
♦ Weed, Environmental Weed
♦ 3, 152, 191
♦ cultivated, herbal. Origin: India.

Eleusine aegyptiaca Desf.
Poaceae
♦ Weed
♦ 87, 88
♦ G, herbal.

Eleusine africana Kenn.-O'Byrne
Poaceae
= *Eleusine coracana* (L.) Gaertn. ssp.
africana (Kenn.-O'Byrne) Hilu & de
Wet
♦ African goosegrass
♦ Weed, Quarantine Weed
♦ 51, 76, 87, 88, 203, 220
♦ G, arid, herbal.

**Eleusine compressa (Forssk.) Asch. &
Schwienf.**
Poaceae
Eleusine flagellifera Nees
♦ Weed, Quarantine Weed
♦ 76, 87, 88, 203, 220, 221
♦ G.

Eleusine coracana (L.) Gaertn.
Poaceae
Cynosurus coracanus L.
♦ African finger millet, African millet,
dagussa, finger millet, gagussa,
garindi, Indian millet, korakan, millet,
poko grass, rapoko grass, Indian
millet, ragi, dragon's claw millet,
African goosegrass
♦ Weed, Naturalised, Introduced
♦ 7, 88, 91, 98, 101, 121, 203, 228
♦ aG, arid, cultivated, herbal. Origin:
Africa.

**Eleusine coracana (L.) Gaertn. ssp.
africana (Kenn.-O'Byrne) Hilu & de Wet**
Poaceae
Eleusine africana Kenn.-O'Byrne (see)
♦ African goosegrass, rapoko grass,
African finger millet

- ♦ Weed, Naturalised
- ♦ 88, 101, 158
- ♦ aG. Origin: southern Africa.

Eleusine indica (L.) Gaertn
Poaceae
Cynosurus indicus L., *Eleusine japonica* Steud.
- ♦ goosegrass, wiregrass, goosefoot, crow's foot, bullgrass, umog, reh takai, manienie ali'i, fahitalo, te uteute, deskim, keteketarmalk, kavoronaisivi, vorovoroisivi, mahkwekwe, yaa teen kaa, wild finger millet, silver crabgrass, pie de gallina, grama carraspera, capim pé de galinha
- ♦ Weed, Quarantine Weed, Naturalised, Introduced
- ♦ 3, 6, 7, 12, 14, 15, 23, 30, 32, 34, 38, 39, 50, 53, 55, 68, 70, 86, 87, 88, 91, 93, 98, 101, 107, 134, 157, 161, 170, 174, 176, 179, 180, 185, 186, 198, 203, 204, 206, 207, 209, 211, 212, 218, 230, 235, 236, 237, 238, 239, 240, 241, 243, 245, 249, 253, 255, 257, 258, 261, 262, 263, 269, 270, 273, 274, 275, 276, 280, 281, 286, 295, 297
- ♦ aG, cultivated, herbal, toxic. Origin: Africa.

Eleusine indica (L.) Gaertn. ssp. indica
Poaceae
- ♦ Indian goosegrass, ox grass, crabgrass, crow's foot grass, goosegrass, landgrass, rapoko grass, wiregrass, yard grass
- ♦ Weed, Native Weed
- ♦ 88, 121, 158
- ♦ a/pG. Origin: Africa.

Eleusine multiflora Hochst. ex A.Rich.
Poaceae
Eragrostis kwaiensis Peter
- ♦ goosegrass
- ♦ Weed
- ♦ 243
- ♦ G. Origin: Africa.

Eleusine tristachya (Lam.) Lam.
Poaceae
- ♦ threespike goosegrass, American crow's foot grass
- ♦ Weed, Quarantine Weed, Naturalised, Introduced, Casual Alien
- ♦ 39, 42, 76, 86, 87, 88, 98, 101, 121, 198, 203, 228, 237, 241, 280, 295, 300
- ♦ aG, arid, cultivated, toxic. Origin: South America.

Eleutheranthera ruderalis (Sw.) Sch.Bip.
Asteraceae
Eleutheranthera ovata Poit. ex Steud., *Eleutheranthera prostrata* (L.) Sch.Bip.
- ♦ ogiera
- ♦ Weed, Naturalised
- ♦ 6, 13, 28, 86, 87, 88, 93, 101, 170, 191, 206, 243, 261, 276
- ♦ aH. Origin: tropical America.

Eleutherococcus lasiogyne (Harms) S.Y.Hu
Araliaceae
- ♦ Quarantine Weed
- ♦ 220
- ♦ cultivated.

Eleutherococcus pentaphyllus (Sieb. & Zucc.) Nakai
Araliaceae
- ♦ ginseng
- ♦ Naturalised
- ♦ 101

Eleutherococcus sieboldianus (Makino) Koidz.
Araliaceae
- ♦ hime ukogi, ukogi
- ♦ Quarantine Weed
- ♦ 220
- ♦ S, cultivated, herbal. Origin: east Asia.

Elionurus muticus (Spreng.) Kunth
Poaceae
Elionurus argenteus Nees, *Lycurus muticus* Spreng.
- ♦ lemon grass, lemon scented grass, matrass grass, silky grass, Simon grass, sour grass, wine grass, wiregrass
- ♦ Native Weed
- ♦ 121
- ♦ pG, arid. Origin: Africa.

Ellisia nyctelea (L.) L.
Hydrophyllaceae
- ♦ nyctelea, Aunt Lucy, waterpod, ellisia
- ♦ Weed, Native Weed
- ♦ 49, 87, 88, 161, 174, 210, 218
- ♦ herbal. Origin: North America.

Elodea Michx. spp.
Hydrocharitaceae
- ♦ waterweed
- ♦ Quarantine Weed, Naturalised
- ♦ 198, 220
- ♦ wH.

Elodea callitrichoides (Rich.) Casp.
Hydrocharitaceae
- ♦ elodea
- ♦ Weed
- ♦ 295
- ♦ wH, cultivated.

Elodea canadensis Michx.
Hydrocharitaceae
Anacharis canadensis (Michx.) Planch. (see), *Udora canadensis* Nutt.
- ♦ Canadian pondweed, American elodea, oxygen weed, waterweed, elodea, vesirutto, common waterweed
- ♦ Weed, Sleeper Weed, Quarantine Weed, Noxious Weed, Naturalised, Garden Escape, Environmental Weed
- ♦ 18, 23, 40, 42, 62, 63, 70, 72, 76, 86, 87, 88, 93, 98, 121, 126, 147, 152, 159, 161, 165, 171, 176, 191, 198, 200, 203, 208, 217, 218, 225, 241, 246, 269, 272, 280, 283, 287, 291, 296, 300
- ♦ wpH, cultivated, herbal. Origin: North America.

Elodea densa (Planch.) Casp.
Hydrocharitaceae
= *Egeria densa* Planch.
- ♦ South American waterweed, densa waterweed
- ♦ Weed
- ♦ 45, 88, 146, 218
- ♦ wpH, cultivated.

Elodea longivaginata St.John
Hydrocharitaceae
= *Elodea bifoliata* St.John (NoR)
- ♦ Canada waterweed, water thyme, ditch moss
- ♦ Noxious Weed
- ♦ 299
- ♦ wH, herbal.

Elodea nuttallii (Planch.) St.John
Hydrocharitaceae
Anacharis nuttallii Planch., *Helodea nuttallii* (Planch) St.John
- ♦ Nuttall's pondweed
- ♦ Weed, Naturalised, Environmental Weed
- ♦ 40, 87, 88, 126, 152, 286, 287
- ♦ wpH, cultivated, herbal.

Elsholtzia ciliata (Thunb.) Hyl.
Lamiaceae
Elsholtzia cristata Willd. (see), *Elsholtzia patrinii* (Lepech.) Garcke, *Sideritis ciliata* Thunb.
- ♦ elsholtzia, helttaminttu, crested latesummer mint, Vietnamese balm, rau kinh gio'i
- ♦ Weed, Naturalised, Casual Alien
- ♦ 23, 42, 88, 94, 101, 133, 195, 204, 224, 272, 275, 286, 297
- ♦ aH, cultivated, herbal.

Elsholtzia cristata Willd.
Lamiaceae
= *Elsholtzia ciliata* (Thunb.) Hyl.
- ♦ Weed
- ♦ 87, 88
- ♦ herbal.

Elsholtzia densa Benth.
Lamiaceae
- ♦ denseflower elsholtzia
- ♦ Weed
- ♦ 297
- ♦ aH, cultivated. Origin: Asia.

Elsholtzia nipponica Ohwi
Lamiaceae
- ♦ Weed
- ♦ 286

Elsholtzia patrini (Lepech.) Garcke
Lamiaceae
Mentha patrini Lepech.
- ♦ Weed
- ♦ 88, 204, 243

Elsholtzia rugulosa Hemsl.
Lamiaceae
- ♦ rugulose elsholtzia
- ♦ Weed
- ♦ 297

Elsholtzia stauntonii Benth.
Lamiaceae
- ♦ staunton elsholtzia, mint balm
- ♦ Weed
- ♦ 297
- ♦ cultivated, herbal.

Elvira biflora (L.) DC.
Asteraceae
= *Delilia biflora* (L.) Kuntze
- ♦ espoleta
- ♦ Weed
- ♦ 157, 255
- ♦ aH. Origin: Americas.

Elymus L. spp.
Poaceae
♦ wildrye, wild rice, lyme grass
♦ Weed
♦ 39, 272
♦ G, toxic.

Elymus arenarius L.
Poaceae
= *Leymus arenarius* (L.) Hochst
♦ blue lyme grass, lyme grass
♦ Weed
♦ 23, 88
♦ pG, cultivated, herbal.

Elymus breviaristatus (Hitchc.) Á.Löve
Poaceae
= *Elymus hitchcockii* Davidse (NoR)
♦ Quarantine Weed
♦ 220
♦ G.

Elymus canadensis L.
Poaceae
♦ kanadanrantavehnä, Canada wildrye
♦ Weed, Casual Alien
♦ 42, 161
♦ pG, arid, cultivated, herbal.

Elymus caninus (L.) L.
Poaceae
Agropyron caninum (L.) Beauv. (see), *Agropyron biflorum* (Brign.) Schult., *Agropyron donianum* F.B.White, *Brachypodium caninum* (L.) Lindm., *Goulardia canina* (L.) Husn., *Elytrigia canina* (L.) Drobow, *Roegneria behmii* Melderis, *Roegneria canina* (L.) Nevski, *Roegneria doniana* (F.B.White) Melderis, *Triticum biflorum* Brign., *Triticum caninum* L., *Triticum rupestre* Link
♦ bearded wheatgrass, fibrous wheatgrass, bearded couchgrass, koiranvehnä
♦ Weed, Naturalised
♦ 70, 101
♦ G, cultivated.

Elymus caput-medusae L.
Poaceae
= *Taeniatherum caput-medusae* (L.) Nevski
♦ Weed
♦ 88, 218
♦ G, herbal.

Elymus dahuricus Turcz. ex Griseb.
Poaceae
♦ bunchgrass, wildrye, Dahurian lyme grass
♦ Weed
♦ 114
♦ G.

Elymus elongatus (P.Beauv.) Runem.
Poaceae
= *Elytrigia elongata* (Host) Nev.
♦ Weed, Naturalised, Environmental Weed
♦ 86, 98, 203
♦ G, herbal.

Elymus elymoides (Raf.) Swezey
Poaceae
♦ squirreltail barley, bottlebrush squirreltail, squirreltail
♦ Weed
♦ 161, 212
♦ pG, herbal.

Elymus farctus (Viv.) Runemark ex Melderis ssp. *boreali-atlanticus* (Simonet & Guin.) Melderis
Poaceae
♦ Naturalised
♦ 86
♦ G.

Elymus hispidus (Opiz) Melderis
Poaceae
= *Elytrigia intermedia* (Host) Nevski ssp. *intermedia* (NoR) [see *Agropyron intermedium* (Host) Beauv., *Thinopyrum intermedium* (Host) Barkworth & D.R.Dewey, *Elymus hispidus* (Opiz) Melderis ssp. *barbulatus* (Schur) Melderis]
♦ sea couchgrass, intermediate wheatgrass
♦ Weed
♦ 272
♦ G, cultivated, herbal.

Elymus hispidus (Opiz) Melderis ssp. *barbulatus* (Schur) Melderis
Poaceae
= *Elytrigia intermedia* (Host) Nevski ssp. *intermedia* (NoR) [see *Agropyron intermedium* (Host) Beauv., *Thinopyrum intermedium* (Host) Barkworth & D.R.Dewey, *Elymus hispidus* (Opiz) Meld.]
♦ Introduced
♦ 228
♦ G.

Elymus hoffmannii K.B.Jensen & K.H.Asay
Poaceae
♦ RS wheatgrass
♦ Naturalised
♦ 101
♦ G.

Elymus humidus Osada
Poaceae
♦ Weed
♦ 286
♦ G.

Elymus mayebaranus (Honda) S.L.Chen
Poaceae
= *Elymus humidus* Osada × *Elymus tsukushiensis* Honda [some view as a natural hybrid]
♦ Weed
♦ 286
♦ G.

Elymus multisetus (J.G.Sm.) Burtt Davy
Poaceae
Sitanion jubatum J.G.Sm.
♦ big squirreltail
♦ Weed
♦ 161, 180
♦ pG.

Elymus patagonicus Speg.
Poaceae
♦ Introduced
♦ 228
♦ G, arid.

Elymus racemifer (Steud.) Tzvelev
Poaceae
♦ Weed
♦ 286
♦ G.

Elymus rectisetus (Nees) Á.Löve & Connor
Poaceae
♦ Naturalised
♦ 280
♦ G.

Elymus repens (L.) Gould
Poaceae
= *Elytrigia repens* (L.) Nev.
♦ English couchgrass, twitch, quackgrass, quitch grass, dog grass, scutch, quickgrass, couchgrass, quitch
♦ Weed, Noxious Weed, Naturalised, Introduced
♦ 28, 70, 86, 97, 101, 174, 222, 229, 243, 272, 286, 287
♦ pG, cultivated, herbal. Origin: Eurasia.

Elymus repens Beauv. var. *aristatum* Baumg.
Poaceae
♦ Naturalised
♦ 287
♦ G.

Elymus semicostatum (Nees ex Steud.) Á.Löve
Poaceae
Agropyron semicostatum (Nees ex Steud.) Boiss. (see), *Agropyron striatum* (Nees ex Steud.) Hook.f., *Triticum semicostatum* Nees ex Steud.
♦ drooping wildrye
♦ Naturalised
♦ 101
♦ G, arid.

Elymus sibiricus L.
Poaceae
♦ squirreltail barley, Siberian wildrye
♦ Weed, Casual Alien
♦ 42, 88, 114, 243, 297
♦ G, herbal.

Elymus subsecundus (Link) Á. & D.Löve
Poaceae
♦ kaarivehnä, bearded wheatgrass
♦ Casual Alien
♦ 42
♦ G.

Elymus trachycaulus (Link) Gould ex Shinners
Poaceae
♦ hoikkavehnä, slender wheatgrass
♦ Casual Alien
♦ 42
♦ pG, cultivated, herbal.

Elymus triticoides (Nutt.) Buckl.
Poaceae
= *Leymus triticoides* (Buckl.) Pilg.
♦ alkali ryegrass, beardless wildrye, creeping wildrye, wild ryegrass
♦ Weed
♦ 87, 88, 121, 180, 218
♦ pG, promoted, herbal. Origin: western North America.

***Elymus tsukusiensis* Honda var. *transiens* Osada**
 Poaceae
 ♦ Weed
 ♦ 286
 ♦ G.

***Elytraria crenata* Vahl**
 Acanthaceae/Nelsoniaceae
 Elytraria lyrata Vahl
 ♦ Weed, Quarantine Weed
 ♦ 76, 87, 88, 203, 220

***Elytraria imbricata* (Vahl) Pers.**
 Acanthaceae/Nelsoniaceae
 Elytraria squamosa (Jacq.) Lindau
 ♦ purple scalystem
 ♦ Weed, Naturalised
 ♦ 157, 257
 ♦ pH, arid, herbal.

***Elytrigia elongata* (Host) Nev.**
 Poaceae
 Agropyron elongatum (Host) Beauv.
 (see), *Agropyron ruthenicum* (Griseb.)
 Prok., *Elymus elongatus* (P.Beauv.)
 Runem. (see), *Elymus elongatus* ssp.
 ponticus (Podp.) Melderis, *Elytrigia
 prokudinii* Druleva ex O.N.Dubovik,
 Elytrigia ruthenica (Griseb.) Prokudin,
 Lophopyrum elongatum (Host) Á.Löve
 (see), *Triticum ponticum* Tzvelev,
 Triticum rigidum Schrad.
 ♦ tall wheatgrass
 ♦ Weed, Introduced
 ♦ 34, 80, 228
 ♦ pG, arid.

***Elytrigia intermedia* (Host) Nevski**
 Poaceae
 Agropyron hispidum Opiz, *Agropyron
 intermedium* (Host) Beauv. (see), *Elymus
 hispidus* (Opiz) Meld. (see), *Thinopyrum
 intermedium* (Host) Barkworth &
 D.R.Dewey (see)
 ♦ intermediate wheatgrass
 ♦ Weed, Introduced
 ♦ 21, 80
 ♦ pG, herbal. Origin: Europe.

***Elytrigia pontica* (Podp.) Holub ssp.
pontica**
 Poaceae
 [see *Lophopyrum ponticum* (Подр.)
 Á.Löve]
 ♦ rush wheatgrass
 ♦ Introduced
 ♦ 34
 ♦ pG.

***Elytrigia pungens* (Pers.) Tutin**
 Poaceae
 Agropyron littorale (Host) Dur. (see),
 Agropyron pungens (Pers.) Roem. &
 Schult., *Elymus pungens* (Pers.) Meld.,
 Triticum littorale Host, *Triticum pungens*
 Pers.
 ♦ sea couchgrass
 ♦ Weed, Naturalised
 ♦ 86, 98, 198, 203
 ♦ G, cultivated.

***Elytrigia pycnantha* (Godr.) Á.Löve**
 Poaceae
 ♦ Naturalised

 ♦ 280
 ♦ G.

***Elytrigia repens* (L.) Nev.**
 Poaceae
 Agropyron firmum J.Presl, *Agropyron
 repens* (L.) Beauv. (see), *Agropyron
 repens* var. *bromiforme* Schur, *Agropyron
 repens* var. *glaucescens* Peterm., *Elymus
 repens* (L.) Gould (see), *Triticum firmum*
 (J.Presl) Link, *Triticum repens* L.
 ♦ quackgrass, couchgrass, quickgrass,
 twitch grass, dog grass
 ♦ Weed, Noxious Weed, Naturalised,
 Introduced, Environmental Weed
 ♦ 15, 26, 34, 35, 86, 98, 151, 161, 176,
 195, 198, 203, 211, 212, 228, 241, 243,
 253, 266, 280, 300
 ♦ pG, arid, cultivated, herbal. Origin:
 Mediterranean.

***Elytropappus rhinocerotis* (L.f.) Less.**
 Asteraceae
 ♦ rhenoster bush, rhinoceros bush
 ♦ Weed, Quarantine Weed, Native
 Weed
 ♦ 76, 87, 88, 121, 203, 220
 ♦ pS. Origin: southern Africa.

***Elytrophorus articulatus* Beauv.**
 Poaceae
 = *Elytrophorus spicatus* (Willd.) Camus
 ♦ Weed
 ♦ 87, 88
 ♦ G.

***Elytrophorus spicatus* (Willd.) Camus**
 Poaceae
 Dactylis spicata Willd., *Elytrophorus
 articulatus* Beauv. (see)
 ♦ Weed
 ♦ 88, 238
 ♦ aG, arid.

***Embothrium coccineum* J.R. & G.Forst.**
 Proteaceae
 ♦ Chilean firetree, Chilean firebush
 ♦ Casual Alien
 ♦ 280
 ♦ cultivated, herbal. Origin: Chile.

***Emex australis* Steinh.**
 Polygonaceae
 ♦ spiny emex, three cornered jack,
 devil's thorn, doublegee, prickly jack,
 bull's head, cape spinach, cat's head,
 goat's head, southern threecorner jack
 ♦ Weed, Quarantine Weed, Noxious
 Weed, Naturalised, Native Weed,
 Introduced, Environmental Weed
 ♦ 7, 39, 50, 51, 53, 55, 62, 67, 72, 76, 86,
 87, 88, 93, 98, 101, 121, 140, 147, 158,
 165, 171, 176, 178, 198, 203, 205, 218,
 228, 229, 243, 246, 269, 280
 ♦ aH, arid, toxic. Origin: South Africa.

***Emex spinosa* (L.) Campd.**
 Polygonaceae
 Rumex spinosus L.
 ♦ prickly dock, spiny threecorner jack,
 spiny emex, lesser jack, doublegee,
 devil's thorn
 ♦ Weed, Quarantine Weed, Noxious
 Weed, Naturalised, Introduced
 ♦ 7, 34, 38, 62, 67, 76, 86, 87, 88, 98, 101,

 140, 161, 185, 198, 203, 218, 221, 228,
 229, 237, 243, 269, 287, 295, 300
 ♦ aH, cultivated, herbal. Origin:
 Mediterranean.

***Emilia coccinea* (Sims) G.Don**
 Asteraceae
 Emilia saggittata (Vahl) DC.
 ♦ scarlet tasselflower, tasselflower,
 pincel, serralha mirim
 ♦ Weed
 ♦ 32, 88, 255
 ♦ aH, cultivated, herbal. Origin:
 Africa.

***Emilia fosbergii* D.H.Nicols.**
 Asteraceae
 ♦ Cupid's shaving brush, Florida
 tasselflower, clavelillo de cafetal
 ♦ Weed, Naturalised
 ♦ 101, 161, 179, 249, 261
 ♦ herbal.

***Emilia javanica* (Burm.f.) C.B.Rob.**
 Asteraceae
 = *Emilia sonchifolia* (L.) DC. var. *javanica*
 (Burm.f.) Mattf.
 ♦ Weed
 ♦ 87, 88
 ♦ cultivated, herbal.

***Emilia praetermissa* Milne-Redh.**
 Asteraceae
 ♦ Weed
 ♦ 87, 88

***Emilia prenanthoidea* DC.**
 Asteraceae
 ♦ Weed
 ♦ 276

***Emilia sagittata* (Vahl) DC.**
 Asteraceae
 ♦ tasselflower
 ♦ Naturalised
 ♦ 287
 ♦ aH, promoted.

***Emilia sonchifolia* (L.) DC.**
 Asteraceae
 ♦ red tasselflower, emilia, purple sow
 thistle, red groundsel, lilac tasselflower
 ♦ Weed, Naturalised, Introduced
 ♦ 6, 7, 12, 13, 14, 66, 87, 88, 93, 98, 101,
 157, 179, 203, 204, 206, 209, 217, 218,
 230, 235, 243, 245, 255, 261, 274, 275,
 276, 286, 297
 ♦ aH, promoted, herbal. Origin:
 tropical Asia.

***Emilia sonchifolia* (L.) DC. var. *javanica*
(Burm.f.) Mattf.**
 Asteraceae
 Emilia javanica (Burm.f.) C.B.Rob. (see)
 ♦ lilac tasselflower
 ♦ Weed, Naturalised
 ♦ 86, 101, 273
 ♦ Origin: Asia.

***Emilia sonchifolia* (L.) DC. var.
*sonchifolia***
 Asteraceae
 ♦ lilac tasselflower
 ♦ Naturalised, Environmental Weed
 ♦ 86
 ♦ Origin: tropical Africa, Americas.

Eminium spiculatum (Blume) Kuntze
Araceae
♦ Weed
♦ 221
♦ pH, promoted.

Empetrum rubrum Vahl ex Willd.
Empetraceae/Ericaceae
= *Empetrum eamesii* Fern. & Wieg. ssp.
eamesii (NoR)
♦ murtilla
♦ Weed
♦ 295
♦ S, cultivated, herbal.

Enarthrocarpus lyratus (Forssk.) DC.
Brassicaceae
Raphanus lyratus Forssk.
♦ tormokki
♦ Weed, Casual Alien
♦ 42, 87, 88, 185, 221

Encelia farinosa A.Gray ex Torr.
Asteraceae
♦ brittle bush, goldenhills
♦ Introduced
♦ 228
♦ S, arid, cultivated, herbal.

Encelia mexicana Mart. ex DC. *nom. inval.*
Asteraceae
= *Simsia amplexicaulis* (Cav.) Pers.
♦ Weed, Quarantine Weed
♦ 76, 87, 88, 203, 220

Encephalocereus Berger spp.
Cactaceae
= *Pelecyphora* Ehrenb. spp.
♦ Weed, Quarantine Weed
♦ 88, 203, 220

Encyclia rufa (Lindl.) Britt. & Millsp.
Orchidaceae
♦ rufous butterfly orchid
♦ Naturalised
♦ 101

Endymion non-scriptus (L.) Garcke
Liliaceae
= *Hyacinthoides non-scripta* (L.)
Chouard ex Rothm.
♦ bluebell
♦ Weed, Naturalised
♦ 23, 88, 198
♦ cultivated, herbal.

Enicostema verticillatum (L.) Engl. ex Gilg
Gentianaceae
♦ whitehead
♦ Weed
♦ 87, 88
♦ herbal.

Enneapogon avenaceus (Lindl.) C.E.Hubb.
Poaceae
Pappophorum avenaceum Lindl.,
Pappophorum commune F.Muell. var.
avenaceum (Lindl.) Maiden & Betcke
♦ Naturalised
♦ 86
♦ G, arid, cultivated. Origin: Australia.

Enneapogon borealis (Griseb.) Honda
Poaceae

♦ short eared enneapogon
♦ Weed
♦ 297
♦ G.

Enneapogon cenchroides (Roem. & Schult.) C.E.Hubb.
Poaceae
♦ common nine awned grass, furgrass, Sabi grass, sour grass
♦ Native Weed
♦ 121
♦ a/pG. Origin: southern Africa.

Enneapogon desvauxii Desv. ex Beauv.
Poaceae
Pappophorum wrightii S.Watson
♦ nineawn pappusgrass
♦ Weed, Introduced
♦ 88, 228
♦ pG, arid, herbal.

Enneapogon mollis Lehm.
Poaceae
♦ soft feather pappusgrass
♦ Naturalised
♦ 101
♦ G.

Enneapogon scoparius Stapf
Poaceae
♦ wiry nine awned grass
♦ Native Weed
♦ 121
♦ pG, herbal. Origin: southern Africa.

Ensete ventricosum (Welw.) Cheesman
Musaceae
♦ Abyssinian banana, Ethiopian banana, butembe
♦ Introduced, Casual Alien
♦ 228, 280
♦ pH, arid, cultivated.

Entada phaseoloides (L.) Merr.
Fabaceae/Mimosaceae
♦ St. Thomas bean, tupe
♦ Naturalised
♦ 39, 101
♦ H, cultivated, herbal, toxic. Origin: Asia, Australia.

Entada polystachia (L.) DC.
Fabaceae/Mimosaceae
Mimosa polystachia L.
♦ Introduced
♦ 228
♦ arid, herbal.

Enterolobium contortisiliquum (Vell.) Morong
Fabaceae/Mimosaceae
Enterolobium timbouva Mart., *Mimosa contortisiliqua* Vell.
♦ earpod tree, timbó, elephant's ear tree, orelha de macaco, tamboril do campo, pacara earpod tree
♦ Weed, Naturalised, Introduced
♦ 80, 87, 88, 101, 112, 228
♦ arid, cultivated.

Enterolobium cyclocarpum (Jacq.) Griseb.
Fabaceae/Mimosaceae
Feuillea cyclocarpa Kuntze, *Inga cyclocarpa* Willd., *Mimosa cyclocarpa* Jacq., *Mimosa parota* Sessé & Moç.,

Pithecellobium cyclocarpum (Jacq.) C.Mart.
♦ ear tree, guanacaste, monkey soap, elephant's ear, oreja de mono
♦ Weed, Naturalised, Introduced, Environmental Weed, Casual Alien
♦ 3, 39, 87, 88, 101, 179, 191, 228, 261
♦ arid, cultivated, herbal, toxic. Origin: continental tropical America.

Enterolobium saman (Jacq.) Prain ex King
Fabaceae/Mimosaceae
= *Samanea saman* (Jacq.) Merr.
♦ Weed
♦ 22
♦ T.

Enteromorpha intestinalis (L.) Grev.
Ulvaceae
♦ green alga
♦ Weed
♦ 87, 88, 197
♦ algae, herbal.

Enteropogon acicularis (Lindl.) Lazarides
Poaceae
Chloris moorei F.Muell.
♦ curly windmill grass
♦ Introduced
♦ 228
♦ G, arid, cultivated.

Enteropogon prieurii (Kunth) Clayton
Poaceae
Chloris prieurii Kunth (see)
♦ Prieur's umbrellagrass
♦ Naturalised
♦ 101
♦ G, arid.

Entolasia marginata (R.Br.) Hughes
Poaceae
♦ Australian panicgrass
♦ Naturalised
♦ 101, 280
♦ G, cultivated. Origin: Australia.

Enydra fluctuans Lour.
Asteraceae
Enydra anagallis Gardner, *Meyera fluctuans* (Lour.) Spreng.
♦ Weed
♦ 13, 87, 88, 170, 191, 262
♦ wpH, herbal.

Enydra sessilis (Sw.) DC.
Asteraceae
Eclipta sessilis Sw., *Meyera sessilis* Sw.
♦ smallray swampwort
♦ Weed, Quarantine Weed
♦ 14, 76, 87, 88, 203, 220

Epacris purpurascens R.Br.
Epacridaceae/Ericaceae
♦ Naturalised
♦ 280
♦ cultivated.

Epaltes australis Less.
Asteraceae
= *Sphaeromorphaea australis* (Less.) Kit. (NoR)
♦ large golden rail
♦ Weed
♦ 274
♦ cultivated. Origin: Australia.

Epaltes cunninghamii (Hook.) Benth.
Asteraceae
Ethulia cunninghamii Hook., *Ethuliopsis cunninghamii* (Hook.) F.Muell., *Ethuliopsis dioica* F.Muell.
♦ Weed
♦ 87, 88
♦ Origin: Australia.

Ephedra alata Decne.
Ephedraceae
♦ Weed
♦ 221
♦ arid, herbal.

Ephedra americana Humb. & Bonpl. ex Willd.
Ephedraceae
♦ Introduced
♦ 228
♦ arid, herbal.

Ephedra aphylla Forssk.
Ephedraceae
♦ Weed
♦ 221

Ephedra gerardiana Wall. ex Stapf
Ephedraceae
♦ Gerard jointfir, ma huang, efedra
♦ Introduced
♦ 228
♦ S, arid, cultivated, herbal.

Ephedra likiangensis Florin
Ephedraceae
♦ Quarantine Weed
♦ 220

Ephedra saxatilis Royle ex Florin
Ephedraceae
Ephedra gerardiana Wall. ex C.A.Mey. var. *saxatilis* Stapf
♦ Quarantine Weed
♦ 220

Ephedra sinica Stapf
Ephedraceae
♦ Chinese ephedra, ma huang
♦ Weed
♦ 39, 275, 297
♦ S, promoted, herbal, toxic.

Ephedra torreyana S.Wats.
Ephedraceae
♦ Torrey ephedra, Mexican tea, Torrey's jointfir
♦ Weed
♦ 87, 88, 218
♦ S, promoted, herbal.

Ephedra triandra Tull.
Ephedraceae
♦ tramontana
♦ Weed
♦ 295
♦ S, promoted.

Ephedra trifurca Torr.
Ephedraceae
♦ longleaf ephedra, longleaf jointfir
♦ Weed
♦ 87, 88, 218
♦ S, promoted, herbal.

Ephemerum serratum (Hedw.) Hampe
Ephemeraceae
♦ serrate ephemerum moss, serrated earth moss
♦ Naturalised
♦ 280
♦ moss.

Epidendrum carpophorum B.Rodrigues
Orchidaceae
♦ Naturalised
♦ 101

Epidendrum obrienianum Rolfe
Orchidaceae
= *Epidendrum* × *obrienianum* Rolfe
♦ Naturalised
♦ 98

Epidendrum × obrienianum Rolfe
Orchidaceae
= *Epidendrum evectum* Hook.f. × *Epidendrum ibaguaense* Kunth [see *Epidendrum obrienianum* Rolfe]
♦ O'Brien's star orchid
♦ Naturalised
♦ 86, 101

Epidendrum radicans Pav. ex Lindl.
Orchidaceae
♦ Weed, Naturalised, Cultivation Escape
♦ 101, 179, 261
♦ cultivated, herbal.

Epilobium L. spp.
Onagraceae
♦ willow herb
♦ Weed
♦ 243, 272
♦ herbal.

Epilobium adenocaulon Hauss.
Onagraceae
= *Epilobium ciliatum* Raf.
♦ American willow herb, northern willow herb, Amerikanhorsma
♦ Weed, Casual Alien
♦ 42, 68, 70, 88, 161, 180, 243
♦ herbal.

Epilobium adnatum Griseb.
Onagraceae
= *Epilobium tetragonum* L. ssp. *tetragonum*
♦ Weed
♦ 68, 88

Epilobium angustifolium L.
Onagraceae
Chamaenerion angustifolium (L.) Scop. (see), *Chamerion angustifolium* (L.) Holub, *Epilobium spicatum* Lam. (see)
♦ fireweed, rosebay willow herb, willow herb, flowering willow, burnt weed, Indian wickup, blooming Sally
♦ Weed
♦ 23, 44, 70, 87, 88, 94, 136, 161, 212, 218, 272, 286, 291, 297
♦ pH, cultivated, herbal.

Epilobium billardierianum Ser.
Onagraceae
♦ aboriginal willow herb
♦ Naturalised, Native Weed
♦ 101, 269
♦ cultivated, herbal. Origin: Australia.

Epilobium billardierianum Ser. ssp. cinereum (A.Rich.) Raven & Engelhorn
Onagraceae
♦ aboriginal willow herb
♦ Naturalised
♦ 101

Epilobium brachycarpum C.Presl
Onagraceae
♦ annual fireweed, autumn willowweed, tall annual willow herb
♦ Naturalised
♦ 300
♦ aH, herbal.

Epilobium brunnescens (Cockayne) Raven & Engelhorn
Onagraceae
♦ New Zealand willow herb
♦ Naturalised
♦ 40
♦ cultivated. Origin: Australia.

Epilobium ciliatum Raf.
Onagraceae
Epilobium adenocaulon Hauss. (see)
♦ tall willow herb, glandular willow herb, willow herb, American willow herb
♦ Weed, Naturalised, Environmental Weed
♦ 7, 15, 20, 40, 42, 72, 86, 88, 98, 165, 176, 198, 203, 208, 243, 253, 280, 290
♦ aH, aqua, herbal. Origin: North and South America.

Epilobium collinum C.C.Gmel.
Onagraceae
♦ mäkihorsma
♦ Weed
♦ 88, 94
♦ herbal.

Epilobium dodonaei Vill.
Onagraceae
Chamaenerion dodonaei (Vill.) Schur, *Chamaenerion palustre* auct. mult. non (L.) Scop., *Chamaenerion rosmarinifolium* (Haenke) Moench, *Chamaenerion argustissimum* (Weber) Sosn., *Epilobium angustissimum* Weber, *Epilobium rosmarinifolium* Haenke
♦ Quarantine Weed
♦ 220
♦ cultivated, herbal.

Epilobium glaberrimum Barbey
Onagraceae
♦ glaucus willow herb, smooth willowweed
♦ Quarantine Weed
♦ 220
♦ pH, herbal.

Epilobium glandulosum Lehm.
Onagraceae
= *Epilobium ciliatum* Raf. ssp. *glandulosum* (Lehm.) P.Hoch & Raven (NoR)
♦ lännenhorsma
♦ Naturalised
♦ 42
♦ herbal.

Epilobium halleanum Hausskn.
Onagraceae
♦ glandular willow herb
♦ Weed
♦ 23, 88
♦ pH, herbal.

Epilobium hirsutum L.
Onagraceae
♦ hairy willow herb, willow herb,
hairy willowweed, great willow herb,
karvahorsma, codlins and cream
♦ Weed, Quarantine Weed,
Naturalised, Introduced, Garden
Escape, Environmental Weed
♦ 42, 44, 54, 70, 72, 76, 80, 86, 87, 88,
101, 133, 136, 185, 195, 198, 218, 221,
222, 253, 272, 275, 297
♦ pH, cultivated, herbal. Origin:
Eurasia, Africa.

Epilobium hypericifolium Tausch
Onagraceae
♦ kuismahorsma
♦ Naturalised
♦ 42

Epilobium komarovianum Lév.
Onagraceae
♦ uudenseelanninhorsma, bronzy
willow herb
♦ Cultivation Escape, Casual Alien
♦ 40, 42
♦ cultivated.

Epilobium lanceolatum Sebast. & Mauri
Onagraceae
♦ suikealehtihorsma, spear leaved
willow herb
♦ Weed, Casual Alien
♦ 42, 272
♦ cultivated.

Epilobium montanum L.
Onagraceae
♦ broad leaved willow herb,
lehtohorsma
♦ Weed, Naturalised
♦ 23, 44, 70, 88, 253, 272, 280
♦ cultivated, herbal.

Epilobium nerterioides A.Cunn.
Onagraceae
♦ Weed
♦ 23, 88

Epilobium nummulariifolium R.Cunn.
Onagraceae
♦ creeping willow herb
♦ Weed
♦ 165
♦ cultivated.

Epilobium obscurum Schreb.
Onagraceae
♦ dwarf willow herb, tummahorsma,
dull leaved willow herb, short fruited
willow herb, thin runner willow herb
♦ Naturalised
♦ 86, 101, 176, 241, 280, 300
♦ cultivated. Origin: Europe.

Epilobium palustre L.
Onagraceae
♦ marsh willow herb, bog willow herb
♦ Weed

♦ 23, 87, 88, 272, 275
♦ pH, cultivated, herbal.

**Epilobium paniculatum Nutt. ex Torr. &
Gray**
Onagraceae
= *Epilobium brachycarpum* C.Presl
(NoR)
♦ panicle willowweed
♦ Weed
♦ 87, 88, 161, 180, 218, 243
♦ cultivated, herbal.

Epilobium parviflorum Schreb.
Onagraceae
♦ smallflower hairy willow herb,
willow herb
♦ Weed, Quarantine Weed,
Naturalised
♦ 15, 24, 44, 87, 88, 101, 220, 253, 272,
280
♦ pH, cultivated, herbal.

Epilobium pubescens Roth
Onagraceae
♦ Weed
♦ 23, 88

**Epilobium pygmaeum (Speg.) Hoch &
P.H.Raven**
Onagraceae
♦ smooth spike primrose
♦ Naturalised
♦ 300
♦ aH.

**Epilobium pyrricholophum Franch. &
Savat.**
Onagraceae
♦ akabana
♦ Weed
♦ 87, 88, 263, 286
♦ pH, promoted, herbal. Origin: east
Asia.

Epilobium roseum Schr.
Onagraceae
♦ pale willow herb, small flowered
willow herb, pedicelled willow herb
♦ Weed, Naturalised
♦ 42, 70, 87, 88, 253, 272
♦ cultivated.

Epilobium rosmarinifolium Pursh
Onagraceae
= *Epilobium leptophyllum* Raf. (NoR)
♦ Quarantine Weed
♦ 220

Epilobium rotundifolium G.Forst.
Onagraceae
♦ Naturalised
♦ 86, 176
♦ Origin: New Zealand.

Epilobium spicatum Lam.
Onagraceae
= *Epilobium angustifolium* L.
♦ Weed
♦ 87, 88

Epilobium tetragonum L.
Onagraceae
Epilobium adnatum Griseb. (see)
♦ square stalked willow herb, willow
herb, särmähorsma
♦ Weed, Naturalised, Casual Alien

♦ 7, 15, 42, 68, 70, 88, 94, 98, 203, 253,
272, 280
♦ pH, cultivated, herbal.

**Epilobium tetragonum L. ssp. lamyi
(F.W.Schultz) Nyman**
Onagraceae
♦ southern willow herb
♦ Naturalised
♦ 300

**Epilobium tetragonum L. ssp. laymi
(F.Schultz) Nyman**
Onagraceae
Epilobium lamyi F.Schultz
♦ Naturalised
♦ 241

Epilobium tetragonum L. ssp. tetragonum
Onagraceae
Epilobium adnatum Griseb. (see)
♦ square stalked willow herb
♦ Naturalised, Environmental Weed
♦ 86
♦ cultivated. Origin: Eurasia, North
Africa.

Epilobium watsonii Barbey
Onagraceae
= *Epilobium ciliatum* Raf. ssp. *watsonii*
(Barbey) P.Hoch & Raven (NoR)
♦ Watson's willow herb
♦ Weed
♦ 136
♦ herbal.

Epimedium acuminatum Franch.
Berberidaceae/Podophyllaceae
♦ Quarantine Weed
♦ 220

Epimedium chlorandrum Stearn
Berberidaceae/Podophyllaceae
♦ Quarantine Weed
♦ 220

Epimedium epsteinii Stearn
Berberidaceae/Podophyllaceae
♦ Quarantine Weed
♦ 220

Epimedium × omiense W.T.Stearn
Berberidaceae/Podophyllaceae
= *Epimedium acuminatum* Franch. ×
Epimedium fangii W.T.Stearn
♦ Quarantine Weed
♦ 220
♦ Origin: south central China.

Epimedium pauciflorum K.C.Yen
Berberidaceae/Podophyllaceae
♦ Quarantine Weed
♦ 220

Epimedium rhizomatosum Stearn
Berberidaceae/Podophyllaceae
♦ Quarantine Weed
♦ 220

**Epimedium sagittatum (Siebold & Zucc.)
Maxim.**
Berberidaceae/Podophyllaceae
Aceranthus sagittatus Siebold & Zucc.
♦ Quarantine Weed
♦ 220
♦ pH, promoted, herbal.

Epimedium sempervirens **Nakai ex Maek.**
Berberidaceae/Podophyllaceae
♦ Quarantine Weed
♦ 220
♦ herbal.

Epimedium × *youngianum* **Fisch. & C.A.Mey. (***pro* **sp.)**
Berberidaceae/Podophyllaceae
= *Epimedium grandiflorum* C.Morrison × *Epimedium diphyllum* (Morr. & Decne.) Lodd.
♦ Young's epimedium
♦ Naturalised
♦ 101
♦ cultivated.

Epipactis atrorubens **(O.Hoffm.) Bess.**
Orchidaceae
♦ royal helleborine, dark red helleborine, tummaneidonvaippa
♦ Naturalised
♦ 101

Epipactis helleborine **(L.) Crantz**
Orchidaceae
♦ broad helleborine, helleborine orchid, broad leaved helleborine, helleborine
♦ Weed, Naturalised
♦ 79, 80, 101, 195
♦ pH, cultivated, herbal.

Epipactis palustris **(L.) Crantz**
Orchidaceae
♦ marsh orchid, marsh helleborine, suoneidonvaippa
♦ Naturalised
♦ 101
♦ herbal.

Epiphyllum hookeri **(Link & Otto) Haw.**
Cactaceae
Epiphyllum phyllanthus (L.) Haw. var. *hookeri* (Haw.) Kimn. (see), *Epiphyllum stenopetalum* (C.F.Först.) Britt. & Rose, *Epiphyllum strictum* (Lem.) Britt. & Rose, *Phyllocactus hookeri* Salm-Dyck
♦ flor de baile, flor de retreta
♦ Introduced
♦ 261
♦ Origin: Trinidad, Tobago, Venezuela.

Epiphyllum oxypetalum **(DC.) Haw.**
Cactaceae
Phyllocactus oxypetalus Link ex Walp.
♦ Dutchman's pipe cactus, night blooming cereus
♦ Naturalised, Introduced
♦ 101, 261
♦ cultivated, herbal. Origin: Mexico, Central America, northern South America.

Epiphyllum phyllanthus **(L.) Haw.**
Cactaceae
Phyllocactus phyllanthus Link.
♦ climbing cactus
♦ Weed, Naturalised
♦ 54, 88, 98, 101, 203
♦ cultivated, herbal.

Epiphyllum phyllanthus **(L.) Haw. var. *hookeri* (Haw.) Kimn.**
Cactaceae
= *Epiphyllum hookeri* (Link & Otto) Haw.

♦ climbing cactus
♦ Weed, Naturalised
♦ 86, 101, 179
♦ cultivated.

Epipremnum aureum **(Lindl. & André) Bunt.**
Araceae
= *Epipremnum pinnatum* (L.) Engl.
♦ pothos, golden pothos, devil's ivy, amapalo amarillo, trepapalo amarillo, golden Ceylon creeper, golden hunter's robe, hunter's robe, ivy arum, malanga trepadora, Solomon Island ivy, taro vine, variegated philodendron
♦ Cultivation Escape
♦ 39, 154, 189, 247, 261
♦ cultivated, toxic. Origin: Solomon Islands.

Epipremnum pinnatum **(L.) Engl.**
Araceae
Epipremnum aureum (Linden & André) Bunt. (see), *Pothos aurea* Linden & André, *Pothos pinnata* L., *Philodendron nechodomii* Britt., *Scindapsus aureus* (Lindl. & André) Engl. (see)
♦ pothos, siempre viva, air plant, centipede tongavine, taro vine
♦ Weed, Naturalised, Environmental Weed, Cultivation Escape
♦ 3, 80, 88, 101, 112, 122, 179, 191, 261
♦ H, cultivated, herbal, toxic. Origin: south-east Asia.

Episcia cupreata **(Hook.) Hanst.**
Gesneriaceae
♦ flame violet
♦ Naturalised, Introduced
♦ 101, 261
♦ cultivated. Origin: Brazil, Colombia, Venezuela.

Equisetum **L. spp.**
Equisetaceae
♦ horsetail, scouringrush, common horsetail
♦ Weed, Quarantine Weed, Noxious Weed, Naturalised, Garden Escape, Environmental Weed
♦ 39, 62, 76, 86, 88, 155, 171, 191, 243, 246
♦ pH, cultivated, herbal, toxic. Origin: northern hemisphere.

Equisetum arvense **L.**
Equisetaceae
♦ field horsetail, scouringrush, western horsetail, horsetail, foxtail, rush, horsetail fern, meadow pine, pine grass, foxtail rush, bottle brush, horse pipes, snake grass, mare's tail, shave grass, coda cavallina
♦ Weed, Quarantine Weed, Noxious Weed, Naturalised, Native Weed, Garden Escape, Environmental Weed
♦ 8, 15, 39, 44, 45, 52, 54, 62, 70, 76, 86, 87, 88, 136, 147, 159, 161, 165, 169, 174, 180, 186, 189, 203, 204, 210, 211, 212, 218, 220, 225, 229, 232, 240, 243, 246, 253, 263, 272, 275, 286, 289, 293, 296, 297, 299
♦ pH, aqua, cultivated, herbal, toxic. Origin: temperate northern hemisphere.

Equisetum bogotense **Kunth**
Equisetaceae
♦ Weed
♦ 87, 88
♦ cultivated, herbal.

Equisetum debile **Roxb. ex Vaucher**
Equisetaceae
♦ Weed
♦ 243
♦ herbal.

Equisetum diffusum **D.Don**
Equisetaceae
♦ Weed
♦ 275
♦ pH.

Equisetum fluviatile **L.**
Equisetaceae
♦ water horsetail, swamp horsetail, järvikorte
♦ Weed
♦ 70, 87, 88, 161, 218
♦ pH, aqua, cultivated, herbal, toxic.

Equisetum giganteum **L.**
Equisetaceae
Equisetum martii Milde, *Equisetum pyramidale* Goldm., *Equisetum xylochaetum* Mett.
♦ cavalinha, cauda de reposa, pinheirinho, arvore de Natal, rabo de cavalo, erva canudo, lixa vegetal
♦ Weed
♦ 87, 88, 255, 295
♦ herbal, toxic. Origin: Americas.

Equisetum hyemale **L.**
Equisetaceae
Equisetum hiemale L.
♦ scouringrush, greater horsetail, rough horsetail, kangaskorte, horsetail, common horsetail
♦ Weed, Sleeper Weed, Quarantine Weed, Noxious Weed, Naturalised, Environmental Weed, Garden Escape
♦ 8, 39, 86, 87, 88, 159, 161, 218, 225, 246, 272, 290
♦ pH, aqua, cultivated, herbal, toxic. Origin: Europe, Asia, North America.

Equisetum hyemale **L. var. *affine* (Engelm.) A.A.Eat.**
Equisetaceae
♦ scouringrush horsetail
♦ Weed
♦ 179

Equisetum laevigatum **A.Braun**
Equisetaceae
♦ Braun's horsetail, Braun's scouringrush, horsetail, rush, smooth horsetail, smooth scouringrush
♦ Weed
♦ 87, 88, 159, 161, 180, 212, 243
♦ pH, cultivated, herbal, toxic.

Equisetum maximum **auct.**
Equisetaceae
= *Equisetum telmateia* Ehrh.
♦ great horsetail
♦ Weed
♦ 44
♦ cultivated, herbal.

Equisetum palustre **L.**
Equisetaceae
♦ marsh horsetail, horsetail
♦ Weed, Quarantine Weed
♦ 39, 44, 70, 76, 87, 88, 161, 186, 203, 218, 220, 243, 263, 272
♦ pH, aqua, cultivated, herbal, toxic.

Equisetum pratense **Ehrh.**
Equisetaceae
♦ meadow horsetail, lehtokorte
♦ Weed
♦ 87, 88, 272
♦ pH, promoted, herbal.

Equisetum ramosissimum **Desf.**
Equisetaceae
♦ branched scouringrush, drill grass, horsetail, mare's tail, scouringrush, Transvaal horsetail
♦ Weed, Quarantine Weed, Noxious Weed, Naturalised
♦ 39, 70, 76, 86, 87, 88, 121, 135, 161, 191, 203, 220, 221, 248, 253, 272, 275, 297
♦ pH, herbal, toxic. Origin: Madagascar.

Equisetum ramosissimum **Desf. ssp. debile** (Roxb.) **Hauke**
Equisetaceae
♦ inudokusa
♦ Weed
♦ 273

Equisetum sylvaticum **L.**
Equisetaceae
♦ sylvan horsetail, wood horsetail
♦ Weed
♦ 70, 87, 88, 161, 218, 243, 272
♦ pH, cultivated, herbal, toxic.

Equisetum telmateia **Ehrh.**
Equisetaceae
Equisetum maximum auct. (see)
♦ great horsetail, giant horsetail
♦ Weed, Noxious Weed
♦ 45, 70, 80, 87, 88, 136, 161, 212, 218, 229, 253, 272
♦ cultivated, herbal, toxic.

Equisetum telmateia **Ehrh. var. braunii** (Milde) **Milde**
Equisetaceae
♦ giant horsetail
♦ Noxious Weed
♦ 229

Eragrostis **Wolf spp.**
Poaceae
♦ lovegrass
♦ Quarantine Weed
♦ 88, 220, 243
♦ G.

Eragrostis abyssinica (Jacq.) **Link**
Poaceae
= *Eragrostis tef* (Zucc.) Trotter
♦ teff
♦ Naturalised
♦ 287
♦ G.

Eragrostis acuminata **Döll**
Poaceae
♦ Weed

♦ 255
♦ G. Origin: tropical America.

Eragrostis aethiopica **Chiov.**
Poaceae
♦ Weed, Quarantine Weed
♦ 76, 87, 88, 203, 220
♦ G, arid.

Eragrostis airoides **Nees**
Poaceae
♦ darnel lovegrass
♦ Weed, Quarantine Weed, Naturalised
♦ 76, 87, 88, 101, 203, 220, 255
♦ pG. Origin: South America.

Eragrostis amabilis (L.) **Wight & Arn. ex Nees**
Poaceae
Cynodon amabilis (L.) Raspail, *Eragrostis amabilis* var. *plumosa* (Retz.) E.G.Camus & A.Camus, *Eragrostis ciliaris* (L.) R.Br. var. *patens* Chapm. ex Beal, *Eragrostis plumosa* (Retz.) Link, *Eragrostis tenella* (L.) P.Beauv. ex Roem. & Schult. (see), *Eragrostis tenella* (L.) P.Beauv. ex Roem. & Schult. var. *plumosa* (Retz.) Stapf, *Erochloe amabilis* Raf. ex B.D.Jacks., *Erochloe spectabilis* (Pursh) Raf. ex B.D.Jacks., *Megastachya amabilis* (L.) P.Beauv., *Poa amabilis* L., *Poa plumosa* Retz., *Poa tenella* L.
♦ feather lovegrass, shohmaleh
♦ Weed, Naturalised, Introduced
♦ 32, 38, 88, 179, 218, 228, 273, 274, 275
♦ aG, arid, herbal. Origin: tropical Africa, Asia, Madagascar.

Eragrostis arenicola **C.E.Hubb.**
Poaceae
♦ Weed, Native Weed
♦ 87, 88, 121
♦ aG. Origin: southern Africa.

Eragrostis aspera (Jacq.) **Nees**
Poaceae
Poa aspera Jacq.
♦ rough lovegrass
♦ Weed, Native Weed
♦ 51, 87, 88, 91, 121, 221
♦ aG, arid. Origin: Africa.

Eragrostis atrovirens (Desf.) **Trin.**
Poaceae
♦ thalia lovegrass, wiry lovegrass
♦ Weed, Quarantine Weed, Naturalised, Introduced
♦ 38, 76, 86, 87, 88, 91, 101, 179, 203, 220
♦ pG. Origin: tropical Africa and Asia.

Eragrostis bahiensis (Schrad. ex J.A.Schult.) **J.A.Schult.**
Poaceae
Eragrostis bahiensis fo. *riparia* Burkart, *Eragrostis bahiensis* var. *contracta* Döll, *Eragrostis bahiensis* var. *laxiuscula* Döll, *Eragrostis blepharophylla* Jedwabn., *Eragrostis elatior* Stapf, *Eragrostis expansa* Link, *Eragrostis firma* Trin., *Eragrostis macra* Jedwabn., *Eragrostis microstachya* (Link) Link, *Eragrostis pilosa* (L.) P.Beauv. var. *bahiensis* (Schrad. ex Schult.) Kuntze, *Eragrostis*

psammodes Trin., *Eragrostis psammodes* var. *microstachya* (Link) Döll, *Poa brasiliensis* Raddi, *Poa expansa* (Link) Kunth, *Poa microstachya* Link, *Poa psammodes* (Trin.) Kunth
♦ Bahia lovegrass
♦ Weed, Naturalised
♦ 86, 87, 88, 98, 101, 179, 203
♦ G. Origin: South America.

Eragrostis barrelieri **Dav.**
Poaceae
♦ Mediterranean lovegrass, pitted lovegrass
♦ Weed, Naturalised, Introduced, Environmental Weed
♦ 86, 87, 88, 91, 93, 98, 101, 179, 203, 205, 218, 228, 253
♦ aG, arid, cultivated, herbal. Origin: Mediterranean.

Eragrostis benthamii **Mattei**
Poaceae
♦ bay grass
♦ Weed, Naturalised
♦ 15, 280
♦ G.

Eragrostis bergiana (Kunth) **Trin.**
Poaceae
♦ kwagga lovergrass
♦ Native Weed
♦ 121
♦ pG. Origin: southern Africa.

Eragrostis bicolor **Nees**
Poaceae
♦ vlei lovegrass
♦ Native Weed
♦ 121
♦ pG. Origin: southern Africa.

Eragrostis biflora **Hack. ex Schinz**
Poaceae
♦ Native Weed
♦ 121
♦ aG. Origin: southern Africa.

Eragrostis blepharostachya **Schum.**
Poaceae
♦ Weed
♦ 87, 88
♦ G.

Eragrostis brownii (Kunth) **Nees ex Wight**
Poaceae
♦ Naturalised
♦ 134, 280
♦ G, cultivated.

Eragrostis capensis (Thunb.) **Trin.**
Poaceae
♦ cape lovegrass, small heartseed grass
♦ Native Weed
♦ 121
♦ pG, cultivated. Origin: southern Africa.

Eragrostis chariis (Schult.) **Hitchc.**
Poaceae
= *Eragrostis nutans* (Retz.) Nees ex Steud. (NoR)
♦ Naturalised
♦ 287
♦ G.

Eragrostis chloromelas Steud.
Poaceae
= *Eragrostis curvula* (Schrad.) Nees
♦ blue lovegrass
♦ Weed, Native Weed
♦ 87, 88, 121
♦ pG, cultivated, herbal. Origin:
southern Africa.

Eragrostis cilianensis (All.) Vig. ex Janch.
Poaceae
Eragrostis major Host, *Eragrostis
megastachya* (Koel.) Link (see),
Eragrostis multiflora (Forssk.) Aschieri
var. *insularis* Chiov., *Eragrostis
multiflora* var. *subbiloba* Chiov.,
Eragrostis multiflora var. *glandulifera*
Chiov., *Eragrostis pappii* Gand.,
Eragrostis polysperma Peter, *Eragrostis
schweinfurthiana* Jedwabn.
♦ stinking eragrostis, stink lovegrass,
stink eragrostis, stinkgrass,
tanakkaröllinurmikka, candy grass,
lovegrass, spreading lovegrass
♦ Weed, Naturalised, Native Weed,
Introduced, Environmental Weed,
Casual Alien, Cultivation Escape
♦ 7, 34, 38, 39, 42, 51, 55, 70, 80, 86, 87,
88, 98, 101, 121, 158, 93, 161, 174, 176,
179, 185, 198, 203, 205, 210, 212, 218,
221, 228, 237, 240, 243, 252, 253, 257,
269, 272, 275, 280, 286, 297, 300
♦ aG, arid, cultivated, herbal, toxic.
Origin: Africa.

Eragrostis ciliaris (L.) R.Br.
Poaceae
Eragrostis pulchella Parl., *Poa ciliaris* L.
♦ mutema lovegrass, woolly lovegrass,
gophertail lovegrass
♦ Weed, Naturalised, Native Weed
♦ 32, 87, 88, 121, 161, 179, 221, 243, 249,
255, 257, 287, 300
♦ aG, arid, herbal. Origin: Africa.

Eragrostis cumingii Steud.
Poaceae
♦ Cuming's lovegrass
♦ Weed, Naturalised
♦ 101, 179, 235
♦ G.

Eragrostis curvula (Schrad.) Nees
Poaceae
Eragrostis chloromelas Steud. (see),
Eragrostis jeffreysii Hack., *Poa curvula*
Schrad.
♦ African lovegrass, Boer lovegrass,
weeping lovegrass, Ermelo lovegrass,
weeping grass, wiregrass
♦ Weed, Sleeper Weed, Quarantine
Weed, Noxious Weed, Naturalised,
Native Weed, Introduced,
Environmental Weed, Cultivation
Escape
♦ 7, 9, 20, 72, 79, 80, 86, 87, 88, 91, 98,
101, 121, 147, 155, 158, 169, 171, 176,
179, 195, 198, 199, 203, 225, 228, 232,
246, 261, 269, 280, 286, 287, 289, 296,
300
♦ pG, arid, cultivated, herbal. Origin:
South Africa.

Eragrostis cylindriflora Hochst.
Poaceae
Eragrostis horizontalis Peter
♦ cylinderflower lovegrass
♦ Weed, Naturalised, Native Weed
♦ 101, 121
♦ a/pG. Origin: southern Africa.

Eragrostis diarrhena (Schult.) Steud.
Poaceae
= *Eragrostis japonica* (Thunb.) Trin.
♦ Weed, Quarantine Weed
♦ 76, 87, 88, 203, 220
♦ G.

Eragrostis diffusa Buckl.
Poaceae
= *Eragrostis pectinacea* (Michx.) Nees ex
Jedwabn.
♦ diffuse lovegrass, spreading
lovegrass, lovegrass
♦ Weed, Casual Alien
♦ 180, 243, 280
♦ G, herbal.

Eragrostis diplachnoides Steud.
Poaceae
= *Eragrostis japonica* (Thunb.) Trin.
♦ Weed
♦ 87, 88, 242
♦ G, arid.

Eragrostis echinochloidea Stapf
Poaceae
♦ African lovegrass
♦ Naturalised
♦ 101
♦ G.

Eragrostis elongata (Willd.) J.F.Jacq.
Poaceae
♦ long lovegrass
♦ Weed, Naturalised
♦ 87, 88, 101
♦ G, arid, cultivated, herbal.

Eragrostis fascicularis Trin.
Poaceae
♦ Weed
♦ 87, 88
♦ G.

Eragrostis ferruginea (Thunb.) Beauv.
Poaceae
♦ kazekusa
♦ Weed, Quarantine Weed
♦ 23, 76, 87, 88, 203, 220, 263, 275, 286,
297
♦ pG.

Eragrostis frankii C.A.Mey. ex Steud.
Poaceae
♦ sandbar lovegrass
♦ Weed
♦ 87, 88, 218
♦ G, herbal.

Eragrostis gangetica (Roxb.) Steud.
Poaceae
Eragrostis cambessediana (Kunth) Steud.,
Eragrostis stenophylla Hochst. ex Miq.,
Poa gangetica Roxb.
♦ slimflower lovegrass
♦ Weed, Naturalised
♦ 87, 88, 101, 179
♦ G, arid.

Eragrostis glomerata (Walter) Dewey
Poaceae
= *Eragrostis japonica* (Thunb.) Trin.
♦ pond lovegrass
♦ Weed, Naturalised
♦ 87, 88, 287
♦ G, arid.

Eragrostis glutinosa (Sw.) Trin.
Poaceae
♦ sticky lovegrass
♦ Naturalised
♦ 101
♦ G.

Eragrostis gummiflua Nees
Poaceae
♦ gum grass, stickystem lovegrass
♦ Native Weed
♦ 121
♦ pG, herbal. Origin: southern Africa.

**Eragrostis hypnoides (Lam.) Britton,
Sterns & Pogg.**
Poaceae
♦ teal lovegrass
♦ Weed
♦ 87, 88
♦ aG, herbal.

Eragrostis infecunda J.M.Black
Poaceae
♦ southern canegrass
♦ Weed, Naturalised
♦ 86, 93, 191, 205
♦ pG, cultivated. Origin: Australia.

Eragrostis intermedia Hitchc.
Poaceae
Eragrostis lugens non Nees
♦ plains lovegrass
♦ Weed, Naturalised
♦ 199, 287
♦ G, arid, herbal.

Eragrostis interrupta (Lam.) Döll
Poaceae
= *Eragrostis japonica* (Thunb.) Trin.
♦ Weed
♦ 87, 88
♦ G.

Eragrostis japonica (Thunb.) Trin.
Poaceae
Catabrosa micrantha Hochst. ex A.Rich.,
Diandrochloa diplachnoides (Steud.)
A.N.Henry, *Diandrochloa glomerata*
(Walt.) Burk. (see), *Diandrochloa
japonica* (Thunb.) A.N.Henry,
Diandrochloa namaquensis (Nees)
De Winter (see), *Eragrostis aturensis*
(Kunth) Trin. ex Steud., *Eragrostis
aurea* Steud., *Eragrostis brasiliana*
Nees, *Eragrostis conferta* (Elliott) Trin.,
Eragrostis diplachnoides Steud. (see),
Eragrostis diarrhena (Schult.) Steud.
(see), *Eragrostis elegans* Nees, *Eragrostis
elegans* var. *laxiflora* Arechav., *Eragrostis
glomerata* (Walt.) Dewey (see),
Eragrostis hapalantha Trin., *Eragrostis
interrupta* (Lam.) Döll (see), *Eragrostis
interrupta* var. *tenuissima* (Schrad. ex
Nees) Stapf, *Eragrostis koenigii* (Kunth)
Steud., *Eragrostis namaquensis* Nees ex
Schrad. (see), *Eragrostis namaquensis*

var. *diplachnoides* (Steud.) Clayton,
Eragrostis pallida Vasey, *Eragrostis
tenellula* (Kunth) Steud., *Eragrostis
tenuissima* Schrad. ex Nees, *Glyceria
micrantha* Steud., *Megastachya aturensis*
(Kunth) Roem. & Schult., *Megastachya
brasiliensis* Schult., *Megastachya
glomerata* (Walter) Schult., *Poa aturensis*
Kunth, *Poa glomerata* Walter, *Poa
interrupta* Lam., *Poa japonica* Thunb.,
Poa sporoboloides A.Rich., *Poa tenella*
R.Br., *Poa tenellula* Kunth, *Roshevitzia
diplachnoides* (Steud.) Tzvelev,
Roshevitzia glomerata (Walter) Tzvelev,
Roshevitzia japonica (Thunb.) Tzvelev,
Sporobolus confertiflorus A.Rich.,
Sporobolus verticillatus Nees, *Vilfa
confertiflora* (A.Rich.) Steud.
♦ pond lovegrass
♦ Weed, Naturalised, Introduced
♦ 32, 87, 88, 228, 274, 275, 286
♦ aG, arid. Origin: Madagascar.

Eragrostis lacunaria F.Muell. ex Benth.
Poaceae
♦ Introduced
♦ 228
♦ G, arid, cultivated. Origin: Australia.

Eragrostis lehmanniana Nees
Poaceae
Eragrostis chaunantha Pilg.
♦ Lehmann lovegrass, Atherstone
lovegrass, Cochise, eastern province
vlei grass, landgrass
♦ Weed, Naturalised, Native Weed,
Introduced
♦ 80, 88, 101, 121, 228, 228
♦ pG, arid, cultivated, herbal. Origin:
southern Africa.

Eragrostis leptocarpa Benth.
Poaceae
♦ Introduced
♦ 228
♦ G, arid.

Eragrostis leptostachya (R.Br.) Steud.
Poaceae
Poa leptostachya R.Br.
♦ Australian lovegrass
♦ Weed, Naturalised, Introduced,
Casual Alien
♦ 86, 87, 88, 101, 228, 280
♦ G, arid, cultivated. Origin: Australia.

Eragrostis lugens Nees
Poaceae
♦ mourning lovegrass, lovegrass,
pasto pelillo
♦ Weed, Quarantine Weed
♦ 76, 87, 88, 203, 220, 236, 237, 295
♦ G, herbal.

Eragrostis malayana Stapf
Poaceae
♦ Weed, Quarantine Weed
♦ 76, 87, 88, 203, 220
♦ G.

Eragrostis megastachya (Koel.) Link
Poaceae
= *Eragrostis cilianensis* (All.) Vig. ex
Janch.
♦ stinkgrass, snake grass, candy grass,

spreading lovegrass, grey lovegrass
♦ Weed
♦ 30, 49, 87, 88, 91, 242, 295
♦ aG, arid, herbal.

Eragrostis mexicana (Hornem.) Link
Poaceae
♦ Mexican lovegrass, lovegrass,
Orcutt's virescens lovegrass
♦ Weed, Naturalised, Casual Alien
♦ 86, 87, 88, 91, 93, 98, 157, 161, 191,
198, 203, 205, 218, 243, 280
♦ aG, arid, cultivated, herbal. Origin:
North and South America.

**Eragrostis mexicana (Hornem.) Link ssp.
mexicana**
Poaceae
Eragrostis neomexicana Vasey ex
L.H.Dewey (see)
♦ Mexican lovegrass, New Mexico
lovegrass
♦ Weed, Naturalised
♦ 34, 198
♦ aG.

**Eragrostis mexicana (Hornem.) Link ssp.
virescens (J.Presl) S.D.Koch & Sánchez**
Poaceae
Eragrostis orcuttiana Vasey, *Eragrostis
virescens* Presl (see)
♦ Mexican lovegrass, Orcutt lovegrass
♦ Weed, Naturalised
♦ 180, 198, 243
♦ aG.

Eragrostis minor Host
Poaceae
Eragrostis poaeoides P.Beauv. *nom. illeg.*
(see), *Poa eragrostis* L.
♦ smaller stinkgrass, little lovegrass
♦ Weed, Naturalised, Introduced,
Casual Alien
♦ 7, 34, 42, 79, 86, 88, 91, 93, 98, 101,
161, 198, 203, 253, 272, 280, 300
♦ aG, cultivated, herbal. Origin:
Eurasia.

Eragrostis montana Balansa
Poaceae
♦ Weed
♦ 87, 88
♦ G.

Eragrostis multicaulis Steud.
Poaceae
= *Eragrostis pilosa* (L.) Beauv.
♦ niwahokori
♦ Weed
♦ 87, 88, 235, 263, 286
♦ aG.

Eragrostis multispicula Kit.
Poaceae
= *Eragrostis pilosa* (L.) Beauv.
♦ Weed
♦ 286
♦ G.

Eragrostis namaquensis Nees ex Schrad.
Poaceae
= *Eragrostis japonica* (Thunb.) Trin.
♦ Weed
♦ 87, 88
♦ G.

**Eragrostis neomexicana Vasey ex
L.H.Dewey**
Poaceae
= *Eragrostis mexicana* (Hornem.) Link
ssp. *mexicana*
♦ New Mexican lovegrass
♦ Weed, Naturalised
♦ 87, 88, 98, 203, 218, 237, 295
♦ G, herbal.

Eragrostis nigra Nees ex Steud.
Poaceae
♦ Weed
♦ 80, 87, 88, 244, 275
♦ pG.

Eragrostis niwahokori Honda
Poaceae
♦ Weed
♦ 87, 88, 204
♦ G.

Eragrostis pallens Hack.
Poaceae
♦ Weed
♦ 88
♦ G, arid.

Eragrostis paniciformis (A.Braun) Steud.
Poaceae
♦ Naturalised
♦ 86
♦ G. Origin: tropical Africa.

**Eragrostis papposa (Roem. & Schult.)
Steud.**
Poaceae
Eragrostis aulacosperma Steud. var.
perennis Schweinf., *Megastachya papposa*
Roem. & Schult.
♦ Weed
♦ 88
♦ G, arid.

Eragrostis parviflora (R.Br.) Trin.
Poaceae
♦ smallflower lovegrass
♦ Naturalised
♦ 101
♦ G, cultivated.

Eragrostis patens Oliv.
Poaceae
♦ annual lovegrass, spreading annual
lovegrass
♦ Native Weed
♦ 121
♦ aG. Origin: southern Africa.

**Eragrostis pectinacea (Michx.) Nees ex
Jedwabn.**
Poaceae
Eragrostis brizoides Schult., *Eragrostis
caroliniana* (Biehler) Scribn., *Eragrostis
cognata* Steud., *Eragrostis delicatula*
Trin., *Eragrostis diffusa* Buckl. (see),
Eragrostis diffusa Buckl. var. *diffusa,
Eragrostis nuttalliana* Steud., *Eragrostis
pennsylvanica* Scheele, *Eragrostis pilosa*
(L.) P.Beauv. var. *caroliniana* (Biehler)
Farw., *Eragrostis purshii* A.Gray,
Eragrostis purshii Schrad., *Eragrostis
purshii* var. *delicatula* Munro ex
Scribn., *Eragrostis purshii* var. *diffusa*
(Buckley) Vasey, *Eragrostis purshii* var.
pauciflora E.Fourn., *Eragrostis spectabilis*

(Pursh) Steud. var. *sparsihirsuta* Farw.,
Eragrostis tracyi Hitchc., *Eragrostis
unionis* Steud., *Poa diandra* Schrad., *Poa
eragrostis* Elliott, *Poa nuttallii* Kunth,
Poa pectinacea Michx., *Poa tenella* Nutt.
♦ Carolina lovegrass, pectinate
lovegrass, tufted lovegrass
♦ Weed, Naturalised
♦ 87, 88, 91, 98, 161, 199, 203, 218, 243
♦ aG, arid, herbal.

Eragrostis pectinacea (Michx.) Nees var. *pectinacea*
Poaceae
♦ purple lovegrass, tufted lovegrass
♦ Weed
♦ 34
♦ aG.

Eragrostis pilosa (L.) Beauv.
Poaceae
Eragrostis multicaulis Steud. (see),
Eragrostis multispicula Kit. (see),
Eragrostis tenuiflora Steud., *Poa pilosa* L.
♦ India lovegrass, soft lovegrass,
Indian lovegrass, pilose
eragrostis, small tufted lovegrass,
speargrass, slender meadowgrass,
hoikkaröllinurmikka
♦ Weed, Naturalised, Introduced,
Environmental Weed, Casual Alien
♦ 32, 42, 86, 87, 88, 91, 93, 98, 121, 179,
191, 198, 203, 217, 218, 221, 228, 230,
245, 253, 255, 257, 272, 275, 280, 287,
300
♦ aG, aqua, cultivated, herbal. Origin:
Eurasia.

Eragrostis pilosa (L.) Beauv. var. *imberbis* Franch
Poaceae
♦ Weed
♦ 275
♦ aG.

Eragrostis plana Nees
Poaceae
♦ fan lovegrass, ox grass, tough
lovegrass
♦ Weed, Quarantine Weed,
Naturalised, Native Weed
♦ 88, 101, 121, 158, 172, 255, 280
♦ pG, herbal. Origin: southern Africa.

Eragrostis planiculmis Nees
Poaceae
♦ broom lovegrass
♦ Native Weed
♦ 121
♦ pG, herbal. Origin: southern Africa.

Eragrostis poaeoides P.Beauv. *nom. illeg.*
Poaceae
= *Eragrostis minor* Host
♦ little lovegrass, low eragrostis
♦ Weed, Naturalised
♦ 87, 88, 180, 218, 235, 243, 248, 275,
286, 287, 297
♦ aG, herbal.

Eragrostis prolifera (Sw.) Steud.
Poaceae
♦ Dominican lovegrass
♦ Weed
♦ 179

♦ G.

Eragrostis pseudosclerantha Chiov.
Poaceae
♦ footpath lovegrass
♦ Weed, Native Weed
♦ 88, 121, 158
♦ pG. Origin: southern Africa.

Eragrostis pusilla Hack.
Poaceae
♦ Weed
♦ 87, 88
♦ G.

Eragrostis racemosa (Thunb.) Steud.
Poaceae
Eragrostis chalcantha Trin.
♦ narrow heart lovegrass
♦ Weed, Native Weed
♦ 88, 121, 158
♦ pG. Origin: southern Africa.

Eragrostis rigidior Pilg.
Poaceae
♦ curly leaved lovegrass
♦ Native Weed
♦ 121
♦ pG. Origin: southern Africa.

Eragrostis scaligera Salzm. ex Steud.
Poaceae
♦ tender lovegrass
♦ Weed, Naturalised
♦ 101, 179
♦ G.

Eragrostis schimperi Benth.
Poaceae
Harpachne schimperi Hochst.
♦ Weed
♦ 87, 88
♦ G.

Eragrostis setifolia Nees
Poaceae
Eragrostis chaetophylla Steud., *Poa
diandra* F.Muell. *non* R.Br.
♦ bristleleaf lovegrass, bristly
lovegrass
♦ Naturalised
♦ 101
♦ G, arid, cultivated.

Eragrostis silveana Swallen
Poaceae
♦ Silveus' lovegrass
♦ Naturalised
♦ 287
♦ G.

Eragrostis spectabilis (Pursh) Steud.
Poaceae
♦ purple lovegrass
♦ Native Weed
♦ 161, 174
♦ G, cultivated, herbal. Origin: North
America.

Eragrostis suaveolens Becker ex Claus
Poaceae
♦ candy lovegrass
♦ Naturalised
♦ 101
♦ G.

Eragrostis subsecunda (Lam.) Fourn.
Poaceae

♦ Naturalised
♦ 86
♦ G. Origin: China.

Eragrostis superba Peyr.
Poaceae
Eragrostis platystachys Franch.
♦ Wilman lovegrass, Massai lovegrass,
flatseed lovegrass, heartseed grass,
heartseed lovegrass, lovegrass,
sawtooth love grass, weeluiseragrostis,
herzstraubgras, pasto avena
♦ Weed, Naturalised, Native Weed
♦ 86, 87, 88, 91, 98, 101, 121, 158, 198,
203, 287
♦ pG, arid, cultivated. Origin: Africa.

Eragrostis tef (Zucc.) Trotter
Poaceae
Eragrostis abyssinica (Jacq.) Link (see),
Eragrostis pilosa (L.) P.Beauv. var. *tef*
Zucc., *Eragrostis pilosa* ssp. *abyssinica*
Jacq., *Eragrostis abyssinica* (Jacq.),
Cynodon abyssinicus (Jacq.), *Poa cerealis*
Salisb., *Poa abyssinica* Jacq., *Poa tef*
Zucc.
♦ tef, tafi, taf, teff, Williams lovegrass,
chimanganga, ndzungulu
♦ Weed, Naturalised, Introduced
♦ 86, 98, 101, 121, 198, 203, 228
♦ aG, arid, cultivated, herbal. Origin:
Africa.

Eragrostis tenella (L.) Beauv. ex Roem. & Schult.
Poaceae
= *Eragrostis amabilis* (L.) Wight & Arn.
ex Nees
♦ bug's egg grass, feather lovegrass,
Japanese lovegrass, feathery lovegrass,
lovegrass, feathery eragrostis
♦ Weed, Naturalised, Native Weed
♦ 6, 86, 87, 88, 91, 93, 101, 121, 170, 204,
209, 261, 262, 286, 287, 297
♦ aG, cultivated, herbal. Origin:
tropical Africa, Asia.

Eragrostis tenuifolia (A.Rich.) Steud.
Poaceae
♦ elastic grass
♦ Weed, Quarantine Weed,
Naturalised, Introduced
♦ 38, 53, 76, 86, 87, 88, 91, 93, 98, 101,
176, 199, 203, 205, 228, 276
♦ pG, arid. Origin: Madagascar.

Eragrostis tremula (Lam.) Steud.
Poaceae
Poa tremula Lam.
♦ Weed, Quarantine Weed, Introduced
♦ 76, 87, 88, 91, 203, 220, 228
♦ G, arid, cultivated, herbal.

Eragrostis trichophora Coss. & Dur.
Poaceae
Eragrostis atherstonei Stapf
♦ Atherstone's grass, lovegrass,
hairyflower lovegrass, blousaadgras,
Cochise
♦ Weed,
Naturalised, Native Weed
♦ 88, 121, 158
♦ pG, arid. Origin: Africa.

Eragrostis turgina (Schum.) De Wild.
Poaceae
♦ Weed
♦ 87, 88
♦ G.

Eragrostis unioloides (Retz.) Nees ex
Steud.
Poaceae
Eragrostis rubens Hochst., *Poa unioloides*
Retz.
♦ Chinese lovegrass
♦ Weed, Naturalised
♦ 32, 87, 88, 91, 101, 170, 179, 191
♦ a/pG. Origin: tropical Africa, Asia,
Australia.

Eragrostis virescens Presl
Poaceae
= *Eragrostis mexicana* (Hornem.) Link
ssp. *virescens* (J.Presl) S.D.Koch &
Sánchez
♦ Chilean lovegrass, Orcutt's
lovegrass, gramilla de huerta
♦ Weed, Naturalised
♦ 51, 87, 88, 91, 98, 121, 203, 236, 237,
295
♦ aG. Origin: South America.

Eragrostis viscosa (Retz.) Trin.
Poaceae
Poa viscosa Retz.
♦ sticky lovegrass, viscid lovegrass
♦ Weed, Quarantine Weed, Native
Weed, Introduced
♦ 76, 87, 88, 121, 203, 220, 228
♦ aG, arid. Origin: southern Africa.

Eragrostis xerophila Domin
Poaceae
♦ Introduced
♦ 228
♦ G, arid, cultivated.

Eragrostis xylanica Hack.
Poaceae
♦ Weed, Quarantine Weed
♦ 76, 87, 88, 203, 220
♦ G.

Eranthemum pulchellum Andrews
Acanthaceae
♦ blue sage
♦ Naturalised
♦ 101
♦ cultivated.

Eranthemum viscidum Bl.
Acanthaceae
♦ Weed
♦ 13
♦ herbal.

Eranthis hyemalis (L.) Salisb.
Ranunculaceae
♦ winter aconite
♦ Naturalised, Casual Alien
♦ 39, 101, 280
♦ cultivated, herbal, toxic.

Erdisia Britt. & Rose spp.
Cactaceae
= *Corryocactus* spp.
♦ Weed, Quarantine Weed
♦ 76, 88, 203

Erechtites arguta (A.Rich.) DC.
Asteraceae
= *Erechtites glomerata* (Desf. ex Poir.)
DC.
♦ Australian fireweed
♦ Weed
♦ 87, 88
♦ herbal.

Erechtites atkinsoniae F.Muell.
Asteraceae
♦ Weed
♦ 87, 88

Erechtites glomerata (Poir.) DC.
Asteraceae
Erechtites arguta (A.Rich.) DC. (see)
♦ Australian fireweed, New Zealand
fireweed, cutleaf burnweed
♦ Weed, Noxious Weed, Naturalised,
Environmental Weed
♦ 35, 78, 80, 88, 101, 116, 151, 161, 231
♦ a/pH, herbal.

Erechtites hieraciifolia (L.) Raf. ex DC.
Asteraceae
Senecio hieraciifolius L. (see)
♦ American burnweed, pilewort,
fireweed, burnweed, capiçoba,
capeçoba
♦ Weed, Naturalised, Introduced,
Casual Alien
♦ 13, 75, 87, 88, 157, 161, 218, 249, 255,
257, 263, 280, 286, 287, 295
♦ aH, cultivated, herbal. Origin: Brazil.

Erechtites hieraciifolia (L.) Raf. ex DC.
var. *cacalioides* (Fisch. ex Spreng.)
Griseb.
Asteraceae
♦ achicoria de cabra, American
burnweed
♦ Weed, Naturalised
♦ 237, 261, 286, 287

Erechtites hieraciifolia (L.) Raf. ex DC.
var. *hieraciifolia*
Asteraceae
♦ achicoria de cabra, American
burnweed
♦ Weed
♦ 261
♦ a/pH.

Erechtites minima (Poir.) DC.
Asteraceae
Erechtites prenanthoides (A.Rich) DC.
(see), *Senecio minimus* Poir. (see)
♦ Australian fireweed, coastal
burnweed, little fireweed, toothed
coast burnweed
♦ Weed, Noxious Weed, Naturalised,
Environmental Weed
♦ 34, 35, 78, 80, 87, 88, 101, 116, 146,
231
♦ a/pH, herbal.

Erechtites prenanthoides (A.Rich) DC.
Asteraceae
= *Erechtites minima* (Poir.) DC.
♦ Australian burnweed
♦ Weed
♦ 87, 88, 218
♦ herbal.

Erechtites quadridentata DC.
Asteraceae
♦ Weed
♦ 87, 88

Erechtites scaberula Hook.f.
Asteraceae
♦ Weed
♦ 87, 88

Erechtites valerianifolia (Wolf) DC.
Asteraceae
Senecio valerianaefolius Wolf.
♦ Brazilian fireweed, Commonwealth
weed, tropical burnweed, capiçoba
♦ Weed, Naturalised
♦ 13, 28, 86, 87, 88, 98, 101, 134, 165,
203, 206, 243, 255, 269, 274, 276, 280,
287
♦ aH, arid, herbal. Origin: Central and
South America.

Eremobium aegyptiacum (Spreng.) Boiss.
Brassicaceae
Malcolmia aegyptiaca Spreng.
♦ Weed
♦ 221
♦ arid.

Eremocarpus setigerus (Hook.) Benth.
Euphorbiaceae
= *Croton setigerus* Hook. (NoR)
♦ turkey mullein, doveweed,
grayweed, woolly white drought
weed, yerba del pescado
♦ Weed, Quarantine Weed, Noxious
Weed, Naturalised
♦ 7, 39, 76, 86, 87, 88, 98, 121, 136, 147,
161, 180, 198, 203, 212, 218, 241, 243,
300
♦ a/bH, cultivated, herbal, toxic.
Origin: western North America.

Eremochloa ciliaris (L.) Merr.
Poaceae
♦ fringed centipede grass
♦ Weed, Naturalised
♦ 87, 88, 101
♦ G. Origin: Australia.

Eremochloa ophiuroides (Munro) Hack.
Poaceae
Ischaemum ophiuroides Munro
♦ centipede grass, cienpiés
♦ Weed, Naturalised, Cultivation
Escape
♦ 23, 80, 88, 101, 179, 235, 243, 261, 287
♦ pG, cultivated, herbal. Origin: south-
east Asia.

Eremocitrus glauca Swingle
Rutaceae
= *Citrus glauca* (Lindl.) Burkill (NoR)
♦ Australian desert lime
♦ Weed, Introduced
♦ 87, 88, 228
♦ arid, cultivated. Origin: Australia.

Eremodaucus lehmanni Bunge
Apiaceae
♦ Weed
♦ 87, 88

Eremophila debilis (Andrews) Chinnock
Myoporaceae

♦ amulla
♦ Naturalised
♦ 280
♦ cultivated.

Eremophila maculata F.Muell.
Myoporaceae
♦ native fuchsia
♦ Weed, Native Weed
♦ 87, 88, 269
♦ S, cultivated. Origin: Australia.

Eremophila mitchellii Benth.
Myoporaceae
♦ budda, sandalwood
♦ Weed, Native Weed
♦ 39, 87, 88, 177, 269
♦ cultivated, toxic. Origin: Australia.

Eremopogon foveolatus (Delile) Stapf
Poaceae
♦ Weed
♦ 221
♦ G.

Eremopyrum bonaepartis (Spreng.)
Nevski
Poaceae
♦ tapertip false wheatgrass
♦ Naturalised
♦ 101, 243
♦ G.

Eremopyrum orientale (L.) Jaub. & Spach
Poaceae
♦ oriental false wheatgrass, idänvaivaisvehnä
♦ Naturalised, Casual Alien
♦ 42, 101
♦ G.

Eremopyrum triticeum (Gaertn.) Nevski
Poaceae
Agropyron triticeum Gaertn. (see)
♦ annual wheatgrass, arovaivaisvehnä
♦ Weed, Naturalised, Casual Alien
♦ 42, 98, 101, 203, 275
♦ aG.

Eremostachys loasaefolia Benth.
Lamiaceae
♦ Weed
♦ 243

Eremurus persicus Boiss.
Liliaceae/Asphodelaceae
♦ Weed
♦ 243

Erepsia heteropetala (Haw.) Schwantes
Aizoaceae/Mesembryanthemaceae
Mesembryanthemum heteropetalum Haw.
♦ lesser sea fig
♦ Naturalised
♦ 40
♦ cultivated. Origin: South Africa.

Erianthemum dregei (Eckl. & Zeyh.)
V.Tiegh.
Loranthaceae
Loranthus dregei Eckl. & Zeyh.
♦ mistletoe
♦ Native Weed
♦ 121
♦ pH parasitic. Origin: southern Africa.

Erianthus angustifolius Nees
Poaceae
Erianthus saccharoides Willd. var. *neesii* Hack.
♦ Weed
♦ 255
♦ pG. Origin: South America.

Erianthus arundinaceus (Retz.) Jeswiet
Poaceae
♦ Weed
♦ 12, 209
♦ G, herbal.

Erianthus formosanus Stapf var. *pollinioides* Ohwi
Poaceae
♦ Naturalised
♦ 287
♦ G.

Erianthus rufipilus (Steud.) Griseb.
Poaceae
♦ redhair plumegrass
♦ Weed
♦ 297
♦ G.

Erica L. spp.
Ericaceae
♦ heath
♦ Weed, Naturalised
♦ 198, 272

Erica andromedaeflora Andrews
Ericaceae
♦ Naturalised
♦ 86, 176
♦ cultivated. Origin: South Africa.

Erica arborea L.
Ericaceae
♦ tree heath, briar root
♦ Weed, Naturalised, Garden Escape, Environmental Weed
♦ 72, 86, 87, 88, 98, 155, 176, 198, 203, 251, 280, 290
♦ S, cultivated, herbal. Origin: Mediterranean, east Africa, Middle East, Canary and Madeira Islands.

Erica australis L.
Ericaceae
♦ Spanish heath
♦ Weed
♦ 70
♦ cultivated, herbal.

Erica baccans L.
Ericaceae
♦ berry flower heath, berry heath
♦ Weed, Naturalised, Garden Escape, Environmental Weed
♦ 7, 72, 86, 88, 98, 155, 176, 198, 203, 280, 289
♦ S, cultivated. Origin: South Africa.

Erica caffra L.
Ericaceae
♦ scented heath
♦ Weed, Naturalised
♦ 86, 98, 176, 203, 280
♦ cultivated.
Origin: South Africa.

Erica ciliaris L.
Ericaceae
♦ Dorset heath, ciliate heath
♦ Weed
♦ 70
♦ cultivated.

Erica cinerea (Sincl.) Zabel
Ericaceae
♦ cross leaved heath, bog heather, Scotch heath, purple heather, bell heather
♦ Weed, Quarantine Weed, Naturalised, Environmental Weed
♦ 80, 101, 246, 280
♦ cultivated, herbal.

Erica herbacea L.
Ericaceae
= *Erica carnea* L. (NoR)
♦ spring heath
♦ Weed
♦ 272
♦ cultivated.

Erica lusitanica Rudolphi
Ericaceae
♦ Spanish heath, Portuguese heath, heath
♦ Weed, Noxious Weed, Quarantine Weed, Naturalised, Garden Escape, Environmental Weed
♦ 15, 35, 40, 72, 76, 78, 86, 88, 98, 101, 116, 152, 155, 165, 176, 198, 203, 225, 246, 280, 289, 296
♦ pS, cultivated, herbal. Origin: south-west Europe.

Erica melanthera L.
Ericaceae
♦ heath
♦ Weed, Naturalised, Garden Escape, Environmental Weed, Casual Alien
♦ 54, 72, 86, 88, 155, 280
♦ S, cultivated. Origin: South Africa.

Erica quadrangularis Salisb.
Ericaceae
♦ angled heath, erica
♦ Weed, Naturalised, Garden Escape, Environmental Weed
♦ 54, 72, 86, 88, 155, 198
♦ S, cultivated. Origin: South Africa.

Erica scoparia L.
Ericaceae
♦ Weed, Naturalised, Environmental Weed
♦ 54, 86, 88, 98, 176, 203
♦ cultivated. Origin: Mediterranean, Canary Islands.

Erica tetralix L.
Ericaceae
♦ cross leaved heath, heather
♦ Weed, Naturalised
♦ 70, 80, 101, 272
♦ S, cultivated, herbal.

Erica vagans L.
Ericaceae
♦ Cornish heath
♦ Weed, Naturalised
♦ 80, 101, 280
♦ S, cultivated, herbal.

Erica × *willmorei* **Knowles & Westc.**
Ericaceae
♦ Naturalised
♦ 86, 176

Ericameria discoidea **(Nutt.) Nesom**
Asteraceae
♦ whitestem goldenbush, whitestem heathgoldenrod, macronema goldenbush
♦ Quarantine Weed
♦ 220
♦ S.

Ericameria laricifolia **(Gray) Shinners**
Asteraceae
Haplopappus laricifolius A.Gray (see)
♦ turpentine bush
♦ Quarantine Weed
♦ 220
♦ S, herbal.

Ericameria monactis **McClatchie**
Asteraceae
= *Aplopappus monactis* A.Gray (NoR)
♦ Quarantine Weed
♦ 220

Ericameria nauseosa **(Pall. ex Pursh) Nesom & Baird**
Asteraceae
Chrysothamnus nauseosa (Pall. ex Pursh) Britt.
♦ gray rabbitbrush, rubber rabbitbrush
♦ Weed
♦ 88, 213

Ericameria nelsonii **Blake**
Asteraceae
Bigelowia nelsonii Fernald (see)
♦ Quarantine Weed
♦ 220

Erigeron **L. spp.**
Asteraceae
♦ fleabane, przymiotno, buva voadeira
♦ Weed
♦ 23, 39, 88, 245, 272
♦ herbal, toxic.

Erigeron acer **L.**
Asteraceae
♦ blue fleabane
♦ Weed
♦ 70, 88, 94, 272
♦ b/pH, cultivated, herbal.

Erigeron acris **L.**
Asteraceae
♦ bitter fleabane
♦ Weed
♦ 87, 88, 243
♦ aH, cultivated, herbal.

Erigeron alpinus **L.**
Asteraceae
Stenactis alpinus Cassin., *Tessenia alpina* Bub.
♦ alpine fleabane, turica alpínska
♦ Weed
♦ 272
♦ cultivated, herbal.

Erigeron annuus **(L.) Pers.**
Asteraceae
♦ annual fleabane
♦ Weed, Quarantine Weed,

Naturalised, Cultivation Escape
♦ 42, 70, 76, 87, 88, 94, 106, 161, 203, 204, 207, 218, 220, 243, 253, 256, 263, 272, 275, 280, 297
♦ a/pH, cultivated, herbal.

Erigeron asteroides **Roxb.**
Asteraceae
♦ Weed
♦ 87, 88

Erigeron bonariensis **L.**
Asteraceae
♦ flax leaved fleabane, horseweed, buva
♦ Weed
♦ 23, 50, 51, 88, 235, 245, 274
♦ herbal.

Erigeron canadensis **L.**
Asteraceae
= *Conyza canadensis* (L.) Cronquist
♦ horseweed, horseweed fleabane
♦ Weed, Noxious Weed, Naturalised
♦ 23, 39, 44, 51, 52, 88, 204, 218, 235, 263, 274, 286, 287, 299
♦ aH, herbal, toxic.

Erigeron canadensis **L. var.** *glabratus* **Gray**
Asteraceae
= *Conyza canadensis* (L.) Cronquist var. *glabrata* (Gray) Cronquist (NoR)
♦ Weed, Naturalised
♦ 286, 287

Erigeron delphinifolius **Willd.**
Asteraceae
♦ Weed
♦ 199

Erigeron divaricatus **Michx.**
Asteraceae
= *Conyza ramosissima* Cronquist
♦ dwarf fleabane
♦ Weed
♦ 87, 88, 218

Erigeron floribundus **(H.B.K.) Sch.Bip.**
Asteraceae
= *Conyza bonariensis* (L.) Cronquist var. *leiotheca* (Blake) Cuatrec.
♦ tall fleabane, fleabane, lukaja
♦ Weed
♦ 50, 51, 87, 88

Erigeron galeottii **(A.Gray) Greene**
Asteraceae
♦ Weed
♦ 199

Erigeron glaucus **Ker Gawl.**
Asteraceae
♦ beach aster, seaside daisy, seaside fleabane
♦ Naturalised
♦ 40
♦ pH, cultivated, herbal.

Erigeron × *hybridus* **Bergmans**
Asteraceae
= *Erigeron atticus* Vill. × *Erigeron aurantiacus* Reg.
♦ jalokallioinen
♦ Cultivation Escape
♦ 42
♦ cultivated.

Erigeron jamaicense **L.**
Asteraceae
♦ Weed
♦ 14

Erigeron karvinskianus **DC.**
Asteraceae
♦ Mexican daisy, seaside daisy, Latin American fleabane, Santa Barbara daisy, fleabane
♦ Weed, Quarantine Weed, Naturalised, Introduced, Garden Escape, Environmental Weed, Cultivation Escape
♦ 3, 7, 15, 38, 40, 70, 86, 87, 88, 98, 101, 134, 155, 165, 176, 181, 198, 203, 225, 233, 241, 246, 280, 286, 287, 289, 300
♦ pH, cultivated, herbal. Origin: Mexico.

Erigeron linifolius **Willd.**
Asteraceae
= *Conyza bonariensis* (L.) Cronquist
♦ Weed
♦ 87, 88, 204

Erigeron longipes **DC.**
Asteraceae
Erigeron scaposus DC.
♦ Weed
♦ 157, 199
♦ pH.

Erigeron mucronatus **DC.**
Asteraceae
♦ fleabane
♦ Naturalised
♦ 287
♦ cultivated, herbal.

Erigeron philadelphicus **L.**
Asteraceae
♦ Philadelphia fleabane, daisy fleabane
♦ Weed, Naturalised, Casual Alien
♦ 24, 40, 52, 68, 87, 88, 161, 218, 243, 263, 280, 286, 287
♦ pH, promoted, herbal, toxic.

Erigeron philadelphicus **L. fo.** *angustatus* **Vict. & Rousseau**
Asteraceae
♦ fleabane
♦ Naturalised
♦ 287

Erigeron philadelphicus **L. fo.** *purpureus* **Farw.**
Asteraceae
♦ fleabane
♦ Naturalised
♦ 287

Erigeron philadelphicus **L. fo.** *scaturicola* **Cronquist**
Asteraceae
♦ fleabane
♦ Naturalised
♦ 287

Erigeron philadelphicus **L. var.** *glaber* **Henry**
Asteraceae
♦ fleabane
♦ Naturalised
♦ 287

Erigeron pseudo-annuus **Makino**
Asteraceae
♦ Naturalised
♦ 287

Erigeron pulchellus **Michx.**
Asteraceae
♦ Robin's plantain, poor Robin's plantain
♦ Weed
♦ 87, 88, 218
♦ cultivated, herbal.

Erigeron pusillus **Nutt.**
Asteraceae
= *Conyza canadensis* (L.) Cronquist var. *pusilla* (Nutt.) Cronquist
♦ Weed, Naturalised
♦ 286, 287
♦ herbal.

Erigeron quercifolius **Lam.**
Asteraceae
♦ southern fleabane. oakleaf fleabane
♦ Weed
♦ 161, 249
♦ herbal.

Erigeron speciosus **(Lindl.) DC.**
Asteraceae
♦ Oregon fleabane, Aspen fleabane, turica ozdobná, showy fleabane
♦ Weed
♦ 87, 88, 218
♦ cultivated, herbal.

Erigeron spiculosus **Hook. & Arn.**
Asteraceae
♦ Weed
♦ 87, 88

Erigeron strigosus **Muhl. ex Willd.**
Asteraceae
♦ prairie fleabane, rough fleabane, daisy fleabane
♦ Weed, Native Weed, Introduced
♦ 24, 34, 52, 87, 88, 161, 174, 210, 218, 249
♦ a/pH, herbal. Origin: North America.

Erigeron strigosus **Muhl. fo.** *discoideus* **Fernald**
Asteraceae
♦ fleabane
♦ Naturalised
♦ 287

Erigeron sumatrensis **Retz.**
Asteraceae
= *Conyza albida* Willd. ex Spreng.
♦ fleabane
♦ Weed
♦ 13, 68, 87, 88, 170, 191, 204, 243, 276
♦ pH, herbal.

Erigeron tenuis **Torr. & Gray**
Asteraceae
♦ fleabane, slenderleaf fleabane
♦ Weed
♦ 161
♦ herbal.

Erinna gilliesioides **R.A.Phil.**
Liliaceae/Alliaceae
♦ Quarantine Weed
♦ 220

Erinus alpinus **L.**
Scrophulariaceae
♦ fairy foxglove, alpine balsam
♦ Naturalised
♦ 40
♦ cultivated, herbal.

Eriobotrya japonica **(Thunb.) Lindl.**
Rosaceae
Mespilus japonicus Thunb., *Photinia japonica* (Thunb.) A.Gray
♦ loquat, Japanese plum, Japanese medlar
♦ Weed, Noxious Weed, Naturalised, Introduced, Garden Escape, Environmental Weed, Cultivation Escape
♦ 3, 7, 15, 22, 39, 73, 86, 88, 98, 101, 107, 121, 134, 155, 161, 179, 198, 203, 225, 228, 230, 233, 246, 247, 261, 280, 283
♦ T, arid, cultivated, herbal, toxic. Origin: Japan, China.

Eriocaulon benthamii **Kunth**
Eriocaulaceae
♦ Bentham's pipewort
♦ Naturalised
♦ 101

Eriocaulon buergerianum **Körn.**
Eriocaulaceae
Eriocaulon pachypetalum Hay.
♦ Weed
♦ 87, 88, 275, 297
♦ aH, aqua, herbal.

Eriocaulon cinereum **R.Br.**
Eriocaulaceae
Eriocaulon formosanum Hay.
♦ ashy pipewort
♦ Weed, Naturalised
♦ 87, 88, 101, 170, 191, 286
♦ cultivated.

Eriocaulon cinereum **R.Br. var.** *sieboldiana* **(Sieb. & Zucc.) T.Koyama**
Eriocaulaceae
♦ Weed
♦ 235

Eriocaulon decemflorum **Maxim.**
Eriocaulaceae
Eriocaulon nipponicum Maxim.
♦ itoinunohige
♦ Weed
♦ 87, 88, 263

Eriocaulon decemflorum **Maxim. var.** *nipponicum* **Nakai**
Eriocaulaceae
♦ Weed
♦ 286

Eriocaulon echinulatum **Mart.**
Eriocaulaceae
♦ Weed
♦ 87, 88

Eriocaulon gracile **Mart.**
Eriocaulaceae
♦ Weed
♦ 87, 88

Eriocaulon heterolepis **Steud.**
Eriocaulaceae
♦ Weed
♦ 88

Eriocaulon heterolepis **Steud. var.** *nigricans* **Koern.**
Eriocaulaceae
♦ Weed
♦ 170, 191

Eriocaulon hondoense **Satake**
Eriocaulaceae
♦ Nippon inunohige
♦ Weed
♦ 87, 88, 286

Eriocaulon longifolium **Nees ex Kunth**
Eriocaulaceae
♦ pipewort
♦ Weed
♦ 88, 170, 191
♦ Origin: Australia.

Eriocaulon luzulaefolium **Mart. in Wall.**
Eriocaulaceae
♦ Weed, Quarantine Weed
♦ 76, 87, 88, 203, 220

Eriocaulon miquelianum **Körn.**
Eriocaulaceae
♦ Weed
♦ 87, 88, 263, 286

Eriocaulon odoratum **Dalz.**
Eriocaulaceae
♦ Weed
♦ 87, 88

Eriocaulon quinquangulare **L.**
Eriocaulaceae
♦ Weed
♦ 87, 88

Eriocaulon robustius **(Maxim.) Makino**
Eriocaulaceae
♦ hirohainunohige
♦ Weed
♦ 87, 88, 263, 275, 286, 297
♦ aH, aqua.

Eriocaulon septangulare **With.**
Eriocaulaceae
= *Eriocaulon aquaticum* (Hill) Druce (NoR)
♦ Weed
♦ 23, 88
♦ herbal.

Eriocaulon setaceum **L.**
Eriocaulaceae
♦ eriocaulon
♦ Weed
♦ 87, 88
♦ aqua, cultivated.

Eriocaulon sexangulare **L.**
Eriocaulaceae
Eriocaulon hexangularis Kunth, *Eriocaulon sieboldianum* Sieb. & Zucc.
♦ Weed
♦ 87, 88
♦ aqua, cultivated, herbal.

Eriocaulon sieboldtianum **Sieb. & Zucc.**
Eriocaulaceae
♦ Weed
♦ 88, 204, 263, 274
♦ aqua.

Eriocaulon truncatum Buch. -Ham. ex Mart.
Eriocaulaceae
♦ Weed, Quarantine Weed
♦ 76, 87, 88, 135, 170, 191, 203, 220

Eriocephalus africanus L.
Asteraceae
Eriocephalus umbellulatus Cass.
♦ Cape of Good Hope shrub, rosemary, wild rosemary
♦ Weed, Naturalised, Native Weed
♦ 86, 98, 121, 203
♦ pS, arid, cultivated. Origin: southern Africa.

Eriocephalus racemosus L.
Asteraceae
♦ wild rosemary
♦ Weed, Quarantine Weed, Naturalised
♦ 76, 86, 88, 220
♦ cultivated.

Eriocereus (Berg.) Riccob. spp.
Cactaceae
= *Harrisia* Britt. spp.
♦ harrisia cactus
♦ Weed, Quarantine Weed, Noxious Weed, Naturalised
♦ 86, 203

Eriocereus martinii (Lab.) Riccob.
Cactaceae
= *Harrisia martinii* (Lab.) Britt. & Rose
♦ harrisia cactus, moonlight cactus, snake cactus
♦ Weed, Quarantine Weed, Noxious Weed, Naturalised
♦ 62, 76, 87, 88, 93, 98, 147, 191, 203, 269
♦ Origin: South America.

Eriocereus regelii (Weing.) Backeb.
Cactaceae
= *Harrisia regelii* (Weing.) Borg (NoR)
♦ Weed
♦ 87, 88
♦ cultivated.

Eriocereus tortuosus (Otto & A.Dietr.) Riccob.
Cactaceae
= *Harrisia tortuosa* (J.Forbes ex Otto & A.Dietr.) Britt. & Rose
♦ harrisia cactus, moonlight cactus, snake cactus
♦ Weed, Noxious Weed, Naturalised
♦ 86, 87, 88, 98, 147, 191, 203
♦ cultivated. Origin: South America.

Eriochloa acuminata (Presl) Hitchc. var. acuminata
Poaceae
Eriochloa gracilis (Fourn.) Hitchc. (see)
♦ tapertip cupgrass
♦ Weed
♦ 34
♦ aG.

Eriochloa contracta A.S.Hitchc.
Poaceae
♦ prairie cupgrass, cupgrass, wiregrass, summergrass, watergrass, silkkihirssi
♦ Weed, Naturalised, Casual Alien

♦ 42, 161, 179, 180, 243, 287
♦ aG, herbal.

Eriochloa crebra S.T.Blake
Poaceae
♦ Introduced
♦ 228
♦ G, arid. Origin: Australia.

Eriochloa fatmensis (Hochst. & Steud.) W.D.Clayton
Poaceae
Eriochloa acrotricha (Steud.) Thell., *Eriochloa fouchei* Stent, *Eriochloa nubica* (Steud.) Hack. & Stapf ex Thell. (see), *Eriochloa procera* (Retz.) C.E.Hubb. (see), *Eriochloa punctata* (L.) Ham. var. *acrotricha* (Steud.) Schum., *Helopus acrotrichus* Steud., *Helopus nubicus* Steud., *Panicum annulatum* A.Rich., *Panicum fatmense* Hochst. & Steud.
♦ tropical cupgrass
♦ Naturalised
♦ 101
♦ G, arid, cultivated.

Eriochloa gracilis (Fourn.) Hitchc.
Poaceae
= *Eriochloa acuminata* (Presl) Hitchc. var. *acuminata*
♦ south-western cupgrass, cupgrass, summergrass, wiregrass
♦ Weed
♦ 87, 88, 90, 161, 180, 212, 218, 243
♦ G, herbal.

Eriochloa leersioides (Munro) Hack.
Poaceae
♦ sharp cupgrass
♦ Naturalised
♦ 101
♦ G.

Eriochloa lemmoni Vasey & Scribn.
Poaceae
♦ canyon cupgrass
♦ Weed
♦ 161
♦ G.

Eriochloa meyeriana (Nees) Pilg.
Poaceae
♦ Weed, Naturalised
♦ 86, 98, 203
♦ G.

Eriochloa montevidensis Griseb.
Poaceae
♦ Naturalised
♦ 241, 300
♦ G.

Eriochloa nubica (Steud.) Hack. & Stapf ex Thell.
Poaceae
= *Eriochloa fatmensis* (Hochst. & Steud.) Clayton
♦ Weed
♦ 242
♦ G, arid.

Eriochloa pacifica Mez
Poaceae
♦ Weed
♦ 87, 88
♦ G, arid.

Eriochloa polystachya Kunth
Poaceae
Eriochloa subglabra (Nash) Hitchc.
♦ carib grass
♦ Weed, Quarantine Weed
♦ 76, 88, 135, 170, 179, 191, 203, 220
♦ G, aqua.

Eriochloa procera (Retz.) C.E.Hubb.
Poaceae
= *Eriochloa fatmensis* (Hochst. & Steud.) Clayton
♦ Weed, Naturalised, Introduced
♦ 12, 32, 38, 87, 88, 228, 286
♦ G, arid, herbal.

Eriochloa pseudoacrotricha (Stapf ex Thell.) C.E.Hubb. ex S.T.Blake
Poaceae
♦ perennial cupgrass
♦ Naturalised
♦ 101
♦ G, arid, cultivated.

Eriochloa punctata (L.) Desv. ex Ham.
Poaceae
Milium punctatum L., *Helopus punctatus* (L.) Nees
♦ Louisiana cupgrass, pasto amargo
♦ Weed, Naturalised
♦ 68, 87, 88, 90, 241, 255, 271, 295, 300
♦ pG, aqua, cultivated. Origin: tropical and subtropical America.

Eriochloa villosa (Thunb.) Kunth
Poaceae
♦ hairy cupgrass, villahirssi, woolly cupgrass, sparrow's millet
♦ Weed, Noxious Weed, Naturalised, Introduced, Casual Alien
♦ 42, 101, 161, 174, 210, 229, 275, 286, 297
♦ aG.

Eriochrysis cayanensis Beauv.
Poaceae
♦ Weed
♦ 87, 88
♦ G.

Eriodictyon angustifolium Nutt.
Hydrophyllaceae
♦ narrowleaf yerbasanta
♦ Weed
♦ 87, 88, 218
♦ S, herbal.

Eriodictyon californicum (Hook. & Arn.) Greene
Hydrophyllaceae
Eriodictyon glutinosum Benth.
♦ California yerbasanta, eriodicto, yerba santa
♦ Weed
♦ 87, 88, 218
♦ S, cultivated, herbal.

Eriodictyon crassifolium Benth.
Hydrophyllaceae
♦ woolly yerbasanta, thickleaf yerbasanta
♦ Weed
♦ 87, 88
♦ S, cultivated, herbal.

Eriodictyon crassifolium Benth. var.
niveum Brandeg.
Hydrophyllaceae
♦ woolly yerbasanta
♦ Weed
♦ 218

Eriogonum annuum Nutt.
Polygonaceae
♦ umbrella plant, annual eriogonum,
annual buckwheat, wild buckwheat
♦ Native Weed
♦ 161, 174
♦ herbal. Origin: North America.

Eriogonum caespitosum Nutt.
Polygonaceae
♦ matted buckwheat
♦ Quarantine Weed
♦ 220
♦ herbal.

Eriogonum compositum Douglas ex
Benth.
Polygonaceae
♦ arrowleaf buckwheat
♦ Weed
♦ 23, 88
♦ cultivated, herbal.

Eriogonum deflexum Torr.
Polygonaceae
♦ skeletonweed eriogonum, flat
topped buckwheat, flatcrown
buckwheat
♦ Weed
♦ 87, 88, 161, 218
♦ aH, herbal.

Eriogonum elatum Douglas ex Benth.
Polygonaceae
♦ tall woolly buckwheat
♦ Weed
♦ 23, 88
♦ pH, herbal.

Eriogonum fasciculatum Benth.
Polygonaceae
♦ California buckwheat, eastern
Mojave buckwheat
♦ Weed
♦ 116
♦ S, arid, cultivated, herbal.

Eriogonum stellatum Benth.
Polygonaceae
= *Eriogonum umbellatum* Torr. var.
ellipticum (Nutt.) Reveal (NoR)
♦ Weed
♦ 23, 88
♦ herbal.

Eriope tumidicaulis Harley
Lamiaceae
♦ Weed
♦ 255
♦ aH. Origin: Brazil.

Eriophorum angustifolium Honck.
Cyperaceae
♦ common cottongrass, cottongrass,
luhtavilla
♦ Weed
♦ 23, 70, 88, 272
♦ pG, aqua, cultivated, herbal.

Eriophorum gracile Koch
Cyperaceae
♦ slender cottongrass, cottongrass,
hoikkavilla
♦ Weed
♦ 70, 272
♦ pG, aqua, cultivated, herbal.

Eriophorum latifolium Hoppe
Cyperaceae
♦ broad leaved cottongrass
♦ Weed
♦ 70, 272, 297
♦ G, aqua, cultivated, herbal.

Eriophorum scheuchzeri Hoppe
Cyperaceae
♦ Scheuchzer's cottongrass, white
cottongrass, töppövilla
♦ Weed
♦ 70
♦ G, herbal.

Eriophyllum confertiflorum (DC.) Gray
Asteraceae
♦ golden yarrow, yellow yarrow
♦ Quarantine Weed
♦ 220
♦ S, cultivated, herbal.

Eriophyllum lanatum (Pursh) Forbes
Asteraceae
♦ common woolly sunflower, woolly
eriophyllum, woolly sunflower
♦ Weed
♦ 23, 88
♦ pH, cultivated, herbal.

Eriophyllum lanatum (Pursh) Forbes var.
grandiflorum (A.Gray) Jeps.
Asteraceae
♦ common woolly sunflower, woolly
sunflower, sunflower
♦ Weed
♦ 161, 180, 243
♦ pH.

Eriophyllum nevinii Gray
Asteraceae
♦ Nevin's woolly sunflower
♦ Quarantine Weed
♦ 220
♦ S, cultivated, herbal.

Eriosema cordatum E.Mey.
Fabaceae/Papilionaceae
♦ Native Weed
♦ 121
♦ pH, arid, herbal. Origin: southern
Africa.

Eriosema crinitum (Kunth) G.Don.
Fabaceae/Papilionaceae
♦ island sand pea
♦ Weed
♦ 14

Eriosema psoraleoides (Lam.) G.Don
Fabaceae/Papilionaceae
Eriosema cajanoides Hook.
♦ yellow seed
♦ Native Weed
♦ 121
♦ pS, arid, herbal, toxic. Origin:
southern Africa.

Eriospermum burchellii Bak.
Liliaceae/Eriospermaceae
♦ Native Weed
♦ 121
♦ pH. Origin: southern Africa.

Erlangea cordifolia (Benth. ex Oliv.)
S.Moore
Asteraceae
Erlangea marginata (Oliv. & Hiern)
S.Moore (see), *Gutenbergia cordifolia*
Benth. ex Oliv. (see), *Vernonia marginata*
Oliv. & Hiern *nom. illeg.*
♦ Weed
♦ 87, 88

Erlangea laxa (N.E.Br.) S.Moore
Asteraceae
♦ Weed, Quarantine Weed, Native
Weed
♦ 76, 87, 88, 121, 203, 220
♦ aH. Origin: southern Africa.

Erlangea marginata (Oliv. & Hiern)
S.Moore
Asteraceae
= *Erlangea cordifolia* (Benth. ex Oliv.)
S.Moore
♦ Weed
♦ 87, 88

Erodium L'Hér. ex Aiton spp.
Geraniaceae
♦ stork's bill
♦ Weed
♦ 18, 88, 116, 243, 272

Erodium angustilobum Carolin
Geraniaceae
♦ Naturalised
♦ 86
♦ Origin: Australia.

Erodium arborescens (Desf.) Willd.
Geraniaceae
♦ Weed
♦ 221

Erodium aureum Carolin
Geraniaceae
♦ Weed, Naturalised, Environmental
Weed
♦ 7, 86, 93, 98, 203, 205
♦ aH. Origin: Asia.

Erodium botrys (Cav.) Bertol.
Geraniaceae
♦ broadleaf filaree, filaree, long beaked
filaree, long stork's bill, big heron's
bill, longbeak stork's bill
♦ Weed, Naturalised, Environmental
Weed
♦ 7, 9, 23, 72, 86, 87, 88, 93, 98, 101, 161,
176, 180, 198, 203, 205, 218, 241, 253,
269, 280, 287, 300
♦ aH, cultivated, herbal. Origin:
Mediterranean.

Erodium brachycarpum (Godr.) Thell.
Geraniaceae
♦ shortfruit stork's bill, hairy pit
heron's bill, foothill filaree, heron's bill
♦ Weed, Naturalised
♦ 86, 98, 101, 198, 203
♦ aH. Origin: North Africa.

Erodium bryoniaefolium Boiss.
Geraniaceae
♦ Weed
♦ 221

Erodium carvifolium Boiss & Reut.
Geraniaceae
♦ Quarantine Weed
♦ 220

Erodium castellanum (Pau) Guitt.
Geraniaceae
♦ Weed, Quarantine Weed
♦ 76, 88, 220

Erodium cazorlanum Heywood
Geraniaceae
♦ Quarantine Weed
♦ 220

Erodium chium (L.) Willd.
Geraniaceae
♦ Weed
♦ 88, 158, 221
♦ Origin: Africa.

Erodium ciconium (L.) L'Hér. ex Aiton
Geraniaceae
♦ common stork's bill
♦ Weed, Naturalised
♦ 88, 94, 101, 221, 253, 272
♦ cultivated, herbal.

Erodium cicutarium (L.) L'Hér. ex Aiton
Geraniaceae
♦ redstem filaree, alfilaree, alfidalaria, pin clover, pin grass, stork's bill, heron's bill, filaree, peltokurjennokka, musk heron's bill, pin weed
♦ Weed, Noxious Weed, Naturalised, Introduced, Garden Escape, Environmental Weed
♦ 7, 9, 15, 21, 24, 34, 39, 44, 45, 51, 70, 72, 80, 86, 87, 88, 93, 94, 98, 101, 121, 136, 146, 161, 162, 165, 176, 180, 198, 203, 205, 211, 212, 218, 221, 228, 229, 236, 237, 241, 243, 249, 253, 269, 272, 280, 286, 287, 295, 299, 300
♦ a/bH, arid, cultivated, herbal, toxic. Origin: Eurasia.

Erodium cicutarium (L.) L'Hér. ex Aiton ssp. bipinnatum Tourlet
Geraniaceae
♦ redstem stork's bill, redstem filaree
♦ Noxious Weed, Naturalised
♦ 101, 229

Erodium cicutarium (L.) L'Hér. ex Aiton ssp. cicutarium
Geraniaceae
♦ redstem filaree, redstem stork's bill
♦ Noxious Weed, Naturalised
♦ 101, 229, 280

Erodium cicutarium (L.) L'Hér. ex Aiton ssp. jacquinianum Fisch., (C.A.Mey. & Avé-Lall.) Briq.
Geraniaceae
♦ redstem filaree, redstem stork's bill
♦ Noxious Weed, Naturalised
♦ 101, 229

Erodium cicutarium L'Hér. var. pimpinellifolium Sm.
Geraniaceae
♦ Naturalised

♦ 287

Erodium crinitum Carolin
Geraniaceae
♦ blue crow's foot, blue stork's bill
♦ Weed, Naturalised, Native Weed
♦ 55, 269, 287
♦ aH, cultivated. Origin: Australia.

Erodium cygnorum Nees
Geraniaceae
♦ Australian stork's bill
♦ Weed, Naturalised
♦ 87, 88, 101
♦ aH. Origin: Australia.

Erodium geoides St.-Hil.
Geraniaceae
♦ Naturalised
♦ 300

Erodium glaucophyllum (L.) L'Hér.
Geraniaceae
♦ Weed
♦ 221

Erodium gruinum (L.) L'Hér. ex Aiton
Geraniaceae
♦ Iranian stork's bill, long beaked stork's bill, hygrometer plant
♦ Weed, Naturalised
♦ 87, 88, 101, 221
♦ a/bH, cultivated. Origin: Mediterranean.

Erodium hirtum Willd.
Geraniaceae
♦ Weed, Introduced
♦ 221, 228
♦ pH, arid, cultivated.

Erodium laciniatum (Cav.) Willd.
Geraniaceae
♦ cutleaf stork's bill
♦ Weed, Naturalised, Introduced
♦ 101, 221, 228
♦ arid, cultivated.

Erodium laciniatum (Cav.) Willd. var. bovei (Delile) Murb.
Geraniaceae
♦ cutleaf stork's bill
♦ Naturalised
♦ 101

Erodium malacoides (L.) L'Hér. ex Aiton
Geraniaceae
Erodium aragonense Loscos, *Erodium subtrilobum* Jord. (see), *Geranium malacoides* L.
♦ Mediterranean stork's bill, oval heron's bill
♦ Weed, Naturalised, Introduced
♦ 70, 86, 87, 88, 94, 98, 101, 111, 176, 198, 203, 221, 228, 237, 243, 253, 280, 295
♦ aH, arid, cultivated, herbal. Origin: Mediterranean.

Erodium malacoides (L.) L'Hér. ex Aiton var. malacoides
Geraniaceae
♦ Naturalised
♦ 241, 300

Erodium malacoides (L.) L'Hér. ex Aiton var. ribifolium (Jacq.) DC.

Geraniaceae
♦ Naturalised
♦ 300

Erodium manescavii Coss.
Geraniaceae
♦ Weed, Quarantine Weed
♦ 76, 88, 220
♦ cultivated.

Erodium moschatum (L.) L'Hér. ex Aiton
Geraniaceae
Herodium moschatum Reich.
♦ whitestem filaree, musk filaree, filaree, stork's bill, musky stork's bill, myskikurjennokka, musk heron's bill, heron's bill, musk heron's bill, musk clover, white stemmed filaree
♦ Weed, Naturalised, Environmental Weed
♦ 7, 9, 15, 42, 51, 70, 86, 87, 88, 94, 98, 101, 121, 134, 158, 161, 165, 176, 180, 198, 199, 203, 218, 221, 241, 243, 250, 253, 269, 280, 287, 300
♦ a/bH, arid, cultivated, herbal. Origin: Eurasia.

Erodium moschatum (L.) L'Hér. ex Aiton var. moschatum
Geraniaceae
♦ musky stork's bill
♦ Naturalised
♦ 101

Erodium moschatum (L.) L'Hér. ex Aiton var. praecox Lange
Geraniaceae
♦ musky stork's bill
♦ Naturalised
♦ 101

Erodium reichardii (Murray) DC.
Geraniaceae
♦ alpine geranium
♦ Weed, Quarantine Weed
♦ 88, 220
♦ cultivated.

Erodium romanum (L.) Willd.
Geraniaceae
♦ Weed
♦ 87, 88

Erodium stephanianum Willd.
Geraniaceae
Erodium stephenianum Willd. (see)
♦ stork's bill
♦ Weed
♦ 88, 114, 243, 275, 297
♦ a/pH, promoted.

Erodium stephenianum Willd.
Geraniaceae
= *Erodium stephanianum* Willd.
♦ Stephen's stork's bill
♦ Naturalised
♦ 101

Erodium subtrilobum Jord.
Geraniaceae
= *Erodium malacoides* (L.) L'Hér. ex Ait.
♦ Weed
♦ 221

Erodium trifolium Cav.
Geraniaceae
♦ Weed, Quarantine Weed
♦ 76, 88, 220

Erophila verna (L.) Chevall.
Brassicaceae
Draba verna L. (see), *Draba verna* L. ssp.
verna, *Erophila stenocarpa* Jord., *Erophila vulgaris* DC. *nom. illeg.* (see)
♦ whitlowgrass, common whitlowgrass, spring whitlowgrass
♦ Weed, Naturalised
♦ 44, 70, 88, 94, 98, 176, 198, 203, 241, 243, 253, 272, 280, 287
♦ aH, cultivated, herbal.

Erophila verna (L.) Chevall. ssp. praecox (Stevens) Walters
Brassicaceae
♦ whitlowgrass
♦ Naturalised, Cultivation Escape
♦ 86, 198, 252
♦ cultivated. Origin: Europe.

Erophila verna (L.) Chevall. ssp. verna
Brassicaceae
♦ whitlowgrass
♦ Naturalised, Cultivation Escape
♦ 86, 198, 252
♦ cultivated. Origin: Europe.

Erophila vulgaris DC. nom. illeg.
Brassicaceae
= *Erophila verna* (L.) Chev.
♦ Weed
♦ 87, 88

Eruca cappadocica Reut.
Brassicaceae
♦ Weed
♦ 87, 88

Eruca sativa Mill.
Brassicaceae
= *Eruca vesicaria* (L.) Cav. ssp. *sativa* (Mill.) Thell.
♦ garden rocket, salad rocket, garden eruca
♦ Weed, Naturalised, Introduced
♦ 39, 87, 88, 98, 121, 176, 203, 218, 221, 228, 243, 271, 297
♦ aH, arid, cultivated, herbal, toxic. Origin: Eurasia.

Eruca vesicaria Cav.
Brassicaceae
♦ rocket salad, rocket, garden rocket, garden sativa rocket
♦ Weed, Naturalised, Casual Alien
♦ 1, 42, 87, 88, 94, 101, 146, 161, 237, 241, 272, 280, 300
♦ cultivated, herbal.

Eruca vesicaria (L.) Cav. ssp. sativa (Mill.) Thell.
Brassicaceae
Eruca sativa Mill. (see)
♦ garden rocket, rocket salad, bladder eruca, salad rocket
♦ Weed, Noxious Weed, Naturalised, Introduced, Environmental Weed, Casual Alien
♦ 34, 40, 56, 80, 86, 101, 198, 229, 280
♦ a/bH, cultivated.

Erucaria aleppica Gaertn.
Brassicaceae
= *Erucaria hispanica* (L.) Druce
♦ Weed, Quarantine Weed
♦ 76, 87, 88, 203, 220

Erucaria crassifolia (Forssk.) Delile
Brassicaceae
♦ Weed
♦ 221

Erucaria hispanica (L.) Druce
Brassicaceae
Erucaria aleppica Gaertn. (see), *Erucaria lineariloba* Boiss., *Erucaria myagroides* (L.) Hal., *Erucaria tenuifolia* DC.
♦ pink mustard
♦ Weed, Introduced
♦ 88, 94, 115, 221, 228
♦ arid, cultivated.

Erucaria pinnata (Viv.) Täckh. & Boulos
Brassicaceae
Reboudia pinnata (Viv.) O.Schulz
♦ Weed
♦ 221
♦ arid.

Erucastrum arabicum Fisch. & C.Mey.
Brassicaceae
♦ Weed
♦ 53, 87, 88, 240
♦ aH, arid.

Erucastrum gallicum (Willd.) O.E.Schulz
Brassicaceae
♦ dogmustard, hairy rocket, kaalisinapi, common dogmustard
♦ Weed, Noxious Weed, Naturalised, Casual Alien
♦ 23, 36, 42, 70, 80, 87, 88, 94, 101, 161, 162, 179, 218, 243, 272, 287, 299
♦ aH, cultivated, herbal.

Erucastrum nasturtiifolium (Poir.) O.E.Schulz
Brassicaceae
♦ watercress dogmustard
♦ Weed
♦ 88, 94, 253
♦ cultivated, herbal.

Erucastrum strigosum (Thunb.) O.E.Schulz
Brassicaceae
Sisymbrium strigosum Thunb.
♦ Native Weed
♦ 121
♦ aH, arid. Origin: southern Africa.

Eryngium alpinum L.
Apiaceae
♦ isopiikkiputki, sea holly, alpine eryngo
♦ Cultivation Escape
♦ 42
♦ cultivated, herbal.

Eryngium amethystinum L.
Apiaceae
♦ amethyst eryngo, eryngo, amethyst holly, amethyst
♦ Weed, Naturalised, Casual Alien
♦ 101, 272, 280
♦ cultivated, herbal.

Eryngium campestre L.
Apiaceae
♦ field eryngo
♦ Weed, Naturalised
♦ 70, 86, 87, 88, 94, 98, 101, 203, 221, 272, 280
♦ pH, cultivated, herbal. Origin:

Eurasia, North Africa.

Eryngium coronatum Hook. & Arn.
Apiaceae
♦ cardilla
♦ Weed
♦ 87, 88, 237, 295

Eryngium creticum Lam.
Apiaceae
♦ eryngo
♦ Weed
♦ 87, 88, 272
♦ pH, promoted, herbal.

Eryngium divaricatum Hook. & Arn.
Apiaceae
♦ ballast eryngo
♦ Naturalised
♦ 101

Eryngium ebracteatum Lam.
Apiaceae
♦ caraguatá
♦ Weed
♦ 295
♦ cultivated.

Eryngium eburneum Decne.
Apiaceae
♦ carda caraguatá
♦ Weed
♦ 87, 88, 295
♦ cultivated.

Eryngium elegans Cham. & Schlecht.
Apiaceae
♦ gravaterinho
♦ Weed
♦ 87, 88, 255
♦ pH. Origin: Brazil.

Eryngium foetidum L.
Apiaceae
♦ spiritweed, Mexican coriander, false coriander, phakchee farang
♦ Weed, Naturalised
♦ 87, 88, 101, 179, 206, 238, 243, 275, 297
♦ pH, cultivated, herbal.

Eryngium hookeri Walp.
Apiaceae
♦ Hooker's eryngo
♦ Weed
♦ 87, 88, 218
♦ herbal.

Eryngium horridum Malme
Apiaceae
♦ caraguatá
♦ Weed
♦ 33, 255
♦ S, cultivated. Origin: South America.

Eryngium ilicifolium Lam.
Apiaceae
♦ Weed
♦ 243

Eryngium maritimum L.
Apiaceae
♦ seaside eryngo, sea holly
♦ Weed, Naturalised, Garden Escape, Environmental Weed
♦ 70, 86, 87, 88, 98, 101, 155, 203, 272
♦ pH, cultivated, herbal. Origin: Mediterranean.

Eryngium nudicaule Lam.
Apiaceae
♦ caraguatá
♦ Weed
♦ 295

Eryngium ovinum Cunn.
Apiaceae
♦ blue devil
♦ Native Weed
♦ 269
♦ Origin: Australia.

Eryngium pandanifolium Cham. & Schltdl.
Apiaceae
♦ Weed, Naturalised
♦ 86, 87, 88, 255, 280
♦ pH, cultivated. Origin: south America.

Eryngium pandanifolium Cham. & Schltdl. var. pandanifolium
Apiaceae
♦ Naturalised
♦ 280

Eryngium paniculatum Cav. & Dombey
Apiaceae
♦ chupalla
♦ Weed
♦ 87, 88, 237, 295
♦ cultivated, herbal.

Eryngium plantagineum F.Muell.
Apiaceae
♦ Weed
♦ 87, 88
♦ cultivated. Origin: Australia.

Eryngium planum L.
Apiaceae
♦ plains eryngo, sea holly, sinipiikkiputki
♦ Weed, Naturalised, Casual Alien
♦ 42, 101, 272
♦ pH, cultivated, herbal.

Eryngium prostratum Nutt. ex DC.
Apiaceae
♦ eryngo, creeping eryngo
♦ Weed
♦ 161
♦ herbal.

Eryngium rostratum Cav. status uncertain
Apiaceae
♦ blue devil
♦ Weed
♦ 87, 88

Eryngium sanguisorba Cham. & Schlecht.
Apiaceae
♦ Weed, Quarantine Weed
♦ 76, 87, 88, 203, 220

Eryngium tenue Lam.
Apiaceae
♦ Weed
♦ 70

Eryngium tricuspidatum L.
Apiaceae
♦ Weed
♦ 87, 88
♦ cultivated.

Eryngium triquetrum Vahl
Apiaceae
♦ Weed
♦ 87, 88

Eryngium vaseyi J.Coult. & Rose
Apiaceae
♦ coyote thistle
♦ Weed
♦ 161
♦ pH.

Eryngium yuccifolium Michx.
Apiaceae
♦ rattlesnake master, button snakeroot, button eryngo
♦ Weed, Quarantine Weed
♦ 76, 87, 88, 218
♦ pH, cultivated, herbal.

Erysimum L. spp.
Brassicaceae
♦ wallflower
♦ Weed, Naturalised
♦ 198, 272

Erysimum allionii hort.
Brassicaceae
= *Cheiranthus × allionii* hort. (NoR) [see *Erysimum × allionii* hort. nom. illeg.]
♦ Siberian wallflower
♦ Naturalised
♦ 56
♦ cultivated.

Erysimum × allionii hort. nom. illeg.
Brassicaceae
= *Cheiranthus × allionii* hort. (NoR) [see *Erysimum allionii* hort.]
♦ kesäkultalakka, Siberian wallflower
♦ Cultivation Escape
♦ 42
♦ pH, cultivated. Origin: horticultural hybrid.

Erysimum alpestre Kotschy ex Boiss.
Brassicaceae
♦ Weed
♦ 87, 88

Erysimum asperum (Nutt.) DC.
Brassicaceae
= *Erysimum capitatum* (Douglas) Greene
♦ western wallflower
♦ Weed
♦ 23, 87, 88, 218
♦ herbal.

Erysimum bungei (Kitag.) Kitag.
Brassicaceae
Erysimum aurantiacum (Bunge) Maxim.
♦ orange erysimum
♦ Weed
♦ 297

Erysimum canescens Roth.
Brassicaceae
= *Erysimum diffusum* Ehrh.
♦ Weed
♦ 87, 88
♦ herbal, toxic.

Erysimum capitatum (Douglas) Greene
Brassicaceae
Erysimum asperum (Nutt.) DC. (see), *Cheiranthus capitatus* Douglas ex Hook.

♦ sanddune wallflower, western wallflower, coastal wallflower
♦ Weed
♦ 23, 88
♦ pH, cultivated, herbal.

Erysimum cheiranthoides L.
Brassicaceae
♦ wormseed mustard, wormseed wallflower, wormseed, treacle hedge mustard, peltoukonnauris
♦ Weed, Quarantine Weed, Noxious Weed, Naturalised, Introduced, Environmental Weed
♦ 23, 34, 39, 40, 44, 52, 70, 76, 86, 87, 88, 94, 101, 161, 162, 174, 179, 203, 210, 218, 220, 243, 253, 272, 275, 280, 297, 299
♦ aH, cultivated, herbal, toxic.

Erysimum cheiri (L.) Crantz
Brassicaceae
Cheiranthus cheiri L. (see)
♦ wallflower, wormseed mustard, Aegean wallflower
♦ Weed, Naturalised, Cultivation Escape
♦ 34, 40, 56, 101
♦ pH, cultivated, herbal. Origin: horticultural.

Erysimum crepidifolium Rchb.
Brassicaceae
♦ kelttoukonnauris
♦ Weed, Casual Alien
♦ 39, 42, 272
♦ cultivated, toxic.

Erysimum diffusum Ehrh.
Brassicaceae
Cheiranthus alpinus Jacq., *Erysimum canescens* Roth. (see)
♦ Weed
♦ 272
♦ b/pH, promoted.

Erysimum durum J.& C.Presl
Brassicaceae
Erysimum hieracifolium L. ssp. *durum* (J. & C.Presl) Celak.
♦ hard wallflower
♦ Naturalised
♦ 101

Erysimum hieraciifolium L.
Brassicaceae
Erysimum strictum Gaertn.
♦ tall wormseed mustard, tall wormseed wallflower, European wallflower, rantaukonnauris
♦ Weed, Naturalised
♦ 52, 87, 88, 101, 161, 272
♦ cultivated.

Erysimum inconspicuum (S.Watson) MacMill.
Brassicaceae
♦ shy wallflower, smallflower wallflower
♦ Weed
♦ 87, 88, 161
♦ herbal.

Erysimum odoratum Ehrh.
Brassicaceae
♦ smelly wallflower
♦ Naturalised

♦ 101
♦ cultivated.

Erysimum perofskianum Fisch. & C.A.Mey.
Brassicaceae
♦ Afghan erysimum, erysimum gold shot
♦ Naturalised
♦ 101
♦ a/pH, cultivated, herbal.

Erysimum repandum L.
Brassicaceae
Erysimum patens Loscos
♦ treacle mustard, spreading wallflower, bushy wallflower
♦ Weed, Naturalised, Cultivation Escape, Casual Alien
♦ 34, 42, 44, 86, 87, 88, 94, 98, 101, 161, 198, 203, 237, 252, 272, 287, 295
♦ aH, cultivated, herbal. Origin: Mediterranean.

Erythraea centaurium Pers.
Gentianaceae
= *Centaurium erythraea* Rafn
♦ Weed
♦ 87, 88
♦ herbal.

Erythraea lomae Gilg
Gentianaceae
♦ Introduced
♦ 38

Erythraea ramosissima Pers.
Gentianaceae
♦ Weed
♦ 87, 88

Erythrina americana Mill.
Fabaceae/Papilionaceae
Corallodendron americanum (Mill.) Kuntze, *Erythrina carnea* Aiton, *Erythrina enneandra* DC., *Erythrina fulgens* Loisel
♦ naked coral tree, tzompantle
♦ Introduced
♦ 228
♦ arid, cultivated, herbal.

Erythrina berteriana Urb.
Fabaceae/Papilionaceae
♦ machete
♦ Naturalised
♦ 101

Erythrina burana Chiov.
Fabaceae/Papilionaceae
♦ Introduced
♦ 228
♦ arid.

Erythrina caffra Thunb.
Fabaceae/Papilionaceae
♦ coral tree
♦ Naturalised, Casual Alien
♦ 134, 280
♦ cultivated.

Erythrina crista-galli L.
Fabaceae/Papilionaceae
♦ cockspur coral tree, crybabytree, Indian coral tree, coral tree, ceibo, cresta de gallo, Peruvian national flower

♦ Weed, Quarantine Weed, Noxious Weed, Naturalised, Garden Escape, Environmental Weed
♦ 3, 54, 73, 76, 86, 88, 101, 155, 203, 261, 280
♦ S, cultivated, herbal. Origin: South America.

Erythrina flabelliformis Kearney
Fabaceae/Papilionaceae
♦ western coralbean, coral bean
♦ Weed
♦ 161
♦ arid, cultivated, herbal, toxic. Origin: south-west North America to Mexico.

Erythrina fusca Lour.
Fabaceae/Papilionaceae
♦ bucayo, swamp imortelle, immortelle, pahr
♦ Naturalised, Introduced
♦ 101, 230, 261
♦ S/T, cultivated, herbal. Origin: Madagascar.

Erythrina herbacea L.
Fabaceae/Papilionaceae
Erythrina arborea Small.
♦ eastern coralbean, cardinal spear, Cherokee bean, coral bean, red cardinal
♦ Weed
♦ 87, 88, 161, 218, 247
♦ pH, arid, cultivated, herbal, toxic.

Erythrina leptorhiza DC.
Fabaceae/Papilionaceae
♦ Weed
♦ 199

Erythrina poeppigiana (Walp.) O.F.Cook
Fabaceae/Papilionaceae
Erythrina amasisa Spruce, *Erythrina darienensis* Standl., *Erythrina micropteryx* Poepp., *Erythrina pisamo* Posada-Ar.
♦ mountain immortelle, palo de boya, immortelle
♦ Naturalised, Introduced
♦ 101, 228, 261
♦ T, arid, cultivated. Origin: Peru.

Erythrina rubrinervia Kunth
Fabaceae/Papilionaceae
♦ chochos de arbol
♦ Introduced
♦ 228
♦ T, arid, cultivated.

Erythrina speciosa Andrews
Fabaceae/Papilionaceae
♦ mulungu
♦ Naturalised
♦ 134
♦ cultivated.

Erythrina × sykesii Barneby & Krukoff
Fabaceae/Papilionaceae
= *Erythrina coralloides* DC. × *Erythrina lysistemon* Hutch.
♦ coral tree
♦ Weed, Sleeper Weed, Naturalised, Environmental Weed
♦ 73, 86, 88, 225, 246, 280, 290, 296
♦ cultivated. Origin: India.

Erythrina variegata L.
Fabaceae/Papilionaceae
Erythrina indica Lam.
♦ coral tree, tiger claw, gaogao, wiliwili haole, drala dina
♦ Weed, Naturalised, Introduced
♦ 3, 101, 161, 228
♦ arid, cultivated, herbal, toxic. Origin: Asia, Australia.

Erythrina variegata L. var. orientalis (L.) Merr.
Fabaceae/Papilionaceae
♦ tiger's claw, bucayo Haitiano, pahr
♦ Naturalised, Introduced, Cultivation Escape
♦ 101, 230, 261
♦ T, cultivated, herbal. Origin: eastern Asia, Pacific Islands.

Erythrina velutina Willd.
Fabaceae/Papilionaceae
♦ mulungu
♦ Introduced
♦ 228
♦ T, arid, cultivated.

Erythrina vespertilio Benth.
Fabaceae/Papilionaceae
♦ bean tree, batwing coral tree
♦ Weed
♦ 205
♦ T, arid, cultivated, herbal. Origin: Australia.

Erythrococca oleracea Prain
Euphorbiaceae
♦ Weed
♦ 87, 88,

Erythronium dens-canis L.
Liliaceae
Erythronium maculatum Lam.
♦ dog's tooth violet
♦ Weed
♦ 39, 272
♦ pH, cultivated, herbal, toxic.

Erythronium oregonum Applegate
Liliaceae
♦ dog's tooth violet, giant white fawnlily, white fawnlily
♦ Weed
♦ 39, 161
♦ pH, cultivated, herbal, toxic.

Erythrostictus punctatus Schltdl.
Liliaceae/Colchicaceae
♦ Weed
♦ 221

Erythroxylum P.Browne spp.
Erythroxylaceae
♦ coca
♦ Weed, Quarantine Weed
♦ 203, 220

Erythroxylum chlorostachys (F.Muell.) Bail.
Erythroxylaceae
♦ Weed
♦ 87, 88

Erythroxylum coca **Lam.**
Erythroxylaceae
Erythroxylum bolivianum Burck.,
Erythroxylum peruvianum Mitchel. &
Pascal. ex Steud.
- cocaleaf, coca, Peru coca
- Weed, Quarantine Weed, Noxious
Weed, Naturalised, Garden Escape
- 39, 76, 86, 87, 88, 147, 161, 169, 203,
230
- S, cultivated,
herbal, toxic.

Escallonia bifida **Link & Otto**
Grossulariaceae/Escalloniaceae
- Weed, Naturalised
- 86, 98, 203, 280
- cultivated. Origin: South America.

Escallonia × exoniensis hort.
Grossulariaceae/Escalloniaceae
= *Escallonia rosea* Griseb. × *Escallonia rubra* (Ruiz & Pav.) Pers.
- escallonia
- Casual Alien
- 280
- S, cultivated. Origin: horticultural
hybrid.

Escallonia macrantha **Hook. & Arn.**
Grossulariaceae/Escalloniaceae
= *Escallonia rubra* (Ruiz & Pav.) Pers.
var. *macrantha* (Hook. & Arn.) Reiche
- gum box
- Naturalised
- 40
- cultivated, herbal.

Escallonia rubra **(Ruiz & Pav.) Pers.**
Grossulariaceae/Escalloniaceae
- red claws, red escallonia
- Weed, Naturalised, Environmental
Weed
- 15, 101, 246, 280
- S, cultivated.

Escallonia rubra **(Ruiz & Pav.) Pers. var.**
macrantha (Hook. & Arn.) Rchb.
Grossulariaceae/Escalloniaceae
Escallonia macrantha Hook. & Arn. (see)
- red claws
- Naturalised
- 101, 280

Eschscholzia californica **Cham.**
Papaveraceae
- California poppy, tuliunikko
- Weed, Naturalised, Introduced,
Garden Escape, Environmental Weed,
Cultivation Escape
- 7, 15, 20, 42, 80, 86, 87, 88, 98, 102,
121, 161, 165, 176, 180, 198, 203, 218,
228, 241, 243, 269, 280, 290, 296, 300
- a/pH, arid, cultivated, herbal, toxic.
Origin: western North America.

Escobaria vivipara **(Nutt.) F.Buxb.**
Cactaceae
- purple mammillaria, spinystar
- Weed
- 161
- S, cultivated.

Escontria **Rose spp.**
Cactaceae

- Weed, Quarantine Weed
- 76, 88, 203, 220

Espadaea amoena **A.Rich.**
Goetzeaceae/Solanaceae
- rascabarriga
- Weed
- 14

Espostoa **Britton & Rose spp.**
Cactaceae
- Weed
- 203

Espostoopsis **Buxb. spp.**
Cactaceae
Gerocephalus Ritter spp.
- Weed, Quarantine Weed
- 76, 88, 203, 220

Ethulia conyzoides **L.f.**
Asteraceae
- abu elafein
- Weed, Naturalised
- 86, 98, 203, 221
- arid, herbal. Origin: Africa.

Etlingera cevuga **(Seem.) R.M.Sm.**
Zingiberaceae
- waxflower
- Naturalised
- 101

Etlingera elatior **(Jack) R.M.Sm.**
Zingiberaceae
Alpinia elatior Jack, *Alpinia magnifica*
Roscoe, *Amomum magnificum ined.*,
Elettaria speciosa Blume, *Nicolaia elatior*
(Jack) Horan. (see), *Nicolaia speciosa*
(Blume) Horan., *Phaeomeria magnifica*
(Roscoe) K.Schum. (see), *Phaeomeria*
speciosa (Blume) Koord.
- torch ginger, grand turmeric, pink
torch ginger
- Naturalised
- 101
- cultivated.

Eucalyptus **L'Hér. spp.**
Myrtaceae
- gum trees, gum, eucalyptus, silver
dollar tree
- Weed, Environmental Weed
- 10, 18, 88, 121, 247, 257, 277, 279
- S/T, herbal, toxic. Origin: Australia.

Eucalyptus alba **Reinw. ex Blume**
Myrtaceae
- white eucalyptus, white gum
- Introduced
- 228
- arid, cultivated.

Eucalyptus albens **Benth.**
Myrtaceae
- whitebox, white box gum
- Naturalised, Introduced
- 101, 228
- arid, cultivated. Origin: Australia.

Eucalyptus amygdalina **Labill.**
Myrtaceae
- black peppermint
- Naturalised
- 101
- cultivated, herbal. Origin: Australia.

Eucalyptus annulata **Benth.**
Myrtaceae
- open fruited mallee
- Introduced
- 228
- arid, cultivated.

Eucalyptus astringens **(Maiden) Maiden**
Myrtaceae
- brown mallet
- Introduced
- 228
- arid, cultivated.

Eucalyptus blakelyi **Maiden**
Myrtaceae
- Blakely's red gum
- Introduced
- 228
- arid, cultivated. Origin: Australia.

Eucalyptus botryoides **Sm.**
Myrtaceae
- southern mahogany, blue gum,
bangalay
- Weed, Sleeper Weed, Naturalised,
Native Weed, Introduced, Garden
Escape, Environmental Weed
- 7, 72, 86, 88, 101, 134, 225, 228, 246,
280
- T, arid, cultivated, herbal. Origin:
Australia.

Eucalyptus bridgesiana **R.Bak.**
Myrtaceae
- applebox
- Naturalised
- 101
- cultivated. Origin: Australia.

Eucalyptus brockwayi **C.Gardner**
Myrtaceae
- dundas mahogany
- Introduced
- 228
- arid, cultivated. Origin: Australia.

Eucalyptus calophylla **R.Br.**
Myrtaceae
- red gum
- Naturalised, Introduced
- 86, 101, 228
- arid, cultivated, herbal. Origin:
Australia.

Eucalyptus calycogona **Turcz.**
Myrtaceae
- gooseberry mallee
- Introduced
- 228
- arid, cultivated.

Eucalyptus camaldulensis **Dehnh.**
Myrtaceae
Eucalyptus rostrata Schldl.
- Murray red gum, red gum, river red
gum, rostrata gum, rooibloekom
- Weed, Noxious Weed, Naturalised,
Introduced, Garden Escape,
Environmental Weed, Cultivation
Escape
- 22, 34, 39, 63, 80, 86, 88, 95, 101, 116,
121, 151, 152, 216, 228, 261, 283
- T, arid, cultivated, herbal, toxic.
Origin: Australia.

Eucalyptus cambageana Maiden
Myrtaceae
♦ Coowarra box
♦ Weed
♦ 87, 88
♦ cultivated. Origin: Australia.

Eucalyptus cinerea F.Muell. ex Benth.
Myrtaceae
♦ Argyle apple, silver dollar eucalyptus, silver dollar gum
♦ Weed, Naturalised
♦ 101, 279, 280
♦ cultivated, herbal. Origin: Australia.

Eucalyptus citriodora Hook.
Myrtaceae
= *Corymbia citriodora* (Hook.) K.D.Hill & L.A.S.Johnson (NoR)
♦ lemon scented gum, lemon gum, citron scented gum
♦ Naturalised, Native Weed, Introduced, Garden Escape, Environmental Weed
♦ 7, 86, 101, 228, 230, 261
♦ T, arid, cultivated, herbal. Origin: Australia.

Eucalyptus cladocalyx F.Muell.
Myrtaceae
♦ sugar gum, suikerbloekom
♦ Weed, Noxious Weed, Naturalised, Native Weed, Introduced, Garden Escape, Environmental Weed, Cultivation Escape
♦ 7, 63, 72, 86, 88, 95, 101, 121, 198, 228, 283, 296
♦ T, arid, cultivated, herbal, toxic. Origin: Australia.

Eucalyptus cloeziana Muell.
Myrtaceae
♦ iron gum, yellow messmate, Gympie messmate
♦ Weed
♦ 279
♦ cultivated, herbal.

Eucalyptus conferruminata D. & S.Carr
Myrtaceae
♦ Bald Island marlock
♦ Weed, Naturalised, Native Weed, Environmental Weed
♦ 7, 72, 86, 88
♦ T, cultivated. Origin: Australia.

Eucalyptus cornuta Labill.
Myrtaceae
♦ yate
♦ Naturalised
♦ 101
♦ cultivated.

Eucalyptus crebra F.Muell.
Myrtaceae
Eucalyptus racemosa Cav. var. *longiflora* Blakely
♦ narrowleaf red ironbark, narrow leaved ironbark
♦ Naturalised, Introduced
♦ 101, 228
♦ cultivated, herbal. Origin: Australia.

Eucalyptus cypellocarpa L.A.S.Johnson
Myrtaceae

♦ mountain grey gum
♦ Casual Alien
♦ 280
♦ cultivated.

Eucalyptus deanei Maiden
Myrtaceae
♦ roundleaf gum
♦ Naturalised
♦ 101
♦ cultivated. Origin: Australia.

Eucalyptus deglupta Blume
Myrtaceae
♦ Mindanao gum, Indonesian gum
♦ Weed, Naturalised, Introduced
♦ 22, 101, 230, 261
♦ T, cultivated, herbal. Origin: southeast Asia, Pacific Islands.

Eucalyptus delegatensis R.T.Bak.
Myrtaceae
♦ alpine ash
♦ Naturalised
♦ 280
♦ cultivated.

Eucalyptus delegatensis R.T.Bak. ssp. *delegatensis*
Myrtaceae
♦ Naturalised
♦ 280

Eucalyptus diversicolor F.Muell.
Myrtaceae
♦ karri
♦ Weed, Noxious Weed, Cultivation Escape
♦ 88, 95, 283
♦ cultivated, herbal. Origin: Australia.

Eucalyptus dives Schauer
Myrtaceae
♦ broad leaved peppermint, peppermint gum
♦ Introduced
♦ 228
♦ arid, cultivated, herbal. Origin: Australia.

Eucalyptus dundasi Maiden
Myrtaceae
♦ Introduced
♦ 228
♦ arid.

Eucalyptus elata Dehnh.
Myrtaceae
♦ river peppermint
♦ Casual Alien
♦ 280
♦ cultivated.

Eucalyptus eremophila (Diels) Maiden
Myrtaceae
♦ tall sand mallee
♦ Introduced
♦ 228
♦ arid, cultivated.

Eucalyptus eugenioides Spreng.
Myrtaceae
♦ thin leaved stringybark
♦ Naturalised
♦ 280
♦ cultivated.

Eucalyptus fastigata Deane & Maiden
Myrtaceae
♦ cut tail gum, brown barrel gum, brown barrel
♦ Weed, Naturalised, Introduced
♦ 15, 228, 279, 280
♦ cultivated. Origin: Australia.

Eucalyptus ferruginea Schauer
Myrtaceae
♦ rusty bloodwood
♦ Weed
♦ 87, 88
♦ cultivated.

Eucalyptus fibrosa F.Muell.
Myrtaceae
♦ broad leaved red ironbark
♦ Naturalised
♦ 134
♦ cultivated. Origin: Australia.

Eucalyptus ficifolia F.Muell.
Myrtaceae
= *Corymbia ficifolia* (F.Muell.) K.D.Hill & L.A.S.Johnson
♦ redflower gum, scarlet gum, scarlet flowered gum
♦ Weed, Naturalised, Casual Alien
♦ 15, 101, 280
♦ cultivated, herbal.

Eucalyptus forrestiana Diels
Myrtaceae
♦ fuschia mallee
♦ Introduced
♦ 228
♦ arid, cultivated, herbal.

Eucalyptus globoidea Blakely
Myrtaceae
♦ white stringybark
♦ Weed
♦ 57
♦ cultivated. Origin: Australia.

Eucalyptus globulus Labill.
Myrtaceae
♦ Tasmanian blue gum, blue gum eucalyptus, blue gum, common eucalyptus
♦ Weed, Sleeper Weed, Noxious Weed, Naturalised, Native Weed, Introduced, Garden Escape, Environmental Weed
♦ 3, 7, 35, 39, 78, 80, 88, 101, 116, 142, 151, 225, 230, 231, 246, 280
♦ T, cultivated, herbal, toxic. Origin: Australia.

Eucalyptus globulus Labill. ssp. *globulus*
Myrtaceae
♦ Tasmanian blue gum
♦ Naturalised
♦ 101, 280
♦ cultivated, herbal.

Eucalyptus globulus Labill. ssp. *maidenii* (F.Muell.) J.B.Kirkp.
Myrtaceae
♦ Tasmanian blue gum, Maiden's gum
♦ Naturalised, Introduced
♦ 101, 228
♦ cultivated, herbal.

Eucalyptus

Eucalyptus

Eucalyptus gomphocephala DC.
Myrtaceae
♦ tuart, tuart gum
♦ Weed, Naturalised, Native
Weed, Introduced, Garden Escape,
Environmental Weed
♦ 72, 86, 88, 101, 228
♦ T, arid, cultivated, herbal. Origin:
Australia.

Eucalyptus goniocalyx F.Muell. ex Miq.
Myrtaceae
♦ mountain gray gum, long leaved
box
♦ Naturalised
♦ 101
♦ cultivated.
Origin: Australia.

Eucalyptus gracilis F.Muell.
Myrtaceae
♦ saligna gum, yorrell
♦ Weed, Introduced
♦ 87, 88, 228
♦ arid, cultivated.

Eucalyptus grandis W.Hill ex Maid.
Myrtaceae
♦ grand eucalyptus, blue gum,
rose gum, saligna, flooded gum,
salignabloekom, saligna gum
♦ Weed, Noxious Weed, Naturalised,
Introduced, Cultivation Escape
♦ 63, 95, 101, 121, 179, 228, 279, 280,
283
♦ T, arid, cultivated. Origin: Australia.

**Eucalyptus gummifera (Sol. ex Gaertn.)
Hochr.**
Myrtaceae
= *Corymbia gummifera* (Gaertn.)
K.D.Hill & L.A.S.Johnson (NoR)
♦ red bloodwood
♦ Naturalised
♦ 101
♦ T, cultivated, herbal. Origin:
Australia.

Eucalyptus gunnii Hook.f.
Myrtaceae
Eucalyptus divaricata McAulay & Brett
♦ cider gum, apple eucalyptus
♦ Naturalised, Introduced
♦ 228, 280
♦ T, arid, cultivated, herbal. Origin:
Australia.

Eucalyptus intertexta R.Bak.
Myrtaceae
♦ gum barked coolibah
♦ Introduced
♦ 228
♦ arid, cultivated.

Eucalyptus largiflorens F.Muell.
Myrtaceae
Eucalyptus bicolor Cunn. ex Hook.
♦ black box
♦ Weed, Introduced
♦ 57, 228
♦ T, arid, cultivated, herbal. Origin: Australia.

Eucalyptus lehmannii (Schauer) Benth.
Myrtaceae
♦ bush yate, Lehmann's gum, spider

gum, spinnekopbloekom, mallee yate
♦ Weed, Noxious Weed, Naturalised,
Native Weed, Garden Escape,
Environmental Weed, Cultivation
Escape
♦ 63, 72, 86, 88, 95, 121, 198, 283
♦ T, cultivated, herbal. Origin:
Australia.

Eucalyptus leucoxylon F.Muell.
Myrtaceae
♦ yellow gum, white ironbark
♦ Weed, Naturalised, Native
Weed, Introduced, Garden Escape,
Environmental Weed
♦ 72, 86, 88, 228
♦ T, arid, cultivated, herbal. Origin:
Australia.

**Eucalyptus longicornis (F.Muell.)
F.Muell. ex Maiden**
Myrtaceae
Eucalyptus oleosa F.Muell. ex Miq. var.
longicornis (F.Muell.) C.Gardner
♦ red morell
♦ Introduced
♦ 228
♦ arid, cultivated.

Eucalyptus longifolia Link
Myrtaceae
♦ woollybutt
♦ Introduced
♦ 228
♦ arid, cultivated. Origin: Australia.

Eucalyptus loxophleba Benth.
Myrtaceae
♦ York gum
♦ Introduced
♦ 228
♦ arid, cultivated.

**Eucalyptus macarthurii H.Deane &
Maiden**
Myrtaceae
♦ Camden woollybutt
♦ Naturalised, Introduced, Casual
Alien
♦ 198, 228, 280
♦ arid, cultivated, herbal. Origin:
Australia.

Eucalyptus maculata Hook.
Myrtaceae
= *Corymbia maculata* (Hook.) K.D.Hill
& L.A.S.Johnson (NoR)
♦ spotted gum, spotted iron gum
♦ Weed, Naturalised, Native
Weed, Introduced, Garden Escape,
Environmental Weed
♦ 7, 72, 86, 88, 228
♦ T, arid, cultivated, herbal. Origin:
Australia.

Eucalyptus marginata Donn ex Sm.
Myrtaceae
♦ jarrah
♦ Weed, Naturalised
♦ 87, 88, 101
♦ cultivated, herbal.

Eucalyptus melanophloia F.Muell.
Myrtaceae
♦ silver leaved ironbark

♦ Introduced
♦ 228
♦ arid, cultivated. Origin: Australia.

Eucalyptus melliodora Cunn. ex Schauer
Myrtaceae
♦ yellow box, honey eucalyptus
♦ Introduced
♦ 228
♦ arid, cultivated, herbal. Origin:
Australia.

Eucalyptus microcorys F.Muell.
Myrtaceae
♦ Australian tallow wood, tallow
wood
♦ Naturalised, Native Weed,
Introduced
♦ 7, 101, 228
♦ T, arid, cultivated, herbal. Origin:
Australia.

Eucalyptus microtheca F.Muell.
Myrtaceae
♦ Weed, Introduced
♦ 57, 228
♦ arid, cultivated, herbal, toxic. Origin:
Australia.

Eucalyptus miniata A.Cunn. ex Schauer
Myrtaceae
♦ Weed
♦ 87, 88
♦ cultivated.

**Eucalyptus × mortoniana Kinney (pro
sp.)**
Myrtaceae
= *Eucalyptus globulus* Labill. ×
Eucalyptus viminalis Labill.
♦ Morton eucalyptus
♦ Naturalised
♦ 101

Eucalyptus muelleriana A.Howitt
Myrtaceae
♦ yellow stringybark
♦ Naturalised
♦ 7, 86
♦ cultivated. Origin: Australia.

Eucalyptus nicholii Maiden & Blakely
Myrtaceae
♦ narrow leaved peppermint, willow
peppermint, peppermint eucalyptus
♦ Introduced
♦ 228
♦ arid, cultivated, herbal. Origin:
Australia.

**Eucalyptus nitens (Deane & Maiden)
Maiden**
Myrtaceae
Eucalyptus goniocalyx F.Muell. ex Miq.
var. *nitens* Deane & Maiden
♦ shining gum
♦ Weed, Naturalised, Introduced
♦ 15, 228, 280
♦ arid, cultivated, herbal. Origin:
Australia.

**Eucalyptus nova-anglica Deane &
Maiden**
Myrtaceae
♦ New England peppermint
♦ Introduced

♦ 228
♦ arid, cultivated, herbal. Origin: Australia.

Eucalyptus obliqua L'Hér.
Myrtaceae
Eucalyptus oblique L'Hér. var. *discocarpa* Blakely
♦ messmate stringybark, messmate
♦ Naturalised, Introduced
♦ 228, 280
♦ T, arid, cultivated, herbal. Origin: Australia.

Eucalyptus occidentalis Endl.
Myrtaceae
♦ flat topped yate
♦ Introduced
♦ 228
♦ arid, cultivated.

Eucalyptus ochrophloia F.Muell.
Myrtaceae
♦ yapunyah
♦ Introduced
♦ 228
♦ arid, cultivated. Origin: Australia.

Eucalyptus oleosa F.Muell. ex Miq.
Myrtaceae
Eucalyptus oleosa F.Muell. ex Miq. var. *obtusa* C.Gardner, *Eucalyptus oleosa* F.Muell. ex Miq. var. *angustifolia* Maiden
♦ eucalyptus
♦ Introduced
♦ 228
♦ arid, cultivated, herbal.

Eucalyptus ovata Labill.
Myrtaceae
♦ swamp gum
♦ Naturalised, Introduced
♦ 228, 280
♦ arid, cultivated. Origin: Australia.

Eucalyptus oxymitra Blakely
Myrtaceae
♦ sharp capped mallee
♦ Introduced
♦ 228
♦ arid, cultivated.

Eucalyptus pachyphylla F.Muell.
Myrtaceae
♦ redbud mallee
♦ Introduced
♦ 228
♦ T, arid, cultivated.

Eucalyptus paniculata Sm.
Myrtaceae
♦ gray ironbark, grysysterbasbloekom
♦ Weed, Noxious Weed, Naturalised, Introduced, Cultivation Escape
♦ 63, 101, 228, 279, 283
♦ arid, cultivated, herbal. Origin: Australia.

Eucalyptus papuana F.Muell.
Myrtaceae
= *Corymbia papuana* (F.Muell.) K.D.Hill & L.A.S.Johnson (NoR)
♦ ghost gum
♦ Introduced
♦ 228

♦ arid, cultivated, herbal. Origin: New Guinea, Australia.

Eucalyptus parvifolia Cambage
Myrtaceae
♦ small leaved gum
♦ Introduced
♦ 228
♦ arid, cultivated. Origin: Australia.

Eucalyptus pilularis J.E.Sm.
Myrtaceae
♦ blackbutt
♦ Weed, Naturalised, Introduced
♦ 87, 88, 101, 228, 280
♦ arid, cultivated, herbal. Origin: Australia.

Eucalyptus piperita Sm.
Myrtaceae
♦ Sydney peppermint, peppermint stringybark
♦ Casual Alien
♦ 280
♦ T, cultivated, herbal.

Eucalyptus platypus Hook.
Myrtaceae
♦ coastal moort, red flowering coastal moort, thicket moort, moort
♦ Introduced
♦ 228
♦ arid, cultivated.

Eucalyptus polyanthemos Schauer
Myrtaceae
♦ redbox, silver dollar gum
♦ Weed, Naturalised, Native Weed, Introduced, Environmental Weed
♦ 7, 34, 80, 86, 88, 101, 151, 228
♦ T, arid, cultivated, herbal. Origin: Australia.

Eucalyptus populnea F.Muell.
Myrtaceae
Eucalyptus populifolia Hook.
♦ poplar box
♦ Weed, Introduced
♦ 87, 88, 228
♦ arid, cultivated. Origin: Australia.

Eucalyptus propinqua Deane & Maiden
Myrtaceae
♦ small fruited grey gum
♦ Introduced
♦ 228
♦ arid, cultivated. Origin: Australia.

Eucalyptus pulchella Desf.
Myrtaceae
♦ white peppermint gum, white peppermint
♦ Naturalised
♦ 40, 280
♦ cultivated, herbal. Origin: Australia.

Eucalyptus pulverulenta Sims
Myrtaceae
♦ silverleaf mountain gum, money tree, powdered gum
♦ Naturalised, Introduced
♦ 101, 228
♦ T, arid, cultivated, herbal. Origin: Australia.

Eucalyptus punctata DC.
Myrtaceae
♦ grey gum
♦ Introduced
♦ 228
♦ T, arid, cultivated. Origin: Australia.

Eucalyptus radiata Sieber ex DC.
Myrtaceae
Eucalyptus australiana R.Bak. & H.G.Sm., *Eucalyptus phellandra* R.Bak. & H.G.Sm.
♦ peppermint eucalyptus
♦ Introduced
♦ 228
♦ arid, cultivated, herbal. Origin: Australia.

Eucalyptus raveretiana F.Muell.
Myrtaceae
♦ black ironbox
♦ Naturalised
♦ 101
♦ cultivated. Origin: Australia.

Eucalyptus redunca Schauer
Myrtaceae
Eucalyptus redunca Schauer var. *melanophloia* Benth.
♦ black marlock
♦ Introduced
♦ 228
♦ arid, cultivated.

Eucalyptus regnans F.Muell.
Myrtaceae
♦ mountain ash
♦ Naturalised
♦ 280
♦ T, cultivated, herbal.

Eucalyptus resinifera Sm.
Myrtaceae
♦ red mahogany
♦ Naturalised, Introduced, Casual Alien
♦ 101, 228, 261, 280
♦ arid, cultivated, herbal. Origin: Australia.

Eucalyptus robusta Sm.
Myrtaceae
Eucalyptus multiflora Poir.
♦ swamp mahogany
♦ Weed, Naturalised, Introduced
♦ 22, 101, 179, 228, 228, 280
♦ T, arid, cultivated, herbal. Origin: Australia.

Eucalyptus rubida Deane & Maiden
Myrtaceae
♦ candlebark
♦ Introduced
♦ 228
♦ arid, cultivated. Origin: Australia.

Eucalyptus rudis Sm.
Myrtaceae
♦ Western Australian flooded gum, flooded gum
♦ Naturalised, Introduced
♦ 101, 228
♦ arid, cultivated.

Eucalyptus saligna Sm.
Myrtaceae
Eucalyptus saligna Sm. var. *protrosa*
Blakely & McKie
♦ blue gum, saligna, Sydney blue gum
♦ Weed, Sleeper Weed, Naturalised, Native Weed, Introduced, Environmental Weed
♦ 7, 86, 101, 121, 225, 228, 246, 280
♦ T, arid, cultivated, herbal. Origin: Australia.

Eucalyptus salmonophloia F.Muell.
Myrtaceae
♦ salmon gum
♦ Introduced
♦ 228
♦ arid, cultivated, herbal.

Eucalyptus sargentii Maiden
Myrtaceae
♦ salt river gum
♦ Introduced
♦ 228
♦ cultivated.

Eucalyptus siderophloia Benth.
Myrtaceae
Eucalyptus decepta Blakely, *Eucalyptus nanglei* R.Bak.
♦ ironbark
♦ Introduced
♦ 228
♦ arid, cultivated. Origin: Australia.

Eucalyptus sideroxylon A.Cunn.
Myrtaceae
♦ red ironbark, swartysterbasbloekom, black ironbark
♦ Weed, Noxious Weed, Naturalised, Cultivation Escape
♦ 63, 101, 283
♦ T, cultivated, herbal. Origin: Australia.

Eucalyptus sideroxylon Cunn. ex Woolls ssp. sideroxylon
Myrtaceae
♦ Introduced
♦ 228

Eucalyptus sieberi L.A.S.Johnson
Myrtaceae
♦ silvertop ash
♦ Casual Alien
♦ 280
♦ cultivated.

Eucalyptus spathulata Hook.
Myrtaceae
♦ swamp mallet
♦ Introduced
♦ 228
♦ arid, cultivated.

Eucalyptus stricklandii Maiden
Myrtaceae
♦ Strickland's gum
♦ Introduced
♦ 228
♦ arid, cultivated.

Eucalyptus tenuiramis Miq.
Myrtaceae
♦ silver peppermint
♦ Naturalised

♦ 280
♦ cultivated.

Eucalyptus tereticornis Sm.
Myrtaceae
Eucalyptus umbellata (Gaertn.) Domin
♦ forest red gum, Queensland blue gum, horncap eucalyptus
♦ Weed, Naturalised, Introduced, Environmental Weed
♦ 80, 88, 101, 151, 228, 280
♦ T, arid, cultivated. Origin: Australia.

Eucalyptus terminalis F.Muell.
Myrtaceae
= *Corymbia terminalis* (F.Muell.) K.D.Hill & L.A.S.Johnson (NoR)
♦ inland bloodwood
♦ Introduced
♦ 228
♦ arid, cultivated, herbal.

Eucalyptus tessellaris F.Muell.
Myrtaceae
♦ Moreton Bay ash
♦ Introduced
♦ 228
♦ arid, cultivated. Origin: Australia.

Eucalyptus tetradonta F.Muell.
Myrtaceae
♦ Weed
♦ 87, 88

Eucalyptus thozetiana F.Muell. ex R.Bak.
Myrtaceae
♦ mountain yapunyah
♦ Introduced
♦ 228
♦ arid, cultivated, herbal. Origin: Australia.

Eucalyptus torelliana F.Muell.
Myrtaceae
= *Corymbia torelliana* (F.Muell.) K.D.Hill & L.A.S.Johnson
♦ cadaga, cadargi
♦ Weed, Environmental Weed
♦ 22, 260
♦ T, cultivated. Origin: Australia.

Eucalyptus torquata Luehm.
Myrtaceae
♦ coral gum
♦ Naturalised, Introduced
♦ 101, 228
♦ arid, cultivated, herbal. Origin: Australia.

Eucalyptus transcontinentalis Maiden
Myrtaceae
Eucalyptus oleosa F.Muell. ex Miq. var. *glauca* Maiden
♦ redwood
♦ Introduced
♦ 228
♦ arid, cultivated.

Eucalyptus triantha Link
Myrtaceae
Eucalyptus acmenioides Schauer
♦ Introduced
♦ 228
♦ arid.

Eucalyptus viminalis Labill.
Myrtaceae

Eucalyptus huberiana Naudin, *Eucalyptus mannifera* Mudie
♦ manna gum
♦ Naturalised, Introduced
♦ 39, 101, 228, 280
♦ T, arid, cultivated, herbal, toxic. Origin: Australia.

Eucalyptus viridis R.Bak.
Myrtaceae
♦ green mallee
♦ Introduced
♦ 228
♦ arid, cultivated, herbal. Origin: Australia.

Eucalyptus wandoo Blakely
Myrtaceae
Eucalyptus redunca Schauer var. *elata* Benth.
♦ wandoo
♦ Introduced
♦ 228
♦ arid, cultivated.

Eucharis grandiflora Planch. & Linden
Liliaceae/Amaryllidaceae
♦ Amazon lily
♦ Naturalised, Cultivation Escape
♦ 101, 261
♦ cultivated, herbal. Origin: Colombia.

Eucheuma denticulatum (Burm.) Collins & Herv.
Solieriaceae
♦ red alga
♦ Weed
♦ 197
♦ algae. Origin: east African coast.

Eucheuma isiforme (C.Agardh) J.Agardh
Solieriaceae
♦ stiff spineweed, red alga
♦ Weed
♦ 197
♦ algae.

Euchiton audax (Drury) Anderb.
Asteraceae
Gnaphalium audax Drury
♦ Naturalised
♦ 86
♦ Origin: New Zealand.

Euchiton gymnocephalus (DC.) A.Anderb.
Asteraceae
Gnaphalium collinum Labill. (see)
♦ creeping cudweed
♦ Naturalised
♦ 101

Euchiton japonicus (Thunb.) A.Anderb.
Asteraceae
Gnaphalium japonicum Thunb. (see)
♦ father and child plant
♦ Naturalised
♦ 101

Euchiton sphaericus (Willd.) A.Anderb.
Asteraceae
♦ cudweed
♦ Weed, Naturalised
♦ 55, 101

Euclea divinorum Hiern
Ebenaceae

♦ Weed, Quarantine Weed
♦ 76, 87, 88, 203, 220
♦ cultivated.

Euclea keniensis R.E.Fr.
Ebenaceae
♦ Weed
♦ 87, 88

Euclea schimperi (A.DC.) Dandy
Ebenaceae
♦ Weed
♦ 87, 88, 221

Euclea undulata Thunb.
Ebenaceae
♦ common guarri, fire fighter's blessing, guarri bush, guarri wood
♦ Native Weed
♦ 121
♦ S/T, cultivated. Origin: southern Africa.

Euclidium syriacum (L.) R.Br.
Brassicaceae
Myagrum syriacum Lam.
♦ Syrian mustard, linnunnokka
♦ Weed, Naturalised, Cultivation Escape, Casual Alien
♦ 34, 42, 86, 87, 88, 98, 101, 161, 203, 243, 248, 252, 272
♦ aH, cultivated. Origin: Eurasia.

Eucomis comosa (Houtt.) hort. ex Wehrh.
Liliaceae/Hyacinthaceae
♦ pineapple lily, pineapple flower
♦ Naturalised
♦ 198
♦ cultivated.

Eugenia L. spp.
Myrtaceae
♦ jambos, stopper
♦ Weed
♦ 18

Eugenia apiculata DC.
Myrtaceae
= *Luma apiculata* (DC.) Burret
♦ shortleaf stopper, arrayan, temu
♦ Naturalised
♦ 101
♦ cultivated.

Eugenia axillaris (Sw.) Willd.
Myrtaceae
♦ white stopper
♦ Weed
♦ 14
♦ cultivated, herbal.

Eugenia caryophyllus (Spr.) Bull. & Harr.
Myrtaceae
= *Syzygium aromaticum* (L.) Merr. & L.M.Perry (NoR)
♦ Introduced
♦ 230
♦ T, herbal.

Eugenia cumini (L.) Druce
Myrtaceae
= *Syzygium cumini* (L.) Skeels
♦ Weed
♦ 3, 80
♦ herbal.

Eugenia jambolana Lam.
Myrtaceae
= *Syzygium cumini* (L.) Skeels
♦ Java plum, jambul
♦ Weed
♦ 88, 218
♦ herbal.

Eugenia jambos L.
Myrtaceae
= *Syzygium jambos* (L.) Alston
♦ rose apple, iou en wai
♦ Weed, Introduced, Environmental Weed
♦ 3, 39, 80, 88, 152, 218, 230, 257
♦ T, cultivated, herbal, toxic. Origin: south-east Asia.

Eugenia javanica Lam.
Myrtaceae
= *Syzygium samarangense* (Bl.) Merr. & Perr. (NoR)
♦ apel en wai
♦ Introduced
♦ 230
♦ T.

Eugenia ligustrina (Sw.) Willd.
Myrtaceae
♦ privet stopper
♦ Weed
♦ 14

Eugenia malaccensis L.
Myrtaceae
= *Syzygium malaccense* (L.) Merr. & Perry
♦ Malay apple, apel
♦ Introduced, Environmental Weed
♦ 230, 257
♦ T, cultivated, herbal.

Eugenia maleolens Pers.
Myrtaceae
= *Eugenia monticola* (Sw.) DC. (NoR)
♦ Weed
♦ 14

Eugenia myrtoides Poir.
Myrtaceae
= *Eugenia foetida* Pers. (NoR)
♦ boxleaf eugenia
♦ Weed
♦ 87, 88, 218

Eugenia punicifolia (Kunth) DC.
Myrtaceae
♦ Weed
♦ 14
♦ T.

Eugenia uniflora L.
Myrtaceae
♦ Surinam cherry, red Brazil cherry, kafika, kafika paplagi
♦ Weed, Sleeper Weed, Noxious Weed, Naturalised, Introduced, Garden Escape, Environmental Weed, Cultivation Escape
♦ 3, 22, 73, 80, 86, 88, 98, 101, 112, 122, 134, 151, 155, 179, 191, 203, 230, 261, 283
♦ S, cultivated, herbal. Origin: central South America.

Eulalia amaura (Büse ex Miq.) Ohwi
Poaceae
= *Polytrias indica* (Houtt.) Veldkamp (NoR) [see *Polytrias amaura* (Büse) Kuntze *nom. illeg.*, *Polytrias praemorsa* (Nees) Hack.]
♦ Weed
♦ 179
♦ G.

Eulalia villosa (Thunb.) Nees
Poaceae
♦ golden velvetgrass
♦ Native Weed
♦ 121
♦ pG. Origin: southern Africa.

Eulophia graminea Lindl.
Orchidaceae
♦ Weed
♦ 19, 285
♦ cultivated.

Eulychnia Phil. spp.
Cactaceae
Philippicereus Backeb. spp. (see)
♦ Weed, Quarantine Weed
♦ 76, 88, 203, 220

Euonymus L. spp.
Celastraceae
♦ spindle tree, burningbush, strawberry bush, wahoo, euonymus
♦ Weed, Naturalised, Garden Escape, Environmental Weed
♦ 54, 72, 86, 88, 161, 247, 252
♦ cultivated, toxic.

Euonymus alata (Thunb.) Sieb.
Celastraceae
= *Euonymus alatus* (Thunb.) Sieb.
♦ winged burningbush, burningbush, winged euonymus
♦ Weed, Naturalised, Introduced, Garden Escape, Environmental Weed
♦ 88, 101, 102, 142, 151, 222
♦ cultivated, herbal.

Euonymus alatus (Thunb.) Sieb.
Celastraceae
Celastrus alatus Thunb., *Euonymus alata* (Thunb.) Sieb. (see), *Euonymus alatus* fo. *subtriflorus* (Blume) Ohwi, *Euonymus alatus* var. *apterus* Regel, *Euonymus alatus* var. *ciliato-dentatus* Franch. & Sav., *Euonymus alatus* var. *pubescens* Maxim., *Euonymus alatus* var. *subtriflorus* (Blume) Franch. & Sav., *Euonymus subtriflorus* Blume, *Euonymus thunbergianus* Blume
♦ winged euonymus, winged spindle tree, burningbush, winged wahoo, Japanese spindle tree
♦ Weed, Environmental Weed
♦ 80, 133, 182, 195, 224
♦ S, cultivated, herbal. Origin: China, Japan.

Euonymus americanus L.
Celastraceae
♦ strawberry bush
♦ Weed
♦ 8, 39, 161
♦ S, promoted, herbal, toxic.

Euonymus aptera **Regel**
Celastraceae
♦ corky spindletree
♦ Naturalised
♦ 101

Euonymus atropurpureus **Jacq.**
Celastraceae
♦ eastern wahoo, spindle tree, wahoo
♦ Weed
♦ 39, 87, 88, 161, 218
♦ S, cultivated, herbal, toxic.

Euonymus bungeanum **Maxim.**
Celastraceae
♦ winterberry
♦ Naturalised
♦ 101

Euonymus europaea **L.**
Celastraceae
= *Euonymus europaeus* L.
♦ European spindletree
♦ Naturalised
♦ 39, 101
♦ cultivated, herbal, toxic.

Euonymus europaeus **L.**
Celastraceae
Euonymus bulgarica Velen., *Euonymus europaea* L. (see), *Euonymus europaeus* fo. *atrorubens* (C.K.Schneid.) Hegi, *Euonymus floribundus* Stev., *Euonymus mediorossica* Klok., *Euonymus suberosus* Klok., *Euonymus vulgaris* Mill.
♦ spindle tree, skewer wood, spindle berry, European euonymus, sorvarinpensas
♦ Weed, Naturalised, Environmental Weed, Cultivation Escape
♦ 15, 39, 42, 70, 80, 161, 165, 181, 225, 246, 272, 280
♦ S, cultivated, herbal, toxic. Origin: southern Europe.

Euonymus fortunei **(Turcz.) Hand.-Mazz.**
Celastraceae
Euonymus radicans Sieb. ex Miq.
♦ winter creeper, climbing euonymus
♦ Weed, Naturalised, Introduced, Garden Escape, Casual Alien
♦ 80, 88, 101, 102, 129, 133, 142, 195, 222, 280
♦ pC, cultivated, herbal.

Euonymus fortunei **(Turcz.) Hand.-Mazz. var.** *fortunei*
Celastraceae
♦ winter creeper
♦ Naturalised
♦ 101

Euonymus fortunei **(Turcz.) Hand.-Mazz. var.** *radicans* **(Sieb. ex Miq.) Rehd.**
Celastraceae
♦ winter creeper
♦ Naturalised
♦ 101

Euonymus hamiltoniana **Wall.**
Celastraceae
♦ Hamilton's spindletree
♦ Naturalised
♦ 101
♦ cultivated.

Euonymus hamiltoniana **Wall. ssp.** *maackii* **(Rupr.) Komarov**
Celastraceae
♦ Hamilton's spindletree
♦ Naturalised
♦ 101

Euonymus hamiltoniana **Wall. ssp.** *sieboldianus* **(Blume) Hara**
Celastraceae
♦ Hamilton's spindletree
♦ Naturalised
♦ 101

Euonymus japonicus **Thunb.**
Celastraceae
Euonymus japonica Thunb.
♦ Japanese spindletree
♦ Weed, Naturalised, Environmental Weed
♦ 15, 101, 225, 246, 280
♦ S, cultivated, herbal.

Euonymus kiautschovica **Loes.**
Celastraceae
♦ creeping strawberry bush
♦ Naturalised
♦ 101
♦ cultivated.

Euonymus nana **Bieb.**
Celastraceae
♦ Weed
♦ 80
♦ cultivated.

Euonymus nanus **Bieb.**
Celastraceae
♦ kääpiösorvarinpensas
♦ Cultivation Escape
♦ 42
♦ cultivated.

Euonymus occidentalis **Nutt. ex Torr.**
Celastraceae
♦ western wahoo, burningbush, western burningbush
♦ Weed
♦ 39, 161
♦ S, herbal, toxic.

Euonymus pendulus **Wall.**
Celastraceae
♦ Casual Alien
♦ 280

Euonymus phellomana **Loes. ex Diels**
Celastraceae
= *Euonymus phellomanus* Loes. ex Diels
♦ corktree
♦ Naturalised
♦ 101

Euonymus phellomanus **Loes. ex Diels**
Celastraceae
Euonymus phellomana Loes. ex Diels (see)
♦ Casual Alien
♦ 280

Eupatoriadelphus purpureus **(L.) R.M.King & H.Rob.**
Asteraceae
= *Eupatorium purpureum* L. var. *purpureum* (NoR)
♦ Naturalised
♦ 280

♦ herbal.

Eupatorium **L. spp.**
Asteraceae
♦ dogfennel, thoroughwort, purple boneset, white snakeroot, joepyeweed
♦ Weed, Environmental Weed
♦ 80, 80, 88, 151
♦ herbal, toxic.

Eupatorium adenophorum **Spreng.**
Asteraceae
= *Ageratina adenophora* (Spreng.) King & Rob.
♦ pamakani, crofton weed, saap maa
♦ Weed
♦ 18, 67, 88, 106, 121, 161, 199, 218, 238, 239, 244, 256
♦ pH, herbal, toxic. Origin: Americas.

Eupatorium altissimum **L.**
Asteraceae
♦ tall joepyeweed, tall thoroughwort, tall boneset
♦ Native Weed
♦ 161, 174
♦ arid, promoted, herbal. Origin: North America.

Eupatorium ballotaefolium **H.B.K.**
Asteraceae
♦ picao roxo
♦ Weed
♦ 87, 88, 255
♦ Origin: Brazil.

Eupatorium betonicaeforme **Bak.**
Asteraceae
♦ Weed
♦ 87, 88

Eupatorium buniifolium **Hook. & Arn.**
Asteraceae
♦ chilca, chirca
♦ Weed
♦ 237, 255, 295
♦ Origin: Brazil.

Eupatorium cannabinum **L.**
Asteraceae
♦ hemp agrimony, punalatva, canapa acquatica
♦ Weed, Quarantine Weed, Naturalised, Environmental Weed
♦ 15, 23, 39, 70, 76, 87, 88, 94, 101, 220, 246, 272, 280
♦ pH, arid, cultivated, herbal, toxic.

Eupatorium capillifolium **(Lam.) Small**
Asteraceae
♦ dogfennel, summer cedar, hogweed
♦ Weed
♦ 14, 87, 88, 161, 211, 218, 243, 249
♦ pH, cultivated, herbal.

Eupatorium chinense **L.**
Asteraceae
♦ Chinese eupatorium
♦ Weed
♦ 297
♦ pH, cultivated, herbal.

Eupatorium chinense **L. var.** *oppositifolium* **Murata & H.Koyama**
Asteraceae
♦ Weed
♦ 286

Eupatorium clematideum Griseb.
Asteraceae
= *Praxelis clematidea* (Griseb.) R.M.King
& H.Rob.
♦ eupatorio clematídeo
♦ Weed
♦ 237, 295

Eupatorium coelestinum L.
Asteraceae
= *Conoclinium coelestinum* (L.) DC.
(NoR)
♦ mistflower
♦ Weed
♦ 161, 207, 275, 297
♦ pH, cultivated, herbal. Origin: North
America.

Eupatorium compositifolium Walter
Asteraceae
♦ yankeeweed
♦ Weed
♦ 87, 88, 218
♦ pH, promoted, herbal.

Eupatorium fistulosum Barratt
Asteraceae
♦ joepyeweed, trumpetweed
♦ Weed
♦ 161, 207
♦ bH, cultivated, herbal. Origin: North
America.

Eupatorium formosanum Hayata
Asteraceae
♦ Weed
♦ 87, 88

Eupatorium glandulosum H.B.K.
Asteraceae
= *Ageratina adenophora* (Spreng.) King
& Rob.
♦ sticky eupatorium
♦ Weed
♦ 87, 88, 218
♦ cultivated.

Eupatorium hirsutum Bak.
Asteraceae
♦ Weed
♦ 87, 88

Eupatorium inulaefolium H.B.K.
Asteraceae
= *Austroeupatorium inulaefolium*
(H.B.K.) R.M.King & H.Rob.
♦ Weed
♦ 13
♦ S, herbal.

Eupatorium japonicum Thunb.
Asteraceae
♦ Weed, Quarantine Weed
♦ 76, 87, 88, 203, 220
♦ herbal.

Eupatorium laevigatum Lam.
Asteraceae
= *Chromolaena laevigata* (Lam.)
R.M.King & H.Rob.
♦ cambará falso
♦ Weed
♦ 255, 295
♦ Origin: Americas.

Eupatorium liatrideum DC.
Asteraceae

Eupatorium squarrulosum Hook. & Arn.
♦ Weed, Quarantine Weed
♦ 76, 87, 88, 203, 220

Eupatorium lindleyanum DC.
Asteraceae
♦ joepyeweed, boneset, eupatorium,
sawahiyodori
♦ Weed
♦ 286, 297
♦ pH, aqua, promoted, herbal. Origin:
China, Japan, Korea, Manchuria.

Eupatorium macrocephalum Lees.
Asteraceae
= *Campuloclinium macrocephalum* (Less.)
DC.
♦ eupatorium, triffid weed
♦ Weed, Quarantine Weed
♦ 76, 87, 88, 121, 203, 220, 255, 295
♦ pH. Origin: South America.

Eupatorium macrophyllum L.
Asteraceae
= *Hebeclinium macrophyllum* (L.) DC.
(NoR)
♦ largeleaf thoroughwort
♦ Weed
♦ 87, 88
♦ cultivated, herbal.

Eupatorium maculatum L.
Asteraceae
♦ joepyeweed, spotted joepyeweed
♦ Weed, Quarantine Weed
♦ 76, 87, 88, 218
♦ pH, cultivated, herbal.

Eupatorium maxmilianii Schrad.
Asteraceae
Eupatorium conyzoides Vahl var.
maxmilianii Bak.
♦ mato pasto
♦ Weed
♦ 255
♦ Origin: Americas.

Eupatorium microstemon Cass.
Asteraceae
= *Fleischmannia microstemon* (Cass.)
R.M.King & H.Rob.
♦ tropical thoroughwort
♦ Weed
♦ 87, 88

Eupatorium odoratum L.
Asteraceae
= *Chromolaena odorata* (L.) King &
Robins.
♦ bitter bush, Siam weed, saap suea
♦ Weed, Naturalised, Environmental
Weed
♦ 3, 14, 51, 88, 157, 170, 204, 216, 238,
239, 256, 262, 275, 276, 287, 297
♦ pH, herbal, toxic.

Eupatorium pallescens DC.
Asteraceae
♦ Weed, Quarantine Weed
♦ 76, 87, 88, 203, 220

Eupatorium pauciflorum Kunth
Asteraceae
= *Praxelis pauciflora* (Kunth) R.M.King
& H.Rob. (NoR)

♦ chirca, eupatório
♦ Weed
♦ 88, 255
♦ Origin: Brazil.

Eupatorium perfoliatum L.
Asteraceae
♦ boneset, thoroughwort, common
boneset
♦ Weed, Quarantine Weed
♦ 87, 88, 218, 220
♦ pH, cultivated, herbal.

Eupatorium purpureum L.
Asteraceae
♦ sweet joepyeweed, gravel root
♦ Weed, Quarantine Weed,
Naturalised
♦ 23, 76, 86, 88
♦ pH, cultivated, herbal.

Eupatorium pycnocephalum Less.
Asteraceae
= *Fleischmannia pycnocephala* (Less.)
R.M.King & H.Rob. (NoR)
♦ lavender thoroughwort
♦ Weed
♦ 87, 88
♦ herbal.

Eupatorium riparium Regel.
Asteraceae
= *Ageratina riparia* (Regel) R.M.King
& H.Rob.
♦ river eupatorium
♦ Weed
♦ 88, 218, 268
♦ herbal.

Eupatorium rotundifolium L.
Asteraceae
♦ roundleaf thoroughwort,
joepyeweed
♦ Weed
♦ 23, 88
♦ cultivated, herbal.

Eupatorium rugosum Houtt.
Asteraceae
= *Ageratina altissima* (L.) R.M.King &
H.Rob.
♦ white snakeroot
♦ Weed, Quarantine Weed,
Naturalised
♦ 88, 121, 161, 163, 207, 210, 218, 220,
247, 287
♦ pH, cultivated, herbal, toxic. Origin:
North America.

Eupatorium serotinum Michx.
Asteraceae
♦ late eupatorium, lateflowering
thoroughwort, late boneset
♦ Weed, Naturalised
♦ 86, 87, 88, 98, 161, 203, 210, 218
♦ pH, herbal. Origin: North America.

Eupatorium squalidum DC.
Asteraceae
= *Chromolaena squalida* (DC.) R.M.King
& H.Rob. (NoR)
♦ cambará roxo, chilca
♦ Weed
♦ 255

Eupatorium triplinerve Vahl
Asteraceae
Ayapana triplinervis (Vahl) King &
H.E.Rob.
♦ yapana
♦ Naturalised, Cultivation Escape
♦ 101, 261
♦ cultivated, herbal. Origin: tropical
America.

Eupatorium triste DC.
Asteraceae
♦ Weed
♦ 87, 88

Eupatorium variable Makino
Asteraceae
♦ Quarantine Weed
♦ 220

Eupatorium villosum Sw.
Asteraceae
= *Koanophyllon villosum* (Sw.) King &
H.E.Rob. (NoR)
♦ bitter bush
♦ Weed
♦ 14, 87, 88
♦ herbal.

Euphorbia L. spp.
Euphorbiaceae
♦ euphorbia, milkweed, spurge
♦ Weed,
Quarantine Weed
♦ 18, 39, 88, 154, 220, 221, 237, 243, 272
♦ herbal, toxic.

Euphorbia acalyphoides Hochst. ex Boiss.
Euphorbiaceae
♦ Weed, Quarantine Weed
♦ 76, 87, 88, 203, 220, 242
♦ arid.

Euphorbia aegyptiaca Boiss.
Euphorbiaceae
Euphorbia forsskalii Jay., *Euphorbia
thymifolia* Forssk.
♦ Weed, Quarantine Weed
♦ 76, 87, 88, 203, 220, 242
♦ arid, herbal.

Euphorbia agraria M.Bieb.
Euphorbiaceae
♦ urban spurge
♦ Weed, Naturalised
♦ 39, 80, 101, 272
♦ toxic.

Euphorbia albomarginata Torr. & Gray
Euphorbiaceae
= *Chamaesyce albomarginata* (Torr. &
Gray) Small
♦ whitemargin spurge
♦ Weed
♦ 87, 88, 161, 218
♦ herbal.

Euphorbia aleppica L.
Euphorbiaceae
♦ spurge
♦ Weed, Quarantine Weed
♦ 39, 76, 87, 88, 203, 220, 272
♦ herbal, toxic.

**Euphorbia ambovombensis Rauh &
Razaf.**
Euphorbiaceae

♦ Quarantine Weed
♦ 220
♦ Origin: Madagascar.

Euphorbia amygdaloides L.
Euphorbiaceae
♦ wood spurge, purple spurge, spurge
♦ Weed, Naturalised
♦ 243, 272, 280
♦ cultivated, herbal, toxic.

Euphorbia anychioides Boiss.
Euphorbiaceae
♦ Weed
♦ 199

Euphorbia arguta Banks & Sol.
Euphorbiaceae
♦ Weed
♦ 87, 88, 185

Euphorbia atoto Forst.f.
Euphorbiaceae
= *Chamaesyce atoto* (Forst.f.) Croizat
(NoR)
♦ Weed
♦ 87, 88
♦ herbal.

Euphorbia avasmontana Dinter
Euphorbiaceae
♦ Quarantine Weed
♦ 220
♦ cultivated.

**Euphorbia barnardii White, Dyer &
Sloane**
Euphorbiaceae
♦ Quarantine Weed
♦ 220
♦ arid, cultivated.

Euphorbia bothae Lotsy & Goddijn
Euphorbiaceae
♦ Native Weed
♦ 121
♦ pS, arid. Origin: South Africa.

Euphorbia brasiliensis Lam.
Euphorbiaceae
= *Chamaesyce hyssopifolia* (L.) Small
♦ Weed
♦ 87, 88, 270
♦ herbal.

Euphorbia caerulescens Haw.
Euphorbiaceae
♦ Quarantine Weed
♦ 220

Euphorbia californica Benth.
Euphorbiaceae
♦ Quarantine Weed
♦ 220
♦ herbal.

Euphorbia canariensis L.
Euphorbiaceae
♦ spurge
♦ Weed, Quarantine Weed
♦ 39, 76, 88, 220
♦ arid, cultivated, herbal, toxic.

Euphorbia capitellata Engelm.
Euphorbiaceae
= *Chamaesyce capitellata* (Engelm.)
Millsp.
♦ Weed
♦ 87, 88

♦ herbal.

Euphorbia chamaepeplus Boiss. & Gaill.
Euphorbiaceae
♦ Weed
♦ 221

Euphorbia chamaesyce L.
Euphorbiaceae
= *Chamaesyce prostrata* (Ait.) Small
♦ groundfig spurge, blue weed, hairy
creeping milkweed, hairy prostrate
euphorbia, prostrate spurge, red
caustic creeper
♦ Weed, Naturalised
♦ 39, 51, 87, 88, 93, 94, 121, 218, 272,
286, 287, 295
♦ aH, herbal, toxic. Origin: tropical
America.

**Euphorbia chamaesyce L. ssp.
massiliensis Thell.**
Euphorbiaceae
♦ Naturalised
♦ 287

Euphorbia characias L.
Euphorbiaceae
♦ large Mediterranean spurge, wulffen
spurge
♦ Weed, Naturalised
♦ 15, 39, 54, 70, 88, 280
♦ pH, cultivated,
herbal, toxic.

**Euphorbia characias L. ssp. wulfenii
(Hoppe ex W.D.J.Koch) Radcl.-Sm.**
Euphorbiaceae
Euphorbia wulfenii Hoppe ex
W.D.J.Koch (see), *Euphorbia veneta*
Willd. (see)
♦ Quarantine Weed, Naturalised
♦ 76, 86
♦ cultivated. Origin: east
Mediterranean.

Euphorbia coerulescens Haw.
Euphorbiaceae
♦ Weed, Quarantine Weed, Native
Weed
♦ 121, 220
♦ pS, arid, cultivated. Origin: South
Africa.

Euphorbia colletioides Benth.
Euphorbiaceae
♦ Quarantine Weed
♦ 220
♦ herbal.

Euphorbia collina Phil. var. collina
Euphorbiaceae
♦ lechetrés
♦ Weed
♦ 295

**Euphorbia collina Phil. var. patagonica
(Hieron.) Subils**
Euphorbiaceae
Euphorbia patagonica Hieron. (see)
♦ Weed
♦ 237

Euphorbia colorata Engelm.
Euphorbiaceae
♦ Quarantine Weed
♦ 220

Euphorbia cornigera Boiss.
Euphorbiaceae
♦ Quarantine Weed
♦ 220
♦ cultivated. Origin: India, Pakistan.

Euphorbia corollata L.
Euphorbiaceae
Euphorbia paniculata Ell.
♦ flowering spurge, large flowering
spurge, blooming spurge, milk
purslane, snake milk, white purslane,
wild spurge
♦ Weed, Quarantine Weed,
Naturalised
♦ 39, 76, 86, 87, 88, 109, 161, 203, 210,
218, 220, 247
♦ pH, cultivated, herbal, toxic.

Euphorbia cotinifolia L.
Euphorbiaceae
♦ Mexican shrubby spurge, spurge
♦ Weed, Quarantine Weed,
Naturalised, Garden Escape
♦ 39, 76, 87, 88, 101, 220, 261
♦ cultivated, herbal, toxic.

Euphorbia cuneata Vahl
Euphorbiaceae
♦ Weed
♦ 221

Euphorbia cuphosperma (Engelm.) Boiss.
Euphorbiaceae
♦ Quarantine Weed
♦ 220

Euphorbia cyathophora J.A.Murray
Euphorbiaceae
Euphorbia heterophylla L. var.
cyathophora (Murr.) Griseb. (see)
♦ dwarf poinsettia, fire on the
mountain, Mexican fire plant, painted
spurge, poinsettia
♦ Weed, Naturalised, Introduced,
Garden Escape, Environmental Weed
♦ 7, 39, 86, 88, 93, 98, 161, 201, 203, 205,
230, 261, 300
♦ aH, cultivated, herbal, toxic. Origin:
tropical America.

Euphorbia cybirensis Boiss.
Euphorbiaceae
♦ Weed
♦ 111, 243

Euphorbia cyparissias L.
Euphorbiaceae
Tithymalus cyparissias (L.) Scop.
♦ cypress spurge, salvers spurge,
quack salvers grass, graveyard weed
♦ Weed, Noxious Weed, Naturalised,
Introduced, Environmental Weed,
Cultivation Escape
♦ 23, 39, 40, 42, 44, 49, 52, 70, 80, 86, 87,
88, 94, 101, 133, 151, 161, 162, 176, 195,
210, 218, 222, 224, 229, 243, 247, 252,
272, 280, 287, 291, 299
♦ cultivated, herbal, toxic. Origin:
Europe.

Euphorbia davidii Subils
Euphorbiaceae
Euphorbia dentata Michx. (see)
♦ David's spurge
♦ Naturalised

♦ 86, 101
♦ Origin: Argentina.

Euphorbia dendroides L.
Euphorbiaceae
♦ tree spurge, spurge
♦ Weed, Naturalised
♦ 7, 39, 86, 98, 101, 203
♦ cultivated, herbal, toxic. Origin:
Mediterranean.

Euphorbia densa Schrenk
Euphorbiaceae
♦ Weed
♦ 243

Euphorbia dentata Michx.
Euphorbiaceae
= *Euphorbia davidii* Subils
♦ toothed spurge, toothedleaf
poinsettia, lecherón
♦ Weed, Noxious Weed, Naturalised
♦ 39, 49, 87, 88, 98, 101, 161, 174, 203,
212, 219, 229, 236, 270
♦ aH, herbal, toxic.

Euphorbia dentata Michx. var. dentata
Euphorbiaceae
♦ toothed spurge
♦ Noxious Weed
♦ 229

Euphorbia dentata Michx. var. lasiocarpa Boiss.
Euphorbiaceae
♦ toothed spurge
♦ Noxious Weed, Naturalised
♦ 101, 229

Euphorbia depauperata A.Rich.
Euphorbiaceae
♦ Weed, Naturalised
♦ 98, 203, 280
♦ cultivated.

Euphorbia depauperata A.Rich. var. pubescens Pax
Euphorbiaceae
♦ Naturalised
♦ 86, 280
♦ Origin: Africa.

Euphorbia didiereoides Denis ex Leandri
Euphorbiaceae
♦ Weed, Quarantine Weed
♦ 76, 88, 220
♦ arid. Origin: Madagascar.

Euphorbia discolor Ledeb.
Euphorbiaceae
= *Euphorbia esula* L.
♦ leafy spurge
♦ Weed
♦ 88, 114, 243

Euphorbia donii Oudejans
Euphorbiaceae
♦ Quarantine Weed
♦ 220

Euphorbia dracunculoides Lam.
Euphorbiaceae
♦ Weed, Quarantine Weed
♦ 76, 87, 88, 203, 220, 243
♦ herbal.

Euphorbia drummondii Boiss.
Euphorbiaceae

= *Chamaesyce drummondii* (Boiss.)
D.C.Hassall
♦ caustic creeper, caustic weed, spurge
♦ Weed
♦ 39, 55, 87, 88, 109
♦ aH, cultivated, herbal, toxic.

Euphorbia edulis Lour.
Euphorbiaceae
♦ Weed
♦ 87, 88

Euphorbia engelmannii Boiss.
Euphorbiaceae
♦ Naturalised
♦ 300

Euphorbia epithymoides L.
Euphorbiaceae
♦ cushion spurge
♦ Weed, Naturalised
♦ 80, 101, 272
♦ cultivated, herbal.

Euphorbia eremophila A.Cunn. ex Hook.
Euphorbiaceae
♦ Weed
♦ 39, 87, 88
♦ toxic.

Euphorbia esula L.
Euphorbiaceae
Euphorbia discolor Ledeb. (see),
Euphorbia esula var. *cyparioides* Boiss.,
Euphorbia esula var. *latifolia* Ledeb.,
Euphorbia glomerulans Prokh., *Euphorbia
gmelinii* Steud., *Euphorbia jaxartica*
Prokh., *Euphorbia kaleniczenkii* Czern.
ex Trautv., *Euphorbia lunulata* Bunge
(see), *Euphorbia mandshurica* Maxim.,
Euphorbia minxianensis W.T.Wang,
Euphorbia subcordata C.A.Mey. ex
Ledeb., *Euphorbia tarokoensis* Hayata,
Euphorbia uralensis Fisch. ex Link,
Galarrhoeus esula (L.) Rydb., *Tithymalus
esula* (L.) Hill, *Tithymalus glomerulans*
Prokh., *Tithymalus jaxarticus* Prokh.
♦ leafy spurge, Hungarian spurge,
wolf's milk, Faitour's grass
♦ Weed, Quarantine Weed, Noxious
Weed, Naturalised, Introduced,
Environmental Weed, Casual Alien
♦ 1, 4, 16, 23, 26, 34, 35, 36, 39, 40, 42,
48, 52, 70, 76, 78, 80, 87, 88, 94, 101, 103,
103, 104, 105, 116, 129, 130, 133, 136,
138, 139, 141, 146, 151, 161, 162, 172,
174, 195, 210, 211, 212, 218, 219, 220,
222, 224, 229, 231, 243, 264, 266, 267,
272, 293, 297, 299
♦ pH, cultivated, herbal, toxic. Origin:
Eurasia.

Euphorbia esula L.var. esula
Euphorbiaceae
Euphorbia pseudovirgata (Schur) Soó
♦ leafy spurge
♦ Noxious Weed, Naturalised
♦ 101, 229

Euphorbia esula L. var. orientalis Boiss.
Euphorbiaceae
♦ leafy spurge, oriental leafy spurge
♦ Noxious Weed, Naturalised
♦ 101, 229

Euphorbia esula L. var. *uralensis* (Fisch. ex Link) Dorn
Euphorbiaceae
Euphorbia virgata Waldst. & Kit. (see)
♦ leafy spurge, Russian leafy spurge
♦ Noxious Weed, Naturalised
♦ 101, 229

Euphorbia exigua L.
Euphorbiaceae
♦ dwarf spurge, pikkutyräkki
♦ Weed, Naturalised, Environmental Weed, Casual Alien
♦ 23, 39, 42, 44, 70, 86, 87, 88, 94, 98, 101, 176, 203, 243, 253, 272, 280
♦ cultivated, herbal, toxic. Origin: Eurasia.

Euphorbia falcata L.
Euphorbiaceae
♦ sickle spurge, sirppityräkki
♦ Weed, Naturalised, Casual Alien
♦ 39, 42, 70, 86, 87, 88, 94, 98, 101, 203, 241, 253, 272, 300
♦ herbal, toxic. Origin: Eurasia.

Euphorbia ferox Marloth
Euphorbiaceae
♦ Weed, Quarantine Weed
♦ 76, 88, 220
♦ arid.

Euphorbia fimbriata Scop.
Euphorbiaceae
♦ Weed, Quarantine Weed
♦ 76, 88, 220
♦ arid, cultivated.

Euphorbia fischeriana Steud.
Euphorbiaceae
♦ Fischer euphorbia
♦ Weed
♦ 297

Euphorbia forsskalii J.Gray
Euphorbiaceae
= *Euphorbia aegyptiaca* Boiss.
♦ Weed
♦ 88, 185, 221

Euphorbia fulgens Karw. ex Klotzsch
Euphorbiaceae
♦ scarlet plume
♦ Weed
♦ 87, 88
♦ cultivated, herbal.

Euphorbia gaillardoti Boiss. & Blanche
Euphorbiaceae
♦ Weed, Quarantine Weed
♦ 76, 87, 88, 203, 220

Euphorbia geniculata Ortega
Euphorbiaceae
= *Euphorbia heterophylla* L.
♦ painted euphorbia, painted milkweed, painted spurge, yaa yaang
♦ Weed
♦ 51, 66, 121, 221, 239, 276
♦ aH, arid, herbal. Origin: tropical North America.

Euphorbia glomerifera (Millsp.) Wheeler
Euphorbiaceae
= *Chamaesyce hypericifolia* (L.) Millsp.
♦ Weed, Quarantine Weed
♦ 76, 87, 88, 203, 220
♦ herbal.

Euphorbia glyptosperma Engelm.
Euphorbiaceae
= *Chamaesyce glyptosperma* (Engelm.) Small
♦ ridgeseed spurge
♦ Weed, Noxious Weed
♦ 39, 87, 88, 136, 161, 212, 299
♦ aH, herbal, toxic.

Euphorbia graminea Jacq.
Euphorbiaceae
♦ grassleaf spurge, lechosa
♦ Weed, Naturalised
♦ 87, 88, 101, 179, 199, 281
♦ herbal.

Euphorbia grandidens Haw.
Euphorbiaceae
♦ big toothed euphorbia, large toothed euphorbia
♦ Weed, Native Weed
♦ 39, 121
♦ T, arid, cultivated, herbal, toxic. Origin: southern Africa.

Euphorbia graniticola Leach
Euphorbiaceae
♦ Quarantine Weed
♦ 220
♦ Origin: Mozambique, Zimbabwe.

Euphorbia granulata Forssk.
Euphorbiaceae
♦ Weed
♦ 87, 88, 185, 243
♦ herbal.

Euphorbia granulata Forssk. var. *glabrata* (Gay) Boiss.
Euphorbiaceae
♦ Weed
♦ 88

Euphorbia handiensis Burchard
Euphorbiaceae
♦ Weed, Quarantine Weed
♦ 76, 88, 220
♦ cultivated.

Euphorbia hedyotoides N.E.Br.
Euphorbiaceae
♦ Weed, Quarantine Weed
♦ 76, 88, 220
♦ arid.

Euphorbia helioscopia L.
Euphorbiaceae
♦ sun spurge, madwoman's milk, wart spurge, wart weed, wart grass, cat's milk, sun euphorbia, umbrella milkweed, viisisädetyräkki
♦ Weed, Naturalised, Garden Escape, Environmental Weed
♦ 7, 15, 23, 24, 34, 39, 44, 51, 70, 80, 86, 87, 88, 94, 98, 101, 115, 118, 121, 158, 161, 176, 185, 198, 203, 217, 217, 218, 221, 237, 241, 243, 253, 269, 272, 275, 280, 286, 295, 297, 300
♦ aH, cultivated, herbal, toxic. Origin: Eurasia.

Euphorbia hernariifolia Willd.
Euphorbiaceae
♦ spurge
♦ Naturalised
♦ 101

Euphorbia heterophylla L.
Euphorbiaceae
Euphorbia geniculata Orteg. (see), *Euphorbia prunifolia* Jacq. (see), *Poinsettia geniculata* (Ortega) Klotzsch & Garcke, *Poinsettia heterophylla* (L.) Klotzsch & Garcke
♦ painted spurge, fire plant, wild poinsettia, various leaved euphorbia, milkweed, desert spurge, Mexican fire plant, yellow spurge, amendoim bravo, leiteiro
♦ Weed, Quarantine Weed, Noxious Weed, Naturalised, Introduced
♦ 6, 7, 12, 14, 39, 55, 68, 76, 86, 87, 88, 93, 98, 121, 147, 157, 158, 161, 185, 199, 203, 209, 217, 228, 229, 240, 242, 243, 245, 255, 258, 262, 271, 276, 281, 295
♦ a/bH, arid, cultivated, herbal, toxic. Origin: Central America.

Euphorbia heterophylla L. var. *cyathophora* (Murr.) Griseb.
Euphorbiaceae
= *Euphorbia cyathophora* J.A.Murray
♦ Weed, Naturalised
♦ 286, 287

Euphorbia hirta L.
Euphorbiaceae
= *Chamaesyce hirta* (L.) Millsp.
♦ garden spurge, pillpod spurge, asthma weed, nam nom raatchasee, asthma plant, hairy spurge, Queensland asthma weed, red euphorbia, red milkweed, snakeweed, golondrina
♦ Weed, Naturalised, Native Weed, Introduced, Garden Escape, Environmental Weed
♦ 12, 28, 50, 51, 53, 66, 86, 87, 88, 93, 98, 121, 161, 170, 185, 186, 199, 203, 205, 206, 209, 218, 228, 235, 239, 240, 242, 243, 263, 274, 275, 276, 280, 281, 295, 297
♦ aH, arid, cultivated, herbal, toxic. Origin: tropical Africa.

Euphorbia hirta L. var. *hirta*
Euphorbiaceae
♦ Naturalised
♦ 300

Euphorbia hirta L. var. *ophtalmica* (Pers) Allem & Irgang
Euphorbiaceae
♦ Naturalised
♦ 300

Euphorbia hirta L. var. *procumbens* (DC.) N.E.Br.
Euphorbiaceae
= *Chamaesyce ophthalmica* (Pers.) Burch
♦ Weed, Naturalised
♦ 199, 241, 270

Euphorbia hispida Boiss.
Euphorbiaceae
♦ Weed
♦ 87, 88

Euphorbia hormorrhiza Radcl.-Sm.
Euphorbiaceae
♦ Quarantine Weed
♦ 220

Euphorbia horombensis **Ursch & Leandri**
Euphorbiaceae
♦ Weed, Quarantine Weed
♦ 76, 88, 220
♦ arid, cultivated.

Euphorbia humifusa **Willd.**
Euphorbiaceae
Euphorbia pseudochamaesyce Fisch. &
Mey.
♦ rentotyräkki
♦ Weed, Casual Alien
♦ 42, 87, 88, 94, 243, 272, 275, 297
♦ aH, promoted, herbal.

Euphorbia humifusa **Willd. var.**
pseudochamaesyce **Murata**
Euphorbiaceae
♦ Weed
♦ 263, 286
♦ aH.

Euphorbia humistrata **Engelm.**
Euphorbiaceae
= *Chamaesyce humistrata* (Engelm.)
Small
♦ spreading spurge, prostrate spurge
♦ Weed
♦ 80, 88, 102, 161, 243
♦ herbal, toxic.

Euphorbia hypericifolia **L.**
Euphorbiaceae
= *Chamaesyce hypericifolia* (L.) Millsp.
♦ milkweed, large spotted spurge,
garden spurge, black purslane, milk
purslane, eyebright, flux weed, spurge,
golondrina
♦ Weed, Naturalised
♦ 87, 88, 109, 170, 191, 241, 281, 287,
295
♦ arid, herbal.

Euphorbia hyssopifolia **L.**
Euphorbiaceae
= *Chamaesyce hyssopifolia* (L.) Small
♦ hyssop spurge, painted spurge
♦ Weed, Naturalised
♦ 28, 55, 87, 88, 93, 98, 161, 191, 199,
203, 205, 218, 243, 245, 287
♦ aH, herbal, toxic.

Euphorbia inaequilatera **Sond.**
Euphorbiaceae
= *Chamaesyce inaequilatera* (Sond.) Sojak
♦ prostrate spurge, smooth creeping
milkweed, smooth prostrate euphorbia
♦ Weed, Native Weed
♦ 50, 51, 87, 88, 121, 158, 243
♦ aH, herbal. Origin: southern Africa.

Euphorbia indica **Lam.**
Euphorbiaceae
♦ Weed
♦ 88, 185, 242, 243
♦ arid.

Euphorbia ingens **E.Mey. ex Boiss.**
Euphorbiaceae
♦ cactus euphorbia, candelabra
euphorbia, candelabra tree, common
tree euphorbia, tree euphorbia
♦ Weed, Native Weed
♦ 39, 121
♦ T, arid, cultivated, herbal, toxic.
Origin: southern Africa.

Euphorbia kalaharica **Marloth**
Euphorbiaceae
♦ Quarantine Weed
♦ 220

Euphorbia kansui **Liou ex S.B.Ho**
Euphorbiaceae
♦ Weed
♦ 275
♦ pH, herbal.

Euphorbia lactea **Haw.**
Euphorbiaceae
♦ mottled spurge, candelabra cactus
♦ Weed, Naturalised, Garden Escape
♦ 39, 101, 161, 179, 189, 247, 261
♦ S/T, arid, cultivated, herbal, toxic.

Euphorbia lagascae **Spreng.**
Euphorbiaceae
♦ Introduced
♦ 228
♦ arid.

Euphorbia lamarckii **Sweet**
Euphorbiaceae
♦ Quarantine Weed
♦ 220

Euphorbia lanata **Sieber ex Spreng.**
Euphorbiaceae
♦ Weed
♦ 87, 88

Euphorbia lasiocarpa **Klotzsch**
Euphorbiaceae
= *Chamaesyce lasiocarpa* (Klotzsch)
Arthur
♦ Weed
♦ 87, 88
♦ arid, herbal.

Euphorbia lathyris **L.**
Euphorbiaceae
Euphorbia lathyrus L. (see), *Tithymalus
lathyris* (L.) Hill
♦ caper spurge
♦ Weed, Quarantine Weed,
Naturalised, Introduced, Garden
Escape
♦ 15, 24, 34, 39, 40, 70, 76, 80, 87, 88,
101, 109, 160, 161, 165, 228, 247, 272,
280, 287, 295, 300
♦ a/pH, arid, cultivated, herbal, toxic.
Origin: southern Europe to China.

Euphorbia lathyrus **L.**
Euphorbiaceae
= *Euphorbia lathyris* L.
♦ caper spurge
♦ Weed, Quarantine Weed, Noxious
Weed, Naturalised, Garden Escape,
Environmental Weed
♦ 7, 23, 72, 86, 98, 116, 147, 176, 198,
203, 220, 241, 269
♦ pH, cultivated, herbal. Origin:
southern Europe, west Asia.

Euphorbia ledienii **Berger**
Euphorbiaceae
♦ Native Weed
♦ 121
♦ pS, arid, cultivated, toxic. Origin:
South Africa.

Euphorbia leptocaula **Boiss.**
Euphorbiaceae

♦ Weed
♦ 87, 88

Euphorbia leucocephala **Lotsy**
Euphorbiaceae
♦ pascuita, little Christmas flower
♦ Naturalised, Garden Escape
♦ 101, 261
♦ cultivated.

Euphorbia longifolia **D.Don**
Euphorbiaceae
♦ Quarantine Weed
♦ 220
♦ herbal.

Euphorbia lorentzii **Müll.Arg.**
Euphorbiaceae
♦ Weed
♦ 237, 270
♦ herbal.

Euphorbia lucida **Waldst. & Kit.**
Euphorbiaceae
♦ shining spurge
♦ Weed, Naturalised
♦ 87, 88, 101, 218, 272
♦ cultivated.

Euphorbia lunulata **Bunge**
Euphorbiaceae
= *Euphorbia esula* L.
♦ Weed
♦ 275
♦ pH.

Euphorbia maculata **L.**
Euphorbiaceae
= *Chamaesyce maculata* (L.) Small
♦ spotted spurge, prostrate spurge,
täplätyräkki
♦ Weed, Naturalised, Environmental
Weed, Casual Alien
♦ 39, 42, 80, 86, 87, 88, 94, 98, 155, 161,
198, 203, 204, 210, 211, 212, 218, 237,
247, 263, 272, 280, 286, 287, 295, 300
♦ aH, herbal, toxic.

Euphorbia makinoi **Hayata**
Euphorbiaceae
♦ Weed
♦ 235

Euphorbia marginata **Pursh**
Euphorbiaceae
Dichrophyllum marginatum Klotzch &
Garcke, *Euphorbia variegata* Sims (see),
Leptadenia marginata (Pursh) Niewland
♦ snow on the mountain, white
margined spurge, variegated spurge,
smoke on the prairie
♦ Weed, Naturalised, Native Weed,
Introduced, Garden Escape
♦ 7, 23, 39, 49, 86, 87, 88, 98, 109, 161,
174, 189, 203, 210, 218, 228, 247
♦ aH, arid, cultivated, herbal, toxic.
Origin: North America.

Euphorbia mauritanica **L.**
Euphorbiaceae
♦ jackal's food, yellow milkbush,
spurge
♦ Weed, Native Weed, Casual Alien
♦ 39, 121, 280
♦ pS, cultivated, herbal, toxic. Origin:
southern Africa.

Euphorbia medicaginea **Boiss.**
Euphorbiaceae
♦ Weed
♦ 87, 88

Euphorbia mercurialina **Michx.**
Euphorbiaceae
♦ mercury spurge
♦ Weed
♦ 39, 179
♦ herbal, toxic.

Euphorbia micromera **Boiss. ex Engelm.**
Euphorbiaceae
= *Chamaesyce micromera* (Boiss.) Woot. & Standl.
♦ littleleaf spurge
♦ Weed
♦ 87, 88, 161, 218
♦ herbal.

Euphorbia microphylla **B.Heyne ex Roth**
Euphorbiaceae
♦ Weed, Quarantine Weed
♦ 76, 87, 88, 203, 220

Euphorbia milii **Des Moul.**
Euphorbiaceae
Euphorbia splendens Bojer ex Hook.
♦ Christ plant, crown of thorns, corona de cristo
♦ Weed, Naturalised, Garden Escape
♦ 39, 86, 87, 88, 101, 161, 194, 247, 261
♦ S, cultivated,
herbal, toxic. Origin: Madagascar.

Euphorbia multiceps **Berger**
Euphorbiaceae
♦ Quarantine Weed
♦ 220
♦ cultivated.

Euphorbia myrsinites **L.**
Euphorbiaceae
♦ myrtle spurge, spurge
♦ Weed, Noxious Weed, Naturalised
♦ 39, 79, 80, 101, 229, 247, 272
♦ cultivated, herbal, toxic.

Euphorbia nematocypha **Hand.-Mazz.**
Euphorbiaceae
♦ Quarantine Weed
♦ 220

Euphorbia neriifolia **L.**
Euphorbiaceae
♦ Indian spurgetree
♦ Naturalised, Garden Escape
♦ 101, 261, 287
♦ arid, cultivated, herbal.

Euphorbia nubica **N.E.Br.**
Euphorbiaceae
♦ Weed
♦ 221
♦ arid. Origin: Africa.

Euphorbia nutans **Lag.**
Euphorbiaceae
= *Chamaesyce nutans* (Lag.) Small
♦ nodding spurge
♦ Weed, Naturalised
♦ 88, 94, 98, 161, 199, 203, 272, 280
♦ cultivated, herbal, toxic.

Euphorbia oblongata **Griseb.**
Euphorbiaceae
♦ oblong spurge, eggleaf spurge

♦ Weed, Noxious Weed, Naturalised
♦ 35, 88, 101, 229
♦ pH, cultivated.

Euphorbia obovata **Decne.**
Euphorbiaceae
♦ Weed
♦ 221

Euphorbia obtusifolia **Poir.**
Euphorbiaceae
♦ Quarantine Weed
♦ 220
♦ cultivated.

Euphorbia oerstediana **(Klotzsch & Garcke) Boiss.**
Euphorbiaceae
♦ West Indian spurge
♦ Weed, Quarantine Weed
♦ 76, 87, 88, 203, 220
♦ herbal.

Euphorbia ovalifolia **Engelm. ex Klotzsch & Garcke**
Euphorbiaceae
♦ Weed
♦ 87, 88
♦ herbal.

Euphorbia pachypodioides **Boiteau**
Euphorbiaceae
♦ Weed, Quarantine Weed
♦ 76, 88, 220
♦ arid.

Euphorbia palustris **L.**
Euphorbiaceae
Pityussa grandis Thal, *Tithymalus fruticosus* Gilib., *Tithymalus palustris* Hill
♦ rantatyräkki
♦ Weed
♦ 39, 87, 88, 272
♦ a/pH, cultivated, herbal, toxic.

Euphorbia papillosa **St.-Hil.**
Euphorbiaceae
♦ spurge
♦ Weed
♦ 87, 88, 237, 295
♦ herbal.

Euphorbia paralias **L.**
Euphorbiaceae
♦ sea spurge, spurge
♦ Weed, Quarantine Weed, Naturalised, Environmental Weed
♦ 7, 39, 72, 86, 88, 101, 176, 198, 203, 220, 221, 289, 296
♦ pH, herbal, toxic. Origin: Europe, North Africa.

Euphorbia patagonica **Hieron.**
Euphorbiaceae
= *Euphorbia collina* Phil. var. *patagonica* (Hieron.) Subils
♦ Weed
♦ 87, 88

Euphorbia peplis **L.**
Euphorbiaceae
♦ purple spurge
♦ Weed
♦ 39, 70
♦ herbal, toxic. Origin: Mediterranean.

Euphorbia peplus **L.**
Euphorbiaceae
Euphorbia hyrcana Grossh., *Galarhoeus peplus* (L.) Haw. ex Small, *Tithymalus peplus* (L.) Hill
♦ petty spurge, milkweed, radium plant, cancer weed, stinging milkweed
♦ Weed, Quarantine Weed, Naturalised, Introduced, Garden Escape, Environmental Weed
♦ 7, 9, 15, 23, 34, 38, 39, 44, 51, 70, 80, 86, 87, 88, 93, 94, 98, 101, 115, 118, 121, 134, 136, 158, 161, 165, 176, 180, 185, 198, 199, 203, 205, 218, 220, 221, 228, 237, 241, 243, 253, 269, 270, 272, 280, 287, 295, 300
♦ aH, arid, cultivated, herbal, toxic. Origin: Europe, Middle East.

Euphorbia perrieri **Drake**
Euphorbiaceae
♦ Quarantine Weed
♦ 220
♦ cultivated.

Euphorbia pilulifera **L.**
Euphorbiaceae
= *Chamaesyce hirta* (L.) Millsp.
♦ pill bearing spurge, snakeweed, cat's hair, Queensland asthma weed, flowery headed spurge
♦ Weed, Naturalised
♦ 39, 88, 109, 245, 286, 287
♦ arid, herbal, toxic.

Euphorbia pinea **L.**
Euphorbiaceae
♦ Weed
♦ 87, 88

Euphorbia pinetorum **(Small) G.L.Webster**
Euphorbiaceae
♦ pineland spurge
♦ Quarantine Weed
♦ 220

Euphorbia pithyusa **L.**
Euphorbiaceae
♦ sea spurge
♦ Quarantine Weed
♦ 39, 220
♦ cultivated, herbal, toxic. Origin: Mediterranean.

Euphorbia platyphyllos **L.**
Euphorbiaceae
♦ broad leaved spurge, leveälehtityräkki, broad spurge
♦ Weed, Naturalised, Cultivation Escape
♦ 39, 42, 44, 70, 86, 87, 88, 94, 101, 161, 198, 241, 252, 253, 272, 280, 300
♦ cultivated, herbal, toxic. Origin: Eurasia.

Euphorbia polychroma **A.Kern.**
Euphorbiaceae
♦ kultatyräkki
♦ Cultivation Escape
♦ 42
♦ cultivated, herbal.

Euphorbia polycnemoides **Hochst. ex Boiss.**
Euphorbiaceae

♦ Weed
♦ 87, 88

Euphorbia portulacoides L. var.
portulacoides emend Spreng
Euphorbiaceae
♦ pichoa
♦ Weed
♦ 295

Euphorbia preslii Guss.
Euphorbiaceae
= *Chamaesyce nutans* (Lag.) Small
♦ spotted spurge
♦ Weed
♦ 39, 87, 88, 210
♦ herbal, toxic.

Euphorbia primulifolia Bak.
Euphorbiaceae
♦ Quarantine Weed
♦ 220
♦ Origin: Madagascar.

Euphorbia prostrata Aiton
Euphorbiaceae
= *Chamaesyce prostrata* (Ait.) Small
♦ blue weed, ground spurge, red
caustic weed, golondrina
♦ Weed, Naturalised
♦ 39, 50, 87, 88, 93, 98, 134, 161, 185,
191, 199, 203, 205, 212, 242, 243, 245,
276, 281
♦ aH, arid, herbal, toxic.

Euphorbia prunifolia Jacq.
Euphorbiaceae
= *Euphorbia heterophylla* L.
♦ spurge, wild spurge, painted
euphorbia
♦ Weed, Naturalised
♦ 67, 87, 88, 161, 170, 191, 217, 286, 287

Euphorbia pseudocactus A.Berger
Euphorbiaceae
♦ Quarantine Weed
♦ 220
♦ arid, cultivated.

Euphorbia pseudodendroides H.Lindb.
Euphorbiaceae
= *Euphorbia lamarckii* Sweet var.
pseudodendroides (H.Lindb.) Oudejans
(NoR)
♦ Quarantine Weed
♦ 220

Euphorbia pterococca Brot.
Euphorbiaceae
♦ Weed
♦ 70, 87, 88

Euphorbia pulcherrima Willd. ex
Klotzsch
Euphorbiaceae
♦ poinsettia, Christmas flower,
Christmas star, Easter flower
♦ Weed, Naturalised, Garden Escape,
Environmental Weed
♦ 39, 80, 101, 154, 161, 189, 194, 247,
257, 261
♦ cultivated, herbal, toxic.

Euphorbia radians Benth.
Euphorbiaceae
♦ sun spurge
♦ Weed, Quarantine Weed

♦ 199, 220
♦ cultivated.

Euphorbia regis-jubae Webb & Berthel.
Euphorbiaceae
= *Tithymalus regis-jubae* (Webb &
Berthel.) Klotzsch & Garcke
♦ Quarantine Weed
♦ 39, 220
♦ cultivated, toxic.

Euphorbia reinwardtiana Steud.
Euphorbiaceae
= *Euphorbia serrulata* Thuill.
♦ Weed
♦ 87, 88

Euphorbia restiacea Benth.
Euphorbiaceae
♦ Quarantine Weed
♦ 220

Euphorbia retusa Forssk.
Euphorbiaceae
♦ Weed
♦ 221

Euphorbia rossica P.Smirn.
Euphorbiaceae
♦ Weed
♦ 87, 88

Euphorbia royleana Boiss.
Euphorbiaceae
♦ churee
♦ Quarantine Weed
♦ 39, 220, 247
♦ cultivated, herbal, toxic.

Euphorbia sagittaria Marloth
Euphorbiaceae
♦ Quarantine Weed
♦ 220

Euphorbia salicifolia Host
Euphorbiaceae
♦ Weed
♦ 272
♦ cultivated.

Euphorbia sanguinea Hochst. & Steud.
Euphorbiaceae
♦ Weed
♦ 87, 88

Euphorbia schimperiana Scheele
Euphorbiaceae
♦ Weed
♦ 88, 240
♦ aH.

Euphorbia schinzii Pax.
Euphorbiaceae
♦ Native Weed
♦ 121
♦ S, arid, cultivated. Origin: southern
Africa.

Euphorbia schlechtendalii Boiss.
Euphorbiaceae
♦ Quarantine Weed
♦ 220
♦ herbal.

Euphorbia scordifolia Jacq.
Euphorbiaceae
♦ Weed
♦ 87, 88, 221

Euphorbia segetalis L.
Euphorbiaceae
♦ shortstem carnation weed, meadow
spurge, grainfield spurge
♦ Weed, Quarantine Weed,
Naturalised, Environmental Weed
♦ 7, 39, 70, 76, 86, 87, 88, 94, 98, 101,
203, 253, 280
♦ cultivated, toxic. Origin:
Mediterranean.

Euphorbia seguierana Neck.
Euphorbiaceae
Euphorbia gerardiana Jacq.
♦ Weed
♦ 272
♦ cultivated.

Euphorbia serpens H.B.K.
Euphorbiaceae
= *Chamaesyce serpens* (Kunth) Small
♦ menoa
♦ Weed, Quarantine Weed, Casual
Alien
♦ 39, 76, 87, 88, 199, 203, 220, 237, 280,
295
♦ herbal, toxic.

Euphorbia serpens H.B.K. var. *montev*
Kunth
Euphorbiaceae
♦ meona
♦ Weed
♦ 236

Euphorbia serpens H.B.K. var. *serpens*
Kunth
Euphorbiaceae
♦ Weed
♦ 270

Euphorbia serpyllifolia Pers.
Euphorbiaceae
= *Chamaesyce serpyllifolia* (Pers.) Small
ssp. *serpyllifolia*
♦ Weed
♦ 39, 87, 88, 243
♦ aH, promoted, herbal, toxic.

Euphorbia serrata L.
Euphorbiaceae
♦ toothed spurge, serrate spurge
♦ Weed, Noxious Weed, Naturalised
♦ 35, 39, 87, 88, 94, 101, 121, 161, 218,
229, 253
♦ pH, toxic. Origin: Eurasia.

Euphorbia serrula Engelm.
Euphorbiaceae
= *Chamaesyce serrula* (Engelm.) Woot.
& Standl.
♦ sawtooth spurge
♦ Weed
♦ 88, 218
♦ herbal.

Euphorbia serrulata Thuill.
Euphorbiaceae
Euphorbia stricta L. (see), *Euphorbia
reinwardtiana* Steud. (see)
♦ upright spurge
♦ Weed
♦ 272
♦ cultivated, herbal.

Euphorbia spathulata Lam.
Euphorbiaceae
♦ netseed spurge, spurge, warty spurge
♦ Weed
♦ 161
♦ aH, herbal.

Euphorbia stellata Willd.
Euphorbiaceae
♦ Weed, Quarantine Weed, Naturalised
♦ 76, 86, 88, 220
♦ arid, cultivated.

Euphorbia stevenii Bailey
Euphorbiaceae
♦ bottletree spurge
♦ Naturalised
♦ 86
♦ cultivated. Origin: Australia.

Euphorbia striata Thunb.
Euphorbiaceae
♦ milkweed, milkwood, spurge
♦ Native Weed
♦ 121
♦ pH, herbal. Origin: southern Africa.

Euphorbia stricta L.
Euphorbiaceae
= *Euphorbia serrulata* Thuill.
♦ upright spurge
♦ Weed, Naturalised
♦ 87, 88, 280
♦ cultivated.

Euphorbia strigosa Hook. & Arn.
Euphorbiaceae
♦ Quarantine Weed
♦ 220

Euphorbia sulcata De Lens ex Loisel.
Euphorbiaceae
♦ Weed
♦ 87, 88, 94

Euphorbia supina Raf.
Euphorbiaceae
= *Chamaesyce maculata* (L.) Small.
♦ prostrate spurge
♦ Weed, Quarantine Weed, Naturalised
♦ 76, 87, 88, 93, 136, 191, 203, 205, 218, 220, 263, 270, 286, 287, 297
♦ aH, herbal.

Euphorbia tarapacana Phil.
Euphorbiaceae
♦ Weed
♦ 87, 88

Euphorbia taurinensis All.
Euphorbiaceae
Euphorbia graeca Boiss. & Sprun.
♦ Weed
♦ 272

Euphorbia terracina L.
Euphorbiaceae
♦ false caper, Geraldton carnation weed, terracina spurge
♦ Weed, Quarantine Weed, Noxious Weed, Naturalised
♦ 7, 9, 76, 86, 87, 88, 98, 101, 147, 171, 198, 199, 203, 220
♦ Origin: Mediterranean.

Euphorbia thymifolia L.
Euphorbiaceae
= *Chamaesyce thymifolia* (L.) Millsp.
♦ spurge, Nam nom raatchasee lek
♦ Weed, Naturalised
♦ 39, 66, 87, 88, 98, 203, 209, 235, 239, 262, 274, 275, 276, 286, 297
♦ aH, herbal, toxic.

Euphorbia tinctoria Boiss. & É.Huet.
Euphorbiaceae
= *Euphorbia macroclada* Boiss. (NoR)
♦ Weed
♦ 87, 88
♦ pH, herbal. Origin: Middle East.

Euphorbia tirucalli L.
Euphorbiaceae
♦ Indian tree spurge, naked lady, pencil tree, milk bush, mnyala
♦ Weed, Naturalised, Introduced, Garden Escape, Environmental Weed
♦ 39, 54, 86, 88, 101, 161, 179, 228, 247, 257, 261
♦ arid, cultivated, herbal, toxic. Origin: Africa.

Euphorbia trichotoma Kunth
Euphorbiaceae
♦ sanddune spurge
♦ Weed
♦ 14

Euphorbia turcomanica Boiss.
Euphorbiaceae
♦ Weed
♦ 87, 88

Euphorbia vachellii Hook. & Arn.
Euphorbiaceae
♦ Weed
♦ 286

Euphorbia variegata Sims
Euphorbiaceae
= *Euphorbia marginata* Pursh
♦ Weed, Quarantine Weed
♦ 88, 220
♦ herbal.

Euphorbia veneta Willd.
Euphorbiaceae
= *Euphorbia characias* L. ssp. *wulfenii* (Hoppe ex W.D.J.Koch) Radcl.-Sm.
♦ Weed, Quarantine Weed, Naturalised
♦ 76, 86, 88
♦ cultivated. Origin: east Mediterranean.

Euphorbia vermiculata Raf.
Euphorbiaceae
= *Chamaesyce vermiculata* (Raf.) House
♦ hairy spurge
♦ Weed
♦ 87, 88, 161, 218
♦ herbal, toxic.

Euphorbia viguieri Denis
Euphorbiaceae
♦ Weed, Quarantine Weed
♦ 76, 88, 220
♦ arid, cultivated. Origin: Madagascar.

Euphorbia villosa Waldst. & Kit.
Euphorbiaceae
♦ hairy spurge

♦ Weed
♦ 272
♦ cultivated.

Euphorbia virgata Waldst. & Kit.
Euphorbiaceae
= *Euphorbia esula* L. var. *uralensis* (Fisch. ex Link) Dorn
♦ leafy spurge, tithymal, Faitour's grass
♦ Weed, Quarantine Weed
♦ 49, 87, 88, 220
♦ herbal.

Euphorbia wulfenii Hoppe ex W.D.J.Koch
Euphorbiaceae
= *Euphorbia characias* L. ssp. *wulfenii* (Hoppe ex W.D.J.Koch) Radcl.-Sm.
♦ Dalmatian spurge
♦ Quarantine Weed
♦ 39, 220
♦ cultivated, herbal, toxic. Origin: east Mediterranean.

Euphrasia arctica Lange
Scrophulariaceae
Euphrasia brevipila auct.
♦ greater eyebright
♦ Weed
♦ 272
♦ herbal.

Euphrasia disjuncta Fern. & Wieg.
Scrophulariaceae
♦ polar eyebright
♦ Naturalised
♦ 101
♦ herbal.

Euphrasia micrantha Rchb.
Scrophulariaceae
♦ northern eyebright, common slender eyebright, nummisilmäruoho
♦ Naturalised
♦ 101

Euphrasia minima Jacq. ex DC.
Scrophulariaceae
♦ eufrasia minima
♦ Weed
♦ 272
♦ herbal.

Euphrasia nemorosa (Pers.) Wallr.
Scrophulariaceae
♦ common eyebright, tanakkasilmäruoho
♦ Naturalised
♦ 280
♦ herbal.

Euphrasia odontites L.
Scrophulariaceae
= *Odontites verna* (Bell.) Dumort.
♦ red bartsia
♦ Weed
♦ 44, 243

Euphrasia officinalis L. nom. ambig.
Scrophulariaceae
= *Euphrasia rostkoviana* Hayne
♦ eyebright, common eyebright, drug eyebright
♦ Weed
♦ 272
♦ aH parasitic, promoted, herbal.

Euphrasia rostkoviana Hayne
Scrophulariaceae
Euphrasia officinalis L. *nom. ambig.* (see)
♦ large flowered sticky eyebright, eyebright
♦ Weed
♦ 44, 272
♦ cultivated, herbal.

Euphrasia stricta D.Wolff ex J.F.Lehm.
Scrophulariaceae
Euphrasia tavastiensis W.Becker, *Euphrasia brevipila* Burnat & Gremli, *Euphrasia asturica* Pugsley, *Euphrasia brevipila* Burnat & Gremli ssp. *brevipila*, *Euphrasia condensata* Jord., *Euphrasia parviflora sensu* Lange *p.p. non* Fr., *Euphrasia majalis* Jord., *Euphrasia ericetorum* Jord. ex Reut., *Euphrasia reuteri* Wettst., *Euphrasia pumila* A.Kern., *Euphrasia tatarica auct. non* Fisch. ex Spreng.
♦ drug eyebright, ketosilmäruoho
♦ Weed, Naturalised
♦ 101, 272
♦ cultivated, herbal.

Euphrasia tetraquetra (Brèb) Arrond.
Scrophulariaceae
♦ maritime eyebright, broad leaved eyebright
♦ Naturalised
♦ 101

Euryops abrotanifolius (L.) DC.
Asteraceae
♦ euryops, geelmagriet
♦ Weed, Naturalised, Garden Escape, Environmental Weed, Casual Alien
♦ 72, 86, 88, 98, 176, 198, 203, 280, 289
♦ S, cultivated. Origin: southern Africa.

Euryops annae Phill.
Asteraceae
♦ Native Weed
♦ 121
♦ pS, herbal. Origin: southern Africa.

Euryops anthemoides B.Nord.
Asteraceae
♦ Native Weed
♦ 121
♦ S. Origin: South Africa.

Euryops brachypodus (DC.) B.Nord.
Asteraceae
♦ Native Weed
♦ 121
♦ pS. Origin: South Africa.

Euryops brevipapposus M.D.Hend.
Asteraceae
♦ Native Weed
♦ 121
♦ pS, cultivated, toxic. Origin: South Africa.

Euryops caespitosus Mark.
Asteraceae
♦ Quarantine Weed
♦ 220

Euryops candollei Harv.
Asteraceae
Jacobaeastrum candollei Kuntze (see)
♦ Quarantine Weed
♦ 220

Euryops chrysanthemoides (DC.) R.Nord.
Asteraceae
♦ daisybush, bull's eye
♦ Weed, Naturalised, Native Weed, Casual Alien
♦ 86, 98, 101, 121, 179, 203, 280
♦ pS, cultivated. Origin: southern Africa.

Euryops empetrifolius DC.
Asteraceae
Euryops sulcatus (Thunb.) Harv. var. *densifolius* Sond. ex Harv., *Euryops sulcatus auct. non* (Thunb.) Harv.
♦ resin bush
♦ Native Weed
♦ 121
♦ S, cultivated. Origin: South Africa.

Euryops floribundus N.E.Br.
Asteraceae
♦ Kamdeboo resin bush, resin bush, rosin bush, pimple bush
♦ Weed, Native Weed
♦ 51, 87, 88, 121
♦ pS. Origin: South Africa.

Euryops galpinii Phill.
Asteraceae
♦ Quarantine Weed
♦ 220

Euryops imbricatus (Thunb.) DC.
Asteraceae
Euryops lateriflorus (L.f.) DC. var. *imbricatus* (Thunb.) Harv., *Othonna imbricata* Thunb.
♦ Quarantine Weed
♦ 220

Euryops lateriflorus (L.f.) DC.
Asteraceae
♦ soetharpuisbos
♦ Native Weed
♦ 121
♦ pS, cultivated. Origin: southern Africa.

Euryops multifidus (Thunb.) DC.
Asteraceae
♦ sweet resin bush, hawk's eye, resin bush
♦ Weed, Naturalised, Native Weed
♦ 80, 101, 121
♦ pS, cultivated. Origin: southern Africa.

Euryops multinervis N.E.Br. ex S.Moore
Asteraceae
♦ Quarantine Weed
♦ 220

Euryops neptunicus S.Moore
Asteraceae
♦ Quarantine Weed
♦ 220

Euryops oligoglossus DC. ssp. oligoglossus
Asteraceae

♦ resin bush
♦ Native Weed
♦ 121
♦ pS. Origin: southern Africa.

Euryops othonnoides (DC.) B.Nord.
Asteraceae
Ruckeria euryopoides Drège ex Harv., *Ruckeria euryopsidis* DC., *Ruckeria othonnoides* DC.
♦ Quarantine Weed
♦ 220
♦ cultivated.

Euryops pectinatus (L.) Cass.
Asteraceae
♦ grey leaved euryops
♦ Quarantine Weed
♦ 220
♦ cultivated.

Euryops setiloba N.E.Br.
Asteraceae
Euryops setilobus N.E.Br.
♦ Quarantine Weed
♦ 220

Euryops spathaceus DC.
Asteraceae
♦ Weed
♦ 23, 88

Euryops speciosissimus DC.
Asteraceae
Euryops athanasiae (L.f.) Harv. (Less.)
♦ resin bush, Clanwilliam euryops
♦ Native Weed
♦ 121
♦ pS, cultivated. Origin: southern Africa.

Euryops striata N.E.Br.
Asteraceae
Euryops striatus N.E.Br.
♦ Quarantine Weed
♦ 220

Euryops subcarnosus DC. ssp. vulgaris B.Nord.
Asteraceae
♦ sweet resin bush
♦ Weed
♦ 26
♦ cultivated.

Euryops tenuissimus (L.) DC.
Asteraceae
♦ resin bush
♦ Native Weed
♦ 121
♦ pS, cultivated, toxic. Origin: southern Africa.

Euryops transvaalensis Klatt
Asteraceae
♦ Quarantine Weed
♦ 220
♦ cultivated.

Eustachys caribaea (Spreng.) Herter
Poaceae
♦ Caribbean fingergrass
♦ Naturalised
♦ 101
♦ G.

Eustachys distichophylla (Lag.) Nees
Poaceae
Chloris distichophylla Lag. (see)
♦ weeping fingergrass, evergreen
chloris
♦ Weed, Naturalised
♦ 7, 86, 98, 101, 179, 203, 241
♦ G. Origin: South America.

**Eustachys paspaloides (Vahl) Lanza &
Mattei**
Poaceae
♦ fangrass, brown Rhodes grass, red
Rhodes grass
♦ Native Weed
♦ 121
♦ pG, cultivated, toxic. Origin: Africa.

Eustachys petraea (Sw.) Desv.
Poaceae
Chloris petraea Sw. (see)
♦ pinewoods fingergrass, rock
fingergrass
♦ Weed, Introduced
♦ 230, 249
♦ G, herbal.

Eustachys retusa (Lag.) Kunth
Poaceae
♦ Argentine fingergrass
♦ Naturalised
♦ 101
♦ G.

Eusteralis stellata Murata
Lamiaceae
♦ Weed
♦ 286
♦ aqua, cultivated.

Eustoma exaltatum (L.) Salisb. ex G.Don
Gentianaceae
♦ catchfly prairie gentian, catchfly
gentian
♦ Weed
♦ 14
♦ a/pH, cultivated, herbal.

Evax multicaulis DC.
Asteraceae
= *Evax verna* Raf. var. *verna* (NoR)
♦ manystem evax
♦ Weed
♦ 87, 88, 218

Evax pygmaea (L.) Brot.
Asteraceae
Filago pygmaea L.
♦ laate
♦ Weed, Naturalised, Casual Alien
♦ 42, 86, 98, 203
♦ cultivated. Origin: Mediterranean.

Evodia hortensis Forst.
Rutaceae
♦ Introduced
♦ 230
♦ S, herbal.

Evodia ponapensis Kaneh. & Hatus.
Rutaceae
♦ Introduced
♦ 230
♦ S.

Evolvulus alsinoides (L.) L.
Convolvulaceae
♦ slender dwarf morningglory
♦ Weed
♦ 66, 87, 88, 157, 179, 221, 297
♦ pH, cultivated, herbal. Origin:
pantropical.

**Evolvulus alsinoides (L.) L. var. debilis
(Kunth) Ooststr.**
Convolvulaceae
♦ slender dwarf morningglory
♦ Naturalised, Introduced
♦ 101, 261

**Evolvulus alsinoides (L.) L. var. linifolius
(L.) Bak.**
Convolvulaceae
= *Evolvulus alsinoides* (L.) L. var.
angustifolius Torr. (NoR)
♦ Native Weed
♦ 121
♦ pH. Origin: southern Africa.

Evolvulus filipes Mart.
Convolvulaceae
♦ Maryland dwarf morningglory
♦ Naturalised
♦ 101

Evolvulus glomeratus Nees & Mart.
Convolvulaceae
♦ blue daze
♦ Naturalised
♦ 101
♦ cultivated.

**Evolvulus glomeratus Nees & Mart. ssp.
grandiflorus (Parodi) Ooststr.**
Convolvulaceae
♦ Weed, Naturalised
♦ 101, 179

Evolvulus nummularius (L.) L.
Convolvulaceae
Convolvulus nummularius L., *Evolvulus
capreolatus* Mart. ex Choisy, *Evolvulus
dichondroides* Oliv., *Evolvulus
domingensis* Spreng. ex Choisy,
Evolvulus nummularius var. *grandifolia*
Hoehne, *Evolvulus reniformis* Salzm.
ex Choisy, *Evolvulus repens* D.Parodi,
Evolvulus veronicaeifolius Kunth,
Evolvulus yunnanensis S.H.Huang,
Volvulopsis nummularium (L.) Roberty
♦ agracejo rastrero
♦ Weed, Naturalised, Environmental
Weed
♦ 86, 93, 216, 221
♦ Origin: Tropics.

Evolvulus sericeus Sw.
Convolvulaceae
♦ silver dwarf morningglory
♦ Weed
♦ 14, 295
♦ herbal.

Evolvulus tenuis Mart. ex Choisy
Convolvulaceae
♦ skyblue dwarf morningglory
♦ Weed
♦ 87, 88

Exacum pedunculatum L.
Gentianaceae
♦ Weed
♦ 66
♦ cultivated, herbal.

Evolvulus alsinoides (L.) L.
Convolvulaceae

Exochorda racemosa (Lindl.) Rehd.
Rosaceae
♦ pearlbush, common pearlbrush
♦ Weed, Naturalised, Casual Alien
♦ 80, 101, 280
♦ cultivated.

**Exomis microphylla (Thunb.) Aell. var.
axyrioides (Fenzl) Aell.**
Chenopodiaceae
♦ exomis
♦ Native Weed
♦ 121
♦ pS. Origin: southern Africa.

Exotheca chevalieri A.Camus
Poaceae
♦ Weed
♦ 87, 88
♦ G.

F

Facelis apiculata Cass.
Asteraceae
= *Facelis retusa* (Lam.) Sch.Bip.
♦ facelis
♦ Weed
♦ 87, 88, 218

Facelis retusa (Lam.) Sch.Bip.
Asteraceae
Facelis apiculata Cass. (see)
♦ annual trampweed, facelis
♦ Weed, Naturalised
♦ 51, 86, 87, 88, 98, 101, 121, 134, 161, 203, 269, 280, 295
♦ aH, arid, herbal. Origin: South America.

Facheiroa Britt. & Rose spp.
Cactaceae
Zehntnerella Britt. & Rose spp. (see)
♦ Weed, Quarantine Weed
♦ 76, 88, 203, 220

Fadogia monticola Robyns
Rubiaceae
♦ wild date
♦ Native Weed
♦ 121
♦ S. Origin: southern Africa.

Fadogia tetraquetra Krause
Rubiaceae
♦ Native Weed
♦ 121
♦ S, toxic. Origin: southern Africa.

Fagonia arabica L.
Zygophyllaceae
♦ Weed
♦ 221
♦ herbal.

Fagonia bruguieri DC.
Zygophyllaceae
♦ Weed
♦ 221
♦ herbal.

Fagonia glutinosa Delile
Zygophyllaceae
♦ Weed
♦ 221

Fagonia indica Burm.f.
Zygophyllaceae
♦ Weed
♦ 88

Fagonia tristis Sickenb.
Zygophyllaceae
♦ Weed
♦ 221

Fagopyrum cymosum (Trev.) Meisn.
Polygonaceae
= *Fagopyrum dibotrys* (D.Don.) Hara.
♦ phakbung som, buckwheat
♦ Weed, Naturalised
♦ 238, 243, 275, 287
♦ pH, herbal.

Fagopyrum dibotrys (D.Don.) Hara.
Polygonaceae
Fagopyrum cymosum (Trev.) Meisn. (see), *Polygonum chinense* L. (see), *Polygonum cymosum* Trev., *Polygonum dibotrys* D.Don, *Polygonum labordei* Lév. & Vaniot, *Polygonum tristachyum* H.Lév.
♦ perennial buckwheat
♦ Weed, Casual Alien
♦ 88, 280, 297
♦ pH, cultivated.

Fagopyrum esculentum Moench
Polygonaceae
Fagopyrum sagittatum Gilib., *Fagopyrum vulgare* Hill, *Polygonum fagopyrum* L.
♦ buckwheat, viljatatar, fagopyrum
♦ Weed, Naturalised, Cultivation Escape, Casual Alien
♦ 7, 34, 39, 42, 86, 87, 88, 94, 98, 101, 161, 179, 180, 203, 243, 247, 252, 255, 272, 280, 287
♦ aH, cultivated, herbal, toxic. Origin: Eurasia.

Fagopyrum tataricum (L.) Gaertn.
Polygonaceae
Fagopyrum rotundatum Bab., *Polygonum tataricum* L.
♦ Tartary buckwheat, buckwheat, duck wheat, sarrasin
♦ Weed, Noxious Weed, Naturalised, Casual Alien
♦ 23, 42, 87, 88, 94, 101, 114, 161, 162, 218, 243, 272, 287, 292, 297, 299
♦ aH, cultivated, herbal.

Fagraea berterana A.Gray ex Benth. var. sair (Gilg & Benedict) F.R.Fosberg
Loganiaceae/Potaliaceae
Fagraea sair Gilg & Benedict
♦ seir pwur
♦ Introduced
♦ 230
♦ T.

Fagraea racemosa Jack ex Wall.
Loganiaceae/Potaliaceae
♦ Weed
♦ 87, 88
♦ cultivated, herbal. Origin: Australia.

Fagus grandifolia Ehrh.
Fagaceae
♦ American beech
♦ Weed
♦ 39, 87, 88, 218
♦ T, cultivated, herbal, toxic.

Fagus orientalis Lipsky
Fagaceae
♦ oriental beech
♦ Weed
♦ 80
♦ T, cultivated, herbal.

Fagus sylvatica L.
Fagaceae
♦ beech, European beech, common beech
♦ Weed, Naturalised, Cultivation Escape
♦ 39, 40, 42, 80, 101, 161
♦ T, cultivated, herbal, toxic.

Fagus sylvatica L. var. cuprea Oliv.
Fagaceae
♦ copper beech
♦ Naturalised
♦ 40

Falcaria vulgaris Bernh.
Apiaceae
Falcaria rivini Host, *Falcaria sioides* (Wib.) Asch., *Prionitis falcaria* Dum.
♦ sickleweed, longleaf, sirppiputki
♦ Weed, Quarantine Weed, Naturalised, Casual Alien
♦ 40, 42, 70, 76, 86, 87, 88, 94, 101, 161, 203, 220, 243, 253, 272
♦ cultivated.

Falcataria moluccana (Miq.) Barneby & Grimes
Fabaceae/Mimosaceae
♦ peacocksplume
♦ Naturalised
♦ 101

Falkia repens L.f.
Convolvulaceae
♦ Native Weed
♦ 121
♦ pH. Origin: southern Africa.

Fallopia aubertii (Henry) Holub
Polygonaceae
= *Polygonum aubertii* Henry
♦ Russian vine, pohánkovec ãínsky
♦ Weed, Naturalised
♦ 15, 280
♦ cultivated.

Fallopia baldschuanica (Regel) Holub
Polygonaceae
= *Polygonum baldschuanicum* Regel
♦ Russian vine, bokharabinda
♦ Naturalised
♦ 40
♦ cultivated, herbal.

Fallopia × bohemica (Chrtek & Chrtková) J.Bailey.
Polygonaceae
= *Fallopia japonica* (Houtt.) Ronse Decr. × *Fallopia sachalinensis* (F.Schmidt) Ronse Decr.
♦ hörtsätatar
♦ Naturalised
♦ 40
♦ pH, herbal.

Fallopia convolvulus (L.) Á.Löve
Polygonaceae
Bilderdykia convolvulus (L.) Dumort. (see), *Fagopyrum convolvulus* (L.) H.Gross, *Polygonum convolvulus* L. (see), *Polygonum convolvulus* L. var. *convolvulus* (see), *Polygonum convolvulus* L. var. *subulatum* Lej. & Court. (see), *Reynoutria convolvulus* (L.) Shinners

♦ climbing knotweed, black knotweed, wild buckwheat, climbing buckwheat, cornbind, convolvolo nero, black bindweed
♦ Weed, Naturalised, Introduced
♦ 7, 15, 38, 53, 55, 68, 86, 88, 98, 118, 134, 158, 165, 176, 198, 203, 243, 253, 269, 280, 286, 287, 300
♦ aH, cultivated, herbal. Origin: Asia, Europe, North Africa.

Fallopia dentato-alata (F.Schmidt) Holub
Polygonaceae
Polygonum dentato-alatum F.Schmidt ex Maxim (see)
♦ Naturalised
♦ 287

Fallopia dumetorum (L.) Holub
Polygonaceae
♦ pensaikkotatar, heckenknoterich, copse bindweed, poligono delle siepi
♦ Naturalised
♦ 287
♦ cultivated, herbal.

Fallopia japonica (Houtt.) Ronse Decr.
Polygonaceae
Polygonum cuspidatum Sieb. & Zucc. (see), *Reynoutria japonica* Houtt. (see)
♦ Japanese knotweed
♦ Weed, Quarantine Weed, Noxious Weed, Naturalised, Environmental Weed, Cultivation Escape
♦ 18, 54, 76, 86, 88, 132, 139, 151, 176, 184, 184, 243, 252
♦ pC, cultivated, herbal, toxic. Origin: east Asia.

Fallopia japonica (Houtt.) Ronse Decr. var. *compacta* (Hook.f.) J.P.Bailey
Polygonaceae
Polygonum cuspidatum Sieb. & Zucc. var. *compactum* (Hook.f.) L.H.Bailey
♦ dwarf Japanese knotweed
♦ Naturalised
♦ 40
♦ herbal.

Fallopia japonica (Houtt.) Ronse Decr.var. *japonica*
Polygonaceae
♦ Japanese knotweed
♦ Naturalised
♦ 40

Fallopia sachalinensis (Schmidt) Ronse Decr.
Polygonaceae
Polygonum sachalinense Schmidt. (see), *Reynoutria sachalinensis* (Schmidt) Nakai (see)
♦ giant knotweed
♦ Naturalised, Introduced
♦ 40, 150, 198
♦ H, cultivated, herbal. Origin: northern Japan.

Famatina maulensis Ravenna
Liliaceae/Amaryllidaceae
♦ Quarantine Weed
♦ 220

Famatina saxatillis Ravenna
Liliaceae/Amaryllidaceae

♦ Quarantine Weed
♦ 220
♦ Origin: Argentina.

Farfugium japonicum (L.) Kitam.
Asteraceae
Ligularia kaempferi (DC.) Sieb. & Zucc., *Ligularia tussilaginea* (Burm.) Makino., *Senecio kaempferi* DC., *Tussilago japonica* L.
♦ Weed
♦ 87, 88
♦ pH, cultivated, herbal, toxic.

Farsetia aegyptia Turra
Brassicaceae
Farsetia edgeworthii Hook.f. & Thoms
♦ Weed
♦ 87, 88, 221
♦ arid.

Farsetia jacquemontii Hook.f. & Thoms.
Brassicaceae
♦ Weed
♦ 87, 88
♦ herbal.

Farsetia longisiliqua Decne.
Brassicaceae
♦ Weed
♦ 221, 242
♦ arid.

Farsetia ramosissima Hochst. ex Fourn.
Brassicaceae
♦ Weed
♦ 221
♦ herbal.

Farsetia stenoptera Hochst.
Brassicaceae
♦ Weed, Quarantine Weed
♦ 76, 87, 88, 203, 220

Fatoua villosa (Thunb.) Nakai
Moraceae
♦ hairy crabweed
♦ Weed, Naturalised
♦ 80, 87, 88, 101, 102, 179, 261, 263, 286
♦ aH. Origin: Asia.

Fatsia japonica (Thunb.) Decne. & Planch.
Araliaceae
♦ Japanese aralia, paper plant
♦ Weed, Naturalised
♦ 15, 280
♦ cultivated, herbal.

Fedia caput-bovis Pomel
Valerianaceae
♦ Weed
♦ 87, 88

Fedia cornucopiae (L.) Gaertn.
Valerianaceae
Fedia scorpioides Dufr.
♦ African valerian, house of plenty
♦ Weed
♦ 87, 88, 94
♦ aH, cultivated, herbal.

Fedia graciliflora Fisch. & C.A.Mey.
Valerianaceae
♦ Weed, Naturalised
♦ 86, 98, 203
♦ cultivated. Origin: North Africa.

Felicia amelloides (L.) Voss
Asteraceae
♦ Casual Alien
♦ 280
♦ cultivated, herbal.

Felicia bergeriana (Spreng.) Hoffm. ex A.Zahlbr.
Asteraceae
Cineraria bergeriana Spreng.
♦ kingfisher daisy
♦ Weed
♦ 23, 88
♦ aH, cultivated, herbal.

Felicia filifolia (Vent.) Burtt Davy
Asteraceae
♦ wild aster
♦ Native Weed
♦ 121
♦ S, cultivated, toxic. Origin: southern Africa.

Felicia fruticosa (L.) H.Nicholson
Asteraceae
♦ Naturalised
♦ 280
♦ cultivated.

Felicia muricata (Thunb.) Nees
Asteraceae
♦ wild aster
♦ Native Weed
♦ 121
♦ pS. Origin: southern Africa.

Felicia petiolata (Harv.) N.E.Br.
Asteraceae
♦ felicia
♦ Weed, Naturalised, Casual Alien
♦ 54, 86, 88, 198, 280
♦ cultivated. Origin: South Africa.

Felicia rotundifolia G.C.Taylor
Asteraceae
♦ Weed
♦ 23, 88
♦ cultivated.

× *Ferobergia* Glass spp.
Cactaceae
= *Ferocactus* Britton & Rose spp. × *Leuchtenbergia* Hook. spp.
♦ Quarantine Weed
♦ 220

Ferraria crispa Burm.
Iridaceae
♦ black flag, spinnekopblom
♦ Weed, Naturalised, Garden Escape, Environmental Weed
♦ 7, 9, 72, 88, 98, 134, 203
♦ pH, cultivated. Origin: South Africa.

Ferraria crispa Burm. ssp. *crispa*
Iridaceae
♦ black flag
♦ Naturalised, Environmental Weed
♦ 86, 198
♦ Origin: South Africa.

Ferula assa-foetida L.
Apiaceae
Ferula erubescens Boiss., *Ferula pseudalliaca* Rech.f., *Ferula rubicaulis* Boiss., *Narthex polakii* Stapf &n Wettst.
♦ asafetida

♦ Quarantine Weed
♦ 39, 220
♦ pH, arid, cultivated, herbal, toxic.

Ferula bungeana **Kitag.**
Apiaceae
♦ Weed
♦ 275
♦ pH.

Ferula communis **L.**
Apiaceae
Ferula brevifolia Hoff. & Link, *Ferula nodiflora* L., *Ferula linkii* Webb & Berthel.
♦ giant fennel, African ammoniacum
♦ Weed, Naturalised
♦ 39, 70, 86, 87, 88, 98, 203, 272
♦ pH, cultivated, herbal, toxic. Origin: Mediterranean, Middle East.

Ferula sinaica **Boiss.**
Apiaceae
♦ Weed
♦ 221

Ferulago campestris **(Besser) Grec.**
Apiaceae
Ferula ferulago L., *Ferulago galbanifera* Koch
♦ ferula finocchiazzo
♦ Weed
♦ 272
♦ cultivated, herbal.

Festuca **L. spp.**
Poaceae
♦ fescue
♦ Weed
♦ 15, 39, 80, 243, 272
♦ G, herbal, toxic.

Festuca airoides **Lam.**
Poaceae
♦ tufted fescue
♦ Naturalised
♦ 101
♦ G.

Festuca altissima **All.**
Poaceae
♦ wood fescue, kostrava lesná
♦ Weed
♦ 70
♦ G, cultivated, herbal.

Festuca ampla **Hack.**
Poaceae
♦ Weed
♦ 70
♦ G.

Festuca argentina **(Speg.) Parodi**
Poaceae
♦ coirón
♦ Weed
♦ 295
♦ G.

Festuca arundinacea **Schreb.**
Poaceae
= *Lolium arundinaceum* (Schreb.) S.J.Darbysh.
♦ tall fescue, alata fescue, reed fescue, coarse fescue, New Zealand tall fescue
♦ Weed, Noxious Weed, Naturalised, Introduced, Garden Escape,

Environmental Weed
♦ 7, 15, 34, 35, 39, 70, 72, 78, 86, 87, 88, 91, 98, 116, 121, 142, 146, 151, 161, 176, 181, 198, 199, 203, 211, 212, 225, 228, 243, 246, 252, 253, 272, 280, 286, 287, 300
♦ pG, arid, cultivated, herbal, toxic. Origin: Eurasia.

Festuca arundinacea **Schreb.** × *Lolium perenne* **L.**
Poaceae
♦ Weed
♦ 15
♦ G.

Festuca arvernensis **Auq. Kerg. & Markgr.-Dann.**
Poaceae
Festuca ovina L. var. *glauca* (Lam.) W.D.J.Koch (see)
♦ field fescue
♦ Naturalised
♦ 101
♦ G.

Festuca bromoides **L.**
Poaceae
= *Vulpia bromoides* (L.) Gray
♦ Weed
♦ 80
♦ G, herbal.

Festuca elatior **L.**
Poaceae
= *Lolium arundinaceum* (Schreb.) S.J.Darbysh.
♦ tall fescue, alta fescue, meadow fescue
♦ Weed
♦ 17, 80, 88, 93, 191, 195, 205
♦ pG, cultivated, herbal.

Festuca filiformis **Pourr.**
Poaceae
Festuca tenuifolia Sibth. (see)
♦ fineleaf sheep fescue
♦ Naturalised
♦ 101, 280, 300
♦ G, cultivated.

Festuca gigantea **(L.) Vill.**
Poaceae
= *Lolium giganteum* (L.) S.J.Darbysh.
♦ giant fescue, tall brome, lehtonata, giant green fescue
♦ Weed, Naturalised
♦ 23, 70, 88, 272, 287
♦ G, cultivated, herbal.

Festuca heteromalla **Pourr.**
Poaceae
♦ varioushair fescue, kostrava mnohokvetá
♦ Naturalised
♦ 101
♦ G.

Festuca heterophylla **Lam.**
Poaceae
♦ variousleaf fescue, kostrava rôznolistá
♦ Weed, Naturalised
♦ 101, 272, 286, 287
♦ G, cultivated, herbal.

Festuca juncifolia **St.-Amans**
Poaceae
♦ rush leaved fescue
♦ Naturalised
♦ 300
♦ G, cultivated.

Festuca kashmiriana **Stapf**
Poaceae
♦ dog fescue
♦ Naturalised
♦ 101
♦ G.

Festuca litorea **Wahlenb.**
Poaceae
♦ Weed
♦ 23, 88
♦ G.

Festuca longifolia **Thuill.**
Poaceae
♦ hard fescue, blue fescue
♦ Weed, Naturalised
♦ 80, 98, 203
♦ G, cultivated.

Festuca lugens **(E.Fourn.) Hitchc. ex Hern.-Xol.**
Poaceae
♦ Weed
♦ 199
♦ G.

Festuca megalura **Nutt.**
Poaceae
= *Vulpia myuros* (L.) C.Gmel.
♦ foxtail fescue
♦ Weed
♦ 23, 87, 88, 180, 218, 243
♦ G, arid, cultivated, herbal.

Festuca myuros **L.**
Poaceae
= *Vulpia myuros* (L.) C.Gmel.
♦ rat's tail fescue
♦ Weed
♦ 87, 88, 218
♦ G, herbal.

Festuca nigrescens **Lam.**
Poaceae
♦ Chewing's fescue
♦ Weed, Naturalised
♦ 15, 86, 176
♦ G, cultivated. Origin: Europe.

Festuca octoflora **Walter**
Poaceae
= *Vulpia octoflora* (Walter) Rydb.
♦ sixweeks fescue
♦ Weed
♦ 23, 87, 88, 218
♦ G, herbal.

Festuca ovina **L.**
Poaceae
Avena ovina Salisb., *Festuca saximontana* Rydb., *Festuca vulgaris* (W.D.J.Koch) Hayek
♦ sheep's fescue
♦ Weed, Naturalised, Introduced, Environmental Weed
♦ 21, 70, 80, 87, 88, 91, 101, 151, 272, 280, 286
♦ pG, arid, cultivated, herbal.

Festuca ovina L. ssp. *hirtula* (W.G.Travis) M.J.Wilk.
Poaceae
♦ Naturalised
♦ 280
♦ G.

Festuca ovina L. var. *glauca* (Lam.) W.D.J.Koch
Poaceae
= *Festuca arvernensis* Auq. Kerg. & Markgr.-Dann.
♦ blue fescue
♦ Weed
♦ 116
♦ G, herbal.

Festuca pallescens (St.-Yves) Parodi
Poaceae
♦ Introduced
♦ 228
♦ G, arid.

Festuca parvigluma Steud.
Poaceae
♦ toboshigara
♦ Weed
♦ 286
♦ G.

Festuca pratensis Huds.
Poaceae
Lolium pratense (Huds.) S.J.Darbysh. (see)
♦ tall fescue, ryegrass, nurminata, meadow fescue
♦ Weed, Naturalised, Introduced, Environmental Weed, Cultivation Escape
♦ 7, 21, 23, 34, 80, 86, 88, 98, 102, 146, 151, 195, 198, 203, 228, 252, 272, 286, 287
♦ pG, arid, cultivated, herbal. Origin: Eurasia.

Festuca rubra L.
Poaceae
♦ red fescue, creeping fescue, Chewing's fescue, fine fescue, creeping red fescue
♦ Weed, Naturalised, Environmental Weed
♦ 7, 39, 70, 72, 80, 87, 88, 91, 98, 151, 155, 176, 181, 198, 199, 203, 241, 243, 272, 280, 286, 300
♦ pG, cultivated, herbal, toxic.

Festuca rubra L. fo. *vivipara* S.Kawano
Poaceae
♦ mukagoooushinokegusa
♦ Naturalised
♦ 287
♦ G.

Festuca rubra L. ssp. *commutata* Gaud.
Poaceae
♦ Chewing's fescue
♦ Naturalised
♦ 198, 280
♦ G, cultivated.

Festuca rubra L. ssp. *rubra*
Poaceae
♦ creeping fescue, creeping red fescue
♦ Weed, Naturalised, Environmental Weed

♦ 15, 86, 198, 280
♦ G, cultivated. Origin: Eurasia, North America.

Festuca rubra L. var. *musashiensis* Ohwi
Poaceae
♦ Naturalised
♦ 287
♦ G.

Festuca rupicola Heuff.
Poaceae
Festuca sulcata (Hack.) Nyman.
♦ Weed
♦ 272
♦ G, cultivated.

Festuca sulcata (Hack.) Beck
Poaceae
♦ Weed
♦ 23, 88
♦ G.

Festuca tenuifolia Sibth.
Poaceae
= *Festuca filiformis* Pourr.
♦ awnless sheep's fescue, fine leaved fescue
♦ Weed, Naturalised
♦ 241, 272
♦ G, cultivated.

Festuca trachyphylla (Hack.) Krajina
Poaceae
♦ hard fescue
♦ Naturalised
♦ 101
♦ pG.

Festuca valesiaca Schleich. ex Gaudin
Poaceae
Festuca ovina L. var. *vallesiaca* Koch
♦ Volga fescue
♦ Weed, Naturalised
♦ 272
♦ G, cultivated.

× Festulolium holmbergii (Dörfl.) P.Fourn.
Poaceae
= *Festuca* L. sp. × *Lolium* L. sp.
♦ Naturalised
♦ 280
♦ G.

Fevillea cordifolia L.
Cucurbitaceae
♦ antidote caccoon
♦ Weed
♦ 39, 87, 88
♦ herbal, toxic.

Fibigia clypeata (L.) Medik.
Brassicaceae
Farsetia clypeata (L.) R.Br.
♦ Roman shields
♦ Weed, Quarantine Weed
♦ 87, 88, 220
♦ pH, cultivated, herbal.

Ficaria grandiflora Robert
Ranunculaceae
♦ Weed
♦ 87, 88

Ficaria verna Huds.
Ranunculaceae
= *Ranunculus ficaria* L.
♦ lesser celandine

♦ Weed, Naturalised
♦ 54, 86, 88, 198
♦ herbal. Origin: Europe.

Ficindica St.-Lag. spp.
Cactaceae
= *Opuntia* Mill. spp.
♦ Weed, Quarantine Weed
♦ 76, 88, 220

Ficinia filiformis (Lam.) Schrad.
Cyperaceae
♦ stargrass
♦ Native Weed
♦ 121
♦ pG. Origin: southern Africa.

Ficinia indica (Lam.) Pfeiffer
Cyperaceae
♦ stargrass
♦ Native Weed
♦ 121
♦ pG. Origin: southern Africa.

Ficus L. spp.
Moraceae
♦ ornamental figs, fig
♦ Weed, Environmental Weed
♦ 88, 151, 154
♦ herbal, toxic.

Ficus altissima Blume
Moraceae
♦ false banyan, council tree, banyan tree, lofty fig
♦ Weed, Quarantine Weed, Noxious Weed, Naturalised, Introduced, Environmental Weed
♦ 3, 39, 80, 88, 101, 112, 122, 179, 191, 220, 228
♦ arid, cultivated, toxic.

Ficus aurantiaca Griff.
Moraceae
Ficus callicarpa Miq.
♦ Weed
♦ 87, 88

Ficus benghalensis L.
Moraceae
Urostigma benghalense Gaspar
♦ banyan fig, Indian banyan, vada tree, bar
♦ Weed, Noxious Weed, Naturalised, Introduced, Garden Escape
♦ 3, 86, 101, 122, 142, 179, 191, 228, 252
♦ T, arid, cultivated, herbal. Origin: southern Asia.

Ficus benjamina L.
Moraceae
Ficus comosa Roxb., *Ficus exotica* hort., *Ficus nitida* Thunb., *Ficus waringiana* auct., *Urostigma benjaminum* (L.) Miq.
♦ weeping fig, baka, benjamin tree, Java fig, Java tree, tropic laurel, waringin, weeping fig, weeping laurel, wariengien
♦ Weed, Naturalised, Introduced
♦ 3, 7, 80, 86, 88, 101, 112, 194, 247, 261
♦ T, cultivated, herbal, toxic. Origin: south-east Asia, Australia.

Ficus carica L.
Moraceae
♦ common ficus, fig tree, edible fig, common fig, aitoviikuna, laurel de

mysore
♦ Weed, Noxious Weed, Naturalised, Introduced, Garden Escape, Environmental Weed, Casual Alien, Cultivation Escape
♦ 7, 15, 34, 35, 38, 39, 42, 78, 80, 86, 88, 98, 101, 116, 142, 151, 161, 198, 203, 221, 231, 257, 261, 280, 290
♦ T, cultivated, herbal, toxic. Origin: Mediterranean, Middle East.

Ficus drupacea Thunb.
Moraceae
♦ brown woolly fig
♦ Naturalised, Introduced
♦ 101, 261
♦ cultivated. Origin: tropical Asia.

Ficus elastica Roxb. ex Hornem.
Moraceae
♦ India rubber tree, rubber plant, komunoki, komunokí, rapah, gak'iynigoma
♦ Weed, Naturalised, Introduced, Environmental Weed
♦ 3, 38, 73, 80, 88, 101, 107, 201, 230, 259, 290
♦ T, cultivated, herbal.

Ficus exasperata Vahl
Moraceae
♦ lwago, lukenga, jolo, bolala
♦ Weed
♦ 88

Ficus fistulosa Reinw. ex Bl.
Moraceae
♦ Weed
♦ 87, 88
♦ herbal.

Ficus hirta Vahl
Moraceae
♦ Weed
♦ 87, 88

Ficus lacor Buch.-Ham.
Moraceae
♦ Quarantine Weed
♦ 220
♦ herbal.

Ficus lyrata Warb.
Moraceae
♦ fiddleleaf fig
♦ Naturalised, Introduced
♦ 101, 261
♦ cultivated, herbal. Origin: tropical Africa.

Ficus macrophylla Desf. ex Pers.
Moraceae
♦ Moreton Bay fig
♦ Sleeper Weed, Environmental Weed, Casual Alien
♦ 225, 246, 280
♦ cultivated, herbal. Origin: Australia.

Ficus macrophylla Desf. ex Pers. ssp. *macrophylla*
Moraceae
♦ Moreton Bay fig
♦ Naturalised
♦ 198

Ficus microcarpa L.f.
Moraceae
♦ laurel fig, Chinese banyan, Malayan banyan, Indian laurel
♦ Weed, Sleeper Weed, Noxious Weed, Introduced, Garden Escape, Environmental Weed, Cultivation Escape
♦ 22, 80, 88, 112, 122, 142, 179, 225, 233, 246, 261
♦ T, cultivated, herbal. Origin: Australasia.

Ficus natalensis Hochst.
Moraceae
♦ Natal fig, kahumo
♦ Quarantine Weed
♦ 220
♦ cultivated. Origin: Africa.

Ficus nekbudu Warb.
Moraceae
♦ African cloth bark tree
♦ Introduced
♦ 261
♦ cultivated. Origin: Africa.

Ficus nota (Blanco) Merr.
Moraceae
♦ tibig
♦ Naturalised
♦ 101

Ficus palmata Forssk.
Moraceae
♦ Punjab fig, wild fig
♦ Naturalised
♦ 101
♦ T, cultivated, herbal.

Ficus pseudosycomorus Decne.
Moraceae
♦ Weed
♦ 221

Ficus pumila L.
Moraceae
♦ climbing fig, creeping rubber plant, fig, creeping fig
♦ Weed, Sleeper Weed, Naturalised, Introduced, Garden Escape, Environmental Weed
♦ 39, 86, 87, 88, 98, 101, 203, 225, 246, 261, 280
♦ pC, cultivated, herbal, toxic. Origin: Asia.

Ficus religiosa L.
Moraceae
♦ bo tree, peepul tree, sacred bo tree, alamo
♦ Weed, Naturalised, Introduced
♦ 22, 80, 88, 101, 112, 261
♦ T, cultivated, herbal. Origin: tropical Asia.

Ficus retusa L.
Moraceae
Ficus nitida Bl. *nom. illeg.*
♦ Weed, Introduced
♦ 38, 87, 88
♦ T, cultivated, herbal.

Ficus rubiginosa Desf ex Vent.
Moraceae
Ficus australis Willd. *non hort.*
♦ Port Jackson fig, rusty fig, rusty leaved fig
♦ Weed, Sleeper Weed, Naturalised,

Environmental Weed, Cultivation Escape
♦ 80, 101, 225, 233, 246, 280
♦ T, cultivated. Origin: Australia.

Ficus salicifolia Vahl
Moraceae
♦ Weed
♦ 221

Ficus stricta Miq.
Moraceae
♦ Quarantine Weed
♦ 220

Ficus sur Forssk.
Moraceae
Ficus capensis Thunb.
♦ broom cluster fig, bush fig, Cape fig, fire sticks, kooman, wild fig, kafumo, besemtrosvy
♦ Native Weed
♦ 121
♦ T, cultivated. Origin: southern Africa.

Ficus sycomorus L.
Moraceae
Ficus damarensis Engl., *Ficus graphalocarpa* (Miq.) A.Rich.
♦ sycamore fig, kamsaongo
♦ Introduced
♦ 228
♦ arid, cultivated, herbal. Origin: Madagascar.

Ficus thonningii Blume
Moraceae
♦ Chinese banyan, common wild fig, jolo, kafula, gafula, kajimonsole
♦ Quarantine Weed, Naturalised
♦ 101, 220
♦ cultivated.

Ficus triangularis Warb.
Moraceae
= *Ficus natalensis* Hochst. ssp. *leprieurii* (Miq.) Berg (NoR)
♦ Quarantine Weed
♦ 220
♦ cultivated.

Ficus villosa Bl.
Moraceae
♦ Weed
♦ 87, 88

Filaginella uliginosa (L.) Opiz
Asteraceae
= *Gnaphalium uliginosum* L.
♦ marsh cudweed
♦ Weed
♦ 70, 88, 94, 272
♦ herbal.

Filago L. spp.
Asteraceae
♦ cudweed, cottonrose
♦ Weed, Naturalised
♦ 198, 272

Filago arvensis L.
Asteraceae
= *Logfia arvensis* (L.) Holub
♦ ketotuulenlento, field cottonrose
♦ Weed
♦ 44, 87, 88, 243
♦ herbal.

Filago gallica L.
Asteraceae
= *Logfia gallica* (L.) Coss. & Germ.
♦ filago, narrow leaved filago, narrowleaf cottonrose, narrow leaved cudweed, narrow cudweed, ranskantuulenlento
♦ Weed, Naturalised, Introduced, Casual Alien
♦ 7, 34, 42, 44, 87, 88, 241
♦ aH, herbal.

Filago germanica L.
Asteraceae
= *Filago vulgaris* Lam.
♦ cudweed, common cudweed
♦ Weed
♦ 44, 87, 88
♦ promoted, herbal.

Filago lutescens Jord.
Asteraceae
♦ red tipped cudweed
♦ Weed
♦ 88, 94

Filago minima (Sm.) Pers.
Asteraceae
= *Logfia minima* (Sm.) Dumort.
♦ slender cudweed, small cudweed, pikkutuulenlento, little cottonrose
♦ Weed, Casual Alien
♦ 42, 87, 88

Filago pyramidata L.
Asteraceae
♦ broadleaf cottonrose, broad leaved cudweed
♦ Weed, Naturalised, Casual Alien
♦ 42, 70, 86, 88, 94, 98, 101, 198, 203, 243, 253, 272, 280
♦ cultivated, herbal. Origin: southern Europe, Mediterranean, south-west Asia.

Filago spathulata Presl
Asteraceae
♦ spathulate cudweed, broad leaved cudweed
♦ Weed
♦ 87, 88

Filago vulgaris Lam.
Asteraceae
Filago germanica L. (see)
♦ cudweed, creeping cudweed, Jersey cudweed, marsh cudweed, slender cudweed, saksantuulenlento
♦ Weed, Naturalised, Casual Alien
♦ 42, 70, 88, 94, 101, 253, 272, 280

Filicium decipiens (Wight & Arn.) Thwaites ex Hook.f.
Sapindaceae
♦ fern tree
♦ Cultivation Escape
♦ 233
♦ cultivated.

Filipendula kamtschatica (Pall.) Maxim.
Rosaceae
♦ jättiangervo, Kamchatkan meadowsweet
♦ Cultivation Escape
♦ 42
♦ pH, cultivated.

Filipendula palmata (Pall.) Maxim.
Rosaceae
♦ palmate meadowsweet
♦ Weed
♦ 297
♦ cultivated.

Filipendula ulmaria (L.) Maxim
Rosaceae
♦ meadowsweet, queen of the meadow, sirea ulmaria
♦ Weed, Naturalised, Casual Alien
♦ 70, 87, 88, 101, 195, 272, 280
♦ pH, cultivated, herbal.

Filipendula ulmaria (L.) Maxim. ssp. *denudata* (J.& C.Presl) Hayek
Rosaceae
♦ queen of the meadow
♦ Naturalised
♦ 101

Filipendula ulmaria (L.) Maxim. ssp. *ulmaria*
Rosaceae
♦ queen of the meadow
♦ Naturalised
♦ 101

Filipendula vulgaris Moench
Rosaceae
Filipendula hexapetala Gilib., *Spiraea filipendula* L., *Ulmaria filipendula* Kostel.
♦ dropwort
♦ Weed, Naturalised
♦ 15, 70, 101, 272, 280
♦ pH, cultivated, herbal.

Fimbristylis Vahl spp.
Cyperaceae
♦ fimbristylis, globe fringe rush, barba de indio
♦ Weed, Quarantine Weed
♦ 88, 220, 236
♦ G.

Fimbristylis acuminata Vahl
Cyperaceae
♦ pointed fimbristylis
♦ Weed
♦ 87, 88, 126, 170
♦ pG, aqua, cultivated.

Fimbristylis aestivalis (Retz.) Vahl
Cyperaceae
♦ summer fimbry
♦ Weed, Naturalised
♦ 87, 88, 101, 126, 170, 191, 263, 275, 286, 297
♦ aG, aqua, cultivated, herbal. Origin: Asia, Australia.

Fimbristylis albo-viridis C.B.Clarke
Cyperaceae
= *Fimbristylis albovirides* C.B.Clarke
♦ Weed
♦ 88, 170, 191
♦ G.

Fimbristylis albovirides C.B.Clarke
Cyperaceae
Fimbristylis albo-viridis C.B.Clarke (see)
♦ Weed, Naturalised
♦ 86, 93
♦ G.

Fimbristylis annua (All.) Roem. & Schult.
Cyperaceae
Fimbristylis baldwiniana (Schult.) Torr.
♦ cortadera coqueta, annual fimbry, coyolillo
♦ Weed, Naturalised
♦ 86, 153, 272, 281
♦ G. Origin: Europe.

Fimbristylis aphylla Steud.
Cyperaceae
Fimbristylis globulosa (Retz.) Kunth var. *aphylla* Miq.
♦ Weed
♦ 88, 170, 191
♦ G.

Fimbristylis atollensis St.John
Cyperaceae
= *Fimbristylis spathacea* Roth
♦ Weed
♦ 191
♦ G.

Fimbristylis autumnalis (L.) Roem & Schult.
Cyperaceae
♦ slender fimbristylis, autumn rush, slender fimbry
♦ Weed, Quarantine Weed
♦ 76, 87, 88, 126, 203, 218, 255
♦ a/pG, aqua, herbal. Origin: tropical America.

Fimbristylis bisumbellata (Forssk.) Bub.
Cyperaceae
Scirpus bisumbellatus Forssk.
♦ Weed, Native Weed
♦ 66, 87, 88, 121, 170, 185, 191, 221
♦ aG. Origin: obscure.

Fimbristylis complanata (Retz.) Link
Cyperaceae
♦ Puerto Rico fimbry
♦ Weed
♦ 87, 88, 286
♦ G. Origin: Australia.

Fimbristylis cymosa R.Br.
Cyperaceae
♦ tropical fimbry
♦ Weed
♦ 6, 87, 88
♦ G, herbal.

Fimbristylis dichotoma (L.) Vahl
Cyperaceae
Fimbristylis annua (All.) Roem. & Schult. var. *diphylla* (Retz.) Kük., *Fimbristylis brizoides* Nees & Meyen, *Fimbristylis diphylla* (Retz.) Vahl (see), *Fimbristylis laxa* Vahl, *Fimbristylis longispica* Steud., *Fimbristylis polymorpha* Boeck., *Iria polymorpha* (Boeck.) Kuntze, *Scirpus annuus* All., *Scirpus dichotomus* L., *Scirpus diphyllus* Retz.
♦ two leaved fimbristylis, tall fringe rush, lesser fimbristylis, forked fimbry, yaa niu nuu
♦ Weed, Native Weed
♦ 6, 87, 88, 121, 126, 170, 186, 204, 239, 255, 262, 263, 273, 275, 276, 286, 297
♦ a/pG, aqua, cultivated, herbal.

Origin: obscure.

Fimbristylis dichotoma (L.) Vahl var. *floribunda* (Miq.) T.Koyama
Cyperaceae
Fimbristylis diphylla (Retz.) Vahl var. *floribunda* Miq.
♦ Weed
♦ 88, 191, 204
♦ G.

Fimbristylis diphylla (Retz.) Vahl
Cyperaceae
= *Fimbristylis dichotoma* (L.) Vahl
♦ twoleaf fimbristylis
♦ Weed
♦ 88, 218
♦ G, herbal.

Fimbristylis diphylloides Makino
Cyperaceae
♦ Weed
♦ 286
♦ G.

Fimbristylis ferruginea (L) Vahl
Cyperaceae
Fimbristylis arvensis Vahl
♦ West Indian fimbry
♦ Weed
♦ 12, 87, 88, 126, 243
♦ a/pG, aqua, cultivated. Origin: Madagascar.

Fimbristylis globulosa (Retz.) Kunth
Cyperaceae
Fimbristylis efoliata Steud., *Fimbristylis umbellaris* (Lam.) Vahl (see), *Scirpus globulosus* Retz.
♦ globular fimbristylis, globe fimbry
♦ Weed, Quarantine Weed
♦ 76, 87, 88, 126, 170, 191, 203, 209
♦ pG, aqua, herbal.

Fimbristylis griffithii Boeck.
Cyperaceae
♦ Weed
♦ 88, 170, 191
♦ G.

Fimbristylis hispidula (Vahl) Kunth
Cyperaceae
Bulbostylis hispidula (Vahl) R.Haines (see), *Fimbristylis exilis* (H.B.K.) Roem. & Schult.
♦ slender biesie, slender sedge
♦ Weed, Native Weed
♦ 88, 121, 126, 158, 240
♦ a/pG, aqua. Origin: southern Africa.

Fimbristylis koidzumiana Ohwi.
Cyperaceae
♦ Weed
♦ 87, 88
♦ G.

Fimbristylis littoralis Gaudich.
Cyperaceae
Fimbristylis miliacea (L.) Vahl (see)
♦ fimbry, likarak en wel
♦ Weed, Naturalised
♦ 12, 32, 87, 88, 204, 257
♦ G.

Fimbristylis miliacea (L.) Vahl
Cyperaceae
= *Fimbristylis littoralis* Gaudich.

♦ lesser fimbristylis, grasslike fimbristylis, yaa rat khiat, cortadera coqueta, globe fringe rush
♦ Weed
♦ 6, 23, 28, 68, 88, 126, 153, 161, 170, 179, 186, 191, 204, 239, 243, 255, 262, 263, 274, 275, 286, 297
♦ aG, aqua, herbal. Origin: tropical America.

Fimbristylis ovata (Burm.f.) Kern
Cyperaceae
= *Abildgaardia ovata* (Burm.f.) Kral
♦ Weed
♦ 87, 88, 170, 191
♦ G. Origin: Australia.

Fimbristylis puberula (Michx.) Vahl
Cyperaceae
Fimbristylis anomala Boeck., *Fimbristylis drummondii* (Torr. & Hook.) Boeck., *Fimbristylis puberula* var. *interior* (Britton) Kral, *Fimbristylis puberula* var. *puberula* (Michx.) Vahl
♦ hairy fimbry
♦ Weed
♦ 28, 199, 243
♦ G, herbal.

Fimbristylis quinquangularis (Vahl) Kunth
Cyperaceae
Scirpus quinquangularis Vahl
♦ five angled fimbry
♦ Weed, Quarantine Weed, Naturalised
♦ 76, 87, 88, 101, 203
♦ G.

Fimbristylis schoenoides (Retz.) Vahl
Cyperaceae
Fimbristylis bispicata Nees & Mey., *Scirpus schoenoides* Retz.
♦ ditch fimbry
♦ Weed, Naturalised
♦ 87, 88, 101, 170, 179, 191
♦ G, herbal. Origin: Australia.

Fimbristylis spathacea Roth
Cyperaceae
Fimbristylis atollensis St.John (see)
♦ hurricanegrass
♦ Weed, Naturalised
♦ 101, 157, 161, 249
♦ pG.

Fimbristylis squarrosa Vahl
Cyperaceae
♦ Weed, Native Weed
♦ 87, 88, 121, 235, 286
♦ aG. Origin: southern Africa.

Fimbristylis stauntonii Deb. & Franch
Cyperaceae
♦ Weed
♦ 275
♦ G.

Fimbristylis subbispicata Nees & Mey.
Cyperaceae
♦ Weed
♦ 87, 88, 286
♦ G.

Fimbristylis tenera Roem & Schult.
Cyperaceae

♦ Weed
♦ 87, 88
♦ G.

Fimbristylis tetragona R.Br.
Cyperaceae
♦ Weed
♦ 87, 88
♦ G.

Fimbristylis thonningiana Boeck
Cyperaceae
♦ Weed
♦ 87, 88
♦ G.

Fimbristylis tomentosa Vahl
Cyperaceae
Fimbristylis annua (All.) Roem. & Schult. fo. *tomentosa* (Vahl) Kük., *Fimbristylis dichotoma* (L.) Vahl fo. *tomentosa* (Vahl) Ohwi, *Fimbristylis diphylla* (Retz.) Vahl var. *tomentosa* (Vahl) Benth.
♦ woolly fimbry
♦ Weed
♦ 88, 170, 191
♦ G. Origin: Australia.

Fimbristylis tristachya R.Br.
Cyperaceae
♦ fimbry
♦ Weed
♦ 87, 88
♦ G. Origin: Australia.

Fimbristylis umbellaris (L.) Vahl
Cyperaceae
= *Fimbristylis globulosa* (Retz.) Kunth
♦ globular fimbristylis
♦ Weed, Quarantine Weed
♦ 88, 135, 191
♦ G.

Fingerhuthia africana Lehm.
Poaceae
♦ Zulu fescue
♦ Naturalised
♦ 101
♦ G.

Fioria vitifolia (L.) Mattei
Malvaceae
♦ tropical fanleaf
♦ Weed, Naturalised
♦ 101, 261

Firmiana simplex (L.) W.Wight
Sterculiaceae
Firmiana plantanifolia Schott & Endl., *Sterculia platanifolia* L.f.
♦ Chinese parasol tree, Chinese bottletree, Japanese varnish tree, phoenix tree
♦ Weed, Naturalised, Environmental Weed
♦ 80, 88, 101, 151, 287
♦ T, cultivated.

Fissidens bryoides Hedw.
Fissidentaceae
♦ bryoid fissidens moss, lesser pocket moss
♦ Naturalised
♦ 280
♦ moss.

Fissidens exilis Hedw.
Fissidentaceae
- ♦ fissidens moss, slender pocket moss
- ♦ Naturalised
- ♦ 280
- ♦ moss.

Fissidens taxifolius Hedw.
Fissidentaceae
- ♦ fissidens moss, common pocket moss
- ♦ Naturalised
- ♦ 280
- ♦ moss.

Fitchia speciosa Cheeseman
Asteraceae
- ♦ burr daisytree
- ♦ Naturalised
- ♦ 101

Flacourtia indica (Burm.f.) Merr.
Flacourtiaceae
Flacourtia hirtiuscula Oliv., *Flacourtia latifolia* T.Cooke, *Flacourtia ramontchi* L'Hér., *Flacourtia sepiaria* Roxb., *Gmelina indica* Burm.f.
- ♦ Governor's plum, lusungunimba, Madagascar plum
- ♦ Weed, Noxious Weed, Naturalised, Introduced, Environmental Weed, Cultivation Escape
- ♦ 3, 22, 39, 80, 86, 87, 88, 101, 112, 151, 179, 228, 261
- ♦ S/T, arid, cultivated, herbal, toxic. Origin: Africa, Madagascar.

Flacourtia inermis Roxb.
Flacourtiaceae
- ♦ batoko plum, louvi
- ♦ Naturalised, Cultivation Escape
- ♦ 101, 261
- ♦ cultivated. Origin: tropical Africa, southern Asia, Pacific Islands.

Flacourtia jangomas (Lour.) Raeusch.
Flacourtiaceae
Flacourtia cataphracta Roxb ex Willd.
- ♦ Indian plum, coffee palm
- ♦ Weed, Naturalised, Environmental Weed
- ♦ 3, 22, 86, 98, 152, 191, 203
- ♦ S, cultivated, herbal. Origin: Assam, Burma.

Flagellaria indica L.
Flagellariaceae
- ♦ idanwel
- ♦ Weed
- ♦ 87, 88
- ♦ cultivated, herbal. Origin: Australia.

Flaveria australasica Hook.
Asteraceae
- ♦ Australian yellow weed, speedy weed, yellow twin stem
- ♦ Weed
- ♦ 50, 55, 87, 88, 205
- ♦ aH, cultivated.

Flaveria bidentis (L.) Kuntze
Asteraceae
Flaveria contrayerba Pers. (see)
- ♦ coastal plain yellowtops, smelter's bush, yellowtop, valda, smeltersbossie
- ♦ Weed, Naturalised, Environmental

Weed
- ♦ 51, 87, 88, 121, 158, 236, 237, 257, 261, 270, 287, 295
- ♦ aH. Origin: tropical America.

Flaveria contrayerba Pers.
Asteraceae
= *Flaveria bidentis* (L.) Kuntze
- ♦ Weed, Quarantine Weed
- ♦ 50, 76, 87, 88, 203, 220

Flaveria linearis Lag.
Asteraceae
- ♦ yellowtop, narrowleaf yellowtops
- ♦ Weed
- ♦ 161, 249

Flaveria trinervia (Spreng.) C.Mohr
Asteraceae
- ♦ clustered yellowtops, clusterflower
- ♦ Weed, Introduced
- ♦ 14, 34, 66, 87, 88, 161, 199, 240, 243
- ♦ aH, herbal.

Fleischmannia microstemon (Cass.) R.M.King & H.Rob.
Asteraceae
Eupatorium microstemon Cass. (see)
- ♦ tropical thoroughwort
- ♦ Weed, Quarantine Weed
- ♦ 76, 88, 203, 220

Fleischmannia pratensis (Klatt) King & Robins.
Asteraceae
- ♦ Naturalised
- ♦ 257

Flemingia macrophylla (Willd.) Merr.
Acanthaceae
- ♦ Naturalised
- ♦ 86
- ♦ cultivated. Origin: tropical Africa, Asia.

Flemingia strobilifera (L.) Aiton & W.T.Aiton
Acanthaceae
Moghania strobilifera (L.) St.-Hil. ex Kuntze (see)
- ♦ wild hops, besungelaiei
- ♦ Weed, Naturalised, Cultivation Escape
- ♦ 3, 32, 101, 191, 261
- ♦ cultivated, herbal. Origin: East Indies.

Fleurya aestuans (L.) Gaudich. ex Miq.
Urticaceae
= *Laportea aestuans* (L.) Chew
- ♦ lufulo
- ♦ Weed
- ♦ 88

Fleurya cuneata (A.Rich) Wedd.
Urticaceae
= *Fleurya umbellata* Wedd. (NoR)
- ♦ Weed
- ♦ 14

Fleurya ovalifolia (Schum. & Thonn.) Dandy
Urticaceae
= *Laportea ovalifolia* (Schum.) Chew (NoR)
- ♦ Weed, Quarantine Weed
- ♦ 76, 88, 203, 220

Flindersia brayleyana F.Muell.
Rutaceae/Flindersiaceae
- ♦ Queensland maple, silk wood
- ♦ Weed, Naturalised
- ♦ 3, 22, 101
- ♦ T, cultivated, herbal. Origin: Australia.

Floresia Krainz & F.Ritter ex Backeb. spp.
Cactaceae
= *Haageocereus* Backeb. spp.
- ♦ Weed, Quarantine Weed
- ♦ 76, 88, 220

Floscopa scandens Lour.
Commelinaceae
- ♦ Weed
- ♦ 12
- ♦ herbal.

Flourensia cernua DC.
Asteraceae
Helianthus cernuus (DC.) Benth. & Hook.
- ♦ tarbush, American tarwort
- ♦ Weed
- ♦ 39, 87, 88, 161, 218
- ♦ arid, herbal, toxic.

Flueggea acidoton (L.) G.L.Webster
Euphorbiaceae
- ♦ simpleleaf bushweed
- ♦ Naturalised
- ♦ 101

Flueggea virosa (Roxb. ex Willd.) Voigt
Euphorbiaceae
- ♦ flueggea
- ♦ Weed, Environmental Weed, Cultivation Escape
- ♦ 80, 88, 112, 179, 261
- ♦ cultivated. Origin: Madagascar.

Foeniculum vulgare Mill.
Apiaceae
Foeniculum officinale All.
- ♦ fennel, aniseed, dill, anise, sweet anise, sweet fennel, hinojo, venkoli, wild fennel
- ♦ Weed, Noxious Weed, Naturalised, Introduced, Garden Escape, Environmental Weed, Cultivation Escape
- ♦ 7, 8, 15, 24, 34, 35, 38, 40, 42, 70, 72, 78, 80, 86, 87, 88, 94, 97, 98, 101, 116, 121, 146, 147, 151, 158, 161, 165, 167, 176, 180, 198, 203, 218, 228, 231, 236, 237, 241, 243, 253, 261, 269, 272, 280, 287, 295, 296, 300
- ♦ pH, arid, cultivated, herbal. Origin: Eurasia.

Foeniculum vulgare Mill. var. *vulgare*
Apiaceae
- ♦ fennel, aniseed
- ♦ Environmental Weed
- ♦ 289
- ♦ cultivated, herbal.

Fontanesia phillyreoides Labill.
Oleaceae
- ♦ Naturalised
- ♦ 101
- ♦ cultivated.

Fontanesia phillyreoides Labill. ssp.
fortunei (Carr.) Yaltirik
Oleaceae
♦ Naturalised
♦ 101

Forchhammeria polyandra (Griseb.)
Alain
Capparaceae
♦ manystamen forchhammeria
♦ Naturalised
♦ 101

Forestiera acuminata (Michx.) Poir.
Oleaceae
♦ swamp privet, eastern swamp privet
♦ Weed
♦ 87, 88, 218
♦ S, promoted, herbal.

Forestiera neomexicana Gray
Oleaceae
= *Forestiera pubescens* Nutt. var.
pubescens (NoR)
♦ New Mexico forestiera, wild olive
♦ Weed
♦ 87, 88, 218
♦ cultivated, herbal.

Forestiera pubescens Nutt.
Oleaceae
♦ elbowbush, desert olive, stretchberry
♦ Weed
♦ 87, 88, 218
♦ S, herbal.

Forrestia hispida A.Rich.
Commelinaceae
= *Amischotolype hispida* (Less. &
A.Rich.) D.Y.Hong (NoR)
♦ Weed
♦ 87, 88

Forsskaolea tenacissima L.
Urticaceae
Forsskaolea cossoniana Webb
♦ Weed
♦ 221

Forsskaolea viridis Ehrenb. ex Webb
Urticaceae
♦ Weed
♦ 221
♦ cultivated.

Forsythia × *intermedia* Zabel
Oleaceae
= *Forsythia suspensa* (Thunb.) Vahl ×
Forsythia viridissima Lindl.
♦ showy forsythia, border forsythia
♦ Naturalised
♦ 101
♦ S, cultivated, herbal. Origin:
horticultural hybrid.

Forsythia ovata Nakai
Oleaceae
♦ early forsythia, Korean forsythia
♦ Naturalised
♦ 101
♦ cultivated.

Forsythia suspensa (Thunb.) Vahl
Oleaceae
Syringa suspensa Thunb.
♦ weeping forsythia, golden bell, lian
qiao, forsythia
♦ Naturalised, Casual Alien

♦ 39, 101, 280
♦ S, cultivated, herbal, toxic.

Forsythia suspensa (Thunb.) Vahl var.
sieboldii Zabel
Oleaceae
♦ Casual Alien
♦ 280

Forsythia viridissima Lindl.
Oleaceae
♦ greenstem forsythia
♦ Naturalised
♦ 39, 101
♦ cultivated, herbal, toxic.

Fortunella crassifolia Swingle
Rutaceae
♦ meiwa kumquat
♦ Naturalised
♦ 287
♦ cultivated.

Fortunella japonica (Thunb.) Swingle
Rutaceae
♦ round kumquat
♦ Naturalised
♦ 101
♦ S, cultivated, herbal.

Fortunella margarita (Lour.) Swingle
Rutaceae
♦ oval kumquat
♦ Naturalised, Introduced
♦ 101, 261
♦ S, cultivated. Origin: China.

Fouquieria splendens Engelm.
Fouquieriaceae
♦ ocotillo, vine cactus, coach whip,
candlewood
♦ Quarantine Weed
♦ 76
♦ arid, cultivated, herbal.

Fragaria L. spp.
Rosaceae
♦ strawberry
♦ Weed
♦ 88
♦ herbal.

Fragaria × *ananassa* Duchesne (*pro sp.*)
Rosaceae
= *Fragaria virginiana* Mill. × *Fragaria
chiloensis* (L.) Mill. [most probable
combination]
♦ garden strawberry, tarhamansikka,
strawberry, jahoda ananásová
♦ Naturalised, Cultivation Escape,
Casual Alien
♦ 40, 42, 280
♦ pH, cultivated, herbal. Origin:
horticultural hybrid.

Fragaria chiloensis (L.) Mill.
Rosaceae
♦ beach strawberry, wild strawberry,
sand strawberry
♦ Naturalised
♦ 101
♦ pH, cultivated, herbal.

Fragaria chiloensis (L.) Mill. ssp. *lucida*
(Vilm.) Staudt
Rosaceae
♦ beach strawberry
♦ Naturalised

♦ 101

Fragaria indica Andrews
Rosaceae
= *Duchesnea indica* (Andr.) Focke
♦ Weed
♦ 87, 88
♦ cultivated, herbal.

Fragaria moschata Duchesne
Rosaceae
Fragaria elatior Ehrh., *Fragaria magna*
Thuill., *Fragaria muricata* Mill. (see)
♦ ukkomansikka, hautbois strawberry
♦ Cultivation Escape
♦ 42
♦ pH, cultivated, herbal. Origin:
central Europe.

Fragaria muricata Mill.
Rosaceae
= *Fragaria moschata* Duchesne
♦ hautbois strawberry
♦ Naturalised
♦ 40

Fragaria vesca L.
Rosaceae
♦ European strawberry, woodland
strawberry, wild strawberry
♦ Weed, Naturalised
♦ 87, 88, 165, 218, 272, 280, 287
♦ pH, cultivated, herbal. Origin:
Eurasia.

Fragaria vesca L. ssp. *americana* (Porter)
Staudt
Rosaceae
♦ American strawberry, woodland
strawberry
♦ Weed
♦ 218

Fragaria virginiana Duchesne
Rosaceae
♦ Virginia strawberry, wild strawberry,
thick leaved wild strawberry, jahoda
virgínska, scarlet strawberry, mountain
strawberry
♦ Weed
♦ 87, 88, 161, 211, 218
♦ pH, cultivated, herbal.

Fragaria viridus Duchesne
Rosaceae
♦ small wild strawberry
♦ Weed
♦ 272

Francoeuria crispa (Forssk.) Cass.
Asteraceae
= *Pulicaria crispa* (Forssk.) Oliv.
♦ Weed
♦ 221, 242
♦ arid.

Frangula alnus Mill.
Rhamnaceae
Rhamnus frangula L. (see)
♦ European buckthorn, fen buckthorn,
tall hedge buckthorn, glossy
buckthorn, columnar buckthorn, alder
buckthorn, berry bearing alder
♦ Weed, Naturalised, Introduced,
Garden Escape, Environmental Weed
♦ 70, 101, 103, 133, 142, 161, 222, 224
♦ S/T, cultivated, herbal, toxic.

Frangula californica (Eschsch.) Gray
Rhamnaceae
♦ California coffeeberry, California buckthorn
♦ Weed
♦ 161
♦ toxic.

Frangula purshiana (DC.) Cooper
Rhamnaceae
Rhamnus purshiana DC. (see)
♦ cascara, Pursh's buckthorn
♦ Weed, Naturalised
♦ 161, 280
♦ toxic.

Frankenia hirsuta L.
Frankeniaceae
Frankenia revoluta Forssk. (see)
♦ Weed
♦ 87, 88

Frankenia pulverulenta L.
Frankeniaceae
♦ European sea heath, Mediterranean sea heath
♦ Weed, Naturalised
♦ 86, 98, 101, 134, 198, 203, 221
♦ cultivated, herbal. Origin: Europe.

Frankenia revoluta Forssk.
Frankeniaceae
= *Frankenia hirsuta* L.
♦ Weed
♦ 221

Frankenia salina (Mol.) I.M.Johnst.
Frankeniaceae
♦ alkali heath, alkali sea heath
♦ Weed
♦ 87, 88
♦ pH, aqua, herbal.

Franseria acanthicarpa (Hook.) Coville
Asteraceae
= *Ambrosia acanthicarpa* Hook.
♦ annual bursage
♦ Weed
♦ 23, 39, 87, 88, 218
♦ herbal, toxic.

Franseria confertiflora (DC.) Rydb.
Asteraceae
= *Ambrosia confertiflora* DC.
♦ slimleaf bursage
♦ Weed
♦ 87, 218
♦ herbal.

Franseria discolor Nutt.
Asteraceae
= *Ambrosia tomentosa* Nutt.
♦ skeletonleaf bursage
♦ Weed
♦ 87, 88, 218
♦ herbal.

Franseria strigulosa Rydb.
Asteraceae
= *Ambrosia confertiflora* DC.
♦ Weed
♦ 87

Franseria tenuifolia Harv. & Gray
Asteraceae
= *Ambrosia tenuifolia* Spreng.
♦ Weed

♦ 87, 88
♦ herbal.

Franseria tomentosa A.Gray
Asteraceae
= *Ambrosia grayi* (A.Nels.) Shinners
♦ woollyleaf bursage
♦ Weed
♦ 87, 88, 218

Fraxinus americana L.
Oleaceae
♦ white ash, American ash, Biltmore ash, Biltmore white ash
♦ Weed
♦ 87, 88, 121, 211, 218, 243
♦ T, cultivated, herbal. Origin: eastern North America.

Fraxinus angustifolia Vahl
Oleaceae
Fraxinus angustifolia var. *lentiscifolia* (Desf.) A.Henry, *Fraxinus lentiscifolia* Desf., *Fraxinus rotundifolia* Mill. (see)
♦ desert ash, narrow leaved ash
♦ Naturalised, Garden Escape, Environmental Weed
♦ 20, 198
♦ T, cultivated.

Fraxinus angustifolia Vahl ssp. angustifolia
Oleaceae
♦ desert ash
♦ Naturalised, Environmental Weed
♦ 86, 198, 289, 296
♦ cultivated. Origin: Mediterranean, south-west Asia.

Fraxinus angustifolia Vahl ssp. oxycarpa (M.Bieb. ex Willd.) Franco & Rocha Afonso
Oleaceae
Fraxinus oxycarpa M.Bieb. ex Willd., *Fraxinus oxyphylla* M.Bieb.
♦ Syrian ash
♦ Naturalised
♦ 198, 198

Fraxinus caroliniana Mill.
Oleaceae
♦ Carolina ash, water ash
♦ Weed
♦ 87, 88, 218
♦ cultivated, herbal.

Fraxinus excelsior L.
Oleaceae
♦ ash, European ash, saarni
♦ Weed, Naturalised
♦ 15, 39, 70, 101, 280
♦ T, cultivated, herbal, toxic. Origin: Eurasia.

Fraxinus griffithii
Oleaceae
♦ flowering ash, guang la shu
♦ Environmental weed
♦ 201
♦ T, Cultivated.

Fraxinus latifolia Benth.
Oleaceae
Fraxinus oregona Nutt. (see)
♦ Oregon ash
♦ Weed

♦ 87, 88, 218
♦ T, cultivated, herbal.

Fraxinus nigra Marsh.
Oleaceae
♦ black ash, black swamp ash, swamp ash, water ash
♦ Weed
♦ 87, 88, 218
♦ T, cultivated, herbal.

Fraxinus oregona Nutt.
Oleaceae
= *Fraxinus latifolia* Benth.
♦ Weed
♦ 87, 88

Fraxinus ornus L.
Oleaceae
Ornus europaea Pers.
♦ flowering ash, manna ash, manna
♦ Weed, Naturalised, Garden Escape, Environmental Weed
♦ 15, 54, 72, 86, 88, 198, 280
♦ T, cultivated, herbal. Origin: Eurasia.

Fraxinus pennsylvanica Marsh.
Oleaceae
Fraxinus lanceolata Borkh., *Fraxinus pubescens* Lam.
♦ green ash, red ash
♦ Weed
♦ 87, 88, 121, 211, 218, 243
♦ T, cultivated, herbal. Origin: eastern North America.

Fraxinus quadrangulata Michx.
Oleaceae
♦ blue ash
♦ Weed
♦ 87, 88, 218
♦ T, cultivated, herbal.

Fraxinus rotundifolia Mill.
Oleaceae
= *Fraxinus angustifolia* Vahl
♦ desert ash
♦ Weed, Naturalised, Environmental Weed
♦ 7, 72, 86, 88, 98
♦ T, cultivated.

Fraxinus tomentosa Michx.f.
Oleaceae
= *Fraxinus profunda* (Bush) Bush (NoR)
♦ pumpkin ash
♦ Weed
♦ 87, 88, 218
♦ herbal.

Fraxinus uhdei (Wenz.) Lingelsh.
Oleaceae
♦ shamel ash
♦ Weed, Naturalised, Environmental Weed, Cultivation Escape
♦ 22, 80, 88, 101, 151, 261
♦ T, cultivated, herbal. Origin: Mexico to Guatemala.

Fraxinus velutina Torr.
Oleaceae
♦ velvet ash, Arizona ash
♦ Weed
♦ 87, 88, 218
♦ T, cultivated, herbal.

Freesia alba (G.L.Mey.) Gumbl.
Iridaceae
♦ white freesia, wild freesia
♦ Naturalised, Garden Escape
♦ 198
♦ cultivated.

Freesia alba × leichtlinii (G.L.Mey.) Gumbl. & Klatt
Iridaceae
Freesia × hybrid *sensu* (see)
♦ freesia, common freesia, wild freesia
♦ Naturalised, Garden Escape, Environmental Weed
♦ 86, 176, 198, 252, 289
♦ cultivated.

Freesia corymbosa (Burm.f.) N.E.Br.
Iridaceae
♦ common freesia
♦ Naturalised, Garden Escape
♦ 101
♦ cultivated.

Freesia × hybrid *sensu*
Iridaceae
= *Freesia alba* (G.L.Mey.) Gumbl. × *Freesia leichtlinii* Klatt [in Australia, a complex hybrid elsewhere]
♦ freesias
♦ Garden Escape, Environmental Weed
♦ 296

Freesia laxa (Thunb.) Goldblatt & J.C.Manning
Iridaceae
Anomatheca laxa (Thunb.) Goldblatt (see)
♦ false freesia
♦ Naturalised, Garden Escape
♦ 101

Freesia leichtlinii Klatt
Iridaceae
♦ freesia
♦ Weed, Naturalised, Garden Escape, Environmental Weed
♦ 7, 9, 72, 86, 88, 203
♦ pH, cultivated. Origin: South Africa.

Freesia leichtlinii × refracta Klatt & (Jacq.) Klatt
Iridaceae
♦ Naturalised, Garden Escape
♦ 98
♦ cultivated.

Freesia refracta (Jacq.) Eckl. ex Klatt
Iridaceae
Gladiolus refractus Jacq.
♦ freesia, common freesia
♦ Weed, Naturalised, Garden Escape, Cultivation Escape
♦ 15, 34, 280
♦ pH, cultivated, herbal. Origin: South Africa.

Freycinetia ponapensis Martelli
Pandanaceae
♦ rahrah, freycinetia
♦ Introduced
♦ 230
♦ H.

Fritillaria atropurpurea Nutt.
Liliaceae
♦ tiger lily, chocolate lily, purple fritillary, spotted fritillary, spotted missionbells
♦ Weed
♦ 161
♦ pH, cultivated, herbal, toxic.

Fritillaria imperialis L.
Liliaceae
♦ crown imperial lily
♦ Weed
♦ 39, 80
♦ pH, cultivated, herbal, toxic.

Fritillaria meleagris L.
Liliaceae
♦ kirjopikarililja, snake's head fritillary, snake's head bulb, fritillary
♦ Weed, Naturalised
♦ 39, 42, 161, 272
♦ pH, cultivated, herbal, toxic.

Froelichia floridana (Nutt.) Moq.
Amaranthaceae
♦ cottontails, Florida snake cotton, plains snake cotton, prairie froelichia, cottonweed, field snakecotton
♦ Weed, Quarantine Weed, Noxious Weed, Naturalised, Native Weed
♦ 76, 86, 88, 147, 161, 174, 203
♦ cultivated, herbal. Origin: eastern north America.

Froelichia gracilis (Hook.) Moq.
Amaranthaceae
♦ cottonweed, slender snakecotton
♦ Weed, Naturalised
♦ 54, 86, 88, 98, 133, 195, 203, 224, 287
♦ cultivated, herbal. Origin: North America.

Froelichia tomentosa (Mart.) Moq.
Amaranthaceae
= *Froelichia interrupta* (L.) Moq. (NoR)
♦ froelichia
♦ Weed
♦ 87, 88, 237, 295

Fuchsia boliviana Carr.
Onagraceae
♦ Bolivian fuchsia, fuchsia, lady's eardrops
♦ Weed, Naturalised, Environmental Weed, Cultivation Escape
♦ 3, 22, 101, 152, 233, 280
♦ S, cultivated. Origin: Bolivia.

Fuchsia boliviana Carr. var. *luxurians* I.M.Johnst.
Onagraceae
♦ Naturalised
♦ 280

Fuchsia hybrida hort. ex Sieb. & Voss
Onagraceae
♦ hybrid fuchsia
♦ Naturalised
♦ 101
♦ cultivated, herbal. Origin: horticultural.

Fuchsia magellanica Lam.
Onagraceae
♦ fuchsia, hardy fuchsia, lady's eardrops, earing flower, kulapepeiao
♦ Weed, Naturalised, Garden Escape,
Environmental Weed, Cultivation Escape
♦ 3, 7, 40, 72, 86, 87, 88, 98, 132, 152, 155, 176, 198, 203, 233, 280, 290, 296
♦ pS, cultivated, herbal. Origin: Chile, Argentina.

Fuchsia magellanica Lam. var. *gracilis* L.H.Bailey
Onagraceae
♦ Naturalised
♦ 40
♦ pS.

Fuchsia magellanica Lam. var. *macrostema* (Ruiz & Pav.) Munz
Onagraceae
♦ fuchsia
♦ Naturalised
♦ 40
♦ pS.

Fuchsia paniculata Lindl.
Onagraceae
♦ shrubby fuchsia, fuchsia, lady's eardrops
♦ Naturalised, Cultivation Escape, Casual Alien
♦ 101, 233, 280
♦ S, cultivated.

Fuirena caerulescens Steud.
Cyperaceae
♦ Naturalised
♦ 101
♦ G.

Fuirena ciliaris (L.) Roxb.
Cyperaceae
Scirpus ciliarus L.
♦ umbrella grass, yaa khom baang klom
♦ Weed
♦ 88, 126, 170, 204, 209, 239, 286
♦ aG, aqua.

Fuirena glomerata Lam.
Cyperaceae
= *Fuirena umbel!ata* Rottb.
♦ Weed, Quarantine Weed
♦ 76, 87, 88, 203, 220
♦ G.

Fuirena scirpoidea Michx.
Cyperaceae
♦ southern umbrella sedge, leafless fuirena
♦ Quarantine Weed
♦ 258
♦ G, cultivated, herbal.

Fuirena simplex Vahl
Cyperaceae
♦ western umbrella sedge
♦ Weed
♦ 14, 157
♦ pG, herbal.

Fuirena umbellata Rottb.
Cyperaceae
Fuirena camptotricha C.Wr., *Fuirena glomerata* Lam. (see), *Fuirena hildebrandti* Boeck., *Fuirena mahouxii* Cherm., *Fuirena mauritiana* Nees, *Fuirena paniculata* L.f., *Fuirena quinquangularis* Hassk., *Fuirena seriata*

C.B.Clarke, *Fuirena tereticulmis* J. &
C.Presl, *Fuirena thouarsiana* Kunth,
Scirpus fuirena T.Koyama, *Scirpus
umbellatus* (Rottb.) Kuntze
- ♦ bosesansinga, yefen
- ♦ Weed
- ♦ 87, 88, 126, 170, 255
- ♦ pG, aqua. Origin: obscure.

Fumaria L. spp.
Papaveraceae/Fumariaceae
- ♦ fumitory
- ♦ Weed, Naturalised
- ♦ 198, 243, 272

Fumaria agraria Lag.
Papaveraceae/Fumariaceae
- ♦ field fumitory
- ♦ Weed, Quarantine Weed,
Naturalised, Introduced
- ♦ 38, 70, 76, 87, 88, 94, 203, 220, 237,
241, 295, 300

Fumaria asepala Boiss.
Papaveraceae/Fumariaceae
- ♦ Weed
- ♦ 87, 88

Fumaria bastardii Boreau
Papaveraceae/Fumariaceae
- ♦ bastard's fumitory, tall ramping
fumitory
- ♦ Weed, Naturalised
- ♦ 7, 86, 98, 176, 198, 203, 269, 280
- ♦ aH. Origin: Europe.

Fumaria capreolata L.
Papaveraceae/Fumariaceae
- ♦ white ramping fumitory, ramping
fumitory, rampant fumitory, climbing
fumitory
- ♦ Weed, Naturalised, Introduced,
Environmental Weed, Casual Alien
- ♦ 7, 9, 15, 38, 42, 70, 72, 86, 87, 88, 94,
98, 101, 198, 203, 237, 241, 253, 271, 272,
280, 287, 295, 300
- ♦ aH, arid, cultivated, herbal. Origin:
Eurasia.

Fumaria capreolata L. ssp. *capreolata*
Papaveraceae/Fumariaceae
- ♦ white flowered fumitory
- ♦ Weed, Naturalised
- ♦ 269, 280
- ♦ aH. Origin: Europe.

Fumaria densiflora DC.
Papaveraceae/Fumariaceae
Fumaria micrantha Lag. (see)
- ♦ fumitory, narrowleaf fumitory, dense
flowered fumitory
- ♦ Weed, Naturalised, Introduced,
Environmental Weed
- ♦ 7, 38, 68, 86, 87, 88, 93, 94, 98, 176,
185, 198, 203, 205, 221, 243, 253, 269,
280
- ♦ aH. Origin: Europe.

Fumaria indica (Hausskn.) Pugsley
Papaveraceae/Fumariaceae
- ♦ Indian fumitory, American fumitory
- ♦ Weed, Naturalised
- ♦ 86, 87, 88, 93, 98, 198, 203, 243, 248
- ♦ herbal. Origin: central and south-
west Asia.

Fumaria media Loisel.
Papaveraceae/Fumariaceae
- ♦ Weed
- ♦ 87, 88

Fumaria micrantha Lag.
Papaveraceae/Fumariaceae
= *Fumaria densiflora* DC.
- ♦ Weed
- ♦ 87, 88

Fumaria muralis Sond. ex Koch
Papaveraceae/Fumariaceae
Fumaria officinalis L. var. *capensis* Harv.
- ♦ drug fumitory, fumitory, wall
fumitory, common ramping fumitory,
scrambling fumitory
- ♦ Weed, Naturalised
- ♦ 7, 15, 70, 87, 88, 94, 98, 121, 134, 158,
165, 176, 203, 243, 253, 269, 272, 280
- ♦ aH. Origin: Eurasia.

Fumaria muralis Sond. ex Koch ssp.
boraei (Jordan) Pugsley
Papaveraceae/Fumariaceae
- ♦ common ramping fumitory, few
flowered fumitory, fumeterre des
murailles
- ♦ Naturalised, Environmental Weed
- ♦ 86
- ♦ Origin: Europe, North Africa.

Fumaria muralis Sond. ex Koch ssp.
muralis
Papaveraceae/Fumariaceae
- ♦ wall fumitory
- ♦ Naturalised, Environmental Weed
- ♦ 86, 198, 280
- ♦ Origin: west Europe.

Fumaria occidentalis Pugsley
Papaveraceae/Fumariaceae
- ♦ western fumitory, Cornish fumitory,
western ramping fumitory
- ♦ Naturalised
- ♦ 101

Fumaria officinalis L.
Papaveraceae/Fumariaceae
- ♦ fumitory, peltoemäkki, common
fumitory, red flowering fumitory, drug
fumitory
- ♦ Weed, Naturalised
- ♦ 7, 23, 24, 34, 39, 44, 51, 55, 70, 87, 88,
94, 98, 101, 115, 118, 133, 161, 176, 179,
195, 203, 217, 218, 236, 237, 241, 243,
253, 255, 272, 280, 287, 295
- ♦ aH, cultivated, herbal, toxic. Origin:
Europe.

Fumaria officinalis L. ssp. *officinalis*
Papaveraceae/Fumariaceae
- ♦ drug fumitory, common fumitory
- ♦ Naturalised, Environmental Weed
- ♦ 86, 101
- ♦ Origin: Europe, Mediterranean.

Fumaria officinalis L. ssp. *wirtgenii*
(W.D.J.Koch) Arcang.
Papaveraceae/Fumariaceae
- ♦ drug fumitory
- ♦ Naturalised, Environmental Weed
- ♦ 86, 101
- ♦ Origin: Europe, Mediterranean.

Fumaria parviflora Lam.
Papaveraceae/Fumariaceae
Fumaria caespitosa Losc. ex Willk. &
Lange
- ♦ narrowleaf fumitory, small white
fumitory, fine leaved fumitory, small
flowered fumitory
- ♦ Weed, Naturalised, Introduced
- ♦ 7, 38, 44, 55, 70, 87, 88, 94, 98, 101,
115, 185, 199, 203, 205, 221, 241, 243,
253, 272, 300
- ♦ aH, arid, cultivated, herbal.

Fumaria parviflora Lam. var. *parviflora*
Papaveraceae/Fumariaceae
- ♦ fumitory, small flowered fumitory
- ♦ Weed, Naturalised
- ♦ 86, 93, 198

Fumaria rostellata Knaf
Papaveraceae/Fumariaceae
- ♦ Weed
- ♦ 88, 94, 272

Fumaria schleicheri Soy.-Will.
Papaveraceae/Fumariaceae
- ♦ Weed
- ♦ 44, 87, 88, 94, 272

Fumaria sepium Boiss. & Reut.
Papaveraceae/Fumariaceae
- ♦ Weed
- ♦ 70

Fumaria vaillantii Lois.
Papaveraceae/Fumariaceae
- ♦ few flowered fumitory, earthsmoke,
pikkuemäkki
- ♦ Weed, Naturalised
- ♦ 44, 87, 88, 94, 98, 101, 203, 243, 253,
272
- ♦ cultivated, herbal.

Funastrum clausum (Jacq.) Schltr.
Asclepiadaceae/Apocynaceae
= *Sarcostemma clausum* (Jacq.) Schult.
(NoR)
- ♦ white twinevine
- ♦ Weed
- ♦ 87, 88
- ♦ herbal.

Funtumia elastica (Preuss) Stapf
Apocynaceae
- ♦ African rubber tree, silk rubber
- ♦ Weed, Naturalised, Cultivation
Escape
- ♦ 3, 101, 191, 261
- ♦ cultivated, herbal. Origin: west
tropical Africa.

Furcraea Vent.spp.
Agavaceae
- ♦ furcraea
- ♦ Weed
- ♦ 18, 88

Furcraea bedinghausii K.Koch
Agavaceae
- ♦ Introduced
- ♦ 228
- ♦ arid, cultivated.

Furcraea cabuya Trel.
Agavaceae

♦ Central American sisal
♦ Weed
♦ 80

Furcraea cubensis (Jacq.) Vent.
Agavaceae
Furcraea hexapetala (Jacq.) Urb. (see)
♦ Cuban hemp, hemp, cubuya
♦ Weed, Environmental Weed
♦ 3, 152, 191, 257
♦ Origin: Central and South America.

Furcraea foetida (L.) Haw.
Agavaceae
Furcraea gigantea Vent.
♦ Mauritius hemp
♦ Weed, Sleeper Weed, Naturalised, Garden Escape, Environmental Weed, Cultivation Escape
♦ 3, 7, 80, 86, 98, 101, 152, 179, 225, 233, 246, 280
♦ cultivated, herbal. Origin: Cuba, South America.

Furcraea hexapetala (Jacq.) Urb.
Agavaceae
= *Furcraea cubensis* (Jacq.) Vent.
♦ Cuban hemp
♦ Weed
♦ 22
♦ pH/S, cultivated, herbal.

Furcraea longaeva Karw. & Zucc.
Agavaceae
♦ Naturalised
♦ 280

Furcraea selloa K.Koch
Agavaceae
♦ wild sisal
♦ Weed, Naturalised, Cultivation Escape
♦ 86, 98, 179, 203, 252
♦ cultivated. Origin: Colombia.

Furcraea tuberosa (Mill.) Aiton
Agavaceae
♦ female karata
♦ Weed
♦ 87, 88
♦ herbal.

G

Gadellia lactiflora (Bieb.) Schulkina
Campanulaceae
♦ milky bellflower
♦ Naturalised
♦ 101

Gagea Salisb. spp.
Liliaceae
♦ star of Bethlehem, gagea
♦ Weed
♦ 272

Gagea arvensis (Pers.) Dum.
Liliaceae
= *Gagea villosa* (M.Bieb.) Duby
♦ hairy star of Bethlehem
♦ Weed
♦ 70, 87, 88, 126, 272

Gagea fistulosa Ker Gawl.
Liliaceae
♦ star of Bethlehem
♦ Naturalised
♦ 101

Gagea lutea (L.) Ker Gawl.
Liliaceae
Gagea silvatica Loud., *Ornithogalum luteum* L., *Ornithogalum majus* Gilib.
♦ yellow star of Bethlehem
♦ Weed
♦ 272
♦ pH, promoted, herbal.

Gagea minima (L.) Ker Gawl.
Liliaceae
Gagea baumgarteniana Schur, *Ornithogalum callosum* Kit., *Ornithogalum minus* Gilib.
♦ little star of Bethlehem, pikkukäenrieska
♦ Weed
♦ 272

Gagea pratensis (Pers.) Dumort.
Liliaceae
♦ puistokäenrieska
♦ Weed, Casual Alien
♦ 42, 272

Gagea pusilla (Schmidt) Schult. & Schult.f.
Liliaceae
♦ small star of Bethlehem
♦ Weed
♦ 272

Gagea reticulata (Pall.) Schult.f.
Liliaceae
♦ Weed
♦ 221

Gagea villosa (M.Bieb.) Duby
Liliaceae

Gagea arvensis (Pers.) Dum. (see)
♦ Naturalised
♦ 101
♦ H, herbal. Origin: North Africa, Middle East, Europe.

Gaillardia aristata Pursh
Asteraceae
♦ common gaillardia, blanketflower, gajlardia
♦ Weed, Naturalised
♦ 23, 86, 88, 98, 203
♦ pH, cultivated, herbal. Origin: North America.

Gaillardia × grandiflora hort. ex Van Houtte
Asteraceae
= *Gaillardia aristata* Pursh × *Gaillardia pulchella* Foug.
♦ Naturalised
♦ 86, 101, 280
♦ cultivated. Origin: North America.

Gaillardia pulchella Foug.
Asteraceae
♦ rosering gaillardia, firewheel, Indian blanket, nadobna
♦ Weed, Naturalised, Cultivation Escape
♦ 34, 87, 88, 98, 161, 179, 203, 218, 249, 287
♦ aH, cultivated, herbal, toxic.

Gaillardia pulchella Foug. var. picta (Sweet) Gray
Asteraceae
♦ firewheel
♦ Naturalised
♦ 86
♦ Origin: North America.

Gaillardia pulchella Foug. var. pulchella
Asteraceae
♦ firewheel
♦ Naturalised
♦ 86
♦ cultivated. Origin: North America.

Galactia jussiaeana Kunth
Fabaceae/Papilionaceae
♦ Weed
♦ 14

Galactia striata (Jacq.) Urb.
Fabaceae/Papilionaceae
Galactia tenuiflora (Willd.) Wight & Arn. (see)
♦ Florida hammock milkpea
♦ Weed
♦ 14
♦ cultivated.

Galactia tenuiflora (Willd.) Wight & Arn.
Fabaceae/Papilionaceae
= *Galactia striata* (Jacq.) Urb.
♦ Weed, Naturalised
♦ 87, 88, 257

Galactites elegans (All.) Nyman ex Soldano
Asteraceae
Galactites pumila Porta
♦ milk thistle, sädeohdake
♦ Weed
♦ 253

Galactites tomentosa **Moench**
Asteraceae
- milk thistle, sädeohdake
- Weed, Environmental Weed, Casual Alien
- 42, 70, 87, 88, 290
- a/bH, cultivated. Origin: Mediterranean, S.W. Europe.

Galanthus elewesii **Hook.f.**
Liliaceae/Amaryllidaceae
= *Galanthus elwesii* Hook.f.
- giant snowdrop
- Naturalised
- 101

Galanthus elwesii **Hook.f.**
Liliaceae/Amaryllidaceae
Galanthus elwesii Hook.f. (see)
- giant snowdrop
- Naturalised
- 125
- cultivated, herbal.

Galanthus nivalis **L.**
Liliaceae/Amaryllidaceae
- lumikello, snowdrop
- Weed, Naturalised, Cultivation Escape
- 39, 40, 42, 70, 101, 154, 161, 247, 272
- pH, cultivated, herbal, toxic.

Galaxia fugacissima **(L.f.) Druce**
Iridaceae
- galaxia
- Weed, Naturalised, Cultivation Escape
- 86, 98, 198, 203, 252
- cultivated.

Galega officinalis **L.**
Fabaceae/Papilionaceae
- goat's rue, professor weed, French lilac
- Weed, Noxious Weed, Naturalised, Environmental Weed, Cultivation Escape
- 15, 38, 39, 40, 67, 80, 87, 88, 101, 140, 161, 165, 212, 218, 229, 237, 241, 246, 272, 280, 295, 300
- pH, cultivated, herbal, toxic. Origin: Eurasia.

Galega orientalis **Lam.**
Fabaceae/Papilionaceae
- rehuvuohenherne, Caucasian goat's rue
- Casual Alien
- 39, 42
- pH, cultivated, toxic.

Galenia africana **L. var. *africana***
Aizoaceae
- yellowbush
- Native Weed
- 121
- S, toxic. Origin: southern Africa.

Galenia pubescens **(Eckl. & Zeyh.) Druce**
Aizoaceae
- coastal galenia, galenia
- Weed, Naturalised, Garden Escape, Environmental Weed
- 7, 72, 88, 98, 101, 176, 198, 203, 251, 290, 300
- pH, cultivated.

Galenia pubescens **(Eckl. & Zeyh.) Druce**
var. *pubescens*
Aizoaceae
- galenia
- Weed, Noxious Weed, Naturalised
- 86, 93, 205
- pH.

Galenia secunda **(L.f.) Sond.**
Aizoaceae
- galenia, one sided galenia
- Weed, Naturalised, Native Weed, Introduced
- 86, 98, 101, 121, 163, 198, 203, 228
- p, arid, cultivated. Origin: southern Africa.

Galeobdolon luteum **Huds**
Lamiaceae
Lamiastrum galeobdolon (L.) L.
- aluminium plant, yellow archangel
- Weed, Sleeper Weed, Naturalised, Garden Escape, Environmental Weed
- 15, 165, 225, 246, 280
- cultivated. Origin: Eurasia.

Galeobdoion luteum **Huds cv.
'Variegatum'**
Lamiaceae
- aluminium plant, yellow archangel
- Naturalised
- 280
- cultivated. Origin: horticultral.

Galeola ponapensis **(Kaneh. & Yamamoto) Tuyama**
Orchidaceae
- lamahk
- Introduced
- 230
- H.

Galeopsis **L. spp.**
Lamiaceae
- hempnettle
- Weed
- 243
- H.

Galeopsis angustifolia **Ehrh. ex Hoffm.**
Lamiaceae
- narrow leaved hempnettle, red hempnettle, kaitapillike
- Weed, Naturalised, Casual Alien
- 42, 70, 87, 88, 94, 101, 253, 272
- herbal.

Galeopsis bifida **Boenn.**
Lamiaceae
- splitlip hempnettle, peltopillike, bifid hempnettle
- Weed, Quarantine Weed, Naturalised
- 44, 70, 76, 88, 94, 101, 114, 203, 220, 243, 253, 272, 275, 286, 297
- aH, promoted, herbal.

Galeopsis ladanum **L.**
Lamiaceae
Galeopsis intermedia Vill., *Galeopsis latifolia* Hoffm., *Galeopsis segetum* S.F.Gray
- red hempnettle, pehmytpillike, canapetta violacea, breitblättriger, galéopsis intermédiaire, canapetta a foglie strette, smalbladige raai

- Weed, Naturalised
- 39, 44, 70, 87, 88, 94, 101, 243, 253, 272, 280
- aH, cultivated, herbal, toxic.

Galeopsis ladanum **L. var. *ladanum***
Lamiaceae
- red hempnettle
- Naturalised
- 101

Galeopsis ladanum **L. var. *latifolia* (Hoffm.) Wallr.**
Lamiaceae
- red hempnettle
- Naturalised
- 101

Galeopsis pubescens **Bess.**
Lamiaceae
- downy hempnettle, karvapillike
- Weed, Casual Alien
- 42, 70, 87, 88, 94, 243, 253, 272
- cultivated, herbal.

Galeopsis segetum **Neck.**
Lamiaceae
Galeopsis dubia Leers, *Galeopsis ochroleuca* Lam.
- yellow hempnettle, downy hempnettle, myllypillike, gelber hohlzahn, galeopside
- Weed, Quarantine Weed, Casual Alien
- 42, 44, 70, 76, 88, 94, 203, 253, 272
- aH, cultivated, herbal.

Galeopsis speciosa **Mill.**
Lamiaceae
Galeopsis versicolor Curt. (see)
- large flowered hempnettle, kirjopillike, hempnettle
- Weed, Quarantine Weed, Naturalised
- 44, 70, 76, 86, 87, 88, 94, 203, 243, 253, 272
- aH, cultivated.

Galeopsis tetrahit **L.**
Lamiaceae
- hempnettle, common hempnettle, dog nettle, bee nettle, wild hemp, flowering nettle, ironweed, brittlestem hempnettle, canapetta
- Weed, Quarantine Weed, Noxious Weed, Naturalised
- 23, 44, 52, 68, 70, 76, 80, 86, 87, 88, 94, 101, 118, 161, 162, 165, 195, 203, 218, 243, 253, 272, 280, 287, 292, 299, 300
- aH, cultivated, herbal. Origin: Eurasia.

Galeopsis versicolor **Curt.**
Lamiaceae
= *Galeopsis speciosa* Mill.
- Weed
- 87, 88

Galinsoga **Ruiz & Pav. spp.**
Asteraceae
- quickweed, galinsoga, gallant soldier
- Weed, Naturalised, Environmental Weed
- 88, 151, 198, 206

Galinsoga caracasana (DC.) Sch.Bip.
Asteraceae
= *Galinsoga quadriradiata* Ruiz & Pav.
♦ Weed
♦ 87, 88, 157
♦ aH.

Galinsoga ciliata (Raf.) Blake
Asteraceae
= *Galinsoga quadriradiata* Ruiz & Pav.
♦ Peruvian daisy, hairy galinsoga, ciliate galinsoga, galinsoga, quickweed, shaggy soldier, ripsisaurikki
♦ Weed, Naturalised
♦ 23, 42, 44, 52, 68, 70, 80, 87, 88, 94, 121, 157, 158, 161, 180, 211, 218, 263, 272, 297
♦ aH, herbal. Origin: South America.

Galinsoga parviflora Cav.
Asteraceae
♦ galinsoga, smallflower galinsoga, potato weed, gallant soldier, yellow weed, joey hooker, tarhasaurikki, small flowered quickweed, galinsoga weed, chickweed, kew weed
♦ Weed, Naturalised, Introduced, Garden Escape, Environmental Weed
♦ 7, 13, 15, 23, 40, 42, 44, 50, 51, 53, 55, 70, 80, 86, 87, 88, 94, 98, 101, 121, 134, 158, 161, 165, 170, 176, 186, 198, 203, 210, 218, 228, 236, 237, 238, 240, 243, 245, 253, 255, 261, 269, 270, 272, 275, 276, 280, 286, 287, 293, 295
♦ aH, arid, cultivated, herbal. Origin: tropical America.

Galinsoga parviflora Cav. var. parviflora
Asteraceae
♦ Weed
♦ 34, 167
♦ aH.

Galinsoga quadriradiata Ruiz & Pav.
Asteraceae
Galinsoga caracasana (DC.) Sch.Bip. (see), *Galinsoga ciliata* (Raf.) Blake (see), *Galinsoga urticaefolia* Benth. (see)
♦ fringed quickweed, shaggy soldier, hairy galinsoga, quickweed
♦ Weed, Naturalised, Introduced
♦ 40, 44, 80, 101, 174, 179, 195, 207, 243, 253, 255, 261, 273, 280, 286, 287, 293
♦ aH, arid, herbal. Origin: tropical America.

Galinsoga urticaefolia Benth.
Asteraceae
= *Galinsoga quadriradiata* Ruiz & Pav.
♦ olla nueva
♦ Weed, Naturalised
♦ 87, 88, 257, 281

Galium L. spp.
Rubiaceae
♦ cleavers, bedstraw
♦ Weed
♦ 243, 272
♦ herbal.

Galium album Mill.
Rubiaceae
♦ paimenmatara, upright hedge bedstraw, white bedstraw

♦ Naturalised
♦ 42, 101
♦ herbal.

Galium aparine L.
Rubiaceae
Galium vaillantii DC. *Galium aparine* (see), *Galium spurium* L. var. *echinospermum* (Wallr.) Hayek (see)
♦ cleavers, goosegrass, scratch grass, grip grass, catchweed bedstraw, white hedge, bedstraw, sticky willy, velcro plant, robin run over the hedge, attaccamano, gallio, pega pega
♦ Weed, Quarantine Weed, Noxious Weed, Naturalised, Native Weed, Introduced, Garden Escape, Environmental Weed
♦ 7, 9, 15, 23, 34, 36, 39, 44, 62, 70, 72, 76, 86, 87, 88, 94, 97, 98, 115, 118, 136, 159, 161, 162, 165, 174, 176, 181, 186, 198, 203, 204, 207, 210, 211, 212, 218, 236, 237, 241, 243, 249, 253, 269, 272, 280, 289, 291, 295, 299, 300
♦ aH, arid, cultivated, herbal, toxic. Origin: Eurasia.

Galium aparine L. var. tenerum (Gren. & Godr.) Rchb.
Rubiaceae
♦ Weed
♦ 275, 297
♦ aH.

Galium aristatum L.
Rubiaceae
♦ awned bedstraw
♦ Naturalised
♦ 101

Galium asperuloides Edgew.
Rubiaceae
♦ Weed
♦ 248

Galium asprellum Michx.
Rubiaceae
♦ rough bedstraw
♦ Weed
♦ 39, 87, 88, 218
♦ herbal, toxic.

Galium bifolium Wats.
Rubiaceae
♦ twinleaf bedstraw
♦ Weed, Naturalised
♦ 23, 88, 287
♦ aH, herbal.

Galium boreale L.
Rubiaceae
♦ northern bedstraw, ahomatara
♦ Weed
♦ 23, 70, 87, 88, 161, 218
♦ pH, cultivated, herbal.

Galium canescens Kunth
Rubiaceae
♦ Naturalised
♦ 257

Galium debile Desv.
Rubiaceae
♦ slender marsh bedstraw, pond bedstraw
♦ Naturalised

♦ 280
♦ herbal.

Galium divaricatum Pourr. ex Lam.
Rubiaceae
♦ Lamarck's bedstraw, slender bedstraw
♦ Weed, Naturalised, Introduced, Environmental Weed
♦ 7, 9, 15, 34, 86, 98, 101, 165, 176, 198, 203, 253, 280, 287, 300
♦ aH, cultivated, herbal. Origin: Europe, Asia.

Galium glaucum L.
Rubiaceae
Asperula glauca (L.) Besser
♦ harmaamatara, waxy bedstraw
♦ Naturalised, Casual Alien
♦ 42, 101
♦ cultivated.

Galium gracilens (A.Gray) Makino
Rubiaceae
♦ himeyotsubamugura
♦ Weed
♦ 87, 88

Galium hamatum L.
Rubiaceae
♦ Weed
♦ 88, 240
♦ aH.

Galium humifusum Bieb.
Rubiaceae
Asperula humifusa (Bieb.) Bess. (see)
♦ spreading bedstraw
♦ Weed, Naturalised
♦ 80, 101, 272, 280

Galium lutchuense Nakai
Rubiaceae
♦ Weed
♦ 235

Galium mollugo L.
Rubiaceae
Galium mollugo L. ssp. *erectum* (Huds.) Briq. (see)
♦ field madder, smooth bedstraw, great hedge bedstraw, white bedstraw, wild madder, false baby's breath
♦ Weed, Naturalised, Introduced, Environmental Weed, Casual Alien
♦ 4, 23, 34, 42, 44, 70, 80, 87, 88, 101, 104, 133, 161, 195, 218, 237, 272, 280, 287, 295
♦ pH, cultivated, herbal.

Galium mollugo L. ssp. erectum (Huds.) Briq.
Rubiaceae
= *Galium mollugo* L.
♦ upright hedge bedstraw
♦ Weed
♦ 253

Galium murale (L.) All.
Rubiaceae
♦ tiny bedstraw, yellow wall bedstraw, small bedstraw
♦ Weed, Naturalised, Introduced, Environmental Weed
♦ 7, 9, 34, 72, 86, 88, 98, 101, 176, 198, 203, 241, 280, 300
♦ aH, cultivated. Origin: Europe.

Galium odoratum **(L.) Scop.**
Rubiaceae
Asperula odorata L. (see)
♦ sweet scented bedstraw, tuoksumatara, woodruff, sweet woodruff
♦ Weed, Naturalised
♦ 39, 70, 101
♦ pH, cultivated, herbal, toxic.

Galium palustre **L.**
Rubiaceae
♦ marsh bedstraw, rantamatara, common marsh bedstraw
♦ Weed, Naturalised, Environmental Weed
♦ 15, 54, 70, 80, 86, 87, 88, 176, 272, 280
♦ cultivated, herbal. Origin: Europe.

Galium parisiense **L.**
Rubiaceae
♦ wall bedstraw, cleavers, bedstraw, goosegrass, ranskanmatara
♦ Weed, Naturalised, Casual Alien
♦ 34, 42, 87, 88, 94, 101, 161, 180, 241, 243, 253, 272, 300
♦ aH, cultivated, herbal.

Galium pedemontanum **(Bellardi) All.**
Rubiaceae
= *Cruciata pedemontana* (Bellardi) Ehrend.
♦ piedmont bedstraw
♦ Weed
♦ 80
♦ herbal.

Galium × *pomeranicum* **Retz.**
Rubiaceae
= *Galium mollugo* L. × *Galium vernum* Scop.
♦ bedstraw, hybrid bedstraw
♦ Naturalised
♦ 101

Galium prusense **Koch.**
Rubiaceae
♦ Weed
♦ 111, 243

Galium pseudo-asprellum **Makino**
Rubiaceae
♦ oobanyaemugura
♦ Weed
♦ 87, 88, 286

Galium pumilum **Murray**
Rubiaceae
♦ nurmimatara, slender bedstraw
♦ Weed, Naturalised
♦ 42, 272
♦ cultivated.

Galium rotundifolium **L.**
Rubiaceae
♦ round leaved bedstraw
♦ Weed
♦ 87, 88, 272
♦ herbal.

Galium rubioides **L.**
Rubiaceae
♦ European bedstraw
♦ Weed, Naturalised
♦ 101, 272
♦ cultivated.

Galium saccharatum **All.**
Rubiaceae
♦ Weed
♦ 87, 88

Galium saxatile **L.**
Rubiaceae
Galium harcynicum Weigel
♦ heath bedstraw, nummimatara
♦ Weed, Naturalised
♦ 70, 101, 272
♦ pH, cultivated.

Galium schultesii **Vest**
Rubiaceae
♦ Schultes's bedstraw
♦ Naturalised
♦ 101
♦ pH, cultivated.

Galium setaceum **Lam.**
Rubiaceae
♦ Weed
♦ 221

Galium sinaicum **(Delile ex Decne.) Boiss.**
Rubiaceae
♦ Weed
♦ 221

Galium spurium **L.**
Rubiaceae
♦ false cleavers, sticky willy, Marin County bedstraw, peltomatara
♦ Weed, Quarantine Weed, Noxious Weed, Naturalised, Introduced
♦ 16, 34, 36, 44, 68, 76, 86, 87, 88, 94, 98, 101, 158, 162, 203, 204, 243, 272, 287, 291
♦ aH, cultivated, herbal. Origin: Europe to west Asia, Africa.

Galium spurium **L. ssp.** *africanum* **Verdc.**
Rubiaceae
♦ catchweed, catchweed bedstraw, goosegrass
♦ Weed, Native Weed
♦ 53, 88, 121
♦ aC.

Galium spurium **L. var.** *echinospermum* **(Wallr.) Hayek**
Rubiaceae
= *Galium aparine* L.
♦ false cleavers, yaemugura
♦ Weed
♦ 235, 243, 263, 286
♦ aH.

Galium sylvaticum **L.**
Rubiaceae
♦ Scotch mist
♦ Weed, Naturalised
♦ 101, 272
♦ cultivated, herbal.

Galium tinctorium **(L.) Scop.**
Rubiaceae
Asperula tinctoria L. (see)
♦ stiff marsh bedstraw, marsh bedstraw
♦ Weed
♦ 161, 249
♦ aqua, herbal.

Galium trachyspermum **A.Gray**
Rubiaceae
♦ yotsubamugura
♦ Weed
♦ 87, 88, 286

Galium tricorne **Stokes**
Rubiaceae
= *Galium tricornutum* Dandy
♦ Weed
♦ 23, 87, 88, 275, 297
♦ aH.

Galium tricornutum **Dandy**
Rubiaceae
Galium tricorne Stokes (see)
♦ bedstraw, rough bedstraw, roughfruit corn bedstraw, sarvimatara, three horned bedstraw, corn cleavers
♦ Weed, Quarantine Weed, Noxious Weed, Naturalised, Casual Alien
♦ 7, 34, 42, 44, 62, 70, 76, 86, 88, 94, 98, 101, 115, 198, 203, 241, 243, 248, 253, 269, 272, 287, 300
♦ aH, cultivated, herbal. Origin: Africa, Europe, Middle East.

Galium trifidum **L.**
Rubiaceae
♦ three petal bedstraw, pikkumatara
♦ Weed
♦ 87, 88
♦ pH, herbal.

Galium uliginosum **L.**
Rubiaceae
♦ fen bedstraw, luhtamatara
♦ Weed, Naturalised
♦ 87, 88, 272, 280
♦ herbal.

Galium vaillantii **DC.**
Rubiaceae
= *Galium aparine* L.
♦ Weed
♦ 87, 88

Galium verrucosum **Huds.**
Rubiaceae
♦ warty bedstraw
♦ Weed, Naturalised
♦ 44, 70, 88, 94, 101, 243
♦ cultivated.

Galium verum **L.**
Rubiaceae
♦ yellow bedstraw, lady's bedstraw, keltamatara
♦ Weed, Naturalised
♦ 70, 80, 87, 88, 101, 161, 195, 218, 272, 280, 286, 287, 297
♦ C, cultivated, herbal.

Galium verum **L. fo.** *nikkoense* **(Nakai) Ohwi**
Rubiaceae
♦ Weed
♦ 286

Galium verum **L. var.** *asiaticum* **Nakai**
Rubiaceae
♦ kibanakawaramatsuba
♦ Weed
♦ 286

Galium viscosum **Vahl**
Rubiaceae

♦ Weed
♦ 87, 88

Galium wirtgenii F.W.Schultz
Rubiaceae
♦ Wirtgen's bedstraw
♦ Naturalised
♦ 101
♦ cultivated.

Galphimia brasiliensis (L.) A.Juss.
Malpighiaceae
♦ Weed
♦ 87, 88

Galphimia gracilis Bartl.
Malpighiaceae
♦ slender goldshower, rain of gold
♦ Weed, Naturalised, Casual Alien
♦ 101, 179, 261
♦ cultivated. Origin: Mexico.

Gamochaeta americana (Mill.) Wedd.
Asteraceae
Gnaphalium americanum Mill. (see)
♦ woolly cudweed, American everlasting
♦ Weed, Naturalised
♦ 86, 87, 88, 295
♦ Origin: Americas.

Gamochaeta antillarum (Urb.) Anderb.
Asteraceae
Gnaphalium antillarum Urb.
♦ Naturalised
♦ 86

Gamochaeta calviceps (Fern.) Cabrera
Asteraceae
= *Gamochaeta falcata* (Lam.) Cabrera (NoR) [see *Gnaphalium calviceps* Fern. & *Gnaphalium falcatum* Lam.]
♦ grey cudweed
♦ Weed, Naturalised, Environmental Weed
♦ 86, 198, 295
♦ Origin: South America.

Gamochaeta coarctata (Willd.) Kerguélen
Asteraceae
♦ Weed
♦ 237

Gamochaeta pensylvanica (Willd.) Cabrera
Asteraceae
Gnaphalium pensylvanicum Willd. (see), *Gnaphalium peregrinum* Fern. (see)
♦ woolly cudweed, wandering cudweed, Pennsylvania everlasting, cudweed
♦ Weed, Naturalised
♦ 55, 86, 198, 295
♦ Origin: tropical Americas.

Gamochaeta purpurea (L.) Cabrera
Asteraceae
Gnaphalium purpureum L. (see)
♦ spoonleaf purple everlasting
♦ Naturalised
♦ 86, 134, 176
♦ a/pH, arid, herbal.

Gamochaeta sphacilata (Kunth) Cabrera
Asteraceae
♦ owl's crown

♦ Naturalised
♦ 101

Gamochaeta spicata (Lam.) Cabrera
Asteraceae
Gnaphalium coarctatum (Willd.) Ker. (see), *Gnaphalium spicatum* Lam. (see)
♦ peludilla
♦ Weed
♦ 236, 245, 295
♦ Origin: America.

Gamochaeta subfalcata (Cabrera) Cabrera
Asteraceae
Gnaphalium subfalcatum Cabrera (see)
♦ Naturalised
♦ 86

Gamolepis caudata Klatt ex Burtt Davy & Pott-Leendertz
Asteraceae
♦ Quarantine Weed
♦ 220

Gamolepis chrysanthemoides DC.
Asteraceae
♦ Weed, Naturalised
♦ 54, 86, 88
♦ cultivated.

Garaventia graminifolia (F.Phil. ex Phil.) Looser
Liliaceae/Alliaceae
♦ Quarantine Weed
♦ 220

Garcia nutans Vahl
Euphorbiaceae
♦ false tung oil tree
♦ Introduced
♦ 39, 228
♦ arid, herbal, toxic.

Garcinia dulcis (Roxb.) Kurz
Clusiaceae
♦ gourka, mangosteen, garcinia
♦ Naturalised, Cultivation Escape
♦ 101, 261
♦ cultivated, herbal. Origin: Indo Malay region.

Garcinia livingstonei T.Anderson
Clusiaceae
Garcinia baikieana Vesque, *Garcinia ferrandii* Chiov.
♦ Livingstone's garcinia, imbe, pembe fruit
♦ Introduced
♦ 228
♦ arid, cultivated.

Garcinia mangostana L.
Clusiaceae
♦ mangosteen, purple mangosteen
♦ Naturalised, Introduced
♦ 101, 230, 261
♦ T, cultivated, herbal. Origin: Indonesia, Malaysia.

Garcinia ponapensis Lauterb.
Clusiaceae
♦ kehnpuil
♦ Introduced
♦ 230
♦ T.

Gardenia angusta (L.) Merr.
Rubiaceae
= *Gardenia jasminoides* J.Ellis (NoR) [see *Gardenia augusta* (L.) Merr. *nom. illeg.*]
♦ cape jasmine
♦ Naturalised
♦ 101

Gardenia augusta (L.) Merr. nom. illeg.
Rubiaceae
= *Gardenia jasminoides* J.Ellis (NoR) [see *Gardenia angusta* (L.) Merr.]
♦ iosef sarawi, cape jasmine, tulipa
♦ Introduced, Cultivation Escape
♦ 230, 261
♦ S, cultivated, herbal. Origin: China.

Garnotia stricta Brongn.
Poaceae
♦ blue lawngrass
♦ Naturalised
♦ 101
♦ G. Origin: Australia.

Garrya elliptica Douglas ex Lindl.
Garryaceae
♦ tree silktassel, coast silktassel, wavyleaf silktassel
♦ Weed
♦ 87, 88, 218
♦ S, cultivated, herbal.

Garrya fadyenia Hook.
Garryaceae
♦ Fadyen's silktassel
♦ Naturalised
♦ 101

Garrya flavescens S.Wats.
Garryaceae
♦ boxleaf silktassel, ashy silktassel, pale tasselbush
♦ Weed
♦ 87, 88
♦ S, cultivated, herbal.

Garrya flavescens S.Wats. var. buxifolia (A.Gray) Jeps.
Garryaceae
♦ boxleaf silktassel
♦ Weed
♦ 218

Garrya fremontii Torr.
Garryaceae
♦ Fremont silktassel, bearbrush, fever bush
♦ Weed
♦ 39, 87, 88, 218
♦ S, cultivated, herbal, toxic.

Garrya × issaquahensis E.C.Nelson
Garryaceae
= *Garrya elliptica* Lindl. × *Garrya fremontii* Torr.
♦ hybrid coast silktassel
♦ Quarantine Weed
♦ 220
♦ cultivated. Origin: northwest North America.

Garrya veatchii Kellogg
Garryaceae
♦ Veitch silktassel, canyon silktassel
♦ Quarantine Weed
♦ 220
♦ S, cultivated, herbal.

Gastridium phleoides (Nees & Meyen) C.E.Hubb.
Poaceae
Gastridium ventricosum (Gouan) Schinz & Thell. (see)
♦ nutgrass, nit grass
♦ Weed, Naturalised
♦ 7, 86, 98, 198, 203, 241, 300
♦ G, cultivated.

Gastridium ventricosum (Gouan) Schinz & Thell.
Poaceae
= *Gastridium phleoides* (Nees & Meyen) C.E.Hubb.
♦ nit grass, nyyläheinä
♦ Weed, Naturalised, Introduced, Casual Alien
♦ 34, 38, 42, 86, 87, 88, 98, 161, 176, 203, 218, 241, 280, 300
♦ aG, herbal. Origin: Europe, North Africa, west Asia.

Gastrocotyle hispida (Forssk.) Bunge
Boraginaceae
Anchusa hispida Forssk.
♦ Weed
♦ 88

Gastrolobium callistachys Meissn.
Fabaceae/Papilionaceae
♦ Weed
♦ 39, 87, 88
♦ cultivated, herbal, toxic.

Gastrolobium grandiflorum F.Muell.
Fabaceae/Papilionaceae
♦ desert poison bush, heartleaf poison, wallflower poison
♦ Weed, Noxious Weed, Naturalised
♦ 86, 87, 88, 147, 203
♦ cultivated.

Gastrolobium parviflorum Benth.
Fabaceae/Papilionaceae
♦ Weed
♦ 87, 88
♦ cultivated.

Gastrolobium villosum Benth.
Fabaceae/Papilionaceae
♦ crinkle leaved poison
♦ Weed
♦ 39, 87, 88
♦ cultivated, toxic.

Gaudichaudia albida Cham. & Schlecht.
Malpighiaceae
♦ Weed
♦ 157
♦ pC.

Gaudinia fragilis (L.) Beauv.
Poaceae
Avena fragilis L.
♦ haprakaura, fragile oat, French oatgrass
♦ Weed, Naturalised, Casual Alien
♦ 40, 42, 70, 86, 87, 88, 101, 176, 198
♦ G, cultivated, herbal. Origin: Eurasia.

Gaultheria L. spp.
Ericaceae
♦ snowberry
♦ Quarantine Weed

♦ 220
♦ herbal.

Gaultheria mucronata Remy
Ericaceae
= *Gaultheria eriophylla* (Pers.) Sleumer ex Burtt var. *mucronata* (Remy) Luteyn (NoR)
♦ prickly heath, pernettya
♦ Weed, Naturalised
♦ 40, 161, 280
♦ S, cultivated, toxic.

Gaultheria shallon Pursh
Ericaceae
♦ salal, shallon
♦ Weed, Naturalised
♦ 87, 88, 218, 280, 291
♦ S, cultivated, herbal.

Gaura L. spp.
Onagraceae
♦ clockweed, beeblossom
♦ Weed, Quarantine Weed, Naturalised
♦ 76, 86, 88, 116, 203
♦ cultivated.

Gaura biennis L.
Onagraceae
♦ biennial gaura, gaura
♦ Weed, Naturalised
♦ 87, 88, 218, 287
♦ a/bH, cultivated, herbal.

Gaura coccinea Nutt. ex Pursh
Onagraceae
Gaura odorata auct. non Sessé ex Lag.
♦ scarlet gaura, scarlet beeblossom
♦ Weed, Noxious Weed, Native Weed
♦ 34, 35, 87, 88, 161, 174, 218, 229
♦ a/pH, cultivated, herbal. Origin: North America.

Gaura drummondii (Spach) Torr. & Gray
Onagraceae
Gaura odorata Sessé ex Lag. (see)
♦ scented gaura, Drummond's beeblossom
♦ Weed, Noxious Weed
♦ 35, 88, 229
♦ pH.

Gaura gracilis Wooton & Standl.
Onagraceae
= *Gaura hexandra* Ortega ssp. *gracilis* (Woot. & Standl.) Raven & Gregory (NoR)
♦ sirokesäkynttilä
♦ Casual Alien
♦ 42
♦ herbal.

Gaura hexandra Ortega ssp. *hexandra*
Onagraceae
♦ Weed
♦ 199

Gaura lindheimeri Engelm. & Gray
Onagraceae
♦ Lindheimer's beeblossom, the bride
♦ Weed, Naturalised, Garden Escape
♦ 86, 98, 121, 203, 287
♦ pH, cultivated, herbal. Origin: southern North America.

Gaura longiflora Spach
Onagraceae

♦ biennial gaura, longflower beeblossom
♦ Weed
♦ 161
♦ herbal.

Gaura mutabilis Cav.
Onagraceae
♦ Weed
♦ 199

Gaura odorata Sessè ex Lag.
Onagraceae
= *Gaura drummondii* (Spach) Torr. & A.Gray
♦ scented gaura
♦ Weed
♦ 87, 88, 161, 218
♦ herbal.

Gaura parviflora Douglas
Onagraceae
♦ smallflower gaura, velvet weed, gaura, butterfly weed
♦ Weed, Naturalised, Introduced
♦ 34, 38, 49, 86, 87, 88, 98, 161, 203, 212, 218, 287
♦ a/pH, cultivated, herbal. Origin: southern North America.

Gaura sinuata Nutt. ex Ser.
Onagraceae
♦ wavyleaf gaura, wavyleaf beeblossom
♦ Weed, Noxious Weed
♦ 34, 35, 87, 88, 121, 161, 218, 229
♦ pH, herbal.

Gaura villosa Torr.
Onagraceae
♦ hairy gaura, woolly beeblossom
♦ Weed
♦ 87, 88, 218
♦ herbal.

Gaya guerkeana K.Schum.
Malvaceae
♦ malva, guaxima
♦ Weed
♦ 255
♦ pH. Origin: tropical America.

Gaya pilosa K.Schum.
Malvaceae
♦ guanxuma, guaxima
♦ Weed
♦ 255
♦ aH, arid. Origin: Brazil.

Gaylussacia frondosa (L.) Torr. & Gray ex Torr.
Ericaceae
Vaccinium frondosum L., *Vaccinium venustum* Ait.
♦ dangleberry, blue huckleberry
♦ Weed
♦ 87, 88, 218
♦ S, promoted, herbal.

Gazania Gaertn. spp.
Asteraceae
♦ treasure flower, gazania
♦ Weed, Naturalised, Environmental Weed
♦ 116, 198, 246, 290

Gazania krebsiana Less. ssp. *serrulata* (DC.) Roessl.

Asteraceae
- butter flower, common gazania
- Native Weed
- 121
- pH. Origin: southern Africa.

Gazania linearis (Thunb.) Druce
Asteraceae
- treasure flower, gazania
- Weed, Naturalised, Garden Escape, Environmental Weed
- 7, 15, 34, 54, 72, 86, 88, 98, 101, 198, 203, 251, 280, 290, 296
- pH, cultivated. Origin: South Africa.

Gazania rigens (L.) Gaertn.
Asteraceae
- coastal gazania, treasure flower, gazania
- Weed, Naturalised, Garden Escape, Environmental Weed
- 15, 40, 86, 93, 155, 198, 205, 280, 290, 296
- p, cultivated. Origin: southern Africa.

Geanthus humilis Phil.
Zingiberaceae
= *Speea humilis* Loes.
- Quarantine Weed
- 220

Geigeria alata (DC.) Oliv. & Hiern
Asteraceae
Diplostemma alatum DC., *Geigeria macdougalii* S.Moore
- Weed
- 221

Geigeria aspera Harv. var. *aspera*
Asteraceae
- Native Weed
- 121
- S, toxic. Origin: southern Africa.

Geigeria burkei Harv. ssp. *burkei* var. *burkei*
Asteraceae
- Native Weed
- 121
- S, toxic. Origin: southern Africa.

Geigeria burkei Harv. ssp. *burkei* var. *zeyheri* (Harv.) Merxm.
Asteraceae
- Native Weed
- 121
- S, toxic. Origin: southern Africa.

Geigeria ornativa O.Hoffm.
Asteraceae
- Weed, Native Weed
- 39, 121
- pH, toxic. Origin: southern Africa.

Geijera parviflora Lindl.
Rutaceae
- wilga
- Introduced
- 228
- arid, cultivated. Origin: Australia.

Geissaspis cristata Wight & Arn.
Fabaceae/Papilionaceae
- Weed
- 87, 88

Geitonoplesium cymosum (R.Br.) A.Cunn. ex R.Br.
Smilacaceae/Luzuriagaceae/ Eustrephaceae
- scrambling lily
- Quarantine Weed, Environmental Weed
- 246
- pC, cultivated, herbal. Origin: Australia.

Gelidium vagum Okamura
Gelidiaceae
- red alga, agar weed
- Weed
- 197
- algae.

Gelsemium sempervirens (L.) J.St-Hil.
Loganiaceae/Gelsemiaceae
Bignonia sempervirens L., *Gelsemium lucidum* Poir., *Gelsemium nitidum* Michx., *Lisianthus sempervirens* Mill. ex Steud.
- yellow jessamine, evening trumpetflower, Carolina yellow jessamine, Carolina jasmine, false jasmine
- Weed, Naturalised
- 8, 39, 86, 87, 88, 98, 154, 161, 189, 203, 218, 247
- pC, cultivated, herbal, toxic. Origin: south-eastern North America.

Gendarussa vulgaris Nees
Acanthaceae
Justicia gendarussa Burm.f.
- Weed
- 13
- herbal.

Geniostoma stenurum Gig. & Benedict
Loganiaceae/Geniostomaceae
- kehnmant
- Introduced
- 230
- T.

Genista L. spp.
Fabaceae/Papilionaceae
Teline Medik. spp. (see)
- broom, ginster, woad waxen, janowiec
- Weed, Naturalised, Environmental Weed
- 54, 86, 198, 272, 290
- herbal.

Genista acanthoclada DC.
Fabaceae/Papilionaceae
- Weed
- 272

Genista barnadesii Graells
Fabaceae/Papilionaceae
= *Echinospartum barnadesii* (Graells) Rothm.
- Quarantine Weed
- 220

Genista canariensis L.
Fabaceae/Papilionaceae
Cytisus canariensis (L.) Masf., *Cytisus canariensis* var. *ramosissimus* (Poir.) Briq., *Cytisus hillebrandii* (H.Christ) Briq., *Cytisus ramosissimus* Poir., *Genista*

hillebrandii H.Christ, *Teline canariensis* (L.) Webb & Berthel., *Teline hillebrandii* (Christ) Kunkel (see)
- canarybroom, broom
- Weed, Naturalised, Introduced, Garden Escape, Environmental Weed
- 7, 34, 86, 98, 101, 203
- S, cultivated. Origin: Canary Islands.

Genista germanica L.
Fabaceae/Papilionaceae
Cytisus germanicus Vis.
- saksanherne
- Weed, Casual Alien
- 42, 272
- S, cultivated, herbal.

Genista horrida (Vahl) DC.
Fabaceae/Papilionaceae
- Weed, Naturalised, Cultivation Escape
- 7, 86, 98, 203, 252
- cultivated

Genista linifolia L.
Fabaceae/Papilionaceae
Cytisus linifolius (L.) Lam., *Teline linifolia* (L.) Webb & Berthel. (see)
- Mediterranean broom, flax leaved broom, flax broom, flaxleaf broom, dyer's broom, greenwold
- Weed, Noxious Weed, Naturalised, Introduced, Garden Escape, Environmental Weed
- 7, 34, 39, 72, 80, 86, 87, 88, 98, 101, 116, 147, 155, 198, 203, 289, 296
- pS, arid, cultivated, toxic. Origin: Mediterranean.

Genista maderensis (Webb & Berthel.) Lowe
Fabaceae/Papilionaceae
- Madeira dyer's greenweed, Madeira broom
- Weed, Naturalised
- 39, 98, 101, 203
- S, cultivated, toxic.

Genista monosperma Lam.
Fabaceae/Papilionaceae
= *Retama monosperma* (L.) Boiss.
- white weeping broom
- Naturalised, Garden Escape, Environmental Weed
- 39, 86
- cultivated, herbal, toxic.

Genista monspessulana (L.) L.A.S.Johnson
Fabaceae/Papilionaceae
Cytisus candicans (L.) DC., *Cytisus candicans* (L.) Lam., *Cytisus kunzeanus* Willk., *Cytisus monspessulanus* L. (see), *Cytisus monspessulanus* var. *umbellulatus* (Webb) Briq., *Cytisus pubescens* Moench, *Cytisus syriacus* Boiss. & Blanche, *Genista candicans* L., *Genista eriocarpa* Kunze, *Genista syriaca* Boiss. & Blanche, *Teline candicans* (L.) Presl, *Teline candicans* var. *umbellulatus* Webb, *Teline medicagoides* Medik., *Teline monspessulana* (L.) K.Koch (see)
- French broom, soft broom, canary broom, Montpellier broom, Madeira

broom, cape broom, broom, retamo liso
♦ Weed, Quarantine Weed, Noxious Weed, Naturalised, Introduced, Garden Escape, Environmental Weed
♦ 34, 35, 39, 72, 76, 78, 80, 86, 88, 98, 101, 116, 137, 147, 151, 155, 171, 176, 198, 203, 228, 229, 231, 232, 269, 289, 296
♦ pS, arid, cultivated, toxic. Origin: Mediterranean, Middle East.

Genista pilosa L.
Fabaceae/Papilionaceae
Genista repens Lam., *Spartium pilosum* Roth, *Telinaria pilosa* Presl
♦ hairy greenweed
♦ Weed
♦ 272
♦ cultivated, herbal.

Genista polyanthos Roem.
Fabaceae/Papilionaceae
♦ Weed
♦ 70

Genista racemosa hort. hybrid
Fabaceae/Papilionaceae
= *Genista canariensis* L. × *Genista stenopetala* Webb & Berth. [*Cytisus × spachianus* Kuntze]
♦ fragrant broom, florist's broom
♦ Naturalised, Environmental Weed
♦ 198, 290

Genista stenopetala Webb & Berthel.
Fabaceae/Papilionaceae
Cytisus everestianus Carrière, *Cytisus racemosus* Marnock, *Cytisus racemosus* var. *everestianus* (Carrière) Rehder, *Genista spachiana* Webb, *Cytisus stenopetalus* (Webb & Berthel.) Masf., *Teline stenopetala* (Webb & Berthel.) Webb & Berthel. (see)
♦ leafy broom
♦ Naturalised, Introduced, Cultivation Escape
♦ 34, 86, 101, 176, 252
♦ S, cultivated. Origin: Africa, Spain, Canary Islands.

Genista tinctoria L.
Fabaceae/Papilionaceae
♦ dyer's greenweed, woadwaxen, dyer's broom, pensasväriherne
♦ Weed, Naturalised, Cultivation Escape
♦ 39, 42, 54, 80, 86, 87, 88, 101, 218, 272, 280
♦ S, cultivated, herbal, toxic. Origin: Europe, Middle East.

Genlisea A.St.-Hil. spp.
Lentibulariaceae
♦ Quarantine Weed
♦ 76, 220

Genlisea aurea St.-Hil.
Lentibulariaceae
♦ Quarantine Weed
♦ 220
♦ cultivated.

Gentiana L. spp.
Gentianaceae
♦ gentian

♦ Weed
♦ 272
♦ herbal.

Gentiana amarella L.
Gentianaceae
♦ autumn gentian
♦ Quarantine Weed
♦ 220
♦ herbal.

Gentiana andrewsii Griseb.
Gentianaceae
♦ closed bottle gentian, closed gentian, bottle gentian
♦ Weed
♦ 23, 88
♦ pH, cultivated, herbal.

Gentiana crinita Froel.
Gentianaceae
= *Gentianopsis crinita* (Froel.) Ma (NoR)
♦ Weed
♦ 23, 88
♦ herbal.

Gentiana cruciata L.
Gentianaceae
Hippion cruciatum Schmidt, *Tretorrhiza cruciata* Opiz
♦ star gentian, cross gentian, cross leaved gentian
♦ Naturalised
♦ 101
♦ pH, cultivated, herbal.

Gentiana dahurica Fisch.
Gentianaceae
♦ dahuri gentian
♦ Weed
♦ 297
♦ pH, cultivated.

Gentiana lutea L.
Gentianaceae
Asterias lutea Borkh., *Swertia lutea* Vest.
♦ yellow gentian, great yellow gentian
♦ Weed
♦ 272
♦ pH, cultivated, herbal.

Gentiana macrophylla Pall.
Gentianaceae
♦ largeleaf gentian
♦ Weed
♦ 297
♦ pH, cultivated, herbal. Origin: China, Siberia.

Gentiana nutans Bunge
Gentianaceae
♦ tundra gentian
♦ Naturalised
♦ 101

Gentiana pneumonanthe L.
Gentianaceae
Ciminalis pneumonanthe Borkh., *Pneumonanthe vulgaris* Schmidt
♦ marsh gentian, kellokatkero
♦ Weed, Cultivation Escape
♦ 23, 42, 88
♦ pH, cultivated, herbal.

Gentiana purdomii C.Marquand
Gentianaceae
♦ Przewalsk gentiana

♦ Weed
♦ 297

Gentiana purpurea L.
Gentianaceae
♦ punakatkero, purple gentian
♦ Cultivation Escape
♦ 42
♦ pH, cultivated, herbal.

Gentiana septemfida Pall.
Gentianaceae
♦ crested gentian
♦ Naturalised
♦ 101
♦ cultivated, herbal.

Gentiana verna L.
Gentianaceae
Ericoila verna Borkh., *Hippion vernum* (L.) F.W.Schmidt
♦ spring gentian
♦ Weed
♦ 23, 39, 88
♦ cultivated, herbal, toxic.

Gentianella amarella (L.) Börner
Gentianaceae
Gentiana amarella. L., *Gentiana axillaris* (F.W.Schmidt) Rchb.
♦ autumn dwarf gentian, felwort, horkkakatkero
♦ Quarantine Weed
♦ 220
♦ bH, promoted, herbal.

Gentianella ciliata (L.) Borkh.
Gentianaceae
Gentiana ciliata L., *Hippion ciliatum* Schmidt
♦ fringed gentian
♦ Weed
♦ 272

Gentianopsis grandis (H.Sm.) Ma
Gentianaceae
♦ largeflower gentianopsis
♦ Weed
♦ 297

Geoffroea decorticans (Gill. ex Hook. & Arn.) Burkart
Fabaceae/Papilionaceae
Gourliea decorticans Gill. ex Hook. & Arn. (see)
♦ bastard cabbage tree, Chilean palo verde, chañar
♦ Weed, Quarantine Weed, Introduced
♦ 87, 88, 220, 228, 236, 237, 295
♦ arid, cultivated.

Geoffroea spinosa Jacq.
Fabaceae/Papilionaceae
♦ Quarantine Weed, Introduced, Environmental Weed
♦ 220, 228, 257
♦ arid.

Geophila repens (L.) I.M.Johnst.
Rubiaceae
♦ corrida yerba de guava, togo, tono
♦ Weed, Naturalised
♦ 86, 98, 203
♦ S, herbal. Origin: Madagascar.

Geranium L. spp.
Geraniaceae

- geranio selvatico, geranium, crane's bill
- Weed
- 118, 221, 243, 272
- H.

Geranium arabicum Forssk.
Geraniaceae
- Weed
- 87, 88

Geranium bicknellii Britt.
Geraniaceae
- Bicknell's geranium, Bicknell's crane's bill
- Noxious Weed
- 299
- aH, promoted, herbal. Origin: northern North America.

Geranium caffrum Eckl. & Zeyh.
Geraniaceae
- crane's bill
- Native Weed
- 121
- a/pH, cultivated. Origin: southern Africa.

Geranium carolinianum L.
Geraniaceae
- Carolina geranium, wild geranium, Carolina crane's bill, crane's bill, Amerikankurjenpolvi
- Weed, Quarantine Weed, Naturalised, Introduced, Casual Alien
- 42, 76, 84, 87, 88, 136, 161, 180, 207, 210, 211, 218, 228, 243, 249, 286, 287, 297
- aH, arid, cultivated, herbal. Origin: North America.

Geranium columbinum L.
Geraniaceae
Geranium schrenkianum Trautv. ex Krylov
- longstalk geranium, longstalk crane's bill, pakost holubí, dove's foot crane's bill, kivikkokurjenpolvi
- Weed, Naturalised
- 44, 70, 87, 88, 94, 101, 218, 241, 253, 272, 300
- aH, cultivated, herbal.

Geranium core-core Steud.
Geraniaceae
- core core
- Weed
- 87, 88

Geranium dissectum L.
Geraniaceae
- cutleaf geranium, wrinkle seeded crane's bill, liuskakurjenpolvi
- Weed, Naturalised, Introduced, Environmental Weed
- 7, 15, 23, 24, 38, 44, 70, 86, 87, 88, 94, 98, 101, 134, 136, 161, 165, 176, 198, 203, 218, 237, 243, 250, 253, 269, 272, 280, 287, 295, 300
- aH, cultivated, herbal. Origin: Mediterranean, west Asia.

Geranium dissectum L. var. dissectum
Geraniaceae
- Naturalised
- 241

Geranium dissectum L. var. glabratum Hook.f.
Geraniaceae
- Naturalised
- 300

Geranium divaricatum Ehrh.
Geraniaceae
- harakurjenpolvi, fanleaf geranium
- Weed, Naturalised, Casual Alien
- 42, 101, 272
- cultivated.

Geranium endressii J.Gay
Geraniaceae
- French crane's bill
- Naturalised
- 40
- cultivated.

Geranium fremontii Torr. ex A.Gray
Geraniaceae
= *Geranium caespitosum* James var. *fremontii* (Torr. ex Gray) Dorn (NoR)
- Introduced
- 228
- arid, cultivated, herbal.

Geranium homeanum Turcz.
Geraniaceae
Geranium carolinianum L. var. *australe* (Benth.) Fosberg, *Geranium glabratum* (Hook.) Small
- Australasian geranium
- Weed, Naturalised
- 101, 238
- pH, cultivated, herbal. Origin: Australia.

Geranium ibericum Cav.
Geraniaceae
- Iberian geranium
- Naturalised
- 101
- cultivated, herbal.

Geranium lozani Rose
Geraniaceae
- Weed
- 199

Geranium lucidum L.
Geraniaceae
Robertium lucidum Picard
- shining crane's bill, shining geranium, kiiltokurjenpolvi
- Weed, Quarantine Weed, Noxious Weed, Naturalised
- 76, 80, 88, 101, 137
- aH, cultivated, herbal.

Geranium macrorrhizum L.
Geraniaceae
Robertium macrorrhizum Picard
- bigroot crane's bill, bigroot geranium, Bulgarian geranium, rock crane's bill, tuoksukurjenpolvi
- Weed, Naturalised, Garden Escape, Cultivation Escape
- 40, 42, 241, 272
- pH, cultivated, herbal.

Geranium maculatum L.
Geraniaceae
- spotted geranium, wild crane's bill, wild geranium

- Weed
- 87, 88, 218
- pH, cultivated, herbal.

Geranium maderense Yeo
Geraniaceae
- giant herb Robert
- Weed, Naturalised
- 15, 40, 280
- a/pH, cultivated.

Geranium × magnificum Hyl.
Geraniaceae
= *Geranium ibericum* Cav. × *Geranium platypetalum* Fisch. & Mey.
- purple crane's bill
- Naturalised
- 40
- cultivated.

Geranium meeboldii Briq.
Geraniaceae
- idänkurjenpolvi
- Cultivation Escape
- 42
- cultivated.

Geranium molle L.
Geraniaceae
- woodland geranium, dovefoot geranium, pehmytkurjenpolvi, soft crane's bill, crane's bill
- Weed, Naturalised, Garden Escape, Environmental Weed
- 7, 9, 15, 23, 24, 34, 44, 70, 72, 80, 87, 88, 94, 98, 101, 121, 136, 161, 165, 176, 198, 203, 218, 237, 241, 243, 250, 253, 272, 280, 287, 295, 300
- a/pH, cultivated, herbal. Origin: Eurasia.

Geranium molle L. ssp. molle
Geraniaceae
- dove's foot, dove's foot crane's bill
- Weed, Naturalised, Environmental Weed
- 86, 269
- aH, cultivated. Origin: Europe.

Geranium nepalense Sweet
Geraniaceae
- Nepalese crane's bill, sweet Nepalese crane's bill, bhanda
- Weed
- 133, 195, 224, 243
- pH, cultivated, herbal. Origin: Afghanistan, western China, Bhutan, India, Myanmar, Nepal, Pakistan, Sri Lanka.

Geranium × oxonianum Yeo
Geraniaceae
= *Geranium endressii* J.Gay × *Geranium versicolor* L.
- Druce's crane's bill
- Naturalised
- 40
- cultivated.

Geranium palmatum Cav.
Geraniaceae
Geranium anemonifolium L'Hér. ex Ait.
- Canary Island geranium
- Weed, Naturalised
- 54, 86, 88, 101
- cultivated.

Geranium palustre L.
Geraniaceae
♦ marsh crane's bill, long stalked
crane's bill, longstalk geranium
♦ Weed
♦ 272
♦ cultivated.

Ceranium phaeum L.
Geraniaceae
♦ tummakurjenpolvi, dusky crane's
bill
♦ Weed, Naturalised, Casual Alien
♦ 40, 42, 272, 280
♦ cultivated.

Geranium pilosum Cav.
Geraniaceae
= *Geranium solanderi* Carolin
♦ traveller's geranium
♦ Weed
♦ 87, 88, 218
♦ pH, promoted, herbal.

**Geranium platypetalum Fisch. &
C.A.Mey.**
Geraniaceae
♦ kaukasiankurjenpolvi
♦ Cultivation Escape
♦ 42
♦ cultivated.

Geranium potentillaefolium DC.
Geraniaceae
♦ Weed
♦ 199

Geranium potentilloides L'Hér. ex DC.
Geraniaceae
♦ cinquefoil geranium, native carrot
♦ Naturalised
♦ 101
♦ pH, cultivated. Origin: Australia.

Geranium pratense L.
Geraniaceae
♦ small geranium, meadow geranium,
small geranium, meadow crane's bill
♦ Weed, Naturalised, Cultivation
Escape
♦ 23, 42, 70, 87, 88, 101, 218, 272, 280
♦ pH, cultivated, herbal.

Geranium purpureum Vill.
Geraniaceae
♦ little robin, lesser herb Robert
♦ Weed, Naturalised
♦ 86, 98, 203, 280, 300
♦ cultivated.

Geranium pusillum L.
Geraniaceae
♦ small geranium, small flowered
crane's bill, pihakurjenpolvi
♦ Weed, Quarantine Weed,
Naturalised
♦ 23, 34, 44, 70, 76, 80, 87, 88, 94, 101,
218, 243, 253, 272, 280, 287, 300
♦ aH, cultivated, herbal.

Geranium pyrenaicum Burm.f.
Geraniaceae
♦ hedgerow geranium, mountain
crane's bill, pyreneittenkurjenpolvi,
hedgerow crane's bill
♦ Weed, Quarantine Weed,
Naturalised, Casual Alien

♦ 42, 70, 76, 88, 94, 101, 272, 300
♦ cultivated, herbal.

Geranium retrorsum L'Hér. ex DC.
Geraniaceae
♦ New Zealand geranium
♦ Naturalised, Introduced
♦ 101, 228
♦ pH, arid, herbal.

Geranium robertianum L.
Geraniaceae
Robertiella robertianum Hanks
♦ herb Robert, Robert's geranium,
stinky Bob, erba Roberta,
haisukurjenpolvi
♦ Weed, Noxious Weed, Naturalised,
Introduced, Garden escape
♦ 1, 15, 24, 34, 45, 70, 80, 86, 87, 88, 94,
137, 139, 146, 165, 198, 241, 272, 280,
300
♦ aH, cultivated, herbal. Origin:
temperate Eurasia.

Geranium rotundifolium L.
Geraniaceae
♦ round leaved crane's bill,
pyöreälehtikurjenpolvi
♦ Weed, Naturalised, Casual Alien
♦ 42, 44, 86, 87, 88, 94, 101, 253, 272
♦ aH, cultivated, herbal. Origin:
Eurasia.

Geranium rubescens Yeo
Geraniaceae
♦ Weed, Quarantine Weed,
Naturalised
♦ 54, 76, 86, 88, 280
♦ cultivated.

Geranium sanguineum L.
Geraniaceae
♦ bloody crane's bill, verikurjenpolvi,
bloody geranium
♦ Weed, Naturalised
♦ 70, 80, 86, 101, 272
♦ cultivated, herbal. Origin:
Mediterranean, Middle East.

Geranium sibiricum L.
Geraniaceae
♦ siperiankurjenpolvi, Siberian
geranium
♦ Weed, Naturalised, Casual Alien
♦ 42, 86, 101, 272, 275, 287
♦ pH, cultivated, herbal. Origin:
eastern Europe, Russia, Middle East.

**Geranium sibiricum L. var. *glabrius*
(Hara) Ohwi**
Geraniaceae
♦ ichigefuuro
♦ Weed
♦ 286

Geranium simense Hochst. ex A.Rich.
Geraniaceae
Geranium compar R.Br., *Geranium
emirnense* Bojer, *Geranium frigidum*
Hochst., *Geranium latistiulatum* Hochst.
♦ Weed
♦ 87, 88

Geranium sinense R.Knuth
Geraniaceae
♦ Weed, Quarantine Weed

♦ 76, 87, 88, 203, 220
♦ cultivated.

Geranium solanderi Carolin
Geraniaceae
Geranium pilosum Cav. (see)
♦ Solander's geranium, crane's bill
♦ Weed, Naturalised
♦ 15, 101, 134
♦ pH, cultivated. Origin: Australia.

Geranium sylvaticum L.
Geraniaceae
♦ wood crane's bill, woodland
geranium, metsäkurjenpolvi
♦ Weed
♦ 70
♦ pH, cultivated, herbal.

Geranium thunbergii Sieb. & Zucc.
Geraniaceae
♦ Thunberg's geranium
♦ Weed, Quarantine Weed,
Naturalised
♦ 76, 87, 88, 101, 204, 286
♦ pH, cultivated, herbal.

Geranium tuberosum L.
Geraniaceae
♦ tuberous crane's bill, crane's bill
♦ Weed, Quarantine Weed,
Naturalised
♦ 76, 86, 87, 88, 115, 203, 220, 243, 272
♦ pH, cultivated.

Geranium versicolor L.
Geraniaceae
♦ veiny geranium, streaked crane's
bill, pencilled crane's bill
♦ Naturalised
♦ 40, 101
♦ cultivated.

Geranium viscosissimum Fisch. & Mey.
Geraniaceae
♦ sticky purple geranium, sticky
geranium
♦ Weed
♦ 23, 88
♦ pH, cultivated, herbal.

Geranium wilfordii Maxim.
Geraniaceae
♦ mitsubafuuro
♦ Weed
♦ 243
♦ pH, cultivated.

Geranium yemense Deflers
Geraniaceae
♦ Weed
♦ 88

Geranium yeoi Aedo & Muñoz Garm.
Geraniaceae
♦ greater herb Robert
♦ Naturalised
♦ 198
♦ cultivated. Origin: Madeira.

Gerbera anandria (L.) Sch.Bip.
Asteraceae
♦ common gerbera, ghostly daisy
♦ Weed
♦ 297
♦ herbal.

Gerbera jamesonii **Bolus ex Hook.f.**
Asteraceae
♦ Barberton daisy
♦ Naturalised
♦ 101
♦ a/pH, cultivated.

Geropogon glaber **L.**
Asteraceae
= *Tragopogon hybridus* L.
♦ Weed
♦ 87, 88

Geropogon glabrum **L.**
Asteraceae
= *Tragopogon crocifolius* L. (NoR)
♦ Weed
♦ 221

Gethyum atropupureum **Phil**
Liliaceae/Alliaceae
♦ Quarantine Weed
♦ 220

Geum aleppicum **Jacq.**
Rosaceae
Geum aleppicum Jacq. var. *strictum* (Ait.)
Fern. (see), *Geum strictum* Ait.
♦ yellow avens, idänkellukka
♦ Weed, Naturalised
♦ 23, 42, 87, 88, 161, 199, 272, 280, 286, 297
♦ pH, cultivated, herbal. Origin: Eurasia to Australia.

Geum aleppicum **Jacq. var.** *strictum*
(Aiton) Fern.
Rosaceae
= *Geum aleppicum* Jacq.
♦ yellow avens
♦ Weed
♦ 218
♦ herbal.

Geum japonicum **Thunb.**
Rosaceae
♦ daikonsou
♦ Weed
♦ 286
♦ pH, cultivated, herbal. Origin: North America, east Asia.

Geum macrophyllum **Willd.**
Rosaceae
♦ largeleaf avens, yellow avens, Japaninkellukka
♦ Weed, Cultivation Escape
♦ 42, 87, 88, 218
♦ pH, cultivated, herbal.

Geum peckii **Pursh**
Rosaceae
♦ mountain avens
♦ Weed
♦ 23, 88

Geum rivale **L.**
Rosaceae
Caryophyllata rivalis Scop.
♦ water avens, purple avens, ojakellukka
♦ Weed
♦ 23, 70, 88, 272
♦ pH, cultivated, herbal.

Geum urbanum **L.**
Rosaceae
Caryophyllata urbana Scop.

♦ wood avens, herb bennet
♦ Weed, Naturalised
♦ 23, 70, 87, 88, 101, 272, 280
♦ pH, cultivated, herbal.

Ghinia curassavica **(L.) Mill.**
Verbenaceae
♦ Weed
♦ 14

Gibasis geniculata **(Jacq.) Rohw.**
Commelinaceae
Tradescantia geniculata Jacq., *Aneilema geniculatum* (Jacq.) Woods.
♦ Tahitian bridal veil
♦ Naturalised
♦ 287
♦ cultivated. Origin: South America.

Gibasis pellucida **(M.Martens & Galeotti)**
D.R.Hunt
Commelinaceae
♦ dotted bridal veil, bridal veil
♦ Weed, Naturalised
♦ 101, 179
♦ herbal.

Gibasis schiediana **(Kunth) D.R.Hunt**
Commelinaceae
♦ Casual Alien
♦ 280

Gilia **Ruiz & Pav. spp.**
Polemoniaceae
♦ gilia
♦ Weed
♦ 88, 218

Gilia capitata **Sims**
Polemoniaceae
♦ pallerokiurunkukka, bluehead gilia, blue field gilia, globe gilia
♦ Casual Alien
♦ 40, 42
♦ aH, cultivated, herbal.

Gilia laciniata **Ruiz & Pav.**
Polemoniaceae
♦ liuskakiurunkukka
♦ Casual Alien
♦ 39, 42
♦ arid, toxic.

Gilia tricolor **Benth.**
Polemoniaceae
♦ bird's eye, bird's eye gilia, tricolor gilia
♦ Naturalised
♦ 86
♦ aH, cultivated, herbal. Origin: California.

Ginkgo biloba **(Van Geert) Beissn.**
Ginkgoaceae
Salisburia adiantifolia Sm., *Salisburya biloba* Hoffmanns.
♦ maidenhair tree, ginkgo, white nut
♦ Weed, Naturalised
♦ 39, 80, 101, 247
♦ T, cultivated, herbal, toxic.

Girardinia suborbiculata **C.J.Chen**
Urticaceae
Girardinia cuspidata Wedd.
♦ suborbiculata girardinia
♦ Weed
♦ 297

Gisekia africana **(Lour.) Kuntze**
Phytolaccaceae/Gisekiaceae
♦ Weed, Quarantine Weed
♦ 76, 87, 88, 203, 220

Gisekia africana **(Lour.) Kuntze var.**
africana
Phytolaccaceae/Gisekiaceae
♦ gisekia
♦ Weed, Quarantine Weed, Native Weed
♦ 121, 220
♦ H. Origin: southern Africa.

Gisekia africana **(Lour.) Kuntze var.**
pedunculata **(Oliv.) Brenan**
Phytolaccaceae/Gisekiaceae
♦ gisekia
♦ Weed
♦ 50

Gisekia pharnaceoides **L.**
Phytolaccaceae/Gisekiaceae
= *Gisekia pharnacioides* L.
♦ old maid, gisekia
♦ Weed, Introduced
♦ 221, 228, 243
♦ arid, herbal.

Gisekia pharnacioides **L.**
Phytolaccaceae/Gisekiaceae
Gisekia pharnaceoides L. (see)
♦ old maid, gisekia
♦ Weed, Naturalised, Native Weed
♦ 39, 50, 87, 88, 101, 121, 158, 179
♦ aH, herbal, toxic. Origin: southern Africa.

Gladiolus **L. spp.**
Iridaceae
♦ gladiolus, peacock orchid, sword lily
♦ Weed, Naturalised, Garden Escape, Environmental Weed
♦ 72, 86, 198, 203
♦ cultivated, herbal.

Gladiolus alatus **L.**
Iridaceae
♦ Weed, Naturalised, Cultivation Escape
♦ 7, 86, 98, 203, 252
♦ cultivated. Origin: South Africa.

Gladiolus angustus **L.**
Iridaceae
♦ long tubed painted lady
♦ Weed, Naturalised, Garden Escape, Environmental Weed
♦ 7, 9, 86, 98, 203
♦ cultivated. Origin: South Africa.

Gladiolus byzantinus **Mill.**
Iridaceae
= *Gladiolus communis* L. ssp. *byzantinus* (Mill.) A.P.Ham.
♦ Weed
♦ 87, 88
♦ cultivated, herbal.

Gladiolus cardinalis **Curtis**
Iridaceae
♦ waterfall gladiolus
♦ Weed, Naturalised, Garden Escape, Environmental Weed
♦ 7, 54, 86, 88, 98, 203
♦ cultivated. Origin: South Africa.

***Gladiolus carneus* D.Delaroche**
Iridaceae
Gladiolus blandus Aiton
♦ painted lady
♦ Weed, Naturalised, Cultivation
Escape
♦ 7, 15, 86, 98, 203, 252
♦ cultivated. Origin: South Africa.

***Gladiolus caryophyllaceus* (Burm.f.) Poir.**
Iridaceae
♦ wild gladiolus, sandveldlelie
♦ Weed, Naturalised, Garden Escape,
Environmental Weed
♦ 7, 9, 86, 98, 203
♦ cultivated. Origin: southern Africa.

***Gladiolus* × *colvillei* Sweet**
Iridaceae
= *Gladiolus cardinalis* Curtis × *Gladiolus tristis* L.
♦ Colville's gladiolus
♦ Naturalised
♦ 101
♦ cultivated.

***Gladiolus communis* L.**
Iridaceae
♦ Byzantine gladiolus, cornflag,
etelänmiekkalilja
♦ Weed, Naturalised, Garden Escape,
Environmental Weed, Casual Alien
♦ 7, 39, 42, 72, 88, 98, 101, 176, 203
♦ pH, cultivated, herbal, toxic.

***Gladiolus communis* L. ssp. *byzantinus* (Mill.) A.P.Ham.**
Iridaceae
Gladiolus byzantinus Mill. (see)
♦ Byzantine gladiolus, cornflag
♦ Naturalised, Environmental Weed
♦ 40, 86, 101, 176, 198
♦ cultivated. Origin: western
Mediterranean.

Gladiolus communis* L. ssp. *communis
Iridaceae
♦ cornflag
♦ Naturalised
♦ 101

***Gladiolus cuspidatus* Jacq.**
Iridaceae
♦ Weed
♦ 87, 88
♦ cultivated.

***Gladiolus floribundus* Jacq.**
Iridaceae
♦ Weed, Naturalised, Cultivation
Escape
♦ 86, 98, 203, 252
♦ cultivated. Origin: South Africa.

***Gladiolus* × *gandavensis* Van Houtte**
Iridaceae
= *Gladiolus dalenii* Van Geel (NoR)
× *Gladiolus oppositiflorus* Herb. [see
Gladiolus × *hortulanus* L.H.Bailey]
♦ gladiolus
♦ Naturalised
♦ 86, 98, 101
♦ cultivated. Origin: horticultural
hybrid.

***Gladiolus gueinzii* Kunze**
Iridaceae

♦ coastal gladiolus
♦ Weed, Naturalised, Cultivation
Escape
♦ 86, 98, 203, 252
♦ cultivated. Origin: South Africa.

***Gladiolus* × *hortulanus* L.H.Bailey**
Iridaceae
= *Gladiolus* × *gandavensis* Van Houtte
♦ Naturalised
♦ 134
♦ cultivated.

***Gladiolus* × *hybridus* hort.**
Iridaceae
♦ Naturalised
♦ 287

***Gladiolus illyricus* Koch**
Iridaceae
♦ wild gladiolus
♦ Weed
♦ 70, 272
♦ cultivated.

***Gladiolus imbricatus* L.**
Iridaceae
♦ idänmiekkalilja
♦ Weed, Casual Alien
♦ 42, 272
♦ cultivated.

***Gladiolus italicus* Mill.**
Iridaceae
Gladiolus segetum Ker Gawl. (see)
♦ Italian gladiolus, wild gladiolus,
common sword lily, Italianmiekkalilja,
corn gladiolus, cornfield gladiolus,
gladiolo de campo, glaïeul des
moissons, gladiolo dei campi,
espadana das searas
♦ Weed, Naturalised, Introduced,
Casual Alien
♦ 34, 42, 70, 101, 126, 221, 253, 272
♦ pH, cultivated, herbal. Origin:
Mediterranean.

Gladiolus natalensis* (Eckl.) Hook.f. *nom. illeg.
Iridaceae
= *Gladiolus dalenii* Van Geel (NoR) [see
Gladiolus × *gandavensis* Van Houtte]
♦ Weed, Naturalised, Cultivation
Escape
♦ 54, 86, 88, 252, 280
♦ pH, cultivated.

***Gladiolus palustris* Gaud.**
Iridaceae
Gladiolus pratensis Dietr.
♦ marsh gladiolus
♦ Weed
♦ 272
♦ cultivated.

***Gladiolus papilio* Hook.f.**
Iridaceae
♦ goldblotch gladiolus
♦ Naturalised
♦ 101
♦ cultivated.

***Gladiolus scullyi* Bak.**
Iridaceae
♦ Native Weed
♦ 121

♦ pH, cultivated. Origin: South Africa.

***Gladiolus segetum* Ker Gawl.**
Iridaceae
= *Gladiolus italicus* Mill.
♦ Weed, Quarantine Weed
♦ 30, 39, 70, 76, 87, 88, 203, 220
♦ cultivated, herbal, toxic.

***Gladiolus tristis* L.**
Iridaceae
♦ evening flower gladiolus, ever
flowering gladiolus, gladiolus, marsh
Afrikaner
♦ Weed, Naturalised, Garden Escape,
Environmental Weed
♦ 7, 72, 86, 88, 98, 101, 176, 198, 203,
289
♦ pH, cultivated. Origin: South Africa.

***Gladiolus undulatus* L.**
Iridaceae
♦ wild gladiolus, gladiolus
♦ Weed, Naturalised, Garden Escape,
Environmental Weed
♦ 7, 72, 86, 88, 98, 165, 176, 198, 203,
280, 289
♦ pH, cultivated. Origin: South Africa.

***Glandularia bipinnatifida* (Nutt.) Nutt.**
Verbenaceae
♦ Dakota mock vervain
♦ Weed
♦ 243
♦ herbal.

***Glandularia dissecta* (Willd. ex Spreng.) Schnack & Covas**
Verbenaceae
Verbena dissecta Willd. ex Spreng.
(see), *Verbena matthewsii* Briq., *Verbena pulchella* Sweet fo. *latiloba* Moldenke
♦ Weed
♦ 87, 88

***Glandularia hispida* Ruiz & Pav.**
Verbenaceae
Verbena hispida Ruiz & Pav. (see)
♦ hispid mock vervain
♦ Naturalised
♦ 101

***Glandularia* × *hybrida* (Grönland & Rümpler) Nesom & Pruski**
Verbenaceae
= *Glandularia peruviana* (L.) Druce ×
Glandularia incisa (Hook.) Tronc. (see) ×
Glandularia phlogiflora (Cham.) Schnack
& Covas × *Glandularia platensis*
(Spreng.) Schnack & Covas [complex
hybrid involving some or all]
♦ garden verbena
♦ Naturalised, Garden Escape
♦ 101
♦ cultivated.

***Glandularia incisa* (Hook.) Tronc.**
Verbenaceae
♦ crisped mock vervain
♦ Naturalised
♦ 101

***Glandularia lacinata* (L.) Schnack & Covas**
Verbenaceae
♦ Weed
♦ 87, 88

Glandularia peruviana **(L.) Small**
Verbenaceae
Erinus peruvianus L., *Verbena peruviana* (L.) Britton (see)
♦ Peruvian mock vervain
♦ Weed, Naturalised
♦ 87, 88, 101, 237, 295

Glandularia pulchella **(Sweet) Tronc.**
Verbenaceae
♦ South American mock vervain
♦ Weed, Naturalised
♦ 101, 179

Glandularia tenera **(Spreng.) Cabrera**
Verbenaceae
Glandularia laciniata auct. non (L.) Schnack & Covas, *Verbena tenera* Spreng. (see)
♦ Latin American mock vervain
♦ Weed, Naturalised
♦ 101, 237, 295

Glastaria deflexa **Boiss.**
Brassicaceae
Texiera glastifolia (DC.) Jaub. & Spach
♦ Weed
♦ 87, 88, 243

Glaucium arabicum **Fresen.**
Papaveraceae
♦ Weed
♦ 221

Glaucium corniculatum **(L.) Rudolph**
Papaveraceae
Chelidonium corniculatum L., *Glaucium grandiflorum sensu* Hayek *non* Boiss. & Huet, *Glaucium phoeniceum* Crantz, *Glaucium rubrum* Sibth. & Sm.
♦ bristly horned poppy, blackspot hornpoppy, red horned poppy, horned poppy
♦ Weed, Naturalised, Casual Alien
♦ 7, 39, 40, 42, 70, 86, 87, 88, 93, 94, 98, 101, 161, 198, 203, 205, 237, 243, 272, 280, 295
♦ aH, cultivated, herbal, toxic. Origin: Europe.

Glaucium flavum **Crantz**
Papaveraceae
Glaucium luteum Scop.
♦ horned poppy, yellow horned poppy, yellow hornpoppy, sea poppy, keltaneidonunikko
♦ Weed, Naturalised, Garden Escape, Environmental Weed, Casual Alien
♦ 23, 34, 39, 42, 70, 72, 80, 86, 87, 88, 94, 98, 101, 176, 181, 195, 198, 203, 272, 280, 287
♦ pH, cultivated, herbal, toxic. Origin: southern Europe.

Glaucium vitellinum **Boiss. & Buhse**
Papaveraceae
♦ Quarantine Weed
♦ 220
♦ cultivated.

Glaux maritima **L.**
Primulaceae
♦ sea milkwort, saltwort, black saltwort
♦ Weed
♦ 23, 88, 297

♦ pH parasitic, aqua, cultivated, herbal.

Glechoma hederacea **L.**
Lamiaceae
Nepeta glechoma Benth., *Nepeta hederacea* (L.) Trev. (see)
♦ ground ivy, haymaids, gill over the ground, creeping Charlie, maahumala, alehoof, cat's foot, field balm, lierre terrestre
♦ Weed, Quarantine Weed, Noxious Weed, Naturalised, Introduced, Garden Escape, Environmental Weed
♦ 8, 15, 17, 23, 34, 36, 39, 44, 52, 70, 76, 80, 86, 87, 88, 94, 101, 133, 136, 151, 155, 161, 165, 174, 194, 195, 207, 210, 211, 218, 222, 224, 241, 243, 249, 266, 272, 280, 300
♦ pC, aqua, cultivated, herbal, toxic. Origin: Eurasia.

Glechoma hederacea **L. ssp. *grandis* Hara**
Lamiaceae
♦ ground ivy
♦ Weed
♦ 263
♦ pH.

Glechoma hederacea **L. var. *grandis* (A.Gray) Kudô**
Lamiaceae
♦ Weed
♦ 286

Glechoma hirsuta **Waldst. & Kit.**
Lamiaceae
♦ Weed
♦ 272

Glechoma longituba **(Nakai) Kupr.**
Lamiaceae
♦ longtube ground ivy
♦ Weed
♦ 297

Glechon ciliata **Benth.**
Lamiaceae
♦ Weed
♦ 87, 88

Gleditsia amorphoides **(Gr.) Taub.**
Fabaceae/Caesalpiniaceae
Garugandra amorphoides Gr.
♦ espina de corona
♦ Weed
♦ 39, 295
♦ cultivated, toxic.

Gleditsia triacanthos **L.**
Fabaceae/Caesalpiniaceae
Gleditsia heterophylla Raf.
♦ common honey locust, honey locust, honey shuck, sweet locust, Amerikaanse driedoring, soetpeulboom
♦ Weed, Noxious Weed, Native Weed, Naturalised, Native Weed, Introduced, Environmental Weed, Cultivation Escape, Casual Alien
♦ 7, 8, 39, 63, 71, 80, 86, 87, 88, 98, 121, 161, 174, 203, 211, 218, 228, 243, 279, 280, 283, 290
♦ T, arid, cultivated, herbal, toxic. Origin: eastern North America.

Gleichenia laevigata **(Willd.) Hook.**
Gleicheniaceae
♦ Weed
♦ 87, 88

Gleichenia linearis **(Burm.) Clarke**
Gleicheniaceae
= *Dicranopteris linearis* (Burm.) Underw.
♦ Weed
♦ 87, 88
♦ herbal.

Gleichenia polypodioides **(L.) J.E.Sm.**
Gleicheniaceae
♦ creepingfern
♦ Native Weed
♦ 121
♦ pH. Origin: Madagascar.

Gleichenia weatherbyi **Fosb.**
Gleicheniaceae
Dicranopteris weatherbyi (Fosb.) Glassman
♦ Introduced
♦ 230
♦ H.

Glinus dahomensis **A.Cheval.**
Molluginaceae
♦ Weed
♦ 87, 88

Glinus lotoides **L.**
Molluginaceae
Mollugo hirta Thunb., *Mollugo lotoides* Wight & Arn. ex Clarke, *Tryphera prostrata* Blume
♦ hairy glinus, hairy carpetweed, lotus sweetjuice
♦ Weed, Naturalised
♦ 87, 88, 101, 170, 185, 198, 221
♦ aH, cultivated, herbal.

Glinus oppositifolius **(L.) DC.**
Molluginaceae
Mollugo oppositifolia L., *Mollugo spergula* L., *Mollugo verticillata* Roxb. *non* L.
♦ slender carpetweed
♦ Weed, Naturalised
♦ 86, 87, 88, 170, 198, 209
♦ herbal. Origin: Australia.

Gliricidia sepium **(Jacq.) Kunth ex Walp.**
Fabaceae/Papilionaceae
Gliricidia maculata (Kunth) Steud.
♦ mother of cocoa, mother of cacao, madre de cacao, quickstick, Mexican lilac
♦ Weed, Quarantine Weed, Naturalised, Introduced, Garden Escape, Environmental Weed, Cultivation Escape
♦ 3, 7, 32, 39, 76, 86, 87, 88, 101, 155, 179, 203, 228, 230, 261
♦ T, arid, cultivated, herbal, toxic. Origin: Mexico, Central America.

Globba parviflora **Presl**
Zingiberaceae
♦ Weed
♦ 87, 88

Globularia arabica **Jaub. & Spach**
Globulariaceae
♦ Weed
♦ 221

Globularia vulgaris L.
Globulariaceae
- ♦ Quarantine Weed
- ♦ 220
- ♦ pH, cultivated, herbal.

Glochidion littorale Bl.
Euphorbiaceae
- ♦ Weed
- ♦ 87, 88
- ♦ herbal.

Glochidion ponapense Hosok.
Euphorbiaceae
- ◊ mwehkenand
- ◊ Introduced
- ♦ 230
- ♦ S.

Glochidion puberulum Hosok.
Euphorbiaceae
- ♦ mwehkenand
- ♦ Introduced
- ♦ 230
- ♦ S.

Glochidion senyavinianum Glassman
Euphorbiaceae
- ♦ mwehkenand
- ♦ Introduced
- ♦ 230
- ♦ S.

Glomera carolinensis Williams
Orchidaceae
- ♦ Introduced
- ♦ 230
- ♦ H.

Gloriosa rothschildiana O'Brien
Liliaceae/Colchicaceae
= *Gloriosa superba* L. cv.
'Rothschildiana'
- ♦ gloriosa lily
- ♦ Weed
- ♦ 39, 161
- ♦ cultivated, herbal, toxic.

Gloriosa superba L.
Liliaceae/Colchicaceae
Gloriosa rothschildiana O'Brien (see)
- ♦ gloriosa lily, glory lily, climbing lily, Rhodesian flame lily, flame lily
- ♦ Weed, Naturalised, Garden Escape, Environmental Weed
- ♦ 3, 39, 54, 73, 86, 87, 88, 98, 101, 155, 161, 201, 203, 209, 247, 269, 290
- ♦ aC, cultivated, herbal, toxic. Origin: Africa.

Glossonema nubicum Decne.
Asclepiadaceae/Apocynaceae
= *Glossonema boveanum* (Decne.) Decne. ssp. *nubicum* (Decne.) Bullock (NoR)
- ♦ Weed
- ♦ 221

Glossostigma diandrum (L.) Kunze
Scrophulariaceae
- ♦ mud mat
- ♦ Weed, Naturalised
- ♦ 101, 197
- ♦ wH.

Gloxinia perennis (L.) Fritsch
Gesneriaceae
- ♦ Canterbury bells

- ♦ Weed, Naturalised, Cultivation Escape
- ♦ 32, 101, 261
- ♦ cultivated. Origin: South America.

Glyceria R.Br. spp.
Poaceae
- ♦ mannagrass
- ♦ Weed, Quarantine Weed, Naturalised
- ♦ 76, 86, 88, 220
- ♦ G.

Glyceria acutiflora Torr.
Poaceae
- ♦ creeping mannagrass
- ♦ Weed
- ♦ 286
- ♦ pG, aqua, promoted. Origin: China, Japan, eastern North America.

Glyceria canadensis (Michx.) Trin.
Poaceae
- ♦ rattlesnake grass, rattlesnake mannagrass
- ♦ Weed
- ♦ 80
- ♦ G, cultivated, herbal.

Glyceria declinata Breb.
Poaceae
- ♦ sweetgrass, waxy mannagrass, mannagrass, hybrid sweetgrass, glaucous sweetgrass, floating sweetgrass
- ♦ Weed, Quarantine Weed, Naturalised, Introduced, Environmental Weed
- ♦ 15, 70, 72, 76, 86, 87, 88, 98, 101, 176, 198, 203, 228, 280
- ♦ wpG, cultivated. Origin: Europe, North America.

Glyceria fluitans (L.) R.Br.
Poaceae
Desvauxia fluitans (L.) P.Beauv. ex Kunth, *Festuca fluitans* L., *Glyceria fluitans* var. *acutiflora* Döll, *Hydrochloa fluitans* (L.) Hartm., *Melica fluitans* (L.) Raspail, *Panicularia brachyphylla* Nash, *Panicularia fluitans* (L.) Kuntze, *Poa fluitans* (L.) Scop.
- ♦ water mannagrass, ojasorsimo, flote grass, floating sweetgrass
- ♦ Weed, Naturalised, Environmental Weed
- ♦ 23, 70, 86, 87, 88, 98, 101, 176, 203, 208, 218, 225, 241, 246, 272, 280, 300
- ♦ pG, aqua, cultivated, herbal. Origin: Eurasia.

Glyceria grandis S.Watson
Poaceae
- ♦ mannagrass, American mannagrass
- ♦ Weed, Introduced
- ♦ 159, 228
- ♦ pG, arid/aqua, herbal.

Glyceria ischyroneura Steud.
Poaceae
- ♦ dojoutsunagi
- ♦ Weed
- ♦ 286
- ♦ G.

Glyceria maxima (Hartm.) Holmb.
Poaceae
Glyceria aquatica (L.) Wahlb., *Molinia maxima* Hartm., *Panicularia aquatica* (L.) Kuntze, *Poa aquatica* L. (see)
- ♦ tall mannagrass, reed sweetgrass, English watergrass, great mannagrass, variegated watergrass
- ♦ Weed, Sleeper Weed, Noxious Weed, Naturalised, Introduced, Environmental Weed, Cultivation Escape
- ♦ 4, 7, 9, 15, 23, 39, 42, 70, 72, 80, 86, 87, 88, 98, 101, 104, 133, 147, 152, 176, 181, 195, 197, 198, 200, 203, 208, 225, 228, 246, 252, 272, 280, 296
- ♦ wpG, cultivated, herbal, toxic. Origin: Eurasia.

Glyceria × *pedicellata* F.Towns.
Poaceae
= *Glyceria fluitans* (L.) R.Br. × *Glyceria notata* Chevall.
- ♦ hybrid sweetgrass
- ♦ Naturalised
- ♦ 280
- ♦ G.

Glyceria plicata (Fr.) Fr.
Poaceae
Molinia plicata Hartmann
- ♦ plicate sweetgrass, savisorsimo
- ♦ Weed, Naturalised, Environmental Weed
- ♦ 54, 86, 88, 176, 198, 272, 280
- ♦ pG, aqua, promoted, herbal. Origin: Mediterranean, Middle East.

Glyceria striata (Lam.) Hitchc.
Poaceae
Glyceria nervata (Willd.) Trin., *Panicularia nervata* (Willd.) Kuntze
- ♦ fowl mannagrass
- ♦ Weed, Naturalised
- ♦ 15, 23, 39, 88, 161, 280
- ♦ pG, aqua, cultivated, herbal, toxic.

Glyceria triflora (Korsn.) Komar.
Poaceae
- ♦ Weed
- ♦ 275, 297
- ♦ pG.

Glycine hedysaroides Willd.
Fabaceae/Papilionaceae
- ♦ Weed
- ♦ 87, 88

Glycine max (L.) Merr.
Fabaceae/Papilionaceae
Dolichos soja L., *Glycine gracilis* Skvortzov, *Glycine hispida* (Moench) Maxim., *Glycine hispida* var. *brunnea* Skvortzov, *Glycine hispida* var. *lutea* Skvortzov, *Glycine soja* (L.) Merr. nom. illeg., *Soja hispida* Moench, *Soja max* (L.) Piper
- ♦ soija, soybean, sojabna, soijapapu
- ♦ Weed, Naturalised, Casual Alien, Cultivation Escape
- ♦ 39, 42, 86, 98, 101, 179, 203, 257, 261
- ♦ aH, cultivated, herbal, toxic.

Glycine max (L.) Merr. ssp. *soja* (Siebold & Zucc.) H.Ohashi

Fabaceae/Papilionaceae
= *Glycine soja* Sieb. & Zucc.
♦ Weed
♦ 286

Glycine microphylla (Benth.) Tind.
Fabaceae/Papilionaceae
♦ Naturalised
♦ 134
♦ Origin: Australia.

Glycine soja Sieb. & Zucc.
Fabaceae/Papilionaceae
Glycine formosana Hosok., *Glycine max*
(L.) Merr. ssp. *soja* (Siebold & Zucc.)
H.Ohashi (see), *Glycine ussuriensis*
Regel & Maack
♦ wild soybean, tsurumame, reseeding
soybean
♦ Weed, Naturalised
♦ 87, 88, 101, 275, 297
♦ cultivated.

Glycine tabacina (Labill.) Benth.
Fabaceae/Papilionaceae
Desmodium novo-hollandicum F.Muell.,
Kennedya tabacina Labill., *Leptolobium
elongatum* Benth., *Leptolobium
tabacinum* Benth.
♦ glycine, vanilla glycine
♦ Introduced
♦ 228
♦ pC, arid, cultivated, herbal.

Glycine wightii (Wight & Arn.) Verdc.
Fabaceae/Papilionaceae
= *Neonotonia wightii* (Wight &
Arn.) Verdc. var. *wightii* (NoR) [see
Neonotonia wightii (Wight & Arn.)
Lackey]
♦ soja perene
♦ Weed
♦ 80, 240, 255
♦ pH, cultivated. Origin: Asia.

Glycine wightii (Wight & Arn.) Verdc.
ssp. *wightii* var. *longicauda* (Schweinf.)
Verdc.
Fabaceae/Papilionaceae
Glycine javanica auct. non L.
♦ soya bean
♦ Native Weed
♦ 121
♦ pC. Origin: southern Africa.

Glycosmis parviflora (Sims) Little
Rutaceae
♦ flower axistree
♦ Weed, Naturalised
♦ 22, 101, 179
♦ S.

Glycosmis pentaphylla (Retz.) DC.
Rutaceae
♦ Jamaica mandarin orange
♦ Quarantine Weed
♦ 220
♦ cultivated, herbal. Origin: tropical
and temperate Asia.

Glycyrrhiza L. spp.
Fabaceae/Papilionaceae
♦ licorice
♦ Weed
♦ 272
♦ herbal.

Glycyrrhiza acanthocarpa (Lindl.)
J.M.Black
Fabaceae/Papilionaceae
♦ southern licorice
♦ Weed, Native Weed
♦ 87, 88, 269
♦ arid, cultivated. Origin: Australia.

Glycyrrhiza astragalina Gill.
Fabaceae/Papilionaceae
♦ orozú
♦ Weed
♦ 87, 88, 237, 295

Glycyrrhiza echinata L.
Fabaceae/Papilionaceae
♦ Russian licorice, wild liquorice
♦ Weed, Quarantine Weed
♦ 76, 87, 88, 94, 203, 220, 272
♦ pH, cultivated, herbal.

Glycyrrhiza foetida Desf.
Fabaceae/Papilionaceae
♦ Weed
♦ 87, 88

Glycyrrhiza glabra L.
Fabaceae/Papilionaceae
Liquiritia officinalis Moench
♦ cultivated liquorice, liquorice,
licorice
♦ Weed, Naturalised, Garden Escape,
Environmental Weed, Cultivation
Escape
♦ 34, 39, 51, 86, 87, 88, 98, 101, 115, 121,
155, 198, 203, 272
♦ pH, cultivated, herbal, toxic. Origin:
Eurasia.

Glycyrrhiza inflata Batal.
Fabaceae/Papilionaceae
♦ Weed
♦ 275
♦ pH, cultivated.

Glycyrrhiza lepidota (Nutt.) Pursh
Fabaceae/Papilionaceae
♦ wild licorice, American liquorice,
licorice
♦ Weed, Noxious Weed, Native Weed
♦ 49, 87, 88, 161, 174, 210, 212, 218, 219,
229, 264
♦ pH, arid, cultivated, herbal. Origin:
North America.

Glycyrrhiza pallidiflora Maxim.
Fabaceae/Papilionaceae
♦ pricklyfruit licorice
♦ Weed
♦ 297

Glycyrrhiza uralensis Fisch. ex DC.
Fabaceae/Papilionaceae
♦ Chinese licorice, gan zao, Asian
liquorice
♦ Weed
♦ 39, 275, 297
♦ pH, arid, cultivated, herbal, toxic.

Glycyrrhiza yunnanensis S.S.Cheng &
L.K.Tai
Fabaceae/Papilionaceae
♦ Yunnan licorice
♦ Weed
♦ 297

Gmelina arborea Roxb.
Lamiaceae/Verbenaceae
♦ white teak, Malay bush beech,
snapdragon tree, gumhar
♦ Weed, Naturalised, Casual Alien
♦ 22, 86, 93, 121, 261
♦ T, cultivated, herbal. Origin: Eurasia.

Gmelina asiatica L.
Lamiaceae/Verbenaceae
Gmelina coromandelina Burm., *Gmelina
integrifolia* Hunter, *Gmelina parvifolia*
Roxb., *Gmelina tomentosa* Fletcher
♦ badhara bush, Asiatic beechberry,
oval leaved gmelina
♦ Weed, Quarantine Weed, Noxious
Weed, Naturalised, Garden Escape
♦ 76, 86, 88, 101, 147, 191, 203, 261
♦ cultivated, herbal. Origin: Africa,
China, India, Sri Lanka, south-east
Asia.

Gmelina elliptica Sm.
Lamiaceae/Verbenaceae
♦ Naturalised
♦ 86
♦ herbal. Origin: Asia, Pacific Islands.

Gnaphalium L. spp.
Asteraceae
♦ cudweed
♦ Weed
♦ 243
♦ H, herbal.

Gnaphalium adnatum (Wall. ex DC.)
Kitam.
Asteraceae
♦ Naturalised
♦ 287

Gnaphalium affine D.Don
Asteraceae
Gnaphalium luteo-album L. var. *affine*
(D.Don) Kost.
♦ Jersey cudweed
♦ Weed
♦ 235, 238, 263, 274, 275, 286, 297
♦ aH, promoted, herbal.

Gnaphalium americanum Mill.
Asteraceae
= *Gamochaeta americana* (Mill.) Wedd.
♦ Weed, Naturalised
♦ 98, 203, 280

Gnaphalium baicalense Kirp.
Asteraceae
♦ cudweed
♦ Weed
♦ 114

Gnaphalium calviceps Fern.
Asteraceae
= *Gamochaeta falcata* (Lam.) Cabrera
(NoR) [see *Gamochaeta calviceps* (Fern.)
Cabrera & *Gnaphalium falcatum* Lam.]
♦ linearleaf cudweed, silky cudweed
♦ Weed, Naturalised
♦ 15, 87, 88, 98, 203, 218, 280, 286, 287

Gnaphalium chartaceum Greenm.
Asteraceae
♦ Weed
♦ 199

Gnaphalium cheiranthifolium Lam.
Asteraceae
Gnaphalium citrinum Hook. & Arn.
♦ marcela
♦ Weed
♦ 87, 88, 237, 295

Gnaphalium chilense Spreng.
Asteraceae
= *Pseudognaphalium stramineum*
(Kunth) W.A.Weber (NoR)
♦ cotton batting cudweed
♦ Weed
♦ 23, 88, 136, 161
♦ herbal.

Gnaphalium claviceps Fern.
Asteraceae
♦ Weed, Naturalised
♦ 7, 269
♦ aH. Origin: South America.

Gnaphalium coarctatum (Willd.) Ker.
Asteraceae
= *Gamochaeta spicata* (Lam.) Cabrera
♦ purple cudweed, cudweed
♦ Weed, Naturalised
♦ 7, 15, 165, 280
♦ Origin: South America.

Gnaphalium collinum Labill.
Asteraceae
= *Euchiton gymnocephalus* (DC.)
A.Anderb.
♦ creeping cudweed
♦ Weed
♦ 34
♦ pH, cultivated.

Gnaphalium conoideum H.K.B.
Asteraceae
♦ Weed
♦ 199
♦ herbal.

Gnaphalium falcatum Lam.
Asteraceae
= *Gamochaeta falcata* (Lam.) Cabrera
(NoR) [see *Gamochaeta calviceps* (Fern.)
Cabrera and *Gnaphalium calviceps*
Fern.]
♦ narrowleaf cudweed
♦ Weed
♦ 87, 88, 249
♦ herbal.

Gnaphalium gaudichaudianum DC.
Asteraceae
Gnaphalium mendocinum Phil.
♦ cudweed, vira vira
♦ Weed
♦ 236, 237, 295

Gnaphalium glomerulatum Sond. ex Harv.
Asteraceae
♦ Native Weed
♦ 121
♦ aH. Origin: southern Africa.

Gnaphalium hypoleucum DC.
Asteraceae
♦ Weed
♦ 238, 275, 297
♦ aH, promoted.

Gnaphalium indicum auct. non L.
Asteraceae
= *Gnaphalium polycaulon* Pers.
♦ Weed
♦ 87, 88, 221
♦ aH, promoted.

Gnaphalium inornatum DC.
Asteraceae
♦ Weed
♦ 199

Gnaphalium involucratum G.Forst.
Asteraceae
♦ creeping cudweed
♦ Weed
♦ 165
♦ Origin: New Zealand, Taiwan, Java,
Philippines, Australia.

Gnaphalium japonicum Thunb.
Asteraceae
= *Euchiton japonicus* (Thunb.)
A.Anderb.
♦ father and child plant
♦ Weed
♦ 34, 87, 88, 191, 204, 235, 263, 286
♦ aH, cultivated, herbal.

Gnaphalium luteo-album L.
Asteraceae
= *Pseudognaphalium luteo-album* (L.)
Hilliard & Burtt
♦ cudweed, Jersey cudweed,
valkojäkkärä, everlasting cudweed,
gnaphale jaunâtre
♦ Weed, Naturalised, Introduced,
Casual Alien
♦ 23, 34, 42, 50, 51, 70, 87, 88, 94, 121,
179, 185, 199, 221, 241, 243, 272, 286,
287, 300
♦ aH, cultivated, herbal, toxic. Origin:
Eurasia.

**Gnaphalium luteo-album L. ssp. affine
(D.Don.) Kost.**
Asteraceae
♦ Jersey cudweed
♦ Weed
♦ 273

Gnaphalium macounii Greene
Asteraceae
= *Pseudognaphalium macounii* (Greene)
Kartesz *comb. nov. ined.* (NoR)
♦ clammy cudweed
♦ Weed
♦ 87, 88, 218
♦ herbal.

Gnaphalium microcephalum Nutt.
Asteraceae
= *Pseudognaphalium canescens* (DC.)
W.A.Weber ssp. *microcephalum* (Nutt.)
Kartesz *comb. nov. ined.* (NoR)
♦ Weed
♦ 23, 88
♦ herbal.

Gnaphalium multiceps DC.
Asteraceae
♦ Weed
♦ 88, 204
♦ bH, herbal.

Gnaphalium obtusifolium L.
Asteraceae

= *Pseudognaphalium obtusifolium* (L.)
Hilliard & Burtt ssp. *obtusifolium* (NoR)
♦ fragrant cudweed, sweet everlasting,
rabbit tobacco, fragrant everlasting
♦ Weed
♦ 23, 87, 88, 161, 210, 218
♦ bH, herbal.

**Gnaphalium oligandrum (DC.) Hilliard
& Burtt**
Asteraceae
♦ undulate cudweed
♦ Native Weed
♦ 121
♦ aH. Origin: southern Africa.

Gnaphalium palustre Nutt.
Asteraceae
♦ lowland cudweed, western marsh
cudweed
♦ Weed, Noxious Weed
♦ 136, 161, 212, 299
♦ aH, herbal.

Gnaphalium pensylvanicum Willd.
Asteraceae
= *Gamochaeta pensylvanica* (Willd.)
Cabrera
♦ gnaphalium, wandering cudweed
♦ Weed, Naturalised
♦ 7, 87, 88, 98, 121, 158, 161, 203, 249,
255, 263, 269, 280, 286, 287
♦ aH. Origin: Americas.

Gnaphalium peregrinum Fern.
Asteraceae
= *Gamochaeta pensylvanica* (Willd.)
Cabrera
♦ wandering cudweed
♦ Weed
♦ 87, 88, 218
♦ herbal.

Gnaphalium polycaulon Pers.
Asteraceae
Gnaphalium indicum auct. non L. (see)
♦ manystem cudweed
♦ Weed, Naturalised
♦ 7, 86, 297
♦ Origin: India.

Gnaphalium pulvinatum Delile
Asteraceae
♦ Weed
♦ 87, 88, 185, 221

Gnaphalium purpureum L.
Asteraceae
= *Gamochaeta purpurea* (L.) Cabrera
♦ purple cudweed, cudweed, chafe
weed, cat's foot, rabbit tobacco,
everlasting, punajäkkärä, narrowleaf
falcatum cudweed
♦ Weed, Naturalised, Environmental
Weed, Casual Alien
♦ 15, 23, 34, 42, 72, 86, 87, 88, 98, 161,
179, 180, 203, 211, 218, 249, 255, 257,
273, 274, 280, 286, 287
♦ a/pH, cultivated, herbal. Origin:
Americas.

Gnaphalium purpureum L. var. purpureum
Asteraceae
♦ Naturalised
♦ 280

Gnaphalium purpureum L. var. *ustulatum* (Nutt.) B.Boivin
Asteraceae
= *Gamochaeta ustulata* (Nutt.) Nesom (NoR)
♦ Naturalised
♦ 280

Gnaphalium semiamplexicaule DC.
Asteraceae
♦ Weed
♦ 199
♦ herbal.

Gnaphalium simplicicaule Spreng.
Asteraceae
♦ Naturalised
♦ 280

Gnaphalium sphacilathum H.B.K.
Asteraceae
♦ Weed
♦ 199

Gnaphalium sphaericum Willd.
Asteraceae
♦ common cudweed
♦ Native Weed
♦ 269
♦ aH. Origin: Australia.

Gnaphalium spicatum Lam.
Asteraceae
= *Gamochaeta spicata* (Lam.) Cabrera
♦ purple cudweed, shiny cudweed
♦ Weed, Naturalised, Native Weed
♦ 121, 161, 249, 255, 287
♦ a/bH, herbal.

Gnaphalium stagnale I.M.Johnst.
Asteraceae
♦ Weed
♦ 199

Gnaphalium subfalcatum Cabrera
Asteraceae
= *Gamochaeta subfalcata* (Cabrera) Cabrera
♦ Weed, Naturalised
♦ 98, 203, 280

Gnaphalium sylvaticum L.
Asteraceae
= *Omalotheca sylvatica* (L.) Sch.Bip. & Schultz
♦ heath cudweed, wood cudweed, ahojäkkärä
♦ Weed, Naturalised
♦ 87, 88, 287
♦ cultivated, herbal.

Gnaphalium uliginosum L.
Asteraceae
Filaginella uliginosa (L.) Opiz (see)
♦ low cudweed, marsh cudweed, savijäkkärä
♦ Weed, Quarantine Weed, Naturalised, Casual Alien
♦ 23, 44, 76, 87, 88, 101, 136, 161, 203, 218, 220, 243, 253, 280, 286
♦ aH, cultivated, herbal.

Gnaphalium undulatum L.
Asteraceae
= *Pseudognaphalium undulatum* (L.) Hilliard & Burtt

♦ undulate cudweed, cape cudweed, undulated cudweed
♦ Weed, Naturalised
♦ 23, 40, 51, 87, 88, 121
♦ herbal. Origin: South Africa.

Gnaphalium vira-vira Molina
Asteraceae
♦ Naturalised
♦ 257

Gnidia burchellii (Meisn.) Gilg
Thymelaeaceae
♦ Weed, Native Weed
♦ 39, 121
♦ pS, toxic. Origin: southern Africa.

Gnidia capitata L.f.
Thymelaeaceae
♦ Native Weed
♦ 121
♦ pH, toxic. Origin: southern Africa.

Gnidia cuneata Meisn.
Thymelaeaceae
♦ Native Weed
♦ 121
♦ pS. Origin: southern Africa.

Gnidia polycephala (C.A.Mey.) Gilg
Thymelaeaceae
♦ Weed, Native Weed
♦ 39, 121
♦ S, toxic. Origin: southern Africa.

Gnidia squarrosa (L.) Druce
Thymelaeaceae
♦ Naturalised
♦ 86
♦ cultivated. Origin: Africa.

Godetia amoena (Lehm.) G.Don
Onagraceae
= *Clarkia amoena* (Lehm.) A.Nelson & J.F.Macbr.
♦ sirosilkkikukka, atlasblomma
♦ Cultivation Escape
♦ 42
♦ cultivated, herbal.

Godetia tenuifolia (Cav.) Spach
Onagraceae
= *Clarkia tenella* (Cav.) F.H. & M.R.Lewis ssp. *tenuifolia* (Cav.) D.M.Moore & Lewis (NoR)
♦ Weed
♦ 87, 88

Goebelia alopecuroides (L.) Bunge
Fabaceae/Papilionaceae
= *Sophora alopecuroides* L.
♦ Weed
♦ 87, 88

Goebelia pachycarpa (Schrenk ex C.A.Mey.) Bunge ex Boiss.
Fabaceae/Papilionaceae
= *Sophora pachycarpa* Schrenk ex C.A.Mey.
♦ Weed
♦ 87, 88

Goldbachia laevigata (M.Bieb.) DC.
Brassicaceae
♦ Weed
♦ 248, 272

Goldmania foetida (Jacq.) Standl.
Fabaceae/Mimosaceae
♦ Quarantine Weed
♦ 220

Gomphia serrata (Gaertn.) Kanis
Ochnaceae
♦ Weed
♦ 12
♦ herbal.

Gomphocarpus cancellatus (Burm.f.) Bruyns
Asclepiadaceae/Apocynaceae
Asclepias cancellata Burm.f., *Asclepias rotundifolia* Mill. (see)
♦ broadleaf cottonbush
♦ Naturalised
♦ 86, 198
♦ Origin: Africa.

Gomphocarpus fruticosus (L.) W.T.Aiton
Asclepiadaceae/Apocynaceae
Asclepias fruticosa L. (see)
♦ swan plant, narrowleaf cottonbush, balloon cotton, cape cotton, duck bush, milkweed, wild cotton
♦ Weed, Quarantine Weed, Noxious Weed, Naturalised, Introduced, Garden Escape, Environmental Weed
♦ 7, 23, 62, 76, 86, 87, 88, 147, 158, 176, 198, 203, 215, 246, 269, 280
♦ pH, cultivated, herbal, toxic. Origin: southern Africa to Ethiopia.

Gomphocarpus physocarpus E.Mey.
Asclepiadaceae/Apocynaceae
Asclepias physocarpa (E.Mey.) Schlecht. (see), *Asclepias semilunata* (A.Rich.) N.E.Br.
♦ gomphocarpus, balloon cottonbush
♦ Weed, Naturalised, Casual Alien
♦ 7, 86, 87, 88, 98, 134, 203, 218, 269, 280
♦ pH, arid, cultivated, herbal. Origin: southern Africa.

Gomphocarpus sinaicus Boiss.
Asclepiadaceae/Apocynaceae
Asclepias sinaica (Boiss.) Muschl.
♦ Weed
♦ 221
♦ arid.

Gomphrena celosioides C.Mart.
Amaranthaceae
Gomphrena alba Peter, *Gomphrena celosioides* Mart. fo. *villosa* Suess., *Gomphrena decumbens* (non Jacq.) auct., *Gomphrena globosa* L. ssp. *africana* Stuchlik
♦ bachelor's button, globe amaranth, prostrate globe amaranth, baan mai ruu roi paa, gomphrena weed, soft khakiweed, white eye
♦ Weed, Naturalised, Introduced, Environmental Weed, Casual Alien
♦ 39, 50, 51, 66, 86, 87, 88, 93, 98, 121, 158, 170, 203, 204, 205, 209, 228, 237, 239, 255, 256, 262, 269, 273, 276, 280, 286, 287, 295
♦ pH, arid, cultivated, herbal, toxic. Origin: tropical America.

Gomphrena decumbens Jacq.
Amaranthaceae
= *Gomphrena serrata* L.
♦ arrasa con todo
♦ Weed
♦ 23, 87, 88, 199
♦ aH, cultivated, herbal.

Gomphrena dispersa Standl.
Amaranthaceae
= *Gomphrena serrata* L.
♦ arrasa con todo
♦ Weed
♦ 87, 88, 157
♦ pH, herbal.

Gomphrena globosa L.
Amaranthaceae
♦ common globe amaranth, globe
amaranth, gomfréna hlávkatá, pahwis,
amaranto
♦ Weed, Naturalised, Introduced,
Cultivation Escape
♦ 32, 87, 88, 93, 101, 205, 228, 230, 261,
262
♦ aH, arid, cultivated, herbal.

Gomphrena martiana Gill. ex Moq.
Amaranthaceae
♦ Weed, Naturalised
♦ 101, 237, 270

Gomphrena serrata L.
Amaranthaceae
Gomphrena decumbens Jacq. (see),
Gomphrena dispersa Standl. (see)
♦ arrasa con todo
♦ Weed
♦ 179, 261

Goniocaulon indicum (Klein ex Willd.)
C.B.Clarke
Asteraceae
♦ Weed
♦ 66

Gonocarpus chinensis (Lour.) Orchard
Haloragaceae
♦ Chinese raspwort
♦ Naturalised
♦ 101
♦ herbal. Origin: Australia.

Gonocarpus chinensis (Lour.) Orchard
ssp. *verrucosus* (Maiden & Betcke)
Orchard
Haloragaceae
♦ Chinese raspwort
♦ Naturalised
♦ 101

Gonolobus carolinensis (Jacq.) Schult.
Asclepiadaceae/Apocynaceae
= *Matelea carolinensis* (Jacq.) Woods.
(NoR)
♦ hairy milkvine
♦ Weed
♦ 87, 88, 218

Gonolobus gonocarpos (Walter) Perry
Asclepiadaceae/Apocynaceae
= *Matelea gonocarpos* (Walt.) Shinners
(NoR)
♦ anglepod milkvine
♦ Weed
♦ 87, 88, 218

Gonolobus stenanthus (Standl.) Woodson
Asclepiadaceae/Apocynaceae
♦ Weed
♦ 87, 88

Gonospermum gomerae Bolle
Asteraceae
♦ Quarantine Weed
♦ 220

Gonostegia hirta (Hassk.) Miq.
Urticaceae
Pouzolzia hirta Hassk.
♦ Weed
♦ 87, 88, 286

Gonzalagunia spicata (Lam.) G.Maza
Rubiaceae
= *Gonzalagunia hirsuta* (Jacq.) K.Schum.
(NoR) [see *Duggena hirsuta* (Jacq.) Britt.
ex Britt. & Wilson]
♦ Weed
♦ 87, 88

Goodenia koningsbergeri (Back.) Back.
ex Bold
Goodeniaceae
Selliera koningsbergeri Back.
♦ Weed
♦ 13, 88, 170, 191
♦ aqua.

Goodia lotifolia Salisb.
Fabaceae/Papilionaceae
♦ golden tip clover tree, golden tip
♦ Naturalised
♦ 280
♦ cultivated, herbal.

Gorteria affinis DC.
Asteraceae
♦ Quarantine Weed
♦ 220

Gorteria calendulaceae DC.
Asteraceae
♦ Quarantine Weed
♦ 220

Gorteria diffusa Thunb.
Asteraceae
♦ beetle daisy
♦ Quarantine Weed
♦ 220
♦ cultivated.

Gorteria personata L.
Asteraceae
♦ gorteria
♦ Weed, Quarantine Weed, Noxious
Weed, Naturalised
♦ 7, 76, 86, 88, 98, 147, 203
♦ cultivated. Origin: South Africa.

Gossypium barbadense L.
Malvaceae
Gossypium barbadense var. *acuminatum*
(Roxb. ex G.Don) Triana & Planch.
(see), *Gossypium brasiliense* Macfad.
♦ sea island cotton, Creole cotton,
koatun
♦ Weed, Naturalised, Introduced
♦ 39, 86, 101, 161, 230, 241, 300
♦ pH, cultivated, herbal, toxic. Origin:
tropical America.

Gossypium barbadense L. var.
acuminatum (Roxb. ex G.Don) Triana &
Planch.
Malvaceae
= *Gossypium barbadense* L.
♦ Casual Alien
♦ 261

Gossypium barbadense L. var. *barbadense*
Malvaceae
♦ algodón, sea island cotton
♦ Casual Alien
♦ 261

Gossypium hirsutum L.
Malvaceae
♦ upland cotton, cotton
♦ Weed, Naturalised
♦ 7, 22, 39, 86, 93, 98, 203
♦ S, cultivated, herbal, toxic. Origin:
Americas.

Gossypium hirsutum L. var. *punctatum*
(K.Schum. & Thonn.) Roberty
Malvaceae
= *Gossypium hirsutum* L. var. *hirsutum*
(NoR)
♦ algodón, algodón silvestre
♦ Naturalised
♦ 261

Gossypium thurberi Tod.
Malvaceae
♦ Thurber's cotton, wild desert cotton
♦ Naturalised
♦ 86
♦ cultivated, herbal. Origin: North
America.

Gossypium tomentosum Nutt.
Malvaceae
♦ Hawaiian cotton
♦ Weed
♦ 87, 88, 218
♦ cultivated.

Gouania lupuloides Urb.
Rhamnaceae
Banisteria lupuloides L., *Gouania
domingensis* L., *Rhamnus domingensis*
Jacq.
♦ whiteroot
♦ Weed
♦ 87, 88
♦ S/T, arid, herbal.

Gouania polygama (Jacq.) Urb.
Rhamnaceae
Gouania tomentosa Jacq., *Rhamnus
polygama* Jacq.
♦ liane savon
♦ Weed, Naturalised
♦ 14, 257
♦ arid, herbal.

Gourliea chilensis Clos
Fabaceae/Papilionaceae
♦ Quarantine Weed
♦ 220

Gourliea decorticans Gillies ex Hook.
& Arn.
Fabaceae/Papilionaceae
= *Geoffroea decorticans* (Gill. ex Hook. &
Arn.) Burkart
♦ Quarantine Weed
♦ 220

Gourliea spinosa (Molina) Skeels
Fabaceae/Papilionaceae
- Quarantine Weed
- 220
- arid.

Grabowskia duplicata Arn.
Solanaceae
Ehretia duplicata Nees
- Introduced
- 228
- arid, cultivated.

Gracilaria Grev. spp.
Gracilariaceae
- red alga
- Weed
- 197
- algae, herbal.

Gracilaria epihippisora Hoyle
Gracilariaceae
- red alga
- Weed
- 197
- algae.

Gracilaria eucheumatoides Harv.
Gracilariaceae
- red alga
- Weed
- 197
- algae.

Gracilaria salicornia (C.Agardh) Dawson
Gracilariaceae
- red alga
- Weed
- 197, 282
- algae.

Gracilaria tikvahiae McLachlan
Gracilariaceae
- graceful redweed, red alga
- Weed
- 197
- algae. Origin: eastern Atlantic coast to Caribbean.

Graderia scabra (L.f.) Benth.
Scrophulariaceae
- Native Weed
- 121
- pH parasitic. Origin: southern Africa.

Grammatotheca bergiana (Cham.) C.Presl
Campanulaceae/Lobeliaceae
- Naturalised
- 86
- Origin: South Africa.

Grammitis blechnoides Grev.
Grammitidaceae
= *Ctenopteris blechnoides* (Grev.) W.H.Wagner & Grether (NoR)
- Introduced
- 230
- H.

Grammitis ponapensis Copel.
Grammitidaceae
- Introduced
- 230
- H.

Grammitis scleroglossoides Copel.
Grammitidaceae

- Introduced
- 230
- H.

Grangea maderaspatana (L.) Poir.
Asteraceae
Artemisia maderaspatana L., *Cotula maderaspatana* Willd., *Cotula sphaeranthus* Link, *Grangea ceruanoides* Cass., *Grangea hispida* Humbert, *Grangea procumbens* DC., *Grangea sphaeranthus* (Link) K.Koch, *Tanacetum aegytiacum* Juss. ex Jacq.
- marcella
- Weed
- 13, 87, 88, 170, 191
- herbal. Origin: Africa, Madagascar, India, China, south-east Asia.

Graptopetalum paraguayense (N.E.Br.) E.Walther
Crassulaceae
- Naturalised
- 86
- cultivated.

Graptophyllum pictum (L.) Griff.
Acanthaceae
- caricature plant
- Naturalised, Introduced
- 230
- H, cultivated, herbal. Origin: Australia.

Grateloupia doryphora (Mont.) Howe
Cryptonemiaceae
Halymenia doryphora Montague, *Halymenia lanceola* J.Agardh, *Grateloupia lanceola* (J.Agardh) J.Agardh
- red alga
- Weed
- 197, 282, 288
- algae. Origin: Pacific Ocean.

Grateloupia filicina (Wulfen) J.Agardh var. luxurians A. & E.Gepp
Cryptonemiaceae
- Weed
- 288
- algae.

Gratiola japonica Miq.
Scrophulariaceae
- ooabunome
- Weed
- 87, 88, 204, 263, 275, 286, 297
- aH.

Gratiola juncea Roxb.
Scrophulariaceae
= *Dopatrium junceum* (Roxb.) Buch.-Ham.
- Weed
- 88, 191, 204

Gratiola neglecta Torr.
Scrophulariaceae
- kuntio, clammy hedgehyssop
- Naturalised
- 42
- aH, herbal.

Gratiola officinalis L.
Scrophulariaceae
- hedgehyssop
- Weed

- 39, 272
- pH, cultivated, herbal, toxic.

Grayia spinosa (Hook.) Moq.
Chenopodiaceae
- spiny hopsage, hopsage
- Weed
- 23, 88
- S, arid, cultivated, herbal.

Greenovia aurea (C.A.Sm.) Webb & Berthel.
Crassulaceae
- Casual Alien
- 280
- cultivated.

Grevillea arenaria R.Br.
Proteaceae
- grey grevillea
- Weed, Naturalised, Environmental Weed
- 72, 86, 88, 198
- cultivated. Origin: Australia.

Grevillea aspleniifolia R.Br. ex Salisb.
Proteaceae
- Naturalised
- 280
- cultivated.

Grevillea banksii R.Br
Proteaceae
- scarlet grevillea, red flowered silky oak, kahili flower, red silky oak, Bank's grevillea, haiku
- Weed, Noxious Weed, Naturalised
- 3, 22, 39, 80, 87, 88, 101, 121, 161, 218, 229
- S/T, cultivated, toxic. Origin: Australia.

Grevillea floribunda R.Br. ssp. floribunda
Proteaceae
- rusty grevillea
- Naturalised
- 86, 198
- Origin: Australia.

Grevillea juniperina R.Br.
Proteaceae
- prickly spiderflower
- Naturalised
- 198
- cultivated. Origin: Australia.

Grevillea juniperina × victoriae R.Br. & F.Muell.
Proteaceae
- Naturalised
- 198

Grevillea leucopteris Meisn.
Proteaceae
- white plumed grevillea
- Naturalised, Native Weed, Garden Escape, Cultivation Escape
- 7
- cultivated.

Grevillea robusta A.Cunn. ex R.Br.
Proteaceae
- Australian silky oak, silky oak
- Weed, Noxious Weed, Naturalised, Introduced, Garden Escape, Environmental Weed, Cultivation Escape

♦ 3, 22, 39, 80, 86, 88, 95, 101, 121, 134,
151, 152, 161, 179, 198, 226, 228, 230,
233, 261, 279, 280, 283
♦ T, arid, cultivated, herbal, toxic.
Origin: Australia.

Grevillea rosmarinifolia A.Cunn.
Proteaceae
♦ rosemary grevillea
♦ Weed, Naturalised, Native Weed,
Garden Escape, Environmental Weed
♦ 72, 86, 88, 198, 290
♦ pS, cultivated. Origin: Australia.

Grewia asiatica L.
Tiliaceae
Grewia subinaequalis DC.
♦ grewia, phalsa
♦ Weed, Sleeper Weed, Naturalised,
Introduced, Environmental Weed
♦ 3, 39, 86, 155, 191, 228
♦ S/T, arid, cultivated, herbal, toxic.
Origin: central Asia.

Grewia bicolor Juss.
Tiliaceae
Grewia disticha Dinter & Burret, *Grewia*
kwebensis N.E.Br., *Grewia miniata* Mast.
ex Hiern
♦ bastard brandybush, bastard raisin
bush, false brandybush, twocolour
grewia, basterosyntjie, bastard raisin
♦ Weed, Native Weed
♦ 10, 63, 121
♦ S/T, arid, cultivated. Origin:
southern Africa.

Grewia biloba G.Don
Tiliaceae
♦ bilobed grewia
♦ Naturalised
♦ 101
♦ cultivated.

Grewia caffra Meisn.
Tiliaceae
♦ climbing raisin, climbing grewia
♦ Sleeper Weed, Naturalised
♦ 86
♦ cultivated.

Grewia carpinifolia Juss.
Tiliaceae
♦ Introduced
♦ 228
♦ arid, cultivated, herbal.

Grewia flava DC.
Tiliaceae
♦ brandybush, wild currant, wild
raisin, fluweelrosyntjie, wilderosyntjie
♦ Weed, Native Weed
♦ 10, 63, 121
♦ S/T, cultivated. Origin: southern
Africa.

Grewia flavescens Juss.
Tiliaceae
♦ donkeyberry, rough leaved raisin,
skurwerosyntjie, sandpaper raisin
♦ Weed, Native Weed
♦ 10, 63, 121
♦ S/T. Origin: southern Africa.

Grewia occidentalis L.
Tiliaceae
♦ crossberry, four corners

♦ Native Weed
♦ 121
♦ S/T, cultivated, herbal. Origin:
southern Africa.

Grewia tenax (Forssk.) Fiori
Tiliaceae
Chadara tenax Forssk., *Grewia betulifolia*
Schinz, *Grewia populifolia* Vahl
♦ Weed
♦ 221
♦ arid, herbal.

Grewia villosa Willd.
Tiliaceae
♦ Weed
♦ 221
♦ herbal.

Grindelia Willd. spp.
Asteraceae
♦ gumweed
♦ Weed
♦ 136
♦ herbal.

Grindelia brachystephana Griseb.
Asteraceae
♦ Weed
♦ 87, 88

Grindelia camporum Greene
Asteraceae
♦ gumplant, great valley grindelia,
grindelia, gumweed, tarweed,
common gumplant
♦ Weed, Naturalised, Introduced
♦ 98, 161, 180, 203, 228, 243
♦ pH, arid, cultivated, herbal. Origin:
Australia.

Grindelia camporum Greene var.
australis Steyerm.
Asteraceae
♦ Naturalised
♦ 86
♦ cultivated.

Grindelia inuloides Willd. var. glandulosa
(Greenm.) Steyerm.
Asteraceae
♦ Weed
♦ 199

Grindelia inuloides Willd. var. inuloides
Asteraceae
♦ Weed
♦ 199

Grindelia nana Nutt.
Asteraceae
♦ gumweed, smallflowered, Idaho
gumweed, Idaho resinweed, low
gumweed
♦ Weed
♦ 23, 88, 161
♦ pH, cultivated, herbal.

Grindelia rubricaulis DC.
Asteraceae
♦ Weed
♦ 87, 88

Grindelia squarrosa (Pursh) Dunal
Asteraceae
♦ gumweed, gumplant, broad leaved
gumplant, gumweed, rosinweed,
tarweed, scaly grindelia, curlycup

gumweed, rilpiö
♦ Weed, Noxious Weed, Native Weed,
Introduced, Casual Alien
♦ 23, 23, 39, 40, 42, 49, 87, 88, 161, 174,
210, 212, 218, 228, 229, 272
♦ b/pH, arid, cultivated, herbal, toxic.
Origin: North America.

Grindelia squarrosa (Pursh) Dunal var.
quasiperennis Lunell
Asteraceae
♦ gumweed, curlycup gumweed
♦ Noxious Weed
♦ 229

Grindelia squarrosa (Pursh) Dunal var.
serrulata (Rydb.) Steyerm.
Asteraceae
♦ gumweed, Great Plains resinweed,
curlycup gumweed
♦ Noxious Weed
♦ 229
♦ pH.

Grindelia squarrosa (Pursh) Dunal var.
squarrosa
Asteraceae
♦ gumweed, curlycup gumweed
♦ Noxious Weed
♦ 229

Groenlandia densa (L.) Fourr.
Potamogetonaceae
Potamogeton densus L.
♦ opposite leaved pondweed,
äervenaäka hustolistá
♦ Weed
♦ 70, 272
♦ herbal.

Grusonia Rchb. ex Britt. & Rose spp.
Cactaceae
= *Opuntia* Mill. spp.
♦ Weed, Quarantine Weed
♦ 76, 88, 203

Guadua angustifolia Kunth
Poaceae
= *Bambusa guadua* Kunth
♦ Weed
♦ 255
♦ pG, cultivated. Origin: South
America.

Guadua angustifolia Kunth ssp.
angustifolia
Poaceae
Bambusa aculeata (Rupr.) Hitchc.,
Guadua aculeata Rupr. ex Fourn.
♦ Introduced
♦ 228
♦ G.

Guadua trinii (Nees) Nees ex Rupr.
Poaceae
Arundarbor trinii (Nees) Kuntze,
Bambusa ribbentropii Herter, *Bambusa*
riograndensis Dutra, *Bambusa tacuara*
Arechav., *Bambusa tomentosa* (Hack. &
Lindm.) McClure, *Bambusa trinii* Nees,
Chusquea heterophylla Griseb., *Guadua*
riograndensis (Dutra) Herter, *Guadua*
tomentosa Hack. & Lindm., *Guadua*
trinii (Nees) Nees ex Rupr. var. *scabra*
Döll

- Weed
- 87, 88
- G.

Guaiacum officinale L.
Zygophyllaceae
Guajacum officinale L. (see)
- lignum vitae
- Weed, Introduced
- 87, 88, 228
- arid, cultivated, herbal.

Guajacum officinale L.
Zygophyllaceae
= *Guaiacum officinale* L.
- lignum vitae, guajakkipuu, guaiaco, legno santo
- Weed
- 39, 179
- herbal, toxic.

Guarea trichilioides L.
Meliaceae
= *Guarea guidonia* (L.) Sleumer (NoR)
- American muskwood
- Weed
- 87, 88
- herbal.

Guazuma ulmifolia Lam.
Sterculiaceae
Guazuma guazuma Cockerell, *Guazuma polybotrya* Cav., *Guazuma tomentosa* Kunth, *Theobroma guazuma* L.
- bastard cedar, guacimo, West Indian elm
- Weed, Quarantine Weed, Introduced
- 14, 76, 88, 203, 228
- T, arid, cultivated, herbal.

Gueldenstaedtia diversifolia Maxim.
Fabaceae/Papilionaceae
- diversifolious gueldenstaedtia
- Weed
- 297

Gueldenstaedtia multiflora Bunge
Fabaceae/Papilionaceae
- Weed
- 275, 297
- pH, herbal.

Gueldenstaedtia stenophylla Bunge
Fabaceae/Papilionaceae
- Weed
- 275, 297
- pH.

Guettarda calyptrata A.Rich.
Rubiaceae
- Weed
- 14

Guettarda platypoda DC.
Rubiaceae
- Weed
- 87, 88

Guettarda speciosa L.
Rubiaceae
- puapua, ihd
- Weed
- 6, 88
- T, cultivated, herbal. Origin: Madagascar.

Guettarda valenzuelana A.Rich.
Rubiaceae
- cucubano de monte
- Weed
- 14

Guilandina bonduc L.
Fabaceae/Caesalpiniaceae
= *Caesalpinia bonduc* (L.) Roxb.
- Weed
- 3, 191

Guilleminea densa (Humb. & Bonpl. ex Willd.) Moq.
Amaranthaceae
Brayulinea densa (H.& B.) Small (see), *Guilleminea illecebroides* Kunth, *Illecebrum densum* Roem. & Schult.
- carrot weed, small matweed
- Weed, Naturalised, Introduced
- 86, 88, 98, 158, 161, 199, 203, 228
- arid, herbal. Origin: Central and South America.

Guizotia abyssinica (L.f.) Cass.
Asteraceae
Guizotia oleifera DC., *Polymnia abyssinica* L.f.
- ramtilla, niger seed, niger, keltiö
- Weed, Naturalised, Introduced, Environmental Weed, Casual Alien
- 34, 40, 42, 86, 98, 101, 203, 228, 280, 287
- aH, arid, cultivated, herbal. Origin: tropical Africa.

Guizotia scabra (Vis.) Chiov.
Asteraceae
- Weed
- 240
- aH.

Guizotia villosa Sch.Bip. ex A.Rich.
Asteraceae
- Weed, Quarantine Weed
- 76, 87, 88, 203, 220

Gundelia tournefortii L.
Asteraceae
- Tournefort's gundelia
- Weed, Quarantine Weed
- 87, 88, 220, 221, 243
- pH, arid, promoted.

Gunnera tinctoria (Molina) Mirb.
Gunneraceae
Gunnera chilensis Lam., *Gunnera scabra* Ruiz & Pav., *Panke tinctoria* Molina
- Chilean 'ape'ape, Chilean gunnera, Chilean rhubarb
- Weed, Sleeper Weed, Quarantine Weed, Naturalised, Environmental Weed, Cultivation Escape
- 15, 34, 40, 101, 181, 225, 246, 280, 290
- wpH, cultivated, herbal. Origin: Chile.

Gutenbergia cordifolia Benth. ex Oliv.
Asteraceae
= *Erlangea cordifolia* (Benth. ex Oliv.) S.Moore
- Weed
- 53, 88
- aH.

Gutierrezia alamanii A.Gray
Asteraceae
- Weed
- 199

Gutierrezia dracunculoides (DC.) Blake
Asteraceae
= *Amphiachyris dracunculoides* (DC.) Nutt. (NoR)
- common broomweed
- Weed
- 87, 88, 161, 210, 218
- herbal.

Gutierrezia microcephala (DC.) Gray
Asteraceae
- threadleaf snakeweed, sticky snakeweed
- Weed, Native Weed
- 87, 88, 161, 218, 267
- S, herbal, toxic. Origin: south-western USA.

Gutierrezia sarothrae (Pursh) Britt. & Rusby
Asteraceae
Brachyachyris euthamiae (Nutt.) Spreng., *Brachyris divaricata* Nutt., *Brachyris euthamiae* Nutt., *Galinsoga linearifolia* (Lag.) Spreng., *Gutierrezia corymbosa* A.Nelson, *Gutierrezia digyna* S.F.Blake, *Gutierrezia divaricata* (Nutt.) Torr. & A.Gray, *Gutierrezia diversifolia* Greene, *Gutierrezia euthamiae* (Nutt.) Torr. & A.Gray, *Gutierrezia fasciculata* Greene, *Gutierrezia filifolia* Greene, *Gutierrezia fulva* Lunell, *Gutierrezia furfuracea* Greene, *Gutierrezia globosa* A.Nelson, *Gutierrezia goldmanii* Greene, *Gutierrezia greenei* Lunell, *Gutierrezia haenkei* Sch.Bip., *Gutierrezia ionensis* Lunell, *Gutierrezia juncea* Greene, *Gutierrezia laricina* Greene, *Gutierrezia lepidota* Greene, *Gutierrezia linearifolia* Lag., *Gutierrezia linearis* Rydb., *Gutierrezia linoides* Greene, *Gutierrezia longifolia* Greene, *Gutierrezia longipappa* S.F.Blake, *Gutierrezia myriocephala* A.Nelson, *Gutierrezia pomariensis* (S.L.Welsh) S.L.Welsh, *Gutierrezia sarothrae* var. *pomariensis* S.L.Welsh, *Gutierrezia scoparia* Rydb., *Gutierrezia tenuis* Greene, *Solidago sarothrae* Pursh, *Xanthocephalum digynum* (S.F.Blake) Shinners, *Xanthocephalum longipappum* (S.F.Blake) Shinners, *Xanthocephalum petradoria* S.L.Welsh & Goodrich, *Xanthocephalum sarothrae* (Pursh) Shinners, *Xanthocephalum sarothrae* var. *pomariense* (S.L.Welsh) S.L.Welsh, *Xanthocephalum tenue* (Greene) Shinners
- broomweed, turpentine weed, broom snakeweed, stinkweed, perennial snakeweed, yellowtop, matchbrush
- Weed, Native Weed
- 39, 80, 87, 88, 136, 161, 174, 212, 218, 267
- S, cultivated, herbal, toxic. Origin: south-western USA.

***Gutierrezia texana* Torr. & Gray**
Asteraceae
♦ Texas broomweed
♦ Weed
♦ 87, 88, 218
♦ herbal.

***Gymnadenia conopsea* (L.) R.Br.**
Orchidaceae
♦ fragrant orchid, sweet scented orchid, kirkiruoho
♦ Weed, Naturalised
♦ 70, 101, 272
♦ pH, cultivated, herbal.

***Gymnadenia conopsea* (L.) R.Br. ssp. *borealis* (Druce) F.Rose**
Orchidaceae
Habenaria gymnadenia Druce var. *borealis* Druce
♦ Cultivation Escape
♦ 40
♦ cultivated.

***Gymnadenia odoratissima* (L.) Rich.**
Orchidaceae
♦ marsh fragrant orchid, short spurred fragrant orchid
♦ Weed
♦ 70
♦ cultivated, herbal.

***Gymnanthocereus* Backeb. spp.**
Cactaceae
= *Browningia* Britt. & Rose spp. (NoR)
[see *Gymnocereus* Backeb. spp.]
♦ Weed, Quarantine Weed
♦ 76, 88, 203

***Gymnarrhena micrantha* Desf.**
Asteraceae
♦ Weed
♦ 221

***Gymnaster koraiensis* (Nakai) Kitam.**
Asteraceae
♦ Naturalised
♦ 287

***Gymnocarpos decandrum* Forssk.**
Caryophyllaceae/Illecebraceae
♦ Weed
♦ 221

***Gymnocereus* Backeb. spp.**
Cactaceae
= *Browningia* Britt. & Rose spp. (NoR)
[see *Gymnanthocereus* Backeb. spp.]
♦ Weed, Quarantine Weed
♦ 76, 88, 203

***Gymnocladus dioicus* (L.) K.Koch**
Fabaceae/Caesalpiniaceae
♦ Kentucky coffee tree
♦ Weed
♦ 8, 39, 87, 88, 161, 218, 247
♦ cultivated, herbal, toxic.

***Gymnocoronis spilanthoides* DC.**
Asteraceae
♦ Senegal tea plant, temple plant, spadeleaf plant
♦ Weed, Quarantine Weed, Noxious Weed, Naturalised, Garden Escape, Environmental Weed
♦ 3, 54, 62, 76, 86, 87, 88, 98, 147, 155,

169, 171, 191, 200, 203, 220, 225, 232, 246, 251, 280, 290
♦ wpH, cultivated. Origin: tropical Americas.

***Gymnopetalum cochinchinense* (Lour.) Kurz**
Cucurbitaceae
♦ Weed
♦ 13
♦ herbal.

***Gymnopetalum leucostictum* Miq.**
Cucurbitaceae
♦ Weed
♦ 13, 87, 88
♦ herbal.

***Gymnostyles anthemifolia* Juss.**
Asteraceae
Soliva anthemifolia (Juss.) R.Br. ex Less. (see)
♦ button burrweed
♦ Naturalised
♦ 101

***Gymnostyles stolonifera* (Brot.) Tutin**
Asteraceae
Soliva nasturtiifolia (Juss.) DC. (see), *Soliva stolonifera* (Brot) Loudon (see)
♦ carpet burrweed
♦ Naturalised
♦ 101

***Gynandriris setifolia* (L.f.) R.C.Foster**
Iridaceae
Iris setifolia L.
♦ thread iris
♦ Weed, Naturalised, Garden Escape, Environmental Weed, Casual Alien
♦ 7, 72, 86, 88, 98, 198, 203, 280
♦ pH, cultivated.

***Gynandropsis gynandra* (L.) Briq.**
Capparaceae/Cleomaceae
= *Cleome gynandra* L.
♦ spiderflower, nyeve, tsuma, ulude, lyaka, African spiderflower
♦ Weed, Introduced
♦ 50, 51, 53, 88, 185, 218, 221, 228, 240, 243
♦ aH, arid, promoted, herbal.

***Gynatrix pulchella* (Willd.) Alef.**
Malvaceae
♦ hemp bush
♦ Naturalised
♦ 300
♦ S, cultivated. Origin: Australia.

***Gynerium sagittatum* (Aubl.) Beauv.**
Poaceae
Gynerium parviflorum Nees, *Gynerium saccharoides* Humb. & Bonpl., *Saccharum sagittatum* Aubl.
♦ wild cane
♦ Weed, Naturalised, Introduced
♦ 87, 88, 228, 241, 300
♦ G, arid, herbal.

***Gynostemma pentaphyllum* (Thunb.) Makino**
Cucurbitaceae
♦ sweet tea vine, amachazuru
♦ Weed
♦ 286

♦ a/pH, promoted, herbal. Origin: China, Japan, Korea.

***Gynura aurantiaca* (Blume) DC.**
Asteraceae
♦ velvetplant, purple passion
♦ Weed, Naturalised
♦ 101, 179
♦ cultivated, herbal.

***Gynura bicolor* (Roxb. ex Willd.) DC.**
Asteraceae
♦ Naturalised
♦ 287

***Gynura crepidioides* Benth.**
Asteraceae
= *Crassocephalum crepidioides* (Benth.) S.Moore
♦ Weed
♦ 3, 191, 275
♦ aH, herbal.

***Gynura japonica* Juel**
Asteraceae
♦ Naturalised
♦ 287
♦ herbal.

***Gynura procumbens* (Lour.) Merr.**
Asteraceae
♦ Weed
♦ 13
♦ cultivated, herbal.

***Gynura pseudo-china* (L.) DC.**
Asteraceae
♦ Weed
♦ 87, 88
♦ herbal.

***Gypsophila acutifolia* Steven ex Spreng.**
Caryophyllaceae
♦ idänharso, sharpleaf baby's breath
♦ Naturalised, Casual Alien
♦ 42, 101
♦ pH, promoted.

***Gypsophila alsinoides* Bunge**
Caryophyllaceae
♦ Weed
♦ 243

***Gypsophila australis* (Schltdl.) A.Gray**
Caryophyllaceae
= *Gypsophila tubulosa* (Jaub. & Spach) Boiss.
♦ Naturalised
♦ 280

***Gypsophila capillaris* (Forssk.) C.Chr.**
Caryophyllaceae
♦ Weed
♦ 221

***Gypsophila elegans* Bieb.**
Caryophyllaceae
♦ kesäharso, showy baby's breath, baby's breath
♦ Naturalised, Cultivation Escape
♦ 42, 101
♦ aH, cultivated, herbal.

***Gypsophila muralis* L.**
Caryophyllaceae
Saponaria muralis (L.) Lam.
♦ wall gypsophila, low baby's breath, ketoraunikki

♦ Weed, Naturalised
♦ 44, 70, 87, 88, 94, 101, 243, 253, 272, 287
♦ aH, cultivated.

Gypsophila oldhamiana Miq.
Caryophyllaceae
♦ Oldham's baby's breath, Manchurian baby's breath
♦ Naturalised
♦ 101
♦ pH, cultivated, herbal.

Gypsophila paniculata L.
Caryophyllaceae
Arrostia paniculata (L.) Raf.
♦ baby's breath, baby's breath gypsophila, morsiusharso
♦ Weed, Noxious Weed, Naturalised, Garden Escape, Environmental Weed, Cultivation Escape
♦ 1, 21, 35, 39, 42, 80, 88, 101, 136, 142, 146, 151, 156, 161, 195, 212, 229, 247, 272, 280, 294, 299
♦ pH, cultivated, herbal, toxic. Origin: Eurasia.

Gypsophila paniculata L. var. paniculata
Caryophyllaceae
♦ baby's breath
♦ Introduced
♦ 34
♦ pH.

Gypsophila perfoliata L.
Caryophyllaceae
♦ perfoliate baby's breath, gypsomilka prerastená
♦ Naturalised
♦ 101

Gypsophila pilosa Huds.
Caryophyllaceae
Gypsophila porrigens (L.) Boiss. (see)
♦ karvaraunikki, Turkish baby's breath
♦ Naturalised, Casual Alien
♦ 40, 42, 101, 243

Gypsophila porrigens (L.) Boiss.
Caryophyllaceae
= *Gypsophila pilosa* Huds.
♦ Weed
♦ 87, 88

Gypsophila repens L.
Caryophyllaceae
Saponaria diffusa Lam.
♦ creeping baby's breath, gipsowka
♦ Naturalised
♦ 101
♦ cultivated, herbal.

Gypsophila scorzonerifolia Ser.
Caryophyllaceae
♦ garden baby's breath
♦ Naturalised
♦ 101
♦ pH, cultivated.

Gypsophila tubulosa (Jaub. & Spach) Boiss.
Caryophyllaceae
Gypsophila australis (Schltdl.) A.Gray (see)
♦ chalkwort

♦ Weed, Naturalised, Environmental Weed
♦ 7, 86, 93, 98, 176, 198, 203, 205
♦ aH. Origin: Eurasia.

Gypsophila vaccaria (L.) Sibth. & Sm.
Caryophyllaceae
♦ Weed
♦ 23, 88

H

Haageocactus Backeb. spp.
Cactaceae
= *Haageocereus* Backeb. spp.
♦ Quarantine Weed
♦ 76, 88, 220

Haageocereus Backeb. spp.
Cactaceae
Floresia Krainz & F.Ritter ex Backeb. spp. (see), *Haageocactus* Backeb. spp. (see), *Lasiocereus* Ritter spp. (see), *Peruvocereus* Akers. spp. (see), *Pygmaeocereus* Johnson & Backeb. spp. (see)
♦ Weed, Quarantine Weed
♦ 76, 88, 203, 220

Habenaria blephariglottis (Willd.) Hook.
Orchidaceae
= *Platanthera blephariglottis* (Willd.) Lindl. var. *blephariglottis* (NoR)
♦ Weed
♦ 196
♦ herbal.

Habenaria carolinensis Schltr.
Orchidaceae
Peristylis carolinensis Tuyama
♦ Introduced
♦ 230
♦ H.

Hablitzia thamnoides Bieb.
Chenopodiaceae
♦ köynnöspinaatti
♦ Cultivation Escape
♦ 42
♦ cultivated.

Habranthus tubispathus (L'Hér.) Traub
Liliaceae/Amaryllidaceae
♦ Rio Grande copperlily
♦ Weed
♦ 179
♦ cultivated, herbal.

Hackelia floribunda (Lehm.) I.M.Johnst.
Boraginaceae
Lappula floribunda (Lehm.) Greene
♦ western stickseed, manyflower stickseed, large flowered stickseed, western sheepbur
♦ Weed, Quarantine Weed
♦ 49, 76, 87, 88, 218
♦ pH, herbal.

Hackelochloa granularis **(L.) Kuntze**
Poaceae
Manisuris granularis (L.) Naezén
♦ pitscale grass, lizardtail grass
♦ Weed, Naturalised, Native Weed,
Introduced
♦ 32, 38, 66, 87, 88, 101, 121, 157, 170,
179, 191
♦ aG, herbal, toxic.

Haemanthus **L. spp.**
Liliaceae/Amaryllidaceae
♦ blood lilies
♦ Weed
♦ 161
♦ cultivated, toxic.

Haematoxylum campechianum **L.**
Fabaceae/Caesalpiniaceae
Cymbosepalum baronii Bak.,
Haematoxylon campechianum L.
♦ logwood, bloodwood tree,
campeachy wood, tinto mey, bois
campeche, campeche, campechier,
campeggio, kampes agaci, palo de
campeche
♦ Weed, Quarantine Weed,
Naturalised, Introduced,
Environmental Weed
♦ 3, 86, 98, 101, 191, 226, 227, 228
♦ arid, cultivated, herbal. Origin:
Central America, West Indies.

Haemodorum corymbosum **Vahl**
Haemodoraceae
♦ Weed, Naturalised
♦ 98, 203
♦ cultivated.

Hagenia abyssinica **J.F.Gmel.**
Rosaceae
♦ musuzi, kousso
♦ Casual Alien
♦ 280
♦ cultivated, herbal.

Hainardia cylindrica **(Willd.) Greuter**
Poaceae
Lepturus cylindricus (Willd.) Trin. (see),
Monerma cylindrica (Willd.) Coss. &
Dur. (see)
♦ common barbgrass, barbgrass,
thintail, piiskaheinä
♦ Weed, Naturalised, Introduced,
Environmental Weed, Casual Alien
♦ 7, 9, 42, 72, 86, 88, 98, 101, 161, 176,
198, 203, 228, 272, 280
♦ aG, arid/aqua, cultivated, herbal.
Origin: Mediterranean, Middle East,
Azores, Madeira and Canary Islands.

Hakea **Schrad. spp.**
Proteaceae
♦ needle bush
♦ Weed
♦ 18, 88, 181

Hakea costata **Meissner**
Proteaceae
♦ Naturalised, Native Weed
♦ 7, 86
♦ cultivated. Origin: Australia.

Hakea drupacea **(C.F.Gaertn.) Roem. &
Schult.**
Proteaceae

Hakea suaveolens R.Br. (part) (see)
♦ sweet hakea, soethakea
♦ Weed, Noxious Weed, Naturalised,
Garden Escape, Environmental Weed,
Cultivation Escape
♦ 63, 86, 88, 95, 198, 277, 283, 290, 296
♦ cultivated. Origin: Australia.

Hakea elliptica **(Sm.) R.Br.**
Proteaceae
♦ bronzy hakea, oval leaved hakea
♦ Weed, Naturalised, Native Weed,
Environmental Weed
♦ 72, 86, 88
♦ S, cultivated.

Hakea francisiana **F.Muell.**
Proteaceae
♦ narukalja
♦ Naturalised, Native Weed
♦ 7, 86
♦ cultivated. Origin: Australia.

Hakea gibbosa **(Sm.) Cav.**
Proteaceae
♦ rock hakea, downy hakea, harige
hakea
♦ Weed, Noxious Weed, Naturalised,
Environmental Weed, Cultivation
Escape
♦ 10, 51, 63, 87, 88, 95, 121, 152, 158,
181, 198, 225, 246, 277, 278, 280, 283
♦ S, cultivated. Origin: Australia.

Hakea laurina **R.Br.**
Proteaceae
♦ pincushion hakea
♦ Weed, Naturalised, Native Weed,
Garden Escape, Environmental Weed
♦ 72, 86, 88, 198
♦ S, cultivated, herbal. Origin:
Australia.

Hakea pycnoneura **Meissner**
Proteaceae
♦ Naturalised, Native Weed
♦ 7, 86
♦ cultivated. Origin: Australia.

Hakea salicifolia **(Vent.) B.L.Burtt**
Proteaceae
♦ willowleaf hakea, willowleaved
hakea, willow hakea
♦ Weed, Naturalised, Native Weed,
Garden Escape, Environmental Weed,
Cultivation Escape
♦ 15, 72, 86, 88, 134, 165, 198, 225, 246,
280, 283, 289
♦ S, cultivated. Origin: eastern
Australia.

Hakea saligna **(Andrews) Knight**
Proteaceae
♦ hedge hakea, willow hakea
♦ Weed
♦ 39, 121
♦ pS, cultivated, herbal, toxic. Origin:
Australia.

Hakea sericea **Schrad. & J.C.Wendl.**
Proteaceae
Hakea tenuifolia (Salisb.) Domin (see)
♦ silky hakea, syerige hakea, silky
wattle, needle bush, prickly hakea,
needle hakea
♦ Weed, Noxious Weed, Naturalised,

Native Weed, Garden Escape,
Environmental Weed, Cultivation
Escape
♦ 10, 63, 72, 86, 87, 88, 95, 121, 134, 152,
158, 165, 181, 198, 225, 246, 277, 278,
280, 283
♦ S, cultivated. Origin: eastern
Australia.

Hakea suaveolens **R.Br.**
Proteaceae
= *Hakea drupacea* (C.F.Gaertn.) Roem.
& Schult.
♦ sweet hakea, fork leaved hakea,
scented hakea
♦ Weed, Noxious Weed, Naturalised,
Native Weed, Garden Escape,
Environmental Weed
♦ 51, 72, 86, 87, 88, 121, 152, 158, 278,
280
♦ S, cultivated. Origin: Western
Australia.

Hakea tenuifolia **(Salisb.) Domin.**
Proteaceae
= *Hakea sericea* Schrad. & J.C.Wendl.
♦ silky hakea
♦ Weed
♦ 51, 87, 88

Halenia corniculata **(L.) Cornaz**
Gentianaceae
♦ corniculate spurgentian, hanaikari
♦ Weed
♦ 297

Halenia elliptica **D.Don**
Gentianaceae
♦ ellipticleaf spurgentian, spurred
gentian
♦ Weed
♦ 297
♦ pH, cultivated. Origin: Himalayas.

Halerpestes ruthenica **(Jacq.) Ovcz.**
Ranunculaceae
♦ longleaf halerpestes
♦ Weed
♦ 297

Halerpestes sarmentosa **(Adams.) Kom.**
Ranunculaceae
♦ Weed
♦ 275, 297
♦ pH.

Halimione portulacoides **(L.) Aellen**
Chenopodiaceae
Atriplex portulacoides L.
♦ sea purslane
♦ Weed, Introduced
♦ 221, 228
♦ S, arid, promoted, herbal.

Halimodendron halodendron **(Pall.) Voss**
Fabaceae/Papilionaceae
Caragana argentea Lam., *Halimodendron
argenteum* (Lam.) DC. *nom. illeg.*,
Robinia halodendron Pall.
♦ Russian salttree, common salttree
♦ Weed, Noxious Weed, Naturalised
♦ 35, 88, 101, 229
♦ S, promoted.

Halleria lucida **L.**
Scrophulariaceae

♦ hilarious Lucy, tree fuchsia, white olive, wild fuchsia, African honeysuckle
♦ Weed, Native Weed, Casual Alien
♦ 121, 280
♦ S/T, cultivated, herbal. Origin: southern Africa, Madagascar.

Halocnemum strobilaceum (Pall.) Bieb.
Chenopodiaceae
♦ Weed
♦ 221
♦ arid.

Halogeton alopecuroides (Delile) Moq.
Chenopodiaceae
♦ Weed
♦ 221

Halogeton arachnoideus Moq.
Chenopodiaceae
♦ Weed
♦ 275
♦ aH.

Halogeton glomeratus (Bieb.) C.A.Mey.
Chenopodiaceae
Anabasis glomerata Bieb.
♦ halogeton, salt lover, barilla
♦ Weed, Quarantine Weed, Noxious Weed, Naturalised
♦ 23, 26, 35, 36, 39, 76, 78, 80, 87, 88, 101, 116, 130, 136, 146, 161, 212, 218, 219, 220, 229, 231, 264, 275
♦ aH, herbal, toxic. Origin: Asia.

Haloragis aspera Lindl.
Haloragaceae
♦ raspweed
♦ Weed, Naturalised, Native Weed
♦ 55, 269, 280
♦ pH, cultivated. Origin: Australia.

Haloragis erecta (Banks ex Murr.) Oken
Haloragaceae
♦ erect seaberry, shrubby haloragis
♦ Weed, Naturalised, Introduced
♦ 34, 101, 165
♦ S, cultivated, herbal.

Haloragis heterophylla Brongn.
Haloragaceae
♦ variable raspwort
♦ Native Weed
♦ 39, 269
♦ a/pH, cultivated, toxic. Origin: Australia.

Haloragis micrantha (Thunb.) R.Br.
Haloragaceae
♦ arinotougusa
♦ Weed
♦ 88, 204, 286

Haloragis stricta Benth.
Haloragaceae
♦ raspweed
♦ Weed
♦ 55
♦ pH. Origin: Australia.

Haloxylon recurvum Bunge ex Boiss.
Chenopodiaceae
Caroxylon indicum Wight, *Caroxylon recurvum* Moq.
♦ Weed
♦ 243
♦ arid, herbal.

Hamamelis macrophylla Pursh
Hamamelidaceae
= *Hamamelis virginiana* L.
♦ southern witchhazel
♦ Weed
♦ 87, 88, 218
♦ herbal.

Hamamelis virginiana L.
Hamamelidaceae
Hamamelis macrophylla Pursh (see)
♦ witchhazel, American witchhazel, Virginian witchhazel
♦ Weed
♦ 87, 88, 218
♦ S, cultivated, herbal.

Hamatocactus setispinus Britt. & Rose
Cactaceae
= *Thelocactus setispinus* (Engelm.) E.F.Anderson (NoR)
♦ Quarantine Weed
♦ 220
♦ cultivated.

Hamelia patens Jacq.
Rubiaceae
Hamelia erecta Jacq.
♦ scarletbush, scarlet firebush
♦ Weed
♦ 39, 87, 88
♦ S/T, cultivated, herbal, toxic.

Hammada elegans (Bunge) Botsch.
Chenopodiaceae
♦ Weed
♦ 221

Hammada scoparia (Pomel) Iljin
Chenopodiaceae
♦ Weed
♦ 221

Hancornia speciosa Gomes
Apocynaceae
♦ mangabeira
♦ Introduced
♦ 228
♦ S/T, arid.

Haplocarpha lyrata Harv.
Asteraceae
♦ harp onefruit
♦ Weed, Naturalised, Native Weed
♦ 39, 101, 121
♦ pH, toxic. Origin: southern Africa.

Haplocarpha scaposa Harv.
Asteraceae
♦ tonteldoosbossie
♦ Quarantine Weed
♦ 220
♦ cultivated, herbal.

Haplocarpha thunbergii Less.
Asteraceae
♦ Quarantine Weed
♦ 220

Haplopappus arborescens (A.Gray) H.M.Hall
Asteraceae
= *Ericameria arborescens* (A.Gray) Greene (NoR)
♦ fleece goldenweed
♦ Weed
♦ 87, 88, 218

♦ herbal.

Haplopappus bloomeri A.Gray
Asteraceae
= *Ericameria bloomeri* (A.Gray) J.F.Macbr. (NoR)
♦ rabbitbrush goldenweed
♦ Weed
♦ 87, 88, 218
♦ herbal.

Haplopappus coronopifolius DC.
Asteraceae
= *Haplopappus glutinosus* Cass. (NoR)
♦ Weed, Quarantine Weed
♦ 88, 220

Haplopappus discoideus DC.
Asteraceae
♦ Quarantine Weed
♦ 220

Haplopappus heterophyllus (Gray) Blake
Asteraceae
= *Isocoma pluriflora* (Torr. & Gray) Greene (NoR) [see *Haplopappus pluriflorus* (Torr. & Gray) H.M.Hall]
♦ rayless goldenrod
♦ Weed
♦ 39, 161
♦ herbal, toxic.

Haplopappus laricifolius A.Gray
Asteraceae
= *Ericameria laricifolia* (Gray) Shinners
♦ turpentine brush
♦ Weed, Quarantine Weed
♦ 87, 88, 218, 220
♦ arid, herbal.

Haplopappus macronema A.Gray
Asteraceae
= *Bigelowia macronema* (A.Gray) M.E.Jones
♦ Quarantine Weed
♦ 220
♦ herbal.

Haplopappus monactis A.Gray
Asteraceae
♦ Quarantine Weed
♦ 220

Haplopappus pluriflorus (Torr. & Gray) H.M.Hall
Asteraceae
= *Isocoma pluriflora* (Torr. & Gray) Greene (NoR) [see *Haplopappus heterophyllus* (Gray) Blake]
♦ jimmyweed
♦ Weed
♦ 87, 88, 218

Haplopappus tenuisectus (Greene) Blake
Asteraceae
= *Isocoma tenuisecta* Greene (NoR)
♦ burroweed
♦ Weed
♦ 87, 88, 161, 218
♦ arid, toxic.

Haplopappus venetus (Kunth) S.F.Blake
Asteraceae
♦ coast goldenbush
♦ Weed
♦ 199
♦ herbal, toxic.

Haplophyllum buxbaumii (Poir.) Boiss.
Rutaceae
♦ Weed
♦ 87, 88

Haplophyllum perforatum (M.Bieb.) Kar. & Kir.
Rutaceae
♦ Weed
♦ 243

Haplophyllum suaveolens (DC.) Don
Rutaceae
Haplophyllum biebersteinii Spach
♦ Weed
♦ 272

Haplophyllum tuberculatum (Forssk.) A.Juss.
Rutaceae
Ruta tuberculata Forssk.
♦ Weed
♦ 88, 221

Hardenbergia comptoniana (Andrews) Benth.
Fabaceae/Papilionaceae
♦ native lilac
♦ Naturalised
♦ 86
♦ cultivated, herbal. Origin: Australia.

Harpagonella palmeri Gray
Boraginaceae
♦ Palmer's grapplinghook
♦ Weed, Quarantine Weed
♦ 76, 88
♦ aH, cultivated, herbal.

Harpagophytum procumbens (Burch.) DC. ex Meissn.
Pedaliaceae
Uncaria procumbens Burch.
♦ devil's claw, grapple plant, grapple thorn, quickthorn, sandbur, veldspider, wool spider, artiglio dei diavolo, sengaparile
♦ Weed, Quarantine Weed, Native Weed
♦ 51, 76, 87, 88, 121, 220
♦ pH, cultivated, herbal. Origin: southern Africa.

Harpagophytum zeyheri Decne.
Pedaliaceae
♦ Quarantine Weed
♦ 220
♦ cultivated.

Harpephyllum caffrum Bernh. ex Krauss
Anacardiaceae
♦ amagwenya, ingwenya, mmedibibi, ungwenya, wild plum, wildepruim
♦ Introduced
♦ 228
♦ arid, cultivated, herbal.

Harpullia arborea (Blanco) Radlk.
Sapindaceae
♦ puas
♦ Weed, Naturalised
♦ 39, 101, 179
♦ cultivated, herbal, toxic. Origin: Asia, Australia.

Harrisia Britt. spp.
Cactaceae

Eriocereus (Berg.) Riccob. spp. (see), *Roseocereus* Backeb. spp. (see)
♦ apple cactus
♦ Weed, Quarantine Weed, Naturalised
♦ 76, 86, 88, 203, 220
♦ cultivated.

Harrisia bonplandii (Parm.) Britt. & Rose
Cactaceae
= *Harrisia pomanensis* (Weber) Britton & Rose
♦ Naturalised
♦ 86
♦ cultivated.

Harrisia eriophora (Pfeiffer.) Britt.
Cactaceae
♦ Weed
♦ 14
♦ cultivated.

Harrisia guelichii (Speg.) Britton & Rose
Cactaceae
Cereus guelichii Speg., *Eriocereus guelichii* Berger
♦ Introduced
♦ 228
♦ arid, cultivated.

Harrisia martinii (Lab.) Britt. & Rose
Cactaceae
Cereus martinii Labour., *Eriocereus martinii* (Lab.) Riccob. (see)
♦ moon cactus, harrisia cactus, moonlight cactus, snake cactus, toukaktus, harrisia kaktus
♦ Weed, Noxious Weed, Naturalised, Garden Escape
♦ 10, 63, 86, 86, 88, 95, 101, 121, 158, 229, 278, 279, 283
♦ pC, cultivated. Origin: South America.

Harrisia pomanensis (Weber) Britton & Rose
Cactaceae
Cereus bonplandii Parm., *Eriocereus bonplandii* (Pfeiff.) Riccob., *Eriocereus pomanensis* Berger, *Harrisia bonplandii* (Parm.) Britt & Rose (see)
♦ apple cactus
♦ Naturalised
♦ 101
♦ cultivated. Origin: South America.

Harrisia tortuosa (J.Forbes ex Otto & A.Dietr.) Britt. & Rose
Cactaceae
Cereus tortuosus J.Forbes ex Otto & A.Dietr., *Eriocereus arendtii* (Schum.) Ritter, *Eriocereus tortuosus* (Otto & A.Dietr.) Riccob. (see)
♦ Naturalised
♦ 86
♦ cultivated. Origin: South America.

Harungana madagascariensis Lam. ex Poir.
Clusiaceae/Hypericaceae
♦ botonongolo, djene, harungana, mtunu, mutungulu
♦ Weed, Sleeper Weed, Naturalised, Environmental Weed
♦ 3, 22, 86, 155, 191

♦ S/T, cultivated, herbal. Origin: Madagascar, Mauritius, tropical Africa.

Haumania danckelmaniana (J.Braun & K.Schum.) Milne-Redh.
Marantaceae
♦ basele
♦ Weed, Quarantine Weed
♦ 76, 88, 220

Hazardia cana (Gray) Greene
Asteraceae
♦ Guadeloupe hazardia, San Clemente Island hazardia, island hazardia
♦ Quarantine Weed
♦ 220
♦ S, herbal. Origin: south-western North America.

Hebe barkeri (Cockayne) Cockayne
Scrophulariaceae
♦ Barker's hebe
♦ Naturalised
♦ 40
♦ cultivated.

Hebe brachysiphon Summerh.
Scrophulariaceae
♦ Hooker's hebe
♦ Naturalised
♦ 40
♦ S, cultivated.

Hebe dieffenbachii (Benth.) Cockayne & Allan
Scrophulariaceae
♦ Dieffenbach's hebe
♦ Naturalised
♦ 40
♦ S, cultivated.

Hebe elliptica (G.Forst.) Pennell
Scrophulariaceae
♦ hebe
♦ Naturalised, Garden Escape, Environmental Weed
♦ 86
♦ cultivated. Origin: New Zealand.

Hebe × franciscana (Eastw.) Souster
Scrophulariaceae
= *Hebe elliptica* (G.Forst.) Pennell × *Hebe speciosa* (R.Cunn. ex A.Cunn.) Andersen
♦ Francisco hebe, hebe, hedge veronica
♦ Naturalised, Introduced
♦ 34, 40, 101
♦ S, cultivated.

Hebe × kirkii (J.B.Armstr.) Cockayne & Allan
Scrophulariaceae
Veronica kirkii J.B.Armstr., *Veronica salicifolia* G.Forst. var. *kirkii* (J.B.Armstr.) Cheeseman
♦ Cultivation Escape
♦ 40
♦ cultivated.

Hebe × lewisii (J.B.Armstr.) Cockayne & Allan
Scrophulariaceae
= *Hebe salicifolia* (Forst.) Penn. × *Hebe elliptica* (Forst.) Penn. [most probable combination]
♦ Lewis's hebe
♦ Naturalised
♦ 40

Hebe parviflora (Vahl) Cockayne & Allan
Scrophulariaceae
♦ Naturalised
♦ 86
♦ cultivated. Origin: New Zealand.

Hebe salicifolia (G.Forst.) Pennell
Scrophulariaceae
Veronica salicifolia G.Forst.
♦ koromiko
♦ Naturalised
♦ 40
♦ S, cultivated, herbal.

Hebe speciosa (Cunn.) Cockayne & Allan
Scrophulariaceae
Veronica speciosa R.Cunn. ex A.Cunn.
♦ hebe, New Zealand hebe, titirangi, napuka
♦ Naturalised, Introduced
♦ 34, 86, 101
♦ S, cultivated. Origin: New Zealand.

Hebenstretia dentata L.
Scrophulariaceae/Selaginaceae
♦ Weed, Naturalised
♦ 86, 98, 203
♦ cultivated, herbal. Origin: South Africa.

Hebenstretia repens Jaroscz
Scrophulariaceae/Selaginaceae
♦ Native Weed
♦ 121
♦ H. Origin: southern Africa.

Hebestigma cubense (Kunth) Urb.
Fabaceae/Papilionaceae
♦ Introduced
♦ 261
♦ Origin: Cuba.

Hedeoma hispida Pursh
Lamiaceae
♦ rough pennyroyal, mock pennyroyal, rough false pennyroyal
♦ Weed, Native Weed
♦ 161, 174
♦ herbal. Origin: North America.

Hedeoma pulegioides (L.) Pers.
Lamiaceae
♦ American pennyroyal, American false pennyroyal
♦ Weed
♦ 39, 87, 88, 218
♦ aH, cultivated, herbal, toxic.

Hedera L. spp.
Araliaceae
♦ Boston ivy, English ivy, ivy
♦ Weed, Naturalised
♦ 154, 181, 198
♦ toxic.

Hedera canariensis Willd.
Araliaceae
♦ Algerian ivy, canary ivy
♦ Weed
♦ 39, 116, 161, 247
♦ cultivated, toxic.

Hedera colchica (K.Koch) K.Koch
Araliaceae
♦ colchis ivy
♦ Naturalised, Casual Alien
♦ 101, 280
♦ cultivated.

Hedera helix L.
Araliaceae
♦ English ivy, ivy, needlepoint ivy, ripple ivy, common ivy, murgrna
♦ Weed, Noxious Weed, Naturalised, Introduced, Garden Escape, Environmental Weed, Cultivation Escape
♦ 4, 7, 20, 22, 34, 35, 38, 39, 45, 70, 72, 78, 80, 86, 87, 88, 98, 101, 102, 104, 105, 116, 133, 137, 142, 146, 151, 155, 161, 165, 176, 179, 181, 194, 195, 198, 203, 218, 222, 225, 231, 243, 246, 247, 272, 280, 289, 296
♦ pC, cultivated, herbal, toxic. Origin: Eurasia.

Hedera helix L. ssp. canariensis (Willd.) Cout.
Araliaceae
♦ canary ivy
♦ Weed, Naturalised
♦ 15, 280

Hedera helix L. ssp. helix
Araliaceae
♦ ivy
♦ Weed, Naturalised
♦ 15, 280
♦ cultivated.

Hedera rhombea (Miq.) Bean
Araliaceae
♦ kizuta
♦ Weed
♦ 286

Hedychium Koenig spp.
Zingiberaceae
♦ garland lily
♦ Environmental Weed
♦ 257
♦ cultivated.

Hedychium coccineum Buch.-Ham.
Zingiberaceae
♦ red ginger lily
♦ Weed, Noxious Weed, Garden Escape, Environmental Weed
♦ 88, 95, 226, 283
♦ cultivated, herbal. Origin: Asia.

Hedychium coronarium J.König
Zingiberaceae
♦ white ginger butterfly lily, ginger lily, garland flower, awapuhi ke'oke'o, white ginger lily, butterfly lily, sinser
♦ Weed, Noxious Weed, Naturalised, Introduced, Garden Escape, Environmental Weed, Cultivation Escape, Casual Alien
♦ 3, 80, 86, 87, 88, 95, 98, 101, 107, 151, 155, 179, 191, 203, 226, 230, 233, 255, 261, 283, 287
♦ pH, cultivated, herbal. Origin: eastern India.

Hedychium flavescens Carey ex Roscoe
Zingiberaceae
Hedychium emeiense Z.Y.Zhu, *Hedychium panzhuum* Z.Y.Zhu
♦ yellow ginger, awapuhi melemele, cream ginger, wild ginger, cream ginger lily, cream garland lily, longoze
♦ Weed, Quarantine Weed, Noxious

Weed, Naturalised, Garden Escape, Environmental Weed, Cultivation Escape
♦ 3, 15, 76, 80, 88, 95, 101, 107, 132, 152, 191, 225, 233, 246, 280, 283
♦ pH, cultivated, herbal. Origin: Himalayas.

Hedychium flavum Roxb.
Zingiberaceae
♦ yellow ginger, ginger lily
♦ Naturalised
♦ 261
♦ cultivated. Origin: tropical Asia.

Hedychium gardnerianum Sheppard ex Ker Gawl.
Zingiberaceae
♦ kahili ginger, yellow ginger lily, cevuga dromodromo, sinter weitahta, ginger lily
♦ Weed, Quarantine Weed, Noxious Weed, Naturalised, Introduced, Garden Escape, Environmental Weed, Cultivation Escape
♦ 3, 15, 18, 73, 80, 86, 88, 95, 98, 101, 107, 132, 151, 152, 165, 191, 203, 225, 226, 230, 233, 246, 280, 283, 289, 296
♦ pH, cultivated, herbal. Origin: India.

Hedyotis auricularia L.
Rubiaceae
Oldenlandia auricularia (L.) F.Muell.
♦ Weed, Quarantine Weed, Naturalised
♦ 13, 76, 86, 87, 88, 203, 220, 297
♦ cultivated, herbal.

Hedyotis biflora (L.) Lam.
Rubiaceae
Oldenlandia biflora L.
♦ Weed, Introduced
♦ 13, 87, 88, 230
♦ H, herbal. Origin: Australia.

Hedyotis capitellata Wall. ex G.Don
Rubiaceae
♦ Weed
♦ 12
♦ herbal.

Hedyotis commutata J.H. & J.A.Schult.
Rubiaceae
= *Oldenlandia lancifolia* (Schum.) DC.
♦ Weed
♦ 261
♦ Origin: tropical Africa, Madagascar.

Hedyotis corymbosa (L.) Lam.
Rubiaceae
= *Oldenlandia corymbosa* L.
♦ old world diamond flower
♦ Weed, Naturalised, Introduced, Environmental Weed
♦ 13, 86, 87, 88, 161, 170, 209, 230, 235, 249, 261, 262, 273, 276, 297
♦ aH, herbal. Origin: Australia.

Hedyotis costata (Roxb.) Kurz
Rubiaceae
Hedyotis capituliflora Miq., *Hedyotis vestita* R.Br. (see), *Metabolus lineatus* Bartl., *Spermacoce costata* Roxb.
♦ Weed
♦ 88
♦ herbal.

Hedyotis diffusa Willd.
Rubiaceae
Hedyotis herbacea Lour., *Oldenlandia*
diffusa (Willd.) Roxb., *Oldenlandia*
brachypoda DC., *Oldenlandia herbacea*
(L.) Roxb. var. *uniflora* Benth.
♦ futabamugura
♦ Weed
♦ 13, 87, 88, 170, 191, 262, 273, 274, 275,
286, 297
♦ aH, aqua, herbal.

Hedyotis diffusa Willd. var. *longipes*
Nakai
Rubiaceae
♦ Weed
♦ 286

Hedyotis glabra R.Br.
Rubiaceae
♦ Weed
♦ 12
♦ herbal.

Hedyotis herbacea L.
Rubiaceae
= *Oldenlandia herbacea* (L.) Roxb.
♦ Weed
♦ 13, 87, 88
♦ herbal.

Hedyotis hispida Retz.
Rubiaceae
♦ Weed
♦ 275
♦ pH, herbal.

Hedyotis lindleyana Hook.
Rubiaceae
♦ Weed
♦ 87, 88

Hedyotis lindleyana Hook. var. *hirsuta*
(L.f.) Hara
Rubiaceae
♦ Weed
♦ 286

Hedyotis lineata Roxb.
Rubiaceae
♦ Weed
♦ 87, 88

Hedyotis nigricans (Lam.) Fosberg
Rubiaceae
♦ narrowleaf bluet, diamond flower
♦ Weed
♦ 161
♦ herbal.

Hedyotis nitida Wight & Arn.
Rubiaceae
♦ Weed
♦ 12

Hedyotis ovatifolia Cav.
Rubiaceae
Oldenlandia ovatifolia (Cav.) DC.
♦ Weed
♦ 13

Hedyotis pinifolia (Wall. ex G.Don)
K.Schum.
Rubiaceae
Oldenlandia pinifolia Wall. ex G.Don
♦ Weed

♦ 13, 87, 88
♦ herbal.

Hedyotis ponapensis (Val.) Kaneh.
Rubiaceae
♦ hedyotis
♦ Introduced
♦ 230
♦ H.

Hedyotis pseudocorymbosa Bakh.f.
Rubiaceae
Oldenlandia burmanniana (Wall.) G.Don
♦ Weed
♦ 13

Hedyotis pterita Bl.
Rubiaceae
Oldenlandia pterita (Bl.) Miq.
♦ Weed
♦ 13, 87, 88

Hedyotis racemosa Lam.
Rubiaceae
♦ Weed
♦ 88

Hedyotis tenellifora Bl.
Rubiaceae
♦ Weed
♦ 87, 88

Hedyotis umbellata (L.) Lam.
Rubiaceae
Oldenlandia umbellata L.
♦ Weed
♦ 87, 88
♦ herbal.

Hedyotis uncinelloides (Val.) Hosok.
Rubiaceae
♦ Introduced
♦ 230
♦ H.

Hedyotis verticillata Lam.
Rubiaceae
♦ Weed
♦ 12, 13
♦ herbal.

Hedyotis vestita R.Br.
Rubiaceae
= *Hedyotis costata* (Roxb.) Kurz
♦ Weed
♦ 12, 13
♦ herbal.

Hedypnois cretica (L.) Dum.Cours.
Asteraceae
Hedypnois polymorpha DC., *Hedypnois*
rhagadioloides (L.) Schmidt. ssp. *cretica*
(L.) Hayek (see)
♦ Cretan weed, Cretan hedypnois,
hedypnois, vuohensilmä
♦ Weed, Naturalised, Environmental
Weed, Casual Alien
♦ 34, 42, 70, 72, 86, 87, 88, 93, 94, 101,
116, 198, 205, 237, 241, 272, 295, 300
♦ aH, cultivated, herbal. Origin:
Mediterranean.

Hedypnois rhagadioloides (L.) Willd.
Asteraceae
Hedypnois persica M.Bieb., *Hyoseris*
rhagadioloides L.
♦ Cretan weed

♦ Weed, Naturalised
♦ 7, 9, 98, 176, 203, 253

Hedypnois rhagadioloides (L.) Willd. ssp.
cretica (L.) Schmidt (L.) Hayek
Asteraceae
= *Hedypnois cretica* (L.) Dum.Cours.
♦ Cretan weed
♦ Weed, Naturalised, Environmental
Weed
♦ 86, 269
♦ arid. Origin: Mediterranean.

Hedysarum boreale (Rydb.) Rollins
Fabaceae/Papilionaceae
♦ MacKenzie northern sweetvetch,
sweetvetch, boreal sweetvetch
♦ Weed
♦ 161
♦ pH, cultivated, herbal, toxic.

Hedysarum brachypterum Bunge
Fabaceae/Papilionaceae
♦ shortwing sweetvetch
♦ Weed
♦ 297

Hedysarum carnosum Desf.
Fabaceae/Papilionaceae
♦ Quarantine Weed
♦ 220
♦ arid.

Hedysarum coronarium L.
Fabaceae/Papilionaceae
♦ French honeysuckle, sulla, sekernica
vencová
♦ Weed, Naturalised, Introduced
♦ 7, 86, 88, 94, 98, 176, 203, 228, 280
♦ arid, cultivated, herbal. Origin:
Mediterranean.

Hedysarum flexuosum L.
Fabaceae/Papilionaceae
Hedysarum algeriense Pomel
♦ Quarantine Weed
♦ 220
♦ arid.

Hedysarum laeve Maxim.
Fabaceae/Papilionaceae
♦ laeve sweetvetch
♦ Weed
♦ 297

Heimia myrtifolia Cham. & Schltdl.
Lythraceae
♦ Weed
♦ 121
♦ pS, cultivated. Origin: South
America.

Heimia salicifolia (Kunth) Link & Otto
Lythraceae
♦ shrubby yellowcrest, hachinal,
sinicuiche
♦ Weed, Naturalised
♦ 86, 161, 295
♦ S, cultivated, herbal, toxic. Origin:
Central and South America.

Helenium L. spp.
Asteraceae
♦ sneezeweed
♦ Weed, Quarantine Weed,
Naturalised

♦ 76, 86, 88, 203, 220, 247
♦ cultivated, herbal, toxic.

Helenium amarum (Raf.) H.Rock
Asteraceae
♦ bitter sneezeweed, yellowdicks, bitterweed
♦ Weed, Naturalised
♦ 14, 23, 34, 39, 86, 87, 88, 98, 161, 203, 207, 210, 218, 249
♦ aH, cultivated, herbal, toxic. Origin: eastern north America.

Helenium aromaticum Bailey
Asteraceae
♦ Weed
♦ 87, 88
♦ arid.

Helenium autumnale L.
Asteraceae
♦ common sneezeweed, bitterweed, false sunflower, sneezeweed
♦ Weed, Quarantine Weed, Native Weed, Casual Alien
♦ 8, 23, 39, 76, 87, 88, 161, 174, 210, 218, 256, 280
♦ pH, cultivated, herbal, toxic. Origin: North America.

Helenium autumnale L. var. *parviflorum* (Nutt.) Fern.
Asteraceae
= *Helenium autumnale* L. var. *autumnale* (NoR)
♦ smallflower sneezeweed
♦ Weed
♦ 218

Helenium bigelovii A.Gray
Asteraceae
♦ Bigelow's sneezeweed, sneezeweed
♦ Weed
♦ 161, 180, 243
♦ pH, cultivated, herbal.

Helenium flexuosum Raf.
Asteraceae
Helenium nudiflorum Nutt. (see)
♦ purplehead sneezeweed
♦ Weed
♦ 161
♦ cultivated, herbal, toxic.

Helenium hoopesii Gray
Asteraceae
= *Hymenoxys hoopesii* (A.Gray) Bierner (NoR)
♦ western sneezeweed, orange sneezeweed
♦ Weed, Quarantine Weed
♦ 39, 76, 87, 88, 161, 218
♦ pH, cultivated, herbal, toxic.

Helenium microcephalum DC.
Asteraceae
♦ smallhead sneezeweed
♦ Weed
♦ 39, 161
♦ arid, herbal, toxic.

Helenium nudiflorum Nutt.
Asteraceae
= *Helenium flexuosum* Raf.
♦ purplehead sneezeweed

♦ Weed
♦ 39, 87, 88, 218
♦ herbal, toxic.

Helenium puberulum DC.
Asteraceae
♦ rosilla
♦ Naturalised
♦ 280
♦ pH, arid, promoted, herbal.

Helenium tenuifolium Nutt.
Asteraceae
= *Helenium amarum* (Raf.) H.Rock var. *amarum* (NoR)
♦ Weed, Naturalised
♦ 39, 87, 88, 287
♦ herbal, toxic.

Heleocharis Lestib. spp.
Cyperaceae
= *Eleocharis* R.Br. spp.
♦ Weed
♦ 221
♦ G.

Heleocharis palustris (L.) R.Br.
Cyperaceae
= *Eleocharis palustris* (L.) Roem. & Schult.
♦ Weed
♦ 221
♦ G.

Heleochloa Host ex Roem. spp.
Poaceae
♦ Weed
♦ 221
♦ G.

Heleochloa schoenoides (L.) Host ex Roem.
Poaceae
= *Crypsis schoenoides* (L.) Lam.
♦ swamp timothy
♦ Weed
♦ 88, 218, 221
♦ G, herbal.

Heliabravoa Backeb. spp.
Cactaceae
= *Polaskia* Backeb. spp.
♦ Weed, Quarantine Weed
♦ 76, 88, 203, 220

Heliabravoa Backeb. × *Heliaporus* Rowley spp.
Cactaceae
= *Polaskia* Backeb. spp. × [*Heliocereus* (Berg.) Britt. & Rose spp. × *Aporocactus* Lem. spp.]
♦ Weed, Quarantine Weed
♦ 76, 88

Helianthemum Mill. spp.
Cistaceae
♦ rockrose
♦ Weed
♦ 221, 272
♦ herbal.

Helianthemum lippii (L.) Dum.-Cours.
Cistaceae
♦ Weed
♦ 221
♦ cultivated.

Helianthemum nummularium (L.) Mill.
Cistaceae
♦ common rockrose, rockrose
♦ Weed
♦ 272
♦ S, cultivated, herbal.

Helianthemum salicifolium (L.) Mill.
Cistaceae
♦ willowleaf frostweed
♦ Weed, Naturalised
♦ 87, 88, 101
♦ cultivated.

Helianthemum sancti-antonii Boiss.
Cistaceae
♦ Weed
♦ 221

Helianthemum ventosum Boiss.
Cistaceae
♦ Weed
♦ 221

Helianthemum vulgare Gaertn.
Cistaceae
♦ rockrose
♦ Weed
♦ 87, 88
♦ herbal.

Helianthus L. spp.
Asteraceae
♦ sunflower
♦ Weed, Naturalised
♦ 15, 198
♦ herbal.

Helianthus angustifolius L.
Asteraceae
♦ swamp sunflower
♦ Weed
♦ 23, 88
♦ cultivated, herbal.

Helianthus annuus L.
Asteraceae
Helianthus aridus Rydb. (see)
♦ sunflower, common sunflower, wild sunflower, annual sunflower, wild artichoke, isoauringonkukka
♦ Weed, Naturalised, Native Weed, Garden Escape, Cultivation Escape, Casual Alien
♦ 7, 23, 34, 39, 40, 42, 49, 68, 86, 87, 88, 93, 94, 98, 121, 136, 161, 174, 179, 180, 198, 199, 203, 205, 210, 212, 218, 229, 236, 237, 243, 261, 269, 270, 272, 280, 295
♦ a/bH, arid, cultivated, herbal, toxic. Origin: North America.

Helianthus argophyllus Torr. & Gray
Asteraceae
♦ silverleaf sunflower
♦ Weed, Naturalised
♦ 86, 98, 121, 203, 287
♦ aH, arid, cultivated, herbal. Origin: southern North America.

Helianthus aridus Rydb.
Asteraceae
= *Helianthus annuus* L.
♦ Weed
♦ 23, 88

Helianthus californicus DC.
Asteraceae
♦ California sunflower
♦ Weed
♦ 87, 88, 161, 218
♦ pH, aqua.

Helianthus ciliaris DC.
Asteraceae
♦ Texas blueweed, blue weed
♦ Weed, Quarantine Weed, Noxious Weed, Naturalised
♦ 1, 26, 34, 35, 76, 86, 87, 88, 98, 146, 161, 203, 212, 218, 229, 243, 269
♦ a/pH, cultivated, herbal. Origin: southern North America, Mexico.

Helianthus debilis Nutt.
Asteraceae
Helianthus cucumerifolius Torr. & Gray
♦ cucumberleaf sunflower, polyheaded sunflower, dune sunflower
♦ Weed, Naturalised
♦ 7, 54, 86, 88, 98, 203, 287
♦ aH, cultivated, herbal. Origin: southern North America.

Helianthus giganteus L.
Asteraceae
Helianthus altissimus L.
♦ jättiauringonkukka, giant sunflower
♦ Casual Alien
♦ 42
♦ pH, cultivated, herbal.

Helianthus grosseserratus Mart.
Asteraceae
♦ sawtooth sunflower
♦ Weed, Native Weed
♦ 87, 88, 161, 174, 218
♦ cultivated. Origin: North America.

Helianthus laciniatus A.Gray
Asteraceae
♦ alkali sunflower
♦ Weed
♦ 199

Helianthus × laetiflorus Pers.
Asteraceae
= *Helianthus pauciflorus* Nutt. × *Helianthus tuberosus* L.
♦ mountain sunflower, cheerful sunflower, perennial sunflower
♦ Naturalised
♦ 280
♦ herbal.

Helianthus laevigatus Torr. & Gray
Asteraceae
♦ smooth sunflower
♦ Naturalised
♦ 287
♦ herbal.

Helianthus laevis L.
Asteraceae
= *Bidens laevis* (L.) Britton, Sterns & Poggenb.
♦ Naturalised
♦ 287

Helianthus maximiliani Schrad.
Asteraceae
♦ Maximilian sunflower, perennial

sunflower
♦ Weed
♦ 87, 88, 161, 210, 218
♦ pH, cultivated, herbal.

Helianthus nuttallii Torr. & Gray
Asteraceae
♦ Nuttall's sunflower
♦ Weed
♦ 161, 212
♦ pH, cultivated, herbal.

Helianthus petiolaris Nutt.
Asteraceae
♦ prairie sunflower, pikkuauringonkukka
♦ Weed, Casual Alien
♦ 23, 42, 49, 87, 88, 161, 210, 218
♦ aH, promoted, herbal.

Helianthus petiolaris Nutt. ssp. petiolaris
Asteraceae
♦ prairie sunflower
♦ Weed
♦ 34
♦ pH.

Helianthus rigidus (Cass.) Desf.
Asteraceae
♦ stiff sunflower, preeria auringonkukka
♦ Weed, Casual Alien
♦ 42, 161
♦ herbal.

Helianthus salicifolius A.Dietr.
Asteraceae
♦ willowleaf sunflower
♦ Naturalised
♦ 280
♦ cultivated, herbal.

Helianthus simulans E.Watson
Asteraceae
♦ muck sunflower
♦ Weed
♦ 179

Helianthus strumosus L.
Asteraceae
♦ rough sunflower, paleleaf woodland sunflower, woodland sunflower
♦ Weed, Naturalised
♦ 286, 287
♦ pH, cultivated, herbal. Origin: North America.

Helianthus subrhomboideus Rydb.
Asteraceae
= *Helianthus pauciflorus* Nutt. ssp. *subrhomboideus* (Rydb.) O.Spring & E.E.Schill. (NoR)
♦ soikkoauringonkukka
♦ Casual Alien
♦ 42

Helianthus tuberosus L.
Asteraceae
Helianthus mollissimus E.E.Watson
♦ Jerusalem artichoke, girasole, earth apple, maa artisokka
♦ Weed, Noxious Weed, Naturalised, Native Weed, Garden Escape, Environmental Weed, Casual Alien
♦ 15, 39, 40, 42, 72, 86, 87, 88, 98, 132, 152, 161, 174, 180, 195, 203, 210, 211,

218, 229, 241, 243, 253, 263, 269, 272, 280, 286, 287, 291, 300
♦ pH, cultivated, herbal, toxic. Origin: Americas.

Helichrysum Mill. spp.
Asteraceae
♦ everlasting
♦ Weed
♦ 272
♦ herbal.

Helichrysum anomalum Less.
Asteraceae
♦ Native Weed
♦ 121
♦ S. Origin: southern Africa.

Helichrysum arenarium (L.) Moench
Asteraceae
♦ everlasting flower, yellow everlasting
♦ Weed
♦ 272
♦ pH, cultivated, herbal.

Helichrysum argyrophyllum DC.
Asteraceae
♦ amatola weed, everlasting weed, golden guinea everlasting, Moe's gold
♦ Native Weed
♦ 121
♦ S, cultivated, herbal. Origin: southern Africa.

Helichrysum argyrosphaerum DC.
Asteraceae
♦ wild everlasting
♦ Weed, Native Weed
♦ 39, 121
♦ H, cultivated, toxic. Origin: southern Africa.

Helichrysum athrixiifolium (Kuntze) Moeser
Asteraceae
♦ Native Weed
♦ 121
♦ S. Origin: southern Africa.

Helichrysum aureonitens Sch.Bip.
Asteraceae
♦ Native Weed
♦ 121
♦ pH, cultivated. Origin: southern Africa.

Helichrysum bracteatum (Vent.) Andrews
Asteraceae
= *Bracteantha bracteata* (Vent.) Anderb. & Haegi
♦ bracted strawflower, strawflower, jättiolkikukka, golden everlasting, paperflower, yellow paper daisy
♦ Naturalised, Cultivation Escape
♦ 42, 101, 261, 280
♦ aH, cultivated, herbal. Origin: Australia.

Helichrysum cerastioides DC.
Asteraceae
♦ Native Weed
♦ 121
♦ pH. Origin: southern Africa.

Helichrysum cooperi Harv.
Asteraceae

- Weed
- 88, 158
- Origin: southern Africa.

Helichrysum cylindriflorum (L.) Hilliard & Burtt
Asteraceae
- Native Weed
- 121
- S. Origin: southern Africa.

Helichrysum cymosum (L.) D.Don
Asteraceae
- Casual Alien
- 280
- cultivated.

Helichrysum foetidum (L.) Cass.
Asteraceae
Gnaphalium foetidum L.
- stinking strawflower, tuoksuolkikukka
- Weed, Naturalised
- 70, 101
- bH, cultivated, herbal.

Helichrysum kraussii Sch.Bip.
Asteraceae
- straw everlasting
- Native Weed
- 121
- pS. Origin: southern Africa.

Helichrysum mixtum (Kuntze) Moeser
Asteraceae
- Native Weed
- 121
- pH. Origin: southern Africa.

Helichrysum odoratissimum (L.) Sweet
Asteraceae
- Native Weed
- 121
- pH, herbal. Origin: southern Africa.

Helichrysum oxyphyllum DC.
Asteraceae
- Native Weed
- 121
- pH. Origin: southern Africa.

Helichrysum petiolare Hilliard & Burtt
Asteraceae
Helichrysum petiolatum auct. non (L.) DC., *Gnaphalium lanatum* hort.
- helichrysum, licorice plant
- Weed, Noxious Weed, Naturalised, Environmental Weed
- 35, 78, 80, 88, 101, 116, 231, 280
- cultivated, herbal. Origin: South Africa.

Helichrysum rosum (Berg.) Less.
Asteraceae
- strawflower
- Native Weed
- 121
- pS. Origin: southern Africa.

Helichrysum ruderale Hilliard & Burtt
Asteraceae
- Weed
- 88, 158
- Origin: southern Africa.

Helichrysum rugulosum Less.
Asteraceae
- Native Weed

- 121
- pH. Origin: southern Africa.

Helichrysum tenax M.D.Hend.
Asteraceae
- Native Weed
- 121
- pS. Origin: southern Africa.

Heliconia bihai (L.) L.
Heliconiaceae/Strelitziaceae
- macaw flower, bastard plantain, firebird, macaw
- Weed, Naturalised
- 87, 88, 101
- cultivated, herbal.

Heliconia cannoidea A.Rich.
Heliconiaceae/Strelitziaceae
= *Heliconia psittacorum* L.f.
- Weed
- 87, 88

Heliconia latispatha Benth.
Heliconiaceae/Strelitziaceae
- expanded lobsterclaw, gold heliconia
- Naturalised
- 101
- cultivated, herbal.

Heliconia metallica Planch. & Linden ex Hook.
Heliconiaceae/Strelitziaceae
Bihai metallica (Planch. & Linden ex Hook.) Kuntze, *Heliconia nana* G.Rodr., *Heliconia nitens* hort., *Heliconia osaensis* Cufod. var. *rubescens* Stiles, *Heliconia vinosa* Ender
- shining bird of paradise
- Naturalised
- 101
- cultivated.

Heliconia psittacorum L.f.
Heliconiaceae/Strelitziaceae
Heliconia cannoidea A.Rich. (see)
- parakeetflower, parrot's flower
- Weed, Quarantine Weed, Naturalised, Cultivation Escape
- 76, 87, 88, 101, 203, 261
- cultivated, herbal. Origin: tropical South America.

Heliconia subulata Ruiz & Pav.
Heliconiaceae/Strelitziaceae
- Guatemalan bird of paradise
- Naturalised, Cultivation Escape
- 101, 261
- pH, cultivated, herbal. Origin: South America.

Helicteres isora L.
Sterculiaceae
- Weed, Introduced
- 209, 228
- arid, cultivated, herbal. Origin: Australia.

Helicteres jamaicensis Jacq.
Sterculiaceae
- screwtree
- Weed
- 87, 88

Helictotrichon longifolium (Nees) Schwieck.
Poaceae

- Native Weed
- 121
- pG. Origin: southern Africa.

Helictotrichon longum (Stapf) Schweick.
Poaceae
- Native Weed
- 121
- pG. Origin: southern Africa.

Helictotrichon natalense (Stapf) Schweick.
Poaceae
- Native Weed
- 121
- pG. Origin: southern Africa.

Helictotrichon pubescens (Huds.) Bess. ex Pilg.
Poaceae
Avena pubescens Huds., *Avenula pubescens* (Huds.) Dumort. (see), *Avenochloa pubescens* (Huds.) Holub
- downy alpine oatgrass, downy oatgrass, hairy oatgrass
- Naturalised, Casual Alien
- 101, 280
- G.

Helictotrichon turgidulum (Stapf) Schwieck.
Poaceae
- small cat grass
- Native Weed
- 121
- pG. Origin: southern Africa.

× *Heliocactus* Janse spp.
Cactaceae
= *Heliocereus* (A.Berger) Britt. & Rose spp. × *Phyllocactus* Link spp.
- Weed, Quarantine Weed
- 76, 88, 220

Heliocarpus americanus L.
Tiliaceae
Heliocarpus popayanensis Kunth (see), *Heliocarpus tomentosus* Turcz.
- Introduced
- 228
- T, arid.

Heliocarpus popayanensis Kunth
Tiliaceae
= *Heliocarpus americanus* L.
- white moho
- Weed, Naturalised
- 3, 22, 80, 101, 191
- T, herbal.

Heliocereus (Berger) Britt. & Rose spp.
Cactaceae
Mediocactus Britt. & Rose spp. (see)
- Weed, Quarantine Weed, Naturalised
- 76, 86, 88, 203, 220
- cultivated.

Heliocereus (Berg.) Britt. & Rose × *Heliochia* Rowley spp.
Cactaceae
= *Heliocereus* (Berg.) Britt. & Rose. spp. × [*Heliocereus* (Berg.) Britt. & Rose. spp. × *Disocactus* Lindl. spp.]
- Weed, Quarantine Weed
- 76, 88, 220

Heliocereus (Berg.) Britt. & Rose ×
Helioselenius Rowley spp.
   Cactaceae
   = *Heliocereus* (Berg.) Britt. & Rose. spp.
   × [*Heliocereus* (Berg.) Britt. & Rose spp.
   × *Selenicereus* (Berg.) Britt. & Rose spp.]
   ♦ Weed, Quarantine Weed
   ♦ 76, 88, 220

Heliocereus (Berg.) Britt. & Rose ×
Heliphyllum Rowley spp.
   Cactaceae
   = *Heliocereus* (Berg.) Britt. & Rose. spp.
   × [*Heliocereus* (Berg.) Britt. & Rose spp.
   × *Epiphyllum* Haw. spp.]
   ♦ Weed, Quarantine Weed
   ♦ 76, 88, 220

× *Heliochia* Rowley spp.
   Cactaceae
   = *Heliocereus* (Berg.) Britt. & Rose spp.
   × *Nopalxochia* Britt. & Rose spp.
   ♦ Weed, Quarantine Weed
   ♦ 76, 88, 220

Heliophila coronopifolia L.
   Brassicaceae
   ♦ Casual Alien
   ♦ 280
   ♦ aH, cultivated.

Heliophila pusilla L.f.
   Brassicaceae
   ♦ heliophila
   ♦ Weed, Naturalised, Garden Escape,
   Environmental Weed
   ♦ 7, 9, 86, 98, 203
   ♦ cultivated. Origin: southern Africa.

Heliopsis helianthoides (L.) Sweet
   Asteraceae
   ♦ oxeye, oxeye sunflower, heliopsis
   sunflower, smooth oxeye
   ♦ Native Weed
   ♦ 161, 174
   ♦ cultivated, herbal. Origin: North
   America.

× *Helioselenius* Rowley spp.
   Cactaceae
   = *Heliocereus* (Berger) Britt. & Rose spp.
   × *Selenicereus* (Berg.) Britt. & Rose spp.
   ♦ Weed, Quarantine Weed
   ♦ 76, 88, 220

Heliotropium L. spp.
   Boraginaceae
   ♦ heliotrope
   ♦ Weed
   ♦ 39, 154, 221, 247, 272
   ♦ herbal, toxic.

Heliotropium aegyptiacum Lehm.
   Boraginaceae
   Heliotropium cinerascens Steud. ex DC.
   & A.DC., *Heliotropium pallens* Delile
   ♦ Weed
   ♦ 88
   ♦ Origin: east Africa, Middle East.

Heliotropium amplexicaule Vahl
   Boraginaceae
   Cochranea anchusaefolia (Poir.)
   Gürke, *Heliotropium anchusaefolium*
   Poir., *Heliotropium anchusifolium*
   Poir., *Heliotropium anchusifolium*

var. *angustifolium* DC., *Heliotropium*
bolivianum Rusby, *Heliotropium*
lithospermifolium Speg., *Tournefortia*
heliotropioides Hook.
   ♦ blue heliotrope, wild verbena,
   clasping heliotrope, purpletop,
   turnsole, wild heliotrope, verveine
   sauvage
   ♦ Weed, Quarantine Weed, Noxious
   Weed, Naturalised, Introduced,
   Garden Escape
   ♦ 34, 39, 55, 76, 86, 87, 88, 98, 101, 121,
   147, 198, 203, 218, 261, 269, 270, 295
   ♦ pH, cultivated, herbal, toxic. Origin:
   South America.

Heliotropium angiospermum Murr.
   Boraginaceae
   Heliotropium parviflorum L. (see)
   ♦ scorpion's tail, cotorrilla
   ♦ Weed, Naturalised
   ♦ 14, 28, 39, 88, 161, 206, 241, 243, 300
   ♦ arid, herbal, toxic.

Heliotropium anomalum Hook. & Arn.
   Boraginaceae
   ♦ silvery heliotrope, Polynesian
   heliotrope
   ♦ Weed
   ♦ 87, 88, 218
   ♦ herbal.

Heliotropium arbainense Fresen.
   Boraginaceae
   ♦ Weed
   ♦ 221

Heliotropium arborescens L.
   Boraginaceae
   Heliotropium arborescens var. *grisellum*
   I.M.Johnst., *Heliotropium corymbosum*
   Ruiz & Pav., *Heliotropium peruvianum*
   L.
   ♦ garden heliotrope, cherry pie plant,
   white heliotrope, fragrant heliotrope,
   heliotrope
   ♦ Naturalised, Casual Alien
   ♦ 39, 101, 261
   ♦ S, arid, cultivated, herbal, toxic.
   Origin: Peru.

Heliotropium bacciferum Forssk.
   Boraginaceae
   Heliotropium undulatum Vahl
   ♦ Weed
   ♦ 221, 242
   ♦ arid.

Heliotropium bovei Boiss.
   Boraginaceae
   ♦ Weed, Quarantine Weed
   ♦ 76, 87, 88, 203, 220

Heliotropium bracteatum R.Br.
   Boraginaceae
   ♦ Weed
   ♦ 87, 88
   ♦ Origin: Australia.

Heliotropium crispum Desf.
   Boraginaceae
   ♦ Weed
   ♦ 88

Heliotropium curassavicum L.
   Boraginaceae

   ♦ seaside heliotrope, salt heliotrope,
   smooth heliotrope, lännenheliotrooppi,
   alkali oculatum heliotrope
   ♦ Weed, Naturalised, Environmental
   Weed, Casual Alien
   ♦ 7, 14, 39, 42, 86, 87, 88, 98, 121, 161,
   203, 218, 287
   ♦ pH, arid, cultivated, herbal, toxic.
   Origin: South America.

Heliotropium curassavicum L. var.
oculatum (Heller) I.M.Johnst.
   Boraginaceae
   ♦ alkali heliotrope, wild heliotrope,
   salt heliotrope, heliotrope, whiteweed,
   Chinese pusley, devilweed, quail plant,
   yerba del torojo, seaside heliotrope
   ♦ Weed
   ♦ 180

Heliotropium dasycarpum Ledeb. ex
Eichw.
   Boraginaceae
   ♦ Weed
   ♦ 243

Heliotropium digynum (Forssk.) Asch. ex
C.Christ.
   Boraginaceae
   ♦ Weed
   ♦ 221

Heliotropium eduardii Martelli
   Boraginaceae
   ♦ Weed
   ♦ 87, 88

Heliotropium eichwaldi Steud.
   Boraginaceae
   ♦ Weed
   ♦ 87, 88
   ♦ herbal.

Heliotropium ellipticum Ledeb.
   Boraginaceae
   ♦ Weed
   ♦ 272, 275
   ♦ aH.

Heliotropium elongatum Hoffm.
   Boraginaceae
   Heliotropium decipiens Back.
   ♦ Weed
   ♦ 13
   ♦ herbal.

Heliotropium europaeum L.
   Boraginaceae
   ♦ barooga weed, bishop's beard,
   caterpillar weed, heliotrope,
   potato weed, European heliotrope,
   rikkaheliotrooppi
   ♦ Weed, Quarantine Weed, Noxious
   Weed, Naturalised, Environmental
   Weed, Casual Alien
   ♦ 7, 39, 42, 62, 70, 72, 76, 86, 87, 88, 94,
   98, 101, 115, 147, 161, 180, 203, 217, 221,
   243, 253, 269, 272
   ♦ aH, arid, cultivated, herbal, toxic.
   Origin: Mediterranean.

Heliotropium filiforme H.B.K.
   Boraginaceae
   ♦ Weed
   ♦ 87, 88

Heliotropium foliosissimum **McBride**
Boraginaceae
♦ Weed
♦ 199

Heliotropium fruticosum **L.**
Boraginaceae
♦ Key West heliotrope
♦ Weed
♦ 87, 88

Heliotropium indicum **L.**
Boraginaceae
Tiaridium indicum (L.) Lehm.
♦ Indian heliotrope, white clary, wild clary, devilweed, turnsole, yaa nguang chaang
♦ Weed, Naturalised, Environmental Weed
♦ 7, 12, 13, 14, 32, 39, 86, 87, 88, 93, 98, 101, 121, 157, 161, 170, 186, 203, 209, 218, 239, 243, 255, 261, 262, 269, 275, 286, 287, 295, 297
♦ aH, aqua, cultivated, herbal, toxic. Origin: Eurasia.

Heliotropium lanceolatum **Ruiz & Pav.**
Boraginaceae
♦ sete sangrias
♦ Weed
♦ 255
♦ arid. Origin: South America.

Heliotropium lasiocarpum **Fisch. & Mey.**
Boraginaceae
Heliotropium eichwaldii Steud. var. *lasiocarpum* (Fisch. & C.A.Mey.) C.B.Clarke, *Heliotropium ellipticum* Ledeb. var. *lasiocarpum* (Fisch. & C.A.Mey.) Popov, *Heliotropium europaeum* L. var. *lasiocarpum* (Fisch. & C.A.Mey.) Kazmi, *Heliotropium europaeum* L. var. *tenuiflorum* L. (Guss.) Boiss.
♦ Weed
♦ 88, 185
♦ H. Origin: Middle East, India, Pakistan. China.

Heliotropium lineare **(A.DC.) C.H.Wright**
Boraginaceae
♦ Native Weed
♦ 121
♦ pH. Origin: southern Africa.

Heliotropium longiflorum **(Hochst. & Steud. ex A.DC.) Jaub. & Spach**
Boraginaceae
♦ Weed
♦ 88

Heliotropium ovalifolium **Forssk.**
Boraginaceae
♦ forget me not, greyleaf heliotrope
♦ Weed, Native Weed
♦ 50, 87, 88, 121, 242
♦ pH, arid, herbal. Origin: Madagascar.

Heliotropium parviflorum **L.**
Boraginaceae
= *Heliotropium angiospermum* Murr.
♦ Weed
♦ 87, 88
♦ herbal.

Heliotropium procumbens **Mill.**
Boraginaceae
♦ fourspike heliotrope
♦ Weed, Naturalised
♦ 87, 88, 179, 237, 241, 255
♦ herbal. Origin: tropical America.

Heliotropium pterocarpum **(DC. & A.DC.) Hochst.**
Boraginaceae
♦ Weed
♦ 221

Heliotropium ramosissimum **(Lehm.) DC.**
Boraginaceae
♦ wavy heliotrope
♦ Naturalised
♦ 101

Heliotropium rufipilum **(Benth.) Johnst.**
Boraginaceae
♦ Naturalised
♦ 257

Heliotropium scabrum **Retz.**
Boraginaceae
♦ Weed
♦ 87, 88

Heliotropium steudneri **Vatke**
Boraginaceae
♦ Weed, Native Weed
♦ 87, 88, 121
♦ S. Origin: southern Africa.

Heliotropium strigosum **(L.) Willd.**
Boraginaceae
♦ Weed
♦ 87, 88, 221
♦ herbal.

Heliotropium suaveolens **Bieb.**
Boraginaceae
♦ Weed
♦ 272

Heliotropium sudanicum **F.W.Andrews**
Boraginaceae
Heliotropium europaeum (*non* L.) Broun & Massey
♦ Weed, Quarantine Weed
♦ 76, 87, 88, 203, 220, 242
♦ H, arid.

Heliotropium supinum **L.**
Boraginaceae
♦ dwarf heliotrope, prostrate heliotrope, creeping heliotrope
♦ Weed, Naturalised, Environmental Weed
♦ 7, 66, 72, 86, 87, 88, 93, 98, 101, 185, 198, 203, 205, 221, 242, 269
♦ aH, arid, herbal. Origin: Mediterranean.

Heliotropium ternatum **Vahl**
Boraginaceae
♦ bushy heliotrope
♦ Weed
♦ 157

Heliotropium undulatifolium **Turrill**
Boraginaceae
♦ Weed
♦ 87, 88
♦ arid.

Heliotropium veronicifolium **Griseb.**
Boraginaceae
♦ Weed
♦ 237

Heliotropium zeylanicum **(Burm.f.) Lam.**
Boraginaceae
♦ Weed
♦ 221, 240
♦ aH.

× *Heliphyllum* **Rowley spp.**
Cactaceae
= *Epiphyllum* Haw. spp. × *Heliocereus* (Berger) Britt. & Rose spp.
♦ Weed, Quarantine Weed
♦ 76, 88, 220

Helipterum roseum **(Hook.) Benth.**
Asteraceae
Acroclinium roseum Hook.
♦ rusoikikukka, acroclinium
♦ Cultivation Escape
♦ 42
♦ aH, cultivated, herbal.

Helleborus cyclophyllus **Boiss.**
Ranunculaceae
♦ stinking hellebore
♦ Weed
♦ 272
♦ cultivated.

Helleborus dumetorum **Waldst. & Kit.**
Ranunculaceae
Helleborus pallidus Host
♦ shrubby hellebore
♦ Weed
♦ 272
♦ cultivated.

Helleborus foetidus **L.**
Ranunculaceae
Helleboraster foetidus (L.) Moench
♦ stinking hellebore, bear's foot, bear's foot hellebore
♦ Weed
♦ 70, 161, 247
♦ pH, cultivated, herbal, toxic.

Helleborus lividus **Aiton**
Ranunculaceae
♦ Corsican hellebore
♦ Weed
♦ 161, 215
♦ cultivated, toxic.

Helleborus niger **L.**
Ranunculaceae
♦ black hellebore, Christmas rose
♦ Weed, Naturalised
♦ 39, 101, 161, 189, 194, 247, 272
♦ pH, cultivated, herbal, toxic.

Helleborus odorus **Waldst. & Kit.**
Ranunculaceae
♦ Weed
♦ 39, 272
♦ cultivated, toxic.

Helleborus orientalis **Lam.**
Ranunculaceae
♦ Lenten rose, winter rose
♦ Weed, Naturalised
♦ 15, 39, 40, 161, 247
♦ cultivated, herbal, toxic.

Helleborus viridis **L.**
Ranunculaceae
♦ green hellebore
♦ Weed, Naturalised
♦ 39, 101, 161
♦ pH, cultivated, herbal, toxic.

Helleborus viridis **L. ssp.** *occidentalis*
(Reut.) Schiffn.
Ranunculaceae
♦ green hellebore
♦ Naturalised
♦ 40
♦ cultivated. Origin: southern Europe.

Helminthostachys zeylanica **(L.) Hook.**
Ophioglossaceae
♦ flowering fern, kamraj
♦ Weed
♦ 12, 87, 88
♦ cultivated, herbal.

Helminthotheca echioides **(L.) Holub**
Asteraceae
= *Picris echioides* L.
♦ oxtongue
♦ Naturalised, Environmental Weed
♦ 7, 86, 176, 198, 296
♦ a/bH. Origin: Mediterranean,
Middle East.

Hemarthria altissima **(Poir.) Stapf &**
C.E.Hubb.
Poaceae
Hemarthria compressa (L.f.) R.Br. var.
fasciculata (Lam.) Keng, *Hemarthria
compressa* (L.f.) R.Br. ssp. *altissima*
(Poir.) Maire, *Hemarthria fasciculata*
(Lam.) Kunth, *Manisuris altissima*
(Poir.) A.Hitchc., *Manisuris fasciculata*
(Lam.) A.Hitchc., *Rottboellia altissima*
Poir., *Rottboellia compressa* L.f. var.
fasciculata (Lam.) Hack., *Rottboellia
fasciculata* Lam., *Rottboellia heterochroa*
Gaud.
♦ red swampgrass, Batavian
quickgrass, couchgrass, red vlei grass,
swamp couch, limpograss
♦ Weed, Naturalised, Native Weed,
Introduced, Casual Alien
♦ 32, 38, 87, 88, 101, 121, 158, 179, 221,
275, 280, 297
♦ pG, arid, cultivated, herbal. Origin:
southern Africa.

Hemarthria compressa **(L.f.) R.Br.**
Poaceae
Rottboellia compressa L.f.
♦ whipgrass
♦ Weed
♦ 87, 88, 90
♦ pG, aqua.

Hemarthria longiflora **(Hook f.) A.Camus**
Poaceae
♦ Weed
♦ 87, 88
♦ G.

Hemarthria protensa **Nees ex Steud.**
Poaceae
♦ Weed
♦ 87, 88
♦ G.

Hemarthria sibirica **(Gand.) Ohwi**
Poaceae
Hemarthria japonica (Hack.) Roshev.
♦ ushinoshippei
♦ Weed
♦ 87, 88, 286
♦ G.

Hemerocallis fulva **(L.) L.**
Liliaceae/Hemerocallidaceae
Hemerocallis fulva (L.) L. var. *kwanso*
Regel (see)
♦ daylily, orange daylily, tawny
daylily, pale daylily, rusopäivänlilja,
common daylily
♦ Weed, Naturalised, Introduced,
Environmental Weed, Garden Escape,
Cultivation Escape
♦ 4, 8, 40, 42, 80, 87, 88, 101, 133, 151,
195, 198, 218, 222, 280
♦ pH, cultivated, herbal, toxic. Origin:
obscure.

Hemerocallis fulva **(L.) L. var.** *kwanso*
Regel
Liliaceae/Hemerocallidaceae
= *Hemerocallis fulva* (L.) L.
♦ Weed
♦ 286

Hemerocallis fulva **(L.) L. var.**
sempervirens **M.Hotta**
Liliaceae/Hemerocallidaceae
♦ Naturalised
♦ 287

Hemerocallis lilioasphodelus **L.**
Liliaceae/Hemerocallidaceae
♦ keltapäivänlilja, yellow daylily
♦ Naturalised, Garden Escape,
Cultivation Escape, Casual Alien
♦ 40, 42, 101
♦ pH, cultivated, herbal. Origin:
obscure, possibly China.

Hemerocallis minor **Mill.**
Liliaceae/Hemerocallidaceae
= *Hemerocallis flava* (L.) L. var. *minor*
(Mill.) M.Hotta (NoR)
♦ small daylily, grassleaf daylily,
dwarf yellow daylily
♦ Naturalised
♦ 101
♦ pH, cultivated, herbal.

Hemicarpha micrantha **(Vahl) Britt.**
Cyperaceae
♦ Naturalised
♦ 257
♦ G, herbal.

Hemidiodia ocimifolia **(Willd. ex Roem.**
& Schult.) K.Schum.
Rubiaceae
= *Diodia ocymifolia* (Willd. ex Roem. &
Schult.) Bremek.
♦ Weed
♦ 87, 88
♦ arid.

Hemigraphis alternata **(Burm.f.)**
T.Anderson
Acanthaceae
Blechum cordatum Leonard, *Ruellia
alternata* Burm.f.
♦ red ivy, cemetery plant, red flame

ivy
♦ Weed, Naturalised, Cultivation
Escape
♦ 32, 101, 179, 261
♦ cultivated, herbal. Origin: tropical
Asia.

Hemigraphis brunelloides **(Lam.) Brem.**
Acanthaceae
Hemigraphis hirsuta (Vahl) T.Anderson
♦ Weed
♦ 13

Hemigraphis hirta **T.Anderson**
Acanthaceae
♦ Weed
♦ 87, 88

Hemigraphis javanica **Brem.**
Acanthaceae
Hemigraphis decaisneana non
T.Anderson
♦ Weed
♦ 13

Hemigraphis primulaefolia **Villar**
Acanthaceae
♦ Weed
♦ 87, 88

Hemigraphis repanda **(L.) Hallier f.**
Acanthaceae
♦ hemigraphis
♦ Weed
♦ 243
♦ pH, aqua, cultivated.

Hemigraphis reptans **(Forst.) T.Anderson**
Acanthaceae
♦ red flame
♦ Weed, Naturalised
♦ 101, 179, 261
♦ cultivated. Origin: tropical Africa.

Hemigraphis serpens **(Nees) Boerl.**
Acanthaceae
♦ Weed
♦ 13

Hemisteptia lyrata **Bunge**
Asteraceae
Saussurea affinis Spreng. ex DC.
♦ kitsuneazami, lyrate hemistepta
♦ Weed, Naturalised
♦ 68, 86, 87, 88, 235, 273, 274, 275, 286,
297
♦ bH. Origin: Australia.

Hemizonia clementina **Brandegee**
Asteraceae
♦ Catalina tarweed
♦ Quarantine Weed
♦ 220
♦ S, cultivated, herbal.

Hemizonia congesta **DC.**
Asteraceae
♦ common tarweed, hayfield tarweed
♦ Weed
♦ 87, 88, 218
♦ aH, herbal.

Hemizonia kelloggii **Greene**
Asteraceae
♦ Kellogg's tarweed, tarweed
♦ Weed
♦ 161, 180, 243
♦ aH, cultivated, herbal.

Hemizonia pungens **(Hook. & Arn.) Torr. & Gray**
Asteraceae
♦ spikeweed, common spikeweed, common tarweed
♦ Weed, Noxious Weed, Naturalised, Environmental Weed, Casual Alien
♦ 1, 40, 80, 86, 88, 98, 146, 151, 161, 180, 203, 212, 229, 243
♦ aH, herbal. Origin: California.

Hemizonia pungens **(Hook. & Arn.) Torr. & Gray ssp.** *laevis* **Keck**
Asteraceae
♦ spikeweed, common tarweed, smooth tarplant
♦ Noxious Weed
♦ 229
♦ aH, cultivated.

Hemizonia pungens **(Hook. & Arn.) Torr. & Gray ssp.** *maritima* **(Greene) Keck**
Asteraceae
♦ spikeweed, common tarweed, maritime spikeweed
♦ Noxious Weed
♦ 229
♦ aH.

Hemizonia pungens **(Hook. & Arn.) Torr. & Gray ssp.** *pungens*
Asteraceae
♦ spikeweed, common tarweed
♦ Noxious Weed
♦ 229
♦ aH.

Hemizonia pungens **(Hook. & Arn.) Torr. & Gray ssp.** *septentrionalis* **Keck**
Asteraceae
♦ spikeweed, common tarweed
♦ Noxious Weed
♦ 229
♦ aH.

Hemizygia canescens **(Guerke) Ashby**
Lamiaceae
♦ Native Weed
♦ 121
♦ pH, cultivated. Origin: southern Africa.

Hemizygia transvaalensis **(Schltr.) Ashby**
Lamiaceae
♦ Quarantine Weed
♦ 220
♦ cultivated.

Hemizygia welwitschii **(Rolfe) Ashby**
Lamiaceae
♦ Weed
♦ 87, 88

Heracleum douglasii **DC.**
Apiaceae
♦ Quarantine Weed
♦ 220

Heracleum lanatum **Michx.**
Apiaceae
= *Heracleum sphondylium* L. ssp. *montanum* (Schleich. ex Gaudin) Briq. (NoR) [see *Heracleum maximum* Bartr.]
♦ cow parsnip, common cow parsnip
♦ Weed, Quarantine Weed
♦ 8, 39, 161, 180, 212, 220

♦ pH, arid, cultivated, herbal, toxic.

Heracleum lehmannianum **Bunge**
Apiaceae
♦ Weed, Quarantine Weed
♦ 76, 88, 220
♦ cultivated.

Heracleum mantegazzianum **Sommier & Levier**
Apiaceae
Heracleum giganteum (Hornem.) hort. (*non* Fisch.), *Heracleum villosum* hort. (*non* Fisch.)
♦ giant hogweed, cartwheel flower, wild parsnip, wild rhubarb, kaukasianjättiputki
♦ Weed, Quarantine Weed, Noxious Weed, Naturalised, Environmental Weed, Cultivation Escape
♦ 1, 24, 24, 39, 40, 42, 67, 70, 76, 79, 80, 86, 88, 99, 101, 105, 119, 139, 140, 143, 146, 151, 152, 161, 165, 184, 184, 195, 229, 252, 280, 289
♦ b/pH, aqua, cultivated, herbal, toxic. Origin: eastern Europe.

Heracleum maximum **Bartr.**
Apiaceae
= *Heracleum sphondylium* ssp. *montanum* (Schleich. ex Gaudin) Briq. (NoR) [see *Heracleum lanatum* Michx.]
♦ common cow parsnip
♦ Weed, Quarantine Weed
♦ 23, 87, 88, 218, 220
♦ herbal.

Heracleum moellendorffii **Hance**
Apiaceae
Heracleum nipponicum Kitag.
♦ Moellendorffii cow parsnip
♦ Weed
♦ 297

Heracleum montanum **Schleich. ex Gaudin**
Apiaceae
♦ Quarantine Weed
♦ 220

Heracleum persicum **Desf. ex Fisch.**
Apiaceae
♦ persianjättiputki
♦ Cultivation Escape
♦ 42
♦ cultivated.

Heracleum sphondylium **L.**
Apiaceae
♦ hogweed cow parsnip, cow parsnip, hogweed, keck, ukonputki
♦ Weed, Quarantine Weed, Naturalised
♦ 23, 39, 44, 87, 88, 94, 101, 218, 220, 243, 253, 272, 280
♦ b/pH, cultivated, herbal, toxic.

Heracleum sphondylium **L. ssp.** *sibiricum* **(L.) Simonk.**
Apiaceae
♦ eltrot
♦ Naturalised
♦ 101

Heracleum sphondylium **L. ssp.** *sphondylium*

Apiaceae
♦ hogweed
♦ Weed
♦ 70, 243

Herbertia lahue **(Molina) Goldblatt**
Iridaceae
Alophia lahue (Molina) Espinosa, *Ferraria lahue* Molina
♦ prairie nymph
♦ Weed, Naturalised, Cultivation Escape
♦ 86, 98, 203, 252
♦ cultivated. Origin: southern north America.

Herissantia crispa **(L.) Brizicky**
Malvaceae
Abutilon crispum (L.) Medik.
♦ bladder mallow, mela bode, malva de lava prato
♦ Weed
♦ 157, 199, 255
♦ a/pH, arid, herbal. Origin: tropical America.

Herissantia tiubae **(K.Schum.) Brizicky**
Malvaceae
♦ guamxuma branca, malva branca
♦ Weed
♦ 255
♦ pH. Origin: Brazil.

Hermannia candicans **Ait.**
Sterculiaceae
♦ Quarantine Weed
♦ 220

Hermannia depressa **N.E.Br.**
Sterculiaceae
♦ Native Weed
♦ 121
♦ pH, herbal. Origin: southern Africa.

Hermannia modesta **(Ehrenb.) Mast.**
Sterculiaceae
Trichanthera modesta Ehrenb.
♦ Weed
♦ 221

Hermannia paucifolia **Turcz.**
Sterculiaceae
♦ Native Weed
♦ 121
♦ pH, toxic. Origin: southern Africa.

Hermbstaedtia linearis **Schinz**
Amaranthaceae
Celosia linearis (Schinz) Schinz
♦ woolflower
♦ Native Weed
♦ 121
♦ pH. Origin: southern Africa.

Hermbstaedtia odorata **(Burch.) T.Cooke**
Amaranthaceae
♦ guinea flower
♦ Naturalised
♦ 101

Hermbstaedtia odorata **(Burch.) T.Cooke var.** *odorata*
Amaranthaceae
♦ Native Weed
♦ 121
♦ pH. Origin: southern Africa.

Hermodactylus tuberosus (L.) Mill.
Iridaceae
- snake's head iris, widow iris
- Naturalised
- 40
- pH, cultivated, herbal.

Herniaria L. spp.
Caryophyllaceae/Illecebraceae
- rupturewort
- Weed, Naturalised
- 198, 272

Herniaria cinerea DC.
Caryophyllaceae/Illecebraceae
= *Herniaria hirsuta* L. ssp. *cinerea* (DC.)
Cout.
- gray herniaria, herniaria
- Weed, Naturalised, Environmental Weed
- 86, 87, 88, 180, 198, 241, 300
- herbal. Origin: Mediterranean.

Herniaria erckertii Herm. ssp. *erckertii*
var. *dewetii* Herm.
Caryophyllaceae/Illecebraceae
- Native Weed
- 121
- pH. Origin: southern Africa.

Herniaria glabra L.
Caryophyllaceae/Illecebraceae
- smooth rupturewort, smooth herniary, rupture wort
- Weed, Naturalised, Casual Alien
- 42, 70, 88, 94, 101, 243, 253, 272, 280, 287
- b/pH, cultivated, herbal.

Herniaria hemistemon J.Gay
Caryophyllaceae/Illecebraceae
- Weed
- 221

Herniaria hirsuta L.
Caryophyllaceae/Illecebraceae
Herniaria besseri Fisch., *Herniaria incana*
L., *Herniaria macrocarpa* Sibth.
- hairy rupturewort, gray cinerea hernaria
- Weed, Naturalised, Introduced, Environmental Weed, Casual Alien
- 7, 40, 42, 72, 86, 88, 94, 98, 101, 161, 203, 228, 253, 272, 280
- aH, arid, promoted, herbal.

Herniaria hirsuta L. ssp. *cinerea* (DC.)
Cout.
Caryophyllaceae/Illecebraceae
Herniaria cinerea DC. (see)
- hairy rupturewort
- Naturalised, Introduced
- 34, 101
- aH.

Herniaria hirsuta L. ssp. *hirsuta*
Caryophyllaceae/Illecebraceae
- hairy rupturewort, herniaria
- Naturalised
- 101
- aH.

Herniaria incana Lam.
Caryophyllaceae/Illecebraceae
- gray rupturewort, rupturewort
- Weed, Naturalised

- 272

Herniaria polygama Gay
Caryophyllaceae/Illecebraceae
- tuoksutyräruoho
- Casual Alien
- 42

Hertia intermedia Kuntze
Asteraceae
= *Othonna media* C.Jeffrey (NoR) [see
Othonnopsis intermedia Boiss.]
- Weed
- 243

Hertia pallens (DC.) Kuntze
Asteraceae
- springbokbush
- Weed, Native Weed
- 39, 121
- pS, toxic. Origin: southern Africa.

Hertrichocereus Backeb. spp.
Cactaceae
= *Stenocereus* (Berger) Riccob. spp.
- Weed, Quarantine Weed
- 76, 88, 220

Hesperantha falcata (L.f.) Ker Gawl.
Iridaceae
- hesperantha
- Weed, Naturalised, Garden Escape, Environmental Weed
- 7, 9, 86, 98, 203
- cultivated. Origin: southern Africa.

Hesperis L. spp.
Brassicaceae
- rocket
- Weed
- 272

Hesperis laciniata All.
Brassicaceae
- Naturalised
- 56

Hesperis matronalis L.
Brassicaceae
Hesperis nivea Baumg.
- Dame's violet, illakko, Dame's rocket, sweet rocket, rocket
- Weed, Noxious Weed, Naturalised, Introduced, Garden Escape, Environmental Weed, Cultivation Escape
- 4, 23, 24, 34, 40, 42, 56, 80, 87, 88, 94, 101, 102, 104, 133, 138, 142, 151, 159, 161, 174, 195, 218, 222, 224, 229, 241, 272, 280, 287, 299, 300
- b/pH, cultivated, herbal.

Hesperis tristis L.
Brassicaceae
- Weed
- 272

Hesperocnide sandwicensis Wedd.
Urticaceae
- stinging weed, Hawai'i stinging nettle
- Weed, Quarantine Weed
- 76, 87, 88, 203, 218, 220

Hesperocnide tenella Torr.
Urticaceae
- western stinging nettle, western nettle

- Weed
- 161
- aH, herbal, toxic.

Hetaeria Bl. spp.
Orchidaceae
- Introduced
- 230

Heteranthemis viscidehirta Schott
Asteraceae
Chrysanthemum viscidehirtum (Schott)
Thell.
- sticky oxeye
- Weed, Naturalised
- 70, 101

Heteranthera Ruiz & Pav. spp.
Pontederiaceae
- mud plantain
- Weed, Quarantine Weed
- 76, 88, 220

Heteranthera callifolia Rchb. ex Kunth
Pontederiaceae
- Weed
- 88

Heteranthera dubia (Jacq.) MacMill.
Pontederiaceae
Zosterella dubia (Jacq.) Small (see)
- water stargrass, mud plantain, grassleaf mud plantain
- Weed, Quarantine Weed
- 23, 76, 87, 88, 126, 161, 203, 218
- wpH, cultivated, herbal.

Heteranthera limosa (Sw.) Willd.
Pontederiaceae
Pontederia limosa Sw.
- duck salad, blue mud plantain, waterlily
- Weed, Naturalised
- 87, 88, 126, 157, 161, 179, 180, 218, 253, 255, 263, 286, 287
- wpH, herbal. Origin: tropical America.

Heteranthera reniformis Ruiz & Pav.
Pontederiaceae
Heteranthera peduncularis Benth.
- mud plantain, kidneyleaf mud plantain, roundleaf mud plantain
- Weed
- 87, 88, 126, 161, 218, 253, 255
- pH, aqua, cultivated, herbal. Origin: tropical America.

Heterocaryum subsessile Vatke
Boraginaceae
- Weed
- 248

Heterocentron elegans (Schltdl.) Kuntze
Melastomataceae
- Spanish shawl
- Casual Alien
- 280
- cultivated.

Heterocentron subtriplinervium (Link &
Otto) A.Braun & C.D.Bouche
Melastomataceae
- pearl flower
- Weed, Naturalised, Cultivation Escape
- 3, 101, 191, 233
- cultivated, herbal.

Heterodendrum oleaefolium Desf.
Sapindaceae
♦ Weed
♦ 87, 88

Heteroderis stocksiana Boiss.
Asteraceae
♦ Weed
♦ 243

Heterolepis aliena (L.f.) Druce
Asteraceae
♦ daisybush
♦ Quarantine Weed
♦ 220
♦ cultivated.

Heteromeles arbutifolia (Lindl.) M.Roem.
Rosaceae
Photinia arbutifolia (Aiton) Lindl.,
Photinia salicifolia (Decne.)
C.K.Schneid., *Crataegus arbutifolia* Ait.
non Lam.
♦ toyon, Christmas berry
♦ Weed
♦ 161
♦ S, cultivated, herbal, toxic.

Heteropappus altaicus (Willd.) Novopokr.
Asteraceae
♦ Weed
♦ 275, 297
♦ pH, cultivated.

Heteropappus hispidus (Thunb.) Less.
Asteraceae
♦ Weed
♦ 87, 88, 297
♦ pH, promoted, herbal.

Heteropogon contortus (L.) Roem. & Schult.
Poaceae
Andropogon allioni Lam. & DC.,
Andropogon contortus L.
♦ assegai fix, common speargrass, kusal grass, piercing grass, speargrass, stickgrass, tanglehead, yaa nuat ruesee
♦ Weed, Noxious Weed, Native Weed, Introduced
♦ 35, 39, 51, 87, 88, 90, 121, 158, 209, 228, 229, 239, 297
♦ pG, arid, cultivated, herbal, toxic. Origin: southern Africa.

Heteropteris beecheyana Andrews Juss.
Malpighiaceae
♦ Weed
♦ 179

Heteropteris laurifolia (L.) Juss.
Malpighiaceae
♦ dragon withe
♦ Weed
♦ 14
♦ cultivated.

Heteropteris purpurea (L.) Kunth
Malpighiaceae
Banisteria purpurea L.
♦ bull withe
♦ Weed
♦ 87, 88

Heterospathe elata Scheff.
Arecaceae
♦ sagisi palm, palma brava, asbo,

demailéi, ebouch, buag bbuag
♦ Weed
♦ 3, 191
♦ cultivated.

Heterosperma diversifolium Kunth
Asteraceae
♦ Weed
♦ 237, 295

Heterotheca grandiflora Nutt.
Asteraceae
♦ telegraph weed, telegraph plant
♦ Weed, Naturalised, Environmental Weed
♦ 3, 86, 87, 88, 98, 155, 161, 180, 191, 203, 218, 243, 269, 287
♦ a/pH, arid, herbal. Origin: Central America.

Heterotheca inuloides Cass.
Asteraceae
♦ fancy false goldenaster
♦ Weed
♦ 199
♦ herbal.

Heterotheca latifolia Buckley
Asteraceae
= *Heterotheca subaxillaris* (Lam.) Britt. & Rusby
♦ Weed
♦ 237
♦ herbal.

Heterotheca subaxillaris (Lam.) Britt. & Rusby
Asteraceae
Heterotheca latifolia Buckl. (see)
♦ camphor weed, goldenaster
♦ Weed, Naturalised, Native Weed
♦ 59, 87, 88, 161, 174, 212, 218, 287
♦ a/bH, cultivated, herbal. Origin: tropical Americas.

Heterotheca villosa (Pursh) Shinners
Asteraceae
♦ hairy goldenaster, hairy false goldenaster
♦ Weed
♦ 161
♦ pH, herbal.

Heterotrichum cymosum (Wendl.) Urb.
Melastomataceae
♦ camasey terciopelo
♦ Weed
♦ 87, 88

Heuchera americana L.
Saxifragaceae
♦ American alumroot, alumroot, rock geranium
♦ Weed
♦ 23, 88
♦ pH, cultivated, herbal.

Heuchera sanguinea Engelm.
Saxifragaceae
♦ alumroot, coralbells
♦ Casual Alien
♦ 280
♦ pH, cultivated, herbal.

Hevea Aubl. spp.
Euphorbiaceae
♦ hevea, para rubber

♦ Weed, Quarantine Weed, Naturalised
♦ 76, 86, 88, 220
♦ cultivated, herbal.

Hevea brasiliensis (Willd. ex A.Juss.) Müll.Arg.
Euphorbiaceae
♦ rubber, para rubber, para rubber tree
♦ Weed, Naturalised, Introduced, Environmental Weed
♦ 3, 86, 191, 230, 259
♦ T, cultivated, herbal. Origin: Americas.

Hewittia sublobata (L.f.) Kuntze
Convolvulaceae
♦ ng'ubisigo
♦ Weed, Native Weed
♦ 13, 88, 121
♦ pC, cultivated. Origin: southern Africa.

Hexaglottis lewisiae Goldblatt
Iridaceae
= *Moraea lewisiae* (Goldblatt) Goldblatt (NoR)
♦ yellow hexaglottis
♦ Weed, Naturalised, Garden Escape, Environmental Weed
♦ 7, 72, 86, 88, 98, 198, 203
♦ pH, cultivated.

Hibbertia cuneiformis (Labill.) Sm.
Dilleniaceae
♦ cutleaf guinea flower
♦ Naturalised, Cultivation Escape
♦ 7
♦ cultivated.

Hibiscus L. spp.
Malvaceae
♦ flower of an hour, rosemallow, mallow
♦ Weed
♦ 243
♦ herbal.

Hibiscus abelmoschus L.
Malvaceae
= *Abelmoschus moschatus* Medik.
♦ metei, fau tagaloa
♦ Weed, Introduced
♦ 12, 87, 88, 230
♦ S, herbal.

Hibiscus acetosella Welw. ex Hiern
Malvaceae
♦ African rosemallow, red leaved hibiscus, false roselle
♦ Weed, Naturalised, Introduced, Garden Escape
♦ 101, 179, 228, 261
♦ a/pH, arid, cultivated, herbal. Origin: tropical Africa.

Hibiscus altissimus Hornby
Malvaceae
♦ Native Weed
♦ 121
♦ pC. Origin: southern Africa.

Hibiscus articulatus Hochst. ex A.Rich.
Malvaceae
♦ Weed
♦ 87, 88

Hibiscus asper Hook.f.
Malvaceae
♦ Weed
♦ 88, 223
♦ pH, cultivated. Origin: West Africa.

Hibiscus calyphyllus Cav.
Malvaceae
♦ hibiscus, Pondoland hibiscus, lemonyellow rosemallow, ntobotobo
♦ Native Weed
♦ 121
♦ pS, cultivated. Origin: southern Africa.

Hibiscus cannabinus L.
Malvaceae
Hibiscus aspera Hook.f.
♦ kenaf, Indian hemp, lukelekese, wild stockrose, deccan hemp, brown Indian hemp, stockrose, ambari hemp, bastard jute, bimli jute, deckaner hemp, gambo hemp, kenaf hibiscus, kenaf seed oil, wild hollyhock, wild hibiscus
♦ Weed, Naturalised, Garden Escape
♦ 50, 87, 88, 101, 121, 158, 179, 221, 261
♦ aH, cultivated, herbal. Origin: obscure.

Hibiscus costatus A.Rich.
Malvaceae
♦ Weed
♦ 14

Hibiscus diversifolius Jacq.
Malvaceae
♦ swamp hibiscus
♦ Weed, Naturalised, Garden Escape, Environmental Weed
♦ 7, 39, 86, 98, 203, 257
♦ S, cultivated, toxic. Origin: Australia.

Hibiscus elatus Sw.
Malvaceae
= *Hibiscus tiliaceus* L.
♦ Cuban bast, mahoe
♦ Weed, Naturalised, Cultivation Escape
♦ 87, 88, 101, 261
♦ cultivated, herbal.

Hibiscus esculentus L.
Malvaceae
= *Abelmoschus esculentus* (L.) Moench
♦ okra, gobo
♦ Weed, Introduced
♦ 87, 88, 230
♦ S, herbal.

Hibiscus ficulneus L.
Malvaceae
♦ Weed
♦ 87, 88
♦ herbal.

Hibiscus lasiocarpus Cav.
Malvaceae
♦ woolly rosemallow, rosemallow
♦ Weed
♦ 87, 88, 218
♦ pH, aqua, cultivated.

Hibiscus lunariifolius Willd.
Malvaceae
♦ hibiscus

♦ Introduced
♦ 228
♦ arid.

Hibiscus macrophyllus Roxb. ex Hornem.
Malvaceae
♦ largeleaf rosemallow
♦ Naturalised
♦ 101

Hibiscus mastersianus Hiern
Malvaceae
♦ Weed, Native Weed
♦ 87, 88, 121
♦ aH. Origin: Africa.

Hibiscus meeusei Exell
Malvaceae
♦ Native Weed
♦ 121
♦ H. Origin: southern Africa.

Hibiscus micranthus L.f.
Malvaceae
♦ Weed
♦ 87, 88, 221
♦ herbal.

Hibiscus moscheutos L.
Malvaceae
♦ common rosemallow, swamp rosemallow, mallow rose, swamp hibiscus, crimson eyed rosemallow, southern bell hibiscus
♦ Weed
♦ 80
♦ pH, aqua, cultivated, herbal.

Hibiscus mutabilis L.
Malvaceae
♦ Dixie rosemallow, Confederate rose, cottonrose
♦ Weed, Naturalised, Introduced, Casual Alien
♦ 86, 98, 101, 203, 230, 280
♦ S, cultivated, herbal. Origin: China, Japan.

Hibiscus obtusilobus Garcke
Malvaceae
♦ Weed, Quarantine Weed
♦ 76, 87, 88, 203, 220

Hibiscus palustris L.
Malvaceae
= *Hibiscus moscheutos* L. ssp. *moscheutos* (NoR)
♦ swamp rosemallow
♦ Weed
♦ 87, 88, 218
♦ herbal.

Hibiscus panduraeformis Burm.f.
Malvaceae
♦ Weed
♦ 87, 88
♦ herbal.

Hibiscus pedunculatus L.f.
Malvaceae
♦ wild hibiscus
♦ Naturalised
♦ 134
♦ cultivated.

Hibiscus physaloides Guill. & Perr.
Malvaceae

♦ Native Weed
♦ 121
♦ pH. Origin: southern Africa.

Hibiscus pusillus Thunb.
Malvaceae
♦ Native Weed
♦ 121
♦ pH, cultivated. Origin: southern Africa.

Hibiscus radiatus Cav.
Malvaceae
♦ monarch rosemallow
♦ Weed, Naturalised
♦ 101, 179
♦ cultivated, herbal.

Hibiscus rosa-sinensis L.
Malvaceae
♦ Chinese hibiscus, shoebackplant, kinaros, aute
♦ Weed, Naturalised, Environmental Weed
♦ 80, 86, 87, 88, 101, 257
♦ S/T, cultivated, herbal. Origin: China.

Hibiscus rosa-sinensis L. var. rosa-sinensis
Malvaceae
♦ keleu en wai, amapola, candelada, carta abierta
♦ Weed, Introduced, Cultivation Escape
♦ 179, 230, 261
♦ S, cultivated. Origin: tropical Asia.

Hibiscus rosa-sinensis L. var. schizopetalus Mast.
Malvaceae
= *Hibiscus schizopetalus* (Mast.) Hook.f.
♦ lira
♦ Weed, Casual Alien
♦ 179, 261
♦ cultivated. Origin: tropical east Asia.

Hibiscus rugosus Roxb. ex Steud.
Malvaceae
♦ Weed
♦ 87, 88

Hibiscus sabdariffa L.
Malvaceae
Abelmoschus cruentus Bertol., *Hibiscus cordofanus* Turcz., *Hibiscus cruentus* Bertol., *Hibiscus digitatus* Cav., *Hibiscus fraternus* L., *Hibiscus palmatilobus* Baill., *Sabdariffa rubra* Kostel.
♦ rosella, red sorrel, Jamaica sorrel, karkadè, Indian sorrel, roselle, sorrel, oseille de Guinée, Malventee, acedera de Guinea, rosa de Jamaica, serení
♦ Weed, Naturalised, Garden Escape, Environmental Weed, Casual Alien
♦ 7, 86, 87, 88, 93, 98, 101, 203, 221, 261
♦ a/pH, cultivated, herbal. Origin: Old World Tropics.

Hibiscus schizopetalus (Mast.) Hook.f.
Malvaceae
Hibiscus rosa-sinensis L. var. *schizopetalus* Mast. (see)
♦ fringed rosemallow, Japanese lantern, coral hibiscus

♦ Weed, Naturalised, Introduced
♦ 22, 101, 230, 262
♦ S/T, cultivated, herbal. Origin:
Africa.

Hibiscus surattensis L.
Malvaceae
♦ sisangulu
♦ Native Weed
♦ 121
♦ pC, cultivated, herbal. Origin:
southern Africa.

Hibiscus syriacus L.
Malvaceae
Ketmia syriaca (L.) Scop.
♦ shrub althea, rose of Sharon,
common hibiscus
♦ Weed, Naturalised, Casual Alien
♦ 40, 80, 88, 101, 102, 280, 287
♦ S, cultivated, herbal. Origin: obscure,
possibly east Asia.

Hibiscus tetraphyllus Roxb.
Malvaceae
♦ fau tagaloa
♦ Weed
♦ 87, 88
♦ herbal.

Hibiscus tiliaceus L.
Malvaceae
Hibiscus azanzae DC., *Hibiscus
bracteosus* DC., *Hibiscus elatus* Sw. (see),
Kydia calycina Roxb., *Paritium tiliaceum*
(L.) A.L.Juss.
♦ mahoe, sea hibiscus, keleu, fau,
cottonwood
♦ Weed, Noxious Weed, Naturalised,
Introduced, Environmental Weed
♦ 22, 80, 87, 88, 101, 112, 122, 151, 179,
228, 230, 280
♦ T, arid, cultivated, herbal. Origin:
Australia, Africa, Polynesia.

Hibiscus trionum L.
Malvaceae
Hibiscus africanus Mill., *Hibiscus
hispidus* Mill., *Hibiscus ternatus* Cav.,
Hibiscus trionum var. *ternatus* DC.,
Hibiscus vesicarius Cav., *Trionum
annuum* Medik., *Ketmia trionum* (L.)
Scop.
♦ Venice mallow, bladder ketmia,
flower of an hour, rosemallow,
modesty, shoofly, bladder hibiscus,
bladder weed, black eyed Susan,
ajannäyttäjä
♦ Weed, Noxious Weed, Naturalised,
Native Weed, Introduced,
Environmental Weed, Casual Alien
♦ 1, 7, 23, 34, 40, 42, 50, 51, 55, 80, 87,
88, 94, 98, 101, 121, 146, 151, 158, 161,
165, 174, 176, 180, 185, 203, 207, 210,
211, 212, 217, 218, 221, 228, 229, 240,
242, 243, 253, 256, 269, 272, 275, 280,
286, 287, 297, 299, 300
♦ a/pH, arid, cultivated, herbal.
Origin: Africa.

Hibiscus trionum L. var. trionum
Malvaceae
♦ bladder ketmia
♦ Naturalised, Environmental Weed

♦ 86, 198
♦ Origin: Eastern Europe.

Hibiscus vitifolius L.
Malvaceae
♦ tropical rosemallow
♦ Weed, Introduced
♦ 87, 88, 221, 228
♦ arid, cultivated.

Hibiscus vitifolius L. ssp. vitifolius
Malvaceae
♦ Native Weed
♦ 121
♦ pH. Origin: southern Africa.

Hieracium L. spp.
Asteraceae
♦ hawkweed
♦ Weed, Quarantine Weed,
Environmental Weed
♦ 18, 70, 80, 88, 181, 220, 225, 246, 272

Hieracium argillaceum group Jordan
Asteraceae
♦ Naturalised
♦ 280

**Hieracium × atramentarium (Nägeli &
Peter) Zahn ex Engl. (pro sp.)**
Asteraceae
= *Hieracium aurantiacum* L. × *Hieracium
piloselloides* Vill.
♦ hawkweed
♦ Naturalised
♦ 101

Hieracium atratum Fr.
Asteraceae
♦ polar hawkweed
♦ Weed, Noxious Weed
♦ 1, 88, 139

Hieracium aurantiacum L.
Asteraceae
Pilosella aurantiaca (L.) F.W.Schultz &
Sch.Bip. (see)
♦ orange hawkweed, fox and cubs,
devil's paintbrush, missionary weed
♦ Weed, Sleeper Weed, Quarantine
Weed, Noxious Weed, Naturalised,
Introduced, Garden Escape,
Environmental Weed
♦ 1, 21, 52, 76, 80, 86, 87, 88, 98, 101,
136, 139, 146, 156, 161, 165, 176, 179,
195, 203, 210, 212, 218, 219, 222, 229,
241, 267, 280, 286, 287, 289, 298, 299,
300
♦ pH, cultivated, herbal. Origin:
northern, central and eastern Europe.

**Hieracium aurantiacum L. ssp.
carpathicola Nägeli & Peter**
Asteraceae
Hieracium scandicum (Nägeli &
Peter) Omang, *Pilosella aurantiaca*
(L.) F.W.Schultz & Sch.Bip. ssp.
brunneocrocea (Pugsley) P.D.Sell &
C.West, *Hieracium brunneocroceum*
Pugsley
♦ Naturalised
♦ 280
♦ Origin: Europe.

Hieracium bauhini Schwägr. ex Schrank
Asteraceae

♦ Weed
♦ 87, 88

Hieracium × brachiatum Berthel. ex DC.
Asteraceae
= *Hieracium pilosella* L. × *Hieracium
piloselloides* Vill.
♦ Naturalised
♦ 101

Hieracium caespitosum Dumort.
Asteraceae
Hieracium pratense Tausch (see),
Pilosella caespitosa (Dumort.) P.D.Sell
& C.West
♦ meadow hawkweed, yellow
hawkweed, field hawkweed
♦ Weed, Noxious Weed, Naturalised,
Environmental Weed
♦ 1, 80, 88, 101, 139, 146, 229, 246, 267,
272, 280, 298
♦ herbal. Origin: northern, central and
eastern Europe.

Hieracium canadense Michx.
Asteraceae
♦ Canada hawkbeard, Canadian
hawkweed, fireweed, yellow
hawkweed, hawkweed
♦ Weed, Native Weed
♦ 136, 222
♦ cultivated, herbal.

Hieracium cheriense Jord.
Asteraceae
♦ Naturalised
♦ 40

Hieracium echioides Lumn.
Asteraceae
♦ Weed
♦ 272

Hieracium eriosphaerophorum Zahn
Asteraceae
♦ Naturalised
♦ 241

Hieracium × flagellare Willd. (pro sp.)
Asteraceae
= *Hieracium caespitosum* Dumort. ×
Hieracium pilosella L., *Pilosella flagellaris*
(Willd.) P.D.Sell & C.West
♦ hawkweed
♦ Naturalised
♦ 101, 300

**Hieracium × flagellare Willd. (pro sp.)
var. cernuiforme (Nägeli & Peter) Lepage
(pro nm.)**
Asteraceae
♦ hawkweed
♦ Naturalised
♦ 101

**Hieracium × flagellare Willd. (pro sp.)
var. flagellare**
Asteraceae
♦ hawkweed
♦ Naturalised
♦ 101

Hieracium florentinum All.
Asteraceae
♦ kingdevil hawkweed
♦ Weed
♦ 52, 87, 88, 161, 218
♦ herbal.

**Hieracium × floribundum Wimm. & Grab.
(pro sp.)**
Asteraceae
= *Hieracium caespitosum* Dumort. ×
Hieracium lactucella Wallr.
♦ yellow devil hawkweed, hawkweed,
kingdevil hawkweed
♦ Naturalised, Weed, Noxious Weed
♦ 1, 79, 87, 88, 101, 139, 156, 161, 218
♦ herbal.

**Hieracium × fuscatrum Nägeli & Peter
(pro sp.)**
Asteraceae
= *Hieracium aurantiacum* L. × *Hieracium
caespitosum* Dumort.
♦ hawkweed
♦ Naturalised
♦ 101

Hieracium kalmii L.
Asteraceae
♦ Canada hawkweed, Kalm's
hawkweed
♦ Weed
♦ 195

Hieracium lachenalii C.C.Gmel.
Asteraceae
Hieracium vulgatum Fr. (see)
♦ common hawkweed, hawkweed
♦ Weed, Naturalised
♦ 101, 195
♦ cultivated, herbal.

Hieracium laevigatum Willd.
Asteraceae
♦ smooth hawkweed
♦ Weed, Noxious Weed
♦ 1, 88, 139
♦ cultivated.

Hieracium lepidulum Stenstr.
Asteraceae
♦ hawkweed, tussock hawkweed
♦ Naturalised, Environmental Weed
♦ 40, 246, 280, 298

Hieracium maculatum Sm.
Asteraceae
♦ spotted hawkweed, hawkweed
♦ Naturalised
♦ 101
♦ cultivated.

Hieracium murorum L.
Asteraceae
♦ wall hawkweed, hawkweed
♦ Naturalised
♦ 101, 280, 300
♦ cultivated, herbal.

Hieracium pilosella L.
Asteraceae
Pilosella officinarum Schultz & Sch.Bip.
♦ mouse ear hawkweed
♦ Weed, Sleeper Weed, Quarantine
Weed, Noxious Weed, Naturalised,
Garden Escape, Environmental Weed
♦ 1, 15, 70, 76, 80, 86, 87, 88, 101, 139,
146, 152, 161, 165, 172, 181, 203, 218,
229, 243, 246, 272, 280, 298
♦ pH, cultivated, herbal. Origin:
Eurasia.

**Hieracium pilosella L. ssp. euronotum
Nägeli & Peter**

Asteraceae
♦ Naturalised
♦ 241, 300

**Hieracium pilosella L. var. niveum
Müll.Arg.**
Asteraceae
♦ mouse ear hawkweed
♦ Noxious Weed, Naturalised
♦ 101, 229

Hieracium pilosella L. var. pilosella
Asteraceae
♦ mouse ear hawkweed
♦ Noxious Weed, Naturalised
♦ 101, 229

Hieracium piloselloides Vill.
Asteraceae
♦ kingdevil, tall hawkweed, yellow
flowered hawkweed
♦ Weed, Naturalised
♦ 79, 80, 88, 101, 156
♦ cultivated, herbal.

Hieracium pollichiae Sch.Bip.
Asteraceae
♦ Naturalised
♦ 280

Hieracium praealtum Vill. ex Gochnat
Asteraceae
♦ tall kingdevil hawkweed, kingdevil
♦ Weed, Naturalised, Environmental
Weed
♦ 87, 88, 101, 152, 165, 246, 272, 280,
298, 300
♦ Origin: Europe.

**Hieracium praealtum Vill. ex Gochnat
var. decipiens W.D.J.Koch**
Asteraceae
♦ tall kingdevil hawkweed, kingdevil
♦ Weed, Naturalised
♦ 101, 218

Hieracium pratense Tausch
Asteraceae
= *Hieracium caespitosum* Dumort.
♦ yellow hawkweed, yellow kingdevil,
field hawkweed
♦ Weed, Quarantine Weed,
Naturalised
♦ 76, 79, 80, 87, 88, 156, 161, 203, 210,
211, 212, 218, 219, 287
♦ cultivated, herbal.

Hieracium racemosum Waldst. & Kit.
Asteraceae
Hieracium barbatum Tausch
♦ sparviere racemoso
♦ Weed
♦ 272
♦ herbal.

Hieracium sabaudum L.
Asteraceae
♦ New England hawkweed, broad
leaved hawkweed
♦ Weed, Naturalised
♦ 101, 272, 280
♦ cultivated, herbal.

Hieracium × stoloniflorum Waldst. & Kit.
Asteraceae
♦ hawkweed
♦ Naturalised
♦ 280

Hieracium umbellatum L.
Asteraceae
♦ narrow leaved hawkbeard,
hawkweed, narrowleaf hawkweed,
yanagitanpopo
♦ Weed
♦ 23, 80, 87, 88, 136, 272, 286
♦ cultivated, herbal.

Hieracium vulgatum Fr.
Asteraceae
= *Hieracium lachenalii* C.C.Gmel.
♦ common hawkweed, wood
hawkweed
♦ Weed
♦ 23, 87, 88, 218
♦ cultivated, herbal.

**Hierochloe borealis Roem. & Schult.
nom. illeg.**
Poaceae
= *Hierochloe odorata* (L.) P.Beauv.
♦ Weed
♦ 39, 87, 88
♦ G, herbal, toxic.

Hierochloe bungeana Trin.
Poaceae
♦ Weed
♦ 286
♦ G.

Hierochloe glabra Trin.
Poaceae
♦ glabrous sweetgrass
♦ Weed
♦ 297
♦ G.

Hierochloe odorata (L.) P.Beauv.
Poaceae
Hierochloe borealis Roem. & Schult. *nom.
illeg.* (see), *Holcus odoratus* L., *Torresia
odorata* (L.) Hitchc.
♦ sweetgrass, vanilla grass, holy
grass, mannagrass, seneca grass,
duftmariengras
♦ Weed, Naturalised
♦ 87, 88, 218, 275, 287, 297
♦ pG, cultivated, herbal. Origin:
Eurasia.

Hilaria cenchroides H.B.K.
Poaceae
♦ Weed
♦ 199
♦ G.

Hilaria rigida (Thurb.) Scribn.
Poaceae
= *Pleuraphis rigida* Thurb. (NoR)
♦ galleta grass
♦ Weed
♦ 39, 161
♦ G, arid, cultivated, herbal, toxic.

Hildewintera Ritt. spp.
Cactaceae
= *Cleistocactus* Lem. spp.
♦ Weed, Quarantine Weed
♦ 76, 88, 203

Hillia parasitica Jacq.
Rubiaceae
♦ tibey trepador
♦ Weed

- 87, 88
- pH.

Himalayacalamus falconeri (Hook.f. ex Munro) Keng.f.
Poaceae
Thamnocalamus falconeri Hook.f. ex Munro
- fountain bamboo
- Weed, Naturalised
- 15, 280
- G.

Hippeastrum puniceum (Lam.) Kuntze
Liliaceae/Amaryllidaceae
Amaryllis belladona auct. nom., *Amaryllis punicea* Lam., *Amaryllis equestris* Aiton, *Hippeastrum equestre* (Aiton) Herb., *Hippeastrum purpureum* (Lam.) Kuntze
- Barbados lily, Easter lily, amaryllis
- Weed, Naturalised
- 87, 88, 101, 161
- herbal, toxic.

Hippobroma longiflora (L.) G.Don
Campanulaceae/Lobeliaceae
Isotoma longiflora Presl (see), *Laurentia longiflora* (L.) Peterm. (see)
- madam fate, horse poison
- Weed, Naturalised, Introduced
- 28, 39, 86, 98, 179, 199, 230, 243
- H, cultivated, toxic. Origin: Central and South America.

Hippobromus pauciflorus (L.f) Radlk.
Sapindaceae
- bastard horsewood, false horsewood, horsewood
- Native Weed
- 121
- S/T, toxic. Origin: southern Africa.

Hippocratea pallens Planch. ex Oliv.
Celastraceae/Hippocrateaceae
Apodostigma pallens (Planch. ex Oliv.) R.Wilczek
- Weed
- 88

Hippocrepis L. spp.
Fabaceae/Papilionaceae
- horseshoe vetch, hippocrepis
- Weed
- 272

Hippocrepis bicontorta Loisel.
Fabaceae/Papilionaceae
- Weed
- 221

Hippocrepis biflora Spreng.
Fabaceae/Papilionaceae
- Weed
- 221

Hippocrepis comosa L.
Fabaceae/Papilionaceae
Hippocrepis perennis Lam.
- horseshoe vetch, podkova chochlatá
- Naturalised
- 101
- cultivated, herbal.

Hippocrepis constricta Kunze
Fabaceae/Papilionaceae
- Weed
- 221

Hippocrepis unisiliquosa L.
Fabaceae/Papilionaceae
- horseshoe vetch
- Weed
- 70
- cultivated.

Hippomane mancinella L.
Euphorbiaceae
- manzanillo, manchineel tree
- Weed
- 14, 39, 82, 96, 109, 161
- herbal, toxic.

Hippophae rhamnoides L.
Elaeagnaceae
Elaeagnus rhamnoides (L.) A.Nelson, *Hippophae angustifolia* Lodd., *Hippophae littoralis* Salisb., *Hippophae rhamnoideum* St.-Lag., *Hippophae sibirica* Lodd., *Hippophae stourdziana* J.Szabó, *Osyris rhamnoides* Scop., *Rhamnoides hippophae* Moench
- sea buckthorn, tyrni, argasse, argousier, grisset, sanddorn, espino armarillo, espino falso
- Naturalised
- 101
- S, cultivated, herbal.

Hippuris vulgaris L.
Hippuridaceae
Hippuris tetraphylla L.f.
- mare's tail, common mare's tail, vesikuusi
- Weed, Noxious Weed
- 87, 88, 272, 297, 299
- pH, aqua, cultivated, herbal.

Hiptage Gaertn. spp.
Malpighiaceae
- Weed
- 18, 88

Hiptage benghalensis (L.) Kurz
Malpighiaceae
Hiptage madablota Gaertn. (see), *Triopteris jamaicensis* L. (see)
- hiptage, liane de cerf
- Weed, Naturalised, Garden Escape, Environmental Weed, Cultivation Escape
- 3, 22, 39, 80, 86, 88, 112, 131, 132, 152, 155, 233
- pC, cultivated, herbal, toxic. Origin: south-east Asia.

Hiptage madablota Gaertn.
Malpighiaceae
= *Hiptage benghalensis* (L.) Kurz
- hiptage
- Weed, Naturalised
- 3, 86, 191
- cultivated.

Hirpicium bechuanense (S.Moore) Roessl.
Asteraceae
- Native Weed
- 121
- pH. Origin: Africa.

Hirpicium linearifolium (H. Bol.) Roessl.
Asteraceae
- Quarantine Weed
- 220

Hirschfeldia incana (L.) Lagr.-Foss.
Brassicaceae
Brassica adpressa (Moench) Boiss. (see), *Brassica geniculata* (Desf.) Benth., *Brassica incana* Meigen, *Crucifera hirschfeldia* E.H.L.Krause, *Hirschfeldia adpressa* Moench, *Raphanus incanus* Crantz, *Sinapis geniculata* Desf., *Sinapis incana* L. (see)
- buchan weed, hoary mustard, shortpod mustard, hairy brassica, hairy mustard, Mediterranean mustard, short podded mustard
- Weed, Quarantine Weed, Noxious Weed, Naturalised, Introduced, Environmental Weed, Casual Alien
- 34, 35, 40, 42, 70, 76, 78, 86, 87, 88, 94, 98, 101, 115, 116, 144, 147, 161, 165, 167, 171, 176, 180, 198, 203, 218, 228, 237, 243, 253, 269, 272, 280, 300
- aH, arid, cultivated. Origin: southern Europe.

Hoffmannseggia densiflora Benth.
Fabaceae/Caesalpiniaceae
= *Hoffmannseggia glauca* (Ortega) Eifert
- Weed
- 87, 88, 218

Hoffmannseggia falcaria Cav.
Fabaceae/Caesalpiniaceae
= *Hoffmannseggia glauca* (Ortega) Eifert
- algarrobilla fina
- Weed
- 87, 88, 295

Hoffmannseggia glauca (Ortega) Eifert
Fabaceae/Caesalpiniaceae
Hoffmannseggia densiflora Benth. (see), *Hoffmannseggia falcaria* Cav. (see), *Larrea densiflora* (Benth. ex A.Gray) Britton
- hogpotato, Indian rushpea, pignut
- Weed, Noxious Weed
- 161, 212, 229, 237
- pH, arid, herbal.

Hohenbergia penduliflora (A.Rich) Mez
Bromeliaceae
- Weed
- 14
- cultivated.

Hoheria populnea A.Cunn
Malvaceae
- lacebark, houhere
- Weed
- 88, 131
- T, cultivated, herbal. Origin: New Zealand.

Holarrhena curtisii King & Gamble
Apocynaceae
- Weed
- 12

Holcus lanatus L.
Poaceae
Aira holcus-lanata Vill., *Avena pallida* Salisb.
- Yorkshire fog, velvetgrass, tufted softgrass, meadow softgrass, common velvetgrass, karvamesiheinä, mesquite, bambagione pubescente, mesquite grass

♦ Weed, Naturalised, Introduced, Garden Escape, Environmental Weed, Cultivation Escape

♦ 7, 9, 15, 18, 20, 23, 30, 34, 38, 39, 42, 44, 70, 72, 80, 86, 87, 88, 91, 98, 101, 118, 121, 146, 151, 152, 159, 161, 176, 180, 181, 198, 203, 204, 211, 212, 218, 228, 241, 243, 250, 263, 269, 272, 280, 286, 287, 289, 292, 295, 296, 300

♦ aG, arid, cultivated, herbal, toxic. Origin: Eurasia.

Holcus mollis L.
Poaceae
Notholcus mollis (L.) Hitchc.
♦ creeping velvetgrass, German velvetgrass, pehmytmesiheinä, creeping fog, creeping softgrass, soft foggrass, bambagione aristato
♦ Weed, Quarantine Weed, Naturalised, Environmental Weed
♦ 23, 30, 42, 44, 70, 76, 80, 86, 87, 88, 91, 98, 101, 118, 121, 146, 151, 161, 176, 198, 203, 212, 218, 243, 250, 253, 272, 280, 287
♦ pG, cultivated, herbal. Origin: Eurasia.

Holcus setiger Nees
Poaceae
♦ annual fog
♦ Weed, Naturalised, Environmental Weed
♦ 7, 9, 86, 98, 203
♦ G. Origin: South Africa.

Holcus setosus Trin.
Poaceae
♦ annual fog
♦ Weed, Naturalised, Environmental Weed
♦ 72, 86, 88, 98, 198, 203
♦ aG, cultivated. Origin: Mediterranean.

Holmskioldia sanguinea Retz.
Lamiaceae/Verbenaceae
♦ Chinese hat plant, cup and saucer plant, Mandarin's hat, platillo
♦ Weed, Naturalised, Introduced
♦ 101, 179, 261
♦ cultivated. Origin: northern India.

Holocalyx glaziovii Taub.
Fabaceae/Papilionaceae
♦ Weed
♦ 87, 88

Holocarpha virgata (Gray) Keck
Asteraceae
♦ pitgland tarweed, narrow tarplant, yellowflower tarweed, virgate tarweed
♦ Weed
♦ 161
♦ aH.

Holosteum umbellatum L.
Caryophyllaceae
Cerastium umbellatum (L.) Crantz
♦ jagged chickweed, umbellate chickweed, umbrella spurrey
♦ Weed, Naturalised, Introduced
♦ 70, 87, 88, 94, 101, 161, 228, 248, 253, 272, 280
♦ arid, cultivated, herbal.

Holosteum umbellatum L. ssp. *umbellatum*
Caryophyllaceae
♦ jagged chickweed
♦ Introduced
♦ 34
♦ aH.

Homalanthus populifolius Graham
Euphorbiaceae
♦ Queensland poplar, bleedingheart tree
♦ Weed, Sleeper Weed, Quarantine Weed, Naturalised, Environmental Weed
♦ 15, 22, 121, 225, 246, 280
♦ S/T, cultivated, herbal. Origin: Australia.

Homalocladium platycladum (F.J.Muell.) Bailey
Polygonaceae
Muehlenbeckia platyclada (F.J.Muell.) Lind. (see)
♦ ribbon bush, centipede plant, helecho chino
♦ Naturalised
♦ 101, 261
♦ cultivated. Origin: Polynesia.

Homeria Vent. spp.
Iridaceae
= *Moraea* Mill. spp. (NoR)
♦ cape tulip
♦ Weed, Quarantine Weed, Noxious Weed, Naturalised
♦ 76, 88, 140, 171, 198

Homeria breyniana (L.) Lewis
Iridaceae
= *Moraea collina* Thunb.(NoR) [see *Homeria collina* (Thunb.) Vent.]
♦ Weed
♦ 39, 87, 88
♦ toxic.

Homeria britteniae Bolus
Iridaceae
= *Moraea britteniae* (Bolus) Goldblatt (NoR)
♦ Native Weed
♦ 121
♦ pH, toxic. Origin: southern Africa.

Homeria collina (Thunb.) Vent.
Iridaceae
= *Moraea collina* Thunb. (NoR) [see *Homeria breyniana* (L.) Lewis]
♦ aprico homeria, apricot tulip, cape tulip, cape tulp, oneleaf cape tulip, red tulip, red tulp, salmon homeria, yellow tulp
♦ Weed, Sleeper Weed, Noxious Weed, Naturalised, Native Weed, Environmental Weed, Casual Alien
♦ 39, 40, 86, 121, 225, 246, 280
♦ pH, cultivated, toxic. Origin: southern Africa.

Homeria flaccida Sweet
Iridaceae
= *Moraea flaccida* (Sweet) Steud.
♦ oneleaf cape tulip, cape tulip
♦ Weed, Quarantine Weed, Noxious

Weed, Naturalised, Native Weed, Garden Escape, Environmental Weed
♦ 7, 9, 39, 62, 72, 76, 86, 88, 98, 121, 134, 147, 176, 198, 203, 269, 290
♦ pH, cultivated, toxic. Origin: South Africa.

Homeria miniata (Andrews) Sweet
Iridaceae
= *Moraea miniata* Andrews
♦ twoleaf cape tulip, poison bulb, red tulip, red tulp, cape tulip, rootulup
♦ Weed, Quarantine Weed, Noxious Weed, Naturalised, Native Weed, Garden Escape, Environmental Weed
♦ 7, 39, 62, 72, 76, 86, 87, 88, 98, 121, 147, 158, 198, 203, 269, 290
♦ pH, cultivated, toxic. Origin: South Africa.

Homeria ochroleuca Salisb.
Iridaceae
= *Moraea ochroleuca* (Salisb.) Drapiez (NoR)
♦ yellow flowered cape tulip, cape tulip, white cape tulip
♦ Weed, Naturalised, Garden Escape, Environmental Weed
♦ 7, 39, 72, 86, 88, 98, 198, 203
♦ pH, cultivated, toxic.

Homeria pallida Bak.
Iridaceae
= *Moraea pallida* (Bak.) Goldblatt (NoR)
♦ Transvaal yellow tulp, Natal yellow tulp, yellow homeria, yellow tulip, yellow tulp
♦ Weed, Native Weed
♦ 39, 88, 121, 158
♦ pH, cultivated, herbal, toxic. Origin: southern Africa.

Homoglossum watsonium (Thunb.) N.E.Br.
Iridaceae
♦ homogiossum
♦ Weed, Naturalised, Environmental Weed
♦ 7, 86, 98, 203
♦ cultivated.

Homolepis aturensis (Kunth) Chase
Poaceae
♦ torourco pasto amargo
♦ Weed, Naturalised
♦ 32, 87, 88, 153, 157
♦ pG.

Homopholis proluta (F.Muell.) R.D.Webster
Poaceae
Panicum prolutum F.Muell.
♦ Introduced
♦ 228
♦ G, arid, cultivated.

Hordelymus europaeus (L.) Harz
Poaceae
Cuviera europaea (L.) Koeler, *Elymus europaeus* L.
♦ wood barley
♦ Weed
♦ 70, 272
♦ G, cultivated.

Hordeum L. spp.
Poaceae
♦ barley
♦ Weed, Environmental Weed
♦ 88, 151
♦ G.

Hordeum arizonicum Covas
Poaceae
♦ Arizona barley, Arizona foxtail
♦ Quarantine Weed
♦ 76
♦ aG.

Hordeum berteroanum E.Desv. ex C.Gay
Poaceae
♦ Weed
♦ 87, 88
♦ G.

Hordeum bogdanii Wilensky
Poaceae
Critesion bogdanii (Wilensky) Á.Löve,
Hordeum secalinum Schreb. var. *bogdanii*
(Wilensky) Roshev.
♦ Quarantine Weed
♦ 76
♦ G.

Hordeum brachyantherum Nevski
Poaceae
♦ meadow barley, idänohra
♦ Weed, Naturalised, Casual Alien
♦ 42, 87, 88, 161, 218, 287
♦ pG, cultivated, herbal.

Hordeum brachyantherum Nevski (Covas & Stebbins) ssp. *californicum* V.Bothmer N.Jacobsen & Seberg
Poaceae
♦ California barley, California meadow barley, meadow barley
♦ Weed
♦ 167
♦ pG.

Hordeum brevisubulatum (Trin.) Link
Poaceae
♦ Quarantine Weed
♦ 76
♦ G.

Hordeum bulbosum L.
Poaceae
Hordeum nodosum Ucria *non* L.
♦ bulbous barley, bulbous barleygrass
♦ Weed, Naturalised, Introduced
♦ 87, 88, 98, 101, 111, 203, 228, 243, 272
♦ pG, arid, cultivated.

Hordeum capense Thunb.
Poaceae
♦ cape wild barley
♦ Native Weed
♦ 121
♦ pG. Origin: southern Africa.

Hordeum comosum J. & C.Presl
Poaceae
♦ Introduced
♦ 228
♦ G, arid.

Hordeum distichon L.
Poaceae
= *Hordeum vulgare* L.
♦ tworow barley, barley, cebada cervecera
♦ Weed, Naturalised, Casual Alien
♦ 15, 40, 80, 176, 236, 280, 287
♦ G, cultivated, herbal. Origin: obscure.

Hordeum geniculatum All.
Poaceae
= *Hordeum marinum* Huds. ssp. *gussoneanum* (Parl.) Thell.
♦ Mediterranean barleygrass
♦ Weed, Naturalised
♦ 7, 39, 80
♦ G, herbal, toxic.

Hordeum glaucum Steud.
Poaceae
= *Hordeum murinum* L. ssp. *glaucum* (Steud.) Tzvelev
♦ barleygrass, northern barleygrass, wall barley
♦ Weed, Naturalised, Environmental Weed
♦ 7, 68, 86, 87, 88, 243, 248, 269, 280
♦ aG, herbal. Origin: Mediterranean, west Asia.

Hordeum hystrix Roth
Poaceae
= *Hordeum marinum* Huds. ssp. *gussoneanum* (Parl.) Thell.
♦ Mediterranean barley, pörröohra
♦ Weed, Naturalised, Casual Alien
♦ 23, 42, 86, 87, 88, 161, 218, 269, 280, 287
♦ aG, cultivated. Origin: Mediterranean, Eurasia.

Hordeum jubatum L.
Poaceae
Critesion jubatum (L.) Nevski (see), *Hordeum caespitosum* Scribn. ex Pammel, *Hordeum jubatum* L. var. *caespitosum* (Scribn. ex Pammel) A.S.Hitchc.
♦ foxtail barley, wild barley, squirreltail grass, skunktail grass, ticklegrass, partaohra, flickertail
♦ Weed, Quarantine Weed, Noxious Weed, Naturalised, Native Weed, Casual Alien
♦ 23, 39, 40, 42, 49, 52, 76, 80, 87, 88, 91, 136, 161, 174, 180, 203, 210, 212, 218, 220, 243, 256, 272, 280, 287, 294, 299, 300
♦ pG, cultivated, herbal, toxic. Origin: North America.

Hordeum leporinum Link
Poaceae
= *Hordeum murinum* L. ssp. *leporinum* (Link) Arcang.
♦ barleygrass, wild barley, hare barley, wild barley, mouse barley, foxtail, common foxtail, farmer's foxtail, wall barley
♦ Weed, Naturalised, Environmental Weed
♦ 7, 9, 23, 39, 68, 80, 86, 87, 88, 91, 111, 161, 180, 212, 218, 221, 237, 243, 269, 287, 295

♦ aG, herbal, toxic. Origin: Mediterranean, Eurasia.

Hordeum marinum Huds.
Poaceae
♦ sea barleygrass, seaside barley, squirreltail grass, salt barleygrass, meriohra
♦ Weed, Naturalised, Environmental Weed, Casual Alien
♦ 7, 42, 70, 86, 87, 88, 101, 176, 181, 221, 269, 272, 280, 287
♦ aG, cultivated. Origin: Mediterranean, Eurasia.

Hordeum marinum Huds. ssp. gussoneanum (Parl.) Thell.
Poaceae
Critesion hystrix (Roth) Á.Löve (see), *Hordeum hystrix* Roth (see), *Hordeum geniculatum* All. (see), *Hordeum marinum* Huds. ssp. *gussonianum* (Parl.) Thell. (see)
♦ Mediterranean barley
♦ Naturalised
♦ 300
♦ aG.

Hordeum marinum Huds. ssp. gussonianum (Parl.) Thell.
Poaceae
= *Hordeum marinum* Huds. ssp. *gussoneanum* (Parl.) Thell.
♦ Mediterranean barley
♦ Weed, Naturalised
♦ 34, 101, 176
♦ G.

Hordeum marinum Huds. ssp. marinum
Poaceae
Critesion marinum (Huds.) Á.Löve (see)
♦ seaside barley
♦ Naturalised
♦ 101, 176, 300
♦ G.

Hordeum murinum (Steud.) Tzvelev
Poaceae
Critesion murinum (L.) Á.Löve (see), *Zeocrithon murinum* Pall.
♦ wild barley, wild barley, barleygrass, mouse barley, false barley, ball barleygrass
♦ Weed, Naturalised, Introduced, Casual Alien
♦ 30, 38, 39, 42, 51, 70, 80, 86, 87, 88, 91, 101, 115, 118, 121, 158, 176, 217, 218, 237, 243, 253, 280, 286, 287, 295
♦ aG, arid, cultivated, herbal, toxic. Origin: Eurasia.

Hordeum murinum L. ssp. glaucum (Steud.) Tzvelev
Poaceae
Critesion glaucum (Steud.) Á.Löve (see), *Critesion murinum* (L.) Á.Löve ssp. *glaucum* (Steud.) W.A.Weber (see), *Hordeum glaucum* Steud. (see), *Hordeum stebbinsii* Covas (see)
♦ blue foxtail, smooth barley
♦ Weed, Naturalised
♦ 34, 101, 134, 167, 176
♦ aG.

**Hordeum murinum L. ssp. *leporinum*
(Link) Arcang.**
Poaceae
Critesion murinum (L.) Á.Löve ssp.
leporinum (Link) Á.Löve (see), *Hordeum
leporinum* Link (see)
♦ foxtail, foxtail barley, leporinum
barley, mouse barley
♦ Weed, Naturalised
♦ 15, 34, 101, 134, 176, 243, 250, 272,
280, 300
♦ aG.

Hordeum murinum L. ssp. *murinum*
Poaceae
Critesion murinum (L.) Á.Löve (see)
♦ wall barley
♦ Weed, Naturalised
♦ 15, 101, 176, 272, 280, 300
♦ aG.

Hordeum pusillum Nutt.
Poaceae
♦ little barley, little wild barley,
pikkuohra
♦ Weed, Naturalised, Native Weed,
Casual Alien
♦ 42, 87, 88, 161, 174, 210, 218, 243, 249,
287
♦ aG, herbal. Origin: Americas.

Hordeum secalinum Schreb.
Poaceae
Critesion secalinum (Schreb.) Á.Löve
(see), *Hordeum nodosum* L.
♦ meadow barley
♦ Weed, Naturalised
♦ 70, 86, 87, 88, 243, 272, 280, 300
♦ G, cultivated, herbal. Origin:
Mediterranean.

Hordeum spontanum Koch
Poaceae
♦ Weed
♦ 87, 88
♦ G.

Hordeum stebbinsii Covas
Poaceae
= *Hordeum murinum* L. ssp. *glaucum*
(Steud.) Tzvelev
♦ Weed
♦ 23, 88
♦ G.

Hordeum stenostachys Godr.
Poaceae
♦ centenillo
♦ Weed
♦ 295
♦ G.

Hordeum violaceum Boiss. & Hohen.
Poaceae
= *Hordeum brevisubulatum* (Trin.)
Link ssp. *violaceum* (Boiss. & Hohen.)
Tzvelev (NoR)
♦ sinipunaohra
♦ Casual Alien
♦ 42
♦ G, arid, cultivated.

Hordeum vulgare L.
Poaceae
Frumentum hordeum Krause,
Hordeum distichon L. (see), *Hordeum*

lagunculiforme (Bachteev) Bachteev ex
Nikif., *Hordeum sativum* Pers., *Hordeum
sinojaponicum* Vav. & Bachteev,
Hordeum tetrastichum Körn., *Hordeum
zeocriton* L.
♦ barley, steptoe barley, six rowed
barley, orzo mondo frutti, common
barley, barleygrass, orzo coltivato
♦ Weed, Naturalised, Cultivation
Escape, Casual Alien
♦ 7, 15, 39, 40, 42, 80, 87, 88, 98, 101,
176, 198, 199, 203, 237, 241, 243, 280,
295
♦ aG, cultivated, herbal, toxic. Origin:
obscure.

**Hordeum vulgare L. ssp. *distichon* (L.)
Körn**
Poaceae
♦ tworow barley
♦ Naturalised
♦ 86, 198, 280
♦ G, cultivated. Origin: Europe.

Hordeum vulgare L. ssp. *vulgare*
Poaceae
♦ barley
♦ Naturalised, Casual Alien
♦ 86, 198, 280
♦ G, cultivated. Origin: Europe.

Hormuzakia aggregata (Lehm.) Gusul.
Boraginaceae
♦ Weed
♦ 221

Hornungia petraea (L.) Rchb.
Brassicaceae
Astylus petraea Dulac, *Lepidium
petraeum* L., *Teesdalia petraea* Rchb.,
Thlaspi petraeum Moritzi
♦ hutchinsia, rock hutchinsia,
kivikrassi
♦ Weed, Casual Alien
♦ 23, 42, 88
♦ cultivated, herbal.

Horsfordia newberryi (S.Wats.) Gray
Malvaceae
♦ Newberry's velvetmallow
♦ Naturalised
♦ 287
♦ pH, herbal.

Hosta lancifolia (Thunb.) Engl.
Lamiaceae/Verbenaceae
♦ narrowleaf plantain lily
♦ Naturalised
♦ 101
♦ cultivated, herbal.

Hosta montana F.Maek.
Lamiaceae/Verbenaceae
♦ oba giboshi
♦ Weed
♦ 88, 204
♦ pH, cultivated.

Hosta plantaginea (Lam.) Asch.
Lamiaceae/Verbenaceae
♦ fragrant plantain lily
♦ Naturalised
♦ 101
♦ cultivated, herbal.

**Hosta sieboldiana Engl. var. *gigantea*
Kitam.**

Lamiaceae/Verbenaceae
♦ oobagiboushi
♦ Weed
♦ 286

Hosta ventricosa (Salisb.) Stearn
Lamiaceae/Verbenaceae
Bryocles ventricosa Salisb., *Funkia
caerulea* Andrews, *Funkia ovata* Spreng.,
Hemerocallis caerulea Andrews, *Hosta
japonica* (Thunb.) Voss var. *caerulea*
Makino, *Hosta coerulea* Tratt., *Hosta
ovata* Spreng.
♦ blue plantain lily, murasaki giboshi
♦ Naturalised
♦ 101
♦ pH, cultivated, herbal.

Hottonia palustris L.
Primulaceae
Hottonia millefolium Gilib.
♦ water violet
♦ Weed
♦ 23, 87, 88, 272
♦ pH, aqua, cultivated, herbal.

Houstonia caerulea L.
Rubiaceae
♦ azure bluet, bluets
♦ Naturalised
♦ 287
♦ cultivated, herbal.

Houstonia longifolia Gaertn.
Rubiaceae
♦ longleaf summer bluet, slender
leaved bluet
♦ Weed
♦ 161
♦ herbal.

Houstonia micrantha (Shinners) Terrell
Rubiaceae
♦ southern bluet
♦ Weed
♦ 161
♦ herbal.

Houstonia pusilla Schoepf
Rubiaceae
♦ tiny bluet, small bluet
♦ Weed
♦ 161
♦ herbal.

Houttuynia cordata Thunb.
Saururaceae
Gymnotheca chinensis Decne., *Polypara
cochinchinensis* Lour.
♦ vap ca, tsi
♦ Weed, Quarantine Weed,
Naturalised, Environmental Weed
♦ 76, 87, 88, 101, 204, 220, 246, 275, 286,
290, 297
♦ pH, aqua, cultivated, herbal.

**Houttuynia cordata Thunb. cv.
'Chameleon'**
Saururaceae
♦ Sleeper Weed
♦ 225
♦ pH, aqua, cultivated, herbal. Origin:
horticultural.

Hovenia dulcis Thunb.
Rhamnaceae
Hovenia acerba Lindl.

- ◆ Japanese raisintree
- ◆ Weed, Naturalised
- ◆ 22, 54, 86, 88, 98, 101, 203
- ◆ T, cultivated, herbal. Origin: China, Korea, Japan.

Howeia forsteriana (C.Moore & F.Mill.) Becc.
Arecaceae
- ◆ sentrypalm
- ◆ Introduced
- ◆ 230
- ◆ T.

Hoya carnosa (L.f.) R.Br.
Asclepiadaceae/Apocynaceae
- ◆ porcelain flower, wax plant, mata de cera
- ◆ Naturalised, Garden Escape
- ◆ 101, 261
- ◆ cultivated. Origin: southern China.

Hoya schneei Schltr.
Asclepiadaceae/Apocynaceae
- ◆ takituk
- ◆ Introduced
- ◆ 230
- ◆ H, cultivated.

Hoya serpens Hook f.
Asclepiadaceae/Apocynaceae
- ◆ Naturalised
- ◆ 86
- ◆ cultivated. Origin: India.

Hulthemia persica (Gmel.) Bornm.
Rosaceae
- ◆ Weed
- ◆ 87, 88

Humulus japonicus Siebold & Zucc.
Cannabaceae
Humulus scandens auct. non (Lour.) Merr. (see)
- ◆ Japanese hops, hop
- ◆ Weed, Naturalised, Environmental Weed
- ◆ 80, 87, 88, 101, 133, 151, 195, 224, 286
- ◆ pC, aqua, cultivated, herbal.

Humulus lupulus L.
Cannabaceae
Cannabis lupulus Scop., *Humulus americanus* Nutt.
- ◆ hops, common hop, luppolo, wild hop, hop bine
- ◆ Weed, Naturalised, Garden Escape, Environmental Weed, Cultivation Escape
- ◆ 7, 15, 34, 39, 70, 72, 86, 87, 88, 98, 101, 161, 195, 203, 225, 246, 247, 272, 280
- ◆ pC, cultivated, herbal, toxic. Origin: Eurasia.

Humulus lupulus L. var. lupulus
Cannabaceae
- ◆ common hop
- ◆ Naturalised
- ◆ 101

Humulus scandens auct. non (Lour.) Merr.
Cannabaceae
= *Humulus japonicus* Siebold & Zucc.
- ◆ Japanese hope
- ◆ Weed
- ◆ 263, 273, 275, 297

- ◆ pH, herbal.

Hunnemannia fumariifolia Sweet
Papaveraceae
- ◆ Mexican tulip poppy
- ◆ Weed, Naturalised
- ◆ 3, 101, 191
- ◆ a/pH, cultivated.

Hura crepitans L.
Euphorbiaceae
- ◆ sand box tree, ajuapar, monkey pistol, javillo
- ◆ Weed, Sleeper Weed, Quarantine Weed, Naturalised, Environmental Weed
- ◆ 3, 14, 39, 54, 76, 86, 88, 109, 155, 161, 179, 191, 203
- ◆ T, cultivated, herbal, toxic. Origin: Central and South America, West Indies.

Hutchinsia procumbens (L.) Desv.
Brassicaceae
= *Hymenolobus procumbens* (L.) Nutt. ex Schinz & Thell.
- ◆ slenderweed, prostrate hutchinsia
- ◆ Weed
- ◆ 161
- ◆ aH, herbal.

Hyacinthoides hispanica (Mill.) Rothm.
Liliaceae/Hyacinthaceae
Scilla campanulata Aiton, *Scilla hispanica* Mill.
- ◆ Spanish squill, Spanish bluebell, hispanic hyacinthoides
- ◆ Weed, Naturalised
- ◆ 40, 54, 70, 86, 88, 101, 198
- ◆ cultivated, herbal. Origin: Portugal, Spain.

Hyacinthoides non-scripta (L.) Chouard ex Rothm.
Liliaceae/Hyacinthaceae
Endymion non-scriptus (L.) Garcke (see), *Scilla non-scripta* (L.) Hoffmanns. & Link (see)
- ◆ English bluebell, bluebell
- ◆ Weed, Naturalised
- ◆ 39, 54, 70, 86, 88, 101, 161, 280
- ◆ pH, cultivated, toxic. Origin: Europe.

Hyacinthus orientalis L.
Liliaceae/Hyacinthaceae
- ◆ garden hyacinth, hyacinth
- ◆ Weed, Naturalised, Introduced, Casual Alien
- ◆ 39, 101, 154, 161, 228, 247, 280
- ◆ pH, arid, cultivated, herbal, toxic.

Hyalis argentea D.Don
Asteraceae
Plazia argentea (D.Don) Kuntze
- ◆ olivillo
- ◆ Weed, Introduced
- ◆ 228, 295
- ◆ arid.

Hyalis argentea D.Don var. latisquama Cabr.
Asteraceae
- ◆ Weed
- ◆ 237

Hybanthus attenuatus (Humb. & Bonpl. ex Schult.) G.K.Schulze

Violaceae
- ◆ western greenviolet
- ◆ Weed, Quarantine Weed
- ◆ 88, 135, 157, 170, 191, 191, 243
- ◆ aH.

Hybanthus capensis (Thunb.) Engl.
Violaceae
- ◆ Quarantine Weed
- ◆ 220

Hybanthus communis (A.St.-Hil.) Taub.
Violaceae
- ◆ bandeira branca
- ◆ Weed
- ◆ 255
- ◆ S. Origin: South America.

Hybanthus enneaspermus (L.) F.Muell.
Violaceae
- ◆ pink ladies slipper
- ◆ Weed, Native Weed
- ◆ 87, 88, 121, 242
- ◆ pH, arid, cultivated, herbal.

Hybanthus havanensis Jacq.
Violaceae
- ◆ Weed
- ◆ 14

Hybanthus humilis Standl.
Violaceae
- ◆ Weed
- ◆ 87, 88

Hybanthus monopetalus (Schult.) Domin
Violaceae
- ◆ Casual Alien
- ◆ 280
- ◆ cultivated.

Hybanthus parviflorus (L.f.) Baill.
Violaceae
Ionidium glutinosum Vent., *Ionidium parviflorum* Vent.
- ◆ violetilla
- ◆ Weed
- ◆ 87, 88, 121, 295
- ◆ pH. Origin: South America.

Hydrangea arborescens L.
Hydrangeaceae
- ◆ smooth hydrangea, wild hydrangea, seven barks, arborescent hydrangea
- ◆ Weed
- ◆ 8, 39, 87, 88, 161, 218
- ◆ S, cultivated, herbal, toxic.

Hydrangea macrophylla (Thunb.) Ser. in DC.
Hydrangeaceae
Hortensia opuloides Lam., *Hydrangea chungii* Rehder, *Hydrangea hortensia* Siebold, *Hydrangea hortensia* var. *otaksa* A.Gray, *Hydrangea hortensis* Sm., *Hydrangea macrophylla* fo. *otaksa* E.H.Wilson, *Hydrangea macrophylla* var. *otaksa* (E.H.Wilson) L.H.Bailey, *Hydrangea opuloides* var. *hortensia* Dippel, *Hydrangea opuloides* var. *plena* Rehder, *Hydrangea otaksa* Siebold & Zucc., *Viburnum macrophyllum* Thunb.
- ◆ hydrangea, largeleaf hydrangea, lace cap hydrangea, French hydrangea, garden hydrangea
- ◆ Weed, Naturalised, Cultivation Escape

♦ 15, 38, 39, 42, 161, 280
♦ S, cultivated, herbal, toxic.

Hydrangea paniculata Siebold
Hydrangeaceae
♦ panicled hydrangea, hortenzia metlinatá
♦ Weed, Naturalised
♦ 80, 101
♦ S, cultivated, herbal.

Hydrangea quercifolia Bartr.
Hydrangeaceae
♦ oakleaf hydrangea
♦ Weed
♦ 39, 87, 88, 161, 218
♦ cultivated, herbal, toxic.

Hydrastis canadensis L.
Ranunculaceae/Hydrastidaceae
♦ goldenseal, ground raspberry, orange root
♦ Weed
♦ 39, 161, 247
♦ pH, cultivated, herbal, toxic.

Hydrilla verticillata (L.f.) Royle
Hydrocharitaceae
Elodea verticillata (L.f.) F.Muell., *Hottonia serrata* Willd., *Hydrilla angustifolia* Bl., *Hydrilla dentata* Casp., *Hydrilla lithuanica* (Andrz. ex Besser) Dandy, *Hydrilla ovalifolia* Rich., *Hydrilla wrightii* Planch., *Lepanthes verticillatus* Wight., *Serpicula verticillata* L.f., *Vallisneria verticillata* Roxb.
♦ hydrilla, water thyme, Florida elodea, wasserquirl
♦ Weed, Quarantine Weed, Noxious Weed, Naturalised, Native Weed, Garden Escape, Environmental Weed, Cultivation Escape, Casual Alien
♦ 1, 3, 6, 17, 18, 26, 35, 35, 67, 76, 78, 80, 84, 86, 87, 88, 101, 102, 105, 112, 116, 126, 133, 137, 139, 140, 142, 146, 147, 151, 152, 161, 170, 177, 179, 193, 195, 197, 203, 204, 209, 217, 220, 224, 225, 229, 231, 233, 246, 252, 258, 264, 268, 269, 275, 280, 286, 297
♦ wpH, cultivated, herbal. Origin: Africa, Asia, Australia.

Hydrilla verticillata (L.f.) Royle var. roxburghii Casp.
Hydrocharitaceae
♦ 274
♦ wH.

Hydrocera triflora (L.) Wight & Arn.
Balsaminaceae
Impatiens triflora L., *Impatiens natans* Willd., *Impatiens angustifolia* Bl., *Hydrocera anguistifolia* Bl., *Tytonia natans* G.Don.
♦ marsh henna, tien nam
♦ Weed
♦ 88, 239, 262
♦ pH, aqua.

Hydrocharis dubia (Bl.) Backer
Hydrocharitaceae
♦ frogbit
♦ Weed
♦ 87, 88, 204, 209, 275, 286, 297

♦ wH, cultivated, herbal. Origin: Australia.

Hydrocharis morsus-ranae L.
Hydrocharitaceae
♦ European frogbit, kilpukka, frogbit, floating frogbit, petit nénuphar, morène aquatique
♦ Weed, Naturalised, Environmental Weed
♦ 4, 80, 87, 88, 101, 103, 104, 133, 195, 197, 272
♦ wH, cultivated, herbal. Origin: Australia.

Hydrochloa caroliniensis Beauv.
Poaceae
= *Luziola fluitans* (Michx.) Terrell & H.Robins.
♦ southern watergrass
♦ Weed, Quarantine Weed
♦ 87, 88, 218, 258
♦ G.

Hydrocleys nymphoides (Humb. & Bonpl. ex Willd.) Buch.
Limnocharitaceae
Hydrocleys azurea Schult.f. ex Seub., *Hydrocleys commersonii* Rich., *Hydrocleys humboldtii* (Rich.) Endl. (see), *Limnocharis commersonii* (Rich.) Spreng., *Sagittaria ranunculoides* Arráb. ex Vell., *Stratiotes nymphoides* Willd., *Vespuccia humboldtii* (Rich.) Parl.
♦ water poppy
♦ Weed, Quarantine Weed, Naturalised, Environmental Weed
♦ 54, 72, 76, 86, 88, 98, 101, 134, 179, 191, 197, 198, 200, 220, 246, 280
♦ wpH, cultivated. Origin: South America.

Hydrocotyle L. spp.
Apiaceae
♦ pennywort, dollarweed, hydrocotyle
♦ Weed, Quarantine Weed
♦ 161, 249, 258

Hydrocotyle americana L.
Apiaceae
= *Hydrocotyle sibthorpioides* Lam.
♦ marshpennywort, American marshpennywort, navelwort, water pennywort
♦ Weed
♦ 87, 88, 121, 158, 218
♦ pH, aqua, cultivated, herbal. Origin: obscure.

Hydrocotyle batrachium Hance
Apiaceae
♦ Weed
♦ 273

Hydrocotyle bonariensis Comm. ex Lam.
Apiaceae
Hydrocotyle petiolaris DC., *Hydrocotyle polystachya* A.Rich. var. *quinqueradiata* Thouars ex A.Rich.
♦ pennywort, Kurnell curse, largeleaf pennywort, paragüita
♦ Weed, Naturalised, Environmental Weed
♦ 3, 70, 72, 86, 88, 98, 155, 198, 203, 236, 237, 249, 255, 269, 290, 295, 296

♦ pH, arid, cultivated, herbal. Origin: South America.

Hydrocotyle bowlesioides Mathias & Constance
Apiaceae
♦ largeleaf marshpennywort
♦ Naturalised
♦ 101

Hydrocotyle brasiliense Scheidw. ex Otto & F.Dietr.
Apiaceae
♦ Naturalised
♦ 198

Hydrocotyle exigua (Urb.) Malme
Apiaceae
♦ Weed
♦ 87, 88

Hydrocotyle formosana Masam.
Apiaceae
♦ Weed
♦ 87, 88

Hydrocotyle heteromeria A.Rich.
Apiaceae
♦ hydrocotyle
♦ Weed
♦ 243
♦ H. Origin: Australia.

Hydrocotyle japonica Makino
Apiaceae
♦ Weed
♦ 87, 88

Hydrocotyle laxiflora DC.
Apiaceae
♦ Weed
♦ 87, 88
♦ cultivated. Origin: Australia.

Hydrocotyle leucocephala Cham. & Schlecht.
Apiaceae
♦ Brazilian pennywort
♦ Weed
♦ 87, 88, 237, 295
♦ aqua, cultivated.

Hydrocotyle maritima Honda
Apiaceae
♦ nichidome
♦ Weed
♦ 87, 88, 286

Hydrocotyle mexicana Cham. & Schlecht.
Apiaceae
♦ Weed
♦ 87, 88

Hydrocotyle moschata G.Forst.
Apiaceae
♦ musky marshpennywort, hydrocotyle, hairy pennywort
♦ Weed, Naturalised, Casual Alien
♦ 40, 101, 165

Hydrocotyle nepalensis Hook.
Apiaceae
♦ Weed
♦ 235
♦ cultivated.

Hydrocotyle novae-zeelandiae DC.
Apiaceae
♦ New Zealand pennywort

- ♦ Weed, Naturalised
- ♦ 40, 87, 88

Hydrocotyle poeppigi DC.
Apiaceae
- ♦ Weed
- ♦ 87, 88

Hydrocotyle ramiflora Maxim.
Apiaceae
- ♦ oochidome
- ♦ Weed
- ♦ 87, 88, 204, 286

Hydrocotyle ranunculoides L.f.
Apiaceae
- ♦ water pennywort, hydrocotyle, floating marshpennywort
- ♦ Weed, Quarantine Weed, Noxious Weed, Naturalised, Garden Escape, Environmental Weed
- ♦ 3, 7, 54, 62, 76, 86, 87, 88, 98, 155, 171, 200, 203
- ♦ wpH, cultivated, herbal. Origin: North to South America.

Hydrocotyle sibthorpioides Lam.
Apiaceae
Hydrocotyle americana L. (see), *Hydrocotyle americana* var. *monticola* (Hook.f.) Hiern, *Hydrocotyle confusa* H.Wolff, *Hydrocotyle monticola* Hook.f., *Hydrocotyle rotundifolia* Roxb.
- ♦ lawn pennywort, lawn marshpennywort
- ♦ Weed, Naturalised
- ♦ 87, 88, 101, 170, 191, 218, 273, 286
- ♦ wpH, cultivated, herbal. Origin: Africa, Asia, Australia.

Hydrocotyle taxiflora DC.
Apiaceae
- ♦ sticking pennywort, shitweed
- ♦ Native Weed
- ♦ 269
- ♦ pH. Origin: Australia.

Hydrocotyle tripartita R.Br. ex A.Rich.
Apiaceae
- ♦ Australian hydrocotyle
- ♦ Weed, Naturalised, Native Weed
- ♦ 87, 88, 165, 269, 280
- ◊ Origin: easten Australia.

Hydrocotyle umbellata L.
Apiaceae
- ♦ water pennywort, manyflower marshpennywort
- ♦ Weed, Quarantine Weed
- ♦ 39, 87, 88, 161, 218, 249, 258
- ♦ wpH, herbal, toxic.

Hydrocotyle verticillata Thunb.
Apiaceae
- ♦ whorled pennywort, hydrocotyl, whorled marshpennywort, umbrella plant, shield pennywort, whorled umbrella
- ♦ Weed, Quarantine Weed, Noxious Weed, Naturalised, Environmental Weed
- ♦ 62, 76, 87, 88, 218, 246, 249, 287
- ♦ wpH, cultivated, herbal. Origin: Australia.

Hydrocotyle vulgaris L.
Apiaceae

- ♦ marshpennywort, ecuelle d'eau, pennywort, umbrella plant, common pennywort
- ♦ Naturalised
- ♦ 39, 287
- ♦ pH, aqua, cultivated, herbal, toxic. Origin: Europe.

Hydrodictyon reticulatum Lagerh.
Hydrodictyaceae
- ♦ water net
- ♦ Weed, Environmental Weed
- ♦ 87, 88, 204, 225, 246, 286, 297
- ♦ algae.

Hydrolea glabra Schum. & Thonn.
Hydrophyllaceae
Hydrolea guineensis Choi
- ♦ Weed, Quarantine Weed
- ♦ 76, 87, 88, 203, 220

Hydrolea graminifolia A.W.Benn.
Hydrophyllaceae
- ♦ Weed, Quarantine Weed
- ♦ 76, 87, 88, 203, 220

Hydrolea macrosepala A.W.Benn.
Hydrophyllaceae
- ♦ Weed
- ♦ 87, 88

Hydrolea quadrivalvis Walter
Hydrophyllaceae
- ♦ waterpod
- ♦ Weed
- ♦ 87, 88

Hydrolea spinosa L.
Hydrophyllaceae
- ♦ spiny false fiddleleaf
- ♦ Weed
- ♦ 13, 14, 87, 88, 170, 179, 191
- ♦ herbal.

Hydrolea uniflora Raf.
Hydrophyllaceae
- ♦ oneflower hydrolea, oneflower false fiddleleaf
- ♦ Weed
- ♦ 87, 88, 218
- ♦ herbal.

Hydrolea zeylanica (L.) Vahl
Hydrophyllaceae
Beloanthera oppositifolia Hassk., *Hydrolea inermis* Lour., *Hydrolea javanica* Blume
- ♦ Dee plaa lai
- ♦ Weed
- ♦ 13, 87, 88, 170, 191, 238
- ♦ herbal. Origin: Asia, Australia.

Hydrophyllum capitatum Douglas ex Benth.
Hydrophyllaceae
- ♦ ballhead waterleaf, cat's breeches
- ♦ Weed
- ♦ 23, 88
- ♦ pH, promoted, herbal.

Hydrostemma kunstleri (Ridl.) B.C.Stone
Nymphaeaceae/Barclayaceae
= *Barclaya kunstleri* (King) Ridl.
- ♦ Weed, Quarantine Weed
- ♦ 76, 88
- ♦ wH.

Hydrostemma motleyi (Hook.f.) Mabb.
Nymphaeaceae/Barclayaceae
= *Barclaya motleyi* Hook.f.
- ♦ Weed, Quarantine Weed
- ♦ 76, 88
- ♦ wH.

Hydrothrix Hook.f. spp.
Pontederiaceae
- ♦ Weed, Quarantine Weed
- ♦ 76, 88, 220

Hygrophila R.Br. spp.
Acanthaceae
- ♦ swampweed
- ♦ Quarantine Weed
- ♦ 220

Hygrophila angustifolia R.Br.
Acanthaceae
= *Hygrophila salicifolia* (Vahl) Nees (NoR)
- ♦ Weed
- ♦ 87, 88
- ♦ aqua, cultivated, herbal.

Hygrophila aristata Nees
Acanthaceae
- ♦ Weed
- ♦ 87, 88

Hygrophila auriculata (Schum.) Heine
Acanthaceae
- ♦ Weed
- ♦ 66
- ♦ cultivated, herbal.

Hygrophila chevalieri Benoist
Acanthaceae
- ♦ Weed
- ♦ 87, 88

Hygrophila corymbosa (Blume) Lindau
Acanthaceae
- ♦ starhorn, temple plant, giant hygro
- ♦ Weed, Naturalised
- ♦ 101, 179
- ♦ aqua, cultivated.

Hygrophila costata Nees
Acanthaceae
Hygrophila atricheta Bridar., *Hygrophila brasiliensis* (Spreng.) Lindau, *Hygrophila conferta* Nees, *Hygrophila guianensis* Nees, *Hygrophila lacustris* Morong & Britton, *Hygrophila longifolia* (Mart.) Nees, *Hygrophila pubescens* Nees, *Hygrophila rivularis* (Schltdl.) Nees, *Hygrophila verticillata* (Spreng.) Cabrera & G.Dawson, *Ruellia brasiliensis* Spreng., *Ruellia rivularis* Schltdl., *Ruellia verticillata* Spreng.
- ♦ yerba de hicotea, hygrophila
- ♦ Noxious Weed, Naturalised
- ♦ 86, 229
- ♦ Origin: Central and South America.

Hygrophila difformis (L.f.) Blume
Acanthaceae
- ♦ water wisteria, Asian wisteria
- ♦ Weed, Sleeper Weed, Naturalised, Environmental Weed
- ♦ 86, 155, 191
- ♦ aqua, cultivated. Origin: India, Malayan Peninsula.

Hygrophila erecta (Burm.f.) Hochr.
Acanthaceae
- Weed
- 13, 32, 209, 262
- aH, aqua.

Hygrophila phlomoides Nees
Acanthaceae
- Weed, Quarantine Weed
- 76, 87, 88, 203
- herbal.

Hygrophila pobeguini Benoist
Acanthaceae
- Weed, Quarantine Weed
- 76, 87, 88, 203

Hygrophila polysperma (Roxb.) T.Anderson
Acanthaceae
- green hygro, Indian swampweed, Miramar weed, hygro, dwarf hygrophila
- Weed, Quarantine Weed, Noxious Weed, Naturalised, Environmental Weed
- 67, 80, 84, 88, 101, 112, 140, 151, 161, 179, 193, 197, 229, 246
- aqua, cultivated.

Hygrophila quadrivalvis Nees
Acanthaceae
- Weed
- 87, 88
- herbal.

Hygrophila salicifolia (Vahl) Nees
Acanthaceae
- willowleaf hygro, willowleaf
- Weed
- 286
- aqua, cultivated, herbal. Origin: Asia.

Hygrophila serpyllum (Nees) T.Anderson
Acanthaceae
- Weed
- 66

Hygrophila spinosa T.Anderson
Acanthaceae
Asteracantha longifolia (L.) Nees
- Weed, Quarantine Weed
- 76, 87, 88, 203
- herbal.

Hygrophila triflora (Roxb.) Fosberg & Sachet
Acanthaceae
- hygrophila
- Weed, Naturalised
- 86, 93, 191
- aqua. Origin: India.

Hygroryza aristata (Retz.) Nees
Poaceae
- Weed
- 87, 88
- G, herbal.

Hylocereus (Berger) Britt. & Rose spp.
Cactaceae
- night blooming cactus
- Weed, Quarantine Weed
- 76, 88, 203

Hylocereus costaricensis (Weber) Britt. & Rose
Cactaceae
- Costa Rica night blooming cactus
- Naturalised
- 101

Hylocereus undatus (Haw.) Britton & Rose
Cactaceae
Cereus triangularis Haw. var. *major* DC., *Cereus triangularis* var. *aphyllus* Jacq., *Cereus tricostatus* Gosselin, *Cereus trigonus* Haw. var. *guatemalensis* Eichlam, *Cereus undatus* Haw. (see)
- night blooming cereus, Pa nani o ka, dragon fruit, queen of the night, night blooming cactus
- Weed, Naturalised, Introduced, Environmental Weed
- 73, 86, 88, 98, 145, 179, 203, 228, 261
- pC, arid, cultivated, herbal. Origin: tropical America.

Hylotelephium erythrostictum (Miq.) H.Ohba
Crassulaceae
- garden stonecrop
- Naturalised
- 101

Hylotelephium spectabile (Boreau) H.Ohba
Crassulaceae
Sedum spectabile Boreau. (see)
- showy stonecrop
- Naturalised
- 101
- cultivated.

Hylotelephium telephium (L.) H.Ohba
Crassulaceae
- orphine, witch's moneybags, stonecrop, liveforever
- Quarantine Weed, Naturalised
- 76, 101
- cultivated.

Hylotelephium telephium (L.) H.Ohba ssp. *fabaria* (W.D.J.Koch) H.Ohba
Crassulaceae
Sedum telephium L. ssp. *fabaria* (W.D.J.Koch) Schinz & R.Keller
- witch's moneybags
- Naturalised
- 101
- cultivated.

Hylotelephium telephium (L.) H.Ohba ssp. *maximum* (L.) H.Ohba
Crassulaceae
Sedum maximum (L.) Suter
- Weed
- 243

Hylotelephium telephium (L.) H.Ohba ssp. *telephium*
Crassulaceae
Sedum purpureum (L.) J.A.Schult. (see), *Sedum telephium* L. (see)
- witch's moneybags
- Naturalised
- 101

Hymenachne acutigluma (Steud.) Gill.
Poaceae
= *Hymenachne amplexicaulis* (Rudge) Nees
- hymenachne
- Weed
- 88, 170, 262
- pG, aqua, cultivated. Origin: Asia, Australia.

Hymenachne amplexicaulis (Rudge) Nees
Poaceae
Agrostis monostachya Poir., *Hymenachne acutigluma* (Steud.) Gill. (see), *Hymenachne myosurus* (Rich.) Nees, *Panicum acutiglumum* Steud., *Panicum amplexicaule* Rudge (see), *Panicum amplexicaule* var. *deflexa* Döll, *Panicum amplexicaule* var. *erecta* Döll, *Panicum hymenachne* Desv., *Panicum perdensum* Steud.
- West Indian marsh grass, water stargrass, olive hymenachne, West Indian grass, trompetilla
- Weed, Quarantine Weed, Naturalised, Environmental Weed
- 3, 14, 54, 76, 80, 86, 87, 88, 93, 112, 151, 155, 179, 191, 197, 203, 220, 255
- pG, aqua, cultivated. Origin: South and Central tropical America.

Hymenachne pseudointerrupta C.Muell.
Poaceae
- yaa plong
- Weed
- 88, 191, 204, 239
- pG, aqua.

Hymenaea courbaril L.
Fabaceae/Caesalpiniaceae
- copal, stinking toe
- Quarantine Weed
- 39, 220
- cultivated, herbal, toxic. Origin: Central America.

Hymenaea verrucosa Gaertn.
Fabaceae/Caesalpiniaceae
Trachylobium verrucosum (Gaertn.) Oliv.
- East African copal
- Introduced
- 228
- arid. Origin: Madagascar.

Hymenocallis Salisb. spp.
Liliaceae/Amaryllidaceae
- spiderlily
- Weed
- 161, 247
- toxic.

Hymenocallis caribaea (L.) Herb.
Liliaceae/Amaryllidaceae
- Caribbean spiderlily, spiderlily
- Weed, Naturalised, Garden Escape, Environmental Weed, Cultivation Escape
- 32, 54, 86, 88, 98, 155, 203
- cultivated, herbal. Origin: West Indies.

Hymenocallis caroliniana (L.) Herb.
Liliaceae/Amaryllidaceae
- Carolina spiderlily, spiderlily
- Weed
- 161
- aqua, cultivated, herbal, toxic.

Hymenocallis liriosme (Raf.) Shinners
Liliaceae/Amaryllidaceae

♦ spring spiderlily, Texan spiderlily
♦ Weed
♦ 161
♦ toxic.

Hymenocallis littoralis (Jacq.) Salisb.
Liliaceae/Amaryllidaceae
♦ beach spiderlily, American spiderlily, spiderlily, kiup
♦ Weed, Naturalised, Introduced, Cultivation Escape
♦ 32, 101, 161, 230
♦ H, cultivated, herbal, toxic.

Hymenocardia acida Tul.
Euphorbiaceae/Hymenocardiaceae
♦ kampalaga
♦ Weed
♦ 87, 88
♦ herbal.

Hymenocarpos circinnata Savi
Fabaceae/Papilionaceae
♦ Weed
♦ 87, 88

Hymenoclea monogyra Torr. & Gray
Asteraceae
♦ burro brush, leafy burrobush, singlewhorl burrobush
♦ Weed
♦ 88, 218
♦ S, arid, herbal.

Hymenolobus procumbens (L.) Nutt. ex Schinz & Thell.
Brassicaceae
Hutchinsia procumbens (L.) Desv. (see)
♦ oval purse
♦ Weed, Naturalised, Garden Escape, Environmental Weed
♦ 7, 86, 98, 176, 203, 280
♦ cultivated, herbal. Origin: temperate northern hemisphere.

Hymenomonas roseola F.Stein.
Hymenomonadaceae
♦ coccolithophorid, stonewort, brown algae
♦ Weed
♦ 197
♦ algae.

Hymenophysa pubescens C.A.Mey.
Brassicaceae
= *Cardaria pubescens* (C.A.Mey.) Jarm.
♦ Weed
♦ 295

Hymenoxys anthemoides (Juss.) Cass.
Asteraceae
♦ South American rubberweed
♦ Naturalised
♦ 101

Hymenoxys chrysathemoides (H.B.K.) DC.
Asteraceae
♦ Weed
♦ 199

Hymenoxys lemmoni (Greene) Cockerell
Asteraceae
♦ goldfields
♦ Weed
♦ 39, 161
♦ arid, toxic.

Hymenoxys odorata DC.
Asteraceae
Hymenoxys chrysanthemoides DC. var. *excurrens* Cockerell, *Hymenoxys davidsonii* (E.Greene) Cockerell
♦ bitter rubberweed
♦ Weed
♦ 39, 87, 88, 161, 218
♦ aH, arid, herbal, toxic.

Hymenoxys richardsonii (Hook.) Cock.
Asteraceae
Actinea richardsonii (Hook.) Kuntze, *Hymenoxys floribunda* (A.Gray) Cockerell, *Picradenia richardsonii* Hook.
♦ pingue, Colorado rubberweed, pingue hymenoxys, pingue rubberweed
♦ Weed
♦ 39, 87, 88, 161
♦ pH, promoted, herbal, toxic.

Hymenoxys richardsonii (Hook.) Cockl. var. *floribunda* (Gray) Parker
Asteraceae
♦ pingue, Colorado rubberweed
♦ Weed
♦ 218
♦ herbal.

Hymenoxys tweediei Hook. & Arn.
Asteraceae
♦ Weed
♦ 87, 88

Hyoscyamus L. spp.
Solanaceae
♦ henbane
♦ Weed
♦ 221, 272

Hyoscyamus agrestis Kit.
Solanaceae
♦ Weed
♦ 23, 88

Hyoscyamus albus L.
Solanaceae
♦ white henbane, Russian henbane, capseta, gangeet, shirraband, zairiya
♦ Weed, Naturalised, Introduced
♦ 39, 70, 86, 88, 94, 98, 101, 176, 203, 215, 221, 228, 272
♦ a/bH, arid, cultivated, herbal, toxic. Origin: Mediterranean.

Hyoscyamus bohemicus Schmidt.
Solanaceae
♦ Weed
♦ 275, 297
♦ aH.

Hyoscyamus desertorum (Asch. ex Boiss.) V.Tackh.
Solanaceae
♦ Weed
♦ 221

Hyoscyamus muticus L.
Solanaceae
♦ Egyptian henbane, bange, imbiras, sakaran, sekkoran, henbane
♦ Weed, Introduced
♦ 23, 39, 88, 221, 228
♦ arid, herbal, toxic.

Hyoscyamus niger L.
Solanaceae
♦ black henbane, black woody henbane, henbane
♦ Weed, Noxious Weed, Naturalised
♦ 1, 23, 35, 39, 44, 70, 80, 86, 87, 88, 94, 98, 101, 136, 146, 154, 161, 203, 212, 218, 219, 229, 243, 247, 264, 272, 280, 297, 299
♦ a/bH, arid, cultivated, herbal, toxic. Origin: Europe.

Hyoscyamus pusillus L.
Solanaceae
♦ Weed
♦ 221
♦ herbal.

Hyoscyamus reticulatus L.
Solanaceae
♦ turkinhullukaali
♦ Weed, Casual Alien
♦ 39, 42, 87, 88, 221, 243
♦ herbal, toxic.

Hyparrhenia anamesa Clayton
Poaceae
♦ Native Weed
♦ 121
♦ pG. Origin: southern Africa.

Hyparrhenia dregeana (Nees) Stapf
Poaceae
Hyparrhenia pilosissima (Hack.) J.G.Anders.
♦ Drege's deckgrass, tambuki grass, silky thatching grass
♦ Weed, Naturalised, Native Weed
♦ 101, 121
♦ pG. Origin: southern Africa.

Hyparrhenia filipendula (Hochst.) Stapf
Poaceae
♦ bluegrass, fine thatching grass, tambookie, thatching grass
♦ Weed
♦ 121
♦ pG, cultivated. Origin: Africa, Asia, Australia, Madagascar.

Hyparrhenia gazensis (Rendle) Stapf
Poaceae
♦ Weed, Quarantine Weed, Native Weed
♦ 76, 87, 88, 121, 203, 220
♦ pG. Origin: southern Africa.

Hyparrhenia hirta (L.) Stapf
Poaceae
Andropogon hirtus L.
♦ tambookie grass, thatching grass, bluegrass, common thatching grass, thatch grass, coolati grass
♦ Weed, Naturalised, Native Weed, Introduced, Environmental Weed
♦ 7, 38, 86, 87, 88, 98, 101, 121, 158, 198, 203, 221, 228, 272, 296
♦ pG, arid, cultivated. Origin: Africa.

Hyparrhenia involucrata Stapf
Poaceae
♦ Weed
♦ 88
♦ G.

Hyparrhenia pilgerana **C.E.Hubb.**
Poaceae
♦ Native Weed
♦ 121
♦ pG. Origin: southern Africa.

Hyparrhenia quarrei **Robyns**
Poaceae
♦ thatching grass
♦ Weed, Naturalised, Native Weed
♦ 98, 121, 203
♦ pG. Origin: southern Africa.

Hyparrhenia rufa **(Nees) Stapf**
Poaceae
Andropogon rufus (Nees) Kunth,
Cymbopogon rufus (Nees) Rendle,
Trachypogon rufus Nees
♦ thatch grass, nkuku, jaragua grass,
jaragua
♦ Weed, Naturalised, Environmental
Weed, Cultivation Escape
♦ 3, 80, 87, 88, 101, 152, 157, 179, 191,
218, 255, 261
♦ pG, cultivated, herbal. Origin:
tropical Africa.

Hyparrhenia rufa **(Nees) Stapf ssp.**
altissima **(Stapf) B.K.Simon**
Poaceae
♦ Naturalised
♦ 86
♦ G. Origin: Africa.

Hyparrhenia rufa **(Nees) Stapf ssp.** *rufa*
Poaceae
♦ Naturalised
♦ 86
♦ G. Origin: Africa.

Hyparrhenia tamba **(Steud.) Stapf**
Poaceae
♦ blue thatching grass
♦ Weed
♦ 88, 158
♦ pG. Origin: southern Africa.

Hypecoum aegyptiacum **(Forssk.) Asch.**
& Schweinf.
Papaveraceae/Fumariaceae
♦ Weed
♦ 221

Hypecoum deuteroparviflorum **Fedde**
Papaveraceae/Fumariaceae
♦ Weed
♦ 221

Hypecoum erectum **L.**
Papaveraceae/Fumariaceae
♦ Weed
♦ 275, 297
♦ a/bH.

Hypecoum grandiflorum **Benth.**
Papaveraceae/Fumariaceae
= *Hypecoum imberbe* Sibth. & Sm.
♦ Persian poppy
♦ Weed, Quarantine Weed
♦ 76, 87, 88, 203, 220

Hypecoum imberbe **Sibth. & Sm.**
Papaveraceae/Fumariaceae
Hypecoum grandiflorum Benth. (see)
♦ Persian poppy, sicklefruit hypecoum
♦ Weed, Naturalised
♦ 101, 243, 272

♦ cultivated.

Hypecoum leptocarpum **Hook.f. &**
Thomson
Papaveraceae/Fumariaceae
♦ thinfruit hypecoum
♦ Weed
♦ 297

Hypecoum pendulum **L.**
Papaveraceae/Fumariaceae
♦ Persian poppy, nodding hypecoum,
nuokkuliuskio
♦ Weed, Naturalised, Casual Alien
♦ 42, 86, 88, 94, 98, 101, 198, 203, 221,
243, 248, 272
♦ Origin: Middle East.

Hypecoum procumbens **L.**
Papaveraceae/Fumariaceae
♦ Persian poppy, rentoliuskio
♦ Weed, Casual Alien
♦ 39, 42, 87, 88, 94, 243
♦ cultivated, toxic.

Hypecoum trilobum **Trautv.**
Papaveraceae/Fumariaceae
♦ Weed
♦ 87, 88

Hypericum **L. spp.**
Clusiaceae/Hypericaceae
♦ St. John's wort
♦ Quarantine Weed
♦ 220
♦ herbal.

Hypericum aethiopicum **Thunb.**
Clusiaceae/Hypericaceae
♦ St. John's wort
♦ Weed, Native Weed
♦ 39, 121
♦ pH, cultivated, herbal, toxic. Origin:
southern Africa.

Hypericum androsaemum **L.**
Clusiaceae/Hypericaceae
Androsaemum officinale All.,
Androsaemum vulgare Gaertn.
♦ tutsan, sweet amber
♦ Weed, Quarantine Weed, Noxious
Weed, Naturalised, Garden Escape,
Environmental Weed, Cultivation
Escape
♦ 3, 7, 15, 20, 62, 72, 76, 79, 80, 86, 87,
88, 98, 101, 147, 155, 165, 176, 198, 203,
220, 225, 241, 246, 269, 280, 289, 296,
300
♦ S, cultivated, herbal, toxic. Origin:
southern and western Europe, North
Africa and Middle East.

Hypericum androsaemum **L. × inodorum**
Mill.
Clusiaceae/Hypericaceae
♦ Quarantine Weed
♦ 220

Hypericum angustifolium **Lam.**
Clusiaceae/Hypericaceae
♦ Weed
♦ 87, 88

Hypericum ascyron **L.**
Clusiaceae/Hypericaceae
♦ giant St. John's wort, tomoesou
♦ Weed

♦ 297
♦ pH, cultivated, herbal, toxic.

Hypericum brasiliense **Choisy**
Clusiaceae/Hypericaceae
♦ Weed
♦ 87, 88

Hypericum calycinum **L.**
Clusiaceae/Hypericaceae
♦ rose of Sharon, large flowered St.
John's wort, Aaron's beard, creeping
St. John's wort
♦ Weed, Quarantine Weed,
Naturalised, Garden Escape,
Environmental Weed
♦ 40, 70, 72, 86, 88, 101, 116, 176, 198,
220, 280
♦ pS, cultivated, herbal. Origin: east
Europe.

Hypericum canadense **L.**
Clusiaceae/Hypericaceae
♦ lesser Canadian St. John's wort, Irish
St. John's wort
♦ Quarantine Weed
♦ 220
♦ herbal.

Hypericum canariense **L.**
Clusiaceae/Hypericaceae
♦ Canary Island St. John's wort
♦ Introduced
♦ 34
♦ S, cultivated, herbal.

Hypericum chinense **L. var.** *salicifolium*
Y.Kimura
Clusiaceae/Hypericaceae
♦ Naturalised
♦ 287

Hypericum concinnum **Benth.**
Clusiaceae/Hypericaceae
♦ gold wire
♦ Weed
♦ 161
♦ pH, herbal, toxic.

Hypericum coris **L.**
Clusiaceae/Hypericaceae
♦ St. John's wort, Aaron's beard
♦ Weed
♦ 116
♦ cultivated.

Hypericum crispum **L. nom. illeg.**
Clusiaceae/Hypericaceae
= *Hypericum triquetrifolium* Turra
♦ Weed, Quarantine Weed
♦ 39, 87, 88, 220
♦ herbal, toxic.

Hypericum elodes **Huds.**
Clusiaceae/Hypericaceae
♦ marsh St. John's wort, bog St. John's
wort, marsh hypericum
♦ Weed, Naturalised, Garden Escape,
Environmental Weed
♦ 86, 87, 88
♦ aqua, cultivated. Origin: Europe.

Hypericum erectum **Thunb.**
Clusiaceae/Hypericaceae
♦ otogirisou
♦ Weed, Quarantine Weed
♦ 23, 87, 88, 220, 286

♦ pH, cultivated, herbal. Origin: east Asia.

Hypericum fragile Heldr. & Sartori ex Boiss.
Clusiaceae/Hypericaceae
♦ dwarf St. John's wort
♦ Quarantine Weed
♦ 220
♦ cultivated, herbal.

Hypericum frondosum Michx.
Clusiaceae/Hypericaceae
♦ cedarglade St. John's wort
♦ Quarantine Weed
♦ 220
♦ cultivated, herbal.

Hypericum gramineum G.Forst.
Clusiaceae/Hypericaceae
♦ grassy St. John's wort
♦ Naturalised
♦ 39, 101
♦ cultivated, toxic.

Hypericum grandifolium Choisy
Clusiaceae/Hypericaceae
♦ Canary Island St. John's wort
♦ Weed, Naturalised
♦ 54, 86, 88, 198
♦ cultivated. Origin: Canary Islands.

Hypericum henryi H.Lév. & Vaniot
Clusiaceae/Hypericaceae
♦ Naturalised
♦ 280
♦ cultivated.

Hypericum henryi H.Lév. & Vaniot ssp. henryi
Clusiaceae/Hypericaceae
♦ Naturalised
♦ 280
♦ cultivated.

Hypericum hircinum L.
Clusiaceae/Hypericaceae
Androsaemum foetidum Thal,
Androsaemum hircinum (L.) Spach,
Hypericum cambessedesii Coss. ex Marès & Vigin.
♦ stinking tutsan
♦ Naturalised
♦ 241
♦ cultivated.

Hypericum hircinum L. ssp. majus (Aiton) N.K.B.Robson
Clusiaceae/Hypericaceae
♦ stinking tutsan
♦ Naturalised
♦ 40

Hypericum hirsutum L.
Clusiaceae/Hypericaceae
Hypericum villosum Crantz
♦ hairy St. John's wort, karvakuisma
♦ Weed
♦ 272
♦ cultivated, herbal.

Hypericum humifusum L.
Clusiaceae/Hypericaceae
♦ trailing St. John's wort, matalakuisma, creeping St. John's wort
♦ Weed, Naturalised, Casual Alien
♦ 39, 42, 44, 86, 88, 94, 101, 198, 243,

253, 272, 280
♦ b/pH, cultivated, herbal, toxic. Origin: Europe.

Hypericum inodorum Mill.
Clusiaceae/Hypericaceae
= *Hypericum × inodorum* Mill.
♦ Quarantine Weed
♦ 220
♦ cultivated.

Hypericum × inodorum Mill.
Clusiaceae/Hypericaceae
= *Hypericum hircinum* L. × *Hypericum androsaemum* L. [see *Hypericum inodorum* Mill.]
♦ tall St. John's wort, tall tutsan, tutsan
♦ Weed, Sleeper Weed, Quarantine Weed, Naturalised, Noxious Weed, Garden Escape
♦ 62, 76, 86, 88, 198, 220, 280, 300
♦ cultivated. Origin: southern Europe.

Hypericum japonicum Thunb.
Clusiaceae/Hypericaceae
Brathys japonica (Thunb.) Wight,
Brathys japonica var. *acutisepala* Miq.,
Brathys laxa Blume, *Hypericum cavaleriei* H.Lév., *Hypericum chinense* Osbeck,
Hypericum japonicum var. *calyculatum* R.Keller, *Hypericum japonicum* var.
cavaleriei (H.Lév.) Koidz., *Hypericum japonicum* var. *maximowiczii* R.Keller,
Hypericum japonicum var. *thunbergii* (Franch. & Sav.) R.Keller, *Hypericum japonicum* var. *typicum* Hochr.,
Hypericum laxum (Bl.) Koidz. (see),
Hypericum nervatum Hance, *Hypericum thunbergii* Franch. & Sav., *Sarothra japonica* (Thunb.) Y.Kimura, *Sarothra laxa* (Bl.) Y.Kimura
♦ Japanese St. John's wort, matted St. John's wort
♦ Weed
♦ 87, 88, 170, 235, 274, 275, 286, 297
♦ a/pH, aqua, cultivated, herbal. Origin: Asia, Australasia.

Hypericum kalmianum L.
Clusiaceae/Hypericaceae
♦ Kalm's St. John's wort
♦ Quarantine Weed
♦ 220
♦ cultivated, herbal.

Hypericum kouytchense H.Lév.
Clusiaceae/Hypericaceae
♦ Naturalised
♦ 280
♦ cultivated. Origin: China.

Hypericum laxum (Blume) Koidz.
Clusiaceae/Hypericaceae
= *Hypericum japonicum* Thunb.
♦ Weed
♦ 87, 88, 286

Hypericum leschenaultii Choisy
Clusiaceae/Hypericaceae
♦ Quarantine Weed
♦ 220
♦ cultivated.

Hypericum linarifolium Vahl
Clusiaceae/Hypericaceae
♦ flax leaved St. John's wort, narrow

leaved St. John's wort, slender St. John's wort, toadflax leaved St. John's wort
♦ Weed, Naturalised
♦ 23, 70, 88, 280
♦ cultivated.

Hypericum maculatum Cran.
Clusiaceae/Hypericaceae
Hypericum dubium Leers *p.p.*,
Hypericum fallax Grimm
♦ imperforate St. John's wort, särmäkuisma
♦ Weed
♦ 39, 70, 87, 88, 272
♦ cultivated, herbal, toxic.

Hypericum majus (Gray) Britt.
Clusiaceae/Hypericaceae
♦ large St. John's wort
♦ Naturalised
♦ 287
♦ herbal.

Hypericum montanum L.
Clusiaceae/Hypericaceae
♦ mountain St. John's wort, pale St. John's wort, vuorikuisma
♦ Naturalised
♦ 280
♦ cultivated, herbal.

Hypericum montbretii Spach
Clusiaceae/Hypericaceae
♦ Weed
♦ 87, 88

Hypericum × moserianum André
Clusiaceae/Hypericaceae
= *Hypericum calycinum* L. × *Hypericum patulum* Thunb.
♦ St. John's wort
♦ Naturalised, Garden Escape
♦ 86, 269
♦ cultivated. Origin: Europe.

Hypericum mutilum L.
Clusiaceae/Hypericaceae
♦ dwarf St. John's wort
♦ Naturalised
♦ 280
♦ pH, aqua, herbal. Origin: North America.

Hypericum mutilum L. var. mutilum
Clusiaceae/Hypericaceae
♦ Introduced
♦ 38
♦ aqua.

Hypericum nitidum Lam.
Clusiaceae/Hypericaceae
♦ Carolina St. John's wort
♦ Weed
♦ 14

Hypericum oliganthum Fr. & Sav.
Clusiaceae/Hypericaceae
♦ Weed
♦ 286

Hypericum olympicum L.
Clusiaceae/Hypericaceae
♦ Quarantine Weed
♦ 220
♦ cultivated, herbal.

Hypericum parvulum Greene
Clusiaceae/Hypericaceae
♦ Sierra Madre St. John's wort
♦ Naturalised
♦ 101

Hypericum patulum Thunb. ex Murray
Clusiaceae/Hypericaceae
Komana patula (Thunb.) Y.Kimura ex Honda, *Norysca patula* (Thunb.) J.Voigt
♦ Quarantine Weed, Naturalised
♦ 220, 287
♦ S, cultivated, herbal. Origin: China, Japan, Himalayas.

Hypericum perfoliatum L.
Clusiaceae/Hypericaceae
♦ Weed
♦ 70
♦ cultivated.

Hypericum perforatum L.
Clusiaceae/Hypericaceae
Hypericum officinale Gater. ex Steud., *Hypericum vulgare* Bauhin
♦ common St. John's wort, goatweed, perforate St. John's Wort, St. John's wort, iperico, tipton weed, gammock, goat's beard, goatweed, herb John, Klamath weed, penny John, rosin rose, St. John's grass, tipton weed, touch and heal
♦ Weed, Quarantine Weed, Noxious Weed, Naturalised, Introduced, Garden Escape, Environmental Weed, Cultivation Escape
♦ 1, 4, 7, 15, 18, 20, 21, 21, 23, 34, 35, 39, 44, 45, 51, 52, 62, 63, 70, 72, 76, 80, 86, 87, 88, 94, 95, 97, 98, 101, 102, 104, 116, 121, 136, 138, 139, 146, 147, 151, 158, 161, 165, 169, 174, 176, 177, 180, 195, 198, 203, 210, 212, 218, 219, 219, 220, 222, 225, 229, 241, 243, 246, 253, 267, 269, 272, 275, 278, 280, 283, 287, 289, 292, 295, 296, 300
♦ pH, cultivated, herbal, toxic. Origin: Eurasia.

Hypericum perforatum L. var. angustifolium DC.
Clusiaceae/Hypericaceae
♦ Weed, Naturalised
♦ 286, 287

Hypericum polyphyllum Boiss. & Balansa
Clusiaceae/Hypericaceae
Hypericum olympicum L. misapplied in cultivation
♦ Quarantine Weed
♦ 220
♦ cultivated. Origin: Eurasia.

Hypericum prolificum L.
Clusiaceae/Hypericaceae
♦ shrubby St. John's wort
♦ Weed
♦ 161
♦ cultivated, herbal.

Hypericum pubescens Boiss.
Clusiaceae/Hypericaceae
♦ villakuisma
♦ Casual Alien
♦ 42

Hypericum pulchrum L.
Clusiaceae/Hypericaceae
♦ slender St. John's wort, beautiful St. John's wort
♦ Naturalised
♦ 280
♦ cultivated.

Hypericum punctatum Lam.
Clusiaceae/Hypericaceae
♦ spotted St. John's wort
♦ Weed
♦ 87, 88, 161, 218
♦ cultivated, herbal, toxic.

Hypericum quadrangulum L.
Clusiaceae/Hypericaceae
♦ Weed
♦ 23, 87, 88
♦ cultivated.

Hypericum repens L.
Clusiaceae/Hypericaceae
♦ Quarantine Weed
♦ 220
♦ cultivated.

Hypericum revolutum Vahl
Clusiaceae/Hypericaceae
Hypericum lancelatum Lam., *Hypericum leucoptychodes* Steud. ex A.Rich
♦ curry bush, forest primrose, St. John's wort
♦ Weed, Native Weed
♦ 39, 121
♦ pS, cultivated, herbal, toxic. Origin: Africa.

Hypericum richeri Vill.
Clusiaceae/Hypericaceae
Hypericum fimbriatum Lam.
♦ alpine St. John's wort
♦ Weed
♦ 272
♦ cultivated, herbal.

Hypericum scouleri Hook.
Clusiaceae/Hypericaceae
♦ pale St. John's wort, Scouler's St. John's wort
♦ Weed
♦ 161
♦ herbal.

Hypericum styphelioides A.Rich.
Clusiaceae/Hypericaceae
♦ Weed
♦ 14

Hypericum tetrapterum Fr.
Clusiaceae/Hypericaceae
Hypericum acutum Moench
♦ St. Peter's wort, square stemmed hypericum, square stemmed St. John's wort, winged St. John's wort
♦ Weed, Quarantine Weed, Noxious Weed, Naturalised, Environmental Weed
♦ 72, 76, 86, 87, 88, 147, 176, 198, 203, 220, 272, 280
♦ pH, cultivated, herbal, toxic. Origin: Mediterranean.

Hypericum tricolour nom. illeg.
Clusiaceae/Hypericaceae
= *Hypericum calycinum* L. × *Hypericum patulum* Thunb.

♦ Quarantine Weed
♦ 220

Hypericum triquetrifolium Turra
Clusiaceae/Hypericaceae
Hypericum crispum L. *nom. illeg.* (see)
♦ wavyleaf St. John's wort, tangled hypericum, curled leaved St. John's wort
♦ Weed, Quarantine Weed, Noxious Weed, Naturalised
♦ 76, 86, 88, 115, 147, 198, 203, 220, 272
♦ cultivated, toxic. Origin: Mediterranean, eastern Europe.

Hypericum uliginosum Kunth
Clusiaceae/Hypericaceae
♦ Weed
♦ 157
♦ pH, arid.

Hypertelis bowkeriana Sond.
Molluginaceae
♦ Native Weed
♦ 121
♦ pH. Origin: southern Africa.

Hypertelis salsoloides (Burch.) Adamson
Molluginaceae
♦ braksuring
♦ Native Weed
♦ 121
♦ pH, cultivated. Origin: southern Africa.

Hyperthelia dissoluta (Nees ex Steud.) Clayton
Poaceae
♦ thatching grass, yellow thatching grass
♦ Native Weed
♦ 121
♦ pG, cultivated. Origin: Madagascar.

Hyphaene thebaica (L.) Mart.
Arecaceae
♦ gingerbread palm, doum nut, vegetable ivory palm, Egyptian doum palm, akoka, corozo, doam, dom palm, gingerbread tree, ivory nut, mana, nabidh, tageyt, vegetable ivory, doum palm
♦ Weed, Introduced
♦ 39, 221, 228
♦ arid, cultivated, herbal, toxic.

Hypnea musciformis (Wulfen) Lamour.
Hypneaceae
♦ red alga
♦ Weed
♦ 197, 282
♦ algae.

Hypobathrum microcarpum (Bl.) Bakh.f.
Rubiaceae
Petunga microcarpum (Bl.) DC.
♦ Weed
♦ 13

Hypocalyptus sophoroides (Bergius) Druce
Fabaceae/Papilionaceae
♦ Quarantine Weed
♦ 220
♦ S/T, cultivated.

Hypochaeris L. spp.
Asteraceae

♦ cat's ear
♦ Weed, Naturalised, Environmental Weed
♦ 198, 272, 289

Hypochaeris brasiliensis (Less.) Griseb.
Asteraceae
Hypochoeris brasiliensis (Lees.) Benth. & Hook.f.
♦ cat's ear, Brazilian cat's ear
♦ Weed, Naturalised
♦ 87, 88, 101, 121, 237, 243, 245, 255, 295
♦ pH. Origin: South America.

Hypochaeris brasiliensis (Less.) Griseb. var. *tweedyi* (Hook. & Arn.) Bak.
Asteraceae
♦ Tweedy's cat's ear
♦ Naturalised
♦ 101
♦ Origin: South America.

Hypochaeris ciliata (Thunb.) Makino
Asteraceae
♦ common hypochaeris
♦ Weed
♦ 297

Hypochaeris glabra L.
Asteraceae
Hypochoeris glabra L.
♦ smooth cat's ear, glabrous cat's ear, annual flatweed, flatweed, cat's ear, kaljuhäränsilmä, porcelle glabre
♦ Weed, Naturalised, Environmental Weed
♦ 7, 9, 15, 23, 34, 42, 70, 72, 80, 86, 87, 88, 93, 94, 98, 101, 121, 134, 161, 167, 176, 180, 198, 203, 205, 218, 241, 243, 250, 253, 269, 272, 280, 286, 287, 290, 300
♦ a/pH, arid, cultivated, herbal. Origin: Eurasia.

Hypochaeris maculata L.
Asteraceae
♦ spotted cat's ear, spotted hawkweed
♦ Weed
♦ 272
♦ pH, cultivated, herbal.

Hypochaeris microcephala (Sch.Bip.) Cabrera
Asteraceae
♦ smallhead cat's ear
♦ Weed, Naturalised
♦ 98, 101, 203
♦ herbal.

Hypochaeris microcephala (Sch.Bip.) Cabrera var. *albiflora* (Kuntze) Cabrera
Asteraceae
♦ smallhead cat's ear, flatweed
♦ Weed, Naturalised
♦ 86, 101, 269
♦ pH. Origin: Europe, North Africa.

Hypochaeris microcephala (Sch.Bip.) Cabrera var. *microcephala*
Asteraceae
♦ smallhead cat's ear
♦ Naturalised
♦ 86
♦ Origin: South America.

Hypochaeris radicata L.
Asteraceae
Hypochoeris radicata L.
♦ flatweed, cat's ear, hairy wild lettuce, spotted cat's ear, hairy cat's ear, rough cat's ear, smooth cat's ear, cat's ear dandelion, almeirao do campo
♦ Weed, Noxious Weed, Naturalised, Introduced, Environmental Weed, Casual Alien
♦ 1, 3, 7, 15, 23, 39, 42, 51, 52, 55, 70, 72, 80, 86, 87, 88, 93, 94, 98, 101, 121, 134, 136, 146, 152, 158, 161, 165, 176, 178, 181, 198, 203, 204, 212, 218, 228, 229, 237, 241, 243, 249, 250, 253, 255, 263, 269, 272, 280, 286, 287, 290, 293, 295, 296, 300
♦ pH, arid, cultivated, herbal, toxic. Origin: Europe, North Africa.

Hypochaeris tweediei (Hook. & Arn.) Cabrera
Asteraceae
Hypochoeris tweediei (Hook. & Arn.) Cabr.
♦ radicheta salvaje
♦ Weed, Casual Alien
♦ 87, 88, 237, 280, 295

Hypoestes antennifera S.Moore
Acanthaceae
♦ Naturalised
♦ 86
♦ cultivated.

Hypoestes aristata (Vahl) Sol.
Acanthaceae
♦ ribbon plant
♦ Naturalised
♦ 86
♦ cultivated. Origin: Africa.

Hypoestes cancellata Nees
Acanthaceae
♦ Weed
♦ 88

Hypoestes decaisneana Nees
Acanthaceae
Hypoestes rosea Decne.
♦ Weed
♦ 13

Hypoestes phyllostachya Bak.
Acanthaceae
Hypoestes sanguinolenta hort.
♦ polkadot plant, freckle face
♦ Weed, Naturalised, Garden Escape, Environmental Weed, Cultivation Escape
♦ 86, 98, 101, 134, 155, 203, 261
♦ a/pH, cultivated. Origin: Madagascar.

Hypoestes polythyrsa Miq.
Acanthaceae
♦ Weed
♦ 13

Hypoestes verticillaris (L.f.) Sol. ex Roem. & Schult.
Acanthaceae
Hypoestes forskalei L.f.
♦ Weed
♦ 87, 88

♦ pH, arid.

Hypolepis dicksonioides (Endl.) Hook.
Dennstaedtiaceae
♦ Naturalised
♦ 86
♦ cultivated. Origin: New Zealand, Pacific.

Hypolepis punctata (Thunb.) Mett.
Dennstaedtiaceae
♦ dotted beadfern
♦ Weed
♦ 286
♦ cultivated, herbal.

Hypolepis rugosula (Labill.) J.Sm.
Dennstaedtiaceae
Polypodium rugulosum Labill.
♦ fern
♦ Naturalised, Environmental Weed
♦ 86
♦ cultivated. Origin: Australia.

Hypolytrum nemorum (Vahl) Spreng.
Cyperaceae
♦ Weed
♦ 12

Hypoxis decumbens L.
Liliaceae/Hypoxidaceae
= *Hypoxis hirsuta (L.) Coville*
♦ falsa tiririca
♦ Weed
♦ 87, 88, 237, 255, 295
♦ pH, herbal. Origin: South America.

Hypoxis hirsuta (L.) Coville
Liliaceae/Hypoxidaceae
Hypoxis decumbens L. (see)
♦ common goldstar, yellow stargrass
♦ Weed
♦ 161
♦ cultivated, herbal.

Hypoxis hookeri Geerinck
Liliaceae/Hypoxidaceae
♦ Naturalised
♦ 280

Hypoxis obtusa Burch. ex Edwards
Liliaceae/Hypoxidaceae
Hypoxis nitida Verd.
♦ Native Weed
♦ 121
♦ pH, cultivated. Origin: southern Africa.

Hypoxis rigidula Bak. var. *pilosissima* Bak.
Liliaceae/Hypoxidaceae
♦ Native Weed
♦ 121
♦ pH. Origin: southern Africa.

Hypserpa ponapensis (Kaneh.) Kaneh.
Menispermaceae
♦ Introduced
♦ 230
♦ C.

Hyptis atrorubens Poit.
Lamiaceae
♦ marubio oscuro
♦ Weed
♦ 87, 88, 255
♦ pH, herbal. Origin: tropical America.

Hyptis brevipes **Poit.**
Lamiaceae
♦ lesser roundweed
♦ Weed, Quarantine Weed, Naturalised
♦ 12, 13, 76, 87, 88, 135, 153, 170, 191, 203, 220, 243, 255, 287
♦ aH, cultivated, herbal. Origin: tropical America.

Hyptis capitata **Jacq.**
Lamiaceae
♦ botones, batunes, t'aiegarabao, knobweed
♦ Weed, Quarantine Weed, Noxious Weed, Naturalised, Introduced
♦ 3, 13, 76, 86, 87, 88, 93, 107, 147, 153, 157, 170, 203, 206, 230, 243
♦ pH, herbal. Origin: Central America.

Hyptis gaudichaudii **Benth.**
Lamiaceae
♦ Weed
♦ 87, 88

Hyptis lanceolata **Poir.**
Lamiaceae
♦ Weed, Quarantine Weed
♦ 76, 87, 88, 203, 220

Hyptis lophantha **Mart. ex Benth.**
Lamiaceae
♦ catirina
♦ Weed
♦ 245, 255
♦ aH. Origin: Brazil.

Hyptis mutabilis **(Rich.) Briq.**
Lamiaceae
♦ tropical bushmint
♦ Weed, Naturalised
♦ 87, 88, 101, 179, 237, 255
♦ aH. Origin: Americas.

Hyptis pectinata **(L.) Poit.**
Lamiaceae
♦ comb hyptis, mint weed, purpletop, comb bushmint, mumutun lahe, mumutun palaoan, mumutan ademelon, fausse menthe, tamoli ni vavalangi, timothi ni vavalangi, wavuwavu, ndamoli, ben tulsia
♦ Weed, Quarantine Weed, Noxious Weed, Naturalised, Environmental Weed
♦ 3, 6, 13, 54, 76, 86, 87, 88, 155, 179, 191, 203, 206, 229, 243, 255, 262, 276
♦ aH, arid, herbal. Origin: tropical America.

Hyptis pedalipes **Griseb.**
Lamiaceae
Hyptis eriocauloides A.Rich
♦ Weed
♦ 14

Hyptis recurvata **Poit.**
Lamiaceae
♦ Weed
♦ 157
♦ pH.

Hyptis rhomboidea **M.Martens & Galeotti**
Lamiaceae
♦ capitate bushmint, knobweed

♦ Weed, Naturalised
♦ 3, 191, 235, 257, 262, 274, 276, 287
♦ a/pH.

Hyptis sidifolia **(L'Hér.) Briq.**
Lamiaceae
♦ Naturalised
♦ 257
♦ arid.

Hyptis spicigera **Lam.**
Lamiaceae
♦ marubio
♦ Weed
♦ 87, 88, 179
♦ Origin: Central and South America.

Hyptis suaveolens **Poit.**
Lamiaceae
Ballota suaveolens L., *Bystropogon suaveolens* (L.) L'Hér., *Hyptis congesta* Leonard, *Hyptis shaferi* Britt., *Mesosphaerum suaveolens* (L.) Kuntze, *Schaueria suaveolens* (L.) Hassk.
♦ wild spikenard, maeng lak khaa, pignut, hyptis, mint weed, horehound, mumutun
♦ Weed, Quarantine Weed, Noxious Weed, Naturalised, Environmental Weed
♦ 3, 7, 12, 13, 14, 55, 76, 86, 87, 88, 93, 98, 147, 203, 209, 216, 218, 229, 239, 245, 255, 261, 276, 287
♦ aH, arid, cultivated, herbal. Origin: tropical America.

Hyptis urticoides **H.B.K.**
Lamiaceae
♦ Weed
♦ 157
♦ pH.

Hyptis verticillata **Jacq.**
Lamiaceae
♦ John Charles
♦ Weed
♦ 14, 87, 88, 179
♦ herbal.

Hyssopus officinalis **L.**
Lamiaceae
♦ iisoppi, hyssop
♦ Naturalised, Introduced, Cultivation Escape
♦ 39, 42, 101, 228
♦ S, arid, cultivated, herbal, toxic.

Hysterionica jasionoides **Willd.**
Asteraceae
♦ Naturalised
♦ 241

I

Ibatia maritima **(Jacq.) Decne.**
Asclepiadaceae/Apocynaceae
= *Matelea maritima* (Jacq.) Woodson (NoR)
♦ Weed
♦ 87, 88
♦ herbal.

Iberis acutiloba **Bertol.**
Brassicaceae
= *Iberis odorata* L.
♦ Weed
♦ 243

Iberis amara **L.**
Brassicaceae
Thlaspi amarum Crantz
♦ wild candytuft, rocket candytuft, katkerasaippo, iberka horká, candytuft, bitter candytuft, annual candytuft
♦ Weed, Naturalised, Cultivation Escape
♦ 42, 44, 70, 87, 88, 94, 101, 243, 280
♦ a/bH, cultivated, herbal.

Iberis crenata **Lam.**
Brassicaceae
♦ Weed, Naturalised
♦ 86, 98, 203
♦ Origin: Spain, Portugal.

Iberis gibraltarica **L.**
Brassicaceae
♦ Gibraltar candytuft
♦ Naturalised
♦ 101
♦ cultivated, herbal.

Iberis odorata **L.**
Brassicaceae
Iberis acutiloba Bertol. (see)
♦ tuoksusaippo
♦ Cultivation Escape
♦ 42
♦ cultivated.

Iberis pinnata **L.**
Brassicaceae
Crucifera pinnata Krause
♦ Weed
♦ 70, 88, 94
♦ cultivated.

Iberis pruitii **Tineo**
Brassicaceae
♦ Pruit's candytuft
♦ Naturalised
♦ 101
♦ cultivated.

Iberis semperflorens L.
Brassicaceae
♦ Weed
♦ 88, 94
♦ cultivated.

Iberis sempervirens L.
Brassicaceae
♦ evergreen candytuft, candytuft, edging candytuft
♦ Weed, Naturalised, Cultivation Escape
♦ 56, 88, 94, 101, 116
♦ cultivated, herbal.

Iberis umbellata L.
Brassicaceae
Thlaspi umbellatum Crantz
♦ sarjasaippo, globe candytuft, garden candytuft, candytuft
♦ Naturalised, Cultivation Escape, Casual Alien
♦ 40, 42, 101, 280, 287
♦ aH, cultivated, herbal.

Ibicella lutea (Lindl.) Van Eselt.
Pedaliaceae/Martyniaceae
Martynia lutea (Sm.) Stapf, *Martynia montevidensis* Cham., *Proboscidea lutea* (Lindl.) Stapf (see)
♦ yellowflower devil's claw, elephant tusks, goat's head, unicorn plant, cuernos del diablo
♦ Weed, Quarantine Weed, Noxious Weed, Naturalised
♦ 51, 76, 86, 87, 88, 98, 101, 121, 147, 161, 180, 191, 198, 203, 220, 236, 237, 269, 270, 295
♦ aH, cultivated. Origin: South America.

Icacina senegalensis Juss.
Icacinaceae
♦ false yam
♦ Weed, Quarantine Weed
♦ 76, 87, 88, 203, 220
♦ herbal, toxic.

Icacina trichantha Oliv.
Icacinaceae
♦ Weed
♦ 88

Ichnanthus pallens (Sw.) Munro
Poaceae
♦ caruzo
♦ Weed
♦ 28, 206, 243
♦ G.

Idesia polycarpa Maxim.
Flacourtiaceae
♦ iigiri tree
♦ Naturalised
♦ 280
♦ T, cultivated.

Ifloga fontansii Cass.
Asteraceae
♦ Weed
♦ 87, 88

Ifloga spicata (Forssk.) Sch.Bip.
Asteraceae
♦ zenaymch
♦ Weed

♦ 221, 243
♦ arid.

Ilex L. spp.
Aquifoliaceae
♦ holly
♦ Weed, Naturalised
♦ 161, 198
♦ herbal, toxic.

Ilex × *altaclerensis* (Loudon) Dallim.
Aquifoliaceae
= *Ilex aquifolium* L. × *Ilex perado* Aiton
♦ highclere holly
♦ Cultivation Escape
♦ 40
♦ cultivated.

Ilex ambigua (Michx.) Torr.
Aquifoliaceae
♦ Carolina holly
♦ Weed
♦ 87, 88, 218
♦ cultivated, herbal.

Ilex aquifolium L.
Aquifoliaceae
Ilex balearica Desf.
♦ English holly, holly, variegated holly
♦ Weed, Noxious Weed, Quarantine Weed, Naturalised, Garden Escape, Environmental Weed
♦ 15, 20, 22, 35, 39, 45, 72, 76, 78, 80, 86, 88, 98, 101, 116, 137, 142, 146, 151, 155, 176, 198, 203, 246, 272, 280, 289, 296
♦ S/T, cultivated, herbal, toxic. Origin: Mediterranean, Middle East.

Ilex cassine L.
Aquifoliaceae
♦ dahoon holly, cassine, dahoon
♦ Garden Escape
♦ 39, 261
♦ T, cultivated, herbal, toxic. Origin: tropical Africa.

Ilex cornuta Lindl. & Paxton
Aquifoliaceae
♦ Chinese holly, horned holly
♦ Naturalised
♦ 101
♦ S, cultivated, herbal.

Ilex crenata Thunb.
Aquifoliaceae
♦ Japanese holly
♦ Weed, Naturalised
♦ 101, 195
♦ S, cultivated, herbal.

Ilex decidua Walter
Aquifoliaceae
♦ deciduous yaupon, possumhaw
♦ Weed
♦ 87, 88, 218
♦ cultivated, herbal.

Ilex glabra (L.) Gray
Aquifoliaceae
Prinos glaber L.
♦ gallberry, inkberry
♦ Weed
♦ 39, 87, 88, 218
♦ S, cultivated, herbal, toxic.

Ilex opaca Aiton
Aquifoliaceae

Ilex quercifolia Meerb.
♦ American holly, holly berry
♦ Weed
♦ 8, 39, 87, 88, 218, 247
♦ S, cultivated, herbal, toxic.

Ilex paraguariensis St.-Hil.
Aquifoliaceae
Ilex paraguayensis St.-Hil. (see)
♦ Paraguay tea, mate, matebaum
♦ Weed
♦ 22
♦ T, cultivated, herbal.

Ilex paraguayensis St.-Hil.
Aquifoliaceae
= *Ilex paraguariensis* St.-Hil.
♦ Paraguay tea, yerba mate
♦ Naturalised
♦ 101
♦ herbal.

Ilex volkensiana (Loes.) Kaneh. & Hatus.
Aquifoliaceae
♦ holly
♦ Introduced
♦ 230
♦ H.

Ilex vomitoria Aiton
Aquifoliaceae
♦ yaupon, Appalachian holly, yaupon holly
♦ Weed
♦ 8, 39, 87, 88, 218, 247
♦ S, cultivated, herbal, toxic.

Illecebrum verticillatum L.
Caryophyllaceae/Illecebraceae
Paronychia verticillata Lam.
♦ coral necklace, knotgrass, rustokki'
♦ Weed, Naturalised, Casual Alien
♦ 42, 88, 94, 101, 272, 280
♦ cultivated. Origin: Europe.

Illicium anisatum L.
Illiciaceae
Illicium religiosum Sieb. & Zucc.
♦ Japanese anise, star anise
♦ Weed
♦ 39, 161
♦ S, cultivated, herbal, toxic.

Illicium floridanum J.Ellis
Illiciaceae
♦ Florida anisetree, aniseed tree
♦ Weed
♦ 39, 161
♦ S, cultivated, herbal, toxic.

Ilysanthes ciliata Kuntze
Scrophulariaceae
= *Lindernia ciliata* (Colsm.) Pennell
♦ Weed
♦ 87, 88

Ilysanthes parviflora Benth.
Scrophulariaceae
♦ Weed
♦ 87, 88
♦ aqua, cultivated.

Ilysanthes veronicaefolia Urb.
Scrophulariaceae
♦ Weed
♦ 87, 88

Impatiens L. spp.
Balsaminaceae
♦ touch me not, balsams
♦ Weed, Environmental Weed
♦ 18, 73, 88, 243

Impatiens balfourii Hook.f.
Balsaminaceae
♦ Balfour's touch me not, Kashmir balsam
♦ Weed, Naturalised
♦ 80, 101
♦ pH, cultivated, herbal.

Impatiens balsamina L.
Balsaminaceae
Balsamina hortensis Desp.
♦ garden balsam, balsam, spotted snapweed, china
♦ Weed, Naturalised, Introduced, Cultivation Escape, Casual Alien
♦ 23, 87, 88, 101, 218, 230, 243, 261, 262, 280
♦ aH, cultivated, herbal. Origin: southern Asia.

Impatiens biflora Walt.
Balsaminaceae
= *Impatiens capensis* Meerb.
♦ jewelweed
♦ Naturalised
♦ 39, 287
♦ aH, herbal, toxic.

Impatiens capensis Meerb.
Balsaminaceae
Impatiens biflora Walt. (see), *Impatiens fulva* Nutt.
♦ spotted snapweed, orange balsam, jewelweed, spotted touch me not, spotted jewelweed
♦ Weed, Cultivation Escape
♦ 42, 87, 88, 218
♦ aH, cultivated, herbal.

Impatiens chinensis L.
Balsaminaceae
♦ Weed
♦ 87, 88

Impatiens ecalcarata Blank.
Balsaminaceae
♦ touch me not, spurless touch me not
♦ Weed
♦ 159
♦ aH, promoted, herbal.

Impatiens glandulifera Royle
Balsaminaceae
Impatiens glanduligera Lindl., *Impatiens roylei* Walp.
♦ policeman's helmet, ornamental jewelweed, jewelweed, Himalayan balsam, Indian balsam, purple jewelweed, touch me not
♦ Weed, Naturalised, Environmental Weed, Cultivation Escape
♦ 4, 24, 40, 42, 79, 80, 88, 101, 125, 132, 133, 152, 159, 184, 184, 195, 246, 280
♦ aH, cultivated, herbal, toxic. Origin: Eurasia.

Impatiens noli-tangere L.
Balsaminaceae
♦ western touch me not, touch me not, touch me not balsam, wild balsam

♦ Weed
♦ 23, 39, 70, 80, 88, 243, 272
♦ aH, cultivated, herbal, toxic.

Impatiens oliveri C.Wright ex W.Wats.
Balsaminaceae
♦ Oliver's touch me not
♦ Naturalised
♦ 101
♦ cultivated.

Impatiens pallida Nutt.
Balsaminaceae
♦ pale snapweed, yellow jewelweed, pale touch me not
♦ Weed
♦ 87, 88, 218
♦ aH, promoted, herbal.

Impatiens parviflora DC.
Balsaminaceae
♦ smallflower touch me not, small yellow balsam, small balsam
♦ Weed, Naturalised
♦ 23, 42, 88, 132, 272
♦ aH, cultivated.

Impatiens platypetala Lindl.
Balsaminaceae
♦ Weed
♦ 87, 88
♦ herbal.

Impatiens poilanei Tardieu
Balsaminaceae
♦ Weed
♦ 87, 88

Impatiens sodenii Engl. & Warb.
Balsaminaceae
♦ shrub balsam
♦ Sleeper Weed, Naturalised, Environmental Weed
♦ 225, 246, 280
♦ cultivated.

Impatiens sultani Hook.f.
Balsaminaceae
= *Impatiens walleriana* J.D.Hook.
♦ impatiens
♦ Weed
♦ 80, 87, 88
♦ cultivated, herbal.

Impatiens textori Miq.
Balsaminaceae
♦ tsurifunesou
♦ Weed
♦ 286
♦ aH, cultivated. Origin: Japan, Korea.

Impatiens walleriana Hook.f.
Balsaminaceae
Impatiens sultani Hook.f. (see)
♦ balsam, busy Lizzy, buzzy Lizzy
♦ Weed, Naturalised, Garden Escape, Environmental Weed, Cultivation Escape
♦ 3, 73, 86, 88, 98, 101, 155, 157, 179, 201, 203, 255, 261, 280, 290
♦ a/pH, cultivated. Origin: Brazil.

Imperata arundinacea Cirillo
Poaceae
= *Imperata cylindrica* (L.) Beauv.
♦ cogon grass, cogon
♦ Weed

♦ 3, 191
♦ G.

Imperata brasiliensis Trin.
Poaceae
Saccharum sape St.-Hil., *Imperata arundinacea* Cirillo var. *americana* Andrs.
♦ cogon grass, Brazilian satintail
♦ Weed, Quarantine Weed, Noxious Weed, Naturalised
♦ 14, 67, 76, 87, 88, 90, 101, 140, 161, 203, 220, 229, 237, 255, 258, 295
♦ pG, herbal. Origin: Central and South America.

Imperata brevifolia Vasey
Poaceae
♦ satintail, California satintail, satintail grass
♦ Weed, Noxious Weed
♦ 35, 88, 229
♦ pG, cultivated, herbal.

Imperata conferta (J.S.Presl) Ohwi
Poaceae
= *Imperata cylindrica* (L.) Beauv.
♦ bladygrass
♦ Weed, Environmental Weed
♦ 3, 6, 88, 152, 191
♦ G. Origin: south-east Asia.

Imperata contracta (Kunth) Hitchc.
Poaceae
♦ guayanilla
♦ Weed
♦ 14, 237, 295
♦ G.

Imperata cylindrica (L.) Beauv.
Poaceae
Calamagrostis lagurus Koeler, *Imperata arundinacea* Cirillo (see), *Imperata conferta* (J.S.Presl) Ohwi (see), *Imperata koenigii* (Retz.) P.Beauv., *Imperata koenigii* var. *major* Nees, *Lagurus cylindricus* L., *Saccharum cylindricum* (L.) Lam., *Saccharum koenigii* Retz., *Saccharum laguroides* Pourr.
♦ woolly grass, bai mao gen, bladygrass, cogon grass, swordgrass, alang alang grass, silver spike, cottonwool grass, mnyaki motomoto, yaa khaa, lalang, alang alang, Japanese blood grass, impérate, chi gaya, fushige chi gaya
♦ Weed, Quarantine Weed, Noxious Weed, Naturalised, Native Weed, Garden Escape, Environmental Weed
♦ 3, 12, 17, 30, 53, 67, 77, 80, 84, 87, 88, 90, 101, 112, 115, 121, 129, 140, 142, 151, 158, 161, 179, 185, 186, 204, 209, 218, 221, 229, 238, 239, 243, 249, 253, 258, 259, 262, 272, 280
♦ pG, arid, cultivated, herbal. Origin: south-east Asia.

Imperata cylindrica (L.) Beauv. var. koenigii Durand & Schinz
Poaceae
♦ cogon grass
♦ Weed
♦ 263, 286
♦ pG.

Imperata cylindrica (L.) Beauv. var. *major* (Nees) C.E.Hubb.
Poaceae
- cogon grass, bladygrass
- Weed, Sleeper Weed, Naturalised, Environmental Weed
- 170, 225, 246, 269, 273, 274, 275, 280, 297
- pG.

Imperata exaltata Brongn.
Poaceae
- Weed
- 87, 88
- G.

Imperata tenuis Hack.
Poaceae
- Weed
- 87, 88
- G.

Incarvillea sinensis Lam.
Bignoniaceae
- Weed
- 275, 297
- aH, cultivated, herbal.

Indigofera adenoides Bak.f.
Fabaceae/Papilionaceae
- Native Weed
- 121
- pH. Origin: southern Africa.

Indigofera arrecta A.Rich.
Fabaceae/Papilionaceae
- Natal indigo
- Weed, Native Weed
- 87, 88, 121
- pS, cultivated. Origin: Madagascar.

Indigofera articulata Gouan
Fabaceae/Papilionaceae
Indigofera glauca Lam.
- Weed
- 221
- arid, herbal.

Indigofera astragalina DC.
Fabaceae/Papilionaceae
- Weed, Introduced
- 66, 228
- arid.

Indigofera australis Willd.
Fabaceae/Papilionaceae
- Australian indigo
- Weed
- 39, 87, 88
- arid, cultivated, herbal, toxic.

Indigofera bracteola DC.
Fabaceae/Papilionaceae
- Weed
- 87, 88

Indigofera circinella Bak.f.
Fabaceae/Papilionaceae
- Naturalised
- 86
- Origin: Africa.

Indigofera colutea (Burm.f.) Lam.
Fabaceae/Papilionaceae
Indigofera viscosa Lam.
- rusty indigo
- Weed, Naturalised
- 87, 88, 101, 179, 221

Indigofera cordifolia Heyne ex Roth
Fabaceae/Papilionaceae
- bekar, bekario
- Weed
- 66
- arid, herbal, toxic.

Indigofera decora Lindl.
Fabaceae/Papilionaceae
- Weed, Naturalised
- 54, 86, 88, 98, 203, 280
- S, cultivated, herbal. Origin: China, Japan.

Indigofera dimidiata Vogel ex Walp.
Fabaceae/Papilionaceae
- Native Weed
- 121
- pH. Origin: southern Africa.

Indigofera fastigiata E.Mey.
Fabaceae/Papilionaceae
- Native Weed
- 121
- pH, herbal. Origin: southern Africa.

Indigofera filipes Benth. ex Harv.
Fabaceae/Papilionaceae
- Native Weed
- 121
- H. Origin: southern Africa.

Indigofera glandulosa Willd.
Fabaceae/Papilionaceae
- Weed, Naturalised
- 66, 86, 87, 88, 93, 191
- herbal. Origin: India, Indonesia.

Indigofera guatemalensis Moç., Sessè & Cerv. ex Backer
Fabaceae/Papilionaceae
- Guatemalan indigo
- Weed
- 157

Indigofera hendecaphylla Jacq.
Fabaceae/Papilionaceae
Indigofera endecaphylla Jacq. ex Lam. orth. var., *Indigofera neglecta* N.E.Br., *Indigofera spicata* auct. p.p. major
- creeping indigo, spicate indigo, trailing indigo, añil rastrero
- Introduced
- 38
- toxic. Origin: Africa, Asia, Pacific.

Indigofera heterantha Brandis
Fabaceae/Papilionaceae
- indigobush
- Casual Alien
- 280
- S, cultivated.

Indigofera heterotricha DC.
Fabaceae/Papilionaceae
- Native Weed
- 121
- S. Origin: southern Africa.

Indigofera hilaris Eckl. & Zeyh.
Fabaceae/Papilionaceae
- Native Weed
- 121
- pH, cultivated. Origin: southern Africa.

Indigofera hirsuta L.
Fabaceae/Papilionaceae
- roughhairy indigo, hairy indigo
- Weed, Naturalised, Introduced, Cultivation Escape
- 38, 55, 86, 87, 88, 93, 101, 179, 228, 255, 261, 287
- aH, arid, cultivated, herbal. Origin: pantropical.

Indigofera hochstetteri Bak.
Fabaceae/Papilionaceae
Indigofera anabaptista Steud., *Indigofera arenaria* A.Rich., *Indigofera jaubertiana* Schweinf., *Indigofera ornithopoides* Hochst. & Steud. ex Jaub. & Spach
- Weed
- 88, 221, 242
- arid.

Indigofera kirilowii Maxim. ex Palib.
Fabaceae/Papilionaceae
Indigofera macrostachya Bunge non Vent., *Indigofera koreana* Ohwi.
- Kirilow's indigo
- Naturalised
- 101
- S, promoted.

Indigofera linifolia (L.f.) Retz.
Fabaceae/Papilionaceae
Hedysarum linifolium L.
- bakereya, bhur bhura, lambio bekario, leel, sankhahuli, sidio
- Weed, Introduced
- 66, 87, 88, 228, 243
- a/pH, arid, cultivated, herbal.

Indigofera linnaei Ali
Fabaceae/Papilionaceae
Indigofera dominii H.Eichler, *Indigofera enneaphylla* L. nom. illeg.
- Weed, Introduced
- 39, 87, 88, 228
- arid, cultivated, toxic.

Indigofera macrophylla Schum & Thonn.
Fabaceae/Papilionaceae
- Weed
- 87, 88

Indigofera mucronata Spreng. ex DC. nom. illeg.
Fabaceae/Papilionaceae
= *Indigofera trita* L.f. ssp. *scabra* (Roth) de Kort & Thijsse (NoR)
- Weed
- 87, 88

Indigofera oblongifolia Forssk.
Fabaceae/Papilionaceae
Indigofera desmodioides Bak., *Indigofera lotoides* Lam., *Indigofera paucifolia* Del.
- Weed, Naturalised, Environmental Weed
- 7, 54, 86, 88, 221
- arid, cultivated, herbal. Origin: North Africa to India.

Indigofera oxytropis Benth. ex Harv.
Fabaceae/Papilionaceae
- Native Weed
- 121
- pH. Origin: southern Africa.

Indigofera parviflora Heyne
Fabaceae/Papilionaceae
♦ smallflower indigo
♦ Weed, Naturalised
♦ 87, 88, 101
♦ arid.

Indigofera parviflora Heyne ex Wight & Arn. var. *parviflora*
Fabaceae/Papilionaceae
♦ woolly finger bush
♦ Native Weed
♦ 121
♦ pH. Origin: southern Africa.

Indigofera pilosa Poir.
Fabaceae/Papilionaceae
♦ soft hairy indigo
♦ Weed, Naturalised, Introduced
♦ 101, 179, 228
♦ arid, cultivated.

Indigofera pseudo-tinctoria Matsum.
Fabaceae/Papilionaceae
♦ Weed
♦ 87, 88, 286
♦ S, promoted, herbal.

Indigofera sanguinea N.E.Br.
Fabaceae/Papilionaceae
♦ Native Weed
♦ 121
♦ pH. Origin: southern Africa.

Indigofera schimperi Jaub. & Spach
Fabaceae/Papilionaceae
♦ Weed
♦ 240
♦ S, arid.

Indigofera spicata Forssk.
Fabaceae/Papilionaceae
Indigofera bolusii N.E.Br., *Indigofera hendecaphylla* Jacq., *Indigofera neglecta* N.E.Br., *Indigofera parkeri* Bak., *Indigofera parvula sensu* Robyns
♦ creeping indigo, trailing indigo
♦ Weed, Naturalised, Introduced
♦ 28, 80, 86, 87, 88, 98, 101, 161, 179, 203, 206, 230, 243, 249, 261, 273, 287
♦ H, arid, toxic. Origin: Africa, temperate Asia.

Indigofera spinosa Forssk.
Fabaceae/Papilionaceae
Indigofera intricata sensu auct., Indigofera intricata Boiss. *sensu* Hutch. & Bruce
♦ Weed
♦ 221
♦ arid.

Indigofera subulata Vahl
Fabaceae/Papilionaceae
= *Indigofera trita* L.f. ssp. *subulata* (Vahl ex Poir.) Ali (NoR)
♦ Weed
♦ 87, 88
♦ Origin: Australia.

Indigofera suffruticosa Mill.
Fabaceae/Papilionaceae
Indigofera angolensis D.Dietr., *Indigofera anil* L., *Indigofera divaricata* Jacq., *Indigofera micrantha* Desv. *non* Bunge, *Indigofera truxillensis* Kunth (see), *Indigofera uncinata* G.Don
♦ anil de pasto, indigo, aniles, iniko,

akauveli, indigo dye plant
♦ Weed, Naturalised, Introduced
♦ 3, 6, 14, 22, 39, 86, 87, 88, 98, 134, 179, 191, 203, 228, 241, 255, 287, 295
♦ pH, arid, cultivated, herbal, toxic. Origin: Brazil.

Indigofera tenuisiliqua Schweinf.
Fabaceae/Papilionaceae
♦ Weed
♦ 221

Indigofera tinctoria L.
Fabaceae/Papilionaceae
Indigofera sumatrana Gaertn.
♦ true indigo, indigo, indaco del bengala
♦ Weed, Naturalised, Cultivation Escape
♦ 32, 38, 39, 86, 87, 88, 93, 101, 179, 191, 235, 252, 261, 287
♦ cultivated, herbal, toxic. Origin: Old World Tropics.

Indigofera tomentosa L.
Fabaceae/Papilionaceae
♦ Weed
♦ 87, 88

Indigofera trifoliata L.
Fabaceae/Papilionaceae
♦ threeleaf indigo
♦ Naturalised
♦ 101
♦ herbal. Origin: Asia, Australia.

Indigofera trita L.f.
Fabaceae/Papilionaceae
♦ Asian indigo
♦ Weed
♦ 87, 88
♦ pH, cultivated, herbal.

Indigofera truxillensis Kunth
Fabaceae/Papilionaceae
= *Indigofera suffruticosa* Mill.
♦ indigo
♦ Weed, Naturalised
♦ 255, 300
♦ pH, arid. Origin: South America.

Indigofera viscidissima Bak.
Fabaceae/Papilionaceae
♦ Weed
♦ 87, 88

Indosasa sinica C.D.Chu & C.S.Chao
Poaceae
♦ Quarantine Weed
♦ 220
♦ G.

Inga edulis Mart.
Fabaceae/Mimosaceae
♦ icecream bean
♦ Environmental Weed
♦ 257
♦ T, cultivated, herbal.

Inga fastuosa (Jacq.) Willd.
Fabaceae/Mimosaceae
♦ guaba peluda, guaba venezolana
♦ Introduced
♦ 261
♦ Origin: Colombia, Venezuela.

Inga ingoides (Rich.) Willd.
Fabaceae/Mimosaceae

♦ icecream bean
♦ Cultivation Escape
♦ 261
♦ S/T, cultivated.

Inga laurina (Sw.) Willd.
Fabaceae/Mimosaceae
Mimosa laurina Sw.
♦ sacky sac bean
♦ Introduced
♦ 228
♦ S/T, arid.

Inga nobilis Willd.
Fabaceae/Mimosaceae
♦ Naturalised
♦ 101
♦ T.

Inga nobilis Willd. ssp. quaternata (Poepp. & Endl.) T.D.Penn.
Fabaceae/Mimosaceae
Inga quaternata Poepp. & Endl. (see)
♦ guamá venezolano
♦ Naturalised
♦ 101

Inga quaternata Poepp. & Endl.
Fabaceae/Mimosaceae
= *Inga nobilis* Willd. ssp. *quaternata* (Poepp. & Endl.) T.D.Penn.
♦ guamá venezolano
♦ Cultivation Escape
♦ 261
♦ S/T, cultivated. Origin: Central and South America.

Inga schimpffii Harms
Fabaceae/Mimosaceae
= *Inga spectabilis* (Vahl) Willd. var. *schimpffii* (Harms) Little (NoR)
♦ Naturalised
♦ 257
♦ T.

Inocarpus fagiferus (Park.) Fosberg
Fabaceae/Papilionaceae
♦ mwuropw
♦ Weed, Introduced
♦ 22, 230
♦ T.

Inula L. spp.
Asteraceae
♦ yellowhead
♦ Weed
♦ 272
♦ herbal.

Inula britannica L.
Asteraceae
Aster britannicus All., *Conyza britannica* (L.) Moris ex Rupr., *Inula britannica* var. *tymiensis* Kudô, *Inula hispanica* Pau, *Inula japonica* Thunb. (see), *Inula serrata* Gilib., *Inula tymiensis* Kudô
♦ British yellowhead, vanukehirvenjuuri, xuan fu hua, British elecampane
♦ Weed, Naturalised
♦ 42, 87, 88, 101, 272, 297
♦ pH, cultivated, herbal, toxic.

Inula cappa (Ham. ex D.Don) DC.
Asteraceae
♦ naat kham

♦ Weed
♦ 238
♦ S, herbal.

***Inula caspica* Blume**
Asteraceae
♦ Weed
♦ 272

***Inula conyza* DC.**
Asteraceae
= *inula conyzae* (Griess.) Meikle
♦ ploughman's spikenard,
hirvenjuurilaji
♦ Weed, Quarantine Weed,
Naturalised
♦ 76, 86, 88, 203, 220
♦ b/pH, cultivated, herbal.

***Inula conyzae* (Griess.) Meikle**
Asteraceae
Aster conyzae Griess., *Inula conyza* DC.
(see), *Inula vulgaris* Trevis.
♦ ploughman's spikenard
♦ Naturalised
♦ 39, 280
♦ cultivated, herbal, toxic.

***Inula crithmoides* L.**
Asteraceae
♦ golden samphire
♦ Weed
♦ 70, 88, 94, 221
♦ pH, arid, cultivated, herbal.

***Inula dysenterica* L.**
Asteraceae
= *Pulicaria dysenterica* (L.) Bernh.
♦ common fleabane
♦ Weed
♦ 87, 88
♦ herbal.

***Inula ensifolia* L.**
Asteraceae
Aster ensifolius Scop., *Inula linifolia*
Wend.
♦ narrow leaved fleabane
♦ Weed
♦ 272
♦ cultivated, herbal.

***Inula germanica* L.**
Asteraceae
♦ Weed
♦ 272
♦ cultivated.

***Inula graveolens* (L.) Desf.**
Asteraceae
= *Dittrichia graveolens* (L.) Greuter
♦ cape khakiweed, camphor inula,
caledonbos, khakibush, khakiweed,
stinkweed
♦ Weed
♦ 39, 51, 87, 88, 121, 158
♦ pS, herbal, toxic. Origin: Eurasia.

***Inula helenium* L.**
Asteraceae
Aster helenium Scop., *Corvisartia
helenium* Mérat
♦ elecampane, elecampane inula,
enula campana, isohirvenjuuri
♦ Weed, Naturalised, Introduced,
Cultivation Escape

♦ 34, 39, 40, 42, 87, 88, 101, 161, 218,
272, 280
♦ pH, cultivated, herbal, toxic.

***Inula heterolepis* Boiss.**
Asteraceae
♦ Weed
♦ 87, 88

***Inula hirta* L.**
Asteraceae
Aster hirtus Scop., *Inula hirsuta* Suffr.,
Pulicaria hirta Presl
♦ hairy fleabane, downy elecampane
♦ Weed
♦ 272
♦ cultivated, herbal.

***Inula indica* L.**
Asteraceae
♦ Weed, Quarantine Weed
♦ 76, 87, 88, 203, 220

***Inula japonica* Thunb.**
Asteraceae
= *Inula britannica* L.
♦ Weed
♦ 275, 286, 297
♦ pH, herbal.

***Inula oculus-christi* L.**
Asteraceae
♦ Weed
♦ 272
♦ cultivated.

***Inula salicina* L.**
Asteraceae
Aster salicinus Scop., *Conyza salicina*
Rupre.
♦ willowleaf yellowhead, Irish
fleabane, rantahirvenjuuri
♦ Weed, Naturalised
♦ 101, 272
♦ cultivated, herbal.

***Inula salsoloides* (Turcz.) Ostenf.**
Asteraceae
♦ Weed
♦ 275, 297
♦ pH.

***Inula viscosa* (L.) Aiton**
Asteraceae
= *Dittrichia viscosa* (L.) Greuter
♦ Weed
♦ 87, 88, 221
♦ arid, cultivated, herbal.

***Iochroma grandiflorum* Benth.**
Solanaceae
♦ Weed, Naturalised
♦ 15, 280
♦ S/T.

***Iondraba auriculata* (L.) Webb & Berthel.**
Brassicaceae
= *Biscutella auriculata* L.
♦ Weed, Quarantine Weed
♦ 76, 87, 88, 203, 220
♦ cultivated.

***Ionidium suffruticosum* Ging.**
Violaceae
♦ Weed
♦ 87, 88

***Ionopsidium acaule* (Desf.) Rchb.**
Brassicaceae

Cochlearia acaulis Desf.
♦ diamond flower, false
diamondflower, violet cress
♦ Weed, Naturalised, Cultivation
Escape, Casual Alien
♦ 34, 40, 101
♦ aH, cultivated.

***Ipheion uniflorum* (Lindl.) Raf.**
Liliaceae/Alliaceae
Tristagma uniflorum (Lindl.) Traub (see)
♦ spring starflower, springstar
♦ Weed, Naturalised, Garden Escape,
Environmental Weed, Cultivation
Escape
♦ 7, 34, 72, 86, 88, 98, 161, 198, 203, 249,
280, 287
♦ pH, cultivated, herbal. Origin:
Argentina, Uruguay.

***Iphiona mucronata* (Forssk.) Asch. &
Schweinf.**
Asteraceae
♦ Weed
♦ 221

***Iphiona scabra* DC.**
Asteraceae
♦ zafrah
♦ Weed
♦ 221
♦ arid.

***Ipomoea* L. spp.**
Convolvulaceae
♦ weir vine, morningglory, fue sina i,
bejuco, mtuntumba, sifu, quinamul,
batatilla
♦ Weed, Quarantine Weed, Noxious
Weed, Naturalised
♦ 3, 26, 86, 88, 147, 191, 198, 203, 204,
236, 245, 258, 281
♦ herbal.

***Ipomoea acuminata* Roem & Schult.**
Convolvulaceae
= *Ipomoea indica* (Burm.f.) Merr.
♦ corda de viola corriola
♦ Weed
♦ 87, 88, 201, 245
♦ herbal.

***Ipomoea adenioides* Schinz**
Convolvulaceae
♦ trumpet flower
♦ Native Weed
♦ 121
♦ pS, arid, cultivated. Origin: southern
Africa.

***Ipomoea alba* L.**
Convolvulaceae
Ipomoea bona-nox L., *Calonyction
aculeatum* (L.) House (see)
♦ moonflower, tropical white
morningglory, white morningglory,
moonvine, lonya
♦ Weed, Noxious Weed, Naturalised,
Garden Escape, Environmental Weed
♦ 14, 54, 73, 80, 86, 87, 88, 95, 98, 134,
155, 161, 203, 229, 255, 257, 279, 280,
283, 287
♦ pC, arid, cultivated, herbal, toxic.
Origin: tropical America.

Ipomoea amnicola Morong
Convolvulaceae
♦ redcenter morningglory
♦ Noxious Weed, Naturalised
♦ 101, 229

Ipomoea amoena Choisy
Convolvulaceae
♦ Weed
♦ 87, 88

Ipomoea angustifolia Jacq.
Convolvulaceae
= *Xenostegia tridentata* (L.) Austin &
Staples (NoR) [see *Merremia tridentata*
(L.) Hallier f., *Merremia tridentata*
(L.) Hallier f. ssp. *angustifolia* (Jacq.)
Ooststr.]
♦ Weed
♦ 87, 88

Ipomoea aquatica Forssk.
Convolvulaceae
Ipomoea reptans (L.) Poir.
♦ water spinach, swamp
morningglory, kangkong, phak
bung, pink convolvulus, potato vine,
creeping swamp morningglory
♦ Weed, Noxious Weed, Naturalised,
Introduced, Environmental Weed
♦ 13, 38, 67, 80, 87, 88, 101, 112, 140,
151, 161, 179, 193, 197, 204, 209, 217,
218, 229, 239, 246, 262, 263, 286, 287
♦ wpH, cultivated, herbal.

**Ipomoea aristolochiaefolia G.Don.
Austin**
Convolvulaceae
♦ corda de viola corriola
♦ Weed
♦ 88, 237, 245, 295

**Ipomoea asarifolia (Desr.) Roem. &
Schult.**
Convolvulaceae
♦ batatarana, salsa do rio
♦ Weed, Naturalised
♦ 14, 88, 101, 179, 255
♦ arid. Origin: tropical America.

Ipomoea asperifolia Hallier f.
Convolvulaceae
♦ Weed
♦ 87, 88

Ipomoea barbatisepala Gray
Convolvulaceae
♦ canyon morningglory
♦ Noxious Weed
♦ 229
♦ herbal.

Ipomoea barbigera Sweet
Convolvulaceae
= *Ipomoea hederacea* (L.) Jacq.
♦ southern morningglory
♦ Weed
♦ 87, 88, 218

Ipomoea batatas (L.) Lam.
Convolvulaceae
Convolvulus tiliaceus auct. non Willd.,
Convolvulus batatas L., *Ipomoea tiliacea*
(Willd.) Choisy (see)
♦ sweet potato, darkeye morningglory,
pehdede, limenge, ilando kiazi kitamu

♦ Weed, Noxious Weed, Naturalised,
Introduced, Cultivation Escape
♦ 13, 32, 39, 86, 87, 88, 98, 101, 161, 179,
203, 229, 230, 256, 257, 261, 280
♦ C, cultivated, herbal, toxic. Origin:
Central and South America.

Ipomoea biflora (L.) Pers.
Convolvulaceae
♦ white woodrose
♦ Weed
♦ 273

**Ipomoea blepharosepala Hochst. ex
A.Rich.**
Convolvulaceae
♦ Weed, Quarantine Weed
♦ 76, 87, 88, 203, 220

Ipomoea cairica (L.) Sweet
Convolvulaceae
Ipomoea palmata Forssk., *Convolvulus
caricus* L.
♦ Cairo morningglory, coast
morningglory, Messina creeper,
mile a minute, mile a minute vine,
lunsengansenga
♦ Weed, Noxious Weed, Naturalised,
Native Weed, Garden Escape,
Environmental Weed
♦ 7, 73, 86, 87, 88, 98, 101, 121, 134, 158,
179, 201, 203, 218, 221, 229, 255, 262,
269, 274, 286, 287, 290, 295, 296, 297
♦ pC, cultivated, herbal. Origin: South
America.

Ipomoea calantha Griseb.
Convolvulaceae
♦ moonvine
♦ Noxious Weed
♦ 229

Ipomoea calobra W.Hill & F.Muell.
Convolvulaceae
♦ weir vine
♦ Weed, Quarantine Weed
♦ 76, 87, 88
♦ cultivated.

Ipomoea caloneura Meissn.
Convolvulaceae
♦ Weed
♦ 87, 88

Ipomoea capillacea (Kunth) G.Don
Convolvulaceae
♦ purple morningglory
♦ Noxious Weed
♦ 229
♦ herbal.

Ipomoea cardiophylla Gray
Convolvulaceae
♦ heartleaf morningglory
♦ Noxious Weed
♦ 229
♦ herbal.

Ipomoea cardiosepala Meissn.
Convolvulaceae
= *Ipomoea sinensis* (Desr.) Choisy
♦ Weed
♦ 87, 88

Ipomoea carnea Jacq.
Convolvulaceae
♦ gloria de la manana, shrubby

fistulosa morningglory, bush
morningglory
♦ Weed, Naturalised, Environmental
Weed
♦ 39, 93, 98, 101, 161, 203, 216
♦ S, arid, cultivated, herbal, toxic.

**Ipomoea carnea Jacq. ssp. *fistulosa* (Mart.
ex Choisy) D.Austin**
Convolvulaceae
Ipomoea fistulosa Mart. ex Choisy (see)
♦ gloria de la manana, bush
morningglory, canudo, capa bode
♦ Weed, Noxious Weed, Naturalised,
Introduced, Cultivation Escape
♦ 22, 86, 101, 179, 229, 255, 261, 283
♦ aqua, cultivated, toxic. Origin:
tropical America.

Ipomoea cholulensis H.B.K.
Convolvulaceae
♦ Weed
♦ 157
♦ aC.

Ipomoea chryseides Ker Gawl.
Convolvulaceae
♦ Weed
♦ 87, 88

Ipomoea coccinea L.
Convolvulaceae
Quamoclit coccinea (L.) Moench (see)
♦ scarlet morningglory, red
morningglory, starglory, redstar
♦ Weed, Noxious Weed, Naturalised
♦ 80, 87, 88, 101, 161, 212, 218, 229, 286,
287
♦ aH, arid, cultivated, herbal, toxic.

Ipomoea congesta R.Br.
Convolvulaceae
= *Ipomoea indica* (Burm.f.) Merr.
♦ blue morningglory, morningglory
♦ Weed
♦ 14, 87, 88, 95, 218
♦ cultivated, herbal. Origin: tropical
America.

**Ipomoea coptica (L.) Roth ex Roem. &
Schult.**
Convolvulaceae
Ipomoea dissecta Willd.
♦ alamovine
♦ Weed, Noxious Weed
♦ 87, 88, 229
♦ cultivated, herbal.

Ipomoea cordatotriloba Dennst.
Convolvulaceae
♦ sharppod morningglory, tievine
♦ Noxious Weed
♦ 161, 229

**Ipomoea cordatotriloba Dennst. var.
australis O'Donell**
Convolvulaceae
Ipomoea trichocarpa Elliot var. *australis*
O'Donell (see), *Ipomoea triloba* auct.
non L.
♦ Weed
♦ 237

**Ipomoea cordatotriloba Dennst. var.
*cordatotriloba***
Convolvulaceae

Ipomoea trichocarpa Ell. (see)
♦ tievine
♦ Noxious Weed
♦ 229

Ipomoea cordatotriloba Dennst. var. torreyana (Gray) D.Austin
Convolvulaceae
Ipomoea trichocarpa Ell. var. *torreyana* (Gray) Shinners (see)
♦ Torrey's tievine, cotton morningglory
♦ Weed, Noxious Weed
♦ 161, 229

Ipomoea cordifolia Carey ex Voight
Convolvulaceae
♦ morningglory, heartleaf morningglory
♦ Noxious Weed, Naturalised
♦ 101, 229

Ipomoea cordofana Choisy
Convolvulaceae
♦ Weed, Quarantine Weed
♦ 76, 87, 88, 203, 220, 242
♦ arid.

Ipomoea coscinosperma Hochst. ex Choisy
Convolvulaceae
♦ Weed, Native Weed
♦ 51, 87, 88, 121, 158, 242
♦ aH, arid. Origin: southern Africa.

Ipomoea costellata Torr.
Convolvulaceae
♦ crestrib morningglory
♦ Noxious Weed
♦ 229
♦ herbal.

Ipomoea crassifolia Cav.
Convolvulaceae
♦ Weed
♦ 87, 88
♦ arid, herbal.

Ipomoea crassipes Hook.
Convolvulaceae
♦ Native Weed
♦ 121
♦ pH, herbal, toxic. Origin: southern Africa.

Ipomoea cristulata Hallier f.
Convolvulaceae
Quamoclit gracilis Hallier f.
♦ TransPecos morningglory
♦ Noxious Weed
♦ 229
♦ herbal.

Ipomoea cymosa Roem & Schult.
Convolvulaceae
♦ Weed
♦ 87, 88

Ipomoea cynanchifolia Meisn.
Convolvulaceae
♦ Sabi morningglory
♦ Weed, Quarantine Weed
♦ 50, 76, 87, 88, 203, 220
♦ Origin: Brazil, Guyana.

Ipomoea digitata L.
Convolvulaceae
♦ Weed, Naturalised

♦ 87, 88, 209, 287
♦ herbal. Origin: Dominican Republic, Haiti.

Ipomoea dumetorum Willd. ex Roem. & Schult.
Convolvulaceae
♦ railwaycreeper
♦ Noxious Weed, Naturalised
♦ 229, 300
♦ arid. Origin: South America.

Ipomoea eggersiana Peter
Convolvulaceae
♦ jumby potato
♦ Noxious Weed
♦ 229

Ipomoea eriocarpa R.Br.
Convolvulaceae
Ipomoea hispida (Vahl) Roem. & Schult. *nom. illeg.* (see)
♦ morningglory
♦ Weed, Noxious Weed, Naturalised
♦ 13, 87, 88, 229, 240, 242, 287
♦ aH, arid, herbal. Origin: Madagascar.

Ipomoea fimbriosepala Choisy
Convolvulaceae
♦ jitirana, enredaderia
♦ Weed
♦ 255
♦ Origin: Madagascar.

Ipomoea fistulosa Mart. ex Choisy
Convolvulaceae
= *Ipomoea carnea* Jacq. ssp. *fistulosa* (Mart. ex Choisy) D.Austin
♦ tree morningglory, standing morningglory
♦ Weed
♦ 13, 14, 22, 87, 88, 218, 295
♦ S, herbal.

Ipomoea gossypioides Parodi
Convolvulaceae
♦ Weed
♦ 87, 88

Ipomoea gracilis auct. non R.Br.
Convolvulaceae
= *Ipomoea littoralis* Blume
♦ Weed
♦ 87, 88, 209
♦ herbal.

Ipomoea grandifolia (Dammer) O'Donell
Convolvulaceae
♦ corda de viola corriola, campainha, corriola
♦ Weed
♦ 237, 245, 255, 270, 295
♦ Origin: Brazil.

Ipomoea hardwickii Sweet
Convolvulaceae
= *Ipomoea sinensis* (Desr.) Choisy
♦ Weed, Quarantine Weed
♦ 76, 87, 88, 203, 220

Ipomoea hederacea Jacq.
Convolvulaceae
Ipomoea barbigera Sweet (see), *Ipomoea desertorum* House, *Ipomoea hederacea* (L.) Jacq. var. *integriuscula* A.Gray (see), *Ipomoea hirsutala* auct. non Jacq.f., *Pharbitis barbigera* (Sweet) G.Don,

Pharbitis githaginea Hochst., *Pharbitis hederacea* (Jacq.) Choisy, *Pharbitis hispida* A.Rich., *Pharbitis purpurea* Asch.
♦ ivyleaf morningglory, Mexican morningglory, entireleaf morningglory, liuskaelämänlanka
♦ Weed, Quarantine Weed, Noxious Weed, Naturalised, Introduced, Casual Alien
♦ 23, 42, 76, 80, 86, 87, 88, 98, 101, 161, 174, 180, 203, 207, 210, 211, 212, 218, 220, 229, 243, 263, 286, 287, 299
♦ aC, cultivated, herbal, toxic. Origin: tropical America.

Ipomoea hederacea (L.) Jacq. var. integriuscula A.Gray
Convolvulaceae
= *Ipomoea hederacea* (L.) Jacq.
♦ Weed, Naturalised
♦ 286, 287

Ipomoea hederifolia L.
Convolvulaceae
Ipomoea angulata Mart. ex Choisy
♦ scarletcreeper, red flowered bellvine, red convolvulus
♦ Weed, Quarantine Weed, Noxious Weed, Naturalised
♦ 13, 55, 66, 76, 84, 86, 87, 88, 93, 98, 203, 229, 245, 255, 276
♦ aH, cultivated, herbal. Origin: tropical and subtropical America.

Ipomoea hirsutula Jacq.f.
Convolvulaceae
= *Ipomoea purpurea* (L.) Roth
♦ woolly morningglory
♦ Weed
♦ 87, 88, 218
♦ herbal.

Ipomoea hispida (Vahl) Roem. & Schult. nom. illeg.
Convolvulaceae
= *Ipomoea eriocarpa* R.Br.
♦ Weed
♦ 88, 243

Ipomoea hochstetteri House
Convolvulaceae
♦ Native Weed
♦ 121
♦ pC. Origin: southern Africa.

Ipomoea horsfalliae Hook.f.
Convolvulaceae
♦ Lady Doorly's morningglory, princess vine
♦ Noxious Weed, Naturalised
♦ 229, 287
♦ cultivated.

Ipomoea imperati (Vahl) Griseb.
Convolvulaceae
Ipomoea stolonifera (Cyrillo) Gmel. (see)
♦ beach morningglory
♦ Noxious Weed
♦ 229
♦ herbal.

Ipomoea indica (Burm.) Merr.
Convolvulaceae
Convolvulus acuminatus Vahl, *Convolvulus congestus* (R.Br.) Spreng., *Convolvulus indicus* Burm, *Ipomoea*

acuminata (Vahl) Roem. & Schult. (see),
Ipomoea amoena Bl., *Ipomoea cataractae*
Endl., *Ipomoea cathartica* Poir., *Ipomoea
congesta* R.Br. (see), *Ipomoea indica var.
acuminata* (Burm.f.) Merr (see), *Ipomoea
insularis* (Choisy) Steud., *Ipomoea
kiuninsularis* Masam., *Ipomoea learii*
Knight ex Paxton, *Ipomoea mutabilis*
Lindl., *Parasitipomoea formosana*
Hayata, *Pharbitis acuminata* (Vahl)
Choisy, *Pharbitis acuminata var. congesta*
(R.Br.) Choisy, *Pharbitis cathartica*
(Poir.) Choisy, *Pharbitis indica* (Burm.)
R.C.Fang, *Pharbitis insularis* Choisy,
Pharbitis learii (Knight ex Paxton)
Lindl.
♦ blue morningglory, purple
morningglory, ocean blue
morningglory, fue moa, purperwinde,
morningglory
♦ Weed, Quarantine Weed, Noxious
Weed, Naturalised, Garden Escape,
Environmental Weed
♦ 7, 15, 63, 72, 73, 86, 88, 98, 101, 134,
165, 198, 201, 203, 225, 229, 246, 263,
269, 280, 283, 286, 289, 296
♦ C, arid, cultivated, herbal. Origin:
pantropical.

**Ipomoea indica (Burm.f.) Merr. var.
acuminata (Vahl) Fosb.**
Convolvulaceae
= *Ipomoea indica* (Burm.) Merr.
♦ bejuco de gloria
♦ Weed
♦ 261

Ipomoea indivisa H.Hallier
Convolvulaceae
Convolvulus indivisus Vell.
♦ Weed
♦ 87, 88, 255
♦ Origin: South America.

Ipomoea involucrata Beauv.
Convolvulaceae
♦ fulu, ilandala, bololo mwasi, mgubi
♦ Weed
♦ 87, 88
♦ cultivated, herbal.

Ipomoea krugii Urb.
Convolvulaceae
♦ Krug's white morningglory
♦ Noxious Weed
♦ 229

Ipomoea lacunosa L.
Convolvulaceae
♦ pitted morningglory, whitestar,
kaljuelämänlanka, small flowered
white morningglory
♦ Weed, Noxious Weed, Naturalised,
Casual Alien
♦ 42, 84, 87, 88, 161, 179, 207, 211, 229,
243, 263, 286, 287
♦ aH, cultivated, herbal. Origin:
tropical north America.

Ipomoea lacunosa L. fo. purpurata Fern.
Convolvulaceae
♦ Naturalised
♦ 287

Ipomoea leari Paxt.
Convolvulaceae
♦ Weed
♦ 87, 88

Ipomoea leptophylla Torr.
Convolvulaceae
♦ bush morningglory, bush
moonflower
♦ Noxious Weed
♦ 39, 229
♦ pH, cultivated, herbal, toxic.

Ipomoea × leucantha Jacq. (pro sp.)
Convolvulaceae
= *Ipomoea cordatotriloba* Dennst. ×
Ipomoea lacunosa L.
♦ morningglory
♦ Noxious Weed
♦ 229

Ipomoea lindheimeri Gray
Convolvulaceae
♦ Lindheimer's morningglory
♦ Noxious Weed
♦ 229
♦ herbal.

Ipomoea littoralis Blume
Convolvulaceae
Ipomoea gracilis auct. (*non* R.Br.) (see)
♦ whiteflower beach morningglory,
omp, lau tagamimi
♦ Noxious Weed, Naturalised,
Introduced
♦ 101, 229, 230
♦ C, cultivated, herbal. Origin: Asia.

Ipomoea lonchophylla J.M.Black
Convolvulaceae
♦ cowvine
♦ Weed
♦ 55
♦ cultivated.

Ipomoea longifolia Benth.
Convolvulaceae
♦ pinkthroat morningglory, wild
morningglory
♦ Noxious Weed
♦ 161, 229
♦ aC, arid, herbal, toxic.

Ipomoea macrantha Roem. & Schult.
Convolvulaceae
= *Ipomoea violacea* L.
♦ Weed
♦ 6, 88
♦ cultivated, herbal.

Ipomoea macrorhiza Michx.
Convolvulaceae
♦ largeroot morningglory
♦ Weed, Noxious Weed
♦ 179, 229

Ipomoea magnusiana Schinz
Convolvulaceae
♦ Weed
♦ 88

Ipomoea marginisepala O'Donell
Convolvulaceae
♦ bejuco
♦ Weed
♦ 237, 295

Ipomoea mauritiana Jacq.
Convolvulaceae
Ipomoea digitata auct., *Ipomoea insignis*
Ker Gawl.
♦ likam
♦ Weed, Quarantine Weed,
Naturalised, Introduced
♦ 76, 86, 88, 203, 230
♦ pC, cultivated. Origin: pantropical.

Ipomoea maxima G.Don ex Sweet
Convolvulaceae
Ipomoea sepiaria Koen. ex Roxb. (see)
♦ Weed
♦ 87, 88, 262
♦ herbal.

Ipomoea meyeri (Spreng.) G.Don
Convolvulaceae
♦ Meyer's morningglory
♦ Noxious Weed
♦ 229

Ipomoea microdactyla Griseb.
Convolvulaceae
♦ bejuco colorado, calcareous
morningglory, wild potato
♦ Noxious Weed
♦ 229

Ipomoea muelleri Benth.
Convolvulaceae
♦ poison morningglory
♦ Weed
♦ 39, 87, 88
♦ cultivated, herbal, toxic.

Ipomoea × multifida (Raf.) Shinners
Convolvulaceae
= *Ipomoea coccinea* L. × *Ipomoea
quamoclit* L.
♦ morningglory, cardinal climber
♦ Noxious Weed
♦ 229

Ipomoea muricata (L.) Jacq.
Convolvulaceae
= *Ipomoea turbinata* Lag.
♦ iteleka
♦ Weed, Naturalised, Introduced
♦ 87, 88, 228, 287
♦ arid, herbal.

Ipomoea nil (L.) Roth
Convolvulaceae
Convolvulus hederaceus L., *Convolvulus
nil* L., *Ipomoea hederacea auct. non* Jacq.,
Pharbitis limbata Lindl., *Pharbitis nil* (L.)
Choisy (see)
♦ whiteedge morningglory, mbwi
♦ Weed, Noxious Weed, Naturalised
♦ 14, 39, 66, 86, 87, 88, 98, 157, 203, 229,
237, 255, 257, 261, 270, 271, 274, 286,
287, 295
♦ a/pC, cultivated, herbal, toxic.
Origin: pantropical.

Ipomoea obscura (L.) Ker Gawl.
Convolvulaceae
♦ obscure morningglory, small white
morningglory
♦ Weed, Noxious Weed, Naturalised
♦ 13, 87, 88, 98, 101, 203, 218, 229, 287
♦ pC, herbal. Origin: Madagascar.

**Ipomoea obscura (L.) Ker Gawl. var.
fragilis (Choisy) A.Meeuse**

Convolvulaceae
- wild petunia
- Native Weed
- 121
- pC. Origin: southern Africa.

Ipomoea ochracea (Lindl.) G.Don
Convolvulaceae
- fence morningglory, lisisusamgubi
- Noxious Weed, Naturalised, Cultivation Escape
- 86, 101, 229, 261
- cultivated. Origin: possibly tropical Africa.

Ipomoea ochracea (Lindl.) G.Don var. curtisii (House) Stearn
Convolvulaceae
- fence morningglory, African white morningglory
- Noxious Weed
- 229

Ipomoea ochracea (Lindl.) G.Don var. ochracea
Convolvulaceae
- fence morningglory
- Noxious Weed, Naturalised
- 101, 229

Ipomoea pandurata (L.) G.Mey.
Convolvulaceae
- bigroot morningglory, bigfoot morningglory, man under ground, man of the earth, wild potato vine
- Weed, Noxious Weed, Naturalised
- 39, 86, 87, 88, 161, 207, 210, 218, 229, 287
- pC, cultivated, herbal, toxic. Origin: North America.

Ipomoea peltata (L.) Choisy
Convolvulaceae
- = *Merremia peltata* (L.) Merr.
- Weed
- 3, 191

Ipomoea pes-caprae (L.) Sweet
Convolvulaceae
- beach morningglory, bayhops, railroad vine, fue moa, fue vili
- Weed, Noxious Weed
- 14, 87, 88, 209, 218, 221, 229, 286
- pH, cultivated, herbal.

Ipomoea pes-caprae (L.) R.Br. ssp. brasiliensis (L.) Ooststr.
Convolvulaceae
- Brazilian bayhops, sonsol, bayhops, bejuco de playa
- Weed, Noxious Weed, Introduced
- 13, 229, 230, 261
- pC.

Ipomoea pes-tigridis L.
Convolvulaceae
- morningglory, tiger foot morningglory
- Weed, Noxious Weed, Naturalised, Environmental Weed
- 13, 86, 87, 88, 93, 98, 203, 209, 229, 287
- aH, cultivated, herbal. Origin: tropical Africa and Asia.

Ipomoea pileata Roxb.
Convolvulaceae

- Naturalised
- 287

Ipomoea plebeia R.Br.
Convolvulaceae
- bellvine
- Weed, Native Weed
- 39, 55, 87, 88, 269, 276
- cultivated, toxic. Origin: Australia.

Ipomoea plebeia R.Br. ssp. africana A.Meeuse
Convolvulaceae
- Native Weed
- 121
- aC. Origin: southern Africa.

Ipomoea plummerae Gray
Convolvulaceae
- Huachuca mountain morningglory
- Noxious Weed
- 229
- herbal.

Ipomoea plummerae Gray var. cuneifolia (Gray) J.F.Macbr.
Convolvulaceae
- Huachuca mountain morningglory
- Noxious Weed
- 229

Ipomoea plummerae Gray var. plummerae
Convolvulaceae
- Huachuca mountain morningglory
- Noxious Weed
- 229

Ipomoea polyantha Roem. & Schult.
Convolvulaceae
- Weed
- 87, 88

Ipomoea polymorpha Roem. & Schult.
Convolvulaceae
- *Ipomoea heterophylla* R.Br.
- Weed
- 13
- bH.

Ipomoea pulchella Roth
Convolvulaceae
- Environmental Weed
- 257

Ipomoea purga (Wender.) Hayne
Convolvulaceae
- jalap, gialappa vera
- Noxious Weed
- 39, 229
- promoted, herbal, toxic.

Ipomoea purpurea (L.) Roth
Convolvulaceae
- *Convolvulus purpureus* L., *Ipomoea hirsutula* Jacq.f. (see), *Pharbitis purpurea* (L.) Voigt (see)
- tall morningglory, common morningglory, morningglory, purperwinde, wilec purpurowy
- Weed, Quarantine Weed, Noxious Weed, Naturalised, Introduced, Garden Escape, Environmental Weed, Cultivation Escape, Casual Alien
- 34, 39, 42, 51, 55, 63, 73, 76, 80, 86, 87, 88, 95, 98, 101, 121, 157, 158, 161, 174, 180, 198, 203, 207, 210, 211, 212, 218, 221, 228, 229, 236, 237, 243, 245, 255,

261, 269, 270, 279, 280, 283, 286, 287, 295, 300
- aC, arid, cultivated, herbal, toxic. Origin: tropical America.

Ipomoea quamoclit L.
Convolvulaceae
- *Quamoclit pinnata* (Desv.) Boj.
- cypress vine morningglory, cypress vine, morningglory, star of Bethlehem, cardinal climber, Cupid's flower
- Weed, Noxious Weed, Naturalised, Introduced, Garden Escape, Environmental Weed
- 7, 13, 84, 86, 87, 88, 93, 98, 101, 155, 161, 179, 203, 218, 229, 230, 245, 255, 261, 262, 286, 287
- aC, cultivated, herbal. Origin: tropical America.

Ipomoea quinata R.Br.
Convolvulaceae
- Naturalised
- 287

Ipomoea ramosissima (Poir.) Choisy
Convolvulaceae
- *Convolvulus ramosissimus* Poir.
- campainha, corriola
- Weed
- 255
- Origin: Mexico, South America.

Ipomoea repanda Jacq.
Convolvulaceae
- bejuco colorado
- Noxious Weed
- 229

Ipomoea rupicola House
Convolvulaceae
- cliff morningglory
- Noxious Weed
- 229
- herbal.

Ipomoea sagittata Poir.
Convolvulaceae
- saltmarsh morningglory
- Noxious Weed
- 229
- herbal.

Ipomoea sepiaria Roxb.
Convolvulaceae
- = *Ipomoea maxima* G.Don ex Sweet
- Weed
- 13
- cultivated.

Ipomoea setifera Poir.
Convolvulaceae
- bejuco de puerco
- Weed, Noxious Weed
- 87, 88, 229

Ipomoea setosa Ker Gawl.
Convolvulaceae
- Brazilian morningglory
- Noxious Weed, Naturalised
- 101, 229

Ipomoea shumardiana (Torr.) Shinners
Convolvulaceae
- narrowleaf morningglory
- Noxious Weed
- 229

Ipomoea sinensis (Desr.) Choisy
Convolvulaceae
Convolvulus calycinus Roxb.,
Convolvulus hardwickii Spreng. *nom. illeg.*, *Ipomoea calycina* (Roxb.) Benth. ex C.B.Clarke, *Ipomoea cardiosepala* Meissn. (see), *Ipomoea hardwickii* Sweet (see)
♦ Weed
♦ 88, 158, 221, 274
♦ herbal. Origin: southern Africa.

Ipomoea sinensis (Desr.) Choisy ssp. blepharosepala (Hochst. ex A.Rich.) Verdc. ex A.Meeuse
Convolvulaceae
♦ Weed, Native Weed
♦ 121, 242
♦ aC, arid. Origin: southern Africa.

Ipomoea sloteri (House) Ooststr.
Convolvulaceae
Quamoclit sloteri House
♦ Naturalised
♦ 287

Ipomoea stans Cav.
Convolvulaceae
♦ Weed
♦ 199
♦ herbal.

Ipomoea steudelii Millsp.
Convolvulaceae
♦ Steudel's morningglory
♦ Noxious Weed
♦ 229

Ipomoea stolonifera (Cyrillo) Gmel.
Convolvulaceae
= *Ipomoea imperati* (Vahl) Griseb.
♦ seafoam morningglory
♦ Weed
♦ 87, 88, 218, 221
♦ herbal. Origin: Australia.

Ipomoea tenuiloba Torr.
Convolvulaceae
♦ spiderleaf
♦ Noxious Weed
♦ 229
♦ herbal.

Ipomoea tenuiloba Torr. var. lemmonii (Gray) Yatskievych & Mason
Convolvulaceae
♦ spiderleaf
♦ Noxious Weed
♦ 229

Ipomoea tenuiloba Torr. var. tenuiloba
Convolvulaceae
♦ spiderleaf
♦ Noxious Weed
♦ 229

Ipomoea tenuissima Choisy
Convolvulaceae
♦ rockland morningglory
♦ Noxious Weed
♦ 229

Ipomoea thurberi Gray
Convolvulaceae
♦ Thurber's morningglory
♦ Noxious Weed
♦ 229

Ipomoea tiliacea (Willd.) Choisy
Convolvulaceae
= *Ipomoea batatas* (L.) Lam.
♦ darkeye morningglory
♦ Weed
♦ 14, 28, 87, 88, 206, 243
♦ Origin: Central and South America.

Ipomoea trichocarpa Ell.
Convolvulaceae
= *Ipomoea cordatotriloba* Dennst. var. *cordatotriloba*
♦ cotton morningglory
♦ Weed
♦ 84, 87, 88
♦ cultivated, herbal.

Ipomoea trichocarpa Elliot var. australis O'Donell
Convolvulaceae
= *Ipomoea cordatotriloba* Dennst. var. *australis* O'Donell
♦ bejuco
♦ Weed
♦ 295

Ipomoea trichocarpa Ell. var. torreyana (Gray) Shinners
Convolvulaceae
= *Ipomoea cordatotriloba* Dennst. var. *torreyana* (Gray) D.Austin
♦ cotton morningglory
♦ Weed
♦ 218

Ipomoea tricolor Cav.
Convolvulaceae
Convolvulus pauciflorus Willd. ex Roem. & Schult., *Convolvulus pulchellus* Kunth, *Convolvulus rubrocaeruleus* (Hook.) Dietr., *Convolvulus venustus* Spreng., *Ipomoea hookeri* G.Don, *Ipomoea oligantha* Choisy, *Ipomoea pulchella* (Kunth) G.Don, *Ipomoea rubrocaerulea* Hook., *Ipomoea rubro-caerulea* Hook., *Ipomoea violacea* auct. L., *Pharbitis rubrocaeruleus* (Hook.) Planch., *Pharbitis tricolor* (Cav.) Chitt., *Quamoclit mutica* Choisy
♦ morningglory, heavenly blue, pearly gates, flying saucers, wedding bells, summer skies, blue water, grannyvine, multicoloured morningglory, heavenly blue morningglory, liane douce
♦ Weed, Quarantine Weed, Noxious Weed, Naturalised
♦ 39, 76, 88, 101, 161, 189, 203, 220, 229, 247
♦ pC, cultivated, herbal, toxic. Origin: South America.

Ipomoea trifida (Kunth) G.Don
Convolvulaceae
♦ threefork morningglory
♦ Weed, Quarantine Weed
♦ 14, 76, 87, 88, 203, 220

Ipomoea triloba L.
Convolvulaceae
♦ threelobe morningglory, pink convolvulus, potato vine, yaa dok khon, little bell
♦ Weed, Quarantine Weed, Noxious Weed, Naturalised, Introduced,

Environmental Weed
♦ 13, 14, 23, 26, 34, 38, 64, 64, 64, 67, 76, 86, 87, 88, 93, 161, 170, 179, 203, 204, 217, 218, 230, 239, 258, 261, 262, 269, 276, 286, 287
♦ aC, herbal. Origin: tropical America.

Ipomoea tuba (Schltdl.) G.Don
Convolvulaceae
= *Ipomoea violacea* L.
♦ Weed
♦ 87, 88
♦ herbal.

Ipomoea tuberosa L.
Convolvulaceae
= *Merremia tuberosa* (L.) Rendle
♦ Weed
♦ 3, 39
♦ herbal, toxic.

Ipomoea tuboides Deg. & Ooststr.
Convolvulaceae
♦ Hawai'i morningglory
♦ Weed, Noxious Weed
♦ 87, 88, 229
♦ herbal.

Ipomoea turbinata Lag.
Convolvulaceae
Ipomoea muricata (L.) Jacq. (see)
♦ lilac bell, purple moonflower
♦ Weed, Noxious Weed, Naturalised
♦ 101, 161, 179, 229

Ipomoea tyrianthina Lindl.
Convolvulaceae
♦ Weed
♦ 199

Ipomoea vagans Bak.
Convolvulaceae
♦ Weed
♦ 88

Ipomoea violacea L.
Convolvulaceae
Convolvulus tuba Schltdl., *Ipomoea macrantha* Roem. & Schult. (see), *Ipomoea tuba* (Schltdl.) G.Don (see)
♦ heavenly blue morningglory, morningglory, coral de sabana
♦ Weed, Quarantine Weed, Noxious Weed
♦ 76, 88, 154, 220, 229, 261
♦ cultivated, herbal, toxic.

Ipomoea wrightii Gray
Convolvulaceae
♦ Wright's morningglory, palmleaf morningglory
♦ Noxious Weed, Naturalised, Introduced
♦ 38, 101, 161, 229

Ipomopsis pinnata (Cav.) V.Grant
Polemoniaceae
♦ San Luis Mountain ipomopsis
♦ Weed
♦ 199

Iresine calea (Ibanez) Standl.
Amaranthaceae
♦ Weed
♦ 87, 88, 157
♦ S, herbal.

Iresine celosia L.
Amaranthaceae
= *Iresine diffusa* Humb. & Bonp. ex
Willd.
♦ pata de paloma
♦ Weed
♦ 87, 88, 281
♦ herbal.

Iresine diffusa Humb. & Bonp. ex Willd.
Amaranthaceae
Celosia paniculata L., *Iresine celosia*
L. (see), *Iresine celosioides* L., *Iresine*
paniculata (L.) Kuntze, *Iresine*
polymorpha Mart.
♦ Juba's bush, neve da montanha,
paina, pluma
♦ Weed
♦ 157, 255
♦ pH. Origin: Americas.

Iresine herbstii Hook. ex Lindl.
Amaranthaceae
♦ Herbst's bloodleaf, beefsteak
plant, beef plant, blood feast, chicken
gizzard, bloodleaf
♦ Naturalised
♦ 101
♦ cultivated, herbal, toxic.

Iresine lindenii Van Houtte
Amaranthaceae
♦ Linden's bloodleaf
♦ Naturalised
♦ 101
♦ aqua, cultivated.

Iresine rhizomatosa Standl.
Amaranthaceae
♦ bloodleaf, Juda's bush
♦ Weed
♦ 87, 88, 218
♦ herbal.

Iris L. spp.
Iridaceae
♦ iris, flag iris
♦ Weed, Naturalised
♦ 154, 161, 198, 247
♦ herbal, toxic.

Iris aphylla L.
Iridaceae
♦ stool iris
♦ Noxious Weed, Naturalised
♦ 101, 229
♦ cultivated.

Iris bracteata S.Wats.
Iridaceae
♦ Siskiyou iris
♦ Noxious Weed
♦ 229
♦ pH, cultivated, herbal.

Iris brevicaulis Raf.
Iridaceae
♦ zigzag iris
♦ Noxious Weed
♦ 229
♦ cultivated.

Iris chrysophylla T.J.Howell
Iridaceae
♦ yellowleaf iris, yellow flowered iris
♦ Noxious Weed

♦ 229
♦ pH, cultivated, herbal.

Iris cristata Aiton
Iridaceae
♦ dwarf crested iris, crested iris
♦ Noxious Weed
♦ 229
♦ pH, cultivated, herbal.

Iris dichotoma Pall.
Iridaceae
♦ vesper iris
♦ Weed
♦ 297

Iris douglasiana Herb.
Iridaceae
♦ Douglas iris, Pacific coast iris, Marin
iris
♦ Weed, Quarantine Weed, Noxious
Weed
♦ 35, 76, 88, 229
♦ pH, cultivated, herbal.

Iris ensata Thunb.
Iridaceae
Iris kaempferi Sieb. ex Lem.
♦ Russian iris, Japanese water iris,
Japanese iris
♦ Noxious Weed, Naturalised
♦ 101, 229, 297
♦ pH, arid, cultivated, herbal.

Iris fernaldii R.C.Foster
Iridaceae
♦ Fernald's iris
♦ Noxious Weed
♦ 229
♦ pH, cultivated.

Iris flavescens Delile
Iridaceae
♦ lemonyellow iris
♦ Noxious Weed, Naturalised
♦ 101, 229

Iris foetidissima L.
Iridaceae
♦ stinking iris, roast beef plant,
gladwyn, gladdon iris, stinking
galdwyn
♦ Weed, Noxious Weed, Naturalised,
Garden Escape, Environmental Weed
♦ 15, 39, 70, 86, 87, 88, 98, 161, 165, 176,
198, 203, 225, 229, 246, 251, 280
♦ pH, cultivated, herbal, toxic. Origin:
southern Europe.

Iris fulva Ker Gawl.
Iridaceae
♦ copper iris
♦ Noxious Weed
♦ 229
♦ cultivated, herbal.

Iris × fulvala Dykes
Iridaceae
= *Iris fulva* Ker Gawl. × *Iris brevicaulis*
Raf.
♦ iris
♦ Noxious Weed, Naturalised
♦ 101, 229
♦ cultivated.

Iris × germanica L.
Iridaceae

Iris germanica L. (see)
♦ German iris, fleur de lis, bearded
iris, purple flag, saksankurjenmiekka,
ireos, orris root
♦ Naturalised, Garden Escape,
Environmental Weed
♦ 39, 86, 252
♦ cultivated, herbal, toxic. Origin:
obscure, probably the Mediterranean.

Iris germanica L.
Iridaceae
= *Iris × germanica* L.
♦ German iris, fleur de lis, bearded
iris, purple flag, saksankurjenmiekka,
ireos
♦ Weed, Noxious Weed, Naturalised,
Environmental Weed, Cultivation
Escape, Casual Alien
♦ 7, 39, 40, 42, 70, 98, 101, 161, 176, 198,
203, 229, 252, 280, 290
♦ pH, cultivated, herbal, toxic. Origin:
obscure.

Iris germanica L. var. florentina (L.) Dykes
Iridaceae
♦ German iris, orris root, white
German iris
♦ Noxious Weed
♦ 229
♦ cultivated, toxic.

Iris giganticaerulea Small
Iridaceae
♦ giant blue iris
♦ Noxious Weed
♦ 229
♦ herbal.

Iris hartwegii Bak.
Iridaceae
♦ rainbow iris, Pacific coast iris
♦ Noxious Weed
♦ 229
♦ pH, cultivated, herbal.

Iris hartwegii Bak. ssp. australis (Parish) Lenz
Iridaceae
♦ rainbow iris
♦ Noxious Weed
♦ 229
♦ pH.

Iris hartwegii Bak. ssp. columbiana Lenz
Iridaceae
♦ rainbow iris, tuolumne iris
♦ Noxious Weed
♦ 229
♦ pH.

Iris hartwegii Bak. ssp. hartwegii
Iridaceae
♦ Hartweg's iris, rainbow iris
♦ Noxious Weed
♦ 229
♦ pH.

Iris hartwegii Bak. ssp. pinetorum (Eastw.) Lenz
Iridaceae
♦ rainbow iris
♦ Noxious Weed
♦ 229
♦ pH.

Iris helenae Barbey ex Boiss.
Iridaceae
♦ Weed
♦ 221

Iris hexagona Walter
Iridaceae
♦ Dixie iris, blue flag iris
♦ Noxious Weed
♦ 229
♦ cultivated, herbal.

**Iris hexagona Walter var. *flexicaulis*
(Small) R.C.Foster**
Iridaceae
♦ Dixie iris
♦ Noxious Weed
♦ 229

Iris hexagona Walter var. *hexagona*
Iridaceae
♦ Dixie iris
♦ Noxious Weed
♦ 229

**Iris hexagona Walter var. *savannarum*
(Small) R.C.Foster**
Iridaceae
♦ savanna iris
♦ Noxious Weed
♦ 229

Iris innominata Hend.
Iridaceae
♦ Del Norte County iris
♦ Noxious Weed
♦ 229
♦ pH, cultivated, herbal.

Iris japonica Thunb. × *confusa* Sealy
Iridaceae
♦ Weed
♦ 15

Iris lactea Pall. var. *chinensis* Koidz.
Iridaceae
♦ Chinese iris
♦ Weed
♦ 297

Iris lacustris Nutt.
Iridaceae
♦ dwarf lake iris
♦ Noxious Weed
♦ 229

Iris laevigata Fisch.
Iridaceae
♦ iris, Japanese water iris, rabbitear
iris
♦ Noxious Weed, Naturalised, Casual
Alien
♦ 101, 229, 280
♦ aqua, cultivated, herbal.

Iris latifolia (Mill.) Voss
Iridaceae
♦ English iris
♦ Cultivation Escape
♦ 40
♦ cultivated.

Iris lutescens Lam.
Iridaceae
Iris biflora L.
♦ Crimean iris
♦ Weed, Quarantine Weed, Noxious
Weed

♦ 70, 76, 229
♦ cultivated.

Iris macrosiphon Torr.
Iridaceae
♦ bowltube iris, ground iris
♦ Noxious Weed
♦ 229
♦ pH, cultivated, herbal.

Iris missouriensis Nutt.
Iridaceae
♦ Rocky Mountain iris, western blue
flag, western iris, wild iris
♦ Weed, Quarantine Weed, Noxious
Weed
♦ 35, 39, 76, 87, 88, 159, 161, 212, 218,
229
♦ pH, cultivated, herbal, toxic.

Iris munzii R.C.Foster
Iridaceae
♦ Munz's iris
♦ Noxious Weed
♦ 229
♦ pH, cultivated, herbal.

Iris orientalis Mill.
Iridaceae
Iris sanguinea Hornem. ex Donn (see)
♦ yellowband iris, Turkish iris
♦ Weed, Noxious Weed, Naturalised,
Introduced, Garden Escape,
Environmental Weed, Cultivation
Escape
♦ 15, 40, 54, 86, 88, 98, 101, 155, 203,
229, 280
♦ cultivated, herbal. Origin: eastern
Mediterranean.

Iris pallida Lam.
Iridaceae
♦ Dalmatian iris, sweet iris, orris
♦ Noxious Weed, Naturalised
♦ 101, 229
♦ pH, cultivated, herbal. Origin:
Mediterranean.

Iris planifolia (Mill.) Fiori & Paol.
Iridaceae
♦ Weed
♦ 70
♦ cultivated.

Iris prismatica Pursh ex Ker Gawl.
Iridaceae
♦ slender blue iris
♦ Weed, Noxious Weed
♦ 23, 39, 88, 229
♦ cultivated, herbal, toxic.

Iris pseudacorus L.
Iridaceae
♦ yellow flag, yellow water iris, iris,
pale yellow iris, flag iris, iris jaune,
giaggiolo d'acqua, keltakurjenmiekka,
wasserschwertlilie
♦ Weed, Quarantine Weed, Noxious
Weed, Naturalised, Garden Escape,
Environmental Weed
♦ 4, 23, 34, 39, 45, 54, 70, 72, 76, 80, 86,
87, 88, 101, 104, 133, 137, 146, 151, 155,
156, 161, 195, 197, 198, 218, 224, 225,
229, 237, 246, 272, 280, 286, 287, 289,
295, 296, 300
♦ wpH, cultivated, herbal, toxic.

Origin: Europe, Middle East.

Iris pumila L.
Iridaceae
♦ kääpiökurjenmiekka, dwarf iris
♦ Noxious Weed, Naturalised,
Cultivation Escape
♦ 42, 101, 229
♦ cultivated.

Iris purdyi Eastw.
Iridaceae
♦ Purdy's iris
♦ Noxious Weed
♦ 229
♦ pH, cultivated.

Iris × robusta E.Anders.
Iridaceae
= *Iris versicolor* L. × *Iris virginica* L.
♦ iris
♦ Noxious Weed
♦ 229
♦ cultivated.

Iris sancti-cyrii Rouss.
Iridaceae
♦ sanctimonious iris
♦ Noxious Weed
♦ 229

Iris sanguinea Hornem. ex Donn
Iridaceae
= *Iris orientalis* Mill.
♦ Japanese iris
♦ Noxious Weed, Naturalised
♦ 101, 229
♦ pH, cultivated.

Iris setosa Pall. ex Link
Iridaceae
♦ beachhead iris, Arctic iris, iris
♦ Noxious Weed
♦ 229
♦ pH, cultivated, herbal.

**Iris setosa Pall. ex Link var. *canadensis*
M.Foster ex B.L.Robins. & Fern.**
Iridaceae
♦ Canada beachhead iris
♦ Noxious Weed
♦ 229

**Iris setosa Pall. ex Link var. *interior*
E.Anders.**
Iridaceae
♦ wild flag
♦ Noxious Weed
♦ 229

Iris setosa Pall. ex Link var. *setosa*
Iridaceae
♦ beachhead iris
♦ Noxious Weed
♦ 229

Iris sibirica L.
Iridaceae
Xiphium sibiricum Schrank, *Xiridion
sibiricum* Klatt
♦ Siberian iris, siperiankurjenmiekka
♦ Weed, Noxious Weed, Naturalised,
Cultivation Escape
♦ 23, 39, 42, 88, 101, 229, 272
♦ pH, cultivated, herbal, toxic.

Iris sintenisii Janka
Iridaceae

♦ Weed
♦ 272
♦ cultivated.

Iris sisyrinchium L.
Iridaceae
♦ Weed
♦ 221, 248

Iris spuria L.
Iridaceae
♦ seashore iris, iris, butterfly iris, blue iris
♦ Weed, Noxious Weed, Naturalised, Garden Escape, Environmental Weed
♦ 54, 72, 86, 88, 101, 229, 280
♦ pH, cultivated, herbal. Origin: Eurasia.

Iris spuria L. ssp. *ochroleuca* (L.) Dykes
Iridaceae
♦ wild iris, seashore iris
♦ Noxious Weed, Naturalised
♦ 101, 229

Iris tectorum Maxim.
Iridaceae
♦ wild iris, Japanese roof iris, roof iris, wall iris
♦ Noxious Weed, Naturalised
♦ 101, 229
♦ pH, cultivated, herbal.

Iris tenax Douglas ex Lindl.
Iridaceae
♦ wild iris, Oregon iris, toughleaf iris
♦ Noxious Weed
♦ 229
♦ pH, cultivated, herbal.

Iris tenax Douglas ex Lindl. ssp. klamathensis Lenz
Iridaceae
♦ wild iris, Klamath iris, Orleans iris
♦ Noxious Weed
♦ 229
♦ pH.

Iris tenax Douglas ex Lindl. ssp. *tenax*
Iridaceae
♦ wild iris, toughleaf iris
♦ Noxious Weed
♦ 229

Iris tenuis S.Wats.
Iridaceae
♦ Clackamas iris
♦ Noxious Weed
♦ 229
♦ cultivated, herbal.

Iris tenuissima Dykes
Iridaceae
♦ longtube iris, iris
♦ Noxious Weed
♦ 229
♦ pH, cultivated, herbal.

Iris tenuissima Dykes ssp. *purdyiformis* (R.C.Foster) Lenz
Iridaceae
♦ longtube iris
♦ Noxious Weed
♦ 229
♦ pH.

Iris tenuissima Dykes ssp. *tenuissima*
Iridaceae

♦ longtube iris
♦ Noxious Weed
♦ 229
♦ pH.

Iris thompsonii R.C.Foster
Iridaceae
♦ Thompson's iris
♦ Noxious Weed
♦ 229

Iris tingitana Boiss. & Reut.
Iridaceae
♦ Morocco iris
♦ Noxious Weed
♦ 229
♦ cultivated.

Iris tingitana Boiss. & Reut. × *xiphium* L.
Iridaceae
♦ Spanish iris
♦ Weed
♦ 15

Iris tridentata Pursh
Iridaceae
♦ savannah iris
♦ Noxious Weed
♦ 229
♦ cultivated.

Iris unguicularis Poir.
Iridaceae
♦ winter flowering iris, winter iris, Algerian iris
♦ Weed, Naturalised, Garden Escape, Cultivation Escape
♦ 86, 98, 203, 252
♦ cultivated, herbal. Origin: Mediterranean.

Iris variegata L.
Iridaceae
♦ Hungarian iris
♦ Noxious Weed, Naturalised
♦ 39, 101, 229
♦ cultivated, herbal, toxic.

Iris verna L.
Iridaceae
♦ dwarf violet iris
♦ Noxious Weed
♦ 229
♦ herbal.

Iris verna L. var. *smalliana* Fern. ex M.E.Edwards
Iridaceae
♦ dwarf violet iris
♦ Noxious Weed
♦ 229

Iris verna L. var. *verna*
Iridaceae
♦ dwarf violet iris
♦ Noxious Weed
♦ 229

Iris versicolor L.
Iridaceae
♦ blue flag iris, harlequin blue flag, purple iris, wild iris, blue flag
♦ Weed, Noxious Weed
♦ 8, 23, 39, 87, 88, 161, 218, 229
♦ pH, cultivated, herbal, toxic.

Iris vinicolor Small (*pro* sp.)
Iridaceae

♦ winecolor iris
♦ Noxious Weed
♦ 229
♦ cultivated.

Iris virginica L.
Iridaceae
♦ Virginia iris, southern blue flag
♦ Quarantine Weed, Noxious Weed
♦ 229, 258
♦ cultivated, herbal.

Iris virginica L. var. *shrevei* (Small) E.Anders.
Iridaceae
♦ Shreve's iris
♦ Noxious Weed
♦ 229

Iris virginica L. var. *virginica*
Iridaceae
♦ Virginia iris
♦ Noxious Weed
♦ 229

Iris xiphium L.
Iridaceae
Iris hispanica hort. ex Steud., *Iris lusitanica* Ker Gawl., *Iris taitii* Foster
♦ Spanish iris
♦ Weed, Noxious Weed, Naturalised
♦ 40, 54, 70, 86, 88, 101, 229
♦ cultivated, herbal. Origin: North Africa, southern Europe.

Isachne carolinensis Ohwi
Poaceae
♦ isachne
♦ Introduced
♦ 230
♦ G.

Isachne chevalieri A.Camus.
Poaceae
♦ Weed
♦ 87, 88
♦ G.

Isachne debilis Rendle
Poaceae
♦ Weed
♦ 235
♦ G.

Isachne dispar Trin.
Poaceae
= *Isachne globosa* (Thunb.) Kuntze
♦ Weed
♦ 87, 88
♦ G.

Isachne globosa (Thunb.) Kuntze
Poaceae
Isachne australis R.Br., *Isachne dispar* Trin (see)
♦ chigozasa
♦ Weed
♦ 87, 88, 170, 191, 263, 275, 286, 297
♦ pG, aqua, cultivated. Origin: Australasia.

Isachne kunthiana (Wight & Arn.) Miq.
Poaceae
Isachne schmidii Hack.
♦ Weed, Quarantine Weed
♦ 76, 87, 88, 203, 220
♦ G.

Isachne myosotis Nees
Poaceae
♦ Weed
♦ 87, 88
♦ G.

Isachne ponapensis Hosok.
Poaceae
♦ Introduced
♦ 230
♦ G.

Isachne pulchella Roem. & Schult.
Poaceae
Isachne miliacea Roth
♦ isachne
♦ Weed, Naturalised
♦ 7, 86, 88, 98, 170, 203
♦ G, aqua. Origin: tropical Asia, Australia.

Isatis lusitanica L.
Brassicaceae
♦ Weed
♦ 221, 243
♦ a/bH, cultivated, herbal.

Isatis tinctoria L.
Brassicaceae
♦ dyer's woad, woad, morsinko, Marlahan mustard
♦ Weed, Quarantine Weed, Noxious Weed, Naturalised, Introduced, Environmental Weed
♦ 1, 26, 34, 35, 56, 70, 76, 78, 80, 87, 88, 94, 101, 116, 136, 138, 146, 161, 180, 203, 212, 218, 219, 220, 229, 241, 243, 264, 267, 272, 300
♦ b/pH, cultivated, herbal. Origin: south-eastern Russia.

Ischaemum L. spp.
Poaceae
♦ muraina grass
♦ Weed
♦ 6, 88, 191
♦ G.

Ischaemum afrum (J.F.Gmel.) Dandy
Poaceae
Ischaemum brachyatherum (Hochst.) Fenzl ex Hack. (see), *Ischaemum glaucostachyum* Stapf
♦ Weed, Quarantine Weed, Naturalised, Native Weed
♦ 76, 86, 87, 88, 121, 203, 220, 242
♦ pG, arid. Origin: Africa.

Ischaemum aristatum L.
Poaceae
Andropogon crassipes Steud., *Ischaemum crassipes* (Steud.) Thell. (see)
♦ Weed, Quarantine Weed
♦ 76, 87, 88, 203, 220
♦ G.

Ischaemum barbatum Retz.
Poaceae
♦ Weed
♦ 12, 88, 191, 204
♦ G. Origin: Australia.

Ischaemum brachyatherum (Hochst.) Fenzl ex Hack.
Poaceae
= *Ischaemum afrum* (J.F.Gmel.) Dandy

♦ Weed
♦ 87, 88
♦ G.

Ischaemum crassipes (Steud.) Thell.
Poaceae
= *Ischaemum aristatum* L.
♦ Weed
♦ 87, 88
♦ G.

Ischaemum digitatum Brongn. var. polystachyum (Presl) Hack.
Poaceae
= *Ischaemum polystachyum* J.Presl
♦ Weed
♦ 3, 191
♦ G.

Ischaemum fasciculatum Brongn.
Poaceae
Ischaemum arcuatum (Nees) Stapf
♦ border grass
♦ Native Weed
♦ 121
♦ pG. Origin: southern Africa.

Ischaemum indicum (Houtt.) Merr.
Poaceae
Ischaemum aristatus auct. non L., *Ischaemum ciliare* Retz., *Phleum indicum* Houtt.
♦ smutgrass, Batiki bluegrass, Indian murainagrass
♦ Weed, Naturalised, Introduced
♦ 87, 88, 90, 228
♦ a/pG, arid, cultivated.

Ischaemum muticum L.
Poaceae
♦ seashore, centipede grass, droughtgrass
♦ Weed
♦ 87, 88, 90
♦ pG, herbal. Origin: Asia, Australia.

Ischaemum pilosum (Klein ex Willd.) Wight
Poaceae
♦ Weed
♦ 87, 88
♦ G.

Ischaemum polystachyum Presl
Poaceae
Ischaemum digitatum Brongn. var. *polystachyum* (Presl) Hack. (see)
♦ paddle grass, reh padil, mah
♦ Weed
♦ 3, 107, 191
♦ G.

Ischaemum rugosum Salisb.
Poaceae
Andropogon arnottianus (Nees) Steud., *Andropogon griffithsiae* (Nees ex Steud.) Steud., *Andropogon segetum* (Trin.) Steud., *Andropogon tong-dong* Steud., *Colladoa distachia* Cav., *Ischaemum alkoense* Honda, *Ischaemum colladoa* Spreng., *Ischaemum rugosum* Salisb. var. *rugosum*, *Ischaemum rugosum* var. *segetum* (Trin.) Hack., *Ischaemum segetum* Trin., *Meoschium arnottianum* Nees, *Meoschium griffithii* Nees & Arn., *Meoschium griffithsiae* Nees ex Steud.,

Meoschium royleanum Nees ex Steud., *Meoschium rugosum* (Salisb.) Nees, *Meoschium rugosum* Wall., *Meoschium wightianum* Nees
♦ wrinklegrass, wrinkled grass, wrinkle duck beak, yaa daeng, saramolla grass, ribbed murainagrass, muraina grass, tho muraina, co muraina, saramatta grass, falsa caminadora
♦ Weed, Quarantine Weed, Noxious Weed, Naturalised, Introduced
♦ 3, 14, 30, 32, 38, 67, 68, 76, 87, 88, 90, 101, 140, 161, 170, 186, 191, 204, 229, 239, 255, 263, 281
♦ a/pG, aqua. Origin: tropical Asia.

Ischaemum timorense Kunth
Poaceae
♦ waidoi grass, centipede grass, stalkleaf murainagrass
♦ Weed, Quarantine Weed, Naturalised
♦ 3, 76, 87, 88, 90, 101, 135, 170, 191, 203, 220
♦ pG.

Ischnosiphon leucophoeus Körn.
Marantaceae
♦ Weed
♦ 87, 88

Iseilema laxum Hack.
Poaceae
♦ Weed
♦ 66
♦ G, cultivated.

Iseilema membranaceum (Lindl.) Domin
Poaceae
♦ small Flinders grass
♦ Introduced
♦ 228
♦ G, arid, cultivated.

Iseilema vaginiflorum Domin
Poaceae
♦ Weed
♦ 87, 88
♦ G, arid, cultivated.

Isnardia alternifolia DC.
Onagraceae
♦ Quarantine Weed
♦ 220

Isocarpha bilbergiana Less.
Asteraceae
♦ Weed
♦ 87, 88

Isocarpha oppositifolia (L.) Cass.
Asteraceae
♦ Rio Grande pearlhead
♦ Weed, Naturalised
♦ 101, 157
♦ aH, herbal.

Isocoma acradenia (Greene) Greene
Asteraceae
♦ paleleaf goldenweed, desert isocoma, alkali goldenbush, alkali jimmyweed
♦ Weed
♦ 161
♦ S, herbal, toxic.

Isocoma coronopifolia (Gray) Greene
Asteraceae
♦ common goldenbush, common goldenweed
♦ Weed
♦ 161
♦ arid, toxic.

Isodon adenanthus (Diels) Kudo
Lamiaceae
♦ glandularflower rabdosia
♦ Weed
♦ 297

Isodon glaucocalyx (Maxim.) Kudo
Lamiaceae
♦ Japanese rabdosia
♦ Weed
♦ 297

Isolatocereus Backeb. spp.
Cactaceae
♦ Weed, Quarantine Weed
♦ 76, 88, 220

Isolepis antarctica Nees
Cyperaceae
Scirpus antarcticus L.
♦ sedge
♦ Native Weed
♦ 121
♦ aG. Origin: southern Africa.

Isolepis cernua (Vahl) Roem. & Schult.
Cyperaceae
Scirpus cernuus Vahl (see)
♦ slender clubrush, low bulrush, nodding scirpus
♦ Native Weed
♦ 121
♦ a/pG, cultivated.

Isolepis fluitans (L.) R.Br.
Cyperaceae
Scirpus fluitans L. (see)
♦ floating clubrush
♦ Native Weed
♦ 121
♦ pG, aqua, cultivated.

Isolepis hystrix (Thunb.) Nees
Cyperaceae
♦ bottlebrush bulrush, awned clubrush, awned club sedge
♦ Weed, Naturalised, Environmental Weed
♦ 7, 9, 72, 86, 88, 98, 101, 176, 198, 203
♦ aG, cultivated. Origin: southern Africa.

Isolepis marginata (Thunb.) Diels
Cyperaceae
♦ Weed, Naturalised, Native Weed
♦ 7, 15, 86, 280
♦ G. Origin: southern Africa.

Isolepis natans (Thunb.) Dietr.
Cyperaceae
Scirpus subprolifer Boeck.
♦ Native Weed
♦ 121
♦ pG. Origin: southern Africa.

Isolepis nodosa (Rottb.) R.Br.
Cyperaceae
♦ knobby clubrush

♦ Weed
♦ 181
♦ G, cultivated.

Isolepis platycarpa (S.T.Blake) Soják
Cyperaceae
♦ Naturalised
♦ 280
♦ G, cultivated.

Isolepis prolifera (Rottb.) R.Br.
Cyperaceae
Scirpus prolifer Rottb.
♦ proliferous club sedge
♦ Weed, Naturalised, Native Weed, Environmental Weed
♦ 7, 86, 98, 121, 198, 200, 203
♦ wpG. Origin: southern Africa.

Isolepis pseudosetacea Daveau
Cyperaceae
♦ espanjanluikka
♦ Casual Alien
♦ 42
♦ G.

Isolepis sepulcralis Steud.
Cyperaceae
Scirpus chlorostachyus Levyns
♦ African club sedge
♦ Weed, Naturalised, Native Weed
♦ 15, 86, 121, 198, 280
♦ pG. Origin: southern Africa.

Isolepis setacea (L.) R.Br.
Cyperaceae
Scirpus setaceus L. (see)
♦ sukaluikka, bristleleaf bulrush, bristle scirpus, bristle clubrush
♦ Weed, Naturalised, Native Weed, Casual Alien
♦ 15, 42, 86, 101, 121, 176, 280
♦ pG, aqua, cultivated, herbal. Origin: Europe, Africa, west Asia.

Isolepis stellata (C.B.Clarke) K.L.Wilson
Cyperaceae
♦ Naturalised
♦ 86
♦ G, cultivated. Origin: Australia.

Isoplexis isabelliana (Webb) Masf.
Scrophulariaceae
♦ Quarantine Weed
♦ 220
♦ cultivated.

Isotoma longiflora Presl
Campanulaceae/Lobeliaceae
= *Hippobroma longiflora* (L.) G.Don
♦ shrub sharebell
♦ Weed, Quarantine Weed, Naturalised
♦ 39, 76, 86, 88, 218, 220
♦ cultivated, herbal, toxic.

Iva angustifolia Nutt. ex DC.
Asteraceae
♦ narrowleaf sumpweed, narrowleaf marsh elder
♦ Weed
♦ 161
♦ herbal, toxic.

Iva annua L.
Asteraceae

♦ annual marsh elder
♦ Weed
♦ 161, 207
♦ herbal. Origin: North America.

Iva axillaris Pursh
Asteraceae
Iva axillaris ssp. *robustior* (Hook.) Bassett (see)
♦ lesser marsh elder, small flowered marsh elder, iva povertyweed, povertyweed, death weed, devil's weed, mouse ear povertyweed, poverty sumpweed
♦ Weed, Quarantine Weed, Noxious Weed, Naturalised, Native Weed
♦ 23, 35, 49, 52, 76, 87, 88, 98, 147, 161, 174, 203, 210, 212, 218, 220, 229, 243, 269, 294, 299
♦ pH, herbal. Origin: North America.

Iva axillaris Pursh ssp. *robustior* (Hook.) Bassett
Asteraceae
= *Iva axillaris* Pursh
♦ povertyweed, death weed, devil's weed, mouse ear povertyweed, poverty sumpweed
♦ Noxious Weed, Naturalised
♦ 86, 198
♦ pH, aqua, herbal. Origin: North America.

Iva cheiranthifolia Kunth
Asteraceae
♦ fly marsh elder, bush iva
♦ Weed, Naturalised
♦ 14, 101

Iva ciliata Willd.
Asteraceae
= *Iva annua* L. var. *annua* (NoR)
♦ rough sumpweed
♦ Weed, Casual Alien
♦ 42, 87, 88, 218
♦ herbal.

Iva xanthifolia Nutt.
Asteraceae
Cyclachaena xanthifolia (Nutt.) Fresen.
♦ marsh elder, giant marsh elder, horseweed, false ragweed, burrweed marsh elder, giant sumpweed, karheaiiva
♦ Weed, Noxious Weed, Naturalised, Native Weed, Casual Alien
♦ 23, 39, 40, 42, 49, 52, 87, 88, 136, 161, 174, 210, 212, 218, 229, 272, 287
♦ cultivated, herbal, toxic. Origin: North America.

Ixeris chinensis (Thunb.) Nakai
Asteraceae
♦ rabbit milkweed
♦ Weed
♦ 235, 273, 274, 275, 297
♦ pH, promoted, herbal.

Ixeris chinensis (Thunb.) Nakai ssp. *strigosa* Kitam.
Asteraceae
♦ Weed
♦ 286

Ixeris debilis (Thunb.) A.Gray
Asteraceae
♦ oojishibari
♦ Weed
♦ 263, 286
♦ pH.

Ixeris dentata (Thunb.) Nakai
Asteraceae
Lactuca dentata (Thunb.) Robins.,
Lactuca thunbergii (A.Gray) Maxim.
♦ niga na, lettuce
♦ Weed
♦ 87, 88, 204, 263, 286
♦ pH, promoted.

Ixeris japonica (Burm.) Nakai
Asteraceae
Lactuca debilis (Thunb.) Benth
♦ Weed
♦ 87, 88, 235
♦ pH, promoted.

Ixeris laevigata (Bl.) Sch.Bip.
Asteraceae
♦ Weed
♦ 87, 88

Ixeris polycephala Cass.
Asteraceae
Lactuca polycephala (Cass.) Benth.
♦ Weed
♦ 87, 88, 275, 286, 297
♦ bH, promoted.

Ixeris sonchifolia (Bunge) Hance
Asteraceae
♦ Weed
♦ 275, 297
♦ pH.

Ixeris stolonifera A.Gray
Asteraceae
Lactuca stolonifera (A.Gray) Benth. (see)
♦ creeping lettuce, iwanigana
♦ Weed, Quarantine Weed,
Naturalised
♦ 76, 87, 88, 101, 203, 220, 263, 286
♦ pH, promoted.

Ixia campanulata Houtt.
Iridaceae
♦ red cornlily, bellflower African
cornlily
♦ Naturalised
♦ 40, 101
♦ cultivated.

Ixia flexuosa L.
Iridaceae
♦ koringblommetjie, ixia
♦ Weed, Naturalised, Garden Escape,
Environmental Weed
♦ 72, 86, 88, 98, 176, 203, 251
♦ pH, cultivated. Origin: South Africa.

Ixia hybrida Ker Gawl.
Iridaceae
♦ Weed, Naturalised
♦ 98, 203

Ixia longituba N.E.Br.
Iridaceae
♦ longflower ixia, ixia
♦ Weed, Sleeper Weed, Naturalised,
Environmental Weed

♦ 54, 86, 88, 155, 198
♦ Origin: South Africa.

Ixia maculata L.
Iridaceae
♦ spotted African cornlily,
geelkalossie, African cornlily, yellow
ixia
♦ Weed, Naturalised, Garden Escape,
Environmental Weed
♦ 7, 15, 72, 86, 88, 98, 101, 176, 198, 203,
280, 289
♦ pH, cultivated, herbal. Origin: South
Africa.

Ixia paniculata D.Delaroche
Iridaceae
♦ tubular cornlily, ixia
♦ Weed, Naturalised, Environmental
Weed, Casual Alien
♦ 7, 40, 72, 86, 88, 98, 198, 203, 280
♦ pH, cultivated. Origin: South Africa.

Ixia polystachya L.
Iridaceae
Ixia leucantha Jacq.
♦ African cornlily, cornlily, variable
ixia, koringblommetjie
♦ Weed, Naturalised, Native Weed,
Garden Escape, Environmental Weed
♦ 7, 72, 86, 88, 98, 121, 176, 198, 203,
251, 280
♦ pH, cultivated. Origin: South Africa.

Ixia viridiflora Lam.
Iridaceae
♦ green ixia
♦ Weed, Naturalised, Garden Escape,
Environmental Weed
♦ 54, 72, 86, 88, 98, 203, 251
♦ pH, cultivated. Origin: South Africa.

Ixiolirion montanum Herb.
Liliaceae/Ixioliriaceae
♦ Weed
♦ 87, 88, 243

Ixophorus unisetus (J.Presl) Schltdl.
Poaceae
Urochloa uniseta Presl
♦ pasto Honduras, Mexican grass,
zacate leche
♦ Weed, Naturalised
♦ 28, 68, 87, 88, 101, 157, 199, 243, 281
♦ G, cultivated.

Ixora acuminata Roxb.
Rubiaceae
♦ bola de nieve, nevado, sharpleaf
ixora
♦ Naturalised, Introduced
♦ 101, 261
♦ Origin: Himalayas.

Ixora chinensis Lam.
Rubiaceae
♦ Naturalised
♦ 287
♦ cultivated, herbal.

Ixora coccinea L.
Rubiaceae
♦ jungle geranium, scarlet jungleflame,
amor ardiente, bola de coral
♦ Weed, Naturalised, Introduced,
Cultivation Escape

♦ 32, 101, 179, 230, 261
♦ T, cultivated, herbal. Origin: India.

Ixora paludosa (Bl.) Kurz
Rubiaceae
♦ Weed
♦ 13

Ixora parviflora Vahl nom. illeg.
Rubiaceae
= *Ixora pavetta* Andrews
♦ torch tree
♦ Weed
♦ 179
♦ cultivated, herbal.

Ixora pavetta Andrews
Rubiaceae
Ixora parviflora Vahl nom. illeg. (see)
♦ torch tree
♦ Naturalised
♦ 101
♦ cultivated.

Ixora thwaitesii Hook.f.
Rubiaceae
♦ white jungleflame, bola de nieve,
white ixora
♦ Naturalised, Introduced
♦ 101, 261
♦ Origin: Sri Lanka.

J

Jaborosa bergii Hieron.
Solanaceae
♦ jaborosa
♦ Weed
♦ 237, 295

Jaborosa integrifolia Lam.
Solanaceae
♦ spring blossom
♦ Weed, Naturalised
♦ 87, 88, 101, 237, 295

Jaborosa runcinata Lam.
Solanaceae
♦ flor de sapo
♦ Weed, Quarantine Weed
♦ 76, 87, 88, 203, 220, 237, 295

Jacaranda Juss. spp.
Bignoniaceae
♦ green ebony, jacaranda
♦ Quarantine Weed
♦ 220
♦ herbal.

Jacaranda mimosaefolia D.Don
Bignoniaceae
= *Jacaranda mimosifolia* D.Don
♦ jacaranda
♦ Casual Alien
♦ 280
♦ cultivated.

Jacaranda mimosifolia D.Don
Bignoniaceae
Jacaranda mimosaefolia D.Don (see)
♦ jacaranda tree, black poui, jacaranda,
blue Brazilian, Brazilian rosewood,
fern tree
♦ Weed, Noxious Weed, Naturalised,
Garden Escape, Environmental Weed,
Cultivation Escape
♦ 22, 63, 73, 86, 88, 95, 101, 121, 152,
158, 179, 260, 261, 277, 279, 283
♦ T, cultivated, herbal. Origin: South
America.

Jacaranda oxyphylla Cham.
Bignoniaceae
♦ Weed
♦ 87, 88
♦ herbal.

Jacaratia corumbensis Kuntze
Caricaceae
Jacaratia hassleriana Chodat, *Jacaratia
heptaphylla* (Vell.) A.DC. fo. *inermis*
Kuntze
♦ Introduced
♦ 228
♦ arid.

Jacobaeastrum candollei Kuntze
Asteraceae
= *Euryops candollei* Harv.
♦ Quarantine Weed
♦ 220

Jacobinia carnea (Lindl.) G.Nicholson
Acanthaceae
= *Justicia carnea* Lindl.
♦ pink plume flower, pink jacobinia
♦ Weed
♦ 3, 191
♦ cultivated.

**Jacquemontia agrestis (Mart. ex Choisy)
Meisn.**
Convolvulaceae
♦ midnightblue clustervine
♦ Naturalised
♦ 101

**Jacquemontia heterantha (Nees & Mart.)
Hallier f.**
Convolvulaceae
Dufourea heterantha Nees & Mart
♦ campainha, corda de viola, corriola
♦ Weed
♦ 255
♦ Origin: tropical America.

Jacquemontia martii Choisy
Convolvulaceae
♦ Weed
♦ 87, 88

**Jacquemontia paniculata (Burm.f.)
Hallier f.**
Convolvulaceae
♦ Weed
♦ 13
♦ C, cultivated.

Jacquemontia pentantha (Jacq.) G.Don
Convolvulaceae
♦ Weed
♦ 87, 88
♦ cultivated.

Jacquemontia pycnocephala Benth.
Convolvulaceae
♦ Weed
♦ 87, 88

Jacquemontia sandwicensis A.Gray
Convolvulaceae
= *Jacquemontia ovalifolia* (Vahl ex West)
Hallier.f. ssp. *sandwicensis* (Gray)
Robertson (NoR)
♦ beach jacquemontia
♦ Weed
♦ 87, 88, 218
♦ herbal.

Jacquemontia tamnifolia (L.) Griseb.
Convolvulaceae
♦ smallflower morningglory, hairy
clustervine
♦ Weed, Quarantine Weed,
Naturalised
♦ 14, 76, 84, 87, 88, 157, 161, 203, 221,
206, 218, 220, 243, 287
♦ aC, herbal.

Jacquinia aculeata (L.) Mez.
Theophrastaceae
♦ Weed
♦ 14

Jacquinia arborea Vahl
Theophrastaceae
= *Jacquinia armillaris* Jacq. (NoR)
♦ braceletwood
♦ Weed
♦ 39, 179
♦ toxic.

Jacquinia macrocarpa Cav.
Theophrastaceae
Jacquinia arenicola Brandegee, *Jacquinia
aurantiaca* W.T.Aiton var. *latifolia* Mez,
Jacquinia aurantiaca W.T.Aiton, *Jacquinia
axillaris* Oerst., *Jacquinia cuneata*
Standl., *Jacquinia leptopoda* Lundell,
Jacquinia liebmannii Mez, *Jacquinia
luzonensis* C.Presl, *Jacquinia mexicana
hort.* ex Regel, *Jacquinia racemosa*
A.DC., *Jacquinia schiedeana* Mez
♦ Weed
♦ 179
♦ herbal.

Jaegeria hirta (Lag) Less.
Asteraceae
♦ botao de ouro
♦ Weed
♦ 255
♦ Origin: South America.

**Jarava plumosa (Spreng.) S.L.W.Jacobs
& J.Everett**
Poaceae
Stipa papposa Nees (see)
♦ South American ricegrass
♦ Naturalised, Environmental Weed
♦ 86, 101, 290
♦ G. Origin: South America.

Jasione montana L.
Campanulaceae
♦ sheep's bit, vuorimunkki
♦ Weed, Naturalised, Casual Alien
♦ 23, 88, 94, 101, 272, 280
♦ a/bH, cultivated, herbal.

Jasminocereus Britton & Rose spp.
Cactaceae
♦ Weed, Quarantine Weed
♦ 76, 88, 203, 220

Jasminum L. spp.
Oleaceae
♦ jasmine, Gold Coast jasmine, Azores
jasmine
♦ Weed, Environmental Weed
♦ 73, 88, 151, 161
♦ toxic.

Jasminum azoricum auct. non L.
Oleaceae
= *Jasminum fluminense* Vell.
♦ Weed
♦ 87, 88
♦ cultivated.

Jasminum beesianum Forrest & Diels
Oleaceae
♦ Naturalised
♦ 280
♦ cultivated.

Jasminum bifarium Wall ex G.Don
Oleaceae
♦ Weed
♦ 12
♦ herbal.

Jasminum dichotomum Vahl
Oleaceae
♦ Gold Coast jasmine
♦ Weed, Noxious Weed, Naturalised,
Garden Escape, Environmental Weed
♦ 3, 80, 88, 101, 112, 122, 142, 179, 191
♦ cultivated.

Jasminum floribundum R.Br. ex Fresen.
Oleaceae
♦ Abyssinia jasmine
♦ Weed
♦ 39, 221
♦ cultivated, herbal, toxic.

Jasminum fluminense Vell.
Oleaceae
Jasminum azoricum auct. non L. (see)
♦ Brazilian jasmine, jazmin de trapo
♦ Weed, Noxious Weed, Naturalised,
Garden Escape, Environmental Weed,
Cultivation Escape
♦ 3, 80, 88, 101, 112, 122, 142, 179, 191,
233, 261
♦ cultivated. Origin: Africa.

Jasminum fruticans L.
Oleaceae
Jasminum humile hort.
♦ jasmine
♦ Weed
♦ 87, 88
♦ cultivated, herbal.

Jasminum grandiflorum L.
Oleaceae
= *Jasminum officinale* L.
♦ Spanish jasmine, luktjasmin,
aitojasmiini, jasmin
♦ Introduced, Casual Alien
♦ 38, 228, 261
♦ arid, cultivated, herbal. Origin:
south-west Asia.

Jasminum humile L.
Oleaceae
Jasminum bignoniaceum Wall.
♦ yellow jasmine, Italian yellow
jasmine
♦ Quarantine Weed, Naturalised,
Introduced, Environmental Weed,
Cultivation Escape
♦ 38, 225, 246, 280, 283
♦ S, cultivated, herbal. Origin: China
to the Himalayas.

Jasminum mesnyi Hance
Oleaceae
♦ primrose jasmine, Japanese jasmine
♦ Weed, Naturalised
♦ 15, 86, 101, 179, 198, 280
♦ cultivated.

**Jasminum multiflorum (Burm.f.)
Andrews**
Oleaceae
♦ star jasmine, jazmín de papel
♦ Weed, Naturalised, Introduced,
Cultivation Escape
♦ 38, 80, 101, 179, 261
♦ cultivated, herbal. Origin: India,
Nepal, Pakistan.

Jasminum multipartitum Hochst.
Oleaceae
♦ jasmine, many petalled jasmine,

wild jasmine, wild jessamine
♦ Native Weed
♦ 121
♦ pS, cultivated. Origin: southern
Africa.

Jasminum nitidum Skan
Oleaceae
♦ angelwing jasmine, royal jasmine
♦ Weed, Naturalised
♦ 101, 179
♦ cultivated.

Jasminum nudiflorum Lindl.
Oleaceae
Jasminum sieboldianum Blume
♦ winter jasmine, winter flowering
jasmin
♦ Naturalised
♦ 101
♦ S, cultivated, herbal.

Jasminum officinale L.
Oleaceae
Jasminum grandiflorum L. (see)
♦ poet's jasmine, common jasmine,
true jasmione, jessamine, gelsomino,
common white jasmine
♦ Weed, Naturalised
♦ 39, 101, 179, 280
♦ pC, cultivated, herbal, toxic.

Jasminum polyanthum Franch.
Oleaceae
♦ winter jasmine, jasmine
♦ Weed, Naturalised, Garden Escape,
Environmental Weed, Cultivation
Escape
♦ 15, 86, 155, 165, 225, 246, 280, 290
♦ pC, cultivated. Origin: west China.

Jasminum sambac (L.) Aiton
Oleaceae
♦ Arabian jasmine
♦ Weed, Naturalised, Introduced,
Environmental Weed, Cultivation
Escape
♦ 38, 39, 80, 88, 101, 112, 179, 261
♦ cultivated, herbal, toxic. Origin:
tropical Asia.

Jasminum subtriplinerve Bl.
Oleaceae
♦ Weed
♦ 87, 88

Jatropha angustifolia Griseb.
Euphorbiaceae
♦ Weed
♦ 14

Jatropha cathartica Terán & Berland.
Euphorbiaceae
♦ Berlandier's nettlespurge
♦ Weed
♦ 39, 161
♦ toxic.

Jatropha curcas L.
Euphorbiaceae
Curcas curcas (L.) Britton
♦ physic nut, Barbados nut, curcas
bean, purge nut, purging nut, pinhao
de purga, pinhao do Paraguai, pinhao
bravo, pourghere, bed bug plant, big
purge nut, black vomit nut, Brazilian
stinging nut, Cuban physic nut, hell

oil, purging nut tree
♦ Weed, Quarantine Weed, Noxious
Weed, Naturalised, Introduced,
Environmental Weed, Cultivation
Escape
♦ 22, 39, 62, 76, 86, 87, 88, 93, 98, 101,
121, 147, 161, 179, 191, 203, 228, 257,
261, 279
♦ S/T, arid, cultivated, herbal, toxic.
Origin: South America.

Jatropha dioica Sessè ex Cerv.
Euphorbiaceae
Jatropha spathulata (Ort.) Müll.Arg.
(see), *Mozinna spathulata* Ortega
♦ leatherstem
♦ Weed
♦ 161
♦ cultivated, herbal, toxic.

Jatropha glandulifera Roxb.
Euphorbiaceae
♦ Weed
♦ 39, 87, 88
♦ toxic.

Jatropha gossypiifolia L.
Euphorbiaceae
Adenoropium gossypifolium (L.) Pohl.
(see)
♦ bellyache bush, cottonleaf physic
nut
♦ Weed, Quarantine Weed, Noxious
Weed, Naturalised, Environmental
Weed
♦ 6, 7, 11, 12, 14, 22, 39, 62, 76, 86, 87,
88, 93, 98, 101, 147, 155, 161, 179, 203,
205, 209, 258
♦ S, cultivated, herbal, toxic. Origin:
Caribbean, tropical America.

**Jatropha gossypiifolia L. var. *elegans*
(Pohl) Müll.Arg.**
Euphorbiaceae
♦ bellyache bush
♦ Naturalised
♦ 101

Jatropha integerrima Jacq.
Euphorbiaceae
♦ peregrina
♦ Weed, Naturalised, Introduced
♦ 39, 101, 161, 179, 261
♦ arid, cultivated, herbal, toxic. Origin:
Cuba.

**Jatropha integerrima Jacq. var. *hastata*
(Jacq.) Fosberg**
Euphorbiaceae
♦ peregrina
♦ Naturalised
♦ 101

**Jatropha integerrima Jacq. var.
*integerrima***
Euphorbiaceae
♦ peregrina
♦ Naturalised
♦ 101

Jatropha macrorhiza Benth.
Euphorbiaceae
♦ jicamilla, ragged nettlespurge
♦ Weed
♦ 39, 161
♦ herbal, toxic.

Jatropha messinica E.A.Bruce
Euphorbiaceae
♦ Native Weed
♦ 121
♦ pS, toxic. Origin: southern Africa.

Jatropha multifida L.
Euphorbiaceae
♦ coralbush, coral plant
♦ Weed, Naturalised, Native Weed
♦ 39, 101, 161, 179, 261
♦ cultivated, herbal, toxic. Origin:
Puerto Rico.

Jatropha podagrica Hook.
Euphorbiaceae
♦ goutystalk nettlespurge, tartogo,
gout plant, Australian bottle plant
♦ Weed, Naturalised, Cultivation
Escape
♦ 39, 86, 101, 161, 261
♦ arid, cultivated, toxic. Origin:
Mexico.

Jatropha spathulata (Ort.) Muell.& Arg.
Euphorbiaceae
= *Jatropha dioica* Sessè ex Cerv.
♦ sangre de drago
♦ Weed
♦ 39, 161
♦ herbal, toxic.

Jatropha spinosa Vahl
Euphorbiaceae
♦ Weed
♦ 88

Jatropha urens L.
Euphorbiaceae
= *Cnidoscolus urens* (L.) Arthur
♦ Weed
♦ 199
♦ herbal.

Jatropha villosa (Forssk.) Müll.Arg.
Euphorbiaceae
Croton villosus Forssk.
♦ Weed
♦ 88
♦ herbal.

Josephinia eugeniae F.Muell.
Pedaliaceae
♦ Weed
♦ 87, 88
♦ cultivated. Origin: Australia.

Jovibarba heuffelii (Schott) Á.& D.Löve
Crassulaceae
♦ hen and chickens
♦ Naturalised
♦ 101
♦ cultivated.

Jovibarba sobolifera (Sims) Opiz
Crassulaceae
Sempervivum soboliferum auct. non Sims
♦ mehiparta, hen and chickens
houseleek
♦ Cultivation Escape
♦ 42
♦ pH, cultivated.

Juglans ailanthifolia Carr.
Juglandaceae
= *Juglans ailanthifolia* Carr.
♦ Japanese walnut
♦ Naturalised

♦ 101
♦ cultivated.

Juglans ailantifolia Carr.
Juglandaceae
Juglans ailanthifolia Carr. (see), *Juglans
allardiana* Dode, *Juglans lavellei* Dode,
Juglans mirabunda Koidz., *Juglans
sachalinensis* Komatsu, *Juglans
sieboldiana* Maxim.
♦ Japanese walnut, heartnut
♦ Weed, Naturalised, Environmental
Weed
♦ 15, 80, 225, 246, 280
♦ T, cultivated, herbal. Origin: Japan.

Juglans × bixbyi Rehd.
Juglandaceae
= *Juglans cinerea* L. × *Juglans cordiformis*
Maxim. var. *ailantifolia* (Carriére)
Rehder [most probable combination]
♦ buartnut, Bixby walnut
♦ Naturalised
♦ 101
♦ T, cultivated. Origin: horticultural
hybrid.

Juglans cinerea L.
Juglandaceae
♦ butternut, white walnut
♦ Weed
♦ 39, 87, 88, 161, 218
♦ T, cultivated, herbal, toxic.

Juglans × intermedia Carr.
Juglandaceae
= *Juglans nigra* L. × *Juglans regia* L.
♦ intermediate walnut
♦ Naturalised
♦ 101
♦ cultivated.

Juglans microcarpa Berland.
Juglandaceae
Juglans rupestris Engelm. ex Torr.,
Juglans subrupestris Dode
♦ river walnut, little walnut, Texas
walnut
♦ Weed
♦ 87, 88, 218
♦ T, cultivated, herbal.

Juglans neotropica Diels
Juglandaceae
♦ Andean walnut
♦ Environmental Weed
♦ 257
♦ T, cultivated.

Juglans nigra L.
Juglandaceae
Pericarya nigra Dochn., *Wallia nigra*
Alef.
♦ black walnut, walnut
♦ Weed
♦ 8, 39, 87, 88, 161, 218
♦ T, cultivated, herbal, toxic.

**Juglans × quadrangulata (Carr.) Rehd.
(pro sp.)**
Juglandaceae
= *Juglans cinerea* L. × *Juglans regia* L.
♦ Naturalised
♦ 101

Juglans regia L.
Juglandaceae

♦ English walnut, wallflower, walnut,
Persian walnut, Madeira nut
♦ Weed, Naturalised
♦ 39, 54, 86, 88, 101, 161, 198, 280
♦ T, cultivated, herbal, toxic. Origin:
eastern Europe to northern Asia.

Julocroton Mart. spp.
Euphorbiaceae
= *Croton* L. spp.
♦ Weed
♦ 270

**Juncellus alopecuroides (Rottb.)
C.B.Clarke**
Cyperaceae
= *Cyperus alopecuroides* Rottb.
♦ mat sedge
♦ Weed
♦ 126
♦ pG, aqua.

Juncellus laevigatus (L.) C.B.Clarke
Cyperaceae
= *Cyperus laevigatus* L.
♦ Weed
♦ 126
♦ pG, aqua.

Juncellus serotinus (Rottb.) C.B.Clarke
Cyperaceae
= *Cyperus serotinus* Rottb.
♦ Weed
♦ 88, 126, 253, 275, 297
♦ pG, aqua.

Juncus L. spp.
Juncaceae
♦ rush, wiregrass
♦ Weed, Quarantine Weed
♦ 181, 208, 212, 220, 243
♦ G, herbal.

Juncus acuminatus Mich.
Juncaceae
♦ prickly rush, tapertip rush
♦ Weed, Naturalised, Environmental
Weed
♦ 15, 54, 86, 88, 98, 176, 198, 199, 203,
280
♦ pG, herbal. Origin: North and South
America.

Juncus acutiflorus Ehrh.
Juncaceae
Juncus sylvaticus auct. mult. (*non* Reich.
Wahl.)
♦ sharp flowered rush, otavihvilä
♦ Weed, Sleeper Weed, Naturalised,
Environmental Weed, Casual Alien
♦ 42, 70, 86, 155, 176, 272, 280
♦ pG, herbal. Origin: Eurasia, North
Africa.

Juncus acutus L.
Juncaceae
♦ spiny rush, sharp rush, sharp
pointed rush, sharp sea rush
♦ Weed, Noxious Weed, Naturalised,
Native Weed, Introduced,
Environmental Weed
♦ 7, 15, 23, 70, 72, 87, 88, 98, 121, 147,
176, 203, 221, 225, 228, 246, 269, 280,
290, 295, 296
♦ pG, aqua, cultivated, herbal. Origin:
obscure.

Juncus acutus L. ssp. *acutus*
Juncaceae
♦ spiny rush, sharp rush, sharp pointed rush
♦ Weed, Noxious Weed, Naturalised
♦ 86, 93, 198
♦ pG.

Juncus alatus Franch. & Sav.
Juncaceae
♦ Weed
♦ 87, 88, 126, 286
♦ pG, aqua.

Juncus amabilis Edgar
Juncaceae
♦ Naturalised
♦ 86, 280
♦ G, cultivated. Origin: Australia.

Juncus ambiguus Guss.
Juncaceae
♦ rush, toad rush, frog rush
♦ Naturalised
♦ 280
♦ aG.

Juncus aridicola L.A.S.Johnson
Juncaceae
♦ Naturalised
♦ 86
♦ G, cultivated. Origin: Australia.

Juncus articulatus L.
Juncaceae
Juncus lampocarpus Ehrh. (see)
♦ jointed rush, Baltic rush, solmuvihvilä
♦ Weed, Naturalised, Environmental Weed
♦ 7, 9, 15, 20, 23, 44, 70, 72, 86, 87, 88, 98, 126, 134, 176, 181, 198, 200, 203, 208, 225, 246, 269, 272, 280, 297
♦ wpG, cultivated, herbal. Origin: Europe.

Juncus articulatus L. × holoschoenus R.Br.
Juncaceae
♦ jointed rush, jointleaf rush
♦ Naturalised
♦ 198
♦ G.

Juncus balticus Willd.
Juncaceae
♦ Baltic rush, merivihvilä
♦ Weed
♦ 87, 88, 218
♦ pG, aqua, cultivated, herbal.

Juncus brachycarpus Engelm.
Juncaceae
♦ whiteroot rush
♦ Naturalised
♦ 280
♦ G, herbal.

Juncus bufonius L.
Juncaceae
♦ toad rush, common toad rush, grass rush, rush, konnanvihvilä, annual rush
♦ Weed, Naturalised, Environmental Weed
♦ 7, 9, 15, 23, 30, 44, 70, 86, 87, 88, 93, 121, 126, 134, 159, 161, 180, 185, 205,

212, 218, 221, 243, 253, 269, 272, 280, 286
♦ aG, arid, cultivated, herbal, toxic. Origin: cosmopolitan.

Juncus bufonius L. var. *bufonius*
Juncaceae
♦ toad rush
♦ Naturalised
♦ 280
♦ aG, herbal.

Juncus bufonius L. var. *congestus* Wahlenb.
Juncaceae
♦ toad rush
♦ Naturalised
♦ 280
♦ aG.

Juncus bufonius L. var. *parviflorus* Asch. & Graebn.
Juncaceae
♦ Naturalised
♦ 280
♦ G.

Juncus bulbosus L.
Juncaceae
♦ bulbous rush, rentovihvilä
♦ Weed, Naturalised, Environmental Weed
♦ 72, 86, 88, 98, 176, 198, 203, 225, 246, 272, 280, 300
♦ wpG, cultivated, herbal. Origin: Eurasia, North Africa.

Juncus canadensis Gay
Juncaceae
♦ Canadian rush
♦ Weed, Sleeper Weed, Naturalised, Environmental Weed
♦ 15, 86, 98, 155, 203, 280
♦ pG, herbal. Origin: North America.

Juncus capensis Thunb.
Juncaceae
♦ Weed, Naturalised
♦ 86, 98, 203
♦ G. Origin: South Africa.

Juncus capillaceus Lam.
Juncaceae
♦ string rush, rush
♦ Weed, Naturalised, Environmental Weed
♦ 72, 86, 88, 98, 198, 203
♦ pG. Origin: temperate South America.

Juncus capitatus Weigel
Juncaceae
♦ capitate rush, dwarf rush, leafybract dwarf rush, mykerövihvilä, sitina hlaviäkatá
♦ Weed, Naturalised, Introduced, Environmental Weed, Casual Alien
♦ 7, 9, 34, 42, 72, 86, 88, 98, 101, 176, 198, 203, 243, 272, 280
♦ aG, cultivated, herbal. Origin: Eurasia, Africa.

Juncus cognatus Kunth
Juncaceae
♦ Weed, Naturalised
♦ 86, 98, 203

♦ G, cultivated. Origin: south America.

Juncus compressus Jacq.
Juncaceae
♦ roundfruit rush, tannervihvilä
♦ Weed, Naturalised
♦ 23, 87, 88, 101, 126, 272, 297
♦ a/pG, aqua, cultivated, herbal.

Juncus conglomeratus L.
Juncaceae
= *Juncus effusus* L. var. *conglomeratus* (L.) Engelm. (NoR) [see *Juncus effusus* var. *compactus* Lej. & Courtois]
♦ compact rush, sitina klbkatá, common rush, jonc aggloméré
♦ Weed, Naturalised, Environmental Weed
♦ 23, 44, 54, 70, 86, 87, 88, 176, 272, 280
♦ pG, aqua, cultivated, herbal. Origin: Eurasia.

Juncus continuus L.A.S.Johnson
Juncaceae
♦ Naturalised
♦ 280
♦ G, cultivated.

Juncus cyperoides Laharpe
Juncaceae
♦ Forbestown rush
♦ Naturalised
♦ 101
♦ pG.

Juncus dichotomus Ell.
Juncaceae
♦ forked rush
♦ Weed, Naturalised
♦ 15, 280
♦ G, herbal.

Juncus dregeanus Kunth
Juncaceae
♦ Naturalised
♦ 280
♦ G.

Juncus dudleyi Wieg.
Juncaceae
♦ Dudley's rush
♦ Naturalised
♦ 287
♦ pG, cultivated, herbal.

Juncus effusus L.
Juncaceae
Juncus bogotensis Kunth, *Juncus communis* E.Mey., *Juncus communis* var. *effusus* (L.) Mey., *Juncus effusus* var. *solutus* Fernald & Wiegand, *Juncus glomeratus* Thunb. (see), *Juncus mauritianus* Bojer
♦ soft rush, Japanese mat rush, common rush, bogrush, jonc à lier, jonc épars, flatterbinse, junco, junquera
♦ Weed, Quarantine Weed, Naturalised, Garden Escape, Environmental Weed
♦ 15, 20, 23, 39, 70, 72, 76, 86, 87, 88, 98, 108, 126, 161, 176, 191, 198, 203, 204, 218, 225, 246, 258, 272, 280, 289, 297
♦ pG, aqua, cultivated, herbal, toxic. Origin: temperate Northern Hemisphere.

Juncus effusus L. var. compactus Lej. & Courtois
Juncaceae
= *Juncus effusus* L. var. *conglomeratus* (L.) Engelm. (NoR) [see *Juncus conglomeratus* L.]
♦ Naturalised
♦ 280
♦ G.

Juncus effusus L. var. decipiens Buchenau
Juncaceae
♦ lamp rush
♦ Weed
♦ 263, 286
♦ pG.

Juncus effusus L. var. effusus
Juncaceae
♦ common rush
♦ Naturalised
♦ 280
♦ G.

Juncus effusus L. var. pacificus Fern. & Wieg.
Juncaceae
♦ soft rush, bogrush, common rush, rush, Pacific rush
♦ Weed
♦ 180, 243
♦ pG.

Juncus ensifolius Wikstr.
Juncaceae
♦ swordleaf rush, rush, dwarf rush, miekkavihvilä
♦ Weed, Naturalised, Casual Alien
♦ 42, 159, 280
♦ pG, aqua, cultivated, herbal. Origin: North America, eastern Russia, Japan.

Juncus filicaulis Buchenau
Juncaceae
♦ Naturalised
♦ 280
♦ G, cultivated.

Juncus filiformis L.
Juncaceae
♦ thread rush, jouhivihvilä
♦ Weed
♦ 23, 87, 88, 272
♦ pG, cultivated, herbal.

Juncus flavidus L.A.S.Johnson
Juncaceae
♦ Naturalised
♦ 280
♦ G.

Juncus fockei Buchenau
Juncaceae
♦ Weed, Naturalised
♦ 15, 280
♦ G, cultivated. Origin: Australia.

Juncus fontanesii Gay
Juncaceae
♦ rush
♦ Weed, Naturalised, Environmental Weed
♦ 54, 72, 88, 98, 203
♦ pG.

Juncus fontanesii J.Gay ex Laharpe ssp. fontanesii
Juncaceae
♦ spring rush
♦ Naturalised, Environmental Weed
♦ 86, 198
♦ G. Origin: Mediterranean.

Juncus fontanesii Gay ssp. pyramidatus (Laharpe) Snogerup
Juncaceae
Juncus pyramidatus Laharpe
♦ Weed
♦ 185
♦ G, aqua.

Juncus gerardii Loisel.
Juncaceae
Juncus gerardi Loisel.
♦ rush, salt mud rush, salt meadow rush, saltmarsh rush, suolavihvilä, mud rush
♦ Weed, Naturalised, Environmental Weed
♦ 72, 87, 88, 218, 280
♦ pG, herbal.

Juncus gerardii Loisel. ssp. gerardii
Juncaceae
♦ mud rush
♦ Naturalised
♦ 86, 198
♦ G. Origin: North America.

Juncus glaucus Sibth.
Juncaceae
= *Juncus inflexus* L.
♦ Weed
♦ 87, 88
♦ G.

Juncus glomeratus Thunb.
Juncaceae
♦ clustered rush
♦ Naturalised
♦ 86
♦ G.

Juncus homalocaulis F.Muell.
Juncaceae
♦ Naturalised
♦ 280
♦ G, cultivated.

Juncus hybridus Brot.
Juncaceae
♦ hybrid rush
♦ Weed
♦ 88, 185
♦ G, aqua.

Juncus imbricatus Laharpe
Juncaceae
♦ folded rush
♦ Weed, Naturalised
♦ 7, 86, 98, 198, 203, 280
♦ G. Origin: temperate South America.

Juncus imbricatus Laharpe var. chamissonis (Kunth) Buchenau
Juncaceae
♦ Naturalised
♦ 280
♦ G.

Juncus indescriptus Steud.
Juncaceae
♦ Weed, Naturalised, Environmental Weed

♦ 54, 86, 88, 176
♦ G. Origin: South Africa.

Juncus inflexus L.
Juncaceae
Juncus glaucus Sibth. (see)
♦ hard rush, sea green rush, European meadow rush, blue rush, sinivihvilä
♦ Weed, Naturalised, Environmental Weed, Casual Alien
♦ 23, 39, 42, 54, 70, 72, 86, 87, 88, 101, 126, 155, 161, 198, 221, 272, 280
♦ pG, aqua, cultivated, herbal, toxic. Origin: Eurasia.

Juncus koidzumii Satake
Juncaceae
♦ Weed
♦ 87, 88
♦ G.

Juncus krameri Fr. & Sav.
Juncaceae
♦ tachikougaizekishou
♦ Weed
♦ 286
♦ G.

Juncus kraussii Hochst.
Juncaceae
♦ Naturalised
♦ 280
♦ G, cultivated.

Juncus kraussii Hochst. ssp. australiensis (Buchenau) Snogerup
Juncaceae
♦ Naturalised
♦ 280
♦ G.

Juncus lampocarpus Ehrh.
Juncaceae
= *Juncus articulatus* L.
♦ Weed
♦ 30, 87, 88, 275
♦ pG.

Juncus leschenaultii J.Gay ex Laharpe
Juncaceae
= *Juncus prismatocarpus* R.Br.
♦ kougaizekishou
♦ Weed
♦ 87, 88, 235, 263, 286
♦ pG.

Juncus lomatophyllus Spreng.
Juncaceae
♦ Naturalised
♦ 280
♦ G.

Juncus marginatus Rostk.
Juncaceae
♦ shore rush, grassleaf rush
♦ Weed
♦ 87, 88, 146, 218
♦ G, herbal.

Juncus maritimus Lam.
Juncaceae
♦ sea rush
♦ Weed, Naturalised
♦ 23, 70, 87, 88, 101, 126, 272
♦ pG, aqua, cultivated, herbal.

Juncus microcephalus Kunth
Juncaceae
♦ rush, tiny headed rush, junquinho, junco
♦ Weed, Naturalised, Environmental Weed
♦ 7, 15, 72, 86, 87, 88, 98, 176, 198, 203, 255, 280
♦ wpG. Origin: tropical America.

Juncus monticola Steud.
Juncaceae
♦ komochizekishou
♦ Weed
♦ 286
♦ G.

Juncus nodosus L.
Juncaceae
♦ jointed rush, knotted rush
♦ Weed, Naturalised
♦ 54, 86, 88
♦ pG, aqua, herbal.

Juncus oxycarpus E.Mey. ex Kunth
Juncaceae
♦ spinyfruit rush, rush
♦ Weed, Naturalised, Native Weed, Environmental Weed
♦ 7, 72, 86, 88, 98, 121, 198, 203
♦ pG. Origin: southern Africa.

Juncus papillosus Franch. & Sav.
Juncaceae
♦ aokougaizekishou
♦ Weed
♦ 87, 88, 126, 263, 286
♦ pG, aqua.

Juncus phaeocephalus Engelm.
Juncaceae
♦ brownhead rush
♦ Weed
♦ 87, 88, 161, 218
♦ pG, herbal, toxic.

Juncus planifolius R.Br.
Juncaceae
♦ broadleaf rush
♦ Weed, Naturalised
♦ 80, 101
♦ G, cultivated, herbal. Origin: Australia.

Juncus polyanthemos Buchen.
Juncaceae
= *Juncus polyanthemus* Buchen.
♦ manyflower rush
♦ Weed, Naturalised, Native Weed
♦ 7, 80, 86, 101
♦ G, cultivated.

Juncus polyanthemus Buchen.
Juncaceae
Juncus polyanthemos Buchen. (see)
♦ Weed, Naturalised
♦ 9, 87, 88
♦ G. Origin: Australia.

Juncus prismatocarpus R.Br.
Juncaceae
Juncus leschenaultii J.Gay ex Laharpe (see)
♦ Weed
♦ 39, 87, 88, 126, 170, 191, 238, 274

♦ pG, aqua, cultivated, toxic. Origin: Asia, Australasia.

Juncus procerus E.Mey.
Juncaceae
♦ Weed, Naturalised
♦ 87, 88, 280
♦ pG, aqua, cultivated. Origin: Australia.

Juncus punctorius L.f.
Juncaceae
♦ Weed
♦ 221
♦ G.

Juncus rigidus Desf.
Juncaceae
♦ Weed
♦ 221
♦ G.

Juncus roemerianus Scheele
Juncaceae
♦ needle rush, needlegrass rush, black rush
♦ Weed
♦ 87, 88, 218
♦ G, cultivated, herbal.

Juncus rugosus Steud.
Juncaceae
♦ Weed
♦ 87, 88
♦ G.

Juncus scirpoides Lam.
Juncaceae
♦ needlepod rush
♦ Naturalised
♦ 287
♦ G, herbal.

Juncus setchuensis Buchen. var. *effusoides* Buchen.
Juncaceae
♦ Weed
♦ 286
♦ G.

Juncus sphaerocarpus Nees
Juncaceae
♦ Weed
♦ 272
♦ G, herbal.

Juncus squarrosus L.
Juncaceae
Juncus sprengelii Willd.
♦ heath rush, mosquito rush, harjasvihvilä
♦ Weed, Naturalised, Environmental Weed
♦ 42, 54, 70, 86, 88, 176, 181, 225, 246, 272, 280
♦ G, cultivated. Origin: Europe.

Juncus subnodulosus Schrank
Juncaceae
♦ bluntflower rush, giunco subnodoso
♦ Weed, Naturalised
♦ 15, 23, 86, 88, 98, 101, 203, 272, 280
♦ pG, cultivated, herbal. Origin: Northern Hemisphere.

Juncus subsecundus N.A.Wakef.
Juncaceae

♦ Naturalised
♦ 280
♦ G, cultivated.

Juncus subulatus Forssk.
Juncaceae
♦ Weed
♦ 221
♦ G.

Juncus tenageia Ehrh. ex L.f.
Juncaceae
♦ Weed
♦ 272
♦ G.

Juncus tenuis Willd.
Juncaceae
Juncus tenuis Willd. var. *anthelatus* (Wieg.) (see)
♦ slender rush, path rush, field rush, slender yard rush, wiregrass, poverty rush, soft rush, sitina tenká, nurmivihvilä, jonc grêle
♦ Weed, Naturalised, Casual Alien
♦ 23, 40, 42, 44, 70, 86, 87, 88, 98, 126, 161, 176, 198, 203, 210, 211, 218, 243, 249, 261, 272, 280, 286
♦ pG, cultivated, herbal. Origin: North and South America.

Juncus tenuis Willd. var. anthelatus Wieg.
Juncaceae
= *Juncus tenuis* Willd.
♦ Weed, Naturalised
♦ 15, 280, 287
♦ G.

Juncus tenuis Willd. var. nakaii Satake
Juncaceae
♦ Naturalised
♦ 287
♦ G.

Juncus tenuis Willd. var. tenuis
Juncaceae
♦ poverty rush
♦ Weed, Naturalised
♦ 15, 280
♦ G.

Juncus usitatus L.A.S.Johnson
Juncaceae
♦ common rush
♦ Naturalised, Native Weed
♦ 7, 86, 269
♦ pG, aqua, cultivated. Origin: Australia.

Juncus wallichianus Laharpe
Juncaceae
♦ harikougaizekishou
♦ Weed
♦ 286
♦ G.

Juncus yokoscensis (Fr. & Sav.) Satake
Juncaceae
♦ rush
♦ Weed
♦ 87, 88
♦ G.

Junellia tridens (Kuntze) Moldenke
Verbenaceae
Verbena tridens Lag. (see)

♦ Weed
♦ 237

***Jungia hirsuta* Cuatr.**
Asteraceae
♦ Naturalised
♦ 257

***Juniperus* L. spp.**
Cupressaceae
♦ juniper
♦ Weed, Naturalised
♦ 23, 86, 88, 154, 198
♦ T, herbal, toxic.

***Juniperus ashei* Buchh.**
Cupressaceae
♦ Ashe's juniper
♦ Weed
♦ 87, 88, 218
♦ T, cultivated, herbal.

***Juniperus bermudiana* L.**
Cupressaceae
♦ Environmental Weed
♦ 246
♦ cultivated, herbal.

***Juniperus communis* L.**
Cupressaceae
♦ common juniper, genevrier, ginepro, enebro, baccae juniperi, dwarf juniper
♦ Weed
♦ 8, 39, 70, 87, 88, 218, 247, 272
♦ S, cultivated, herbal, toxic. Origin: northern hemisphere.

***Juniperus deppeana* Steud.**
Cupressaceae
♦ alligator juniper
♦ Weed
♦ 80, 87, 88, 161, 218
♦ T, cultivated, herbal.

***Juniperus horizontalis* Moench**
Cupressaceae
♦ creeping juniper, blue rug juniper
♦ Weed
♦ 87, 88, 218
♦ S, cultivated, herbal.

***Juniperus monosperma* (Engelm.) Sarg.**
Cupressaceae
♦ oneseed juniper, cherrystone juniper
♦ Weed
♦ 87, 88, 161, 218
♦ T, cultivated, herbal.

***Juniperus occidentalis* Hook.**
Cupressaceae
♦ western juniper
♦ Weed
♦ 80, 87, 88, 218
♦ T, promoted, herbal.

***Juniperus osteosperma* (Torr.) Little**
Cupressaceae
Juniperus californica Carr. var. *utahensis* (Engelm.), *Juniperus utahensis* (Engelm.) Lemmon
♦ Utah juniper, desert juniper
♦ Weed
♦ 87, 88, 161, 218
♦ S, cultivated, herbal.

***Juniperus oxycedrus* L. ssp. *macrocarpa* (Sibth. & Sm.) Ball**
Cupressaceae

♦ prickly juniper
♦ Naturalised
♦ 198

***Juniperus phoenicea* L.**
Cupressaceae
♦ Phoenician juniper
♦ Weed, Introduced
♦ 221, 228
♦ S/T, arid, cultivated, herbal.

***Juniperus pinchotii* Sudw.**
Cupressaceae
♦ redberry juniper, Pinchot's juniper
♦ Weed
♦ 87, 88, 218
♦ T, cultivated, herbal.

***Juniperus scopulorum* Sarg.**
Cupressaceae
Juniperus virginiana L. ssp. *scopulorum* (Sarg.) E.Murr., *Juniperus virginiana* L. var. *scopulorum* (Sarg.) Lemmon
♦ Rocky Mountain juniper
♦ Weed
♦ 87, 88, 218
♦ T, cultivated, herbal.

***Juniperus silicicola* (Small) Bailey**
Cupressaceae
= *Juniperus virginiana* L. var. *silicicola* (Small) J.Silba (NoR)
♦ southern red cedar
♦ Weed
♦ 87, 88, 218
♦ T, cultivated, herbal.

***Juniperus virginiana* L.**
Cupressaceae
♦ eastern red cedar, red cedar juniper, pencil cedar, red cedar, eastern red cedar, borievka virgínska
♦ Weed, Noxious Weed, Naturalised, Native Weed, Cultivation Escape
♦ 8, 39, 54, 80, 86, 87, 88, 95, 137, 161, 174, 195, 218, 247, 283
♦ T, cultivated, herbal, toxic. Origin: North America.

***Jurinea alata* Cass.**
Asteraceae
♦ Quarantine Weed
♦ 220

***Jurinea mollis* (L.) Rchb.**
Asteraceae
♦ Weed
♦ 272
♦ cultivated.

***Jussiaea californica* (Wats.) Jeps.**
Onagraceae
= *Ludwigia peploides* (Kunth) Raven
♦ California water primrose
♦ Weed
♦ 87, 88, 218

***Jussiaea decurrens* (Walter) DC.**
Onagraceae
= *Ludwigia decurrens* Walter
♦ winged water primrose
♦ Weed
♦ 88, 218
♦ herbal.

***Jussiaea erecta* L.**
Onagraceae

= *Ludwigia erecta* (L.) H.Hara
♦ clavito
♦ Weed
♦ 153

***Jussiaea grandiflora* Michx. *non* Ruiz & Pav.**
Onagraceae
= *Ludwigia peruviana* (L.) H.Hara
♦ Weed
♦ 3, 191

***Jussiaea leptocarpa* Nutt.**
Onagraceae
= *Ludwigia leptocarpa* (Nutt.) H.Hara
♦ Weed, Quarantine Weed
♦ 203, 220

***Jussiaea linifolia* Vahl**
Onagraceae
= *Ludwigia hyssopifolia* (G.Don) Exell
♦ water primrose, thian naa
♦ Weed
♦ 23, 88, 191, 204, 239
♦ aqua, herbal.

***Jussiaea michauxiana* Fern.**
Onagraceae
= *Ludwigia uruguayensis* (Camb.) H.Hara
♦ perennial water primrose
♦ Weed
♦ 87, 88, 218

***Jussiaea peruviana* L.**
Onagraceae
= *Ludwigia peruviana* (L.) H.Hara
♦ Weed
♦ 3, 191
♦ herbal.

***Jussiaea repens* L.**
Onagraceae
= *Ludwigia peploides* (Kunth) Raven ssp. *glabrescens* (Kuntze) Raven (NoR) [see *Jussiaea repens* L. var. *glabrescens* Ktze.]
♦ creeping water primrose, phak phaeng phuai
♦ Weed
♦ 88, 185, 191, 204, 221, 239
♦ wpH, herbal.

***Jussiaea repens* L. var. *glabrescens* Ktze.**
Onagraceae
= *Ludwigia peploides* (Kunth) Raven ssp. *glabrescens* (Kuntze) Raven (NoR) [see *Jussiaea repens* L.]
♦ creeping water primrose
♦ Weed
♦ 218

***Jussiaea suffruticosa* L. Him**
Onagraceae
= *Ludwigia octovalvis* (Jacq.) Raven
♦ Weed
♦ 88
♦ herbal.

***Justicia americana* (L.) Vahl**
Acanthaceae
♦ waterwillow, American waterwillow
♦ Weed
♦ 87, 88, 218
♦ aqua, cultivated, herbal.

Justicia betonica L.
Acanthaceae
♦ squirrel's tail
♦ Weed, Naturalised, Cultivation
Escape
♦ 32, 86, 101, 191, 252
♦ cultivated, herbal. Origin: Asia and
tropical Africa.

Justicia brandegeana Wassh. & L.B.Sm.
Acanthaceae
♦ shrimp plant
♦ Weed, Naturalised
♦ 101, 179
♦ cultivated.

Justicia carnea Lindl.
Acanthaceae
Cyrtanthera magnifica Nees, *Cyrtanthera
pohliana* Nees, *Jacobinia carnea* (Lindl.)
G.Nicholson (see), *Jacobinia magnifica*
(Nees) Lindau, *Jacobinia pohliana* (Nees)
Lindau, *Jacobinia pohliana* var. *velutina*
Nees, *Jacobinia velutina* (Nees) Voss,
Justicia magnifica Pohl ex Nees
♦ Brazilian plume flower, Brazilian
plume
♦ Weed, Casual Alien
♦ 3, 15, 191, 280
♦ cultivated. Origin: Brazil.

Justicia diffusa Willd.
Acanthaceae
♦ Weed
♦ 87, 88
♦ herbal.

Justicia exigua S.Moore
Acanthaceae
♦ Weed
♦ 87, 88

Justicia flava Vahl
Acanthaceae
♦ Weed, Quarantine Weed
♦ 76, 87, 88, 203, 220
♦ pH.

Justicia heterocarpa T.Anderson
Acanthaceae
♦ Weed
♦ 221

Justicia insularis T.Anders.
Acanthaceae
♦ Weed, Quarantine Weed
♦ 76, 87, 88, 203
♦ cultivated.

Justicia palustris (Hochst) Anders.
Acanthaceae
♦ Weed
♦ 242
♦ arid.

Justicia pectoralis Jacq.
Acanthaceae
Dianthera pectoralis (Jacq.) Murr. (see)
♦ freshcut, yacu piri piri
♦ Weed
♦ 87, 88, 179
♦ cultivated, herbal.

Justicia procumbens L.
Acanthaceae
Rostellularia sundana Bremek. (see)
♦ waterwillow

♦ Weed
♦ 87, 88, 235, 273, 286
♦ aH, cultivated, herbal.

Justicia prostrata Gamble
Acanthaceae
♦ Weed
♦ 66, 87, 88
♦ cultivated.

Justicia schimperi (Hochst.) Dandy
Acanthaceae
♦ Weed
♦ 240

Justicia simplex D.Don
Acanthaceae
♦ Weed, Quarantine Weed
♦ 76, 87, 88, 203, 220, 243

Justicia spicigera Schlect.
Acanthaceae
♦ mohintli, Mexican indigo
♦ Naturalised
♦ 101
♦ cultivated.

Justicia striata (Klotzsch) Bullock
Acanthaceae
♦ Weed
♦ 87, 88

K

Kaempferia L. spp.
Zingiberaceae
♦ kaempferia
♦ Weed, Naturalised
♦ 98, 203

Kaempferia atrovirens N.E.Br.
Zingiberaceae
♦ peacock plant
♦ Naturalised
♦ 287
♦ cultivated.

Kaempferia pulchra Ridl.
Zingiberaceae
♦ Weed
♦ 12
♦ cultivated.

Kaempferia rotunda L.
Zingiberaceae
♦ resurrection lily, duende violeta,
llangilang de tierra, lirio misterioso
♦ Naturalised, Cultivation Escape
♦ 39, 101, 261
♦ cultivated, herbal, toxic. Origin: East
Indies.

Kalanchoe Adans. spp.
Crassulaceae
♦ widow's thrill, palm beachbells,
kalanchoe
♦ Weed, Environmental Weed
♦ 80, 88, 151
♦ herbal.

Kalanchoe beauverdii Raym.-Hamet
Crassulaceae
= *Bryophyllum beauverdii* (Raym.-
Hamet) A.Berger
♦ Beauverd's widow's thrill
♦ Naturalised
♦ 101
♦ cultivated.

Kalanchoe blossfeldiana Poelln.
Crassulaceae
♦ Madagascar widow's thrill
♦ Naturalised, Garden Escape
♦ 101, 261
♦ cultivated. Origin: Madagascar.

**Kalanchoe daigremontiana Raym.-Hamet
& H.Perrier**
Crassulaceae
= *Bryophyllum daigremontianum*
(Raym.-Hamet & H.Perrier) A.Berger
♦ devil's backbone
♦ Weed, Naturalised, Casual Alien
♦ 101, 179, 261
♦ pH, cultivated, toxic. Origin:
Madagascar.

Kalanchoe delagoensis Eckl. & Zeyh.
Crassulaceae
= *Bryophyllum delagoense* (Eckl. &
Zeyh.) Schinz
♦ chandelier plant
♦ Quarantine Weed, Naturalised,
Cultivation Escape
♦ 101, 220, 233
♦ cultivated.

Kalanchoe fedtschenkoi Hamet & Perrier
Crassulaceae
= *Bryophyllum fedtschenkoi* (Hamet &
Perrier) Cheng
♦ lavender scallops
♦ Weed, Naturalised
♦ 101, 179
♦ cultivated. Origin: Australia.

**Kalanchoe gastonis-bonnieri Hamet &
Perrier**
Crassulaceae
♦ palm beachbells
♦ Weed, Naturalised
♦ 101, 179
♦ cultivated.

Kalanchoe grandiflora Wight & Arn.
Crassulaceae
Kalanchoe nyikae Engl.
♦ Naturalised
♦ 280
♦ cultivated.

Kalanchoe integra (Medik.) Kuntze
Crassulaceae
♦ neverdie, madre de bruja roja
♦ Naturalised, Casual Alien
♦ 101, 261
♦ cultivated, herbal. Origin: tropical
Africa.

**Kalanchoe integra (Medik.) Kuntze var.
crenata (Andrews) Cufod.**
Crassulaceae
♦ neverdie
♦ Naturalised
♦ 101

**Kalanchoe integra (Medik.) Kuntze var.
verea (Jacq.) Cufod.**
Crassulaceae
♦ neverdie
♦ Naturalised
♦ 101

Kalanchoe laciniata (L.) DC.
Crassulaceae
♦ Christmas tree plant
♦ Weed, Naturalised
♦ 101, 179
♦ cultivated, herbal.

Kalanchoe lanceolata (Forssk.) Pers.
Crassulaceae
♦ lanceleaf air plant
♦ Weed, Native Weed
♦ 39, 121, 161
♦ H, cultivated, toxic. Origin:
Madagascar.

Kalanchoe lateritia Engl. var. lateritia
Crassulaceae
♦ Naturalised
♦ 86
♦ Origin: east Africa.

Kalanchoe longiflora Schltr.
Crassulaceae
♦ Weed, Naturalised
♦ 86, 98, 203
♦ cultivated.

Kalanchoe pinnata (Lam.) Pers.
Crassulaceae
= *Bryophyllum pinnatum* (Lam.) Oken
♦ life plant, air plant, resurrection
plant, Canterbury bells, Mexican love
plant, bulatawamudu, cathedral bells,
oliwa ku kahakai
♦ Weed, Naturalised, Environmental
Weed, Cultivation Escape
♦ 3, 18, 80, 87, 88, 98, 101, 107, 122, 152,
161, 179, 191, 203, 233, 257
♦ pH, cultivated, herbal, toxic. Origin:
Africa, Madagascar, India, Indian
Ocean Islands.

Kalanchoe rotundifolia (Haw.) Haw.
Crassulaceae
♦ roundleaf air plant
♦ Weed, Native Weed
♦ 39, 121, 161
♦ pH, cultivated, toxic. Origin:
southern Africa.

Kalanchoe tubiflora (Haw.) Raym.-Hamet
Crassulaceae
= *Bryophyllum delagoense* (Eckl. &
Zeyh.) Schinz
♦ Weed, Quarantine Weed,
Naturalised
♦ 179, 220, 287
♦ cultivated, herbal.

Kalanchoe verticillata Scott-Elliot
Crassulaceae
= *Bryophyllum delagoense* (Eckl. &
Zeyh.) Schinz
♦ Weed, Naturalised
♦ 87, 88, 220, 287

Kalimeris incisa (Fisch.) DC.
Asteraceae
Boltonia incisa (Fisch.) Benth.
♦ Weed
♦ 286
♦ pH, cultivated. Origin: China, Japan,
Korea, Siberia.

Kalimeris indica (L.) Sch.Bip.
Asteraceae
♦ Indian aster
♦ Weed, Naturalised
♦ 101, 275, 286
♦ pH.

Kalimeris integrifolia Turcz. ex DC.
Asteraceae
♦ integrifolius kalimeris
♦ Weed
♦ 275, 297
♦ pH.

Kalimeris pinnatifida (Maxim) Kitam.
Asteraceae
Boltonia cantoniensis (Blume.) Franch.
& Savat.
♦ Weed
♦ 87, 88, 286
♦ pH, promoted. Origin: China, Japan.

Kalimeris yomena Kitam.
Asteraceae
♦ Weed
♦ 87, 88, 286
♦ pH, cultivated.

**Kalimeris yomena Kitam. var. dentatus
Hara**
Asteraceae
♦ Weed
♦ 286

Kallstroemia californica (Wats.) Vail
Zygophyllaceae/Tribulaceae
♦ California caltrop
♦ Weed
♦ 87, 88, 161, 218
♦ aH, herbal.

Kallstroemia caribaea Rydb.
Zygophyllaceae/Tribulaceae
= *Kallstroemia pubescens* (G.Don) Dandy
♦ Weed
♦ 87, 88

Kallstroemia grandiflora Torr.
Zygophyllaceae/Tribulaceae
♦ orange caltrop, Arizona poppy,
Arizona caltrop
♦ Weed
♦ 87, 88, 161, 218
♦ aH, cultivated, herbal.

Kallstroemia hirsutissima Vail
Zygophyllaceae/Tribulaceae
♦ hairy caltrop, carpetweed
♦ Weed
♦ 87, 88, 161, 218
♦ herbal, toxic.

Kallstroemia maxima (L.) Hook. & Arn.
Zygophyllaceae/Tribulaceae
Kallstroemia tribuloides (Mart.) Steud.
(see)
♦ big caltrop
♦ Weed, Quarantine Weed
♦ 14, 76, 87, 88, 157, 203, 220
♦ aH, herbal.

Kallstroemia pubescens (G.Don) Dandy
Zygophyllaceae/Tribulaceae
Kallstroemia caribaea Rydb. (see)
♦ Caribbean caltrop
♦ Weed
♦ 87, 88

Kallstroemia tribuloides (Mart.) Steud.
Zygophyllaceae/Tribulaceae
= *Kallstroemia maxima* (L.) Hook. &
Arn.
♦ rabo de calango
♦ Weed
♦ 255
♦ aH. Origin: Brazil.

Kalmia angustifolia L.
Ericaceae
♦ sheep laurel, lambkill, calfkill, dwarf
laurel
♦ Weed
♦ 8, 39, 87, 88, 161, 189, 218, 247, 294
♦ S, cultivated, herbal, toxic.

Kalmia latifolia L.
Ericaceae
Kalmia lucida Koch. nom. inval.
♦ mountain laurel, calico bush, mountain laurel, ivy bush, kalmi, American laurel
♦ Weed
♦ 8, 39, 87, 88, 154, 161, 218, 247
♦ S, cultivated, herbal, toxic.

Kalmia microphylla (Hook.) Heller
Ericaceae
♦ bog laurel, alpine laurel
♦ Weed
♦ 39, 161
♦ cultivated, herbal, toxic.

Kalmia polifolia Wang
Ericaceae
Kalmia glauca Aiton
♦ pale laurel, swamp laurel, bog laurel, mountain laurel
♦ Weed
♦ 39, 87, 88, 218
♦ S, cultivated, herbal, toxic.

Kalopanax septemlobus (Thunb. ex A.Murr.) Koidz.
Araliaceae
Kalopanax pictus (Thunb.) Nakai., *Kalopanax ricinifolium* (Siebold & Zucc.) Miq., *Acanthopanax ricinifolium* (Siebold & Zucc.) Seem., *Acer pictum* Thunb. ex Murray, *Acer septemlobum* Thunb. ex Murray, *Panax ricinifolium* Siebold & Zucc.
♦ castor aralia, tree aralia
♦ Naturalised
♦ 101
♦ T, cultivated.

Kappaphycus alvarezii (M.S.Doty) M.S.Doty ex P.C.Silva
Solieriaceae
Eucheuma alvarezii M.S.Doty
♦ red alga
♦ Weed
♦ 197
♦ algae.

Kappaphycus striatum (F.Schmitz) M.S.Doty ex P.C.Silva
Solieriaceae
Eucheuma striatum F.Schmitz
♦ red alga, brown licorice
♦ Weed
♦ 197, 282
♦ algae.

Karelinia caspia (Pall.) Less.
Asteraceae
= *Pluchea caspia* (Pall.) O.Hoffm. ex Paulsen (NoR)
♦ Weed
♦ 272, 275
♦ pH.

Karroochloa purpurea (L.f.) Conert & Türpe
Poaceae
♦ South African oatgrass
♦ Naturalised
♦ 101
♦ G.

Karwinskia humboldtiana (Roem. & Schult.) Zucc.
Rhamnaceae
Karwinskia affinis Schldl., *Karwinskia biniflora* Schldl., *Karwinskia glandulosa* Zucc., *Karwinskia subcordata* Schldl., *Rhamnus biniflorus* DC., *Rhamnus humboldtiana* Roem. & Schult., *Rhamnus maculata* Sessé & Moç.
♦ coyotillo, buckthorn, tullidora, cacachila, cachila
♦ Weed, Quarantine Weed
♦ 39, 76, 87, 88, 154, 161, 189, 203, 218, 220
♦ arid, herbal, toxic.

Kedrostis foetidissima (Jacq.) Cogn.
Cucurbitaceae
♦ Naturalised
♦ 32
♦ herbal.

Kennedia Vent. spp.
Fabaceae/Papilionaceae
♦ Weed
♦ 23
♦ herbal.

Kennedia nigricans Lindl.
Fabaceae/Papilionaceae
♦ black coral pea, tiger snake vine
♦ Naturalised, Native Weed, Garden Escape, Cultivation Escape
♦ 7, 86, 252
♦ cultivated, herbal. Origin: Australia.

Kennedia rubicunda Vent.
Fabaceae/Papilionaceae
♦ dusky coral pea, coral pea
♦ Weed, Naturalised, Native Weed, Garden Escape, Environmental Weed
♦ 72, 86, 88, 176, 280
♦ pC, cultivated. Origin: Australia.

Kerria japonica (L.) DC.
Rosaceae
♦ Japanese rose, bachelor's button, Jew's mallow, kéria japonská
♦ Weed, Naturalised, Garden Escape
♦ 80, 101, 280
♦ S, cultivated, herbal.

Khaya nyasica Stapf ex Bak.
Meliaceae
♦ Nyasaland mahogany, African mahogany
♦ Naturalised, Cultivation Escape
♦ 101, 261
♦ cultivated, herbal. Origin: tropical Africa.

Khaya senegalensis (Desr.) A.Juss.
Meliaceae
♦ African mahogany, Senegal mahogany, quinquina du Senegal
♦ Weed, Sleeper Weed, Naturalised, Environmental Weed, Cultivation Escape
♦ 3, 86, 155, 191, 261
♦ arid, cultivated, herbal. Origin: tropical Africa.

Kickxia acerbiana (Boiss.) V.Täckh. & Boulos
Apocynaceae
♦ Weed

♦ 221

Kickxia aegyptiaca (L.) Nabelek
Apocynaceae
♦ Weed
♦ 221

Kickxia cirrhosa (L.) Fritsch
Apocynaceae
♦ hentonielukukka
♦ Casual Alien
♦ 42

Kickxia commutata (Bernh. ex Rchb.) Fritsch
Apocynaceae
♦ Weed, Naturalised
♦ 88, 94, 98, 203
♦ Origin: Mediterranean.

Kickxia commutata (Bernh. ex Rchb.) Fritsch ssp. graeca (Bory & Chaub.) R.Fern.
Apocynaceae
♦ Naturalised
♦ 86
♦ Origin: Mediterranean.

Kickxia elatine (L.) Dumort.
Apocynaceae
Elatinoides elatine Wetts., *Linaria elatine* (L.) Mill. (see)
♦ sharppoint fluvellin, sharpleaf cancerwort, sharp leaved fluvellin, keihäsnielukukka, sharppoint fluellin
♦ Weed, Naturalised, Introduced, Casual Alien
♦ 7, 34, 42, 70, 87, 88, 94, 98, 101, 161, 176, 198, 199, 203, 212, 218, 241, 243, 253, 272, 280, 287, 300
♦ pH, cultivated, herbal.

Kickxia elatine (L.) Dumort. ssp. crinita (Mabille) W.Greuter
Apocynaceae
♦ sharp leaved fluvellin
♦ Naturalised, Environmental Weed
♦ 86, 198

Kickxia elatine (L.) Dumort. ssp. elatine
Apocynaceae
♦ sharp leaved fluvellin
♦ Naturalised
♦ 86, 198
♦ Origin: North Africa, south-west Asia.

Kickxia heterophylla (Schousb.) Dandy
Apocynaceae
♦ Weed
♦ 221

Kickxia lanigera Hand.-Mazz.
Apocynaceae
♦ Weed
♦ 87, 88, 94

Kickxia macilenta (Decne.) Danin
Apocynaceae
♦ Weed
♦ 221

Kickxia spuria (L.) Dum.
Apocynaceae
Antirrhinum spurium L., *Cymbalaria spuria* (L.) Baumg., *Elatinoides spuria* Wetts., *Linaria spuria* (L.) Mill. (see)
♦ round leaved fluellen, roundleaf

cancerwort, sarvinielukukka, female fluvellin, blunt leaved fluellen
♦ Weed, Naturalised, Introduced, Casual Alien
♦ 7, 34, 42, 70, 86, 87, 88, 94, 98, 101, 161, 176, 180, 198, 203, 212, 221, 243, 253, 272, 280
♦ pH, cultivated, herbal.

Kigelia africana **(Lam.) Benth.**
Bignoniaceae
Bignonia africana Lam., *Crescentia pinnata* Jacq., *Kigelia abyssinica* A.Rich., *Kigelia acutifolia* Engl. ex Spreng., *Kigelia aethiopum* (Fenzl) Dandy, *Kigelia aethopica* Decne., *Kigelia elliottii* Sprague, *Kigelia elliptica* Sprague, *Kigelia impressa* Sprague, *Kigelia pinnata* (Jacq.) DC., *Kigelia spragueana* Wernham, *Kigelia talbotii* Hutch. & Dalziel, *Kigelia tristis* A.Chev., *Sotor aethiopiumm* Fenzl, *Tanaecium pinnatum* (Jacq.) Willd.
♦ sausage tree, jillahi, palo de salchichón
♦ Naturalised, Introduced
♦ 101, 228, 261
♦ T, arid, cultivated, herbal. Origin: tropical Africa.

Kindbergia praelonga **(Hedw.) Ochyra**
Brachytheciaceae
= *Eurhynchium praelongum* (Hedw.) Schimp. (NoR)
♦ Naturalised
♦ 280
♦ moss.

Kirkia acuminata **Oliv.**
Kirkiaceae/Simaroubaceae
♦ white syringa
♦ Weed
♦ 87, 88
♦ cultivated.

Kleinhovia hospita **L.**
Sterculiaceae
♦ guest tree, fu'afu'a
♦ Naturalised, Introduced
♦ 101, 228, 261
♦ arid, cultivated, herbal. Origin: south-east Asia.

Kleinia longiflora **DC.**
Asteraceae
= *Senecio longiflorus* (DC.) Sch.Bip.
♦ succulent daisy
♦ Quarantine Weed
♦ 220
♦ arid, cultivated.

Knautia **L. spp.**
Dipsacaceae
♦ scabiosa
♦ Weed
♦ 272
♦ herbal.

Knautia arvensis **(L.) Coult.**
Dipsacaceae
Scabiosa arvensis L. (see)
♦ field scabious, ruusuruoho, blue buttons
♦ Weed, Noxious Weed, Naturalised

♦ 23, 44, 70, 87, 88, 101, 156, 159, 161, 162, 218, 243, 253, 272, 299
♦ pH, cultivated, herbal.

Knautia integrifolia **(L.) Bertol.**
Dipsacaceae
♦ ambretta annuale
♦ Weed, Naturalised
♦ 253, 272, 300
♦ herbal.

Knautia macedonica **Griseb.**
Dipsacaceae
♦ Weed
♦ 272
♦ cultivated, herbal.

Knautia sylvatica **(L.) Duby** *nom. ambig.*
Dipsacaceae
= *Knautia dipsacifolia* Kreutzer (NoR)
♦ Weed
♦ 87, 88

Kniphofia galpinii **Bak.**
Aloeaceae/Asphodelaceae
♦ Naturalised
♦ 40
♦ cultivated.

Kniphofia praecox **Bak.**
Aloeaceae/Asphodelaceae
♦ greater red hot poker
♦ Naturalised
♦ 40

Kniphofia uvaria **(L.) Hook.f.**
Aloeaceae/Asphodelaceae
Aletris sarmentosa Andrews, *Aloe uvaria* L., *Kniphofia alooides* Moench, *Tritoma uvaria* (L.) Ker Gawl.
♦ red hot poker, torch lily
♦ Weed, Naturalised, Garden Escape, Environmental Weed
♦ 54, 72, 86, 88, 98, 101, 155, 198
♦ pH, cultivated, herbal. Origin: South Africa.

Knowltonia vesicatoria **Sims**
Ranunculaceae
♦ Quarantine Weed
♦ 39, 220
♦ toxic.

Kochia aphylla **R.Br.**
Chenopodiaceae
= *Maireana aphylla* (R.Br.) P.G.Wilson (NoR)
♦ cottonbush
♦ Weed
♦ 87, 88
♦ arid.

Kochia indica **Wight**
Chenopodiaceae
= *Bassia indica* (Wight) A.J.Scott (NoR)
♦ Weed, Introduced
♦ 23, 87, 88, 228
♦ arid, herbal.

Kochia laniflora **(S.G.Gmel.) Borbás**
Chenopodiaceae
= *Bassia laniflora* (S.G.Gmel.) A.J.Scott (NoR)
♦ Weed
♦ 88, 94, 272

Kochia prostrata **(L.) Schrad.**
Chenopodiaceae
= *Bassia prostrata* (L.) A.J.Scott (NoR)
♦ prostrate summer cypress
♦ Naturalised
♦ 101
♦ arid.

Kochia pubescens **Moq.**
Chenopodiaceae
♦ Native Weed
♦ 121
♦ pS. Origin: southern Africa.

Kochia scoparia **(L.) Schrad.**
Chenopodiaceae
= *Bassia scoparia* (L.) Scott
♦ kochia, summer cypress, Mexican burningbush, mock cypress, Mexican fireweed, fireweed, common kochia, belvedere, red belvedere, belvedere cypress, fireball, firebush
♦ Weed, Quarantine Weed, Noxious Weed, Naturalised, Introduced, Environmental Weed, Casual Alien, Cultivation Escape
♦ 1, 7, 23, 34, 39, 42, 49, 52, 54, 68, 70, 76, 80, 87, 94, 101, 121, 133, 136, 138, 139, 146, 151, 155, 161, 169, 171, 174, 180, 195, 203, 210, 211, 212, 218, 220, 224, 228, 229, 232, 236, 263, 264, 272, 275, 280, 295, 297, 299
♦ aH, arid, cultivated, herbal, toxic. Origin: Eurasia.

Kochia scoparia **(L.) Schrad.** **var.** *culta* **Farw.**
Chenopodiaceae
= *Bassia scoparia* (L.) Scott
♦ Naturalised
♦ 287

Kochia scoparia **Schrad.** **var.** *scoparia*
Chenopodiaceae
♦ Naturalised
♦ 287

Kochia villosa **Lindl.**
Chenopodiaceae
= *Maireana villosa* (Lindl.) P.G.Wilson (NoR)
♦ Weed
♦ 87, 88

Koeberlinia spinosa **Zucc.**
Capparaceae/Koeberliniaceae
♦ allthorn, crown of thorns
♦ Weed
♦ 87, 88, 218
♦ arid, herbal.

Koeleria **Pers. spp.**
Poaceae
♦ junegrass
♦ Weed
♦ 272
♦ G.

Koeleria capensis **(Steud.) Nees**
Poaceae
♦ crested koeleria, koeleria grass, junegrass
♦ Native Weed
♦ 121
♦ pG. Origin: southern Africa.

Koeleria glauca (Schrad.) DC.
Poaceae
♦ sinitoppo, koeleria grass
♦ Quarantine Weed, Casual Alien
♦ 42, 220
♦ G, cultivated.

Koeleria macrantha (Ledeb.) Schult.
Poaceae
Aira cristata L., *Aira macrantha* Ledeb., *Koeleria albescens* auct., *Koeleria cristata* auct., *Koeleria gracilis* Pers., *Koeleria mukdenensis* Domin, *Koeleria nitida* Nutt., *Koeleria pyramidata* auct. Amer. (see)
♦ prairie junegrass, junegrass, hentotoppo, crested hairgrass, minoboro
♦ Weed, Naturalised, Casual Alien
♦ 42, 86, 98, 203, 272
♦ pG, arid, cultivated, herbal. Origin: Eurasia.

Koeleria phleoides (Vill.) Pers.
Poaceae
= *Rostraria cristata* (L.) Tzvelev
♦ annual junegrass, annual cat's tail, annual koeleria, crested polgrass
♦ Weed, Introduced
♦ 34, 87, 88, 91, 111, 121, 243
♦ aG, herbal, toxic. Origin: Eurasia.

Koeleria pyramidata (Lam.) Beauv.
Poaceae
♦ röyhötoppo, prairie junegrass, crested cat's tail, crested hairgrass
♦ Weed, Casual Alien
♦ 42, 91, 272
♦ pG, cultivated, herbal.

Koelpinia linearis Pali.
Asteraceae
♦ Weed
♦ 272

Koelreuteria elegans (Seem.) A.C.Sm.
Sapindaceae
♦ Chinese rain tree, flame gold, golden rain tree
♦ Weed, Naturalised, Environmental Weed
♦ 3, 86, 88, 101, 112, 155, 191, 201
♦ cultivated.

Koelreuteria elegans (Seem.) A.C.Sm. ssp. formosana (Hayata) F.G.Mey.
Sapindaceae
♦ flame gold, golden rain tree
♦ Weed, Naturalised
♦ 101, 179

Koelreuteria paniculata Laxm.
Sapindaceae
♦ Chinese rain tree, flame gold, golden rain tree, golden shower tree, golden chain tree
♦ Weed, Naturalised
♦ 3, 39, 80, 101, 191
♦ T, cultivated, herbal, toxic.

Koenigia islandica L.
Polygonaceae
♦ island purslane, kurjentatar
♦ Naturalised
♦ 241, 300
♦ herbal.

Kohautia Cham. & Schlechtd. spp.
Rubiaceae
♦ Weed
♦ 240
♦ aH.

Kohautia amatymbica Eckl. & Zeyh.
Rubiaceae
♦ Native Weed
♦ 121
♦ pH, herbal. Origin: southern Africa.

Kohautia aspera (Heyne ex Roth.) Bremek.
Rubiaceae
♦ Weed
♦ 242
♦ arid.

Kohautia caespitosa Schnizl.
Rubiaceae
♦ Weed, Native Weed
♦ 121, 221
♦ H. Origin: southern Africa.

Kohautia grandiflora DC.
Rubiaceae
♦ Weed
♦ 87, 88

Kohautia lasiocarpa Klotzsch
Rubiaceae
♦ Native Weed
♦ 121
♦ aH. Origin: southern Africa.

Kohautia senegalensis Cham. & Schlecht.
Rubiaceae
♦ Weed
♦ 87, 88

Kohautia virgata (Willd.) Brem.
Rubiaceae
♦ Native Weed
♦ 121
♦ pH. Origin: southern Africa.

Kohlrauschia prolifera (L.) Kunth
Caryophyllaceae
= *Petrorhagia prolifera* (L.) P.Ball & Heywood
♦ Weed
♦ 87, 88

Kolkwitzia amabilis Graebn.
Linnaeaceae/Caprifoliaceae
♦ beautybush
♦ Naturalised
♦ 101, 280
♦ cultivated, herbal.

Kolobopetalum chevalieri (Hutch. & Dalziel) Troupin
Menispermaceae
♦ bokaso
♦ Weed
♦ 87, 88

Kopsia fruticosa
Apocynaceae
♦ pink kopsia
♦ Weed
♦ 12
♦ herbal.

Kopsia ramosa (L.) Dumort.
Apocynaceae
= *Orobanche ramosa* L.

♦ Weed
♦ 87, 88

Kosteletzkya virginica (L.) Presl
Malvaceae
♦ seashore mallow, Virginia saltmarsh mallow, saltmarsh mallow
♦ Weed
♦ 87, 88, 218
♦ herbal.

Krameria grayi Rose & Painter
Krameriaceae
♦ white ratany, white rhatany
♦ Weed, Quarantine Weed
♦ 76, 88
♦ S parasitic, herbal.

Krameria iluca Phil.
Krameriaceae
♦ Introduced
♦ 228
♦ arid.

Krapovickasia physaloides (Presl) Fryxell
Malvaceae
♦ Naturalised
♦ 101

Krascheninnikovia ceratoides (L.) Gueldenst.
Chenopodiaceae
= *Ceratoides latens* (J.F.Gmel.) Rev. & Holmgren
♦ Weed
♦ 221
♦ arid.

Krigia cespitosa (Raf.) K.L.Chambers
Asteraceae
Krigia caespitosa (Raf.) Chambers, *Krigia oppositifolia* Raf.
♦ dwarf dandelion
♦ Weed
♦ 161
♦ herbal.

Krigia dandelion (L.) Nutt.
Asteraceae
♦ potato dandelion, potato dwarf dandelion
♦ Weed
♦ 161
♦ herbal.

Krigia virginica (L.) Willd.
Asteraceae
♦ Virginia dwarf dandelion, dwarf dandelion
♦ Weed
♦ 87, 88, 161, 218, 249
♦ herbal.

Kuhnia eupatorioides L.
Asteraceae
= *Brickellia eupatorioides* L. var. *eupatorioides* (NoR) [see *Kuhnia glutinosa* Ell.]
♦ boneset, false boneset
♦ Weed
♦ 161
♦ pH, cultivated, herbal.

Kuhnia glutinosa Ell.
Asteraceae
= *Brickellia eupatorioides* L. var. *eupatorioides* (NoR) [see *Kuhnia*

eupatorioides L.]
- Weed
- 23, 88
- herbal.

Kummerowia stipulacea (Maxim.) Makino
Fabaceae/Papilionaceae
Lespedeza stipulacea Maxim. (see)
- Korean clover, Korean lespedeza
- Weed, Naturalised
- 80, 87, 88, 101, 102, 275, 286, 297
- aH, promoted, herbal.

Kummerowia striata (Thunb.) Schindl.
Fabaceae/Papilionaceae
Hedysarum striatum Thunb., *Lespedeza striata* (Thunb.) Hook. & Arn. (see), *Microlespedeza striata* Makino
- Japanese clover, yahazusou
- Weed, Naturalised
- 80, 87, 88, 98, 101, 102, 179, 191, 203, 204, 235, 275, 286, 297
- aH, promoted, herbal. Origin: China, Japan, Korea, Manchuria.

Kundmannia sicula DC.
Apiaceae
- Weed
- 87, 88

Kunzea baxteri (Klotzsch) Schauer
Myrtaceae
- scarlet gold tip kunzea
- Naturalised, Native Weed, Garden Escape, Cultivation Escape
- 7
- S, cultivated. Origin: Australia.

Kunzea ericoides (A.Rich.) J.Thomps.
Myrtaceae
= *Leptospermum ericoides* Rich.
- kanuka, burgan
- Weed
- 3, 88, 131
- cultivated. Origin: Australia, New Zealand.

Kyllinga alba Nees
Cyperaceae
- white sedge, lulimbu
- Weed, Native Weed
- 88, 121, 158
- pG. Origin: southern Africa.

Kyllinga aurata (Nees) Nees
Cyperaceae
- Weed
- 87, 88
- G.

Kyllinga brevifolia Rottb.
Cyperaceae
Cyperus brevifolius (Rottb.) Hassk. (see), *Kyllinga colorata* (L.) Druce (see)
- green kyllinga, shortleaf spikesedge
- Weed, Naturalised
- 6, 34, 80, 88, 121, 126, 134, 161, 191, 218, 235, 273, 274, 297
- pG, arid, herbal, toxic. Origin: Africa.

Kyllinga brevifolia Rottb. var. *leiolepis* (Franch. & Sav.) Hara
Cyperaceae
- Weed

- 275
- pG.

Kyllinga brevifolioides (Thieret & Delahoussaye) G.Tucker
Cyperaceae
= *Kyllinga gracillima* Miq. (NoR) [see *Cyperus brevifolius* (Rottb.) Endl. ex Hassk. var. *leiolepis* (Franch. & Savigny) T.Koyama]
- Weed
- 80
- G.

Kyllinga bulbosa P.Beauv.
Cyperaceae
- Weed
- 88
- G.

Kyllinga cephalotes (Jacq.) Druce
Cyperaceae
= *Kyllinga nemoralis* (J.R. & G.Forst.) Dandy ex Hutch. & Dalziel
- Weed
- 3, 191
- G.

Kyllinga colorata (L.) Druce
Cyperaceae
= *Kyllinga brevifolia* Rottb.
- Weed
- 275
- pG.

Kyllinga erecta Schum.
Cyperaceae
Cyperus erectus (Schum.) Mattf. & Kuek., *Kyllinga elata* Steud., *Kyllinga polyphylla* Willd. ex Kunth (see)
- white kyllinga, greater kyllinga, greater kyllinga, white sedge, kieye
- Weed, Native Weed
- 51, 87, 88, 121, 126, 158
- pG. Origin: Madagascar.

Kyllinga monocephala Rottb.
Cyperaceae
= *Kyllinga nemoralis* (J.R. & G.Forst.) Dandy ex Hutch. & Dalziel
- white kyllinga
- Weed
- 3, 28, 39, 88, 126, 191, 243
- pG, herbal, toxic.

Kyllinga nemoralis (J.R. & G.Forst.) Dandy ex Hutch. & Dalziel
Cyperaceae
Cyperus kyllingia Endl. (see), *Kyllinga cephalotes* (Jacq.) Druce (see), *Kyllinga monocephala* Rottb. (see)
- white kyllinga, kili'o'opu, mo'u upo'o, tuise, pakopako, pakopako 'ae kuma, whitehead spikesedge
- Weed, Naturalised
- 3, 6, 88, 101, 107, 191, 273
- pG, herbal.

Kyllinga odorata Vahl
Cyperaceae
= *Cyperus sesquiflorus* (Torr.) Kük.
- annual kyllinga, fragrant spikesedge
- Weed, Native Weed
- 80, 87, 88, 121, 161
- pG.

Kyllinga polyphylla Willd. ex Kunth
Cyperaceae
= *Kyllinga erecta* Schum.
- navua sedge
- Weed
- 3, 6, 87, 88, 191
- pG.

Kyllinga pumila Michx.
Cyperaceae
- low spikesedge
- Weed
- 28, 87, 88, 206, 243, 281
- G, herbal.

Kyllinga sesquiflora Torr.
Cyperaceae
= *Cyperus sesquiflorus* (Torr.) Kük.
- cortadera
- Weed
- 153
- G.

Kyllinga squamulata Thonn. ex Vahl
Cyperaceae
Cyperus metzii (Hochst. ex Steud.) Mattf. & Kük. ex Kük. (see)
- Asian spikesedge
- Weed, Naturalised
- 88, 101, 179
- G, herbal.

Kyphocarpa angustifolia (Moq.) Lopr.
Amaranthaceae
Kyphocarpa zeyheri (Moq.) Lopr. (see)
- silky burrweed
- Native Weed
- 121
- pH. Origin: southern Africa.

Kyphocarpa zeyheri (Moq.) Lopr.
Amaranthaceae
= *Kyphocarpa angustifolia* (Moq.) Lopr.
- silky burrweed
- Weed
- 50

L

Lablab purpureus (L.) Sweet
Fabaceae/Papilionaceae
Dolichos lablab L. (see), *Dolichos purpureus* L., *Lablab niger* Medik., *Lablab vulgaris* (L.) Savi
♦ lablab bean, hyacinth bean, dolichos bean, bonavista bean, hyacinth bean vine, chícharos de jardín, sweet pea
♦ Weed, Naturalised, Environmental Weed, Casual Alien
♦ 7, 86, 98, 101, 134, 161, 179, 203, 243, 257, 261
♦ pC, arid, cultivated, herbal, toxic. Origin: Madagascar.

Lablab purpureus (L.) Sweet ssp. purpureus
Fabaceae/Papilionaceae
♦ bonavista bean, Egyptian bean, hyacinth bean, Indian bean, lablab bean, lubia bean
♦ Weed
♦ 121
♦ pC, cultivated. Origin: Africa.

Laburnum alpinum (Mill.) Bercht. & J.Presl
Fabaceae/Papilionaceae
Cytisus alpinus Mill.
♦ kultasade, Scotch laburnum
♦ Cultivation Escape
♦ 42
♦ T, cultivated, herbal.

Laburnum anagyroides Medik.
Fabaceae/Papilionaceae
Cytisus laburnum L., *Genista laburnum* Krause
♦ etelänkultasade, laburnum, golden chain, golden chain, golden rain, bean tree
♦ Weed, Naturalised, Cultivation Escape
♦ 15, 39, 42, 87, 88, 154, 161, 189, 247, 280
♦ S/T, cultivated, herbal, toxic.

Laburnum × watereri (Kirchn.) Dipp.
Fabaceae/Papilionaceae
= *Laburnum alpinum* (Mill.) Bercht. & J.Presl × *Laburnum anagyroides* Medic.
♦ golden chain tree
♦ Weed
♦ 45
♦ S/T, cultivated, herbal.

Lachenalia aloides (L.f.) hort. ex Engl.
Liliaceae/Hyacinthaceae
♦ lachenalia, cape cowslip
♦ Weed, Naturalised, Garden Escape,

Environmental Weed
♦ 7, 72, 86, 88, 98, 203
♦ pH, cultivated. Origin: South Africa.

Lachenalia aloides (L.f.) hort. ex Engl. var. aurea (J.W.Loudon & Wooster) Engl.
Liliaceae/Hyacinthaceae
♦ Environmental Weed
♦ 155
♦ Origin: South Africa.

Lachenalia bulbifera (Cirillo) Asch. & Graebn.
Liliaceae/Hyacinthaceae
♦ rooinaltjie
♦ Weed, Sleeper Weed, Naturalised, Garden Escape, Environmental Weed, Cultivation Escape
♦ 7, 54, 86, 88, 98, 155, 203
♦ cultivated. Origin: South Africa.

Lachenalia mutabilis Sweet
Liliaceae/Hyacinthaceae
♦ lachenalia
♦ Weed, Sleeper Weed, Naturalised, Garden Escape, Environmental Weed, Cultivation Escape
♦ 7, 54, 86, 88, 155
♦ cultivated. Origin: South Africa.

Lachenalia reflexa Thunb.
Liliaceae/Hyacinthaceae
♦ cape cowslip, lachenalia
♦ Weed, Naturalised, Garden Escape, Environmental Weed
♦ 7, 9, 86, 98, 155, 203
♦ pH, cultivated. Origin: South Africa.

Lachnanthes caroliana (Lam.) Dandy
Haemodoraceae
Lachnanthes tinctoria (J.F.Gmel.) Ell. (see)
♦ redroot, Carolina redroot
♦ Weed, Quarantine Weed
♦ 161, 258
♦ pH, promoted, herbal, toxic.

Lachnanthes tinctoria (J.F.Gmel.) Ell.
Haemodoraceae
= *Lachnanthes caroliana* (Lam.) Dandy
♦ redroot, lachnanthes
♦ Weed
♦ 87, 88, 218
♦ herbal.

Lachnocaulon Kunth spp.
Eriocaulaceae
♦ bogbutton, lachnocaulon
♦ Quarantine Weed
♦ 258

Lacmellia Karst. spp.
Apocynaceae
♦ Quarantine Weed
♦ 220

Lactuca altaica Fisch. & C.A.Mey.
Asteraceae
♦ Weed
♦ 272

Lactuca biennis (Moench) Fern.
Asteraceae
♦ biennial lettuce, tall blue lettuce, wild lettuce
♦ Weed
♦ 8, 23, 87, 88, 161, 218

♦ pH, cultivated, herbal, toxic.

Lactuca canadensis L.
Asteraceae
♦ tall lettuce, tall wild lettuce, wild lettuce, Canada lettuce
♦ Weed
♦ 8, 34, 87, 88, 161, 210, 218, 249
♦ pH, promoted, herbal, toxic.

Lactuca capensis Thunb.
Asteraceae
♦ Weed, Quarantine Weed, Native Weed
♦ 23, 76, 87, 88, 121, 203, 220
♦ pH, arid, promoted, herbal. Origin: southern Africa.

Lactuca dissecta D.Don
Asteraceae
♦ Weed
♦ 87, 88, 243

Lactuca dregeana DC.
Asteraceae
= *Lactuca serriola* L.
♦ wild lettuce
♦ Weed
♦ 23, 88, 121
♦ H. Origin: southern Africa.

Lactuca floridana (L.) Gaertn.
Asteraceae
♦ Florida wild lettuce, woodland lettuce
♦ Weed
♦ 161
♦ herbal.

Lactuca formosana Maxim.
Asteraceae
♦ Formosan lettuce
♦ Weed
♦ 87, 88, 274
♦ H, promoted.

Lactuca indica L.
Asteraceae
Lactuca laciniata (Houtt.) Makino
♦ Indian lettuce, akinonogeshi, wild lettuce
♦ Weed
♦ 87, 88, 204, 235, 263, 274, 275, 286, 297
♦ a/pH, promoted, herbal. Origin: Asia.

Lactuca indica L. fo. indivisa Hara
Asteraceae
♦ Weed
♦ 286

Lactuca intybacea Jacq.
Asteraceae
= *Launaea intybacea* (Jacq.) Beauv.
♦ Weed
♦ 87, 88
♦ herbal.

Lactuca ludoviciana (Nutt.) Riddell
Asteraceae
♦ biannual lettuce, wild lettuce
♦ Weed
♦ 23, 88
♦ pH, promoted, herbal.

Lactuca muralis (L.) Gaertn.
Asteraceae

= *Mycelis muralis* (L.) Dumort.
♦ wall lettuce
♦ Weed
♦ 23, 80, 87, 88, 146, 218
♦ herbal.

Lactuca orientalis Boiss.
Asteraceae
Scariola orientalis (Boiss.) Sojak
♦ Weed
♦ 87, 88, 221
♦ arid.

Lactuca perennis L.
Asteraceae
♦ blue lettuce, perennial lettuce
♦ Weed
♦ 88, 94, 253
♦ pH, cultivated, herbal.

Lactuca pulchella (Pursh) DC.
Asteraceae
= *Lactuca tatarica* (L.) C.A.Mey. var.
pulchella (Pursh) Breitung (NoR)
♦ blue flowering lettuce, blue lettuce,
showy lettuce, large flowered blue
lettuce
♦ Weed, Quarantine Weed,
Naturalised
♦ 49, 52, 76, 80, 87, 88, 136, 161, 210,
212, 218, 220, 287
♦ pH, promoted, herbal.

Lactuca runcinata DC.
Asteraceae
♦ Weed
♦ 87, 88

Lactuca saligna L.
Asteraceae
♦ willowleaf lettuce, least lettuce, wild
lettuce
♦ Weed, Naturalised, Environmental
Weed
♦ 7, 34, 55, 80, 86, 87, 88, 98, 101, 161,
176, 198, 203, 218, 221, 269, 272, 280
♦ aH, cultivated, herbal. Origin:
Europe.

Lactuca sativa L.
Asteraceae
♦ lettuce, garden lettuce, ruoka
salaatti, insalata, kopfsalat
♦ Weed, Naturalised, Environmental
Weed, Cultivation Escape, Casual
Alien
♦ 40, 42, 86, 93, 101, 205, 241, 252, 257,
261, 280, 300
♦ a/bH, cultivated, herbal. Origin:
obscure possibly horticultural.

Lactuca scariola L.
Asteraceae
= *Lactuca serriola* L.
♦ prickly lettuce, true prickly lettuce,
compass plant, milk thistle
♦ Weed, Naturalised
♦ 8, 23, 49, 52, 87, 88, 218, 286, 287
♦ a/bH, herbal, toxic.

Lactuca scariola L. fo. integrifolia L.
Asteraceae
♦ Naturalised
♦ 287

Lactuca serriola L.
Asteraceae
Lactuca scariola L. (see), *Lactuca
dregeana* DC. (see)
♦ prickly lettuce, wild lettuce, China
lettuce, compass plant, milk thistle,
horse thistle, wild opium
♦ Weed, Noxious Weed, Naturalised,
Introduced, Garden Escape,
Environmental Weed, Casual Alien
♦ 7, 21, 34, 39, 40, 42, 44, 51, 55, 68, 70,
72, 80, 86, 87, 88, 93, 94, 98, 101, 115,
121, 136, 146, 151, 158, 161, 167, 174,
176, 180, 185, 195, 198, 199, 203, 205,
207, 210, 211, 212, 221, 237, 241, 243,
248, 250, 253, 269, 272, 280, 295, 299,
300
♦ a/bH, arid, cultivated, herbal, toxic.
Origin: Eurasia.

Lactuca sibirica (L.) Maxim.
Asteraceae
♦ siperiansinivalvatti
♦ Weed
♦ 272, 297
♦ pH, promoted, herbal.

Lactuca stolonifera (Gray) Maxim.
Asteraceae
= *Ixeris stolonifera* A.Gray
♦ Weed
♦ 88, 204

**Lactuca taraxacifolia (Willd.) Schum. &
Thonn. ex Hornem.**
Asteraceae
= *Launaea cornuta* (Oliv. & Hiern)
C.Jeffrey
♦ wild lettuce
♦ Weed, Quarantine Weed
♦ 39, 76, 87, 88, 203, 220
♦ arid, toxic.

Lactuca tatarica (L.) C.A.Mey.
Asteraceae
= *Mulgedium tataricum* (L.) DC.
♦ blue lettuce, tataarisininvalvatti
♦ Weed, Naturalised
♦ 39, 42, 87, 88, 272, 275, 297
♦ pH, herbal, toxic.

Lactuca versicolor Sch.Bip.
Asteraceae
♦ Weed
♦ 87, 88

Lactuca viminea (L.) J. & C.Presl
Asteraceae
Chondrilla viminea Lam.
♦ Weed
♦ 87, 88, 243, 272
♦ arid, cultivated.

Lactuca virosa L.
Asteraceae
♦ great lettuce, prickly lettuce,
compass plant, milk thistle, bitter
lettuce, acrid lettuce
♦ Weed, Naturalised, Casual Alien
♦ 7, 15, 24, 39, 40, 49, 70, 98, 101, 161,
165, 203, 243, 253, 272, 280, 287, 300
♦ pH, cultivated, herbal, toxic. Origin:
Mediterranean, west Asia.

Laennecia coulteri (Gray) Nesom
Asteraceae

♦ Coulter's conyza, conyza
♦ Weed
♦ 161
♦ toxic.

Lafoensia pacari St.-Hil.
Lythraceae
♦ Weed
♦ 87, 88

Lagarosiphon Harv. spp.
Hydrocharitaceae
♦ African elodea, lagarosiphon,
oxygen weed
♦ Weed, Quarantine Weed, Noxious
Weed
♦ 10, 62, 67, 76, 203
♦ wH.

Lagarosiphon cordofanus Casp.
Hydrocharitaceae
Largosiphon crispus Rendle
♦ Weed
♦ 121
♦ wH.

Lagarosiphon major (Ridl.) Moss
Hydrocharitaceae
Elodea crispa hort.
♦ lagarosiphon, oxygen weed, curly
water thyme, coarse oxygen weed
♦ Weed, Quarantine Weed, Noxious
Weed, Naturalised, Garden Escape,
Environmental Weed
♦ 15, 40, 54, 67, 76, 86, 87, 88, 93, 98,
126, 140, 147, 152, 155, 121, 165, 169,
171, 176, 191, 193, 200, 203, 208, 225,
229, 232, 246, 280, 290
♦ wpH, cultivated. Origin: tropical
and southern Africa.

Lagarosiphon muscoides Harv.
Hydrocharitaceae
♦ fine oxygen weed
♦ Weed
♦ 121
♦ wH, cultivated.

Lagarosiphon roxburghii Benth.
Hydrocharitaceae
♦ Weed, Quarantine Weed
♦ 76, 87, 88, 203
♦ wH.

Lagarosiphon verticillifolius Oberm.
Hydrocharitaceae
♦ Weed
♦ 121
♦ wH.

Lagascea decipiens Hemsl.
Asteraceae
= *Nocca decipiens* (Hemsl.) Kuntze
♦ doll's head
♦ Quarantine Weed
♦ 220
♦ cultivated, herbal.

Lagascea mollis Cav.
Asteraceae
♦ acuate, silk leaf, yaa kammayee
♦ Weed, Noxious Weed, Naturalised
♦ 14, 66, 87, 88, 101, 157, 191, 204, 218,
229, 239
♦ aH.

Lagenaria siceraria (Molina) Standl.
Cucurbitaceae
Adenopus abyssinicus Hook.f. var.
somaliensis Chiov., *Cucurbita lagenaria*
L., *Cucurbita siceraria* Molina, *Lagenaria*
leucantha (Duch.) Rusby, *Lagenaria*
vulgaris Ser.
♦ bottle gourd, amabhanga, calabash,
calabash bottle, calabash gourd, cheho,
club gourd, digo, dudhiva, fagufagu,
gardu, ghiba, ikhomane, inshubaba,
iselwa, kadu, kalabas, kalbas,
kalbaspatat, ladu, lau, lauki, leraka,
maraga, moraka, ntshubaba, sego,
seho, sehoana, smooth gourd, thotse,
tshikumbu, white pumpkin, white
flowered gourd
♦ Weed, Naturalised, Introduced,
Cultivation Escape
♦ 13, 32, 39, 86, 98, 101, 179, 203, 228,
243, 261
♦ aC, arid, cultivated, herbal, toxic.
Origin: tropical Africa, Asia.

Lagenaria siceraria (Molina) Standl. ssp.
asiatica (Kobiakova) Heiser
Cucurbitaceae
♦ sinek
♦ Introduced
♦ 230
♦ C.

Lagerstroemia indica L.
Lythraceae
♦ crape myrtle, crepe myrtle
♦ Weed, Naturalised, Introduced,
Environmental Weed, Cultivation
Escape
♦ 80, 88, 101, 151, 179, 230, 261
♦ S, cultivated, herbal. Origin: east and
southern Asia.

Lagerstroemia speciosa (L.) Pers.
Lythraceae
♦ pride of India, queen's crape myrtle
♦ Weed, Naturalised, Cultivation
Escape
♦ 32, 101, 261
♦ cultivated, herbal. Origin: Asia.

Laggera aurita Sch.Bip. ex C.B.Clarke
Asteraceae
Blumea aurita DC. (see)
♦ Weed
♦ 87, 88
♦ herbal.

Laggera pterodonta (DC.) Sch.Bip. ex
Oliv.
Asteraceae
Blumea pterodonta DC.
♦ naat liam
♦ Weed
♦ 238, 297
♦ aH.

Lagonychium farctum (Banks & Sol.)
Bobrov.
Fabaceae/Mimosaceae
= *Prosopis farcta* (Banks & Sol.)
J.F.Macbr.
♦ Weed
♦ 87, 88, 221

Lagopsis supina (Stephan ex Willd.)
Ikonn.-Gal. ex Knorr.
Lamiaceae
♦ Weed
♦ 275, 297
♦ pH.

Lagoseris sancta Maly
Asteraceae
Hieracium sanctum L.
♦ Weed
♦ 221

Lagunaria patersonia (Andr.) G.Don
Malvaceae/Bombacaceae
= *Lagunaria patersonii* (Andr.) G.Don
♦ Norfolk Island hibiscus
♦ Naturalised, Casual Alien
♦ 7, 280
♦ cultivated, herbal. Origin: Australia.

Lagunaria patersonia (Andrews) G.Don
ssp. *patersonia*
Malvaceae/Bombacaceae
♦ Casual Alien
♦ 280

Lagunaria patersonii (Andr.) G.Don
Malvaceae/Bombacaceae
Lagunaria patersonia (Andr.) G.Don
(see)
♦ white wood, Norfolk Island
hibiscus, pyramid tree, Queensland
pyramid tree
♦ Weed, Naturalised
♦ 86, 121
♦ S/T, cultivated, herbal, toxic. Origin:
Australia.

Lagurus ovatus L.
Poaceae
♦ hare's tail grass, jänönhäntä,
bunnie's tails, hare's foot, rabbit tail
grass
♦ Weed, Naturalised, Garden Escape,
Environmental Weed, Cultivation
Escape, Casual Alien
♦ 7, 9, 15, 34, 40, 42, 51, 72, 86, 87, 88,
91, 98, 101, 121, 158, 161, 176, 198, 203,
241, 272, 280, 287, 289, 300
♦ aG, cultivated, herbal. Origin:
Eurasia.

Lallemantia canescens (L.) Fisch. &
C.A.Mey.
Lamiaceae
Dracocephalum canescens L.
♦ harmaakormikki
♦ Casual Alien
♦ 42
♦ cultivated.

Lallemantia iberica (Bieb.) Fisch. & Mey.
Lamiaceae
Lallemantia sulphurea (Bieb.) Fisch. &
Mey., *Dracocephalum ibericum* M.Bieb.
♦ espanjankormikki
♦ Weed, Introduced, Casual Alien
♦ 42, 87, 88, 228
♦ a/bH, arid, cultivated, herbal.

Lallemantia peltata (L.) Fisch. &
C.A.Mey.
Lamiaceae
♦ kilpikormikki, lion's heart
♦ Naturalised, Casual Alien

♦ 42, 101
♦ cultivated.

Lallemantia royleana (Benth.) Benth.
Lamiaceae
♦ Weed
♦ 248

Lamarckia aurea (L.) Moench
Poaceae
♦ goldentop, goldentop grass,
kultaheinä
♦ Weed, Naturalised, Introduced,
Environmental Weed, Casual Alien
♦ 7, 34, 38, 42, 72, 86, 87, 88, 91, 98, 101,
167, 198, 203, 218, 241, 300
♦ aG, arid, cultivated, herbal. Origin:
Mediterranean, Madeira and Canary
Islands.

Lamiastrum galeobdolon (L.) Ehrend. &
Polat.
Lamiaceae
= *Lamium galeobdolon* (L.) L.
♦ yellow deadnettle, yellow archangel,
polatschek nukulat
♦ Weed, Naturalised, Cultivation
Escape
♦ 42, 70, 80, 101, 272
♦ cultivated, herbal.

Lamiastrum galeobdolon (L.) Ehrend. &
Polat. ssp. *argentatum* (Smejkal) Stace
Lamiaceae
Galeobdolon argentatum Smejkal
♦ variegated yellow archangel
♦ Naturalised
♦ 40

Lamium L. spp.
Lamiaceae
♦ deadnettle
♦ Weed, Naturalised
♦ 39, 198, 243
♦ H, toxic.

Lamium album L.
Lamiaceae
♦ white deadnettle, valkopeippi
♦ Weed, Naturalised
♦ 39, 44, 70, 87, 88, 101, 272, 280, 297
♦ pH, cultivated, herbal, toxic.

Lamium album L. var. *barbatum* Franch.
& Sav.
Lamiaceae
♦ odorikosou
♦ Weed
♦ 263
♦ pH.

Lamium amplexicaule L.
Lamiaceae
Galeobdolon amplexicaule Moench,
Lamium stepposum Kossko ex Klok.,
Lamiopsis amplexicaulis (L.) Opiz,
Pollichia amplexicaulis (L.) Willd.
♦ henbit, henbit deadnettle,
sepiväpeippi, blind nettle, bee nettle,
giraffe head, deadnettle, falsa ortica
reniforme, erba ruota
♦ Weed, Noxious Weed, Naturalised,
Introduced
♦ 7, 15, 23, 34, 38, 39, 44, 55, 70, 80, 86,
87, 88, 93, 94, 98, 101, 115, 118, 121, 136,
161, 162, 165, 174, 176, 179, 180, 185,

198, 203, 204, 207, 210, 211, 212, 217,
217, 218, 221, 235, 236, 237, 240, 241,
243, 249, 250, 253, 263, 269, 271, 272,
280, 286, 295, 297, 299, 300
♦ a/bH, cultivated, herbal, toxic.
Origin: Eurasia.

Lamium barbatum Sieb. & Zucc.
Lamiaceae
♦ Weed
♦ 286
♦ herbal.

Lamium bifidum Cyr.
Lamiaceae
♦ Weed
♦ 272

Lamium confertum Fr.
Lamiaceae
♦ northern deadnettle, välipeippi
♦ Naturalised
♦ 280

Lamium galeobdolon (L.) L.
Lamiaceae
Galeobdolon luteum Huds (see),
Galeopsis galeobdolon L., *Lamiastrum
galeobdolon* (L.) Ehrend. & Polat. (see)
♦ aluminium plant, yellow archangel
♦ Weed
♦ 88
♦ pH, cultivated, herbal. Origin:
Eurasia.

**Lamium galeobdolon (L.) L. fo.
argentatum (Smejkal) Mennema**
Lamiaceae
♦ aluminium plant, yellow archangel
♦ Weed, Naturalised
♦ 54, 86, 198
♦ Origin: Eurasia.

Lamium hybridum Vill.
Lamiaceae
Lamium incisum Willd.
♦ cut leaved deadnettle, deadnettle
hybrid
♦ Weed, Noxious Weed, Naturalised
♦ 1, 44, 70, 80, 88, 94, 136, 139, 161, 253,
272, 280, 287

Lamium macrodon Boiss.
Lamiaceae
♦ Weed
♦ 87, 88

Lamium maculatum L.
Lamiaceae
♦ spotted deadnettle, spotted henbit,
täplälehtipeippi, red deadnettle
♦ Weed, Naturalised, Casual Alien
♦ 15, 40, 42, 80, 87, 88, 101, 195, 218,
272, 280
♦ cultivated, herbal.

Lamium moluccellifolium Fr.
Lamiaceae
♦ northern deadnettle, intermediate
deadnettle
♦ Weed
♦ 70, 88, 94, 253, 272

Lamium purpureum L.
Lamiaceae
♦ purple deadnettle, red deadnettle,
falsa ortica purpurea

♦ Weed, Naturalised, Introduced
♦ 15, 23, 34, 44, 70, 80, 86, 87, 88, 94,
98, 101, 118, 136, 161, 165, 176, 198, 199,
203, 212, 218, 243, 249, 253, 263, 272,
280, 286, 287
♦ aH, cultivated, herbal. Origin:
Eurasia.

**Lamium purpureum L. var. incisum
(Willd.) Pers.**
Lamiaceae
Lamium hybridum auct. *non* Vill.
♦ hybrid deadnettle, purple deadnettle
♦ Noxious Weed, Naturalised
♦ 101, 229

Lamium purpureum L. var. purpureum
Lamiaceae
♦ purple deadnettle
♦ Naturalised
♦ 101

Lampranthus coccineus (Haw.) N.E.Br.
Aizoaceae/Mesembryanthemaceae
♦ redflush
♦ Naturalised
♦ 86, 101
♦ cultivated.

Lampranthus falciformis N.E.Br.
Aizoaceae/Mesembryanthemaceae
Mesembryanthemum falciforme Haw.
♦ sickle leaved dewplant
♦ Naturalised
♦ 40

Lampranthus glaucus (L.) N.E.Br.
Aizoaceae/Mesembryanthemaceae
♦ lampranthus
♦ Weed, Naturalised, Environmental
Weed
♦ 7, 72, 86, 88, 98, 176, 203, 246, 280
♦ S, cultivated. Origin: South Africa.

**Lampranthus multiradiatus (Jacq.)
N.E.Br.**
Aizoaceae/Mesembryanthemaceae
♦ Naturalised
♦ 7, 86
♦ cultivated. Origin: South Africa.

Lampranthus spectabilis (Haw.) N.E.Br.
Aizoaceae/Mesembryanthemaceae
♦ trailing iceplant
♦ Weed, Naturalised
♦ 116, 280
♦ cultivated.

Lampranthus tegens (F.Muell.) N.E.Br.
Aizoaceae/Mesembryanthemaceae
♦ little noonflower
♦ Weed, Naturalised, Environmental
Weed
♦ 72, 86, 88, 98, 198, 203
♦ S, cultivated. Origin: South Africa.

Lamprocapnos spectabilis (L.) Fukuhara
Papaveraceae/Fumariaceae
Dicentra spectabilis (L.) Lem. (see)
♦ bleedingheart
♦ Naturalised
♦ 101

Landolphia dulcis (Sabine) Pichon
Apocynaceae
Carpodinus dulcis Sabine (see)
♦ Quarantine Weed

♦ 220
♦ herbal.

**Landolphia tenuifolia (Pierre ex Stapf)
Pichon**
Apocynaceae
Carpodinus tenuifolia Pierre ex Stapf
(see)
♦ Quarantine Weed
♦ 220

Lannea coromandelica (Houtt.) Merr.
Anacardiaceae
♦ djavaran, golra, jiyal, kaju djaran,
kaju kuda, kayu kuda, kedongdong,
reo, wodier
♦ Introduced
♦ 228
♦ arid, cultivated, herbal.

**Lannea schweinfurthii (Engl.) Engl. var.
stuhlmannii (Engl.) Kokwaro**
Anacardiaceae
Lannea stuhlmannii (Engl.) Engl.
♦ bastard marula, false marula, tree
grape
♦ Native Weed
♦ 121
♦ T. Origin: southern Africa.

Lansium domesticum Corr.
Meliaceae
♦ langsat
♦ Introduced
♦ 39, 230
♦ T, cultivated, herbal, toxic.

Lantana L. spp.
Verbenaceae
♦ lantana
♦ Weed, Quarantine Weed,
Naturalised
♦ 76, 86, 88, 220
♦ cultivated, herbal.

Lantana camara L.
Verbenaceae
Camara vulgaris Benth, *Lantana aculeata*
L., *Lantana armata* Schauer, *Lantana
armata* var. *guianensis* Moldenke,
Lantana camara fo. *mista* (L.) Moldenke,
Lantana camara var. *aculeata* (L.)
Moldenke (see), *Lantana camara* var.
mista (L.) L.H.Bailey, *Lantana camara*
var. *moritziana* (Otto & A.Dietr.) López-
Pal., *Lantana camara* var. *nivea* (Vent.)
L.H.Bailey, *Lantana glandulosissima*
Hayek, *Lantana hybrida* hort. (see),
Lantana mista L., *Lantana moritziana*
Otto & A.Dietr. (see), *Lantana nivea*
Vent.
♦ lantana, yellow sage, red flowered
sage, tickberry, wild sage, prickly
lantana, white sage, chiPoniwe,
mikinolia hihiu, curse of India,
landana, lanitana, rantana, rahndana,
tukasuweth, te kaibuaka, talatala,
kauboica, latora moa, tatara moa, ros
fonacni, latana, lakana, talatala, talatala
talmoa, te kaibuaja, taramoa, migiroa,
kaumboitha, mbonambulumakau,
mbona ra mbulumakau, tokalau,
waiwai, taratara hamoa
♦ Weed, Noxious Weed, Naturalised,

Introduced, Garden Escape, Environmental Weed, Cultivation Escape
♦ 3, 6, 7, 10, 12, 13, 14, 15, 18, 22, 37, 38, 39, 50, 51, 53, 63, 73, 80, 85, 86, 87, 88, 93, 95, 98, 107, 112, 121, 122, 132, 134, 147, 151, 152, 153, 154, 157, 158, 161, 165, 178, 179, 186, 186, 189, 194, 201, 203, 209, 216, 218, 233, 247, 255, 256, 257, 262, 268, 269, 274, 275, 276, 277, 278, 279, 280, 283, 286, 287, 289, 295, 296
♦ S, cultivated, herbal, toxic. Origin: Central and South America.

Lantana camara L. var. *aculeata* (L.) Moldenke
Verbenaceae
= *Lantana camara* L.
♦ randana, lantana
♦ Quarantine Weed, Naturalised, Introduced, Environmental Weed
♦ 225, 230, 246, 280
♦ S.

Lantana camara L. var. *crocea* L.
Verbenaceae
♦ Naturalised
♦ 98

Lantana canescens Kunth
Verbenaceae
♦ cambarazinho, hammock shrubverbena
♦ Weed
♦ 255
♦ S. Origin: South America.

Lantana crocea Jacq.
Verbenaceae
♦ Weed
♦ 87, 88

Lantana cujabensis Shau.
Verbenaceae
♦ Weed
♦ 87, 88

Lantana fucata Lindl.
Verbenaceae
♦ erva de grilo
♦ Weed
♦ 255
♦ S. Origin: South America.

Lantana hybrida hort.
Verbenaceae
= *Lantana camara* L.
♦ lantana, tickberry
♦ Weed
♦ 63

Lantana involucrata L.
Verbenaceae
♦ button sage, involucred lantana
♦ Weed
♦ 87, 88, 161
♦ herbal, toxic.

Lantana montevidensis (Spreng.) Briq.
Verbenaceae
Lantana sellowiana Link & Otto
♦ trailing shrubverbena, creeping lantana, purple lantana, small lantana, trailing lantana, weeping lantana
♦ Weed, Noxious Weed, Naturalised,

Introduced, Garden Escape
♦ 34, 39, 80, 86, 87, 88, 93, 98, 101, 116, 147, 161, 179, 201, 203, 261, 269, 280
♦ C, cultivated, herbal, toxic. Origin: South America.

Lantana moritziana Otto & Dietr.
Verbenaceae
= *Lantana camara* L.
♦ Weed
♦ 87, 88

Lantana ovata Hayek
Verbenaceae
♦ Weed
♦ 87, 88

Lantana ovatifolia auct. *p.p. non* Britt.
Verbenaceae
= *Lantana depressa* Small (NoR)
♦ ovalleaf lantana
♦ Weed
♦ 39, 161
♦ herbal, toxic.

Lantana reticulata Pers.
Verbenaceae
Lantana fucata Lindl. var. *antillana* Moldenke
♦ netted shrubverbena
♦ Weed
♦ 14

Lantana salvifolia Jacq.
Verbenaceae
♦ Weed
♦ 87, 88, 221

Lantana tiliifolia Cham.
Verbenaceae
♦ Weed, Naturalised
♦ 98, 203
♦ arid.

Lantana trifolia L.
Verbenaceae
♦ threeleaf shrubverbena
♦ Weed
♦ 13, 14, 87, 88, 121, 255
♦ pS, herbal. Origin: South America.

Lapeirousia laxa (Thunb.) N.E.Br.
Iridaceae
= *Freesia laxa* (Thunb.) Goldblatt & J.C.Manning
♦ Casual Alien
♦ 280
♦ cultivated.

Laportea aestuans (L.) Chew
Urticaceae
Fleurya aestuans (L.) Gaudich. ex Miq. (see)
♦ West Indian woodnettle
♦ Weed
♦ 28, 39, 87, 88, 157, 206, 243
♦ aH, herbal, toxic. Origin: Madagascar.

Laportea alatipes N.E.Br.
Urticaceae
Fleurya alatipes (Hook.f.) N.E.Br.
♦ nettle
♦ Native Weed
♦ 121
♦ pH. Origin: southern Africa.

Laportea canadensis (L.) Wedd.
Urticaceae
Urtica canadensis L.
♦ Canadian woodnettle, woodnettle
♦ Weed
♦ 39, 161, 247
♦ pH, cultivated, herbal, toxic.

Laportea gigas Wedd.
Urticaceae
♦ Weed
♦ 87, 88

Laportea interrupta (L.) Chew
Urticaceae
Fleurya interrupta (L.) Gaudich.
♦ Hawai'i woodnettle, ogoogo
♦ Weed, Naturalised
♦ 87, 88, 101, 276
♦ aH, herbal. Origin: Australia.

Laportea peduncularis (Wedd.) Chew
Urticaceae
Fleurya mitis Wedd.
♦ stinging nettle
♦ Native Weed
♦ 121
♦ pH. Origin: southern Africa.

Lappula barbata (Bieb.) Gürke
Boraginaceae
♦ Weed
♦ 272

Lappula echinata Gilib.
Boraginaceae
= *Lappula squarrosa* (Retz.) Dumort.
♦ European sticktight, blue burr, European stickseed, stickseed
♦ Weed, Noxious Weed, Naturalised
♦ 23, 36, 51, 52, 87, 88, 121, 161, 162, 210, 218, 275, 287
♦ a/bH, herbal. Origin: Eurasia.

Lappula marginata (Bieb.) Gürke
Boraginaceae
Lappula patula (Lehm.) Menyh.
♦ rikkasirkunjyvä, margined stickseed
♦ Weed, Naturalised, Casual Alien
♦ 42, 101, 272

Lappula myosotis Moench.
Boraginaceae
= *Lappula squarrosa* (Retz.) Dumort.
♦ European stickseed
♦ Weed
♦ 297
♦ herbal.

Lappula occidentalis (S.Wats.) Greene
Boraginaceae
♦ western sticktight, flatspine stickseed, hairy stickseed
♦ Weed, Native Weed
♦ 49, 87, 88, 161, 174, 212, 218
♦ herbal. Origin: North America.

Lappula redowskii (Hornem.) Greene
Boraginaceae
= *Lappula occidentalis* (S.Wats.) Greene var. *occidentalis* (NoR) [see *Lappula redowskii* (Hornem.) Greene var. *occidentalis* (Wats.) Rydb]
♦ Redowski's stickseed
♦ Weed
♦ 87, 88
♦ aH, herbal.

**Lappula redowskii (Hornem.) Greene
var. occidentalis (Wats.) Rydb.**
Boraginaceae
= *Lappula occidentalis* (S.Wats.) Greene
var. *occidentalis* (NoR) [see *Lappula
redowskii* (Hornem.) Greene]
♦ stinkseed
♦ Weed, Noxious Weed
♦ 36

Lappula spinocarpos (Forssk.) Kuntze
Boraginaceae
♦ Weed
♦ 221

Lappula squarrosa (Retz.) Dumort.
Boraginaceae
Echinospermum fabrei Sennen,
Echinospermum lappula (L.) Lehm. (see),
Lappula echinata Gilib. (see), *Myosotis
lappula* L. (see), *Lappula myosotis*
Moench (see), *Myosotis squarrosa* Retz.
♦ European stickseed, stickseed, burr
forget me not
♦ Weed, Noxious Weed, Naturalised,
Introduced, Casual Alien
♦ 34, 40, 42, 88, 94, 101, 272, 292, 299
♦ aH, cultivated, herbal.

Lapsana apogonoides Maxim.
Asteraceae
= *Lapsanastrum apogonoides* (Maxim.)
J.H.Pak & K.Bremer
♦ Japanese nipplewort
♦ Weed, Quarantine Weed
♦ 76, 87, 88, 203, 220, 263, 286, 297
♦ aH, herbal.

Lapsana communis L.
Asteraceae
♦ nipplewort, succory dock
nipplewort, ballogan
♦ Weed, Naturalised
♦ 15, 23, 34, 44, 70, 80, 86, 87, 88, 94, 98,
101, 136, 161, 165, 176, 195, 198, 199,
203, 212, 218, 237, 241, 243, 253, 272,
280, 287, 295, 300
♦ aH, cultivated, herbal. Origin:
Europe, North Africa.

Lapsana humilis (Thunb.) Makino
Asteraceae
♦ yabutabirako
♦ Weed
♦ 87, 88, 286

Lapsana stellata L.
Asteraceae
= *Rhagadiolus stellatus* (L.) Gaertn.
♦ Weed
♦ 87, 88

**Lapsanastrum apogonoides (Maxim.)
J.H.Pak & K.Bremer**
Asteraceae
Lapsana apogonoides Maxim. (see)
♦ Japanese nipplewort
♦ Naturalised
♦ 101

Larix Mill. spp.
Pinaceae
♦ larch, golden larch
♦ Weed
♦ 39, 181
♦ toxic.

Larix decidua Mill.
Pinaceae
Larix europaea DC., *Pinus larix* L.
♦ European larch, larch
♦ Weed, Sleeper Weed, Naturalised,
Environmental Weed, Cultivation
Escape
♦ 41, 42, 88, 101, 151, 225, 246, 280
♦ T, cultivated, herbal.

Larix kaempferi (Lam.) Carr.
Pinaceae
♦ Japanese larch
♦ Naturalised
♦ 101
♦ T, cultivated.

Larix laricina (Du Roi) K.Koch
Pinaceae
♦ tamarack, American larch, black
larch
♦ Weed
♦ 87, 88, 218
♦ T, cultivated, herbal.

Larix occidentalis Nutt.
Pinaceae
♦ western larch
♦ Weed
♦ 87, 88, 218
♦ T, cultivated, herbal.

Larix sibirica Ledeb.
Pinaceae
♦ siperianlehtikuusi
♦ Cultivation Escape
♦ 42
♦ cultivated, herbal.

Larrea cuneifolia Cav.
Fabaceae/Caesalpiniaceae
♦ Introduced
♦ 228
♦ arid.

Larrea divaricata Cav.
Fabaceae/Caesalpiniaceae
♦ creosote bush
♦ Weed, Quarantine Weed, Introduced
♦ 39, 76, 87, 88, 203, 220, 228, 295
♦ arid, cultivated, herbal, toxic.

Larrea nitida Cav.
Fabaceae/Caesalpiniaceae
♦ Introduced
♦ 228
♦ arid, herbal.

Larrea tridentata (DC.) Coville
Fabaceae/Caesalpiniaceae
Larrea divaricata auct. non Cav., *Larrea
mexicana* Moric.
♦ creosote bush, chaparral, South
American creosote bush
♦ Weed
♦ 23, 39, 87, 88, 161, 212, 218
♦ S, cultivated, herbal, toxic.

Laser trilobum (L.) Borkh.
Apiaceae
Siler trilobum Crantz.
♦ gladich
♦ Weed
♦ 272
♦ pH, cultivated.

Laserpitium latifolium L.
Apiaceae

♦ sermountain, bastard lovage,
laserwort, karvasputki
♦ Weed
♦ 272
♦ pH, cultivated, herbal.

Lasiacis ligulata A.S.Hitchc. & Chase
Poaceae
♦ thicket tribisee
♦ Weed
♦ 87, 88
♦ G.

Lasiacis ruscifolia (Kunth) Hitchc.
Poaceae
♦ climbing tribisee
♦ Weed, Naturalised
♦ 87, 88, 101
♦ G.

Lasiocereus Ritter spp.
Cactaceae
= *Haageocereus* Backeb. spp.
♦ Weed, Quarantine Weed
♦ 76, 88, 203, 220

Lasiochloa echinata (Thunb.) Adamson
Poaceae
= *Tribolium echinatum* (Thunb.)
Renvoize
♦ Native Weed
♦ 121
♦ aG. Origin: southern Africa.

Lasiochloa longifolia (Schrad.) Kunth
Poaceae
♦ haregrass
♦ Native Weed
♦ 121
♦ pG. Origin: southern Africa.

Lasiopogon muscoides (Desf.) DC.
Asteraceae
♦ Weed
♦ 221

**Lasiospermum bipinnatum (Thunb.)
Druce**
Asteraceae
♦ cocoon head
♦ Weed, Naturalised, Native Weed
♦ 39, 86, 98, 101, 121, 176, 203
♦ pH, arid, cultivated, herbal, toxic.
Origin: southern Africa.

Lasiurus hirsutus (Vahl) Boiss. nom. illeg.
Poaceae
♦ Weed
♦ 88, 221
♦ G, cultivated, herbal.

**Lasthenia chrysantha (Greene ex Gray)
Greene**
Asteraceae
Crockeria chrysantha Greene ex Gray
(see)
♦ crockeria, alkali sink goldfields
♦ Weed
♦ 161
♦ aH.

Lastrea thelypteris (L.) Bory.
Thelypteridaceae
= *Thelypteris palustris* Schott
♦ himeshida
♦ Weed
♦ 88, 204

Latania loddigesii **Mart.**
Arecaceae
♦ blue latan palm
♦ Introduced
♦ 230
♦ T, cultivated.

Latania lontaroides **(Gaertn.) H.E.Moore**
Arecaceae
♦ red latan palm
♦ Introduced
♦ 230
♦ T, cultivated.

Lathraea clandestina **L.**
Scrophulariaceae
♦ purple toothwort
♦ Naturalised
♦ 40
♦ pH parasitic, cultivated.

Lathraea squamaria **L.**
Scrophulariaceae
♦ toothwort, suomukka, latrea comune
♦ Weed
♦ 70, 272
♦ herbal.

Lathyrus **L. spp.**
Fabaceae/Papilionaceae
♦ sweet pea, caley pea, wild vetchling, peavine
♦ Weed, Naturalised
♦ 198, 221, 243, 247, 272
♦ herbal, toxic.

Lathyrus angulatus **L.**
Fabaceae/Papilionaceae
♦ angled pea, angled peavine, angular pea
♦ Weed, Naturalised
♦ 39, 86, 87, 88, 98, 101, 198, 203, 250, 253, 272
♦ C, cultivated, herbal, toxic. Origin: Eurasia.

Lathyrus annuus **L.**
Fabaceae/Papilionaceae
Lathyrus chius Boiss. & Orph., *Lathyrus hierosolymitanus* Boiss. var. *grandiflorus* Boiss.
♦ annual yellow vetchling, fodder pea
♦ Weed, Casual Alien
♦ 40, 42, 87, 88, 253, 272
♦ aH, arid, cultivated, herbal.

Lathyrus aphaca **L.**
Fabaceae/Papilionaceae
Aphaca vulgaris Presl, *Lathyrus affinis* Guss., *Orobus aphaca* Doell.
♦ yellow vetchling, yellow flowered pea, yellow peavine, korvakenätkelmä
♦ Weed, Quarantine Weed, Naturalised, Introduced, Casual Alien
♦ 34, 39, 42, 44, 70, 76, 86, 87, 88, 94, 98, 101, 115, 185, 203, 220, 221, 243, 253, 272, 280, 287
♦ aH, cultivated, herbal, toxic.

Lathyrus articulatus **L.**
Fabaceae/Papilionaceae
♦ Weed
♦ 87, 88, 243
♦ cultivated, herbal.

Lathyrus blepharicarpus **Boiss.**
Fabaceae/Papilionaceae

♦ Weed
♦ 87, 88
♦ cultivated.

Lathyrus cicera **L.**
Fabaceae/Papilionaceae
♦ red vetchling, red peavine, flat pod peavine, vetchling, chickling vetch, etelännätkelmä
♦ Weed, Naturalised, Introduced, Casual Alien
♦ 39, 40, 42, 70, 87, 88, 94, 101, 161, 228, 241, 243, 253, 272, 300
♦ aH, arid, cultivated, herbal, toxic.

Lathyrus clymenum **L.**
Fabaceae/Papilionaceae
♦ Spanish vetchling, Spanish vetch
♦ Weed, Naturalised
♦ 39, 70, 88, 94, 161, 287
♦ aH, cultivated, herbal, toxic.

Lathyrus erectus **Lag.**
Fabaceae/Papilionaceae
♦ Weed
♦ 87, 88

Lathyrus grandiflorus **Sibth. & Sm.**
Fabaceae/Papilionaceae
♦ two flowered pea
♦ Naturalised
♦ 280
♦ cultivated.

Lathyrus heterophyllus **L.**
Fabaceae/Papilionaceae
Pisum heterophyllum Krause
♦ pallenätkelmä
♦ Weed, Casual Alien
♦ 42, 87, 88
♦ cultivated.

Lathyrus hirsutus **L.**
Fabaceae/Papilionaceae
Pisum hirsutum Krause
♦ rough podded vetchling, singletary peavine, pea, hairy vetchling, rough peavine, caley pea
♦ Weed, Naturalised, Cultivation Escape, Casual Alien
♦ 34, 39, 40, 42, 70, 87, 88, 94, 101, 161, 185, 241, 243, 253, 272
♦ aH, cultivated, herbal, toxic.

Lathyrus inconspicuus **L.**
Fabaceae/Papilionaceae
♦ vähänätkelmä, inconspicuous peavine, inconspicuous pea
♦ Weed, Naturalised, Casual Alien
♦ 42, 101, 272, 287

Lathyrus japonicus **Willd.**
Fabaceae/Papilionaceae
♦ beach pea, sea peavine, sea pea, purple beach pea, Japanese beach pea
♦ Weed
♦ 23, 88
♦ pH, cultivated, herbal.

Lathyrus japonicus **Willd. var. *aleuticus* (Greene) Fernald**
Fabaceae/Papilionaceae
♦ Naturalised
♦ 300

Lathyrus japonicus **Willd. var. *japonicus***
Fabaceae/Papilionaceae

♦ beach pea, sea peavine
♦ Naturalised
♦ 241, 300

Lathyrus latifolius **L.**
Fabaceae/Papilionaceae
Lathyrus megalanthus Steud., *Lathyrus membranaceus* C.Presl, *Lathyrus silvester* L. *p.*, *Pisum latifolium* Krause
♦ perennial sweet pea, everlasting pea, broad leaved everlasting, perennial pea
♦ Weed, Naturalised, Introduced, Garden Escape, Environmental Weed, Cultivation Escape
♦ 4, 15, 34, 39, 40, 72, 82, 86, 87, 88, 98, 101, 116, 161, 165, 195, 198, 203, 212, 272, 280, 287
♦ pH, cultivated, herbal, toxic. Origin: southern Europe.

Lathyrus montanus **Bernh.**
Fabaceae/Papilionaceae
♦ bitter vetch, groszek
♦ Weed
♦ 70
♦ cultivated, herbal.

Lathyrus niger **(L.) Bernh.**
Fabaceae/Papilionaceae
Orobus niger L., *Pisum nigrum* Krause
♦ black bitter vetch, mustalinnunherne, black pea
♦ Weed
♦ 272
♦ cultivated, herbal.

Lathyrus nissolia **L.**
Fabaceae/Papilionaceae
Anurus nissolia E.Mey., *Orobus nissolia* Doell.
♦ grass vetchling, peavine, grass leaved pea, grass pea
♦ Weed, Naturalised
♦ 39, 44, 70, 86, 87, 88, 94, 98, 101, 121, 176, 203, 243, 253, 272, 280
♦ aH, cultivated, herbal, toxic. Origin: Eurasia.

Lathyrus ochroleucus **Hook.**
Fabaceae/Papilionaceae
♦ yellow vetchling, cream peavine, cream pea
♦ Weed
♦ 161
♦ herbal, toxic.

Lathyrus ochrus **(L.) DC.**
Fabaceae/Papilionaceae
Lathyrus thirkeanus Koch, *Pisum ochrus* L.
♦ Cyprus vetch, louvana
♦ Weed, Naturalised, Introduced
♦ 70, 87, 88, 94, 228, 243, 253, 287
♦ aH, arid, cultivated, herbal.

Lathyrus odoratus **L.**
Fabaceae/Papilionaceae
Lathyrus maccaguenii Tod. ex Nyman, *Pisum odoratum* (L.) E.H.L.Krause
♦ sweet pea, sweetpea peavine, tuoksuherne, garden sweet pea
♦ Weed, Naturalised, Cultivation Escape
♦ 34, 39, 42, 54, 86, 88, 101, 154, 161,

198, 252, 280
♦ aH, cultivated, herbal, toxic. Origin:
Italy, Sicily.

Lathyrus palustris L.
Fabaceae/Papilionaceae
Lathyrus occidentalis Torr. & A.Gray,
Lathyrus incurvus Reichb., *Lathyrus
macranthus* (T.White) Rydb., *Lathyrus
myrtifolius* Willd., *Lathyrus paluster
sensu auct.*, *Lathyrus pilosus* Cham.,
Orobus myrtifolius (Willd.) Hall, *Orobus
myrtifolius* Alef., *Pisum palustre* Krause
♦ marsh peavine, marsh vetchling,
blue marsh vetchling, slenderstem
peavine, rantanätkelmä
♦ Weed
♦ 87, 88, 161, 218, 272
♦ pH, aqua, cultivated, herbal, toxic.

Lathyrus polymorphus Nutt.
Fabaceae/Papilionaceae
♦ singletary incanus pea, manystem
pea, manystem peavine
♦ Weed
♦ 161
♦ pH, promoted, herbal, toxic.

Lathyrus pratensis L.
Fabaceae/Papilionaceae
Orobus pratensis Doell.
♦ meadow peavine, yellow meadow
vetchling, meadow vetchling,
niittynätkelmä
♦ Weed, Quarantine Weed,
Naturalised
♦ 44, 70, 76, 87, 88, 101, 218, 272, 280,
287
♦ pH, cultivated, herbal.

Lathyrus pusillus Ellis
Fabaceae/Papilionaceae
♦ tiny peavine, tiny pea, low peavine
♦ Weed
♦ 39, 161
♦ herbal, toxic.

Lathyrus quadrimarginatus Bory & Chaub.
Fabaceae/Papilionaceae
♦ Weed
♦ 87, 88

Lathyrus quinquenervius (Miq.) Litv. ex Kom.
Fabaceae/Papilionaceae
♦ fivevein vetchling
♦ Weed
♦ 275, 297
♦ pH, promoted. Origin: China, Japan,
Korea.

Lathyrus rotundifolius Willd.
Fabaceae/Papilionaceae
Lathyrus drummondii hort., *Lathyrus
litvinovii* Iljin
♦ Persian everlasting pea
♦ Introduced
♦ 100
♦ cultivated.

Lathyrus rotundifolius Willd. ssp. *miniatus* (M.Bieb. ex Steven) P.Davis
Fabaceae/Papilionaceae
Lathyrus miniatus M.Bieb. ex Steven
♦ Introduced

♦ 192
♦ Origin: Iran, Iraq.

Lathyrus sativus L.
Fabaceae/Papilionaceae
Lathyrus asiaticus (Zalk.) Kudrj.,
Lathyrus sativas L., *Lathyrus sativus* L.
var. *stenophyllus* Boiss.
♦ white peavine, grass pea, chickling
pea, Indian vetch, almorta, khesari,
batura, alverjas, gilban, guaya,
matri, gesette, pisello bretonne, blue
vetchling, Indian pea, khasari, dhal,
saat platterbse, teora, giant lentil,
ajilbane, baqia, blue vetchling, djiben
el biod, djilben, gasse, commune,
gilbaan, gwaya, hamikou, hurtumann,
jarosse, julban, kerfala, khesari,
khessary pea, pois carré, quarfala,
sa'eyda, sabbäre
♦ Weed, Quarantine Weed,
Naturalised, Introduced, Casual Alien
♦ 39, 42, 54, 70, 86, 87, 88, 94, 98, 100,
101, 121, 161, 198, 203, 221, 228, 241,
272, 300
♦ aC, arid, cultivated, herbal, toxic.
Origin: Eurasia.

Lathyrus sphaericus Retz.
Fabaceae/Papilionaceae
Orobus sphaericus Alef.
♦ grass pea, grass peavine, round
peavine
♦ Weed, Naturalised
♦ 23, 39, 86, 87, 88, 98, 101, 161, 203,
243, 272, 280
♦ aH, cultivated, herbal, toxic. Origin:
Europe, North Africa to central Asia.

Lathyrus splendens Kellogg
Fabaceae/Papilionaceae
♦ pride of California, Campo pea,
sweet pea
♦ Weed
♦ 39, 161
♦ pH, cultivated, toxic.

Lathyrus sylvestris L.
Fabaceae/Papilionaceae
♦ flat peavine, narrow leaved
everlasting pea, wild pea,
metsänätkelmä, flat pea
♦ Weed, Naturalised
♦ 7, 23, 39, 70, 87, 88, 98, 101, 161, 195,
203, 218, 272, 280
♦ arid, cultivated, herbal, toxic.

Lathyrus tingitanus L.
Fabaceae/Papilionaceae
♦ Tangier pea, Tangier scarlet pea,
Tangier peavine, sweet pea
♦ Weed, Naturalised, Environmental
Weed, Cultivation Escape
♦ 7, 15, 39, 70, 86, 87, 88, 98, 101, 136,
161, 176, 198, 203, 252, 280, 290, 296
♦ aH, cultivated, herbal, toxic. Origin:
south-west Europe.

Lathyrus tuberosus L.
Fabaceae/Papilionaceae
Pisum tuberosum (L.) E.H.L.Krause
♦ tuberous sweetpea, tuberous
vetchling, mukulanätkelmä
♦ Weed, Quarantine Weed,
Naturalised, Casual Alien

♦ 40, 42, 44, 70, 76, 86, 87, 88, 94, 101,
161, 203, 220, 243, 253, 272
♦ pC, cultivated, herbal, toxic.

Lathyrus venetus (Mill.) Wohlf.
Fabaceae/Papilionaceae
Lathyrus variegatus (Ten.) Gren. &
Godr., *Orobus venetus* Mill.
♦ cicerchia veneta
♦ Weed
♦ 272
♦ cultivated, herbal.

Lathyrus venosus Muhl. ex Willd.
Fabaceae/Papilionaceae
♦ bushy vetchling, veiny peavine,
veiny pea, vetchling
♦ Weed
♦ 161
♦ cultivated, herbal, toxic.

Lathyrus vestitus Nutt.
Fabaceae/Papilionaceae
♦ Pacific pea, chicharillo, tight aiefeldii
peavine
♦ Weed
♦ 161
♦ pH, arid, cultivated, herbal, toxic.

Latipes senegalensis Kunth
Poaceae
♦ Weed
♦ 221
♦ G.

Latua pubiflora (Griseb.) Bail.
Solanaceae
♦ Weed
♦ 39, 87, 88
♦ cultivated, toxic.

Launaea Cass. spp.
Asteraceae
♦ launaea
♦ Weed
♦ 221

Launaea angustifolia (Desf.) Kuntze
Asteraceae
♦ Weed
♦ 221

Launaea asplenifolia Hook f.
Asteraceae
♦ Weed, Quarantine Weed
♦ 76, 87, 88, 203, 220
♦ herbal.

Launaea capitata (Spreng.) Dandy
Asteraceae
Launaea glomerata (Cass.) Hook.f.
♦ Weed
♦ 221
♦ arid.

Launaea cassiniana (Jaub. & Spach) Burkill
Asteraceae
♦ Weed
♦ 221

Launaea cornuta (Oliv. & Hiern) C.Jeffrey
Asteraceae
Lactuca taraxacifolia (Willd.) Schum.
& Thonn. ex Hornem. (see), *Sonchus
bipontini* (*non* Asch.) Broun & Massey
p.p, *Sonchus cornutus* Hochst. ex Oliv.

& Hiern, *Sonchus exauriculatus* (Oliv. &
Hiern) O.Hoffm. (see)
♦ wild lettuce
♦ Weed
♦ 53, 88, 240
♦ pH, arid.

Launaea fallax (Jaub. & Spach) Kuntze
Asteraceae
Microrhynchus fallax Jaub. & Spach
♦ Weed
♦ 221

Launaea intybacea (Jacq.) Beauv.
Asteraceae
Lactuca intybacea Jacq. (see)
♦ achicoria azul
♦ Weed
♦ 179, 261

Launaea massauensis (Fresen.) Kuntze
Asteraceae
♦ Weed
♦ 221

Launaea nudicaulis (L.) Hook.f.
Asteraceae
♦ walha
♦ Weed
♦ 87, 88, 221
♦ arid, herbal.

Launaea procumbens (Roxb.) Ram. & Raj.
Asteraceae
♦ Weed
♦ 66, 88, 221, 243

**Launaea rarifolia (Oliv. & Hiern)
L.Boulos**
Asteraceae
Sonchus rarifolius Oliv. & Hiern
♦ Native Weed
♦ 121
♦ pH. Origin: southern Africa.

**Launaea sarmentosa (Willd.) Sch.Bip. ex
Kuntze**
Asteraceae
Launaea pinnatifida Cass.
♦ Weed, Naturalised
♦ 7, 86, 87, 88
♦ herbal. Origin: Australia.

Launaea spinosa (Forssk.) Sch.Bip.
Asteraceae
♦ Weed
♦ 221

Launaea tenuiloba (Boiss.) Kuntze
Asteraceae
♦ Weed
♦ 221

Laurentia longiflora (L.) Peterm.
Campanulaceae/Lobeliaceae
= *Hippobroma longiflora* (L.) G.Don
♦ madam fate
♦ Weed, Quarantine Weed,
Naturalised
♦ 13, 14, 76, 86, 87, 88, 203, 220
♦ aH, cultivated, herbal.

Laurus nobilis L.
Lauraceae
Laurus undulata Mill.
♦ sweet bay, bay, laurel, sweet laurel,
alloro lauro
♦ Weed, Naturalised

♦ 15, 40, 86, 101, 280
♦ T, cultivated, herbal. Origin: Eurasia.

Lavandula angustifolia Mill.
Lamiaceae
Lavandula officinalis Chaix, *Lavandula
spica* Cav. (see)
♦ English lavender, common lavender
♦ Weed, Naturalised
♦ 7, 86, 98, 101, 203
♦ S, cultivated, herbal. Origin:
Mediterranean.

**Lavandula angustifolia Mill. ssp.
angustifolia**
Lamiaceae
Lavandula vera DC.
♦ English lavender, lavender, true
lavender
♦ Weed
♦ 121
♦ pS. Origin: Eurasia.

Lavandula dentata L.
Lamiaceae
Stoechas dentata Mill.
♦ Spanish lavender, toothed lavender
♦ Weed, Naturalised, Garden Escape
♦ 7, 15, 86, 98, 203, 269, 280
♦ S, cultivated, herbal. Origin: North
Africa, Spain.

Lavandula multifida L.
Lamiaceae
♦ cut leaved lavender, fernleaf
lavender, oregano scented lavender
♦ Naturalised
♦ 86
♦ aH, cultivated, herbal. Origin:
southern Europe.

Lavandula pubescens Decne.
Lamiaceae
♦ Weed
♦ 221

Lavandula spica Cav.
Lamiaceae
= *Lavandula angustifolia* Mill.
♦ lavender, lavanda
♦ Weed
♦ 87, 88
♦ cultivated, herbal.

Lavandula stoechas L.
Lamiaceae
Stoechas arabica Garsault, *Stoechas
officinarum* Mill.
♦ topped lavender, bush lavender,
French lavender, Italian lavender,
Spanish lavender
♦ Weed, Noxious Weed, Naturalised,
Garden Escape, Environmental Weed,
Casual Alien
♦ 7, 42, 70, 72, 86, 87, 88, 98, 147, 155,
198, 203, 269, 280, 289, 296
♦ pS, cultivated, herbal. Origin:
Mediterranean.

Lavandula stricta Delile
Lamiaceae
= *Lavandula coronipifolia* Poir. (NoR)
♦ Weed
♦ 221

Lavatera arborea L.
Malvaceae

= *Malva dendromorpha* M.F.Ray (NoR)
♦ tree mallow, bushmallow
♦ Weed, Naturalised, Introduced,
Garden Escape, Environmental Weed,
Cultivation Escape
♦ 7, 15, 38, 72, 86, 88, 98, 101, 121, 158,
165, 176, 198, 203, 272, 280, 283
♦ bH, cultivated, herbal, toxic. Origin:
Eurasia.

Lavatera assurgentiflora Kellogg
Malvaceae
= *Malva assurgentiflora* (Kellogg)
M.F.Ray
♦ malva rosa, island mallow
♦ Weed, Naturalised
♦ 86, 98, 203, 241, 280
♦ S, cultivated, herbal.

Lavatera cretica L.
Malvaceae
= *Malva linnaei* M.F.Ray (NoR)
♦ small tree mallow, rikkamalvikki,
lesser tree mallow, Cretan hollyhock,
Cornish mallow
♦ Weed, Naturalised, Environmental
Weed, Cultivation Escape
♦ 7, 15, 42, 70, 86, 87, 88, 98, 101, 111,
121, 176, 198, 203, 243, 252, 253, 272,
280
♦ a/bH, cultivated, herbal. Origin:
Eurasia.

Lavatera mauritanica Durieu
Malvaceae
♦ Weed, Naturalised
♦ 98, 203

Lavatera olbia L.
Malvaceae
♦ hyeres tree mallow
♦ Naturalised
♦ 280
♦ cultivated.

Lavatera plebeia Sims
Malvaceae
♦ Australian hollyhock
♦ Introduced
♦ 228
♦ arid, cultivated. Origin: Australia.

Lavatera punctata All.
Malvaceae
♦ annual tree mallow
♦ Weed, Casual Alien
♦ 40, 87, 88
♦ cultivated.

Lavatera thuringiaca L.
Malvaceae
Malva thuringiaca (L.) Vis.
♦ tree lavatera, gay mallow,
harmaamalvikki, lavatera
♦ Weed, Naturalised, Cultivation
Escape
♦ 42, 87, 88, 101, 241, 272
♦ pH, cultivated, herbal.

Lavatera trimestris L.
Malvaceae
♦ annual mallow, annual lavatera,
royal mallow
♦ Weed, Naturalised, Cultivation
Escape, Casual Alien
♦ 7, 40, 42, 86, 87, 88, 94, 98, 101, 203,

252, 280
♦ aH, cultivated, herbal. Origin:
Mediterranean.

Lavigeria macrocarpa Pierre
Icacinaceae
♦ Quarantine Weed
♦ 220

Lawsonia inermis L.
Lythraceae
Lawsonia alba Lam.
♦ henna, mignonette tree, hennè
♦ Naturalised, Introduced, Cultivation
Escape
♦ 101, 228, 261
♦ arid, cultivated, herbal. Origin:
Madagascar.

Leandra longicoma Cogn.
Melastomataceae
♦ Weed
♦ 87, 88

Lechenaultia biloba Lindl.
Goodeniaceae
♦ blue lechenaultia
♦ Naturalised, Native Weed, Garden
Escape, Cultivation Escape
♦ 7
♦ pH, cultivated.

Ledebouria cooperi (Hook.f.) Jessop
Liliaceae/Hyacinthaceae
Scilla adlamii Bak., *Scilla cooperi* Hook.f.,
Scilla minima Bak.
♦ Cooper's squill, wild squill
♦ Native Weed
♦ 121
♦ pH, cultivated, toxic. Origin:
southern Africa.

Ledebouria revoluta (L.f.) Jessop
Liliaceae/Hyacinthaceae
Scilla lanceaefolia (Jacq.) Bak.
♦ Native Weed
♦ 121
♦ pH, toxic. Origin: southern Africa.

Ledum columbianum Piper
Ericaceae
= *Ledum glandulosum* Nutt. var.
columbianum (Piper) C.L.Hitchc. (NoR)
♦ Pacific Labrador tea
♦ Weed
♦ 161
♦ S, cultivated, toxic.

Ledum glandulosum Nutt.
Ericaceae
♦ western Labrador tea, Labrador tea
♦ Weed
♦ 39, 161
♦ S, cultivated, herbal, toxic.

Ledum groenlandicum Oeder
Ericaceae
Ledum palustre L. ssp. *groenlandicum*
(Oeder) Hultén, *Ledum palustre* var.
latifolium (Jacq.) Michx., *Rhododendron
groenlandicum* (Oeder) K.A.Kron &
Judd
♦ Labrador tea, bog Labrador tea
♦ Weed, Garden Escape
♦ 39, 87, 88, 161, 218
♦ S, cultivated, herbal, toxic.

Ledum palustre L.
Ericaceae
♦ Labrador tea, crystal tea, wild
rosemary, marsh Labrador tea,
suopursu, bagno
♦ Weed
♦ 39, 70, 272
♦ S, cultivated, herbal, toxic.

Leersia hexandra Sw.
Poaceae
Asprella australis (R.Br.) Roem. &
Schult., *Asprella hexandra* (Sw.)
P.Beauv., *Asprella hexandra* (Sw.) Roem.
& Schult., *Asprella mexicana* (Kunth)
Roem. & Schult., *Homalocenchrus
angustifolius* Kuntze, *Homalocenchrus
gouinii* (E.Fourn.) Kuntze,
Homalocenchrus hexandrus (Sw.)
Kuntze, *Hygroryza ciliata* (Retz.) Nees
ex Steud., *Leersia abyssinica* Hochst.
ex A.Rich., *Leersia aegyptiaca* Fig. &
De Not., *Leersia angustifolia* Munro ex
Prod., *Leersia australis* R.Br., *Leersia
capensis* Müll.Hal., *Leersia ciliaris* Griff.,
Leersia ciliata (Retz.) Roxb., *Leersia
contracta* Nees, *Leersia dubia* F.Aresch.,
Leersia elongata Willd. ex Trin.,
Leersia glaberrima Trin., *Leersia gouinii*
E.Fourn., *Leersia gracilis* Willd. ex Trin.,
Leersia griffithiana Müll.Stuttg., *Leersia
hexandra* ssp. *grandiflora* (Döll) Roseng.,
Leersia luzonensis J.Presl, *Leersia
mauritiaca* Salzm. ex Trin., *Leersia
mexicana* Kunth, *Leersia parviflora* Desv.,
Oryza australis A.Braun ex Schweinf.,
Oryza hexandra (Sw.) Döll, *Oryza
hexandra* var. *grandiflora* Döll, *Oryza
hexandra* var. *hexandra* (Sw.) Döll, *Oryza
mexicana* (Kunth) Döll, *Pharus ciliatus*
Retz., *Pseudoryza ciliata* (Retz.) Griff.
♦ ricegrass, southern cutgrass, swamp
ricegrass, yaa sai, tiger's tongue grass,
cutgrass, white grass, wild ricegrass
♦ Weed, Native Weed
♦ 14, 87, 88, 91, 121, 158, 170, 186, 191,
204, 218, 221, 237, 239, 243, 255, 274,
275, 286, 295, 297
♦ wpG, cultivated, herbal. Origin:
tropical America.

Leersia japonica Honda ex Honda
Poaceae
♦ ashikaki
♦ Weed
♦ 87, 88, 204, 275, 286, 297
♦ pG, aqua.

Leersia oryzoides (L.) Sw.
Poaceae
Homalocenchrus oryzoides Pollich, *Oryza
oryzoides* (L.) Brand., *Phalaris oryzoides*
Sw.
♦ cutgrass, rice cutgrass,
ezonosayanukagusa, riso selvatico
♦ Weed, Naturalised, Environmental
Weed
♦ 70, 72, 86, 87, 88, 91, 98, 161, 198, 203,
218, 253, 263, 272, 280, 286
♦ wpG, cultivated, herbal. Origin:
northern hemisphere.

Leersia sayanuka Ohwi
Poaceae
♦ cutgrass
♦ Weed
♦ 88, 204, 263, 286
♦ pG, aqua.

Legazpia polygonoides (Benth.) T.Yamaz.
Scrophulariaceae
♦ Naturalised
♦ 287

Legousia Dura. spp.
Campanulaceae
♦ legousia
♦ Weed
♦ 272

Legousia falcata (Ten.) Fritsch
Campanulaceae
♦ Weed
♦ 88, 272

Legousia hybrida (L.) Del.
Campanulaceae
Campanula hybrida L., *Prismatocarpus
hybridus* L'Hér., *Specularia hybrida* DC.
♦ Venus's lookingglass
♦ Weed
♦ 44, 70, 87, 88, 94, 243, 253, 272
♦ herbal.

Legousia pentagonia (L.) Druce
Campanulaceae
Specularia pentagonia (L.) DC.
♦ large Venus's lookingglass
♦ Weed
♦ 272
♦ aH, promoted.

Legousia speculum-veneris (L.) Chaix
Campanulaceae
Campanula speculum-veneris L.,
Specularia speculum-veneris (L.) A.DC.
(see)
♦ greater Venus's lookingglass, large
Venus's lookingglass, European
Venus's lookingglass, venuksenpeili
♦ Weed, Naturalised, Casual Alien
♦ 42, 44, 70, 88, 94, 101, 243, 253, 272
♦ aH, cultivated, herbal. Origin:
Eurasia.

Lemaireocereus Britt. & Rose spp.
Cactaceae
= *Pachycereus* (Berg.) Britt. & Rose spp.
♦ Weed, Quarantine Weed
♦ 88, 220

Lembotropis nigricans (L.) Griseb.
Fabaceae/Papilionaceae
Cytisus nigricans L., *Laburnum nigricans*
Presl
♦ mustuvapensasapila
♦ Weed, Cultivation Escape
♦ 42, 272
♦ cultivated, herbal.

Lemna L. spp.
Lemnaceae
♦ duck's meat, duckweed, frog
buttons
♦ Weed, Native Weed, Cultivation
Escape
♦ 121, 208, 233, 237, 262, 295
♦ wH, cultivated. Origin: tropical,
temperate cosmopolitan.

380

357802579

Lemna aequinoctialis Welw.
Lemnaceae
Lemna paucicostata Hegelm. (see),
Lemna perpusilla Torr. (see)
♦ duckweed, lesser duckweed
♦ Native Weed
♦ 121
♦ wpH. Origin: South America.

Lemna aoukikusa Beppu & Murata
Lemnaceae
♦ Weed
♦ 286
♦ wpH.

Lemna gibba L.
Lemnaceae
Lemna trichorrhiza Thuill., *Telmatophace gibba* Schleid.
♦ swollen duckweed, fat duckweed, gibbous duckweed, duck's meat, duckweed, frog buttons, kupulimaska, thick duckweed, eendekroos, lentille d'eau gibbeuse
♦ Weed, Noxious Weed, Naturalised, Native Weed
♦ 23, 70, 87, 88, 121, 158, 221, 253, 255, 272, 278, 287
♦ wpH, cultivated, herbal. Origin: cosmopolitan.

Lemna minima Philip. ex Hegelm. non Thuill. ex Beauv.
Lemnaceae
= *Lemna minuta* Kunth (NoR)
♦ Naturalised
♦ 287
♦ wpH, cultivated, herbal.

Lemna minor L.
Lemnaceae
Lemna cyclostasa Elliott
♦ common duckweed, water lentil, pikkulimaska, duckweed, lesser duckweed, floating duckweed, European duckweed, petite lentille d'eau, koukikusa
♦ Weed, Naturalised
♦ 23, 45, 70, 86, 87, 88, 98, 165, 191, 198, 203, 204, 217, 218, 253, 272, 275, 286, 297
♦ wpH, cultivated, herbal. Origin: North America, Europe, Asia, North Africa.

Lemna paucicostata Hegelm.
Lemnaceae
= *Lemna aequinoctialis* Welw.
♦ duckweed
♦ Weed
♦ 88, 204, 263, 274
♦ wpH.

Lemna perpusilla Torr.
Lemnaceae
= *Lemna aequinoctialis* Welw.
♦ minute duckweed
♦ Weed, Quarantine Weed, Naturalised
♦ 76, 87, 88, 203, 209, 221, 287
♦ wpH, cultivated, herbal.

Lemna polyrhiza L.
Lemnaceae
= *Spirodela polyrrhiza* (L.) Schleid.

♦ greater duckweed
♦ Weed
♦ 88, 204
♦ wpH.

Lemna trisulca L.
Lemnaceae
♦ star duckweed, ivy duckweed, ivy leaved duckweed, ristilimaska, floating duckweed
♦ Weed, Noxious Weed
♦ 87, 88, 218, 272, 297, 299
♦ wpH, cultivated, herbal. Origin: Eurasia.

Lemna valdiviana Phil.
Lemnaceae
♦ valdivia duckweed
♦ Quarantine Weed, Naturalised
♦ 258, 287
♦ wpH, cultivated, herbal.

Lennoa madreporoides La Llave & Lex.
Lennoaceae
Lennoa madrepoides Steud., *Lennoa madreporoides* ssp. *australis* Steyerm.
♦ Weed
♦ 199
♦ H parasitic.

Lens culinaris Medik.
Fabaceae/Papilionaceae
Cicer lens Willd., *Ervum lens* L., *Lens esculenta* Moench.
♦ lentil
♦ Weed, Naturalised, Cultivation Escape, Casual Alien
♦ 34, 42, 101, 280
♦ aH, cultivated, herbal. Origin: obscure, possibly Mediterranean.

Lens orientalis (Boiss.) Popov
Fabaceae/Papilionaceae
♦ Weed
♦ 272

Leocereus Britt. & Rose spp.
Cactaceae
♦ Weed, Quarantine Weed
♦ 39, 76, 88, 203, 220
♦ toxic.

Leonotis africana Th. & Dur.
Lamiaceae
♦ Weed
♦ 87, 88

Leonotis cardiaca L.
Lamiaceae
♦ Naturalised
♦ 287

Leonotis intermedia (Burm.f.) Iwarsson var. natalensis Lindl.
Lamiaceae
♦ Native Weed
♦ 121
♦ pH. Origin: southern Africa.

Leonotis leonurus (L.) R.Br.
Lamiaceae
♦ lion's ear, cape hemp, lion's ear, lion's tail, minaret flower, red dacha, red dagga, wild dagga, wild hemp
♦ Weed, Naturalised, Garden Escape, Environmental Weed
♦ 7, 86, 87, 88, 98, 101, 158, 203, 290

♦ cultivated, herbal. Origin: southern Africa.

Leonotis leonurus (L.) R.Br. var. leonurus
Lamiaceae
♦ cape hemp, lion's ear, lion's tail, minaret flower, red dacha, red dagga, wild dagga, wild hemp
♦ Native Weed
♦ 121
♦ pH. Origin: southern Africa.

Leonotis microphylla Skan
Lamiaceae
♦ wild dagga
♦ Native Weed
♦ 121
♦ pH. Origin: southern Africa.

Leonotis mollis Benth.
Lamiaceae
♦ balm of Gilead
♦ Native Weed
♦ 121
♦ pH, herbal. Origin: southern Africa.

Leonotis mollissima Guercke
Lamiaceae
♦ Weed
♦ 23, 87, 88

Leonotis nepetaefolia (L.) R.Br.
Lamiaceae
= *Leonotis nepetifolia* (L.) R.Br.
♦ lion's ear
♦ Weed
♦ 14, 209
♦ herbal.

Leonotis nepetifolia (L.) R.Br.
Lamiaceae
Leonotis kwebensis N.E.Br., *Leonotis nepetaefolia* (L.) R.Br. (see), *Phlomis nepetifolia* L.
♦ lion's tail, lion's ear, Christmas candlestick, lulyolwasebe, baldhead, bird honey, Johnny Collins, gros bouton, gros tête, pompon soldat, rubim, tolonga
♦ Weed, Noxious Weed, Naturalised, Native Weed, Garden Escape, Environmental Weed, Casual Alien
♦ 3, 7, 13, 86, 87, 88, 93, 98, 101, 121, 147, 155, 179, 199, 203, 255, 261, 280, 287, 290, 295
♦ aH, cultivated, herbal. Origin: South America.

Leonotis ocymifolia (L.) R.Br.
Lamiaceae
♦ rock dagga
♦ Weed, Naturalised
♦ 88, 158, 280
♦ cultivated. Origin: southern Africa.

Leontice leontopetalum L.
Berberidaceae/Leonticaceae
♦ leontice, rakaf
♦ Weed, Quarantine Weed
♦ 39, 76, 87, 88, 203, 220, 221, 272
♦ pH, arid, promoted, herbal, toxic.

Leontodon L. spp.
Asteraceae
♦ hawkbit
♦ Weed, Naturalised

♦ 272

Leontodon autumnalis L.
Asteraceae
Apargia autumnalis Hoff., *Hedypnois autumnalis* Huds.
♦ fall hawkbit, fall dandelion, smooth hawkbit, syysmaitiainen, autumn hawkbit, autumnal hawkbit
♦ Weed, Quarantine Weed, Naturalised
♦ 23, 45, 52, 70, 76, 86, 87, 88, 94, 101, 161, 195, 203, 218, 220, 243, 272, 280, 295, 300
♦ pH, cultivated, herbal.

Leontodon autumnalis L. ssp. autumnalis
Asteraceae
♦ fall dandelion
♦ Naturalised
♦ 101, 280

Leontodon autumnalis L. ssp. pratensis (Link) Arcang.
Asteraceae
♦ fall dandelion
♦ Naturalised
♦ 101

Leontodon hirtus L.
Asteraceae
Apargia nudicaulis (L.) Britt., *Leontodon nudicaulis* (L.) Banks ex Schinz & R.Keller (see)
♦ rough hawkbit
♦ Weed, Quarantine Weed, Naturalised, Environmental Weed
♦ 76, 88, 98, 101, 151, 300

Leontodon hispidulus (Delile) Boiss.
Asteraceae
♦ Weed
♦ 221

Leontodon hispidus L.
Asteraceae
Leontodon proteiformis Vill.
♦ rough hawkbit, bristly hawkbit, kesämaitiainen, léontondon changeant, dente di leone comune
♦ Weed, Naturalised, Environmental Weed
♦ 70, 87, 88, 101, 272, 290
♦ pH, cultivated, herbal.

Leontodon hispidus L. ssp. danubialis (Jacq.) Simonk.
Asteraceae
♦ bristly hawkbit
♦ Naturalised
♦ 101
♦ cultivated.

Leontodon hispidus L. ssp. hispidus
Asteraceae
♦ bristly hawkbit
♦ Naturalised
♦ 101

Leontodon leysseri (Wallr.) G.Beck
Asteraceae
= *Leontodon taraxacoides* (Vill.) Mérat ssp. *taraxacoides*
♦ rough hawkbit
♦ Weed, Naturalised
♦ 161, 287
♦ herbal.

Leontodon nudicaulis (L.) Banks ex Schinz & R.Keller
Asteraceae
= *Leontodon hirtus* L.
♦ rough hawkbit
♦ Weed
♦ 23, 87, 88, 218, 243

Leontodon saxatilis Lam.
Asteraceae
= *Leontodon taraxacoides* (Vill.) Mérat
♦ lesser hawkbit
♦ Weed, Naturalised
♦ 9, 98, 203, 300
♦ cultivated.

Leontodon taraxacoides (Vill.) Mérat
Asteraceae
Leontodon saxatilis Lam. (see)
♦ lesser hawkbit, hawkbit, autumnal hawkbit, hairy hawkbit, rough hawkbit, smooth hawkbit
♦ Weed, Naturalised, Environmental Weed, Casual Alien
♦ 7, 15, 34, 42, 72, 88, 165, 176, 181, 198, 241, 269, 280, 287
♦ pH. Origin: Eurasia.

Leontodon taraxacoides (Vill.) Mérat ssp. longirostris Finch & Sell
Asteraceae
♦ lesser hawkbit
♦ Weed, Naturalised
♦ 34, 101
♦ pH.

Leontodon taraxacoides (Vill.) Mérat ssp. taraxacoides
Asteraceae
Leontodon leysseri (Wallr.) G.Beck (see)
♦ lesser hawkbit, hairy hawkbit
♦ Weed, Naturalised
♦ 34, 86, 101
♦ pH. Origin: Eurasia.

Leontopodium leontopodioides (Willd.) Beauv.
Asteraceae
♦ common edelweiss
♦ Weed
♦ 297

Leontopodium longifolium Ling
Asteraceae
♦ longleaf edelweiss
♦ Weed
♦ 297

Leonurus cardiaca L.
Lamiaceae
♦ motherwort, common motherwort, nukula
♦ Weed, Naturalised, Introduced, Environmental Weed, Cultivation Escape
♦ 8, 23, 39, 42, 70, 80, 87, 88, 101, 121, 151, 161, 174, 195, 210, 218, 272, 280
♦ pH, cultivated, herbal, toxic. Origin: Eurasia.

Leonurus heterophyllus Sweet
Lamiaceae
♦ Weed
♦ 275
♦ a/bH, promoted, herbal.

Leonurus indicus L.
Lamiaceae
♦ Naturalised
♦ 86, 98

Leonurus japonicus Houtt.
Lamiaceae
Leonurus artemisia auct. non Lour., *Leonurus sibiricus* L. (see), *Stachys artemisia* Lour.
♦ honeyweed, Chinese motherwort, yi mu cao
♦ Weed, Naturalised
♦ 32, 286, 297
♦ herbal. Origin: east Asia.

Leonurus marrubiastrum L.
Lamiaceae
= *Chaiturus marrubiastrum* (L.) Rchb.
♦ false motherwort
♦ Weed
♦ 272
♦ cultivated, herbal.

Leonurus sibiricus L.
Lamiaceae
= *Leonurus japonicus* Houtt.
♦ honeyweed, Siberian motherwort, Chinese motherwort
♦ Weed, Quarantine Weed, Naturalised, Introduced
♦ 13, 38, 76, 86, 87, 88, 98, 101, 203, 235, 237, 245, 255, 261, 274, 295, 297
♦ a/bH, cultivated, herbal. Origin: Siberia, China.

Leonurus villosus Desf. ex D'Urv.
Lamiaceae
♦ Weed
♦ 87, 88

Lepechinia caulescens (Ortega) Epling
Lamiaceae
♦ Weed
♦ 199
♦ cultivated.

Lepianthes umbellata (L.) Raf.
Piperaceae
= *Lepianthes peltata* (L.) Raf. (NoR) [see *Piper umbellatum* L., *Pothomorphe dombeyana* Miq., *Pothomorphe peltata* (L.) Miq., *Pothomorphe umbellata* (L.) Miq.]
♦ baquina
♦ Weed
♦ 179

Lepidagathis javanica Bl.
Acanthaceae
♦ Weed
♦ 13

Lepidagathis javanica Bl. var. parviflora (Bl.) Brem.
Acanthaceae
Lepidagathis parviflora Bl.
♦ Weed
♦ 13

Lepidium L. spp.
Brassicaceae
♦ peppergrass, whitetop, pepperweed
♦ Weed, Noxious Weed
♦ 80, 141, 272

Lepidium africanum (Burm.f.) DC.
Brassicaceae
♦ common peppercress, rubble peppercress, peppercress, pepperweed, birdseed, cape peppercress, peppergrass, pepperwort, tongue grass, African pepperwort
♦ Weed, Naturalised, Garden Escape, Environmental Weed
♦ 7, 15, 51, 55, 72, 86, 87, 88, 93, 98, 101, 158, 165, 176, 198, 203, 205, 243, 269, 280
♦ aH, cultivated. Origin: South Africa.

Lepidium africanum (Burm.f.) DC. ssp. africanum
Brassicaceae
♦ birdseed, cape peppercress, peppergrass, pepperweed, pepperwort, tongue grass
♦ Native Weed
♦ 121
♦ a/bH. Origin: southern Africa.

Lepidium apetalum Willd.
Brassicaceae
Lepidium micranthum Lebour.
♦ Weed, Naturalised
♦ 275, 286, 287, 297
♦ a/bH, promoted, herbal.

Lepidium auriculatum Regel & Körn.
Brassicaceae
♦ Weed
♦ 87, 88
♦ arid.

Lepidium bipinnatifidum Desv.
Brassicaceae
♦ Weed, Quarantine Weed
♦ 76, 87, 88, 203, 220

Lepidium bonariense L.
Brassicaceae
♦ birdseed, peppercress, pepperweed, Argentine pepperweed, Argentiinankrassi, Argentine cress, mastuerzo loco
♦ Weed, Naturalised, Garden Escape, Environmental Weed, Casual Alien
♦ 7, 15, 40, 42, 51, 55, 86, 87, 88, 98, 101, 121, 134, 158, 198, 203, 237, 241, 280, 286, 287, 295, 300
♦ H, arid, cultivated. Origin: South America.

Lepidium campestre (L.) R.Br.
Brassicaceae
Crucifera lepidium E.H.L.Krause, *Thlaspi campestre* L.
♦ fieldcress, kenttäkrassi, pepperwort, field pepperwort, field pepperwort, field peppergrass, downy peppergrass
♦ Weed, Noxious Weed, Naturalised, Introduced, Environmental Weed, Cultivation Escape
♦ 21, 23, 34, 36, 44, 70, 80, 86, 87, 88, 94, 98, 101, 136, 161, 176, 198, 203, 207, 210, 211, 218, 243, 252, 253, 272, 280, 287
♦ a/pH, cultivated, herbal. Origin: Europe.

Lepidium capense Thunb.
Brassicaceae
♦ cape peppercress

♦ Native Weed
♦ 121
♦ pH, herbal. Origin: southern Africa.

Lepidium cartilagineum (J.Mey.) Thell.
Brassicaceae
♦ Weed
♦ 248

Lepidium chalepense L.
Brassicaceae
= *Cardaria draba* (L.) Desv. ssp. *chalepensis* (L.) Schulz
♦ Weed, Quarantine Weed
♦ 76, 87, 88, 203, 220, 295

Lepidium densiflorum Schrad.
Brassicaceae
Crucifera apetala Krause
♦ dense peppergrass, peppergrass, greenflower pepperweed, common pepperweed, ratakrassi
♦ Weed, Noxious Weed, Naturalised, Native Weed, Introduced, Environmental Weed
♦ 21, 23, 42, 52, 56, 80, 87, 88, 94, 161, 174, 180, 210, 218, 256, 272, 280, 287, 299
♦ aH, promoted, herbal. Origin: North America.

Lepidium desvauxii Thell.
Brassicaceae
♦ Naturalised
♦ 280

Lepidium divaricatum W.T.Aiton
Brassicaceae
♦ Casual Alien
♦ 280

Lepidium draba L.
Brassicaceae
= *Cardaria draba* (L.) Desv.
♦ hoary cress, hoary pepperwort, whitetop, devil's cabbage, perennial peppergrass, whiteweed, thanet cress
♦ Weed, Noxious Weed, Naturalised
♦ 39, 40, 44, 49, 80, 88, 199, 283
♦ b/pH, herbal, toxic. Origin: Mediterranean, Eurasia.

Lepidium draba L. ssp. chalepense (L.) Schulz
Brassicaceae
= *Cardaria draba* (L.) Desv. ssp. *chalepensis* (L.) Schulz
♦ Naturalised
♦ 40

Lepidium draba L. ssp. draba
Brassicaceae
♦ hoary cress
♦ Cultivation Escape
♦ 40
♦ cultivated.

Lepidium fremontii S.Watson
Brassicaceae
♦ desert pepperweed, peppergrass
♦ Introduced
♦ 228
♦ pH, arid, cultivated, herbal.

Lepidium graminifolium L.
Brassicaceae
Crucifera graminifolia Krause, *Iberis*

graminifolia Roth
♦ grassleaf pepperweed, tall pepperwort
♦ Weed, Naturalised, Introduced
♦ 56, 87, 88, 94, 101, 272
♦ pH, cultivated, herbal.

Lepidium heterophyllum Benth.
Brassicaceae
♦ Smith's pepperwort, Smith's cress, purpleanther field pepperweed, variable leaved pepperwort, nurmikrassi
♦ Weed, Naturalised
♦ 42, 86, 98, 101, 176, 203, 280
♦ aH, cultivated, herbal. Origin: Europe.

Lepidium hirtum (L.) Sm.
Brassicaceae
♦ Mediterranean pepperweed
♦ Naturalised
♦ 101

Lepidium hyssopifolium Desv.
Brassicaceae
Lepidium ambiguum F.Muell., *Lepidium desvauxii* Thell. var. *hookeri* Thell., *Lepidium desvauxii* Thell. var. *gracilescens* Thell., *Lepidium dubium* Thell., *Lepidium hyssopifolium* Desv. var. *tasmanicum* (Thell.) Domin, *Lepidium tasmanicum* Thell.
♦ hyssopleaf pepperweed
♦ Weed, Naturalised
♦ 87, 88, 101, 280
♦ a/bH, arid, cultivated. Origin: Australia.

Lepidium lasiocarpum Nutt.
Brassicaceae
♦ shaggyfruit pepperweed, sand pepperweed
♦ Weed
♦ 23, 88, 161
♦ herbal.

Lepidium latifolium L.
Brassicaceae
Cardaria latifolia (L.) Spach, *Crucifera latifolia* Krause
♦ perennial pepperweed, tall whitetop, broad leaved pepperweed, peppergrass, tall pepperwort, dittander
♦ Weed, Quarantine Weed, Noxious Weed, Naturalised, Environmental Weed, Cultivation Escape
♦ 1, 34, 35, 39, 70, 76, 78, 80, 86, 87, 88, 98, 101, 116, 130, 133, 138, 139, 141, 146, 151, 156, 161, 176, 180, 195, 199, 203, 212, 218, 219, 224, 229, 231, 243, 252, 264, 272, 287
♦ wpH, cultivated, herbal, toxic. Origin: Eurasia.

Lepidium latifolium L. var. affine (Ledeb.) C.A.Mey.
Brassicaceae
♦ Weed
♦ 275, 297
♦ pH.

Lepidium neglectum Thell.
Brassicaceae
= *Lepidium densiflorum* Schrad. var.

densiflorum (NoR)
- least pepperwort, rikkakrassi
- Weed, Casual Alien
- 42, 88, 94

Lepidium nitidum Nutt.
Brassicaceae
- tongue pepperweed, shining pepperweed, common peppergrass
- Weed
- 87, 88, 161, 218
- aH, promoted, herbal.

Lepidium perfoliatum L.
Brassicaceae
Crucifera diversifolia Krause
- clasping pepperweed, yellow pepperweed, yellowflower pepperweed, perfoliate pepperwort, peppergrass, sepiväkrassi
- Weed, Naturalised, Introduced, Casual Alien, Cultivation Escape
- 21, 23, 34, 40, 42, 56, 80, 86, 87, 88, 94, 98, 101, 136, 161, 203, 212, 218, 243, 252, 264, 272, 287
- aH, cultivated, herbal. Origin: east Europe, west Asia.

Lepidium pinnatifidum Ledeb.
Brassicaceae
- featherleaf pepperweed
- Weed, Naturalised
- 101, 272
- a/pH.

Lepidium pseudohyssopifolium Hewson
Brassicaceae
- Casual Alien
- 280

Lepidium pseudotasmanicum Thell.
Brassicaceae
- narrow leaved cress
- Weed, Naturalised
- 15, 280

Lepidium pubescens Desv.
Brassicaceae
- matted peppercress
- Weed, Naturalised, Cultivation Escape
- 86, 98, 198, 203, 252
- cultivated.

Lepidium ramosissimum A.Nelson
Brassicaceae
- haarakrassi, many branched pepperweed, branched pepperweed
- Weed, Casual Alien
- 42, 161
- herbal.

Lepidium repens (Schrenk) Boiss.
Brassicaceae
= *Cardaria draba* (L.) Desv. ssp. *chalepensis* (L.) Schulz
- lens peppergrass, whiteweed, hoary cress
- Weed
- 49, 87, 88, 218

Lepidium ruderale L.
Brassicaceae
Crucifera ruderalis Krause
- stinking pepperweed, roadside pepperweed, pihakrassi, narrow

leaved pepperwort, pepper grass, fetid peppergrass
- Weed, Naturalised
- 23, 40, 44, 70, 80, 87, 88, 94, 101, 272, 280, 287
- a/bH, cultivated, herbal.

Lepidium sagittulatum Thell.
Brassicaceae
- Weed
- 87, 88

Lepidium sativum L.
Brassicaceae
Cardamon sativum Fourr., *Crucifera nasturtium* Krause, *Lepidium hortense* Forssk.
- garden cress, vihanneskrassi, gardencress pepperweed, cress, habb al rrshad, cresson alénois
- Weed, Naturalised, Introduced, Cultivation Escape, Casual Alien
- 39, 40, 42, 70, 80, 86, 87, 88, 94, 98, 101, 176, 185, 198, 203, 221, 228, 241, 243, 252, 272, 280, 287, 300
- aH, arid, cultivated, herbal, toxic. Origin: Eurasia.

Lepidium schinzii Thell.
Brassicaceae
- Schinz's pepperweed, pepperwort
- Weed, Quarantine Weed, Naturalised, Native Weed
- 76, 87, 88, 101, 121, 203, 220
- aH, herbal. Origin: southern Africa.

Lepidium sordidum A.Gray
Brassicaceae
- sordid pepperweed
- Weed
- 199

Lepidium virginicum L.
Brassicaceae
Clypeola caroliniana Walt., *Crucifera virginica* Krause, *Dilephium virginicum* Raf.
- Virginia pepperweed, poor man's pepper, peppergrass, Virginian peppercress, virginiankrassi
- Weed, Naturalised, Introduced, Casual Alien, Cultivation Escape
- 14, 23, 28, 32, 34, 38, 40, 42, 56, 68, 86, 87, 88, 94, 98, 136, 161, 198, 203, 206, 210, 211, 218, 228, 243, 245, 249, 252, 255, 256, 259, 261, 263, 272, 274, 275, 280, 286, 287, 297
- aH, arid, cultivated, herbal. Origin: North America.

Lepidophorum repandum (L.) DC.
Asteraceae
- Weed
- 70

Lepidospartum squamatum (Gray) Gray
Asteraceae
- California broomsage, California broomshrub, scale broom
- Weed
- 161
- S, herbal, toxic.

Lepilaena J.Drumm. ex Harv. spp.
Potamogetonaceae/Zannichelliaceae
- Weed, Quarantine Weed

- 76, 88, 220

Lepilaena australis Harv.
Potamogetonaceae/Zannichelliaceae
- Weed
- 200
- aqua, cultivated.

Lepilaena biloculata Kirk
Potamogetonaceae/Zannichelliaceae
- Weed
- 87, 88

Lepinia ponapensis Hosok.
Apocynaceae
- Introduced
- 230
- S.

Lepironia articulata Domin.
Cyperaceae
- Naturalised
- 287
- pG, aqua.

Lepistemon binectariferum (Wall) Kuntze
Convolvulaceae
Lepistemon flavescens Bl.
- Weed
- 13
- C.

Leptadenia heterophylla (Delile) Decne.
Asclepiadaceae/Apocynaceae
Cynanchum heterophyllum Delile
- Weed
- 221

Leptadenia pyrotechnica (Forssk.) Decne.
Asclepiadaceae/Apocynaceae
Cynanchum pyrotechnicum Forssk., *Leptadenia spartium* Wight & Arn.
- Weed
- 221
- arid, cultivated.

Leptadenia reticulata Wight
Asclepiadaceae/Apocynaceae
- Weed
- 87, 88
- cultivated, herbal. Origin: Madagascar.

Leptaleum filifolium (Willd.) DC.
Brassicaceae
- Weed
- 221, 243, 272

Leptilon pusillum (Nutt.) Britt.
Asteraceae
= *Conyza canadensis* (L.) Cronquist var. *pusilla* (Nutt.) Cronquist
- Weed
- 87, 88

Leptocarpus disjunctus Mast.
Restionaceae
- Weed
- 87, 88
- G.

Leptocarydion vulpiastrum (De Not.) Stapf
Poaceae
- spade grass
- Native Weed
- 121
- aG. Origin: southern Africa.

Leptocereus (Berger) Britt. & Rose spp.
Cactaceae
Neoabbottia Britt. & Rose spp. (see)
- leptocereus
- Weed, Quarantine Weed
- 76, 88, 203, 220

Leptochloa P.Beauv. spp.
Poaceae
- sprangletop
- Weed
- 88, 249
- G.

Leptochloa chinensis (L.) Nees
Poaceae
Leptochloa decipiens (R.Br.) Druce, *Poa chinensis* L.
- red sprangletop, feathergrass, yaa yon huu, Chinese sprangletop, Asian sprangletop
- Weed, Quarantine Weed, Noxious Weed
- 11, 67, 76, 87, 88, 91, 135, 140, 170, 186, 191, 203, 204, 220, 229, 239, 243, 262, 263, 274, 275, 286, 297
- a/pG, aqua.

Leptochloa chloridiformis (Hack. ex Stuck.) Parodi
Poaceae
- Argentine sprangletop
- Naturalised
- 101
- G.

Leptochloa coerulescens Steud.
Poaceae
- Weed, Quarantine Weed
- 76, 87, 88, 203, 220
- G.

Leptochloa decipiens (R.Br.) Stapf ex Maiden
Poaceae
- Australian sprangletop
- Naturalised
- 101
- G, cultivated. Origin: Australia.

Leptochloa decipiens (R.Br.) Stapf ex Maiden ssp. peacockii Maiden & Betche
Poaceae
- peacock sprangletop
- Naturalised
- 101
- G.

Leptochloa digitata (R.Br.) Domin
Poaceae
- finger sprangletop, umbrella canegrass
- Naturalised
- 101
- pG, aqua, cultivated.

Leptochloa divaricatissima S.T.Blake
Poaceae
- spreading sprangletop
- Naturalised
- 101
- G. Origin: Australia.

Leptochloa dubia (Kunth) Nees
Poaceae
Chloris dubia Kunth, *Leptochloa pringlei* Beal

- green sprangletop
- Weed, Naturalised, Introduced
- 86, 199, 228
- G, arid, cultivated, herbal. Origin: southern North America to Argentina.

Leptochloa fascicularis (Lam.) Gray
Poaceae
= *Leptochloa fusca* (L.) Kunth ssp. *fascicularis* (Lam.) N.Snow (NoR) [see *Diplachne fascicularis* (Lam.) P.Beauv.]
- bearded sprangletop, clustered saltgrass, loose flowered sprangletop, raygrass, sprangletop, spreading millet
- Weed
- 30, 87, 88, 161, 180, 218, 243
- aG, herbal.

Leptochloa filiformis (Lam.) Beauv.
Poaceae
= *Leptochloa panicea* (Retz.) Ohwi ssp. *brachiata* (Steudl.) N.Snow (NoR)
- red sprangletop, pasto morado, plumilla, paja mona
- Weed, Naturalised
- 30, 88, 91, 98, 157, 161, 203, 218, 236, 237, 255, 257, 281, 295
- aG, herbal. Origin: Asia.

Leptochloa fusca (L.) Kunth
Poaceae
= *Leptochloa malabarica* (L.) Veldkamp (NoR) [see *Diplachne fusca* (L.) P.Beauv. ex Roem. & Schult. & *Diplachne malabarica* (L.) Merr.]
- Malabar sprangletop
- Weed
- 88, 185
- G.

Leptochloa fusca (L.) Kunth ssp. uninervia (J.Presl) N.Snow
Poaceae
Diplachne uninervia (J.Presl) Parodi (see), *Leptochloa uninervia* (Presl) Hitchc. & Chase (see), *Megastachya uninervia* J.Presl
- Mexican sprangletop, clustered saltgrass, ryegrass, sprangletop, spreading millet
- Weed, Sleeper Weed, Naturalised
- 86, 93, 243
- aG. Origin: Americas.

Leptochloa mucronata (Michx.) Kunth
Poaceae
= *Leptochloa panicea* (Retz.) Ohwi ssp. *mucronata* (Michx.) Nowack (NoR)
- mucronate sprangletop, cola de zorra, red sprangletop
- Weed, Naturalised
- 28, 86, 199, 243
- G, arid.

Leptochloa panicea (Retz.) Ohwi
Poaceae
Leptochloa filiformis auct. non (Lam.) Beauv.
- sprangletop, mucronate sprangletop, red sprangletop, thread sprangletop
- Weed, Quarantine Weed
- 76, 87, 88, 91, 135, 186, 191, 203, 220, 235, 275, 286, 297
- aG, arid/aqua.

Leptochloa panicoides (J.Presl) Hitchc.
Poaceae
- Amazon sprangletop
- Weed
- 161, 261
- G.

Leptochloa procera Nees
Poaceae
- Weed
- 270
- G.

Leptochloa scabra Nees
Poaceae
- rough sprangletop
- Weed, Quarantine Weed
- 76, 87, 88, 91, 203, 220, 271
- G, aqua.

Leptochloa uninervia (Presl) Hitchc. & Chase
Poaceae
= *Leptochloa fusca* (L.) Kunth ssp. *uninervia* (J.Presl) N.Snow
- Mexican sprangletop, clustered saltgrass, ryegrass, sprangletop, spreading millet
- Weed, Quarantine Weed
- 76, 87, 88, 161, 180, 203, 212, 218, 220, 243
- aG, arid, herbal.

Leptochloa virgata (L.) Beauv.
Poaceae
Cynosurus virgatus L.
- Judd's grass, tropical sprangletop
- Weed, Naturalised
- 14, 87, 88, 91, 255, 257, 271, 295
- pG.

Leptocoryphium lanatum (Kunth) Nees
Poaceae
- lanilla
- Weed
- 14
- G.

Leptopyrum fumarioides Reichb.
Ranunculaceae
- common leptopyrum
- Weed
- 297
- cultivated.

Leptospermum ericoides Rich.
Myrtaceae
Kunzea ericoides (A.Rich.) J.Thomps. (see)
- tree manuka, tree manuba, kanuka
- Weed
- 3
- S, cultivated, herbal.

Leptospermum erubescens Schauer
Myrtaceae
- pink teatree
- Naturalised
- 9
- cultivated. Origin: Australia.

Leptospermum flavescens Sm.
Myrtaceae
- common teatree
- Naturalised
- 101
- cultivated, herbal.

Leptospermum laevigata (Sol. ex Gaertn.) F.Muell.
Myrtaceae
= *Leptospermum laevigatum* (Gaertn.) F.Muell.
♦ coastal teatree
♦ Naturalised
♦ 101
♦ Origin: Australia.

Leptospermum laevigatum (Gaertn.) F.Muell.
Myrtaceae
Leptospermum laevigata (Sol. ex Gaertn.) F.Muell. (see)
♦ Australian myrtle, Victorian teatree, coast teatree, Australian teatree
♦ Weed, Noxious Weed, Naturalised, Native Weed, Garden Escape, Environmental Weed, Cultivation Escape
♦ 7, 10, 15, 63, 72, 86, 88, 95, 101, 121, 152, 158, 277, 278, 280, 283, 289
♦ pS, cultivated, herbal. Origin: Australia.

Leptospermum lanigerum (Sol. ex Aiton) Sm.
Myrtaceae
♦ woolly teatree, silky teatree
♦ Naturalised
♦ 40
♦ S, cultivated. Origin: Australia.

Leptospermum petersonii F.M.Bailey
Myrtaceae
Leptospermum citratum Challinor, Cheel & Penfold
♦ lemon scented teatree, lemon teatree
♦ Weed, Naturalised, Native Weed, Garden Escape, Environmental Weed
♦ 72, 86, 88, 198, 228
♦ S/T, cultivated, herbal. Origin: Australia.

Leptospermum rotundifolium (Maiden & Betche) F.Rodway ex Cheel
Myrtaceae
♦ teatree, roundleaf teatree
♦ Naturalised, Native Weed
♦ 7
♦ cultivated. Origin: Australia.

Leptospermum scoparium J.R. & G.Forst.
Myrtaceae
Leptospermum nichollsii Dorr.Sm.
♦ broom teatree, manuka, New Zealand tea manuka, kahikatoa, teatree, rose flowered teatree
♦ Weed, Naturalised, Cultivation Escape
♦ 3, 22, 40, 80, 87, 88, 101, 131, 165, 181, 233
♦ S, cultivated, herbal. Origin: Australia, New Zealand.

Leptotaenia dissecta Nutt.
Apiaceae
= *Lomatium dissectum* (Nutt.) Mathias & Constance
♦ carrotleaf leptotaenia
♦ Weed
♦ 87, 88

Leptotaenia dissecta Nutt. var. *multifida* (Nutt.) Jeps.
Apiaceae
= *Lomatium dissectum* (Nutt.) Mathias & Constance
♦ carrotleaf leptotaenia
♦ Weed
♦ 218

Lepturus cylindricus (Willd.) Trin.
Poaceae
= *Hainardia cylindrica* (Willd.) Greuter
♦ Weed
♦ 87, 88
♦ G.

Lepyrodiclis holosteoides (Mey.) Fenzl ex Fisch.
Caryophyllaceae
♦ lepyrodicilis, false jagged chickweed, rennokki
♦ Weed, Noxious Weed, Naturalised, Casual Alien
♦ 1, 42, 80, 88, 101, 146, 161, 229, 243, 287, 297
♦ cultivated.

Lepyrodiclis paniculata Stapf
Caryophyllaceae
♦ Weed
♦ 248

Lespedeza bicolor Turcz.
Fabaceae/Papilionaceae
♦ bicolor lespedeza, bushclover, shrub bushclover, shrubby lespedeza, Siberian bushclover
♦ Weed, Naturalised
♦ 77, 80, 88, 101, 102, 286
♦ S, cultivated, herbal. Origin: China, Japan, Korea.

Lespedeza capitata Michx.
Fabaceae/Papilionaceae
♦ round headed bushclover, roundhead lespedeza, lespedeza
♦ Quarantine Weed
♦ 220
♦ S, cultivated, herbal.

Lespedeza caraganae Bunge
Fabaceae/Papilionaceae
♦ Weed
♦ 275

Lespedeza cuneata (Dum.Cours.) G.Don
Fabaceae/Papilionaceae
Lespedeza sericea (Thunb.) Miq.
♦ Chinese lespedeza, Himalayan bushclover, silky bushclover, perennial lespedeza, sericea lespedeza, bushclover
♦ Weed, Naturalised, Environmental Weed
♦ 17, 80, 87, 88, 101, 102, 121, 129, 133, 151, 161, 195, 218, 243, 275, 286, 297
♦ pH, cultivated, herbal. Origin: Asia.

Lespedeza cuneata G.Don var. *serpens* (Nakai) Ohwi
Fabaceae/Papilionaceae
♦ Weed
♦ 286

Lespedeza cyrtobotrya Miq.
Fabaceae/Papilionaceae

♦ leafy lespedeza, maruba hagi
♦ Naturalised
♦ 101
♦ S, promoted.

Lespedeza daurica (Laxm.) Schindl.
Fabaceae/Papilionaceae
♦ Dahurian lespedeza, Dahurian bushclover
♦ Weed, Naturalised
♦ 101, 275, 287, 297

Lespedeza formosa (Vogel) Koehne
Fabaceae/Papilionaceae
♦ oriental lespedeza
♦ Naturalised
♦ 101

Lespedeza grandis Koidz.
Fabaceae/Papilionaceae
♦ Quarantine Weed
♦ 76, 220

Lespedeza hirta (L.) Hornem.
Fabaceae/Papilionaceae
♦ hairy lespedeza
♦ Weed
♦ 179
♦ herbal.

Lespedeza juncea (L.f.) Pers.
Fabaceae/Papilionaceae
♦ Naturalised
♦ 287
♦ S, cultivated. Origin: Himalayas, China, Japan.

Lespedeza liukiuensis Hatsus.
Fabaceae/Papilionaceae
♦ Naturalised
♦ 287

Lespedeza pilosa (Thunb.) Sieb. & Zucc.
Fabaceae/Papilionaceae
♦ Weed
♦ 87, 88, 286
♦ pH, promoted. Origin: China, Japan, Korea.

Lespedeza stipulacea Maxim.
Fabaceae/Papilionaceae
= *Kummerowia stipulacea* (Maxim.) Makino
♦ Korean lespedeza, Korean bushclover, Korean clover
♦ Weed
♦ 39, 80, 88, 161, 195, 218, 243
♦ aH, cultivated, herbal, toxic.

Lespedeza striata (Thunb.) Hook. & Arn.
Fabaceae/Papilionaceae
= *Kummerowia striata* (Thunb.) Schindl.
♦ common lespedeza, Japanese clover, Japanese lespedeza, bushclover
♦ Weed, Naturalised
♦ 80, 86, 87, 88, 161, 195, 218, 243, 249, 263
♦ aH, herbal. Origin: China, Japan.

Lespedeza stuevei Nutt.
Fabaceae/Papilionaceae
♦ tall lespedeza
♦ Naturalised
♦ 287
♦ herbal.

Lespedeza thunbergii (DC.) Nakai
Fabaceae/Papilionaceae
♦ Thunberg's lespedeza, miyagino
hagi, shrub lespedeza
♦ Weed, Quarantine Weed,
Naturalised
♦ 76, 86, 88, 101, 220
♦ S, cultivated, herbal. Origin:
horticultural.

Lespedeza tomentosa (Thunb.) Sieb.
Fabaceae/Papilionaceae
♦ Weed
♦ 275, 286
♦ pH, promoted.

Lespedeza violacea (L.) Pers.
Fabaceae/Papilionaceae
♦ violet lespedeza
♦ Weed
♦ 87, 88, 218
♦ herbal.

Lespedeza virgata (Thunb.) DC.
Fabaceae/Papilionaceae
♦ wand lespedeza
♦ Naturalised
♦ 101

Lesquerella gordonii (Gray) Wats.
Brassicaceae
Alyssum gordonii (A.Gray) Kuntze,
Vesicaria gordonii A.Gray
♦ Gordon's bladderpod, bladder pod
♦ Weed
♦ 23, 87, 88, 161, 218
♦ aH, arid, cultivated, herbal.

Lessertia diffusa R.Br.
Fabaceae/Papilionaceae
= *Galega dubia* (R.Br.) Jacq. (NoR)
♦ Quarantine Weed
♦ 220
♦ cultivated.

Lessertia perennans (Jacq.) DC.
Fabaceae/Papilionaceae
Colutea perennans (Medik.) Jacq.
♦ Quarantine Weed
♦ 220
♦ cultivated.

Leucadendron argenteum (L.) R.Br.
Proteaceae
♦ silver tree
♦ Naturalised
♦ 280
♦ cultivated.

Leucadendron ericifolium R.Br.
Proteaceae
Leucadendron uniflorum Phill.
♦ erica leaved yellowbush
♦ Native Weed
♦ 121
♦ pS. Origin: southern Africa.

Leucadendron rubrum Burm.f.
Proteaceae
Leucadendron parviflorum (L.) Druce
♦ spinning conebush
♦ Native Weed
♦ 121
♦ pS, cultivated. Origin: southern
Africa.

Leucaena Benth. spp.
Fabaceae/Mimosaceae
♦ leadtree
♦ Environmental weed
♦ 201
♦ S/T, herbal.

Leucaena esculenta (DC.) Benth.
Fabaceae/Mimosaceae
Acacia esculenta Sessé & Moç., *Mimosa
esculenta* Sessé & Moç.
♦ Introduced
♦ 228
♦ arid, herbal.

Leucaena glabrata Rose
Fabaceae/Mimosaceae
♦ Introduced
♦ 228
♦ arid.

Leucaena glauca (L.) Benth.
Fabaceae/Mimosaceae
= *Leucaena leucocephala* (Lam.) de Wit
♦ leadtree, horse tamarind, wild
tamarind
♦ Weed, Noxious Weed
♦ 3, 23, 39, 88, 122, 191, 218
♦ S/T, cultivated, herbal, toxic.

Leucaena leucocephala (Lam.) de Wit
Fabaceae/Mimosaceae
Albizia julibrissin sensu Blakelock,
Leucaena glauca auct. *non* (L.) Benth.
(see), *Mimosa leucocephala* Lam.
♦ leadtree, koa haole, reuse wattel,
leucaena, wild tamarind, ipil ipil,
white leadtree, haole koa, coffee bush,
leadtree, acacia negra, faux acacia, faux
mimosa, koa haole, tangantangan,
ganitnityuwan tangantan, telentund,
namas, vaivai, tuhngantuhngan,
rohbohtin, lopa samoa, pepe, siale
mohemohe, nito, cassis, te kaitetua,
balori
♦ Weed, Noxious Weed, Naturalised,
Introduced, Environmental Weed,
Cultivation Escape
♦ 3, 6, 7, 18, 22, 37, 39, 63, 80, 87, 88, 93,
95, 98, 107, 112, 121, 122, 132, 151, 152,
155, 158, 161, 179, 203, 216, 228, 230,
255, 257, 259, 260, 268, 283, 287
♦ S/T, arid, cultivated, herbal, toxic.
Origin: Central and South America.

Leucaena leucocephala (Lam.) de Wit ssp.
glabrata (Rose) Zárate
Fabaceae/Mimosaceae
♦ Naturalised
♦ 86
♦ S/T. Origin: Mexico.

Leucaena leucocephala (Lam.) de Wit ssp.
leucocephala
Fabaceae/Mimosaceae
♦ leucaena
♦ Naturalised
♦ 86
♦ S/T. Origin: Mexico.

Leucaena pulverulenta (Schltdl.) Benth.
Fabaceae/Mimosaceae
Acacia pulverulenta Schldl.
♦ great leadtree
♦ Introduced

♦ 228
♦ arid.

Leucaena retusa Benth.
Fabaceae/Mimosaceae
♦ littleleaf leadtree
♦ Introduced
♦ 228
♦ arid, cultivated, herbal.

Leucanthemella serotina (L.) Tzvelev
Asteraceae
♦ giant daisy
♦ Naturalised
♦ 101
♦ cultivated.

Leucanthemum lacustre (Brot.) Samp.
Asteraceae
♦ Portuguese daisy
♦ Weed, Naturalised
♦ 70, 101

Leucanthemum maximum (Ramond) DC.
Asteraceae
Chrysanthemum maximum Ramond
(see)
♦ shasta daisy, isopäivänkakkara, max
chrysanthemum
♦ Weed, Naturalised, Garden Esacpe,
Environmental Weed, Cultivation
Escape
♦ 42, 72, 86, 88, 98, 101, 198, 203, 252,
280, 289
♦ a/pH, cultivated, herbal. Origin:
south-west Europe.

Leucanthemum myconis (L.) Giraud
Asteraceae
♦ Weed, Quarantine Weed
♦ 76, 87, 88, 203, 220

Leucanthemum nipponicum Franch. ex
Maxim.
Asteraceae
♦ Nippon daisy
♦ Naturalised
♦ 101

Leucanthemum paludosum (Poir.) Bonnet
& Barratte
Asteraceae
Chrysanthemum paludosum Poir. (see)
♦ peikonkakkara
♦ Weed, Cultivation Escape, Casual
Alien
♦ 42, 87, 88, 280
♦ cultivated.

Leucanthemum segetum (L.) Stankov
Asteraceae
♦ Weed
♦ 87, 88

Leucanthemum × superbum (J.W.Ingram)
Berg. ex Kent
Asteraceae
= *Leucanthemum lacustre* (Brot.) Samp.
× *Leucanthemum maximum* (Raymond)
DC. [see *Chrysanthemum × superbum*
J.W.Ingram]
♦ shasta daisy
♦ Naturalised
♦ 40, 101
♦ herbal.

Leucanthemum sylvaticum (Hoffm. & Link) Nyman
Asteraceae
♦ Weed
♦ 70

Leucanthemum vulgare Lam.
Asteraceae
Chrysanthemum leucanthemum L. (see), *Chrysanthemum leucanthemum* L. var. *pinnatifidum* Lecoq & Lam. (see), *Leucanthemum leucanthemum* (L.) Rydb.
♦ oxeye daisy, päivänkakkara, dog daisy, margriet, marguerite daisy, moon daisy, white daisy, yellow daisy, margaréta biela
♦ Weed, Noxious Weed, Naturalised, Introduced, Environmental Weed, Cultivation Escape
♦ 1, 7, 15, 20, 21, 34, 35, 70, 72, 78, 80, 86, 87, 88, 94, 98, 101, 139, 146, 147, 151, 165, 176, 181, 198, 203, 222, 229, 231, 237, 241, 243, 252, 256, 269, 272, 280, 290, 300
♦ pH, cultivated, herbal. Origin: Eurasia.

Leucas aspera (Willd.) Link
Lamiaceae
♦ Weed
♦ 13, 87, 88, 276, 297
♦ aH, cultivated, herbal.

Leucas biflora (Vahl) R.Br.
Lamiaceae
♦ Weed
♦ 66

Leucas cephalotes Spreng.
Lamiaceae
♦ Weed, Naturalised
♦ 86, 87, 88
♦ herbal.

Leucas ciliata Benth.
Lamiaceae
♦ ciliate leucas
♦ Weed
♦ 297

Leucas decemdentata (Willd.) Sm.
Lamiaceae
♦ ogoogo tea
♦ Naturalised
♦ 86
♦ herbal.

Leucas decurvata Bak. ex Hiern
Lamiaceae
♦ Weed
♦ 87, 88

Leucas glabrata R.Br.
Lamiaceae
♦ Weed, Quarantine Weed
♦ 76, 87, 88, 203, 220
♦ herbal.

Leucas javanica Benth.
Lamiaceae
♦ Weed
♦ 13, 87, 88

Leucas lanata Benth.
Lamiaceae
♦ Weed
♦ 87, 88
♦ herbal.

Leucas lavandulaefolia J.E.Sm.
Lamiaceae
Leucas lavandulifolia Sm. (see), *Leucas linifolia* (Roth) Spreng. (see)
♦ Weed
♦ 87, 88, 170, 191
♦ herbal.

Leucas lavandulifolia Sm.
Lamiaceae
Leucas lavandulaefolia J.E.Sm. (see)
♦ Weed
♦ 13
♦ herbal.

Leucas linifolia (Roth) Spreng.
Lamiaceae
= *Leucas lavandulaefolia* J.E.Sm.
♦ Naturalised
♦ 86
♦ Origin: south-east Asia.

Leucas martinicensis (Jacq.) R.Br.
Lamiaceae
♦ bobbin weed, tumbleweed, tolbossie, whitewort
♦ Weed, Naturalised, Native Weed
♦ 50, 51, 86, 87, 88, 98, 121, 203, 240, 255
♦ aH, herbal. Origin: pantropical.

Leucas mollissima Wall.
Lamiaceae
♦ Weed
♦ 87, 88, 235

Leucas neufliseana Courb.
Lamiaceae
♦ Weed
♦ 87, 88

Leucas sexdentata Skan
Lamiaceae
♦ Native Weed
♦ 121
♦ aH. Origin: southern Africa.

Leucas urticaefolia R.Br.
Lamiaceae
= *Leucas urticifolia* (Vahl) R.Br.
♦ Weed, Quarantine Weed
♦ 76, 87, 88, 203, 220

Leucas urticifolia (Vahl) R.Br.
Lamiaceae
Leucas urticaefolia R.Br. (see)
♦ Weed
♦ 66, 242
♦ arid.

Leucas zeylanica R.Br.
Lamiaceae
♦ Weed, Naturalised
♦ 13, 86, 88
♦ herbal.

Leuchtenbergia Hook. spp.
Cactaceae
♦ Weed, Quarantine Weed, Naturalised
♦ 76, 86, 88, 220
♦ cultivated.

Leucocrinum montanum Nutt. ex Gray
Liliaceae/Anthericaceae
♦ mountain lily, common starlily, sand lily
♦ Weed
♦ 39, 161

♦ pH, promoted, herbal, toxic.

Leucojum aestivum L.
Liliaceae/Amaryllidaceae
Nivaria aestivalis Moench, *Nivaria monadelpha* Medik.
♦ summer snowflake, London lily, snowflake
♦ Weed, Naturalised, Garden Escape, Environmental Weed
♦ 7, 15, 39, 86, 87, 88, 98, 101, 195, 198, 203, 272, 280
♦ pH, cultivated, herbal, toxic. Origin: Europe.

Leucojum aestivum L. ssp. aestivum
Liliaceae/Amaryllidaceae
♦ summer snowflake
♦ Naturalised
♦ 101

Leucojum aestivum L. ssp. pulchellum (Salisb.) Briq.
Liliaceae/Amaryllidaceae
♦ summer snowflake
♦ Naturalised
♦ 101

Leucojum autumnale L.
Liliaceae/Amaryllidaceae
♦ autumn snowflake
♦ Weed
♦ 272
♦ cultivated.

Leucojum vernum L.
Liliaceae/Amaryllidaceae
♦ kevätkello, spring snowflake
♦ Naturalised, Cultivation Escape
♦ 39, 42, 101
♦ pH, cultivated, herbal, toxic.

Leucosidea sericea Eckl. & Zeyh.
Rosaceae
♦ chechebush, oldwood, ouhout
♦ Weed, Native Weed
♦ 10, 63, 121
♦ S/T, cultivated, herbal. Origin: southern Africa.

Leucosphaera bainesii (Hook.f.) Gilg
Amaranthaceae
♦ Native Weed
♦ 121
♦ S. Origin: southern Africa.

Leucothoe axillaris (Lam.) D.Don
Ericaceae
♦ coastal doghobble, drooping leucothoe
♦ Weed
♦ 161, 247
♦ S, cultivated, herbal, toxic.

Leucothoe davisiae Torr.
Ericaceae
♦ Sierra laurel
♦ Weed
♦ 39, 161
♦ S, cultivated, herbal, toxic.

Leucothoe fontanesiana (Steud.) Sleumer
Ericaceae
♦ doghobble, drooping leucothoe, fetterbush, highland doghobble
♦ Weed
♦ 161, 247
♦ cultivated, herbal, toxic.

Leucothoe racemosa (L.) Gray
Ericaceae
♦ sweet bells, swamp doghobble
♦ Weed
♦ 39, 161
♦ cultivated, herbal, toxic.

Leucothoe recurva (Buckley) A.Gray
Ericaceae
♦ redtwig doghobble, recurved fetterbush
♦ Weed
♦ 161
♦ cultivated, herbal, toxic.

Leuzea conifera (L.) DC.
Asteraceae
Centaurea conifera L.
♦ Weed, Quarantine Weed
♦ 76, 88, 220
♦ cultivated.

Levisticum officinale Koch
Apiaceae
♦ liperi, lovage, garden lovage
♦ Naturalised, Cultivation Escape
♦ 42, 101, 300
♦ pH, cultivated, herbal.

Leycesteria formosa Wall
Caprifoliaceae
♦ Himalayan honeysuckle, flowering nutmeg
♦ Weed, Naturalised, Garden Escape, Environmental Weed, Cultivation Escape
♦ 15, 20, 40, 72, 86, 88, 98, 101, 155, 165, 176, 198, 203, 225, 246, 280, 289, 296
♦ S, cultivated. Origin: Himalayas, western China, India, Nepal, Myanmar.

Leymus angustus (Trin.) Pilg.
Poaceae
♦ Altai wildrye
♦ Naturalised
♦ 101
♦ G.

Leymus arenarius (L.) Hochst.
Poaceae
Elymus arenarius L. (see)
♦ sand ryegrass, lyme grass, rantavehnä
♦ Weed, Naturalised
♦ 101, 176, 241, 272, 280, 300
♦ G, cultivated. Origin: Europe.

Leymus chinensis Tzvel.
Poaceae
Agropyron berezovcanum Prodán, *Agropyron chinense* (Trin.) Ohwi, *Agropyron pseudoagropyrum* (Trin. ex Griseb.) Franch., *Agropyron uninerve* Candargy, *Aneurolepidium chinense* (Trin.) Kitag., *Aneurolepidium pseudoagropyrum* (Trin. ex Griseb.) Nevski, *Elymus chinensis* (Trin.) Keng, *Elymus pseudoagropyrum* (Trin. ex Greseb.) Turcz., *Leymus pseudoagropyrum* (Trin. ex Griseb.) Tzvelev, *Triticum chinense* Trin., *Triticum pseudoagropyrum* Trin. ex Griseb.
♦ Chinese leymus

♦ Weed
♦ 114, 243, 297
♦ G, arid.

Leymus multicaulis (Kar. & Kir.) Tzvelev
Poaceae
♦ manystem wildrye, Siberian wildrye
♦ Weed, Naturalised
♦ 98, 101, 198, 203, 272
♦ G.

Leymus racemosus (Lam.) Tzvelev
Poaceae
Elymus giganteus Vahl
♦ mammoth wildrye
♦ Naturalised
♦ 280
♦ G, arid.

Leymus ramosus (Trin.) Tzvelev
Poaceae
Agropyron ramosum (Trin.) K.Richt. (see), *Aneurolepidium ramosum* (Trin.) Nevski, *Elymus trinii* Melderis, *Triticum ramosum* Trin. (see)
♦ Weed
♦ 272
♦ G. Origin: temperate Asia, eastern Europe.

Leymus secalinus (Georgi) Tzvel.
Poaceae
♦ common leymus, blue wild ryegrass, ryegrass
♦ Weed
♦ 297
♦ G, cultivated. Origin: Asia.

Leymus triticoides (Buckl.) Pilg.
Poaceae
Elymus triticoides (Nutt.) Buckl. (see)
♦ beardless wildrye
♦ Weed
♦ 161
♦ pG, arid.

Liatris punctata Hook.
Asteraceae
Laciniaria punctata (Hook.) Kuntze
♦ dotted gayfeather, dotted blazing star, snakeroot
♦ Weed
♦ 23, 88, 161
♦ pH, cultivated, herbal.

Liatris spicata (L.) Willd.
Asteraceae
♦ gayfeather, dense blazing star, button snakewort
♦ Casual Alien
♦ 280
♦ pH, cultivated, herbal.

Libertia formosa Graham
Iridaceae
♦ snowy mermaid
♦ Naturalised, Introduced
♦ 34, 40, 101
♦ pH, cultivated, herbal.

Libocedrus decurrens Torr.
Cupressaceae
= *Calocedrus decurrens* (Torr.) Florin (NoR)
♦ incense cedar
♦ Weed
♦ 87, 88, 218

♦ cultivated, herbal.

Licuala grandis (hort. ex W.Bull) H.Wendl.
Arecaceae
♦ ruffled fan palm, palmier cuillère, fan palm
♦ Weed
♦ 3
♦ cultivated.

Ligaria cuneifolia (Ruiz & Pav.) Tiegh.
Loranthaceae
Psittacanthus cuneifolius (Ruiz & Pav.) Blume ex Schult.f., *Loranthus cuneifolius* Ruiz & Pav., *Loranthus montevidensis* Spreng., *Ligaria coronata* Tiegh.
♦ liga
♦ Weed
♦ 237, 295
♦ pS parasitic.

Ligularia dentata (A.Gray) Hara
Asteraceae
♦ kallionauhus
♦ Cultivation Escape
♦ 42
♦ cultivated, herbal.

Ligularia fischeri (Ledeb.) Turcz.
Asteraceae
♦ otakarakou
♦ Weed, Quarantine Weed
♦ 76, 88, 220
♦ cultivated.

Ligularia intermedia Nakai
Asteraceae
♦ narrowbract goldenray
♦ Weed
♦ 297
♦ pH, promoted. Origin: northern China, Japan, Korea.

Ligularia japonica (Thunb.) Less.
Asteraceae
Arnica japonica Thunb., *Senecio japonicus* (Thunb.) Sch.Bip.
♦ Quarantine Weed
♦ 220
♦ pH, cultivated, herbal.

Ligularia sibirica (L.) Cass.
Asteraceae
♦ Weed, Quarantine Weed
♦ 76, 88, 220
♦ pH, cultivated. Origin: Eurasia.

Ligularia stenocephala (Maxim.) Matsum. & Koidz.
Asteraceae
Senecio stenocephalus Maxim.
♦ Weed, Quarantine Weed
♦ 76, 88, 220
♦ cultivated.

Ligularia tangutica (Maxim.) Mattf. ex Rehder & Kobuski
Asteraceae
♦ Quarantine Weed
♦ 220
♦ herbal.

Ligularia vorobievii Worosch.
Asteraceae
♦ Quarantine Weed
♦ 220

Ligularia wilsoniana (Hemsl.) Greenm.
Asteraceae
Ligularia polycephala (Hemsl.) Nakai,
Ligularia sibirica (L.) Cass. var.
polycephalus (Hemsl.) Diels, *Senecillis wilsoniana* (Hemsl.) Kitam., *Senecio cacaliifolius* Sch.Bip. var. *polycephalus* (Hemsl.) Franch., *Senecio iochanense* H.Lév., *Senecio ligularia* Hook.f. var. *polycephalus* Hemsl., *Senecio wilsonianus* Hemsl.
♦ giant groundsel
♦ Quarantine Weed
♦ 220
♦ cultivated.

Ligustrum L. spp.
Oleaceae
♦ privet
♦ Weed, Naturalised, Garden Escape, Environmental Weed
♦ 3, 18, 80, 86, 88, 151, 198
♦ cultivated, herbal.

Ligustrum amurense Carr.
Oleaceae
♦ Amur privet
♦ Weed, Naturalised
♦ 80, 101
♦ cultivated, herbal.

Ligustrum indicum (Lour.) Merr.
Oleaceae
Phillyrea indica Lour.
♦ Weed
♦ 80
♦ S, cultivated, herbal. Origin: east Asia.

Ligustrum japonicum Thunb.
Oleaceae
♦ waxleaf privet, Japanese privet, Japanese liguster, Japanese wax leaved privet
♦ Weed, Quarantine Weed, Noxious Weed, Naturalised, Introduced, Garden Escape, Environmental Weed
♦ 39, 63, 76, 77, 80, 88, 95, 101, 108, 112, 142, 161, 247, 261, 279, 283
♦ S/T, cultivated, herbal, toxic. Origin: China, Korea, Japan.

Ligustrum lucidum W.T.Aiton
Oleaceae
Esquirolia sinensis H.Lév., *Ligustrum compactum* (Wall. ex G.Don) Hook.f. & Thomson ex Brandis var. *latifolium* W.C.Cheng, *Ligustrum esquirolii* H.Lév.
♦ glossy privet, tree privet, largeleaf privet, broadleaf privet, Chinese wax leaved privet, Chinese privet, Nepal privet, privet, waxleaf privet, white wax tree
♦ Weed, Noxious Weed, Quarantine Weed, Naturalised, Garden Escape, Environmental Weed, Cultivation Escape
♦ 15, 35, 39, 63, 72, 73, 76, 78, 80, 86, 87, 88, 95, 98, 101, 112, 116, 121, 134, 152, 155, 161, 165, 179, 198, 201, 203, 225, 233, 246, 269, 280, 283, 287, 289, 296
♦ S/T, cultivated, herbal, toxic. Origin: China, Korea.

Ligustrum obtusifolium Sieb. & Zucc.
Oleaceae
♦ border privet, Amur river privet
♦ Weed, Naturalised
♦ 39, 80, 101, 133, 195, 224
♦ S, cultivated, herbal, toxic. Origin: east Asia, China, Japan.

Ligustrum ovalifolium Hassk.
Oleaceae
♦ Californian privet, Kaliforniese liguster, privet, garden privet, hedge privet, oval leaved privet, troène des haies
♦ Weed, Noxious Weed, Naturalised, Introduced, Garden Escape, Environmental Weed
♦ 15, 39, 40, 63, 80, 88, 95, 101, 133, 161, 198, 224, 261, 280, 283, 290
♦ S, cultivated, herbal, toxic. Origin: Japan.

Ligustrum quihoui Carr.
Oleaceae
♦ waxyleaf privet, privet
♦ Weed, Naturalised
♦ 80, 101
♦ cultivated, herbal.

Ligustrum robustum (Roxb.) Blume
Oleaceae
♦ Weed
♦ 3, 22, 39, 88
♦ T, cultivated, herbal, toxic.

Ligustrum robustum (Roxb.) Blume ssp. walkeri (Decne.) P.S.Green
Oleaceae
Ligustrum walkeri Decne., *Ligustrum ceylanicum* Decne.
♦ privet
♦ Weed, Environmental Weed
♦ 132, 152
♦ Origin: India, Sri Lanka.

Ligustrum sinense Lour.
Oleaceae
Ligustrum villosum May
♦ Chinese privet, hedge privet, small leaved privet, privet, Chinese liguster
♦ Weed, Quarantine Weed, Noxious Weed, Naturalised, Garden Escape, Environmental Weed, Cultivation Escape
♦ 15, 17, 39, 63, 72, 73, 76, 77, 80, 86, 88, 95, 98, 101, 102, 112, 134, 142, 152, 155, 165, 179, 198, 201, 203, 225, 233, 246, 261, 269, 279, 280, 283, 290, 296
♦ S, cultivated, herbal, toxic. Origin: China.

Ligustrum vulgare L.
Oleaceae
Olea humilis Salisb., *Ligustrum italicum* Mill
♦ common privet, wild privet, golden privet, gewone liguster, European privet, aitalikusteri
♦ Weed, Noxious Weed, Naturalised, Introduced, Garden Escape, Environmental Weed, Cultivation Escape
♦ 4, 15, 27, 39, 42, 63, 72, 80, 86, 88, 95, 98, 101, 102, 133, 142, 154, 155, 161, 176,

195, 198, 203, 222, 224, 279, 280, 283, 289
♦ S/T, cultivated, herbal, toxic. Origin: Mediterranean.

Lilaea scilloides (Poir.) Hauman
Juncaginaceae/Lilaeaceae
♦ awlleaf lilaea, flowering quillwort, lilaea
♦ Weed, Naturalised, Environmental Weed
♦ 72, 86, 88, 98, 198, 203
♦ wH, cultivated, herbal. Origin: North and South America.

Lilium bulbiferum L.
Liliaceae
♦ ruskolilja, orange lily, fire lily
♦ Naturalised, Cultivation Escape
♦ 42, 101
♦ pH, cultivated, herbal.

Lilium candidum L.
Liliaceae
♦ Madonna lily
♦ Naturalised, Garden Escape
♦ 101
♦ pH, cultivated, herbal.

Lilium formosanum Wallace
Liliaceae
Lilium longiflorum Thunb. var. *formosanum* Bak., *Lilium philippinense* Bak. var. *formosanum* (Wallace) E.H.Wilson, *Lilium yoshidai* Leichtlin
♦ lily, Formosa lily, Taiwan lily
♦ Weed, Noxious Weed, Naturalised, Garden Escape, Environmental Weed
♦ 15, 72, 73, 86, 88, 98, 101, 121, 155, 198, 203, 246, 251, 269, 280, 283, 287, 290, 296
♦ pH, cultivated, herbal. Origin: Taiwan.

Lilium lancifolium Thunb.
Liliaceae
Lilium tigrinum Ker Gawl. (see)
♦ tiger lily
♦ Weed, Naturalised, Garden Escape, Environmental Weed
♦ 54, 86, 88, 101, 252, 286
♦ pH, cultivated, herbal. Origin: China, Japan, Korea.

Lilium leichtlinii Hook.f. var. tigrinum (Regel) G.Nicholson
Liliaceae
♦ Weed
♦ 286

Lilium longiflorum Thunb.
Liliaceae
♦ Easter lily, trumpet lily, white trumpet lily
♦ Naturalised
♦ 101
♦ pH, cultivated, herbal.

Lilium martagon L.
Liliaceae
♦ martagon lily, Turk's cap lily, varjolilja
♦ Weed, Cultivation Escape
♦ 42, 70
♦ pH, cultivated, herbal.

Lilium maximowiczii **Regel**
Liliaceae
 ♦ Weed
 ♦ 88, 191, 204

Lilium philippinense **Bak.**
Liliaceae
 ♦ Philippine lily
 ♦ Naturalised
 ♦ 101
 ♦ cultivated.

Lilium pumilum **DC.**
Liliaceae
 ♦ low lily, coral lily
 ♦ Weed
 ♦ 297
 ♦ pH, cultivated. Origin: east China to Siberia.

Lilium pyrenaicum **Gouan**
Liliaceae
 ♦ Pyrenean lily, straw coloured Turk's cap
 ♦ Naturalised
 ♦ 40
 ♦ cultivated, herbal.

Lilium tigrinum **Ker Gawl.**
Liliaceae
 = *Lilium lancifolium* Thunb.
 ♦ tiger lily
 ♦ Weed, Naturalised, Garden Escape, Environmental Weed
 ♦ 72, 86, 88, 198, 280
 ♦ pH, cultivated, herbal.

Limeum fenestratum **(Fenzl) Heimerl var. fenestratum**
Molluginaceae
 ♦ Native Weed
 ♦ 121
 ♦ aH. Origin: southern Africa.

Limeum linifolium **Fenzl**
Molluginaceae
 ♦ Weed
 ♦ 87, 88

Limeum pterocarpum **(Gay) Heimerl var. pterocarpum**
Molluginaceae
 ♦ Native Weed
 ♦ 121
 ♦ aH. Origin: southern Africa.

Limeum sulcatum **(Klotzsch) Hutch. var. sulcatum**
Molluginaceae
 ♦ Native Weed
 ♦ 121
 ♦ H. Origin: southern Africa.

Limeum viscosum **(Gay) Fenzl ssp. viscosum var. kraussii Friedr.**
Molluginaceae
 ♦ Native Weed
 ♦ 121
 ♦ aH. Origin: southern Africa.

Limnanthes douglasii **R.Br.**
Limnanthaceae
 ♦ poached eggs, meadowfoam, poached egg plant, Douglas's meadowfoam, common meadowfoam
 ♦ Weed, Naturalised, Casual Alien

 ♦ 15, 40, 280
 ♦ aH, cultivated, herbal.

Limnobium boscii **Rich**
Hydrocharitaceae
 ♦ Weed
 ♦ 87, 88
 ♦ wH.

Limnobium laevigatum **(Humb. & Bonpl. ex Willd.) Heine**
Hydrocharitaceae
 Limnobium spongia (Bosc) Steud. ssp. *laevigatum* (Humb. & Bonpl. ex Willd.) Lowden (see), *Limnobium stoloniferum* (G.Mey.) Griseb. (see)
 ♦ Amazon frogbit, South America frogbit
 ♦ Naturalised
 ♦ 241
 ♦ wH, cultivated.

Limnobium spongia **(Bosc) L.C.Rich. ex Steud.**
Hydrocharitaceae
 Hydrocharis spongia Bosc
 ♦ frogbit, American frogbit, American spongeplant
 ♦ Weed, Quarantine Weed
 ♦ 88, 161, 197, 218, 258
 ♦ wH, cultivated, herbal.

Limnobium spongia **(Bosc) Steud. ssp. laevigatum (Humb. & Bonpl. ex Willd.) Lowden**
Hydrocharitaceae
 = *Limnobium laevigatum* (Humb. & Bonpl. ex Willd.) Heine
 ♦ Naturalised
 ♦ 300
 ♦ wH.

Limnobium stoloniferum **(G.Mey.) Griseb.**
Hydrocharitaceae
 = *Limnobium laevigatum* (Humb. & Bonpl. ex Willd.) Heine
 ♦ Naturalised
 ♦ 287
 ♦ wH, cultivated.

Limnocharis emarginata **Humb. & Bonpl.**
Limnocharitaceae
 = *Limnocharis flava* (L.) Buch.
 ♦ Weed
 ♦ 87, 88
 ♦ wH.

Limnocharis flava **(L.) Buch.**
Limnocharitaceae
 Limnocharis emarginata Humb. & Bonp., *Alisma flava* L. (see)
 ♦ yellow burrhead, burrhead, yellow sawah lettuce, taalapat ruesee
 ♦ Weed, Quarantine Weed, Noxious Weed, Naturalised
 ♦ 11, 68, 76, 86, 87, 88, 135, 170, 191, 203, 209, 220, 239, 255, 262, 263
 ♦ wpH, cultivated. Origin: Caribbean, Mexico to Paraguay.

Limnocharis humboldtii **(Rich.) Endl.**
Limnocharitaceae
 = *Hydrocleys nymphoides* (Humb. & Bonpl. ex Willd.) Buch.
 ♦ water poppy

 ♦ Weed, Quarantine Weed
 ♦ 76, 88
 ♦ wH.

Limnophila **R.Br. spp.**
Scrophulariaceae
 ♦ tamole vai, limnophila
 ♦ Quarantine Weed
 ♦ 220
 ♦ herbal.

Limnophila aromatica **(Lam.) Merr.**
Scrophulariaceae
 ♦ Asian limno, rau ngo, limnophila
 ♦ Weed
 ♦ 13, 87, 88
 ♦ aqua, cultivated, herbal. Origin: Australia.

Limnophila ceratophylloides **(Hiern) Skan**
Scrophulariaceae
 ♦ Native Weed
 ♦ 121
 ♦ wH. Origin: southern Africa.

Limnophila chinensis **(Osbeck) Merr.**
Scrophulariaceae
 ♦ Weed
 ♦ 262
 ♦ aH.

Limnophila chinensis **(Osbeck) Merr. ssp. aromatica (Lam.) T.Yamaz.**
Scrophulariaceae
 ♦ Weed
 ♦ 286
 ♦ Origin: Asia.

Limnophila conferta **Benth.**
Scrophulariaceae
 ♦ Weed, Quarantine Weed
 ♦ 76, 87, 88, 203

Limnophila dasyantha **Skan**
Scrophulariaceae
 ♦ Weed
 ♦ 87, 88

Limnophila erecta **Benth.**
Scrophulariaceae
 ♦ Weed
 ♦ 13, 88, 170, 191

Limnophila gratioloides **R.Br.**
Scrophulariaceae
 ♦ Weed
 ♦ 87, 88
 ♦ aqua, cultivated.

Limnophila heterophylla **Benth.**
Scrophulariaceae
 ♦ Weed, Quarantine Weed
 ♦ 76, 87, 88, 203
 ♦ aqua, cultivated.

Limnophila heterophylla **Benth. var. reflexa Hook.f.**
Scrophulariaceae
 ♦ saaraai chat
 ♦ Weed
 ♦ 239
 ♦ aqua.

Limnophila indica **(L.) Druce**
Scrophulariaceae
 ♦ Indian marshweed, Indian ambulia, ambulia

♦ Weed, Quarantine Weed,
Naturalised
♦ 76, 87, 88, 101, 197, 262, 286
♦ wH, cultivated, herbal.

Limnophila laotica **Bonati**
Scrophulariaceae
♦ naang rak tung
♦ Weed
♦ 238

Limnophila × ludoviciana **Thieret**
Scrophulariaceae
= *Limnophila indica* (L.) Druce ×
Limnophila sessiliflora (Vahl) Blume
♦ marshweed
♦ Weed
♦ 197
♦ aqua.

Limnophila micrantha **Benth.**
Scrophulariaceae
♦ Weed, Quarantine Weed
♦ 76, 87, 88, 203

Limnophila rugosa (Roth) **Merr.**
Scrophulariaceae
♦ Weed
♦ 13
♦ herbal.

Limnophila sessiliflora (Vahl) **Blume**
Scrophulariaceae
♦ Asian marshweed, dwarf ambulia,
ambulia, limnophila
♦ Weed, Quarantine Weed, Noxious
Weed, Naturalised
♦ 67, 68, 76, 84, 87, 88, 101, 140, 161,
179, 193, 197, 204, 229, 263, 286
♦ wpH, cultivated, toxic.

Limnophila villosa **Bl.**
Scrophulariaceae
Limnophila javanica A.DC.
♦ Weed
♦ 88, 170, 191

Limoniastrum monopetalum (L.) **Boiss.**
Plumbaginaceae
♦ Weed
♦ 221
♦ cultivated.

Limonium angustifolium (Tausch) **Turrill**
Plumbaginaceae
♦ Weed
♦ 221

Limonium arborescens (Brouss.) **Kuntze**
Plumbaginaceae
♦ tree limonium
♦ Naturalised
♦ 101
♦ pH, cultivated.

Limonium axillare (Forssk.) **Kuntze**
Plumbaginaceae
♦ Weed
♦ 221

Limonium bicolor (Bunge) **Kuntze**
Plumbaginaceae
♦ Weed
♦ 275, 297
♦ pH.

Limonium binervosum (Sm.) **Salmon**
Plumbaginaceae
♦ rock sea lavender

♦ Weed, Naturalised
♦ 86, 98, 203
♦ cultivated. Origin: Atlantic coastal
Europe.

Limonium bonduellii (F.Lestib.) **Kuntze**
Plumbaginaceae
♦ Naturalised
♦ 280

Limonium companyonis (Gren. & Billot) **Kuntze**
Plumbaginaceae
♦ sea lavender, Riviera sea lavender
♦ Weed, Naturalised, Environmental
Weed, Casual Alien
♦ 7, 72, 86, 88, 98, 198, 203, 280
♦ pH. Origin: southern Europe.

Limonium hyblaeum **Brullo**
Plumbaginaceae
♦ Sicilian sea lavender
♦ Naturalised
♦ 86, 198
♦ Origin: Sicily.

Limonium leptostachyum (Boiss.) **Kuntz**
Plumbaginaceae
♦ statice
♦ Naturalised
♦ 101

Limonium lobatum (L.f.) **Kuntze**
Plumbaginaceae
♦ winged sea lavender
♦ Weed, Naturalised
♦ 7, 86, 98, 198, 203
♦ Origin: Mediterranean.

Limonium myrianthum (Schrenk) **Kuntze**
Plumbaginaceae
♦ Weed, Naturalised
♦ 98, 203

Limonium otolepis (Schrenk) **Kuntze**
Plumbaginaceae
♦ saltmarsh sea lavender
♦ Weed, Naturalised, Cultivation
Escape
♦ 34, 86, 98, 101, 203, 252, 275
♦ pH, cultivated. Origin: temperate
Asia.

Limonium perezii (Stapf) **Hubb.**
Plumbaginaceae
♦ Perez's sea lavender, sea lavender,
statice, marsh rosemary
♦ Naturalised, Introduced
♦ 34, 101
♦ pH, cultivated, herbal.

Limonium pruinosum (L.) **Kuntze**
Plumbaginaceae
♦ Weed
♦ 221
♦ cultivated.

Limonium ramosissimum (Poir.) **Maire**
Plumbaginaceae
♦ malephora crocea
♦ Weed
♦ 35, 78, 88, 116

Limonium sinense (Girard) **Kuntze**
Plumbaginaceae
♦ Weed
♦ 275
♦ pH, cultivated.

Limonium sinuatum (L.) **Mill.**
Plumbaginaceae
♦ wavyleaf sea lavender, winged
sea lavender, notchleaf sea lavender,
perennial sea lavender, statice
♦ Weed, Naturalised, Garden Escape,
Environmental Weed, Casual Alien
♦ 7, 86, 98, 101, 116, 198, 203, 280
♦ pH, aqua, cultivated, herbal. Origin:
Mediterranean.

Limonium thouinii (Voiv.) **Kuntze**
Plumbaginaceae
♦ Weed, Naturalised
♦ 98, 203
♦ pH, cultivated.

Limonium tubiflorum (Delile) **Kuntze**
Plumbaginaceae
♦ Weed
♦ 221

Limosella aquatica **L.**
Scrophulariaceae
♦ water mudwort, mudwort,
mutayrtti, kitamisou
♦ Weed
♦ 39, 272
♦ aH, aqua, cultivated, herbal, toxic.

Linanthus androsaceus (Benth.) **Greene**
Polemoniaceae
♦ pellavikko, false babystars
♦ Cultivation Escape
♦ 42
♦ aH, cultivated, herbal.

Linaria **Mill. spp.**
Scrophulariaceae
♦ toadflax
♦ Weed, Noxious Weed, Naturalised
♦ 36, 54, 86, 88, 272

Linaria aeruginea (Gouan) **Cav.**
Scrophulariaceae
♦ roadside toadflax
♦ Naturalised
♦ 101
♦ herbal.

Linaria algarviana **Chav.**
Scrophulariaceae
♦ Weed
♦ 70

Linaria alpina (L.) **Mill.**
Scrophulariaceae
♦ alpine toadflax
♦ Casual Alien
♦ 280
♦ cultivated, herbal.

Linaria amethystea (Lam.) **Hoffm. & Link**
Scrophulariaceae
♦ ametistikannusruoho
♦ Cultivation Escape
♦ 42
♦ aH, cultivated.

Linaria angustissima (Loisel.) **Borbás**
Scrophulariaceae
Linaria italica Trev.
♦ Italian toadflax
♦ Weed, Naturalised
♦ 88, 94, 101, 272
♦ cultivated.

Linaria arvensis (L.) Desf.
Scrophulariaceae
♦ field linaria, corn toadflax
♦ Weed, Naturalised
♦ 15, 70, 86, 88, 94, 98, 198, 203, 243, 253, 272, 280
♦ Origin: Mediterranean, west Asia.

Linaria aucheri Boiss.
Scrophulariaceae
♦ Weed
♦ 87, 88

Linaria biebersteinii Bess.
Scrophulariaceae
♦ Weed
♦ 87, 88

Linaria bipartita Willd.
Scrophulariaceae
♦ clovenlip toadflax
♦ Weed, Naturalised, Cultivation Escape
♦ 34, 101, 287
♦ aH, cultivated, herbal.

Linaria canadensis (L.) Dum.Cours.
Scrophulariaceae
♦ oldfield toadflax, blue toadflax, smaller blue toadflax, toadflax, kanadankannusruoho
♦ Weed, Naturalised, Casual Alien
♦ 42, 87, 88, 161, 180, 199, 218, 237, 249, 286, 287
♦ a/pH, arid, promoted, herbal.

Linaria canadensis (L.) Dum.Cours. var. texana (Scheele) Pennell
Scrophulariaceae
Linaria texana Scheele (see)
♦ blue toadflax
♦ Naturalised
♦ 287
♦ a/pH.

Linaria chalepensis (L.) Mill.
Scrophulariaceae
♦ Weed
♦ 88, 94

Linaria commutata Bernh. ex Rchb.
Scrophulariaceae
♦ Weed
♦ 87, 88

Linaria corifolia Desf.
Scrophulariaceae
♦ Weed
♦ 87, 88

Linaria dalmatica (L.) Mill.
Scrophulariaceae
♦ Dalmatian toadflax, Balkan toadflax, broad leaved toadflax
♦ Weed, Quarantine Weed, Noxious Weed, Naturalised, Garden Escape, Environmental Weed
♦ 1, 23, 26, 40, 48, 76, 80, 87, 88, 98, 101, 130, 136, 147, 151, 162, 203, 218, 219, 220, 229, 267, 286, 287, 299
♦ herbal, toxic. Origin: Mediterranean.

Linaria dalmatica (L.) Mill. ssp. dalmatica
Scrophulariaceae
Linaria genistifolia (L.) Mill. ssp. *dalmatica* (L.) Maire & Petitm. (see)

♦ Dalmatian toadflax
♦ Weed, Noxious Weed, Naturalised, Introduced, Environmental Weed
♦ 21, 101, 139, 229
♦ pH, toxic. Origin: Mediterranean.

Linaria dalmatica (L.) Mill. ssp. macedonica (Griseb.) D.A.Sutton
Scrophulariaceae
♦ Dalmatian toadflax
♦ Noxious Weed, Naturalised
♦ 101, 229
♦ Origin: Mediterranean.

Linaria elatine (L.) Mill.
Scrophulariaceae
= *Kickxia elatine* (L.) Dumort.
♦ sharp pointed toadflax, sharppoint fluvellin
♦ Weed
♦ 44, 87, 88, 243
♦ herbal.

Linaria genistifolia (L.) Mill.
Scrophulariaceae
♦ Dalmatica toadflax, broomleaf toadflax, härmäkannusruoho
♦ Weed, Noxious Weed, Naturalised, Casual Alien
♦ 35, 42, 88, 98, 101, 105, 146, 161, 203, 272, 280
♦ cultivated, herbal. Origin: south-eastern Europe.

Linaria genistifolia (L.) Mill. ssp. dalmatica (L.) Maire & Petitm.
Scrophulariaceae
= *Linaria dalmatica* (L.) Mill. ssp. *dalmatica* (Griseb.) D.A.Sutton
♦ Dalmatian toadflax, broad leaved toadflax, toadflax, narrow leaved toadflax
♦ Weed, Noxious Weed, Naturalised, Environmental Weed
♦ 86, 138, 152, 212, 280
♦ pH, cultivated, herbal. Origin: south-eastern Europe.

Linaria haelava (Forssk.) Delile
Scrophulariaceae
♦ Weed
♦ 221

Linaria hirta (L.) Moench
Scrophulariaceae
♦ Weed
♦ 88, 94

Linaria incarnata (Vent.) Spreng.
Scrophulariaceae
♦ tarhakannusruoho, crimson toadflax
♦ Weed, Naturalised, Casual Alien
♦ 42, 86, 88, 94, 98, 101, 203
♦ Origin: south-west Europe, north-west Africa.

Linaria lamarckii Rouy
Scrophulariaceae
♦ Weed
♦ 70

Linaria latifolia Desf.
Scrophulariaceae
♦ Weed
♦ 87, 88, 94

Linaria maroccana Hook.f.
Scrophulariaceae

plants cultivated under this name are probably complex hybrids which differ from the wild Moroccan plants
♦ Moroccan toadflax, garden linaria, sirkoannusruoho, annual toadflax
♦ Weed, Naturalised, Cultivation Escape, Casual Alien
♦ 15, 34, 40, 42, 86, 98, 101, 161, 180, 203, 243, 252, 280, 287
♦ aH, cultivated, herbal. Origin: North Africa.

Linaria micrantha (Cav.) Hoffmans. & Link
Scrophulariaceae
♦ Weed
♦ 88, 94

Linaria minor (L.) Desf.
Scrophulariaceae
= *Chaenorrhinum minus* (L.) Lange
♦ small toadflax
♦ Weed
♦ 23, 44, 87, 88, 218, 243

Linaria munbyana Boiss. & Reut.
Scrophulariaceae
♦ etelänkannusruoho
♦ Casual Alien
♦ 42

Linaria odora (Bieb.) Fisch.
Scrophulariaceae
♦ tuoksukannusruoho
♦ Casual Alien
♦ 42

Linaria pelisseriana (L.) Mill.
Scrophulariaceae
♦ Jersey toadflax, Pelisser's toadflax
♦ Weed, Naturalised, Casual Alien
♦ 40, 86, 88, 94, 98, 198, 203, 280
♦ Origin: Mediterranean.

Linaria pinifolia (Poir.) Thell.
Scrophulariaceae
♦ pineneedle toadflax
♦ Weed, Naturalised, Cultivation Escape
♦ 34, 101
♦ aH, cultivated, herbal.

Linaria platycalyx Boiss.
Scrophulariaceae
♦ Casual Alien
♦ 280

Linaria purpurea (L.) Mill.
Scrophulariaceae
♦ purple toadflax, purple linaria, perennial snapdragon
♦ Weed, Naturalised, Introduced
♦ 15, 34, 40, 101, 165, 280
♦ pH, cultivated, herbal. Origin: Italy.

Linaria reflexa (L.) Desf.
Scrophulariaceae
♦ Weed
♦ 88, 94

Linaria repens (L.) Mill.
Scrophulariaceae
Antirrhinum repens L.
♦ pale toadflax, striped toadflax, juovakannusruoho, creeping toadflax
♦ Weed, Naturalised
♦ 42, 88, 94, 101, 243, 253, 280

♦ a/pH, cultivated, herbal.

Linaria reticulata (Sm.) Desf.
Scrophulariaceae
♦ purplenet toadflax
♦ Naturalised
♦ 101
♦ aH, herbal.

Linaria ricardoi Cout.
Scrophulariaceae
♦ Weed
♦ 88, 94

Linaria simplex (Willd.) DC.
Scrophulariaceae
Antirrhinum simplex Willd.
♦ Weed
♦ 253

Linaria spartea (L.) Chaz.
Scrophulariaceae
♦ ballast toadflax
♦ Weed, Naturalised
♦ 88, 94, 101, 250, 253

Linaria spuria (L.) Mill.
Scrophulariaceae
= *Kickxia spuria* (L.) Dumort.
♦ round leaved toadflax
♦ Weed
♦ 44, 87, 88, 243
♦ herbal.

Linaria supina (L.) Chaz.
Scrophulariaceae
Antirrhinum supinum L., *Linaria ambigua* E.Huet, *Linaria multicaulis* (L.) Mill., *Linaria masedae* Merino, *Linaria maritima* DC., *Linaria pyrenaica* DC., *Linaria haenseleri sensu* Merino *non* Boiss. & Reut.
♦ lesser butter and eggs, rentokannusruoho, prostrate toadflax
♦ Weed, Naturalised, Introduced, Casual Alien
♦ 34, 42, 87, 88, 94, 101, 163, 253
♦ aH, cultivated.

Linaria texana Scheele
Scrophulariaceae
= *Linaria canadensis* (L.) Dum.Cours. var. *texana* (Scheele) Pennell
♦ Weed, Naturalised
♦ 241, 295, 300
♦ herbal.

Linaria triornithophora (L.) Willd.
Scrophulariaceae
♦ three birds flying
♦ Casual Alien
♦ 280
♦ cultivated.

Linaria triphylla (L.) Mill.
Scrophulariaceae
♦ Weed, Quarantine Weed
♦ 76, 87, 88, 94, 203, 220
♦ cultivated.

Linaria vulgaris Mill.
Scrophulariaceae
♦ common toadflax, butter and eggs, yellow toadflax, ramsted, flaxweed, wild snapdragon, eggs and bacon, Jacob's ladder, toadflax
♦ Weed, Quarantine Weed, Noxious Weed, Naturalised, Introduced, Garden Escape, Environmental Weed, Cultivation Escape
♦ 1, 16, 21, 21, 23, 34, 39, 44, 52, 70, 76, 80, 86, 87, 88, 94, 98, 101, 121, 130, 136, 138, 139, 146, 155, 159, 161, 162, 174, 176, 195, 198, 203, 210, 211, 212, 218, 219, 229, 241, 243, 253, 264, 272, 275, 280, 286, 287, 297, 299, 300
♦ aH, cultivated, herbal, toxic. Origin: Eurasia.

Linaria vulgaris L. fo. leucantha Fernald
Scrophulariaceae
♦ Naturalised
♦ 287

Lindenbergia abyssinica Hochst. ex Benth.
Scrophulariaceae
♦ Weed
♦ 221

Lindenbergia sinaica (Decne.) Benth.
Scrophulariaceae
♦ Weed
♦ 221
♦ Origin: Middle East.

Lindera benzoin (L.) Blume
Lauraceae
♦ common spicebush, northern spicebush, spicebush
♦ Weed
♦ 39, 88, 218
♦ S, cultivated, herbal, toxic.

Lindera strychnifolia (Blume) Fern.-Vill.
Lauraceae
♦ Naturalised
♦ 287
♦ S, promoted, herbal. Origin: China.

Lindernia anagallidea (Michx.) Pennell
Scrophulariaceae
= *Lindernia dubia* (L.) Pennell var. *anagallidea* (Michx.) Cooperrider (NoR)
♦ Weed, Naturalised
♦ 286, 287
♦ herbal.

Lindernia anagallis (Burm.f.) Pennell
Scrophulariaceae
♦ heart vandellia
♦ Weed
♦ 13, 87, 88, 170, 191, 204, 273, 275, 286
♦ aH, aqua. Origin: Australia.

Lindernia anagallis (Burm.f.) Pennell var. verbenefolia Yamaz.
Scrophulariaceae
Illysanthes serrata Makino.
♦ Weed
♦ 274

Lindernia angustifolia (Benth.) Wettst.
Scrophulariaceae
♦ Weed, Quarantine Weed
♦ 76, 87, 88, 191, 203, 204, 220, 275, 286
♦ aH.

Lindernia antipoda (L.) Alston
Scrophulariaceae
Bonnaya antipoda (L.) Druce, *Bonnaya veronicifolia* (Retz.) Spreng., *Gratiola veronicifolia* Retz., *Ilysanthes antipoda* (L.) Merr., *Lindernia veronicifolia* (Retz.) F.Muell., *Ruellia antipoda* L., *Vandellia veronicifolia* (Retz.) Haines.
♦ sparrow false pimpernel
♦ Weed, Naturalised, Introduced
♦ 13, 88, 101, 170, 191, 230, 262, 273, 275, 286, 297
♦ aH, herbal.

Lindernia ciliata (Colsm.) Pennell
Scrophulariaceae
Bonnaya brachiata Link & Otto, *Gratiola ciliata* Colsm., *Ilysanthes ciliata* Kuntze (see), *Ilysanthes serrata* (Roxb.) Urb.
♦ fringed false pimpernel, phak hom ho paa
♦ Weed, Naturalised
♦ 13, 87, 88, 101, 170, 191, 239, 262
♦ pH. Origin: Australia.

Lindernia cordifolia (Colsm.) Merr.
Scrophulariaceae
♦ Weed, Quarantine Weed
♦ 76, 87, 88, 203, 220, 274
♦ herbal.

Lindernia crustacea (L.) F.Muell.
Scrophulariaceae
Capraria crustacea L., *Gratiola lucida* Vahl, *Lindernia minuta* Koord., *Mimulus javanicus* Bl., *Pyxidaria crustacea* (L.) Kuntze, *Torenia crustacea* (L.) Cham. & Schltdl., *Torenia minuta* Bl., *Vandellia bodinieri* H.Lév., *Vandellia crustacea* (L.) Benth. (see), *Vandellia minuta* Miq.
♦ Malaysian false pimpernel, yaa kaaphoi tua mia
♦ Weed, Naturalised, Introduced
♦ 12, 13, 28, 86, 86, 87, 88, 93, 153, 170, 179, 206, 209, 230, 239, 243, 262, 273, 274, 275, 276, 286, 297
♦ aH, herbal. Origin: Asia.

Lindernia diffusa (L.) Wettst.
Scrophulariaceae
♦ spreading false pimpernel
♦ Weed
♦ 87, 88, 261
♦ herbal.

Lindernia dubia (L.) Pennell
Scrophulariaceae
♦ false pimpernel, moistbank pimpernel, yellowseed false pimpernel, low false pimpernel
♦ Weed
♦ 68, 70, 87, 88, 253, 263
♦ aH, herbal.

Lindernia dubia (L.) Pennell ssp. major Pennell
Scrophulariaceae
♦ Weed, Naturalised
♦ 286, 287

Lindernia dubia (L.) Pennell ssp. typica Pennell
Scrophulariaceae
♦ Weed, Naturalised
♦ 286, 287

Lindernia dubia (L.) Pennell var. major (Pursh) Pennell
Scrophulariaceae
= *Lindernia dubia* (L.) Pennell var. *dubia* (NoR)
♦ low false pimpernel
♦ Weed
♦ 68

Lindernia hyssopioides (L.) Haines
Scrophulariaceae
Bonnaya hyssopioides (L.) Benth.,
Gratiola hyssopioides L., *Ilysanthes hyssopioides* (L.) Benth.
♦ Weed
♦ 87, 88, 170, 191
♦ aqua.

Lindernia micrantha D.Don
Scrophulariaceae
♦ azetogarashi
♦ Weed
♦ 68

Lindernia multiflora (Roxb.) Mukherjee
Scrophulariaceae
Lindernia glandulifera (Bl.) Back.
♦ Weed
♦ 13

Lindernia parviflora (Roxb.) Haines
Scrophulariaceae
♦ Weed
♦ 87, 88

Lindernia procumbens (Knock.) Borb.
Scrophulariaceae
Anagalloides procumbens Krock.,
Lindernia gratioloides sensu Hayek
non J.Lloyd, *Lindernia pyxidaria sensu*
Pennell. *non* L. (see), *Vandellia erecta*
Benth.
♦ prostrate false pimpernel
♦ Weed, Quarantine Weed, Naturalised
♦ 13, 68, 76, 87, 88, 101, 170, 191, 203, 204, 220, 262, 263, 272, 275, 286, 297
♦ aH, herbal.

Lindernia pusilla (Thumb.) Merr.
Scrophulariaceae
♦ Introduced
♦ 230
♦ H.

Lindernia pyxidaria sensu Pennell. non L.
Scrophulariaceae
= *Lindernia procumbens* (Knock.) Borb.
♦ false pimpernel
♦ Weed
♦ 88, 191, 204, 274

Lindernia rotundifolia (L.) Standl. & L.O.Williams
Scrophulariaceae
♦ Naturalised
♦ 32
♦ aqua, cultivated.

Lindernia ruellioides (Colsm.) Pennell
Scrophulariaceae
Bonnaya reptans (Roxb.) Spreng.,
Gratiola reptans Roxb., *Gratiola ruellioides* Colsm., *Ilysanthes ruellioides* (Colsm.) Kuntze
♦ Weed
♦ 87, 88
♦ cultivated.

Lindernia sessiliflora (Bth.) Wettst.
Scrophulariaceae
♦ Weed
♦ 13

Lindernia viscosa (Hornem.) Bold
Scrophulariaceae

♦ Weed
♦ 13

Linociera intermedia Wight
Oleaceae
♦ Weed
♦ 22, 80

Linum L. spp.
Linaceae
♦ flax
♦ Weed
♦ 272
♦ herbal, toxic.

Linum anatolicum Boiss.
Linaceae
♦ Weed
♦ 87, 88

Linum austriacum L.
Linaceae
♦ Asian flax
♦ Weed
♦ 87, 88, 272
♦ cultivated.

Linum bienne Mill.
Linaceae
♦ narrowleaf flax, pale flax
♦ Weed, Naturalised, Environmental Weed
♦ 15, 34, 54, 86, 88, 101, 165, 176, 280
♦ aH, cultivated, herbal. Origin: western Europe, Mediterranean, Middle East.

Linum catharticum L.
Linaceae
Cathartolinum pratense Reich.
♦ fairy flax, purging flax, white flax, ahopellava
♦ Weed, Naturalised, Garden Escape, Environmental Weed
♦ 15, 39, 86, 101, 176, 272, 280, 300
♦ aH, cultivated, herbal, toxic. Origin: Eurasia.

Linum flavum L.
Linaceae
♦ golden flax
♦ Weed, Naturalised
♦ 87, 88, 287
♦ cultivated, herbal.

Linum grandiflorum Desf.
Linaceae
♦ garden flax, flowering flax, punapellava, flax, scarlet flax
♦ Weed, Naturalised, Cultivation Escape, Casual Alien
♦ 34, 42, 101
♦ aH, cultivated, herbal.

Linum hirsutum L.
Linaceae
♦ hairy flax
♦ Weed
♦ 272
♦ cultivated, herbal.

Linum hologynum Rchb.
Linaceae
♦ Weed
♦ 272

Linum humile Mill.
Linaceae

= *Linum usitatissimum* L.
♦ Weed
♦ 221
♦ aH, promoted.

Linum lewisii Pursh
Linaceae
= *Linum perenne* L.
♦ prairie flax, western blue flax
♦ Weed
♦ 161
♦ pH, cultivated, herbal, toxic.

Linum marginale A.Cunn.
Linaceae
♦ native flax
♦ Naturalised
♦ 39, 134
♦ pH, cultivated, toxic.

Linum nelsonii Rose
Linaceae
♦ Weed
♦ 157
♦ pH.

Linum neomexicanum Greene
Linaceae
♦ New Mexico flax, yellow pine flax
♦ Weed
♦ 39, 87, 88, 161, 218
♦ herbal, toxic.

Linum nodiflorum L.
Linaceae
♦ Weed
♦ 272

Linum perenne L.
Linaceae
Linum lewisii Pursh (see)
♦ blue flax, sinipellava, perennial flax
♦ Weed, Naturalised, Casual Alien
♦ 42, 101, 272
♦ pH, cultivated, herbal.

Linum peyroni Post
Linaceae
♦ Weed
♦ 87, 88

Linum rigidum Pursh
Linaceae
♦ stiffstem flax, yellow flax
♦ Weed
♦ 39, 161
♦ herbal, toxic.

Linum rzedowskii Arreguín
Linaceae
♦ Weed
♦ 199

Linum striatum Walt.
Linaceae
♦ ridged yellow flax
♦ Naturalised
♦ 287
♦ herbal.

Linum strictum L.
Linaceae
♦ upright yellow flax
♦ Weed, Naturalised
♦ 98, 203, 272
♦ aH, cultivated, herbal.

Linum strictum L. ssp. strictum
Linaceae

♦ upright yellow flax
♦ Naturalised
♦ 86, 198
♦ Origin: Mediterranean, Middle East.

Linum tenuifolium L.
Linaceae
Cathartolinum tenuifolium Reichb.,
Linum cilicicum Fenzl
♦ white flax, narrow leaved flax
♦ Weed
♦ 272
♦ cultivated, herbal.

Linum trigynum L.
Linaceae
♦ yellow flax, French flax, lino spinato
♦ Weed, Naturalised, Garden Escape,
Environmental Weed
♦ 7, 9, 15, 70, 86, 98, 101, 134, 165, 176,
198, 203, 269, 272, 280
♦ aH, cultivated, herbal. Origin:
Eurasia.

Linum usitatissimum L.
Linaceae
Linum humile Mill. (see)
♦ common flax, lino, linseed, field flax,
flax
♦ Weed, Naturalised, Casual Alien
♦ 7, 39, 40, 42, 86, 87, 88, 98, 101, 161,
176, 179, 198, 199, 203, 218, 241, 247,
269, 280, 287, 295, 300
♦ aH, cultivated, herbal, toxic. Origin:
obscure, possibly Europe.

Linum usitatissimum L. var. bienne Mill.
Linaceae
♦ Naturalised
♦ 287

Linum virginianum L.
Linaceae
♦ yellow flax, woodland flax
♦ Weed, Naturalised
♦ 286, 287
♦ cultivated, herbal.

Liparis odorata Lindl.
Orchidaceae
♦ Introduced
♦ 230
♦ H, herbal.

Lipocarpha aristulata (Coville) G.Tucker
Cyperaceae
Lipocarpha microcephala (R.Br.) Kunth
(see)
♦ awned halfchaff sedge
♦ Weed
♦ 179
♦ aG.

Lipocarpha chinensis (Osb.) Kern
Cyperaceae
Cyperus lipocarpha T.Koyama, *Cyperus
submaculatus* Koyama, *Hypaelyptum
albidum* Willd. ex Kunth, *Hypaelyptum
argenteum* Vahl, *Hypaelyptum
senegalense* (Lam.) K.Schum.,
Hypolytrum argenteum (Vahl) Kunth,
Hypolytrum laevigatum (Roxb.) Spreng.,
Hypolytrum senegalemse (Lam.) Rich.
ex Pers., *Kyllinga albescens* Steud.,
Lipocarpha argentea (Vahl) R.Br. ex
Nees, *Lipocarpha chinensis* (Osbeck)

Tang & Wang, *Lipocarpha debilis* Ridl.,
Lipocarpha laevigata (Roxb.) Nees,
Lipocarpha senegalensis (Lam.) Dandy,
Lipocarpha senegalensis (Lam.) T.Durand
& H.Durand, *Schoenus laevigatus* Roxb.
ex Nees, *Scirpus chinensis* Osbeck,
Scirpus senegalensis Lam., *Scirpus
squarrosus* L., *Tunga laevigata* Roxb.
♦ goose tongue sedge
♦ Weed, Native Weed
♦ 87, 88, 121, 126, 170, 191
♦ wpG. Origin: Madagascar.

Lipocarpha microcephala (R.Br.) Kunth
Cyperaceae
= *Lipocarpha aristulata* (Cov.) G.Tucker
♦ smallhead halfchaff sedge
♦ Weed
♦ 87, 88, 286
♦ G, cultivated.

Lipocarpha rehmanii (Ridl.) Goetgh.
Cyperaceae
♦ Rehman's halfchaff sedge
♦ Naturalised
♦ 101
♦ G.

**Lippia alba (Mill.) N.E.Br. ex Britt. &
Wilson**
Verbenaceae
Lippia geminata Kunth (see)
♦ bushy lippia, anise verbena
♦ Weed
♦ 87, 88, 93, 179, 191, 255
♦ pH, arid, herbal. Origin: Brazil.

**Lippia alba (Mill.) N.E.Br. ex Britt. &
P.Wilson var. alba**
Verbenaceae
♦ Naturalised
♦ 86
♦ Origin: Central and South America.

Lippia aristata Schauer
Verbenaceae
= *Lantana aristata* (Schauer) Briq. (NoR)
♦ Introduced
♦ 228
♦ arid.

Lippia asperifolia Rich.
Verbenaceae
♦ Weed
♦ 87, 88

Lippia cuneifolia (Torr.) Steud.
Verbenaceae
= *Phyla cuneifolia* (Torr.) E.Greene
♦ wedgeleaf fogfruit
♦ Weed
♦ 49, 87, 88, 218
♦ herbal.

Lippia filiformis Schrad.
Verbenaceae
♦ Weed
♦ 87, 88

Lippia geminata Kunth
Verbenaceae
= *Lippia alba* (Mill.) N.E.Br. ex Britt. &
Wilson
♦ Weed
♦ 87, 88
♦ herbal.

Lippia javanica (Burm.f.) Spreng.
Verbenaceae
♦ common lippia, fever tea, fever tree,
wild sage, wild tea
♦ Weed, Native Weed
♦ 87, 88, 121
♦ pS, toxic. Origin: east and southern
Africa.

Lippia nodiflora (L.) Michx.
Verbenaceae
= *Phyla nodiflora* (L.) Greene
♦ mat lippia
♦ Weed
♦ 13, 88, 218, 221, 243, 286
♦ bH, arid, cultivated, herbal.

Lippia rehmannii H.Pearson
Verbenaceae
♦ Rehmann lippia
♦ Weed, Native Weed
♦ 39, 121
♦ pS, toxic. Origin: southern Africa.

Lippia reptans Kunth
Verbenaceae
= *Phyla nodiflora* (L.) Greene
♦ Weed
♦ 87, 88

Lippia scaberrima Sond.
Verbenaceae
♦ beukessboss
♦ Native Weed
♦ 121
♦ pS, herbal, toxic. Origin: southern
Africa.

Lippia strigulosa Martens & Galeotti
Verbenaceae
= *Phyla strigulosa* (Mart. & Gal.)
Moldenke var. *strigulosa* (NoR)
♦ Weed
♦ 157
♦ pH.

Lippia turbinata Griseb.
Verbenaceae
♦ poleo
♦ Weed
♦ 295

Lippia ukambensis Vatke
Verbenaceae
♦ Weed
♦ 87, 88

Liquidambar styraciflua L.
Hamamelidaceae/Altingiaceae
♦ sweet gum, liquidamber
♦ Weed, Naturalised, Casual Alien
♦ 87, 88, 198, 218, 280
♦ T, cultivated, herbal.

Liriodendron tulipifera L.
Magnoliaceae
♦ tuliptree
♦ Weed, Casual Alien
♦ 39, 87, 88, 218, 280
♦ T, cultivated, herbal, toxic.

Liriope muscari (Decne.) Bailey
Liliaceae/Convallariaceae
Liriope platyphylla F.Wang & T.Tang
♦ big blue lilyturf, lilyturf, yaburan
♦ Naturalised
♦ 101
♦ pH, cultivated, herbal.

Liriope spicata (Thunb.) Lour.
Liliaceae/Convallariaceae
Liriope spicatum Lour. (see)
♦ creeping lilyturf, lilyturf
♦ Weed
♦ 80
♦ pH, cultivated, herbal.

Liriope spicatum Lour.
Liliaceae/Convallariaceae
= *Liriope spicata* (Thunb.) Lour.
♦ creeping lilyturf
♦ Naturalised
♦ 101

Lisaea heterocarpa (DC.) Boiss.
Apiaceae
♦ Weed
♦ 87, 88

Lisaea syriaca Boiss.
Apiaceae
♦ Weed
♦ 87, 88

Lisianthus alatus Aubl.
Gentianaceae
= *Irlbachia alata* (Aubl.) Maas ssp. *alata*
(NoR)
♦ Weed
♦ 87, 88

Lithocarpus densiflorus (Hook. & Arn.) Rehd.
Fagaceae
Pasania densiflora Oerst., *Quercus densiflora* Hook. & Arn.
♦ tanoak, tanbark oak
♦ Weed
♦ 87, 88, 218
♦ S, cultivated, herbal.

Lithocarpus densiflorus (Hook. & Arn.) Rehder var. echinoides (R.Br. Campst.) Abrams
Fagaceae
♦ scrub tanoak, serpentine bush tanoak, tanoak
♦ Weed
♦ 218
♦ S.

Lithospermum L. spp.
Boraginaceae
♦ stoneseed
♦ Weed
♦ 272
♦ toxic.

Lithospermum arvense L.
Boraginaceae
= *Buglossoides arvensis* (L.) I.M.Johnst.
♦ corn gromwell, bastard alkanet, gromwell, red root, wheat thief, stoneseed, puccoon, pigeon weed, field gromwell, peltorusojuuri
♦ Weed, Naturalised, Introduced
♦ 23, 34, 44, 87, 98, 121, 161, 203, 207, 212, 218, 237, 243, 275, 280, 286, 287, 295
♦ a/bH, cultivated, herbal. Origin: Eurasia.

Lithospermum erythrorhizon Siebold & Zucc.
Boraginaceae
♦ lithospermum

♦ Quarantine Weed
♦ 220
♦ pH, promoted, herbal.

Lithospermum officinale L.
Boraginaceae
Margarospermum officinale Dec.
♦ gromwell, European stoneseed, rohtorusojuuri, pearl gromwell
♦ Weed, Naturalised
♦ 42, 86, 87, 88, 101, 121, 161, 176, 218, 237, 272, 287
♦ pH, cultivated, herbal. Origin: Eurasia.

Lithospermum purpurocaeruleum L.
Boraginaceae
♦ blue gromwell, purple gromwell
♦ Naturalised
♦ 280

Lithospermum ruderale Douglas
Boraginaceae
♦ western gromwell, Columbia puccoon
♦ Weed
♦ 23, 87, 88, 218
♦ pH, promoted, herbal.

Lithospermum strictum Lehm.
Boraginaceae
♦ Weed
♦ 199

Lithospermum zollingera A.DC.
Boraginaceae
♦ Weed
♦ 87, 88

Lithrea molleoides (Vell.) Engl.
Anacardiaceae
♦ aroeira blanca
♦ Naturalised
♦ 101
♦ cultivated.

Litsea Lam. spp.
Lauraceae
♦ litsea
♦ Weed
♦ 18, 88

Litsea cubena (Lour.) Pers.
Lauraceae
Litsea citrata Bl.
♦ Weed
♦ 87, 88

Litsea glutinosa (Lour.) C.B.Rob.
Lauraceae
Litsea sebifera Pers.
♦ Indian laurel, Indiese lourier
♦ Weed, Noxious Weed, Garden Escape
♦ 22, 63, 87, 88, 95, 283
♦ S/T, cultivated, herbal, toxic. Origin: tropical Asia.

Litsea glutinosa (Lour.) C.B.Rob. var. brideliifolia (Hayata) Merr.
Lauraceae
♦ Indian laurel
♦ Weed
♦ 121
♦ S/T, toxic. Origin: Eurasia.

Litsea monopetala (Roxb.) Pers.
Lauraceae

♦ Weed
♦ 22
♦ T, herbal.

Litwinowia tenuissima (Pall.) Woronow
Brassicaceae
♦ Weed
♦ 272

Livistona australis (R.Br.) Mart.
Arecaceae
Livistona inermis R.Br., *Corypha australis* R.Br.
♦ cabbage tree palm
♦ Weed, Naturalised
♦ 15, 280
♦ T, cultivated, herbal. Origin: Australia.

Livistona carinensis (Chiov.) J.Dransf.
Arecaceae
Hyphaene carinensis Chiov., *Wissmannia carinensis* (Chiov.) Burret
♦ Introduced
♦ 228
♦ arid.

Livistona chinensis (Jacq.) R.Br. ex Mart.
Arecaceae
♦ fountain palm, Chinese fan palm, Chinese fan
♦ Weed, Naturalised, Environmental Weed
♦ 101, 179
♦ cultivated, herbal.

Livistona rotundifolia (Lam.) Mart.
Arecaceae
♦ footstool palm, footstool
♦ Introduced
♦ 230
♦ cultivated.

Lobelia affinis Wall.
Campanulaceae/Lobeliaceae
♦ Weed, Quarantine Weed
♦ 87, 88, 220

Lobelia angulata Forst.
Campanulaceae/Lobeliaceae
♦ Weed
♦ 87, 88
♦ herbal.

Lobelia berlandieri A.DC.
Campanulaceae/Lobeliaceae
♦ Berland's cardinalflower, Berlandier's lobelia
♦ Weed
♦ 39, 161
♦ herbal, toxic.

Lobelia cardinalis L.
Campanulaceae/Lobeliaceae
♦ cardinalflower, propinqua cardinalflower, scarlet lobelia, Indian pink
♦ Weed, Quarantine Weed
♦ 8, 23, 39, 87, 88, 161, 218, 220
♦ pH, aqua, cultivated, herbal, toxic.

Lobelia chinensis Lour.
Campanulaceae/Lobeliaceae
Lobelia radicans Thunb. (see)
♦ Chinese lobelia
♦ Weed, Quarantine Weed, Naturalised

♦ 76, 87, 88, 101, 203, 204, 220, 243, 263, 273, 275, 286, 297
♦ pH, herbal.

Lobelia cliffortiana L.
Campanulaceae/Lobeliaceae
♦ cardenala azul
♦ Weed, Quarantine Weed
♦ 14, 76, 87, 88, 203, 220
♦ herbal.

Lobelia comosa Cav.
Campanulaceae/Lobeliaceae
= *Siphocampylus comosus* (Cav.) G.Don (NoR)
♦ Quarantine Weed
♦ 220
♦ cultivated.

Lobelia coronopifolia L.
Campanulaceae/Lobeliaceae
♦ Quarantine Weed
♦ 220
♦ cultivated.

Lobelia dortmanna L.
Campanulaceae/Lobeliaceae
♦ water lobelia, nuottaruoho, Dortmann's cardinalflower
♦ Weed
♦ 23, 88, 161
♦ pH, aqua, cultivated, herbal, toxic.

Lobelia erinus L.
Campanulaceae/Lobeliaceae
♦ lobelia, edging lobelia, sinilobelia, fountains lobelia
♦ Weed, Naturalised, Cultivation Escape
♦ 15, 40, 42, 54, 86, 88, 98, 101, 198, 203, 252, 280
♦ a/pH, cultivated, herbal. Origin: South Africa.

Lobelia erinus L. var. erinus
Campanulaceae/Lobeliaceae
♦ edging lobelia, garden lobelia, wild lobelia
♦ Native Weed
♦ 121
♦ aH. Origin: southern Africa.

Lobelia fenestralis Cav.
Campanulaceae/Lobeliaceae
♦ fringeleaf lobelia
♦ Weed
♦ 199
♦ herbal.

Lobelia × gerardii Sauv.
Campanulaceae/Lobeliaceae
= *Lobelia cardinais* L. × *Lobelia siphilitica* L.
♦ hybridlobelia, tarhalobelia
♦ Casual Alien
♦ 280
♦ herbal.

Lobelia gruina Cav. var. gruina
Campanulaceae/Lobeliaceae
♦ Weed
♦ 199

Lobelia inflata L.
Campanulaceae/Lobeliaceae
♦ Indian tobacco, lobelia
♦ Weed, Naturalised

♦ 8, 23, 39, 87, 88, 161, 218, 287
♦ aH, cultivated, herbal, toxic.

Lobelia kalmii L.
Campanulaceae/Lobeliaceae
♦ kalminlobelia, Ontario lobelia
♦ Casual Alien
♦ 39, 42
♦ herbal, toxic.

Lobelia laxiflora Kunth
Campanulaceae/Lobeliaceae
♦ Sierra Madre lobelia, Mexican cardinalflower
♦ Weed, Naturalised, Garden Escape
♦ 86, 157, 161
♦ pH, cultivated, herbal, toxic. Origin: Central and South America.

Lobelia laxiflora Kunth var. angustifolia A.DC.
Campanulaceae/Lobeliaceae
♦ Sierra Madre lobelia
♦ Weed
♦ 199

Lobelia nuda Hemsl.
Campanulaceae/Lobeliaceae
Lobelia natalensis E.Wimm.
♦ Native Weed
♦ 121
♦ aH. Origin: southern Africa.

Lobelia pinifolia L.
Campanulaceae/Lobeliaceae
♦ Quarantine Weed
♦ 220
♦ cultivated, herbal.

Lobelia pratioides Benth.
Campanulaceae/Lobeliaceae
♦ Weed
♦ 39, 87, 88
♦ toxic.

Lobelia purpurascens R.Br.
Campanulaceae/Lobeliaceae
♦ Weed
♦ 39, 87, 88
♦ herbal, toxic.

Lobelia radicans Thunb.
Campanulaceae/Lobeliaceae
= *Lobelia chinensis* Lour.
♦ Chinese lobelia
♦ Weed, Quarantine Weed
♦ 76, 88, 274
♦ pH, promoted, herbal.

Lobelia siphilitica L.
Campanulaceae/Lobeliaceae
♦ great lobelia, blue cardinalflower
♦ Weed, Quarantine Weed
♦ 8, 23, 39, 87, 88, 161, 218, 220
♦ pH, cultivated, herbal, toxic.

Lobelia spicata Lam.
Campanulaceae/Lobeliaceae
♦ pale spike lobelia, pale spike
♦ Weed
♦ 8, 39, 161
♦ a/pH, cultivated, herbal, toxic.

Lobelia tomentosa L.f.
Campanulaceae/Lobeliaceae
♦ Quarantine Weed
♦ 220

Lobelia trigona Roxb.
Campanulaceae/Lobeliaceae
♦ Weed, Quarantine Weed
♦ 87, 88, 220

Lobelia trinitensis Griseb.
Campanulaceae/Lobeliaceae
♦ Weed, Quarantine Weed
♦ 87, 88, 220

Lobularia arabica (Boiss.) Muschl.
Brassicaceae
♦ Weed
♦ 221

Lobularia libyca (Viv.) Webb & Berthel.
Brassicaceae
♦ Weed
♦ 221

Lobularia maritima (L.) Desv.
Brassicaceae
Alyssum maritimum (L.) Lam (see), *Alyssum minimum* L., *Clypeola maritima* L.
♦ sweet Alison, sea alyssum, alyssum
♦ Weed, Naturalised, Introduced, Garden Escape, Environmental Weed, Cultivation Escape
♦ 7, 15, 34, 40, 42, 56, 86, 88, 94, 98, 101, 116, 121, 134, 161, 165, 176, 179, 198, 203, 228, 241, 280, 295, 300
♦ pH, arid, cultivated, herbal. Origin: Eurasia.

Lochnera pusilla K.Schum.
Apocynaceae
♦ Weed, Quarantine Weed
♦ 76, 87, 88, 203, 220

Loeflingia hispanica L.
Caryophyllaceae
♦ espanjanruoho
♦ Casual Alien
♦ 42

Loeselia ciliata L.
Polemoniaceae
♦ Weed
♦ 157
♦ pH.

Loeselia mexicana (Lam.) Brand
Polemoniaceae
♦ Mexican false calico
♦ Weed
♦ 199
♦ herbal.

Logfia arvensis (L.) Holub
Asteraceae
Filago arvensis L. (see)
♦ least cudweed, field cottonrose
♦ Weed, Naturalised
♦ 70, 88, 94, 101, 253, 272
♦ herbal.

Logfia gallica (L.) Coss. & Germ.
Asteraceae
Filago gallica L. (see)
♦ narrowleaf cottonrose, narrow leaved cudweed, logfia
♦ Weed, Naturalised, Environmental Weed
♦ 86, 88, 94, 98, 101, 176, 198, 203, 250, 253, 280, 300
♦ Origin: Europe.

Logfia minima (Sm.) Dumort.
Asteraceae
Filago minima (Sm.) Pers. (see)
♦ small cudweed, little cottonrose
♦ Weed, Naturalised
♦ 70, 88, 94, 253, 272, 280

Lola lubrica Setch. & Gard.
Cladophoraceae
♦ green alga
♦ Weed
♦ 197
♦ algae.

Lolium L. spp.
Poaceae
♦ ryegrass
♦ Weed, Naturalised, Environmental Weed
♦ 39, 88, 181, 198, 203, 221, 243, 290
♦ G, toxic.

Lolium arundinaceum (Schreb.) S.J.Darbysh.
Poaceae
Bromus arundinaceus (Schreb.) Roth, *Festuca arundinacea* Schreb. (see), *Festuca elatior* L. (see), *Festuca elatior* ssp. *arundinacea* (Schreb.), *Festuca elatior* var. *arundinacea* (Schreb.) Wimm., *Schedonorus arundinaceus* (Schreb.) Dumort.
♦ tall fescue, reed fescue, Alta fescue, coarse fescue
♦ Naturalised, Introduced
♦ 101, 174
♦ G.

Lolium giganteum (L.) S.J.Darbysh.
Poaceae
Festuca gigantea (L.) Vill. (see)
♦ giant fescue
♦ Naturalised
♦ 101
♦ G.

Lolium × hubbardii Jansen & Wacht. ex B.K.Simon
Poaceae
= *Lolium multiflorum* Lam. × *Lolium rigidum* Gaud.
♦ Naturalised
♦ 86, 98
♦ G. Origin: Queensland.

Lolium × hybridum Hausskn.
Poaceae
= *Lolium multiflorum* Lam. × *Lolium perenne* L.
♦ hybrid ryegrass
♦ Naturalised
♦ 86, 98, 198, 287
♦ G, cultivated. Origin: Europe.

Lolium loliaceum (Bory & Chaub.) Hand-Mazz.
Poaceae
Lolium rigidum Gaudin var. *loliaceum* (Bory & Chaub.) Hal., *Lolium rigidum* Gaudin var. *rottboellioides* Heldr. ex Boiss. (see), *Rottboellia loliacea* Bory & Chaub.
♦ darnel, Swiss ryegrass, stiff ryegrass, rigid ryegrass
♦ Weed, Naturalised

♦ 86, 93, 98, 121, 176, 198, 203, 269
♦ aG, arid, cultivated. Origin: Eurasia.

Lolium multiflorum Lam.
Poaceae
Lolium gaudinii Parl. *nom. inval.*, *Lolium italicum* A.Braun. *nom. inval.*, *Lolium perenne* L. ssp. *multiflorum* (Lam.) Husn. (see), *Lolium scabrum* J.S.Presl ex C.Presl
♦ Italian ryegrass, annual ryegrass, Australian ryegrass, ryegrass, westerwolds ryegrass, ivraie multiflore, raygrass d'Italie, italienisches raygras, welsches weidelgras, westerwoldisches weidelgras, azevém, ballico Italiano, raygras Italiano
♦ Weed, Naturalised, Introduced, Environmental Weed, Cultivation Escape
♦ 7, 9, 15, 34, 38, 39, 40, 42, 44, 51, 68, 70, 72, 86, 87, 88, 91, 98, 111, 118, 121, 146, 158, 161, 176, 180, 185, 198, 199, 203, 211, 212, 218, 221, 228, 236, 237, 241, 243, 245, 250, 252, 253, 255, 256, 263, 272, 275, 280, 286, 287, 295, 300
♦ a/bG, arid, cultivated, herbal, toxic. Origin: Eurasia.

Lolium multiflorum Lam. fo. ramosum Guss.
Poaceae
♦ Naturalised
♦ 287
♦ G.

Lolium multiflorum Lam. × rigidum Gaud.
Poaceae
♦ Naturalised
♦ 86, 198
♦ G.

Lolium perenne L.
Poaceae
Festuca anglica E.H.L.Krause, *Lolium cristatum* Pers, *Lolium vulgare* Host, *Lolium linicolum* A.Br.
♦ English ryegrass, Italian ryegrass, perennial ryegrass, raigrás perenne, ryegrass perenne, ryegrass, loglio comune, loglio perenne, Englanninraiheinä, lyme grass, strand wheat
♦ Weed, Naturalised, Introduced, Garden Escape, Environmental Weed, Cultivation Escape
♦ 7, 15, 21, 34, 38, 39, 42, 44, 68, 70, 72, 86, 87, 88, 91, 93, 98, 101, 111, 118, 121, 134, 146, 158, 161, 176, 179, 181, 185, 195, 198, 199, 203, 205, 221, 225, 228, 236, 241, 243, 246, 253, 263, 271, 272, 280, 286, 287, 296, 297, 300
♦ pG, arid, cultivated, herbal, toxic. Origin: Eurasia.

Lolium perenne L. ssp. multiflorum (Lam.) Husn.
Poaceae
= *Lolium multiflorum* Lam.
♦ Italian ryegrass
♦ Naturalised
♦ 101
♦ G.

Lolium perenne L. ssp. perenne
Poaceae
Lolium perenne L. var. *cristatum* (Pers. ex B.D.Jacks.) (see)
♦ perennial ryegrass, English ryegrass
♦ Weed, Naturalised
♦ 101, 180
♦ G.

Lolium perenne L. var. cristatum Pers. ex B.D.Jacks.
Poaceae
= *Lolium perenne* L. ssp. *perenne*
♦ perennial ryegrass
♦ Naturalised
♦ 86, 198
♦ G, cultivated. Origin: Europe, Mediterranean, temperate Asia.

Lolium perenne L. var. perenne
Poaceae
♦ perennial ryegrass
♦ Naturalised
♦ 86, 198
♦ G, cultivated. Origin: Europe, Mediterranean, temperate Asia.

Lolium perenne L. × rigidum Gaud.
Poaceae
♦ Naturalised
♦ 86, 98, 198
♦ G. Origin: Mediterranean.

Lolium persicum Boiss & Hohen.
Poaceae
♦ Persian ryegrass, Persian darnel
♦ Weed, Noxious Weed, Naturalised, Introduced, Casual Alien
♦ 36, 42, 68, 87, 88, 91, 101, 161, 162, 228, 256, 299
♦ G, aqua.

Lolium pratense (Huds.) S.J.Darbysh.
Poaceae
= *Festuca pratensis* Huds.
♦ meadow ryegrass
♦ Naturalised
♦ 101
♦ G.

Lolium remotum Sch.
Poaceae
Lolium linicola A.Braun
♦ hardy ryegrass, pellavaraiheinä
♦ Weed, Naturalised
♦ 23, 30, 44, 70, 87, 88, 91, 98, 203, 272, 280
♦ G, cultivated, herbal. Origin: Europe, west Asia.

Lolium rigidum Gaud.
Poaceae
Lolium perenne L. ssp. *rigidum* (Gaudin) Á.& D.Löve, *Lolium rigidum* Gaudin ssp. *lepturoides* (Boiss.) (see), *Lolium strictum* auct. non J.Presl, *Lolium subulatum* Vis. (see)
♦ annual ryegrass, ryegrass, Wimmera ryegrass, Swiss ryegrass, rigid ryegrass
♦ Weed, Naturalised, Introduced, Environmental Weed, Casual Alien
♦ 7, 9, 23, 30, 39, 42, 68, 70, 72, 86, 87, 88, 91, 93, 98, 101, 111, 115, 176, 198, 203, 205, 221, 228, 243, 250, 253, 269, 272, 280, 286, 287, 300

- aG, arid, cultivated, toxic. Origin: Mediterranean to India.

Lolium rigidum Gaudin ssp. *lepturoides* (Boiss.) Senn & Maur.
Poaceae
= *Lolium rigidum* Gaud.
- Wimmera ryegrass
- Naturalised
- 241
- G.

Lolium rigidum Gaud. var. *rigidum*
Poaceae
- Naturalised
- 134
- G.

Lolium rigidum Gaud. var. *rottboellioides* Heldr. ex Boiss.
Poaceae
= *Lolium loliaceum* (Bory & Chaub.) Hand-Mazz.
- Naturalised
- 134
- G.

Lolium subulatum Vis.
Poaceae
= *Lolium rigidum* Gaud.
- Naturalised
- 241, 287
- G.

Lolium temulentum L.
Poaceae
Bromus temulentus Bernh., *Craepalia temulenta* Schrank, *Lolium annuum* Gilib., *Lolium temulentum* fo. *arvense* (With.) Junge (see), *Lolium temulentum* L. var. *arvense* (With.) Lilj. (see), *Lolium temulentum* L. var. *leptochaeton* A.Br. (see)
- darnel, poison ryegrass, bearded ryegrass, annual darnel, myrkkyraiheinä, cheat, darnel ryegrass, dragge, drawke, drunk, poison darnel, poison raygrass, sturdy ryle, Virginian oat
- Weed, Naturalised, Introduced, Garden Escape, Environmental Weed, Casual Alien
- 7, 23, 30, 34, 39, 40, 42, 44, 51, 70, 72, 86, 87, 88, 91, 98, 101, 106, 121, 158, 161, 176, 185, 186, 198, 203, 210, 218, 221, 228, 237, 241, 243, 253, 256, 269, 272, 275, 280, 286, 287, 295, 297, 300
- aG, arid, cultivated, herbal, toxic. Origin: Mediterranean.

Lolium temulentum L. fo. *arvense* (With.) Junge
Poaceae
= *Lolium temulentum* L.
- Weed, Naturalised, Introduced
- 38, 176, 185
- G.

Lolium temulentum L. fo. *leptochaeton* A.Br.
Poaceae
- Weed
- 185
- G.

Lolium temulentum L. fo. *macrochaeton* (A.Braun) Junge
Poaceae
- Weed
- 185
- G.

Lolium temulentum L. fo. *temulentum* L.
Poaceae
- Weed, Introduced
- 38, 185
- G.

Lolium temulentum L. var. *arvense* (With.) Lilj.
Poaceae
= *Lolium temulentum* L.
- bearded ryegrass
- Weed, Naturalised, Casual Alien
- 40, 86, 198, 256, 275
- G, cultivated. Origin: Mediterranean.

Lolium temulentum L. var. *leptochaeton* A.Br.
Poaceae
= *Lolium temulentum* L.
- Naturalised
- 287
- G.

Lolium temulentum L. var. *longiaristatum* Parn.
Poaceae
- Weed
- 256
- G.

Lolium temulentum L. var. *temulentum*
Poaceae
- bearded ryegrass
- Naturalised
- 86, 198
- G, cultivated. Origin: Mediterranean.

Lomandra longifolia Labill.
Xanthorrhoeaceae/Dasypogonaceae
- longleaf mat rush, spiny headed mat rush, mat rush, spinyhead matrush
- Native Weed
- 269
- pH, cultivated, herbal. Origin: Australia.

Lomandra multiflora (R.Br.) Britten ssp. *multiflora*
Xanthorrhoeaceae/Dasypogonaceae
- many flowered matrush
- Native Weed
- 269
- cultivated. Origin: Australia.

Lomatium dissectum (Nutt.) Mathias & Constance
Apiaceae
Ferula multifida (Nutt.) A.Gray, *Leptotaenia dissecta* Nutt. (see), *Leptotaenia dissecta* Nutt. var. *multifida* (Nutt.) Jeps. (see), *Leptotaenia multifida* Nutt.
- fernleaf biscuitroot, fernleaf lomatium, fernleaf mountain parsley, coughroot, lomatium
- Weed

- 23, 88
- pH, arid, cultivated, herbal.

Lomatium grayi (Coult. & Rose) Coult. & Rose
Apiaceae
- Gray's biscuitroot
- Weed
- 23, 88
- pH, herbal.

Lomatium leptocarpum (Nutt.) Coult. & Rose
Apiaceae
= *Lomatium bicolor* (S.Watson) J.Coult. & Rose var. *leptocarpum* (Torr. & A.Gray) M.Schlessman (NoR)
- bicolor biscuitroot
- Weed
- 87, 88, 218

Lomatium nudicaule (Pursh) Coult. & Rose
Apiaceae
- pestle parsnip, barestem biscuitroot, pestle lomatium
- Weed
- 23, 88
- pH, cultivated, herbal.

Lomentaria clavellosa (Turner) Gaillon
Lomentariaceae
Fucus clavellosus Turner, *Chrysymenia clavellosa* Agardh
- red alga
- Weed
- 197
- algae.

Lomentaria hakodatensis Yendo
Lomentariaceae
- red alga
- Weed
- 197
- algae.

Lomoplis ceratonia (L.) Raf.
Fabaceae/Mimosaceae
Mimosa ceratonia L.
- black ambret
- Weed
- 87, 88

Lonas annua (L.) Vines & Druce
Asteraceae
- ikikulta, African daisy, yellow ageratum
- Casual Alien
- 42
- aH, cultivated, herbal.

Lonchocarpus capassa Rolfe
Fabaceae/Papilionaceae
Capassa violacea Klotzsch, *Derris violacea* Harms, *Lonchocarpus violaceus* Oliv.
- apple leaf, lance tree, rain tree, lancepod
- Weed, Native Weed
- 87, 88, 121
- T, arid, cultivated. Origin: southern Africa.

Lonchocarpus punctatus Kunth
Fabaceae/Papilionaceae
- dotted lancepod
- Weed, Naturalised
- 101, 179

Lonchocarpus sericeus (Poir.) Kunth ex DC.
Fabaceae/Papilionaceae
Derris sericea (Poir.) Kunth,
Lonchocarpus cruentus Lundell,
Lonchocarpus pyxidarius DC., *Robinia sericea* Poir.
♦ lancepod, cabelouro da caatinga, garrapato, ingazeira, juimai, jumay, jumayo, savonette poilue, savonnette rivisre, vainillo
♦ Weed, Introduced
♦ 14, 228
♦ arid, cultivated, herbal.

Lonicera L. spp.
Caprifoliaceae
♦ bush honeysuckle, honeysuckle
♦ Weed, Naturalised, Environmental Weed
♦ 7, 80, 88, 151, 198, 243
♦ toxic.

Lonicera × americana C.Koch
Caprifoliaceae
= *Lonicera caprifolium* L. × *Lonicera etrusca* Santi
♦ Naturalised
♦ 86, 280
♦ cultivated. Origin: North America.

Lonicera × bella Zabel
Caprifoliaceae
= *Lonicera morrowii* A.Gray × *Lonicera tatarica* L.
♦ Bell's honeysuckle, hybrid pretty honeysuckle, whitebell honesuckle, bella honeysuckle
♦ Weed, Naturalised, Introduced
♦ 80, 101, 129, 133, 195, 222, 224
♦ aqua, cultivated, toxic. Origin: Eurasia.

Lonicera caprifolium L.
Caprifoliaceae
♦ Italian woodbine, perfoliate honeysuckle, tuoksuköynnöskuusama, caprifoglio
♦ Weed, Naturalised, Cultivation Escape, Casual Alien
♦ 40, 42, 80, 101
♦ pC, cultivated, herbal. Origin: Europe.

Lonicera etrusca Santi
Caprifoliaceae
Caprifolium etruscum Roem. & Schult.
♦ Etruscan honeysuckle, honeysuckle, zapletina
♦ Weed, Naturalised, Introduced
♦ 34, 80, 101
♦ S, cultivated, herbal.

Lonicera fragrantissima Lindl. & Paxton
Caprifoliaceae
♦ January jasmine, sweet breath of spring, fragrant honeysuckle
♦ Weed, Naturalised, Garden Escape
♦ 80, 86, 88, 98, 101, 102, 129, 198, 203, 251
♦ cultivated, herbal, toxic. Origin: Eurasia.

Lonicera × heckrottii Rehd.
Caprifoliaceae
= *Lonicera sempervirens* L. × *Lonicera × americana* C.Koch
♦ goldflame honeysuckle, honeysuckle
♦ Naturalised
♦ 101
♦ cultivated.

Lonicera hirsuta Eaton
Caprifoliaceae
♦ hairy honeysuckle
♦ Weed
♦ 87, 88, 218
♦ herbal.

Lonicera involucrata (Richards.) Banks ex Spreng.
Caprifoliaceae
Xylosteum involucratum (Banks ex Spreng.) Richards.
♦ kehtokuusama, twinberry honeysuckle, twinberry
♦ Cultivation Escape
♦ 39, 42
♦ S, cultivated, herbal, toxic.

Lonicera japonica Thunb.
Caprifoliaceae
Caprifolium japonicum (Thunb.) Dum.Cours., *Nintooa japonica* (Thunb.) Sweet
♦ Japanese honeysuckle, Chinese honeysuckle, honekakala, honeysuckle
♦ Weed, Quarantine Weed, Naturalised, Introduced, Garden Escape, Environmental Weed, Cultivation Escape
♦ 3, 4, 7, 8, 15, 17, 22, 34, 39, 40, 72, 73, 77, 80, 86, 87, 88, 98, 101, 102, 112, 133, 134, 142, 151, 152, 155, 161, 165, 179, 181, 195, 198, 201, 203, 204, 211, 218, 222, 224, 225, 233, 237, 243, 246, 247, 255, 261, 269, 280, 286, 289, 295, 296
♦ pC, arid, cultivated, herbal, toxic. Origin: east Asia.

Lonicera japonica Thunb. var. halliana Nichols
Caprifoliaceae
Lonicera flexuosa Thunb. var. *halliana* Dipp.
♦ Japanese honeysuckle, Hall's honeysuckle, Japanese kanferfoelie
♦ Garden Escape
♦ 283
♦ cultivated. Origin: China, Korea, Japan.

Lonicera maackii (Rupr.) Herder
Caprifoliaceae
♦ Amur honeysuckle, Amur bush honeysuckle
♦ Weed, Naturalised, Introduced, Garden Escape
♦ 80, 88, 101, 102, 129, 133, 142, 195, 222, 224
♦ cultivated, herbal, toxic. Origin: Eurasia.

Lonicera × minutiflora Zabel
Caprifoliaceae
= *Lonicera morrowii* A.Gray × (*Lonicera tatarica* L. × *Lonicera xylosteum* L.)
♦ smallflower honeysuckle
♦ Naturalised
♦ 101

Lonicera morrowii Gray
Caprifoliaceae
♦ Morrow's honeysuckle, fly honeysuckle, Morrow's bush honeysuckle
♦ Weed, Naturalised, Introduced, Garden Escape
♦ 17, 80, 88, 101, 102, 129, 133, 142, 195, 222, 224
♦ aqua, cultivated, herbal, toxic. Origin: Eurasia.

Lonicera nitida E.H.Wilson
Caprifoliaceae
♦ hedge honeysuckle, boxleaf honeysuckle, Wilson's honeysuckle
♦ Weed, Naturalised
♦ 15, 40, 80, 280
♦ S, cultivated.

Lonicera × notha Zabel
Caprifoliaceae
= *Lonicera ruprechtiana* Reg. × *Lonicera tatarica* L.
♦ honeysuckle
♦ Naturalised
♦ 101

Lonicera periclymenum L.
Caprifoliaceae
Caprifolium periclymenum Roem. & Schult.
♦ woodbine, honeysuckle, European honeysuckle, ruotsinköynnöskuusama
♦ Weed, Naturalised, Cultivation Escape
♦ 39, 42, 80, 86, 98, 101, 161, 176, 203, 252, 280
♦ pC, cultivated, herbal, toxic. Origin: Mediterranean.

Lonicera ruprechtiana Regel
Caprifoliaceae
♦ Manchurian honeysuckle
♦ Naturalised
♦ 101
♦ cultivated.

Lonicera sempervirens L.
Caprifoliaceae
♦ trumpet honeysuckle
♦ Weed
♦ 80, 87, 88, 218
♦ S, cultivated, herbal.

Lonicera standishii Jacques
Caprifoliaceae
♦ Standish's honeysuckle
♦ Weed, Naturalised
♦ 17, 39, 88, 101, 129, 129
♦ cultivated, toxic. Origin: Eurasia.

Lonicera subspicata Hook. & Arn.
Caprifoliaceae
♦ moronel honeysuckle, southern honeysuckle
♦ Weed
♦ 87, 88, 218
♦ S, herbal.

Lonicera tatarica L.
Caprifoliaceae
♦ Tartarian honeysuckle, bush honeysuckle, garden fly honey suckle, rusokuusama
♦ Weed, Naturalised, Introduced,

Garden Escape, Environmental Weed, Cultivation Escape
- 4, 39, 42, 80, 87, 88, 101, 104, 129, 133, 142, 159, 161, 195, 211, 218, 222, 224
- S, cultivated, herbal, toxic. Origin: Eurasia.

Lonicera × xylosteoides Tausch
Caprifoliaceae
= *Lonicera tatarica* L. × *Lonicera xylosteum* L.
- fly honeysuckle
- Naturalised
- 101
- cultivated.

Lonicera xylosteum L.
Caprifoliaceae
Caprifolium xylosteum Gaertn.
- fly honeysuckle, bush honeysuckle, lehtokuusama, woody honeysuckle, European fly honeysuckle, dwarf honeysuckle
- Weed, Naturalised, Casual Alien
- 39, 40, 80, 101, 129, 133, 161, 195, 224
- cultivated, herbal, toxic. Origin: Eurasia.

Lopezia mexicana Jacq.
Onagraceae
= *Lopezia racemosa* Cav.
- Weed, Quarantine Weed
- 76, 87, 88, 203, 220

Lopezia racemosa Cav.
Onagraceae
Lopezia hirsuta Jacq., *Lopezia mexicana* Jacq. (see)
- pink brush, pretty rose
- Weed
- 157, 243
- a/pH, cultivated, herbal.

Lophatherum gracile Brongn.
Poaceae
- common lophantherum
- Weed
- 297
- G, herbal.

Lophiocarpus guyanensis Micheli
Phytolaccaceae/Petiveriaceae
- Weed
- 87, 88

Lophiocarpus tenuissimus Hook.f.
Phytolaccaceae/Petiveriaceae
- Native Weed
- 121
- aH. Origin: southern Africa.

Lophocereus (Berg.) Britt. & Rose spp.
Cactaceae
= *Pachycereus* (Berg.) Britt. & Rose spp.
- Weed, Quarantine Weed
- 76, 88, 220

Lophochloa cristata (L.) Hyl.
Poaceae
= *Rostraria cristata* (L.) Tzvelev
- tähkiötoppo
- Weed, Naturalised, Casual Alien
- 42, 241, 272, 286
- G, herbal.

Lophochloa phleoides (Vill.) Rchb.
Poaceae

= *Rostraria cristata* (L.) Tzvelev
- lopocloa
- Weed
- 237, 295
- G.

Lophochloa pumila (Desf.) Bor
Poaceae
= *Rostraria pumila* (Desf.) Tzvelev
- roughtail
- Naturalised
- 7
- G.

Lophochloa smyrnacea Trin.
Poaceae
= *Rostraria smyrnacea* (Trin.) H.Scholz (NoR)
- cat's tail
- Weed
- 68, 88
- G.

Lopholaena coriifolia (Sond.) Phillips & C.A.Sm.
Asteraceae
- chiGunguru, pluisbossie, lopholaena
- Weed
- 50, 63

Lophophora Coult. spp.
Cactaceae
- peyote, lophophora
- Quarantine Weed
- 220
- toxic.

Lophophora diffusa (Croizat) Bravo
Cactaceae
Lophophora lutea (Rouhier) Backeb.
- peyote
- Quarantine Weed
- 76
- cultivated, herbal.

Lophophora williamsii (Lem. ex Salm-Dyck) Coult.
Cactaceae
Anhalonium williamsii (Lem.) Lem., *Echinocactus rapa* Fisch. & C.Mey., *Echinocactus williamsii* Lem., *Lophophora echinata* Croizat, *Lophophora fricii* Haberm., *Lophophora jourdaniana* Haberm., *Mammillaria williamsii* (Lem.) J.M.Coult.
- peyote, dumpling cactus, mescal, mescal buttons
- Weed, Quarantine Weed
- 39, 76, 88, 161, 189, 203, 247
- arid, cultivated, herbal, toxic.

Lophopyrum elongatum (Host) Á.Löve
Poaceae
= *Elytrigia elongata* (Host) Nev.
- elongate wheatgrass, tall wheatgrass
- Weed, Naturalised, Environmental Weed
- 72, 86, 88, 176, 198
- pG, cultivated, herbal. Origin: southern Europe.

Lophopyrum ponticum (Podp.) Á.Löve
Poaceae
= *Elytrigia pontica* (Podp.) Holub (NoR) [see *Elytrigia pontica* (Podp.) Holub ssp. *pontica*]

- tall wheatgrass
- Environmental Weed
- 296
- G.

Lophospermum erubescens D.Don
Scrophulariaceae
Asarina erubescens (D.Don) Pennell, *Lophospermum scandens* Sessé & Moç. ex D.Don, *Maurandya erubescens* (D.Don) A.Gray
- Mexican twist, creeping gloxinia
- Weed, Naturalised, Cultivation Escape
- 86, 98, 101, 203, 261, 280
- cultivated. Origin: Mexico.

Lophostemon confertus (R.Br.) Wilson & Waterhouse
Myrtaceae
Tristania conferta R.Br.
- brush box, Brisbane box, vinegar tree
- Weed, Naturalised, Environmental Weed
- 3, 7, 86, 88, 101, 151, 191
- cultivated, herbal. Origin: Australia.

Loranthus acaciae Zucc.
Loranthaceae
- Weed
- 221
- parasitic, herbal.

Loranthus curviflorus Benth. ex Oliv.
Loranthaceae
= *Plicosepalus curviflorus* (Benth. ex Oliv.) Tiegh. (NoR)
- Weed
- 221
- parasitic.

Loranthus elasticus Desr.
Loranthaceae
- Weed
- 87, 88
- parasitic.

Loranthus longiflorus Desr.
Loranthaceae
- Weed
- 87, 88
- parasitic.

Loranthus pentandrus L.
Loranthaceae
= *Dendrophthoe pentandra* (L.) Miq. (NoR)
- Weed
- 209
- parasitic, herbal.

Loranthus pulverulentus Wall.
Loranthaceae
- Weed
- 87, 88
- parasitic.

Lotononis bainesii Bak.
Fabaceae/Papilionaceae
- lotononis
- Weed, Naturalised
- 86, 98, 101, 203
- cultivated. Origin: South Africa.

Lotononis listii Polhill
Fabaceae/Papilionaceae
Listia heterophylla E.Mey.
- Native Weed
- 121
- pH. Origin: southern Africa.

Lotononis platycarpa (Viv.) Pic.Serm.
Fabaceae/Papilionaceae
- Weed
- 221
- toxic.

Lotus L. spp.
Fabaceae/Papilionaceae
- bird's foot trefoil, trefoil, deervetch
- Weed
- 39, 272
- herbal, toxic.

Lotus angustissimus L.
Fabaceae/Papilionaceae
- slender bird's foot trefoil, long fruited bird's foot trefoil, slender lotus
- Weed, Naturalised, Introduced, Environmental Weed, Casual Alien
- 7, 9, 15, 34, 39, 42, 86, 98, 101, 134, 176, 198, 203, 253, 272, 280
- aH, cultivated, herbal, toxic. Origin: Eurasia, Mediterranean.

Lotus arabicus L.
Fabaceae/Papilionaceae
Lotus glinoides sensu Bak. *non* Del., *Lotus mossamedensis* Welw. ex Bak., *Lotus roseus* Forssk.
- Weed
- 39, 87, 88, 185, 221, 242
- arid, herbal, toxic.

Lotus arenarius Brot.
Fabaceae/Papilionaceae
Lotus aurantiacus Boiss. (see)
- Quarantine Weed
- 220

Lotus aurantiacus Boiss.
Fabaceae/Papilionaceae
= *Lotus arenarius* Brot.
- Quarantine Weed
- 220

Lotus australis Andrews
Fabaceae/Papilionaceae
- austral trefoil
- Native Weed
- 39, 269
- arid, cultivated, toxic. Origin: Asia, Australia.

Lotus canescens Ktze.
Fabaceae/Papilionaceae
- Quarantine Weed
- 220

Lotus corniculatus L.
Fabaceae/Papilionaceae
- bird's foot trefoil, crowtoes, bloomfell, birdfoot deervetch, ground honeysuckle, cat's clover, broadleaf bird's foot trefoil
- Weed, Naturalised, Introduced, Garden Escape, Environmental Weed
- 34, 39, 70, 72, 80, 86, 87, 88, 98, 101, 136, 142, 146, 151, 155, 161, 174, 176, 180, 185, 195, 198, 203, 211, 218, 221,

228, 241, 269, 272, 280, 286, 287, 293, 295, 297, 300
- a/pH, arid, cultivated, herbal, toxic. Origin: Eurasia.

Lotus corniculatus L. var. corniculatus
Fabaceae/Papilionaceae
- bird's foot trefoil
- Naturalised
- 86, 176, 198
- cultivated. Origin: Eurasia, Mediterranean.

Lotus corniculatus L. var. japonicus Regel
Fabaceae/Papilionaceae
Lotus japonicus (Regel) K.Larsen
- Japanese bird's foot trefoil, miyakogusa
- Weed
- 286

Lotus corniculatus L. var. tenuifolius L.
Fabaceae/Papilionaceae
= *Lotus tenuis* Waldst. & Kit. ex Willd.
- narrow bird's foot trefoil
- Naturalised
- 86, 176, 198
- cultivated. Origin: Eurasia, Mediterranean.

Lotus creticus L.
Fabaceae/Papilionaceae
Lotus commutatus Guss., *Lotus salzmannii* Boiss. & Reut.
- southern bird's foot trefoil, Cretan trefoil, lotus, esshb oshb, grain, hueta, lotier de Crète, lotier de maritime, silvery bird's foot trefoil, zghiga
- Weed, Naturalised, Environmental Weed
- 54, 70, 72, 86, 88, 198, 221
- pH, arid, cultivated. Origin: Mediterranean.

Lotus cruentus Court
Fabaceae/Papilionaceae
- redflower lotus, red flowered trefoil
- Native Weed
- 39, 269
- cultivated, toxic. Origin: Australia.

Lotus glaber Mill. nom. rej. prop.
Fabaceae/Papilionaceae
= *Lotus tenuis* Waldst. & Kit. ex Willd.
- narrow leaved bird's foot trefoil, lotier glabre
- Naturalised
- 300

Lotus glinoides Delile
Fabaceae/Papilionaceae
- Weed
- 221

Lotus halophilus Boiss. & Spruner
Fabaceae/Papilionaceae
Lotus aucheri Boiss. & Spruner, *Lotus pusillus* Viv., *Lotus pusillus* Viv. var. *majus* Boiss., *Lotus villosus* Forssk.
- greater bird's foot trefoil
- Weed
- 221
- pH, arid, cultivated.

Lotus hamatus E.Greene
Fabaceae/Papilionaceae
- San Diego bird's foot trefoil
- Introduced
- 228
- aH, arid, herbal.

Lotus hispidus Desf. ex DC.
Fabaceae/Papilionaceae
= *Lotus parviflorus* Desf.
- hairy bird's foot trefoil
- Weed, Naturalised
- 87, 88, 98, 203, 253

Lotus lalambensis Schweinf.
Fabaceae/Papilionaceae
- Weed
- 221

Lotus ornithopodioides L.
Fabaceae/Papilionaceae
- Weed
- 221
- cultivated.

Lotus palustris Willd.
Fabaceae/Papilionaceae
- Quarantine Weed
- 220

Lotus parviflorus Desf.
Fabaceae/Papilionaceae
Lotus hispidus Desf. ex DC. (see)
- hairy bird's foot trefoil, smallflower bird's foot trefoil
- Quarantine Weed, Naturalised
- 220

Lotus pedunculatus Cav.
Fabaceae/Papilionaceae
Lotus granadensis Zertová, *Lotus major* Sm.
- lotus, big trefoil, greater bird's foot trefoil, large bird's foot trefoil
- Weed, Naturalised, Introduced, Environmental Weed
- 15, 39, 80, 98, 101, 165, 181, 203, 225, 228, 246, 253, 280
- arid, cultivated, herbal, toxic. Origin: Mediterranean, west Asia.

Lotus preslii Ten.
Fabaceae/Papilionaceae
- bird's foot trefoil
- Naturalised
- 86, 198
- Origin: Mediterranean.

Lotus purpureus Webb
Fabaceae/Papilionaceae
- Weed, Quarantine Weed
- 76, 88, 220
- Origin: Cape Verde Islands.

Lotus purshianus (Benth.) Clem. & E.G.Clem. var. purshianus
Fabaceae/Papilionaceae
- Spanish clover
- Weed
- 180, 243
- aH, cultivated.

Lotus scoparius (Nutt.) Oxley
Fabaceae/Papilionaceae
- broom deervetch, deerweed, wild broom, common deerweed
- Weed

- 87, 88, 218
- pH, arid, cultivated, herbal.

Lotus suaveolens Pers.
Fabaceae/Papilionaceae
- hairy bird's foot trefoil, Boyd's clover
- Weed, Naturalised, Environmental Weed
- 7, 9, 15, 39, 86, 98, 165, 176, 198, 203, 280
- toxic. Origin: Mediterranean.

Lotus subbiflorus Lag.
Fabaceae/Papilionaceae
- kankeakarvamaite, hairy bird's foot trefoil, lotus
- Weed, Naturalised, Environmental Weed, Casual Alien
- 42, 72, 86, 88, 98, 101, 158, 203, 287
- pH, cultivated, herbal.

Lotus tenuis Waldst. & Kit. ex Willd.
Fabaceae/Papilionaceae
Lotus glaber Mill. *nom. rej. prop.* (see),
Lotus corniculatus L. var. *tenuifolius* L. (see)
- hentomaite, narrow trefoil
- Weed, Naturalised
- 15, 42, 87, 88, 98, 101, 161, 203, 218, 241, 272, 275, 280, 287
- aH, cultivated, herbal, toxic.

Lotus tetragonolobus L.
Fabaceae/Papilionaceae
= *Tetragonolobus purpureus* Moench
- winged pea, spargelerbse
- Weed, Naturalised, Casual Alien
- 80, 86, 280
- aH, cultivated, herbal. Origin: Mediterranean.

Lotus uliginosus Schkuhr
Fabaceae/Papilionaceae
- isomaite, greater lotus, greater bird's foot trefoil, large bird's foot trefoil, marsh bird's foot trefoil
- Weed, Naturalised, Introduced, Environmental Weed
- 7, 20, 34, 42, 70, 72, 86, 88, 98, 176, 198, 203, 241, 269, 272, 286, 287, 300
- pH, cultivated, herbal. Origin: Europe, northern Africa.

Lotus unifoliolatus (Hook.) Benth.
Fabaceae/Papilionaceae
Lotus americanus (Nutt.) Bisch. *nom. illeg.*, *Lotus unifoliolatus* var. *unifoliolatus* (Hook.) Benth., *Hosackia americana* (Nutt.) Piper, *Hosackia purshiana* Benth. *nom. illeg.*, *Lotus purshianus* Clem. & E.G.Clem., *Lotus sericeus* Pursh. *nom. illeg.*, *Trigonella americana* Nutt.
- American bird's foot trefoil, Spanish clover, deervetch
- Weed
- 161

Loudetia arundinacea (Hochst. ex A.Rich.) Steud.
Poaceae
- lujange
- Weed
- 88
- pG.

Loudetia simplex (Nees) C.E.Hubb
Poaceae
- russet grass, lwejwe
- Native Weed
- 121
- pG, cultivated. Origin: Madagascar.

Loxanthocereus Backeb. spp.
Cactaceae
= *Cleistocactus* Lem. spp.
- Weed, Quarantine Weed
- 76, 88, 203

Lucuma spinosa Molina
Sapotaceae
- Quarantine Weed
- 220

Ludwigia L. spp.
Onagraceae
- water primrose, primrose willow
- Weed, Quarantine Weed
- 181, 220, 258
- H/S, aqua.

Ludwigia abyssinica A.Rich
Onagraceae
- Weed
- 88
- wa/pH.

Ludwigia adscendens (L.) H.Hara
Onagraceae
Jussiaea diffusa Forsk.
- water primrose, creeping water primrose, red ludwigia
- Weed, Environmental Weed
- 87, 88, 170, 191, 209, 216, 217, 262, 275, 276, 297
- wpH, cultivated, herbal. Origin: Asia, Australia.

Ludwigia affinis (DC.) H.Hara
Onagraceae
Jussiaea affinis DC.
- Weed
- 87, 88
- wH.

Ludwigia alternifolia L.
Onagraceae
- rattlebox, seedbox
- Quarantine Weed
- 258
- wH, cultivated, herbal.

Ludwigia decurrens Walter
Onagraceae
Jussiaea decurrens (Walt.) DC. (see)
- wingleaf primrose willow, lamparita
- Weed, Naturalised
- 87, 88, 157, 281, 286, 287
- waH, herbal.

Ludwigia elegans (Cambess.) H.Hara
Onagraceae
- eruz de malta
- Weed
- 255
- wa/pH. Origin: South America.

Ludwigia epilobioides Maxim.
Onagraceae
- Weed
- 263, 286
- waH.

Ludwigia erecta (L.) H.Hara
Onagraceae
Jussiaea erecta L. (see)
- yerba de jicotea
- Weed, Quarantine Weed
- 14, 76, 87, 88, 203
- wH.

Ludwigia hexapetala (Hook. & Arn.) Zardini, H.Gu & P.H.Raven
Onagraceae
= *Ludwigia uruguayensis* (Camb.) Hara
- Uruguay seedbox, sixpetal water primrose
- Weed
- 197
- wpH, cultivated.

Ludwigia hyssopifolia (G.Don) Exell
Onagraceae
Jussiaea linifolia Vahl (see)
- seedbox, telurik, linearleaf water primrose
- Weed
- 12, 87, 88, 170, 209, 275, 276, 297
- waH. Origin: pantropical.

Ludwigia leptocarpa (Nutt.) H.Hara
Onagraceae
Jussiaea leptocarpa Nutt. (see)
- anglestem primrose willow, cruz de malta
- Weed, Quarantine Weed
- 76, 87, 88, 203, 255
- wa/pH, herbal. Origin: Americas.

Ludwigia linearis Walt.
Onagraceae
- narrowleaf primrose willow
- Naturalised
- 287
- wH, herbal.

Ludwigia longifolia (DC.) H.Hara
Onagraceae
Jussiaea longifolia (DC.) Hara
- longleaf primrose willow, primrose willow
- Weed, Sleeper Weed, Quarantine Weed, Naturalised, Environmental Weed
- 3, 54, 76, 86, 87, 88, 101, 155, 191, 203
- wpS, cultivated. Origin: Brazil, Argentina.

Ludwigia micrantha (Kunze) Hara
Onagraceae
- Naturalised
- 287
- wH.

Ludwigia microcarpa Michx.
Onagraceae
- smallfruit primrose willow
- Quarantine Weed
- 220
- wH, herbal.

Ludwigia octovalvis (Jacq.) Raven
Onagraceae
Jussiaea angustifolia Lam., *Jussiaea blumeana* DC., *Jussiaea calycina* C.Presl, *Jussiaea costata* C.Presl, *Jussiaea frutescens* Jacq.f. ex DC., *Jussiaea haenkeana* Steud., *Jussiaea hirsuta* Mill., *Jussiaea ligustrifolia* Kunth, *Jussiaea*

linearis Hochst., *Jussiaea occidentalis*
Nutt. ex Torr. & A.Gray, *Jussiaea octofila*
DC., *Jussiaea octonervia* Lam., *Jussiaea
octovalvis* (Jacq.) Sw., *Jussiaea parviflora*
Cambess., *Jussiaea persicariifolia*
Schltdl. fo. *minor* Schltdl., *Jussiaea
persicariifolia* Schltdl. fo. *major* Schltdl.,
Jussiaea persicariifolia Schltdl., *Jussiaea
peruviana* L. var. *octofila* Bertoni,
Jussiaea pubescens L., *Jussiaea sagreana*
A.Rich., *Jussiaea salicifolia* Kunth,
Jussiaea suffruticosa L. Him. (see),
Jussiaea tetragona Spreng., *Jussiaea
venosa* C.Presl, *Ludwigia angustifolia*
(Lam.) M.Gómez, *Ludwigia octovalvis*
(Jacq.) P.H.Raven var. *octofila* (Bertoni)
Alain, *Ludwigia octovalvis* (Jacq.)
P.H.Raven var. *ligustrifolia* (Kunth)
Alain, *Ludwigia pubescens* (L.) H.Hara,
Ludwigia pubescens (L.) H.Hara var.
linearifolia (Hassl.) A.Fern.& R.Fern.,
Ludwigia pubescens (L.) H.Hara var.
ligustrifolia (Kunth) H.Hara, *Ludwigia
sagreana* (A.Rich) M.Gómez, *Oenothera
octovalvis* Jacq.
  ♦ willow primrose, false primrose,
  yellow willow herb, primrose willow,
  water primrose, Mexican primrose
  willow, telurik, yaa rak na
  ♦ Weed
  ♦ 6, 14, 39, 87, 88, 157, 170, 217, 235,
  239, 255, 262, 274, 276
  ♦ wpS, cultivated, herbal, toxic.
  Origin: tropical America.

**Ludwigia octovalvis (Jacq.) Raven var.
sessiliflora (M.Micheli) Shinners**
  Onagraceae
  ♦ Weed
  ♦ 286
  ♦ wH.

Ludwigia palustris (L.) Elliott
  Onagraceae
  ♦ water purslane, marsh seedbox,
  marsh ludwigia, false loosestrife,
  Hampshire purslane, broadleaf
  ludwigia
  ♦ Weed, Naturalised, Garden Escape,
  Environmental Weed
  ♦ 15, 72, 86, 87, 88, 98, 121, 155, 165,
  181, 198, 203, 208, 218, 246, 272, 280,
  286, 287
  ♦ wpH, cultivated, herbal. Origin:
  northern hemisphere.

Ludwigia parviflora Roxb.
  Onagraceae
  = *Ludwigia perennis* L.
  ♦ Weed
  ♦ 88, 191, 204
  ♦ wH.

Ludwigia peploides (Kunth) Raven
  Onagraceae
  Jussiaea repens sensu Munz *non* L.,
  Jussiaea californica (Wats.) Jeps. (see),
  Jussiaea gomezii Ram.Goyena, *Jussiaea
  patibilcensis* Kunth, *Jussiaea peploides*
  Kunth, *Jussiaea polygonoides* Kunth,
  Jussiaea repens var. *peploides* (Kunth)
  Griseb.
  ♦ creeping water primrose, California

water primrose, water primrose,
yellow water primrose, primrose
willow
  ♦ Weed, Quarantine Weed,
  Naturalised, Garden Escape,
  Environmental Weed
  ♦ 15, 87, 88, 98, 152, 157, 161, 180, 203,
  207, 208, 237, 246, 280, 295
  ♦ wpH, cultivated, herbal. Origin:
  tropical America.

**Ludwigia peploides (Kunth) P.H.Raven
ssp. montevidensis (Spreng.) Raven**
  Onagraceae
  ♦ water primrose, floating primrose
  willow, floating water primrose
  ♦ Weed, Quarantine Weed,
  Naturalised, Environmental Weed
  ♦ 76, 86, 269, 280
  ♦ wpH, cultivated.

**Ludwigia peploides (Kunth) Raven ssp.
stipulacea (Ohwi) Raven**
  Onagraceae
  Ludwigia stipulacea (Ohwi) Ohwi (see)
  ♦ Weed
  ♦ 274
  ♦ wH.

Ludwigia perennis L.
  Onagraceae
  Ludwigia gracilis Miq., *Ludwigia
  leucorrhiza* Bl., *Ludwigia lythroides* Bl.,
  Ludwigia parviflora Roxb. (see), *Jussiaea
  perennis* Brenan
  ♦ Weed
  ♦ 87, 88, 170
  ♦ wpH, herbal.

Ludwigia peruviana (L.) H.Hara
  Onagraceae
  Jussiaea grandiflora L. (see), *Jussiaea
  peruviana* L. (see)
  ♦ ludwigia, Peruvian primrose bush,
  water primrose
  ♦ Weed, Quarantine Weed, Noxious
  Weed, Naturalised, Garden Escape,
  Environmental Weed
  ♦ 3, 14, 54, 76, 86, 87, 88, 98, 101, 147,
  155, 170, 191, 200, 203, 246, 251, 269,
  290, 296
  ♦ wpS, cultivated. Origin: tropical
  South America.

Ludwigia prostrata Roxb.
  Onagraceae
  Jussiaea prostrata Lev.
  ♦ choujitade, climbing seedbox
  ♦ Weed, Quarantine Weed
  ♦ 76, 87, 88, 191, 203, 204, 274, 275, 297
  ♦ waH, herbal.

Ludwigia repens J.R.Forst.
  Onagraceae
  ♦ creeping primrose willow, creeping
  water primrose, creeping ludwigia, red
  ludwigia
  ♦ Quarantine Weed, Naturalised
  ♦ 258, 287
  ♦ wpH, cultivated, herbal.

Ludwigia sericea (Cambess.) H.Hara
  Onagraceae
  ♦ cruz de malta
  ♦ Weed

  ♦ 255
  ♦ wpH. Origin: South America.

Ludwigia stenorraphe (Brenan) Hara
  Onagraceae
  Jussiaea stenorraphe Brenan
  ♦ Weed, Quarantine Weed
  ♦ 76, 87, 88, 203
  ♦ wH.

Ludwigia stipulacea (Ohwi) Ohwi
  Onagraceae
  = *Ludwigia peploides* (Kunth) Raven
  ssp. *stipulacea* (Ohwi) Raven
  ♦ Weed
  ♦ 286
  ♦ wH.

**Ludwigia stolonifera (Guill. & Perr.)
Raven**
  Onagraceae
  ♦ willow herb
  ♦ Weed
  ♦ 121
  ♦ wa/pH. Origin: obscure.

Ludwigia tomentosa (Cambess.) H.Hara
  Onagraceae
  ♦ cruz de malta
  ♦ Weed
  ♦ 255
  ♦ wpS. Origin: tropical America.

Ludwigia uruguayensis (Camb.) Hara
  Onagraceae
  Ludwigia hexapetala (Hook. & Arn.)
  Zardini, H.Gu & P.H.Raven (see),
  Jussiaea michauxiana Fern. (see), *Jussiaea
  uruguayensis* Camb.
  ♦ Uruguay water primrose, hairy
  water primrose, water primrose,
  Uruguayan primrose willow
  ♦ Weed, Quarantine Weed, Noxious
  Weed, Naturalised
  ♦ 67, 76, 87, 88, 101, 102, 116, 161, 203,
  229, 255
  ♦ wpH, herbal. Origin: South America.

Luehea divaricata Mart.
  Tiliaceae
  ♦ Introduced
  ♦ 228
  ♦ arid, cultivated.

Luehea speciosa Willd.
  Tiliaceae
  Luehea tarapotina J.F.Macbr. (see)
  ♦ Naturalised, Cultivation Escape
  ♦ 101, 261
  ♦ T, cultivated. Origin: Cuba.

Luehea tarapotina J.F.Macbr.
  Tiliaceae
  = *Luehea speciosa* Willd.
  ♦ Weed
  ♦ 87, 88

Luetzelburgia auriculata Allemão
  Fabaceae/Papilionaceae
  ♦ Introduced
  ♦ 228
  ♦ arid.

Luffa acutangula (L.) Roxb.
  Cucurbitaceae
  Cucumis acutangulus L., *Cucurbita
  umbellata* Klein ex Willd, *Luffa*

acutangula (L.) Roxb. var. *amara* (Roxb.) C.B.Clarke, *Luffa forskalii* Schweinf. ex Harms (see), *Luffa umbellata* (Klein ex Willd.) M.Roem.
♦ sinkwa towelsponge, angular loofah, ara torui, dishcloth gourd, dishrag gourd, fluted loofah, jhinga, karavi tori, liane torchon, loofah, luffa sponge, papangaie, pipangaie, ribbed gourd, ribbed luffa, snake gourd, vegetable sponge, Chinese okra
♦ Naturalised, Introduced, Casual Alien
♦ 39, 101, 228, 261
♦ arid, cultivated, herbal, toxic. Origin: Old World Tropics.

Luffa aegyptiaca Mill.
Cucurbitaceae
= *Luffa cylindrica* (L.) M.Roem.
♦ sponge gourd, loofah
♦ Weed, Naturalised, Cultivation Escape
♦ 39, 87, 88, 101, 179, 255, 261
♦ pH, cultivated, herbal, toxic. Origin: Eurasia.

Luffa cylindrica (L.) M.Roem.
Cucurbitaceae
Cucumis fricatorius Sessé & Moç., *Luffa acutangula* (L.) Roxb. var. *subangulata* (Miq.) Cogn., *Luffa aegyptiaca* Mill. (see), *Luffa subangulata* Miq., *Melothria touchanensis* H.Lév., *Momordica cylindrica* L., *Momordica luffa* L.
♦ ndodoki mdodoki, ki raci, dishcloth gourd, estropajo, ghin torai, loofah, smooth loofah, sponge gourd, vegetable sponge
♦ Weed, Introduced
♦ 32, 87, 88, 228, 243
♦ arid, cultivated, herbal. Origin: Australia.

Luffa cylindrica (L.) Roem. var. *insularum* (A.Gray) Cogn.
Cucurbitaceae
♦ smooth loofah, vegetable sponge
♦ Weed
♦ 3
♦ C.

Luffa forskalii Schweinf. ex Harms
Cucurbitaceae
= *Luffa acutangula* (L.) Roxb.
♦ Weed
♦ 88

Luffa operculata (L.) Cogn.
Cucurbitaceae
♦ luffa, wild luffa
♦ Weed
♦ 39, 87, 88
♦ cultivated, herbal, toxic.

Luma apiculata (DC.) Burret
Myrtaceae
Eugenia affinis Gillies ex Hook. & Arn., *Eugenia apiculata* DC. (see), *Eugenia apiculata* var. *arnyan* Hook.f., *Eugenia barneoudii* O.Berg, *Eugenia cuspidata* Phil., *Eugenia ebracteata* Phil., *Eugenia gilliesi* Hook. & Arn., *Eugenia hookeri* Steud., *Eugenia luma* O.Berg, *Eugenia*

modesta Phil., *Eugenia mucronata* Phil., *Eugenia palenae* Phil., *Eugenia proba* O.Berg, *Eugenia spectabilis* Phil., *Luma gilliesi* (Hook. & Arn.) Burret, *Luma hookeri* (Steud.) Burret, *Luma spectablis* (Phil.) Burret, *Myrceugenella apiculata* (DC.) Kausel, *Myrceugenella apiculata* var. *australis* Kausel, *Myrceugenella apiculata* var. *genuina* Kausel, *Myrceugenella apiculata* var. *nahuelhuapensis* Kausel, *Myrceugenella apiculata* var. *spectabilis* (Phil.) Kausel, *Myrceugenella grandjotii* Kausel, *Myrceugenia apiculata* (DC.) Nied., *Myrtus chequenilla* Kuntze
♦ temu, arrayan, Chilean myrtle, arrayán, collimamol, palo colorado
♦ Introduced
♦ 34
♦ S, cultivated.

Lumnitzera racemosa Willd.
Combretaceae
♦ Weed
♦ 179
♦ cultivated, herbal.

Lunaria annua L.
Brassicaceae
Lunaria biennis Moench, *Lunaria inodora* Lam.
♦ honesty, money flower, moonwort, satin flower, kuuruoho
♦ Weed, Naturalised, Cultivation Escape
♦ 4, 15, 24, 34, 40, 42, 56, 86, 88, 98, 101, 195, 203, 252, 280, 287
♦ a/pH, cultivated, herbal. Origin: south-east Europe.

Lunaria annua L. ssp. annua
Brassicaceae
♦ Naturalised
♦ 280

Lunaria rediviva L.
Brassicaceae
Lunaria alpina Berg.
♦ perennial honesty, money plant, European honesty
♦ Weed, Naturalised, Casual Alien
♦ 40, 101, 195
♦ cultivated, herbal.

Lunularia cruciata (L.) Dumort.
Lunulariaceae
♦ mesiacovka kríľovitá, crescent cup liverwort
♦ Weed, Naturalised
♦ 211, 243, 287
♦ liverwort.

Lupinus L. spp.
Fabaceae/Papilionaceae
♦ lupin
♦ Weed, Naturalised
♦ 18, 88, 154, 181, 198, 247, 272
♦ toxic.

Lupinus albus L.
Fabaceae/Papilionaceae
Lupinus graecus Boiss. & Spruner, *Lupinus jugoslavicus* Kazim. & Nowacki, *Lupinus termis* Forssk.

♦ white lupin, lupino
♦ Weed, Naturalised, Introduced, Environmental Weed, Casual Alien
♦ 7, 39, 70, 86, 87, 88, 98, 101, 203, 228, 280, 300
♦ aH, arid, cultivated, herbal, toxic. Origin: southern Balkans, Aegean.

Lupinus alpestris A.Nelson
Fabaceae/Papilionaceae
♦ silvery lupine
♦ Weed
♦ 161
♦ toxic.

Lupinus angustifolius L.
Fabaceae/Papilionaceae
Lupinus linifolius Roth, *Lupinus reticulatus* Desv.
♦ sinilupiini, narrowleaf lupine, New Zealand blue lupin, bitter lupin, blue lupin
♦ Weed, Naturalised, Garden Escape, Environmental Weed, Casual Alien
♦ 7, 15, 39, 42, 70, 86, 87, 88, 94, 98, 101, 176, 198, 203, 272, 280, 290, 300
♦ aH, cultivated, herbal, toxic. Origin: Mediterranean, Middle East.

Lupinus arboreus Sims
Fabaceae/Papilionaceae
Lupinus macrocarpus Hook. & Arn. (see), *Lupinus propinquus* E.Greene, *Lupinus rivulars* Dougl. ex Lindl.
♦ tree lupin, yellowbush lupine, bush lupin, coastal bush lupine
♦ Weed, Quarantine Weed, Noxious Weed, Naturalised, Introduced, Garden Escape, Environmental Weed
♦ 15, 35, 40, 72, 76, 78, 80, 86, 87, 88, 98, 116, 146, 151, 152, 155, 165, 176, 181, 198, 203, 225, 228, 231, 241, 246, 280, 290, 296, 300
♦ pS, arid, cultivated, herbal, toxic. Origin: south-western North America.

Lupinus argenteus Pursh
Fabaceae/Papilionaceae
♦ silvery lupine, Wyeth lupine, silver lupine
♦ Weed, Noxious Weed
♦ 36, 39, 87, 88, 161, 212, 218
♦ pH, herbal, toxic.

Lupinus bicolor Lindl.
Fabaceae/Papilionaceae
♦ Lindley's annual lupine, bicolor lupine, fairy lupin, miniature lupine, pygmy leaved lupin, summer pipersmithii lupine
♦ Weed
♦ 161, 180, 243
♦ a/pH, cultivated, herbal.

Lupinus bilineatus Benth.
Fabaceae/Papilionaceae
= *Lupinus mexicanus* Cerv. ex Lag.
♦ Weed
♦ 199

Lupinus campestris Cham. &. Schltdl.
Fabaceae/Papilionaceae
♦ Weed
♦ 199

Lupinus caudatus **Kell.**
Fabaceae/Papilionaceae
♦ tailcup lupine
♦ Weed
♦ 39, 87, 88, 161, 218
♦ cultivated, herbal, toxic.

Lupinus cosentinii **Guss.**
Fabaceae/Papilionaceae
♦ Western Australian blue lupin, sand plain lupin
♦ Weed, Naturalised, Garden Escape, Environmental Weed
♦ 7, 9, 39, 86, 88, 98, 134, 203
♦ cultivated, toxic. Origin: Mediterranean.

Lupinus diffusus **Nutt.**
Fabaceae/Papilionaceae
♦ skyblue lupine, Oak Ridge lupine
♦ Weed
♦ 161
♦ toxic.

Lupinus formosus **Greene**
Fabaceae/Papilionaceae
♦ Lindley's robustus lupine, summer lupine, western lupine
♦ Weed
♦ 161
♦ pH, herbal.

Lupinus formosus **E.Greene** var. *robustus* **C.P.Sm.**
Fabaceae/Papilionaceae
♦ summer lupine, late lupine, lupine
♦ Weed
♦ 180
♦ pH.

Lupinus hartwegii **Lindl.**
Fabaceae/Papilionaceae
♦ Hartweg's bluebonnet
♦ Weed
♦ 199
♦ aH, cultivated.

Lupinus hirsutus **L. nom. rej.**
Fabaceae/Papilionaceae
= *Lupinus pilosus* L.
♦ lupine
♦ Weed
♦ 87, 88
♦ aH, promoted, herbal.

Lupinus hybridus **Lem.**
Fabaceae/Papilionaceae
♦ hybrid lupine
♦ Naturalised
♦ 101
♦ herbal.

Lupinus kingii **S.Wats.**
Fabaceae/Papilionaceae
♦ king's lupine
♦ Weed
♦ 87, 88, 161, 218
♦ herbal.

Lupinus latifolius **Lindl. ex Agardh.**
Fabaceae/Papilionaceae
♦ bigleaf lupine, broadleaf lupine
♦ Weed
♦ 161
♦ pH, cultivated, herbal, toxic.

Lupinus laxiflorus **Douglas ex Lindl.** *non* **Amer.** *auct.*
Fabaceae/Papilionaceae
= *Lupinus argenteus* Pursh ssp. *argenteus* var. *laxiflorus* (Dougl. ex Lindl.) Dorn (NoR)
♦ grassland lupine, Douglas spurred lupine
♦ Weed
♦ 87, 88, 161, 218
♦ toxic.

Lupinus leucophyllus **Douglas**
Fabaceae/Papilionaceae
♦ velvet lupine
♦ Weed
♦ 39, 87, 88, 161, 218
♦ pH, herbal, toxic.

Lupinus luteus **L.**
Fabaceae/Papilionaceae
♦ lupin, keltalupiini, yellow lupin, lupina Íltá, European yellow lupine, yellow sweet lupin
♦ Weed, Naturalised, Garden Escape, Environmental Weed, Casual Alien
♦ 7, 39, 42, 70, 86, 87, 88, 94, 98, 101, 121, 203, 250, 280
♦ aH, cultivated, herbal, toxic. Origin: Eurasia.

Lupinus macrocarpus **Hook. & Arn.**
Fabaceae/Papilionaceae
= *Lupinus arboreus* Sims
♦ Weed
♦ 87, 88

Lupinus mexicanus **Cerv. ex Lag.**
Fabaceae/Papilionaceae
Lupinus bilineatus Benth. (see), *Lupinus ehrenbergii* Schlecht.
♦ Weed
♦ 199

Lupinus micranthus **Guss.**
Fabaceae/Papilionaceae
♦ hairy lupin
♦ Weed
♦ 70
♦ herbal.

Lupinus multiflorus **Desv.**
Fabaceae/Papilionaceae
Lupinus albescens Hook. & Arn., *Lupinus incanus* Graham
♦ Introduced
♦ 228
♦ arid.

Lupinus mutabilis **Sweet**
Fabaceae/Papilionaceae
Lupinus tauris Hook.
♦ tarwi, pearl lupin, lupina menlivá
♦ Weed, Naturalised, Introduced
♦ 39, 98, 203, 228
♦ aH, arid, cultivated, herbal, toxic.

Lupinus nootkatensis **Donn ex Sims**
Fabaceae/Papilionaceae
♦ nootka lupine
♦ Weed
♦ 161
♦ pH, promoted, herbal, toxic.

Lupinus onustus **S.Watson**
Fabaceae/Papilionaceae

♦ plumas lupine
♦ Weed
♦ 39, 161
♦ pH, toxic.

Lupinus perennis **L.**
Fabaceae/Papilionaceae
♦ perennial lupine, blue bean, sundial lupin, wild lupine, lupina trváca
♦ Weed
♦ 8, 23, 87, 88, 161, 218
♦ pH, cultivated, herbal, toxic.

Lupinus pilosus **L.**
Fabaceae/Papilionaceae
Lupinus hirsutus L. *nom. rej.* (see)
♦ Russell lupin
♦ Weed, Naturalised, Cultivation Escape
♦ 7, 39, 86, 87, 88, 98, 252
♦ cultivated, herbal, toxic. Origin: southern Europe, Middle East.

Lupinus plattensis **S.Watson**
Fabaceae/Papilionaceae
♦ Platte lupine, Nebraska lupine
♦ Native Weed
♦ 161, 174
♦ Origin: North America.

Lupinus polyphyllus **Lindl.**
Fabaceae/Papilionaceae
Lupinus adscendens Rydb., *Lupinus ammophilus* Greene, *Lupinus amplus* Greene, *Lupinus biddlei* Hend. ex C.P.Sm., *Lupinus burkei* S.Watson, *Lupinus crassus* Payson, *Lupinus elongatus* Greene ex A.Heller, *Lupinus grandifolius* Lindl. ex J.Agardh, *Lupinus holmgrenianus* C.P.Sm., *Lupinus humicola* A.Nelson, *Lupinus polyphyllus* var. *saxosus* (Howell) Barneby, *Lupinus procerus* Greene ex A.Heller, *Lupinus prunophilus* M.E.Jones, *Lupinus* 'Russell hybrid' (see), *Lupinus subsericeus* B.L.Rob. ex Piper, *Lupinus superbus* A.Heller, *Lupinus tooelensis* C.P.Sm.
♦ komealupiini, Russell lupin, large leaved lupine, garden lupin, bluepod lupin, bigleaf lupine
♦ Weed, Naturalised, Garden Escape, Environmental Weed, Cultivation Escape
♦ 15, 39, 42, 54, 72, 86, 88, 152, 155, 161, 198, 225, 241, 246, 280
♦ pH, cultivated, herbal, toxic. Origin: North America.

Lupinus pusillus **Pursh**
Fabaceae/Papilionaceae
♦ low lupine, rusty lupine, small lupine
♦ Weed, Native Weed
♦ 39, 87, 88, 161, 174, 218
♦ herbal, toxic. Origin: North America.

Lupinus rivularis **Douglas ex Lindl.**
Fabaceae/Papilionaceae
♦ stream lupine, riverbank lupine
♦ Weed
♦ 87, 88, 218
♦ pH, cultivated, herbal.

Lupinus **'Russell hybrid' Lindl.**
Fabaceae/Papilionaceae

= *Lupinus polyphyllus* Lindl. × *Lupinus arboreus* Sims. × annual *Lupinus* L. spp.
♦ lupin, Russell lupin
♦ Naturalised, Garden Escape, Environmental Weed
♦ 20, 86, 252, 289
♦ cultivated. Origin: horticultural hybrid.

Lupinus sericeus Pursh
Fabaceae/Papilionaceae
♦ silky lupine
♦ Weed
♦ 39, 87, 88, 161, 218
♦ cultivated, herbal, toxic.

Lupinus sparsiflorus Benth.
Fabaceae/Papilionaceae
♦ Coulter's lupine, Mojave lupine
♦ Weed
♦ 161
♦ aH, cultivated, herbal, toxic.

Lupinus wyethii S.Wats.
Fabaceae/Papilionaceae
= *Lupinus polyphyllus* Lindl. var. *humicola* (A.Nelson) Barneby (NoR)
♦ Wyeth's lupine
♦ Weed
♦ 161, 212
♦ herbal, toxic.

Luziola fluitans (Michx.) Terrell & H.Robins.
Poaceae
Hydrochloa caroliniensis Beauv. (see)
♦ southern watergrass
♦ Weed, Quarantine Weed
♦ 161, 258
♦ G.

Luziola peruviana J.F.Gmel.
Poaceae
Luziola leiocarpa Lind., *Luziola mexicana* Kunth
♦ Peruvian watergrass
♦ Weed
♦ 255, 295
♦ pG, aqua. Origin: South America.

Luziola spruceana Benth. ex Döll
Poaceae
= *Luziola subintegra* Swallen (NoR)
♦ Weed, Quarantine Weed
♦ 76, 87, 88, 203, 220
♦ G.

Luzula albida DC.
Juncaceae
= *Luzula luzuloides* (Lam.) Dandy & Wilmott
♦ erba lucciola bianca
♦ Weed
♦ 23, 88
♦ G, herbal.

Luzula campestris (L.) DC.
Juncaceae
Juncus campestris L.
♦ woodrush, field woodrush, ketopiippo, sweeps brush
♦ Weed, Naturalised
♦ 15, 23, 70, 86, 87, 88, 161, 176, 272, 280
♦ G, cultivated, herbal. Origin: Europe, North Africa, North America.

Luzula canariensis Poir.
Juncaceae
♦ Quarantine Weed
♦ 220
♦ G, cultivated.

Luzula capitata (Miq.) Miq.
Juncaceae
♦ suzumenoyari
♦ Weed
♦ 286
♦ pG, promoted. Origin: China, Japan.

Luzula congesta (Thuill.) Lej.
Juncaceae
♦ woodrush, heath woodrush
♦ Weed, Naturalised
♦ 15, 86, 101, 176, 280
♦ G. Origin: Europe.

Luzula flaccida (Buchenau) Edgar
Juncaceae
♦ Naturalised
♦ 280
♦ G, cultivated.

Luzula luzuloides (Lam.) Dandy & Wilmott
Juncaceae
Juncoides nemorosum (Pollard) Kuntze, *Luzula albida* DC. (see), *Luzula angustifolia* (Wulfen) Wender., *Luzula nemorosa* (Pollich) E.Mey. *non* Hornem.
♦ variksenkukka, oakforest woodrush, white woodrush
♦ Naturalised, Casual Alien
♦ 40, 42, 101
♦ G, cultivated, herbal.

Luzula maxima (Reichard) DC.
Juncaceae
= *Luzula sylvatica* (Huds.) Gaud.
♦ greater woodrush
♦ Quarantine Weed
♦ 220
♦ G, cultivated.

Luzula multiflora (Ehrh.) Lej.
Juncaceae
♦ many headed woodrush, nurmipiippo, common woodrush, woodrush, nurmipiippo, yamasuzumenohie
♦ Weed, Naturalised
♦ 15, 23, 86, 88, 176, 204, 272, 280, 286
♦ G, cultivated, herbal. Origin: Europe.

Luzula pilosa (L.) Willd.
Juncaceae
= *Luzula acuminata* Raf. var. *acuminata* (NoR)
♦ hairy woodrush, kevätpiippo
♦ Weed
♦ 70
♦ G, cultivated, herbal.

Luzula sylvatica (Huds.) Gaud.
Juncaceae
Juncus silvaticus Huds., *Luzula maxima* (Reichard) DC. (see)
♦ great woodrush
♦ Weed, Quarantine Weed
♦ 70, 220
♦ pG, cultivated, herbal.

Lychnis alba Mill.
Caryophyllaceae

= *Silene latifolia* Poir. ssp. *alba* (Mill.) Greut. & Burd.
♦ white cockle, white campion
♦ Weed, Noxious Weed
♦ 23, 36, 52, 80, 87, 88, 136, 162, 210, 218
♦ b/pH, herbal.

Lychnis chalcedonica L.
Caryophyllaceae
♦ Maltese cross, palavarakkaus
♦ Weed, Naturalised, Cultivation Escape, Casual Alien
♦ 39, 42, 80, 86, 98, 101, 203, 241, 252, 280
♦ cultivated, herbal, toxic. Origin: central and eastern Russia.

Lychnis coronaria (L.) Desr.
Caryophyllaceae
Agrostemma coronaria L., *Silene coronaria* (L.) Clairv. (see)
♦ rose campion, mullein pink
♦ Weed, Naturalised, Cultivation Escape
♦ 7, 15, 23, 34, 40, 80, 86, 88, 98, 101, 136, 198, 203, 241, 252, 280, 287, 300
♦ pH, cultivated, herbal. Origin: Europe.

Lychnis dioica L.
Caryophyllaceae
= *Silene dioica* (L.) Clairv.
♦ red campion
♦ Weed
♦ 39, 80, 87, 88, 218
♦ herbal, toxic.

Lychnis flos-cuculi L.
Caryophyllaceae
Coronaria flos-cuculi (L.) A.Braun, *Silene flos-cuculi* (L.) Clairv. (see)
♦ ragged robin, käenkukka, meadow campion
♦ Weed, Naturalised
♦ 23, 39, 70, 80, 87, 88, 101, 133, 195, 218, 224, 272, 280
♦ pH, cultivated, herbal, toxic.

Lychnis fulgens Fisch. ex Sims
Caryophyllaceae
♦ brilliant campion, ezosennou
♦ Weed, Naturalised
♦ 101, 297
♦ pH, cultivated. Origin: China, Japan, Korea, Manchuria, Siberia.

Lychnis githago (L.) Scop.
Caryophyllaceae
= *Agrostemma githago* L.
♦ Weed
♦ 80
♦ herbal.

Lychnis viscaria L.
Caryophyllaceae
Silene viscaria (L.) Jess. (see), *Steris viscaria* (L.) Raf., *Viscaria viscosa* (Scop.) Asch.
♦ clammy campion, German catchfly, red catchfly, mäkitervakko, sticky catchfly
♦ Weed, Naturalised, Casual Alien
♦ 23, 88, 101, 272, 280
♦ cultivated, herbal.

Lycianthes asarifolia (Kunth & Bouché)
Bitter
   Solanaceae
- gingerleaf
- Weed, Quarantine Weed,
Naturalised
- 76, 88, 101

Lycianthes rantonnettii (Carrière ex
Lescuy.) Bitter
   Solanaceae
- blue potatobush
- Naturalised
- 101

Lycium afrum L.
   Solanaceae
- kaffir boxthorn
- Naturalised, Environmental Weed
- 86, 198
- S, cultivated. Origin: southern
Africa.

Lycium barbarum L.
   Solanaceae
   Jasminoides flaccida Veill., *Lycium*
   halimifolium Mill. (see)
- matrimony vine, Duke of Argyll's
teaplant, Chinese boxthorn, morali,
murali, boxthorn
- Weed, Naturalised, Introduced,
Environmental Weed, Cultivation
Escape
- 15, 34, 39, 40, 86, 87, 88, 98, 101, 132,
174, 176, 198, 203, 228, 252, 269, 272,
280
- S, arid, cultivated, herbal, toxic.
Origin: Eurasia.

Lycium carolinianum Walter
   Solanaceae
- salt matrimony vine, Carolina desert
thorn, Christmas berry
- Weed
- 39, 161
- S, promoted, herbal, toxic.

Lycium chilense Miers ex Bertero
   Solanaceae
   Lycium grevilleanum Gillet ex
   Miers, *Lycium lasiopetalum* Speg.,
   Lycium patagonicum Miers, *Lycium*
   pulverulentum Skottsb., *Lycium*
   scoparium Miers
- coralillo, yaullìn
- Introduced
- 228
- arid, cultivated.

Lycium chinense Mill.
   Solanaceae
- Chinese desert thorn, matrimony
vine, Duke of Argyll's teatree, Chinese
matrimony vine, Chinese boxthorn,
boxthorn
- Weed, Naturalised, Introduced
- 34, 101, 275, 286
- S, cultivated, herbal.

Lycium cinereum Thunb. (*sens. lat.*)
   Solanaceae
   Lycium arenicola Miers, *Lycium*
   caespitosum Dinter & Dammer, *Lycium*
   colletioides Dammer, *Lycium echinatum*
   Dunal, *Lycium kraussii* Dunal, *Lycium*

leptacanthum C.H.Wright, *Lycium*
minutiflorum Dammer, *Lycium*
omahekense Dammer, *Lycium oxycladum*
Miers, *Lycium pendulinum* Miers,
Lycium pumilum Dammer, *Lycium*
roridum Miers, *Lycium tenue* Willd.,
Lycium tetrandrum Thunb.
- honey thorn
- Weed
- 10

Lycium europaeum L.
   Solanaceae
   Lycium intricatum Boiss.
- ad gorad, aushaz, ekakebekete, fub,
fursaa, fursh, lokei, ol okii, pkata
- Weed
- 87, 88, 221, 272
- S, arid, cultivated, herbal.

Lycium ferocissimum Miers
   Solanaceae
   Lycium campanulatum E.Mey. ex
   C.H.Wright, *Lycium macrocalyx* Domin
- African boxthorn, boxthorn, cape
boxthorn
- Weed, Quarantine Weed, Noxious
Weed, Naturalised, Native Weed,
Garden Escape, Environmental Weed
- 7, 15, 39, 67, 72, 76, 86, 87, 88, 93, 98,
121, 134, 140, 147, 161, 165, 169, 171,
176, 181, 198, 203, 205, 225, 229, 246,
269, 280, 289, 296
- pS, arid, cultivated, toxic. Origin:
southern Africa.

Lycium halimifolium Mill.
   Solanaceae
   = *Lycium barbarum* L.
- matrimony vine, Duke of Argyll's
teaplant
- Weed
- 39, 87, 88, 161, 218
- herbal, toxic.

Lycium shawii Roem. & Schult.
   Solanaceae
   Lycium albiflorum Phil., *Lycium*
   albiflorum Dammer, *Lycium arabicum*
   Schweinf. ex Boiss., *Lycium cufodonfii*
   Lanza, *Lycium ellenbeckii* Dammer,
   Lycium jaegeri Dammer, *Lycium*
   javallense Lanza, *Lycium merkeri*
   Dammer, *Lycium orientale* Miers,
   Lycium somalense Dammer, *Lycium*
   tenuiramosum Dammer, *Lycium*
   withaniifolium Dammer
- awsaj
- Weed
- 221
- arid.

Lycopersicon esculentum Mill.
   Solanaceae
   Lycopersicon lycopersicum (L.) Karst.
   nom. rej. (see), *Solanum lycopersicum* L.
   var. *lycopersicum* (see)
- tomato, garden tomato, tomat,
tomaatti, tomate, pomidor
- Weed, Naturalised, Environmental
Weed, Cultivation Escape, Casual
Alien
- 7, 15, 34, 39, 40, 42, 86, 98, 154, 161,
179, 198, 247, 252, 257, 280

- a/pH, cultivated, herbal, toxic.
Origin: obscure.

Lycopersicon esculentum Mill. var.
cerasiforme (Dunal) Gray
   Solanaceae
   Lycopersicon cerasiforme Dunal,
   Lycopersicon lycopersicum L. var.
   cerasiforme auct., *Solanum lycopersicum*
   L. var. *cerasiforme* (Dunal) Spoon.
   J.Anders. & R.K.Jansen (see)
- kirsikkatomaatti
- Weed, Naturalised
- 93, 287
- herbal. Origin: obscure.

Lycopersicon lycopersicum (L.) Karst.
nom. rej.
   Solanaceae
   = *Lycopersicon esculentum* Mill.
- tomato
- Weed
- 39, 80, 87, 88
- cultivated, herbal, toxic.

Lycopersicon pennellii (Correll) D'Arcy
   Solanaceae
   Solanum pennellii Correll
- Introduced
- 228
- arid.

Lycopersicon peruvianum Mill.
   Solanaceae
- Peruvian tomato
- Weed
- 87, 88
- a/pH, arid, promoted.

Lycopersicon pimpinellifolium (Jusl.)
Mill.
   Solanaceae
   Solanum pimpinellifolium Jusl. (see)
- cherry tomato, currant tomato
- Weed
- 87, 88
- arid, cultivated, herbal.

Lycopodium cernuum L.
   Lycopodiaceae
   = *Lycopodiella cernua* (L.) Pic.Serm. var.
   cernua (NoR)
- mizusugi
- Weed
- 14, 87, 88
- H, cultivated, herbal.

Lycopodium clavatum L.
   Lycopodiaceae
- common clubmoss, running
clubmoss, ground pine, stag's horn
clubmoss, clubmoss, katinlieko, shen
jin cao, licopodio
- Weed
- 8, 39, 272
- pH, aqua, cultivated, herbal, toxic.

Lycopodium sabinifolium Willd.
   Lycopodiaceae
- purple loosestrife, savinleaf
groundpine
- Noxious Weed
- 229
- herbal.

Lycopsis arvensis L.
Boraginaceae
= *Anchusa arvensis* (L.) Bieb.
♦ small bugloss, bugloss
♦ Weed
♦ 87, 88, 218, 237, 243, 269, 295
♦ pH, cultivated, herbal. Origin:
Eurasia.

Lycopsis orientalis L.
Boraginaceae
♦ Weed, Casual Alien
♦ 40, 275, 297
♦ aH.

Lycopus americanus Muhl. ex W.Bart.
Lamiaceae
♦ American bugleweed, water
horehound, American water
horehound
♦ Weed, Native Weed
♦ 87, 88, 161, 174, 218
♦ pH, cultivated, herbal. Origin: North
America.

Lycopus asper Greene
Lamiaceae
Lycopus lucidus auct. p.p. non Turcz. ex
Benth. (see)
♦ rough bugleweed
♦ Weed
♦ 87, 88, 218
♦ pH, promoted, herbal.

Lycopus europaeus L.
Lamiaceae
Lycopus aquaticus Moench
♦ European bugleweed, gypsywort,
rantayrtti
♦ Weed, Naturalised
♦ 23, 70, 87, 88, 101, 218, 272, 280, 300
♦ pH, aqua, cultivated, herbal.

Lycopus exaltatus L.
Lamiaceae
Lycopus pinnatifidus Pall.
♦ erba sega maggiore
♦ Weed
♦ 272
♦ cultivated, herbal.

**Lycopus lucidus auct. p.p. non Turcz. ex
Benth.**
Lamiaceae
= *Lycopus asper* Greene
♦ shirone
♦ Weed
♦ 87, 88, 286, 297
♦ pH, promoted, herbal. Origin: east
Asia.

Lycopus maackianus (Maxim.) Makino
Lamiaceae
♦ himeshirone
♦ Weed
♦ 286
♦ pH, promoted. Origin: China, Japan.

Lycopus uniflorus Michx.
Lamiaceae
Lycopus parviflorus Maxim.
♦ oneflower bugleweed, northern
bugleweed, bugleweed
♦ Weed
♦ 87, 88, 218
♦ pH, promoted, herbal.

Lycoris aurea (L'Hér.) Herb.
Liliaceae/Amaryllidaceae
♦ golden spiderlily
♦ Weed
♦ 161, 247
♦ pH, cultivated, herbal, toxic.

Lycoris radiata (L'Hér.) Herb.
Liliaceae/Amaryllidaceae
♦ red spiderlily, spiderlily
♦ Weed, Naturalised
♦ 39, 101, 161, 247, 286
♦ pH, cultivated, herbal, toxic.

Lycoris sanguinea Maxim.
Liliaceae/Amaryllidaceae
♦ spiderlily
♦ Weed
♦ 286
♦ pH, cultivated, herbal. Origin:
China, Japan.

Lycoris squamigera Maxim.
Liliaceae/Amaryllidaceae
♦ resurrection lily, magic lily
♦ Weed, Naturalised
♦ 39, 101, 161, 247, 287
♦ pH, cultivated, herbal, toxic.

Lycurus phleoides Kunth
Poaceae
♦ common wolfstail
♦ Introduced
♦ 228
♦ G, arid, herbal.

Lygeum spartum Loefl. ex L.
Poaceae
♦ lygeum
♦ Weed, Introduced
♦ 221, 228
♦ pG, arid, cultivated.

Lygodesmia juncea (Pursh) D.Don
Asteraceae
♦ skeleton weed, rush skeletonweed,
rush skeletonplant, rushlike
lygodesmia, skeleton pink
♦ Weed, Native Weed
♦ 49, 87, 88, 161, 174, 210, 212, 218, 243
♦ pH, promoted, herbal, toxic. Origin:
North America.

Lygodium circinnatum Sw.
Schizaeaceae
♦ Weed
♦ 87, 88
♦ herbal.

Lygodium flexuosum (L.) Sw.
Schizaeaceae
♦ climbing fern
♦ Weed
♦ 12, 87, 88, 209
♦ cultivated, herbal.

Lygodium japonicum (Thunb.) Sw.
Schizaeaceae
♦ Japanese climbing fern, climbing
fern
♦ Weed, Noxious Weed, Naturalised,
Garden Escape, Environmental Weed
♦ 77, 80, 86, 87, 88, 101, 112, 122, 142,
151, 152, 179, 286, 297
♦ cultivated, herbal. Origin: Japan to
Australia.

Lygodium microphyllum (Cav.) R.Br.
Schizaeaceae
♦ old world climbing fern, climbing
maidenhair fern
♦ Weed, Naturalised, Environmental
Weed
♦ 80, 88, 101, 112, 151, 179, 246
♦ cultivated, herbal.

Lygodium polymorphum (Cav.) H.B.K.
Schizaeaceae
♦ Weed
♦ 87, 88

Lygodium scandens (L.) Sw.
Schizaeaceae
♦ Weed, Quarantine Weed
♦ 76, 87, 88, 203, 220
♦ cultivated, herbal.

Lygos monosperma (L.) Heywood
Fabaceae/Papilionaceae
= *Retama monosperma* (L.) Boiss.
♦ Naturalised, Environmental Weed
♦ 86

Lygos raetam (Forssk.) Heywood
Fabaceae/Papilionaceae
= *Retama raetam* (Forssk.) Webb &
Berthel.
♦ Weed
♦ 221

Lyonia ligustrina (L.) DC.
Ericaceae
♦ maleberry
♦ Weed
♦ 161
♦ cultivated, herbal, toxic.

Lyonia mariana (L.) D.Don.
Ericaceae
♦ stagger bush, piedmont staggerbush
♦ Weed
♦ 8, 161
♦ cultivated, herbal, toxic.

**Lyonothamnus floribundus A.Gray ssp.
asplenifolius (Greene) P.H.Raven**
Rosaceae
♦ Santa Cruz Island ironwood,
Catalina ironwood, fernleaf ironwood,
fern leaved Catalina ironwood
♦ Weed
♦ 116
♦ T, cultivated.

**Lysichiton americanus Hultén &
H.St.John**
Araceae
♦ American skunk cabbage, yellow
skunk cabbage
♦ Weed, Naturalised
♦ 40, 161
♦ cultivated, herbal, toxic.

Lysiloma acapulcensis (Kunth) Benth.
Fabaceae/Mimosaceae
♦ Quarantine Weed
♦ 220
♦ cultivated, herbal.

Lysiloma bahamensis Benth.
Fabaceae/Mimosaceae
♦ wild tamarind
♦ Weed
♦ 32
♦ cultivated.

Lysiloma divaricatum (Jacq.) J.F.Macbr.
Fabaceae/Mimosaceae
Acacia divaricata (Jacq.) Willd.,
Lysiloma australis Britton & Rose,
Lysiloma calderonii Britton & Rose,
Lysiloma cayacensis M.E.Jones, *Lysiloma
chiapensis* Britton & Rose, *Lysiloma
divaricata* (Jacq.) J.F.Macbr. (see),
Lysiloma kellermanii Britton & Rose,
Lysiloma salvadorensis Britton & Rose,
Lysiloma schiedeana Benth., *Lysiloma
seemannii* Britton & Rose, *Mimosa
divaricata* Jacq.
♦ Quarantine Weed, Introduced
♦ 220, 228
♦ arid.

Lysiloma sabicu Benth.
Fabaceae/Mimosaceae
♦ horseflesh mahogany, wild
tamarind, horse flesh
♦ Weed, Naturalised
♦ 101, 179
♦ cultivated.

Lysiloma tergemina Benth.
Fabaceae/Mimosaceae
♦ Quarantine Weed
♦ 220

Lysimachia atropurpurea L.
Primulaceae
♦ Weed
♦ 272
♦ cultivated.

Lysimachia barystachys Bunge
Primulaceae
♦ Manchurian yellow loosestrife, bog
loosestrife
♦ Weed, Naturalised
♦ 101, 297
♦ pH, cultivated, herbal.

Lysimachia candida Lindl.
Primulaceae
♦ Weed
♦ 275, 297
♦ a/bH.

Lysimachia christinae Hance
Primulaceae
♦ Christina loosestrife
♦ Weed
♦ 297
♦ pH, promoted, herbal. Origin:
China.

Lysimachia clethroides Duby
Primulaceae
♦ gooseneck yellow loosestrife,
gooseneck loosestrife, loosestrife
♦ Weed, Naturalised
♦ 80, 87, 88, 101, 204, 286, 297
♦ pH, aqua, cultivated, herbal.

Lysimachia fortunei Maxim
Primulaceae
♦ numatoranoo
♦ Weed
♦ 87, 88, 286
♦ pH, aqua, promoted, herbal. Origin:
east Asia.

Lysimachia grammica Hance
Primulaceae

♦ striate loosestrife
♦ Weed
♦ 297

Lysimachia japonica Thunb.
Primulaceae
♦ Japanese yellow loosestrife
♦ Weed, Naturalised
♦ 86, 87, 88, 101, 204, 286
♦ cultivated, herbal. Origin: Asia.

**Lysimachia japonica Thunb. fo.
subsessilis Murata**
Primulaceae
♦ Weed
♦ 286

Lysimachia leucantha Miq.
Primulaceae
♦ Weed
♦ 235

Lysimachia mauritiana Lam.
Primulaceae
♦ spoonleaf yellow loosestrife
♦ Weed
♦ 87, 88

Lysimachia nemorum L.
Primulaceae
Ephemerum nemorosum Schur, *Lerouxia
nemorum* (L.) Mérat
♦ yellow pimpernel
♦ Weed
♦ 70
♦ cultivated, herbal.

Lysimachia nummularia L.
Primulaceae
Ephemerum nummularia Schur, *Lerouxia
nummularia* (L.) Á.Löve
♦ moneywort, creeping loosestrife,
yellow myrtle, creeping Jenny,
creeping Charlie, herb twopence,
twopenny grass, creeping Penny,
moneywory
♦ Weed, Naturalised, Introduced,
Environmental Weed, Cultivation
Escape
♦ 4, 39, 40, 42, 70, 80, 86, 87, 88, 101,
102, 104, 133, 151, 161, 176, 195, 211,
218, 222, 224, 252, 272, 280, 287
♦ wpH, cultivated, herbal, toxic.
Origin: Eurasia.

Lysimachia pentapetala Bunge
Primulaceae
♦ fivepetal loosestrife
♦ Weed
♦ 297

Lysimachia punctata L.
Primulaceae
Lysimachia quadrifolia Mill., *Lysimachia
verticillaris* Spreng.
♦ large yellow loosestrife, tarha alpi,
dotted loosestrife, garden loosestrife
♦ Weed, Naturalised, Cultivation
Escape
♦ 40, 42, 80, 87, 88, 101
♦ cultivated, herbal.

Lysimachia ruhmeriana Vatke
Primulaceae
♦ Weed
♦ 240

♦ aH.

**Lysimachia terrestris (L.) Britton, Sterns
& Pogg.**
Primulaceae
♦ Amerikanalpi, earth loosestrife
♦ Weed, Casual Alien
♦ 42, 80
♦ cultivated, herbal.

Lysimachia thyrsiflora L.
Primulaceae
Naumburgia thyrsiflora (L.) Rchb.
♦ tufted loosestrife, terttualpi,
loosestrife
♦ Weed
♦ 23, 87, 88, 159, 243
♦ pH, herbal.

Lysimachia vulgaris L.
Primulaceae
♦ yellow loosestrife, ranta alpi, garden
loosestrife, common loosestrife
♦ Weed, Noxious Weed, Naturalised
♦ 1, 23, 70, 80, 87, 88, 98, 101, 133, 139,
146, 161, 195, 197, 198, 203, 224, 229,
272, 280, 287
♦ pH, aqua, cultivated, herbal.

**Lysimachia vulgaris L. var. davurica
(Ledeb.) Knuth**
Primulaceae
♦ yellow loosestrife
♦ Naturalised
♦ 86, 198
♦ cultivated. Origin: Eurasia.

Lysimachia vulgaris L. var. vulgaris
Primulaceae
♦ yellow loosestrife
♦ Naturalised
♦ 86, 198
♦ cultivated. Origin: Eurasia.

Lysiphyllum hookeri (F.Muell.) Pedley
Fabaceae/Caesalpiniaceae
♦ Hooker's bauhinia
♦ Introduced
♦ 228
♦ arid, cultivated. Origin: Australia.

Lythrum L. spp.
Lythraceae
♦ loosestrife
♦ Weed
♦ 67, 272

Lythrum acutangulum Lag.
Lythraceae
♦ Weed
♦ 87, 88

Lythrum alatum Pursh
Lythraceae
♦ loosestrife, winged lythrum
♦ Weed, Noxious Weed
♦ 159, 229
♦ herbal.

Lythrum alatum Pursh var. alatum
Lythraceae
♦ winged lythrum
♦ Noxious Weed
♦ 229

**Lythrum alatum Pursh var. lanceolatum
(Ell.) Torr. & Gray ex Rothr.**
Lythraceae

- winged lythrum
- Noxious Weed
- 229

Lythrum album H.B.K.
Lythraceae
- Naturalised
- 241, 300
- herbal.

Lythrum anceps (Kohne) Makino
Lythraceae
- spiked loosestrife, purple loosestrife, black blood
- Weed
- 87, 88, 263, 286
- pH, cultivated, herbal.

Lythrum californicum Torr. & Gray
Lythraceae
- California loosestrife
- Noxious Weed
- 229
- pH, herbal.

Lythrum curtissii Fern.
Lythraceae
- Curtis's loosestife
- Noxious Weed
- 229
- herbal.

Lythrum flagellare Shuttlw. ex Chapman
Lythraceae
- Florida loosestrife
- Noxious Weed
- 229
- herbal.

Lythrum hyssopifolia L.
Lythraceae
Lythrum adsurgens Greene, *Lythrum hyssopifolium* L. (see)
- hyssop lythrum, hyssop loosestrife, hyssop leaved loosestrife, grass poly, grass roly poly
- Weed, Noxious Weed, Naturalised, Environmental Weed, Cultivation Escape, Casual Alien
- 7, 9, 15, 42, 70, 86, 87, 88, 94, 98, 101, 134, 161, 165, 200, 203, 208, 218, 229, 237, 241, 243, 253, 269, 272, 280, 287, 295
- aH, aqua, cultivated, herbal. Origin: Mediterranean, west Asia.

Lythrum hyssopifolium L.
Lythraceae
= *Lythrum hyssopifolia* L.
- hyssop loosestrife, grass poly, hyssop lythrum, loosestrife
- Weed, Naturalised
- 80, 121, 180, 221, 300
- a/pH. Origin: Europe.

Lythrum intermedium Ledeb. ex Colla
Lythraceae
- Weed
- 275
- pH.

Lythrum junceum Banks & Sol.
Lythraceae
- Mediterranean loosestrife, rose loosestrife, false grass poly, kimppurantakukka'

- Weed, Naturalised, Environmental Weed, Casual Alien
- 40, 42, 72, 86, 87, 88, 94, 98, 198, 203, 208, 253, 280
- pH, aqua, cultivated. Origin: Mediterranean.

Lythrum lineare L.
Lythraceae
- wand lythrum
- Noxious Weed
- 229
- herbal.

Lythrum maritimum Kunth
Lythraceae
- pukamole
- Noxious Weed, Naturalised
- 101, 229, 241, 300
- arid, herbal.

Lythrum ovalifolium Koehne
Lythraceae
- low loosestrife
- Noxious Weed
- 229

Lythrum portula (L.) Webb
Lythraceae
Peplis portula L. (see)
- broadleaf loosestrife, spatulaleaf loosestrife, water purslane
- Weed, Noxious Weed, Naturalised, Introduced
- 34, 70, 101, 229, 272, 280, 300
- aH, aqua, cultivated, herbal.

Lythrum salicaria L.
Lythraceae
Lythrum palustre Salisb., *Lythrum spicatum* S.F.Gray, *Lythrum spiciforme* Dulac
- purple loosestrife, spiked loosestrife, rantakukka, rainbow weed, salicaire, swamp loosestrife
- Weed, Sleeper Weed, Quarantine Weed, Noxious Weed, Naturalised, Native Weed, Introduced, Garden Escape, Environmental Weed, Cultivation Escape
- 1, 4, 15, 17, 18, 21, 23, 26, 34, 35, 35, 45, 48, 52, 63, 70, 78, 80, 86, 87, 88, 94, 101, 102, 103, 104, 105, 116, 129, 130, 133, 136, 137, 138, 139, 141, 142, 146, 151, 152, 159, 161, 162, 174, 195, 197, 211, 212, 218, 219, 222, 224, 225, 229, 231, 246, 252, 267, 269, 272, 275, 280, 283, 286, 291, 297, 299, 300
- pH, aqua, cultivated, herbal. Origin: Eurasia.

Lythrum thymifolia L.
Lythraceae
- timjamirantakukka, thymeleaf loosestrife
- Noxious Weed, Naturalised, Casual Alien
- 42, 101, 229

Lythrum tribracteatum Salzm. ex Spreng.
Lythraceae
- threebract loosestrife
- Noxious Weed, Naturalised
- 101, 229
- aH, herbal.

Lythrum virgatum L.
Lythraceae
Lythrum austriacum Jacq.
- European wand loosestrife, loosestrife wand, purple loosestrife
- Weed, Noxious Weed, Naturalised
- 1, 17, 80, 88, 94, 101, 139, 146, 161, 229, 272
- cultivated, herbal.

Lythrum virgatum L. × alatum Pursh
Lythraceae
- loosestrife
- Weed
- 159

M

Macadamia integrifolia Maiden & Betche
Proteaceae
♦ macadamia nut, Queensland nut, smooth macadamia
♦ Naturalised
♦ 39, 101, 261
♦ T, cultivated, toxic. Origin: Australia.

Macadamia tetraphylla L.A.S.Johnson
Proteaceae
♦ macadamia nut, Queensland nut, macadamia, rough macadamia
♦ Naturalised
♦ 39, 280
♦ T, cultivated, toxic.

Macaranga carolinensis Volk. var. grandiflora Pax & Hoff.
Euphorbiaceae
♦ apwid
♦ Introduced
♦ 230
♦ T.

Macaranga harveyana Müll.Arg.
Euphorbiaceae
♦ lau pata
♦ Weed, Quarantine Weed
♦ 76, 87, 88, 203, 220
♦ herbal.

Macaranga mappa (L.) Müll.Arg.
Euphorbiaceae
♦ pengua
♦ Weed, Naturalised
♦ 80, 101

Macaranga tanarius (L.) Müll.Arg.
Euphorbiaceae
♦ parasolleaf tree
♦ Weed, Naturalised
♦ 80, 101
♦ cultivated, herbal. Origin: Australia.

Macaranga triloba (Reinw.) Müll.Arg.
Euphorbiaceae
♦ Weed, Quarantine Weed
♦ 76, 87, 88, 203, 220
♦ herbal.

Macfadyena unguis-cati (L.) A.H.Gentry
Bignoniaceae
Bignonia unguis-cati L. (see), *Batocydia unguis* (L. ex DC.) Mart. ex DC., *Bignonia acutistipula* Schltdl., *Bignonia inflata* Griseb., *Bignonia tweediana* Lindl., *Bignonia unguis* L. ex DC., *Doxantha acutistipula* (Schltdl.) Miers, *Doxantha unguis-cati* (L.) Miers ex Rehder, *Doxantha unguis* (L. ex DC.) Miers, *Bignonia exoleta* Vell. (see)

♦ cat's claw vine, claw vine, cat's claw creeper, cat's claw trumpet, funnel creeper, katteklouranker
♦ Weed, Quarantine Weed, Noxious Weed, Naturalised, Garden Escape, Environmental Weed, Casual Alien
♦ 3, 63, 73, 76, 80, 86, 88, 95, 98, 101, 112, 151, 155, 179, 191, 201, 203, 220, 246, 255, 269, 279, 280, 283, 290, 296
♦ pC, cultivated. Origin: Brazil.

Machaeocereus Britt. & Rose spp.
Cactaceae
= *Stenocereus* (Berger) Riccob. spp.
♦ Weed, Quarantine Weed
♦ 76, 88, 203

Machaeranthera canescens (Pursh) Gray
Asteraceae
Aster canescens Pursh
♦ purple aster, hoary tansyaster, hoary aster
♦ Weed
♦ 23, 88, 161, 212
♦ pH, herbal.

Machaeranthera pinnatifida (Hook.) Shinners
Asteraceae
Haplopappus spinulosus (Pursh.) DC.
♦ lacy tansyaster, cutleaf ironplant
♦ Weed
♦ 161
♦ herbal.

Machaerina falcata (Nees) T.Koyama
Cyperaceae
Cladium ponapense Ohwi
♦ twigrush
♦ Introduced
♦ 230
♦ G.

Machaerium nictitans (Vell. Conc.) Benth.
Fabaceae/Papilionaceae
♦ Introduced
♦ 228
♦ arid.

Mackaya bella Harv.
Acanthaceae
♦ forest bell bush, kapbuske makaija
♦ Weed, Casual Alien
♦ 15, 280
♦ cultivated, herbal.

Macleaya cordata (Willd.) R.Br. ex G.Don
Papaveraceae
♦ plume poppy, tree celandine
♦ Weed, Naturalised
♦ 80, 87, 88, 101, 102, 286
♦ pH, cultivated, herbal.

Maclura pomifera (Raf.) C.K.Schneid.
Moraceae
♦ osage orange, bowwood
♦ Weed, Naturalised, Environmental Weed, Cultivation Escape
♦ 8, 39, 80, 86, 87, 88, 98, 121, 151, 161, 195, 198, 203, 218, 280
♦ S/T, cultivated, herbal, toxic. Origin: North America.

Macrocystis pyrifera (L.) C.Agardh
Lessoniaceae

♦ brown alga, California giant kelp, kelp
♦ Weed
♦ 197
♦ algae, herbal.

Macronema discoidea Nutt.
Asteraceae
= *Ericameria discoidea* (Nutt.) G.L.Nesom (NoR)
♦ Quarantine Weed
♦ 220

Macropiper melchior W.R.Sykes
Piperaceae
♦ Quarantine Weed
♦ 220
♦ cultivated. Origin: New Zealand, south-west Pacific.

Macroptilium atropurpureum (DC.) Urb.
Fabaceae/Papilionaceae
Phaseolus atropurpureus DC. (see), *Phaseolus canescens* M.Martens & Galeotti, *Phaseolus dysophyllus* Benth., *Phaseolus vestitus* Hook.
♦ siratro, phasey bean, pini, cowpea, purple bushbean
♦ Weed, Naturalised, Introduced, Environmental Weed
♦ 3, 7, 86, 93, 98, 155, 179, 203, 228, 255, 261
♦ pH, arid, cultivated, herbal. Origin: tropical America.

Macroptilium gibbosifolium (Ortega) A.Delgado
Fabaceae/Papilionaceae
♦ variableleaf bushbean
♦ Naturalised
♦ 101
♦ herbal.

Macroptilium lathyroides (L.) Urb.
Fabaceae/Papilionaceae
Phaseolus lathyroides L. (see)
♦ cowpea, phasey bean, wild bushbean
♦ Weed, Naturalised, Environmental Weed
♦ 6, 7, 14, 86, 88, 93, 98, 179, 203, 255, 287
♦ aH, arid, cultivated, herbal. Origin: tropical America.

Macroptilium longipedunculatus (C.Mart. ex Benth.) Urb.
Fabaceae/Papilionaceae
Phaseolus campestris Benth., *Phaseolus longepedunculatus* C.Mart. ex Benth.
♦ Introduced
♦ 228
♦ arid.

Macroptilium prostratum (Benth.) Urb.
Fabaceae/Papilionaceae
Phaseolus prostratus Benth.
♦ porotillo del campo
♦ Weed
♦ 295

Macrothelypteris polypodioides (Hook.) Holttum
Thelypteridaceae
Lastrea leucolepis Presl
♦ Weed

♦ 87, 88
♦ cultivated. Origin: Australia.

Macrothelypteris torresiana (Gaud.) Ching
Thelypteridaceae
♦ swordfern
♦ Weed, Naturalised
♦ 101, 121, 179
♦ pH, cultivated, herbal. Origin: North America.

Macrotyloma axillare (E.Mey.) Verdc.
Fabaceae/Papilionaceae
Dolichos axillaris E.Mey.
♦ perennial horse gram
♦ Weed, Naturalised, Native Weed
♦ 86, 98, 121, 203
♦ pC, cultivated. Origin: Africa, Madagascar.

Macrotyloma geocarpum (Harms) Maréchal & Baudet
Fabaceae/Papilionaceae
Kerstingiella geocarpa Harms
♦ geocarpa groundnut
♦ Introduced
♦ 228
♦ arid.

Macrotyloma maranguense (Taub.) Verdc.
Fabaceae/Papilionaceae
Dolichos taubertii Bak.f.
♦ Native Weed
♦ 121
♦ pC. Origin: southern Africa.

Macrozamia communis (L.) L.A.S.Johnson
Zamiaceae
♦ Australian cycad
♦ Weed, Naturalised
♦ 39, 87, 88, 198
♦ cultivated, herbal, toxic. Origin: Australia.

Macrozamia spiralis (Salisb.) Miq.
Zamiaceae
♦ burrawong
♦ Weed
♦ 39, 87, 88
♦ cultivated, toxic. Origin: Australia.

Madia capitata Nutt.
Asteraceae
= *Madia sativa* Molina
♦ Casual Alien
♦ 280
♦ herbal.

Madia elegans D.Don ex Lindl.
Asteraceae
Madaria elegans (D.Don ex Lindl.) DC.
♦ showy tarweed, common madia, tarweed
♦ Weed
♦ 87, 88, 161, 218
♦ aH, cultivated, herbal.

Madia exigua (Sm.) Gray
Asteraceae
♦ little tarweed, small tarweed, threadstem tarweed
♦ Weed
♦ 23, 88
♦ aH, herbal.

Madia glomerata Hook.
Asteraceae
♦ cluster tarweed, mountain tarweed
♦ Weed
♦ 23, 87, 88, 136, 161, 218
♦ aH, promoted, herbal.

Madia sativa Molina
Asteraceae
Madia capitata Nutt. (see), *Madia sativa* Molina var. *congesta* Torr. & Gray (see)
♦ Chilean tarweed, madia oil, madi, coast tarweed
♦ Weed, Noxious Weed, Naturalised, Casual Alien
♦ 23, 34, 40, 86, 87, 88, 98, 161, 165, 198, 203, 212, 229, 237, 280, 295
♦ aH, cultivated, herbal. Origin: Chile, Argentina.

Madia sativa Molina var. congesta Torr. & Gray
Asteraceae
= *Madia sativa* Molina
♦ Chilean tarweed
♦ Weed
♦ 218

Maerua crassifolia Forssk.
Capparaceae
Maerua rigida R.Br. ex G.Don, *Maerua senegalensis* R.Br. ex G.Don
♦ maeru
♦ Weed
♦ 221
♦ arid, herbal. Origin: North Africa, Middle East.

Maesa tenera Mez.
Myrsinaceae
♦ Weed
♦ 87, 88

Maesopsis eminii Engl.
Rhamnaceae
♦ musizi, mumgunguleLushogo, bosongu, umbrella tree
♦ Weed, Naturalised, Environmental Weed, Cultivation Escape
♦ 22, 37, 101, 152, 261
♦ T, cultivated. Origin: west tropical Africa.

Magnolia acuminata L.
Magnoliaceae
♦ cucumbertree, cucumber magnolia
♦ Weed
♦ 88, 218
♦ T, cultivated, herbal.

Magnolia grandiflora L.
Magnoliaceae
♦ bull bay, magnolia, southern magnolia
♦ Introduced
♦ 39, 139, 261
♦ T, cultivated, herbal, toxic. Origin: southern North America.

Magnolia macrophylla Michx.
Magnoliaceae
♦ bigleaf magnolia, great leaved magnolia
♦ Weed
♦ 88, 218
♦ T, cultivated, herbal.

Magnolia × soulangeana Soul.-Bod.
Magnoliaceae
= *Magnolia heptapeta* (Buc'hoz) Dandy × *Magnolia quinquepeta* (Buc'hoz) Dandy
♦ Chinese magnolia
♦ Naturalised
♦ 101
♦ cultivated, herbal.

Magnolia stellata (Sieb. & Zucc.) Maxim.
Magnoliaceae
♦ star magnolia
♦ Naturalised
♦ 101
♦ S, cultivated, herbal.

Magnolia tripetala (L.) L.
Magnoliaceae
♦ umbrella tree, magnolia parasolowata
♦ Weed
♦ 195
♦ cultivated, herbal.

Magnolia virginiana L.
Magnoliaceae
♦ sweetbay magnolia, laurel magnolia, sweet bay, white laurel
♦ Weed
♦ 88, 218
♦ T, cultivated, herbal.

Mahoberberis C.K.Schneid. spp.
Berberidaceae
♦ Quarantine Weed
♦ 220

Mahonia Nutt. spp.
Berberidaceae
♦ Oregon grape, holly barber
♦ Quarantine Weed
♦ 220

Mahonia aquifolium (Pursh) Nutt.
Berberidaceae
Berberis aquifolium Pursh (see)
♦ Oregon grape, mountain grape, mahonia, holly leaved barberry, Oregon grapeholly
♦ Weed, Naturalised, Cultivation Escape
♦ 42, 86, 88, 218, 252, 280
♦ S, cultivated, herbal. Origin: western North America.

Mahonia bealei (Fortune) Carr.
Berberidaceae
♦ Beale's barberry, leatherleaf mahonia
♦ Weed, Naturalised
♦ 80, 101
♦ S, cultivated, herbal.

Mahonia haematocarpa (Woot.) Fedde
Berberidaceae
Berberis haematocarpa Woot. (see)
♦ red mahonia, Mexican barberry
♦ Weed
♦ 88, 218
♦ S, cultivated.

Mahonia japonica (Thunb.) DC.
Berberidaceae
♦ Naturalised
♦ 280
♦ S, cultivated, herbal.

Mahonia leschenaultii (Wall.) Takeda
Berberidaceae
♦ mahonia
♦ Weed, Naturalised, Garden Escape, Environmental Weed, Cultivation Escape
♦ 54, 86, 88, 98, 155, 203
♦ cultivated. Origin: India, east Asia.

Mahonia pinnata (Lag.) Fedde
Berberidaceae
♦ wavyleaf barberry
♦ Naturalised, Cultivation Escape
♦ 86, 252
♦ S, cultivated.

Mahonia trifoliolata (Moric.) Fedde
Berberidaceae
Berberis ilicifolia G.Forst., *Berberis trifoliata* Moric., *Berberis trifoliolata* Moric. (see)
♦ agarito, algerita, Mexican barberry
♦ Weed
♦ 88, 218
♦ S, promoted.

Maianthemum dilatatum (Wood) Nelson & Macbr.
Liliaceae/Convallariaceae
Maianthemum kamtschaticum (J.F.Gmel. ex Cham.) Nakai
♦ twoleaf Solomon seal, false lily of the valley, twoleaf false Solomon's seal
♦ Quarantine Weed
♦ 220
♦ pH, promoted, herbal.

Maihuenia (Phil. ex Weber) Schumm. spp.
Cactaceae
♦ Weed, Quarantine Weed
♦ 76, 88, 203, 220

Maihueniopsis Speg. spp.
Cactaceae
= *Opuntia* Mill. spp.
♦ Weed, Quarantine Weed
♦ 76, 88, 203

Maireana brevifolia (R.Br.) P.G.Wilson
Chenopodiaceae
Kochia brevifolia R.Br., *Kochia tamariscina* (Lindl.) J.Black, *Kochia thymifolia* Lindl.
♦ yanga bush, smallleaf bluebush
♦ Naturalised
♦ 300
♦ arid, cultivated.

Malabaila suaveolens Coss.
Apiaceae
♦ Weed
♦ 221
♦ Origin: Egypt.

Malachium aquaticum (L.) Fr.
Caryophyllaceae
= *Myosoton aquaticum* (L.) Moench
♦ ushihakobe
♦ Weed
♦ 88, 275
♦ pH.

Malachra alceifolia Jacq.
Malvaceae
= *Malachra capitata* (L.) L.
♦ yellow leafbract, malachra, malva de caballo

♦ Weed, Noxious Weed, Environmental Weed
♦ 28, 88, 157, 199, 229, 243, 257, 261
♦ pH, arid, herbal.

Malachra capitata (L.) L.
Malvaceae
Malachra alceaefolia Jacq., *Malachra alceifolia* Jacq. (see), *Sida capitata* L.
♦ malva de caballo, malachra, malva
♦ Weed, Quarantine Weed, Naturalised
♦ 14, 76, 86, 87, 88, 93, 179, 191, 203, 218, 220, 261
♦ cultivated, herbal. Origin: tropical America.

Malachra fasciata Jacq.
Malvaceae
♦ roadside leafbract, malachra
♦ Weed, Quarantine Weed, Naturalised
♦ 76, 86, 87, 88, 93, 191, 199, 203, 220, 243, 261, 262
♦ aH. Origin: Central and South America.

Malachra radiata (L.) L.
Malvaceae
♦ tropical leafbract, guamxuma espinhenta
♦ Weed
♦ 255
♦ pH. Origin: tropical America.

Malcolmia africana (L.) R.Br.
Brassicaceae
♦ African mustard, Malcolm stock, Afrikanillakko, Virginia stock, malkolmia Africká
♦ Weed, Quarantine Weed, Naturalised, Casual Alien
♦ 42, 54, 76, 86, 87, 88, 94, 101, 156, 161, 203, 220, 243, 248, 275, 297
♦ aH, cultivated, herbal. Origin: Mediterranean, Middle East.

Malcolmia exacoides Spreng
Brassicaceae
Malcolmia coringioides Boiss.
♦ Weed
♦ 87, 88

Malcolmia maritima (L.) R.Br.
Brassicaceae
♦ Virginia stock, meri illakko
♦ Naturalised, Cultivation Escape, Casual Alien
♦ 40, 42, 56, 98, 101, 280
♦ cultivated, herbal.

Malcolmia scorpioides (Bunge) Boiss.
Brassicaceae
♦ Weed
♦ 248

Malephora crocea (Jacq.) Schwantes
Aizoaceae/Mesembryanthemaceae
♦ coppery mesemb, coppery mesembryanthemum, iceplant, croceumice plant, copper vygie, fingerkanna, coppery mesemb
♦ Weed, Noxious Weed, Naturalised, Native Weed, Cultivation Escape
♦ 34, 35, 88, 101, 116, 121
♦ pH, cultivated, herbal. Origin: southern Africa.

Malephora lutea (Haw.) Schwantes
Aizoaceae/Mesembryanthemaceae
♦ Naturalised
♦ 86
♦ cultivated. Origin: South Africa.

Mallotus barbatus Müll.Arg.
Euphorbiaceae
♦ Weed
♦ 12
♦ herbal.

Mallotus japonicus (Thunb.) Müll.Arg.
Euphorbiaceae
♦ Weed
♦ 88, 191, 204, 286
♦ cultivated, herbal.

Mallotus oppositifolius (Geissl.) Müll.Arg.
Euphorbiaceae
♦ Weed
♦ 88
♦ herbal.

Mallotus philippensis (Lam.) Müll.Arg.
Euphorbiaceae
♦ kamala tree
♦ Naturalised
♦ 101
♦ cultivated, herbal. Origin: Australia.

Mallotus tiliaefolius Müll.Arg.
Euphorbiaceae
♦ Introduced
♦ 230
♦ S.

Malope trifida Cav.
Malvaceae
♦ maloppi, annual malope
♦ Weed, Cultivation Escape, Casual Alien
♦ 40, 42, 87, 88
♦ aH, cultivated.

Malpighia cubensis Kunth
Malpighiaceae
♦ Weed
♦ 14

Malpighia emarginata DC.
Malpighiaceae
♦ Barbados cherry, acerola
♦ Weed, Naturalised, Casual Alien
♦ 101, 179, 261
♦ S/T, cultivated.

Malpighia glabra L.
Malpighiaceae
♦ Barbados cherry, wild crape myrtle, acerola
♦ Introduced
♦ 228
♦ S/T, arid, cultivated, herbal.

Malus baccata (L.) Borkh.
Rosaceae
Pyrus baccata L.
♦ Siberian crabapple, marjaomenapuu, Chinese crab
♦ Naturalised, Cultivation Escape
♦ 42, 101
♦ T, cultivated, herbal.

Malus × domestica Borkh.
Rosaceae
= *Malus domestica* Borkh.
♦ apple, tarhaomenapuu

♦ Weed, Naturalised, Environmental
Weed, Cultivation Escape
♦ 15, 42, 72, 86, 88, 176, 198, 252, 280,
290
♦ T, cultivated.

Malus domestica Borkh.
Rosaceae
Malus × domestica Borkh. (see), *Malus
malus* (L.) Britton *nom. inval.*, *Malus
pumila auct.*, *Malus sylvestris auct.*,
Malus sylvestris Mill. var. *domestica*
(Borkh.) Mansf., *Pyrus malus* L. (see)
♦ cultivated apple, apple
♦ Weed, Naturalised
♦ 39, 40, 88, 98, 203
♦ cultivated, herbal, toxic. Origin:
Asia.

Malus floribunda Sieb. ex Van Houtte
Rosaceae
Pyrus pulcherrima Asch. & Graebn.
♦ Japanese flowering crabapple,
Japanese crabapple
♦ Weed, Naturalised
♦ 88, 101
♦ T, cultivated, herbal.

Malus fusca (Raf.) Schneid.
Rosaceae
Malus rivularis Douglas ex Hook.
♦ crabapple, Oregon crabapple
♦ Weed, Environmental Weed
♦ 88, 151
♦ S, cultivated, herbal.

Malus glabrata Rehd.
Rosaceae
♦ Biltmore crabapple
♦ Naturalised
♦ 101
♦ T, promoted.

Malus ioensis (Wood) Britt.
Rosaceae
Pyrus ioensis (Wood) Bailey (see)
♦ prairie crabapple, flowering
crabapple
♦ Weed, Naturalised
♦ 86, 88
♦ T, cultivated, herbal.

Malus × magdeburgensis Spach
Rosaceae
= *Malus pumila* Mill. × *Malus spectabilis*
Hartwig
♦ malus
♦ Naturalised
♦ 101

Malus prunifolia (Willd.) Borkh.
Rosaceae
Pyrus prunifolia Willd.
♦ plumleaf crabapple, Chinese apple,
plumleaf crabapple, plum leaved
Chinese apple, siperianomenapuu
♦ Weed, Naturalised, Environmental
Weed, Cultivation Escape
♦ 42, 80, 88, 101, 151
♦ T, cultivated, herbal.

Malus pumila Mill.
Rosaceae
Malus communis Poir., *Malus dasyphylla*
Borkh., *Malus dasyphylla* var. *domestica*
Koidz., *Malus niedzwetzkyana* Dieck,

Malus paradisiaca (L.) Medik., *Malus
pumila* var. *domestica* C.K.Schneid.,
Malus pumila var. *niedzwetzkyana*
(Dieck) C.K.Schneid., *Malus pumila*
var. *paradisiaca* (L.) C.K.Schneid, *Malus
silvestris* Mill. ssp. *mitis* Mansf., *Malus
sylvestris* Mill. var. *niedzwetskyana*
(Dieck) L.H.Bailey, *Pyrus malus* L. var.
paradisiaca L., *Pyrus malus* var. *pumila*
Henry, *Pyrus niedzwetzkyana* (Dieck)
Hemsl.
♦ paradise apple, common apple,
cultivated apple
♦ Weed, Naturalised
♦ 80, 101, 243
♦ T, cultivated, herbal.

Malus sieboldii (Regel) Rehd.
Rosaceae
♦ toringa crabapple
♦ Naturalised
♦ 101
♦ T, cultivated, herbal.

**Malus sieboldii (Regel) Rehd. var.
sieboldii**
Rosaceae
♦ toringa crabapple
♦ Naturalised
♦ 101

**Malus sieboldii (Regel) Rehd. var. zumi
(Matsum.) Asami**
Rosaceae
♦ toringa crabapple
♦ Naturalised
♦ 101

Malus sylvestris Mill.
Rosaceae
Malus acerba Mérat, *Malus communis*
Poir. var. *sylvestris* (Mill.) DC., *Malus
praecox* (Pall.) Borkh., *Malus sylvestris*
ssp. *praecox* (Pall.) Soó
♦ crabapple, European crabapple,
omenapuu, wild crab
♦ Weed, Naturalised, Environmental
Weed, Cultivation Escape
♦ 34, 39, 80, 86, 88, 98, 101, 151, 203,
247, 252
♦ T, cultivated, herbal, toxic. Origin:
Europe.

Malva L. spp.
Malvaceae
♦ mallow, malva
♦ Weed, Naturalised
♦ 39, 198, 243, 245, 272
♦ H, herbal, toxic.

Malva alcea L.
Malvaceae
Alcea palmata Gilib.
♦ ruusumalva, vervain mallow
♦ Weed, Naturalised, Cultivation
Escape
♦ 42, 101, 272
♦ pH, cultivated, herbal.

Malva assurgentiflora (Kellogg) M.F.Ray
Malvaceae
Lavatera assurgentiflora Kellogg (see)
♦ Naturalised
♦ 300
♦ Origin: California to western South
America.

Malva crispa (L.) L.
Malvaceae
= *Malva verticillata* L. var. *crispa* L.
♦ curly mallow
♦ Weed, Naturalised
♦ 87, 88, 101
♦ cultivated, herbal.

Malva dendromorpha M.F.Ray
Malvaceae
♦ Naturalised
♦ 300
♦ Origin: Mediterranean.

Malva hispanica L.
Malvaceae
♦ Spanish mallow
♦ Weed
♦ 70, 87, 88, 94

Malva mohileviensis Downar
Malvaceae
♦ mallow weed
♦ Weed
♦ 88, 114, 243
♦ aH.

Malva montana auct.
Malvaceae
= *Malva nicaeensis* All.
♦ Weed
♦ 88

Malva moschata L.
Malvaceae
♦ musk mallow, myskimalva
♦ Weed, Naturalised, Cultivation
Escape
♦ 42, 80, 86, 87, 88, 98, 101, 136, 161,
176, 195, 198, 203, 218, 241, 252, 280,
286, 287, 300
♦ pH, cultivated, herbal. Origin:
southern Europe.

Malva neglecta Wallr.
Malvaceae
Malva rotundifolia auct. non L., *Malva
vulgaris* Fr.
♦ common mallow, dwarf
mallow, round leaved mallow,
katinjuustomalva, cheeseweed,
cheeses, cheese mallow, running
mallow, malice, round dock,
buttonweed, low mallow
♦ Weed, Naturalised, Introduced,
Casual Alien
♦ 15, 24, 34, 39, 42, 44, 52, 70, 80, 86, 87,
88, 94, 98, 101, 136, 161, 174, 176, 195,
198, 203, 207, 210, 211, 212, 218, 228,
243, 253, 272, 280, 286, 287, 300
♦ a/pH, arid, cultivated, herbal, toxic.
Origin: Europe, North Africa, west
Asia.

Malva nicaeensis All.
Malvaceae
Malva arvensis J. & C.Presl, *Malva
montana auct.* (see)
♦ bull mallow, French mallow,
Rivieranmalva, malva común
♦ Weed, Naturalised, Casual Alien
♦ 15, 34, 40, 42, 86, 87, 88, 94, 98, 101,
115, 161, 176, 198, 199, 203, 218, 241,
280, 295, 300
♦ aH, cultivated, herbal. Origin:
Mediterranean.

Malva parviflora L.
Malvaceae
♦ little mallow, small flowered mallow, marshmallow, myllymalva, small mallow, bread and cheese, least mallow, cheeseweed mallow
♦ Weed, Naturalised, Introduced, Environmental Weed, Casual Alien
♦ 7, 15, 34, 38, 39, 40, 42, 51, 55, 80, 86, 87, 88, 93, 94, 98, 101, 115, 116, 121, 134, 158, 161, 165, 176, 180, 185, 198, 203, 205, 218, 228, 236, 237, 241, 243, 248, 253, 257, 269, 280, 286, 287, 295, 300
♦ aH, arid, cultivated, herbal, toxic. Origin: Eurasia.

Malva pusilla Sm.
Malvaceae
♦ small mallow, kylämalva
♦ Weed, Naturalised, Casual Alien
♦ 40, 44, 70, 87, 88, 94, 121, 272, 286, 287, 291
♦ H, cultivated. Origin: Eurasia.

Malva rotundifolia L.
Malvaceae
♦ dwarf mallow, common mallow, cheeses, running mallow, round dock
♦ Weed, Noxious Weed, Naturalised
♦ 23, 49, 80, 87, 88, 101, 161, 162, 210, 218, 243, 275, 287, 299
♦ pH, herbal.

Malva sylvestris L.
Malvaceae
Malva ambigua Guss., *Malva erecta* Gilib., *Malva sylvestris* L. var. *mauritiana* (L.) Boiss. (see)
♦ high mallow, large flowered mallow, common mallow, cheeses, high malva
♦ Weed, Naturalised, Introduced, Cultivation Escape
♦ 15, 23, 34, 38, 39, 42, 44, 70, 80, 86, 87, 88, 94, 98, 101, 115, 121, 165, 176, 203, 228, 243, 250, 252, 253, 272, 280, 287, 295, 300
♦ aH, arid, cultivated, herbal, toxic. Origin: Eurasia.

Malva sylvestris L. var. *mauritiana* (L.) Boiss.
Malvaceae
= *Malva sylvestris* L.
♦ zebrina mallow
♦ Weed, Naturalised
♦ 286, 287
♦ cultivated, herbal.

Malva sylvestris L. var. *sylvestris*
Malvaceae
♦ tall mallow
♦ Naturalised
♦ 198

Malva verticillata L.
Malvaceae
Althaea verticillata (L.) Alef., *Malva mohileviensis* Downar, *Malva montana* Forssk.
♦ cluster mallow, curled mallow, whorled mallow, mallow
♦ Weed, Naturalised, Casual Alien
♦ 42, 53, 80, 86, 87, 88, 101, 272, 280, 287, 297

♦ a/bH, cultivated, herbal. Origin: North Africa, west Asia.

Malva verticillata L. var. *crispa* (L.) L.
Malvaceae
Malva crispa (L.) L. (see)
♦ Naturalised
♦ 287
♦ cultivated, herbal.

Malvastrum americanum (L.) Torr.
Malvaceae
Malvastrum spicatum (L.) A.Gray
♦ malvastrum, spiked malvastrum, Indian valley false mallow
♦ Weed, Naturalised, Introduced, Environmental Weed
♦ 7, 55, 86, 87, 88, 93, 98, 203, 205, 228, 257, 261
♦ a/p, arid, cultivated, herbal.

Malvastrum capense Gray & Harv.
Malvaceae
♦ Naturalised
♦ 101
♦ cultivated, herbal.

Malvastrum corchorifolium (Desv.) Britt. ex Small
Malvaceae
♦ false mallow
♦ Weed
♦ 88, 261

Malvastrum coromandelianum (L.) Garcke
Malvaceae
Malvastrum carpinifolium A.Gray, *Malvastrum tricuspidatum* (Ait.) A.Gray
♦ prickly malvastrum, threelobe false mallow, escoba dura
♦ Weed, Naturalised, Environmental Weed
♦ 14, 55, 66, 86, 87, 88, 93, 101, 121, 134, 158, 179, 199, 236, 237, 241, 255, 256, 257, 261, 274, 275, 276, 286, 287, 295, 297
♦ a/pH, cultivated, herbal. Origin: tropical New World.

Malvastrum peruvianum (L.) A.Gray
Malvaceae
= *Urocarpidium peruvianum* (Gray) Krapov. (NoR)
♦ Weed, Quarantine Weed
♦ 76, 87, 88, 203, 220
♦ herbal.

Malvastrum scabrum (Cav.) Gray
Malvaceae
= *Malvastrum tomentosum* (L.) S.R.Hill ssp. *tomentosum* (NoR) [see *Malvastrum scoparium* (L'Hér.) A.Gray]
♦ Weed
♦ 87, 88

Malvastrum scoparium (L'Hér.) A.Gray
Malvaceae
= *Malvastrum tomentosum* (L.) S.R.Hill ssp. *tomentosum* (NoR) [see *Malvastrum scabrum* (Cav.) Gray]
♦ Weed, Naturalised
♦ 87, 88, 257

Malvaviscus arboreus Dill. ex Cav.
Malvaceae

♦ Turk's cap, fire dart, wax mallow
♦ Weed, Naturalised, Cultivation Escape
♦ 15, 87, 88, 101, 121, 261, 280
♦ pS, cultivated, herbal. Origin: South America.

Malvaviscus arboreus Dill. ex Cav. var. *arboreus*
Malvaceae
♦ wax mallow, Turk's cap
♦ Sleeper Weed, Naturalised, Environmental Weed
♦ 101, 225, 246

Malvaviscus arboreus Dill. ex Cav. var. *drummondii* (Torr. & A.Gray) Schery
Malvaceae
♦ wax mallow
♦ Weed
♦ 179
♦ cultivated.

Malvaviscus penduliflorus DC.
Malvaceae
♦ mazapan, sleepy mallow
♦ Weed, Naturalised, Cultivation Escape
♦ 80, 101, 179, 261
♦ cultivated, herbal.

Malvella leprosa (Ortega) Krapov.
Malvaceae
Malva leprosa Ort., *Sida leprosa* (Ort.) K.Schum. (see), *Sida leprosa* (Ort.) Shum. var. *hederacea* (Dougl. ex Hook) Shum. (see)
♦ alkali sida, ivyleaf sida, alkali mallow, whiteweed
♦ Weed, Quarantine Weed, Noxious Weed, Naturalised
♦ 35, 76, 86, 88, 147, 171, 198, 203, 229, 243
♦ pH, arid, herbal. Origin: western north America.

Malvella leprosa (Ortega) Krapov. var. *hederacea* (Hook.) Schumann
Malvaceae
Sida hederacea (Dougl. ex Hook.) Torr. ex Gray (see)
♦ alkali sida, alkali mallow, creeping mallow, star mallow, white mallow, whiteweed
♦ Weed
♦ 180, 243
♦ toxic.

Mammillaria vivipara (Nutt.) Haw.
Cactaceae
= *Escobaria vivipara* (Nutt.) Buxb. var. *vivipara* (NoR)
♦ purple mammillaria
♦ Weed
♦ 87, 88, 218
♦ herbal.

Mandevilla laxa (Ruiz & Pav.) Woodson
Apocynaceae
Mandevilla suaveolens Lindl.
♦ Chilean jasmine
♦ Weed, Naturalised
♦ 86, 98, 203
♦ cultivated, herbal. Origin: South America.

Mandragora autumnalis Bertol.
Solanaceae
♦ mandrake, autumn mandrake
♦ Weed
♦ 39, 87, 88
♦ cultivated, herbal, toxic.

Manettia cordifolia Mart.
Rubiaceae
Guagnebina ignita Vell., *Manettia ignita*
(Vell.) K.Schum.
♦ poeja do mato, hummingbird vine,
firecracker vine
♦ Weed
♦ 255
♦ S. Origin: South America.

Manettia hispida Poepp. & Endl.
Rubiaceae
♦ Weed
♦ 87, 88
♦ pC.

Mangifera indica L.
Anacardiaceae
♦ mango, am, amba
♦ Weed, Naturalised, Introduced,
Garden Escape, Environmental Weed,
Cultivation Escape
♦ 3, 7, 22, 39, 80, 86, 87, 88, 93, 98, 101,
121, 151, 152, 161, 179, 203, 228, 230,
257, 261, 279
♦ T, arid, cultivated, herbal, toxic.
Origin: Asia.

Mangifera minor Bl.
Anacardiaceae
♦ kehngid
♦ Introduced
♦ 230
♦ T.

Manihot Mill. spp.
Euphorbiaceae
♦ cassava, manioca, tapioca, manihot,
maniok
♦ Weed
♦ 121
♦ pH, herbal, toxic. Origin: South
America.

Manihot dichotoma Ule
Euphorbiaceae
♦ manicoba rubber, likajeba
♦ Introduced
♦ 228
♦ arid.

Manihot dulcis (Gmel.) Pax
Euphorbiaceae
= *Manihot esculenta* Crantz
♦ Introduced
♦ 228
♦ arid.

Manihot esculenta Crantz
Euphorbiaceae
Manihot dulcis (Gmel.) Pax (see)
♦ tapioca, Yokohauna bean, manioka,
manioc, kehptuhke, cassava
♦ Weed, Naturalised, Introduced,
Cultivation Escape
♦ 39, 86, 88, 98, 101, 161, 179, 203, 230,
247, 261
♦ pH, cultivated, herbal, toxic. Origin:
tropical South America.

Manihot glaziovii Müll.Arg.
Euphorbiaceae
♦ ceara rubber tree
♦ Weed, Naturalised, Introduced
♦ 22, 39, 86, 101, 228, 230
♦ T, arid, cultivated, herbal, toxic.
Origin: Brazil.

Manihot grahamii Hook
Euphorbiaceae
♦ Graham's manihot
♦ Weed, Naturalised
♦ 54, 86, 88, 101, 179
♦ cultivated, herbal. Origin: South
America.

Manilkara hexandra (Roxb.) Dubard
Sapotaceae
Mimusops hexandra Roxb.
♦ khirni, kirni
♦ Introduced
♦ 228
♦ arid, cultivated, herbal.

Manilkara roxburghiana (Wight) Dubard
Sapotaceae
♦ Weed
♦ 179
♦ cultivated.

Manilkara zapota (L.) van Royen
Sapotaceae
Achras zapota L. nom. illeg. (see), *Achras
zapota* L. var. *zapotilla* Jacq., *Manilkara
zapotilla* (Jacq.) Gilly, *Sapota achras*
Mill., *Sapota zapotilla* (Jacq.) Cov.
♦ sapodilla, nispero
♦ Weed, Naturalised, Introduced,
Environmental Weed
♦ 22, 32, 80, 87, 88, 101, 151, 161, 179,
228, 230, 261
♦ T, arid, cultivated, herbal, toxic.
Origin: southern Mexico, Central
America.

Manniophyton fulvum Müll.Arg.
Euphorbiaceae
♦ lokosa, kusa, kosa
♦ Weed, Quarantine Weed
♦ 76, 88, 203, 220
♦ pC.

Mansoa alliacea (Lam.) A.H.Gentry
Bignoniaceae
♦ garlic vine, bejuco de ajo, mata de
ajo
♦ Naturalised, Introduced
♦ 101, 261
♦ T, cultivated. Origin: northern South
America.

Mansoa hymenaea (DC.) A.H.Gentry
Bignoniaceae
♦ membranous garlic vine, bejuco de
ajo, mata de ajo, lavender garlic vine
♦ Naturalised, Cultivation Escape
♦ 101, 261
♦ T, cultivated. Origin: Mexico to
Brazil.

Mantisalca salmantica (L.) Briq. & Cav.
Asteraceae
Centaurea salmantica L. (see),
Microlonchus salmanticus (L.) DC. (see)
♦ dagger flower, mantisalca

♦ Weed, Naturalised
♦ 86, 88, 94, 98, 101, 158, 198, 203
♦ Origin: Mediterranean, Middle East.

Marah fabaceus (Naudin.) Greene
Cucurbitaceae
♦ coast wild cucumber, bigroot,
common manroot, California manroot
♦ Weed
♦ 161
♦ pH, cultivated, herbal.

**Marah oreganus (Torr. & S.Wats.)
T.J.Howell**
Cucurbitaceae
Echinocystis oregana (Torr. & Gray)
Cogn. (see)
♦ western wild cucumber, bigroot,
manroot, old man in the ground, coast
manroot
♦ Weed
♦ 161, 212
♦ pH, herbal, toxic.

Maranta arundinacea L.
Marantaceae
♦ arrowroot, obedience plant,
amaranta, maranta, yuquilla
♦ Weed, Naturalised, Cultivation
Escape
♦ 101, 179, 261
♦ cultivated, herbal. Origin: South
America.

Marchantia polymorpha L.
Marchantiaceae
♦ common liverwort
♦ Weed, Naturalised
♦ 211, 243, 280, 286
♦ liverwort.

**Marchantia polymorpha L. var. aquatica
Nees**
Marchantiaceae
♦ Naturalised
♦ 280
♦ liverwort.

Marenopuntia Backeb. spp.
Cactaceae
= *Opuntia* Mill. spp.
♦ Weed, Quarantine Weed
♦ 76, 88, 203

Maresia pygmaea (Delile) O.E.Schulz
Brassicaceae
♦ Weed
♦ 221

Marginatocereus (Backeb.) Backeb. spp.
Cactaceae
= *Pachycereus* (Berg.) Britt. & Rose spp.
♦ Weed, Quarantine Weed
♦ 76, 88, 203

Mariscus albescens Gaud.
Cyperaceae
♦ Weed
♦ 87, 88
♦ G.

Mariscus alternifolius Vahl
Cyperaceae
= *Cyperus cyperoides* (L.) Kuntze
♦ Weed
♦ 88
♦ G.

Mariscus aristatus (Rottb.) Ts.Tang & F.T.Wang
Cyperaceae
♦ Weed
♦ 88
♦ G.

Mariscus capensis (Steud.) Schrad.
Cyperaceae
♦ monkey bulb
♦ Native Weed
♦ 121
♦ pG. Origin: southern Africa.

Mariscus coloratus Nees.
Cyperaceae
♦ Weed
♦ 87, 88
♦ G.

Mariscus compactus (Retz.) Bold.
Cyperaceae
= *Cyperus compactus* Retz.
♦ Weed
♦ 88, 126, 297
♦ pG, aqua.

Mariscus congestus (Vahl) C.B.Clarke
Cyperaceae
= *Cyperus congestus* Vahl
♦ Weed, Naturalised, Native Weed
♦ 121, 287
♦ pG, herbal. Origin: southern Africa.

Mariscus dregeanus Kunth
Cyperaceae
♦ Weed, Native Weed
♦ 87, 88, 121
♦ pG. Origin: southern Africa.

Mariscus dubius (Rottb.) Kük. ex G.E.C.Fisch.
Cyperaceae
♦ Native Weed
♦ 121
♦ pG. Origin: southern Africa.

Mariscus durus (Kunth) C.B.Clarke
Cyperaceae
♦ Native Weed
♦ 121
♦ pG. Origin: southern Africa.

Mariscus flabelliformis Kunth var. *flabelliformis*
Cyperaceae
♦ Weed
♦ 88
♦ G.

Mariscus flavus Vahl
Cyperaceae
= *Cyperus aggregatus* (Willd.) Endl.
♦ Weed
♦ 126
♦ pG.

Mariscus indecorus (Kunth) Podlech
Cyperaceae
Mariscus albomarginatus C.B.Clarke
♦ sedge
♦ Native Weed
♦ 121
♦ pG. Origin: southern Africa.

Mariscus keniensis (Kük.) Hooper
Cyperaceae
♦ Native Weed

♦ 121
♦ pG. Origin: southern Africa.

Mariscus ligularis (L.) Urb.
Cyperaceae
= *Cyperus ligularis* L.
♦ Weed
♦ 87, 88
♦ G.

Mariscus longibracteatus Cherm.
Cyperaceae
♦ Weed, Quarantine Weed
♦ 76, 88, 203, 220
♦ G.

Mariscus macer Kunth ssp. *magaliesmontanum* P.J.Vorster
Cyperaceae
♦ Native Weed
♦ 121
♦ pG. Origin: southern Africa.

Mariscus mutisii Kunth
Cyperaceae
= *Cyperus mutisii* (Kunth) Griseb.
♦ Weed
♦ 87, 88
♦ G.

Mariscus radians (Nees & Meyen) Tang & F.T.Wang
Cyperaceae
= *Cyperus radians* Nees
♦ short stemmed cyperus
♦ Weed
♦ 126
♦ pG, aqua.

Mariscus rehmannianus C.B.Clarke
Cyperaceae
♦ Native Weed
♦ 121
♦ pG. Origin: southern Africa.

Mariscus rufus H.B.K.
Cyperaceae
= *Cyperus ligularis* L.
♦ Weed, Quarantine Weed
♦ 76, 87, 88, 203, 220
♦ G.

Mariscus squarrosus (L.) C.B.Clarke
Cyperaceae
= *Cyperus squarrosus* L.
♦ Weed
♦ 126
♦ aG, aqua.

Mariscus sumatrensis (Retz.) Raynal
Cyperaceae
= *Cyperus cyperoides* (L.) Kuntze
♦ Weed
♦ 121, 126
♦ pG.

Mariscus thunbergii (Vahl) Schrad.
Cyperaceae
♦ Native Weed
♦ 121
♦ pG. Origin: southern Africa.

Mariscus umbellatus (Rottb.) Vahl
Cyperaceae
= *Cyperus cyperoides* (L.) Kuntze
♦ Weed, Quarantine Weed
♦ 76, 87, 88, 203, 220, 275, 297
♦ pG.

Mariscus ustulatus C.B.Clarke
Cyperaceae
♦ Weed, Quarantine Weed
♦ 76, 87, 88, 203, 220
♦ G.

Maritimocereus Ackers & Buining spp.
Cactaceae
= *Cleistocactus* Lem. spp.
♦ Weed, Quarantine Weed
♦ 76, 88, 203

Marrubium alysson L.
Lamiaceae
♦ Weed
♦ 221

Marrubium cuneatum Sol.
Lamiaceae
♦ Weed
♦ 87, 88

Marrubium parviflorum Fisch. & Mey.
Lamiaceae
♦ Weed
♦ 87, 88

Marrubium peregrinum L.
Lamiaceae
♦ branched horehound
♦ Weed
♦ 272
♦ cultivated, herbal.

Marrubium radiatum Delile ex Benth.
Lamiaceae
♦ Weed
♦ 87, 88

Marrubium supinam (Steph. ex Willd.) Hu
Lamiaceae
♦ Weed
♦ 87, 88

Marrubium vulgare L.
Lamiaceae
Marrubium album Gilib., *Prasium marrubium* Krause
♦ white horehound, common horehound, horehound, houndsbane, marrube, marvel
♦ Weed, Quarantine Weed, Noxious Weed, Naturalised, Introduced, Garden Escape, Environmental Weed, Casual Alien
♦ 7, 15, 20, 23, 34, 39, 42, 62, 72, 76, 80, 86, 87, 88, 97, 98, 101, 116, 121, 134, 147, 151, 161, 165, 167, 169, 171, 176, 180, 198, 203, 212, 218, 221, 228, 237, 241, 243, 248, 269, 272, 280, 287, 289, 295, 296, 300
♦ pH, arid, cultivated, herbal, toxic. Origin: Eurasia, North Africa.

Marsdenia rostrata R.Br.
Asclepiadaceae/Apocynaceae
♦ Weed
♦ 39, 87, 88
♦ cultivated, toxic. Origin: Australia.

Marshallocereus Backeb. spp.
Cactaceae
= *Stenocereus* (Berger) Riccob. spp.
♦ Weed, Quarantine Weed
♦ 76, 88, 203

Marsilea L. spp.
Marsileaceae
♦ waterclover
♦ Quarantine Weed
♦ 220

Marsilea aegyptiaca Willd.
Marsileaceae
♦ nardoo
♦ Weed
♦ 87, 88, 185, 221

Marsilea crenata Presl
Marsileaceae
♦ waterclover, clover fern, phak waen, dwarf fourleaf clover
♦ Weed
♦ 87, 88, 170, 197, 204, 209, 239, 286
♦ aqua, cultivated. Origin: tropical Asia, Australia.

Marsilea drummondii A.Braun
Marsileaceae
♦ clover fern, common nardoo
♦ Weed, Native Weed
♦ 39, 87, 88, 269
♦ aqua, cultivated, herbal, toxic. Origin: Australia.

Marsilea hirsuta R.Br.
Marsileaceae
♦ rough waterclover
♦ Naturalised
♦ 39, 101
♦ aqua, cultivated, toxic.

Marsilea macropoda Engelm. ex A.Braun
Marsileaceae
♦ bigfoot waterclover
♦ Weed
♦ 179

Marsilea minuta L.
Marsileaceae
♦ dwarf waterclover
♦ Weed, Naturalised
♦ 87, 88, 101, 221

Marsilea mucronata A.Br.
Marsileaceae
= *Marsilea vestita* Hook. & Grev. ssp. *vestita* (NoR)
♦ pepperwort
♦ Weed
♦ 88, 218

Marsilea mutica Mett.
Marsileaceae
♦ nardoo
♦ Naturalised, Garden Escape, Environmental Weed
♦ 39, 86, 176, 246
♦ aqua, cultivated, toxic. Origin: Australia.

Marsilea quadrifolia L.
Marsileaceae
Marsilia quadrifolia L.
♦ waterclover, water shamrock, European pepperwort, pepperwort, European waterclover, Australian four leaved clover
♦ Weed, Quarantine Weed, Naturalised
♦ 70, 76, 87, 88, 101, 133, 195, 197, 203, 204, 217, 224, 263, 272, 274, 275, 286, 297

♦ pH, aqua, cultivated, herbal.

Marsilea vestita Hook. & Grev.
Marsileaceae
♦ hairy waterclover
♦ Weed
♦ 87, 88
♦ herbal.

Marsypianthes chamaedrys (Vahl) Kuntze
Lamiaceae
♦ ortela, vassoura
♦ Weed
♦ 87, 88, 157, 255
♦ aH. Origin: Americas.

Martynia annua L.
Pedaliaceae/Martyniaceae
Carpoceras angulata A.Rich., *Craniolaria annua* L., *Disteira angulosa* (Lam.) Raf., *Martynia angulosa* Lam., *Martynia diandra* Gloxin *nom. illeg.*
♦ devil's claw, smallfruit devil's claw, tiger's claw, iceplant, escorzoera
♦ Weed, Quarantine Weed, Noxious Weed, Naturalised
♦ 62, 66, 76, 86, 87, 88, 93, 98, 147, 191, 203, 220, 261, 269
♦ aH, cultivated, herbal. Origin: Americas.

Martynia louisiana Mill.
Pedaliaceae/Martyniaceae
= *Proboscidea louisianica* (Mill.) Thell.
♦ Quarantine Weed
♦ 220

Mascagnia concinna Morton
Malpighiaceae
♦ Weed, Quarantine Weed
♦ 76, 88, 203, 220

Mascagnia pubiflora Griseb.
Malpighiaceae
♦ Weed, Quarantine Weed
♦ 76, 87, 88, 203, 220

Mascarena lagenicaulis Bailey
Arecaceae
♦ Introduced
♦ 230
♦ T.

Mascarena verschaffeltii (Wendl.) Bailey
Arecaceae
♦ Introduced
♦ 230
♦ T.

Matelea decumbens W.D.Stevens
Asclepiadaceae/Apocynaceae
♦ Weed
♦ 199

Matricaria L. spp.
Asteraceae
Chamomilla S.F.Gray spp. (see)
♦ chamomile, mayweed
♦ Weed, Naturalised
♦ 198, 203, 243, 272
♦ H, cultivated, herbal.

Matricaria aurea (L.) Sch.Bip.
Asteraceae
Chamomilla aurea (L.) J.Gay ex Coss. & Kral. (see)
♦ Weed

♦ 221
♦ herbal.

Matricaria chamomilla L.
Asteraceae
= *Tripleurospermum perforatum* (Mérat) Laínz
♦ German chamomile, Hungarian chamomile, wild chamomile, scented mayweed
♦ Weed, Quarantine Weed, Naturalised
♦ 44, 76, 86, 87, 88, 118, 161, 217, 236, 243, 271, 286, 287, 295
♦ aH, arid, cultivated, herbal.

Matricaria courrantiana DC.
Asteraceae
= *Matricaria recutita* L.
♦ crown mayweed
♦ Naturalised
♦ 101

Matricaria discoidea DC.
Asteraceae
= *Matricaria matricarioides* (Less.) Porter
♦ rayless chamomile, disc mayweed, pineapple weed
♦ Weed, Naturalised, Introduced
♦ 5, 40, 87, 88, 101, 165, 174, 176, 195, 243, 253, 280
♦ aH, cultivated, herbal.

Matricaria inodora L. nom. illeg.
Asteraceae
= *Tripleurospermum perforatum* (Mérat) Laínz
♦ maruna bezwonna
♦ Weed, Naturalised
♦ 87, 88, 243, 263, 286, 287
♦ aH, herbal.

Matricaria maritima L.
Asteraceae
= *Tripleurospermum maritima* (L.) W.D.J.Koch ssp. *maritima* (NoR)
♦ scentless chamomile, false chamomile
♦ Weed, Quarantine Weed, Noxious Weed
♦ 23, 36, 70, 76, 80, 87, 88, 156, 161, 162, 210, 272
♦ cultivated, herbal.

Matricaria maritima L. var. *agrestis* (Knaf) Wilmott
Asteraceae
= *Tripleurospermum perforatum* (Mérat) Laínz
♦ false chamomile
♦ Weed
♦ 218

Matricaria matricarioides (Less.) Porter
Asteraceae
Artemisia matricarioides Less., *Chamomilla suaveolens* (Pursh) Rydb. (see), *Matricaria discoidea* DC. (see), *Matricaria suaveolens* (Pursh) Buch. *non* L. (see), *Santolina suaveolens* Pursh
♦ pineapple weed, rayless chamomile, rounded chamomile, pineapple mayweed, rayless dogfennel
♦ Weed, Noxious Weed, Naturalised
♦ 21, 39, 42, 52, 68, 80, 86, 87, 88, 98,

136, 146, 161, 180, 198, 203, 207, 210, 211, 212, 218, 243, 286, 287, 299, 300
♦ aH, arid, cultivated, herbal, toxic. Origin: North America, eastern Russia, Japan.

Matricaria nigellifolia **DC.**
Asteraceae
♦ bovine staggers plant, staggers weed
♦ Weed
♦ 39, 88, 158
♦ toxic. Origin: southern Africa.

Matricaria nigellifolia **DC. var.** *nigellifolia*
Asteraceae
♦ bovine staggers plant, staggers weed
♦ Native Weed
♦ 121
♦ pH, toxic. Origin: southern Africa.

Matricaria perforata **Mérat**
Asteraceae
= *Tripleurospermum perforatum* (Mérat) Laínz
♦ scentless mayweed, scentless chamomile, wild chamomile, mayweed, false chamomile
♦ Weed, Noxious Weed, Naturalised, Introduced
♦ 1, 16, 21, 68, 70, 86, 88, 94, 98, 146, 161, 203, 241, 243, 253, 272, 291, 299
♦ a/bH, cultivated, herbal. Origin: Eurasia, North Africa.

Matricaria recutita **L.**
Asteraceae
Chamomilla recutita (L.) Rauschert (see), *Chamomilla courrantiana* C.Koch, *Matricaria chamomilla* auct. non L., *Matricaria courrantiana* DC. (see), *Matricaria suaveolens* L. (see)
♦ wild chamomile, chamomile, German chamomile, mayweed, kamomillasaunio, scented mayweed
♦ Weed, Naturalised, Garden Escape, Environmental Weed, Casual Alien
♦ 86, 87, 88, 98, 101, 155, 185, 199, 203, 221, 237, 243, 253, 280, 300
♦ aH, cultivated, herbal. Origin: Europe, western Asia.

Matricaria suaveolens **L.**
Asteraceae
= *Matricaria recutita* L.
♦ rayless mayweed, pineapple weed
♦ Weed
♦ 44, 87, 88
♦ herbal.

Matricaria suaveolens **(Pursh) Buch.** *non* **L.**
Asteraceae
= *Matricaria matricarioides* (Less.) Porter
♦ rayless mayweed, pineapple weed
♦ Weed
♦ 44
♦ aH, herbal.

Matricaria tenella **DC.**
Asteraceae
♦ stinkweed
♦ Native Weed
♦ 121
♦ aH, toxic. Origin: southern Africa.

Matricaria trichophylla **(Boiss.) Boiss.**
Asteraceae
Matricaria tenuifolia (Kit.) Sim., *Tripeurospermum tenuifolium* (Kit.) Freyn
♦ Weed
♦ 88, 94, 272

Matricaria tridentata **Ball**
Asteraceae
♦ Weed
♦ 221

Matthiola arabica **Boiss.**
Brassicaceae
♦ Weed
♦ 221
♦ arid.

Matthiola elliptica **R.Br.**
Brassicaceae
♦ Weed
♦ 221

Matthiola incana **(L.) R.Br.**
Brassicaceae
Cheiranthus incanus L.
♦ tenweeks stock, stock, hoary stock, gilliflower, queen stock, tarhaleukoija, brompton stock, evening stock, fiala sivá
♦ Weed, Naturalised, Cultivation Escape
♦ 7, 15, 34, 42, 86, 98, 101, 134, 198, 203, 252, 272, 280
♦ pH, cultivated, herbal. Origin: Mediterranean.

Matthiola livida **(Delile) DC.**
Brassicaceae
♦ Weed
♦ 221
♦ arid.

Matthiola longipetala **(Vent.) DC.**
Brassicaceae
Matthiola longipetala ssp. *bicornis* (Sibth. & Sm.) Ball (see)
♦ night scented stock, evening scented stock
♦ Weed, Naturalised, Cultivation Escape, Casual Alien
♦ 42, 98, 101, 203, 243, 280
♦ aH, cultivated, herbal.

Matthiola longipetala **(Vent.) DC. ssp.** *bicornis* **(Sibth. & Sm.) Ball**
Brassicaceae
= *Matthiola longipetala* (Vent.) DC.
♦ nightstock, night scented stock
♦ Naturalised, Casual Alien
♦ 40, 86, 98, 198, 280
♦ cultivated. Origin: eastern Europe.

Matthiola parviflora **(Schousb.) R.Br.**
Brassicaceae
♦ Weed
♦ 87, 88

Matthiola sinuata **(L.) R.Br.**
Brassicaceae
♦ sea stock
♦ Weed
♦ 272
♦ bH, cultivated.

Maurandya antirrhiniflora **Humb. &**

Bonpl. ex Willd.
Scrophulariaceae
= *Maurandella antirrhiniflora* (Humb. & Bonpl. ex Willd.) Rothm. (NoR)
♦ Weed
♦ 179, 199
♦ cultivated, herbal.

Maurandya barclaiana **Lindl.**
Scrophulariaceae
Asarina barclaiana (Lindl.) Pennell
♦ Mexican viper
♦ Weed, Sleeper Weed, Naturalised, Environmental Weed, Casual Alien
♦ 7, 54, 86, 88, 101, 155, 199, 280
♦ pC, cultivated, herbal. Origin: Mexico.

Mauria heterophylla **Kunth var.** *contracta* **Loes.**
Anacardiaceae
= *Mauria heterophylla* Kunth (NoR)
♦ Introduced
♦ 38

Mayaca fluviatilis **Aubl.**
Mayacaceae
Mayaca aubletii Michx., *Mayaca vandellii* Schott & Endl.
♦ stream bogmoss, floating moo herb
♦ Quarantine Weed
♦ 76, 258
♦ aqua, cultivated. Origin: south-east USA and Central America.

Maytenus boaria **Molina**
Celastraceae
Maytenus chilensis DC.
♦ mayten tree
♦ Weed, Naturalised
♦ 15, 280
♦ T, cultivated, herbal.

Maytenus heterophylla **(Eckl. & Zeyh.) N.Robson**
Celastraceae
Maytenus cymosa (Sol.) Exell
♦ common spikethorn, quickthorn, spikethorn
♦ Weed, Quarantine Weed, Native Weed
♦ 10, 121, 220
♦ S, cultivated, herbal. Origin: southern Africa.

Maytenus mossambicensis **(Klotzch) Blakelock var.** *mossambicensis*
Celastraceae
♦ blackforest spikethorn, long spined maytenus, red forest spikethorn
♦ Native Weed
♦ 121
♦ S/T. Origin: southern Africa.

Maytenus nemorosa **(Eckl. & Zeyh.) Marais**
Celastraceae
♦ forest maytenus, white forest spikethorn
♦ Native Weed
♦ 121
♦ S/T. Origin: southern Africa.

Maytenus oleoides **(Lam.) Loes.**
Celastraceae
♦ rock candlewood, mountain

maytenus
- ♦ Native Weed
- ♦ 121
- ♦ S/T, cultivated. Origin: southern Africa.

Maytenus senegalensis (Lam.) Exell
Celastraceae
- ♦ confetti tree, red spikethorn, mwesya, rooipendoring, red spikethorn
- ♦ Weed, Native Weed
- ♦ 63, 121, 221
- ♦ S/T, cultivated, herbal. Origin: southern Africa.

Maytenus spinosa (Griseb.) Lourteig & O'Don.
Celastraceae
Gymnosporia establei Herter, *Gymnosporia spinosa* (Griseb.) Loes., *Moya spinosa* Griseb.
- ♦ Introduced
- ♦ 228
- ♦ arid.

Maytenus vitis-idaea Griseb.
Celastraceae
- ♦ Introduced
- ♦ 228
- ♦ arid.

Mazus fauriei Bonati
Scrophulariaceae
- ♦ Weed
- ♦ 235

Mazus japonicus (Thunb.) Kuntze
Scrophulariaceae
= *Mazus pumilus* (Burm.f.) Steen.
- ♦ tokiwahaze
- ♦ Weed
- ♦ 88, 263, 274, 275, 286
- ♦ aH, promoted, herbal. Origin: east Asia.

Mazus miquelii Makino
Scrophulariaceae
Mazus stolonifer Makino (see)
- ♦ Miquel's mazus
- ♦ Weed, Quarantine Weed, Naturalised
- ♦ 76, 87, 88, 101, 203, 220, 263, 275, 286, 297
- ♦ aH, cultivated.

Mazus pumilus (Burm.f.) Steen.
Scrophulariaceae
Mazus japonicus (Thunb.) Kuntze (see), *Mazus rugosus* Lour.
- ♦ Japanese mazus, Asian mazus
- ♦ Weed, Quarantine Weed, Naturalised
- ♦ 76, 87, 88, 101, 161, 179, 203, 220, 235, 243, 273, 297

Mazus stachydifolius (Turcz.) Maxim.
Scrophulariaceae
- ♦ betonyleaf mazus
- ♦ Weed
- ♦ 297
- ♦ promoted. Origin: China, Korea.

Mazus stolonifer Makino
Scrophulariaceae
= *Mazus miquelii* Makino

- ♦ Weed
- ♦ 88, 204

Mecardonia dianthera (Sw.) Pennell
Scrophulariaceae
= *Mecardonia procumbens* (Mill.) Small
- ♦ Weed, Quarantine Weed
- ♦ 76, 87, 88, 203, 220

Mecardonia procumbens (Mill.) Small
Scrophulariaceae
Bacopa chamaedryoides (Kunth) Wettst., *Bacopa chamaidryoides* Wettst., *Bacopa dianthera* (Sw.) Descole & Borsini (see), *Bacopa procumbens* (Mill.) Greenm. (see), *Erinus procumbens* Mill., *Herpestis caprarioides* Kunth, *Herpestis chamaedryoides* Kunth, *Herpestis colubrina* Kunth, *Herpestis peduncularis* Benth., *Herpestis procumbens* (Mill.) Urb., *Lindernia dianthera* Sw., *Mecardonia dianthera* (Sw.) Pennell (see), *Mecardonia tenuis* Small, *Mecardonia viridis* Small, *Moniera dianthera* (Sw.) Millsp., *Monniera dianthera* Millps., *Pagesia diathera* (Sw.) Pennell, *Pagesia peduncularis* (Benth.) Pennell, *Pagesia procumbens* (Mill.) Pennell
- ♦ baby jumpup
- ♦ Weed, Naturalised
- ♦ 11, 88, 300
- ♦ herbal.

Meconopsis cambrica (L.) Vig.
Papaveraceae
- ♦ keltavaleunikko, Welsh poppy
- ♦ Casual Alien
- ♦ 42, 280
- ♦ cultivated, herbal.

Medemia argun (Mart.) Wurtt. ex H.A.Wendl.
Arecaceae
- ♦ Weed
- ♦ 221
- ♦ Origin: Egypt, Sudan.

Medicago L. spp.
Fabaceae/Papilionaceae
- ♦ medic, alfalfa
- ♦ Weed, Naturalised
- ♦ 198, 243, 272
- ♦ H, herbal.

Medicago aculeata Willd. nom. illeg.
Fabaceae/Papilionaceae
= *Medicago doliata* Carmign. var. *muricata* Heyn (NoR)
- ♦ Weed
- ♦ 88, 94

Medicago agrestis Ten.
Fabaceae/Papilionaceae
- ♦ Weed
- ♦ 87, 88

Medicago arabica (L.) Huds.
Fabaceae/Papilionaceae
Medicago maculata Sibth. (see), *Medicago polymorpha* L. var. *arabica* L.
- ♦ laikkumailanen, spotted medick, spotted burr medick, burr clover, luzerne tachée
- ♦ Weed, Naturalised, Introduced,

Casual Alien
- ♦ 7, 15, 42, 80, 86, 87, 88, 94, 98, 101, 161, 176, 198, 203, 218, 228, 237, 241, 249, 253, 272, 280, 286, 287, 295, 300
- ♦ aH, arid, cultivated, herbal. Origin: Europe, Africa, south-west Asia.

Medicago arborea L.
Fabaceae/Papilionaceae
- ♦ tree medick, moon trefoil
- ♦ Weed, Naturalised, Introduced, Garden Escape, Environmental Weed
- ♦ 54, 86, 88, 98, 176, 203, 228, 241, 251, 280, 290, 300
- ♦ S, arid, cultivated, herbal. Origin: Mediterranean and Canary Islands.

Medicago aschersoniana Urb.
Fabaceae/Papilionaceae
- ♦ Weed
- ♦ 221

Medicago carstiensis Jacq.
Fabaceae/Papilionaceae
- ♦ Weed, Naturalised
- ♦ 272, 287
- ♦ cultivated.

Medicago ciliaris (L.) All.
Fabaceae/Papilionaceae
Medicago intertexta (L.) Mill. var. *ciliaris* (L.) Heyn (see), *Medicago polymorpha* L. var. *ciliaris* L.
- ♦ Weed, Quarantine Weed, Naturalised
- ♦ 76, 86, 87, 88, 94, 203
- ♦ herbal. Origin: Mediterranean.

Medicago coronata (L.) Bartal.
Fabaceae/Papilionaceae
- ♦ Weed
- ♦ 272

Medicago dentatus Pers.
Fabaceae/Papilionaceae
- ♦ Weed, Quarantine Weed
- ♦ 76, 87, 88, 203, 220

Medicago denticulata Willd.
Fabaceae/Papilionaceae
= *Medicago polymorpha* L.
- ♦ burr clover
- ♦ Weed
- ♦ 87, 88, 243
- ♦ toxic.

Medicago disciformis DC.
Fabaceae/Papilionaceae
- ♦ Weed
- ♦ 87, 88

Medicago falcata L.
Fabaceae/Papilionaceae
= *Medicago sativa* L. ssp. *falcata* (L.) Arcang.
- ♦ sickle medick, sirppimailanen, yarrow, yellow medic, yellow sickle medic, lucerne yellowflower
- ♦ Weed, Naturalised
- ♦ 23, 42, 80, 86, 87, 88, 98, 114, 176, 203
- ♦ cultivated, herbal.

Medicago glomerata Balb.
Fabaceae/Papilionaceae
- ♦ Naturalised
- ♦ 280

Medicago granatensis Willd.
Fabaceae/Papilionaceae
♦ Weed
♦ 87, 88

Medicago heldreichii E.Small
Fabaceae/Papilionaceae
♦ Heldreich's alfalfa
♦ Naturalised
♦ 101

Medicago hispida Gaertn.
Fabaceae/Papilionaceae
= *Medicago polymorpha* L.
♦ rough medic, burr clover, silver medic
♦ Weed, Naturalised
♦ 51, 80, 87, 88, 212, 243, 255, 287, 297
♦ arid, herbal.

Medicago hybrida (Pourr.) Trautv.
Fabaceae/Papilionaceae
♦ hybrid alfalfa
♦ Naturalised
♦ 101

Medicago intertexta (L.) Mill.
Fabaceae/Papilionaceae
Medicago polymorpha L. var. *intertexta* L.
♦ calvary medic
♦ Weed, Naturalised, Introduced, Environmental Weed
♦ 7, 70, 86, 87, 88, 98, 198, 203, 221, 228
♦ arid. Origin: western Mediterranean.

Medicago intertexta (L.) Mill. var. *ciliaris* (L.) Heyn
Fabaceae/Papilionaceae
= *Medicago ciliaris* (L.) All.
♦ Weed
♦ 185

Medicago italica (Mill.) Fiori
Fabaceae/Papilionaceae
Medica italica Mill., *Medicago corrugata* Durieu, *Medicago helix* Willd., *Medicago helix* var. *spinulosa* Moris, *Medicago obscura* Retz. (see), *Medicago obscura* ssp. *helix* (Willd.) Urb., *Medicago obscura* var. *aculeata* Guss., *Medicago obscura* var. *rugulosa* Ser., *Medicago polymorpha* L. var. *tornata* L., *Medicago striata* T.Bastard, *Medicago tornata* (L.) Mill. (see), *Medicago tornata* var. *aculeata* (Guss.) Heyn, *Medicago tornata* var. *rugulosa* (Ser.) Heyn *nom. illeg.*, *Medicago tornata* var. *spinulosa* (Moris) Heyn, *Medicago tornata* var. *striata* (T.Bastard) K.A. & I.Lesins
♦ Weed, Naturalised
♦ 86, 93
♦ aH.

Medicago laciniata (L.) Mill.
Fabaceae/Papilionaceae
Medicago aschersonianan Urb., *Medicago polymorpha* L. var. *laciniata* L.
♦ burr clover, little burrweed, veld shamrock, cutleaf medic
♦ Weed, Naturalised, Introduced
♦ 7, 88, 98, 101, 121, 158, 203, 228, 243, 287
♦ aH, arid, herbal. Origin: Eurasia.

Medicago laciniata (L.) Mill.var. laciniata

Fabaceae/Papilionaceae
♦ cutleaf medic
♦ Naturalised
♦ 86, 198
♦ Origin: North Africa, Middle East.

Medicago littoralis Rohde ex Loisel.
Fabaceae/Papilionaceae
♦ rantamailanen, water medic, strand medic
♦ Weed, Naturalised, Casual Alien
♦ 39, 42, 86, 98, 101, 203, 272
♦ arid, cultivated, toxic. Origin: Mediterranean.

Medicago lupulina L.
Fabaceae/Papilionaceae
Lupulina aureata Noulet, *Medicago lupulina* L. var. *willdenowii* (Mérat) Urb., *Medicago willdenowii* (Mérat) Urb., *Melilotus lupulinus* Trautv.
♦ yellow trefoil, nonesuch, black medic, hop medic, hop clover, trefoil, black clover, nurmimailanen
♦ Weed, Noxious Weed, Naturalised, Introduced, Environmental Weed
♦ 7, 15, 21, 23, 24, 34, 42, 49, 52, 70, 80, 84, 86, 87, 88, 94, 98, 101, 134, 154, 161, 165, 174, 176, 179, 198, 203, 207, 210, 211, 212, 218, 228, 235, 236, 237, 241, 243, 249, 253, 272, 275, 280, 286, 287, 293, 295, 297, 299, 300
♦ a/pH, arid, cultivated, herbal. Origin: Mediterranean, west Asia.

Medicago maculata Sibth.
Fabaceae/Papilionaceae
= *Medicago arabica* (L.) Huds.
♦ Weed
♦ 87, 88

Medicago marina L.
Fabaceae/Papilionaceae
♦ sea medick
♦ Weed
♦ 70

Medicago minima (L.) Bartal.
Fabaceae/Papilionaceae
Medicago polymorpha L. var. *minima* L., *Medicago polymorpha* var. *recta* Desf., *Medicago recta* (Desf.) Willd.
♦ pikkumailanen, woolly burr medic, small medic, little medic, little burr clover, burr medic, small burr medic
♦ Weed, Naturalised, Introduced, Environmental Weed, Casual Alien
♦ 7, 39, 42, 72, 86, 87, 88, 93, 94, 98, 101, 161, 176, 198, 203, 205, 218, 228, 237, 241, 253, 272, 275, 280, 287, 295, 297, 300
♦ aH, arid, cultivated, herbal, toxic. Origin: Europe, Asia, Africa.

Medicago monantha (C.A.Mey.) Trautv.
Fabaceae/Papilionaceae
Trigonella incisa Benth. (see), *Trigonella monantha* C.A.Mey. (see), *Trigonella monantha* ssp. *incisa* (Benth.) Ali, *Trigonella monantha* ssp. *noeana* (Boiss.) Hub.-Mor., *Trigonella noeana* Boiss.
♦ medick
♦ Naturalised
♦ 101

Medicago monspeliaca (L.) Trautv.
Fabaceae/Papilionaceae
Trigonella monspeliaca L. (see), *Trigonella monspeliaca* ssp. *subacaulis* Feinbrun
♦ hairy medick
♦ Naturalised
♦ 101

Medicago murex Willd. var. *aculeata* Urb. var. *sphaerica* Urb.
Fabaceae/Papilionaceae
♦ Naturalised
♦ 287

Medicago nigra (L.) Krock.
Fabaceae/Papilionaceae
Medicago hysterix Ten., *Medicago pentacycla* DC.
♦ Weed, Naturalised
♦ 250, 280

Medicago obscura Retz.
Fabaceae/Papilionaceae
= *Medicago italica* (Mill.) Fiori
♦ Weed
♦ 87, 88
♦ aH.

Medicago orbicularis (L.) Bartal.
Fabaceae/Papilionaceae
Medicago applaudata Willd., *Medicago polymorpha* L. var. *orbicularis* L.
♦ buttonclover, blackdisk medick, kiekkomailanen, snail medick
♦ Weed, Naturalised, Introduced, Casual Alien
♦ 7, 34, 40, 42, 86, 87, 88, 94, 98, 101, 198, 203, 228, 241, 253, 272, 287, 300
♦ aH, arid, cultivated, herbal. Origin: Mediterranean, Middle East.

Medicago polymorpha L.
Fabaceae/Papilionaceae
Medicago denticulata Willd. (see), *Medicago hispida* Gaertn. (see), *Medicago lappacea* Desr., *Medicago polycarpa* Willd., *Medicago polymorpha* L. var. *brevispina* (Benth.) Heyn (see), *Medicago polymorpha* L. var. *vulgaris* (Benth.) Shinn. (see), *Medicago terebellum* Willd.
♦ toothed medick, California burr clover, burr clover, medic, piikkimailanen, hairy medick, burr medic, spiny burr medic
♦ Weed, Noxious Weed, Naturalised, Introduced, Environmental Weed, Casual Alien
♦ 7, 9, 15, 26, 39, 42, 70, 72, 78, 86, 87, 88, 94, 98, 101, 115, 116, 121, 134, 158, 161, 167, 176, 179, 180, 185, 198, 203, 205, 221, 228, 229, 235, 240, 243, 253, 272, 286, 287, 295, 300
♦ aH, arid, cultivated, herbal, toxic. Origin: Eurasia.

Medicago polymorpha L. var. *brevispina* (Benth.) Heyn
Fabaceae/Papilionaceae
= *Medicago polymorpha* L.
♦ Weed
♦ 93

Medicago polymorpha L. var. *confinis* (Koch) Oostr. & Reichg.
Fabaceae/Papilionaceae

Medicago hispida Gaertn. var. *confinis*
(Koch) Burk.
- Weed, Naturalised
- 237, 287

Medicago polymorpha L. var. *lapponica*
Burnet
Fabaceae/Papilionaceae
- Naturalised
- 287

Medicago polymorpha L. var. *microdon*
Ehr.
Fabaceae/Papilionaceae
- Naturalised
- 287

Medicago polymorpha L. var. *polymorpha*
Fabaceae/Papilionaceae
- Naturalised
- 241

Medicago polymorpha L. var. *vulgaris*
(Benth.) Shinn.
Fabaceae/Papilionaceae
= *Medicago polymorpha* L.
- California burr clover
- Weed, Introduced
- 93, 218, 228, 237
- arid.

Medicago praecox DC.
Fabaceae/Papilionaceae
- Mediterranean medick
- Weed, Naturalised, Introduced
- 7, 34, 86, 98, 101, 203
- aH, arid, cultivated. Origin:
Mediterranean, Middle East.

Medicago rigidula (L.) All.
Fabaceae/Papilionaceae
Medicago gerardi Waldst. & Kit. ex
Willd.
- tifton burr clover, lucerna tvrdá
- Weed, Naturalised, Introduced
- 101, 228, 253, 272
- arid, herbal.

Medicago rugosa Desr.
Fabaceae/Papilionaceae
Medicago elegans Jacq. ex Willd.
- wrinkled medick, gama medic,
medic
- Weed, Naturalised, Introduced
- 70, 86, 101, 198, 228
- arid. Origin: Mediterranean.

Medicago sativa L.
Fabaceae/Papilionaceae
Medica sativa Lam.
- lucerne, alfalfa, sinimailanen,
sickle medick, purple lucerne, purple
medick, common medick
- Weed, Naturalised, Introduced,
Garden Escape, Environmental Weed,
Cultivation Escape, Casual Alien
- 4, 7, 15, 34, 39, 42, 80, 87, 88, 93, 94,
98, 101, 104, 121, 136, 158, 161, 165, 176,
179, 195, 203, 205, 215, 218, 228, 241,
256, 261, 280, 286, 287, 299, 300
- pH, arid, cultivated, herbal, toxic.
Origin: Eurasia.

Medicago sativa L. ssp. *caerulea* (Less. ex
Ledeb.) Schmalh.
Fabaceae/Papilionaceae
- Naturalised
- 86

- cultivated. Origin: Russia, central
Asia.

Medicago sativa L. ssp. *falcata* (L.)
Arcang.
Fabaceae/Papilionaceae
Medicago borealis Grossh., *Medicago
difalcata* Sinskaya, *Medicago falcata* L.
(see), *Medicago falcata* var. *romanica*
(Prodán) O.Schwarz & Klink., *Medicago
quasifalcata* Sinskaya, *Medicago
romanica* Prodán, *Medicago tenderiensis*
Opperman ex Klokov
- yellow alfalfa, sickle medick, sickle
alfalfa, yellow lucerne, yellowflower
alfalfa, luzerne de suède, luzerne
jaune, sichelklee, sichelluzerne,
luzerna de sequeiro, alfalfa amarilla,
alfalfa sueca
- Weed, Naturalised
- 101, 272
- cultivated. Origin: Eurasia, Africa.

Medicago sativa L. ssp. *sativa*
Fabaceae/Papilionaceae
Medicago sativa L. ssp. *varia* (Martyn)
Arcang. (see)
- lucerne, alfalfa, medic
- Weed, Naturalised, Environmental
Weed
- 40, 86, 101, 198, 272
- cultivated. Origin: Europe, central
Asia, North Africa.

Medicago sativa L. ssp. *varia* (Martyn)
Arcang.
Fabaceae/Papilionaceae
= *Medicago sativa* L. ssp. *sativa*
- sand lucerne
- Naturalised
- 40

Medicago scutellata (L.) Mill.
Fabaceae/Papilionaceae
Medicago inermis L., *Medicago
polymorpha* L. var. *scutellata* L.
- snail medic
- Weed, Naturalised, Introduced,
Casual Alien
- 7, 70, 86, 87, 88, 93, 94, 98, 101, 176,
198, 203, 205, 228, 253, 280
- aH, arid, cultivated. Origin:
Mediterranean.

Medicago tornata (L.) Mill.
Fabaceae/Papilionaceae
= *Medicago italica* (Mill.) Fiori
- disc medic
- Weed, Naturalised
- 86, 93, 98, 203, 205
- aH, arid, cultivated. Origin:
Mediterranean.

Medicago tribuloides Desr.
Fabaceae/Papilionaceae
= *Medicago truncatula* Gaertn.
- Weed
- 87, 88

Medicago truncatula Gaertn.
Fabaceae/Papilionaceae
Medicago tribuloides Desr. (see)
- barrel medic, palikkamailanen,
strong spined medick
- Weed, Naturalised, Introduced,
Environmental Weed, Casual Alien

- 7, 39, 40, 42, 72, 86, 88, 93, 94, 98, 198,
203, 205, 228, 253
- aH, arid, cultivated, toxic. Origin:
Mediterranean.

Medicago truncatula Gaertn. var.
truncatula
Fabaceae/Papilionaceae
Medicago tribuloides Desr. var.
truncatula (Gaertn.) Koch
- Introduced
- 228
- arid.

Medicago turbinata (L.) All.
Fabaceae/Papilionaceae
Medicago polymorpha L. var. *turbinata*
L., *Medicago tuberculata* Willd.
- southern medick
- Weed, Naturalised, Introduced
- 87, 88, 101, 228, 241, 300
- arid, cultivated.

Medicago × varia Martyn
Fabaceae/Papilionaceae
= *Medicago sativa* L. ssp. *falcata* (L.)
Arcang. × *Medicago sativa* L. ssp. *sativa*
- ristimailanen, lucerna menlivá,
bastard lucerne
- Casual Alien
- 42
- promoted, herbal.

Medinilla cumingii Naud.
Melastomataceae
- medinilla
- Cultivation Escape
- 233
- cultivated.

Medinilla venosa (Blume) Blume
Melastomataceae
- holdtight, medinilla
- Weed, Noxious Weed, Naturalised,
Cultivation Escape
- 3, 101, 191, 229, 233
- cultivated.

Mediocactus Britt. & Rose spp.
Cactaceae
= *Heliocereus* (Berger) Britt. & Rose spp.
- Weed, Quarantine Weed
- 76, 88, 203, 220

Mediocalcar ponapense Hawkes
Orchidaceae
- Introduced
- 230
- H.

Melaleuca L. spp.
Myrtaceae
- melaleuca
- Weed, Quarantine Weed,
Environmental Weed
- 18, 88, 226, 258

Melaleuca armillaris (Sol. ex Gaertn.)
Sm.
Myrtaceae
- giant honey myrtle, bracelet honey
myrtle
- Weed, Naturalised, Native Weed,
Environmental Weed, Cultivation
Escape
- 72, 86, 88, 252, 296
- S/T, cultivated. Origin: Australia.

Melaleuca bracteata F.Muell.
Myrtaceae
♦ white cloud tree
♦ Introduced
♦ 228
♦ S, arid, cultivated, herbal.

Melaleuca decussata R.Br.
Myrtaceae
♦ totem poles, niaouli cajeput, crossed leaved honey myrtle
♦ Weed, Naturalised, Native Weed, Garden Escape, Environmental Weed
♦ 72, 86, 88
♦ S, cultivated, herbal. Origin: Australia.

Melaleuca diosmifolia R.Br.
Myrtaceae
♦ green honey myrtle
♦ Weed, Naturalised, Native Weed, Garden Escape, Environmental Weed, Cultivation Escape
♦ 7, 72, 86, 88, 198
♦ S, cultivated. Origin: Australia.

Melaleuca halmaturorum F.Muell. ex Miq.
Myrtaceae
♦ salt paperbark
♦ Introduced
♦ 228
♦ arid, cultivated. Origin: Australia.

Melaleuca hypericifolia Sm.
Myrtaceae
♦ hillock bush, cajuput, red honey myrtle, teatree
♦ Weed, Naturalised, Native Weed, Garden Escape, Environmental Weed
♦ 72, 86, 88, 198
♦ S, cultivated, herbal. Origin: Australia.

Melaleuca lanceolata Otto
Myrtaceae
Melaleuca pubescens Schauer
♦ moonah
♦ Naturalised, Native Weed, Introduced
♦ 7, 86, 228
♦ arid, cultivated.

Melaleuca leucadendra (L.) L.
Myrtaceae
Melaleuca leucadendron (L.) L. (see)
♦ broad leaved paperbark, cajeput
♦ Weed
♦ 3, 87, 88
♦ cultivated, herbal.

Melaleuca leucadendron (L.) L.
Myrtaceae
= *Melaleuca leucadendra* (L.) L.
♦ melaleuca, punk tree, cajeput
♦ Weed, Noxious Weed
♦ 22, 122
♦ T, cultivated, herbal.

Melaleuca linariifolia Sm.
Myrtaceae
♦ cajeput tree, snow in summer
♦ Weed, Naturalised
♦ 39, 101, 179
♦ S, cultivated, herbal, toxic. Origin: Australia.

Melaleuca nesophila F.Muell.
Myrtaceae
♦ mauve honey myrtle, showy honey myrtle
♦ Weed, Naturalised, Native Weed, Garden Escape, Environmental Weed
♦ 72, 86, 88
♦ S, cultivated. Origin: Australia.

Melaleuca parvistaminea Brynes
Myrtaceae
♦ rough paperbark
♦ Weed, Naturalised, Native Weed, Environmental Weed
♦ 72, 86, 88
♦ S, cultivated. Origin: Australia.

Melaleuca pentagona Labill.
Myrtaceae
♦ Naturalised, Native Weed
♦ 7, 86
♦ cultivated. Origin: Australia.

Melaleuca quinquenervia (Cav.) T.Blake
Myrtaceae
Melaleuca leucadendron auct. non (L.) L., *Metrosideros quinquenervia* Cav.
♦ melaleuca, paperbark tree, swamp teatree, punk tree, niaouli, cajeput, paperbark, cajeput tree, broadleaf paperbark
♦ Weed, Noxious Weed, Naturalised, Native Weed, Garden Escape, Environmental Weed, Cultivation Escape
♦ 3, 6, 7, 22, 35, 37, 67, 78, 80, 88, 101, 105, 107, 112, 116, 122, 129, 140, 142, 151, 152, 161, 179, 193, 197, 200, 229, 233, 246, 261
♦ T, aqua, cultivated, herbal, toxic. Origin: Australasia.

Melaleuca styphelioides Sm.
Myrtaceae
♦ prickly paperbark
♦ Naturalised, Casual Alien
♦ 198, 280
♦ cultivated. Origin: Australia.

Melaleuca viminalis (Gaertn.) Byrnes
Myrtaceae
♦ Weed
♦ 179
♦ Origin: Australia.

Melampodium L. spp.
Asteraceae
♦ blackfoot
♦ Weed
♦ 88
♦ cultivated.

Melampodium arvense Rob.
Asteraceae
♦ Weed, Quarantine Weed
♦ 76, 87, 88, 203, 220

Melampodium divaricatum (Rich) DC.
Asteraceae
♦ boton de oro, flor amarilla, florecilla
♦ Weed, Quarantine Weed
♦ 14, 76, 87, 88, 157, 199, 203, 220, 243, 255, 281
♦ aH, arid, herbal. Origin: South America.

Melampodium longifolium Cerv.
Asteraceae
♦ Weed
♦ 199

Melampodium paludosum Kunth
Asteraceae
= *Melampodium leucanthum* Torr. & A.Gray (NoR)
♦ Weed, Quarantine Weed
♦ 76, 88
♦ aH, cultivated.

Melampodium paniculatum Gardner
Asteraceae
Melampodium brachyglossum Donn.Sm.
♦ botao de ouro
♦ Weed
♦ 255
♦ Origin: Brazil.

Melampodium perfoliatum (Cav.) Kunth
Asteraceae
Alcina perfoliata Cav.
♦ perfoliate blackfoot, estrelinha, botao de cachorro
♦ Weed, Naturalised
♦ 101, 157, 243, 255
♦ aH.

Melampodium sericeum Lag.
Asteraceae
♦ rough blackfoot
♦ Naturalised
♦ 101

Melampyrum L. spp.
Scrophulariaceae
♦ cow wheat
♦ Weed
♦ 272

Melampyrum arvense L.
Scrophulariaceae
♦ field cow wheat, field cowpea, peltomaitikka
♦ Weed, Quarantine Weed
♦ 39, 44, 70, 76, 88, 94, 203, 220, 243, 253, 272
♦ cultivated, herbal, toxic.

Melampyrum barbatum Waldst. & Kit. ex Willd.
Scrophulariaceae
♦ bearded cow wheat
♦ Weed
♦ 88, 94, 272

Melampyrum cristatum L.
Scrophulariaceae
♦ crested cow wheat, tähkämaitikka
♦ Weed
♦ 23, 88, 272
♦ herbal.

Melampyrum lineare Desr.
Scrophulariaceae
♦ narrowleaf cow wheat
♦ Weed
♦ 23, 88
♦ herbal.

Melampyrum pratense L.
Scrophulariaceae
♦ common cow wheat, kangasmaitikka
♦ Weed

♦ 70, 272
♦ herbal.

Melandrium album (Mill.) Garcke
Caryophyllaceae
= *Silene latifolia* Poir. ssp. *alba* (Mill.)
Greut. & Burd.
♦ white campion
♦ Weed
♦ 44, 243
♦ herbal.

Melandrium noctiflorum (L.) Fr.
Caryophyllaceae
= *Silene noctiflora* L.
♦ night flowering campion, night
flowering catchfly, clammy cockle,
sticky cockle
♦ Weed
♦ 44, 87, 88, 243
♦ cultivated, herbal.

**Melanocenchris abyssinica (R.Br. ex
Fresen.) Hochst.**
Poaceae
♦ Weed
♦ 221
♦ G.

**Melanoselinum decipiens (Schrad. &
J.C.Wendl) Hoffm.**
Apiaceae
♦ parsnip palm
♦ Weed, Naturalised, Environmental
Weed
♦ 15, 86, 280
♦ cultivated.

Melanthera aspera (Jacq.) Small
Asteraceae
♦ yerba de cabra
♦ Weed
♦ 157
♦ pH.

Melanthera confusa Britton
Asteraceae
= *Melanthera aspera* (Jacq.) Steud.
ex Small var. *glabriuscula* (Kuntze)
J.C.Parks (NoR)
♦ Weed
♦ 87, 88

Melanthera deltoidea L.C.Rich. ex Michx.
Asteraceae
= *Melanthera aspera* (Jacq.) Steud. ex
Small var. *aspera* (NoR)
♦ Weed
♦ 14

Melanthera nivea (L.) Small
Asteraceae
Bidens nivea L.
♦ snow squarestem, botoncillo
♦ Weed
♦ 87, 88, 281
♦ herbal. Origin: Americas.

**Melanthera scandens (Schum. & Thonn.)
Brenan**
Asteraceae
♦ Weed, Quarantine Weed
♦ 76, 87, 88, 203, 220

Melanthium latifolium Desr.
Liliaceae/Melanthiaceae
♦ slender bunchflower, hybrid

bunchflower
♦ Weed
♦ 161
♦ herbal, toxic.

Melanthium parviflorum (Michx.) S.Wats.
Liliaceae/Melanthiaceae
♦ Appalachian bunchflower, small
flowered hellebore
♦ Weed
♦ 161
♦ herbal, toxic.

Melanthium virginicum L.
Liliaceae/Melanthiaceae
♦ bunchflower, Virginia bunchflower
♦ Weed
♦ 39, 161
♦ pH, cultivated, herbal, toxic.

**Melanthium woodii (J.W.Robbins ex
Wood) Bodkin**
Liliaceae/Melanthiaceae
♦ Indian poke, Wood's bunchflower
♦ Weed
♦ 161
♦ herbal, toxic.

Melastoma affine D.Don
Melastomataceae
Melastoma polyanthum Bl.
♦ Weed
♦ 87, 88, 170, 191
♦ cultivated, herbal. Origin: Australia.

Melastoma candidum D.Don
Melastomataceae
Melastoma septemnervium Lour. *non*
Jacq.
♦ Indian rhododendron, Asian
melastome
♦ Weed, Noxious Weed, Naturalised,
Cultivation Escape
♦ 3, 80, 87, 88, 101, 191, 229, 233
♦ cultivated.

Melastoma decifidium Roxb.
Melastomataceae
♦ Weed
♦ 22
♦ S.

Melastoma malabathricum L.
Melastomataceae
♦ Bank's melastoma, Malabar
melastome, moegalo, Singapore
rhododendron
♦ Weed, Noxious Weed, Naturalised
♦ 12, 22, 28, 67, 80, 87, 88, 101, 140, 217,
218, 229, 243
♦ S, cultivated, herbal.

Melastoma sanguineum Sims.
Melastomataceae
Melastoma decemfidum Roxb.
♦ fox tongued melastoma, red
melastome, pink tibouchina
♦ Weed, Quarantine Weed,
Naturalised, Cultivation Escape
♦ 76, 87, 88, 101, 203, 220, 233
♦ cultivated.

Melhania denhamii R.Br.
Sterculiaceae
♦ Weed
♦ 221

Melhania forbesii Planch. ex Mast.
Sterculiaceae
♦ Native Weed
♦ 121
♦ S. Origin: southern Africa.

Melia azedarach L.
Meliaceae
♦ Chinaberry, white cedar, cape lilac,
tulip cedar, syringa, Indian bead tree,
Persian lilac, maksering, bessieboom,
Chinaberry tree, margosa tree,
azedarach, bead tree, berrytree, cape
syringa, China tree, Chinese umbrella,
Indian lilac, Japanese bead tree,
paradise tree, pride of China, pride
of Persia, red seringea, South African
syringa, Syrian bead tree, Texas
umbrella tree
♦ Weed, Noxious Weed, Naturalised,
Introduced, Garden Escape,
Environmental Weed, Cultivation
Escape, Casual Alien
♦ 3, 7, 10, 22, 34, 39, 63, 77, 80, 86, 87,
88, 93, 95, 101, 102, 107, 112, 121, 134,
142, 151, 152, 154, 158, 161, 179, 190,
205, 218, 228, 230, 233, 247, 252, 257,
261, 277, 279, 280, 283, 295
♦ T, arid, cultivated, herbal, toxic.
Origin: south-east Asia.

Melia baccifera Roth
Meliaceae
♦ Quarantine Weed
♦ 220

Melia volkensii Gürke
Meliaceae
♦ tree of knowledge
♦ Quarantine Weed
♦ 220

Melianthus comosus Vahl
Melianthaceae
♦ tufted honeyflower
♦ Weed, Quarantine Weed, Noxious
Weed, Naturalised, Native Weed,
Garden Escape
♦ 39, 76, 86, 87, 88, 98, 121, 147, 161,
198, 203, 220, 269
♦ pS, cultivated, herbal, toxic. Origin:
southern Africa.

Melianthus major L.
Melianthaceae
♦ honeyflower, large honeyflower,
cape honeyflower
♦ Weed, Naturalised, Native Weed,
Garden Escape, Environmental Weed
♦ 7, 15, 39, 86, 88, 98, 121, 161, 165, 198,
203, 225, 246, 269, 280, 289
♦ pS, cultivated, herbal, toxic. Origin:
southern Africa.

Melica altissima L.
Poaceae
♦ isohelmikkä, Siberian melicgrass
♦ Quarantine Weed, Naturalised,
Casual Alien
♦ 42, 101, 220
♦ G, cultivated.

Melica ciliata L.
Poaceae
Beckeria montana Bernh., *Melica glauca*
F.Schultz
♦ silky melic, tähkähelmikkä
♦ Weed, Quarantine Weed,
Naturalised
♦ 86, 220, 272
♦ G, cultivated, herbal. Origin:
Eurasia.

Melica decumbens Thunb.
Poaceae
♦ staggers grass
♦ Weed, Native Weed
♦ 39, 121
♦ pG, toxic. Origin: southern Africa.

Melica minuta L.
Poaceae
♦ Naturalised
♦ 280
♦ G, cultivated.

Melica nutans L.
Poaceae
Melica montana Huds.
♦ mountain melick, nodding melick,
nodding pearl, nuokkuhelmikkä
♦ Weed
♦ 70, 272, 286
♦ G, cultivated, herbal.

Melica racemosa Thunb.
Poaceae
♦ melic grass
♦ Native Weed
♦ 121
♦ pG. Origin: southern Africa.

Melica scabrosa Trin.
Poaceae
♦ Weed
♦ 275, 297
♦ pG.

Melica uniflora Retz.
Poaceae
♦ wood melick, rönsyhelmikkä
♦ Weed
♦ 70, 272
♦ G, cultivated, herbal.

Melicoccus bijugatus Jacq.
Sapindaceae
♦ Spanish lime, genip, mamoncillo,
quenepa
♦ Weed, Naturalised, Environmental
Weed, Cultivation Escape
♦ 22, 32, 88, 101, 151, 179, 261
♦ T, cultivated. Origin: northern South
America.

Melicope ponapensis Lauterb.
Rutaceae
♦ kahmet painte
♦ Introduced
♦ 230
♦ T.

Melilotus Mill. spp.
Fabaceae/Papilionaceae
♦ sweetclover, melilot, yellow
sweetclover, white sweetclover
♦ Weed, Naturalised, Environmental
Weed
♦ 80, 104, 198, 272

♦ H, herbal, toxic.

Melilotus alba Medik.
Fabaceae/Papilionaceae
= *Melilotus albus* Medik.
♦ white sweetclover, white melilot,
bokhara clover, bukhara clover, clover,
hubam clover, sweetclover
♦ Weed, Noxious Weed, Naturalised,
Introduced, Cultivation Escape, Casual
Alien
♦ 4, 7, 21, 34, 39, 42, 51, 52, 70, 80, 87,
88, 94, 98, 102, 121, 136, 146, 158, 161,
195, 203, 207, 212, 218, 222, 228, 241,
243, 261, 269, 272, 286, 287, 294, 299
♦ a/bH, arid, cultivated, herbal, toxic.
Origin: Eurasia.

Melilotus albus Medik.
Fabaceae/Papilionaceae
Melilotus alba Medik. (see), *Melilotus
albus* var. *annuus* H.S.Coe, *Melilotus
leucanthus* W.D.J.Koch ex DC.
♦ sweetclover, bokhara clover,
komonica biela
♦ Weed, Naturalised
♦ 15, 40, 86, 176, 179, 198, 237, 255, 256,
275, 280, 295, 297, 300
♦ a/bH, cultivated, herbal. Origin:
Eurasia.

Melilotus altissima Thuill.
Fabaceae/Papilionaceae
Melilotus altissimus Thuill. (see),
Trifolium altissimum (Thuill.) Loisel.,
Sertula altissima (Thuill.) Kuntze
♦ tall yellow sweetclover, isomesikkä,
tall melilot, yellow melilot
♦ Weed, Naturalised
♦ 39, 42, 70, 87, 88, 218, 241, 272, 287
♦ b/pH, cultivated, herbal, toxic.

Melilotus altissimus Thuill.
Fabaceae/Papilionaceae
= *Melilotus altissima* Thuill.
♦ tall yellow sweetclover, tall melilot
♦ Naturalised
♦ 40, 101, 300
♦ H, cultivated, herbal.

Melilotus dentata (Waldst. & Kit.) Pers.
Fabaceae/Papilionaceae
♦ rantamesikkä
♦ Casual Alien
♦ 42
♦ H.

Melilotus elegans Salz.
Fabaceae/Papilionaceae
♦ elegant sweetclover
♦ Weed
♦ 70
♦ aH, promoted.

Melilotus indica (L.) All.
Fabaceae/Papilionaceae
= *Melilotus indicus* (L.) All.
♦ sour clover, small melilot,
Indian sweetclover, annual yellow
sweetclover, bitter clover, Indian
melilot, sweetclover, yellow melilot
♦ Weed, Naturalised, Introduced,
Environmental Weed, Casual Alien,
Cultivation Escape
♦ 7, 39, 42, 51, 70, 72, 88, 94, 98, 115,

121, 158, 161, 167, 180, 203, 212, 218,
228, 241, 243, 253, 269, 272, 286, 287
♦ aH, arid, cultivated, herbal, toxic.
Origin: Eurasia.

Melilotus indicus (L.) All.
Fabaceae/Papilionaceae
Melilotus bonplandii Ten., *Melilotus
indica* (L.) All. (see), *Melilotus indicus*
(L.) All. ssp. *permixtus* (Jordan) Rouy,
Melilotus melilotus-indica Asch. &
Graebn., *Melilotus melilotus-indicus*
Asch. & Graebn., *Melilotus officinalis
sensu* Bojer, *Melilotus parviflora* Desf.
(see), *Melilotus parviflorus* Desf.,
Melilotus permixtus Jordan, *Melilotus
tommasinii* Jordan, *Trifolium indica* L.,
Trifolium indicum L.
♦ Hexham scent, annual yellow
sweetclover, sour clover, senji, Indian
sweetclover, small melilot, mélilot des
Indes, kleinblütiger steinklee, trevo de
cheiro
♦ Weed, Naturalised, Environmental
Weed, Casual Alien
♦ 9, 15, 40, 55, 80, 86, 87, 93, 101, 134,
165, 176, 179, 185, 198, 205, 221, 235,
237, 243, 271, 280, 295, 300
♦ aH, herbal, toxic. Origin:
Mediterranean to India.

Melilotus infesta Guss.
Fabaceae/Papilionaceae
♦ Weed, Naturalised
♦ 98, 203
♦ H.

Melilotus italica (L.) Lam.
Fabaceae/Papilionaceae
♦ Italian melilot
♦ Weed, Introduced
♦ 70, 228
♦ H, arid.

Melilotus messanensis (L.) All.
Fabaceae/Papilionaceae
= *Melilotus sicula* Vitm.
♦ Mediterranean melilot
♦ Weed, Naturalised, Environmental
Weed
♦ 7, 70, 72, 86, 87, 88, 98, 185, 198, 203
♦ aH, cultivated.

Melilotus neapolitanus Ten.
Fabaceae/Papilionaceae
= *Melilotus sulcatus* Desf.
♦ European sweetclover
♦ Naturalised
♦ 101
♦ H.

Melilotus officinalis Lam.
Fabaceae/Papilionaceae
Melilotus albus Medik. (see), *Melilotus
arvensis* Wallr., *Melilotus graveolens*
Bunge, *Melilotus officinalis* fo. *suaveolens*
(Ledeb.) H.Ohashi & Tateishi, *Melilotus
officinalis* Lam. var. *micranthus*
O.E.Schulz (see), *Melilotus suaveolens*
Ledeb. (see), *Melilotus vulgaris* Hill,
Trifolium officinale L.
♦ yellow sweetclover, common
melilot, ribbed melilot, rohtomesikkä,
field melilot, yellow melilot, mélilot

jaune, mélilot officinal, gelber
Steinklee, meliloto giallo, trevo
cheiroso, cornilla real, meliloto
amarillo, trébol de olor
♦ Weed, Naturalised, Introduced,
Environmental Weed
♦ 4, 7, 15, 21, 34, 39, 39, 40, 42, 52, 70,
78, 80, 86, 87, 88, 94, 98, 101, 102, 116,
136, 146, 151, 161, 174, 176, 195, 198,
203, 212, 218, 222, 237, 241, 243, 253,
272, 280, 287, 294, 295, 300
♦ a/bH, arid, cultivated, herbal, toxic.
Origin: Eurasia.

**Melilotus officinalis Lam. var. *micranthus*
O.E.Schulz**
Fabaceae/Papilionaceae
= *Melilotus officinalis* Lam.
♦ Naturalised
♦ 287
♦ H.

Melilotus parviflora Desf.
Fabaceae/Papilionaceae
= *Melilotus indicus* (L.) All.
♦ Weed
♦ 243
♦ T, arid.

Melilotus segetalis (Brot.) Ser.
Fabaceae/Papilionaceae
♦ corn melilot
♦ Weed
♦ 70, 87, 88, 94, 253
♦ H, arid.

Melilotus sicula Vitm.
Fabaceae/Papilionaceae
Melilotus messanensis (L.) All. (see),
Trifolium messanense L., *Trifolium
siculum* Turra
♦ Weed
♦ 87, 88
♦ aH. Origin: North Africa, Middle
East, southern Europe.

Melilotus suaveolens Ledeb.
Fabaceae/Papilionaceae
= *Melilotus officinalis* Lam.
♦ sweetclover
♦ Weed, Naturalised
♦ 275, 286, 287, 297
♦ a/bH, promoted. Origin: temperate
Asia.

Melilotus sulcata Desf.
Fabaceae/Papilionaceae
= *Melilotus sulcatus* Desf.
♦ uurremesikkä
♦ Weed, Naturalised, Casual Alien
♦ 42, 87, 88, 94, 98, 203, 253
♦ H.

Melilotus sulcatus Desf.
Fabaceae/Papilionaceae
Melilotus neapolitanus Ten. (see),
Melilotus sulcata Desf. (see), *Sertula
sulcata* (Desf.) Kuntze
♦ Mediterranean sweetclover,
furrowed melilot, grooved melilot,
mélilote des moissons, gefurchter
Steinklee, anafa, anafe, trébol amarillo,
trébol real de olor
♦ Weed, Naturalised
♦ 101, 243
♦ H.

Melilotus wolgica Poir.
Fabaceae/Papilionaceae
Melilotus ruthenica (Bieb.) Ser.
♦ volganmesikä
♦ Casual Alien
♦ 42
♦ bH, cultivated.

Melinis Beauv. spp.
Poaceae
♦ stinkgrass, mollasses grass
♦ Quarantine Weed, Naturalised
♦ 198, 220
♦ G.

Melinis minutiflora Beauv.
Poaceae
Panicum minutiflora (P.Beauv.) Rasp.
(see), *Panicum melinis* Trin. (see)
♦ molasses grass, wynne grass,
puakatau, Brazilian stinkgrass,
dordura grass, efwatakala grass,
gordura grass, honey grass
♦ Weed, Naturalised, Native Weed,
Introduced, Environmental Weed,
Cultivation Escape
♦ 3, 7, 18, 38, 80, 86, 87, 88, 98, 101, 112,
121, 134, 151, 152, 157, 179, 203, 226,
241, 245, 246, 255, 257, 261, 287
♦ pG, cultivated, herbal. Origin:
tropical Africa.

Melinis repens (Willd.) Zizka
Poaceae
Melinis rosea (Nees) Hack., *Monachyron
roseum* (Nees) Parl., *Monachyron
tonsum* (Nees) Parl., *Panicum braunii*
Steud., *Panicum roseum* (Nees) Steud.,
Panicum tonsum (Nees) Steud.,
Rhynchelytrum dregeanum Nees,
Rhynchelytrum dregeanum var. *annuum*
Chiov., *Rhynchelytrum dregeanum* var.
intermedium Chiov., *Rhynchelytrum
repens* (Willd.) C.E.Hubb. (see),
Rhynchelytrum repens var. *roseum*
(Nees) Chiov., *Rhynchelytrum roseum*
(Nees) Stapf & C.E.Hubb. ex Bews
(see), *Rhynchelytrum tonsum* (Nees)
Lanza & Mattei, *Saccharum sphacelatum*
(Benth.) Walp., *Saccharum repens* Willd.,
Tricholaena dregeana (Nees) T.Durand
& Schinz, *Tricholaena fragilis* A.Braun,
Tricholaena repens (Willd.) Hitchc. (see),
Tricholaena rosea Nees (see), *Tricholaena
sphacelata* Benth., *Tricholaena tonsa*
Nees, *Tricholaena tonsa* var. *submutica*
Schweinf.
♦ Natal redtop, red Natal grass, rose
Natal grass
♦ Weed, Naturalised, Introduced,
Environmental Weed, Cultivation
Escape, Casual Alien
♦ 7, 55, 86, 88, 93, 101, 158, 198, 205,
228, 261, 280
♦ a/pG, arid, cultivated. Origin:
Africa.

Meliosma myriantha Siebold & Zucc.
Sabiaceae/Meliosmaceae
♦ Quarantine Weed
♦ 220
♦ cultivated.

Melissa officinalis L.
Lamiaceae
Melissa graveolens Host, *Thymus melissa*
E.H.L.Krause
♦ melissa balm, lemon balm, balm,
bee balm, common balm, cedronella,
melissa vera, citronella, erba limona
♦ Weed, Noxious Weed, Naturalised,
Introduced, Garden Escape,
Environmental Weed
♦ 15, 34, 40, 70, 72, 80, 88, 97, 101, 137,
165, 176, 181, 198, 203, 228, 241, 243,
280, 300
♦ pH, arid, cultivated, herbal. Origin:
Eurasia.

Melissa officinalis L. ssp. *officinalis*
Lamiaceae
♦ lemon balm
♦ Naturalised, Environmental Weed
♦ 86
♦ cultivated. Origin: Mediterranean.

Melissitus ruthenicus (L.) C.W.Chang
Fabaceae/Papilionaceae
= *Medicago ruthenica* (L.) Trautv. (NoR)
♦ Russian fenugreek
♦ Weed
♦ 297

Melittis melissophyllum L.
Lamiaceae
Melittis grandiflora Sm., *Melittis
sylvestris* Lam.
♦ bastard balm
♦ Weed
♦ 70
♦ pH, cultivated, herbal.

Melochia compacta Hochr.
Sterculiaceae
= *Melochia umbellata* (Houtt.) Stapf
♦ kotol
♦ Weed
♦ 3, 191
♦ T.

Melochia concatenata L.
Sterculiaceae
= *Melochia corchorifolia* L.
♦ Weed
♦ 23, 87, 88, 262
♦ aH.

Melochia corchorifolia L.
Sterculiaceae
Riedlea corchorifolia (L.) DC., *Visenia
corchorifolia* (L.) Spreng.
♦ redweed, chocolateweed, wire bush,
seng lek
♦ Weed, Naturalised
♦ 88, 101, 161, 170, 179, 204, 209, 218,
235, 239, 276, 286
♦ pH, cultivated, herbal. Origin:
Madagascar.

Melochia indica Kurz
Sterculiaceae
= *Melochia umbellata* (Houtt.) Stapf
♦ Weed
♦ 3, 191

Melochia lupulina Sw.
Sterculiaceae
♦ Weed
♦ 87, 88, 157
♦ pH.

Melochia manducata Wright
Sterculiaceae
Melochia glandulifera Standl.
♦ Weed
♦ 157
♦ aH.

Melochia melissaefolia Benth.
Sterculiaceae
♦ Weed
♦ 87, 88

**Melochia mollis (Kunth) Triana &
Planch.**
Sterculiaceae
♦ Weed
♦ 87, 88
♦ S.

Melochia nodiflora Sw.
Sterculiaceae
♦ bretonica prieta
♦ Weed
♦ 14, 262
♦ aH, cultivated.

Melochia pyramidata L.
Sterculiaceae
♦ pyramid flower, guanxuma rosa
♦ Weed, Naturalised, Introduced,
Environmental Weed
♦ 28, 86, 87, 88, 98, 157, 199, 203, 228,
243, 255
♦ pH, arid, cultivated. Origin: Brazil.

Melochia tomentosa L.
Sterculiaceae
♦ teabush
♦ Weed
♦ 87, 88
♦ herbal.

Melochia umbellata (Houtt.) Staf.
Sterculiaceae
Melochia compacta Hochr. (see),
Melochia indica Kurz. (see)
♦ hierba del soldado, melochia
♦ Weed, Naturalised
♦ 3, 22, 80, 101, 107
♦ S/T, herbal.

Melochia villosa (Mill) Fawc. & Rendle
Sterculiaceae
= *Melochia spicata* (L.) Fryxell (NoR)
♦ Weed
♦ 14, 87, 88
♦ herbal.

**Melolobium candicans (E.Mey) Eckl. &
Zeyh.**
Fabaceae/Papilionaceae
♦ honey bush
♦ Native Weed
♦ 121
♦ S. Origin: southern Africa.

Melolobium humile Eckl. & Zeyh.
Fabaceae/Papilionaceae
♦ Native Weed
♦ 121
♦ S. Origin: southern Africa.

Melothria fluminensis Gardn.
Cucurbitaceae
= *Melothria pendula* L.
♦ Weed

♦ 87, 88
♦ herbal.

Melothria formosana Hayata
Cucurbitaceae
♦ Weed
♦ 87, 88

Melothria guadalupensis (Spreng.) Cogn.
Cucurbitaceae
= *Melothria pendula* L.
♦ Weed
♦ 87, 88

Melothria heterophylla (Lour.) Cogn.
Cucurbitaceae
♦ Weed
♦ 87, 88
♦ herbal.

Melothria japonica (Thunb.) Maxim.
Cucurbitaceae
♦ Weed
♦ 87, 88, 286

Melothria maderaspatana (L.) Cogn.
Cucurbitaceae
= *Mukia maderaspatana* (L.) M.J.Roem.
♦ Weed
♦ 13, 87, 88
♦ herbal.

Melothria mucronata (Bl.) Cogn.
Cucurbitaceae
♦ Weed
♦ 87, 88

Melothria pendula L.
Cucurbitaceae
Melothria fluminensis Gardn. (see),
Melothria guadalupensis (Spreng.) Cogn.
(see)
♦ Guadeloupe cucumber, creeping
cucumber, mini cukemelon
♦ Naturalised
♦ 247, 287
♦ cultivated, herbal, toxic.

Melothria tridactyla Hook.f.
Cucurbitaceae
♦ Weed
♦ 87, 88

Memecylon floribundum Blume
Melastomataceae/Memecylaceae
♦ Weed, Environmental Weed
♦ 3, 152, 191
♦ Origin: Indonesia.

Memecylon myrsinoides Bl.
Melastomataceae/Memecylaceae
♦ Weed
♦ 12

Memora peregrina (Miers) Sandwith
Bignoniaceae
♦ cipo arame, ciganinha
♦ Weed
♦ 255
♦ Origin: Brazil.

Menispermum canadense L.
Menispermaceae
♦ common moonseed, moonseed,
Canada moonseed, Texas sasparilla,
yellow sasparilla, yellow parilla
♦ Weed
♦ 8, 39, 87, 88, 154, 161, 189, 218, 247
♦ pC, cultivated, herbal, toxic.

Menispermum dauricum DC.
Menispermaceae
♦ aasiankilpikierto, koumorikazura
♦ Weed, Cultivation Escape
♦ 42, 297
♦ cultivated, herbal.

Menonvillea alata Rollins
Brassicaceae
♦ Naturalised
♦ 241
♦ arid.

Mentha L. spp.
Lamiaceae
♦ mint
♦ Weed, Naturalised
♦ 7, 39, 80, 221, 243, 272
♦ herbal, toxic.

Mentha aquatica L.
Lamiaceae
♦ mint, water mint, vesiminttu
♦ Weed, Naturalised
♦ 7, 8, 23, 70, 80, 86, 87, 88, 98, 101, 203,
241, 272, 287, 300
♦ pH, aqua, cultivated, herbal, toxic.
Origin: Eurasia.

Mentha arvensis L.
Lamiaceae
Mentha arvensis var. *canadensis*
(L.) Kuntze (see), *Mentha austriaca*
Jacq., *Mentha gentilis* L. (see),
Mentha lapponica Wahlenb., *Mentha
parietariifolia* J.Beck.
♦ mint, corn mint, Japanese mint,
field mint, American wild mint, menta
campestre
♦ Weed, Quarantine Weed,
Naturalised, Introduced
♦ 38, 44, 70, 76, 80, 86, 87, 88, 94, 118,
161, 174, 210, 218, 243, 253, 272, 280,
287
♦ pH, cultivated, herbal. Origin:
Eurasia.

**Mentha arvensis L. var. canadensis (L.)
Kuntze**
Lamiaceae
= *Mentha arvensis* L.
♦ Naturalised
♦ 287

**Mentha arvensis L. var. javanica (Bl.)
Hook.**
Lamiaceae
♦ Weed
♦ 13

**Mentha arvensis L. var. piperascens
Malinv. ex L.H.Bailey**
Lamiaceae
= *Mentha canadensis* L. (NoR)
♦ Japaninminttu, Japanische minze,
Japanese mint
♦ Weed
♦ 286
♦ herbal.

Mentha australis R.Br.
Lamiaceae
♦ river mint
♦ Naturalised
♦ 39, 280
♦ pH, cultivated, toxic.

Mentha cardiaca (S.F.Gray) Gerarde ex Bak.
  Lamiaceae
  = *Mentha × gracilis* Sole (*pro* sp.)
  ♦ Naturalised
  ♦ 287
  ♦ cultivated, herbal.

Mentha × dalmatica L.
  Lamiaceae
  = *Mentha arvensis* L. × *Mentha longifolia* (L.) Huds.
  ♦ karjalanminttu
  ♦ Cultivation Escape
  ♦ 42
  ♦ cultivated, herbal.

Mentha gentilis L.
  Lamiaceae
  = *Mentha arvensis* L.
  ♦ red mint
  ♦ Weed
  ♦ 87, 88, 218
  ♦ cultivated, herbal.

Mentha × gracilis Sole (*pro* sp.)
  Lamiaceae
  = *Mentha arvensis* L. × *Mentha spicata* L. [see *Mentha cardiaca* (S.F.Gray) Gerarde ex Bak.]
  ♦ small leaved mint, gingermint, bushy mint, mäta jemná, mint
  ♦ Weed, Naturalised
  ♦ 80, 101
  ♦ cultivated, herbal. Origin: Europe.

Mentha haplocalyx Briq.
  Lamiaceae
  ♦ Chinese mint
  ♦ Weed
  ♦ 275, 297
  ♦ pH, herbal.

Mentha lavandulacea Willd.
  Lamiaceae
  = *Mentha longifolia* (L.) Huds.
  ♦ Introduced
  ♦ 38

Mentha longifolia (L.) Huds.
  Lamiaceae
  Mentha alaica Boriss., *Mentha asiatica* Boriss., *Mentha darvasica* Boriss., *Mentha kopetdaghensis* Boriss., *Mentha lavandulacea* Willd. (see), *Mentha longifolia* ssp. *caucasica* (Gand.) Briq., *Mentha pamiroalaica* Boriss., *Mentha spicata* L. ssp. *longifolia* (L.) Tacik ex Towpasz, *Mentha sylvestris* L., *Mentha vagans* Boriss.
  ♦ horsemint, downy mint, mint
  ♦ Weed, Naturalised, Casual Alien
  ♦ 39, 42, 86, 87, 88, 94, 198, 248, 272, 287
  ♦ pH, cultivated, herbal, toxic. Origin: Eurasia.

Mentha longifolia (L.) Huds. ssp. typhoides (Birq.) Harley
  Lamiaceae
  ♦ Weed
  ♦ 185

Mentha microphylla Koch
  Lamiaceae

♦ Weed
♦ 87, 88, 221

Mentha nemorosa Willd. ex L.
  Lamiaceae
  = *Mentha spicata* L. × *Mentha suaveolens* Ehrh. [see *Mentha × villosa* Huds.]
  ♦ yerba buena
  ♦ Introduced
  ♦ 261
  ♦ herbal.

Mentha × piperita L.
  Lamiaceae
  = *Mentha aquatica* L. × *Mentha spicata* L. [*Mentha nigricans* Mill.]
  ♦ peppermint, white peppermint, piparminttu, eau de Cologne mint, black peppermint, agua florida, menta piperita
  ♦ Weed, Naturalised, Garden Escape, Environmental Weed, Cultivation Escape
  ♦ 8, 42, 72, 80, 86, 87, 88, 98, 101, 102, 134, 161, 167, 176, 198, 218, 241, 257, 261, 280, 286, 287, 295, 300
  ♦ pH, cultivated, herbal, toxic. Origin: Great Britain.

Mentha × piperita L. var. citrata (Ehrh.) Briq.
  Lamiaceae
  = *Mentha aquatica* L. × *Mentha spicata* L.
  ♦ lemon mint, eau de Cologne mint, lime mint, bergamot mint
  ♦ Weed, Naturalised
  ♦ 15, 198, 208, 280
  ♦ aqua, herbal. Origin: Europe.

Mentha × piperita L. var. piperita
  Lamiaceae
  = *Mentha aquatica* L. × *Mentha spicata* L.
  ♦ peppermint
  ♦ Weed, Naturalised
  ♦ 165, 198, 280
  ♦ aqua, cultivated, herbal. Origin: Europe.

Mentha pulegium L.
  Lamiaceae
  Pulegium vulgare Mill.
  ♦ peppermint, pennyroyal, European pennyroyal, pennyroyal mint, puolanminttu
  ♦ Weed, Noxious Weed, Naturalised, Garden Escape, Environmental Weed, Casual Alien
  ♦ 7, 15, 35, 39, 42, 70, 72, 78, 80, 86, 87, 88, 94, 97, 98, 101, 116, 147, 151, 154, 161, 165, 176, 198, 200, 203, 221, 231, 241, 253, 269, 272, 280, 286, 287, 296, 300
  ♦ pH, aqua, cultivated, herbal, toxic. Origin: Eurasia.

Mentha requienii Benth.
  Lamiaceae
  ♦ Corsican mint, mint
  ♦ Naturalised
  ♦ 40
  ♦ pH, cultivated, herbal.

Mentha × rotundifolia (L.) Huds.
  Lamiaceae

= *Mentha longifolia* (L.) Huds. × *Mentha suaveolens* Ehrh.
  ♦ roundleaf mint, apple mint, apple scented mint
  ♦ Weed, Naturalised
  ♦ 80, 86, 87, 88, 98, 101, 198, 287, 295
  ♦ cultivated, herbal. Origin: Europe.

Mentha satureiodes R.Br.
  Lamiaceae
  ♦ Weed
  ♦ 87, 88

Mentha spicata L.
  Lamiaceae
  Mentha crispa L., *Mentha crispata* Schrad., *Mentha niliaca* auct. non Juss. ex Jacq., *Mentha cordifolia* auct. non Opiz, *Mentha longifolia* auct. non (L.) Huds., *Mentha spicata* L. var. *crispa* (Benth.) Danert (see), *Mentha sylvestris* auct. non L., *Mentha viridis* (L.) L. (see)
  ♦ spearmint, garden mint, pea mint, mint, lambmint
  ♦ Weed, Naturalised, Introduced, Garden Escape, Environmental Weed, Cultivation Escape
  ♦ 8, 21, 40, 42, 72, 80, 86, 87, 88, 94, 98, 101, 102, 116, 134, 155, 165, 176, 198, 203, 208, 218, 272, 280, 287, 296
  ♦ pH, aqua, cultivated, herbal, toxic. Origin: south and central Europe.

Mentha spicata L.ssp. spicata
  Lamiaceae
  ♦ spearmint
  ♦ Weed, Naturalised
  ♦ 15, 280

Mentha spicata L. ssp. tomentosa (Briq.) Harley
  Lamiaceae
  = *Mentha spicata* ssp. *condensata* (Briq.) Greuter & Burdet (NoR) [see *Mentha tomentosa* D'Urv. non Sm. *nom. illeg.*]
  ♦ hairy spearmint
  ♦ Weed, Naturalised
  ♦ 15, 280

Mentha spicata L. var. crispa (Benth.) Danert
  Lamiaceae
  = *Mentha spicata* L.
  ♦ curled mint
  ♦ Weed, Naturalised
  ♦ 286, 287
  ♦ herbal.

Mentha spicata L. var. spicata
  Lamiaceae
  ♦ field mint, spearmint
  ♦ Introduced
  ♦ 34
  ♦ pH.

Mentha suaveolens Ehrh.
  Lamiaceae
  ♦ round leaved mint, apple mint, mint
  ♦ Weed, Naturalised, Introduced, Casual Alien
  ♦ 15, 34, 42, 70, 80, 88, 94, 98, 101, 203, 250, 253, 280, 300
  ♦ pH, cultivated, herbal.

Mentha tomentosa **D'Urv.** *non* **Sm.** *nom. illeg.*
Lamiaceae
= *Mentha spicata* L. ssp. *condensata* (Briq.) Greuter & Burdet (NoR) [see *Mentha spicata* ssp. *tomentosa* (Briq.) Harley]
♦ Weed
♦ 87, 88

Mentha × verticillata **L.**
Lamiaceae
= *Mentha arvensis* L. × *Mentha aquatica* L.
♦ whorled mint, kiehkuraminttu
♦ Weed, Naturalised
♦ 80, 101, 272

Mentha × villosa **Huds.**
Lamiaceae
= *Mentha spicata* L. × *Mentha suaveolens* Ehrh. [see *Mentha nemorosa* Willd. ex L.]
♦ large apple mint
♦ Naturalised
♦ 40, 101
♦ pH, promoted, herbal.

Mentha viridis **L.**
Lamiaceae
= *Mentha spicata* L.
♦ menta dolce, spearmint
♦ Weed, Naturalised
♦ 86, 87, 88, 98
♦ cultivated, herbal.

Mentzelia albicaulis **(Douglas ex Hook.) Douglas ex Torr. & A.Gray**
Loasaceae
♦ prairie lily, whitestem blazingstar, whitestem stickleaf
♦ Weed
♦ 87, 88, 161
♦ aH, cultivated, herbal.

Mentzelia aspera **L.**
Loasaceae
♦ tropical blazingstar
♦ Weed
♦ 87, 88
♦ arid, herbal.

Mentzelia decapetala **(Pursh ex Sims) Urb. & Gilg ex Gilg**
Loasaceae
♦ evening starflower, tenpetal mentzelia, tenpetal stickleaf, tenpetal blazingstar, prairie stickleaf
♦ Native Weed
♦ 161, 174
♦ cultivated, herbal. Origin: North America.

Mentzelia laevicaulis **(Douglas) Torr. & Gray**
Loasaceae
♦ blazingstar, smoothstem blazingstar
♦ Weed
♦ 161, 180
♦ pH, cultivated, herbal.

Mentzelia lindleyi **Torr. & Gray**
Loasaceae
♦ Lindley's blazingstar, blazingstar
♦ Weed, Cultivation Escape

♦ 34
♦ aH, cultivated, herbal.

Mentzelia nuda **(Pursh) Torr. & A.Gray**
Loasaceae
♦ stiff mentzelia, bractless mentzelia, stiff nuttallia, branched nuttallia, blazingstar, stickleaf, evening star
♦ Weed
♦ 49
♦ cultivated, herbal.

Menyanthes trifoliata **L.**
Menyanthaceae
♦ common bogbean, buckbean, bogbean, tnfoglio fibrino, vachta trojlistá, marsh trefoil
♦ Weed, Quarantine Weed, Naturalised, Environmental Weed
♦ 23, 39, 70, 87, 88, 218, 246, 272, 280
♦ pH, aqua, cultivated, herbal, toxic.

Menziesia ferruginea **Sm.**
Ericaceae
♦ mock azalea, rusty menziesia, false huckleberry
♦ Weed
♦ 39, 39, 161
♦ S, promoted, herbal, toxic.

Mercurialis annua **L.**
Euphorbiaceae
♦ annual mercury, mercury, mercury weed, rikkasinijuuri, dog's mercury, mercorella comune
♦ Weed, Naturalised, Garden Escape, Environmental Weed, Casual Alien
♦ 7, 15, 23, 34, 39, 42, 44, 70, 86, 87, 88, 94, 98, 101, 109, 115, 118, 121, 161, 203, 221, 243, 253, 272, 280, 300
♦ a/bH, cultivated, herbal, toxic. Origin: Eurasia.

Mercurialis perennis **L.**
Euphorbiaceae
Mercurialis nemoralis Salisb., *Mercurialis longifolia* Host.
♦ dog's mercury, lehtosinijuuri
♦ Weed
♦ 39, 272
♦ pH, cultivated, herbal, toxic.

Merremia aegyptia **(L.) Urb.**
Convolvulaceae
Ipomoea aegyptia L.
♦ hairy woodrose
♦ Weed, Naturalised
♦ 11, 28, 86, 87, 88, 93, 179, 206, 221, 237, 243, 255
♦ aC, arid, herbal. Origin: tropical America.

Merremia angustifolia **Hallier f.**
Convolvulaceae
♦ Weed
♦ 87, 88

Merremia bracteata **Blume**
Convolvulaceae
♦ ambui
♦ Weed
♦ 3, 191

Merremia cissoides **(Lam.) Hallier f.**
Convolvulaceae
♦ roadside woodrose

♦ Weed
♦ 87, 88, 179, 255
♦ Origin: tropical America.

Merremia dissecta **(Jacq.) Hallier f.**
Convolvulaceae
♦ white convolvulus creeper, noyau vine
♦ Weed, Naturalised, Garden Escape, Environmental Weed
♦ 7, 86, 87, 88, 93, 98, 203, 205, 255
♦ pC, cultivated, herbal. Origin: tropical America.

Merremia distillatoria **(Blanco) Merr.**
Convolvulaceae
♦ Weed
♦ 87, 88
♦ herbal.

Merremia emarginata **(Burm.f.) Hallier f.**
Convolvulaceae
♦ Weed
♦ 13, 87, 88, 242
♦ bH, arid, herbal. Origin: Asia, Australia.

Merremia gemella **(Burm.f.) Hallier f.**
Convolvulaceae
♦ Weed
♦ 13, 87, 88
♦ C, cultivated.

Merremia hederacea **(Burm.f.) Hallier f.**
Convolvulaceae
♦ ivy woodrose
♦ Weed, Naturalised
♦ 87, 88, 204, 287
♦ herbal.

Merremia hirta **(L.) Merr.**
Convolvulaceae
♦ Weed
♦ 3, 13, 191
♦ C. Origin: Asia.

Merremia macrocalyx **(Ruiz & Pav.) O'Donell**
Convolvulaceae
Convolvulus macrocalyx Ruiz & Pav.
♦ jitirana, batata de purga
♦ Weed
♦ 255
♦ Origin: tropical America.

Merremia nymphaeifolia **(Dietr.) Hallier f.**
Convolvulaceae
= *Merremia peltata* (L.) Merr.
♦ Weed
♦ 3, 191

Merremia pacifica **(L.) Merr.**
Convolvulaceae
♦ Weed
♦ 3, 191

Merremia peltata **(L.) Merr.**
Convolvulaceae
Merremia nymphaeifolia (Dietr.) Hallier f. (see), *Ipomoea peltata* (L.) Choisy (see), *Operculina peltata* (L.) Hallier f. (see)
♦ merrimia, lohl, yol, kebeas, lagon, lagun, pala, fue, fue vao, fue kula, iol, puhlah, fue lautetele, fue mea, abui, grobihi, arosumou, wa mbula, wa bula, wa damu, wa ndamu, viliyawa,

wiliviwa, veliyana, wiliao
- ♦ Weed, Introduced
- ♦ 3, 6, 87, 88, 107, 230
- ♦ C, herbal. Origin: Asia, Australia.

Merremia quinquefolia (L.) Hallier f.
Convolvulaceae
Ipomoea quinquefolia L.
- ♦ merremia, merremia vine
- ♦ Weed, Quarantine Weed, Naturalised
- ♦ 76, 86, 87, 88, 157, 179, 203
- ♦ aC, arid. Origin: tropical America.

Merremia tridentata (L.) Hallier f.
Convolvulaceae
= *Xenostegia tridentata* (L.) Austin & Staples (NoR) [see *Ipomoea angustifolia* Jacq., *Merremia tridentata* (L.) Hallier f. ssp. *angustifolia* (Jacq.) Ooststr.]
- ♦ Weed
- ♦ 87, 88
- ♦ cultivated, herbal.

Merremia tridentata (L.) Hallier f. ssp. *angustifolia* (Jacq.) Ooststr.
Convolvulaceae
= *Xenostegia tridentata* (L.) Austin & Staples (NoR) [see *Ipomoea angustifolia* Jacq., *Merremia tridentata* (L.) Hallier f.]
- ♦ Weed, Native Weed
- ♦ 88, 121
- ♦ a/pC, toxic.

Merremia tridentata (L.) Hallier f. ssp. *hastata* Ooststr.
Convolvulaceae
- ♦ Weed
- ♦ 13
- ♦ C.

Merremia tuberosa (L.) Rendle
Convolvulaceae
Ipomoea tuberosa L. (see)
- ♦ wood rose, Spanish arborvine, merremia
- ♦ Weed, Sleeper Weed, Noxious Weed, Naturalised, Garden Escape, Environmental Weed
- ♦ 3, 22, 80, 86, 88, 112, 122, 142, 151, 179, 191, 261
- ♦ pC, cultivated, herbal. Origin: tropical America.

Merremia umbellata (L.) Hallier f.
Convolvulaceae
- ♦ hogvine
- ♦ Weed, Naturalised
- ♦ 13, 14, 87, 88, 157, 179, 255, 257, 261
- ♦ pC, herbal. Origin: Americas.

Merremia vitifolia (Burm.f.) Hallier f.
Convolvulaceae
- ♦ Weed
- ♦ 13, 209
- ♦ C, herbal.

Mertensia fusiformis Greene
Boraginaceae
= *Mertensia oblongifolia* (Nutt.) G.Don (NoR)
- ♦ spindleroot bluebells
- ♦ Weed
- ♦ 23, 88
- ♦ herbal.

Mertensia maritima (L.) S.F.Gray
Boraginaceae
Pneumaria maritima (L.) Hill, *Pulmonaria maritima* L.
- ♦ oyster leaf, oyster plant, austernpflanze, Asiatic lungwort, sea lungwort
- ♦ Weed
- ♦ 23, 88
- ♦ pH, cultivated, herbal. Origin: Europe.

Merxmuellera disticha (Nees) Conert
Poaceae
Danthonia disticha Nees
- ♦ wiry danthonia
- ♦ Native Weed
- ♦ 121
- ♦ pG. Origin: southern Africa.

Meryta senfftiana Volk.
Araliaceae
- ♦ Introduced
- ♦ 230
- ♦ H.

Mesechites rosea (A.DC.) Miers
Apocynaceae
- ♦ Weed
- ♦ 14

Mesembryanthemum aitonis Jacq.
Aizoaceae/Mesembryanthemaceae
Cryophytum aitonis (Thunb.) N.E.Br
- ♦ sea spinach, angled iceplant
- ♦ Weed, Naturalised, Native Weed, Environmental Weed
- ♦ 7, 86, 98, 121, 198, 203
- ♦ a/bH. Origin: southern Africa.

Mesembryanthemum crystallinum L.
Aizoaceae/Mesembryanthemaceae
Cryophytum crystallinum (L.) N.E.Br., *Gasoul crystallinum* (L.) Rothm.
- ♦ crystalline iceplant, iceplant, common iceplant
- ♦ Weed, Noxious Weed, Naturalised, Introduced, Garden Escape, Environmental Weed, Casual Alien
- ♦ 7, 35, 72, 78, 80, 86, 88, 98, 101, 116, 142, 152, 176, 198, 203, 228, 231, 241, 280, 296, 300
- ♦ aH, arid, cultivated, herbal. Origin: west coast Africa.

Mesembryanthemum forskahlii Hochst. ex Boiss.
Aizoaceae/Mesembryanthemaceae
- ♦ Weed
- ♦ 221

Mesembryanthemum nodiflorum L.
Aizoaceae/Mesembryanthemaceae
- ♦ slenderleaf iceplant, angled iceplant, iceplant
- ♦ Weed, Noxious Weed, Naturalised, Introduced, Garden Escape, Environmental Weed, Casual Alien
- ♦ 7, 35, 42, 78, 86, 88, 98, 101, 116, 198, 203, 228, 300
- ♦ aH, arid, cultivated, herbal. Origin: South Africa.

Mespilus germanica L.
Rosaceae
Mespilus sylvestris Mill., *Pyrus*

germanica J.D.Hook.
- ♦ medlar
- ♦ Naturalised
- ♦ 40
- ♦ T, cultivated, herbal.

Messerschmidia argentea (L.f.) Johnst.
Boraginaceae
= *Tournefortia argentea* L.f.
- ♦ Weed
- ♦ 88
- ♦ cultivated, herbal.

Messerschmidia sibirica L. var. *angustoir* (DC.) Kitag.
Boraginaceae
- ♦ Siberian messerschmidia
- ♦ Weed
- ♦ 297

Metalasia muricata (L.) D.Don
Asteraceae
- ♦ blombos
- ♦ Weed, Quarantine Weed, Native Weed
- ♦ 121, 220
- ♦ pS, arid, cultivated, herbal. Origin: southern Africa.

Metaplexis japonica (Thunb.) Makino
Asclepiadaceae/Apocynaceae
Metaplexis chinensis Decne., *Metaplexis stauntonii* Schult., *Pergularia japonica* Thunb., *Urostelma chinense* Bunge
- ♦ rough potato
- ♦ Weed, Naturalised
- ♦ 87, 88, 101, 191, 204, 263, 275, 286, 297
- ♦ pH, cultivated, herbal.

Metopium toxiferum (L.) Krug & Urb.
Anacardiaceae
- ♦ poison wood, Florida poisontree
- ♦ Weed
- ♦ 14, 39, 82, 161
- ♦ herbal, toxic.

Metrodorea flavida K.Krause
Rutaceae
- ♦ Introduced
- ♦ 38

Metrosideros excelsa Sol. ex Gaertn.
Myrtaceae
Metrosideros tomentosa A.Rich.
- ♦ New Zealand bottlebrush, pohutukawa, Nieu Seelandse perdestert
- ♦ Weed, Noxious Weed, Cultivation Escape
- ♦ 15, 63, 88, 95, 131, 283
- ♦ cultivated, herbal. Origin: New Zealand.

Metrosideros kermadecensis W.R.B.Oliv.
Myrtaceae
- ♦ Naturalised
- ♦ 134
- ♦ cultivated.

Metroxylon americarum (Wendl.) Becc.
Arecaceae
- ♦ oahs
- ♦ Introduced
- ♦ 230
- ♦ T.

Metroxylon sagu **Rottb.**
Arecaceae
♦ sago palm
♦ Introduced
♦ 230
♦ T, cultivated, herbal.

Meum athamanticum **Jacq.**
Apiaceae
♦ karhunjuuri, spignel, bald money
♦ Casual Alien
♦ 42
♦ pH, cultivated, herbal.

Meyna grisea **W.Robyns**
Rubiaceae
Vangueria spinosa Roxb.
♦ Weed
♦ 13

Mibora minima **(L.) Desv.**
Poaceae
Agrostis minima L., *Chamagrostis minima*
Borkh., *Sturmia minima* Hoppe
♦ early sandgrass, sand bent, keriheinä
♦ Weed, Naturalised, Environmental
Weed, Casual Alien
♦ 7, 30, 42, 86, 88, 98, 101, 203, 243, 253
♦ G. Origin: western Europe,
Mediterranean.

Michelia champaca **L.**
Magnoliaceae
♦ michelia, fragrant champaca,
champaca
♦ Introduced
♦ 39, 228
♦ arid, cultivated, herbal, toxic.

Miconia **Ruiz & Pav. spp.**
Melastomataceae
♦ miconia, johnnyberry
♦ Noxious Weed, Naturalised
♦ 86

Miconia affinis **DC.**
Melastomataceae
♦ saquiyac
♦ Noxious Weed
♦ 229

Miconia calvescens **DC.**
Melastomataceae
Miconia magnifica Triana (see)
♦ bush currant, miconia, purple
plague, velvetleaf
♦ Weed, Quarantine Weed, Noxious
Weed, Naturalised, Environmental
Weed
♦ 3, 6, 18, 22, 37, 76, 80, 86, 88, 101, 105,
129, 132, 151, 152, 155, 191, 246
♦ S, cultivated. Origin: tropical
America.

Miconia chamissois **Naudin**
Melastomataceae
♦ Weed
♦ 87, 88

Miconia foveolata **Cogn.**
Melastomataceae
♦ Puerto Rico johnnyberry
♦ Noxious Weed
♦ 229

Miconia impetiolaris **(Sw.) D.Don ex DC.**
Melastomataceae

♦ camasey de costilla
♦ Noxious Weed
♦ 229

Miconia laevigata **(L.) D.Don**
Melastomataceae
♦ smooth johnnyberry
♦ Weed, Noxious Weed
♦ 87, 88, 229

Miconia lanata **(DC.) Triana**
Melastomataceae
♦ hairy johnnyberry
♦ Noxious Weed
♦ 229

Miconia lateriflora **Cogn.**
Melastomataceae
♦ Weed
♦ 87, 88

Miconia magnifica **Triana**
Melastomataceae
= *Miconia calvescens* DC.
♦ Weed
♦ 3, 191
♦ cultivated.

Miconia mirabilis **(Aubl.) L.O.Williams**
Melastomataceae
♦ camasey cuatrocanales
♦ Noxious Weed
♦ 229

Miconia nervosa **(Sm.) Triana**
Melastomataceae
♦ Weed
♦ 87, 88
♦ S.

Miconia pachyphylla **Cogn.**
Melastomataceae
♦ camasey racimoso
♦ Noxious Weed
♦ 229

Miconia prasina **(Sw.) DC.**
Melastomataceae
♦ granadillo bobo
♦ Noxious Weed
♦ 229

Miconia punctata **(Desr.) D.Don ex DC.**
Melastomataceae
♦ auquey
♦ Noxious Weed
♦ 229

Miconia pycnoneura **Urb.**
Melastomataceae
♦ ridge johnnyberry
♦ Noxious Weed
♦ 229

Miconia racemosa **(Aubl.) DC.**
Melastomataceae
♦ camasey felpa
♦ Noxious Weed
♦ 229

Miconia rubiginosa **(Bonpl.) DC.**
Melastomataceae
♦ peraleio
♦ Noxious Weed
♦ 229

Miconia serrulata **(DC.) Naud.**
Melastomataceae
♦ jau jau

♦ Noxious Weed
♦ 229

Miconia sintenisii **Cogn.**
Melastomataceae
♦ mountain johnnyberry
♦ Noxious Weed
♦ 229

Miconia stenostachya **(Schr. & Mart.)
DC.**
Melastomataceae
♦ Weed
♦ 87, 88

Miconia subcorymbosa **Britt.**
Melastomataceae
♦ forest johnnyberry
♦ Noxious Weed
♦ 229

Miconia tetrandra **(Sw.) D.Don**
Melastomataceae
♦ rajador
♦ Noxious Weed
♦ 229

Miconia tetrastoma **Naud.**
Melastomataceae
♦ graceful johnnyberry
♦ Noxious Weed
♦ 229

Miconia thomasiana **DC.**
Melastomataceae
♦ camasey tomaso
♦ Noxious Weed
♦ 229

Micranthemum umbrosum **(Walter ex
Gmel.) Blake**
Scrophulariaceae
♦ shade mudflower
♦ Quarantine Weed
♦ 258
♦ aqua, cultivated, herbal.

Micrargeria filiformis **(Schumach. &
Thonn.) Hutch. & Dalziel**
Scrophulariaceae
♦ Casual Alien
♦ 280

Microcarpaea minima **(K.D.Koenig)
Merr.**
Scrophulariaceae
Microcarpaea alternifiora Bl.,
Microcarpaea muscosa R.Br., *Paederota
minima* K.D.Koenig
♦ suzumenohakobe
♦ Weed
♦ 87, 88, 170, 191, 286
♦ Origin: tropical Asia.

Microchloa caffra **Nees**
Poaceae
♦ pinchusion grass
♦ Native Weed
♦ 121
♦ pG. Origin: southern Africa.

Microchloa indica **(L.f.) Beauv.**
Poaceae
Microchloa indica (L.f.) Kuntze,
Microchloa indica var. *gracilis* Rendle,
Microchloa kunthii Desv., *Microchloa
setacea* (Roxb.) R.Br., *Nardus indica* L.f.,
Rottboellia setacea Roxb.

- ♦ Weed, Naturalised
- ♦ 32, 241, 300
- ♦ G. Origin: Australia.

Micrococca mercurialis **(L.) Benth.**
Euphorbiaceae
- ♦ Weed, Naturalised
- ♦ 86, 87, 88, 93, 98, 191, 203
- ♦ cultivated. Origin: tropical Africa, India.

Microcystis toxica **Stephens**
Chroococcaceae
- ♦ Weed, Quarantine Weed, Naturalised
- ♦ 76, 86, 87, 88, 203, 220

Microlaena stipoides **(Labill.) R.Br.**
Poaceae
= *Ehrharta stipoides* Labill.
- ♦ meadow ricegrass
- ♦ Weed, Naturalised
- ♦ 3, 80, 87, 88, 191, 241
- ♦ G, cultivated, herbal. Origin: Asia, Australia, New Zealand.

Microlepia speluncae **(L.) Moore**
Dennstaedtiaceae
- ♦ limpleaf fern
- ♦ Weed, Naturalised
- ♦ 32, 101
- ♦ pH, cultivated, herbal.

Microloma tenuifolium **K.Schum.**
Asclepiadaceae/Apocynaceae
- ♦ coral creeper, red wax creeper, wax twiner
- ♦ Native Weed
- ♦ 121
- ♦ pC, cultivated. Origin: southern Africa.

Microlonchus salmanticus **(L.) DC.**
Asteraceae
= *Mantisalca salmantica* (L.) Briq. & Cav.
- ♦ Weed
- ♦ 121
- ♦ aH, herbal. Origin: Eurasia.

Micromelum pubescens **Blume**
Rutaceae
- ♦ Weed
- ♦ 87, 88

Micromeria brownei **(Sw.) Benth.**
Lamiaceae
= *Clinopodium brownei* (Sw.) Kuntze (NoR)
- ♦ Browne's savory
- ♦ Weed
- ♦ 87, 88
- ♦ herbal.

Micromeria sinaica **Benth.**
Lamiaceae
- ♦ Weed
- ♦ 221

Micromeria viminea **(L.) Urb.**
Lamiaceae
- ♦ Jamaican mint
- ♦ Weed
- ♦ 87, 88
- ♦ cultivated, herbal.

Micropterum papulosum **(L.f.) Schwantes**
Aizoaceae
Micropterum papillosum nom. illeg.

- ♦ Weed, Naturalised, Environmental Weed
- ♦ 7, 86, 98, 203
- ♦ cultivated. Origin: South Africa.

Micropuntia **Daston spp.**
Cactaceae
- ♦ Weed, Quarantine Weed
- ♦ 76, 88, 203

Micropus supinus **L.**
Asteraceae
- ♦ haunio
- ♦ Casual Alien
- ♦ 42

Microsechium helleri **(Peyr.) Cogn.**
Cucurbitaceae
- ♦ Weed
- ♦ 199
- ♦ herbal.

Microseris troximoides **Gray**
Asteraceae
= *Nothocalais troximoides* (Gray) Greene (NoR)
- ♦ Weed
- ♦ 23, 88
- ♦ herbal.

Microsorum heterophyllum **(L.) A.D.Hawkes**
Polypodiaceae
Polypodium exiguum Hews.
- ♦ Weed
- ♦ 87, 88

Microsorum scolopendria **(Burm.f.) Copel.**
Polypodiaceae
= *Phymatosorus scolopendria* (Burm.f.) Pic.Serm.
- ♦ wart fern, monarch fern, lau 'autâ
- ♦ Weed, Naturalised, Native Weed
- ♦ 101, 121
- ♦ pH, cultivated, herbal.

Microstegium japonicum **(Miq.) Koidz.**
Poaceae
- ♦ Weed
- ♦ 286
- ♦ G.

Microstegium nudum **(Trin.) A.Camus**
Poaceae
- ♦ Weed, Naturalised
- ♦ 98, 203
- ♦ G.

Microstegium spectabile **(Trin.) Camus**
Poaceae
- ♦ Introduced
- ♦ 230
- ♦ G.

Microstegium spectabile **(Trin.) Camus fo. *cryptochaetum* Ohwi f.**
Poaceae
- ♦ Introduced
- ♦ 230
- ♦ G.

Microstegium vimineum **(Trin.) A.Camus**
Poaceae
Eulalia viminea (Trin.) Kuntze
- ♦ Japanese stilt grass, Nepal grass, Nepalese browntop, stiltgrass, Nepal microstegium, flexible seagrass,

Chinese packing grass
- ♦ Weed, Naturalised, Environmental Weed
- ♦ 17, 77, 80, 88, 90, 101, 102, 129, 132, 133, 151, 161, 195, 224, 249, 275, 286, 297
- ♦ pG, herbal. Origin: Japan, Korea, China, Malaysia, India.

Microstegium vimineum **(Trin.) A.Camus var. *polystachyum* (Franch. & Savat.) Ohwi**
Poaceae
- ♦ Weed
- ♦ 286
- ♦ G.

Microtatorchis hosokawae **Fuk.**
Orchidaceae
- ♦ Introduced
- ♦ 230
- ♦ H.

Microtea debilis **Sw.**
Phytolaccaceae/Petiveriaceae
- ♦ weak jumby pepper
- ♦ Weed
- ♦ 87, 88
- ♦ herbal.

Microtea paniculata **Moq.**
Phytolaccaceae/Petiveriaceae
- ♦ Weed
- ♦ 255
- ♦ pH. Origin: Brazil.

Microthlaspi perfoliatum **(L.) F.K.Mey.**
Brassicaceae
= *Thlaspi perfoliatum* L.
- ♦ claspleaf pennycress
- ♦ Naturalised
- ♦ 101

Microtrichia perrotteti **DC.**
Asteraceae
- ♦ Weed
- ♦ 87, 88

Microula sikkimensis **(C.B.Clarke) Hemsl.**
Boraginaceae
- ♦ sikkim microula
- ♦ Weed
- ♦ 297

Miersia chilensis **Lindl.**
Liliaceae/Alliaceae
- ♦ Quarantine Weed
- ♦ 220
- ♦ arid.

Miersia myodes **Bert.**
Liliaceae/Alliaceae
- ♦ Quarantine Weed
- ♦ 220

Miersia rusbyi **Britton**
Liliaceae/Alliaceae
- ♦ Quarantine Weed
- ♦ 220
- ♦ Origin: South America.

Mikania **Willd. spp.**
Asteraceae
- ♦ mile a minute, mikania, hempvine
- ♦ Noxious Weed, Naturalised, Environmental Weed
- ♦ 86, 155

Mikania congesta DC.
Asteraceae
Mikania micrantha Kunth var. *congesta*
(DC.) Robins. (see)
♦ guaco
♦ Weed
♦ 87, 88

Mikania cordata (Burm.f.) B.L.Robins.
Asteraceae
♦ African mile a minute, mile a
minute, heartleaf hempvine, bololo
mobali
♦ Weed, Quarantine Weed, Noxious
Weed, Naturalised, Environmental
Weed
♦ 3, 12, 13, 33, 67, 76, 87, 88, 135, 140,
161, 186, 191, 203, 216, 220, 229, 243,
273, 287
♦ pC, herbal. Origin: south-east Asia,
east Africa.

Mikania cordifolia (L.f.) Willd.
Asteraceae
♦ Florida Keys hempvine
♦ Weed
♦ 87, 88, 255
♦ herbal. Origin: Brazil.

Mikania micrantha Kunth
Asteraceae
♦ mile a minute weed, Chinese
creeper, liane américaine, kwalo
koburu, fue saina, fou laina, wa
mbosuthu, wa mbosuvu, wa mbutako,
wa ndamele, ovaova, climbing
hempvine, bittervine
♦ Weed, Quarantine Weed, Noxious
Weed, Naturalised, Garden Escape,
Environmental Weed
♦ 3, 6, 18, 67, 76, 86, 87, 88, 106, 135,
140, 152, 161, 170, 186, 191, 203, 206,
220, 229, 239, 243, 258, 259, 276, 295
♦ pC, cultivated, herbal. Origin:
Central and South America.

Mikania micrantha Kunth var. *congesta* (DC.) Robins.
Asteraceae
= *Mikania congesta* DC.
♦ Weed
♦ 14

Mikania periplocifolia Hook. & Arn.
Asteraceae
Mikania scandens Willd. var.
periplocifolia (Hook. & Arn.) Bak.
♦ guaco
♦ Weed
♦ 237, 295

Mikania ranunculifolia A.Rich.
Asteraceae
♦ Weed
♦ 14

Mikania scandens (L.) Willd.
Asteraceae
♦ climbing hempweed, climbing
boneset, mile a minute
♦ Weed, Quarantine Weed, Noxious
Weed, Environmental Weed
♦ 3, 22, 76, 87, 88, 152, 161, 186, 191,
203, 218, 220, 229
♦ pC, cultivated, herbal. Origin: North
America.

Milium effusum L.
Poaceae
♦ wood millet, American milletgrass,
tesma
♦ Weed, Casual Alien
♦ 23, 39, 70, 88, 272, 280
♦ G, cultivated, herbal, toxic.

Milium vernale Bieb.
Poaceae
♦ milium, spring milletgrass, early
millet, spring millet
♦ Weed, Noxious Weed, Naturalised
♦ 80, 101, 161, 219, 229, 272
♦ aG.

Milleria quinqueflora L.
Asteraceae
♦ Weed
♦ 14, 88, 157
♦ aH.

Millettia dura Dunn.
Fabaceae/Papilionaceae
♦ shungurhi
♦ Weed
♦ 22
♦ T, cultivated.

Millettia extensa (Benth.) Benth. ex Bak.
Fabaceae/Papilionaceae
Millettia auriculata Brandis
♦ Introduced
♦ 228
♦ arid, herbal.

Millettia grandis (E.Mey.) Skeels
Fabaceae/Papilionaceae
Millettia caffra Meissner
♦ umzimbeet
♦ Introduced
♦ 228
♦ arid, cultivated.

Millettia leucantha Vatke
Fabaceae/Papilionaceae
♦ Introduced
♦ 228
♦ arid.

Mimosa L. spp.
Fabaceae/Mimosaceae
♦ sensitive plant, mimosa
♦ Weed
♦ 88, 204
♦ herbal.

Mimosa acantholoba (Willd.) Poir.
Fabaceae/Mimosaceae
♦ Weed, Naturalised
♦ 22, 257
♦ S.

Mimosa albida Rudd.
Fabaceae/Mimosaceae
♦ la vergonzosa
♦ Weed, Naturalised
♦ 22, 257
♦ T, arid, cultivated.

Mimosa asperata L.
Fabaceae/Mimosaceae
= *Mimosa pigra* L.
♦ Puerto Rico sensitive briar
♦ Quarantine Weed
♦ 220

Mimosa bimucronata (DC.) Kuntze
Fabaceae/Mimosaceae
♦ silva, espinheiro
♦ Weed, Cultivation Escape
♦ 32, 255
♦ T, cultivated. Origin: Brazil.

Mimosa biuncifera Benth.
Fabaceae/Mimosaceae
= *Mimosa aculeaticarpa* Ortega var.
biuncifera (Benth.) Barneby (NoR)
♦ waitaminutebush, cat's claw mimosa
♦ Weed
♦ 87, 88, 218
♦ arid, cultivated, herbal.

Mimosa brachyloba Muhlenb. ex Steud.
Fabaceae/Mimosaceae
♦ Quarantine Weed
♦ 220

Mimosa caesalpiniifolia Benth.
Fabaceae/Mimosaceae
♦ sabi
♦ Introduced
♦ 228
♦ arid.

Mimosa casta L.
Fabaceae/Mimosaceae
♦ graceful mimosa
♦ Weed, Quarantine Weed
♦ 32, 87, 88, 258
♦ herbal.

Mimosa contortuplicata Zucc.
Fabaceae/Mimosaceae
♦ Quarantine Weed
♦ 220

Mimosa debilis Humb. & Bonpl. ex Willd.
Fabaceae/Mimosaceae
♦ dormideira
♦ Weed
♦ 255
♦ Origin: Brazil.

Mimosa diplotricha C.Wright ex Sauvalle
Fabaceae/Mimosaceae
Mimosa invisa Mart. nom. illeg.
♦ giant sensitive plant, singbiguin
sasa, giant false sensitive plant
♦ Weed, Noxious Weed, Naturalised
♦ 3, 88, 101, 140, 191, 229
♦ S.

Mimosa diplotricha C.Mart. ex Colla var. *diplotricha*
Fabaceae/Mimosaceae
♦ giant sensitive plant
♦ Weed, Quarantine Weed, Noxious
Weed, Naturalised
♦ 6, 76, 86, 88, 98, 147, 170, 218
♦ S, herbal. Origin: Brazil.

Mimosa dulcis Roxb.
Fabaceae/Mimosaceae
= *Pithecellobium dulce* (Roxb.) Benth.
♦ Weed, Quarantine Weed
♦ 3, 76, 88, 191, 220

Mimosa flavescens Splitg.
Fabaceae/Mimosaceae
♦ Weed
♦ 87, 88

Mimosa glandulosa Michx.
Fabaceae/Mimosaceae
= *Acuan glandulosa* A.A.Heller
♦ Quarantine Weed
♦ 220

Mimosa illinoensis Michx.
Fabaceae/Mimosaceae
= *Desmanthus illinoensis* (Michx.)
MacMill. ex B.L.Rob. & Fern.
♦ Quarantine Weed
♦ 220

Mimosa invisa C.Mart. ex Colla
Fabaceae/Mimosaceae
♦ giant sensitive plant, grande
sensitive, sensitive gèante, singbiguin
sasa, mechiuaiu, vao fefe palagi,
la'au fefe tele, la'au fefe palagi,
wa ngandrongandro levu, wa
ngandrongandro ni wa ngalelevu,
limemeihr laud, co gadrogadro,
prickly mimosa
♦ Weed, Quarantine Weed, Noxious
Weed, Introduced, Environmental
Weed
♦ 3, 11, 22, 62, 67, 76, 87, 88, 93, 107,
152, 157, 161, 186, 203, 209, 220, 230,
238, 239, 255, 262, 268, 269, 276, 295,
297
♦ pH. Origin: Brazil.

**Mimosa invisa C.Mart. ex Colla var.
inermis Adelb.**
Fabaceae/Mimosaceae
♦ spineless Brazilian sensitive plant
♦ Weed
♦ 297

Mimosa lebbeck L.
Fabaceae/Mimosaceae
= *Albizia lebbeck* (L.) Benth.
♦ Weed
♦ 3, 191

Mimosa nilotica L.
Fabaceae/Mimosaceae
= *Acacia nilotica* (L.) Willd. ex Delile
♦ Weed
♦ 22
♦ T, herbal.

**Mimosa pellita Humb. & Bonpl. ex
Willd.**
Fabaceae/Mimosaceae
= *Mimosa pigra* L.
♦ lollipop mimosa
♦ Weed
♦ 261
♦ arid.

Mimosa pigra L.
Fabaceae/Mimosaceae
Mimosa asperata L. (see), *Mimosa
asperata* var. *pigra* Willd., *Mimosa
berlandieri* A.Gray ex Torr., *Mimosa
brasiliensis* Niederl., *Mimosa canescens*
Willd., *Mimosa ciliata* Willd., *Mimosa
hispida* Willd., *Mimosa pellita* Humb.
& Bonpl. ex Willd. (see), *Mimosa
pigra* var. *berlandieri* (A.Gray ex Torr.)
B.L.Turner, *Mimosa polyacantha* Willd.
♦ cat's claw mimosa, mimosa, giant
sensitive plant, giant sensitive tree,
miyaraap ton, giant mimosa, thorny

sensitive plant
♦ Weed, Quarantine Weed, Noxious
Weed, Naturalised, Native Weed,
Introduced, Garden Escape,
Environmental Weed
♦ 3, 6, 14, 18, 22, 37, 62, 63, 67, 76, 80,
84, 86, 87, 88, 93, 98, 101, 112, 121, 122,
140, 147, 151, 152, 161, 177, 179, 191,
199, 200, 203, 209, 220, 221, 228, 229,
238, 239, 243, 257, 262, 268, 269, 277,
283
♦ S/T, aqua, cultivated, herbal. Origin:
Central and South America.

Mimosa pudica L.
Fabaceae/Mimosaceae
♦ sensitive plant, shameplant,
touch me not, shame lady, mimosa,
shamebush, action plant, humble
plant, live and die, shame weed,
vergonzosa
♦ Weed, Quarantine Weed, Noxious
Weed, Naturalised, Introduced,
Environmental Weed, Garden Escape
♦ 3, 6, 14, 23, 62, 76, 86, 87, 88, 93, 98,
107, 147, 153, 161, 170, 179, 186, 191,
194, 199, 203, 209, 216, 218, 230, 239,
243, 255, 256, 261, 262, 273, 274, 275,
276, 286, 287, 297
♦ pH, cultivated, herbal, toxic. Origin:
tropical America.

Mimosa pudica L. var. hispida Brenan
Fabaceae/Mimosaceae
♦ action plant, humble plant, live
and die, sensitive plant, shame plant,
shame weed, touch me not
♦ Weed, Naturalised, Introduced
♦ 38, 86, 121
♦ H, cultivated. Origin: south-west
Mexico.

**Mimosa pudica L. var. tetrandra (Humb.
& Bonpl. ex Willd.) DC.**
Fabaceae/Mimosaceae
♦ common sensitive plant
♦ Naturalised
♦ 86
♦ cultivated. Origin: tropical America.

**Mimosa pudica L. var. unijuga (Walp. &
Duchass.) Griseb.**
Fabaceae/Mimosaceae
♦ common sensitive plant, shameplant
♦ Naturalised
♦ 86
♦ cultivated. Origin: Caribbean,
Central and South America.

Mimosa ramosissima Benth.
Fabaceae/Mimosaceae
♦ juqueri
♦ Weed
♦ 255
♦ S. Origin: Brazil.

Mimosa rubicaulis Lam.
Fabaceae/Mimosaceae
♦ Introduced
♦ 228
♦ arid, herbal.

Mimosa scabrella Benth.
Fabaceae/Mimosaceae
Mimosa bracaatinga Hoehne

♦ bracaatinga
♦ Quarantine Weed, Introduced
♦ 220, 228
♦ arid, cultivated, herbal.

Mimosa sensitiva L.
Fabaceae/Mimosaceae
♦ Weed
♦ 87, 88
♦ arid, herbal.

Mimosa sepiaria Benth.
Fabaceae/Mimosaceae
♦ Weed
♦ 22, 87, 88
♦ T.

Mimosa setosa Benth.
Fabaceae/Mimosaceae
♦ malicia, dormideira
♦ Weed
♦ 255
♦ S. Origin: Brazil.

**Mimosa somnians Humb. & Bonpl. ex
Willd.**
Fabaceae/Mimosaceae
♦ Weed
♦ 87, 88

Mimosa strigillosa Torr. & Gray
Fabaceae/Mimosaceae
♦ mimosa vine, powder puff
♦ Weed
♦ 161, 249
♦ pH, herbal.

Mimosa tenuiflora (Willd.) Poir.
Fabaceae/Mimosaceae
Acacia tenuiflora Willd., *Mimosa cabrera*
H.Karst., *Mimosa hostilis* (Mart.) Benth.,
Mimosa limana Rizzini, *Mimosa nigra*
Huber
♦ calumbi, jurema preta, carbon chele
♦ Quarantine Weed
♦ 220
♦ arid, cultivated, herbal.

Mimosa velloziana Mart.
Fabaceae/Mimosaceae
♦ Introduced
♦ 38

Mimulus cardinalis Benth.
Scrophulariaceae
= *Mimulus eastwoodiae* Rydb. (NoR)
♦ scarlet monkeyflower, crimson
monkeyflower
♦ Weed
♦ 23, 88
♦ pH, aqua, cultivated, herbal.

Mimulus guttatus DC.
Scrophulariaceae
Mimulus langsdorfii Donn.
♦ seep monkeyflower, monkeyflower,
common monkeyflower, yellow
monkeyflower, täpläapinankukka,
monkeymusk, common large
monkeyflower
♦ Weed, Naturalised, Environmental
Weed
♦ 15, 23, 40, 42, 86, 88, 98, 161, 165, 180,
203, 208, 225, 246, 280
♦ wa/pH, cultivated, herbal. Origin:
western North America.

Mimulus lewisii Pursh
  Scrophulariaceae
  ♦ purple monkeyflower, great purple monkeyflower
  ♦ Weed
  ♦ 23, 88
  ♦ pH, aqua, cultivated, herbal.

Mimulus luteus L.
  Scrophulariaceae
  ♦ glabrous monkeyflower, blood drop emlets, monkeymusk, yellow monkeymusk, blotched monkeyflower
  ♦ Naturalised, Cultivation Escape, Casual Alien
  ♦ 42, 198, 280, 287
  ♦ pH, aqua, cultivated. Origin: Chile.

Mimulus moniliformis Greene
  Scrophulariaceae
  = *Mimulus moschaius* Dougl. ex Lindl. var. *moniliformis* (Greene) Munz (NoR)
  ♦ Naturalised
  ♦ 287

Mimulus moschatus Douglas ex Lindl.
  Scrophulariaceae
  ♦ musk monkeyflower, musk flower, musk, tahma apinankukka
  ♦ Weed, Naturalised, Garden Escape, Environmental Weed, Cultivation Escape
  ♦ 15, 20, 40, 42, 72, 86, 88, 98, 176, 198, 203, 208, 280, 290
  ♦ wpH, cultivated, herbal. Origin: western North America.

Mimulus nepalensis Benth.
  Scrophulariaceae
  ♦ Weed
  ♦ 87, 88

Mimulus orbicularis Wall.
  Scrophulariaceae
  ♦ phak taptao
  ♦ Weed, Quarantine Weed
  ♦ 76, 87, 88, 203, 220, 239, 262
  ♦ aH, aqua.

Mimulus ringens L.
  Scrophulariaceae
  ♦ ringen monkeyflower, Allegheny monkeyflower, square stemmed monkeyflower, lavender musk
  ♦ Weed
  ♦ 23, 88
  ♦ pH, aqua, cultivated, herbal.

Mimusops L. spp.
  Sapotaceae
  ♦ mimusops
  ♦ Weed, Cultivation Escape
  ♦ 32
  ♦ cultivated.

Mimusops balata (Aubl.) Gaertn.f.
  Sapotaceae
  imperfectly known sp. may be *Mimusops coriacea* (A.DC.) Miq. or *Mimusops maxima* (Poir.) R.E.Vaughan
  ♦ Weed
  ♦ 179
  ♦ cultivated.

Mimusops laurifolia (Forssk.) I.Friis
  Sapotaceae

Mimusops schimperi Hochst. ex A.Rich.
  ♦ Introduced
  ♦ 228
  ♦ arid.

Minuartia L. spp.
  Caryophyllaceae
  ♦ sandwort, stitchwort
  ♦ Weed, Naturalised
  ♦ 198, 272

Minuartia hybrida (Vill.) Schischk.
  Caryophyllaceae
  Alsine tenuifolia Crantz, *Minuartia tenuifolia* Hiern
  ♦ fine leaved sandwort
  ♦ Weed, Naturalised
  ♦ 7, 9, 88, 94, 98, 203, 253, 280
  ♦ herbal.

Minuartia mediterranea (Link) K.Maly
  Caryophyllaceae
  ♦ fine leaved sandwort
  ♦ Naturalised, Environmental Weed
  ♦ 86, 176, 198
  ♦ Origin: Mediterranean.

Minuartia picta (Sibth. & Sm.) Bornm.
  Caryophyllaceae
  ♦ Weed
  ♦ 221

Minuartia viscosa (Schreb.) Schinz & Thell.
  Caryophyllaceae
  Alsine viscosa Schreb., *Arenaria pentandra* Dufour, *Alsinella viscosa* (Schreb.) Hartm.
  ♦ sticky sandwort
  ♦ Weed
  ♦ 272

Mirabilis hirsuta (Pursh) MacMill.
  Nyctaginaceae
  ♦ hairy four o'clock
  ♦ Weed
  ♦ 87, 88, 218
  ♦ cultivated, herbal.

Mirabilis jalapa L.
  Nyctaginaceae
  ♦ prairie four o'clock, peteli, marvel of Peru, common four o'clock, beauty of the night, ihmekukka, false jalap, jalapa falsa, jalapa bastarda
  ♦ Weed, Naturalised, Environmental Weed, Garden Escape, Cultivation Escape
  ♦ 7, 15, 32, 34, 40, 42, 80, 86, 87, 88, 98, 101, 121, 154, 157, 161, 179, 189, 198, 203, 215, 218, 237, 247, 252, 255, 256, 257, 261, 269, 280, 286, 287, 295
  ♦ pH, cultivated, herbal, toxic. Origin: South America.

Mirabilis linearis (Pursh) Heimerl
  Nyctaginaceae
  ♦ narrowleaf four o'clock
  ♦ Weed, Cultivation Escape
  ♦ 34, 87, 88, 218
  ♦ pH, cultivated, herbal.

Mirabilis multiflora (Torr.) Gray
  Nyctaginaceae
  ♦ glandulosa mirabilis, Colorado four o'clock, desert four o'clock, high desert

four o'clock
  ♦ Weed
  ♦ 161
  ♦ pH, cultivated, herbal.

Mirabilis nyctaginea (Michx.) MacMill.
  Nyctaginaceae
  Oxybaphus nyctagineus (Michx.) Sweet
  ♦ heartleaf four o'clock, wild four o'clock, heart leaved umbrella wort, four o'clock
  ♦ Weed, Noxious Weed, Native Weed, Introduced
  ♦ 1, 8, 34, 80, 87, 88, 133, 146, 161, 174, 195, 210, 218, 229
  ♦ pH, cultivated, herbal, toxic. Origin: North America.

Miscanthus floridulus (Labill.) Schum. & Laut.
  Poaceae
  Miscanthus japonicus Anderss. (see), *Saccharum floridulum* Labill. (see)
  ♦ Chinese silvergrass, Japanese silvergrass, eulalia, Chinese fairygrass, Pacific island silvergrass, swordgrass, sawgrass, reedgrass, nete, tupon nette, nette, mah, sapala, sapeleng, sapalang, aset, banga ruchel, medecherecher bokso, pagaluel, ngasau
  ♦ Weed, Noxious Weed, Naturalised
  ♦ 3, 6, 87, 88, 90, 101, 107, 191, 229, 238, 286
  ♦ pG, promoted, herbal.

Miscanthus × giganteus J.M.Greef & Dueter ex Hodk. & Renvoize *nom. inval.*
  Poaceae
  = *Miscanthus sinensis* Anders. × *Miscanthus sacchariflorus* (Maxim.) Hack. [most probable combination]
  ♦ Introduced
  ♦ 5
  ♦ G.

Miscanthus japonicus Anderss.
  Poaceae
  = *Miscanthus floridulus* (Labill.) Schum. & Laut.
  ♦ Japanese silvergrass
  ♦ Weed, Quarantine Weed
  ♦ 3, 76, 87, 88, 191, 203, 218, 220
  ♦ G, herbal.

Miscanthus nepalensis (Trin.) Hack.
  Poaceae
  ♦ Himalayan fairygrass
  ♦ Weed, Sleeper Weed, Quarantine Weed, Naturalised, Environmental Weed
  ♦ 15, 225, 246, 280
  ♦ G, cultivated.

Miscanthus sacchariflorus (Maxim.) Hack.
  Poaceae
  Imperata sacchariflora Maxim., *Miscanthus saccharifer* Benth.
  ♦ silver banner grass, pampas grass, Amur silvergrass, silver plume grass
  ♦ Weed, Naturalised
  ♦ 80, 101, 263, 275, 286, 297
  ♦ pG, cultivated. Origin: east Asia, China.

Miscanthus sinensis Anders.
Poaceae
Eulalia japonica Trin., *Miscanthus sinensis* fo. *glaber* Honda, *Miscanthus sinensis* var. *gracillimus* Hitchc., *Miscanthus sinensis* var. *variegatus* Beal, *Miscanthus sinensis* var. *zebrinus* Beal (see), *Saccharum japonicum* Thunb.
♦ zebra grass, silvergrass, eulalia grass, Chinese silvergrass, eulalia, Chinese fairygrass, susuki
♦ Weed, Naturalised, Garden Escape, Environmental Weed
♦ 4, 7, 15, 80, 87, 88, 98, 101, 102, 133, 142, 151, 195, 203, 204, 224, 246, 263, 280, 286, 300
♦ pG, cultivated, herbal. Origin: China, Japan, Korea, Russia, Taiwan, Indonesia, Philippines.

Miscanthus sinensis Anders. ssp. sinensis
Poaceae
♦ eulalia, Japanese silvergrass
♦ Naturalised
♦ 86
♦ G. Origin: Asia.

Miscanthus sinensis Anders. var. zebrinus Beal
Poaceae
= *Miscanthus sinensis* Anders. cv. 'Zebrinus'
♦ Naturalised
♦ 86
♦ G, cultivated. Origin: east Asia.

Misopates orontium (L.) Raf.
Scrophulariaceae
Antirrhinum orontium L. (see)
♦ weasel's snout, linearleaf snapdragon, pikkuleijonankita, lesser snapdragon
♦ Weed, Naturalised, Environmental Weed, Casual Alien
♦ 7, 42, 70, 86, 94, 98, 101, 134, 161, 198, 203, 243, 250, 253, 272, 300
♦ aH, cultivated, herbal. Origin: Eurasia.

Mitracarpus frigidus K.Schum.
Rubiaceae
♦ Weed
♦ 87, 88

Mitracarpus hirtus (L.) DC.
Rubiaceae
Diodia villosa Moç. & Sessé ex DC., *Mitracarpus breviflorus* A.Gray, *Mitracarpus simplex* Rusby, *Mitracarpus villosus* (Sw.) Cham. & Schltdl. (see), *Spermacoce hirta* L., *Spermacoce villosa* Sw.
♦ herbe a macornet, tropical girdlepod
♦ Weed, Naturalised
♦ 55, 86, 88, 93, 98, 179, 191, 203, 255
♦ aH. Origin: Americas.

Mitracarpus scaber Zucc.
Rubiaceae
Mitracarpum verticillatum Vatke
♦ Weed, Quarantine Weed
♦ 76, 87, 88, 203, 220
♦ herbal.

Mitracarpus villosus (Sw.) Cham. & Schlecht. ex DC.
Rubiaceae
= *Mitracarpus hirtus* (L.) DC.
♦ Weed
♦ 12, 88, 157, 170, 191, 238
♦ aH.

Miyamayomena koraiensis Kitam.
Asteraceae
Aster koraiensis Nakai
♦ Naturalised
♦ 287

Modiola caroliniana (L.) G.Don
Malvaceae
Mondolia reptans St.-Hil.
♦ red flowered mallow, creeping mallow, Carolina bristlemallow, Carolina mallow, bristly mallow, mauve
♦ Weed, Naturalised, Introduced, Environmental Weed
♦ 7, 15, 34, 39, 55, 72, 86, 87, 88, 98, 121, 134, 158, 161, 165, 176, 198, 203, 218, 237, 241, 249, 269, 280, 287, 295, 300
♦ pH, cultivated, herbal, toxic. Origin: tropical America.

Modiolastrum malvifolium (Griseb.) K.Schmann
Malvaceae
Modiola malvifolia Griseb.
♦ Weed
♦ 87, 88

Moehringia muscosa L.
Caryophyllaceae
Alsine moehringia Crantz, *Arenaria muscosa* (L.) Medik., *Stellaria muscosa* (L.) Jess.
♦ mossy sandwort, meringia
♦ Weed
♦ 272
♦ cultivated, herbal.

Moehringia trinervia (L.) Clairv.
Caryophyllaceae
Alsine trinervia (L.) Crantz, *Arenaria trinervia* L.
♦ three veined sandwort, lehtoarho
♦ Weed, Naturalised, Casual Alien
♦ 101, 272, 280
♦ cultivated, herbal.

Moenchia erecta (L.) Gaertn., B.Mey. & Scherb.
Caryophyllaceae
Sagina erecta L.
♦ upright chickweed, erect chickweed, upright moenchia
♦ Weed, Naturalised, Environmental Weed
♦ 7, 72, 86, 88, 98, 101, 176, 198, 203, 280
♦ aH, cultivated, herbal. Origin: Europe.

Moenchia erecta (L.) Gaertn., Mey. & Scherb. ssp. erecta
Caryophyllaceae
♦ Weed
♦ 121
♦ aH. Origin: Eurasia.

Moenchia mantica (L.) Bartl.
Caryophyllaceae
Cerastium manticum L., *Stellaria mantica* (L.) DC.
♦ peverina di mantico
♦ Weed
♦ 272
♦ herbal.

Moerenhoutia leucantha Schltr. var. glabrata Schltr.
Orchidaceae
♦ Introduced
♦ 230
♦ H.

Moerenhoutia leucantha Schltr. var. leucantha
Orchidaceae
♦ Introduced
♦ 230
♦ H.

Moerenhoutia leucantha Schltr. var. minor Tuyama
Orchidaceae
♦ Introduced
♦ 230
♦ H.

Moghania lineata (L.) Ktze.
Fabaceae/Papilionaceae
Flemingia lineata (L.) Ktze.
♦ Weed
♦ 87, 88

Moghania strobilifera (L.) St.-Hil. ex Kuntze
Fabaceae/Papilionaceae
= *Flemingia strobilifera* (L.) Aiton & W.T.Aiton
♦ wild hops
♦ Weed, Environmental Weed
♦ 87, 88, 226, 227
♦ herbal.

Molineria capitulata (Lour.) Herb.
Liliaceae/Hypoxidaceae
Curculigo capitulata (Lour.) Kuntze (see)
♦ weevil grass
♦ Weed
♦ 12

Molineriella minuta (L.) Rouy
Poaceae
Aira minuta L., *Periballia minuta* (L.) Asch. & Graebn. (see)
♦ small hairgrass
♦ Weed, Naturalised, Environmental Weed
♦ 7, 72, 86, 88, 98, 176, 198, 203
♦ aG. Origin: Mediterranean.

Molinia caerulea (L.) Moench
Poaceae
Aira caerulea L.
♦ purple moorgrass, siniheinä
♦ Weed, Quarantine Weed, Naturalised
♦ 39, 70, 101, 220, 272, 287
♦ G, cultivated, herbal, toxic.

Mollugo berteriana Ser.
Molluginaceae
= *Mollugo verticillata* L.
♦ Weed
♦ 87, 88

Mollugo cerviana (L.) Ser.
Molluginaceae
♦ carpetweed, threadstem carpetweed
♦ Weed, Naturalised, Introduced
♦ 34, 87, 88, 101
♦ aH, herbal.

Mollugo cerviana (L.) Ser. var. *cerviana*
Molluginaceae
♦ Native Weed
♦ 121
♦ aH.

Mollugo gracillima Anderss.
Molluginaceae
♦ slender carpetweed
♦ Naturalised
♦ 101

Mollugo nudicaulis Lam.
Molluginaceae
♦ nakedstem carpetweed
♦ Weed, Quarantine Weed
♦ 76, 87, 88, 121, 203, 220, 240
♦ aH, herbal.

Mollugo pentaphylla L.
Molluginaceae
Mollugo stricta L. (see)
♦ mollugo
♦ Weed
♦ 12, 66, 87, 88, 170, 191, 235, 263, 275, 276, 286, 297
♦ aH, cultivated, herbal. Origin: Australia.

Mollugo stricta L.
Molluginaceae
= *Mollugo pentaphylla* L.
♦ Weed
♦ 88, 204
♦ herbal.

Mollugo verticillata L.
Molluginaceae
Mollugo berteriana Ser. (see)
♦ carpetweed, verdolaga alfombra, Indian chickweed, whorled chickweed, devil's grip, green carpetweed
♦ Weed, Quarantine Weed, Noxious Weed, Naturalised, Native Weed
♦ 34, 76, 86, 88, 157, 161, 174, 179, 180, 198, 207, 210, 211, 218, 236, 243, 245, 249, 255, 261, 270, 286, 287, 295, 299, 300
♦ a/pH, arid, cultivated, herbal. Origin: tropical America.

Molopospermum peloponnesiacum (L.) Koch
Apiaceae
Cicutaria peloponnesiaca Kuntze
♦ vuorikirveli
♦ Cultivation Escape
♦ 42
♦ cultivated.

Moltkia coerulea (Willd.) Lehm.
Boraginaceae
♦ Weed
♦ 87, 88

Moltkiopsis ciliata (Forssk.) I.M.Johnst.
Boraginaceae
Lithospermum callosum Vahl, *Moltkia callosa* (Vahl) Wettst., *Moltkia ciliata* Forssk.

♦ Weed
♦ 221
♦ arid.

Moluccella laevis L.
Lamiaceae
♦ shell flower, bells of Ireland, molucca balm
♦ Weed, Naturalised, Cultivation Escape
♦ 7, 34, 86, 87, 88, 98, 101, 198, 203, 252
♦ aH, cultivated, herbal. Origin: western Mediterranean, Middle East.

Momordica balsamina L.
Cucurbitaceae
♦ balsam apple, African cucumber, barh karelo, bave, ingaca, mbaua, mbave, mucaca, mucacana, ncaca, ncacana, nganga, southern balsam pear
♦ Weed, Naturalised, Native Weed, Introduced
♦ 38, 39, 87, 88, 93, 101, 121, 161, 179, 191, 218, 228
♦ pC, arid, cultivated, herbal, toxic. Origin: Africa, Asia, Australasia.

Momordica charantia L.
Cucurbitaceae
Momordica muricata Willd.
♦ balsam pear, balsam apple, bitter melon, melega saga, cerasee, bitter gourd, peria, squirting cucumber, atmagoso, markoso, kerala, jaiva, carilla gourd, concombre africain, margose, momordique, balsambirne, bittergurke, karela, bálsamo, balsamito, cundeamor
♦ Weed, Naturalised, Introduced
♦ 3, 6, 13, 14, 39, 84, 86, 87, 88, 98, 101, 107, 154, 161, 179, 199, 203, 206, 217, 230, 243, 247, 255, 257, 261, 271, 276, 281, 295
♦ aC, cultivated, herbal, toxic. Origin: Africa, Asia, Australasia.

Momordica cochinchinensis (Lour.) Spreng.
Cucurbitaceae
Muricia cochinchinensis Lour.
♦ spiny bitter cucumber, balsampear, bhat karela, Chinese bitter cucumber, Chinese cucumber, spiny bitter cucumber, cundeamor
♦ Weed
♦ 39, 262
♦ aH, cultivated, herbal, toxic. Origin: Asia, Australia.

Momordica tuberosa (Roxb.) Cogn.
Cucurbitaceae
= *Momordica cymbalaria* Hook.f. (NoR)
♦ Weed, Quarantine Weed
♦ 76, 87, 88, 203, 220, 242
♦ arid, herbal.

Monadenia bracteata (Sw.) T.Durand & Schinz
Orchidaceae
= *Disa bracteata* Sw.
♦ South African orchid weed, South African orchid
♦ Weed, Naturalised, Environmental Weed, Cultivation Escape

♦ 7, 9, 74, 86, 98, 155, 203, 252, 296
♦ cultivated.

Monadenium ritchiei Bally
Euphorbiaceae
♦ Quarantine Weed
♦ 220

Monadenium schubei (Pax) N.E.Br.
Euphorbiaceae
♦ Weed, Quarantine Weed
♦ 76, 88, 220
♦ cultivated.

Monadenium yattanum Bally
Euphorbiaceae
♦ Weed, Quarantine Weed
♦ 76, 88, 220

Monarda citriodora Cerv. ex Lag.
Lamiaceae
♦ lemon beebalm, lemon bergamot, common lemon mint
♦ Weed
♦ 87, 88, 218
♦ a/pH, cultivated, herbal.

Monarda didyma L.
Lamiaceae
♦ väriminttu, bergamot, bee balm, oswego tea
♦ Cultivation Escape
♦ 42
♦ pH, cultivated, herbal.

Monarda fistulosa L.
Lamiaceae
= *Monarda bradburiana* Beck (NoR)
♦ wild bergamot
♦ Weed, Naturalised, Native Weed
♦ 23, 87, 88, 161, 174, 218, 287
♦ pH, cultivated, herbal. Origin: North America.

Monarda pectinata Nutt.
Lamiaceae
♦ spotted beebalm, plains beebalm, pony beebalm, horsemint, plains lemon monarda
♦ Native Weed
♦ 161, 174
♦ aH, promoted, herbal. Origin: North America.

Monarda punctata L.
Lamiaceae
♦ spotted beebalm, horsemint, American horsemint, dotted horsemint
♦ Weed, Naturalised
♦ 87, 88, 218, 287
♦ a/pH, cultivated, herbal. Origin: North America.

Monarda punctata L. var. *occidentalis* Palmer & Steyerm.
Lamiaceae
♦ spotted beebalm, American horsemint, horsemint
♦ Naturalised
♦ 287

Monardella lanceolata Gray
Lamiaceae
♦ mustangmint, mustang mountainbalm, mustang monardella, pennyroyal
♦ Weed

♦ 161
♦ aH, cultivated, herbal.

Monardella odoratissima Benth.
Lamiaceae
♦ coyote mint, mountain monardella, Pacific monardella, mountain pennyroyal
♦ Quarantine Weed
♦ 220
♦ pH, cultivated, herbal.

Mondia whitei (Hook.f.) Skeels
Asclepiadaceae/Apocynaceae/Periplocaceae
♦ lufute lwa matwi
♦ Quarantine Weed
♦ 220

Monechma ciliatum (Jacq.) Milne-Redh.
Acanthaceae
Justicia ciliata Jacq.
♦ Weed
♦ 88

Monechma debile (Forssk.) Nees
Acanthaceae
♦ Native Weed
♦ 121
♦ H. Origin: southern Africa.

Monerma cylindrica (Willd.) Coss. & Dur.
Poaceae
= *Hainardia cylindrica* (Willd.) Greuter
♦ common barbgrass
♦ Weed, Naturalised
♦ 87, 88, 121, 287, 300
♦ aG. Origin: Eurasia.

Monerma filiformis Trin.
Poaceae
♦ Weed
♦ 87, 88
♦ G.

Monnina amplibracteata Ferr.
Polygalaceae
♦ Weed
♦ 87, 88

Monochaetum floribundum (Schlecht.) Naudin
Melastomataceae
♦ Weed
♦ 157
♦ S.

Monochoria elata Ridl. var. eleta
Pontederiaceae
♦ Weed
♦ 209

Monochoria hastata (L.) Solms
Pontederiaceae
Monochoria hastaefolia Presl, *Pontederia hastata* L.
♦ arrowleaf false pickerelweed, monochoria, phak top thai, pickerelweed
♦ Weed, Noxious Weed
♦ 6, 67, 87, 88, 140, 161, 170, 186, 191, 193, 209, 229, 239, 262
♦ pH, aqua, cultivated, herbal. Origin: tropical Asia and Africa.

Monochoria korsakowii Regel & Maack
Pontederiaceae
♦ mizuaoi

♦ Weed
♦ 68, 87, 88, 263, 272, 275, 286, 297
♦ aqua.

Monochoria vaginalis (Burm.f.) Presl ex Kunth
Pontederiaceae
Monochoria vaginalis (Burm.f.) Presl ex Kunth var. *pauciflora* (Bl.) Merr. (see), *Pontederia vaginalis* Burm.f.
♦ pickerelweed, heartshape false pickerelweed, monochoria, ninlabon, ovalleaf monochoria, ovalleaf pondweed
♦ Weed, Noxious Weed, Naturalised
♦ 34, 67, 87, 88, 101, 126, 140, 161, 170, 180, 186, 193, 204, 209, 229, 239, 262, 263, 274, 275, 297
♦ wpH, herbal.

Monochoria vaginalis (Burm.f.) Presl ex Kunth var. pauciflora (Bl.) Merr.
Pontederiaceae
= *Monochoria vaginalis* (Burm.f.) Presl ex Kunth
♦ Weed
♦ 275
♦ aqua.

Monochoria vaginalis (Burm.f.) Presl var. plantaginea Solms
Pontederiaceae
♦ monochoria, pickerelweed, phak khiat
♦ Weed
♦ 239, 286
♦ aH, aqua.

Monocosmia corrigioloides Fenzl
Portulacaceae
= *Talinum monandrum* Ruiz & Pav. (NoR)
♦ Weed
♦ 87, 88

Monocymbium ceresiiforme (Nees) Stapf
Poaceae
♦ oatgrass, wild oatgrass, wild oat
♦ Native Weed
♦ 121
♦ pG. Origin: southern Africa.

Monolepis nuttalliana (J.A.Schult.) Greene
Chenopodiaceae
Blitum chenopodioides Nutt., *Blitum nuttallianum* Schult., *Chenopodium trifidum* Trev., *Monolepis chenopodioides* Moq.
♦ povertyweed, monolepis, patata, Nuttall's povertyweed, Nuttall's monolepis, papago spinach, patote, patata, suolasavikka
♦ Weed, Noxious Weed, Naturalised, Native Weed, Introduced, Casual Alien
♦ 23, 42, 87, 88, 161, 174, 174, 180, 218, 228, 287, 299
♦ aH, arid, herbal, toxic. Origin: North America.

Monolepis spathulata A.Gray
Chenopodiaceae
♦ beaver povertyweed, beaver monolepis
♦ Weed, Naturalised
♦ 54, 86, 88, 98, 203

♦ aH, herbal. Origin: North America.

Monopsis debilis (L.) C.Presl
Campanulaceae/Lobeliaceae
Monopsis monantha Wimm., *Monopsis simplex* (L.) Wimm. (see)
♦ Naturalised, Environmental Weed
♦ 9, 86
♦ cultivated.

Monopsis lutea (L.) Urb. var. lutea
Campanulaceae/Lobeliaceae
♦ yellow lobelia
♦ Native Weed
♦ 121
♦ pH. Origin: southern Africa.

Monopsis simplex (L.) Wimm.
Campanulaceae/Lobeliaceae
= *Monopsis debilis* (L.) C.Presl
♦ monopsis
♦ Weed, Naturalised
♦ 7, 98, 198, 203

Monotagma plurispicatum (Körn.) K.Schum.
Marantaceae
♦ Weed
♦ 87, 88

Monsonia angustifolia E.Mey. ex A.Rich.
Geraniaceae
♦ crane's bill
♦ Native Weed
♦ 121
♦ aH, cultivated. Origin: southern Africa.

Monsonia burkeana Planch. ex Harv.
Geraniaceae
Monsonia biflora DC.
♦ crane's bill, dysentery herb
♦ Native Weed
♦ 121
♦ S. Origin: southern Africa.

Monsonia glauca Knuth
Geraniaceae
Monsonia ovata Cav.
♦ dysentery herb
♦ Native Weed
♦ 121
♦ pH. Origin: southern Africa.

Monsonia heliotropoides Boiss.
Geraniaceae
♦ Weed
♦ 221

Monsonia nivea (Decne.) Decne. ex Webb
Geraniaceae
Erodium niveum Decne.
♦ Weed
♦ 221

Monsonia senegalensis Guill. & Perr.
Geraniaceae
♦ Weed
♦ 221

Monstera deliciosa Liebm.
Araceae
♦ splitleaf philodendron, cheese plant, taro vine, ceriman, casiman
♦ Weed, Cultivation Escape
♦ 39, 179, 194, 247, 261
♦ cultivated, herbal, toxic. Origin: Mexico, Central America.

***Monstera obliqua* Miq.**
Araceae
Monstera expilata Schott, *Monstera falcifolia* Engl., *Monstera falcifolia* var. *latifolia* K.Krause, *Monstera fendleri* Engl., *Monstera killipii* K.Krause, *Monstera microstachys* Schott, *Monstera obliqua* var. *expilata* (Schott) Engl., *Monstera snethlagei* K.Krause
♦ Weed
♦ 88
♦ pH.

***Montanoa bipinnatifida* (Kunth) K.Koch**
Asteraceae
Uhdea bipinnatifida Kunth
♦ tree daisy
♦ Weed, Naturalised
♦ 86, 98, 121, 203
♦ pS, cultivated. Origin: Mexico.

***Montanoa hibiscifolia* Benth.**
Asteraceae
♦ tree daisy, montanoa
♦ Weed, Quarantine Weed, Noxious Weed, Naturalised, Garden Escape, Cultivation Escape
♦ 3, 76, 86, 87, 88, 95, 98, 101, 121, 191, 203, 229, 233, 252, 283
♦ S/T, cultivated. Origin: Central America.

***Montia fontana* L.**
Portulacaceae
♦ blinks, hetekaali, water blinks, annual water miner's lettuce, water chickweed
♦ Weed
♦ 70, 88, 94, 272
♦ aH, aqua, cultivated, herbal. Origin: Australia.

***Montia fontana* L. ssp. *chondrosperma* (Fenzl) Walter**
Portulacaceae
Montia minor auct., *Montia minor* C.C.Gmel., *Montia verna* auct., *Montia verna* Neck.
♦ dwarf montia, annual water miner's lettuce
♦ Weed, Naturalised
♦ 165, 253, 280
♦ aH. Origin: central and southern Europe.

***Montia perfoliata* (Donn) Howell**
Portulacaceae
= *Claytonia perfoliata* Donn. ex Willd.
♦ miner's lettuce, winter purslane, spring beauty
♦ Weed, Quarantine Weed, Naturalised
♦ 24, 70, 76, 87, 88, 94, 212, 218, 287
♦ aH, cultivated, herbal, toxic.

***Montrichardia arborescens* (L.) Schott**
Araceae
♦ yautia madera
♦ Weed
♦ 39, 87, 88
♦ herbal, toxic.

***Monttea aphylla* Hauman**
Scrophulariaceae
♦ Introduced

♦ 228
♦ arid.

***Monvillea* Britt. & Rose spp.**
Cactaceae
= *Acanthocereus* (Engelm. ex Berg.) Britt. & Rose spp.
♦ Weed, Quarantine Weed
♦ 76, 88, 203

***Moraea aristata* (D.Delaroche) Asch. & Graebn.**
Iridaceae
♦ blou ooguintje
♦ Weed, Naturalised, Cultivation Escape
♦ 54, 86, 88, 98, 203, 252
♦ cultivated. Origin: South Africa.

***Moraea bellendenii* (Sweet) N.E.Br.**
Iridaceae
♦ moraea
♦ Weed, Naturalised, Garden Escape, Environmental Weed
♦ 39, 72, 86, 88, 98, 203
♦ pH, cultivated, toxic. Origin: South Africa.

***Moraea bipartita* L.Bol.**
Iridaceae
Moraea polyanthos sensu Goldbl.
♦ blue tulip, cape tulip, bloutulp
♦ Weed, Native Weed
♦ 39, 121
♦ pH, cultivated, toxic. Origin: southern Africa.

***Moraea flaccida* (Sweet) Steud.**
Iridaceae
Homeria flaccida Sweet (see)
♦ oneleaf cape tulip
♦ Environmental Weed
♦ 296

***Moraea fugax* (D.Delaroche) Jacq.**
Iridaceae
Moraea edulis (L.f.) Ker Gawl.
♦ peacock lily, uintjie
♦ Weed, Naturalised, Garden Escape, Environmental Weed
♦ 7, 86, 98, 203
♦ pH, cultivated. Origin: South Africa.

***Moraea graminicola* Oberm.**
Iridaceae
♦ butterfly iris, moraea, Natal lily
♦ Weed, Native Weed
♦ 39, 121
♦ pH, toxic. Origin: southern Africa.

***Moraea miniata* Andrews**
Iridaceae
Homeria miniata (Andrews) Sweet (see)
♦ twoleaf cape tulip
♦ Environmental Weed
♦ 296

***Moraea pavonia* (Andrews) Ker Gawl.**
Iridaceae
♦ Weed, Naturalised
♦ 7, 86, 98, 203
♦ Origin: southern Africa.

***Moraea polystachya* (Thunb.) Ker Gawl.**
Iridaceae
♦ blue moraea, blue tulip, cape blue tulip

♦ Weed, Native Weed
♦ 39, 88, 121, 158
♦ pH, cultivated, toxic. Origin: southern Africa.

***Moraea setifolia* (L.f.) Druce**
Iridaceae
♦ thread iris, throw thread iris
♦ Garden Escape, Environmental Weed
♦ 251, 290
♦ cultivated. Origin: South Africa.

***Moraea spathulata* (L.f.) Klatt**
Iridaceae
Moraea spathacea (Thunb.) Ker Gawl.
♦ large yellow moraea, mountain moraea, yellow tulip
♦ Native Weed
♦ 121
♦ pH, cultivated, herbal, toxic. Origin: southern Africa.

***Moraea vegeta* L.**
Iridaceae
♦ moraea
♦ Weed, Naturalised, Garden Escape, Environmental Weed
♦ 7, 72, 86, 88, 98, 203
♦ pH, cultivated. Origin: South Africa.

***Morangaya* Rowley spp.**
Cactaceae
= *Echinocereus* Engelm. spp. (NoR)
♦ Weed, Quarantine Weed
♦ 76, 88, 203

***Morawetzia* Backeb. spp.**
Cactaceae
= *Oreocereus* (Berg.) Riccob. spp. (NoR)
♦ Weed, Quarantine Weed
♦ 76, 88, 203

***Morella faya* (Aiton) Wilbur**
Myricaceae
Myrica faya Ait. (see)
♦ firetree, fayatree, candleberry myrtle
♦ Noxious Weed, Naturalised
♦ 101, 229

***Morettia canescens* Boiss.**
Brassicaceae
♦ Weed
♦ 221

***Morettia philaeana* (Del.) DC.**
Brassicaceae
♦ Weed
♦ 221
♦ arid.

***Moricandia arvensis* (L.) DC.**
Brassicaceae
Brassica arvensis L. *non* Amer. *auct.*
♦ violet cabbage, purple mistress, moricandia
♦ Weed, Naturalised
♦ 88, 94, 101, 272, 287
♦ aH, arid, cultivated.

***Moricandia sinaica* (Boiss.) Boiss.**
Brassicaceae
Brassica sinaica Boiss.
♦ Weed
♦ 221
♦ arid.

464

Morinda citrifolia L.
Rubiaceae
♦ Indian mulberry, nonu
♦ Weed, Naturalised, Garden Escape
♦ 22, 39, 87, 88, 101, 179, 261
♦ T, cultivated, herbal, toxic. Origin:
tropical Asia, Australia.

Morinda citrifolia L. var. citrifolia
Rubiaceae
♦ weipwul
♦ Introduced
♦ 230
♦ T.

Morinda royoc L.
Rubiaceae
♦ redgal
♦ Weed
♦ 87, 88
♦ herbal.

Morinda tomentosa Roth
Rubiaceae
Morinda tinctoria Roxb.
♦ Weed
♦ 13

Moringa oleifera Lam.
Moringaceae
Guilandina moringa L., *Moringa erecta*
Salisb., *Moringa myrepsica* Thell.,
Moringa nux-eben Desf., *Moringa
octogona* Stokes, *Moringa polygona* DC.,
Moringa pterygosperma Gaertn. *nom.
illeg.* (see), *Moringa zeylanica* Pers.
♦ horseradish tree, ben nut, drumstick
tree, sprokiesboom
♦ Weed, Naturalised, Introduced,
Environmental Weed, Casual Alien
♦ 22, 88, 101, 151, 179, 228, 230, 261
♦ T, arid, cultivated, herbal. Origin:
India.

Moringa peregrina (Forssk.) Fiori
Moringaceae
♦ Weed
♦ 221
♦ herbal.

**Moringa pterygosperma Gaertn. nom.
illeg.**
Moringaceae
= *Moringa oleifera* Lam.
♦ horseradish tree
♦ Weed, Naturalised
♦ 86, 98, 203, 279
♦ cultivated, herbal. Origin: India.

Moringa stenopetala (Bak.f.) Cufod.
Moringaceae
♦ Quarantine Weed
♦ 220

Morrenia odorata (Hook. & Arn.) Lindl.
Asclepiadaceae/Apocynaceae
♦ latexplant
♦ Weed, Naturalised
♦ 87, 88, 101, 179, 295
♦ herbal.

Morus alba L.
Moraceae
Morus indica L., *Morus multicaulis* Perr.
♦ white mulberry, mulberry,
witmoerbei, gewone moerbei, common
mulberry

♦ Weed, Noxious Weed, Naturalised,
Introduced, Garden Escape,
Environmental Weed, Cultivation
Escape
♦ 4, 34, 38, 39, 63, 73, 80, 86, 87, 88, 95,
98, 101, 104, 121, 133, 146, 151, 161, 179,
195, 203, 211, 218, 222, 228, 243, 247,
277, 279, 283
♦ T, arid, cultivated, herbal, toxic.
Origin: Eurasia.

Morus australis Poir.
Moraceae
Morus acidosa Griff., *Morus alba* L. var.
stylosa Griff.
♦ Korean mulberry
♦ Weed
♦ 87, 88
♦ T, cultivated.

Morus nigra L.
Moraceae
Morus laciniata Mill., *Morus siciliana*
Mill.
♦ black mulberry, mulberry
♦ Weed, Naturalised, Environmental
Weed
♦ 39, 87, 88, 101, 121, 151, 218, 261
♦ T, cultivated, herbal, toxic. Origin:
Eurasia.

Morus rubra L.
Moraceae
♦ red mulberry, mulberry
♦ Weed
♦ 39, 87, 88, 161, 218, 247
♦ T, cultivated, herbal, toxic.

Moschosma polystachyon (L.) Benth.
Lamiaceae
= *Basilicum polystachyon* (L.) Moench
♦ Weed
♦ 87, 88
♦ herbal.

Mosla dianthera (Ham.) Maxim.
Lamiaceae
Mosla grosseserrata Maxim., *Lycopus
dianthera* Buch.-Ham. ex Roxb.,
Orthodon dianthera (Buch.-Ham. ex
Roxb.) Hand.-Mazz. (see), *Orthodon
grosseserratum* (Maxim.) Kud.
♦ miniature beefsteak, miniature
beefsteak plant
♦ Weed, Naturalised
♦ 68, 80, 87, 88, 101, 102, 191, 204, 286
♦ aH, promoted.

Mosla formosana Maxim.
Lamiaceae
♦ Weed
♦ 235

Mosla japonica (Oliv.) Maxim.
Lamiaceae
♦ yamajiso
♦ Weed
♦ 88, 204
♦ herbal.

Mosla leucantha Hayata
Lamiaceae
♦ Weed
♦ 235

Mosla lysimachiiflora Hayata
Lamiaceae
♦ Weed
♦ 235

Mosla punctulata (J.F.Gmel.) Nakai
Lamiaceae
♦ inukouju
♦ Weed
♦ 87, 88, 286, 297
♦ aH, promoted. Origin: east Asia.

Mosla scabra (Thunb.) C.Y.Wu & H.W.Li
Lamiaceae
♦ Weed
♦ 275
♦ aH.

Mouriri valenzuelana A.Rich.
Melastomataceae/Memecylaceae
♦ Weed
♦ 14

Mucuna albertisii B.L.Rob.
Fabaceae/Papilionaceae
♦ ularat
♦ Weed, Environmental Weed
♦ 3, 191, 259

Mucuna coriacea Bak.
Fabaceae/Papilionaceae
♦ lwase, upupu
♦ Weed, Quarantine Weed
♦ 76, 87, 88, 203, 220

**Mucuna coriacea Bak. ssp. irritans (Burtt
Davy) Verdc.**
Fabaceae/Papilionaceae
♦ buffalo bean, firebean, hell fire bean
♦ Native Weed
♦ 121
♦ pC. Origin: southern Africa.

Mucuna gigantea (Willd.) DC.
Fabaceae/Papilionaceae
♦ sea bean, small seabean, oxeye bean,
gayetan, bayogo dikike, kikiki gaogao,
bayogon dailaili, akankan, bayogon
dikiké, dikiki gaogao, gaggao dálalai,
kakatea, feteka uli, bayogo, keldellel,
ka'e'e, tutae pua'a, tupe, pa'anga 'ae
kuma, valai, wa kore, wa kurikuri, fue
vai
♦ Weed
♦ 3, 39, 87, 88
♦ C, arid, cultivated, herbal, toxic.

Mucuna pruriens (L.) DC.
Fabaceae/Papilionaceae
Dolichos pruriens L., *Strizolobium
pruriens* (L.) Medik., *Strizolobium
deeringianum* Bort., *Strizolobium
aterimum* Piper & Tracey, *Strizolobium
niveum* Kuntze
♦ cow itch, horseeye bean, velvet
bean, cowhage, maamui, lwase upupu,
Florida velvet bean, Bengal bean
♦ Weed, Quarantine Weed, Garden
Escape, Environmental Weed
♦ 3, 11, 14, 39, 76, 80, 87, 88, 135, 155,
157, 161, 179, 191, 203, 206, 209, 220,
239, 243
♦ aH, cultivated, herbal, toxic. Origin:
India.

Mucuna pruriens (L.) DC. var. *pruriens*
Fabaceae/Papilionaceae
Stizolobium pruritum (Wight) Piper
(see)
♦ buffalo bean, cow itch, cowhage,
nettle, pica pica, velvet bean, vine
gungo pea
♦ Introduced
♦ 228
♦ arid. Origin: Madagascar.

Mucuna pruriens (L.) DC. var. *utilis*
(Wight) Burck
Fabaceae/Papilionaceae
Mucuna deeringiana (Bort) Merr.
♦ Bengal bean, Mauritius bean,
Portuguese coffee, cafe Brazilii, go
bouwo, kaf, velvet bean, cowhage, cow
itch
♦ Weed, Naturalised, Introduced
♦ 54, 86, 228
♦ arid, cultivated. Origin: Asia.

Mucuna rostrata Benth.
Fabaceae/Papilionaceae
♦ mucuna
♦ Naturalised
♦ 257

Mucuna sloanei Fawc. & Rendle
Fabaceae/Papilionaceae
Mucuna urens auct.
♦ horseeye bean
♦ Weed
♦ 87, 88
♦ arid, herbal.

Muehlenbeckia australis (Forst.) Meissn.
Polygonaceae
♦ pohuehue, large leaved
muehlenbeckia
♦ Weed
♦ 165, 181
♦ pC, cultivated.

Muehlenbeckia axillaris (Hook.f.) Walp.
Polygonaceae
= *Muehlenbeckia complexa* (A.Cunn.)
Meisn.
♦ wireplant
♦ Naturalised
♦ 101
♦ pC, cultivated. Origin: Australia.

Muehlenbeckia complexa (A.Cunn.)
Meissn.
Polygonaceae
Muehlenbeckia axillaris (Hook.f.) Walp.
(see)
♦ maidenhair vine, wireplant, scrub
pohuehue
♦ Weed, Quarantine Weed,
Naturalised
♦ 7, 40, 76, 86, 87, 88, 101, 181, 203, 220
♦ C, cultivated, herbal. Origin: New
Zealand.

Muehlenbeckia cunninghamii (Meissner)
F.Muell.
Polygonaceae
♦ Weed
♦ 87, 88
♦ cultivated.

Muehlenbeckia diclina (F.Muell.) F.Muell.
Polygonaceae

♦ Weed
♦ 87, 88
♦ cultivated. Origin: Australia.

Muehlenbeckia ephedroides Hook.f.
Polygonaceae
♦ Weed, Quarantine Weed,
Naturalised
♦ 76, 86, 88, 203, 220
♦ cultivated.

Muehlenbeckia florulenta Meissn.
Polygonaceae
Muehlenneckia cunninghamii (Meissn.)
Muell.
♦ lignum
♦ Weed, Native Weed
♦ 200, 269
♦ S, aqua. Origin: Australia.

Muehlenbeckia hastatula (Sm.)
I.M.Johnst.
Polygonaceae
♦ wirevine, hastate muehlenbeckia
♦ Naturalised
♦ 101
♦ C, cultivated.

Muehlenbeckia platyclada (F.J.Muell.)
Lindl.
Polygonaceae
= *Homalocladium platycladum*
(F.J.Muell.) Bailey
♦ Weed
♦ 262
♦ cultivated.

Muehlenbeckia sagittifolia Meissn.
Polygonaceae
Coccoloba sagittifolia Ort.
♦ zarzaparrila colorada
♦ Weed
♦ 87, 88, 237, 295

Muhlenbergia asperifolia (Nees & Mey.)
Parodi
Poaceae
♦ alkali muhly, scratch grass
♦ Weed
♦ 87, 88, 159, 218
♦ pG, herbal.

Muhlenbergia frondosa (Poir.) Fern.
Poaceae
♦ wirestem muhly, Mexican dropseed,
satin grass, wood grass, knot root
grass
♦ Weed, Noxious Weed
♦ 87, 88, 161, 210, 211, 218, 229, 243
♦ pG, herbal.

Muhlenbergia implicata (H.B.K.) Kunth
Poaceae
♦ Weed
♦ 199
♦ G.

Muhlenbergia japonica Steud.
Poaceae
♦ nezumigaya
♦ Weed
♦ 87, 88, 286
♦ G, cultivated.

Muhlenbergia macroura (Kunth) Hitchc.
Poaceae
♦ muhly
♦ Weed

♦ 199
♦ G.

Muhlenbergia mexicana (L.) Trin.
Poaceae
♦ Mexican muhly
♦ Weed
♦ 87, 88, 161, 218
♦ pG, cultivated, herbal.

Muhlenbergia microsperma (DC.) Kunth
Poaceae
Agrostis debilis (Kunth) Spreng.,
Agrostis microcarpa hort. ex Steud.,
Agrostis microsperma Lag., *Agrostis
setosa* (Kunth) Spreng., *Muhlenbergia
debilis* (Kunth) Kunth, *Muhlenbergia
fasciculata* Trin., *Muhlenbergia
microsperma* Trin., *Muhlenbergia
purpurea* Nutt., *Muhlenbergia
ramosissima* Vasey ex S.Watson,
Muhlenbergia setosa (Kunth) Kunth,
Podosemum debile Kunth, *Podosemum
setosum* Kunth, *Trichochloa debilis*
(Kunth) Roem. & Schult., *Trichochloa
microsperma* DC., *Trichochloa setosa*
(Kunth) Roem. & Schult.
♦ annual muhly, littleseed muhly
♦ Weed
♦ 199
♦ aG, arid, herbal.

Muhlenbergia racemosa (Michx.) Britton,
Sterns & Pogg.
Poaceae
♦ green muhly, marsh muhly
♦ Native Weed
♦ 161, 174
♦ G, cultivated, herbal. Origin: North
America.

Muhlenbergia ramulosa (Kunth) Swallen
Poaceae
♦ green muhly
♦ Weed
♦ 199
♦ G.

Muhlenbergia rigens (Benth.) A.S.Hitchc.
Poaceae
♦ deergrass
♦ Weed
♦ 161, 180
♦ pG, cultivated, herbal.

Muhlenbergia schreberi G.F.Gmel
Poaceae
♦ nimblewill, wiregrass, dropseed,
nimblewill muhly
♦ Weed, Noxious Weed, Naturalised,
Native Weed
♦ 35, 87, 88, 91, 161, 174, 207, 210, 211,
218, 229, 243, 249, 287
♦ pG, aqua, herbal. Origin: North
America.

Muhlenbergia tenuifolia (Kunth) Trin.
Poaceae
♦ slimflower muhly
♦ Weed
♦ 199
♦ G.

Mukia maderaspatana (L.) M.J.Roem.
Cucurbitaceae
Bryonia cordifolia L., *Bryonia scabrella*

L., *Coccinia cordifolia* (L.) Cogn. (see), *Cucumis maderaspatanus* L., *Melothria maderaspatana* (L.) Cogn. (see), *Mukia scabrella* (L.) Arn.
♦ heen kekiri, kakobakansimba
♦ Native Weed
♦ 121
♦ pC, arid, cultivated, toxic. Origin: Africa, Asia, Australasia.

Mulgedium tataricum (L.) DC.
Asteraceae
Agathyrsus tataricus D.Don, *Crepis charbonnelii* H.Lév., *Lactuca multipes* H.Lév. & Vaniot, *Lactuca tatarica* (L.) C.A.Mey. (see), *Lagedium tataricum* Soják, *Mulgedium runcinatum* Cass., *Sonchus tataricus* L.
♦ Weed
♦ 87, 88

Mulinum spinosum Pers.
Apiaceae
♦ neneo
♦ Weed, Introduced
♦ 228, 295
♦ arid.

Mundulea sericea (Willd.) Chev.
Fabaceae/Papilionaceae
Mundulea suberosa (DC.) Benth.
♦ silver bush, cork bush
♦ Introduced
♦ 39, 228
♦ arid, cultivated, herbal, toxic. Origin: Madagascar.

Muntingia calabura L.
Tiliaceae/Elaeocarpaceae/ Muntingiaceae
♦ Japanese cherry, Singapore cherry, ornamental cherry, calabur, Panama berry, sirsen, strawberry tree
♦ Weed, Quarantine Weed, Naturalised, Environmental Weed, Cultivation Escape
♦ 3, 22, 76, 86, 88, 101, 151, 179, 203, 259, 261
♦ T, cultivated, herbal. Origin: Caribbean, Mexico to Peru.

Muraltia heisteria (L.) DC.
Polygalaceae
♦ furze muraltia
♦ Weed, Naturalised, Environmental Weed
♦ 86, 98, 198, 203, 290, 296
♦ cultivated. Origin: South Africa.

Murdannia blumei (Hassk.) Brenan
Commelinaceae
Aneilema blumei (Hassk.) Bakh.f., *Aneilema hamiltonianum* Wall., *Dichoespermum blumei* Hassk.
♦ Weed
♦ 88, 170, 191
♦ aqua, cultivated.

Murdannia keisak (Hassk.) Hand.-Mazz.
Commelinaceae
Aneilema japonicum auct. non (Thunb.) Kunth, *Aneilema keisak* Hassk. (see)
♦ marsh dewflower, wart removing herb, Asian spiderwort, aneilima
♦ Weed, Quarantine Weed, Naturalised

♦ 17, 76, 80, 88, 101, 102, 126, 197, 263, 286
♦ aH, aqua, herbal.

Murdannia loriformis (Hassk.) R.Rao & Kammathy
Commelinaceae
♦ severalflower dewflower
♦ Weed
♦ 286

Murdannia nudiflora (L.) Brenan
Commelinaceae
Aneilema nudiflorum (L.) R.Br. (see), *Aneilema malabarica* (L.) Merr., *Commelina nudiflora* L. (see)
♦ common spiderwort, spreading dayflower, doveweed, nakedstem dewflower, kinkung noi, trapoerabinha
♦ Weed, Naturalised, Environmental Weed
♦ 12, 86, 87, 88, 93, 101, 126, 161, 179, 186, 204, 239, 243, 249, 255, 261, 276
♦ wpH. Origin: tropical Africa and Asia.

Murdannia simplex (Vahl) Brenan
Commelinaceae
♦ itesa
♦ Weed
♦ 235, 273, 274
♦ Origin: Australia.

Murdannia spirata (L.) Brückn.
Commelinaceae
Commelina spirata L., *Aneilema nanum* (Roxb.) Kunth, *Aneilema spirata* (L.) Wall, *Aneilema spiratum* (L.) Sweet (see)
♦ Asiatic dewflower
♦ Weed, Naturalised
♦ 88, 101, 126, 170, 179, 191
♦ aqua.

Murraya exotica L.
Rutaceae
= *Murraya paniculata* (L.) Jack
♦ murraya, Chinese box, orange jessamine
♦ Weed, Naturalised, Garden Escape, Environmental Weed, Casual Alien
♦ 73, 86, 88, 101, 201, 261
♦ cultivated, herbal. Origin: south-east Asia.

Murraya koenigii (L.) Spreng.
Rutaceae
♦ curryleaf tree, curryleaf
♦ Naturalised
♦ 86
♦ cultivated, herbal. Origin: China, India, south-east Asia.

Murraya paniculata (L.) Jack
Rutaceae
Chalcas exotica (L.) Millsp., *Chalcas paniculata* L., *Murraya exotica* L. (see), *Murraya omphalocarpa* Hayata, *Murraya paniculata* var. *omphalocarpa* Tanaka
♦ orange jasmine, murraya, Chinese box, cosmetic bark tree, Hawaiian mock orange, jasmine orange, orange jessamine, satinwood
♦ Weed, Naturalised, Environmental Weed, Cultivation Escape
♦ 3, 54, 73, 80, 86, 88, 112, 122, 179, 252
♦ S, cultivated, herbal. Origin: east and

south-east Asia to Australia.

Musa Colla × hybrids
Musaceae
= *Musa acuminata* Colla × *Musa balbisiana* Colla
♦ banana, soa'a
♦ Casual Alien
♦ 261

Musa acuminata Colla
Musaceae
Musa malaccensis Ridl.
♦ banana, plantain
♦ Weed, Naturalised
♦ 28, 86, 87, 88, 101, 179, 243
♦ pH, cultivated, herbal. Origin: China, India, Sri Lanka, Myanmar, south-east Asia.

Musa acuminata Colla × balbisiana Colla
Musaceae
Musa × paradisiaca L. (see)
♦ Naturalised
♦ 86

Musa basjoo Siebold & Zucc. ex Iinuma
Musaceae
♦ Japanese banana
♦ Naturalised
♦ 287
♦ pH, cultivated, herbal. Origin: east Asia.

Musa liukiuensis (Matsum.) Makino
Musaceae
♦ Naturalised
♦ 287

Musa × paradisiaca L.
Musaceae
= *Musa acuminata* Colla × *Musa balbisiana* Colla [see *Musa × sapientum* L.]
♦ uht, French plantain
♦ Weed, Naturalised, Introduced
♦ 87, 88, 101, 230
♦ pH, cultivated.

Musa × sapientum L.
Musaceae
= *Musa acuminata* Colla × *Musa balbisiana* Colla [see *Musa × paradisiaca* L.]
♦ uht
♦ Introduced
♦ 230
♦ T.

Musa textilis Nees
Musaceae
♦ Manila hemp, abaca, uhtisel
♦ Introduced
♦ 230
♦ T, cultivated, herbal.

Musa tikap Warb.
Musaceae
♦ tikap
♦ Introduced
♦ 230
♦ T.

Musa troglodytarum L.
Musaceae
♦ fe'i banana, soa'a
♦ Naturalised
♦ 101, 230
♦ cultivated, herbal.

Musa uranoscopos Lour.
Musaceae
- Naturalised
- 287

Musa velutina H.Wendl. & Drude
Musaceae
- hairy banana, pink velvet banana
- Naturalised, Cultivation Escape
- 101, 261
- cultivated. Origin: India.

Muscari Mill. spp.
Liliaceae/Hyacinthaceae
- grape hyacinth, hyacinth
- Weed, Naturalised
- 70, 198, 272

Muscari armeniacum Leichtlin ex Bak.
Liliaceae/Hyacinthaceae
Muscari cyaneo-violaceum Turrill,
Muscari colchicum Grossh.
- Armenian grape hyacinth, grape hyacinth
- Weed, Naturalised, Garden Escape, Environmental Weed
- 15, 40, 86, 98, 101, 198, 203, 251, 280, 290
- cultivated. Origin: Balkans, Turkey.

Muscari atlanticum Boiss. & Reut.
Liliaceae/Hyacinthaceae
= *Muscari neglectum* Guss. ex Ten.
- grape hyacinth
- Weed
- 80, 88, 102
- herbal.

Muscari botryoides (L.) Mill.
Liliaceae/Hyacinthaceae
Botryanthus vulgaris Kunth, *Muscari strangwaysii* Ten.
- hentohelmililjat, grape hyacinth, common grape hyacinth, Italian grape hyacinth
- Weed, Quarantine Weed, Naturalised, Garden Escape, Cultivation Escape
- 42, 76, 80, 87, 88, 101, 102, 126, 161, 218, 272
- pH, cultivated, herbal. Origin: Europe.

Muscari comosum (L.) Mill.
Liliaceae/Hyacinthaceae
Bellevalia comosa Kunth, *Leopoldia comosa* (L.) Parl.
- tassel hyacinth, feather hyacinth
- Weed, Naturalised, Introduced
- 7, 30, 40, 70, 80, 86, 87, 88, 98, 101, 102, 126, 203, 228, 253, 272
- pH, arid, cultivated, herbal. Origin: Europe, North Africa, Asia.

Muscari neglectum Ten.
Liliaceae/Hyacinthaceae
Bothryanthus atlanticus (Boiss. & Reut.) Nyman, *Bothryanthus commutatus* (Guss.) Kunth, *Bothryanthus mordoanus* (Heldr.) Nyman, *Bothryanthus neglectus* (Guss.) Kunth, *Hyacinthus racemosus* L. nom. ambig., *Leopoldia neumayrii* Heldr., *Muscari atlanticum* Boiss. & Reut. (see), *Muscari granetense* Freyn, *Muscari*

leucostomum Woronow, *Muscari mordoanum* Heldr., *Muscari neumayrii* (Heldr.) Boiss., *Muscari racemosum* (L.) Lam. & DC., *Muscari speciosum* Marches., *Muscari skorpilii* Velen., *Muscari vandasii* Velen.
- common grape hyacinth, starch grape hyacinth, modrica nebadaná, grape hyacinth
- Weed, Naturalised
- 30, 40, 88, 101, 126, 253, 272
- pH, cultivated. Origin: North Africa, Middle East, Europe.

Muscari racemosum Mill. nom. confus.
Liliaceae/Hyacinthaceae
= *Muscari muscarimi* Medik. (NoR)
- grape hyacinth, starch grape hyacinth
- Weed, Quarantine Weed
- 30, 76, 87, 88, 161, 203, 218, 220
- cultivated, herbal.

Muscari tenuiflorum Tausch
Liliaceae/Hyacinthaceae
- Weed
- 272
- pH, cultivated.

Mussaenda cambodiana Pierre ex Pit.
Rubiaceae
- Weed
- 12
- herbal.

Mussaenda flava (Verde.) Bakh.f.
Rubiaceae
- Weed
- 262

Mussaenda frondosa L.
Rubiaceae
- Weed
- 13
- S, herbal, toxic.

Mutisia spinosa Ruiz & Pav.
Asteraceae
- Quarantine Weed
- 220

Myagrum L. spp.
Brassicaceae
- myagrum
- Weed, Naturalised
- 198, 243
- H.

Myagrum perfoliatum L.
Brassicaceae
- musk weed, mitre cress, bird's eye cress, myagrum, miagro liscio, ontervio
- Weed, Quarantine Weed, Noxious Weed, Naturalised, Casual Alien
- 42, 70, 76, 86, 87, 88, 94, 98, 101, 118, 147, 198, 203, 243, 253, 269, 272
- aH, cultivated, herbal. Origin: Eurasia.

Mycelis muralis (L.) Dumort.
Asteraceae
Chondrilla muralis Lam., *Lactuca muralis* (L.) Gaertn. (see), *Prenanthes muralis* L.
- wall lettuce, jänönsalaatti
- Weed, Naturalised, Casual Alien

- 15, 40, 87, 88, 101, 165, 181, 272, 280
- pH, cultivated, herbal. Origin: Mediterranean, west Asia.

Myoporum acuminatum R.Br.
Myoporaceae
- water bush
- Weed
- 39, 87, 88, 121
- S/T, cultivated, toxic. Origin: Australia.

Myoporum deserti Cunn. ex Benth.
Myoporaceae
- turkey bush, Ellangowan poison bush
- Weed, Native Weed
- 39, 87, 88, 269
- S, cultivated, toxic. Origin: Australia.

Myoporum insulare R.Br.
Myoporaceae
- Australian ngaio, boobyalla
- Sleeper Weed, Naturalised, Environmental Weed
- 225, 246, 280
- S, cultivated, herbal.

Myoporum laetum G.Forst.
Myoporaceae
- myoporum, ngaio tree
- Weed, Noxious Weed, Naturalised, Garden Escape, Environmental Weed
- 35, 39, 78, 80, 88, 101, 116, 131, 142, 151, 161, 231
- S/T, cultivated, herbal, toxic. Origin: New Zealand.

Myoporum serratum R.Br.
Myoporaceae
Pogonia tetrandra Labill.
- manitoka
- Weed
- 39, 121
- S/T, cultivated, toxic. Origin: Australia.

Myoporum tenuifolium G.Forst.
Myoporaceae
- chinnock, manatoka
- Weed
- 63, 88, 95
- cultivated. Origin: Australia.

Myoporum tenuifolium G.Forst ssp. montanum (R.Br.) Chinnock
Myoporaceae
Myoporum monatnum R.Br.
- manatoka
- Noxious Weed, Cultivation Escape
- 283
- cultivated, toxic. Origin: Australia.

Myosotis L. spp.
Boraginaceae
- forget me not
- Weed, Environmental Weed
- 272, 296
- herbal.

Myosotis alpestris F.W.Schmidt
Boraginaceae
- alpine forget me not
- Weed
- 23, 88
- pH, cultivated, herbal.

Myosotis arvensis (L.) Hill
Boraginaceae
Myosotis heteropoda Trautv., *Myosotis intermedia* Link
♦ field forget me not, nontiscordardime, peltolemmikki, common forget me not
♦ Weed, Naturalised, Introduced, Environmental Weed
♦ 15, 23, 24, 44, 70, 80, 86, 87, 88, 94, 98, 101, 118, 165, 176, 198, 243, 253, 272, 280, 286, 287, 300
♦ waH, cultivated, herbal. Origin: Eurasia.

Myosotis asiatica (Vesterg.) Schischk. & Serg.
Boraginaceae
♦ Asian forget me not
♦ Naturalised
♦ 101
♦ herbal.

Myosotis azorica H.C.Wats. ex Hook.
Boraginaceae
♦ Azores forget me not
♦ Naturalised
♦ 101, 241
♦ herbal.

Myosotis baltica Sam.
Boraginaceae
Myosotis laxa auct. non Lehm., *Myosotis laxa* Lehm. ssp. *baltica* (Sam. ex Lindm.) Hyl. ex Nordh., *Myosotis palustris* (L.) L. var. *strigulosa* (Rchb.) Mert. & Koch
♦ Naturalised
♦ 287

Myosotis caespitosa Schultz
Boraginaceae
Myosotis laxa Lehm. ssp. *caespitosa* (Schultz) Hyl. ex Nordh. (see)
♦ lesser water forget me not
♦ Weed, Naturalised
♦ 87, 88, 98, 203

Myosotis collina Hoffm.
Boraginaceae
= *Myosotis discolor* Pers.
♦ Weed
♦ 87, 88

Myosotis discolor Pers.
Boraginaceae
Myosotis collina Hoffm. (see), *Myosotis collina* Hoffm. ssp. *collina*, *Myosotis filiformis* Schleich., *Myosotis versicolor* (Pers.) Sm., *Myosotis versicolor* (Pers.) Sm. ssp. *versicolor*
♦ yellow forget me not, blue forget me not, grassland forget me not, changing forget me not, forget me not, kirjolemmikki, nezábudka pestrá
♦ Weed, Naturalised, Casual Alien
♦ 15, 42, 80, 86, 88, 94, 98, 101, 176, 198, 203, 253, 272, 280, 286, 287
♦ waH, herbal. Origin: Europe, north-west Africa, west Asia.

Myosotis discolor Pers. ssp. *canariensis* (Pit.) Grau
Boraginaceae
♦ Naturalised
♦ 300

Myosotis discolor Pers. ssp. *discolor* (Pit.) Grau
Boraginaceae
Myosotis discolor Pers. ssp. *versicolor* (Pers.) Hyl.
♦ Naturalised
♦ 241, 300

Myosotis lappula L.
Boraginaceae
= *Lappula squarrosa* (Retz.) Dumort.
♦ Weed
♦ 87, 88

Myosotis latifolia Poir.
Boraginaceae
♦ broadleaf forget me not, forget me not
♦ Weed, Naturalised, Cultivation Escape
♦ 34, 101, 116, 241, 300
♦ pH, cultivated, herbal.

Myosotis laxa Lehm.
Boraginaceae
♦ bay forget me not, forget me not, tufted forget me not, water forget me not, rantalemmikki
♦ Weed, Naturalised, Environmental Weed
♦ 72, 88, 98, 159, 176, 181, 203, 241, 272, 280, 300
♦ pH, herbal.

Myosotis laxa Lehm. ssp. *caespitosa* (Schultz) Hyl. ex Nordh.
Boraginaceae
= *Myosotis caespitosa* Schultz
♦ water forget me not
♦ Weed, Naturalised, Environmental Weed
♦ 15, 86, 176, 198, 280
♦ Origin: Europe.

Myosotis micrantha Pall. ex Lehm.
Boraginaceae
= *Myosotis stricta* Link
♦ forget me not
♦ Weed, Introduced
♦ 34, 80, 87, 88, 159, 243
♦ waH, herbal.

Myosotis palustris Hill nom. illeg.
Boraginaceae
= *Myosotis scorpioides* L.
♦ Weed
♦ 87, 88
♦ aqua, cultivated, herbal.

Myosotis ramosissima Roch.
Boraginaceae
Myosotis hispida Schlech., *Myosotis collina* auct. plur. non Hoffm.
♦ early forget me not, mäkilemmikki
♦ Weed, Naturalised
♦ 44, 70, 88, 94, 253, 272, 300
♦ herbal.

Myosotis scorpioides L.
Boraginaceae
Myosotis palustris Hill *nom. illeg.* (see)
♦ true forget me not, yelloweye forget me not, forget me not, luhtalemmikki, common forget me not
♦ Weed, Noxious Weed, Naturalised, Introduced, Cultivation Escape
♦ 34, 70, 80, 86, 88, 101, 133, 137, 159, 176, 195, 222, 224, 241, 252, 272, 280, 286, 287, 300
♦ wpH, cultivated, herbal. Origin: Eurasia.

Myosotis scorpioides L. fo. *gracilis* Benningh
Boraginaceae
♦ Naturalised
♦ 287

Myosotis sicula Guss.
Boraginaceae
♦ Jersey forget me not
♦ Weed
♦ 87, 88

Myosotis sparsiflora Mikan
Boraginaceae
= *Strophiostoma sparsiflorum* (Mikan) Turcz.
♦ harsulemmikki
♦ Naturalised
♦ 42

Myosotis stricta Link
Boraginaceae
Myosotis arenaria Schrad., *Myosotis micrantha* Pall. ex Lehm. (see)
♦ upright forget me not, strict forget me not, hietalemmikki, grassland forget me not
♦ Weed, Naturalised
♦ 15, 23, 44, 70, 88, 94, 101, 241, 253, 272, 280, 300
♦ herbal.

Myosotis sylvatica Hoffm.
Boraginaceae
♦ woodland forget me not, puistolemmikki, wood forget me not, ezomurasaki
♦ Weed, Naturalised, Garden Escape, Environmental Weed, Cultivation Escape
♦ 7, 15, 42, 72, 86, 87, 88, 98, 101, 155, 176, 198, 203, 272, 280, 289
♦ pH, cultivated, herbal. Origin: Europe, Russia, northern Asia.

Myosotis verna Nutt.
Boraginaceae
Myosotis virginica auct. non (L.) Britton, Sterns & Pogg. (see)
♦ forget me not, spring forget me not
♦ Weed
♦ 34
♦ a/pH, herbal.

Myosotis virginica auct. non (L.) Britton, Sterns & Pogg.
Boraginaceae
= *Myosotis verna* Nutt.
♦ scorpion grass
♦ Weed
♦ 161
♦ herbal.

Myosoton aquaticum (L.) Moench
Caryophyllaceae
Alsine uliginosa Vill., *Cerastium aquaticum* L., *Malachium aquaticum* (L.) Fr. (see), *Stellaria aquatica* (L.) Scop. (see)
♦ water chickweed, vata, giant

chickweed, parrot's feather, great
chickweed
♦ Weed, Naturalised, Casual Alien
♦ 40, 80, 88, 94, 101, 243, 272, 286, 297
♦ a/pH, aqua, cultivated, herbal.

Myosurus minimus L.
Ranunculaceae
♦ mousetail, little mousetail, tiny
mousetail
♦ Weed, Naturalised, Introduced,
Environmental Weed
♦ 23, 70, 86, 88, 94, 161, 176, 228, 243,
253, 272
♦ aH, arid, cultivated, herbal.

Myriactis nepalensis Less.
Asteraceae
♦ Nepalese myriactis
♦ Weed
♦ 297

Myrica californica Cham. & Schlecht.
Myricaceae
= *Morella californica* (Cham. &
Schlecht.) Wilbur (NoR)
♦ Pacific wax myrtle, Californian
bayberry, wax myrtle, Pacific bayberry
♦ Weed
♦ 87, 88, 218
♦ S, cultivated, herbal.

Myrica cerifera L.
Myricaceae
= *Morella cerifera* (L.) Small (NoR)
♦ southern wax myrtle, bayberry, wax
myrtle, candleberry
♦ Weed
♦ 39, 87, 88, 218
♦ S, cultivated, herbal, toxic.

Myrica faya Aiton
Myricaceae
= *Morella faya* (Ait.) Wilbur
♦ firetree, fayatree, firebush
♦ Weed, Sleeper Weed, Quarantine
Weed, Naturalised, Environmental
Weed
♦ 22, 37, 76, 80, 86, 87, 88, 129, 151, 152,
155, 203, 225, 246
♦ S/T, cultivated, herbal. Origin:
Madeira, Canary Islands, Azores.

Myrica pensylvanica Mirb.
Myricaceae
= *Morella pensylvanica* (Mirb.) Kartesz
comb. nov. ined. (NoR)
♦ northern bayberry, bayberry
♦ Weed
♦ 8, 87, 88, 218, 294
♦ S, cultivated, herbal, toxic.

Myricaria germanica (L.) Desv.
Tamaricaceae
Tamariscus germanicus Scop., *Tamarix
germanica* Scop.
♦ false tamarisk, pensaskanerva,
myricaria
♦ Quarantine Weed, Naturalised,
Environmental Weed
♦ 246, 280
♦ S, cultivated, herbal.

Myriophyllum L. spp.
Haloragaceae
♦ parrot's feather, watermilfoil

♦ Weed, Quarantine Weed, Noxious
Weed
♦ 88, 137, 208, 220, 258
♦ wH.

Myriophyllum alterniflorum DC.
Haloragaceae
♦ alternate flowered watermilfoil
♦ Weed, Quarantine Weed
♦ 23, 76, 87, 88, 203, 220, 272
♦ wH, cultivated, herbal.

Myriophyllum aquaticum (Vell.) Verdc.
Haloragaceae
Enydria aquatica Vell., *Myriophyllum
brasiliense* Camb. (see), *Myriophyllum
proserpinacoides* Gillies ex Hook. & Arn.
♦ parrot's feather, watermilfoil, thread
for life, waterduisendblaar, Brazilian
watermilfoil, parrot's feather
♦ Weed, Quarantine Weed, Noxious
Weed, Naturalised, Garden Escape,
Environmental Weed, Cultivation
Escape
♦ 1, 7, 15, 34, 35, 40, 62, 63, 70, 72, 76,
78, 80, 86, 88, 95, 98, 101, 102, 116, 121,
132, 133, 139, 147, 152, 158, 161, 165,
170, 176, 179, 180, 181, 191, 195, 197,
198, 200, 203, 208, 220, 224, 225, 229,
231, 237, 246, 255, 258, 278, 280, 283,
290, 295, 296
♦ wpH, cultivated, herbal. Origin:
South America.

Myriophyllum brasiliense Camb.
Haloragaceae
= *Myriophyllum aquaticum* (Vell.) Verdc.
♦ parrot's feather, Brasilian milfoil
♦ Weed, Quarantine Weed,
Naturalised, Environmental Weed
♦ 23, 80, 87, 88, 151, 191, 204, 218, 286,
287
♦ wpH, cultivated, herbal.

Myriophyllum crispatum Orch.
Haloragaceae
♦ common watermilfoil
♦ Naturalised, Native Weed,
Environmental Weed
♦ 86, 176, 269
♦ wpH, cultivated. Origin: Australia.

Myriophyllum elatinoides Gaud.
Haloragaceae
= *Myriophyllum quitense* Kunth
♦ gambarusa
♦ Weed, Quarantine Weed
♦ 76, 87, 88, 220, 295
♦ wH, cultivated.

Myriophyllum exalbescens Fern.
Haloragaceae
= *Myriophyllum sibiricum* Kom. (NoR)
♦ northern watermilfoil, watermilfoil
♦ Weed, Noxious Weed
♦ 87, 88, 195, 218, 299
♦ wH, herbal.

Myriophyllum heterophyllum Michx.
Haloragaceae
♦ broadleaf watermilfoil, twoleaf
watermilfoil, variable watermilfoil
♦ Weed
♦ 80, 87, 88, 133, 161, 195, 197, 218, 224
♦ wpH, cultivated, herbal.

Myriophyllum indicum Willd.
Haloragaceae
♦ Weed, Quarantine Weed
♦ 76, 87, 88, 203, 220
♦ wH.

Myriophyllum papillosum Orch.
Haloragaceae
Myriophyllum propinquum A.Cunn.
(see)
♦ common watermilfoil
♦ Weed, Native Weed
♦ 200, 269
♦ wpH, cultivated. Origin: Australia.

Myriophyllum pinnatum (Walter) Britton, Sterns & Pogg.
Haloragaceae
= *Myriophyllum scabratum* Michx.
(NoR)
♦ watermilfoil, cutleaf watermilfoil
♦ Weed
♦ 80
♦ wH, cultivated, herbal.

Myriophyllum propinquum A.Cunn.
Haloragaceae
= *Myriophyllum papillosum* Orch.
♦ common watermilfoil
♦ Weed, Naturalised, Native Weed
♦ 86, 87, 88, 181, 220, 269
♦ wpH, cultivated. Origin: Australia.

Myriophyllum quitense Kunth
Haloragaceae
Myriophyllum elatiniodes Gaud. (see)
♦ Andean watermilfoil
♦ Weed
♦ 80, 237
♦ wH, herbal.

Myriophyllum salsugineum Orch.
Haloragaceae
♦ watermilfoil
♦ Weed
♦ 200
♦ wpH, cultivated.

Myriophyllum scandens Kunth
Haloragaceae
♦ myrisphyllum
♦ Weed, Environmental Weed
♦ 72, 88
♦ wpH, cultivated. Origin: South
Africa.

Myriophyllum simulans Orch.
Haloragaceae
♦ common watermilfoil
♦ Native Weed, Naturalised
♦ 269, 280
♦ wpH, cultivated. Origin: Australia.

Myriophyllum spicatum L.
Haloragaceae
Myriophyllum exalbescens Fern.
♦ Eurasian watermilfoil, spiked
watermilfoil, watermilfoil,
myriophylle en epi, tähkä ärviä
♦ Weed, Quarantine Weed, Noxious
Weed, Naturalised, Native Weed,
Introduced, Garden Escape,
Environmental Weed
♦ 1, 4, 23, 35, 45, 63, 67, 76, 78, 80, 86,
87, 88, 95, 101, 102, 103, 104, 105, 112,
116, 121, 129, 133, 135, 139, 146, 151,

152, 159, 161, 162, 191, 195, 197, 200, 203, 217, 218, 220, 222, 224, 229, 231, 246, 258, 272, 275, 283, 286, 293, 297
- wpH, cultivated, herbal. Origin: Eurasia.

Myriophyllum variifolium **Hook.f.**
Haloragaceae
- common watermilfoil
- Native Weed
- 269
- wpH, cultivated. Origin: Australia.

Myriophyllum verrucosum **Lindl.**
Haloragaceae
- red watermilfoil
- Weed, Quarantine Weed, Naturalised, Native Weed
- 86, 87, 88, 200, 220, 269
- wpH, cultivated. Origin: Australia.

Myriophyllum verticillatum **L.**
Haloragaceae
Myriophyllum pectinatum DC.
- whorlleaf watermilfoil, watermilfoil, myriad leaf, kiehkuraärviä
- Weed
- 87, 88, 159, 263, 272, 275, 286, 297
- wpH, cultivated, herbal.

Myristica fragrans **Houtt.**
Myristicaceae
- nutmeg, macis, noce moscata, nuez moscada
- Weed, Naturalised, Introduced
- 39, 87, 88, 154, 261
- cultivated, herbal, toxic. Origin: East Indies.

Myrocarpus frondosus **Allemão**
Fabaceae/Papilionaceae
- Introduced
- 228
- T, arid, cultivated.

Myrosma cannifolia **L.f.**
Marantaceae
- cannaleaf myrosma
- Naturalised, Cultivation Escape
- 101, 261
- cultivated.

Myrospermum frutescens **Jacq.**
Fabaceae/Papilionaceae
Calusia emarginata Bertero ex Klotzsch, *Myrospermum emarginatum* (Bertero ex Klotzsch) Klotzsch, *Myroxylon frutescens* (Jacq.) Willd.
- cereipo, cercipo
- Cultivation Escape
- 261
- arid, cultivated, herbal.

Myrospermum sousanum **A.Delgado & M.C.Johnst.**
Fabaceae/Papilionaceae
- palo nuevo
- Quarantine Weed
- 220
- cultivated.

Myroxylon balsamum **(L.) Harms**
Fabaceae/Papilionaceae
Myroxylon toluiferum Kunth (see), *Toluifera balsamum* L.
- Tolu balsam tree, balsam of Tolu

- Weed, Introduced
- 228, 230, 268
- T, arid, cultivated, herbal.

Myroxylon pereirae **Klotszch**
Fabaceae/Papilionaceae
= *Myroxylon balsamum* (L.) Harms var. *pereirae* (Royle) Harms (NoR)
- Introduced
- 39, 228
- arid, herbal, toxic.

Myroxylon toluiferum **Kunth**
Fabaceae/Papilionaceae
= *Myroxylon balsamum* (L.) Harms
- Environmental Weed
- 39, 152
- T, toxic. Origin: tropical America.

Myrrhis odorata **(L.) Scop.**
Apiaceae
Chaerophyllum odoratum Lam., *Scandix odorata* L.
- saksankirveli, anise, sweet cicely, myrrh
- Naturalised, Cultivation Escape
- 42, 101
- pH, cultivated, herbal.

Myrsine africana **L.**
Myrsinaceae
Myrsine retusa Sol.
- African boxwood, cape beech, cape myrtle, wild myrtle
- Weed, Native Weed
- 39, 121, 161
- pS, cultivated, herbal, toxic. Origin: Africa to east Asia.

Myrsine carolinensis **(Mez.) Fosb. & Sachet**
Myrsinaceae
Rapanea carolinensis Mez.
- Introduced
- 230
- T.

Myrsiphyllum **Willd. spp.**
Liliaceae/Asparagaceae
= *Asparagus* L. spp.
- bridal creeper
- Noxious Weed
- 171

Myrsiphyllum asparagoides **(L.) Willd.**
Liliaceae/Asparagaceae
= *Asparagus asparagoides* (L.) W.Wight
- bridal creeper
- Weed, Noxious Weed, Naturalised, Environmental Weed
- 7, 88, 98, 147, 152, 176, 203
- cultivated. Origin: South Africa.

Myrsiphyllum declinatum **(L.) Oberm.**
Liliaceae/Asparagaceae
= *Asparagus declinatus* L. (NoR)
- Weed, Naturalised
- 7, 98, 203
- cultivated.

Myrsiphyllum scandens **(Thunb.) Oberm.**
Liliaceae/Asparagaceae
= *Asparagus scandens* Thunb.
- Weed, Naturalised
- 98, 176, 203
- cultivated.

Myrtillocactus **Console spp.**
Cactaceae
- Weed, Quarantine Weed
- 76, 88, 220

Myrtillocactus **Console × *Heliaporus* Rowley spp.**
Cactaceae
= *Myrtillocactus* Console spp. × [*Heliocereus* (Berg.) Britt. & Rose spp. × *Aporocactus* Lem. spp.]
- Weed, Quarantine Weed
- 76, 88, 220

Myrtillocereus **Fric & Kreuz. spp.**
Cactaceae
= *Cereus* Mill. spp.
- Quarantine Weed
- 220

Myrtus communis **L.**
Myrtaceae
- myrtle, common myrtle, mirto
- Weed, Naturalised, Introduced
- 39, 70, 86, 228
- S, arid, cultivated, herbal, toxic. Origin: North Africa, Mediterranean.

N

Nabalus trifoliolatus Cass.
Asteraceae
= *Prenanthes trifoliolata* (Cass.) Fern.
♦ Quarantine Weed
♦ 220

Nageia rospigliosii (Pilg.) Laubenf.
Podocarpaceae
♦ Introduced
♦ 38

Najas L. spp.
Najadaceae/Hydrocharitaceae
♦ naiad, naiad pondweed, waternymph
♦ Weed, Quarantine Weed, Cultivation Escape
♦ 88, 161, 204, 220, 233
♦ wH, cultivated.

Najas armata H.Lindb.
Najadaceae/Hydrocharitaceae
♦ Weed
♦ 221
♦ wH.

Najas filifolia Haynes
Najadaceae/Hydrocharitaceae
♦ south-eastern naiad, needleleaf waternymph
♦ Weed
♦ 197
♦ wH.

Najas flexilis (Willd.) Rostk. & Schmidt
Najadaceae/Hydrocharitaceae
Caulinia flexilis Willd.
♦ slender naiad, notkeanäkinruoho, nodding waternymph, flexible naiad
♦ Weed
♦ 23, 87, 88, 218
♦ waH, cultivated, herbal.

Najas graminea Del.
Najadaceae/Hydrocharitaceae
♦ ricefield waternymph, bushy pondweed, saaraai, ricefield naiad
♦ Weed, Naturalised
♦ 87, 88, 101, 197, 217, 239, 272, 274, 275, 286, 297
♦ waH, cultivated, herbal. Origin: Africa, Middle East, Asia.

Najas guadalupensis (Spreng.) Magnus
Najadaceae/Hydrocharitaceae
♦ najas microdon, southern naiad, southern waternymph
♦ Weed, Quarantine Weed
♦ 76, 87, 203, 220
♦ waH, cultivated, herbal.

Najas malesiana de Wilde
Najadaceae/Hydrocharitaceae

Najas falciculata A.Br.
♦ Weed
♦ 88, 170, 191
♦ wH. Origin: Australia.

Najas marina L.
Najadaceae/Hydrocharitaceae
Najas major All., *Najas monosperma* Willd
♦ hollyleaf naiad, spiny naiad, bushy pondweed, marine naiad, waternymph
♦ Weed, Quarantine Weed, Noxious Weed, Environmental Weed
♦ 23, 70, 87, 88, 179, 217, 218, 229, 246, 272, 275, 297
♦ waH, promoted, herbal.

Najas minor All.
Najadaceae/Hydrocharitaceae
Caulinia minor Coss., *Fluvialis minor* Pers., *Ittnera minor* Gmel.
♦ slender leaved naiad, waternymph, European waternymph, brittle naiad, spiny naiad, eutrophic waternymph
♦ Weed, Quarantine Weed, Naturalised
♦ 23, 70, 76, 80, 87, 88, 101, 133, 195, 197, 203, 224, 253, 272, 275, 286, 297
♦ wH, herbal.

Najas oguraensis Miki
Najadaceae/Hydrocharitaceae
♦ Weed
♦ 286
♦ wH.

Najas pectinata (Parl.) Magnus
Najadaceae/Hydrocharitaceae
♦ sawgrass, saw weed
♦ Weed, Quarantine Weed, Native Weed
♦ 76, 87, 88, 121, 203
♦ wpH, cultivated. Origin: southern Africa.

Najas wrightiana A.Braun
Najadaceae/Hydrocharitaceae
♦ Wright's waternymph
♦ Naturalised
♦ 101
♦ wH.

Nama dichotomum (Ruiz & Pav.) Choisy
Hydrophyllaceae
♦ wishbone fiddleleaf
♦ Weed
♦ 157
♦ aH, herbal.

Nama jamaicense L.
Hydrophyllaceae
♦ Jamaican weed
♦ Weed
♦ 14, 87, 88, 237, 295

Nama undulatum Kunth
Hydrophyllaceae
♦ whitewhisker fiddleleaf
♦ Weed
♦ 199

Nandina domestica Thunb.
Berberidaceae/Nandinaceae
♦ nandina, heavenly bamboo, sacred bamboo
♦ Weed, Naturalised, Garden Escape,

Environmental Weed, Cultivation Escape
♦ 39, 54, 80, 86, 88, 101, 112, 142, 151, 161, 247, 252
♦ S, cultivated, herbal, toxic. Origin: India, China, Japan.

Nanocnide japonica Bl.
Urticaceae
♦ Japanese nanocnide
♦ Weed
♦ 297

Narcissus L. spp.
Liliaceae/Amaryllidaceae
♦ narcissus, daffodil, jonquil
♦ Weed, Naturalised, Environmental Weed
♦ 154, 161, 198, 247, 290
♦ herbal, toxic.

Narcissus assoanus Dufour
Liliaceae/Amaryllidaceae
♦ rushleaf jonquil, tiny jonquil
♦ Naturalised
♦ 101
♦ cultivated.

Narcissus bulbocodium L.
Liliaceae/Amaryllidaceae
♦ petticoat daffodil, hoop petticoat daffodil
♦ Naturalised, Casual Alien
♦ 101, 280
♦ cultivated, herbal.

Narcissus bulbocodium L. ssp. obesus (Salisb.) Maire
Liliaceae/Amaryllidaceae
♦ hoop daffodil
♦ Weed
♦ 70

Narcissus × incomparabilis Mill.
Liliaceae/Amaryllidaceae
= *Narcissus poeticus* L. × *Narcissus pseudonarcissus* L.
♦ tähtinarsissi, nonesuch daffodil
♦ Naturalised, Cultivation Escape
♦ 40, 42, 101, 280
♦ cultivated.

Narcissus jonquiila L.
Liliaceae/Amaryllidaceae
♦ jonquil, common jonquil
♦ Naturalised
♦ 7, 39, 86, 101, 154
♦ pH, cultivated, herbal, toxic. Origin: Spain, Portugal.

Narcissus × medioluteus Mill.
Liliaceae/Amaryllidaceae
= *Narcissus tazetta* L. × *Narcissus poeticus* L.
♦ primrose peerless
♦ Naturalised
♦ 40, 86, 101, 280
♦ Origin: France.

Narcissus × odorus L.
Liliaceae/Amaryllidaceae
= *Narcissus jonquilla* L. × *Narcissus pseudonarcissus* L.
♦ hybrid jonquil, Campernelle jonquil
♦ Naturalised, Casual Alien
♦ 40, 101, 280

Narcissus papyraceus Ker Gawl.
Liliaceae/Amaryllidaceae
♦ paperwhite narcissus, paperwhite
♦ Weed, Naturalised
♦ 7, 86, 98, 101, 203
♦ cultivated. Origin: southern Europe.

Narcissus papyraceus Ker Gawl. ssp. panizzianus (Parl.) Arc.
Liliaceae/Amaryllidaceae
♦ papery narcissus
♦ Weed
♦ 70

Narcissus poeticus L.
Liliaceae/Amaryllidaceae
Narcissus angustifolius Curtis
♦ daffodil, narcis biely, poet's narcissus, valkonarsissi, pheasant's eye narcissus
♦ Weed, Naturalised, Cultivation Escape
♦ 15, 39, 40, 42, 101, 161, 280
♦ pH, cultivated, herbal, toxic.

Narcissus poeticus L. ssp. poeticus
Liliaceae/Amaryllidaceae
♦ pheasant's eye daffodil
♦ Cultivation Escape
♦ 40
♦ cultivated.

Narcissus pseudonarcissus L.
Liliaceae/Amaryllidaceae
♦ wild daffodil, daffodil
♦ Weed, Naturalised, Garden Escape, Cultivation Escape
♦ 7, 23, 39, 42, 54, 86, 87, 88, 98, 101, 161, 176, 198, 203, 252, 280
♦ pH, cultivated, herbal, toxic. Origin: western Europe.

Narcissus pseudonarcissus L. ssp. major (Curtis) Bak.
Liliaceae/Amaryllidaceae
♦ Spanish daffodil
♦ Naturalised
♦ 40

Narcissus pseudonarcissus L. ssp. pseudonarcissus
Liliaceae/Amaryllidaceae
♦ daffodil
♦ Naturalised
♦ 40

Narcissus serotinus L.
Liliaceae/Amaryllidaceae
♦ Weed
♦ 70
♦ cultivated, herbal.

Narcissus tazetta L.
Liliaceae/Amaryllidaceae
♦ polyanthus narcissus, jonquil, cream narcissus, daffodil
♦ Weed, Naturalised, Garden Escape, Environmental Weed
♦ 7, 15, 39, 40, 70, 86, 87, 88, 98, 101, 154, 161, 198, 203, 221, 280
♦ pH, cultivated, herbal, toxic. Origin: Mediterranean, Middle East.

Narcissus tazetta L. ssp. tazetta
Liliaceae/Amaryllidaceae
♦ Weed
♦ 253

Narcissus tazetta L. var. chinensis Roem.
Liliaceae/Amaryllidaceae
♦ Naturalised
♦ 287
♦ herbal.

Narcissus tazetta L. var. plenus Nakai
Liliaceae/Amaryllidaceae
♦ Naturalised
♦ 287

Nardus stricta L.
Poaceae
♦ matgrass, psica tuhá
♦ Weed, Noxious Weed, Naturalised
♦ 15, 23, 70, 79, 80, 86, 87, 88, 98, 101, 161, 176, 203, 219, 229, 272, 280
♦ pG, cultivated, herbal. Origin: Europe, west Asia.

Narenga porphyrocoma (Hance ex Trimen) Bor
Poaceae
Eriochrysis porphyrocoma Hance
♦ Weed
♦ 262
♦ pG.

Narthecium ossifragum (L.) Huds.
Liliaceae/Melanthiaceae/Nartheciaceae
♦ bog asphodel
♦ Weed
♦ 39, 87, 88
♦ aqua, cultivated, herbal, toxic.

Nassauvia glomerulosa D.Don
Asteraceae
♦ Introduced
♦ 228
♦ arid.

Nassella (Trin.) Desv. spp.
Poaceae
♦ needlegrass, tussockgrass
♦ Weed, Naturalised
♦ 18, 88, 198
♦ G.

Nassella brachychaeta (Godr.) Barkworth
Poaceae
= *Achnatherum brachychaetum* (Godr.) Barkworth
♦ punagrass
♦ Weed
♦ 161
♦ G.

Nassella cernua (Steb. & Á.Löve) Barkworth
Poaceae
Stipa cernua Steb. & Á.Löve (see)
♦ California needlegrass, nodding stipa, tussockgrass, nodding tussockgrass
♦ Quarantine Weed
♦ 76
♦ pG, cultivated.

Nassella charruana (Arechav.) Barkworth
Poaceae
♦ Uruguayan needlegrass, lobed needlegrass
♦ Weed, Noxious weed, Naturalised, Environmental Weed
♦ 54, 86, 88, 198, 232, 290
♦ G. Origin: South America.

Nassella chilensis (Trin.) Desv.
Poaceae
♦ Chilean tussockgrass
♦ Naturalised
♦ 101
♦ G, arid.

Nassella formicarum (Delile) Barkworth
Poaceae
♦ tropical tussockgrass
♦ Naturalised
♦ 101
♦ pG.

Nassella hyalina (Nees) Barkworth
Poaceae
♦ cane needlegrass, speargrass
♦ Weed, Naturalised, Environmental Weed
♦ 72, 86, 88, 198, 290
♦ pG. Origin: South America.

Nassella inconspicua (J.Presl) Barkworth
Poaceae
♦ Introduced
♦ 38
♦ G.

Nassella leucotricha (Trin. & Rupr.) Pohl
Poaceae
Stipa leucotricha Trin. & Rupr. (see)
♦ Texas needlegrass
♦ Naturalised, Environmental Weed
♦ 86, 198, 290
♦ G. Origin: southern North America.

Nassella megapotamia (Spreng. ex Trin.) Barkworth
Poaceae
♦ Naturalised
♦ 86
♦ G. Origin: South America.

Nassella neesiana (Trin. & Rupr.) Barkworth
Poaceae
Stipa neesiana Trin. & Rupr. (see)
♦ Chilean needlegrass, speargrass, Uruguayan tussockgrass
♦ Weed, Quarantine Weed, Noxious weed, Naturalised, Environmental Weed
♦ 39, 72, 76, 86, 88, 101, 158, 169, 198, 203, 232, 246, 280, 289, 296
♦ pG, toxic. Origin: South America.

Nassella neesiana (Trin. & Rupr.) Barkworth var. neesiana
Poaceae
♦ Naturalised
♦ 280
♦ G.

Nassella tenuissima (Trin.) Barkworth
Poaceae
Stipa cirrosa E.Fourn. ex Hemsl. (see), *Stipa geniculata* Phil. (see), *Stipa mendocina* Phil. (see), *Stipa oreophila* Speg. (see), *Stipa subulata* E.Fourn. ex Hemsl. (see), *Stipa tenuissima* Trin. (see), *Stipa tenuissima* var. *oreophila* (Speg.) Speg., *Stipa tenuissima* var. *planicola* Speg.
♦ white tussock, tussockgrass, finestem tussockgrass, witpolgras
♦ Weed, Quarantine Weed, Noxious

Weed, Garden Escape, Environmental Weed, Casual Alien
- ♦ 63, 76, 88, 95, 158, 220, 246, 251, 280, 283, 289
- ♦ pG, cultivated. Origin: South America.

Nassella trichotoma (Nees) Hack.
Poaceae
Stipa trichotoma Nees (see)
- ♦ serrated tussock, nassella tussock, Yass River tussock, nassella polgras
- ♦ Weed, Quarantine Weed, Noxious Weed, Naturalised, Garden Escape, Environmental Weed
- ♦ 20, 26, 51, 63, 67, 72, 76, 86, 87, 88, 95, 98, 140, 147, 152, 158, 161, 169, 171, 176, 177, 198, 203, 220, 229, 232, 246, 251, 269, 280, 283, 289, 296
- ♦ pG, cultivated. Origin: South America.

Nasturtium R.Br. spp.
Brassicaceae
- ♦ water arum
- ♦ Environmental Weed
- ♦ 246

Nasturtium aquaticum Wahlenb.
Brassicaceae
= *Nasturtium officinale* R.Br.
- ♦ Weed, Quarantine Weed
- ♦ 76, 88, 220
- ♦ herbal.

Nasturtium indicum (L.) DC.
Brassicaceae
= *Rorippa indica* (L.) Hiern
- ♦ Weed
- ♦ 88, 204
- ♦ herbal.

Nasturtium madagascariensis DC.
Brassicaceae
- ♦ Weed
- ♦ 87, 88

Nasturtium microphyllum Boenn. ex Rchb.
Brassicaceae
Nasturtium officinale R.Br. var. *olgae* (Regel & Schmalh.) N.Busch, *Rorippa microphylla* (Boenn. ex Rchb.) Hyl. ex Á. & D.Löve (see)
- ♦ watercress, one rowed watercress
- ♦ Weed, Naturalised
- ♦ 23, 88, 208, 280, 287, 300
- ♦ pH, aqua, promoted, herbal. Origin: Eurasia.

Nasturtium montanum Wall.
Brassicaceae
- ♦ Weed
- ♦ 87, 88
- ♦ herbal.

Nasturtium officinale R.Br.
Brassicaceae
Nasturtium nasturtium-aquaticum (L.) H.Karst. *nom. inval*, *Radicula nasturtium* Cav. *nom. illeg.*, *Radicula nasturtium-aquaticum* (L.) Rendle & Britten, *Rorippa nasturtium* Beck *nom. illeg.*, *Rorippa nasturtium-aquaticum* (L.) Hayek (see), *Sisymbrium nasturtium* Thunb. *nom. illeg.*, *Sisymbrium*

nasturtium-aquaticum L.
- ♦ watercress, green watercress, cresson d'eau, cresson de fontaine, brunnenkresse, selada air, mizu garashi, oranda garashi, agrião, berro
- ♦ Weed, Naturalised, Cultivation Escape
- ♦ 23, 42, 70, 80, 87, 88, 102, 133, 159, 161, 195, 197, 208, 218, 224, 272, 275, 280, 286, 287, 297, 300
- ♦ pH, aqua, cultivated, herbal. Origin: Eurasia.

Nasturtium officinale R.Br. var. officinale
Brassicaceae
- ♦ Naturalised
- ♦ 241

Nasturtium palustre (L.) DC.
Brassicaceae
= *Rorippa palustris* (L.) Besser
- ♦ Weed
- ♦ 88, 204
- ♦ herbal.

Navarretia intertexta (Benth.) Hook.
Polemoniaceae
- ♦ woolly gilia, interwoven navarretia, needleleaf navarretia
- ♦ Weed, Casual Alien
- ♦ 42, 87, 88, 161, 218
- ♦ aH, herbal.

Navarretia squarrosa (Esch.) Hook. & Arn.
Polemoniaceae
- ♦ skunkweed gilia, skunk bush, Californian stinkweed
- ♦ Weed, Naturalised, Environmental Weed
- ♦ 7, 86, 87, 88, 98, 165, 176, 198, 203, 218, 246, 269, 280
- ♦ aH, promoted, herbal. Origin: western North America.

Neatostema apulum (L.) Johnst.
Boraginaceae
- ♦ gromwell
- ♦ Weed, Naturalised, Environmental Weed
- ♦ 86, 98, 198, 203
- ♦ Origin: Mediterranean, Canary Islands, Middle East.

Nechamandra alternifolia (Roxb.) Thw.
Hydrocharitaceae
= *Lagarosiphon alternifolia* (Roxb.) Druce (NoR)
- ♦ Weed
- ♦ 87, 88
- ♦ wH, cultivated.

Nelumbo lutea (Willd.) Pers.
Nelumbonaceae
Nelumbo pentapetala (Walter) Willd., *Nelumbo pentaphylla* (Walter) Fern. (see)
- ♦ American lotus, water chinquapin, yellow lotus
- ♦ Weed, Quarantine Weed, Naturalised
- ♦ 14, 76, 86, 88, 133, 161, 203, 218, 220, 224, 258
- ♦ wpH, cultivated, herbal. Origin: North America.

Nelumbo nucifera Gaertn.
Nelumbonaceae
Nelumbium nuciferum Gaertn., *Nelumbium speciosum* Willd.
- ♦ lotus, sacred lotus, Indian lotus
- ♦ Weed, Naturalised
- ♦ 87, 88, 101, 179, 191, 197, 204, 262, 287
- ♦ wpH, cultivated, herbal. Origin: Asia, Australia.

Nelumbo pentaphylla (Walter) Fern.
Nelumbonaceae
= *Nelumbo lutea* (Willd.) Pers.
- ♦ Weed
- ♦ 87, 88
- ♦ wpH.

Nemacystus decipiens (Itomozuku) Masakuni
Spermatochnaceae
- ♦ brown alga, mozuku
- ♦ Weed
- ♦ 197
- ♦ algae.

Nemesia floribunda Lehm.
Scrophulariaceae
- ♦ Naturalised
- ♦ 280
- ♦ aH, cultivated.

Nemesia strumosa Benth.
Scrophulariaceae
- ♦ nemesia, leeubekkie, kohtalonkukka
- ♦ Weed, Naturalised, Native Weed, Garden Escape, Cultivation Escape
- ♦ 42, 86, 121, 252, 280
- ♦ pH, cultivated, herbal. Origin: southern Africa.

Nemophila aphylla (L.) Brummitt
Hydrophyllaceae
- ♦ baby blue eyes, smallflower baby blue eyes
- ♦ Weed
- ♦ 161
- ♦ herbal.

Nemophila menziesii Hook. & Arn.
Hydrophyllaceae
- ♦ sinisievikki, baby blue eyes, Menzies' baby blue eyes
- ♦ Cultivation Escape
- ♦ 42
- ♦ aH, cultivated, herbal.

Neoabbottia Britt. & Rose spp.
Cactaceae
= *Leptocereus* (Berger) Britt. & Rose spp.
- ♦ Weed, Quarantine Weed
- ♦ 76, 203

Neobouteloua lophostachya (Griseb.) Gould
Poaceae
Bouteloua lophostachya Griseb.
- ♦ Introduced
- ♦ 228
- ♦ G, arid.

Neoevansia Marshall spp.
Cactaceae
= *Peneocereus* Britt. & Rose spp. (NoR)
- ♦ Weed, Quarantine Weed
- ♦ 76, 88, 220

Neohickenia Fric spp.
Cactaceae
= *Parodia* Speg. spp. (NoR)
♦ Weed, Quarantine Weed
♦ 76, 88, 220

Neolamarckia cadamba (Roxb.) Bosser
Rubiaceae
♦ kadam
♦ Naturalised, Cultivation Escape
♦ 101, 261
♦ cultivated. Origin: south-east Asia.

Neomarica caerulea (Ker Gawl.) Sprague
Iridaceae
♦ fan iris
♦ Weed, Cultivation Escape
♦ 32
♦ cultivated.

Neomarica northiana (Schneev.) Sprague
Iridaceae
Moraea northiana Schneev.
♦ North's false flag
♦ Naturalised, Garden Escape
♦ 101, 261
♦ cultivated. Origin: Brazil.

Neonotonia wightii (Wight & Arn.) Lackey
Fabaceae/Papilionaceae
Notonia wightii Arn.
♦ perennial soybean, tropical legume, Tinaroo glycine, glycine
♦ Weed, Sleeper Weed, Naturalised, Environmental Weed
♦ 3, 55, 86, 88, 98, 101, 151, 155, 191, 203
♦ cultivated. Origin: tropical Africa and Asia.

Neoporteria Britt. & Rose spp.
Cactaceae
Islaya Backeb. spp., *Pyrrhocactus* (Berg.) Backeb. spp., *Horridocactus* Backeb. spp., *Neochilenia* Doelz. spp., *Thelocephala* Ito spp., *Delaetia* Backeb. spp.
♦ Weed, Quarantine Weed
♦ 203

Neotinea maculata (Desf.) Stearn.
Orchidaceae
♦ dense flowered orchid
♦ Weed
♦ 272

Nepeta cataria L.
Lamiaceae
Cataria vulgaris Moench, *Glechoma cataria* (L.) Kuntze
♦ catnip, catmint, catwort, field balm, aitokissanminttu
♦ Weed, Naturalised, Introduced, Cultivation Escape
♦ 15, 23, 24, 34, 39, 42, 86, 87, 88, 98, 101, 161, 174, 176, 180, 195, 198, 203, 210, 218, 252, 272, 280, 287
♦ pH, cultivated, herbal, toxic. Origin: south and eastern Europe.

Nepeta × faassenii Bergmans ex Stearn
Lamiaceae
= *Nepeta racemosa* Lam. × *Nepeta nepetella* L.
♦ garden cat mint, Faassen's catnip,

mirrinminttu
♦ Naturalised, Cultivation Escape
♦ 40, 42
♦ cultivated, herbal. Origin: sterile horticultural hybrid.

Nepeta grandiflora Bieb.
Lamiaceae
♦ isokissanminttu, Caucasus catmint
♦ Naturalised, Cultivation Escape
♦ 42, 101
♦ cultivated, herbal.

Nepeta hederacea (L.) Trev.
Lamiaceae
= *Glechoma hederacea* L.
♦ Weed
♦ 87, 88
♦ cultivated.

Nepeta micrantha Bunge
Lamiaceae
Nepeta meyeri Benth.
♦ Weed
♦ 243

Nepeta mussinii Spreng. ex Henckel
Lamiaceae
♦ kollinminttu, Persian catmint
♦ Naturalised, Cultivation Escape
♦ 42, 101, 280
♦ cultivated, herbal.

Nepeta nuda L.
Lamiaceae
Nepeta pannonica L.
♦ Weed
♦ 272
♦ cultivated.

Nepeta pilinux P.H.Davis
Lamiaceae
♦ Weed
♦ 87, 88

Nepeta racemosa Lam.
Lamiaceae
♦ raceme catnip, katinminttu, dwarf catnip
♦ Naturalised
♦ 101
♦ cultivated, herbal.

Nepeta septemcrenata Ehrenb. ex Benth.
Lamiaceae
♦ Weed
♦ 221

Nephelium lappaceum L.
Sapindaceae
♦ rambutan
♦ Introduced
♦ 39, 230
♦ T, cultivated, herbal, toxic.

Nephrolepis auriculata (L.) Trimen
Nephrolepidaceae/Oleandraceae
♦ Weed
♦ 87, 88
♦ cultivated.

Nephrolepis × averyi C.E.Nauman
Nephrolepidaceae/Oleandraceae
= *Nephrolepis biserrata* (Sw.) Schott × *Nephrolepis exaltata* (L.) Schott
♦ Avery's swordfern
♦ Weed
♦ 179

Nephrolepis biserrata (Sw.) Schott
Nephrolepidaceae/Oleandraceae
♦ giant swordfern, large swordfern, broad swordfern
♦ Weed, Native Weed, Environmental Weed
♦ 12, 87, 88, 121, 254, 259
♦ pH, cultivated, herbal.

Nephrolepis cordifolia (L.) C.Presl
Nephrolepidaceae/Oleandraceae
♦ swordfern, narrow swordfern, erect swordfern, ladder fern, tuberous swordfern, fishbone fern
♦ Weed, Noxious Weed, Naturalised, Garden Escape, Environmental Weed
♦ 7, 15, 73, 80, 86, 88, 101, 112, 151, 179, 201, 225, 246, 290, 296
♦ cultivated, herbal. Origin: obscure.

Nephrolepis exaltata (L.) Schott
Nephrolepidaceae/Oleandraceae
Nephrolepis bostoniensis hort.
♦ Boston fern, swordfern, Boston swordfern
♦ Weed, Noxious Weed, Garden Escape
♦ 87, 88, 121, 283
♦ pH, cultivated, herbal. Origin: pantropical and subtropical.

Nephrolepis falcata (Cav.) C.Chr.
Nephrolepidaceae/Oleandraceae
♦ fishtail swordfern, fishtail fern
♦ Weed, Naturalised
♦ 101, 254
♦ cultivated.

Nephrolepis falcata (Cav.) C.Chr. fo. furcans (T.Moore) Proctor
Nephrolepidaceae/Oleandraceae
♦ Weed
♦ 179

Nephrolepis hirsutula (Forst.) Presl
Nephrolepidaceae/Oleandraceae
♦ scaly swordfern
♦ Weed
♦ 6, 80, 87, 88, 179, 191, 254
♦ cultivated, herbal.

Nephrolepis laurifolia (Christ) Proctor
Nephrolepidaceae/Oleandraceae
♦ laurelleaf swordfern
♦ Naturalised
♦ 101

Nephrolepis multiflora (Roxb.) Jarrett ex Morton
Nephrolepidaceae/Oleandraceae
♦ Asian swordfern
♦ Weed, Naturalised, Environmental Weed
♦ 28, 80, 80, 88, 101, 112, 179, 206, 227, 243, 259, 261
♦ cultivated, herbal. Origin: India, tropical Asia.

Nephrolepis rivularis (Vahl) Mett. ex Krug
Nephrolepidaceae/Oleandraceae
Nephrolepis undulata (Afzel. & Sw.) J.Sm.
♦ streamside swordfern
♦ Weed, Quarantine Weed
♦ 76, 87, 88, 203, 220

Nepsera aquatica **(Aubl.) Naud.**
Melastomataceae
♦ altea
♦ Weed
♦ 87, 88

Neptunia gracilis **Benth.**
Fabaceae/Mimosaceae
♦ native sensitive plant, sensitive plant
♦ Weed, Introduced
♦ 55, 228
♦ arid, cultivated. Origin: Australia.

Neptunia monosperma **F.Muell. ex Benth.**
Fabaceae/Mimosaceae
♦ sensitive plant
♦ Weed
♦ 93, 205
♦ pH, cultivated.

Neptunia natans **(L.f.) Druce**
Fabaceae/Mimosaceae
Neptunia oleracea Lour. (see)
♦ Weed, Quarantine Weed
♦ 76, 87, 88, 209, 220

Neptunia oleracea **Lour.**
Fabaceae/Mimosaceae
= *Neptunia natans* (L.f.) Druce
♦ Weed, Quarantine Weed
♦ 88, 262
♦ aH, cultivated, herbal.

Neptunia plena **(L.) Benth.**
Fabaceae/Mimosaceae
Acacia lycopodioides Desv., *Acacia punctata* (L.) Desf., *Desmanthus comosus* A.Rich., *Desmanthus plenus* (L.) Willd., *Desmanthus polyphyllus* DC., *Desmanthus punctatus* (L.) Willd., *Mimosa adenanthera* Roxb., *Mimosa lycopodioides* Desf., *Mimosa plena* L., *Mimosa punctata* L., *Neptunia polyphylla* (DC.) Benth., *Neptunia surinamensis* Steud.
♦ water dead and awake
♦ Quarantine Weed, Introduced
♦ 220, 228
♦ aqua, cultivated.

Neptunia triquetra **Benth.**
Fabaceae/Mimosaceae
♦ Naturalised
♦ 287
♦ herbal.

Nerine **Herb. spp.**
Liliaceae/Amaryllidaceae
♦ nerine
♦ Weed
♦ 161
♦ toxic.

Nerine filifolia **Bak.**
Liliaceae/Amaryllidaceae
♦ Weed, Naturalised, Casual Alien
♦ 54, 86, 88, 280
♦ cultivated. Origin: South Africa.

Nerium oleander **L.**
Apocynaceae
Nerium indicum Mill., *Nerium odoratum* Lam., *Nerium odorum* Sol., *Nerium verecundum* Salisb.
♦ oleander, te orian, selonsroos, Ceylon rose, dogbane, double oleander, laurier rose, rose bay, rose laurel, rose of Ceylon, south sea rose, sweet scented oleander, adelfa, alhelí
♦ Weed, Noxious Weed, Naturalised, Introduced, Garden Escape, Environmental Weed, Cultivation Escape
♦ 3, 7, 32, 39, 63, 70, 78, 80, 86, 87, 88, 95, 101, 116, 121, 151, 152, 154, 155, 161, 179, 189, 194, 215, 228, 247, 261, 269, 277, 280, 283
♦ S/T, arid, cultivated, herbal, toxic. Origin: Eurasia.

Nesaea verticillata **H.B.K.**
Lythraceae
♦ Quarantine Weed
♦ 39, 220
♦ toxic.

Neslia apiculata **Fisch., C.A.Mey. & Avé-Lall.**
Brassicaceae
= *Neslia paniculata* (L.) Desv. ssp. *thracica* (Velen.) Bornm.
♦ Weed
♦ 243, 248

Neslia paniculata **(L.) Desv.**
Brassicaceae
Myagrum paniculatum L., *Vogelia paniculata* (L.) Hornem. (see)
♦ ball mustard, ohraruoho, repinka metlinatá
♦ Weed, Quarantine Weed, Noxious Weed, Naturalised, Introduced, Casual Alien
♦ 23, 40, 42, 44, 52, 56, 68, 70, 76, 86, 87, 88, 94, 98, 101, 161, 162, 198, 203, 210, 218, 228, 237, 243, 253, 280, 287, 295, 299
♦ aH, arid, cultivated, herbal. Origin: Europe.

Neslia paniculata **(L.) Desv. ssp. paniculata**
Brassicaceae
♦ ball mustard
♦ Weed, Casual Alien
♦ 272, 280

Neslia paniculata **(L.) Desv. ssp. thracica (Velen.) Bornm.**
Brassicaceae
Neslia apiculata Fisch., C.A.Mey. & Avé-Lall. (see), *Alyssum paniculatum* Willd., *Vogelia apiculata* (Fisch., C.A.Mey. & Avé-Lall.) Vierh. (see)
♦ Weed
♦ 272

Nestegis apetala **(Vahl) L.A.S.Johnson**
Oleaceae
♦ Quarantine Weed
♦ 220

Neurada procumbens **L.**
Neuradaceae/Tiliaceae
♦ Weed
♦ 221
♦ aH.

Neurolaena lobata **(L.) R.Br.**
Asteraceae
Conyza lobata L., *Conyza symphytifolia* Mill., *Pluchea symphytifolia* (Mill.) Gillis (see)

♦ sepi
♦ Weed
♦ 14, 87, 88
♦ promoted, herbal.

Newbouldia laevis **(Beauv.) Seem. ex Bureau**
Bignoniaceae
♦ boundary tree
♦ Weed
♦ 88
♦ arid, herbal.

Neyraudia arundinacea **(L.) Henr.**
Poaceae
♦ Madagascar grass, kaswegenda
♦ Naturalised
♦ 101
♦ G. Origin: Madagascar.

Neyraudia reynaudiana **L.**
Poaceae
Neyraudia madagascariensis (Kunth) Hook.f. var. *zollingeri* (Büse) Hook.f.
♦ Burma reed, canegrass, silk reed, yoong, false reed
♦ Weed, Naturalised, Environmental Weed
♦ 3, 80, 88, 101, 112, 129, 151, 179, 191, 209, 238
♦ pG, toxic.

Nicandra physalodes **(L.) Gaertn.**
Solanaceae
Atropa physaloides L., *Nicandra physaloides* (L.) Gaertn. (see), *Physalodes peruviana* Kuntze
♦ apple of Peru, shoofly plant, wild hops, Peru apple, Chinese lantern, wild gooseberry, rakkokoiso, belladona, manzana de Peru, toloache, tomate de burro
♦ Weed, Naturalised, Introduced, Casual Alien, Cultivation Escape
♦ 13, 23, 34, 39, 40, 42, 50, 51, 53, 55, 86, 87, 98, 101, 134, 157, 158, 161, 165, 179, 198, 203, 217, 218, 228, 241, 247, 252, 257, 269, 280, 295
♦ aH, arid, cultivated, herbal, toxic. Origin: South America.

Nicandra physaloides **(L.) Gaertn.**
Solanaceae
= *Nicandra physalodes* (L.) Gaertn.
♦ apple of Peru, rivabe, shoofly plant, farolito
♦ Weed, Naturalised
♦ 15, 88, 121, 199, 236, 237, 245, 255, 270, 272, 275, 286, 287, 297
♦ aH, cultivated, herbal. Origin: South America.

Nicodemia madagascariensis **(Lam.) R.Parker**
Loganiaceae/Buddlejaceae
= *Buddleja madagascariensis* Lam.
♦ Weed, Naturalised
♦ 98, 203
♦ cultivated.

Nicolaia elatior **(Jack.) Horan.**
Zingiberaceae
= *Etlingera elatior* (Jack.) R.M.Sm.
♦ flor de cera
♦ Garden Escape

- 261
- cultivated. Origin: Malaysia.

Nicotiana L. spp.
Solanaceae
- tobacco
- Quarantine Weed
- 220
- herbal, toxic.

Nicotiana acuminata (Graham) Hook. var. multiflora (Phil.) Rchb.
Solanaceae
- many flowered tobacco
- Introduced
- 34
- aH, herbal.

Nicotiana alata Link & Otto
Solanaceae
Nicotiana affinis Moore, *Nicotiana persica* Lindl.
- jasmine tobacco, white flowering tobacco, sweet scented tobacco, koristetupakka
- Weed, Naturalised, Cultivation Escape, Casual Alien
- 39, 40, 42, 87, 88, 101, 108, 255, 280
- pH, cultivated, herbal, toxic. Origin: Central and South America.

Nicotiana attenuata Torr. ex S.Watson
Solanaceae
- coyote tobacco, mountain tobacco
- Weed
- 39, 161
- aH, cultivated, herbal, toxic.

Nicotiana bigelovii (Torr.) S.Watson
Solanaceae
= *Nicotiana quadrivalvis* Pursh var. *bigelovii* (Torr.) DeWolf (NoR)
- Indian tobacco, wild tobacco, Bigelow's tobacco
- Weed
- 39, 180
- aH, cultivated, herbal, toxic.

Nicotiana glauca Graham
Solanaceae
Siphaulax glabra Raf.
- tree tobacco, mustard tree, wild tobacco, wildetabak, Mexican tobacco, coneton, San Juan tree, tobacco plant, akkue musa, cestrum, corneton, free tobacco, jantwak, le tabaque glauque, mahasatpurush, masseyss, palau pazau, satpurush, tabaco cimarron, tabaco de arbol, tabakboom, tabaqueira, tobacco bush, tobacco tree, tombak el gerey
- Weed, Noxious Weed, Naturalised, Introduced, Garden Escape, Environmental Weed, Cultivation Escape
- 7, 34, 35, 39, 42, 50, 51, 63, 72, 78, 79, 80, 86, 87, 88, 93, 95, 98, 101, 116, 121, 151, 152, 158, 161, 180, 189, 198, 199, 203, 218, 221, 228, 241, 243, 247, 269, 277, 279, 280, 283, 295, 300
- S/T, arid, cultivated, herbal, toxic. Origin: South America.

Nicotiana glutinosa L.
Solanaceae

- tobacco
- Weed
- 39, 87, 88
- arid, toxic.

Nicotiana longiflora Cav.
Solanaceae
- longflower tobacco
- Weed, Naturalised, Casual Alien
- 39, 87, 88, 101, 121, 161, 237, 280, 295
- H, cultivated, toxic. Origin: South America.

Nicotiana paniculata L.
Solanaceae
- tobacco
- Weed
- 87, 88
- arid, cultivated.

Nicotiana plumbaginifolia Viv.
Solanaceae
- Tex Mex tobacco
- Weed, Naturalised
- 14, 101, 179
- T.

Nicotiana quadrivalvis Pursh
Solanaceae
- Indian bigelovii tobacco, Indian tobacco
- Weed
- 39, 161
- aH, herbal, toxic.

Nicotiana rustica L.
Solanaceae
Nicotiana rugosa Mill.
- palturitupakka, wild tobacco, palturitupakka, Aztec tobacco
- Weed, Naturalised, Casual Alien
- 39, 42, 101, 161, 280
- aH, cultivated, herbal, toxic. Origin: obscure, possibly Americas.

Nicotiana × sanderae hort. ex W.Wats. (pro sp.)
Solanaceae
= *Nicotiana alata* Link & Otto × *Nicotiana forgetiana* hort. ex Hemsl.
- Sander's tobacco
- Naturalised, Introduced, Cultivation Escape, Casual Alien
- 40, 101, 280
- aH, cultivated.

Nicotiana suaveolens Lehm.
Solanaceae
- dead shot, native tobacco, native tobacco bush, scented tobacco, tobacco bush, tobacco weed, wild tobacco, Australian tobacco
- Weed, Naturalised
- 39, 87, 88, 101
- arid, cultivated, toxic. Origin: Australia.

Nicotiana sylvestris Speg. & Comes
Solanaceae
- South American tobacco
- Weed, Naturalised, Introduced
- 15, 34, 101, 280, 300
- pH, cultivated, herbal.

Nicotiana tabacum L.
Solanaceae
Nicotiana chinensis Fisch. ex Lehm.,

Nicotiana mexicana Schltdl., *Nicotiana mexicana* var. *rubriflora* Dunal, *Nicotiana pilosa* Dunal
- tobacco, Virginiantupakka, tipaker, cultivated tobacco, tabac commun, tabak, tabaco
- Weed, Noxious Weed, Naturalised, Introduced, Environmental Weed, Cultivation Escape, Casual Alien
- 8, 32, 39, 42, 86, 87, 88, 98, 101, 134, 147, 161, 179, 198, 203, 230, 241, 247, 252, 257, 261, 280, 287
- aH, cultivated, herbal, toxic. Origin: tropical America.

Nicotiana trigonophylla Dun.
Solanaceae
= *Nicotiana obtusifolia* Mert. & Galeotti var. *obtusifolia* (NoR)
- desert tobacco
- Weed, Naturalised
- 39, 87, 88, 161, 218, 243, 287
- pS, cultivated, herbal, toxic.

Nicotiana undulata Ruiz & Pav.
Solanaceae
- mata tabaco
- Weed
- 295

Nidorella anomala Steetz
Asteraceae
Nidorella angustifolia O.Hoffm.
- Native Weed
- 121
- H. Origin: southern Africa.

Nidorella hottentotica DC.
Asteraceae
- Native Weed
- 121
- H. Origin: southern Africa.

Nidorella resedifolia DC.
Asteraceae
- stinkkruid
- Weed
- 88, 158, 243
- Origin: southern Africa.

Nidorella resedifolia DC. ssp. resedifolia
Asteraceae
- Native Weed
- 121
- aH. Origin: southern Africa.

Nierembergia frutescens Durieu
Solanaceae
- tall cupflower
- Naturalised
- 101

Nierembergia hippomanica Miers
Solanaceae
- dwarf cupflower
- Weed, Naturalised
- 39, 86, 87, 88, 98, 101, 203, 295
- a/pH, cultivated, toxic. Origin: South America.

Nierembergia hippomanica Miers var. caerulea (Miers) Millán
Solanaceae
- dwarf cupflower
- Naturalised
- 101

Nierembergia linariaefolia Graham var. *linariaefolia*
Solanaceae
♦ Weed
♦ 237

Nierembergia repens Ruiz & Pav.
Solanaceae
♦ Casual Alien
♦ 280
♦ cultivated.

Nierembergia scoparia Sendtn.
Solanaceae
♦ broom cupflower
♦ Naturalised
♦ 101
♦ cultivated.

Nigella arvensis L.
Ranunculaceae
Nigella foeniculacea DC.
♦ wild fennel, field nigella, rikkaneito
♦ Weed, Quarantine Weed, Casual Alien
♦ 42, 70, 76, 87, 88, 94, 203, 243, 272
♦ aH, cultivated, herbal.

Nigella assyriaca Boiss.
Ranunculaceae
♦ Weed
♦ 221

Nigella damascena L.
Ranunculaceae
♦ love in a mist, nigella, devil in the bush, tarhaneito, fanciullacce, damigella scapigliata
♦ Weed, Naturalised, Cultivation Escape
♦ 39, 42, 70, 86, 87, 88, 94, 98, 101, 203, 252, 253, 272, 280
♦ aH, cultivated, herbal, toxic. Origin: Mediterranean, Middle East.

Nigella hispanica L.
Ranunculaceae
♦ Spanish fennel
♦ Weed, Quarantine Weed, Naturalised
♦ 70, 76, 86, 87, 88, 94
♦ aH, cultivated, herbal.

Nigella sativa L.
Ranunculaceae
Melanthium indica Roxb. ex Flem.
♦ black cumin, fennel flower
♦ Weed, Naturalised
♦ 39, 88, 94, 101, 272
♦ aH, cultivated, herbal, toxic.

Nigella segetalis Bieb.
Ranunculaceae
♦ Weed
♦ 272

Nipa fruticans Thunb.
Arecaceae
♦ nipa palm, pahrem
♦ Weed
♦ 19
♦ T, herbal.

Nitella C.A.Agardh spp.
Characeae
♦ stonewort, musk grass, algae, nitela
♦ Weed

♦ 255
♦ algae.

Nitella batrachosperma Ag.
Characeae
♦ Weed
♦ 87, 88
♦ wH.

Nitella flexilis (L.) Ag.
Characeae
♦ Weed
♦ 87, 88
♦ wH.

Nitella hyalina (DC.) Agardh
Characeae
♦ stonewort
♦ Weed, Quarantine Weed
♦ 76, 87, 88, 203, 220
♦ wH.

Nitellopsis obtusa (Desv.) J.Groves
Characeae
Chara stelligera Bauer
♦ stonewort, starry stonewort, tähtimukulaparta, stjärnslinke
♦ Weed
♦ 197
♦ wH.

Nitraria billardierei DC.
Zygophyllaceae/Nitrariaceae
♦ nitre bush, dillon bush
♦ Native Weed, Introduced
♦ 228, 269
♦ cultivated. Origin: Australia.

Nitraria retusa (Forssk.) Asch.
Zygophyllaceae/Nitrariaceae
Nitraria tridentata Desf.
♦ Weed
♦ 221
♦ arid, cultivated, herbal.

Noaea mucronata (Forssk.) Asch. & Schweinf.
Chenopodiaceae
♦ Weed
♦ 221
♦ arid.

Nocca decipiens (Hemsl.) Kuntze
Asteraceae
Lagascea decipiens Hemsl. (see)
♦ Quarantine Weed
♦ 220

Nolana crassulifolia Poepp. ssp. *revoluta* (Ruiz & Pav.) Mesa
Nolanaceae/Solanaceae
= *Nolana revoluta* Ruiz & Pav. (NoR)
♦ Introduced
♦ 38

Nolana prostrata L.
Nolanaceae/Solanaceae
= *Nolana humifusa* (Gouan) I.M.Johnst. (NoR)
♦ perunkoisokki
♦ Casual Alien
♦ 42
♦ cultivated.

Nolina microcarpa S.Wats.
Agavaceae/Dracaenaceae/Nolinaceae
♦ sacahuista, beargrass
♦ Weed

♦ 39, 87, 88, 161, 218
♦ cultivated, herbal, toxic.

Nolina texana S.Wats.
Agavaceae/Dracaenaceae/Nolinaceae
♦ Texas sacahuista, beargrass, bunchgrass, sacahuista
♦ Weed
♦ 39, 87, 88, 161, 218
♦ cultivated, herbal, toxic.

Nolletia ciliaris (DC.) Steetz
Asteraceae
♦ Native Weed
♦ 121
♦ S, herbal. Origin: southern Africa.

Noltea africana (L.) Rchb. ex Harv. & Sond.
Rhamnaceae
♦ Naturalised
♦ 39, 86
♦ cultivated, herbal, toxic. Origin: South Africa.

Nomaphila parishii T.And.
Acanthaceae
♦ Weed
♦ 12

Nomaphila stricta (Vahl) Nees
Acanthaceae
♦ giant hygrophila
♦ Weed, Naturalised
♦ 13, 101
♦ aqua, cultivated.

Nonea alba DC.
Boraginaceae
♦ Weed
♦ 87, 88

Nonea lutea (Desr.) Rchb. ex A.DC.
Boraginaceae
♦ yellow monkswort, yellow alkanet
♦ Weed, Naturalised, Environmental Weed
♦ 86, 87, 88, 101, 176, 198, 272, 287
♦ cultivated. Origin: eastern Europe.

Nonea nigricans DC.
Boraginaceae
♦ Weed
♦ 87, 88

Nonea picta Sweet
Boraginaceae
♦ Weed
♦ 87, 88

Nonea pulla (L.) DC.
Boraginaceae
♦ nonnea, rusonunna
♦ Weed, Casual Alien
♦ 42, 87, 88, 94, 114, 243, 272
♦ cultivated.

Nonea rosea (Bieb.) Link
Boraginaceae
♦ rose monkswort
♦ Naturalised
♦ 101

Nonea versicolor (Steven) Sweet
Boraginaceae
♦ kirjonunna
♦ Cultivation Escape
♦ 42
♦ cultivated.

Nonea vesicaria **(L.) Rchb.**
Boraginaceae
♦ red monkswort
♦ Weed, Naturalised
♦ 88, 94, 101

Nopalea **Salm-Dyck spp.**
Cactaceae
= *Opuntia* Mill. spp.
♦ Weed, Quarantine Weed
♦ 76, 88, 203

Nopalea cochenillifera **(L.) Salm-Dyck**
Cactaceae
= *Opuntia cochenillifera* (L.) Mill.
♦ Introduced
♦ 228
♦ arid, herbal.

Normanbokea **Kladiwa & Bux. spp.**
Cactaceae
= *Turbinocarpus* (Backeb.) Bux. &
Backeb. spp. (NoR)
♦ Weed, Quarantine Weed
♦ 76, 88, 220

Noronhia emarginata **(Lam.) Thouars**
Oleaceae
♦ Madagascar olive
♦ Weed
♦ 179
♦ cultivated.

Nothofagus antarctica **(G.Forst.) Oerst.**
Nothofagaceae/Fagaceae
♦ Antarctic beech
♦ Casual Alien
♦ 280
♦ cultivated.

Nothoscordum bivalve **(L.) Britt.**
Liliaceae/Alliaceae
♦ garlic, crowpoison
♦ Weed, Naturalised
♦ 161, 243, 287
♦ pH, arid, cultivated, herbal.

Nothoscordum borbonicum **Kunth**
Liliaceae/Alliaceae
♦ fragrant false garlic
♦ Weed, Naturalised
♦ 7, 101, 134, 253, 300

Nothoscordum fragrans **(Vent.) Kunth**
Liliaceae/Alliaceae
= *Nothoscordum gracile* (Aiton) Stearn
♦ Weed, Naturalised
♦ 286, 287

Nothoscordum gracile **(Aiton) Stearn**
Liliaceae/Alliaceae
Nothoscordum inodorum (Ait.) auct. non
(Sol. ex Ait.) Nichols., *Nothoscordum*
fragrans (Vent.) Kunth (see)
♦ fragrant false garlic, onion weed,
slender false garlic
♦ Weed, Naturalised, Garden Escape,
Environmental Weed
♦ 86, 88, 101, 158, 176, 198, 237, 269,
280
♦ pH, cultivated. Origin: Americas.

Nothoscordum inodorum **(Aiton)**
Nicholson
Liliaceae/Alliaceae
= *Allium neapolitanum* Cirillo
♦ viirulaukka, fragrant false garlic,

Gowie's curse, Gowie weed, onion
weed, wild onion
♦ Weed, Noxious Weed, Introduced,
Garden Escape, Cultivation Escape
♦ 15, 34, 35, 42, 51, 70, 87, 88, 121, 126,
180, 255, 295
♦ pH, arid, cultivated. Origin: South
America.

Nothosmyrnium japonicum **Miq.**
Apiaceae
♦ straw weed
♦ Naturalised
♦ 287
♦ herbal.

Notobasis syriaca **(L.) Cass.**
Asteraceae
Cirsium syriacum (L.) Gaertn. (see)
♦ Syrian thistle
♦ Weed, Naturalised
♦ 86, 87, 88, 198
♦ arid, cultivated. Origin: Eurasia.

Notoceras bicorne **(Sol.) Caruel**
Brassicaceae
♦ Weed
♦ 221

Notospartium glabrescens **Petrie**
Fabaceae/Papilionaceae
♦ Quarantine Weed
♦ 220
♦ cultivated.

Notothylas orbicularis **(Schwein.) Sull.**
Notothyladaceae
♦ round notothylas
♦ Quarantine Weed
♦ 220
♦ moss.

Nuphar **Sm. spp.**
Nymphaeaceae
♦ pond lily
♦ Weed, Quarantine Weed,
Naturalised
♦ 76, 86, 88, 220
♦ wH, cultivated.

Nuphar advena **(Aiton) W.T.Aiton f.**
Nymphaeaceae
= *Nuphar lutea* (L.) Sm. ssp. *advena*
(Aiton) Kartesz & Gandhi (NoR)
♦ spatterdock, yellow waterlily,
common spatterdock
♦ Weed
♦ 87, 88, 218
♦ wpH, cultivated, herbal.

Nuphar japonicum **DC.**
Nymphaeaceae
♦ koohone
♦ Weed
♦ 87, 88, 286
♦ wpH, cultivated, herbal. Origin: east
Asia.

Nuphar lutea **(L.) Sibth. & Sm.**
Nymphaeaceae
Nuphar luteum (L.) Sibth. & Sm. (see),
Nymphaea lutea L.
♦ yellow waterlily, yellow pond lily,
spatterdock
♦ Weed, Quarantine Weed,
Naturalised, Environmental Weed

♦ 23, 70, 80, 87, 88, 246, 272, 280
♦ wpH, cultivated, herbal.

Nuphar luteum **(L.) Sibth. & Sm.**
Nymphaeaceae
= *Nuphar lutea* (L.) Sibth. & Sm.
♦ yellow waterlily, spatterdock, yellow
pond lily, dwarf waterlily
♦ Weed, Quarantine Weed
♦ 8, 161, 258
♦ wpH, cultivated, herbal, toxic.

Nuphar polysepala **Engelm.**
Nymphaeaceae
= *Nuphar lutea* (L.) Sm. ssp. *polysepala*
(Engelm.) E.O.Beal (NoR)
♦ yellow waterlily
♦ Weed
♦ 159
♦ wpH, promoted, herbal.

Nuphar pumila **(Timm) DC.**
Nymphaeaceae
= *Nuphar lutea* (L.) Sm. ssp. *pumila*
(Timm) E.O.Beal (NoR)
♦ spatterdock, konnanulpukka, least
waterlily
♦ Weed
♦ 23, 88
♦ wpH, cultivated.

Nuphar variegatum **Engelm.**
Nymphaeaceae
= *Nuphar variegata* Durand (NoR)
♦ Weed
♦ 23, 88
♦ wpH, herbal.

Nuttallanthus canadensis **(L.) D.A.Sutton**
Scrophulariaceae
Antirrhinum canadense L., *Linaria*
canadensis Dum.Cours. var. *canadensis*
♦ Canada toadflax
♦ Naturalised
♦ 86
♦ Origin: North America.

Nyctocereus **(Berg) Britt. & Rose spp.**
Cactaceae
= *Peniocereus* (Berg) Britt. & Rose spp.
♦ Quarantine Weed
♦ 76, 88, 203

Nyctocereus serpentinus **(Lag. & Rodr.)**
Britt. & Rose
Cactaceae
= *Peniocereus serpentinus* (Lag. & Rodr.)
N.P.Taylor
♦ Weed, Naturalised
♦ 98, 203
♦ cultivated.

Nymphaea **L. spp.**
Nymphaeaceae
♦ waterlily
♦ Weed, Naturalised, Introduced,
Cultivation Escape
♦ 198, 221, 230, 233
♦ wH, cultivated.

Nymphaea alba **L.**
Nymphaeaceae
Castalia alba (L.) Woodv. & Wood,
Castalia minoriflora Simonk.,
Castalia speciosa Salisb. *nom. illeg.*,
Leuconymphaea alba (L.) Kuntze,

Nymphaea alba fo. *csepelensis* Soó,
Nymphaea alba fo. *limosa* Soó, *Nymphaea alba* ssp. *occidentalis* (Ostenf.) Hyl.,
Nymphaea alba var. *melocarpa* Casp.,
Nymphaea erythrocarpa Hentze,
Nymphaea exumbonata Rupr., *Nymphaea melocarpa* (Casp.) Asch. & Graebn.,
Nymphaea minoriflora (Simonk.)
E.D.Wissjul., *Nymphaea occidentalis* (Ostenf.) Moss, *Nymphaea officinalis* Gaterau nom. illeg., *Nymphaea parviflora* Hentze, *Nymphaea polystigma* E.H.L.Krause, *Nymphaea rotundifolia* Hentze, *Nymphaea splendens* Hentze, *Nymphaea suaveolens* Dumort.,
Nymphaea urceolata Hentze, *Nymphaea venusta* Hentze
♦ waterlily, white waterlily, European white lily
♦ Weed, Naturalised, Garden Escape, Environmental Weed
♦ 23, 39, 70, 72, 86, 87, 88, 98, 165, 198, 203, 241, 246, 272, 280, 300
♦ wpH, cultivated, herbal, toxic. Origin: Mediterranean, Middle East.

Nymphaea amazonum Mart. & Zucc.
Nymphaeaceae
Castalia amazonum (Mart. & Zucc.) Britton & P.Wilson, *Leuconymphaea amazonum* (Mart. & Zucc.) Kuntze
♦ Amazon waterlily
♦ Weed, Quarantine Weed
♦ 76, 87, 88, 203, 220
♦ wpH, cultivated.

Nymphaea ampla DC.
Nymphaeaceae
Castalia ampla Salisb., *Nymphaea ampla* var. *plumieri* Planch. nom. illeg.,
Nymphaea leiboldiana Lehm.
♦ dotleaf waterlily
♦ Weed
♦ 197, 255
♦ wpH, cultivated, herbal. Origin: tropical America.

Nymphaea blanda G.F.W.Mey. nom. illeg.
Nymphaeaceae
= *Nymphaea glandulifera* Rodschied (NoR)
♦ sleeping beauty waterlily, waterlily
♦ Weed
♦ 197
♦ wpH, cultivated.

Nymphaea caerulea Sav.
Nymphaeaceae
Castalia caerulea (Sav.) Tratt.,
Leuconymphaea caerulea (Sav.) Kuntze,
Nymphaea abbreviata Guill. & Perr.,
Nymphaea ampla Kotschy ex Casp. nom. inval., *Nymphaea caerulea* var. *albiflora* Casp., *Nymphaea caerulea* var. *versicolor* T.Durand & H.Durand,
Nymphaea coerulea Sav. (see), *Nymphaea calliantha* Conard, *Nymphaea calliantha* var. *nelsonii* Burtt Davy, *Nymphaea calliantha* var. *tenuis* Conard, *Nymphaea cyclophylla* R.E.Fr., *Nymphaea discolor* Steud. ex Lehm., *Nymphaea engleri* Gilg, *Nymphaea maculata* Schumach. & Thonn. (see), *Nymphaea magnifica*

Gilg, *Nymphaea mildbraedii* Gilg,
Nymphaea muschleriana Gilg, *Nymphaea muschleriana* var. *megaphylla* Hauman nom. inval., *Nymphaea nouchali* Burm.f. var. *caerulea* (Savigny) Verdc.,
Nymphaea nouchali var. *mutandaensis* Verdc., *Nymphaea nubica* Lehm.,
Nymphaea poecila Lehm., *Nymphaea radiata* Bercht. & Opiz
♦ blue lotus, blue waterlily, Egyptian lotus, waterlily
♦ Weed, Quarantine Weed, Naturalised, Native Weed
♦ 39, 101, 121, 220
♦ wpH, cultivated, toxic. Origin: southern Africa.

Nymphaea caerulea Sav. ssp. zanzibarensis (Casp.) S.W.L.Jacobs
Nymphaeaceae
= *Nymphaea capensis* Thunb.
♦ Naturalised
♦ 86
♦ wpH, cultivated. Origin: southern Africa.

Nymphaea capensis Thunb.
Nymphaeaceae
Castalia capensis (Thunb.) J.Schust.,
Castalia scutifolia Salisb., *Castalia zanzibariensis* (Casp.) Britton,
Leuconymphaea berneriana (Planch.) Kuntze, *Leuconymphaea emirnensis* (Planch.) Kuntze, *Leuconymphaea zanzibariensis* (Casp.) Kuntze,
Nymphaea bernieriana Planch.,
Nymphaea caerulea Sav. ssp. *zanzibariensis* (Casp.) S.W.L.Jacobs (see), *Nymphaea capensis* var. *katangensis* Hauman nom. inval., *Nymphaea capensis* var. *madagascariensis* (DC.) Conard,
Nymphaea capensis var. *zanzibariensis* (Casp.) Conard (see), *Nymphaea colorata* Peter, *Nymphaea colorata* var. *parviflora* Peter, *Nymphaea emirnensis* Planch.,
Nymphaea grandiflora Peter, *Nymphaea madagascariensis* DC., *Nymphaea nouchali* Burm.f. var. *zanzibariensis* (Casp.) Verdc., *Nymphaea polychroma* Peter, *Nymphaea purpurascens* Peter,
Nymphaea scutifolia (Salisb.) DC.,
Nymphaea spectabilis Gilg, *Nymphaea sphaerantha* Peter, *Nymphaea stellata* Willd. var. *zanzibariensis* (Casp.) Hook.f., *Nymphaea zanzibariensis* Casp.,
Nymphaea zanzibariensis var. *pallida* Peter
♦ blue waterlily, cape blue waterlily, lotus lily
♦ Weed, Naturalised, Native Weed
♦ 98, 101, 121, 197, 203
♦ wpH, cultivated. Origin: Madagascar.

Nymphaea capensis Thunb. var. zanzibariensis (Casp.) Conard
Nymphaeaceae
= *Nymphaea capensis* Thunb.
♦ cape blue waterlily
♦ Weed, Naturalised
♦ 101, 179
♦ wpH, cultivated.

Nymphaea coerulea Sav.
Nymphaeaceae
= *Nymphaea caerulea* Sav.
♦ skyblue waterlily
♦ Weed, Quarantine Weed
♦ 76, 87, 88, 203, 221
♦ wpH, cultivated, herbal.

Nymphaea × daubenyana W.T.Baxter ex Daubeny
Nymphaeaceae
= *Nymphaea caerulea* Sav. × *Nymphaea micrantha* Guill. & Perr. [most probable combination]
♦ Weed
♦ 179, 197
♦ wpH, cultivated.

Nymphaea gigantea Hook.
Nymphaeaceae
Castalia gigantea (Hook.) Britten,
Leuconymphaea gigantea (Hook.) Kuntze, *Nymphaea gigantea* fo. *hudsonii* (anon.) K.C.Landon, *Nymphaea gigantea* var. *hudsoniana* F.Henkel et al.,
Nymphaea gigantea var. *hudsonii* anon.,
Nymphaea gigantea var. *media* F.Henkel et al., *Nymphaea gigantea* var. *neorosea* K.C.Landon, *Victoria fitzroyana* hort. ex Loudon nom. inval.
♦ giant waterlily
♦ Naturalised
♦ 7, 86
♦ wpH, cultivated. Origin: Australia.

Nymphaea hybrida Peck
Nymphaeaceae
Nuphar advena Ait. var. *hybrida* Peck
♦ Naturalised
♦ 287
♦ wH.

Nymphaea jamesoniana Planch.
Nymphaeaceae
Castalia gibertii Morong, *Castalia jamesoniana* (Planch.) Britton & P.Wilson, *Leuconymphaea gibertii* (Morong) Conard. nom. inval.,
Leuconymphaea jamesoniana (Planch.) Kuntze, *Nymphaea gibertii* (Morong) Conard, *Nymphaea sagittariaefolia* Lehm.
♦ James waterlily
♦ Weed
♦ 197
♦ wH, cultivated, herbal.

Nymphaea lotus L.
Nymphaeaceae
Castalia lotus (L.) Woodv. & Wood,
Castalia mystica Salisb. nom. illeg.,
Castalia thermalis (DC.) Simonk.,
Leuconymphaea lotus (L.) Kuntze,
Nymphaea acutidens Peter, *Nymphaea aegyptiaca* Opiz, *Nymphaea dentata* Schumach. & Thonn., *Nymphaea hypotricha* Peter, *Nymphaea leucantha* Peter, *Nymphaea liberiensis* A.Chev. nom. inval., *Nymphaea lotus* fo. *thermalis* (DC.) Tuzson, *Nymphaea lotus* var. *dentata* (Schumach. & Thonn.) G.Nicholson, *Nymphaea lotus* var. *monstrosa* C.A.Barber, *Nymphaea lotus* var. *parviflora* Peter, *Nymphaea*

ortgiesiana Planch., *Nymphaea reichardiana* F.Hoffm., *Nymphaea thermalis* DC., *Nymphaea boucheana* Planch., *Nymphaea zenkeri* Gilg (see)
♦ white Egyptian lotus, Egyptian lily, Egyptian waterlily, lotus, sacred lotus, white waterlily, tiger lotus
♦ Weed, Naturalised, Native Weed
♦ 69, 87, 88, 101, 121, 179, 197, 221
♦ wpH, cultivated, herbal. Origin: southern Africa.

Nymphaea maculata Schum. & Thonn.
Nymphaeaceae
= *Nymphaea caerulea* Sav.
♦ African tiger lotus
♦ Weed
♦ 87, 88
♦ wpH, cultivated.

Nymphaea mexicana Zucc.
Nymphaeaceae
Castalia flava (Leitn. ex A.Gray) Greene, *Castalia mexicana* (Zucc.) J.M.Coult., *Leuconymphaea flava* (Leitn. ex A.Gray) Kuntze, *Leuconymphaea mexicana* (Zucc.) Kuntze, *Nymphaea flava* Leitn. ex A.Gray, *Nymphaea lutea* Treat. *nom. illeg.*, *Nymphaea planchonii* Casp. ex Conard *nom. inval.*
♦ banana waterlily, yellow waterlily, Mexican waterlily
♦ Weed, Quarantine Weed, Noxious Weed, Naturalised, Introduced, Garden Escape, Environmental Weed
♦ 7, 35, 86, 87, 88, 98, 161, 197, 198, 200, 203, 218, 229, 246, 258, 280
♦ wpH, cultivated. Origin: North America.

Nymphaea nouchali Burm.f.
Nymphaeaceae
Castalia acutiloba (DC.) Hand.-Mazz., *Castalia stellaris* Salisb. *nom. illeg.*, *Castalia stellata* (Willd.) Blume, *Leuconymphaea stellata* (Willd.) Kuntze, *Nymphaea acutiloba* DC., *Nymphaea cahlara* Donn *nom. inval.*, *Nymphaea cyanea* Roxb., *Nymphaea edgeworthii* Lehm., *Nymphaea henkeliana* Rehnelt, *Nymphaea hookeriana* Lehm., *Nymphaea malabarica* Poir., *Nymphaea membranacea* Wall. ex Casp. *nom. inval.*, *Nymphaea minima* F.M.Bailey, *Nymphaea punctata* Edgew. *nom. illeg.*, *Nymphaea rhodantha* Lehm., *Nymphaea stellata* Willd. (see), *Nymphaea stellata* var. *cyanea* (Roxb.) Hook.f. & Thomson, *Nymphaea stellata* var. *parviflora* Hook.f. & Thomson, *Nymphaea stellata* var. *versicolor* (Sims) Hook.f. & Thomson, *Nymphaea versicolor* Sims, *Nymphaea voalefoka* Lat.-Marl. ex W.Watson *nom. nudum*
♦ Weed, Naturalised, Introduced
♦ 85, 87, 88, 209, 228
♦ wpH, cultivated, herbal.

Nymphaea odorata Aiton
Nymphaeaceae
Nymphaea maximiliani Lehm., *Nymphaea rosea* Raf. *nom. illeg.*
♦ waterlily, fragrant waterlily, American white waterlily
♦ Weed, Quarantine Weed,

Naturalised, Garden Escape, Environmental Weed
♦ 7, 8, 86, 87, 88, 98, 161, 197, 203, 218, 258
♦ wpH, cultivated, herbal, toxic. Origin: North America, West Indies.

Nymphaea polysepela (Engelm.) Greene
Nymphaeaceae
= *Nuphar polysepala* Engelm. (NoR)
♦ waterlily
♦ Weed, Noxious Weed
♦ 88, 137
♦ wpH.

Nymphaea pubescens Willd.
Nymphaeaceae
Castalia edulis Salisb., *Castalia magnifica* Salisb., *Castalia pubescens* (Willd.) Woodv. & Wood, *Castalia sacra* Salisb., *Leuconymphaea lotus* (L.) Kuntze var. *pubescens* (Willd.) Kuntze, *Leuconymphaea rubra* (Roxb. ex Andrews) Kuntze, *Nymphaea coteka* Roxb. ex Salisb. *nom. inval.*, *Nymphaea devoniensis* Hook., *Nymphaea edulis* (Salisb.) DC., *Nymphaea esculenta* Roxb., *Nymphaea lotus* L. var. *pubescens* (Willd.) Hook.f. & Thomson, *Nymphaea magnifica* (Salisb.) Conard *nom. illeg.*, *Nymphaea nouchali* auct. *non*, *Nymphaea purpurea* Rehnelt & F.Henkel, *Nymphaea rosea* (Sims) Sweet, *Nymphaea rubra* Roxb. ex Andrews, *Nymphaea rubra* var. *rosea* Sims, *Nymphaea sagittata* Edgew., *Nymphaea semisterilis* Lehm., *Nymphaea spontanea* K.C.Landon *nom. nudum*
♦ Weed
♦ 87, 88, 262
♦ wpH, cultivated. Origin: Asia, Australia.

Nymphaea stellata Willd.
Nymphaeaceae
= *Nymphaea nouchali* Burm.f.
♦ blue waterlily, red waterlily
♦ Weed, Quarantine Weed, Naturalised
♦ 12, 25, 76, 88, 203, 220
♦ wpH, cultivated, herbal.

Nymphaea tetragona Georgi
Nymphaeaceae
Castalia crassifolia Hand.-Mazz., *Castalia tetragona* (Georgi) G.Lawson, *Leuconymphaea tetragona* (Georgi) Kuntze, *Nymphaea crassifolia* (Hand.-Mazz.) Nakai, *Nymphaea esquirolii* H.Lév. & Vaniot, *Nymphaea fennica* Mela, *Nymphaea tetragona* var. *crassifolia* (Hand.-Mazz.) Y.C.Chu, *Nymphaea tetragona* var. *lata* Casp., *Nymphaea tetragona* var. *wenzelii* (Maack) Vorosch., *Nymphaea wenzelii* Maack
♦ white waterlily, pygmy waterlily, suomenlumme
♦ Weed
♦ 159, 297
♦ wpH, cultivated, herbal. Origin: Europe, Asia, North America.

Nymphaea tuberosa Paine
Nymphaeaceae
= *Nymphaea odorata* Aiton ssp. *tuberosa*

(Paine) Wiersema & Hellq. (NoR)
♦ white waterlily, tuberous waterlily, magnolia waterlily
♦ Weed
♦ 23, 39, 87, 88, 161, 218
♦ wpH, cultivated, herbal, toxic.

Nymphaea zenkeri Gilg
Nymphaeaceae
= *Nymphaea lotus* L.
♦ skyblue waterlily
♦ Quarantine Weed
♦ 76
♦ wpH, cultivated.

Nymphoides Hill spp.
Menyanthaceae
♦ floating heart
♦ Quarantine Weed
♦ 220

Nymphoides aquatica (J.F.Gmel.) Kuntze
Menyanthaceae
♦ big floating heart, underwater banana, banana plant
♦ Naturalised
♦ 287
♦ aqua, cultivated, herbal.

Nymphoides aquaticum Fern.
Menyanthaceae
♦ Weed, Quarantine Weed
♦ 87, 88

Nymphoides coreana (H.Lév.) H.Hara
Menyanthaceae
♦ Weed
♦ 286

Nymphoides crenata (F.Muell.) Kuntze
Menyanthaceae
♦ wavy marshwort
♦ Weed
♦ 200
♦ aqua, cultivated.

Nymphoides cristata (Roxb.) Kuntze
Menyanthaceae
Limnanthemum cristatum (Roxb.) Griseb.
♦ white water snowflake, white water
♦ Weed
♦ 87, 88
♦ aqua, cultivated.

Nymphoides geminata (R.Br.) Kuntze
Menyanthaceae
♦ Quarantine Weed, Naturalised, Environmental Weed
♦ 246, 280
♦ cultivated. Origin: Australia.

Nymphoides humboldtianum (H.B.K.) Kuntze
Menyanthaceae
♦ Weed, Quarantine Weed
♦ 76, 87, 88, 203

Nymphoides indica (L.) Kuntze
Menyanthaceae
Limnanthemum indicum (L.) Griseb.
♦ water snowflake, bua ba
♦ Weed, Naturalised
♦ 87, 88, 101, 170, 209, 239, 255, 262, 286, 297
♦ aqua, cultivated. Origin: tropical Asia, Australia.

Nymphoides indica (L.) Kuntze ssp.
occidentalis A.Raynal
Menyanthaceae
♦ floating heart, water snowflake,
yellow pond lily
♦ Native Weed
♦ 121
♦ wpH. Origin: southern Africa.

Nymphoides peltata (Gmel.) Kuntze
Menyanthaceae
Limnanthemum nymphoides Hoffm. &
Link, *Limnanthemum peltatum* Gmel.
♦ yellow floating heart, yellow fringe,
water fringe, floating heart, fringed
waterlily
♦ Weed, Quarantine Weed,
Naturalised, Environmental Weed,
Casual Alien
♦ 4, 76, 80, 87, 88, 101, 104, 133, 195,
197, 246, 272, 275, 280, 286, 297
♦ wpH, cultivated, herbal.

Nymphoides thunbergiana (Griseb.)
Kuntze
Menyanthaceae
♦ Native Weed
♦ 121
♦ wpH. Origin: southern Africa.

Nypa fruticans Wurmb
Arecaceae
♦ nipa palm, mangrove palm, nipa
♦ Weed
♦ 87, 88
♦ cultivated, herbal. Origin: Asia,
Australia.

Nyssa aquatica L.
Cornaceae/Nyssaceae
Nyssa uniflora Wangenh.
♦ water tupelo, cotton gum
♦ Weed
♦ 87, 88, 218
♦ T, aqua, cultivated, herbal.

Nyssa sylvatica Marsh.
Cornaceae/Nyssaceae
Nyssa caroliniana Poir., *Nyssa multiflora*
Wangenh., *Nyssa sylvatica* Marsh. var.
dilatata Fernald
♦ blackgum, black tupelo
♦ Weed
♦ 87, 88, 218
♦ T, cultivated, herbal.

Nyssa sylvatica Marsh. var. *biflora*
(Walter) Sarg.
Cornaceae/Nyssaceae
= *Nyssa biflora* Walt. (NoR)
♦ swamp tupelo
♦ Weed
♦ 218

Nyssanthes diffusa R.Br.
Amaranthaceae
♦ barbwire weed
♦ Native Weed
♦ 269
♦ pH. Origin: Australia.

O

Oberonia losokawae Fuk.
Orchidaceae
♦ Introduced
♦ 230
♦ H.

Oberonia ponapensis Tuyama
Orchidaceae
♦ Introduced
♦ 230
♦ H.

Ochagavia Phil. spp.
Bromeliaceae
♦ bromeliad
♦ Weed
♦ 15
♦ cultivated.

Ochna jabotapita L.
Ochnaceae
♦ bird's eye bush
♦ Naturalised, Cultivation Escape
♦ 101, 261
♦ cultivated, herbal. Origin: Sri Lanka.

Ochna mossambicensis Klotsch
Ochnaceae
♦ Mozambique ochna, kabukobuko
♦ Cultivation Escape
♦ 261
♦ cultivated. Origin: Mozambique.

Ochna natalitia (Meisn.) Walp.
Ochnaceae
♦ coast boxwood, coast redwood,
Natal plane, showy ochna
♦ Weed, Quarantine Weed, Native
Weed
♦ 76, 88, 121, 203, 220
♦ S/T. Origin: southern Africa.

Ochna pulchra Hook.
Ochnaceae
♦ peeling bark ochna, peeling plane,
wild pear, wild plum
♦ Weed, Quarantine Weed, Native
Weed
♦ 76, 88, 121, 203, 220
♦ S/T, cultivated. Origin: southern
Africa.

Ochna serrulata (Hochst.) Walp.
Ochnaceae
Ochna multiflora hort. not DC., *Ochna
japonica* hort., *Ochna serratifolia* hort.
♦ ochna, Mickey Mouse plant, bird's
eye bush
♦ Weed, Noxious Weed, Naturalised,
Garden Escape, Environmental Weed
♦ 3, 73, 86, 88, 98, 101, 134, 155, 201,
203, 269, 280, 290, 296

♦ cultivated. Origin: South Africa.

Ochna thomasiana Engl. & Gilg
Ochnaceae
♦ Mickey Mouse plant, ochna, ochna
kirkii
♦ Naturalised, Cultivation Escape
♦ 101, 233
♦ cultivated.

Ochradenus baccatus Delile
Resedaceae
♦ Weed
♦ 221

Ochroma lagopus Sw.
Bombacaceae
= *Ochroma pyramidale* (Cav. ex Lam.)
Urb.
♦ balsa, corkwood
♦ Weed
♦ 3, 191
♦ cultivated, herbal.

Ochroma pyramidale (Cav. ex Lam.) Urb.
Bombacaceae
Ochroma lagopus Sw. (see)
♦ West Indian balsa, balsa
♦ Weed, Environmental Weed
♦ 3, 22, 191, 257
♦ T, herbal.

Ochrosia elliptica Labill.
Apocynaceae
Ochrosia parviflora G.Don (see)
♦ elliptic yellowwood, ochrosia plum,
pokosola, gu cheng mei gui shu
♦ Weed, Naturalised, Environmental
Weed
♦ 101, 161, 179
♦ cultivated, herbal, toxic. Origin:
Australasia.

Ochrosia parviflora G.Don
Apocynaceae
= *Ochrosia elliptica* Labill.
♦ kopsia
♦ Weed
♦ 80, 88, 112

Ocimum adscendens Willd.
Lamiaceae
♦ Weed
♦ 87, 88
♦ cultivated.

Ocimum americanum L.
Lamiaceae
Ocimum canum Sims (see)
♦ hoary basil, sweet basil, lemon basil,
American basil
♦ Weed
♦ 13, 66, 87, 88, 158
♦ a/pH, cultivated, herbal. Origin:
tropical Africa and Asia.

Ocimum americanum L. var. *americanum*
Lamiaceae
♦ Naturalised
♦ 86
♦ Origin: tropical Africa and Asia.

Ocimum basilicum L.
Lamiaceae
♦ sweet basil, common basil,
lukalanga, la'au sauga, basil, basilico,
basilic, basilienkraut, basilikum,
basilico, alfavaca, albahaca

- ♦ Weed, Naturalised, Introduced, Garden Escape, Environmental Weed
- ♦ 7, 38, 86, 87, 88, 98, 101, 121, 179, 203, 228, 230, 240, 242, 261, 280
- ♦ a/bH, arid, cultivated, herbal. Origin: obscure, possibly Asia.

Ocimum basilicum L. var. *pilosum* (Willd.) Benth.
Lamiaceae
- ♦ sweet basil
- ♦ Weed
- ♦ 297

Ocimum canum Sims
Lamiaceae
= *Ocimum americanum* L.
- ♦ hoary basil, basil
- ♦ Weed, Native Weed
- ♦ 88, 121, 243
- ♦ H, herbal.

Ocimum × *citriodorum* Vis.
Lamiaceae
= *Ocimum basilicum* L. × *Ocimum americanum* L.
- ♦ Naturalised
- ♦ 86
- ♦ Origin: Sudan, Arabia, Iran, India, China.

Ocimum gratissimum L.
Lamiaceae
Ocimum viride Willd. (see)
- ♦ African basil, East Indian basil, wild basil, la'au sauga
- ♦ Weed, Quarantine Weed, Naturalised
- ♦ 6, 13, 32, 76, 87, 88, 101, 203
- ♦ cultivated, herbal. Origin: Madagascar.

Ocimum gratissimum L. var. *suave* (Willd.) Hook f.
Lamiaceae
- ♦ Weed
- ♦ 275

Ocimum menthaefolium Hochst. ex Benth.
Lamiaceae
= *Ocimum basilicum* L.
- ♦ Weed
- ♦ 221
- ♦ cultivated.

Ocimum micranthum Willd.
Lamiaceae
= *Ocimum campechianum* Mill. (NoR)
- ♦ alfavaca
- ♦ Weed
- ♦ 14, 87, 88, 255
- ♦ pH, herbal. Origin: Americas.

Ocimum sanctum L.
Lamiaceae
= *Ocimum tenuiflorum* L.
- ♦ sacred basil, tulsi, holy basil, la'au sauga, katerihn
- ♦ Weed, Naturalised, Introduced
- ♦ 13, 87, 88, 98, 203, 230
- ♦ H, cultivated, herbal.

Ocimum tenuiflorum L.
Lamiaceae
Ocimum sanctum L. (see)

- ♦ holy basil, sacred basil
- ♦ Weed, Naturalised
- ♦ 86, 101, 261, 262
- ♦ pH, cultivated, herbal. Origin: Old World Tropics.

Ocimum viride Willd.
Lamiaceae
= *Ocimum gratissimum* L.
- ♦ West African basil, mosquito plant
- ♦ Weed
- ♦ 87, 88
- ♦ herbal.

Octodon setosum Hiern
Rubiaceae
Borreria setosa (Hiern) K.Schum.
- ♦ Weed
- ♦ 87, 88

Odontites lutea (L.) Clairv.
Scrophulariaceae
Euphrasia lutea L.
- ♦ yellow bartsia, yellow odontites
- ♦ Weed
- ♦ 272
- ♦ cultivated, herbal.

Odontites rubra Pers. ex Besser *nom. inval.*
Scrophulariaceae
= *Odontites verna* (Bell.) Dumort.
- ♦ Weed
- ♦ 87, 88, 243
- ♦ herbal.

Odontites serotina (Lam.) Dum.
Scrophulariaceae
= *Odontites verna* (Bellardi) Dumort. ssp. *serotina* (Dumort.) Corb. (NoR)
- ♦ red bartsia
- ♦ Weed, Noxious Weed
- ♦ 23, 36, 88, 162, 297
- ♦ herbal.

Odontites verna (Bell.) Dumort.
Scrophulariaceae
Bartsia odontites (L.) Huds. (see), *Euphrasia odontites* L. (see), *Odontites rubra* Pers. ex Besser *nom. inval.* (see), *Odontites serotina* Reichen. ssp. *verna* (Bell.) Hayek, *Odontites vernus* (Bell.) Dumort. (see)
- ♦ red bartsia, red rattle, peltosänkiö, acker zahntrost, algarabia, odontitès du printemps, perlina di primavera
- ♦ Weed
- ♦ 70, 87, 88, 94, 253, 272
- ♦ cultivated. Origin: Europe.

Odontites vernus (Bell.) Dumort.
Scrophulariaceae
= *Odontites verna* (Bell.) Dumort.
- ♦ red bartsia
- ♦ Naturalised
- ♦ 101

Odontites vernus (Bellardi) Dumort. ssp. *serotinus* (Dumort.) Corb.
Scrophulariaceae
- ♦ red bartsia, common red bartsia
- ♦ Naturalised
- ♦ 101

Odontites vulgaris Moench
Scrophulariaceae
- ♦ punasänkiö

- ♦ Naturalised
- ♦ 101
- ♦ cultivated.

Odontonema callistachyum (Schltr. & Cham.) Kuntze
Acanthaceae
= *Odontonema tubaeforme* (Nees) Kuntze
- ♦ Weed
- ♦ 3
- ♦ cultivated.

Odontonema cuspidatum (Nees) Kuntze
Acanthaceae
- ♦ mottled toothedthread, coral de jardín
- ♦ Naturalised, Cultivation Escape
- ♦ 101, 261
- ♦ cultivated. Origin: Central America.

Odontonema strictum (Nees) Kuntze
Acanthaceae
= *Odontonema tubaeforme* (Nees) Kuntze
- ♦ firespike, scarlet flame
- ♦ Weed
- ♦ 3, 163
- ♦ cultivated, herbal.

Odontonema tubaeforme (Nees) Kuntze
Acanthaceae
Odontonema callistachyum (Schltr. & Cham.) Kuntze (see), *Odontonema tubiforme* (Bertol.) Kuntze (see), *Odontonema strictum* (Nees) Kuntze (see)
- ♦ firespike
- ♦ Weed, Quarantine Weed, Naturalised, Environmental Weed
- ♦ 3, 12, 76, 82, 86, 88, 191
- ♦ cultivated. Origin: Central America.

Odontonema tubiforme (Bertol.) Kuntze
Acanthaceae
= *Odontonema tubaeforme* (Nees) Kuntze
- ♦ firespike
- ♦ Weed, Naturalised
- ♦ 101, 179

Odontosoria aculeata (L.) J.Sm.
Dennstaedtiaceae
- ♦ thicket creepingfern
- ♦ Weed
- ♦ 87, 88

Odontosoria chinensis (L.) J.Sm.
Dennstaedtiaceae
Sphenomeris chinensis (L.) Maxon (see)
- ♦ Chinese creepingfern
- ♦ Naturalised
- ♦ 101
- ♦ cultivated.

Odontosoria wrightiana Maxon.
Dennstaedtiaceae
- ♦ Weed
- ♦ 14

Odontostomum hartwegii Torr.
Liliaceae/Tecophilaeaceae
- ♦ Hartweg's doll's lily, Hartweg's odontostomum, odontostomum
- ♦ Quarantine Weed
- ♦ 220
- ♦ pH, cultivated, herbal.

Oeceoclades maculata (Lindl.) Lindl.
Orchidaceae
- ground orchid, monk orchid
- Weed, Quarantine Weed, Environmental Weed
- 76, 80, 88, 112, 179
- cultivated.

Oenanthe L. spp.
Apiaceae
- water dropwort
- Weed
- 272

Oenanthe aquatica (L.) Poir.
Apiaceae
Oenanthe phellandrium Lam.
- fineleaf water dropwort, pahaputki, horse bane
- Weed, Naturalised
- 39, 87, 88, 101, 208, 208, 272, 280
- pH, aqua, cultivated, herbal, toxic.

Oenanthe banatica Heuff.
Apiaceae
- Weed
- 272

Oenanthe benghalensis (Roxb.) Kurz
Apiaceae
- Weed
- 275, 297

Oenanthe crocata L.
Apiaceae
- water dropwort, hemlock water dropwort
- Weed
- 39, 70, 161
- cultivated, herbal, toxic.

Oenanthe fistulosa L.
Apiaceae
Phellandrium fistulosum Clairv., *Selinum fistulosum* E.H.L.Krause
- tubular water dropwort, common water dropwort, halucha dutá
- Weed
- 39, 87, 88, 272
- herbal, toxic.

Oenanthe javanica (Blume) DC.
Apiaceae
Oenanthe stolonifera (Roxb.) DC.
- phak an, water dropwort, Java water dropwort
- Weed
- 87, 88, 204, 238, 263, 274, 275, 286, 297
- pH, aqua, cultivated. Origin: Asia, Australia.

Oenanthe lachenalii Gmel.
Apiaceae
- parsley water dropwort
- Weed
- 39, 70
- cultivated, toxic.

Oenanthe pimpinelloides L.
Apiaceae
- water dropwort, meadow parsley, parsley dropwort, corkyfruit water dropwort
- Weed, Noxious Weed, Naturalised, Environmental Weed

- 39, 54, 86, 87, 88, 101, 155, 165, 171, 272, 280
- pH, cultivated, herbal, toxic. Origin: Mediterranean.

Oenanthe sarmentosa C.Presl ex DC.
Apiaceae
- water parsely, water dropwort
- Weed, Naturalised
- 23, 88, 161, 280
- pH, promoted, herbal, toxic.

Oenanthe silaifolia Bieb.
Apiaceae
- narrow leaved water dropwort
- Weed
- 39, 272
- cultivated, herbal, toxic.

Oenothera L. spp.
Onagraceae
- evening primroses, aandblomme
- Naturalised, Garden Escape, Environmental Weed
- 198, 283, 296
- cultivated, herbal. Origin: Americas.

Oenothera acaulis Cav.
Onagraceae
- Weed, Naturalised
- 86, 98, 203, 287
- cultivated. Origin: Chile.

Oenothera affinis Camb.
Onagraceae
- longflower evening primrose
- Weed, Naturalised, Environmental Weed
- 7, 55, 72, 86, 88, 98, 101, 134, 198, 203, 269, 280, 295
- pH. Origin: South America.

Oenothera albicaulis Pursh
Onagraceae
Anogra albicaulis (Pursh) Britt.
- prairie evening primrose
- Weed
- 87, 88, 218
- a/bH, promoted, herbal.

Oenothera biennis L.
Onagraceae
Oenothera muricata L. (see), *Onagra biennis* (L.) Scop.
- common evening primrose, yellow evening primrose, evening primrose, helokki, German rampion
- Weed, Naturalised, Native Weed
- 23, 23, 24, 40, 42, 51, 52, 70, 87, 88, 94, 98, 121, 136, 158, 161, 174, 207, 210, 218, 272, 280, 286, 287, 291, 299, 300
- pH, arid, cultivated, herbal. Origin: North America.

Oenothera biennis L. var. canescens Torr. & Gray
Onagraceae
= *Oenothera villosa* Thunb. ssp. *villosa* (NoR)
- western yellow evening primrose
- Weed
- 218

Oenothera cambrica Rostanski
Onagraceae
- small flowered evening primrose

- Naturalised
- 40

Oenothera cordata J.W.Loudon
Onagraceae
- heartleaf evening primrose
- Naturalised
- 101

Oenothera coronifera Renner
Onagraceae
- hohtohelokki
- Casual Alien
- 42

Oenothera curvifolia H.Fisch.
Onagraceae
- Weed
- 23, 88

Oenothera dentata Cav.
Onagraceae
- contorted primrose
- Weed
- 161, 180, 243
- herbal.

Oenothera drummondii Hook.
Onagraceae
Raimannia drummondii (Hook.) Rose ex Sprague & Riley
- beach evening primrose, coastal evening primrose
- Weed, Naturalised, Introduced
- 7, 98, 179, 203, 228, 269
- a/pH, arid, cultivated, herbal.

Oenothera drummondii Hook. ssp. drummondii
Onagraceae
- Naturalised, Environmental Weed
- 86
- Origin: south-east North America.

Oenothera elata Kunth
Onagraceae
- Hooker's evening primrose
- Weed
- 161, 199
- pH, cultivated, herbal.

Oenothera erythrosepala Borbás
Onagraceae
= *Oenothera glazioviana* Micheli
- large flowered evening primrose
- Weed, Naturalised
- 87, 88, 121, 215, 286, 287
- H, cultivated, herbal.

Oenothera fallax Renner & Rostanski
Onagraceae
- intermediate evening primrose
- Cultivation Escape
- 40
- cultivated.

Oenothera fennoscandica Rostanski
Onagraceae
- pohjanhelokki
- Casual Alien
- 42

Oenothera flava (A.Nels.) Garrett
Onagraceae
- yellow evening primrose
- Weed
- 199
- herbal.

Oenothera fruticosa L.
Onagraceae
♦ narrowleaf evening primrose
♦ Naturalised
♦ 287
♦ cultivated, herbal.

Oenothera glauca Michx.
Onagraceae
♦ Naturalised
♦ 287

Oenothera glazioviana Micheli
Onagraceae
= *Oenothera grandiflora* L'Hér. ×
Oenothera elata Kunth [most probable
combination – see *Oenothera
erythrosepala* Borbás]
♦ redsepal evening primrose, reddish
evening primrose, evening primrose,
onagre à grandes fleurs
♦ Weed, Naturalised, Environmental
Weed, Cultivation Escape
♦ 7, 15, 34, 40, 72, 86, 88, 98, 101, 165,
176, 198, 203, 252, 280, 300
♦ pH, cultivated, herbal. Origin:
horticultural hybrid.

Oenothera grandiflora L'Hér.
Onagraceae
Oenothera biennis L. var. *grandiflora*
(L'Hér.) Torr. & A.Gray, *Oenothera
fusiformis* Munz & I.M.Johnst., *Onagra
grandiflora* (L'Hér.) Cockerell & Atkins,
Oenothera lamarckiana Ser.
♦ godetia evening primrose,
largeflower evening primrose
♦ Weed, Naturalised
♦ 23, 51, 87, 88, 121, 286, 287
♦ bH. Origin: North America.

Oenothera hookeri Torr. & Gray
Onagraceae
♦ Hooker's evening primrose
♦ Weed
♦ 180, 243
♦ cultivated, herbal.

Oenothera humifusa Nutt.
Onagraceae
♦ seabeach evening primrose, seaside
evening primrose
♦ Naturalised
♦ 287
♦ cultivated, herbal.

Oenothera indecora Cambess.
Onagraceae
♦ smallflower evening primrose,
evening primrose
♦ Weed, Naturalised
♦ 7, 55, 87, 88, 98, 158, 203, 280, 295
♦ b/pH.

**Oenothera indecora Nutt. ssp.
bonariensis Dietr.**
Onagraceae
♦ smallflower evening primrose
♦ Weed, Naturalised
♦ 86, 269
♦ Origin: South America.

**Oenothera indecora Nutt. ssp. bonariensis
W.Dietr. × stricta Ledeb. ex Link**
Onagraceae

♦ Naturalised
♦ 86

**Oenothera indecora Cambess. ssp.
indecora**
Onagraceae
♦ evening primrose
♦ Weed
♦ 121
♦ a/bH. Origin: South America.

Oenothera insignis Bartl.
Onagraceae
♦ outohelokki
♦ Casual Alien
♦ 42

Oenothera jamesii Torr. & Gray
Onagraceae
♦ giant evening primrose, trumpet
evening primrose
♦ Weed
♦ 51, 87, 88, 121, 158
♦ pH, herbal. Origin: North and
Central America.

Oenothera kunthiana (Spach) Munz
Onagraceae
♦ Kunth's evening primrose
♦ Quarantine Weed
♦ 220
♦ cultivated, herbal.

Oenothera laciniata Hill
Onagraceae
Oenothera albicans Lam., *Oenothera
laciniata* var. *pubescens* (Willd. ex
Spreng.) Munz, *Oenothera mexicana*
Spach, *Oenothera prostrata* Ruiz &
Pav. (see), *Oenothera pubescens* Willd.
ex Spreng., *Oenothera sinuata* L.,
Raimannia laciniata (Hill) Rose
♦ cutleaf evening primrose, laciniate
evening primrose, liuskahelokki,
ragged evening primrose
♦ Weed, Naturalised, Introduced,
Casual Alien
♦ 34, 40, 42, 87, 88, 98, 121, 161, 180,
198, 203, 211, 218, 243, 249, 273, 286,
287
♦ a/bH, arid, cultivated, herbal.
Origin: southern and eastern North
America.

Oenothera laciniata Hill ssp. laciniata
Onagraceae
♦ Naturalised
♦ 86
♦ Origin: eastern North America.

**Oenothera laciniata Hill var. grandiflora
(S.Wats.) B.L.Robins.**
Onagraceae
= *Oenothera grandis* (Britt.) Smyth
(NoR)
♦ Weed, Naturalised
♦ 286, 287

Oenothera lamarckiana Ser.
Onagraceae
♦ Weed
♦ 87, 88

Oenothera linifolia Nutt.
Onagraceae
♦ flaxleaf evening primrose, threadleaf
evening primrose, pellavahelokki

♦ Weed, Casual Alien
♦ 42, 87, 88, 218
♦ herbal.

Oenothera longiflora L.
Onagraceae
♦ Weed, Naturalised
♦ 86, 98, 203
♦ herbal. Origin: South America.

Oenothera missouriensis Sims
Onagraceae
= *Oenothera macrocarpa* Nutt. ssp.
macrocarpa (NoR)
♦ Ozark sundrops
♦ Naturalised
♦ 287
♦ cultivated, herbal.

Oenothera mollissima L.
Onagraceae
♦ Argentine evening primrose
♦ Weed, Naturalised
♦ 7, 86, 87, 88, 98, 101, 203
♦ Origin: eastern South America.

Oenothera muricata L.
Onagraceae
= *Oenothera biennis* L.
♦ Weed
♦ 272

Oenothera nuttallii Sweet
Onagraceae
♦ Nuttall's evening primrose, white
evening primrose
♦ Weed
♦ 87, 88, 161
♦ herbal.

Oenothera odorata Jacq.
Onagraceae
Raimannia odorata Sprague & Riley
♦ fragrant evening primrose
♦ Weed
♦ 87, 88
♦ pH, cultivated, herbal. Origin:
Australia.

Oenothera pallida Lindl.
Onagraceae
♦ pale evening primrose, jasmine
evening primrose
♦ Weed
♦ 87, 88, 218
♦ cultivated, herbal.

Oenothera parodiana Munz
Onagraceae
♦ Weed
♦ 87, 88

**Oenothera parodiana Jacq. ssp.
parodiana Munz**
Onagraceae
♦ Weed
♦ 121
♦ pH. Origin: South America.

Oenothera parviflora L.
Onagraceae
♦ northern evening primrose, small
flowered evening primrose, pupalka
malokvetá
♦ Weed, Naturalised
♦ 87, 88, 161, 204, 272, 280, 286, 287
♦ herbal.

Oenothera perennis **L.**
Onagraceae
♦ perennial sundrops, little evening
primrose, sundrops
♦ Weed, Naturalised
♦ 87, 88, 161, 218, 287
♦ cultivated, herbal.

Oenothera prostrata **Ruiz & Pav.**
Onagraceae
= *Oenothera laciniata* Hill
♦ Weed
♦ 87, 88

Oenothera renneri **H.Scholz**
Onagraceae
♦ rennerinhelokki
♦ Casual Alien
♦ 42

Oenothera rosea **L'Hér. ex Aiton**
Onagraceae
Hartmannia affinis Spach, *Hartmannia
gauroides* Spach, *Hartmannia rosea*
(L'Hér. ex Aiton) G.Don, *Hartmannia
virgata* (Ruiz & Pav.) Spach, *Oenothera
psychrophila* Ball, *Oenothera purpurea*
Lam., *Oenothera rubra* Cav., *Oenothera
virgata* Ruiz & Pav., *Xylopleurum roseum*
(L'Hér. ex Aiton) Raim.
♦ rose evening primrose, evening
primrose
♦ Weed, Naturalised, Introduced,
Casual Alien
♦ 34, 40, 50, 51, 86, 87, 88, 98, 121, 134,
158, 198, 203, 241, 243, 275, 280, 286,
287, 295, 297, 300
♦ pH, arid, cultivated, herbal. Origin:
Central America.

Oenothera rubricaulis **Kleb.**
Onagraceae
♦ täplähelokki
♦ Casual Alien
♦ 42

Oenothera speciosa **Nutt.**
Onagraceae
Oenothera speciosa Nutt. var. *childsii*
(Bailey) Munz (see)
♦ white evening primrose, pink ladies,
komeahelokki, showy primrose,
showy evening primrose
♦ Weed, Naturalised, Cultivation
Escape, Casual Alien
♦ 7, 34, 42, 86, 87, 88, 98, 161, 179, 198,
203, 218, 249, 252, 280, 286, 287
♦ pH, cultivated, herbal. Origin: North
America.

Oenothera speciosa **Nutt. var.** *childsii*
(Bailey) Munz
Onagraceae
= *Oenothera speciosa* Nutt.
♦ Naturalised
♦ 287

Oenothera striata **Link**
Onagraceae
♦ Weed
♦ 87

Oenothera stricta **Ledeb. ex Link**
Onagraceae
♦ common evening primrose, Chilean
evening primrose, sand primrose,

scented evening primrose, fragrant
evening primrose
♦ Weed, Naturalised, Introduced,
Garden Escape, Environmental Weed
♦ 3, 7, 15, 51, 72, 88, 98, 101, 121, 155,
158, 176, 203, 228, 251, 269, 280, 286,
287, 290
♦ pH, arid, cultivated, herbal. Origin:
South America.

Oenothera stricta **Ledeb. ex Link ssp.**
stricta
Onagraceae
♦ common evening primrose
♦ Naturalised, Introduced,
Environmental Weed
♦ 34, 86, 93, 134, 198
♦ pH. Origin: Chile, Argentina.

Oenothera stricta **Ledeb. ex Link var.**
stricta
Onagraceae
♦ Naturalised
♦ 241

Oenothera strigosa **(Rydb.) Mack. &**
Bush
Onagraceae
= *Oenothera villosa* Thunb. ssp. *strigosa*
(Rydb.) W.Dietr. & P.H.Raven (NoR)
♦ karheahelokki
♦ Weed, Casual Alien
♦ 23, 42, 88
♦ herbal.

Oenothera tetragona **Roth**
Onagraceae
= *Oenothera fruticosa* L. ssp. *glauca*
(Michx.) Straley (NoR)
♦ golden sundrops
♦ Weed
♦ 87, 88
♦ cultivated, herbal.

Oenothera tetraptera **Cav.**
Onagraceae
♦ evening primrose, white evening
primrose, fourwing evening primrose
♦ Weed, Naturalised, Casual Alien
♦ 51, 86, 87, 88, 98, 121, 134, 158, 203,
280
♦ H, herbal. Origin: Central America.

Oenothera triloba **Nutt.**
Onagraceae
♦ stemless evening primrose
♦ Weed, Naturalised
♦ 86, 98, 203
♦ cultivated, herbal. Origin: southern
North America.

Oenothera wratislawiensis **Rostanski**
Onagraceae
♦ puolanhelokki
♦ Casual Alien
♦ 42

Oldenburgia arbuscula **DC.**
Asteraceae
♦ Weed
♦ 23, 39, 88
♦ cultivated, toxic.

Oldenlandia aspera **DC.**
Rubiaceae
♦ Weed

♦ 87, 88

Oldenlandia callitrichoides **Griseb.**
Rubiaceae
= *Oldenlandiopsis callitrichoides* (Griseb.)
Terrell & W.H.Lewis
♦ Weed
♦ 14

Oldenlandia capensis **L.f.**
Rubiaceae
♦ Weed, Quarantine Weed
♦ 76, 87, 88, 203, 220

Oldenlandia corymbosa **L.**
Rubiaceae
Hedyotis corymbosa (L.) Lam. (see)
♦ flattop mille graines
♦ Weed, Naturalised, Introduced
♦ 38, 88, 93, 243, 257
♦ H, herbal.

Oldenlandia corymbosa **L. var.** *caespitosa*
(Benth.) Verdc.
Rubiaceae
♦ Naturalised, Environmental Weed
♦ 86

Oldenlandia corymbosa **L. var.**
corymbosa
Rubiaceae
♦ Naturalised, Environmental Weed
♦ 86

Oldenlandia decumbens **Hiern**
Rubiaceae
♦ Weed
♦ 87, 88

Oldenlandia herbacea **(L.) Roxb.**
Rubiaceae
Hedyotis herbacea L. (see)
♦ false spurrey
♦ Weed, Native Weed
♦ 50, 88, 121
♦ aH, cultivated.

Oldenlandia lancifolia **(Schum.) DC.**
Rubiaceae
Hedyotis commutata J.H. & J.A.Schult.
(see)
♦ calycose mille graines
♦ Weed, Quarantine Weed,
Naturalised
♦ 32, 76, 87, 88, 101, 203, 220
♦ S.

Oldenlandia praecox **Pierre ex Pit.**
Rubiaceae
♦ Weed
♦ 87, 88

Oldenlandia prostrata **Blume**
Rubiaceae
♦ Weed
♦ 87, 88

Oldenlandia salzmannii **(DC.) Benth. &**
Hook.f. ex B.D.Jacks.
Rubiaceae
♦ Salzmann's mille graines
♦ Naturalised
♦ 101

Oldenlandia strumosa **Hiern**
Rubiaceae
♦ Weed
♦ 87, 88

Oldenlandia thesiifolia **K.Schum.**
Rubiaceae
♦ Weed
♦ 87, 88

Oldenlandiopsis callitrichoides **(Griseb.) Terrell & W.H.Lewis**
Rubiaceae
Oldenlandia callitrichoides Griseb. (see)
♦ creeping bluet
♦ Weed
♦ 32

Olea africana **Mill.**
Oleaceae
= *Olea europaea* L. ssp. *africana* (Mill.) P.Green
♦ African olive
♦ Environmental Weed
♦ 155
♦ cultivated, herbal.

Olea chrysophylla **Lam.**
Oleaceae
= *Olea europaea* L. ssp. *africana* (Mill.) P.Green
♦ Weed
♦ 221

Olea europaea **L.**
Oleaceae
Olea officinarum Crantz
♦ olive, European olive
♦ Weed, Noxious Weed, Naturalised, Garden Escape, Environmental Weed, Cultivation Escape
♦ 7, 22, 72, 80, 86, 88, 98, 101, 147, 151, 155, 171, 203, 233, 280
♦ S/T, cultivated, herbal. Origin: southern Europe, Mediterranean.

Olea europaea **L. ssp. *africana* (Mill.) P.Green**
Oleaceae
Olea africana Mill. (see), *Olea chrysophylla* Lam. (see), *Olea europaea* L. var. *nubica* Schweinf. ex Bak.
♦ African olive, anireju, badda, brown olive, ejars, ejass, ejerssa, emdit, emitiet, emitiot, ethelei, euriepei, ilnyirei, jerso, jiemdet, kang'g, kango, korosiondet, larak, lorien, molialundi, mutero, muthamayu, muthata, muthatha, mutheru, ol oirien, ol orien, olerenit, wild olive, yemdid, yemit, yemtit, yemut, yernit, zertun
♦ Weed, Noxious Weed, Naturalised, Introduced, Environmental Weed
♦ 3, 73, 86, 98, 101, 152, 228, 280
♦ arid, cultivated. Origin: tropical and southern Africa.

Olea europaea **L. ssp. *cuspidata* (Wall. ex G.Don) Cif.**
Oleaceae
♦ African olive
♦ Naturalised, Environmental Weed
♦ 134, 225, 246, 296
♦ Origin: Africa, Madagascar.

Olea europaea **L. ssp. *europaea***
Oleaceae
♦ European olive, olive
♦ Noxious Weed, Naturalised, Environmental Weed

♦ 86, 98, 101, 198, 280, 289, 296
♦ S/T, cultivated. Origin: Mediterranean, southwest Asia.

Olearia elliptica **DC.**
Asteraceae
♦ sticky daisybush
♦ Weed
♦ 87, 88
♦ cultivated. Origin: Australia.

Olgaea leucophylla **Iljin**
Asteraceae
♦ whiteleaf olgaea
♦ Weed
♦ 297

Oligomeris linifolia **(Vahl) Macbr. Matuda**
Resedaceae
♦ lineleaf whitepuff
♦ Weed
♦ 221
♦ aH, herbal.

Olyra latifolia **L.**
Poaceae
♦ carrycillo, lintentwa
♦ Weed
♦ 87, 88
♦ pG, cultivated, herbal. Origin: Madagascar.

Omalanthus populifolius **Graham**
Euphorbiaceae
♦ bleedingheart, native poplar, bleedingheart tree
♦ Naturalised, Environmental Weed
♦ 39, 86
♦ cultivated, herbal, toxic. Origin: Australia.

Omalotheca sylvatica **(L.) Sch.Bip. & Schultz**
Asteraceae
Gnaphalium sylvaticum L. (see)
♦ heath cudweed, woodland Arctic cudweed
♦ Weed
♦ 70, 272

Omphalocarpum elatum **Miers**
Sapotaceae
♦ bate
♦ Quarantine Weed
♦ 220

Omphalodes linifolia **(L.) Moench**
Boraginaceae
Cynoglossum linifolium L.
♦ kesäkaihonukka, whiteflower navelwort, navelwort
♦ Naturalised, Cultivation Escape
♦ 42, 101, 300
♦ aH, cultivated.

Omphalodes verna **Moench**
Boraginaceae
Cynoglossum omphalodes L., *Picotia verna* Roem. & Schult.
♦ kevätkaihonkukka, blue eyed Mary
♦ Cultivation Escape, Casual Alien
♦ 40, 42
♦ cultivated.

Oncidium ensatum **Lindl.**
Orchidaceae

♦ Latin American orchid
♦ Naturalised
♦ 101

Oncidium variegatum **Sw.**
Orchidaceae
= *Tolumnia variegata* (Sw.) G.J.Braem (NoR)
♦ harlequin dancinglady orchid
♦ Weed
♦ 87, 88

Oncoba echinata **Oliv.**
Flacourtiaceae
= *Caloncoba echinata* (Oliv.) Gilg (NoR)
♦ gorli oncoba
♦ Cultivation Escape
♦ 261
♦ cultivated, herbal. Origin: tropical Africa.

Oncosiphon grandiflorum **(Thunb.) Källersjö**
Asteraceae
Tanacetum grandiflorum Thunb.
♦ pentzia pincushion
♦ Weed, Quarantine Weed
♦ 76, 88
♦ cultivated.

Oncosiphon piluliferum **(L.f.) Källersjö**
Asteraceae
Matricaria globifera (Thunb.) Fenzl ex Harv., *Pentzia globifera* (Thunb.) Hutch. (see), *Pentzia pilulifera* (L.f.) Fourc. (see)
♦ globe chamomile, stinknet
♦ Naturalised
♦ 86, 101, 198
♦ H. Origin: South Africa.

Oncosiphon suffruticosum **(L.) Källersjö**
Asteraceae
♦ pentzia, shrubby mayweed
♦ Naturalised
♦ 86, 101, 198
♦ Origin: South Africa.

Onobrychis aequidentata **(Sibth. & Sm.) D'Urv.**
Fabaceae/Papilionaceae
Onobrychis foveolata DC.
♦ sainfoin
♦ Weed
♦ 272

Onobrychis caput-galli **(L.) Lam.**
Fabaceae/Papilionaceae
Hedysarum caput-galli L.
♦ cock's comb sainfoin
♦ Weed
♦ 272
♦ herbal.

Onobrychis crista-galli **Lam.**
Fabaceae/Papilionaceae
Onobrychis squamosa Viv.
♦ Weed
♦ 221
♦ arid.

Onobrychis ptolemaica **(Del.) DC.**
Fabaceae/Papilionaceae
Hedysarum ptolemaica Del.
♦ Weed
♦ 221
♦ arid.

Onobrychis viciaefolia Scop.
Fabaceae/Papilionaceae
= *Onobrychis viciifolia* Scop.
♦ sanfion
♦ Introduced, Cultivation Escape
♦ 21
♦ cultivated, herbal.

Onobrychis viciifolia Scop.
Fabaceae/Papilionaceae
Hedysarum onobrychis Neck., *Onobrychis sativa* Lam., *Onobrychis viciaefolia* Scop. (see), *Onobrychis vulgaris* Hill.
♦ esparsetti, sainfoin, common sainfoin, esparcet, holy clover
♦ Weed, Naturalised, Casual Alien
♦ 42, 86, 98, 101, 121, 176, 203, 272, 280
♦ pH, arid, cultivated, herbal. Origin: Eurasia.

Onoclea sensibilis L.
Dryopteridaceae/Woodsiaceae
♦ sensitive fern, meadow brake, polypody brake
♦ Weed
♦ 39, 87, 88, 161, 218
♦ H, cultivated, herbal, toxic.

Onoclea sensibilis L. var. interrupta Maxim.
Dryopteridaceae/Woodsiaceae
♦ kouyawarabi
♦ Weed
♦ 286

Ononis L. spp.
Fabaceae/Papilionaceae
♦ restharrow
♦ Weed, Naturalised
♦ 198, 272
♦ cultivated, herbal.

Ononis alopecuroides L.
Fabaceae/Papilionaceae
♦ restharrow
♦ Quarantine Weed, Naturalised, Environmental Weed
♦ 76, 86, 182
♦ cultivated. Origin: western Mediterranean region, northern Africa, southern Europe.

Ononis arvensis L.
Fabaceae/Papilionaceae
♦ field restharrow, restharrow
♦ Weed, Naturalised
♦ 101, 272
♦ cultivated, herbal.

Ononis biflora Desf.
Fabaceae/Papilionaceae
♦ Weed
♦ 87, 88

Ononis campestris G.Koch & Ziz
Fabaceae/Papilionaceae
Ononis spinosa L. (see)
♦ restharrow, prickly restharrow, spiny restharrow
♦ Quarantine Weed, Naturalised
♦ 76, 101, 280

Ononis maweana Ball
Fabaceae/Papilionaceae
♦ etelänorakko
♦ Casual Alien
♦ 42

Ononis natrix L.
Fabaceae/Papilionaceae
♦ yellow restharrow, large yellow restharrow
♦ Weed, Quarantine Weed
♦ 70, 220, 221
♦ cultivated, herbal.

Ononis procurrens Wallr.
Fabaceae/Papilionaceae
= *Ononis repens* L.
♦ Weed
♦ 87, 88

Ononis pusilla L.
Fabaceae/Papilionaceae
Ononis apula Ten., *Ononis parviflora* Lam., *Ononis subocculta* Vill.
♦ ononide piccina
♦ Weed
♦ 272
♦ herbal.

Ononis reclinata L.
Fabaceae/Papilionaceae
♦ small restharrow
♦ Weed
♦ 88, 94, 221

Ononis repens L.
Fabaceae/Papilionaceae
Ononis miniana Plan., *Ononis procurrens* Wallr. (see)
♦ common restharrow, restharrow
♦ Weed, Naturalised, Casual Alien
♦ 42, 70, 86, 88, 94, 98, 101, 198, 272, 280
♦ pH, cultivated. Origin: Europe.

Ononis serrata Forssk.
Fabaceae/Papilionaceae
♦ Weed
♦ 221

Ononis sicula Guss.
Fabaceae/Papilionaceae
♦ Quarantine Weed
♦ 76, 220

Ononis spinosa L.
Fabaceae/Papilionaceae
♦ spring restharrow, restharrow, prickly restharrow, spiny restharrow, piikkiorakko, dornige hauhechel
♦ Weed, Naturalised, Cultivation Escape, Casual Alien
♦ 42, 70, 86, 87, 88, 98, 119, 176, 198, 252, 272
♦ pH, cultivated, herbal. Origin: Eurasia.

Ononis viscosa L.
Fabaceae/Papilionaceae
♦ Weed, Quarantine Weed
♦ 25, 220

Onopordum L. spp.
Asteraceae
♦ farting donkey, cotton thistle, onopordum thistles
♦ Weed, Noxious Weed, Naturalised
♦ 35, 88, 198, 272
♦ H, herbal.

Onopordum acanthium L.
Asteraceae
♦ Scotch thistle, cotton thistle, heraldic thistle, silver thistle, woolly thistle, Scotch cotton thistle
♦ Weed, Quarantine Weed, Noxious Weed, Naturalised, Introduced, Garden Escape, Environmental Weed, Casual Alien
♦ 1, 20, 26, 39, 40, 42, 70, 76, 80, 86, 87, 88, 94, 98, 101, 130, 136, 138, 139, 141, 146, 147, 151, 156, 161, 165, 169, 174, 176, 180, 195, 198, 203, 210, 212, 219, 220, 229, 236, 237, 241, 264, 267, 269, 272, 280, 287, 290, 295, 300
♦ a/bH, arid, cultivated, herbal, toxic. Origin: Eurasia.

Onopordum acaulon L.
Asteraceae
♦ stemless thistle, horse thistle, stemless onopordum
♦ Weed, Quarantine Weed, Noxious Weed, Naturalised, Environmental Weed
♦ 7, 62, 72, 76, 86, 87, 88, 98, 147, 198, 203, 243, 269
♦ aH. Origin: western Mediterranean.

Onopordum alexandrinum Boiss.
Asteraceae
♦ Weed
♦ 221
♦ H.

Onopordum anatolicum (Boiss.) Eig.
Asteraceae
♦ turkinkruunuohdake
♦ Casual Alien
♦ 42
♦ H.

Onopordum anisacanthum Boiss.
Asteraceae
♦ Weed
♦ 87, 88
♦ H.

Onopordum candidum Nab.
Asteraceae
♦ valkokruunuohdake
♦ Casual Alien
♦ 42
♦ H.

Onopordum carduelinum Bolle
Asteraceae
♦ Quarantine Weed
♦ 220
♦ H.

Onopordum heteracanthum C.A.Mey.
Asteraceae
♦ Weed
♦ 87, 88
♦ H.

Onopordum illyricum L.
Asteraceae
♦ Illyrian thistle, Illyrian cotton thistle, cotton thistle
♦ Weed, Quarantine Weed, Noxious Weed, Naturalised, Environmental Weed
♦ 76, 86, 87, 88, 98, 101, 147, 155, 198, 203, 220, 269, 272, 287
♦ bH, promoted. Origin: Mediterranean.

Onopordum jordanicolum Eig.
Asteraceae
- ♦ Weed
- ♦ 243
- ♦ H.

Onopordum leptolepis DC.
Asteraceae
- ♦ Weed
- ♦ 87, 88
- ♦ H.

Onopordum macrocanthum Schousb.
Asteraceae
- ♦ Weed
- ♦ 87, 88
- ♦ H.

Onopordum nervosum Boiss.
Asteraceae
- ♦ Moor's cotton thistle, cotton thistle
- ♦ Weed, Quarantine Weed, Naturalised, Garden Escape, Cultivation Escape
- ♦ 76, 86, 88, 220
- ♦ bH, cultivated, herbal.

Onopordum nogalesii Svent.
Asteraceae
- ♦ Quarantine Weed
- ♦ 220
- ♦ H.

Onopordum sibthorpianum Boiss.& Heldr.
Asteraceae
- ♦ Weed
- ♦ 88
- ♦ H.

Ono *pordum tauricum* Willd.
Asteraceae
- ♦ Turkish thistle, taurian thistle, taurus cotton thistle, bull cotton thistle, Scotch thistle
- ♦ Weed, Noxious Weed, Naturalised, Introduced, Casual Alien
- ♦ 34, 86, 101, 198, 229, 267, 272, 280
- ♦ pH. Origin: southern Europe.

Onosma arenaria Waldst. & Kit.
Boraginaceae
- ♦ purple goldendrop
- ♦ Weed
- ♦ 272
- ♦ cultivated.

Onosma visianii Clem.
Boraginaceae
- ♦ rumenica visianiho
- ♦ Weed
- ♦ 272

Onosmodium virginianum (L.) A.DC.
Boraginaceae
- ♦ wild Job's tears
- ♦ Weed
- ♦ 179
- ♦ herbal.

Onychium melanolepis (Decne.) Kunze
Pteridaceae/Adiantaceae
- ♦ Weed
- ♦ 221

Operculina peltata (L.) Hallier f.
Convolvulaceae
= *Merremia peltata* (L.) Merr.

- ♦ Weed
- ♦ 3, 191

Operculina turpethum (L.) S.Manso
Convolvulaceae
- ♦ St. Thomas lidpod
- ♦ Weed, Naturalised, Introduced
- ♦ 13, 32, 39, 87, 88, 101, 287
- ♦ C, herbal, toxic. Origin: Africa, Asia, Australia.

Operculina turpethum (L.) Silva Manso var. *ventricosa* (Bertero) Staples & Austin
Convolvulaceae
Operculina ventricosa (Bert.) Peter (see)
- ♦ St. Thomas lidpod
- ♦ Naturalised, Introduced, Cultivation Escape
- ♦ 32, 101, 261
- ♦ cultivated.

Operculina ventricosa (Bertero) Peter
Convolvulaceae
= *Operculina turpethum* (L.) Silva Manso var. *ventricosa* (Bertero) Staples & Austin
- ♦ wood rose, alalag, palulu, fue hina
- ♦ Weed, Environmental Weed
- ♦ 3, 152, 191
- ♦ herbal. Origin: tropical America.

Ophioglossum pendulum L.
Ophioglossaceae
- ♦ ribbon fern, old world adder's tongue
- ♦ Weed
- ♦ 179
- ♦ H, cultivated, herbal. Origin: Australia.

Ophioglossum petiolatum Hook.
Ophioglossaceae
- ♦ longstem adder's tongue
- ♦ Weed
- ♦ 273
- ♦ cultivated, herbal.

Ophioglossum vulgatum L.
Ophioglossaceae
- ♦ adder's tongue fern, adder's tongue, käärmeenkieli, southern adder's tongue, English adder's tongue
- ♦ Weed
- ♦ 272
- ♦ H, promoted, herbal.

Ophiopogon jaburan (Sieb.) Lodd.
Liliaceae/Convallariaceae
- ♦ lilyturf, jaburan lilyturf
- ♦ Naturalised
- ♦ 101
- ♦ aqua, cultivated.

Ophiopogon japonicus (L.f.) Ker Gawl.
Liliaceae/Convallariaceae
- ♦ dwarf lilyturf, mondo grass, janohige
- ♦ Weed
- ♦ 286
- ♦ pH, aqua, cultivated, herbal.

Ophrys L. spp.
Orchidaceae
- ♦ Weed
- ♦ 272

Ophrys apifera Huds.
Orchidaceae
- ♦ bee orchid
- ♦ Weed
- ♦ 70
- ♦ pH, cultivated, herbal.

Ophrys bombyliflora Link
Orchidaceae
- ♦ bumblee orchid
- ♦ Weed
- ♦ 70
- ♦ pH, cultivated, herbal.

Ophrys fusca Link
Orchidaceae
- ♦ brown bee orchid, sombre bee orchid
- ♦ Weed
- ♦ 70
- ♦ pH, promoted, herbal.

Ophrys lutea (Gouan) Cav.
Orchidaceae
- ♦ yellow bee orchid
- ♦ Weed
- ♦ 70
- ♦ pH, promoted.

Ophrys monophyllos Pav. ex Lindl.
Orchidaceae
= *Malaxis monophyllos* (Pav ex Lindl.) Sw. (NoR)
- ♦ Introduced
- ♦ 38

Ophrys scolopax Cav.
Orchidaceae
- ♦ woodcock orchid
- ♦ Weed
- ♦ 70
- ♦ pH, promoted.

Opizia stolonifera J.Presl
Poaceae
- ♦ Acapulco grass
- ♦ Weed, Naturalised
- ♦ 101, 179
- ♦ G.

Oplismenus burmannii (Retz.) Beauv.
Poaceae
Oplismenus affinis J.Presl, *Oplismenus affinis* Schult., *Oplismenus affinis* var. *humboldtianus* U.Scholz, *Oplismenus africanus* P.Beauv., *Oplismenus albus* (Poir.) Roem. & Schult., *Oplismenus bromoides* (Lam.) P.Beauv., *Oplismenus burmannii* var. *mulisetus* (Hochst. ex A.Rich.) U.Scholz, *Oplismenus cristatus* J.Presl, *Oplismenus humboldtianus* Nees, *Oplismenus humboldtianus* var. *nudicaulis* Vasey, *Oplismenus indicus* Duthie, *Oplismenus multisetus* Hochst. ex A.Rich., *Oplismenus preslii* Kunth, *Orthopogon albus* (Poir.) Nees ex Steud., *Orthopogon burmannii* (Retz.) Trin., *Panicum album* Poir., *Panicum bromoides* Lam., *Panicum burmannii* Retz., *Panicum hirtellum* Burm., *Panicum japonicum* Steud., *Panicum multisetum* (Hochst. ex A.Rich.) Steud.
- ♦ zacatillo
- ♦ Weed
- ♦ 66, 87, 88, 90, 157, 243, 281
- ♦ aG, herbal. Origin: cosmopolitan.

Oplismenus compositus (L.) Beauv.
Poaceae
♦ running mountaingrass, armgrass
♦ Weed, Naturalised
♦ 87, 88, 90, 101, 240, 243, 273
♦ pG, cultivated, herbal. Origin:
cosmopolitan.

Oplismenus hirtellus (L.) Beauv.
Poaceae
Oplismenus setarius (Lam.) Roem. &
Schult. (see)
♦ basketgrass, bristle basketgrass
♦ Weed
♦ 87, 88, 90
♦ pG, cultivated, herbal. Origin:
cosmopolitan.

**Oplismenus setarius (Lam.) Roem. &
Schult.**
Poaceae
= *Oplismenus hirtellus* (L.) Beauv.
♦ shortleaf basketgrass
♦ Weed, Naturalised
♦ 87, 88, 257
♦ G, herbal.

Oplismenus undulatifolius (Ard.) Beauv.
Poaceae
Panicum undulatifolium Ard.,
Hoplismenus undulatifolius Landolt
♦ wavyleaf basketgrass, spicate
armgrass
♦ Weed, Naturalised
♦ 87, 88, 90, 101, 286, 297
♦ pG, cultivated. Origin: Africa,
Europe, Asia, Australia.

**Oplismenus undulatifolius (Ard.) Roem.
& Schult. var. japonicus (Steud.) Koidz.**
Poaceae
♦ Weed
♦ 286
♦ G.

Opopanax chironium (L.) Koch
Apiaceae
Malabaila opoponax Baill.
♦ opoponax
♦ Weed
♦ 272
♦ pH, arid, cultivated, herbal.

Opophytum aquosum (L.Bolus) N.E.Br.
Aizoaceae
= *Mesembryanthemum aquosum* L.Bolus
(NoR)
♦ Native Weed
♦ 121
♦ aH, cultivated. Origin: southern
Africa.

Opuntia Mill. spp.
Cactaceae
Austrocylindropuntia Backeb. spp.
(see), *Brasiliopuntia* (Schumm.) Berg.
spp., *Chaffeyopuntia* A.V.Fric spp. (see),
Consolea Lem. spp. (see), *Corynopuntia*
F.M.Knuth. spp. (see), *Cumulopuntia*
F.Ritter spp., *Cylindropuntia* (Engelm.)
F.M.Knuth spp. (see), *Ficindica* St.-Lag.
spp. (see), *Grusonia* Reichen. ex Britt.
& Rose spp. (see), *Maihueniopsis* Speg.
spp. (see), *Marenopuntia* Backeb. spp.
(see), *Nopalea* Salm-Dyck spp. (see),

Platyopuntia (Engelm.) Ritt. spp. (see),
Pseudotephrocactus Fric & Schelle spp.
(see), *Puna* Kiesling spp., *Salmiopuntia*
Fric spp. (see), *Tephrocactus* Lem. spp.
(see), *Tunas* Lunell spp. (see)
♦ cholla, prickly pear, opuncja
♦ Weed, Quarantine Weed, Noxious
Weed, Naturalised, Garden Escape,
Environmental Weed
♦ 18, 23, 62, 76, 86, 88, 163, 191, 198,
203, 210, 220, 247, 289, 296
♦ cultivated, herbal, toxic.

Opuntia aurantiaca Lindl.
Cactaceae
♦ jointed cactus, tiger pear,
litjieskaktus, jointed prickly pear
♦ Weed, Quarantine Weed, Noxious
Weed, Naturalised, Garden Escape,
Environmental Weed
♦ 10, 51, 63, 67, 72, 76, 86, 87, 88, 95, 98,
121, 140, 147, 152, 158, 198, 203, 229,
269, 278, 283
♦ pS, cultivated. Origin: South
America.

Opuntia bergeriana A.Weber ex A.Berger
Cactaceae
♦ redflower prickly pear, prickly pear
♦ Weed, Noxious Weed, Naturalised
♦ 54, 86, 88, 198
♦ cultivated.

Opuntia bigelovii Engelm.
Cactaceae
Cylindropuntia bigelowii (Engelm.)
F.M.Knuth
♦ jumping cholla, teddy bear cholla
♦ Weed, Quarantine Weed
♦ 76, 88
♦ S, cultivated, herbal.

Opuntia borinquensis Britton & Rose
Cactaceae
♦ olaga
♦ Weed
♦ 87, 88

Opuntia brasiliensis (Willd.) Haw.
Cactaceae
Brasiliopuntia bahiensis Berg.,
Brasiliopuntia brasiliensis (Willd.) Haw.,
Brasiliopuntia subacarpa Rizzini &
A.Mattos, *Opuntia bahiensis* Britt. &
Rose
♦ Brazilian prickly pear
♦ Weed, Naturalised, Environmental
Weed, Cultivation Escape
♦ 88, 101, 151, 261
♦ cultivated. Origin: South America.

Opuntia cochenillifera (L.) Mill.
Cactaceae
Nopalea cochenillifera (L.) Salm-Dyck
(see)
♦ cochineal nopal cactus, cochineal
cactus
♦ Weed, Naturalised, Casual Alien
♦ 101, 179, 261
♦ cultivated. Origin: Central America.

Opuntia compressa (Salisb.) Macbr.
Cactaceae
= *Opuntia ficus-indica* (L.) Mill.

♦ prickly pear, Indian fig, prickly pear
cactus
♦ Noxious Weed, Naturalised
♦ 86, 98
♦ pH, cultivated, herbal.

Opuntia cordobensis Speg.
Cactaceae
Platyopuntia cordobensis (Speg.) F.Ritter
♦ Argentine prickly pear
♦ Introduced
♦ 228
♦ arid, cultivated.

Opuntia cylindrica (Lam.) DC.
Cactaceae
Austrocylindropuntia cylindrica (Lam.)
Backeb., *Austrocylindropuntia intermedia*
Rauh & Backeb., *Cylindropuntia
intermedia* (Rauh & Backeb.) Rauh &
Backeb.
♦ San Pedro cactus, cane cactus,
prickly pear
♦ Weed, Noxious Weed, Naturalised,
Casual Alien
♦ 86, 98, 198, 203, 280
♦ cultivated, toxic. Origin: Central and
South America.

Opuntia dejecta Salm-Dyck
Cactaceae
Nopalea dejecta (Salm-Dyck) Salm-Dyck
♦ Weed, Naturalised
♦ 98, 203
♦ cultivated.

Opuntia dillenii (Ker Gawl.) Harv.
Cactaceae
= *Opuntia stricta* (Haw.) Haw.
♦ Dillen prickly pear, pipestem prickly
pear, prickly pear
♦ Weed, Quarantine Weed, Noxious
Weed, Introduced
♦ 14, 51, 76, 87, 88, 121, 203, 228, 258,
278, 297
♦ pS, arid, cultivated, herbal. Origin:
Americas.

Opuntia distans Britton & Rose
Cactaceae
♦ Introduced
♦ 228
♦ arid.

Opuntia elatior Mill.
Cactaceae
Opuntia nigricans Par. ex Foerst.
♦ prickly pear
♦ Weed, Noxious Weed, Naturalised,
Introduced, Garden Escape,
Environmental Weed
♦ 86, 87, 88, 93, 98, 155, 203, 205, 228
♦ pH, arid, cultivated, herbal. Origin:
tropical South America.

**Opuntia engelmannii Salm-Dyck ex
Engelm.**
Cactaceae
♦ Engelmann prickly pear, cactus
apple, prickly pear
♦ Weed
♦ 87, 88, 218
♦ arid, cultivated, herbal. Origin:
southern North America.

Opuntia erinacea Engelm. & Bigelow ex Engelm.
Cactaceae
- old man prickly pear, grizzlybear prickly pear, prickly pear
- Weed, Noxious Weed, Naturalised
- 54, 86, 88, 98, 203
- S, cultivated, herbal.

Opuntia exaltata Berg.
Cactaceae
= *Opuntia subulata* (Muehlenpf.) Engelm.
- long spine cactus, langdoringkaktus
- Weed, Quarantine Weed, Noxious Weed, Cultivation Escape
- 63, 76, 88, 121, 203, 278, 283
- S/T, cultivated. Origin: South America.

Opuntia ficus-indica (L.) Mill.
Cactaceae
Cactus ficus-indica L., *Opuntia castillae* Griffiths, *Opuntia compressa* (Salisb.) Macbr. (see), *Opuntia incarnadilla* Griffiths, *Opuntia megacantha* Salm-Dyck (see), *Opuntia occidentalis* Engelm., *Opuntia vulgaris* Mill. (see)
- Indian fig, tuna cactus, sweet prickly pear, mission prickly pear, prickly pear, spineless cactus, Boereturksvy, grootdoringturksvy, spiny pest pear
- Weed, Quarantine Weed, Noxious Weed, Naturalised, Introduced, Garden Escape, Environmental Weed, Cultivation Escape
- 10, 34, 51, 63, 72, 76, 86, 87, 88, 95, 98, 101, 121, 151, 152, 158, 198, 203, 228, 261, 269, 272, 278, 279, 283, 287, 300
- pS, arid, cultivated, herbal, toxic. Origin: Central America.

Opuntia fragilis (Nutt.) Haw.
Cactaceae
- brittle prickly pear, pygmy tuna, little prickly pear, jumping cactus, fragile cactus
- Native Weed
- 161, 174
- S, promoted, herbal. Origin: North America.

Opuntia fulgida Engelm.
Cactaceae
Cylindropuntia fulgida (Engelm.) Knuth.
- jumping cholla
- Weed, Noxious Weed, Garden Escape
- 87, 88, 218, 283
- cultivated, herbal. Origin: south-west North America, Mexico.

Opuntia humifusa (Raf.) Raf.
Cactaceae
Cactus humifusus Raf., *Opuntia allairei* Griffiths, *Opuntia calcicola* Wherry, *Opuntia compressa* auct., *Opuntia cumulicola* Small, *Opuntia fusco-atra* Engelm., *Opuntia impedata* Small, *Opuntia nemoralis* Griffiths, *Opuntia rafinesquei* Engelm., *Opuntia rubiflora* Engelm., *Opuntia vulgaris* auct. non.
- large flowered prickly pear, creeping prickly pear, devil's tongue, spreading prickly pear, prickly pear cactus
- Weed, Noxious Weed, Naturalised, Garden Escape
- 8, 63, 87, 88, 98, 161, 203, 218, 249, 269, 283
- cultivated, herbal, toxic. Origin: eastern North America.

Opuntia imbricata (Harv.) DC.
Cactaceae
Cylindropuntia imbricata (Haw.) Knuth (see)
- imbricate prickly pear, imbrikaatkaktus, kabelturksvy, imbricate cactus, imbricate prickly pear, devil's rope pear, chain link cactus, tree cholla, walkingstick cholla
- Weed, Quarantine Weed, Noxious Weed, Naturalised, Garden Escape, Environmental Weed
- 10, 51, 63, 72, 76, 86, 87, 88, 93, 95, 98, 121, 152, 155, 158, 203, 205, 218, 278, 279, 283
- pS, cultivated, herbal. Origin: Central America.

Opuntia imbricata (Haw.) DC. var. imbricata
Cactaceae
- devil's rope, tree cholla
- Naturalised
- 198

Opuntia inermis (DC.) DC.
Cactaceae
= *Opuntia stricta* (Haw.) Haw.
- Weed
- 87, 88

Opuntia joconostle A.Weber
Cactaceae
- Introduced
- 228
- arid.

Opuntia leptocaulis DC.
Cactaceae
Cylindropuntia brittonii (G.Ortega) Backeb., *Cylindropuntia leptocaulis* DC.
- tasajillo, Christmas cactus, Christmas cholla
- Weed, Noxious Weed, Naturalised
- 86, 87, 88, 218
- cultivated, herbal. Origin: southern North America.

Opuntia leucotricha DC.
Cactaceae
Opuntia fulvispina Salm-Dyck, *Opuntia leucotricha* DC. var. *fulvispina* (Salm-Dyck) F.A.C.Weber
- arborescent prickly pear, prickly pear, semaphore cactus
- Weed, Noxious Weed, Naturalised, Introduced
- 54, 86, 88, 98, 101, 198, 203, 228
- arid, cultivated. Origin: Mexico.

Opuntia lindheimeri Engelm.
Cactaceae
Opuntia aciculata Griffiths, *Opuntia engelmannii* Salm-Dyck var. *aciculata* (Griffiths) Bravo, *Opuntia tardospina* Griffiths (see)
- Lindheimer prickly pear, cholla, small round leaved prickly pear, prickly pear, Klein rondeblaarturksvy, Texas prickly pear
- Weed, Noxious Weed, Naturalised, Garden Escape, Environmental Weed
- 54, 63, 87, 88, 93, 95, 98, 121, 155, 198, 203, 205, 218, 278, 283
- pS, arid, cultivated, herbal. Origin: North and Central America.

Opuntia lindheimeri Engelm. var. lindheimeri
Cactaceae
- Texas prickly pear, prickly pear
- Noxious Weed, Naturalised
- 86, 198

Opuntia lindheimeri Engelm. var. linguiformis (Griffiths) L.Benson
Cactaceae
= *Opuntia engelmannii* Salm-Dyck var. *linguiformis* (Griffiths) Parfitt & Pinkava (NoR)
- cow's tongue, prickly pear
- Noxious Weed, Naturalised
- 86, 198

Opuntia lubrica Griffiths
Cactaceae
= *Opuntia rufida* Engelm. (NoR)
- prickly pear
- Noxious Weed, Naturalised
- 86
- cultivated.

Opuntia megacantha Salm-Dyck
Cactaceae
= *Opuntia ficus-indica* (L.) Mill.
- mission prickly pear, prickly pear
- Weed
- 50, 51, 87, 88, 218
- arid, cultivated, herbal.

Opuntia microdasys (Lehm.) Pfeiffer
Cactaceae
Opuntia macrocalyx Griffiths
- prickly pear, bunny ears
- Weed, Noxious Weed, Naturalised
- 39, 86, 98, 203
- pH, arid, cultivated, toxic. Origin: Mexico.

Opuntia monacantha (Willd.) Haw.
Cactaceae
Platyopuntia brunneogemmia Ritter, *Platyopuntia vulgaris* (Mill.) Ritter
- common prickly pear, suurturksvy, luisiesturksvy, cochineal prickly pear, drooping prickly pear
- Weed, Noxious Weed, Naturalised, Introduced, Cultivation Escape
- 63, 101, 228, 283
- arid, cultivated. Origin: South America.

Opuntia pachypus Schum.
Cactaceae
Austrocylindropuntia pachypus (K.Schum.) Backeb.
- Weed, Naturalised
- 54, 88, 98, 203
- cultivated.

Opuntia paraguayensis K.Schum.
Cactaceae
Opuntia bonaerensis Spreg., *Opuntia chakensis* Speg.
♦ Riverina pear, prickly pear
♦ Weed, Noxious Weed, Naturalised
♦ 86, 98, 198, 203, 269
♦ cultivated. Origin: South America.

Opuntia phaeacantha Engelm.
Cactaceae
♦ prickly pear, tulip prickly pear, bastard fig
♦ Weed, Naturalised
♦ 54, 88, 98, 203
♦ S, arid, cultivated, herbal.

Opuntia phaeacantha Engelm. var. discata (Griffiths) Benson & Walk.
Cactaceae
= *Opuntia engelmannii* Salm-Dyck var. *engelmannii*
♦ prickly pear
♦ Noxious Weed, Naturalised
♦ 86
♦ cultivated.

Opuntia polyacantha Haw.
Cactaceae
Opuntia nicholii L.Benson, *Opuntia arenaria* Engelm.
♦ plains prickly pear, prickly pear
♦ Weed, Native Weed
♦ 87, 88, 136, 161, 174, 212, 218
♦ pH, cultivated, herbal. Origin: North America.

Opuntia puberula Pfeiff.
Cactaceae
Opuntia heliae Matuda, *Opuntia maxonii* Ortega
♦ blind prickly pear, prickly pear
♦ Noxious Weed, Naturalised
♦ 86, 198
♦ cultivated. Origin: Mexico.

Opuntia quimilo Schum.
Cactaceae
Platyopuntia quimilo (Schum.) F.Ritter
♦ Weed
♦ 237, 295
♦ cultivated.

Opuntia robusta H.L.Wendl. ex Pfeiff.
Cactaceae
Opuntia guerrana Griffiths
♦ wheel pear, wheel cactus, bartolona, camuesa, nopal camueso, nopal tápon, sweet purple cactus
♦ Weed, Quarantine Weed, Noxious Weed, Naturalised, Garden Escape, Environmental Weed
♦ 72, 76, 86, 88, 98, 147, 198, 203, 269
♦ S, arid, cultivated. Origin: Mexico.

Opuntia rosea DC.
Cactaceae
Cylindropuntia rosea (DC.) Backeb., *Opuntia pallida* Rose
♦ rosea cactus, Douglas pest, roseakaktus
♦ Weed, Quarantine Weed, Noxious Weed, Environmental Weed
♦ 10, 63, 76, 88, 95, 121, 152, 203, 278, 279

♦ pS, cultivated. Origin: Central America.

Opuntia schickendantzii Weber
Cactaceae
♦ lion's tongue, prickly pear
♦ Weed, Noxious Weed, Naturalised
♦ 54, 86, 88, 198
♦ cultivated. Origin: Argentina.

Opuntia schumannii Weber
Cactaceae
♦ Schumann prickly pear
♦ Weed, Quarantine Weed, Native Weed
♦ 51, 76, 87, 88, 121, 203
♦ pS. Origin: South America.

Opuntia soehrensis Britton & Rose
Cactaceae
♦ Introduced
♦ 228
♦ arid.

Opuntia spinosior (Engelm. & Bigel.) Toumey
Cactaceae
Cylindropuntia spinosior (Toumey) F.M.Knuth
♦ spiny cholla, walkingstick cactus
♦ Weed
♦ 87, 88, 218
♦ cultivated, herbal. Origin: North America.

Opuntia spinulifera Salm-Dyck
Cactaceae
Opuntia candelabriformis C.Mart., *Opuntia heliabravoana* Mart.
♦ large round leaved prickly pear, saucepan cactus, blouturksvy, groot rondeblaar turksvy
♦ Weed, Quarantine Weed, Noxious Weed, Native Weed, Garden Escape
♦ 51, 63, 76, 87, 88, 121, 203, 278, 283
♦ pS, cultivated. Origin: Mexico.

Opuntia streptacantha Lem.
Cactaceae
Opuntia cardona F.A.C.Weber
♦ prickly pear, white spined prickly pear
♦ Weed, Noxious Weed, Naturalised, Introduced
♦ 86, 87, 88, 98, 203, 228, 269
♦ arid, cultivated. Origin: Mexico.

Opuntia stricta (Haw.) Haw.
Cactaceae
Cactus strictus Haw., *Opuntia anahuacensis* Griffiths, *Opuntia atrocapensis* Small, *Opuntia bahamana* Britt. & Rose, *Opuntia dillenii* (Ker Gawl.) Haw. (see), *Opuntia inermis* (DC.) DC. (see), *Opuntia keyensis* Britt. ex Small, *Opuntia magnifica* Small, *Opuntia melanosperma* Svenson, *Opuntia nitens* Small, *Opuntia zebrina* Small
♦ erect prickly pear, common prickly pear, Araluen pear, common pest pear, Gayndah pear, spiny pest pear, sour prickly pear, suurturksvy, pest pear of Australia, Australian pest pear
♦ Weed, Quarantine Weed, Noxious

Weed, Naturalised, Garden Escape, Environmental Weed
♦ 10, 14, 63, 72, 76, 86, 87, 88, 95, 98, 121, 147, 152, 203, 268, 278, 283
♦ pS, cultivated, herbal. Origin: tropical America.

Opuntia stricta (Haw.) Haw. var. dillenii (Ker Gawl.) L. Benson
Cactaceae
♦ erect prickly pear, common prickly pear, Araluen pear, common pest pear, Gayndah pear, spiny pest pear
♦ Noxious Weed, Naturalised
♦ 86
♦ herbal.

Opuntia stricta (Haw.) Haw. var. stricta
Cactaceae
♦ common prickly pear, erect prickly pear, Araluen pear, common pest pear, Gayndah pear, spiny pest pear
♦ Weed, Noxious Weed, Naturalised, Introduced
♦ 86, 93, 198, 205, 228, 269
♦ S, arid. Origin: Americas.

Opuntia subulata (Muhlenpf.) Engelm.
Cactaceae
Austrocylindropuntia exaltata (A.Berger) Backeb., *Austrocylindropuntia subulata* (Engelm.) Backeb. (see), *Opuntia exaltata* Berg. (see)
♦ prickly pear
♦ Weed, Noxious Weed, Naturalised
♦ 86, 98, 203
♦ cultivated. Origin: South America.

Opuntia sulphurea G.Don ex Loudon
Cactaceae
Opuntia vulpina A.Web. (see), *Platyopuntia sulphurea* (G.Don) F.Ritter
♦ prickly pear
♦ Weed, Noxious Weed, Naturalised
♦ 86, 98, 203
♦ cultivated. Origin: South America.

Opuntia tardospina Griffiths
Cactaceae
= *Opuntia lindheimeri* Engelm.
♦ small round leaved prickly pear
♦ Weed
♦ 51, 87, 88

Opuntia tomentosa Salm-Dyck
Cactaceae
Opuntia hernandezii DC., *Opuntia macdougaliana* Rose, *Opuntia sarca* Griffiths ex Scheinvar
♦ woollyjoint prickly pear, prickly pear, velvet tree pear
♦ Weed, Noxious Weed, Naturalised, Introduced, Garden Escape
♦ 86, 87, 88, 98, 101, 203, 228, 269
♦ arid, cultivated. Origin: Mexico.

Opuntia tuna (L.) Mill.
Cactaceae
♦ elephantear prickly pear, tuna prickly pear
♦ Weed, Introduced
♦ 87, 88, 228
♦ arid, cultivated, herbal.

Opuntia tunicata (Lehm.) Link & Otto
Cactaceae

Cylindropuntia tunicata (Lehm.) Knuth
(see)
- thistle cholla
- Weed, Naturalised
- 54, 88, 98, 203
- arid, cultivated, herbal. Origin: North America.

Opuntia tunicata (Lehm.) Link & Otto var. tunicata
Cactaceae
- chain link cactus, prickly pear, thistle cholla
- Noxious Weed, Naturalised
- 86, 198

Opuntia versicolor Engelm.
Cactaceae
Cylindropuntia versicolor (Engelm.) Knuth
- staghorn cholla
- Weed
- 87, 88, 218
- cultivated, herbal.

Opuntia vulgaris Mill.
Cactaceae
= *Opuntia ficus-indica* (L.) Mill.
- drooping prickly pear, Barbary fig, drooping tree pear, smooth tree pear, spiny prickly pear, spreading prickly pear, tuna
- Weed, Quarantine Weed, Noxious Weed, Naturalised, Garden Escape, Environmental Weed
- 15, 72, 76, 86, 87, 88, 98, 121, 147, 152, 179, 198, 203, 269, 278, 280
- pS, cultivated, herbal. Origin: South America.

Opuntia vulpina A.Web.
Cactaceae
= *Opuntia sulphurea* G.Don
- Weed
- 87, 88

Orbea variegata (L.) Haw.
Asclepiadaceae/Apocynaceae
Stapelia variegata L. (see)
- African carrion flower, toad cactus, toad plant, starfish cactus, starfish plant, orbea
- Naturalised, Environmental Weed
- 86
- pH, arid, cultivated. Origin: South Africa.

Orchis L. spp.
Orchidaceae
- Weed
- 272

Orchis coriophora L.
Orchidaceae
Orchis coriophorus L.
- bug orchid, bug orchis
- Weed
- 39, 70, 272
- pH, promoted, herbal, toxic.

Orchis laxiflora Lam.
Orchidaceae
- marsh orchis, lax flowered orchid, loose flowered orchid, Jersey orchid
- Weed
- 87, 88
- pH, promoted, herbal.

Orchis militaris L.
Orchidaceae
- soikkokämmekkä, military orchid, military orchis, soldier orchid
- Weed, Naturalised
- 42, 272
- pH, cultivated, herbal.

Orchis morio L.
Orchidaceae
Orchis crenulatus Gilib.
- green winged orchid, green winged meadow orchid
- Weed
- 272
- pH, cultivated, herbal.

Orchis palustris Jacq.
Orchidaceae
- vstavaä moäiarny
- Weed
- 87, 88
- herbal.

Origanum bilgeri P.H.Davis
Lamiaceae
- Weed
- 87, 88

Origanum heracleoticum L.
Lamiaceae
- winter majoram, Greek oregano
- Weed
- 272
- pH, cultivated, herbal.

Origanum majorana L.
Lamiaceae
Majorana hortensis Moench, *Majorana majorana* (L.) Karst.
- sweet marjoram, maggiorana, knotted majoram
- Naturalised
- 39, 101
- pH, cultivated, herbal, toxic.

Origanum syriacum L.
Lamiaceae
- bible hyssop
- Weed
- 111, 221, 243
- pH, promoted, herbal.

Origanum vulgare L.
Lamiaceae
- wild marjoram, pot oregano, wintersweet, organdy, origano, mäkimeirami
- Weed, Naturalised, Environmental Weed
- 4, 15, 39, 70, 86, 87, 88, 101, 104, 165, 218, 272, 280
- pH, cultivated, herbal, toxic. Origin: Eurasia.

Orlaya Hoffm. spp.
Apiaceae
- orlaya
- Weed
- 272

Orlaya daucoides (L.) Greuter
Apiaceae
Caucalis daucoides L. (see)
- Weed
- 44, 243

Orlaya grandiflora (L.) Hoff.
Apiaceae
Caucalis grandiflora L., *Daucus grandiflora* (L.) Scop. (*non* Desf.), *Selinum grandiflorum* Krause
- orlaya
- Weed
- 44, 70, 88, 94, 243, 272
- cultivated, herbal.

Orlaya kochii Heywood
Apiaceae
- Weed
- 87, 88, 94
- herbal.

Orlaya platycarpos (L.) Koch
Apiaceae
Caucalis polycarpos auct. non L.
- orlaya
- Weed
- 87, 88, 111, 243

Ormenis mixta Dum.
Asteraceae
= *Chamaemelum mixtum* (L.) All.
- Weed
- 87, 88
- cultivated.

Ormenis praecox (Link) Briq. & Cavill.
Asteraceae
- Weed
- 87, 88

Ormocarpum trichocarpum (Taub.) Engl.
Fabaceae/Papilionaceae
Ormocarpum setosum Burtt Davy
- caterpillar bush
- Weed, Quarantine Weed, Native Weed
- 10, 76, 87, 88, 121, 203, 220
- S/T, arid. Origin: southern Africa.

Ornithogalum L. spp.
Liliaceae/Hyacinthaceae
- ornithogalum, chincherinchee, star of Bethlehem, wonder flower
- Weed, Naturalised
- 198, 272
- toxic.

Ornithogalum angustifolium Bor.
Liliaceae/Hyacinthaceae
- star of Bethlehem
- Naturalised
- 40, 176

Ornithogalum arabicum L.
Liliaceae/Hyacinthaceae
- star of Bethlehem, lesser cape lily
- Weed, Naturalised, Garden Escape, Environmental Weed
- 7, 86, 98, 161, 198, 203
- cultivated, toxic. Origin: Mediterranean, Middle East.

Ornithogalum boucheanum (Kunth) Asch.
Liliaceae/Hyacinthaceae
Albucea chlorantha Reichb., *Ornithogalum undulatum* Bouché, *Ornithogalum chloranthum* Saut.
- bledavka boucheova
- Weed
- 272

Ornithogalum caudatum W.T.Aiton
Liliaceae/Hyacinthaceae
♦ false sea onion, star of Bethlehem,
sea onion
♦ Weed, Naturalised
♦ 80, 98, 101, 161, 203
♦ a/pH, cultivated, toxic.

Ornithogalum comosum L.
Liliaceae/Hyacinthaceae
♦ Weed
♦ 272
♦ cultivated.

Ornithogalum conicum Jacq.
Liliaceae/Hyacinthaceae
♦ chink, chinkerinchee
♦ Weed, Native Weed
♦ 39, 121
♦ pH, toxic. Origin: southern Africa.

Ornithogalum longibracteatum Jacq.
Liliaceae/Hyacinthaceae
♦ false sea onion, pregnant onion, wild
onion, sea onion
♦ Weed, Naturalised, Native Weed
♦ 7, 54, 86, 88, 121, 198
♦ pH, cultivated, herbal. Origin:
southern Africa.

Ornithogalum narbonense L.
Liliaceae/Hyacinthaceae
♦ star of Bethlehem
♦ Weed
♦ 70, 87, 88, 126, 253, 272
♦ pH, cultivated, herbal.

Ornithogalum nutans L.
Liliaceae/Hyacinthaceae
Albucea nutans Reich., *Myogalum
nutans* Link
♦ nuokkutähdikki, drooping star
of Bethlehem, bledavka ovisnutá,
nodding star of Bethlehem
♦ Weed, Naturalised, Cultivation
Escape
♦ 30, 39, 42, 70, 80, 87, 88, 101, 126, 161
♦ cultivated, herbal, toxic.

**Ornithogalum ornithogaloides (Kunth)
Oberm.**
Liliaceae/Hyacinthaceae
♦ grass chinkerinchee, vlei
chinckerinchee
♦ Weed, Native Weed
♦ 39, 121
♦ pH, toxic. Origin: southern Africa.

Ornithogalum orthophyllum Ten.
Liliaceae/Hyacinthaceae
♦ Weed
♦ 272
♦ cultivated.

Ornithogalum prasinum Lindl.
Liliaceae/Hyacinthaceae
♦ Weed, Native Weed
♦ 39, 121
♦ pH, cultivated, toxic. Origin:
southern Africa.

Ornithogalum pyramidale L.
Liliaceae/Hyacinthaceae
♦ Weed, Naturalised
♦ 86, 87, 88, 98, 203, 272
♦ cultivated. Origin: Europe.

Ornithogalum pyrenaicum L.
Liliaceae/Hyacinthaceae
♦ spiked star of Bethlehem, Pyrenees
star of Bethlehem, bath asparagus,
latte di gallina a fiori giallastri
♦ Weed, Naturalised
♦ 86, 87, 88, 98, 101, 203, 272
♦ pH, cultivated, herbal. Origin:
southern Europe.

Ornithogalum refractum Kit.
Liliaceae/Hyacinthaceae
♦ Weed
♦ 272
♦ cultivated.

Ornithogalum saundersiae Bak.
Liliaceae/Hyacinthaceae
♦ chincherinchee, giant chinkerinchee,
Transvaal chinkerinchee
♦ Weed, Native Weed
♦ 39, 121
♦ pH, cultivated, toxic. Origin:
southern Africa.

Ornithogalum sibthorpii Greut.
Liliaceae/Hyacinthaceae
Ornithogalum nanum Sibth. & Sm.
♦ Weed
♦ 272

Ornithogalum sphaerocarpum Kern.
Liliaceae/Hyacinthaceae
♦ Weed
♦ 272

**Ornithogalum tenuifolium Guss. nom.
illeg.**
Liliaceae/Hyacinthaceae
= *Ornithogalum gussonei* Ten. (NoR)
♦ Naturalised
♦ 287
♦ cultivated.

Ornithogalum thyrsoides Jacq.
Liliaceae/Hyacinthaceae
♦ African wonder flower,
chincherinchee, chinkerinchee,
common chinkerinchee, star of
Bethlehem, wonder flower, cape lily,
black eyed Susan
♦ Weed, Naturalised, Native Weed,
Garden Thug
♦ 7, 39, 86, 98, 121, 126, 161, 203, 247
♦ pH, cultivated, toxic. Origin:
southern Africa.

Ornithogalum trichophyllum Bak.
Liliaceae/Hyacinthaceae
♦ Weed
♦ 221

Ornithogalum umbellatum L.
Liliaceae/Hyacinthaceae
Scilla campestris Savi, *Stellaris corymbosa*
Moench
♦ sarjatähdikki, star of Bethlehem,
summer snowflake, starflower, sleepy
dick, dove's dung, pigeon dung,
bird's milk, cape lily, common star of
Bethlehem, snowdrops, nap at noon
♦ Weed, Quarantine Weed,
Naturalised, Garden Escape,
Environmental Weed, Cultivation
Escape

♦ 30, 39, 42, 70, 76, 80, 86, 87, 88, 98,
101, 102, 126, 133, 154, 161, 176, 195,
198, 203, 207, 211, 218, 224, 247, 249,
253, 272, 280, 287
♦ pH, cultivated, herbal, toxic. Origin:
Europe, Mediterranean.

Ornithoglossum viride (L.f.) Aiton
Liliaceae/Colchicaceae
Ornithoglossum glaucum Salisb.
♦ cape poison onion, cape slangkop,
yellow slangkop
♦ Weed, Native Weed
♦ 39, 121
♦ pH, cultivated, toxic. Origin:
southern Africa.

Ornithopus compressus L.
Fabaceae/Papilionaceae
♦ yellow serradella, keltalinnunjalka
♦ Weed, Naturalised, Environmental
Weed, Casual Alien
♦ 7, 9, 40, 42, 70, 86, 87, 88, 94, 98, 101,
198, 203, 241, 250, 253, 272, 287, 300
♦ cultivated. Origin: Mediterranean,
Middle East.

Ornithopus perpusillus L.
Fabaceae/Papilionaceae
Ornithopus subumbellatus Gilib.
♦ kääpiölinnunjalka, little white bird's
foot, wild serradella, bird's foot
♦ Weed, Naturalised, Environmental
Weed, Casual Alien
♦ 42, 86, 88, 94, 101, 165, 198, 253, 280,
287
♦ Origin: Europe, western Russia.

Ornithopus pinnatus (Mill.) Druce
Fabaceae/Papilionaceae
♦ yellow serradella, orange bird's foot,
sand bird's foot, slender serradella
♦ Weed, Naturalised, Environmental
Weed
♦ 7, 9, 15, 70, 86, 87, 88, 94, 98, 101, 176,
198, 203, 250, 280
♦ a/pH. Origin: Europe,
Mediterranean.

Ornithopus sativus Brot.
Fabaceae/Papilionaceae
Coronilla serradella Krause
♦ bird's foot, ornithopus, common
bird's foot, rusolinnunjalka, French
serradella
♦ Weed, Naturalised, Introduced,
Casual Alien
♦ 7, 34, 39, 40, 42, 88, 94, 98, 101, 203,
228, 241, 280, 287, 300
♦ aH, arid, cultivated, toxic.

**Ornithopus sativus Brot. ssp.
isthmocarpus (Coss.) Dost.**
Fabaceae/Papilionaceae
♦ serradella, common bird's foot
♦ Weed, Naturalised
♦ 70, 101

Ornithopus sativus Brot. ssp. sativus
Fabaceae/Papilionaceae
♦ French serradella, common bird's
foot
♦ Naturalised, Environmental Weed
♦ 86, 101, 198
♦ Origin: south-west Europe.

Orobanche L. spp.
Orobanchaceae
♦ broomrape
♦ Weed, Quarantine Weed, Noxious Weed
♦ 39, 62, 67, 76, 88, 171, 172, 203, 220, 221, 240, 243, 258, 272
♦ H parasitic, toxic.

Orobanche aegyptiaca Pers.
Orobanchaceae
Kopsia aegyptiaca (Pers.) Caruel, *Phelypaea aegyptiaca* (Pers.) Walp. (see), *Phelypaea longiflora* C.A.Mey. (see)
♦ Egyptian broomrape
♦ Weed, Quarantine Weed
♦ 76, 87, 88, 115, 172, 185, 203, 221, 243, 272, 275
♦ aH parasitic, herbal.

Orobanche arenaria Borkh.
Orobanchaceae
Kopsia borkhausenii (Andrz. ex Besser) Caruel, *Phelypaea arenaria* (Borkh.) Walp., *Orobanche laevis* L. *p.p.*
♦ sand broomrape
♦ Weed
♦ 70
♦ H parasitic.

Orobanche australiana F.Muell.
Orobanchaceae
♦ Weed
♦ 87, 88
♦ H parasitic.

Orobanche boninsimae (Maxim.) Tuyama
Orobanchaceae
♦ Weed
♦ 87, 88
♦ H parasitic.

Orobanche brassicae Novopokr.
Orobanchaceae
= *Orobanche ramosa* L. ssp. *mutelii* (F.W.Schultz) Cout.
♦ Weed
♦ 87, 88
♦ H parasitic.

Orobanche bulbosa (Gray) G.Beck
Orobanchaceae
♦ chaparral broomrape
♦ Weed
♦ 67
♦ pH parasitic, herbal.

Orobanche calendulae Pom.
Orobanchaceae
Orobanche mauretanica Beck
♦ Weed
♦ 70
♦ H parasitic.

Orobanche californica Cham. & Schlecht.
Orobanchaceae
♦ California broomrape
♦ Weed
♦ 67
♦ pH parasitic, promoted, herbal.

Orobanche cernua Loefl.
Orobanchaceae
Orobanche bicolor C.A.Mey., *Orobanche cumana* Loefl. (see), *Orobanche sarmatica* Kotov
♦ nodding broomrape, drooping broomrape

♦ Weed, Quarantine Weed, Noxious Weed
♦ 23, 87, 88, 115, 140, 172, 221, 229, 243, 272
♦ aH parasitic, cultivated. Origin: Africa, Eurasia.

Orobanche coerulescens Stephan
Orobanchaceae
♦ hamautsubo
♦ Weed
♦ 275, 297
♦ H parasitic.

Orobanche cooperi (Gray) Heller
Orobanchaceae
♦ Cooper's broomrape, desert broomrape
♦ Weed, Noxious Weed
♦ 67, 87, 88, 229
♦ pH parasitic, herbal.

Orobanche cooperi (Gray) Heller ssp. cooperi
Orobanchaceae
♦ Cooper's broomrape, desert broomrape
♦ Noxious Weed
♦ 229
♦ H parasitic.

Orobanche cooperi (Gray) Heller ssp. latiloba (Munz) Collins comb. nov. ined.
Orobanchaceae
♦ Cooper's broomrape, desert broomrape
♦ Noxious Weed
♦ 229
♦ H parasitic.

Orobanche corymbosa (Rydb.) Ferris
Orobanchaceae
♦ flattop broomrape
♦ Weed
♦ 67
♦ aH parasitic.

Orobanche crenata Forssk.
Orobanchaceae
Orobanche speciosa DC. (see)
♦ bean broomrape, scalloped broomrape
♦ Weed, Quarantine Weed, Noxious Weed
♦ 70, 76, 87, 88, 115, 140, 172, 185, 203, 221, 229, 243, 253
♦ aH parasitic, cultivated, herbal.

Orobanche cumana Wallr.
Orobanchaceae
= *Orobanche cernua* Loefl.
♦ broomrape
♦ Weed
♦ 87, 88, 161, 275, 297
♦ aH parasitic.

Orobanche dugesii (S.Watson) Munz
Orobanchaceae
♦ Duges' broomrape
♦ Weed
♦ 67
♦ H parasitic.

Orobanche elatior Sut.
Orobanchaceae
♦ knapweed broomrape, tall broomrape

♦ Weed, Quarantine Weed
♦ 70, 76, 88, 203, 272
♦ H parasitic, herbal.

Orobanche fasciculata Nutt.
Orobanchaceae
♦ clustered broomrape, cancer root
♦ Weed
♦ 67
♦ pH parasitic, cultivated, herbal.

Orobanche foetida Poir.
Orobanchaceae
♦ fetid broomrape
♦ Weed
♦ 70
♦ H parasitic.

Orobanche gracilis Sm.
Orobanchaceae
Orobanche grandiuscula Moris, *Orobanche spruneri* F.W.Schultz, *Orobanche cruenta* Bertol.
♦ záraza útla
♦ Weed
♦ 44, 70, 87, 88, 272
♦ a/pH parasitic, herbal.

Orobanche hederae Duby
Orobanchaceae
Orobanche salisii Req. ex Coss., *Orobanche laurina* Rchb.f. *non* Bertol., *Orobanche balearica* Sennen & Pau, *Orobanche glaberrima* Guss. ex Reut.
♦ ivy broomrape
♦ Weed
♦ 23, 70, 87, 88
♦ H parasitic, cultivated, herbal.

Orobanche indica Buch.-Ham.
Orobanchaceae
♦ Weed
♦ 87, 88
♦ H parasitic.

Orobanche loricata Recihb.
Orobanchaceae
Orobanche ambigua Moris, *Orobanche artemisiae-campestris* Gaudin, *Orobanche centaurina* Bertol., *Orobanche santolinae* Loscos, *Orobanche picridis* F.W.Schultz ex W.D.J.Koch (see)
♦ oxtongue broomrape, picris broomrape
♦ Weed, Naturalised
♦ 86, 87, 88
♦ H parasitic. Origin: Europe.

Orobanche ludoviciana Nutt.
Orobanchaceae
♦ Louisiana broomrape, Cooper's broomrape, broomrape
♦ Weed, Noxious Weed
♦ 23, 35, 67, 87, 88, 161, 199, 218
♦ H parasitic, promoted, herbal.

Orobanche lutea Baumg.
Orobanchaceae
Orobanche concreta (Beck) Rouy, *Orobanche fragrantissima* Bertol., *Orobanche hians* Steven, *Orobanche rubens* Wallr. (see)
♦ yellow broomrape
♦ Weed, Quarantine Weed
♦ 44, 70, 76, 87, 88, 203, 272
♦ b/pH parasitic, herbal.

Orobanche minor Sm.
Orobanchaceae
Orobanche euglossa Rchb.f., *Orobanche hyalina* Spruner ex Reut., *Orobanche crithmi* Bertol., *Orobanche laurina* Bertol., *Orobanche livida* Sendtn. ex Freyn, *Orobanche pyrrha* Rchb.f., *Orobanche concolor* Duby, *Orobanche yuccae* Pa.Savi ex Bertol., *Orobanche unicolor* Boreau
♦ lesser broomrape, clover broomrape, klawerbesemraap, bremraap, hemp broomrape, branched broomrape, hellroot, small broomrape
♦ Weed, Noxious Weed, Naturalised, Environmental Weed
♦ 7, 9, 15, 23, 39, 44, 51, 63, 70, 72, 80, 86, 87, 88, 98, 101, 121, 134, 140, 161, 165, 176, 198, 203, 217, 218, 229, 241, 269, 272, 278, 280, 283, 286, 287, 300
♦ aH parasitic, cultivated, herbal, toxic. Origin: Eurasia.

Orobanche minor Sm. **var.** *flava* Regel
Orobanchaceae
♦ Naturalised
♦ 287
♦ H parasitic.

Orobanche multicaulis Brandeg.
Orobanchaceae
♦ spiked broomrape
♦ Weed
♦ 67
♦ H parasitic.

Orobanche mutelii F.Schultz
Orobanchaceae
= *Orobanche ramosa* L. ssp. *mutelii* (F.W.Schultz) Cout.
♦ Weed
♦ 87, 88
♦ H parasitic.

Orobanche nicotianae Wight
Orobanchaceae
♦ Weed
♦ 87, 88
♦ H parasitic.

Orobanche pallidiflora Wimm. & Grab.
Orobanchaceae
= *Orobanche reticulata* Wallr. (NoR)
♦ Weed
♦ 87, 88
♦ H parasitic.

Orobanche parishii (Jeps.) Heckard
Orobanchaceae
♦ Parish's broomrape
♦ Weed
♦ 67
♦ pH parasitic, herbal.

Orobanche picridis F.W.Schultz ex W.D.J.Koch
Orobanchaceae
= *Orobanche loricata* Rchb.
♦ picris broomrape
♦ Weed
♦ 87, 88
♦ H parasitic.

Orobanche pinorum Geyer ex Hook.
Orobanchaceae

♦ conifer broomrape, forest broomrape
♦ Weed
♦ 67
♦ pH parasitic, promoted, herbal.

Orobanche pycnostachya Hance
Orobanchaceae
♦ yellowflower broomrape
♦ Weed
♦ 297
♦ H parasitic.

Orobanche ramosa L.
Orobanchaceae
Kopsia ramosa (L.) Dumort. (see), *Phelypaea ramosa* (L.) C.A.Mey. (see)
♦ branched broomrape, hemp broomrape, yellow broomrape, orobanche, clover broomrape, blue broomrape
♦ Weed, Quarantine Weed, Noxious Weed, Naturalised, Environmental Weed
♦ 14, 23, 26, 34, 35, 44, 51, 62, 70, 76, 86, 87, 88, 101, 115, 121, 140, 155, 161, 180, 185, 203, 217, 218, 229, 242, 243, 253, 275, 289, 300
♦ aH parasitic, arid, cultivated, herbal. Origin: Eurasia.

Orobanche ramosa L. **ssp.** *mutelii* (F.W.Schultz) Cout.
Orobanchaceae
Orobanche brassicae Novopokr. (see), *Orobanche mutelii* F.Schultz. (see), *Phelypaea ramosa* (L.) Mey.
♦ Weed
♦ 272
♦ H parasitic.

Orobanche ramosa L. **ssp.** *ramosa*
Orobanchaceae
♦ hemp broomrape
♦ Weed
♦ 272
♦ H parasitic.

Orobanche rapum-genistae Thu.
Orobanchaceae
Orobanche bracteata Viv., *Orobanche major* L. *p.p.*, *Orobanche rapum* auct., *Orobanche benthamii* Timb.-Lagr.
♦ greater broomrape
♦ Weed
♦ 70, 87, 88
♦ H parasitic, herbal.

Orobanche riparia Collins sp. nov. ined.
Orobanchaceae
♦ river broomrape
♦ Noxious Weed
♦ 140, 229
♦ H parasitic.

Orobanche rubens Wallr.
Orobanchaceae
= *Orobanche lutea* Baumg.
♦ Weed
♦ 87, 88
♦ H parasitic.

Orobanche speciosa DC.
Orobanchaceae
= *Orobanche crenata* Forssk.

♦ Weed
♦ 23, 87, 88
♦ H parasitic.

Orobanche uniflorum L.
Orobanchaceae
♦ Weed
♦ 67
♦ H parasitic.

Orobanche valida Jeps.
Orobanchaceae
♦ Rock Creek broomrape
♦ Weed
♦ 67
♦ pH parasitic.

Orobanche vallicola (Jeps.) Heckard
Orobanchaceae
♦ hillside broomrape
♦ Weed
♦ 67
♦ aH parasitic.

Orontium aquaticum L.
Araceae
♦ goldenclub
♦ Weed, Quarantine Weed
♦ 23, 87, 88, 161, 218, 247, 258
♦ pH, aqua, cultivated, herbal, toxic.

Oroxylum indicum (L.) Vent.
Bignoniaceae
Bignonia indica L., *Calosanthes indica* Blume
♦ midnight horror
♦ Introduced
♦ 228
♦ arid, cultivated, herbal.

Orthocarpus barbatus Cotton
Scrophulariaceae
♦ cotton, Grand Coulee owl's clover
♦ Weed
♦ 23, 88

Orthocarpus purpurascens Benth.
Scrophulariaceae
= *Castilleja exserta* (Heller) Chuang & Heckard ssp. *exserta*
♦ owl's clover, escobita, common owl's clover
♦ Weed
♦ 180, 243
♦ cultivated, herbal. Origin: Australia.

Orthocarpus tenuifolius (Pursh) Benth.
Scrophulariaceae
♦ thin leaved owl's clover
♦ Weed
♦ 23, 88
♦ herbal.

Orthoclada laxa (Rich.) Beauv.
Poaceae
♦ Weed
♦ 87, 88
♦ G.

Orthodon dianthera (Buch.-Ham. ex Roxb.) Hand.-Mazz.
Lamiaceae
= *Mosla dianthera* (Ham.) Maxim.
♦ Weed
♦ 80

Orthodontium lineare Schwäegr.
Bryaceae
♦ cape thread moss
♦ Environmental Weed
♦ 152
♦ moss. Origin: Southern Hemisphere.

Orthopappus angustifolius (Sw.) Gleason
Asteraceae
= *Elephantopus angustifolius* Sw. (NoR)
♦ lingua de vaca
♦ Weed
♦ 88, 255
♦ Origin: South America.

Orthoraphium coreanum Ohwi
Poaceae
Stipa coreana Honda ex Nakai
♦ Naturalised
♦ 287
♦ G.

Orthosiphon aristatus (Blume) Miq.
Lamiaceae
♦ alis en kaht
♦ Weed, Introduced
♦ 13, 230
♦ H, cultivated, herbal. Origin: Asia, Australia.

Orthosiphon pallidus Royle ex Benth.
Lamiaceae
♦ Weed
♦ 87, 88

Orthrosanthus chimboracensis (Kunth) Bak.
Iridaceae
Moraea chimboracensis Kunth
♦ chimborazo dawn flower
♦ Weed
♦ 87, 88
♦ cultivated.

Orychophragmus violaceus (L.) Schultz.
Brassicaceae
♦ Weed, Naturalised
♦ 87, 88, 287, 297
♦ a/bH, cultivated.

Orychophragmus violaceus O.E.Schulz var. lasiocarpus Migo
Brassicaceae
♦ Naturalised
♦ 287

Oryctanthus occidentalis (L.) Eichler
Loranthaceae
♦ Weed
♦ 87, 88
♦ parasitic.

Oryza alta Swallen
Poaceae
♦ Weed
♦ 87, 88
♦ G.

Oryza barthii A.Chev.
Poaceae
Oryza breviligulata A.Chev. & Roehr., *Oryza longistaminata* A.Chev. & Roehr. (see)
♦ wild rice, red rice, Barth's rice
♦ Weed, Quarantine Weed
♦ 76, 87, 88, 91, 203, 217, 220
♦ aG, aqua, herbal.

Oryza latifolia Desv.
Poaceae
♦ broadleaf rice
♦ Weed
♦ 157
♦ pG, herbal.

Oryza longistaminata A.Chev. & Roehr.
Poaceae
= *Oryza barthii* A.Chev.
♦ red rice, longstamen rice
♦ Weed, Quarantine Weed, Noxious Weed
♦ 67, 76, 88, 140, 203, 220, 229
♦ G. Origin: Madagascar.

Oryza minuta Presl
Poaceae
Oryza latifolia Back.
♦ Weed
♦ 87, 88
♦ G.

Oryza officinalis Wall. ex Watt.
Poaceae
♦ wild rice, red rice
♦ Weed
♦ 88, 217
♦ pG, aqua.

Oryza perennis Moench nom. dub.
Poaceae
♦ Weed
♦ 87, 88
♦ G.

Oryza punctata Kotschy ex Steud.
Poaceae
♦ wild rice, red rice
♦ Weed, Quarantine Weed, Noxious Weed
♦ 67, 76, 87, 88, 91, 140, 203, 217, 220, 229
♦ G, aqua, herbal. Origin: Madagascar.

Oryza ridleyi Hook.f.
Poaceae
♦ Weed
♦ 88, 191, 204
♦ G.

Oryza rufipogon Griff.
Poaceae
Oryza fatua J.König ex Trin. *nom. nudum*, *Oryza sativa* L. var. *fatua* Prain
♦ red rice, brownbeard rice, wild rice, wild red rice
♦ Weed, Quarantine Weed, Noxious Weed, Naturalised
♦ 32, 35, 67, 76, 80, 86, 87, 88, 91, 101, 140, 161, 170, 179, 203, 217, 229, 271
♦ pG, aqua.

Oryza sativa L.
Poaceae
Oryza glutinosa Lour., *Oryza nivara* S.D.Sharma & Shastry
♦ rice, wild rice, red rice, upland rice, domestic rice, paddy rice, jing mi, arroz rojo
♦ Weed, Naturalised, Introduced, Garden Escape, Environmental Weed, Cultivation Escape
♦ 7, 14, 23, 32, 39, 86, 87, 88, 93, 98, 101, 161, 197, 203, 217, 218, 230, 241, 245,

255, 261, 262, 263, 281, 295, 300
♦ aG, aqua, cultivated, herbal, toxic. Origin: Asia.

Oryza sativa L. var. savannae Körn
Poaceae
♦ Weed
♦ 237
♦ G.

Oryza sativa L. var. sundensis Körn.
Poaceae
♦ Weed
♦ 237
♦ G.

Oryzopsis coerulescens (Desf.) Hack.
Poaceae
♦ Mediterranean ricegrass
♦ Weed
♦ 80
♦ G.

Oryzopsis hymenoides (Roem. & Schult.) Ricker.
Poaceae
= *Achnatherum hymenoides* (Roem. & Schult.) Barkworth (NoR)
♦ Indian ricegrass, Indian millet
♦ Weed
♦ 88, 218
♦ pG, arid, cultivated, herbal.

Oryzopsis miliacea (L.) Benth. & Hook.f. ex Asch. & C.Schweinf.
Poaceae
= *Piptatherum miliaceum* (L.) Coss.
♦ smilo grass, rice millet, bamboo grass, smilo ricegrass, mountain rice, San Diego grass
♦ Weed, Naturalised
♦ 87, 88, 91, 121, 180, 221, 241, 243, 287
♦ pG, herbal. Origin: Eurasia.

Oryzopsis nymenoides (Roem. & Schult.) Ricker
Poaceae
♦ Weed
♦ 87, 88
♦ G.

Oscularia Schwantes spp.
Aizoaceae/Mesembryanthemaceae
♦ iceplant
♦ Weed
♦ 116

Oscularia caulescens (Mill.) Schwantes
Aizoaceae/Mesembryanthemaceae
= *Lampranthus deltoides* (L.) Glen (NoR)
♦ Casual Alien
♦ 280

Oscularia deltoides Schwantes
Aizoaceae/Mesembryanthemaceae
♦ deltoid leaved dewplant
♦ Naturalised
♦ 40
♦ cultivated.

Osmorhiza aristata (Thunb.) Rydb.
Apiaceae
♦ yabuninjin
♦ Weed
♦ 286
♦ pH, promoted. Origin: China, Japan.

Osmunda cinnamomea L.
Osmundaceae
♦ cinnamon fern
♦ Weed
♦ 87, 88, 218
♦ H, cultivated, herbal.

Osmunda japonica Thunb.
Osmundaceae
♦ flowering fern, royal fern
♦ Weed
♦ 87, 88, 286
♦ H, cultivated.

Osmunda regalis L.
Osmundaceae
♦ royal fern
♦ Weed, Quarantine Weed,
Environmental Weed
♦ 70, 225, 246, 272
♦ H, cultivated, herbal.

Ossaea DC. spp.
Melastomataceae
♦ Weed
♦ 18, 88

Ossaea marginata (Desr.) Triana
Melastomataceae
♦ Weed, Environmental Weed
♦ 3, 22, 152, 191
♦ T. Origin: Brazil.

Osteospermum barberae (Harv.) Norl.
Asteraceae
♦ Naturalised
♦ 198
♦ cultivated.

Osteospermum calendulaceum L.f.
Asteraceae
♦ stinking Roger
♦ Weed, Naturalised, Native Weed,
Environmental Weed
♦ 7, 86, 98, 101, 121, 203
♦ aH, toxic. Origin: southern Africa.

Osteospermum caulescens Harv.
Asteraceae
♦ Introduced
♦ 228
♦ arid.

Osteospermum clandestinum (Less.) Norl.
Asteraceae
♦ daisy, tripteris, osteospermum
♦ Weed, Naturalised, Native Weed,
Environmental Weed
♦ 7, 9, 72, 86, 88, 98, 121, 158, 198, 203
♦ aH, cultivated, toxic. Origin:
southern Africa.

Osteospermum ecklonis (DC.) Norl.
Asteraceae
Dimorphotheca ecklonis DC. (see)
♦ blue daisybush, white daisybush,
African daisy, marigold, Sunday's
river daisy, Van Staden's daisy,
Van Staden osteospermum, white
daisybush
♦ Weed, Naturalised, Native Weed,
Cultivation Escape
♦ 34, 39, 86, 101, 121, 252
♦ pS, arid, cultivated, toxic. Origin:
southern Africa.

Osteospermum fruticosum (L.) Norl.
Asteraceae

♦ shrubby daisybush, trailing African
daisy, freeway daisy, dimorphotheca
♦ Weed, Naturalised, Cultivation
Escape
♦ 54, 86, 88, 98, 101, 116, 161, 165, 203,
252, 280
♦ cultivated, toxic. Origin: southern
Africa.

Osteospermum jucundum (Phill.) T.Norl.
Asteraceae
♦ dimorphotheca
♦ Weed, Naturalised
♦ 15, 39, 280
♦ cultivated, toxic.

Osteospermum moniliferum L.
Asteraceae
= *Chrysanthemoides monilifera* (L.) Norl.
♦ Quarantine Weed
♦ 220
♦ S.

Osteospermum muricatum E.Mey. ex DC.
Asteraceae
♦ Weed, Naturalised, Introduced
♦ 86, 93, 191, 205, 228
♦ pH, arid. Origin: Africa.

**Osteospermum muricatum E.Mey. ex DC.
ssp. *muricatum***
Asteraceae
♦ Native Weed
♦ 121
♦ pH. Origin: southern Africa.

Osteospermum sinuatum (DC.) Norl.
Asteraceae
♦ glandular cape marigold
♦ Introduced
♦ 228
♦ arid, cultivated.

Osteospermum vaillantii (Decne.) Norl.
Asteraceae
♦ Weed
♦ 88, 221
♦ arid.

Ostrya virginiana (Mill.) C.Koch
Betulaceae/Corylaceae/Carpinaceae
♦ eastern hophornbeam, ironwood,
hophornbeam
♦ Weed
♦ 87, 88, 218
♦ T, cultivated, herbal.

Otholobium decumbens (Aiton) C.H.Stirt.
Fabaceae/Papilionaceae
Psoralea decumbens Ait. (see)
♦ Weed, Quarantine Weed
♦ 76, 88, 220

Othonna capensis L.H.Bailey
Asteraceae
Othonna crassifolia Harv.
♦ little pickles
♦ Naturalised
♦ 280
♦ cultivated.

Othonnopsis intermedia Boiss.
Asteraceae
= *Othonna media* C.Jeffrey (NoR) [see
Hertia intermedia Kuntze]
♦ Weed
♦ 87, 88

Otospermum glabrum (Lag.) Willk.
Asteraceae
♦ Weed
♦ 87, 88

Otostegia fruticosa (Forssk.) Briq.
Lamiaceae
♦ Weed
♦ 221

Ottelia Pers. spp.
Hydrocharitaceae
♦ ottelia
♦ Weed, Quarantine Weed,
Naturalised
♦ 76, 86, 88, 220
♦ wH, cultivated.

Ottelia alismoides (L.) Pers.
Hydrocharitaceae
Damasonium indicum Willd., *Ottelia
condorensis* Gagnep., *Ottelia dioecia*
Yan, *Ottelia japonica* Miq. (see), *Ottelia
javanica* Miq., *Stratiotes alismoides* L.
♦ waterplantain, duck lettuce,
santawaa, waterplantain ottelia
♦ Weed, Quarantine Weed, Noxious
Weed, Naturalised, Environmental
Weed
♦ 87, 88, 101, 126, 140, 161, 170, 191,
197, 204, 209, 221, 229, 239, 246, 262
♦ wpH, cultivated, herbal. Origin:
tropical Africa, Asia, Australia.

Ottelia japonica Miq.
Hydrocharitaceae
= *Ottelia alismoides* (L.) Pers.
♦ Weed
♦ 286
♦ wpH.

Ottelia ovalifolia (R.Br.) Rich.
Hydrocharitaceae
♦ swamplily
♦ Weed, Sleeper Weed, Naturalised,
Environmental Weed
♦ 15, 165, 208, 225, 246, 280
♦ wpH, cultivated. Origin: Australia.

Ottelia ulvaefolia Planch.
Hydrocharitaceae
♦ Weed
♦ 87, 88
♦ wH.

Ottochloa nodosa (Kunth) Dandy
Poaceae
Digitaria divulsa Mez, *Digitaria
urochloides* Büse, *Hemigymnia multinodis*
Stapf, *Ichnanthus oblongus* Hughes,
Ottochloa arnottiana (Nees ex Steud.)
Dandy, *Panicum aequabile* Domin,
Panicum arnottianum Nees ex Steud.,
Panicum multinode J.Presl, *Panicum
nodosum* Kunth, *Panicum urochloides*
(Büse) Boerl.
♦ slender panicgrass, short glumed
panic
♦ Weed
♦ 12, 28, 87, 88, 90, 243
♦ pG. Origin: Asia, Australia.

Oxalis L. spp.
Oxalidaceae
♦ woodsorrel, oxalis, sorrel, shamrock,
lucky clover, good luck plant, trébol

♦ Quarantine Weed, Naturalised, Garden Escape, Environmental Weed
♦ 76, 86, 88, 154, 155, 220, 247, 281
♦ pH, cultivated, herbal, toxic.

Oxalis acetosella L.
Oxalidaceae
Oxys acetosella Scop., *Oxalis alba* Gilib.
♦ woodsorrel, shamrock plant, acetosella, käenkaali, ketunleipä
♦ Weed
♦ 39, 70, 87, 88, 154, 272
♦ pH, cultivated, herbal, toxic.

Oxalis adenophylla Gillies ex Hook. & Arn.
Oxalidaceae
♦ sauer klee
♦ Quarantine Weed
♦ 220
♦ cultivated.

Oxalis albicans Kunth
Oxalidaceae
♦ radishroot woodsorrel, white oxalis
♦ Weed
♦ 199
♦ pH, herbal.

Oxalis anthelmintica A.Rich.
Oxalidaceae
♦ Weed
♦ 87, 88

Oxalis articulata Savigny
Oxalidaceae
♦ pink oxalis, sour grass, woodsorrel, jointed woodsorrel, shamrock oxalis, pink sorrel, rubra woodsorrel
♦ Weed, Noxious Weed, Naturalised, Introduced, Garden Escape, Environmental Weed
♦ 15, 40, 72, 86, 87, 88, 98, 161, 165, 176, 198, 203, 215, 237, 269, 280, 286, 287, 295
♦ pH, cultivated. Origin: South America.

Oxalis articulata Savigny var. *articulata*
Oxalidaceae
♦ Naturalised
♦ 280

Oxalis articulata Savigny var. *hirsuta* Progel
Oxalidaceae
♦ Naturalised
♦ 280

Oxalis bahiensis Progel
Oxalidaceae
♦ Weed
♦ 87, 88

Oxalis barrelieri L.
Oxalidaceae
Acetosella amazonica (Progel) Kuntze, *Acetosella barrelieri* (L.) Kuntze, *Lotoxalis barrelieri* (L.) Small, *Oxalis amazonica* Progel
♦ Barrelier's woodsorrel
♦ Weed, Introduced
♦ 28, 87, 88, 170, 206, 230, 243, 261
♦ pH, promoted, herbal.

Oxalis bifurca G.Lodd.
Oxalidaceae
♦ Weed, Naturalised, Cultivation Escape
♦ 86, 98, 203, 252, 269
♦ cultivated. Origin: South Africa.

Oxalis bowieana G.Lodd.
Oxalidaceae
♦ Weed, Naturalised
♦ 286, 287
♦ Origin: South Africa.

Oxalis bowiei Herb. ex Lindl.
Oxalidaceae
♦ David Bowie woodsorrel, Bowie woodsorrel
♦ Weed, Naturalised, Garden Escape, Environmental Weed, Casual Alien
♦ 7, 72, 86, 87, 88, 98, 101, 198, 203, 269, 280
♦ pH, cultivated.

Oxalis brasiliensis Lodd.
Oxalidaceae
♦ Weed, Naturalised, Cultivation Escape
♦ 86, 98, 203, 252, 269, 287
♦ cultivated. Origin: South America.

Oxalis caprina Thunb.
Oxalidaceae
♦ oxalis
♦ Weed, Naturalised, Garden Escape, Environmental Weed
♦ 7, 86, 98, 203
♦ cultivated. Origin: South Africa.

Oxalis cernua Thunb.
Oxalidaceae
= *Oxalis pes-caprae* L.
♦ Weed, Naturalised
♦ 87, 88, 241

Oxalis chrysantha Progel
Oxalidaceae
= *Oxalis conorrhiza* Jacq.
♦ vinagrillo
♦ Weed
♦ 295

Oxalis compressa L.f.
Oxalidaceae
♦ Weed, Quarantine Weed, Naturalised, Cultivation Escape
♦ 76, 86, 87, 88, 98, 203, 252
♦ cultivated. Origin: South Africa.

Oxalis conorrhiza Jacq.
Oxalidaceae
Acetosella caespitosa (A.St.-Hil.) Kuntze, *Acetosella chrysantha* (Progel) Kuntze, *Acetosella cineraceae* (A.St.-Hil.) Kuntze, *Acetosella commersonii* (Pers.) Kuntze, *Acetosella conorrhiza* (Jacq.) Kuntze, *Acetosella megapotamica* (Spreng.) Kuntze, *Oxalis andicola* Gillies ex Hook. & Arn., *Oxalis andicola* Gillies ex Hook. & Arn. var. *wallichiana* Stuck., *Oxalis brevipes* Fredr., *Oxalis caespitosa* A.St.-Hil., *Oxalis chrysantha* Progel (see), *Oxalis chrysantha* Progel var. *pusilla* Progel, *Oxalis cineracea* A.St.-Hil., *Oxalis commersonii* Pers., *Oxalis conorhixa* Larrañaga, *Oxalis cordobensis* Kunth (see), *Oxalis cordobensis* Kunth var. *humilior* R.Knuth, *Oxalis corniculata* var. *serpens* R.Knuth, *Oxalis hassleriana* Chodat, *Oxalis linneaformis* R.Knuth,

Oxalis megapotamica Spreng., *Oxalis repens* Thunb. fo. *uniflora* Hieron. & Lorentz ex R.Knuth, *Oxalis sexenata* Savigny, *Xanthoxalis chrysantha* (Progel) Holub, *Xanthoxalis cordobensis* (R.Knuth) Holub
♦ Weed
♦ 237

Oxalis cordobensis Kunth
Oxalidaceae
= *Oxalis conorrhiza* Jacq.
♦ Weed
♦ 87, 88

Oxalis corniculata L.
Oxalidaceae
Oxalis corniculata L. var. *atropurpurea* (Planch.) (see), *Oxalis repens* Thunb. (see), *Xanthoxalis corniculata* (L.) Small
♦ yellow woodsorrel, procumbent yellow sorrel, Indian sorrel, sheep sorrel, sour grass, creeping woodsorrel, creeping sorrel, vinagrillo, tarhakäenkaali, oxalis, woodsorrel, yellow oxalis, creeping lady's sorrel, creeping oxalis
♦ Weed, Noxious Weed, Naturalised, Introduced, Garden Escape, Casual Alien
♦ 6, 7, 8, 14, 15, 23, 32, 34, 39, 42, 44, 50, 51, 66, 70, 84, 87, 88, 93, 94, 98, 115, 121, 157, 158, 161, 165, 167, 170, 176, 180, 185, 186, 203, 204, 205, 206, 209, 212, 218, 221, 228, 235, 236, 237, 238, 241, 243, 253, 255, 257, 262, 263, 266, 269, 272, 273, 274, 275, 276, 280, 286, 292, 295, 297, 299, 300
♦ a/pH, arid, cultivated, herbal, toxic. Origin: obscure.

Oxalis corniculata L. ssp. *corniculata*
Oxalidaceae
♦ creeping woodsorrel, yellow woodsorrel, creeping oxalis
♦ Weed, Naturalised
♦ 86, 198, 261, 280

Oxalis corniculata L. var. *atropurpurea* Planch.
Oxalidaceae
= *Oxalis corniculata* L.
♦ Naturalised
♦ 40

Oxalis corymbosa DC.
Oxalidaceae
= *Oxalis debilis* Kunth var. *corymbosa* (DC.) Lourt.
♦ violet woodsorrel, azedinha, trevo azedo
♦ Weed, Naturalised
♦ 7, 87, 88, 94, 98, 203, 255, 256, 257, 263, 273, 275, 286, 287, 297
♦ pH, cultivated, herbal. Origin: tropical America.

Oxalis debilis Kunth
Oxalidaceae
♦ pink woodsorrel, pink shamrock, large flowered pink sorrel
♦ Weed, Naturalised, Casual Alien
♦ 40, 80, 87, 88, 101, 134, 165, 280
♦ arid, herbal. Origin: Central America.

Oxalis debilis Kunth var. corymbosa (DC.) Lourteig
Oxalidaceae
Acetosella martiana (Zucc.) Kunze,
Ionoxalis martiana (Zucc.) Small, *Oxalis bipunctata* Graham ex Hook., *Oxalis bulbillifera* Herter, *Oxalis corymbosa* DC. (see), *Oxalis grandifolia* DC., *Oxalis japonica* Franch. & Sav., *Oxalis macrophylla* Kunth, *Oxalis martiana* Zucc. (see), *Oxalis multibulbosa* Turcz., *Oxalis urbica* A.St.-Hil.
♦ pink shamrock, pink woodsorrel
♦ Weed, Naturalised, Garden Escape, Cultivation Escape
♦ 28, 86, 93, 101, 179, 198, 205, 206, 237, 243, 252, 261, 269, 295
♦ pH, cultivated. Origin: South America.

Oxalis dembeyi St.-Hil.
Oxalidaceae
♦ Weed
♦ 87, 88

Oxalis depressa Eckl. & Zeyh.
Oxalidaceae
♦ Weed, Naturalised, Native Weed, Cultivation Escape
♦ 86, 121, 252
♦ pH, cultivated, herbal, toxic. Origin: southern Africa.

Oxalis dillenii Jacq.
Oxalidaceae
= *Oxalis stricta* L.
♦ Dillen's oxalis
♦ Weed
♦ 44, 87, 88
♦ herbal.

Oxalis dillenii Jacq. ssp. dillenii
Oxalidaceae
♦ Dillen's oxalis
♦ Weed
♦ 292

Oxalis dillenii Jacq. ssp. filipes (Small) Eiten
Oxalidaceae
= *Oxalis stricta* L.
♦ Dillen's oxalis
♦ Weed
♦ 292

Oxalis eggersii Urb.
Oxalidaceae
= *Oxalis latifolia* Kunth
♦ Eggers' woodsorrel
♦ Weed
♦ 14

Oxalis europaea Jord.
Oxalidaceae
= *Oxalis stricta* L.
♦ European woodsorrel, upright yellow sorrel, yellow woodsorrel, lady's sorrel
♦ Weed
♦ 44, 87, 88, 94, 161, 218
♦ pH, promoted, herbal.

Oxalis exilis A.Cunn.
Oxalidaceae
♦ pikkukäenkaali, least yellow sorrel
♦ Casual Alien

♦ 40, 42
♦ cultivated. Origin: Australia.

Oxalis flava L.
Oxalidaceae
♦ yellow oxalis
♦ Weed, Naturalised, Garden Escape, Environmental Weed
♦ 7, 86, 87, 88, 98, 203, 269
♦ cultivated. Origin: South Africa.

Oxalis floribunda Lehm. ssp. floribunda
Oxalidaceae
♦ Naturalised
♦ 300

Oxalis florida Salisb.
Oxalidaceae
♦ yellow woodsorrel, Florida yellow woodsorrel
♦ Weed
♦ 84, 87, 88, 161, 218
♦ herbal.

Oxalis fontana Bunge
Oxalidaceae
♦ pystykäenkaali
♦ Weed, Naturalised, Casual Alien
♦ 42, 253, 280, 286
♦ cultivated, herbal.

Oxalis glabra Thunb.
Oxalidaceae
♦ fingerleaf oxalis
♦ Weed, Naturalised, Garden Escape, Environmental Weed
♦ 7, 9, 86, 98, 203
♦ cultivated. Origin: South Africa.

Oxalis hirta L.
Oxalidaceae
♦ tropical woodsorrel, hairy woodsorrel
♦ Weed, Naturalised, Cultivation Escape
♦ 7, 86, 98, 101, 198, 203, 252, 269, 280
♦ pH, cultivated. Origin: South Africa.

Oxalis incarnata L.
Oxalidaceae
♦ pale woodsorrel, pale pink sorrel, lilac oxalis, crimson woodsorrel
♦ Weed, Noxious Weed, Naturalised, Introduced, Garden Escape, Environmental Weed
♦ 7, 15, 34, 40, 72, 86, 88, 98, 101, 176, 198, 203, 269, 280
♦ pH, cultivated, herbal. Origin: South Africa.

Oxalis intermedia A.Rich.
Oxalidaceae
= *Oxalis latifolia* Kunth
♦ Cuban purple woodsorrel, West Indian woodsorrel
♦ Weed
♦ 161, 249

Oxalis lactea Hook.
Oxalidaceae
♦ Weed, Naturalised
♦ 98, 203, 269
♦ cultivated.

Oxalis latifolia Kunth
Oxalidaceae
Acetosella violacea (L.) Kuntze ssp.

latifolia (Kunth) Kuntze, *Acetosella violacea* var. *albida* Kuntze, *Acetosella violacea* var. *rosea* Kuntze, *Ionoxalis attenuata* Small, *Ionoxalis calcaria* Small, *Ionoxalis intermedia* (A.Rich.) Small, *Ionoxalis latifolia* (Kunth) Rose, *Ionoxalis stipitata* Rose, *Ionoxalis tenuiloba* Rose, *Ionoxalis vallicola* Rose, *Ionoxalis vespertilionis* (Zucc.) Rose, *Oxalis araucana* Reiche, *Oxalis atroglandulosa* R.Knuth, *Oxalis binervis* Regel, *Oxalis calcaria* (Small) R.Knuth, *Oxalis chiriquensis* Woodson, *Oxalis eggersii* Urb. (see), *Oxalis elegans* Knuth var. *karwinskii* Progel ex R.Knuth, *Oxalis intermedia* A.Rich. (see), *Oxalis lilacina* Klotzsch, *Oxalis mauritiana* Lodd., *Oxalis morelosensis* R.Knuth, *Oxalis multipes* R.Knuth, *Oxalis ramonensis* R.Knuth, *Oxalis schraderiana* Kunth, *Oxalis stipulata* (Rose) Rose ex R.Knuth, *Oxalis stylosa* Klotzsch ex R.Knuth, *Oxalis tenuiloba* (Rose) R.Knuth, *Oxalis vallicola* (Rose) R.Knuth (see), *Oxalis vespertilionis* Zucc.
♦ purple flowered oxalis, shamrock, red garden sorrel, fishtail oxalis, broadleaf woodsorrel, garden sorrel, kolmiokäenkaali, Mexican oxalis
♦ Weed, Quarantine Weed, Noxious Weed, Naturalised, Introduced, Garden Escape, Environmental Weed, Casual Alien
♦ 15, 23, 34, 39, 40, 42, 51, 53, 72, 76, 86, 87, 88, 94, 98, 121, 147, 158, 165, 176, 179, 198, 199, 203, 206, 217, 240, 243, 253, 255, 261, 269, 280
♦ pH, arid, cultivated, herbal, toxic. Origin: tropical South America.

Oxalis laxa Hook. & Arn.
Oxalidaceae
= *Oxalis radicosa* A.Rich.
♦ dwarf woodsorrel
♦ Introduced
♦ 34
♦ aH, arid.

Oxalis luteola Jacq.
Oxalidaceae
♦ pink sorrel
♦ Native Weed
♦ 121
♦ pH, toxic. Origin: southern Africa.

Oxalis mallobolba Cav.
Oxalidaceae
= *Oxalis perdicaria* (Molina) Bertero
♦ macachín
♦ Weed
♦ 87, 88, 295
♦ Origin: South America.

Oxalis martiana Zucc.
Oxalidaceae
= *Oxalis debilis* Kunth var. *corymbosa* (DC.) Lourt.
♦ pink woodsorrel
♦ Weed
♦ 87, 88, 204, 218, 237, 245, 274
♦ arid, herbal.

Oxalis megalorrhiza Jacq.
Oxalidaceae

♦ fleshy yellow sorrel
♦ Naturalised
♦ 40
♦ arid, cultivated.

Oxalis micrantha **Bertero ex Colla**
Oxalidaceae
= *Oxalis radicosa* A.Rich.
♦ Weed
♦ 87, 88
♦ arid.

Oxalis montana **Raf.**
Oxalidaceae
♦ mountain woodsorrel
♦ Weed
♦ 80
♦ pH, promoted, herbal.

Oxalis neaei **DC.**
Oxalidaceae
♦ Weed
♦ 87, 88, 157
♦ pH.

Oxalis obliquifolia **Steud. ex A.Rich**
Oxalidaceae
♦ Weed, Native Weed
♦ 87, 88, 121
♦ pH, herbal, toxic. Origin: southern Africa.

Oxalis obtusa **Jacq.**
Oxalidaceae
♦ primrose woodsorrel, suring
♦ Weed, Naturalised, Garden Escape, Environmental Weed
♦ 54, 72, 86, 88, 198
♦ Ph, cultivated. Origin: South Africa.

Oxalis perdicaria **(Molina) Bertero**
Oxalidaceae
Oxalis mallobolba Cav. (see)
♦ Chilean woodsorrel, Chilean oxalis
♦ Weed, Naturalised, Cultivation Escape
♦ 86, 98, 203, 252, 269
♦ cultivated. Origin: South America.

Oxalis perennans **Haw.**
Oxalidaceae
♦ woody root oxalis
♦ Weed, Naturalised
♦ 15, 280
♦ cultivated.

Oxalis pes-caprae **L.**
Oxalidaceae
Bolboxalis cernua (Thunb.) Small (see), *Oxalis cernua* Thunb. (see)
♦ soursob, Bermuda buttercup, African woodsorrel, buttercup oxalis, cape cowslip, oxalis, sorrel, sour grass, yellow flowered oxalis, yellow sorrel, soursop, wild sorrel, woodsorrel
♦ Weed, Noxious Weed, Naturalised, Native Weed, Introduced, Garden Escape, Environmental Weed
♦ 9, 15, 34, 39, 40, 51, 70, 72, 78, 86, 87, 88, 93, 94, 98, 101, 115, 116, 121, 147, 158, 161, 165, 171, 176, 178, 180, 198, 199, 203, 205, 218, 221, 228, 243, 253, 269, 272, 280, 287, 289, 296, 300
♦ pH, arid, cultivated, herbal, toxic. Origin: southern Africa.

Oxalis polyphylla **Jacq.**
Oxalidaceae
♦ Weed, Naturalised, Garden Escape, Environmental Weed, Casual Alien
♦ 7, 86, 98, 203, 280
♦ cultivated.

Oxalis polyphylla **Jacq. var. *pentaphylla*** (Sims) **T.M.Salter**
Oxalidaceae
♦ oxalis
♦ Casual Alien
♦ 280
♦ Origin: South Africa.

Oxalis polyphylla **Jacq. var. *polyphylla***
Oxalidaceae
♦ finger sorrel
♦ Native Weed
♦ 121
♦ pH, toxic. Origin: southern Africa.

Oxalis purpurata **Jacq.**
Oxalidaceae
♦ Weed
♦ 39, 87, 88
♦ toxic.

Oxalis purpurea **L.**
Oxalidaceae
♦ purple oxalis, purple woodsorrel, grand duchess, sorrel, largeflower woodsorrel, four o'clock
♦ Weed, Noxious Weed, Naturalised, Garden Escape, Environmental Weed, Cultivation Escape
♦ 7, 9, 15, 34, 70, 72, 86, 87, 88, 98, 101, 158, 176, 198, 269, 280, 289
♦ pH, cultivated. Origin: southern Africa.

Oxalis radicosa **A.Rich.**
Oxalidaceae
Oxalis laxa Hook. & Arn. (see), *Oxalis micrantha* Bert. ex Colla (see)
♦ dwarf woodsorrel
♦ Naturalised
♦ 101
♦ pH.

Oxalis repens **Thunb.**
Oxalidaceae
= *Oxalis corniculata* L.
♦ Weed
♦ 87, 88

Oxalis rosea **Jacq.**
Oxalidaceae
♦ annual pink sorrel
♦ Naturalised
♦ 40, 241

Oxalis rubra **St.-Hil.**
Oxalidaceae
♦ windowbox woodsorrel, rose woodsorrel, red oxalis
♦ Weed, Naturalised, Introduced
♦ 34, 80, 87, 88, 101, 218
♦ pH, cultivated, herbal.

Oxalis semiloba **Sond.**
Oxalidaceae
♦ Transvaal sorrel, sorrel, oxalis
♦ Weed, Quarantine Weed, Native Weed, Casual Alien
♦ 40, 50, 76, 87, 88, 121, 203

♦ pH, herbal, toxic. Origin: southern Africa.

Oxalis sepium **St.-Hil**
Oxalidaceae
♦ Weed
♦ 87, 88

Oxalis stricta **L.**
Oxalidaceae
Oxalis dillenii Jacq. (see), *Oxalis dillenii* Jacq. ssp. *filipes* (Small) Eiten (see), *Oxalis europaea* Jord. (see), *Xanthoxalis cymosa* (Small) Small, *Xanthoxalis stricta* (L.) Small
♦ common yellow woodsorrel, upright oxalis, yellow woodsorrel, toad sorrel, sheep sorrel, sour grass, common yellow oxalis
♦ Weed, Quarantine Weed, Naturalised, Native Weed, Casual Alien
♦ 23, 40, 42, 70, 76, 87, 88, 94, 161, 174, 203, 207, 210, 211, 218, 243, 249, 266, 272, 286, 287, 292
♦ aH, cultivated, herbal. Origin: Obscure.

Oxalis tetraphylla **Link & Otto**
Oxalidaceae
♦ good luck plant, four leaved pink sorrel
♦ Weed, Naturalised
♦ 40, 98, 203, 269, 287
♦ pH, cultivated, herbal.

Oxalis thompsoniae **B.J.Conn & P.G.Richards**
Oxalidaceae
♦ Naturalised
♦ 280

Oxalis tuberosa **Molina**
Oxalidaceae
♦ oca, truffete acide, oxalide tubéreuse, oka, aleluya tuberosa
♦ Naturalised
♦ 280
♦ pH, cultivated. Origin: Colombia, Peru, Chile.

Oxalis vallicola **(Rose) R.Knuth**
Oxalidaceae
= *Oxalis latifolia* Kunth
♦ Naturalised
♦ 280

Oxalis variabilis **Jacq.**
Oxalidaceae
♦ Naturalised
♦ 287

Oxalis versicolor **L.**
Oxalidaceae
♦ Naturalised
♦ 280
♦ cultivated. Origin: South America.

Oxalis violacea **L.**
Oxalidaceae
Ionoxalis violacea (L.) Small
♦ violet woodsorrel
♦ Weed, Naturalised, Cultivation Escape
♦ 7, 87, 88, 98, 161, 203, 261
♦ pH, cultivated, herbal. Origin: eastern North America.

Oxalis zonata Lieb.
Oxalidaceae
♦ Naturalised
♦ 241

Oxanthera neocaledonica (Guillaumin) Tanaka
Rutaceae
♦ Quarantine Weed
♦ 220

Oxychloe andina Phil.
Juncaceae/Cyperaceae
♦ Introduced
♦ 38
♦ G.

Oxychloris scariosa (F.Muell.) Lazarides
Poaceae
Chloris scariosa F.Muell.
♦ Introduced
♦ 228
♦ G, arid.

Oxydendrum arboreum (L.) DC.
Ericaceae
♦ sour wood, sorrel tree
♦ Weed
♦ 39, 87, 88, 218
♦ T, cultivated, herbal, toxic.

Oxygonum atriplicifolium (Meissn.) Mart.
Polygonaceae
Ceratogonon atriplicifolium Meisn., *Oxygonum somalense* Chiov., *Polygonum owenii* Bojer
♦ Weed, Quarantine Weed
♦ 76, 87, 88, 203, 220, 221, 242
♦ arid.

Oxygonum dregeanum Meisn. var. canescens (Sond.) R.A.Grah.
Polygonaceae
Oxygonum zeyheri Sond.
♦ Native Weed
♦ 121
♦ pH. Origin: southern Africa.

Oxygonum sinuatum (Meisn.) Dammer
Polygonaceae
Ceratogonum cordofanum Meisn., *Ceratogonum sinuatum* Hochst. & Steud. ex Meisn., *Oxygonum elongatum* Dammer, *Oxygonum somalense* Chiov. var. *pterocarpum* Chiov.
♦ oxygonum, double thorn
♦ Weed, Quarantine Weed
♦ 53, 76, 87, 88, 158, 203, 220, 240
♦ aH. Origin: east and southern Africa.

Oxylobium lanceolatum (Vent.) Druce
Fabaceae/Papilionaceae
= *Callistachys lanceolata* Vent.
♦ oxylobium
♦ Weed, Naturalised, Native Weed, Garden Escape, Environmental Weed
♦ 72, 86, 88, 225, 246, 280
♦ S, cultivated.

Oxypetalum caeruleum (D.Don ex Sweet) Decne.
Asclepiadaceae/Apocynaceae
= *Oxypetalum coeruleum* (D.Don ex Sweet) Decne.
♦ Casual Alien

♦ 280
♦ a/pH, cultivated.

Oxypetalum coeruleum (D.Don ex Sweet) Decne.
Asclepiadaceae/Apocynaceae
Gothofreda coerulea (D.Don ex Sweet) Kuntze, *Oxypetalum caeruleum* (D.Don ex Sweet) Decne. (see), *Tweedia coerulea* D.Don ex Sweet (see)
♦ Naturalised
♦ 86
♦ cultivated. Origin: Argentina, Brazil, Uruguay.

Oxypetalum solanoides Hook. & Arn.
Asclepiadaceae/Apocynaceae
♦ plumerillo
♦ Weed
♦ 87, 88, 237, 295

Oxyrhynchus volubilis Brandeg.
Fabaceae/Papilionaceae
♦ twining bluehood
♦ Naturalised
♦ 101

Oxyspora paniculata (D.Don) DC.
Melastomataceae
♦ oxyspora, bristletips
♦ Weed, Noxious Weed, Naturalised
♦ 3, 80, 101, 191, 229

Oxystelma alpini Decne.
Asclepiadaceae/Apocynaceae
= *Oxystelma esculentum* (L.f.) Sm. var. *alpini* (Decne.) R.Br. (NoR)
♦ Weed
♦ 88, 185

Oxystelma esculentum (L.f.) Sm.
Asclepiadaceae/Apocynaceae
♦ Weed
♦ 87, 88, 221
♦ herbal.

Oxytenanthera abyssinica (A.Rich.) Munro
Poaceae
Bambusa abyssinica A.Rich.
♦ Abyssinia oxytenanthera, lulonje mwanzi
♦ Introduced
♦ 228
♦ pG, arid.

Oxytenia acerosa Nutt.
Asteraceae
= *Iva acerosa* (Nutt.) R.C.Jacks. (NoR)
♦ copperweed
♦ Weed
♦ 39, 87, 88, 161, 218
♦ herbal, toxic.

Oxytropis aciphylla Ledeb.
Fabaceae/Papilionaceae
♦ spinyleaf crazyweed
♦ Weed
♦ 297
♦ arid.

Oxytropis bicolor Bunge
Fabaceae/Papilionaceae
♦ Weed
♦ 275, 297
♦ pH.

Oxytropis campestris (L.) DC.
Fabaceae/Papilionaceae
Astragalus campestris L., *Phaca campestris* Wahlenb., *Spiesia campestris* Kuntze
♦ yellow milkvetch, field locoweed, yellow oxytropis, cold mountain crazyweed
♦ Weed
♦ 272
♦ cultivated, herbal.

Oxytropis coerulea (Pall.) DC.
Fabaceae/Papilionaceae
♦ skyblueflower crazyweed
♦ Weed
♦ 297

Oxytropis hirta Bunge
Fabaceae/Papilionaceae
♦ hirsute crazyweed
♦ Weed
♦ 297

Oxytropis kansuensis Bunge
Fabaceae/Papilionaceae
♦ Kansu crazyweed
♦ Weed
♦ 297

Oxytropis lambertii Pursh
Fabaceae/Papilionaceae
Aragallus angustatus Rydb., *Aragallus aven-nelsonii* Lunell, *Aragallus falcatus* Greene, *Aragallus formosus* Greene, *Aragallus involutus* A.Nelson, *Aragallus lambertii* (Pursh) Greene, *Astragalus lambertii* (Pursh) Spreng. (see), *Oxytropis angustata* (Rydb.) A.Nelson, *Oxytropis aven-nelsonii* (Lunell) A.Nelson, *Oxytropis bushii* Gand., *Oxytropis falcata* (Greene) A.Nelson, *Oxytropis hookeriana* Nutt. ex Torr. & A.Gray, *Oxytropis involuta* (A.Nelson) K.Schum., *Oxytropis lambertii* fo. *mixta* Gand., *Oxytropis lambertii* var. *bigelovii* A.Gray, *Oxytropis plattensis* Nutt. ex Torr. & A.Gray, *Spiesia lambertii* (Pursh) Kuntze (see)
♦ Lambert crazyweed, Lambert loco, whitepoint locoweed, white woollyloco, purple locoweed, stemless loco, crazyweed
♦ Weed, Quarantine Weed, Native Weed
♦ 39, 87, 88, 161, 174, 212, 218, 220, 264
♦ pH, cultivated, herbal, toxic. Origin: North America.

Oxytropis macounii (Greene) Rydb.
Fabaceae/Papilionaceae
= *Oxytropis campestris* (L.) DC. var. *spicata* Hook.
♦ spike crazyweed, early yellow locoweed
♦ Weed, Noxious Weed
♦ 36, 87, 88, 218

Oxytropis moellendorffii Bunge ex Maxim
Fabaceae/Papilionaceae
♦ Moellendorff crazyweed
♦ Weed
♦ 297

Oxytropis monticola A.Grey
Fabaceae/Papilionaceae
♦ late yellow locoweed, crazyweed, yellowflower locoweed
♦ Noxious Weed
♦ 299
♦ herbal, toxic.

Oxytropis nana Nutt.
Fabaceae/Papilionaceae
♦ Wyoming locoweed
♦ Weed
♦ 161
♦ toxic.

Oxytropis pilosa (L.) DC.
Fabaceae/Papilionaceae
Astragalus pilosus L., *Spiesia pilosa* Kuntze
♦ woolly milkvetch
♦ Weed
♦ 272
♦ cultivated.

Oxytropis psammocharis Hance.
Fabaceae/Papilionaceae
♦ Weed
♦ 275, 297
♦ pH.

Oxytropis saximontana Nels.
Fabaceae/Papilionaceae
♦ Rocky Mountain crazyweed
♦ Weed
♦ 39, 87, 88, 218
♦ toxic.

Oxytropis sericea Nutt. ex Torr. & Gray
Fabaceae/Papilionaceae
♦ silky crazyweed, locoweed, white locoweed
♦ Weed, Native Weed
♦ 39, 161, 212, 264
♦ cultivated, herbal, toxic. Origin: North America.

Oxytropis splendens Douglas ex Hook.
Fabaceae/Papilionaceae
♦ showy crazyweed, showy locoweed
♦ Weed
♦ 87, 88, 218
♦ cultivated, herbal.

Ozoroa paniculosa (Sond.) R. & A.Fern.
Anacardiaceae
♦ common resin tree
♦ Weed
♦ 10

P

× Pachgerocereus Moran. spp. not validly published
Cactaceae
= *Bergerocactus* Britt. & Rose spp. × *Pachycereus* (Berg.) Britt. & Rose spp.
♦ Quarantine Weed
♦ 220

Pachira aquatica Aubl.
Bombacaceae
♦ Guyana chestnut, Malabar chestnut, saba nut, waterchestnut, ceiba de agua
♦ Introduced
♦ 230, 261
♦ T, cultivated, herbal. Origin: Mexico, Peru, Brazil.

Pachira insignis (Sw.) Sw. ex Savigny
Bombacaceae
♦ wild chestnut, shaving brush tree
♦ Naturalised, Introduced
♦ 101, 261
♦ cultivated. Origin: South America.

Pachyanthus cubensis A.Rich.
Melastomataceae
♦ Weed
♦ 14

Pachycarpus campanulatus (Harv.) N.E.Br.
Asclepiadaceae/Apocynaceae
Gomphocarpus campanulatus Harv.
♦ Quarantine Weed
♦ 220
♦ cultivated.

Pachycarpus grandiflorus (L.f.) E.Mey.
Asclepiadaceae/Apocynaceae
Asclepias grandiflora L.f.
♦ Quarantine Weed
♦ 220
♦ cultivated.

Pachycereus (Berg.) Britt. & Rose spp.
Cactaceae
Anisocereus Backeb. spp. (see), *Lemaireocereus* Britt. & Rose spp. (see), *Lophocereus* (Berg.) Britt. & Rose spp. (see), *Marginatocereus* (Backeb.) Backeb. spp. (see), *Mitrocereus* (Backeb.) Backeb. spp., *Backbergia* Bravo spp., *Pterocereus* MacDoug. & Miranda spp., *Pseudomitrocereus* Bravo & Bux. spp.
♦ pachycereus
♦ Weed, Quarantine Weed
♦ 203, 220
♦ herbal.

Pachygone ledermannii Deils
Menispermaceae

♦ sel en eihr
♦ Introduced
♦ 230
♦ C.

Pachyphragma macrophyllum (Hoffm.) N.Buch.
Brassicaceae
♦ Weed, Naturalised
♦ 23, 88, 163
♦ cultivated.

Pachyrhizus erosus (L.) Urb.
Fabaceae/Papilionaceae
♦ yam bean, jicama, bangkwang
♦ Weed, Naturalised, Introduced
♦ 27, 39, 86, 98, 161, 179, 228
♦ arid, cultivated, herbal, toxic. Origin: tropical America.

Pachyrhizus tuberosus (Lam.) Spreng.
Fabaceae/Papilionaceae
♦ jicama, yam bean, sincamas, ajipo
♦ Cultivation Escape
♦ 39, 261
♦ pC, cultivated, herbal, toxic. Origin: Amazon Region.

Pachysandra terminalis Sieb. & Zucc.
Buxaceae
♦ pachysandra, Japanese pachysandra, spurge
♦ Weed, Naturalised, Environmental Weed
♦ 88, 101, 133, 151, 195
♦ S, cultivated, herbal.

Pachystachys spicata (Ruiz & Pav.) Wassh.
Acanthaceae
♦ Cardinal's guard
♦ Naturalised, Cultivation Escape
♦ 101, 261
♦ cultivated. Origin: northern South America.

Pachystigma pygmaeum (Schltr.) Robyns
Rubiaceae
♦ gousiekte bush, hairy gousiekte bossie, Transvaal gousiektebossie, western Transvaal gousiektebossie
♦ Weed, Native Weed
♦ 39, 121
♦ pH, toxic. Origin: southern Africa.

Pachystigma thamnus Robyns
Rubiaceae
♦ Natal gousiektebos, smooth gousiektebossie
♦ Weed, Native Weed
♦ 39, 121
♦ pH, toxic. Origin: southern Africa.

Packera glabella (Poir.) C.Jeffrey
Asteraceae
Senecio glabellus Poir. (see)
♦ cressleaf, butterweed
♦ Noxious Weed
♦ 229

Pacouria dulcis (Sabine) Roberty
Apocynaceae
♦ Quarantine Weed
♦ 220

Paederia chinensis **Hance**
Rubiaceae
♦ Weed
♦ 87, 88
♦ herbal.

Paederia cruddasiana **Prain**
Rubiaceae
♦ sewer vine, onion vine
♦ Weed, Naturalised, Environmental
Weed
♦ 80, 88, 101, 112, 179

Paederia foetida **L.**
Rubiaceae
Gentiana scandens Lour., *Paederia
magnifica* Noronha *nom. nudum,
Paederia scandens* (Lour.) Merr. (see),
Paederia tomentosa Blume
♦ skunk vine, maile pilau, lesser
Malayan stinkwort, stinkvine
♦ Weed, Quarantine Weed,
Naturalised, Environmental Weed
♦ 3, 39, 76, 80, 87, 88, 101, 112, 135, 151,
179, 191, 203, 220, 243, 273
♦ pC, herbal, toxic. Origin: Asia.

Paederia lanuginosa **Wall.**
Rubiaceae
♦ Weed
♦ 262
♦ pH.

Paederia linearis **Hook.f.**
Rubiaceae
♦ fever vine, tot muu tot maa
♦ Weed
♦ 239

Paederia pilifera **Hook.f.**
Rubiaceae
♦ Weed
♦ 209

Paederia scandens **(Lour.) Merr.**
Rubiaceae
= *Paederia foetida* L.
♦ stink vine
♦ Weed
♦ 3, 13, 87, 88, 191, 204, 263, 274, 286
♦ pC, cultivated, herbal.

Paeonia lactiflora **Pall.**
Paeoniaceae
Paeonia albiflora Pall.
♦ jalopionit, Chinese peony
♦ Naturalised, Cultivation Escape
♦ 42, 101
♦ pH, cultivated, herbal.

Paeonia officinalis **L.**
Paeoniaceae
♦ common peony
♦ Naturalised
♦ 39, 101, 154
♦ pH, cultivated, herbal, toxic.

Pagesia dianthera **(Sw.) Pennell**
Scrophulariaceae
♦ Weed
♦ 87, 88

Palafoxia arida **B.L.Turner & Morris**
Asteraceae
♦ desert palafox, Spanish needle
♦ Weed
♦ 243

♦ aH, herbal.

Palafoxia linearis **(Cav.) Lag.**
Asteraceae
♦ palafox
♦ Weed
♦ 23, 88
♦ herbal.

Palafoxia rosea **(Bush) Cory**
Asteraceae
♦ rosy palafox
♦ Naturalised
♦ 86
♦ herbal. Origin: Central America,
southern North America.

Palaquium karrak **Kaneh.**
Sapotaceae
♦ kalak
♦ Introduced
♦ 230
♦ T.

Palicourea macrobotrys **(Ruiz & Pav.)
DC.**
Rubiaceae
♦ Weed
♦ 87, 88

Palicourea marcgravii **St.-Hil.**
Rubiaceae
♦ cafezinho
♦ Weed
♦ 87, 88, 255
♦ pH. Origin: Brazil.

Palicourea paraensis **(Müll.Arg.) Standl.**
Rubiaceae
= *Palicourea crocea* (Sw.) Roem. &
Schult. (NoR)
♦ Weed
♦ 87, 88

Palicourea triphylla **DC.**
Rubiaceae
♦ Weed
♦ 39, 87, 88
♦ S, toxic.

Palisota hirsuta **(Thunb.) Schum.**
Commelinaceae
♦ kamokamko, mangabo, liteletele
♦ Weed
♦ 88
♦ herbal.

Paliurus spina-christi **Mill.**
Rhamnaceae
Zizyphus paliurus Willd.
♦ Jerusalem thorn, Christ's thorn
♦ Naturalised, Introduced
♦ 101, 228
♦ arid, cultivated, herbal.

Pallenis spinosa **(L.) Cass.**
Asteraceae
Asteriscus spinosus Sch.Bip. (see)
♦ asterisco spinoso
♦ Weed, Naturalised
♦ 70, 86, 87, 88, 94, 272
♦ cultivated, herbal. Origin:
Mediterranean.

Pancratium arabicum **Sickenb.**
Liliaceae/Amaryllidaceae
♦ Weed
♦ 221

Pancratium maritimum **L.**
Liliaceae/Amaryllidaceae
♦ sea daffodil
♦ Weed, Naturalised, Introduced
♦ 39, 70, 86, 221, 228
♦ pH, arid, cultivated, herbal, toxic.
Origin: Mediterranean.

Pancratium sickenbergeri **Asch. &
Schweinf.**
Liliaceae/Amaryllidaceae
♦ Weed
♦ 221

Pancratium tortuosum **Herb.**
Liliaceae/Amaryllidaceae
♦ Weed
♦ 221

Pandanus baptistii hort. **Veitch ex
Misonne**
Pandanaceae
♦ Weed, Cultivation Escape
♦ 179, 261
♦ cultivated. Origin: New Britian
Island.

Pandanus odoratissimus **L.f.** *nom. illeg.*
Pandanaceae
= *Pandanus odorifer* (Forssk.) Kuntze
nom. dub. (NoR)
♦ breadfruit, screwpine
♦ Introduced
♦ 39, 261
♦ cultivated, herbal, toxic. Origin:
south-east Asia.

Pandanus pacificus **H.J.Veitch ex
M.T.Mast.**
Pandanaceae
♦ pandano
♦ Cultivation Escape
♦ 261
♦ cultivated. Origin: New Guinea,
Molucas and Mariana Islands.

Pandanus patina **Mart.**
Pandanaceae
♦ peet, pandanus
♦ Introduced
♦ 230
♦ T.

Pandanus tectorius **Parkinson ex Zucc.**
Pandanaceae
♦ Tahitian screwpine, kipar en wel,
fala, screwpine
♦ Weed, Naturalised, Introduced
♦ 22, 101, 261
♦ T, cultivated, herbal. Origin: Tahiti.

Pandanus utilis **Bory**
Pandanaceae
♦ common screwpine, screwpine,
mongo
♦ Naturalised, Introduced
♦ 101, 261
♦ cultivated, herbal. Origin:
Madagascar.

Pandanus veitchii hort. **Veitch ex Mast.
& T.Moore**
Pandanaceae
♦ Veitch's screwpine
♦ Naturalised, Introduced
♦ 101, 261

♦ cultivated. Origin: Polynesia.

Pandorea jasminoides (Lindl.) Schum.
Bignoniaceae
♦ bower vine, bower plant, pink bower plant
♦ Sleeper Weed, Environmental Weed, Casual Alien
♦ 225, 246, 280
♦ pC, cultivated. Origin: Australia.

Pandorea pandorana (Andrews) Steenis
Bignoniaceae
♦ wonga wonga vine
♦ Naturalised, Environmental Weed
♦ 225, 246, 280
♦ pC, cultivated, herbal.

Pangium edulae L.
Flacourtiaceae
♦ football tree, sis, pangi, duhrien
♦ Weed
♦ 3, 191

Panicum L. spp.
Poaceae
♦ panicgrass, sefa
♦ Noxious Weed
♦ 36, 39
♦ G, herbal, toxic.

Panicum adspersum Trin.
Poaceae
= *Urochloa adspersa* (Trin.) R.Webster
♦ broadleaf panicum
♦ Weed
♦ 87, 88, 161, 249
♦ G.

Panicum aequinerve Nees
Poaceae
♦ Native Weed
♦ 121
♦ a/pG. Origin: southern Africa.

Panicum agrostoides Spreng
Poaceae
= *Panicum rigidulum* Bosc ex Nees var. *rigidulum* (NoR)
♦ Weed
♦ 87, 88
♦ G, herbal.

Panicum amarum Elliott
Poaceae
♦ beachgrass, bitter panicgrass, panicgrass
♦ Weed, Environmental Weed
♦ 88, 151, 195
♦ G, cultivated, herbal.

Panicum ambiguum Trin.
Poaceae
= *Brachiaria paspaloides* (J.Presl ex C.Presl) C.E.Hubb.
♦ Weed
♦ 87, 88
♦ G.

Panicum amplexicaule Rudge
Poaceae
= *Hymenachne amplexicaulis* (Rudge) Nees
♦ Weed, Quarantine Weed
♦ 76, 87, 88, 203, 220
♦ G.

Panicum anceps Michx.
Poaceae
♦ panicgrass, beaked panicgrass
♦ Weed
♦ 161
♦ G, herbal.

Panicum antidotale Retz.
Poaceae
♦ giant panic, blue panic, gramana, blue panicgrass
♦ Weed, Noxious Weed, Naturalised, Introduced
♦ 7, 30, 35, 39, 86, 88, 90, 93, 98, 101, 161, 203, 205, 228, 229
♦ pG, arid, cultivated, herbal, toxic. Origin: Middle East to India.

Panicum arizonicum Scribn. & Merr.
Poaceae
= *Urochloa arizonica* (Scribn. & Merr.) Morrone & Zuloaga (NoR) [see *Brachiaria arizonica* (Scribn. & Merr.) S.T.Blake]
♦ Arizona panicum
♦ Weed
♦ 87, 88, 218
♦ G, herbal.

Panicum attenuatum Willd.
Poaceae
♦ Weed
♦ 87, 88
♦ G.

Panicum auritum Presl
Poaceae
♦ Weed
♦ 87, 88
♦ G.

Panicum austroasiaticum Ohwi
Poaceae
Panicum humile Nees ex Steud., *Panicum vescum* R.R.Stewart
♦ Weed, Quarantine Weed
♦ 76, 87, 88, 203, 220
♦ G.

Panicum barbinode Trin.
Poaceae
= *Urochloa mutica* (Forssk.) Nguyen
♦ Weed
♦ 3, 191
♦ G.

Panicum bergii Arechav.
Poaceae
♦ Berg's panicgrass, paja voladora
♦ Weed, Quarantine Weed, Naturalised
♦ 76, 87, 88, 101, 236, 237, 295
♦ G.

Panicum bisulcatum Thunb.
Poaceae
♦ Japanese panicgrass, chaff panic, blackseed panic
♦ Weed, Naturalised
♦ 86, 87, 88, 90, 98, 101, 198, 203, 275, 286, 297
♦ aG, aqua, cultivated. Origin: India, China, Japan, Indonesia, Australia.

Panicum boliviense Hack.
Poaceae

= *Panicum polygonatum* Schrad.
♦ Bolivian panicgrass
♦ Naturalised
♦ 101
♦ G.

Panicum brevifolium L.
Poaceae
Isachne tricarinata Roth, *Panicum arborescens* L., *Panicum biflorum* Lam., *Panicum brevifolium* var. *hirtifolium* (Ridl.) Jansen, *Panicum dubium* Lam., *Panicum gladiatum* Wawra, *Panicum guineense* Desv. ex Poir., *Panicum hirtifolium* Ridl., *Panicum litigosum* Steud., *Panicum ovalifolium* Poir., *Panicum tricarinatum* (Roth) Steud., *Panicum trichopioides* Mez
♦ short leaved panic, botai
♦ Weed
♦ 87, 88, 90
♦ aG. Origin: central and eastern Asia.

Panicum bulbosum Kunth
Poaceae
♦ bulb panicgrass, bulbous panic
♦ Naturalised, Introduced
♦ 98, 228
♦ G, arid, cultivated, herbal. Origin: southern North America to northern South America.

Panicum buncei F.Muell. ex Benth.
Poaceae
♦ Naturalised, Introduced
♦ 86, 228
♦ G, arid. Origin: Australia.

Panicum cambogiense Bal.
Poaceae
♦ Weed, Naturalised
♦ 12, 88, 98, 191, 203, 204
♦ G.

Panicum campestre Nees
Poaceae
♦ capim câiana
♦ Weed
♦ 245
♦ G.

Panicum capillare L.
Poaceae
Panicum barbipulvinatum Nash, *Panicum capillare* L. var. *brevifolium* Vasey ex Rydb. & Shear (see), *Panicum capillare* L. var. *occidentale* Rydb. (see)
♦ witchgrass, old witchgrass, ticklegrass, witches hair, tumbleweed grass, fool hay, food hay, mousseline, panicgrass, tumbleweed, windmill grass
♦ Weed, Noxious Weed, Naturalised, Native Weed, Environmental Weed, Casual Alien
♦ 7, 15, 23, 24, 30, 34, 40, 42, 49, 52, 68, 72, 86, 87, 88, 90, 98, 118, 161, 174, 176, 180, 198, 203, 210, 211, 212, 218, 236, 241, 243, 253, 272, 280, 286, 287, 295, 299, 300
♦ aG, arid, cultivated, herbal, toxic. Origin: North America.

Panicum capillare L. var. *brevifolium*
Vasey ex Rydb. & Shear
  Poaceae
  = *Panicum capillare* L.
  ♦ witchgrass
  ♦ Naturalised
  ♦ 86, 198
  ♦ G.

Panicum capillare L. var. *capillare*
  Poaceae
  ♦ witchgrass
  ♦ Naturalised
  ♦ 86, 176, 198
  ♦ G. Origin: North America.

Panicum capillare L. var. *occidentale*
Rydb.
  Poaceae
  = *Panicum capillare* L.
  ♦ Naturalised
  ♦ 86, 176
  ♦ G. Origin: North America.

Panicum caudiglume **Hack.**
  Poaceae
  ♦ Weed
  ♦ 87, 88
  ♦ G.

Panicum chloroticum **Nees**
  Poaceae
  = *Panicum dichotomiflorum* Michx.
  ♦ Weed, Quarantine Weed
  ♦ 76, 87, 88, 203, 220
  ♦ G.

Panicum ciliatifolium **Kunth**
  Poaceae
  ♦ Naturalised
  ♦ 86
  ♦ G.

Panicum ciliatum **Ell.**
  Poaceae
  = *Dichanthelium strigosum* (Muhl. ex
  Ell.) Freckmann var. *leucoblepharis*
  (Trin.) Freckmann (NoR)
  ♦ Weed, Quarantine Weed
  ♦ 76, 87, 88, 203, 220
  ♦ G.

Panicum clandestinum **L.**
  Poaceae
  Dichanthelium clandestinum (L.) Gould
  (see)
  ♦ deertongue dichanthelium
  ♦ Weed
  ♦ 90
  ♦ G, herbal.

Panicum coloratum **L.**
  Poaceae
  ♦ makarikari grass, small buffalograss,
  small panicum, white buffalograss,
  coolah grass, blue panicgrass, Klein
  grass, bambatsi panicgrass, Klein grass
  ♦ Weed, Naturalised, Native Weed,
  Introduced, Environmental Weed
  ♦ 39, 72, 80, 86, 88, 98, 101, 121, 185,
  198, 203, 221, 228
  ♦ pG, arid, cultivated, toxic. Origin:
  Africa.

Panicum coloratum **L. var.** *colaratum*
  Poaceae

Panicum swynnertonii Rendle
  ♦ coloured guinea grass, Klein grass,
  white buffalograss, hijé, buntes
  Guineagras, capim macaricam, pasto
  colorado
  ♦ Naturalised
  ♦ 86
  ♦ G.

Panicum coloratum **L. var.**
makarikariense **Gooss.**
  Poaceae
  ♦ Naturalised
  ♦ 86
  ♦ G. Origin: southern Africa.

Panicum decompositum **R.Br.**
  Poaceae
  ♦ Australian native millet, native
  millet
  ♦ Introduced
  ♦ 39, 228, 243
  ♦ pG, arid, cultivated, toxic. Origin:
  Australia.

Panicum deustum **Thunb.**
  Poaceae
  ♦ buffalograss, reed panicum
  ♦ Native Weed
  ♦ 121
  ♦ pG. Origin: southern Africa.

Panicum dichotomiflorum **Michx.**
  Poaceae
  Panicum chloroticum Nees (see)
  ♦ fall panicgrass, smooth witchgrass,
  western witchgrass, sprouting
  crabgrass, panico delle risaie, giavane
  Americano, fall panicum, capim do
  banhado
  ♦ Weed, Noxious Weed, Naturalised,
  Native Weed, Environmental Weed,
  Casual Alien
  ♦ 15, 30, 34, 42, 68, 87, 88, 90, 118, 151,
  161, 174, 207, 210, 211, 212, 218, 229,
  236, 237, 243, 249, 253, 255, 263, 272,
  280, 286, 287, 295, 300
  ♦ aG, cultivated, herbal. Origin: North
  America.

Panicum dichotomiflorum **Michx. var.**
bartowense **(Scribn. & Merr.) Fern.**
  Poaceae
  ♦ fall panicum, fall panicgrass
  ♦ Noxious Weed
  ♦ 229
  ♦ G.

Panicum dichotomiflorum **Michx. var.**
dichotomiflorum
  Poaceae
  ♦ fall panicum, fall panicgrass
  ♦ Noxious Weed
  ♦ 229
  ♦ G.

Panicum dichotomiflorum **Michx. var.**
puritanorum **Svens.**
  Poaceae
  ♦ fall panicum, fall panicgrass
  ♦ Noxious Weed
  ♦ 229
  ♦ G.

Panicum dilatatum **Steud.**
  Poaceae

  ♦ paspalum
  ♦ Weed
  ♦ 269
  ♦ pG. Origin: South America.

Panicum effusum **R.Br.**
  Poaceae
  ♦ Weed, Introduced
  ♦ 39, 87, 88, 228, 269
  ♦ G, arid, cultivated, toxic. Origin:
  Australia.

Panicum elephantipes **Nees**
  Poaceae
  ♦ elephant panicgrass, camalote
  ♦ Weed
  ♦ 14, 87, 88, 236, 237, 295
  ♦ G, aqua.

Panicum erectum **Pollacci**
  Poaceae
  ♦ Weed, Quarantine Weed
  ♦ 76, 87, 88, 203, 220
  ♦ G.

Panicum fasciculatum **Sw.**
  Poaceae
  = *Urochloa fasciculata* (Sw.) Webster
  ♦ browntop panicum, pajilla
  ♦ Weed, Quarantine Weed
  ♦ 14, 30, 76, 87, 88, 157, 161, 203, 220,
  281
  ♦ G, herbal.

Panicum fasciculatum **Sw. var.**
reticulatum **(Torr.) Beal**
  Poaceae
  = *Urochloa fasciculata* (Sw.) Webster
  ♦ browntop panicum
  ♦ Weed
  ♦ 218
  ♦ G.

Panicum flavidum **Retz.**
  Poaceae
  = *Paspalidium flavidum* (Retz.) A.Camus
  ♦ Weed
  ♦ 23, 88
  ♦ G, herbal.

Panicum gattingeri **Nash**
  Poaceae
  ♦ Gattinger panicum
  ♦ Weed
  ♦ 87, 88, 90, 218
  ♦ G, herbal.

Panicum geminatum **Forssk.**
  Poaceae
  = *Setaria geminata* (Forssk.) Veldkamp
  (NoR)
  ♦ Naturalised
  ♦ 287
  ♦ G.

Panicum gilvum **Launert**
  Poaceae
  Panicum laevifolium Hack. var.
  contractum Pilg.
  ♦ panic
  ♦ Weed, Naturalised, Introduced,
  Environmental Weed
  ♦ 86, 93, 98, 155, 176, 191, 198, 203, 228,
  269
  ♦ aG, arid, cultivated. Origin: southern
  Africa.

Panicum glabrescens Steud.
Poaceae
= *Panicum subalbidum* Kunth
♦ Weed
♦ 87, 88
♦ G.

Panicum gouini Fourn.
Poaceae
♦ Weed, Quarantine Weed
♦ 76, 87, 88, 203, 220
♦ G.

Panicum guadaloupense Spreng. ex Steud.
Poaceae
= *Urochloa mutica* (Forssk.) Nguyen
♦ Weed
♦ 3, 191
♦ G.

Panicum hallii Vasey
Poaceae
♦ Hall's panicgrass
♦ Weed, Introduced
♦ 87, 88, 228
♦ G, arid, herbal.

Panicum hemitomon Schult.
Poaceae
♦ maidencane
♦ Weed, Quarantine Weed
♦ 87, 88, 161, 258
♦ G, cultivated, herbal.

Panicum heterostachyum Hack.
Poaceae
♦ Introduced
♦ 228
♦ G, arid.

Panicum hians Ell.
Poaceae
= *Steinchisma hians* (Ell.) Nash (NoR)
♦ Weed
♦ 121
♦ aG, herbal. Origin: North America.

Panicum hillmanii Chase
Poaceae
♦ witchpanic, Hillman's panicgrass
♦ Weed, Quarantine Weed, Naturalised, Environmental Weed
♦ 72, 76, 86, 88, 176, 198
♦ aG. Origin: southern North America.

Panicum hirticaule J.Presl
Poaceae
♦ Mexican panicgrass
♦ Weed, Introduced
♦ 28, 199, 228, 243
♦ aG, arid, herbal.

Panicum huachucae Ashe
Poaceae
= *Panicum acuminatum* Sw. var. *fasciculatum* (Torr.) Beetle (NoR)
♦ Naturalised
♦ 280
♦ G.

Panicum hygrocharis Steud.
Poaceae
♦ Weed, Quarantine Weed
♦ 76, 87, 88, 203, 220, 242
♦ G, arid.

Panicum incomtum Trin.
Poaceae
♦ yaa khaihao, scandent panic
♦ Weed
♦ 238
♦ pG. Origin: Australia.

Panicum kerstingii Mez
Poaceae
♦ Weed
♦ 87, 88
♦ G.

Panicum laevifolium Hack.
Poaceae
= *Panicum schinzii* Hack.
♦ sweetgrass, vlei panicum, landsgrass
♦ Weed, Naturalised
♦ 51, 87, 88, 90, 280
♦ aG, aqua, herbal.

Panicum laevinode Lindl.
Poaceae
Panicum whitei J.M.Black
♦ Introduced
♦ 228
♦ G, arid.

Panicum lancearum Trin.
Poaceae
♦ Weed
♦ 87, 88
♦ G.

Panicum lanuginosum Elliott *non* Bosc ex Spreng.
Poaceae
= *Panicum acuminatum* Sw. var. *acuminatum* (NoR)
♦ Naturalised
♦ 287
♦ G, herbal.

Panicum laxiflorum Lam.
Poaceae
= *Dichanthelium laxiflorum* (Lam.) Gould (NoR)
♦ Naturalised
♦ 287
♦ G, herbal.

Panicum laxum Sw.
Poaceae
♦ lax panicgrass, panico
♦ Weed, Quarantine Weed
♦ 76, 87, 88, 153, 203, 206, 220, 243
♦ G, cultivated.

Panicum lindheimeri Nash
Poaceae
Dichanthelium lindheimeri (Nash) Gould, *Panicum funstonii* Scribn. & Merr., *Panicum lanuginosum* Elliott var. *lindheimeri* (Nash) Fernald, *Panicum lindheimeri* var. *typicum* Fernald
♦ Casual Alien
♦ 280
♦ G, herbal.

Panicum luzonense Presl
Poaceae
♦ Weed, Naturalised
♦ 87, 88, 98, 203
♦ G.

Panicum maximum Jacq.
Poaceae

Panicum maximum Jacq. var. *trichoglume* Robyns (see), *Panicum hirsutissimum* Steud.
♦ guinea grass, green panic, buffalograss, saafa, herbe de Guinéa, panic élevé, capime guiné, fataque, katengalujinga, barbe grass, browntop buffel grass, bush buffalograss, common buffalograss, purpletop buffalograss, purpletop buffelsgras, rainbow grass, unabe grass, hamil grass, pasto guinea, Gatton panic
♦ Weed, Quarantine Weed, Naturalised, Native Weed, Environmental Weed
♦ 3, 6, 7, 12, 14, 30, 32, 39, 53, 55, 80, 87, 88, 90, 98, 107, 121, 134, 152, 157, 158, 161, 179, 186, 203, 218, 226, 236, 245, 246, 249, 255, 257, 262, 280, 281, 286, 287, 295
♦ pG, cultivated, herbal, toxic. Origin: Africa.

Panicum maximum Jacq. var. *coloratum* C.T.White
Poaceae
♦ Naturalised
♦ 86
♦ G. Origin: Africa.

Panicum maximum Jacq. var. *maximum*
Poaceae
♦ guinea grass
♦ Weed, Naturalised
♦ 86, 269
♦ G. Origin: Africa.

Panicum maximum Jacq. var. *trichoglume* Robyns
Poaceae
= *Panicum maximum* Jacq.
♦ green panicgrass
♦ Weed, Naturalised
♦ 55, 86, 280
♦ G. Origin: Africa.

Panicum melinis Trin.
Poaceae
= *Melinis minutiflora* Beauv.
♦ Weed
♦ 3, 191
♦ G.

Panicum mertensii Roth
Poaceae
Panicum megiston Schult.
♦ Introduced
♦ 228
♦ G, arid.

Panicum meyerianum Nees
Poaceae
♦ Weed
♦ 87, 88
♦ G.

Panicum miliaceum L.
Poaceae
Milium panicum Mill.
♦ proso millet, wild proso millet, panicum, broomcorn millet, hog millet, panic millet, French millet, proso, broomcorn, yellow millet, proso siate
♦ Weed, Noxious Weed, Naturalised, Introduced, Casual Alien, Cultivation Escape

♦ 7, 30, 34, 39, 40, 42, 70, 80, 86, 87, 88,
93, 90, 98, 101, 114, 121, 161, 174, 176,
179, 198, 199, 203, 205, 210, 211, 212,
218, 219, 228, 229, 243, 252, 253, 272,
280, 287, 295, 299, 300
♦ aG, arid, cultivated, herbal, toxic.
Origin: Eurasia.

Panicum miliaceum L. ssp. *miliaceum*
Poaceae
♦ wild proso millet, broomcorn millet,
panicum
♦ Noxious Weed, Naturalised
♦ 101, 229
♦ G.

Panicum miliaceum L. ssp. *ruderale*
(Kitag.) Tzvelev
Poaceae
Panicum miliaceum L. var. *ruderale*
Kitag. (see)
♦ wild proso millet, broomcorn millet,
panicum
♦ Noxious Weed, Naturalised
♦ 101, 229
♦ G.

Panicum miliaceum L. var. *ruderale* Kitag.
Poaceae
= *Panicum ruderale* (Kitag.) Liou (NoR)
♦ Weed
♦ 275
♦ aG.

Panicum millegrana Poir.
Poaceae
♦ Weed
♦ 87, 88
♦ G.

Panicum minutiflora (P.Beauv.) Rasp.
Poaceae
= *Melinis minutiflora* Beauv.
♦ Weed
♦ 3, 191
♦ G.

Panicum monostachyum H.B.K.
Poaceae
♦ Weed
♦ 87, 88
♦ G.

Panicum montanum Roxb.
Poaceae
♦ Weed
♦ 87, 88
♦ G.

Panicum muticum Forssk.
Poaceae
= *Urochloa mutica* (Forssk.) Nguyen
♦ Weed
♦ 3, 191
♦ G.

Panicum napaliense Davidse
Poaceae
♦ Napali panicgrass
♦ Naturalised
♦ 101
♦ G.

Panicum natalense Hochst.
Poaceae
♦ Natal buffalograss
♦ Native Weed
♦ 121

♦ pG. Origin: southern Africa.

Panicum notatum Retz.
Poaceae
♦ Weed
♦ 12, 275
♦ pG.

Panicum novemnerve Stapf
Poaceae
♦ Weed, Naturalised, Native Weed
♦ 86, 98, 121, 203
♦ aG. Origin: southern Africa.

Panicum obseptum Trin.
Poaceae
♦ white water panic
♦ Naturalised
♦ 86, 198
♦ G, cultivated. Origin: Australia.

Panicum obtusum Kunth
Poaceae
Brachiaria obtusa (Kunth) Nash
♦ vine mesquite, obtuse panicgrass
♦ Weed, Introduced
♦ 87, 88, 90, 199, 218, 228
♦ pG, aqua, promoted, herbal.

Panicum palmifolium König
Poaceae
= *Setaria palmifolia* (König) Stapf
♦ Weed
♦ 3, 191
♦ G.

Panicum paludosum Roxb.
Poaceae
Panicum proliferum Lam. var. *paludosun*
(Roxb.) Stapf
♦ Chesapeake panicgrass
♦ Weed, Naturalised
♦ 87, 88, 101, 170, 191, 275
♦ pG, cultivated. Origin: Asia,
Australia.

Panicum phyllopogon Stapf
Poaceae
= *Echinochloa oryzoides* (Ard.) Fritsch
♦ Weed, Quarantine Weed
♦ 76, 87, 88, 203, 220
♦ G.

Panicum pilipes Nees & Arn. ex Büse
Poaceae
= *Cyrtococcum oxyphyllum* (Hochst. ex
Steud.) Stapf
♦ Weed
♦ 87, 88
♦ G.

Panicum pilosum Sw.
Poaceae
Setaria pilosa Kunth (see)
♦ Weed
♦ 88, 153
♦ G, cultivated.

Panicum polygonatum Schrad.
Poaceae
Panicum boliviense Hack. (see)
♦ Weed
♦ 87, 88
♦ G.

Panicum prionitis Nees
Poaceae
♦ Weed, Introduced

♦ 87, 88, 228
♦ G, arid.

Panicum psilopodium Trin.
Poaceae
♦ barefoot panicgrass
♦ Weed, Quarantine Weed,
Naturalised
♦ 66, 76, 88, 101, 203, 220
♦ G, cultivated, herbal.

Panicum purpurascens Raddi
Poaceae
= *Urochloa mutica* (Forssk.) Nguyen
♦ paragrass, zacate pará
♦ Weed, Environmental Weed
♦ 3, 14, 88, 191, 218, 257, 281
♦ pG.

Panicum queenslandicum Domin
Poaceae
♦ Weed, Introduced
♦ 87, 88, 228
♦ G, arid, cultivated. Origin: south-
east Asia, Australia.

Panicum racemosum (P.Beauv.) Spreng.
Poaceae
Monachne racemosum P.Beauv.,
Saccharum reptans Lam., *Thalasium
montevidense* Spreng.
♦ branched panic
♦ Weed, Naturalised, Introduced
♦ 54, 86, 88, 198, 228, 237, 295
♦ G, arid. Origin: South America.

Panicum repens Linn.
Poaceae
Panicum gouinii auct. non Fourn.
♦ torpedo grass, couch, creeping
panicgrass, panic rampant, yaa
channakaat, creeping panic, couch
panicum, wainaku grass
♦ Weed, Quarantine Weed, Noxious
Weed, Naturalised, Native Weed,
Environmental Weed
♦ 3, 26, 30, 70, 80, 86, 87, 88, 90, 98, 112,
121, 151, 161, 170, 179, 185, 186, 191,
197, 203, 209, 218, 221, 229, 237, 239,
243, 246, 249, 253, 255, 272, 273, 274,
286, 295, 297
♦ pG, aqua, cultivated, herbal. Origin:
Brazil.

Panicum reptans L.
Poaceae
= *Urochloa reptans* (L.) Stapf
♦ Weed
♦ 199
♦ G.

Panicum rivulare Trin.
Poaceae
Agrostis pernamucensis Spreng.
♦ palha branca
♦ Weed
♦ 255
♦ pG. Origin: Brazil.

Panicum sabulorum Lam.
Poaceae
♦ Weed, Quarantine Weed
♦ 76, 87, 88, 203, 220
♦ G.

Panicum sarmentosum Roxb.
Poaceae

♦ scrambling panicgrass
♦ Weed
♦ 87, 88, 90
♦ pG, herbal.

Panicum schinzii Hack.
Poaceae
Panicum laevifolium Hack.
♦ bluegrass, buffalograss, landgrass, old landgrass, sweetgrass, sweet buffalograss, vlei panicum
♦ Weed, Naturalised, Native Weed, Environmental Weed
♦ 86, 88, 98, 121, 158, 203, 280
♦ aG. Origin: southern Africa.

Panicum scoparium Lam.
Poaceae
= *Dichanthelium scoparium* (Lam.) Gould
♦ Naturalised
♦ 287
♦ G, herbal.

Panicum sonorum Beal
Poaceae
♦ sauwi
♦ Introduced
♦ 228
♦ G, arid, promoted, herbal.

Panicum spectabile Nees
Poaceae
= *Echinochloa polystachya* (Nees ex Trin.) Mart. var. *spectabilis* Crov. (NoR)
♦ Weed
♦ 87, 88
♦ G.

Panicum sphaerocarpon Elliott
Poaceae
Dichanthelium sphaerocarpon (Elliott) Gould, *Panicum auburne* Ashe, *Panicum dichotomum* (Elliott) Alph. var. *sphaerocarpum* Wood, *Panicum heterophyllum* Sw. ex Wikstr., *Panicum inflatum* Scribn. & J.G.Sm., *Panicum kalmii* Sw. ex Wikstr., *Panicum microcarpon* Muhl. ex Elliott var. *sphaerocarpon* (Elliott) Vasey, *Panicum nitidum* Lam. var. *crassifolium* A.Gray, *Panicum sphaerocarpon* ssp. *inflatum* (Scribn. & J.G.Sm.) Hitchc., *Panicum vicarium* E.Fourn.
♦ Naturalised
♦ 280
♦ G, herbal.

Panicum stapfianum Fourc.
Poaceae
Panicum minus Stapf
♦ Introduced
♦ 228
♦ G, arid.

Panicum subalbidum Kunth
Poaceae
Panicum glabrescens Steud., *Panicum ingens* Peter. *Panicum kermesinum* Mez, *Panicum longijubatum* (Stapf) Stapf, *Panicum longiramum* Peter, *Panicum proliferum* Lam. var. *longijubatum* Stapf
♦ elbow buffalograss
♦ Weed, Native Weed
♦ 88, 121
♦ a/pG, arid. Origin: southern Africa.

Panicum subquadriparum Trin.
Poaceae
= *Urochloa subquadripara* (Trin.) R.D.Webster
♦ Weed
♦ 3, 191
♦ G.

Panicum subxerophilum Domin
Poaceae
♦ Introduced
♦ 228
♦ G, arid, cultivated. Origin: Australia.

Panicum texanum Buckl.
Poaceae
= *Urochloa texana* (Buckley) R.D.Webster
♦ Texas panicum, Texas millet, Colorado grass
♦ Weed
♦ 30, 87, 88, 161, 180, 218
♦ G.

Panicum toridum Gaudich.
Poaceae
♦ Weed
♦ 87, 88
♦ G.

Panicum trichocladum Hack. ex K.Schum.
Poaceae
♦ lukoka
♦ Weed, Quarantine Weed, Naturalised
♦ 32, 76, 87, 88, 203, 220
♦ pG, arid, cultivated.

Panicum trichoides Sw.
Poaceae
♦ ticklegrass, tropical panicgrass, zacate ilusión
♦ Weed
♦ 28, 30, 87, 88, 90, 157, 199, 206, 243, 281
♦ aG.

Panicum trypheron Schult.
Poaceae
♦ Weed
♦ 87, 88
♦ G.

Panicum turgidum Forssk.
Poaceae
Panicum nubicum Fig. & De Not.
♦ Weed
♦ 90, 221
♦ G, arid, cultivated, herbal.

Panicum umbellatum Trin.
Poaceae
♦ Weed
♦ 87, 88
♦ G.

Panicum urvilleanum Kunth
Poaceae
Monachne urvilleana (Kunth) Herter, *Panicum megastachyum* J.Presl, *Panicum patagonicum* Hieron., *Panicum preslii* Kunth
♦ desert panicgrass
♦ Weed, Introduced
♦ 87, 88, 228, 237, 295
♦ pG, arid, cultivated.

Panicum venezuelae Hack.
Poaceae
♦ Venezuelan panicgrass
♦ Naturalised
♦ 101
♦ G.

Panicum virgatum L.
Poaceae
Chasea virgata (L.) Nieuwl., *Eatonia purpurascens* Raf., *Ichnanthus glaber* Link ex Steud., *Milium virgatum* (L.) Lunell, *Milium virgatum* var. *elongatum* (Vasey) Lunell, *Panicum coloratum* Walter, *Panicum giganteum* Scheele, *Panicum glaberrimum* Steud., *Panicum ichnanthoides* E.Fourn., *Panicum kunthii* E.Fourn., *Panicum pruinosum* Bernh. ex Trin.
♦ switchgrass, Blackwell switchgrass, fall panicgrass, Heller's rosette grass, roundseed rosette grass, tapered rosette grass, wand panicgrass, luutahirssi
♦ Weed, Quarantine Weed, Introduced, Casual Alien
♦ 30, 42, 76, 87, 88, 90, 161, 218, 220, 228
♦ pG, arid, cultivated, herbal, toxic.

Panicum volutans J.G.Anders.
Poaceae
♦ rolling grass, tumbleweed
♦ Native Weed
♦ 121
♦ aG. Origin: southern Africa.

Panicum vulgaris L.
Poaceae
♦ panicgrass
♦ Noxious Weed
♦ 36
♦ G.

Panicum zizanoides Kunth
Poaceae
♦ Weed
♦ 88
♦ G.

Panzeria alaschanica Kupr.
Lamiaceae
♦ alashan panzeria
♦ Weed
♦ 297

Papaver L. spp.
Papaveraceae
♦ poppy
♦ Weed, Naturalised
♦ 23, 88, 154, 198, 243
♦ toxic.

Papaver aculeatum Thunb.
Papaveraceae
♦ wild poppy, thorny poppy, Californian poppy, red poppy, bristle poppy, Iranian poppy
♦ Weed, Naturalised, Native Weed, Environmental Weed, Casual Alien
♦ 39, 51, 86, 87, 88, 98, 121, 158, 198, 203, 269, 280
♦ aH, cultivated, herbal, toxic. Origin: southern Africa.

Papaver apulum Ten.
Papaveraceae
Papaver argemonoides Ces.
♦ Weed
♦ 23, 88

Papaver arenarium Bieb.
Papaveraceae
♦ Weed, Casual Alien
♦ 40, 272

Papaver argemone L.
Papaveraceae
♦ prickly poppy, prickly long headed poppy, long pricklyhead poppy, pinnate poppy, hietaunikko
♦ Weed, Quarantine Weed, Naturalised, Introduced, Garden Escape, Cultivation Escape, Casual Alien
♦ 23, 34, 42, 44, 70, 76, 86, 87, 88, 94, 98, 101, 115, 118, 161, 176, 198, 203, 243, 252, 253, 269, 272, 280
♦ aH, cultivated, herbal.

Papaver atlanticum (Ball) Coss.
Papaveraceae
♦ Atlas poppy
♦ Naturalised
♦ 280
♦ cultivated, herbal.

Papaver bracteatum Lindl.
Papaveraceae
♦ blood poppy, bracteate poppy, oriental poppy
♦ Weed, Quarantine Weed
♦ 76, 88, 220
♦ cultivated, herbal, toxic.

Papaver commutatum Fisch. & C.Mey.
Papaveraceae
♦ ladybird poppy
♦ Naturalised
♦ 287
♦ aH, cultivated. Origin: Armenia, Azerbaijan, Iran, Turkey.

Papaver croceum Ledeb.
Papaveraceae
♦ siperianunikko, ice poppy
♦ Cultivation Escape
♦ 42
♦ cultivated.

Papaver decaisnei Hochst. & Steud. ex Elkan
Papaveraceae
♦ Weed
♦ 221

Papaver dubium L.
Papaveraceae
Papaver lamottei Bor.
♦ poppy, long headed poppy, ruisunikko, field poppy, blindeyes, long smooth headed poppy, papavero a clava
♦ Weed, Quarantine Weed, Naturalised, Garden Escape, Cultivation Escape, Casual Alien
♦ 23, 42, 44, 70, 76, 80, 86, 87, 88, 94, 98, 101, 102, 118, 176, 198, 203, 218, 243, 252, 253, 272, 280, 286, 287, 300
♦ aH, cultivated, herbal. Origin: Eurasia.

Papaver glaucum Boiss. & Hausskn.
Papaveraceae
♦ tulip poppy
♦ Naturalised
♦ 101
♦ cultivated.

Papaver gorodkovii Tolm. & Petrovsky
Papaveraceae
♦ Arctic poppy
♦ Naturalised
♦ 101

Papaver hybridum L.
Papaveraceae
♦ rough poppy, round prickly headed poppy, round rough headed poppy, karvaunikko
♦ Weed, Naturalised, Garden Escape, Environmental Weed, Cultivation Escape
♦ 7, 42, 44, 70, 86, 87, 88, 93, 94, 98, 101, 118, 176, 198, 203, 205, 243, 253, 269, 272, 280, 287, 300
♦ aH, cultivated, herbal. Origin: Europe.

Papaver lecoqii Lamotte
Papaveraceae
♦ Babington's poppy, yellow juiced poppy
♦ Weed
♦ 23, 88, 94

Papaver macrostomum Boiss. & E.Huet
Papaveraceae
♦ Weed
♦ 272

Papaver nudicaule L.
Papaveraceae
♦ Icelandic poppy, Arctic poppy, Iceland poppy
♦ Weed, Casual Alien
♦ 39, 161, 280, 297
♦ pH, cultivated, herbal, toxic.

Papaver orientale L.
Papaveraceae
♦ oriental poppy, idänunikko
♦ Weed, Naturalised, Cultivation Escape, Casual Alien
♦ 40, 42, 101, 161
♦ pH, cultivated, herbal, toxic.

Papaver pavoninum Fisch. & C.A.Mey.
Papaveraceae
♦ Turkestaninunikko
♦ Weed, Casual Alien
♦ 42, 243
♦ cultivated.

Papaver pinnatifidum Mor.
Papaveraceae
♦ Weed
♦ 70, 88, 94

Papaver pseudorientale (Fedde) Medw.
Papaveraceae
♦ Casual Alien
♦ 280

Papaver rhoeas L.
Papaveraceae
Papaver insignitum Jord., *Papaver intermedium* Beck, *Papaver roubiaei* Vig., *Papaver tenuissimum* Fedde, *Papaver*

trilobum Wallr., *Papaver tumidulum* Klokov, *Papaver strigosum* (Boenn.) Schur (see), *Papaver tenuissimum* Fedde, *Papaver trilobum* Wallr., *Papaver tumidulum* Klokov
♦ field poppy, common red poppy, corn poppy, Shirley poppy, silkkiunikko, papavero comune, Flanders poppy, papavero fiori rossi, coquelicot, amapola
♦ Weed, Naturalised, Introduced, Garden Escape, Casual Alien, Cultivation Escape
♦ 7, 15, 23, 24, 34, 39, 42, 44, 68, 70, 80, 86, 87, 88, 94, 98, 101, 115, 118, 121, 158, 161, 165, 176, 198, 203, 217, 218, 221, 228, 243, 252, 253, 256, 269, 272, 280, 287, 300
♦ aH, arid, cultivated, herbal, toxic. Origin: Eurasia.

Papaver setigerum DC.
Papaveraceae
= *Papaver somniferum* L. ssp. *setigerum* (DC.) Corb.
♦ Weed, Naturalised
♦ 39, 98, 203
♦ herbal, toxic.

Papaver somniferum L.
Papaveraceae
Papaver album Crantz, *Papaver nigrum* Crantz
♦ opium poppy, breadseed poppy, poppy, oopiumiunikko
♦ Weed, Quarantine Weed, Noxious Weed, Naturalised, Garden Escape, Cultivation Escape
♦ 7, 23, 24, 34, 39, 40, 42, 70, 80, 87, 88, 93, 98, 101, 134, 154, 161, 169, 176, 198, 203, 205, 220, 229, 247, 280, 287, 300
♦ aH, cultivated, herbal, toxic. Origin: Eurasia.

Papaver somniferum L. ssp. setigerum (DC.) Corb.
Papaveraceae
Papaver setigerum DC. (see)
♦ smallflower opium poppy
♦ Weed, Noxious Weed, Naturalised
♦ 86, 147, 176, 243, 269, 280, 287
♦ aH, cultivated, toxic. Origin: Mediterranean.

Papaver somniferum L. ssp. somniferum Mansf.
Papaveraceae
♦ opium poppy
♦ Weed, Noxious Weed, Naturalised
♦ 15, 86, 147, 176, 243, 280
♦ cultivated. Origin: Eurasia.

Papaver × strigosum (Boenn.) Schur (pro sp.)
Papaveraceae
= *Papaver dubium* L. × *Papaver rhoeas* L.
♦ strigose poppy, silosilkkiunikko
♦ Naturalised, Casual Alien
♦ 42, 101
♦ aH.

Pappea capensis Eckl. & Zeyh.
Sapindaceae
♦ indaba tree, jacket plum, kaffir

plum, wild amandel, wild cherry, wild plum tree
♦ Weed, Native Weed
♦ 87, 88, 121
♦ T, cultivated, herbal. Origin: southern Africa.

Pappophorum pappiferum (Lam.) Kuntze
Poaceae
♦ limestone pappusgrass
♦ Naturalised
♦ 32
♦ G, arid.

Paracaryum intermedium (Fresen.) Lipsky
Boraginaceae
♦ Weed
♦ 221

Paracaryum rugulosum (DC.) Boiss.
Boraginaceae
♦ Weed
♦ 221

Paracaryum strictum Boiss.
Boraginaceae
♦ Weed
♦ 87, 88

Paragenipa lancifolia (Boj. ex Bak.) D.D.Tirveng. & E.Robbr.
Rubiaceae
Pyrostria lancifolia Boj. ex Bak.
♦ Quarantine Weed
♦ 220

Paramacrolobium coeruleum (Taub.) J.Leonard
Fabaceae/Caesalpiniaceae
♦ bimba
♦ Quarantine Weed
♦ 220

Parameria barbata (Blume) K.Schum.
Apocynaceae
♦ Quarantine Weed
♦ 220
♦ herbal.

Parapholis incurva (L.) C.E.Hubb.
Poaceae
Aegilops incurva L., *Aegilops incurvata* L., *Lepturus incurvatus* (L.) Trin., *Lepturus incurvus* (L.) Druce, *Pholiurus incurvatus* (L.) A.Hitchc., *Pholiurus incurvus* (L.) Schinz & Thell.
♦ coast barbgrass, curved sicklegrass, sickle grass, curved hardgrass
♦ Weed, Naturalised, Introduced, Environmental Weed
♦ 7, 15, 72, 36, 87, 88, 98, 101, 121, 161, 176, 198, 203, 228, 241, 280, 287, 300
♦ aG, arid/aqua, cultivated, herbal. Origin: Eurasia.

Parapholis strigosa (Dumort.) C.E.Hubb.
Poaceae
Lepiurus strigosus Dumort.
♦ slender barbgrass, strigose sicklegrass, sea hardgrass, hardgrass
♦ Weed, Naturalised, Environmental Weed, Casual Alien
♦ 42, 72, 86, 88, 98, 101, 176, 198, 203, 241, 280, 300

♦ aG, cultivated, herbal. Origin: Mediterranean.

Parapiptadenia rigida (Benth.) Brenan
Fabaceae/Mimosaceae
Acacia angico Mart., *Piptadenia rigida* Benth., *Piptadenia rigida* Benth. var. *grandis* Lindm.
♦ Introduced
♦ 228
♦ arid, cultivated.

Paraserianthes falcataria (L.) I.C.Nielson
Fabaceae/Mimosaceae
Albizia falcataria (L.) Fosb. (see)
♦ molucca albizia, tuhke kerosene, tuhkehn karisihn, ukall ra ngebard, tamaligi palagi, tuhkenkerosin, batai wood, peacocksplume
♦ Weed, Introduced, Environmental Weed
♦ 3, 22, 80, 107, 152, 191, 230
♦ T, cultivated. Origin: South Pacific islands

Paraserianthes lophantha (Willd.) I.C.Nielsen
Fabaceae/Mimosaceae
Acacia distachya (Vent.) J.F.Macbr., *Albizia lophantha* (Willd.) Benth. (see), *Paraserianthes lophantha* (Willd.) I.C.Nielsen ssp. *lophantha* (see)
♦ stinkbean, brush wattle, cape wattle, plume albizia, crested wattle, Australiese albizia, stinkboon, Australian albizia
♦ Weed, Noxious Weed, Naturalised, Native Weed, Introduced, Garden Escape, Environmental Weed
♦ 3, 15, 63, 72, 80, 88, 95, 101, 134, 158, 165, 176, 181, 198, 225, 228, 246, 277, 278, 280, 283, 296
♦ S/T, arid, cultivated, toxic. Origin: Western Australia.

Paraserianthes lophantha (Willd.) I.C.Nielsen ssp. *lophantha*
Fabaceae/Mimosaceae
= *Paraserianthes lophantha* (Willd.) I.C.Nielsen
♦ cape wattle, crested wattle
♦ Naturalised, Environmental Weed
♦ 86, 289
♦ S/T, cultivated. Origin: Australia.

Parastrephia lepidophylla (Wedd.) Cabrera
Asteraceae
Dolichogyne lepidophylla Wedd.
♦ Introduced
♦ 228
♦ T, arid.

Parathesis cubana (A.DC.) Molt. & Maza.
Myrsinaceae
♦ Weed
♦ 14

Parentucellia latifolia (L.) Caruel
Scrophulariaceae
♦ southern red bartsia, red bartsia, broadleaf glandweed
♦ Weed, Naturalised, Environmental Weed
♦ 7, 9, 72, 86, 88, 98, 101, 116, 176, 203,

269, 280, 300
♦ aH, cultivated, herbal. Origin: Europe.

Parentucellia latifolia (L.) Caruel ssp. *latifolia*
Scrophulariaceae
♦ common bartsia
♦ Naturalised
♦ 198

Parentucellia viscosa (L.) Caruel
Scrophulariaceae
♦ yellow glandweed, tarweed, yellow bartsia, sticky parentucellia, parentucellia, yellow parentucellia
♦ Weed, Noxious Weed, Naturalised, Environmental Weed, Casual Alien
♦ 7, 9, 15, 34, 35, 42, 72, 78, 80, 86, 88, 94, 98, 101, 116, 136, 146, 151, 161, 165, 176, 198, 203, 241, 269, 280, 287, 300
♦ aH, cultivated, herbal. Origin: Europe, Mediterranean.

Parietaria alsinifolia Delile
Urticaceae
♦ Weed
♦ 221

Parietaria cretica L.
Urticaceae
♦ Weed
♦ 272
♦ herbal.

Parietaria debilis Forst.f.
Urticaceae
♦ Florida pellitory
♦ Weed
♦ 87, 88, 295
♦ arid, cultivated, herbal.

Parietaria diffusa Mert. & Koch
Urticaceae
= *Parietaria judaica* L.
♦ pellitory of the wall
♦ Weed, Naturalised
♦ 272, 287
♦ pH, cultivated, herbal.

Parietaria floridana Nutt.
Urticaceae
♦ Florida pellitory
♦ Weed
♦ 87, 88, 161, 218, 249
♦ herbal.

Parietaria judaica L.
Urticaceae
Parietaria diffusa Mert. & Koch (see)
♦ pellitory, wall pellitory, spreading pellitory, sticky weed, pellitory of the wall, muuriyrtti
♦ Weed, Noxious Weed, Naturalised, Garden Escape, Environmental Weed, Casual Alien
♦ 7, 34, 42, 86, 87, 88, 98, 101, 147, 155, 198, 203, 269, 280, 289, 300
♦ pH, cultivated, herbal. Origin: Mediterranean.

Parietaria lusitanica L.
Urticaceae
♦ Weed
♦ 70
♦ cultivated.

Parietaria officinalis L.
Urticaceae
Parietaria erecta Mert. & Koch
♦ pellitory, wall pellitory, pellitory of the wall, upright pellitory
♦ Weed, Quarantine Weed, Noxious Weed, Naturalised
♦ 70, 76, 86, 87, 88, 101, 147, 203, 272, 280, 295
♦ cultivated, herbal.

Parietaria pensylvanica Muhl. ex Willd.
Urticaceae
♦ Pennsylvania pellitory
♦ Weed, Naturalised, Native Weed
♦ 87, 88, 161, 174, 218, 287
♦ aH, herbal. Origin: North America.

Parietaria punctata Willd.
Urticaceae
♦ Weed
♦ 221

Parinari capensis Harv. ssp. *capensis*
Chrysobalanaceae
♦ bosapple, dwarf mobola, sand apple
♦ Native Weed
♦ 121
♦ S. Origin: southern Africa.

Paris polyphylla Sm.
Liliaceae/Trilliaceae
♦ herb Paris, rhizoma paridis
♦ Quarantine Weed
♦ 220
♦ herbal. Origin: China, India.

Paris quadrifolia L.
Liliaceae/Trilliaceae
Paris quadrifolius L.
♦ herb Paris, sudenmarja
♦ Weed
♦ 39, 272
♦ pH, cultivated, herbal, toxic.

Parkia bicolor A.Chev.
Fabaceae/Mimosaceae
Parkia agboensis A.Chev., *Parkia klainei* A.Chev., *Parkia zenkeri* Harms
♦ bicolor parkia, lilembe, lulele
♦ Introduced
♦ 228
♦ arid.

Parkia korom Kaneh.
Fabaceae/Mimosaceae
♦ kurum
♦ Introduced
♦ 230
♦ T.

Parkia platycephala Benth.
Fabaceae/Mimosaceae
♦ visgueiro
♦ Introduced
♦ 228
♦ arid.

Parkinsonia aculeata L.
Fabaceae/Caesalpiniaceae
♦ Jerusalem thorn, parkinsonia, horse bean, retama, Mexican palo verde
♦ Weed, Quarantine Weed, Noxious Weed, Naturalised, Introduced, Garden Escape, Environmental Weed, Cultivation Escape

♦ 3, 7, 22, 34, 62, 76, 86, 87, 88, 93, 98, 107, 121, 147, 152, 171, 179, 191, 203, 205, 218, 228, 279, 283, 290, 295
♦ T, arid, cultivated, herbal, toxic. Origin: southern US, Central and northern South America.

Parmentiera aculeata (Kunth) Seem.
Bignoniaceae
Parmentiera edulis DC. (see)
♦ cuachilote, guajalote
♦ Naturalised, Garden Escape
♦ 86, 101, 261
♦ cultivated. Origin: Mexico, Central America.

Parmentiera cereifera Seem.
Bignoniaceae
♦ candle tree, arbol de cera, palo de vela
♦ Naturalised, Cultivation Escape
♦ 86, 101, 261
♦ cultivated, herbal. Origin: Central and southern North America.

Parmentiera edulis DC.
Bignoniaceae
= *Parmentiera aculeata* (Kunth) Seem.
♦ cuajilote
♦ Naturalised
♦ 86, 98
♦ cultivated, herbal.

Parnassia palustris L.
Saxifragaceae/Parnassiaceae
♦ grass of Parnassus, vilukko, marsh grass of Parnassus, northern grass of Parnassus
♦ Weed
♦ 23, 88, 272, 297
♦ pH, cultivated, herbal.

Parochetus communis Buch.-Ham. ex D.Don
Fabaceae/Papilionaceae
♦ shamrock pea
♦ Weed, Naturalised
♦ 15, 280
♦ cultivated.

Paronychia arabica (L.) DC.
Caryophyllaceae/Illecebraceae
♦ Weed
♦ 221

Paronychia argentea Lam.
Caryophyllaceae/Illecebraceae
♦ Algerian tea, mattovuohenpolvi
♦ Weed, Naturalised, Casual Alien
♦ 42, 70, 86, 88, 94, 98, 203, 221
♦ pH, cultivated, herbal. Origin: Mediterranean.

Paronychia brasiliana DC.
Caryophyllaceae/Illecebraceae
Paronychia braziliana DC. (see)
♦ nailwort
♦ Weed, Naturalised
♦ 86, 87, 88, 98, 134, 176, 198, 203, 269, 280
♦ pH. Origin: South America.

Paronychia braziliana DC.
Caryophyllaceae/Illecebraceae
= *Paronychia brasiliana* DC.
♦ Brazilian paronychia, chickweed, nailwort

♦ Weed
♦ 51, 121
♦ aH. Origin: South America.

Paronychia cymosa (L.) DC.
Caryophyllaceae/Illecebraceae
♦ pikkuvuohenpolvi
♦ Casual Alien
♦ 42

Paronychia desertorum Boiss.
Caryophyllaceae/Illecebraceae
♦ Weed
♦ 221

Paronychia echinulata Chater
Caryophyllaceae/Illecebraceae
♦ Eurasian nailwort
♦ Weed, Naturalised
♦ 70, 88, 94, 101

Paronychia franciscana Eastw.
Caryophyllaceae/Illecebraceae
♦ San Francisco nailwort, nailwort
♦ Weed, Naturalised
♦ 86, 98, 198, 203
♦ pH, herbal. Origin: Chile.

Paronychia mexicana Hemsl.
Caryophyllaceae/Illecebraceae
♦ Weed
♦ 199

Paronychia nivea DC.
Caryophyllaceae/Illecebraceae
= *Paronychia capitata* (L.) Lam. (NoR)
♦ Weed
♦ 88

Parthenium argentatum Gray
Asteraceae
♦ guayule, kumipensas
♦ Weed, Introduced
♦ 23, 39, 88, 228
♦ S, arid, promoted, herbal, toxic. Origin: southern North America, Mexico.

Parthenium hysterophorus L.
Asteraceae
Argyrochaeta bipinnatifida Cav., *Villanova bipinnatifida* Ort., *Echetrosis pentaspermum* Phil.
♦ ragweed parthenium, parthenium weed, ragweed, Santa Maria feverfew, bitterweed, carrot grass, congress grass, false ragweed, ragweed parthenium, whitetop, demoina weed, ajenjo cimarrón, yerba amarga
♦ Weed, Quarantine Weed, Noxious Weed, Naturalised, Environmental Weed
♦ 6, 14, 18, 39, 55, 62, 63, 66, 68, 76, 86, 87, 88, 93, 98, 101, 121, 147, 158, 161, 169, 171, 178, 179, 191, 199, 203, 218, 220, 232, 237, 240, 241, 243, 245, 255, 256, 258, 261, 268, 269, 283, 287, 290, 295, 297
♦ pH, herbal, toxic. Origin: tropical America, Caribbean.

Parthenocissus Planch. spp.
Vitaceae
♦ Virginia creeper, creeper
♦ Weed
♦ 243
♦ pC.

Parthenocissus inserta (Kern.) Fritsch
Vitaceae
= *Parthenocissus quinquefolia* (L.)
Planch.
♦ false Virginia creeper, Virginia
creeper
♦ Naturalised, Cultivation Escape
♦ 40, 42, 280
♦ cultivated, herbal.

Parthenocissus quinquefolia (L.) Planch.
Vitaceae
Parthenocissus inserta (Kern.) Fritsch
(see)
♦ Virginia creeper, Boston ivy,
Japanese ivy
♦ Weed, Naturalised, Cultivation
Escape, Garden Escape
♦ 8, 32, 42, 86, 88, 98, 161, 194, 203, 209,
211, 218, 243, 247, 252, 287
♦ pC, cultivated, herbal, toxic. Origin:
eastern North America.

Parthenocissus tricuspidata (Sieb. & Zucc.) Planch.
Vitaceae
Ampelopsis tricuspidata Siebold & Zucc.,
Cissus thunbergii Siebold & Zucc.,
Parthenocissus thunbergii (Siebold
& Zucc.) Nakai, *Psedera thunbergii*
(Siebold & Zucc.) Nakai, *Psedera
tricuspidata* (Siebold & Zucc.) Rehder,
Quinaria tricuspidata Koehne, *Vitis
inconstans* Miq., *Vitis taquetii* H.Lév.
♦ Virginia creeper, Boston ivy,
Japanese ivy
♦ Weed, Naturalised, Garden Escape,
Environmental Weed, Casual Alien
♦ 39, 80, 101, 116, 155, 161, 280, 286
♦ pC, cultivated, herbal, toxic.

Parthenocissus vitacea (Knerr) Hitchc.
Vitaceae
♦ Virginia creeper, woodbine
♦ Weed
♦ 161
♦ S, herbal, toxic.

Parviopuntia Soulaire spp.
Cactaceae
= *Opuntia* Mill. spp.
♦ Weed, Quarantine Weed
♦ 76, 88, 220

Pascalia glauca Ortega
Asteraceae
Wedelia glauca (Ortega) O.Hoffm. ex
Hicken (see)
♦ beach creeping oxeye
♦ Naturalised
♦ 101

Pascopyrum smithii (Rydb.) Á.Löve
Poaceae
Agropyron smithii Rydb. (see), *Elymus
smithii* (Rydb.) Gould
♦ western wheatgrass
♦ Weed
♦ 161
♦ pG, arid, herbal.

Pasithea coerulea (Ruiz & Pav.) D.Don
Liliaceae/Phormiaceae
♦ Quarantine Weed
♦ 220
♦ arid.

Paspalidium distans (Trin.) Hughes
Poaceae
♦ spreading panicgrass
♦ Naturalised
♦ 101
♦ G.

Paspalidium flavidum (Retz.) A.Camus
Poaceae
Panicum flavidum Retz.
♦ Weed
♦ 87, 88, 90, 262, 275
♦ pG, aqua, herbal. Origin: Australia.

Paspalidium geminatum (Forssk.) Stapf
Poaceae
Panicum appressum (Lam.) Doell.,
Panicum carnosum Salzm. ex Steud.,
Paspalum appressum Lam., *Paspalum
geminatum* Forssk. (see)
♦ Egyptian panicgrass, watergrass
♦ Weed, Quarantine Weed
♦ 76, 87, 88, 90, 185, 203, 220, 221
♦ G, arid.

Paspalidium gracile (R.Br.) Hughes
Poaceae
♦ Introduced
♦ 228
♦ G, arid, cultivated.

Paspalidium jubiflorum (Trin.) Hughes
Poaceae
♦ Introduced
♦ 228
♦ G, arid, cultivated.

Paspalidium obtusifolium (Del.) Simps.
Poaceae
♦ Weed
♦ 87, 88
♦ G.

Paspalidium paludivagum (Hitchc. & Chase) Par.
Poaceae
= *Paspalidium geminatum* (Forsk.) Stapf
var. *paludivagum* (Hitchc. & Chase)
Gould (NoR)
♦ paspalidio
♦ Weed
♦ 237, 295
♦ G.

Paspalidium philippianum Parodi
Poaceae
♦ Philippine watercrown grass
♦ Naturalised
♦ 101
♦ G.

Paspalidium punctatum (Burm.) Camus
Poaceae
Panicum punctatum Burm.f.
♦ Weed
♦ 90
♦ pG, aqua.

Paspalum L. spp.
Poaceae
♦ crowngrass, Dallas grass
♦ Weed, Naturalised
♦ 18, 88, 198
♦ G.

Paspalum acuminatum Raddi
Poaceae
♦ brook crowngrass
♦ Weed

♦ 88, 179
♦ G.

Paspalum almum Chase
Poaceae
♦ Comb's crowngrass
♦ Naturalised, Introduced
♦ 101, 228
♦ G, arid.

Paspalum atratum Swallen
Poaceae
♦ atra paspalum
♦ Weed, Quarantine Weed
♦ 76, 88
♦ G, cultivated.

Paspalum boscianum Flüggé
Poaceae
Paspalum amazonicum Trin., *Paspalum
brunneum* Bosc ex Flüggé, *Paspalum
confertum* J.Le Conte, *Paspalum
purpurascens* Elliott, *Paspalum virgatum*
Walter var. *purpurascens* (Elliott)
A.W.Wood, *Paspalum virgatum* Walter
♦ bull crowngrass, bull paspalum
♦ Quarantine Weed
♦ 76
♦ G, herbal.

Paspalum candidum (Humb. & Bonpl. ex Flüggé) Kunth
Poaceae
♦ Weed
♦ 157
♦ aG, arid.

Paspalum cartilagineum Presl
Poaceae
= *Paspalum scrobiculatum* L.
♦ Weed
♦ 88, 170, 191
♦ G.

Paspalum ciliatifolium Michx.
Poaceae
= *Paspalum setaceum* Michx.
♦ fringeleaf paspalum
♦ Weed, Naturalised
♦ 86, 88, 90, 218
♦ pG, aqua, cultivated, herbal.

Paspalum commersonii Lam.
Poaceae
= *Paspalum scrobiculatum* L.
♦ koda grass, scrobic
♦ Weed
♦ 87, 88, 170, 191
♦ G.

Paspalum conjugatum Berg.
Poaceae
Paspalum ciliatum Trin.
♦ buffalograss, hilo grass, sour grass,
thruston grass, sour paspalum, yellow
grass, caraboa grass, tororuco, yaa
nom non, kandanda, rehn wai, grama,
antenita
♦ Weed, Naturalised, Introduced,
Environmental Weed
♦ 3, 6, 7, 12, 14, 80, 86, 87, 88, 90, 93, 98,
107, 151, 152, 153, 157, 170, 186, 199,
203, 204, 218, 230, 235, 238, 243, 245,
255, 257, 262, 269, 273, 274, 275, 276,
280, 281, 286, 287, 295, 297
♦ a/pG, herbal. Origin: tropical
America.

Paspalum conjugatum Berg. var. conjugatum
Poaceae
- Weed
- 206
- G.

Paspalum conspersum Schrad. ex Schult.
Poaceae
Paspalum virgatum Walter var. *conspersum* (Schrad. ex Schult.) Döll
- capim milhã do brejo
- Weed
- 243, 245, 255
- pG. Origin: tropical America.

Paspalum convexum Humb. & Bonpl. ex Flueggé
Poaceae
- Latin American crowngrass
- Weed, Naturalised, Introduced
- 101, 199, 228
- G, arid.

Paspalum coryphaeum Trin.
Poaceae
- emperor crowngrass
- Naturalised
- 101
- G.

Paspalum dasypleurum Desv.
Poaceae
- Weed, Naturalised
- 98, 203
- G.

Paspalum dilatatum Poir.
Poaceae
Digitaria dilatata (Poir.) Coste
- paspalum, watergrass, water paspalum, pasto miel, caterpillar grass, Leichhardt grass, bastard milletgrass, common paspalum, golden crowngrass, dallis grass, hairy flowered paspalum, large watergrass, large waterseed paspalum, yerba dalis, grama comprida
- Weed, Naturalised, Introduced, Environmental Weed, Cultivation Escape, Casual Alien
- 3, 6, 7, 9, 15, 30, 34, 39, 42, 72, 79, 80, 86, 87, 88, 90, 93, 98, 101, 121, 158, 161, 176, 179, 180, 186, 198, 203, 204, 205, 211, 212, 218, 228, 236, 237, 243, 244, 245, 249, 253, 255, 256, 261, 263, 272, 280, 286, 287, 289, 295, 296
- aG, aqua, cultivated, herbal, toxic. Origin: South America.

Paspalum distachyon Willd. ex Döll
Poaceae
= *Paspalum notatum* Flüggé
- Weed
- 87, 88
- G.

Paspalum distichum L.
Poaceae
Digitaria paspaloides Michx., *Paspalum distichum* L. var. *indutum* Shinners (see), *Panicum paspaliforme* J.Presl, *Paspalum paspalodes* (Michx.) Scribn. (see), *Paspalum paspaloides* (Michx.) Scribn. (see)
- knotgrass, water couch, mercer grass, couch paspalum, water fingergrass, paspalum, buffalo quick paspalum
- Weed, Naturalised, Introduced, Environmental Weed
- 7, 15, 40, 51, 72, 80, 86, 87, 88, 98, 157, 158, 161, 170, 176, 180, 181, 198, 199, 200, 203, 204, 217, 218, 225, 228, 236, 237, 241, 243, 246, 253, 262, 263, 269, 273, 274, 275, 280, 286, 287, 295, 297
- wpG, cultivated, herbal. Origin: cosmopolitan.

Paspalum distichum L. var. indutum Shinners
Poaceae
= *Paspalum distichum* L.
- Naturalised
- 287
- G.

Paspalum exaltatum J.Presl
Poaceae
- Weed, Naturalised
- 86, 98, 203
- G. Origin: South America.

Paspalum fasciculatum Willd. ex Flueggé
Poaceae
- Mexican crowngrass
- Weed, Naturalised, Cultivation Escape
- 7, 87, 88, 98, 101, 199, 203, 243, 261
- G, aqua, cultivated. Origin: tropical South and Central America.

Paspalum fimbriatum Kunth
Poaceae
- Panama paspalum, fimbriate paspalum, Colombia grass
- Weed, Naturalised, Introduced
- 3, 32, 87, 88, 90, 179, 191, 218, 228, 256, 287
- aG, arid, herbal.

Paspalum fluitans (Elliott) Kunth
Poaceae
= *Paspalum repens* P.J.Bergius var. *fluitans* (Elliott) Wipff & S.D.Jones (NoR)
- water paspalum, horsetail paspalum, floating grass
- Weed
- 87, 88, 90, 218
- pG, aqua, herbal.

Paspalum gardnerianum Nees
Poaceae
- Introduced
- 228
- G, arid.

Paspalum geminatum Forssk.
Poaceae
= *Paspalidium geminatum* (Forssk.) Stapf
- Weed
- 87, 88
- G.

Paspalum haenkeanum Presl
Poaceae
- Weed
- 87, 88
- G.

Paspalum hydrophilum Henr.
Poaceae
- water paspalum
- Naturalised
- 101
- G.

Paspalum intermedium Munro ex Morong & Britton
Poaceae
- intermediate paspalum
- Naturalised, Introduced
- 101, 228
- G, arid.

Paspalum laeve Michx.
Poaceae
Paspalum alternans Steud., *Paspalum angustifolium* J.Le Conte, *Paspalum angustifolium* var. *tenue* A.W.Wood, *Paspalum australe* Nash, *Paspalum laeve* var. *angustifolium* (J.Le Conte) Vasey, *Paspalum laeve* var. *australe* (Nash) Nash ex Hitchc., *Paspalum laeve* var. *brevifolium* Vasey, *Paspalum laeve* var. *undulosum* (J.Le Conte) A.W.Wood, *Paspalum lecomteanum* Schult., *Paspalum punctulatum* Bertol., *Paspalum tenue* Darby, *Paspalum undulosum* J.Le Conte
- field paspalum
- Weed, Quarantine Weed
- 76, 87, 88, 90, 161, 203, 207, 218, 220, 249
- pG, herbal. Origin: North America.

Paspalum lividum Trin.
Poaceae
Paspalum hieronymi Hack., *Paspalum proliferum* Arech. (see)
- longtom
- Weed
- 87, 88, 90, 237, 295
- pG, arid.

Paspalum longifolium Roxb.
Poaceae
Paspalum cognatum Steud., *Paspalum flexuosum* Klein ex J.S.Presl, *Paspalum houttuynii* Van Hall ex De Vriese, *Paspalum longifolium* var. *hirsutum* Boerl., *Paspalum longifolium* var. *trichocoleum* Hack., *Paspalum orbiculare* G.Forst. var. *otobedii* Fosberg & Sachet, *Paspalum platycoleum* Ridl., *Paspalum scrobiculatum* L. var. *longifolium* (Roxb.) Domin, *Paspalum scrobiculatum* var. *philippinense* Merr., *Paspalum sumatrense* Roth ex Roem. & Schult.
- long leaved paspalum, longleaf paspalum
- Weed, Naturalised
- 88, 90, 170, 191, 287
- pG, aqua. Origin: Asia, Australia.

Paspalum macrophyllum Kunth
Poaceae
- bigleaf paspalum
- Naturalised
- 101
- G.

Paspalum maculosum Trin.
Poaceae

- ◆ Weed, Quarantine Weed
- ◆ 76, 87, 88, 203, 220
- ◆ G.

Paspalum malacophyllum Trin.
Poaceae
- ◆ ribbed paspalum
- ◆ Naturalised
- ◆ 101
- ◆ G.

Paspalum maritimum Trin.
Poaceae
- ◆ coastal sand paspalum
- ◆ Weed
- ◆ 87, 88, 255
- ◆ pG, arid. Origin: tropical America.

Paspalum melanospermum Desv. ex Poir.
Poaceae
- ◆ Weed
- ◆ 87, 88
- ◆ G.

Paspalum millegrana Schrad. ex Schult.
Poaceae
Paspalum karwinskyi Fourn., *Paspalum underwoodii* Nash
- ◆ paja brava
- ◆ Weed, Quarantine Weed
- ◆ 76, 87, 88, 203, 220
- ◆ G, arid.

Paspalum minus E.Fourn.
Poaceae
- ◆ matted paspalum
- ◆ Naturalised
- ◆ 287
- ◆ G.

Paspalum modestum Mez.
Poaceae
- ◆ capim do brejo
- ◆ Weed
- ◆ 255
- ◆ a/pG, aqua. Origin: South America.

Paspalum nicorae Parodi
Poaceae
- ◆ Brunswick grass
- ◆ Weed, Quarantine Weed, Naturalised
- ◆ 76, 86, 87, 88, 101, 179, 237, 295
- ◆ G, cultivated. Origin: South America.

Paspalum notatum Flügge
Poaceae
Paspalum cromyorhizon Trin. ex Döll, *Paspalum distachyon* Willd. ex Döll (see), *Paspalum notatum* var. *cromyorhizon* (Trin. ex Döll) Herter, *Paspalum notatum* var. *eriorhizon* Griseb., *Paspalum notatum* var. *latiflorum* Döll (see), *Paspalum notatum* var. *saurae* Parodi (see), *Paspalum saltense* Arechav., *Paspalum saurae* (Parodi) Parodi, *Paspalum taphrophyllum* Steud., *Paspalum tephophyllum* Steud., *Paspalum uruguayense* Arechav.
- ◆ Bahia grass, water couch, notatum grass, lawn paspalum
- ◆ Weed, Naturalised, Environmental Weed

- ◆ 14, 39, 80, 86, 87, 88, 90, 98, 101, 121, 151, 153, 158, 161, 191, 203, 204, 237, 243, 249, 255, 286, 287, 295
- ◆ pG, aqua, cultivated, herbal, toxic. Origin: South America.

Paspalum notatum Flueggé var. latiflorum Döll
Poaceae
= *Paspalum notatum* Flüggé
- ◆ Bahia grass
- ◆ Naturalised
- ◆ 101
- ◆ G.

Paspalum notatum Flueggé var. notatum
Poaceae
- ◆ Bahia grass
- ◆ Weed
- ◆ 179
- ◆ G.

Paspalum notatum Flueggé var. saurae Parodi
Poaceae
= *Paspalum notatum* Flüggé
- ◆ Bahia grass
- ◆ Naturalised
- ◆ 101, 287
- ◆ G.

Paspalum nutans Lam.
Poaceae
- ◆ Weed
- ◆ 28, 87, 88, 206, 243
- ◆ G.

Paspalum orbiculare Forst.
Poaceae
= *Paspalum scrobiculatum* L.
- ◆ ricegrass paspalum
- ◆ Weed, Naturalised
- ◆ 67, 87, 88, 191, 204, 218, 274, 275, 280, 286, 297
- ◆ pG, herbal.

Paspalum orbiculare Forst. var. orbiculare
Poaceae
- ◆ reh nta
- ◆ Introduced
- ◆ 230
- ◆ G.

Paspalum paniculatum L.
Poaceae
- ◆ Russell river grass, galmarra grass, arrocillo, grama touceira
- ◆ Weed, Naturalised, Casual Alien
- ◆ 3, 6, 28, 86, 87, 88, 98, 191, 203, 206, 243, 245, 255, 280, 281, 287
- ◆ pG. Origin: South America.

Paspalum paspalodes (Michx.) Scribn.
Poaceae
= *Paspalum distichum* L.
- ◆ knotgrass, mercer grass, water couch, buffalo quick paspalum, couch paspalum
- ◆ Weed, Native Weed
- ◆ 39, 70, 121, 185, 208, 250, 272
- ◆ pG, aqua, toxic. Origin: tropical America.

Paspalum paspaloides (Michx.) Scribn.
Poaceae
= *Paspalum distichum* L.

- ◆ knotgrass, couch paspalum, jointgrass, water couch, mercer grass
- ◆ Weed
- ◆ 87, 88, 90, 181, 221, 243
- ◆ pG, aqua, cultivated, herbal.

Paspalum pilosum Lam.
Poaceae
- ◆ Weed
- ◆ 88
- ◆ G.

Paspalum plicatulum Michx.
Poaceae
Paspalum oligostachyum Salz. ex Steud. var. *pilosum* Salz. ex Doell., *Paspalum saxatile* Salz. ex Doell., *Paspalum texanum* Swallen
- ◆ brownseed paspalum, circular fruited paspalum
- ◆ Weed, Naturalised, Introduced
- ◆ 14, 86, 87, 88, 90, 93, 98, 157, 203, 228, 245, 255
- ◆ pG, arid, cultivated, herbal. Origin: Brazil.

Paspalum polystachyum R.Br.
Poaceae
= *Paspalum scrobiculatum* L.
- ◆ Weed
- ◆ 87, 88
- ◆ G.

Paspalum proliferum Arech.
Poaceae
= *Paspalum lividum* Trin.
- ◆ Weed
- ◆ 87, 88
- ◆ G.

Paspalum prostratum Scribn. & Merr.
Poaceae
- ◆ Weed
- ◆ 199
- ◆ G.

Paspalum pubiflorum Rupr. ex Galeotti
Poaceae
- ◆ smooth scaled glabrum paspalum, hairyseed paspalum
- ◆ Weed, Casual Alien
- ◆ 15, 161, 280
- ◆ G, herbal.

Paspalum pulchellum Kunth
Poaceae
- ◆ grand paspalum
- ◆ Naturalised
- ◆ 101
- ◆ G.

Paspalum pumilum Nees
Poaceae
- ◆ Weed, Quarantine Weed
- ◆ 76, 87, 88, 203, 220
- ◆ G.

Paspalum quadrifarium Lam.
Poaceae
- ◆ tussock paspalum, paja mansa, paspalum
- ◆ Weed, Noxious Weed, Naturalised
- ◆ 86, 87, 88, 98, 198, 203, 236, 237, 295
- ◆ G, cultivated. Origin: South America.

Paspalum racemosum Lam.
Poaceae
- Peruvian paspalum
- Naturalised
- 87, 88, 101
- G, arid, cultivated.

Paspalum regnellii Mez
Poaceae
- Naturalised
- 86
- G. Origin: South America.

Paspalum repens P.J.Bergius
Poaceae
Paspalum gracile Rudge.
- canarana rasteira
- Weed
- 237, 255, 295
- G, aqua. Origin: tropical America.

Paspalum scrobiculatum L.
Poaceae
Paspalum borbatum Schum., *Paspalum borbonicum* Steud., *Paspalum cartilagineum* Presl (see), *Paspalum commersonii* Lam. (see), *Paspalum jardinii* Steud., *Paspalum ledermannii* Mez, *Paspalum orbiculare* Forst. (see), *Paspalum polystachyum* R.Br. (see)
- ricegrass paspalum, kodomillet, ditch millet, kodra, kodo, Indian paspalum, kado millet, native millet, scrobic, water couchgrass, wild paspalum
- Weed, Noxious Weed, Naturalised, Native Weed
- 12, 39, 86, 87, 88, 90, 101, 121, 140, 204, 229, 275
- pG, aqua, cultivated, herbal, toxic. Origin: Africa, India, China, south-east Asia.

Paspalum setaceum (Nash) D.Banks
Poaceae
Paspalum ciliatifolium Michx. (see), *Paspalum debile* Michx., *Paspalum rigidifolium* Nash, *Paspalum separatum* Shinn.
- thin paspalum, bull paspalum, thin paspalum, fall panicgrass, slender crowngrass, downy lens grass, hairy beadgrass, straw coloured hairy beadgrass
- Weed, Quarantine Weed, Introduced
- 161, 228, 249
- pG, arid, herbal.

Paspalum thunbergii Kunth ex Steud.
Poaceae
- Japanese paspalum, Korean paspalum
- Weed
- 68, 87, 88, 90, 204, 263, 275, 286, 297
- pG.

Paspalum urvillei Steud.
Poaceae
Paspalum larrangai Arechav.
- vaseygrass, pasto vasey, giant paspalum, upright paspalum, tall paspalum
- Weed, Naturalised, Introduced, Environmental Weed, Cultivation Escape

- 3, 7, 34, 80, 86, 87, 88, 90, 98, 101, 121, 161, 176, 179, 198, 203, 218, 228, 237, 245, 249, 255, 261, 280, 286, 287, 295, 300
- pG, aqua, cultivated, herbal. Origin: South America.

Paspalum vaginatum Sw.
Poaceae
Paspalum distichom L. var. *vaginatum* (Sweet) Griseb., *Paspalum distichum* auct. non L., *Paspalum sqamatum* Steud.
- seashore paspalum, saltwater couch, water couchgrass
- Weed, Naturalised, Native Weed, Environmental Weed
- 6, 7, 86, 87, 88, 90, 93, 121, 170, 198, 205, 237, 262, 280, 286, 295, 300
- pG, aqua, cultivated, herbal. Origin: Australia.

Paspalum virgatum L.
Poaceae
Paspalum secans Hitchc. & Chase
- talquezal, remolina, cabezona, navajuela, upright paspalum, water couch, camalote blanco, marciega
- Weed, Naturalised, Introduced
- 14, 54, 86, 87, 88, 90, 153, 191, 228, 281
- pG, aqua. Origin: Americas.

Paspalum wettsteinii Hack.
Poaceae
- broadleaf paspalum
- Weed, Naturalised
- 86, 98, 203
- G, cultivated. Origin: South America.

Passerina L. spp.
Thymelaeaceae
- gonna
- Native Weed
- 121
- pS. Origin: southern Africa.

Passerina ericoides L.
Thymelaeaceae
Chymococca empetroides Meisn. (see)
- Quarantine Weed
- 220

Passerina glomerata Thunb.
Thymelaeaceae
- Native Weed
- 121
- pS. Origin: southern Africa.

Passerina montana Thoday
Thymelaeaceae
- Native Weed
- 121
- pS, herbal. Origin: southern Africa.

Passiflora L. spp.
Passifloraceae
- granadillas, passionflower
- Environmental Weed
- 279
- herbal.

Passiflora alba Link & Otto
Passifloraceae
- Weed
- 87, 88

Passiflora antioquiensis H.Karst.
Passifloraceae
- passionflower, banana passionfruit
- Casual Alien
- 280
- cultivated.

Passiflora bicornis Mill.
Passifloraceae
Passiflora pulchella Kunth (see)
- wingleaf passionfruit
- Noxious Weed, Naturalised
- 101, 229

Passiflora biflora Lam.
Passifloraceae
- twoflower passionflower
- Weed, Naturalised, Environmental Weed
- 87, 88, 101, 179
- cultivated.

Passiflora caerulea L.
Passifloraceae
Passiflora coerulea auct. L. (see)
- blue passionflower, common granadilla, grenadilla, passionflower, passionfruit, Brazilian passionflower
- Weed, Noxious Weed, Naturalised, Garden Escape, Environmental Weed
- 3, 39, 72, 86, 88, 95, 98, 101, 121, 198, 203, 246, 279, 280, 283
- pC, cultivated, herbal, toxic. Origin: South America.

Passiflora cincinnata Mast.
Passifloraceae
- crato passionvine
- Weed
- 255
- pC, cultivated, herbal. Origin: tropical America.

Passiflora cinnabarina Lindl.
Passifloraceae
- red passionflower
- Weed, Naturalised, Native Weed, Garden Escape, Environmental Weed
- 39, 72, 86, 88, 176
- pC, cultivated, herbal, toxic. Origin: Australia.

Passiflora coccinea Aubl.
Passifloraceae
- scarlet passionflower, red passionflower
- Weed, Naturalised
- 39, 87, 88, 101
- pC, cultivated, toxic. Origin: tropical South America.

Passiflora coerulea auct. L.
Passifloraceae
= *Passiflora caerulea* L.
- siergrenadella, blue passionflower
- Weed
- 63, 295

Passiflora edulis Sims
Passifloraceae
- purple granadilla, passionfruit, yellow passionfruit, purple passionfruit, liliko'i, qarandila, vaine tonga, pasio, pompom en wai
- Weed, Naturalised, Introduced, Garden Escape, Environmental Weed,

Cultivation Escape
♦ 3, 15, 18, 22, 39, 63, 73, 86, 87, 88, 95, 98, 101, 107, 121, 134, 151, 179, 191, 198, 203, 225, 230, 246, 257, 261, 279, 280, 283
♦ pC, cultivated, herbal, toxic. Origin: South America.

Passiflora filamentosa **Cav.**
Passifloraceae
♦ Weed, Naturalised
♦ 7, 86, 98, 203
♦ Origin: Americas.

Passiflora foetida **Vell.**
Passifloraceae
♦ stinking passionflower, love in a mist, mossy passionflower, scarlet fruited passionflower, wild passionfruit, running pop, ka thok rok, dulce, kudamono, pasio vao, vaine 'ae kuma, pohapoha, tea biku, sou, loliloli ni kalavo, wild water lemon, fetid passionflower
♦ Weed, Naturalised, Environmental Weed
♦ 3, 6, 7, 12, 39, 73, 80, 87, 88, 93, 98, 101, 107, 112, 121, 157, 170, 179, 203, 204, 209, 217, 218, 239, 262, 269, 276, 286, 287, 297, 300
♦ C, arid, cultivated, herbal, toxic. Origin: tropical America.

Passiflora foetida **L. var.** *foetida*
Passifloraceae
♦ fetid passionflower
♦ Naturalised, Environmental Weed
♦ 86
♦ Origin: South America, West Indies.

Passiflora foetida **L. var.** *gossypifolia* **(Desv. ex Ham.) Mast.**
Passifloraceae
♦ Naturalised, Environmental Weed
♦ 86
♦ Origin: South America, West Indies.

Passiflora foetida **L. var.** *isthmia* **Killip**
Passifloraceae
♦ scarletfruit passionflower
♦ Naturalised
♦ 101

Passiflora foetida **L. var.** *riparia* **(C.Wright) Killip**
Passifloraceae
= *Passiflora ciliata* Ait. var. *riparia* C.Wright (NoR)
♦ Naturalised, Environmental Weed
♦ 86
♦ Origin: South America, West Indies.

Passiflora gracilis **Jacq. ex Link**
Passifloraceae
♦ crinkled passionflower, annual passionflower
♦ Naturalised
♦ 101
♦ cultivated.

Passiflora incarnata **L.**
Passifloraceae
♦ maypop passionflower, purple passionflower, apricot vine, maypop
♦ Weed

♦ 39, 84, 87, 88, 161, 207, 218, 243
♦ pC, cultivated, herbal, toxic. Origin: North America.

Passiflora laurifolia **Linn.**
Passifloraceae
♦ yellow granadilla, belle apple, pasio, yellow water lemon, bellapple, golden bellapple, water lemon
♦ Weed, Naturalised, Cultivation Escape
♦ 3, 39, 86, 191, 233, 252
♦ cultivated, herbal, toxic. Origin: South America, West Indies.

Passiflora ligularis **Juss.**
Passifloraceae
♦ sweet granadilla, yellow passionfruit
♦ Weed, Naturalised, Environmental Weed
♦ 3, 22, 80, 101, 191, 257
♦ pC, cultivated, herbal.

Passiflora lutea **Ruiz & Pav. ex Mast.**
Passifloraceae
♦ yellow passionflower
♦ Weed
♦ 161
♦ cultivated, herbal.

Passiflora maliformis **L.**
Passifloraceae
♦ conch apple, sweet calabash, apple fruited granadilla
♦ Weed
♦ 3, 6, 88, 191
♦ cultivated, herbal. Origin: South America.

Passiflora manicata **(Juss.) Pers.**
Passifloraceae
♦ red passionflower
♦ Naturalised
♦ 101
♦ cultivated.

Passiflora mixta **L.**
Passifloraceae
♦ banana passionfruit, passionflower, northern banana passionfruit
♦ Weed, Quarantine Weed, Naturalised, Environmental Weed
♦ 3, 191, 225, 246, 280
♦ pC, cultivated.

Passiflora mollissima **(Kunth) L.H.Bailey**
Passifloraceae
= *Passiflora tarminiana* Coppens & Barney sp. *nov.* (NoR) [where *Passiflora mollissima* (Kunth) L.H.Bailey misapp.]
♦ banana poka, banana passionfruit, bananaadilla, pink banana passionfruit
♦ Weed, Quarantine Weed, Noxious Weed, Naturalised, Garden Escape, Environmental Weed
♦ 3, 15, 18, 22, 37, 72, 80, 86, 88, 95, 98, 101, 151, 152, 155, 165, 176, 191, 198, 203, 225, 229, 243, 246, 280, 283, 289, 296
♦ pC, cultivated, herbal. Origin: tropical South America.

Passiflora morifolia **Mast.**
Passifloraceae
♦ woodland passionflower
♦ Weed, Naturalised

♦ 86, 98, 101, 203
♦ cultivated. Origin: South America.

Passiflora × pfordtii hort. **ex O.Deg.**
Passifloraceae
= *Passiflora alata* C.Curtis × *Passiflora caerulea* L.
♦ Weed
♦ 179

Passiflora pinnatistipula **Cav.**
Passifloraceae
♦ gulupa, tin tin
♦ Naturalised
♦ 280
♦ cultivated.

Passiflora pulchella **Kunth**
Passifloraceae
= *Passiflora bicornis* Mill.
♦ wingleaf passionflower, two lobed passionflower
♦ Weed, Quarantine Weed
♦ 3, 76, 87, 88, 191, 203, 218, 220
♦ cultivated, herbal.

Passiflora quadrangularis **L.**
Passifloraceae
♦ granadilla, giant granadilla, parapotina maata, palatini, vine fua lalahi, tinitini, pasione, pasio, kudamono, passionfruit granadilla
♦ Weed, Naturalised, Introduced, Environmental Weed, Cultivation Escape
♦ 3, 32, 39, 86, 98, 101, 107, 191, 203, 230, 252, 257, 261
♦ pC, cultivated, herbal, toxic. Origin: Central America.

Passiflora × rosea **(H.Karst.) Killip**
Passifloraceae
Poggendorffia rosea Karst., *Tacsonia rosea* (Karst.) Sodiro
♦ Casual Alien
♦ 280

Passiflora rubra **L.**
Passifloraceae
♦ passionflower, Dutchman's laudanum
♦ Weed, Environmental Weed
♦ 3, 39, 87, 88, 152, 191
♦ cultivated, herbal, toxic. Origin: tropical America.

Passiflora sanguinolenta **Mast. & Linden**
Passifloraceae
♦ Naturalised
♦ 86
♦ cultivated.

Passiflora sexflora **A.Juss.**
Passifloraceae
♦ goat's foot, six flowered passionflower
♦ Weed
♦ 87, 88
♦ cultivated.

Passiflora suberosa **L.**
Passifloraceae
♦ devil's pumpkin, indigo berry, huehue haole, wild passionfruit, corky passionflower, corky passionfruit, pointed leaf passionfruit

♦ Weed, Quarantine Weed, Noxious Weed, Naturalised, Garden Escape, Environmental Weed
♦ 3, 14, 22, 39, 73, 76, 80, 86, 87, 88, 93, 95, 98, 121, 151, 152, 191, 201, 203, 269, 283, 287
♦ pC, arid, cultivated, herbal, toxic. Origin: South America.

Passiflora subpeltata Ortega
Passifloraceae
♦ granadina, white passionflower, wild grenadella, wild grenadilla
♦ Weed, Noxious Weed, Naturalised, Garden Escape, Environmental Weed
♦ 3, 39, 73, 86, 87, 88, 95, 98, 101, 121, 158, 191, 198, 203, 269, 283
♦ pC, cultivated, herbal, toxic. Origin: South America.

Passiflora tuberosa Jacq.
Passifloraceae
♦ tuberous passionflower
♦ Naturalised
♦ 101
♦ cultivated.

Passiflora violacea Vell.
Passifloraceae
♦ Naturalised
♦ 86
♦ cultivated.

Passiflora vitifolia Kunth
Passifloraceae
♦ perfumed passionflower, red passionflower
♦ Naturalised
♦ 101
♦ cultivated, herbal.

Pastinaca sativa L.
Apiaceae
Anethum pastinaca Wibel, *Peucedanum sativum* S.Watson
♦ wild parsnip, bird's nest, hart's eye, madnip, hogweed
♦ Weed, Noxious Weed, Naturalised, Introduced, Environmental Weed
♦ 7, 15, 23, 34, 39, 44, 52, 70, 80, 86, 87, 88, 94, 98, 101, 102, 121, 151, 161, 165, 174, 176, 195, 198, 203, 207, 210, 218, 222, 229, 237, 241, 247, 269, 272, 280, 287, 295, 300
♦ a/pH, cultivated, herbal, toxic. Origin: Europe.

Patrinia heterophylla Bunge
Valerianaceae
♦ Weed
♦ 275, 297
♦ pH.

Patrinia scabiosaefolia Fisch.
Valerianaceae
♦ Dahurian patrinia, mountain parsley, ominaeshi
♦ Weed
♦ 297
♦ pH, aqua, cultivated, herbal. Origin: east Asia.

Patrinia scabra Bunge
Valerianaceae
♦ scabrous patrinia
♦ Weed

♦ 297

Paullinia cupana Kunth
Sapindaceae
♦ guaranà
♦ Quarantine Weed
♦ 39, 220
♦ herbal, toxic.

Paullinia densiflora Sm.
Sapindaceae
♦ Weed
♦ 87, 88

Paullinia fuscescens Kunth
Sapindaceae
♦ mouldy bread and cheese
♦ Weed
♦ 14
♦ herbal.

Paullinia pinnata L.
Sapindaceae
♦ bread and cheese, mpatwe, pbamampbo
♦ Weed, Introduced
♦ 39, 87, 88, 228
♦ arid, herbal, toxic. Origin: Madagascar.

Paulownia tomentosa (Thunb.) Steud.
Scrophulariaceae
Paulownia imperialis Sieber & Zucc., *Paulownia recurva* Rehder, *Bignonia tomentosa* Thunb.
♦ empress tree, princess tree, royal paulownia, paulownia, powton, Chinese empress tree
♦ Weed, Naturalised, Garden Escape, Environmental Weed
♦ 3, 8, 15, 80, 86, 87, 88, 101, 102, 129, 133, 142, 151, 155, 195, 218, 224, 280, 290
♦ S/T, cultivated, herbal, toxic. Origin: western and central China.

Pavetta harborii S.Moore
Rubiaceae
♦ pavetta
♦ Weed, Native Weed
♦ 39, 121
♦ S, toxic. Origin: southern Africa.

Pavetta indica Linn.
Rubiaceae
♦ Weed
♦ 12
♦ cultivated, herbal.

Pavetta schumanniana F.Hoffm. ex K.Schum.
Rubiaceae
♦ gousiekte tree, poison bride's bush, poison pavetta, tree gousiekte
♦ Native Weed
♦ 121
♦ S/T, toxic. Origin: southern Africa.

Pavonia burchellii (DC.) R.A.Dyer
Malvaceae
Pavonia patens (Andr.) Chiov.
♦ Native Weed
♦ 121
♦ H. Origin: southern Africa.

Pavonia cancellata (L.) Cav.
Malvaceae

Hibiscus anonimus Mart. ex Colla, *Hibiscus cancellatus* L., *Malache cancellata* (L.) Kuntze, *Malache deltoides* (Mart.) Kuntze, *Malache modesta* (Mart.) Kuntze, *Pavonia cancellata* fo. *montana* Huber, *Pavonia cancellata* var. *cordata* Hassl., *Pavonia cancellata* var. *crassivenosa* Gürke, *Pavonia cancellata* var. *deltoidea* (Mart.) A.St.-Hil. & Naudin, *Pavonia cancellata* var. *modesta* (Mart.) Garcke, *Pavonia deltoidea* Mart., *Pavonia guanacastensis* Standl., *Pavonia hirta* Klotzsch ex Schlecht., *Pavonia modesta* Mart., *Pavonia procumbens* Casar.
♦ malva rasteira
♦ Weed
♦ 255
♦ aH, herbal. Origin: tropical America.

Pavonia coccinea Cav.
Malvaceae
♦ Naturalised
♦ 86
♦ Origin: Caribbean.

Pavonia communis A.St.-Hil.
Malvaceae
♦ arranca estrepe
♦ Weed
♦ 255
♦ pH. Origin: South America.

Pavonia coxii Tad. & Jacob.
Malvaceae
♦ Weed
♦ 87, 88

Pavonia fruticosa (Mill.) Fawc. & Rendle
Malvaceae
Typhalea fruticosa (Mill.) Britton (see)
♦ anamu
♦ Weed
♦ 88
♦ S, herbal.

Pavonia hastata Cav.
Malvaceae
♦ spearleaf swampmallow
♦ Weed, Naturalised
♦ 7, 80, 86, 98, 101, 134, 203, 269, 280, 295
♦ S, cultivated. Origin: South America.

Pavonia lasiopetala Scheele
Malvaceae
Pavonia wrightii Gray (see)
♦ Texas swampmallow
♦ Weed, Quarantine Weed
♦ 76, 88, 220
♦ cultivated.

Pavonia malvacea (Vell.) Krapov. & Cristóbal
Malvaceae
= *Pavonia sepium* St.-Hil.
♦ malvavisco de cerco
♦ Weed
♦ 295

Pavonia procumbens Boiss.
Malvaceae
♦ Weed
♦ 87, 88

Pavonia rosea **Schltdl.**
Malvaceae
= *Pavonia schiedeana* Steud. (NoR)
♦ Weed
♦ 88

Pavonia sepium **St.-Hil.**
Malvaceae
Pavonia malvacea (Vell.) Krapov. &
Cristóbal (see), *Sida malvacea* Vell.
♦ Weed
♦ 237
♦ Origin: South America.

Pavonia sidifolia **Kunth**
Malvaceae
♦ vassoura
♦ Weed
♦ 255
♦ pC. Origin: Americas.

Pavonia spinifex **(L.) Cav.**
Malvaceae
♦ gingerbush
♦ Weed
♦ 80, 87, 88, 261
♦ cultivated, herbal.

Pavonia triloba **Cav.**
Malvaceae
♦ Weed
♦ 221

Pavonia urens **Cav.**
Malvaceae
♦ Weed
♦ 87, 88

Pavonia wrightii **Gray**
Malvaceae
= *Pavonia lasiopetala* Scheele
♦ Weed, Quarantine Weed
♦ 76, 88, 220

Pavonia zeylanica **Cav.**
Malvaceae
♦ Weed
♦ 39, 87, 88, 221
♦ cultivated, herbal, toxic.

Pechuel-loeschea leubnitziae **(Kuntze)
O.Hoffm.**
Asteraceae
Piptocarpha leubnitziae Kuntze, *Pluchea
leubnitziae* (Kuntze) N.E.Br.
♦ stinkbush, bitteros, edimba, edimba
lendume, omadimba mandume,
bitterossie
♦ Weed
♦ 121
♦ pS, arid. Origin: southern Africa.

Pectis ciliaris **L.**
Asteraceae
♦ donkeyweed
♦ Weed
♦ 87, 88
♦ herbal.

Pectis floribunda **A.Rich.**
Asteraceae
= *Pectis elongata* Kunth var. *floribunda*
(A.Rich.) D.J.Keil (NoR)
♦ Weed
♦ 14

Pectis glaucescens **(Cass.) D.J.Keil**
Asteraceae

♦ stalked chickenweed, sanddune
cinchweed
♦ Weed
♦ 161, 249

Pectis humifusa **Sw.**
Asteraceae
♦ yerba de San Juan
♦ Weed
♦ 179

Pectis linifolia **L.**
Asteraceae
♦ romero macho
♦ Naturalised
♦ 257

Pectis papposa **Harr. & Gray ex Gray**
Asteraceae
♦ cinchweed, fetid marigold, many
bristle cinchweed, cinchweed fetid
marigold
♦ Weed
♦ 23, 88, 243
♦ aH, arid, promoted, herbal.

Pectis prostrata **Cav.**
Asteraceae
♦ spreading cinchweed
♦ Weed
♦ 14, 157
♦ herbal.

Pedalium murex **L.**
Pedaliaceae
♦ burra gookeroo
♦ Weed
♦ 87, 88
♦ cultivated, herbal.

Pedicularis **L. spp.**
Scrophulariaceae
♦ lousewort
♦ Weed
♦ 23, 39, 88
♦ herbal, toxic.

Pedicularis canadensis **L.**
Scrophulariaceae
♦ wood betony, common lousewort,
lousewort, Canadian lousewort
♦ Weed
♦ 23, 88
♦ pH, cultivated, herbal.

Pedicularis comosa **L.**
Scrophulariaceae
♦ crested lousewort
♦ Weed
♦ 87, 88, 272

Pedicularis comptoniaefolia **Franch. ex
Maxim.**
Scrophulariaceae
Pedicularis comptoniifolia Franch. ex
Maxim.
♦ Weed, Quarantine Weed
♦ 76, 88, 220

Pedicularis confertiflora **Prain**
Scrophulariaceae
♦ Weed, Quarantine Weed
♦ 76, 88, 220

Pedicularis foliosa **L.**
Scrophulariaceae
♦ Quarantine Weed
♦ 220

♦ promoted, herbal.

Pedicularis groenlandica **Retz.**
Scrophulariaceae
Elephantella groenlandica (Retz.) Rydb.
(see)
♦ elephant's head lousewort,
elephant's head, bog elephant's head
♦ Quarantine Weed
♦ 220
♦ pH, cultivated, herbal.

Pedicularis kaufmannii **Pinzger**
Scrophulariaceae
♦ arokuusio
♦ Casual Alien
♦ 42

Pedicularis palustris **L.**
Scrophulariaceae
♦ marsh lousewort, red rattle,
luhtakuusio
♦ Weed
♦ 39, 70, 272
♦ cultivated, toxic.

Pedicularis resupinata **L.**
Scrophulariaceae
♦ resupinate woodbetony,
shiogamagiku
♦ Weed
♦ 297
♦ pH, promoted, herbal.

Pedicularis sceptrum-carolinum **L.**
Scrophulariaceae
♦ kaarlenvaltikka
♦ Quarantine Weed
♦ 220

Pedicularis spicata **Pall.**
Scrophulariaceae
♦ spicate wood betony,
hozakishiogama
♦ Weed
♦ 297

Pedicularis striata **Pall.**
Scrophulariaceae
♦ redstriate wood betony
♦ Weed
♦ 297

Pedicularis surrecta **Benth.**
Scrophulariaceae
♦ Quarantine Weed
♦ 220

Pedicularis sylvatica **L.**
Scrophulariaceae
♦ metsäkuusio, lousewort, dwarf red
rattle
♦ Casual Alien
♦ 39, 42
♦ herbal, toxic.

Pedicularis tatarinowii **Maxim.**
Scrophulariaceae
♦ tatarinow woodbetony
♦ Weed
♦ 297

Pedilanthus macrocarpus **Benth.**
Euphorbiaceae
♦ redbird cactus
♦ Weed
♦ 161
♦ cultivated, toxic.

Pedilanthus tithymaloides Poit.
Euphorbiaceae
♦ slipper flower, Christmas candle, redbird flower
♦ Weed
♦ 39, 154, 161, 247
♦ cultivated, herbal, toxic.

Pedilanthus tithymaloides (L.) Poit. ssp. parasiticus (Klotzsch & Garcke) Dressler
Euphorbiaceae
♦ redbird flower
♦ Cultivation Escape
♦ 261
♦ cultivated. Origin: Mexico, British Honduras.

Pedilanthus tithymaloides (L.) Poit. ssp. smallii (Millsp.) Dressler
Euphorbiaceae
♦ Small's redbird flower
♦ Naturalised
♦ 86
♦ Origin: Mexico to Brazil, West Indies.

Pedilanthus tithymaloides (L.) Poit. ssp. tithymaloides
Euphorbiaceae
♦ fiddle flower, ipecacuana, itamo real, redbird flower
♦ Cultivation Escape
♦ 261
♦ cultivated. Origin: Mexico to Surinam.

Pediomelum argophyllum (Pursh) J.W.Grimes
Fabaceae/Papilionaceae
♦ silver leaved scurfpea, silverleaf Indian breadroot
♦ Weed
♦ 161
♦ toxic.

Peganum harmala L.
Zygophyllaceae/Peganaceae
♦ African rue, peganum, Syrian rue, harmal, harmala, harmara, harmel, hermal, hurmur, isbendlahouri, khokrana, wild rue
♦ Weed, Quarantine Weed, Noxious Weed, Naturalised, Introduced
♦ 1, 26, 35, 39, 62, 76, 79, 80, 86, 87, 88, 98, 101, 121, 130, 146, 147, 161, 171, 198, 203, 212, 218, 221, 228, 229, 243, 264, 272, 275, 297
♦ pH, arid, cultivated, herbal, toxic. Origin: North Africa, Asia.

Peganum nigellastrum Bunge
Zygophyllaceae/Peganaceae
♦ Weed
♦ 33
♦ toxic.

Pegolettia senegalensis Cass.
Asteraceae
♦ Weed
♦ 221
♦ herbal.

Pelargonium L'Hér. ex Aiton spp.
Geraniaceae
♦ geranium, garden geranium
♦ Weed, Naturalised, Garden Escape, Environmental Weed
♦ 7, 72, 86, 88, 116, 203, 247
♦ cultivated, herbal, toxic.

Pelargonium alchemilloides (L.) L'Hér.
Geraniaceae
♦ Naturalised, Environmental Weed
♦ 7, 54, 86, 88
♦ cultivated, herbal. Origin: South Africa.

Pelargonium × asperum Ehrh. ex Willd.
Geraniaceae
♦ rose oil geranium, rose geranium
♦ Naturalised, Environmental Weed
♦ 86, 98, 176, 198, 280
♦ cultivated, herbal. Origin: Mediterranean.

Pelargonium australe Willd.
Geraniaceae
♦ ivy geranium, austral stork's bill
♦ Naturalised
♦ 134
♦ pH, cultivated.

Pelargonium capitatum (L.) L'Hér.
Geraniaceae
ornamental plants cultivated under this name may be of hybrid origin
♦ rose scented geranium, wild pelargonium, atomic snowflake geranium
♦ Weed, Naturalised, Native Weed, Environmental Weed, Casual Alien
♦ 7, 9, 72, 86, 88, 98, 101, 121, 198, 203, 280
♦ pS, cultivated, herbal. Origin: southern Africa.

Pelargonium crispum (L.) L'Hér.
Geraniaceae
♦ lemon geranium, geranium
♦ Casual Alien
♦ 280
♦ S, cultivated, herbal.

Pelargonium × domesticum L.H.Bailey
Geraniaceae
= *Pelargonium grandiflorum* (Andrews) Willd. × *Pelargonium cucullatum* (L.) L'Hér. [complex hybrid including others]
♦ garden geranium, regal pelargonium
♦ Naturalised, Environmental Weed
♦ 86, 98, 101, 176, 198, 280
♦ cultivated. Origin: horticultural hybrid.

Pelargonium × fragrans Willd.
Geraniaceae
= *Pelargonium exstipulatum* (Cav.) L'Hér. × *Pelargonium odoratissimum* (L.) L'Hér. ex Aiton [covers a group of ornamental cultivars]
♦ nutmeg geranium, geranium
♦ Weed, Naturalised, Introduced
♦ 86, 98, 203, 228, 280
♦ S, arid, cultivated, herbal. Origin: South Africa.

Pelargonium graveolens L'Hér. ex Aiton
Geraniaceae
♦ sweet scented geranium, rose geranium, geranium
♦ Naturalised, Garden Escape

♦ 101, 261
♦ S, cultivated, herbal. Origin: South Africa.

Pelargonium grossularioides (L.) L'Hér. ex Aiton
Geraniaceae
♦ gooseberry geranium, coconut geranium
♦ Weed, Naturalised
♦ 34, 101
♦ a/pH, cultivated, herbal.

Pelargonium × hortorum Bailey (pro sp.)
Geraniaceae
= *Pelargonium inquinans* (L.) L'Hér. ex Aiton × *Pelargonium zonale* (L.) L'Hér. ex Aiton
♦ zonal geranium
♦ Naturalised
♦ 101, 280
♦ aH, cultivated.

Pelargonium inodorum Willd.
Geraniaceae
♦ scentless geranium
♦ Naturalised
♦ 101
♦ cultivated. Origin: Australia.

Pelargonium inquinans (L.) L'Hér. ex Aiton
Geraniaceae
♦ scarlet geranium
♦ Naturalised
♦ 101
♦ S, cultivated.

Pelargonium odoratissimum (L.) L'Hér. ex Aiton
Geraniaceae
♦ apple geranium
♦ Naturalised, Garden Escape
♦ 101, 261
♦ pH, arid, cultivated, herbal. Origin: South Africa.

Pelargonium panduriforme Eckl. & Zeyh.
Geraniaceae
♦ oakleaf garden geranium, geranium
♦ Naturalised, Garden Escape
♦ 101
♦ S, cultivated.

Pelargonium peltatum (L.) L'Hér. ex Aiton
Geraniaceae
♦ ivy geranium, Austrian geranium, ivyleaf geranium
♦ Weed, Naturalised, Garden Escape, Cultivation Escape
♦ 15, 34, 39, 101, 261, 280
♦ pH, cultivated, toxic. Origin: South Africa.

Pelargonium quercifolium (L.f.) L'Hér. ex Aiton
Geraniaceae
♦ oakleaf pelargonium, pelargonium, staghorn oak geranium, almond geranium
♦ Weed, Naturalised, Environmental Weed
♦ 54, 72, 86, 88, 101
♦ pS, cultivated, herbal. Origin: South Africa.

Pelargonium tomentosum Jacq.
Geraniaceae
- peppermint scented geranium, geranium
- Naturalised
- 40, 280
- S, cultivated, herbal.

Pelargonium vitifolium (L.) L'Hér. ex Aiton
Geraniaceae
- grapeleaf geranium, balm scented geranium
- Naturalised, Introduced
- 34, 101, 280
- pH, cultivated, herbal.

Pelargonium zonale (L.) L'Hér. ex Aiton
Geraniaceae
- horseshoe geranium
- Naturalised, Garden Escape
- 101, 261
- S, cultivated. Origin: South Africa.

Pellaea viridis (Forssk.) Prantl
Pteridaceae/Adiantaceae
- green cliffbrake
- Naturalised
- 86
- cultivated, herbal. Origin: Africa, Madagascar.

Peltandra virginica (L.) Schott & Endl.
Araceae
- arrow arum, green arrow arum
- Weed, Quarantine Weed
- 23, 87, 88, 218, 258
- pH, aqua, cultivated, herbal.

Peltophorum africanum Sond.
Fabaceae/Caesalpiniaceae
- African blackwood, weeping wattle
- Weed, Quarantine Weed, Introduced
- 76, 87, 88, 203, 220, 228
- arid, cultivated, herbal.

Peltophorum dubia (Spreng.) Taub.
Fabaceae/Caesalpiniaceae
- horse bush
- Weed, Naturalised
- 101, 179

Peltophorum pterocarpum (DC.) Backer ex K.Heyne
Fabaceae/Caesalpiniaceae
Peltophorum ferrugineum Benth., *Peltophorum inerme* (Roxb.) Naves ex Villar, *Peltophorum roxburghii* (G.Don) Deg., *Poinciana roxburghii* G.Don, *Inga pterocarpa* DC.
- peltophorum, yellow poinciana
- Weed, Naturalised, Introduced, Garden Escape
- 86, 101, 179, 228, 261
- arid, cultivated, herbal. Origin: tropical Asia, Australia.

Peniocereus (Berg.) Britton & Rose spp.
Cactaceae
Nyctocereus (Berg.) Britton & Rose spp. (see)
- peniocereus
- Weed, Quarantine Weed, Naturalised
- 76, 86, 88, 203, 220

Peniocereus serpentinus (Lag. & Rodr.) N.P.Taylor
Cactaceae
Nyctocereus castellanosii Scheinvar, *Nyctocereus serpentinus* (Lag. & Rodr.) Britton & Rose (see)
- Naturalised
- 86
- cultivated. Origin: Mexico.

Pennisetum L.C.Rich. ex Pers. spp.
Poaceae
- fountaingrass
- Weed, Quarantine Weed, Noxious Weed, Naturalised, Environmental Weed
- 88, 137, 198, 221, 246
- G.

Pennisetum alopecuroides (L.) Spreng.
Poaceae
Alopecurus hordeiformis L., *Panicum alopecuroides* L., *Pennisetum compressum* R.Br., *Pennisetum hordeiforme* (Thunb.) Spreng., *Pennisetum japonicum* Trin. ex Spreng. (see)
- swamp foxtail grass, Chinese pennisetum, Chinese fountaingrass
- Weed, Quarantine Weed, Naturalised, Native Weed, Garden Escape, Environmental Weed
- 15, 72, 80, 86, 87, 88, 90, 98, 101, 155, 198, 203, 204, 246, 251, 263, 269, 275, 280, 290, 296, 297
- pG, cultivated, herbal. Origin: southeast Asia, Australia.

Pennisetum alopecuroides (L.) Spreng. fo. *purpurascens* Ohwi
Poaceae
- Weed
- 286
- G.

Pennisetum americanum (L.) Leeke
Poaceae
= *Pennisetum glaucum* (L.) R.Br.
- pearl millet, bulrush millet, American fountaingrass, spiked millet, bajra
- Weed
- 90, 255
- aG, cultivated, herbal. Origin: Africa.

Pennisetum centrasiaticum Tzvel.
Poaceae
= *Pennisetum flaccidum* Hochst ex Steud.
- flaccid pennisetum
- Weed
- 297
- G.

Pennisetum chilense (Desv.) Jacks.
Poaceae
- esporal
- Weed, Introduced
- 87, 88, 228, 295
- G, arid.

Pennisetum ciliare (L.) Link
Poaceae
= *Cenchrus ciliaris* L.
- buffel grass, bufle
- Weed, Naturalised

- 80, 101
- G, herbal.

Pennisetum ciliare (L.) Link var. *ciliare*
Poaceae
- buffel grass
- Naturalised
- 101
- G.

Pennisetum ciliare (L.) Link var. *setigerum* (Vahl) Leeke
Poaceae
- cow sandbur
- Naturalised
- 101
- G.

Pennisetum clandestinum Hochst. ex Chiov.
Poaceae
Pennisetum inclusum Pilg., *Pennisetum longistylum* Hochst., *Pennisetum longistylum* var. *clandestinum* (Hochst. ex Chiov.) Leeke
- kikuyu grass, yaa kikuyu, kikuyu, yerba kikuyo, capim quicuio, pasto africano
- Weed, Noxious Weed, Naturalised, Introduced, Garden Escape, Environmental Weed, Cultivation Escape
- 3, 7, 9, 15, 18, 23, 26, 30, 34, 35, 39, 53, 67, 72, 78, 80, 86, 88, 90, 93, 95, 98, 101, 116, 121, 134, 140, 152, 157, 158, 161, 176, 180, 181, 186, 198, 199, 203, 205, 212, 218, 225, 228, 229, 238, 243, 246, 251, 255, 257, 261, 269, 280, 283, 289, 295, 296, 300
- pG, arid, cultivated, herbal, toxic. Origin: tropical east Africa.

Pennisetum divisum (L.) Link
Poaceae
Gymnotrix longiglumis Munro ex Hook., *Panicum dichotomum* Forssk., *Panicum divisum* J.Gmel., *Pennisetum dichotomum* (Forssk.) Del., *Pennisetum elatum* Hochst. ex Steud. (see)
- Weed, Introduced
- 221, 228
- G, arid.

Pennisetum elatum Hochst. ex Steud.
Poaceae
= *Pennisetum divisum* (L.) Link
- Weed
- 221
- G.

Pennisetum flaccidum Hochst ex Steud.
Poaceae
Gymnothrix flaccida (Griseb.) Munro ex Ait., *Pennisetum centrasiaticum* Tzvel. (see)
- kikuyu grass
- Weed
- 88, 275
- pG, arid, cultivated.

Pennisetum fructescens Leeke
Poaceae
- Weed
- 87, 88
- G.

Pennisetum glaucum (L.) R.Br.
Poaceae
Panicum glaucum L., *Pennisetum americanum* (L.) Leeke. (see), *Pennisetum typhoides* (Burm.) Stapf & C.E.Hubb., *Setaria glauca* (L.) Beauv. (see)
♦ yellow foxtail, pearl millet, pigeongrass, yellow foxtail, yellow bristlegrass, wild millet, bulrush millet, African millet, babala, Indian millet, kaffir millet, poko grass, pussy grass
♦ Weed, Noxious Weed, Naturalised, Introduced, Casual Alien
♦ 7, 86, 93, 98, 101, 121, 174, 179, 203, 229, 280, 287
♦ aG, arid, cultivated, herbal.

Pennisetum hordeoides (Lam.) Steud.
Poaceae
♦ Weed
♦ 87, 88
♦ G.

Pennisetum japonicum Trin.
Poaceae
= *Pennisetum alopecuroides* (L.) Spreng.
♦ Weed, Quarantine Weed
♦ 23, 76, 87, 88, 203, 220
♦ G.

Pennisetum latifolium Spreng.
Poaceae
♦ Uruguay pennisetum, Uruguayan fountaingrass
♦ Weed, Naturalised
♦ 15, 101, 280, 287
♦ G.

Pennisetum macrourum Trin.
Poaceae
♦ African feathergrass, beddingrass
♦ Weed, Quarantine Weed, Noxious Weed, Naturalised, Garden Escape, Environmental Weed
♦ 7, 67, 72, 76, 86, 87, 88, 98, 101, 140, 147, 155, 171, 176, 198, 203, 225, 229, 246, 251, 280, 289
♦ pG, cultivated. Origin: southern Africa.

Pennisetum nervosum (Nees) Trin.
Poaceae
♦ bentspike fountaingrass
♦ Naturalised
♦ 101
♦ G.

Pennisetum orientale L.C.Rich.
Poaceae
Pennisetum fasciculatum Trin., *Pennisetum triflorum* Nees ex Steud.
♦ oriental pennisetum
♦ Weed
♦ 221
♦ G, cultivated, herbal. Origin: North Africa, Middle East, India.

Pennisetum orientale L.C.Rich. var. triflorum Stapf
Poaceae
♦ Naturalised
♦ 287
♦ G.

Pennisetum pedicellatum Trin.
Poaceae
Eriochaeta secundiflora Fig. & De Not., *Pennisetum amoenum* A.Rich., *Pennisetum densiflorum* (Fig. & De Not.) T.Durand & Schinz, *Pennisetum dillonii* Steud., *Pennisetum implicatum* Steud., *Pennisetum lanuginosum* Hochst.
♦ kyasuwa grass, feather pennisetum, yaa khachyon chop
♦ Weed, Quarantine Weed, Noxious Weed, Naturalised, Introduced
♦ 7, 12, 67, 76, 87, 88, 90, 98, 101, 140, 161, 179, 186, 203, 204, 228, 229, 239
♦ aG, arid, cultivated.

Pennisetum pedicellatum Trin. ssp. pedicellatum
Poaceae
♦ deenanth grass
♦ Weed, Naturalised, Environmental Weed
♦ 86, 93
♦ G. Origin: Africa.

Pennisetum pedicellatum Trin. ssp. unispiculum Brunken
Poaceae
♦ kyasuma grass, deenanth grass
♦ Weed, Noxious Weed, Naturalised, Environmental Weed
♦ 86, 93, 101, 140, 205, 229
♦ G. Origin: Africa, India.

Pennisetum petiolare (Hochst.) Chiov.
Poaceae
♦ petioled fountaingrass
♦ Naturalised
♦ 101
♦ G.

Pennisetum polystachion (L.) Schult.
Poaceae
Cenchrus retusus Sw., *Panicum barbatum* Roxb., *Panicum cauda-ratti* Schumach., *Panicum cenchroides* Rich., *Panicum densispica* Poir., *Panicum longisetum* Poir., *Panicum polystachion* L., *Panicum triticoides* Poir., *Pennisetum alopecuroides* Desv. ex Ham., *Pennisetum amethystinum* P.Beauv., *Pennisetum borbonicum* Kunth, *Pennisetum breve* Nees, *Pennisetum cauda-ratti* (Schumach.) Franch., *Pennisetum ciliatum* Parl. ex Hook., *Pennisetum dasistachyum* Desv., *Pennisetum erubescens* (Willd.) Desv. ex Ham., *Pennisetum gabonense* Franch., *Pennisetum multiflorum* E.Fourn., *Pennisetum nicaraguense* E.Fourn., *Pennisetum polystachyon* (L.) Schult. (see), *Pennisetum richardii* Kunth, *Pennisetum setosum* (Sw.) Rich. var. *breve* (Nees) Döll, *Pennisetum tenuispiculatum* Steud., *Pennisetum triticoides* (Poir.) Roem. & Schult.
♦ mission grass, feathery pennisetum, mission grass, yaa khachorn chop dok lek
♦ Weed, Quarantine Weed, Noxious Weed, Naturalised, Environmental Weed
♦ 6, 11, 76, 86, 98, 147, 152, 155, 179, 191, 203, 209, 238, 243
♦ a/pG. Origin: tropical Africa.

Pennisetum polystachion (L.) Schult. ssp. polystachion
Poaceae
♦ mission grass
♦ Weed
♦ 93
♦ G.

Pennisetum polystachion (L.) Schult. var. polystachion
Poaceae
♦ Introduced
♦ 230
♦ G.

Pennisetum polystachyon (L.) Schult.
Poaceae
= *Pennisetum polystachion* (L.) Schult.
♦ missiongrass, feathery pennisetum, queue de chat, pwokso, o tamata, feather pennisetum, yaa khachyon chop, elephant grass, West Indian pennisetum, thin napier grass
♦ Weed, Noxious Weed, Naturalised
♦ 3, 12, 67, 87, 88, 90, 101, 107, 140, 161, 170, 186, 204, 229, 239, 262
♦ a/pG, arid, cultivated.

Pennisetum polystachyon (L.) Schult. ssp. setosum (Sw.) Brunken
Poaceae
= *Pennisetum setosum* (Sw.) Rich.
♦ missiongrass, thin napier grass
♦ Noxious Weed, Naturalised
♦ 101, 140, 229
♦ G.

Pennisetum purpureum Schumach.
Poaceae
Gymnotrix nitens Anders., *Pennisetum benthamii* Steud., *Pennisetum flavicomum* Leeke, *Pennisetum flexispica* K.Schum., *Pennisetum macrostachyum* Benth., *Pennisetum merkerim* Leeke, *Pennisetum nitens* (Anders.) Hack., *Pennisetum palescens* Leeke, *Pennisetum pruinosum* Leeke
♦ elephant grass, napier grass, merker grass, bokso, puk soh, acfucsracsracsr, herbe éléphant, fausse canne à sucre, ilengesongo, iswe bingobingo, napier fodder
♦ Weed, Sleeper Weed, Quarantine Weed, Noxious Weed, Naturalised, Introduced, Environmental Weed, Cultivation Escape
♦ 3, 6, 7, 18, 38, 39, 80, 86, 87, 88, 90, 95, 98, 101, 107, 112, 121, 122, 132, 134, 152, 179, 186, 197, 203, 209, 218, 225, 228, 230, 245, 246, 255, 257, 258, 261, 263, 269, 274, 280, 283, 286, 287
♦ pG, aqua, cultivated, herbal, toxic. Origin: tropical Africa.

Pennisetum ruppelii Steud.
Poaceae
= *Pennisetum setaceum* (Forssk.) Chiov.
♦ fountaingrass
♦ Weed
♦ 3, 87, 88, 191, 218
♦ G, cultivated.

Pennisetum setaceum (Forssk.) Chiov.
Poaceae
Cenchrus asperifolius Desf., *Pennisetum asperifolium* auct., *Pennisetum ruppelii* Steud. (see), *Pennisetum ruppelianum* hort. ex Mez, *Phalaris setacea* Forsk.
♦ fountaingrass, crimson fountaingrass, pronkgras, African fountaingrass
♦ Weed, Quarantine Weed, Noxious Weed, Naturalised, Introduced, Garden Escape, Environmental Weed, Cultivation Escape
♦ 3, 7, 15, 18, 34, 35, 63, 78, 79, 80, 86, 87, 88, 90, 93, 95, 98, 101, 116, 121, 129, 151, 152, 155, 158, 161, 179, 180, 198, 203, 221, 225, 228, 229, 231, 233, 246, 280, 283, 289, 296
♦ pG, arid, cultivated, herbal. Origin: Africa, west Asia.

Pennisetum setosum (Sw.) Rich.
Poaceae
Cenchrus setosus Sw. (see), *Gymnothrix geniculata* Schult., *Panicum erubescens* Willd., *Pennisetum flavescens* J.Presl, *Pennisetum hamiltonii* Steud., *Pennisetum hirsutum* Nees, *Pennisetum indicum* Murr. var. *purpurascens* (Kunth) Kuntze, *Pennisetum pallidum* Nees, *Pennisetum polystachion* (L.) Schult. ssp. *setosum* (Sw.) Brunken, *Pennisetum polystachyon* ssp. *setosum* (Sw.) Brunken (see), *Pennisetum purpurascens* Kunth, *Pennisetum richardii* Kunth, *Pennisetum sieberi* Kunth, *Pennisetum uniflorum* Kunth, *Setaria cenchroides* (Rich.) Roem. & Schult., *Setaria erubescens* (Willd.) P.Beauv.
♦ West Indies pennisetum, yaa khachyon chop dok luang
♦ Weed, Naturalised
♦ 3, 12, 87, 88, 191, 218, 239, 245, 255, 256, 287
♦ pG. Origin: Madagascar.

Pennisetum subangustum Stapf & C.E.Hubb.
Poaceae
♦ Weed
♦ 88
♦ G.

Pennisetum thunbergii Kunth
Poaceae
♦ Weed, Naturalised
♦ 86, 98, 203
♦ G. Origin: tropical Africa.

Pennisetum villosum R.Br. ex Fresen.
Poaceae
♦ feathertop, longstyle feathergrass, feathergrass, long styled feathergrass, white foxtail, veergras, zacate plumosa
♦ Weed, Noxious Weed, Naturalised, Introduced, Garden Escape, Environmental Weed, Cultivation Escape
♦ 7, 34, 38, 63, 72, 86, 87, 88, 90, 95, 98, 101, 116, 121, 147, 158, 161, 176, 180, 198, 199, 203, 228, 243, 252, 269, 280, 283, 289, 295, 300

♦ pG, arid, cultivated, herbal. Origin: Africa.

Pennisetum violaceum (Lam.) Rich.
Poaceae
♦ Weed
♦ 88
♦ G.

Penstemon angustifolius Nutt. ex Pursh
Scrophulariaceae
♦ broadbeard beardtongue
♦ Weed
♦ 23, 88
♦ cultivated, herbal.

Penstemon Schmidel spp.
Scrophulariaceae
♦ beardtongue, beardtongue penstemon
♦ Weed
♦ 161
♦ herbal.

Penstemon campanulatus (Cav.) Willd.
Scrophulariaceae
♦ bellflower beardtongue
♦ Weed
♦ 199
♦ cultivated.

Penstemon cobaea Nutt. × hartwegii Benth.
Scrophulariaceae
♦ Casual Alien
♦ 280

Penstemon digitalis Nutt. ex Sims
Scrophulariaceae
♦ digitalis penstemon, foxglove beardtongue, false foxglove, talus slope penstemon
♦ Weed
♦ 23, 87, 88, 195, 218
♦ cultivated, herbal.

Penstemon gracilis Nutt.
Scrophulariaceae
♦ slender penstemon, lilac penstemon
♦ Weed
♦ 87, 88, 218
♦ cultivated, herbal.

Penstemon procerus Douglas ex R.Grah.
Scrophulariaceae
♦ littleflower penstemon, small flowered penstemon
♦ Weed
♦ 23, 88
♦ pH, cultivated, herbal.

Penstemon rydbergii A.Nelson
Scrophulariaceae
♦ Rydberg's penstemon
♦ Weed
♦ 87, 88, 218
♦ pH, cultivated, herbal.

Penstemon speciosus Douglas ex Lindl.
Scrophulariaceae
♦ royal penstemon
♦ Weed
♦ 23, 88
♦ pH, cultivated, herbal.

Penstemon wilcoxii Rydb.
Scrophulariaceae
♦ Wilcox's penstemon

♦ Weed
♦ 23, 88
♦ herbal.

Pentaglottis sempervirens (L.) Tausch ex L.Bailey
Boraginaceae
♦ pentaglottis, evergreen bugloss, alkanet, green alkanet
♦ Weed, Naturalised
♦ 15, 40, 70, 86, 101, 176, 198, 280
♦ pH, cultivated, herbal. Origin: Europe.

Pentalinon luteum (L.) B.F.Hansen & Wunderlin
Apocynaceae
Urechites lutea (L.) Britt. (see)
♦ yellow nightshade, hammock viper's tail
♦ Weed
♦ 161
♦ toxic.

Pentapetes phoenicea L.
Sterculiaceae
♦ flor impia, midday flower, baan thiang
♦ Weed, Naturalised, Cultivation Escape
♦ 87, 88, 101, 191, 204, 209, 239, 261, 262
♦ aH, aqua, cultivated, herbal. Origin: southern Asia.

Pentaphalangium solomonense A.C.Small
Clusiaceae
♦ Introduced
♦ 230
♦ S.

Pentapogon quadrifidus (Labill.) Baill.
Poaceae
♦ Naturalised
♦ 280
♦ G, cultivated.

Pentas lanceolata (Forssk.) Deflers
Rubiaceae
Pentas carnea Benth.
♦ starcluster, Egyptian starcluster, pentas
♦ Naturalised, Introduced
♦ 101, 134, 261
♦ cultivated, herbal. Origin: tropical Africa, Middle East.

Pentaschistis airoides (Nees) Stapf
Poaceae
♦ falsehair grass
♦ Weed, Naturalised, Environmental Weed
♦ 7, 9, 72, 86, 88, 98, 198, 203
♦ aG, arid, cultivated, herbal. Origin: South Africa.

Pentaschistis pallida (Thunb.) H.P.Linder
Poaceae
♦ pussy tail
♦ Naturalised
♦ 7, 86, 198
♦ G. Origin: South Africa.

Pentaschistis thunbergii (Kunth) Stapf
Poaceae
♦ dune grass
♦ Weed, Naturalised, Native Weed, Environmental Weed
♦ 7, 9, 86, 98, 121, 203
♦ pG. Origin: southern Africa.

Pentatropis microphylla Wight & Arn.
Asclepiadaceae/Apocynaceae
♦ Weed
♦ 87, 88

Penthorum chinense Pursh.
Saxifragaceae/Penthoraceae
♦ Weed
♦ 275, 297
♦ pH.

Penthorum sedoides L.
Saxifragaceae/Penthoraceae
♦ Virginian stonecrop, ditch stonecrop
♦ Weed
♦ 23, 88
♦ pH, aqua, cultivated, herbal.

Pentodon pentandrus (K.Schum.) Vatke
Rubiaceae
♦ Hale's pentodon
♦ Weed
♦ 88

Pentzia albida (DC.) Hutch. var. *annua* (DC.) Merxm. & Eberle
Asteraceae
Pentzia annua DC.
♦ Native Weed
♦ 121
♦ aH. Origin: southern Africa.

Pentzia cooperi Harv.
Asteraceae
♦ Native Weed
♦ 121
♦ pS. Origin: southern Africa.

Pentzia globifera (Thunb.) Hutch.
Asteraceae
= *Oncosiphon piluliferum* (L.f.) Källersjö
♦ Weed, Naturalised
♦ 7, 98, 203
♦ H.

Pentzia globosa Less.
Asteraceae
♦ bitter karoo bush, hair karroo
♦ Native Weed
♦ 121
♦ S. Origin: southern Africa.

Pentzia grandiflora (Thunb.) Hutch.
Asteraceae
♦ matricaria, stinkweed, stinkkruid
♦ Native Weed
♦ 121
♦ aH, cultivated. Origin: southern Africa.

Pentzia incana (Thunb.) Kuntze
Asteraceae
Pentzia virgata Less.
♦ common karro, good karoo bush, karroo bush, sheepbush, sweet karoo, African sheepbush
♦ Weed, Naturalised, Native Weed
♦ 86, 98, 101, 121, 203

♦ S, arid, cultivated. Origin: southern Africa.

Pentzia pilulifera (L.f.) Fourc.
Asteraceae
= *Oncosiphon piluliferum* (L.f.) Källersjö
♦ cattle bush
♦ Native Weed
♦ 121
♦ H, cultivated, toxic. Origin: southern Africa.

Pentzia sphaerocephala DC.
Asteraceae
Pentzia cinerascens DC., *Pentzia grisea* Muschl. ex Dinter
♦ large karoo bush
♦ Native Weed
♦ 121
♦ S, arid. Origin: southern Africa.

Pentzia suffruticosa (L.) Hutch. ex Merxm.
Asteraceae
= *Pentzia spinescens* Less. (NoR)
♦ calomba daisy, chamomile, matricaria, mayweed, stinking weed, yellowtop, yellow weed, karoo bush, matricaria
♦ Weed, Quarantine Weed, Noxious Weed, Naturalised, Native Weed, Environmental Weed
♦ 7, 76, 86, 88, 98, 121, 147, 171, 203, 243, 290
♦ aH, cultivated. Origin: southern Africa.

Peperomia amplexicaulis (Sw.) A.Dietr.
Piperaceae/Peperomiaceae
♦ Jackie's saddle
♦ Naturalised
♦ 101

Peperomia breviramula C.DC.
Piperaceae/Peperomiaceae
♦ Introduced
♦ 230
♦ H.

Peperomia pellucida (L.) Kunth
Piperaceae/Peperomiaceae
Micropiper pellucidum (L.) Miq., *Peperomia concinna* A.Dietr., *Peperomia pellucida* var. *minor* Miq., *Peperomia pellucida* var. *pygmaea* Kunth, *Peperomia translucens* Trel., *Piper concinnum* Haw., *Piper pellucidum* L.
♦ peperomia, vao vai, man to man
♦ Weed, Naturalised, Introduced, Garden Escape, Environmental Weed, Cultivation Escape
♦ 3, 28, 54, 86, 87, 88, 93, 98, 155, 157, 179, 191, 203, 206, 209, 230, 243, 262, 276, 297
♦ aH, cultivated, herbal. Origin: pantropical.

Peperomia ponapensis C.DC. var. *ponapensis*
Piperaceae/Peperomiaceae
♦ Introduced
♦ 230
♦ H.

Peplis portula L.
Lythraceae
= *Lythrum portula* (L.) D.A.Webber
♦ ojakaali
♦ Naturalised
♦ 241

Pera bumeliifolia Griseb.
Euphorbiaceae/Peraceae
♦ jiqi
♦ Cultivation Escape
♦ 261
♦ cultivated.

Pereskia Mill.spp.
Cactaceae
Rhodocactus (Berg.) Kunth spp. (see), *Rittereocereus* Backeb. spp. (see)
♦ pereskia
♦ Weed, Quarantine Weed
♦ 76, 88, 203, 220

Pereskia aculeata Mill.
Cactaceae
Pereskia godseffiana Sander, *Pereskia pereskia* (L.) Karst.
♦ Barbados gooseberry, leafy cactus, lemon vine, pereskia creeper, primitive cactus, Spanish gooseberry
♦ Weed, Sleeper Weed, Quarantine Weed, Noxious Weed, Naturalised, Introduced, Environmental Weed, Cultivation Escape
♦ 3, 10, 63, 76, 86, 86, 88, 95, 98, 101, 121, 152, 155, 158, 179, 191, 203, 228, 261, 278, 279, 283
♦ pC, arid, cultivated. Origin: South America.

Pereskia grandifolia Haw.
Cactaceae
Rhodocactus grandifolius (Haw.) Kunth, *Rhodocactus tampicanus* (Weber.) Backeb., *Pereskia ochnacarpa* Miq., *Pereskia tampicana* F.A.C.Weber
♦ large flowered Barbados gooseberry, large leaved Barbados gooseberry, rose cactus, grootblaar, Barbadosstekelbessie
♦ Weed, Noxious Weed, Naturalised, Introduced, Cultivation Escape
♦ 88, 101, 121, 158, 179, 228, 261, 278
♦ S/T, arid, cultivated, herbal. Origin: tropical South America.

Pereskia sacharosa Griseb.
Cactaceae
Pereskia moorei Britt. & Rose, *Pereskia saipinensis* Cárdenas, *Pereskia sparsiflora* Ritter., *Rhodocactus sacharosa* (Griseb.) Backeb., *Rhodocactus saipinensis* (Cárdenas) Backeb.
♦ Introduced
♦ 228
♦ arid, cultivated.

Pereskiopsis Britt. & Rose spp.
Cactaceae
♦ Weed, Quarantine Weed
♦ 76, 88, 203, 220

Pergularia daemia (Forssk.) Chiov.
Asclepiadaceae/Apocynaceae
Asclepias daemia Forssk., *Pergularia extensa* (R.Br.) N.E.Br.

♦ mkoboso, pergularia
♦ Weed
♦ 221
♦ arid, herbal.

Pergularia tomentosa L.
Asclepiadaceae/Apocynaceae
♦ Weed
♦ 221
♦ arid, herbal.

Periballia minuta (L.) Asch. & Graebn.
Poaceae
= *Molineriella minuta* (L.) Rouy
♦ Naturalised, Environmental Weed
♦ 86
♦ G.

Pericallis cuneata (L'Hér.) Bolle
Asteraceae
Senecio hybridus (Willd.) Regel (see)
♦ common ragwort
♦ Naturalised
♦ 101

Pericallis × hybrida B.Nord.
Asteraceae
= *Pericallis lanata* (L'Hér.) B.Nord. ×
Pericallis cruenta (Masson ex L'Hér.)
Bolle [possibly includes other species]
♦ cineraria
♦ Weed, Naturalised
♦ 15, 40, 280

Perilla frutescens (L.) Britton
Lamiaceae
Ocimum frutescens L., *Perilla ocymoides*
L.
♦ beefsteak mint, perilla, shiso,
beefsteak plant, mint perilla
♦ Weed, Naturalised
♦ 8, 39, 80, 87, 88, 101, 161, 195, 207,
235, 275, 286, 287, 297
♦ aH, cultivated, herbal, toxic. Origin:
eastern Asia.

Perilla frutescens (L.) Britton fo. *viridis*
Makino
Lamiaceae
♦ Weed
♦ 286
♦ toxic.

Perilla frutescens (L.) Britton var. *crispa*
(Benth.) Deane
Lamiaceae
♦ beefsteak plant
♦ Naturalised
♦ 101
♦ herbal, toxic.

Perilla frutescens (L.) Britton var. *crispa*
(Benth.) Deane fo. *purpurea* Makino
Lamiaceae
♦ Weed
♦ 286
♦ toxic.

Perilla frutescens (L.) Britton var.
frutescens
Lamiaceae
♦ beefsteak plant
♦ Weed, Naturalised
♦ 101, 243
♦ toxic.

Periploca angustifolia Labill.
Asclepiadaceae/Apocynaceae/
Periplocaceae
♦ Weed
♦ 221

Periploca aphylla Decne.
Asclepiadaceae/Apocynaceae/
Periplocaceae
♦ Weed
♦ 221
♦ arid, herbal.

Periploca graeca L.
Asclepiadaceae/Apocynaceae/
Periplocaceae
♦ silk vine
♦ Weed, Naturalised
♦ 39, 101, 272
♦ pC, cultivated, herbal, toxic.

Periploca laevigata Ait.
Asclepiadaceae/Apocynaceae/
Periplocaceae
♦ Quarantine Weed
♦ 220
♦ arid, cultivated. Origin: Cape Verde,
Canary Islands, Mediterranean.

Periploca linearifolia Dill. & A.Rich.
Asclepiadaceae/Apocynaceae/
Periplocaceae
♦ kanondo nondo kadogo, bugaga,
lufute lwa mnyala
♦ Weed
♦ 221

Periploca sepium Bunge
Asclepiadaceae/Apocynaceae/
Periplocaceae
♦ Weed
♦ 275, 297
♦ herbal.

Peristrophe bicalyculata Nees
Acanthaceae
♦ Weed
♦ 87, 88, 221
♦ herbal.

Peristrophe paniculata (Forssk.)
Brummitt
Acanthaceae
♦ Weed
♦ 66

Peristrophe roxburghiana (Roem. &
Schult.) Bremek.
Acanthaceae
Peristrophe tinctoria Nees.
♦ Weed
♦ 262
♦ pH, cultivated.

Peristrophe roxurghiana (Schult.) Bre.
Acanthaceae
Peristrophe bivalvis (L.) Merr.
♦ Weed
♦ 13

Perotis hordeiformis Nees ex Hook.
Poaceae
♦ Weed
♦ 12

Perotis indica (L.) Kuntze
Poaceae

♦ Weed
♦ 87, 88
♦ G, cultivated. Origin: Asia.

Perotis patens Gand.
Poaceae
♦ bottlebrush grass, purplespike
perotis, purple spike cat's tail
♦ Weed, Native Weed
♦ 88, 121, 158
♦ a/pG. Origin: Madagascar.

Persea americana Mill.
Lauraceae
♦ avocado pear, alligator pear,
aguacate, palta, saboka, avioka,
apokado
♦ Weed, Naturalised, Introduced,
Environmental Weed, Casual Alien
♦ 3, 22, 32, 39, 42, 86, 88, 101, 151, 152,
161, 179, 189, 194, 230, 257, 261, 280
♦ T, cultivated, herbal, toxic. Origin:
Mexico, Central America.

Persea americana Mill. var. *americana*
Lauraceae
Laurus persea L., *Persea gratissima*
Gaertn., *Persea leiogyna* S.F.Blake
♦ avocado
♦ Introduced
♦ 228
♦ arid.

Persea americana Mill. var. *drymifolia*
(Schltdl. & Cham.) S.F.Blake
Lauraceae
Persea drymifolia Schldl. & Cham.
♦ Mexican avocado
♦ Introduced
♦ 228
♦ arid.

Persea borbonia (L.) Spreng
Lauraceae
♦ redbay, red tip bay
♦ Weed
♦ 87, 88, 218
♦ T, promoted, herbal.

Persicaria aestiva Ohwi
Polygonaceae
♦ Weed
♦ 286

Persicaria blumei H.Gross
Polygonaceae
♦ Weed
♦ 88, 204

Persicaria bungeana (Turcz.) Nakai ex
T.Mori
Polygonaceae
= *Polygonum bungeanum* Turcz.
♦ Naturalised
♦ 287

Persicaria campanulata (Hook.f.) Ronse
Decr.
Polygonaceae
= *Polygonum campanulatum* Hook.f.
♦ lesser knotweed
♦ Naturalised
♦ 40

Persicaria capitata (Buch.-Ham. ex D.Don) H.Gross
Polygonaceae
Persicaria capitatum Hamlt. (see),
Polygonum capitatum Buch.-Ham. ex D.Don (see)
♦ persicaria, nuppitatar
♦ Weed, Naturalised, Garden Escape, Environmental Weed
♦ 7, 86, 98, 203, 269
♦ cultivated, herbal. Origin: Asia.

Persicaria capitatum Hamlt.
Polygonaceae
= *Persicaria capitata* (Buch.-Ham. ex D.Don) H.Gross
♦ Naturalised
♦ 287

Persicaria chinensis Nakai
Polygonaceae
♦ Weed
♦ 286

Persicaria conspicua (Nakai) Nakai
Polygonaceae
♦ Weed
♦ 286

Persicaria decipiens (R.Br.) K.L.Wilson
Polygonaceae
= *Polygonum salicifolium* Brouss.
♦ slender knotweed
♦ Native Weed
♦ 269
♦ aqua, cultivated. Origin: Australia.

Persicaria erecto-minor (Makino) Nakai
Polygonaceae
♦ Weed
♦ 286

Persicaria hydropiper (L.) Spach
Polygonaceae
= *Polygonum hydropiper* L.
♦ waterpepper
♦ Weed
♦ 88, 204, 269, 286

Persicaria japonica (Meisn.) H.Gross
Polygonaceae
♦ Weed
♦ 286

Persicaria lapathifolia (L.) S.F.Gray
Polygonaceae
= *Polygonum lapathifolium* L.
♦ pale persicaria
♦ Weed, Naturalised
♦ 86, 176, 269, 286
♦ herbal. Origin: Eurasia.

Persicaria lapathifolia S.F.Gray ssp. *lanigera* (Danser) Sugimoto
Polygonaceae
♦ Naturalised
♦ 287

Persicaria longiseta (de Bruyn) Kitag.
Polygonaceae
♦ Weed
♦ 286

Persicaria longiseta (de Bruyn) Kitag. fo. *albiflorum* Makino
Polygonaceae
♦ Weed
♦ 286

Persicaria maackiana (Regel) Nakai
Polygonaceae
♦ Weed
♦ 286

Persicaria maculosa S.F.Gray
Polygonaceae
Persicaria maculata (Raf.) Á. & D.Löve,
Persicaria vulgaris Webb & Moq. (see),
Polgonum maculatum Raf., *Polygonum persicaria* L. (see)
♦ willowweed, persicaria, lady's thumb, spotted lady's thumb, redshank, willowweed, lady's thumb smartweed
♦ Weed, Naturalised, Environmental Weed
♦ 72, 86, 88, 98, 176, 198, 203, 243
♦ aH. Origin: Europe.

Persicaria microcephala (D.Don) Sasaki
Polygonaceae
♦ Quarantine Weed
♦ 220

Persicaria nepalensis (Meisn.) H.Gross
Polygonaceae
♦ Weed
♦ 88, 286

Persicaria nipponensis (Makino) H.Gross
Polygonaceae
♦ Weed
♦ 286

Persicaria orientalis (L.) Spach
Polygonaceae
= *Polygonum orientale* L.
♦ purpurpilrt, purppuratatar
♦ Weed, Naturalised
♦ 86, 98, 203, 269, 287
♦ cultivated, herbal. Origin: Australia.

Persicaria pensylvanica (L.) M.Gómez
Polygonaceae
= *Polygonum pensylvanicum* L.
♦ Pennsylvania smartweed
♦ Naturalised
♦ 287
♦ Origin: North America.

Persicaria perfoliata (L.) H.Gross
Polygonaceae
= *Polygonum perfoliatum* L.
♦ Weed
♦ 286

Persicaria pilosa (Roxb.) Kitag.
Polygonaceae
♦ Weed, Naturalised
♦ 286, 287

Persicaria praetermissa (Hook.f.) Hara
Polygonaceae
Polygonum praetermissum Hook.f.
♦ Weed
♦ 269, 286

Persicaria prostrata (R.Br.) Sojak.
Polygonaceae
Polygonum prostratum R.Br. (see)
♦ Weed
♦ 269

Persicaria pubescens (Blume) Hara
Polygonaceae
♦ Weed
♦ 286

Persicaria runcinata (Buch.-Ham. ex D.Don) H.Gross
Polygonaceae
Polygonum runcinatum Buch.-Ham. ex D.Don
♦ Weed
♦ 88

Persicaria scabra (Moench) Mold.
Polygonaceae
♦ Weed
♦ 286

Persicaria senticosa (Meisn.) H.Gross ex Nakai
Polygonaceae
♦ Weed
♦ 286

Persicaria serrulata (Lag.) Webb & Moq.
Polygonaceae
♦ Weed
♦ 88

Persicaria sieboldii (Meisn.) Ohki
Polygonaceae
Polygonum sieboldii Meisn. (see)
♦ Weed
♦ 286

Persicaria strigosa (R.Br.) Gross
Polygonaceae
♦ Weed, Naturalised
♦ 98, 203, 269
♦ cultivated.

Persicaria thunbergii (Sieb. & Zucc.) H.Gross
Polygonaceae
♦ Weed
♦ 286

Persicaria tinctoria (Ait.) H.Gross
Polygonaceae
♦ Naturalised
♦ 287

Persicaria viscofera (Makino) H.Gross ex Nakai
Polygonaceae
♦ Weed
♦ 286

Persicaria viscosa (Buch.-Ham. ex D.Don) H.Gross ex Nakai
Polygonaceae
Polygonum viscosum Buch.-Ham. ex D.Don (see)
♦ Weed, Naturalised
♦ 286, 287

Persicaria vulgaris Webb & Moq.
Polygonaceae
= *Persicaria maculosa* S.F.Gray
♦ Weed
♦ 286

Persicaria wallichii W.Greuter & Burdet
Polygonaceae
♦ Himalayan knotweed
♦ Naturalised
♦ 40

Pertya robusta Beauverd
Asteraceae
♦ Quarantine Weed
♦ 220

Pertya scandens **Sch.Bip.**
Asteraceae
♦ Quarantine Weed
♦ 220

Peruvocereus **Akers. spp.**
Cactaceae
= *Haageocereus* Backeb. spp.
♦ Weed, Quarantine Weed
♦ 76, 88, 203

Perymenium berlandieri **DC.**
Asteraceae
♦ Weed
♦ 243

Perymenium subsquarrosum **Rob. & Greenm.**
Asteraceae
♦ Weed, Quarantine Weed
♦ 76, 87, 88, 203, 220

Peschiera fuchsiaefolia **(A.DC.) Miers**
Apocynaceae
Tabernaemontana fuchsiaefolia A.DC. (see)
♦ leiteiro, leiteira
♦ Weed
♦ 255
♦ Origin: Brazil.

Petalidium engleranum **(Schinz) C.B.Clarke**
Acanthaceae
♦ Native Weed
♦ 121
♦ pS. Origin: southern Africa.

Petalostemon occidentale **(Heller) Fern.**
Fabaceae/Papilionaceae
= *Dalea candida* Michx. ex Willd. var. *oligophylla* (Torr.) Shinners (NoR)
♦ Weed
♦ 23, 88

Petasites albus **(L.) Gaertn.**
Asteraceae
Tussilago alba L.
♦ white butterbur, butterbur
♦ Weed
♦ 272
♦ pH, cultivated, herbal.

Petasites fragrans **(Vill.) C.Presl**
Asteraceae
Nardosmia denticulata Cass., *Tussilago suaveolens* Desf.
♦ winter heliotrope, farfaraccio vaniglione
♦ Weed, Naturalised, Garden Escape, Environmental Weed
♦ 15, 40, 70, 72, 86, 88, 98, 155, 176, 198, 203, 280
♦ pH, cultivated, herbal. Origin: western Mediterranean.

Petasites hybridus **(L.) Gaertn., Mey. & Scherb.**
Asteraceae
Petasites officinalis Moench. (see), *Petasites ovatus* Hill, *Petasites vulgaris* Hill, *Tussilago petasites* L., *Tussilago hybrida* L.
♦ butterbur, pestilence wort, etelänruttojuuri, bog rhubarb, bog rhubarb butterbur

♦ Weed, Naturalised, Cultivation Escape
♦ 42, 70, 87, 88, 101, 272
♦ pH, aqua, cultivated, herbal. Origin: Europe.

Petasites japonicus **(Siebold & Zucc.) Maxim.**
Asteraceae
Nardosmia japonica Siebold & Zucc.
♦ giant butterbur, Japanese sweet colt's foot, sweet colt's foot
♦ Weed, Naturalised
♦ 40, 87, 88, 101, 204, 286
♦ pH, cultivated, herbal.

Petasites japonicus **(Siebold & Zucc.) Maxim. ssp.** *giganteus* **(F.Schmidt ex Trautv.) Kitam.**
Asteraceae
♦ akitabuki
♦ Weed
♦ 286

Petasites officinalis **Moench**
Asteraceae
= *Petasites hybridus* (L.) Gaertn. Mey. & Scherb.
♦ Weed
♦ 39, 87, 88
♦ herbal, toxic.

Petasites spurius **(Retz.) Rchb.**
Asteraceae
♦ rantaruttojuuri
♦ Weed, Naturalised
♦ 42, 272

Petitia domingensis **Jacq.**
Lamiaceae/Verbenaceae
♦ bastard stopper
♦ Weed
♦ 179

Petiveria alliacea **L.**
Phytolaccaceae/Petiveriaceae
♦ guinea henweed, Congo root, gully root
♦ Weed
♦ 14, 39, 87, 88, 157, 261
♦ pH, cultivated, herbal, toxic.

Petrea kohautiana **C.Presl**
Verbenaceae
= *Petrea volubilis* L.
♦ Introduced
♦ 261
♦ Origin: tropical America.

Petrea volubilis **L.**
Verbenaceae
Petrea kohautiana C.Presl (see)
♦ queen's wreath, petrea, purple wreath
♦ Weed, Naturalised, Garden Escape
♦ 101, 179, 261
♦ a/pH, cultivated.

Petrorhagia dubia **(Raf.) López & Romo**
Caryophyllaceae
Petrorhagia velutina (Guss.) Ball & Heywood (see), *Tunica velutina* (Guss.) Fisch. & Mey. (see)
♦ wilding pink, hairypink
♦ Weed, Naturalised
♦ 34, 101, 300
♦ aH. Origin: southern Europe.

Petrorhagia nanteuilii **(Burnat) Ball & Heywood**
Caryophyllaceae
♦ childing pink, proliferous pink
♦ Weed, Naturalised, Environmental Weed
♦ 7, 86, 98, 101, 176, 198, 203, 241, 269, 286, 287
♦ aH, cultivated. Origin: Europe.

Petrorhagia prolifera **(L.) P.Ball & Heywood**
Caryophyllaceae
Dianthus prolifer L. (see), *Kohlrauschia prolifera* (L.) Kunth (see), *Tunica prolifera* (L.) Scop.
♦ childing pink, proliferous pink, tunika prerastená
♦ Weed, Naturalised, Environmental Weed
♦ 72, 88, 98, 101, 203, 241, 253, 280, 286, 287, 300
♦ aH, cultivated, herbal.

Petrorhagia saxifraga **(L.) Link**
Caryophyllaceae
Dianthus saxifragus L., *Tunica saxifraga* (L.) Scop.
♦ saxifrage pink, tunic flower
♦ Naturalised
♦ 101
♦ cultivated, herbal.

Petrorhagia velutina **(Guss.) Ball & Heywood**
Caryophyllaceae
= *Petrorhagia dubia* (Raf.) López & Romo
♦ velvet pink, hairy pink
♦ Weed, Naturalised, Environmental Weed
♦ 7, 9, 86, 98, 134, 176, 198, 203, 280
♦ herbal.

Petroselinum crispum **(Mill.) Nyman ex A.W.Hill**
Apiaceae
♦ parsley, wild parsley, persilja, prezzemolo, garden parsley
♦ Weed, Naturalised, Introduced, Cultivation Escape, Casual Alien
♦ 7, 34, 38, 40, 42, 86, 88, 94, 98, 101, 121, 165, 176, 198, 203, 252, 257, 261, 280, 287
♦ b/pH, cultivated, herbal. Origin: Eurasia.

Petroselinum segetum **(L.) Koch**
Apiaceae
Apium segetum (L.) Dumort., *Sison segetum* L.
♦ corn caraway, corn parsley
♦ Weed
♦ 70, 87, 88, 94, 243, 253
♦ bH, promoted, herbal. Origin: Europe.

Petunia × *atkinsiana* **D.Don ex Loudon**
Solanaceae
= *Petunia axillaris* (Lam.) Britt. Sterns & Pogg. (see) × *Petunia integrifolia* (Hook.) Schinz & Thell. (see)
♦ Naturalised
♦ 101

Petunia axillaris (Lam.) Britton, Sterns & Pogg.
Solanaceae
- large white petunia
- Weed, Naturalised
- 86, 87, 88, 101, 237, 269, 270, 295
- pH, cultivated. Origin: South America.

Petunia × hybrida hort. ex E.Vilm.
Solanaceae
= *Petunia axillaris* (Lam.) Britton, Sterns & Pogg. × *Petunia integrifolia* (Hook.) Schinz & Thell. [see *Petunia × atkinsiana* D.Don ex Loud.]
- common garden petunia, petunia, garden petunia, tarhapetunia
- Weed, Naturalised, Cultivation Escape
- 42, 121, 134, 280, 287
- aH, cultivated. Origin: South America.

Petunia integrifolia (Hook.) Schinz & Thell.
Solanaceae
Petunia violacea Lindl. (see)
- violetflower petunia
- Naturalised
- 101
- cultivated, herbal.

Petunia parviflora Juss.
Solanaceae
= *Calibrachoa parviflora* (Juss.) D'Arcy
- wild petunia, fewflower petunia, petunia
- Weed, Naturalised, Casual Alien
- 98, 203, 241, 280, 300
- aH, cultivated, herbal.

Petunia violacea Lindl.
Solanaceae
= *Petunia integrifolia* (Hook.) Schinz & Thell.
- petunia morada
- Casual Alien
- 261
- cultivated, herbal. Origin: tropical and subtropical South America.

Peucedanum carvifolia Vill.
Apiaceae
Cervaria rivini Gaertn., *Schlosseria chabrei* Schloss. & Vuk., *Seseli carvifolia* L.
- Weed
- 272

Peucedanum cervaria (L.) Lapeyr.
Apiaceae
Selinum cervaria L., *Athamantha cervaria* L., *Cervaria nigra* Bern.
- broad leaved spignel
- Weed
- 272
- pH, promoted, herbal.

Peucedanum galbanum (L.) Benth. & Hook.f.
Apiaceae
- blister bush, blistering bush, wild celery
- Native Weed
- 121

- pH, cultivated, herbal. Origin: southern Africa.

Peucedanum kerstenii Engl.
Apiaceae
- Quarantine Weed
- 220

Peucedanum lancifolium Lang.
Apiaceae
- Weed
- 70

Peucedanum officinale L.
Apiaceae
Selinum officinale Vest, *Selinum peucedanum officinale* Crantz
- hog's fennel, sea hog's fennel
- Weed
- 70, 272
- pH, promoted, herbal.

Peucedanum oreoselinum (L.) Moench
Apiaceae
Oreoselinum nigrum Delarb., *Selinum oreoselinum* Crantz
- imperatoria apio montano
- Weed
- 272
- pH, cultivated, herbal.

Peucedanum ostruthium (L.) W.D.J.Koch
Apiaceae
Imperatoria ostruthium L., *Selinum imperatoria* Crantz
- masterwort, imperatoria
- Naturalised, Introduced
- 39, 101, 125
- pH, cultivated, herbal, toxic.

Peucedanum palustre (L.) Moench
Apiaceae
Calestania palustris (L.) Koso-Pol., *Selinum palustre* L. (see), *Thysselinum palustre* Hoffm.
- hogfennel, marsh hog's fennel, milk parsley
- Weed, Naturalised
- 87, 88, 101, 272
- bH, promoted, herbal.

Pfaffia glomerata (Spreng.) Pedersen
Amaranthaceae
Pfaffia stenophylla (Spreng.) Stuchlik (see)
- Weed
- 237, 295

Pfaffia iresinoides (Kunth) A.Spreng.
Amaranthaceae
- Weed
- 87, 88
- herbal.

Pfaffia sericea (Spreng.) Mart.
Amaranthaceae
= *Pfaffia tuberosa* (Spreng.) Hicken
- Weed
- 87, 88

Pfaffia stenophylla (Spreng.) Stuchlik.
Amaranthaceae
= *Pfaffia glomerata* (Spreng.) Pedersen
- Weed
- 87, 88

Pfaffia tuberosa (Spreng.) Hicken
Amaranthaceae

Pfaffia sericea (Spreng.) Mart. (see)
- batatilla
- Weed
- 237, 295
- herbal.

Phacelia artemisioides Griseb.
Hydrophyllaceae
- Weed
- 121
- aH. Origin: South America.

Phacelia campanularia Gray
Hydrophyllaceae
- desert bells, California bluebell, desert bluebells
- Weed, Quarantine Weed
- 39, 76, 88, 220
- aH, cultivated, herbal, toxic.

Phacelia ciliata Benth.
Hydrophyllaceae
- great valley phacelia, great valley scorpionweed
- Weed
- 161, 180
- aH, cultivated.

Phacelia coulteri Greenm.
Hydrophyllaceae
- Weed
- 199

Phacelia crenulata Torr. ex S.Watson
Hydrophyllaceae
- scorpion flower, cleftleaf wild heliotrope
- Weed
- 39, 161
- aH, cultivated, herbal, toxic.

Phacelia grandiflora (Benth.) A.Gray
Hydrophyllaceae
Phacelia whitlavia Gray
- giant flowered phacelia, largeflower scorpionweed
- Quarantine Weed
- 76
- aH, cultivated, herbal.

Phacelia heterophylla Pursh
Hydrophyllaceae
- varileaf phacelia
- Weed
- 23, 88
- herbal.

Phacelia hirsuta Nutt.
Hydrophyllaceae
- hairy phacelia, fuzzy phacelia
- Weed
- 161
- herbal.

Phacelia leucophylla Torr.
Hydrophyllaceae
= *Phacelia hastata* Dougl. ex Lehm. var. *hastata* (NoR)
- Weed
- 23, 88
- herbal.

Phacelia linearis (Pursh) Holz.
Hydrophyllaceae
- threadleaf phacelia
- Weed

♦ 23, 88
♦ aH, herbal.

Phacelia minor (Harv.) Thell.
Hydrophyllaceae
♦ suppilohunajakukka, wild
Canterbury bells, California bluebell
♦ Casual Alien
♦ 39, 42
♦ aH, cultivated, herbal, toxic.

Phacelia pedicellata Gray
Hydrophyllaceae
♦ scorpionweed, pedicellate phacelia
♦ Weed
♦ 161
♦ aH, herbal, toxic.

Phacelia platycarpa (Cav.) Spreng.
Hydrophyllaceae
♦ Weed
♦ 199

Phacelia purshii Buckl.
Hydrophyllaceae
♦ pursh phacelia, Miami mist
♦ Weed
♦ 87, 88, 218
♦ herbal.

Phacelia tanacetifolia Benth.
Hydrophyllaceae
♦ tansy phacelia, lacy scorpionweed,
fiddleneck, phacelia, aitohunajakukka
♦ Weed, Naturalised, Cultivation
Escape, Casual Alien
♦ 7, 15, 40, 42, 86, 87, 88, 98, 176, 203,
218, 252, 272, 280
♦ aH, cultivated, herbal. Origin:
eastern North America.

Phaenocoma prolifera D.Don
Asteraceae
Xeranthemum proliferum L.
♦ pink everlasting
♦ Weed, Quarantine Weed,
Naturalised
♦ 76, 86, 88
♦ cultivated.

Phaeomeria magnifica (Roscoe) K.Schum.
Zingiberaceae
= *Etlingera elatior* (Jack) R.M.Sm.
♦ torch ginger
♦ Weed
♦ 3, 191
♦ herbal.

Phaeopappus scoparius Boiss.
Asteraceae
= *Centaurea scoparia* Sieber ex DC.
(NoR)
♦ Weed
♦ 221

Phaeoptilum spinosum Radlk.
Nyctaginaceae
♦ Native Weed
♦ 121
♦ pS. Origin: southern Africa.

Phagnalon Cass. spp.
Asteraceae
♦ Weed
♦ 221

**Phagnalon barbeyanum Asch. &
Schweinf.**
Asteraceae
♦ Weed
♦ 221

Phagnalon nitidum Fresen.
Asteraceae
♦ Weed
♦ 221

Phagnalon rupestre (L.) DC.
Asteraceae
♦ Weed
♦ 221
♦ arid, cultivated.

Phagnalon saxatile (L.) Cass.
Asteraceae
♦ Weed
♦ 70
♦ arid, cultivated.

Phagnalon sinaicum Bornm. & Kneuck.
Asteraceae
♦ Weed
♦ 221

**Phaius tancarvilleae (Banks ex L'Hér.)
Blume**
Orchidaceae
♦ nun's hood orchid
♦ Naturalised
♦ 101

**Phalacrachena inuloides (Fisch. ex Janka)
Iljin**
Asteraceae
♦ suolakaunokki
♦ Casual Alien
♦ 42

Phalaris L. spp.
Poaceae
♦ reed canarygrass, canarygrass
♦ Weed, Noxious Weed, Naturalised
♦ 88, 137, 198, 243
♦ G, herbal.

Phalaris angusta Nees ex Trin.
Poaceae
♦ kapeatähkähelpi, timothy
canarygrass
♦ Weed, Naturalised, Casual Alien
♦ 7, 30, 40, 42, 86, 87, 88, 98, 121, 203,
237, 243, 280, 295
♦ aG. Origin: South America.

Phalaris aquatica L.
Poaceae
Phalaris bulbosa L., *Phalaris nodosa*
(L.) Murray, *Phalaris tuberosa* L. (see),
Phalaris tuberosa L. var. *stenoptera*
(Hack.) A.S.Hitchc. (see)
♦ harding grass, tuberous canarygrass,
Toowoomba canarygrass, phalaris,
canarygrass
♦ Weed, Noxious Weed, Naturalised,
Introduced, Environmental Weed,
Casual Alien
♦ 7, 15, 34, 35, 38, 39, 40, 70, 72, 78, 80,
86, 87, 88, 91, 98, 101, 116, 121, 146, 158,
161, 176, 198, 203, 228, 231, 241, 269,
280, 287, 289, 296, 300
♦ pG, aqua, cultivated, herbal, toxic.
Origin: Eurasia.

Phalaris aquatica L. × arundinacea R.Br.
Poaceae
♦ Naturalised, Weed
♦ 86, 98, 203
♦ G.

Phalaris arundinacea L.
Poaceae
Arundo colorata Ait., *Arundo riparia*
Salisb., *Baldingera arundinacea* (L.)
Dumort., *Baldingera arundinacea*
var. *picta* (L.) Nyman, *Baldingera
arundinacea* var. *rotgesii* Foucaud
& Mandon ex Husn., *Baldingera
colorata* P.Gaertn., B.Mey. & Scherb.,
Calamagrostis colorata (Ait.) Sibth.,
Calamagrostis variegata With., *Digraphis
americana* Elliott ex Loud., *Digraphis
arundinacea* (L.) Trin. (see), *Endallex
arundinacea* Raf., *Endallex arundinaceae*
Raf. ex B.D.Jacks., *Phalaridantha
arundinacea* (L.) St.-Lag., *Phalaris
arundinacea* var. *picta* L. (see), *Phalaris
caesia* Nees, *Phalaris hispanica* Coincy,
Phalaris japonica Steud., *Phalaroides
arundinacea* (L.) Rausch., *Typhoides
arundinacea* (L.) Moench
♦ reed canarygrass, lady grass, spires,
doggers, swordgrass, ladies' laces,
bride's laces, London lace, gardener's
garters, ribbon grass, variegated grass,
alpiste roseau, Rohrglanzgras, kusa-
yoshi, caniço malhado, hierba cinta,
pasto cinto
♦ Weed, Noxious Weed, Naturalised,
Native Weed, Introduced, Garden
Escape, Environmental Weed
♦ 1, 4, 7, 15, 23, 30, 38, 39, 45, 70, 72, 80,
88, 91, 98, 102, 104, 121, 132, 133, 136,
138, 139, 142, 146, 151, 159, 161, 174,
176, 195, 198, 203, 204, 208, 208, 211,
212, 218, 222, 224, 228, 229, 272, 280,
286, 297, 300
♦ wpG, cultivated, herbal, toxic.
Origin: Eurasia.

Phalaris arundinacea L. var. arundinacea
Poaceae
♦ reed canarygrass
♦ Naturalised, Environmental Weed
♦ 86, 98, 280
♦ G, cultivated. Origin: North Africa,
Asia, Europe, North America.

Phalaris arundinacea L. var. picta L.
Poaceae
= *Phalaris arundinacea* L.
♦ variegated reed canarygrass, ribbon
grass, reed canarygrass
♦ Weed, Naturalised, Environmental
Weed
♦ 15, 40, 86, 98, 280, 287
♦ G, cultivated. Origin: Europe.

Phalaris brachystachys Link
Poaceae
♦ shortspike canarygrass, falaridi,
scagliole
♦ Weed, Quarantine Weed,
Naturalised
♦ 70, 76, 87, 88, 91, 101, 118, 203, 220,
243, 253, 272
♦ aG, herbal.

Phalaris canariensis L.
Poaceae
♦ canarygrass, common canarygrass, birdseed grass, alpiste, lesknica kanárska
♦ Weed, Naturalised, Introduced, Environmental Weed, Casual Alien
♦ 7, 30, 34, 38, 40, 42, 80, 86, 87, 88, 91, 98, 101, 102, 116, 121, 158, 176, 198, 199, 203, 218, 228, 236, 237, 243, 263, 280, 286, 287, 295, 300
♦ aG, arid, cultivated, herbal. Origin: Eurasia.

Phalaris caroliniana Walter
Poaceae
♦ lännenhelpi, Carolina canarygrass
♦ Weed, Naturalised, Casual Alien
♦ 42, 161, 241, 300
♦ aG, herbal.

Phalaris coerulescens Desf.
Poaceae
♦ sunolgrass, blue canarygrass, phalaris
♦ Weed, Quarantine Weed, Naturalised
♦ 61, 70, 76, 86, 98, 198, 203, 253, 272
♦ G, arid, cultivated. Origin: Mediterranean.

Phalaris lemmonii Vasey
Poaceae
♦ phalaris, Lemmon's canarygrass
♦ Weed, Naturalised
♦ 86, 98, 198, 203
♦ aG, herbal. Origin: California.

Phalaris minor Retz.
Poaceae
♦ littleseed canarygrass, Mediterranean canarygrass, pikkuhelpi, small canarygrass, falaridi, scagliole, canarygrass
♦ Weed, Naturalised, Introduced, Environmental Weed, Casual Alien
♦ 7, 9, 15, 30, 34, 39, 40, 42, 68, 70, 72, 80, 86, 87, 88, 91, 98, 101, 115, 118, 121, 134, 157, 158, 161, 176, 179, 180, 185, 198, 203, 212, 218, 221, 228, 237, 241, 243, 253, 269, 272, 280, 286, 287, 295, 300
♦ aG, arid, cultivated, herbal, toxic. Origin: Eurasia.

Phalaris paradoxa L.
Poaceae
Phalaris paradoxa var. *praemorsa* (Lam.) Coss. & Durieu (see)
♦ paradoxa grass, hood canarygrass, bristle spiked canarygrass, paradoxial canarygrass, rikkahelpi, annual canarygrass
♦ Weed, Naturalised, Introduced, Environmental Weed, Casual Alien
♦ 7, 34, 40, 42, 55, 68, 72, 86, 87, 88, 91, 98, 101, 111, 115, 118, 176, 185, 198, 203, 218, 221, 228, 237, 243, 253, 269, 272, 280, 287, 295
♦ aG, arid, cultivated, herbal. Origin: Mediterranean.

Phalaris paradoxa L. var. praemorsa (Lam.) Coss. & Durieu
Poaceae
= *Phalaris paradoxa* Lam.
♦ Naturalised, Casual Alien
♦ 40, 287
♦ G.

Phalaris platensis Henrard ex Wacht.
Poaceae
♦ Weed
♦ 237, 295
♦ G.

Phalaris tuberosa L.
Poaceae
= *Phalaris aquatica* L.
♦ bulbous canarygrass, harding grass
♦ Weed
♦ 87, 88
♦ G, herbal.

Phalaris tuberosa L. var. stenoptera (Hack.) A.S.Hitchc.
Poaceae
= *Phalaris aquatica* L.
♦ harding grass, canarygrass, tuberous canarygrass
♦ Weed
♦ 180
♦ G, herbal.

Pharbitis nil (L.) Choisy
Convolvulaceae
= *Ipomoea nil* (L.) Roth
♦ Weed
♦ 275, 297
♦ aH, herbal.

Pharbitis purpurea (L.) Voigt
Convolvulaceae
= *Ipomoea purpurea* (L.) Roth
♦ Weed
♦ 256, 275, 297
♦ aH, cultivated.

Phaseolus aconitifolius Jacq.
Fabaceae/Papilionaceae
= *Vigna aconitifolia* (Jacq.) Maréchal
♦ Weed
♦ 39, 87, 88
♦ toxic.

Phaseolus acutifolius A.Gray
Fabaceae/Papilionaceae
♦ tepary bean
♦ Introduced
♦ 228
♦ arid, cultivated, herbal.

Phaseolus acutifolius A.Gray var. latifolius F.L.Freeman
Fabaceae/Papilionaceae
Phaseolus latifolius Freeman, *Phaseolus tenuifolius* Wooton & Standl.
♦ tepary bean
♦ Introduced
♦ 228
♦ arid. Origin: Central and North America.

Phaseolus adenanthus G.F.W.Mey.
Fabaceae/Papilionaceae
= *Vigna adenantha* (G.Mey.) Maréchal, Mascherpa & Stainier
♦ phaseolus
♦ Weed
♦ 87, 88, 295
♦ arid, cultivated, herbal.

Phaseolus atropurpureus DC.
Fabaceae/Papilionaceae
= *Macroptilium atropurpureum* (DC.) Urb.
♦ siratro
♦ Weed
♦ 87, 88, 157, 239
♦ pH.

Phaseolus aureus Roxb.
Fabaceae/Papilionaceae
= *Vigna radiata* (L.) R.Wilczek
♦ Weed
♦ 87, 88
♦ aC, herbal.

Phaseolus coccineus L.
Fabaceae/Papilionaceae
Phaseolus multiflorus Lam.
♦ scarlet runner, runner bean, scarlet runner bean
♦ Weed, Naturalised, Casual Alien
♦ 39, 40, 101, 161, 262, 280
♦ pH, cultivated, herbal, toxic.

Phaseolus lathyroides L.
Fabaceae/Papilionaceae
= *Macroptilium lathyroides* (L.) Urb.
♦ phasey bean, thua phee
♦ Weed, Naturalised
♦ 87, 88, 209, 239, 257, 262, 287
♦ pH, arid, herbal.

Phaseolus linearis Kunth
Fabaceae/Papilionaceae
= *Vigna linearis* (Kunth) Maréchal, Mascherpa & Stainier (NoR)
♦ Weed
♦ 87, 88

Phaseolus lunatus L.
Fabaceae/Papilionaceae
♦ sieva bean, lima bean, broad bean, butter bean, duffin bean, haba, haba lima, limabna
♦ Weed, Naturalised, Introduced, Casual Alien
♦ 39, 86, 87, 88, 98, 101, 161, 179, 203, 228, 247, 261, 280
♦ pC, arid, cultivated, herbal, toxic. Origin: tropical America.

Phaseolus trilobus auct.
Fabaceae/Papilionaceae
= *Vigna trilobata* (L.) Verdc.
♦ Weed, Quarantine Weed
♦ 76, 87, 88, 203, 220
♦ herbal.

Phaseolus vulgaris L.
Fabaceae/Papilionaceae
♦ tarhapapu, pensaspapu, salkopapu, green bean, kidney bean, runner bean, common bean, haricot bean, navy bean, habichuela, fagiolo baccello
♦ Weed, Naturalised, Cultivation Escape, Casual Alien
♦ 39, 42, 101, 179, 261
♦ aH, cultivated, herbal, toxic. Origin: obscure.

Phaulopsis angolana S.Moore
Acanthaceae
♦ Weed
♦ 87, 88

***Phellodendron amurense* Rupr.**
Rutaceae
♦ Amur corktree, huang bai
♦ Weed, Naturalised, Environmental Weed
♦ 80, 101, 182
♦ T, cultivated, herbal. Origin: east Asia, northern China, Manchuria.

***Phellodendron japonicum* Maxim.**
Rutaceae
♦ corktree, Japanese corktree
♦ Weed, Naturalised
♦ 101, 133, 195
♦ cultivated.

***Phelypaea aegyptiaca* Walp.**
Orobanchaceae
= *Orobanche aegyptiaca* Pers.
♦ Weed
♦ 87, 88
♦ herbal.

***Phelypaea longiflora* C.A.Mey.**
Orobanchaceae
= *Orobanche aegyptiaca* Pers.
♦ Weed
♦ 87, 88

***Phelypaea ramosa* (L.) C.A.Mey.**
Orobanchaceae
= *Orobanche ramosa* L.
♦ Weed
♦ 87, 88

***Phenax sonneratii* (Poir.) Wedd.**
Urticaceae
♦ fura parede
♦ Weed, Naturalised, Introduced
♦ 28, 32, 87, 88, 206, 243, 255, 261
♦ aH. Origin: Brazil.

***Philadelphus cordifolius* Lange**
Hydrangeaceae/Philadelphaceae
♦ heartleaf mock orange
♦ Naturalised
♦ 101

***Philadelphus coronarius* L.**
Hydrangeaceae/Philadelphaceae
Philadelphus pallidus Hayek
♦ pihajasmike, mock orange, sweet mock orange
♦ Naturalised, Cultivation Escape
♦ 39, 42, 101
♦ S, cultivated, herbal, toxic.

***Philadelphus* × *cymosus* Rehder**
Hydrangeaceae/Philadelphaceae
♦ Casual Alien
♦ 280

***Philadelphus karwinskianus* Koehne**
Hydrangeaceae/Philadelphaceae
♦ evergreen mock orange, mock orange philadelphus, syringa
♦ Naturalised, Cultivation Escape
♦ 101, 233
♦ cultivated.

***Philadelphus lewisii* Pursh**
Hydrangeaceae/Philadelphaceae
♦ tähtijasmike, Lewis's mock orange, mock orange
♦ Cultivation Escape
♦ 39, 42
♦ S, cultivated, herbal, toxic.

***Philadelphus mexicanus* Schltdl.**
Hydrangeaceae/Philadelphaceae
♦ Naturalised
♦ 280
♦ cultivated, herbal.

***Philadelphus pubescens* Loisel.**
Hydrangeaceae/Philadelphaceae
♦ hovijasmike, hoary mock orange
♦ Cultivation Escape
♦ 42
♦ S, cultivated, herbal.

***Philippicereus* Backeb. spp.**
Cactaceae
= *Eulychnia* Phil. spp.
♦ Weed, Quarantine Weed
♦ 76, 88, 203

***Phillyrea angustifolia* L.**
Oleaceae
♦ narrowleaf jasmine box, narrow leaved phillyrea
♦ Weed, Naturalised
♦ 70, 198
♦ cultivated, herbal.

***Phillyrea latifolia* L.**
Oleaceae
♦ broadleaf jasmine box, mock privet
♦ Naturalised
♦ 86, 198
♦ T, cultivated, herbal. Origin: Mediterranean.

***Philodendron* Schott spp.**
Araceae
♦ philodendron, malanguilla
♦ Weed
♦ 154, 161, 247, 281
♦ toxic.

***Philoxerus portulacoides* A.St.-Hil.**
Amaranthaceae
♦ philoxerus
♦ Introduced
♦ 228
♦ arid.

***Philoxerus vermicularis* (L.) R.Br. ex Sm.**
Amaranthaceae
= *Blutaparon vermiculare* (L.) Mears (NoR)
♦ silverhead
♦ Weed
♦ 87, 88
♦ cultivated, herbal.

***Philydrum lanuginosum* Banks & Sol. ex Gaertn.**
Philydraceae
♦ frogmouth
♦ Weed
♦ 39, 87, 88, 200
♦ aqua, cultivated, toxic.

***Phleum* L. spp.**
Poaceae
♦ timothy grass, timothy
♦ Weed, Naturalised
♦ 198, 272
♦ G.

***Phleum alpinum* L.**
Poaceae
Phleum commutatum Gaudin
♦ alpine timothy, mountain timothy, pohjantähkiö, alpine cat's tail

♦ Naturalised
♦ 241
♦ pG, cultivated, herbal.

***Phleum arenarium* L.**
Poaceae
Chilochloa arenaria Beauv.
♦ hietatähkiö, sand timothy, sand cat's tail
♦ Naturalised, Casual Alien
♦ 7, 42, 86, 101
♦ G, cultivated. Origin: southern Europe.

***Phleum paniculatum* Huds.**
Poaceae
Chilochloa aspera Beauv., *Phleum asperum* Jacq.
♦ British timothy
♦ Weed, Naturalised
♦ 87, 88, 101, 243, 272, 275
♦ G, cultivated, herbal.

***Phleum phleoides* (L.) Karst.**
Poaceae
Chilochloa boehmeri P.Beauv., *Phleum boehmeri* Wib.
♦ purple cat's ear, timotejka tuhá, helpitähkiö, purple stalked cat's tail, purplestem cat's tail
♦ Weed
♦ 272
♦ G, cultivated, herbal.

***Phleum pratense* L.**
Poaceae
Phleum ciliatum Gilib., *Phleum parnassicum* Boiss., *Stelephusos pratensis* Lunell
♦ timothy, timothy grass, cat's tail grass, herd's grass
♦ Weed, Naturalised, Introduced, Garden Escape, Environmental Weed, Cultivation Escape
♦ 7, 9, 15, 20, 21, 30, 34, 39, 72, 80, 86, 87, 88, 98, 101, 121, 146, 151, 155, 161, 176, 179, 195, 198, 199, 203, 211, 218, 241, 243, 253, 256, 272, 280, 286, 287, 290, 295, 296, 300
♦ pG, cultivated, herbal, toxic. Origin: Eurasia.

***Phleum pratense* L. ssp. *bertolonii* (DC.) Born.**
Poaceae
Phleum nodosum L., *Phleum bertolonii* DC.
♦ smaller cat's tail, small timothy, turf grass
♦ Weed
♦ 70, 243
♦ G, cultivated.

Phleum pratense* L. ssp. *pratense
Poaceae
♦ timothy
♦ Weed
♦ 70, 243
♦ G.

***Phleum pratense* L. ssp. *serotinum* (Jordan) Berher**
Poaceae
♦ Weed
♦ 253
♦ G.

Phleum subulatum (Savi) Asch. & Graebn.
Poaceae
Phleum tenue Schrad.
♦ Italian timothy
♦ Weed, Naturalised
♦ 86, 98, 101, 203, 221, 272
♦ G. Origin: southern Europe, Middle East.

Phlomis aurea Decne.
Lamiaceae
♦ Weed
♦ 221

Phlomis floccosa D.Don
Lamiaceae
♦ Weed
♦ 221

Phlomis fruticosa L.
Lamiaceae
♦ shrubby Jerusalem sage, Jerusalem sage
♦ Weed, Naturalised, Casual Alien
♦ 40, 101, 272, 280
♦ S, cultivated, herbal.

Phlomis herba-venti L.
Lamiaceae
♦ Weed
♦ 87, 88, 272
♦ cultivated.

Phlomis kurdica K.H.Rech.
Lamiaceae
♦ Weed
♦ 87, 88

Phlomis lychnitis L.
Lamiaceae
♦ lampwick plant
♦ Weed
♦ 70
♦ cultivated, herbal.

Phlomis orientalis Mill.
Lamiaceae
♦ Weed
♦ 87, 88
♦ cultivated.

Phlomis russeliana (Sims) Benth.
Lamiaceae
♦ Naturalised
♦ 280
♦ pH, cultivated, herbal.

Phlomis tuberosa L.
Lamiaceae
♦ mukulapaloyrtti, tuberous Jerusalem sage
♦ Weed, Naturalised, Casual Alien
♦ 42, 101, 272
♦ pH, cultivated.

Phlomis umbrosa Turcz.
Lamiaceae
♦ shady Jerusalem sage
♦ Weed
♦ 297

Phlox L. spp.
Polemoniaceae
♦ phlox, sweet William
♦ Environmental Weed
♦ 39, 257
♦ cultivated, toxic.

Phlox carolina L.
Polemoniaceae
♦ mountain phlox, thickleaf phlox
♦ Weed
♦ 161
♦ cultivated, herbal.

Phlox divaricata L.
Polemoniaceae
♦ wild blue phlox, blue phlox
♦ Weed
♦ 161
♦ cultivated, herbal.

Phlox drummondii Hook.
Polemoniaceae
♦ annual phlox, garden phlox
♦ Weed, Naturalised
♦ 179, 280
♦ cultivated, herbal.

Phlox paniculata L.
Polemoniaceae
♦ syysleimu, fall phlox, garden phlox, perennial phlox
♦ Naturalised, Cultivation Escape
♦ 42, 280
♦ pH, cultivated, herbal.

Phlox pilosa L.
Polemoniaceae
♦ sand prairie phlox, downy phlox, prairie phlox
♦ Weed
♦ 161
♦ aH, cultivated, herbal.

Phlox × procumbens Lehm. (pro sp.)
Polemoniaceae
= *Phlox stolonifera* Sims. × *Phlox subulata* L.
♦ Naturalised
♦ 101

Phlox subulata L.
Polemoniaceae
♦ sammalleimu, moss phlox
♦ Cultivation Escape, Casual Alien
♦ 42, 280
♦ cultivated, herbal.

Phoenix L. spp.
Arecaceae
♦ date palm
♦ Weed, Naturalised
♦ 116, 198
♦ herbal.

Phoenix canariensis hort. ex Chabaud
Arecaceae
♦ Canary Island date palm, phoenix palm
♦ Weed, Sleeper Weed, Naturalised, Introduced, Garden Escape, Environmental Weed, Casual Alien
♦ 15, 54, 72, 86, 88, 101, 179, 198, 225, 230, 246, 280
♦ T, cultivated, herbal. Origin: Canary Islands.

Phoenix dactylifera L.
Arecaceae
♦ date palm, arrak, taatelipalmu, dadelpalm, taateli, dattel, date
♦ Weed, Naturalised, Introduced, Garden Escape, Environmental Weed,

Casual Alien
♦ 7, 22, 39, 42, 86, 93, 98, 101, 179, 203, 205, 228, 230
♦ S/T, arid, cultivated, herbal, toxic. Origin: obscure.

Phoenix reclinata Jacq.
Arecaceae
Phoenix leonensis Lodd. ex Kunth *nom. nud.*, *Phoenix spinosa* Schumach.
♦ Senegal date palm, reclining date palm, lusanda mkindu
♦ Weed, Naturalised, Environmental Weed
♦ 80, 80, 88, 101, 112, 151, 179
♦ T, cultivated, herbal. Origin: Madagascar.

Phoenix roebelenii O'Brien
Arecaceae
♦ pygmy date palm
♦ Introduced
♦ 230
♦ T, cultivated. Origin: east Asia.

Phoenix sylvestris (L.) Roxb.
Arecaceae
♦ wild date palm, date sugar palm, wild date
♦ Naturalised
♦ 86
♦ cultivated, herbal. Origin: India, Nepal.

Pholiurus pannonicus (Host) Trin.
Poaceae
♦ Weed, Naturalised
♦ 86, 98, 203
♦ G. Origin: Eurasia.

Phoradendron affine Nutt.
Viscaceae
♦ erva de passarinho
♦ Weed
♦ 255
♦ T parasitic.

Phoradendron randiae (Bello) Britt.
Viscaceae
= *Phoradendron quadrangulare* (Kunth) Krug & Urb. (NoR)
♦ Weed
♦ 14
♦ H parasitic.

Phoradendron villosum (Nutt.) Nutt.
Viscaceae
♦ oak mistletoe, Pacific mistletoe, mistletoe
♦ Weed
♦ 39, 154, 161
♦ S parasitic, herbal, toxic.

Phormium cookianum Le Jol.
Agavaceae/Phormiaceae
♦ lesser New Zealand flax, mountain flax
♦ Naturalised
♦ 40
♦ cultivated.

Phormium tenax J.R. & G.Forst.
Agavaceae/Phormiaceae
♦ New Zealand flax, harakeke, flax
♦ Weed, Naturalised, Garden Escape, Environmental Weed

♦ 3, 40, 54, 72, 80, 86, 88, 101, 131, 155, 176, 191, 198
♦ pH, aqua, cultivated, herbal. Origin: New Zealand.

Photinia arbutifolia (Aiton) Lindl.
Rosaceae
= *Heteromeles arbutifolia* (Lindl.) M.Roem.
♦ Christmas berry
♦ Weed
♦ 87, 88, 218
♦ S, herbal.

Photinia davidiana (Decne.) Cardot
Rosaceae
Stranvaesia undulata Decne., *Stranvaesia davidiana* Decne. (see), *Stranvaesia davidiana* var. *undulata* (Decne.) Rehder & E.H.Wilson
♦ Chinese photinia, photinia
♦ Naturalised, Cultivation Escape, Casual Alien
♦ 40, 101, 233
♦ S, cultivated. Origin: China, Indonesia, Malaysia, Vietnam.

Photinia davidsoniae Rehder & E.H.Wilson
Rosaceae
♦ Naturalised
♦ 280
♦ T, cultivated.

Photinia glabra (Thunb.) Maxim.
Rosaceae
Crataegus glabra Thunb.
♦ Japanese photinia
♦ Naturalised
♦ 86, 101, 198
♦ S, cultivated, herbal. Origin: China, Japan.

Photinia serratifolia (Desf.) Kalkman
Rosaceae
Photinia serrulata Lindl. (see), *Crataegus serratifolia* Desf.
♦ Taiwanese photinia, Chinese hawthorn
♦ Weed, Sleeper Weed, Naturalised, Garden Escape
♦ 54, 86, 88, 101
♦ S, cultivated. Origin: China, Taiwan, Philippines.

Photinia serrulata Lindl.
Rosaceae
= *Photinia serratifolia* (Desf.) Kalkman
♦ Naturalised
♦ 198
♦ cultivated, herbal.

Photinia villosa (Thunb.) DC.
Rosaceae
Crataegus villosa Thunb., *Photinia villosa* var. *longipes* ined., *Pourthiaea villosa* (Thunb.) Decne., *Pourthiaea villosa* var. *longipes* Nakai
♦ oriental photinia
♦ Naturalised
♦ 101
♦ S, cultivated, herbal.

Phragmites australis (Cav.) Trin. ex Steud.
Poaceae

Arundo australis Cav., *Arundo phragmites* L., *Phragmites australis* (Cav.) Trin. ex Steud. var. *stenophylla* (Boiss.) Bor, *Phragmites communis* Trin. (see), *Phragmites communis* Trin. var. *stenophylla* Boiss., *Phragmites maxima* (Forssk.) Blatt. & McCann, *Phragmites vulgaris* (Lam.) Crep., *Trichoon phragmites* (L.) Rendle
♦ common reed, giant reed, phragmites, common reedgrass, canegrass, giant reedgrass, ditch reed, reedgrass, carricillo
♦ Weed, Quarantine Weed, Noxious Weed, Naturalised, Native Weed, Introduced, Garden Escape, Environmental Weed
♦ 7, 10, 17, 70, 76, 80, 86, 87, 88, 91, 102, 103, 121, 133, 151, 158, 159, 161, 178, 185, 186, 195, 200, 208, 211, 220, 221, 224, 228, 229, 237, 246, 248, 253, 258, 269, 272, 280, 286, 295
♦ pG, aqua, cultivated, herbal. Origin: cosmopolitan.

Phragmites communis Trin.
Poaceae
= *Phragmites australis* (Cav.) Trin.
♦ common reed, carrizo, reedgrass, trzcina
♦ Weed, Naturalised
♦ 4, 23, 30, 88, 204, 218, 243, 263, 275, 280, 297
♦ pG, aqua, cultivated, herbal.

Phragmites karka (Retz.) Trin. ex Steud.
Poaceae
Arundo karka Retz., *Phragmites roxburghii* (Kunth) Steud., *Phragmites nepalensis* Nees ex Steud.
♦ tall reed, lirau, reed, yaa khaem, flute reed
♦ Weed, Naturalised
♦ 87, 88, 101, 186, 238, 286
♦ pG, aqua, cultivated.

Phragmites mauritianus Kunth
Poaceae
Phragmites communis Trin. var. *mauritianus* (Kunth) Bak., *Phragmites communis* Trin. var. *mossambicensis* Anders., *Phragmites laxiflorus* Steud., *Phragmites pungens* Hack., *Phragmites vulgaris* Crép. var. *mauritianus* (Kunth) T.Dyrand & Schinz, *Phragmites vulgaris* Crép. var. *mossambicensis* (Anders.) T.Durand & Schinz
♦ lowveld reed, ibano tete
♦ Weed, Native Weed
♦ 10, 87, 88, 121, 158
♦ pG, aqua. Origin: southern Africa.

Phragmites vallatoria (L.) Veldk.
Poaceae
♦ Weed
♦ 262
♦ pG.

Phreatia ponapensis Schltr.
Orchidaceae
♦ Introduced
♦ 230
♦ H.

Phreatia pseudo-thompsonsii Tuyama
Orchidaceae
♦ Introduced
♦ 230
♦ H.

Phtheirospermum japonicum (Thunb.) Kanitz
Scrophulariaceae
♦ Japanese phtheirospermum, koshiogama
♦ Weed
♦ 297

Phthirusa bicolor (Krug & Urb.) Engl.
Loranthaceae
= *Dendropemon bicolor* Krug & Urb. (NoR)
♦ Puerto Rico leechbush
♦ Weed
♦ 87, 88
♦ parasitic.

Phthirusa pauciflora Eichl.
Loranthaceae
= *Dendropemon pauciflorus* (Sw.) Tiegh. (NoR)
♦ Weed
♦ 87, 88
♦ parasitic.

Phthirusa purpurea (L.) Engl.
Loranthaceae
= *Dendropemon purpureus* (L.) Krug & Urb. (NoR)
♦ Weed
♦ 87, 88
♦ parasitic.

Phygelius capensis E.Mey ex Benth
Scrophulariaceae
♦ cape fuchsia
♦ Weed, Naturalised
♦ 15, 163, 280
♦ cultivated, herbal.

Phyla betulifolia (Kunth) Greene
Verbenaceae
= *Lippia betulifolia* Kunth (NoR)
♦ Weed
♦ 87, 88

Phyla canescens (Kunth) Greene
Verbenaceae
Lippia canescens Kunth, *Lippia repens* Spreng., *Lippia uncinuligera* Nees ex Walp., *Phyla nodiflora* (L.) Greene var. *canescens* (Kunth) Moldenke, *Phyla nodiflora* var. *pusilla* (Briq.) Moldenke, *Phyla nodiflora* var. *rosea* (D.Don) Moldenke, *Zapania canescens* Gilbert
♦ fogfruit, lippia
♦ Weed, Noxious Weed, Naturalised
♦ 86, 198, 237, 241, 295, 300
♦ cultivated.

Phyla cuneifolia (Torr.) E.Greene
Verbenaceae
Lippia cuneifolia (Torr.) Steud. (see)
♦ fogfruit, wedgeleaf
♦ Weed
♦ 161
♦ arid, herbal.

Phyla incisa **Small**
Verbenaceae
= *Phyla nodiflora* (L.) Greene
♦ Naturalised
♦ 287
♦ herbal.

Phyla lanceolata **(Michx.) Greene**
Verbenaceae
Lippia lanceolata Michx.
♦ lanceleaf fogfruit, northern fogfruit
♦ Weed
♦ 161
♦ pH, arid/aqua, herbal.

Phyla nodiflora **(L.) Greene**
Verbenaceae
Blairia nodiflora (L.) Gaertn., *Lippia nodiflora* (L.) Michx. (see), *Lippia nodiflora* fo. *brevipes* Kuntze, *Lippia nodiflora* var. *normalis* Kuntze, *Lippia nodiflora* var. *repens* (Bertol.) Schauer, *Lippia nodiflora* var. *sarmentosa* (Willd.) Schauer, *Lippia reptans* Kunth (see), *Phyla chinensis* Lour., *Phyla incisa* Small (see), *Verbena nodiflora* L., *Zapania nodiflora* (L.) Lam., *Zapania nodiflora* Pers. ex Bak., *Zappania nodiflora* var. *rosea* D.Don
♦ matgrass, creeping vervain, lippia, frogfruit, carpetweed, condamine couch, no mow, turkey tangle fogfruit
♦ Weed, Noxious Weed, Naturalised, Garden Escape, Cultivation Escape
♦ 7, 34, 35, 66, 78, 80, 86, 87, 88, 98, 116, 161, 170, 185, 199, 203, 249, 257, 269, 275, 280, 297
♦ pH, cultivated, herbal. Origin: Australia.

Phyla nodiflora **(L.) Greene var.** *nodiflora*
Verbenaceae
♦ capeweed, daisylawn, frogfruit, matgrass, turkey tangle
♦ Weed, Naturalised, Native Weed
♦ 121, 241
♦ pH. Origin: tropical America.

Phyla reptans **(H.B.K.) Greene**
Verbenaceae
♦ Naturalised
♦ 241, 300

Phyla scaberrima **(Juss. ex Pers.) Moldenke**
Verbenaceae
♦ rough fogfruit
♦ Weed
♦ 14
♦ herbal.

Phyla stoechadifolia **(L.) Small**
Verbenaceae
♦ southern fogfruit
♦ Weed
♦ 14
♦ herbal.

Phyla strigulosa **(Mart. & Gal.) Moldenke**
Verbenaceae
♦ diamondleaf fogfruit
♦ Weed
♦ 179

Phyllanthus **L. spp.**
Euphorbiaceae
♦ fua lili'i, leafflower
♦ Quarantine Weed
♦ 220
♦ herbal.

Phyllanthus abnormis **Baill.**
Euphorbiaceae
♦ leafflower, Drummond's leafflower
♦ Weed
♦ 39, 161
♦ herbal, toxic.

Phyllanthus acidus **(L.) Skeels**
Euphorbiaceae
Cicca disticha L., *Phyllanthus distichus* Muell., *Averrhoa acida* L.
♦ Tahitian gooseberry tree, gooseberry tree, otaheite gooseberry
♦ Weed, Naturalised, Introduced, Cultivation Escape
♦ 22, 101, 179, 228, 261
♦ T, arid, cultivated, herbal. Origin: Brazil.

Phyllanthus acuminatus **Vahl**
Euphorbiaceae
♦ Jamaican gooseberry tree
♦ Cultivation Escape
♦ 261
♦ cultivated, herbal.

Phyllanthus amarus **Schum. & Thonn.**
Euphorbiaceae
♦ limeirpwong, fua lili'i, carry me seed, niruri, luuk tai bai
♦ Weed, Naturalised, Casual Alien
♦ 6, 12, 14, 86, 87, 88, 93, 101, 239, 261, 273, 280
♦ pH, cultivated, herbal. Origin: tropical America.

Phyllanthus angulatus **Schum. & Thonn.**
Euphorbiaceae
Fluggea microcarpa Bl.
♦ Weed
♦ 88

Phyllanthus angustifolius **(Sw.) Sw.**
Euphorbiaceae
♦ foliage flower
♦ Weed, Naturalised
♦ 101, 179
♦ cultivated.

Phyllanthus carolinensis **Walter**
Euphorbiaceae
♦ Carolina leafflower
♦ Weed
♦ 28, 199, 243

Phyllanthus carolinensis **Walter var.** *saxicola* **(Small) Webst.**
Euphorbiaceae
♦ Carolina leafflower
♦ Weed
♦ 14

Phyllanthus corcovadensis **Müll.Arg.**
Euphorbiaceae
♦ Weed, Quarantine Weed
♦ 76, 87, 88, 203, 220
♦ herbal.

Phyllanthus debilis **Klein ex Willd.**
Euphorbiaceae

♦ lagoon spurge, niruri
♦ Weed, Naturalised, Introduced
♦ 32, 86, 88, 101, 170, 191, 230, 261, 286, 287
♦ H, herbal, toxic. Origin: tropical Asia, Sri Lanka.

Phyllanthus diffusus **Klotzsch**
Euphorbiaceae
= *Phyllanthus stipulatus* (Raf.) Webst.
♦ Weed, Quarantine Weed
♦ 76, 87, 88, 203, 220

Phyllanthus emblica **L.**
Euphorbiaceae
Emblica officinalis Gaertn.
♦ Indian gooseberry, myrobalan, emblic myrobalan, emblic
♦ Weed, Sleeper Weed, Naturalised, Introduced, Garden Escape, Environmental Weed
♦ 3, 39, 86, 93, 101, 155, 191, 228, 261
♦ arid, cultivated, herbal, toxic. Origin: India, Sri Lanka.

Phyllanthus fraternus **Webster**
Euphorbiaceae
♦ gulf leafflower
♦ Weed, Naturalised, Introduced
♦ 66, 87, 88, 101, 228
♦ arid, herbal.

Phyllanthus galeottianus **Baill.**
Euphorbiaceae
♦ Weed
♦ 87, 88

Phyllanthus graminicola **Britton**
Euphorbiaceae
♦ Weed
♦ 87, 88

Phyllanthus grandifolius **L.**
Euphorbiaceae
♦ Weed
♦ 87, 88

Phyllanthus lacerilobus **Croizat**
Euphorbiaceae
♦ Introduced
♦ 38

Phyllanthus lathyroides **Kunth**
Euphorbiaceae
= *Phyllanthus niruri* L. ssp. *lathyroides* (H.B.K.) G.L.Webster
♦ Weed
♦ 87, 88

Phyllanthus maderaspatensis **L.**
Euphorbiaceae
♦ canoe weed
♦ Weed, Native Weed
♦ 50, 87, 88, 121, 221, 242
♦ p, arid, herbal. Origin: Madagascar.

Phyllanthus matsumurae **Hayata**
Euphorbiaceae
♦ Weed
♦ 87, 88, 286

Phyllanthus minutiflorus **F.Muell. ex Müll.Arg.**
Euphorbiaceae
♦ Weed
♦ 87, 88
♦ cultivated.

Phyllanthus niruri L.
Euphorbiaceae
Diasperus niruri (L.) Kuntze,
Phyllanthus asperulatus Hutch.,
Phyllanthus filiformis Pav. ex Bail.,
Phyllanthus niruri var. *genuinus*
Müll.Arg.
♦ gale of the wind, chanca piedra,
phyllanthus, lagoon spurge, carry me
seed, quinine weed, seed underleaf,
quebra pedra, flor escondida,
tamarindillo
♦ Weed
♦ 39, 87, 88, 153, 209, 217, 218, 242, 243,
245, 255, 262, 276, 281, 286
♦ aH, arid, promoted, herbal, toxic.
Origin: West Indies.

Phyllanthus niruri L. ssp. *amarus*
Leandri
Euphorbiaceae
♦ Weed, Naturalised
♦ 286, 287

Phyllanthus niruri L. ssp. *lathyroides*
(H.B.K.) G.L.Webster
Euphorbiaceae
Phyllanthus lathyroides Müll.Arg. (see)
♦ sarandicito, gale of the wind
♦ Weed
♦ 295

Phyllanthus nummulariifolius Poir.
Euphorbiaceae
♦ Weed
♦ 87, 88
♦ pH.

Phyllanthus odontadenius Müll.Arg.
Euphorbiaceae
♦ Weed
♦ 87, 88

Phyllanthus psuedo-conami Müll.Arg.
Euphorbiaceae
♦ Weed
♦ 87, 88

Phyllanthus pulcher Wall. ex Müll.Arg.
Euphorbiaceae
♦ tropical leafflower
♦ Weed, Naturalised
♦ 12, 101
♦ herbal.

Phyllanthus reticulatus Poir.
Euphorbiaceae
♦ potatobush, kajibajiba
♦ Weed, Naturalised, Native Weed
♦ 121, 287
♦ S/T, cultivated, herbal.

Phyllanthus rotundifolius Klein ex Willd.
Euphorbiaceae
Phyllanthus aspericaulis Pax
♦ Weed
♦ 221

Phyllanthus simplex Retz.
Euphorbiaceae
= *Phyllanthus virgatus* Forst.f.
♦ Weed
♦ 275
♦ aH, herbal.

Phyllanthus stipulatus (Raf.) Webst.
Euphorbiaceae

Phyllanthus diffusus Klotzsch (see)
♦ stipulate leafflower
♦ Weed
♦ 28, 87, 88, 206, 243, 261
♦ herbal.

Phyllanthus sublanatus Schum. &
Thonn.
Euphorbiaceae
♦ Weed
♦ 87, 88

Phyllanthus tenellus Roxb.
Euphorbiaceae
♦ Mascarene Island leafflower, long
stalked phyllanthus
♦ Weed, Naturalised, Introduced
♦ 28, 32, 86, 87, 88, 98, 101, 134, 161,
179, 203, 206, 230, 243, 249, 255, 261
♦ pH. Origin: Brazil.

Phyllanthus urinaria L.
Euphorbiaceae
♦ chamber bitter, niruri
♦ Weed, Naturalised
♦ 32, 39, 87, 88, 101, 161, 170, 179, 235,
249, 261, 273, 274, 276, 286, 297
♦ H, herbal, toxic. Origin: Asia.

Phyllanthus virgatus Forst.f.
Euphorbiaceae
Phyllanthus simplex Retz. (see)
♦ avasâ, moimoi
♦ Weed
♦ 87, 88, 170, 297
♦ herbal.

Phyllitis scolopendrium (L.) Newman
Aspleniaceae
= *Asplenium scolopendrium* L. var.
scolopendrium (NoR)
♦ Hart's tongue
♦ Weed
♦ 87, 88, 272
♦ cultivated, herbal.

Phyllocereus Knebel spp.
Cactaceae
= *Epiphyllum* Haw. spp.
♦ Weed, Quarantine Weed
♦ 76, 88, 220

Phyllodium elegans (Lour.) Desv.
Fabaceae/Papilionaceae
♦ elegant phyllodium
♦ Weed
♦ 297

Phyllopodium cordatum (Thunb.)
O.M.Hilliard
Scrophulariaceae
♦ Naturalised, Environmental Weed
♦ 7, 86
♦ Origin: southern Africa.

Phyllostachys Sieb. & Zucc. spp.
Poaceae
♦ bamboo, black bamboo, Japanese
bamboo, oriental bamboo, tall bamboo
♦ Weed, Noxious Weed, Naturalised,
Garden Escape, Environmental Weed
♦ 80, 86, 88, 151, 195, 198
♦ G, cultivated, herbal.

Phyllostachys aurea Rivière & C.Rivière
Poaceae
Bambusa aurea hort. ex Rivière &

C.Rivière, *Bambusa aurea* Sieber,
Phyllostachys bambusoides Siebold &
Zucc. var. *aurea* (Rivière & C.Rivière)
Makino, *Phyllostachys formosana*
Hayata, *Phyllostachys meyeri* McClure
var. *aurea* (Rivière & C.Rivière) Pilip.,
Phyllostachys reticulata (Rupr.) K.Koch
var. *aurea* (Rivière & C.Rivière) Makino
♦ golden bamboo, fishpole bamboo,
bamboo
♦ Weed, Noxious Weed, Naturalised,
Garden Escape
♦ 15, 54, 86, 88, 98, 101, 203, 280
♦ G, cultivated, herbal. Origin: China.

Phyllostachys aureosulcata McClure
Poaceae
♦ yellow grove bamboo
♦ Naturalised
♦ 101
♦ G, cultivated.

Phyllostachys bambusoides Siebold &
Zucc.
Poaceae
♦ Japanese timber bamboo, giant
timber bamboo
♦ Weed, Noxious Weed, Naturalised,
Introduced, Cultivation Escape
♦ 15, 40, 86, 101, 252, 280
♦ G, cultivated. Origin: China.

Phyllostachys dulcis McClure
Poaceae
♦ sweetshoot bamboo
♦ Naturalised
♦ 101
♦ G, promoted.

Phyllostachys edulis (Carrière) Houz.
Poaceae
Bambusa edulis Carrière
♦ tortoiseshell bamboo, moso bamboo,
moso chiku, tall bamboo
♦ Naturalised
♦ 101
♦ G, cultivated.

Phyllostachys flexuosa A. & C.Rivière
Poaceae
♦ drooping timber bamboo, zigzag
bamboo
♦ Naturalised
♦ 101
♦ G, cultivated.

Phyllostachys henionis Mitford
Poaceae
= *Phyllostachys nigra* (Lodd. ex Lindl.)
Munro
♦ Weed
♦ 19
♦ G.

Phyllostachys meyeri McClure
Poaceae
♦ Meyer's bamboo
♦ Naturalised
♦ 101
♦ G, cultivated.

Phyllostachys mitis A. & C.Rivière
Poaceae
♦ Weed, Quarantine Weed
♦ 76, 87, 88, 203, 220
♦ G.

Phyllostachys nigra (Lodd. ex Lindl.) Munro
Poaceae
Bambusa nigra Lodd. ex Lindl., *Bambusa puberula* Miq., *Phyllostachys boryana* Mitford, *Phyllostachys henionis* Mitford (see), *Phyllostachys nigra* fo. *boryana* (Mitford) Makino, *Phyllostachys nigra* var. *henonis* (Mitford) Stapf ex Rendle (see), *Phyllostachys puberula* (Miq.) Munro, *Phyllostachys puberula* var. *boryana* Makino, *Phyllostachys puberula* var. *nigra* (Lodd. ex Lindl.) J.Houz.
♦ black bamboo
♦ Weed, Quarantine Weed, Noxious Weed, Naturalised, Garden Escape, Environmental Weed
♦ 73, 76, 80, 86, 88, 98, 101, 155, 203, 280
♦ pG, cultivated, herbal. Origin: southern China.

Phyllostachys nigra (Lodd. ex Lindl.) Munro var. *henonis* (Mitford) Stapf ex Rendle
Poaceae
= *Phyllostachys nigra* (Lodd. ex Lindl.) Munro
♦ bamboo, ha chiku
♦ Weed, Naturalised
♦ 15, 280
♦ G, cultivated.

Phyllostachys nigra (Lodd. ex Lindl.) Munro var. *nigra*
Poaceae
♦ black bamboo
♦ Weed, Naturalised
♦ 15, 280
♦ G.

Phyllostachys rubromarginata McClure
Poaceae
♦ reddish bamboo
♦ Naturalised
♦ 101
♦ G, cultivated.

Phyllostachys viridiglaucescens (Carrière) A. & C.Rivière
Poaceae
♦ greenwax golden bamboo
♦ Naturalised
♦ 101
♦ G, cultivated.

Phymatosorus scolopendria (Burm.f.) Pic.Serm.
Polypodiaceae
Microsorum scolopendria (Burm.f.) Copel. (see), *Phymatodes scolopendria* (Burm.f.) Ching, *Pleopeltis phymatodes* (L.) T.Moore, *Polypodium phymatodes* L., *Polypodium scolopendria* Burm.f. (see)
♦ Weed
♦ 179
♦ cultivated, herbal. Origin: east Africa.

Physalis acutifolia (Miers) Sandw.
Solanaceae
Physalis wrightii A.Gray (see), *Saracha acutifolia* Miers
♦ sharpleaf groundcherry, Wright groundcherry, Rydberg twinpod,

sharpleaf twinpod, southern twinpod
♦ Weed
♦ 34, 161, 180, 243
♦ a/pH, arid, promoted, herbal. Origin: south-western North America.

Physalis alkekengi L.
Solanaceae
Physalis franchetii Mast., *Physalis hyemalis* Salisb.
♦ Chinese lantern plant, wintercherry, alkekengi
♦ Weed, Naturalised, Cultivation Escape
♦ 23, 39, 42, 86, 87, 88, 101, 198, 218, 252, 272
♦ pH, cultivated, herbal, toxic. Origin: Europe to India, China.

Physalis alkekengi L. var. franchetii hort.
Solanaceae
♦ Weed, Naturalised
♦ 286, 287

Physalis angulata L.
Solanaceae
Physalis angulata var. *pendula* (Rydb.) Waterf., *Physalis angulata* var. *lanceifolia* (Nees) Waterf. (see), *Physalis angulata* var. *capsicifolia* (Dunal) Griseb., *Physalis capsicifolia* Dunal, *Physalis esquirolii* H.Lév. & Vaniot, *Physalis lanceifolia* Nees (see), *Physalis linkiana* Nees, *Physalis pendula* Rydb. (see), *Physalis ramosissima* Mill.
♦ cutleaf groundcherry, groundcherry, lanceleaf groundcherry, purplevein groundcherry, south-west groundcherry, alquequenje, sacabuche
♦ Weed, Naturalised, Introduced, Casual Alien
♦ 6, 13, 39, 42, 50, 51, 70, 84, 86, 87, 88, 121, 153, 158, 161, 170, 191, 217, 228, 236, 237, 242, 243, 255, 261, 262, 270, 273, 274, 275, 276, 286, 287, 295, 297
♦ aH, arid, promoted, herbal, toxic. Origin: southern North America.

Physalis angulata L. var. angulata
Solanaceae
♦ Introduced
♦ 230
♦ H.

Physalis angulata L. var. lanceifolia (Nees) Waterf.
Solanaceae
= *Physalis angulata* L.
♦ Introduced
♦ 230
♦ H.

Physalis angulata L. var. villosa Bonati
Solanaceae
♦ villous groundcherry
♦ Weed
♦ 297

Physalis chenopodiifolia Lam.
Solanaceae
♦ Weed
♦ 199

Physalis cordata Mill.
Solanaceae
♦ heartleaf groundcherry, alquequenje,

sacabuche
♦ Weed, Naturalised
♦ 257, 261

Physalis divaricata D.Don
Solanaceae
♦ Weed
♦ 87, 88

Physalis floridana Rydb.
Solanaceae
= *Physalis pubescens* L.
♦ Weed
♦ 88

Physalis greenei Vasey & Rose
Solanaceae
= *Physalis crassifolia* Benth. var. *crassifolia* (NoR)
♦ Naturalised
♦ 287
♦ aH, promoted. Origin: southern North America.

Physalis heterophylla Nees.
Solanaceae
Physalis ambigua (A.Gray) Britton, *Physalis nyctaginea* Dunal, *Physalis sinuata* Rybd.
♦ clammy groundcherry
♦ Weed, Quarantine Weed, Naturalised, Native Weed
♦ 8, 39, 76, 87, 88, 161, 174, 207, 210, 218, 287
♦ pH, arid, cultivated, herbal, toxic. Origin: North America.

Physalis ignota Britton
Solanaceae
♦ Weed
♦ 157
♦ aH.

Physalis ixocarpa Brot. ex DC.
Solanaceae
Physalis aequata J.Jacq. ex Nees
♦ groundcherry, tomatillo groundcherry, tomatillo, tomate verde
♦ Weed, Quarantine Weed, Naturalised, Cultivation Escape
♦ 55, 76, 86, 87, 88, 98, 161, 180, 203, 218, 243, 269
♦ pH, arid, cultivated, herbal, toxic. Origin: Mexico.

Physalis lagascae Roem. & Schult.
Solanaceae
= *Physalis minima* L.
♦ Weed
♦ 87, 88, 157
♦ aH.

Physalis lanceifolia Nees
Solanaceae
= *Physalis angulata* L.
♦ lanceleaf groundcherry, groundcherry, narrowleaf tomatillo
♦ Weed, Naturalised
♦ 34, 86, 98, 161, 180, 198, 203, 212
♦ aH, toxic. Origin: southern North America, Mexico.

Physalis lanceolata Michx.
Solanaceae
♦ lanceleaf groundcherry
♦ Weed
♦ 39, 87, 88

♦ pH, promoted, herbal, toxic.

Physalis lobata Torr.
Solanaceae
= *Quincula lobata* (Torr.) Raf.
♦ purpleflower groundcherry, lobed groundcherry, purple groundcherry
♦ Weed
♦ 87, 88, 161, 218
♦ pH, arid, herbal.

Physalis longifolia Nutt.
Solanaceae
♦ longleaf groundcherry
♦ Weed, Introduced
♦ 23, 34, 39, 87, 88, 161, 218
♦ pH, herbal, toxic.

Physalis longifolia Nutt. var. longifolia
Solanaceae
♦ smooth groundcherry, longleaf groundcherry
♦ Noxious Weed
♦ 229

Physalis macrophysa Rydb.
Solanaceae
= *Physalis longifolia* Nutt. var. *subglabrata* (Mack. & Bush) Cronquist (NoR) [see *Physalis subglabrata* Mack. & Bush.]
♦ Weed
♦ 87, 88
♦ pH, promoted, herbal.

Physalis mendocina Phil.
Solanaceae
♦ camambu
♦ Weed
♦ 87, 88, 236

Physalis micrantha Link
Solanaceae
♦ Weed
♦ 87, 88
♦ Origin: Africa.

Physalis minima L.
Solanaceae
Physalis lagascae Roem. & Schult. (see)
♦ gooseberry, wild gooseberry, pygmy groundcherry, native gooseberry, chirphoti, chirpotoka, chirpotyo, papotan, pipat, Chinese lantern plant, thong theng
♦ Weed, Naturalised, Introduced, Garden Escape, Environmental Weed
♦ 12, 13, 39, 55, 66, 86, 87, 88, 93, 121, 209, 228, 239, 269, 286, 287, 297
♦ pH, arid, cultivated, herbal, toxic. Origin: Australia.

Physalis neesiana Sendtn.
Solanaceae
♦ pocotillo
♦ Weed
♦ 295

Physalis nicandroides Schlecht.
Solanaceae
♦ Weed
♦ 87, 88
♦ herbal.

Physalis orizabae Dunal
Solanaceae
♦ Weed
♦ 199

Physalis pendula Rydb.
Solanaceae
= *Physalis angulata* L.
♦ Naturalised
♦ 287

Physalis peruviana L.
Solanaceae
Alkekengi pubescens Moench, *Physalis esculenta* Salisb.
♦ Peruvian groundcherry, cape gooseberry, Barbados gooseberry, cherry tomato, gooseberry tomato, groundcherry, love apple, poha, strawberry tomato, wild gooseberry, wintercherry
♦ Weed, Naturalised, Native Weed, Environmental Weed
♦ 7, 15, 39, 50, 72, 86, 87, 88, 98, 101, 121, 134, 161, 165, 176, 198, 203, 218, 241, 257, 280, 287, 300
♦ pH, cultivated, herbal, toxic. Origin: tropical America.

Physalis philadelphica Lam.
Solanaceae
Physalis chenopodifolia Willd., *Physalis ixocarpa sensu* auct., *Physalis laevigata* G.Martens & Galeotti
♦ purple gooseberry, Mexican groundcherry, tomate, tomate verde, tomatillo, tomatillo groundcherry
♦ Weed, Naturalised, Introduced, Cultivation Escape
♦ 7, 34, 53, 86, 88, 98, 101, 167, 203, 228, 240, 243, 252, 272, 280
♦ aH, arid, cultivated. Origin: North America.

Physalis philadelphica Lam. var. immaculata Waterf.
Solanaceae
♦ Mexican groundcherry
♦ Naturalised
♦ 101

Physalis pruinosa L.
Solanaceae
♦ strawberry tomato, Cossack pineapple
♦ Naturalised
♦ 287
♦ aH, cultivated, herbal. Origin: eastern North America.

Physalis pubescens L.
Solanaceae
Physalis barbadensis Jacq., *Physalis floridana* Rydb. (see), *Physalis pruinosa* auct. non L., *Physalis turbinata* Medik. (see), *Physalis villosa* Mill.
♦ downy groundcherry, husk tomato, low hair groundcherry, miltomate, tomate, tomatillo, tomato fesadilla
♦ Weed, Naturalised, Introduced
♦ 7, 14, 23, 34, 70, 86, 87, 88, 98, 203, 218, 228, 237, 241, 255, 256, 261, 272, 275, 280
♦ pH, arid, promoted, herbal. Origin: tropical America.

Physalis pubescens L. var. grisea Waterf.
Solanaceae
= *Physalis grisea* (Waterf.) M.Martínez (NoR)

♦ gray groundcherry
♦ Introduced
♦ 34
♦ aH.

Physalis solanacae Mert. ex Roth
Solanaceae
♦ Weed
♦ 87, 88

Physalis sordida Fern.
Solanaceae
♦ Weed
♦ 199

Physalis subglabrata Mack. & Bush.
Solanaceae
= *Physalis longifolia* Nutt. var. *subglabrata* (Mack. & Bush) Cronquist (NoR) [see *Physalis macrophysa* Rydb.]
♦ smooth groundcherry, husk tomato
♦ Weed
♦ 23, 87, 88, 161, 210, 211, 218, 243
♦ pS, cultivated, herbal, toxic. Origin: eastern North America.

Physalis turbinata Medik.
Solanaceae
= *Physalis pubescens* L.
♦ thicket groundcherry, alquequenje, sacabuche
♦ Weed
♦ 87, 88, 261

Physalis virginiana Mill.
Solanaceae
Physalis intermedia Rydb., *Physalis lanceolata* auct. p.p. non Michx., *Physalis lanceolata* Michx. var. *laevigata* A.Gray, *Physalis lanceolata* Michx. var. *longifolia* (Nutt.) Trel., *Physalis longifolia* (Nutt.) Trel., *Physalis longifolia* (Nutt.) Trel. var. *sonorae* (Torr.) Waterf., *Physalis monticola* C.Mohr, *Physalis pumila* Nutt. var. *sonorae* Torr., *Physalis rigida* Pollard & Ball, *Physalis virginiana* Mill. var. *sonorae* (Torr.) Waterf.
♦ Texas groundcherry, Virginia groundcherry, perennial groundcherry, smooth groundcherry, lanceleaf groundcherry
♦ Weed, Noxious Weed, Naturalised, Native Weed
♦ 23, 35, 39, 55, 86, 87, 88, 98, 161, 174, 203, 212, 269
♦ pH, arid, promoted, herbal, toxic. Origin: North America.

Physalis viscosa L.
Solanaceae
Physalis fuscomaculata Rouville ex Dunal
♦ sticky gooseberry, sticky physalis, wild gooseberry, prairie groundcherry, tomato weed, sticky cape gooseberry, sticky groundcherry, wild tomato, camambu
♦ Weed, Quarantine Weed, Noxious Weed, Naturalised, Introduced
♦ 7, 34, 35, 51, 76, 86, 87, 88, 98, 121, 147, 158, 198, 203, 228, 229, 236, 237, 269, 270, 295, 300
♦ pH, arid, promoted, herbal, toxic. Origin: North to South America.

***Physalis viscosa* L. var. *cinerascens* (Dunal) Waterf.**
Solanaceae
= *Physalis cinerascens* (Dunal) A.S.Hitchc. var. *cinerascens* (NoR)
♦ Weed
♦ 199

***Physalis wrightii* A.Gray**
Solanaceae
= *Physalis acutifolia* (Miers) Sandw.
♦ Wright groundcherry
♦ Weed
♦ 87, 88, 212, 218
♦ aH, promoted. Origin: south-western North America.

***Physcomitrella californica* Crum & Anderson**
Funariaceae
= *Physcomitrella readeri* (Crum & Anderson) Stone & Scott (NoR)
♦ Naturalised
♦ 287
♦ moss.

***Physocarpus opulifolius* (L.) Maxim.**
Rosaceae
Physocarpus stellatus (Rydb. ex Small) Rehder, *Spiraea opulifolia* L., *Neillia opulifolia* Benth. & Hook.f.
♦ lännenheisiangervo, ninebark, eastern ninebark, common ninebark
♦ Weed, Cultivation Escape, Casual Alien
♦ 8, 42, 195, 280
♦ S, cultivated, herbal, toxic. Origin: central and eastern North America.

***Physospermum cornubiense* (L.) DC.**
Apiaceae
♦ bladderseed
♦ Weed
♦ 272
♦ cultivated, herbal.

***Physostegia virginiana* (L.) Benth.**
Lamiaceae
Dracocephalum virginianum L.
♦ obedient plant, false dragonhead
♦ Weed, Naturalised
♦ 87, 88, 287
♦ cultivated, herbal.

***Physostigma mesoponticum* Taub.**
Fabaceae/Papilionaceae
♦ Weed
♦ 87, 88
♦ herbal.

***Physostigma virginiana* (L.) Benth.**
Fabaceae/Papilionaceae
♦ Weed
♦ 87, 88

***Phyteuma nigrum* F.W.Schmidt**
Campanulaceae
♦ tummatähkämunkki
♦ Casual Alien
♦ 42
♦ cultivated.

***Phyteuma spicatum* L.**
Campanulaceae
♦ vaaleatähkämunkki, spiked rampion
♦ Casual Alien

♦ 42
♦ pH, cultivated, herbal.

***Phyteuma tetramerum* Schur**
Campanulaceae
♦ romaniantähkämunkki
♦ Casual Alien
♦ 42

***Phytolacca* L. spp.**
Phytolaccaceae
♦ inkweed, pokeweed, alkiermes
♦ Weed, Naturalised
♦ 88, 198
♦ herbal.

***Phytolacca americana* L.**
Phytolaccaceae
Phytolacca decandra L. (see)
♦ pokeweed, inkweed, pokeberry, common pokeweed, American pokeweed, Virginia poke, scoke, garget, inkberry, red inkplant, coakum, American cancer, American nightshade, cancer jalap, cancer root, chongras, common pokeberry, crowberry, jalap, kermes bush, poke, pokeroot, pocan, pigeon berry, redweed, scoke berry, shoke
♦ Weed, Quarantine Weed, Naturalised, Native Weed
♦ 8, 23, 34, 39, 76, 86, 87, 88, 94, 98, 121, 154, 161, 174, 180, 189, 203, 207, 210, 211, 215, 218, 243, 247, 255, 256, 263, 269, 272, 280, 286, 287, 297
♦ pS, cultivated, herbal, toxic. Origin: North America.

***Phytolacca bogotensis* Kunth**
Phytolaccaceae
♦ pokeweed
♦ Naturalised
♦ 101
♦ cultivated.

***Phytolacca clavigera* W.W.Sm.**
Phytolaccaceae
♦ Naturalised
♦ 280
♦ cultivated.

***Phytolacca decandra* L.**
Phytolaccaceae
= *Phytolacca americana* L.
♦ poke, pokeweed, pocan, red inkplant
♦ Weed
♦ 97
♦ herbal, toxic.

***Phytolacca dioica* (L.) Moq.**
Phytolaccaceae
♦ belhambra, bella ombre, omboe, ombu, umbo, umbra tree, bella sombra, elephant tree
♦ Weed, Noxious Weed, Naturalised, Environmental Weed, Cultivation Escape
♦ 39, 85, 86, 98, 121, 260, 283
♦ T, cultivated, herbal, toxic. Origin: South America.

***Phytolacca dodecandra* L'Hér.**
Phytolaccaceae
♦ pokeweed, ntembotelemya, lisingo
♦ Weed, Quarantine Weed

♦ 76, 87, 88, 203, 220
♦ cultivated, herbal. Origin: Madagascar.

***Phytolacca esculenta* Van Houtte**
Phytolaccaceae
♦ Asian pokeweed
♦ Naturalised
♦ 287
♦ pH, cultivated, herbal. Origin: China.

***Phytolacca heptandra* Retz.**
Phytolaccaceae
♦ inkberry, wild sweet potato
♦ Weed, Native Weed
♦ 88, 121, 158
♦ pH, herbal, toxic. Origin: southern Africa.

***Phytolacca heteropetala* H.Walter**
Phytolaccaceae
♦ Mexican pokeweed
♦ Naturalised
♦ 101

***Phytolacca icosandra* L.**
Phytolaccaceae
= *Phytolacca octandra* L.
♦ Weed
♦ 39, 157, 199
♦ pH, herbal, toxic.

***Phytolacca octandra* L.**
Phytolaccaceae
Phytolacca icosandra L. (see)
♦ eightstamen pokeweed, red inkplant, dyeberry, inkweed, forest inkberry, inkberry, phytolacca
♦ Weed, Naturalised, Environmental Weed
♦ 7, 15, 39, 72, 86, 87, 88, 98, 121, 134, 158, 165, 181, 198, 203, 218, 225, 246, 269, 280, 289, 296
♦ pS, herbal, toxic. Origin: tropical America.

***Phytolacca polyandra* Batal.**
Phytolaccaceae
♦ manystemon pokeberry
♦ Weed
♦ 297
♦ cultivated.

***Phytolacca rigida* Small**
Phytolaccaceae
= *Phytolacca americana* L. var. *rigida* (Small) Caulkins & Wyatt (NoR)
♦ Weed
♦ 87, 88

***Phytolacca rivinoides* Kunth & Bouche**
Phytolaccaceae
♦ Venezuelan pokeweed
♦ Weed
♦ 39, 87, 88
♦ herbal, toxic.

***Picea* A.Dietr. spp.**
Pinaceae
♦ spruce
♦ Weed, Naturalised
♦ 54, 86, 88

***Picea abies* (L.) Karst.**
Pinaceae
Abies excelsa Lam. & DC. *non* Link,

Picea excelsa (Lam.) Link
- Norway spruce, Norwegian spruce
- Weed, Naturalised, Environmental Weed
- 4, 39, 40, 80, 88, 101, 151, 280
- T, cultivated, herbal, toxic. Origin: Europe to Siberia.

Picea glauca (Moench) Voss
Pinaceae
Abies canadensis Mill., *Picea alba* (Aiton) Link, *Picea alba* var. *albertiana* (S.Br.) Beissn., *Picea albertiana* S.Br., *Picea canadensis* (Mill.) Britton, Sterns & Poggenb., *Picea canadensis* var. *glauca* (Moench) Sudw., *Picea glauca* fo. *aurea* (J.Nelson) Rehder, *Picea glauca* var. *albertiana* (S.Br.) Sarg., *Picea glauca* var. *conica* Rehder, *Picea glauca* var. *densata* Bailey, *Picea glauca* var. *porsildii* Raup, *Pinus alba* Aiton, *Pinus glauca* Moench, *Pinus laxa* Ehrh.
- white spruce, dwarf Alberta spruce
- Weed
- 87, 88, 218
- T, cultivated, herbal.

Picea mariana (Mill.) Britton, Sterns & Pogg.
Pinaceae
Picea nigra Mill., *Abies mariana* (L.) Karst., *Pinus nigra* Arnold
- black spruce
- Weed
- 87, 88, 218
- T, cultivated, herbal.

Picea pungens Engelm.
Pinaceae
- blue spruce, Colorado spruce
- Weed
- 87, 88, 218
- T, cultivated, herbal. Origin: southwestern North America.

Picea rubens Sarg.
Pinaceae
Picea rubra (Du Roi) Link.
- red spruce
- Weed
- 87, 88, 218
- T, cultivated, herbal.

Picea sitchensis (Bong.) Carr.
Pinaceae
Abies falcata Raf., *Abies menziesii* (Douglas ex D.Don) Lindl., *Picea falcata* (Raf.) Suringar, *Picea menziesii* (Douglas ex D.Don) Carriére, *Pinus sitchensis* Bong.
- sitka spruce, sitkafichte
- Weed, Naturalised
- 80, 87, 88, 218, 280
- T, cultivated, herbal. Origin: North America.

Picnomon acarna (L.) Cass.
Asteraceae
Cirsium acarna (L.) Moench (see), *Cnicus acarna* (L.) L.
- soldier thistle, yellow plumed thistle, siiliohdake
- Weed, Quarantine Weed, Noxious Weed, Naturalised, Casual Alien

- 42, 76, 86, 88, 94, 98, 147, 171, 203, 220, 243, 269, 272
- aH, herbal. Origin: Eurasia.

Picradeniopsis oppositifolia (Nutt.) Rydb. ex Britt.
Asteraceae
Bahia oppositifolia (Nutt.) Gray (see)
- oppositeleaf bahia, plains bahia
- Weed
- 161
- herbal, toxic.

Picraena excelsa Lindl.
Simaroubaceae
- quassia
- Weed
- 39, 87, 88
- herbal, toxic.

Picramnia antidesma Sw.
Picramniaceae/Simaroubaceae
- amarga cascara
- Weed
- 87, 88
- herbal.

Picridium vulgare Desf.
Asteraceae
- picridium
- Weed
- 87, 88, 218

Picris L. spp.
Asteraceae
- oxtongue
- Weed
- 272

Picris abyssinica Sch.Bip.
Asteraceae
- Weed
- 88

Picris echioides L.
Asteraceae
Helminthia echioides (L.) Gaertn., *Helminthotheca echioides* (L.) Holub (see)
- bristly oxtongue, oxtongue bugloss, bugloss picris, picride fausse vipérine
- Weed, Naturalised, Introduced, Environmental Weed, Casual Alien
- 15, 34, 40, 42, 51, 70, 72, 78, 80, 86, 87, 88, 94, 98, 101, 116, 121, 158, 161, 165, 180, 199, 203, 218, 228, 237, 241, 243, 253, 269, 272, 287, 295, 300
- a/bH, arid, cultivated, herbal. Origin: Eurasia.

Picris hieracioides L.
Asteraceae
- hawkweed oxtongue, hawkweed picris, keltanokitkerö, tao
- Weed, Noxious Weed, Naturalised, Introduced, Environmental Weed, Casual Alien
- 1, 23, 39, 70, 72, 80, 86, 87, 88, 94, 98, 101, 134, 146, 161, 195, 203, 204, 218, 228, 229, 253, 272, 280
- b, arid, cultivated, herbal, toxic.

Picris hieracioides L. ssp. *hieracioides*
Asteraceae
- hawkweed oxtongue, hawkweed
- Weed, Noxious Weed, Naturalised
- 101, 229, 269

- aH. Origin: Europe.

Picris hieracioides L. ssp. *japonica* Krylov
Asteraceae
- kouzorina
- Weed
- 275
- bH.

Picris hieracioides L. ssp. *kamtschatica* (Ledeb.) Hultén
Asteraceae
- hawkweed oxtongue
- Noxious Weed
- 229

Picris hieracioides L. var. *glabrescens* (Regel) Ohwi
Asteraceae
- hawkweed picris
- Weed
- 263, 286
- bH.

Picris japonica Thunb.
Asteraceae
- Japanese oxtongue
- Weed
- 297

Picris pauciflora Willd.
Asteraceae
- smallflower oxtongue
- Naturalised
- 101

Picris radicata (Forssk.) Less.
Asteraceae
- Weed
- 221

Picris sprengeriana (L.) Poir.
Asteraceae
- bitterweed
- Weed, Naturalised
- 101, 221

Picrosia longifolia D.Don.
Asteraceae
- picrosia
- Weed
- 237, 295

Pieris floribunda (Pursh) Benth. & Hook.f.
Ericaceae
- fetterbush, mountain fetterbush
- Weed
- 39, 161, 247
- cultivated, herbal, toxic.

Pieris japonica (Thunb. ex Murray) D.Don ex G.Don
Ericaceae
Andromeda japonica Thunb., *Pieris taiwanensis* Hayata
- Japanese pieris, lily of the valley bush, asebi
- Weed
- 39, 85, 161, 247
- S, cultivated, herbal, toxic.

Pieris mariana (L.) Benth. & Hook.f.
Ericaceae
- Weed
- 23, 39, 88
- toxic.

Pikea californica Harv.
Dumontiaceae
Pikea pinnata Setch.
♦ Introduced
♦ 288
♦ algae.

Pilayella Bory spp.
Phaeophyceae/Ectocarpaceae
♦ algae
♦ Weed
♦ 282
♦ algae.

Pilea depressa (Sw.) Blume
Urticaceae
♦ depressed clearweed
♦ Naturalised, Cultivation Escape
♦ 101, 261
♦ cultivated. Origin: Jamaica, Hispaniola.

Pilea involucrata (Sims) Urb.
Urticaceae
♦ friendship plant
♦ Naturalised
♦ 101
♦ cultivated, herbal.

Pilea microphylla (L.) Liebm.
Urticaceae
♦ rock weed, beldroega
♦ Weed, Naturalised, Introduced
♦ 11, 86, 87, 88, 98, 134, 203, 230, 243, 255, 257, 276, 286, 287
♦ a/pH, cultivated, herbal. Origin: tropical America.

Pilea nummulariifolia (Sw.) Wedd.
Urticaceae
♦ creeping Charlie
♦ Weed
♦ 87, 88, 179, 206, 243, 255
♦ pH, cultivated. Origin: tropical America.

Pilea pumila (L.) Gray
Urticaceae
♦ Canadian clearweed, clearweed
♦ Weed
♦ 286
♦ cultivated, herbal.

Pilea serpyllifolia (Poir.) Wedd.
Urticaceae
= *Pilea trianthemoides* (Sw.) Lindl.
♦ Naturalised
♦ 287

Pilea trianthemoides (Sw.) Lindl.
Urticaceae
Pilea serpyllifolia (Poir.) Wedd. (see)
♦ artillery plant
♦ Naturalised
♦ 101

Pilinella californica Hollenb.
Chaetophoraceae
♦ green alga
♦ Weed
♦ 197
♦ algae.

Piliostigma reticulatum (DC.) Hochst.
Fabaceae/Caesalpiniaceae
Bauhinia reticulata DC.
♦ nama tene

♦ Introduced
♦ 228
♦ arid, cultivated.

Piliostigma thonningii (Schumach.) Milne-Redh.
Fabaceae/Caesalpiniaceae
Bauhinia thonningii Schum.
♦ camel's foot, monkey bread, abu khameira, barandé, barkalléhi, bmabmahi, diamara, eko nammon, epamambo, faa mho, fara diambanmésô'hõ, gnamahon, kabâb baut, kammeelspoor, kurukuru, kárgoó, madazagi, nammarehi, ngigis bambuk, opitipata, pouúndquè, sì farun, tafatafa, waku, yafe, yorokoye
♦ Weed, Quarantine Weed
♦ 76, 87, 88, 203, 220
♦ arid, cultivated, herbal.

Pilosella aurantiaca (L.) F.W.Schultz & Sch.Bip.
Asteraceae
= *Hieracium aurantiacum* L.
♦ orange hawkweed, fox and cubs
♦ Weed, Quarantine Weed, Naturalised
♦ 40, 76, 88, 220
♦ cultivated.

Pilosocereus Byles & Rowley spp.
Cactaceae
Pseudopilocereus Bux. spp. (see)
♦ tree cactus
♦ Weed, Quarantine Weed
♦ 76, 88, 203, 220

Pimelea linifolia Sm.
Thymelaeaceae
♦ Weed
♦ 39, 87, 88
♦ cultivated, toxic. Origin: Australia.

Pimelea prostrata (J.R. & G.Forst.) Willd.
Thymelaeaceae
Banksia prostrata J.R. & G.Forst.
♦ riceflower
♦ Weed
♦ 161
♦ S, cultivated, toxic. Origin: New Zealand.

Pimelea simplex Muell.
Thymelaeaceae
♦ desert riceflower
♦ Native Weed
♦ 269
♦ aH, cultivated, toxic. Origin: Australia.

Pimenta dioica (L.) Merr.
Myrtaceae
Pimenta officinalis Lindl.
♦ allspice
♦ Weed, Naturalised, Cultivation Escape
♦ 101, 179, 233, 261
♦ cultivated, herbal. Origin: Mexico, Central America.

Pimenta racemosa (Mill.) J.W.Moore
Myrtaceae
♦ bayrum tree, bay rum malagueta, bay tree
♦ Cultivation Escape

♦ 233
♦ cultivated, herbal.

Pimpinella anisum L.
Apiaceae
Anisum officinarum Moench, *Anisum vulgare* Gaertn.
♦ anisruoho, aniseed, anise, anise burnet saxifrage, common anise
♦ Weed, Naturalised, Introduced, Casual Alien
♦ 42, 101, 228, 272
♦ aH, arid, cultivated, herbal.

Pimpinella major (L.) Huds.
Apiaceae
Carum magnum Baill., *Tragium majus* Lam.
♦ hollowstem burnet saxifrage, greater burnet saxifrage, isopukinjuuri
♦ Naturalised
♦ 101
♦ pH, cultivated, herbal.

Pimpinella peregrina L.
Apiaceae
♦ tragoselino calcatrippa
♦ Weed
♦ 272
♦ cultivated, herbal.

Pimpinella saxifraga L.
Apiaceae
Apium saxifragum Calest., *Tragoselinum saxifragum* Moench
♦ lesser burnet, solidstem burnet saxifrage, burnet saxifrage, pukinjuuri
♦ Weed, Naturalised
♦ 23, 39, 87, 88, 101, 243, 272
♦ pH, cultivated, herbal, toxic.

Pimpinella saxifraga L. ssp. nigra (Mill.) Gaudin
Apiaceae
♦ solidstem burnet saxifrage
♦ Naturalised
♦ 101

Pimpinella saxifraga L. ssp. saxifraga
Apiaceae
♦ solidstem burnet saxifrage
♦ Naturalised
♦ 101

Pimpinella schweinfurthii Asch.
Apiaceae
♦ Weed
♦ 221
♦ herbal.

Pinellia pedatisecta Schott
Araceae
Pinellia tuberifera Ten. var. *pedatisecta* (Schott) Engl.
♦ pedate pinellia
♦ Weed
♦ 275, 297
♦ pH, promoted. Origin: north and west China.

Pinellia ternata (Thunb.) Makino
Araceae
Arisaema cochinchinense Blume, *Arum dracontium* Lour., *Arum ternatum* Thunb., *Pinellia cochinchinense* (Blume) W.Wight, *Pinellia ternata* (Thunb.) Druce, *Pinellia tuberifera* Ten., *Pinellia*

wawrae Engl.
♦ crowdipper
♦ Weed, Quarantine Weed, Naturalised
♦ 76, 80, 87, 88, 101, 203, 204, 220, 275, 286, 297
♦ pH, cultivated, herbal.

Pinus L. spp.
Pinaceae
♦ pine, wilding pine
♦ Weed, Naturalised, Garden Escape, Environmental Weed
♦ 18, 39, 80, 86, 88, 116, 181, 198, 201, 279
♦ cultivated, herbal, toxic.

Pinus attenuata Lemmon
Pinaceae
♦ knobcone pine
♦ Weed
♦ 87, 88, 218
♦ T, cultivated, herbal.

Pinus banksiana Lamb.
Pinaceae
Pinus divaricata (Aiton) Sudw., *Pinus sylvestris* L. var. *divaricata* Aiton
♦ jack pine, pin gris
♦ Weed, Naturalised
♦ 87, 88, 218, 280
♦ T, cultivated, herbal.

Pinus brutia Ten.
Pinaceae
♦ Turkish pine, pine
♦ Naturalised, Cultivation Escape
♦ 86, 252
♦ T, cultivated. Origin: eastern Mediterranean.

Pinus canariensis C.Sm.
Pinaceae
♦ Canary pine, Canary Island pine
♦ Weed, Noxious Weed, Naturalised, Introduced, Garden Escape, Cultivation Escape
♦ 54, 86, 88, 95, 228, 251, 283
♦ T, arid, cultivated. Origin: Canary Islands.

Pinus caribaea Morelet
Pinaceae
Pinus taeda L. var. *heterophylla* Elliott
♦ Caribbean pine, pitch pine
♦ Weed, Naturalised
♦ 22, 80, 86, 261
♦ T, cultivated, herbal. Origin: Central America.

Pinus caribaea Morelet var. *hondurensis* (Sénécl.) W.H.G.Barrett & Golfari
Pinaceae
♦ Caribbean pine
♦ Environmental Weed
♦ 260
♦ T.

Pinus clausa (Chapm. ex Engelm.) Vasey ex Sarg.
Pinaceae
♦ sand pine, Choctawhatchee sand pine
♦ Weed
♦ 87, 88, 218
♦ T, cultivated, herbal.

Pinus contorta Douglas ex Loudon
Pinaceae
♦ lodgepole pine, shore pine, kontortamänty, beach pine
♦ Weed, Quarantine Weed, Naturalised, Garden Escape, Environmental Weed, Cultivation Escape
♦ 15, 42, 54, 86, 87, 88, 152, 181, 218, 225, 246, 280
♦ T, cultivated, herbal. Origin: western North America.

Pinus densiflora Siebold & Zucc.
Pinaceae
♦ Japanese red pine
♦ Weed
♦ 286
♦ T, cultivated, herbal. Origin: northeast China, Japan, Korea.

Pinus echinata Mill.
Pinaceae
♦ shortleaf pine, yellow pine, hard pine
♦ Weed
♦ 87, 88, 218
♦ T, cultivated, herbal.

Pinus edulis Engelm.
Pinaceae
Pinus cembroides (Engelm.) Voss.
♦ pinyon pine, twoneedle pinyon, twoneedle nut pine, Rocky Mountain pinyon, Colorado pinyon
♦ Weed
♦ 87, 88, 218
♦ T, cultivated, herbal.

Pinus elliottii Engelm.
Pinaceae
♦ slash pine, pine tree, basden
♦ Weed, Noxious Weed, Naturalised, Garden Escape, Environmental Weed, Cultivation Escape
♦ 19, 54, 63, 80, 86, 87, 88, 95, 121, 218, 277, 279, 283
♦ T, cultivated, herbal. Origin: southeast North and Central America, West Indies.

Pinus elliottii Engelm. var. *elliottii*
Pinaceae
♦ Honduras pine
♦ Environmental Weed
♦ 260
♦ T.

Pinus halepensis Mill.
Pinaceae
♦ Aleppo pine, halepensis pine, Jerusalem pine
♦ Weed, Noxious Weed, Naturalised, Garden Escape, Environmental Weed, Cultivation Escape
♦ 63, 86, 88, 95, 98, 101, 121, 155, 171, 198, 203, 260, 277, 280, 283, 296
♦ T, cultivated, herbal. Origin: Eurasia.

Pinus jeffreyi Grev. & Balf. ex A.Murray
Pinaceae
♦ Jeffrey pine
♦ Weed
♦ 87, 88, 218
♦ T, cultivated, herbal.

Pinus lambertiana Douglas
Pinaceae
♦ sugar pine
♦ Weed
♦ 87, 88, 218
♦ T, cultivated, herbal.

Pinus luchuensis Mayer
Pinaceae
♦ Weed
♦ 22
♦ cultivated.

Pinus monticola Douglas ex D.Don
Pinaceae
♦ western white pine
♦ Weed
♦ 87, 88, 218
♦ T, cultivated, herbal.

Pinus mugo Turra
Pinaceae
Pinus montana Mill. ssp. *uncinata* (Mill.) Domin, *Pinus rotundata* Link
♦ mugo pine, Swiss mountain pine, mountain pine, dwarf pine, wilding pine, vuorimänty, dwarf mountain pine
♦ Naturalised, Garden Escape, Environmental Weed, Cultivation Escape
♦ 42, 101, 225, 246, 251, 280
♦ T, cultivated, herbal. Origin: central to south-east Europe.

Pinus muricata D.Don
Pinaceae
♦ bishop pine
♦ Naturalised
♦ 280
♦ T, cultivated, herbal.

Pinus nigra J.F.Arnold
Pinaceae
♦ black pine, Austrian pine, Corsican pine, schwarzkiefer
♦ Weed, Naturalised, Garden Escape, Environmental Weed
♦ 15, 72, 80, 88, 98, 101, 151, 152, 181, 203, 280
♦ T, cultivated, herbal. Origin: Europe.

Pinus nigra J.F.Arnold ssp. *laricio* (Poir.) Maire
Pinaceae
♦ Corsican pine
♦ Naturalised, Environmental Weed
♦ 246, 280
♦ T.

Pinus nigra J.F.Arnold ssp. *nigra*
Pinaceae
Pinus asutriaca Hoess., *Pinus nigricans* Host
♦ Austrian pine
♦ Naturalised, Environmental Weed
♦ 40, 246, 280
♦ T, cultivated.

Pinus nigra J.F.Arnold var. *corsicana* (Loudon) Hyl.
Pinaceae
♦ Corsican pine
♦ Naturalised, Environmental Weed
♦ 86, 198
♦ T, cultivated. Origin: Corsica.

Pinus occidentalis Sw.
Pinaceae
♦ pino
♦ Introduced
♦ 261
♦ herbal. Origin: Cuba, Hispaniola.

Pinus palustris Mill.
Pinaceae
Pinus australis F.Michx., *Pinus australis*
var. *filius* Michx., *Pinus longifolia* Salisb.
♦ longleaf pine, pitch pine, longleaf
yellow pine, southern yellow pine
♦ Weed
♦ 8, 87, 88, 218
♦ T, cultivated, herbal, toxic. Origin:
south-east North America.

Pinus patula Scheide ex Schltdl. & Cham.
Pinaceae
♦ patula pine, Mexican yellow pine,
spreading leaved pine, Mexican
weeping pine, jelecote pine, pino triste,
treurden
♦ Weed, Noxious Weed, Naturalised,
Environmental Weed, Cultivation
Escape
♦ 22, 37, 63, 80, 86, 88, 95, 121, 151, 152,
158, 233, 252, 260, 277, 279, 280, 283
♦ T, cultivated. Origin: Mexico.

Pinus pinaster Aiton
Pinaceae
Pinus maritima Poir., *Pinus sylvestris*
non L.
♦ cluster pine, maritime pine, trosden,
wilding pine
♦ Weed, Noxious Weed, Naturalised,
Garden Escape, Environmental Weed,
Cultivation Escape
♦ 7, 10, 15, 40, 41, 63, 72, 80, 86, 88, 95,
98, 101, 121, 152, 155, 158, 198, 203, 225,
246, 277, 278, 280, 283, 290, 296
♦ T, cultivated, herbal. Origin: Eurasia.

Pinus pinea L.
Pinaceae
♦ Italian stone pine, umbrella pine,
stone pine, Ponderosa pine, western
yellow pine
♦ Weed, Naturalised, Garden Escape,
Environmental Weed, Cultivation
Escape
♦ 54, 80, 86, 88, 95, 101, 121, 151, 198,
251, 277, 283
♦ T, cultivated, herbal. Origin: Eurasia.

Pinus ponderosa Douglas ex P.Lawson & Lawson
Pinaceae
♦ Ponderosa pine, western yellow
pine, bull pine
♦ Weed, Naturalised, Environmental
Weed
♦ 39, 86, 87, 88, 98, 161, 181, 203, 218,
246, 280
♦ T, cultivated, herbal, toxic. Origin:
Mexico to Canada.

Pinus radiata D.Don
Pinaceae
Pinus insignis Douglas ex Loudon
♦ radiata pine, Monterey pine, wilding

pine, insignis, radiata
♦ Weed, Noxious Weed, Naturalised,
Garden Escape, Environmental Weed,
Cultivation Escape
♦ 7, 9, 15, 20, 35, 63, 72, 78, 80, 86, 88,
95, 98, 116, 121, 152, 176, 181, 198, 203,
225, 241, 246, 257, 260, 277, 280, 283,
296
♦ T, cultivated, herbal. Origin:
California.

Pinus radiata D.Don var. radiata
Pinaceae
♦ radiata pine
♦ Naturalised, Environmental Weed
♦ 198, 289
♦ T. Origin: California, Mexico.

Pinus resinosa Aiton
Pinaceae
♦ red pine
♦ Weed
♦ 87, 88, 218
♦ T, cultivated, herbal.

Pinus rigida Mill.
Pinaceae
♦ pitch pine, northern pitch pine
♦ Weed
♦ 87, 88, 218
♦ T, cultivated, herbal.

Pinus roxburghii Sarg.
Pinaceae
Pinus longifolia Roxb.
♦ chir pine, emodi pine, long leaved
Indian pine
♦ Weed, Noxious Weed, Cultivation
Escape
♦ 121, 283
♦ T, cultivated, herbal. Origin:
subtropical Himalayas.

Pinus sabiniana Douglas ex D.Don.
Pinaceae
♦ digger pine, California foothill pine,
foothill pine
♦ Weed, Naturalised, Cultivation
Escape
♦ 86, 87, 88, 198, 218, 252
♦ T, cultivated, herbal. Origin:
California.

Pinus serotina Michx.
Pinaceae
Pinus rigida Mill. ssp. *serotina* (Michx.)
Clausen
♦ eastern white pine, pond pine
♦ Weed
♦ 87, 88, 218
♦ T, cultivated, herbal.

Pinus strobus L.
Pinaceae
♦ white pine, Weymouth pine, eastern
white pine
♦ Weed, Naturalised
♦ 87, 88, 280
♦ T, cultivated, herbal.

Pinus sylvestris L.
Pinaceae
♦ Scotch pine, Scot's pine
♦ Weed, Naturalised, Garden Escape,
Environmental Weed
♦ 4, 39, 41, 80, 87, 88, 101, 104, 151, 218,

246, 251, 280
♦ T, cultivated, herbal, toxic. Origin:
Europe.

Pinus taeda L.
Pinaceae
♦ loblolly pine, loblollyden
♦ Weed, Noxious Weed, Naturalised,
Environmental Weed, Cultivation
Escape
♦ 19, 63, 80, 86, 87, 88, 121, 161, 218,
260, 279, 280, 283
♦ T, cultivated, herbal, toxic. Origin:
south-east North America.

Pinus thunbergiana Franco nom. illeg.
Pinaceae
= *Pinus thunbergii* Parl.
♦ Japanese black pine
♦ Weed, Naturalised
♦ 101, 195
♦ T, cultivated, herbal.

Pinus thunbergii Parl.
Pinaceae
Pinus thunbergiana Franco *nom. illeg.*
(see)
♦ black pine, Japanese black pine
♦ Weed
♦ 80
♦ T, cultivated, herbal.

Pinus virginiana Mill.
Pinaceae
♦ Virginia pine, scrub pine, spruce
pine
♦ Weed
♦ 87, 88, 195, 218
♦ T, cultivated, herbal.

Piper L. spp.
Piperaceae
♦ pepper
♦ Quarantine Weed
♦ 220
♦ herbal.

Piper aduncum L.
Piperaceae
♦ spiked pepper, yaqona ni onolulu,
false kava, false matico, higuillo de
hoja menuda
♦ Weed, Quarantine Weed, Noxious
Weed
♦ 3, 14, 22, 76, 87, 88, 135, 179, 191, 203,
218, 220, 229, 243, 255
♦ pH, herbal. Origin: tropical America.

Piper amalago L.
Piperaceae
♦ higuillo de limon, rough leaved
pepper
♦ Weed
♦ 87, 88
♦ herbal.

Piper auritum Kunth
Piperaceae
Artanthe aurita (Kunth) Miq., *Artanthe*
seemanniana Miq., *Piper alstonii*
Trel., *Piper auritilaminum* Trel., *Piper*
auritilimbum Trel., *Piper heraldi* Trel. var.
cocleanum Trel., *Piper perlongipes* Trel.,
Schilleria aurita (Kunth) Kunth
♦ eared pepper, anise piper, hoja santa,
anisillo, hinojo, sabalero, hoja de la

estrella, Hawaiian sakau, false sakau, false kava, Vera Cruz pepper, cowfoot, sacred pepper, Mexican pepper leaves, Hawaiian sakau, bolhoof, cowfoot
♦ Weed, Quarantine Weed, Naturalised
♦ 76, 101, 107, 179
♦ cultivated, herbal. Origin: tropical America.

Piper betle L.
Piperaceae
♦ betel pepper, betelpippuri, betelpfeffer, betel
♦ Weed
♦ 87, 88
♦ cultivated, herbal.

Piper betle L. fo. *densum* Bl.f. var. *trukensis* (Yunck.) Fosb.
Piperaceae
♦ Introduced
♦ 230
♦ C.

Piper caninum Bl.
Piperaceae
♦ Weed
♦ 87, 88
♦ cultivated. Origin: Australia.

Piper dilatatum L.C.Rich
Piperaceae
♦ higuillo
♦ Weed
♦ 28, 206, 243

Piper guianense (Klotzsch) C.DC.
Piperaceae
♦ Weed
♦ 87, 88

Piper hispidum Sw.
Piperaceae
♦ Jamaican pepper
♦ Weed
♦ 39, 87, 88
♦ herbal, toxic.

Piper jamaicense D.DC.
Piperaceae
♦ Weed
♦ 87, 88

Piper methysticum G.Forst.
Piperaceae
♦ kava, sakau, ava
♦ Naturalised, Introduced
♦ 39, 101, 230
♦ S, cultivated, herbal, toxic.

Piper nigrinodum C.DC.
Piperaceae
♦ Weed
♦ 87, 88

Piper nigrum L.
Piperaceae
♦ black pepper, peppar, pippuri, peper, pfeffer, poivre, pepper, white pepper, hu jiao, poivre, poivre blanc, poivre noir, pepe, kosho, pimenta, pimienta
♦ Weed, Introduced
♦ 32, 39, 230
♦ C, cultivated, herbal, toxic.

Piper ponapense C.DC.
Piperaceae
♦ pepper, konok
♦ Introduced
♦ 230
♦ C.

Piper tuberculatum Jacq.
Piperaceae
= *Piper arboreum* Aubl. ssp. *tuberculatum* (Jacq.) Tebbs (NoR)
♦ Weed, Quarantine Weed
♦ 76, 87, 88, 203, 220

Piper umbellatum L.
Piperaceae
= *Lepianthes peltata* (L.) Raf. (NoR) [see *Lepianthes umbellata* (L.) Raf., *Pothomorphe dombeyana* Miq., *Pothomorphe peltata* (L.) Miq., *Pothomorphe umbellata* (L.) Miq.]
♦ umbelled pepper, mombenju, ndembelembe, lilombolombo, ilendelyakenyinamwami
♦ Weed
♦ 39, 87, 88
♦ pH, cultivated, herbal, toxic.

Piptadenia peregrina (L.) Benth.
Fabaceae/Mimosaceae
= *Anadenanthera peregrina* (L.) Speg.
♦ cohoba, coxoba, yoke
♦ Weed, Quarantine Weed
♦ 76, 88, 203, 220
♦ herbal, toxic.

Piptanthocereus (Berg.) Riccob. spp.
Cactaceae
= *Cereus* Mill. spp.
♦ Weed, Quarantine Weed
♦ 76, 88, 220

Piptatherum miliaceum (L.) Coss.
Poaceae
Agrostis miliacea L., *Oryzopsis miliacea* (L.) Benth. & Hook.f. ex Asch. & C.Schweinf. (see)
♦ smilo grass, bamboo grass, milo, ricegrass, rice millet
♦ Weed, Noxious Weed, Naturalised, Introduced, Environmental Weed
♦ 7, 15, 34, 35, 70, 72, 78, 80, 86, 88, 98, 101, 116, 161, 176, 198, 203, 228, 253, 272, 280, 300
♦ pG, arid, cultivated. Origin: Mediterranean.

Piptocarpha poeppigiana (DC.) Bak.
Asteraceae
♦ Weed
♦ 87, 88

Piptochaetium bicolor (Vahl) Desv.
Poaceae
♦ Weed
♦ 295
♦ G.

Piptochaetium montevidense (Spreng.) Parodi
Poaceae
Caryochloa montevidensis Spreng., *Caryochloa montevidensis* var. *brasiliensis* (Trin. & Rupr.) Döll ex

Ekman, *Caryochloa montevidensis* var. *montevidensis* Spreng., *Oryzopsis montevidensis* (Spreng.) Hauman, *Oryzopsis montevidensis* (Spreng.) Speg., *Oryzopsis montevidensis* fo. *brasiliensis* Speg., *Oryzopsis montevidensis* fo. *trachycarpa* Speg., *Oryzopsis montevidensis* fo. *typica* Speg., *Oryzopsis montevidensis* var. *brasiliensis* (Trin. & Rupr.) Speg., *Oryzopsis tuberculata* (E.Desv.) Speg., *Oryzopsis verrucosa* (Phil.) Speg., *Oryzopsis verruculosa* (Phil.) Speg., *Piptochaetium granulatum* Phil., *Piptochaetium humile* Phil., *Piptochaetium leiocarpum* (Speg.) Hack. fo. *subpapillosa* Hack. ex Stuck., *Piptochaetium moelleri* Phil., *Piptochaetium montevidense* (Spreng.) Herter, *Piptochaetium panicoides* (Lam.) E.Desv. fo. *subpapillosum* (Hack. ex Stuck.) Parodi, *Piptochaetium subnudum* Phil., *Piptochaetium tuberculatum* E.Desv., *Piptochaetium verrucosum* Phil., *Stipa panicoides* Nees, *Urachne depressa* Steud., *Urachne panicoides* Trin. ex Nees, *Urachne panicoides* var. *brasiliensis* Trin. & Rupr.
♦ Uruguayan ricegrass, pasto pampa
♦ Weed, Naturalised, Environmental Weed
♦ 54, 72, 86, 88, 198, 290, 295
♦ pG. Origin: South America.

Piptochaetium napostaense (Speg.) Hack.
Poaceae
Oryzopsis napotaensis Speg., *Stipa capillifolia* Hack.
♦ Introduced
♦ 228
♦ G, arid.

Piptochaetium setosum (Trin.) Arechav.
Poaceae
♦ bristly speargrass
♦ Naturalised
♦ 101
♦ pG.

Piptochaetium stipoides Hack.
Poaceae
♦ purple speargrass
♦ Weed
♦ 295
♦ G.

Piriqueta racemosa (Jacq.) Sweet ex Steud.
Turneraceae
♦ rigid stripeseed
♦ Weed
♦ 12

Piscidia piscipula (L.) Sarg.
Fabaceae/Papilionaceae
Erythrina piscipula L., *Ichthyomethia communis* S.T.Blake, *Piscidia communis* (S.T.Blake) Harms
♦ Florida fishpoison tree, Jamaica dogwood
♦ Weed
♦ 39, 87, 88, 96
♦ arid, cultivated, herbal, toxic.

Pisonia aculeata L.
Nyctaginaceae
Pisonia helleri Standl., *Pisonia yaguapinda* D.Parodi
♦ pullback, pullback and hold
♦ Weed
♦ 14, 88
♦ S, cultivated, herbal. Origin: humid tropical and subtropical regions.

Pisonia grandis R.Br.
Nyctaginaceae
♦ grand devil's claws, catchbird tree
♦ Naturalised
♦ 101
♦ cultivated, herbal. Origin: Australia.

Pistacia atlantica Desf.
Anacardiaceae/Pistaciaceae
♦ Mt. Atlas mastic tree, betoum, butum, halibah
♦ Naturalised, Introduced
♦ 101, 228
♦ T, arid, cultivated.

Pistacia chinensis Bunge
Anacardiaceae/Pistaciaceae
♦ Chinese pistache, pistachio nut tree, Chinese pistachio
♦ Weed, Sleeper Weed, Naturalised, Garden Escape, Environmental Weed
♦ 54, 86, 88, 101, 128, 155
♦ T, cultivated, herbal, toxic. Origin: Afghanistan, China, India, Nepal, Philippines, Pakistan, Taiwan.

Pistacia khinjuk Stocks
Anacardiaceae/Pistaciaceae
♦ Weed
♦ 39, 221
♦ herbal, toxic.

Pistacia lentiscus L.
Anacardiaceae/Pistaciaceae
♦ mastic tree, lentisc, mastix strauch, mastixbaum, arbre au mastique, charneca, lentisco, lentisque, pistachier lentisque, gomma mastice, mastice greco, chios mastic tree, mastiksipistaasi
♦ Weed
♦ 70
♦ S, arid, cultivated, herbal.

Pistacia terebinthus L.
Anacardiaceae/Pistaciaceae
Pistacia palaestina Boiss., *Pistacia terebinthus* L. ssp. *palaestina* (Boiss.) Engl.
♦ turpentine tree, chiang turpentine tree, Cyprus turpentine tree, terebinth
♦ Introduced
♦ 228
♦ T, arid, cultivated, herbal.

Pistacia vera L.
Anacardiaceae/Pistaciaceae
♦ pistachio, pistachio nut, pistache
♦ Introduced
♦ 228
♦ T, arid, cultivated, herbal.

Pistia stratiotes L.
Araceae
Apiospermum ocidentalis Blume, *Limonesis commutata* Klotzsch,

Limonesis friedrichsthaliana Klotzsch, *Pistia aethiopica* Fenzl ex Klotzsch, *Pistia africana* C.Presl, *Pistia commutata* Schleid., *Pistia crispata* Blume, *Pistia occidentalis* Blume
♦ water lettuce, tropical duckweed, laitue d'eau, pistie, lechuguita de agua, repollo de agua, apon apon, apoe apoe, beo cai, chawk, Nile cabbage, waterslaai, shell flower, water fern, floating aroid, chok, repollito de agua
♦ Weed, Quarantine Weed, Noxious Weed, Naturalised, Native Weed, Garden Escape, Environmental Weed, Cultivation Escape, Casual Alien
♦ 3, 6, 7, 10, 18, 62, 63, 76, 80, 86, 87, 88, 93, 95, 98, 112, 121, 126, 147, 151, 152, 158, 161, 169, 170, 179, 186, 191, 197, 200, 203, 220, 221, 229, 232, 233, 237, 239, 241, 246, 247, 255, 258, 262, 269, 275, 278, 280, 283, 295, 297, 300
♦ wpH, cultivated, herbal, toxic. Origin: pantropical.

Pistia stratiotes L. var. *cuneata* Engl.
Araceae
♦ water lettuce
♦ Weed, Naturalised
♦ 286, 287

Pisum elatius Steven ex M.Bieb.
Fabaceae/Papilionaceae
♦ wild pea
♦ Weed
♦ 87, 88
♦ cultivated.

Pisum fulvum Sibth. & Sm.
Fabaceae/Papilionaceae
♦ tawny pea
♦ Naturalised
♦ 101

Pisum sativum L.
Fabaceae/Papilionaceae
Pisum arvense L., *Pisum umbellatum* Mill.
♦ garden pea, wild pea, pea, hernefield pea
♦ Weed, Naturalised, Garden Escape, Cultivation Escape, Casual Alien
♦ 34, 39, 39, 40, 42, 86, 88, 93, 94, 98, 101, 198, 203, 205, 252, 272, 280
♦ aH, cultivated, herbal, toxic.

Pisum sativum L. ssp. *elatius* (Steven ex M.Bieb.) Asch. & Graebn.
Fabaceae/Papilionaceae
♦ Weed
♦ 94

Pisum sativum L. ssp. *sativum*
Fabaceae/Papilionaceae
♦ garden pea
♦ Weed
♦ 94

Pitcairnia campii L.B.Sm.
Bromeliaceae
♦ Quarantine Weed
♦ 220

Pitcairnia sceptrigera Mez
Bromeliaceae
♦ Quarantine Weed
♦ 220

Pithecellobium carbonarium (Britt.) Niez. & Nevl.
Fabaceae/Mimosaceae
= *Albizia carbonaria* Britt.
♦ carbonero
♦ Cultivation Escape
♦ 261
♦ cultivated. Origin: Colombia.

Pithecellobium dulce (Roxb.) Benth.
Fabaceae/Mimosaceae
Mimosa dulcis Roxb. (see)
♦ Madras thorn, Manila tamarind, monkey pod, camachili, kamachile, kamatire, kamatsíri 'opiuma, kataiya, Victorian box, Victorian laurel, guama americano
♦ Weed, Quarantine Weed, Noxious Weed, Naturalised, Introduced, Garden Escape, Environmental Weed, Cultivation Escape
♦ 3, 11, 22, 76, 80, 86, 87, 88, 101, 107, 155, 179, 191, 203, 220, 228, 230, 259, 261
♦ T, arid, cultivated, herbal. Origin: Americas.

Pithecellobium filamentosum Benth.
Fabaceae/Mimosaceae
♦ Weed
♦ 87, 88

Pithecellobium graciliflorum Blake
Fabaceae/Mimosaceae
♦ Guadeloupe blackbead
♦ Weed, Naturalised
♦ 101, 179

Pithecellobium lanceolatum (Humb. & Bonpl. ex Willd.) Benth.
Fabaceae/Mimosaceae
Inga lanceolata Humb. & Bonpl. ex Willd., *Mimosa ligustrina* Jacq., *Pithecellobium ligustrinum* (Jacq.) Klotzsch ex Benth. *nom. illeg.* (see)
♦ Weed
♦ 87, 88

Pithecellobium ligustrinum (Jacq.) Klotzsch ex Benth. *nom. illeg.*
Fabaceae/Mimosaceae
= *Pithecellobium lanceolatum* (Humb. & Bonpl. ex Willd.) Benth.
♦ Weed
♦ 87, 88
♦ herbal.

Pithecellobium pachypus Pittier
Fabaceae/Mimosaceae
♦ Weed
♦ 87, 88

Pithecellobium saman (Jacq.) Benth.
Fabaceae/Mimosaceae
= *Samanea saman* (Jacq.) Merr.
♦ Weed
♦ 3, 14, 191
♦ T, cultivated, herbal.

Pithecellobium unguis-cati (L.) Benth.
Fabaceae/Mimosaceae
Mimosa unguis-cati L.
♦ cat's claw, bread and cheese, campeche, cat's claw blackbead
♦ Weed, Introduced
♦ 22, 228
♦ T, arid, herbal.

Pithecoctenium crucigerum (L.) A.H.Gentry
Bignoniaceae
♦ monkey's comb
♦ Weed, Naturalised
♦ 101, 179

Pithecoctenium cynanchoides DC.
Bignoniaceae
♦ Weed, Naturalised
♦ 54, 86, 88
♦ cultivated. Origin: South America.

Pithophora Wittr. spp.
Cladophoraceae
♦ algae, borra, agua podre, limo
♦ Weed
♦ 255
♦ algae.

Pithophora zelli Wittr.
Cladophoraceae
♦ Weed
♦ 286
♦ algae.

Pitraea cuneato-ovata (Cav.) Caro
Verbenaceae
Castelia cuneato-ovata Cav., *Priva cuneato-ovata* (Cav.) Rusby, *Priva laevis* Juss. (see)
♦ papilla
♦ Weed
♦ 237, 270, 295

Pittosporum Banks spp.
Pittosporaceae
♦ pittosporum, Australian laurel
♦ Weed
♦ 116, 161
♦ toxic.

Pittosporum crassifolium Banks & Sol. ex A.Cunn.
Pittosporaceae
♦ karo pittosporum, karo, stiffleaf cheesewood, thick leaved box, dwarf karo
♦ Weed, Naturalised, Garden Escape, Environmental Weed
♦ 39, 40, 54, 72, 86, 88, 98, 101, 131, 134, 198, 290
♦ S/T, cultivated, herbal, toxic. Origin: New Zealand.

Pittosporum eugenioides A.Cunn.
Pittosporaceae
♦ tarata, lemonwood
♦ Weed, Naturalised, Garden Escape, Environmental Weed
♦ 39, 54, 72, 86, 88, 198
♦ T, cultivated, toxic. Origin: New Zealand.

Pittosporum ferrugineum Aiton
Pittosporaceae
♦ kamal
♦ Weed
♦ 22, 39
♦ T, cultivated, herbal, toxic. Origin: Australia.

Pittosporum pentandrum (Blanco) Merr.
Pittosporaceae
♦ pittosporum, Taiwanese cheesewood, mamalis
♦ Weed, Naturalised, Cultivation Escape

♦ 3, 80, 88, 101, 112, 179, 191, 233
♦ cultivated.

Pittosporum phylliraeoides DC.
Pittosporaceae
♦ butterbush
♦ Introduced
♦ 228
♦ arid, cultivated.

Pittosporum tenuifolium Gaertn.
Pittosporaceae
♦ kohuhu, black matipo, tawhiwhi
♦ Weed, Naturalised, Garden Escape, Environmental Weed
♦ 40, 54, 72, 86, 88, 101
♦ T, cultivated. Origin: New Zealand.

Pittosporum tobira (Thunb.) Aiton
Pittosporaceae
♦ tobira, Japanese pittosporum, pittosporum, mock orange, Wheeler's mock orange, Japanese cheesewood
♦ Weed, Naturalised, Cultivation Escape
♦ 3, 39, 80, 88, 101, 112, 122, 191
♦ S/T, cultivated, herbal, toxic.

Pittosporum undulatum Vent.
Pittosporaceae
♦ Australian cheesewood, Victorian box, mock orange, sweet pittosporum, New Zealand daphne, Victorian laurel, orange pittosporum, wild coffee, Australiese kasuur, soet pittosporum
♦ Weed, Noxious Weed, Naturalised, Native Weed, Garden Escape, Environmental Weed, Cultivation Escape
♦ 3, 7, 22, 67, 39, 63, 72, 80, 86, 87, 88, 95, 101, 116, 121, 134, 152, 176, 226, 229, 233, 280, 283, 289, 296
♦ S/T, cultivated, herbal, toxic. Origin: Australasia.

Pittosporum undulatum Vent. ssp. undulatum
Pittosporaceae
♦ Naturalised
♦ 176

Pittosporum undulatum Vent. ssp. × emmettii hybrid
Pittosporaceae
= *Pittosporum undulatum* Vent. × *Pittosporum bicolor* Hook.
♦ sweet pittosporum
♦ Naturalised, Environmental Weed
♦ 86
♦ cultivated. Origin: Australia.

Pittosporum viridiflorum Sims
Pittosporaceae
♦ cape cheesewood, cape pittosporum
♦ Weed, Naturalised, Cultivation Escape
♦ 22, 101, 233
♦ T, cultivated, herbal. Origin: Madagascar.

Pittosporum viridiflorum Sims var. viridiflorum
Pittosporaceae
Pittosporum abyssinicum Del. var. *angolensis* Oliv., *Pittosporum commutatum* Putterl., *Pittosporum*

floribundum Wight & Arn., *Pittosporum vosseleri* Engl.
♦ Introduced
♦ 228
♦ arid.

Pituranthos tortuosus (DC.) Benth. & Hook.
Apiaceae
Bubon tortuosum Desf., *Deverra tortuosa* (Desf.) DC.
♦ Weed
♦ 221
♦ arid.

Pituranthos triradiatus (Hochst. ex Boiss.) Asch. & Schweinf.
Apiaceae
Deverra triradiata Hochst. ex Boiss.
♦ Weed
♦ 221

Pityrodia teckiana (F.Muell.) E.Pritz.
Lamiaceae/Verbenaceae/ Chloanthaceae
♦ Weed, Naturalised, Cultivation Escape
♦ 7
♦ cultivated.

Pityrogramma austroamericana Domin
Pteridaceae/Adiantaceae
Pityrogramma calomelanos (L.) Link var. *aureoflava* (Hook.) Weath. ex Bailey (see), *Pityrogramma calomelanos* (L.) Link var. *austroamericana* (Domin) Farw. (see)
♦ leatherleaf goldback fern
♦ Naturalised
♦ 98

Pityrogramma calomelanos (L.) Link
Pteridaceae/Adiantaceae
♦ silver fern, Dixie silverback fern, jefame
♦ Weed, Naturalised
♦ 28, 87, 88, 93, 98, 191, 203, 206, 243, 254, 287
♦ pH, herbal.

Pityrogramma calomelanos (L.) Link var. aureoflava (Hook.) Weath. ex Bailey
Pteridaceae/Adiantaceae
= *Pityrogramma austroamericana* Domin
♦ gold fern, golden fern
♦ Weed
♦ 121
♦ pH. Origin: South America.

Pityrogramma calomelanos (L.) Link var. austroamericana (Domin) Farw.
Pteridaceae/Adiantaceae
= *Pityrogramma austroamericana* Domin
♦ silver fern
♦ Naturalised
♦ 86
♦ Origin: South America.

Pityrogramma calomelanos (L.) Link var. calomelanos
Pteridaceae/Adiantaceae
♦ silver fern
♦ Naturalised
♦ 86
♦ Origin: Central and South America.

Placea arzae **Phil.**
Liliaceae/Amaryllidaceae
♦ Quarantine Weed
♦ 220

Placea germaini **Phil.**
Liliaceae/Amaryllidaceae
♦ Quarantine Weed
♦ 220

Placea grandiflora **Lem.**
Liliaceae/Amaryllidaceae
♦ Quarantine Weed
♦ 220

Placea lutea **Phil.**
Liliaceae/Amaryllidaceae
♦ Quarantine Weed
♦ 220

Placea ornata **Miers ex Lindl.**
Liliaceae/Amaryllidaceae
♦ Quarantine Weed
♦ 220

Plagiobothrys canescens **Benth.**
Boraginaceae
♦ valley popcorn flower, grey popcorn flower
♦ Weed, Quarantine Weed, Naturalised, Native Weed
♦ 76, 86, 88, 98, 161, 180, 198, 203, 269
♦ aH, herbal. Origin: California.

Plagiobothrys mollis **(Gray) I.M.Johnst.**
Boraginaceae
♦ popcorn flower, soft popcorn flower
♦ Weed
♦ 159
♦ pH.

Plagiobothrys scouleri **(Hook. & Arn.) I.M.Johnst.**
Boraginaceae
♦ popcorn flower, Scouler's popcorn flower, mierokki
♦ Weed, Casual Alien
♦ 42, 159
♦ herbal.

Plagiochloa uniolae **(L.f.) Adamson & Sprague**
Poaceae
= *Tribolium uniolae* (L.f.) Renvoize
♦ Weed, Naturalised, Native Weed
♦ 98, 121, 203
♦ pG. Origin: southern Africa.

Planera aquatica **(Walter) J.F.Gmel.**
Ulmaceae
♦ planetree
♦ Weed
♦ 87, 88, 218
♦ herbal.

Plantago **L. spp.**
Plantaginaceae
♦ plantaiga, plantain, plantagos
♦ Weed, Environmental Weed
♦ 106, 272, 296
♦ herbal.

Plantago afra **L.**
Plantaginaceae
Plantago afrum (L.) Mirb.
♦ psyllium
♦ Weed, Introduced, Casual Alien

♦ 88, 94, 221, 228, 243, 280
♦ arid, cultivated, herbal.

Plantago aitchisonii **Pilg.**
Plantaginaceae
♦ Weed
♦ 243

Plantago albicans **L.**
Plantaginaceae
♦ Weed, Naturalised
♦ 70, 86, 98, 203, 221
♦ Origin: Mediterranean.

Plantago alismatifolia **Pilg.**
Plantaginaceae
♦ Weed
♦ 199

Plantago altissima **L.**
Plantaginaceae
♦ piantaggine palustre
♦ Weed
♦ 272
♦ cultivated, herbal.

Plantago amplexicaulis **Cav.**
Plantaginaceae
♦ ispaghula
♦ Weed, Naturalised
♦ 23, 88, 221, 248, 287
♦ aH, cultivated, herbal.

Plantago arabica **Boiss.**
Plantaginaceae
♦ Weed
♦ 221

Plantago arenaria **Waldst. & Kit.**
Plantaginaceae
Plantago indica L. *nom. illeg.* (see),
Plantago psyllium L. *nom. ambig.* (see),
Plantago scabra Moench *nom. illeg.* (see)
♦ sand plantain, branched plantain, haarovaratamo
♦ Weed, Naturalised, Casual Alien
♦ 40, 42, 54, 86, 88, 94, 198, 272, 287
♦ aH, cultivated, herbal. Origin: Eurasia.

Plantago aristata **Michx.**
Plantaginaceae
♦ large bracted plantain, western buck's horn, bristly buck's horn, western ripple grass, bracted plantain, otaratamo
♦ Weed, Naturalised, Casual Alien
♦ 23, 34, 42, 49, 80, 87, 88, 161, 210, 218, 249, 280, 287
♦ a/pH, herbal.

Plantago asiatica **L.**
Plantaginaceae
♦ oobako
♦ Weed, Quarantine Weed
♦ 87, 88, 204, 220, 263, 273, 275, 286, 297
♦ pH, promoted, herbal. Origin: east Asia.

Plantago australis **Lam.**
Plantaginaceae
Plantago hirtella Kunth
♦ Mexican plantain, swamp plantain, southern plantain
♦ Weed, Naturalised, Environmental Weed

♦ 15, 54, 86, 86, 88, 98, 157, 176, 198, 203, 280
♦ pH, promoted, herbal. Origin: South America.

Plantago australis **Lam. ssp.** *hirtella* **(H.B.K.) Rahn**
Plantaginaceae
♦ Mexican plantain
♦ Weed
♦ 199

Plantago bellardii **All.**
Plantaginaceae
♦ hairy plantain, bellardinratamo, silky plantain
♦ Weed, Naturalised, Environmental Weed, Casual Alien
♦ 42, 72, 86, 88, 98, 198, 203, 272
♦ aH. Origin: southern Europe.

Plantago bicallosa **Decne.**
Plantaginaceae
♦ Weed
♦ 87, 88

Plantago camtschatica **Cham.**
Plantaginaceae
♦ Kamchatkan plantain
♦ Weed
♦ 87, 88, 286
♦ pH, cultivated.

Plantago ciliata **Desf.**
Plantaginaceae
♦ Weed
♦ 221
♦ herbal.

Plantago coronopus **L.**
Plantaginaceae
Asterogeum laciniatum S.F.Gray,
Coronopus vulgaris Fourr., *Plantago coronopus* ssp. *commutata* (Guss.) Pilg. (see)
♦ buck's horn plantain, liuskaratamo
♦ Weed, Naturalised, Garden Escape, Environmental Weed, Casual Alien
♦ 7, 15, 23, 42, 70, 72, 87, 88, 94, 98, 101, 165, 176, 181, 198, 203, 221, 241, 269, 272, 280, 287, 300
♦ pH, aqua, cultivated, herbal. Origin: Eurasia.

Plantago coronopus **L. ssp.** *commutata* **(Guss.) Pilg.**
Plantaginaceae
= *Plantago coronopus* L.
♦ buck's horn plantain
♦ Naturalised, Environmental Weed
♦ 86, 198
♦ cultivated. Origin: Mediterranean.

Plantago coronopus **L. ssp.** *coronopus*
Plantaginaceae
♦ buck's horn plantain
♦ Naturalised, Environmental Weed
♦ 86, 198
♦ cultivated. Origin: Eurasia.

Plantago coronopus **L. var.** *coronopus*
Plantaginaceae
♦ buck's horn plantain
♦ Weed
♦ 93, 205
♦ pH.

Plantago cretica L.
Plantaginaceae
♦ Weed, Naturalised
♦ 7, 86, 98, 203
♦ Origin: eastern Mediterranean.

Plantago cunninghamii Decne.
Plantaginaceae
♦ sago weed
♦ Weed
♦ 55
♦ Origin: Australia.

Plantago cylindrica Forssk.
Plantaginaceae
♦ Weed
♦ 221

Plantago debilis R.Br.
Plantaginaceae
♦ weak plantain
♦ Naturalised
♦ 101, 134, 280
♦ Origin: Australia.

Plantago depressa Willd.
Plantaginaceae
♦ Weed, Naturalised
♦ 275, 287, 297
♦ aH, promoted, herbal.

Plantago elongata Pursh
Plantaginaceae
♦ longleaf plantain, slender plantain, plantain
♦ Weed
♦ 159, 161
♦ aH, herbal.

Plantago fastigiata Morris
Plantaginaceae
= *Plantago ovata* Forssk.
♦ Weed
♦ 23, 88
♦ herbal.

Plantago firma Kunze ex Walp.
Plantaginaceae
♦ Chilean plantain
♦ Naturalised
♦ 101

Plantago heterophylla Nutt.
Plantaginaceae
♦ slender plantain
♦ Naturalised
♦ 287
♦ herbal.

Plantago hirtelle H.B.K.
Plantaginaceae
♦ Weed
♦ 87, 88

Plantago hybrida W.Bart.
Plantaginaceae
= *Plantago pusilla* Nutt.
♦ dwarf plantain
♦ Weed
♦ 161
♦ herbal.

Plantago indica L. nom. illeg.
Plantaginaceae
= *Plantago arenaria* Waldst. & Kit.
♦ whorled plantain, Indian plantain
♦ Weed, Naturalised

♦ 87, 88, 98, 203, 218, 221
♦ aH, cultivated, herbal.

Plantago insularis Eastw.
Plantaginaceae
= *Plantago ovata* Forssk.
♦ plantain
♦ Quarantine Weed
♦ 220
♦ cultivated, herbal.

Plantago intermedia Godr.
Plantaginaceae
Plantago biebersteinii Opiz, *Plantago major* L. ssp. *plejosperma* Pilg., *Plantago scopulorum* (Fr.) Pavlova, *Plantago uliginosa* F.W.Schmidt
♦ Weed
♦ 44

Plantago lagopus L.
Plantaginaceae
♦ round headed plantain, plantain, jänönratamo
♦ Weed, Casual Alien
♦ 40, 42, 68, 70, 87, 88, 94, 185, 221, 243, 272
♦ cultivated, herbal.

Plantago lanceolata L.
Plantaginaceae
Arnoglossum lanceolatum (L.) S.F.Gray, *Plantago lanceaefolia* Salisb.
♦ lance leaved plantain, buck's horn, ribwort plantain, ribwort, English plantain, narrow leaved plantain, buck's horn plantain, rib grass, rat tail, heinäratamo, German psyllium, lamb's tongue, small plantian, wild sago
♦ Weed, Noxious Weed, Naturalised, Introduced, Garden Escape, Environmental Weed, Casual Alien
♦ 7, 8, 9, 14, 15, 23, 34, 36, 38, 44, 49, 50, 51, 52, 70, 72, 80, 86, 87, 88, 93, 94, 97, 98, 101, 115, 121, 134, 136, 158, 161, 165, 174, 176, 179, 180, 186, 195, 198, 203, 204, 205, 207, 210, 211, 212, 218, 229, 237, 240, 241, 243, 249, 250, 253, 261, 263, 266, 269, 271, 272, 275, 280, 286, 287, 289, 293, 295, 297, 300
♦ pH, arid, cultivated, herbal. Origin: Eurasia.

Plantago lanceolata L. fo. *composita* Farw.
Plantaginaceae
♦ Naturalised
♦ 287

Plantago lanceolata L. var. *mediterranea* Pilg.
Plantaginaceae
♦ Naturalised
♦ 287

Plantago linearis Kunth var. *mexicana* (Lam.) Brand
Plantaginaceae
♦ Weed
♦ 199

Plantago major L.
Plantaginaceae
Plantago asiatica auct. non L., *Plantago*

major L. ssp. *intermedia* (DC.) Arcang. (see)
♦ broadleaf plantain, yaa en yued, great plantain, ribwort plantain, llantén, large plantain, common plantain, dooryard plantain, whiteman's foot, grand plantain, ribwort, ribgrass, narrow leaved plantain, buck's horn plantain, cart track plan, wild sagot
♦ Weed, Noxious Weed, Naturalised, Introduced, Garden Escape, Environmental Weed
♦ 7, 9, 15, 23, 34, 38, 44, 49, 51, 52, 70, 72, 86, 87, 88, 94, 97, 98, 121, 134, 136, 157, 158, 161, 165, 176, 179, 180, 185, 186, 195, 198, 203, 204, 210, 211, 212, 218, 221, 235, 236, 237, 238, 241, 243, 248, 249, 253, 255, 257, 261, 262, 266, 269, 272, 275, 276, 280, 286, 287, 294, 295, 299, 300
♦ pH, arid, cultivated, herbal. Origin: Eurasia.

Plantago major L. ssp. *intermedia* (DC.) Arcang.
Plantaginaceae
= *Plantago major* L.
♦ Cultivation Escape
♦ 40
♦ cultivated.

Plantago major L. var. *japonica* Miyabe
Plantaginaceae
♦ Weed
♦ 286

Plantago major L. var. *kimurae* Yamamoto
Plantaginaceae
♦ rippleseed plantain, dooryard weed
♦ Weed
♦ 274

Plantago maritima L.
Plantaginaceae
♦ goose tongue, maritime plantain, sea plantain
♦ Naturalised
♦ 241
♦ pH, aqua, cultivated, herbal.

Plantago media L.
Plantaginaceae
♦ hoary plantain, soikkoratamo, lamb's tongue, mittelwegerich
♦ Weed, Noxious Weed, Naturalised
♦ 23, 44, 52, 70, 87, 88, 94, 101, 161, 218, 243, 253, 272, 299
♦ pH, cultivated, herbal.

Plantago minor L.
Plantaginaceae
= *Plantago tenuiflora* Waldst. & Kit. (NoR)
♦ common plantain
♦ Weed
♦ 114
♦ herbal.

Plantago myosuros Lam.
Plantaginaceae
♦ Weed, Naturalised
♦ 98, 121, 203

Plantago myosuros **Lam. ssp.** *myosuros*
Plantaginaceae
♦ mouse plantain
♦ Naturalised
♦ 86, 198
♦ Origin: South America.

Plantago notata **Lag.**
Plantaginaceae
♦ Weed
♦ 221

Plantago ovata **Forssk.**
Plantaginaceae
Plantago decumbens Forssk., *Plantago fastigiata* Morris (see), *Plantago insularis* Eastw. (see), *Plantago insularis* Eastw. var. *fastigiata* (Morris) Jeps., *Plantago ispaghula* Roxb. ex Fleming
♦ desert Indian wheat, blond psyllium
♦ Weed, Introduced
♦ 221, 228
♦ aH, arid, cultivated, herbal.

Plantago patagonica **Jacq.**
Plantaginaceae
Plantago purshii Roem. & Schult. (see), *Plantago spinulosa* Decne. (see)
♦ woolly plantain, woolly Indian wheat, Patagonia plantain, karvaratamo
♦ Weed, Native Weed, Casual Alien
♦ 42, 87, 88, 161, 174, 212, 237, 295
♦ aH, cultivated, herbal. Origin: North America.

Plantago psyllium **L. nom. ambig.**
Plantaginaceae
= *Plantago arenaria* Waldst. & Kit.
♦ sand plantain, fleawort, Spanish psyllium, whorled plantain
♦ Weed, Naturalised
♦ 87, 88, 101, 121
♦ pH, cultivated, herbal. Origin: Eurasia.

Plantago pumila **L.**
Plantaginaceae
♦ Weed
♦ 221

Plantago purshii **Roem. & Schult.**
Plantaginaceae
= *Plantago patagonica* Jacq.
♦ woolly plantain
♦ Weed
♦ 23, 87, 88, 210, 218
♦ herbal.

Plantago pusilla **Nutt.**
Plantaginaceae
Plantago hybrida W.Bart. (see)
♦ slender plantain, dwarf plantain, little plantain
♦ Weed
♦ 87, 88, 218
♦ aH, herbal.

Plantago rhodosperma **Decne.**
Plantaginaceae
♦ Virginia plantain, redseed plantain
♦ Weed
♦ 34
♦ a/pH, herbal.

Plantago rugelii **Decne.**
Plantaginaceae
♦ blackseed plantain, broadleaf plantain, dooryard plantain, common plantain, Rugel's plantain
♦ Weed, Native Weed
♦ 23, 49, 52, 87, 88, 161, 174, 210, 211, 218, 243, 266, 294
♦ a/pH, cultivated, herbal. Origin: North America.

Plantago scabra **Moench nom. illeg.**
Plantaginaceae
= *Plantago arenaria* Waldst. & Kit.
♦ Weed, Naturalised
♦ 86, 98, 203, 280
♦ Origin: Europe, North Africa to central Asia.

Plantago serraria **L.**
Plantaginaceae
♦ Weed
♦ 88, 94

Plantago spinulosa **Decne.**
Plantaginaceae
= *Plantago patagonica* Jacq.
♦ Naturalised
♦ 287

Plantago stepposa **Kuprian.**
Plantaginaceae
♦ Weed
♦ 23, 88

Plantago tomentosa **Lam.**
Plantaginaceae
♦ tanchagem
♦ Weed
♦ 87, 88, 255
♦ pH, herbal. Origin: South America.

Plantago turrifera **Briggs, Carolin & Pulley**
Plantaginaceae
♦ Naturalised
♦ 86
♦ Origin: Australia.

Plantago tweedyi **A.Gray**
Plantaginaceae
♦ Tweedy's plantain
♦ Weed
♦ 23, 88
♦ herbal.

Plantago varia **R.Br.**
Plantaginaceae
♦ Weed
♦ 87, 88
♦ cultivated. Origin: Australia.

Plantago virginica **L.**
Plantaginaceae
♦ paleseed plantain, dwarf plantain, sand plantain, lännenratamo
♦ Weed, Naturalised, Casual Alien
♦ 23, 34, 42, 87, 88, 121, 161, 167, 218, 249, 256, 286, 287, 297
♦ a/pH, herbal. Origin: Americas.

Platanthera bifolia **(L.) Rich.**
Orchidaceae
Gymnadenia bifolia Mey., *Lysias bifolia* Salisb., *Orchis bifolia* L.
♦ lesser butterfly orchid
♦ Weed

♦ 272
♦ pH, cultivated, herbal.

Platanus × acerifolia **(Aiton) Willd.**
Platanaceae
= *Platanus orientalis* L. × *Platanus occidentalis* L. [see *Platanus × hispanica* Mill ex Münchh.]
♦ London plane
♦ Weed, Naturalised, Environmental Weed, Casual Alien
♦ 54, 86, 88, 280
♦ herbal.

Platanus × hispanica **Mill ex Münchh.**
Platanaceae
= *Platanus × acerifolia* (Aiton) Willd.
♦ London plane
♦ Weed, Naturalised, Environmental Weed
♦ 86
♦ cultivated.

Platanus × hybrida **Brot.**
Platanaceae
= *Platanus occidentalis* L. × *Platanus orientalis* L.
♦ planetree, London planetree, London plane
♦ Weed, Naturalised, Garden Escape, Environmental Weed
♦ 72, 80, 86, 88
♦ T, cultivated. Origin: horticultural hybrid.

Platanus occidentalis **L.**
Platanaceae
♦ American planetree, American plane, American sycamore, sycamore, buttonball, eastern sycamore, buttonwood
♦ Weed
♦ 39, 87, 88, 179, 218
♦ T, cultivated, herbal, toxic. Origin: North America.

Plathymenia reticulata **Benth.**
Fabaceae/Mimosaceae
♦ vinhatico
♦ Introduced
♦ 228
♦ arid.

Platostoma africanum **Beauv.**
Lamiaceae
♦ Weed, Quarantine Weed
♦ 76, 87, 88, 203, 220
♦ a/pH, herbal.

Platycapnos spicata **(L.) Bernh.**
Papaveraceae/Fumariaceae
♦ Weed, Naturalised
♦ 86, 87, 88, 94, 98, 203

Platycerium bifurcatum **(Cav.) C.Chr.**
Polypodiaceae
♦ elkhorn fern, common staghorn fern, staghorn fern
♦ Cultivation Escape
♦ 233
♦ cultivated, herbal. Origin: Australia.

Platycladus orientalis **(L.) Franco**
Cupressaceae
Platycladus stricta Spach, *Biota orientalis* (L.) Endl., *Thuja orientalis* L. (see)
♦ oriental arborvitae

♦ Naturalised
♦ 101
♦ T, cultivated, herbal.

Platycodon grandiflorum (Jacq.) A.DC.
Campanulaceae
♦ balloon flower, kikyou
♦ Weed, Naturalised
♦ 101, 297
♦ cultivated, herbal, toxic.

Platygyna hexandra (Jacq.) Müll.Arg.
Euphorbiaceae
♦ Weed
♦ 14, 39
♦ toxic.

Platyopuntia (Engelm.) Ritt. spp.
Cactaceae
= *Opuntia* Mill. spp.
♦ Weed, Quarantine Weed
♦ 76, 88, 203

Plectranthus amboinicus (Lour.) Spreng.
Lamiaceae
Coleus amboinicus Lour. (see)
♦ Mexican mint, Cuban oregano, Spanish thyme, variegated Spanish thyme
♦ Naturalised, Cultivation Escape
♦ 86, 101, 261
♦ cultivated, herbal. Origin: Africa.

Plectranthus barbatus Andrews
Lamiaceae
Coleus barbatus (Andrews) Benth., *Coleus forskohlii* auct., *Plectranthus forskohlii* auct.
♦ Weed
♦ 121
♦ pS, cultivated. Origin: tropical Asia.

Plectranthus behrii Compton
Lamiaceae
♦ Casual Alien
♦ 280

Plectranthus caninus Roth
Lamiaceae
Coleus caninus (Roth) Vatke, *Coleus omahekense* Dinter
♦ Naturalised
♦ 86
♦ cultivated. Origin: Africa to India.

Plectranthus ciliatus E.Mey.
Lamiaceae
♦ African spurflower, spurflower
♦ Weed, Sleeper Weed, Quarantine Weed, Naturalised, Garden Escape, Environmental Weed
♦ 54, 76, 86, 88, 98, 165, 198, 203, 225, 246, 251, 280, 290
♦ cultivated, herbal. Origin: south-east Africa.

Plectranthus comosus Sims
Lamiaceae
Coleus grandis Cramer
♦ Abyssinian coleus, woolly plectranthus, Abessiniese coleus
♦ Noxious Weed, Garden Escape
♦ 283
♦ cultivated.

Plectranthus ecklonii Benth.
Lamiaceae

♦ blue spurflower, spurflower
♦ Weed, Sleeper Weed, Naturalised, Garden Escape, Environmental Weed
♦ 54, 86, 88, 198, 225, 246, 251, 280
♦ cultivated. Origin: South Africa.

Plectranthus flaccidus Guerke
Lamiaceae
♦ Weed
♦ 87, 88

Plectranthus grandis (Cramer) Willems
Lamiaceae
♦ plectranthus
♦ Sleeper Weed, Naturalised, Environmental Weed
♦ 225, 246, 280

Plectranthus neochilus Schltr.
Lamiaceae
Coleus carnosus A.Chev., *Coleus neochilus* (Schltr.) Codd, *Coleus pentheri* Gürke, *Coleus schinzii* Gürke
♦ dogbane
♦ Naturalised
♦ 86
♦ cultivated. Origin: southern Africa.

Plectranthus nummularius Briq.
Lamiaceae
= *Plectranthus verticillatus* (L.f.) Druce
♦ Garden Escape
♦ 261
♦ cultivated. Origin: South Africa.

Plectranthus ornatus Codd
Lamiaceae
♦ Naturalised, Casual Alien
♦ 86, 280
♦ cultivated. Origin: east Africa.

Plectranthus parviflorus Willd.
Lamiaceae
= *Basilicum polystachyon* (L.) Moench
♦ little spurflower
♦ Naturalised
♦ 101
♦ cultivated, herbal. Origin: Asia, Australia.

Plectranthus scutellarioides (L.) R.Br.
Lamiaceae
= *Coleus scutellarioides* (L.) Benth.
♦ koramahd, oleus, nazareno, tocador
♦ Weed, Introduced, Cultivation Escape
♦ 3, 191, 230, 261
♦ S, cultivated, herbal. Origin: east Asia.

Plectranthus stocksii Hook.f.
Lamiaceae
♦ Weed
♦ 66

Plectranthus verticillatus (L.f.) Druce
Lamiaceae
Plectranthus nummularius Briq. (see)
♦ whorled plectranthus
♦ Naturalised
♦ 86, 101
♦ cultivated. Origin: south-east Africa.

Plectranthus zatarhendi (Forssk.) E.A.Bruce var. woodii (Guerke) Codd
Lamiaceae

Plectranthus woodii Guerke
♦ Native Weed
♦ 121
♦ pH. Origin: southern Africa.

Plectrocarpa tetracantha Gillies ex Hook.
Zygophyllaceae
♦ Introduced
♦ 228
♦ arid.

Plectronia parviflora Bedd.
Oliniaceae
♦ Weed
♦ 87, 88

Pleioblastus Nakai spp.
Poaceae
= *Arundinaria* Michx. spp. (NoR)
♦ dwarf bamboo
♦ Weed
♦ 88, 204
♦ G.

Pleioblastus auricomus (Mitford) D.C.McClint.
Poaceae
= *Arundinaria auricoma* Mitford (NoR)
♦ kamuro zasa
♦ Naturalised
♦ 280
♦ G, cultivated.

Pleioblastus chino (Franch. & Sav.) Makino
Poaceae
= *Arundinaria chino* (Franch. & Sav.) Makino (NoR)
♦ adzuma nezasa
♦ Weed, Naturalised
♦ 15, 280, 286
♦ G, cultivated.

Pleioblastus chino (Franch. & Sav.) Makino var. viridis (Makino) S.Suzuki
Poaceae
= *Arundinaria variegata* (Siebold ex Miq.) Makino var. *viridis* Makino (NoR)
♦ Weed
♦ 286
♦ G.

Pleioblastus gramineus (Bean) Nakai
Poaceae
= *Arundinaria graminea* (Bean) Makino (NoR)
♦ Weed, Naturalised
♦ 15, 280
♦ G, promoted.

Pleioblastus hindsii (Munro) Nakai
Poaceae
= *Arundinaria hindsii* Munro (NoR)
♦ Weed, Naturalised
♦ 15, 280
♦ G, cultivated.

Pleioblastus linearis (Hack.) Nakai
Poaceae
= *Arundinaria linearis* Hack. (NoR)
♦ Weed
♦ 88, 204
♦ G, cultivated.

Pleioblastus shibuyanus **Makino & Nakai fo.** *pubescens* **S.Suzuki**
Poaceae
= *Arundinaria variegata* (Siebold ex Miq.) Makino fo. *pubescens* Makino (NoR)
♦ Weed
♦ 286
♦ G.

Pleioblastus variegatus **(Siebold ex Miq.) Makino**
Poaceae
= *Arundinaria variegata* (Siebold ex Miq.) Makino (NoR)
♦ dwarf white striped bamboo
♦ Weed, Naturalised
♦ 15, 87, 88, 280
♦ G, cultivated. Origin: Japan.

Pleuropterus multiflorus **(Thunb.) Turcz.**
Polygonaceae
Polygonum multiflorum Thunb. ex Murray (see)
♦ Weed, Naturalised
♦ 286, 287

Pleurothallis lancilabris **(Rchb.f.) Schltr. var.** *oxyglossa* **(Schltdl.) C.Schweinf.**
Orchidaceae
= *Platystele oxyglossa* (Schltr.) Garay (NoR)
♦ Introduced
♦ 38

Pleurothallis vittata **Lindl.**
Orchidaceae
= *Pleurothallis hemirhoda* Lindl. & Paxton (NoR)
♦ Introduced
♦ 38

Pluchea carolinensis **(Jacq.) G.Don.**
Asteraceae
♦ cure for all
♦ Weed
♦ 14
♦ herbal.

Pluchea dioscoridis **(L.) DC.**
Asteraceae
Baccharis dioscoridis L., *Conyza dioscoridis* Desf. (see)
♦ ploughman's spikenard
♦ Weed
♦ 88, 185

Pluchea indica **(L.) Less.**
Asteraceae
♦ Indian pulchea, Indian fleabane, Indian camphorweed
♦ Weed, Naturalised, Introduced, Environmental Weed
♦ 3, 6, 13, 22, 80, 87, 88, 101, 209, 228, 259, 262, 287
♦ S, arid, herbal. Origin: Australia.

Pluchea lanceolata **(DC.) Oliv. & Hiern**
Asteraceae
♦ Weed, Quarantine Weed
♦ 76, 87, 88, 203, 220, 243
♦ arid, herbal.

Pluchea odorata **(L.) Cass.**
Asteraceae
Conyza cortesii Kunth, *Conyza odorata*

L., *Pluchea cortesii* DC., *Pluchea purpurascens* (Sw.) DC. (see)
♦ saltmarsh fleabane, sweet scent
♦ Weed, Quarantine Weed, Naturalised
♦ 3, 22, 76, 80, 87, 88, 191, 203, 220, 287
♦ a/pH, arid, herbal.

Pluchea purpurascens **(Sw.) DC.**
Asteraceae
= *Pluchea odorata* (L.) Cass.
♦ marsh fleabane
♦ Weed
♦ 87, 88
♦ cultivated, herbal.

Pluchea rosea **Godfrey**
Asteraceae
♦ rosy camphorweed
♦ Weed
♦ 14
♦ herbal.

Pluchea sagittalis **(Lam.) Cabrera**
Asteraceae
Pluchea quitoc DC.
♦ wingstem camphorweed, lucera, madrecravo
♦ Weed, Quarantine Weed
♦ 76, 87, 88, 203, 220, 255, 273, 295
♦ herbal. Origin: Americas.

Pluchea sericea **(Nutt.) Cov.**
Asteraceae
Pluchea borealis A.Gray, *Polypappus sericeus* Nutt., *Tessaria borealis* Torr. & A.Gray ex A.Gray, *Tessaria sericea* (Nutt.) Shinners
♦ arrowweed, arrowwood
♦ Weed, Quarantine Weed
♦ 23, 76, 87, 88, 161, 203, 218, 220
♦ S, arid, cultivated, herbal.

Pluchea symphytifolia **(Mill.) Gillis**
Asteraceae
= *Neurolaena lobata* (L.) R.Br.
♦ sour bush
♦ Weed, Quarantine Weed
♦ 3, 76, 88, 191, 203, 220
♦ herbal.

Plumbagella micrantha **(Ledeb.) Spach**
Plumbaginaceae
♦ littleflower plumbagella
♦ Weed
♦ 297
♦ cultivated.

Plumbago auriculata **Lam.**
Plumbaginaceae
Plumbago capensis Thunb.
♦ cape leadwort, leadwort, plumbago
♦ Weed, Naturalised, Native Weed, Introduced
♦ 86, 101, 121, 161, 179, 261
♦ pC, cultivated, herbal, toxic. Origin: South Africa.

Plumbago europaea **L.**
Plumbaginaceae
♦ European plumbago, plumbago
♦ Weed
♦ 272
♦ pH, promoted, herbal.

Plumbago indica **L.**
Plumbaginaceae

♦ whorled plantain, scarlet leadwort, plumbago, zapatitos de la virgen
♦ Weed, Naturalised, Introduced
♦ 38, 101, 161, 261
♦ cultivated, herbal, toxic. Origin: south-east Asia.

Plumbago scandens **L.**
Plumbaginaceae
♦ doctorbush
♦ Weed
♦ 14, 39, 87, 88, 157
♦ arid, cultivated, herbal, toxic.

Plumbago zeylanica **L.**
Plumbaginaceae
♦ leadwort, wild leadwort, lautafifi
♦ Weed, Garden Escape
♦ 39, 87, 88
♦ pH, cultivated, herbal, toxic.

Plumeria **L. spp.**
Apocynaceae
♦ frangipani
♦ Naturalised, Garden Escape, Environmental Weed, Cultivation Escape
♦ 86, 155
♦ cultivated, toxic.

Plumeria acutifolia **Poir.**
Apocynaceae
= *Plumeria rubra* L. fo. *acutifolia* (Poir.) Woodson (NoR)
♦ Introduced
♦ 228
♦ arid, cultivated, herbal.

Plumeria rubra **L.**
Apocynaceae
♦ temple tree, frangipani, pua fiti, pomeria, red paucipan
♦ Weed, Naturalised, Introduced, Cultivation Escape
♦ 32, 39, 101, 161, 230, 261
♦ T, cultivated, herbal, toxic. Origin: Mexico, Central America.

Poa **L. spp.**
Poaceae
♦ meadowgrass, bluegrass
♦ Weed
♦ 272
♦ G.

Poa acroleuca **Steud.**
Poaceae
♦ mizoichigotsunagi
♦ Weed
♦ 286
♦ G.

Poa alpina **L.**
Poaceae
Poa divaricata Vill.
♦ alpine meadowgrass, alpine bluegrass
♦ Weed, Casual Alien
♦ 272, 280
♦ G, cultivated, herbal.

Poa ampla **Merr.**
Poaceae
= *Poa secunda* J.S.Presl (NoR)
♦ big bluegrass, western bluegrass
♦ Naturalised

♦ 300
♦ pG, herbal.

Poa angustifolia L.
Poaceae
= *Poa pratensis* L. ssp. *angustifolia* (L.)
Lej. (NoR)
♦ narrow leaved meadowgrass, lipnica
úzkolistá
♦ Naturalised
♦ 300
♦ G, herbal.

Poa annua L.
Poaceae
Poa annua L. var. *reptans* Hausskn.
(see), *Poa triangularis* Gilib.
♦ annual bluegrass, winter grass,
kylänurmikka, fienarola annuale,
annual poa, annual meadowgrass, low
speargrass, sixweeks grass, goosegrass,
pasto de invierno, walkgrass
♦ Weed, Noxious Weed, Naturalised,
Introduced, Garden Escape,
Environmental Weed
♦ 7, 9, 15, 23, 30, 34, 44, 45, 51, 68, 70,
80, 86, 87, 88, 91, 93, 98, 101, 118, 121,
134, 152, 158, 161, 167, 174, 176, 179,
180, 185, 195, 198, 203, 204, 205, 210,
211, 212, 217, 218, 228, 235, 236, 237,
238, 241, 243, 249, 250, 253, 255, 261,
263, 269, 272, 273, 275, 280, 286, 290,
293, 295, 297, 299, 300
♦ aG, arid, cultivated, herbal. Origin:
Eurasia.

Poa annua L. var. reptans Hausskn.
Poaceae
= *Poa annua* L.
♦ Naturalised
♦ 287
♦ G.

Poa aquatica L.
Poaceae
= *Glyceria maxima* (Hartm.) Holmb.
♦ Weed
♦ 87, 88
♦ G.

Poa barrosiana Parodi
Poaceae
♦ Introduced
♦ 228
♦ G, arid.

Poa bonariensis (Lam.) Kunth
Poaceae
♦ poa bonaerense
♦ Weed
♦ 295
♦ G, arid.

Poa bulbosa L.
Poaceae
Poa crispa Thuill., *Poa prolifera* Schmidt
♦ bulbous bluegrass, bulbous
meadowgrass, bulbous poa
♦ Weed, Naturalised, Introduced,
Environmental Weed
♦ 7, 44, 72, 80, 87, 88, 91, 98, 101, 136,
161, 176, 195, 198, 203, 212, 218, 228,
243, 272, 280, 287
♦ pG, arid, cultivated, herbal.

Poa bulbosa L. var. bulbosa
Poaceae
♦ bulbous meadowgrass
♦ Naturalised, Environmental Weed
♦ 86, 198
♦ G. Origin: Europe, North Africa,
west Asia, Canary and Madeira
Islands.

Poa bulbosa L. var. vivipara Koeler
Poaceae
♦ bulbous meadowgrass
♦ Weed, Naturalised
♦ 86, 198, 248, 287
♦ G.

Poa chaixii Vill.
Poaceae
♦ broad leaved meadowgrass, lipnica
chaixova, broadleaf bluegrass
♦ Weed, Naturalised, Casual Alien
♦ 40, 70, 101, 272
♦ G, cultivated.

Poa cita Edgar
Poaceae
♦ silver tussock
♦ Weed
♦ 181
♦ G, cultivated.

Poa colensoi Hook.f.
Poaceae
♦ kiwi snow grass
♦ Naturalised
♦ 198
♦ G.

Poa compressa L.
Poaceae
Poa subcompressa Parn.
♦ Canada bluegrass, flattened poa,
wiregrass, flat stalked meadowgrass,
flattened meadowgrass
♦ Weed, Naturalised, Introduced,
Environmental Weed
♦ 4, 15, 21, 80, 86, 87, 88, 91, 98, 101,
104, 133, 146, 176, 195, 198, 203, 218,
222, 224, 272, 280, 287, 300
♦ pG, cultivated, herbal. Origin:
Eurasia, North Africa.

Poa conglomerata Rupr.
Poaceae
♦ Weed
♦ 199
♦ G.

Poa crassinervis Honda
Poaceae
♦ Weed
♦ 286
♦ G.

Poa glauca Vahl
Poaceae
Poa balfourii Parn., *Poa caesia* Sm.,
Poa glauca var. *glaucanthos* (Gaudin)
Lindm., *Poa glaucanthos* Gaudin
♦ glaucous meadowgrass,
takaneichigotsunagi, glaucous
bluegrass, glaucantha bluegrass,
blaugrünes Rispengras
♦ Naturalised
♦ 300
♦ G, herbal.

Poa holciformis J.Presl
Poaceae
♦ Weed
♦ 295
♦ G.

Poa infirma Kunth
Poaceae
♦ Scilly Isles meadowgrass, weak
bluegrass, early meadowgrass,
bluegrass
♦ Weed, Naturalised, Environmental
Weed
♦ 7, 86, 98, 101, 176, 198, 203, 253, 280
♦ aG, cultivated. Origin: Europe.

Poa labillardieri Steud.
Poaceae
♦ silver tussockgrass, large
tussockgrass, poa tussock,
tussockgrass
♦ Naturalised
♦ 280
♦ G, cultivated. Origin: Australia.

Poa labillardieri Steud. var. labillardieri
Poaceae
♦ poa tussock, tussockgrass
♦ Native Weed
♦ 269
♦ pG, arid, cultivated. Origin:
Australia.

Poa lanuginosa Poir.
Poaceae
Festuca lanata Spreng.
♦ Introduced
♦ 228
♦ G, arid.

Poa ligularis Nees ex Steud.
Poaceae
♦ Introduced
♦ 228
♦ G, arid.

Poa nemoralis L.
Poaceae
Paneion nemorale Lunell
♦ wood meadowgrass, wood poa,
wood bluegrass, lipnica hájna
♦ Weed, Naturalised, Introduced
♦ 15, 34, 70, 241, 272, 280, 300
♦ pG, cultivated, herbal.

Poa nipponica Koidz.
Poaceae
♦ ooichigotsunagi
♦ Weed
♦ 286
♦ G. Origin: east Asia.

Poa palustris L.
Poaceae
♦ fowl bluegrass, marsh
meadowgrass, swamp meadowgrass
♦ Weed, Naturalised, Introduced
♦ 15, 80, 228, 241, 272, 280, 287, 300
♦ pG, arid, cultivated, herbal.

Poa pratensis L.
Poaceae
Paneion pratense Lunell
♦ Kentucky bluegrass, smooth
meadowgrass, niittynurmikka,
meadowgrass, smooth stalked

meadowgrass, English meadowgrass, junegrass, speargrass, winter grass
♦ Weed, Naturalised, Introduced, Environmental Weed
♦ 4, 7, 15, 21, 38, 44, 70, 72, 80, 86, 87, 88, 91, 93, 98, 101, 104, 118, 121, 134, 136, 146, 151, 159, 161, 174, 176, 181, 195, 198, 203, 205, 218, 222, 241, 243, 261, 263, 272, 280, 286, 287, 295, 300
♦ pG, aqua, cultivated, herbal. Origin: Eurasia.

Poa pratensis L. ssp. *irrigata* (Lindm.) Lindb.f.
Poaceae
♦ spreading bluegrass
♦ Naturalised
♦ 101
♦ G.

Poa pratensis L. var. *angustifolia* Sm.
Poaceae
♦ Naturalised
♦ 287
♦ G.

Poa pratensis L. var. *hirsuta* Asch. & Graebn.
Poaceae
♦ Naturalised
♦ 287
♦ G.

Poa remota Forselles
Poaceae
♦ Casual Alien
♦ 280
♦ G.

Poa sieberiana Spreng.
Poaceae
♦ Naturalised
♦ 280
♦ G, cultivated.

Poa sinaica Steud.
Poaceae
♦ Weed
♦ 221
♦ G, arid.

Poa sphondylodes Trin.
Poaceae
♦ ichigotsunagi
♦ Weed, Quarantine Weed
♦ 76, 87, 88, 203, 220, 275, 286, 297
♦ pG.

Poa stenantha Trin.
Poaceae
♦ northern bluegrass
♦ Naturalised
♦ 300
♦ G, herbal.

Poa subcaerulea Sm.
Poaceae
♦ speading meadowgrass, spreading bluegrass
♦ Naturalised
♦ 287
♦ G.

Poa supina Schrad.
Poaceae
♦ bluegrass, juurtonurmikka
♦ Weed

♦ 44
♦ pG, herbal.

Poa trivialis L.
Poaceae
Aira semineutra Waldst. & Kit., *Phalaris semineutra* Roem. & Schult.
♦ rough meadowgrass, roughstalk bluegrass, rough bluegrass, rough stalk meadowgrass, karheanurmikka, fienarola comune
♦ Weed, Naturalised, Introduced
♦ 15, 30, 38, 44, 70, 80, 87, 88, 91, 98, 101, 118, 176, 198, 203, 211, 241, 243, 253, 272, 280, 286, 287, 300
♦ pG, cultivated, herbal.

Poa trivialis L. ssp. *sylvicola* (Guss.) H.Lindb.
Poaceae
♦ rough meadowgrass
♦ Naturalised
♦ 86, 198
♦ G. Origin: south-east Europe.

Poa trivialis L. ssp. *trivialis*
Poaceae
♦ rough meadowgrass
♦ Naturalised
♦ 86, 198
♦ G, cultivated. Origin: Eurasia, North Africa.

Poacynum hendersonii (Hook.f.) Woodson
Apocynaceae
♦ Weed
♦ 275
♦ pH.

Podalyria sericea R.Br.
Fabaceae/Papilionaceae
Sophora sericea Andrews *nom. illeg.*
♦ silky podalyria
♦ Weed, Naturalised, Garden Escape, Environmental Weed
♦ 7, 15, 54, 72, 86, 88, 198, 280, 290
♦ S, cultivated. Origin: South Africa.

Podocarpus cf. *blumei* Endl.
Podocarpaceae
♦ Introduced
♦ 230
♦ T.

Podocarpus macrophyllus (Thunb.) Sweet
Podocarpaceae
♦ yew plum pine, southern yew, Japanese yew pine, Bhuddist pine, Japanese yew, kusamaki
♦ Weed, Naturalised
♦ 101, 161, 247
♦ T, cultivated, toxic. Origin: Asia.

Podocarpus macrophyllus (Thunb.) Sweet var. *maki* Endl.
Podocarpaceae
♦ yew plum pine, maki, podocarpo
♦ Naturalised, Introduced
♦ 101, 261
♦ Origin: China.

Podophyllum peltatum L.
Berberidaceae/Podophyllaceae
♦ mayapple, American mandrake,

behen, devil's apple, hog apple, Indian apple, raccoon berry, umbrella leaf, wild jalap, wild lemon
♦ Weed
♦ 8, 39, 133, 161, 189, 195, 247
♦ pH, cultivated, herbal, toxic.

Podospermum resedifolium (L.) DC.
Asteraceae
Scorzonera resedifolia L.
♦ Weed, Naturalised
♦ 86, 98, 203
♦ herbal.

Podranea ricasoliana (Tanfani) Sprague
Bignoniaceae
♦ Zimbabwe creeper, Port St. John vine, pink trumpet vine, millonaria
♦ Weed, Sleeper Weed, Naturalised, Environmental Weed, Cultivation Escape
♦ 15, 101, 225, 246, 261, 280
♦ pC, cultivated. Origin: South Africa.

Pogonarthria squarrosa (Licht. ex Roem. & Schult.) Pilg.
Poaceae
♦ crossgrass, herringbone grass, sickle grass
♦ Native Weed
♦ 121
♦ b/pG. Origin: southern Africa.

Pogonatherum crinitum (Thunb.) Kunth
Poaceae
♦ Weed
♦ 87, 88, 286
♦ G, herbal.

Pogostemon auricularia (L.) El Gazzar & L.Watson
Lamiaceae
Dysophylla auricularia (L.) Bl. (see), *Mentha auricularia* L.
♦ Weed
♦ 88, 170, 191
♦ herbal.

Pogostemon stellatus (Lour.) Kuntze
Lamiaceae
Dysophylla stellata (Lour.) Bth.
♦ niam dong
♦ Weed
♦ 238
♦ aH, cultivated. Origin: Australia.

Poinciana gilliesii Wall. ex Hook.
Fabaceae/Caesalpiniaceae
= *Caesalpinia gilliesii* (Wall. ex Hook.) Dietr.
♦ bird of paradise bush, bird of paradise
♦ Naturalised
♦ 86, 194
♦ cultivated, herbal, toxic.

Poinciana regia Bojer ex Hook.
Fabaceae/Caesalpiniaceae
= *Delonix regia* (Bojer ex Hook.) Raf.
♦ Weed
♦ 3
♦ herbal.

Polanisia dodecandra (L.) DC.
Capparaceae/Cleomaceae
♦ clammyweed, redwhisker

clammyweed, western trachysperma
clammyweed
- Weed
- 161
- aH, cultivated, herbal.

Polanisia graveolens **Raf.**
Capparaceae/Cleomaceae
= *Polanisia dodecandra* (L.) DC. ssp.
dodecandra (NoR)
- clammyweed
- Weed
- 87, 88, 218

Polanisia trachysperma **Torr. & Gray**
Capparaceae/Cleomaceae
= *Polanisia dodecandra* (L.) DC. ssp.
trachysperma (Torr. & A.Gray) Iltis
(NoR)
- western clammyweed
- Weed
- 23, 87, 88, 218
- herbal.

Polaskia **Backeb. spp.**
Cactaceae
Chichipia Backeb. spp., *Heliabravoa*
Backeb. spp. (see)
- Weed, Quarantine Weed
- 76, 88, 203, 220

Polemonium caeruleum **L.**
Polemoniaceae
Polemonium cashmerianum nom. inval.
(see)
- Jacob's ladder, charity, lehtosinilatva
- Weed, Naturalised
- 23, 39, 88, 280
- pH, cultivated, herbal, toxic.

Polemonium cashmerianum nom. inval.
Polemoniaceae
= *Polemonium caeruleum* L.
- Quarantine Weed
- 220
- cultivated.

Polemonium cuspidatum **Sieb. & Zucc.**
Polemoniaceae
- stiffpoint Jacob's ladder
- Naturalised
- 101

Polemonium delicatum **Rydb.**
Polemoniaceae
= *Polemonium pulcherrimum* Hook. ssp.
delicatum (Rydb.) Brand (NoR)
- Weed, Quarantine Weed
- 76, 88, 220
- herbal.

Polemonium liniflorum **V.Vassil.**
Polemoniaceae
- Weed, Quarantine Weed
- 76, 88, 220
- cultivated.

Polemonium micranthum **Benth.**
Polemoniaceae
- annual polemonium
- Weed
- 87, 88, 161
- aH, herbal.

Polemonium pauciflorum **S.Watson**
Polemoniaceae

- yellow Jacob's ladder, pine trumpets,
fewflower Jacob's ladder
- Weed, Quarantine Weed
- 76, 88
- cultivated, herbal.

Polemonium pulcherrimum **Hook.**
Polemoniaceae
- Jacob's ladder, skunkleaf
polemonium, sky pilot, skyblue Jacob's
ladder
- Weed, Quarantine Weed
- 76, 88, 220
- pH, cultivated, herbal.

Polemonium reptans **L.**
Polemoniaceae
- niitysinilatva, Greek valerian,
Jacob's ladder, abcess root
- Cultivation Escape
- 39, 42
- pH, cultivated, herbal, toxic.

Polemonium yezoense **Kitam.**
Polemoniaceae
- Weed, Quarantine Weed
- 76, 88, 220

Polianthes tuberosa **L.**
Agavaceae
- tuberos, tuberosa, tuberose double
pearl, tuberose tub, reuse tuberoza
wonna
- Introduced
- 39, 228
- pH, arid, cultivated, herbal, toxic.

Pollia japonica **Thunb.**
Commelinaceae
- Quarantine Weed
- 220
- cultivated, herbal.

Polycarena heterophylla **(L.f.) Levyns**
Scrophulariaceae
- Weed, Naturalised, Environmental
Weed
- 7, 86, 98, 203

Polycarena leipoldtii **Hiern**
Scrophulariaceae
- Weed, Naturalised
- 86, 98, 203

Polycarpaea corymbosa **(L.) Lam.**
Caryophyllaceae
- old man's cap
- Weed, Naturalised, Native Weed
- 87, 88, 101, 121, 179
- aH, herbal.

Polycarpaea repens **(Forssk.) Asch. &
Schweinf.**
Caryophyllaceae
- Weed
- 221

Polycarpon indicum **(Retz.) Merr.**
Caryophyllaceae
- Weed, Quarantine Weed
- 76, 87, 88, 203, 220

Polycarpon prostratum **(Forssk.) Pax**
Caryophyllaceae
- Weed
- 87, 88, 221
- herbal.

Polycarpon tetraphyllum **(L.) L.**
Caryophyllaceae
Alsine polycarpa Crantz, *Mollugo
tetraphylla* L.
- four leaved allseed, fourleaf
manyseed, kynsiyrtti, allseed
- Weed, Naturalised, Environmental
Weed, Casual Alien
- 7, 15, 34, 42, 51, 72, 86, 87, 88, 93, 94,
98, 101, 121, 134, 165, 176, 191, 198, 203,
205, 241, 253, 269, 272, 280, 287, 295,
300
- aH, cultivated, herbal. Origin:
Eurasia.

Polycarpon tetraphyllum **(L.) L. ssp.
alsinifolium (Biv.) P.W.Ball**
Caryophyllaceae
- fourleaf manyseed
- Naturalised
- 101

Polycarpon tetraphyllum **(L.) L. ssp.
*tetraphyllum***
Caryophyllaceae
- fourleaf manyseed
- Naturalised
- 101

Polycephalium poggei **Engl.**
Icacinaceae
- Weed
- 87, 88

Polycnemum arvense **L.**
Chenopodiaceae
- otarustoruoho
- Weed, Casual Alien
- 42, 70, 88, 94, 253, 272
- herbal.

Polycnemum majus **A.Braun**
Chenopodiaceae
- giant needleleaf, piikkirustoruoho
- Weed, Naturalised, Casual Alien
- 42, 88, 94, 101, 272
- cultivated, herbal.

Polygala **L. spp.**
Polygalaceae
- milkwort, polygala
- Weed
- 272

Polygala acuminata **Willd.**
Polygalaceae
- Weed
- 87, 88
- S.

Polygala adenophylla **St.-Hil.**
Polygalaceae
- Weed
- 87, 88

Polygala ambigua **Nutt.**
Polygalaceae
- whorled milkwort
- Naturalised
- 287
- herbal.

Polygala arenaria **Willd.**
Polygalaceae
- Weed
- 87, 88

Polygala arvensis Willd.
Polygalaceae
♦ Weed
♦ 66

Polygala chinensis L.
Polygalaceae
Polygala polyfolia Presl
♦ Weed, Naturalised, Introduced
♦ 86, 87, 88, 230
♦ H, herbal. Origin: Asia.

Polygala comosa Schk.
Polygalaceae
♦ tufted milkwort, tupsulinnunruoho
♦ Weed
♦ 272
♦ herbal.

Polygala cornuta Kellogg
Polygalaceae
♦ milkwort, Sierra milkwort
♦ Weed
♦ 161
♦ pH, herbal.

Polygala duartena St.-Hil.
Polygalaceae
♦ Weed, Naturalised
♦ 86, 98, 203

Polygala erioptera DC.
Polygalaceae
♦ Weed
♦ 87, 88, 221, 242
♦ arid.

Polygala glochidiata Kunth
Polygalaceae
♦ tropical milkwort
♦ Weed
♦ 87, 88
♦ H.

Polygala glomerata Lour.
Polygalaceae
♦ Weed
♦ 87, 88
♦ herbal.

Polygala guineensis Willd.
Polygalaceae
♦ Weed
♦ 87, 88

Polygala hottentotta Presl
Polygalaceae
♦ Native Weed
♦ 121
♦ pH, cultivated. Origin: southern Africa.

Polygala japonica Houtt.
Polygalaceae
♦ himehagi
♦ Weed
♦ 87, 88, 204, 286
♦ pH, cultivated, herbal. Origin: Australasia.

Polygala luteo-alba Gagnep.
Polygalaceae
♦ Weed
♦ 87, 88

Polygala monspeliaca L.
Polygalaceae
♦ annual milkwort

♦ Weed, Naturalised
♦ 86, 98, 198, 203
♦ Origin: Mediterranean.

Polygala myrtifolia L.
Polygalaceae
♦ sweet pea bush, myrtleleaf milkwort
♦ Weed, Quarantine Weed, Naturalised, Garden Escape, Environmental Weed
♦ 3, 72, 86, 88, 98, 134, 165, 176, 198, 203, 225, 246, 280, 289, 296
♦ S, cultivated. Origin: South Africa.

Polygala nicaeensis Risso ex Koch
Polygalaceae
♦ Nicean milkwort
♦ Weed
♦ 272
♦ herbal.

Polygala paniculata L. Folsom
Polygalaceae
Polygala variabilis Kunth
♦ island snakeroot, root beer plant, namupululola, airoinituraga, teketekeniulumatua, senikuila, kisinpwil, orosne
♦ Weed, Naturalised, Environmental Weed
♦ 3, 86, 87, 88, 98, 155, 157, 170, 191, 203, 204, 276, 287
♦ aH, herbal. Origin: tropical America.

Polygala persicariifolia DC.
Polygalaceae
♦ Weed
♦ 242
♦ arid. Origin: Australia.

Polygala sanguinea L.
Polygalaceae
Polygala viridescens L.
♦ blood milkwort, blood polygala
♦ Weed, Naturalised
♦ 39, 161, 287
♦ herbal, toxic.

Polygala scoparia Kunth
Polygalaceae
♦ Weed
♦ 221
♦ herbal.

Polygala senega L.
Polygalaceae
♦ seneca snakeroot
♦ Weed
♦ 39, 161
♦ pH, cultivated, herbal, toxic.

Polygala serpyllifolia J.A.C.Hose
Polygalaceae
Polygala serpyllacea Weihe
♦ heath milkwort, nummilinnunruoho, thyme leaved milkwort
♦ Naturalised, Casual Alien
♦ 42, 280

Polygala sibirica L.
Polygalaceae
♦ Siberian milkwort
♦ Weed
♦ 297
♦ pH, promoted, herbal.

Polygala tatarinowii Regel.
Polygalaceae
♦ tatarinow milkwort
♦ Weed
♦ 297

Polygala tenuifolia Willd.
Polygalaceae
♦ Weed
♦ 39, 275, 297
♦ pH, promoted, herbal, toxic.

Polygala verticillata L.
Polygalaceae
♦ whorled milkwort
♦ Naturalised
♦ 280, 287
♦ herbal.

Polygala violaceae Aubl.
Polygalaceae
♦ roxinha
♦ Weed
♦ 255
♦ aH. Origin: Americas.

Polygala virgata Thunb.
Polygalaceae
♦ polygala, purple broom, kalimbesokola
♦ Weed, Naturalised, Garden Escape, Environmental Weed
♦ 7, 72, 86, 88, 98, 155, 198, 203, 280, 290
♦ S, cultivated. Origin: southern Africa.

Polygala vulgaris L.
Polygalaceae
♦ common milkwort, milkwort, isolinnunruoho
♦ Weed, Naturalised
♦ 15, 86, 101, 176, 198, 272, 280
♦ pH, cultivated, herbal. Origin: Europe.

Polygonatum biflorum (Walter) Ellis
Liliaceae/Convallariaceae
♦ smooth Solomon's seal, Solomon's seal, smooth commutatum Solomon's seal, American Solomon's seal
♦ Weed, Native Weed
♦ 23, 39, 88, 161, 174
♦ pH, cultivated, herbal, toxic. Origin: North America.

Polygonatum falcatum A.Gray
Liliaceae/Convallariaceae
♦ narukoyuri
♦ Weed
♦ 286
♦ pH, cultivated, herbal. Origin: China, Japan, Korea.

Polygonatum hirsutum (Bosc ex Poir.) Pursh
Liliaceae/Convallariaceae
Polygonatum latifolium (Jacq.) Desf. (see)
♦ broadleaf Solomon's seal
♦ Naturalised
♦ 101

Polygonatum × hybridum Brügger
Liliaceae/Convallariaceae
= *Polygonatum multiflorum* (L.) All. ×

Polygonatum odoratum (Mill.) Druce
- ♦ garden Solomon's seal
- ♦ Naturalised
- ♦ 40, 280
- ♦ cultivated, herbal.

Polygonatum latifolium (Jacq.) Desf.
Liliaceae/Convallariaceae
= *Polygonatum hirsutum* (Bosc ex Poir.) Pursh
- ♦ broad leaved Solomon's seal
- ♦ Weed
- ♦ 272
- ♦ cultivated, herbal.

Polygonatum odoratum (Mill.) Druce
Liliaceae/Convallariaceae
- ♦ scented Solomon's seal, angular Solomon's seal, kalliokielo, Solomon's seal, fragrant Solomon's seal, yu zhu
- ♦ Weed
- ♦ 70, 297
- ♦ pH, cultivated, herbal.

Polygonatum verticillatum (L.) All.
Liliaceae/Convallariaceae
Convallaria verticillata L., *Evallaria verticillata* Neck.
- ♦ whorled Solomon's seal
- ♦ Weed
- ♦ 272
- ♦ pH, cultivated, herbal.

Polygonum L. spp.
Polygonaceae
- ♦ willowweed, polygonum, smartweeds, fleeceflower, American smartweed
- ♦ Weed, Quarantine Weed
- ♦ 88, 181, 237, 243, 258, 272
- ♦ H, herbal, toxic.

Polygonum achoreum Blake
Polygonaceae
- ♦ striated knotweed, leathery knotweed, striate knotweed
- ♦ Weed, Native Weed
- ♦ 52, 87, 88, 161, 174
- ♦ herbal. Origin: North America.

Polygonum acre H.B.K.
Polygonaceae
= *Polygonum punctatum* Ell.
- ♦ bitter smartweed
- ♦ Weed
- ♦ 23, 39, 87, 88, 218
- ♦ H, herbal, toxic.

Polygonum acuminatum Kunth
Polygonaceae
- ♦ tapertip smartweed
- ♦ Weed, Naturalised
- ♦ 87, 88, 237, 255, 295, 300
- ♦ pH. Origin: Americas.

Polygonum aequale Lindm.
Polygonaceae
= *Polygonum arenastrum* Boreau
- ♦ Weed
- ♦ 88, 191, 204

Polygonum afghanicum Meissn.
Polygonaceae
- ♦ Weed
- ♦ 243

Polygonum alatum Ham. ex D.Don
Polygonaceae
= *Polygonum nepalense* Meissn.
- ♦ Weed
- ♦ 87, 88

Polygonum amphibium L.
Polygonaceae
Persicaria amphibia (L.) Gray
- ♦ water smartweed, amphibious bistort, willow grass, water knotweed, vesitatar, knotweed
- ♦ Weed
- ♦ 23, 44, 70, 87, 88, 94, 159, 161, 218, 243, 253, 272, 275, 297
- ♦ wpH, cultivated, herbal.

Polygonum amphibium L. var. *emersum* Michx.
Polygonaceae
Polygonum coccineum Muhl. ex Willd. (see)
- ♦ swamp smartweed, devil's shoestring, kelp, marsh smartweed, swamp knotweed, swamp persicaria, tanweed, longroot smartweed, water smartweed
- ♦ Weed, Noxious Weed, Native Weed
- ♦ 174, 180, 229, 243
- ♦ pH, aqua. Origin: North America.

Polygonum amphibium L. var. *terrestre* Leyss.
Polygonaceae
- ♦ Weed
- ♦ 275
- ♦ pH.

Polygonum aquisetiferme Sibth. & Sm.
Polygonaceae
- ♦ Weed
- ♦ 87, 88

Polygonum arenarium Waldst. & Kit.
Polygonaceae
- ♦ European knotweed
- ♦ Weed, Naturalised
- ♦ 101, 272
- ♦ cultivated.

Polygonum arenastrum Jord. ex Boreau
Polygonaceae
Polygonum acetosellum Klokov, *Polygonum aequale* Lindm. (see), *Polygonum aviculare* auct., *Polygonum littorale* auct., *Polygonum microspermum* Jord. ex Boreau, *Polygonum propinquum* Ledeb.
- ♦ matweed, ovalleaf knotweed, doorweed, prostrate knotweed, common knotweed, small leaved wireweed, tannertatar, wireweed
- ♦ Weed, Noxious Weed, Naturalised, Introduced, Environmental Weed
- ♦ 7, 15, 21, 34, 70, 72, 86, 88, 94, 98, 101, 161, 167, 174, 176, 198, 203, 241, 243, 272, 280, 286, 287, 299
- ♦ aH, aqua, herbal. Origin: Europe.

Polygonum argyrocoleon Kunze
Polygonaceae
Persicaria argyrocoleum Steud.
- ♦ silversheath knotweed
- ♦ Weed, Naturalised
- ♦ 7, 34, 80, 87, 88, 98, 101, 161, 180, 199,
203, 212, 218, 243, 272
- ♦ aH, herbal.

Polygonum aubertii Henry
Polygonaceae
Fallopia aubertii (Henry) Holub (see)
- ♦ Russian vine, Chinese fleecevine, silver lace vine
- ♦ Weed, Naturalised
- ♦ 23, 80, 88, 101, 195
- ♦ cultivated, herbal.

Polygonum aviculare L.
Polygonaceae
Polygonum aequale Lindm. (see), *Polygonum agreste* Sumn., *Polygonum aphyllum* Krock., *Polygonum araraticum* Kom., *Polygonum aviculare* ssp. *aequale* (Lindm.) Asch. & Graebn., *Polygonum berteroi* Phil., *Polygonum calcatum* Lindm. (see), *Polygonum geniculatum* Poir., *Polygonum heterophyllum* Lindm. nom. illeg. (see), *Polygonum monspeliense* Pers. (see), *Polygonum retinerve* Worosch., *Polygonum striatum* K.Koch, *Polygonum uruguense* H.Gross
- ♦ wireweed, prostrate knotweed, knotweed, hogweed, ironweed, knotgrass, Ray's knotgrass, sea knotgrass, small leaved knotgrass, doorweed, matgrass, pink weed, bird grass, stone grass, way grass, goosegrass, sanguinaria, Cien nudos
- ♦ Weed, Naturalised, Introduced, Environmental Weed
- ♦ 7, 15, 23, 38, 39, 44, 49, 51, 52, 53, 55, 68, 70, 80, 86, 87, 88, 93, 94, 98, 101, 114, 115, 118, 121, 136, 157, 158, 161, 165, 176, 179, 180, 198, 203, 204, 205, 207, 210, 211, 212, 217, 218, 221, 228, 236, 237, 240, 241, 243, 248, 249, 250, 253, 263, 269, 271, 272, 275, 280, 286, 295, 297, 300
- ♦ a/bH, arid, cultivated, herbal, toxic. Origin: Eurasia.

Polygonum aviculare L. var. *condensatum* Beck.
Polygonaceae
- ♦ Naturalised
- ♦ 287

Polygonum aviculare L. var. *monospeliense* Thibaud
Polygonaceae
- ♦ Naturalised
- ♦ 287

Polygonum baldschuanicum Regel
Polygonaceae
Fallopia baldschuanica (Regel) Holub (see)
- ♦ mile a minute, bukhara fleeceflower, galaxy a month, Russian disaster vine
- ♦ Quarantine Weed, Naturalised
- ♦ 76, 101
- ♦ pC, cultivated, herbal.

Polygonum barbatum L.
Polygonaceae
- ♦ knotgrass, smartweed
- ♦ Weed
- ♦ 39, 87, 88, 170, 240, 275, 276
- ♦ aH, aqua, promoted, herbal, toxic.

Polygonum bellardii All.
Polygonaceae
Polygonum agrestinum Jord. ex
Boreau, *Polygonum aviculare* L. var.
angustissimum (Meisn.), *Polygonum
aviculare* ssp. *rectum* Chrtek, *Polygonum
aviculare* var. *virgatum* Peterm.,
Polygonum kitaibelianum Sadler,
Polygonum neglectum Bess. (see),
Polygonum nervosum Wallr., *Polygonum
patulum* M.Bieb. ssp. *kitaibelianum*
(Sadler) Asch. & Graebn., *Polygonum
procumbens* Gilib., *Polygonum rurivagum*
Jord. ex Boreau (see), *Polygonum rectum*
(Chrtek) H.Scholz, *Polygonum rectum*
ssp. *virgatum* (Peterm.) H.Scholz
♦ narrowleaf knotweed, tree hogweed
♦ Weed, Quarantine Weed,
Naturalised
♦ 76, 87, 88, 98, 101, 221, 243

Polygonum bistorta L.
Polygonaceae
♦ snakeweed, meadow bistort,
common bistort, bistort, snakeroot
♦ Weed, Naturalised, Cultivation
Escape
♦ 39, 42, 70, 87, 88, 101, 272, 297
♦ pH, cultivated, herbal, toxic.

Polygonum bistorta L. var. bistorta
Polygonaceae
♦ meadow bistort
♦ Naturalised
♦ 101

Polygonum bistortoides Pursh
Polygonaceae
♦ American bistort
♦ Weed
♦ 87, 88, 218
♦ pH, cultivated, herbal.

Polygonum blumei Meisn. ex Miq.
Polygonaceae
♦ Weed
♦ 88, 204
♦ herbal.

Polygonum brasiliense Koch.
Polygonaceae
Polygonum camporum Meisn.
♦ Weed, Naturalised
♦ 237, 295, 300

Polygonum bungeanum Turcz.
Polygonaceae
Persicaria bungeana (Turcz.) Nakai ex
T.Mori (see)
♦ Bunge's smartweed, idäntatar
♦ Weed, Quarantine Weed,
Naturalised, Casual Alien
♦ 42, 76, 101, 243, 275, 297
♦ aH, promoted. Origin: China.

Polygonum caespitosum Blume
Polygonaceae
Polygonum cespitosum Blume (see)
♦ bunchy knotweed, oriental lady's
thumb, tufted longisetum knotweed
♦ Weed, Naturalised
♦ 23, 80, 87, 88, 101, 102, 161, 275
♦ aH.

**Polygonum caespitosum Blume var.
caespitosum**

Polygonaceae
♦ oriental lady's thumb
♦ Naturalised
♦ 101

**Polygonum caespitosum Blume var.
longisetum (Bruijn) A.N.Stewart**
Polygonaceae
Polygonum longisetum De Bruyn (see)
♦ oriental lady's thumb, tufted
knotweed, smartweed
♦ Weed
♦ 249

Polygonum calcatum Lindm.
Polygonaceae
= *Polygonum aviculare* L.
♦ sunajimichiyanagi
♦ Weed
♦ 88, 94

Polygonum campanulatum Hook.f.
Polygonaceae
Persicaria campanulata (Hook.f.) Ronse
Decr. (see)
♦ bellflower smartweed
♦ Naturalised
♦ 101, 300

**Polygonum capitatum Buch.-Ham. ex
D.Don**
Polygonaceae
= *Persicaria capitata* (Buch.-Ham. ex
D.Don) H.Gross
♦ pinkhead smartweed, pink
pinheads, knotweed
♦ Weed, Naturalised, Environmental
Weed, Cultivation Escape
♦ 15, 34, 101, 121, 280, 290, 297
♦ pH, cultivated, herbal. Origin:
Eurasia.

Polygonum cespitosum Blume
Polygonaceae
= *Polygonum caespitosum* Blume
♦ oriental lady's thumb
♦ Weed
♦ 17, 133, 195, 224
♦ herbal.

Polygonum chinense L.
Polygonaceae
= *Fagopyrum dibotrys* (D.Don) Hara
♦ Chinese knotweed, phayaa dong,
red bush
♦ Weed, Quarantine Weed,
Naturalised, Environmental Weed
♦ 76, 87, 88, 101, 203, 220, 226, 235, 238,
273, 274, 297
♦ cultivated, herbal.

Polygonum cilinode Michx.
Polygonaceae
♦ blackfringe knotweed
♦ Weed
♦ 23, 87, 88, 161, 218
♦ herbal.

Polygonum coccineum Muhl. ex Willd.
Polygonaceae
= *Polygonum amphibium* L. var. *emersum*
Michx.
♦ swamp smartweed, kelp
♦ Weed, Noxious Weed
♦ 23, 35, 45, 87, 88, 161, 210, 218, 299
♦ pH, aqua, promoted, herbal.

Polygonum cognatum Meissn.
Polygonaceae
Polygonum ammanioides Jaub. & Spach,
Polygonum myriophyllum H.Gross,
Polygonum rupestre Kar. & Kir.
♦ Weed
♦ 243

Polygonum conspicuum (Nakai) Nakai
Polygonaceae
♦ sakuratade
♦ Weed
♦ 87, 88, 263
♦ pH, promoted. Origin: east Asia.

Polygonum convolvulus L.
Polygonaceae
= *Fallopia convolvulus* (L.) Á.Löve
♦ wild buckwheat, black bindweed,
knot bindweed, bear bind, ivy
bindweed, climbing bindweed,
climbing buckwheat, cornbind,
Enredadera anual, cornbind, dullseed
cornbind
♦ Weed, Noxious Weed, Naturalised,
Introduced
♦ 21, 23, 34, 39, 44, 45, 49, 51, 52, 68, 80,
87, 88, 94, 101, 114, 136, 161, 162, 174,
180, 186, 199, 210, 211, 212, 218, 229,
236, 237, 243, 255, 271, 275, 293, 295,
297, 299
♦ aH, cultivated, herbal, toxic. Origin:
Eurasia.

**Polygonum convolvulus L. var.
convolvulus**
Polygonaceae
= *Fallopia convolvulus* (L.) Á.Löve
♦ wild buckwheat, black bindweed
♦ Noxious Weed, Naturalised
♦ 101, 229

**Polygonum convolvulus L. var.
subulatum Lej. & Court.**
Polygonaceae
= *Fallopia convolvulus* (L.) Á.Löve
♦ wild buckwheat, black bindweed
♦ Noxious Weed, Naturalised
♦ 101, 229

Polygonum coriarium Grigorj
Polygonaceae
♦ Weed
♦ 23, 88
♦ cultivated.

Polygonum cuspidatum Siebold & Zucc.
Polygonaceae
= *Fallopia japonica* (Houtt.) Ronse Decr.
♦ Japanese knotweed, Mexican
bamboo, Japanese bamboo, Japanese
grass, Japanese polygonum,
fleeceflower
♦ Weed, Noxious Weed, Naturalised,
Garden Escape, Environmental Weed,
Cultivation Escape
♦ 1, 4, 17, 18, 23, 34, 35, 39, 79, 80, 87,
88, 101, 102, 103, 129, 133, 136, 137, 139,
142, 146, 151, 156, 161, 195, 204, 207,
211, 212, 218, 224, 229, 263
♦ pH, aqua, cultivated, herbal, toxic.
Origin: eastern Asia.

Polygonum decipiens R.Br.
Polygonaceae
= *Polygonum salicifolium* Brouss.
♦ swamp willowweed
♦ Weed
♦ 208
♦ aqua.

Polygonum densiflorum Meisn.
Polygonaceae
♦ denseflower knotweed
♦ Weed
♦ 14, 87, 88
♦ herbal.

Polygonum dentato-alatum F.Schmidt ex Maxim
Polygonaceae
= *Fallopia dentato-alata* (F.Schmidt) Holub
♦ dentate winged fruit knotweed, ootsuruitadori
♦ Weed
♦ 297

Polygonum divaricatum L.
Polygonaceae
♦ knotgrass, röyhytatar
♦ Weed, Cultivation Escape
♦ 42, 88, 114, 275, 297
♦ pH, cultivated, herbal.

Polygonum douglassi Greene
Polygonaceae
♦ Weed
♦ 23, 88

Polygonum dumetorum L.
Polygonaceae
= *Fallopia dumetorum* (L.) Holub (NoR) [see *Bilderdykia dumetorum* (L.) Dumort.]
♦ copse buckwheat, copse bindweed
♦ Weed
♦ 23, 87, 88, 237, 243, 295
♦ aH, promoted, herbal.

Polygonum equisetiforme Sibth. & Sm.
Polygonaceae
♦ horsetail knotgrass, kortetatar
♦ Weed, Casual Alien
♦ 42, 70, 88, 185, 221, 250, 272
♦ pH, cultivated.

Polygonum erecto-minus Makino
Polygonaceae
♦ himetade
♦ Weed
♦ 87, 88

Polygonum erectum L.
Polygonaceae
♦ erect knotweed, Russian knotgrass
♦ Weed
♦ 23, 87, 88, 161, 210, 212, 218, 243
♦ herbal.

Polygonum flaccidum Meissn.
Polygonaceae
♦ Weed
♦ 39, 87, 88
♦ herbal, toxic.

Polygonum franchetii Vorosch.
Polygonaceae
Polygonum aviculare L. var. *minutiflorum* Franch., *Polygonum plebejum* auct.

♦ Weed
♦ 243

Polygonum fugax Small
Polygonaceae
= *Polygonum viviparum* L. (NoR)
♦ Weed
♦ 87, 88
♦ pH, promoted.

Polygonum glabrum Willd.
Polygonaceae
♦ Weed, Naturalised
♦ 86, 87, 88
♦ arid, herbal. Origin: Australia.

Polygonum hartwrightii Gray
Polygonaceae
= *Polygonum amphibium* L. var. *stipulaceum* (Coleman) (NoR)
♦ Weed
♦ 199

Polygonum hastato-sagittatum Makino
Polygonaceae
♦ nagabanounagitsukami
♦ Weed
♦ 87, 88, 235

Polygonum herniarioides Delile
Polygonaceae
♦ knotweed
♦ Naturalised
♦ 101

Polygonum heterophyllum Lindm. nom. illeg.
Polygonaceae
= *Polygonum aviculare* L.
♦ Weed
♦ 88, 94, 243
♦ herbal.

Polygonum higegaweri Steud.
Polygonaceae
♦ Weed, Quarantine Weed
♦ 76, 87, 88, 203, 220

Polygonum hydropiper L.
Polygonaceae
Persicaria hydropiper (L.) Spach (see)
♦ waterpepper, redleaf, marshpepper, smartweed, red shank, marshpepper knotweed, mild waterpepper
♦ Weed, Naturalised
♦ 15, 23, 24, 39, 44, 68, 70, 80, 87, 88, 94, 101, 159, 161, 165, 191, 204, 208, 217, 218, 243, 253, 263, 272, 274, 275, 280, 297, 300
♦ waH, cultivated, herbal, toxic. Origin: Eurasia.

Polygonum hydropiperoides Michx.
Polygonaceae
Polygonum persicarioides Kunth (see), *Polygonum opelousanum* Riddell ex Small
♦ mild smartweed, swamp smartweed, waterpepper
♦ Weed, Naturalised
♦ 23, 39, 87, 88, 161, 218, 241, 243, 255, 295, 300
♦ pH, arid, herbal, toxic. Origin: Eurasia.

Polygonum japonicum Meissn.
Polygonaceae

♦ Japanese knotweed
♦ Weed
♦ 87, 88, 263, 274, 275, 297
♦ pH, cultivated, herbal.

Polygonum lacerum Kunth
Polygonaceae
♦ knotweed
♦ Naturalised
♦ 101

Polygonum lanigerum Bak. & C.H.Wright var. africanum Meisn.
Polygonaceae
♦ Weed
♦ 88

Polygonum lapathifolium L.
Polygonaceae
Persicaria incarnata (Ell.) Small, *Persicaria lapathifolia* (L.) S.F.Gray (see), *Persicaria tomentosa* (Schrank) Bickn., *Polygonum incarnatum* Ell., *Polygonum incanum* F.W.Schmidt, *Polygonum lapathifolium* L. ssp. *lapathifolium* (see), *Polygonum lapathifolium* ssp. *pallidum* (With.) Fr. (see), *Polygonum lapathifolium* L. var. *salicifolium* Sibth. (see), *Polygonum nodosum* Pers. (see), *Polygonum oneillii* Brenckle, *Polygonum pensylvanicum* L. ssp. *oneillii* (Brenckle) Hultén, *Polygonum scabrum* Moench (see), *Polygonum tenuiflorum* Presl (see), *Polygonum tomentosum* Willd. (see)
♦ pale smartweed, common knotweed, pale persicaria, smartweed, willow smartweed, willowweed, wireweed, poligono nodoso, ukontatar, pale persicaria, curltop knotweed, curltop lady's thumb
♦ Weed, Noxious Weed, Quarantine Weed, Naturalised, Native Weed
♦ 23, 44, 52, 68, 76, 80, 87, 88, 94, 118, 136, 158, 159, 161, 174, 179, 180, 203, 208, 212, 217, 218, 235, 241, 243, 250, 253, 255, 262, 263, 272, 273, 274, 275, 280, 295, 297, 299, 300
♦ aH, aqua, cultivated, herbal. Origin: Europe.

Polygonum lapathifolium L. ssp. lapathifolium
Polygonaceae
= *Polygonum lapathifolium* L.
♦ bulbous persicaria
♦ Weed
♦ 70
♦ cultivated.

Polygonum lapathifolium L. ssp. maculatum (S.F.Gray) T.-Dyer & Trim.
Polygonaceae
♦ spotted knotweed
♦ Native Weed
♦ 121
♦ aH. Origin: southern Africa.

Polygonum lapathifolium L. ssp. pallidum (With.) Fr.
Polygonaceae
= *Polygonum lapathifolium* L.
♦ pale persicaria
♦ Weed
♦ 70

**Polygonum lapathifolium L. var.
salicifolium Sibth.**
Polygonaceae
= *Polygonum lapathifolium* L.
♦ curlytop knotweed
♦ Weed
♦ 275, 297
♦ aH.

Polygonum limbatum Meisn.
Polygonaceae
♦ Weed, Native Weed
♦ 121, 221
♦ H. Origin: southern Africa.

Polygonum linicola Sutulov
Polygonaceae
♦ Weed
♦ 87, 88

Polygonum longisetum de Bruyn
Polygonaceae
= *Polygonum caespitosum* Blume var.
longisetum (de Bruyn) A.N.Steward
♦ tufted knotweed
♦ Weed, Quarantine Weed
♦ 23, 76, 87, 88, 203, 220, 235, 263
♦ aH, promoted.

Polygonum maackianum Regel
Polygonaceae
♦ sadekusa
♦ Weed
♦ 87, 88, 275, 297
♦ aH, promoted. Origin: east Asia.

Polygonum maritimum L.
Polygonaceae
♦ sea knotgrass
♦ Weed, Naturalised
♦ 87, 88, 221, 237, 241, 295, 300
♦ herbal.

Polygonum milletii (H.Lév.) H.Lév.
Polygonaceae
Bistorta milletii H.Lév. (see)
♦ Quarantine Weed
♦ 220
♦ cultivated.

Polygonum minus Huds.
Polygonaceae
♦ small waterpepper, mietotatar, lesser
persicaria
♦ Weed
♦ 44, 87, 88, 94, 243, 272
♦ aH, cultivated, herbal. Origin:
Australia.

Polygonum mite Schrank
Polygonaceae
♦ tasteless waterpepper, lax flowered
persicaria, marsh persicaria
♦ Weed, Naturalised
♦ 44, 87, 88, 94, 241, 243, 272
♦ cultivated, herbal.

Polygonum molle D.Don
Polygonaceae
♦ syystatar
♦ Cultivation Escape, Casual Alien
♦ 42, 280
♦ pH, cultivated, herbal.

Polygonum monspeliense Pers.
Polygonaceae
= *Polygonum aviculare* L.

♦ Weed
♦ 80, 88, 94, 221, 243, 248

**Polygonum multiflorum Thunb. ex
Murray**
Polygonaceae
= *Pleuropterus multiflorus* (Thunb.)
Turcz.
♦ fo ti
♦ Quarantine Weed
♦ 220
♦ pC, promoted, herbal. Origin:
eastern Asia.

Polygonum neglectum Bess.
Polygonaceae
= *Polygonum bellardii* All.
♦ liettuanpihatatar
♦ Weed, Casual Alien
♦ 42, 87, 88

Polygonum nepalense Meissn.
Polygonaceae
Polygonum alatum Ham. ex D.Don.
(see)
♦ Nepalese smartweed
♦ Weed, Quarantine Weed,
Naturalised
♦ 76, 87, 88, 101, 121, 203, 220, 238, 240,
243, 275, 276, 297
♦ aH, cultivated. Origin: China, Japan,
Korea, Himalayas.

Polygonum nipponense Makino
Polygonaceae
♦ yanonegusa
♦ Weed
♦ 87, 88, 263
♦ aH.

Polygonum nodosum Pers.
Polygonaceae
= *Polygonum lapathifolium* L.
♦ spotted persicaria
♦ Weed
♦ 23, 88, 204, 243
♦ herbal.

Polygonum orientale L.
Polygonaceae
Persicaria orientalis (L.) Spach (see)
♦ prince's feather, shui hong,
purppuratatar, kiss me over the garden
gate
♦ Weed, Naturalised
♦ 23, 39, 80, 87, 88, 101, 102, 218, 235,
241, 272, 275, 276, 280, 297, 300
♦ aH, cultivated, herbal, toxic.

**Polygonum oxyspermum C.A.Mey. &
Bun.**
Polygonaceae
♦ Ray's knotgrass, sharpfruit
knotweed, meritatar
♦ Weed
♦ 272

**Polygonum paronychia Cham. &
Schlecht.**
Polygonaceae
♦ beach knotweed
♦ Weed
♦ 23, 88
♦ pH, herbal.

Polygonum patulum M.Bieb.
Polygonaceae

♦ tree hogweed, Bellard's smartweed,
eteläntatar
♦ Weed, Quarantine Weed,
Naturalised, Casual Alien
♦ 42, 76, 86, 88, 98, 101, 198, 203, 220,
272
♦ aH, aqua, herbal. Origin: Europe.

Polygonum pensylvanicum L.
Polygonaceae
Persicaria pensylvanica (L.) M.Gómez
(see)
♦ Pennsylvania smartweed,
Pennsylvania knotweed, purple weed,
swamp persicary, glandular persicary,
purple head, pink weed, hearts ease
♦ Weed, Quarantine Weed, Noxious
Weed, Native Weed
♦ 23, 49, 68, 76, 87, 88, 161, 174, 203,
207, 210, 211, 218, 220, 229, 243
♦ aH, cultivated, herbal. Origin: North
America.

Polygonum perfoliatum L.
Polygonaceae
Ampelygonum perfoliatum (L.) Roberty
& Vautier, *Echinocaulon perfoliatum* (L.)
Meisn. ex Hassk., *Persicaria perfoliata*
(L.) H.Gross (see), *Tracaulon perfoliatum*
(L.) Greene, *Truellum perfoliatum* (L.)
Soják
♦ tearthumb, mile a minute vine,
devil's tail tearthumb, giant climbing
tearthumb, som khom khiang, Asiatic
tearthumb
♦ Weed, Quarantine Weed, Noxious
Weed, Naturalised, Environmental
Weed
♦ 17, 67, 76, 80, 87, 88, 101, 103, 105,
123, 129, 133, 151, 161, 195, 211, 220,
224, 229, 238, 243, 263, 273, 274, 275,
297
♦ a/pC, cultivated, herbal.

Polygonum persicaria L.
Polygonaceae
= *Persicaria maculosa* S.F.Gray
♦ lady's thumb, redshank,
willowweed, common persicaria,
persicaria, spotted smartweed,
heartweed, spotted knotweed, red
shank, lovers pride
♦ Weed, Noxious Weed, Naturalised
♦ 15, 23, 34, 39, 44, 52, 68, 70, 80, 87, 88,
94, 101, 102, 118, 121, 136, 159, 161, 162,
165, 179, 208, 210, 212, 217, 218, 229,
235, 236, 237, 241, 245, 253, 255, 270,
272, 275, 280, 295, 300
♦ aH, aqua, cultivated, herbal, toxic.
Origin: Eurasia.

Polygonum persicarioides Kunth
Polygonaceae
= *Polygonum hydropiperoides* Michx.
♦ Weed
♦ 157
♦ pH.

Polygonum plebeium R.Br.
Polygonaceae
♦ small knotweed
♦ Weed, Naturalised
♦ 86, 87, 88, 101, 185, 221, 235, 273, 274,
275, 280, 297
♦ aH, herbal.

Polygonum polycnemoides Jaub. & Spach
Polygonaceae
- manyleg knotweed
- Naturalised
- 101

Polygonum polystachyum Wall. ex Meissn.
Polygonaceae
- Indian knotweed, Himalayan knotweed, cultivated knotweed
- Weed, Noxious Weed, Naturalised
- 34, 35, 80, 101, 146, 156, 165, 229, 280
- pH, cultivated, herbal. Origin: temperate Himalayas.

Polygonum prostratum R.Br.
Polygonaceae
= *Persicaria prostrata* (R.Br.) Soják
- Naturalised
- 280
- cultivated.

Polygonum pubescens Bl.
Polygonaceae
- phak phai naam
- Weed
- 87, 88, 235, 238
- pH, promoted, herbal.

Polygonum punctatum Ell.
Polygonaceae
Polygonum acre H.B.K. (see), *Persicaria punctacta* Small
- dotted smartweed, common water smartweed, punctate smartweed
- Weed, Quarantine Weed, Naturalised
- 14, 23, 39, 76, 86, 87, 88, 203, 208, 218, 220, 237, 245, 280, 295
- pH, aqua, cultivated, herbal, toxic.

Polygonum ramosissimum Michx.
Polygonaceae
- bushy knotweed, tall knotweed
- Weed, Naturalised, Native Weed
- 34, 161, 174, 287
- aH, herbal. Origin: North America.

Polygonum robustius (Small) Fern.
Polygonaceae
- stout smartweed
- Weed
- 23, 88

Polygonum rurivagum Jordan ex Boreau
Polygonaceae
= *Polygonum bellardii* All.
- pikkupihatatar, cornfield knotgrass
- Casual Alien
- 42
- herbal.

Polygonum sachalinense Schmidt.
Polygonaceae
= *Fallopia sachalinensis* (Schmidt) Ronse Decr.
- giant knotweed, spreading knotweed, sakhalin knotweed
- Weed, Noxious Weed, Naturalised, Cultivation Escape
- 34, 35, 45, 80, 87, 88, 101, 102, 156, 195, 218, 229, 263
- pH, cultivated, herbal.

Polygonum sagittatum L.
Polygonaceae
Polygonum sagittatum L. var. *gracilentum* Fern.
- arrowleaf tearthumb, false buckwheat
- Weed
- 23, 88
- aH, promoted, herbal.

Polygonum sagittifolium Levl. & Vant.
Polygonaceae
- Weed
- 275
- pH.

Polygonum salicifolium Brouss.
Polygonaceae
Persicaria decipiens (R.Br.) K.L.Wilson (see), *Persicaria salicifolia* (Brouss. ex Willd.) Assenov *nom. illeg.*, *Polygonum decipiens* R.Br. (see)
- knotweed, willow leaved knotweed, lumpululu, snakeroot
- Weed
- 87, 88, 158, 185, 221
- aH, cultivated, herbal. Origin: Africa, Eurasia, Australasia.

Polygonum scabrum Moench
Polygonaceae
= *Polygonum lapathifolium* L.
- green smartweed
- Weed, Quarantine Weed
- 52, 76, 87, 88, 203, 218, 220, 243, 263
- aH, herbal.

Polygonum scandens L.
Polygonaceae
Bilderdykia scandens (L.) Greene
- hedge smartweed, climbing false buckwheat
- Weed
- 23, 87, 88, 161, 218
- herbal.

Polygonum segetum Kunth
Polygonaceae
- field smartweed
- Weed, Quarantine Weed, Introduced
- 76, 87, 88, 203, 220, 261
- Origin: tropical America.

Polygonum senegalense Meisn.
Polygonaceae
- snakeroot, mushomorangoko
- Weed, Native Weed
- 87, 88, 121, 185, 221
- pH. Origin: Madagascar.

Polygonum senticosum (Meissn.) Fr. & Sav.
Polygonaceae
- mamakono shirinugui
- Weed, Quarantine Weed
- 76, 87, 88, 203, 220, 235, 297

Polygonum serrulatum Lag.
Polygonaceae
- Weed
- 87, 88

Polygonum setaceum Baldw.
Polygonaceae
- bristly smartweed, bog smartweed
- Weed

- 87, 88, 218
- herbal.

Polygonum sibiricum Laxm.
Polygonaceae
- dock weed, buckwheat
- Weed
- 88, 114, 275, 297
- pH, promoted.

Polygonum sieboldii Meisn.
Polygonaceae
= *Persicaria sieboldii* (Meisn.) Ohwi
- Weed
- 87, 88, 297

Polygonum sieboldii Meisn. var. aestivum Ohwi
Polygonaceae
- Weed
- 263
- aH.

Polygonum stelligerum Cham.
Polygonaceae
Polygonum bonaerense Speg.
- lirio de agua
- Weed
- 237, 295

Polygonum strigosum R.Br.
Polygonaceae
- rough knotweed
- Weed, Naturalised
- 275, 276, 280
- pH, herbal, toxic.

Polygonum taipaishanense H.W.Kung
Polygonaceae
- Quarantine Weed
- 220

Polygonum taquettii Lev.
Polygonaceae
Polygonum minutulum Makino
- Weed
- 87, 88
- aH, promoted.

Polygonum tenuiflorum Presl
Polygonaceae
= *Polygonum lapathifolium* L.
- Weed
- 23, 88

Polygonum thunbergii Sieb. & Zucc.
Polygonaceae
- mizosoba
- Weed, Quarantine Weed
- 76, 87, 88, 203, 220, 263, 297
- aH, promoted. Origin: east Asia.

Polygonum tomentosum Willd.
Polygonaceae
= *Polygonum lapathifolium* L.
- knotweed, phak phai nam
- Weed
- 88, 94, 209, 239, 243, 262
- aqua, herbal.

Polygonum trigonocarpum (Makino) Kudô & Masam.
Polygonaceae
Polygonum minus Huds. fo. *trigonocarpum* Makino
- Weed
- 235

Polygonum virginianum **L.**
Polygonaceae
♦ jumpseed
♦ Weed
♦ 23, 88
♦ pH, cultivated, herbal.

Polygonum viscosum **Ham.**
Polygonaceae
= *Persicaria viscosa* (Buch.-Ham. ex D.Don) H.Gross ex Nakai
♦ nioitade
♦ Weed
♦ 87, 88, 275, 297
♦ aH.

Polymeria longifolia **Lindl.**
Convolvulaceae
♦ polymeria, Peak Downs curse, polymeria take all, take all, clumped bindweed, erect bindweed
♦ Weed, Native Weed
♦ 55, 87, 88, 269
♦ pH, cultivated. Origin: Australia.

Polymeria pusilla **R.Br.**
Convolvulaceae
♦ Weed
♦ 55
♦ Origin: Australia.

Polymnia sonchifolia **Poepp. & Endl.**
Asteraceae
♦ apple of the earth, jicama, yacón
♦ Introduced
♦ 228
♦ arid, cultivated.

Polypodium scolopendria **Burm.f.**
Polypodiaceae
= *Phymatosorus scolopendria* (Burm.f.) Pic.Serm.
♦ kitieu, wart fern
♦ Introduced
♦ 261
♦ H, cultivated, herbal. Origin: southeast Asia.

Polypodium vulgare **L.**
Polypodiaceae
♦ wall fern, kallioimarre, polypody
♦ Weed
♦ 39, 87, 88, 272
♦ pH, cultivated, herbal, toxic.

Polypogon australis **Brongn.**
Poaceae
♦ Chilean rabbit's foot grass, Chilean beardgrass, Chilean polypogon
♦ Weed, Naturalised
♦ 34, 87, 88, 101
♦ pG, herbal.

Polypogon chilensis **(Kunth) Pilg.**
Poaceae
♦ Weed, Naturalised
♦ 54, 86, 88
♦ G. Origin: South America.

Polypogon fugax **Nees**
Poaceae
♦ Asia Minor bluegrass
♦ Weed, Naturalised
♦ 15, 88, 101, 235, 275, 280, 286, 297
♦ G.

Polypogon interruptus **Kunth**
Poaceae
♦ ditch polypogon, ditch beardgrass, ditch rabbit's foot grass
♦ Weed, Naturalised
♦ 87, 88, 101, 161, 199, 218
♦ pG, arid, herbal.

Polypogon maritimus **Willd.**
Poaceae
♦ Mediterranean beardgrass, Mediterranean polypogon, coastal beardgrass, Mediterranean rabbit's foot grass, merisukaheinä
♦ Weed, Naturalised, Casual Alien
♦ 7, 42, 98, 101, 176, 203, 241, 272, 300
♦ aG, cultivated, herbal.

Polypogon maritimus **Willd. var. subspathaceus (Req.) Bonnier & Layens**
Poaceae
♦ coast beardgrass
♦ Naturalised, Environmental Weed
♦ 86, 198
♦ G. Origin: Mediterranean, Middle East.

Polypogon monspeliensis **(L.) Desf.**
Poaceae
Agrostis alopecuroides Lam., *Agrostis crinita* (Schreb.) Moench, *Alopecurus aristatus* Gouan var. *monspeliensis* (L.) Huds., *Alopecurus monspeliensis* L., *Phalaris aristata* Gouan ex P.Beauv., *Phalaris crinita* Forssk., *Phalaris cristata* Forssk., *Phleum crinitum* Schreb., *Phleum monspliense* (L.) Koeler, *Polypogon crinitus* (Schreb.) Nutt., *Polypogon flavescens* J.Presl, *Polypogon monspeliensis* fo. *argentinus* Hack., *Polypogon monspeliensis* fo. *nana* Stuck., *Santia monspeliensis* (L.) Parl.
♦ rabbit's foot grass, rabbit's foot polypogon, annual beardgrass, beardgrass, tawny beardgrass, ranskansukaheinä
♦ Weed, Naturalised, Introduced, Environmental Weed, Casual Alien
♦ 7, 9, 15, 23, 34, 42, 68, 70, 72, 86, 87, 88, 91, 93, 98, 101, 121, 158, 161, 176, 179, 180, 185, 198, 200, 203, 205, 212, 218, 228, 241, 243, 253, 269, 272, 274, 275, 280, 286, 300
♦ aG, aqua, cultivated, herbal. Origin: Eurasia.

Polypogon semiverticillatus **(Forssk.) Hyl.**
Poaceae
= *Polypogon viridis* (Gouan) Breistr.
♦ Weed, Naturalised
♦ 88, 241
♦ G, herbal.

Polypogon tenellus **R.Br.**
Poaceae
♦ Naturalised
♦ 7
♦ G.

Polypogon viridis **(Gouan) Breistr.**
Poaceae
Agrostis semiverticillata (Forssk.) Christ., *Agrostis verticillata* Vill. (see),

Agrostis viridis Gouan (see), *Phalaris semiverticillata* (Forssk.) Christ. (see), *Polypogon semiverticillatus* (Forssk.) Hyl. (see)
♦ water bentgrass, water bent, kiehkurasukaheinä, beardless rabbit's foot grass
♦ Weed, Naturalised, Casual Alien
♦ 7, 40, 42, 86, 88, 91, 98, 101, 121, 176, 185, 198, 203, 253, 261, 272, 280, 300
♦ pG, aqua, cultivated. Origin: Eurasia.

Polypremum procumbens **L.**
Loganiaceae
♦ polypremen, juniper leaf, rustweed
♦ Weed
♦ 87, 88, 161, 218, 249
♦ herbal. Origin: Americas.

Polyscias **J.R. & G.Forst. spp.**
Araliaceae
♦ coffee tree, aralia
♦ Weed
♦ 161, 247
♦ toxic.

Polyscias cumingiana **(C.Presl) Fern.-Vill.**
Araliaceae
♦ cubano, polyscias
♦ Naturalised, Introduced
♦ 101, 261
♦ Origin: Pacific islands.

Polyscias fruticosa **(L.) Harms**
Araliaceae
Nothopanax fruticosum (L.) Miq.
♦ ming aralia
♦ Naturalised, Introduced
♦ 101, 230, 261
♦ H, cultivated, herbal. Origin: India to Polynesia.

Polyscias guilfoylei **(W.Bull) L.Bailey**
Araliaceae
♦ geranium aralia, gallego, frosted aralia
♦ Naturalised, Introduced
♦ 101, 261
♦ cultivated. Origin: Polynesia.

Polyscias pinnata **J.R. & G.Forst.**
Araliaceae
= *Polyscias scutellaria* (Burm.f.) Fosberg
♦ Introduced
♦ 230, 261
♦ H, herbal. Origin: Malaysia.

Polyscias sambucifolia **(Sieber ex DC.) Harms**
Araliaceae
♦ elderberry panax
♦ Naturalised, Garden Escape, Environmental Weed
♦ 86, 176
♦ S, cultivated. Origin: Australia.

Polyscias scutellaria **(Burm.f.) Fosberg**
Araliaceae
Nothopanax scutellarium (Burm.f.) Merr., *Polyscias pinnata* J.R. & G.Forst. (see)
♦ shield aralia, tagitagi, geraniumleaf aralia
♦ Naturalised, Introduced

- 101, 230, 261
- cultivated, herbal. Origin: New Caledonia.

Polyscias tricochleata (Miq.) Fosb.
Araliaceae
Nothopanax tricochleatum Miq.
- Weed, Naturalised, Introduced
- 98, 203, 230
- H.

Polysiphonia breviarticulata (C.Agardh) Zanardini
Rhodomelaceae
- red alga
- Weed
- 197
- algae.

Polysiphonia denudata (Dillwyn) Grev. ex Harv.
Rhodomelaceae
Polysiphonia variegata (Agardh) Zanardini, *Polysiphonia divergens* Agardh, *Polysiphonia variegata* (Agardh) Zanardini fo. *divergens* (Agardh) De Toni
- red alga
- Weed
- 197
- algae.

Polysiphonia harveyi Bailey
Rhodomelaceae
Polysiphonia insidiosa P. & H.Crouan
- red alga, algae
- Weed
- 282, 288
- algae.

Polystichum aculeatum (L.) Roth
Dryopteridaceae
Aspidium lobatum (Huds.) Sw.
- hard shield fern, piikkihärkylä
- Weed
- 272
- H, cultivated, herbal.

Polystichum munitum (Kaulf.) Presl
Dryopteridaceae
- western swordfern, swordfern, giant hollyfern
- Weed
- 87, 88, 218
- pH, cultivated, herbal.

Polytrias amaura (Büse) Kuntze *nom. illeg.*
Poaceae
= *Polytrias indica* (Houtt.) Veldkamp (NoR) [see *Eulalia amaura* (Büse ex Miq.) Ohwi, *Polytrias praemorsa* (Nees) Hack.]
- Java grass
- Weed, Naturalised
- 87, 88, 101, 170, 191
- G.

Polytrias praemorsa (Nees) Hack.
Poaceae
= *Polytrias indica* (Houtt.) Veldkamp (NoR) [see *Eulalia amaura* (Büse ex Miq.) Ohwi, *Polytrias amaura* (Büse) Kuntze *nom. illeg.*]
- grama de Java, yerba de Java, Javanese grass

- Cultivation Escape
- 261
- G, cultivated. Origin: Java.

Polyzygus tuberosus Dalz.
Apiaceae
- Weed
- 87, 88

Pomaderris aspera DC.
Rhamnaceae
- hazel pomaderris
- Weed, Naturalised
- 15, 280
- cultivated. Origin: Australia.

Poncirus trifoliata (L.) Raf.
Rutaceae
Aegle sepiaria DC., *Citrus trifoliata* L., *Pseudaegle trifoliata* Makino, *Limonia trichocarpa* Hance, *Limonia trifoliata* L.
- Japanese orange, hardy orange, bitter orange, trifoliate orange
- Weed, Naturalised, Environmental Weed
- 22, 39, 80, 88, 101, 151, 161, 247, 287
- S, cultivated, herbal, toxic.

Pongamia pinnata (L.) Pierre
Fabaceae/Papilionaceae
= *Millettia pinnata* (L.) Panigrahi (NoR) [see *Derris indica* (Lam.) Bennet]
- Indian beech, milletia pinnata
- Weed, Introduced, Garden Escape
- 22, 179, 228, 261
- T, arid, cultivated, herbal. Origin: tropical Asia.

Pontederia cordata L.
Pontederiaceae
Pontederia lanceolata Nutt. (see)
- pickerelweed, pontederia
- Weed, Quarantine Weed, Noxious Weed, Naturalised, Garden Escape, Environmental Weed
- 7, 23, 39, 40, 63, 72, 87, 88, 98, 161, 198, 200, 203, 218, 255, 258, 283, 295
- wpH, cultivated, herbal, toxic. Origin: Americas.

Pontederia cordata L. var. *cordata*
Pontederiaceae
- pickerelweed
- Naturalised, Environmental Weed
- 86
- cultivated. Origin: Americas.

Pontederia lanceolata Nutt.
Pontederiaceae
= *Pontederia cordata* L.
- pickerelweed
- Weed
- 14
- aqua, cultivated.

Pontederia rotundifolia L.f.
Pontederiaceae
- tropical pickerelweed, rainha dos lagos
- Weed, Quarantine Weed, Noxious Weed, Naturalised
- 76, 86, 88, 203, 220, 229, 255, 295
- pH, aqua. Origin: Central America.

Populus L. spp.
Salicaceae
- poplar, aspen, cottonwood

- Weed, Quarantine Weed, Naturalised
- 39, 88, 198, 203
- herbal, toxic.

Populus alba L.
Salicaceae
Populus nivea hort.
- white poplar, silver leaved poplar, abele, silver poplar, poplar
- Weed, Noxious Weed, Naturalised, Introduced, Garden Escape, Environmental Weed, Cultivation Escape
- 4, 7, 15, 34, 42, 63, 70, 72, 80, 86, 87, 88, 95, 98, 101, 102, 104, 121, 129, 133, 142, 146, 151, 155, 165, 195, 198, 203, 218, 222, 224, 225, 228, 246, 279, 280, 283, 296
- T, arid, cultivated, herbal. Origin: Eurasia.

Populus alba L. cv. 'Nivea'
Salicaceae
- white poplar
- Naturalised
- 280
- T, cultivated, herbal. Origin: horticultral.

Populus alba L. cv. 'Pyramidalis'
Salicaceae
Populus alba var. *croatica* L., *Populus alba* var. *bolleana* L.
- white poplar
- Naturalised
- 280
- T, cultivated, herbal. Origin: horticultral.

Populus balsamifera L.
Salicaceae
- balsam poplar, balm of Gilead, tacamahac, eastern balsam poplar, palsamipoppeli
- Weed, Cultivation Escape
- 39, 42, 87, 88, 218
- T, cultivated, herbal, toxic.

Populus × *berolinensis* (C.Koch) Dipp.
Salicaceae
= *Populus laurifolia* Ledeb. × *Populus nigra* L. cv. 'Italica'
- berliininpoppeli
- Cultivation Escape
- 42
- cultivated.

Populus × *canadensis* Moench
Salicaceae
= *Populus deltoides* Marsh. × *Populus nigra* L.
- Carolina poplar, Italian poplar, poplar, Canadian poplar
- Weed, Naturalised, Environmental Weed
- 15, 88, 151, 280
- T, cultivated, herbal. Origin: North America.

Populus × *canescens* (Aiton) J.E.Sm.
Salicaceae
= *Populus alba* L. × *Populus tremula* L. [*Populus hybrida* M.Bieb.]
- grey poplar, poplar

♦ Weed, Quarantine Weed, Noxious Weed, Naturalised, Environmental Weed, Cultivation Escape
♦ 63, 70, 76, 80, 86, 88, 95, 98, 101, 121, 151, 158, 198, 203, 277, 279, 280, 283
♦ T, cultivated, herbal. Origin: Eurasia.

Populus deltoides Marsh.
Salicaceae
♦ eastern cottonwood, cottonwood, broad leaved poplar, Carolina poplar, match poplar, necklace poplar, vuurhoutjiepolpulier
♦ Weed, Naturalised, Environmental Weed, Cultivation Escape
♦ 63, 87, 88, 121, 211, 218, 243, 279, 280, 283
♦ T, cultivated, herbal. Origin: southeast North America.

Populus deltoides Bartram ex Marshall ssp. *wislizenii* (Watson) Eckenw.
Salicaceae
Populus wislizenii Sarg. (see)
♦ Rio Grande cottonwood
♦ Weed
♦ 218

Populus euphratica Olivier
Salicaceae
♦ Weed, Introduced
♦ 221, 228
♦ T, arid, cultivated, herbal.

Populus fremontii S.Wats.
Salicaceae
♦ Fremont cottonwood, cottonwood
♦ Weed
♦ 87, 88, 218
♦ T, cultivated, herbal.

Populus × gileadensis Rouleau
Salicaceae
= *Populus deltoides* W.Bartram ex Marshall × *Populus balsamifera* L. [most probable combination]
♦ Naturalised
♦ 280

Populus grandidentata Michx.
Salicaceae
♦ bigtooth aspen, Canadian aspen
♦ Weed
♦ 87, 88, 195, 218
♦ T, cultivated, herbal.

Populus heterophylla L.
Salicaceae
♦ swamp cottonwood
♦ Weed
♦ 87, 88, 218
♦ T, cultivated, herbal.

Populus laurifolia Ledeb.
Salicaceae
♦ laakeripoppeli
♦ Cultivation Escape
♦ 42
♦ cultivated.

Populus × jackii Sarg.
Salicaceae
= *Populus deltoides* Marsh. × *Populus balsamifera* L.
♦ balm of Gilead
♦ Weed, Naturalised
♦ 54, 86, 88

♦ T, promoted. Origin: obscure.

Populus nigra L.
Salicaceae
Populus nigra L. var. *italica* Du Roi (see)
♦ black poplar, Lombardy poplar, black cherry
♦ Weed, Naturalised, Introduced, Garden Escape, Environmental Weed, Cultivation Escape
♦ 7, 42, 72, 80, 86, 87, 88, 98, 101, 151, 155, 198, 203, 228, 280
♦ T, arid, cultivated, herbal.

Populus nigra L. cv. 'Italica'
Salicaceae
Populus nigra L. var. *italica* Du Roi (see), *Populus nigra* L. var. *pyramidalis* (Rozier) Spach
♦ Lombardy poplar, black cherry
♦ Naturalised, Environmental Weed
♦ 280, 296
♦ T, cultivated, herbal. Origin: Italy.

Populus nigra L. var. *italica* Du Roi
Salicaceae
= *Populus nigra* L. cv. 'Italica'
♦ black poplar, Lombardy poplar
♦ Weed, Naturalised, Cultivation Escape
♦ 4, 88, 121, 218, 279, 283, 287
♦ T, cultivated. Origin: Eurasia.

Populus × petrowskiana R.I.Schröd. ex Regel
Salicaceae
♦ tsaarinpoppeli
♦ Cultivation Escape
♦ 42
♦ cultivated.

Populus × rasumowskiana (R.I.Schröd. ex Regel) Dippel
Salicaceae
♦ ruhtinaanpoppeli
♦ Cultivation Escape
♦ 42
♦ cultivated.

Populus sargentii Dode
Salicaceae
= *Populus deltoides* Bartr. ex Marsh. ssp. *monilifera* (Ait.) Eckenw. (NoR)
♦ plains cottonwood
♦ Weed
♦ 87, 88, 218
♦ herbal.

Populus simoni Carrière
Salicaceae
♦ Introduced
♦ 228
♦ arid.

Populus tomentosa Carrière
Salicaceae
♦ Chinese white poplar
♦ Naturalised
♦ 101

Populus tremula L.
Salicaceae
♦ aspen, European aspen
♦ Weed, Naturalised
♦ 54, 70, 80, 86, 88, 101, 198, 280
♦ T, cultivated, herbal. Origin: Europe, North Africa to China.

Populus tremula L. × tremuloides Michx.
Salicaceae
♦ hybridihaapa
♦ Cultivation Escape
♦ 42
♦ cultivated.

Populus tremuloides Michx.
Salicaceae
Populus trepida Willd.
♦ trembling aspen, quaking aspen, poplar, American aspen
♦ Weed
♦ 80, 87, 88, 195, 218
♦ T, arid, cultivated, herbal.

Populus trichocarpa Torr. & Gray
Salicaceae
Populus balsamifera L. ssp. *trichocarpa* (Torr. & A.Gray) Brayshaw
♦ black cottonwood, western balsam poplar, balsam poplar
♦ Weed, Quarantine Weed, Naturalised
♦ 76, 87, 88, 218, 280
♦ T, cultivated, herbal.

Populus tristis Fisch.
Salicaceae
♦ tummapoppeli
♦ Cultivation Escape
♦ 42
♦ cultivated.

Populus wislizenii Sarg.
Salicaceae
= *Populus deltoides* Bartram ex Marshall ssp. *wislizenii* (Watson) Eckenw.
♦ poplar, valley cottonwood
♦ Weed
♦ 121
♦ T, herbal. Origin: North America.

Populus yunnanensis Dode
Salicaceae
♦ Yunnan poplar, Chinese rustfree poplar
♦ Weed, Naturalised
♦ 15, 280
♦ cultivated.

Porana paniculata Roxb.
Convolvulaceae
= *Poranopsis paniculata* (Roxb.) Roberty
♦ white corallita
♦ Weed
♦ 179
♦ herbal.

Porana volubilis Burm.f.
Convolvulaceae
♦ Weed
♦ 13
♦ herbal.

Poranopsis paniculata (Roxb.) Roberty
Convolvulaceae
Porana paniculata Roxb. (see)
♦ bridal bouquet, velo de novia
♦ Naturalised, Garden Escape
♦ 101, 261
♦ cultivated. Origin: China, India, Pakistan.

Porlieria angustifolia (Engelm.) Gray
Zygophyllaceae

= *Guajacum angustifolium* Engelm.
(NoR)
♦ guayacan
♦ Weed
♦ 87, 88, 218
♦ herbal.

Porophyllum lanceolatum DC.
Asteraceae
♦ yerba del venado
♦ Weed
♦ 87, 88, 237, 295

Porophyllum panctatum Blake
Asteraceae
♦ Weed
♦ 87, 88

Porophyllum ruderale (Jacq.) Cass.
Asteraceae
Porophyllum ellipticum Cass.
♦ papalo, papaloquelite, yerba porosa
♦ Weed, Quarantine Weed,
Naturalised, Environmental Weed
♦ 76, 87, 88, 170, 191, 220, 255, 257, 270,
300
♦ aH, promoted, herbal. Origin: Brazil.

Porphyra C.Agardh spp.
Lamiaceae/Verbenaceae
♦ nori, red alga
♦ Weed
♦ 197
♦ algae, herbal.

Portulaca albiflora hort.
Portulacaceae
♦ Weed
♦ 262
♦ pH.

Portulaca amilis Speg.
Portulacaceae
♦ Paraguayan purslane, broadleaf
pink purslane
♦ Weed, Naturalised
♦ 101, 161, 179, 249

Portulaca australis Endl.
Portulacaceae
♦ Introduced
♦ 39, 230
♦ H, toxic. Origin: Australia.

Portulaca cryptopetala Speg.
Portulacaceae
♦ Weed
♦ 87, 88

Portulaca formosana Poelln.
Portulacaceae
♦ Weed, Quarantine Weed
♦ 76, 87, 88, 203, 220

Portulaca grandiflora Hook.
Portulacaceae
♦ rose moss, Don Diego gigante
♦ Weed, Naturalised, Cultivation
Escape, Casual Alien
♦ 86, 87, 88, 98, 101, 179, 203, 237, 261,
280, 295
♦ aH, cultivated, herbal. Origin: South
America.

**Portulaca lanceolata Engelm. ex A.Gray
nom. illeg.**
Portulacaceae
= *Portulaca coronata* Small (NoR)

♦ Weed
♦ 87, 88

Portulaca oleracea L.
Portulacaceae
Portulaca officinarum Crantz
♦ purselane, wild protulaca, pussley,
pursley, duckweed, akulikuli kula,
phak bia yai, green purslane, little
hogweed, pigweed, verdolaga, aiyo,
beldroega, beldroeginha de ovelha,
chingongo, common purslane,
ekalitete, gatamatonga, graviol,
jiabara, kalunda, kulfe ka sag, lunak,
luni, lyawa, mama luni, marare,
masonde, mataga atsanu, natola,
nkhotchwe, nunar, ojipa, onjoroleo,
paxlac, pourpier, pusky, rigla, salunak,
salunak, serepe, siachamubili,
sirrussirso, ssezzira, tebere, titiuhuire,
tsotsope, verdolaga, xukul
♦ Weed, Noxious Weed, Naturalised,
Native Weed, Introduced, Garden
Escape
♦ 6, 7, 12, 14, 15, 23, 24, 26, 34, 39, 40,
44, 49, 50, 51, 52, 53, 55, 66, 68, 70, 80,
86, 87, 88, 94, 111, 115, 118, 121, 134,
136, 153, 157, 158, 161, 165, 170, 174,
179, 180, 185, 186, 204, 206, 207, 210,
211, 212, 218, 221, 228, 229, 235, 236,
237, 238, 239, 240, 241, 242, 243, 245,
249, 250, 253, 255, 257, 261, 263, 269,
270, 272, 273, 274, 275, 276, 280, 281,
286, 293, 295, 297, 299, 300
♦ aH, arid, cultivated, herbal, toxic.
Origin: Eurasia.

**Portulaca oleracea L. var. granulato-
stellulata (Poelln.) A.Danin & H.G.Baker**
Portulacaceae
♦ Introduced
♦ 230
♦ H.

Portulaca pilosa L.
Portulacaceae
♦ kiss me quick
♦ Weed, Naturalised, Environmental
Weed
♦ 14, 86, 87, 88, 93, 98, 157, 203, 205,
235, 269, 286, 287
♦ aH, arid, cultivated, herbal, toxic.
Origin: tropics.

**Portulaca pilosa L. ssp. grandiflora
(Hook.) Geesink**
Portulacaceae
Portulaca caryophylloides hort. ex Vilm.,
Portulaca gilliesii Engelm., *Portulaca
hilaireana* G.Don, *Portulaca immerso-
stellata* Poelln., *Portulaca megalantha*
Steud., *Portulaca mendocinensis* Gill.
ex Rohrb., *Portulaca multistaminata*
Poelln., *Portulaca pilosa* L. var.
grandiflora Kuntze, *Portulaca pilosa* L.
var. *microphylla* Kuntze, *Portulaca pilosa*
L. ssp. *cisplatina* D.Legrand
♦ common rose moss, lunia, sun plant
♦ Introduced
♦ 228

Portulaca quadrifida L.
Portulacaceae

Portulaca foliosa Bak.f., *Portulaca
geniculata* Royle, *Portulaca meridiana* L.,
Portulaca quadrifida L. var. *tormosana*
Hayata, *Portulaca walteriana* Poelln.
♦ pusley, wild purslane, chicken weed,
bordinte, bwanda, chota, lunkibuti,
melkhena, nunya, nunyka sag,
purslane, sanimarumbi
♦ Weed, Quarantine Weed,
Naturalised
♦ 76, 86, 87, 88, 101, 121, 158, 203, 220,
242
♦ pH, arid, cultivated, herbal.

Portulaca umbraticola Kunth
Portulacaceae
Portulaca denudata Poelln., *Portulaca
lanceolata* Engelm., *Portulaca plano-
operculata* Kuntze
♦ wingpod purslane
♦ Weed
♦ 237, 295
♦ herbal.

Portulacaria afra (L.) Jacq.
Portulacaceae
♦ elephant's food, gya nese,
iGwanishe, inDibili enkula,
isAmbilane, isiCococo, isiDonwane,
sala ni marumbi, spekboom, elephant
bush, jade plant
♦ Introduced
♦ 228
♦ arid, cultivated.

Posidonia oceania (L.) Delile
Posidoniaceae
♦ Mediterranean tapeweed
♦ Weed, Naturalised
♦ 101, 183, 221
♦ herbal.

Potamogeton L. spp.
Potamogetonaceae
♦ pondweed
♦ Weed, Sleeper Weed, Quarantine
Weed
♦ 10, 23, 88, 191, 204, 208, 221, 225, 237,
258, 272
♦ aqua, cultivated.

**Potamogeton acutifolius Link ex Roem.
& Schult.**
Potamogetonaceae
♦ sharp leaved pondweed
♦ Weed, Naturalised
♦ 98, 203
♦ cultivated. Origin: Europe.

Potamogeton alpinus Balb.
Potamogetonaceae
Potamogeton rufescens Schrad.
♦ red pondweed, alpine pondweed
♦ Weed
♦ 272
♦ cultivated, herbal.

Potamogeton amplifolius Tuckerm.
Potamogetonaceae
♦ largeleaf pondweed, broad leaved
pondweed
♦ Weed
♦ 87, 88, 218
♦ pH, aqua, herbal.

Potamogeton berteroanus Phil.
Potamogetonaceae
Potamogeton pusillus auct. arg. non L.
♦ Weed
♦ 295

Potamogeton cheesemanii A.Benn.
Potamogetonaceae
♦ pondweed
♦ Weed
♦ 208
♦ pH, aqua, promoted.

Potamogeton crispus L.
Potamogetonaceae
Potamogeton serratus Huds.
♦ curlyleaf pondweed, curled
pondweed, poimuvita, crisp leaved
pondweed, poimuvita
♦ Weed, Sleeper Weed, Quarantine
Weed, Naturalised, Native Weed,
Introduced, Environmental Weed
♦ 4, 15, 34, 35, 70, 76, 78, 80, 86, 87, 88,
101, 102, 104, 116, 121, 133, 161, 165,
195, 197, 208, 217, 218, 220, 221, 222,
224, 225, 246, 272, 274, 275, 280, 286,
292, 297
♦ wpH, cultivated, herbal. Origin:
Eurasia, Africa, Australia, North
America.

Potamogeton cristatus Regel & Maack
Potamogetonaceae
♦ kobanohirumushiro
♦ Weed
♦ 87, 88, 274, 275, 286, 297
♦ pH, aqua, promoted. Origin: east
Asia.

Potamogeton distinctus A.Benn.
Potamogetonaceae
♦ hirumushiro
♦ Weed, Quarantine Weed
♦ 76, 87, 88, 203, 204, 220, 263, 274, 275,
286, 297
♦ pH, aqua, promoted. Origin: east
Asia.

Potamogeton diversifolius Raf.
Potamogetonaceae
♦ waterthread pondweed, diverse
leaved pondweed
♦ Weed, Quarantine Weed
♦ 76, 87, 88, 203, 218, 220
♦ wpH, herbal.

Potamogeton epihydrus Raf.
Potamogetonaceae
♦ ribbonleaf pondweed, American
pondweed
♦ Weed
♦ 87, 88, 218
♦ herbal.

Potamogeton ferrugineus Hagstr.
Potamogetonaceae
♦ Weed
♦ 295
♦ aqua.

Potamogeton filiformis Pers.
Potamogetonaceae
= *Stuckenia filiformis* (Pers.) Börner ssp.
filiformis (NoR)
♦ fineleaf pondweed, leafy pondweed,
slender leaved pondweed, merivita

♦ Weed
♦ 87, 88, 218
♦ pH, aqua, herbal.

Potamogeton fluitans Roth
Potamogetonaceae
= *Potamogeton nodosus* Poir.
♦ Weed
♦ 87, 88

Potamogeton foliosus Raf.
Potamogetonaceae
♦ leafy pondweed
♦ Weed, Quarantine Weed
♦ 76, 87, 88, 161, 203, 218, 220
♦ wpH, herbal.

Potamogeton gayii A.Benn.
Potamogetonaceae
♦ Weed, Quarantine Weed
♦ 76, 88, 220, 295

Potamogeton gramineus L.
Potamogetonaceae
♦ pondweed, variableleaf pondweed,
heinävita
♦ Weed
♦ 159, 272
♦ pH, aqua, cultivated, herbal.

Potamogeton illinoensis Morong
Potamogetonaceae
♦ Illinois pondweed, shining
pondweed
♦ Weed, Quarantine Weed
♦ 87, 88, 218, 258
♦ pH, aqua, herbal.

Potamogeton indicus Roxb.
Potamogetonaceae
♦ Weed
♦ 87, 88

Potamogeton lucens L.
Potamogetonaceae
♦ shining pondweed, välkevita
♦ Weed
♦ 87, 88, 272, 297
♦ aqua, herbal. Origin: Australia.

Potamogeton malaianus Miq.
Potamogetonaceae
♦ pondweed, nae paak pet, sasabamo
♦ Weed
♦ 239, 275, 286, 297
♦ aqua, cultivated.

Potamogeton natans L.
Potamogetonaceae
♦ floatingleaf pondweed, uistinvita,
broad leaved pondweed, floating
pondweed
♦ Weed, Quarantine Weed,
Naturalised
♦ 70, 76, 86, 87, 88, 159, 203, 217, 218,
220, 272, 275
♦ wpH, cultivated, herbal.

Potamogeton nodosus Poir.
Potamogetonaceae
Potamogeton americanus Chamis. &
Schlech., *Potamogeton fluitans* Roth
(see)
♦ American pondweed, Loddon
pondweed, longleaf pondweed
♦ Weed, Quarantine Weed
♦ 23, 76, 87, 88, 161, 203, 218, 220, 221,

253, 272
♦ wpH, herbal.

Potamogeton ochreatus Raoul
Potamogetonaceae
♦ blunt pondweed
♦ Weed, Quarantine Weed
♦ 76, 87, 88, 220
♦ aqua, cultivated.

Potamogeton octandrus Poir.
Potamogetonaceae
♦ pondweed, hosobamizuhikimo
♦ Weed
♦ 121, 286
♦ wpH, cultivated.

Potamogeton oxyphyllus Miq.
Potamogetonaceae
♦ yanagimo
♦ Weed
♦ 286
♦ pH, aqua, cultivated. Origin: China,
Japan, Korea.

Potamogeton pectinatus L.
Potamogetonaceae
Stuckenia pectinatus (L.) Boerner
♦ sago pondweed, fennel leaved
pondweed, sago pondweed, bushy
pondweed, fennel pondweed, brasca
delle lagune
♦ Weed, Quarantine Weed,
Naturalised
♦ 23, 70, 76, 86, 87, 88, 121, 159, 161,
217, 218, 220, 221, 258, 272, 297
♦ wpH, cultivated, herbal. Origin:
cosmopolitan.

Potamogeton pectinatus L. var. *striatus*
(Ruiz & Pav.) Hagstr.
Potamogetonaceae
Potamogeton striatus Ruiz & Pav. (see)
♦ potamogeton estriado
♦ Weed
♦ 295

Potamogeton perfoliatus L.
Potamogetonaceae
♦ perfoliate pondweed, claspingleaf
pondweed, clasped pondweed,
ahvenvita
♦ Weed, Quarantine Weed,
Environmental Weed
♦ 87, 88, 246, 272, 297
♦ aqua, cultivated, herbal. Origin:
Australia.

Potamogeton polygonifolius Pour.
Potamogetonaceae
Potamogeton oblongus Viv.
♦ bog pondweed, tatarvita,
cinnamonspot pondweed
♦ Weed
♦ 272

Potamogeton praelongus Wulf.
Potamogetonaceae
♦ whitestem pondweed, pitkälehtivita,
white stemmed pondweed, long
stalked pondweed
♦ Weed
♦ 87, 88, 218
♦ pH, aqua, herbal. Origin: Europe,
North America.

Potamogeton pusillus L.
Potamogetonaceae
♦ small pondweed, lesser pondweed, hentovita, potamot fluet
♦ Weed
♦ 87, 88, 121, 161, 218, 272, 274, 286
♦ waH, cultivated, herbal.

Potamogeton richardsonii (A.Benn.) Rydb.
Potamogetonaceae
♦ Richardson pondweed
♦ Weed, Noxious Weed
♦ 87, 88, 161, 218, 299
♦ pH, aqua, herbal.

Potamogeton robbinsii Oakes
Potamogetonaceae
♦ flatleaf pondweed, Robbins' pondweed
♦ Weed
♦ 87, 88, 218
♦ pH, aqua, herbal.

Potamogeton schweinfurthii A.Benn.
Potamogetonaceae
♦ pondweed
♦ Weed, Native Weed
♦ 87, 88, 121
♦ wpH. Origin: southern Africa.

Potamogeton striatus Ruiz & Pav.
Potamogetonaceae
= *Potamogeton pectinatus* L. var. *striatus* (Ruiz & Pav.) Hagstr.
♦ Weed
♦ 87, 88
♦ aqua.

Potamogeton strictifolius A.Benn.
Potamogetonaceae
♦ narrowleaf pondweed
♦ Weed
♦ 87, 88, 218
♦ herbal.

Potamogeton thunbergii Cham. & Schlechtd.
Potamogetonaceae
♦ floating pondweed, pondweed
♦ Native Weed
♦ 121
♦ wpH. Origin: southern Africa.

Potamogeton tricarinatus F.Muell. & A.Benn.
Potamogetonaceae
♦ floating pondweed
♦ Weed, Quarantine Weed, Naturalised
♦ 86, 87, 88, 200, 220
♦ aqua, cultivated.

Potamogeton trichoides Cham. & Schlechtd.
Potamogetonaceae
♦ hairlike pondweed, brasca capillare
♦ Weed
♦ 121
♦ wpH, herbal.

Potamogeton uruguayensis Benn. & Graebn.
Potamogetonaceae
♦ Weed
♦ 295

Potamogeton vaginatus Turcz.
Potamogetonaceae
= *Stuckenia vaginatus* (Turcz.) Holub (NoR)
♦ giant pondweed, tuppivita, sheathed pondweed
♦ Weed
♦ 87, 88, 218
♦ herbal.

Potamogeton zosterifolius Schum.
Potamogetonaceae
Potamogeton zosteraefolius Schum., *Spirillus zosteraefolius* Nieuwl.
♦ flatstem pondweed
♦ Weed
♦ 218

Potamogeton zosteriformis Fern.
Potamogetonaceae
♦ flatstem pondweed, eelgrass pondweed
♦ Weed
♦ 87, 88
♦ aH, aqua, herbal.

Potentilla L. spp.
Rosaceae
♦ fingerkraut, cinquefoil
♦ Weed, Naturalised
♦ 272

Potentilla alba L.
Rosaceae
Potentilla cordata Schrank
♦ white cinquefoil, cinquefoil
♦ Weed, Naturalised
♦ 101, 272
♦ cultivated.

Potentilla amurensis Maxim.
Rosaceae
♦ cinquefoil
♦ Naturalised
♦ 287

Potentilla anglica Laichard.
Rosaceae
♦ creeping cinquefoil, trailing tromentil, English cinquefoil, lännenhanhikki
♦ Weed, Naturalised
♦ 15, 86, 101, 176, 280, 287
♦ cultivated, herbal.

Potentilla anserina L.
Rosaceae
= *Argentina anserina* (L.) Rydb.
♦ silverweed cinquefoil, silverweed
♦ Weed, Naturalised
♦ 23, 44, 70, 86, 87, 88, 94, 98, 114, 161, 176, 198, 203, 218, 241, 243, 272, 291, 297, 300
♦ pH, cultivated, herbal.

Potentilla anserinoides Raoul
Rosaceae
♦ silverweed
♦ Weed, Native Weed
♦ 165

Potentilla argentea L.
Rosaceae
Potentilla tomentosa Gilib.
♦ silvery cinquefoil, hopeahanhikki
♦ Weed, Naturalised

♦ 23, 52, 80, 87, 88, 101, 161, 195, 210, 218, 272, 280, 294
♦ cultivated, herbal.

Potentilla argentea L. var. argentea
Rosaceae
♦ silver cinquefoil
♦ Naturalised
♦ 101

Potentilla argentea L. var. pseudocalabra T.Wolf
Rosaceae
♦ silver cinquefoil
♦ Naturalised
♦ 101

Potentilla arguta Pursh
Rosaceae
♦ white cinquefoil, tall cinquefoil
♦ Weed
♦ 23, 87, 88, 218
♦ cultivated, herbal.

Potentilla atrosanguinea Lodd. ex D.Don
Rosaceae
♦ vesimeloni, ruby cinquefoil, cinquefoil
♦ Cultivation Escape
♦ 42
♦ cultivated.

Potentilla bifurca L.
Rosaceae
♦ kaksiliuskahanhikki
♦ Weed, Naturalised
♦ 42, 87, 88, 272, 297

Potentilla canadensis L.
Rosaceae
♦ common cinquefoil, dwarf cinquefoil
♦ Weed
♦ 23, 87, 88, 161, 218
♦ herbal.

Potentilla canescens Bess.
Rosaceae
= *Potentilla inclinata* Vill.
♦ Weed
♦ 87, 88
♦ cultivated.

Potentilla caulescens L.
Rosaceae
Potentilla sororia Wend.
♦ kalliohanhikki
♦ Cultivation Escape
♦ 42
♦ cultivated.

Potentilla chinensis Ser.
Rosaceae
♦ Chinese cinquefoil
♦ Weed
♦ 87, 88, 275, 286, 297
♦ pH, promoted, herbal.

Potentilla collina Wibel
Rosaceae
Potentilla argentea L. × *verna* Hegi, *Potentilla jaeggiana* Siegfr.
♦ palmleaf cinquefoil
♦ Naturalised
♦ 101
♦ cultivated.

Potentilla crantzii (Crantz) G.Beck ex Fritsch
Rosaceae
= *Potentilla neumanniana* Rchb. (NoR)
[see *Potentilla tabernaemontani* Asch.]
♦ alpine cinquefoil, keväthanhikki
♦ Weed, Casual Alien
♦ 23, 88, 280
♦ cultivated, herbal.

Potentilla cuneata Wall.
Rosaceae
♦ cuneate cinquefoil
♦ Weed
♦ 297

Potentilla discolor Bunge
Rosaceae
♦ Weed
♦ 235, 297
♦ pH, promoted, herbal. Origin: China, Japan, Korea.

Potentilla diversifolia Lehm.
Rosaceae
♦ blueleaf cinquefoil, varileaf cinquefoil
♦ Weed
♦ 87, 88, 218
♦ cultivated, herbal.

Potentilla erecta (L.) Rausch.
Rosaceae
Fragaria tormentilla Crantz, *Potentilla sylvestris* Neck., *Potentilla tormentilla* (Stokes) Neck. (see), *Tormentilla erecta* L
♦ erect cinquefoil, tormentil, common tormentil, rätvänä
♦ Weed, Naturalised, Garden Escape
♦ 70, 101, 142, 272
♦ pH, cultivated, herbal.

Potentilla erecta (L.) Raeusch. ssp. strictissima (Zimmeter) A.J.Richards
Rosaceae
Potentilla strictissima Zimmeter
♦ Cultivation Escape
♦ 40
♦ cultivated.

Potentilla etomentosa Rydb.
Rosaceae
= *Potentilla gracilis* Dougl. ex Hook. var. *fastigiata* (Nutt.) S.Wats. (NoR)
♦ Naturalised
♦ 287

Potentilla flabelliformis Lehm.
Rosaceae
= *Potentilla gracilis* Hook. var. *flabelliformis* (Lehm.) Torr. & A.Gray (NoR)
♦ Weed
♦ 23, 88

Potentilla flagellaris Willd.
Rosaceae
♦ Weed
♦ 275, 297
♦ pH.

Potentilla fragarioides L.
Rosaceae
♦ dewberry cinquefoil
♦ Weed
♦ 297
♦ pH, promoted, herbal.

Potentilla fragarioides L. var. major Maxim.
Rosaceae
♦ kijimushiro
♦ Weed
♦ 286

Potentilla fragiformis Willd. ex Schlecht.
Rosaceae
♦ strawberry cinquefoil
♦ Naturalised
♦ 101
♦ cultivated.

Potentilla freyniana Bornm.
Rosaceae
♦ cinquefoil
♦ Weed
♦ 88, 204, 286

Potentilla fruticosa L.
Rosaceae
Dasiphora fruticosa (L.) Rydb., *Pentaphylloides fruticosa* (L.) O.Schwarz, *Potentilla fruticosa* var. *farreri* Besant, *Potentilla fruticosa* var. *pyrenaica* Willd. ex Schltdl.
♦ shrubby cinquefoil, pensashanhikki, bush cinquefoil
♦ Weed, Cultivation Escape
♦ 23, 42, 87, 88, 218
♦ S, cultivated, herbal.

Potentilla glandulosa Lindl.
Rosaceae
♦ sticky cinquefoil, common cinquefoil, gland cinquefoil
♦ Weed
♦ 23, 88, 136
♦ pH, cultivated, herbal.

Potentilla gracilis Douglas ex Hook.
Rosaceae
♦ slender cinquefoil, grace cinquefoil, north-west cinquefoil
♦ Weed
♦ 136
♦ pH, cultivated, herbal.

Potentilla hirta L.
Rosaceae
♦ hairy cinquefoil
♦ Weed
♦ 272
♦ cultivated, herbal.

Potentilla inclinata Vill.
Rosaceae
Potentilla canescens Bess. (see)
♦ harmaahanhikki, ashy cinquefoil
♦ Naturalised, Casual Alien
♦ 42, 101

Potentilla intermedia L.
Rosaceae
Potentilla diffusa Rchb., *Potentilla visurgina* Weihe
♦ downy cinquefoil, huhtahanhikki, Russian cinquefoil
♦ Weed, Naturalised, Casual Alien
♦ 23, 40, 87, 88, 101, 218
♦ cultivated, herbal.

Potentilla kleiniana Wight & Arn.
Rosaceae
Potentilla anemonefolia Lehm., *Potentilla*

bodinieri H.Lév.
♦ Weed
♦ 87, 88, 204
♦ pH, promoted, herbal.

Potentilla longifolia Willd. ex Schlecht
Rosaceae
♦ longleaf cinquefoil
♦ Weed
♦ 297

Potentilla millegrana Engelm. ex Lehm.
Rosaceae
= *Potentilla rivalis* Nutt. var. *millegrana* (Engelm. ex Lehm.) S.Wats. (NoR)
♦ Weed, Naturalised
♦ 23, 88, 287

Potentilla multifida L.
Rosaceae
♦ liuskahanhikki, staghorn cinquefoil
♦ Naturalised
♦ 42
♦ pH, promoted, herbal.

Potentilla nepalensis Hook.
Rosaceae
♦ Weed
♦ 80
♦ pH, cultivated, herbal.

Potentilla norvegica L.
Rosaceae
Potentilla grossa Douglas ex Hook., *Potentilla monspeliensis* L., *Potentilla norvegica* ssp. *monspeliensis* (L.) Asch. & Graebn.
♦ rough cinquefoil, peltohanhikki, Norwegian cinquefoil, ternate leaved cinquefoil
♦ Weed, Noxious Weed, Environmental Weed, Casual Alien
♦ 23, 40, 52, 87, 88, 161, 162, 210, 218, 294, 299
♦ a/pH, cultivated, herbal.

Potentilla palustris (L.) Scop.
Rosaceae
= *Comarum palustre* L.
♦ purple marshlocks, marsh cinquefoil, kurjenjalka, potentilla
♦ Weed
♦ 23, 88
♦ pH, aqua, cultivated, herbal.

Potentilla pensylvanica L.
Rosaceae
♦ siperianhanhikki, Pennsylvania cinquefoil
♦ Casual Alien
♦ 42
♦ pH, cultivated, herbal.

Potentilla ranunculoides Humb. & Bon.
Rosaceae
♦ Weed
♦ 199

Potentilla recta L.
Rosaceae
Potentilla corymbosa Moench, *Potentilla sulphurea* Lam. & DC.
♦ erect cinquefoil, sulphur cinquefoil, upright cinquefoil, roughfruit cinquefoil, Austrian fieldcress
♦ Weed, Noxious Weed, Naturalised,

Introduced, Environmental Weed,
Casual Alien
♦ 1, 23, 34, 40, 42, 52, 80, 86, 87, 88, 98,
101, 136, 139, 146, 151, 161, 165, 174,
176, 195, 198, 203, 210, 212, 218, 219,
229, 267, 272, 280, 287, 294
♦ pH, cultivated, herbal. Origin:
eastern Mediterranean.

Potentilla reptans L.
Rosaceae
Fragaria reptans Crantz
♦ creeping cinquefoil, suikerohanhikki
♦ Weed, Naturalised
♦ 44, 70, 86, 87, 88, 94, 98, 101, 176, 179,
198, 203, 218, 253, 272, 280, 300
♦ pH, cultivated, herbal. Origin:
Eurasia, North Africa.

Potentilla reptans L. var. *sericophylla* Franch.
Rosaceae
♦ Weed
♦ 275, 297
♦ pH.

Potentilla rupestris L.
Rosaceae
Fragaria rupestris Crantz
♦ valkohanhikki, rock cinquefoil
♦ Casual Alien
♦ 42
♦ pH, cultivated, herbal.

Potentilla simplex Michx.
Rosaceae
♦ oldfield cinquefoil, old field five
fingers
♦ Weed
♦ 87, 88, 161, 211, 218, 249
♦ pH, promoted, herbal.

Potentilla sundaica Kuntze var. *robusta* Kitag.
Rosaceae
♦ Weed
♦ 286

Potentilla supina L.
Rosaceae
Argentina supina Lam., *Comarum supinum* Alef.
♦ cinquefoil, spreading cinquefoil,
rentohanhikki
♦ Weed, Naturalised, Casual Alien
♦ 42, 87, 88, 94, 98, 185, 203, 221, 272,
275, 286, 287, 297
♦ a/pH, cultivated, herbal. Origin:
Europe, North Africa to China.

Potentilla tabernaemontani Asch.
Rosaceae
= *Potentilla neumanniana* Rchb. (NoR)
[see *Potentilla crantzii* (Crantz) G.Beck
ex Fritsch]
♦ spring cinquefoil
♦ Weed
♦ 23, 88
♦ cultivated, herbal.

Potentilla tanacetifolia Willd. ex Schltdl.
Rosaceae
♦ tansyleaf cinquefoil
♦ Weed
♦ 297

Potentilla thuringiaca Bernh. ex Link
Rosaceae
♦ saksanhanhikki, European
cinquefoil
♦ Naturalised
♦ 42
♦ cultivated.

Potentilla tormentilla (Stokes) Neck.
Rosaceae
= *Potentilla erecta* (L.) Raeusch.
♦ tormentil
♦ Weed
♦ 87, 88
♦ cultivated, herbal.

Potentilla tridentata Aiton
Rosaceae
= *Sibbaldiopsis tridentata* (Aiton) Rydb.
(NoR)
♦ three toothed cinquefoil
♦ Weed
♦ 23, 88
♦ cultivated, herbal.

Poterium magnolii Spach
Rosaceae
♦ Weed
♦ 87, 88

Poterium polygamum Waldst. & Kit.
Rosaceae
= *Sanguisorba minor* Scop. ssp. *muricata*
(Spach) Briq.
♦ Weed, Naturalised
♦ 87, 88, 98, 203, 248

Poterium sanguisorba L.
Rosaceae
= *Sanguisorba minor* Scop. ssp. *minor*
(NoR)
♦ salad burnet
♦ Weed
♦ 87, 88
♦ cultivated, herbal.

Poterium spinosum L.
Rosaceae
♦ Weed
♦ 87, 88, 111, 243

Poterium villosum Sibth. & Sm.
Rosaceae
♦ Weed
♦ 87, 88

Pothomorphe dombeyana Miq.
Piperaceae
= *Lepianthes peltata* (L.) Raf. (NoR) [see
Lepianthes umbellata (L.) Raf., *Piper
umbellatum* L., *Pothomorphe peltata* (L.)
Miq., *Pothomorphe umbellata* (L.) Miq.]
♦ Weed
♦ 87, 88

Pothomorphe peltata (L.) Miq.
Piperaceae
= *Lepianthes peltata* (L.) Raf. (NoR) [see
Lepianthes umbellata (L.) Raf., *Piper
umbellatum* L., *Pothomorphe dombeyana*
Miq., *Pothomorphe umbellata* (L.) Miq.]
♦ Weed, Environmental Weed
♦ 39, 87, 88, 257
♦ pH, cultivated, herbal, toxic.

Pothomorphe umbellata (L.) Miq.
Piperaceae

= *Lepianthes peltata* (L.) Raf. (NoR) [see
Lepianthes umbellata (L.) Raf., *Piper
umbellatum* L., *Pothomorphe dombeyana*
Miq., *Pothomorphe peltata* (L.) Miq.]
♦ pariparoba
♦ Weed
♦ 255
♦ S, herbal. Origin: South America.

Pouteria campechiana L.
Sapotaceae
Lucuma nervosa DC., *Lucuma rivicoa*
Gaertn.f.
♦ canistel, egg fruit tree, yellow sapote
♦ Weed, Naturalised, Environmental
Weed, Cultivation Escape
♦ 3, 80, 88, 101, 122, 151, 179, 191, 261
♦ cultivated. Origin: Mexico to
Panama.

Pouteria dominigensis (Gaertn.f.) Baehni
Sapotaceae
♦ jacana, Dominican pouteria
♦ Weed, Naturalised, Environmental
Weed
♦ 88, 101, 151

Pouteria sapota (Jacq.) H.E.Moore & Stearn
Sapotaceae
♦ naseberry, mamey sapote, sapote,
sapota
♦ Naturalised
♦ 101
♦ cultivated, herbal.

Pouzolzia guineensis Benth.
Urticaceae
♦ Weed
♦ 87, 88

Pouzolzia zeylanica (L.) Benn.
Urticaceae
Pouzolzia indica (L.) Gaudich.
♦ graceful pouzolzsbush
♦ Weed, Naturalised
♦ 12, 87, 88, 101, 179, 235
♦ herbal.

Praecitrullus fistulosus (Stocks) Pangalo
Cucurbitaceae
Citrullus fistulosus Stocks, *Citrullus
lanatus* (Thunb.) Matsum. & Nakai
var. *fistulosus* (Stocks) Duthie & Fuller,
Citrullus vulgaris Schrad. var. *fistulosus*
(Stocks) Stewart, *Colocynthis citrullus*
(L.) Kuntze var. *fistulosus* (Stocks)
Chakrav.
♦ Stock's squash melon, round melon,
squash melon, tinda, tinsi
♦ Introduced
♦ 228
♦ arid.

Prangos trifida (Mill.) Herrnst. & Heyn
Apiaceae
♦ Quarantine Weed
♦ 220
♦ cultivated.

Prasium majus L.
Lamiaceae
♦ Weed
♦ 221

Pratia begoniifolia (Wall.) Lindl.
Campanulaceae/Lobeliaceae
= *Pratia nummularia* (Lam.) A.Braun &
Asch. (NoR)
♦ Weed
♦ 275
♦ pH.

Pratia concolor (R.Br.) Druce
Campanulaceae/Lobeliaceae
♦ poison pratia
♦ Weed, Native Weed
♦ 87, 88, 269
♦ pH, cultivated, toxic. Origin:
Australia.

Pratia pedunculata (R.Br.) Benth.
Campanulaceae/Lobeliaceae
♦ Naturalised
♦ 280
♦ cultivated.

Pratia purpurascens (R.Br.) E.Wimm.
Campanulaceae/Lobeliaceae
♦ Naturalised
♦ 134, 280
♦ cultivated. Origin: Australia.

**Praxelis clematidea (Griseb.) R.M.King
& H.Rob.**
Asteraceae
Eupatorium clematideum Griseb. (see)
♦ praxelis
♦ Weed, Naturalised, Environmental
Weed
♦ 3, 54, 86, 88, 155, 191
♦ aH. Origin: South America.

Premna obtusifolia R.Br.
Lamiaceae/Verbenaceae
= *Premna serratifolia* L.
♦ premna
♦ Weed
♦ 6, 88
♦ herbal. Origin: Madagascar.

Premna odorata Blanco
Lamiaceae/Verbenaceae
♦ fragrant premna
♦ Weed, Naturalised
♦ 101, 179
♦ herbal. Origin: Australia.

Premna resinosa (Hochst.) Schauer
Lamiaceae/Verbenaceae
♦ Weed
♦ 221

Premna serratifolia L.
Lamiaceae/Verbenaceae
Premna obtusifolia R.Br. (see)
♦ aloalo
♦ Weed
♦ 191
♦ cultivated, herbal. Origin: Australia.

Prenanthes purpurea L.
Asteraceae
Chondrilla purpurea Lam.
♦ Weed, Quarantine Weed
♦ 70, 220
♦ cultivated, herbal.

Prenanthes trifoliolata (Cass.) Fern.
Asteraceae
Nabalus trifoliolatus Cass. (see)
♦ gall of the earth

♦ Quarantine Weed
♦ 220
♦ herbal.

Prenia pallens N.E.Br.
Aizoaceae/Mesembryanthemaceae
♦ Quarantine Weed
♦ 220

**Prestonia acutifolia (Benth. ex
Müll.Arg.) Schum.**
Apocynaceae
♦ Weed
♦ 87, 88

Prestonia quinquangularis (Jacq.) Spreng.
Apocynaceae
♦ Weed
♦ 87, 88

Priestleya myrtifolia DC.
Fabaceae/Papilionaceae
♦ Quarantine Weed
♦ 220
♦ cultivated.

Primula L. spp.
Primulaceae
♦ primrose
♦ Weed
♦ 39, 154, 194, 272
♦ herbal, toxic.

Primula anisodora Balf.f. & G.Forrest
Primulaceae
♦ anise primrose, primrose
♦ Naturalised
♦ 101
♦ cultivated.

Primula boveana Decne.
Primulaceae
♦ Weed
♦ 221
♦ cultivated.

Primula elatior (L.) Hill
Primulaceae
♦ etelänkevätesikko, oxlip, polyanthus
♦ Weed, Naturalised
♦ 42, 272
♦ pH, cultivated, herbal.

Primula japonica Gray
Primulaceae
♦ Japanese cowslip, Japanese primrose
♦ Naturalised
♦ 40, 101
♦ cultivated, herbal.

Primula malacoides Franch.
Primulaceae
♦ fairy primrose
♦ Weed, Naturalised
♦ 15, 280
♦ cultivated.

Primula obconica Hance
Primulaceae
♦ German primula
♦ Weed
♦ 39, 161, 247
♦ cultivated, herbal, toxic.

Primula × polyantha Mill.
Primulaceae
= *Primula veris* L. × *Primula vulgaris*
Huds.
♦ elatior hybrid primrose, false oxlip,

polyanthus
♦ Naturalised
♦ 101

Primula scotica Hook.
Primulaceae
♦ Scottish primrose
♦ Weed
♦ 23, 88
♦ cultivated.

Primula veris L.
Primulaceae
Primula officinalis (L.) Hill
♦ cowslip primrose, cowslip,
kevätesikko
♦ Weed, Naturalised
♦ 23, 39, 88, 101, 272
♦ pH, cultivated, herbal, toxic.

Primula vulgaris Huds.
Primulaceae
Primula acaulis (L.) Hill, *Primula veris* L.
var. *acaulis* L.
♦ kääpiöesikko, primrose, English
primrose
♦ Weed, Naturalised, Cultivation
Escape
♦ 42, 70, 280
♦ pH, cultivated, herbal.

Prismatomeris malayana Ridl.
Rubiaceae
♦ Weed
♦ 12
♦ herbal.

Pritchardia pacifica Seem. & Wendl.
Arecaceae
♦ Fiji fan palm
♦ Introduced
♦ 230
♦ T, cultivated.

Priva bahiensis DC.
Verbenaceae
♦ carrapicho
♦ Weed
♦ 255
♦ aH. Origin: Brazil.

Priva cordifolia (L.f.) Druce
Verbenaceae
♦ Weed
♦ 87, 88
♦ herbal.

**Priva cordifolia (L.f.) Druce var.
abyssinica (Jaub. & Spach) Moldenke**
Verbenaceae
♦ Native Weed
♦ 121
♦ pH. Origin: southern Africa.

Priva laevis Juss.
Verbenaceae
= *Pitraea cuneato-ovata* (Cav.) Caro
♦ Weed
♦ 87, 88
♦ T.

Priva lappulacea (L.) Pers.
Verbenaceae
♦ cat's tongue
♦ Weed, Naturalised
♦ 14, 28, 87, 88, 157, 199, 243, 257
♦ a/pH, herbal.

Priva leptostachya Juss.
Verbenaceae
♦ Weed
♦ 87, 88

Priva meyeri Jaub. & Spach var. *meyeri*
Verbenaceae
♦ Native Weed
♦ 121
♦ pH. Origin: southern Africa.

Proboscidea fragrans (Lindl.) Decne.
Pedaliaceae/Martyniaceae
= *Proboscidea louisianica* (Mill.) Thell.
ssp. *fragrans* (Lindl.) Bretting
♦ sweet unicorn plant, ram's horn,
Chihuahua wild devil's claw
♦ Weed, Naturalised
♦ 86, 98, 203, 269
♦ aH, cultivated, herbal. Origin: south-western North America.

Proboscidea jussieui Medik.
Pedaliaceae/Martyniaceae
= *Proboscidea louisianica* (Mill.) Thell.
ssp. *fragrans* (Lindl.) Bretting
♦ Weed
♦ 87, 88

Proboscidea louisianica (Mill.) Thell.
Pedaliaceae/Martyniaceae
Martynia louisiana Mill. (see), *Martynia
louisianica* Mill., *Martynia proboscidea*
Glox., *Proboscidea jussieui* Medik. (see)
♦ unicorn plant, devil's claw,
purpleflower devil's claw, elephant
tusks, goat's head, unicorn plant,
ram's horn, aphid trap
♦ Weed, Quarantine Weed, Noxious
Weed, Naturalised, Native Weed
♦ 1, 7, 39, 62, 76, 80, 86, 87, 88, 98, 146,
147, 161, 174, 179, 191, 198, 203, 210,
218, 229, 243, 269
♦ aH, cultivated, herbal, toxic. Origin:
North America.

**Proboscidea louisianica (Mill.) Thell.
ssp. *fragrans* (Lindl.) Bretting**
Pedaliaceae/Martyniaceae
Proboscidea fragrans (Lindl.) Decne.
(see)
♦ unicorn plant, ram's horn
♦ Noxious Weed
♦ 229
♦ Origin: south-western North
America.

**Proboscidea louisianica (Mill.) Thell.
ssp. *louisianica***
Pedaliaceae/Martyniaceae
♦ Louisiana ram's horn, common
devil's claw, ram's horn, unicorn plant
♦ Noxious Weed, Introduced
♦ 34, 229
♦ aH. Origin: southern North America.

Proboscidea lutea (Lindl.) Stapf
Pedaliaceae/Martyniaceae
= *Ibicella lutea* (Lindl.) Van Eselt.
♦ yellow devil's claw
♦ Weed
♦ 34, 87, 88
♦ aH.

**Proboscidea parviflora (Woot.) Woot. &
Standl.**

Pedaliaceae/Martyniaceae
♦ doubleclaw
♦ Weed, Quarantine Weed
♦ 76, 88
♦ herbal.

**Prosopidastrum globosum (Gillies ex
Hook. & Arn.) Burk.**
Fabaceae/Mimosaceae
♦ Introduced
♦ 228
♦ arid.

Prosopis L. spp.
Fabaceae/Mimosaceae
♦ mesquite
♦ Weed, Quarantine Weed, Noxious
Weed, Naturalised, Garden Escape,
Environmental Weed
♦ 3, 18, 62, 76, 86, 88, 155, 171, 191, 198,
203, 277, 279
♦ cultivated, herbal.

Prosopis affinis Spreng.
Fabaceae/Mimosaceae
Prosopis algarobilla Griseb. (see),
Prosopis algarobilla var. *nandubay*
(Lorentz) Hassl., *Prosopis nandubey*
Lorentz ex Griseb.
♦ Introduced
♦ 228
♦ S/T, arid. Origin: South America.

Prosopis alapataco R.A.Phil.
Fabaceae/Mimosaceae
♦ Weed
♦ 67, 88

Prosopis alba Griseb.
Fabaceae/Mimosaceae
♦ algarrobo blanco
♦ Weed, Quarantine Weed, Introduced
♦ 76, 88, 203, 220, 228, 295
♦ T, arid, cultivated, herbal. Origin:
Argentina, Paraguay.

Prosopis algarobilla Griseb.
Fabaceae/Mimosaceae
= *Prosopis affinis* Spreng.
♦ ñandubay
♦ Weed
♦ 295

Prosopis alpataco Phil.
Fabaceae/Mimosaceae
Prosopis stenoloba Phil.
♦ mesquite
♦ Noxious Weed, Introduced
♦ 140, 228, 229
♦ arid.

Prosopis argentina Burkart
Fabaceae/Mimosaceae
♦ mesquite
♦ Weed, Noxious Weed, Introduced
♦ 67, 88, 140, 228, 229
♦ arid.

Prosopis articulata S.Wats.
Fabaceae/Mimosaceae
= *Prosopis velutina* Woot.
♦ Weed
♦ 67, 88
♦ arid, cultivated.

Prosopis burkartii Muñoz
Fabaceae/Mimosaceae

♦ mesquite
♦ Weed, Noxious Weed
♦ 67, 88, 140, 229

Prosopis caldenia Burkart
Fabaceae/Mimosaceae
Prosopis dulcis Hook.
♦ mesquite, calden
♦ Weed, Noxious Weed, Introduced
♦ 67, 88, 140, 228, 229
♦ arid.

Prosopis calingastana Burkart
Fabaceae/Mimosaceae
♦ mesquite, cusqui
♦ Weed, Noxious Weed
♦ 67, 88, 140, 229

Prosopis campestris Griseb
Fabaceae/Mimosaceae
♦ mesquite
♦ Weed, Noxious Weed
♦ 67, 87, 88, 140, 229

Prosopis castellanosii Burkart
Fabaceae/Mimosaceae
♦ mesquite
♦ Weed, Noxious Weed
♦ 67, 88, 140, 229

Prosopis chilensis (Molina) Stuntz
Fabaceae/Mimosaceae
Ceratonia chilensis Molina, *Prosopis
schinopoma* Stuck., *Prosopis siliquastrum*
(Lag.) DC.
♦ algarrobo
♦ Weed, Quarantine Weed, Introduced
♦ 76, 87, 88, 203, 220, 228
♦ arid, cultivated, herbal.

Prosopis cineraria (L.) Druce
Fabaceae/Mimosaceae
Mimosa cineraria L., *Prosopis spicata*
Burm., *Prosopis spicigera* L. (see)
♦ jand, khejri
♦ Weed, Quarantine Weed
♦ 76, 88, 203, 220
♦ arid, cultivated, herbal.

Prosopis denudans Benth.
Fabaceae/Mimosaceae
♦ mesquite
♦ Weed, Noxious Weed, Introduced
♦ 67, 88, 140, 228, 229
♦ arid.

Prosopis elata (Burkart) Burkart
Fabaceae/Mimosaceae
Prosopis campestris Griseb. var. *elata*
Burkart
♦ mesquite
♦ Weed, Noxious Weed, Introduced
♦ 67, 88, 140, 228, 229
♦ arid.

Prosopis farcta (Banks & Sol.) J.F.Macbr.
Fabaceae/Mimosaceae
Lagonychium farctum (Banks & Sol.)
Bobrov (see), *Mimosa farcta* Sol. ex
Russell, *Prosopis stephaniana* (Bieb.)
Kunth ex Spreng. (see)
♦ mesquite, Syrian mesquite, yanbout
♦ Weed, Noxious Weed, Naturalised
♦ 67, 87, 88, 101, 115, 140, 161, 229
♦ arid, herbal.

Prosopis ferox Griseb.
Fabaceae/Mimosaceae
♦ mesquite
♦ Weed, Noxious Weed, Introduced
♦ 67, 88, 140, 228, 229
♦ arid.

Prosopis fiebrigii Harms
Fabaceae/Mimosaceae
♦ mesquite
♦ Weed, Noxious Weed
♦ 67, 88, 140, 229

Prosopis flexuosa DC.
Fabaceae/Mimosaceae
Acacia flexuosa Lag. *nom. illeg.*
♦ Quilpie mesquite
♦ Weed, Quarantine Weed, Noxious
Weed, Naturalised
♦ 76, 86, 88, 98, 147, 203

Prosopis glandulosa Torr.
Fabaceae/Mimosaceae
Prosopis chilensis auct. non (Mol.)
Stuntz, *Prosopis juliflora* auct. non (Sw.)
DC.
♦ honey mesquite, mesquite, prosopis
♦ Weed, Quarantine Weed, Noxious
Weed, Naturalised, Environmental
Weed
♦ 7, 10, 39, 76, 87, 88, 95, 98, 121, 152,
158, 161, 203, 212, 269, 278, 279
♦ S/T, arid, cultivated, herbal, toxic.
Origin: South America.

Prosopis gladulosa Torr. × velutina Woot.
Fabaceae/Mimosaceae
♦ mesquite
♦ Noxious Weed, Naturalised
♦ 86

Prosopis glandulosa Torr. var. glandulosa
Fabaceae/Mimosaceae
Prosopis chilensis (Molina) Stuntz var.
glandulosa (Torr.) Standl., *Prosopis*
juliflora (Sw.) DC. var. *glandulosa* (Torr.)
Cockerell (see)
♦ honey mesquite
♦ Weed, Noxious Weed, Naturalised,
Introduced
♦ 86, 147, 228
♦ arid. Origin: southern North
America.

Prosopis glandulosa Torr. var. torreyana (L.D.Benson) M.C.Johnst.
Fabaceae/Mimosaceae
Prosopis juliflora (Sw.) DC. var. *torreyana*
Benson, *Prosopis oderata* Torr. & Frém.
♦ western honey mesquite, honey
mesquite, mesquite, heuningprosopis
♦ Weed, Noxious Weed, Cultivation
Escape
♦ 63, 212, 218, 283
♦ S, arid, cultivated, toxic. Origin:
North and Central America.

Prosopis hassleri Harms ex Hassl.
Fabaceae/Mimosaceae
♦ mesquite
♦ Weed, Noxious Weed
♦ 67, 88, 140, 229

Prosopis humilis Gillies ex Hook. & Arn.
Fabaceae/Mimosaceae
♦ mesquite

♦ Weed, Noxious Weed
♦ 67, 87, 88, 140, 229

Prosopis juliflora (Sw.) DC.
Fabaceae/Mimosaceae
♦ mesquite
♦ Weed, Quarantine Weed, Noxious
Weed, Naturalised, Introduced
♦ 3, 7, 22, 39, 76, 87, 88, 98, 101, 147,
161, 198, 203, 218, 228, 229, 258, 268,
269
♦ S/T, arid, cultivated, herbal, toxic.
Origin: tropical and subtropical
Americas.

Prosopis juliflora (Sw.) DC. var. glandulosa (Torr.) Cockerell
Fabaceae/Mimosaceae
= *Prosopis glandulosa* Torr. var.
glandulosa
♦ honey mesquite
♦ Weed
♦ 23, 218

Prosopis juliflora (Sw.) DC.var. juliflora
Fabaceae/Mimosaceae
Mimosa juliflora Sw., *Neltuma juliflora*
(Sw.) Raf.
♦ mesquite
♦ Noxious Weed, Naturalised
♦ 86
♦ Origin: Mexico, Central America,
northern South America.

Prosopis juliflora (Sw.) DC. var. velutina (Woot.) Sarg.
Fabaceae/Mimosaceae
= *Prosopis velutina* Woot.
♦ velvet mesquite
♦ Weed
♦ 23, 218

Prosopis juliflora (Sw.) DC. × velutina Woot.
Fabaceae/Mimosaceae
♦ mesquite
♦ Weed, Quarantine Weed, Noxious
Weed, Naturalised
♦ 76, 86, 88, 147

Prosopis kuntzei Harms
Fabaceae/Mimosaceae
Prosopis barba-tigridis Stuck., *Prosopis*
casadensis Penz.
♦ mesquite, itín
♦ Weed, Noxious Weed, Introduced
♦ 67, 88, 140, 228, 229, 295
♦ arid.

Prosopis laevigata (Humb. & Bonpl. ex Willd.) M.C.Johnst.
Fabaceae/Mimosaceae
Acacia laevigata Humb. & Bonpl. ex
Willd. (see), *Algarobia dulcis* Benth.,
Mimosa laevigata (Humb. & Bonpl. ex
Willd.) Poir., *Mimosa rotundata* Sessé &
Moç., *Neltuma attenuata* Britton & Rose,
Neltuma laevigata (Humb. & Bonpl.
ex Willd.) Britton & Rose, *Neltuma*
michoacana Britton & Rose, *Neltuma*
pallescens Britton & Rose, *Neltuma*
palmeri Britton & Rose, *Prosopis dulcis*
Kunth
♦ smooth mesquite
♦ Weed, Quarantine Weed,

Naturalised, Introduced
♦ 76, 88, 101, 203, 220, 228
♦ T, arid.

Prosopis limensis Benth.
Fabaceae/Mimosaceae
= *Prosopis pallida* (Humb. & Bon. ex
Willd.) Kunth
♦ algaroba
♦ Noxious Weed, Naturalised
♦ 86, 98
♦ arid.

Prosopis nigra (Griseb.) Hieron.
Fabaceae/Mimosaceae
Prosopis algarobilla Griseb. var. *nigra*
Griseb., *Prosopis dulcis* Kunth var.
australis Benth.
♦ algarrobo negro
♦ Weed, Quarantine Weed, Introduced
♦ 76, 88, 203, 220, 228, 295
♦ arid, herbal.

Prosopis pallida (Humb. & Bonpl. ex Willd.) Kunth
Fabaceae/Mimosaceae
Acacia pallida Willd., *Mimosa pallida*
Poir., *Prosopis limensis* Benth. (see)
♦ kiawe, mesquite, algaroba
♦ Weed, Quarantine Weed, Noxious
Weed, Naturalised, Introduced,
Environmental Weed
♦ 3, 7, 22, 67, 76, 80, 86, 87, 88, 93, 98,
101, 140, 147, 152, 161, 203, 228, 229
♦ T, arid, cultivated, herbal. Origin:
southern US, central and northern
South America.

Prosopis palmeri S.Wats.
Fabaceae/Mimosaceae
♦ mesquite
♦ Weed, Noxious Weed
♦ 67, 88, 140, 229
♦ arid.

Prosopis pubescens Benth.
Fabaceae/Mimosaceae
Prosopis emoryi Torr., *Prosopis odorata*
Torr. & Fremont
♦ screwbean mesquite, screwbean
mesquite, tornillo, Fremont screwbean
♦ Weed, Quarantine Weed, Introduced
♦ 87, 88, 218, 220, 228
♦ S, arid, cultivated, herbal.

Prosopis reptans Benth.
Fabaceae/Mimosaceae
♦ tornillo
♦ Weed, Noxious Weed
♦ 88, 140, 229

Prosopis reptans Benth. var. cinerascens (Gray) Burkart
Fabaceae/Mimosaceae
♦ tornillo
♦ Noxious Weed
♦ 140, 229

Prosopis reptans Benth. var. reptans
Fabaceae/Mimosaceae
♦ tornillo
♦ Weed
♦ 67

Prosopis rojasiana Burkart
Fabaceae/Mimosaceae
♦ mesquite

- Weed, Noxious Weed
- 67, 88, 140, 229

Prosopis ruizlealii Burkart
Fabaceae/Mimosaceae
- mesquite
- Weed, Noxious Weed
- 67, 88, 140, 229

Prosopis ruscifolia Griseb.
Fabaceae/Mimosaceae
- mesquite
- Weed, Quarantine Weed, Noxious Weed, Introduced
- 39, 67, 76, 87, 88, 140, 203, 220, 228, 229, 237, 295
- arid, herbal, toxic.

Prosopis sericantha Gill. ex Hook. & Arn.
Fabaceae/Mimosaceae
- mesquite
- Weed, Noxious Weed
- 67, 88, 140, 229

Prosopis spicigera L.
Fabaceae/Mimosaceae
= *Prosopis cineraria* (L.) Druce
- Weed
- 87, 88
- cultivated.

Prosopis stephaniana (Bieb.) Kunth ex Spreng.
Fabaceae/Mimosaceae
= *Prosopis farcta* (Banks & Sol.) J.F.Macbr.
- Weed, Quarantine Weed
- 23, 76, 87, 88, 203, 220
- herbal.

Prosopis strombulifera (Lam.) Benth.
Fabaceae/Mimosaceae
- creeping mesquite, Argentine screwbean, algarrobilla
- Weed, Noxious Weed, Naturalised, Introduced, Cultivation Escape
- 34, 35, 67, 88, 101, 140, 161, 228, 229
- S/T, arid, cultivated.

Prosopis tamarugo F.Phil.
Fabaceae/Mimosaceae
- velvet mesquite, tamarugo
- Weed, Quarantine Weed
- 76, 88
- arid, cultivated.

Prosopis torquata (Cav. ex Lag.) DC.
Fabaceae/Mimosaceae
Acacia torquata Cav., *Prosopis adesmioides* Griseb.
- mesquite
- Weed, Noxious Weed
- 67, 88, 140, 229
- arid.

Prosopis velutina Woot.
Fabaceae/Mimosaceae
Prosopis articulata S.Wats. (see), *Prosopis juliflora* (Sw.) DC. var. *velutina* (Woot.) Sarg. (see)
- velvet mesquite, fluweelprosopis
- Weed, Quarantine Weed, Noxious Weed, Naturalised, Introduced, Garden Escape, Environmental Weed, Cultivation Escape
- 34, 63, 76, 86, 88, 95, 98, 121, 140, 147,

152, 161, 203, 212, 228, 229, 269, 279, 283
- S/T, arid, cultivated, herbal, toxic. Origin: southern US, Central and northern South America.

Prosopis vinalillo Stuck.
Fabaceae/Mimosaceae
- Introduced
- 228
- arid.

Protasparagus aethiopicus (L.) Oberm.
Liliaceae/Asparagaceae
= *Asparagus densiflorus* (Kunth) Jessop
- asparagus fern, sprengeris fern, fern asparagus, ground asparagus
- Weed, Quarantine Weed, Noxious Weed, Environmental Weed, Garden Escape
- 73, 76, 88, 147, 203

Protasparagus africanus (Lam.) Oberm.
Liliaceae/Asparagaceae
= *Asparagus africanus* Lam.
- climbing asparagus
- Weed, Naturalised, Environmental Weed
- 73, 88, 98, 201, 203

Protasparagus cooperii (Bak.) Oberm.
Liliaceae/Asparagaceae
= *Asparagus cooperi* Bak. (NoR)
- katbos
- Weed
- 10

Protasparagus densiflorus (Kunth) Oberm.
Liliaceae/Asparagaceae
= *Asparagus densiflorus* (Kunth) Jessop
- protasparagus
- Weed, Naturalised, Garden Escape, Environmental Weed
- 7, 72, 88, 98, 201, 203
- pH, cultivated. Origin: South Africa.

Protasparagus laricinus (Burch) Oberm.
Liliaceae/Asparagaceae
= *Asparagus laricinus* Burch. (NoR)
- wild asparagus
- Weed
- 88, 158
- cultivated. Origin: southern Africa.

Protasparagus plumosus (Bak.) Oberm.
Liliaceae/Asparagaceae
= *Asparagus plumosus* Bak.
- ferny asparagus, climbing asparagus fern
- Weed, Quarantine Weed, Noxious Weed, Naturalised, Environmental Weed
- 73, 76, 88, 98, 147, 201, 203
- cultivated.

Protasparagus setaceus (Kunth) Oberm.
Liliaceae/Asparagaceae
= *Asparagus setaceus* (Kunth) Jessop
- Weed, Naturalised
- 98, 203
- cultivated.

Protasparagus virgatus (Bak.) Oberm.
Liliaceae/Asparagaceae
= *Asparagus virgatus* Bak.
- Weed, Naturalised, Environmental

Weed
- 98, 203, 290
- cultivated.

Prunella grandiflora (L.) Scholler
Lamiaceae
Brunella alpina Tim.-Lag.
- isoniittyhumala, loveliness
- Weed, Quarantine Weed, Naturalised, Casual Alien
- 42, 76, 86, 88, 220, 272
- pH, cultivated, herbal.

Prunella hyssopifolia L.
Lamiaceae
- Weed, Quarantine Weed
- 76, 88, 220
- cultivated.

Prunella laciniata (L.) L.
Lamiaceae
- cutleaf self heal
- Weed, Naturalised, Casual Alien
- 86, 98, 101, 176, 198, 203, 272, 280
- cultivated, herbal. Origin: Mediterranean.

Prunella vulgaris L.
Lamiaceae
Brunella officinalis Crantz, *Brunella vulgaris* L. var. *indivisa* Neilr., *Prunella reptans* Dumort.
- heal all, self heal, carpenters weed, common self heal
- Weed, Noxious Weed, Naturalised, Native Weed, Garden Escape, Environmental Weed
- 7, 15, 23, 34, 36, 44, 70, 86, 87, 88, 94, 98, 121, 136, 159, 161, 165, 174, 176, 181, 195, 198, 203, 210, 211, 218, 235, 241, 243, 249, 253, 269, 272, 275, 280, 295, 300
- pH, cultivated, herbal. Origin: Eurasia.

Prunella vulgaris L. ssp. asiatica (Nakai) Hara
Lamiaceae
- utsubogusa
- Weed
- 286

Prunus L. spp.
Rosaceae
- plum, peach, cherry, apricot, evergreen cherry
- Weed, Naturalised, Environmental Weed
- 88, 154, 181, 198, 296
- herbal, toxic.

Prunus americana Marsh.
Rosaceae
Cerasus americana Hook., *Prunus latifolia* Moench
- American plum, Canada plum
- Weed
- 39, 87, 88, 218
- T, cultivated, herbal, toxic.

Prunus angustifolia Marsh.
Rosaceae
- chickasaw plum
- Weed
- 39, 87, 88, 218
- T, cultivated, herbal, toxic.

Prunus angustifolia Marsh. var. *watsonii*
(Sarg.) Waugh
Rosaceae
♦ sand plum, Watson's plum
♦ Weed
♦ 218

Prunus armeniaca L.
Rosaceae
Prunus tiliaefolia Salisb.
♦ apricot, Chinese almond
♦ Weed, Naturalised, Casual Alien
♦ 39, 80, 86, 98, 101, 161, 198, 203, 247, 279, 280
♦ T, cultivated, herbal, toxic. Origin: Asia.

Prunus avium L.
Rosaceae
Cerasus avium (L.) Moench, *Cerasus nigra* Mill.
♦ bird cherry, wild cherry, sweet cherry, gean
♦ Weed, Noxious Weed, Naturalised, Environmental Weed, Cultivation Escape
♦ 39, 42, 79, 80, 87, 88, 98, 101, 137, 146, 151, 161, 195, 203, 218, 225, 246, 247, 280
♦ T, arid, cultivated, herbal, toxic.

Prunus campanulata Maxim.
Rosaceae
♦ Taiwan cherry
♦ Sleeper Weed, Naturalised, Environmental Weed
♦ 225, 246, 280, 287
♦ T, cultivated, herbal.

Prunus caroliniana (Mill.) Aiton
Rosaceae
Laurocerasus caroliniana (Mill.) Roem.
♦ Carolina laurelcherry, American cherry laurel
♦ Weed
♦ 39, 161, 247
♦ S, cultivated, herbal, toxic.

Prunus cerasifera Ehrh.
Rosaceae
Prunus domestica L. var. *myrobalana* L., *Prunus myrobalana* (L.) Loisel.
♦ cherry plum, myrobalan plum, thundercloud cherry, myrobalan, purpleleaf cherryplum
♦ Weed, Noxious Weed, Naturalised, Garden Escape, Environmental Weed
♦ 15, 20, 72, 80, 86, 88, 98, 101, 137, 198, 203, 280, 289
♦ T, cultivated, herbal. Origin: Eurasia.

Prunus cerasus L.
Rosaceae
Cerasus vulgaris Mill., *Prunus vulgaris* Schur
♦ sour cherry, morello cherry, tart cherry, pie cherry, maraschino cherry
♦ Weed, Naturalised, Environmental Weed, Cultivation Escape
♦ 39, 40, 42, 80, 86, 87, 88, 98, 101, 108, 146, 151, 203, 218, 252, 280
♦ T, cultivated, herbal, toxic. Origin: Eurasia.

Prunus domestica L.
Rosaceae
= *Prunus spinosa* L. (see) × *Prunus cerasifera* Ehrh. ssp. *divaricata* (Ledeb.) L.H.Bailey [*Druparia prunus* Clairv., *Prunus communis non* L., *Prunus sativa* Rouy & Camus]
♦ European plum, damson plum, bullace plum, domestic plum, luumupuu, plum, slivka domáca
♦ Weed, Naturalised, Environmental Weed, Cultivation Escape
♦ 7, 39, 40, 42, 80, 88, 98, 101, 151, 176, 203, 280
♦ T, cultivated, herbal, toxic.

Prunus domestica L. ssp. *domestica*
Mansf.
Rosaceae
Prunus domestica L. ssp. *oeconomica* (Borkh.) C.K.Schneid., *Prunus oeconomica* Borkh., *Prunus pruna* Crantz
♦ garden plum, plum
♦ Naturalised
♦ 40, 86
♦ cultivated. Origin: Eurasia.

Prunus domestica L. ssp. *insititia* (L.)
Schneid.
Rosaceae
= *Prunus domestica* L. var. *insititia* (L.) Fiori & Paol.
♦ bullace, bullace plum, damson plum, haferpflaume, kriechenpflaume
♦ Naturalised
♦ 40, 86, 176
♦ cultivated.

Prunus domestica L. var. *domestica*
Rosaceae
♦ European plum
♦ Naturalised
♦ 101

Prunus domestica L. var. *insititia* (L.)
Fiori & Paol.
Rosaceae
Prunus insititia L., *Prunus italica* Borkh., *Prunus domestica* L. ssp. *insititia* (L.) Schneid. (see), *Prunus domestica* L. ssp. *italica* (Borkh.) Hegi
♦ European plum
♦ Naturalised
♦ 101

Prunus dulcis (Mill.) D.A.Webb
Rosaceae
Amygdalus communis L., *Amygdalus dulcis* Mill., *Prunus amygdalus* Batsch
♦ domestic almond, almond, sweet almond, manteli
♦ Weed, Naturalised, Cultivation Escape, Casual Alien
♦ 7, 34, 42, 86, 98, 101, 161, 203, 252
♦ T, cultivated, herbal, toxic. Origin: east Mediterranean to central Asia.

Prunus emarginata (Douglas ex Hook.)
Walp.
Rosaceae
♦ bitter cherry
♦ Weed
♦ 87, 88, 218
♦ S, promoted, herbal.

Prunus fruticosa Pall.
Rosaceae
Cerasus fruticosa (Pall.) Woronow, *Prunus intermedia* Poir.
♦ European dwarf cherry, European groundcherry, groundcherry, steppe cherry, Mongolian cherry
♦ Weed, Naturalised
♦ 80, 101
♦ S, cultivated.

Prunus glandulosa Thunb.
Rosaceae
♦ flowering almond, Korean cherry
♦ Weed, Naturalised
♦ 80, 101
♦ S, cultivated, herbal.

Prunus grisea (Blume ex C.Muell.)
Kalkman
Rosaceae
♦ Weed, Naturalised
♦ 98, 203

Prunus ilicifolia (Nutt.) Walp. ssp. *lyonii*
(Eastw.) Raven
Rosaceae
♦ Catalina cherry, Catalina cherry, island cherry, hollyleaf cherry
♦ Weed, Cultivation Escape
♦ 34
♦ T, cultivated.

Prunus laurocerasus L.
Rosaceae
Cerasus laurocerasus (L.) Loisel., *Laurocerasus officinalis* Roem., *Padus laurocerasus* Mill.
♦ cherry laurel, common cherry laurel, Portuguese laurel
♦ Weed, Sleeper Weed, Noxious Weed, Naturalised, Garden Escape, Environmental Weed
♦ 15, 39, 40, 45, 72, 80, 86, 88, 98, 101, 137, 146, 155, 161, 181, 198, 203, 225, 246, 247, 280, 289
♦ T, cultivated, herbal, toxic. Origin: east Europe, west Asia.

Prunus lusitanica L.
Rosaceae
Laurocerasus lusitanica (L.) M.Roem., *Padus lusitanica* L.
♦ Portugal laurel
♦ Weed, Sleeper Weed, Noxious Weed, Naturalised, Garden Escape, Environmental Weed
♦ 39, 54, 72, 80, 86, 88, 98, 101, 137, 198, 203, 225, 246, 280
♦ T, cultivated, herbal, toxic. Origin: south-west Europe, Morocco, Azores, Madeira and Canary Islands.

Prunus lusitanica L. ssp. *lusitanica*
Rosaceae
♦ Naturalised
♦ 280

Prunus mahaleb L.
Rosaceae
Cerasus mahaleb Mill.
♦ perfumed cherry, mahaleb cherry, St. Lucie cherry
♦ Weed, Naturalised, Environmental Weed

♦ 39, 54, 80, 86, 88, 98, 101, 146, 151, 195, 203, 280
♦ T, cultivated, herbal, toxic. Origin: Eurasia, north-west Africa.

Prunus munsoniana **W.Wight & Hedrick**
Rosaceae
♦ wild goose plum
♦ Weed, Naturalised
♦ 86, 98, 203
♦ T, cultivated, herbal. Origin: North America.

Prunus nigra **Aiton**
Rosaceae
Cerasus nigra Loisel. ex Duhamel
♦ Canada plum, bitter cherry
♦ Weed
♦ 87, 88, 218
♦ T, cultivated, herbal.

Prunus padus **L.**
Rosaceae
Padus avium Mill., *Padus racemosa* Gilib., *Prunus racemosa* Lam
♦ European bird cherry, bird cherry, tuomi
♦ Weed, Naturalised, Environmental Weed
♦ 80, 88, 101, 151
♦ T, cultivated, herbal.

Prunus pensylvanica **L.f.**
Rosaceae
♦ pin cherry
♦ Weed
♦ 39, 87, 88, 161, 218, 247
♦ T, cultivated, herbal, toxic.

Prunus persica **(L.) Batsch**
Rosaceae
Persica nucipersica Borkh., *Persica vulgaris* Mill.
♦ peach, nectarine, persikka
♦ Weed, Naturalised, Garden Escape, Environmental Weed, Casual Alien
♦ 15, 39, 42, 80, 88, 98, 101, 121, 151, 161, 198, 203, 243, 247, 251, 279, 280
♦ T, cultivated, herbal, toxic. Origin: Eurasia.

Prunus persica **(L.) Batsch var. *nectarina* (Aiton) Maxim.**
Rosaceae
= *Prunus persica* (L.) Batsch var. *nucipersica* (Suckow) C.Schneid. (NoR)
♦ nektarin, nektariini, nektarynka, nectarine
♦ Naturalised
♦ 86
♦ cultivated, herbal.

Prunus persica **(L.) Batsch var. *persica***
Rosaceae
♦ peach
♦ Naturalised
♦ 86
♦ cultivated. Origin: China.

Prunus salicina **Lindl.**
Rosaceae
Prunus triflora Roxb.
♦ Japanese plum, plum
♦ Weed, Naturalised
♦ 86, 98, 203
♦ T, cultivated, herbal. Origin:

east Asia.

Prunus serotina **Ehrh.**
Rosaceae
Cerasus serotina Loisel. ex Duhamel, *Padus serotina* (Ehrh.) Borkh.
♦ black cherry, rum cherry, wild cherry, kiiltotuomi
♦ Weed, Naturalised, Garden Escape, Environmental Weed, Casual Alien
♦ 8, 39, 86, 87, 88, 98, 152, 155, 161, 203, 211, 218, 243, 247, 279, 280, 293
♦ T, cultivated, herbal, toxic. Origin: North America.

Prunus serotina **Ehrh. ssp. *serotina***
Rosaceae
♦ Introduced
♦ 38

Prunus serrulata **Lindl.**
Rosaceae
Cerasus serrulata (Lindl.) G.Don, *Prunus tenuiflora* Koehne
♦ Japanese flowering cherry, yamazakura, oriental cherry
♦ Sleeper Weed, Naturalised, Environmental Weed, Cultivation Escape
♦ 40, 101, 225, 246, 280
♦ T, cultivated, herbal.

Prunus spinosa **L.**
Rosaceae
Druparia spinosa Clair.
♦ sloe, black thorn, épine noire, prunellier, schlehdorn, schlehe, schwarzdorn, abrunheiro, ciruelo silvestre, endrino, espino negro
♦ Weed, Naturalised, Garden Escape, Environmental Weed, Casual Alien
♦ 39, 54, 72, 80, 86, 88, 98, 101, 151, 176, 198, 203, 280
♦ S, cultivated, herbal, toxic. Origin: Eurasia.

Prunus subhirtella **Miq.**
Rosaceae
Cerasus subhirtella (Miq.) S.Ya.Sokolov
♦ higan cherry, rosebud cherry
♦ Naturalised
♦ 101
♦ T, cultivated.

Prunus tomentosa **Thunb.**
Rosaceae
♦ Nanking cherry
♦ Weed, Naturalised
♦ 80, 101
♦ S, cultivated, herbal.

Prunus triloba **Lindl.**
Rosaceae
Amygdalus petzoldii (K.Koch) Ricker, *Amygdalus triloba* (Lindl.) Ricker, *Prunus petzoldii* K.Koch, *Prunus triloba* var. *petzoldii* (K.Koch) L.H.Bailey, *Prunus ulmifolia* Franch.
♦ flowering plum, flowering almond, mandelbäumchen
♦ Naturalised
♦ 101
♦ S, cultivated. Origin: China.

Prunus turneriana **(Bailey) Kalkman**
Rosaceae

♦ Weed, Naturalised
♦ 98, 203
♦ cultivated.

Prunus umbellata **Ell.**
Rosaceae
♦ flatwoods plum, hog plum, black sloe
♦ Weed
♦ 87, 88, 218
♦ T, promoted, herbal.

Prunus virginiana **L.**
Rosaceae
Padus virginiana (L.) Mill.
♦ common chokecherry, western demissa chokecherry, wild cherry
♦ Weed, Native Weed
♦ 8, 39, 87, 88, 161, 174, 211, 218, 243, 293
♦ S, cultivated, herbal, toxic. Origin: North America.

Prunus virginiana **L. var. *demissa* (Nutt.) Torr.**
Rosaceae
♦ western chokecherry, chokecherry
♦ Weed
♦ 218
♦ S, cultivated.

Prunus virginiana **L. var. *melanocarpa* (A.Nels.) Sarg.**
Rosaceae
♦ black chokecherry, Rocky Mountain chokecherry
♦ Weed
♦ 161, 218

Psacadocalymma comatum **(Sw.) Bremek.**
Acanthaceae
♦ Weed
♦ 276
♦ pH. Origin: tropical America.

Psammochloa villosa **(Trin.) Bor**
Poaceae
Psammochloa mongolica A.Hitchc.
♦ Mongolian psamochloa
♦ Weed
♦ 297
♦ G, arid.

Psathyrostachys fragilis **(Boiss.) Nevski**
Poaceae
Elymus fragilis (Boiss.) Griseb., *Hordeum fragile* Boiss.
♦ Quarantine Weed
♦ 220
♦ G, arid. Origin: Middle East.

Psathyrostachys juncea **(Fisch.) Nevski**
Poaceae
Elymus alberti Regel, *Elymus caespitosus* Sukaczev, *Elymus cretaceus* Zing. ex Nevski, *Elymus desertorum* Karav. & Kir., *Elymus desertorum* var. *angustifolius* Karav., *Elymus desertorum* var. *desertorum* Karav. & Kir., *Elymus desertorum* var. *latifolius* Kar. & Kir., *Elymus hyalanthus* Rupr., *Elymus junceus* Fisch., *Elymus junceus* var. *caespitosus* (Sukaczev) Reverd., *Elymus junceus* Fisch. var. *junceus*, *Elymus junceus* var. *typica* Trautv., *Elymus*

junceus var. *villosus* Drob., *Elymus kokczetavicus* Drob., *Psathyrostachys caespitosa* (Sukaczev) G.A.Peshkova, *Psathyrostachys hyalantha* (Rupr.) Tzvelev, *Psathyrostachys juncea* ssp. *hyalantha* (Rupr.) Tzvelev, *Psathyrostachys perennis* Keng, *Triticum junccellum* F.Herm.
♦ Russian wildrye
♦ Naturalised, Introduced
♦ 21, 101
♦ G, herbal.

Psathyrotes annua (Nutt.) Gray
Asteraceae
♦ psathyrotes, annual psathyrotes
♦ Weed
♦ 39, 161
♦ aH, herbal, toxic.

Pseudabutilon stuckertii R.E.Fr.
Malvaceae
♦ velvetleaf Indian mallow
♦ Naturalised
♦ 101

Pseudanamomis umbellulifera (Kunth) Kausel
Myrtaceae
♦ ciruelas
♦ Naturalised, Garden Escape
♦ 101, 261
♦ cultivated. Origin: Central and South America.

Pseudechinolaena polystachya (Kunth) Stapf
Poaceae
= *Echinochloa polystachya* (Kunth) Hitchc.
♦ Weed
♦ 87, 88, 157, 243
♦ aG.

Pseudelephantopus Rohr spp.
Asteraceae
♦ lengua de perro, dog's tongue, pseudelephantopus
♦ Weed
♦ 153

Pseudelephantopus funckii (Turcz.) Philipson
Asteraceae
= *Pseudelephantopus spiralis* (Less.) Cronquist
♦ Weed
♦ 87, 88

Pseudelephantopus spicatus (Juss. ex Aubl.) C.F.Bak.
Asteraceae
Elephantopus spicatus Juss. ex Aubl. (see)
♦ spike elephantopus, false elephant's foot, vao malini, vao elefane, yasawa tobacco weed, faux tabac des Samoa, dog's tongue, matapasto
♦ Weed, Quarantine Weed, Naturalised
♦ 3, 6, 13, 14, 76, 86, 87, 88, 101, 153, 157, 179, 191, 203, 257
♦ pH, herbal. Origin: Central and northern South America.

Pseudelephantopus spiralis (Less.) Cronquist
Asteraceae
Chaetospira funckii (Philipson) S.F.Blake, *Chaetospira spiralis* (Less.) Aspl. & S.F.Blake, *Distreptus spiralis* Less., *Pseudelephantopus funckii* (Turcz.) Philipson (see), *Spirochaeta funckii* Turcz.
♦ Naturalised
♦ 257
♦ pH.

Pseuderanthemum acuminatissimum (Miq.) Benoist.
Acanthaceae
♦ Introduced
♦ 230
♦ H.

Pseuderanthemum carruthersii (Seem.) Guill.
Acanthaceae
♦ Naturalised, Introduced
♦ 101, 261
♦ Origin: Polynesia.

Pseuderanthemum carruthersii (Seem.) Guill. var. *atropurpureum* (Bull.) Fosb.
Acanthaceae
Pseuderanthemum atropurpureum (Bull.) L.
♦ Introduced
♦ 230
♦ H.

Pseuderanthemum diversifolium (Bl.) Radlk.
Acanthaceae
♦ Weed
♦ 13

Pseuderanthemum fasciculatum (Oerst.) Léonard
Acanthaceae
♦ false face
♦ Weed
♦ 179, 261
♦ Origin: Mexico.

Pseuderanthemum praecox (Benth.) Léonard
Acanthaceae
♦ night and afternoon
♦ Naturalised
♦ 101

Pseuderia micronesiaca Schltr.
Orchidaceae
♦ Introduced
♦ 230
♦ H.

Pseudobrachiaria deflexa (Schum.) Launert
Poaceae
= *Urochloa deflexa* (Schum.) H.Scholz (NoR) [see *Pseudobrachiaria deflexa* (Schum.) Launert]
♦ false signalgrass
♦ Native Weed
♦ 121
♦ aG. Origin: southern Africa.

Pseudocanthocereus Ritt. spp.
Cactaceae

= *Acanthocereus* (Engelm. ex Berg.) Britt. & Rose spp.
♦ Weed, Quarantine Weed
♦ 76, 88, 203

Pseudocydonia sinensis (Dum.Cours.) Schneid.
Rosaceae
Chaenomeles sinensis (Thouin) Koehne, *Cydonia sinensis* Thouin (see)
♦ Chinese quince
♦ Naturalised
♦ 101
♦ T, cultivated, herbal.

Pseudofumaria alba (Mill.) Lidén
Papaveraceae/Fumariaceae
♦ Naturalised
♦ 176

Pseudofumaria alba (Mill.) Lidén ssp. alba
Papaveraceae/Fumariaceae
♦ Naturalised, Environmental Weed
♦ 86
♦ Origin: Europe.

Pseudofumaria lutea (L.) Borkh.
Papaveraceae/Fumariaceae
Capnoides luteum (L.) Gaertn., *Corydalis lutea* (L.) DC. (see), *Fumaria lutea* L.
♦ yellow corydalis, rock fumewort
♦ Naturalised
♦ 40, 101
♦ Origin: Italy, Switzerland.

Pseudogaltonia clavata (Mast. ex Bak.) Phill.
Liliaceae/Hyacinthaceae
Galtonia clavata Mast., *Lindneria clavata* (Mast.) Speta
♦ cape hyacinth, south west Africa slangkop
♦ Native Weed
♦ 121
♦ pH, cultivated, toxic. Origin: southern Africa.

Pseudognaphalium attenuatum (DC.) A.Anderb.
Asteraceae
♦ tapered cudweed
♦ Naturalised
♦ 101

Pseudognaphalium elegans (Kunth) Kartesz comb. nov. ined.
Asteraceae
♦ royal cudweed
♦ Naturalised
♦ 101

Pseudognaphalium luteo-album (L.) Hilliard & Burtt
Asteraceae
Gnaphalium luteo-album L. (see)
♦ cudweed, Jersey cudweed
♦ Weed, Naturalised, Introduced, Environmental Weed
♦ 7, 55, 86, 88, 101, 158, 165, 228, 243, 269
♦ aH, arid, cultivated. Origin: Africa, Eurasia, Australia.

Pseudognaphalium roseum (Kunth) A.Anderb.

Asteraceae
- rosy cudweed
- Naturalised
- 101

Pseudognaphalium undulatum (L.) Hilliard & Burtt
Asteraceae
Gnaphalium undulatum L. (see), *Helichrysum montosicolum* Gand.
- cudweed
- Weed
- 88, 158
- Origin: southern Africa.

Pseudogynoxys chenopodioides (Kunth) Cabrera
Asteraceae
- Mexican flame vine
- Weed, Naturalised, Cultivation Escape
- 101, 161, 179, 261
- cultivated, toxic.

Pseudogynoxys oerstedii (Benth.) Cuatrec.
Asteraceae
Senecio chinotegensis Klatt
- Weed
- 157
- pC.

Pseudopanax crassifolius (Sol. ex A.Cunn.) K.Koch × lessonii (DC.) K.Koch
Araliaceae
- hybrid five fingers
- Weed
- 15

Pseudopilocereus Bux. spp.
Cactaceae
= *Pilosocereus* Byles & Rowley spp.
- Weed, Quarantine Weed
- 76, 88, 220

Pseudorchis albida (L.) Á. & D.Löve
Orchidaceae
Orchis albida (L.) Scop., *Leucorchis albida* (L.) E.Mey. ssp. *albida*, *Habenaria albida* (L.) R.Br., *Leucorchis albida* (L.) E.Mey., *Gymnadenia albida* (L.) Rich., *Bicchia albida* (L.) Parl.
- small white orchid
- Weed
- 70

Pseudorlaya pumila (L.) Grande
Apiaceae
- Weed
- 221

Pseudosasa japonica (Siebold & Zucc. ex Steud.) Makino ex Nakai
Poaceae
- running bamboo, arrow bamboo, metake
- Weed, Naturalised, Environmental Weed
- 15, 40, 88, 101, 151, 280
- G, cultivated, herbal.

Pseudoscleropodium purum (Hedw.) Fleisch. ex Broth.
Entodontaceae
- pseudoscleropodium moss

- Naturalised, Environmental Weed
- 198, 280, 289
- moss.

Pseudotephrocactus Fric & Schelle spp.
Cactaceae
= *Opuntia* Mill. spp.
- Weed, Quarantine Weed
- 76, 88, 220

Pseudotsuga menziesii (Mirb.) Franco
Pinaceae
Abies californica hort., *Pinus taxifolia* (Lambert) Britton, *Tsuga douglasii* Carr.
- Douglas fir
- Weed, Naturalised, Environmental Weed
- 15, 54, 86, 87, 88, 181, 198, 218, 225, 246, 280
- T, cultivated, herbal. Origin: western north America.

Psiadia arabica Jaub. & Spach
Asteraceae
- Weed, Quarantine Weed
- 76, 87, 88, 203, 220
- arid.

Psidium cattleianum Sabine
Myrtaceae
Psidium cattleianum Sabine var. *littorale* (Raddi) Mattos (see), *Psidium littorale* Raddi var. *longipes* (O.Berg) Fosb. (see)
- strawberry guava, cherry guava, cattley guava, Chinese guava, kuahpa, red cherry guava, waiawi, ngguava, goyavier de Chine, tuava tinito
- Weed, Sleeper Weed, Naturalised, Noxious Weed, Naturalised, Environmental Weed, Garden Escape, Cultivation Escape
- 3, 18, 22, 73, 80, 87, 88, 98, 101, 107, 112, 121, 129, 129, 132, 151, 152, 155, 179, 191, 203, 218, 225, 233, 246, 261, 280, 283
- S/T, cultivated, herbal, toxic. Origin: South America.

Psidium cattleianum Sabine fo. lucidum Deg.f.
Myrtaceae
- kuahpa
- Introduced
- 230

Psidium cattleianum Sabine var. cattleianum
Myrtaceae
- strawberry guava
- Naturalised
- 134
- cultivated.

Psidium cattleianum Sabine var. littorale (Raddi) Mattos
Myrtaceae
Psidium littorale Raddi (see)
- strawberry guava
- Naturalised, Environmental Weed
- 86, 134, 155
- cultivated. Origin: Brazil.

Psidium × durbanensis Baijnath ined.
Myrtaceae
= *Psidium guajava* L. × *Psidium guineense* Sw.

- Durban guava
- Weed, Noxious Weed, Garden Escape
- 88, 95, 283
- cultivated. Origin: South Africa.

Psidium guajava L.
Myrtaceae
- apple guava, guava, abas, apas, bonongu, guabang, kuabang, guahva, quwawa, koejawel, kuahpa, kuava, amrut, kautoga, ku'ava, kuhfahfah, kautonga, kuawa, goyavier, ku'avu, tu'avu, te kuawa, kuwawa, mpela, nguava, ngguava ni India, yellow guava
- Weed, Sleeper Weed, Noxious Weed, Naturalised, Introduced, Garden Escape, Environmental Weed, Cultivation Escape
- 3, 6, 10, 14, 18, 22, 37, 63, 73, 80, 86, 87, 88, 95, 98, 101, 107, 112, 121, 122, 132, 134, 151, 152, 153, 155, 158, 179, 203, 218, 225, 230, 243, 246, 257, 259, 277, 279, 280, 283, 287
- S/T, cultivated, herbal. Origin: tropical America.

Psidium guineense Sw.
Myrtaceae
- Brazilian guava, guinea guava, Brasiliaanse koejawel, ocker berry
- Weed, Noxious Weed, Naturalised, Garden Escape
- 3, 63, 86, 88, 95, 98, 191, 203, 283
- S/T, cultivated. Origin: Brazil.

Psidium littorale Raddi
Myrtaceae
= *Psidium cattleianum* Sabine var. *littorale* (Raddi) Mattos
- strawberry guava, cherry guava, yellow strawberry guava
- Weed, Naturalised, Cultivation Escape
- 3, 80, 88, 95, 122, 191, 287
- S, cultivated. Origin: South America.

Psidium littorale Raddi var. longipes (O.Berg) Fosb.
Myrtaceae
= *Psidium cattleianum* Sabine
- aarbeikoejawel, strawberry guava
- Weed
- 63
- cultivated.

Psilocaulon rogersiae L. Bol.
Aizoaceae/Mesembryanthemaceae
- prenia vygie
- Native Weed
- 121
- H. Origin: southern Africa.

Psilocaulon tenue (Haw.) Schwantes
Aizoaceae/Mesembryanthemaceae
Mesembryanthemum tenue (Haw.) Haw.
- wiry noonflower
- Weed, Naturalised, Environmental Weed
- 72, 86, 88, 98, 198, 203
- S, arid.

Psilostrophe cooperi (Gray) Greene
Asteraceae
♦ whitestem paperflower, paperflower, paper daisy
♦ Weed
♦ 161
♦ S, cultivated, herbal, toxic.

Psilostrophe gnaphalodes DC.
Asteraceae
♦ paperflower
♦ Weed
♦ 39, 161
♦ toxic.

Psilostrophe sparsifolia A.Nelson
Asteraceae
♦ greenstem paperflower
♦ Weed
♦ 161
♦ toxic.

Psilostrophe tagetina (Nutt.) Greene
Asteraceae
♦ woolly paperflower
♦ Weed
♦ 161
♦ herbal, toxic.

Psilothonna speciosa (Pillans) Phillips
Asteraceae
♦ wild coffee
♦ Weed
♦ 23

Psilurus incurvus (Gouan) Schinz & Thell.
Poaceae
♦ bristle tail grass, setolina
♦ Weed, Naturalised
♦ 86, 98, 198, 203
♦ G, herbal. Origin: Mediterranean.

Psittacanthus cordatus (Hoffm.) Blume
Loranthaceae
♦ erva de passarinho
♦ Weed
♦ 255
♦ pS parasitic, herbal.

Psittacanthus robustus (Mart.) Mart.
Loranthaceae
♦ erva de passarinho
♦ Weed
♦ 255
♦ pH parasitic. Origin: tropical America.

Psoralea americana L.
Fabaceae/Papilionaceae
= *Cullen americana* (L.) Rydb.
♦ Weed, Quarantine Weed
♦ 76, 87, 88, 220

Psoralea bituminosa L.
Fabaceae/Papilionaceae
= *Bituminaria bituminosa* (L.) Stirt.
♦ pitch trefoil, scurfy pea
♦ Weed, Quarantine Weed
♦ 70, 76, 87, 88, 220, 221, 272
♦ cultivated, herbal.

Psoralea corylifolia L.
Fabaceae/Papilionaceae
= *Cullen corylifolia* (L.) Medik.
♦ Malay tea
♦ Weed, Quarantine Weed,

Naturalised
♦ 76, 87, 88, 220, 287
♦ pH, arid, promoted, herbal, toxic.

Psoralea decumbens Ait.
Fabaceae/Papilionaceae
= *Otholobium decumbens* (Aiton) C.H.Stirt.
♦ Weed, Quarantine Weed
♦ 76, 88, 220
♦ arid.

Psoralea floribunda Nutt. ex Torr. & Gray
Fabaceae/Papilionaceae
= *Psoralidium tenuiflorum* (Pursh) Rydb.
♦ Weed, Quarantine Weed
♦ 76, 88, 220
♦ herbal.

Psoralea lanceolata Pursh
Fabaceae/Papilionaceae
= *Psoralidium lanceolatum* (Pursh) Rydb.
♦ lemon scurfpea
♦ Weed, Quarantine Weed
♦ 76, 87, 88, 161, 220
♦ pH, arid, promoted, herbal.

Psoralea macrostachya DC.
Fabaceae/Papilionaceae
= *Hoita macrostachya* (DC.) Rydb. (NoR)
♦ leather root, large leather root
♦ Introduced
♦ 39, 228
♦ pH, arid, cultivated, herbal, toxic.

Psoralea obtusifolia Torr. & Gray
Fabaceae/Papilionaceae
= *Psoralidium tenuiflorum* (Pursh) Rydb.
♦ Weed, Quarantine Weed
♦ 76, 88, 220

Psoralea pinnata L.
Fabaceae/Papilionaceae
♦ blue psoralea, African scurfpea, dally pine, fountain bush, taylorina
♦ Weed, Naturalised, Garden Escape, Environmental Weed
♦ 7, 72, 86, 88, 98, 155, 165, 176, 198, 203, 225, 246, 280, 289, 296
♦ S, cultivated. Origin: southern Africa.

Psoralea plicata Del.
Fabaceae/Papilionaceae
♦ Weed
♦ 221
♦ arid.

Psoralea tenuiflora Pursh
Fabaceae/Papilionaceae
= *Psoralidium tenuiflorum* (Pursh) Rydb.
♦ scurfy pea
♦ Weed, Quarantine Weed
♦ 39, 76, 87, 88, 220
♦ pH, arid, cultivated, herbal, toxic.

Psoralidium batesii Rydb.
Fabaceae/Papilionaceae
= *Psoralidium tenuiflorum* (Pursh) Rydb.
♦ Weed, Quarantine Weed
♦ 76, 88, 220

Psoralidium lanceolatum (Pursh) Rydb.
Fabaceae/Papilionaceae
Psoralea lanceolata Pursh (see)
♦ lemon scurfpea, scurfpea
♦ Weed, Quarantine Weed

♦ 76, 88, 220
♦ pH, herbal.

Psoralidium tenuiflorum (Pursh) Rydb.
Fabaceae/Papilionaceae
Psoralea floribunda Nutt. ex Torr. & Gray (see), *Psoralea obtusifolia* Torr. & Gray (see), *Psoralea tenuiflora* Pursh (see), *Psoralidium batesii* Rydb. (see)
♦ slimflower scurfpea, slender scurfpea
♦ Weed, Quarantine Weed
♦ 76, 88, 161, 220
♦ herbal, toxic.

Psychine stylosa Desf.
Brassicaceae
♦ Weed
♦ 87, 88
♦ cultivated.

Psychotria alba Ruiz & Pav.
Rubiaceae
= *Psychotria carthagenensis* Jacq. (NoR)
♦ Weed
♦ 87, 88
♦ cultivated.

Psychotria curviflora Wall.
Rubiaceae
Chasalia curviflora (Wall.) Thw.
♦ Weed
♦ 13

Psychotria hombroniana (Baill.) Fosberg var. hirtella (Val.) Fosb.
Rubiaceae
♦ Introduced
♦ 230
♦ S.

Psychotria lasianthoides Val.
Rubiaceae
♦ Introduced
♦ 230
♦ S.

Psychotria merrillii Kaneh.
Rubiaceae
♦ Introduced
♦ 230
♦ S.

Psychotria nervosa Sw.
Rubiaceae
Psychotria granadensis Benth., *Psychotria hirta* Humb. & Bonpl. ex Roem. & Schult., *Psychotria nervosa* ssp. *rufescens* (Kunth) Steyerm., *Psychotria nervosa* var. *rufescens* (Kunth) L.O.Williams, *Psychotria quiinifolia* Dwyer, *Psychotria rufescens* Kunth, *Psychotria rufescens* var. *haenkeana* DC., *Psychotria undata* Jacq., *Uragoga granadensis* (Benth.) Kuntze
♦ wild coffee, seminole balsamo
♦ Weed
♦ 14
♦ cultivated, herbal.

Psychotria pubescens Sw.
Rubiaceae
♦ hairy wild coffee
♦ Weed
♦ 14

Psychotria punctata Vatke
Rubiaceae

- dotted wild coffee
- Weed, Naturalised
- 101, 179

Psychotria rhombocarpoides Hosok.
Rubiaceae
- Introduced
- 230
- S.

Psychotria ruelliaefolia Müll.Arg.
Rubiaceae
Cephaelis ruelliaefolia Cham. & Schlect.
- Weed, Quarantine Weed
- 76, 87, 88, 203, 220

Psylliostachys spicata (Willd.) Nevski
Plumbaginaceae
- unkarinsyreeni, punatähkämö
- Cultivation Escape
- 42
- cultivated.

Ptelea trifoliata L.
Rutaceae
- hoptree, water ash, common hoptree
- Weed
- 8, 39, 87, 88, 161, 218
- S/T, cultivated, herbal, toxic.

Pteridium Gled. ex Scop. spp.
Dennstaedtiaceae
- brackenfern, brake fern, eagle fern
- Weed
- 18, 88
- toxic.

Pteridium aquilinum (L.) Kuhn
Dennstaedtiaceae
Eupteris aquilina Newman, *Pteridium lanuginosum* Clute, *Pteridium latiusculum* (Desv.) Maxon, *Pteris aquilina* L.
- bracken, western brackenfern, kelikundjuku, brackenfern, brake, eagle fern, hog pasture brake, pasture brake, lusilisilu, samambaia
- Weed, Noxious Weed, Native Weed, Environmental Weed
- 8, 10, 23, 39, 70, 87, 88, 121, 136, 154, 157, 157, 158, 161, 204, 210, 217, 226, 229, 238, 245, 247, 254, 255, 272, 294
- pH, arid, cultivated, herbal, toxic. Origin: Europe and Africa.

Pteridium aquilinum (L.) Kuhn var. decompositum (Gaud.) R.Tryon
Dennstaedtiaceae
- bracken, decomposition brackenfern
- Noxious Weed
- 229

Pteridium aquilinum (L.) Kuhn var. latiusculum (Desv.) Underwood ex Heller
Dennstaedtiaceae
- eastern bracken, western brackenfern, bracken
- Weed, Noxious Weed
- 218, 229, 286, 297

Pteridium aquilinum (L.) Kuhn var. pseudocaudatum (Clute) Heller
Dennstaedtiaceae
- bracken, western brackenfern
- Noxious Weed
- 229

Pteridium aquilinum (L.) Kuhn var. pubescens Underwood
Dennstaedtiaceae
- western brackenfern, brackenfern, bracken, hairy brackenfern
- Weed, Noxious Weed
- 212, 218, 229
- pH.

Pteridium caudatum (L.) Maxon
Dennstaedtiaceae
- southern brackenfern
- Weed
- 14

Pteridium esculentum (Forst.) Nakai
Dennstaedtiaceae
- bracken
- Weed, Native Weed
- 39, 87, 88, 181, 269
- cultivated, herbal, toxic. Origin: Australia.

Pteridium revolutum (Blume) Nakai
Dennstaedtiaceae
- Weed
- 39, 87, 88
- toxic. Origin: Australia.

Pteris cretica L.
Pteridaceae
- Cretan fern, Cretan brake
- Weed, Naturalised
- 15, 40, 101
- pH, cultivated, herbal. Origin: Madagascar.

Pteris cretica L. var. albolineata Hook.
Pteridaceae
- Cretan brake
- Weed, Naturalised
- 101, 179

Pteris cretica L. var. cretica
Pteridaceae
- Cretan brake
- Naturalised
- 101

Pteris ensiformis Burm.f.
Pteridaceae
- slender brake, sword brake
- Weed, Naturalised
- 86, 87, 88, 101
- cultivated, herbal. Origin: Asia, Australia.

Pteris grandifolia L.
Pteridaceae
- elephantleaf brake
- Weed
- 179
- H, herbal.

Pteris multifida Poir.
Pteridaceae
- spiderbrake
- Weed, Naturalised
- 101, 179, 297
- cultivated, herbal.

Pteris plumula Desv.
Pteridaceae
- striped brake
- Weed
- 179

Pteris semipinnata L.
Pteridaceae
- Naturalised
- 86
- cultivated. Origin: east and south-east Asia.

Pteris tremula R.Br.
Pteridaceae
- Australian brake, brake fern, brake
- Naturalised
- 86, 101
- pH, cultivated. Origin: Australia.

Pteris tripartita Sw.
Pteridaceae
- giant brake, giant bracken
- Weed, Naturalised
- 32, 101, 179
- cultivated, herbal. Origin: Australia.

Pteris vittata L.
Pteridaceae
- ladder brake
- Weed, Naturalised, Native Weed, Cultivation Escape
- 32, 34, 86, 87, 88, 101, 121, 179, 252, 255
- pH, cultivated, herbal. Origin: Africa, Asia, Australia.

Pterisanthes stonei A.Latiff
Vitaceae
- Quarantine Weed
- 220
- Origin: Malaysia.

Pterocactus Schumm. spp.
Cactaceae
- Quarantine Weed
- 220

Pterocarpus erinaceus Poir.
Fabaceae/Papilionaceae
Pterocarpus zenkeri Harms
- African gum, kino tree, African rosewood, African teak, cornwood, Gambia gum, kinotree gum, lancewood, rosewood, Molompi woodtree, Senegal rosewood, West African rosewood, barwood, African kino
- Quarantine Weed
- 76
- arid, herbal.

Pterocarpus indicus Willd.
Fabaceae/Papilionaceae
- angsana, padauk, Burmese rosewood, liki, sang dragon, pterocarpus
- Weed, Introduced, Garden Escape, Environmental Weed
- 3, 191, 228, 259, 261
- arid, cultivated, herbal. Origin: tropical Asia.

Pterocarpus lucens Lepr. ex Guill. & Perr.
Fabaceae/Papilionaceae
Pterocarpus abyssinicus Hochst. ex A.Rich., *Pterocarpus lucens* Lepr. ex Guill. & Perr. var. *simplicifolius* (Bak.) A.Chev., *Pterocarpus simplicifolius* Bak.
- Introduced
- 228
- arid, cultivated, herbal.

Pterocarpus macrocarpus Kurz
Fabaceae/Papilionaceae
♦ Garden Escape
♦ 261
♦ cultivated. Origin: southern Asia.

Pterocarpus marsupium Roxb.
Fabaceae/Papilionaceae
♦ Malabar kino, bastard teak, rosewood, Indian kino tree, bijasal
♦ Introduced
♦ 39, 228
♦ arid, cultivated, herbal, toxic.

Pterocarpus officinalis Jacq.
Fabaceae/Papilionaceae
♦ pall de pollo, dragon's blood tree
♦ Weed
♦ 261
♦ herbal.

Pterocarpus rotundifolius (Sond.) Druce
Fabaceae/Papilionaceae
♦ round leaved teak
♦ Weed
♦ 10
♦ cultivated.

Pterocarpus rotundifolius (Sond.) Druce ssp. rotundifolius
Fabaceae/Papilionaceae
♦ round leaved bloodwood, round leaved kiaat, round leaved teak
♦ Native Weed
♦ 121
♦ T. Origin: southern Africa.

Pterocarya × rehderiana C.K.Schneid.
Juglandaceae
= *Pterocarya fraxinifolia* (Lam.) Spach × *Pterocarya stenoptera* C.DC.
♦ Casual Alien
♦ 280

Pterocarya stenoptera C.DC.
Juglandaceae
♦ Chinese wingnut
♦ Naturalised
♦ 101, 287
♦ T, cultivated, herbal.

Pterocaulon lanatum Kuntze
Asteraceae
♦ branqueja, verbasco
♦ Weed
♦ 255
♦ Origin: Brazil.

Pterocaulon redolens (Willd.) Fern.-Vill.
Asteraceae
♦ ragweed
♦ Native Weed
♦ 269
♦ Origin: Australia.

Pterocaulon rugosum (Vahl) Malme
Asteraceae
♦ Weed, Quarantine Weed
♦ 76, 87, 88, 203, 220

Pterocaulon sphacelatum (Labill.) F.Muell.
Asteraceae
♦ scented daisy, fruit salad plant
♦ Weed, Native Weed
♦ 87, 88, 269
♦ arid, cultivated. Origin: Australia.

Pterocaulon virgatum (L.) DC.
Asteraceae
♦ wand blackroot, barbasco, verbasco do Brasil
♦ Weed
♦ 87, 88, 255
♦ Origin: Americas.

Pterocephalus Adans. spp.
Dipsacaceae
♦ Weed
♦ 221

Pterocephalus sanctus Decne.
Dipsacaceae
♦ Weed
♦ 221

Pterocypsela indica (L.) C.Shih
Asteraceae
♦ India lettuce, wild lettuce
♦ Weed
♦ 273

Pterogyne nitens Tul.
Fabaceae/Caesalpiniaceae
♦ amendoim
♦ Weed, Introduced
♦ 87, 88, 228, 255, 295
♦ T, arid, cultivated. Origin: Brazil.

Pterolepis glomeratum Miq.
Melastomataceae
♦ Weed
♦ 87, 88

Pterolepis pumila (Bonpl.) Cogn.
Melastomataceae
= *Pterolepis trichotoma* (Rottb.) Cogn. var. *pumila* (Bonpl.) Renner. *ined.* (NoR)
♦ Weed
♦ 157
♦ pH.

Pterolobium stellatum (Forssk.) Brenan
Fabaceae/Caesalpiniaceae
Cantuffa stellata (Forssk.) Chiov., *Pterolobium abyssinicum* (A.Rich.) A.Rich., *Pterolobium exosum* (J.Gmel.) Bak.f.
♦ katdoring, rank wan n bietjie, vlam wag n bietjie
♦ Native Weed
♦ 121
♦ pS, arid. Origin: southern Africa.

Pteronia incana (Burm.) DC.
Asteraceae
♦ bluebush
♦ Native Weed
♦ 121
♦ pS, arid. Origin: southern Africa.

Pteronia pallens L.f.
Asteraceae
♦ scholtz bush
♦ Weed, Native Weed
♦ 39, 121
♦ S, toxic. Origin: southern Africa.

Ptilostemon afer (Jacq.) Greuter
Asteraceae
♦ ivory thistle
♦ Naturalised
♦ 280
♦ cultivated.

Ptychosperma aff. lineare (Burret) Burret
Arecaceae
♦ Introduced
♦ 230
♦ T.

Ptychosperma elegans (R.Br.) Blume
Arecaceae
♦ solitaire palm
♦ Weed, Naturalised
♦ 101, 179
♦ cultivated. Origin: Australia.

Ptychosperma hosinoi (Kaneh.) Moore & Fosb.
Arecaceae
♦ kedei
♦ Introduced
♦ 230
♦ T.

Ptychosperma ledermaniana (Becc.) Moore & Fosb.
Arecaceae
♦ kedei
♦ Introduced
♦ 230
♦ T.

Puccinellia airoides (Nutt.) Wats. & Coult.
Poaceae
= *Puccinellia nuttalliana* (Schult.) A.Hitchc. (NoR)
♦ Nuttall alkaligrass
♦ Weed
♦ 87, 88, 218
♦ G, herbal.

Puccinellia angusta (Nees) Sm. & Hubb
Poaceae
♦ finch alkaligrass
♦ Native Weed
♦ 121
♦ pG. Origin: southern Africa.

Puccinellia ciliata Bor.
Poaceae
♦ puccinellia
♦ Weed, Sleeper Weed, Naturalised, Environmental Weed
♦ 7, 86, 98, 155, 203
♦ G, arid. Origin: Turkey.

Puccinellia distans (Jacq.) Parl.
Poaceae
Atropis distans Rupr., *Festuca distans* (Wahlb.) Kunth, *Glyceria distans* Wahlb.
♦ weeping alkaligrass, reflexed saltmarsh grass, reflexed poa, kujasorsimo
♦ Weed, Naturalised, Introduced, Environmental Weed
♦ 86, 87, 88, 98, 176, 203, 218, 228, 280, 287
♦ pG, arid, cultivated, herbal. Origin: Eurasia, Mediterranean.

Puccinellia fasciculata (Torr.) Bickn.
Poaceae
Festuca borreri Bab., *Glyceria borreri* (Bab.) Bab., *Sclerochloa borreri* (Bab.) Bab.
♦ saltmarsh alkaligrass, Borrer's saltmarsh grass, tufted saltmarsh grass

♦ Weed, Naturalised, Environmental
Weed
♦ 72, 86, 88, 98, 198, 203, 280
♦ pG, cultivated, herbal. Origin:
Eurasia, Mediterranean.

Puccinellia lemmoni (Vasey) Scribn.
Poaceae
♦ Lemmon alkaligrass
♦ Weed
♦ 87, 88, 218
♦ G.

Puccinellia maritima (Huds.) Parl.
Poaceae
Glyceria maritima (Huds.) Wahlenb.,
Poa maritima Huds. *non* Savi,
Sclerochloa maritima (Huds.) Lindl. ex
Bab.
♦ seaside alkaligrass, common
saltmarsh grass, sea poa
♦ Naturalised
♦ 101
♦ G, cultivated, herbal.

Puccinellia rupestris (With.) Fern. &
Weath.
Poaceae
Glyceria rupestris (Wither.) Marsh., *Poa
rupestris* Wither., *Sclerochloa rupestris*
(Wither.) Britt. & Rendle
♦ British alkaligrass, procumbent poa,
stiff saltmarsh grass
♦ Naturalised
♦ 101, 280
♦ G, cultivated.

Puccinellia stricta (Hook.f.) Blom
Poaceae
♦ Weed
♦ 87, 88
♦ G, cultivated.

Puccinellia tenuiflora (Griseb.) Scribn.
& Merr.
Poaceae
♦ Weed
♦ 275, 297
♦ a/pG.

Pueraria harmsii Rech.
Fabaceae/Papilionaceae
♦ Weed
♦ 3, 191

Pueraria lobata (Willd.) Owhi
Fabaceae/Papilionaceae
= *Pueraria montana* (Lour.) Merr. var.
lobata (Willd.) Maesen & Almeida
♦ kudzu vine, Japanese arrowroot,
kudzu, acha, nepalem, aka, a'a, yaka,
wa yaka, nggariaka
♦ Weed, Noxious Weed, Naturalised,
Environmental Weed, Cultivation
Escape, Casual Alien
♦ 3, 10, 17, 18, 63, 86, 88, 95, 98, 102,
121, 133, 134, 152, 161, 191, 195, 204,
207, 211, 218, 224, 262, 263, 280, 283,
286, 297
♦ pC, arid, cultivated, herbal. Origin:
Asia.

Pueraria montana (Lour.) Merr.
Fabaceae/Papilionaceae
Dolichos montanus Lour., *Glycine
javanica* L., *Pachyrhizus montanus*

(Lour.) DC., *Pueraria thunbergiana*
(Siebold & Zucc.) Benth. var. *formosana*
Hosok., *Pueraria tonkinensis* Gagnep.
♦ kudzu
♦ Weed, Quarantine Weed,
Naturalised
♦ 76, 77, 80, 88, 101, 112, 243
♦ pC. Origin: China, Japan.

Pueraria montana (Lour.) Merr. var.
lobata (Willd.) Maesen & Almeida
Fabaceae/Papilionaceae
Dolichos lobatus Willd., *Pueraria hirsuta*
(Thunb.) Matsum. *nom. illeg.*, *Pueraria
lobata* (Willd.) Ohwi (see), *Pueraria
thunbergiana* (Siebold & Zucc.) Benth.
(see), *Pueraria triloba* (Lour.) Makino
(see)
♦ kudzu, ye ge
♦ Weed, Noxious Weed, Naturalised,
Environmental Weed
♦ 101, 105, 129, 140, 151, 179, 229
♦ pC.

Pueraria novo-guineensis Warb.
Fabaceae/Papilionaceae
♦ Weed
♦ 3, 191

Pueraria phaseoloides (Roxb.) Benth
Fabaceae/Papilionaceae
Pueraria javanica (Benth.) Benth.
♦ tropical kudzu, puero
♦ Weed, Noxious Weed, Naturalised,
Environmental Weed, Cultivation
Escape
♦ 3, 32, 86, 87, 88, 101, 191, 229, 257,
261
♦ pC, cultivated, herbal. Origin:
tropical Asia.

Pueraria thomsonii Benth.
Fabaceae/Papilionaceae
= *Pueraria montana* (Lour.) Merr.
var. *thomsonii* (Benth.) Wiersema ex
D.B.Ward (NoR)
♦ Weed, Naturalised
♦ 87, 88, 287
♦ pC, promoted.

Pueraria thunbergiana (Siebold & Zucc.)
Benth.
Fabaceae/Papilionaceae
= *Pueraria montana* (Lour.) Merr. var.
lobata (Willd.) Maesen & Almeida
♦ kudzu, Thunberg's kudzu bean
♦ Weed
♦ 3, 88, 164, 191
♦ cultivated, herbal.

Pueraria triloba (Lour.) Makino
Fabaceae/Papilionaceae
= *Pueraria montana* (Lour.) Merr. var.
lobata (Willd.) Maesen & Almeida
♦ kudzu vine, kudzu
♦ Weed, Quarantine Weed
♦ 3, 76, 87, 88, 203

Pugionium cornutum (L.) Gaertn.
Brassicaceae
♦ cornuted pugionium
♦ Weed
♦ 297
♦ arid.

Pulicaria Gaertn. spp.
Asteraceae
♦ false fleabane
♦ Weed
♦ 221

Pulicaria arabica (L.) Cass.
Asteraceae
Inula arabica L., *Vicoa auriculata* auct.
non Cass. (see)
♦ ladies' false fleabane
♦ Weed, Naturalised
♦ 87, 88, 101, 185, 221

Pulicaria crispa (Forssk.) Benth. &
Hook.f. ex Oliv. & Hiern
Asteraceae
Francoeuria crispa (Forssk.) Cass. (see),
Aster crispus Forssk.
♦ Weed, Quarantine Weed
♦ 76, 87, 88, 203, 220
♦ arid, cultivated, herbal.

Pulicaria dysenterica (L.) Bernh.
Asteraceae
Inula dysenterica L. (see)
♦ fleabane, alpine fleabane, blue
fleabane, Canadian fleabane, lesser
fleabane, small fleabane, meadow false
fleabane, common fleabane
♦ Weed, Naturalised, Casual Alien
♦ 42, 70, 87, 88, 101, 272, 280
♦ pH, cultivated, herbal.

Pulicaria gnaphalodes (Vent) Boiss.
Asteraceae
♦ Weed
♦ 243

Pulicaria inuloides (Poir.) DC.
Asteraceae
♦ Weed
♦ 221

Pulicaria odora (L.) Rchb.
Asteraceae
♦ Weed
♦ 70, 272

Pulicaria orientalis Jaub. & Spach
Asteraceae
♦ Weed
♦ 88
♦ arid.

Pulicaria paludosa Link
Asteraceae
♦ Spanish false fleabane
♦ Weed, Naturalised
♦ 34, 70, 101
♦ a/pH.

Pulicaria prostrata (Gilib.) Asch.
Asteraceae
Aster pulicarius (L.) Scop., *Diplopappus
pulicarius* (L.) Bluff & Fingerh., *Inula
prostrata* Gilib., *Inula pulicaria* L.,
Inula undulata L., *Pulicaria undulata*
(L.) C.A.Mey. (see), *Pulicaria vulgaris*
Gaertn. (see)
♦ Weed
♦ 87, 88

Pulicaria scabra (Thunb.) Druce
Asteraceae
♦ Native Weed
♦ 121
♦ H, herbal. Origin: southern Africa.

Pulicaria undulata (L.) C.A.Mey.
Asteraceae
= *Pulicaria prostrata* (Gilib.) Asch.
♦ Weed
♦ 221
♦ herbal.

Pulicaria vulgaris Gaertn.
Asteraceae
= *Pulicaria prostrata* (Gilib.) Asch.
♦ lesser fleabane, small fleabane
♦ Weed
♦ 88, 94, 272
♦ cultivated, herbal.

Pulmonaria angustifolia L.
Boraginaceae
Pulmonaria media Host
♦ lungwort, cowslip
♦ Weed
♦ 272
♦ cultivated, herbal.

Pulmonaria officinalis L.
Boraginaceae
Pulmonaria maculosa Hayne
♦ lungwort, polmonaria, Jerusalem
sage, common lungwort
♦ Weed, Naturalised
♦ 40, 101, 272
♦ pH, cultivated, herbal.

Pulmonaria saccharata Mill.
Boraginaceae
♦ Bethlehem lungwort, Jerusalem sage
♦ Naturalised
♦ 101
♦ pH, cultivated, herbal.

Pulsatilla chinensis (Bunge) Regel
Ranunculaceae
♦ Chinese pulsatilla
♦ Weed
♦ 297
♦ pH, promoted, herbal.

Pulsatilla patens (L.) Mill.
Ranunculaceae
Anemone patens L.
♦ wild multifida crocus, American
pasqueflower, eastern pasqueflower,
kylmänkukka
♦ Weed
♦ 161
♦ pH, cultivated, herbal, toxic.

Pulsatilla pratensis (L.) Mill.
Ranunculaceae
Anemone pratensis L.
♦ small pasqueflower, pasqueflower
♦ Weed
♦ 272
♦ pH, cultivated.

Pulsatilla vulgaris Mill.
Ranunculaceae
Anemone pulsatilla L. (see)
♦ pasqueflower, European
pasqueflower
♦ Weed
♦ 161, 272
♦ pH, cultivated, herbal, toxic.

Pultenaea daphnoides J.C.Wendl.
Fabaceae/Papilionaceae
♦ largeleaf pea bush

♦ Naturalised
♦ 280
♦ cultivated.

Punica granatum L.
Lythraceae/Punicaceae
Granatum punicum St.-Lag.
♦ pomegranate, humma, melograno
♦ Weed, Naturalised, Introduced,
Garden Escape, Casual Alien
♦ 22, 39, 80, 86, 98, 101, 203, 228, 261,
280
♦ S, arid, cultivated, herbal, toxic.
Origin: Mediterranean to east Asia.

Pupalia atropurpurea (Lam.) Moq.
Amaranthaceae
♦ Weed
♦ 87, 88

Pupalia lappacea (L.) Juss.
Amaranthaceae
♦ sweet hearts, mamata, sawnee
♦ Weed, Naturalised, Environmental
Weed
♦ 7, 51, 86, 87, 88, 98, 121, 155, 203, 221,
243
♦ pH, arid, cultivated, herbal. Origin:
Eurasia.

Purshia tridentata (Pursh) DC.
Rosaceae
♦ bitterbrush, antelope bitterbrush,
desert bitterbrush, buckbrush,
greasewood
♦ Weed, Quarantine Weed
♦ 39, 76, 87, 88, 203, 218, 220
♦ S, cultivated, herbal, toxic.

Puya alpestris (Poepp.) C.Gay
Bromeliaceae
Pourretia alpestris Poepp., *Puya whytei*
Hook.f.
♦ Weed, Quarantine Weed
♦ 76, 88, 220
♦ cultivated.

Puya chilensis Molina
Bromeliaceae
Puya coarctata Fisch.
♦ Weed, Quarantine Weed
♦ 76, 88, 220
♦ pH, cultivated, herbal.

Puya coerulea Lindl.
Bromeliaceae
Pitcairnia coerulea (Lindl.) Benth. ex
Bak., *Pourretia coerulea* Miers, *Puya
rubricaulis* Steud.
♦ Weed, Quarantine Weed
♦ 76, 88, 220
♦ cultivated.

Puya venusta Phil.
Bromeliaceae
Pitcairnia sphaerocephala Bak., *Pitcairnia
venusta* Bak., *Puya gaudichaudii* Mez
♦ Weed, Quarantine Weed
♦ 76, 88, 220
♦ cultivated.

**Pycnanthemum flexuosum (Walter) Britt.,
Sterns & Pogg.**
Lamiaceae
♦ Appalachian mountain mint,
mountain mint

♦ Quarantine Weed, Naturalised
♦ 220, 287
♦ pH, cultivated, herbal.

Pycnocycla tomentosa Decne.
Apiaceae
♦ Weed
♦ 221

Pycreus albo-marginatus Nees
Cyperaceae
Cyperus albo-marginatus Mart. &
Schrad. ex Nees
♦ Weed
♦ 87, 88
♦ G.

Pycreus decumbens T.Koyama
Cyperaceae
♦ Weed
♦ 255
♦ pG. Origin: Brazil.

Pycreus ferrugineus C.B.Clarke
Cyperaceae
= *Pycreus intactus* (Vahl) Raynal
♦ Weed
♦ 87, 88
♦ G.

Pycreus flavescens (L.) Rchb.
Cyperaceae
= *Cyperus flavescens* L.
♦ Weed, Native Weed
♦ 87, 88, 121
♦ a/pG. Origin: Africa.

Pycreus globosus (All.) Rchb.
Cyperaceae
Cyperus flavidus Retz. (see), *Cyperus
globosus* All. (see), *Cyperus vulgaris*
Sieb. ex Kunth
♦ globose flatsedge
♦ Weed
♦ 274, 275, 297
♦ G, aqua.

Pycreus intactus (Vahl) Raynal
Cyperaceae
Pycreus ferrugineus C.B.Clarke (see)
♦ Weed
♦ 121
♦ pG.

Pycreus lanceolatus C.B.Clarke
Cyperaceae
♦ Weed, Quarantine Weed
♦ 76, 87, 88, 203, 220
♦ G.

Pycreus mundtii Nees
Cyperaceae
♦ Weed, Quarantine Weed
♦ 76, 87, 88, 203, 220
♦ G.

Pycreus nitens (Retz.) Nees
Cyperaceae
= *Cyperus pumilus* Nees
♦ Weed
♦ 87, 88
♦ G.

Pycreus nitidus (Lam.) J.Raynal
Cyperaceae
Pycreus lanceus Turrill
♦ Native Weed
♦ 121
♦ pG. Origin: southern Africa.

Pycreus odoratus (L.) Urb.
Cyperaceae
♦ Weed
♦ 235
♦ G.

Pycreus polystachyos (Rottb.) Beauv.
Cyperaceae
Chlorocyperus polystachys Rikli,
Cyperus holosericeus Link, *Cyperus
paniculatus* Rottb., *Cyperus polystachyos*
Rottb. (see), *Cyperus polystachyos* ssp.
holosericeus (Link) T.Koyama, *Cyperus
polystachyos* var. *leptostachyus* Boeck.
♦ bunchy sedge, field sedge
♦ Weed, Naturalised
♦ 88, 126, 134, 274
♦ a/pG, aqua, herbal. Origin:
Madagascar.

Pycreus polystachyos (Rottb.) Beauv. var. *polystachyos*
Cyperaceae
♦ Weed
♦ 121
♦ pG.

Pycreus propinquus Nees
Cyperaceae
Cyperus fasciculatus Ell.
♦ Weed
♦ 87, 88
♦ G.

Pycreus sanguinolentus (Vahl) Nees
Cyperaceae
Cyperus sanguinolentus (Vahl) Nees
(see)
♦ Weed
♦ 126, 275, 297
♦ a/pG, aqua. Origin: East Africa,
Asia, Australia.

Pycreus tremulus (Poir.) C.B.Clarke
Cyperaceae
♦ Weed, Quarantine Weed
♦ 76, 87, 88, 203, 220
♦ G.

Pygmaeocereus Johnson & Backeb. spp.
Cactaceae
= *Haageocereus* Backeb. spp.
♦ Quarantine Weed
♦ 220

Pygmaeothamnus zeyheri Sond. var. *zeyheri*
Rubiaceae
Fadogia zeyheri (Sond.) Hiern
♦ sand apple
♦ Native Weed
♦ 121
♦ pH. Origin: southern Africa.

Pygmea ciliolata Hook.f.
Scrophulariaceae
♦ Weed, Naturalised
♦ 98, 203

Pyracantha Roem. spp.
Rosaceae
♦ firethorn, pyracantha
♦ Weed, Naturalised, Garden Escape,
Environmental Weed
♦ 86, 116, 161, 198, 277, 279
♦ cultivated, herbal, toxic.

Pyracantha angustifolia (Franch.) C.K.Schneid.
Rosaceae
Cotoneaster angustifolia Franch.
♦ orange firethorn, firethorn, yellow
firethorn, geelbranddoring, narrowleaf
firethorn
♦ Weed, Quarantine Weed, Noxious
Weed, Naturalised, Garden Escape,
Environmental Weed, Cultivation
Escape
♦ 20, 34, 63, 72, 76, 80, 86, 88, 95, 98,
101, 121, 155, 198, 203, 225, 233, 246,
269, 279, 280, 283, 287, 289, 296
♦ S/T, cultivated, herbal, toxic. Origin:
Asia.

Pyracantha coccinea M.Roem.
Rosaceae
Cotoneaster pyracantha (L.) Spach,
Crataegus pyracantha Voss, *Mespilus
pyracantha* L.
♦ firethorn, red firethorn, scarlet
firethorn
♦ Weed, Naturalised, Introduced,
Environmental Weed
♦ 38, 101, 121, 279, 290
♦ S/T, cultivated, herbal. Origin:
Eurasia.

Pyracantha crenatoserrata (Hance) Rehder
Rosaceae
♦ firethorn
♦ Weed, Naturalised, Environmental
Weed
♦ 15, 280, 290
♦ cultivated.

Pyracantha crenulata (D.Don) M.Roem.
Rosaceae
♦ Himalayan firethorn, Nepalese
whitethorn, rooivuurdoring, firethorn
♦ Weed, Quarantine Weed, Noxious
Weed, Naturalised, Garden Escape,
Environmental Weed, Cultivation
Escape
♦ 15, 63, 72, 76, 86, 88, 95, 98, 101, 198,
203, 269, 280, 283, 290
♦ S, cultivated, herbal, toxic. Origin:
China, Nepal, Bhutan, India.

Pyracantha fortuneana (Maxim) H.L.Li
Rosaceae
♦ Chinese firethorn, firethorn,
broadleaf firethorn
♦ Weed, Quarantine Weed,
Naturalised, Garden Escape,
Environmental Weed
♦ 76, 86, 88, 98, 101, 198, 203, 269, 289
♦ S, cultivated, herbal. Origin: China.

Pyracantha koidzumii (Hayata) Rehder
Rosaceae
♦ Formosa firethorn, firethorn
♦ Weed, Quarantine Weed,
Naturalised, Garden Escape,
Environmental Weed
♦ 54, 76, 86, 88, 98, 101, 203, 290
♦ S, cultivated. Origin: Taiwan.

Pyracantha rogersiana (A.B.Jacks) Bean
Rosaceae
♦ firethorn
♦ Weed, Naturalised, Garden Escape,

Environmental Weed
♦ 86, 98, 203, 290
♦ S, cultivated. Origin: China.

Pyrethrum santolinoides DC.
Asteraceae
= *Tanacetum santolinoides* (DC.)
Feinbrun & Fertig (NoR)
♦ Weed
♦ 221

Pyrostegia ignea (Vell.) C.Presl
Bignoniaceae
= *Pyrostegia venusta* (Ker Gawl.) Miers
♦ golden shower vine, flame vine,
flame flower
♦ Weed, Environmental Weed
♦ 73, 87, 88
♦ cultivated.

Pyrostegia venusta (Ker Gawl.) Miers
Bignoniaceae
Bignonia ignea Vell., *Bignonia
tecomaeflora* Rusby, *Bignonia venusta*
Ker Gawl., *Pyrostegia dichotoma* Miers
ex Schum., *Pyrostegia ignea* (Vell.)
C.Presl (see), *Pyrostegia intaminata*
Miers, *Pyrostegia pallida* Miers,
Pyrostegia parvifolia Miers, *Pyrostegia
reticulata* Miers, *Tecoma venusta* Lem.
♦ flame vine, flame flower, golden
shower vine
♦ Weed, Naturalised, Introduced,
Garden Escape, Environmental Weed
♦ 3, 19, 73, 86, 87, 88, 98, 101, 155, 179,
203, 246, 255, 261
♦ pC, cultivated, herbal. Origin: Brazil.

Pyrrhopappus carolinianus (Walter) DC.
Asteraceae
♦ Carolina false dandelion, false
dandelion, Carolina desert chicory
♦ Weed
♦ 87, 88, 161, 218, 243, 249
♦ a/bH, herbal.

Pyrrhopappus pauciflorus (D.Don) DC.
Asteraceae
♦ smallflower desert chicory, desert
chicory
♦ Weed
♦ 179

Pyrus angustifolia Aiton
Rosaceae
= *Malus angustifolia* (Ait.) Michx. var.
angustifolia (NoR)
♦ southern crabapple
♦ Weed
♦ 87, 88, 218
♦ herbal.

Pyrus arbutifolia (L.) L.f.
Rosaceae
= *Photinia pyrifolia* (Lam.) Robertson &
Phipps (NoR)
♦ red chokeberry
♦ Weed
♦ 87, 88, 218

Pyrus calleryana Decne.
Rosaceae
♦ Callery pear, Bradford pear
♦ Naturalised, Environmental Weed
♦ 86, 101, 155
♦ T, cultivated, herbal.

Pyrus communis L.
Rosaceae
Pyrus domestica Medik., *Pyrus sativa*
Lam. & DC.
♦ wild pear, pear, common pear,
päärynäpuu
♦ Weed, Naturalised, Environmental
Weed, Garden Escape, Cultivation
Escape
♦ 42, 86, 88, 98, 101, 151, 198, 203, 252,
280
♦ T, cultivated, herbal. Origin:
probably of hybrid origin.

Pyrus coronaria L.
Rosaceae
= *Malus coronaria* (L.) Mill. var.
coronaria (NoR)
♦ sweet crabapple
♦ Weed
♦ 87, 88, 218
♦ herbal.

Pyrus ioensis (Wood) Bailey
Rosaceae
= *Malus ioensis* (Wood) Britt.
♦ prairie crabapple
♦ Weed
♦ 87, 88, 218
♦ herbal.

Pyrus malus L.
Rosaceae
= *Malus domestica* Borkh.
♦ apple
♦ Weed
♦ 87, 88, 218
♦ herbal.

Pyrus melanocarpa (Michx.) Willd.
Rosaceae
= *Photinia melanocarpa* (Michx.)
Robertson & Phipps (NoR)
♦ black chokeberry
♦ Weed
♦ 87, 88, 218, 294
♦ herbal.

Pyrus pyrifolia (Burm.f.) Nakai
Rosaceae
Ficus pyrifolia Burm.f., *Pyrus serotina*
Rehd.
♦ Chinese pear, oriental pear, nashi
pear, sand pear
♦ Naturalised
♦ 101
♦ T, cultivated, herbal.

Pyrus ussuriensis Maxim.
Rosaceae
Pyrus simonii Carr.
♦ Manchurian pear, Chinese pear
♦ Sleeper Weed, Naturalised,
Environmental Weed
♦ 86, 155
♦ T, cultivated.

Q

Quamoclit coccinea (L.) Moench
Convolvulaceae
= *Ipomoea coccinea* L.
♦ star glory
♦ Weed
♦ 87, 88
♦ cultivated, herbal.

Quassia amara L.
Simaroubaceae
♦ quassia wood, bitterwood, quassia
♦ Naturalised, Introduced, Casual
Alien
♦ 39, 101, 230, 261
♦ S, cultivated, herbal, toxic. Origin:
tropical America.

Quercus L. spp.
Fagaceae
♦ oak
♦ Weed, Naturalised
♦ 88, 154, 198, 247
♦ herbal, toxic.

Quercus acutissima Carruth.
Fagaceae
♦ sawtooth oak
♦ Weed, Naturalised
♦ 80, 101
♦ T, cultivated, herbal.

Quercus agrifolia Nee
Fagaceae
♦ California live oak, encina, coast live
oak, Californian field oak, liveforever
♦ Weed
♦ 87, 88, 218
♦ T, cultivated, herbal.

Quercus alba L.
Fagaceae
♦ white oak
♦ Weed
♦ 8, 39, 87, 88, 218
♦ T, cultivated, herbal, toxic.

Quercus arizonica Sarg.
Fagaceae
♦ Arizona white oak
♦ Weed
♦ 87, 88, 218
♦ cultivated, herbal.

Quercus bicolor Willd.
Fagaceae
♦ swamp white oak
♦ Weed
♦ 87, 88, 218
♦ T, cultivated, herbal.

Quercus canariensis Willd.
Fagaceae
♦ Algerian oak
♦ Weed, Naturalised
♦ 54, 86, 88
♦ cultivated. Origin: Spain, Portugal,
Morocco.

Quercus cerris L.
Fagaceae
Quercus aegilops Scop.
♦ turkey oak, European turkey oak
♦ Weed, Naturalised
♦ 40, 87, 88, 101, 280
♦ T, cultivated, herbal.

Quercus chrysolepis Liebm.
Fagaceae
♦ canyon live oak, live oak
♦ Weed
♦ 87, 88, 218
♦ T, cultivated, herbal.

Quercus coccinea Muenchh.
Fagaceae
♦ scarlet oak
♦ Weed
♦ 39, 87, 88, 161, 218
♦ T, cultivated, herbal, toxic.

Quercus douglasii Hook. & Arn.
Fagaceae
♦ blue oak
♦ Weed
♦ 87, 88, 161, 218
♦ T, cultivated, herbal, toxic.

Quercus dumosa Nutt.
Fagaceae
♦ California scrub oak, coastal sage
scrub oak
♦ Weed
♦ 87, 88, 218
♦ S, cultivated, herbal.

Quercus durandii Buckl.
Fagaceae
= *Quercus sinuata* Walt. var. *sinuata*
(NoR)
♦ Weed
♦ 88

**Quercus durandii Buckl. var. breviloba
(Torr.) Palmer**
Fagaceae
= *Quercus sinuata* Walt. var. *breviloba*
(Torr.) Muller (NoR)
♦ bigleaf shin oak
♦ Weed
♦ 218

Quercus durata Jeps.
Fagaceae
♦ leather oak
♦ Weed
♦ 87, 88, 218
♦ S, cultivated, herbal.

Quercus ellipsoidalis E.J.Hill
Fagaceae
♦ northern pin oak
♦ Weed
♦ 87, 88, 218
♦ T, cultivated, herbal.

Quercus emoryi Torr.
Fagaceae
- Emory's oak, black oak
- Weed
- 87, 88, 161, 218
- T, cultivated, herbal, toxic.

Quercus falcata Michx.
Fagaceae
- southern red oak
- Weed
- 87, 88, 218
- T, cultivated, herbal.

Quercus gambelii Nutt.
Fagaceae
Quercus utahensis (A.DC.) Rydb.
- Gambel's oak, shin oak
- Weed
- 87, 88, 161, 218
- S, cultivated, herbal, toxic.

Quercus garryana Douglas ex Hook.
Fagaceae
- Oregon white oak, western oak, Oregon oak, Garry oak
- Weed
- 87, 88, 218
- T, cultivated, herbal.

Quercus havardii Rydb.
Fagaceae
- sand shinnery oak, Havard's oak, shinnery oak
- Weed
- 39, 87, 88, 161, 218
- herbal, toxic.

Quercus hypoleucoides A.Camus
Fagaceae
- silverleaf oak
- Weed
- 87, 88, 218
- cultivated, herbal.

Quercus ilex L.
Fagaceae
Quercus smilax L.
- holly oak, Holm oak, bellotas, ballota, evergreen oak
- Naturalised, Garden Escape, Environmental Weed
- 40, 86, 280
- T, cultivated, herbal. Origin: Mediterranean.

Quercus ilicifolia Wangenh.
Fagaceae
- bear oak
- Weed
- 87, 88, 218
- cultivated, herbal.

Quercus imbricaria Michx.
Fagaceae
- shingle oak
- Weed
- 87, 88, 218
- T, cultivated, herbal.

Quercus incana Bartr.
Fagaceae
- bluejack oak
- Weed
- 87, 88, 218
- cultivated, herbal.

Quercus kelloggii Newb.
Fagaceae
Quercus californica (Torr.) Cooper
- California black oak
- Weed
- 87, 88, 161, 218
- T, cultivated, herbal, toxic.

Quercus laevis Walter
Fagaceae
Quercus catesbaei Michx.
- turkey oak, American turkey oak
- Weed
- 87, 88, 218
- T, cultivated, herbal.

Quercus laurifolia Michx.
Fagaceae
- laurel oak, laurel leaved oak
- Weed
- 87, 88, 218
- cultivated, herbal.

Quercus lobata Nee
Fagaceae
Quercus hindsii Benth.
- California white oak, valley oak
- Weed
- 87, 88, 218
- T, cultivated, herbal.

Quercus lyrata Walter
Fagaceae
- overcup oak
- Weed
- 87, 88, 218
- T, cultivated, herbal.

Quercus macrocarpa Michx.
Fagaceae
- burr oak
- Weed
- 87, 88, 218
- T, cultivated, herbal.

Quercus marilandica Muenchh.
Fagaceae
- blackjack oak, blackjack
- Weed
- 39, 87, 88, 161, 218
- T, cultivated, herbal, toxic.

Quercus michauxii Nutt.
Fagaceae
- swamp chestnut oak
- Weed
- 87, 88, 218
- T, cultivated, herbal.

Quercus muehlenbergii Engelm.
Fagaceae
- chinquapin oak, chinkapin oak, yellow chestnut oak
- Weed
- 87, 88, 218
- T, cultivated, herbal.

Quercus myrtifolia Willd.
Fagaceae
- myrtle oak
- Weed
- 87, 88, 218
- cultivated, herbal.

Quercus nigra L.
Fagaceae
Quercus marylandica Du Roi

- water oak
- Weed
- 87, 88, 218
- T, cultivated, herbal.

Quercus oblongifolia Torr.
Fagaceae
- Mexican blue oak, live oak
- Weed
- 87, 88, 218
- S, cultivated, herbal.

Quercus palustris Muenchh.
Fagaceae
- pin oak, Spanish oak, oak
- Weed
- 87, 88, 181, 218
- T, cultivated, herbal. Origin: northeastern and central North America.

Quercus phellos L.
Fagaceae
- willow oak
- Weed
- 87, 88, 218
- T, cultivated, herbal.

Quercus prinoides Willd.
Fagaceae
- dwarf chinquapin oak, dwarf chinkapin oak, chinquapin oak
- Weed
- 87, 88, 218
- S, cultivated, herbal.

Quercus prinus L.
Fagaceae
- chestnut oak, basket oak, rock chestnut oak
- Weed
- 39, 87, 88, 161, 218
- T, cultivated, herbal, toxic.

Quercus pubescens Willd.
Fagaceae
- downy oak
- Weed
- 87, 88
- T, cultivated, herbal.

Quercus pungens Liebm.
Fagaceae
- sandpaper oak, pungent oak
- Weed
- 87, 88, 218
- S, promoted, herbal.

Quercus robur L.
Fagaceae
Quercus femina Mill, *Quercus fructipendula* Schrank
- English oak, pedunculate oak, common oak, truffle oak, oak tree
- Weed, Naturalised, Garden Escape, Environmental Weed, Cultivation Escape
- 15, 39, 80, 86, 88, 95, 98, 101, 121, 151, 161, 195, 198, 203, 280, 283
- T, cultivated, herbal, toxic. Origin: Eurasia.

Quercus robur L. × cerris L.
Fagaceae
- hybrid turkey oak
- Cultivation Escape
- 40
- cultivated.

Quercus rubra L.
Fagaceae
Quercus borealis Michx.
♦ northern red oak, red oak
♦ Weed, Naturalised
♦ 8, 39, 87, 88, 161, 218, 280
♦ T, cultivated, herbal, toxic.

Quercus sessiliflora Salisb.
Fagaceae
= *Quercus petraea* (Matt.) Liebl. (NoR)
♦ durmast oak, sessile oak
♦ Weed
♦ 87, 88
♦ herbal.

Quercus shumardii Buckl.
Fagaceae
♦ Shumard's red oak, Shumard's oak
♦ Weed
♦ 87, 88, 218
♦ T, cultivated, herbal.

Quercus sinuata Walter
Fagaceae
♦ bastard oak, bastard breviloba oak
♦ Weed
♦ 161
♦ herbal, toxic.

Quercus stellata Wangenh.
Fagaceae
♦ post oak
♦ Weed
♦ 39, 87, 88, 161, 218
♦ T, cultivated, herbal, toxic.

Quercus suber L.
Fagaceae
♦ cork oak
♦ Weed, Naturalised
♦ 54, 86, 88, 198
♦ T, cultivated, herbal. Origin:
southern Europe to North Africa.

Quercus turbinella Greene
Fagaceae
Quercus subturbinella Trel.
♦ shrub live oak, Sonoran scrub oak,
grey oak
♦ Weed
♦ 87, 88, 218
♦ S, arid, herbal.

Quercus vaccinifolia Kellogg
Fagaceae
♦ huckleberry oak
♦ Weed
♦ 87, 88, 218
♦ S, herbal.

Quercus velutina Lam.
Fagaceae
♦ black oak, yellow barked oak
♦ Weed
♦ 39, 87, 88, 161, 218
♦ T, cultivated, herbal, toxic.

Quercus virginiana Mill.
Fagaceae
♦ live oak, southern live oak
♦ Weed
♦ 87, 88, 218
♦ T, cultivated, herbal.

Quercus wislizenii A.DC.
Fagaceae

♦ interior live oak, live oak
♦ Weed
♦ 87, 88, 218
♦ S, cultivated, herbal.

Quiabentia Britt. & Rose spp.
Cactaceae
♦ Weed, Quarantine Weed
♦ 76, 88, 203, 220

Quincula lobata (Torr.) Raf.
Solanaceae
Physalis lobata Torr. (see)
♦ purple flowered groundcherry,
Chinese lantern
♦ Weed
♦ 49
♦ herbal.

Quisqualis indica L.
Combretaceae
♦ Rangoon creeper
♦ Weed, Sleeper Weed, Naturalised,
Introduced, Garden Escape,
Environmental Weed, Cultivation
Escape
♦ 3, 39, 86, 93, 98, 101, 155, 203, 228,
261, 287
♦ pC, arid, cultivated, herbal, toxic.

R

Rabdosiella calycina (Benth.) Codd
Lamiaceae
♦ Quarantine Weed
♦ 220
♦ cultivated.

**Racosperma baileyanum (F.Muell.)
Pedley**
Fabaceae/Mimosaceae
= *Acacia baileyana* F.Muell.
♦ Naturalised
♦ 280

Racosperma dealbatum (Link) Pedley
Fabaceae/Mimosaceae
= *Acacia dealbata* Link
♦ Naturalised
♦ 280

Racosperma decurrens (Willd.) Pedley
Fabaceae/Mimosaceae
= *Acacia decurrens* (J.C.Wendl.) Willd.
♦ Naturalised
♦ 280

Racosperma elatum (Benth.) Pedley
Fabaceae/Mimosaceae
= *Acacia elata* A.Cunn. ex Benth.
♦ Naturalised
♦ 280

Racosperma floribundum (Vent.) Pedley
Fabaceae/Mimosaceae
= *Acacia floribunda* (Vent.) Willd.
♦ Naturalised
♦ 280

**Racosperma longifolium (Andrews)
C.Mart.**
Fabaceae/Mimosaceae
= *Acacia longifolia* (Andrews) Willd.
♦ Naturalised
♦ 280

Racosperma mearnsii (De Wild.) Pedley
Fabaceae/Mimosaceae
= *Acacia mearnsii* De Wild.
♦ black wattle
♦ Weed, Naturalised
♦ 165, 280
♦ Origin: south-eastern Australia.

Racosperma melanoxylon (R.Br.) Mart.
Fabaceae/Mimosaceae
= *Acacia melanoxylon* R.Br.
♦ Tasmanian blackwood
♦ Sleeper Weed, Naturalised
♦ 225, 280
♦ T.

Racosperma paradoxum (DC.) Mart.
Fabaceae/Mimosaceae
= *Acacia paradoxa* DC.

♦ Naturalised
♦ 280
♦ Origin: Australia.

Racosperma parramattense **(Tindale) Pedley**
Fabaceae/Mimosaceae
= *Acacia parramattensis* Tindale
♦ Naturalised
♦ 280

Racosperma podalyriifolium **(G.Don) Pedley**
Fabaceae/Mimosaceae
= *Acacia podalyriifolia* A.Cunn. ex G.Don
♦ Casual Alien
♦ 280
♦ S/T.

Racosperma sophorae **(Labill.) Mart.**
Fabaceae/Mimosaceae
= *Acacia sophorae* (Labill.) R.Br.
♦ Naturalised
♦ 280

Racosperma strictum **(Andrews) Mart.**
Fabaceae/Mimosaceae
= *Acacia stricta* (Andrews) Willd.
♦ Casual Alien
♦ 280

Racosperma verticillatum **(L'Hér.) Mart.**
Fabaceae/Mimosaceae
= *Acacia verticillata* (L'Hér.) Willd.
♦ Naturalised
♦ 280

Radermachera pentandra **Hemsl.**
Bignoniaceae
♦ Casual Alien
♦ 280

Radermachera sinica **(Hance) Hemsl.**
Bignoniaceae
♦ China doll
♦ Environmental weed
♦ 201
♦ T, Cultivated. Origin: China.

Radiola linoides **Roth**
Linaceae
♦ allseed
♦ Naturalised
♦ 101, 243
♦ herbal.

Randia aculeata **L.**
Rubiaceae
Randia mitis L. (see)
♦ white indigoberry, indigo berry
♦ Weed
♦ 14, 39
♦ cultivated, herbal, toxic.

Randia armata **(Sw.) DC.**
Rubiaceae
Gardenia armata Sw., *Mussaenda spinosa* Jacq., *Randia spinosa* Karst.
♦ Introduced
♦ 228
♦ arid, cultivated, herbal.

Randia formosa **(Jacq.) K.Schum.**
Rubiaceae
Mussaenda formosa Jacq., *Randia mussaenda* (L.f.) DC.
♦ jasmin de rosa, blackberry jam,

blackberry jam fruit
♦ Naturalised, Garden Escape
♦ 101, 261
♦ cultivated. Origin: South America.

Randia mitis **L.**
Rubiaceae
= *Randia aculeata* L.
♦ Weed
♦ 87, 88
♦ herbal.

Randia spinifex **(Roem. & Schult.) Standl.**
Rubiaceae
♦ Weed
♦ 14

Ranunculus **L. spp.**
Ranunculaceae
♦ buttercup, crow's foot, ranunkler
♦ Weed
♦ 39, 154, 189, 243, 247, 272
♦ herbal, toxic.

Ranunculus abortivus **L.**
Ranunculaceae
♦ smallflower buttercup, early woodbuttercup, smallflower crow's foot
♦ Weed, Native Weed
♦ 39, 87, 88, 161, 174, 207, 210, 218, 249
♦ a/bH, cultivated, herbal, toxic. Origin: North America.

Ranunculus acer **L.**
Ranunculaceae
♦ meadow buttercup, tall field buttercup
♦ Weed
♦ 44, 87, 88

Ranunculus aconitifolius **L.**
Ranunculaceae
♦ white batchelor's button
♦ Weed
♦ 272
♦ cultivated, herbal.

Ranunculus acris **L.**
Ranunculaceae
Ranunculus acer auct. non L.
♦ tall buttercup, meadow buttercup, upright meadow crow's foot, giant buttercup, common buttercup, acrid buttercup
♦ Weed, Quarantine Weed, Noxious Weed, Naturalised, Introduced
♦ 15, 23, 34, 39, 52, 68, 70, 76, 80, 86, 87, 88, 101, 136, 136, 161, 162, 165, 176, 195, 198, 203, 210, 210, 212, 218, 229, 272, 280, 287, 299
♦ pH, cultivated, herbal, toxic. Origin: Europe, northern Asia.

Ranunculus acris **L. var.** *acris*
Ranunculaceae
♦ tall buttercup, showy buttercup
♦ Noxious Weed, Naturalised
♦ 101, 229

Ranunculus acris **L. var.** *frigidus* **Regel**
Ranunculaceae
♦ tall buttercup
♦ Noxious Weed
♦ 229

Ranunculus alismifolius **Geyer ex Benth.**
Ranunculaceae
Ranunculus alismaefolius Geyer ex Benth.
♦ plantainleaf buttercup, alisma leaved buttercup
♦ Weed
♦ 87, 88, 218
♦ pH, herbal.

Ranunculus amansii **Jord.**
Ranunculaceae
= *Ranunculus serpens* Schrank
♦ Quarantine Weed
♦ 220

Ranunculus apiifolius **Pers.**
Ranunculaceae
♦ apio del diablo
♦ Weed
♦ 295
♦ herbal.

Ranunculus aquatilis **L.**
Ranunculaceae
Batrachium aquatile (L.) Dumort., *Ranunculus radians* Revel
♦ white water crow's foot, water crow's foot, ojasätkin, common water crow's foot
♦ Weed
♦ 39, 70, 87, 88, 159, 272
♦ a/pH, aqua, cultivated, herbal, toxic.

Ranunculus arvensis **L.**
Ranunculaceae
♦ corn crow's foot, field buttercup, hunger weed, corn buttercup, peltoleinikki, buttercup
♦ Weed, Naturalised, Environmental Weed
♦ 23, 34, 39, 44, 70, 80, 86, 87, 88, 94, 98, 101, 115, 161, 203, 218, 221, 243, 253, 256, 272, 280, 287, 300
♦ aH, cultivated, herbal, toxic. Origin: Eurasia, North Africa.

Ranunculus asiaticus **L.**
Ranunculaceae
♦ turban buttercup, ranunculus
♦ Weed
♦ 39, 221, 243
♦ a/pH, cultivated, herbal, toxic.

Ranunculus auricomus **L.**
Ranunculaceae
♦ Greenland buttercup, Goldilocks buttercup, Goldilocks, kevätleinikki
♦ Weed
♦ 39, 70, 87, 88, 272
♦ cultivated, toxic.

Ranunculus bonariensis **Poir.**
Ranunculaceae
♦ Carter's buttercup
♦ Naturalised
♦ 101

Ranunculus bonariensis **Poir. var.** *trisepalus* **(Gillies ex Hook. & Arn.) Lourteig**
Ranunculaceae
♦ Carter's buttercup, vernal pool buttercup
♦ Naturalised
♦ 101
♦ aH.

Ranunculus bulbosus L.
Ranunculaceae
Ranunculus bulbifer Jordan, *Ranunculus valdepubens* Jordan
♦ bulbous buttercup, St. Anthony's turnip, mäkileinikki, bottone d'oro, bulbous crow's foot, yellow weed, blister flower, gowan
♦ Weed, Naturalised, Introduced
♦ 15, 23, 34, 39, 44, 70, 80, 87, 88, 101, 161, 211, 218, 253, 272, 280, 287
♦ pH, cultivated, herbal, toxic.

Ranunculus bullatus L.
Ranunculaceae
♦ Weed
♦ 70
♦ herbal.

Ranunculus californicus Benth.
Ranunculaceae
Ranunculus californicus Benth. var. *rugulosus* (Greene) L.Benson (see)
♦ California buttercup
♦ Weed
♦ 161
♦ pH, promoted, herbal, toxic.

Ranunculus californicus Benth. var. rugulosus (Greene) L.Benson
Ranunculaceae
= *Ranunculus californicus* Benth.
♦ Weed
♦ 180, 243

Ranunculus calthaefolius (Guss.) Jord.
Ranunculaceae
= *Ranunculus ficaria* L. ssp. *ficariiformis* (F.W.Schultz) Rouy & Foucaud (NoR)
♦ Weed, Quarantine Weed, Naturalised
♦ 76, 86, 87, 88

Ranunculus cantoniensis DC.
Ranunculaceae
♦ Weed
♦ 87, 88, 235, 263, 286
♦ pH, herbal.

Ranunculus canuti Coss. ex Ardoino
Ranunculaceae
♦ Quarantine Weed
♦ 220

Ranunculus chinensis Bunge
Ranunculaceae
♦ kokitsunenobotan
♦ Weed
♦ 275, 297
♦ pH, promoted.

Ranunculus circinatus Sibth.
Ranunculaceae
Batrachium circinatum (Sibth.) Spach, *Batrachium foeniculaceum auct. non* (Gilib.) V.I.Krecz., *Ranunculus capillaceus* Thuill., *Batrachium circinatum* (Sibth.) Fr., *Ranunculus divaricatus sensu* H.J.Coste *non* Schrank
♦ circular leaved crow's foot, buttercup, fan leaved water crow's foot, pyörösätkin
♦ Weed
♦ 87, 88, 272
♦ herbal.

Ranunculus cornutus DC.
Ranunculaceae
♦ Weed
♦ 111, 243

Ranunculus cortusifolius Willd.
Ranunculaceae
Ranunculus cortusaefolius Willd.
♦ Quarantine Weed
♦ 220
♦ cultivated.

Ranunculus cymbalaria Pursh
Ranunculaceae
♦ suolaleinikki, alkali buttercup, shore buttercup
♦ Weed, Naturalised
♦ 39, 42, 161
♦ arid, cultivated, herbal, toxic.

Ranunculus falcatus L.
Ranunculaceae
= *Ceratocephala falcata* (L.) Pers.
♦ Weed
♦ 87, 88
♦ herbal.

Ranunculus fascicularis Muhl. ex Bigelow
Ranunculaceae
♦ early buttercup
♦ Weed
♦ 161
♦ herbal, toxic.

Ranunculus ficaria L.
Ranunculaceae
Ficaria degenii Hervier, *Ficaria nudicaulis* A.Kern., *Ficaria ranunculoides* Roth, *Ficaria verna* Huds. (see), *Ficaria vulgaris* A.St.-Hil.
♦ lesser celandine, pilewort, fig buttercup, mukulaleinikki, lesser crow's foot, buttercup
♦ Weed, Naturalised, Environmental Weed, Casual Alien
♦ 17, 39, 44, 70, 80, 87, 88, 98, 101, 129, 133, 151, 161, 195, 203, 224, 253, 272, 280
♦ pH, aqua, cultivated, herbal, toxic. Origin: Europe.

Ranunculus ficaria L. ssp. bulbilifer Lambinon
Ranunculaceae
♦ Naturalised
♦ 280
♦ Origin: north and central Europe.

Ranunculus ficaria L. ssp. calthifolius (Rchb.) Arcang.
Ranunculaceae
Ficaria calthifolia Rchb., *Ranunculus calthifolius* Rchb.
♦ Casual Alien
♦ 280
♦ Origin: south, central and eastern Europe.

Ranunculus ficaria L. ssp. ficariiformis Rouy & Foucaud
Ranunculaceae
♦ celadine
♦ Weed, Naturalised
♦ 15, 280
♦ Origin: southern Europe.

Ranunculus ficaria L. var. bulbifera Marsden-Jones
Ranunculaceae
♦ fig buttercup
♦ Naturalised
♦ 101

Ranunculus flabellaris Raf.
Ranunculaceae
♦ yellow water buttercup, water buttercup
♦ Weed
♦ 87, 88, 218
♦ pH, herbal.

Ranunculus flammula L.
Ranunculaceae
♦ lesser spearwort, spearwort, spearwort buttercup, water buttercup, greater creeping spearwort
♦ Weed, Naturalised, Garden Escape, Environmental Weed
♦ 15, 39, 70, 72, 87, 88, 161, 165, 176, 181, 272, 280
♦ wpH, cultivated, herbal, toxic. Origin: southern Europe, Mediterranean.

Ranunculus flammula L. ssp. flammula
Ranunculaceae
♦ lesser spearwort
♦ Naturalised, Environmental Weed
♦ 86, 198
♦ cultivated.

Ranunculus fluitans Lam.
Ranunculaceae
Batrachium fluitans (Lam.) Wimm.
♦ river crow's foot, water buttercup, river water crow's foot
♦ Weed
♦ 23, 87, 88, 208
♦ aqua, cultivated.

Ranunculus friesianus Jord.
Ranunculaceae
Ranunculus acris L. ssp. *friesianus* (Jordan) Rouy & Fouc., *Ranunculus acris* ssp. *scandinavicus* (Orlova) Á. & D.Löve, *Ranunculus scandinavicus* Orlova, *Ranunculus silvaticus* Fr.
♦ Weed
♦ 44
♦ toxic.

Ranunculus garganicus Ten.
Ranunculaceae
♦ Quarantine Weed
♦ 220

Ranunculus grandifolius C.A.Mey.
Ranunculaceae
♦ Quarantine Weed
♦ 220

Ranunculus hederaceus L.
Ranunculaceae
Batrachyum hederaceum (L.) S.F.Gray, *Ranunculus asarifolius* Diard
♦ ivy leaved crow's foot, ivy buttercup, ivy leaved water crow's foot
♦ Weed
♦ 23, 88

Ranunculus heucheraefolius Presl
Ranunculaceae

♦ Quarantine Weed
♦ 220

Ranunculus illyricus L.
Ranunculaceae
♦ Weed
♦ 272

Ranunculus inundatus R.Br. ex DC.
Ranunculaceae
♦ river buttercup
♦ Native Weed
♦ 39, 269
♦ pH, cultivated, toxic. Origin: Australia.

Ranunculus japonicus Thunb.
Ranunculaceae
♦ umanoashigata
♦ Weed
♦ 87, 88, 235, 275, 286, 297
♦ pH, promoted, herbal. Origin: east Asia.

Ranunculus lanuginosus L.
Ranunculaceae
♦ woolly buttercup, villaleinikki
♦ Weed
♦ 39, 272
♦ cultivated, herbal, toxic.

Ranunculus lappaceus Sm.
Ranunculaceae
♦ common buttercup, native buttercup, buttercup
♦ Native Weed
♦ 39, 269
♦ pH, cultivated, toxic. Origin: Australia.

Ranunculus lingua L.
Ranunculaceae
♦ greater spearwort, jokileinikki, great spearwort
♦ Weed
♦ 39, 272
♦ aqua, cultivated, herbal, toxic.

Ranunculus lomatocarpus Fisch. & Mey.
Ranunculaceae
♦ Weed
♦ 87, 88

Ranunculus macranthus Scheele
Ranunculaceae
♦ large buttercup
♦ Weed
♦ 199

Ranunculus marginatus d'Urv.
Ranunculaceae
♦ margined buttercup, St. Martin's buttercup
♦ Naturalised
♦ 40
♦ cultivated.

Ranunculus marginatus d'Urv. var. trachycarpus (Fisch. & C.A.Mey.) Arn.
Ranunculaceae
♦ margined buttercup
♦ Naturalised
♦ 101

Ranunculus megaphyllus Steud.
Ranunculaceae
♦ Quarantine Weed
♦ 220

Ranunculus meyerianus Rupr.
Ranunculaceae
= *Ranunculus polyanthemos* L.
♦ Quarantine Weed
♦ 220

Ranunculus micranthus Nutt.
Ranunculaceae
♦ rock buttercup, small flowered crow's foot
♦ Weed
♦ 161
♦ herbal, toxic.

Ranunculus millefoliatus Vahl
Ranunculaceae
♦ Quarantine Weed
♦ 220
♦ cultivated.

Ranunculus mixtus Jord.
Ranunculaceae
= *Ranunculus serpens* Schrank
♦ Quarantine Weed
♦ 220

Ranunculus montanus Willd.
Ranunculaceae
Ranunculus geraniifolius auct.,
Ranunculus nivalis Crantz
♦ mountain buttercup, vuorileinikki
♦ Weed, Casual Alien
♦ 42, 70, 272
♦ cultivated.

Ranunculus multifidus Forssk.
Ranunculaceae
♦ buttercup, wild buttercup, botterblom
♦ Weed, Native Weed
♦ 39, 121
♦ pH, cultivated, herbal, toxic. Origin: Madagascar.

Ranunculus muricatus L.
Ranunculaceae
♦ roughseed buttercup, sharp buttercup, buttercup, spinyfruit buttercup, prickle fruited buttercup, Scilly buttercup
♦ Weed, Naturalised, Introduced, Garden Escape, Environmental Weed
♦ 7, 34, 38, 40, 70, 86, 87, 88, 94, 98, 101, 134, 161, 176, 180, 198, 203, 218, 241, 253, 263, 269, 272, 280, 286, 287, 295, 300
♦ a/pH, cultivated, herbal, toxic. Origin: Mediterranean to India.

Ranunculus natans C.A.Mey.
Ranunculaceae
♦ water buttercup, nodding buttercup
♦ Weed
♦ 159
♦ herbal.

Ranunculus neapolitanus Ten.
Ranunculaceae
♦ Weed
♦ 272

Ranunculus nemorosus DC.
Ranunculaceae
= *Ranunculus serpens* Schrank
♦ pyökkileinikki
♦ Weed, Quarantine Weed, Casual Alien

♦ 42, 220, 272
♦ cultivated, herbal.

Ranunculus ophioglossifolius Vill.
Ranunculaceae
♦ adder's tongue spearwort, snake tongue buttercup
♦ Weed, Naturalised
♦ 86, 87, 88, 198, 272, 280
♦ aH, cultivated, herbal. Origin: Eurasia.

Ranunculus parviflorus L.
Ranunculaceae
♦ small flowered buttercup, smallflower buttercup, short buttercup, few flowered buttercup, buttercup
♦ Weed, Quarantine Weed, Naturalised
♦ 15, 34, 39, 70, 76, 80, 86, 88, 94, 98, 101, 134, 161, 176, 198, 203, 241, 243, 253, 280, 300
♦ aH, herbal, toxic. Origin: south-west Europe, Morocco, Azores, Madeira and Canary Islands.

Ranunculus pensylvanicus L.f.
Ranunculaceae
♦ Pennsylvania buttercup, Amerikanleinikki
♦ Casual Alien
♦ 42
♦ herbal.

Ranunculus petiolaris Kunth ex DC. ssp. arsenei H.B.K.
Ranunculaceae
♦ Weed
♦ 199

Ranunculus petiolaris Kunth ex DC. var. trahens H.B.K.
Ranunculaceae
♦ Weed
♦ 199

Ranunculus platensis Spreng.
Ranunculaceae
♦ prairie buttercup
♦ Naturalised
♦ 101
♦ herbal.

Ranunculus plebeius R.Br. ex DC.
Ranunculaceae
♦ common Australian buttercup
♦ Naturalised
♦ 101
♦ Origin: Australia.

Ranunculus polyanthemoides Schur
Ranunculaceae
= *Ranunculus serpens* Schrank ssp. *polyanthemoides* (Boreau) M.Kerguélen & J.Lambinon (NoR)
♦ Quarantine Weed
♦ 220

Ranunculus polyanthemophyllus W.Koch & H.Hess
Ranunculaceae
= *Ranunculus polyanthemos* L. ssp. *polyanthemophyllus* (W.Koch & H.Hess) M.Baltisb. (NoR)
♦ Quarantine Weed
♦ 220

Ranunculus polyanthemos L.
Ranunculaceae
Ranunculus meyerianus Rupr.
♦ aholeinikki
♦ Weed, Quarantine Weed
♦ 39, 220, 272
♦ cultivated, toxic.

Ranunculus pseudobulbosus Schur
Ranunculaceae
♦ romanianleinikki
♦ Casual Alien
♦ 42

Ranunculus pumilio R.Br. ex DC. var.
***politus* Melville**
Ranunculaceae
♦ Naturalised
♦ 86
♦ Origin: Australia.

Ranunculus quelpaertensis (Lev.) Nakai
Ranunculaceae
♦ Weed
♦ 87, 88
♦ pH, promoted.

Ranunculus radicescens Jord.
Ranunculaceae
♦ Quarantine Weed
♦ 220

Ranunculus recurvatus Poir.
Ranunculaceae
♦ blister wort, hooked buttercup
♦ Weed
♦ 161
♦ herbal, toxic.

Ranunculus repens L.
Ranunculaceae
♦ creeping buttercup, ranuncolo
strisciante, rönsyleinikki
♦ Weed, Quarantine Weed,
Naturalised, Introduced, Garden
Escape, Environmental Weed,
Cultivation Escape
♦ 15, 20, 23, 34, 38, 39, 44, 52, 70, 72, 76,
80, 86, 87, 88, 94, 98, 101, 118, 134, 136,
161, 165, 176, 181, 195, 198, 200, 203,
212, 217, 218, 220, 243, 253, 269, 272,
280, 290, 291, 295, 296, 300
♦ pH, aqua, cultivated, herbal, toxic.
Origin: Eurasia.

Ranunculus repens L. var. *repens*
Ranunculaceae
♦ Naturalised
♦ 241
♦ herbal.

Ranunculus reptans L.
Ranunculaceae
♦ pineleaf buttercup, itokinpouge,
creeping spearwort, rantaleinikki
♦ Weed
♦ 297
♦ herbal.

Ranunculus sardous Crantz
Ranunculaceae
Ranunculus hirsutus Curtis, *Ranunculus
philonotis* Ehrh.
♦ buttercup, hairy buttercup, pale
hairy buttercup, etelänleinikki

♦ Weed, Naturalised, Casual Alien
♦ 34, 42, 70, 80, 86, 87, 88, 94, 101, 161,
165, 176, 198, 243, 249, 253, 263, 272,
280, 286, 287
♦ pH, cultivated, herbal, toxic. Origin:
Europe.

Ranunculus sceleratus L.
Ranunculaceae
♦ celeryleaf buttercup, cursed
buttercup, celery buttercup, poison
buttercup, konnanleinikki, marsh
crow's foot, celery leaved crow's foot
♦ Weed, Quarantine Weed, Noxious
Weed, Naturalised, Introduced
♦ 15, 23, 34, 39, 70, 76, 86, 87, 88, 98,
161, 171, 176, 185, 200, 200, 203, 218,
220, 221, 235, 243, 263, 269, 272, 274,
275, 280, 286, 297, 300
♦ wH, cultivated, herbal, toxic. Origin:
Europe.

Ranunculus sceleratus L. ssp. *sceleratus*
Ranunculaceae
♦ celery buttercup
♦ Naturalised
♦ 86, 198
♦ cultivated.

Ranunculus septentrionalis Prior.
Ranunculaceae
= *Ranunculus hispidus* Michx. var.
nitidus (Chapman) T.Duncan (NoR)
♦ northern buttercup
♦ Weed
♦ 23, 39, 88, 161
♦ herbal, toxic.

Ranunculus serbicus Vis.
Ranunculaceae
♦ serbianleinikki
♦ Casual Alien
♦ 42

Ranunculus serpens Schrank
Ranunculaceae
Ranunculus amansii Jord. (see),
Ranunculus breyninus auct. non
Crantz, *Ranunculus mixtus* Jord. (see),
Ranunculus nemorosus DC. (see),
Ranunculus tuberosus Lapeyr. (see)
♦ Quarantine Weed
♦ 220

Ranunculus sessiliflorus R.Br. ex DC.
Ranunculaceae
♦ Naturalised
♦ 39, 134
♦ cultivated, toxic.

Ranunculus sieboldii Miq.
Ranunculaceae
♦ Cantonese buttercup
♦ Weed
♦ 274, 275, 286, 297
♦ pH.

Ranunculus silerifolius H.Lév.
Ranunculaceae
♦ Weed
♦ 286

Ranunculus teneriffae Pers.
Ranunculaceae
♦ Quarantine Weed
♦ 220

Ranunculus ternatus Thunb.
Ranunculaceae
Ranunculus zuchharinii Miq.
♦ Weed
♦ 87, 88
♦ pH, promoted.

Ranunculus testiculatus Crantz
Ranunculaceae
= *Ceratocephala testiculata* (Crantz) Bess.
♦ testiculate buttercup, burr buttercup,
little burr, hornseed buttercup
♦ Weed
♦ 39, 80, 87, 88, 136, 161, 212, 218
♦ aH, herbal, toxic.

Ranunculus thomasii Ten.
Ranunculaceae
♦ Quarantine Weed
♦ 220

Ranunculus trichophyllus Chaix
Ranunculaceae
Batrachium aquatile (L.) Dumort. var.
trichophyllus (Chaix) Spach
♦ white water buttercup, water
buttercup, white water crow's foot,
threadleaf crow's foot, thread leaved
water crow's foot, purosätkin, dark
hair crow's foot
♦ Weed, Naturalised, Environmental
Weed
♦ 15, 23, 87, 88, 165, 181, 218, 246, 272,
280
♦ aqua, herbal.

Ranunculus trilobus Desf.
Ranunculaceae
♦ threelobe buttercup, large annual
buttercup, buttercup
♦ Weed, Naturalised, Environmental
Weed
♦ 7, 86, 98, 101, 176, 198, 203, 253
♦ cultivated, herbal.

Ranunculus tuberosus Lapeyr.
Ranunculaceae
♦ Quarantine Weed
♦ 220

Ranunculus velutinus Ten.
Ranunculaceae
♦ ranuncolo vellutato
♦ Weed
♦ 272
♦ herbal.

Raphanus maritimus Sm.
Brassicaceae
♦ sea radish
♦ Weed, Naturalised, Cultivation
Escape
♦ 86, 98, 176, 198, 203, 252
♦ cultivated, herbal. Origin: Europe.

Raphanus microcarpus Wilk.
Brassicaceae
♦ Weed
♦ 87, 88

Raphanus raphanistrum L.
Brassicaceae
Raphanistrum arvense Mérat, *Raphanus
silvestris* Lam., *Rapistrum arvense* All.,
Rapistrum raphanistrum Crantz
♦ jointed charlock, wild radish, jointed

charlock, white charlock, jointed radish, wild kale, wild turnip, cadlock, rabizon, runch, ravanello selvatico, peltoretikka, sea radish
♦ Weed, Noxious Weed, Naturalised, Garden Escape
♦ 7, 15, 23, 34, 36, 39, 44, 51, 52, 53, 55, 68, 70, 78, 80, 84, 86, 87, 88, 94, 98, 101, 114, 115, 116, 118, 121, 136, 147, 158, 161, 162, 176, 178, 179, 180, 198, 203, 210, 211, 212, 217, 218, 229, 236, 237, 241, 243, 245, 253, 255, 272, 280, 287, 295, 299, 300
♦ aH, cultivated, herbal, toxic. Origin: Eurasia.

***Raphanus raphanistrum* L. ssp. *maritimus* (Sm.) Thell.**
Brassicaceae
♦ sea radish
♦ Naturalised
♦ 280

Raphanus raphanistrum* L. ssp. *raphanistrum
Brassicaceae
♦ wild radish
♦ Weed, Naturalised
♦ 40, 165, 280
♦ toxic. Origin: Mediterranean, west Asia.

***Raphanus sativus* L.**
Brassicaceae
Raphanus chinensis Mill., *Raphanus officinalis* Crantz, *Raphanus raphanistrum* L. var. *sativus* Beck
♦ radish, wild radish, nabón, garden radish, salad radish, ruokaretikka, jointed charlock, runch, white charlock, wild charlock, wild kale, wild turnip, garden radish, salad radish
♦ Weed, Naturalised, Environmental Weed, Cultivation Escape, Casual Alien
♦ 15, 34, 40, 42, 56, 78, 80, 86, 87, 88, 98, 101, 116, 136, 161, 179, 198, 203, 212, 236, 237, 241, 243, 252, 255, 257, 261, 270, 280, 295, 300
♦ a/bH, cultivated, herbal. Origin: Europe.

***Raphanus sativus* L. fo. *raphanistroides* Mak.**
Brassicaceae
♦ Naturalised
♦ 287

***Raphanus sativus* L. var. *raphanistroides* Mak.**
Brassicaceae
♦ Weed, Naturalised
♦ 286, 287

***Raphia* P.Beauv. spp.**
Arecaceae
♦ Introduced
♦ 230

***Raphia vinifera* P.Beauv.**
Arecaceae
♦ West African piassava palm, mpeke, mosende
♦ Introduced

♦ 230
♦ T, herbal.

***Raphionacme procumbens* Schltr.**
Asclepiadaceae/Apocynaceae/Periplocaceae
♦ Native Weed
♦ 121
♦ pH, toxic. Origin: southern Africa.

***Rapistrum* Cran. spp.**
Brassicaceae
♦ bastard cabbage
♦ Weed, Naturalised
♦ 198, 272

***Rapistrum hispanicum* (L.) Crantz**
Brassicaceae
Myagrum hispanicum L.
♦ Weed
♦ 87, 88

***Rapistrum orientale* DC.**
Brassicaceae
♦ Weed
♦ 87, 88

***Rapistrum perenne* (L.) All.**
Brassicaceae
Cakile perennis L'Hér., *Crucifera rapistra* Krause, *Myagrum biarticulatum* Crantz, *Rapistrum costatum* DC., *Schrankia divaricata* Moench
♦ sileäsaksanretikka, perennial bastard cabbage, steppe cabbage
♦ Weed, Casual Alien
♦ 42, 272
♦ cultivated.

***Rapistrum rugosum* (L.) All.**
Brassicaceae
Caulis rugosus Krause, *Crucifera rugosa* Krause, *Myagrum rugosum* L., *Schrankia rugosa* Medik.
♦ turnip weed, wild turnip, annual bastard cabbage, giant mustard, bastard cabbage, mostacilla, kurttusaksanretikka
♦ Weed, Noxious Weed, Naturalised, Introduced, Casual Alien
♦ 7, 34, 42, 55, 68, 70, 86, 87, 88, 93, 94, 98, 101, 121, 134, 147, 158, 176, 198, 199, 203, 205, 228, 236, 237, 241, 243, 253, 255, 269, 271, 272, 280, 287, 295, 300
♦ aH, arid, cultivated, herbal. Origin: Eurasia.

***Rapistrum rugosum* (L.) All. ssp. *orientale* (L.) Arcang.**
Brassicaceae
♦ oriental bastard cabbage
♦ Naturalised
♦ 101

Rapistrum rugosum* (L.) All. ssp. *rugosum
Brassicaceae
♦ annual bastard cabbage
♦ Naturalised
♦ 101
♦ herbal.

***Rapistrum rugosum* (L.) All. var. *microcarpum* Thell.**
Brassicaceae
♦ Naturalised
♦ 287

***Rathbunia* Britt. & Rose spp.**
Cactaceae
♦ Weed, Quarantine Weed
♦ 76, 88, 203

***Ratibida columnaris* (Sims) D.Don**
Asteraceae
= *Ratibida columnifera* (Nutt.) Woot. & Standl.
♦ upright prairie coneflower, yellow Mexican hat, Mexican hat
♦ Weed
♦ 23
♦ promoted, herbal.

***Ratibida columnifera* (Nutt.) Woot. & Standl.**
Asteraceae
Ratibida columnaris (Sims) D.Don (see), *Lepachys columnifera* (Nutt.) J.F.Macbr.
♦ prairie coneflower, upright prairie coneflower, yellow Mexican hat, Mexican hat, pylväspäivänhattu
♦ Weed, Quarantine Weed, Casual Alien
♦ 42, 87, 88, 161, 218, 220
♦ pH, cultivated, herbal.

***Ratibida pinnata* (Vent.) Barnh.**
Asteraceae
♦ pinnate coneflower, pinnate prairie coneflower, prairie coneflower
♦ Weed
♦ 87, 88, 218
♦ a/pH, cultivated, herbal.

***Rauhocereus* Backeb. spp.**
Cactaceae
= *Weberbauerocereus* Britt. & Rose spp.
♦ Weed, Quarantine Weed
♦ 76, 88, 203

***Raulinoreitzia tremula* (Hook. & Arn.) King & Rob.**
Asteraceae
Eupatorium tremulum Hook. & Arn.
♦ Weed
♦ 87, 88

***Rauvolfia ligustrina* Willd. ex Roem. & Schult.**
Apocynaceae
♦ Weed
♦ 14
♦ herbal.

***Rauvolfia salicifolia* Griseb.**
Apocynaceae
♦ Weed
♦ 14

***Rauvolfia tetraphylla* L.**
Apocynaceae
♦ be still tree
♦ Weed, Naturalised
♦ 14, 86, 157, 179
♦ S/T, cultivated, herbal. Origin: Central and South America, Caribbean.

***Rauvolfia vomitoria* Afzel.**
Apocynaceae
♦ poison devil's pepper, likete
♦ Naturalised
♦ 101, 261
♦ cultivated, herbal. Origin: tropical Africa.

Ravenala madagascariensis **Sonn.**
Musaceae/Strelitziaceae
♦ traveller's tree, traveller's palm, Madagascar traveller's tree, travel palm
♦ Weed, Naturalised, Environmental Weed, Casual Alien
♦ 3, 22, 101, 152, 191, 230, 261
♦ T, cultivated, herbal. Origin: Madagascar.

Reaumuria hirtella **Jaub. & Spach**
Tamaricaceae
♦ Weed
♦ 221
♦ cultivated.

Reaumuria vermiculata **L.**
Tamaricaceae
♦ Weed
♦ 221

Regnellidium diphyllum **Lindm.**
Marsileaceae
♦ waterclover
♦ Weed, Quarantine Weed
♦ 76, 87, 88, 203, 220
♦ aqua, cultivated.

Rehmannia glutinosa **(Gaertn.) Steud.**
Scrophulariaceae
Digitalis glutinosa Gaertn., *Rehmannia chinensis* Libosch. ex Fisch. & C.A.Mey., *Rehmannia glutinosa* fo. *huechingensis* (Chao & Shih) P.G.Xiao, *Rehmannia glutinosa* fo. *purpurea* Matsuda, *Rehmannia glutinosa* var. *hemsleyana* Diels, *Rehmannia glutinosa* var. *huechingensis* Chao & Shih
♦ Chinese foxglove, jiwhang
♦ Weed
♦ 68, 88, 275, 297
♦ pH, promoted, herbal.

Reichardia intermedia **(Sch.Bip.) Cout.**
Asteraceae
♦ Weed
♦ 87, 88, 94

Reichardia orientalis **(L.) Hochr.**
Asteraceae
= *Reichardia tingitana* (L.) Roth
♦ Weed
♦ 221

Reichardia picroides **(L.) Roth**
Asteraceae
Reichardia hypochoeriformis Ginzb., *Reichardia integrifolia* Moench
♦ French scorzonera, common brighteyes
♦ Weed, Naturalised
♦ 87, 88, 94, 101
♦ pH, promoted, herbal.

Reichardia tingitana **(L.) Roth**
Asteraceae
Picridium tingitanum (L.) Desf., *Reichardia orientalis* (L.) Hochr. (see), *Reichardia tingitana* L. var. *orientalis* (L.) Asch. & Schweinf., *Reichardia tingitana* L. var. *arabica* (Hochst. & Steud.) Asch. & Schweinf., *Scorzonera orientalis* L., *Scorzonera tingitana* L., *Sonchus tingitanus* (L.) Lam.

♦ false sow thistle, halawla, huwwa, maknn, murr
♦ Weed, Naturalised, Environmental Weed
♦ 7, 72, 86, 88, 94, 98, 101, 198
♦ aH, arid, cultivated, herbal. Origin: Mediterranean, Middle East.

Reineckea carnea **(Andrews) Kunth**
Liliaceae/Convallariaceae
♦ reineckea
♦ Naturalised
♦ 40
♦ cultivated, herbal.

Reinwardtia indica **Dumort.**
Linaceae
Linum trigynum Roxb., *Linum repens* Buch.-Ham. ex D.Don, *Reinwardtia trigyna* (Roxb.) Planch. (see), *Reinwardtia sinensis* Hemsl., *Tirpitzia sinensis* (Hemsl.) Hall.
♦ kham paa, yellow flax
♦ Weed
♦ 238
♦ cultivated, herbal.

Reinwardtia trigyna **(Roxb.) Planch.**
Linaceae
= *Reinwardtia indica* Dumort.
♦ Naturalised
♦ 287
♦ cultivated.

Relhania genistifolia **(L.) L'Hér.**
Asteraceae
♦ pepperbush, sticky shrublet
♦ Native Weed
♦ 121
♦ S. Origin: southern Africa.

Renealmia antillarum **Gagnep.**
Bromeliaceae
♦ Weed
♦ 87, 88
♦ herbal.

Reseda alba **L.**
Resedaceae
♦ white mignonette, white upright mignonette, wild mignonette, dyer's rocket
♦ Weed, Naturalised, Introduced, Casual Alien, Cultivation Escape
♦ 7, 15, 34, 40, 42, 86, 87, 88, 94, 98, 101, 161, 176, 198, 203, 218, 228, 252, 272, 280
♦ pH, arid, cultivated, herbal. Origin: Mediterranean.

Reseda alba **L. ssp.** *decursiva* **Forssk.**
Resedaceae
♦ Weed
♦ 221

Reseda arabica **Boiss.**
Resedaceae
♦ Weed
♦ 221

Reseda decursiva **Forssk.**
Resedaceae
♦ Weed, Naturalised
♦ 221, 300

Reseda lutea **L.**
Resedaceae

♦ yellow mignonette, cutleaf mignonette, wild mignonette
♦ Weed, Noxious Weed, Naturalised, Introduced, Garden Escape, Casual Alien
♦ 7, 34, 42, 70, 86, 87, 88, 94, 98, 101, 121, 147, 158, 161, 171, 176, 177, 198, 203, 218, 221, 243, 253, 269, 272, 280, 287, 300
♦ a/pH, cultivated, herbal. Origin: Eurasia.

Reseda luteola **L.**
Resedaceae
Luteola resedoides Fuss, *Reseda undulata* Gilib.
♦ wild mignonette, dyer's rocket, dyer's weed, weld, yellow weed
♦ Weed, Noxious Weed, Naturalised, Introduced, Garden Escape, Environmental Weed, Casual Alien
♦ 7, 23, 34, 39, 42, 70, 86, 87, 88, 94, 98, 101, 147, 165, 176, 198, 203, 221, 228, 241, 243, 269, 272, 280, 287, 290, 295
♦ pH, arid, cultivated, herbal, toxic. Origin: Mediterranean, west Asia.

Reseda media **Lag.**
Resedaceae
♦ Weed
♦ 70, 253

Reseda muricata **C.Presl**
Resedaceae
♦ Weed
♦ 221
♦ cultivated.

Reseda odorata **L.**
Resedaceae
Reseda nilgherrensis Müll.Arg.
♦ garden mignonette, mignonette, tuoksureseda, sweet mignonette, common mignonette
♦ Weed, Naturalised, Introduced, Garden Escape, Cultivation Escape, Casual Alien
♦ 34, 40, 42, 98, 101, 203, 228, 241, 280, 300
♦ aH, arid, cultivated, herbal. Origin: south-east Mediterranean.

Reseda orientalis **(Muller) Boiss.**
Resedaceae
♦ Weed
♦ 87, 88, 221

Reseda phyteuma **L.**
Resedaceae
Pectanisia phyteuma (L.) Raf., *Reseda calicinalis* Lam.
♦ rampion mignonette, mignonette, corn mignonette
♦ Weed, Quarantine Weed, Noxious Weed, Naturalised, Casual Alien
♦ 40, 54, 70, 76, 86, 87, 88, 101, 171, 198, 243, 253, 272, 300
♦ a/b, cultivated, herbal. Origin: Mediterranean.

Reseda pruinosa **Delile**
Resedaceae
♦ Weed
♦ 221

Retama monosperma (L.) Boiss.
Fabaceae/Papilionaceae
Genista monosperma (L.) Lam. (see),
Lygos monosperma (L.) Heywood (see)
♦ bridal broom, white weeping broom,
bridal veil broom
♦ Weed, Sleeper Weed, Noxious Weed,
Naturalised
♦ 35, 78, 80, 88, 98, 116, 231
♦ cultivated. Origin: southern Europe.

**Retama raetam (Forssk.) Webb &
Berthel.**
Fabaceae/Papilionaceae
Genista raetam Forssk., *Lygos raetam*
(Forssk.) Heywood (see)
♦ white weeping broom, ratamals,
white broom
♦ Weed, Naturalised, Introduced,
Garden Escape, Environmental Weed
♦ 86, 98, 155, 203, 228
♦ arid, cultivated, herbal. Origin:
Mediterranean.

Reverchonia arenaria Gray
Euphorbiaceae
♦ reverchonia, sand reverchonia
♦ Weed
♦ 39, 161
♦ herbal, toxic.

Reynoutria japonica Houtt.
Polygonaceae
= *Fallopia japonica* (Houtt.) Ronse Decr.
♦ Japanese knotweed, Asiatic
knotweed, Japanintatar
♦ Weed, Quarantine Weed, Noxious
Weed, Naturalised, Environmental
Weed, Cultivation Escape
♦ 42, 70, 86, 98, 138, 152, 165, 203, 225,
246, 252, 272, 280, 286, 290
♦ cultivated, herbal. Origin: Japan,
Korea and northern China.

**Reynoutria sachalinensis (F.Schmidt)
Nakai**
Polygonaceae
= *Fallopia sachalinensis* (Schmidt) Ronse
Decr.
♦ giant knotweed, sacaline, jättitatar,
ostrich fern
♦ Weed, Naturalised, Environmental
Weed, Cultivation Escape
♦ 42, 86, 121, 152, 225, 246, 252, 280,
286
♦ pH, cultivated, herbal. Origin: Asia.

Rhagadiolus cathartica L.
Asteraceae
♦ Weed, Quarantine Weed
♦ 76, 87, 88, 203, 220

Rhagadiolus edulis Gaertn.
Asteraceae
= *Rhagadiolus stellatus* (L.) Gaertn.
♦ Weed
♦ 87, 88
♦ herbal.

Rhagadiolus stellatus (L.) Gaertn.
Asteraceae
Lapsana stellata L. (see), *Rhagadiolus
edulis* Gaertn. (see)
♦ endive daisy
♦ Weed, Naturalised

♦ 87, 88, 94, 101, 243, 272
♦ aH, cultivated, herbal.

Rhagodia hastata (R.Br.) Scott
Chenopodiaceae
= *Einadia hastata* (R.Br.) Scott (NoR)
♦ saloop, berry saltbush
♦ Introduced
♦ 98

Rhagodia nutans R.Br.
Chenopodiaceae
♦ Weed
♦ 121
♦ S, cultivated.

Rhagodia spinescens R.Br.
Chenopodiaceae
♦ Naturalised
♦ 86
♦ S, arid, cultivated, herbal. Origin:
Australia.

Rhamnus alaternus L.
Rhamnaceae
♦ Italian buckthorn, evergreen
buckthorn, Mediterranean buckthorn
♦ Weed, Quarantine Weed,
Naturalised, Introduced, Garden
Escape, Environmental Weed
♦ 3, 7, 9, 15, 39, 72, 86, 88, 98, 165, 176,
198, 203, 225, 228, 246, 280, 289, 296
♦ S, arid, cultivated, herbal, toxic.
Origin: Mediterranean.

Rhamnus alnifolia L'Hér.
Rhamnaceae
♦ alderleaf buckthorn, alderleaf
coffeeberry
♦ Noxious Weed
♦ 229
♦ S, herbal.

Rhamnus arguta Maxim.
Rhamnaceae
♦ buckthorn
♦ Noxious Weed, Naturalised
♦ 101, 229

**Rhamnus arguta Maxim. var. velutina
Hand.-Mazz.**
Rhamnaceae
♦ buckthorn
♦ Noxious Weed, Naturalised
♦ 101, 229

Rhamnus californica Eschsch.
Rhamnaceae
= *Frangula californica* (Eschsch.) Gray
ssp. *californica* (NoR)
♦ California buckthorn, coffeeberry
♦ Weed
♦ 39, 87, 88
♦ S, arid, cultivated, herbal, toxic.

**Rhamnus californica Eschsch. var. ursina
(Greene) McMinn**
Rhamnaceae
= *Frangula californica* (Eschsch.) Gray
ssp. *ursina* (Greene) Kartesz & Gandhi
(NoR)
♦ California buckthorn
♦ Weed
♦ 218

Rhamnus caroliniana Walter
Rhamnaceae

= *Frangula caroliniana* (Walt.) Gray
(NoR)
♦ Carolina buckthorn
♦ Weed
♦ 39, 87, 88, 218
♦ cultivated, herbal, toxic.

Rhamnus cathartica L.
Rhamnaceae
Cervispina cathartica (L.) Moench,
Rhamnus catharticus L. (see)
♦ common buckthorn, European
buckthorn, nerprun cathartique, Hart's
thorn, European waythorn, purging
buckthorn, buckthorn
♦ Weed, Noxious Weed, Naturalised,
Introduced, Garden Escape,
Environmental Weed
♦ 4, 39, 80, 87, 88, 101, 103, 103, 104,
129, 133, 142, 151, 156, 161, 195, 218,
222, 224, 229
♦ S/T, cultivated, herbal, toxic. Origin:
Europe.

Rhamnus catharticus L.
Rhamnaceae
= *Rhamnus cathartica* L.
♦ buckthorn, spincervino,
orapaatsama
♦ Introduced
♦ 228
♦ arid, cultivated, herbal.

Rhamnus crocea Nutt.
Rhamnaceae
♦ hollyleaf buckthorn, redberry
buckthorn, redberry, spiny redberry,
red berried buckthorn
♦ Weed, Noxious Weed
♦ 87, 88, 229
♦ S, arid, cultivated, herbal.

Rhamnus crocea Nutt. ssp. crocea
Rhamnaceae
♦ redberry buckthorn
♦ Noxious Weed
♦ 229

**Rhamnus crocea Nutt. ssp. pilosa (Trel.)
C.B.Wolf**
Rhamnaceae
♦ hollyleaf buckthorn
♦ Noxious Weed
♦ 229

**Rhamnus crocea Nutt. var. ilicifolia
(Kellogg) Greene**
Rhamnaceae
= *Rhamnus ilicifolia* Kellogg (NoR)
♦ hollyleaf buckthorn
♦ Weed
♦ 218

Rhamnus davurica Pall.
Rhamnaceae
♦ Dahurian buckthorn
♦ Weed, Noxious Weed, Naturalised
♦ 101, 195, 229
♦ cultivated, herbal.

Rhamnus davurica Pall. ssp. davurica
Rhamnaceae
♦ Dahurian buckthorn
♦ Noxious Weed, Naturalised
♦ 101, 229

Rhamnus davurica Pall. ssp. *nipponica*
(Makino) Kartesz & Gandhi
Rhamnaceae
♦ Dahurian buckthorn, Japanese
buckthorn
♦ Noxious Weed, Naturalised
♦ 101, 229

Rhamnus disperma Ehrenb. ex Boiss.
Rhamnaceae
♦ Weed
♦ 221

Rhamnus frangula L.
Rhamnaceae
= *Frangula alnus* Mill.
♦ glossy buckthorn, alder buckthorn,
kruszyna
♦ Weed, Environmental Weed
♦ 4, 39, 80, 87, 88, 102, 104, 151, 195,
218
♦ S, cultivated, herbal, toxic.

Rhamnus globosus Bunge
Rhamnaceae
♦ lokao
♦ Noxious Weed
♦ 229
♦ S, cultivated.

Rhamnus japonica Maxim.
Rhamnaceae
♦ Japanese buckthorn
♦ Noxious Weed, Naturalised
♦ 101, 229
♦ herbal.

Rhamnus lanceolata Pursh
Rhamnaceae
♦ lanceleaf buckthorn
♦ Noxious Weed
♦ 229
♦ herbal.

Rhamnus lanceolata Pursh ssp. *glabrata*
(Gleason) Kartesz & Gandhi
Rhamnaceae
♦ lanceleaf buckthorn
♦ Noxious Weed
♦ 229

Rhamnus lanceolata Pursh ssp.
lanceolata
Rhamnaceae
♦ lanceleaf buckthorn
♦ Noxious Weed
♦ 229

Rhamnus pirifolia Greene
Rhamnaceae
♦ island redberry
♦ Noxious Weed
♦ 229
♦ S, cultivated.

Rhamnus prinoides L'Hér.
Rhamnaceae
♦ blinkblaar, gloss leaf, dogwood
♦ Weed
♦ 63
♦ cultivated, herbal.

Rhamnus purshiana DC.
Rhamnaceae
= *Frangula purshiana* (DC.) Cooper
♦ cascara buckthorn, cascara sagrada,
California buckthorn

♦ Weed
♦ 39, 87, 88, 218
♦ S, cultivated, herbal, toxic.

Rhamnus serrata Humb. & Bonpl. ex
Schult.
Rhamnaceae
♦ sawleaf buckthorn
♦ Noxious Weed
♦ 229
♦ herbal.

Rhamnus smithii Greene
Rhamnaceae
♦ Smith's buckthorn
♦ Noxious Weed
♦ 229
♦ herbal.

Rhamnus utilis Decne.
Rhamnaceae
♦ Chinese buckthorn
♦ Noxious Weed, Naturalised
♦ 101, 229
♦ S, promoted.

Rhamphicarpa fistulosa (Hochst.) Benth.
Scrophulariaceae
Macrosiphon fistulosus Hochst.,
Rhamphicarpa australiensis Steenis
♦ Weed
♦ 163, 202, 243
♦ H parasitic. Origin: Madagascar.

Rhamphicarpa longiflora Benth.
Scrophulariaceae
♦ Weed
♦ 87, 88
♦ Origin: Africa.

Rhanteriopsis S.Rauschert spp.
Asteraceae
Postia Boiss. & Blanche spp.
♦ Quarantine Weed
♦ 220

Rhaphiodon echinus (Nees & Mart.)
Schauer
Lamiaceae
♦ falsa menta
♦ Weed
♦ 255
♦ pH. Origin: Brazil.

Rhaphiolepis × *delacourii* André
Rosaceae
= *Rhaphiolepis indica* (L.) Lindl. ×
Rhaphiolepis umbellata (Thunb.) Mak.
♦ Casual Alien
♦ 280

Rhaphiolepis indica (L.) Lindl. ex Ker
Gawl.
Rosaceae
Crataegus indica L.
♦ Indian hawthorn, cherry laurel
♦ Weed, Naturalised, Garden Escape,
Environmental Weed
♦ 3, 86, 98, 155, 191, 198, 201, 203
♦ S, cultivated. Origin: China, India.

Rhaphiolepis umbellata (Thunb.)
Makino
Rosaceae
Laurus umbellata Thunb., *Rhaphiolepis
japonica* Siebold & Zucc.
♦ sexton's bride, yedda hawthorn,

Japanese hawthorn, sharimbai
♦ Weed, Sleeper Weed, Naturalised,
Environmental Weed
♦ 15, 134, 225, 246, 280
♦ S, cultivated.

Rhaphiolepis umbellata (Thunb.)
Makino fo. *ovata* (Briot) C.K.Schneid.
Rosaceae
♦ Naturalised
♦ 280

Rhapis excelsa (Thunb.) Henry ex
Rehder
Arecaceae
♦ lady finger palm, lady palm, large
lady palm
♦ Introduced
♦ 230
♦ T, cultivated.

Rhapis humilis Blume
Arecaceae
♦ slender lady palm, reed palm
♦ Naturalised, Introduced
♦ 230, 287
♦ T, cultivated.

Rhaponticum uniflorum (L.) DC.
Asteraceae
♦ uniflower Swiss centaury
♦ Weed
♦ 297

Rheum palmatum L.
Polygonaceae
♦ turkey rhubarb, koristeraparperi,
Chinese rhubarb, rebarbora okrasná
♦ Cultivation Escape
♦ 39, 42
♦ pH, cultivated, herbal, toxic.

Rheum rhabarbarum L.
Polygonaceae
Rheum undulatum L.
♦ raparperi, rhubarb, garden rhubarb
♦ Weed, Naturalised, Cultivation
Escape
♦ 15, 39, 42, 101, 154, 161, 280
♦ cultivated, herbal, toxic.

Rhigozum brevispinosum Kuntze
Bignoniaceae
♦ western rhigozum, shortthorn
pomegranate
♦ Weed
♦ 10
♦ cultivated.

Rhigozum obovatum Burch.
Bignoniaceae
♦ yellow pomegranate
♦ Quarantine Weed
♦ 220
♦ cultivated.

Rhigozum trichotomum Burch.
Bignoniaceae
♦ threethorn rhigozum, wildegranaat,
driedoring, wild granate
♦ Weed, Native Weed
♦ 10, 63, 121
♦ pS. Origin: southern Africa.

Rhinacanthus nasutus (L.) Kurz
Acanthaceae
♦ Weed

♦ 13
♦ herbal.

Rhinanthus L. spp.
Scrophulariaceae
♦ yellow rattle
♦ Weed
♦ 39, 70
♦ toxic.

Rhinanthus alectorolophus (Scop.) Pollich
Scrophulariaceae
= *Rhinanthus major* L.
♦ European yellow rattle, villalaukku
♦ Weed, Naturalised
♦ 42, 44, 88, 94, 101, 243, 272
♦ herbal.

Rhinanthus angustifolius C.C.Gmel.
Scrophulariaceae
♦ great yellow rattle
♦ Weed
♦ 70, 272
♦ cultivated.

Rhinanthus apterus Ostenf.
Scrophulariaceae
Alectorolophus apterus (Fr.) Ostenf.
♦ Weed
♦ 87, 88

Rhinanthus crista-galli L.
Scrophulariaceae
♦ cock's comb rattleweed, yellow rattle
♦ Weed
♦ 23, 39, 87, 88, 218
♦ herbal, toxic.

Rhinanthus glaber Lam.
Scrophulariaceae
Alectorolophus major (L.) Rchb.,
Alectorolophus songaricus Stern.,
Alectorolophus vernalis N.W.Zinger,
Rhinanthus major Ehrh., *Rhinanthus songaricus* (Stern.) B.Fedtsch.,
Rhinanthus vernalis (N.W.Zinger) Schischk. & Serg.
♦ Weed
♦ 87, 88

Rhinanthus major L.
Scrophulariaceae
Mimulus alectorolophus Scop.,
Rhinanthus alectorolophus (Scop.) Pollich (see)
♦ Weed
♦ 39, 87, 88
♦ toxic.

Rhinanthus minor L.
Scrophulariaceae
Alectorolophus minor (L.) Wim. & Grab., *Alectorolophus parviflorus* Wall.,
Fistularia crista-galli Wetts., *Rhinanthus crista-galli* L. var. *minor* F.L.Walther
♦ lesser yellow rattle, yellow rattle, little yellow rattle
♦ Weed
♦ 39, 70, 272
♦ aH, cultivated, herbal, toxic.

Rhinanthus rumelicus Velen.
Scrophulariaceae
Alectorolophus rumelicus (Velen.) Bord.

♦ Weed
♦ 272

Rhinanthus serotinus (Schönh.) Oborný
Scrophulariaceae
Alectorolophus grandiflorus Wallr.,
Alectorolophus serotinus Schönh.,
Rhinanthus angustifolius auct. non C.C.Gmel., *Rhinanthus grandiflorus* (Wallr.) Bluff & Fingerh., *Rhinanthus grandiflorus* (Wallr.) Soó, *Rhinanthus montanus* Saut.
♦ late flowering yellow rattle, greater hayrattle, isolaukku
♦ Weed, Naturalised
♦ 87, 88, 101
♦ herbal.

Rhizoclonium riparium (Roth) Kütz. ex Harv.
Cladophoraceae
Conferva riparia Roth, *Tiresias riparia* (Roth) Aresch.
♦ common crinkle grass
♦ Weed
♦ 286
♦ algae. Origin: north-east Atlantic, Mediterranean.

Rhizophora mangle L.
Rhizophoraceae
♦ American mangrove, mangrove, togo togo
♦ Weed, Quarantine Weed
♦ 22, 39, 76, 80, 87, 88, 203, 218, 220
♦ T, arid, cultivated, herbal, toxic.

Rhizophora stylosa Griff.
Rhizophoraceae
♦ Weed
♦ 22
♦ T, cultivated.

Rhodanthe chlorocephala (Turcz.) P.G.Wilson ssp. *rosea* (Hook.) P.G.Wilson
Asteraceae
♦ Naturalised, Native Weed, Garden Escape, Cultivation Escape
♦ 7
♦ cultivated.

Rhodiola dumulosa (Franch.) S.H.Fu
Crassulaceae
♦ shrub berry rhodiola
♦ Weed
♦ 297
♦ cultivated.

Rhodocactus (Berg.) Kunth spp.
Cactaceae
= *Pereskia* Mill. spp.
♦ Weed, Quarantine Weed
♦ 76, 88, 203

Rhododendron L. spp.
Ericaceae
♦ azalea, rhododendron, laurel, alpine azalea
♦ Weed
♦ 154, 161, 247
♦ herbal, toxic.

Rhododendron albiflorum Hook.
Ericaceae
♦ cascade azalea, white flowered rhododendron
♦ Weed

♦ 39, 161
♦ cultivated, herbal, toxic.

Rhododendron arboreum Sm.
Ericaceae
♦ Weed
♦ 22, 39
♦ S, cultivated, herbal, toxic.

Rhododendron canadense (L.) Torr.
Ericaceae
♦ Canadian rhododendron, rhodora
♦ Weed
♦ 87, 88, 218
♦ cultivated, herbal.

Rhododendron canescens (Michx.) Sweet
Ericaceae
♦ piedmont azalea, mountain azalea
♦ Weed
♦ 87, 88, 218
♦ cultivated, herbal.

Rhododendron catawbiense Michx.
Ericaceae
♦ Catawba rosebay, mountain rosebay, Catawba rhododendron
♦ Weed
♦ 39, 161
♦ cultivated, herbal, toxic.

Rhododendron japonicum (Gray) Sur.
Ericaceae
= *Rhododendron molle* (Blume) G.Don ssp. *japonicum* (A.Gray) Kron (NoR)
♦ Japanese azalea
♦ Naturalised
♦ 101
♦ S, cultivated, herbal.

Rhododendron macgregoriae F.Muell.
Ericaceae
♦ Weed
♦ 276
♦ cultivated, toxic.

Rhododendron macrophyllum D.Don ex G.Don
Ericaceae
♦ Pacific rhododendron, coast rhododendron, California rosebay, rhododendron, rosebay
♦ Weed
♦ 39, 87, 88, 161, 218
♦ S, cultivated, herbal, toxic.

Rhododendron maximum L.
Ericaceae
♦ rosebay rhododendron, great laurel, white laurel, great rhododendron
♦ Weed
♦ 8, 39, 87, 88, 161, 218
♦ S, cultivated, herbal, toxic.

Rhododendron obtusum (Lindl.) Planch. var. *kaempferi* Wils.
Ericaceae
♦ yamatsutsuji
♦ Weed
♦ 286

Rhododendron occidentale (Torr. & Gray) Gray
Ericaceae
♦ western azalea, azalea
♦ Weed
♦ 39, 87, 88, 161, 194, 218
♦ S/T, cultivated, herbal, toxic.

Rhododendron ponticum L.
Ericaceae
Rhododendron lancifolium Hook.f.
♦ rhododendron, wild rhododendron
♦ Weed, Sleeper Weed, Naturalised, Environmental Weed
♦ 15, 18, 39, 70, 87, 88, 152, 225, 246, 280, 290
♦ S, cultivated, herbal, toxic. Origin: southern Europe.

Rhododendron ponticum L. ssp. ponticum
Ericaceae
♦ rhododendron
♦ Naturalised
♦ 40

Rhodomyrtus macrocarpa Benth.
Myrtaceae
♦ Weed
♦ 39, 87, 88
♦ cultivated, herbal, toxic. Origin: Asia, Australia.

Rhodomyrtus tomentosa (Aiton) Hassk.
Myrtaceae
♦ rose myrtle, downy myrtle, isenberg bush, hill cherry, Ceylon hill cherry, hill guava, downy rose myrtle
♦ Weed, Quarantine Weed, Noxious Weed, Naturalised, Environmental Weed, Cultivation Escape
♦ 3, 12, 22, 47, 76, 80, 83, 86, 87, 88, 101, 112, 135, 151, 179, 191, 203, 218, 220, 233, 229, 252
♦ S/T, cultivated, herbal.

Rhodosciadium tolucense (H.B.K.) Mathias
Apiaceae
♦ Weed
♦ 199

Rhodostachys bicolor Benth. & Hook.
Bromeliaceae
Bromelia bicolor Ruiz & Pav.
♦ Weed
♦ 87, 88

Rhodotypos scandens (Thunb.) Makino
Rosaceae
Corchorus scandens Thunb., *Kerria tetrapetala* Siebold, *Rhodotypos kerrioides* Siebold & Zucc., *Rhodotypos tetrapetala* (Siebold) Makino
♦ jetbead, white kerria, jetberry bush
♦ Weed, Naturalised, Cultivation Escape
♦ 39, 80, 85, 101, 161, 247
♦ cultivated, herbal, toxic.

Rhoeo spathacea (Sw.) Stearn
Commelinaceae
= *Tradescantia spathacea* Sw.
♦ oyster plant, talo talo, Moses in a boat, te ruru ni, boat lily
♦ Weed, Naturalised, Environmental Weed, Cultivation Escape
♦ 3, 80, 88, 112, 122, 151, 191, 247, 287
♦ cultivated, herbal, toxic.

Rhoicissus tomentosa (Lam.) Wild. & R.B.Drumm.
Vitaceae
Cissus tomentosa Lam., *Rhoicissus capensis* Planch., *Vitis capensis* Thunb.

♦ common forest grape, monkey rope, simple leaved grape, wild grape, wild vine
♦ Weed, Native Weed, Introduced
♦ 121, 228
♦ pC, arid, cultivated. Origin: southern Africa.

Rhoicissus tridentata (L.f.) Wild & R.B.Drumm.
Vitaceae
♦ bitter grape, bushman's grape, common forest grape, bobbejaantou, ilyungulyungu
♦ Native Weed
♦ 121
♦ pC, cultivated, herbal. Origin: southern Africa.

Rhus abyssinica Hochst.
Anacardiaceae
♦ Weed
♦ 221

Rhus aromatica Aiton
Anacardiaceae
Rhus canadensis Marsh. *nom. illeg.*
♦ fragrant sumac, lemon sumach, sweet sumach, stinking sumac
♦ Weed
♦ 87, 88, 218
♦ S, cultivated, herbal.

Rhus ciliata Licht. ex Schult.
Anacardiaceae
♦ suurkaree, sour karree
♦ Weed
♦ 63

Rhus copallina L.
Anacardiaceae
♦ shining sumac, dwarf sumac, winged sumac
♦ Weed
♦ 23, 87, 88, 218
♦ S, cultivated, herbal.

Rhus copallina L. var. leucantha DC.
Anacardiaceae
♦ Weed
♦ 14

Rhus coriaria L.
Anacardiaceae
♦ Sicilian sumac, elm leaved sumach
♦ Weed
♦ 39, 221
♦ S, cultivated, herbal, toxic.

Rhus diversiloba Torr. & Gray
Anacardiaceae
= *Toxicodendron diversilobum* (Torr. & A.Gray) Greene
♦ Pacific poison oak, poison oak, western poison oak
♦ Weed, Quarantine Weed
♦ 39, 45, 76, 87, 88, 203, 218, 220
♦ S, cultivated, herbal, toxic.

Rhus glabra L.
Anacardiaceae
Rhus cismontana Greene, *Rhus glabra* var. *cismontana* (Greene) Cockerell
♦ smooth sumac, western sumac, red sumac, scarlet sumac, vinegar tree
♦ Weed, Quarantine Weed, Naturalised, Native Weed, Garden

Escape, Environmental Weed
♦ 8, 23, 39, 76, 86, 87, 88, 136, 151, 161, 174, 195, 203, 218, 220
♦ S, cultivated, herbal, toxic. Origin: North America.

Rhus glauca Thunb.
Anacardiaceae
♦ blinkblaar, suurbessie, kuni bush
♦ Weed
♦ 63
♦ cultivated.

Rhus hirta (L.) Sudw.
Anacardiaceae
Rhus typhina L. (see)
♦ stag's horn sumach, staghorn sumac
♦ Naturalised
♦ 40

Rhus javanica L. var. roxburghii (DC.) Rehder & E.H.Wilson
Anacardiaceae
♦ Weed
♦ 286

Rhus laevigata L.
Anacardiaceae
Rhus mucronata Thunb., *Rhus viminalis* Vahl
♦ dune currant, red currant, dune taaibos, white karree
♦ Weed, Native Weed, Introduced
♦ 121, 228
♦ S/T, arid, cultivated. Origin: southern Africa.

Rhus lancea L.f.
Anacardiaceae
♦ bastard willow, common karee, karoo tree, karree, kareeboom, mosilabele, willow rhus, African sumac, karee, karree
♦ Weed, Native Weed, Introduced
♦ 63, 80, 121, 228
♦ S/T, arid, cultivated. Origin: southern Africa.

Rhus lanceolata (A.Gray) Britt.
Anacardiaceae
♦ prairie sumac
♦ Weed
♦ 87, 88, 218
♦ herbal.

Rhus laurina Nutt.
Anacardiaceae
= *Malosma laurina* (Nutt.) Nutt. ex Abrams (NoR)
♦ laurel sumac
♦ Weed
♦ 87, 88, 218
♦ herbal.

Rhus leptodictya Diels
Anacardiaceae
♦ mountain karree, rock rhus
♦ Native Weed
♦ 121
♦ S/T, cultivated. Origin: southern Africa.

Rhus lucida L.
Anacardiaceae
♦ glossy currant, glossy taaibos, shiny leaved rhus, wild currant, blinktaaibos, besembos

- Weed, Native Weed
- 39, 63, 121
- S/T, cultivated, toxic. Origin: southern Africa.

Rhus marlothii Engl.
Anacardiaceae
- bitter karree
- Weed, Native Weed
- 10, 121
- pS. Origin: southern Africa.

Rhus microphylla Engelm. ex Gray
Anacardiaceae
- littleleaf sumac, desert sumach
- Weed
- 87, 88, 218
- S, cultivated, herbal.

Rhus ovata S.Wats.
Anacardiaceae
- sugar sumac, sugarbush
- Weed
- 23, 87, 88, 218
- S, cultivated, herbal.

Rhus pyroides Burch.
Anacardiaceae
Rhus baurii Schönland
- common wild currant, wild currant, fire thorned rhus, mogadiri, taaibos
- Weed, Introduced
- 10, 87, 88, 228
- arid, cultivated.

Rhus radicans L.
Anacardiaceae
= *Toxicodendron radicans* (L.) Kuntze
- poison ivy
- Weed, Quarantine Weed, Noxious Weed, Naturalised, Environmental Weed
- 39, 52, 76, 86, 87, 88, 136, 155, 203, 207, 210, 218, 220, 294, 299
- pC, cultivated, herbal, toxic. Origin: North America.

Rhus rehmanniana Engl.
Anacardiaceae
- suur taaibos, sour taaibos, blunt leaved taaibos
- Weed
- 63
- cultivated.

Rhus succedanea L.
Anacardiaceae
= *Toxicodendron succedaneum* (L.) Kuntze
- wax tree, Japanese lacquer tree, Japanese wax, kakrasingi
- Weed, Noxious Weed, Introduced, Garden Escape, Environmental Weed, Cultivation Escape, Casual Alien
- 39, 88, 95, 121, 228, 251, 280, 283, 290
- S/T, arid, cultivated, herbal, toxic. Origin: Eurasia.

Rhus succedanea L. var. japonica Engl.
Anacardiaceae
- Introduced
- 230
- S.

Rhus sylvestris Siebold & Zucc.
Anacardiaceae

- Weed
- 286
- T, cultivated, herbal, toxic. Origin: China, Japan, Korea.

Rhus taitensis Guill.
Anacardiaceae
- Weed
- 203
- S, cultivated, herbal. Origin: Asia, Australia.

Rhus toxicodendron L.
Anacardiaceae
= *Toxicodendron pubescens* Mill. (NoR)
- poison oak, hiedra, poison ivy
- Weed
- 39, 87, 88, 121, 218
- pS, cultivated, herbal, toxic. Origin: south-eastern North America.

Rhus transvaalensis Engl.
Anacardiaceae
- Transvaal currant, Transvaal taaibos
- Native Weed
- 121
- S/T, cultivated. Origin: southern Africa.

Rhus trilobata Nutt. ex Torr. & A.Gray
Anacardiaceae
- skunkbush sumac, skunk bush, squawbush, lemonade sumac
- Weed
- 87, 88, 218
- S, cultivated, herbal.

Rhus tripartita (Ucria) DC.
Anacardiaceae
- Weed
- 221

Rhus typhina L.
Anacardiaceae
= *Rhus hirta* (L.) Sudw.
- staghorn sumac, stag's horn sumach, staghorn
- Weed, Naturalised
- 87, 88, 161, 211, 218, 280
- S, cultivated, herbal.

Rhus undulata Jacq.
Anacardiaceae
- kuni bush
- Weed
- 10
- cultivated.

Rhus verniciflua Stokes
Anacardiaceae
= *Toxicodendron vernicifluum* (Stokes.) F.Barkley
- lacquer tree, lakkapuu
- Naturalised
- 287
- T, cultivated, herbal.

Rhus vernix L.
Anacardiaceae
= *Toxicodendron vernix* (L.) Kuntze
- poison sumac
- Weed
- 88, 218
- S, cultivated, herbal.

Rhus virens Lindh. ex Gray
Anacardiaceae

- evergreen sumac
- Weed
- 87, 88, 218
- herbal.

Rhynchanthera grandiflora (Aubl.) DC.
Melastomataceae
- Weed
- 87, 88

Rhynchelytrum nerviglume (Franch.) Chiov.
Poaceae
Rhynchelytrum setifolium (Stapf) Chiov.
- bistle leaved redtop, redtop grass
- Native Weed
- 121
- pG. Origin: southern Africa.

Rhynchelytrum repens (Willd.) Hubb.
Poaceae
= *Melinis repens* (Willd.) Zizka
- Natal grass, Natal redtop, yaa dok shompuu, red Natal grass, Holme's grass, blanket grass, salapona, herbe du Natal, herbe rose, herbe pappangue, fairygrass, Natal redtop grass, redtop grass, Natal ruby grass, rose Natal grass, yaa dok shompuu
- Weed, Naturalised, Introduced, Environmental Weed, Casual Alien
- 3, 12, 14, 32, 34, 38, 80, 87, 88, 90, 98, 112, 121, 134, 151, 179, 199, 203, 209, 218, 239, 255, 262, 269, 273, 280, 286, 287
- pG, arid, cultivated, herbal. Origin: southern Africa.

Rhynchelytrum roseum (Nees) Stapf & C.E.Hubb. ex Bews
Poaceae
= *Melinis repens* (Willd.) Zizka
- capim favorito
- Weed
- 88, 237, 245, 295
- G.

Rhynchocorys elephas (L.) Griseb.
Scrophulariaceae
- Weed
- 87, 88

Rhynchophreatia carolinensis (Schltr.) Fosb. & Sachet
Orchidaceae
Phreatia carolinensis Schltr.
- rhynchophreatia
- Introduced
- 230
- H.

Rhynchosia Lour. spp.
Fabaceae/Papilionaceae
- ikalamalungwe, good luck seeds, snoutbean
- Weed
- 88, 240
- pH.

Rhynchosia caribaea (Jacq.) DC.
Fabaceae/Papilionaceae
- Caribbean snoutbean
- Introduced
- 261
- cultivated, herbal. Origin: Africa.

Rhynchosia discolor **Martens & Galeotti**
Fabaceae/Papilionaceae
♦ Weed
♦ 157
♦ pC, herbal.

Rhynchosia hirta **(Andrews) Meikle & Verdc.**
Fabaceae/Papilionaceae
Cylista albiflora Sims, *Dolichos hirtus* Andr., *Rhynchosia albidiflora* (Sims) Alston, *Rhynchosia albiflora* (Sims) Alston, *Rhynchosia cyanosperma* Bak.
♦ Weed, Native Weed, Introduced
♦ 121, 228
♦ pC, arid, cultivated. Origin: southern Africa.

Rhynchosia memnonia **(Del.) DC.**
Fabaceae/Papilionaceae
= *Rhynchosia minima* (L.) DC. var. *memnonia* (Del.) Cooke
♦ Weed
♦ 87, 88
♦ cultivated.

Rhynchosia minima **(L.) DC.**
Fabaceae/Papilionaceae
Dolicholus minimus L. (see)
♦ least snoutbean, rhyncho
♦ Weed, Naturalised, Introduced
♦ 14, 28, 39, 55, 87, 88, 199, 221, 228, 241, 243, 271, 300
♦ arid, cultivated, herbal, toxic. Origin: Australia.

Rhynchosia minima **(L.) DC. var. memnonia (Del.) Cooke**
Fabaceae/Papilionaceae
Rhynchosia memnonia (Del.) DC. (see)
♦ Weed
♦ 242
♦ arid.

Rhynchosia pentheri **Schltr. ex Zahlbr.**
Fabaceae/Papilionaceae
♦ Native Weed
♦ 121
♦ pH. Origin: southern Africa.

Rhynchosia pyramidalis **(Lam.) Urb.**
Fabaceae/Papilionaceae
♦ pyramid snoutbean
♦ Weed
♦ 161
♦ cultivated, herbal, toxic.

Rhynchosia reticulata **(Sw.) DC.**
Fabaceae/Papilionaceae
♦ habilla
♦ Weed
♦ 14

Rhynchosia sublobata **(Schum.) Meikle**
Fabaceae/Papilionaceae
♦ Weed
♦ 87, 88
♦ cultivated. Origin: Madagascar.

Rhynchosida physocalyx **(Gray) Fryxell**
Malvaceae
Sida physocalyx A.Gray (see)
♦ tuberous sida, buffpetal
♦ Weed
♦ 161
♦ herbal.

Rhynchosinapis cheiranthos **(Vill.) Dandy**
Brassicaceae
= *Coincya monensis* (L.) Greuter & Burdet ssp. *recurvata* (All.) Leadlay
♦ wallflower cabbage
♦ Weed
♦ 70, 88, 94
♦ herbal.

Rhynchosinapis erucastrum **Dandy.**
Brassicaceae
♦ Naturalised
♦ 287

Rhynchospora **Vahl spp.**
Cyperaceae
Dichromena Michx. spp. (see)
♦ beaksedge, beak rushes
♦ Quarantine Weed
♦ 258
♦ G, cultivated.

Rhynchospora aurea **Vahl**
Cyperaceae
= *Rhynchospora corymbosa* (L.) Britt.
♦ Weed
♦ 255
♦ pG. Origin: cosmopolitan.

Rhynchospora caduca **Ell.**
Cyperaceae
♦ anglestem beaksedge
♦ Weed
♦ 80
♦ G, herbal.

Rhynchospora cephalotes **(L.) Vahl**
Cyperaceae
♦ Weed
♦ 157
♦ pG.

Rhynchospora corymbosa **(L.) Britt.**
Cyperaceae
Rhynchospora aurea Vahl (see), *Rhynchospora longflora* Presl, *Scirpus corymbosus* L.
♦ golden beaksedge, matamat
♦ Weed
♦ 87, 88, 126, 170, 191, 237, 295
♦ pG, aqua, herbal. Origin: pan tropics, subtropics.

Rhynchospora kunthii **Nees ex Kunth**
Cyperaceae
♦ Kunth's beaksedge
♦ Naturalised
♦ 101
♦ G.

Rhynchospora nervosa **(Vahl) Boeck.**
Cyperaceae
♦ yerba de estrella
♦ Weed, Naturalised
♦ 88, 255, 257
♦ pG, aqua. Origin: tropical America.

Rhynchospora rubra **(Lour.) Makino**
Cyperaceae
♦ rhynchospora
♦ Weed
♦ 87, 88, 286
♦ G. Origin: Australia.

Rhynchospora rugosa **(Vahl) Gale ssp. lavarum (Gaud.) Koyama**

Cyperaceae
♦ matamat
♦ Introduced
♦ 230
♦ G.

Rhynchospora stellata **(Lam.) Griseb.**
Cyperaceae
= *Rhynchospora colorata* (L.) Pfeiffer (NoR)
♦ Weed
♦ 14
♦ G.

Rhynchospora tenuis **Link**
Cyperaceae
= *Rhynchospora wrightiana* Boeck. (NoR)
♦ quill beaksedge
♦ Weed
♦ 87, 88
♦ G.

Rhytidiadelphus squarrosus **(Hedw.) Warnst.**
Rhytidiaceae
♦ square gooseneck moss, springy turf moss
♦ Naturalised
♦ 280
♦ moss.

Rhytidiadelphus triquetrus **(Hedw.) Warnst.**
Rhytidiaceae
♦ rough gooseneck moss, big shaggy moss
♦ Naturalised
♦ 280
♦ moss.

Rhytidophyllum tomentosum **(L.) Mart. ex G.Don**
Gesneriaceae
Rhytidophyllum tomentosum L., *Gesneria tomentosa* L.
♦ Weed
♦ 87, 88
♦ cultivated.

Ribes alpinum **L.**
Grossulariaceae
Liebichia alpina Opiz, *Ribes doicum* Moench
♦ alpine currant, mountain currant
♦ Naturalised
♦ 101
♦ S, cultivated, herbal.

Ribes americanum **Mill.**
Grossulariaceae
♦ American black currant, black currant, gall, rain in the face
♦ Weed
♦ 87, 88, 218
♦ S, cultivated, herbal.

Ribes aureum **Pursh**
Grossulariaceae
♦ keltaherukka, golden currant
♦ Cultivation Escape
♦ 39, 42
♦ S, cultivated, herbal, toxic.

Ribes binominatum **Heller**
Grossulariaceae

♦ Siskiyou gooseberry, ground gooseberry, trailing gooseberry
♦ Weed
♦ 87, 88, 218
♦ S, herbal.

Ribes bracteosum Douglas ex Hook.
Grossulariaceae
♦ stink currant
♦ Weed
♦ 87, 88, 218
♦ S, cultivated, herbal.

Ribes californicum Hook. & Arn.
Grossulariaceae
Grossularia californica (Hook. & Arn.) Cav. & Brit.
♦ California gooseberry, hillside gooseberry
♦ Weed
♦ 87, 88, 218
♦ S, promoted, herbal.

Ribes cereum Douglas
Grossulariaceae
♦ wax currant, squaw currant
♦ Weed
♦ 39, 87, 88, 218
♦ S, cultivated, herbal, toxic.

Ribes cynosbati L.
Grossulariaceae
Grossularia cynosbati (L.) Mill.
♦ pasture gooseberry, eastern prickly gooseberry, dog berry
♦ Weed
♦ 87, 88, 218
♦ S, cultivated, herbal.

Ribes glandulosum Grauer
Grossulariaceae
Ribes prostratum L'Hér.
♦ skunk currant
♦ Weed
♦ 87, 88, 218
♦ S, cultivated, herbal. Origin: North America.

Ribes glutinosum Benth.
Grossulariaceae
= *Ribes sanguineum* Pursh var. *glutinosum* (Benth.) Loud. (NoR)
♦ nutmeg currant
♦ Weed
♦ 87, 88, 218
♦ cultivated.

Ribes hirtellum Michx.
Grossulariaceae
Grossularia hirtella (Michx.) Cov. & Britt.
♦ hairystem gooseberry, currant gooseberry
♦ Weed
♦ 87, 88, 218
♦ S, promoted, herbal.

Ribes inerme Rydb.
Grossulariaceae
Grossularia inermis (Rydb.) Cov. & Britt.
♦ whitestem gooseberry
♦ Weed
♦ 87, 88, 218
♦ S, promoted, herbal.

Ribes lacustre (Pers.) Poir.
Grossulariaceae

♦ swamp black currant, prickly black currant, prickly currant, swamp currant
♦ Weed
♦ 87, 88, 218
♦ S, promoted, herbal.

Ribes laxiflorum Pursh
Grossulariaceae
♦ trailing black currant
♦ Weed
♦ 87, 88, 218
♦ S, promoted, herbal.

Ribes lobbii A.Gray
Grossulariaceae
Grossularia lobbii (A.Gray) Cov. & Britt.
♦ Lobb's gooseberry, gummy gooseberry
♦ Weed
♦ 87, 88, 218
♦ S, promoted, herbal.

Ribes marshallii Greene
Grossulariaceae
♦ Hupa gooseberry, Marshall's gooseberry
♦ Weed
♦ 87, 88, 218
♦ S, herbal.

Ribes menziesii Pursh
Grossulariaceae
Grossularia menziesii (Pursh.) Cov. & Britt.
♦ Menzies' gooseberry, canyon gooseberry
♦ Weed
♦ 87, 88, 218
♦ S, cultivated, herbal.

Ribes missouriense Nutt. ex Ton. & Gray
Grossulariaceae
♦ Missouri gooseberry
♦ Weed
♦ 87, 88, 218
♦ S, promoted, herbal.

Ribes montigenum McClatchie
Grossulariaceae
Grossularia montigenum McClatchie
♦ mountain gooseberry, gooseberry currant
♦ Weed
♦ 87, 88, 218
♦ S, cultivated, herbal.

Ribes nevadense Kellogg
Grossulariaceae
♦ Sierra currant
♦ Weed
♦ 87, 88, 218
♦ S, cultivated, herbal.

Ribes nigrum L.
Grossulariaceae
Botryocarpum nigrum Opiz, *Grossularia nigra* Rupre., *Ribes olidum* Moench
♦ black currant, red currant, European black currant, mustaherukka
♦ Weed, Naturalised
♦ 39, 40, 80, 101, 280
♦ S, cultivated, herbal, toxic.

Ribes odoratum H.L.Wendl.
Grossulariaceae

= *Ribes aureum* Pursh var. *villosum* DC.
♦ Missouri currant, buffalo currant, clove currant
♦ Weed, Naturalised
♦ 80, 280
♦ S, cultivated, herbal.

Ribes oxyacanthoides L.
Grossulariaceae
Grossularia oxyacanthoides (L.) Cov. & Britt.
♦ northern gooseberry, Canadian gooseberry, American mountain gooseberry
♦ Weed
♦ 87, 88, 218
♦ S, promoted, herbal.

Ribes petiolare Douglas
Grossulariaceae
= *Ribes hudsonianum* Richards var. *petiolare* (Douglas) Jancz. (NoR)
♦ western black currant
♦ Weed
♦ 87, 88, 218
♦ S, promoted.

Ribes × pallidum Otto & Dietr.
Grossulariaceae
= *Ribes petraeum* Wulfen × *Ribes spicatum* Robson ssp. *spicatum*
♦ hollanninpunaherukka
♦ Cultivation Escape
♦ 42
♦ cultivated.

Ribes roezlii Regel
Grossulariaceae
Grossularia roezlii (Regel.) Cov. & Britt.
♦ Sierra gooseberry
♦ Weed
♦ 87, 88, 218
♦ S, cultivated, herbal.

Ribes rubrum L.
Grossulariaceae
Grossularia rubra Scop., *Ribes domesticum* Jancz., *Ribes sativum* (Rchb.) Syme (see)
♦ red currant
♦ Weed, Naturalised, Cultivation Escape
♦ 15, 40, 42, 80, 88, 101, 146, 280
♦ S, cultivated, herbal.

Ribes sanguineum Pursh
Grossulariaceae
♦ red flowered currant, flowering currant
♦ Weed, Sleeper Weed, Naturalised, Garden Escape, Environmental Weed
♦ 40, 54, 86, 87, 88, 165, 176, 218, 225, 246, 280
♦ S, cultivated, herbal. Origin: western North America.

Ribes sativum (Rchb.) Syme
Grossulariaceae
= *Ribes rubrum* L.
♦ garden red currant, white currant, red currant
♦ Weed
♦ 133, 195
♦ S, cultivated, herbal.

Ribes speciosum **Pursh**
Grossulariaceae
♦ fuchsia gooseberry, fuchsia flowered gooseberry
♦ Weed
♦ 87, 88, 218
♦ S, cultivated, herbal.

Ribes triste **Pall.**
Grossulariaceae
♦ swamp red currant, American red currant, red currant
♦ Weed
♦ 87, 88, 218
♦ S, cultivated, herbal.

Ribes tularense **(Cov.) Fedde**
Grossulariaceae
Grossularia tularensis Coville
♦ Tulare gooseberry
♦ Weed
♦ 87, 88, 218
♦ S, herbal. Origin: North America.

Ribes uva-crispa **L.**
Grossulariaceae
Grossularia reclinata Mill., *Oxyacanthus uva-crispa* Chev., *Ribes reclinatum* L.
♦ English gooseberry, European gooseberry, karviainen
♦ Weed, Naturalised, Cultivation Escape
♦ 15, 42, 80, 86, 98, 101, 181, 198, 203, 252, 280
♦ S, cultivated, herbal. Origin: southern Europe.

Ribes uva-crispa **L. var.** *sativum* **DC.**
Grossulariaceae
Ribes grossularia L.
♦ European gooseberry
♦ Naturalised
♦ 101

Ribes velutinum **Greene**
Grossulariaceae
♦ desert gooseberry
♦ Weed
♦ 87, 88, 218
♦ S, herbal.

Ribes viscosissimum **Pursh**
Grossulariaceae
♦ sticky currant, sticky flowering currant
♦ Weed
♦ 87, 88, 218
♦ S, promoted, herbal.

Riccia **L. spp.**
Ricciaceae
♦ riccia
♦ Weed
♦ 243
♦ liverwort.

Riccia bifurca **Hoffm.**
Ricciaceae
♦ lizard crystalwort
♦ Naturalised
♦ 280
♦ liverwort.

Riccia ciliata **Hoffm.**
Ricciaceae
♦ Naturalised

♦ 280
♦ liverwort.

Riccia crystallina **L.**
Ricciaceae
♦ blue crystalwort
♦ Naturalised
♦ 280
♦ liverwort.

Riccia glauca **L.**
Ricciaceae
♦ glaucous crystalwort
♦ Naturalised
♦ 280
♦ liverwort.

Ricciocarpus natans **(L.) Corda**
Ricciaceae
Riccia natans L.
♦ ricciocarpus, purple fringed liverwort
♦ Weed
♦ 87, 88, 204, 263, 286, 295
♦ w liverwort, cultivated.

Richardia brasiliensis **(Moq.) Gomez**
Araceae
Richardia scabra St.-Hil., *Richardsonia rosea* St.-Hil.
♦ Brazil callalily, tropical Mexican clover, tropical richardia, Mexican richardia, white eye, yerba del sapo, poaia branca, Brazil pusley
♦ Weed, Naturalised
♦ 7, 13, 14, 50, 51, 55, 84, 86, 87, 88, 98, 121, 158, 161, 170, 179, 203, 218, 236, 237, 243, 245, 249, 255, 269, 287, 295
♦ pH, arid, cultivated, herbal. Origin: South America.

Richardia grandiflora **(Cham. & Schlecht.) J.A. & J.H.Schult.**
Araceae
♦ largeflower pusley, largeflower Mexican clover
♦ Weed, Naturalised
♦ 101, 161, 179, 249, 255
♦ aH. Origin: Brazil.

Richardia humistrata **(Cham. & Schlechtd.) Steud.**
Araceae
♦ peelton richardia, peelton weed, South American Mexican clover
♦ Weed, Naturalised
♦ 86, 101, 121
♦ pH. Origin: South America.

Richardia scabra **L.**
Araceae
♦ Florida purslane, pursley, yaa thaa phra, rough Mexican clover, white eye, poaia do cerrado, ricardia
♦ Weed, Naturalised
♦ 55, 84, 86, 87, 88, 93, 98, 161, 179, 203, 218, 239, 243, 249, 255, 263, 281, 287
♦ aH, arid, herbal. Origin: tropical America.

Richardia stellaris **(Cham. & Schltdl.) Steud.**
Araceae
♦ field madder
♦ Weed, Naturalised

♦ 86, 98, 203, 269
♦ Origin: Brazil.

Ricinus communis **L.**
Euphorbiaceae
♦ castor bean, castorbean tree, castor oil bush, castor oil plant, castor oil tree, wonder tree, risiini, kasterolieboom, umFude, umHlafuto, muPfuta, palma christi, African coffee tree, agaliya, gelug, maskerekur, uluchula skoki, mbele ni vavalagi, toto ni vavalagi, utouto, lama papalagi, tuitui, tuitui fua ikiiki, koli, lama palagi, lepo, ricin
♦ Weed, Quarantine Weed, Noxious Weed, Naturalised, Introduced, Garden Escape, Environmental Weed, Casual Alien
♦ 3, 6, 7, 8, 15, 22, 34, 35, 39, 42, 50, 50, 51, 63, 70, 72, 73, 78, 80, 86, 87, 88, 93, 95, 98, 101, 116, 121, 122, 134, 147, 151, 152, 154, 157, 158, 161, 165, 167, 179, 189, 198, 203, 212, 215, 218, 220, 228, 231, 237, 241, 242, 247, 255, 257, 261, 262, 269, 279, 280, 283, 286, 287, 289, 295, 296, 300
♦ S/T, arid, cultivated, herbal, toxic. Origin: tropical Africa.

Ridolfia segetum **Moris.**
Apiaceae
♦ corn parsley, false caraway
♦ Weed, Quarantine Weed, Introduced
♦ 38, 70, 76, 87, 88, 94, 115, 203, 220, 243, 253
♦ cultivated, herbal.

Riencourtia latifolia **Gardn.**
Asteraceae
♦ Weed
♦ 32

Riencourtia oblongifolia **Gardn.**
Asteraceae
♦ Weed
♦ 32

Ritchiea reflexa **(Schum. & Thonn.) Gilg & Benedict**
Capparaceae
♦ Introduced
♦ 228
♦ arid.

Ritterocereus **Backeb. spp.**
Cactaceae
= *Stenocereus* (Berger) Riccob. spp.
♦ Weed, Quarantine Weed
♦ 76, 88, 203

Rivea corymbosa **(L.) Hallier f.**
Convolvulaceae
= *Turbina corymbosa* (L.) Raf.
♦ oliliuqui
♦ Weed, Quarantine Weed
♦ 76, 88, 203, 220
♦ herbal.

Rivina humilis **L.**
Phytolaccaceae/Petiveriaceae
♦ baby pepper, bloodberry, coralberry, rouge plant, polo, bloedbessie
♦ Weed, Noxious Weed, Naturalised, Garden Escape, Environmental Weed
♦ 3, 39, 73, 86, 87, 88, 98, 121, 134, 155, 161, 191, 257, 283, 287

♦ a/pH, cultivated, herbal, toxic.
Origin: South America.

Robbairea delileana **Milne-Redh.**
Caryophyllaceae
♦ Weed
♦ 88, 221

Robinia hispida **L.**
Fabaceae/Papilionaceae
♦ bristly locust, rose acacia
♦ Weed, Environmental Weed
♦ 80, 88, 151, 195
♦ T, cultivated, herbal.

Robinia neomexicana **(Wooton & Standl.)**
W.C.Martin & Hutchins ex Peabody
Fabaceae/Papilionaceae
Robinia luxurians Rydb., *Robinia subvelutina* Rydb.
♦ New Mexico locust, desert locust
♦ Weed
♦ 161
♦ S, arid, cultivated, herbal, toxic.

Robinia pseudoacacia **L.**
Fabaceae/Papilionaceae
Pseudacacia odorata Moench, *Robinia acacia* L., *Robinia fragilis* Salisb.
♦ black locust, false acacia, locust tree, yellow locust, witakasia, valeakaasia, robinia
♦ Weed, Sleeper Weed, Noxious Weed, Naturalised, Introduced, Garden Escape, Environmental Weed, Cultivation Escape, Casual Alien
♦ 4, 7, 8, 15, 18, 34, 35, 39, 40, 42, 63, 72, 78, 80, 86, 87, 88, 95, 98, 104, 116, 121, 129, 133, 137, 142, 146, 151, 152, 154, 161, 195, 198, 203, 211, 218, 224, 225, 228, 231, 243, 246, 247, 256, 269, 279, 280, 283, 286, 287, 289, 300
♦ T, arid, cultivated, herbal, toxic.
Origin: North America.

Robinia viscosa **Vent.**
Fabaceae/Papilionaceae
♦ clammy locust
♦ Weed
♦ 39, 161
♦ T, cultivated, herbal, toxic.

Rochefortia spinosa **(Jacq.) Urb.**
Boraginaceae/Ehretiaceae
♦ espino
♦ Weed
♦ 87, 88

Rochelia disperma **(L.f.) C.Koch**
Boraginaceae
Lithospermum dispermum L.f.
♦ rochélia dvojsemenná
♦ Weed
♦ 87, 88, 121, 243
♦ aH. Origin: Eurasia.

Roegneria ciliaris **(Trin.) Nev.**
Poaceae
= *Elymus ciliaris* (Trin.) Tzvel. (NoR)
♦ Weed
♦ 275, 297
♦ pG.

Roegneria kamoji **(Ohwi) Ohwi ex Keng**
Poaceae
= *Elymus tsukushiensis* Honda var.

transiens (Hack.) Osada (NoR) [see
Agropyron kamoji Ohwi, *Agropyron tsukushiense* (Honda) Ohwi var.
transiens (Hack.) Ohwi]
♦ Weed
♦ 275
♦ pG.

Roegneria mayebarana **(Honda) Ohwi**
Poaceae
♦ oriental roegneria
♦ Weed
♦ 297
♦ G.

Roemeria hybrida **(L.) DC.**
Papaveraceae
♦ violet horned poppy, saksanunikko
♦ Weed, Naturalised, Casual Alien
♦ 42, 86, 87, 88, 94, 98, 203, 221, 243, 272
♦ cultivated. Origin: Mediterranean.

Roemeria refracta **DC.**
Papaveraceae
♦ Roemer poppy, horned poppy, spotted Asian poppy
♦ Weed, Naturalised, Casual Alien
♦ 42, 87, 88, 101, 218
♦ cultivated.

Roemeria rhoeadiflora **Boiss.**
Papaveraceae
♦ Weed
♦ 87, 88

Rolandra fruticosa **(L.) Kuntze**
Asteraceae
♦ yerba de plata
♦ Weed
♦ 87, 88
♦ herbal.

Roldana petasitis **(DC.) H.Rob. & Brettell**
Asteraceae
Senecio petasitis (Sims) DC. (see),
Cineraria petasitis Sims
♦ velvet groundsel
♦ Naturalised
♦ 86, 198, 280
♦ cultivated. Origin: Mexico.

Romneya coulteri **Harv.**
Papaveraceae
♦ Coulter's matilija poppy, matilija poppy, Californian tree poppy, matilija
♦ Weed, Naturalised, Casual Alien
♦ 7, 98, 203, 280
♦ pH, cultivated, herbal.

Romneya trichocalyx **Eastw.**
Papaveraceae
♦ bristly matilija poppy, hairy matilija poppy
♦ Weed, Naturalised
♦ 7, 54, 86, 88, 98, 203
♦ pH, cultivated. Origin: western North America.

Romulea bulbocodium **(L.) Sebast. & Mauri**
Iridaceae
♦ romulea
♦ Weed, Naturalised, Garden Escape, Environmental Weed

♦ 72, 88, 198, 251
♦ pH, cultivated.

Romulea flava **(Lam.) M.P.de Vos**
Iridaceae
♦ frutang
♦ Weed, Naturalised, Garden Escape
♦ 7, 9, 98, 203
♦ cultivated.

Romulea flava **(Lam.) M.P.de Vos var. minor (Beg.) M.P.de Vos**
Iridaceae
♦ yellow oniongrass, frutang, yellow flowered oniongrass
♦ Naturalised, Environmental Weed
♦ 86, 198
♦ Origin: southern Africa.

Romulea minutiflora **Klatt**
Iridaceae
♦ frutang, small oniongrass, smallflower oniongrass
♦ Weed, Naturalised, Native Weed, Garden Escape, Environmental Weed
♦ 7, 72, 86, 88, 98, 121, 176, 198, 203
♦ pH, cultivated. Origin: southern Africa.

Romulea obscura **Klatt**
Iridaceae
♦ romulea
♦ Weed, Naturalised, Garden Escape, Environmental Weed
♦ 7, 9, 86, 98, 203
♦ cultivated. Origin: southern Africa.

Romulea pratensis **M.P.de Vos**
Iridaceae
♦ Native Weed
♦ 121
♦ pH, cultivated. Origin: southern Africa.

Romulea rosea **(L.) Eckl.**
Iridaceae
♦ oniongrass, Guildford grass, australis oniongrass, rosy sandcrocus
♦ Weed, Naturalised, Garden Escape, Environmental Weed
♦ 7, 9, 15, 54, 87, 88, 98, 101, 161, 165, 176, 198, 203, 243, 280, 296
♦ cultivated, toxic. Origin: South Africa.

Romulea rosea **(L.) Eckl. var.** *australis* **(Ewart) M.P.de Vos**
Iridaceae
♦ rosy sandcrocus, Guildford grass, frutang, pink romulea, common oniongrass
♦ Weed, Naturalised, Native Weed, Introduced, Environmental Weed
♦ 9, 34, 72, 86, 101, 121, 176, 198, 269, 289
♦ pH, cultivated. Origin: South Africa.

Romulea rosea **(L.) Eckl. var.** *communis* **M.P.de Vos**
Iridaceae
♦ common oniongrass, Guildford grass
♦ Naturalised, Environmental Weed
♦ 86
♦ cultivated. Origin: South Africa.

Romulea rosea (L.) Eckl. var. *reflexa* (Eckl.) Beg.
Iridaceae
♦ largeflower oniongrass, common oniongrass, Guildford grass
♦ Naturalised, Environmental Weed
♦ 72, 86, 198
♦ pH, cultivated. Origin: South Africa.

Rorippa Scop. spp.
Brassicaceae
♦ yellowcress
♦ Weed
♦ 272

Rorippa amphibia (L.) Bess.
Brassicaceae
Armoracia amphibia Peterm., *Myagrum aquaticum* Lam., *Nasturtium amphibium* R.Br.
♦ great watercress, vesinenätti, marshcress, great yellowcress, amphibious yellowcress
♦ Weed, Naturalised, Environmental Weed
♦ 4, 23, 80, 87, 88, 101, 104, 195, 197, 272, 280
♦ pH, aqua, cultivated, herbal.

Rorippa × *anceps* (Wahlenb.) Rchb.
Brassicaceae
= *Rorippa amphibia* (L.) Bess. × *Rorippa sylvestris* (L.) Bess.
♦ kärsänenätti
♦ Naturalised
♦ 42

Rorippa aquatica (Eaton) Palmer & Steyerm.
Brassicaceae
= *Neobeckia aquatica* (Eaton) Greene (NoR)
♦ Weed
♦ 87, 88
♦ aqua, cultivated.

Rorippa × *armoracioides* (Tausch) Fuss
Brassicaceae
= *Rorippa austriaca* (Crantz) Bess. × *Rorippa sylvestris* (L.) Bess.
♦ puistonenätti
♦ Naturalised
♦ 42

Rorippa atrovirens (Hornem.) Ohwi & Hara
Brassicaceae
= *Rorippa indica* (L.) Hiern
♦ indica peppercress
♦ Weed
♦ 263, 274
♦ pH.

Rorippa austriaca (Crantz) Besser
Brassicaceae
Nasturtium austriacum Crantz
♦ Austrian fieldcress, Austrian yellowcress
♦ Weed, Quarantine Weed, Noxious Weed, Naturalised, Introduced
♦ 1, 26, 34, 35, 42, 56, 76, 80, 87, 88, 101, 146, 161, 203, 210, 218, 220, 229, 272, 287
♦ pH, herbal.

Rorippa cantoniensis (Lour.) Ohwi
Brassicaceae
Rorippa microsperma (DC.) L.H.Bailey (see)
♦ Chinese yellowcress, yellowcress
♦ Weed, Naturalised
♦ 87, 88, 101, 235, 273, 274, 286, 297

Rorippa curvisiliqua (Hook.) Bess. ex Britt.
Brassicaceae
♦ yellowcress, curvepod yellowcress, western yellowcress
♦ Weed, Naturalised
♦ 136, 287
♦ a/pH, herbal.

Rorippa dubia (Pers.) H.Hara
Brassicaceae
Cardamine sublyrata Miq., *Nasturtium dubium* (Pers.) Kuntze, *Nasturtium heterophyllum* Blume, *Nasturtium indicum* (L.) DC. var. *apetalum* DC., *Nasturtium indicum* var. *javanum* Bl., *Nasturtium sublyratum* (Miq.) Franch. & Sav., *Rorippa heterophylla* (Bl.) R.O.Williams (see), *Rorippa indica* (L.) Hiern var. *apetala* (DC.) Hochr. (see), *Rorippa sublyrata* (Miq.) H.Hara, *Sisymbrium dubium* Pers.
♦ phak kaat nam
♦ Weed
♦ 32, 87, 88, 238, 286, 297
♦ aH.

Rorippa fluviatilis (E.Mey. ex Sond.) Thell. var. *fluviatilis*
Brassicaceae
♦ Native Weed
♦ 121
♦ pH. Origin: southern Africa.

Rorippa globosa (Turcz. ex Fisch. & C.A.Mey.) Hayek
Brassicaceae
♦ globe yellowcress
♦ Weed, Naturalised
♦ 101, 243, 297
♦ pH.

Rorippa heterophylla (Bl.) R.O.Williams
Brassicaceae
= *Rorippa dubia* (Pers.) H.Hara
♦ Weed
♦ 157, 261
♦ Origin: tropical Asia.

Rorippa hilariana (Walp.) Cabrera
Brassicaceae
♦ Weed
♦ 87, 88

Rorippa indica (L.) Hiern
Brassicaceae
Rorippa atrovirens (Hornem.) Ohwi & Hara (see), *Nasturtium indicum* (L.) DC. (see), *Sinapis patens* Roxb., *Sisymbrium indicum* L.
♦ variableleaf yellowcress
♦ Weed, Quarantine Weed, Naturalised
♦ 76, 87, 88, 101, 170, 191, 203, 220, 235, 248, 261, 262, 273, 275, 286
♦ aH.

Rorippa indica (L.) Hiern var. *apetala* (DC.) Hochr.
Brassicaceae
= *Rorippa dubia* (Pers.) H.Hara
♦ variableleaf yellowcress
♦ Naturalised
♦ 101

Rorippa indica (L.) Hiern var. *indica*
Brassicaceae
Rorippa montana Small (see)
♦ variableleaf yellowcress
♦ Naturalised
♦ 101
♦ aH.

Rorippa islandica (Oeder) Borbás
Brassicaceae
Myagrum palustre Lam., *Radicula islandica* Druce, *Rorippa palustris* Moench, *Sisymbrium islandicum* Oeder
♦ marsh yellowcress, northern marsh yellowcress, marshcress, common yellowcress
♦ Weed
♦ 23, 44, 70, 87, 88, 94, 161, 211, 218, 243, 253, 272, 286, 297
♦ aH, cultivated, herbal.

Rorippa islandica (Oeder ex Murray) Borbás var. *hispida* (Desv.) Butters & Abbe
Brassicaceae
= *Rorippa palustris* (L.) Bess. ssp. *hispida* (Desv.) Jonsell (NoR)
♦ kesukashitagobou
♦ Weed, Naturalised
♦ 261, 287

Rorippa lippizensis (Wulf.) Rchb.
Brassicaceae
Nasturtium lippizense (Wulf.) DC.
♦ Weed
♦ 272

Rorippa microphylla (Boenn. ex Rchb.) Hyl. ex Á. & D.Löve
Brassicaceae
= *Nasturtium microphyllum* Boenn. ex Rchb.
♦ one rowed watercress, brown watercress, narrow fruited watercress
♦ Weed, Naturalised, Garden Escape, Environmental Weed
♦ 15, 86, 98, 101, 176, 198, 203, 280
♦ aqua, cultivated.

Rorippa microsperma (DC.) L.H.Bailey
Brassicaceae
= *Rorippa cantoniensis* (Lour.) Ohwi
♦ Weed
♦ 87, 88

Rorippa montana Small
Brassicaceae
= *Rorippa indica* (L.) Hiern var. *indica*
♦ Weed
♦ 87, 88, 275
♦ aH, herbal.

Rorippa nasturtium-aquaticum (L.) Hayek
Brassicaceae
= *Nasturtium officinale* R.Br.
♦ watercress, great watercress,

bronkors, green watercress, tworow watercress
♦ Weed, Sleeper Weed, Noxious Weed, Naturalised, Introduced, Garden Escape, Environmental Weed, Cultivation Escape
♦ 7, 15, 32, 34, 63, 72, 80, 86, 88, 95, 98, 116, 121, 134, 136, 146, 158, 165, 176, 179, 198, 200, 203, 225, 237, 269, 280, 283, 295, 296
♦ wpH, cultivated, herbal. Origin: Eurasia.

Rorippa obtusa (Nutt.) Britt.
Brassicaceae
= *Rorippa teres* (Michx.) R.Stuckey
♦ Naturalised
♦ 287
♦ herbal.

Rorippa palustris (L.) Besser
Brassicaceae
Cardamine palustris (L.) Kuntze fo. *barbareaefolia* (Del.) Kuntze, *Nasturtium palustre* (L.) DC. (see), *Nasturtium semipinnatifidum* Hook., *Nasturtium terreste* (With.) Aiton, *Nasturtium terreste* (With.) Aiton var. *semipinnatifidum* (Hook.) Hook., *Sisymbrium amphibium* L. var. *palustre* L., *Sisymbrium barbareaefolium* Del., *Sisymbrium palustre* (L.) Leyss., *Sisymbrium terrestre* With.
♦ marshcress, common yellowcress, bog yellowcress, bog marshcress, yellow watercress, marsh yellowcress, rantanenätti
♦ Weed, Naturalised, Native Weed, Garden Escape, Environmental Weed
♦ 32, 72, 86, 88, 98, 159, 174, 176, 198, 203, 263, 275, 296
♦ a/pH, arid, cultivated, herbal. Origin: Northern Hemisphere.

Rorippa palustris (L.) Bess. ssp. *occidentalis* (S.Wats.) Abrams
Brassicaceae
♦ marsh yellowcress, cress, marshcress, yellow watercress, western bog yellowcress
♦ Weed
♦ 180, 243

Rorippa prostrata (Bergeret) Schinz & Thell.
Brassicaceae
Myagrum prostratum Bergeret, *Sisymbrium anceps* Wahlenb.
♦ prostrate yellowcress
♦ Naturalised
♦ 101
♦ herbal.

Rorippa pyrenaica (Lam.) Rchb.
Brassicaceae
Alyssum pyrenaicum Clairv., *Brachiolobos pyrenaicus* All., *Lepidium stylosum* Pers., *Nasturtium pyrenaicum* R.Br.
♦ crescione dei pirenei
♦ Weed
♦ 272
♦ cultivated, herbal.

Rorippa sarmentosa (G.Forst. ex DC.) J.F.Macbr.
Brassicaceae
Cardamine sarmentosa Forst.f. (see)
♦ longrunner, a'atasi
♦ Naturalised
♦ 101
♦ herbal.

Rorippa silvestris (L.) Besser
Brassicaceae
Radicula silvestris Druce
♦ creeping yellowcress, yellowcress
♦ Weed
♦ 44
♦ pH.

Rorippa sinuata (Nutt.) A.S.Hitchc.
Brassicaceae
Radicula sinuata (Nutt.) Greene
♦ spreading yellowcress, yellowcress
♦ Weed, Native Weed
♦ 49, 159, 161, 174
♦ pH, aqua, herbal. Origin: North America.

Rorippa × sterilis Airy Shaw
Brassicaceae
= *Rorippa microphylla* (Boenn.) N.Hyl. ex Á. & D.Löve × *Nasturtium officinale* R.Br.
♦ hybrid watercress, brown watercress
♦ Naturalised
♦ 101
♦ herbal.

Rorippa sylvestris (L.) Besser
Brassicaceae
Cardamine silvestris Kuntze, *Nasturtium silvestre* R.Br., *Sisymbrium silvestre* L.
♦ creeping yellowcress, yellow fieldcress, rikkanenätti
♦ Weed, Noxious Weed, Naturalised
♦ 15, 42, 67, 70, 80, 86, 87, 88, 94, 101, 146, 161, 165, 218, 229, 241, 243, 253, 272, 280, 286, 287, 300
♦ aqua, cultivated, herbal. Origin: temperate Eurasia.

Rorippa teres (Michx.) R.Stuckey
Brassicaceae
Cardamine teres Michx., *Radicula obtusa* (Nutt.) Greene, *Radicula walteri* (Ell.) Greene, *Rorippa obtusa* (Nutt.) Britt. (see), *Rorippa walteri* (Ell.) C.Mohr
♦ southern marsh yellowcress
♦ Weed
♦ 243
♦ H, herbal.

Rosa L. spp.
Rosaceae
♦ rose, ros, ruusu
♦ Weed, Naturalised, Introduced, Environmental Weed, Cultivation Escape
♦ 15, 23, 40, 72, 80, 86, 88, 198, 272
♦ S, cultivated, herbal.

Rosa × alba L. (*pro* sp.)
Rosaceae
= *Rosa arvensis* Huds. × *Rosa gallica* L. or *Rosa corymbifera* Borkh. or *Rosa canina* L. × *Rosa damascena* Mill. [disputed parentage with various possible combinations]
♦ white rose of York
♦ Naturalised, Introduced, Cultivation Escape
♦ 40, 101
♦ cultivated, herbal.

Rosa arabica Crep.
Rosaceae
= *Rosa rubiginosa* L.
♦ Weed
♦ 221

Rosa arkansana Porter
Rosaceae
♦ Arkansas rose, prairie rose, low prairie rose, prairie wild rose, wild rose
♦ Weed, Noxious Weed, Native Weed
♦ 87, 88, 161, 174, 210, 218, 299
♦ S, cultivated, herbal. Origin: North America.

Rosa arvensis Huds.
Rosaceae
Rosa repens Scop., *Rosa serpens* Wib., *Rosa silvestris* Herm.
♦ field rose, trailing rose
♦ Weed
♦ 272
♦ cultivated, herbal.

Rosa blanda Aiton
Rosaceae
♦ rönsyruusu, Labrador rose, meadow rose, smooth rose
♦ Casual Alien
♦ 42
♦ S, cultivated, herbal.

Rosa × borboniana Desp. (*pro* sp.)
Rosaceae
= *Rosa chinensis* Jacq. × *Rosa damascena* Mill. or *Rosa gallica* L.
♦ Bourbon rose
♦ Naturalised
♦ 101

Rosa bracteata J.C.Wendl.
Rosaceae
♦ Macartney rose
♦ Weed, Naturalised, Cultivation Escape
♦ 7, 80, 86, 87, 88, 98, 101, 179, 203, 218, 252
♦ cultivated, herbal. Origin: China.

Rosa californica Cham. & Schlecht.
Rosaceae
♦ California rose, California wild rose
♦ Weed
♦ 87, 88, 218
♦ S, cultivated, herbal.

Rosa canina L.
Rosaceae
Rosa caucasica Pall.
♦ dog rose, briar rose, rosa canina
♦ Weed, Quarantine Weed, Noxious Weed, Naturalised, Introduced, Garden Escape, Environmental Weed
♦ 34, 39, 76, 80, 86, 87, 88, 97, 98, 101, 146, 147, 171, 176, 198, 203, 241, 272, 280, 290, 300
♦ S, cultivated, herbal, toxic. Origin: Eurasia.

Rosa × centifolia L.
Rosaceae
= *Rosa gallica* L. × *Rosa moschata* Herrm.
× *Rosa canina* L. × *Rosa × damascena*
Mill. [possible combination]
♦ Burgundy rose, cabbage rose,
Holland rose, moss rose, pale rose,
Provence rose, rose de Mai, rosier cent
feuilles, Kohlrose, Zentifolie, rosa das
cem folhas, rosa común
♦ Naturalised, Casual Alien
♦ 101, 280
♦ S, cultivated, herbal. Origin:
horticultural hybrid.

Rosa chinensis Jacq.
Rosaceae
Rosa nankinensis Lour., *Rosa sinica* L.
♦ China rose
♦ Weed, Naturalised
♦ 86, 98, 101, 203
♦ S, cultivated, herbal. Origin: China.

Rosa chinensis Jacq. hybrids
Rosaceae
♦ China rose
♦ Casual Alien
♦ 280

Rosa chinensis × moschata Jacq.
Rosaceae
♦ noisette
♦ Naturalised
♦ 7, 98

Rosa chinensis × multiflora Jacq.
Rosaceae
♦ Naturalised
♦ 98

Rosa cinnamomea L.
Rosaceae
= *Rosa majalis* J.Herrm.
♦ cinnamon rose
♦ Weed, Naturalised
♦ 80, 101
♦ cultivated, herbal.

Rosa corymbifera Borkh.
Rosaceae
= *Rosa dumetorum* Thuill.
♦ Weed
♦ 272
♦ S, cultivated.

Rosa × damascena Mill. (*pro* sp.)
Rosaceae
= *Rosa gallica* L. × *Rosa moschata* Herrm.
[*Rosa belgica* Mill., *Rosa calendarum*
Borkh.]
♦ damask rose
♦ Naturalised
♦ 80, 101
♦ S, promoted, cultivated, herbal.

Rosa × dumalis Bechst.
Rosaceae
= *Rosa canina* L. × *Rosa squarrosa* Rau
[probable combination]
♦ hybrid dog rose
♦ Cultivation Escape
♦ 40
♦ cultivated.

Rosa dumetorum Thuill.
Rosaceae

Rosa corymbifera Borkh. (see)
♦ corymb rose
♦ Naturalised
♦ 101
♦ cultivated.

Rosa eglanteria L.
Rosaceae
= *Rosa rubiginosa* L.
♦ sweet briar rose, eglantine, sweetleaf
rose
♦ Weed, Noxious Weed, Naturalised,
Introduced
♦ 4, 34, 80, 87, 88, 95, 101, 121, 137, 146,
195, 218, 279, 295
♦ pS, cultivated, herbal. Origin:
Eurasia.

Rosa gallica L.
Rosaceae
Rosa austriaca Crantz, *Rosa pumila* Jacq.,
Rosa rubra Lam.
♦ French rose
♦ Weed, Naturalised, Garden Escape
♦ 39, 80, 98, 101, 203, 251, 272, 280
♦ S, cultivated, herbal, toxic.

Rosa gallica L. var. gallica
Rosaceae
♦ French rose
♦ Naturalised
♦ 101

Rosa gallica L. var. officinalis Thory
Rosaceae
♦ French rose
♦ Naturalised
♦ 101

Rosa glutinosa Sibth. & Sm.
Rosaceae
♦ Weed
♦ 272
♦ cultivated, herbal.

Rosa gymnocarpa Nutt.
Rosaceae
♦ baldhip rose, wood rose, dwarf rose
♦ Weed
♦ 87, 88, 218
♦ S, cultivated, herbal.

Rosa × harisonii Rivers
Rosaceae
= *Rosa foetida* Herrm. × *Rosa*
spinosissima L.
♦ Naturalised
♦ 101

Rosa indica L.
Rosaceae
♦ cyme rose, China rose, tea rose
♦ Weed, Naturalised, Casual Alien
♦ 15, 101, 261
♦ herbal. Origin: China.

Rosa laevigata Michx.
Rosaceae
♦ Cherokee rose
♦ Weed, Naturalised, Garden Escape
♦ 86, 87, 88, 98, 101, 179, 203, 218, 251
♦ S, cultivated, herbal. Origin: east
Asia.

Rosa majalis J.Herrm.
Rosaceae
Rosa cinnamomea L. (see), *Rosa*

foecundissima Münchh.
♦ double cinnamon rose, metsäruusu,
cinnamon rose
♦ Naturalised
♦ 101
♦ S, promoted, herbal.

Rosa micrantha Borrer ex Sm.
Rosaceae
Rosa viscida Puget
♦ smallflower sweet briar, lesser sweet
briar, rose
♦ Weed, Naturalised
♦ 80, 101, 272, 280
♦ S, promoted, herbal.

Rosa × moschata J.Herrm.
Rosaceae
♦ musk rose, Himalayan musk rose
♦ Naturalised
♦ 101, 241, 280, 300
♦ S, cultivated, herbal. Origin: obscure,
possibly Asia.

Rosa multiflora Thunb. ex Murr
Rosaceae
Rosa multiflora var. *carnea* Thory, *Rosa*
multiflora var. *platyphylla* Thory, *Rosa*
floribunda hort. ex Andrews *nom. illeg.*
♦ multiflora rose, baby rose, Japanese
rose, seven sisters rose, bramble
flowered rose
♦ Weed, Noxious Weed, Naturalised,
Introduced, Garden Escape,
Environmental Weed, Cultivation
Escape
♦ 4, 17, 77, 79, 80, 87, 88, 101, 102, 104,
129, 133, 137, 142, 151, 161, 182, 195,
211, 222, 224, 229, 280, 283, 286
♦ S, cultivated, herbal. Origin: China,
Korea, Japan.

Rosa nutkana Presl
Rosaceae
Rosa fraxinifolia non Borkh.
♦ nootka rose
♦ Weed
♦ 87, 88, 159, 218
♦ S, cultivated, herbal.

Rosa × odorata (Andrews) Sweet
Rosaceae
= *Rosa chinensis* Jacq. × *Rosa gigantea*
Collett [most probable combination]
♦ tea rose
♦ Sleeper Weed, Naturalised, Garden
Escape, Environmental Weed
♦ 39, 86, 101, 155, 251
♦ S, cultivated, herbal, toxic. Origin:
horticultural hybrid.

Rosa palustris Marshall
Rosaceae
♦ swamp rose
♦ Weed, Environmental Weed
♦ 80, 88, 151
♦ cultivated, herbal.

Rosa pendulina L.
Rosaceae
Rosa alpina L., *Rosa rupestris* Crantz
♦ rosa alpina
♦ Weed
♦ 272
♦ cultivated, herbal.

Rosa pimpinellifolia L.
Rosaceae
= *Rosa spinosissima* L.
♦ juhannusruusu, burnet rose
♦ Weed, Naturalised, Cultivation Escape
♦ 42, 272, 280
♦ S, cultivated, herbal.

Rosa pratincola Greene
Rosaceae
= *Rosa arkansana* Porter var. *suffulta* (Greene) Cockerell (NoR)
♦ Weed
♦ 88, 218
♦ herbal.

Rosa × rehderiana Blackb.
Rosaceae
= *Rosa chinensis* Jacq. × *Rosa multiflora* Thunb.
♦ polyantha rose
♦ Naturalised
♦ 101

Rosa roxburghii Tratt.
Rosaceae
♦ chestnut rose
♦ Naturalised, Cultivation Escape, Casual Alien
♦ 86, 252, 280
♦ S, cultivated, herbal. Origin: China.

Rosa rubiginosa L.
Rosaceae
Rosa arabica Crep. (see), *Rosa eglanteria* L. (see)
♦ sweet briar, eglantine, sweet briar rose, wilderoos
♦ Weed, Noxious Weed, Naturalised, Garden Escape, Environmental Weed, Cultivation Escape
♦ 7, 15, 20, 42, 63, 72, 86, 87, 88, 97, 98, 147, 152, 165, 169, 171, 176, 181, 198, 203, 225, 241, 243, 246, 269, 280, 283, 289, 296, 300
♦ S, cultivated, herbal. Origin: Europe.

Rosa rubrifolia Vill.
Rosaceae
♦ redleaf rose
♦ Naturalised
♦ 101
♦ cultivated, herbal.

Rosa rugosa Thunb.
Rosaceae
Rosa ferox Lawr., *Rosa kamtchatica* Vent.
♦ Japanese rose, rugosa rose, ramanas rose, ninebark, saltspray rose, wrinkled rose, beach rose, rose
♦ Weed, Naturalised, Environmental Weed, Cultivation Escape
♦ 4, 40, 42, 80, 88, 101, 133, 151, 195, 224, 280
♦ S, cultivated, herbal.

Rosa sempervirens L.
Rosaceae
♦ evergreen rose
♦ Naturalised, Casual Alien
♦ 101, 261
♦ cultivated, herbal. Origin: Mediterranean.

Rosa sempervirens hybrids L.
Rosaceae
♦ Naturalised
♦ 280

Rosa serafinii Viv.
Rosaceae
♦ Mediterranean rose
♦ Naturalised
♦ 101

Rosa setigera Michx.
Rosaceae
♦ climbing rose, prairie rose
♦ Weed
♦ 161
♦ cultivated, herbal.

Rosa spinosissima L.
Rosaceae
Rosa pimpinellifolia L. (see)
♦ Scotch rose, burnet rose
♦ Weed, Naturalised
♦ 80, 101
♦ cultivated, herbal.

Rosa sulphurea Aiton
Rosaceae
♦ Weed
♦ 87, 88

Rosa tomentosa Sm.
Rosaceae
♦ harsh downy rose, white woolly rose, downy rose
♦ Weed, Naturalised, Casual Alien
♦ 87, 88, 101, 280
♦ S, cultivated.

Rosa villosa L.
Rosaceae
Rosa hispida hort. ex Poir., *Rosa pomifera* Herm.
♦ apple rose, soft leaved rose
♦ Naturalised
♦ 101
♦ S, cultivated, herbal.

Rosa wichuraiana Crép.
Rosaceae
♦ memorial rose
♦ Weed, Naturalised
♦ 15, 80, 88, 101, 179, 204, 286
♦ S, cultivated, herbal.

Rosa wichuraiana hybrids Crép.
Rosaceae
♦ Naturalised
♦ 280

Rosa woodsii Lindl.
Rosaceae
♦ Wood's rose, western wild rose
♦ Weed
♦ 87, 88, 218
♦ S, cultivated, herbal.

Rosa xanthina Lindl.
Rosaceae
♦ rose
♦ Naturalised
♦ 101
♦ cultivated.

Roseocereus Backeb. spp.
Cactaceae
= *Harrisia* Britt. spp.

♦ Weed, Quarantine Weed
♦ 76, 88, 203

Rosmarinus officinalis L.
Lamiaceae
Rosmarinus angustifolius Mill., *Rosmarinus latifolius* Mill., *Salvia rosmarinus* Schleid.
♦ rosemary
♦ Weed, Naturalised, Introduced, Garden Escape, Casual Alien
♦ 38, 39, 86, 97, 98, 101, 203, 261, 280
♦ S, cultivated, herbal, toxic. Origin: Eurasia.

Rostellularia obtusa Nees var. grandifolia Miq.
Acanthaceae
♦ Weed
♦ 13

Rostellularia obtusa Nees var. neesiana Bre.
Acanthaceae
Justicia procumbens L.
♦ Weed
♦ 13

Rostellularia procumbens (L.) Nees
Acanthaceae
♦ Weed
♦ 275, 297
♦ aH.

Rostellularia sundana Bremek.
Acanthaceae
= *Justicia procumbens* L.
♦ Weed
♦ 88, 170, 191

Rostraria cristata (L.) Tzvelev
Poaceae
Bromus cristatus (L.) Spreng., *Festuca cristata* L., *Festuca gerardii* Vill., *Festuca phleoides* Vill., *Koeleria brachystachya* DC., *Koeleria campestris* Phil., *Koeleria cristata* (L.) Pers., *Koeleria gerardii* (Vill.) Shinners, *Koeleria phleoides* (Vill.) Pers. (see), *Lophochloa cristata* (L.) Hyl. (see), *Lophochloa phleoides* (Vill.) Reich. (see), *Trisetaria cristata* (L.) Kerguélen (see), *Trisetaria phleoides* (Vill.) Nevski, *Trisetum cristatum* (L.) Potztal, *Trisetum minutiflorum* Phil.
♦ annual junegrass, annual cat's tail, Mediterranean hairgrass
♦ Weed, Naturalised, Environmental Weed, Casual Alien
♦ 40, 86, 98, 101, 134, 176, 198, 203, 243, 253, 280, 287, 300
♦ G. Origin: Eurasia, Mediterranean.

Rostraria pumila (Desf.) Tzvelev
Poaceae
Avena pumila Desf., *Koeleria pumila* (Desf.) Domin, *Koeleria sinaica* Boiss., *Lophochloa pumila* (Desf.) Bor (see), *Trisetum pumilum* (Desf.) Kunth
♦ roughtail, tiny bristlegrass
♦ Weed, Naturalised, Introduced, Environmental Weed
♦ 86, 98, 198, 203, 228
♦ G, arid. Origin: Spain.

Rotala L. spp.
Lythraceae
♦ rotala
♦ Quarantine Weed
♦ 220

Rotala densiflora (Roth) Koehne
Lythraceae
♦ Weed
♦ 87, 88
♦ Origin: Australia.

Rotala indica (Willd.) Koehne
Lythraceae
Ameletia acutidens Miq., *Ammannia peploides* Spreng. (see), *Peplis indica* Willd.
♦ toothcup, Indian toothcup, Huai chinnasee
♦ Weed, Quarantine Weed, Naturalised
♦ 76, 87, 88, 101, 135, 161, 170, 180, 191, 191, 203, 204, 220, 238, 262, 274, 275, 297
♦ aH, aqua, cultivated, herbal.

Rotala indica Koehne var. uliginosa Koehne
Lythraceae
♦ kikashigusa
♦ Weed
♦ 263, 286
♦ aH.

Rotala leptopetala (Blume) Koehne
Lythraceae
= *Rotala rosea* (Poir.) C.D.K.Cook
♦ Weed
♦ 87, 88

Rotala leptopetala (Blume) Koehne var. littorea (Miq.) Koehne
Lythraceae
♦ Weed
♦ 286

Rotala mexicana Cham. & Schlecht.
Lythraceae
♦ Weed
♦ 87, 88, 170, 262
♦ aqua.

Rotala pusilla Tul.
Lythraceae
♦ Weed
♦ 263, 286
♦ aH, aqua.

Rotala ramosior (L.) Koehne
Lythraceae
Ammannia catholica Cham. & Schltdl., *Ammannia dentifera* A.Gray, *Ammannia humilis* Michx., *Ammannia monoflora* Blanco, *Ammannia occidentalis* DC., *Ammannia ramosa* Hill, *Ammannia ramosior* L., *Boykinia humilis* (Michx.) Raf., *Peplis occidentalis* Spreng., *Rotala catholica* (Cham. & Schltdl.) Leeuwen, *Rotala dentifera* (A.Gray) Koehne, *Rotala ramosior* var. *dentifera* (A.Gray) Lundell
♦ toothcup, lowland rotala
♦ Weed, Quarantine Weed
♦ 76, 87, 88, 157, 170, 191, 203, 218, 220
♦ waH, herbal.

Rotala rosea (Poir.) Cook
Lythraceae

Rotala leptopetala (Blume) Koehne (see),
Ammannia leptopetala Blume, *Ammannia pentandra* Roxb. (see)
♦ Weed
♦ 88, 170, 191

Rotala rotundifolia Koehne
Lythraceae
Ammannia rotundifolia Buch.-Ham. ex Roxb. (see)
♦ dwarf rotala, rotala
♦ Weed, Sleeper Weed, Quarantine Weed, Naturalised, Environmental Weed
♦ 3, 54, 76, 86, 87, 88, 155, 191, 203, 238, 274, 275, 286, 297
♦ pH, aqua, cultivated. Origin: India to Japan.

Rotala uliginosa Miq.
Lythraceae
♦ Weed, Quarantine Weed
♦ 76, 87, 88, 203, 220

Rothia trifoliata (Roth) Pers.
Fabaceae/Papilionaceae
Dillwynia trifoliata Roth
♦ Weed
♦ 87, 88
♦ Origin: Australia.

Rottboellia cochinchinensis (Lour.) Clayton
Poaceae
Manisuris exaltata (L.) Kuntze, *Rottboellia exaltata* (L.) L.f. (see), *Stegosia cochinchinensis* Lour.
♦ itchgrass, raoulgrass, guinea fowl grass, shamva grass, caminadora, peluda
♦ Weed, Quarantine Weed, Noxious Weed, Naturalised, Introduced
♦ 6, 28, 32, 38, 39, 53, 67, 76, 80, 86, 88, 93, 101, 140, 158, 161, 172, 179, 191, 206, 229, 237, 243, 258, 261, 271, 281, 295
♦ aG, toxic. Origin: Africa.

Rottboellia exaltata (L.) L.f.
Poaceae
= *Rottboellia cochinchinensis* (Lour.) Clayton
♦ kokoma grass, guinea fowl grass, corn grass, lisofya, caminadora, shamva grass, raoulgrass, aa prong khaai
♦ Weed, Native Weed
♦ 30, 50, 87, 88, 90, 121, 153, 157, 170, 186, 236, 239, 255, 263, 276
♦ aG, cultivated, herbal. Origin: India.

Rottboellia exaltata L. var. appendiculata Hack.
Poaceae
♦ Weed, Naturalised
♦ 286, 287
♦ G.

Rourea surinamensis Miq.
Connaraceae
♦ Juan Caliente
♦ Weed
♦ 87, 88

Royena sericea Bernh.
Ebenaceae
= *Diospyros lycioides* Desf. ssp. *sericea*

(Bernh.) De Winter
♦ Weed, Quarantine Weed
♦ 76, 87, 88, 203, 220

Roystonea elata (Bart.) Harper
Arecaceae
♦ Florida royal palm
♦ Introduced
♦ 230
♦ T, cultivated, herbal.

Roystonea oleracea (Jacq.) Cook
Arecaceae
♦ Caribbean royal palm
♦ Introduced
♦ 230
♦ T, cultivated.

Roystonea regia (Kunth) Cook
Arecaceae
♦ Cuban royal palm, royal palm, palma de Cubana, yaguas Cubana
♦ Weed, Garden Escape
♦ 179, 261
♦ cultivated, herbal. Origin: Cuba.

Rubia akane Nakai
Rubiaceae
♦ Indian madder
♦ Weed
♦ 263, 273
♦ pC, promoted, herbal.

Rubia argyi Hara
Rubiaceae
♦ Weed
♦ 286

Rubia cordifolia L.
Rubiaceae
♦ Indian madder, lukelangafu, lukera batuzi
♦ Weed
♦ 87, 88, 275, 297
♦ pH, promoted, herbal.

Rubia peregrina L.
Rubiaceae
♦ wild madder, levany madder
♦ Weed
♦ 70, 88, 94, 253
♦ pH, cultivated, herbal.

Rubia tinctoria L.
Rubiaceae
= *Rubia tinctorum* L.
♦ madder
♦ Naturalised
♦ 101
♦ cultivated, herbal.

Rubia tinctorum L.
Rubiaceae
Rubia sylvestris Mill., *Rubia tinctoria* L. (see)
♦ madder, garance des teinturiers, färberröte, krapp, garanca, granza, rubia de tintes
♦ Weed, Naturalised
♦ 88, 94, 241, 272, 287, 300
♦ pH, cultivated, herbal.

Rubus L. spp.
Rosaceae
♦ blackberries, brambles, raspberries, dewberries
♦ Weed, Cultivation Escape,

Environmental Weed
♦ 3, 18, 22, 80, 88, 161, 211, 233, 243, 253, 257, 272, 279
♦ pS, cultivated, herbal.

Rubus affinis **Wight & Arn.**
Rosaceae
♦ blackberry
♦ Weed, Quarantine Weed
♦ 76, 88, 121, 203, 220, 279
♦ pS, promoted.

Rubus alceaefolius **Poir.**
Rosaceae
= *Rubus alceifolius* Poir.
♦ giant bramble
♦ Weed, Quarantine Weed, Noxious Weed, Naturalised
♦ 22, 76, 86, 87, 88, 132, 147, 191, 203
♦ herbal.

Rubus alceifolius **Poir.**
Rosaceae
Rubus alceaefolius Poir. (see), *Rubus fimbriiferus* Focke
♦ Naturalised
♦ 86, 98
♦ cultivated, herbal. Origin: China, Taiwan, south-east Asia.

Rubus allegheniensis **Porter**
Rosaceae
♦ Allegheny blackberry, sow teat blackberry, Allegheny brombeere, Amerikanvatukka
♦ Weed, Naturalised, Introduced, Cultivation Escape
♦ 34, 42, 87, 88, 218, 287
♦ S, cultivated, herbal. Origin: North America.

Rubus amplificatus **E.Lees**
Rosaceae
♦ Casual Alien
♦ 280

Rubus annamensis **Cardot**
Rosaceae
♦ Weed
♦ 87, 88

Rubus apetalus **Poir.**
Rosaceae
♦ raspberry
♦ Weed, Quarantine Weed, Native Weed
♦ 76, 88, 121, 203, 220
♦ pS. Origin: southern Africa.

Rubus arcticus **L.**
Rosaceae
♦ Arctic blackberry, Arctic bramble, mesimarja
♦ Weed
♦ 87, 88
♦ pH, cultivated, herbal.

Rubus argutus **Link**
Rosaceae
Rubus penetrans L.H.Bailey (see)
♦ Florida prickly blackberry, Florida blackberry, sawtooth blackberry, highbush blackberry
♦ Weed, Quarantine Weed, Noxious Weed, Naturalised, Environmental Weed

♦ 22, 76, 80, 86, 88, 151, 152, 203, 220, 229, 280
♦ S, promoted, herbal. Origin: eastern North America.

Rubus armeniacus **Focke**
Rosaceae
= *Rubus procerus* Muell.
♦ Himalayan giant
♦ Naturalised
♦ 40, 86

Rubus bellobatus **L.H.Bailey**
Rosaceae
Rubus alumnus L.H.Bailey
♦ Kittatinny blackberry
♦ Weed, Naturalised, Environmental Weed
♦ 7, 86, 98, 203
♦ S, cultivated. Origin: north-eastern North America.

Rubus bifrons **Vest ex Tratt.**
Rosaceae
♦ bramble, Himalayan berry, ruusukarhunvatukka
♦ Weed, Naturalised, Casual Alien
♦ 42, 80, 101
♦ S, promoted, herbal.

Rubus bogotensis **Kunth**
Rosaceae
♦ Environmental Weed
♦ 257

Rubus brasiliensis **Mart.**
Rosaceae
♦ Weed
♦ 87, 88

Rubus caesius **L.**
Rosaceae
Rubus turkestanicus Pav.
♦ European dewberry, dewberry, sinivatukka, bramble
♦ Weed, Quarantine Weed, Naturalised, Casual Alien
♦ 76, 80, 87, 88, 94, 101, 203, 220, 243, 272, 280, 295
♦ S, promoted, herbal.

Rubus candidans **Weihe ex Rchb.**
Rosaceae
♦ Naturalised
♦ 300

Rubus canescens **DC.**
Rosaceae
♦ rovo tomentoso
♦ Weed
♦ 272
♦ herbal.

Rubus cardiophyllus **Lefèvre & P.J.Müll.**
Rosaceae
♦ Naturalised
♦ 280

Rubus chevalieri **Touss.**
Rosaceae
♦ Weed
♦ 87, 88

Rubus chloocladus **W.C.R.Watson**
Rosaceae
♦ blackberry
♦ Weed, Naturalised
♦ 86, 98, 198, 203

Rubus cissburiensis **Barton & Ridd.**
Rosaceae
♦ blackberry
♦ Weed, Naturalised, Environmental Weed
♦ 72, 86, 88, 98, 198, 203, 280
♦ S, cultivated. Origin: Europe.

Rubus cissburiensis **W.C.Barton & Ridd.** × *ulmifolius* **Schott**
Rosaceae
♦ Naturalised
♦ 280

Rubus collinus **DC.**
Rosaceae
♦ Weed
♦ 111, 243

Rubus constrictus **Muell. & Lef.**
Rosaceae
♦ ostruľina stiahnutá
♦ Naturalised
♦ 241, 300

Rubus coronarius **(Sims) Sweet**
Rosaceae
= *Rubus rosifolius* Sm.
♦ cultivated raspberry
♦ Naturalised, Cultivation Escape
♦ 101, 261
♦ cultivated, herbal.

Rubus couchii **Focke**
Rosaceae
♦ Cultivation Escape
♦ 40
♦ cultivated.

Rubus cuneifolius **Pursh**
Rosaceae
♦ American bramble, Gozard's curse, sand blackberry, sand bramble, Amerikaanse braam
♦ Weed, Quarantine Weed, Noxious Weed, Environmental Weed, Cultivation Escape
♦ 10, 51, 63, 76, 87, 88, 95, 95, 121, 152, 158, 203, 220, 277, 278, 279, 283
♦ pS, cultivated, herbal. Origin: North America.

Rubus discolor **Weihe & Nees**
Rosaceae
= *Rubus ulmifolius* Schott
♦ Himalaya berry, blackberry, Himalayan giant blackberry, Armeniankarhunvatukka
♦ Weed, Noxious Weed, Naturalised, Environmental Weed, Cultivation Escape, Casual Alien
♦ 4, 7, 18, 20, 34, 35, 42, 45, 72, 78, 80, 86, 88, 98, 101, 104, 137, 146, 151, 180, 198, 203, 231, 237, 243, 252, 295
♦ pS, cultivated, herbal. Origin: Eurasia.

Rubus divaricatus **P.J.Mull.**
Rosaceae
Rubus nitidus Weihe & Nees, *Rubus caesius* auct. arg. non L., *Rubus fruticosum* auct. non L.
♦ zarzamora
♦ Weed
♦ 237, 295

Rubus echinatus Lindl.
Rosaceae
♦ Naturalised
♦ 280
♦ Origin: southern Europe.

Rubus ellipticus Sm.
Rosaceae
Rubus flavus Buch.-Ham. ex D.Don
♦ yellow Himalayan raspberry
♦ Weed, Naturalised, Environmental
Weed
♦ 22, 80, 86, 88, 98, 101, 129, 151, 152,
203
♦ S, cultivated, herbal. Origin: India,
Sri Lanka, Burma, tropical China,
Philippines.

Rubus ellipticus Sm. var. *obcordatus*
Focke
Rosaceae
♦ yellow Himalayan raspberry
♦ Noxious Weed, Naturalised
♦ 101, 229

Rubus errabundus W.C.R.Watson
Rosaceae
♦ Casual Alien
♦ 280

Rubus erythrops Weihe. & Nees.
Rosaceae
Rubus rosaceus Weihe (see)
♦ Naturalised
♦ 86, 280
♦ Origin: Europe.

Rubus fioniae Frid. ex Neuman
Rosaceae
♦ tanskankarhunvatukka
♦ Casual Alien
♦ 42

Rubus flagellaris Willd.
Rosaceae
♦ northern dewberry, bramble
♦ Weed, Naturalised, Cultivation
Escape
♦ 87, 88, 95, 218, 280, 283, 287
♦ S, cultivated, herbal. Origin: Europe.

Rubus fraxinifolius Poir.
Rosaceae
♦ Weed, Naturalised
♦ 98, 203

Rubus fruticosus L. agg.
Rosaceae
numerous species in this aggregate
♦ European blackberry, shrubby
blackberry, wild blackberry,
brombeere, bramble, blackberry, braam
♦ Weed, Quarantine Weed, Noxious
Weed, Naturalised, Garden Escape,
Environmental Weed, Cultivation
Escape
♦ 15, 62, 63, 67, 70, 76, 86, 87, 88, 95, 98,
134, 140, 147, 152, 165, 169, 171, 176,
178, 181, 203, 218, 225, 229, 232, 246,
251, 269, 272, 280, 283, 286, 287, 289,
296
♦ S, cultivated, herbal. Origin: Europe.

Rubus geniculatus Kaltenb.
Rosaceae
♦ false Himalayan berry

♦ Naturalised
♦ 101

Rubus glaucus Benth.
Rosaceae
♦ Andean blackberry, Andes berry,
Mora de castilla
♦ Weed
♦ 22, 80
♦ S, cultivated, herbal.

Rubus grayanus Maxim.
Rosaceae
♦ Weed
♦ 88, 204
♦ S, promoted.

Rubus hillii F.Muell.
Rosaceae
♦ blackberry
♦ Naturalised
♦ 7
♦ cultivated, herbal.

Rubus hirsutus Thunb.
Rosaceae
♦ kusaichigo
♦ Weed
♦ 286
♦ S, promoted. Origin: China, Japan,
Korea.

Rubus hirtus Waldst. & Kit.
Rosaceae
♦ Weed
♦ 272

Rubus hispidus L.
Rosaceae
♦ swamp dewberry, bristly dewberry
♦ Weed
♦ 293
♦ S, promoted, herbal. Origin: eastern
North America.

Rubus idaeus L.
Rosaceae
Rubus sericeus Gilib.
♦ red raspberry, wild raspberry,
raspberry, American red raspberry,
lampone
♦ Weed, Naturalised
♦ 70, 86, 87, 88, 98, 198, 203, 218, 272,
280
♦ S, cultivated, herbal. Origin: Eurasia.

Rubus illecebrosus Focke
Rosaceae
♦ strawberry raspberry
♦ Naturalised
♦ 101
♦ S, promoted, herbal.

Rubus koehleri Weihe & Nees
Rosaceae
♦ Weed, Naturalised
♦ 86, 98, 203

Rubus laciniatus F.Muell.
Rosaceae
Rubus vulgaris Weihe & Nees (see)
♦ cut leaved blackberry, evergreen
blackberry, liuskavatukka, Oregon
cutleaf blackberry
♦ Weed, Naturalised, Environmental
Weed, Cultivation Escape
♦ 15, 34, 39, 40, 42, 45, 72, 80, 87, 88, 98,

101, 136, 146, 176, 198, 203, 218, 280
♦ pS, cultivated, herbal, toxic. Origin:
obscure.

Rubus laciniatus Willd. ssp. *laciniatus*
Rosaceae
♦ cutleaf bramble
♦ Noxious Weed, Naturalised
♦ 86, 198
♦ Origin: obscure.

Rubus laciniatus Willd. ssp. *selmeri*
(Lindeb.) Beek
Rosaceae
♦ blackberry, cutleaf blackberry
♦ Noxious Weed, Naturalised
♦ 86, 198

Rubus lasiocarpus Sm.
Rosaceae
= *Rubus niveus* Thunb.
♦ Weed
♦ 87, 88
♦ herbal.

Rubus leightonii Lees & F.M.Leight.
Rosaceae
♦ Weed, Naturalised
♦ 86, 98, 203
♦ Origin: Europe.

Rubus leptothyrsos G.Braun
Rosaceae
♦ Naturalised
♦ 280

Rubus leucodermis Douglas ex Torr. &
Gray
Rosaceae
♦ whitebark raspberry, western
raspberry
♦ Weed
♦ 87, 88, 218
♦ S, promoted, herbal.

Rubus linkianus Ser.
Rosaceae
♦ Link's blackberry
♦ Naturalised
♦ 101
♦ S, promoted.

Rubus loganobaccus L.H.Bailey
Rosaceae
= *Rubus idaeus* L. × *Rubus ursinus*
Cham. & Schldl. [most probable
combination]
♦ loganberry, boysenberry
♦ Weed, Naturalised
♦ 40, 86, 98, 203
♦ S, cultivated. Origin: horticultural
hybrid.

Rubus longepedicellatus (C.E.Gust.)
C.H.Stirt.
Rosaceae
♦ raspberry
♦ Weed, Quarantine Weed, Native
Weed
♦ 76, 88, 121, 203, 220
♦ pS. Origin: southern Africa.

Rubus macrophyllus Weihe & Nees
Rosaceae
♦ largeleaf blackberry
♦ Naturalised
♦ 101

Rubus mollior L.H.Bailey
Rosaceae
♦ softleaf blackberry
♦ Naturalised
♦ 280
♦ herbal.

Rubus moluccanus L.
Rosaceae
♦ Ceylon blackberry, eelkek, molucca raspberry, wild blackberry, broad leaved bramble, molucca bramble, kohkihl, soni, wa sori, wa ngandrongandro, wa votovotoa
♦ Weed, Noxious Weed, Environmental Weed
♦ 3, 22, 67, 80, 87, 88, 140, 152, 229
♦ S, promoted, herbal. Origin: east Asia, Himalayas to Sri Lanka, Australia.

Rubus moluccanus L. var. moluccanus
Rosaceae
♦ Naturalised
♦ 86
♦ Origin: Asia, Australia.

Rubus moluccanus L. var. trilobus A.R.Bean
Rosaceae
♦ Naturalised
♦ 86
♦ Origin: Australia.

Rubus mucronulatus Boreau
Rosaceae
♦ Naturalised
♦ 86, 280

Rubus nemoralis P.J.Müll.
Rosaceae
♦ Naturalised
♦ 280

Rubus nivalis Doug. ex Hook.
Rosaceae
♦ snow bramble, snow dwarf bramble, snow raspberry
♦ Weed
♦ 80
♦ pC, herbal.

Rubus niveus Thunb.
Rosaceae
Rubus lasiocarpus Sm. (see)
♦ hill raspberry, Java bramble, mysore raspberry, snowpeaks raspberry, Ceylon raspberry
♦ Weed, Quarantine Weed, Noxious Weed, Naturalised, Environmental Weed
♦ 18, 22, 76, 88, 101, 121, 179, 203, 220, 229, 257, 279
♦ S, promoted. Origin: Eurasia.

Rubus occidentalis L.
Rosaceae
♦ black raspberry, Virginian raspberry
♦ Weed
♦ 87, 88, 218
♦ S, promoted, herbal.

Rubus odoratus L.
Rosaceae
♦ tuoksuvatukka, purple flowering raspberry, thimbleberry

♦ Cultivation Escape
♦ 42
♦ S, cultivated, herbal.

Rubus ostryifolius Rydb.
Rosaceae
♦ highbush blackberry
♦ Naturalised
♦ 280
♦ herbal.

Rubus palmatus Thunb.
Rosaceae
♦ Weed
♦ 87, 88
♦ S, promoted.

Rubus parviflorus Nutt.
Rosaceae
♦ western thimbleberry, thimbleberry
♦ Weed
♦ 87, 88, 218, 265
♦ S, cultivated, herbal. Origin: western North America.

Rubus parvifolius L.
Rosaceae
Rubus triphyllus Thunb. (see)
♦ Japanese raspberry, western thimbleberry, thimbleberry
♦ Weed, Naturalised, Garden Escape, Environmental Weed
♦ 86, 204, 286
♦ S, cultivated, herbal. Origin: Asia, Australia.

Rubus pascuus Bailey
Rosaceae
♦ Chesapeake blackberry
♦ Weed
♦ 279
♦ S, promoted. Origin: eastern North America.

Rubus penetrans L.H.Bailey
Rosaceae
= *Rubus argutus* Link
♦ Weed
♦ 87, 88
♦ herbal.

Rubus pensilvanicus Poir.
Rosaceae
♦ Pennsylvania blackberry
♦ Weed, Cultivation Escape
♦ 34
♦ S, cultivated, herbal.

Rubus phoenicolasius Maxim.
Rosaceae
♦ wineberry, Japanese wineberry, wine raspberry, raspberry
♦ Weed, Naturalised
♦ 15, 17, 80, 86, 88, 98, 101, 102, 121, 133, 165, 195, 198, 203, 224, 280
♦ pC, cultivated, herbal. Origin: Eurasia.

Rubus pinnatus Willd.
Rosaceae
♦ lutandula, bramble, cape bramble, South African blackberry, South African bramble, South African raspberry, makangawa
♦ Weed, Quarantine Weed, Native Weed
♦ 76, 88, 121, 203, 220

♦ pS. Origin: southern Africa.

Rubus plicatus Weihe & Nees
Rosaceae
♦ oimuvatukka
♦ Naturalised
♦ 42
♦ herbal.

Rubus polyanthemus Lindeb.
Rosaceae
♦ blackberry
♦ Weed, Naturalised, Environmental Weed
♦ 72, 86, 86, 88, 98, 198, 203, 280
♦ S. Origin: Europe.

Rubus procerus Muell.
Rosaceae
Rubus armeniacus Focke (see), *Rubus discolor* auct., *Rubus hedycarpus* Focke var. *armeniacus* (Focke) Focke, *Rubus praecox* Bertol. *nom. dubium*
♦ Himalaya blackberry
♦ Weed, Naturalised
♦ 87, 88, 98, 116, 136, 203, 218, 280
♦ cultivated, herbal.

Rubus × proteus C.H.Stirt.
Rosaceae
= *Rubus cuneifolius* Pursh (Mpumalanga form) × *Rubus longepedicellatus* (C.E.Gust.) C.H.Stirt. (native)
♦ bramble, Amerikaanse braam, American bramble
♦ Weed, Noxious Weed, Cultivation Escape
♦ 63, 88, 95, 283
♦ cultivated. Origin: South Africa.

Rubus pyramidalis Kaltenb.
Rosaceae
♦ Weed, Naturalised
♦ 54, 86, 88, 98, 203

Rubus racemosus Roxb.
Rosaceae
♦ Weed
♦ 22
♦ S, promoted.

Rubus radula Weihe & Boenn.
Rosaceae
♦ Weed, Naturalised
♦ 86, 98, 203

Rubus rigidus Sm.
Rosaceae
♦ bramble
♦ Weed, Quarantine Weed, Native Weed
♦ 76, 88, 121, 158, 203, 220
♦ pS, herbal. Origin: southern Africa.

Rubus roribaccus Rydb.
Rosaceae
♦ dewberry, Lucretia dewberry
♦ Weed, Sleeper Weed, Naturalised, Environmental Weed
♦ 86, 98, 155, 203
♦ Origin: North America.

Rubus roridus Lindl.
Rosaceae
♦ Weed
♦ 87, 88

Rubus rosaceus Weihe
Rosaceae
= *Rubus erythrops* Weihe & Nees.
♦ blackberry
♦ Weed, Naturalised, Environmental
Weed
♦ 72, 86, 88, 98, 198, 203
♦ S.

Rubus rosaefolius Sm.
Rosaceae
= *Rubus rosifolius* Sm.
♦ Mauritius raspberry, ruusuvatukka
♦ Weed, Casual Alien
♦ 42, 87, 88, 158
♦ S, promoted, herbal.

Rubus rosifolius Sm.
Rosaceae
Rubus commersonii Poir., *Rubus coronarius* (Sims) Sweet (see), *Rubus eustephanos* Focke ex Diels var. *coronarius* (Sims) Koidz., *Rubus rosaefolius* Sm. (see), *Rubus rosifolius* var. *coronarius* Sims
♦ West Indian raspberry, strawberry raspberry, roseleaf raspberry, thimbleberry, ola'a, bramble of the Cape, Mauritius raspberry, javanische Himbeere, tokin ibara, frambueso de Africa
♦ Weed, Naturalised, Introduced
♦ 3, 15, 22, 38, 80, 88, 98, 101, 203, 255, 261, 280
♦ pC, cultivated, herbal. Origin: Brazil.

Rubus rugosus Sm.
Rosaceae
♦ keriberry, Himalayan blackberry
♦ Naturalised
♦ 280
♦ Origin: central Asia.

Rubus sanctus Schreb.
Rosaceae
♦ Weed
♦ 221

Rubus sieboldii Blume
Rosaceae
♦ molucca raspberry, palmleaf dewberry
♦ Noxious Weed, Naturalised
♦ 101, 229
♦ S, promoted.

Rubus spectabilis Pursh
Rosaceae
♦ salmonberry
♦ Weed, Naturalised
♦ 40, 87, 88, 218, 265
♦ S, cultivated, herbal.

Rubus strigosus Michx.
Rosaceae
Rubus idaeus L. var. *peramoenus* (Greene) Fernald, *Rubus idaeus* var. *strigosus* (Michx.) Maxim.
♦ American red raspberry
♦ Weed
♦ 265
♦ herbal.

Rubus taiwanianus Matsum.
Rosaceae
♦ Weed

♦ 87, 88

Rubus thyrsoides C.Wimm.
Rosaceae
♦ Great Britain blackberry
♦ Naturalised
♦ 101

Rubus tomentosus Borkh.
Rosaceae
♦ woolly blackberry
♦ Naturalised
♦ 101

Rubus tomentosus Borkh. var. *canescens* Wirtg.
Rosaceae
♦ woolly blackberry
♦ Naturalised
♦ 101

Rubus triphyllus Thunb.
Rosaceae
= *Rubus parvifolius* L.
♦ threeleaf blackberry
♦ Naturalised
♦ 101

Rubus trivialis Michx.
Rosaceae
♦ southern dewberry
♦ Weed
♦ 87, 88, 218
♦ S, promoted, herbal.

Rubus tuberculatus Bab.
Rosaceae
♦ Naturalised
♦ 280

Rubus ulmifolius Schott
Rosaceae
Rubus discolor Weihe & Nees (see)
♦ blackberry, elmleaf blackberry, jalavakarhunvatukka
♦ Weed, Naturalised, Environmental
Weed, Casual Alien
♦ 7, 42, 72, 80, 86, 87, 88, 98, 101, 146, 198, 203, 241, 280, 300
♦ S, cultivated, herbal. Origin: Europe, North Africa, Azores, Madeira and Canary Islands.

Rubus ulmifolius Schott var. *inermis* (Willd.) W.O.Focke
Rosaceae
♦ elmleaf bramble, elmleaf blackberry
♦ Weed, Naturalised, Cultivation Escape
♦ 34, 101
♦ S, cultivated.

Rubus ursinus Cham. & Schltdl.
Rosaceae
Rubus vitifolius Cham. & Schltdl. (see)
♦ California blackberry, loganberry, Pacific dewberry, trailing blackberry
♦ Weed, Naturalised, Environmental
Weed
♦ 86, 98, 203
♦ S, cultivated, herbal.

Rubus velox L.Bailey
Rosaceae
♦ fuzzy dewberry
♦ Weed, Naturalised
♦ 98, 203

Rubus vestitus Weihe
Rosaceae
♦ blackberry, European blackberry
♦ Weed, Naturalised, Environmental
Weed
♦ 72, 80, 86, 88, 98, 101, 146, 198, 203, 280
♦ S, herbal. Origin: Europe.

Rubus villosus Aiton
Rosaceae
♦ American blackberry, blackberry
♦ Weed
♦ 39, 87, 88
♦ S, promoted, herbal, toxic.

Rubus vitifolius Cham. & Schltdl.
Rosaceae
= *Rubus ursinus* Cham. & Schltdl.
♦ grapeleaf blackberry, Pacific blackberry, Pacific dewberry
♦ Weed, Quarantine Weed
♦ 76, 87, 88, 203, 218, 220
♦ S, promoted, herbal.

Rubus vulgaris Weihe & Nees
Rosaceae
= *Rubus laciniatus* Willd.
♦ Weed
♦ 87, 88

Rudbeckia amplexicaulis Vahl
Asteraceae
= *Dracopis amplexicaulis* (Vahl) Cass. (NoR)
♦ clasping coneflower
♦ Weed
♦ 87, 88, 218
♦ herbal.

Rudbeckia fulgida Ait.
Asteraceae
♦ orange coneflower
♦ Naturalised
♦ 287
♦ cultivated, herbal.

Rudbeckia hirta L.
Asteraceae
Rudbeckia gracilis Nutt., *Rudbeckia strigosa* Nutt.
♦ hairy coneflower, black eyed Susan, kesäpäivänhattu
♦ Weed, Naturalised, Casual Alien
♦ 23, 24, 39, 42, 52, 87, 88, 161, 195, 210, 218, 280, 286, 287
♦ a/bH, cultivated, herbal, toxic.

Rudbeckia hirta L. var. *pulcherrima* Farw.
Asteraceae
Rudbeckia serotina Nutt. (see)
♦ black eyed Susan
♦ Weed, Naturalised, Introduced
♦ 34, 161, 286, 287
♦ pH.

Rudbeckia laciniata L.
Asteraceae
♦ green headed coneflower, coneflower, black eyed Susan, cutleaf coneflower
♦ Weed, Naturalised, Garden Escape, Environmental Weed, Cultivation Escape
♦ 39, 42, 87, 88, 152, 161, 210, 272, 280, 286, 287

♦ pH, cultivated, herbal, toxic.

Rudbeckia laciniata L. var. hortensis Bailey
Asteraceae
= *Rudbeckia laciniata* L. var. *laciniata* (NoR)
♦ Weed, Naturalised
♦ 286, 287

Rudbeckia occidentalis Nutt.
Asteraceae
♦ niggerhead, western coneflower
♦ Weed
♦ 39, 87, 88, 161, 218
♦ a/bH, cultivated, herbal, toxic.

Rudbeckia serotina Nutt.
Asteraceae
= *Rudbeckia hirta* L. var. *pulcherrima* Farw.
♦ black eyed Susan
♦ Weed
♦ 87, 88, 218
♦ herbal.

Rudbeckia triloba L.
Asteraceae
♦ brown eyed Susan, branched coneflower
♦ Weed, Naturalised
♦ 87, 88, 161, 218, 287
♦ cultivated, herbal.

Ruellia brevifolia (Pohl) C.Ezcurra
Acanthaceae
Stephanophysum longifolium Pohl (see)
♦ tropical wild petunia
♦ Naturalised
♦ 101

Ruellia brittoniana Léonard
Acanthaceae
♦ Britton's wild petunia
♦ Weed, Naturalised, Environmental Weed, Cultivation Escape
♦ 101, 179, 261, 287
♦ cultivated. Origin: Mexico.

Ruellia chartacea (T.Anders.) Wassh.
Acanthaceae
♦ Naturalised, Cultivation Escape
♦ 101, 261
♦ S/T, cultivated. Origin: Peru, Ecuador, southern Colombia.

Ruellia ciliatiflora Hook.
Acanthaceae
♦ hairyflower wild petunia
♦ Weed, Naturalised
♦ 101, 179
♦ arid.

Ruellia coccinea (L.) Vahl
Acanthaceae
♦ yerba maravilla
♦ Weed
♦ 87, 88

Ruellia devosiana hort. Makoy ex E.Murr.
Acanthaceae
♦ Brazilian wild petunia, ruellia
♦ Naturalised, Cultivation Escape
♦ 101, 233
♦ cultivated.

Ruellia malacosperma Greenm.
Acanthaceae
♦ softseed wild petunia
♦ Weed, Naturalised
♦ 86, 179
♦ Origin: Australia.

Ruellia patula Jacq.
Acanthaceae
♦ Weed
♦ 39, 87, 88, 221, 240
♦ aH, toxic.

Ruellia prostrata Poir.
Acanthaceae
Dipteracanthus prostratus (Poir.) Nees (see)
♦ black weed, bell weed, vao uli, vao uliuli
♦ Weed, Naturalised
♦ 3, 6, 87, 88, 101, 191
♦ herbal.

Ruellia repens L.
Acanthaceae
♦ Introduced
♦ 230
♦ H, herbal.

Ruellia squarrosa (Fenzl) Cufod.
Acanthaceae
♦ ruellia
♦ Sleeper Weed, Naturalised, Garden Escape, Environmental Weed
♦ 86, 155
♦ cultivated. Origin: southern Mexico.

Ruellia tuberosa L.
Acanthaceae
♦ minnieroot, popping pod, toi ting
♦ Weed, Naturalised
♦ 12, 13, 14, 32, 39, 86, 87, 88, 93, 191, 204, 239, 262
♦ pH, cultivated, herbal, toxic. Origin: Central America.

Ruellia tweediana Griseb.
Acanthaceae
♦ Weed
♦ 14, 87, 88

Rugelia repens Nees
Asteraceae
♦ Weed, Quarantine Weed
♦ 76, 87, 88, 203, 220

Rumex L. spp.
Polygonaceae
♦ dock, sorrel
♦ Weed, Noxious Weed, Environmental Weed
♦ 18, 36, 80, 88, 154, 243, 247, 296
♦ H, herbal, toxic.

Rumex abyssinicus Jacq.
Polygonaceae
♦ spinach rhubarb, mubelanaga, kamjola
♦ Weed, Quarantine Weed
♦ 39, 76, 87, 88, 203, 220
♦ pH, promoted, toxic. Origin: Madagascar.

Rumex acetosa L.
Polygonaceae
Acetosa pratensis Mill., *Lapathum acetosa* (L.) Scop., *Rumex acidus* Salisb.

♦ sorrel, sour dock, garden sorrel, meadow sorrel, tall sorrel, green sorrel, large leaved sorrel, niittysuolaheinä, common sorrel
♦ Weed, Quarantine Weed, Naturalised
♦ 23, 39, 44, 70, 76, 80, 86, 87, 88, 94, 101, 161, 218, 241, 253, 263, 272, 275, 280, 286, 297, 300
♦ pH, cultivated, herbal, toxic.

Rumex acetosa L. ssp. acetosa
Polygonaceae
♦ garden sorrel
♦ Naturalised
♦ 101

Rumex acetosa L. ssp. thyrsiflorus (Fingerh.) Hayek
Polygonaceae
Rumex thyrsiflorus Fingerh. (see)
♦ garden sorrel
♦ Naturalised
♦ 101

Rumex acetosella L.
Polygonaceae
Rumex fascilobus Klokov
♦ red sorrel, sheep sorrel, sorrel, small sorrel, field sorrel, horse sorrel, sour weed, sower grass, redtop sorrel, cow sorrel, redweed, mountain sorrel, Indian cane, common sheep sorrel, ahosuolaheinä
♦ Weed, Noxious Weed, Naturalised, Introduced, Environmental Weed
♦ 7, 8, 9, 15, 23, 34, 38, 39, 44, 49, 52, 70, 72, 80, 86, 87, 88, 94, 97, 98, 101, 133, 136, 146, 157, 161, 165, 174, 195, 203, 204, 210, 211, 212, 217, 218, 224, 229, 236, 237, 241, 243, 247, 249, 253, 255, 263, 266, 269, 272, 280, 286, 287, 295, 300
♦ pH, cultivated, herbal, toxic. Origin: Eurasia.

Rumex acetosella L. ssp. angiocarpus (Murb.) Murb.
Polygonaceae
Rumex angiocarpus Murb. (see)
♦ sheep sorrel
♦ Weed
♦ 88, 158
♦ herbal. Origin: Europe.

Rumex × acutus L. (pro sp.)
Polygonaceae
= *Rumex crispus* L. × *Rumex obtusifolius* L. [see *Rumex × pratensis* Mert. & Koch]
♦ Naturalised, Weed, Quarantine Weed
♦ 76, 87, 88, 101, 203, 220
♦ herbal.

Rumex alpinus L.
Polygonaceae
Acetosa alpina (L.) Moench, *Lapathum alpinum* (L.) Lam.
♦ monk's rhubarb, munk's rhubarb, alppihierakka, alpine dock, mountain dock
♦ Weed, Naturalised
♦ 42, 44, 70, 87, 88, 101, 272
♦ pH, cultivated, herbal. Origin: Middle East, central and eastern Europe.

Rumex altissimus A.W.Wood
Polygonaceae
♦ pale dock, smooth dock
♦ Weed, Noxious Weed, Native Weed
♦ 23, 87, 88, 161, 174, 210, 218, 229
♦ herbal. Origin: North America.

Rumex angiocarpus Murb.
Polygonaceae
= *Rumex acetosella* L. ssp. *angiocarpus*
(Murb.) Murb.
♦ dock, field sorrel, sheep sorrel
♦ Weed
♦ 39, 50, 51, 70, 87, 88, 97, 121, 243, 250
♦ pH, promoted, herbal, toxic. Origin:
Mediterranean.

Rumex aquaticus L.
Polygonaceae
Lapathum aquaticum (L.) Scop, *Rumex*
hippolapathum Fr.
♦ Scottish dock, red dock,
vesihierakka, Scottish water dock,
western dock
♦ Weed, Naturalised
♦ 272, 287
♦ pH, aqua, cultivated, herbal.

Rumex bequaerti De Wild.
Polygonaceae
♦ Weed, Quarantine Weed
♦ 76, 87, 88, 203, 220, 240

Rumex brownei Campd.
Polygonaceae
= *Rumex brownii* Campd.
♦ Browne's dock
♦ Naturalised
♦ 101

Rumex brownii Campd.
Polygonaceae
Rumex brownei Campd. (see)
♦ swamp dock, brown dock, hooked
dock, slender dock
♦ Weed, Noxious Weed, Naturalised,
Garden Escape, Environmental Weed
♦ 7, 9, 39, 55, 86, 87, 88, 93, 134, 147,
203, 205, 269, 280, 287
♦ p, cultivated, toxic. Origin:
Australia.

Rumex bucephalophorus L.
Polygonaceae
♦ red dock, horned dock,
häränpäähierakka, dock
♦ Weed, Naturalised, Casual Alien
♦ 7, 42, 70, 86, 87, 88, 94, 98, 101, 203,
243, 253, 272
♦ a/pH, promoted, herbal. Origin:
Mediterranean.

Rumex bucephalophorus L. ssp. gallicus
(Steinh.) Rech.f.
Polygonaceae
♦ Weed
♦ 250

Rumex chalepensis Mill.
Polygonaceae
♦ Weed
♦ 248

Rumex confertus Willd.
Polygonaceae
♦ idänhierakka, Asiatic dock

♦ Weed, Naturalised
♦ 42, 272
♦ cultivated. Origin: Eastern Europe.

Rumex × confusus Simonk.
Polygonaceae
= *Rumex crispus* L. × *Rumex patientia* L.
♦ Naturalised
♦ 101

Rumex conglomeratus Murray
Polygonaceae
Rumex glomeratus Schreb., *Rumex*
nemolapathum Ehrh. *p.p.*, *Rumex*
paludosus Wither.
♦ dock, sharp dock, cluster dock,
saksanhierakka, green dock
♦ Weed, Noxious Weed, Naturalised,
Native Weed, Garden Escape,
Environmental Weed, Casual Alien
♦ 7, 15, 39, 42, 44, 70, 72, 86, 87, 88, 98,
101, 121, 134, 147, 161, 165, 176, 180,
198, 203, 218, 237, 243, 250, 253, 263,
269, 272, 280, 286, 287, 295, 300
♦ pH, cultivated, herbal, toxic. Origin:
Eurasia.

Rumex conglomeratus Murr. fo.
atropurporeus Asch.
Polygonaceae
♦ Naturalised
♦ 287

Rumex crispus L.
Polygonaceae
Lapathum crispum (L.) Scop., *Rumex*
elongatus Guss.
♦ curly dock, curled dock, curlyleaf
dock, narrowleaf dock, sour dock,
yellow dock, lengua de vaca
♦ Weed, Noxious Weed, Naturalised,
Introduced, Garden Escape,
Environmental Weed
♦ 7, 9, 23, 34, 38, 44, 49, 52, 55, 70, 72,
80, 84, 86, 87, 88, 93, 94, 97, 98, 101, 115,
116, 118, 121, 136, 146, 147, 157, 158,
161, 165, 167, 174, 176, 179, 180, 186,
195, 198, 200, 203, 205, 207, 210, 211,
212, 218, 228, 229, 236, 237, 241, 243,
247, 249, 250, 253, 255, 257, 261, 266,
269, 271, 272, 275, 276, 280, 286, 287,
295, 297, 299, 300
♦ aH, aqua, cultivated, herbal, toxic.
Origin: Eurasia.

Rumex crispus L. ssp. crispus
Polygonaceae
♦ curly dock
♦ Noxious Weed, Naturalised
♦ 101, 229

Rumex crispus L. ssp. fauriei (Rech.f.)
S.L.Mosyakin & W.L.Wagner
Polygonaceae
♦ curly dock
♦ Noxious Weed, Naturalised
♦ 101, 229

Rumex crispus L. var. japonicus Meissn.
Polygonaceae
Rumex japonicus Houtt. (see)
♦ Japanese curly dock
♦ Weed
♦ 274

Rumex cristatus DC.
Polygonaceae
Rumex graecus Boiss.
♦ Greek dock
♦ Naturalised
♦ 101, 300

Rumex cuneifolius Campd.
Polygonaceae
= *Rumex frutescens* Thouars
♦ lengua de vaca
♦ Weed
♦ 87, 88, 237, 295

Rumex dentatus L.
Polygonaceae
♦ toothed dock, Indian sorrel,
dentated dock, Aegean dock,
hammashierakka
♦ Weed, Naturalised, Casual Alien
♦ 34, 40, 87, 88, 101, 185, 221, 243, 275,
297
♦ a/pH, cultivated, herbal.

Rumex × dissimilis Rech.f. (pro sp.)
Polygonaceae
= *Rumex crispus* L. × *Rumex orbiculatus*
Gray
♦ Naturalised
♦ 101

Rumex domesticus Hartm.
Polygonaceae
= *Rumex longifolius* DC.
♦ Weed, Quarantine Weed
♦ 23, 76, 80, 87, 88, 203, 220
♦ herbal.

Rumex dumosus Cunn. ex Meiss.
Polygonaceae
♦ wiry dock
♦ Native Weed
♦ 269
♦ pH. Origin: Australia.

Rumex dumosus Cunn. ex Meiss. ssp.
dumosiformis Meissner
Polygonaceae
♦ lesser wiry dock
♦ Native Weed
♦ 55

Rumex frutescens Thouars
Polygonaceae
Rumex cuneifolius Campd. (see)
♦ wedgeleaf dock, Argentine dock
♦ Weed, Naturalised
♦ 7, 40, 86, 98, 101, 203, 280

Rumex fueginus Phil.
Polygonaceae
= *Rumex maritimus* L.
♦ tulihierakka
♦ Casual Alien
♦ 42
♦ herbal.

Rumex hastatulus Baldw.
Polygonaceae
♦ heartwing sorrel
♦ Weed
♦ 84, 87, 88, 161, 218
♦ herbal.

Rumex hastatus D.Don
Polygonaceae
Rumex dissectus H.Lév.

- ♦ Weed
- ♦ 87, 88, 297
- ♦ pH, promoted, herbal. Origin: Nepal.

Rumex hydrolapathum Huds.
Polygonaceae
Lapathum giganteum Opiz, *Rumex aquaticus* Poll., *Rumex britannica* Huds.
- ♦ water dock, great water dock
- ♦ Weed, Naturalised
- ♦ 23, 87, 88, 272, 287
- ♦ pH, aqua, cultivated, herbal.

Rumex hymenosepalus Torr.
Polygonaceae
Rumex arizonicus Britt.
- ♦ canaigre, canaigre dock, wild rhubarb
- ♦ Weed, Naturalised, Introduced
- ♦ 39, 87, 88, 98, 161, 203, 218, 228
- ♦ pH, arid, cultivated, herbal, toxic.

Rumex japonicus Houtt.
Polygonaceae
= *Rumex crispus* L. var. *japonicus* Meissn.
- ♦ Weed, Quarantine Weed
- ♦ 76, 87, 88, 203, 204, 220, 263, 286, 297
- ♦ pH, promoted, herbal.

Rumex kerneri Borbás
Polygonaceae
- ♦ unkarinhierakka, Kerner's dock
- ♦ Naturalised, Casual Alien
- ♦ 42, 101
- ♦ pH, cultivated.

Rumex lanceolatus Thunb.
Polygonaceae
- ♦ common dock, smaller dock, smooth dock
- ♦ Weed, Native Weed
- ♦ 88, 121, 158
- ♦ pH, herbal. Origin: southern Africa.

Rumex longifolius DC.
Polygonaceae
Rumex domesticus Hart. (see)
- ♦ long leaved dock, northern dock, dooryard dock, hevonhierakka
- ♦ Weed, Quarantine Weed, Naturalised
- ♦ 70, 76, 87, 88, 94, 101, 161, 203, 220, 272, 300
- ♦ pH, cultivated, herbal.

Rumex maritimus L.
Polygonaceae
Lapathum maritimum (L.) Moench, *Rumex fueginus* Phil. (see)
- ♦ golden dock, keltahierakka
- ♦ Weed
- ♦ 23, 88, 161, 235, 243, 272, 297
- ♦ a/pH, aqua, cultivated, herbal.

Rumex marschallianus Rchb.
Polygonaceae
- ♦ jokihierakka
- ♦ Weed, Casual Alien
- ♦ 42, 297

Rumex mexicanus Meisn.
Polygonaceae
= *Rumex salicifolius* Weinm. var. *mexicanus* (Meisn.) C.L.Hitchc.

(NoR) [see *Rumex salicifolius* Weinm. ssp. *triangulivalvis* Danser, *Rumex triangulivalvis* (Danser) Rech.f.]
- ♦ Mexican dock, willow leaved dock, white dock, pale dock
- ♦ Weed
- ♦ 23, 49, 87, 88, 199, 218, 243
- ♦ pH, promoted, herbal.

Rumex nepalensis Spreng.
Polygonaceae
- ♦ Weed, Introduced
- ♦ 88, 228, 275, 297
- ♦ pH, arid, promoted, herbal.

Rumex nipponicus Franch. & Sav.
Polygonaceae
- ♦ Weed
- ♦ 235, 286

Rumex obovatus Danser
Polygonaceae
- ♦ lapahierakka, tropical dock
- ♦ Weed, Naturalised, Casual Alien
- ♦ 42, 101, 179

Rumex obtusifolius L.
Polygonaceae
Lapathum obtusifolium Moench., *Lapathum silvestre* Lam.
- ♦ bitter dock, broad leaved dock, round leaved dock, tylppälehtihierakka
- ♦ Weed, Noxious Weed, Naturalised, Introduced, Garden Escape
- ♦ 7, 15, 23, 34, 38, 39, 42, 44, 49, 70, 80, 87, 88, 94, 98, 101, 118, 147, 161, 165, 167, 176, 179, 186, 203, 204, 210, 218, 237, 241, 243, 253, 255, 263, 269, 272, 280, 286, 287, 295, 300
- ♦ pH, cultivated, herbal, toxic. Origin: Europe.

Rumex obtusifolius L. ssp. obtusifolius
Polygonaceae
- ♦ broadleaf dock
- ♦ Noxious Weed, Naturalised
- ♦ 86, 176, 198
- ♦ cultivated. Origin: Eurasia.

Rumex obtusifolius L. ssp. transiens (Simonk.) Rech.f.
Polygonaceae
- ♦ Naturalised
- ♦ 176

Rumex occidentalis S.Wats.
Polygonaceae
= *Rumex aquaticus* L. var. *fenestratus* (Greene) Hultén (NoR)
- ♦ western dock
- ♦ Weed
- ♦ 87, 88, 218
- ♦ pH, promoted, herbal.

Rumex orbiculatus Gray
Polygonaceae
- ♦ greater water dock
- ♦ Weed
- ♦ 23, 88
- ♦ pH, herbal.

Rumex palustris Sm.
Polygonaceae
- ♦ mutahierakka, marsh dock, yellow marsh dock

- ♦ Weed, Casual Alien
- ♦ 42, 272
- ♦ herbal.

Rumex paraguayensis Parodi
Polygonaceae
- ♦ Paraguayan dock
- ♦ Weed, Naturalised
- ♦ 87, 88, 101, 237, 295

Rumex patientia L.
Polygonaceae
Lapathum hortense Moench, *Rumex olympicus* Boiss.
- ♦ patience dock, spinach dock
- ♦ Weed, Naturalised
- ♦ 70, 80, 87, 88, 101, 161, 218, 272, 275, 297
- ♦ pH, cultivated, herbal.

Rumex pictus Forssk.
Polygonaceae
- ♦ Weed
- ♦ 221

Rumex × pratensis Mert. & Koch
Polygonaceae
= *Rumex crispus* L. × *Rumex obtusifolius* L. [see *Rumex × acutus* L. (*pro* sp.)]
- ♦ Naturalised
- ♦ 86, 287
- ♦ Origin: South Africa.

Rumex pseudonatronatus Borb.
Polygonaceae
- ♦ field dock, suomenhierakka
- ♦ Weed, Naturalised
- ♦ 87, 88, 101, 161
- ♦ cultivated.

Rumex pulcher L.
Polygonaceae
Rumex pulcher ssp. *divaricatus* (L.) Murb. (see)
- ♦ fiddleleaf dock, soreahierakka, red dock, fiddle dock
- ♦ Weed, Noxious Weed, Naturalised, Casual Alien
- ♦ 7, 9, 15, 34, 42, 70, 86, 87, 88, 94, 98, 101, 111, 147, 161, 176, 179, 203, 218, 221, 237, 241, 243, 253, 269, 272, 280, 287, 295, 300
- ♦ pH, cultivated, herbal.

Rumex pulcher L. ssp. divaricatus (L.) Murb.
Polygonaceae
= *Rumex pulcher* L.
- ♦ Weed, Naturalised, Introduced
- ♦ 38, 86, 250
- ♦ cultivated.

Rumex pulcher L. ssp. pulcher
Polygonaceae
- ♦ fiddle dock
- ♦ Weed, Naturalised
- ♦ 86, 176, 198, 199
- ♦ cultivated. Origin: Mediterranean.

Rumex pulcher L. ssp. woodsii (De Not.) Arcang.
Polygonaceae
- ♦ Naturalised
- ♦ 86
- ♦ cultivated.

Rumex rhodesius Rech.f.
Polygonaceae
♦ Native Weed
♦ 121
♦ pH. Origin: southern Africa.

Rumex rugosus Campd.
Polygonaceae
Acetosa rugosa (Campd.) Holub, *Rumex ambiguus* Gren. & Godr.
♦ wrinkled dock, wrinkled sorrel
♦ Weed
♦ 87, 88, 161
♦ cultivated, herbal.

Rumex sachalinensis Nakai
Polygonaceae
♦ Weed
♦ 23, 88

Rumex sagittatus Thunb.
Polygonaceae
= *Acetosa saggittata* (Thunb.)
L.A.S.Johnson & B.G.Briggs
♦ climbing dock, red sorrel
♦ Weed, Naturalised, Native Weed, Garden Escape, Environmental Weed
♦ 7, 15, 72, 86, 88, 98, 121, 165, 203, 225, 246, 280
♦ pC, cultivated, herbal, toxic. Origin: southern Africa.

Rumex salicifolius J.A.Weinm.
Polygonaceae
♦ willow dock, willow leaved Mexicanus dock
♦ Weed
♦ 23, 87, 88, 161, 167, 218
♦ pH, cultivated, herbal.

Rumex salicifolius Weinm. ssp. triangulivalvis Danser
Polygonaceae
= *Rumex salicifolius* Weinm. var. *mexicanus* (Meisn.) C.L.Hitchc. (NoR)
[see *Rumex mexicanus* Meisn., *Rumex triangulivalvis* (Danser) Rech.f.]
♦ willow leaved dock
♦ Naturalised
♦ 40

Rumex sanguineus L.
Polygonaceae
Rumex nemorosus Schrad.
♦ redvein dock, blood veined dock, wood dock, verihierakka
♦ Weed, Naturalised, Casual Alien
♦ 23, 42, 80, 87, 88, 101, 241, 272, 287, 300
♦ cultivated, herbal.

Rumex sibiricus Hultén
Polygonaceae
♦ Siberian dock
♦ Naturalised
♦ 101
♦ herbal.

Rumex stenophyllus Ledeb.
Polygonaceae
Rumex biformis (Menyh.) Borbás, *Rumex odontocarpus* Sándor ex Borbás
♦ narrowleaf dock, kapealehtihierakka
♦ Weed, Naturalised, Introduced, Casual Alien

♦ 34, 42, 87, 88, 94, 101, 161
♦ pH, cultivated, herbal.

Rumex tenuifolius (Wallr.) Á.Löve
Polygonaceae
= *Rumex acetosella* L. ssp. *acetosella* var. *tenuifolius* Wallr. (NoR)
♦ fine leaved sorrel
♦ Weed, Naturalised
♦ 272, 280
♦ pH, promoted, herbal.

Rumex thyrsiflorus Fingerh.
Polygonaceae
= *Rumex acetosa* L. ssp. *thyrsiflorus* (Fingerh.) Hayek
♦ compact dock, thyrse sorrel, tulisuolaheinä
♦ Weed
♦ 70, 87, 88, 161, 218, 272
♦ cultivated.

Rumex triangulivalvis (Danser) Rech.f.
Polygonaceae
= *Rumex salicifolius* Weinm. var. *mexicanus* (Meisn.) C.L.Hitchc. (NoR)
[see *Rumex mexicanus* Meisn., *Rumex salicifolius* Weinm. ssp. *triangulivalvis* Danser]
♦ meksikonhierakka
♦ Naturalised
♦ 42
♦ cultivated, herbal.

Rumex tuberosus L.
Polygonaceae
♦ Weed
♦ 272

Rumex venosus Pursh
Polygonaceae
♦ veiny dock, sour greens, winged dock, wild begonia
♦ Weed
♦ 23, 88, 161, 218
♦ pH, cultivated, herbal.

Rumex verticillatus L.
Polygonaceae
♦ swamp dock
♦ Weed
♦ 23, 88
♦ herbal.

Rumex vesicarius L.
Polygonaceae
= *Acetosa vesicaria* (L.) Á.Löve
♦ bladder dock
♦ Weed, Naturalised, Environmental Weed, Casual Alien
♦ 7, 72, 86, 88, 98, 203, 221, 280
♦ aH, cultivated, herbal.

Rumfordia media Blake
Asteraceae
♦ Weed
♦ 87, 88

Rumohra adiantiformis (G.Forst.) Ching
Dryopteridaceae
♦ iron fern, leather fern, leatherleaf, nahkasanikka
♦ Weed
♦ 179
♦ cultivated, herbal. Origin: Australia.

Rungia blumeana Val.
Acanthaceae
♦ Weed
♦ 13

Rungia elegans Dalz.
Acanthaceae
♦ Weed
♦ 66

Rungia pectinata (Linn.) Nees
Acanthaceae
♦ Weed
♦ 12
♦ herbal.

Rungia repens Nees
Acanthaceae
♦ Weed, Quarantine Weed
♦ 76, 87, 88, 203, 220
♦ herbal.

Ruppia cirrhosa (Petag.) Grande
Potamogetonaceae/Ruppiaceae
Ruppia spiralis L. ex Dum.
♦ ditchgrass, spiral ditchgrass, spiral tasselweed
♦ Native Weed
♦ 121
♦ wpH, herbal. Origin: southern Africa.

Ruppia maritima L.
Potamogetonaceae/Ruppiaceae
♦ widgeongrass, beaked tasselweed, ditchgrass, merihapsikka, beak tassel pondweed
♦ Weed, Naturalised
♦ 23, 86, 87, 88, 98, 203, 218, 221
♦ wpH, cultivated, herbal. Origin: cosmopolitan.

Ruschia canonotata (L.Bol.) Schwant.
Aizoaceae/Mesembryanthemaceae
♦ cattle vygie
♦ Native Weed
♦ 121
♦ S. Origin: southern Africa.

Ruschia caroli (L.Bolus) Schwantes
Aizoaceae/Mesembryanthemaceae
♦ purple dewplant
♦ Naturalised
♦ 40
♦ cultivated.

Ruschia decumbens L.Bolus
Aizoaceae/Mesembryanthemaceae
♦ Weed, Naturalised
♦ 54, 86, 88

Ruschia tumidula (Haw.) Schwantes
Aizoaceae/Mesembryanthemaceae
♦ looseflower pigface
♦ Weed, Naturalised
♦ 7, 86, 98, 198, 203, 280
♦ cultivated. Origin: southern Africa.

Ruscus aculeatus L.
Liliaceae/Ruscaceae
Ruscus flexuosus Mill.
♦ butcher's broom, box holly, Jew's myrtle
♦ Weed
♦ 15, 15, 39, 87, 88, 272
♦ S, arid, cultivated, herbal, toxic.

Ruscus hypoglossum L.
  Liliaceae/Ruscaceae
- ♦ sirokolisna veprina
- ♦ Weed
- ♦ 272
- ♦ cultivated, herbal.

Russelia equisetiformis Schltr. & Cham.
  Scrophulariaceae
- ♦ firecracker plant, fountain bush, coral de Italia, fountain plant, Madeira plant
- ♦ Weed, Naturalised, Casual Alien
- ♦ 80, 86, 101, 179, 261
- ♦ S, cultivated. Origin: Mexico.

Ruta chalepensis L.
  Rutaceae
  Ruta bracteosa DC.
- ♦ fringed rue, Egyptian rue, ruda
- ♦ Weed, Naturalised, Introduced, Casual Alien
- ♦ 34, 38, 101, 161, 228, 241, 261, 272, 295, 300
- ♦ pH, arid, cultivated, herbal, toxic. Origin: Mediterranean.

Ruta graveolens L.
  Rutaceae
  Ruta altera Mill., *Ruta hortensis* Mill., *Ruta officinalis* Pall.
- ♦ common rue, countryman's treacle, garden rue, herb of grace, herb of repentance, rue, herby grass
- ♦ Weed, Naturalised, Garden Escape, Environmental Weed, Cultivation Escape, Casual Alien
- ♦ 39, 42, 51, 54, 72, 86, 88, 101, 121, 158, 161, 241, 247, 272, 287, 300
- ♦ pS, cultivated, herbal, toxic. Origin: Eurasia.

Rytidosperma auriculatum (J.M.Black) Connor & Edgar
  Poaceae
- ♦ Naturalised
- ♦ 280
- ♦ G.

Rytidosperma caespitosum (Gaudich.) Connor & Edgar
  Poaceae
- ♦ Naturalised
- ♦ 280
- ♦ G.

Rytidosperma erianthum (Lindl.) Connor & Edgar
  Poaceae
- ♦ Naturalised
- ♦ 280
- ♦ G.

Rytidosperma geniculatum (J.M.Black) Connor & Edgar
  Poaceae
- ♦ Naturalised
- ♦ 280
- ♦ G.

Rytidosperma laeve (Vickery) Connor & Edgar
  Poaceae
- ♦ Naturalised
- ♦ 280
- ♦ G.

Rytidosperma penicillatum (Labill.) Connor & Edgar
  Poaceae
- ♦ Naturalised
- ♦ 280
- ♦ G.

Rytidosperma pilosum (R.Br.) Connor & Edgar
  Poaceae
- ♦ hairy wallaby grass
- ♦ Naturalised
- ♦ 101, 280
- ♦ G.

Rytidosperma racemosum (R.Br.) Connor & Edgar
  Poaceae
- = *Danthonia racemosa* R.Br. (NoR)
- ♦ danthonia
- ♦ Weed, Naturalised
- ♦ 15, 280
- ♦ G.

Rytidosperma semiannulare (Labill.) Connor & Edgar
  Poaceae
- ♦ Tasmanian wallaby grass
- ♦ Naturalised
- ♦ 101
- ♦ G.

Rytidosperma tenuius (Steud.) A.Hansen & Sunding
  Poaceae
- ♦ Naturalised
- ♦ 280
- ♦ G.

S

Sabal mexicana Mart.
  Arecaceae
  Inodes exul Cook, *Inodes mexicana* (Mart.) Standl., *Inodes texana* Cook, *Sabal exul* (Cook) L.H.Bailey, *Sabal texana* (Cook) Becc.
- ♦ Mexican palmetto, Oaxaca palmetto, Rio Grande palmetto, Texas palm, Texas palmetto, bayal, bouxaan, cabbage tree, guano, micheros, otoomal, palma de micharo, palma di Micheros, palma huichira, palma llanera, palma real, palma redonda, palma rustica, soyate, xaan, botan
- ♦ Weed, Quarantine Weed, Introduced
- ♦ 87, 88, 203, 220, 228
- ♦ T, arid, cultivated, herbal.

Sabal minor (Jacq.) Pers.
  Arecaceae
- ♦ dwarf palmetto, bush palmetto
- ♦ Weed
- ♦ 87, 88, 218
- ♦ S, cultivated, herbal.

Sabal palmetto (Walter) Schult. & Schult.f.
  Arecaceae
- ♦ cabbage palmetto, cabbage palm
- ♦ Weed, Naturalised, Introduced
- ♦ 86, 87, 88, 218, 230
- ♦ T, cultivated, herbal. Origin: south-eastern North America.

Sabal umbraculifera (Jacq.) Mart.
  Arecaceae
- ♦ Introduced
- ♦ 230
- ♦ T, cultivated.

Saccharum L. spp.
  Poaceae
- ♦ sugarcane
- ♦ Quarantine Weed
- ♦ 220
- ♦ G.

Saccharum arundinaceum Retz.
  Poaceae
- ♦ Weed, Naturalised
- ♦ 87, 88, 287
- ♦ G, arid.

Saccharum bengalense Retz.
  Poaceae
  Erianthus munja (Roxb.) Jesw., *Saccharum munja* Roxb.
- ♦ munj sweetcane
- ♦ Weed, Quarantine Weed, Naturalised, Introduced
- ♦ 76, 87, 88, 203, 220, 228, 261
- ♦ G, arid.

Saccharum floridulum Labill.
Poaceae
= *Miscanthus floridulus* (Labill.) Schum. & Laut.
♦ Weed
♦ 3, 191
♦ G.

Saccharum narenga (Nees ex Steud.) Buch.-Ham. ex Hack.
Poaceae
♦ Weed
♦ 87, 88
♦ G.

Saccharum officinarum L.
Poaceae
♦ sugarcane, sehu, cana de açucar, soqueira de cana, tolo
♦ Weed, Quarantine Weed, Naturalised, Introduced, Casual Alien
♦ 86, 87, 88, 98, 101, 121, 179, 203, 220, 230, 241, 245, 261, 280
♦ pG, cultivated, herbal. Origin: Asia.

Saccharum procerum Roxb.
Poaceae
♦ yaa khamong
♦ Weed
♦ 209, 238
♦ pG.

Saccharum ravennae (L.) L.
Poaceae
Andropogon ravennae L., *Erianthus ravennae* (L.) Beauv.
♦ ravennagrass
♦ Weed, Naturalised
♦ 101, 179, 272
♦ pG, cultivated.

Saccharum spontaneum L.
Poaceae
♦ wild sugarcane, serio grass, wild cane, ahlek, ahlec, banga ruchel
♦ Weed, Quarantine Weed, Noxious Weed, Naturalised, Introduced, Environmental Weed, Casual Alien
♦ 3, 33, 67, 87, 88, 101, 107, 140, 152, 161, 191, 209, 217, 220, 221, 228, 229, 243, 258, 261, 297
♦ pG, arid, cultivated. Origin: tropical Asia.

Sacciolepis africana C.E.Hubb. & Snowden
Poaceae
♦ Weed
♦ 88
♦ G, herbal. Origin: Madagascar.

Sacciolepis angusta (Trin.) Stapf
Poaceae
= *Sacciolepis indica* (L.) A.Chase
♦ Weed
♦ 87, 88
♦ G.

Sacciolepis indica (L.) A.Chase
Poaceae
Sacciolepis angusta (Trin.) Stapf (see), *Hymenachne indica* Büse, *Panicum myuros* Kunth *non* Lam.
♦ glenwoodgrass
♦ Weed, Naturalised
♦ 32, 80, 87, 88, 101, 170, 179, 280, 286,

297
♦ G, herbal.

Sacciolepis indica (L.) Chase var. oryzetorum (Makino) Ohwi
Poaceae
♦ Weed
♦ 235, 286
♦ G.

Sacciolepis insulicola (Steud.) Ohwi
Poaceae
♦ Weed
♦ 87, 88
♦ G.

Sacciolepis interrupta (Willd.) Stapf
Poaceae
Hymenachne interrupta (Willd.) Büse, *Panicum indicum* Hack., *Panicum interruptum* Willd., *Panicum inundatum* Kunth, *Panicum turritum* Thunb., *Panicum uliginosum* Roth ex Roem. & Schult.
♦ Weed, Quarantine Weed
♦ 76, 87, 88, 135, 170, 191, 203, 220
♦ G, herbal.

Sacciolepis myosuroides (R.Br.) A.Camus
Poaceae
Panicum myosuroides R.Br.
♦ Weed
♦ 87, 88, 170
♦ G.

Sacciolepis myurus (Lam.) A.Chase
Poaceae
Hymenachne myurus (Lam.) Beauv.
♦ Weed
♦ 87, 88
♦ G.

Sacciolepis polymorpha A.Chase ex E.G. & A.Camus
Poaceae
♦ Weed
♦ 87, 88
♦ G.

Sageretia brandrethiana Aitch.
Rhamnaceae
♦ Weed
♦ 221

Sageretia thea (Osbeck) M.C.Johnst.
Rhamnaceae
♦ pauper's tea, mock buckthorn
♦ Naturalised
♦ 101
♦ herbal.

Sagina apetala Ard.
Caryophyllaceae
Sagina apetala Ard. var. *barbata* Fenzl (see)
♦ dwarf pearlwort, common pearlwort, pearlwort, ciliate pearlwort, annual pearlwort
♦ Weed, Naturalised, Environmental Weed
♦ 7, 9, 15, 23, 34, 44, 86, 87, 88, 93, 94, 98, 101, 134, 136, 161, 176, 198, 203, 205, 218, 243, 253, 269, 272, 280, 287, 300
♦ aH, arid, herbal. Origin: Europe.

Sagina apetala Ard. var. apetala
Caryophyllaceae

♦ sagina
♦ Weed
♦ 295

Sagina apetala Ard. var. barbata Fenzl
Caryophyllaceae
= *Sagina apetala* Ard.
♦ dwarf pearlwort, pearlwort, sticky pearlwort
♦ Weed
♦ 180

Sagina ciliata Ard.
Caryophyllaceae
♦ ciliate pearlwort
♦ Weed
♦ 44

Sagina decumbens (Ell.) Torr. & Gray
Caryophyllaceae
♦ pearlwort, beach pearlwort, trailing pearlwort
♦ Weed
♦ 161
♦ herbal.

Sagina japonica (Sw.) Ohwi
Caryophyllaceae
♦ Japanese pearlwort
♦ Weed, Quarantine Weed, Naturalised
♦ 76, 87, 88, 101, 203, 220, 263, 286, 297
♦ aH, cultivated.

Sagina maritima Don
Caryophyllaceae
♦ sea pearlwort, merihaarikko
♦ Weed, Naturalised, Environmental Weed
♦ 7, 9, 86, 98, 176, 198, 203, 272
♦ Origin: Mediterranean.

Sagina maxima Gray
Caryophyllaceae
♦ stickystem pearlwort
♦ Weed
♦ 88, 204, 235, 286
♦ a/bH, promoted, herbal.

Sagina nodosa (L.) Fenzl
Caryophyllaceae
Alsine nodosa (L.) Crantz, *Moehringia nodosa* (L.) Crantz, *Spergula nodosa* L.
♦ knotted pearlwort, nyylähaarikko
♦ Weed, Naturalised
♦ 87, 88, 94, 101, 272
♦ cultivated, herbal.

Sagina nodosa (L.) Fenzl ssp. nodosa
Caryophyllaceae
♦ knotted pearlwort
♦ Naturalised
♦ 101

Sagina procumbens L.
Caryophyllaceae
Alsine procumbens (L.) Crantz, *Sagina procumbens* Linn. var. *compacta* Lange (see)
♦ bird's eye pearlwort, procumbent pearlwort, rentohaarikko, spreading pearlwort, bird's eye pearlwort, bird's eye, Arctic pearlwort
♦ Weed, Naturalised
♦ 7, 15, 44, 70, 86, 87, 88, 94, 98, 101, 136, 161, 165, 176, 198, 203, 211, 218, 241, 243, 253, 269, 272, 280, 287, 300

♦ pH, cultivated, herbal. Origin: Eurasia.

Sagina procumbens **Linn. var.** *compacta* **Lange**
Caryophyllaceae
= *Sagina procumbens* L.
♦ procumbent pearlwort
♦ Weed
♦ 238
♦ a/pH.

Sagina subulata **(Sw.) Presl**
Caryophyllaceae
Alsine subulata (Sw.) Jess., *Phaloe subulata* (Sw.) Dumort., *Spergula subulata* Sw.
♦ heath pearlwort, awlleaf pearlwort, Corsican pearlwort
♦ Weed, Naturalised
♦ 88, 94, 101, 272, 280
♦ cultivated, herbal.

Sagittaria **L. spp.**
Alismataceae
♦ sagittaria, arrowhead
♦ Weed, Quarantine Weed, Naturalised, Environmental Weed
♦ 14, 198, 246, 258
♦ wH.

Sagittaria aginashi **Makino**
Alismataceae
♦ aginashi
♦ Weed, Quarantine Weed
♦ 76, 87, 88, 191, 203, 204, 220, 238, 263, 286
♦ wpH, promoted.

Sagittaria calycina **Engelm.**
Alismataceae
♦ California arrowhead, hooded arrowhead
♦ Weed
♦ 87, 88, 218
♦ wH, herbal.

Sagittaria chilensis **Cham. & Schlecht.**
Alismataceae
♦ Weed, Quarantine Weed
♦ 76, 87, 88, 203, 220
♦ wH.

Sagittaria cuneata **Sheldon**
Alismataceae
Sagittaria arifolia Nutt. ex J.G.Sm.
♦ arumleaf arrowhead, arrowhead, wappato, tule potato
♦ Weed
♦ 23, 87, 88, 159
♦ wpH, promoted, herbal.

Sagittaria engelmanniana **J.G.Sm. ssp.** *brevirostrata* **(Mack. & Bush) Bogin**
Alismataceae
♦ arrowhead
♦ Naturalised
♦ 86, 198
♦ wH. Origin: North America.

Sagittaria falcata **Pursh**
Alismataceae
= *Sagittaria lancifolia* L.
♦ coastal arrowhead
♦ Weed
♦ 87, 88, 218
♦ wH, herbal.

Sagittaria graminea **Michx.**
Alismataceae
Sagittaria sinensis Sims.
♦ sagittaria, arrowhead, delta arrowhead, slender arrowhead, grassy arrowhead, Chinese arrowhead, arrowhead
♦ Weed, Quarantine Weed, Noxious Weed, Naturalised, Environmental Weed
♦ 72, 76, 86, 88, 98, 147, 161, 171, 191, 198, 203, 258, 287, 296
♦ wpH, cultivated, herbal. Origin: North America.

Sagittaria graminea **Michx. ssp.** *platyphylla* **Engelm.**
Alismataceae
♦ sagittaria
♦ Weed
♦ 200
♦ wpH.

Sagittaria graminea **Michx. var.** *graminea*
Alismataceae
♦ grassleaf arrowhead, grassy arrowhead
♦ Weed
♦ 197
♦ wpH, cultivated.

Sagittaria guayanensis **Kunth**
Alismataceae
Echinodorus guianensis (Kunth) Griseb.
♦ Guyanese arrowhead, arrowhead, swamp potato, arrowhead lily, aguape, sagitaria, flecha
♦ Weed, Naturalised
♦ 87, 88, 101, 126, 255
♦ wa/pH. Origin: Americas.

Sagittaria guayanensis **Kunth ssp.** *lappula* **H.B.K.**
Alismataceae
Lophotocarpus formosanus Hayata, *Lophotocarpus guayanensis* (Kunth) Micheli ex J.G.Sm., *Sagittaria lappula* D.Don
♦ Weed
♦ 170, 191
♦ wa/pH.

Sagittaria guayanensis **Kunth var.** *guayanensis*
Alismataceae
♦ Guyana arrowhead
♦ Weed
♦ 197
♦ wa/pH.

Sagittaria lancifolia **L.**
Alismataceae
Sagittaria falcata Pursh (see)
♦ arrowhead, narrow leaved sagittaria
♦ Quarantine Weed
♦ 258
♦ wH, cultivated, herbal.

Sagittaria latifolia **Willd.**
Alismataceae
♦ common arrowhead, wappato, tule potato, broadleaf arrowhead, duck potato
♦ Weed, Quarantine Weed, Naturalised

♦ 23, 76, 86, 87, 88, 161, 203, 218, 220, 258
♦ wpH, cultivated, herbal.

Sagittaria longiloba **Engelm. ex J.G.Sm.**
Alismataceae
♦ Gregg's arrowhead, long lobed arrowhead, longbarb arrowhead
♦ Weed
♦ 161, 180
♦ wpH, cultivated.

Sagittaria montevidensis **Cham. & Schltdl.**
Alismataceae
♦ arrowhead, sagittaria, California arrowhead, giant arrowhead, long lobed arrowhead, arrowhead sagittaria
♦ Weed, Sleeper Weed, Quarantine Weed, Noxious Weed, Naturalised, Environmental Weed
♦ 62, 68, 76, 86, 87, 88, 98, 101, 147, 161, 171, 191, 197, 200, 203, 220, 225, 236, 237, 243, 246, 255, 269, 280, 295
♦ wpH, cultivated, herbal. Origin: Americas.

Sagittaria montevidensis **Cham. & Schlecht. ssp.** *calycina* **(Engelm.) Bogin**
Alismataceae
= *Sagittaria calycina* Engelm. var. *calycina* (NoR)
♦ California arrowhead, hooded arrowhead, arrowhead, sagittaria
♦ Weed, Noxious Weed, Naturalised
♦ 86, 180
♦ wpH. Origin: North America.

Sagittaria montevidensis **Cham. & Schltdl. ssp.** *montevidensis*
Alismataceae
♦ arrowhead, sagittaria, California arrowhead
♦ Noxious Weed, Naturalised
♦ 86
♦ wpH. Origin: South America.

Sagittaria platyphylla **(Engelm.) J.G.Sm.**
Alismataceae
♦ delta arrowhead, sagittaria, giant sag
♦ Weed, Quarantine Weed, Noxious Weed, Environmental Weed
♦ 23, 62, 87, 88, 170, 191, 218, 246
♦ wpH, cultivated, herbal. Origin: Central and North America.

Sagittaria pygmaea **Miq.**
Alismataceae
♦ pygmy arrowhead, dwarf arrowhead
♦ Weed, Quarantine Weed, Noxious Weed, Naturalised
♦ 76, 86, 87, 88, 126, 203, 204, 220, 263, 274, 275, 286, 297
♦ wa/pH, cultivated.

Sagittaria rigida **Pursh**
Alismataceae
♦ stiff arrowhead, Canadian arrowhead, deep water duck potato, sessilefruit arrowhead
♦ Weed
♦ 197
♦ wpH, cultivated, herbal.

Sagittaria sagittifolia L.
Alismataceae
Sagittaria leucopetala Miq.
♦ arrowhead, Chinese arrowhead, old world arrowhead, pystykeiholehti, giant sagittaria, Hawai'i arrowhead
♦ Weed, Quarantine Weed, Noxious Weed, Naturalised, Environmental Weed, Casual Alien
♦ 40, 67, 70, 87, 88, 98, 126, 140, 161, 193, 203, 217, 229, 246, 262, 272, 275
♦ wpH, cultivated, herbal.

Sagittaria sagittifolia L. var. *longiloba* Turcz.
Alismataceae
♦ Weed
♦ 275, 297
♦ wpH.

Sagittaria subulata (L.) Buchenau
Alismataceae
Alisma subulatum L., *Sagittaria filiformis* J.G.Sm., *Sagittaria lorata* (Chapm.) Small, *Sagittaria natans* Michx., *Sagittaria natans* var. *lorata* Chapm., *Sagittaria natans* var. *pusilla* (Nutt.) Chapm., *Sagittaria pusilla* Nutt.
♦ dwarf arrowhead, awlleaf arrowhead
♦ Weed, Quarantine Weed, Environmental Weed
♦ 87, 88, 218, 246
♦ wH, cultivated.

Sagittaria trifolia L.
Alismataceae
♦ phak khaang kai, old world arrowhead, threeleaf arrowhead, omodaka
♦ Weed, Quarantine Weed
♦ 76, 87, 88, 203, 204, 220, 238, 263, 274, 286, 297
♦ wpH, promoted. Origin: east Asia.

Sagittaria trifolia L. fo. *longifolia* Makino
Alismataceae
♦ Weed
♦ 286
♦ wpH.

Sagittaria variabilis Engelm.
Alismataceae
♦ Weed
♦ 23, 88
♦ wH, cultivated.

Saintpaulia ionantha H.Wendl.
Gesneriaceae
♦ African violet
♦ Naturalised
♦ 287
♦ cultivated, herbal.

Salacca edulis Reinw.
Arecaceae
♦ salac palm, salak
♦ Introduced
♦ 230
♦ T.

Salicornia europaea L.
Chenopodiaceae
♦ slender grasswort, suolayrtti, glasswort, common glasswort
♦ Weed, Introduced
♦ 228, 275
♦ aH, arid/aqua, promoted, herbal.

Salicornia fruticosa (L.) L.
Chenopodiaceae
= *Sarcocornia fruticosa* (L.) A.J.Scott (NoR)
♦ Weed
♦ 221

Salicornia quinqueflora Bunge ex Ung.
Chenopodiaceae
Salicornia australis Sol., *Sarcocornia quinqueflora* (Bunge ex Ung.-Sternb.) A.J.Scott
♦ chicken claws
♦ Weed
♦ 87, 88
♦ H, promoted.

Salicornia rubra Nelson
Chenopodiaceae
♦ saltwort, red swampfire, slender grasswort
♦ Weed
♦ 159
♦ aH, promoted.

Salicornia virginica L.
Chenopodiaceae
♦ Virginia glasswort, pickleweed
♦ Naturalised
♦ 287
♦ pH, aqua, promoted, herbal. Origin: west Europe, North America.

Salix L. spp.
Salicaceae
♦ willow
♦ Weed, Quarantine Weed, Noxious Weed, Naturalised, Garden Escape, Environmental Weed
♦ 15, 18, 39, 73, 86, 88, 155, 161, 181, 198, 203, 221, 258, 272, 289
♦ T, cultivated, herbal, toxic. Origin: northern hemisphere.

Salix alba L.
Salicaceae
Nectolis vitellina Raf., *Salix alba* var. *vitellina* (L.) Stokes (see), *Salix flexibilis* Gilib.
♦ white willow, willow
♦ Weed, Noxious Weed, Naturalised, Garden Escape, Environmental Weed, Cultivation Escape
♦ 42, 54, 54, 72, 80, 86, 87, 88, 98, 101, 151, 176, 195, 198, 203, 218, 280
♦ pS, cultivated, herbal. Origin: Europe, North Africa to China.

Salix alba L. var. *alba*
Salicaceae
♦ white willow
♦ Noxious Weed, Naturalised
♦ 86, 198, 280
♦ cultivated. Origin: Mediterranean.

Salix alba L. var. *vitellina* (L.) Stokes
Salicaceae
= *Salix alba* L.
♦ golden willow, white willow
♦ Weed, Noxious weed, Naturalised
♦ 15, 86, 198, 232, 280
♦ T, cultivated. Origin: Europe.

Salix alba L. × *babylonica* L.
Salicaceae
Salix × *sepulcralis* Simonk. (see)
♦ white willow
♦ Noxious Weed, Naturalised
♦ 86

Salix alba L. × *fragilis* L.
Salicaceae
Salix × *rubens* Schrank (see)
♦ kujapaju
♦ Naturalised, Cultivation Escape
♦ 42, 98
♦ cultivated.

Salix alba L. × *matsudana* Koidz.
Salicaceae
♦ willow
♦ Weed, Noxious Weed, Naturalised, Environmental Weed
♦ 86, 88, 232

Salix amygdaloides Anderss.
Salicaceae
♦ peachleaf willow
♦ Weed
♦ 87, 88, 218
♦ T, promoted, herbal. Origin: North America.

Salix atrocinerea Brot.
Salicaceae
= *Salix cinerea* L. ssp. *oleifolia* (Sm.) Macreight
♦ willow, saule roux
♦ Weed, Naturalised
♦ 98, 203
♦ cultivated. Origin: Europe.

Salix aurita L.
Salicaceae
Salix spathulata Willd.
♦ eared willow, virpapaju
♦ Naturalised
♦ 101
♦ S, promoted. Origin: Europe.

Salix babylonica L.
Salicaceae
♦ weeping willow, willow tea, Babylon weeping willow, saule de Babylone, saule pleureur, trauerweide, shidare yanagi, sauce de Babilonia, sauce llorón
♦ Weed, Noxious Weed, Naturalised, Introduced, Garden Escape, Environmental Weed, Cultivation Escape
♦ 7, 63, 72, 80, 86, 87, 88, 95, 98, 121, 151, 176, 179, 195, 198, 200, 203, 228, 261, 277, 279, 280, 283, 300
♦ T, aqua, cultivated, herbal. Origin: China.

Salix babylonica L. fo. *rokkoku* Kimura
Salicaceae
♦ Naturalised
♦ 287

Salix babylonica L. fo. *seiko* Kimura
Salicaceae
♦ Naturalised
♦ 287

Salix babylonica L. var. *lavalle* Dode
Salicaceae

♦ Naturalised
♦ 287

Salix bebbiana Sarg.
Salicaceae
♦ Bebb willow, gray willow
♦ Weed, Quarantine Weed
♦ 76, 87, 88, 203, 218, 220
♦ T, aqua, cultivated, herbal. Origin: North America.

Salix × calodendron Wimm.
Salicaceae
= *Salix caprea* L. × *Salix atrocinerea* Brot. (see) × *Salix viminalis* L. [most probable combination]
♦ calodendron
♦ Naturalised
♦ 280
♦ Origin: Europe.

Salix caprea L.
Salicaceae
Salix lanata Lightf.
♦ goat willow, great sallow, pussy willow, raita, saule marsault
♦ Weed, Noxious Weed, Naturalised
♦ 70, 80, 86, 87, 88, 98, 101, 198, 203, 300
♦ T, cultivated, herbal. Origin: Europe.

Salix caroliniana Michx.
Salicaceae
♦ ward willow, coastal plain willow
♦ Weed
♦ 87, 88, 218
♦ herbal.

Salix caudata (Nutt.) Heller
Salicaceae
= *Salix lucida* Muhlenb. ssp. *caudata* (Nutt.) E.Murray (NoR)
♦ whiplash willow
♦ Weed
♦ 87, 88, 218
♦ herbal.

Salix chilensis Molina
Salicaceae
= *Salix humboldtiana* Willd.
♦ pencil willow
♦ Naturalised
♦ 198
♦ T, cultivated.

Salix × chrysocoma Dode
Salicaceae
= *Salix babylonica* L. × *Salix alba* L. var. *vitellina* (L.) Stokes
♦ willow
♦ Noxious Weed, Naturalised
♦ 86, 280
♦ cultivated. Origin: horticultural hybrid.

Salix cinerea L.
Salicaceae
Salix acuminata Mill.
♦ grey sallow, gray willow, large gray willow, common willow, tuhkapaju, fen sallow
♦ Weed, Quarantine Weed, Noxious weed, Naturalised, Garden Escape, Environmental Weed
♦ 15, 20, 72, 80, 86, 87, 88, 98, 101, 165,

176, 181, 198, 203, 225, 232, 246, 280
♦ S/T, cultivated, herbal. Origin: Europe, west Asia, North Africa.

Salix cinerea L. ssp. cinerea
Salicaceae
♦ grey sallow, large gray willow
♦ Naturalised, Environmental Weed
♦ 86, 101, 198, 296
♦ cultivated. Origin: Eurasia, Mediterranean.

Salix cinerea L. ssp. oleifolia (Sm.) Macreight
Salicaceae
Salix atrocinerea Brot. (see)
♦ rusty sallow, grey sallow, large gray willow
♦ Naturalised, Environmental Weed
♦ 86, 101, 198, 296
♦ cultivated. Origin: Mediterranean.

Salix daphnoides Vill.
Salicaceae
Salix bigemmis Hoff.
♦ härmäpaju
♦ Naturalised, Cultivation Escape
♦ 42, 280
♦ T, cultivated.

Salix daphnoides Vill. ssp. acutifolia (Willd.) Andersson
Salicaceae
♦ Naturalised
♦ 280

Salix × dasyclados Wimm.
Salicaceae
= *Salix caprea* L. × *Salix cinerea* L. × *Salix viminalis* L. [most probable combination]
♦ vannepaju
♦ Cultivation Escape
♦ 42
♦ cultivated.

Salix discolor Muhl.
Salicaceae
♦ pussy willow
♦ Weed, Noxious Weed, Naturalised
♦ 86, 87, 88
♦ cultivated, herbal. Origin: North America.

Salix × ehrhartiana Sm.
Salicaceae
= *Salix alba* L. × *Salix pentandra* L.
♦ Ehrhart's willow
♦ Naturalised
♦ 101

Salix elaeagnos Scop.
Salicaceae
♦ Elaeagnus willow
♦ Naturalised
♦ 101, 280
♦ cultivated.

Salix exigua Nutt.
Salicaceae
Salix interior Rowlee (see)
♦ sandbar willow, coyote willow, river willow, narrowleaf willow
♦ Weed, Quarantine Weed
♦ 76, 87, 88, 159, 203, 218, 220
♦ S/T, promoted, herbal. Origin:

North America.

Salix fragilis L.
Salicaceae
♦ brittle willow, crack willow, fragile willow
♦ Weed, Noxious weed, Naturalised, Garden Escape, Environmental Weed, Cultivation Escape
♦ 15, 42, 54, 54, 63, 70, 80, 87, 88, 95, 98, 101, 151, 152, 159, 165, 176, 181, 195, 198, 203, 225, 232, 246, 277, 280, 283, 296
♦ T, cultivated, herbal. Origin: Eurasia.

Salix fragilis L. var. fragilis
Salicaceae
♦ crack willow
♦ Weed, Naturalised, Environmental Weed
♦ 86, 88, 198
♦ cultivated. Origin: western Europe.

Salix fragilis L. var. fragilis × matsudana Koidz.
Salicaceae
♦ crack willow
♦ Naturalised, Environmental Weed
♦ 86

Salix fragilis L. var. fragilis × nigra Marshall
Salicaceae
♦ crack willow
♦ Naturalised, Environmental Weed
♦ 86
♦ cultivated.

Salix fragilis L. var. furcata Gaudin
Salicaceae
♦ forked catkin, crack willow
♦ Naturalised, Environmental Weed
♦ 86, 198
♦ cultivated. Origin: western Europe.

Salix glaucophylloides Fernald
Salicaceae
= *Salix myricoides* Muhl. var. *myricoides* (NoR)
♦ Noxious weed, Naturalised
♦ 232, 280
♦ herbal.

Salix gracilistyla Miq.
Salicaceae
♦ rosegold pussy willow, nekoyanagi
♦ Naturalised
♦ 280
♦ S, cultivated.

Salix humboldtiana Willd.
Salicaceae
Salix chilensis Molina (see)
♦ Humboldt's willow, mimbre, sauce
♦ Weed, Naturalised, Introduced
♦ 54, 86, 88, 101, 261
♦ T, cultivated, herbal. Origin: South America.

Salix interior Rowlee
Salicaceae
= *Salix exigua* Nutt.
♦ ditchbank willow, sandbar willow
♦ Weed
♦ 87, 88, 218
♦ herbal.

Salix koriyanagi **Kimura ex Goerz**
Salicaceae
♦ Naturalised
♦ 287
♦ S, promoted, herbal. Origin: China, Japan, Korea.

Salix laevigata **Bebb**
Salicaceae
♦ red willow, willow
♦ Weed
♦ 87, 88, 218
♦ T, herbal.

Salix lasiandra **Benth.**
Salicaceae
= *Salix lucida* Muhlenb. ssp. *lasiandra* (Benth.) E.Murray (NoR)
♦ Pacific willow, yellow willow
♦ Weed
♦ 87, 88, 218
♦ T, promoted, herbal.

Salix lutea **Nutt.**
Salicaceae
♦ yellow willow
♦ Weed
♦ 87, 88, 218
♦ S, herbal.

Salix matsudana **Koidz.**
Salicaceae
♦ tortured willow, corkscrew willow
♦ Weed, Noxious weed, Naturalised
♦ 54, 54, 86, 88, 101, 198, 232, 280, 287
♦ cultivated, herbal. Origin: China, Korea.

Salix matsudana **Koidz. cv. 'Tortuosa' Rehd.**
Salicaceae
♦ Naturalised, Noxious weed, Garden Escape, Environmental Weed
♦ 287
♦ cultivated, herbal. Origin: horticultural.

Salix matsudana **Koidz. × *chrysochroma* Dode**
Salicaceae
= *Salix matsudana* Koidz. × *Salix sepulcralis* Simonk. var. *chrysocoma* (Dode) Meikle [probable parents]
♦ Weed, Noxious Weed, Naturalised
♦ 86, 88
♦ cultivated.

Salix nigra **Marshall**
Salicaceae
♦ black willow, black American willow
♦ Weed, Noxious weed, Naturalised, Garden Escape, Environmental Weed
♦ 20, 54, 86, 87, 88, 198, 218, 232, 296
♦ T, cultivated, herbal. Origin: North America.

Salix × pendulina **Wender.**
Salicaceae
= *Salix fragilis* L. × *Salix babylonica* L.
♦ weeping willow, Wisconsin weeping willow, pendulous willow
♦ Weed, Noxious Weed, Naturalised
♦ 54, 86, 88, 101, 198
♦ cultivated. Origin: horticultural hybrid from Germany.

Salix pentandra **L.**
Salicaceae
Lusekia laurina Opiz
♦ bay willow, laurel willow, halava
♦ Weed, Naturalised
♦ 80, 101
♦ T, cultivated, herbal. Origin: Eurasia.

Salix petiolaris **J.E.Sm.**
Salicaceae
♦ meadow willow
♦ Weed
♦ 87, 88, 218
♦ S, promoted, herbal. Origin: eastern North America.

Salix phylicifolia **L.**
Salicaceae
♦ tea leaved willow, kiiltopaju
♦ Weed
♦ 87, 88
♦ cultivated, herbal.

Salix purpurea **L.**
Salicaceae
Knafia purpurea Opiz
♦ basket willow, purple osier, purple osier willow, purple willow
♦ Weed, Noxious weed, Naturalised, Cultivation Escape
♦ 15, 42, 54, 80, 86, 88, 101, 146, 198, 232, 252, 280
♦ T, cultivated, herbal. Origin: Europe, North Africa to Japan.

Salix × reichardtii **A.Kern.**
Salicaceae
= *Salix caprea* L. × *Salix cinerea* L.
♦ pussy willow
♦ Weed, Noxious Weed, Naturalised
♦ 15, 54, 86, 88, 198, 280
♦ Origin: Eurasia.

Salix × rubens **Schrank**
Salicaceae
= *Salix alba* L. × *Salix fragilis* L. (see)
♦ basket willow, crack willow, hybrid crack willow, rubens willow
♦ Weed, Noxious weed, Naturalised, Environmental Weed
♦ 15, 72, 86, 88, 98, 101, 176, 198, 232, 280, 296
♦ T, cultivated. Origin: Europe.

Salix × rubra **Huds.**
Salicaceae
= *Salix purpurea* L. × *Salix viminalis* L.
♦ Weed, Noxious Weed, Naturalised
♦ 54, 86, 88
♦ S, cultivated. Origin: horticultural hybrid.

Salix × sepulcralis **Simonk.**
Salicaceae
= *Salix alba* L. × *Salix babylonica* L. [see *Salix alba* L. × *babylonica* L.]
♦ weeping willow, kemp willow, sepulcral willow
♦ Weed, Naturalised
♦ 15, 54, 54, 88, 101, 198, 280
♦ cultivated.

Salix × sepulcralis **Simonk. var. *chrysocoma* (Dode) Meikle**
Salicaceae
= *Salix alba* L. ssp. *vitellina* (L.) Arcang.

× *Salix babylonica* L.
♦ golden weeping willow
♦ Noxious Weed, Naturalised, Cultivation Escape
♦ 40, 86, 198
♦ cultivated. Origin: horticultural hybrid.

Salix × sepulcralis **Simonk. var. *sepulcralis***
Salicaceae
= *Salix alba* L. ssp. *alba* × *Salix babylonica* L. (see)
♦ weeping willow
♦ Noxious Weed, Naturalised
♦ 86, 198
♦ Origin: horticultural hybrid.

Salix × smithiana **Willd.**
Salicaceae
= *Salix cinerea* L. × *Salix viminalis* L. [most probable combination]
♦ silkyleaf osier, Smith's willow
♦ Naturalised
♦ 101
♦ T, promoted. Origin: obscure.

Salix subserrata **Willd.**
Salicaceae
♦ Weed
♦ 221

Salix tetrasperma **Roxb.**
Salicaceae
♦ Weed
♦ 221
♦ herbal.

Salix triandra **L.**
Salicaceae
Salix amygdalifolia Gilib., *Salix auriculata* Mill.
♦ almond willow, almond leaved willow, jokipaju
♦ Weed, Naturalised
♦ 98, 203
♦ T, cultivated, herbal.

Salix triandra **L. × *viminalis* L.**
Salicaceae
♦ vakkapaju
♦ Cultivation Escape
♦ 42
♦ cultivated.

Salix viminalis **L.**
Salicaceae
♦ osier, koripaju, withy, basket willow, common osier
♦ Weed, Noxious weed, Naturalised, Environmental Weed, Cultivation Escape
♦ 15, 42, 80, 86, 88, 98, 101, 151, 203, 232, 252, 280, 300
♦ T, cultivated, herbal. Origin: Europe to China.

Salmonopuntia **P.V.Heath spp.**
Cactaceae
= *Opuntia* Mill. spp.
♦ Weed, Quarantine Weed
♦ 76, 88, 220

Salpichroa origanifolia **(Lam.) Baill.**
Solanaceae
Atropa origanifolia (Lam.) Desf., *Atropa*

rhomboidea Gillies & Hook., *Physalis origanifolia* Lam., *Salpichroa rhomboidea* (Gillies & Hook.) Miers (see)
♦ pampas lily of the valley, lily of the valley vine, cock's eggs
♦ Weed, Quarantine Weed, Noxious Weed, Naturalised, Environmental Weed, Garden Escape, Casual Alien
♦ 7, 15, 34, 40, 72, 86, 87, 88, 98, 101, 147, 161, 165, 176, 179, 180, 198, 203, 220, 237, 243, 269, 270, 280, 290, 295, 296
♦ pH, cultivated, herbal. Origin: South America.

Salpichroa rhomboidea (Gillies & Hook.) Miers
Solanaceae
= *Salpichroa origanifolia* (Lam.) Baill.
♦ Naturalised
♦ 287

Salpiglossis sinuata Ruiz & Pav.
Solanaceae
♦ painted tongue
♦ Naturalised
♦ 280
♦ aH, cultivated, herbal.

Salsola L. spp.
Chenopodiaceae
♦ Russian thistle
♦ Weed
♦ 80, 221

Salsola aphylla L.
Chenopodiaceae
Salsola caffra Sparrm., *Salsola caryoxylon* Moq.
♦ lye bush, lye ganna
♦ Native Weed
♦ 121
♦ pS, arid. Origin: southern Africa.

Salsola australis R.Br.
Chenopodiaceae
= *Salsola tragus* L.
♦ common Russian thistle
♦ Weed, Noxious Weed
♦ 35, 88
♦ herbal.

Salsola baryosma (Schult.) Dandy
Chenopodiaceae
Salsola foetida Delile ex Spreng. (see)
♦ Weed
♦ 221
♦ arid, herbal.

Salsola collina Pall.
Chenopodiaceae
Salsola rutescens Schrad.
♦ spineless Russian thistle, slender Russian thistle, Russian thistle, tumbleweed
♦ Weed, Noxious Weed, Naturalised
♦ 35, 80, 87, 88, 101, 161, 229, 243, 272, 275, 297
♦ aH, arid, promoted, herbal.

Salsola delileana Botsch.
Chenopodiaceae
♦ Weed
♦ 221

Salsola foetida Delile ex Spreng.
Chenopodiaceae

= *Salsola baryosma* (Schult.) Dandy
♦ Weed
♦ 87, 88
♦ herbal.

Salsola iberica auct.
Chenopodiaceae
= *Salsola tragus* L.
♦ Russian thistle, common saltwort, Russian tumbleweed, saltwort, tumbleweed, tumbling weed, windwitch, witchweed
♦ Weed, Noxious Weed, Introduced, Environmental Weed
♦ 21, 68, 80, 88, 138, 146, 151, 161, 180, 212, 243
♦ arid, herbal.

Salsola inermis Forssk.
Chenopodiaceae
♦ Weed
♦ 221

Salsola kali L.
Chenopodiaceae
♦ soft roly poly, prickly saltwort, saltwort, Russian thistle, prickly glasswort, Russian cactus, Russian tumbleweed, tumbling thistle, windwitch, tumbleweed, roly poly, tartor thistle
♦ Weed, Quarantine Weed, Noxious Weed, Naturalised, Native Weed
♦ 23, 39, 44, 55, 70, 86, 87, 88, 94, 95, 98, 101, 121, 136, 158, 198, 203, 210, 217, 220, 221, 229, 237, 241, 243, 263, 272, 280, 283, 287, 299, 300
♦ aH, arid, cultivated, herbal, toxic. Origin: Asia.

Salsola kali L. ssp. *kali*
Chenopodiaceae
♦ Russian thistle, prickly Russian thistle
♦ Noxious Weed, Naturalised
♦ 101, 229

Salsola kali L. ssp. *pontica* (Pall.) Mosyakin
Chenopodiaceae
♦ Russian thistle
♦ Weed, Noxious Weed, Naturalised
♦ 101, 179, 229

Salsola kali L. ssp. *ruthenica* (Iljin) Soó
Chenopodiaceae
= *Salsola tragus* L.
♦ prickly saltwort
♦ Weed, Naturalised, Introduced
♦ 198, 228, 243

Salsola kali L. ssp. *tragus* (L.) Celak.
Chenopodiaceae
= *Salsola tragus* L.
♦ coast saltwort, prickly Russian thistle
♦ Naturalised
♦ 198

Salsola kali L. var. *kali*
Chenopodiaceae
♦ soft roly poly, prickly saltwort, buckbush
♦ Native Weed
♦ 269
♦ aH, arid. Origin: Australia.

Salsola kali L. var. *leptophylla* Benth.
Chenopodiaceae
♦ slender saltwort
♦ Naturalised
♦ 198

Salsola kali L. var. *tenuifolia* Tausch
Chenopodiaceae
= *Salsola tragus* L.
♦ Russian tumbleweed, Russian thistle
♦ Weed, Naturalised
♦ 51, 88, 199, 218, 287

Salsola longifolia Forssk.
Chenopodiaceae
Salsola sieberi Presl, *Salsola zygophylla* Batt.
♦ Weed
♦ 221
♦ arid.

Salsola paulsenii Litv.
Chenopodiaceae
♦ barbwire Russian thistle
♦ Weed, Noxious Weed, Naturalised
♦ 35, 80, 88, 101, 161, 229
♦ a/pH, herbal.

Salsola pestifer Nelson
Chenopodiaceae
= *Salsola tragus* L.
♦ Russian thistle, Russian tumbleweed, tumbling thistle, saltwort
♦ Weed, Noxious Weed
♦ 39, 49, 52, 87, 88, 162, 275, 292
♦ aH, promoted, herbal, toxic.

Salsola ruthenica Iljin
Chenopodiaceae
= *Salsola tragus* L.
♦ harihijiki
♦ Weed, Casual Alien
♦ 87, 88, 280
♦ herbal.

Salsola soda L.
Chenopodiaceae
♦ oppositeleaf Russian thistle, glasswort, barilla plant
♦ Weed, Noxious Weed, Naturalised
♦ 35, 78, 80, 87, 88, 94, 101, 116, 237, 272
♦ aH, promoted, herbal.

Salsola tetrandra Forssk.
Chenopodiaceae
♦ Weed
♦ 221

Salsola tragus L.
Chenopodiaceae
Salsola australis R.Br. (see), *Salsola iberica* auct. (see), *Salsola kali* L. ssp. *ruthenica* (Iljin) Soó (see), *Salsola kali* L. ssp. *tragus* (L.) Celak. (see), *Salsola kali* L. var. *tenuifolia* Tausch (see), *Salsola pestifer* Nels. (see), *Salsola ruthenica* Iljin (see)
♦ Russian thistle, prickly Russian thistle, common Russian thistle, tumbleweed, coast saltwort, tumbling thistle, cardo ruso
♦ Weed, Noxious Weed, Naturalised, Introduced
♦ 78, 80, 101, 116, 167, 174, 229, 236, 243
♦ aH, herbal.

Salsola tuberculatiformis Botsch.
Chenopodiaceae
♦ cauliflower saltwort
♦ Weed, Native Weed
♦ 39, 121
♦ pS, arid, toxic. Origin: southern Africa.

Salsola vermiculata L.
Chenopodiaceae
♦ wormleaf salsola, shrubby Russian thistle, Mediterranean saltwort
♦ Weed, Noxious Weed, Naturalised
♦ 35, 67, 88, 101, 140, 161, 221, 229
♦ cultivated.

Salsola volkensii Schweinf. & Asch.
Chenopodiaceae
♦ Weed
♦ 221

Salvadora persica L.
Salvadoraceae
Galenia asiatica Burm.f., *Salvadora indica* Wight
♦ saltbush, mustard tree, mithi jal, toothbrush tree
♦ Weed, Introduced
♦ 39, 221, 228
♦ arid, cultivated, herbal, toxic.

Salvia L. spp.
Lamiaceae
♦ sage, clary
♦ Weed
♦ 243, 272
♦ herbal.

Salvia acetabulosa L.
Lamiaceae
♦ Weed
♦ 87, 88, 221
♦ cultivated.

Salvia aegyptiaca L.
Lamiaceae
♦ Weed
♦ 221
♦ herbal.

Salvia aethiopis L.
Lamiaceae
Salvia kochiana Kuntze, *Sclarea aethiopis* Mill., *Sclarea lanata* Moench
♦ Mediterranean sage, African sage, clary
♦ Weed, Quarantine Weed, Noxious Weed, Naturalised, Introduced, Casual Alien
♦ 1, 23, 34, 35, 40, 76, 78, 80, 86, 87, 88, 98, 101, 116, 138, 139, 146, 161, 203, 212, 218, 229, 267, 272
♦ pH, cultivated, herbal. Origin: Mediterranean, northern Africa.

Salvia algeriensis Desf.
Lamiaceae
♦ Weed
♦ 87, 88

Salvia apiana Jeps.
Lamiaceae
♦ white sage, bee sage, white sage
♦ Weed
♦ 87, 88, 218
♦ S, arid, cultivated, herbal.

Salvia argentea L.
Lamiaceae
Salvia patula Desf.
♦ silver sage, hopea salvia
♦ Weed, Naturalised
♦ 70, 87, 88, 101
♦ bH, cultivated, herbal.

Salvia aurea L.
Lamiaceae
♦ golden sage
♦ Weed, Naturalised
♦ 86, 98, 198, 203, 280
♦ cultivated. Origin: South Africa.

Salvia austriaca Jacq.
Lamiaceae
♦ Weed
♦ 272
♦ cultivated.

Salvia azurea Michx. ex Lam.
Lamiaceae
♦ azure blue sage, blue sage, New Mexico sage
♦ Weed, Casual Alien
♦ 161, 280
♦ pH, cultivated, herbal.

Salvia barrelieri Etl.
Lamiaceae
♦ Weed
♦ 87, 88
♦ cultivated.

Salvia brachiata Roxb.
Lamiaceae
♦ Weed
♦ 87, 88

Salvia cinnabarina Martens & Galeotti
Lamiaceae
♦ Weed
♦ 157
♦ pH.

Salvia coccinea Buc'hoz ex Etl.
Lamiaceae
♦ Texas sage, red salvia, scarlet salvia, South American sage, Texas salvia, crimson sage, blood sage, tropical sage
♦ Weed, Naturalised, Environmental Weed
♦ 14, 39, 86, 87, 88, 98, 121, 134, 203, 218, 269
♦ pH, cultivated, herbal, toxic. Origin: tropical America.

Salvia coccinea Buc'hoz ex Etl. var. pseudococcinea (Juss. ex Murr.) Back.
Lamiaceae
♦ Weed
♦ 13

Salvia dasycalyx Fern.
Lamiaceae
♦ Weed
♦ 87, 88

Salvia deserti Decne.
Lamiaceae
♦ Weed
♦ 221

Salvia farinacea Benth.
Lamiaceae
♦ mealycup sage, blue sage
♦ Naturalised

♦ 280
♦ a/pH, cultivated, herbal.

Salvia glutinosa L.
Lamiaceae
Sclarea glutinosa Mill.
♦ Jupiter's distaff, sticky sage
♦ Weed, Naturalised
♦ 87, 88, 101, 272
♦ pH, cultivated, herbal.

Salvia grandiflora Etl.
Lamiaceae
♦ balsamic sage
♦ Weed
♦ 272
♦ cultivated, herbal.

Salvia graveolens Linn. ex Jacks.
Lamiaceae
♦ Weed
♦ 221

Salvia guaranitica A.St.-Hil. ex Benth.
Lamiaceae
♦ Naturalised, Casual Alien
♦ 280, 300
♦ cultivated. Origin: South America.

Salvia hispanica L.
Lamiaceae
♦ Spanish sage, Mexican chia, chia sage, chia
♦ Naturalised, Introduced
♦ 101, 228
♦ aH, arid, cultivated, herbal.

Salvia horminum L.
Lamiaceae
= *Salvia viridis* L.
♦ Weed
♦ 87, 88
♦ cultivated, herbal.

Salvia hyptoides Martens & Galeotti
Lamiaceae
♦ Weed
♦ 157
♦ aH, herbal.

Salvia japonica Thunb.
Lamiaceae
♦ Japanese sage
♦ Weed
♦ 286
♦ pH, promoted, herbal. Origin: China, Japan, Korea.

Salvia lanceolata Brouss.
Lamiaceae
= *Salvia reflexa* Hornem.
♦ lance leaved sage, blue sage, mint weed
♦ Weed
♦ 49

Salvia lanigera Poir.
Lamiaceae
♦ Weed
♦ 221
♦ H, promoted.

Salvia leucantha Cav.
Lamiaceae
♦ purple Mexican bush sage, Mexican sage
♦ Weed, Naturalised
♦ 86, 98, 203

♦ cultivated, herbal. Origin: tropical America.

Salvia leucophylla Greene
Lamiaceae
♦ whiteleaf sage, purple sage, San Luis purple sage
♦ Weed
♦ 87, 88, 218
♦ S, cultivated, herbal.

Salvia longistyla Benth.
Lamiaceae
♦ Mexican sage
♦ Naturalised
♦ 101

Salvia lyrata L.
Lamiaceae
♦ cancer weed, lyreleaf sage
♦ Weed
♦ 161
♦ pH, cultivated, herbal.

Salvia mellifera Greene
Lamiaceae
♦ black sage, Californian black sage
♦ Weed
♦ 87, 88, 218
♦ S, cultivated, herbal.

Salvia micrantha Vahl
Lamiaceae
♦ Yucatan sage
♦ Weed
♦ 88
♦ herbal.

Salvia microphylla Kunth
Lamiaceae
♦ baby sage
♦ Naturalised
♦ 101, 280
♦ cultivated.

Salvia microphylla Kunth var. *neurepia* (Fernald) Epling
Lamiaceae
♦ Naturalised
♦ 280
♦ cultivated.

Salvia miltiorrhiza Bunge
Lamiaceae
♦ dan shen, red sage, tan shen, Chinese sage
♦ Weed
♦ 297
♦ promoted, herbal.

Salvia misella Kunth
Lamiaceae
Salvia riparia Kunth (see)
♦ tropical sage
♦ Naturalised
♦ 86
♦ Origin: tropical America.

Salvia moureti Batt. & Pit.
Lamiaceae
♦ Weed
♦ 87, 88

Salvia napifolia Jacq.
Lamiaceae
♦ Naturalised
♦ 287

Salvia nemorosa L.
Lamiaceae
♦ woodland sage
♦ Weed, Naturalised, Casual Alien
♦ 40, 80, 87, 88, 101, 272, 280
♦ cultivated, herbal.

Salvia nutans L.
Lamiaceae
♦ nodding sage
♦ Naturalised
♦ 101
♦ cultivated.

Salvia occidentalis Sw.
Lamiaceae
♦ West Indian sage
♦ Weed
♦ 14, 87, 88, 157
♦ a/pH, herbal.

Salvia officinalis L.
Lamiaceae
Salvia chromatica Hoff., *Salvia tomentosa* Mill.
♦ sage, red sage, broadleaf sage, narrowleaf sage, kitchen sage
♦ Weed, Naturalised
♦ 39, 70, 80, 101, 154, 272, 280, 287
♦ S, cultivated, herbal, toxic.

Salvia palaestina Benth.
Lamiaceae
♦ Weed
♦ 221

Salvia plebeia R.Br.
Lamiaceae
♦ sage weed
♦ Weed, Naturalised
♦ 87, 88, 98, 203, 235, 275, 276, 286, 297
♦ aH, herbal, toxic.

Salvia pratensis L.
Lamiaceae
♦ meadow clary, introduced sage, niittysalvia
♦ Weed, Noxious Weed, Naturalised, Introduced, Casual Alien
♦ 1, 34, 39, 42, 70, 80, 82, 87, 88, 101, 139, 161, 243, 272
♦ pH, cultivated, herbal, toxic.

Salvia pratensis L. × sylvestris L.
Lamiaceae
♦ Casual Alien
♦ 42

Salvia procurrens Benth.
Lamiaceae
♦ Weed
♦ 87, 88

Salvia reflexa Hornem.
Lamiaceae
Salvia lanceolata Brouss. (see)
♦ Rocky Mountain sage, mint weed, lanceleaf sage, narrowleaf sage, wild mint, nuokkusalvia, white chia, lambsleaf sage, blue sage, sage mint
♦ Weed, Noxious Weed, Naturalised, Native Weed, Casual Alien
♦ 7, 39, 40, 42, 55, 62, 86, 87, 88, 98, 121, 147, 158, 161, 174, 198, 199, 203, 212, 243, 269, 280, 287
♦ aH, cultivated, herbal, toxic. Origin: North America.

Salvia repens Benth.
Lamiaceae
♦ creeping sage
♦ Naturalised
♦ 280
♦ cultivated, herbal.

Salvia riparia Kunth
Lamiaceae
= *Salvia misella* Kunth
♦ Florida Keys sage
♦ Weed, Naturalised
♦ 13, 98, 203

Salvia runcinata L.
Lamiaceae
♦ Native Weed
♦ 121
♦ pH, herbal. Origin: southern Africa.

Salvia rutilans Carrière
Lamiaceae
♦ pineapple sage, pineapple scented sage
♦ Weed, Casual Alien
♦ 15, 280
♦ cultivated, herbal.

Salvia sclarea L.
Lamiaceae
Aethiopis sclarea (L.) Fourr., *Sclarea vulgaris* Mill.
♦ clary, Europe sage, clary sage
♦ Weed, Noxious Weed, Naturalised, Introduced
♦ 1, 24, 80, 87, 88, 101, 121, 139, 221, 228, 272, 280
♦ b/pH, arid, cultivated, herbal. Origin: Eurasia.

Salvia serotina L.
Lamiaceae
♦ little woman
♦ Weed
♦ 87, 88
♦ herbal.

Salvia sonomensis Greene
Lamiaceae
♦ Sonoma sage, creeping sage
♦ Weed
♦ 87, 88, 218
♦ pH, cultivated, herbal.

Salvia spinosa L.
Lamiaceae
♦ Weed
♦ 87, 88, 221

Salvia splendens Sellow ex Roem. & Schult.
Lamiaceae
♦ scarlet sage
♦ Naturalised, Introduced, Garden Escape
♦ 38, 101, 261
♦ aH, cultivated, herbal. Origin: Brazil.

Salvia stenophylla Burch. ex Benth.
Lamiaceae
♦ wild sage
♦ Weed, Native Weed
♦ 88, 121, 158
♦ pH, cultivated, herbal. Origin: southern Africa.

Salvia × superba Stapf
Lamiaceae
= *Salvia sylvestris* L. × *Salvia amplexicaulis* Lam.
♦ Naturalised
♦ 101
♦ a/pH, cultivated, herbal.

Salvia × sylvestris L. (pro sp.)
Lamiaceae
= *Salvia nemorosa* L. × *Salvia pratensis* L.
♦ lehtosalvia
♦ Naturalised, Casual Alien
♦ 42, 101
♦ cultivated, herbal.

Salvia syriaca L.
Lamiaceae
♦ Syrian sage
♦ Weed
♦ 87, 88, 115, 243
♦ aH.

Salvia tiliifolia Vahl
Lamiaceae
♦ lindenleaf sage, tarahumara chia
♦ Weed, Naturalised
♦ 101, 121
♦ aH, cultivated, herbal. Origin: South America.

Salvia uliginosa Benth.
Lamiaceae
♦ bog sage
♦ Naturalised
♦ 280
♦ cultivated, herbal.

Salvia urica Epling
Lamiaceae
♦ Weed
♦ 157
♦ pH.

Salvia verbenaca L.
Lamiaceae
Gallitrichum arvale Jordan & Fourr., *Horminum verbenaceum* Mill., *Salvia clandestina* L., *Salvia linnaei* Rouy, *Salvia verbenacea* L. (see)
♦ salvia, vervain salvia, wild sage, wild clary
♦ Weed, Naturalised, Native Weed, Garden Escape, Environmental Weed
♦ 7, 72, 86, 87, 88, 93, 94, 98, 121, 134, 158, 165, 176, 198, 203, 221, 237, 269, 272, 280, 287, 295, 300
♦ pH, cultivated, herbal. Origin: western Europe, Mediterranean.

Salvia verbenaca L. var. verbenaca
Lamiaceae
♦ wild sage
♦ Naturalised
♦ 198

Salvia verbenaca L. var. vernalis Boiss.
Lamiaceae
♦ wild sage
♦ Naturalised
♦ 198

Salvia verbenacea L.
Lamiaceae
= *Salvia verbenaca* L.
♦ verbena sage

♦ Naturalised
♦ 101
♦ pH, herbal.

Salvia verticillata L.
Lamiaceae
Salvia amasiaca Freyn & Bornm., *Salvia uberrima* Rech.
♦ lilac sage, kiehkurasalvia, whorled sage
♦ Weed, Naturalised, Casual Alien
♦ 23, 40, 42, 87, 88, 101, 243, 272
♦ cultivated, herbal.

Salvia virgata Jacq.
Lamiaceae
♦ meadow sage
♦ Weed, Noxious Weed
♦ 35, 88
♦ cultivated.

Salvia viridis L.
Lamiaceae
Salvia horminum L. (see)
♦ annual clary, bluebeard, red topped sage, clary, kirjosalvia, painted sage, kirjosalvia
♦ Weed, Naturalised, Casual Alien
♦ 40, 42, 98, 203
♦ aH, cultivated, herbal.

Salvinia Ség. spp.
Salviniaceae
♦ watervaring, kariba weed, watermoss, salvinia, floating fern
♦ Weed, Quarantine Weed, Noxious Weed, Naturalised, Garden Escape, Cultivation Escape
♦ 63, 67, 76, 86, 88, 220, 233, 237, 258
♦ wpH, cultivated.

Salvinia auriculata Aubl.
Salviniaceae
Salvinia rotundifolia Willd. (see)
♦ water fern, kariba weed, eared watermoss, watervaring, salvinia, butterfly fern, small leaved salvinia
♦ Weed, Noxious Weed, Naturalised
♦ 35, 51, 67, 88, 101, 140, 186, 193, 229, 241, 255, 295, 300
♦ wpH, cultivated. Origin: Brazil.

Salvinia biloba Raddi
Salviniaceae
♦ giant salvinia
♦ Weed, Noxious Weed
♦ 67, 88, 140, 193, 229
♦ wpH.

Salvinia cucullata Roxb.
Salviniaceae
♦ salvinia, small rat's ear, Chok huu nuu
♦ Weed, Quarantine Weed
♦ 76, 87, 88, 135, 170, 191, 203, 209, 239, 262
♦ wpH.

Salvinia hastata Desv.
Salviniaceae
♦ Weed
♦ 87, 88
♦ wpH.

Salvinia herzogii de la Sota
Salviniaceae
♦ giant salvinia

♦ Weed, Noxious Weed
♦ 67, 88, 140, 193, 229
♦ wpH.

Salvinia minima Bak.
Salviniaceae
♦ water spangles, little salvinia
♦ Weed, Noxious Weed
♦ 80, 197, 229
♦ wpH, cultivated.

Salvinia molesta D.Mitch.
Salviniaceae
Salvinia auriculata auct. non Aubl.
♦ water fern, salvinia, kariba weed, African payal, koi kandy, watervaring, African pyle, giant salvinia
♦ Weed, Quarantine Weed, Noxious Weed, Naturalised, Garden Escape, Environmental Weed
♦ 3, 6, 7, 10, 18, 62, 63, 67, 76, 86, 87, 88, 93, 95, 98, 101, 121, 140, 147, 152, 158, 169, 170, 171, 177, 179, 181, 191, 193, 197, 200, 203, 225, 229, 232, 246, 254, 268, 269, 278, 283, 287, 289, 296
♦ wpH, cultivated. Origin: Brazil.

Salvinia natans (L.) All.
Salviniaceae
Marsilea natans L
♦ salvinia, floating watermoss, eared watermoss
♦ Weed, Quarantine Weed, Noxious Weed, Naturalised
♦ 76, 87, 88, 101, 135, 170, 191, 203, 204, 229, 263, 272, 274, 275, 286, 297
♦ wpH, cultivated, herbal.

Salvinia nymphellula Desv.
Salviniaceae
♦ Weed, Quarantine Weed
♦ 76, 87, 88, 203
♦ wpH.

Salvinia rotundifolia Willd.
Salviniaceae
= *Salvinia auriculata* Aubl.
♦ salvinia, kellusaniainen
♦ Weed, Quarantine Weed, Cultivation Escape
♦ 23, 42, 76, 87, 88, 203, 218, 295
♦ wpH, cultivated.

Samaipaticereus Cárdenas spp.
Cactaceae
Yungasocereus Ritt. spp. (see)
♦ Weed, Quarantine Weed
♦ 76, 88, 220

Samanea saman (Jacq.) Merr.
Fabaceae/Mimosaceae
Albizia saman (Jacq.) F.Muell. (see), *Enterolobium saman* (Jacq.) Prain ex King (see), *Pithecellobium saman* (Jacq.) Benth. (see)
♦ monkey pod, rain tree, ohai saman, tronkon mames, gumor ni spanis, vaivai ni vavalangi
♦ Weed, Naturalised, Introduced
♦ 3, 22, 32, 86, 101, 107, 132, 230, 261
♦ T, cultivated, herbal. Origin: continental tropical America.

Sambucus callicarpa Greene
Adoxaceae/Caprifoliaceae/Sambucaceae

= *Sambucus racemosa* L. var. *arborescens*
(Torr. & A.Gray) A.Gray (NoR)
- Pacific red elder
- Weed
- 87, 88, 218
- S, cultivated, herbal.

Sambucus canadensis L.
Adoxaceae/Caprifoliaceae/
Sambucaceae
= *Sambucus nigra* L. ssp. *canadensis* (L.)
R.Bolli (NoR) [see *Sambucus mexicana*
C.Presl ex A.DC., *Sambucus simpsonii*
Rehd. ex Sarg.]
- American elder, elderberry
- Weed, Naturalised, Cultivation
Escape
- 8, 23, 39, 86, 87, 88, 161, 218, 247, 261
- S, cultivated, herbal, toxic. Origin:
Central and North America.

Sambucus cerulea Raf.
Adoxaceae/Caprifoliaceae/
Sambucaceae
= *Sambucus nigra* L. ssp. *cerulea* (Raf.)
R.Bolli (NoR) [see *Sambucus glauca*
Nutt.]
- blue elderberry
- Weed
- 161
- herbal, toxic.

Sambucus chinensis Lindl.
Adoxaceae/Caprifoliaceae/
Sambucaceae
Sambucus formosana Nakai
- Chinese elder
- Weed
- 87, 88, 286
- pH, promoted, herbal.

Sambucus ebulus L.
Adoxaceae/Caprifoliaceae/
Sambucaceae
Ebulus humile Garcke, *Sambucus
herbacea* Gilib., *Sambucus paucijuga*
Steven
- dwarf elder, danewort
- Weed, Naturalised
- 39, 40, 101, 272
- pH, cultivated, herbal, toxic.

Sambucus gaudichaudiana DC.
Adoxaceae/Caprifoliaceae/
Sambucaceae
- Naturalised
- 86
- S, cultivated. Origin: Australia.

Sambucus glauca Nutt.
Adoxaceae/Caprifoliaceae/
Sambucaceae
= *Sambucus nigra* L. ssp. *cerulea* (Raf.)
R.Boll (NoR) [see *Sambucus cerulea*
Raf.]
- blueberry elder, blue elder
- Weed
- 87, 88, 218
- herbal.

Sambucus javanica Reinw. ex Bl.
Adoxaceae/Caprifoliaceae/
Sambucaceae
- Chinese elder
- Weed

- 87, 88
- S, cultivated, herbal.

Sambucus mexicana C.Presl ex A.DC.
Adoxaceae/Caprifoliaceae/
Sambucaceae
= *Sambucus nigra* L. ssp. *canadensis* (L.)
R.Bolli (NoR) [see *Sambucus canadensis*
L., *Sambucus simpsonii* Rehd. ex Sarg.]
- blue elder, Mexican elder
- Weed
- 22, 39, 161
- S, cultivated, herbal, toxic.

Sambucus nigra L.
Adoxaceae/Caprifoliaceae/
Sambucaceae
Sambucus laciniata Mill.
- elder, European black elderberry,
common elder, elderberry, mustaselja,
sambuco
- Weed, Naturalised, Garden Escape,
Environmental Weed, Cultivation
Escape
- 15, 22, 39, 42, 70, 80, 86, 97, 98, 101,
161, 165, 176, 181, 198, 203, 225, 241,
246, 272, 280, 290, 300
- S/T, cultivated, herbal, toxic. Origin:
Mediterranean, west Asia.

Sambucus nigra L. cv. 'Viridis'
Adoxaceae/Caprifoliaceae/
Sambucaceae
- black elder, elder, common elder,
elderberry, European black elderberry,
sambuco
- Naturalised
- 280
- S/T, cultivated, herbal, toxic. Origin:
horticultural.

Sambucus nigra L. ssp. nigra
Adoxaceae/Caprifoliaceae/
Sambucaceae
- European black elderberry
- Naturalised
- 101

Sambucus pubens Michx.
Adoxaceae/Caprifoliaceae/
Sambucaceae
= *Sambucus racemosa* L. ssp. *pubens*
(Michx.) House (NoR)
- American red elder, red berried
elder, scarlet elderberry
- Naturalised
- 39, 247, 280
- S, cultivated, herbal, toxic.

Sambucus racemosa L.
Adoxaceae/Caprifoliaceae/
Sambucaceae
- scarlet elderberry, red elderberry,
red berried elder, black melanocarpa
elderberry, red elder, selja, terttu selja
- Weed, Cultivation Escape
- 39, 42, 87, 88, 161
- S, cultivated, herbal, toxic.

Sambucus racemosa L. ssp. sieboldiana (Miq.) H.Hara
Adoxaceae/Caprifoliaceae/
Sambucaceae
- niwatoko

- Weed
- 286

Sambucus simpsonii Rehd. ex Sarg.
Adoxaceae/Caprifoliaceae/
Sambucaceae
= *Sambucus nigra* L. ssp. *canadensis* (L.)
R.Bolli (NoR) [see *Sambucus canadensis*
L., *Sambucus mexicana* C.Presl ex
A.DC.]
- Weed
- 87, 88
- herbal.

Samolus valerandi L.
Primulaceae
Samolus americanus Spreng., *Samolus
caulescens* Willd. ex Roem. & Schult.,
Samolus floribundus Kunth, *Samolus
parviflorus* Raf., *Samolus valerandi* var.
americanus A.Gray, *Samolus valerandi*
var. *floribundus* (Kunth) R.Knuth
- brookweed, seaside brookweed,
suolapunka
- Weed, Naturalised, Garden Escape,
Environmental Weed, Garden Escape
- 7, 23, 70, 86, 88, 98, 101, 203, 221, 241,
272, 300
- pH, aqua, cultivated, herbal.

Samyda macrantha P.Wils.
Flacourtiaceae
- Weed
- 14

Sanchezia nobilis Hook.f.
Acanthaceae
= *Sanchezia speciosa* Léonard
- Weed
- 3, 191
- cultivated.

Sanchezia parvibracteata Sprague & Hutch.
Acanthaceae
- sanchezia
- Weed, Naturalised, Garden Escape,
Environmental Weed
- 3, 54, 86, 88, 155, 191
- cultivated. Origin: tropical America.

Sanchezia speciosa Léonard
Acanthaceae
Sanchezia nobilis Hook.f. (see)
- sanchezia, shrubby whitevein
- Weed, Naturalised, Cultivation
Escape
- 3, 101, 191, 261
- cultivated. Origin: Ecuador.

Sanguinaria canadensis L.
Papaveraceae
- bloodroot, red puccoon, bloodwort
- Weed
- 8, 39, 161, 247
- pH, cultivated, herbal, toxic.

Sanguisorba canadensis L.
Rosaceae
Poterium canadense A.Gray
- Canadian burnet, American great
burnet
- Weed
- 87, 88, 218
- pH, cultivated, herbal.

Sanguisorba hybrida (L.) Nord.
Rosaceae
♦ Weed
♦ 70

Sanguisorba menziesii Rydb.
Rosaceae
♦ lännenluppio, Menzies' burnet
♦ Cultivation Escape
♦ 42
♦ pH, cultivated, herbal.

Sanguisorba minor Scop.
Rosaceae
Poterium dictyocarpum Spach,
Sanguisorba dictyocarpa Grem.,
Sanguisorba gaillardotii (Boiss.) Hayek
♦ small burnet, sheep's burnet,
salad burnet, garden burnet, little
burnet, petite pimprenelle, kleiner
wiesenknopf, pimpinela menor
♦ Weed, Naturalised
♦ 7, 70, 87, 88, 94, 98, 101, 161, 165, 176,
203, 218, 221, 241, 243, 253, 272, 280,
287, 300
♦ pH, cultivated, herbal. Origin:
Mediterranean, west Asia.

**Sanguisorba Scop. minor ssp. lasiocarpa
(Boiss. & Hausskn.) Nordborg**
Rosaceae
♦ Naturalised
♦ 280

**Sanguisorba minor Scop. ssp. muricata
(Spach) Briq.**
Rosaceae
Poterium polygamum Waldst. & Kit.
(see)
♦ small burnet, fodder burnet, burnet,
garden burnet
♦ Weed, Naturalised, Cultivation
Escape
♦ 34, 40, 86, 101, 198, 252, 280
♦ pH, cultivated, herbal. Origin:
Europe, Middle East.

Sanguisorba occidentalis Nutt.
Rosaceae
♦ western burnet
♦ Weed
♦ 23, 88
♦ a/pH, promoted, herbal.

Sanguisorba officinalis L.
Rosaceae
Pimpinella officinalis Lam., *Poterium
officinale* A.Gray, *Sanguisorba hispanica*
Mill., *Sanguisorba sabauda* Mill.
♦ garden burnet, great burnet, official
burnet, salad burnet, punaluppio
♦ Weed, Quarantine Weed,
Naturalised, Introduced
♦ 23, 38, 42, 76, 87, 88, 204, 218, 241,
272, 286, 297
♦ pH, aqua, cultivated, herbal.

Sanguisorba parviflora Takeda
Rosaceae
♦ whiteflower Siberian burnet
♦ Weed
♦ 297

Sanguisorba tenuifolia Fisch. ex Link
Rosaceae
♦ Weed

♦ 286
♦ pH, cultivated. Origin: China, Japan.

Sanicula bipinnata Hook. & Arn.
Apiaceae
♦ poison sanicle
♦ Weed
♦ 34, 82, 161
♦ pH, herbal, toxic.

Sanicula europaea L.
Apiaceae
Astrantia diapensia Scop., *Caucalis
capiatata* Salisb., *Sanicula trilobata* Gilib.
♦ haavayrtti, sanicle, wood sanicle
♦ Weed
♦ 23, 88
♦ pH, cultivated, herbal.

Sansevieria Thunb. spp.
Agavaceae/Dracaenaceae
♦ sikonje, bowstring hemp, sansevieria
♦ Weed
♦ 80, 161
♦ toxic.

Sansevieria guineensis (L.) Willd.
Agavaceae/Dracaenaceae
♦ Weed, Naturalised, Environmental
Weed
♦ 3, 86, 87, 88, 155, 191
♦ pH, cultivated. Origin: tropical
Africa, Asia.

Sansevieria hyacinthoides (L.) Druce
Agavaceae/Dracaenaceae
♦ bowstring hemp, African bowstring
hemp, iguana tail
♦ Weed, Naturalised, Environmental
Weed, Cultivation Escape
♦ 3, 80, 88, 101, 112, 179, 191, 261
♦ arid, cultivated. Origin: tropical
Africa.

Sansevieria pearsonii N.E.Br.
Agavaceae/Dracaenaceae
Sansevieria desertii N.E.Br.
♦ enghushe, ongushe
♦ Introduced
♦ 228
♦ arid.

Sansevieria trifasciata Prain
Agavaceae/Dracaenaceae
Aloe guineensis Jacq., *Sansevieria
guineensis* auct. non (L.) Willd.
♦ mother in law's tongue, snake plant,
viper's bowstring hemp, lengua de
suegra
♦ Weed, Naturalised, Introduced,
Garden Escape, Environmental Weed
♦ 3, 86, 98, 101, 155, 191, 201, 203, 228,
247, 261
♦ arid, cultivated, herbal, toxic. Origin:
Africa.

Sansevieria zeylanica (L.) Willd.
Agavaceae/Dracaenaceae
♦ Ceylon bowstring hemp
♦ Naturalised
♦ 287
♦ cultivated, herbal.

Santalum album L.
Santalaceae
♦ sandalwood, white sandalwood tree
♦ Weed, Introduced

♦ 39, 87, 88, 179, 230
♦ cultivated, herbal, toxic. Origin:
Asia, Australia.

Santolina chamaecyparissus L.
Asteraceae
Santolina incana Lam., *Santolina
tomentosa* Pers.
♦ lavender cotton, cotton lavender,
santolina, lavender, gray lavender,
cypress cotton
♦ Weed, Naturalised, Introduced,
Casual Alien
♦ 15, 34, 40, 70, 87, 88, 101, 280
♦ S, cultivated, herbal.

Sanvitalia procumbens Lam.
Asteraceae
♦ Mexican creeping zinnia, creeping
zinnia
♦ Naturalised
♦ 101, 287
♦ aH, cultivated, herbal.

Sapindus mukorossi Gaertn.
Sapindaceae
♦ Chinese soapberry, Kashmir
soapberry
♦ Quarantine Weed
♦ 39, 220
♦ cultivated, herbal, toxic.

Sapindus saponaria L.
Sapindaceae
Sapindus amolli Sessé & Moç., *Sapindus
drummondii* Hook. & Arn., *Sapindus
inaequalis* DC., *Sapindus marginatus*
Willd.
♦ soapberry, savonnier, wingleaf
soapberry
♦ Naturalised, Introduced
♦ 39, 228, 300
♦ T, arid, cultivated, herbal, toxic.

Sapindus trifoliatus L.
Sapindaceae
Sapindus emarginatus Vahl, *Sapindus
laurifolius* Vahl
♦ threeleaf soapberry
♦ Introduced
♦ 39, 228
♦ arid, herbal, toxic.

Sapium biloculare (S.Wats.) Pax
Euphorbiaceae
= *Sebastiania bilocularis* S.Wats. (NoR)
♦ Mexican jumping bean
♦ Weed
♦ 161
♦ herbal, toxic.

Sapium discolor (Champ.) Müll.Arg.
Euphorbiaceae
♦ Quarantine Weed
♦ 220
♦ cultivated.

Sapium grahami (Stapf) Prain
Euphorbiaceae
♦ Weed
♦ 87, 88
♦ herbal.

Sapium haematospermum Müll.Arg.
Euphorbiaceae
Sapium bolivianum Pax. & Hoffm.
♦ Weed

♦ 255
♦ T. Origin: Brazil.

Sapium jamaicense Sw. nom. illeg.
Euphorbiaceae
= *Sapium laurifolium* (A.Rich.) Griseb.
(NoR)
♦ tabaiba
♦ Weed
♦ 87, 88

Sapium sebiferum (L.) Roxb.
Euphorbiaceae
= *Triadica sebifera* (L.) Small
♦ tallow tree, Chinese tallow, popcorn
tree, vegetable tallow
♦ Weed, Quarantine Weed, Noxious
Weed, Naturalised, Introduced,
Garden Escape, Environmental Weed
♦ 3, 35, 39, 77, 78, 80, 86, 87, 88, 112,
116, 142, 151, 152, 161, 179, 191, 218,
228, 247, 258, 287
♦ S/T, arid, cultivated, herbal, toxic.
Origin: China, Japan.

Saponaria bellidifolia Sm.
Caryophyllaceae
♦ yellow soapwort
♦ Weed
♦ 272

Saponaria calabrica Guss.
Caryophyllaceae
♦ soapwort, Italian suopayrtti
♦ Weed, Naturalised, Casual Alien
♦ 42, 86, 98, 203
♦ aH, cultivated. Origin: east
Mediterranean.

Saponaria ocymoides L.
Caryophyllaceae
Saponaria repens Lam.
♦ rock soapwort, tumbling ted,
soapwort
♦ Weed, Naturalised
♦ 80, 101
♦ pH, cultivated, herbal.

Saponaria officinalis L.
Caryophyllaceae
Lychnis officinalis (L.) Scop., *Saponaria
hybrida* Mill., *Saponaria vulgaris* Pall.
♦ bouncing Bet, soapwort, sweet
Betty, bladder soapwort, China cockle,
cockle, cow basil, cow cockle, cowfoot,
cowherb, cow soapwort, glong, spring
cockle, suopayrtti, saponaria
♦ Weed, Noxious Weed, Naturalised,
Introduced, Garden Escape,
Environmental Weed, Cultivation
Escape
♦ 8, 15, 23, 34, 39, 40, 42, 49, 70, 80, 86,
87, 88, 94, 98, 101, 121, 138, 161, 174,
176, 195, 198, 203, 207, 210, 212, 218,
228, 229, 241, 247, 269, 272, 280, 286,
287, 300
♦ pH, arid, cultivated, herbal, toxic.
Origin: Eurasia.

Saponaria pumilio (L.) Fenzl ex A.Braun
Caryophyllaceae
♦ pygmy pink
♦ Naturalised
♦ 101
♦ cultivated.

Saponaria vaccaria L.
Caryophyllaceae
= *Vaccaria hispanica* (Mill.) Rausch.
♦ cow cockle, cowherb
♦ Weed, Noxious Weed
♦ 23, 36, 39, 51, 52, 87, 88, 162, 218
♦ herbal, toxic.

**Saposhnikovia divaricata (Turcz.)
Schischk.**
Apiaceae
♦ saposhnikovia
♦ Weed
♦ 297

Sarcobatus vermiculatus (Hook.) Torr.
Sarcobataceae/Chenopodiaceae
Batis vermiculata Hook., *Sarcobatus
maximilliana* Nees
♦ greasewood, black greasewood
♦ Weed, Noxious Weed
♦ 36, 39, 87, 88, 161, 212, 218
♦ S, arid, promoted, herbal, toxic.

**Sarcocaulon patersonii (DC.) Eckl. &
Zeyh.**
Geraniaceae
Monsonia patersonii DC., *Sarcocaulon
l'heritieri* Sweet var. *brevimucronatum*
Schinz, *Sarcocaulon patersonii* (DC.)
Eckl. & Zeyh. ssp. *curvatum* Rehm,
Sarcocaulon rigidum Schinz
♦ Introduced
♦ 228
♦ arid.

Sarcocephalus latifolius (Sm.) Bruce
Rubiaceae/Naucleaceae
Nauclea esculenta (Sabine) Merr.,
Nauclea latifolia Sm., *Sarcocephalus
esculentus* Sabine, *Sarcocephalus
russeggeri* Schweinf.
♦ African cinchona, African peach,
African quinine, doundaké, egbessi
root, guinea peach, kina du Rio Nunez,
njimo, woacroolie root, wuacruli
♦ Introduced
♦ 39, 228
♦ arid, toxic.

Sarcopoterium spinosum (L.) Spach
Rosaceae
♦ thorny burnet
♦ Weed
♦ 272
♦ cultivated.

Sarcostemma australe R.Br.
Asclepiadaceae/Apocynaceae
♦ Weed
♦ 39, 87, 88
♦ cultivated, toxic.

Sarcostemma cynanchoides Decne.
Asclepiadaceae/Apocynaceae
= *Funastrum cynanchoides* (Decne.)
Schltr. ssp. *cynanchoides* (NoR)
♦ climbing milkweed, fringed
twinevine
♦ Weed
♦ 87, 88, 161, 218
♦ herbal.

Sarcostemma viminale (L.) R.Br.
Asclepiadaceae/Apocynaceae

Cynanchum viminale L.
♦ caustic bush, caustic creeper, caustic
vine, spantou
♦ Native Weed
♦ 39, 121
♦ pC, cultivated, herbal, toxic. Origin:
southern Africa.

Sargassum muticum (Yendo) Fensholt
Sargassaceae
♦ Japanese brown alga, Jap weed,
wireweed, strangleweed, algae
♦ Weed
♦ 197, 282, 288
♦ algae.

**Saritaea magnifica (Sprague ex Steenis)
Dugand**
Bignoniaceae
♦ glowvine, purple bignonia, saritaea
♦ Naturalised, Introduced
♦ 86, 101, 261
♦ cultivated. Origin: Central and
South America.

Sarracenia flava L.
Sarraceniaceae
♦ yellow pitcherplant, yellow trumpet
♦ Casual Alien
♦ 39, 280
♦ pH, cultivated, herbal, toxic.

Sartwellia flaveriae Gray
Asteraceae
♦ sartwellia, threadleaf glowwort
♦ Weed
♦ 39, 161
♦ herbal, toxic.

**Sasa kurilensis (Rupr.) Makino &
Shibata**
Poaceae
Arundinaria kurilensis Rupr.
♦ dwarf bamboo, chishimazasa
♦ Weed
♦ 286
♦ G, cultivated. Origin: Japan, Korea.

Sasa nipponica (Maxim.) Mak. & Shib.
Poaceae
Bambusa nipponica Makino
♦ dwarf bamboo, miyakozasa
♦ Weed
♦ 286
♦ G, promoted.

Sasa palmata (hort. ex Burb.) E.G.Camus
Poaceae
Arundinaria palmata (hort. ex Burb.)
Bean, *Bambusa metallica* Mitf.,
Bambusa palmata hort. ex Burb., *Sasa
amplissima* Koidz., *Sasa chimakisasa*
Koidz., *Sasa koshinaiana* Koidz.,
Sasa laevissima Koidz., *Sasa lingulata*
Koidz., *Sasa nakasiretokoensis* Koidz.,
Sasa palmata (hort. ex Burb.) Nakai,
Sasa pseudobrachyphylla Nakai, *Sasa
shikotanensis* Nakai, *Sasa shikotanensis*
var. *pseudobrachyphylla* (Nakai) Koidz.,
Sasa soyensis Nakai
♦ broad leaved bamboo, sasa
♦ Weed, Naturalised
♦ 15, 40, 101, 280, 286
♦ G, cultivated. Origin: Japan.

**Sasa ramosa (Makino) Makino &
Shibata**
Poaceae
♦ Weed
♦ 88, 191, 204
♦ G.

Sasa senanensis (Franch. & Sav.) Rehder
Poaceae
♦ kumaizasa
♦ Weed
♦ 286
♦ G, promoted. Origin: China, Japan.

Sasa veitchii (Carrière) Rehder
Poaceae
Sasa albomarginata (Miq.) Makino &
Shibata, *Bambusa veitchii* Carrière
♦ Veitch's bamboo, kuma zasa
♦ Cultivation Escape
♦ 40
♦ G, cultivated.

Sasaella ramosa (Makino) Makino
Poaceae
♦ hairy bamboo, azuma sasa
♦ Naturalised
♦ 40, 280
♦ G, cultivated.

Sasamorpha borealis (Thunb.) Nakai
Poaceae
♦ suzudake
♦ Weed
♦ 286
♦ G, promoted. Origin: China, Japan,
Korea.

Sassafras albidum (Nutt.) Nees
Lauraceae
♦ sassafras, American sassafras oil
♦ Weed
♦ 87, 88, 218, 247
♦ T, cultivated, herbal, toxic.

Satureja grandiflora (L.) Scheele
Lamiaceae
= *Calamintha grandiflora* (L.) Moench
♦ tarhakäenminttu
♦ Cultivation Escape
♦ 42
♦ cultivated, herbal.

Satureja hortensis L.
Lamiaceae
Clinopodium hortense Kuntze, *Satureja
officinarum* Crantz, *Thymus cunila*
Krause
♦ summer savory, kesäkynteli,
santoreggia
♦ Weed, Naturalised, Cultivation
Escape
♦ 42, 88, 94, 101
♦ aH, cultivated, herbal.

Satureja montana L.
Lamiaceae
Clinopodium montanum Kuntze,
Micromeria montana (L.) Reichen.,
Micromeria pygmaea Reichen.,
Micromeria variegata Reichen.
♦ winter savory, savory, cerea
♦ Weed, Naturalised
♦ 101, 272
♦ cultivated, herbal.

Satureja pseudosimensis Brenan
Lamiaceae
♦ Weed, Quarantine Weed
♦ 76, 87, 88, 203, 220

Satureja vulgare (L.) Fritsch
Lamiaceae
♦ wild basil
♦ Weed
♦ 87, 88, 218

Sauropus androgynus (L.) Merr.
Euphorbiaceae
♦ chekkurmanis
♦ Weed
♦ 87, 88
♦ herbal.

Saururus cernuus L.
Saururaceae
♦ lizardtail, water dragon, swamplily,
American swamplily
♦ Weed, Quarantine Weed
♦ 39, 87, 88, 161, 195, 218, 258
♦ wpH, cultivated, herbal, toxic.

Saururus chinensis (Lour.) Baill.
Saururaceae
♦ Weed
♦ 87, 88, 286
♦ wpH, cultivated, herbal.

Saussurea amara (L.) DC.
Asteraceae
Saussurea glomerata Poir. (see)
♦ saussurea
♦ Weed
♦ 114, 297

Saussurea candicans C.B.Clarke
Asteraceae
♦ Weed
♦ 87, 88

Saussurea glomerata Poir.
Asteraceae
= *Saussurea amara* (L.) DC.
♦ Weed
♦ 275
♦ pH.

Saussurea iodostegia Hance.
Asteraceae
♦ purplebract saussurea
♦ Weed
♦ 297

Saussurea japonica (Thunb.) DC.
Asteraceae
♦ Weed, Quarantine Weed
♦ 220, 275, 297
♦ pH.

Saussurea salsa (Pall.) Spreng
Asteraceae
♦ Weed
♦ 275
♦ pH.

**Saussurea schaginiana (Wydl.) Fisch. ex
Herd.**
Asteraceae
♦ Weed
♦ 114, 243

Saussurea thoroldii Hemsl.
Asteraceae
♦ thorold saussurea

♦ Weed
♦ 297

Sauvagesia brownei Planch.
Ochnaceae
♦ Weed
♦ 87, 88

Sauvagesia erecta L.
Ochnaceae
♦ Creole tea
♦ Weed, Naturalised
♦ 87, 88, 101, 153
♦ H, herbal.

Savignya parviflora (Delile) Webb
Brassicaceae
Farsetia parviflora (Del.) Spreng.,
Lunaria parviflora Del., *Savignya
aegyptiaca* DC.
♦ Weed
♦ 221
♦ arid.

Saxifraga L. spp.
Saxifragaceae
♦ saxifrage
♦ Weed
♦ 23, 88, 272

Saxifraga cespitosa L.
Saxifragaceae
♦ mätäsrikko, tufted alpine saxifrage,
tufted saxifrage
♦ Cultivation Escape
♦ 42
♦ pH, cultivated, herbal.

Saxifraga × geum L.
Saxifragaceae
= *Saxifraga hirsuta* L. × *Saxifraga
umbrosa* L. [intermediate between
parents, see *Saxifraga × urbium*
D.A.Webb.]
♦ scarce London pride
♦ Naturalised
♦ 40
♦ cultivated.

Saxifraga granulata L.
Saxifragaceae
Evaiezoa granulata Raf.
♦ meadow saxifrage, papelorikko
♦ Weed, Naturalised
♦ 23, 40, 88
♦ cultivated, herbal.

Saxifraga hostii Tausch
Saxifragaceae
Saxifraga besleri Sternb., *Saxifraga elatior*
Mert. & Koch
♦ isorikko
♦ Cultivation Escape
♦ 42
♦ cultivated.

Saxifraga hypnoides L.
Saxifragaceae
♦ sammalrikko, mossy saxifrage,
dovedale moss, skalnica
♦ Cultivation Escape
♦ 42
♦ cultivated.

Saxifraga oppositifolia L.
Saxifragaceae
Antiphylla coerulea Haw., *Antiphylla*

oppositifolia Fourr., *Evaiezoa oppositifolia* Raf., *Saxifraga coerulea* Pers.
- purple mountain saxifrage, purple saxifrage, sinirikko
- Weed
- 23, 88
- cultivated, herbal.

Saxifraga rhomboidea Greene
Saxifragaceae
- diamondleaf saxifrage
- Weed
- 23, 88
- herbal.

Saxifraga rosacea Moench
Saxifragaceae
- mattorikko, Irish saxifrage
- Cultivation Escape
- 42
- cultivated.

Saxifraga sibthorpii Boiss.
Saxifragaceae
- yellow saxifrage
- Naturalised
- 39, 101
- toxic.

Saxifraga stolonifera Meerb.
Saxifragaceae
Saxifraga sarmentosa L.
- creeping saxifrage, strawberry geranium, mother of thousands, strawberry saxifrage
- Weed, Naturalised, Cultivation Escape
- 34, 40, 101, 280, 297
- pH, cultivated, herbal.

Saxifraga tridactylites L.
Saxifragaceae
Saxifraga annua Lapeyr., *Saxifraga trifida* Gilib., *Tridactylites annua* Haw.
- rue leaved saxifrage, mäkirikko
- Weed
- 88, 94, 253, 272
- cultivated, herbal.

Saxifraga umbrosa L.
Saxifragaceae
Geum umbrosum Moench, *Hydatica umbrosa* Raf., *Robertsonia umbrosa* Haw.
- varjorikko, London pride, Pyrenees saxifrage
- Naturalised, Cultivation Escape
- 42, 241, 300
- cultivated.

Saxifraga × *urbium* D.A.Webb.
Saxifragaceae
= *Saxifraga umbrosa* L. × *Saxifraga hirsuta* L. [more like *Saxifraga umbrosa*, see *Saxifraga* × *geum* L.]
- London pride
- Naturalised
- 40
- cultivated.

Scabiosa L. spp.
Dipsacaceae
- scabious, pincushions
- Weed, Naturalised
- 198, 272

Scabiosa anthemifolia Eckl. & Zeyh.
Dipsacaceae

- Casual Alien
- 280

Scabiosa arvensis L.
Dipsacaceae
= *Knautia arvensis* (L.) Coult.
- Weed
- 221
- herbal.

Scabiosa atropurpurea L.
Dipsacaceae
Scabiosa maritima L. (see)
- pincushion, mourning bride, pincushion flower, sweet scabious, mournful widow
- Weed, Naturalised, Garden Escape, Environmental Weed, Cultivation Escape
- 7, 34, 40, 86, 87, 88, 98, 101, 176, 198, 203, 237, 241, 252, 269, 280, 295, 300
- aH, cultivated, herbal. Origin: Mediterranean.

Scabiosa canescens Waldst. & Kit.
Dipsacaceae
Asterocephalus suaveolens Spreng., *Scabiosa suaveolens* Desf. (see)
- Weed
- 272
- cultivated.

Scabiosa caucasica Bieb.
Dipsacaceae
- kaukasiantörmäkukka, Caucasian pincushion flower, driakiew
- Cultivation Escape, Casual Alien
- 42, 280
- cultivated.

Scabiosa columbaria L.
Dipsacaceae
- small scabious, dove pincushions, ketotörmäkukka, lesser scabious
- Weed, Naturalised, Casual Alien
- 42, 44, 70, 101, 272, 287
- b/pH, cultivated, herbal.

Scabiosa maritima L.
Dipsacaceae
= *Scabiosa atropurpurea* L.
- Weed
- 87, 88
- herbal.

Scabiosa ochroleuca L.
Dipsacaceae
Asterocephalus ochroleucus Wall., *Scabiosa columbaria* L. var. *ochroleuca* Coult.
- keltatörmäkukka, cream pincushions
- Weed, Naturalised, Casual Alien
- 42, 101, 272
- cultivated.

Scabiosa olivieri Coult.
Dipsacaceae
- Weed
- 243

Scabiosa palaestina L.
Dipsacaceae
Lomelosia palaestina (L.) Raf., *Tremastelma palaestinum* (L.) Janch. (see)
- Balkan pincushions
- Weed, Naturalised
- 87, 88, 101

Scabiosa prolifera L.
Dipsacaceae
- carmel daisy
- Weed
- 87, 88
- aH, cultivated.

Scabiosa rotata Bieb.
Dipsacaceae
- Weed
- 87, 88

Scabiosa semipapposa Salzm. ex DC.
Dipsacaceae
- Weed
- 87, 88

Scabiosa stellata L.
Dipsacaceae
- starflower pincushions, scabious drumstick, pincushion flower, moonflower
- Weed, Naturalised, Cultivation Escape
- 34, 87, 88, 101
- aH, cultivated, herbal.

Scabiosa suaveolens Desf.
Dipsacaceae
= *Scabiosa canescens* Waldst. & Kit.
- Weed
- 23, 88

Scabiosa tschiliensis Grun.
Dipsacaceae
- north China scabious
- Weed
- 297

Scabiosa tschiliensis Grun. var. *superba* (Grun.) S.Y.He
Dipsacaceae
- largeflower scabious
- Weed
- 297

Scabiosa veronica L.
Dipsacaceae
- Weed
- 87, 88

Scaevola plumieri (L.) Vahl
Goodeniaceae
- Florida scaveola, gullfeed
- Environmental Weed
- 152, 261
- cultivated, herbal.

Scaevola sericea Vahl
Goodeniaceae
= *Scaevola taccada* (Gaertn.) Roxb. (NoR) [see *Scaevola taccada* (Gaertn.) Roxb. var. *sericea* (Vahl) St.John]
- beach naupaka, scaevola, half flower, sea lettuce tree
- Weed, Garden Escape, Environmental Weed
- 6, 39, 88, 112, 142, 151, 179
- cultivated, herbal, toxic.

Scaevola taccada (Gaertn.) Roxb. var. *sericea* (Vahl) St.John
Goodeniaceae
- scaevola, half flower
- Weed, Cultivation Escape
- 80, 122
- cultivated.

Scandicium stellatum Thell.
Apiaceae
= *Scandix stellata* Banks & Sol.
♦ Weed
♦ 221

Scandix australis L.
Apiaceae
♦ southern shepherd's needle
♦ Weed
♦ 87, 88, 94, 272

Scandix iberica Bieb.
Apiaceae
♦ kaukasiansarjaputki
♦ Weed, Casual Alien
♦ 42, 87, 88

Scandix pecten-veneris L.
Apiaceae
Chaerophyllum pecten-veneris Crantz,
Myrrhis pecten-veneris All., *Pecten
veneris* Lam., *Pectinaria vulgaris* Bernh.
♦ shepherd's needle, Venus needle,
Venus comb, kampa sarjaputki, acicula
comune, pettine di venere
♦ Weed, Naturalised, Introduced,
Casual Alien
♦ 34, 42, 44, 70, 86, 87, 88, 94, 98, 101,
115, 161, 176, 198, 203, 218, 221, 228,
241, 243, 253, 272, 280, 287, 300
♦ aH, arid, cultivated, herbal. Origin:
Europe to western Himalayas.

Scandix stellata Banks & Sol.
Apiaceae
Scandicium stellatum Thell. (see)
♦ Weed
♦ 88, 94
♦ cultivated.

Schanginia aegyptiaca (Hasselq.) Aellen
Chenopodiaceae
♦ Weed
♦ 221

Schanginia hortensis Moq.
Chenopodiaceae
= *Suaeda hortensis* Forsk. ex Gmel.
♦ Weed
♦ 221

× Schedololium holmbergii (Dörfl.) Holub
Poaceae
= *Lolium* L. sp. × *Schedonorus* P.Beauv.
sp.
♦ Naturalised
♦ 280
♦ G.

Schedonnardus paniculatus (Nutt.) Trel.
Poaceae
Lepturus paniculatus Nutt.
♦ tumblegrass
♦ Weed, Native Weed
♦ 87, 88, 161, 174, 218, 243
♦ G, herbal. Origin: North America.

Schedonorus phoenix (Scop.) Holub
Poaceae
♦ Naturalised
♦ 280
♦ G.

Schefflera actinophylla (Endl.) Harms
Araliaceae
Brassaia actinophylla Endl. (see)

♦ schefflera, Australian umbrella tree,
Queensland umbrella tree, octopus
tree, Australian ivy palm, ivy palm
♦ Weed, Noxious Weed, Naturalised,
Garden Escape, Environmental Weed,
Cultivation Escape
♦ 3, 22, 73, 80, 86, 88, 101, 107, 112, 122,
142, 151, 179, 191, 201, 233, 259, 260
♦ T, cultivated, herbal. Origin: north-
east Australia, southern New Guinea.

Schefflera arboricola (Hayata) Merr.
Araliaceae
Heptapleurum arboriculum Hayata
♦ dwarf brassaia, dwarf schefflera,
umbrella tree
♦ Weed
♦ 3, 179, 191
♦ cultivated, herbal.

Schima superba Gardner & Champ.
Theaceae
♦ Quarantine Weed
♦ 220
♦ cultivated.

**Schimpera arabica Hochst. & Steud. ex
Boiss.**
Brassicaceae
♦ Weed
♦ 221

**Schinopsis quebracho-colorado (Schltdl.)
Barkley & Mey.**
Anacardiaceae
Schinopsis lorentzii (Griseb.) Engl.
♦ quebracho, quebracho colorado,
quebracho colorado santiagueòo, red
quebracho
♦ Introduced
♦ 228
♦ arid, cultivated.

Schinus areira L.
Anacardiaceae
= *Schinus molle* var. *areira* L. (L.) DC.
♦ peppertree, Californian peppertree,
peppercorn tree, pepperina, Brazilian
peppertree
♦ Weed, Naturalised, Garden Escape,
Environmental Weed
♦ 73, 86, 88, 289
♦ cultivated. Origin: northern South
America to Mexico.

Schinus longifolius (Lindl.) Speg.
Anacardiaceae
♦ longleaf peppertree
♦ Naturalised
♦ 101

Schinus molle L.
Anacardiaceae
♦ California peppertree, Peruvian
peppertree, peppertree, Peruvian
mastic tree, aguaribay, false
peppertree, mulli, peppercorn, pirul,
peppercorn tree, pepperina, pirul,
Chilean peppertree
♦ Weed, Naturalised, Introduced,
Environmental Weed, Cultivation
Escape
♦ 3, 7, 22, 39, 72, 80, 87, 88, 95, 98, 101,
116, 121, 155, 161, 198, 201, 203, 218,
228, 233, 257, 261, 277, 279, 280, 283,

296
♦ T, arid, cultivated, herbal, toxic.
Origin: South America.

Schinus molle L. var. areira (L.) DC.
Anacardiaceae
Schinus areira L. (see)
♦ peppertree, Californian peppertree,
aguaribay, mulli, peppercorn
♦ Weed, Introduced
♦ 93, 205, 228
♦ T, arid, cultivated. Origin: South
America.

Schinus polygamus (Cav.) Cabr.
Anacardiaceae
Amyris polygama Cav., *Duvaua cuneata*
Gill., *Duvaua dentata* DC., *Duvaua
dependens* Kunth var. *obovata* Arechav.,
Duvaua dependens var. *ovata* Arechav.,
Duvaua inebrians Gill. ex Hook. &
Arn., *Duvaua ornata* Phil., *Duvaua
ovata* Lindl., *Duvaua polygama* Kunth,
Duvaua praecox (Speg.) Griseb. var.
hyemalis Griseb., *Duvaua praecox* var.
montana Griseb., *Duvaua spinescens*
Ten., *Schinus bonplandianus* Marchand,
Schinus dentatus Andrews, *Schinus
dependens* Ortega, *Schinus dependens*
var. *brevifolia* Fenzl ex Engl., *Schinus
dependens* var. *longifolia* Fenzl. ex
Engl., *Schinus dependens* var. *ovatus*
Engl., *Schinus dependens* var. *ovatus*
(Lindl.) Marchand, *Schinus dependens*
var. *parviflorus* Marchand, *Schinus
dependens* var. *subintegra* Engl., *Schinus
huygan* Ruiz ex Engl., *Schinus huygan*
Kuntze, *Schinus polygamus* fo. *ovatus*
(Lindl.) Cabrera, *Schinus praecox* Speg.
♦ Hardee peppertree, huigen,
peppertree, Peruvian peppertree
♦ Naturalised, Environmental Weed
♦ 101, 182
♦ S, cultivated. Origin: Argentina,
Brazil, Chile, Peru, Uruguay.

Schinus terebinthifolius Raddi
Anacardiaceae
Schinus terebinthifolia Raddi
♦ Brazilian pepper, Christmas berry,
schinus, Florida holly, Brazilian
peppertree, wilelaiki, nani o hilo,
Christmas berrytree, South American
pepper, Brasiliaanse peperboom, faux
poivrier, warui
♦ Weed, Sleeper Weed, Quarantine
Weed, Noxious Weed, Naturalised,
Introduced, Garden Escape,
Environmental Weed, Cultivation
Escape, Casual Alien
♦ 3, 6, 7, 10, 18, 22, 25, 27, 35, 37, 39, 47,
63, 78, 80, 84, 86, 87, 88, 95, 98, 101, 112,
116, 121, 122, 134, 142, 151, 152, 155,
161, 179, 201, 203, 218, 220, 225, 228,
229, 231, 233, 246, 261, 277, 280, 283,
287, 290
♦ T, aqua, cultivated, herbal, toxic.
Origin: South America.

**Schinus terebinthifolius Raddi var.
raddianus Engl.**
Anacardiaceae
♦ peppertree, Brazilian peppertree

♦ Noxious Weed, Naturalised
♦ 101, 229
♦ aqua.

Schismus arabicus Nees
Poaceae
Schismus spectabilis Fig. & De Not.,
Schismus marginatus Thomson,
Danthonia submutica Clarke
♦ Arabian schismus, Mediterranean
grass, schismus, Arabian grass, abu
mashi, split grass, araby grass
♦ Weed, Noxious Weed, Naturalised,
Environmental Weed
♦ 7, 35, 78, 86, 88, 98, 101, 116, 161, 203,
231, 241, 300
♦ aG, herbal. Origin: Mediterranean to
China.

Schismus barbatus (L.) Thell.
Poaceae
Festuca barbata L., *Festuca calycina*
Loefl., *Schismus calycinus* (Loefl.)
K.Koch, *Schismus marginatus* P.Beauv.
♦ Mediterranean grass, Arabian
schismus, schismus, split grass,
common Mediterranean grass, Arabian
grass
♦ Weed, Noxious Weed, Naturalised,
Introduced, Environmental Weed
♦ 7, 34, 35, 72, 78, 80, 86, 87, 88, 93, 98,
101, 116, 161, 198, 203, 205, 218, 221,
228, 231, 241, 300
♦ aG, arid, cultivated, herbal. Origin:
Mediterranean to India.

Schizachyrium brevifolium (Sw.) Nees ex Büse
Poaceae
Andropogon brevifolius Sw. (see)
♦ serillo dulce
♦ Weed, Quarantine Weed
♦ 76, 87, 88, 203, 220, 286
♦ G.

Schizachyrium condensatum (Kunth) Nees
Poaceae
Andropogon condensatus Kunth (see)
♦ Colombian bluestem, bush
beardgrass, little bluestem, rabo de
burro
♦ Weed, Naturalised
♦ 3, 80, 90, 101, 191, 255
♦ pG. Origin: South America.

Schizachyrium microstachyum (Desv.) Roseng., Arril. & Izag.
Poaceae
♦ hierba colorada
♦ Weed
♦ 295
♦ G.

Schizachyrium paniculatum (Kunth) Herter
Poaceae
♦ Weed, Quarantine Weed
♦ 76, 87, 88, 203, 220
♦ G.

Schizachyrium sanguineum (Retz.) Alston
Poaceae
Andropogon domingensis (Schult.)
Hubb., *Andropogon hirtiflorus* (Nees)

Kunth, *Andropogon leptostachys*
Benth., *Andropogon sanguineus* (Retz.)
Merr., *Andropogon scabriflorus* Rupr.
ex Hack., *Andropogon semiberbis*
(Nees) Kunth, *Rottboellia sanguinea*
Retz., *Schizachyrium domingense*
(Schult.) Nash, *Schizachyrium griseum*
Stapf, *Schizachyrium hirtiflorum*
Nees, *Schizachyrium inspersum*
Pilg., *Schizachyrium linoliense* Pilg.,
Schizachyrium scabriflorum (Rupr.)
Camus, *Schizachyrium semiberbe* Nees,
Schizachyrium tenuispicatum Pilg.,
Streptachne domingensis Schult.
♦ crimson bluestem
♦ Naturalised, Introduced
♦ 228, 241, 300
♦ G, arid.

Schizachyrium scoparium (Michx.) Nash
Poaceae
Andropogon scoparius Michx. (see)
♦ little bluestem, bluestem grass,
prairie beardgrass
♦ Weed, Naturalised, Introduced
♦ 87, 88, 90, 228, 241, 300
♦ pG, arid, cultivated, herbal.

Schizanthus pinnatus Ruiz & Pav.
Solanaceae
♦ siro perhoskukka, poor man's
orchid, butterfly flower
♦ Naturalised, Cultivation Escape
♦ 42, 101
♦ aH, cultivated, herbal. Origin: Chile.

Schizanthus × *wisetonensis* hort.
Solanaceae
= *Schizanthus pinnatus* Ruiz & Pav. ×
Schizanthus grahamii Gillies
♦ kirjoperhoskukka, butterfly flower
♦ Cultivation Escape
♦ 42
♦ cultivated.

Schizolobium parahybum (Vell.) Blake
Fabaceae/Caesalpiniaceae
♦ Brazilian firetree
♦ Naturalised, Cultivation Escape
♦ 101, 261
♦ cultivated. Origin: Central and
South America.

Schizonepeta multifida (L.) Briq.
Lamiaceae
♦ common schizonepeta
♦ Weed
♦ 297
♦ cultivated.

Schizostachyum glaucifolium (Rupr.) Munro
Poaceae
Bambusa glaucifolia Rupr.
♦ Polynesian 'ohe, ofe
♦ Naturalised
♦ 101
♦ G, herbal.

Schizostylis coccinea Backh. & Harv.
Iridaceae
♦ crimson flag, kaffir lily
♦ Naturalised, Environmental Weed,
Cultivation Escape

♦ 86, 176, 246, 252, 280
♦ cultivated. Origin: South Africa.

Schkuhria isopappa Benth.
Asteraceae
= *Schkuhria pinnata* (Lam.) Kuntze ex
Thell.
♦ Weed
♦ 87, 88

Schkuhria pinnata (Lam.) Kuntze ex Thell.
Asteraceae
Schkuhria bonariensis Hook & Arn.,
Schkuhria isopappa Benth. (see)
♦ dwarf marigold, schkuhria, starry
skies, mata pulgas, pinnate false
threadleaf, dwarf Mexican marigold,
khakibush, yellow tumbleweed
♦ Weed, Quarantine Weed,
Naturalised, Introduced
♦ 50, 51, 53, 76, 87, 88, 98, 101, 121, 158,
203, 228, 236, 240, 243, 255, 270
♦ aH, arid, cultivated, herbal. Origin:
South America.

Schkuhria pinnata (Lam.) Kuntze ex Thell. var. *abrotanoides* (Roth.) Cabr.
Asteraceae
♦ schkuhria, dwarf marigold
♦ Weed, Naturalised
♦ 86, 198, 237, 269, 287, 295
♦ aH. Origin: South America.

Schkuhria pinnata (Lam.) Kuntze ex Thell. var. *pinnata*
Asteraceae
♦ pinnate false threadleaf
♦ Naturalised
♦ 101

Schkuhria virgata DC.
Asteraceae
= *Schkuhria pinnata* (Lam.) Kuntze ex
Thell. var. *virgata* (La Llave) Heiser
(NoR)
♦ Weed
♦ 87, 88
♦ herbal.

Schlumbergera truncata (Haw.) Moran
Cactaceae
Epiphyllum bridgesii Lem., *Zygocactus
truncatus* (Haw.) K.Schum.
♦ false Christmas cactus, lobster cactus
♦ Naturalised
♦ 101
♦ cultivated.

Schmidtia pappophoroides Steud. ex J.A.Schmidt
Poaceae
Schmidtia bulbosa Stapf
♦ sand quickgrass
♦ Native Weed
♦ 121
♦ pG, arid. Origin: southern Africa.

Schoenefeldia gracilis Kunth
Poaceae
Chloris myosuroides Hook.f., *Chloris
pallida* (Edgew.) Hook.f., *Schoenefeldia
pallida* Edgew.
♦ Weed
♦ 88, 242
♦ G, arid.

Schoenoplectus (Rchb.) Pall. spp.
Cyperaceae
♦ bulrush
♦ Weed, Quarantine Weed, Naturalised
♦ 76, 86, 88, 220
♦ G, cultivated.

Schoenoplectus californicus (C.A.Mey.) Soják
Cyperaceae
Scirpus californicus (C.A.Mey) Steud. (see)
♦ California bulrush
♦ Weed, Sleeper Weed, Naturalised, Environmental Weed
♦ 54, 86, 88, 155, 237, 246, 280
♦ G. Origin: Americas.

Schoenoplectus erectus (Poir.) Palla ex J.Raynal
Cyperaceae
♦ soft club sedge, sharpscale bulrush
♦ Weed, Naturalised
♦ 86, 98, 198, 203
♦ G, cultivated. Origin: Africa.

Schoenoplectus glaucus (Lam.) Kartesz comb. nov. ined.
Cyperaceae
♦ tuberous bulrush
♦ Naturalised
♦ 101
♦ G.

Schoenoplectus juncoides (Roxb.) Pall.
Cyperaceae
Scirpus juncoides Roxb. (see)
♦ rock bulrush
♦ Weed, Quarantine Weed
♦ 88, 135, 191
♦ G. Origin: Madagascar.

Schoenoplectus lacustris (L.) Palla
Cyperaceae
Schoenoplectus lacuster (L.) Pall., *Scirpus altissimus* Gilib., *Scirpus lacustris* L. (see), *Scirpus validus* Vahl
♦ lakeshore bulrush, clubrush, common clubrush, bulrush
♦ Naturalised
♦ 101
♦ G, aqua, cultivated, herbal.

Schoenoplectus leucanthus (Boeck.) Raynal
Cyperaceae
Isolepis supina (L.) R.Br. var. *tenuis* Nees, *Scirpus leucanthus* Boeck., *Scirpus supinus* L. var. *leucosperma* C.B.Clarke, *Scirpus thunbergii* A.Spreng.
♦ Native Weed
♦ 121
♦ a/pG. Origin: southern Africa.

Schoenoplectus lineolatus (Franch. & Sav.) Koyama
Cyperaceae
♦ linear club sedge
♦ Weed, Naturalised
♦ 86, 98, 198, 203
♦ G. Origin: Japan, Taiwan.

Schoenoplectus littoralis (Schrad.) Pall.
Cyperaceae
= *Schoenoplectus subulatus* Vahl (NoR)

[see *Scirpus litoralis* Schrad.]
♦ sedge
♦ Native Weed
♦ 121
♦ pG.

Schoenoplectus mucronatus (L.) Pall.
Cyperaceae
Scirpus glomeratus Scop., *Scirpus mucronatus* L. (see), *Scirpus preslii* A.Dietr., *Scirpus sundanus* Miq.
♦ bog bulrush
♦ Weed, Naturalised
♦ 101, 253
♦ pG, aqua, cultivated, herbal.

Schoenoplectus triqueter (L.) Pall.
Cyperaceae
Heleogiton triquetrum L.Reichen., *Scirpus trigonus* Roth., *Scirpus triqueter* L. (see)
♦ streambank bulrush, triangular clubrush
♦ Naturalised
♦ 101
♦ G, herbal.

Schoenoxiphium lehmannii (Nees) Steud.
Cyperaceae
♦ Native Weed
♦ 121
♦ pG. Origin: southern Africa.

Schoenus apogon Roem. & Schult.
Cyperaceae
♦ smooth bogrush
♦ Weed, Naturalised
♦ 39, 87, 88, 101
♦ G, cultivated, toxic. Origin: Australia.

Schoenus nigricans L.
Cyperaceae
Chaetospora nigricans Kunth
♦ bogrush, black bogrush, black sedge
♦ Weed
♦ 272
♦ pG, cultivated, herbal.

Schotia afra (L.) Thunb.
Fabaceae/Caesalpiniaceae
Schotia angustifolia E.Mey., *Schotia parvifolia* Jacq., *Schotia speciosa* Jacq., *Schotia tamarindifolia* Afzel. ex Sims
♦ karoo Boer bean
♦ Introduced
♦ 228
♦ arid, cultivated.

Schotia brachypetala Sond.
Fabaceae/Caesalpiniaceae
Schotia latifolia sensu Dale, *Schotia rogersii* Burtt Davy, *Schotia semireducta* Merxm.
♦ boerboon, fuchsia tree, huilboerboon, molope, mutanswa, uvovovo, weeping boerbean, weeping schotia, tree fuchsia
♦ Naturalised, Introduced
♦ 86, 228
♦ arid, cultivated, herbal. Origin: Africa.

Schotia latifolia Jacq.
Fabaceae/Caesalpiniaceae
Schotia cuneifolia Gand., *Schotia*

diversifolia Walp.
♦ tree fuchsia
♦ Introduced
♦ 228
♦ arid, cultivated.

Schouwia DC. spp.
Brassicaceae
♦ Weed
♦ 88

Schouwia thebaica Webb
Brassicaceae
♦ Weed
♦ 88, 221

Schrankia leptocarpa DC.
Fabaceae/Mimosaceae
= *Mimosa quadrivalvis* L. var. *leptocarpa* (DC.) Barneby (NoR)
♦ Weed, Quarantine Weed
♦ 76, 87, 88, 203, 220

Schrankia microphylla (Dryand.) Macbr.
Fabaceae/Mimosaceae
= *Mimosa microphylla* Dry. (NoR)
♦ littleleaf sensitive briar
♦ Weed
♦ 87, 88, 218
♦ herbal.

Schrankia nuttallii (DC.) Standl.
Fabaceae/Mimosaceae
= *Mimosa nuttallii* (DC.) B.L.Turner (NoR)
♦ cat's claw sensitive briar
♦ Weed
♦ 87, 88, 218

Schultesia guyanensis Malme
Gentianaceae
♦ Weed
♦ 87, 88

Schwenkia americana L.
Solanaceae
♦ Weed, Quarantine Weed
♦ 76, 87, 88, 203, 220
♦ herbal.

Scilla autumnalis L.
Liliaceae/Hyacinthaceae
Anthericum autumnale Scop., *Ornithogalum autumnale* Lam.
♦ autumn squill
♦ Weed
♦ 39, 272
♦ cultivated, herbal, toxic.

Scilla bifolia L.
Liliaceae/Hyacinthaceae
Anthericum bifolium Scop., *Ornithogalum bifolium* Neck., *Rinopodium bifolium* Salisb., *Scilla dubia* Koch, *Scilla silvatica* Czetz
♦ alpine squill
♦ Weed
♦ 39, 272
♦ cultivated, herbal, toxic.

Scilla hyacinthoides L.
Liliaceae/Hyacinthaceae
♦ Weed, Naturalised
♦ 86, 98, 203
♦ cultivated.

Scilla maritima L.
Liliaceae/Hyacinthaceae

= *Urginea maritima* (L.) Bak.
♦ Weed
♦ 87, 88
♦ herbal.

Scilla natalensis Planch.
Liliaceae/Hyacinthaceae
♦ blue hyacinth, blue squill, wild squill
♦ Weed, Native Weed
♦ 39, 121
♦ pH, cultivated, herbal, toxic. Origin: southern Africa.

Scilla nervosa (Burch.) Jessop
Liliaceae/Hyacinthaceae
Scilla rigidifolia Kunth
♦ sand lily, scilla, wild squill
♦ Native Weed
♦ 121
♦ pH, cultivated, toxic. Origin: southern Africa.

Scilla non-scripta (L.) Hoffmanns. & Link
Liliaceae/Hyacinthaceae
= *Hyacinthoides non-scripta* (L.)
Chouard ex Rothm.
♦ bluebell
♦ Weed
♦ 15
♦ cultivated, herbal.

Scilla peruviana L.
Liliaceae/Hyacinthaceae
♦ Cuban lily, Peruvian lily, Peruvian jacinth, Peruvian squill, squill
♦ Weed, Naturalised, Garden Escape, Environmental Weed, Casual Alien
♦ 15, 39, 40, 86, 98, 161, 176, 198, 203, 280
♦ cultivated, herbal, toxic. Origin: Mediterranean.

Scilla scilloides (Lindl.) Druce
Liliaceae/Hyacinthaceae
Scilla thunbergii Miyabe & Kudo
♦ tsurubo
♦ Weed
♦ 87, 88, 286, 297
♦ pH, cultivated. Origin: east Asia.

Scilla siberica Haw.
Liliaceae/Hyacinthaceae
Orthocallis siberica (Haw.) Speta
♦ idänsinililja, Siberian squill
♦ Weed, Naturalised, Cultivation Escape
♦ 39, 42, 80, 101
♦ cultivated, herbal, toxic.

Scindapsus Schott. spp.
Araceae
♦ devil's ivy
♦ Weed
♦ 161
♦ toxic.

Scindapsus aureus (Lindl. & André) Engl.
Araceae
= *Epipremnum pinnatum* (L.) Engl.
♦ pothos, devil's ivy
♦ Naturalised
♦ 287
♦ cultivated, toxic.

Scirpoides holoschoenus (L.) Soják
Cyperaceae

Scirpus holoschoenus L. (see)
♦ pallokaisla, roundhead bulrush, round headed clubrush
♦ Naturalised, Casual Alien
♦ 42, 101
♦ G, cultivated.

Scirpus L. spp.
Cyperaceae
♦ bulrush
♦ Weed, Quarantine Weed
♦ 88, 191, 204, 221, 258, 272
♦ G.

Scirpus acutus Muhl. ex Bigelow
Cyperaceae
= *Schoenoplectus acutus* (Muhl. ex Bigelow) Á. & D.Löve (NoR)
♦ hardstem bulrush, bull tule, bulrush, tule, viscid bulrush, common tule
♦ Weed
♦ 23, 87, 88, 126, 159, 161, 180, 218
♦ pG, aqua, cultivated, herbal.

Scirpus americanus Pers.
Cyperaceae
= *Schoenoplectus americanus* (Pers.) Volk. ex Schinz & R.Keller (NoR) [see *Scirpus olneyi* Gray ex Engelm. & Gray]
♦ American bulrush, Jersey clubrush, American tule, triangular stemmed sedge, common threesquare
♦ Weed
♦ 23, 39, 87, 88, 161, 218
♦ pG, aqua, cultivated, herbal, toxic.

Scirpus articulatus L.
Cyperaceae
♦ bulrush, song kratiam huawaen
♦ Weed
♦ 87, 88, 170, 191, 204, 239
♦ G, herbal.

Scirpus atrovirens Willd.
Cyperaceae
♦ green bulrush, black bulrush
♦ Weed
♦ 23, 39, 87, 88, 161, 218
♦ G, cultivated, herbal, toxic.

Scirpus australis Murr.
Cyperaceae
= *Scirpoides holoschoenus* (L.) Soják ssp. *australis* (Murray) Soják (NoR)
♦ Weed
♦ 87, 88
♦ G.

Scirpus californicus (C.A.Mey.) Steud.
Cyperaceae
= *Schoenoplectus californicus* (C.A.Mey.) Soják
♦ California bulrush, California tule, giant bulrush, junco
♦ Weed
♦ 87, 88, 218, 236, 295
♦ pG, aqua, cultivated, herbal.

Scirpus cernuus Vahl
Cyperaceae
= *Isolepis cernua* (Vahl) Roem. & Schult.
♦ annual tule, low bulrush, nodding scirpus, slender clubrush
♦ Naturalised
♦ 300
♦ aG, aqua, cultivated, herbal.

Scirpus cespitosus L.
Cyperaceae
= *Trichophorum cespitosum* (L.) Hartm. (NoR)
♦ deergrass
♦ Weed
♦ 23, 88
♦ G, herbal.

Scirpus cubensis Poepp. & Kunth
Cyperaceae
= *Oxycaryum cubense* (Poepp. & Kunth) Lye (NoR)
♦ Cuban bulrush
♦ Weed, Quarantine Weed
♦ 76, 87, 88, 203, 220
♦ G.

Scirpus cyperinus (L.) Kunth
Cyperaceae
Scirpus rubricosus Fern. (see)
♦ woolgrass bulrush, wool grass
♦ Weed
♦ 23, 87, 88, 218
♦ pG, aqua, cultivated, herbal.

Scirpus erectus Poir.
Cyperaceae
= *Schoenoplectus erectus* (Poir.) Palla ex Raynal ssp. *raynalii* (Schuyler) Lye (NoR)
♦ sharpscale bulrush
♦ Weed
♦ 87, 88
♦ G, herbal.

Scirpus fluitans L.
Cyperaceae
= *Isolepis fluitans* (L.) R.Br.
♦ floating clubrush, floating mudrush
♦ Weed
♦ 70
♦ G, herbal.

Scirpus fluviatilis (Torr.) Gray
Cyperaceae
= *Schoenoplectus fluviatilis* (Torr.) Strong (NoR) [see *Bolboschoenus fluviatilis* (Torr.) Sojak]
♦ river bulrush
♦ Weed
♦ 23, 87, 88, 161, 180, 218, 263
♦ wpG, promoted, herbal.

Scirpus georgianus R.M.Harper
Cyperaceae
♦ Georgia bulrush
♦ Naturalised
♦ 280
♦ G, herbal.

Scirpus grossus L.f.
Cyperaceae
♦ kok saamliam
♦ Weed
♦ 87, 88, 170, 191, 204, 209, 239, 262
♦ pG, aqua, herbal, toxic.

Scirpus holoschoenus L.
Cyperaceae
= *Scirpoides holoschoenus* (L.) Soják
♦ cluster headed clubrush, clustered rush, round headed clubrush
♦ Weed
♦ 87, 88, 126, 221, 272
♦ pG, arid, cultivated.

Scirpus hotarui auct. non Ohwi
Cyperaceae
= *Schoenoplectus purshianus* (Fern.)
Strong (NoR)
♦ Weed
♦ 88, 191, 204
♦ G.

Scirpus jacobii C.E.C.Fisch.
Cyperaceae
Scirpus praelongatus auct.
♦ Weed
♦ 88
♦ G.

Scirpus juncoides Roxb.
Cyperaceae
= *Schoenoplectus juncoides* (Roxb.) Pall.
♦ phrong klom yai, chuk nu
♦ Weed, Quarantine Weed
♦ 76, 87, 88, 126, 170, 191, 203, 204, 220,
238, 239, 274, 275, 286, 297
♦ pG, aqua.

**Scirpus juncoides Roxb. ssp. ohwianus
T.Koyama**
Cyperaceae
♦ Weed
♦ 263
♦ pG.

**Scirpus juncoides Roxb. var. ohwianus
T.Koyama**
Cyperaceae
♦ Weed
♦ 286
♦ G.

Scirpus komarovii Roshev.
Cyperaceae
♦ Weed
♦ 286
♦ G.

Scirpus lacustris L.
Cyperaceae
= *Schoenoplectus lacustris* (L.) Palla
♦ common clubrush, bulrush
♦ Weed
♦ 70, 87, 88, 272
♦ pG, aqua, cultivated, herbal.

Scirpus lateriflorus Gmel.
Cyperaceae
Isolepis oryzetorum Steud.,
Schoenoplectus lateriflorus (J.F.Gmel.)
Lye, *Schoenoplectus oryzetorum* (Steud.)
Ohwi, *Schoenoplectus oryzetorum*
(Steud.) V.I.Krecz., *Schoenoplectus
smithii* (Gray) Soják ssp. *lateriflorus*
(J.F.Gmel.) Soják, *Scirpus oryzetorum*
(Steud.) Ohwi, *Scirpus supinus* L. var.
lateriflorus (J.F.Gmel.) T.Koyama
♦ Weed
♦ 87, 88, 170, 191
♦ G.

Scirpus lineatus Michx.
Cyperaceae
♦ drooping bulrush
♦ Weed
♦ 23, 88
♦ G, herbal.

Scirpus lineolatus Franch. & Sav.
Cyperaceae

♦ himehotarui
♦ Weed
♦ 263, 274, 286
♦ pG.

Scirpus litoralis Schrad.
Cyperaceae
= *Schoenoplectus subulatus* Vahl (NoR)
[see *Scheonoplectus littoralis* (Schrad.)
Pall.]
♦ bulrush, coast clubrush
♦ Weed
♦ 87, 88, 126, 185, 262, 272
♦ pG, aqua.

Scirpus maritimus L.
Cyperaceae
= *Bolboschoenus maritimus* (L.) Palla
♦ sea clubrush, saltmarsh bulrush,
puruagrass, prairie rush
♦ Weed, Quarantine Weed,
Naturalised
♦ 23, 30, 70, 86, 87, 88, 126, 135, 191,
204, 217, 263, 272, 300
♦ pG, aqua, cultivated, herbal.

**Scirpus maritimus L. var. paludosus
(Nelson) Kük.**
Cyperaceae
= *Bolboschoenus maritimus* (L.) Palla
♦ Introduced
♦ 228
♦ G, arid.

Scirpus michelianus L.
Cyperaceae
= *Cyperus michelianus* (L.) Delile
♦ Weed
♦ 275
♦ G.

Scirpus microcarpus Presl
Cyperaceae
♦ panicled bulrush
♦ Naturalised
♦ 287
♦ pG, aqua, cultivated, herbal. Origin:
western North America.

Scirpus mucronatus L.
Cyperaceae
= *Schoenoplectus mucronatus* (L.) Pall.
♦ roughseed bulrush, ricefield
bulrush, bog bulrush, roughseed
clubrush, sea scripus
♦ Weed, Naturalised
♦ 34, 68, 70, 87, 88, 126, 161, 170, 180,
217, 218, 263, 272, 300
♦ wpG, cultivated. Origin: Eurasia.

Scirpus nevadensis S.Wats.
Cyperaceae
♦ bulrush, great basin bulrush,
Nevada bulrush
♦ Weed
♦ 159
♦ pG, aqua, promoted, herbal.

Scirpus nipponicus Makino
Cyperaceae
♦ Weed
♦ 263, 286
♦ pG.

Scirpus olneyi Gray ex Engelm. & Gray
Cyperaceae

= *Schoenoplectus americanus* (Pers.)
Volk. ex Schinz & R.Keller (NoR) [see
Scirpus americanus Pers.]
♦ bulrush
♦ Weed
♦ 159
♦ G, herbal.

Scirpus paludosus A.Nels.
Cyperaceae
= *Bolboschoenus maritimus* (L.) Palla
♦ bayonet grass, alkali bulrush
♦ Weed
♦ 23, 88
♦ pG, aqua, cultivated, herbal.

Scirpus palustris L.
Cyperaceae
= *Eleocharis palustris* (L.) Roem. &
Schult.
♦ Weed
♦ 87, 88
♦ G, herbal.

Scirpus pedicellatus Fern.
Cyperaceae
♦ stalked bulrush
♦ Weed
♦ 23, 88
♦ G, herbal.

Scirpus pendulus Muhl.
Cyperaceae
♦ hanging club sedge, rufous bulrush
♦ Weed, Naturalised
♦ 54, 86, 88, 98, 198, 203
♦ pG, cultivated, herbal. Origin: North
America.

Scirpus planiculmis Fr.Schmidt
Cyperaceae
♦ iseukiyagara
♦ Weed
♦ 88, 204, 275, 286, 297
♦ pG, aqua.

Scirpus polystachyus F.Muell.
Cyperaceae
♦ Naturalised
♦ 280
♦ G, cultivated.

Scirpus praelongatus Poir.
Cyperaceae
♦ Weed
♦ 87, 88
♦ G.

Scirpus radicans Schk.
Cyperaceae
Nemocharis radicans Beurl.
♦ juurtokaisla
♦ Weed
♦ 272
♦ G, herbal.

Scirpus robustus Pursh
Cyperaceae
= *Bolboschoenus robustus* (Pursh) Soják
(NoR)
♦ bull tule, alkali bulrush, big bulrush
♦ Weed
♦ 23, 88
♦ pG, aqua, cultivated, herbal.

Scirpus rubricosus Fern.
Cyperaceae

= *Scirpus cyperinus* (L.) Kunth
♦ Weed
♦ 23, 88
♦ G, herbal.

Scirpus setaceus L.
Cyperaceae
= *Isolepis setacea* (L.) R.Br
♦ bristle clubrush
♦ Weed, Introduced
♦ 34, 272
♦ G.

Scirpus smithii A.Gray
Cyperaceae
= *Schoenoplectus smithii* (Gray) Soják
(NoR)
♦ Weed
♦ 23, 87, 88
♦ G, herbal.

Scirpus subterminalis Torr.
Cyperaceae
= *Schoenoplectus subterminalis* (Torr.)
Soják (NoR)
♦ water bulrush
♦ Weed
♦ 23, 88
♦ pG, aqua, promoted, herbal.

Scirpus supinus L.
Cyperaceae
= *Schoenoplectus supinus* (L.) Pall.
(NoR)
♦ Weed
♦ 87, 88, 272
♦ G.

Scirpus sylvaticus L.
Cyperaceae
= *Scirpus expansus* Fern. (NoR)
♦ wood clubrush, korpikaisla, forest
rush
♦ Weed
♦ 23, 70, 87, 88, 272
♦ G, cultivated, herbal.

Scirpus tabernaemontani Gmel.
Cyperaceae
= *Schoenoplectus tabernaemontani*
(Gmel.) Pall. (NoR) [see *Scirpus validus*
Benth.]
♦ mountain bulrush, glaucous
clubrush
♦ Weed
♦ 263, 275, 286, 297
♦ pG, aqua, herbal.

Scirpus triangulatus Roxb.
Cyperaceae
♦ Weed
♦ 87, 88, 286
♦ G.

Scirpus triqueter L.
Cyperaceae
= *Schoenoplectus triqueter* (L.) Pall.
♦ triangular rush, sankakui
♦ Weed
♦ 87, 88, 126, 263, 272, 275, 286, 297
♦ pG, aqua.

Scirpus tuberosus Desf.
Cyperaceae
= *Eleocharis dulcis* (Burm.f.) Trin. ex
Hensch.
♦ sea clubrush, tuberous bulrush

♦ Weed, Introduced
♦ 34, 87, 88, 185, 221
♦ pG, aqua, herbal.

Scirpus validus Benth.
Cyperaceae
= *Schoenoplectus tabernaemontani*
(Gmel.) Pall. (NoR) [see *Scirpus
tabernaemontani* Gmel.]
♦ softstem bulrush, river clubrush
♦ Weed
♦ 23, 87, 88, 218
♦ pG, aqua, cultivated, herbal.

Scirpus wallichii Nees
Cyperaceae
♦ bulrush
♦ Weed, Quarantine Weed
♦ 76, 87, 88, 203, 220, 263, 274, 275, 286
♦ pG.

Scirpus yagara Ohwi
Cyperaceae
♦ Weed
♦ 286, 297
♦ G, aqua.

Scleranthus annuus L.
Caryophyllaceae
Knavel annuum Scop.
♦ knawel, German knotgrass,
annual knawel, annual scleranthus,
viherjäsenruoho
♦ Weed, Noxious Weed, Naturalised,
Environmental Weed
♦ 23, 44, 51, 70, 86, 87, 88, 94, 98, 101,
121, 158, 161, 162, 176, 198, 203, 211,
218, 240, 241, 243, 249, 272, 280, 287,
299, 300
♦ a/bH, cultivated, herbal, toxic.
Origin: Eurasia.

Scleranthus annuus L. ssp. annuus
Caryophyllaceae
♦ scleranthus, knawel, annual knawel
♦ Weed
♦ 34, 253
♦ aH.

Scleranthus fasciculatus (R.Br.) Hook.f.
Caryophyllaceae
♦ Naturalised
♦ 280

Scleranthus perennis L.
Caryophyllaceae
Scleranthus dichotomus Dalla Torre &
Sarnth.
♦ perennial knawel, vaaleajäsenruoho
♦ Weed, Naturalised
♦ 42, 44, 101, 272
♦ herbal.

Scleria bancana Miq.
Cyperaceae
♦ Weed
♦ 87, 88
♦ G.

Scleria barteri Boeck.
Cyperaceae
♦ Weed, Quarantine Weed
♦ 76, 87, 88, 203, 220
♦ G.

Scleria bourgeaui Boeck.
Cyperaceae

♦ Bourgeau's nutrush
♦ Naturalised
♦ 101
♦ G.

Scleria bracteata Cav.
Cyperaceae
♦ bracted nutrush
♦ Weed, Naturalised
♦ 87, 88, 101
♦ G.

Scleria canescens Boeck.
Cyperaceae
= *Scleria scindens* Nees ex Kunth (NoR)
♦ Weed
♦ 87, 88
♦ G.

Scleria ciliaris Nees
Cyperaceae
♦ nutrush
♦ Weed
♦ 88
♦ G, cultivated. Origin: Australia.

Scleria foliosa Hochst. ex A.Rich.
Cyperaceae
♦ Weed
♦ 88
♦ G.

Scleria havanensis Britt.
Cyperaceae
♦ Havana nutrush
♦ Weed
♦ 14
♦ G.

Scleria hebecarpa Nees
Cyperaceae
♦ Weed
♦ 87, 88
♦ G.

Scleria lacustris C.Wright
Cyperaceae
♦ lakeshore nutrush
♦ Weed, Naturalised
♦ 101, 179
♦ G.

Scleria levis Retz.
Cyperaceae
♦ Weed
♦ 87, 88
♦ G, herbal. Origin: Australia.

Scleria lithosperma (L.) Sw.
Cyperaceae
Scleria filiformis Sw., *Scleria puzzoleana*
Schum., *Scirpus lithospermus* L.
♦ Florida Keys nutrush
♦ Weed
♦ 14, 87, 88, 126
♦ pG, aqua, herbal. Origin: Asia,
Australia.

**Scleria melaleuca Rchb. ex Schltdl. &
Cham.**
Cyperaceae
= *Scleria pterota* C.Presl ex C.B.Clarke
var. *melaleuca* (Rchb. ex Schlecht. &
Cham.) Standl. (NoR) [see *Scleria
pterota* C.Presl ex C.B.Clarke]
♦ Weed, Quarantine Weed
♦ 14, 76, 87, 88, 203, 220
♦ G.

Scleria multifoliata **Boeck.**
Cyperaceae
♦ Weed
♦ 87, 88
♦ G.

Scleria myriocarpa **Kunth**
Cyperaceae
♦ Weed
♦ 87, 88
♦ G.

Scleria naumanniana **Boeck.**
Cyperaceae
♦ Weed
♦ 88
♦ G.

Scleria poaeformis **Retz.**
Cyperaceae
Scleria oryzoides Presl
♦ prue
♦ Weed
♦ 87, 88, 239
♦ pG. Origin: Australia.

Scleria polycarpa **Boeck.**
Cyperaceae
Scleria margaritifera Willd.
♦ nutrush
♦ Weed
♦ 87, 88
♦ G. Origin: Australia.

Scleria pterota **C.Presl ex C.B.Clarke**
Cyperaceae
= *Scleria pterota* C.Presl ex C.B.Clarke var. *melaleuca* (Rchb. ex Schlecht. & Cham.) Standl. (NoR) [see *Scleria melaleuca* Rchb. ex Schltdl. & Cham.]
♦ cortadora blanca, cortadera, navajuela
♦ Weed
♦ 87, 88, 153, 255, 281
♦ pG. Origin: Brazil.

Scleria purdiei **C.B.Clarke**
Cyperaceae
♦ Purdie's nutrush
♦ Naturalised
♦ 101
♦ G.

Scleria reflexa **Kunth**
Cyperaceae
= *Scleria secans* (L.) Urb.
♦ Weed
♦ 88
♦ G.

Scleria scabriuscula **Schlecht.**
Cyperaceae
♦ mosquito nutrush
♦ Naturalised
♦ 101
♦ G.

Scleria scrobiculata **Ness & Mey. ex Ness**
Cyperaceae
♦ nutrush
♦ Weed
♦ 87, 88
♦ G, herbal. Origin: Australia.

Scleria secans **(L.) Urb.**
Cyperaceae
Scleria reflexa Kunth (see), *Scleria*

weigeltiana Schrad. ex Boeck.
♦ razor grass
♦ Weed
♦ 88
♦ G.

Scleria setuloso-ciliata **Boeck.**
Cyperaceae
♦ Weed
♦ 14, 157
♦ G.

Scleria sumatrensis **Retz.**
Cyperaceae
♦ nutrush, Sumatran scleria
♦ Weed
♦ 87, 88, 126
♦ pG, aqua, herbal. Origin: Asia, Australia.

Scleria tessellata **Willd.**
Cyperaceae
♦ Weed
♦ 126
♦ G, aqua. Origin: Madagascar.

Scleria verrucosa **Willd.**
Cyperaceae
♦ Weed
♦ 88
♦ G.

Scleroblitum atriplicinum **(F.Muell.) Ulbr.**
Chenopodiaceae
♦ purple goosefoot
♦ Weed
♦ 55, 87, 88
♦ Origin: Australia.

Sclerocarpus africanus **Jacq.**
Asteraceae
♦ African bonebract
♦ Weed, Naturalised
♦ 87, 88, 101

Sclerocarpus coffeaecola **Klatt**
Asteraceae
♦ Weed, Quarantine Weed
♦ 76, 87, 88, 203, 220

Sclerocarpus divaricatus **(Benth.) Benth. & Hook.f. ex Hemsl.**
Asteraceae
♦ Weed
♦ 157
♦ aH.

Sclerocarpus phyllocephalus **Blake**
Asteraceae
♦ Weed
♦ 157
♦ aH.

Sclerocarya birrea **(A.Rich.) Hochst. ssp. caffra (Sond.) Kokwaro**
Anacardiaceae
Sclerocarya caffra Sond.
♦ amaganu, cat thorn, cider tree, didissa, iganu, ikanyi, inkanyi, jelly plum, katetalum, lerula, mafuna, maroela, maroola nut, marula, maroola plum, mufura, muganu, mukwakwa, mupfura, mura, mushomo, mutsomo, muua, nkanyi, ol mangwai, oruluo, pfura, tololokwo, tsua, tsula, ufuongo
♦ Weed, Native Weed
♦ 121, 228

♦ T, cultivated, herbal. Origin: southern Africa.

Sclerocarya birroea **Hochst.**
Anacardiaceae
♦ Weed
♦ 87, 88

Sclerocephalus arabicus **Boiss.**
Caryophyllaceae/Illecebraceae
♦ Weed
♦ 221

Sclerochloa dura **(L.) Beauv.**
Poaceae
Cynosurus durus L., *Eleusine dura* Lam., *Festuca dura* Vill., *Poa dura* (L.) Scop., *Sesleriad dura* Kunth
♦ common hardgrass, hardgrass, tufted hardgrass, hard meadowgrass
♦ Weed, Naturalised, Introduced, Environmental Weed
♦ 34, 86, 87, 88, 98, 101, 161, 176, 198, 203, 212, 272
♦ aG, cultivated, herbal. Origin: Eurasia.

Sclerochloa kengiana **(Ohwi) Tzvelev**
Poaceae
♦ Weed
♦ 243
♦ G.

Sclerolaena bicornis **Lindl.**
Chenopodiaceae
♦ goat's head burr
♦ Native Weed
♦ 269
♦ S, cultivated. Origin: Australia.

Sclerolaena bicornis **Lindl. var. horrida Lindl.**
Chenopodiaceae
♦ Naturalised
♦ 86
♦ Origin: Australia.

Sclerolaena birchii **(F.Muell.) Domin**
Chenopodiaceae
Anisacantha birchii F.Muell., *Bassia birchii* (F.Muell.) F.Muell. (see)
♦ galvanised burr, blue burr, galvanised roly poly, Hermidale lucerne, Woolerino burr
♦ Weed, Quarantine Weed, Noxious Weed, Naturalised, Native Weed
♦ 76, 86, 88, 147, 171, 203, 269
♦ S, arid, cultivated. Origin: Australia.

Sclerolaena calcarata **(Ising) Scott**
Chenopodiaceae
Bassia calcarata Ising
♦ Naturalised
♦ 86
♦ arid. Origin: Australia.

Sclerolaena divaricata **(R.Br.) Domin.**
Chenopodiaceae
Bassia divaricata (R.Br.) Muell.
♦ tangled copper burr
♦ Native Weed
♦ 269
♦ arid, cultivated. Origin: Australia.

Sclerolaena muricata **(Moq.) Domin**
Chenopodiaceae
Anisacantha gracilicuspis Muell.,

Anisacantha muricata Moq., *Anisacantha quinquecuspis* Muell., *Bassia quinquecuspis* (F.Muell.) F.Muell. (see), *Chenolea quinquecuspis* (Muell.) Muell.
♦ five spined saltbush, black roly poly, electric burr, five spined bassia, prickly roly poly, spiny roly poly
♦ Weed, Quarantine Weed, Noxious Weed, Native Weed
♦ 76, 88, 147, 203, 269
♦ arid, cultivated, toxic. Origin: Australia.

Sclerolaena muricata (Moq.) Domin var. *semiglabra* (Ising) A.J.Scott
Chenopodiaceae
♦ five spined saltbush, black roly poly, electric burr, five spined bassia, prickly roly poly, spiny roly poly
♦ Noxious Weed, Naturalised
♦ 86
♦ Origin: Australia.

Sclerolaena tricuspis (Muell.) Ulbr.
Chenopodiaceae
Bassia tricuspis (F.Muell.) R.H.Anderson
♦ giant redbur, three spined bassia
♦ Native Weed
♦ 269
♦ cultivated. Origin: Australia.

Scleropoa rigida (L.) Griseb.
Poaceae
= *Catapodium rigidum* (L.) C.E.Hubb.
♦ hardgrass
♦ Weed, Naturalised
♦ 87, 88, 218, 287
♦ G, herbal.

Scolymus hispanicus L.
Asteraceae
Myscolus hispanicus Cass. ex Diet., *Myscolus microcephalus* Cass., *Scolymus congestus* Lam.
♦ golden thistle, Spanish oyster plant, Spanish salsify, espanjanpertiö
♦ Weed, Quarantine Weed, Noxious Weed, Naturalised, Introduced, Casual Alien
♦ 34, 35, 42, 70, 76, 86, 87, 88, 94, 98, 101, 147, 161, 198, 203, 229, 241, 243, 253, 269, 272, 287, 295, 300
♦ aH, arid, cultivated, herbal. Origin: Mediterranean.

Scolymus maculatus L.
Asteraceae
♦ spotted golden thistle, spotted thistle, kirjopertiö
♦ Weed, Quarantine Weed, Noxious Weed, Naturalised, Casual Alien
♦ 42, 70, 76, 86, 87, 88, 94, 98, 101, 147, 203, 221, 243, 253, 269, 272
♦ aH, cultivated. Origin: Mediterranean.

Scoparia dulcis L.
Scrophulariaceae
Scoparia ternata Forssk.
♦ sweet broom, goatweed, licorice weed, kra tai chaam, vacourinha, krot num
♦ Weed, Quarantine Weed,

Naturalised, Introduced, Environmental Weed
♦ 12, 13, 14, 39, 76, 86, 87, 88, 93, 98, 157, 161, 170, 191, 203, 230, 238, 239, 255, 261, 262, 275, 287, 297
♦ aH, arid, cultivated, herbal, toxic. Origin: tropical America.

Scoparia flava Cham. & Schlecht.
Scrophulariaceae
♦ yellow licorice weed, scoparia
♦ Naturalised
♦ 101

Scoparia montevidensis (Spreng.) R.E.Fr.
Scrophulariaceae
♦ broomwort
♦ Weed, Naturalised
♦ 87, 88, 101, 295

Scoparia montevidensis (Spreng.) R.E.Fr. var. *glandulifera* (Fritsch) R.E.Fr.
Scrophulariaceae
♦ broomwort
♦ Naturalised
♦ 101

Scopolia carniolica Jacq.
Solanaceae
Hyoscyamus scopolia L.
♦ skopolia, European scopolia
♦ Cultivation Escape
♦ 42
♦ pH, cultivated, herbal, toxic.

Scorpiurus muricatus L.
Fabaceae/Papilionaceae
Scorpiurus echinatus Lam., *Scorpiurus laevigatus* Lindl., *Scorpiurus oliverii* Palau, *Scorpiurus sulcata* L. (see), *Scorpiurus sulcatus* L., *Scorpiurus subvillosa* L., *Scorpiurus subvillosus* L., *Scorpiurus subvillosus* L. ssp. *sulcatus* (L.) Hayek
♦ prickly scorpion's tail, scorpion's tail, caterpillar plant
♦ Weed, Naturalised, Introduced, Casual Alien
♦ 40, 87, 88, 94, 101, 185, 221, 228, 243, 253, 280
♦ aH, arid, promoted, herbal. Origin: Europe.

Scorpiurus muricatus L. var. *subvillosus* L.
Fabaceae/Papilionaceae
♦ Naturalised
♦ 86
♦ Origin: Mediterranean.

Scorpiurus sulcata L.
Fabaceae/Papilionaceae
= *Scorpiurus muricatus* L.
♦ Weed
♦ 87, 88

Scorpiurus vermiculatus L.
Fabaceae/Papilionaceae
♦ Weed, Quarantine Weed
♦ 70, 87, 88, 94, 220
♦ aH, cultivated.

Scorzonera L. spp.
Asteraceae
♦ scorzonera

♦ Weed, Naturalised
♦ 198, 272

Scorzonera albicaulis Bunge
Asteraceae
♦ whitestem serpentroot
♦ Weed
♦ 297
♦ pH, promoted.

Scorzonera alexandrina Boiss.
Asteraceae
♦ Weed
♦ 221

Scorzonera austriaca Willd.
Asteraceae
Scorzonera glabra Rupr.
♦ Austrian viper's grass
♦ Weed
♦ 272
♦ pH, promoted.

Scorzonera hispanica L.
Asteraceae
Scorzonera montana Mutel, *Scorzonera sativa* Gaertn.
♦ Spanish salsify, black salsify, salsify
♦ Weed, Naturalised, Cultivation Escape
♦ 34, 42, 101, 287
♦ pH, cultivated, herbal.

Scorzonera hispida Forssk.
Asteraceae
♦ Egyptinsikojuuri
♦ Casual Alien
♦ 42

Scorzonera humilis L.
Asteraceae
Scorzonera pannonica Tabern.
♦ common viper's grass, viper's grass, sikojuuri
♦ Weed
♦ 272
♦ cultivated, herbal.

Scorzonera jacquiniana Boiss.
Asteraceae
Scorzonera cana (C.Mey.) O.Hoffm.
♦ Weed
♦ 243

Scorzonera laciniata L.
Asteraceae
Podospermum calcitrapifolium (Vahl) DC., *Podospermum laciniatum* (L.) DC., *Podospermum residifolium* (L.) DC., *Podospermum willkommii* Sch.Bip., *Scorzonera residifolium* L.
♦ cutleaf viper's grass, self salsify, salsify, cutleaf
♦ Weed, Naturalised, Environmental Weed
♦ 70, 72, 79, 86, 88, 94, 98, 101, 156, 161, 176, 198, 203, 253, 272
♦ b, arid.

Scorzonera laciniata L. var. *calcitrapifolia* (Vahl) Bisch. ex Boiss.
Asteraceae
♦ scorzonera
♦ Naturalised, Environmental Weed
♦ 86, 198

Scorzonera laciniata **L. var.** *laciniata*
Asteraceae
♦ scorzonera
♦ Naturalised, Environmental Weed
♦ 86, 198

Scorzonera lanata **(L.) Hoffm.**
Asteraceae
♦ Weed
♦ 87, 88

Scorzonera mollis **Bieb.**
Asteraceae
♦ Weed
♦ 221
♦ pH, promoted.

Scorzonera mongolica **Maxim.**
Asteraceae
♦ Mongolian serpentroot
♦ Weed
♦ 297
♦ pH, promoted.

Scorzonera purpurea **L.**
Asteraceae
♦ purple viper's grass
♦ Weed
♦ 87, 88, 272

Scorzonera sinensis **Lipsch. & Krasch.**
Asteraceae
♦ Chinese serpentroot
♦ Weed
♦ 297

Scrophularia **L. spp.**
Scrophulariaceae
♦ figwort
♦ Weed
♦ 221, 272
♦ herbal.

Scrophularia auriculata **L.**
Scrophulariaceae
Scrophularia aquatica auct., *Scrophularia balbisii* Hornem.
♦ shoreline figwort, water figwort, water betony, figwort, scrofulaire aquatique
♦ Weed, Naturalised
♦ 15, 39, 86, 101, 176, 280, 300
♦ aqua, cultivated, herbal, toxic. Origin: Mediterranean.

Scrophularia californica **Cham. & Schlecht.**
Scrophulariaceae
♦ California figwort
♦ Weed
♦ 161
♦ pH, cultivated, herbal.

Scrophularia canina **L.**
Scrophulariaceae
♦ dog figwort, French figwort
♦ Weed, Naturalised
♦ 101, 272
♦ bH, cultivated, herbal.

Scrophularia canina **L. ssp.** *canina*
Scrophulariaceae
♦ dog figwort
♦ Naturalised
♦ 101

Scrophularia canina **L. ssp.** *hoppii* **(Koch) Fourn.**
Scrophulariaceae
♦ Hopp's figwort
♦ Naturalised
♦ 101

Scrophularia chrysantha **Jaub. & Spach**
Scrophulariaceae
♦ kultasyyläjuuri
♦ Casual Alien
♦ 42
♦ cultivated.

Scrophularia deserti **Delile**
Scrophulariaceae
♦ Weed
♦ 221

Scrophularia lanceolata **Pursh**
Scrophulariaceae
♦ lanceleaf figwort, figwort, mountain figwort
♦ Weed
♦ 23, 87, 88, 218
♦ pH, cultivated, herbal.

Scrophularia marilandica **L.**
Scrophulariaceae
♦ Maryland figwort, carpenter's square, figwort
♦ Weed
♦ 87, 88, 218
♦ promoted, herbal.

Scrophularia nodosa **L.**
Scrophulariaceae
♦ woodland figwort, knotted figwort, common figwort
♦ Weed, Naturalised
♦ 23, 39, 86, 87, 88, 98, 101, 203, 272, 280
♦ pH, cultivated, herbal, toxic. Origin: Eurasia.

Scrophularia peregrina **L.**
Scrophulariaceae
♦ Mediterranean figwort, nettle leaved figwort
♦ Weed, Naturalised
♦ 88, 94, 101
♦ herbal.

Scrophularia umbrosa **Dumort.**
Scrophulariaceae
Scrophularia alata Gilib., *Scrophularia aquatica* L., *Scrophularia erhardtii* Steven
♦ water figwort, water betony, green figwort, scarce water figwort
♦ Naturalised
♦ 101, 241
♦ pH, cultivated, herbal.

Scrophularia vernalis **L.**
Scrophulariaceae
♦ kevätsyyläjuuri, yellow figwort
♦ Cultivation Escape, Casual Alien
♦ 40, 42
♦ cultivated, herbal.

Scrophularia xanthoglossa **Boiss.**
Scrophulariaceae
♦ Weed
♦ 221

Scutellaria alpina **L.**
Lamiaceae
Scutellaria altaica Fisch., *Scutellaria variegata* Spreng.
♦ alppivuohennokka, alpine skullcap
♦ Cultivation Escape
♦ 42
♦ cultivated.

Scutellaria altissima **L.**
Lamiaceae
Scutellaria commutata Guss.
♦ tall skullcap
♦ Weed, Naturalised
♦ 23, 88, 101
♦ cultivated, herbal.

Scutellaria baicalensis **Georgi**
Lamiaceae
Scutellaria lanceolaria Miq., *Scutellaria macrantha* Fisch.
♦ Baikal skullcap, huang qin
♦ Weed, Naturalised
♦ 287, 297
♦ pH, cultivated, herbal. Origin: east Asia.

Scutellaria dependens **Maxim.**
Lamiaceae
♦ himenamiki
♦ Weed
♦ 87, 88, 297

Scutellaria galericulata **L.**
Lamiaceae
Cassida galericulata Moench, *Cassida major* Gilib.
♦ marsh skullcap, skullcap, common skullcap
♦ Weed
♦ 23, 39, 87, 88, 272
♦ pH, aqua, cultivated, herbal, toxic.

Scutellaria hastifolia **L.**
Lamiaceae
Cassida hastifolia Scop.
♦ spear leaved skullcap, keihäsvuohennokka
♦ Weed
♦ 272
♦ herbal.

Scutellaria indica **L.**
Lamiaceae
♦ Weed
♦ 235, 286
♦ pH, cultivated.

Scutellaria lateriflora **L.**
Lamiaceae
♦ skullcap, side flowering skullcap, hoodwort, mad dog skullcap, quaker bonnet, blue skullcap, Virginian skullcap
♦ Weed
♦ 23, 39, 88
♦ pH, aqua, cultivated, herbal, toxic.

Scutellaria minor **Huds.**
Lamiaceae
♦ lesser skullcap
♦ Naturalised
♦ 280
♦ cultivated, herbal.

Scutellaria parvula **Michx.**
Lamiaceae
♦ small southern australis skullcap, small skullcap
♦ Weed

♦ 161
♦ cultivated, herbal.

Scutellaria racemosa Pers.
Lamiaceae
♦ skullcap, South American skullcap
♦ Weed, Naturalised
♦ 86, 98, 101, 121, 179, 203
♦ pH. Origin: South America.

Scutellaria rivularis Wall. ex Benth.
Lamiaceae
♦ Weed
♦ 235

Scutellaria scordifolia Fisch. ex Schrank
Lamiaceae
♦ twinflower skullcap
♦ Weed
♦ 297

Scutellaria strigillosa Hemsl.
Lamiaceae
♦ namikisou
♦ Weed
♦ 87, 88

Scutia myrtina (Burm.f.) Kurz
Rhamnaceae
♦ cat thorn
♦ Weed, Native Weed
♦ 87, 88, 121
♦ S/T, arid, cultivated, herbal. Origin: southern Africa.

Sebastiania Spreng. spp.
Euphorbiaceae
♦ Mexican jumping beans, Sebastian bush
♦ Weed, Quarantine Weed
♦ 76, 88

Sebastiania chamaelea (L.) Müll.Arg.
Euphorbiaceae
♦ creeping sebastiana
♦ Weed
♦ 87, 88, 297
♦ cultivated.

Sebastiania corniculata (Vahl) Müll.Arg.
Euphorbiaceae
Tragia corniculata Vahl
♦ hato tejas
♦ Weed
♦ 87, 88, 255, 261
♦ aH. Origin: Americas.

Secale L. spp.
Poaceae
♦ ryegrass
♦ Weed, Naturalised
♦ 181, 198
♦ G.

Secale cereale L.
Poaceae
Frumentum secale Krause var. *cereale* Krause, *Secale aestivum* Uspen., *Secale montanum* Guss. (see), *Secale turkestanicum* Ben., *Secale vernum* Poir., *Tririum secale* Link, *Triticum cereale* Asch. & Graebn.
♦ cereal rye, rye, volunteer rye, winter rye, ruis
♦ Weed, Noxious Weed, Naturalised, Cultivation Escape, Casual Alien
♦ 1, 34, 39, 40, 42, 80, 86, 87, 88, 98, 101, 138, 146, 161, 176, 179, 198, 203, 212, 229, 237, 241, 252, 280, 287, 295
♦ aG, cultivated, herbal, toxic. Origin: Russia, Middle East.

Secale cereale L. ssp. cereale
Poaceae
♦ Naturalised
♦ 280
♦ G.

Secale montanum Guss.
Poaceae
= *Secale cereale* L.
♦ mountain rye
♦ Weed
♦ 87, 88
♦ G, arid, cultivated.

Sechium edule (Jacq.) Sw.
Cucurbitaceae
♦ chayote, choko, cayhua
♦ Weed, Naturalised, Introduced
♦ 86, 98, 203, 228, 280, 287
♦ arid, cultivated, herbal. Origin: tropical America.

Securidaca diversifolia (L.) Blake
Polygalaceae
♦ Easter flower
♦ Naturalised, Cultivation Escape
♦ 101, 261
♦ cultivated, herbal.

Securidaca virgata Sw.
Polygalaceae
♦ bejuco de sopla
♦ Weed
♦ 87, 88

Securigera securidaca (L.) Degen & Dörfl.
Fabaceae/Papilionaceae
Coronilla cretica L. (see)
♦ scorpion vetch
♦ Weed, Quarantine Weed, Casual Alien
♦ 40, 76, 87, 88, 94, 272
♦ herbal.

Securigera varia (L.) Lassen
Fabaceae/Papilionaceae
Coronilla varia L. (see)
♦ crownvetch
♦ Naturalised
♦ 40
♦ arid. Origin: Europe, Middle East.

Securigera virosa Thunb.
Fabaceae/Papilionaceae
♦ Weed
♦ 87, 88

Securinega virosa (Roxb. ex Willd.) Baill.
Euphorbiaceae
♦ Weed
♦ 221
♦ herbal.

Seddera latifolia Hochst. & Steud. ex Hochst.
Convolvulaceae
♦ Weed
♦ 221

Sedum L. spp.
Crassulaceae
♦ sedum, stonecrop, roseroot

♦ Weed, Naturalised, Environmental Weed
♦ 72, 86, 88, 247, 272
♦ herbal, toxic.

Sedum acre L.
Crassulaceae
♦ mossy stonecrop, stonecrop, wall pepper, yellow stonecrop, yellow sedum, keltamaksaruoho, goldmoss stonecrop, biting stonecrop
♦ Weed, Naturalised, Garden Escape, Environmental Weed
♦ 39, 80, 86, 87, 88, 98, 101, 161, 165, 176, 181, 195, 198, 203, 218, 225, 241, 246, 272, 280, 287, 300
♦ pH, cultivated, herbal, toxic. Origin: Mediterranean, west Asia.

Sedum aizoon L.
Crassulaceae
♦ siperianmaksaruoho, aizoon stonecrop
♦ Weed, Naturalised, Garden Escape, Cultivation Escape
♦ 42, 101, 275, 297
♦ pH, cultivated.

Sedum album L.
Crassulaceae
Oreosedum album (L.) Grulich, *Sedum athoum* DC.
♦ white stonecrop, valkomaksaruoho, small houseleek, green stonecrop, orpin blanc
♦ Weed, Naturalised
♦ 15, 39, 40, 70, 101, 161, 272, 280
♦ pH, cultivated, herbal, toxic. Origin: Europe, Middle East.

Sedum album L. ssp. micranthum L.
Crassulaceae
♦ Naturalised
♦ 40

Sedum bulbiferum Makino
Crassulaceae
♦ Weed
♦ 87, 88, 286

Sedum caespitosum (Cav.) DC.
Crassulaceae
♦ tiny stonecrop
♦ Naturalised
♦ 86, 198
♦ Origin: Europe.

Sedum confusum Hemsl.
Crassulaceae
♦ lesser Mexican stonecrop, stonecrop, sedum
♦ Naturalised
♦ 40
♦ H. Origin: Mexico.

Sedum dasyphyllum L.
Crassulaceae
♦ thick leaved stonecrop
♦ Naturalised, Casual Alien
♦ 280, 287
♦ cultivated, herbal.

Sedum decumbens R.T.Clausen
Crassulaceae
♦ Naturalised
♦ 280

Sedum dendroideum DC.
Crassulaceae
♦ sedum, tree stonecrop
♦ Weed, Naturalised, Garden Escape,
Environmental Weed
♦ 72, 86, 88
♦ S, cultivated, herbal.

Sedum ewersii Ledeb.
Crassulaceae
= *Hylotelephium ewersii* (Ledeb.)
H.Ohba (NoR)
♦ Turkestaninmaksaruoho
♦ Cultivation Escape
♦ 42
♦ cultivated.

Sedum formosanum N.E.Br.
Crassulaceae
♦ Weed
♦ 87, 88, 274

Sedum forsterianum Sm.
Crassulaceae
Sedum elegans Lej.
♦ rock stonecrop, Welsh stonecrop
♦ Weed, Naturalised
♦ 54, 86, 88, 176
♦ cultivated. Origin: western Europe,
Azores.

Sedum hispanicum L.
Crassulaceae
♦ espanjanmaksaruoho, Spanish
stonecrop
♦ Naturalised, Cultivation Escape
♦ 42, 101
♦ cultivated, herbal.

Sedum hybridum L.
Crassulaceae
♦ Mongolian maksaruoho, hybrid
stonecrop
♦ Garden Escape, Cultivation Escape
♦ 42
♦ cultivated. Origin: central and
southern Urals.

Sedum kamtschaticum Fisch. & C.A.Mey.
Crassulaceae
♦ orange stonecrop, Kamchatkan
sedum
♦ Naturalised
♦ 101
♦ pH, cultivated, herbal.

**Sedum kamtschaticum Fisch. & C.A.Mey.
ssp. ellacombianum (Praeger) Clausen**
Crassulaceae
♦ orange stonecrop
♦ Naturalised
♦ 101
♦ cultivated.

Sedum liebmannianum Hemsl.
Crassulaceae
♦ Casual Alien
♦ 280

Sedum lineare Thunb.
Crassulaceae
♦ needle stonecrop, stonecrop
♦ Weed, Naturalised
♦ 80, 87, 88, 101
♦ pH, cultivated, herbal.

Sedum lydium Boiss.
Crassulaceae
♦ lyydianmaksaruoho
♦ Cultivation Escape
♦ 42
♦ cultivated.

Sedum makinoi Maxim.
Crassulaceae
♦ Weed
♦ 286
♦ pH, promoted, herbal. Origin:
China, Japan.

Sedum mexicanum Britt.
Crassulaceae
♦ Mexican stonecrop
♦ Naturalised
♦ 101, 280, 287

Sedum ochroleucum Chaix
Crassulaceae
Sedum anopetalum DC.
♦ European stonecrop
♦ Weed, Naturalised
♦ 101, 253
♦ cultivated.

Sedum praealtum DC.
Crassulaceae
♦ sedum, green cock's comb
♦ Weed, Naturalised, Garden Escape,
Environmental Weed
♦ 15, 72, 88, 101, 280
♦ S, cultivated. Origin: Mexico.

Sedum praealtum A.DC. ssp. praealtum
Crassulaceae
♦ shrubby stonecrop
♦ Naturalised, Environmental Weed
♦ 86, 198
♦ Origin: Mexico.

Sedum purpureum (L.) J.A.Schult.
Crassulaceae
= *Hylotelephium telephium* (L.) H.Ohba
ssp. *telephium*
♦ liveforever, orpine
♦ Weed
♦ 87, 88, 218, 247
♦ cultivated, herbal, toxic.

Sedum reflexum L.
Crassulaceae
= *Sedum rupestre* L.
♦ reflexed stonecrop, Jenny's
stonecrop, crooked yellow stonecrop
♦ Weed, Naturalised, Garden Escape,
Environmental Weed
♦ 54, 86, 88, 101, 176, 280
♦ pH, cultivated, herbal. Origin:
Europe.

Sedum rubens L.
Crassulaceae
Aithales rubens Webb., *Crassula rubens*
L.
♦ Weed
♦ 253, 272
♦ cultivated.

Sedum × rubrotinctum R.T.Clausen
Crassulaceae
♦ pork and beans
♦ Casual Alien
♦ 280
♦ cultivated.

Sedum rupestre L.
Crassulaceae
Sedum reflexum L. (see)
♦ reflexed stonecrop, stonecrop,
kalliomaksaruoho
♦ Naturalised
♦ 40, 42
♦ pH, cultivated, herbal. Origin:
Eurasia.

Sedum sarmentosum Bunge
Crassulaceae
♦ stringy stonecrop
♦ Weed, Naturalised, Garden Escape
♦ 101, 275, 286, 287, 297
♦ pH, cultivated, herbal.

Sedum sediforme (Jacq.) Pau
Crassulaceae
Sedum altissimum Poir., *Sedum nicaeense*
All.
♦ Weed, Naturalised
♦ 86, 253
♦ pH, cultivated. Origin:
Mediterranean.

Sedum sexangulare L.
Crassulaceae
Sedum boloniense Loisel., *Sedum mite*
Gilib., *Sedum sempervivum* Grimm
♦ tasteless stonecrop,
särmämaksaruoho
♦ Weed, Naturalised, Environmental
Weed, Casual Alien
♦ 40, 86, 101, 176, 272
♦ cultivated, herbal. Origin: Europe.

Sedum spectabile Boreau
Crassulaceae
= *Hylotelephium spectabile* (Boreau)
Ohba
♦ iceplant, showy stonecrop
♦ Weed, Naturalised
♦ 15, 80, 280
♦ pH, cultivated, herbal.

Sedum spurium Bieb.
Crassulaceae
Sedum spurium var. *coccineum hort.*
♦ Caucasian stonecrop,
tworow stonecrop, rozchodnik,
kaukasianmaksaruho, o benkei so
♦ Naturalised, Garden Escape,
Cultivation Escape
♦ 40, 42, 101, 280
♦ pH, cultivated, herbal.

Sedum stoloniferum Gmel.
Crassulaceae
♦ stolon stonecrop, mat forming
stonecrop
♦ Naturalised
♦ 101
♦ pH, cultivated.

Sedum telephium L.
Crassulaceae
= *Hylotelephium telephium* (L.) H.Ohba
ssp. *telephium*
♦ stonecrop, liveforever, livelong,
orpine, witch's moneybags,
isomaksaruoho
♦ Weed
♦ 23, 88, 161, 195, 272
♦ pH, cultivated, herbal, toxic.

Seetzenia lanata (Willd.) Bullock
Zygophyllaceae
♦ Weed
♦ 221

Seguieria langsdorffi Moq.
Phytolaccaceae/Petiveriaceae
♦ agulheiro, espino agulha
♦ Weed
♦ 255
♦ Origin: Brazil.

Sehima ischaemoides Forssk.
Poaceae
♦ Weed
♦ 87, 88
♦ G.

Selaginella Beauv. spp.
Selaginellaceae
♦ spikemoss, selaginella
♦ Environmental Weed
♦ 246

Selaginella apoda (L.) Spring
Selaginellaceae
♦ meadow spikemoss
♦ Naturalised
♦ 241, 300
♦ cultivated, herbal.

Selaginella belluta Ces.
Selaginellaceae
♦ Weed
♦ 87, 88

Selaginella braunii Bak.
Selaginellaceae
♦ Braun's spikemoss
♦ Naturalised
♦ 101

Selaginella flabellata (L.) Spring
Selaginellaceae
♦ fan spikemoss
♦ Weed
♦ 19
♦ herbal.

Selaginella kanehirae Alst.
Selaginellaceae
♦ spikemoss
♦ Introduced
♦ 230
♦ H.

Selaginella kraussiana (Kunze) A.Braun
Selaginellaceae
♦ garden selaginella, Krauss' spikemoss, mossy clubmoss, selaginella, African club moss
♦ Weed, Quarantine Weed, Naturalised, Garden Escape, Environmental Weed
♦ 15, 40, 72, 86, 88, 98, 101, 134, 165, 181, 198, 203, 225, 246
♦ pH, cultivated, herbal. Origin: South Africa.

Selaginella opaca Warb.
Selaginellaceae
♦ Weed
♦ 87, 88

Selaginella plana (Desv. ex Poir.) Hieron.
Selaginellaceae
♦ Asian spikemoss
♦ Weed, Naturalised

♦ 32, 86, 87, 88, 101
♦ Origin: south-east Asia.

Selaginella umbrosa Lem.
Selaginellaceae
♦ Naturalised
♦ 86
♦ cultivated. Origin: Central and South America.

Selaginella uncinata (Desv. ex Poir.) Spring
Selaginellaceae
Lycopodium uncinatum Desv. ex Poir., *Selaginella aristata* (Roxb.) Scott, *Selaginella caesia* Kunze, *Selaginella eurystachya* Warb.
♦ blue spikemoss, peacock moss, rainbow fern, rainbow moss
♦ Weed, Naturalised
♦ 101, 179, 287
♦ cultivated.

Selaginella vogelii Spring
Selaginellaceae
Selaginella dinklageana Sadeb.
♦ Naturalised
♦ 86

Selaginella willdenovii (Desv.) Bak.
Selaginellaceae
Selaginella willdenowii (Desv. ex Poir.) Bak.
♦ peacock fern, Willdenow's spikemoss
♦ Weed, Naturalised, Garden Escape
♦ 80, 86, 101, 179, 261
♦ cultivated, herbal. Origin: tropical Asia.

Selago corymbosa L.
Scrophulariaceae/Selaginaceae
♦ poverty bush, bitterblombos
♦ Weed, Naturalised, Native Weed
♦ 86, 98, 121, 203
♦ S, cultivated. Origin: southern Africa.

Selago thunbergii Choisy
Scrophulariaceae/Selaginaceae
♦ Naturalised
♦ 86
♦ cultivated. Origin: South Africa.

Selenicereus (Berger) Britt. & Rose spp.
Cactaceae
Deamia Britt. & Rose spp. (see)
♦ moonlight cactus
♦ Weed, Quarantine Weed, Naturalised
♦ 76, 86, 88, 203, 220
♦ cultivated.

Selenicereus coniflorus (Weing.) Britt. & Rose
Cactaceae
Selenicereus pringlei Rose
♦ coneflower moonlight cactus
♦ Weed
♦ 179
♦ cultivated.

Selenicereus grandiflorus (L.) Britt. & Rose
Cactaceae
Cereus grandiflorus Mill., *Selenicereus hallensis* Weing., *Selenicereus kunthianus*

(Otto) Britt. & Rose
♦ queen of the night, reina de la noche
♦ Naturalised, Introduced
♦ 101, 261
♦ cultivated, herbal. Origin: Cuba, Hispaniola, Mexico.

Selenicereus macdonaldiae (Hook.) Britt. & Rose
Cactaceae
♦ Naturalised
♦ 86
♦ cultivated. Origin: Uruguay.

Selenicereus pteranthus (Link ex A.Dietr.) Britt. & Rose
Cactaceae
♦ princess of the night
♦ Weed, Naturalised
♦ 101, 179
♦ cultivated.

× *Seleniphyllum* Rowley spp.
Cactaceae
= *Selenicereus* (A.Berger) Britt. & Rose
× *Epiphyllum* Haw. spp.
♦ Weed, Quarantine Weed
♦ 76, 88, 203, 220

Selinum carvifolia (L.) L.
Apiaceae
Laserpitium bavaricum Schrank, *Ligusticum carvifolia* Caruel, *Mylinum carvifolia* Fourr., *Oreoselinum pseudo-carvifolium* Hoff., *Peucedanum carvifolia* Loisel., *Selinum tenuifolium* Salisb.
♦ Cambridge milk parsley, selinum, särmäputki
♦ Weed, Naturalised
♦ 101, 272
♦ cultivated, herbal.

Selinum lineare Schum.
Apiaceae
♦ Weed
♦ 23, 88

Selinum palustre L.
Apiaceae
= *Peucedanum palustre* (L.) Moench
♦ Weed
♦ 23, 88

Semiarundinaria fastuosa (Mitford) Makino
Poaceae
♦ Narihira bamboo, Narihiradake
♦ Naturalised
♦ 280
♦ G, cultivated.

Sempervivum tectorum L.
Crassulaceae
♦ kattomehitähti, houseleak, common houseleek, kattomehitähti
♦ Naturalised, Cultivation Escape, Casual Alien
♦ 40, 42, 101
♦ pH, cultivated, herbal. Origin: obscure, possibly Europe.

Senebiera didyma Pers.
Brassicaceae
= *Coronopus didymus* (L.) J.E.Sm.
♦ Weed
♦ 87, 88, 243

Senecio L. spp.
Asteraceae
♦ senecio, squaw weed, groundsel, ragwort, cineraria
♦ Weed
♦ 154, 221, 247, 272
♦ herbal, toxic.

Senecio abyssinicus Sch.Bip. ex A.Rich.
Asteraceae
♦ Weed
♦ 87, 88, 240
♦ aH.

Senecio aegyptius L.
Asteraceae
♦ Weed
♦ 87, 88, 185, 221
♦ toxic.

Senecio angulatus L.f.
Asteraceae
Senecio macropodus DC.
♦ climbing groundsel, cape ivy
♦ Weed, Naturalised, Garden Escape, Environmental Weed
♦ 15, 72, 86, 88, 155, 198, 225, 246, 280, 289, 300
♦ C, arid, cultivated. Origin: southern Africa.

Senecio angustifolius (Thunb.) Willd.
Asteraceae
♦ Weed, Quarantine Weed, Native Weed
♦ 76, 88, 121, 203, 220
♦ H, toxic. Origin: southern Africa.

Senecio anonymus Wood
Asteraceae
♦ Small's ragwort, Small's groundsel
♦ Weed
♦ 161
♦ herbal, toxic.

Senecio anteuphorbium (L.) Sch.Bip.
Asteraceae
♦ Weed, Quarantine Weed
♦ 76, 88, 220
♦ cultivated.

Senecio apiifolius (DC.) Benth. ex Hook.f. ex O.Hoffm.
Asteraceae
♦ Native Weed
♦ 121
♦ aH. Origin: southern Africa.

Senecio aquaticus Hill
Asteraceae
Senecio divergens Schultz, *Senecio richteri* Schultz
♦ marsh ragwort, ojavillakko
♦ Weed, Naturalised, Casual Alien
♦ 23, 42, 70, 87, 88, 272, 280
♦ cultivated, herbal.

Senecio aquaticus Hill ssp. barbareifolius (Wimm. & Grab.) Walters
Asteraceae
♦ Naturalised
♦ 241, 300

Senecio argentinus Bak.
Asteraceae
♦ senecio argentino

♦ Weed
♦ 87, 88, 295

Senecio arguensis Turcz.
Asteraceae
♦ Argum groundsel
♦ Weed
♦ 297

Senecio articulatus (L.f.) Sch.Bip.
Asteraceae
♦ Weed, Quarantine Weed, Naturalised
♦ 76, 86, 88, 220
♦ cultivated.

Senecio aschenbornianus Schauer
Asteraceae
♦ Weed
♦ 199

Senecio aureus L.
Asteraceae
= *Packera aurea* (L.) Á. & D.Löve (NoR)
♦ golden ragwort, squaw weed, life root
♦ Weed, Quarantine Weed
♦ 8, 23, 39, 87, 88, 161, 210, 218, 220
♦ herbal, toxic.

Senecio bicolor (Willd.) Tod.
Asteraceae
♦ silver ragwort
♦ Weed, Naturalised
♦ 101, 161
♦ cultivated, herbal, toxic.

Senecio bicolor (Willd.) Tod. ssp. cineraria (DC.) Chater
Asteraceae
Senecio cineraria DC. (see)
♦ silver ragwort
♦ Naturalised
♦ 101

Senecio bipinnatisectus Belcher
Asteraceae
♦ fireweed, Australian fireweed
♦ Weed, Naturalised
♦ 15, 39, 165, 280
♦ toxic. Origin: Australia.

Senecio blochmaniae Greene
Asteraceae
♦ dune ragwort, dune senecio
♦ Naturalised
♦ 287
♦ S.

Senecio bonariensis Hook. & Arn.
Asteraceae
♦ margarita de agua
♦ Weed
♦ 87, 237, 295

Senecio brasiliensis (Spreng.) Less.
Asteraceae
Cineraria brasiliensis Spreng.
♦ maria mole, catiao
♦ Weed
♦ 87, 88, 245, 255, 295
♦ Origin: Brazil.

Senecio brasiliensis (Spreng.) Less. var. tripartitus (DC.) Bak.
Asteraceae
♦ Weed
♦ 237

Senecio bupleuroides DC.
Asteraceae
♦ ragwort
♦ Weed, Quarantine Weed, Native Weed
♦ 76, 88, 121, 203, 220
♦ pH, herbal, toxic. Origin: southern Africa.

Senecio burchellii DC.
Asteraceae
♦ guanobush, Molteno disease plant, Molteno disease senecio, ragwort
♦ Weed, Quarantine Weed, Native Weed
♦ 39, 51, 76, 87, 88, 121, 203, 220
♦ H, arid, toxic. Origin: Lesotho, Namibia, South Africa.

Senecio campestris (Retz.) DC.
Asteraceae
= *Senecio integrifolius* (L.) Clairv.
♦ Weed
♦ 87, 88
♦ herbal.

Senecio canadensis L.
Asteraceae
♦ Weed, Naturalised
♦ 98, 203

Senecio cannabinifolius Hook. & Arn.
Asteraceae
♦ hempleaf ragwort
♦ Naturalised
♦ 101

Senecio chrysocoma Meerb.
Asteraceae
Senecio graminifolius Phil., *Senecio paniculatus* Berg. var. *reclinatus* (L.f.) Harv.
♦ Weed
♦ 88, 158

Senecio cineraria DC.
Asteraceae
= *Senecio bicolor* (Willd.) Tod. ssp. *cineraria* (DC.) Chater
♦ dusty miller, silver ragwort, cineraria, silver ragweed
♦ Weed, Naturalised
♦ 39, 40, 116, 280
♦ S, cultivated, herbal, toxic.

Senecio citriformis G.Rowley
Asteraceae
♦ Weed, Quarantine Weed, Naturalised
♦ 76, 86, 88, 220
♦ cultivated.

Senecio clivorum Maxim.
Asteraceae
♦ summer ragwort
♦ Naturalised
♦ 101

Senecio compactus (A.Gray) Rydb.
Asteraceae
♦ Weed
♦ 87, 88

Senecio congestus (R.Br.) DC.
Asteraceae
Cineraria congesta R.Br., *Othonna palustris* L., *Senecio palustris* (L.) Hook.

nom. illeg., Senecio tubicaulis Mansf., *Tephroseris palustris* (L.) Fourr.
- ♦ kosteikkovillakko, marsh fleabane
- ♦ Casual Alien
- ♦ 42
- ♦ herbal.

Senecio consanguineus DC.
Asteraceae
- ♦ ragwort, starvation bush, starvation senecio
- ♦ Weed, Quarantine Weed, Native Weed
- ♦ 51, 76, 87, 88, 121, 158, 203, 220
- ♦ aH. Origin: southern Africa.

Senecio coronopifolius Desf.
Asteraceae
- ♦ groundsel
- ♦ Weed
- ♦ 87, 88, 243
- ♦ H.

Senecio cotyledonis DC.
Asteraceae
- ♦ Weed, Quarantine Weed
- ♦ 76, 88, 220

Senecio crassiflorus (Poir.) DC.
Asteraceae
- ♦ Weed, Naturalised, Casual Alien
- ♦ 86, 98, 203, 280
- ♦ Origin: South America.

Senecio crassissimus Humbert
Asteraceae
- ♦ vertical leaf
- ♦ Weed, Quarantine Weed
- ♦ 76, 88, 220
- ♦ cultivated. Origin: Madagascar.

Senecio daltonii F.Muell.
Asteraceae
- ♦ Dalton weed, Dalton's groundsel
- ♦ Weed, Native Weed
- ♦ 87, 88, 269
- ♦ pH. Origin: Australia.

Senecio decaryi Humbert
Asteraceae
- ♦ Weed, Quarantine Weed
- ♦ 76, 88, 220

Senecio deltoideus Less.
Asteraceae
Cacalia scandens Thunb., *Eupatorium auriculatum* Lam., *Mikania auriculata* Willd., *Senecio durbanensis* Gand., *Senecio fimbrillifer* (B.Rob.), *Senecio mikaniae* DC., *Senecio mikaniaeformis* DC., *Senecio rooseveltianus* De Wild., *Senecio sarmentosus* O.Hoffm., *Senecio tanzaniensis* Cuf.
- ♦ canary creeper
- ♦ Weed, Quarantine Weed, Native Weed
- ♦ 76, 88, 121, 203, 220
- ♦ pC. Origin: southern Africa.

Senecio desfontainei Druce
Asteraceae
- ♦ Weed
- ♦ 87, 88, 221, 243
- ♦ cultivated.

Senecio diaschides D.Drury
Asteraceae

- ♦ Naturalised, Environmental Weed
- ♦ 7, 86, 280
- ♦ Origin: Australia.

Senecio discifolius Oliv.
Asteraceae
- ♦ Weed
- ♦ 87, 88

Senecio discolor (Sw.) DC.
Asteraceae
- ♦ Weed
- ♦ 87, 88

Senecio doria L.
Asteraceae
Senecio altissimus Mill., *Senecio carnosus* Lam.
- ♦ Weed
- ♦ 272
- ♦ cultivated.

Senecio elegans L.
Asteraceae
Senecio pseudo-elegans Less.
- ♦ purple groundsel, purple ragwort, redpurple ragwort
- ♦ Weed, Naturalised, Environmental Weed, Cultivation Escape
- ♦ 7, 15, 34, 70, 72, 86, 88, 98, 101, 116, 161, 176, 180, 198, 203, 252, 280, 289
- ♦ aH, arid, cultivated, herbal, toxic. Origin: South Africa.

Senecio elliottii Torr. & A.Gray
Asteraceae
- ♦ Quarantine Weed
- ♦ 220

Senecio elongatus Druce
Asteraceae
- ♦ Quarantine Weed
- ♦ 220

Senecio eremophilus Richards.
Asteraceae
- ♦ lännenvillakko, desert ragwort, desert groundsel, cut leaved ragwort
- ♦ Weed, Casual Alien
- ♦ 42, 161
- ♦ herbal, toxic.

Senecio erraticus Bertol.
Asteraceae
Senecio barbaraefolius Krock.
- ♦ Weed
- ♦ 39, 87, 88
- ♦ cultivated, herbal, toxic.

Senecio erucifolius L.
Asteraceae
- ♦ hoary ragwort, hoary groundsel, liuskavillakko
- ♦ Weed, Naturalised, Casual Alien
- ♦ 42, 70, 87, 88, 101, 272
- ♦ pH, cultivated, herbal.

Senecio filaginoides DC.
Asteraceae
- ♦ mata mora
- ♦ Weed
- ♦ 237, 295

Senecio flaccidus Less.
Asteraceae
- ♦ threadleaf ragwort, shrubby douglasii butterweed

- ♦ Weed
- ♦ 161
- ♦ S, toxic.

Senecio flavus (Decne.) Sch.Bip.
Asteraceae
- ♦ Weed
- ♦ 221

Senecio fluviatilis Wallr.
Asteraceae
- ♦ jokivillakko, broad leaved ragwort
- ♦ Weed, Cultivation Escape
- ♦ 42, 272
- ♦ cultivated.

Senecio formosus H.B.K.
Asteraceae
- ♦ Weed
- ♦ 87, 88

Senecio fulgens Rydb.
Asteraceae
Kleinia fulgens Hook.f.
- ♦ scarlet klenia, scarlet knights
- ♦ Weed, Quarantine Weed
- ♦ 76, 88, 220
- ♦ cultivated. Origin: South Africa.

Senecio gallicus Chaix
Asteraceae
- ♦ Weed
- ♦ 70, 88, 94

Senecio glabellus Poir.
Asteraceae
= *Packera glabella* (Poir.) C.Jeffrey
- ♦ cressleaf groundsel, butterweed
- ♦ Weed, Quarantine Weed
- ♦ 39, 76, 87, 88, 161, 203, 207, 218, 220
- ♦ herbal, toxic. Origin: North America.

Senecio glastifolius L.f.
Asteraceae
- ♦ large senecio, pink ragwort, holly leaved senecio, waterdissel
- ♦ Weed, Quarantine Weed, Naturalised, Native Weed, Garden Escape, Environmental Weed
- ♦ 7, 15, 54, 76, 86, 88, 121, 155, 181, 203, 220, 246, 251, 280, 289
- ♦ pH, cultivated. Origin: southern Africa.

Senecio glaucus L. ssp. coronopifolius (Maire) J.C.M.Alexander
Asteraceae
- ♦ Weed
- ♦ 88, 185

Senecio glutinosus Thunb.
Asteraceae
- ♦ Native Weed
- ♦ 121
- ♦ aH. Origin: southern Africa.

Senecio gramineus Harv.
Asteraceae
- ♦ Weed, Quarantine Weed, Native Weed
- ♦ 76, 88, 121, 203, 220
- ♦ pH. Origin: southern Africa.

Senecio grandiflorus Berg.
Asteraceae
- ♦ Weed
- ♦ 23, 88

Senecio grisebachii **Bak.**
Asteraceae
♦ primavera
♦ Weed
♦ 87, 88, 236, 237, 271, 295

Senecio hallianus **Rowley**
Asteraceae
♦ Weed, Quarantine Weed
♦ 76, 88, 220
♦ Origin: South Africa.

Senecio herreianus **Dinter**
Asteraceae
♦ Weed, Quarantine Weed
♦ 76, 88, 220
♦ cultivated. Origin: South Africa.

Senecio heterotrichius **DC.**
Asteraceae
♦ Weed
♦ 87, 88

Senecio hieraciifolius **L.**
Asteraceae
= *Erechtites hieraciifolia* (L.) Raf. ex DC.
♦ Weed, Quarantine Weed
♦ 76, 88, 203, 220

Senecio hybridus **(Willd.) Regel**
Asteraceae
= *Pericallis cuneata* (L'Hér.) Bolle
♦ common ragwort, florist's cineraria
♦ Introduced
♦ 34
♦ pH, cultivated, herbal.

Senecio ilicifolius **L.**
Asteraceae
♦ sprinkaan senecio, ragwort
♦ Weed, Native Weed
♦ 39, 51, 87, 88, 121, 158
♦ H, toxic. Origin: southern Africa.

Senecio inaequidens **DC.**
Asteraceae
♦ Molteno disease senecio, canary weed, Burchell senecio
♦ Weed, Quarantine Weed, Native Weed
♦ 76, 88, 121, 158, 203, 220, 253
♦ pH, arid, cultivated, herbal, toxic. Origin: Lesotho, South Africa, Swaziland.

Senecio incognitus **Cabr.**
Asteraceae
= *Senecio madagascariensis* Poir.
♦ Weed
♦ 87, 88

Senecio inornatus **DC.**
Asteraceae
♦ Weed, Quarantine Weed, Native Weed
♦ 76, 88, 121, 203, 220
♦ pH. Origin: southern Africa.

Senecio integerrimus **Nutt.**
Asteraceae
♦ forest groundsel, lamb's tongue groundsel, lamb's tongue ragwort, tall western senecio
♦ Weed
♦ 39, 161
♦ pH, herbal, toxic.

Senecio integrifolius **(L.) Clairv.**
Asteraceae
Cineraria campestris Retz., *Othonna integrifolia* L., *Senecio campestris* (Retz.) DC. (see), *Senecio pratensis* (Hoppe) DC., *Senecio korabensis* Kümmerle & Jáv.
♦ ehytlehtivillakko, field fleawort
♦ Weed, Casual Alien
♦ 42, 87, 88, 272

Senecio integrifolius **(L.) Clairv. ssp.**
fauriei **Kitam.**
Asteraceae
♦ Weed
♦ 286

Senecio iosensis **G.Rowley**
Asteraceae
♦ Weed, Quarantine Weed
♦ 76, 88, 220

Senecio isatideus **DC.**
Asteraceae
♦ Dan's cabbage, poisonous ragwort
♦ Weed, Quarantine Weed, Native Weed
♦ 39, 76, 88, 121, 203, 220
♦ pH, toxic. Origin: southern Africa.

Senecio jacobaea **L.**
Asteraceae
Jacobaea vulgaris Gaertn.
♦ tansy ragwort, stinking Willy, ragwort, St. James wort, stinking Willie, common ragwort, jaakonvillakko
♦ Weed, Quarantine Weed, Noxious Weed, Naturalised, Garden Escape, Environmental Weed, Cultivation Escape
♦ 1, 7, 15, 23, 26, 35, 35, 36, 39, 42, 45, 52, 62, 70, 72, 76, 78, 80, 86, 87, 88, 98, 101, 116, 136, 139, 146, 147, 161, 165, 171, 176, 180, 181, 195, 198, 203, 212, 218, 219, 219, 225, 229, 231, 243, 246, 253, 267, 269, 272, 280, 287, 289, 291, 296
♦ pH, cultivated, herbal, toxic. Origin: Eurasia.

Senecio junceus **(DC.) Harv.**
Asteraceae
Brachyrhynchos junceus DC.
♦ Weed, Quarantine Weed
♦ 76, 88, 220

Senecio juniperinus **L.f.**
Asteraceae
♦ Weed, Quarantine Weed
♦ 76, 88, 203, 220

Senecio juniperinus **L.f. var.** *juniperinus*
Asteraceae
♦ Weed, Quarantine Weed, Native Weed
♦ 121, 220
♦ S. Origin: southern Africa.

Senecio kirilowii **Turcz.**
Asteraceae
Senecio integrifolius Clairv. var. *spathulatus* Hara
♦ Kirilow's groundsel
♦ Weed
♦ 297

Senecio kleinia **Less.**
Asteraceae
♦ Weed, Quarantine Weed
♦ 76, 88, 220
♦ cultivated.

Senecio latifolius **DC.**
Asteraceae
Senecio sceleratus Schweick.
♦ Dan's cabbage, groundsel, Molteno disease plant, pictou disease, ragwort, Rhodesian ragwort, senecio, staggers bush, winton disease, chiGurangu
♦ Weed, Quarantine Weed, Native Weed
♦ 39, 50, 76, 87, 88, 121, 158, 203, 220
♦ pH, herbal, toxic. Origin: southern Africa.

Senecio lautus **(Forst.f.) Willd.**
Asteraceae
♦ fireweed, variable groundsel
♦ Weed, Naturalised, Native Weed
♦ 7, 87, 88, 269
♦ cultivated. Origin: Australia, New Zealand.

Senecio leucanthemifolius **Poir.**
Asteraceae
♦ Weed, Quarantine Weed
♦ 76, 87, 88, 203, 220

Senecio leucanthemifolius **Poir. ssp.**
vernalis **(Waldst. & Kit.) Alexander**
Asteraceae
♦ Weed
♦ 253

Senecio linearifolius **A.Rich**
Asteraceae
♦ fireweed, fireweed groundsel
♦ Native Weed, Naturalised
♦ 39, 269, 280
♦ pH, cultivated, toxic. Origin: Australia.

Senecio lividus **L.**
Asteraceae
♦ sinivillakko
♦ Weed, Casual Alien
♦ 42, 250

Senecio longiflorus **(DC.) Sch.Bip.**
Asteraceae
Kleinia longiflora DC. (see)
♦ Weed, Quarantine Weed
♦ 76, 88, 220
♦ cultivated.

Senecio longilobus **Benth.**
Asteraceae
= *Senecio flaccidus* Less. var. *flaccidus* (NoR)
♦ threadleaf groundsel
♦ Weed
♦ 39, 87, 88, 161, 212, 218
♦ herbal, toxic.

Senecio macroglossus **DC.**
Asteraceae
♦ cape ivy
♦ Casual Alien
♦ 280
♦ cultivated. Origin: South Africa.

Senecio madagascariensis **Poir.**
Asteraceae

Senecio burchellii auct. *non* (DC.) Cabr., *Senecio incognitus* Cabr. (see)
♦ fireweed, senecio amarillo, Madagascar ragwort
♦ Weed, Quarantine Weed, Noxious Weed, Naturalised, Environmental Weed
♦ 3, 76, 86, 88, 98, 101, 147, 158, 169, 178, 191, 203, 229, 232, 236, 237, 269, 271, 290, 295
♦ aH. Origin: South Africa, Natal, Swaziland.

Senecio mandraliscae (Tineo) Jacobsen
Asteraceae
♦ blue iceplant
♦ Weed, Quarantine Weed
♦ 76, 88, 220
♦ cultivated.

Senecio mikanioides Otto ex Walp.
Asteraceae
= *Delairea odorata* Lem.
♦ German ivy, Italian ivy, cape ivy, murattimikania
♦ Weed, Noxious Weed, Naturalised, Introduced, Environmental Weed, Cultivation Escape
♦ 3, 15, 34, 35, 39, 42, 73, 78, 80, 87, 88, 107, 116, 161, 165, 191, 218, 225, 241, 246, 280, 300
♦ pH, cultivated, herbal, toxic. Origin: South Africa.

Senecio minimus Poir.
Asteraceae
= *Erechtites minima* (Poir.) DC.
♦ fireweed
♦ Weed
♦ 181
♦ herbal.

Senecio moorei R.E.Fr.
Asteraceae
♦ Weed, Quarantine Weed
♦ 76, 87, 88, 203, 220

Senecio mulgediifolius Schauer
Asteraceae
♦ Weed
♦ 199

Senecio mutabilis Greene
Asteraceae
= *Packera neomexicana* (Gray) W.A.Weber & Á.Löve var. *mutabilis* (Greene) W.A.Weber & Á.Löve (NoR)
♦ Weed
♦ 23, 88
♦ herbal.

Senecio nemorensis L.
Asteraceae
Senecio ganpinensis Vaniot, *Senecio jacquinianus* Reichen, *Senecio kematongensis* Vaniot, *Senecio nemorensis* var. *octoglossus* (DC.) Koch ex Ledeb., *Senecio nemorensis* var. *subinteger* Hara, *Senecio nemorensis* var. *taiwanensis* (Hayata) Yamam., *Senecio nemorensis* var. *turczaninowii* (DC.) Kom., *Senecio octoglossus* DC., *Senecio sarracenicus* L., *Senecio sarracenicus* var. *turczaninowii* (DC.) Nakai, *Senecio taiwanensis* Hayata, *Senecio tozanensis* Hayata

♦ lehtovillakko, kion
♦ Weed, Cultivation Escape
♦ 42, 272
♦ pH, cultivated, herbal.

Senecio obovatus Muhl. ex Willd.
Asteraceae
= *Packera obovata* (Muhl. ex Willd.) W.A.Weber & Á.Löve (NoR) [see *Senecio rotundus* (Britt.) Small]
♦ squaw weed, roundleaf ragwort
♦ Quarantine Weed
♦ 161, 220
♦ herbal, toxic.

Senecio oligophyllus Bak.
Asteraceae
♦ Weed, Quarantine Weed
♦ 76, 87, 88, 203, 220

Senecio oxyriaefolius DC.
Asteraceae
= *Senecio oxyriifolius* DC.
♦ Weed, Quarantine Weed
♦ 76, 87, 88, 203, 220
♦ cultivated.

Senecio oxyriifolius DC.
Asteraceae
Senecio orbicularis Sond. ex Harv., *Senecio oxyriaefolius* DC. (see), *Senecio peltatus* DC., *Senecio peltiformis* DC., *Senecio subnudus* DC., *Senecio subpeltatus* Steud.
♦ Native Weed
♦ 121
♦ pH. Origin: southern Africa.

Senecio paludosus L.
Asteraceae
Jacobaea paludosa Gaertn., Mey. & Scherb.
♦ fen ragwort, great fen ragwort, senecione palustre
♦ Weed, Naturalised
♦ 272, 287
♦ herbal.

Senecio papposus (Rchb.) Less.
Asteraceae
Senecio bosniacus Beck
♦ downy fleawort
♦ Weed
♦ 272

Senecio pauperculus Michx.
Asteraceae
= *Packera paupercula* (Michx.) Á. & D.Löve (NoR)
♦ balsam groundsel
♦ Weed
♦ 23, 88, 161
♦ herbal, toxic.

Senecio petasitis (Sims) DC.
Asteraceae
= *Roldana petasitis* (DC.) H.Rob. & Brettell
♦ velvet groundsel
♦ Weed, Quarantine Weed, Naturalised, Environmental Weed
♦ 15, 76, 246, 280
♦ cultivated.

Senecio pinnatifidus Less.
Asteraceae

♦ Weed
♦ 87, 88

Senecio plattensis Nutt.
Asteraceae
= *Packera plattensis* (Nutt.) W.A.Weber & Á.Löve (NoR)
♦ prairie groundsel
♦ Weed
♦ 39, 161
♦ herbal, toxic.

Senecio platylepis DC.
Asteraceae
♦ toothed groundsel
♦ Weed
♦ 87, 88
♦ aH, cultivated. Origin: Australia.

Senecio polyanthemoides Poir.
Asteraceae
♦ fireweed
♦ Weed
♦ 88, 158

Senecio pterophorus DC.
Asteraceae
Senecio pterophorus var. *apterus* DC. (see)
♦ African daisy, South African daisy, winged groundsel
♦ Weed, Quarantine Weed, Noxious Weed, Naturalised, Native Weed, Environmental Weed
♦ 39, 72, 76, 86, 88, 121, 147, 198, 203, 220, 269, 290
♦ pH, arid, toxic. Origin: southern Africa.

Senecio pterophorus DC. var. *apterus* DC.
Asteraceae
= *Senecio pterophorus* DC.
♦ Naturalised
♦ 98

Senecio pyrenaicus L.
Asteraceae
♦ Weed
♦ 70

Senecio quadridentatus Labill.
Asteraceae
♦ cotton fireweed
♦ Weed, Native Weed
♦ 39, 87, 88, 269
♦ pH, cultivated, toxic. Origin: Australia.

Senecio retrorsus DC.
Asteraceae
Senecio barbellatus DC., *Senecio graminicolus* C.A.Sm., *Senecio latifolius* DC. var. *retrorsus* (DC.) Harv., *Senecio latifolius* DC. var. *barbellatus* (DC.) Harv., *Senecio retrorsus* DC. var. *subedentulus* DC.
♦ bushweed, Dan's cabbage, grass staggers weed, Molteno disease plant, poisonous ragwort, staggers bush, staggers senecio
♦ Weed, Quarantine Weed, Native Weed
♦ 39, 76, 88, 121, 203, 220
♦ pH, arid, toxic. Origin: southern Africa.

Senecio riddellii Torr. & Gray
Asteraceae
♦ Riddell's ragwort, Riddell's groundsel
♦ Weed, Native Weed
♦ 39, 87, 88, 161, 174, 212, 218
♦ herbal, toxic. Origin: North America.

Senecio rotundus (Britt.) Small
Asteraceae
= *Packera obovata* (Muhl. ex Willd.) W.A.Weber & Á.Löve (NoR) [see *Senecio obovatus* Muhl. ex Willd.]
♦ Quarantine Weed
♦ 220

Senecio rowleyanus Jacobsen
Asteraceae
♦ string of beads
♦ Weed, Quarantine Weed
♦ 76, 88, 220
♦ cultivated.

Senecio rupestris Waldst. & Kit.
Asteraceae
Senecio laciniatus Bertol.
♦ rock ragwort, senecio
♦ Weed, Naturalised
♦ 87, 88, 101
♦ herbal.

Senecio ruwenzoriensis S.Moore
Asteraceae
♦ Weed
♦ 87, 88

Senecio salignus DC.
Asteraceae
= *Barkleyanthus salicifolius* (Kunth) H.E.Rob. & Brett. (NoR)
♦ willow ragwort
♦ Weed
♦ 199
♦ herbal.

Senecio scandens Buch.-Ham. ex D.Don
Asteraceae
= *Delairea odorata* Lem.
♦ groundsel
♦ Naturalised
♦ 86
♦ pC, cultivated, herbal.

Senecio schimperi Sch.Bip. ex A.Rich.
Asteraceae
♦ Weed
♦ 88

Senecio serpens G.D.Rowley
Asteraceae
Kleinia repens L.
♦ senecio, dusty miller, blue chalksticks
♦ Weed, Quarantine Weed, Casual Alien
♦ 76, 88, 220, 280
♦ cultivated.

Senecio skirrhodon DC.
Asteraceae
♦ gravel groundsel
♦ Weed, Naturalised
♦ 15, 165, 280
♦ Origin: Madagascar, Africa.

Senecio spartioides Torr. & A.Gray
Asteraceae
♦ broom groundsel, broomlike ragwort
♦ Weed
♦ 39, 161
♦ pH, herbal, toxic.

Senecio squalidus L.
Asteraceae
♦ Oxford ragwort, rikkavillakko
♦ Weed, Noxious Weed, Naturalised, Casual Alien
♦ 35, 40, 42, 70, 88, 101, 229
♦ cultivated.

Senecio sylvaticus L.
Asteraceae
Obaejaca silvatica Cass.
♦ woodland groundsel, woodland ragwort, mountain groundsel, wood groundsel, heath groundsel, kalliovillakko
♦ Weed, Quarantine Weed, Naturalised
♦ 15, 23, 34, 76, 80, 87, 88, 101, 136, 161, 203, 212, 218, 220, 241, 250, 280, 300
♦ aH, cultivated, herbal, toxic.

Senecio tamoides DC.
Asteraceae
♦ canary creeper, climbing cineraria
♦ Weed, Naturalised, Native Weed
♦ 7, 86, 121, 198
♦ pC, cultivated. Origin: southern Africa.

Senecio tomentosus Michx.
Asteraceae
= *Packera tomentosa* (Michx.) C.Jeffrey (NoR)
♦ woolly ragwort, southern woolly groundsel
♦ Weed
♦ 161
♦ herbal, toxic.

Senecio triangularis Hook.
Asteraceae
♦ arrowleaf groundsel, arrowleaf ragwort, ragwort
♦ Weed
♦ 87, 88, 218
♦ pH, cultivated, herbal.

Senecio vernalis Waldst. & Kit.
Asteraceae
Jacobaea incana Gilib., *Senecio rapistroides* DC., *Senecio sinuatus* Gilib.
♦ spring groundsel, kevätvillakko, eastern groundsel
♦ Weed, Naturalised, Casual Alien
♦ 40, 42, 70, 87, 88, 94, 243, 272, 287
♦ H, cultivated.

Senecio viravira Hieron.
Asteraceae
♦ dusty miller
♦ Weed
♦ 237
♦ cultivated. Origin: Argentina.

Senecio viscosus L.
Asteraceae
Obaejaca viscosa Cass.
♦ sticky groundsel, sticky ragwort, tahmavillakko
♦ Weed, Noxious Weed, Naturalised

♦ 42, 70, 87, 88, 94, 101, 218, 272, 287, 299
♦ aH, cultivated, herbal.

Senecio vulgaris L.
Asteraceae
♦ ragwort, groundsel, sticky groundsel, stinking groundsel, wood groundsel, senecione, old man in the spring, grimsel, simson, birdseed, peltovillakko, common fireweed
♦ Weed, Noxious Weed, Naturalised, Garden Escape, Environmental Weed
♦ 7, 15, 23, 24, 34, 39, 44, 68, 70, 80, 86, 87, 88, 94, 98, 101, 102, 118, 136, 161, 165, 167, 176, 180, 185, 195, 198, 203, 204, 211, 212, 217, 218, 221, 229, 236, 237, 241, 243, 250, 253, 263, 269, 272, 280, 286, 287, 295, 297, 299, 300
♦ aH, arid, cultivated, herbal, toxic. Origin: Mediterranean, west Asia.

Senegalia westiana (DC.) Britton & Rose
Fabaceae/Mimosaceae
= *Acacia westiana* DC. (NoR)
♦ Weed
♦ 87, 88

Senna alata (L.) Roxb.
Fabaceae/Caesalpiniaceae
Cassia alata L. (see)
♦ candle bush, emperor's candlesticks, ringworm shrub, ringworm bush, ringworm senna, empress candle plant, Christmas candle, seven golden candlesticks, candlestick senna
♦ Weed, Quarantine Weed, Noxious Weed, Naturalised, Garden Escape, Environmental Weed
♦ 3, 22, 76, 86, 88, 93, 107, 179, 191, 191, 220, 255
♦ S, cultivated, herbal. Origin: South America.

Senna alexandrina Mill.
Fabaceae/Caesalpiniaceae
Cassia senna L. (see)
♦ Alexandrian senna, Alexandrian wild sensitive plant
♦ Naturalised
♦ 39, 101
♦ arid, cultivated, herbal, toxic.

Senna aphylla (Cav.) Irwin & Barneby
Fabaceae/Caesalpiniaceae
Cassia aphylla Cav.
♦ Introduced
♦ 228
♦ arid.

Senna artemisioides (Gaud. ex DC.) Randell
Fabaceae/Caesalpiniaceae
♦ silver senna, desert cassia, senna
♦ Naturalised, Introduced
♦ 101, 228
♦ arid, cultivated.

Senna artemisioides (Gaud. ex DC.) Randell ssp. coriacea (Benth.) Randell
Fabaceae/Caesalpiniaceae
Cassia eremophila Vogel var. *coriacea* Benth. *p.p.*, *Cassia nemophila* Cunn. ex J.Vogel var. *coriacea* (Benth.) Symon, *Cassia sturtii* R.Br. var. *coriacea* Benth.
♦ Introduced

♦ 228
♦ arid.

Senna artemisioides (Gaud. ex DC.) Randell ssp. *filifolia* Randell
Fabaceae/Caesalpiniaceae
Cassia eremophila Cunn ex Vogel var. *eremophila*
♦ punty bush
♦ Native Weed
♦ 269
♦ S, cultivated. Origin: Australia.

Senna artemisioides (Gaud. ex DC.) Randell ssp. *sturtii* (R.Br.) Randell
Fabaceae/Caesalpiniaceae
Cassia sturtii R.Br.
♦ Introduced
♦ 228

Senna artemisioides (Gaud. ex DC.) Randell var. *platypoda* (R.Br.) Benth
Fabaceae/Caesalpiniaceae
♦ punty bush
♦ Native Weed
♦ 269
♦ S. Origin: Australia.

Senna atomaria (L.) Irwin & Barneby
Fabaceae/Caesalpiniaceae
♦ flor de San Jose, frijolillo
♦ Naturalised
♦ 101
♦ cultivated.

Senna auriculata (L.) Roxb.
Fabaceae/Caesalpiniaceae
Cassia auriculata L. (see), *Cassia densistipulata* Taub.
♦ anwal, anwala, avaram, matara tea, tanner's cassia, tarwar
♦ Introduced
♦ 228
♦ arid, cultivated, herbal.

Senna barclayana (Sweet) Randell
Fabaceae/Caesalpiniaceae
♦ pepperleaf senna
♦ Naturalised, Native Weed
♦ 7, 86, 269
♦ S, cultivated. Origin: Australia.

Senna bauhinioides (Gray) Irwin & Barneby
Fabaceae/Caesalpiniaceae
Cassia bauhinioides A.Gray (see)
♦ twoleaf desert senna, twinleaf senna
♦ Weed
♦ 161
♦ herbal.

Senna bicapsularis (L.) Roxb.
Fabaceae/Caesalpiniaceae
Cassia bicapsularis L. (see), *Cassia emarginata* L. (see), *Cassia manzanilloana* Rose, *Cassia ovalifolia* M.Martens & Galeotti
♦ Christmas bush, rambling cassia
♦ Noxious Weed, Naturalised, Introduced, Garden Escape
♦ 38, 228, 283, 300
♦ arid, cultivated. Origin: West Indies, South America.

Senna birostris (Vogel) H.S.Irwin & Barneby var. *hookeriana* (Hook.) H.S.Irwin & Barneby
Fabaceae/Caesalpiniaceae
Cassia hookeriana Gill. ex Hook & Arn. (see)
♦ Introduced
♦ 228
♦ arid.

Senna corymbosa (Lam.) Irwin & Barneby
Fabaceae/Caesalpiniaceae
Cassia bonariensis Colla, *Cassia corymbosa* Lam. (see)
♦ Argentine senna, Argentine wild sensitive plant
♦ Weed, Naturalised, Introduced, Garden Escape
♦ 86, 98, 101, 179, 203, 228, 283
♦ arid, cultivated, toxic. Origin: north Argentina, Uruguay, southern Brazil.

Senna covesii (Gray) Irwin & Barneby
Fabaceae/Caesalpiniaceae
♦ desert senna, Coves's cassia, Coves's senna
♦ Weed
♦ 161
♦ pH, cultivated.

Senna crassiramea (Benth.) Irwin & Barneby
Fabaceae/Caesalpiniaceae
Cassia crassiramea Benth.
♦ Introduced
♦ 228
♦ arid.

Senna didymobotrya (Fresen.) Irwin & Barneby
Fabaceae/Caesalpiniaceae
Cassia didymobotrya Fresen. (see), *Cassia nairobensis* L.Bailey
♦ wild senna, peanut butter cassia, African wild sensitive plant, candelabra tree, African senna, popcorn bush, grondboontjiebotter-kassia
♦ Weed, Noxious Weed, Naturalised, Introduced, Cultivation Escape
♦ 22, 86, 88, 95, 101, 158, 161, 179, 228, 283
♦ S, arid, cultivated, toxic. Origin: tropical Africa.

Senna × *floribunda* (Cav.) Irwin & Barneby
Fabaceae/Caesalpiniaceae
= *Senna septemtrionalis* (Viv.) Irwin & Barneby × *Senna multiglandulosa* (Jacq.) Irwin & Barneby [see *Cassia floribunda* Cav.]
♦ smooth cassia, winter cassia
♦ Weed, Naturalised, Environmental Weed
♦ 73, 86, 88, 98, 101, 201, 269
♦ S, cultivated, herbal. Origin: Central and South America.

Senna hirsuta (L.) Irwin & Barneby
Fabaceae/Caesalpiniaceae
♦ slimpod glaberrima senna, woolly senna, woolly wild sensitive plant
♦ Weed, Naturalised, Garden Escape
♦ 86, 161, 255, 283
♦ cultivated, herbal, toxic. Origin: Brazil.

Senna hirsuta (L.) Irwin & Barneby var. *hirsuta*
Fabaceae/Caesalpiniaceae
Cassia hirsuta L. (see), *Cassia tomentosa* Wall. ex Arn., *Ditremexa hirsuta* (L.) Britt. & Rose ex Britt. & Wilson
♦ shower tree senna, woolly senna, woolly wild sensitive plant
♦ Weed
♦ 80, 261
♦ Origin: South America.

Senna holosericea (Fresen.) Greuter
Fabaceae/Caesalpiniaceae
Cassia holosericea Fresen.
♦ Introduced
♦ 228
♦ arid.

Senna italica Mill.
Fabaceae/Caesalpiniaceae
Cassia aschrek Forssk., *Cassia italica* (Mill.) F.W.Andrews, *Cassia obovata* Collad.
♦ Port Royal senna, Port Royal wild sensitive plant
♦ Naturalised
♦ 101
♦ arid, herbal.

Senna ligustrina (L.) Irwin & Barneby
Fabaceae/Caesalpiniaceae
Cassia ligustrina L. (see)
♦ privet senna, privet wild sensitive plant
♦ Introduced
♦ 261

Senna lindheimeriana (Scheele) Irwin & Barneby
Fabaceae/Caesalpiniaceae
♦ velvetleaf wild sensitive plant, velvetleaf senna, Lindheimer's senna
♦ Weed
♦ 161
♦ herbal, toxic.

Senna macranthera (DC. ex Collad.) Irwin & Barneby
Fabaceae/Caesalpiniaceae
♦ Weed
♦ 32

Senna multiglandulosa (Jacq.) Irwin & Barneby
Fabaceae/Caesalpiniaceae
Adipera tomentosa (L.f.) Britton & Rose, *Cassia lutescens* G.Don, *Cassia multiglandulosa* Jacq., *Cassia tomentosa* L.f. (see)
♦ buttercup bush, downy senna
♦ Weed, Naturalised, Introduced, Garden Escape
♦ 15, 34, 86, 98, 198, 203, 228, 280, 283
♦ S, arid, cultivated, toxic. Origin: Central and South America.

Senna multijuga (L.C.Rich.) Irwin & Barneby
Fabaceae/Caesalpiniaceae
♦ false sicklepod, November shower
♦ Naturalised
♦ 101
♦ T, cultivated.

Senna multijuga (L.C.Rich.) Irwin & Barneby ssp. *multijuga*
Fabaceae/Caesalpiniaceae
♦ Garden Escape
♦ 261
♦ cultivated. Origin: continental tropical America.

Senna obtusifolia (L.) Irwin & Barneby
Fabaceae/Caesalpiniaceae
Cassia obtusifolia L. (see), *Cassia tora sensu auct. mult. non* L., *Cassia tora* L. var. *obtusifolia* L.
♦ sicklepod senna, coffeeweed, Chinese senna, sicklepod, Java bean, biche manso, casse puante, cassia fétide, chakunda, dormilón, ejotil, ejotil, foetid cassia, sickle senna, fedegoso, dormidera
♦ Weed, Quarantine Weed, Noxious Weed, Naturalised, Introduced
♦ 62, 76, 80, 86, 88, 93, 98, 102, 191, 203, 206, 228, 243, 255, 261
♦ pH/S, arid, cultivated, herbal. Origin: tropical America.

Senna occidentalis (L.) Link
Fabaceae/Caesalpiniaceae
Cassia occidentalis L. (see), *Ditremexa occidentalis* (L.) Britton & Rose
♦ coffee senna, sickle pod, Nigerian senna, bricho, café des noirs, café nègre, casse café, etiatia, faux kinkéliba, fedegoso, feuille de Paradis, negro coffee, omitiwojika, omutiwojoka, stinking pea, stinking weed, septicweed, stinkweed, stypticweed, wild coffee
♦ Weed, Noxious Weed, Naturalised, Introduced, Cultivation Escape
♦ 6, 7, 80, 86, 88, 93, 98, 101, 102, 179, 203, 228, 237, 240, 255, 261, 283
♦ aH, arid, cultivated, herbal, toxic. Origin: tropical America.

Senna pendula (Willd.) H.S.Irwin & Barneby
Fabaceae/Caesalpiniaceae
♦ climbing cassia, Christmas senna, valamuerto, Easter cassia
♦ Weed, Quarantine Weed, Naturalised, Garden Escape, Environmental Weed
♦ 73, 88, 98, 101, 151, 155, 203, 220, 269, 290
♦ S, cultivated, herbal.

Senna pendula (Humb. & Bonpl. ex Willd.) Irwin & Barneby var. *advena* (Vogel) Irwin & Barneby
Fabaceae/Caesalpiniaceae
♦ valamuerto
♦ Naturalised
♦ 101

Senna pendula (Humb. & Bonpl. ex Willd.) Irwin & Barneby var. *glabrata* (Vogel) Irwin & Barneby
Fabaceae/Caesalpiniaceae
Cassia coluteoides Collad. (see)
♦ Easter cassia, valamuerto
♦ Weed, Noxious Weed, Naturalised, Garden Escape, Environmental Weed

♦ 86, 179, 283, 296
♦ cultivated, toxic. Origin: Brazil.

Senna pleurocarpa F.Muell. var. *pleurocarpa*
Fabaceae/Caesalpiniaceae
Cassia pleurocarpa F.Muell. var. *pleurocarpa*
♦ Introduced
♦ 228
♦ arid.

Senna polyphylla (Jacq.) Irwin & Barneby
Fabaceae/Caesalpiniaceae
♦ retama prieta, twin senna
♦ Weed
♦ 32

Senna roemeriana (Scheele) Irwin & Barneby
Fabaceae/Caesalpiniaceae
♦ twoleaf wild sensitive plant, twoleaf senna
♦ Weed
♦ 161
♦ herbal, toxic.

Senna septemtrionalis (Viv.) Irwin & Barneby
Fabaceae/Caesalpiniaceae
Adipera laevigata sensu Britton & Rose, *Adipera laevigata* (Willd.) Britton & Rose, *Cassia elegans* Kunth, *Cassia floribunda sensu* de Wit, *Cassia floribunda* Cav. var. *elegans* (Kunth) J.Vogel, *Cassia laevigata* Willd., *Cassia laevigata sensu* Collad., *Cassia septentrionalis* Zucc., *Cassia septentrionalis* Sessé & Moç., *Cassia vernicosa* D.Clos, *Chamaecassia laevigata* (Willd.) Link, *Chamaefistula laevigata* (Willd.) G.Don, *Senna aurata* Roxb.
♦ senna, buttercup bush, hedionda macho
♦ Weed, Naturalised, Introduced, Garden Escape, Environmental Weed
♦ 3, 101, 134, 191, 225, 228, 246, 280, 283, 300
♦ S, arid, cultivated, toxic. Origin: Central America, Mexico.

Senna siamea (Lam.) H.S.Irwin & Barneby
Fabaceae/Caesalpiniaceae
Cassia siamea Lam. (see)
♦ Siamese cassia, cassod tree, Bombay blackwood
♦ Weed, Naturalised, Introduced, Garden Escape
♦ 32, 86, 101, 191, 228, 261
♦ arid, cultivated. Origin: tropical Asia.

Senna spectabilis (DC.) Irwin & Barneby
Fabaceae/Caesalpiniaceae
♦ casia amarilla, spectacular cassia
♦ Naturalised
♦ 101
♦ T, cultivated. Origin: Central America.

Senna spectabilis (DC.) Irwin & Barneby var. *excelsa* (Schrad.) Irwin & Barneby
Fabaceae/Caesalpiniaceae

Cassia excelsa Schrad.
♦ Introduced
♦ 228
♦ arid.

Senna spectabilis (DC.) Irwin & Barneby var. *spectabilis*
Fabaceae/Caesalpiniaceae
Cassia spectabilis DC. (see)
♦ algarrobilla
♦ Garden Escape
♦ 261
♦ cultivated. Origin: tropical America.

Senna sulfurea (DC. ex Collad.) Irwin & Barneby
Fabaceae/Caesalpiniaceae
♦ smooth wild sensitive plant, smooth senna
♦ Weed, Naturalised, Environmental Weed
♦ 3, 101, 259

Senna surattensis (Burm.f.) Irwin & Barneby
Fabaceae/Caesalpiniaceae
Cassia surattensis Burm.f. (see)
♦ glossy shower, scrambled egg tree, Singapore shower
♦ Weed, Naturalised, Garden Escape
♦ 86, 101, 179, 261
♦ cultivated, herbal. Origin: East Indies, Australia.

Senna tora (L.) Roxb.
Fabaceae/Caesalpiniaceae
Cassia tora L. (see)
♦ Chinese senna, jue ming zi, foetid cassia, Java bean, sickle senna, sicklepod, biche manso, chakaunda, chakonda, chakowar, chakunda, pawad, takla
♦ Weed, Quarantine Weed, Noxious Weed, Naturalised, Casual Alien
♦ 6, 62, 76, 86, 88, 98, 101, 191, 203, 280
♦ arid, cultivated. Origin: India to China.

Senna uniflora (Mill.) Irwin & Barneby
Fabaceae/Caesalpiniaceae
Cassia uniflora Mill. (see)
♦ oneleaf senna, oneleaf wild sensitive plant
♦ Weed
♦ 261
♦ arid.

Sequoia gigantea (Lindl.) Decne.
Taxodiaceae/Cupressaceae
= *Sequoiadendron giganteum* (Lindl.) Buchholz (NoR)
♦ Californian big tree
♦ Weed
♦ 181

Sequoia sempervirens (D.Don) Endl.
Taxodiaceae/Cupressaceae
Taxodium sempervirens D.Don
♦ redwood, coastal redwood, Californian redwood
♦ Weed, Naturalised, Cultivation Escape
♦ 7, 39, 86, 87, 88, 181, 218, 252, 280
♦ T, cultivated, herbal, toxic. Origin: western North America.

Serapias cordigera L.
Orchidaceae
♦ heart flowered helleborine, heart flowered serapias
♦ Weed
♦ 272
♦ herbal.

Serapias lingua L.
Orchidaceae
♦ tongue orchid, tongue helleborine
♦ Weed
♦ 272
♦ herbal.

Serapias vomeracea (Burm.) Briq.
Orchidaceae
Serapias laxiflora Chaub.
♦ serapide maggiore
♦ Weed
♦ 272
♦ cultivated, herbal.

Serenoa repens (W.Bartram) Small
Arecaceae
Corypha repens W.Bartram, *Sabal serrulata* (Michx.) Nutt. ex Schult. & Schult.f.
♦ saw palmetto, sawtooth palmetto
♦ Weed
♦ 87, 88, 218
♦ S, cultivated, herbal.

Serialbizzia acle (Blanco) Kosterm.
Fabaceae/Mimosaceae
Albizia acle (Blanco) Merr., *Mimosa acle* Blanco
♦ Introduced
♦ 230
♦ T.

Seriola aethnensis L.
Asteraceae
♦ Weed
♦ 87, 88

Serissa foetida (L.f.) Lam.
Rubiaceae
♦ snowrose
♦ Naturalised
♦ 101
♦ cultivated, herbal.

Serjania acoma Radlk.
Sapindaceae
♦ Weed
♦ 87, 88

Serjania diversifolia (Jacq.) Radlk.
Sapindaceae
Serjania subdenata Juss.
♦ fowlsfoot
♦ Weed
♦ 14
♦ herbal.

Serjania exarata Radlk.
Sapindaceae
= *Serjania membranacea* Splitg. (NoR)
♦ Naturalised
♦ 86
♦ Origin: South America.

Serjania polyphylla (L.) Radlk.
Sapindaceae
♦ basketwood
♦ Weed

♦ 87, 88

Serjania rubicaulis (Ruiz & Pav.) Benth.
Sapindaceae
♦ Weed
♦ 87, 88

Serratula centauroides L.
Asteraceae
♦ common sawwort
♦ Weed
♦ 297

Serratula cerinthifolia Sibth. & Sm.
Asteraceae
♦ Weed
♦ 87, 88

Serratula radiata (Waldst. & Kit.) Bieb.
Asteraceae
♦ Weed
♦ 272
♦ cultivated.

Serratula tinctoria L.
Asteraceae
♦ dyer's plumeless sawwort, sawwort
♦ Weed, Naturalised
♦ 23, 88, 101, 272
♦ pH, cultivated, herbal.

Serratula wolfii Andrae
Asteraceae
♦ Quarantine Weed
♦ 220
♦ cultivated.

Sesamoides canescens (L.) Kuntze
Resedaceae
♦ tähtireseda
♦ Casual Alien
♦ 42
♦ cultivated.

Sesamum alatum Thonn.
Pedaliaceae
♦ sesamum
♦ Weed, Native Weed
♦ 87, 88, 121
♦ aH, cultivated, herbal. Origin: southern Africa.

Sesamum indicum L.
Pedaliaceae
= *Sesamum orientale* L.
♦ sesame, sim sim, benniseed, benue oil, gingelly oil plant, gingilli, sem sem oil, sesam, sesame ole, teel oil, tilseed, thunderbolt flower, wild foxglove
♦ Weed, Naturalised, Introduced, Casual Alien
♦ 42, 86, 88, 93, 98, 121, 203, 228
♦ aH, arid, cultivated, herbal. Origin: Eurasia, North Africa.

Sesamum orientale L.
Pedaliaceae
Sesamum indicum L. (see)
♦ African simsim, sesame
♦ Weed, Naturalised, Casual Alien
♦ 13, 32, 87, 88, 101, 179, 261
♦ aH, herbal. Origin: East Indies.

Sesamum radiatum Schumach. & Thonn.
Pedaliaceae
♦ sesamum
♦ Weed
♦ 87, 88

Sesamum schinzianum Asch.
Pedaliaceae
♦ Native Weed
♦ 121
♦ pH.

Sesamum triphyllum Welw. ex Asch.
Pedaliaceae
♦ wild sesame, thunderbolt flower
♦ Weed
♦ 88, 158
♦ Origin: southern Africa.

Sesamum triphyllum Welw. ex Asch. var. triphyllum
Pedaliaceae
♦ wild sesame
♦ Native Weed
♦ 121
♦ aH. Origin: southern Africa.

Sesbania aculeata (Willd.) Poir.
Fabaceae/Papilionaceae
= *Sesbania bispinosa* (Jacq.) Wight
♦ sesbania, sano khaang khok
♦ Weed
♦ 239
♦ aH, aqua, cultivated, herbal.

Sesbania arabica Steud. & Hochst. ex Phillips & Hutch.
Fabaceae/Papilionaceae
♦ Weed
♦ 242
♦ arid.

Sesbania benthamiana Domin
Fabaceae/Papilionaceae
♦ Weed
♦ 87, 88
♦ Origin: Australia.

Sesbania bispinosa (Jacq.) Wight
Fabaceae/Papilionaceae
Aeschynomene bispinosa Jacq., *Sesbania aculeata* (Willd.) Poir. (see)
♦ spiny sesbania, dunchi fiber
♦ Weed, Quarantine Weed, Naturalised, Introduced
♦ 76, 88, 101, 158, 203, 220, 228, 261, 262
♦ a/bH, arid, cultivated, herbal. Origin: Eurasia, Madagascar.

Sesbania bispinosa (Jacq.) Wight var. bispinosa
Fabaceae/Papilionaceae
♦ sesbania, spiny sesbania
♦ Weed
♦ 121
♦ a/bH. Origin: Eurasia.

Sesbania brachycarpa F.Muell.
Fabaceae/Papilionaceae
♦ Weed
♦ 205
♦ pH. Origin: Australia.

Sesbania cannabina (Retz.) Pers.
Fabaceae/Papilionaceae
♦ sesbania, sesbania pea, yellow pea bush
♦ Weed, Naturalised
♦ 23, 55, 87, 88, 287, 297
♦ arid, cultivated, herbal. Origin: Asia, Australia.

Sesbania drummondii (Rydb.) Cory
Fabaceae/Papilionaceae
Daubentonia drummondii Rydb.,
Daubentonia texana Pierce (see)
♦ Drummond's rattlebox, poison bean, rattlebox, rattlebush
♦ Weed
♦ 39, 82, 87, 88, 161, 218
♦ herbal, toxic.

Sesbania emerus (Aubl.) Urb.
Fabaceae/Papilionaceae
♦ rattlebox, bequilla, danglepod
♦ Weed
♦ 80, 80

Sesbania exaltata (Raf.) Rydb. ex A.W.Hill
Fabaceae/Papilionaceae
= *Sesbania herbacea* (Mill.) McVaugh (NoR) [see *Sesbania macrocarpa* Muhl. ex Raf.]
♦ sesbania, Colorado River hemp, hemp sesbania
♦ Weed, Naturalised, Casual Alien
♦ 39, 42, 84, 87, 88, 161, 207, 218, 243, 263, 286, 287
♦ aH, arid, cultivated, herbal, toxic. Origin: North America.

Sesbania exasperata Kunth
Fabaceae/Papilionaceae
♦ mangeriroba
♦ Weed
♦ 87, 88, 255
♦ pH. Origin: Brazil.

Sesbania grandiflora (L.) Pers.
Fabaceae/Papilionaceae
Aeschynomene coccinea L.f.,
Aeschynomene grandiflora (L.) L., *Agati coccinea* (L.f.) Desv., *Agati grandiflora* (L.) Desv., *Agati grandiflora* var. *coccinea* (L.f.) Wight & Arn., *Coronilla coccinea* (L.f.) Willd., *Coronilla grandiflora* (L.) Willd., *Dolichos arboreus* Forssk., *Emerus grandiflorus* (L.) Kuntze, *Resupinaria grandiflora* (L.) Raf., *Robinia grandiflora* L., *Sesban coccinea* (L.f.) Poir., *Sesban grandiflorus* (L.) Poir., *Sesbania coccinea* (L.f.) Pers.
♦ vegetable hummingbird, scarlet wistaria tree, hummingbird tree, sesban, katurai, agati, agathi, cresta de gallo, gallito
♦ Weed, Naturalised, Introduced, Environmental Weed, Casual Alien
♦ 32, 87, 88, 101, 107, 151, 179, 228, 230, 261, 262
♦ S/T, arid, cultivated, herbal. Origin: tropical Asia.

Sesbania javanica Miq.
Fabaceae/Papilionaceae
Sesbania roxburgii Merr.
♦ Weed
♦ 87, 88
♦ Origin: Asia, Australia.

Sesbania macrocarpa Muhl. ex Raf.
Fabaceae/Papilionaceae
= *Sesbania herbacea* (Mill.) McVaugh (NoR) [see *Sesbania exaltata* (Raf.) Cory]
♦ hemp sesbania

♦ Weed
♦ 84, 88, 249
♦ herbal.

Sesbania punicea (Cav.) Benth.
Fabaceae/Papilionaceae
Daubentonia punicea (Cav.) DC. (see)
♦ red sesbania, coffeeweed, rattlepod, rooi sesbania, purple sesbane, false poinciana, rattlebox, Brazilian glory pea, glory pea, red sesbania, tango
♦ Weed, Quarantine Weed, Noxious Weed, Naturalised, Garden Escape, Environmental Weed, Cultivation Escape
♦ 10, 22, 39, 63, 76, 80, 86, 87, 88, 95, 101, 121, 151, 152, 155, 158, 161, 179, 203, 218, 247, 252, 277, 278, 279, 283, 295
♦ pH, cultivated, herbal, toxic. Origin: South America.

Sesbania roxburghii Merr.
Fabaceae/Papilionaceae
♦ Weed
♦ 209
♦ herbal.

Sesbania sericea (Willd.) Link
Fabaceae/Papilionaceae
♦ papagayo
♦ Weed
♦ 32, 179

Sesbania sesban (L.) Merr.
Fabaceae/Papilionaceae
♦ Egyptian riverhemp, river bean
♦ Weed, Naturalised
♦ 14, 22, 32, 86, 87, 88, 101, 221, 243, 262
♦ pH, cultivated, herbal. Origin: Eurasia, Australia.

Sesbania sesban (L.) Merr. var. *nubica* Chiov.
Fabaceae/Papilionaceae
♦ msilye
♦ Introduced
♦ 228
♦ arid. Origin: Africa.

Sesbania tomentosa Hook & Arn.
Fabaceae/Papilionaceae
♦ Oahu riverhemp
♦ Cultivation Escape
♦ 261
♦ cultivated, herbal. Origin: Hawaii.

Sesbania vesicaria (Jacq.) Ell.
Fabaceae/Papilionaceae
= *Glottidium vesicarium* (Jacq.) Harper (NoR)
♦ bagpod sesbania
♦ Weed
♦ 39, 84, 87, 88, 161, 218
♦ herbal, toxic.

Sesbania virgata (Cav.) Pers.
Fabaceae/Papilionaceae
♦ wand riverhemp, angiquinho grande
♦ Weed, Naturalised
♦ 101, 255
♦ pH. Origin: Americas.

Seseli L. spp.
Apiaceae

♦ seseli
♦ Weed
♦ 272

Seseli annuum L.
Apiaceae
Sium annuum Roth, *Seseli elatius* L., *Seseli simplex* Poir.
♦ Weed
♦ 272
♦ cultivated.

Seseli campestre Besser
Apiaceae
♦ arohirvenputki
♦ Casual Alien
♦ 42

Seseli libanotis (L.) Koch
Apiaceae
Athamantha libanotis L., *Bubon libanotis* Dumort, *Libanotis alpina* Schur, *Libanotis libanotis* Karst, *Selinum libanotis* Krause, *Torilis libanotis* Clairv.
♦ mooncarrot, hirvenputki
♦ Naturalised
♦ 101, 241, 300
♦ b/pH, cultivated, herbal.

Seseli tortuosum L.
Apiaceae
♦ Weed
♦ 272
♦ herbal.

Sesleria albicans Schult.
Poaceae
♦ Casual Alien
♦ 280
♦ G, cultivated.

Sesleria autumnalis (Scop.) F.W.Schultz
Poaceae
♦ sesleria d'autunno
♦ Casual Alien
♦ 280
♦ G, cultivated, herbal.

Sessea brasiliensis Toledo
Solanaceae
♦ Weed, Quarantine Weed
♦ 76, 87, 88, 203, 220

Sesuvium crithmoides Welw.
Aizoaceae
♦ tropical sea purslane
♦ Naturalised
♦ 101

Sesuvium portulacastrum (L.) L.
Aizoaceae
Portulaca portulacastrum L.
♦ shoreline sea purslane, sea purslane
♦ Weed, Naturalised, Introduced
♦ 87, 88, 228, 262, 287, 300
♦ arid, cultivated, herbal.

Sesuvium sessile Pers.
Aizoaceae
♦ western sea purslane, lowland purslane
♦ Weed
♦ 161

Setaria Beauv. spp.
Poaceae
♦ foxtail, cola de zorro, capim oferecido, bristlegrass, lovegrass,

pigeongrass
- ◆ Weed, Naturalised
- ◆ 88, 198, 236, 243, 245
- ◆ G.

Setaria acromelaena (Hochst.) Dur. & Schinz
Poaceae
Panicum acromelaena Hochst., *Setaria abyssinica* Hack. var. *annua* Chiov.
- ◆ Weed, Quarantine Weed
- ◆ 76, 87, 88, 203, 220, 242
- ◆ G, arid.

Setaria adhaerens (Forssk.) Chiov.
Poaceae
- ◆ burr bristlegrass, pikkupantaheinä
- ◆ Weed, Naturalised, Casual Alien
- ◆ 42, 68, 88, 101, 300
- ◆ G.

Setaria aequalis Stapf
Poaceae
- ◆ Weed, Quarantine Weed
- ◆ 76, 87, 88, 203, 220
- ◆ G.

Setaria aurea Hochst.
Poaceae
- ◆ Weed
- ◆ 87, 88
- ◆ G.

Setaria barbata (Lam.) Kunth
Poaceae
Panicum barbatum Lam., *Panicum basisetum* Steud., *Panicum lineatum* Schum., *Panicum rhachitrichum* Hochst., *Setaria basiseta* (Steud.) Durand & Schinz, *Setaria rhachitricha* (Hochst.) Rendle
- ◆ East Indian bristlegrass, East Indies foxtail grass, corn grass, bristly foxtail grass
- ◆ Weed, Naturalised, Introduced
- ◆ 28, 32, 86, 87, 88, 90, 98, 101, 179, 203, 206, 243, 261, 286, 287
- ◆ aG, arid. Origin: Madagascar.

Setaria conspersum (L.) Beauv.
Poaceae
- ◆ Weed
- ◆ 87, 88
- ◆ G.

Setaria decipiens Schimp. ex Nyman
Poaceae
= *Setaria verticilliformis* Dumort.
- ◆ ambiguous foxtail grass, foxtail
- ◆ Weed
- ◆ 30, 44, 88
- ◆ G.

Setaria dielsii Herrm.
Poaceae
- ◆ Weed, Naturalised, Introduced
- ◆ 98, 203, 228
- ◆ G, arid.

Setaria faberi Herrm.
Poaceae
- ◆ giant foxtail, nodding foxtail, Chinese foxtail, Chinese millet, giant bristlegrass, Japanese bristlegrass, kiinanpantaheinä, nodding foxtail grass, akinoenokorogusa

- ◆ Weed, Quarantine Weed, Noxious Weed, Naturalised, Introduced, Casual Alien
- ◆ 23, 30, 35, 42, 68, 76, 80, 87, 88, 90, 101, 102, 161, 174, 179, 199, 203, 207, 210, 211, 218, 220, 229, 243, 263, 275, 286, 297, 299
- ◆ aG, cultivated, herbal. Origin: China, Japan.

Setaria geniculata (Lam.) Beauv.
Poaceae
= *Setaria parviflora* (Poir.) Kerguélen
- ◆ knotroot foxtail, slender pigeongrass, knotroot foxtail, bristlegrass, knotroot, yaa haang maa ching chok, perennial pigeongrass, perennial foxtail, bristly foxtail, paitén, cola de zorro, capim rabo de raposa
- ◆ Weed, Naturalised
- ◆ 15, 30, 87, 88, 157, 161, 180, 217, 218, 236, 237, 238, 239, 241, 243, 245, 249, 255, 270, 274, 280, 281, 287, 295
- ◆ pG, herbal. Origin: tropical and subtropical America.

Setaria geniculata (Lam.) Beauv. var. pauciseta (Lam.) Desv.
Poaceae
- ◆ slender pigeongrass
- ◆ Weed
- ◆ 269
- ◆ pG. Origin: Central and South America.

Setaria gigantea (Fr. & Sav.) Makino
Poaceae
- ◆ Weed
- ◆ 87, 88
- ◆ G.

Setaria glauca (L.) Beauv.
Poaceae
= *Pennisetum glaucum* (L.) R.Br.
- ◆ yellow foxtail, bottlegrass, foxtail millet, golden foxtail, pigeongrass, pussy grass, summergrass, wild millet, yellow bristlegrass, pale pigeongrass, mongoose tail
- ◆ Weed, Noxious Weed, Naturalised
- ◆ 23, 24, 30, 39, 44, 49, 68, 87, 88, 90, 161, 180, 211, 212, 217, 218, 221, 243, 261, 263, 272, 275, 286, 293, 297, 299, 300
- ◆ aG, cultivated, herbal, toxic. Origin: Europe.

Setaria gracilis Kunth
Poaceae
= *Setaria parviflora* (Poir.) Kerguélen
- ◆ slender pigeongrass, perennial foxtail, yellow foxtail, knotroot bristlegrass
- ◆ Weed, Naturalised
- ◆ 3, 7, 90, 98, 176, 198, 203, 280
- ◆ pG, cultivated, herbal.

Setaria gracilis Kunth var. gracilis
Poaceae
- ◆ slender pigeongrass
- ◆ Naturalised, Environmental Weed
- ◆ 86, 98, 198
- ◆ G.

Setaria gracilis Kunth var. pauciseta (Desv.) Simon
Poaceae
- ◆ slender pigeongrass
- ◆ Naturalised, Environmental Weed
- ◆ 86, 98, 176, 198
- ◆ G.

Setaria grisebachii Fourn.
Poaceae
- ◆ Grisebach's bristlegrass
- ◆ Weed
- ◆ 28, 87, 88, 199, 243
- ◆ G, arid, herbal.

Setaria homonyma (Steud.) Chiov.
Poaceae
- ◆ Weed, Quarantine Weed
- ◆ 76, 87, 88, 203, 220
- ◆ G.

Setaria incrassata (Hochst.) Hack.
Poaceae
Setaria ciliolata Stapf & C.E.Hubb., *Setaria woodii* Hack.
- ◆ purple pigeongrass, wood bristlegrass
- ◆ Weed, Naturalised, Native Weed
- ◆ 86, 87, 88, 93, 98, 121, 203
- ◆ pG, arid. Origin: Africa.

Setaria intermedia (Roth) Roem. & Schult.
Poaceae
Panicum intermedium (Roem. & Schult.) Roth, *Panicum tomentosum* Roxb., *Setaria tomentosa* (Roxb.) Kunth
- ◆ Weed, Introduced
- ◆ 87, 88, 228
- ◆ G, arid.

Setaria italica (L.) Beauv.
Poaceae
Chaetochloa italica (L.) Scribn., *Chamaeraphis italica* (L.) Kuntze, *Ixophorus italicus* (L.) Nash, *Panicum italicum* L., *Pennisetum italicum* R.Br.
- ◆ Italian millet, foxtail millet, German millet, Hungarian millet, Siberian millet, mohár taliansky, whorled pigeongrass, Italianpantaheinä, foxtail bristlegrass, Japanese millet, Hungarian grass
- ◆ Weed, Naturalised, Casual Alien
- ◆ 7, 15, 24, 30, 32, 40, 42, 80, 86, 87, 88, 90, 93, 98, 101, 102, 121, 176, 198, 203, 218, 272, 280
- ◆ aG, arid, cultivated, herbal. Origin: Eurasia.

Setaria lindenbergiana (Nees) Stapf
Poaceae
- ◆ mountain bristlegrass, tussockgrass
- ◆ Native Weed
- ◆ 121
- ◆ pG. Origin: southern Africa.

Setaria longiseta P.Beauv.
Poaceae
- ◆ Weed
- ◆ 87, 88
- ◆ G.

Setaria lutescens (Weigel) Hubb. *nom. dubium*
Poaceae
= *Setaria pumila* (Poir.) Roem. & Schult. ssp. *pumila*
♦ yellow foxtail, cat's tail grass
♦ Weed
♦ 68, 88, 167, 210
♦ G, herbal.

Setaria megaphylla (Steud.) Dur. & Schinz
Poaceae
Setaria chevalieri Stapf & Hubb.
♦ broad leaved setaria, buffalograss, bush buffalograss, fine swordgrass, forest buffalograss, macopo grass, ribbon bristlegrass, ribbon grass, sclitz grass, solitzgrass, bigleaf bristlegrass, lokokoloko
♦ Weed, Naturalised, Native Weed
♦ 10, 88, 101, 121, 158
♦ pG, cultivated. Origin: southern Africa.

Setaria nigrirostris (Nees) Dur. & Schinz
Poaceae
Chaetochloa nigrirostris (Nees) Skeels, *Panicum nigrirostre* Nees
♦ black bristlegrass, largeseed setaria
♦ Weed, Naturalised, Native Weed
♦ 88, 101, 121, 158, 287
♦ pG, arid. Origin: southern Africa.

Setaria pallide-fusca (Schum.) Stapf & Hubb.
Poaceae
= *Setaria pumila* (Poir.) Roem. & Schult. ssp. *pallide-fusca* (Stapf & Hubb.) Simon
♦ annual timothy, cat's tail, garden bristlegrass, garden setaria, horse grass, red bristlegrass, water setaria, Queensland pigeongrass, cat's tail grass, millet sauvage
♦ Weed, Native Weed
♦ 3, 67, 87, 88, 90, 121, 158, 170, 217, 262, 286
♦ aG, herbal. Origin: Africa.

Setaria palmifolia (J.König) Stapf
Poaceae
Panicum palmifolium J.König (see), *Panicum plicatum* Willd., *Panicum neurodes* Schult.
♦ palmgrass, short pitpit, hailans pitpit, broad leaved bristlegrass, yaa kaap phai
♦ Weed, Naturalised, Environmental Weed
♦ 3, 7, 15, 80, 86, 87, 88, 90, 98, 101, 134, 152, 203, 225, 238, 243, 246, 256, 262, 280, 286
♦ pG, cultivated. Origin: India.

Setaria paniculifera (Steud.) Fourn.
Poaceae
Chaetochloa sulcata (Aubl.) Hitchc., *Panicum paniculiferum* Steud., *Panicum sulcatum* Aubl., *Setaria sulcata* (Aubl.) A.Camus. *nom. illeg.*
♦ Weed
♦ 88, 90
♦ pG.

Setaria parviflora (Poir.) Kerguélen
Poaceae
Cenchrus parviflorus Poir., *Chaetochloa corrugata* (Ell.) Scribn. var. *parviflora* (Poir.) Scribn. & Merr., *Chaetochloa geniculata* (Poir.) Millsp. & Chase, *Chaetochloa ventenatii* (Kunth) Nash, *Panicum dasyurum* Willd. ex Nees, *Panicum flavum* Nees, *Panicum geniculatum* Willd., *Panicum imberbe* Poir., *Panicum laevigatum* Muhl. ex Elliott, *Panicum penicillatum* Willd. ex Nees, *Panicum tejucense* Nees, *Panicum ventenatii* (Kunth) Steud., *Setaria berteroniana* Schult., *Setaria discolor* Hack., *Setaria flava* (Nees) Kunth, *Setaria flava* var. *pumila* E.Fourn., *Setaria floriana* Andersson, *Setaria geniculata* (Lam.) Beauv. (see), *Setaria glauca* (L.) Beauv. fo. *normalis* Büse, *Setaria glauca* var. *imberbis* (Poir.) Griseb., *Setaria glauca* var. *penicillata* (Willd. ex Nees) Griseb., *Setaria gracilis* Kunth (see), *Setaria imberbis* (Poir.) Roem. & Schult., *Setaria montana* Reeder, *Setaria penicillata* (Willd. ex Nees) J.Presl, *Setaria purpurascens* Kunth, *Setaria roemeri* Jansen, *Setaria surgens* Stapf, *Setaria tejucensis* (Nees) Kunth, *Setaria tenella* Desv., *Setaria ventenatii* Kunth
♦ knotroot foxtail, marsh bristlegrass, yellow bristlegrass, perennial pigeongrass, perennial foxtail
♦ Weed, Naturalised
♦ 86, 243, 253, 280
♦ a/pG, aqua. Origin: Americas.

Setaria parviflora (Poir.) Kerguélen var. parviflora
Poaceae
♦ Naturalised
♦ 300
♦ G.

Setaria pilosa Kunth
Poaceae
= *Panicum pilosum* Sw.
♦ capim oferecido
♦ Weed
♦ 245
♦ G.

Setaria plicata (Lam.) Cooke
Poaceae
Panicum excurrens Trin, *Panicum plicatum* Lam.
♦ Weed
♦ 87, 88, 275, 297
♦ pG, herbal.

Setaria poiretiana (Schult.) Kunth
Poaceae
Chaetochloa poiretiana (Schult.) Hitchc., *Panicum elongatum* Poir., *Panicum poiretianum* Schult.
♦ Weed, Quarantine Weed, Naturalised
♦ 54, 76, 86, 87, 88, 90, 191, 203, 220, 255
♦ pG. Origin: Central and South America, Caribbean.

Setaria pumila (Poir.) Roem. & Schult.
Poaceae
Setaria glauca (L.) P.Beauv. var. *pallide-fusca* (Schumach.) Koyama, *Setaria rubiginosa* (Steud.) Miq.
♦ yellow bristlegrass, smooth millet, pale pigeongrass, yellow foxtail
♦ Weed, Naturalised, Introduced, Casual Alien
♦ 7, 34, 38, 40, 42, 53, 66, 70, 80, 87, 88, 98, 101, 102, 179, 185, 195, 198, 203, 228, 243, 253, 272, 280, 300
♦ aG, arid, cultivated, herbal. Origin: Eurasia.

Setaria pumila (Poir.) Roem. & Schult. ssp. pallide-fusca (Stapf & Hubb.) Simon
Poaceae
Panicum pallide-fuscum Schum., *Setaria pallide-fusca* (Schum.) Stapf & Hubb. (see)
♦ yellow bristlegrass, cat's tail grass
♦ Noxious Weed, Naturalised
♦ 86, 101, 134, 140, 229, 280
♦ G.

Setaria pumila (Poir.) Roem. & Schult. ssp. pumila
Poaceae
Chaetochloa lutescens (Weigel) Stuntz *nom. dubium*, *Panicum lutescens* Weigel *nom. dubium*, *Panicum pumilum* Poir., *Setaria glauca auct.*, *Setaria lutescens* (Weigel) Hubb. *nom. dubium* (see)
♦ yellow bristlegrass
♦ Weed, Naturalised
♦ 15, 86
♦ G. Origin: Eurasia.

Setaria rariflora Mikan ex Trin.
Poaceae
= *Setaria setosa* (Sw.) P.Beauv.
♦ Brazilian bristlegrass
♦ Weed, Naturalised
♦ 179, 287
♦ G.

Setaria setosa (Sw.) Beauv.
Poaceae
Chaetochloa rariflora (Mikan & Trin.) Hitchc. & Chase, *Panicum caudatum* Lam., *Panicum onurus* Willd. ex Trin., *Panicum setosum* Sw., *Panicum utriculatum* Steud., *Setaria caespitosa* Hack. & Arechav., *Setaria onurus* (Willd. ex Trin.) Griseb., *Setaria rariflora* Mikan ex Trin. (see), *Setaria vaginata* Spreng.
♦ West Indian bristlegrass
♦ Weed, Naturalised, Introduced
♦ 32, 179, 228
♦ G, arid, cultivated.

Setaria sphacelata (Schumach.) M.B.Moss ex Stapf & C.E.Hubb.
Poaceae
Panicum sphacelatum Schumach.
♦ African bristlegrass, golden setaria, common setaria, South African pigeongrass
♦ Weed, Naturalised, Introduced, Casual Alien

♦ 7, 34, 87, 88, 90, 98, 101, 158, 179, 203, 280, 287

♦ pG, arid, cultivated, herbal. Origin: southern Africa.

Setaria sphacelata (Schumach.) M.B.Moss ex Stapf & C.E.Hubb. var. *sericea* (Stapf) Clayton

Poaceae

Setaria almaspicata de Wit, *Setaria anceps* Stapf ex Massey, *Setaria anceps* var. *sericea* Stapf, *Setaria aurea* A.Br. ssp. *palustris* Vanderyst, *Setaria cana* de Wit, *Setaria flabelliformis* de Wit, *Setaria planiflora* Stapf, *Setaria tenuispica* Stapf & Hubb.

♦ Naturalised, Introduced, Environmental Weed

♦ 86, 228

♦ G, arid. Origin: Africa.

Setaria sphacelata (Schumach.) M.B.Moss ex Stapf & C.E.Hubb. var. *sphacelata*

Poaceae

♦ common bristlegrass, golden millet, golden timothy, landgrass, old landsgrass, Rhodesian timothy grass, South African golden milletgrass

♦ Native Weed

♦ 121

♦ a/pG. Origin: southern Africa.

Setaria sphacelata (Schumach.) M.B.Moss ex Stapf & C.E.Hubb. var. *splendida* (Stapf) Clayton

Poaceae

♦ Naturalised, Environmental Weed

♦ 86

♦ G. Origin: Africa.

Setaria sphacelata (Schumach.) M.B.Moss ex Stapf & C.E.Hubb. var. *torta* Clayton

Poaceae

Setaria flabellata Stapf

♦ small creeping foxtail

♦ Native Weed

♦ 121

♦ pG. Origin: southern Africa.

Setaria verticillata (L.) Beauv.

Poaceae

Chaetochloa verticillata Scribn., *Ixophorus verticillatus* (L.) Nash, *Panicum adhaerens* (Forssk.) Chiov., *Panicum verticillatum* L., *Pennisetum aparine* Steud., *Pennisetum verticillatum* (L.) R.Br., *Setaria carnei* A.Hitchc.

♦ bristly foxtail, whorled pigeongrass, lovegrass, foxtail, rough bristlegrass, burr bristlegrass, pabbio verticillato, hooked bristlegrass

♦ Weed, Noxious Weed, Naturalised, Native Weed, Introduced, Casual Alien

♦ 7, 14, 15, 23, 30, 34, 38, 40, 42, 44, 50, 51, 53, 68, 70, 80, 86, 87, 88, 90, 98, 101, 115, 118, 121, 134, 158, 161, 174, 176, 179, 180, 185, 186, 198, 203, 205, 210, 212, 218, 221, 228, 236, 237, 240, 241, 242, 243, 253, 269, 270, 272, 274, 280, 286, 287, 293, 295, 299, 300

♦ aG, arid, cultivated, herbal. Origin: obscure.

Setaria verticillata (L.) Beauv. var. *ambigua* (Guss.) Parl.

Poaceae

= *Setaria verticilliformis* Dumort.

♦ Naturalised

♦ 287

♦ G.

Setaria verticilliformis Dumort.

Poaceae

Chaetochloa ambigua (Guss.) Scribn. & Merr., *Chamaeraphis italica* (L.) Kuntze var. *ambigua* (Guss.) Kuntze, *Panicum ambiguum* (Guss.) Hausskn., *Panicum verticillatum* L. var. *ambiguum* Guss., *Setaria ambigua* (Guss.) Guss., *Setaria decipiens* Schimp. ex Nyman (see), *Setaria gussonei* Kerguélen, *Setaria verticillata* (L.) Beauv. var. *ambigua* (Guss.) Parl. (see), *Setaria viridis* (L.) Beauv. var. *ambigua* (Guss.) Coss. & Durieu, *Setaria × verticilliformis* Dumort.

♦ ambiguous foxtail grass, barbed bristlegrass

♦ Weed, Naturalised

♦ 90, 101

♦ aG.

Setaria viridis (L.) Beauv.

Poaceae

Chaetochloa viridis Scribn., *Ixophorus viridis* (L.) Nash, *Panicum viride* L.

♦ green pigeongrass, green setaria, bottlegrass, green bristlegrass, green foxtail, wild millet, green panicum, blue foxtail, green bottlegrass, pabbio comune, falso panico, viherpantaheinä, cola de zorro

♦ Weed, Noxious Weed, Naturalised, Introduced, Casual Alien

♦ 15, 23, 30, 34, 40, 44, 49, 52, 68, 70, 80, 86, 87, 88, 90, 98, 101, 102, 114, 118, 136, 161, 162, 174, 176, 179, 180, 185, 186, 198, 203, 204, 210, 211, 212, 218, 221, 229, 235, 236, 237, 241, 243, 253, 272, 275, 280, 286, 292, 295, 297, 299, 300

♦ aG, arid, cultivated, herbal. Origin: Eurasia, North Africa.

Setaria viridis Beauv. fo. *misera* Honda

Poaceae

♦ Weed

♦ 286

♦ G.

Setaria viridis (L.) Beauv. var. *gigantea* Fr. & Sav. Matsum

Poaceae

♦ Weed

♦ 275

♦ aG.

Setaria viridis (L.) Beauv. var. *major* (Gaudin) Posp.

Poaceae

♦ green foxtail, green bristlegrass, giant green foxtail

♦ Weed, Noxious Weed, Naturalised

♦ 68, 101, 229

♦ G.

Setaria viridis (L.) Beauv. var. *robusta-alba* M.M.Schreib.

Poaceae

♦ robust white foxtail

♦ Weed

♦ 68

♦ G.

Setaria viridis (L.) Beauv. var. *viridis*

Poaceae

♦ green foxtail

♦ Noxious Weed, Naturalised

♦ 101, 229

♦ G.

Setaria vulpiseta (Lam.) Roem. & Schult.

Poaceae

Chaetochloa trichorhachis (Hack.) Hitchc., *Chaetochloa vulpiseta* (Lam.) Hitchc. & Chase, *Chamaeraphis composita* (Kunth) Kuntze, *Chamaeraphis setosa* (Sw.) Kuntze var. *vulpiseta* (Lam.) Kuntze, *Panicum amplifolium* Steud., *Panicum compositum* (Kunth) Nees, *Panicum macrostachyum* (Kunth) Döll, *Panicum macrourum* Trin., *Panicum subsphaerocarpum* Salzm. ex Schltdl., *Panicum vulpisetum* Lam., *Setaria alopecurus* Trin. ex Steud., *Setaria composita* Kunth, *Setaria lancifolia* R.A.W.Herrm., *Setaria liebmannii* E.Fourn. fo. *trichorhachis* Hack., *Setaria macrostachya* Kunth, *Setaria polystachya* Schrad. ex Schult., *Setaria trichorhachis* (Hack.) R.C.Foster

♦ pajita tempranera, plains bristlegrass, rabo de ardilla, rabo de raposa

♦ Weed, Naturalised, Introduced, Garden Escape

♦ 228, 255, 257, 261

♦ pG, arid, cultivated. Origin: tropical America.

Setaria welwitschii Rendle

Poaceae

♦ Weed

♦ 87, 88

♦ G, arid.

Seticereus Backeb. spp.

Cactaceae

= *Cleistocactus* Lem. spp.

♦ Weed, Quarantine Weed

♦ 76, 88, 203

Seticleistocactus Backeb. spp.

Cactaceae

= *Cleistocactus* Lem. spp.

♦ Weed, Quarantine Weed

♦ 76, 88, 203

Severinia buxifolia (Poir.) Ten.

Rutaceae

♦ Chinese boxorange

♦ Naturalised

♦ 86

♦ cultivated. Origin: China.

Severinia disticha (Blanco) Swingle

Rutaceae

♦ Quarantine Weed

♦ 220

♦ Origin: south-east Asia.

Severinia monophylla (L.) Tanaka
Rutaceae
♦ Chinese boxorange
♦ Weed, Naturalised
♦ 101, 179

Shepherdia argentea (Pursh) Nutt.
Elaeagnaceae
Hippophae argentea Pursh
♦ silver buffaloberry, buffaloberry
♦ Weed
♦ 87, 88, 218
♦ S, cultivated, herbal.

Shepherdia canadensis (L.) Nutt.
Elaeagnaceae
Hippophae canadensis L.
♦ russet buffaloberry, buffaloberry
♦ Weed
♦ 87, 88, 218
♦ S, cultivated, herbal.

Sherardia arvensis L.
Rubiaceae
♦ field madder, blue field madder, spurwort, herb sherard, meadow bedstraw
♦ Weed, Naturalised, Introduced, Garden Escape, Environmental Weed
♦ 7, 9, 15, 23, 34, 44, 70, 72, 86, 87, 88, 94, 98, 101, 134, 161, 165, 167, 176, 198, 199, 203, 218, 237, 241, 243, 249, 253, 272, 280, 286, 287, 295, 300
♦ aH, cultivated, herbal. Origin: Mediterranean, west Asia.

Shorea roxburghii G.Don
Dipterocarpaceae
♦ Quarantine Weed
♦ 220

Sibara virginica (L.) Rollins
Brassicaceae
♦ sibara, cress rock, Virginia rockcress, Virginia winged rockcress
♦ Weed
♦ 87, 88, 161, 180, 207, 218, 243, 249
♦ a/pH, cultivated, herbal. Origin: North America.

Sicana odorifera (Vell.) Naud.
Cucurbitaceae
♦ casa banana, curuba
♦ Quarantine Weed, Naturalised, Cultivation Escape
♦ 101, 220, 261
♦ cultivated, herbal. Origin: South America.

Sicyos angulatus L.
Cucurbitaceae
♦ burcucumber, one seeded burr cucumber, star cucumber, nimble kate, chocho vine, wall burr cucumber, siankurkku
♦ Weed, Noxious Weed, Naturalised, Native Weed, Casual Alien
♦ 23, 40, 42, 87, 88, 161, 174, 207, 210, 211, 218, 229, 243, 272, 286, 287
♦ aC, cultivated, herbal. Origin: North America.

Sicyos angulatus L. fo. *ohtanus* Asai
Cucurbitaceae
♦ Naturalised
♦ 287

Sicyos deppei G.Don
Cucurbitaceae
♦ Weed
♦ 243

Sicyos laciniatus L.
Cucurbitaceae
♦ cutleaf burr cucumber
♦ Weed
♦ 199, 243
♦ herbal.

Sicyos microphyllus Kunth
Cucurbitaceae
♦ burr cucumber, smallleaf burr cucumber
♦ Weed
♦ 243
♦ herbal.

Sicyos parviflorus Willd.
Cucurbitaceae
♦ smallflower burr cucumber
♦ Weed
♦ 199

Sicyos polyacanthus Cogn.
Cucurbitaceae
♦ túpulo, po de mico, cipo de mico
♦ Weed
♦ 236, 237, 255, 295
♦ Origin: Americas.

Sida abutifolia Mill.
Malvaceae
♦ spreading fanpetals
♦ Naturalised
♦ 101
♦ herbal.

Sida acuminata DC.
Malvaceae
= *Sidastrum multiflorum* (Jacq.) Fryxell (NoR)
♦ Weed
♦ 14

Sida acuta Burm.f.
Malvaceae
Sida carpinifolia L.f. (see)
♦ southern sida, spinyhead sida, cheeseweed, broomweed, sida, yaa khat bai yaao, kaweniotz, sinchipichana
♦ Weed, Quarantine Weed, Noxious Weed, Naturalised, Native Weed, Introduced
♦ 6, 7, 12, 22, 23, 53, 55, 62, 76, 86, 87, 88, 93, 98, 121, 147, 153, 157, 161, 186, 191, 199, 203, 205, 209, 218, 230, 239, 240, 243, 249, 257, 261, 262, 269, 276, 286, 287, 297
♦ pS, cultivated, herbal. Origin: Africa.

Sida alba L.
Malvaceae
= *Sida spinosa* L.
♦ spiny sida, prickly sida, spring sida
♦ Weed, Native Weed
♦ 50, 87, 88, 121, 158, 185, 221, 242
♦ H, arid, herbal.

Sida angustifolia L.
Malvaceae
= *Sida spinosa* L.
♦ Weed, Quarantine Weed

♦ 76, 87, 88, 203, 220

Sida antillensis Urb.
Malvaceae
♦ Antilles fanpetals
♦ Naturalised
♦ 101

Sida carpinifolia L.f.
Malvaceae
= *Sida acuta* Burm.f.
♦ malva baixa
♦ Weed, Naturalised
♦ 87, 88, 134, 255
♦ S. Origin: Brazil.

Sida chrysantha Ulbr.
Malvaceae
♦ Native Weed
♦ 121
♦ S. Origin: southern Africa.

Sida ciliaris L.
Malvaceae
♦ bracted fanpetals
♦ Weed
♦ 14
♦ pH, arid, herbal.

Sida cordifolia L.
Malvaceae
♦ flannel weed, heartleaf sida, sida, flannel sida, white burr, llima, iNama, malva branca
♦ Weed, Quarantine Weed, Noxious Weed, Naturalised, Native Weed, Introduced
♦ 6, 50, 51, 55, 62, 76, 86, 87, 88, 93, 98, 121, 147, 158, 179, 191, 203, 218, 228, 255, 269, 274, 276
♦ pH, arid, cultivated, herbal. Origin: obscure, possibly Africa.

Sida corrugata Lindl.
Malvaceae
♦ corrugated sida
♦ Weed
♦ 87, 88, 269
♦ cultivated. Origin: Australia.

Sida corylifolia Wall. ex Mast.
Malvaceae
♦ Weed
♦ 12

Sida corymbosa R.E.Fr.
Malvaceae
♦ Weed
♦ 87, 88

Sida cuneifolia Gray
Malvaceae
= *Billieturnera helleri* (Rose ex Heller) Fryxell (NoR)
♦ Weed
♦ 87, 88

Sida dregei Burtt Davy
Malvaceae
♦ spiderleg, Sutherland's curse
♦ Native Weed
♦ 121
♦ pS, toxic. Origin: southern Africa.

Sida fallax Walp.
Malvaceae
♦ yellow llima
♦ Weed

- 6, 87, 88, 191
- pH, herbal.

Sida glabra Mill.
Malvaceae
Sida arguta Fisch. ex Link, *Sida arguta* Sw., *Sida cearensis* Ulbr., *Sida endlicheriana* C.Presl, *Sida fasciculata* Willd. ex Spreng., *Sida glutinosa* Comm. ex Cav. (see), *Sida insperata* Standl. & L.O.Williams, *Sida nervosa* DC., *Sida rupicola* Hassl., *Sida ulmifolia* Cav., *Sida verruculata* DC., *Sida viscidula* Blume, *Sida willdenowii* D.Dietr.
- smooth fanpetals
- Weed
- 179

Sida glaziovii K.Schum.
Malvaceae
- guanxuma branca
- Weed
- 255
- pH. Origin: Brazil.

Sida glomerata Cav.
Malvaceae
- clustered fanpetals
- Weed
- 87, 88
- herbal.

Sida glutinosa Comm. ex Cav.
Malvaceae
= *Sida glabra* Mill.
- sticky fanpetals
- Weed, Naturalised
- 87, 88, 257

Sida hederacea (Douglas ex Hook.) Torr. ex Gray
Malvaceae
= *Malvella leprosa* (Ortega) Krapov. var. *hederacea* (Hook.) Schumann
- alkali sida, dollarweed, alkali mallow
- Weed
- 87, 88, 161, 212, 218
- herbal.

Sida jamaicensis Vell. or L.
Malvaceae
- Jamaican fanpetals
- Weed
- 39, 87, 88
- toxic.

Sida javensis Cav.
Malvaceae
Sida veronicaefolia Lam.
- Weed
- 87, 88

Sida leprosa (Ortega) Schum.
Malvaceae
= *Malvella leprosa* (Ortega) Krapov.
- ivy leaved sida, alkali mallow
- Weed, Noxious Weed, Naturalised
- 35, 86, 88
- herbal.

Sida leprosa (Ortega) Shum. var. *hederacea* (Douglas ex Hook.) Shum.
Malvaceae
= *Malvella leprosa* (Ortega) Krapov.
- ivyleaf sida

- Weed
- 269

Sida linifolia Cav.
Malvaceae
- flaxleaf fanpetals
- Weed, Naturalised
- 14, 87, 88, 101, 255
- aH. Origin: South America.

Sida micrantha St.-Hil.
Malvaceae
= *Sidastrum micranthum* (St.-Hil.) Fryxell
- Weed
- 87, 88

Sida mysorensis Wight & Arn.
Malvaceae
- Weed
- 87, 88

Sida ovata Forssk.
Malvaceae
- Weed
- 87, 88
- herbal.

Sida paniculata L.
Malvaceae
= *Sidastrum paniculatum* (L.) Fryxell
- Weed, Naturalised
- 39, 87, 88, 241, 257
- toxic.

Sida physocalyx A.Gray
Malvaceae
= *Rhynchosida physocalyx* (Gray) Fryxell
- tuberous sida
- Weed
- 87, 88, 218
- arid, herbal.

Sida platycalyx F.Muell. ex Benth.
Malvaceae
- lifesaver burr
- Weed
- 87, 88
- cultivated.

Sida potentilloides St.-Hil.
Malvaceae
Sida cordifolia L. var. *potentilloides* Gris.
- malva misionera
- Weed
- 87, 88, 237, 295

Sida pseudocordifolia Hochr.
Malvaceae
- Native Weed
- 121
- pS. Origin: southern Africa.

Sida retusa L.
Malvaceae
= *Sida rhombifolia* L. ssp. *retusa* (L.) Boiss.
- Weed
- 23, 88
- herbal.

Sida rhombifolia L.
Malvaceae
- arrowleaf sida, Pretoria sida, Paddy's lucerne, broomstick, common sida, jelly leaf, Queensland hemp, shrub sida, sida retusa, ruutusiida, Cuban jute, mautofu, te'ehosi, motofu,

balais, mamafu'ai
- Weed, Quarantine Weed, Noxious Weed, Naturalised, Native Weed, Introduced, Garden Escape, Casual Alien
- 3, 6, 14, 34, 42, 51, 55, 70, 76, 86, 87, 88, 93, 98, 121, 134, 147, 153, 158, 161, 165, 170, 172, 198, 199, 203, 217, 218, 228, 236, 237, 245, 255, 257, 262, 269, 270, 276, 280, 286, 287, 295, 300
- pH, arid, cultivated, herbal. Origin: Americas.

Sida rhombifolia L. ssp. *retusa* (L.) Boiss.
Malvaceae
Sida retusa L. (see)
- Naturalised
- 87, 287

Sida rhombifolia L. var. *longipedicellata* L.
Malvaceae
- Introduced
- 230
- H.

Sida rhombifolia L. var. *rhombifolia*
Malvaceae
- koyolung
- Introduced
- 230
- H.

Sida samoensis Rech.
Malvaceae
- Weed
- 87, 88

Sida santaremensis Monteiro
Malvaceae
- moth fanpetals, guaxima
- Weed, Naturalised
- 101, 255
- S. Origin: Brazil.

Sida spinosa L.
Malvaceae
Sida alba L. (see), *Sida angustifolia* Lam. (see)
- spiny sida, prickly sida, false mallow, Indian mallow, thistle mallow, Afata hembra
- Weed, Naturalised, Native Weed, Introduced, Casual Alien
- 42, 55, 68, 86, 87, 88, 93, 121, 158, 161, 179, 205, 207, 210, 211, 218, 230, 236, 237, 243, 255, 263, 269, 270, 286, 287, 295, 299, 300
- pS, arid, cultivated, herbal.

Sida spinosa L. var. *spinosa*
Malvaceae
- Naturalised
- 241

Sida stipulata Cav.
Malvaceae
- Weed, Quarantine Weed
- 76, 87, 88, 203, 220

Sida subspicata F.Muell. ex Benth.
Malvaceae
- spiked sida
- Weed, Naturalised
- 87, 88, 287
- cultivated.

Sida ternata **L.f.**
Malvaceae
♦ Native Weed
♦ 121
♦ pH. Origin: southern Africa.

Sida urens **L.**
Malvaceae
♦ tropical fanpetals, pichana peluda
♦ Weed
♦ 14, 39, 87, 88, 153, 255
♦ a/bH, toxic. Origin: South America.

Sidalcea malviflora **(DC.) A.Gray ex Benth.**
Malvaceae
♦ dwarf checkermallow, dwarf checkerbloom
♦ Casual Alien
♦ 280
♦ cultivated, herbal.

Sidastrum micranthum **(St.-Hil.) Fryxell**
Malvaceae
Sida micrantha St.-Hil. (see)
♦ dainty sandmallow, malva preta
♦ Weed
♦ 255
♦ S. Origin: Americas.

Sidastrum paniculatum **(L.) Fryxell**
Malvaceae
Sida paniculata L. (see)
♦ panicled sandmallow, malva roxa
♦ Weed
♦ 255
♦ aH, arid. Origin: Americas.

Sideritis lanata **L.**
Lamiaceae
♦ hairy ironwort
♦ Naturalised
♦ 101

Sideritis montana **L.**
Lamiaceae
♦ mountain ironwort, vuoriraudakki
♦ Weed, Naturalised, Casual Alien
♦ 42, 87, 88, 94, 101, 272
♦ cultivated. Origin: Europe, Middle East. northern Asia.

Sideritis romana **L.**
Lamiaceae
♦ simplebeak ironwort
♦ Weed, Naturalised
♦ 88, 94, 101, 272
♦ herbal.

Sieglingia decumbens **(L.) Bernh.**
Poaceae
= *Danthonia decumbens* (L.) DC.
♦ heath grass, sieglingie décombante
♦ Weed, Naturalised, Environmental Weed
♦ 15, 72, 86, 88, 198, 280
♦ pG. Origin: Eurasia, Mediterranean.

Sigesbeckia agrestis **Poepp. & Endl.**
Asteraceae
♦ Weed
♦ 87, 88

Sigesbeckia cordifolia **H.B.K.**
Asteraceae
♦ Weed
♦ 87, 88

Sigesbeckia glabrescens **(Makino) Makino**
Asteraceae
♦ Weed
♦ 87, 88
♦ herbal.

Sigesbeckia glabrescens **Makino var.** *leucoclada* **Nakai**
Asteraceae
♦ Naturalised
♦ 287

Sigesbeckia jorullensis **Kunth**
Asteraceae
♦ Weed
♦ 199

Sigesbeckia orientalis **L.**
Asteraceae
Siegesbeckia brachiata Roxb.
♦ St. Paul's wort, common St. Paul's wort, Indian weed, siegesbeckia, botao de ouro
♦ Weed, Naturalised, Environmental Weed
♦ 7, 13, 15, 32, 51, 86, 87, 88, 98, 101, 121, 134, 158, 176, 203, 218, 235, 238, 253, 255, 269, 273, 274, 275, 276, 280, 286, 297
♦ aH, arid, herbal, toxic. Origin: obscure.

Sigesbeckia orientalis **L. ssp.** *glabrescens* **Kitam.**
Asteraceae
♦ Weed
♦ 286

Sigesbeckia orientalis **L. ssp.** *pubescens* **Kitam.**
Asteraceae
= *Siegesbeckia pubescens* (Makino) Makino
♦ Weed
♦ 286

Sigesbeckia pubescens **(Makino) Makino**
Asteraceae
Sigesbeckia orientalis L. ssp. *pubescens* Kitam. (see)
♦ common St. Paul's wort, menamomi
♦ Weed
♦ 87, 88, 191, 204, 263, 275, 297
♦ herbal.

Silaum silaus **(L.) Schinz & Thell.**
Apiaceae
Cnidium silaus Spreng, *Crithmum silaus* Wibel, *Ligusticum silaus* Villar., *Meum silaus* Baill., *Selinum silaus* Crantz, *Sium silaus* Roth
♦ ketunputki, pepper saxifrage
♦ Weed, Casual Alien
♦ 42, 272
♦ pH, cultivated.

Silene **L. spp.**
Caryophyllaceae
♦ campion, catchfly
♦ Weed, Naturalised
♦ 198, 243, 272
♦ herbal.

Silene acaulis **L.**
Caryophyllaceae

Cucubalus acaulis L.
♦ moss campion, tunturikohokki
♦ Weed
♦ 23, 88
♦ pH, cultivated, herbal.

Silene aegyptiaca **L.f.**
Caryophyllaceae
♦ Weed
♦ 88, 111, 243
♦ cultivated.

Silene alba **(Mill) Krause**
Caryophyllaceae
= *Silene latifolia* Poir. ssp. *alba* (Mill.) Greut. & Burd.
♦ white campion, white cockle
♦ Weed, Naturalised
♦ 70, 87, 88, 94, 118, 161, 211, 212, 272, 287, 294
♦ cultivated, herbal.

Silene anglica **L.**
Caryophyllaceae
= *Silene gallica* L.
♦ Weed
♦ 87, 88

Silene antirrhina **L.**
Caryophyllaceae
♦ sleepy catchfly, sleepy silene
♦ Weed, Naturalised
♦ 34, 39, 87, 88, 161, 210, 212, 218, 287, 295
♦ aH, herbal, toxic.

Silene apetala **Willd.**
Caryophyllaceae
♦ mallee catchfly, sand catchfly
♦ Weed, Naturalised, Environmental Weed
♦ 7, 72, 86, 88, 98, 198, 203
♦ aH. Origin: southern Europe.

Silene aprica **Turcz. ex Fisch. & Mey.**
Caryophyllaceae
Melandrium apricum (Turcz.) Rohrb.
♦ sunward silene
♦ Weed
♦ 297
♦ herbal.

Silene arabica **Boiss.**
Caryophyllaceae
♦ Weed
♦ 221

Silene armeria **L.**
Caryophyllaceae
♦ tarhakohokki, sweet William catchfly, none so pretty
♦ Weed, Naturalised, Cultivation Escape, Casual Alien
♦ 40, 42, 86, 101, 241, 252, 272, 280, 286, 287, 300
♦ aH, cultivated, herbal. Origin: Europe.

Silene atocioides **Boiss.**
Caryophyllaceae
♦ Naturalised
♦ 86

Silene biappendiculata **Ehr. ex Rohrb.**
Caryophyllaceae
♦ Weed
♦ 221

Silene caroliniana Walter
Caryophyllaceae
♦ sticky catchfly
♦ Weed
♦ 23, 88
♦ herbal.

Silene cerastoides All.
Caryophyllaceae
♦ Weed
♦ 87, 88

Silene chalcedonica (L.) E.H.L.Krause
Caryophyllaceae
♦ Casual Alien
♦ 280

Silene chlorantha (Willd.) Ehrh.
Caryophyllaceae
♦ yellowgreen catchfly, silene
♦ Naturalised
♦ 101

Silene coeli-rosa (L.) Godr.
Caryophyllaceae
♦ rose silene, rose of heaven, viscaria
♦ Weed, Naturalised, Casual Alien
♦ 15, 40, 101, 280
♦ aH, cultivated.

Silene colorata Poir.
Caryophyllaceae
♦ kirjokohokki
♦ Weed, Casual Alien
♦ 42, 87, 88, 221
♦ cultivated.

Silene conica L.
Caryophyllaceae
Conosilene conica (L.) Fourr., *Pleconax conica* (L.) Sourk.
♦ striped corn catchfly, sand catchfly, hietakohokki, conical catchfly
♦ Weed, Naturalised, Casual Alien
♦ 42, 70, 86, 87, 88, 94, 98, 101, 176, 198, 203, 272, 280, 287
♦ aH, cultivated. Origin: Eurasia, Mediterranean.

Silene conoidea L.
Caryophyllaceae
♦ conoid catchfly, weed silene, cone catchfly
♦ Weed, Naturalised, Casual Alien
♦ 23, 34, 42, 87, 88, 94, 101, 115, 161, 185, 212, 221, 243, 275, 286, 287, 297
♦ aH, promoted, herbal.

Silene coronaria (L.) Clairv.
Caryophyllaceae
= *Lychnis coronaria* (L.) Desr.
♦ Naturalised
♦ 280

Silene crassipes Fenzl
Caryophyllaceae
♦ Weed
♦ 87, 88

Silene cretica L.
Caryophyllaceae
♦ Weed
♦ 88, 94
♦ cultivated, herbal.

Silene cserei Baumg.
Caryophyllaceae
♦ biennial campion, Balkan catchfly,

itämaittenkohokki, European catchfly, smooth catchfly
♦ Weed, Noxious Weed, Naturalised, Casual Alien
♦ 23, 42, 87, 88, 101, 161, 162, 218
♦ Origin: eastern Europe.

Silene cucubalus Wibel nom. illeg.
Caryophyllaceae
= *Silene vulgaris* (Moench) Garcke
♦ bladder campion
♦ Weed, Noxious Weed
♦ 23, 36, 52, 87, 88, 162, 210, 218, 299
♦ herbal.

Silene dichotoma Ehrh.
Caryophyllaceae
♦ hairy catchfly, forked catchfly, dichotoma silene, hankakohokki
♦ Weed, Naturalised, Introduced, Casual Alien
♦ 23, 34, 40, 42, 70, 86, 87, 88, 94, 98, 101, 176, 198, 203, 218, 243, 272, 287
♦ aH, cultivated, herbal. Origin: southern Europe, Middle East.

Silene dioica (L.) Clairv.
Caryophyllaceae
Lychnis dioica L. (see), *Melandrium diurnum* (Sibth.) Fries., *Melandrium rubrum* (Weigel) Garcke
♦ red catchfly, red campion, puna ailakki
♦ Weed, Naturalised
♦ 54, 70, 86, 87, 88, 98, 101, 198, 203, 241, 272, 280
♦ b/pH, cultivated, herbal. Origin: Europe.

Silene disticha Willd.
Caryophyllaceae
♦ Naturalised
♦ 280

Silene firma Siebold & Zucc.
Caryophyllaceae
♦ Weed
♦ 286

Silene firma Sieb. & Zucc. fo. pubescens Ohwi & Ohashi
Caryophyllaceae
♦ Weed
♦ 286

Silene flos-cuculi (L.) Clairv.
Caryophyllaceae
= *Lychnis flos-cuculi* L.
♦ Naturalised
♦ 280

Silene fortunei Vis.
Caryophyllaceae
♦ Weed
♦ 275
♦ pH.

Silene fuscata Link ex Brot.
Caryophyllaceae
♦ Weed
♦ 87, 88, 253
♦ cultivated.

Silene gallica L.
Caryophyllaceae
Silene anglica L. (see), *Silene silvestris* Schott

♦ English catchfly, small catchfly, ranskankohokki, windmill pink, common catchfly, small flowered catchfly, French silene, gunpowder weed, calabacilla, alfinetes da terra, flor roxa
♦ Weed, Naturalised, Introduced, Casual Alien
♦ 7, 9, 15, 34, 38, 42, 51, 70, 87, 88, 94, 98, 101, 121, 134, 158, 161, 165, 167, 176, 180, 198, 203, 218, 236, 237, 241, 243, 245, 250, 253, 255, 271, 272, 280, 286, 287, 295, 300
♦ aH, arid, cultivated, herbal, toxic. Origin: Eurasia.

Silene gallica L. var. gallica
Caryophyllaceae
♦ French catchfly
♦ Naturalised, Environmental Weed
♦ 7, 72, 86, 198
♦ aH. Origin: southern Europe, Middle East.

Silene gallica L. var. giraldii (Guss.) Walters
Caryophyllaceae
♦ Naturalised
♦ 287

Silene gallica L. var. quinquevulnera (L.) Koch
Caryophyllaceae
♦ spotted catchfly, French catchfly
♦ Weed, Naturalised, Environmental Weed
♦ 7, 72, 86, 198, 269, 286, 287
♦ aH. Origin: Mediterranean.

Silene giraldii Guss.
Caryophyllaceae
♦ Naturalised
♦ 287

Silene gonocalyx Boiss.
Caryophyllaceae
♦ Weed
♦ 87, 88

Silene inflata Sm.
Caryophyllaceae
= *Silene vulgaris* (Moench) Garcke
♦ Weed
♦ 87, 88, 243

Silene × hampeana Meusel & K.Werner
Caryophyllaceae
= *Silene latifolia* Poir. × *Silene dioica* L.
♦ pink campion, hybrid campion
♦ Naturalised
♦ 101

Silene italica Pers.
Caryophyllaceae
Cucubalus italicus L.
♦ Italian catchfly
♦ Weed, Naturalised, Casual Alien
♦ 101, 272, 280
♦ cultivated, herbal.

Silene italica Pers. ssp. nemoralis (Waldst. & Kit.) Nyman
Caryophyllaceae
= *Silene nemoralis* Waldst. & Kit. (NoR)
♦ Italian catchfly
♦ Naturalised
♦ 101

Silene latifolia Poir.
Caryophyllaceae
♦ bladder campion, white campion, white cockle, silenka biela
♦ Weed, Naturalised
♦ 1, 86, 88, 101, 146, 195, 198, 253, 280
♦ pH, cultivated, herbal. Origin: Europe, Middle East.

Silene latifolia Poir. ssp. *alba* (Mill.) Greut. & Burd.
Caryophyllaceae
Lychnis alba Mill. (see), *Lychnis vespertina* Sibth., *Melandrium album* (Mill.) Garcke (see), *Silene alba* (Mill.) Krause (see), *Silene pratensis* (Raf.) Godr. & Gren. (see)
♦ white cockle, bladder campion, white campion
♦ Noxious Weed, Naturalised
♦ 101, 229, 243
♦ aH, cultivated, herbal.

Silene linearis Decne.
Caryophyllaceae
♦ Weed
♦ 221

Silene linicola Gm.
Caryophyllaceae
♦ Weed
♦ 70, 88, 94
♦ cultivated.

Silene longicaulis Pourr. ex Lag.
Caryophyllaceae
♦ Portuguese catchfly
♦ Weed, Naturalised, Environmental Weed
♦ 72, 86, 88, 98, 176, 198, 203
♦ a/b. Origin: Spain, Portugal.

Silene maritima With.
Caryophyllaceae
♦ sea campion
♦ Weed
♦ 23, 88
♦ cultivated.

Silene micropetala Lag.
Caryophyllaceae
♦ pieniteräkohokki
♦ Casual Alien
♦ 42

Silene multiflora (Waldst. & Kit.) Pers.
Caryophyllaceae
♦ Weed
♦ 272
♦ cultivated.

Silene muscipula L.
Caryophyllaceae
♦ kärpäskohokki
♦ Weed, Casual Alien
♦ 42, 70, 88, 94

Silene neglecta Ten.
Caryophyllaceae
♦ Weed
♦ 88

Silene noctiflora L.
Caryophyllaceae
Melandrium noctiflorum (L.) Fr. (see)
♦ night flowering catchfly, yöailakki, sticky cockle, night flowering campion,

night flowering silene
♦ Weed, Noxious Weed, Naturalised, Introduced
♦ 23, 36, 42, 52, 70, 87, 88, 94, 98, 101, 161, 162, 174, 203, 210, 218, 229, 243, 253, 272, 280, 287, 293, 299
♦ aH, cultivated, herbal. Origin: Europe.

Silene nocturna L.
Caryophyllaceae
♦ night flowering catchfly, night flowering campion, night flowering silene, Mediterranean catchfly, välimerenkohokki
♦ Weed, Naturalised, Environmental Weed, Casual Alien
♦ 7, 9, 42, 70, 72, 86, 88, 93, 98, 176, 185, 198, 205, 253
♦ aH, herbal. Origin: Mediterranean.

Silene nutans L.
Caryophyllaceae
♦ nodding catchfly, Nottingham catchfly, nuokkukohokki, Eurasian catchfly
♦ Weed, Naturalised
♦ 23, 88, 101, 272, 280, 287
♦ cultivated, herbal.

Silene otites (L.) Wibel
Caryophyllaceae
Cucubalus otites L.
♦ arokohokki, Spanish catchfly
♦ Casual Alien
♦ 42
♦ cultivated, herbal.

Silene palaestina Boiss.
Caryophyllaceae
♦ Weed
♦ 221
♦ cultivated.

Silene pendula L.
Caryophyllaceae
♦ rentokohokki, nodding catchfly
♦ Naturalised, Casual Alien
♦ 42, 86, 101, 280, 287
♦ aH, cultivated. Origin: Italy.

Silene pratensis (Raf.) Godr. & Gren.
Caryophyllaceae
= *Silene latifolia* Poir. ssp. *alba* (Mill.) Greut. & Burd.
♦ Weed, Noxious Weed, Naturalised
♦ 86, 98, 176, 203, 287, 299
♦ cultivated, herbal.

Silene pseudo-atocion Desf.
Caryophyllaceae
♦ Naturalised
♦ 86, 101

Silene pusilla Waldst. & Kit.
Caryophyllaceae
Silene quadridentata Pers.
♦ Weed
♦ 272
♦ cultivated.

Silene repens Patrin
Caryophyllaceae
♦ pink campion
♦ Weed
♦ 272, 297
♦ herbal.

Silene rubella L.
Caryophyllaceae
♦ Weed, Quarantine Weed
♦ 76, 87, 88, 185, 203, 220, 221

Silene schafta Hohen.
Caryophyllaceae
♦ Weed, Naturalised
♦ 98, 203
♦ cultivated.

Silene schimperiana Boiss.
Caryophyllaceae
♦ Weed
♦ 221

Silene soczaviana (Schisch.) Bocquet
Caryophyllaceae
♦ boreal catchfly
♦ Naturalised
♦ 101

Silene succulenta Forssk.
Caryophyllaceae
♦ Weed
♦ 221

Silene tridentata Desf.
Caryophyllaceae
♦ Weed, Naturalised
♦ 54, 86, 88
♦ Origin: Spain, North Africa.

Silene trinervia Sebast. & Mauri
Caryophyllaceae
♦ Weed
♦ 272

Silene uniflora Roth
Caryophyllaceae
♦ sea campion, merikohokki
♦ Naturalised
♦ 86
♦ cultivated.

Silene venosa Asch. nom. illeg.
Caryophyllaceae
= *Silene vulgaris* (Moench) Garcke ssp. *vulgaris*
♦ Weed
♦ 87, 88

Silene villosa Forssk.
Caryophyllaceae
♦ Weed
♦ 221

Silene viscaria (L.) Jess.
Caryophyllaceae
= *Lychnis viscaria* L.
♦ Casual Alien
♦ 280

Silene viscosa (L.) Pers
Caryophyllaceae
♦ white sticky catchfly, tahma ailakki, sticky catchfly
♦ Weed
♦ 39, 272
♦ cultivated, toxic.

Silene vulgaris (Moench) Garcke
Caryophyllaceae
Behen vulgaris Moench, *Cucubalus latifolius* Mill., *Silene cucubalus* Wibel (see), *Silene inflata* Sm. (see), *Silene venosa* Asch. (see)
♦ maiden's tears, bladder campion, nurmikohokki, blue root, rattlebox

♦ Weed, Noxious Weed, Naturalised, Garden Escape, Environmental Weed
♦ 7, 16, 34, 70, 72, 86, 87, 88, 94, 98, 101, 115, 147, 161, 165, 171, 176, 195, 198, 203, 212, 241, 243, 253, 269, 272, 280, 286, 287, 300
♦ pH, arid, cultivated, herbal. Origin: Mediterranean, west Asia.

Silene vulgaris (Moench) Garcke ssp. *macrocarpa* (Marsden) Jones & Turrill
Caryophyllaceae
Silene latifolia (Mill.) Britten & Rendl. *non* Poir.
♦ bladder campion, maiden's tears
♦ Weed
♦ 121
♦ H.

Silene vulgaris (Moench) Garcke ssp. *maritima* (With.) Á. & D.Löve
Caryophyllaceae
♦ Naturalised
♦ 86, 280
♦ cultivated.

Silene vulgaris (Moench) Garcke ssp. *vulgaris*
Caryophyllaceae
Silene venosa Asch. *nom. illeg.* (see)
♦ bladder campion
♦ Naturalised
♦ 86, 198, 280
♦ cultivated. Origin: Eurasia, Mediterranean.

Silphium asperrimum Hook.
Asteraceae
= *Silphium radula* Nutt. (NoR)
♦ showy rosinweed
♦ Weed
♦ 161
♦ herbal.

Silphium integrifolium Michx.
Asteraceae
♦ wholeleaf rosinweed, rosinweed
♦ Weed
♦ 161
♦ cultivated, herbal.

Silphium laciniatum L.
Asteraceae
♦ compass plant
♦ Weed
♦ 161
♦ pH, arid, cultivated, herbal.

Silphium perfoliatum L.
Asteraceae
♦ cup plant, cup rosinweed
♦ Weed
♦ 87, 88, 133, 161, 210, 218, 224
♦ pH, cultivated, herbal.

Silybum Adans. spp.
Asteraceae
♦ milk thistle
♦ Weed, Naturalised
♦ 18, 88, 198
♦ herbal.

Silybum marianum (L.) Gaertn.
Asteraceae
Carduus marianum L., *Carduus marianus* L., *Carthamus maculatus* Lam., *Cirsium maculatum* Scop., *Mariana mariana* Hill,

Silybum maculatum Moench
♦ milk thistle, variegated thistle, blessed milk thistle, cabbage thistle, Gundagai thistle, gundy, holy thistle, lady's thistle, spotted thistle, St. Mary's thistle, maarianohdake, cardo mariano
♦ Weed, Quarantine Weed, Noxious Weed, Naturalised, Introduced, Environmental Weed, Casual Alien
♦ 1, 7, 15, 20, 34, 39, 40, 42, 51, 55, 62, 70, 72, 76, 78, 80, 86, 87, 88, 94, 97, 98, 101, 116, 121, 134, 139, 146, 147, 161, 165, 171, 176, 180, 185, 198, 199, 203, 212, 217, 218, 221, 228, 229, 236, 237, 241, 243, 246, 253, 256, 269, 270, 272, 275, 280, 287, 290, 295, 300
♦ aH, arid, cultivated, herbal, toxic. Origin: Eurasia.

Simarouba glauca DC.
Simaroubaceae
♦ paradise tree
♦ Introduced
♦ 228
♦ arid, cultivated, herbal.

Simmondsia chinensis (Link) Schneid.
Simmondsiaceae
Buxus chinensis Link, *Simmondsia californica* Nutt.
♦ pignut, jojoba, goatnut
♦ Introduced
♦ 228
♦ S, arid, cultivated, herbal.

Simsia amplexicaulis (Cav.) Pers.
Asteraceae
Coreopsis amplexicaulis Cav., *Encelia amplexicaulis* (Cav.) Hemsl. (see), *Encelia cordata* (Kunth) Hemsl., *Encelia heterophylla* (Kunth) Hemsl., *Encelia mexicana* Mart. ex DC. *nom. inval.* (see), *Helianthus amplexicaulis* DC., *Helianthus sericeus* Sessé & Moç., *Helianthus trilobatus* Link, *Simsia amplexicaulis* var. *decipiens* (S.F.Blake) S.F.Blake, *Simsia auriculata* DC., *Simsia cordata* (Kunth) Cass., *Simsia foetida* (Cav.) S.F.Blake var. *decipiens* S.F.Blake, *Simsia heterophylla* (Kunth) DC., *Simsia kunthiana* Cass., *Simsia schaffneri* Sch.Bip. ex A.Gray, *Ximenesia cordata* Kunth, *Ximenesia heterophylla* Kunth, *Ximenesia hirta* Mart. ex DC.
♦ simsia
♦ Weed, Quarantine Weed
♦ 76, 88, 203, 220, 243
♦ aH.

Simsia foetida (Cav.) Blake
Asteraceae
♦ Weed
♦ 157
♦ aH.

Simsia grandiflora Benth.
Asteraceae
♦ Weed
♦ 157
♦ aH.

Sinacalia tangutica (Maxim.) B.Nord.
Asteraceae
♦ Weed, Quarantine Weed

♦ 76, 88, 220
♦ cultivated.

Sinapis L. spp.
Brassicaceae
♦ mustard
♦ Weed, Naturalised
♦ 198, 221, 243
♦ H.

Sinapis alba L.
Brassicaceae
Brassica alba (L.) Rabenh., *Brassica hirta* Moench (see)
♦ white mustard, kedlock, keltasinappi
♦ Weed, Naturalised, Introduced, Casual Alien, Cultivation Escape
♦ 7, 34, 39, 40, 42, 44, 70, 86, 87, 88, 94, 98, 101, 115, 176, 203, 221, 243, 252, 261, 272, 280, 287
♦ aH, arid, cultivated, herbal, toxic. Origin: Mediterranean.

Sinapis alba L. ssp. *alba*
Brassicaceae
Bonnania officinalis Presl, *Crucifera lampsana* Krause, *Eruca alba* Noulet, *Leucosinapis alba* Spach, *Raphanus albus* Crantz, *Sinapis hispida* Ten., *Sinapistrum album* Cheval.
♦ white mustard, mustard
♦ Naturalised, Introduced
♦ 56, 198, 280
♦ cultivated.

Sinapis alba L. ssp. *dissecta* (Lag.) Bonnier
Brassicaceae
♦ Naturalised
♦ 280

Sinapis allionii Jacq.
Brassicaceae
= *Sinapis arvensis* L. ssp. *allionii* (Jacq.) Baillarg. (NoR)
♦ Weed
♦ 88, 185

Sinapis arvensis L.
Brassicaceae
Brassica arvensis (L.) Rabenh (see), *Brassica kaber* (DC.) Wheeler (see), *Brassica kaber* (DC.) Wheeler var. *pinnatifida* (Stokes) Wheeler (see), *Brassica sinapis* Vis, *Brassica sinapistrum* Boiss (see), *Caulis sinapiaster* Krause, *Eruca arvensis* Noulet, *Napus agriasinapis* Schimp. & Spenn., *Sinapis orientalis* L., *Sinapis polymorpha* Geners., *Sinapis schkuhriana* Rchb.
♦ charlock, wild mustard, rikkasinappi, senape selvatica, kaber mustard, rapeseed, canola, charlock mustard, wild mustard, common wild mustard, corn mustard, field kale, kilk, brassocks
♦ Weed, Noxious Weed, Naturalised, Introduced
♦ 7, 32, 34, 36, 39, 44, 49, 52, 68, 70, 86, 88, 94, 98, 101, 111, 115, 118, 162, 174, 176, 179, 185, 198, 203, 221, 229, 243, 253, 255, 263, 269, 272, 280, 287, 294
♦ aH, arid, cultivated, herbal, toxic. Origin: Mediterranean.

Sinapis arvensis L. var. *orientalis* Koch
& Ziz.
Brassicaceae
♦ Naturalised
♦ 287

Sinapis arvensis L. var. *schkuhriana*
(Rchb.) Hageng.
Brassicaceae
♦ wild mustard, mostaza silvestre
♦ Weed
♦ 236, 237, 295

Sinapis incana L.
Brassicaceae
= *Hirschfeldia incana* (L.) Lagr.-Foss.
♦ Naturalised
♦ 287

Sinapis juncea L.
Brassicaceae
= *Brassica juncea* (L.) Czern.
♦ Weed
♦ 87, 88

Sinarundinaria anceps (Mitford)
C.S.Chao & Renvoize
Poaceae
Arundinaria anceps Mitford, *Arundinaria
jaunsarensis* Gamble, *Chimonobambusa
jaunsarensis* (Gamble) Bahadur
& Naithani, *Yushania jaunsarensis*
(Gamble) T.P.Yi
♦ Indian fountain bamboo
♦ Naturalised
♦ 40
♦ G.

Sindora siamensis Teijsm. ex Miq.
Fabaceae/Caesalpiniaceae
♦ Quarantine Weed
♦ 220

Sindora supa Merr.
Fabaceae/Caesalpiniaceae
♦ Naturalised
♦ 86
♦ herbal. Origin: Philippines.

Sinocalamus latiflorus (Munro) McClure
Poaceae
♦ wideleaf bamboo
♦ Naturalised
♦ 101
♦ G.

Siphonostegia chinensis Benth.
Scrophulariaceae
♦ Chinese siphonostegia, hikiyomogi
♦ Weed
♦ 297
♦ aH, promoted. Origin: east Asia.

Sison amomum L.
Apiaceae
Apium amomum (L.) Caruel, *Cicuta
amomum* Crantz, *Seseli amomum* Scop.,
Sium aromaticum Lam.
♦ stone parsley, breakstone, bastard
stone parsley, amomo germanico
♦ Weed, Naturalised
♦ 15, 165, 280
♦ bH, cultivated, herbal. Origin:
Eurasia, Mediterranean.

Sisymbrium L. spp.
Brassicaceae

♦ hedge mustard, mustard
♦ Weed, Naturalised
♦ 198, 243, 272

Sisymbrium altissimum L.
Brassicaceae
Norta altissima (L.) Britt., *Sisymbrium
pannonicum* Jacq., *Sisymbrium
sinapistrum* Crantz
♦ Jim Hill mustard, tumbleweed
mustard, tall mustard,
unkarinpernaruoho, tall
tumblemustard, tall rocket, Nabo
chileno
♦ Weed, Noxious Weed, Naturalised,
Introduced, Casual Alien
♦ 21, 23, 34, 40, 42, 44, 49, 52, 70, 80, 86,
87, 88, 94, 98, 101, 136, 146, 161, 167,
174, 179, 203, 210, 212, 218, 236, 237,
241, 243, 272, 280, 287, 295, 299, 300
♦ aH, cultivated, herbal. Origin:
Eurasia.

Sisymbrium austriacum Jacq.
Brassicaceae
Sisymbrium compressum Moench,
Sisymbrium eckartsbergense Willd.,
Sisymbrium multisiliquosum Hoff.
♦ tonavanpernaruoho, jeweled rocket
♦ Naturalised, Casual Alien
♦ 42, 101, 241, 300
♦ cultivated.

Sisymbrium bolgense Bieb.
Brassicaceae
♦ Weed
♦ 87, 88

Sisymbrium burchellii DC. var. *burchellii*
Brassicaceae
♦ Weed
♦ 121

Sisymbrium capense Thunb.
Brassicaceae
♦ cape mustard, cape wild mustard,
wild mustard
♦ Weed, Native Weed
♦ 88, 121, 158
♦ pH, herbal. Origin: southern Africa.

Sisymbrium columnae Jacq.
Brassicaceae
= *Sisymbrium orientale* L.
♦ Weed
♦ 87, 88

Sisymbrium erysimoides Desf.
Brassicaceae
♦ smooth mustard, Mediterranean
rocket
♦ Weed, Naturalised, Introduced,
Environmental Weed, Casual Alien
♦ 7, 86, 87, 88, 93, 98, 101, 198, 203, 205,
221, 228, 269, 280
♦ aH, arid, cultivated. Origin:
Mediterranean.

Sisymbrium exacoides DC.
Brassicaceae
♦ Weed
♦ 87, 88

Sisymbrium heteromallum C.A.Mey.
Brassicaceae
♦ Weed

♦ 275
♦ a/bH.

Sisymbrium irio L.
Brassicaceae
Descurainia irio Webb & Berth.,
Sisymbrium erysimastrum Lam.,
Sisymbrium heteromallum Fourn,
Sisymbrium latifolium Gray
♦ London rocket, desert mustard,
kiiltopernaruoho
♦ Weed, Naturalised, Introduced,
Environmental Weed, Casual Alien
♦ 7, 34, 42, 55, 56, 80, 86, 87, 88, 93, 98,
101, 115, 161, 167, 176, 179, 180, 185,
198, 203, 205, 212, 218, 221, 228, 237,
241, 243, 253, 269, 272, 280, 286, 287,
295, 300
♦ aH, arid, cultivated, herbal. Origin:
Eurasia.

Sisymbrium loeselii L.
Brassicaceae
Erysimum loeselii Rupre., *Leptocarpaea
loeselii* Rupre., *Nasturtium loeselium*
Krause, *Sisymbrium scholare* Fourn.,
Turritis loeselii R.Br.
♦ tall hedge mustard, small
tumbleweed mustard, loesel
tumblemustard, karvapernaruoho
♦ Weed, Noxious Weed, Naturalised,
Introduced, Casual Alien
♦ 23, 36, 40, 42, 44, 80, 87, 88, 101, 114,
136, 161, 174, 218, 272, 287
♦ aH, cultivated, herbal.

Sisymbrium officinale (L.) Scop.
Brassicaceae
Chamaeplium officinale Wall, *Erysimum
officinale* L., *Erysimum runcinatum*
Gilib., *Sisymbrium officinale* var.
leiocarpum (L.) Scop. (see)
♦ hedge mustard, hedge wild mustard,
hedge weed, rohtopernaruoho,
Erísimo, common hedge mustard
♦ Weed, Naturalised, Introduced,
Garden Escape, Environmental Weed
♦ 7, 15, 21, 23, 34, 38, 39, 44, 55, 70, 80,
86, 87, 88, 94, 98, 101, 121, 134, 136, 161,
165, 167, 176, 198, 199, 203, 210, 211,
218, 228, 236, 237, 241, 243, 253, 269,
271, 272, 280, 286, 287, 295, 297, 300
♦ a/bH, arid, cultivated, herbal, toxic.
Origin: Eurasia.

Sisymbrium officinale (L.) Scop. var.
leiocarpum (L.) Scop.
Brassicaceae
= *Sisymbrium officinale* (L.) Scop.
♦ hedge mustard
♦ Weed, Naturalised
♦ 207, 287
♦ Origin: Europe.

Sisymbrium officinale (L.) Scop. var.
officinale
Brassicaceae
♦ Naturalised
♦ 287

Sisymbrium orientale Thunb.
Brassicaceae
Sisymbrium columnae Jacq. (see),
Sisymbrium flexuosum Dulac,

Sisymbrium villosum Moench
- Indian hedge mustard, eastern rocket, oriental mustard, rocket, hedge mustard, oriental hedge mustard, mustard, cress
- Weed, Naturalised, Introduced, Environmental Weed, Casual Alien
- 7, 15, 34, 38, 40, 42, 55, 56, 68, 79, 80, 86, 87, 88, 93, 94, 98, 101, 121, 134, 158, 161, 176, 180, 198, 203, 205, 221, 228, 241, 243, 253, 269, 272, 280, 286, 287, 300
- aH, arid, cultivated, herbal. Origin: Eurasia.

Sisymbrium polyceratium L.
Brassicaceae
- shortfruit hedge mustard
- Weed, Naturalised, Casual Alien
- 40, 88, 94, 101, 272, 280
- herbal.

Sisymbrium polymorphum (Murray) Roth
Brassicaceae
- kaakonpernaruoho
- Casual Alien
- 42

Sisymbrium runcinatum Lag. ex DC.
Brassicaceae
- Weed, Naturalised
- 7, 86, 87, 88, 98, 203, 269, 300
- cultivated. Origin: Europe, Middle East.

Sisymbrium sophia (L.) Webb
Brassicaceae
= *Descurainia sophia* (L.) Webb ex Prantl
- flixweed
- Weed
- 39, 44, 80, 87, 88
- a/bH, herbal, toxic.

Sisymbrium strictissimum L.
Brassicaceae
Cheirinia strictissima Link, *Norta strictissima* Schur, *Sisymbrium nitidulum* Lag.
- jäykkäpernaruoho, perennial rocket
- Weed, Garden Escape, Cultivation Escape
- 42, 272
- cultivated.

Sisymbrium supinum L.
Brassicaceae
Arabis supina Lam., *Braya supina* Koch., *Chamaeplium supinum* Wall., *Erysimum supinum* Link, *Kibera supina* (L.) Fourr.
- matalapernaruoho
- Casual Alien
- 42

Sisymbrium thellungii Schulz
Brassicaceae
Brassica pachypoda Thell.
- African turnip weed, common wild mustard, hedge mustard, wild mustard, Indian hedge mustard
- Weed, Quarantine Weed, Naturalised, Native Weed
- 51, 55, 68, 76, 86, 87, 88, 98, 121, 158, 203, 243, 269
- a/bH, arid, cultivated, herbal. Origin: southern Africa.

Sisymbrium turczaninowii Sond.
Brassicaceae
- Russian rocket
- Naturalised
- 101
- herbal.

Sisymbrium volgense Bieb. ex E.Fourn.
Brassicaceae
- Russian rocket
- Weed, Casual Alien
- 42, 272

Sisyrinchium L. spp.
Iridaceae
- blue eyed grass, scourweed
- Weed, Naturalised
- 54, 86, 88, 198, 280
- cultivated, herbal.

Sisyrinchium albidum Raf.
Iridaceae
- blue eyed grass, white blue eyed grass
- Weed
- 161
- herbal.

Sisyrinchium atlanticum Bick.
Iridaceae
- eastern blue eyed grass, blue eyed grass
- Weed
- 87, 88
- cultivated, herbal.

Sisyrinchium bermudiana L.
Iridaceae
- blue eyed grass, giglietto
- Weed, Naturalised, Casual Alien
- 40, 98, 203, 287
- cultivated, herbal.

Sisyrinchium californicum (Ker Gawl.) W.T.Aiton
Iridaceae
- golden blue eyed grass, golden eyed grass, yellow eyed grass
- Casual Alien
- 40, 280
- pH, aqua, cultivated, herbal.

Sisyrinchium chilense Hook.
Iridaceae
Sisyrinchium scabrum Schltdl. & Cham. (see)
- swordleaf blue eyed grass
- Weed
- 87, 88

Sisyrinchium convolutum Nocca
Iridaceae
- Weed
- 199
- cultivated.

Sisyrinchium exile Bickn.
Iridaceae
= *Sisyrinchium rosulatum* Bickn.
- Weed, Naturalised
- 7, 9, 98, 203, 287
- cultivated, herbal.

Sisyrinchium graminoides Bickn.
Iridaceae
= *Sisyrinchium angustifolium* Mill. (NoR)

- Naturalised
- 287
- herbal.

Sisyrinchium iridifolium Kunth
Iridaceae
Sisyrinchium laxum Sims (see)
- striped rush leaf, purple eyed grass, blue eyed grass, blue pigroot, spreading blue eyed grass
- Weed, Naturalised, Garden Escape, Environmental Weed
- 7, 15, 39, 72, 86, 87, 88, 98, 101, 165, 176, 198, 203, 280
- pH, cultivated, toxic. Origin: Central and South America.

Sisyrinchium iridifolium Humb. Bonpl. & Kunth var. *laxum* Maek.
Iridaceae
- Weed, Naturalised
- 286, 287

Sisyrinchium laxum Sims
Iridaceae
= *Sisyrinchium iridifolium* Kunth
- cebolinha
- Weed
- 255
- pH. Origin: Brazil.

Sisyrinchium micranthum Cav.
Iridaceae
= *Sisyrinchium rosulatum* Bickn.
- Weed, Naturalised
- 39, 98, 134, 203
- arid, cultivated, herbal, toxic.

Sisyrinchium montanum Greene
Iridaceae
- kuovinkukka, mountain blue eyed grass, strict blue eyed grass
- Casual Alien
- 42
- cultivated, herbal.

Sisyrinchium mucronatum Michx.
Iridaceae
- needletip blue eyed grass
- Naturalised
- 287
- herbal.

Sisyrinchium rosulatum Bickn.
Iridaceae
Sisyrinchium exile Bickn. (see), *Sisyrinchium micranthum* Cav. (see)
- annual blue eyed grass
- Weed, Naturalised
- 157, 161, 179, 249, 286, 287
- pH, herbal.

Sisyrinchium scabrum Schltdl. & Cham.
Iridaceae
= *Sisyrinchium chilense* Hook.
- Weed
- 199
- herbal.

Sisyrinchium striatum Sm.
Iridaceae
- pale yellow eyed grass
- Naturalised
- 40, 280
- cultivated.

Sitanion hystrix **(Nutt.) Sm.**
Poaceae
= *Elymus elymoides* (Raf.) Swezey ssp.
elymoides (NoR)
♦ squirreltail
♦ Weed
♦ 87, 88, 218
♦ G, cultivated, herbal.

Sium angustifolium **L. nom. illeg.**
Apiaceae
= *Berula erecta* (Huds.) Coville
♦ Weed
♦ 23, 88
♦ herbal.

Sium erectum **Huds.**
Apiaceae
= *Berula erecta* (Huds.) Coville
♦ Weed
♦ 39, 87, 88
♦ herbal, toxic.

Sium latifolium **L.**
Apiaceae
Cicuta latifolia Crantz, *Coriandrum latifolium* Crantz, *Drepanophyllum palustre* Hoffm., *Selinum sium* Krause, *Sisarum palustre* Bubani, *Sium lunifolium* Gmel.
♦ water parsnip, narrow leaved water parsnip, sorsanputki, greater water parsnip
♦ Weed, Naturalised
♦ 23, 39, 87, 88, 272, 300
♦ pH, aqua, cultivated, herbal, toxic.

Sium sisarum **L.**
Apiaceae
Pimpinella sisarum Jess., *Seseli sisarum* (L.) Crantz, *Sium sisaroideum* DC.
♦ sokerijuuri, skirret
♦ Quarantine Weed, Casual Alien
♦ 42, 220
♦ pH, cultivated, herbal. Origin: Eurasia.

Sium suave **Walter**
Apiaceae
Sium cicutaefolium Schrank
♦ water parsnip, hemlock water parsnip
♦ Weed
♦ 39, 87, 88, 161, 218, 297
♦ wpH, promoted, herbal, toxic.

Sixalix atropurpurea **(L.) Greuter & Burdet**
Dipsacaceae
= *Scabiosa atropurpurea* L. ssp. *atropurpurea* (NoR) [see *Scabiosa atropurpurea* L.]
♦ Weed
♦ 253

Skeletonema potamos **(C.I.Weber) Hasle.**
Skeletonemataceae
♦ Weed
♦ 197
♦ diatom.

Skimmia japonica **Thunb.**
Rutaceae
♦ skimmia, Japanese skimmia
♦ Weed
♦ 39, 161, 247

♦ S, cultivated, herbal, toxic.

Smilacina japonica **A.Gray**
Liliaceae/Convallariaceae
♦ yukizasa
♦ Quarantine Weed
♦ 220
♦ cultivated.

Smilacina racemosa **(L.) Desf.**
Liliaceae/Convallariaceae
= *Maianthemum racemosum* (L.) Link ssp. *racemosum* (NoR)
♦ false Solomon's seal, false spikenard, large false Solomon's seal
♦ Weed
♦ 23, 88
♦ pH, cultivated, herbal.

Smilacina stellata **(L.) Desf.**
Liliaceae/Convallariaceae
= *Maianthemum stellatum* (L.) Link (NoR)
♦ little false Solomon's seal, starflower
♦ Weed
♦ 23, 88
♦ pH, cultivated, herbal.

Smilax **L. spp.**
Smilacaceae
♦ green briar, asparagus, sarsaparilla
♦ Weed
♦ 243
♦ herbal.

Smilax argyrea **Lindl. & Rod.**
Smilacaceae
♦ Introduced
♦ 38

Smilax aspera **L.**
Smilacaceae
Smilax nigra Willd.
♦ sarsaparilla, smilax, Italian sarsaparilla
♦ Weed
♦ 272
♦ pC, cultivated, herbal.

Smilax balbisiana **Kunth**
Smilacaceae
♦ Weed
♦ 87, 88
♦ herbal.

Smilax bona-nox **L.**
Smilacaceae
♦ saw green briar, green briar
♦ Weed
♦ 87, 88, 218
♦ pC, cultivated, herbal.

Smilax brasiliensis **Spreng.**
Smilacaceae
♦ japecanga
♦ Weed
♦ 255
♦ pH. Origin: Brazil.

Smilax china **L.**
Smilacaceae
♦ China root, China smilax
♦ Weed
♦ 88, 204, 286
♦ pC, cultivated, herbal.

Smilax coriacea **Spreng.**
Smilacaceae

Smilax havanensis auct. *non* Jacq. (see)
♦ everglades green briar
♦ Weed
♦ 87, 88

Smilax cumanensis **Humb. & Bonpl. ex Willd.**
Smilacaceae
♦ Weed
♦ 87, 88

Smilax glauca **Walter**
Smilacaceae
♦ cat green briar, saw briar, wild sarsaparilla
♦ Weed
♦ 87, 88, 218
♦ pC, promoted, herbal.

Smilax havanensis auct. *non* **Jacq.**
Smilacaceae
= *Smilax coriacea* Walt.
♦ everglades green briar
♦ Weed
♦ 14
♦ herbal.

Smilax helferi **A.DC.**
Smilacaceae
♦ Weed
♦ 12
♦ herbal.

Smilax kraussiana **Meisn.**
Smilacaceae
♦ wild sarsaparilla, linselele, bofekifeki
♦ Weed, Native Weed
♦ 88, 121
♦ pC, cultivated, herbal. Origin: southern Africa.

Smilax laurifolia **L.**
Smilacaceae
♦ laurel green briar, laurelleaf green briar
♦ Weed
♦ 87, 88, 218
♦ pC, cultivated, herbal.

Smilax ornata **Lem.**
Smilacaceae
♦ Jamaica sarsaparilla
♦ Weed
♦ 87, 88
♦ herbal.

Smilax riparia **A.DC. var.** *ussuriensis* **(Regel) H.Hara & T.Koyama**
Smilacaceae
♦ Weed
♦ 286

Smilax rotundifolia **L.**
Smilacaceae
♦ roundleaf green briar, common green briar, common cat briar, bull briar, horse briar, Mexican sarsaparilla, green briar, cat briar
♦ Weed
♦ 8, 80, 87, 88, 161, 211, 218, 243
♦ pC, promoted, herbal. Origin: eastern North America.

Smilax walteri **Pursh**
Smilacaceae
♦ redbead green briar, coral green

briar
- Weed
- 87, 88, 218
- herbal.

Smithia sensitiva **W.Ait.**
Fabaceae/Papilionaceae
- Naturalised
- 287
- herbal.

Smyrnium olusatrum **L.**
Apiaceae
- alexanders
- Weed, Naturalised
- 24, 40, 70, 241, 272
- bH, cultivated, herbal.

Smyrnium perfoliatum **L.**
Apiaceae
- biennial alexanders, perfoliate alexanders
- Weed
- 272
- bH, cultivated, herbal.

Smyrnium rotundifolium **Mill.**
Apiaceae
- round leaved alexanders, perfoliate alexanders
- Weed
- 272
- bH.

Snowdenia polystachya **(Fresen.) Pilg.**
Poaceae
- Weed, Quarantine Weed
- 76, 87, 88, 203, 220, 240
- aG.

Soehrensia **Backeb. spp.**
Cactaceae
= *Echinopsis* Zucc. spp. (NoR)
- Weed, Quarantine Weed
- 76, 88, 220

Solandra **Sw. spp.**
Solanaceae
- trumpet flower, solandra, chalice vine
- Weed
- 161, 247
- toxic.

Solandra hartwegii **C.F.Ball**
Solanaceae
= *Solandra maxima* (Sessé & Moç.) P.S.Green
- cup of gold, golden cup
- Weed
- 3, 191

Solandra maxima **(Sessé & Moç.) P.S.Green**
Solanaceae
Datura maxima Sessé & Moç., *Solandra hartwegii* C.F.Ball (see), *Solandra selerae* Dammer
- golden chalice, cup of gold, golden cup, chalice vine, golden chalice vine
- Naturalised, Cultivation Escape
- 39, 134, 233
- pC/S, cultivated, herbal, toxic.

Solandra nitida **Zucc.**
Solanaceae
= *Swartsia nitida* (Zuccagni) Standl.

(NoR)
- Weed
- 3, 191
- cultivated, herbal.

Solanum **L. spp.**
Solanaceae
- nightshade, polo
- Weed, Naturalised
- 18, 88, 198, 272
- herbal, toxic.

Solanum abutiloides **(Griseb.) Bitter & Lillo**
Solanaceae
- Weed, Naturalised
- 54, 86, 88, 98, 191, 203
- Origin: South America.

Solanum aculeastrum **Dun.**
Solanaceae
- apple of Sodom, bitterapple, devil's apple, goat apple, goat bitterapple, poison apple, mu Tura, dungwiza, ntobolobo
- Weed, Native Weed
- 39, 50, 87, 88, 121
- S/T, toxic. Origin: southern Africa.

Solanum aculeatissimum **Jacq.**
Solanaceae
- soda apple nightshade, apple of Sodom, devil's apple, love apple
- Weed, Naturalised, Native Weed
- 39, 87, 88, 106, 121, 161, 218, 286, 287
- pS, cultivated, herbal, toxic.

Solanum aethiopicum **L.**
Solanaceae
- Ethiopian nightshade, mock tomato
- Weed
- 221
- S, arid, promoted, herbal.

Solanum alatum **Moench**
Solanaceae
= *Solanum villosum* (L.) Mill.
- Weed, Quarantine Weed
- 76, 87, 88, 203, 220, 275
- aH.

Solanum albicaule **Kotschy ex Dunal**
Solanaceae
- Weed
- 221
- herbal.

Solanum americanum **Mill.**
Solanaceae
Solanum caribaeum Dun. (see), *Solanum nodiflorum* Jacq. (see), *Solanum nigrum* L. var. *americanum* (Mill.) Schulz (see), *Solanum nodiflorum* Jacq. ssp. *nutans* R.J.F.Hend.
- glossy nightshade, American black nightshade, black nightshade, garden nightshade, nightshade, yerba mora negra, maria preta
- Weed, Naturalised, Environmental Weed, Casual Alien
- 6, 7, 9, 34, 39, 42, 55, 68, 72, 88, 93, 98, 161, 180, 198, 199, 203, 243, 245, 247, 255, 257, 261, 262, 286, 287
- pH, arid, cultivated, herbal, toxic. Origin: Americas.

Solanum americanum **Mill. ssp. nodiflorum (Jacq.) R.J.F.Hend.**
Solanaceae
- glossy nightshade
- Naturalised
- 86

Solanum amygdalifolium **Steud.**
Solanaceae
Solanum angustifolium Lam.
- jazmín de córdoba
- Weed
- 295

Solanum anguivi **Lam.**
Solanaceae
Solanum hermannii Dunal *nom. illeg.* (see), *Solanum indicum auct.*, *Solanum scalare* C.H.Wright, *Solanum sodomeum* L. *nom. rej.* (see)
- Native Weed
- 121
- pS, cultivated, herbal. Origin: Africa, Arabia, Madagascar.

Solanum antillarum **O.E.Schulz**
Solanaceae
= *Solanum nudum* Dunal (NoR)
- forest nightshade
- Weed
- 14

Solanum asperolanatum **Ruiz & Pav.**
Solanaceae
- jurubeba
- Weed
- 255
- S, herbal. Origin: tropical America.

Solanum athroanthum **Dunal**
Solanaceae
- Weed
- 13

Solanum atriplicifolium **Gillies ex Nees**
Solanaceae
= *Solanum excisirhombeum* Bitter (NoR)
- Weed
- 237
- herbal.

Solanum atropurpureum **Schrank**
Solanaceae
- Weed
- 87, 88
- cultivated, herbal.

Solanum auriculatum **Aiton**
Solanaceae
= *Solanum mauritianum* Scop.
- Weed
- 87, 88

Solanum aviculare **G.Forst.**
Solanaceae
- New Zealand nightshade, kangaroo apple, gunyang, koonyang, mayakitch, meakitch, mookitch, poroporo, kohoho, bullibulli
- Weed, Naturalised, Native Weed, Introduced, Garden Escape, Environmental Weed
- 34, 39, 80, 86, 98, 101, 161, 165, 203, 228, 269
- S, arid, cultivated, herbal, toxic. Origin: Australasia.

Solanum balbisii Dun.
Solanaceae
= *Solanum sisymbrifolium* Lam. (NoR)
♦ Weed
♦ 87, 88

Solanum betaceum Cav.
Solanaceae
= *Cyphomandra betacea* (Cav.) Sendtn.
♦ Naturalised
♦ 86, 280
♦ cultivated. Origin: South America.

Solanum biflorum Lour.
Solanaceae
♦ Weed
♦ 87, 88

Solanum bonariense L.
Solanaceae
♦ naranjillo
♦ Weed
♦ 87, 88, 237, 295
♦ cultivated.

Solanum capense L.
Solanaceae
♦ nightshade
♦ Native Weed
♦ 121
♦ S, herbal, toxic. Origin: southern Africa.

Solanum capsicastrum Link ex Schauer
Solanaceae
♦ false Jerusalem cherry, huonekoiso, wintercherry
♦ Weed, Naturalised, Cultivation Escape
♦ 39, 42, 87, 88, 101
♦ cultivated, toxic.

Solanum capsicoides All.
Solanaceae
Solanum ciliatum Lam. (see)
♦ cockroach berry
♦ Weed, Naturalised, Introduced
♦ 32, 38, 86, 98, 203, 255, 269
♦ a/pH, cultivated. Origin: Brazil.

Solanum cardiophyllum Lindl.
Solanaceae
♦ heartleaf horsenettle, heartleaf nightshade
♦ Weed, Noxious Weed, Naturalised
♦ 35, 88, 101, 229

Solanum cardiophyllum Lindl. var. cardiophyllum
Solanaceae
♦ Weed
♦ 199

Solanum caribaeum Dun.
Solanaceae
= *Solanum americanum* Mill.
♦ Weed
♦ 87, 88

Solanum carolinense L.
Solanaceae
♦ horsenettle, bull nettle, apple of Sodom, wild tomato, devil's tomato, devil's potato, sand briar, karoliinankoiso, Carolina horsenettle, sand briar
♦ Weed, Quarantine Weed, Noxious

Weed, Naturalised, Native Weed, Introduced, Environmental Weed, Casual Alien
♦ 8, 23, 26, 34, 35, 36, 39, 42, 76, 80, 84, 87, 88, 154, 161, 174, 207, 210, 211, 218, 219, 220, 228, 229, 243, 246, 247, 249, 263, 266, 280, 286, 287, 292
♦ pH, arid, cultivated, herbal, toxic. Origin: south-eastern North America.

Solanum carolinense L. fo. albiflorum Benk.
Solanaceae
♦ Naturalised
♦ 287

Solanum carolinense L. var. carolinense
Solanaceae
♦ Carolina horsenettle
♦ Noxious Weed
♦ 229

Solanum carolinense L. var. *floridanum* (Shuttlw. ex Dunal) Chapman
Solanaceae
♦ Carolina horsenettle
♦ Noxious Weed
♦ 229

Solanum carolinense L. var. *hirsutum* (Nutt.) Gray
Solanaceae
♦ Carolina horsenettle, horsenettle
♦ Noxious Weed
♦ 229

Solanum chacoense Bitter
Solanaceae
♦ papa silvestre
♦ Weed
♦ 87, 88, 237, 295

Solanum chenopodioides Lam.
Solanaceae
Solanum gracilius Herter, *Solanum sublobatum* Willd. ex Roem. & Schult. (see)
♦ goosefoot nightshade, whitetip nightshade, velvety nightshade, tall nightshade
♦ Weed, Naturalised, Introduced, Casual Alien
♦ 15, 40, 86, 87, 88, 98, 198, 203, 228, 237, 270, 280, 295
♦ arid. Origin: South America.

Solanum chrysocarpum Pers.
Solanaceae
= *Solanum hispidum* Pers.
♦ giant devil's fig
♦ Weed, Quarantine Weed
♦ 76, 88

Solanum ciliatum Lam.
Solanaceae
= *Solanum capsicoides* All.
♦ Weed
♦ 87, 88
♦ herbal.

Solanum cinereum R.Br.
Solanaceae
♦ Narrawa burr
♦ Weed, Naturalised, Native Weed
♦ 39, 86, 87, 88, 198, 269
♦ cultivated, herbal, toxic. Origin: Australia.

Solanum coagulans Forssk.
Solanaceae
♦ wild nightshade, esikele
♦ Weed
♦ 297
♦ arid, herbal.

Solanum coccineum Jacq.
Solanaceae
♦ Native Weed
♦ 121
♦ pS. Origin: southern Africa.

Solanum comitis Dunal
Solanaceae
♦ Weed
♦ 13

Solanum commersonii Dun. ex Poir.
Solanaceae
♦ Commerson's nightshade
♦ Weed
♦ 87, 88, 237, 295

Solanum cornutum Lam. auct.
Solanaceae
= *Solanum rostratum* Dunal
♦ kärsäkoiso
♦ Weed, Casual Alien
♦ 42, 272
♦ herbal.

Solanum corymbosum Jacq.
Solanaceae
♦ Weed
♦ 199

Solanum crispum Ruiz & Pav.
Solanaceae
♦ Naturalised
♦ 39, 280
♦ toxic.

Solanum cyananthum Dunal
Solanaceae
♦ Weed, Quarantine Weed
♦ 76, 88, 220

Solanum deflexum Greenm.
Solanaceae
= *Solanum adscendens* Sendtn. (NoR)
♦ sonoita nightshade
♦ Weed
♦ 157
♦ aH, herbal.

Solanum diflorum Vell.
Solanaceae
♦ Jerusalem cherry, false Jerusalem cherry
♦ Weed, Naturalised
♦ 15, 87, 88, 237, 255, 280, 295
♦ pH, toxic. Origin: tropical America.

Solanum dimidiatum Raf.
Solanaceae
♦ Torrey's nightshade, western horsenettle, robust horsenettle
♦ Weed, Quarantine Weed, Noxious Weed, Naturalised, Introduced
♦ 34, 35, 86, 88, 98, 161, 203, 220, 229
♦ pH, toxic. Origin: south-east North America.

Solanum diphyllum L.
Solanaceae
♦ twinleaf nightshade
♦ Weed, Naturalised, Environmental

Weed
- 80, 88, 101, 112, 179

Solanum donianum **Walp.**
Solanaceae
Solanum verbascifolium sensu L. (see)
- mullein nightshade
- Naturalised
- 101
- herbal.

Solanum douglasii **Dunal**
Solanaceae
- Douglas's nightshade, greenspot nightshade
- Weed, Naturalised
- 86, 161, 198
- pH, herbal, toxic.

Solanum dubium **Fresen.**
Solanaceae
- Weed, Quarantine Weed
- 76, 87, 88, 203, 220, 221, 240, 242
- S, arid, herbal.

Solanum dulcamara **L.**
Solanaceae
Dulcamara flexuosa Moench
- bittersweet nightshade, punakoiso, woody nightshade, bittersweet, deadly nightshade, climbing nightshade, European bittersweet, bitter nightshade, blue nightshade, woody nightshade, poison berry, scarlet berry, blue bindweed, dogwood, fellenwort
- Weed, Noxious Weed, Naturalised, Cultivation Escape
- 1, 7, 8, 15, 23, 34, 39, 44, 45, 52, 70, 80, 86, 87, 88, 94, 98, 101, 102, 133, 136, 139, 146, 161, 165, 176, 189, 194, 195, 203, 207, 210, 210, 211, 212, 218, 224, 229, 243, 247, 252, 253, 272, 280
- pC, aqua, cultivated, herbal, toxic. Origin: Eurasia, North Africa.

Solanum dulcamara **L. var.** *dulcamara*
Solanaceae
- bitter nightshade, climbing nightshade
- Noxious Weed, Naturalised
- 101, 229
- toxic.

Solanum dulcamara **L. var.** *villosissimum* **Desv.**
Solanaceae
- bitter nightshade, climbing nightshade
- Noxious Weed, Naturalised
- 101, 229
- toxic.

Solanum elaeagnifolium **Cav.**
Solanaceae
- silverleaf nightshade, white horsenettle, bitterapple, silverleaf, silverleaf bitterapple, silverleaf nettle, tomato weed, tähtikoiso, revienta caballos, bullnettle, prairie berry, sand briar, silver horsenettle, trompillo, white horsenettle, buena mujer
- Weed, Quarantine Weed, Noxious Weed, Naturalised, Native Weed, Introduced, Environmental Weed, Casual Alien
- 1, 7, 34, 35, 42, 51, 63, 80, 86, 87, 88,

93, 95, 98, 121, 146, 147, 158, 161, 171, 179, 180, 198, 203, 205, 210, 212, 215, 218, 219, 228, 229, 236, 237, 243, 264, 266, 269, 272, 278, 283, 287
- pH, arid, cultivated, herbal, toxic. Origin: Americas.

Solanum elaeagnifolium **Cav. fo.** *albiflorum* **Cockerell**
Solanaceae
- Naturalised
- 287

Solanum elaeagnifolium **Cav. var.** *leprosum* **(Ort.) Dun.**
Solanaceae
- revienta caballos
- Weed
- 295

Solanum ellipticum **R.Br.**
Solanaceae
Solanum lithophilum F.Muell.
- a'leljaka, albaranji, alpirrantja, booka booda, desert raisin, e toonba, hillside flannel bush, immaru, kuilpura, native gooseberry, potatobush, potato weed, randa, ranto, tomato bush, velvet nightshade, velvet potatobush, walki, wangi, wanji, wanki, wild gooseberry, yaliljiriki yaliljiriki
- Weed
- 39, 87, 88
- arid, cultivated, toxic.

Solanum erianthum **D.Don**
Solanaceae
Solanum verbascifolium auct. non L. (see)
- berenjena macho, kala mewa, potato tree, thasau rangman, tobacco tree, verenjena, tobacco nightshade, jye yan ye, turkey berry
- Weed, Naturalised, Introduced
- 86, 88, 93, 98, 191, 203, 228, 255, 261
- pH, arid, cultivated, herbal. Origin: tropical America.

Solanum esuriale **Lindl.**
Solanaceae
Solanum ellipticum R.Br. fo. *inermis* Wawra, *Solanum esuriale* Lindl. var. *sublobatum* Domin, *Solanum pulchellum* F.Muell.
- comya, ntaganta, oondoroo, quena, yak ka berry
- Weed, Native Weed
- 39, 87, 88, 269
- pH, arid, cultivated, herbal, toxic. Origin: Australia.

Solanum ferox **L.**
Solanaceae
- nightshade
- Weed, Naturalised
- 98, 203
- herbal. Origin: Central America.

Solanum ferox **L. var.** *involucratum* **(Bl.) Miq.**
Solanaceae
Solanum involucratum Bl.
- Weed
- 13
- S.

Solanum ficifolium **Ortega**
Solanaceae

= *Solanum torvum* Sw.
- Weed
- 88

Solanum fructo-tecto **Cav.**
Solanaceae
- Weed
- 199

Solanum furcatum **Dunal**
Solanaceae
- forked nightshade, nightshade, broad nightshade, black South America nightshade
- Weed, Naturalised, Environmental Weed, Casual Alien
- 34, 39, 72, 86, 88, 161, 176, 198, 280
- pS, herbal, toxic. Origin: South America.

Solanum gayanum **(Remy) Phil.f.**
Solanaceae
- Chilean nightshade
- Naturalised
- 101

Solanum giganteum **Jacq.**
Solanaceae
- goat bitterapple, healing leaf tree, red bitterapple, red bitter berry, African holly
- Native Weed
- 121
- S/T, cultivated, herbal. Origin: Africa.

Solanum glaucophyllum **Desf.**
Solanaceae
Solanum malacoxylon Sendtn., *Solanum glaucum* Dun. ex DC. (see)
- waxyleaf nightshade
- Weed, Naturalised
- 101, 237, 287, 295

Solanum glaucum **Dun. ex DC.**
Solanaceae
= *Solanum glaucophyllum* Desf.
- Weed, Quarantine Weed
- 76, 87, 88, 203, 220

Solanum gracile **Dunal** *nom. illeg.*
Solanaceae
= *Solanum nigrescens* M.Martens & Galeotti (NoR)
- black berry, black nightshade
- Weed, Quarantine Weed, Native Weed
- 39, 76, 87, 88, 121, 203, 220
- pH, herbal, toxic.

Solanum grandiflorum **Ruiz & Pav.**
Solanaceae
- siuca huito
- Weed
- 39, 153, 255
- S/T, toxic. Origin: Brazil.

Solanum grossedentatum **A.Rich.**
Solanaceae
- Weed, Quarantine Weed
- 76, 87, 88, 203, 220

Solanum hamulosum **C.White**
Solanaceae
- Weed
- 87, 88
- Origin: Australia.

Solanum hermannii Dunal *nom. illeg.*
Solanaceae
= *Solanum anguivi* Lam.
♦ bitterapple, Sodom's apple, Dead Sea apple, apple of Sodom
♦ Weed
♦ 39, 121, 203, 269
♦ pH, arid, cultivated, toxic. Origin: Africa, Mediterranean.

Solanum heterodoxum Dunal
Solanaceae
♦ meksikonkoiso, melonleaf nightshade
♦ Weed, Casual Alien
♦ 42, 272
♦ herbal.

Solanum heterodoxum Dunal var. heterodoxum
Solanaceae
♦ melonleaf nightshade
♦ Weed
♦ 199

Solanum hieronymi Kuntze
Solanaceae
Solanum pocote Hieron. ex Millán
♦ pocote
♦ Weed
♦ 237, 270, 295

Solanum hirtum Vahl
Solanaceae
♦ Weed
♦ 87, 88
♦ herbal.

Solanum hispidum Pers.
Solanaceae
Solanum chrysocarpum Pers. (see)
♦ giant devil's fig
♦ Weed, Naturalised, Environmental Weed
♦ 39, 86, 87, 88, 98, 121, 203
♦ pS, toxic. Origin: Central America.

Solanum hoplopetalum Bitter & Summerh.
Solanaceae
♦ Afghan thistle, porcupine solanum, prickly potato weed
♦ Weed, Noxious Weed, Naturalised
♦ 7, 86, 88, 147, 203, 243
♦ Origin: Australia.

Solanum hystrix R.Br.
Solanaceae
♦ Afghan thistle, porcupine solanum, prickly potato weed
♦ Weed, Quarantine Weed, Noxious Weed, Naturalised
♦ 76, 86, 87, 88, 147, 203, 243
♦ arid. Origin: Australia.

Solanum incanum L.
Solanaceae
♦ bitterapple, um dulukwa, mu nhundurwa mukuru, grey bitterapple, thornapple, nightshade, kasongo, Sodom apple, bwanhula, idigaga, jemokimnerkeny, ochok, umucucu, yohola
♦ Weed, Native Weed
♦ 39, 50, 53, 87, 88, 121, 158, 221, 240

♦ pS, arid, herbal, toxic. Origin: southern Africa.

Solanum indicum L.
Solanaceae
= *Solanum lasiocarpum* Dunal (NoR)
♦ ntunfululu
♦ Weed
♦ 87, 88, 209, 297
♦ cultivated, herbal.

Solanum integrifolium Poir.
Solanaceae
♦ Chinese scarlet eggplant, ruffled tomato, ruffled red eggplant
♦ Naturalised
♦ 287
♦ aH, cultivated, herbal.

Solanum jamaicense Mill.
Solanaceae
♦ Jamaica nightshade
♦ Weed, Environmental Weed
♦ 13, 80, 87, 88, 112, 179
♦ cultivated, herbal.

Solanum japonense Nakai
Solanaceae
♦ Japonense nightshade, yamahoroshi
♦ Weed
♦ 297
♦ cultivated.

Solanum jasminoides Paxton
Solanaceae
♦ jasmine nightshade, potato vine
♦ Weed, Naturalised, Garden Escape, Environmental Weed
♦ 15, 39, 86, 98, 161, 198, 203, 225, 246, 280
♦ pC, cultivated, herbal, toxic. Origin: South America.

Solanum kwebense N.E.Br.
Solanaceae
♦ Weed, Native Weed
♦ 39, 121
♦ pS, toxic. Origin: southern Africa.

Solanum laciniatum Aiton
Solanaceae
♦ poroporo, kangaroo apple, koonyang, meakitch, Tasmanian kangaroo apple
♦ Weed, Naturalised, Introduced, Garden Escape, Environmental Weed, Casual Alien
♦ 7, 39, 40, 86, 87, 88, 131, 228
♦ S, arid, cultivated, herbal, toxic. Origin: Australia, New Zealand.

Solanum lanceolatum Cav.
Solanaceae
= *Solanum ruizii* S.Knapp (NoR)
♦ lanceleaf nightshade, orangeberry nightshade
♦ Weed, Noxious Weed, Naturalised
♦ 35, 88, 101, 199, 229
♦ S.

Solanum lasiostylum (Y.C.Liu & C.H.Ou) Tawada
Solanaceae
♦ Naturalised
♦ 287

Solanum lepidotum Humb. & Bonpl. ex Dun.
Solanaceae
♦ Weed
♦ 87, 88

Solanum linnaeanum Hepper & Jaeger
Solanaceae
Solanum hermannii Dunal *auct.*, *Solanum sodomeum auct.*
♦ apple of Sodom, bitterapple, poison apple, Afghan thistle
♦ Weed, Noxious Weed, Naturalised, Environmental Weed
♦ 7, 62, 72, 86, 88, 134, 147, 165, 198, 203, 225, 246, 280, 287
♦ S, herbal, toxic. Origin: Africa.

Solanum luteum Mill.
Solanaceae
= *Solanum villosum* (L.) Mill.
♦ keltakoiso
♦ Weed, Casual Alien
♦ 42, 70, 88, 94, 272
♦ aH, cultivated, herbal.

Solanum lycocarpum St.-Hil
Solanaceae
♦ fruto de lobo
♦ Weed
♦ 255
♦ S. Origin: Brazil.

Solanum lycopersicum L.
Solanaceae
= *Lycopersicon esculentum* Mill. var. *esculentum*
♦ garden tomato, pomidor
♦ Naturalised
♦ 39, 101, 280
♦ herbal, toxic.

Solanum lycopersicum L. var. cerasiforme (Dunal) Spoon. J.Anders. & R.K.Jansen
Solanaceae
= *Lycopersicon esculentum* Mill. var. *cerasiforme* (Dunal) Gray
♦ garden tomato
♦ Naturalised
♦ 101

Solanum lycopersicum L. var. lycopersicum
Solanaceae
= *Lycopersicon esculentum* Mill.
♦ garden tomato
♦ Naturalised, Introduced
♦ 101, 230
♦ H.

Solanum lyratum Thunb.
Solanaceae
♦ hiyodorijougo
♦ Weed
♦ 286, 297
♦ pC, promoted, herbal. Origin: China, Japan, Korea.

Solanum mammosum L.
Solanaceae
♦ nipplefruit, apple of Sodom
♦ Weed
♦ 13, 39, 87, 88
♦ a/pH, cultivated, herbal, toxic. Origin: South America.

Solanum marginatum L.
Solanaceae
♦ white edged nightshade, purple African nightshade, white margined nightshade
♦ Weed, Quarantine Weed, Noxious Weed, Naturalised, Introduced, Environmental Weed
♦ 34, 35, 76, 80, 86, 87, 88, 98, 101, 147, 165, 176, 198, 199, 203, 229, 241, 246, 280, 300
♦ S, cultivated, herbal. Origin: north-east Africa.

Solanum mauritianum Scop.
Solanaceae
Solanum auriculatum Ait. (see)
♦ bugweed, earleaf nightshade, wild tobacco tree, woolly nightshade, bug berry, bugtree, tobacco bush, luisboom
♦ Weed, Noxious Weed, Naturalised, Garden Escape, Environmental Weed
♦ 3, 6, 10, 15, 19, 22, 39, 51, 63, 72, 73, 86, 87, 88, 95, 98, 101, 121, 134, 152, 158, 165, 191, 198, 203, 225, 243, 246, 251, 269, 277, 278, 279, 280, 283, 289, 295, 296
♦ S, cultivated, toxic. Origin: South America.

Solanum maximowiczii Koidz.
Solanaceae
♦ Weed
♦ 286

Solanum melongena L.
Solanaceae
♦ eggplant, aubergine
♦ Weed, Naturalised, Cultivation Escape
♦ 39, 80, 87, 88, 101, 161, 179, 261
♦ pH, cultivated, herbal, toxic. Origin: south-east Asia.

Solanum melongena L. fo. spontanea L.
Solanaceae
♦ Weed
♦ 13

Solanum memphiticum Mart.
Solanaceae
♦ Naturalised
♦ 287

Solanum miniatum Bernh. ex Willd.
Solanaceae
= *Solanum villosum* (L.) Mill.
♦ Weed
♦ 87, 88

Solanum muricatum W.T.Aiton
Solanaceae
♦ melon pear, pepino
♦ Naturalised
♦ 280
♦ S, cultivated, herbal.

Solanum nigrum L.
Solanaceae
♦ black fruited nightshade, black nightshade, black berry, chiSungubvana, common nightshade, deadly nightshade, duscle, enab el dib, erba morella, garden huckleberry, garden nightshade, harsh, hierba mora, hound's berry, i Xabaxaba,

inkberry, makoy, maniloche, maria preta, ma waeng nok, moralle, muSaka, muSungusungu, muTsungutsungu, native currant, nightshade, petty morel, poison berry, potatobush, stubbleberry, tomato bush, wild currant, wonder berry, woody nightshade
♦ Weed, Noxious Weed, Naturalised, Introduced, Garden Escape, Environmental Weed
♦ 7, 8, 9, 13, 23, 34, 39, 44, 45, 50, 51, 53, 55, 68, 70, 72, 86, 87, 88, 93, 94, 98, 101, 115, 118, 121, 158, 161, 165, 176, 180, 185, 186, 189, 195, 198, 203, 205, 209, 212, 218, 221, 222, 228, 229, 235, 238, 240, 242, 243, 245, 247, 250, 253, 263, 269, 272, 273, 274, 275, 276, 280, 286, 292, 297, 299
♦ a/bH, arid, cultivated, herbal, toxic. Origin: Eurasia.

Solanum nigrum L. var. americanum (Mill.) Schulz
Solanaceae
= *Solanum americanum* Mill.
♦ nahsupi
♦ Introduced
♦ 230
♦ S.

Solanum nigrum L. var. miniatum Hook.f.
Solanaceae
♦ Naturalised
♦ 287

Solanum nigrum L. var. nigrum
Solanaceae
♦ Naturalised
♦ 241

Solanum nigrum L. var. villosum L.
Solanaceae
= *Solanum villosum* Mill.
♦ Naturalised
♦ 287
♦ herbal.

Solanum nodiflorum Jacq.
Solanaceae
= *Solanum americanum* Mill.
♦ black berry, black nightshade
♦ Weed, Native Weed
♦ 14, 87, 88, 121, 157
♦ a/bH, herbal.

Solanum orthocarpum Pic.Serm.
Solanaceae
♦ Weed
♦ 87, 88

Solanum palinacanthum Dunal
Solanaceae
♦ arrabenta cavalo
♦ Weed
♦ 255
♦ aH. Origin: Brazil.

Solanum palitans Morton
Solanaceae
♦ Weed, Naturalised
♦ 98, 203

Solanum panduriforme Drège ex Dun.
Solanaceae
♦ apple of Sodom, bitterapple, poison apple, yellow bitterapple

♦ Weed, Native Weed
♦ 39, 87, 88, 121, 158
♦ pS, toxic. Origin: southern Africa.

Solanum paniculatum L.
Solanaceae
♦ jurubeba
♦ Weed
♦ 87, 88, 255
♦ S, herbal. Origin: Brazil.

Solanum peoppigianum Sendt.
Solanaceae
♦ Weed
♦ 87, 88

Solanum peruvianum L.
Solanaceae
= *Lycopersicon peruvianum* (L.) Mill. var. *peruvianum* (NoR)
♦ Peruvian nightshade
♦ Naturalised
♦ 101

Solanum photeinocarpum Nakam. & Odash.
Solanaceae
♦ terimini inuhoozuki
♦ Weed
♦ 87, 88, 204, 275, 286, 297
♦ aH.

Solanum physalifolium Rusby
Solanaceae
Solanum sarrachoides auct. non Sendt.
♦ hairy nightshade, hoe nightshade
♦ Noxious Weed, Naturalised
♦ 176, 229, 280
♦ herbal.

Solanum physalifolium Rusby var. nitidibaccatum (Bitter) Edmonds
Solanaceae
Solanum nitidibaccatum Bitter
♦ cherry nightshade
♦ Naturalised, Environmental Weed
♦ 86, 176, 198, 253

Solanum pilcomayense Morong
Solanaceae
♦ Weed
♦ 270

Solanum pimpinellifolium Jusl.
Solanaceae
= *Lycopersicon pimpinellifolium* (Jusl.) Mill.
♦ currant tomato
♦ Naturalised
♦ 101

Solanum pseudocapsicum L.
Solanaceae
Solanum capsicastrum Link ex Schauer (see)
♦ false capsicum, Jerusalem cherry, Natal cherry, wintercherry, Madeira wintercherry
♦ Weed, Naturalised, Garden Escape, Environmental Weed
♦ 3, 7, 15, 39, 72, 80, 85, 86, 87, 88, 98, 101, 121, 154, 161, 165, 176, 189, 194, 198, 203, 225, 237, 241, 246, 247, 269, 280, 286, 287, 289, 295, 296, 300
♦ S, cultivated, herbal, toxic. Origin: South America.

Solanum pseudogracile Heiser
Solanaceae
♦ glowing nightshade, graceful nightshade
♦ Weed
♦ 161
♦ herbal, toxic.

Solanum ptycanthum Dun.
Solanaceae
♦ eastern black nightshade, deadly nightshade, poison berry, garden nightshade
♦ Weed
♦ 68, 161, 207, 210, 211, 243, 292
♦ aH, cultivated, herbal, toxic. Origin: eastern North America.

Solanum pygmaeum Cav.
Solanaceae
♦ Weed
♦ 87, 88, 237, 295

Solanum quitoense Lam.
Solanaceae
♦ naranjilla
♦ Naturalised, Cultivation Escape
♦ 101, 257, 261
♦ S, cultivated, herbal. Origin: South America.

Solanum radicans L.f.
Solanaceae
♦ Naturalised
♦ 86, 241, 300
♦ arid. Origin: South America.

Solanum rantonnei Carrière
Solanaceae
= *Lycianthes rantonnei* (Carrière) Bitter (NoR)
♦ Naturalised
♦ 280
♦ cultivated.

Solanum reflexum Schrank.
Solanaceae
♦ Weed
♦ 295

Solanum retroflexum Dun.
Solanaceae
♦ momodi, nasgal, nastergal, nightshade, seshoa bohloko
♦ Weed, Naturalised, Native Weed
♦ 39, 86, 87, 88, 98, 121, 158, 203
♦ aH, arid, toxic. Origin: South Africa.

Solanum rigescens Jacq.
Solanaceae
♦ Native Weed
♦ 121
♦ S, cultivated, toxic. Origin: South Africa.

Solanum robustum Wendl.
Solanaceae
♦ shrubby nightshade
♦ Noxious Weed, Naturalised, Casual Alien
♦ 101, 229, 280
♦ cultivated.

Solanum rostratum Dunal
Solanaceae
Androsera rostrata (Dunal) Rydb., *Nycterium rostratum* (Dunal) Link,

Solanum bejarense Dunal, *Solanum cornutum* Lam. *auct.* (see), *Solanum heterandrum* (Pursh), *Solanum propinquum* M.Martens & Galeotti
♦ buffalo burr, beaked nightshade, Colorado burr, Kansas thistle, Mexican thistle, prickly nightshade, sandbur, Texas thistle
♦ Weed, Noxious Weed, Naturalised, Native Weed, Casual Alien
♦ 1, 7, 23, 24, 34, 39, 40, 49, 51, 80, 86, 87, 88, 98, 121, 147, 161, 174, 180, 198, 203, 210, 212, 218, 219, 229, 243, 264, 269, 280, 287
♦ pH, cultivated, herbal, toxic. Origin: North America.

Solanum sanitwongsei Craib
Solanaceae
♦ Weed
♦ 209
♦ herbal.

Solanum sarrachoides Sendt.
Solanaceae
♦ hairy nightshade, green nightshade, Brasilian koiso, leafy fruited nightshade
♦ Weed, Quarantine Weed, Naturalised, Casual Alien
♦ 34, 40, 42, 45, 76, 80, 87, 88, 94, 98, 136, 161, 180, 203, 212, 237, 243, 270, 286, 287, 292, 295
♦ aH, herbal, toxic. Origin: South America.

Solanum seaforthianum Andrews
Solanaceae
♦ potato creeper, Brazilian nightshade, Italian jasmine, climbing nightshade
♦ Weed, Noxious Weed, Naturalised, Garden Escape, Environmental Weed
♦ 3, 11, 39, 63, 73, 80, 86, 87, 88, 95, 98, 121, 161, 179, 203, 279, 283
♦ pC, cultivated, herbal, toxic. Origin: South America.

Solanum semiarmatum F.Muell.
Solanaceae
♦ Weed
♦ 87, 88
♦ cultivated. Origin: Australia.

Solanum sepicula Dunal
Solanaceae
♦ European goldenrod
♦ Weed
♦ 88

Solanum septemlobum Bunge
Solanaceae
♦ Weed
♦ 275
♦ pH, promoted.

Solanum sessiliflorum Dunal
Solanaceae
♦ cocona
♦ Cultivation Escape
♦ 32
♦ cultivated.

Solanum sisymbriifolium Lam.
Solanaceae
♦ sticky nightshade, dense thorned bitterapple, wild tomato, liuskakoiso,

espina colorada, wildetamatie, doringtamatie, bitterapple, arrabenta cavalo
♦ Weed, Noxious Weed, Naturalised, Casual Alien
♦ 3, 7, 10, 34, 42, 51, 63, 86, 87, 88, 95, 98, 101, 121, 158, 179, 203, 236, 237, 245, 255, 270, 278, 280, 283, 287, 295, 300
♦ pS, cultivated, herbal, toxic. Origin: South America.

Solanum sodomeum L. nom. rej.
Solanaceae
= *Solanum anguivi* Lam.
♦ apple of Sodom nightshade, apple of Sodom
♦ Weed, Naturalised, Environmental Weed
♦ 39, 86, 87, 88, 161, 218
♦ cultivated, herbal, toxic.

Solanum stipulaceum Roem. & Schult.
Solanaceae
♦ fumo bravo
♦ Weed
♦ 255
♦ S. Origin: Brazil.

Solanum stramoniifolium Jacq.
Solanaceae
♦ Weed
♦ 87, 88

Solanum subinerme Jacq.
Solanaceae
♦ Weed
♦ 87, 88

Solanum sublobatum Willd. ex Roem. & Schult.
Solanaceae
= *Solanum chenopodifolium* Lam.
♦ nightshade, hierba mora
♦ Weed
♦ 236, 237, 270, 295

Solanum superficiens Adelb.
Solanaceae
♦ Weed, Naturalised
♦ 98, 203
♦ cultivated.

Solanum surattense Burm.
Solanaceae
♦ nightshade
♦ Weed, Naturalised
♦ 66, 87, 88, 101
♦ cultivated, herbal.

Solanum tampicense Dunal
Solanaceae
♦ aquatic soda apple, nightshade, wetland nightshade, scrambling nightshade
♦ Weed, Quarantine Weed, Noxious Weed, Naturalised, Environmental Weed
♦ 76, 80, 88, 101, 112, 140, 151, 179, 197, 220
♦ aqua.

Solanum torvum Sw.
Solanaceae
Solanum ficifolium Ortega (see)
♦ turkey berry, prickly solanum, devil's fig, terongan, fausse aubergine,

piko, tisaipale, kausoni, soni, kauvoto votua, kaisurisuri, katai, bhankatiya, prickly solanum, berenjena cimarrona
♦ Weed, Quarantine Weed, Noxious Weed, Naturalised, Environmental Weed, Casual Alien
♦ 3, 6, 13, 22, 39, 67, 80, 86, 87, 88, 93, 98, 112, 140, 151, 161, 179, 191, 203, 205, 212, 218, 229, 258, 261, 269, 275, 276, 280, 287, 297
♦ S, cultivated, herbal, toxic.

Solanum triflorum Nutt.
Solanaceae
♦ cutleaf nightshade, three flowered nightshade, small nightshade, kolmikukkakoiso, wild tomato
♦ Weed, Noxious Weed, Naturalised, Casual Alien
♦ 7, 21, 23, 34, 39, 40, 42, 49, 86, 87, 88, 98, 161, 180, 198, 203, 212, 218, 269, 287, 299
♦ aH, promoted, herbal, toxic. Origin: North America.

Solanum tuberosum L.
Solanaceae
Lycopersicon tuberosum (L.) Mill., *Solanum aracatscha* Besser, *Solanum sinense* Blanco
♦ potato, Irish potato, white potato, Zulu potato, cetewayo
♦ Weed, Naturalised, Cultivation Escape, Casual Alien
♦ 15, 39, 40, 42, 80, 87, 88, 98, 101, 121, 154, 161, 203, 247, 261, 280
♦ pH, cultivated, herbal, toxic. Origin: South America.

Solanum unguiculatum A.Rich.
Solanaceae
♦ Weed
♦ 88, 221

Solanum verbascifolium L.
Solanaceae
= *Solanum donianum* Walp.
♦ potato tree, tobacco tree
♦ Weed
♦ 13, 39, 87, 88, 276
♦ herbal, toxic.

Solanum viarum Dunal
Solanaceae
Solanum khasianum C.B.Clarke var. *chatterjeeanum* Sengupta
♦ tropical soda apple, tropical nightshade
♦ Weed, Quarantine Weed, Noxious Weed, Naturalised, Environmental Weed
♦ 3, 26, 67, 76, 80, 88, 101, 102, 105, 112, 122, 140, 151, 161, 172, 179, 191, 220, 229, 243, 255, 258
♦ pH, cultivated. Origin: Brazil.

Solanum villosum (L.) Mill.
Solanaceae
Solanum alatum Moench (see), *Solanum luteum* Mill. (see), *Solanum luteum* ssp. *alatum* (Moench) Dostal, *Solanum miniatum* Bernh. ex Willd. (see), *Solanum nigrum* L. var. *villosum* L. (see)
♦ hairy nightshade, red fruited

nightshade
♦ Weed, Naturalised
♦ 39, 86, 87, 88, 98, 121, 203, 218, 253, 280
♦ aH, herbal, toxic. Origin: southern Europe.

Solanum welwitschii C.H.Wright
Solanaceae
♦ Weed
♦ 87, 88

Solanum wendlandii Hook.f.
Solanaceae
♦ giant potato creeper, Costa Rican nightshade, potato vine
♦ Naturalised, Garden Escape
♦ 39, 101, 261
♦ cultivated, herbal, toxic. Origin: Costa Rica.

Solanum wrightii Benth.
Solanaceae
Solanum macranthum hort. ex Carr.
♦ Brazilian potato tree, potato tree
♦ Weed
♦ 121
♦ T, cultivated, toxic. Origin: South America.

Solanum xanthocarpum Schrad. & Wendl.
Solanaceae
♦ yellow berried nightshade, kantikari
♦ Weed, Quarantine Weed, Naturalised
♦ 39, 76, 86, 87, 88, 203, 220, 275, 297
♦ aH, cultivated, herbal, toxic.

Soldanella montana Willd.
Primulaceae
♦ alppikello
♦ Cultivation Escape
♦ 42
♦ cultivated.

Soleirolia soleirolii (Req.) Dandy
Urticaceae
Helxine soleirolii Req.
♦ baby's tears, mother of thousands, helxine, angel's tears, Corscian carpet plant, Corsican curse, Irish moss, Japanese moss, mind your own business, peace in the home, Pollyanna vine, touch me not
♦ Weed, Naturalised, Introduced, Garden Escape, Environmental Weed
♦ 7, 15, 34, 40, 86, 98, 101, 121, 176, 198, 203, 241, 280, 300
♦ pH, cultivated, herbal. Origin: Eurasia.

Solenogyne dominii L.G.Adams
Asteraceae
♦ Naturalised
♦ 280

Solenogyne gunnii (Hook.f.) Cabrera
Asteraceae
♦ Naturalised
♦ 280

Solenogyne mikadoi (Koidz.) Koidz.
Asteraceae
♦ Casual Alien
♦ 280

Solenopsis laurentia (L.) C.Presl
Campanulaceae/Lobeliaceae
♦ Naturalised
♦ 280

Solenostemma argel Del.
Asclepiadaceae/Apocynaceae
♦ Weed
♦ 39, 221
♦ herbal, toxic.

Solenostemon monostachyus (Beauv.) Briq.
Lamiaceae
♦ Weed
♦ 87, 88

Solidago L. spp.
Asteraceae
♦ goldenrod
♦ Weed, Naturalised
♦ 18, 39, 88, 198, 243
♦ H, herbal, toxic.

Solidago altissima L.
Asteraceae
= *Solidago canadensis* L. var. *scabra* Torr. & Gray
♦ tall goldenrod, korkeapiisku
♦ Weed, Naturalised, Cultivation Escape
♦ 42, 98, 106, 161, 203, 263, 286, 287, 297
♦ pH, arid, cultivated, herbal.

Solidago bicolor L.
Asteraceae
♦ white goldenrod
♦ Weed
♦ 23, 88
♦ cultivated, herbal.

Solidago californica Nutt.
Asteraceae
Aster californicus (Nutt.) Kuntze
♦ California goldenrod, goldenrod
♦ Weed
♦ 161, 180
♦ pH, arid, herbal.

Solidago canadensis L.
Asteraceae
Solidago lepida DC.
♦ Canada goldenrod, kanadanpiisku, goldenrod, dwarf goldenrod
♦ Weed, Naturalised, Introduced, Environmental Weed, Cultivation Escape
♦ 7, 40, 42, 52, 87, 88, 94, 98, 132, 136, 152, 161, 203, 211, 212, 218, 228, 243, 246, 256, 269, 272, 280, 287, 293
♦ aH, arid, cultivated, herbal. Origin: North America.

Solidago canadensis L. var. scabra Torr. & Gray
Asteraceae
Solidago altissima L. (see), *Solidago hirsutissima* Mill., *Solidago lunellii* Rydb.
♦ Canada goldenrod
♦ Naturalised
♦ 86, 98, 198
♦ cultivated. Origin: North America.

Solidago chilensis Meyen
Asteraceae
Solidago microglosa DC. var. *linearifolia* (DC.) Bak.
- rnica, sapé macho
- Weed, Quarantine Weed
- 76, 87, 88, 203, 220, 237, 255, 270, 295
- Origin: South America.

Solidago cutleri Fern.
Asteraceae
- Cutler's alpine goldenrod
- Weed
- 23, 88
- cultivated.

Solidago gigantea Aiton
Asteraceae
Solidago glabra Desf., *Solidago gigantea* Ait. var. *leiophylla* (Fern.) (see), *Solidago gigantea* Ait. ssp. *serotina* (Kuntze) McNeill (see), *Solidago serotina* Ait.
- giant goldenrod, smooth goldenrod
- Weed, Quarantine Weed, Casual Alien
- 76, 87, 88, 94, 132, 161, 218, 272, 280
- pH, aqua, cultivated, herbal.

Solidago gigantea Aiton ssp. serotina (Kuntze) McNeill
Asteraceae
= *Solidago gigantea* Ait.
- early goldenrod
- Naturalised
- 40

Solidago gigantea Aiton var. leiophylla Fern.
Asteraceae
= *Solidago gigantea* Ait.
- late goldenrod
- Weed, Naturalised
- 263, 286, 287
- pH.

Solidago graminifolia (L.) Salisb.
Asteraceae
= *Euthamia graminifolia* (L.) Nutt. var. *graminifolia* (NoR)
- narrowleaf goldenrod, grassleaf goldenrod
- Weed, Naturalised
- 87, 88, 161, 287
- pH, cultivated, herbal.

Solidago graminifolia (L.) Salisb. var. nuttallii (Greene) Fernald
Asteraceae
= *Euthamia graminifolia* (L.) Nutt. var. *nuttallii* (Greene) W.Stone (NoR)
- narrowleaf goldenrod
- Weed
- 218

Solidago juncea Ait.
Asteraceae
- early goldenrod
- Naturalised
- 287
- cultivated, herbal.

Solidago marcrophylla Pursh
Asteraceae
- Weed
- 23, 88

Solidago microglossa DC.
Asteraceae
- Weed, Quarantine Weed
- 76, 87, 88, 203, 220

Solidago missouriensis Nutt.
Asteraceae
Solidago marshallii Rothr.
- Missouri goldenrod, prairie goldenrod
- Weed, Native Weed
- 87, 88, 161, 174, 218
- pH, arid, cultivated, herbal. Origin: North America.

Solidago nemoralis Aiton
Asteraceae
- gray goldenrod, dyer's weed goldenrod
- Weed
- 23, 87, 88, 161, 210, 218
- cultivated, herbal, toxic.

Solidago occidentalis (Nutt.) Torr. & Gray
Asteraceae
= *Euthamia occidentalis* Nutt. (NoR)
- western goldenrod, goldenrod, narrow leaved goldenrod
- Weed
- 87, 88, 161, 180, 218, 243
- arid, herbal.

Solidago odora Aiton
Asteraceae
- anise scented goldenrod, sweet goldenrod, sweet scented goldenrod
- Weed
- 23, 39, 88
- pH, cultivated, herbal, toxic.

Solidago rigida L.
Asteraceae
= *Oligoneuron rigidum* (L.) Small var. *rigidum* (NoR)
- rigid goldenrod, stiff goldenrod, hardleaf goldenrod
- Weed
- 23, 87, 88, 161, 210, 218
- pH, arid, cultivated, herbal.

Solidago rugosa Mill.
Asteraceae
- rough goldenrod, wrinkleleaf goldenrod
- Weed
- 23, 87, 88, 161, 218
- cultivated, herbal.

Solidago sempervirens L.
Asteraceae
- seaside goldenrod, Amerikanische goldrute
- Naturalised
- 287
- cultivated, herbal.

Solidago sempervirens L. var. mexicana (L.) Fern.
Asteraceae
- seaside goldenrod, solidago
- Naturalised, Cultivation Escape
- 261, 287
- cultivated. Origin: North America.

Solidago sessilis Ruiz & Pav.
Asteraceae
- Weed
- 87, 88

Solidago spectabilis (Eaton.) A.Gray
Asteraceae
- basin goldenrod, Nevada goldenrod, showy goldenrod
- Weed
- 39, 161
- pH, promoted, herbal, toxic.

Solidago virgaurea L.
Asteraceae
Solidago virga-aurea L., *Solidago virga-aurea* L. ssp. *asiatica* Kitam. (see), *Solidago taurica* Juz., *Solidago lapponica* With., *Solidago lapponica* ssp. *stenophylla* G.E.Schultz
- goldenrod, tall goldenrod, European goldenrod, kultapiisku
- Weed, Casual Alien
- 23, 39, 87, 88, 272, 280
- pH, cultivated, herbal, toxic. Origin: Africa, Asia, Europe.

Solidago virgaurea L. ssp. asiatica Kitam.
Asteraceae
= *Solidago virgaurea* L.
- akinokirinsou
- Weed
- 263

Solieria chordalis (C.Agardh) J.Agardh
Solieriaceae
Delesseria chordalis C.Agardh
- Introduced
- 288
- algae.

Soliva anthemifolia (Juss.) R.Br. ex Less.
Asteraceae
= *Gymnostyles anthemifolia* Juss.
- dwarf jo jo weed, dwarf jo jo, cuspe de caipira, false corianda
- Weed, Naturalised
- 55, 86, 87, 88, 98, 198, 203, 255, 256, 273, 274, 280, 286, 287, 295
- Origin: South America.

Soliva mutisii Kunth
Asteraceae
- Mutis' burrweed
- Naturalised
- 101

Soliva nasturtiifolia (Juss.) DC.
Asteraceae
= *Gymnostyles stolonifera* (Brot.) Tutin
- cressleaf burrweed
- Weed
- 161

Soliva pterosperma (Juss.) Less.
Asteraceae
= *Soliva sessilis* Ruiz & Pav.
- onehunga, jo jo, lawn burrweed, bindy eye, roseta
- Weed, Quarantine Weed, Noxious Weed, Naturalised
- 39, 55, 76, 80, 87, 88, 121, 134, 147, 161, 203, 249, 255, 269, 273, 295
- pH, herbal, toxic. Origin: South America.

Soliva sessilis **Ruiz & Pav.**
Asteraceae
Gymnostyles pterosperma Juss., *Soliva pterosperma* (Juss.) Less. (see)
♦ jo jo, bindii, field burrweed, lawn burrweed, field soliva, onehunga weed
♦ Weed, Naturalised, Environmental Weed
♦ 1, 7, 15, 80, 86, 98, 101, 165, 176, 179, 198, 203, 237, 280, 287, 295
♦ aH, herbal. Origin: South America.

Soliva stolonifera **(Brot.) Loudon**
Asteraceae
= *Gymnostyles stolonifera* (Brot.) Tutin
♦ carpet burrweed
♦ Weed, Naturalised, Introduced
♦ 38, 86, 98, 198, 203, 237, 295
♦ Origin: South America.

Sollya heterophylla **Lindl.**
Pittosporaceae
♦ bluebell creeper, Australian climbing bluebell
♦ Weed, Naturalised, Native Weed, Garden Escape, Environmental Weed
♦ 72, 86, 88, 101, 176, 198, 289, 296
♦ pC, cultivated, herbal. Origin: Australia.

Sonchus **L. spp.**
Asteraceae
♦ sow thistle
♦ Weed
♦ 88, 243, 272
♦ herbal, toxic.

Sonchus arvensis **L.**
Asteraceae
Hieracium arvense Scop., *Sonchus hispidus* Gilib.
♦ perennial sow thistle, field sow thistle, creeping sow thistle, gutweed, milk thistle, field milk thistle, corn sow thistle, swine thistle, tree sow thistle, dindle, grespino dei campi
♦ Weed, Quarantine Weed, Noxious Weed, Naturalised, Introduced
♦ 1, 13, 16, 23, 26, 34, 35, 36, 44, 49, 52, 70, 76, 80, 86, 87, 88, 94, 98, 101, 114, 118, 136, 138, 146, 161, 162, 170, 195, 203, 204, 211, 212, 217, 218, 220, 229, 238, 243, 253, 267, 272, 280, 286, 287, 291, 295, 299, 300
♦ aH, arid, cultivated, herbal, toxic. Origin: Eurasia.

Sonchus arvensis **L. ssp.** *arvensis*
Asteraceae
♦ perennial sow thistle, field sow thistle
♦ Noxious Weed, Naturalised
♦ 101, 229

Sonchus arvensis **L. ssp.** *uliginosus*
(Bieb.) Nyman
Asteraceae
Sonchus uliginosus Bieb. (see)
♦ marsh sow thistle, perennial sow thistle, sow thistle, moist sow thistle, creeping sow thistle
♦ Weed, Noxious Weed, Naturalised, Introduced, Environmental Weed
♦ 21, 101, 212, 219, 229

Sonchus asper **(L.) Hill**
Asteraceae
Sonchus fallax Wall., *Sonchus spinosus* Lam.
♦ spiny sow thistle, prickly sow thistle, sow thistle, spinyleaf sow thistle, rough sow thistle
♦ Weed, Noxious Weed, Naturalised, Introduced, Environmental Weed
♦ 7, 9, 15, 23, 44, 51, 52, 68, 70, 72, 80, 87, 88, 93, 94, 98, 101, 116, 121, 136, 158, 161, 174, 176, 179, 180, 185, 195, 198, 199, 203, 204, 207, 210, 212, 217, 218, 237, 240, 241, 243, 249, 253, 255, 261, 263, 267, 271, 272, 275, 276, 280, 286, 287, 292, 295, 297, 299, 300
♦ aH, cultivated, herbal. Origin: Eurasia.

Sonchus asper **(L.) Hill ssp.** *asper*
Asteraceae
♦ rough sow thistle, prickly sow thistle, spiny sow thistle
♦ Weed, Naturalised, Environmental Weed
♦ 34, 86, 167, 198
♦ aH.

Sonchus asper **(L.) Hill ssp.** *glaucescens*
(Jord.) Ball
Asteraceae
Sonchus glaucescens Jordan (see), *Sonchus nymanii* Tineo & Guss.
♦ rough sow thistle, prickly sow thistle
♦ Weed, Naturalised, Introduced, Environmental Weed
♦ 86, 198, 228, 269
♦ aH. Origin: Europe.

Sonchus brachyotus **DC.**
Asteraceae
♦ hachijouna
♦ Weed
♦ 87, 88, 275, 286, 297
♦ pH, promoted, herbal. Origin: east Asia.

Sonchus cornutus **Hochst. ex Steud.**
Asteraceae
♦ Weed, Quarantine Weed
♦ 76, 87, 88, 203, 220, 242
♦ arid.

Sonchus dregeanus **DC.**
Asteraceae
Sonchus ecklonianus DC.
♦ sow thistle
♦ Native Weed
♦ 121
♦ pH, arid. Origin: southern Africa.

Sonchus exauriculatus **(Oliv. & Hiern)**
O.Hoffm.
Asteraceae
= *Launaea cornuta* (Oliv. & Hiern) C.Jeffrey
♦ Weed, Quarantine Weed
♦ 76, 87, 88, 203, 220

Sonchus glaucescens **Jord.**
Asteraceae
= *Sonchus asper* (L.) Hill ssp. *glaucescens* (Jord.) Ball
♦ Weed
♦ 87, 88

Sonchus macrocarpus **Boulos & C.Jeffrey**
Asteraceae
♦ Weed
♦ 88, 185

Sonchus maritimus **L. ssp.** *aquatilis*
(Pour.) Nyman
Asteraceae
♦ Weed
♦ 70

Sonchus obtusilobus **R.E.Fr.**
Asteraceae
♦ Weed
♦ 88
♦ Origin: Africa.

Sonchus oleraceus **L.**
Asteraceae
Endiuia agrestis Hard., *Sonchus ciliatus* Lam., *Sonchus laevis* Vill., *Sonchus levis* Gars., *Sonchus roseus* Bess.
♦ annual sow thistle, common sow thistle, hare's lettuce, colewort, milk thistle, kaalivalvatti, grespino comune, cerraja, sow thistle
♦ Weed, Noxious Weed, Naturalised, Introduced, Garden Escape, Environmental Weed
♦ 6, 7, 9, 14, 15, 23, 34, 44, 49, 50, 51, 52, 53, 55, 66, 68, 70, 72, 80, 86, 87, 88, 93, 94, 98, 101, 115, 116, 118, 121, 134, 136, 157, 158, 161, 162, 165, 176, 179, 180, 185, 186, 195, 198, 203, 204, 205, 210, 211, 212, 218, 221, 230, 235, 236, 237, 241, 242, 243, 245, 250, 253, 255, 257, 261, 263, 267, 269, 271, 272, 274, 275, 276, 280, 286, 290, 292, 295, 297, 300
♦ aH, arid, cultivated, herbal, toxic. Origin: Eurasia.

Sonchus oleraceus **L. fo.** *integrifolius*
G.Beck
Asteraceae
♦ sow thistle
♦ Weed
♦ 286

Sonchus palustris **L.**
Asteraceae
Sonchus paluster Land.
♦ marsh sow thistle, fen sow thistle
♦ Weed
♦ 80, 272
♦ cultivated, herbal.

Sonchus tenerrimus **L.**
Asteraceae
♦ clammy sow thistle, sow thistle, slender sow thistle, hentovalvatti
♦ Weed, Naturalised, Environmental Weed, Casual Alien
♦ 7, 34, 42, 70, 86, 87, 88, 94, 98, 101, 198, 203, 241, 253, 300
♦ aH, arid, promoted, herbal. Origin: Mediterranean.

Sonchus uliginosus **Bieb.**
Asteraceae
= *Sonchus arvensis* L. ssp. *uliginosus* (Bieb.) Nyman
♦ marsh sow thistle
♦ Weed
♦ 23, 88, 267
♦ herbal. Origin: Europe.

Sonneratia caseolaris (L.) Engl.
Lythraceae/Sonneratiaceae
Sonneratia acida L.f.
♦ Weed
♦ 87, 88
♦ herbal.

Sophora alopecuroides L.
Fabaceae/Papilionaceae
Goebelia alopecuroides (L.) Bunge (see),
Vexibia alopecuroides (L.) Yakovlev
♦ Weed
♦ 39, 87, 88, 243, 275, 297
♦ arid, toxic. Origin: Middle East,
Pakistan, China, Russia.

Sophora flavescens Aiton
Fabaceae/Papilionaceae
Sophora angustifolia Sieb. & Zucc.
♦ Weed
♦ 88, 204, 286, 297
♦ S, promoted, herbal.

Sophora japonica L.
Fabaceae/Papilionaceae
Ormosia esquirolii Lév., *Sophora mairei*
Lév., *Styphnolobium japonicum* (L.)
Schott
♦ Japanese pagoda tree, Japanese
kowhai, Chinese scholar tree
♦ Weed, Naturalised
♦ 39, 101, 247, 279
♦ T, cultivated, herbal, toxic.

Sophora nuttalliana B.Turner
Fabaceae/Papilionaceae
Sophora sericea Nutt. (see)
♦ silky sophora
♦ Weed
♦ 161
♦ herbal, toxic.

Sophora pachycarpa Schrank ex C.A.Mey.
Fabaceae/Papilionaceae
Goebelia pachycarpa (Schrenk ex
C.A.Mey.) Bunge ex Boiss. (see), *Vexibia
pachycarpa* (Schrenk ex C.A.Mey.)
Yakovlev
♦ Weed, Quarantine Weed
♦ 76, 87, 88, 203, 220
♦ Origin: Afghanistan, Iran,
Kazakhstan, Kyrgyzstan, Tajikistan,
Turkmenistan, Uzbekistan.

Sophora secundiflora (Ortega) Lag.
Fabaceae/Papilionaceae
♦ mescalbean, frijolito
♦ Weed
♦ 39, 87, 88, 161, 218, 247
♦ T, cultivated, herbal, toxic.

Sophora sericea Nutt.
Fabaceae/Papilionaceae
= *Sophora nuttalliana* B.Turner
♦ silky sophora
♦ Weed
♦ 39, 87, 88, 218
♦ pH, promoted, toxic. Origin: south-
western North America.

Sophora tetraptera J.F.Mill.
Fabaceae/Papilionaceae
Edwardsia grandiflora Salisb.
♦ kowhai
♦ Naturalised
♦ 198

♦ T, cultivated, herbal. Origin:
Australia, New Zealand.

Sophora tomentosa L.
Fabaceae/Papilionaceae
♦ silver bush, yellow necklacepod,
necklacepod, yellow sophora
♦ Weed, Introduced
♦ 39, 161, 228
♦ arid, cultivated, herbal, toxic. Origin:
Australia.

Sopubia delphinifolia (L.) Don.
Scrophulariaceae
♦ Weed
♦ 66
♦ herbal.

Sorbaria arborea Schneid.
Rosaceae
♦ giant false spiraea, sorbaria
♦ Naturalised
♦ 101
♦ cultivated.

Sorbaria sorbifolia (L.) A.Braun
Rosaceae
♦ false spirea, Ural false spirea
♦ Weed, Naturalised, Garden Escape,
Cultivation Escape
♦ 42, 80, 101, 133, 195
♦ S, cultivated, herbal.

Sorbaria tomentosa (Lindl.) Rehder
Rosaceae
♦ Naturalised
♦ 280
♦ cultivated.

Sorbus americana Marsh.
Rosaceae
Pyrus americana (Marsh.) DC.
♦ American mountain ash
♦ Weed
♦ 87, 88, 218
♦ T, cultivated, herbal.

Sorbus aucuparia L.
Rosaceae
Aucuparia silvestris Medik., *Mespilus
aucuparia* All., *Pirus aucuparia* Gaertn.,
Pyrenia aucuparia Clairv., *Pyrus
aucuparia* Gaertn.
♦ European mountain ash, mountain
ash, rowan
♦ Weed, Naturalised, Garden Escape,
Environmental Weed, Cultivation
Escape
♦ 4, 15, 34, 70, 80, 86, 88, 101, 146, 151,
181, 195, 225, 246, 280, 290
♦ T, cultivated, herbal. Origin: Eurasia.

Sorbus domestica L.
Rosaceae
Cornus domestica Spach, *Malus sorbus*
Borkh., *Mespilus domestica* All., *Pirus
sorbus* Gaertn., *Pyrenia sorbus* Clairv.
♦ service tree, service tree arran
♦ Weed, Naturalised, Garden Escape,
Environmental Weed
♦ 86, 98, 203
♦ T, cultivated, herbal. Origin: Eurasia,
Mediterranean.

Sorbus hybrida L.
Rosaceae
Crataegus fennica Kalm, *Pyrus*

pinnatifida Ehrh., *Sorbus fennica* (Kalm)
Fr.
♦ oakleaf mountain ash,
suomenpihlaja, Swedish servicetree
♦ Naturalised
♦ 101
♦ T, cultivated, herbal. Origin:
northern Europe.

Sorbus × latifolia (Lam.) Pers.
Rosaceae
♦ Naturalised
♦ 280

Sorbus sambucifolia (Cham. & Schlecht.) M.Roem.
Rosaceae
Pyrus sambucifolia Cham. & Schlecht.
♦ Siberian mountain ash
♦ Naturalised
♦ 101
♦ S, cultivated.

Sorghastrum nutans (L.) Nash
Poaceae
Andropogon albescens E.Fourn.,
Andropogon arenaceus Raf., *Andropogon
avenaceum* Michx., *Andropogon
avenaceus* Michx., *Andropogon ciliatus*
Elliott, *Andropogon confertus* Trin.
ex E.Fourn., *Andropogon linnaeanus*
(Hack.) Scribn. & Kearney, *Andropogon
nutans* L., *Andropogon nutans* var.
avenaceus (Michx.) Hack. ex Stuck.,
Chalcoelytrum nutans (L.) Lunell,
Chrysopogon avenaceus (Michx.)
Benth., *Chrysopogon nutans* (L.) Benth.,
Chrysopogon nutans var. *linnaeanus*
(Hack.) C.Mohr, *Chrysopogon nutans*
var. *avenaceus* (Michx.) Trel. ex
Branner & Coville, *Digitaria nutans*
(L.) Beetle, *Holcus nutans* (L.) Kuntze
ex Stuck., *Holcus nutans* var. *avenaceus*
(Michx.) Hack., *Poranthera cilata* Raf.
ex B.D.Jacks., *Poranthera nutans* Raf.
ex B.D.Jacks., *Sorghastrum albescens*
(E.Fourn.) Beetle, *Sorghastrum
avenaceum* (Michx.) Nash, *Sorghastrum
flexuosum* Swallen, *Sorghastrum
linnaeanum* (Hack.) Nash, *Sorghastrum
stipoides* (Kunth) Nash (see),
Sorghastrum viride Swallen, *Sorghum
avenaceum* (Michx.) Chapm., *Sorghum
nutans* (L.) A.Gray, *Sorghum nutans* ssp.
linnaeanus (Hack.) Hack., *Stipa villosa*
Walter, *Trichachne nutans* (L.) B.R.Baum
♦ yellow Indian grass, Indian grass,
wood grass
♦ Weed, Naturalised
♦ 7, 86, 161
♦ G, arid, cultivated, herbal.

Sorghastrum pellitum (Hack.) Parodi
Poaceae
Sorghastrum nutans (L.) Nash ssp.
pellitum (Hack.) Burk.
♦ pasto de vaca
♦ Weed
♦ 295
♦ G.

Sorghastrum stipoides (Kunth) Nash
Poaceae
= *Sorghastrum nutans* (L.) Nash

Sorghum

- ◆ needle Indian grass
- ◆ Naturalised
- ◆ 101
- ◆ G.

Sorghum Moench spp.
Poaceae
- ◆ Columbus grass, forage sorghum, sorghum, perennial sorghum, sorgo
- ◆ Weed
- ◆ 80, 203
- ◆ G, cultivated, herbal, toxic.

Sorghum affine (Presl) E.G. & A.Camus
Poaceae
- ◆ Weed
- ◆ 87, 88
- ◆ G.

Sorghum × *almum* Parodi
Poaceae
= *Sorghum bicolor* (L.) Moench ×
Sorghum halepense (L.) Pers.
- ◆ Columbus grass, almum grass, sorghum almum
- ◆ Weed, Quarantine Weed, Noxious Weed, Naturalised, Introduced
- ◆ 7, 39, 76, 86, 87, 88, 93, 101, 121, 141, 147, 161, 218, 228, 229, 237, 295
- ◆ G, arid, cultivated, toxic. Origin: Argentina.

Sorghum arundinaceum (Desv.) Stapf
Poaceae
Andropogon arundinaceus Willd. *nom. illeg.*, *Andropogon sorghum* (L.) Brot. ssp. *vogelianus* Piper, *Andropogon sorghum* var. *effusus* Hack., *Andropogon sorghum* var. *virgatus* Hack., *Andropogon stapfii* Hook.f., *Andropogon verticilliflorus* Steud., *Holcus exiguus* Forssk., *Holcus sorghum* L. var. *effusus* (Hack.) Hitchc., *Holcus sorghum* var. *exiguus* (Forssk.) Hitchc., *Holcus sorghum* var. *verticilliflorus* (Steud.) Hitchc., *Rhaphis arundinacea* Desv., *Sorghum aethiopicum* (Hack.) Rupr. ex Stapf, *Sorghum bicolor* (L.) Moench ssp. *arundinaceum* (Desv.) de Wet & J.R.Harlan (see), *Sorghum lanceolatum* Stapf, *Sorghum macrochaeta* Snowden, *Sorghum pugionifolium* Snowden, *Sorghum stapfii* (Hook.f.) C.E.C.Fisch., *Sorghum usambarense* Snowden, *Sorghum verticilliflorum* (Steud.) Stapf (see), *Sorghum virgatum* (Hack.) Stapf (see), *Sorghum vogelianum* (Piper) Stapf
- ◆ wild sorghum, sorgo selvagem
- ◆ Weed, Quarantine Weed, Naturalised
- ◆ 6, 76, 87, 88, 134, 179, 203, 220, 242, 255
- ◆ a/pG, arid. Origin: Africa.

Sorghum bicolor (L.) Moench
Poaceae
Andropogon bicolor (L.) Roxb., *Andropogon sorghum* (L.) Brot., *Andropogon sorghum* subvar. *rubidus* Burkill ex C.Benson & C.K.Subba Rao, *Andropogon sorghum* var. *agricolarum* Burkill ex C.Benson & C.K.Subba Rao, *Andropogon sorghum* var. *arduinii* Körn., *Andropogon sorghum* var.

caudatus Hack., *Andropogon sorghum* var. *compactus* Burkill ex C.Benson & C.K.Subba Rao, *Andropogon sorghum* var. *ehrenbergianus* Körn., *Andropogon sorghum* var. *elegans* Körn., *Andropogon sorghum* var. *hians* Stapf, *Andropogon sorghum* var. *miliiformis* Hack., *Andropogon sorghum* var. *splendidus* Hack., *Andropogon sorghum* var. *subglobosus* Hack., *Andropogon sorghum* var. *technicus* Körn., *Andropogon subglabrescens* Steud., *Holcus bicolor* L., *Holcus caffrorum* Thunb., *Holcus cernuus* Ard., *Holcus dochna* Forssk., *Holcus durra* Forssk., *Holcus saccharatus* L., *Holcus sorghum* L., *Holcus sudanensis* (Piper) L.H.Bailey, *Milium nigricans* Ruiz & Pav., *Panicum caffrorum* Retz., *Sorghum basutorum* Snowden, *Sorghum bicolor* (L.) Moench var. *arduinii* (Körn.) Snowden, *Sorghum bicolor* var. *subglobosum* (Hack.) Snowden, *Sorghum bicolor* var. *technicum* (Körn.) Stapf ex Holland, *Sorghum caffrorum* (Thunb.) P.Beauv., *Sorghum caffrorum* var. *brunneolum* Snowden, *Sorghum caffrorum* var. *lasiorhachis* (Hack.) Snowden, *Sorghum caudatum* (Hack.) Stapf, *Sorghum cernuum* (Ard.) Host, *Sorghum cernuum* var. *agricolarum* (Burkill ex C.Benson & C.K.Subba Rao) Snowden, *Sorghum cernuum* var. *orbiculatum* Snowden, *Sorghum conspicuum* Snowden, *Sorghum conspicuum* var. *pilosum* Snowden, *Sorghum conspicuum* var. *rubicundum* Snowden, *Sorghum coriaceum* Snowden, *Sorghum coriaceum* var. *subinvolutum* Snowden, *Sorghum dochna* (Forssk.) Snowden, *Sorghum dochna* var. *technicum* (Körn.) Snowden, *Sorghum durra* (Forssk.) Stapf, *Sorghum elegans* (Körn.) Snowden, *Sorghum exsertum* Snowden, *Sorghum gambicum* Snowden, *Sorghum guineense* Stapf, *Sorghum japonicum* (Hack.) Roshev., *Sorghum margaritiferum* Stapf, *Sorghum melaleucum* Stapf, *Sorghum mellitum* Snowden, *Sorghum membranaceum* Chiov., *Sorghum membranaceum* var. *ehrenbergianum* (Körn.) Snowden, *Sorghum miliiforme* (Hack.) Snowden, *Sorghum nervosum* Besser ex Schult. & Schult.f., *Sorghum nervosum* Chiov., *Sorghum nigricans* (Ruiz & Pav.) Snowden, *Sorghum notabile* Snowden, *Sorghum roxburghii* Stapf, *Sorghum roxburghii* var. *hians* (Stapf) Stapf, *Sorghum saccharatum* (L.) Moench (see), *Sorghum saccharatum* var. *bicolor* (L.) Kerguélen, *Sorghum simulans* Snowden, *Sorghum splendidum* (Hack.) Snowden, *Sorghum subglabrescens* (Steud.) Schweinf. & Asch., *Sorghum subglabrescens* var. *compactum* (Burkill ex C.Benson & C.K.Subba Rao) Snowden, *Sorghum subglabrescens* var. *oviforme* Snowden, *Sorghum subglabrescens* var. *rubidum* (Burkill ex C.Benson & C.K.Subba Rao) Snowden, *Sorghum sudanense* (Piper) Stapf (see),

Sorghum technicum Batt. & Trab., *Sorghum vulgare* Pers. *nom. illeg.* (see), *Sorghum vulgare* var. *caffrorum* (Retz.) C.E.Hubb. & Rehder, *Sorghum vulgare* var. *durra* (Forssk.) C.E.Hubb. & Rehder, *Sorghum vulgare* var. *roxburghii* (Stapf) Haines, *Sorghum vulgare* var. *saccharatum* (L.) Boerl.
- ◆ shattercane, sorghum, wild cane, black amber, chicken corn, gooseneck, sorgho, black amber cane, milo, broomcorn, kaffir corn, isaka mtama, kao lin, broomcorn, durra, feterita, forage sorghum, grain sorghum, great millet, shallu, sweet sorghum, gros mil, sato morokoshi, sorgo, daza
- ◆ Weed, Noxious Weed, Naturalised, Introduced, Environmental Weed, Cultivation Escape
- ◆ 7, 23, 34, 39, 68, 80, 87, 88, 90, 93, 98, 101, 151, 161, 174, 179, 199, 203, 210, 211, 212, 218, 229, 241, 243, 261, 263, 280, 287, 300
- ◆ aG, arid, cultivated, herbal, toxic. Origin: tropical Africa.

Sorghum bicolor (L.) Moench ssp. *arundinaceum* (Desv.) de Wet & Harlan
Poaceae
= *Sorghum arundinaceum* (Desv.) Stapf
- ◆ common wild sorghum, shattercane, broomcorn
- ◆ Weed, Noxious Weed, Naturalised
- ◆ 86, 88, 101, 158, 229, 271
- ◆ pG. Origin: southern Africa.

Sorghum bicolor (L.) Moench ssp. *bicolor*
Poaceae
- ◆ shattercane, wild cane, grain sorghum, broomcorn
- ◆ Noxious Weed, Naturalised, Environmental Weed
- ◆ 86, 101, 229
- ◆ G.

Sorghum bicolor (L.) Moench ssp. *drummondii* (Nees ex Steud.) de Wet & Harlan
Poaceae
Andropogon drummondii Nees ex Steud.
- ◆ shattercane, wild grain sorghum, Drummond's broomcorn, Sudan grass, broomcorn, red kaffir corn, shallu, sugar millet, sugar reed, sugar sorghum, sweet cane, sweet reed, sweet sorghum
- ◆ Weed, Noxious Weed, Naturalised
- ◆ 86, 88, 101, 121, 158, 229
- ◆ aG. Origin: southern Africa.

Sorghum brevicarinatum Snowden
Poaceae
- ◆ lunsamba
- ◆ Weed, Naturalised
- ◆ 86, 98, 203
- ◆ G. Origin: central Africa, Indian Ocean Islands.

Sorghum cv. hybrid 'Silk' hort.
Poaceae
= *Sorghum halepense* (L.) Pers. × *Sorghum roxburghii* Stapf cv. 'Krish') × *Sorghum arundinaceum* (Desv.) Stapf
- ◆ silk sorghum

♦ Weed, Naturalised, Quarantine Weed
♦ 76, 86, 88, 147
♦ G, cultivated. Origin: horticultural.

Sorghum halepense (L.) Pers.
Poaceae
Andropogon arundinacea Scop., *Andropogon controversus* Steud., *Andropogon halepensis* (L.) Brot., *Andropogon halepensis* ssp. *anatherus* Piper, *Andropogon halepensis* var. *genuinus* Stapf, *Andropogon miliaceus* Roxb., *Andropogon miliformis* Schult., *Andropogon sorghum* (L.) Brot. ssp. *halepensis* (L.) Hack., *Andropogon sorghum* ssp. *sudanensis* Piper, *Andropogon sorghum* subvar. *genuinus* Hack., *Andropogon sorghum* subvar. *muticus* Hack., *Andropogon sorghum* var. *halepensis* (L.) Hack., *Blumenbachia halepensis* (L.) Koeler, *Holcus halepensis* L., *Holcus halepensis* var. *miliformis* (Schult.) Hitchc., *Holcus sorghum* L. var. *sudanensis* (Piper) Hitchc., *Milium halepense* (L.) Cav., *Sorghum controversum* (Steud.) Snowden, *Sorghum halepense* var. *genuinum* Hack., *Sorghum halepense* var. *muticum* (Hack.) Hayek, *Sorghum miliaceum* (Roxb.) Snowden (see), *Sorghum miliaceum* var. *parvispicula* Snowden
♦ Johnson grass, St. Mary's grass, Aleppo grass, Egyptian millet, false guineagrass, Cuba grass, Syria grass, evergreen millet, maidencane, meansgrass, sorgo selvatico, sorghetta, sorgagna, melghetta, cannereccia, sorgo de Alepo de semilla, sorgho d'Alep, sorgo de Alepo, herbe de Cuba, zacate Johnson, Aleppo milletgrass, Arabian millet, false guinea, Morocco millet, Egyptian grass
♦ Weed, Quarantine Weed, Noxious Weed, Naturalised, Introduced, Environmental Weed, Cultivation Escape, Casual Alien
♦ 1, 3, 6, 7, 14, 17, 23, 30, 34, 35, 36, 38, 39, 40, 42, 49, 49, 51, 55, 66, 68, 70, 76, 77, 80, 86, 87, 88, 90, 93, 98, 101, 102, 105, 107, 111, 115, 116, 118, 121, 138, 141, 146, 147, 151, 152, 157, 158, 161, 167, 174, 179, 180, 186, 195, 198, 199, 203, 207, 210, 211, 212, 218, 219, 221, 228, 229, 236, 237, 243, 245, 247, 249, 253, 255, 256, 258, 261, 263, 264, 266, 269, 270, 272, 275, 280, 281, 283, 286, 287, 290, 293, 295, 299, 300
♦ pG, arid, cultivated, herbal, toxic. Origin: Mediterranean to India.

Sorghum halepense (L.) Pers. fo. *halepense*
Poaceae
♦ Naturalised
♦ 241
♦ G.

Sorghum halepense (L.) Pers. fo. *muticum* Pers.
Poaceae
♦ Weed, Naturalised
♦ 286, 287
♦ G.

Sorghum halepense (L.) Pers. var. *halepense* fo. *muticum* (Hack.) Hubb.
Poaceae
♦ Introduced
♦ 230
♦ G.

Sorghum miliaceum (Roxb.) Snowden
Poaceae
= *Sorghum halepense* (L.) Pers.
♦ Johnson grass
♦ Weed, Naturalised
♦ 86, 87, 88, 98, 203
♦ G. Origin: North Africa to India.

Sorghum nitidum (Vahl) Pers.
Poaceae
♦ yaa haang maa, glossy wild sorghum
♦ Weed
♦ 238
♦ G. Origin: Asia, Australia.

Sorghum plumosum (R.Br.) Beauv.
Poaceae
♦ perennial canegrass, plume sorghum
♦ Weed
♦ 87, 88
♦ G, arid. Origin: Asia, Australia.

Sorghum propinquum (Kunth) Hitchc.
Poaceae
♦ sorghum
♦ Noxious Weed
♦ 87, 88, 229
♦ G.

Sorghum saccharatum (L.) Moench
Poaceae
= *Sorghum bicolor* (L.) Moench
♦ mijo, millo, sorgo
♦ Casual Alien
♦ 261
♦ G, herbal. Origin: Old World.

Sorghum stipodeum (Ewart & Jean White) Gardner & Hubb.
Poaceae
♦ annual native sorghum
♦ Weed, Naturalised
♦ 86, 93
♦ G.

Sorghum sudanense (Piper) Stapf
Poaceae
= *Sorghum bicolor* (L.) Moench
♦ Sudan grass, chicken corn, white kaffir corn
♦ Weed, Naturalised
♦ 3, 6, 7, 30, 39, 68, 80, 87, 88, 98, 121, 203, 218, 241, 256, 295
♦ aG, cultivated, herbal, toxic.

Sorghum versicolor Andersson
Poaceae
♦ blackseed wild sorghum
♦ Weed
♦ 88, 158
♦ aG, toxic. Origin: southern Africa.

Sorghum verticilliflorum (Steud.) Stapf
Poaceae
= *Sorghum arundinaceum* (Desv.) Stapf
♦ common wild sorghum, evergreen millet, Johnson grass, wild grain sorghum
♦ Weed, Naturalised, Native Weed, Cultivation Escape
♦ 30, 39, 87, 88, 98, 121, 203, 261, 276
♦ a/bG, cultivated, toxic. Origin: southern Africa.

Sorghum virgatum (Hack.) Stapf
Poaceae
= *Sorghum arundinaceum* (Desv.) Stapf
♦ Tunis grass
♦ Weed
♦ 87, 88, 185, 221
♦ G.

Sorghum vulgare Pers. nom. illeg.
Poaceae
= *Sorghum bicolor* (L.) Moench
♦ grain sorghum, sorghum, sweet sorghum, milo, broomcorn, durra, kaffir corn, shattercane
♦ Weed
♦ 39, 87, 88
♦ G, herbal, toxic.

Sorindeia madagascariensis Thouars
Anacardiaceae
Sorindeia somalensis (Chiov.) Chiov.
♦ mataanbiyood, matanbioo, mperiperi
♦ Weed, Introduced
♦ 179, 228
♦ arid, cultivated.

Spananthe paniculata Jacq.
Apiaceae
♦ Weed
♦ 87, 88
♦ arid.

Sparattosperma leucanthemum (Vell.) K.Schum.
Bignoniaceae
Bignonia leucantha Vell., *Bignonia subvernicosa* DC., *Spathodea lithontripticum* C.Mart. ex DC., *Spathodea vernicosa* Cham., *Tecoma subvernicosa* DC., *Sparattosperma vernicosa* (Cham.) Bureau & K.Schum.
♦ taruma, cinco folhas, ipe batata
♦ Weed
♦ 255
♦ Origin: Brazil.

Sparaxis bulbifera (L.) Ker Gawl.
Iridaceae
♦ sparaxis, fluweelblom, harlequin flower
♦ Weed, Naturalised, Garden Escape, Environmental Weed
♦ 7, 9, 72, 86, 88, 98, 176, 198, 203, 280, 289, 296
♦ pH, cultivated. Origin: South Africa.

Sparaxis fragrans (Jacq.) Ker Gawl.
Iridaceae
♦ fragrant wandflower
♦ Naturalised
♦ 101
♦ cultivated.

Sparaxis fragrans (Jacq.) Ker Gawl. ssp. grandiflora (D.Delaroche) Goldbl.
Iridaceae
= *Sparaxis grandiflora* (Delaroche) Ker Gawl.
♦ fragrant wandflower

♦ Naturalised
♦ 101

Sparaxis grandiflora (D.Delaroche) Ker Gawl.
Iridaceae
Sparaxis fragrans (Jacq.) Ker Gawl. ssp. *grandiflora* (D.Delaroche) Goldbl. (see)
♦ plain harlequin flower, sparaxis
♦ Naturalised, Garden Escape, Environmental Weed
♦ 7, 40, 86
♦ pH, cultivated. Origin: South Africa.

Sparaxis pillansii L.Bolus
Iridaceae
♦ harlequin flower, wandflower, tricolor harlequin flower
♦ Naturalised, Garden Escape, Environmental Weed
♦ 86
♦ cultivated. Origin: South Africa.

Sparaxis tricolor (Schneev.) Ker Gawl.
Iridaceae
♦ wandflower, harlequin flower, tricolor harlequin flower
♦ Weed, Naturalised, Garden Escape, Environmental Weed
♦ 7, 15, 72, 88, 98, 101, 198, 203, 251, 280, 290, 296
♦ pH, cultivated. Origin: South Africa.

Sparganium americanum Nutt.
Sparganiaceae
♦ threesquare burreed, American burreed
♦ Weed
♦ 23, 87, 88, 218
♦ pH, aqua, cultivated, herbal.

Sparganium angustifolium Michx.
Sparganiaceae
Sparganium affine Schnitz., *Sparganium emersum* Rehm. (see)
♦ narrowleaf burreed, kaitapalpakko, floating burreed
♦ Weed
♦ 23, 87, 88, 218
♦ pH, herbal.

Sparganium chlorocarpum Rydb.
Sparganiaceae
= *Sparganium erectum* L. ssp. *stoloniferum* (Graebn.) Hara
♦ greenfruit burreed
♦ Weed
♦ 23, 87, 88, 218
♦ herbal.

Sparganium emersum Rehm.
Sparganiaceae
= *Sparganium angustifolium* Michx.
♦ unbranched burreed, rantapalpakko, ezomikuri
♦ Weed
♦ 272
♦ aqua, cultivated, herbal.

Sparganium erectum L.
Sparganiaceae
Sparganium neglectum Beeby, *Sparganium polyedrum* (Asch. & Graebn.) Juz, *Sparganium ramosum* Huds. (see), *Sparganium ramosum* ssp. *polyedrum* Asch. & Graebn.

♦ simplestem burreed, haarapalpakko, exotic burreed, common burreed, branched burreed, burreed
♦ Weed, Quarantine Weed, Noxious Weed, Naturalised, Environmental Weed
♦ 23, 67, 70, 76, 87, 88, 98, 140, 161, 193, 198, 203, 220, 229, 246, 272
♦ pH, aqua, cultivated, herbal.

Sparganium erectum L. ssp. *stoloniferum* (Graebn.) Hara
Sparganiaceae
Sparganium chlorocarpum Rydb. (see)
♦ exotic burreed, erect burreed, simplestem burreed
♦ Naturalised, Noxious Weed
♦ 86, 140, 229, 286
♦ pH, aqua.

Sparganium erectum L. var. *macrocarpum* Hara
Sparganiaceae
♦ Weed
♦ 286

Sparganium eurycarpum Engelm. ex Gray
Sparganiaceae
♦ giant burreed, broadfruit burreed
♦ Weed
♦ 23, 87, 88, 218
♦ herbal.

Sparganium fluctuans (Morong) Robins.
Sparganiaceae
♦ water burreed, floating burreed
♦ Weed
♦ 23, 87, 88, 218
♦ herbal.

Sparganium ramosum Huds.
Sparganiaceae
= *Sparganium erectum* L.
♦ Weed
♦ 87, 88
♦ aqua, cultivated, herbal.

Sparganium stoloniferum Buch.-Ham.
Sparganiaceae
♦ Weed
♦ 87, 88, 275, 297
♦ pH, aqua, promoted, herbal.

Sparmannia africana L.f.
Tiliaceae
♦ German linden, African hemp, indoor linden
♦ Weed, Naturalised
♦ 161, 247, 280
♦ cultivated, herbal, toxic.

Sparmannia africana L.f. cv. 'Flore Pleno'
Tiliaceae
♦ African hemp, German linden, indoor linden
♦ Naturalised
♦ 280
♦ cultivated, herbal, toxic.

Spartina Schreb. spp.
Poaceae
♦ cordgrass, spartina, marsh grass
♦ Weed, Naturalised, Garden Escape, Environmental Weed
♦ 18, 86, 88, 198, 203, 246
♦ G, cultivated.

Spartina alterniflora Loisel.
Poaceae
♦ saltwater cordgrass, smooth cordgrass, Atlantic cordgrass, American spartina, American cordgrass
♦ Weed, Quarantine Weed, Noxious Weed, Naturalised, Environmental Weed
♦ 1, 35, 78, 80, 87, 88, 116, 139, 146, 161, 197, 218, 220, 225, 229, 231, 246, 280
♦ wpG, cultivated, herbal.

Spartina anglica C.E.Hubb.
Poaceae
♦ common cordgrass, ricegrass, spartina, cordgrass
♦ Weed, Quarantine Weed, Noxious Weed, Naturalised, Environmental Weed
♦ 1, 15, 35, 72, 76, 78, 79, 80, 80, 86, 88, 98, 101, 106, 116, 139, 146, 152, 161, 176, 197, 198, 203, 225, 229, 231, 246, 256, 280, 289, 296
♦ wpG, promoted. Origin: Great Britain.

Spartina bakeri Merr.
Poaceae
♦ Florida cordgrass, sand cordgrass
♦ Weed
♦ 87, 88, 218
♦ G, cultivated, herbal.

Spartina coarctata Trin.
Poaceae
♦ Introduced
♦ 228
♦ G, arid.

Spartina cynosuroides (L.) Roth
Poaceae
♦ big cordgrass
♦ Weed
♦ 87, 88, 218
♦ G, herbal.

Spartina densiflora Brongn.
Poaceae
♦ dense flowered cordgrass
♦ Weed, Noxious Weed, Naturalised, Environmental Weed
♦ 35, 78, 80, 87, 88, 101, 116, 197, 229, 231
♦ wpG.

Spartina maritima (Curtis) Fern.
Poaceae
Spartina capensis Nees ex Trin., *Spartina stricta* (Ait.) Roth
♦ small cordgrass, cape cordgrass
♦ Weed, Naturalised, Native Weed
♦ 70, 98, 101, 121, 203
♦ pG.

Spartina patens (Aiton) Muhl.
Poaceae
♦ saltmeadow cordgrass
♦ Weed, Noxious Weed, Environmental Weed
♦ 1, 35, 78, 80, 87, 88, 116, 146, 161, 197, 218, 229, 231
♦ wpG, cultivated, herbal.

Spartina × *townsendii* H. & J.Groves
Poaceae
= *Spartina alterniflora* Lois × *Spartina maritima* (Curtis) Fernald
♦ Townsend's cordgrass, spartina hybrid, ricegrass
♦ Weed, Naturalised, Garden Escape, Environmental Weed
♦ 72, 86, 88, 98, 101, 198, 208, 225, 246, 280, 290, 296
♦ wpG, cultivated.

Spartium junceum L.
Fabaceae/Papilionaceae
Genista juncea Lam., *Spartium americanum* Mey.
♦ Spanish broom, gorse, weaver's broom, Spaanse besem
♦ Weed, Sleeper Weed, Noxious Weed, Naturalised, Introduced, Garden Escape, Environmental Weed
♦ 1, 15, 34, 35, 38, 39, 40, 63, 70, 72, 78, 80, 86, 87, 88, 95, 98, 101, 116, 121, 137, 139, 146, 176, 198, 203, 218, 225, 228, 229, 231, 241, 246, 272, 280, 283, 289, 300
♦ pS, arid, cultivated, herbal, toxic. Origin: Eurasia.

Spathiphyllum Schott spp.
Araceae
♦ spathe flower, peace lily, spathe flower
♦ Weed
♦ 161, 247
♦ aqua, cultivated, toxic.

Spathodea campanulata P.Beauv.
Bignoniaceae
♦ African tulip tree, fireball, fountain tree, tulipier du Gabon, pisse pisse, rarningobchey, tuhke dulip, tiulipe, taga mimi, flame tree
♦ Weed, Naturalised, Introduced, Garden Escape, Environmental Weed, Cultivation Escape
♦ 3, 6, 22, 38, 54, 73, 80, 86, 88, 101, 107, 132, 155, 179, 201, 230, 233, 257, 259
♦ T, cultivated, herbal. Origin: obscure.

Spathoglottis plicata Blume
Orchidaceae
♦ Philippine ground orchid, terrestrial orchid
♦ Weed, Naturalised
♦ 19, 87, 88, 101
♦ cultivated, herbal.

Specularia biflora (Ruiz & Pav.) Fisch. & Mey.
Campanulaceae
= *Triodanis perfoliata* (L.) Nieuwl. var. *biflora* (Ruiz & Pav.) Bradl.
♦ Weed
♦ 87, 88

Specularia perfoliata (L.) A.DC.
Campanulaceae
= *Triodanis perfoliata* (L.) Nieuwl.
♦ Venus's lookingglass
♦ Weed
♦ 87, 88, 218, 286
♦ herbal.

Specularia speculum-veneris (L.) A.DC.
Campanulaceae
= *Legousia speculum-veneris* (L.) Chaix
♦ Venus's lookingglass
♦ Weed
♦ 23, 88
♦ herbal.

Speea humilis (Phil.) Loes.
Liliaceae/Alliaceae
Geanthus humilis Phil. (see)
♦ Quarantine Weed
♦ 220

Spenceria ramalana Trimen
Rosaceae
♦ common spenceria
♦ Weed
♦ 297

Speranskia tuberculata (Bunge) Baill.
Euphorbiaceae
♦ tuberculate speranskia
♦ Weed
♦ 297

Spergula arvensis L.
Caryophyllaceae
Alsine arvensis (L.) Crantz, *Arenaria arvensis* (L.) Cambess., *Spergula decandra* Gilib., *Spergula arvensis* L. var. *sativa* (Boenn.) Koch (see), *Spergula sativa* Boenn., *Spergula vulgaris* Boenn.
♦ corn spurrey, spurrey, renaiola comune, peltohatikka, stickwort, starwort, spurrey
♦ Weed, Noxious Weed, Naturalised, Introduced, Environmental Weed, Cultivation Escape
♦ 7, 9, 15, 23, 38, 44, 51, 52, 53, 70, 86, 87, 88, 94, 98, 101, 118, 121, 136, 157, 158, 161, 162, 165, 176, 186, 198, 203, 210, 212, 218, 236, 237, 241, 243, 245, 250, 253, 255, 269, 272, 280, 286, 287, 295, 299, 300
♦ aH, cultivated, herbal. Origin: Europe.

Spergula arvensis L. ssp. *arvensis*
Caryophyllaceae
♦ spurrey
♦ Weed
♦ 34
♦ aH.

Spergula arvensis L. var. *maxima* Koch
Caryophyllaceae
Spergula maxima Weihe
♦ Naturalised
♦ 287

Spergula arvensis L. var. *sativa* (Boenn.) Koch
Caryophyllaceae
= *Spergula arvensis* L.
♦ Weed, Naturalised
♦ 286, 287

Spergula linicola Bor. ex Nym.
Caryophyllaceae
♦ Weed
♦ 87, 88

Spergula morisonii Boreau
Caryophyllaceae

♦ Morison's spurrey, kalliohatikka
♦ Weed, Naturalised
♦ 101, 272

Spergula pentandra L.
Caryophyllaceae
Alsine marginata Schreb., *Alsine pentandra* (L.) Crantz, *Spergula arvensis* L. var. *marginata* (Schreb.) Moris
♦ fivestamen corn spurrey, wingstem spurrey, fiveanther spurrey
♦ Weed, Naturalised
♦ 70, 86, 88, 94, 98, 101, 198, 203, 272
♦ Origin: Europe.

Spergula platensis (Camb.) Shinn.
Caryophyllaceae
Alsine platensis (Cambess.) House, *Balardia platensis* Cambess., *Buda platensis* (Cambess.) Kuntze, *Spergularia platensis* (Camb.) Fenzl (see), *Spergularia platensis* var. *septentrionalis* (Hassl.) Hauman & Irigoyen, *Tissa platensis* (Cambess.) Hassl., *Tissa platensis* ssp. *septentrionalis* Hassl.
♦ Weed
♦ 295

Spergularia (Pers.) Presl spp.
Caryophyllaceae
♦ sand spurrey
♦ Weed
♦ 272

Spergularia bocconei (Scheele) Asch. & Graebn.
Caryophyllaceae
= *Spergularia bocconii* (Scheele) Asch. & Graebn.
♦ Greek sea spurrey, Boccon's sand spurrey
♦ Naturalised
♦ 280, 300
♦ aH.

Spergularia bocconii (Scheele) Asch. & Graebn.
Caryophyllaceae
Spergularia bocconei (Scheele) Asch. & Graebn. (see)
♦ Boccone's sand spurrey, Bocconi's sand spurrey, sand spurrey
♦ Weed, Naturalised
♦ 86, 101, 121, 161, 176, 180, 198, 287
♦ herbal. Origin: Mediterranean.

Spergularia diandra (Guss.) Heldr. & Sartori
Caryophyllaceae
♦ diandra sand spurrey, lesser sand spurrey, small sand spurrey
♦ Weed, Naturalised, Environmental Weed
♦ 7, 86, 93, 98, 101, 191, 198, 203, 221
♦ Origin: Eurasia.

Spergularia echinosperma (Celak.) Asch. & Graebn.
Caryophyllaceae
♦ bristleseed sand spurrey
♦ Naturalised
♦ 101
♦ herbal.

Spergularia levis Cambess.
Caryophyllaceae
♦ Weed, Naturalised
♦ 86, 98, 203
♦ Origin: South America.

Spergularia marina (L.) Griseb.
Caryophyllaceae
♦ sand spurrey, sea spurrey, perennial sea spurrey, lesser sea spurrey
♦ Weed, Naturalised, Garden Escape, Environmental Weed
♦ 7, 72, 86, 87, 88, 93, 98, 176, 185, 191, 203, 241, 275, 280, 300
♦ wa/pH, cultivated, herbal.

Spergularia maritima (All.) Chiov.
Caryophyllaceae
= *Spergularia media* (L.) C.Presl
♦ merisolmukki, media sand spurrey, sea spurrey
♦ Naturalised, Casual Alien
♦ 42, 101
♦ cultivated.

Spergularia media (L.) C.Presl
Caryophyllaceae
Arenaria maritima All., *Arenaria media* L., *Spergularia maritima* (All.) Chiov. (see)
♦ media sand spurrey, kittel, greater sea spurrey, greater sand spurrey
♦ Weed, Naturalised
♦ 86, 87, 88, 98, 176, 198, 203, 241, 300
♦ a/pH, herbal. Origin: Eurasia, North Africa.

Spergularia mexicana Hemsl.
Caryophyllaceae
♦ Weed
♦ 199

Spergularia platensis (Camb.) Fenzl
Caryophyllaceae
= *Spergula platensis* (Camb.) Shinn.
♦ La Plata sand spurrey
♦ Weed, Naturalised
♦ 87, 88, 101, 300
♦ aH, cultivated.

Spergularia purpurea (Pers.) Don
Caryophyllaceae
♦ sand spurrey, etelänsolmukki
♦ Weed, Naturalised, Casual Alien
♦ 42, 70, 88, 94, 101, 250, 253

Spergularia rubra (L.) J.& K.Presl
Caryophyllaceae
Arenaria campestris auct., *Arenaria rubra* (L.) Pers., *Buda rubra* (L.) Dumort, *Lepigonum rubrum* (L.) Wahlenb., *Tissa campestris* (L.) Pax, *Tissa rubra* (L.) Britt.
♦ red sand spurrey, sand spurrey, cliff sand spurrey
♦ Weed, Naturalised, Introduced, Environmental Weed
♦ 7, 9, 15, 21, 23, 34, 70, 72, 86, 87, 88, 94, 98, 101, 136, 165, 176, 203, 218, 241, 243, 253, 272, 280, 286, 287, 300
♦ a/pH, arid, cultivated, herbal. Origin: Europe.

Spergularia segetalis (L.) G.Don
Caryophyllaceae
Alsine segetalis L., *Alsine unilateralis* Moench, *Delia segetalis* (L.) Dumort,

Lepigonum segetale (L.) Koch., *Spergula segetalis* (L.) Vill., *Spergularia semidecandra* Kitt.
♦ autumn spurrey
♦ Weed
♦ 70, 88, 94, 272

Spergularia villosa (Pers.) Camb.
Caryophyllaceae
♦ hairy sand spurrey
♦ Weed, Naturalised
♦ 101, 167
♦ pH, arid/aqua, herbal. Origin: Chile, Peru.

Spermacoce alata Aubl.
Rubiaceae
Borreria alata (Aubl.) DC. (see)
♦ borreria, broadleaf buttonweed, winged false buttonweed
♦ Noxious Weed
♦ 140, 229

Spermacoce articulatis (L.f.) F.Williams
Rubiaceae
♦ Weed, Naturalised
♦ 98, 203

Spermacoce assurgens Ruiz & Pav.
Rubiaceae
= *Borreria assurgens* (Ruiz & Pav.) Griseb. (NoR)
♦ woodland false buttonweed, buttonweed
♦ Weed
♦ 11, 88, 161, 206, 243, 249
♦ herbal.

Spermacoce auriculata F.Muell.
Rubiaceae
♦ Weed
♦ 93, 191
♦ Origin: Australia.

Spermacoce capitata Ruiz & Pav.
Rubiaceae
= *Borreria capitata* (Ruiz & Pav.) DC. fo. *capitata* (NoR)
♦ baldhead false buttonweed, poaia da praia
♦ Weed, Naturalised
♦ 39, 101, 255
♦ aH, toxic. Origin: South America.

Spermacoce confusa Rendle ex Gillis
Rubiaceae
♦ river false buttonweed
♦ Weed, Naturalised
♦ 28, 206, 243, 257

Spermacoce eryngioides (Cham. & Schl.) Kuntze
Rubiaceae
♦ yerba de garro
♦ Weed
♦ 261

Spermacoce exilis (L.O.Williams) Adams
Rubiaceae
♦ Pacific false buttonweed
♦ Naturalised
♦ 101

Spermacoce glabra Michx.
Rubiaceae
♦ smooth false buttonweed
♦ Naturalised

♦ 287
♦ herbal.

Spermacoce hispida L.
Rubiaceae
Borreria rotundifolia Val.
♦ Weed, Introduced
♦ 93, 191, 230
♦ H, cultivated, herbal. Origin: Asia, Australia.

Spermacoce laevis Lam.
Rubiaceae
= *Spermacoce tenuior* L.
♦ buttonplant
♦ Weed
♦ 88, 218

Spermacoce latifolia Aubl.
Rubiaceae
= *Borreria latifolia* (Aubl.) Schum.
♦ ovalleaf false buttonweed, buttonweed, poaia do campo
♦ Weed, Naturalised, Introduced
♦ 28, 88, 93, 101, 191, 206, 230, 243, 255, 273
♦ aH. Origin: Brazil.

Spermacoce mauritiana Gideon
Rubiaceae
♦ Mauritius false buttonweed
♦ Weed, Quarantine Weed
♦ 88, 135, 191, 273

Spermacoce ovalifolia (Mart. & Gal.) Hemsl.
Rubiaceae
♦ broadleaf false buttonweed
♦ Naturalised
♦ 101

Spermacoce pilosa DC.
Rubiaceae
♦ Weed
♦ 87, 88

Spermacoce prostrata Aubl.
Rubiaceae
= *Borreria prostrata* (Aubl.) Miq. (NoR)
♦ prostrate false buttonweed
♦ Weed
♦ 28, 88, 206, 243

Spermacoce pusilla Wall.
Rubiaceae
♦ Weed
♦ 66
♦ cultivated.

Spermacoce tenuior L.
Rubiaceae
Borreria laevis (Lam.) Griseb. (see), *Spermacoce laevis* Lam. (see), *Spermacoce riparia* Cham. & Schlecht.
♦ slender false buttonweed
♦ Weed
♦ 14, 87, 88

Spermacoce verticillata L.
Rubiaceae
Borreria terminalis Small, *Borreria verticillata* (L.) Mey. (see)
♦ shrubby false buttonweed, whitehead broom, vassourinha
♦ Weed, Naturalised
♦ 80, 88, 101, 161, 179, 249, 255
♦ pH. Origin: Americas.

Sphacelaria fluviatilis **Jao**
Sphacelariaceae
- brown alga
- Weed
- 197
- algae.

Sphacelaria lacustris **sp. nov.**
Sphacelariaceae
- brown alga
- Weed
- 197
- algae.

Sphaeralcea angustifolia **(Cav.) Don**
Malvaceae
- copper globemallow, narrowleaf globemallow
- Weed
- 161
- S, cultivated, herbal.

Sphaeralcea bonariensis **(Cav.) Griseb.**
Malvaceae
Malva bonariensis Cav., *Sphaeralcea cisplatina* St.-Hil
- malva blanca
- Weed
- 121, 237, 295
- pS, herbal. Origin: South America.

Sphaeralcea coccinea **(Nutt.) Rydb.**
Malvaceae
Malvastrum coccineum (Nutt.) Gray
- scarlet globemallow, red false mallow, copper mallow
- Weed, Native Weed
- 49, 161, 174, 212
- cultivated, herbal. Origin: North America.

Sphaeralcea coulteri **(S.Wats.) Gray**
Malvaceae
- Coulter's globemallow
- Weed
- 161
- aH, herbal.

Sphaeralcea emoryi **Torr. ex A.Gray**
Malvaceae
- Emory's globemallow
- Weed
- 161
- herbal.

Sphaeralcea fulva **Greene**
Malvaceae
- sphaeralcea
- Weed
- 243

Sphaeralcea laxa **Woot. & Standl.**
Malvaceae
- caliche globemallow
- Weed
- 161
- herbal.

Sphaeralcea miniata **(Cav.) Spach**
Malvaceae
- malvisco
- Weed
- 295

Sphaeralcea orcuttii **Rose**
Malvaceae
- Carrizo Creek globemallow, Carrizo

mallow, Orcutt's globemallow
- Weed
- 161
- a/pH.

Sphaeralcea parvifolia **A.Nels.**
Malvaceae
- smallflower globemallow, littleleaf globemallow
- Weed
- 161
- herbal.

Sphaeranthus africanus **L.**
Asteraceae
Sphaeranthus microcephalus Willd.
- yaa khon klong
- Weed
- 13, 87, 88, 170, 191, 239, 262
- aH, herbal.

Sphaeranthus bullatus **Mattf.**
Asteraceae
- Weed, Quarantine Weed
- 53, 76, 87, 88, 203, 220
- aH.

Sphaeranthus indicus **L.**
Asteraceae
Sphaeranthus hirtus Willd., *Sphaeranthus mollis* Roxb.
- East Indian globe thistle, matom suea
- Weed
- 13, 87, 88, 170, 191, 204, 238
- cultivated, herbal. Origin: Asia, Australia.

Sphaeranthus senegalensis **DC.**
Asteraceae
- Weed
- 87, 88
- herbal.

Sphaeranthus suaveolens **(Forssk.) DC.**
Asteraceae
- Weed
- 87, 88, 221
- herbal.

Sphaerocarpos texanus **Austin**
Sphaerocarpaceae
- Texas balloonwort
- Naturalised
- 198
- liverwort.

Sphaerocoma hookeri **T.Anderson**
Caryophyllaceae/Illecebraceae
- Weed
- 221

Sphaeromariscus microcephalus **(J. & C.Presl) E.G.Camus**
Cyperaceae
- Weed
- 87, 88
- G.

Sphaerophysa salsula **(Pall.) DC.**
Fabaceae/Papilionaceae
Phaca salsula Pall., *Swainsona salsula* (Pall.) Taub. (see)
- Austrian peaweed, alkali swainson pea, swainsonia
- Weed, Noxious Weed, Naturalised, Introduced

- 1, 34, 35, 49, 80, 88, 101, 146, 161, 212, 228, 229, 264, 275
- pH, arid, herbal. Origin: Asia.

Sphaeropteris cooperi **(F.Muell.) R.M.Tryon**
Cyatheaceae
= *Cyathea cooperi* (Hook. ex F.Muell.) Domin
- Australian tree fern, tree fern
- Weed, Naturalised, Native Weed, Cultivation Escape
- 3, 7, 86, 233, 252
- cultivated.

Sphaerostephanos invisus **(Forst.) Holttum**
Thelypteridaceae
= *Thelypteris forsteri* Morton (NoR)
- Weed
- 6, 88, 191
- herbal.

Sphaerostephanos unitus **(L.) Holttum**
Thelypteridaceae
- Weed
- 6, 88, 191, 254
- Origin: Australia.

Sphagneticola trilobata **(L.C.Rich.) Pruski**
Asteraceae
Acmella brasiliensis Spreng., *Buphthalmum repens* Lam., *Complaya trilobata* (L.) Strother, *Seruneum trilobatum* (L.) Kuntze, *Silphium trilobatum* L., *Sphagneticola ulei* O.Hoffm., *Stemmodontia trilobata* (L.) Small, *Thelechitonia trilobata* (L.) H.Rob. & Cuatrec. (see), *Wedelia brasillensis* (Spreng.) S.F.Blake, *Wedelia carnosa* Rich. ex Pers., *Wedelia crenata* Rich. ex Pers., *Wedelia paludosa* DC., *Wedelia trilobata* (L.) A.Hitchc. (see)
- Bay Biscayne creeping oxeye, wedelia, margaridao
- Weed, Naturalised, Cultivation Escape
- 101, 233, 243, 255
- cultivated. Origin: tropical America.

Sphagnum subnitens **Russow & Warnst.**
Sphagnaceae
- sphagnum, lustrous bog moss
- Naturalised
- 280
- moss.

Sphallerocarpus gracilis **(Bess. ex Trev.) Pol.**
Apiaceae
- Weed
- 88, 114, 243

Sphenoclea zeylanica **Gaertn.**
Sphenocleaceae/Campanulaceae
Rapina herbacea Lour.
- gooseweed, phak pot, chicken spike
- Weed, Quarantine Weed, Naturalised
- 13, 39, 68, 76, 87, 88, 101, 161, 170, 179, 186, 191, 203, 204, 218, 220, 239, 243, 255, 261, 262, 263, 274, 286, 287, 297
- aH, aqua, herbal, toxic. Origin: Asia.

Sphenomeris chinensis (L.) Maxon
Dennstaedtiaceae
= *Odontosoria chinensis* (L.) J.Sm.
♦ Weed
♦ 87, 88
♦ cultivated, herbal.

Sphenopholis obtusata (Michx.) Scribn.
Poaceae
♦ prairie wedgescale, wedgegrass
♦ Weed, Naturalised
♦ 243, 287
♦ pG, herbal.

Sphenopus divaricatus (Gouan) Rchb.
Poaceae
♦ Weed, Naturalised
♦ 98, 203
♦ G.

Sphenosciadium capitellatum Gray
Apiaceae
♦ whiteheads, woollyhead parsnip, swamp whiteheads
♦ Weed
♦ 161
♦ pH, cultivated, herbal, toxic.

Sphondylium lanatum (Michx.) Greene
Apiaceae
♦ Quarantine Weed
♦ 220

Spiesia lambertii (Pursh) Kuntze
Fabaceae/Papilionaceae
= *Oxytropis lambertii* Pursh
♦ Quarantine Weed
♦ 220

Spigelia anthelmia L.
Loganiaceae/Spigeliaceae
♦ West Indian pinkroot, worm grass, pink spigelia, Indian pink, lombricera, waterweed, yerba lombricera
♦ Weed, Quarantine Weed
♦ 14, 39, 76, 87, 88, 161, 170, 191, 203, 206, 220, 243, 255, 261
♦ aH, herbal, toxic. Origin: tropical America.

Spigelia humboldtiana Cham. & Schlecht
Loganiaceae/Spigeliaceae
= *Spigelia scabra* Cham. & Schltdl. (NoR)
♦ Weed
♦ 157
♦ pH.

Spigelia marilandica (L.) L.
Loganiaceae/Spigeliaceae
♦ pink root, woodland pinkroot, Indian pink
♦ Weed
♦ 39, 161, 247
♦ herbal, toxic.

Spilanthes acmella (L.) L.
Asteraceae
♦ toothache plant, para cress
♦ Weed, Quarantine Weed, Naturalised
♦ 39, 87, 88, 209, 220, 255, 257
♦ arid, herbal, toxic. Origin: South America.

Spilanthes americana (Mutis ex L.f.) Hieron.
Asteraceae
= *Acmella oppositifolia* (Lam.) R.K.Jansen (NoR)
♦ botoncillo
♦ Weed, Naturalised
♦ 87, 88, 157, 281, 286, 287
♦ pH, herbal.

Spilanthes calva DC.
Asteraceae
♦ Weed
♦ 87, 88
♦ herbal.

Spilanthes decumbens (Sm.) A.H.Moore
Asteraceae
Ceratocephalus decumbens (Sm.) Kuntze, *Rudbeckia decumbens* Sm., *Spilanthes eurycarena* A.H.Moore
♦ spilanthes
♦ Weed
♦ 88, 158

Spilanthes filicaulis (Schumach. & Thonn.) C.D.Adams
Asteraceae
Eclipta filicaulis Schumach. & Thonn.
♦ Weed
♦ 88

Spilanthes grandiflora Turcz.
Asteraceae
♦ Weed
♦ 276
♦ pH, herbal, toxic.

Spilanthes iabadicensis A.H.Moore
Asteraceae
♦ Weed, Naturalised, Introduced
♦ 13, 88, 170, 191, 230, 287
♦ H, aqua.

Spilanthes limonica Moore
Asteraceae
Spilanthes insipida Jacq.
♦ Weed
♦ 14

Spilanthes macraei Hook. & Arn.
Asteraceae
♦ Weed
♦ 87, 88

Spilanthes ocymifolia (Lamk.) A.H.Moore
Asteraceae
♦ Weed
♦ 13
♦ herbal.

Spilanthes paniculata Wall. ex DC.
Asteraceae
= *Acmella paniculata* (Wall. ex DC.) R.K.Jansen (NoR) [see *Spilanthes paniculata* Wall. ex DC. fo. *bicolor* Kost.]
♦ phak phet, para cress
♦ Weed, Naturalised
♦ 13, 28, 88, 170, 191, 238, 243, 262, 276, 287, 297
♦ aH, herbal.

Spilanthes paniculata Wall. ex DC. fo. bicolor Kost.
Asteraceae

= *Acmella paniculata* (Wall. ex DC.) R.K.Jansen (NoR) [see *Spilanthes paniculata* Wall. ex DC.]
♦ Naturalised
♦ 287

Spilanthes uliginosa Sw.
Asteraceae
= *Acmella uliginosa* (Sw.) Cass. (NoR)
♦ Weed
♦ 87, 88
♦ herbal.

Spilanthes urens Jacq.
Asteraceae
♦ pigeoncoop
♦ Weed
♦ 87, 88

Spiloxene capensis (L.) Garside
Liliaceae/Hypoxidaceae
♦ spiloxene
♦ Naturalised
♦ 86, 198
♦ cultivated.

Spinacia oleracea L.
Chenopodiaceae
Spinacia domestica Borkh.
♦ spinach, pinaatti
♦ Weed, Naturalised, Casual Alien
♦ 39, 42, 98, 101, 203, 243
♦ aH, cultivated, herbal, toxic. Origin: obscure.

Spinifex littoreus (Burm.f.) Merr.
Poaceae
♦ littoral spinifex
♦ Weed
♦ 297
♦ G, cultivated.

Spinifex sericeus R.Br.
Poaceae
♦ Naturalised
♦ 7, 86, 134
♦ G, cultivated. Origin: Australia.

Spiraea alba Du Roi
Rosaceae
♦ narrowleaf meadowsweet, valkopajuangervo, white meadowsweet, meadowsweet
♦ Weed, Cultivation Escape
♦ 42, 52, 87, 88, 218
♦ S, cultivated, herbal.

Spiraea × arguta Zabel
Rosaceae
= *Spiraea multiflora* Zabel × *Spiraea thunbergii* Siebold ex Blume
♦ garland spiraea, morsiusangervo
♦ Weed, Cultivation Escape
♦ 42, 80
♦ S, cultivated. Origin: horticultural hybrid.

Spiraea × billiardii Hérincq (pro sp.)
Rosaceae
= *Spiraea douglasii* Hook. × *Spiraea salicifolia* L.
♦ rusopajuangervo, Billiard's spirea
♦ Cultivation Escape, Casual Alien
♦ 42, 280
♦ cultivated.

Spiraea × bumalda Burv.
Rosaceae
= *Spiraea albiflora* (Miq.) Zabel × *Spiraea japonica* L.
♦ Japanese spiraea, ruusuangervo
♦ Weed, Naturalised, Cultivation Escape
♦ 42, 80, 101
♦ cultivated.

Spiraea cantoniensis Lour.
Rosaceae
♦ cape may, maybush, Reeve's spirea, may, Reeve's meadowsweet
♦ Weed, Naturalised
♦ 22, 54, 86, 88, 101, 121, 198, 280
♦ pS, cultivated, herbal. Origin: Eurasia.

Spiraea chamaedryfolia L.
Rosaceae
♦ virpiangervo, germander meadowsweet
♦ Naturalised, Cultivation Escape
♦ 42, 101
♦ cultivated.

Spiraea chamaedryfolia L. var. chamaedryfolia
Rosaceae
♦ germander meadowsweet
♦ Naturalised
♦ 101

Spiraea chamaedryfolia L. var. ulmifolia (Scop.) Maxim.
Rosaceae
♦ germander meadowsweet
♦ Naturalised
♦ 101

Spiraea douglasii Hook.
Rosaceae
♦ Douglas's spirea, hardhack, punapajuangervo, rose spirea
♦ Weed, Cultivation Escape, Casual Alien
♦ 42, 87, 88, 218, 280
♦ S, cultivated, herbal.

Spiraea hypericifolia L.
Rosaceae
♦ Iberian spirea, spirea
♦ Naturalised
♦ 39, 101
♦ cultivated, toxic.

Spiraea japonica L.f.
Rosaceae
♦ Japanese spiraea, Japaninangervo, Japanese meadowsweet
♦ Weed, Naturalised, Garden Escape, Environmental Weed, Cultivation Escape
♦ 17, 39, 42, 80, 87, 88, 101, 102, 129, 142, 151, 218, 246, 280
♦ S, cultivated, herbal, toxic.

Spiraea japonica L.f. var. fortunei (Planch.) Rehd.
Rosaceae
♦ fortune meadowsweet
♦ Naturalised
♦ 101

Spiraea latifolia (Aiton) Borkh.
Rosaceae
= *Spiraea alba* Du Roi var. *latifolia* (Aiton) Dippel (NoR)
♦ meadowsweet, kaljupajuangervo
♦ Weed, Cultivation Escape
♦ 42, 87, 88, 218, 294
♦ cultivated, herbal.

Spiraea × macrothyrsa Dipp.
Rosaceae
= *Spiraea douglasii* Hook. × *Spiraea latifolia* (Ait.) Borkh.
♦ isopajuangervo
♦ Cultivation Escape
♦ 42
♦ cultivated. Origin: horticultural hybrid.

Spiraea prunifolia Siebold & Zucc.
Rosaceae
♦ bridal wreath spirea
♦ Weed, Naturalised
♦ 80, 98, 101, 203
♦ S, cultivated, herbal.

Spiraea × rubella Dipp.
Rosaceae
♦ purppura angervo
♦ Cultivation Escape
♦ 42
♦ cultivated.

Spiraea salicifolia L.
Rosaceae
♦ bridewort, willowleaf meadowsweet, pajuangervo
♦ Naturalised, Cultivation Escape
♦ 40, 42, 101
♦ S, cultivated.

Spiraea × sanssouciana K.Koch
Rosaceae
= *Spiraea japonica* L. × *Spiraea douglasii* Hook.
♦ keisarinangervo
♦ Cultivation Escape
♦ 42
♦ cultivated. Origin: horticultural hybrid from Japan.

Spiraea thunbergii Siebold ex Blume
Rosaceae
♦ Thunberg's meadowsweet
♦ Weed, Naturalised
♦ 80, 101
♦ S, cultivated, herbal.

Spiraea tomentosa L.
Rosaceae
♦ hardhack, steeplebush
♦ Weed
♦ 39, 87, 88, 218
♦ S, cultivated, herbal, toxic.

Spiraea trilobata L.
Rosaceae
♦ Asian meadowsweet
♦ Naturalised
♦ 101
♦ cultivated.

Spiraea × vanhouttei (Briot) Zabel
Rosaceae
= *Spiraea cantoniensis* Lour. × *Spiraea trilobata* L.

♦ kinosangervo, Vanhoutte spirea
♦ Weed, Naturalised, Cultivation Escape
♦ 42, 80, 101
♦ cultivated, herbal.

Spiranthes aestivalis (Poir.) Rich.
Orchidaceae
Ophrys aestivalis Lam., *Tussacia aestivalis* Desv.
♦ summer lady's tresses
♦ Weed
♦ 272
♦ cultivated, herbal.

Spiranthes sinensis (Pers.) Ames
Orchidaceae
Neottia sinensis Pers., *Spiranthes australis* (R.Br.) Lindl., *Spiranthes lancea* (Thunb.) Bakh.f. & V.Steenis
♦ ladies tresses
♦ Weed, Quarantine Weed
♦ 76, 88, 179, 191, 204, 235, 297
♦ cultivated. Origin: Australia.

Spiranthes sinensis (Pres.) Ames var. amoena Hara
Orchidaceae
♦ Weed
♦ 286

Spirodela Schleid. spp.
Lemnaceae
♦ spirodela, duckweed, stor andmat, isolimaskat, duckmeat
♦ Quarantine Weed
♦ 220
♦ wH, herbal.

Spirodela intermedia W.Koch
Lemnaceae
♦ intermediate duckweed, erva de pato
♦ Weed
♦ 237, 255, 295
♦ wpH.

Spirodela oligorrhiza (Kurz.) Hegelm.
Lemnaceae
= *Spirodela punctata* (G.F.W.Mey.) Thomps.
♦ greater duckweed
♦ Weed, Naturalised
♦ 87, 88, 274, 286, 287
♦ wpH.

Spirodela polyrhiza (L.) Schleid.
Lemnaceae
= *Spirodela polyrrhiza* (L.) Schleid.
♦ giant duckweed, greater duckweed, great duckweed, common duckmeat
♦ Weed, Quarantine Weed, Naturalised
♦ 76, 87, 88, 98, 203, 204, 209, 217, 218, 220, 258, 262, 263, 272, 275, 286, 297
♦ wpH, cultivated, herbal.

Spirodela polyrrhiza (L.) Schleid.
Lemnaceae
Lemna major Mey., *Lemna orbicularis* Kit., *Lemna orbiculata* Roxb., *Lemna polyrhiza* L. (see), *Lemna thermalis* Beauv., *Telmatophace orbicularis* Schur., *Spirodela polyrhiza* (L.) Schleid. (see), *Telmatophace polyrrhiza* Godr.

♦ duckweed, great duckweed, water flaxseed, common duckweed, common duckmeat
♦ Weed
♦ 121, 221
♦ wpH, herbal. Origin: Africa, Europe, Asia.

Spirodela punctata (G.F.W.Mey) Thomps.
Lemnaceae
Spirodela oligorrhiza (Kurz.) Hegelm. (see)
♦ duckweed, dotted duckmeat, dotted duckweed
♦ Weed, Naturalised, Environmental Weed
♦ 121, 179, 197, 246, 280
♦ wpH, cultivated. Origin: Eurasia.

Spirogyra Link spp.
Zygnemataceae
♦ spirogyra
♦ Weed, Quarantine Weed
♦ 76, 88, 218, 220

Spirogyra arcla Kütz.
Zygnemataceae
♦ Weed
♦ 88, 204, 286
♦ algae.

Spirogyra crassa Kütz.
Zygnemataceae
♦ shui mian
♦ Weed
♦ 263, 297
♦ aqua.

Spirostachys africana Sond.
Euphorbiaceae
♦ African mahogany tree, African sandalwood, cape sandalwood, jumping bean seed, sandaleen wood, tamboti
♦ Native Weed
♦ 121
♦ T, toxic. Origin: southern Africa.

Spodiopogon sibiricus Trin.
Poaceae
♦ o aburasusuki
♦ Weed
♦ 88, 191, 204, 286
♦ G, cultivated.

Spondias cytherea Sonn.
Anacardiaceae
= *Spondias dulcis* Sol. ex Parkinson
♦ citara, jobo de la India, ambarella, hog plum
♦ Cultivation Escape
♦ 261
♦ cultivated.

Spondias dulcis Sol. ex Parkinson
Anacardiaceae
Spondias cytherea Sonn. (see)
♦ Jewish plum, oatahette apple, hog plum, ambarella
♦ Naturalised, Introduced
♦ 101, 230
♦ T, cultivated, herbal.

Spondias mombin L.
Anacardiaceae
Spondias lutea L., *Spondias mexicana* S.Watson

♦ abal, atoya xocotl, biaxhi, capuaticacao, chiabal, chupandilla, circuelo obo, ciruelo, coztilxocotl, cupu, hobo, hog plum, jobillo, jobito, jobo, jobo espino, jocote, jocote de jobo, palo de mulato, piets ten, pompoqua, sismoyo, ten mi viad, tzrrobmal, xobo, yellow mombin, Jamaica plum, mombin, prunier mombin, gelbe mombinpflaume, gelbpflaume, imbu, tepereba, ubos
♦ Weed, Introduced, Cultivation Escape
♦ 14, 87, 88, 228, 261
♦ T, arid, cultivated, herbal. Origin: Mexico, Central and South America.

Spondias pinnata (L.f.) Kurz
Anacardiaceae
Mangifera pinnata L.f., *Spondias mangifera* Willd.
♦ ambra, amra
♦ Introduced
♦ 228, 230
♦ T, arid, cultivated, herbal.

Spondias purpurea L.
Anacardiaceae
Spondias cirouella Tussac
♦ purple mombin
♦ Weed, Naturalised, Environmental Weed, Cultivation Escape
♦ 80, 88, 101, 151, 179, 257, 261
♦ T, cultivated, herbal. Origin: Mexico, Central America, Peru, Brazil.

Sporobolus africanus (Poir.) Robyns & Tourn.
Poaceae
= *Sporobolus indicus* (L.) R.Br. var. *capensis* Engelm.
♦ Parramatta grass, African dropseed grass, rat's tail grass, tussockgrass, rat's tail dropseed, rush grass, tough dropseed
♦ Weed, Noxious Weed, Quarantine Weed, Naturalised, Native Weed, Environmental Weed
♦ 15, 76, 86, 87, 88, 93, 121, 134, 147, 158, 191, 203, 246, 269, 280, 289, 296
♦ pG, herbal. Origin: Africa.

Sporobolus airoides (Torr.) Torr.
Poaceae
Agrostis airoides Torr., *Sporobolus diffusissimus* Buckley
♦ alkali sacaton
♦ Weed
♦ 87, 88, 91, 218
♦ pG, arid, cultivated, herbal.

Sporobolus capensis (P.Beauv.) Kunth nom. illeg.
Poaceae
= *Sporobolus indicus* (L.) R.Br. var. *capensis* Engelm.
♦ rat's tail dropseed
♦ Weed
♦ 88, 218
♦ G.

Sporobolus caroli Mez.
Poaceae
♦ Introduced

♦ 228
♦ pG, arid, cultivated.

Sporobolus compositus (Poir.) Merr. var. compositus
Poaceae
Sporobolus asper (P.Beauv.) Kunth
♦ composite dropseed, dropseed
♦ Introduced
♦ 228
♦ G, arid.

Sporobolus coromandelianus (Retz.) Kunth
Poaceae
Sporobolus pyramidatus (Lam.) Hitchc. (see)
♦ dropseed, Madagascar dropseed
♦ Weed, Naturalised
♦ 86, 87, 88, 93, 191
♦ G, arid. Origin: southern Africa, eastern Asia.

Sporobolus creber De Nardi
Poaceae
♦ slender rat's tail grass
♦ Introduced, Native Weed
♦ 228, 269
♦ pG, arid, cultivated. Origin: Australia.

Sporobolus cryptandrus (Torr.) A.Gray
Poaceae
♦ sand dropseed
♦ Weed, Naturalised, Native Weed, Introduced, Casual Alien
♦ 87, 88, 161, 174, 218, 228, 280, 287
♦ pG, arid, cultivated, herbal. Origin: North America.

Sporobolus diander (Retz.) P.Beauv.
Poaceae
= *Sporobolus indicus* (L.) R.Br. var. *flaccidus* (Roem. & Schult.) Veldkamp (NoR) [see *Sporobolus diandrus* (Retz.) P.Beauv.]
♦ Indian dropseed, lesser dropseed, two anthered smutgrass, tussocky sporobolus, tussock dropseed
♦ Weed, Naturalised, Introduced
♦ 87, 88, 91, 101, 218, 230, 275, 286
♦ a/pG, arid, herbal. Origin: Australia.

Sporobolus diandrus (Retz.) P.Beauv.
Poaceae
= *Sporobolus indicus* (L.) R.Br. var. *flaccidus* (Roem. & Schult.) Veldkamp (NoR) [see *Sporobolus diander* (Retz.) P.Beauv.]
♦ Casual Alien
♦ 280
♦ G.

Sporobolus domingensis (Trin.) Kunth
Poaceae
♦ coral dropseed
♦ Weed
♦ 161, 249
♦ pG.

Sporobolus elongatus R.Br.
Poaceae
♦ elongate dropseed
♦ Weed, Naturalised, Native Weed
♦ 101, 269, 273, 280
♦ pG. Origin: Australia.

Sporobolus fertilis (Steud.) W.D.Clayton
Poaceae
Agrostis fertilis Steud., *Sporobolus indicus* (L.) R.Br. var. *major* (Büse) Baaijens
♦ Australian smutgrass
♦ Weed
♦ 235, 238, 244, 274, 286, 297
♦ pG. Origin: tropical Asia.

Sporobolus fimbriatus (Trin.) Nees
Poaceae
♦ common dropseed, bushveld dropseed, fringed dropseed, lulele
♦ Weed, Naturalised, Native Weed
♦ 88, 121, 158
♦ pG, herbal. Origin: southern Africa.

Sporobolus heterolepis (A.Gray) A.Gray
Poaceae
♦ prairie dropseed
♦ Weed
♦ 87, 88, 218
♦ G, cultivated, herbal.

Sporobolus indicus (L.) R.Br.
Poaceae
Sporobolus bertteroanus (Trin.) Hitch. & Chase, *Sporobolus poiretii* (Roem. & Schult.) Hitchc. (see)
♦ smutgrass, West Indian dropseed
♦ Weed, Naturalised, Environmental Weed
♦ 3, 14, 72, 87, 88, 91, 98, 101, 157, 161, 176, 199, 203, 237, 241, 249, 255, 295, 300
♦ pG, arid, cultivated, herbal.

Sporobolus indicus (L.) R.Br. var. africanus (Poir.) Jovet & Guédès
Poaceae
♦ Naturalised
♦ 98
♦ G.

Sporobolus indicus (L.) R.Br. var. capensis Engelm.
Poaceae
Agrostis capensis Willd. *nom. illeg.*, *Sporobolus africanus* (Poir.) Robyns & Tourn. (see), *Sporobolus capensis* (P.Beauv.) Kunth *nom. illeg.* (see)
♦ Parramatta grass, rat's tail, tufty grass, rat's tail grass
♦ Weed, Quarantine Weed, Naturalised, Environmental Weed
♦ 76, 86, 93, 101, 176, 198, 205
♦ pG. Origin: Africa.

Sporobolus indicus (L.) R.Br. var. diandrus (Retz.) Jovet & Guédès
Poaceae
Agrostis diandra Retz.
♦ Weed
♦ 88
♦ G.

Sporobolus indicus (L.) R.Br. var. fertilis (Steud.) Jovet & Guédès
Poaceae
♦ Weed
♦ 88
♦ G. Origin: Australia.

Sporobolus indicus (L.) R.Br. var. indicus
Poaceae
♦ smutgrass
♦ Weed
♦ 179
♦ G.

Sporobolus indicus (L.) R.Br. var. major (Büse) Baaijens
Poaceae
♦ rat's tail grass
♦ Naturalised, Environmental Weed
♦ 7, 86
♦ G.

Sporobolus indicus (L.) R.Br. var. purpurea-suffusus (Ohwi) Koyama
Poaceae
♦ Weed
♦ 275
♦ pG.

Sporobolus indicus (L.) R.Br. var. pyramidalis (Beauv.) Veldkamp
Poaceae
Sporobolus jacquemontii Kunth (see), *Sporobolus pyramidalis* Beauv. (see)
♦ West Indian dropseed
♦ Weed, Naturalised
♦ 101, 179
♦ G. Origin: Madagascar.

Sporobolus ioclados (Trin.) Nees
Poaceae
Sporobolus genalensis Chiov., *Sporobolus gillii* Stent, *Sporobolus ioclados* var. *usitatus* Nees (Stent) Chippind., *Sporobolus laetevirens* Coss., *Sporobolus marginatus* A.Rich., *Sporobolus marginatus* A.Rich. var. *anceps* Chiov., *Sporobolus marginatus* A.Rich. var. *scabrifolius* Chiov., *Sporobolus pallidus* (Trin.) Boiss., *Sporobolus seineri* Mez, *Sporobolus smutsii* Stent, *Sporobolus usitatus* Stent, *Vilfa ioclados* Trin., *Vilfa marginata* (A.Rich.) Steud., *Vilfa scabrifolia* Hochst. & Edgew.
♦ Weed
♦ 240
♦ pG, arid.

Sporobolus jacquemontii Kunth
Poaceae
= *Sporobolus indicus* (L.) R.Br. var. *pyramidalis* (Beauv.) Veldkamp
♦ West Indian dropseed, smutgrass
♦ Weed, Naturalised
♦ 80, 86, 98, 203
♦ G. Origin: tropical America.

Sporobolus mitchelli (Trin.) Hubb. ex Blake
Poaceae
Sporobolus benthamii F.M.Bailey
♦ Introduced
♦ 228
♦ G, arid.

Sporobolus natalensis (Steud.) Dur. & Schinz
Poaceae
♦ giant rat's tail grass
♦ Weed, Quarantine Weed, Noxious Weed, Naturalised, Native Weed
♦ 76, 86, 93, 121, 191
♦ pG. Origin: central and southern Africa.

Sporobolus neglectus Nash
Poaceae
♦ annual dropseed, puffsheath dropseed
♦ Weed
♦ 87, 88, 91, 161, 210, 218
♦ aG, herbal.

Sporobolus piliferus (Trin.) Kunth
Poaceae
♦ Barundi dropseed
♦ Naturalised
♦ 101
♦ G.

Sporobolus poiretii (Roem. & Schult.) Hitchc.
Poaceae
= *Sporobolus indicus* (L.) R.Br.
♦ smutgrass
♦ Weed
♦ 87, 88, 218
♦ G, herbal.

Sporobolus purpurascens (Sw.) Ham.
Poaceae
♦ purple dropseed
♦ Introduced
♦ 38
♦ G.

Sporobolus pyramidalis Beauv.
Poaceae
= *Sporobolus indicus* (L.) R.Br. var. *pyramidalis* (Beauv.) Veldkamp
♦ West Indies smutgrass, Parramatta grass, cat's tail grass
♦ Weed, Quarantine Weed, Naturalised, Native Weed, Introduced
♦ 76, 87, 88, 91, 98, 121, 132, 158, 203, 205, 220, 228
♦ pG, arid. Origin: Africa.

Sporobolus pyramidalis Beauv. var. pyramidalis
Poaceae
♦ giant rat's tail, giant rat's tail grass
♦ Weed, Noxious Weed, Naturalised
♦ 86, 93, 191
♦ G.

Sporobolus pyramidatus (Lam.) Hitchc.
Poaceae
= *Sporobolus coromandelianus* (Retz.) Kunth
♦ whorled dropseed
♦ Weed
♦ 199, 242
♦ G, arid, herbal.

Sporobolus rigens (Trin.) Desv.
Poaceae
Epicampes arundinaceus (Griseb.) Hack., *Sporobolus arundinaceus* (Griseb.) Kunth, *Vilfa rigens* Trin.
♦ unco
♦ Weed, Introduced
♦ 228, 237, 295
♦ G, arid.

Sporobolus robustus Kunth
Poaceae
♦ Weed
♦ 87, 88
♦ G.

Sporobolus scabridus S.T.Blake
Poaceae
- Weed, Naturalised
- 86, 93, 191, 205
- pG.

Sporobolus spicatus (Vahl) Kunth
Poaceae
- Weed
- 221
- G, arid, cultivated.

Sporobolus tenuissimus (Schrank) Kuntze
Poaceae
- tropical dropseed
- Weed
- 87, 88, 179
- G.

Sporobolus tremulus (Willd.) Kunth
Poaceae
- Weed, Quarantine Weed
- 76, 87, 88, 203, 220
- G.

Sporobolus vaginiflorus (Torr.) Wood
Poaceae
- poverty dropseed, poverty grass
- Weed, Naturalised, Native Weed
- 87, 88, 91, 161, 174, 218, 287
- aG, herbal. Origin: North America.

Sporobolus virginicus (L.) Kunth
Poaceae
Agrostis virginicus L., *Vilfa virginica* (L.) Beauv.
- saltwater smutgrass, sandcouch, seashore dropseed, seaside rush
- Weed, Naturalised
- 87, 88, 91, 241, 300
- pG, arid, cultivated, herbal.

Stachys L. spp.
Lamiaceae
- wormwood, stachys
- Weed, Naturalised
- 39, 198, 272
- herbal, toxic.

Stachys aegyptiaca Person
Lamiaceae
- Weed
- 221

Stachys affinis Bunge
Lamiaceae
Stachys sieboldii Miq. (see), *Stachys tuberifera* Naud.
- artichoke betony, Chinese artichoke
- Naturalised
- 101
- pH, cultivated, herbal.

Stachys agraria Schlecht. & Cham.
Lamiaceae
= *Stachys crenata* Raf. (NoR)
- Weed
- 87, 88
- herbal.

Stachys annua (L.) L.
Lamiaceae
Betonica annua L., *Stachys annuus* L., *Stachys micrantha* Koch, *Stachys neglecta* Klok. ex Kossko, *Stachys nervosa* Gater.
- hedgenettle betony, annual

woundwort, annual hedgenettle, keltapähkämö, annual yellow woundwort
- Weed, Quarantine Weed, Naturalised, Casual Alien
- 40, 42, 70, 76, 87, 88, 94, 101, 203, 218, 220, 243, 253, 272, 280
- a/pH, cultivated, herbal.

Stachys arvensis (L.) L.
Lamiaceae
Cardiaca arvensis Lam., *Glechoma arvensis* L., *Glechoma marrubiastrum* Vill., *Sideritis cordi* Thal., *Trixago arvensis* Hoffman. & Link., *Trixago colorata* Presl, *Trixago cordifolia* Moench
- fieldnettle betony, stagger weed, tolanga, rikkapähkämö, annual hedgenettle, field woundwort
- Weed, Naturalised, Introduced, Environmental Weed, Casual Alien
- 7, 15, 34, 38, 39, 42, 55, 70, 86, 87, 88, 94, 98, 101, 134, 161, 165, 176, 198, 203, 218, 236, 237, 250, 253, 255, 269, 272, 280, 286, 287, 295, 300
- aH, arid, herbal, toxic. Origin: Europe.

Stachys baicalensis Fisch. ex Benth.
Lamiaceae
- Baikal betony
- Weed
- 297
- pH, promoted, herbal.

Stachys bullata Benth.
Lamiaceae
- hedgenettle, California hedgenettle
- Weed
- 161
- pH, cultivated, herbal. Origin: south-western North America.

Stachys byzantina K.Koch
Lamiaceae
Eriostomum lanatum Hoffman., *Stachys lanata* Jacq. (see), *Stachys olympica* Poir., *Stachys sublanata* Fleischm.
- woolly lamb's ear, woolly hedgenettle, woolly betony, nukkapähkämö, lamb's ear
- Weed, Naturalised, Cultivation Escape, Casual Alien
- 40, 42, 86, 98, 101, 203, 252, 280
- cultivated, herbal. Origin: Eurasia.

Stachys chinensis Bunge ex Benth.
Lamiaceae
- Chinese betony
- Weed
- 297

Stachys durandiana Coss.
Lamiaceae
- Weed
- 87, 88

Stachys elliptica H.B.K.
Lamiaceae
- Weed
- 87, 88

Stachys floridana Shuttlw. ex Benth.
Lamiaceae
- Florida hedgenettle, hedgenettle,

Florida betony, rattlesnake weed
- Weed, Noxious Weed
- 67, 80, 84, 87, 88, 102, 161, 179, 218, 229, 249
- herbal.

Stachys germanica L.
Lamiaceae
Eriostomum polystachyum Presl, *Stachys argentea* Tausch., *Stachys biennis* Roth., *Stachys pannonica* Lang., *Stachys polystachia* Ten., *Stachys tomentosa* Gater.
- German hedgenettle, downy woundwort, true woundwort
- Weed, Naturalised, Casual Alien
- 86, 101, 272, 280
- pH, cultivated, herbal. Origin: Europe.

Stachys grandidentata Lindl.
Lamiaceae
- Weed
- 87, 88
- arid.

Stachys grandiflora (Willd.) Benth.
Lamiaceae
Betonica grandiflora Stephan ex Willd.
- big sage
- Naturalised
- 101
- cultivated, herbal.

Stachys hyssopoides Burch. ex Benth.
Lamiaceae
- Native Weed
- 121
- pH. Origin: southern Africa.

Stachys keerlii Benth.
Lamiaceae
- Weed
- 199

Stachys lanata Jacq.
Lamiaceae
= *Stachys byzantina* K.Koch ex Scheele
- lamb's tongue, lamb's tail, lamb's ear, woolly betong
- Weed, Introduced
- 38, 87, 88
- cultivated, herbal.

Stachys longispicata Boiss.
Lamiaceae
- longspike hedgenettle, hedgenettle
- Naturalised
- 101

Stachys milanii Pet.
Lamiaceae
- Weed
- 272

Stachys nivea Labill.
Lamiaceae
- Weed
- 87, 88
- cultivated.

Stachys ocymastrum (L.) Briq.
Lamiaceae
- hedgenettle
- Weed, Naturalised
- 70, 87, 88, 101
- cultivated.

Stachys officinalis (L.) Trev.
Lamiaceae
Stachys betonica Benth., *Betonica officinalis* L. (see)
♦ rohtopähkämö, wood betony, betony, lus bheathag, epiaire vulgaire, bishop wort, betoniye, betonica, betoine, pourpre, bathenien, common hedgenettle
♦ Weed, Naturalised, Cultivation Escape
♦ 42, 101, 272
♦ pH, cultivated, herbal. Origin: Europe.

Stachys palustris L.
Lamiaceae
Stachys maeotica Postr., *Stachys paluster* L., *Stachys wolgensis* Wilensky
♦ marsh betony, marsh woundwort, marsh hedgenettle, peltopähkämö, hedgenettle
♦ Weed, Noxious Weed, Naturalised
♦ 15, 39, 70, 86, 87, 88, 94, 159, 176, 218, 243, 253, 272, 280, 287, 299
♦ H, aqua, cultivated, herbal, toxic. Origin: Europe.

Stachys parviflora Benth.
Lamiaceae
♦ Weed
♦ 87, 88

Stachys petiolosa Briq.
Lamiaceae
♦ Weed
♦ 87, 88

Stachys pubescens Ten.
Lamiaceae
♦ Weed
♦ 87, 88

Stachys recta L.
Lamiaceae
Betonica hirta Gouan, *Sideritis hirsuta* Gouan
♦ perennial yellow woundwort, yellow woundwort, seaport hedgenettle
♦ Weed
♦ 272
♦ cultivated, herbal.

Stachys sieboldii Miq.
Lamiaceae
= *Stachys affinis* Bunge
♦ artichoke betony
♦ Weed
♦ 87, 88, 218, 297
♦ cultivated, herbal.

Stachys sylvatica L.
Lamiaceae
Stachys canariensis Jacq., *Stachys canescens* Muss.Puschk. ex Spreng.
♦ wood woundwort, woundwort, whitespot, hedge woundwort, lehtopähkämö
♦ Weed, Naturalised
♦ 15, 23, 70, 88, 101, 165, 272, 280
♦ pH, cultivated, herbal. Origin: Eurasia.

Stachytarpheta × adulterina Urb. & E.Ekman
Verbenaceae
♦ Naturalised, Casual Alien
♦ 86, 280
♦ Origin: Australia (hybrid origin).

Stachytarpheta angustifolia (Mill.) Vahl
Verbenaceae
Stachytarpheta elatior Schrad. ex Schult. & Roem. (see), *Stachytarpheta surinamensis* Miq.
♦ Weed, Quarantine Weed
♦ 76, 87, 88, 203, 220
♦ herbal.

Stachytarpheta australis Mold.
Verbenaceae
= *Stachytarpheta cayennensis* (Rich.) Vahl
♦ snakeweed
♦ Weed, Naturalised, Environmental Weed
♦ 86, 87, 88, 93, 191
♦ Origin: South America.

Stachytarpheta cayennensis (Rich.) Vahl
Verbenaceae
Stachytarpheta australis Mold. (see), *Stachytarpheta dichotoma* (Ruiz & Pav.) Vahl (see)
♦ cayenne snakeweed, blue snakeweed, rough leaved false vervain, blue rat's tail, dark blue snakeweed, snakeweed, verbena negra
♦ Weed, Noxious Weed, Naturalised
♦ 6, 7, 86, 87, 88, 93, 98, 147, 153, 191, 191, 203, 255, 257
♦ pH/S, herbal. Origin: tropical America.

Stachytarpheta dichotoma (Ruiz & Pav.) Vahl
Verbenaceae
= *Stachytarpheta cayennensis* (Rich.) Vahl
♦ branched porterweed
♦ Weed, Naturalised
♦ 98, 101, 203, 286, 287
♦ herbal.

Stachytarpheta elatior Schrad. ex Schult. & Roem.
Verbenaceae
= *Stachytarpheta angustifolia* (Mill.) Vahl
♦ erva de grilo
♦ Weed
♦ 255
♦ aH. Origin: tropical America.

Stachytarpheta × gracilis Danser
Verbenaceae
♦ Naturalised
♦ 101

Stachytarpheta indica (L.) Vahl
Verbenaceae
= *Stachytarpheta jamaicensis* (L.) Vahl
♦ nettleleaf vervain, jimica vervain, phan nguu khieo
♦ Weed, Naturalised
♦ 13, 23, 39, 87, 88, 121, 170, 191, 218, 239, 262, 287

♦ pH, herbal, toxic. Origin: tropical America.

Stachytarpheta × intercedens Danser
Verbenaceae
= *Stachytarpheta jamaicensis* (L.) Vahl × *Stachytarpheta urticifolia* Sims
♦ Naturalised
♦ 101

Stachytarpheta jamaicensis (L.) Vahl
Verbenaceae
Stachytarpheta indica (L.) Vahl (see)
♦ light blue snakeweed, bastard vervain, Brazil tea, Jamaica vervain, blue porterweed, blue snakeweed, mautofu tala, joee
♦ Weed, Quarantine Weed, Noxious Weed, Naturalised
♦ 6, 13, 14, 76, 86, 87, 88, 93, 98, 147, 170, 191, 203, 209, 217, 276, 286, 287, 297
♦ pH, cultivated, herbal. Origin: tropical America.

Stachytarpheta mutabilis (Jacq.) Vahl
Verbenaceae
♦ pink snakeweed, changeable velvetberry
♦ Weed, Quarantine Weed, Noxious Weed, Naturalised
♦ 76, 86, 87, 88, 98, 101, 147, 191, 203
♦ herbal. Origin: tropical and subtropical America.

Stachytarpheta mutabilis (Jacq.) Vahl × jamaicensis (L.) Vahl
Verbenaceae
♦ Naturalised
♦ 98

Stachytarpheta × trimeni Rech.
Verbenaceae
= *Stachytarpheta mutabilis* (Jacq.) Vahl × *Stachytarpheta urticifolia* (Salisb.) Sims
♦ Naturalised
♦ 86, 98, 101

Stachytarpheta urticaefolia (Salisb.) Sims
Verbenaceae
= *Stachytarpheta urticifolia* (Salisb.) Sims
♦ Weed, Naturalised
♦ 286, 287

Stachytarpheta urticifolia (Salisb.) Sims
Verbenaceae
Stachytarpheta urticaefolia (Salisb.) Sims (see)
♦ dark blue snakeweed, nettleleaf vervain, blue rat's tail, false verbena, louch beluu, mautofutala, mautofu tala, mautofu Samoa, iku'i kuma, hiku 'i kuma, mautofu vao, matofu fualanumanoa, te uti, turulakaka, tumbutumbu, serakawa, lavenia
♦ Weed, Quarantine Weed, Noxious Weed, Naturalised, Sleeper Weed
♦ 3, 6, 76, 86, 87, 88, 98, 101, 107, 147, 179, 203, 261, 276
♦ pH, herbal. Origin: tropical Asia, Pacific region.

Stanleya pinnata (Pursh) Britt.
Brassicaceae
Cleome pinnata Pursh, *Stanleya
pinnatifida* Nutt.
♦ desert prince's plume, prince's
plume, desert plume
♦ Weed
♦ 39, 87, 88, 218
♦ S, arid, cultivated, herbal, toxic.

Stapelia gigantea N.E.Br.
Asclepiadaceae/Apocynaceae
♦ Zulu giant, carrion flower, starfish
flower, giant toad plant, giant stapelia
♦ Naturalised, Cultivation Escape
♦ 101, 233
♦ cultivated.

Stapelia grandiflora Masson
Asclepiadaceae/Apocynaceae
♦ Weed, Naturalised
♦ 98, 203
♦ cultivated.

Stapelia variegata L.
Asclepiadaceae/Apocynaceae
= *Orbea variegata* (L.) Haw.
♦ Weed, Naturalised
♦ 98, 203
♦ pH, arid, cultivated. Origin: South
Africa.

Stauntonia hexaphylla (Thunb.) Decne.
Lardizabalaceae
♦ Japanese staunton vine
♦ Weed
♦ 286
♦ pC, cultivated, herbal. Origin:
Burma, Japan, Korea.

Staurogyne spathulata (Bl.) Koord.
Acanthaceae/Nelsoniaceae
♦ Weed
♦ 13

Stellaria L. spp.
Caryophyllaceae
♦ starwort
♦ Weed
♦ 272
♦ herbal.

Stellaria alsine Grimm
Caryophyllaceae
Stellaria brevifolia Gilib., *Stellaria
lateriflora* Krock, *Stellaria uliginosa*
Murr. (see)
♦ bog stitchwort, bog chickweed
♦ Weed, Naturalised
♦ 15, 23, 88, 204, 241, 280, 287, 300
♦ b/pH, promoted, herbal.

**Stellaria alsine Grimm var. undulata
Ohwi**
Caryophyllaceae
♦ slender sandwort
♦ Weed
♦ 235, 263, 286
♦ aH.

Stellaria aquatica (L.) Scop.
Caryophyllaceae
= *Myosoton aquaticum* (L.) Moench
♦ water chickweed, water mouse ear
chickweed
♦ Weed, Quarantine Weed

♦ 44, 76, 87, 88, 203, 204, 220, 235, 238,
243, 263, 273, 274
♦ pH, herbal.

Stellaria chinensis Regel
Caryophyllaceae
♦ Chinese starwort
♦ Weed
♦ 297

Stellaria cuspidata Willd. ex Schlecht.
Caryophyllaceae
♦ Mexican starwort
♦ Weed
♦ 87, 88
♦ arid, herbal.

Stellaria dichotoma L.
Caryophyllaceae
♦ chickweed, cornuculated chickweed
♦ Weed
♦ 88, 114, 243
♦ pH, promoted.

Stellaria graminea L.
Caryophyllaceae
Larbrea graminea (L.) Fuss, *Stellaria
grandiflora* Gilib., *Stellaria paniculata*
Pall., *Stellaria scapigera* Willd.
♦ little starwort, grass leaved
stichwort, grassy starwort, lesser
stitchwort, heinätähtimö
♦ Weed, Naturalised, Environmental
Weed
♦ 15, 23, 34, 44, 70, 80, 86, 87, 88, 94, 98,
101, 151, 161, 165, 176, 195, 198, 203,
211, 218, 243, 272, 280, 287
♦ pH, cultivated, herbal. Origin:
Eurasia.

Stellaria hebecalyx Fenzl
Caryophyllaceae
♦ idäntähtimö
♦ Casual Alien
♦ 42

Stellaria holostea L.
Caryophyllaceae
Alsine holostea (L.) Britt., *Cerastium
holosteum* (L.) Crantz, *Stellaria connata*
Dulac, *Stellaria scabra* Stokes
♦ greater stitchwort, addersmeat,
kevättähtimö
♦ Weed, Quarantine Weed,
Naturalised, Casual Alien
♦ 23, 76, 88, 101, 243, 272, 280, 287
♦ cultivated, herbal.

Stellaria kotschyana Fenzl ex Boiss.
Caryophyllaceae
♦ Weed
♦ 248

Stellaria lanceolata Poir.
Caryophyllaceae
♦ Naturalised
♦ 241

Stellaria media (L.) Cirillo
Caryophyllaceae
Alsine avicularum Lam., *Alsine bipartita*
Gilib., *Alsine media* L., *Alsinella
wallichiana* Benth, *Cerastium medium*
(L.) Crantz, *Holosteum alsine* Sw.
♦ chickweed, common chickweed,
starwort, starweed, winter weed,

satin flower, mouse eared chickweed,
pihatähtimö, esparguta, capiquí,
bindweed, tongue grass, white bird's
eye
♦ Weed, Noxious Weed, Naturalised,
Introduced, Garden Escape,
Environmental Weed
♦ 7, 9, 15, 23, 24, 34, 36, 44, 51, 52, 53,
55, 68, 70, 72, 80, 86, 87, 88, 93, 94, 97,
98, 101, 114, 115, 118, 121, 134, 136, 151,
158, 161, 162, 162, 165, 167, 174, 176,
179, 180, 186, 198, 203, 204, 205, 207,
210, 211, 212, 218, 221, 236, 237, 241,
243, 245, 249, 250, 253, 255, 257, 263,
269, 271, 272, 275, 280, 286, 287, 293,
295, 297, 299, 300
♦ aH, aqua, cultivated, herbal, toxic.
Origin: Eurasia.

Stellaria media (L.) Vill. ssp. *media*
Caryophyllaceae
♦ common chickweed
♦ Naturalised
♦ 101, 280
♦ cultivated.

**Stellaria media (L.) Vill. ssp. *neglecta*
(Weihe) Murb.**
Caryophyllaceae
= *Stellaria neglecta* Weihe ex Bluff &
Fingerh.
♦ common chickweed
♦ Naturalised
♦ 101

**Stellaria media (L.) Vill. ssp. *pallida*
(Dumort.) Asch. & Graebn.**
Caryophyllaceae
= *Stellaria pallida* (Dum.) Piré
♦ common chickweed
♦ Naturalised
♦ 101

**Stellaria neglecta Weihe ex Bluff &
Fingerh.**
Caryophyllaceae
Alsine neglecta (Weihe) Á. & D.Löve,
Stellaria diversiflora Maxim. var.
gymnandra Franch., *Stellaria media* (L.)
Cirillo ssp. *neglecta* (Weihe) Murb.
(see), *Stellaria media* var. *decandra* Fenzl,
Stellaria media var. *procera* Klett &
Richt., *Stellaria octandra* Pobed.
♦ greater chickweed, pyökkitähtimö
♦ Weed, Casual Alien
♦ 42, 70, 87, 88, 286
♦ aH, promoted, herbal.

Stellaria nemorum L.
Caryophyllaceae
Alsine nemorum (L.) Schreb., *Stellaria
montana* Pierrat
♦ wood stitchwort, lehtotähtimö
♦ Weed
♦ 272
♦ cultivated, herbal.

Stellaria ovata Willd. ex Schlecht.
Caryophyllaceae
♦ Weed
♦ 157
♦ pH, arid.

Stellaria pallida (Dum.) Piré
Caryophyllaceae
Alsine pallida Dumort., *Stellaria abortiva*
Naudin, *Stellaria media* (L.) Vill. ssp.
pallida (Dumort.) Asch. & Graebn. (see)
♦ lesser chickweed, chickweed
♦ Weed, Naturalised, Environmental
Weed
♦ 70, 86, 87, 88, 93, 98, 176, 185, 198,
203, 205, 221, 243, 272, 286, 287, 300
♦ aH. Origin: Europe.

Stellaria palustris (Murr.) Retz.
Caryophyllaceae
Larbrea palustris (Retz.) Fuss, *Stellaria
glauca* Wither., *Stellaria persica* Boiss.,
Stellaria stricta Koch
♦ marsh stitchwort, meadow starwort,
European chickweed, luhtatähtimö
♦ Weed, Naturalised
♦ 86, 101, 272
♦ Origin: Australia.

Stellaria parva Pedersen
Caryophyllaceae
♦ pygmy starwort
♦ Naturalised
♦ 101

Stellaria radians L.
Caryophyllaceae
♦ ciliatepetal starwort,
ezoooyamahakobe
♦ Weed
♦ 297
♦ pH, promoted. Origin: Japan to
Siberia.

Stellaria uliginosa Murr.
Caryophyllaceae
= *Stellaria alsine* Grimm
♦ bog stitchwort, lähdetähtimö
♦ Weed, Quarantine Weed
♦ 76, 87, 88, 203, 220

Stellaria verticillata (Mill.) Bold.
Caryophyllaceae
Stellaria parvifolia Ait.
♦ Weed
♦ 87, 88

Stellaria viscosa Roxb.
Caryophyllaceae
♦ Weed
♦ 87, 88

Stellera chamaejasme L.
Thymelaeaceae
♦ Chinese stellera
♦ Weed
♦ 297
♦ pH, promoted, herbal, toxic. Origin:
Himalayas to China.

Stemodia durantifolia (L.) Sw.
Scrophulariaceae
♦ white woolly twintip
♦ Weed
♦ 14, 179
♦ pH, herbal.

Stemodia trifoliata (Link.) Reichb.
Scrophulariaceae
♦ mentinha
♦ Weed
♦ 255

♦ aH. Origin: Americas.

Stemodia verticillata (Mill.) Hassl.
Scrophulariaceae
♦ whorled twintip
♦ Weed, Naturalised, Introduced
♦ 230, 257, 261
♦ H.

Stemona curtisii Hook.f.
Stemonaceae
♦ Weed
♦ 12

Stenactis annuus Cass.
Asteraceae
♦ Weed, Naturalised
♦ 286, 287

Stenactis annuus Cass. fo. *discoideus*
Asteraceae
♦ Naturalised
♦ 287

Stenactis strigosus DC.
Asteraceae
♦ Weed, Naturalised
♦ 286, 287

Stenocarpus sinuatus (Loudon) Endl.
Proteaceae
♦ firewheel tree
♦ Casual Alien
♦ 280
♦ cultivated.

Stenocereus (Berger) Riccob. spp.
Cactaceae
Hertrichocereus Backeb. spp. (see),
Machaeocereus Britt. & Rose spp. (see),
Marshallocereus Backeb. spp. (see)
♦ stenocereus
♦ Weed, Quarantine Weed
♦ 76, 88, 203, 220

Stenocereus griseus (Haw.) Buxb.
Cactaceae
Lemaireocereus griseus (Haw.) Britton
& Rose, *Ritterocereus deficiens* (Otto &
A.Dietr.) Backeb., *Stenocereus deficiens*
(Otto & Dietr.) F.Buxb.
♦ Introduced
♦ 228
♦ arid, cultivated.

Stenochlaena palustris (Burn.) Bedd.
Blechnaceae
♦ Weed
♦ 12, 87, 88
♦ cultivated, herbal.

Stenochlaena tenuifolia (Desv.) Moore
Blechnaceae
♦ bracken, climbing fern, giant vine
fern
♦ Weed, Naturalised, Native Weed
♦ 101, 121
♦ pC, cultivated. Origin: southern
Africa.

Stenomesson pauciflorum Lindl. ex Hook.
var. *pauciflorum*
Liliaceae/Amaryllidaceae
♦ Introduced
♦ 38

Stenosolenium saxatile (Pall.) Turcz.
Boraginaceae
♦ Weed

♦ 275, 297
♦ pH.

Stenotaphrum dimidiatum (L.) Brongn.
Poaceae
Panicum dimidiatum L., *Stenotaphrum
madagascariense* Kunth
♦ Weed, Introduced
♦ 87, 88, 228
♦ G, arid, herbal. Origin: Madagascar.

Stenotaphrum secundatum (Walter)
Kuntze
Poaceae
Diastemanthe platystachys Steud.,
Ischaemum secundatum Walter,
Rottboellia dimidiata Thunb., *Rottboellia
stolonifera* Poir., *Rottboellia tripsacoides*
Lam., *Stenotaphrum americanum*
Schrank, *Stenotaphrum compressum*
Druce, *Stenotaphrum dimidiatum* (L.)
Brongn. var. *americanum* (Schrank)
Hack. ex Stuck., *Stenotaphrum
dimidiatum* var. *secundatum* (Walter)
Domin, *Stenotaphrum glabrum* Trin.,
Stenotaphrum glabrum var. *americanum*
(Schrank) Döll, *Stenotaphrum
sarmentosum* Nees, *Stenotaphrum
secundatum* var. *variegatum* Hitchc.,
Stenotaphrum swartzianum Nees
♦ St. Augustine grass, buffalograss,
crabgrass, pimento grass, cape kweek,
cape quickgrass, carpetgrass, coarse
couchgrass, coarse quickgrass, coastal
buffalograss, coast kweek, couchgrass,
grove kweek, mission grass,
quickgrass, ramsammy grass, seaside
quickgrass
♦ Weed, Sleeper Weed, Naturalised,
Native Weed, Environmental Weed,
Cultivation Escape
♦ 7, 9, 15, 34, 72, 80, 86, 87, 88, 90, 98,
121, 134, 158, 176, 198, 203, 218, 225,
241, 246, 252, 280, 287, 289, 295, 296,
300
♦ pG, cultivated, herbal. Origin:
obscure, possibly Americas.

Stephanandra incisa (Thunb.) Zabel
Rosaceae
Stephanandra flexuosa Siebold & Zucc.,
Spiraea incisa Thunb.
♦ cutleaf stephanandra
♦ Naturalised
♦ 101
♦ S, cultivated.

Stephania cephalantha Hayata
Menispermaceae
♦ Weed
♦ 87, 88
♦ herbal.

Stephania elegans Hook.f. & Thoms.
Menispermaceae
♦ Weed
♦ 87, 88

Stephania japonica (Murr.) Miers
Menispermaceae
Stephania hernandifolia Walp.
♦ Weed
♦ 87, 88
♦ cultivated, herbal. Origin: Australia.

Stephanodiscus binderanus (Kütz) W.Krieg.
Bacillariophyceae
- Weed
- 197
- diatom.

Stephanodiscus subtilis Goor
Bacillariophyceae
- Weed
- 197
- diatom.

Stephanomeria exigua Nutt.
Asteraceae
- small wirelettuce, small stephanomeria
- Weed
- 161
- aH, herbal.

Stephanomeria exigua Nutt. ssp. coronaria (E.Greene) Gottlieb
Asteraceae
- whiteplume wirelettuce
- Weed
- 180, 243
- aH.

Stephanomeria tenuifolia (Raf.) M.H.Hall
Asteraceae
= *Stephanomeria minor* (Hook.) Nutt. var. *minor* (NoR)
- narrowleaf wirelettuce, slender wirelettuce
- Weed
- 87, 88, 161
- pH, herbal.

Stephanomeria virgata Benth.
Asteraceae
- tall stephanomeria, rod wirelettuce
- Weed
- 34
- aH, herbal.

Stephanophysum longifolium Pohl
Acanthaceae
= *Ruellia brevifolia* (Pohl) C.Ezcurra
- Weed, Sleeper Weed, Naturalised, Environmental Weed
- 3, 86, 87, 88, 155, 191
- cultivated. Origin: Brazil.

Steptorhamphus tuberosus (Jacq.) Grossh.
Asteraceae
Lactuca cretica Desf.
- Weed
- 272

Sterculia africana (Lour.) Fiori
Sterculiaceae
- mopopaja tree
- Weed
- 221
- cultivated.

Sterculia apetala (Jacq.) Karst.
Sterculiaceae
- Panama tree
- Naturalised, Cultivation Escape
- 39, 101, 261
- cultivated, herbal, toxic.

Sterculia foetida L.
Sterculiaceae
- Indian almond, hazel sterculia, Java olive, anacagüita
- Weed, Introduced, Garden Escape
- 161, 228, 261
- arid, cultivated, herbal, toxic. Origin: tropical Asia, Australia.

Sterculia ponapensis Kaneh.
Sterculiaceae
- Introduced
- 230
- T.

Sternbergia lutea (L.) Ker Gawl. ex Spreng.
Liliaceae/Amaryllidaceae
- winter daffodil, sternbergis
- Weed, Naturalised
- 86, 98, 101, 203
- cultivated, herbal. Origin: Mediterranean.

Stetsonia Britt. & Rose spp.
Cactaceae
- Weed, Quarantine Weed
- 76, 88, 203, 220

Stevia eupatoria (Spreng.) Willd.
Asteraceae
- stevia, Kempton's weed
- Weed, Quarantine Weed, Noxious Weed, Naturalised
- 76, 86, 88, 98, 147, 198, 203, 220, 269
- pH, herbal. Origin: Mexico.

Stiburus alopecuroides (Hack.) Stapf
Poaceae
- pongwa grass
- Native Weed
- 121
- pG. Origin: southern Africa.

Stictocardia tiliifolia (Desr.) Hallier f.
Convolvulaceae
- spotted heart
- Weed
- 6, 87, 88, 179, 191, 261
- herbal. Origin: Asia, Australia.

Stigmaphyllon blanchetii C.F.Andrews
Malpighiaceae
Stygmaphyllon blanchetii C.F.Andrews
- rabo de rato
- Weed
- 255
- pH. Origin: Brazil.

Stigmaphyllon sagraeanum A.Juss.
Malpighiaceae
- Weed
- 14
- cultivated.

Stigmaphyllon tomentosum A.Juss.
Malpighiaceae
Stigmaphyllon affine A.Juss. var. *paulinum* Nied., *Stigmaphyllon eriocardium* Nied., *Stigmaphyllon psilocardium* Nied.
- Weed
- 87, 88

Stillingia sylvatica Garden ex L.
Euphorbiaceae
- queen's delight
- Weed
- 39, 87, 88, 218
- herbal, toxic.

Stillingia treculeana (Müll.Arg.) Johnst.
Euphorbiaceae
- queen's delight
- Weed
- 39, 161
- herbal, toxic.

Stipa L. spp.
Poaceae
- needlegrass, feathergrass
- Weed, Environmental Weed
- 39, 237, 246, 272
- G, herbal, toxic.

Stipa aristiglumis F.Muell.
Poaceae
Stipa fusiformis Hughes
- Introduced
- 228
- G, arid, cultivated. Origin: Australia.

Stipa avenacea L.
Poaceae
= *Piptochaetium avenaceum* (L.) Parodi (NoR)
- blackseed needlegrass
- Weed
- 87, 88, 218
- G, herbal.

Stipa bormanii Hauman
Poaceae
= *Anatherostipa bomanii* (Hauman) Peñail. (NoR)
- vizcachera hembra
- Weed
- 295
- G.

Stipa brachychaeta Godr.
Poaceae
= *Achnatherum brachychaetum* (Godr.) Barkworth
- espartillo
- Weed, Quarantine Weed, Noxious Weed, Naturalised
- 86, 87, 88, 98, 147, 191, 203, 220, 236, 237, 295
- G.

Stipa bromoides (L.) Dörfl.
Poaceae
Achnatherum bromoides (L.) P.Beauv., *Agrostis bromoides* L.
- slender feathergrass
- Weed
- 272
- G.

Stipa bungeana Trin.
Poaceae
- Weed
- 275
- pG, arid.

Stipa calamagrostis (L.) Wahlenb.
Poaceae
= *Achnatherum calamagrostis* (L.) P.Beauv.
- Quarantine Weed
- 220
- G, cultivated.

Stipa capensis **Thunb.**
Poaceae
Stipa tortilis Desf. (see)
♦ Weed
♦ 221
♦ G, arid, herbal.

Stipa capillata **L.**
Poaceae
Stipa juncea Jacq.
♦ hairy feathergrass, lace veil, needlegrass
♦ Weed
♦ 39, 272
♦ G, arid, cultivated, herbal, toxic.

Stipa caudata **Trin.**
Poaceae
= *Achnatherum caudatum* (Trin.) Jacobs & Everett
♦ espartillo
♦ Weed, Noxious Weed, Naturalised
♦ 86, 88, 98, 147, 203, 295
♦ G. Origin: South America.

Stipa cernua **Steb. & Á.Löve**
Poaceae
= *Nassella cernua* (Steb. & Á.Löve) Barkworth
♦ California needlegrass, nodding stipa, nodding needlegrass
♦ Weed, Quarantine Weed
♦ 87, 88, 161, 180, 203, 220, 243
♦ G, cultivated, herbal.

Stipa charruana **Arechav.**
Poaceae
♦ stipa charrúa
♦ Weed
♦ 237, 295
♦ G.

Stipa cirrosa **E.Fourn. ex Hemsl.**
Poaceae
= *Nassella tenuissima* (Trin.) Barkworth
♦ Quarantine Weed
♦ 220
♦ G.

Stipa clandestina **Hack.**
Poaceae
= *Achnatherum clandestinum* (Hack.) Barkworth
♦ Weed
♦ 121
♦ pG. Origin: South America.

Stipa comata **Trin. & Rupr.**
Poaceae
= *Hesperostipa comata* (Trin. & Rupr.) Barkworth ssp. *comata* (NoR)
♦ needle and thread, needlegrass
♦ Weed
♦ 87, 88, 161, 218
♦ G, arid, cultivated, herbal.

Stipa eminens **Cav.**
Poaceae
= *Achnatherum eminens* (Cav.) Barkworth (NoR)
♦ Weed
♦ 199
♦ G, herbal.

Stipa geniculata **Phil.**
Poaceae

= *Nassella tenuissima* (Trin.) Barkworth
♦ Quarantine Weed
♦ 220
♦ G.

Stipa gigantea **Link**
Poaceae
Lasiagrostis gigantea (Link) Trin. & Rupr.
♦ giant feathergrass, golden oats
♦ Weed, Environmental Weed
♦ 70, 290
♦ pG, cultivated.

Stipa hyalina **Nees**
Poaceae
♦ flechilla mansa
♦ Weed, Naturalised
♦ 87, 88, 98, 203, 237, 295
♦ G.

Stipa hypogona **Hack.**
Poaceae
♦ Introduced
♦ 228
♦ G, arid.

Stipa ibari **Phil.**
Poaceae
♦ Introduced
♦ 228
♦ G, arid.

Stipa ichu **(Ruiz & Pav.) Kunth**
Poaceae
♦ Weed
♦ 157
♦ pG, arid.

Stipa lagascae **Roem. & Schult.**
Poaceae
Stipa fontanesii Parl., *Stipa gigantea* Lag., *Stipa holosericea* Trin. & Rupr.
♦ aadame, addam, feathergrass, gawther, ghawther
♦ Weed
♦ 221
♦ G, arid.

Stipa leptostachya **Griseb.**
Poaceae
Stipa capilliseta Hitch.
♦ vizcachera hembra
♦ Weed
♦ 295
♦ G.

Stipa leucotricha **Trin. & Rupr.**
Poaceae
= *Nassella leucotricha* (Trin. & Rupr.) Pohl
♦ Weed, Naturalised
♦ 98, 203
♦ G, cultivated, herbal.

Stipa megapotamia **Spreng.**
Poaceae
♦ Weed, Naturalised
♦ 98, 203
♦ G.

Stipa melanosperma **Presl**
Poaceae
♦ Weed
♦ 295
♦ G.

Stipa mendocina **Phil.**
Poaceae
= *Nassella tenuissima* (Trin.) Barkworth
♦ Quarantine Weed
♦ 220
♦ G.

Stipa neesiana **Trin. & Rupr.**
Poaceae
= *Nassella neesiana* (Trin. & Rupr.) Barkworth
♦ Chilean needlegrass
♦ Weed, Noxious Weed, Naturalised, Environmental Weed
♦ 68, 86, 87, 88, 98, 121, 203, 225, 237, 295
♦ pG. Origin: South America.

Stipa oreophila **Speg.**
Poaceae
= *Nassella tenuissima* (Trin.) Barkworth
♦ Quarantine Weed
♦ 220
♦ G.

Stipa papposa **Nees**
Poaceae
= *Jarava plumosa* (Spreng.) S.L.W.Jacobs & J.Everett
♦ eibe
♦ Weed
♦ 121, 295
♦ pG. Origin: South America.

Stipa parviflora **Desf.**
Poaceae
♦ Weed
♦ 221
♦ G, arid.

Stipa pennata **L.**
Poaceae
♦ höyhenheinä, feathergrass, needlegrass
♦ Weed, Casual Alien
♦ 39, 42, 272
♦ pG, cultivated, herbal, toxic.

Stipa retorta **Cav.**
Poaceae
♦ Weed
♦ 87, 88
♦ G.

Stipa robusta **(Vasey) Scribn.**
Poaceae
= *Achnatherum robustum* (Vasey) Barkworth (NoR)
♦ sleepy grass
♦ Weed
♦ 39, 161
♦ G, arid, cultivated, herbal, toxic.

Stipa scabra **Lindl. ssp.** *falcata* **(Hughes) Vickery**
Poaceae
Stipa falcata Hughes
♦ Introduced
♦ 228
♦ G.

Stipa spartea **Trin.**
Poaceae
= *Hesperostipa spartea* (Trin.) Barkworth (NoR)
♦ porcupine grass, needlegrass

♦ Weed, Naturalised
♦ 87, 88, 161, 210, 218, 287
♦ pG, cultivated, herbal.

Stipa subulata E.Fourn. ex Hemsl.
Poaceae
= *Nassella tenuissima* (Trin.) Barkworth
♦ Quarantine Weed
♦ 220
♦ G.

Stipa tenacissima L.
Poaceae
Macrochloa tenacissima. (L.) Kunth,
Lasiagrostis tenacissima (L.) Trin & Rupr.
♦ esparto grass, esparto, alfa, alfa
grass, halfa
♦ Weed, Introduced
♦ 70, 163, 228
♦ pG, arid, cultivated.

Stipa tenuis Phil.
Poaceae
Stipa argentina Speg., *Stipa papillosa*
(Hack.) Hitchc., *Stipa puelches* Speg.,
Stipa tenuis Phil. var. *argentina* (Speg.)
Speg.
♦ Introduced
♦ 228
♦ G, arid.

Stipa tenuissima Trin.
Poaceae
= *Nassella tenuissima* (Trin.) Barkworth
♦ white tussock
♦ Weed, Quarantine Weed, Noxious
Weed
♦ 121, 220, 278
♦ pG, cultivated, herbal. Origin: South
America.

Stipa tortilis Desf.
Poaceae
= *Stipa capensis* Thunb.
♦ Weed
♦ 87, 88
♦ G.

Stipa trichotoma Nees
Poaceae
= *Nassella trichotoma* (Nees) Hack.
♦ Australian serrated tussock, nassella
tussock, nassella tussockgrass, New
Zealand tussockgrass, serrated
tussock, serrated tussockgrass,
tumbleweed, Yass River tussock
♦ Weed, Quarantine Weed, Noxious
Weed, Environmental Weed
♦ 10, 121, 181, 220, 225, 278
♦ pG. Origin: South America.

Stipa variabilis Hughes
Poaceae
♦ Weed
♦ 121
♦ pG. Origin: Australia.

Stipa verticillata Spreng.
Poaceae
♦ Weed
♦ 15
♦ G, cultivated. Origin: Australia.

Stipa viridula Trin.
Poaceae
= *Nassella viridula* (Trin.) Barkworth
(NoR)

♦ green needlegrass
♦ Weed
♦ 87, 88, 218
♦ pG, cultivated, herbal.

Stipagrostis Nees spp.
Poaceae
♦ Weed
♦ 221
♦ G.

**Stipagrostis acutiflora (Trin. & Rupr.)
De Winter**
Poaceae
Aristida acutiflora Trin. & Rupr.
♦ Weed
♦ 221
♦ G, arid.

Stipagrostis brevifolia (Nees) De Winter
Poaceae
Aristida brevifolia (Nees) Steud.
♦ twabushman grass
♦ Native Weed
♦ 121
♦ pG. Origin: southern Africa.

Stipagrostis ciliata (Desf.) De Winter
Poaceae
Aristida ciliata Desf.
♦ Weed
♦ 39, 221
♦ G, arid, toxic.

**Stipagrostis ciliata (Desf.) De Winter
var. capensis (Trin. & Rupr.) De Winter**
Poaceae
♦ large bushman grass
♦ Native Weed
♦ 121
♦ pG. Origin: southern Africa.

Stipagrostis lanata (Forssk.) De Winter
Poaceae
Aristida forskahlei Tausch, *Aristida lanata*
Forssk.
♦ Weed
♦ 221
♦ G, arid.

Stipagrostis pennata (Trin.) De Winter
Poaceae
Aristida pennata Trin.
♦ Introduced
♦ 228
♦ G, arid.

**Stipagrostis plumosa (L.) Munro ex
Anderson**
Poaceae
Aristida brachypoda Tausch, *Aristida
plumosa* L.
♦ Weed
♦ 221
♦ G, arid.

Stipagrostis pungens (Desf.) De Winter
Poaceae
Aristida pungens Desf.
♦ Weed, Introduced
♦ 221, 228
♦ G, arid.

**Stipagrostis scoparia (Trin. & Rupr.) De
Winter**
Poaceae
♦ Weed

♦ 221
♦ G.

**Stipagrostis uniplumis (Licht. ex Roem.
& Schult.) De Winter var. uniplumis**
Poaceae
♦ bushman grass
♦ Weed
♦ 121
♦ G.

Stipagrostis zeyheri (Nees) De Winter
Poaceae
♦ Native Weed
♦ 121
♦ pG. Origin: southern Africa.

Stizolobium pruritum (Wight) Piper
Fabaceae/Papilionaceae
= *Mucuna pruriens* (L.) DC. var.
pruriens
♦ Weed, Quarantine Weed
♦ 76, 87, 88, 203, 220

Stobaea grandifolia DC.
Asteraceae
= *Berkheya grandifolia* Willd. (NoR)
♦ Quarantine Weed
♦ 220

Stobaea multijuga DC.
Asteraceae
= *Berkheya multijuga* (DC.) Roessler
♦ Quarantine Weed
♦ 220

Stobaea purpurea DC.
Asteraceae
= *Berkheya purpurea* Benth. & Hook.f.
ex Mast.
♦ Quarantine Weed
♦ 220

Stobaea speciosa DC.
Asteraceae
= *Berkheya speciosa* O.Hoffm.
♦ Quarantine Weed
♦ 220

Stoebe plumosa (L.) Thunb.
Asteraceae
Seriphium plumosum L.
♦ slangbos
♦ Weed, Quarantine Weed
♦ 76, 88, 220
♦ cultivated.

Stoebe spiralis Less.
Asteraceae
♦ Native Weed
♦ 121
♦ S. Origin: southern Africa.

Stoebe vulgaris Levyns
Asteraceae
♦ bankrupt bush, hanya
♦ Weed, Native Weed
♦ 10, 50, 87, 88, 121
♦ pS. Origin: southern Africa.

Stranvaesia davidiana Decne.
Rosaceae
= *Photinia davidiana* (Decne.) Cardot
♦ Casual Alien
♦ 280
♦ cultivated.

Stratiotes aloides L.
Hydrocharitaceae
♦ water soldier, water aloe, crab's claw, sahalehti
♦ Weed, Quarantine Weed, Noxious Weed, Naturalised, Environmental Weed
♦ 67, 76, 86, 87, 88, 193, 200, 203, 220, 229, 246, 272
♦ wpH, cultivated, herbal.

Strelitzia reginae Aiton
Strelitziaceae
♦ bird of paradise flower, bird of paradise
♦ Weed
♦ 161
♦ cultivated, herbal, toxic.

Streptanthus arizonicus Wats.
Brassicaceae
= *Streptanthus carinatus* (S.Wats.) Kruckeb. ssp. *arizonicus* Rodman & Worth. (NoR)
♦ Weed
♦ 23, 88
♦ herbal.

Streptopus amplexifolius (L.) DC.
Liliaceae/Uvulariaceae/Convallariaceae
Streptopus distortus Michx., *Uvularia amplexifolia* L.
♦ claspleaf twistedstalk, liver berry, wild cucumber
♦ Naturalised
♦ 101
♦ pH, cultivated, herbal.

Streptopus amplexifolius (L.) DC. var. papillatus Ohwi
Liliaceae/Uvulariaceae/Convallariaceae
♦ claspleaf twistedstalk
♦ Naturalised
♦ 101

Streptosolen jamesonii (Benth.) Miers
Solanaceae
♦ marmalade bush, streptosolen, orange browallia
♦ Naturalised
♦ 101
♦ cultivated.

Striga Lour. spp.
Scrophulariaceae
♦ witchweed
♦ Weed, Quarantine Weed, Noxious Weed, Naturalised
♦ 26, 67, 76, 86, 88, 172, 203
♦ H parasitic, cultivated.

Striga angustifolia (Don) Saldanha
Scrophulariaceae
♦ witchweed
♦ Weed, Quarantine Weed
♦ 76, 87, 88, 135, 191, 203
♦ H parasitic.

Striga asiatica (L.) Kuntze
Scrophulariaceae
Striga hirsuta Benth., *Striga lutea* Lour. (see), *Buchnera asiatica* L.
♦ witchweed, yaa mae mot, red witchweed, striga, mealie witchweed,

Asiatic witchweed, buri, common mealie witchweed, isona weed, Matabele flower, mealie poison, scarlet lobelia
♦ Weed, Quarantine Weed, Noxious Weed, Naturalised, Native Weed
♦ 50, 51, 53, 76, 88, 101, 121, 135, 140, 158, 161, 170, 191, 209, 229, 239, 240, 243
♦ aH parasitic, cultivated, herbal. Origin: tropics, subtropics of the Old World.

Striga aspera Benth.
Scrophulariaceae
♦ Weed
♦ 87, 88
♦ H parasitic.

Striga bilabiata (Thunb.) Kuntze
Scrophulariaceae
Striga thunbergii Benth. (see)
♦ Native Weed
♦ 121
♦ a/pH parasitic. Origin: southern Africa.

Striga densiflora (Benth.) Benth.
Scrophulariaceae
♦ denseflower witchweed, witchweed
♦ Weed, Quarantine Weed, Noxious Weed
♦ 66, 76, 87, 88, 140, 203, 229
♦ H parasitic.

Striga elegans Benth.
Scrophulariaceae
♦ large mealie witchweed, Matabele flower, witchweed
♦ Weed, Native Weed
♦ 87, 88, 121
♦ pH parasitic. Origin: southern Africa.

Striga forbesii Benth.
Scrophulariaceae
♦ giant mealie witchweed, witchweed
♦ Weed, Quarantine Weed, Native Weed
♦ 50, 51, 76, 87, 88, 121, 203
♦ aH parasitic. Origin: Madagascar.

Striga gesnerioides (Willd.) Vatke
Scrophulariaceae
Striga orobanchoides (R.Br. ex Endl.) Benth.
♦ purple witchweed, tobacco witchweed, iSona, cowpea witchweed
♦ Weed, Noxious Weed, Naturalised, Native Weed
♦ 50, 51, 87, 88, 101, 121, 140, 161, 179, 229
♦ aH parasitic, herbal. Origin: tropics, subtropics of the Old World.

Striga hermonthica (Del.) Benth.
Scrophulariaceae
Buchnera hermonthica Delile, *Striga senegalensis* Benth. (see)
♦ witchweed, purple witchweed, striga
♦ Weed, Quarantine Weed, Noxious Weed
♦ 23, 53, 76, 87, 88, 115, 140, 203, 221, 229, 240, 242

♦ aH parasitic, arid, herbal. Origin: Madagascar.

Striga junodii Schinz
Scrophulariaceae
♦ Native Weed
♦ 121
♦ a/pH parasitic. Origin: southern Africa.

Striga lutea Lour.
Scrophulariaceae
= *Striga asiatica* (L.) Kuntze
♦ witchweed, red witchweed
♦ Weed, Quarantine Weed, Noxious Weed
♦ 23, 35, 87, 88, 186, 218
♦ H parasitic, herbal.

Striga multiflora Benth.
Scrophulariaceae
♦ Weed
♦ 13
♦ aH parasitic.

Striga senegalensis Benth.
Scrophulariaceae
= *Striga hermonthica* (Del.) Benth.
♦ Weed
♦ 87, 88
♦ H parasitic.

Striga thunbergii Benth.
Scrophulariaceae
= *Striga bilabiata* (Thunb.) Kuntze
♦ Weed
♦ 87, 88
♦ H parasitic.

Strobilanthes anisophyllus (Lodd.) T.Anderson
Acanthaceae
♦ Casual Alien
♦ 280
♦ cultivated.

Strobilanthes cusia (Nees) Kuntze
Acanthaceae
♦ Naturalised
♦ 287
♦ herbal.

Strobilanthes japonica Miq.
Acanthaceae
♦ Naturalised
♦ 287

Strophioblachia glandulosa Pax var. pandurifolia Airy Shaw
Euphorbiaceae
♦ Weed
♦ 12

Strophiostoma sparsiflorum (Mikan) Turcz.
Boraginaceae
Myosotis sparsiflora Mikan (see)
♦ Weed
♦ 87, 88

Strophostyles helvula (L.) Ell.
Fabaceae/Papilionaceae
Strophostyles helvola (L.) Ell.
♦ wildbean, trailing fuzzybean
♦ Weed
♦ 243
♦ herbal.

Strophostyles leiosperma (Torr. & A.Gray) Piper
Fabaceae/Papilionaceae
♦ slickseed fuzzybean, smoothseed wildbean
♦ Weed
♦ 161
♦ herbal.

Struchium sparganophorum (L.) Kuntze
Asteraceae
♦ yerba de faja
♦ Weed
♦ 87, 88

Strychnos aculeata Soler.
Loganiaceae/Strychnaceae
♦ Quarantine Weed
♦ 220

Strychnos grayi Griseb.
Loganiaceae/Strychnaceae
♦ Weed
♦ 14

Strychnos madagascariensis Poir.
Loganiaceae/Strychnaceae
Strychnos innocua Del.
♦ black monkeyorange, spineless monkeyorange
♦ Weed, Native Weed
♦ 10, 121
♦ S/T, herbal. Origin: southern Africa.

Strychnos nux-vomica L.
Loganiaceae/Strychnaceae
♦ strychnine, strychnine tree, poison nut, nux vomica, nux vomica tree
♦ Weed, Quarantine Weed, Naturalised
♦ 39, 76, 86, 88, 161, 203, 220
♦ cultivated, herbal, toxic.

Strychnos spinosa Lam.
Loganiaceae/Strychnaceae
♦ Natal orange, spiny monkeyorange
♦ Weed, Quarantine Weed, Naturalised
♦ 39, 76, 88, 101, 179
♦ cultivated, herbal, toxic. Origin: Madagascar.

Stryphnodendron barbatimam Mart.
Fabaceae/Mimosaceae
♦ Weed
♦ 87, 88
♦ herbal.

Stryphnodendron coriaceum Benth.
Fabaceae/Mimosaceae
♦ Introduced
♦ 228
♦ arid.

Stryphnodendron obovatum Benth.
Fabaceae/Mimosaceae
♦ Weed
♦ 87, 88

Stuartina muelleri Sond.
Asteraceae
♦ Naturalised
♦ 280

Stylosanthes erecta Beauv.
Fabaceae/Papilionaceae
Ononis coriifolia Rchb. ex Guill. & Perr.,

Stylosanthes guineensis Schumach.
♦ Nigerian stylo
♦ Naturalised, Introduced
♦ 101, 228
♦ arid.

Stylosanthes fruticosa (Retz.) Alston
Fabaceae/Papilionaceae
Arachis fruticosa Retz., *Hedysarum hamatum* Burm.f., *Stylosanthes bojeri* Vogel, *Stylosanthes mucronata* Willd. nom. illeg.
♦ wild lucerne, shrubby pencilflower
♦ Weed, Naturalised, Native Weed
♦ 101, 121
♦ pH, arid, cultivated, herbal. Origin: Madagascar.

Stylosanthes guianensis (Aubl.) Sw.
Fabaceae/Papilionaceae
♦ stylo, Brazilian lucerne, tropical lucerne
♦ Weed, Naturalised, Introduced, Environmental Weed, Cultivation Escape
♦ 3, 7, 86, 87, 88, 93, 98, 191, 203, 228, 255, 261
♦ a/bH, arid, cultivated. Origin: Central and South America.

Stylosanthes guianensis (Aubl.) Sw. var. intermedia (Vogel) Hassl.
Fabaceae/Papilionaceae
Stylosanthes campestris M.B.Ferreira & Sousa Costa, *Stylosanthes hippocampoides* Mohlenbr., *Stylosanthes montevidensis* Vogel var. *intermedia* Vogel
♦ stylo
♦ Naturalised, Environmental Weed
♦ 86
♦ cultivated. Origin: South America.

Stylosanthes hamata (L.) Taub.
Fabaceae/Papilionaceae
♦ verano stylo, verano, Caribbean stylo, stylo, pencilflower, cheesytoes
♦ Weed, Naturalised, Introduced, Environmental Weed
♦ 14, 54, 86, 87, 88, 93, 161, 228, 249
♦ arid, cultivated. Origin: Central and South America.

Stylosanthes humilis Kunth
Fabaceae/Papilionaceae
Stylosanthes figueroae Mohlenbr., *Stylosanthes mucronata auct. non* Willd. (see), *Stylosanthes sundaica* Taub. (see)
♦ Townsville stylo, Townsville lucerne
♦ Weed, Naturalised, Introduced, Environmental Weed
♦ 7, 86, 93, 98, 203, 228
♦ arid, cultivated. Origin: South America.

Stylosanthes macrocephala Ferreira & Sousa Costa
Fabaceae/Papilionaceae
♦ Introduced
♦ 228
♦ arid.

Stylosanthes mucronata auct. non Willd.
Fabaceae/Papilionaceae

= *Stylosanthes humilis* Kunth
♦ Weed, Naturalised
♦ 7, 86, 98, 203

Stylosanthes scabra Vogel
Fabaceae/Papilionaceae
Stylosanthes diarthra S.F.Blake, *Stylosanthes glioides* S.F.Blake, *Stylosanthes plicata* S.F.Blake
♦ shrubby stylo, pencilflower
♦ Weed, Naturalised, Introduced
♦ 54, 86, 88, 93, 98, 203, 228
♦ arid, cultivated, herbal. Origin: South America.

Stylosanthes sundaica Taub.
Fabaceae/Papilionaceae
= *Stylosanthes humilis* Kunth
♦ Weed
♦ 87, 88

Stylosanthes viscosa Sw.
Fabaceae/Papilionaceae
Stylosanthes prostrata M.E.Jones
♦ poor man's friend, sticky stylo
♦ Weed, Naturalised
♦ 7, 54, 86, 87, 88, 93, 98, 203, 255
♦ pH, arid, cultivated, herbal. Origin: Brazil.

Stypandra glauca R.Br.
Liliaceae/Phormiaceae
♦ Weed
♦ 39, 87, 88
♦ cultivated, toxic.

Styphelia douglasii (Gray) F.Muell. ex Skottsb.
Epacridaceae/Ericaceae
= *Styphelia tameiameiae* (Cham. & Schlecht.) F.Muell.
♦ Weed
♦ 87, 88
♦ herbal.

Styphelia tameiameiae (Cham. & Schlecht.) F.Muell.
Epacridaceae/Ericaceae
Styphelia douglasii (Gray) F.Muell. ex Skottsb. (see)
♦ pukiawe
♦ Weed
♦ 87, 88
♦ herbal.

Styrax camporum Pohl.
Styracaceae
♦ Weed
♦ 87, 88

Styrax japonicus Sieb. & Zucc.
Styracaceae
♦ Japanese snowbell
♦ Naturalised
♦ 101
♦ cultivated, herbal.

Suaeda Forssk. ex Scop. spp.
Chenopodiaceae
♦ Weed
♦ 221

Suaeda aegyptiaca (Hasselq.) Zohary
Chenopodiaceae
♦ Weed, Naturalised
♦ 86, 98, 203

Suaeda altissima (L.) Pall.
Chenopodiaceae
♦ isokilokki
♦ Casual Alien
♦ 42

Suaeda australis (R.Br.) Moq.
Chenopodiaceae
= *Suaeda maritima* (L.) Dumort.
♦ Weed
♦ 87, 88
♦ aH, cultivated.

Suaeda baccifera Pall.
Chenopodiaceae
♦ berry seablite, annual seablite
♦ Weed, Naturalised, Environmental
Weed
♦ 72, 86, 88, 98, 198, 203
♦ aH. Origin: Eastern Europe, Russia.

Suaeda calceoliformis (Hook.) Moq.
Chenopodiaceae
Suaeda depressa (Pursh) S.Watson
♦ pahute weed, Pursh seepweed,
horned seablite, seablite
♦ Weed
♦ 159, 161
♦ aH, arid, herbal.

Suaeda divaricata Moq.
Chenopodiaceae
♦ jume
♦ Weed
♦ 295
♦ arid.

Suaeda fruticosa auct. non Forssk.
Chenopodiaceae
= *Suaeda moquinii* (Torr.) Greene (NoR)
♦ alkali seepweed, inkbush, shrubby
seablite
♦ Weed, Native Weed
♦ 23, 87, 88, 121, 218, 221
♦ pS, arid, promoted, herbal, toxic.
Origin: southern Africa.

Suaeda glauca (Bunge) Bunge
Chenopodiaceae
♦ Weed
♦ 275, 297
♦ aH, promoted.

Suaeda linifolia Pall.
Chenopodiaceae
♦ stalked seablite
♦ Naturalised
♦ 86, 198
♦ Origin: eastern Europe.

Suaeda maritima (L.) Dumort.
Chenopodiaceae
Chenopodina australis (R.Br.)
Moq., *Chenopodina maritima* (L.)
Moq., *Chenopodium australe* R.Br.,
Chenopodium maritimum L., *Lerchea
maritima* (L.) Kuntze, *Salsola indica*
Willd., *Suaeda australis* (R.Br.) Moq.
(see), *Suaeda indica* Moq., *Suaeda
nudiflora* Moq.
♦ herbaceous seepweed, pikkukilokki,
seablite, annual seablite
♦ Weed, Naturalised
♦ 23, 88, 101
♦ aH, arid, cultivated, herbal. Origin:
Australia.

**Suaeda maritima (L.) Dumort ssp.
maritima**
Chenopodiaceae
♦ annual seablite, herbaceous
seepweed
♦ Naturalised
♦ 86, 101, 198
♦ Origin: Eurasia.

Suaeda monoica Forssk. ex Gmel.
Chenopodiaceae
♦ Weed, Introduced
♦ 221, 228, 243
♦ arid.

Suaeda paradoxa (Bunge) Bunge
Chenopodiaceae
♦ Weed
♦ 275

Suaeda rigida Kung. & G.L.Chu
Chenopodiaceae
♦ Weed
♦ 275

Suaeda salsa (L.) Pall.
Chenopodiaceae
♦ Weed
♦ 275, 297
♦ aH, promoted.

Suaeda vera Forssk. ex Gmel.
Chenopodiaceae
♦ shrubby seablite
♦ Weed
♦ 221

Suaeda vermiculata Forssk.
Chenopodiaceae
Suaeda mollis Del.
♦ Weed
♦ 221
♦ arid.

Subpilocereus Backeb. spp.
Cactaceae
= *Cereus* Mill. spp.
♦ Weed, Quarantine Weed
♦ 76, 88, 203

Subularia aquatica L.
Brassicaceae
Crucifera subularia Krause, *Draba
subularia* L.
♦ awlwort, water awlwort, äimäruoho
♦ Weed
♦ 23, 88, 159
♦ aqua, cultivated, herbal.

Succisa australis (Wulf.) Rchb.
Dipsacaceae
= *Succisella inflexa* (Kluk) Beck
♦ devil's bit
♦ Weed
♦ 87, 88, 218

Succisa pratensis Moench
Dipsacaceae
Asterocephalus succisa Wall., *Scabiosa
praemorsa* Gilib., *Scabiosa succisa* L.
♦ devil's bit scabious, devil's bit,
purtojuuri, succise des près
♦ Weed, Naturalised
♦ 23, 39, 70, 88, 101, 272
♦ pH, cultivated, herbal, toxic.

Succisella inflexa (Kluk) Beck
Dipsacaceae

Succisa australis (Wulf.) Reichen. (see)
♦ southern succisella
♦ Naturalised
♦ 101
♦ cultivated, herbal.

Succowia balearica (L.) Medik.
Brassicaceae
♦ Weed, Naturalised, Environmental
Weed
♦ 7, 54, 86, 88, 155
♦ cultivated.

Suckleya suckleyana (Torr.) Rydb.
Chenopodiaceae
♦ poison suckleya
♦ Weed
♦ 39, 82, 87, 88, 161, 218
♦ herbal, toxic.

Sutera caerulea (L.f.) Hiern
Scrophulariaceae
Diascia avasmontana Dinter, *Manulea
caerulea* L.f.
♦ sutera
♦ Weed, Quarantine Weed
♦ 76, 88, 158
♦ Origin: southern Africa.

Sutherlandia frutescens (L.) R.Br.
Fabaceae/Papilionaceae
Colutea frutescens L.
♦ cancer bush, duck plant, kankerbos
♦ Weed, Naturalised
♦ 7, 86, 98, 203
♦ arid, cultivated, herbal. Origin:
South Africa.

**Swainsona formosa (G.Don) Joy
Thomps.**
Fabaceae/Papilionaceae
Clianthus formosus (G.Don) Ford &
Vick., *Donia formosa* G.Don
♦ Sturt's desert pea
♦ Introduced
♦ 228
♦ arid, cultivated, herbal.

Swainsona galegifolia (Andrews) R.Br.
Fabaceae/Papilionaceae
Swainsona coronillifolia Salisb., *Vicia
galegifolia* Andrews
♦ swan flower, smooth Darling pea
♦ Weed
♦ 39, 87, 88, 269
♦ pH, arid, cultivated, toxic. Origin:
Australia.

Swainsona greyana Lindl. ssp. greyana
Fabaceae/Papilionaceae
♦ hairy Darling pea, Darling pea
♦ Native Weed
♦ 269
♦ pH, cultivated. Origin: Australia.

Swainsona luteola Muell.
Fabaceae/Papilionaceae
♦ dwarf Darling pea
♦ Native Weed
♦ 269
♦ arid. Origin: Australia.

Swainsona salsula (Pall.) Taub.
Fabaceae/Papilionaceae
= *Sphaerophysa salsula* (Pall.) DC.
♦ swainson pea
♦ Weed

♦ 23, 87, 88, 218
♦ herbal.

Swainsona swainsonoides (Benth.) Lee ex Black
Fabaceae/Papilionaceae
♦ downy swainsonia, downy Darling pea
♦ Native Weed
♦ 269
♦ pH, cultivated. Origin: Australia.

Sweetia fruticosa Spreng.
Fabaceae/Papilionaceae
Ferreirea spectabilis Allemão
♦ Introduced
♦ 228
♦ arid.

Swietenia macrophylla King
Meliaceae
♦ Honduras mahogany, broad leaved mahogany, mahogany, American mahogany, mogno, águano, araputangá, caoba, cedroaraná
♦ Weed, Quarantine Weed, Naturalised, Introduced, Environmental Weed, Cultivation Escape
♦ 3, 22, 76, 87, 88, 101, 152, 191, 230, 261, 268
♦ T, cultivated, herbal. Origin: Central and South America.

Swietenia mahagoni (L.) Jacq.
Meliaceae
♦ mahogany, West Indian mahogany
♦ Weed
♦ 3, 22, 39, 161, 191
♦ T, cultivated, herbal, toxic.

Syagrus romanzoffiana (Cham.) Glassman
Arecaceae
Arecastrum romanzoffianum (Cham.) Becc. (see), *Arecastrum romanzoffianum* var. *australe* (Mart.) Becc., *Arecastrum romanzoffianum* var. *micropindo* Becc., *Cocos australis* Mart., *Cocos datil* Griseb. & Drude, *Cocos romanzoffiana* Cham., *Syagrus romanzoffianum* (Cham.) Glassman (see)
♦ feathery coconut, queen palm, butia palm
♦ Weed, Introduced
♦ 179, 228
♦ arid, cultivated, herbal. Origin: South America.

Syagrus romanzoffianum (Cham.) Glassman
Arecaceae
= *Syagrus romanzoffiana* (Cham.) Glassman
♦ Cocos palm, queen palm, palma pindo, chirvana
♦ Weed, Naturalised, Garden Escape, Environmental Weed
♦ 73, 86, 88, 155
♦ cultivated.

Symphoricarpos albus (L.) Blake
Caprifoliaceae
♦ common snowberry, snowberry bush, waxberry
♦ Weed, Naturalised

♦ 15, 23, 39, 87, 88, 154, 161, 218, 280
♦ cultivated, herbal, toxic.

Symphoricarpos albus (L.) Blake var. laevigatus (Fern.) Blake
Caprifoliaceae
♦ upright snowberry, common snowberry, snowberry, white snowberry
♦ Naturalised
♦ 40
♦ S.

Symphoricarpos × chenaultii Rehder
Caprifoliaceae
= *Symphoricarpos microphyllus* Kunth × *Symphoricarpos orbiculatus* Moench.
♦ pink snowberry
♦ Naturalised
♦ 40
♦ cultivated.

Symphoricarpos glomeratus C.Koch
Caprifoliaceae
♦ Quarantine Weed
♦ 220

Symphoricarpos occidentalis Hook.
Caprifoliaceae
♦ western snowberry, wolfberry, buckbrush
♦ Weed, Noxious Weed, Native Weed
♦ 39, 87, 88, 161, 174, 218, 299
♦ S, cultivated, herbal, toxic. Origin: North America.

Symphoricarpos orbiculatus Moench
Caprifoliaceae
Symphoricarpos symphoricarpos (L.) MacMill. *nom. inval.* (see), *Symphoricarpos vulgaris* Michx. (see), *Lonicera symphoricarpos* L.
♦ coralberry, buckbrush
♦ Weed, Quarantine Weed, Naturalised
♦ 15, 23, 76, 87, 88, 108, 161, 218, 220, 280
♦ S, cultivated, herbal, toxic. Origin: eastern North America, Mexico.

Symphoricarpos symphoricarpos (L.) MacMill. nom. inval.
Caprifoliaceae
= *Symphoricarpos orbiculatus* Moench
♦ Quarantine Weed
♦ 220

Symphoricarpos vialis A.Gray
Caprifoliaceae
♦ Weed
♦ 87, 88

Symphoricarpos vulgaris Michx.
Caprifoliaceae
= *Symphoricarpos orbiculatus* Moench
♦ Quarantine Weed
♦ 39, 220
♦ toxic.

Symphytum asperum Lepech.
Boraginaceae
Symphytum armeniacum Buckn., *Symphytum asperrimum* Donn., *Symphytum echinatum* Ledeb.
♦ prickly comfrey, rough comfrey, tarharaunioyrtti
♦ Weed, Noxious Weed, Naturalised,

Introduced, Cultivation Escape
♦ 15, 34, 35, 40, 42, 87, 88, 101, 218, 229, 280
♦ pH, cultivated, herbal, toxic.

Symphytum bulbosum Schimp.
Boraginaceae
Symphytum brochum Bory, *Symphytum clusii* Gmel., *Symphytum macrolepis* Gay, *Symphytum punctatum* Gaud., *Symphytum zeyheri* Schimp.
♦ bulbous comfrey
♦ Weed
♦ 88, 94, 272
♦ herbal.

Symphytum grandiflorum A.DC.
Boraginaceae
♦ creeping comfrey, dwarf comfrey, comfrey
♦ Naturalised
♦ 40
♦ pH, cultivated, herbal.

Symphytum officinale L.
Boraginaceae
Symphytum consolida Gueldenst., *Symphytum elatum* Tausch, *Symphytum majus* Bubani, *Symphytum patens* Sibth., *Symphytum uliginosum* Kern.
♦ common comfrey, English comfrey, comfrey, knitbone, rohtorraunioyrtti, consolida maggiore
♦ Weed, Naturalised, Garden Escape, Environmental Weed, Cultivation Escape
♦ 7, 23, 39, 42, 70, 86, 87, 88, 94, 97, 98, 101, 198, 203, 218, 243, 253, 272, 280, 286, 287
♦ pH, cultivated, herbal, toxic. Origin: Eurasia.

Symphytum orientale L.
Boraginaceae
♦ white comfrey
♦ Naturalised
♦ 40
♦ H, cultivated, herbal.

Symphytum tuberosum L.
Boraginaceae
♦ tuberous comfrey
♦ Weed, Naturalised
♦ 101, 272
♦ pH, cultivated, herbal.

Symphytum × uplandicum Nyman (pro sp.)
Boraginaceae
= *Symphytum asperum* Lepech. × *Symphytum officinale* L.
♦ Russian comfrey, ruotsinraunioyrtti
♦ Weed, Naturalised, Garden Escape, Cultivation Escape
♦ 15, 39, 40, 42, 86, 101, 165, 176, 252, 280
♦ cultivated, herbal, toxic.

Symplocarpus foetidus (L.) Nutt.
Araceae
Spathyema foetida (L.) Raf.
♦ skunk cabbage
♦ Weed
♦ 8, 39, 87, 88, 154, 161, 218, 247
♦ pH, cultivated, herbal, toxic.

Synadenium grantii Hook.f.
Euphorbiaceae
♦ African milk bush
♦ Weed, Naturalised
♦ 39, 54, 86, 88, 247
♦ cultivated, herbal, toxic. Origin:
Africa.

Syncarpha gnaphaloides (L.) DC.
Asteraceae
Staehelina gnaphaloides L.
♦ Quarantine Weed
♦ 220
♦ cultivated.

Syncarpha vestita (L.) B.Nord.
Asteraceae
Xeranthemum vestitum L., *Helichrysum
vestitum* (L.) Willd.
♦ cape everlasting, tinder everlasting,
sewejaartjie
♦ Weed, Quarantine Weed
♦ 76, 88, 220
♦ cultivated. Origin: South Africa.

Syncarpia glomulifera (Sw.) Nied.
Myrtaceae
♦ turpentine tree, syncarpia
♦ Weed, Naturalised, Casual Alien
♦ 3, 101, 191, 280
♦ cultivated, herbal. Origin: Australia.

Synedrella nodiflora (L.) Gaertn.
Asteraceae
Verbesina nodiflora L.
♦ synedrella, node weed, porter bush,
fatten barrow, phak khraet, American
weed, botao de ouro, pig grass
♦ Weed, Naturalised, Introduced,
Environmental Weed
♦ 6, 12, 13, 14, 23, 28, 86, 87, 88, 93, 155,
170, 179, 191, 209, 217, 218, 230, 238,
239, 243, 255, 257, 261, 262, 275, 276,
286, 287, 297
♦ aH, cultivated, herbal. Origin:
tropical America.

**Synedrellopsis grisebachii auct. non
Hieron. & Kuntze**
Asteraceae
= *Calyptocarpus vialis* Less.
♦ agriaozinho, agriaao do pasto
♦ Weed, Naturalised
♦ 86, 255
♦ Origin: South America.

Syneilesis aconitifolia (Bunge) Maxim.
Asteraceae
♦ aconiteleaf syneilesis
♦ Weed
♦ 297
♦ cultivated.

Syngonium angustatum Schott
Araceae
Nephthytis triphylla hort. ex L.H.Bailey,
Syngonium albolineatum hort.,
Syngonium gracilis Matuda, *Syngonium
oerstedtianum* Schott, *Syngonium
podophyllum* Schott var. *albolineatum*
(hort.) Engl., *Syngonium podophyllum*
var. *oerstedianum* (Schott) Engl.
♦ five fingers
♦ Weed, Naturalised
♦ 101, 179

Syngonium podophyllum Schott
Araceae
♦ arrowhead vine, goosefoot plant,
arrowhead plant, nephthytis, African
evergreen, conde, comemanos
♦ Weed, Naturalised, Garden Escape,
Environmental Weed, Cultivation
Escape
♦ 3, 80, 86, 87, 88, 101, 107, 112, 122,
155, 179, 194, 247, 261, 281
♦ aqua, cultivated, toxic. Origin:
Central America from Mexico to
Panama.

Synnotia villosa (Burm.f.) N.E.Br.
Iridaceae
♦ perslelie
♦ Weed, Naturalised, Native Weed
♦ 86, 98, 121, 203
♦ pH, cultivated. Origin: southern
Africa.

Syringa josikaea Jacq.f. ex Rchb.
Oleaceae
♦ valkopiippo, Hungarian lilac
♦ Naturalised, Cultivation Escape
♦ 42, 101
♦ S, cultivated.

Syringa × persica L.
Oleaceae
= *Syringa afghanica* C.Schneid. ×
Syringa laciniata Mill. [possible
combination]
♦ Persian lilac
♦ Naturalised
♦ 101

**Syringa reticulata (Blume) H.Hara ssp.
amurensis (Rupr.) Greene & Chang**
Oleaceae
♦ Amur lilac
♦ Naturalised
♦ 101

**Syringa reticulata (Blume) H.Hara ssp.
pekinensis (Rupr.) Greene & Chang**
Oleaceae
♦ Peking tree lilac
♦ Naturalised
♦ 101

**Syringa reticulata (Blume) H.Hara ssp.
reticulata**
Oleaceae
♦ Japanese tree lilac
♦ Naturalised
♦ 101

Syringa villosa Vahl
Oleaceae
♦ villous lilac, late lilac
♦ Weed, Naturalised
♦ 80, 101
♦ cultivated.

Syringa vulgaris L.
Oleaceae
♦ common lilac, lilac, syringa,
pihasyreeni
♦ Weed, Naturalised, Environmental
Weed, Cultivation Escape
♦ 4, 40, 42, 80, 86, 87, 88, 101, 104, 198,
218, 252, 280
♦ S/T, cultivated, herbal. Origin:
eastern Europe.

Syzygium australe (Link) B.Hyland
Myrtaceae
♦ brush cherry
♦ Quarantine Weed, Naturalised,
Environmental Weed
♦ 225, 246, 280
♦ cultivated. Origin: Australia.

Syzygium cumini (L.) Skeels
Myrtaceae
Eugenia cumini (L.) Druce (see), *Eugenia
jambolana* Lam. (see)
♦ jambolan, Java plum, jambolan
plum, duhat, mesegerak, jamelonguier,
kavika ni India, jammun
♦ Weed, Noxious Weed, Naturalised,
Environmental Weed, Garden Escape,
Cultivation Escape
♦ 3, 22, 47, 80, 86, 87, 88, 95, 101, 112,
122, 132, 151, 179, 191, 252, 283
♦ T, cultivated, herbal. Origin: Asia.

Syzygium floribundum F.Muell.
Myrtaceae
= *Waterhousia floribunda* (F.Muell.)
B.Hyland. (NoR)
♦ weeping lilly pilly, weeping myrtle
♦ Weed
♦ 3, 191
♦ cultivated. Origin: Australia.

Syzygium grande (Wight) Wight ex Walp.
Myrtaceae
♦ sea apple
♦ Introduced
♦ 261
♦ cultivated. Origin: south-east Asia.

Syzygium jambos (L.) Alston
Myrtaceae
Eugenia jambos L. (see)
♦ rose apple, Malabar plum, iouen
wai, youenwai, apel en wai, kavika,
kavika ni vavalangi, kavika ni India,
rose apple tree, jamboes
♦ Weed, Sleeper Weed, Noxious
Weed, Naturalised, Garden Escape,
Environmental Weed, Cultivation
Escape
♦ 3, 14, 22, 32, 80, 86, 87, 88, 95, 101,
107, 112, 122, 132, 155, 179, 191, 226,
227, 261, 283, 287
♦ T, cultivated, herbal. Origin: south-
east Asia.

Syzygium malaccense (L.) Merr. & Perry
Myrtaceae
Caryophyllus malaccensis (L.) Stokes,
Eugenia malaccensis L. (see), *Jambosa
malaccensis* (L.) DC.
♦ Malay apple, pomerac, rose apple,
large fruited rose apple, pomerac
jambos, Malay pomarosa, Tahiti apple,
Malaysian apple, nonu fi'afi'a
♦ Weed, Naturalised, Cultivation
Escape
♦ 22, 101, 261
♦ T, cultivated, herbal. Origin:
Malaysia.

Syzygium paniculatum Gaertn.
Myrtaceae
♦ brush cherry, lilly pilly, Australian
brush cherry

♦ Weed, Quarantine Weed,
Naturalised, Environmental Weed,
Cultivation Escape, Casual Alien
♦ 15, 198, 246, 280, 283
♦ cultivated. Origin: Australia.

T

Tabebuia aurea (Manso) Benth. & Hook.f. ex S.Moore
Bignoniaceae
♦ Caribbean trumpet tree
♦ Weed, Naturalised, Introduced
♦ 101, 179, 261
♦ T, cultivated. Origin: South America.

Tabebuia chrysantha (Jacq.) Nichols.
Bignoniaceae
♦ roble amarillo, cortes, yellow poui
♦ Naturalised, Introduced
♦ 101, 261
♦ T, cultivated. Origin: Mexico to Colombia.

Tabebuia donnell-smithii Rose
Bignoniaceae
♦ primavera
♦ Naturalised, Cultivation Escape
♦ 101, 261
♦ cultivated. Origin: southern Mexico to Colombia.

Tabebuia haemantha (Bertol. ex Spreng.) DC.
Bignoniaceae
♦ roble cimarron
♦ Weed
♦ 87, 88

Tabebuia heterophylla (DC.) Britt.
Bignoniaceae
Tabebuia pentaphylla (L.) Hemsl.,
Tabebuia pallida (Lindl.) Miers (see)
♦ pink tecoma, pink trumpet tree, white cedar, calice du pape, pink pui
♦ Weed
♦ 107, 179
♦ cultivated, herbal.

Tabebuia lepidophylla A.Rich.
Bignoniaceae
♦ Weed
♦ 14

Tabebuia lepidota (Kunth) Britt.
Bignoniaceae
♦ one toe
♦ Weed
♦ 14

Tabebuia pallida (Lindl.) Miers
Bignoniaceae
= *Tabebuia heterophylla* (DC.) Britt.
♦ white cedar
♦ Weed
♦ 22
♦ T, cultivated, herbal.

Tabebuia rosea (Bertol.) DC.
Bignoniaceae
♦ pink trumpet tree, pink poui, rosy

trumpet tree, roble venezolano
♦ Naturalised, Garden Escape
♦ 101, 261
♦ cultivated, herbal. Origin: Mexico, Central America.

Tabebuia serratifolia (Vahl) Nichols.
Bignoniaceae
♦ yellow poui
♦ Naturalised, Cultivation Escape
♦ 101, 261
♦ T, cultivated. Origin: southern Antilles, Guyana to Brazil, Bolivia.

Tabernaemontana L. spp.
Apocynaceae
♦ milkwood
♦ Quarantine Weed
♦ 220

Tabernaemontana alba Mill.
Apocynaceae
♦ white milkwood
♦ Weed, Naturalised
♦ 101, 179

Tabernaemontana amblyocarpa Urb.
Apocynaceae
♦ Weed
♦ 14

Tabernaemontana chrysocarpa Blake
Apocynaceae
♦ Weed, Quarantine Weed
♦ 87, 88, 220

Tabernaemontana citrifolia L.
Apocynaceae
♦ milkwood
♦ Weed
♦ 14, 39
♦ herbal, toxic.

Tabernaemontana divaricata (L.) R.Br. ex Roem. & Schult.
Apocynaceae
Ervatamia divaricata (L.) Burkill
♦ crepe jasmine, butterfly gardenia, pinwheel flower
♦ Weed, Naturalised, Introduced, Garden Escape, Cultivation Escape
♦ 32, 38, 39, 101, 161, 230, 261
♦ S, cultivated, herbal, toxic. Origin: India.

Tabernaemontana fuchsiaefolia A.DC.
Apocynaceae
= *Peschiera fuchsiaefolia* (A.DC.) Miers
♦ Weed, Quarantine Weed
♦ 76, 87, 88, 203, 220

Tacazzea apiculata Oliv.
Asclepiadaceae/Apocynaceae/Periplocaceae
♦ lufute katika monga
♦ Native Weed
♦ 121
♦ pS. Origin: southern Africa.

Tacca leontopetaloides (L.) Kuntze
Taccaceae
♦ batflower, kabuga, mâsoâ, Polynesian arrowroot
♦ Weed, Naturalised
♦ 87, 88, 101
♦ pH, cultivated, herbal. Origin: Madagascar.

Tacinga Britt. & Rose spp.
Cactaceae
♦ Weed, Quarantine Weed
♦ 76, 88, 203, 220

Taeniatherum asperum (Sim.) Nevski
Poaceae
= *Taeniatherum caput-medusae* (L.)
Nevski ssp. *asperum* (Simonk.)
Melderis (NoR)
♦ Medusa head
♦ Weed
♦ 23, 80, 87, 88, 218
♦ G, herbal.

Taeniatherum caput-medusae (L.) Nevski
Poaceae
Elymus caput-medusae L. (see)
♦ Medusa head, Medusa head rye
♦ Weed, Quarantine Weed, Noxious
Weed, Naturalised, Environmental
Weed
♦ 34, 35, 76, 78, 80, 86, 88, 98, 101, 116,
141, 146, 151, 161, 172, 203, 212, 219,
220, 221, 229, 231, 267, 272, 300
♦ aG, cultivated. Origin: Eurasia.

Taeniophyllum Blume spp.
Orchidaceae
♦ Introduced
♦ 230
♦ H.

Taeniophyllum petrophilium Schltr.
Orchidaceae
♦ paten en kewelik
♦ Introduced
♦ 230
♦ H.

Taenitis blechnoides (Willd.) Sw.
Pteridaceae/Adiantaceae
♦ Weed
♦ 87, 88
♦ cultivated. Origin: Australia.

Tagetes coronopifolia Willd.
Asteraceae
♦ Weed
♦ 199

Tagetes daucoides Schrad.
Asteraceae
♦ Naturalised
♦ 241

Tagetes erecta L.
Asteraceae
♦ French marigold, African marigold,
makerita, Aztec marigold, ruda de
pasto, isosamettikukka
♦ Weed, Naturalised, Introduced,
Garden Escape, Environmental Weed,
Cultivation Escape
♦ 34, 42, 101, 157, 179, 256, 257, 261,
280
♦ aH, cultivated, herbal. Origin:
Mexico, Central America.

Tagetes filifolia Lag.
Asteraceae
Diglossus variabilis Cass., *Solenotheca
tenella* Nutt., *Tagetes anisata* Lillo,
Tagetes congesta Hook. & Arn., *Tagetes
dichotoma* Turcz., *Tagetes foeniculacea*
Desf., *Tagetes foeniculacea* Poepp. ex
DC., *Tagetes multifida* DC., *Tagetes*

pseudomicrantha Lillo, *Tagetes pusilla*
Kunth (see), *Tagetes scabra* Brandegee,
Tagetes silenoides Meyen & Walp.
♦ Irish lace, Irish lace marigold,
anisillo
♦ Weed
♦ 199
♦ aH, cultivated, herbal. Origin:
Central and South America.

Tagetes foetidissima DC.
Asteraceae
♦ Weed
♦ 199

Tagetes lanulata Ortega
Asteraceae
♦ Weed
♦ 199

Tagetes micrantha Cav.
Asteraceae
Tagetes fragrantissima Sessé & Moç.
♦ licorice marigold
♦ Weed
♦ 243
♦ herbal.

Tagetes minima L.
Asteraceae
♦ miniature marigold
♦ Naturalised
♦ 101

Tagetes minuta L.
Asteraceae
Tagetes glandulifera Schr., *Tagetes
bonariensis* Pers.
♦ wild marigold, stinking Roger,
Mexican marigold, stinkweed, tall
khakiweed, muster John Henry,
chinchilla, pikkusamettikukka,
mbanda, little marigold
♦ Weed, Noxious Weed, Naturalised,
Introduced, Garden Escape,
Environmental Weed, Casual Alien
♦ 7, 23, 34, 35, 39, 42, 50, 51, 53, 80, 86,
87, 88, 98, 101, 121, 134, 158, 198, 203,
217, 218, 228, 229, 236, 237, 240, 243,
255, 269, 270, 280, 286, 287, 295
♦ aH, arid, cultivated, herbal, toxic.
Origin: South America.

Tagetes patula L.
Asteraceae
♦ French marigold, ryhmäsametti-
kukka, clavel de muerto
♦ Weed, Naturalised, Garden Escape,
Cultivation Escape
♦ 42, 87, 88, 101, 256, 261, 280
♦ aH, cultivated, herbal. Origin:
Mexico, Central America.

Tagetes pusilla Kunth
Asteraceae
= *Tagetes filifolia* Lag.
♦ lesser marigold, marigold, sacha
anis
♦ Naturalised
♦ 101
♦ aH, cultivated, herbal.

Tagetes tenuifolia Cav.
Asteraceae
Tagetes signata Bartl.
♦ lemon marigold, signata marigold,

marigold
♦ Weed, Casual Alien
♦ 87, 88, 280
♦ aH, promoted, herbal.

Tagetes terniflora Kunth
Asteraceae
Tagetes cabrerae Ferraro
♦ suidue
♦ Weed
♦ 237, 295
♦ herbal.

Talinum crassifolium (Jacq.) Willd.
Portulacaceae
= *Talinum triangulare* (Jacq.) Willd.
♦ Naturalised
♦ 287

Talinum crispatulum Dinter ex V.Poelln
Portulacaceae
♦ Native Weed
♦ 121
♦ pH. Origin: southern Africa.

Talinum paniculatum (Jacq.) Gaertn.
Portulacaceae
♦ fame flower, jewels of Opar, yuyo
verdolaga, maria gorda
♦ Weed, Naturalised, Casual Alien
♦ 7, 86, 87, 88, 98, 121, 153, 203, 237,
255, 256, 257, 270, 280, 295
♦ pH, cultivated, herbal. Origin:
Americas.

Talinum triangulare (Jacq.) Willd.
Portulacaceae
Calandrinia andrewsii (Sweet) Sweet,
Claytonia triangularis (Jacq.) Kuntze,
Portulaca crassicaulis Jacq., *Portulaca
crassifolia* Jacq., *Portulaca cuneifolia*
Vahl, *Portulaca paniculata* L., *Portulaca
racemosa* L., *Portulaca triangularis* Jacq.,
Talinum andrewsii Sweet, *Talinum
attenuatum* Rose & Standl., *Talinum
confusum* Rose & Standl., *Talinum
crassifolium* (Jacq.) Willd. (see), *Talinum
diffusum* Rose & Standl., *Talinum
fruticosum* Willd., *Talinum paniculatum*
Moench., *Talinum racemosum* (L.)
Rohrb.
♦ bangwelengwele, hierba de cacolote,
lela capé, lustrosa grande
♦ Weed, Introduced
♦ 87, 88, 228, 255, 262, 295
♦ aH, arid, herbal. Origin: tropical
America.

Talisia esculenta (A.St.-Hil.) Radlk.
Sapindaceae
♦ Quarantine Weed
♦ 220
♦ Origin: South America.

Tamarindus indica L.
Fabaceae/Caesalpiniaceae
♦ tamarind, tamarindo, imli, pousga
♦ Weed, Naturalised, Introduced,
Environmental Weed
♦ 7, 22, 80, 86, 88, 93, 98, 101, 151, 152,
179, 203, 228, 230, 257, 261
♦ T, arid, cultivated, herbal. Origin:
tropical Africa.

Tamarix L. spp.
Tamaricaceae

♦ tamarisk, saltcedar
♦ Weed, Quarantine Weed, Noxious Weed, Naturalised, Environmental Weed
♦ 1, 18, 35, 80, 88, 116, 139, 151, 198, 221, 258, 272, 290
♦ herbal.

Tamarix africana Poir.
Tamaricaceae
Tamarix hispanica Boiss.
♦ saltcedar, African tamarisk
♦ Weed, Noxious Weed, Naturalised, Introduced
♦ 80, 101, 228, 229
♦ T, arid, cultivated.

Tamarix aphylla (L.) H.Karst.
Tamaricaceae
Tamarix aphylla (L.) Warb., *Tamarix articulata* Vahl, *Tamarix orientalis* Forssk.
♦ athel, tamarisk, athel pine, athel tree, flowering cypress, athel tamarisk
♦ Weed, Quarantine Weed, Noxious Weed, Naturalised, Introduced, Garden Escape, Environmental Weed, Cultivation Escape
♦ 7, 34, 35, 76, 78, 80, 86, 87, 88, 93, 98, 101, 116, 129, 147, 152, 155, 203, 218, 221, 228, 229, 261, 269, 290
♦ T, arid, cultivated, herbal. Origin: north-east Africa, eastern Europe, Asia.

Tamarix aralensis Bunge
Tamaricaceae
♦ saltcedar, Russian tamarisk
♦ Noxious Weed, Naturalised
♦ 101, 229

Tamarix canariensis Willd.
Tamaricaceae
♦ saltcedar, Canary Island tamarisk, tamarisk
♦ Noxious Weed, Naturalised
♦ 101, 229
♦ T, cultivated, herbal.

Tamarix chinensis Lour.
Tamaricaceae
Tamarix plumosa hort. ex Carrière
♦ tamarisk, Chinese tamarisk, fivestamen tamarisk, saltcedar
♦ Weed, Noxious Weed, Naturalised, Introduced, Garden Escape, Environmental Weed, Cultivation Escape
♦ 34, 63, 78, 80, 101, 105, 129, 142, 159, 228, 229, 231, 264, 280, 283
♦ S, arid, cultivated, herbal. Origin: Eurasia, Africa.

Tamarix dioica Roxb. ex Roth
Tamaricaceae
♦ saltcedar, tamarisk
♦ Noxious Weed
♦ 229
♦ herbal.

Tamarix gallica L.
Tamaricaceae
Tamarix algeriensis hort., *Tamarix anglica* Webb, *Tamarix brachylepis* Sennen, *Tamarix madritensis* Pau & Villar
♦ French tamarisk, saltcedar,

mannaplant, tamarisk
♦ Weed, Quarantine Weed, Noxious Weed, Naturalised, Introduced, Environmental Weed
♦ 40, 76, 78, 80, 87, 88, 101, 129, 218, 228, 229, 231, 261
♦ S/T, arid, cultivated, herbal. Origin: Eurasia, Africa.

Tamarix indica Willd.
Tamaricaceae
Tamarix troupii Hole
♦ Naturalised, Introduced
♦ 86, 228
♦ arid, herbal.

Tamarix nilotica (Ehrenb.) Bunge
Tamaricaceae
♦ Weed
♦ 221
♦ arid.

Tamarix parviflora DC.
Tamaricaceae
Tamarix tetrandra auct. non Pall.
♦ tamarisk, small flowered tamarisk, fourstamen tamarisk, saltcedar
♦ Weed, Noxious Weed, Naturalised, Garden Escape, Environmental Weed, Casual Alien
♦ 34, 78, 80, 101, 105, 129, 138, 142, 161, 212, 229, 231, 280
♦ S/T, cultivated, herbal. Origin: Eurasia, Africa.

Tamarix pentandra Pall. *nom. illeg.*
Tamaricaceae
= *Tamarix ramosissima* Ledeb.
♦ saltcedar
♦ Weed
♦ 23, 87, 88, 146, 218
♦ cultivated, herbal.

Tamarix ramosissima Ledeb.
Tamaricaceae
Tamarix pallasii Desv. var. *brachystachys* Bunge, *Tamarix pentandra* Pall. *nom. illeg.* (see)
♦ tamarisk, saltcedar, perstamarisk, pink tamarisk, athel tree, fivestamen tamarix
♦ Weed, Noxious Weed, Naturalised, Introduced, Garden Escape, Environmental Weed, Cultivation Escape
♦ 54, 63, 78, 80, 86, 88, 95, 101, 105, 129, 138, 142, 152, 161, 198, 212, 228, 229, 231, 269, 277, 283
♦ S/T, arid, cultivated, herbal. Origin: Eurasia, Africa.

Tamarix tetragyna C.Ehrenb.
Tamaricaceae
♦ saltcedar, tamarisk
♦ Noxious Weed, Naturalised
♦ 101, 229

Tamarix usneoides E.Mey. ex Bunge
Tamaricaceae
Tamarix austro-africana Schinz
♦ Introduced
♦ 228
♦ arid.

Tamonea boxiana (Moldenke) Howard
Melastomataceae

♦ crow broom
♦ Naturalised
♦ 101

Tamus communis L.
Dioscoreaceae
♦ black bryony, blackeye root, uva taminia
♦ Weed, Quarantine Weed
♦ 23, 39, 70, 88, 215, 220, 272
♦ pC, cultivated, herbal, toxic.

Tanacetum annuum L.
Asteraceae
♦ annual tansy
♦ Weed
♦ 70

Tanacetum boreale Fisch. ex DC.
Asteraceae
= *Tanacetum vulgare* L.
♦ Weed, Naturalised
♦ 86, 98, 203
♦ Origin: Siberia.

Tanacetum camphoratum Less.
Asteraceae
Tanacetum douglasii DC. (see)
♦ camphor tansy, dune tansy
♦ Quarantine Weed
♦ 220
♦ pH, herbal.

Tanacetum cinerariifolium (Trev.) Sch.Bip.
Asteraceae
Chrysanthemum cinerariifolium (Trev.) Vis. (see), *Chrysanthemum turreanum* Vis., *Pyrethrum cinerariifolium* Trev., *Pyrethrum willemoti* Duches.
♦ pyrethrum, painted daisy, Dalmatianpäivänkakkara, Dalmatian pellitory, Dalmatia pyrethrum, Dalmatian insect flower, pyrethrum, pelitre, piretro
♦ Casual Alien
♦ 42
♦ pH, cultivated, herbal. Origin: Europe.

Tanacetum coccineum (Willd.) Grierson
Asteraceae
Chrysanthemum coccineum Willd. (see), *Chrysanthemum marschallii* Asch. ex O.Hoffm., *Chrysanthemum roseum* Adams, *Pyrethrum carneum* M.Bieb., *Pyrethrum roseum* (Adams) M.Bieb.
♦ punapäivänkakkara, pyrethrum, painted daisy, Persian daisy, Persian insect flower, pyrethrum
♦ Cultivation Escape
♦ 42
♦ pH, cultivated, herbal.

Tanacetum corymbosum (L.) Sch.Bip.
Asteraceae
Chrysanthemum corymbosum L. (see), *Leucanthemum corymbosum* Gren. & Godr., *Matricaria corymbosa* Desr., *Pyrethrum corymbosum* (L.) Scop.
♦ huiskilopäivänkakkara, corymbflower tansy
♦ Weed, Naturalised, Cultivation Escape
♦ 42, 101, 272
♦ cultivated, herbal.

Tanacetum densum (Labill.) Sch.Bip.
Asteraceae
♦ Quarantine Weed
♦ 220
♦ cultivated.

Tanacetum douglasii DC.
Asteraceae
= *Tanacetum camphoratum* Less.
♦ Quarantine Weed
♦ 220
♦ herbal.

Tanacetum haradjanii (Rech.) Grierson
Asteraceae
Chrysanthemum haradjanii Rech.
♦ densum amanum
♦ Weed, Quarantine Weed
♦ 76, 88, 220
♦ cultivated.

Tanacetum huronense Nutt.
Asteraceae
Tanacetum pauciflorum Richards.
♦ Weed
♦ 23, 88
♦ herbal.

Tanacetum macrophyllum (Waldst. & Kit.) Schultz.
Asteraceae
♦ Weed
♦ 272
♦ cultivated.

Tanacetum niveum (Lag.) Sch.Bip.
Asteraceae
Pyrethrum niveum Lag.
♦ silver tansy
♦ Quarantine Weed
♦ 220
♦ cultivated, herbal.

Tanacetum parthenium (L.) Sch.Bip.
Asteraceae
Artemisia tenuifolia Fuchs,
Chrysanthemum parthenium (L.) Bernh.
(see), *Chrysanthemum praealtum*
Vent., *Leucanthemum odoratum* Dulac,
Leucanthemum parthenium Gren. &
Godr., *Matricaria eximia* Voss *nom.
inval.*, *Matricaria febrifuga* Brunfels,
Matricaria latifolium Gilib., *Matricaria
parthenium* L., *Pyrethrum parthenium*
(L.) Sm.
♦ feverfew, featherfew,
reunuspäivänkakkara, grande
camomille, altamisa
♦ Weed, Naturalised, Cultivation
Escape
♦ 24, 34, 40, 42, 86, 98, 101, 176, 198,
203, 241, 252, 272, 280, 287, 300
♦ pH, cultivated, herbal, toxic. Origin:
south-east Europe.

Tanacetum vulgare L.
Asteraceae
Chrysanthemum tanacetum Vis. (see),
Chrysanthemum vulgare (Lam.) Parsa
(see), *Tanacetum boreale* Fisch. ex DC.
(see), *Tanacetum officinarum* Crantz
♦ tansy, common tansy, garden
tansy, pietaryrtti, bitter buttons, hind
head, parsley fern, golden buttons,
barbotine, tanaisie, rainfarn, atanásia,

hierba lombriguera, tanaceto
♦ Weed, Noxious Weed, Naturalised,
Cultivation Escape
♦ 1, 4, 23, 34, 35, 70, 78, 80, 86, 87, 88,
94, 98, 101, 116, 136, 138, 139, 146, 159,
161, 162, 165, 176, 195, 198, 203, 210,
212, 218, 219, 229, 241, 243, 247, 252,
272, 280, 287, 299, 300
♦ pH, cultivated, herbal, toxic. Origin:
Eurasia.

Tanaecium exitiosum Dugand
Bignoniaceae
♦ Weed, Quarantine Weed
♦ 76, 87, 88, 203, 220
♦ herbal.

Tapirira guianensis Aubl.
Anacardiaceae
♦ tapiriri, pau pombo, copiuva, jobo
♦ Weed
♦ 87, 88, 255
♦ T. Origin: Brazil.

Tarasa antofagastana (Phil.) Krapov.
Malvaceae
Malva antofagastana Phil.
♦ Weed
♦ 237, 295

Tarasa geranioides (Cham. & Schltdl.) Krapov.
Malvaceae
♦ Weed
♦ 199

Taraxacum albidum Dahlst.
Asteraceae
♦ Weed
♦ 286
♦ pH, promoted, herbal. Origin: east
Asia.

Taraxacum erythrospermum auct.
Asteraceae
= *Taraxacum laevigatum* (Willd.) DC.
♦ smooth dandelion, dandelion, lesser
dandelion
♦ Weed, Naturalised
♦ 23, 88, 98, 203, 218, 295
♦ herbal.

Taraxacum hondoense Nakai
Asteraceae
♦ ezotanpopo
♦ Weed
♦ 286
♦ pH, promoted. Origin: east Asia.

Taraxacum japonicum Koidz.
Asteraceae
♦ Weed
♦ 286
♦ pH, promoted, herbal. Origin: east
Asia.

Taraxacum kok-saghyz Rodin
Asteraceae
Taraxacum bicorne Boiss.
♦ rubber dandelion
♦ Weed, Naturalised
♦ 23, 86, 88, 98, 176
♦ pH, arid, cultivated, herbal. Origin:
Eurasia.

Taraxacum laevigatum (Willd.) DC. group

Asteraceae
Taraxacum erythrospermum auct. (see)
♦ lesser dandelion, rock dandelion,
red seeded dandelion
♦ Weed, Naturalised
♦ 80, 101, 161, 179, 237, 286, 287
♦ pH, cultivated, herbal.

Taraxacum longe-appendiculatum Nakai
Asteraceae
= *Taraxacum platycarpum* Dahlst. var.
longeappendiculatum (Nakai) T.Morita
(NoR)
♦ Weed
♦ 286

Taraxacum megalorhizon (Forssk.) Hand.-Mazz.
Asteraceae
Leontodon megalorrhizon Forssk.
♦ Weed
♦ 23, 88

Taraxacum mongolicum Hand-Mazz.
Asteraceae
♦ Chinese dandelion
♦ Weed
♦ 275, 297
♦ pH, promoted, herbal.

Taraxacum officinale (L.) Weber ex F.H.Wigg.
Asteraceae
Lentodon taraxacum L., *Taraxacum
dens-leonis* Desf., *Taraxacum retroflexum*
Lindb.f., *Taraxacum taraxacum* Karst.,
Taraxacum vulgare Schrank (see)
♦ common dandelion, English
dandelion, dandelion, little marsh
dandelion, bog dandelion, lesser
dandelion, tarassaco, Diente de león,
lion's tooth, blowball, cankerwort,
doorhead clock, milk witch, puffball,
witch's gowan, yellow gowan
♦ Weed, Noxious Weed, Naturalised,
Native Weed, Introduced, Garden
Escape, Environmental Weed
♦ 7, 15, 21, 23, 34, 44, 45, 49, 51, 52, 70,
80, 86, 87, 88, 93, 94, 97, 101, 114, 121,
134, 136, 157, 158, 161, 162, 165, 167,
174, 176, 179, 180, 181, 195, 203, 204,
205, 210, 211, 212, 217, 218, 236, 237,
241, 243, 245, 248, 249, 253, 255, 261,
266, 269, 271, 272, 280, 286, 287, 295,
299, 300
♦ aH, cultivated, herbal, toxic. Origin:
Eurasia.

Taraxacum officinale G.H.Weber ex Wigg. ssp. officinale
Asteraceae
♦ common dandelion
♦ Naturalised
♦ 101

Taraxacum officinale G.H.Weber ex Wigg. ssp. vulgare (Lam.) Schinz & R.Keller
Asteraceae
♦ common dandelion
♦ Naturalised
♦ 101

Taraxacum ohwianum Kitam.
Asteraceae

♦ Naturalised
♦ 287

Taraxacum palustre (Lyons) Symons
Asteraceae
Hedypnois paludosa Scop., *Leontodon paluster* Sm., *Leontodon raji* Gouan, *Taraxacum lanceolatum* Poir.
♦ marsh dandelion, little marsh dandelion
♦ Naturalised
♦ 101
♦ herbal.

Taraxacum platycarpum Dahlst.
Asteraceae
♦ Weed
♦ 23, 87, 88, 263, 286
♦ pH, promoted.

Taraxacum pseudocalocephalum (Willd.) DC.
Asteraceae
♦ Naturalised
♦ 86

Taraxacum serotinum (Waldst. & Kit.) Poir.
Asteraceae
♦ dandelion
♦ Weed
♦ 51, 87, 88, 121
♦ pH, cultivated. Origin: Eurasia.

Taraxacum spectabile Dahlst.
Asteraceae
♦ showy dandelion, bog dandelion, dandelion, broad leaved marsh dandelion
♦ Naturalised
♦ 101

Taraxacum vulgare Schrank
Asteraceae
= *Taraxacum officinale* F.H.Wigg. group
♦ Weed, Quarantine Weed
♦ 76, 87, 88, 203, 220, 243
♦ herbal.

Tarchonanthus camphoratus L.
Asteraceae
Tarchonanthus minor Less.
♦ African fleabane, camphor bush tree, camphor bush wood, sagewood, wild cotton, wild sage, kanferbos, vaalbos
♦ Weed, Quarantine Weed, Native Weed
♦ 10, 63, 76, 87, 88, 121, 203, 220
♦ S/T, arid, cultivated, herbal. Origin: southern Africa.

Tauschia neglecta Cald. & Const.
Apiaceae
♦ Weed
♦ 199

Taverniera aegyptiaca Boiss.
Fabaceae/Papilionaceae
♦ Weed
♦ 221
♦ arid.

Taxodium distichum (L.) Rich.
Taxodiaceae/Cupressaceae
Cupressus disticha L.
♦ bald cypress, swamp cypress
♦ Weed

♦ 87, 88, 218
♦ T, aqua, cultivated, herbal.

Taxus baccata L.
Taxaceae
♦ English yew, marjakuusi, yew, common yew
♦ Weed, Naturalised
♦ 39, 101, 154, 161, 280
♦ T, cultivated, herbal, toxic.

Taxus brevifolia Nutt.
Taxaceae
♦ Pacific yew
♦ Weed
♦ 39, 87, 88, 161, 218
♦ T, cultivated, herbal, toxic.

Taxus canadensis Marsh.
Taxaceae
♦ Canada yew, American yew
♦ Weed
♦ 8, 39, 154, 161
♦ T, promoted, herbal, toxic.

Taxus cuspidata Sieb. & Zucc.
Taxaceae
♦ Japanese yew, Japaninmarjakuusi
♦ Weed, Naturalised, Garden Escape, Environmental Weed
♦ 39, 80, 88, 101, 133, 142, 151, 161, 195
♦ T, cultivated, herbal, toxic.

Taxus floridana Nutt. ex Chapman
Taxaceae
♦ Florida yew
♦ Weed
♦ 87, 88, 161, 218
♦ herbal, toxic.

Tecoma alata DC.
Bignoniaceae
= *Tecoma guarume* A.DC. (NoR)
♦ cahuato
♦ Naturalised
♦ 86
♦ cultivated. Origin: Peru.

Tecoma capensis (Thunb.) Lindl.
Bignoniaceae
Tecomaria capensis (Thunb.) Spach (see), *Bignonia capensis* Thunb., *Gelseminum capense* (Thunb.) Kuntze, *Tecomaria krebsii* Klotzsch, *Tecomaria petersii* Klotzsch
♦ cape honeysuckle
♦ Weed, Naturalised, Introduced
♦ 54, 86, 88, 101, 179, 261
♦ cultivated, herbal. Origin: South Africa.

Tecoma castanifolia (D.Don) Melch.
Bignoniaceae
♦ chestnutleaf trumpetbush
♦ Weed, Naturalised
♦ 101, 179
♦ cultivated.

Tecoma stans (L.) Juss. ex Kunth
Bignoniaceae
Bignonia stans L., *Stenolobium incisum* Rose & Standl., *Stenolobium quinquejugum* Loes., *Stenolobium stans* (L.) Seem., *Stenolobium tronadora* Loes., *Tecoma mollis* Kunth, *Tecoma stans* L. var. *velutina* DC.

♦ yellow trumpetbush, yellow bells, yellow elder, tagamimi, piti, peeal, trovadora, geelklokkies
♦ Weed, Noxious Weed, Naturalised, Introduced, Garden Escape, Environmental Weed
♦ 3, 6, 7, 22, 39, 63, 86, 87, 88, 95, 98, 107, 155, 179, 201, 203, 218, 228, 255, 259, 279, 283
♦ S/T, arid, cultivated, herbal, toxic.

Tecomaria capensis (Thunb.) Spach
Bignoniaceae
= *Tecoma capensis* (Thunb.) Lindl.
♦ cape honeysuckle, fire flower
♦ Weed, Sleeper Weed, Naturalised, Environmental Weed
♦ 15, 73, 88, 98, 225, 246, 280
♦ pC, cultivated.

Tectaria incisa Cav.
Dryopteridaceae
♦ incised halberd fern
♦ Weed, Environmental Weed
♦ 80, 80, 88, 112, 151
♦ H, cultivated.

Tectona grandis L.f.
Lamiaceae/Verbenaceae
♦ teak, teca
♦ Naturalised, Cultivation Escape
♦ 39, 101, 261
♦ T, cultivated, herbal, toxic. Origin: tropical Asia.

Teesdalia coronopifolia (Berg.) Thell.
Brassicaceae
♦ lesser shepherdscress
♦ Naturalised
♦ 101
♦ aH.

Teesdalia nudicaulis (L.) R.Br.
Brassicaceae
Capsella nudicaulis Prantl., *Crucifera teesdalea* Krause, *Guepinia nudicaulis* Bast., *Iberis nudicaulis* L., *Lepidium scapiferum* Wall., *Teesdalia iberis* DC., *Thlaspi nudicaule* DC. ex Lam.
♦ shepherd's cress, barestem teesdalia, téesdalie
♦ Weed, Naturalised, Garden Escape, Environmental Weed
♦ 23, 54, 70, 86, 88, 94, 98, 101, 176, 203, 253, 272, 300
♦ cultivated, herbal. Origin: central and west Europe.

Telekia speciosa (Schreb.) Baumg.
Asteraceae
Buphthalmum speciosum (Baumg.) Schreb. (see)
♦ heartleaf oxeye, auringontähti
♦ Weed, Quarantine Weed, Cultivation Escape
♦ 42, 76, 88, 220
♦ cultivated, herbal.

Telephium imperati L.
Molluginaceae/Caryophyllaceae
♦ Quarantine Weed
♦ 220
♦ cultivated.

Telfairia pedata (Sims) Hook.
Cucurbitaceae
Fevillea pedata Sims
 ♦ Zanzibar oil vine, castanha de
Inhambane, cungo, cungua, dicungo,
lipeme, meme, oyster nut, umpeme
 ♦ Introduced
 ♦ 39, 228
 ♦ arid, herbal, toxic.

Teline Medik. spp.
Fabaceae/Papilionaceae
= *Genista* L. spp.
 ♦ Weed
 ♦ 18, 88

Teline hillebrandii (Christ) Kunkel
Fabaceae/Papilionaceae
= *Genista canariensis* L.
 ♦ Quarantine Weed
 ♦ 220

Teline linifolia (L.) Webb & Berthel.
Fabaceae/Papilionaceae
= *Genista linifolia* L.
 ♦ Naturalised
 ♦ 280

**Teline linifolia (L.) Webb & Berthel. ssp.
linifolia**
Fabaceae/Papilionaceae
 ♦ Naturalised
 ♦ 280

**Teline microphylla (DC.) P.E.Gibbs &
Dingwall**
Fabaceae/Papilionaceae
= *Genista microphylla* DC. (NoR)
 ♦ Quarantine Weed
 ♦ 220

Teline monspessulana (L.) K.Koch
Fabaceae/Papilionaceae
= *Genista monspessulana* (L.) L.Johnson
 ♦ Montpellier broom
 ♦ Weed, Naturalised, Environmental
Weed
 ♦ 15, 134, 225, 241, 246, 280, 300

**Teline stenopetala (Webb & Berthel.)
Webb & Berthel.**
Fabaceae/Papilionaceae
= *Genista stenopetala* Webb & Berth.
 ♦ Weed, Naturalised
 ♦ 15, 280
 ♦ cultivated.

Teliostachya alopecuroidea (Vahl) Nees
Acanthaceae
Lepidagathis alopecuroidea (Vahl) R.Br.
ex Griseb.
 ♦ pata de gallina
 ♦ Weed
 ♦ 28, 87, 88, 206, 243

**Tellima grandiflora (Pursh) Douglas ex
Lindl.**
Saxifragaceae
 ♦ bigflower tellima, fringe cups
 ♦ Naturalised, Cultivation Escape
 ♦ 40, 42
 ♦ pH, cultivated, herbal.

Telopea oreades F.Muell.
Proteaceae
 ♦ Victorian waratah
 ♦ Naturalised

 ♦ 280
 ♦ cultivated, herbal.

Teloxys aristata Moq.
Chenopodiaceae
= *Chenopodium aristatum* L.
 ♦ seafoam moss
 ♦ Weed
 ♦ 87, 88
 ♦ cultivated, herbal.

Tephrocactus Lem. spp.
Cactaceae
= *Opuntia* Mill. spp.
 ♦ Weed, Quarantine Weed
 ♦ 203, 220

Tephrosia apollinea (Delile) Link
Fabaceae/Papilionaceae
 ♦ Weed
 ♦ 88, 221
 ♦ arid, herbal.

Tephrosia bracteolata Guill. & Perr.
Fabaceae/Papilionaceae
 ♦ Weed
 ♦ 88
 ♦ cultivated, herbal.

Tephrosia candida DC.
Fabaceae/Papilionaceae
 ♦ white hoarypea, tefrosia
 ♦ Weed, Quarantine Weed,
Naturalised, Introduced, Cultivation
Escape
 ♦ 22, 32, 38, 39, 76, 86, 87, 88, 98, 101,
203, 230, 261
 ♦ S, cultivated, herbal, toxic. Origin:
tropical Asia.

Tephrosia cathartica (Sessé & Moç.) Urb.
Fabaceae/Papilionaceae
 ♦ Weed
 ♦ 87, 88

Tephrosia cinerea (L.) Pers.
Fabaceae/Papilionaceae
Cracca cinerea (L.) Morong, *Galega
cinerea* L., *Tephrosia littoralis* (Jacq.)
Pers.
 ♦ anil cenizo, anil bravo, brusca
cimarrona, herbe enivrer, mort aux
poissons, slender goat's rue, ashen
hoarypea
 ♦ Weed, Introduced
 ♦ 14, 39, 87, 88, 228
 ♦ arid, herbal, toxic.

Tephrosia ehrenbergiana Schweinf.
Fabaceae/Papilionaceae
= *Tephrosia villosa* (L.) Pers. ssp.
ehrenbergiana (Schweinf.) Brumm.
 ♦ Weed
 ♦ 87, 88

Tephrosia elegans Schum.
Fabaceae/Papilionaceae
 ♦ Weed, Naturalised
 ♦ 86, 87, 88
 ♦ Origin: Africa.

Tephrosia glomeruliflora Meissner
Fabaceae/Papilionaceae
 ♦ pink tephrosia
 ♦ Weed, Naturalised, Environmental
Weed

 ♦ 3, 86, 98, 155, 191, 203
 ♦ cultivated. Origin: southern Africa.

**Tephrosia grandiflora (L'Hér. ex Aiton)
Pers.**
Fabaceae/Papilionaceae
 ♦ pink pea bush
 ♦ Weed, Naturalised, Environmental
Weed
 ♦ 86, 98, 203
 ♦ cultivated, herbal. Origin: South
Africa.

Tephrosia inandensis F.M.Forbes
Fabaceae/Papilionaceae
 ♦ Weed, Naturalised
 ♦ 54, 86, 88, 98, 203
 ♦ Origin: South Africa.

Tephrosia nana Kotsch.
Fabaceae/Papilionaceae
 ♦ Weed, Naturalised
 ♦ 86, 93, 191
 ♦ Origin: tropical Africa.

Tephrosia noctiflora Bojer ex Bak.
Fabaceae/Papilionaceae
 ♦ Weed, Naturalised, Introduced
 ♦ 86, 98, 203, 261
 ♦ herbal. Origin: Madagascar.

Tephrosia nubica (Boiss.) Bak.
Fabaceae/Papilionaceae
 ♦ Weed
 ♦ 221

**Tephrosia procumbens (Buch.-Ham.)
Benth.**
Fabaceae/Papilionaceae
 ♦ Weed
 ♦ 87, 88
 ♦ herbal.

Tephrosia purpurea (L.) Pers.
Fabaceae/Papilionaceae
 ♦ tephrosia, fishpoison
 ♦ Weed, Naturalised
 ♦ 39, 87, 88, 101, 218
 ♦ arid, cultivated, herbal, toxic. Origin:
Africa, Asia, Australia, Madagascar.

Tephrosia quartiniana Cuf.
Fabaceae/Papilionaceae
Tephrosia vicioides A.Rich.
 ♦ Weed
 ♦ 221
 ♦ arid.

Tephrosia sessiliflora (Poir.) Hassl.
Fabaceae/Papilionaceae
 ♦ sessileflower hoarypea
 ♦ Naturalised
 ♦ 101

Tephrosia sinapou (Buc'hoz) A.Chev.
Fabaceae/Papilionaceae
Cracca multifolia Rose, *Cracca schiedeana*
(Schldl.) Standl., *Cracca toxicaria* (Sw.)
Kuntze, *Tephrosia emarginata* Kunth,
Tephrosia multifolia Rose, *Tephrosia
schiedeana* Schldl., *Tephrosia toxicaria*
(Sw.) Pers.
 ♦ hoarypea
 ♦ Introduced
 ♦ 39, 228
 ♦ S/T, arid, herbal, toxic.

Tephrosia spinosa (L.f.) Pers.
Fabaceae/Papilionaceae
♦ Weed
♦ 87, 88
♦ herbal.

Tephrosia tinctoria Pers.
Fabaceae/Papilionaceae
♦ Weed, Naturalised
♦ 66, 86, 87, 88, 98, 203
♦ Origin: India.

Tephrosia trifoliata DC.
Fabaceae/Papilionaceae
♦ Quarantine Weed
♦ 220

Tephrosia uniflora Pers.
Fabaceae/Papilionaceae
Tephrosia anthylloides Hochst. ex Bak.,
Tephrosia lathyroides Guillem. Perrot.,
Tephrosia transjubens Chiov.
♦ Weed
♦ 87, 88, 242
♦ arid, herbal.

Tephrosia villosa (L.) Pers. ssp. ehrenbergiana (Schweinf.) Brumm.
Fabaceae/Papilionaceae
Tephrosia ehrenbergiana Schweinf. (see)
♦ Introduced
♦ 228
♦ a/pH, arid. Origin: eastern to southern Africa, Madagascar.

Tephrosia virginiana (L.) Pers.
Fabaceae/Papilionaceae
♦ goat's rue, catgut, Virginia tephrosia, hoarypea
♦ Weed
♦ 8, 39, 161, 247
♦ pH, promoted, herbal, toxic.

Tephrosia vogelii Hook.f.
Fabaceae/Papilionaceae
Tephrosia periculosa Bak.
♦ Vogel's tephrosia, pange
♦ Weed, Naturalised, Introduced
♦ 38, 39, 86, 98, 203, 228
♦ arid, cultivated, herbal, toxic. Origin: Africa.

Teramnus labialis (L.f.) Spreng.
Fabaceae/Papilionaceae
♦ blue wiss
♦ Weed, Naturalised
♦ 28, 32, 86, 87, 88, 157, 206, 243
♦ aC, cultivated, herbal. Origin: Madagascar.

Teramnus labialis (L.f.) Spreng. ssp. labialis
Fabaceae/Papilionaceae
Glycine labialis L.f.
♦ Native Weed
♦ 121
♦ pC. Origin: southern Africa.

Teramnus volubilis Sw.
Fabaceae/Papilionaceae
♦ Weed
♦ 87, 88
♦ cultivated.

Terminalia arjuna (Roxb. ex DC.) Wight & Arn.
Combretaceae

Pentaptera glabra Roxb., *Terminalia alata* D.Dietr., *Terminalia ovalifolia* C.B.Clarke
♦ arjuna, maddi, sanmadat, vellamardha, vellamatta, yerramaddi
♦ Weed
♦ 22, 179
♦ T, arid, cultivated, herbal.

Terminalia bellirica (Gaertn.) Roxb.
Combretaceae
Myrobalanus bellirica Gaertn.
♦ ahdan koddai, bahera, bulu, desi badam, tanti, myrobalan
♦ Introduced
♦ 39, 228
♦ arid, cultivated, herbal, toxic.

Terminalia carolinensis Kaneh.
Combretaceae
♦ terminalia, keima
♦ Introduced
♦ 230
♦ T.

Terminalia catappa L.
Combretaceae
♦ tropical almond, Indian almond, false kamani, kamani haole, talie, sea almond, almendra
♦ Weed, Naturalised, Introduced, Environmental Weed, Cultivation Escape
♦ 22, 80, 87, 88, 101, 112, 122, 151, 179, 228, 233, 261
♦ T, arid, cultivated, herbal. Origin: East Indies to Pacific Islands.

Terminalia chebula (Gaertn.) Retz.
Combretaceae
♦ aralu, kadukkay, myrobalan
♦ Introduced
♦ 39, 228
♦ arid, promoted, herbal, toxic.

Terminalia ivorensis Chev.
Combretaceae
♦ Ivory Coast almond, cortesa negra
♦ Naturalised, Cultivation Escape
♦ 101, 261
♦ cultivated, herbal. Origin: west tropical Africa.

Terminalia muelleri Benth.
Combretaceae
♦ Australian almond, West Indian almond
♦ Weed, Naturalised
♦ 101, 179
♦ cultivated. Origin: Australia.

Terminalia myriocarpa Van Heurck & Müll.Arg.
Combretaceae
♦ jhalna, East Indian almond
♦ Weed, Naturalised, Introduced
♦ 3, 22, 101, 191, 261
♦ T, herbal. Origin: India.

Terminalia oblonga (Ruiz & Pav.) Steud.
Combretaceae
♦ Naturalised, Cultivation Escape
♦ 101, 261
♦ T, cultivated. Origin: Peru.

Terminalia oblongata F.Muell.
Combretaceae
♦ Weed

♦ 87, 88
♦ Origin: Australia.

Terminalia prunioides Lawson
Combretaceae
♦ lowveld clusterleaf, lowveld terminalia, purplepod terminalia
♦ Native Weed
♦ 121
♦ S/T, cultivated. Origin: southern Africa.

Terminalia sericea Burch. ex DC.
Combretaceae
Terminalia silozensis Gibbs
♦ assegai wood, sand yellow wood, silver terminalia, silver tree, silver clusterleaf, Transvaal silver tree, wild quince, sandvaalboom, sandgeelhout
♦ Weed, Native Weed
♦ 10, 63, 87, 88, 121
♦ S/T, arid, cultivated, herbal. Origin: southern Africa.

Tessaria absinthioides (Hook. & Arn.) DC.
Asteraceae
Baccharis absinthiodes Hook. & Arn.,
Gynheteria incana Spreng., *Pluchea absinthioides* (Hook. & Arn.) H.Rob. & Cuatrec.
♦ suncho rosado
♦ Weed, Introduced
♦ 87, 88, 228, 295
♦ arid.

Tessaria dodonaefolia (Hook. & Arn.) Cabr.
Asteraceae
Tessaria viscosa Lillo., *Eupatorium dodnaefolium* Hook & Arn.
♦ chilca negra
♦ Weed
♦ 295

Tessaria integrifolia Ruiz & Pav.
Asteraceae
♦ Weed
♦ 87, 88
♦ T.

Tetracera breyniana Schlecht.
Dilleniaceae
♦ Weed
♦ 87, 88
♦ herbal.

Tetracera indica (Christm. & Panz.) Merr.
Dilleniaceae
♦ Weed
♦ 12, 87, 88
♦ herbal.

Tetracera scandens (L.) Merr.
Dilleniaceae
♦ Weed
♦ 87, 88
♦ herbal.

Tetradymia axillaris Nelson
Asteraceae
♦ longspine horsebrush, cotton cat's claw, cottonthorn
♦ Weed
♦ 39, 161
♦ S, cultivated, herbal, toxic.

Tetradymia canescens DC.
Asteraceae
♦ spineless horsebrush, gray
horsebush
♦ Weed
♦ 39, 87, 88, 161, 212, 218
♦ S, herbal, toxic.

Tetradymia glabrata A.Gray
Asteraceae
♦ littleleaf horsebrush, smooth
horsebrush, coal oil brush, spring
rabbitbrush
♦ Weed
♦ 39, 87, 88, 161, 218
♦ S, herbal, toxic.

Tetradymia stenolepis Greene
Asteraceae
♦ Mojave horsebrush, Mojave
cottonthorn
♦ Weed
♦ 161
♦ S, herbal, toxic.

Tetraglochin cristatum (Britton) Rothm.
Rosaceae
Margyricarpus cristatus Britton
♦ Introduced
♦ 228
♦ arid.

Tetragonia caesia Adamson
Aizoaceae/Tetragoniaceae
♦ klappiesbrak
♦ Weed
♦ 88, 158, 243
♦ Origin: southern Africa.

Tetragonia decumbens Mill.
Aizoaceae/Tetragoniaceae
♦ Weed, Naturalised, Environmental
Weed
♦ 7, 9, 86, 98, 203
♦ cultivated. Origin: South Africa.

Tetragonia echinata Aiton
Aizoaceae/Tetragoniaceae
♦ Native Weed
♦ 121
♦ aH, cultivated. Origin: southern
Africa.

Tetragonia macroptera Pax
Aizoaceae/Tetragoniaceae
♦ Native Weed
♦ 121
♦ pS. Origin: southern Africa.

Tetragonia microptera Fenzl var.
microptera
Aizoaceae/Tetragoniaceae
♦ African spinach
♦ Naturalised
♦ 86, 198
♦ Origin: South Africa.

Tetragonia nigrescens Eckl. & Zeyh.
Aizoaceae/Tetragoniaceae
♦ Weed, Naturalised
♦ 86, 98, 203
♦ Origin: South Africa.

Tetragonia tetragonioides (Pall.) Kuntze
Aizoaceae/Tetragoniaceae
= *Tetragonia tetragonoides* (Pall.) Kuntze

♦ kokihi, New Zealand spinach
♦ Weed, Naturalised, Native Weed,
Introduced, Garden Escape, Casual
Alien
♦ 39, 40, 55, 87, 88, 101, 116, 131, 161,
228, 243, 261, 269
♦ aH, arid/aqua, cultivated, herbal,
toxic. Origin: New Zealand.

Tetragonia tetragonoides (Pall.) Kuntze
Aizoaceae/Tetragoniaceae
Demidovia tetragonioides Pall., *Tetragonia
cornuta* Gaertn., *Tetragonia expansa*
Murr., *Tetragonia halimifolia* Forst.f.,
Tetragonia japonica Thunb., *Tetragonia
quadricornis* Stokes, *Tetragonia
tetragonioides* (Pall.) Kuntze (see)
♦ lamopinaatti, New Zealand spinach,
kohiki
♦ Naturalised, Cultivation Escape
♦ 42, 241, 300
♦ aH, cultivated. Origin: New
Zealand.

Tetragonolobus maritimus (L.) Roth
Fabaceae/Papilionaceae
= *Lotus maritimus* L.
♦ dragon's teeth
♦ Weed
♦ 88, 94
♦ pH, cultivated, herbal.

**Tetragonolobus palaestinus Boiss. &
Blanche**
Fabaceae/Papilionaceae
♦ Palestine winged pea
♦ Weed
♦ 88, 115

Tetragonolobus purpureus Moench
Fabaceae/Papilionaceae
Lotus tetragonolobus L. (see), *Scandalida
rubra* Medik., *Scandalida tetragonoloba*
Medik., *Tetragonolobus edulis* Link
♦ winged pea
♦ Weed, Quarantine Weed
♦ 87, 88, 94, 220, 221
♦ cultivated.

Tetramerium hispidum Nees
Acanthaceae
= *Tetramerium nervosum* Nees
♦ Introduced
♦ 228
♦ arid, herbal.

Tetramerium nervosum Nees
Acanthaceae
Tetramerium hispidum Nees (see)
♦ hairy fournwort
♦ Weed
♦ 157
♦ pH, arid.

Tetrapanax papyrifer (Hook.) K.Koch
Araliaceae
Tetrapanax papyriferus (Hook.) K.Koch
(see)
♦ aralia, ricepaper plant
♦ Weed, Naturalised, Garden Escape,
Environmental Weed
♦ 86, 98, 134, 203, 269, 290
♦ S, cultivated. Origin: China, Japan,
Taiwan.

Tetrapanax papyriferus (Hook.) K.Koch
Araliaceae
= *Tetrapanax papyrifer* (Hook.) K.Koch
♦ ricepaper plant
♦ Weed, Naturalised
♦ 15, 101, 165, 280
♦ cultivated, herbal. Origin: south
China, Taiwan.

Tetraselago natalensis (Rolfe) Junell
Scrophulariaceae/Selaginaceae
Selago natalensis Rolfe
♦ Native Weed
♦ 121
♦ pS. Origin: southern Africa.

Tetrastigma umbellata (Hemsl.) Nakai
Vitaceae
♦ Weed
♦ 87, 88

Tetrazygia bicolor (Mill.) Cogn.
Melastomataceae
♦ Florida clover ash
♦ Weed, Cultivation Escape
♦ 14, 233
♦ S/T, cultivated, herbal.

Teucrium botrys L.
Lamiaceae
Chamaedrys botrys Moench, *Scorodonia
botrys* Ser.
♦ cut leaved germander
♦ Weed, Naturalised
♦ 88, 94, 101, 253, 272
♦ cultivated, herbal.

Teucrium canadense L.
Lamiaceae
♦ American germander, Canada
germander, germander, wood sage,
wild basil, creeping germander
♦ Weed, Native Weed
♦ 23, 88, 161, 174, 210
♦ pH, cultivated, herbal. Origin: North
America.

Teucrium chamaedrys L.
Lamiaceae
Chamaedrys officinalis Moench, *Teucrium
officinale* Lam., *Teucrium sinuatum*
Celak.
♦ wall germander, germander
♦ Weed
♦ 39, 70, 272
♦ S, cultivated, herbal, toxic.

Teucrium cubense Jacq.
Lamiaceae
♦ small coastal germander
♦ Weed
♦ 87, 88
♦ herbal.

Teucrium fruticans L.
Lamiaceae
♦ shrubby germander, tree germander
♦ Naturalised
♦ 101
♦ cultivated, herbal.

Teucrium hircanicum L.
Lamiaceae
♦ Naturalised
♦ 280
♦ cultivated.

Teucrium integrifolium Benth.
Lamiaceae
- teucry weed
- Weed
- 55, 87, 88
- Origin: Australia.

Teucrium orientale L.
Lamiaceae
- oriental germander
- Naturalised
- 101

Teucrium polium L.
Lamiaceae
- felty germander, ezovion, bijeli dubacac
- Weed
- 221, 272
- S, cultivated, herbal.

Teucrium resupinatum Desf.
Lamiaceae
- Weed
- 87, 88

Teucrium scordium L.
Lamiaceae
Chamaedrys scordium Moench, *Teucrium arenarium* Gmel., *Teucrium palustre* Lam., *Teucrium petkovii* Urum., *Teucrium serratum* Benth.
- water germander, polio montano, camedrio scordio, iksugawood sage
- Weed, Quarantine Weed, Naturalised
- 39, 70, 76, 86, 87, 88, 203, 220, 272
- pH, cultivated, herbal, toxic.

Teucrium scorodonia L.
Lamiaceae
Monochilon cordifolius Dulac., *Scorodonia heteromalla* Moench, *Teucrium salviaefolium* Salisb., *Teucrium sylvestre* Lam.
- curled woodsage, woodsage, woodland germander, kelta akankaali, germandrée scorodoine
- Weed, Quarantine Weed, Naturalised, Garden Escape, Environmental Weed, Casual Alien
- 42, 54, 76, 86, 88, 101, 176, 280
- pH, cultivated, herbal. Origin: Europe.

Teucrium spinosum L.
Lamiaceae
- Weed
- 87, 88

Thalassiosira guillardii Hasle.
Bacillariophyceae/Thalassiosiraceae
- Weed
- 197
- diatom.

Thalassiosira lacustris (Grunov) Hasle.
Bacillariophyceae/Thalassiosiraceae
- Weed
- 197
- diatom.

Thalassiosira pseudonana Hasle & Heindal
Bacillariophyceae/Thalassiosiraceae

Cyclotella nana Hust.
- Weed
- 197
- diatom.

Thalassiosira weissflogii (Grunov) G.Fryxell & Hasle
Bacillariophyceae/Thalassiosiraceae
Micropodiscus weissflogii Grunov, *Thalassiosira fluviatilis* Hust.
- Weed
- 197
- diatom.

Thalia dealbata Fraser ex Roscoe
Marantaceae
- hardy water canna, powdery alligator flag
- Naturalised
- 287
- aqua, cultivated, herbal.

Thalia geniculata L.
Marantaceae
- bent alligator flag, fire flag, arrowroot
- Weed, Quarantine Weed
- 14, 87, 88, 255, 258, 295
- pH, aqua, cultivated, herbal. Origin: Brazil.

Thalictrum L. spp.
Ranunculaceae
- meadowrue
- Weed
- 39, 272
- herbal, toxic.

Thalictrum aquilegifolium L.
Ranunculaceae
- columbine meadowrue, meadowrue
- Weed, Naturalised, Casual Alien
- 101, 272, 280
- cultivated, herbal.

Thalictrum aquilegifolium L. var. intermedium Nakai
Ranunculaceae
- karamatsusou
- Weed
- 286

Thalictrum flavum L.
Ranunculaceae
- common meadowrue, keltaängelmä
- Weed, Naturalised
- 39, 40, 70, 272
- cultivated, herbal, toxic.

Thalictrum foetidum L.
Ranunculaceae
- saniaisängelmä
- Weed, Quarantine Weed, Casual Alien
- 39, 42, 76, 88, 220
- cultivated, herbal, toxic.

Thalictrum lucidum L.
Ranunculaceae
- kaitaängelmä
- Naturalised
- 42
- cultivated.

Thalictrum minus L.
Ranunculaceae

- small meadowrue, meadowrue, alpine meadowrue, cliff meadowrue, common meadowrue, greater meadowrue, sand meadowrue, lesser meadowrue
- Weed, Cultivation Escape, Casual Alien
- 23, 42, 87, 88, 272, 280
- pH, cultivated, herbal.

Thalictrum minus L. var. hypoleucum (Sieb. & Zucc.) Miq.
Ranunculaceae
- akikaramatsu
- Weed
- 286
- cultivated.

Thalictrum morisonii C.C.Gmel.
Ranunculaceae
Thalictrum exaltatum auct.
- Weed, Quarantine Weed
- 76, 88, 220

Thalictrum petaloideum L.
Ranunculaceae
- petal formed meadowrue
- Weed
- 297
- cultivated.

Thalictrum simplex L.
Ranunculaceae
- hoikkaängelmä
- Weed, Quarantine Weed
- 76, 87, 88, 220, 272
- cultivated, herbal.

Thalictrum sparsiflorum Turcz. ex Fisch. & C.A.Mey.
Ranunculaceae
- fewflower meadowrue
- Weed, Quarantine Weed
- 76, 88, 220
- pH, herbal.

Thalictrum squarrosum Stephan ex Willd.
Ranunculaceae
- Weed, Quarantine Weed
- 76, 88, 220, 297
- cultivated.

Thamnosma texana (Gray) Torr.
Rutaceae
- Dutchman's breeches, rue of the mountains
- Weed
- 161
- herbal, toxic.

Thaumatococcus daniellii (Benn.) Benth. ex Jacks.
Marantaceae
- sweet prayer plant, miracle fruit
- Sleeper Weed, Naturalised, Environmental Weed
- 39, 86, 155
- cultivated, herbal, toxic.

Thea sinensis L.
Theaceae
= *Camellia sinensis* (L.) Kuntze
- Naturalised
- 287
- herbal.

Thelechitonia trilobata (L.) H.Rob. & Cuatrec.
Asteraceae
= *Sphagneticola trilobata* (L.C.Rich.) Pruski
♦ Singapore daisy, Singapoer madeliefie
♦ Weed, Noxious Weed, Garden Escape
♦ 3, 191, 283
♦ cultivated. Origin: tropical America.

Thelesperma megapotamicum (Spreng.) Kuntze
Asteraceae
Bidens megapotamica Spreng. (see)
♦ Hopi tea greenthread
♦ Introduced
♦ 34, 228
♦ pH, arid, herbal.

Theligonum cynocrambe L.
Theligonaceae/Rubiaceae
♦ dog's cabbage
♦ Weed
♦ 70, 111, 243
♦ aH, cultivated.

Thellungiella salsuginea (Pall.) Schulz
Brassicaceae
= *Arabidopsis salsuginea* (Pall.) N.Busch (NoR)
♦ Weed
♦ 275
♦ a/bH.

Thelocactus (K.Schum.) Britt. & Rose spp.
Cactaceae
♦ thelocactus
♦ Weed, Quarantine Weed, Naturalised
♦ 76, 86, 88, 203
♦ cultivated.

Thelypteris dentata (Forssk.) E.St.John
Thelypteridaceae
Cyclosorus dentatus (Forssk.) Ching (see)
♦ downy woodfern, mountain woodfern, downy maiden fern, rabo de gato, samambaia rabo de gato, samambaia do mato
♦ Weed
♦ 28, 179, 206, 243, 255
♦ cultivated, herbal. Origin: tropical America.

Thelypteris gracilescens (Bl.) Ching
Thelypteridaceae
Lastrea gracilescens (Bl.) Moore
♦ Weed
♦ 87, 88

Thelypteris opulenta (Kaulf.) Fosberg
Thelypteridaceae
♦ jeweled maiden fern, swordfern, fern
♦ Weed, Naturalised, Environmental Weed
♦ 32, 179, 227, 261
♦ herbal. Origin: tropical Africa and Asia.

Thelypteris palustris Schott
Thelypteridaceae

Acrostichum thelypteris L., *Dryopteris thelypteris* (L.) A.Gray, *Lastrea thelypteris* (L.) Bory. (see), *Polypodium oreopteris* Ehrh., *Polypodium palustre* Salisb., *Thelypteris thelypterioides* auct.
♦ marsh fern, nevaimarre, eastern marsh fern, felce palustre
♦ Weed
♦ 286
♦ H, cultivated, herbal.

Thelypteris phegopteris (L.) Slos.
Thelypteridaceae
= *Phegopteris connectilis* (F.Michx.) Watt (NoR)
♦ beech fern, korpi imarre
♦ Weed
♦ 70
♦ cultivated, herbal.

Thelypteris poiteana (Bory) Proctor
Thelypteridaceae
♦ dark green maiden fern
♦ Weed
♦ 28, 206, 243

Thelypteris ponapeana (Hosok.) Reed
Thelypteridaceae
Phegoteris ponapeana Hosok.
♦ malkenahna
♦ Introduced
♦ 230
♦ H.

Themeda arguens (L.) Hack.
Poaceae
♦ lesser tanglegrass, Christmas grass
♦ Weed, Naturalised, Cultivation Escape
♦ 86, 87, 88, 90, 93, 101, 261
♦ pG, cultivated, herbal. Origin: Asia, Australia.

Themeda arundinacea (Roxb.) Ridl.
Poaceae
♦ giant themeda
♦ Weed
♦ 297
♦ G.

Themeda australis (R.Br.) Stapf
Poaceae
= *Themeda triandra* Forssk.
♦ kangaroo grass
♦ Weed
♦ 87, 88
♦ G, cultivated.

Themeda avenacea (F.Muell.) Maiden & Betche
Poaceae
Anthistiria avenacea F.Muell.
♦ tall oatgrass
♦ Naturalised
♦ 198
♦ G, arid, cultivated. Origin: Australia.

Themeda gigantea (Cav.) Hack.
Poaceae
♦ Weed, Quarantine Weed
♦ 76, 87, 88, 203, 220
♦ G, arid, cultivated, herbal.

Themeda japonica Tanaka
Poaceae
♦ grass stalk, Japanese kangaroo grass

♦ Weed
♦ 297
♦ G.

Themeda quadrivalvis (L.) Kuntze
Poaceae
Andropogon quadrivalvis L.
♦ grader grass, habana grass, oatgrass, kangaroo grass
♦ Weed, Quarantine Weed, Noxious Weed, Naturalised
♦ 6, 11, 55, 76, 86, 87, 88, 93, 98, 101, 147, 203, 269
♦ aG, cultivated. Origin: India.

Themeda triandra Forssk.
Poaceae
Anthistiria ciliata Nees, *Anthistiria glauca* Desf., *Anthistiria imberbis* Retz., *Anthistiria paleacea* (Vahl) Ball, *Anthistiria punctata* Hochst. ex A.Rich., *Calamina imberbis* (Retz.) Roem. & Schult., *Stipa arguens* Thunb., *Themeda australis* (R.Br.) Stapf (see), *Themeda forskalii* (Kunth) Hack., *Themeda imberbis* (Retz.) Cooke, *Themeda triandra* var. *burchellii* (Hack.) Stapf, *Themeda triandra* var. *hispida* (Nees) Stapf, *Themeda triandra* var. *imberbis* (Retz.) A.Camus, *Themeda triandra* var. *trachyspathea* Gooss., *Themeda triandra* var. *vulgaris* auct. non Hack.
♦ angle grass, bluegrass, kangaroo grass, red grass, red oatgrass, red oat, nkuku
♦ Weed, Native Weed, Naturalised, Introduced
♦ 87, 88, 121, 228, 280
♦ pG, arid, cultivated, herbal. Origin: Africa, Asia, Australia.

Themeda triandra Forssk. var. *japonica* (Willd.) Makino
Poaceae
♦ Weed
♦ 275
♦ pG.

Themeda villosa (Poir.) Camus
Poaceae
♦ greater tanglegrass, Lyon's grass, silky kangaroo grass
♦ Weed, Noxious Weed, Naturalised
♦ 12, 87, 88, 90, 101, 229
♦ aG.

Theobroma cacao L.
Sterculiaceae
♦ cacao, cocoa, koko, cacaotier, kakaowiec, kakao
♦ Naturalised, Introduced, Cultivation Escape
♦ 39, 101, 230, 261
♦ T, cultivated, herbal, toxic. Origin: Central and South America.

Thereianthus minutus (Klatt) G.J.Lewis
Iridaceae
Watsonia minuta Klatt
♦ Weed, Quarantine Weed
♦ 76, 88, 220

Thermopsis caroliniana M.A.Curtis
Fabaceae/Papilionaceae
= *Thermopsis villosa* (Walt.) Fern. &

Schub. (NoR)
- ♦ Carolina lupin
- ♦ Weed, Quarantine Weed
- ♦ 76, 88, 220
- ♦ cultivated, herbal.

Thermopsis lanceolata R.Br.
Fabaceae/Papilionaceae
Sophora lupinoides L., *Thermopsis dahurica* Czefr., *Thermopsis lupinoides* (L.) Link, *Thermopsis sibirica* Czefr.
- ♦ false lupin, golden pea
- ♦ Weed, Quarantine Weed
- ♦ 76, 88, 220, 272, 297
- ♦ pH, arid, cultivated, herbal.

Thermopsis montana Nutt.
Fabaceae/Papilionaceae
- ♦ revonpapu, mountain goldenbanner, mountain thermopsis
- ♦ Weed, Quarantine Weed, Cultivation Escape
- ♦ 42, 76, 88, 161, 212, 220
- ♦ cultivated, herbal.

Thermopsis rhombifolia (Nutt.) Richards
Fabaceae/Papilionaceae
Thermopsis arenosa A.Nels.
- ♦ prairie goldenpea, prairie thermopsis, golden pea, false lupin, golden bean
- ♦ Weed, Noxious Weed
- ♦ 36, 39, 87, 88, 161, 212, 218
- ♦ cultivated, herbal, toxic.

Thesium L. spp.
Santalaceae
- ♦ bastard toadflax
- ♦ Weed
- ♦ 272

Thesium alpinum L.
Santalaceae
- ♦ alpine bastard toadflax
- ♦ Weed
- ♦ 272
- ♦ herbal.

Thesium arvense Horv.
Santalaceae
Thesium ramosum Hayne
- ♦ Weed
- ♦ 272

Thesium australe R.Br.
Santalaceae
- ♦ Weed
- ♦ 87, 88
- ♦ Origin: Australia.

Thesium chinense Turcz.
Santalaceae
- ♦ kanabikisou
- ♦ Weed
- ♦ 286
- ♦ herbal.

Thesium divaricatum Jan.
Santalaceae
- ♦ Weed
- ♦ 272

Thesium humile Vahl
Santalaceae
- ♦ Weed, Quarantine Weed
- ♦ 76, 87, 88, 203, 220, 221, 272

Thesium linophyllon L.
Santalaceae
Thesium intermedium Schrad.
- ♦ flaxleaf
- ♦ Naturalised
- ♦ 101
- ♦ herbal.

Thesium namaquense Schltr.
Santalaceae
- ♦ Namaqua thesium, poison bush
- ♦ Weed, Native Weed
- ♦ 39, 121
- ♦ pH parasitic, toxic. Origin: southern Africa.

Thesium racemosum Bernth.
Santalaceae
- ♦ Native Weed
- ♦ 121
- ♦ pH parasitic, toxic. Origin: southern Africa.

Thespesia garckeana F.Hoffm.
Malvaceae
Azanza garckeana (F.Hoffm.) Exell & Hillc. (see)
- ♦ Weed
- ♦ 87, 88

Thespesia populnea (L.) Sol. ex Correa
Malvaceae
- ♦ seaside mahoe, portia nut, portia tree, pone, milo
- ♦ Weed, Naturalised, Environmental Weed, Cultivation Escape
- ♦ 32, 80, 101, 112, 151, 179
- ♦ T, cultivated, herbal. Origin: Australia.

Thevetia peruviana (Pers.) K.Schum.
Apocynaceae
Cascabela peruviana (Pers.) Raf. (see), *Cascabela thevetia* (L.) Lippoid (see), *Cerbera thevetia* L., *Thevetia neriifolia* Juss. ex Steud.
- ♦ lucky nut, yellow oleander, oléandre jaune, thevetie, oandro amarelo, adelfa amarilla, cabalonga, chirca, snake nut, yellow be still tree, geel oleander
- ♦ Noxious Weed, Naturalised, Introduced, Garden Escape
- ♦ 98, 101, 230, 261, 283
- ♦ S/T, cultivated, herbal, toxic. Origin: tropical America.

Thevetia thevetioides (Kunth) Schum.
Apocynaceae
- ♦ giant thevetia, thevetia
- ♦ Weed
- ♦ 161
- ♦ cultivated, herbal, toxic.

Thinopyrum distichum (Thunb.) Á.Löve
Poaceae
= *Elymus distichus* (Thunb.) Melderis (NoR)
- ♦ Weed, Naturalised, Environmental Weed
- ♦ 86, 98, 203
- ♦ G. Origin: South Africa.

Thinopyrum elongatum (Host) D.R.Dewey
Poaceae

- ♦ Naturalised, Environmental Weed
- ♦ 86
- ♦ G.

Thinopyrum intermedium (Host) Barkworth & D.R.Dewey
Poaceae
= *Elytrigia intermedia* (Host) Nevski ssp. *intermedia* (NoR) [see *Agropyron intermedium* (Host) Beauv., *Elymus hispidus* (Opiz) Melderis ssp. *barbulatus* (Schur) Melderis, *Elymus hispidus* (Opiz) Meld.]
- ♦ intermediate wheatgrass
- ♦ Naturalised, Introduced, Casual Alien
- ♦ 89, 101, 168, 280
- ♦ pG.

Thinopyrum junceiforme (Á. & D.Löve) Á.Löve
Poaceae
= *Elytrigia juncea* (L.) Nevski ssp. *boreali-atlantica* (Simonet & Guin.) Hyl. (NoR)
- ♦ sea wheatgrass, Russian wheatgrass
- ♦ Weed, Naturalised, Environmental Weed
- ♦ 86, 98, 176, 198, 203, 280, 296
- ♦ G. Origin: Europe.

Thinopyrum junceum (L.) Á.Löve
Poaceae
= *Elytrigia juncea* (L.) Nevski ssp. *juncea* (NoR)
- ♦ sea wheatgrass
- ♦ Weed, Naturalised, Environmental Weed
- ♦ 72, 86, 88
- ♦ pG.

Thinopyrum ponticum (Podp.) Barkworth & D.R.Dewey
Poaceae
Triticum ponticum Podp.
- ♦ rush wheatgrass
- ♦ Naturalised
- ♦ 101
- ♦ G.

Thinopyrum pycnanthum (Godr.) Barkworth comb. nov. ined.
Poaceae
Agropyron littorale auct. non (Host) Dur., *Agropyron pungens* auct. non (Pers.) Roem. & Schult., *Agropyron pycnanthum* (Godr.) Godr. & Gren., *Elymus pungens* auct. non (Pers.) Melderis, *Elymus pycnanthus* (Godr.) Á.Löve, *Elytrigia pycnanthes* (Godr.) Á.Löve, *Triticum pungens* auct. non Pers.
- ♦ tick quackgrass
- ♦ Naturalised
- ♦ 101
- ♦ G.

Thladiantha dubia Bunge
Cucurbitaceae
- ♦ goldencreeper, kiinankurkku, Manchu tubergourd
- ♦ Weed, Naturalised, Casual Alien
- ♦ 42, 101, 161, 287
- ♦ pC, cultivated, herbal.

Thlaspi L. spp.
Brassicaceae
♦ pennycress
♦ Weed, Naturalised
♦ 198, 243, 272
♦ H.

Thlaspi alliaceum L.
Brassicaceae
Crucifera thlaspoides Krause
♦ laukkataskuruoho, roadside
pennycress, garlic pennycress
♦ Weed, Casual Alien
♦ 42, 88, 94, 272
♦ cultivated, herbal.

Thlaspi arvense L.
Brassicaceae
Crucifera thlaspi Krause, *Thlaspi
collinum* M.Bieb., *Thlaspidea arvensis*
Opiz
♦ field pennycress, pennycress, French
weed, fanweed, stinkweed, bastard
cress, mithridate mustard, devilweed,
carraspique, fan weed, mithridate
mustard
♦ Weed, Quarantine Weed, Noxious
Weed, Naturalised, Introduced,
Garden Escape
♦ 1, 23, 24, 34, 36, 39, 44, 49, 52, 62, 70,
76, 80, 86, 87, 88, 94, 98, 101, 114, 118,
121, 136, 161, 162, 174, 179, 195, 198,
203, 207, 210, 211, 212, 217, 218, 236,
237, 241, 243, 253, 269, 272, 275, 280,
286, 287, 294, 295, 297, 299, 300
♦ aH, cultivated, herbal, toxic. Origin:
Eurasia.

Thlaspi caerulescens J. & C.Presl
Brassicaceae
Crucifera coerulescens Krause,
Pterotropis gaudiniana Fourr., *Pterotropis
silvestris* Fourr., *Thlaspi gaudinianum*
Jordan, *Thlaspi sylvestre* Jordan
♦ kevättaskuruoho, alpine pennycress
♦ Naturalised
♦ 42

Thlaspi montanum L.
Brassicaceae
Crucifera montana Krause, *Iberis
badensis* Juslen., *Thlaspi beugesiacum*
Jord., *Thlaspi spathulatum* Gater.,
Thlaspi villardensis Jord.
♦ alpine pennycress, mountain
pennycress
♦ Weed
♦ 87, 88
♦ cultivated, herbal.

Thlaspi perfoliatum L.
Brassicaceae
Microthlaspi perfoliatum (L.) F.K.Mey.
(see), *Thlaspi rotundifolium* Tineo,
Thlaspi tinei Nyman
♦ perfoliate pennycress, Cotsworld
pennycress, pennycress, thoroughwort
pennycress, rikkataskuruoho
♦ Weed, Casual Alien
♦ 42, 44, 70, 80, 87, 88, 94, 218, 243, 253,
272
♦ aH, promoted, herbal. Origin: Africa,
west Asia, Europe.

Thlaspi praecox Wulf.
Brassicaceae
Hutchinsia torreana Ten.
♦ Weed
♦ 272

Threlkeldia proceriflora F.Muell.
Chenopodiaceae
♦ Weed
♦ 87, 88

Thrincia tripolitana Sch.Bip. ex Busch
Asteraceae
♦ Weed
♦ 221
♦ Origin: North Africa.

Thrixspermum arachnitiformae Schltr.
Orchidaceae
♦ Introduced
♦ 230
♦ H.

Thrixspermum ponapense Tuyama
Orchidaceae
♦ Introduced
♦ 230
♦ H.

Thryptomene calycina (Lindl.) Stapf
Myrtaceae
♦ Grampians thryptomeme
♦ Weed, Naturalised, Garden Escape,
Environmental Weed
♦ 72, 86, 88
♦ S, cultivated. Origin: Australia.

**Thuarea involuta (Forst.) Roem. &
Schult.**
Poaceae
♦ coastal sasa leaved grass, litoral
creeping grass, kuroiwa grass, thuarea
♦ Weed, Naturalised
♦ 87, 88, 90, 101
♦ pG, aqua, cultivated. Origin: Asia,
Australia, Madagascar.

Thuja occidentalis L.
Cupressaceae
Arbor vitae Clus., *Cupressus arbor-vitae*
Targ., *Thuja obtusa* Moench, *Thuja
odorata* Marsh.
♦ northern white cedar, American
arborvitae, arborvitae, thuja, white
cedar
♦ Weed
♦ 8, 39, 87, 88, 218
♦ T, cultivated, herbal, toxic.

Thuja orientalis L.
Cupressaceae
= *Platycladus orientalis* (L.) Franco
♦ oriental arborvitae, northern white
cedar, Chinese arborvitae
♦ Weed, Introduced, Environmental
Weed
♦ 88, 151, 261
♦ cultivated, herbal. Origin: northern
China, Korea.

Thuja plicata Donn ex D.Don
Cupressaceae
Thuja douglasii Nutt., *Thuja gigantea*
Nutt., *Thuja menziesii* Dougl.
♦ western red cedar, red cedar, tuja
riasnatá
♦ Weed

♦ 87, 88, 218
♦ T, cultivated, herbal.

**Thujopsis dolabrata (L.f.) Siebold &
Zucc.**
Cupressaceae
♦ hibatuija, hib
♦ Cultivation Escape
♦ 42
♦ cultivated, herbal.

Thunbergia alata Bojer ex Sims
Acanthaceae/Thunbergiaceae
♦ black eyed Susan vine,
kakobakansimba, waew taa
♦ Weed, Naturalised, Introduced,
Native Weed, Garden Escape,
Environmental Weed
♦ 3, 13, 14, 22, 38, 73, 86, 87, 88, 98, 101,
121, 155, 157, 179, 201, 203, 238, 255,
261, 269, 276, 280, 287, 290
♦ pC, arid, cultivated, herbal. Origin:
southern Africa.

Thunbergia annua Hochst. ex Nees
Acanthaceae/Thunbergiaceae
♦ thunbergia
♦ Weed, Quarantine Weed, Noxious
Weed, Naturalised
♦ 76, 86, 87, 88, 203, 220, 242
♦ arid.

Thunbergia atriplicifolia E.Mey. ex Nees
Acanthaceae/Thunbergiaceae
♦ Natal primrose
♦ Native Weed
♦ 121
♦ pH. Origin: southern Africa.

Thunbergia coccinea Wall.
Acanthaceae/Thunbergiaceae
♦ Casual Alien
♦ 280
♦ cultivated.

Thunbergia erecta (Benth.) T.Anderson
Acanthaceae/Thunbergiaceae
♦ bush clockvine, king's mantle,
clockbush
♦ Weed, Naturalised, Cultivation
Escape
♦ 101, 179, 261
♦ cultivated. Origin: tropical Africa.

Thunbergia fragrans Roxb.
Acanthaceae/Thunbergiaceae
♦ whitelady, angel wings, flor de
nieve, thunbergia
♦ Weed, Quarantine Weed, Noxious
Weed, Naturalised, Introduced,
Garden Escape, Cultivation Escape
♦ 13, 14, 38, 76, 86, 87, 88, 101, 179, 191,
203, 220, 252, 261
♦ cultivated. Origin: tropical Asia.

**Thunbergia grandiflora (Roxb. ex Rottler)
Roxb.**
Acanthaceae/Thunbergiaceae
♦ blue thunbergia, thunbergia, blue
trumpet vine, Bengal clockvine, blue
skyflower, clockvine, skyflower, sky
vine, large flowered thunbergia
♦ Weed, Sleeper Weed, Quarantine
Weed, Noxious Weed, Naturalised,
Garden Escape, Environmental Weed,
Cultivation Escape

◆ 3, 22, 73, 76, 86, 88, 98, 101, 147, 152, 155, 179, 191, 203, 225, 233, 246, 261
◆ pC, cultivated, herbal. Origin: northern India.

Thunbergia harrisii **Hook.**
Acanthaceae/Thunbergiaceae
= *Thunbergia laurifolia* Lindl.
◆ Weed
◆ 3, 191

Thunbergia laurifolia **Lindl.**
Acanthaceae/Thunbergiaceae
Thunbergia harrisii Hook. (see)
◆ laurel clockvine, blue thunbergia, blue trumpet vine, purple allamanda, skyflower
◆ Weed, Noxious Weed, Naturalised, Garden Escape, Environmental Weed, Cultivation Escape
◆ 3, 54, 86, 88, 101, 155, 233
◆ cultivated, herbal. Origin: India, Malaysia.

Thymelaea hirsuta **(L.) Endl.**
Thymelaeaceae
◆ Weed
◆ 221
◆ cultivated, herbal.

Thymelaea passerina **(L.) Coss. & Germ.**
Thymelaeaceae
Lygia passerina Fasano, *Passerina annua* Wikstr., *Passerina tragi* Camer., *Stellera passerina* L., *Thymelaea annua* Wickstr., *Thymelaea arvensis* L.
◆ thymelaea, mezereon, spurge flax, linaiola
◆ Weed, Naturalised
◆ 80, 86, 87, 88, 94, 98, 101, 198, 203, 272
◆ herbal. Origin: southern Europe, west Asia.

Thymophylla tenuiloba **(DC.) Small**
Asteraceae
Dyssodia tenuiloba (DC.) B.L.Robins., *Hymenatherum tenuilobum* DC.
◆ dahlberg daisy, bristleleaf pricklyleaf, golden fleece
◆ Weed, Quarantine Weed
◆ 76, 88, 179, 220
◆ cultivated, herbal.

Thymus capitatus **(L.) Hoffmanns. & Link**
Lamiaceae
◆ headed savory, thymus
◆ Weed
◆ 272
◆ S, cultivated, herbal.

Thymus cephalotos **L.**
Lamiaceae
◆ Weed
◆ 70

Thymus mongolicus **Rönn**
Lamiaceae
◆ Mongolian thyme
◆ Weed
◆ 297

Thymus pannonicus **All.**
Lamiaceae
◆ Eurasian thyme, thymus

◆ Naturalised
◆ 101
◆ pH, cultivated.

Thymus praecox **Opiz**
Lamiaceae
◆ mother of thyme, wild thyme, thyme, creeping thyme
◆ Naturalised
◆ 101
◆ S, cultivated, herbal.

Thymus praecox **Opiz ssp. *arcticus* (Dur.) Jalas**
Lamiaceae
◆ creeping thyme, mountain thyme, wild thyme
◆ Naturalised
◆ 101
◆ cultivated.

Thymus pulegioides **L.**
Lamiaceae
◆ nurmiajuruoho, lemon thyme, larger wild thyme, wild thyme, creeping thyme, mother of thyme
◆ Weed, Naturalised, Cultivation Escape
◆ 8, 23, 39, 42, 88, 101, 195, 272, 280
◆ S, cultivated, herbal, toxic.

Thymus serpyllum **L.**
Lamiaceae
◆ creeping thyme, wild thyme, common wild thyme, larger wild thyme, hillwort, penny mountain, lemon thyme, mother of thyme, Breckland thyme, kangasajuruoho
◆ Weed, Naturalised
◆ 87, 88, 161, 195, 211, 218, 272, 287
◆ S, cultivated, herbal.

Thymus vulgaris **L.**
Lamiaceae
Thymus webbianus Rouy
◆ thyme, common thyme, garden thyme, timo volgare, culinary thyme
◆ Weed, Naturalised, Cultivation Escape
◆ 40, 42, 70, 86, 87, 88, 98, 101, 121, 161, 165, 203, 252, 280
◆ p, cultivated, herbal, toxic. Origin: Eurasia.

Thysanocarpus curvipes **Hook.**
Brassicaceae
◆ sand fringepod, fringepod
◆ Weed
◆ 161
◆ aH, herbal.

Thysanolaena maxima **(Roxb.) Kuntze**
Poaceae
Agrostis maxima Roxb., *Melica latifolia* Roxb., *Myriachaeta arundinacea* Zoll. & Moritzi, *Myriachaeta glauca* Moritzi ex Steud., *Panicum acariferum* Trin., *Thysanolaena agrostis* Nees, *Thysanolaena assamensis* Gaud., *Thysanolaena birmanica* Gaud., *Thysanolaena sikkimensis* Gand.
◆ kong, tiger grass
◆ Weed, Quarantine Weed
◆ 87, 88, 209, 220, 238
◆ pG, arid, herbal.

Tibouchina **Aubl. spp.**
Melastomataceae
◆ glorytree
◆ Weed
◆ 18, 88

Tibouchina granulosa **(Desr.) Cogn.**
Melastomataceae
◆ glorybush
◆ Naturalised, Introduced
◆ 101, 261
◆ cultivated. Origin: Brazil, Bolivia.

Tibouchina herbacea **(DC.) Cogn.**
Melastomataceae
◆ herbaceous glorytree, glorybush, cane ti, tibouchina, cane tibouchina
◆ Weed, Noxious Weed, Naturalised, Cultivation Escape
◆ 3, 80, 101, 191, 229, 233
◆ cultivated, herbal.

Tibouchina longifolia **(Vahl) Baill. ex Cogn.**
Melastomataceae
◆ longleaf glorytree
◆ Weed, Noxious Weed, Naturalised
◆ 87, 88, 101, 229
◆ herbal.

Tibouchina semidecandra **Cogn.**
Melastomataceae
◆ Weed
◆ 87, 88

Tibouchina urvilleana **(DC.) Cogn.**
Melastomataceae
Tibouchina grandiflora hort.
◆ glorybush, lasiandra, princessflower edwardsii, pleroma, purple glorytree
◆ Weed, Noxious Weed, Naturalised, Introduced, Cultivation Escape
◆ 3, 22, 80, 101, 191, 229, 233, 261, 280
◆ T, cultivated, herbal. Origin: Brazil.

Tibouchina viminea **Cogn.**
Melastomataceae
◆ Weed
◆ 3, 191

Ticanto nuga **(L.) Medik.**
Fabaceae/Caesalpiniaceae
= *Caesalpinia crista* L.
◆ gray nicker
◆ Naturalised
◆ 101

Tidestromia lanuginosa **(Nutt.) Standl.**
Amaranthaceae
Achyranthes lanuginosa Nutt.
◆ woolly tidestromia
◆ Weed, Introduced
◆ 87, 88, 161, 218, 228
◆ aH, arid, herbal.

Tigridia vanhouttei **Roezl ssp. *vanhouttei***
Iridaceae
◆ Weed
◆ 199

Tilia americana **L.**
Tiliaceae
◆ American basswood
◆ Weed
◆ 87, 88, 218
◆ T, cultivated, herbal.

Tilia cordata Mill.
Tiliaceae
Tilia cordifolia Bess., *Tilia microphylla*
Vent., *Tilia silvestris* Desf.
♦ littleleaf linden, small leaved lime
♦ Naturalised
♦ 101
♦ T, cultivated, herbal.

Tilia heterophylla Vent.
Tiliaceae
= *Tilia americana* L. var. *heterophylla*
(Vent.) Loud. (NoR)
♦ white basswood
♦ Weed
♦ 87, 88, 218
♦ T, cultivated, herbal.

Tilia miqueliana Maxim.
Tiliaceae
♦ Naturalised
♦ 287

Tilia mongolica Maxim.
Tiliaceae
♦ Mongolian lime
♦ Quarantine Weed
♦ 220
♦ T, cultivated.

Tilia petiolaris DC.
Tiliaceae
♦ pendent silver linden
♦ Naturalised
♦ 101
♦ cultivated.

Tilia platyphyllos Scop.
Tiliaceae
Tilia europaea L., *Tilia grandifolia* Ehrh.,
Tilia hollandica hort., *Tilia mollis* Spach
♦ isolehtilehmus, largeleaf linden,
large leaved lime, tiglio, alburno di
tiglio
♦ Naturalised, Cultivation Escape
♦ 42, 101
♦ T, cultivated, herbal.

Tilia × vulgaris Hayne
Tiliaceae
= *Tilia cordata* Mill. × *Tilia platyphyllos*
Scop.
♦ common linden, common lime,
puistolehmus
♦ Naturalised, Cultivation Escape
♦ 40, 42, 101
♦ T, cultivated, herbal. Origin:
horticultural hybrid.

**Tiliacora acuminata (Lam.) Hook.
Thomson**
Menispermaceae
Cocculus acuminatus DC., *Tiliacora
acuminata* Miers, *Tiliacora racemosa*
Colebr.
♦ Introduced
♦ 228

Tillandsia L. spp.
Bromeliaceae
♦ air plant
♦ Weed
♦ 14

Tillandsia aeranthos (Loisel.) L.B.Sm.
Bromeliaceae

♦ Weed
♦ 237, 295
♦ cultivated.

Tillandsia bandensis Bak.
Bromeliaceae
♦ clavel del aire
♦ Weed
♦ 237, 295

Tillandsia landbeckii Phil.
Bromeliaceae
♦ Weed
♦ 87, 88
♦ arid, cultivated.

Tillandsia recurvata (L.) L.
Bromeliaceae
Diaphoranthema recurvata (L.) Beer,
Diaphoranthema uniflora (Kunth) Beer,
Tillandsia uniflora Kunth
♦ small ballmoss, ballmoss
♦ Weed, Introduced
♦ 228, 237, 295
♦ arid, cultivated, herbal.

Tillandsia usneoides (L.) L.
Bromeliaceae
Tillandsia trichoides Kunth
♦ camanbaia
♦ Weed
♦ 255
♦ arid, cultivated, herbal. Origin:
South America.

Timonius ledermannii Val.
Rubiaceae
♦ Introduced
♦ 230
♦ T.

Timonius ponapensis Val.
Rubiaceae
♦ kehn
♦ Introduced
♦ 230
♦ T.

Timonius timon (Spreng.) Merr.
Rubiaceae
♦ liberal, sakosia
♦ Weed
♦ 3
♦ cultivated.

Tinantia erecta (Jacq.) Schltdl.
Commelinaceae
♦ canutillo
♦ Weed
♦ 87, 88, 157, 281
♦ arid, cultivated.

Tinantia standleyi Steyerm.
Commelinaceae
♦ Weed
♦ 157

**Tinguarra montana (Webb Ex Christ)
A.Hansen & Kunkel**
Apiaceae
♦ Quarantine Weed
♦ 220
♦ cultivated.

Tinospora cordifolia Miers
Menispermaceae
♦ Weed

♦ 87, 88
♦ cultivated, herbal.

Tipuana tipu (Benth.) Kuntze
Fabaceae/Papilionaceae
♦ rosewood, tipoeboom, tipu tree,
pride of Bolivia
♦ Noxious Weed, Naturalised,
Garden Escape, Environmental Weed,
Cultivation Escape
♦ 63, 86, 260, 279, 283
♦ arid, cultivated. Origin: Brazil,
Bolivia, Argentina.

Tithonia diversifolia (Hemsl.) A.Gray
Asteraceae
♦ tree marigold, Mexican sunflower,
Japanese sunflower, matala,
Mexikaanse sonneblom, girasol
Mexicano
♦ Weed, Noxious Weed, Naturalised,
Introduced, Garden Escape,
Environmental Weed, Cultivation
Escape
♦ 3, 63, 86, 87, 88, 95, 98, 101, 107, 121,
158, 179, 191, 201, 203, 209, 218, 230,
261, 269, 283
♦ pH, cultivated, herbal. Origin:
Central America.

Tithonia rotundifolia (Mill.) S.F.Blake
Asteraceae
Tithonia speciosa (Hook.) Griseb.,
Helianthus speciosus Hook.
♦ rooisonneblom, red sunflower, clavel
de muerto, flor amarilla, girasol
♦ Weed, Noxious Weed, Naturalised,
Garden Escape, Environmental Weed,
Cultivation Escape, Casual Alien
♦ 63, 86, 88, 95, 98, 101, 121, 157, 158,
203, 243, 261, 269, 280, 281, 283, 287
♦ aH, cultivated, herbal. Origin:
Central America.

Tithonia tubaeformis (Jacq.) Cass.
Asteraceae
Helianthus tubaeformis Jacq.
♦ southern sunflower, yuyo Cubano
♦ Weed
♦ 43, 65, 236, 237, 243, 295
♦ arid.

**Tithymalus longifolium (D.Don) Hurus.
& Y.Tanaka**
Euphorbiaceae
♦ Quarantine Weed
♦ 220

**Tithymalus regis-jubae (Webb & Berthel.)
Klotzsch & Garcke**
Euphorbiaceae
Euphorbia regis-jubae Webb & Berthel.
(see)
♦ Quarantine Weed
♦ 220
♦ toxic.

Tococa quadrialata (Naud.) Macbr.
Melastomataceae
♦ Weed
♦ 87, 88

Tofieldia alpina Sm.
Liliaceae/Melanthiaceae/Tofieldiaceae
♦ Quarantine Weed

♦ 220

Tofieldia calyculata Wahl.
Liliaceae/Melanthiaceae/Tofieldiaceae
Anthericum calycinum Braun,
Anthericum calyculatum L., *Heriteria anthericoides* Schrank, *Narthecium flavescens* Wahlenb.
♦ false asphodel
♦ Quarantine Weed
♦ 39, 220
♦ cultivated, toxic.

Tofieldia palustris Huds.
Liliaceae/Melanthiaceae/Tofieldiaceae
♦ Scotch asphodel
♦ Quarantine Weed
♦ 220
♦ herbal.

Tolmiea menziesii Torr. & Gray
Saxifragaceae
Tiarella menziesii Pursh
♦ pick a back plant, youth on age, variegated piggyback plant
♦ Naturalised
♦ 40, 247
♦ pH, cultivated, herbal, toxic.

Tolpis barbata (L.) Gaertn.
Asteraceae
Crepis barbata L., *Tolpis umbellata* Bertol. (see)
♦ tolpis, European umbrella milkwort, partavaunikki
♦ Weed, Naturalised, Environmental Weed, Casual Alien, Cultivation Escape
♦ 7, 34, 42, 70, 86, 87, 88, 98, 101, 176, 198, 203, 250, 252, 253, 280, 300
♦ a/pH, cultivated. Origin: southern Europe.

Tolpis capensis (L.) Sch.Bip.
Asteraceae
Hieracium capense L.
♦ fukuthoane
♦ Native Weed
♦ 121
♦ pH, arid, herbal. Origin: southern Africa.

Tolpis umbellata Bertol.
Asteraceae
= *Tolpis barbata* (L.) Gaertn.
♦ sarjavaunikki
♦ Weed, Naturalised, Casual Alien
♦ 42, 98, 203

Toona ciliata Roem.
Meliaceae
Cedrela toona Roxb. ex Rottl. & Willd. (see)
♦ Australian red cedar, toon tree, toonboom
♦ Weed, Noxious Weed, Naturalised, Introduced, Cultivation Escape
♦ 3, 22, 63, 88, 95, 101, 228, 261, 283
♦ T, arid, cultivated, herbal. Origin: India to China.

Toona ciliata Roem. ssp. ciliata
Meliaceae
♦ Australian red cedar

♦ Naturalised
♦ 101

Toona ciliata Roem. ssp. ciliata var. australis (F.Muell.) Bahadur
Meliaceae
Cedrela toona Roxb. ex Rottl. & Willd. var. *australis* (F.Muell.) C.DC.
♦ Australian red cedar, Australian toon
♦ Naturalised
♦ 101

Toona ciliata Roem. var. australis (F.Muell.) C.DC.
Meliaceae
Toona australis (F.Muell.) Harms
♦ Introduced
♦ 228
♦ arid.

Tordylium aegyptiacum (L.) Lam.
Apiaceae
♦ Egyptian hartwort
♦ Weed, Quarantine Weed
♦ 76, 87, 88, 115, 203, 220, 243
♦ cultivated.

Tordylium apulum L.
Apiaceae
Condylocarpus apulus Hoffm.
♦ Mediterranean hartwort, Roman pimpernel
♦ Weed, Naturalised
♦ 54, 86, 87, 88, 94, 98, 101, 203, 272
♦ pH, promoted, herbal. Origin: Mediterranean.

Tordylium maximum L.
Apiaceae
Caucalis maxima Baumg., *Heracleum tordylium* Spreng., *Tordylium magnum* Brot.
♦ hartwort
♦ Weed
♦ 88, 94, 253, 272
♦ cultivated, herbal.

Tordylium officinale L.
Apiaceae
♦ common hartwort
♦ Weed
♦ 87, 88, 94, 272
♦ herbal.

Torenia asiatica L.
Scrophulariaceae
♦ wishbone flower
♦ Naturalised
♦ 101

Torenia bicolor Dalz.
Scrophulariaceae
♦ Weed
♦ 87, 88

Torenia concolor Lindl.
Scrophulariaceae
♦ Weed
♦ 87, 88

Torenia flava Bth.
Scrophulariaceae
♦ Weed
♦ 13

Torenia fournieri Linden ex Fourn.
Scrophulariaceae
♦ wishbone flower, bluewings
♦ Weed, Naturalised, Cultivation Escape
♦ 32, 101, 179
♦ a/pH, cultivated, herbal. Origin: Australia.

Torenia polygonoides Benth.
Scrophulariaceae
♦ Weed
♦ 12
♦ herbal.

Torenia spicata Engl.
Scrophulariaceae
♦ Weed
♦ 87, 88

Torenia thouarsii (Cham. & Schltdl.) Kuntze
Scrophulariaceae
♦ Weed, Naturalised
♦ 32, 87, 88

Torenia violacea (Azaola ex Blanco) Pennell
Scrophulariaceae
Mimulus violaceus Azaola ex Blanco, *Torenia exappendiculata* Regel, *Torenia peduncularis* Benth., *Torenia violacea* var. *chinensis* T.Yamaz.
♦ Weed
♦ 13, 88, 170, 191, 275
♦ aH.

Torilis Adans. spp.
Apiaceae
♦ hedgeparsley
♦ Weed, Naturalised
♦ 198, 272

Torilis anthriscus (L.) Gmel. nom. illeg.
Apiaceae
= *Torilis japonica* (Houtt.) DC.
♦ Weed
♦ 44, 87, 88

Torilis arvensis (Huds.) Link
Apiaceae
Caucalis arvensis Huds., *Caucalis trifida* Hoffm., *Torilis helvetica* Gmel., *Torilis infesta* Clairv.
♦ hedgeparsley, spreading hedgeparsley, field hedgeparsley
♦ Weed, Quarantine Weed, Noxious Weed, Naturalised, Introduced, Environmental Weed
♦ 1, 15, 34, 44, 54, 70, 72, 76, 80, 86, 87, 88, 94, 101, 102, 121, 139, 146, 161, 198, 229, 243, 253, 272, 280, 300
♦ aH, cultivated, herbal, toxic. Origin: Eurasia.

Torilis arvensis (Huds.) Link ssp. arvensis
Apiaceae
♦ hedgeparsley, spreading hedgeparsley
♦ Noxious Weed, Naturalised, Introduced
♦ 34, 101, 229
♦ aH.

Torilis arvensis (Huds.) Link ssp.
purpurea (Ten.) Hayek
 Apiaceae
 ♦ hedgeparsley, spreading
 hedgeparsley
 ♦ Noxious Weed, Naturalised
 ♦ 101, 229
 ♦ aH.

Torilis glochidiatus Fisch. & Mey.
 Apiaceae
 ♦ Naturalised
 ♦ 287

Torilis japonica (Houtt.) DC.
 Apiaceae
 Caucalis anthriscus Scop., *Caucalis*
 aspera Lam., *Caucalis elata* D.Don,
 Daucus anthriscus Baill., *Selinum torilis*
 Krause, *Tordylium anthriscus* L., *Torilis*
 anthriscus (L.) Gmel. *nom. illeg.* (see),
 Torilis convexa Dulac, *Torilis stricta*
 Wibel
 ♦ Japanese hedgeparsley
 ♦ Weed, Naturalised
 ♦ 15, 23, 44, 87, 88, 101, 161, 218, 235,
 272, 280, 286, 297
 ♦ aH, cultivated, herbal.

Torilis leptophylla (L.) Rchb.f.
 Apiaceae
 Caucalis leptophylla L. (see), *Daucus*
 leptophylla Pomel., *Nigeria parviflora*
 Bubani, *Selinum humile* Krause
 ♦ bristlefruit hedgeparsley
 ♦ Weed, Naturalised
 ♦ 87, 88, 94, 101, 253, 272, 287
 ♦ Origin: North Africa, Middle East,
 southern Europe.

Torilis neglecta Schult.
 Apiaceae
 ♦ Weed
 ♦ 87, 88

Torilis nodosa (L.) Gaertn.
 Apiaceae
 Lappularia nodosa Pomel, *Selinum*
 nodosum Krause, *Tordylium nodosum* L.
 ♦ knotted hedgeparsley, nivelkatko
 ♦ Weed, Naturalised, Introduced,
 Casual Alien
 ♦ 34, 38, 42, 70, 86, 87, 88, 94, 98, 101,
 134, 176, 198, 203, 241, 243, 253, 272,
 280, 287, 295, 300
 ♦ aH, herbal. Origin: Europe, Middle
 East.

Torilis radiata Moench
 Apiaceae
 ♦ Weed
 ♦ 221

Torilis scabra (Thunb.) DC.
 Apiaceae
 Caucalis scabra (Thunb.) Makino
 ♦ rough hedgeparsley
 ♦ Weed, Naturalised
 ♦ 87, 88, 101, 275, 286, 297
 ♦ pH.

Torilis syriaca Boiss. & Blanche
 Apiaceae
 ♦ Weed
 ♦ 87, 88

Torreya californica Torr.
 Taxaceae
 ♦ California nutmeg
 ♦ Weed
 ♦ 87, 88, 218
 ♦ T, cultivated, herbal.

Torularia torulosa (Desf.) O.E.Schulz
 Brassicaceae
 ♦ Weed
 ♦ 243, 248, 272

Torulinium ferax (L.C.Rich) Urb.
 Cyperaceae
 = *Cyperus odoratus* L.
 ♦ Weed
 ♦ 274
 ♦ G.

Torulinium odoratum (L.) Hooper
 Cyperaceae
 = *Cyperus odoratus* L.
 ♦ yellow nutsedge, calingale
 ♦ Weed
 ♦ 126
 ♦ a/pG, aqua, herbal.

Tournefortia argentea L.f.
 Boraginaceae
 Argusia argentea (L.f.) Heine,
 Messerschmidia argentea (L.f.) Johnst.
 (see), *Tournefortia arborea* Blanco
 ♦ velvetleaf soldierbush, titin, tree
 heliotrope
 ♦ Weed, Naturalised
 ♦ 6, 88, 101
 ♦ S, cultivated, herbal. Origin:
 Madagascar.

Tournefortia bicolor Sw.
 Boraginaceae
 ♦ niguita
 ♦ Weed
 ♦ 14, 87, 88
 ♦ herbal.

Tournefortia cuspidata Kunth
 Boraginaceae
 ♦ Weed
 ♦ 87, 88

Tournefortia hirsutissima L.
 Boraginaceae
 ♦ chiggery grapes
 ♦ Weed
 ♦ 87, 88
 ♦ herbal.

Tournefortia maculata Jacq.
 Boraginaceae
 ♦ marmelinho, bejuco de masa
 ♦ Weed
 ♦ 255
 ♦ Origin: Brazil.

Tournefortia poliochros Spreng.
 Boraginaceae
 = *Tournefortia volubilis* L. (NoR)
 ♦ Weed
 ♦ 14
 ♦ herbal.

Toxicodendron diversilobum (Torr. &
A.Gray) Greene
 Anacardiaceae
 Rhus diversiloba Torr. & Gray (see)

 ♦ Pacific poison oak, western poison
 oak
 ♦ Weed
 ♦ 34, 39, 82, 154, 161, 189, 243
 ♦ pS, herbal, toxic.

Toxicodendron radicans (L.) Kuntze
 Anacardiaceae
 Rhus radicans L. (see)
 ♦ poison ivy, markweed, poison
 creeper, three leaved ivy, picry,
 mercury, eastern poison ivy
 ♦ Weed, Quarantine
 Weed, Noxious Weed, Naturalised
 ♦ 8, 39, 76, 82, 86, 88, 147, 154, 161, 171,
 189, 203, 211, 212, 220, 243, 247
 ♦ pC, cultivated, herbal, toxic.

Toxicodendron rydbergii (Small ex Rydb.)
Greene
 Anacardiaceae
 Rhus rydbergii Small
 ♦ western poison ivy, Rydberg's
 poison oak, poison ivy
 ♦ Weed, Native Weed
 ♦ 39, 49, 49, 82, 154, 159, 161, 174, 264
 ♦ S, herbal, toxic. Origin: North
 America.

Toxicodendron succedaneum (L.) Kuntze
 Anacardiaceae
 Rhus succedanea L. (see)
 ♦ rhus, rhus tree, Japanese wax tree,
 scarlet rhus, sumac, wax tree
 ♦ Weed, Quarantine Weed, Noxious
 Weed, Naturalised, Environmental
 Weed
 ♦ 39, 73, 76, 86, 88, 147, 169, 171, 203,
 232
 ♦ cultivated, herbal, toxic. Origin:
 Pakistan to Japan.

Toxicodendron toxicarium (Salisb.) Gillis
 Anacardiaceae
 = *Toxicodendron pubescens* Mill. (NoR)
 ♦ poison oak
 ♦ Weed
 ♦ 161
 ♦ herbal, toxic.

Toxicodendron vernicifluum (Stokes)
F.Barkley
 Anacardiaceae
 Rhus verniciflua Stokes (see)
 ♦ varnish tree, Chinese lacquer,
 Japanese lacquer tree
 ♦ Weed
 ♦ 39, 161
 ♦ cultivated, toxic.

Toxicodendron vernix (L.) Kuntze
 Anacardiaceae
 Rhus vernix L. (see)
 ♦ poison sumac
 ♦ Weed
 ♦ 39, 82, 161, 247
 ♦ herbal, toxic.

Trachelium caeruleum L.
 Campanulaceae
 ♦ blue throatwort, throatwort
 ♦ Naturalised, Casual Alien
 ♦ 86, 280
 ♦ cultivated. Origin: south-west
 Europe.

Trachelospermum asiaticum (Sieb. & Zucc.) Nakai var. *intermedium* Nakai
Apocynaceae
♦ Weed
♦ 286

Trachelospermum jasminoides (Lindl.) Lem.
Apocynaceae
♦ Confederate jasmine, star jasmine
♦ Weed, Naturalised
♦ 101, 179, 247
♦ pC, cultivated, herbal, toxic.

Trachelospermum jasminoides (Lindl.) Lem. var. *pubescens* Makino
Apocynaceae
♦ Weed
♦ 286

Trachyandra divaricata (Jacq.) Kunth
Liliaceae/Asphodelaceae
♦ dune onion weed
♦ Weed, Naturalised, Environmental Weed
♦ 7, 9, 39, 86, 98, 203
♦ cultivated, toxic. Origin: South Africa.

Trachycarpus fortunei (Hook.) H.Wendl.
Arecaceae
Chamaerops fortunei Hook., *Trachycarpus excelsus* (Thunb.) H.Wendl.
♦ fan palm, Chusan fan palm, Chinese windmill palm, Chinese fan palm, hemp palm
♦ Weed, Sleeper Weed, Naturalised, Introduced, Garden Escape, Environmental Weed
♦ 15, 54, 72, 86, 88, 225, 230, 246, 280
♦ T, cultivated, herbal. Origin: central and east China.

Trachymene ochracea L.A.S.Johnson
Apiaceae
♦ white parsnip, wild parsnip
♦ Native Weed
♦ 39, 269
♦ pH, toxic. Origin: Australia.

Trachymene oleracea (Domin) B.L.Burtt
Apiaceae
♦ lace flower
♦ Weed
♦ 87, 88
♦ cultivated.

Trachynia distachya (L.) Link.
Poaceae
= *Brachypodium distachyon* (L.) Beauv.
♦ purple falsebrome
♦ Weed
♦ 91, 111, 221, 243
♦ G, cultivated.

Trachypogon spicatus (L.f) Kuntze
Poaceae
Andropogon plumosus Willd., *Andropogon spicatus* (L.f.) Steud., *Andropogon truncatus* (Nees) Steud., *Heteropogon truncatus* Nees, *Stipa spicata* L.f., *Trachypogon canescens* Nees, *Trachypogon capensis* Trin., *Trachypogon ligularis* Nees, *Trachypogon micans* Andersson, *Trachypogon plumosus* (Willd.) Nees, *Trachypogon thollonii* (Franch.) Stapf, *Trachypogon truncatus* (Nees) Anderss.
♦ giant speargrass, grey tussockgrass, spiked crinkleawn
♦ Native Weed
♦ 121
♦ pG, arid, cultivated. Origin: southern Africa.

Trachyspermum ammi (L.) Sprague ex Turrill
Apiaceae
= *Trachyspermum copticum* (L.) Link
♦ ajowan
♦ Introduced
♦ 228
♦ aH, arid, cultivated, herbal.

Trachyspermum copticum (L.) Link
Apiaceae
Trachyspermum ammi (L.) Sprague ex Turrill (see)
♦ ajowan caraway
♦ Naturalised
♦ 101
♦ herbal.

Trachyspermum strictocarpum (C.B.Clarke) Wolf.
Apiaceae
♦ Weed
♦ 66

Trachystemon orientalis (L.) G.Don f.
Boraginaceae
♦ Abraham Isaac Jacob
♦ Weed, Naturalised
♦ 40, 54, 86, 88, 198
♦ cultivated. Origin: eastern Europe.

Tradescantia L. spp.
Commelinaceae
♦ spiderwort, tradescantia, purple queen tradescantia
♦ Weed, Naturalised
♦ 18, 88, 198, 247
♦ toxic.

Tradescantia albiflora Kunth
Commelinaceae
often confused with *Tradescantia fluminensis* Vell.
♦ wandering Jew, wandering creeper
♦ Weed, Naturalised, Garden Escape, Environmental Weed
♦ 72, 73, 86, 87, 88, 98, 176, 198, 201, 203, 269
♦ pH, cultivated. Origin: South America.

Tradescantia × andersoniana W.Ludw. & Rohweder *nom. inval.*
Commelinaceae
= *Tradescantia virginiana* L. × *Tradescantia ohiensis* Raf. × *Tradescantia subaspera* Ker Gawl. [other species are also associated with garden hybrids under this name]
♦ tradeskancia záhradná
♦ Weed
♦ 80
♦ cultivated.

Tradescantia bracteata Small ex Britt.
Commelinaceae
♦ longbract spiderwort, sticky spiderwort
♦ Weed
♦ 161
♦ herbal.

Tradescantia cerinthoides Kunth
Commelinaceae
♦ Casual Alien
♦ 280
♦ cultivated.

Tradescantia crassifolia Cav.
Commelinaceae
♦ leatherleaf spiderwort, spiderwort
♦ Weed, Quarantine Weed
♦ 39, 76, 87, 88, 199, 203, 220
♦ toxic.

Tradescantia crassula Link & Otto
Commelinaceae
♦ succulent spiderwort
♦ Naturalised
♦ 101

Tradescantia discolor L'Hér.
Commelinaceae
= *Tradescantia spathacea* Sw.
♦ Weed
♦ 262
♦ pH.

Tradescantia fluminensis Vell.
Commelinaceae
Tradescantia albiflora Kunth often misapplied
♦ white flowered wandering Jew, wandering Jew, smallleaf spiderwort, spiderwort
♦ Weed, Quarantine Weed, Naturalised, Introduced, Environmental Weed
♦ 7, 15, 34, 80, 87, 88, 101, 112, 126, 151, 152, 165, 179, 181, 225, 246, 270, 280, 286, 287, 289, 295, 296
♦ pH, cultivated. Origin: Brazil.

Tradescantia gracilis Kunth
Commelinaceae
= *Callisia gracilis* (Kunth) D.R.Hunt (NoR)
♦ Weed
♦ 87, 88

Tradescantia ohiensis Raf.
Commelinaceae
♦ bluejacket, spiderwort
♦ Naturalised
♦ 287
♦ cultivated, herbal.

Tradescantia pallida (Rose) D.R.Hunt
Commelinaceae
Setcreasea pallida Rose, *Setcreasea purpurea* Boom, *Tradescantia purpurea* Boom
♦ purple queen, common spiderwort, widow's tears
♦ Quarantine Weed, Naturalised, Garden Escape
♦ 76, 179, 261
♦ pH, cultivated, toxic. Origin: Mexico.

Tradescantia spathacea Sw.
Commelinaceae
Rhoe spathacea (Sw.) Stearn (see),
Tradescantia discolor L'Hér. (see)
♦ boat lily, oyster plant
♦ Weed, Naturalised, Environmental Weed
♦ 39, 86, 101, 161, 179
♦ cultivated, toxic. Origin: Mexico.

Tradescantia virginiana L.
Commelinaceae
♦ spiderwort, Virginia spiderwort, widow's tears
♦ Casual Alien
♦ 280
♦ pH, cultivated, herbal.

Tradescantia volubis L.
Commelinaceae
♦ Weed
♦ 87, 88

Tradescantia zebrina hort. ex Bosse
Commelinaceae
Zebrina pendula Schnizl. (see)
♦ inchplant, wandering Jew, striped wandering creeper
♦ Weed, Naturalised, Garden Escape, Cultivation Escape
♦ 3, 73, 86, 88, 101, 179
♦ pH, cultivated. Origin: Mexico.

Traganum nudatum Delile
Chenopodiaceae
♦ Weed
♦ 221
♦ arid.

Tragia benthami Bak.
Euphorbiaceae
♦ Weed, Quarantine Weed
♦ 76, 87, 88, 203, 220

Tragia betonicifolia Nutt.
Euphorbiaceae
♦ noseburn, betonyleaf noseburn
♦ Weed
♦ 161
♦ herbal.

Tragia cannabina L.f.
Euphorbiaceae
♦ Weed
♦ 87, 88
♦ herbal.

Tragia involucrata L.
Euphorbiaceae
♦ Weed
♦ 39, 87, 88
♦ cultivated, herbal, toxic.

Tragia mercurialis L.
Euphorbiaceae
♦ Weed
♦ 87, 88

Tragia nepetifolia Cav.
Euphorbiaceae
♦ catnip noseburn
♦ Weed
♦ 87, 88
♦ herbal.

Tragia ramosa Torr.
Euphorbiaceae
Tragia stylaris Müll.Arg. (see)

♦ nettleleaf noseburn, branched noseburn, noseburn
♦ Weed
♦ 161
♦ pH, herbal.

Tragia spathulata Benth.
Euphorbiaceae
♦ Weed
♦ 87, 88

Tragia stylaris Müll.Arg.
Euphorbiaceae
= *Tragia ramosa* Torr.
♦ Weed
♦ 87, 88

Tragia volubilis L.
Euphorbiaceae
♦ fireman
♦ Weed
♦ 14, 39, 87, 88
♦ herbal, toxic.

Tragopogon L. spp.
Asteraceae
♦ goat's beard
♦ Weed, Naturalised
♦ 23, 88, 198, 272

Tragopogon buphthalmoides (DC.) Boiss.
Asteraceae
♦ Weed
♦ 87, 88

Tragopogon collinus DC.
Asteraceae
♦ Weed
♦ 221

Tragopogon × crantzii Dichtl
Asteraceae
= *Tragopogon dubius* Simps. × *Tragopogon pratensis* L.
♦ Naturalised
♦ 101

Tragopogon dubius Scop.
Asteraceae
Tragopogon major Jacq. (see), *Tragopogon pratensis* Hook.f. *non* L.
♦ yellow goat's beard, yellow salsify, western salsify, goat's beard, meadow salsify
♦ Weed, Noxious Weed, Naturalised, Introduced, Environmental Weed
♦ 21, 52, 80, 86, 87, 88, 98, 101, 102, 136, 146, 151, 158, 161, 174, 180, 203, 207, 210, 211, 212, 243, 272, 280, 299
♦ pH, arid, cultivated, herbal. Origin: Europe.

Tragopogon floccosus Waldst. & Kit.
Asteraceae
♦ woolly goat's beard
♦ Naturalised
♦ 101
♦ cultivated.

Tragopogon graminifolius DC.
Asteraceae
♦ Weed
♦ 87, 88

Tragopogon hybridus L.
Asteraceae
Geropogon glaber L. (see)
♦ Weed, Naturalised

♦ 54, 86, 88, 94
♦ Origin: Mediterranean.

Tragopogon latifolius Boiss.
Asteraceae
♦ Weed
♦ 87, 88, 243

Tragopogon longirostris Bisch. ex Sch.Bip.
Asteraceae
♦ Weed
♦ 221

Tragopogon major Jacq.
Asteraceae
= *Tragopogon dubius* Scop.
♦ western salsify
♦ Weed
♦ 23, 87, 88, 218
♦ herbal.

Tragopogon mirabilis Rouy
Asteraceae
♦ Ontario goat's beard
♦ Naturalised
♦ 101

Tragopogon mirus Ownbey
Asteraceae
♦ remarkable goat's beard
♦ Weed
♦ 80

Tragopogon miscellus Ownbey
Asteraceae
♦ Moscow salsify
♦ Weed, Naturalised
♦ 80, 101

Tragopogon × neohybridus Farw.
Asteraceae
= *Tragopogon porrifolius* L. × *Tragopogon pratensis* L.
♦ Naturalised
♦ 101

Tragopogon orientalis L.
Asteraceae
Tragopogon pratensis L. ssp. *orientalis* (L.) Velen.
♦ Weed
♦ 87, 88
♦ cultivated.

Tragopogon porrifolius L.
Asteraceae
Trapopogon eriospermus Ten., *Trapopogon sinuatus* Avé.
♦ common salsify, Jerusalem star, Joseph's flower, oyster plant, persbokbaard, purple goat's beard, purple salsify, salsify, swart wortel, vegetable oyster, wild salsify, wilde skorsenier
♦ Weed, Naturalised, Introduced, Garden Escape, Environmental Weed, Cultivation Escape
♦ 7, 15, 23, 34, 40, 42, 51, 72, 80, 86, 87, 88, 98, 101, 121, 134, 136, 158, 161, 165, 176, 180, 198, 203, 218, 228, 241, 243, 253, 269, 272, 280, 287, 300
♦ a/bH, arid, cultivated, herbal. Origin: Eurasia.

Tragopogon pratensis L.
Asteraceae

♦ meadow salsify, salsify, yellow goat's beard, wild oyster plant, noonflower, pukinparta, meadow goat's beard, Jack go to bed at noon, goat's beard
♦ Weed, Naturalised
♦ 23, 39, 49, 52, 70, 80, 87, 88, 94, 101, 121, 136, 146, 161, 195, 218, 272, 280, 287
♦ pH, cultivated, herbal, toxic. Origin: Eurasia.

Tragus australianus S.T.Blake
Poaceae
♦ Australian burrgrass, small burrgrass
♦ Weed, Naturalised, Native Weed, Introduced
♦ 101, 163, 228, 269
♦ aG, arid, cultivated. Origin: Australia.

Tragus berteronianus Schult.
Poaceae
♦ burrgrass, carrotseed grass, small carrotseed grass, spiked carrotseed grass, spiked burrgrass
♦ Weed, Native Weed
♦ 50, 87, 88, 91, 121, 158, 221, 275, 297
♦ aG, arid, herbal.

Tragus biflorus Schult.
Poaceae
♦ Weed
♦ 87, 88
♦ G.

Tragus heptaneuron W.D.Clayton
Poaceae
♦ Kenya burrgrass
♦ Naturalised
♦ 101
♦ G.

Tragus koelerioides Asch.
Poaceae
♦ creeping carrotseed grass, cushion grass, goat's beardgrass, perennial carrotseed grass
♦ Native Weed
♦ 121
♦ pG. Origin: southern Africa.

Tragus racemosus (L.) All.
Poaceae
Cenchrus racemosus L., *Lappago racemosa* Schreb., *Tragus brevicaulis* Boiss., *Tragus muricatus* Moench, *Tragus paucispinus* Hack.
♦ carrot grass, carrotseed grass, large carrotseed grass, stalked bristlegrass, stalked carrotseed grass, stalked burrgrass
♦ Weed, Naturalised, Native Weed, Introduced
♦ 38, 51, 87, 88, 91, 101, 121, 158, 221, 237, 240, 253, 272, 287, 295
♦ aG, arid, cultivated, herbal.

Trapa L. spp.
Trapaceae
♦ water caltrope, waterchestnut
♦ Weed, Quarantine Weed, Noxious Weed, Naturalised
♦ 67, 76, 86, 88, 220

♦ wH, cultivated.

Trapa bicornis Osbeck
Trapaceae
♦ horn nut, waterchestnut, ling
♦ Weed, Quarantine Weed, Noxious Weed
♦ 76, 88, 203, 229
♦ wpH, promoted.

Trapa bispinosa Roxb.
Trapaceae
= *Trapa natans* L. var. *bispinosa* (Roxb.) Makino
♦ singhara nut
♦ Weed, Quarantine Weed
♦ 76, 88, 203
♦ wpH, cultivated, herbal.

Trapa incisa Siebold & Zucc.
Trapaceae
♦ himebishi
♦ Weed
♦ 286
♦ wH.

Trapa japonica Flerow
Trapaceae
Trapa korshinskyi V.N.Vassil., *Trapa litwinowii* V.N.Vassil.
♦ waterchestnut
♦ Weed
♦ 88, 204, 297
♦ wpH, cultivated.

Trapa maximowiczii Korsh.
Trapaceae
♦ Maximowicz waterchestnut
♦ Weed, Quarantine Weed
♦ 76, 88, 203, 297
♦ wH.

Trapa natans L.
Trapaceae
Trapa astrachanica (Flerow) Winter, *Trapa quadricornis* Stokes, *Trapa quadrispinosa* Roxb.
♦ waterchestnut, waternut, water caltrops, Jesuit nut, horn, bull nut, European waterchestnut
♦ Weed, Quarantine Weed, Noxious Weed, Naturalised, Environmental Weed
♦ 23, 26, 39, 76, 80, 86, 87, 88, 101, 103, 133, 161, 195, 197, 200, 203, 218, 224, 229, 246, 272
♦ wpH, cultivated, herbal, toxic. Origin: Europe, North Africa, Middle East.

Trapa natans L. var. *bispinosa* (Roxb.) Makino
Trapaceae
Trapa bispinosa Roxb. (see)
♦ singhara nut
♦ Noxious Weed
♦ 229, 286
♦ wpH.

Trapa natans L. var. *natans*
Trapaceae
♦ caltrop
♦ Noxious Weed
♦ 229
♦ wpH.

Trapella sinensis F.W.Oliv.
Pedaliaceae
♦ Weed
♦ 23, 88

Trema aspera Bl.
Celtidaceae/Ulmaceae
♦ Weed
♦ 39, 87, 88
♦ cultivated, toxic.

Trema cannabina Lour.
Celtidaceae/Ulmaceae
= *Trema orientalis* (L.) Blume
♦ mâgele ii
♦ Weed
♦ 3, 191
♦ cultivated, herbal.

Trema guineensis (Schum. & Thonn.) Ficalho
Celtidaceae/Ulmaceae
= *Trema orientalis* (L.) Blume
♦ bosesu, jole
♦ Weed
♦ 87, 88, 179
♦ herbal.

Trema micrantha (L.) Blume
Celtidaceae/Ulmaceae
♦ Florida trema
♦ Weed, Environmental Weed
♦ 22, 87, 88, 218, 255, 257
♦ pH, cultivated. Origin: Brazil.

Trema orientale (L.) Blume
Celtidaceae/Ulmaceae
= *Trema orientalis* (L.) Blume
♦ oriental trema
♦ Naturalised
♦ 101

Trema orientalis (L.) Blume
Celtidaceae/Ulmaceae
Trema cannabina Lour. (see), *Trema guineensis* (Schum. & Thonn.) Fical. (see), *Trema orientale* (L.) Blume (see)
♦ charcoal tree, gunpowder tree, agaunai, banahl, elodechoel, uanin, ndrou, ndroundrou, ndrikanaithembe, bulasisi, besesu, pigeonwood, musonsoli, mhehu, Rhodesian elm
♦ Weed, Native Weed
♦ 3, 12, 22, 87, 88, 121, 191
♦ S/T, cultivated, herbal. Origin: Madagascar.

Tremastelma palaestinum (L.) Janch.
Dipsacaceae
= *Scabiosa palaestina* L.
♦ Weed
♦ 272

Trembleya phlogiformis DC.
Melastomataceae
♦ island glorybush
♦ Naturalised
♦ 101

Trembleya phlogiformis DC. var. *parvifolia* Cogn.
Melastomataceae
♦ island glorybush
♦ Naturalised
♦ 101

Triadica sebifera (L.) Small
Euphorbiaceae
Croton sebiferum L., *Sapium sebiferum*
(L.) Roxb. (see), *Stillingia sebifera* (L.)
Michx.
- Chinese tallow tree, tallow tree
- Noxious Weed, Naturalised
- 101, 229
- Origin: China, Japan.

Trianoptiles solitaria (C.B.Clarke) Levyns
Cyperaceae
- subterranean cape sedge
- Weed, Naturalised, Environmental Weed
- 54, 86, 88, 198
- G. Origin: South Africa.

Trianthema L. spp.
Aizoaceae
- trianthema
- Weed
- 39, 243
- H, toxic.

Trianthema australis Melville
Aizoaceae
= *Zaleya galericulata* (Melville)
H.Eichler var. *australis* (Melville)
S.W.L.Jacobs
- Weed
- 87, 88

Trianthema decandra L.
Aizoaceae
- Weed
- 87, 88
- herbal.

Trianthema galericulata Melville
Aizoaceae
= *Zaleya galericulata* (Melville)
H.Eichler
- Weed
- 39, 87, 88
- toxic.

Trianthema monogyna L.
Aizoaceae
= *Trianthema portulacastrum* L.
- Weed, Quarantine Weed
- 76, 87, 88, 203, 220
- herbal.

Trianthema pentandra L.
Aizoaceae
= *Zaleya pentandra* (L.) Jeffr.
- African purslane
- Weed, Quarantine Weed
- 50, 76, 87, 88, 203, 220
- herbal, toxic.

Trianthema portulacastrum L.
Aizoaceae
Portulacastrum monogynum (L.) Medik.,
Trianthema flexuosa Schumach. &
Thonn., *Trianthema littoralis* Cordem.,
Trianthema monanthogyna L., *Trianthema
monogyna* L. (see), *Trianthema
procumbens* Mill.
- horse purslane, verdolaga rastrera, desert purslane, purslane, phak bia hin, giant pigweed, black pigweed, desert horse purslane

- Weed, Naturalised, Environmental Weed
- 32, 39, 55, 86, 87, 88, 93, 98, 157, 161, 180, 203, 205, 217, 218, 236, 237, 239, 242, 243, 269, 276, 295
- a/pH, arid, cultivated, herbal, toxic. Origin: obscure.

Trianthema salsoloides Fenzl ex Oliv.
Aizoaceae
- Weed
- 221
- herbal.

Trianthema triqueta Rottl. ex Willd.
Aizoaceae
- Weed
- 87, 88

Tribolium acutiflorum (Nees) Renv.
Poaceae
- desmazeria, tribolium
- Weed, Naturalised, Environmental Weed
- 72, 86, 86, 88, 98, 198, 203
- pG. Origin: South Africa.

Tribolium echinatum (Thunb.) Renvoize
Poaceae
Alopecurus echinatus Thunb., *Dactylis
ascendens* Schrad., *Lasiochloa ciliaris*
Kunth, *Lasiochloa echinata* (Thunb.)
Adamson (see), *Lasiochloa echinata*
(Thunb.) Henrard
- Weed, Naturalised
- 7, 54, 86, 88, 98, 203
- G. Origin: South Africa.

Tribolium obliterum (Hemsl.) Renvoize
Poaceae
- cape grass
- Naturalised, Environmental Weed
- 86, 182
- pG. Origin: South Africa.

Tribolium uniolae (L.f.) Renvoize
Poaceae
Briza imbricata Steud., *Brizopyrum
alternans* Nees, *Brizopyrum capense*
(Spreng.) Nees, *Brizopyrum capense*
(Trin.) Trin., *Cynosurus paniculatus*
Thunb., *Cynosurus uniolae* L.f.,
Desmazeria alternans (Nees) T.Durand
& Schinz, *Desmazeria capensis* (Spreng.)
E.Phillips, *Desmazeria uniolae* (L.f.)
Kuntze, *Plagiochloa alternans* (Nees)
Adamson & Sprague, *Plagiochloa
uniolae* (L.f.) Adamson & Sprague
(see), *Poa papillosa* Schrad., *Poa uniolae*
(L.f.) Schrad., *Tribolium alternans*
(Nees) Renvoize, *Tribolium amplexum*
Renvoize, *Triticum capense* Spreng.,
Uniola capensis Trin.
- tribolium
- Weed, Naturalised, Environmental Weed
- 7, 86, 98, 155, 163, 203
- G. Origin: South Africa.

Tribulus L. spp.
Zygophyllaceae/Tribulaceae
- puncture vine, caltrop, goat's head
- Weed
- 221
- toxic.

Tribulus cistoides L.
Zygophyllaceae/Tribulaceae
- Jamaica feverplant, puncture vine, burnut, caltrop, te maukinikini, rockrose, carpetweed, false puncture vine, puncture weed
- Weed, Noxious Weed, Naturalised, Introduced, Environmental Weed, Cultivation Escape
- 3, 6, 14, 39, 80, 86, 87, 88, 93, 101, 112, 122, 157, 161, 179, 186, 218, 228, 249, 252, 257, 261, 276, 287
- a/pH, arid, cultivated, herbal, toxic. Origin: tropical America.

Tribulus longipetalus Viv.
Zygophyllaceae/Tribulaceae
- Weed
- 88, 221
- arid.

Tribulus micrococcus Domin
Zygophyllaceae/Tribulaceae
- yellow vine
- Weed, Quarantine Weed
- 76, 88, 203

Tribulus ochroleucus (Maire) Ozenda & Quèzel
Zygophyllaceae/Tribulaceae
- Weed
- 221

Tribulus patens Swallen
Zygophyllaceae/Tribulaceae
- Weed
- 87, 88

Tribulus saccariflora Nees
Zygophyllaceae/Tribulaceae
- Weed
- 87, 88

Tribulus terrestris L.
Zygophyllaceae/Tribulaceae
Tribulus terrester Landolt
- bindy eye, bindii, bull's head, burnut, caltrop, cat's head, common dubbeltjie, devil's thorn, doublegee, dubbeltjie, goat's head, ground burr nut, isiHoho, land caltrop, Maltese cross, Mexican sandbur, puncture vine, puncture weed, roseta francesa, small caltrops, tackweed, Texas sandbur, yellow vine
- Weed, Noxious Weed, Naturalised, Native Weed, Introduced, Casual Alien
- 1, 7, 23, 26, 34, 35, 39, 49, 50, 51, 53, 55, 66, 67, 70, 80, 86, 87, 88, 93, 94, 101, 102, 115, 116, 121, 138, 146, 147, 156, 158, 161, 167, 171, 174, 179, 180, 186, 186, 203, 205, 209, 210, 212, 218, 219, 221, 228, 229, 236, 237, 240, 241, 242, 243, 253, 257, 264, 269, 272, 275, 276, 280, 295, 297, 299, 300
- aH, arid, cultivated, herbal, toxic. Origin: Mediterranean.

Tribulus zeyheri Sond.
Zygophyllaceae/Tribulaceae
- devil thornweed
- Weed, Native Weed
- 87, 88, 121
- pH, toxic. Origin: southern Africa.

Trichachne insularis (L.) Nees
Poaceae
= *Digitaria insularis* (L.) Fedde
♦ sour grass
♦ Weed, Quarantine Weed, Naturalised
♦ 3, 14, 76, 86, 87, 88, 191, 203, 218, 220, 286, 287
♦ G, herbal.

Trichilia emetica Vahl ssp. *emetica*
Meliaceae
Trichilia jubensis Chiov., *Trichilia roka* Chiov. nom. illeg., *Trichilia somalensis* Chiov., *Trichilia umbrifera* Swynn. & Bak.f.
♦ Introduced
♦ 228

Trichilia havanensis Jacq.
Meliaceae
♦ Introduced
♦ 38, 228
♦ T, arid, herbal.

Trichilia hirta L.
Meliaceae
♦ broomstick, wild mahogany
♦ Weed
♦ 14, 179
♦ T, herbal.

Trichloris crinita (Lag.) Parodi
Poaceae
Trichloris mendocina (Phil) Kurtz
♦ trichloris
♦ Weed
♦ 295
♦ G, herbal.

Trichloris pluriflora Fourn.
Poaceae
= *Chloris pluriflora* (E.Fourn.) Clayton
♦ Weed
♦ 237
♦ G.

Trichocaulon marlothii N.E.Br.
Asclepiadaceae/Apocynaceae
♦ Native Weed
♦ 121
♦ pH, toxic. Origin: southern Africa.

Trichocereus (Berg.) Riccob. spp.
Cactaceae
= *Echinopsis* Zucc. spp. (NoR) [see × *Trichoechinopsis* hort. ex Backeb. spp.]
♦ Weed, Quarantine Weed, Naturalised
♦ 76, 86, 88, 203
♦ cultivated.

Trichodesma africanum (L.) Lehm.
Boraginaceae
Borago africana L.
♦ Weed
♦ 87, 88, 221
♦ herbal.

Trichodesma amplexicaule Roth
Boraginaceae
♦ Weed
♦ 87, 88

Trichodesma dekindtianum Guerke
Boraginaceae

♦ Weed
♦ 87, 88

Trichodesma ehrenbergii Boiss.
Boraginaceae
♦ Weed
♦ 221

Trichodesma indicum (L.) Sm.
Boraginaceae
♦ Weed
♦ 87, 88
♦ herbal.

Trichodesma physaloides (Fenzl) A.DC.
Boraginaceae
♦ chocolate bells
♦ Native Weed
♦ 121
♦ pH. Origin: southern Africa.

Trichodesma sedgwickianum S.P.Ban.
Boraginaceae
♦ Weed
♦ 66
♦ herbal.

Trichodesma zeylanicum (Burm.f.) R.Br.
Boraginaceae
♦ late weed, northern bluebell
♦ Weed
♦ 23, 39, 50, 53, 87, 88, 121, 158, 240
♦ aH, cultivated, herbal, toxic. Origin: Eurasia.

× *Trichoechinopsis* hort. ex Backeb. spp.
Cactaceae
= *Echinopsis* Zucc. spp. (NoR) [see *Trichocereus* (Berg.) Riccob. spp.]
♦ Weed, Quarantine Weed
♦ 76, 88, 220

Tricholaena Schult. spp.
Poaceae
♦ Weed
♦ 218
♦ G.

Tricholaena monachne (Trin.) Stapf & C.E.Hubb.
Poaceae
Aira bicolor Schum., *Eremochlamys arenaria* Peter, *Eremochlamys littoralis* Peter, *Melinis glabra* (Stapf) Hack., *Melinis monachne* (Trin.) Pilg., *Melinis trichotoma* Mez, *Panicum gracillimum* Schum., *Panicum madagascariense* Spreng., *Panicum madagascariense* var. *brevispiculum* Rendle, *Panicum monachne* Trin., *Tricholaena arenaria* Nees var. *semiglabra* Hack., *Tricholaena bicolor* (Schumach.) C.E.Hubb., *Tricholaena delicatula* Stapf & C.E.Hubb., *Tricholaena glabra* Stapf, *Tricholaena monachne* (Trin.) Stapf & C.E.Hubb. var. *annua* J.G.Anderson, *Xyochlaena monachne* (Trin.) Stapf
♦ blueseed tricholaena
♦ Weed, Native Weed
♦ 88, 121
♦ pG, arid. Origin: Madagascar, southern Africa.

Tricholaena repens (Willd.) Hitchc.
Poaceae
= *Melinis repens* (Willd.) Zizka
♦ Weed

♦ 3, 88, 191
♦ G.

Tricholaena rosea Nees
Poaceae
= *Melinis repens* (Willd.) Zizka
♦ Weed
♦ 3, 191
♦ G.

Tricholaena teneriffae (L.) Link
Poaceae
Panicum teneriffae (L.f.) R.Br., *Saccharum teneriffae* L.f., *Tricholaena micrantha* Schrad.
♦ Weed
♦ 221
♦ G, arid.

Trichomanes venosum R.Br.
Hymenophyllaceae
♦ Australasian filmy fern
♦ Garden Escape
♦ 40
♦ cultivated.

Trichoneura eleusinoides (Rendle) Ekman
Poaceae
♦ Weed
♦ 121
♦ pG.

Trichoneura grandiglumis (Nees) Ekman
Poaceae
Crossotropis grandiglumis (Nees) Rendle
♦ rolling grass, tumbleweed
♦ Native Weed
♦ 121
♦ pG, herbal. Origin: southern Africa.

Trichosanthes cucumerina L.
Cucurbitaceae
Trichosanthes ambrozii Domin, *Trichosanthes anguina* L., *Trichosanthes brevibrachteata* Kundu (incl. vars), *Trichosanthes pachyrrhachis* Kundu
♦ chichinda, snake gourd
♦ Weed, Introduced
♦ 228, 262
♦ arid, cultivated, herbal. Origin: Asia, Australia.

Trichosanthes cucumeroides (Ser.) Maxim.
Cucurbitaceae
♦ Weed
♦ 286
♦ herbal.

Trichosanthes integrifolia Kurz
Cucurbitaceae
♦ Weed
♦ 12

Trichosanthes kirilowii Maxim.
Cucurbitaceae
♦ Chinese cucumber
♦ Quarantine Weed
♦ 39, 220
♦ pC, cultivated, herbal, toxic.

Trichosanthes kirilowii Maxim. var. *japonica* Kitam.
Cucurbitaceae
♦ kikarasuuri
♦ Weed
♦ 286
♦ cultivated.

Trichostema dichotomum L.
Lamiaceae
♦ blue curls, bastard pennyroyal, forked bluecurls
♦ Weed
♦ 87, 88, 218
♦ herbal.

Trichostema lanceolatum Benth.
Lamiaceae
♦ vinegar weed, blue curls, camphor weed
♦ Weed
♦ 161, 180, 243
♦ aH, cultivated, herbal.

Trichostigma octandrum (L.) Watt.
Phytolaccaceae/Petiveriaceae
♦ hoopvine
♦ Weed
♦ 14
♦ herbal.

Tricyrtis formosana Bak.
Liliaceae/Uvulariaceae/
Convallariaceae/Tricyridaceae
♦ toad lily
♦ Naturalised
♦ 287
♦ cultivated. Origin: Taiwan.

Tricyrtis hirta (Thunb.) Hook.
Liliaceae/Uvulariaceae/
Convallariaceae/Tricyridaceae
Tricyrtis japonica Miq.
♦ toad lily, hototogisu
♦ Naturalised
♦ 101
♦ pH, cultivated, herbal.

Tridax coronopifolia (H.B.K.) Hemsl.
Asteraceae
♦ Weed
♦ 199

Tridax procumbens L.
Asteraceae
♦ tridax daisy, tridax, coat buttons, teen tukkae, Mexican daisy, hierba de toro
♦ Weed, Quarantine Weed, Noxious Weed, Naturalised, Environmental Weed
♦ 6, 7, 12, 13, 14, 67, 86, 87, 88, 93, 98, 101, 121, 140, 157, 158, 161, 170, 179, 199, 203, 209, 217, 229, 239, 243, 249, 255, 257, 258, 261, 273, 274, 275, 276, 281, 286, 287, 297
♦ aH, cultivated, herbal. Origin: Central America.

Tridens flavus (L.) Hitchc.
Poaceae
♦ purpletop, purpletop tridens
♦ Weed
♦ 161
♦ G, herbal.

Trifolium L. spp.
Fabaceae/Papilionaceae
♦ clover
♦ Weed, Naturalised, Environmental Weed
♦ 181, 195, 198, 221, 243, 256, 272, 296
♦ H, herbal, toxic.

Trifolium africanum Ser. var. glabellum (E.Mey.) Harv.
Fabaceae/Papilionaceae
♦ African wild clover, cape clover, red African clover, wild clover, wild pink clover
♦ Native Weed
♦ 121
♦ pH. Origin: southern Africa.

Trifolium agrarium L. p.p.
Fabaceae/Papilionaceae
= *Trifolium aureum* Pollich
♦ hop clover
♦ Weed
♦ 80, 87, 88, 218
♦ H, herbal.

Trifolium alexandrinum L.
Fabaceae/Papilionaceae
Trifolium constantinopolitanum DC.
♦ berseem clover, Egyptian clover
♦ Weed, Naturalised, Introduced, Casual Alien
♦ 39, 40, 70, 86, 98, 176, 203, 221, 228, 248
♦ H, arid, cultivated, toxic.

Trifolium alpestre L.
Fabaceae/Papilionaceae
Lagopus montanus Bernh.
♦ pyökkiapila, purple globe clover
♦ Weed, Casual Alien
♦ 42, 272
♦ H, cultivated.

Trifolium amabile Kunth
Fabaceae/Papilionaceae
♦ Aztec clover
♦ Weed
♦ 87, 88
♦ H, herbal.

Trifolium ambiguum Bieb.
Fabaceae/Papilionaceae
Trifolium ambiguum M.Bieb. var. *majus* Hossain, *Trifolium vaillantii* M.Bieb. ex Fisch.
♦ kura clover, Caucasian clover
♦ Weed, Naturalised, Introduced
♦ 86, 98, 203, 228
♦ H, arid, cultivated. Origin: eastern Europe, Middle East.

Trifolium angustifolium L.
Fabaceae/Papilionaceae
♦ narrowleaf crimson clover, narrow leaved clover
♦ Weed, Naturalised, Introduced, Environmental Weed, Casual Alien
♦ 7, 9, 34, 40, 70, 72, 88, 94, 98, 101, 158, 176, 203, 241, 250, 253, 280, 287, 300
♦ aH, cultivated, herbal.

Trifolium angustifolium L. var. angustifolium
Fabaceae/Papilionaceae
♦ narrowleaf clover
♦ Naturalised, Environmental Weed
♦ 86, 198
♦ H. Origin: southern Europe, North Africa, Middle East.

Trifolium argutum Banks & Sol.
Fabaceae/Papilionaceae

Trifolium xerocephalum Fenzl
♦ Weed, Naturalised
♦ 98, 203
♦ H, cultivated.

Trifolium arvense L.
Fabaceae/Papilionaceae
♦ rabbit's foot clover, stone clover, old field clover, hare's foot clover, jänönapila
♦ Weed, Naturalised, Introduced, Environmental Weed
♦ 7, 9, 15, 24, 34, 44, 70, 72, 80, 87, 88, 94, 98, 101, 161, 165, 176, 179, 203, 211, 218, 228, 241, 243, 249, 250, 253, 272, 280, 287, 300
♦ aH, arid, cultivated, herbal. Origin: Mediterranean, west Asia.

Trifolium arvense L. fo. albiflorum Sylvén
Fabaceae/Papilionaceae
♦ Naturalised
♦ 287
♦ H.

Trifolium arvense L. var. arvense
Fabaceae/Papilionaceae
♦ hare's foot clover
♦ Naturalised, Environmental Weed
♦ 86, 198
♦ H. Origin: Eurasia, North Africa.

Trifolium aureum Pollich
Fabaceae/Papilionaceae
Trifolium agrarium L. *p.p.* (see), *Trifolium fuscum* Desv., *Trifolium strepens* Crantz (see)
♦ yellow hop clover, large trefoil, golden clover, hop clover, kelta apila, large hop trefoil
♦ Weed, Naturalised
♦ 80, 101, 272, 280, 287, 300
♦ aH, cultivated, herbal. Origin: Eurasia.

Trifolium badium Schreb.
Fabaceae/Papilionaceae
♦ rusoapila
♦ Casual Alien
♦ 42
♦ H, cultivated, herbal.

Trifolium burchellianum Ser. ssp. burchellianum
Fabaceae/Papilionaceae
♦ Naturalised
♦ 86
♦ H. Origin: southern Africa.

Trifolium campestre Schreb.
Fabaceae/Papilionaceae
Chrysaspis campestris Desv., *Melilotus agraria* Desf., *Trifolium agrarium sensu auct.*, *Trifolium campestre* Schreb. var. *lagrangei* (Boiss.) Zohary, *Trifolium erythranthum* (Griseb.) Hal., *Trifolium lagrangei* Boiss., *Trifolium procumbens* L. (see), *Trifolium pseudoprocumbens* C.Gmel., *Trifolium pumilum* Hossain, *Trifolium thionanthum* Hausskn.
♦ large hop clover, low clover, yellow clover, hop trefoil, large hop trefoil, rentoapila
♦ Weed, Naturalised, Introduced,

Environmental Weed
♦ 7, 9, 15, 34, 42, 72, 80, 87, 88, 94, 98, 101, 111, 121, 134, 161, 176, 203, 207, 218, 228, 241, 243, 249, 250, 253, 272, 280, 287, 300
♦ a/bH, arid, cultivated, herbal. Origin: Eurasia.

Trifolium campestre Schreb. var. campestre
Fabaceae/Papilionaceae
♦ hop clover
♦ Naturalised, Environmental Weed
♦ 86, 198
♦ H. Origin: Eurasia, North Africa.

Trifolium carolinianum Michx.
Fabaceae/Papilionaceae
♦ Carolina clover
♦ Naturalised
♦ 287
♦ H.

Trifolium cernuum Brot.
Fabaceae/Papilionaceae
♦ drooping flowered clover, nodding clover
♦ Weed, Naturalised, Environmental Weed
♦ 7, 9, 15, 86, 98, 121, 176, 198, 203, 250, 280
♦ aH, arid. Origin: Eurasia.

Trifolium cherleri L.
Fabaceae/Papilionaceae
♦ cupped clover
♦ Weed, Naturalised, Introduced
♦ 7, 86, 88, 94, 98, 203, 228, 253
♦ H, arid. Origin: southern Europe, North Africa, Middle East.

Trifolium clusii Gren. & Godr.
Fabaceae/Papilionaceae
♦ shaftal clover
♦ Weed
♦ 111, 243
♦ H, cultivated.

Trifolium clypeatum L.
Fabaceae/Papilionaceae
♦ helmut clover
♦ Introduced
♦ 228
♦ H, arid.

Trifolium dalmaticum Vis.
Fabaceae/Papilionaceae
♦ Dalmatian clover
♦ Naturalised
♦ 101
♦ H.

Trifolium dubium Sibth.
Fabaceae/Papilionaceae
♦ cowhop clover, Irish shamrock, least hop clover, lesser yellow trefoil, little hop clover, pikkuapila, shamrock, shamrock clover, small hop clover, suckling clover, yellow clover, yellow suckling clover
♦ Weed, Naturalised, Introduced, Environmental Weed
♦ 7, 9, 15, 24, 34, 38, 42, 72, 80, 86, 87, 88, 98, 101, 121, 134, 136, 157, 161, 165, 176, 181, 198, 203, 218, 241, 243, 249,

253, 272, 280, 286, 287, 300
♦ a/bH, arid, cultivated, herbal. Origin: Eurasia.

Trifolium echinatum Bieb.
Fabaceae/Papilionaceae
♦ prickly clover
♦ Weed, Naturalised
♦ 101, 272
♦ H.

Trifolium elegans Savi
Fabaceae/Papilionaceae
= *Trifolium hybridum* L.
♦ Weed
♦ 87, 88
♦ H.

Trifolium filiforme L.
Fabaceae/Papilionaceae
= *Trifolium micranthum* Viv. and *Trifolium campestre* Schreb. in part.
♦ Weed
♦ 87, 88
♦ H.

Trifolium fragiferum L.
Fabaceae/Papilionaceae
Galearia fragifera (L.) Presl, *Trifolium fragiferum* L. ssp. *bonannii* (J. & C.Presl) Soják (see), *Trifolium neglectum* Fisch. & Mey.
♦ strawberry clover, strawberry headed clover, O'Connor's legume, ìatelina jahodovitá, rakkoapila
♦ Weed, Naturalised, Introduced, Environmental Weed
♦ 7, 15, 34, 39, 72, 80, 86, 88, 94, 98, 101, 176, 181, 198, 203, 243, 272, 280, 287
♦ pH, arid, cultivated, herbal, toxic. Origin: Eurasia.

Trifolium fragiferum L. ssp. *bonannii* (J. & C.Presl) Soják
Fabaceae/Papilionaceae
= *Trifolium fragiferum* L.
♦ Naturalised
♦ 241
♦ H.

Trifolium fragiferum L. var. *fragiferum*
Fabaceae/Papilionaceae
♦ strawberry clover
♦ Naturalised, Environmental Weed
♦ 86, 198, 300
♦ H.

Trifolium fragiferum L. var. *pulchellum* Lange
Fabaceae/Papilionaceae
♦ strawberry clover
♦ Naturalised
♦ 86
♦ H. Origin: Eurasia.

Trifolium fucatum Lindl.
Fabaceae/Papilionaceae
♦ bull clover, sour clover
♦ Naturalised
♦ 287
♦ aH, promoted, herbal. Origin: western North America.

Trifolium globosum L.
Fabaceae/Papilionaceae
Trifolium radiosum Wahlenb.

♦ Weed, Naturalised
♦ 86, 98, 203
♦ H. Origin: western Mediterranean.

Trifolium glomeratum L.
Fabaceae/Papilionaceae
♦ flat headed clover, clustered clover, tiheäkukka apila, ball clover
♦ Weed, Naturalised, Introduced, Environmental Weed, Casual Alien
♦ 7, 9, 15, 34, 42, 72, 86, 87, 88, 98, 101, 134, 176, 198, 203, 241, 250, 253, 272, 280, 287, 300
♦ aH, arid, herbal. Origin: south and west Europe, Middle east.

Trifolium hirtum All.
Fabaceae/Papilionaceae
Trifolium hispidum Desf., *Trifolium oxypetasum* Heldr. & Sart. ex Nyman, *Trifolium pictum* Roth.
♦ rose clover, hairy clover
♦ Weed, Naturalised, Introduced, Environmental Weed, Casual Alien
♦ 7, 34, 40, 86, 98, 101, 198, 203, 228, 280, 287
♦ aH, arid, cultivated, herbal. Origin: Eurasia, North Africa.

Trifolium hybridum L.
Fabaceae/Papilionaceae
Trifolium bicolor Moench, *Trifolium elegans* Savi (see), *Trifolium fistulosum* Gilib., *Trifolium intermedium* Lapeyr.
♦ alsike clover, alsikeapila
♦ Weed, Naturalised, Introduced, Environmental Weed
♦ 7, 21, 39, 39, 40, 42, 80, 87, 88, 98, 101, 161, 176, 179, 203, 218, 241, 272, 280, 287, 300
♦ pH, cultivated, herbal, toxic.

Trifolium hybridum L. ssp. *hybridum*
Fabaceae/Papilionaceae
♦ Naturalised
♦ 40
♦ H.

Trifolium hybridum L. var. *hybridum*
Fabaceae/Papilionaceae
♦ alsike clover
♦ Naturalised
♦ 86, 198
♦ H. Origin: south to east Europe.

Trifolium incarnatum L.
Fabaceae/Papilionaceae
Trifolium incarnatum L. var. *elatius* Gibelli & Belli (see), *Trifolium spicatum* Perret
♦ crimson clover, long headed clover, veriapila
♦ Weed, Naturalised, Introduced, Casual Alien
♦ 7, 34, 39, 42, 70, 80, 87, 88, 94, 98, 101, 161, 176, 203, 241, 272, 280, 287, 300
♦ aH, cultivated, herbal, toxic.

Trifolium incarnatum L. ssp. *incarnatum*
Fabaceae/Papilionaceae
♦ crimson clover
♦ Naturalised
♦ 40
♦ H.

***Trifolium incarnatum* L. var. *elatius*
Gibelli & Belli**
Fabaceae/Papilionaceae
= *Trifolium incarnatum* L.
♦ crimson clover
♦ Weed
♦ 218
♦ H.

Trifolium incarnatum* L. var. *incarnatum
Fabaceae/Papilionaceae
♦ crimson clover
♦ Naturalised
♦ 86, 198
♦ H, cultivated. Origin: west and
central Europe.

Trifolium involucratum* Willd. *nom. illeg.
Fabaceae/Papilionaceae
= *Trifolium mucronatum* Willd. ex
Spreng. (NoR)
♦ Naturalised
♦ 241
♦ H.

***Trifolium lappaceum* L.**
Fabaceae/Papilionaceae
♦ burdock clover, burr clover,
takiaisapila
♦ Weed, Naturalised, Casual Alien
♦ 7, 40, 42, 87, 88, 94, 98, 101, 111, 176,
203, 243, 253, 272
♦ H, herbal.

Trifolium lappaceum* L. var. *lappaceum
Fabaceae/Papilionaceae
♦ burdock clover
♦ Naturalised, Environmental Weed
♦ 86, 198
♦ H. Origin: Europe, Middle East.

***Trifolium leucanthum* M.Bieb.**
Fabaceae/Papilionaceae
♦ Weed, Casual Alien
♦ 40, 272
♦ H.

***Trifolium ligusticum* Balb. ex Lois.**
Fabaceae/Papilionaceae
♦ Weed, Naturalised
♦ 7, 86, 98, 203
♦ H. Origin: Mediterranean.

***Trifolium lupinaster* L.**
Fabaceae/Papilionaceae
♦ lupine clover
♦ Weed, Naturalised
♦ 101, 297
♦ H, cultivated.

***Trifolium macraei* Hook. & Arn.**
Fabaceae/Papilionaceae
♦ Chilean clover
♦ Weed
♦ 87, 88
♦ aH, herbal.

***Trifolium medium* L.**
Fabaceae/Papilionaceae
Trifolium flexuosum Jacq., *Trifolium
transsilvanicum* Porc.
♦ zigzag clover, metsäapila
♦ Weed, Naturalised
♦ 86, 87, 88, 98, 101, 176, 203, 218, 272,
280, 287
♦ H, cultivated, herbal. Origin:

Europe, Middle East.

***Trifolium michelianum* Savi**
Fabaceae/Papilionaceae
♦ bigflower clover, balansa clover
♦ Weed, Naturalised
♦ 98, 203, 272
♦ H, cultivated, herbal. Origin:
Mediterranean.

***Trifolium michelianum* Savi var.
*michelianum***
Fabaceae/Papilionaceae
♦ annual white clover
♦ Naturalised
♦ 86, 198
♦ H.

***Trifolium micranthum* Viv.**
Fabaceae/Papilionaceae
Melilotus anomala Ledeb., *Trifolium
capilliforme* Desr., *Trifolium filiforme* L.
p.p. (see)
♦ slender suckling clover, least yellow
trefoil, slender trefoil
♦ Weed, Naturalised
♦ 15, 86, 98, 176,
203, 280
♦ H, herbal. Origin: Madagascar.

***Trifolium microdon* Hook. & Arn.**
Fabaceae/Papilionaceae
♦ thimble clover, valparaiso,
valparaiso clover
♦ Naturalised
♦ 7
♦ aH, herbal. Origin: Americas.

***Trifolium montanum* L.**
Fabaceae/Papilionaceae
Trifolium rupestre Ten.
♦ apilat clover trefoil, mäkiapila
♦ Weed
♦ 87, 88, 272
♦ H, cultivated, herbal.

***Trifolium nervulosum* Boiss. & Heldr.**
Fabaceae/Papilionaceae
♦ Weed
♦ 221
♦ H.

***Trifolium nigrescens* Viv.**
Fabaceae/Papilionaceae
♦ small white clover
♦ Weed, Naturalised, Casual Alien
♦ 39, 40, 87, 88, 101, 253, 272
♦ H, herbal, toxic.

***Trifolium ochroleucon* Huds.**
Fabaceae/Papilionaceae
Trifolium dipsaceum Camus, *Trifolium
ochroleucum* Huds., *Trifolium pallidulum*
Jordan, *Trifolium roseum* J. & C.Presl
♦ kalvasapila, sulphur clover
♦ Weed, Naturalised, Casual Alien
♦ 42, 272, 280
♦ H, cultivated. Origin: Eurasia.

***Trifolium ornithopodioides* L.**
Fabaceae/Papilionaceae
Trigonella ornithopodiodes (L.) DC.
♦ fenugreek, bird's foot clover, bird's
foot fenugreek, bird clover
♦ Weed, Naturalised
♦ 7, 86, 98, 176, 198, 203, 280

♦ H, arid, cultivated. Origin: Eurasia.

***Trifolium pallidum* Waldst. & Kit.**
Fabaceae/Papilionaceae
♦ Weed
♦ 272
♦ H.

***Trifolium pannonicum* Jacq.**
Fabaceae/Papilionaceae
Trifolium alopecuroides Pers.
♦ Hungarian clover, trèfle de Hongrie,
ponnonischer klee
♦ Weed
♦ 272
♦ H, cultivated.

***Trifolium patens* Schreb.**
Fabaceae/Papilionaceae
Trifolium chrysanthum Gaud., *Trifolium
parisiense* DC., *Trifolium savianum*
Willd.
♦ notkea apila
♦ Weed, Casual Alien
♦ 42, 272
♦ H, herbal.

***Trifolium pilulare* Boiss.**
Fabaceae/Papilionaceae
♦ Weed, Naturalised, Introduced,
Casual Alien
♦ 7, 40, 86, 98, 203, 228
♦ H, arid. Origin: western
Mediterranean, Middle East.

***Trifolium polymorphum* Poir.**
Fabaceae/Papilionaceae
Trifolium megalanthum Steud.
♦ peanut clover
♦ Naturalised
♦ 101
♦ H, arid, cultivated.

***Trifolium polymorphum* Poir. var.
*polymorphum***
Fabaceae/Papilionaceae
♦ Naturalised
♦ 241
♦ H.

***Trifolium pratense* L.**
Fabaceae/Papilionaceae
Lagopus pratensis (L.) Bernh, *Trifolium
borysthenicum* Grun., *Trifolium frigidum*
Schur, *Trifolium seravschanicum* Ovcz.
ex Bobr.
♦ red clover, cow grass, trifoglio rosso,
purple clover
♦ Weed, Naturalised, Introduced,
Garden Escape, Environmental Weed,
Cultivation Escape, Casual Alien
♦ 7, 8, 15, 21, 34, 39, 70, 72, 80, 86, 87,
88, 98, 101, 136, 161, 165, 176, 179, 198,
203, 218, 228, 237, 241, 243, 261, 272,
275, 280, 286, 287, 295, 297, 300
♦ pH, arid, cultivated, herbal, toxic.
Origin: Eurasia.

***Trifolium pratense* L. fo. *albiflorum* Alef.**
Fabaceae/Papilionaceae
♦ Naturalised
♦ 287
♦ H.

***Trifolium pratense* L. fo. *sanguineum* Asai**
Fabaceae/Papilionaceae
♦ Naturalised

- 287
- H.

Trifolium procumbens L.
Fabaceae/Papilionaceae
= *Trifolium campestre* Schreb.
- low hop clover
- Weed
- 80, 87, 88, 218
- H, herbal.

Trifolium purpureum Gilib.
Fabaceae/Papilionaceae
- purple clover
- Weed, Naturalised
- 87, 88, 101, 111, 243, 253, 272
- H, cultivated.

Trifolium repens L.
Fabaceae/Papilionaceae
Trifolium anomalum Schrank, *Trifolium limonium* Phil., *Trifolium nigrescens* Schur, *Trifolium nothum* Stev.
- white clover, Dutch clover, honeysuckle clover, white trefoil, purplewort, trébol blanco, white Dutch clover, ladino clover
- Weed, Naturalised, Introduced, Garden Escape, Environmental Weed
- 7, 15, 34, 39, 44, 70, 72, 80, 86, 87, 88, 94, 98, 101, 111, 121, 136, 154, 157, 158, 161, 165, 174, 176, 179, 181, 195, 203, 211, 218, 228, 236, 237, 241, 243, 249, 253, 255, 263, 272, 275, 280, 286, 287, 293, 295, 297, 299, 300
- aH, arid, cultivated, herbal, toxic. Origin: Eurasia.

Trifolium repens L. fo. *giganteum* Lagr.-Foss.
Fabaceae/Papilionaceae
- Naturalised
- 287
- H.

Trifolium repens L. fo. *nigricans* G.Don
Fabaceae/Papilionaceae
- Naturalised
- 287
- H.

Trifolium repens L. fo. *roseum* Peterm.
Fabaceae/Papilionaceae
- Naturalised
- 287
- H.

Trifolium repens L. var. *repens*
Fabaceae/Papilionaceae
- white clover
- Naturalised, Environmental Weed
- 198, 289
- H.

Trifolium resupinatum L.
Fabaceae/Papilionaceae
Trifolium bicorne Forssk., *Trifolium resupinatum* L. var. *majus* Boiss. (see), *Trifolium resupinatum* L. var. *microcephalum* Zohary, *Trifolium suaveolens* Willd.
- reversed clover, shaftal clover, tuoksuapila, Persian clover, bird's eye clover, trèfle renversé, persischer klee, shaftal, trevo da Pérsia, trébol persa

- Weed, Naturalised, Introduced, Casual Alien
- 7, 40, 42, 70, 80, 87, 88, 94, 98, 101, 161, 176, 179, 185, 198, 203, 218, 221, 228, 243, 272, 280, 287
- H, arid, cultivated, herbal, toxic.

Trifolium resupinatum L. var. *majus* Boiss.
Fabaceae/Papilionaceae
= *Trifolium resupinatum* L.
- shaftal clover
- Naturalised
- 86, 198, 287
- H.

Trifolium resupinatum L. var. *resupinatum*
Fabaceae/Papilionaceae
- shaftal clover
- Naturalised
- 86, 198
- H. Origin: southern Europe, Middle East, North Africa.

Trifolium retusum L.
Fabaceae/Papilionaceae
Trifolium parviflorum Ehrh.
- teasel clover
- Weed, Naturalised
- 40, 272, 280
- H.

Trifolium rubens L.
Fabaceae/Papilionaceae
Lagopus glaber Bernh.
- Naturalised
- 86
- H, cultivated, herbal. Origin: centra and east Europe.

Trifolium scabrum L.
Fabaceae/Papilionaceae
- rough clover, rough trefoil, karhea apila, clover
- Weed, Naturalised, Environmental Weed, Casual Alien
- 7, 42, 72, 86, 88, 94, 98, 101, 176, 198, 203, 280
- aH, herbal. Origin: Eurasia, North Africa.

Trifolium spadiceum L.
Fabaceae/Papilionaceae
Trifolium decipiens Hornem., *Trifolium litigiosum* Desv.
- musta apila
- Weed, Naturalised
- 241, 272, 300
- H.

Trifolium spumosum L.
Fabaceae/Papilionaceae
- Mediterranean clover, bladder clover
- Weed, Naturalised, Casual Alien
- 7, 40, 86, 98, 101, 203
- H, cultivated. Origin: southern Europe, Middle East.

Trifolium squamosum L.
Fabaceae/Papilionaceae
- sea clover, pörröapila, teasel headed clover
- Weed, Naturalised, Casual Alien

- 42, 86, 98, 101, 176, 198, 203, 272, 280
- H. Origin: southern Europe, Middle East, North Africa.

Trifolium squarrosum L.
Fabaceae/Papilionaceae
- Weed
- 272
- H.

Trifolium stellatum L.
Fabaceae/Papilionaceae
- starry clover, star clover
- Weed, Naturalised
- 7, 70, 87, 88, 94, 98, 111, 176, 198, 203, 243, 253
- H, cultivated, herbal.

Trifolium stellatum L. var. *stellatum*
Fabaceae/Papilionaceae
- Naturalised, Environmental Weed
- 86
- H. Origin: Eurasia, North Africa.

Trifolium strepens Crantz. *nom. illeg.*
Fabaceae/Papilionaceae
= *Trifolium aureum* Pollich
- Weed
- 87, 88, 243
- H.

Trifolium striatum L.
Fabaceae/Papilionaceae
Trifolium tenuiflorum Ten.
- knotted clover, soft clover, striated clover, soft trefoil, juovikasapila
- Weed, Naturalised, Environmental Weed, Casual Alien
- 7, 15, 42, 72, 86, 87, 88, 94, 98, 101, 176, 198, 203, 241, 272, 280, 300
- aH, cultivated, herbal. Origin: Eurasia, North Africa.

Trifolium subterraneum L.
Fabaceae/Papilionaceae
Calycomorphum subterraneum (L.) Presl, *Trifolium oxaloides* Bunge ex Nyman
- subterranean clover, burrowing clover
- Weed, Naturalised, Introduced, Environmental Weed
- 7, 9, 15, 34, 39, 72, 80, 86, 87, 88, 98, 101, 165, 176, 198, 203, 228, 243, 253, 280, 287, 300
- aH, arid, cultivated, herbal, toxic. Origin: Mediterranean, west Asia.

Trifolium suffocatum L.
Fabaceae/Papilionaceae
- suffocated clover, trèfle étouffé
- Weed, Naturalised
- 7, 86, 98, 176, 198, 203, 241, 280, 300
- H. Origin: Eurasia, North Africa.

Trifolium tomentosum L.
Fabaceae/Papilionaceae
- woolly clover, clover
- Weed, Naturalised, Introduced, Environmental Weed, Casual Alien
- 7, 40, 42, 70, 72, 87, 88, 98, 101, 121, 176, 203, 221, 228, 241, 253, 280, 287, 300
- a/bH, arid, cultivated. Origin: Eurasia.

Trifolium tomentosum **L. var.** *tomentosum*
Fabaceae/Papilionaceae
♦ woolly clover
♦ Naturalised, Environmental Weed
♦ 86, 198
♦ H. Origin: southern Europe, Middle East, North Africa.

Trifolium uniflorum **L.**
Fabaceae/Papilionaceae
♦ oneflower clover
♦ Weed, Naturalised, Garden Escape, Environmental Weed
♦ 54, 86, 88, 98, 176, 203
♦ H, cultivated. Origin: southern Europe, North Africa.

Trifolium vesiculosum **Savi**
Fabaceae/Papilionaceae
♦ arrowleaf clover, pussiapila
♦ Weed, Naturalised, Casual Alien
♦ 42, 54, 88, 101, 272
♦ H, cultivated. Origin: southern Europe, Mediterranean.

Trifolium vesiculosum **Savi var.**
vesiculosum
Fabaceae/Papilionaceae
♦ arrowleaf clover
♦ Naturalised
♦ 86, 198
♦ H.

Triglochin **L. spp.**
Juncaginaceae
♦ arrowgrass
♦ Quarantine Weed
♦ 220

Triglochin concinnum **Burtt Davy**
Juncaginaceae
♦ Utah arrowgrass, common arrowgrass
♦ Weed
♦ 161
♦ herbal, toxic.

Triglochin maritima **L.**
Juncaginaceae
Triglochin maritimum L. (see)
♦ seaside arrowgrass, merisuolake, arrowgrass, goosegrass, sour grass, troscart maritime
♦ Weed, Noxious Weed
♦ 23, 36, 39, 161, 212, 218, 299
♦ wpH, cultivated, herbal, toxic.

Triglochin maritimum **L.**
Juncaginaceae
= *Triglochin maritima* L.
♦ arrowgrass, seaside arrowgrass
♦ Weed, Quarantine Weed, Native Weed
♦ 76, 87, 88, 174, 203, 220, 297
♦ cultivated, herbal.

Triglochin palustre **L.**
Juncaginaceae
= *Triglochin palustris* L.
♦ marsh arrowgrass, hosobanoshibana
♦ Weed
♦ 275, 297
♦ aqua, herbal.

Triglochin palustris **L.**
Juncaginaceae

Triglochin palustre L. (see)
♦ marsh arrowgrass, hentosuolake, troscart des marais
♦ Weed
♦ 23, 39, 87, 88, 161, 218, 272
♦ wpH, promoted, herbal, toxic.

Triglochin procera **R.Br.**
Juncaginaceae
Triglochin procerum R.Br.
♦ water ribbons
♦ Weed, Quarantine Weed, Naturalised
♦ 86, 87, 88, 220
♦ aqua, cultivated. Origin: Australia.

Triglochin striatum **Ruiz & Pav.**
Juncaginaceae
♦ streaked arrowgrass, three ribbed arrowgrass
♦ Weed
♦ 161
♦ aqua, cultivated, herbal, toxic.

Trigonella **L. spp.**
Fabaceae/Papilionaceae
♦ fenugreek, kozieradka
♦ Weed
♦ 221, 272
♦ herbal.

Trigonella arabica **Del.**
Fabaceae/Papilionaceae
♦ Weed
♦ 221
♦ arid.

Trigonella caerulea **(L.) Ser.**
Fabaceae/Papilionaceae
Folliculigera caerulea Pasq.,
Grammocarpus caeruleus Schur,
Melilotus caeruleus Desv., *Teliosma caerulea* Alef., *Telis caerulea* Kuntze,
Trifoliastrum caeruleum Moench,
Trifolium caeruleum Willd, *Trigonella melilotus-coerulea* (L.) Asch. & Graebn.
♦ sinisarviapila, sweet trefoil, curd herb, blue fenugreek
♦ Weed, Naturalised, Introduced, Casual Alien
♦ 40, 42, 88, 94, 101, 179, 228, 272
♦ aH, arid, cultivated, herbal.

Trigonella corniculata **(L.) L.**
Fabaceae/Papilionaceae
♦ etelänsarviapila, cultivated fenugreek
♦ Weed, Naturalised, Casual Alien
♦ 42, 88, 94, 101, 272
♦ aH, promoted, herbal.

Trigonella fischeriana **Ser.**
Fabaceae/Papilionaceae
♦ Quarantine Weed
♦ 220

Trigonella foenum-graecum **L.**
Fabaceae/Papilionaceae
Buceras foenum-graecum (L.) All.,
Foenum-graecum officinale Moench,
Foenum-graecum sativum Medik., *Telis foenumgraecum* Kuntze, *Trigonella jemenensis* (Serp.) Sinsk.
♦ fenugreek, sicklefruit fenugreek, classical fenugreek
♦ Weed, Naturalised, Introduced,

Casual Alien
♦ 39, 40, 70, 87, 88, 94, 101, 228, 272
♦ aH, arid, cultivated, herbal, toxic.

Trigonella grandiflora **Bunge**
Fabaceae/Papilionaceae
♦ komeasarviapila
♦ Casual Alien
♦ 42

Trigonella hamosa **L.**
Fabaceae/Papilionaceae
= *Trigonella hierosolymitana* Boiss. (NoR)
♦ keltasarviapila, Egyptian fenugreek, fenugreek
♦ Weed, Casual Alien
♦ 40, 42, 87, 88, 121, 185, 221, 242
♦ aH, arid.

Trigonella incisa **Benth.**
Fabaceae/Papilionaceae
= *Medicago monantha* (C.A.Mey.) Trautv.
♦ Weed
♦ 87, 88, 243
♦ arid.

Trigonella laciniata **L.**
Fabaceae/Papilionaceae
♦ jagged fenugreek, cutleaf fenugreek, trigonella
♦ Weed, Naturalised
♦ 87, 88, 101, 185, 221, 243

Trigonella medicagnoidea **(Sirj.) Vassilcz.**
Fabaceae/Papilionaceae
♦ Quarantine Weed
♦ 220

Trigonella monantha **C.A.Mey.**
Fabaceae/Papilionaceae
= *Medicago monantha* (C.A.Mey.) Trautv.
♦ turkinsarviapila
♦ Casual Alien
♦ 42

Trigonella monspeliaca **L.**
Fabaceae/Papilionaceae
= *Medicago monspeliaca* (L.) Trautv.
♦ star fruited fenugreek, ranskansarviapila
♦ Weed, Naturalised, Casual Alien
♦ 40, 42, 86, 87, 88, 94, 241, 253, 272, 300
♦ Origin: Mediterranean, Middle East, Russia.

Trigonella orthoceras **Kar. & Kir.**
Fabaceae/Papilionaceae
♦ suorasarviapila
♦ Casual Alien
♦ 42

Trigonella polycerata **L.**
Fabaceae/Papilionaceae
♦ fenugreek
♦ Weed, Quarantine Weed, Casual Alien
♦ 40, 76, 87, 88, 203, 220
♦ herbal.

Trigonella procumbens **(Bess.) Rchb.**
Fabaceae/Papilionaceae
♦ trailing fenugreek, fenugreek
♦ Weed, Naturalised, Casual Alien
♦ 40, 101, 272

Trigonella radiata (L.) Boiss.
Fabaceae/Papilionaceae
= *Medicago radiata* L. (NoR)
♦ Weed
♦ 87, 88

Trigonella ramosa L.
Fabaceae/Papilionaceae
♦ Naturalised
♦ 101

Trigonella stellata Forssk.
Fabaceae/Papilionaceae
♦ Weed
♦ 221

Trigonia nivea Camb.
Trigoniaceae
♦ Weed
♦ 255
♦ pH. Origin: Brazil.

Trigonotis amblyosepala Nakai. & Kitag
Boraginaceae
♦ Weed
♦ 275
♦ pH.

Trigonotis clavata Stev.
Boraginaceae
♦ Weed
♦ 87, 88

Trigonotis peduncularis Benth. ex S.Moore
Boraginaceae
Eritrichium japonicum Miq., *Eritrichium pedunculare* (Trevis.) A.DC., *Myosotis peduncularis* Trevis.
♦ kiurigusa
♦ Weed
♦ 87, 88, 204, 263, 275, 286, 297
♦ aH, promoted, herbal. Origin: east Asia.

Triguera osbekii (L.) Willk.
Solanaceae
♦ Weed
♦ 87, 88

Trillium L. spp.
Liliaceae/Trilliaceae
♦ trillium, wake robin
♦ Weed
♦ 161, 247
♦ toxic.

Trimezia martinicensis (Jacq.) Herb.
Iridaceae
♦ Martinique trimezia, walking iris
♦ Weed, Naturalised, Cultivation Escape
♦ 28, 87, 88, 101, 179, 206, 243, 261
♦ cultivated, herbal.

Trinia glauca (L.) Dumort
Apiaceae
Apium glaucum Börner, *Seseli pumilum* L.
♦ honewort
♦ Weed
♦ 272
♦ cultivated, herbal.

Triodanis Raf. spp.
Campanulaceae
♦ Venus's lookingglass
♦ Weed
♦ 106

Triodanis biflora (Ruiz & Pav.) Greene
Campanulaceae
= *Triodanis perfoliata* (L.) Nieuwl. var. *biflora* (Ruiz & Pav.) Bradl.
♦ clasping Venus's lookingglass, small Venus's lookingglass
♦ Weed, Quarantine Weed, Naturalised
♦ 76, 87, 88, 161, 203, 220, 287, 295
♦ aH, herbal.

Triodanis perfoliata (L.) Nieuwl.
Campanulaceae
Specularia perfoliata (L.) A.DC. (see)
♦ Venus's lookingglass, clasping Venus's lookingglass, common Venus's lookingglass, round leaved triodanis, clasping bellwort
♦ Weed, Naturalised, Native Weed
♦ 86, 98, 161, 174, 199, 203, 211, 249, 286, 287
♦ aH, arid, herbal. Origin: North America.

Triodanis perfoliata (L.) Nieuwl. fo. alba Voigt
Campanulaceae
♦ Naturalised
♦ 287

Triodanis perfoliata (L.) Nieuwl. var. biflora (Ruiz & Pav) Bradl.
Campanulaceae
Triodanis biflora (Ruiz & Pav.) Greene (see), *Specularia biflora* (Ruiz & Pav.) Fisch. & Mey. (see)
♦ clasping Venus's lookingglass
♦ Weed
♦ 237

Triopteris jamaicensis L.
Malpighiaceae
= *Hiptage benghalensis* (L.) Kurz
♦ Weed
♦ 3, 191
♦ herbal.

Triopteris rigida Sw.
Malpighiaceae
♦ Weed
♦ 14

Triphasia trifolia (Burm.f.) P.Wilson
Rutaceae
Limonia trifolia Burm.f., *Triphasia aurantiola* Lour., *Triphasia trifoliata* (L.) DC. (see)
♦ lime berry, limon China, limoncito, lemon China, lemon de China, chinita
♦ Weed, Quarantine Weed, Naturalised, Environmental Weed, Cultivation Escape, Casual Alien
♦ 3, 22, 32, 76, 87, 88, 101, 179, 191, 261
♦ S, cultivated, herbal.

Triphasia trifoliata (L.) DC.
Rutaceae
= *Triphasia trifolia* (Burm.f.) P.Wilson
♦ lime berry
♦ Weed
♦ 80, 112

Triphysaria pusilla (Benth.) Chuang & Heckard
Scrophulariaceae
♦ small owl's clover, dwarf owl's clover, little owl's clover, triphysaria
♦ Naturalised
♦ 86, 198
♦ aH.

Triplaris Loefl. ex L. spp.
Polygonaceae
♦ ant tree
♦ Naturalised, Environmental Weed
♦ 86

Triplaris americana L.
Polygonaceae
♦ triplaris, ant tree
♦ Weed, Noxious Weed, Garden Escape, Introduced
♦ 19, 85, 283
♦ T, cultivated, toxic. Origin: south Panama to south-east Brazil.

Triplaris cumingiana Fisch. & C.A.Mey.
Polygonaceae
♦ triplaria
♦ Introduced
♦ 261
♦ T. Origin: northern South America.

Tripleurospermum auriculatum (Boiss.) Rech.f.
Asteraceae
♦ Weed
♦ 221

Tripleurospermum inodorum Sch.Bip. nom. illeg.
Asteraceae
= *Tripleurospermum perforatum* (Mérat) Laínz
♦ scentless chamomile, scentless mayweed, peltosaunio
♦ Weed, Naturalised, Garden Escape, Environmental Weed
♦ 15, 44, 86, 87, 88, 176, 243, 280
♦ pH, cultivated, herbal. Origin: Australia.

Tripleurospermum limosum (Maxim.) Pobed.
Asteraceae
♦ scentless mayweed
♦ Weed
♦ 297

Tripleurospermum maritimum (L.) Koch
Asteraceae
Chamomilla maritima (L.) Rydb., *Matricaria maritima* L., *Tripleurospermum maritima* (L.) Koch
♦ mayweed, sea mayweed, merisaunio, false mayweed
♦ Weed, Quarantine Weed, Naturalised
♦ 76, 87, 88, 101, 203, 243
♦ aH, cultivated.

Tripleurospermum perforatum (Mérat) Laínz
Asteraceae
Chamomilla inodora (L.) Gilib., *Matricaria chamomilla* L. (see), *Matricaria inodora* L. nom. illeg. (see), *Matricaria maritima* L. var. *agrestis* (Knaf) L. (see), *Matricaria maritima* L. ssp. *inodora* (L.) Clapham, *Matricaria perforata* Mérat (see), *Tripleurospermum inodorum* Sch.Bip. *nom. illeg.* (see),

Tripleurospermum perforata (Mérat) Laínz
- scentless mayweed, scentless false mayweed, inland scentless mayweed, scentless chamomile
- Quarantine Weed, Noxious Weed, Naturalised
- 76, 101, 229, 243, 300

Triplotaxis stellulifera (Benth.) Hutch.
Asteraceae
= *Vernonia stellulifera* (Benth.) C.Jeffrey (NoR)
- Weed
- 87, 88

Tripodanthus acutifolius (Ruiz & Pav.) Tiegh.
Loranthaceae
- erva de passarinho
- Weed
- 255
- pH parasitic. Origin: South America.

Tripogandra cumanensis (Kunth) Woodson
Commelinaceae
- tripa de pollo
- Weed
- 281

Tripogandra disgrega (Kunth) Woodson
Commelinaceae
- Weed
- 157

Tripogandra diuretica (Mart.) Handlos
Commelinaceae
- ondas do mar
- Weed
- 255
- Origin: Americas.

Tripolium pannonicum (Jacq.) Dobrocz.
Asteraceae
Aster tripolium L. (see)
- sea aster
- Naturalised
- 101

Tripolium vulgare Ness.
Asteraceae
- Weed
- 275, 297
- aH.

Tripsacum andersonii J.R.Gray
Poaceae
= *Tripsacum* L. spp. × *Zea* L. spp. [possible intergeneric origin]
- Weed, Introduced
- 32, 38
- G.

Tripsacum dactyloides (L.) L.
Poaceae
- eastern gramagrass, fhakahatchee grass, gamagrass, sesame grass
- Weed, Naturalised
- 98, 203
- pG, cultivated, herbal.

Tripsacum latifolium Hitchc.
Poaceae
- cayenne grass, yerba cayena, wideleaf gamagrass
- Cultivation Escape

- 261
- G, cultivated. Origin: tropical America.

Tripsacum laxum Nash
Poaceae
Tripsacum fasciculatum Trin. ex Asch. *nom. illeg.*
- Guatemala grass
- Weed, Cultivation Escape
- 87, 88, 90, 261
- pG, arid, cultivated. Origin: Mexico, Central America.

Triraphis mollis R.Br.
Poaceae
- Introduced
- 39, 228
- G, arid, cultivated, toxic.

Triraphis pumilio R.Br.
Poaceae
- Weed
- 221
- G.

Trisetaria cristata (L.) Kerguélen
Poaceae
= *Rostraria cristata* (L.) Tzvelev
- Naturalised
- 7
- G.

Trisetobromus hirtus (Trin.) Nev.
Poaceae
= *Bromus berterianus* Colla
- Weed
- 87, 88
- G.

Trisetum aureum (Ten.) Ten.
Poaceae
- golden oatgrass
- Naturalised
- 101
- G.

Trisetum bifidum (Thunb.) Ohwi
Poaceae
- bifid yellow oatgrass
- Weed
- 87, 88, 286
- G.

Trisetum flavescens (L.) Beauv.
Poaceae
Lophochloa flavescens (L.) P.Beauv., *Rebentischia flavescens* Opiz, *Trisetum pratense* Pers.
- yellow oatgrass, keltakaura, avoine dorée
- Weed, Naturalised, Introduced, Casual Alien
- 34, 39, 42, 98, 101, 203, 228, 272, 280, 300
- pG, arid, cultivated, herbal, toxic. Origin: Eurasia, North Africa.

Trisetum paniceum (Lam.) Pers.
Poaceae
- Weed
- 70
- G.

Trisetum spicatum (L.) Richt.
Poaceae
Aira spicata L., *Aira subspicata* L.,

Trisetum pubiflorum Hack., *Trisetum subspicatum* Beauv.
- spike trisetum, tähkäkaura
- Weed
- 23, 88
- pG, arid, cultivated, herbal. Origin: northern hemisphere.

Tristachya leucothrix Nees
Poaceae
Tristachya hispida (L.f.) K.Schum.
- hairy trident grass
- Native Weed
- 121
- pG. Origin: southern Africa.

Tristagma uniflorum (Lindl.) Traub
Liliaceae/Alliaceae
= *Ipheion uniflorum* (Graham) Raf.
- springstar, spring starflower
- Naturalised
- 40, 101

Tristemma mauritianum J.F.Gmel.
Melastomataceae
- Naturalised
- 86
- pH. Origin: Africa, Madagascar.

Triticum L. spp.
Poaceae
- wheat, trigo
- Weed, Naturalised
- 198, 236
- G, herbal.

Triticum aestivum L.
Poaceae
Frumentum triticum Krause, *Triticum aristatum* Krause, *Triticum erinaceum* Krause, *Triticum hystrix* Ser., *Triticum linnaeanum* Lag., *Triticum muticum* Schübl., *Triticum tenax* Schrank, *Triticum vulgare* Lam.
- wheat, pisi ka hola, bread wheat, common wheat, trigo
- Weed, Naturalised, Garden Escape, Environmental Weed, Cultivation Escape, Casual Alien
- 7, 34, 39, 40, 42, 86, 87, 88, 98, 101, 121, 176, 198, 199, 203, 241, 243, 245, 256, 280
- aG, arid, cultivated, herbal, toxic. Origin: Eurasia.

Triticum baeoticum Boiss.
Poaceae
Triticum aegilopoides Balan.
- wild einkorn
- Weed
- 272
- G, cultivated.

Triticum compactum Host
Poaceae
- club wheat, clubbed wheat
- Casual Alien
- 280
- G, promoted. Origin: obscure.

Triticum monococcum L.
Poaceae
Crithodium monococcum (L.) Á.Löve, *Nivieria monococcum* (L.) Ser., *Triticum aestivum* var. *monococcum*

(L.) L.H.Bailey, *Triticum hornemanni*
Clemente
♦ eikorn
♦ Weed
♦ 272
♦ G, cultivated, herbal. Origin:
obscure.

Triticum ramosum Trin.
Poaceae
= *Leymus ramosus* (Trin.) Tzvelev
♦ Weed
♦ 87, 88
♦ G.

Triticum spelta L.
Poaceae
Triticum aestivum L. ssp. *spelta* (L.)
Thell. (NoR)
♦ spelt, spelt wheat
♦ Naturalised
♦ 101
♦ G, cultivated, herbal. Origin:
obscure, possibly Eurasia.

Triticum tauschii (Coss.) Schmalh.
Poaceae
= *Aegilops tauschii* Coss.
♦ Quarantine Weed
♦ 220
♦ G.

Triticum turgidum L.
Poaceae
Triticum aestivum L. ssp. *turgidum*
(L.) Domin, *Triticum compositum* L.,
Triticum vulgare L. ssp. *turgidum* Körn.
♦ kartiovehnä, rivet wheat
♦ Naturalised, Casual Alien
♦ 40, 42, 101
♦ G, cultivated, herbal. Origin:
obscure.

Tritonia crocata (L.) Ker Gawl.
Iridaceae
Tritonia hyalina (L.) Bak.
♦ orange tritonia, mosselbaaikal-
koentjie
♦ Weed, Naturalised, Garden Escape,
Environmental Weed
♦ 7, 15, 72, 86, 88, 98, 198, 203, 280
♦ pH, cultivated. Origin: South Africa.

Tritonia crocosmaeflora Lem.
Iridaceae
= *Crocosmia × crocosmiiflora* (Lem. ex
anon.) N.E.Br.
♦ crocus tritonia
♦ Weed, Naturalised
♦ 88, 218, 286, 287

Tritonia × crocosmiflora (Lem.) Nichols
Iridaceae
= *Crocosmia × crocosmiiflora* (Lem. ex
anon.) N.E.Br.
♦ montbretia
♦ Weed, Naturalised
♦ 87, 88, 241

Tritonia crocosmiflora (Lem.) Nichols
Iridaceae
= *Crocosmia × crocosmiiflora* (Lem. ex
anon.) N.E.Br.
♦ Weed
♦ 199

Tritonia lineata (Salisb.) Ker Gawl.
Iridaceae
♦ lined tritonia, bergkatjietee
♦ Weed, Naturalised, Garden Escape,
Environmental Weed
♦ 7, 15, 72, 86, 88, 98, 134, 176, 198, 203,
280, 289
♦ pH, cultivated. Origin: South Africa.

Tritonia pottsii Benth. & Hook.
Iridaceae
♦ Weed
♦ 87, 88

Tritonia squalida (Sol.) Ker Gawl.
Iridaceae
♦ tritonia, kalkoentjie, pink tritonia
♦ Weed, Naturalised, Garden Escape,
Environmental Weed
♦ 72, 86, 88, 98, 198, 203
♦ pH, cultivated. Origin: South Africa.

**Tritoniopsis caffra (Ker ex Bak.)
Goldblatt**
Iridaceae
♦ Quarantine Weed
♦ 220
♦ cultivated.

Triumfetta acuminata H.B.K.
Tiliaceae
♦ Weed
♦ 87, 88

Triumfetta althaeoides Lam.
Tiliaceae
= *Triumfetta semitriloba* (L.) Jacq.
♦ Weed
♦ 87, 88

Triumfetta angolensis Sprague & Hutch.
Tiliaceae
Triumfetta sonderi Fic. & Hiern
♦ Native Weed
♦ 121
♦ pS. Origin: southern Africa.

Triumfetta annua L.
Tiliaceae
♦ chafunga
♦ Weed, Native Weed
♦ 88, 121, 158
♦ aH. Origin: southern Africa.

Triumfetta bartramia L. nom. illeg.
Tiliaceae
= *Triumfetta rhomboidea* Jacq.
♦ Sacramento burr, barba de boi
♦ Weed
♦ 3, 87, 88, 191, 218, 255, 297
♦ S, herbal. Origin: tropical America.

Triumfetta bogotensis DC.
Tiliaceae
Triumfetta botteriana Turcz., *Triumfetta
dumetorum* Schldl., *Triumfetta lindeniana*
Turcz.
♦ parquet burr
♦ Naturalised, Introduced
♦ 101, 228
♦ arid.

Triumfetta cordifolia A.Rich.
Tiliaceae
♦ cordleaf burrbark, chaunga, ponga,
katonka

♦ Weed, Quarantine Weed
♦ 76, 88, 203, 220
♦ pH, cultivated. Origin: central
Africa.

Triumfetta flavescens Hochst. ex A.Rich.
Tiliaceae
♦ Weed
♦ 87, 88, 221

Triumfetta lappula L.
Tiliaceae
♦ grandcousin
♦ Weed
♦ 87, 88
♦ S/T, herbal.

Triumfetta pentandra A.Rich.
Tiliaceae
♦ fivestamen burrbark
♦ Weed, Naturalised, Native Weed
♦ 86, 87, 88, 93, 121, 179
♦ aH. Origin: Africa.

Triumfetta pilosa Roth
Tiliaceae
♦ burs
♦ Weed, Naturalised
♦ 86, 87, 88, 158
♦ Origin: Africa, Asia.

**Triumfetta pilosa Roth var. effusa (E.Mey.
ex Harv.) Wild**
Tiliaceae
Triumfetta effusa E.Mey. ex Harv.
♦ klitsbossie
♦ Weed, Native Weed
♦ 51, 121
♦ pS. Origin: southern Africa.

**Triumfetta pilosa Roth var. tomentosa
Szyszyl. ex Sprague & Hutch.**
Tiliaceae
♦ burs, klitsbossie
♦ Weed
♦ 51, 121
♦ pS. Origin: Africa.

Triumfetta procumbens Forst.f.
Tiliaceae
♦ mautofu tai
♦ Weed
♦ 87, 88
♦ H, herbal. Origin: Australia.

Triumfetta rhomboidea Jacq.
Tiliaceae
Triumfetta bartramia L. (see)
♦ Chinese burr, paroquet burr, burr
bush, dadangsi, masiksik lahe,
mo'osipo, mosipo, mautofu, qatima,
um Gamgampunja, diamond burrbark
♦ Weed, Quarantine Weed, Noxious
Weed, Naturalised, Native Weed,
Introduced
♦ 3, 6, 50, 66, 76, 86, 87, 88, 107, 121,
134, 158, 179, 191, 203, 229, 230, 276
♦ H, cultivated, herbal. Origin:
obscure.

Triumfetta rotundifolia Lam.
Tiliaceae
♦ Weed
♦ 66, 87, 88
♦ herbal.

Triumfetta semitriloba (L.) Jacq.
Tiliaceae
Triumfetta althaeoides Lam. (see),
Triumfetta rubricaulis Kunth
♦ Sacramento burr, dadangsi, masiksik lahe, Sacramento burrbark, caballusa, burrweed, guanxuma rosa
♦ Weed, Noxious Weed, Naturalised, Introduced, Environmental Weed
♦ 3, 14, 86, 87, 88, 107, 153, 157, 179, 191, 228, 229, 230, 255, 257
♦ H, arid, herbal. Origin: tropical America.

Triumfetta velutina Vahl
Tiliaceae
♦ African burrbark
♦ Weed
♦ 87, 88
♦ Origin: Australia.

Triumfetta welwitschii Mast.
Tiliaceae
♦ Weed
♦ 87, 88

Triumfetta welwitschii Mast. var. *hirsuta* (Sprague & Hutch.) Wild
Tiliaceae
♦ Native Weed
♦ 121
♦ pH. Origin: southern Africa.

Triumfetta welwitschii Mast. var. *welwitschii*
Tiliaceae
♦ Native Weed
♦ 121
♦ pH. Origin: southern Africa.

Trixis inula Crantz
Asteraceae
♦ tropical threefold
♦ Weed
♦ 199

Trixis radiale Lag.
Asteraceae
♦ Weed
♦ 87, 88

Trollius europaeus L.
Ranunculaceae
♦ globe flower, European globeflower, kullero
♦ Weed
♦ 39, 70, 272
♦ pH, cultivated, herbal, toxic. Origin: Europe.

Tropaeolum gracile (Hook. & Arn.) Sparre
Tropaeolaceae
♦ Weed
♦ 87, 88

Tropaeolum majus L.
Tropaeolaceae
Tropaeolum elatum Salisb., *Tropaeolum repandifolium* Stokes, *Tropaeolum schillingii* Vilm.
♦ garden nasturtium, Indian cress, tall nasturtium, nasturtium
♦ Weed, Naturalised, Introduced, Garden Escape, Environmental Weed, Cultivation Escape, Casual Alien
♦ 7, 15, 24, 34, 39, 40, 42, 72, 86, 88, 98, 101, 116, 121, 165, 198, 203, 225, 228, 241, 246, 257, 261, 269, 280, 290, 300
♦ aC, arid, cultivated, herbal, toxic. Origin: South America.

Tropaeolum peltophorum Benth. × majus L.
Tropaeolaceae
♦ hybrid nasturtium
♦ Naturalised
♦ 40

Tropaeolum pentaphyllum Lam.
Tropaeolaceae
♦ Naturalised
♦ 280
♦ cultivated.

Tropaeolum speciosum Poepp. & Endl.
Tropaeolaceae
♦ Chilean flame creeper, flame flower
♦ Weed, Quarantine Weed, Naturalised, Environmental Weed
♦ 15, 225, 246, 280
♦ pC, cultivated.

Tropidocarpum gracile Hook.
Brassicaceae
♦ dobie pod, graceful tropidocarpum
♦ Weed
♦ 161
♦ aH, herbal.

Tsuga canadensis (L.) Carr.
Pinaceae/Abietaceae
Picea canadensis Link, *Pinus canadensis* L., *Tsuga americana* (Mill.) Farw.
♦ eastern hemlock, Canadian hemlock
♦ Weed
♦ 39, 87, 88, 218
♦ T, cultivated, herbal, toxic.

Tsuga heterophylla (Raf.) Sarg.
Pinaceae/Abietaceae
Abies heterophylla Raf., *Tsuga albertiana* Senecl.
♦ western hemlock, Choina zachodnia
♦ Weed
♦ 87, 88, 218
♦ T, cultivated, herbal.

Tsuga mertensiana (Bong.) Carr.
Pinaceae/Abietaceae
♦ mountain hemlock
♦ Weed
♦ 87, 88, 218
♦ T, cultivated, herbal.

Tuberaria guttata (L.) Fourr.
Cistaceae
Helianthemum guttatum (L.) Mill.
♦ spotted rockrose, European frostweed, täplänouto
♦ Weed, Naturalised, Casual Alien
♦ 23, 42, 88, 101, 272
♦ a/pH, cultivated, herbal.

Tuberaria lignosa (Sweet) Samp.
Cistaceae
Tuberaria vulgaris Willk.
♦ Weed
♦ 272
♦ cultivated.

Tubocapsicum anomalum (Franch. & Sav.) Makino
Solanaceae
= *Capsicum anomalum* Franch. & Sav. (NoR)
♦ Weed
♦ 87, 88
♦ cultivated.

Tulbaghia violacea Harv.
Liliaceae/Alliaceae
♦ society garlic
♦ Naturalised, Cultivation Escape, Casual Alien
♦ 86, 252, 280
♦ cultivated, herbal.

Tulipa L. spp.
Liliaceae
♦ tulip
♦ Weed
♦ 154, 161, 247
♦ toxic.

Tulipa clusiana DC.
Liliaceae
Tulipa stellata Hook.f.
♦ lady tulip
♦ Introduced
♦ 215
♦ cultivated.

Tulipa edulis (Miq.) Bak.
Liliaceae
♦ edible tulip
♦ Weed
♦ 286
♦ pH, cultivated, herbal. Origin: east China, south Japan, Korea, Manchuria.

Tulipa gesneriana L.
Liliaceae
♦ Didier's tulip, tarhatulppaani
♦ Naturalised, Cultivation Escape
♦ 39, 42, 101
♦ pH, cultivated, herbal, toxic.

Tulipa lehmanniana Merckl. ex Bunge
Liliaceae
♦ Weed
♦ 243

Tulipa montana Lindl.
Liliaceae
Tulipa wilsoniana Hoog
♦ Weed
♦ 87, 88
♦ pH, cultivated.

Tulipa saxatilis Sieber ex Spreng.
Liliaceae
♦ Cretan tulip
♦ Naturalised
♦ 40
♦ cultivated.

Tulipa sylvestris L.
Liliaceae
Tulipa florentina hort. ex Bak., *Tulipa persica* Willd. ex Kunth *nom. illeg.*
♦ wild tulip
♦ Weed, Naturalised
♦ 39, 101, 253
♦ pH, cultivated, herbal, toxic.

Tunas Lunell spp.
Cactaceae
= *Opuntia* Mill. spp.

♦ Weed, Quarantine Weed
♦ 76, 88, 220

Tunica velutina (Guss.) Fisch. & Mey.
Caryophyllaceae
= *Petrorhagia dubia* (Raf.) López &
Romo
♦ Weed
♦ 87, 88

Tupa portoricensis Vatke
Campanulaceae
= *Lobelia portoricensis* (Vatke) Urb.
(NoR)
♦ Weed
♦ 87, 88

Turbina corymbosa (L.) Raf.
Convolvulaceae
Ipomoea burmanni Choisy, *Ipomoea
sidifolia* Choisy, *Rivea corymbosa* (L.)
Hallier.f. (see)
♦ Christmas vine, ololiuqui, turbina
♦ Weed, Naturalised, Introduced,
Environmental Weed
♦ 3, 14, 39, 86, 155, 179, 191, 228
♦ pC, arid, cultivated, herbal, toxic.
Origin: tropical America.

**Turbina oblongata (E.Mey. ex Choisy)
A.Meeuse**
Convolvulaceae
Ipomoea oblongata E.Mey. ex Choisy
♦ Native Weed
♦ 121
♦ pH, herbal. Origin: southern Africa.

Turczaninovia fastigiata (Fisch.) DC.
Asteraceae
♦ Weed
♦ 275
♦ pH.

Turgenia latifolia (L.) Hoffm.
Apiaceae
Caucalis latifolia L. (see), *Daucus
latifolius* Baill., *Selinum turgenia* Krause,
Tordylium latifolium L.
♦ greater burr parsley, broadleaf false
carrot, lapapurho
♦ Weed, Naturalised, Casual Alien
♦ 42, 70, 87, 88, 94, 101, 243, 253, 272
♦ cultivated, herbal.

Turnera diffusa Willd. ex Schult.
Turneraceae
Bohadschia humifusa C.Presl, *Turnera
aphrodisiaca* Ward, *Turnera humifusa*
Endl., *Turnera microphylla* Desv.,
Turnera pringlei Rose
♦ damiana
♦ Quarantine Weed, Introduced
♦ 220, 228
♦ arid, cultivated, herbal.

Turnera ulmifolia L.
Turneraceae
♦ ramgoat dashalong, yellow alder
♦ Weed, Quarantine Weed,
Naturalised
♦ 7, 76, 86, 87, 88, 93, 157, 179, 220, 255
♦ pH, herbal. Origin: tropical America.

**Turraeanthus africanus (Welw. ex C.DC.)
Pellegr.**
Meliaceae
♦ kisanda

♦ Quarantine Weed
♦ 220

Turricula parryi (Gray) J.F.Macbr.
Hydrophyllaceae
♦ turricula, common turricula, poodle
dog bush
♦ Weed
♦ 161
♦ pH, cultivated, herbal, toxic.

Turritis glabra L.
Brassicaceae
Arabis glabra (L.) Bernh. (see)
♦ tower cress, tower mustard
♦ Weed, Naturalised, Cultivation
Escape
♦ 23, 86, 88, 98, 198, 203, 252
♦ a/bH, cultivated, herbal. Origin:
Europe.

Tussilago farfara L.
Asteraceae
♦ colt's foot, coughwort, ginger root,
clay weed, dove dock, horse hoof,
leskenlehti, tossilaggine, farfara
♦ Weed, Noxious Weed, Naturalised,
Environmental Weed
♦ 4, 8, 23, 39, 44, 70,
80, 87, 88, 94, 101, 102, 118, 133, 161,
195, 218, 224, 225, 229, 243, 246, 253,
272, 280, 287
♦ pH, aqua, cultivated, herbal, toxic.

Tweedia coerulea D.Don ex Sweet
Asclepiadaceae/Apocynaceae
= *Oxypetalum coeruleum* (D.Don ex
Sweet) Decne.
♦ Weed, Naturalised
♦ 98, 203

Tylecodon cacalioides (L.f.) H.Toelken
Crassulaceae
Cotyledon cacalioides L.
♦ nenta
♦ Weed
♦ 161
♦ toxic. Origin: South Africa.

Tylecodon paniculatus (L.f.) Toelken
Crassulaceae
Cotyledon paniculata L.f.
♦ butter bush, botterboom
♦ Native Weed
♦ 121
♦ pS, cultivated, toxic. Origin:
southern Africa.

**Tylecodon reticulatus (L.f.) Toelken ssp.
reticulatus**
Crassulaceae
Cotyledon reticulata L.f.
♦ Native Weed
♦ 121
♦ S, toxic. Origin: southern Africa.

Tylecodon ventricosus (Burm.f.) Toelken
Crassulaceae
Cotyledon ventricosa Burm.f.
♦ nanta, nenta, klipnenta
♦ Weed, Native Weed
♦ 39, 121
♦ S, cultivated, toxic. Origin: southern
Africa.

Tylecodon wallichii (Harv.) Toelken
Crassulaceae

Cotyledon wallichii Harv.
♦ Wallich cotyledon, krimpsiekbos
♦ Weed, Native Weed
♦ 39, 121
♦ S, cultivated, toxic. Origin: southern
Africa.

Tylophora indica (Burm.f.) Merr.
Asclepiadaceae/Apocynaceae
♦ Weed
♦ 87, 88
♦ cultivated, herbal.

Tylophora laevigata Decne.
Asclepiadaceae/Apocynaceae
♦ Weed
♦ 39, 87, 88
♦ toxic.

Tylophora sylvatica Decne.
Asclepiadaceae/Apocynaceae
♦ Weed
♦ 87, 88

Tylophora tenuis Bl.
Asclepiadaceae/Apocynaceae
♦ Weed
♦ 13
♦ C, herbal.

Tylophora villosa Bl.
Asclepiadaceae/Apocynaceae
♦ Weed
♦ 13
♦ C.

Tylosema esculentum (Burch.) A.Schreib.
Fabaceae/Caesalpiniaceae
Bauhinia esculenta Burch.
♦ gemsbok bean, marama bean
♦ Introduced
♦ 228
♦ arid, cultivated.

Typha L. spp.
Typhaceae
♦ cumbungi, bulrush, totora, cat's tail
♦ Weed, Quarantine Weed,
Naturalised, Garden Escape,
Environmental Weed
♦ 76, 86, 88, 203, 220, 236, 237, 258
♦ wH, cultivated, herbal.

Typha angustata Bory & Chaub.
Typhaceae
= *Typha domingensis* Pers.
♦ Weed
♦ 87, 88
♦ wpH, herbal.

Typha angustifolia L.
Typhaceae
♦ narrowleaf cat's tail, narrow leaved
reedmace, small reedmace, lesser
reedmace, bulrush, cumbungi
♦ Weed, Naturalised
♦ 80, 87, 88, 101, 161, 195, 209, 217, 218,
239, 255, 272, 275, 286, 292, 295, 297
♦ wpH, cultivated, herbal. Origin:
South America.

Typha australis Schum. & Thonn.
Typhaceae
♦ Weed
♦ 87, 88
♦ wH, herbal.

Typha capensis (Rohrb.) N.E.Br.
Typhaceae
Typha latifolia L. ssp. *capensis* Rohrb.
- bulrush, cat's tail, common bulrush, common cat's tail, Cossack asparagus, nail rod, poker plant, reedmace, cape bulrush
- Weed, Native Weed
- 10, 87, 88, 121, 158
- wpH, cultivated. Origin: southern Africa.

Typha domingensis Pers.
Typhaceae
Typha angustata Bory & Chaub. (see)
- southern cat's tail, southern reedmace, cumbungi, narrowleaf cumbungi
- Weed, Noxious Weed, Native Weed, Environmental Weed
- 14, 70, 88, 147, 161, 177, 200, 218, 221, 246, 269, 295
- wpH, cultivated, herbal. Origin: cosmopolitan.

Typha elephantina Roxb.
Typhaceae
- Weed
- 87, 88, 221
- wH, herbal.

Typha × glauca Godr. (*pro* sp.)
Typhaceae
= *Typha latifolia* L. × *Typha angustifolia* L.
- blue cat's tail
- Weed
- 87, 88, 218, 292
- wpH, promoted.

Typha javanica Schnitz. ex Zoll.
Typhaceae
- Weed
- 87, 88
- wpH.

Typha latifolia L.
Typhaceae
Typha engelmanni A.Br. ex Rohrb., *Typha gracilis* Raf., *Typha major* Curtis, *Typha spiralis* Raf.
- common cat's tail, reedmace, great reedmace, lesser reedmace, narrow leaved reedmace, great reedmace cat's tail, broad leaved cat's tail, candlestick, cat o' nine tails, soft flag, water torch, leveäosmankäämi, cumbungi
- Weed, Quarantine Weed, Noxious Weed, Naturalised, Native Weed, Environmental Weed
- 8, 23, 39, 70, 72, 76, 86, 87, 88, 98, 147, 161, 174, 176, 180, 189, 195, 198, 203, 204, 212, 217, 218, 246, 269, 272, 275, 286, 292, 295, 297, 299
- wpH, cultivated, herbal, toxic. Origin: Europe.

Typha laxmannii Lepech.
Typhaceae
Typha stenophylla Fisch. & Mey.
- graceful cat's tail
- Weed
- 272
- wpH, cultivated, herbal.

Typha minima Funck
Typhaceae
- dwarf reedmace
- Weed
- 272, 297
- wpH, cultivated, herbal.

Typha muelleri Rohrb.
Typhaceae
= *Typha angustifolia* L. ssp. *muelleri* (Rohrb.) Graebn. (NoR)
- Weed
- 23, 87, 88
- wpH, cultivated, herbal.

Typha orientalis C.Presl
Typhaceae
Typha latifolia L. var. *orientalis* (Presl) Rohrb., *Typha orientalis* var. *brunnea* Skvortsov, *Typha shuttleworthii* Koch & Sond. ssp. *orientalis* (Presl) Graebn.
- raupo, broadleaf cumbungi, cumbungi
- Weed, Noxious Weed, Naturalised, Native Weed, Garden Escape, Environmental Weed
- 86, 87, 88, 147, 177, 200, 208, 269, 286
- wpH, cultivated, herbal. Origin: Australia.

Typhalea fruticosa (Mill.) Britton
Malvaceae
= *Pavonia fruticosa* (Mill.) Fawc. & Rendle
- Weed
- 87, 88

Typhonium blumei Nicolson & Sivad.
Araceae
- Naturalised
- 32

Typhonium divaricatum (L.) Decne.
Araceae
Arum divaricatum L., *Arum diversifolium* Blume, *Arum trilobatum* Thunb.
- Weed
- 87, 88, 273
- cultivated, herbal.

Typhonium trilobatum (L.) Schott
Araceae
Arum orixense Roxb., *Arum trilobatum* L., *Typhonium orixense* Schott, *Typhonium siamense* Engl., *Typhonium triste* Griff.
- Weed
- 87, 88, 170, 191, 209
- herbal.

Typhonodorum lindleyanum Schott
Araceae
Arodendron engleri Werth, *Typhonodorum madagascariensis* Engl. (see)
- via, viha, mangaoka, mangibo
- Quarantine Weed
- 220
- aqua, cultivated.

Typhonodorum madagascariensis Engl.
Araceae
= *Typhonodorum lindleyanum* Schott
- Quarantine Weed
- 220
- Origin: Madagascar.

Tyrimnus leucographus (L.) Cass.
Asteraceae
- Weed
- 253

U

Ugni molinae Turcz.
Myrtaceae
Eugenia ugni Hook. & Arn., *Myrtus molinae* Barnéoud ex Gay, *Myrtus ugni* Mol., *Ugni lanceolata* O.Berg, *Ugni philippii* O.Berg, *Ugni poeppigii* O.Berg, *Ugni ugni* (Molina) Macloskie
♦ Chilean guava, strawberry myrtle
♦ Weed, Naturalised, Environmental Weed
♦ 15, 152, 280
♦ cultivated. Origin: South America.

Ulex europaeus L.
Fabaceae/Papilionaceae
Ulex armoricanus Mabillle, *Ulex compositus* Moench, *Ulex floridus* Salisb., *Ulex grandiflorus* Pour., *Ulex hibernicus* G.Don, *Ulex major* Thore, *Ulex mitis* hort., *Ulex ophistolepis* Webb, *Ulex strictus* Mackay, *Ulex vernalis* Thore
♦ gorse, common gorse, whin, furze, European gorse, piikkiherne, gaspeldoring, ajonc, ajonc d'Europe, bois jonc, jonc marin, vigneau, stechginster, tojo, retama espinosa, toxo
♦ Weed, Quarantine Weed, Noxious Weed, Naturalised, Garden Escape, Environmental Weed, Cultivation Escape, Casual Alien
♦ 1, 7, 15, 18, 20, 22, 35, 35, 37, 39, 42, 62, 63, 72, 76, 78, 80, 86, 87, 88, 95, 98, 101, 116, 121, 136, 139, 146, 147, 151, 152, 161, 165, 169, 171, 176, 178, 181, 198, 203, 212, 217, 218, 225, 229, 231, 232, 237, 241, 246, 255, 268, 269, 280, 283, 287, 289, 295, 296, 300
♦ pS, cultivated, herbal, toxic. Origin: Eurasia.

Ulex minor Roth
Fabaceae/Papilionaceae
Ulex nanus Forst.
♦ small furze, dwarf gorse
♦ Weed, Naturalised
♦ 87, 88, 280
♦ cultivated.

Ullucus Loz. spp.
Basellaceae
♦ olluco, ullucus
♦ Weed
♦ 243
♦ H.

Ulmus alata Michx.
Ulmaceae
♦ winged elm

♦ Weed
♦ 23, 87, 88, 218
♦ T, cultivated, herbal.

Ulmus americana L.
Ulmaceae
Ulmus floridana Chapman
♦ American elm
♦ Weed
♦ 87, 88, 218
♦ T, cultivated, herbal.

Ulmus carpinifolia Gled.
Ulmaceae
= *Ulmus minor* Mill.
♦ smooth elm
♦ Weed
♦ 80
♦ T, cultivated, herbal.

Ulmus crassifolia Nutt.
Ulmaceae
♦ cedar elm
♦ Weed
♦ 87, 88, 218
♦ herbal.

Ulmus davidiana Planch.
Ulmaceae
Ulmus japonica (Rehd.) Sarg.
♦ Japanese elm
♦ Weed
♦ 80
♦ T, cultivated, herbal.

Ulmus glabra Huds.
Ulmaceae
Ulmus batavia hort., *Ulmus campestris* Mill., *Ulmus montana* Sm., *Ulmus scabra* Mill.
♦ wych elm, vuorijalava
♦ Weed, Naturalised, Casual Alien
♦ 39, 80, 101, 280
♦ T, cultivated, herbal, toxic.

Ulmus × hollandica Mill.
Ulmaceae
= *Ulmus minor* Mill. × *Ulmus glabra* Huds.
♦ Dutch elm, suckering elm
♦ Weed, Naturalised, Environmental Weed
♦ 15, 72, 86, 88, 176, 198, 280
♦ T, cultivated. Origin: Europe.

Ulmus minor Mill.
Ulmaceae
Ulmus angustifolia Moench, *Ulmus campestris* (L.) Spach, *Ulmus carpinifolia* Gled. (see), *Ulmus carpinifolia* var. *suberosa* (Moench) Rehder, *Ulmus chinensis* Desf., *Ulmus corylifolia* Host., *Ulmus foliacea* Gilib. nom. inval., *Ulmus foliacea* var. *suberosa* (Moench) Rehder, *Ulmus glabra* Huds. var. *suberosa* (Moench) Gürke, *Ulmus nana* Bork., *Ulmus nemorosa* Bork., *Ulmus nitens* Moench, *Ulmus nuda* Ehrh., *Ulmus sparsa* Dumort., *Ulmus suberosa* Moench, *Ulmus tortuosa* Host., *Ulmus webbiana* Lee ex K.Koch
♦ English elm, smooth leaved elm, smoothleaf elm
♦ Weed
♦ 34, 80

♦ T, cultivated, herbal. Origin: Eurasia, North Africa.

Ulmus parvifolia Jacq.
Ulmaceae
♦ Chinese elm, lacebark elm
♦ Weed, Naturalised, Introduced
♦ 80, 87, 88, 101, 179, 195, 218, 222, 279
♦ T, cultivated, herbal.

Ulmus procera Salisb.
Ulmaceae
♦ English elm, Dutch elm, common elm
♦ Weed, Naturalised, Garden Escape, Environmental Weed
♦ 7, 39, 54, 86, 87, 88, 101, 198, 218, 279
♦ T, cultivated, herbal, toxic. Origin: southern Europe.

Ulmus pumila L.
Ulmaceae
♦ Siberian elm, Chinese elm, dwarf elm
♦ Weed, Naturalised, Introduced, Garden Escape, Environmental Weed
♦ 4, 17, 80, 87, 88, 101, 102, 142, 146, 151, 161, 174, 195, 218
♦ T, arid, cultivated, herbal.

Ulmus rubra Muhl.
Ulmaceae
Ulmus fulva Michx.
♦ slippery elm, punajalava
♦ Weed
♦ 87, 88, 218
♦ T, cultivated, herbal.

Ulmus serotina Sarg.
Ulmaceae
♦ red elm, September elm
♦ Weed
♦ 87, 88, 218

Ulmus thomasi Sarg.
Ulmaceae
♦ rock elm
♦ Weed
♦ 87, 88, 218

Ulva L. spp.
Ulvaceae
♦ Weed
♦ 88, 218, 282
♦ algae.

Umbellularia californica Nutt.
Lauraceae
♦ California laurel, bay tree, California bay, cajeput tree
♦ Weed
♦ 39, 87, 88, 218
♦ T, cultivated, herbal, toxic.

Umbilicus pendulinus DC.
Crassulaceae
♦ Weed
♦ 221
♦ herbal.

Uncarina grandidieri (Baill.) Stapf
Pedaliaceae
Harpagophytum grandidieri Baill.
♦ Weed, Quarantine Weed
♦ 76, 88
♦ cultivated.

Uncinia hamata (Sw.) Urb.
Cyperaceae
♦ birdcatching sedge
♦ Naturalised
♦ 101
♦ G.

Uncinia sinclairii Boott
Cyperaceae
♦ Weed, Naturalised
♦ 98, 203
♦ G.

Undaria pinnatifida (Harv.) Suringer
Alariaceae
♦ Japanese kelp, japweed, seaweed, wakame, blackberry of the sea
♦ Weed, Sleeper Weed, Environmental Weed
♦ 152, 225, 246, 282, 288, 289, 296
♦ Origin: east Asia.

Uniola latifolia Michx.
Poaceae
= *Chasmanthium latifolium* (Michx.) Yates
♦ Quarantine Weed
♦ 220
♦ G, cultivated, herbal.

Uraria crinita Desv.
Fabaceae/Papilionaceae
♦ Weed
♦ 12
♦ herbal.

Uraria lagopodioides (L.) DC.
Fabaceae/Papilionaceae
Doodia lagopodioides (L.) Roxb., *Hedysarum lagopodioides* L., *Lespedeza lagopodioides* (L.) Pers., *Uraria aequilobata* Hosok.
♦ Weed
♦ 87, 88, 170, 191
♦ herbal. Origin: south-east Asia, Australia.

Uraria picta (Jacq.) Desv. ex DC.
Fabaceae/Papilionaceae
♦ Weed
♦ 87, 88
♦ herbal. Origin: Asia, Australia.

Urechites Müll.Arg. spp.
Apocynaceae
= *Pentalinon* Voigt spp. (NoR)
♦ yellow nightshade
♦ Weed
♦ 161
♦ toxic.

Urechites lutea (L.) Britt.
Apocynaceae
= *Pentalinon luteum* (L.) B.F.Hansen & Wunderlin
♦ wild allamanda, yellow nightshade
♦ Weed
♦ 14, 39, 87, 88
♦ cultivated, herbal, toxic.

Urena capitata L.
Malvaceae
♦ Weed
♦ 87, 88

Urena lobata L.
Malvaceae
Urena trilobata Vell. (see), *Urena sinuata* Sw. *non* L.
♦ Caesar's weed, urena weed, pink flowered Chinese burr, hibiscus burr, dadangsi, dadangsi apaka, dadangsi machingat, dádangse, chosuched e kui, karap, korop, nognuk, osuched a rechui, motipo, mautofu, mo'osipo, manutofu, qatima, gataya, jute Africain, nggatima, sachayute, cadillo
♦ Weed, Noxious Weed, Naturalised, Introduced, Garden Escape, Environmental Weed
♦ 3, 6, 12, 14, 23, 28, 80, 86, 87, 88, 107, 112, 153, 206, 218, 229, 230, 243, 255, 261, 262, 269, 275, 276, 297
♦ pH, cultivated, herbal. Origin: Australia.

Urena sinuata L.
Malvaceae
Urena aculeata Mill., *Urena lobata* L. ssp. *sinuata* (L.) Borss.Waalk., *Urena lobata* L. var. *sinuata* (L.) Hochr., *Urena morifolia* DC., *Urena muricata* DC., *Urena paradoxa* Kunth, *Urena swartzii* DC.
♦ burr mallow, cadillo pata de perro
♦ Weed
♦ 3, 87, 88, 191, 261
♦ herbal.

Urena trilobata Vell.
Malvaceae
= *Urena lobata* L.
♦ Weed
♦ 87, 88

Urera baccifera (L.) Gaud.
Urticaceae
♦ scratchbush, urtiga brava
♦ Weed
♦ 39, 87, 88, 255
♦ H, herbal, toxic. Origin: tropical America.

Urera cameroonensis Wedd.
Urticaceae
♦ ikwakasa
♦ Weed
♦ 87, 88

Urera caracasana (Jacq.) Griseb.
Urticaceae
♦ flameberry
♦ Environmental Weed
♦ 257
♦ pC, cultivated, herbal.

Urera hypselodendron (Rich.) Wedd.
Urticaceae
♦ kitumbekakoko
♦ Weed
♦ 87, 88

Urera tenax N.E.Br
Urticaceae
♦ giant nettle, mountain nettle, stinging nettle tree, tree nettle, uluzi
♦ Native Weed
♦ 121
♦ pH. Origin: southern Africa.

Urginea altissima (L.f.) Bak.
Liliaceae/Hyacinthaceae
♦ Weed, Native Weed
♦ 39, 121
♦ pH, cultivated, herbal, toxic. Origin: southern Africa.

Urginea macrocentra Bak.
Liliaceae/Hyacinthaceae
♦ Natal slangkop, poison bulb, snake's head
♦ Weed, Native Weed
♦ 39, 121
♦ pH, herbal, toxic. Origin: southern Africa.

Urginea maritima (L.) Bak.
Liliaceae/Hyacinthaceae
Drimia maritima (L.) Stearn, *Scilla maritima* L. (see), *Urginea scilla* Steinh.
♦ sea onion, red squill, sea squill, white squill
♦ Weed
♦ 39, 87, 88, 215, 221, 247
♦ pH, cultivated, herbal, toxic.

Urginea pusilla (Jacq.) Bak.
Liliaceae/Hyacinthaceae
♦ mountain slangkop
♦ Weed, Native Weed
♦ 39, 121
♦ pH, toxic. Origin: southern Africa.

Urginea sanguinea Schinz
Liliaceae/Hyacinthaceae
Urginea burkei Bak.
♦ Burke's slangkop, red slangkop, Transvaal slangkop
♦ Native Weed
♦ 121
♦ pH, cultivated, toxic. Origin: Africa.

Urochloa Beauv. spp.
Poaceae
♦ signalgrass
♦ Weed, Naturalised
♦ 88, 198
♦ G.

Urochloa adspersa (Trin.) R.Webster
Poaceae
Brachiaria adspersa (Trin.) Parodi, *Panicum adspersum* Trin. (see)
♦ broadleaf panicum, Dominican signalgrass
♦ Weed
♦ 243
♦ G.

Urochloa advena (Vickery) R.Webster
Poaceae
Brachiaria advena Vickery (see)
♦ Weed, Naturalised
♦ 98, 203
♦ G.

Urochloa arrecta (Hack. ex T.Durand & Schinz) Morrone & Zuloaga
Poaceae
Brachiaria arrecta (Hack. ex Dur. & Schinz) Stent (see), *Brachiaria radicans* Napper, *Panicum arrectum* Hack. ex T.Durand & Schinz
♦ African signalgrass
♦ Weed, Naturalised, Introduced

♦ 38, 101, 179
♦ G.

Urochloa bolbodes (Hochst. ex Steud.) Stapf
Poaceae
= *Urochloa oligotricha* (Fig. & De Not.) Henrard
♦ Weed
♦ 87, 88
♦ G.

Urochloa brachyura (Hack.) Stapf
Poaceae
Panicum brachyura Hack.
♦ urochloa
♦ Native Weed
♦ 121
♦ pG, arid. Origin: southern Africa.

Urochloa brizantha (Hochst. ex Rich.) Webster
Poaceae
Brachiaria brizantha (Hochst. ex A.Rich.) Stapf (see), *Brachiaria ruziziensis* R.Germ. & Evrard (see), *Panicum brizanthum* Hochst. ex A.Rich., *Urochloa ruziziensis* (Germ. & Evrard) Morrone & Zuloaga (see)
♦ palisade signalgrass
♦ Weed, Naturalised, Cultivation Escape
♦ 98, 101, 203, 261
♦ G, cultivated. Origin: tropical Africa.

Urochloa decumbens (Stapf) R.Webster
Poaceae
Brachiaria decumbens Stapf (see)
♦ Weed, Naturalised, Introduced, Cultivation Escape
♦ 7, 38, 98, 203, 261
♦ G, cultivated.

Urochloa eminii (Mez) Davidse
Poaceae
♦ milha, milha roxa, surbana
♦ Introduced
♦ 38
♦ G.

Urochloa fasciculata (Sw.) Webster
Poaceae
Brachiaria fasciculata (Sw.) Parodi (see), *Brachiaria fasciculata* var. *reticulata* (Sw.) Parodi (see), *Panicum fasciculatum* Sw. (see), *Panicum fasciculatum* Sw. var. *reticulatum* (Torr.) Beal (see), *Urochloa fasciculata* (Sw.) Webster var. *reticulata* (Torr.) Webster (see)
♦ panizo fasciculado, browntop signalgrass
♦ Weed, Introduced
♦ 199, 228, 243, 271
♦ G, arid.

Urochloa fasciculata (Sw.) Webster var. reticulata (Torr.) Webster
Poaceae
= *Urochloa fasciculata* (Sw.) Webster
♦ Weed
♦ 179
♦ G.

Urochloa maxima (Jacq.) Webster
Poaceae

♦ guinea grass
♦ Weed, Naturalised
♦ 7, 86, 98, 101, 203
♦ G.

Urochloa maxima (Jacq.) R.Webster var. maxima
Poaceae
♦ Weed
♦ 93
♦ G.

Urochloa mosambicensis (Hack.) Dandy
Poaceae
Brachiaria stolonifera Gooss., *Echinochloa notabile* (Hook.f.) Rhind., *Panicum mosambicense* Hack., *Urochloa pullulans* Stapf *nom. illeg.* (see), *Urochloa pullulans* Stapf var. *mosambicensis* (Hack.) Stapf, *Urochloa rhodesiensis* Stent (see), *Urochloa stolonifera* (Gooss.) Chippind.
♦ common urochloa, gonya grass, impunga, Sabi grass, African liverseed grass, bushveld herringbone grass
♦ Weed, Naturalised, Native Weed, Introduced, Environmental Weed, Cultivation Escape
♦ 7, 39, 86, 87, 88, 93, 98, 101, 121, 158, 203, 205, 228, 261
♦ pG, arid, cultivated, toxic. Origin: southern Africa.

Urochloa mutica (Forssk.) Nguyen
Poaceae
Brachiaria glabrinodis (Hack.) Henrard, *Brachiaria mutica* (Forssk.) Stapf (see), *Brachiaria numidiana* (Lam.) Henrard, *Brachiaria purpurascens* (Raddi) Henr. (see), *Panicum amphibium* Steud., *Panicum barbinode* Trin. (see), *Panicum equinum* Salzm. ex Steud., *Panicum glabrinode* Hack., *Panicum guadaloupense* Spreng. ex Steud. (see), *Panicum muticum* Forssk. (see), *Panicum numidianum* Lam., *Panicum pictigluma* Steud., *Panicum purpurascens* Raddi (see)
♦ para grass
♦ Weed, Naturalised, Environmental Weed
♦ 3, 7, 93, 98, 101, 179, 200, 203, 280
♦ pG, aqua.

Urochloa oligobrachiata (Pilg.) Kartesz comb. nov. ined.
Poaceae
Panicum oligobrachiatum Pilg.
♦ weak signalgrass
♦ Naturalised
♦ 101
♦ G.

Urochloa oligotricha (Fig. & De Not.) Henrard
Poaceae
Eriochloa bolbodes (Hochst. ex Steud.) Schweinf., *Helopus bolbodes* Hochst. ex Steud., *Panicum bolbodes* (Hochst. ex Steud.) Asch. & Schweinf., *Panicum oligotrichum* Fig. & De Not., *Urochloa bolbodes* (Hochst. ex Steud.) Stapf (see)
♦ Weed, Naturalised
♦ 86, 98, 203

♦ G, cultivated. Origin: central and south Africa.

Urochloa panicoides Beauv.
Poaceae
Panicum helopus Trin., *Urochloa helopus* (Trin.) Stapf
♦ panic liverseed grass, urochloa grass, liverseed grass, urochloa, garden urochloa, herringbone grass, annual signalgrass, garden grass, herringbone grass, kuri millet, poke
♦ Weed, Noxious Weed, Native Weed, Naturalised, Introduced, Casual Alien
♦ 7, 51, 55, 67, 68, 86, 87, 88, 90, 98, 101, 121, 140, 158, 161, 198, 203, 228, 229, 240, 243, 269, 280
♦ pG, arid, cultivated. Origin: southern Africa.

Urochloa piligera (F.Muell. ex Benth.) Webster
Poaceae
Brachiaria piligera (F.Muell. ex Benth.) D.K.Hughes (see)
♦ wattle signalgrass
♦ Weed, Naturalised
♦ 101, 179
♦ G.

Urochloa plantaginea (Link) R.Webster
Poaceae
Brachiaria plantaginea (Link) A.S.Hitchc. (see), *Panicum plantagineum* Link
♦ Alexander grass, plantain signalgrass
♦ Weed, Naturalised
♦ 101, 179, 271
♦ G.

Urochloa pullulans Stapf nom. illeg.
Poaceae
= *Urochloa mosambicensis* (Hack.) Dandy
♦ Weed, Naturalised
♦ 87, 88, 98, 203
♦ G.

Urochloa ramosa (L.) R.D.Webster
Poaceae
Brachiaria ramosa (L.) Stapf (see), *Panicum ramosum* L.
♦ Dixie signalgrass
♦ Weed, Naturalised
♦ 7, 98, 101, 179, 203
♦ G.

Urochloa reptans (L.) Stapf
Poaceae
Brachiaria reptans (L.) C.A.Gardner & C.E.Hubb. (see), *Panicum reptans* L. (see)
♦ sprawling signalgrass
♦ Weed, Naturalised
♦ 28, 32, 243
♦ G.

Urochloa rhodesiensis Stent
Poaceae
= *Urochloa mosambicensis* (Hack.) Dandy
♦ Weed, Naturalised
♦ 98, 203
♦ G.

Urochloa ruziziensis (Germ. & Evrard) Morrone & Zuloaga
Poaceae
= *Urochloa brizantha* (Hochst. ex Rich.) Webster
♦ Congo grass, Congo signalgrass, ruzi grass
♦ Introduced, Cultivation Escape
♦ 38, 261
♦ G, cultivated. Origin: tropical Africa.

Urochloa subquadripara (Trin.) R.D.Webster
Poaceae
Brachiaria subquadripara (Trin.) Hitchc. (see), *Panicum subquadriparum* Trin. (see)
♦ tropical signalgrass, cori grass, green summergrass
♦ Weed, Naturalised
♦ 7, 101, 179
♦ G.

Urochloa texana (Buckl.) R.D.Webster
Poaceae
Brachiaria texana (Buckl.) Blake (see), *Panicum texanum* Buckl. (see)
♦ Texas signalgrass, buffalograss, Texas millet, Texas panic
♦ Weed, Naturalised
♦ 98, 179, 203
♦ G.

Urochloa trichopus (Hochst.) Stapf
Poaceae
Panicum trichopus Hochst.
♦ gonya grass, roundseed urochloa, rabguk
♦ Weed, Native Weed
♦ 121, 242
♦ aG, arid. Origin: southern Africa.

Urochloa villosa (Lam.) T.Q.Nguyen
Poaceae
Brachiaria distichophylla (Trin.) Stapf (see)
♦ hairy signalgrass
♦ Naturalised
♦ 101
♦ G.

Urospermum dalechampii (L.) Scop. ex Schmidt
Asteraceae
Arnopogon dalechampii Willd.
♦ Weed, Naturalised, Environmental Weed
♦ 86, 88, 94, 176
♦ pH, cultivated. Origin: Mediterranean.

Urospermum picroides (L.) Scop. ex Schmidt
Asteraceae
Tragopogon capensis Jacq., *Tragopogon picroides* L., *Urospermum capense* (Jacq.) Spreng.
♦ prickly goldenfleece, urospermum
♦ Weed, Naturalised, Introduced, Environmental Weed
♦ 7, 9, 34, 86, 87, 88, 94, 98, 101, 111, 121, 185, 198, 203, 221, 241, 243, 272, 300
♦ a/pH, arid, cultivated, herbal. Origin: Eurasia.

Ursinia abrotanifolia (R.Br.) Spreng.
Asteraceae
♦ Native Weed
♦ 121
♦ S, cultivated. Origin: southern Africa.

Ursinia anethoides (DC.) N.E.Br.
Asteraceae
♦ Weed
♦ 23, 88
♦ cultivated.

Ursinia anthemoides (L.) Poir.
Asteraceae
Arctotis anthemoides L., *Sphenogyne anthemoides* (L.) R.Br.
♦ South African marigold
♦ Weed, Naturalised, Environmental Weed, Casual Alien
♦ 7, 9, 86, 98, 203, 280
♦ a/pH, cultivated. Origin: South Africa.

Ursinia anthemoides (L.) Poir. ssp. versicolor (DC.) Prassler
Asteraceae
♦ Casual Alien
♦ 280

Ursinia chrysanthemoides (Less. ex DC.) Harv.
Asteraceae
♦ coral ursinia
♦ Weed, Naturalised, Native Weed
♦ 86, 98, 121, 203
♦ pH, cultivated, toxic. Origin: southern Africa.

Ursinia nana DC.
Asteraceae
♦ dwarf ursinia, yellow Margaret
♦ Weed, Native Weed
♦ 51, 87, 88, 121
♦ aH, cultivated, toxic. Origin: southern Africa.

Ursinia speciosa DC.
Asteraceae
♦ Weed, Naturalised, Environmental Weed
♦ 7, 23, 86, 88, 98, 203
♦ aH, cultivated. Origin: South Africa.

Urtica angustifolia Fisch ex Hornem.
Urticaceae
♦ narrowleaf nettle
♦ Weed
♦ 297
♦ pH, promoted. Origin: China, Japan, Korea.

Urtica ardens Link
Urticaceae
Urtica himalayensis Kunth & Bouché (see), *Urtica parviflora* Roxb., *Urtica virulenta* Wall. (see)
♦ Quarantine Weed
♦ 220

Urtica balaerica L.
Urticaceae
= *Urtica pilulifera* L.
♦ Quarantine Weed
♦ 220

Urtica ballotifolia Wedd.
Urticaceae
♦ nettle
♦ Naturalised
♦ 101

Urtica cannabina L.
Urticaceae
♦ stinging nettle, Kentucky hemp
♦ Weed
♦ 88, 114, 243, 272
♦ aH, cultivated, herbal.

Urtica chamaedryoides Pursh
Urticaceae
♦ southern nettle, heartleaf nettle
♦ Weed
♦ 161
♦ herbal, toxic.

Urtica cordifolia Moench.
Urticaceae
= *Urtica pilulifera* L.
♦ Quarantine Weed
♦ 220

Urtica dioica L.
Urticaceae
Urtica gracilis Ait. (see), *Urtica hispida* DC., *Urtica major* Kanitz
♦ stinging nettle, common stinging nettle, slender nettle, California nettle, tall nettle, perennial nettle, ortica, greater nettle, European perennial nettle, common nettle, Swedish hemp, great stinging nettle
♦ Weed, Noxious Weed, Naturalised, Native Weed, Garden Escape, Environmental Weed
♦ 15, 23, 39, 44, 51, 52, 80, 86, 87, 88, 94, 98, 101, 102, 121, 136, 158, 159, 161, 174, 198, 203, 210, 212, 218, 243, 246, 247, 253, 255, 266, 272, 280, 287, 294, 299
♦ pH, cultivated, herbal, toxic. Origin: Eurasia.

Urtica dioica L. ssp. dioica
Urticaceae
♦ stinging nettle
♦ Naturalised
♦ 101, 280

Urtica dioica L. ssp. gracilis (Aiton) Selander
Urticaceae
♦ American stinging nettle, California nettle, California slender nettle, hoary nettle, giant creek nettle, hedgenettle
♦ Weed, Naturalised
♦ 180, 243, 280
♦ pH.

Urtica dioica L. var. dioica
Urticaceae
♦ Naturalised
♦ 241

Urtica dioica L. var. holosericea (Nutt.) C.L.Hitchc.
Urticaceae
♦ California slender nettle, hoary nettle, giant creek nettle, hedgenettle
♦ Weed
♦ 180, 243

Urtica dioica L. var. procera (Muhl. ex Willd.) Wedd.

Urticaceae
= *Urtica dioica* L. ssp. *gracilis* (Aiton)
Selander (NoR) [see *Urtica lyallii*
S.Wats., *Urtica procera* Muhl.]
♦ stinging nettle, tall nettle, slender
nettle
♦ Weed
♦ 211, 243

Urtica dodartii L.
Urticaceae
= *Urtica pilulifera* L.
♦ Quarantine Weed
♦ 220

Urtica gracilis Ait.
Urticaceae
= *Urtica dioica* L.
♦ slender nettle
♦ Weed
♦ 39, 87, 88, 218
♦ pH, promoted, herbal, toxic.

Urtica himalayensis Kunth & Bouché
Urticaceae
= *Urtica ardens* Link
♦ Quarantine Weed
♦ 220

Urtica incisa Poir.
Urticaceae
♦ scrub nettle
♦ Weed, Naturalised, Native Weed
♦ 87, 88, 98, 203, 269
♦ pH, cultivated. Origin: Australia.

Urtica integrifolia Savigny
Urticaceae
= *Urtica pilulifera* L.
♦ Quarantine Weed
♦ 220

Urtica lyallii S.Wats.
Urticaceae
= *Urtica dioica* L. ssp. *gracilis* (Aiton)
Selander (NoR) [see *Urtica dioica* L. var.
procera (Muhl. ex Willd.) Wedd., *Urtica
procera* Muhl.]
♦ Lyall nettle, stinging nettle
♦ Weed
♦ 87, 88, 161, 218
♦ pH, promoted, herbal, toxic.

Urtica massaica Mildbr.
Urticaceae
♦ Weed
♦ 87, 88

Urtica membranacea Poir. ex Savigny
Urticaceae
♦ Casual Alien
♦ 280

Urtica pilulifera L.
Urticaceae
Urtica balearica L. (see), *Urtica cordifolia*
Moench (see), *Urtica dodartii* L. (see),
Urtica integrifolia Savigny (see)
♦ Roman nettle, pallonokkonen
♦ Weed, Quarantine Weed, Casual
Alien
♦ 39, 42, 87, 88, 161, 220, 221, 272
♦ aH, cultivated, herbal, toxic.

Urtica procera Muhl.
Urticaceae

= *Urtica dioica* L. ssp. *gracilis* (Aiton)
Selander (NoR) [see *Urtica dioica* L. var.
procera (Muhl. ex Willd.) Wedd., *Urtica
lyallii* S.Wats.]
♦ tall nettle, nettle
♦ Weed
♦ 87, 88, 218
♦ pH, promoted, herbal, toxic.

Urtica urens L.
Urticaceae
Urtica intermedia Form., *Urtica minor*
Lam., *Urtica monoica* Gilib., *Urtica
ovalifolia* Stokes, *Urtica quadristipulata*
Dulac.
♦ burning nettle, stinging nettle,
common stinging nettle, small nettle,
dwarf nettle, rautanokkonen, ortica
minore, annual stinging nettle, nettle,
bush nettle, bush stinging nettle
♦ Weed, Naturalised, Introduced,
Garden Escape, Environmental Weed
♦ 7, 15, 23, 34, 38, 39, 44, 51, 72, 86, 87,
88, 93, 94, 97, 98, 101, 115, 118, 121, 134,
158, 161, 165, 176, 180, 185, 198, 199,
203, 205, 217, 218, 221, 228, 236, 237,
241, 243, 253, 269, 272, 280, 287, 295,
300
♦ aH, arid, cultivated, herbal, toxic.
Origin: Eurasia.

Urtica virulenta Wall.
Urticaceae
= *Urtica ardens* Link
♦ Quarantine Weed
♦ 220

Urvillea ulmacea Kunth
Sapindaceae
♦ apaac
♦ Weed
♦ 87, 88

Utricularia L. spp.
Lentibulariaceae
♦ bladderwort
♦ Weed, Quarantine Weed
♦ 161, 221, 258

Utricularia aurea Lour.
Lentibulariaceae
Utricularia blumei (A.DC.) Miq.,
Utricularia flexuosa Vahl (see),
Utricularia reclinata Hassk.
♦ leafy bladderwort, saaraai khaao
nieo
♦ Weed
♦ 87, 88, 170, 191, 204, 209, 239, 262,
275, 286, 297
♦ aqua, cultivated.

Utricularia australis R.Br.
Lentibulariaceae
Utricularia japonica Makino (see),
Utricularia tenuicaulis Miki (see)
♦ bladderwort, utriculaire négligée,
western bladderwort
♦ Weed
♦ 286
♦ cultivated, herbal.

Utricularia bifida L.
Lentibulariaceae
♦ bladderwort

♦ Weed
♦ 13, 87, 88
♦ cultivated, herbal.

Utricularia biflora Lam.
Lentibulariaceae
= *Utricularia gibba* L.
♦ Weed, Naturalised
♦ 98, 203
♦ herbal.

Utricularia flexuosa Vahl
Lentibulariaceae
= *Utricularia aurea* Lour.
♦ Weed
♦ 13
♦ awH.

Utricularia foliosa L.
Lentibulariaceae
♦ leafy bladderwort
♦ Weed, Quarantine Weed
♦ 87, 88, 218, 255, 258
♦ wpH. Origin: Brazil.

Utricularia gibba L.
Lentibulariaceae
Utricularia biflora Lam. (see), *Utricularia
exoleta* R.Br.
♦ floating bladderwort, humped
bladderwort, bladderwort
♦ Weed, Sleeper Weed, Quarantine
Weed, Naturalised, Environmental
Weed
♦ 86, 87, 88, 198, 225, 246, 280
♦ wpH, cultivated, herbal. Origin:
south-west Europe.

Utricularia gibba L. ssp. gibba
Lentibulariaceae
♦ Naturalised
♦ 287

Utricularia inflata Walter
Lentibulariaceae
♦ floating bladderwort, swollen
bladderwort
♦ Weed, Noxious Weed, Naturalised
♦ 87, 88, 143, 197, 218, 286, 287
♦ wpH, cultivated, herbal.

Utricularia inflexa Forssk.
Lentibulariaceae
Hamulia alba Raf., *Utricularia inflexa* var.
inflexa Forssk., *Utricularia inflexa* var.
major Kamienski, *Utricularia inflexa* var.
remota Kamienski, *Utricularia inflexa*
var. *tenuifolia* Kamienski, *Utricularia
oliveri* Kamienski, *Utricularia oliveri*
var. *fimbriata* Kamienski, *Utricularia
oliveri* var. *schweinfurthii* Kamienski,
Utricularia stellaris L.f. (see), *Utricularia
stellaris* var. *inflexa* (Forssk.) C.B.Clarke,
Utricularia thonningii Schumach. &
Thonn. (see), *Utricularia thonningii* var.
laciniata Stapf
♦ Weed
♦ 87, 88
♦ Origin: Australia.

Utricularia japonica Makino
Lentibulariaceae
= *Utricularia australis* R.Br.
♦ Weed
♦ 275
♦ aqua.

Utricularia odorata Pellegr.
Lentibulariaceae
♦ Weed
♦ 87, 88
♦ cultivated. Origin: Australia.

Utricularia purpurea Walter
Lentibulariaceae
♦ purple bladderwort, eastern purple bladderwort
♦ Weed
♦ 87, 88, 218
♦ aqua, cultivated, herbal.

Utricularia stellaris L.f.
Lentibulariaceae
= *Utricularia inflexa* Forssk.
♦ bladderwort, star bladderwort
♦ Weed, Native Weed
♦ 87, 88, 121
♦ wpH, herbal.

Utricularia tenuicaulis Miki
Lentibulariaceae
= *Utricularia australis* R.Br.
♦ Weed
♦ 286

Utricularia thonningii Schum. & Thonn.
Lentibulariaceae
= *Utricularia inflexa* Forssk.
♦ Weed
♦ 87, 88

Utricularia vulgaris L.
Lentibulariaceae
= *Utricularia macrorhiza* Le Conte (NoR)
♦ common bladderwort, greater bladderwort, isovesiherne
♦ Weed
♦ 23, 45, 87, 88, 218, 272
♦ wpH, cultivated, herbal.

V

Vaccaria hispanica (Mill.) Rausch.
Caryophyllaceae
Saponaria vaccaria L. (see), *Vaccaria pyramidata* Medik. (see), *Vaccaria segetalis* Garcke ex Asch. (see), *Vaccaria vulgaris* Host
♦ cow cockle, wheat cockle, spring cockle, pink cockle, cowherb, cow soapwort, toukokukka, cow basil, bladder soapwort
♦ Weed, Naturalised, Introduced, Casual Alien
♦ 7, 34, 40, 42, 49, 86, 88, 93, 98, 101, 176, 179, 185, 191, 198, 203, 205, 228, 243, 280, 287
♦ aH, arid, cultivated, herbal. Origin: Mediterranean, Europe, west Asia.

Vaccaria parviflora Moench
Caryophyllaceae
♦ Weed
♦ 87, 88

Vaccaria pyramidata Medik.
Caryophyllaceae
= *Vaccaria hispanica* (Mill.) Rausch.
♦ cow cockle, cowherb, cockle, cow soapwort
♦ Weed, Noxious Weed, Naturalised
♦ 23, 70, 87, 88, 94, 98, 121, 161, 203, 212, 221, 237, 243, 269, 272, 295, 299
♦ aH, herbal, toxic. Origin: Eurasia.

Vaccaria segetalis Garcke ex Asch.
Caryophyllaceae
= *Vaccaria hispanica* (Mill.) Rausch.
♦ cow cockle
♦ Weed
♦ 136, 210, 210, 275, 297
♦ bH, herbal, toxic.

Vaccinium angustifolium Aiton
Ericaceae
Vaccinium lamarckii Camp.
♦ lowbush blueberry, late lowbush blueberry
♦ Weed
♦ 87, 88, 218
♦ S, promoted, herbal.

Vaccinium arboreum Marsh.
Ericaceae
♦ tree huckleberry, farkleberry
♦ Weed
♦ 87, 88, 218
♦ T, cultivated, herbal.

Vaccinium corymbosum L.
Ericaceae
Vaccinium constablaei A.Gray
♦ blueberry, highbush blueberry,

American blueberry, swamp blueberry, airelle d'Amérique, Amerikanische blueberry, arándano Americano
♦ Environmental Weed, Casual Alien
♦ 225, 246, 280
♦ S, cultivated, herbal.

Vaccinium myrtillus L.
Ericaceae
Myrtillus niger Gilib., *Myrtillus sylvatica* Bubani
♦ bilberry, whortleberry, bog whortleberry, blaeberry, mirtillo nero, mustikka
♦ Weed
♦ 70, 272
♦ S, cultivated, herbal.

Vaccinium ovalifolium Sm.
Ericaceae
♦ ovalleaf blueberry, blue huckleberry, black huckleberry
♦ Weed
♦ 87, 88, 218
♦ S, cultivated, herbal.

Vaccinium ovatum Pursh
Ericaceae
♦ box blueberry, California huckleberry, evergreen huckleberry
♦ Weed
♦ 87, 88, 218
♦ S, cultivated, herbal.

Vaccinium oxycoccus L.
Ericaceae
♦ small cranberry, cranberry, large cranberry
♦ Weed
♦ 87, 88, 218
♦ S, cultivated, herbal.

Vaccinium parvifolium Sm.
Ericaceae
♦ tall red huckleberry, California red huckleberry, red huckleberry, red bilberry
♦ Weed
♦ 87, 88, 218
♦ S, cultivated, herbal.

Vaccinium stamineum L.
Ericaceae
♦ common deerberry, deerberry
♦ Weed
♦ 87, 88, 218
♦ S, cultivated, herbal.

Vaccinium uliginosum L.
Ericaceae
Myrtillus grandis Bub., *Vaccinium ciliatum* Gilib.
♦ bog bilberry, juolukkabog, whortleberry, northern bilberry, bog blueberry
♦ Weed
♦ 272
♦ S, cultivated, herbal.

Vaccinium vitis-idaea L.
Ericaceae
Myrtillus exigua Bubani, *Oxycoccus vitis-idaea* Fries., *Vaccinium buxifolia* Gilib., *Vaccinium jesoense* Miq., *Vaccinium nemorosum* Salisb., *Vaccinium punctatum* Lam., *Vitis-idaea punctata*

Moench
♦ mountain cranberry, lingonberry, puolukka, cowberry, red whortleberry
♦ Weed
♦ 87, 88, 272
♦ S, cultivated, herbal.

Vaccinium vitis-idaea L. var. minus Lodd.
Ericaceae
= *Vaccinium vitis-idaea* L. ssp. *minus* (Lodd.) Hultén (NoR)
♦ mountain cranberry, cowberry
♦ Weed
♦ 218

Vahlia digyna (Retz.) Kuntze
Saxifragaceae/Vahliaceae
Bistella digyna (Retz.) Bullock (see)
♦ Weed
♦ 87, 88

Valantia filiformis Lojac.
Rubiaceae
= *Valantia hispida* L.
♦ Quarantine Weed
♦ 220

Valantia hispida L.
Rubiaceae
Valantia filiformis Lojac. (see)
♦ Quarantine Weed
♦ 220
♦ cultivated.

Valeriana L. spp.
Valerianaceae
♦ valerian, uva grass
♦ Weed
♦ 272
♦ herbal.

Valeriana dioica L.
Valerianaceae
♦ marsh valerian
♦ Weed
♦ 39, 272
♦ pH, cultivated, herbal, toxic.

Valeriana edulis Nutt. ex Torr. & A.Gray ssp. procera (Kunth) F.G.Mey. nom. illeg.
Valerianaceae
= *Valeriana procera* Kunth (NoR)
♦ Weed
♦ 199

Valeriana officinalis L.
Valerianaceae
Valeriana palustris Kreyer, *Valeriana sylvestris* Sadler, *Valeriana vulgaris* Rupr.
♦ garden heliotrope, valerian, setewale, rohtovirmajuuri, lesser valerian, marsh valerian, Pyrenean valerian, red valerian, garden valerian
♦ Weed, Naturalised, Introduced, Environmental Weed
♦ 23, 39, 87, 88, 101, 133, 151, 195, 222, 224, 272
♦ pH, cultivated, herbal, toxic.

Valeriana sorbifolia Kunth
Valerianaceae
♦ pineland valerian
♦ Weed
♦ 157
♦ aH.

Valeriana sorbifolia Kunth var. mexicana (DC.) F.G.Mey.
Valerianaceae
♦ Weed
♦ 199

Valerianella Mill. spp.
Valerianaceae
♦ cornsalad, lamb's lettuce
♦ Weed, Naturalised
♦ 198, 272

Valerianella boissieri Krok
Valerianaceae
♦ Weed
♦ 87, 88

Valerianella carinata Loisel.
Valerianaceae
Fedia carinata Mert. & Koch, *Valerianella praecox* Waldst. & Kit.
♦ keeled fruited cornsalad, European cornsalad, cornsalad, talkavuonankaali
♦ Weed, Naturalised, Casual Alien
♦ 42, 44, 70, 88, 94, 101, 243, 253, 272, 280
♦ a/pH, cultivated.

Valerianella coronata (L.) DC.
Valerianaceae
♦ valeriánka korunkátá
♦ Weed, Naturalised
♦ 44, 88, 94, 243, 253, 272, 287
♦ cultivated, herbal.

Valerianella costata (Stev.) Betcke
Valerianaceae
Valerianella gibbosa (Guss.) DC.
♦ Weed
♦ 272

Valerianella dentata (L.) Poll.
Valerianaceae
Fedia mixta Vahl, *Fedia morisonii* Spreng., *Valerianella morisonii* DC.
♦ narrowfruit cornsalad, rikkavuonankaali
♦ Weed, Naturalised, Casual Alien
♦ 23, 42, 44, 70, 87, 88, 94, 101, 243, 253, 272
♦ cultivated, herbal.

Valerianella discoidea (L.) Loisel.
Valerianaceae
♦ Weed, Naturalised
♦ 86, 87, 88, 94, 98, 203, 253
♦ Origin: Mediterranean, Middle East.

Valerianella echinata (L.) DC.
Valerianaceae
♦ Weed
♦ 88, 94
♦ cultivated.

Valerianella eriocarpa Desv.
Valerianaceae
Valerianella incrassata Nyman
♦ Italian cornsalad, hairy fruited cornsalad
♦ Weed, Naturalised
♦ 44, 86, 88, 94, 98, 176, 198, 203, 243, 253, 300
♦ aH, promoted, herbal. Origin: Europe.

Valerianella locusta (L.) Lat.
Valerianaceae
Fedia locusta Reichen., *Fedia paniculata*

Colla., *Locusta communis* Del., *Valerianella olitoria* (L.) Poll. (see), *Valerianella pusilla* Miég.
♦ lamb's lettuce, cornsalad, European cornsalad, milk grass, common cornsalad, Lewiston cornsalad
♦ Weed, Naturalised, Introduced
♦ 34, 44, 70, 86, 87, 88, 94, 98, 101, 176, 198, 203, 241, 243, 253, 272, 280, 295, 300
♦ aH, cultivated, herbal. Origin: Europe, North Africa, Middle East.

Valerianella muricata (Steven ex Roem. & Schult.) M.Bieb. ex W.H.Baxter & Wooster
Valerianaceae
♦ Naturalised
♦ 86, 98, 241
♦ Origin: Eurasia.

Valerianella olitoria (L.) Poll.
Valerianaceae
= *Valerianella locusta* (L.) Lat.
♦ European cornsalad
♦ Weed, Naturalised
♦ 23, 87, 88, 218, 286, 287
♦ cultivated, herbal.

Valerianella pumila (L.) DC.
Valerianaceae
♦ valeriánka nízka
♦ Weed
♦ 88, 94, 272

Valerianella radiata (L.) Dufr.
Valerianaceae
♦ cornsalad, beaked cornsalad
♦ Weed
♦ 161
♦ aH, promoted, herbal.

Valerianella rimosa Bast.
Valerianaceae
Fedia auricula Roem. & Schult., *Valerianella auricula* DC.
♦ broad fruited cornsalad, uurrevuonankaali, cornsalad
♦ Weed, Naturalised, Environmental Weed, Casual Alien
♦ 23, 42, 44, 70, 86, 87, 88, 94, 98, 176, 203, 243, 253, 272, 300
♦ cultivated, herbal. Origin: Eurasia, North Africa.

Valerianella vesicaria (L.) Moench
Valerianaceae
♦ Weed
♦ 87, 88

Vallaris heynei Spreng.
Apocynaceae
♦ Weed
♦ 87, 88
♦ herbal.

Vallesia glabra (Cav.) Link
Apocynaceae
Rauwolfia glabra Cav., *Rauwolfia oppositiflora* Sessé & Moç., *Vallesia cymbifolia* Ortega, *Vallesia dichotoma* Ruiz & Pav.
♦ pearlberry, huelatave, huetatave, otave
♦ Naturalised, Introduced
♦ 101, 228
♦ S/T, arid, herbal.

Vallisneria L. spp.
Hydrocharitaceae
- eelgrass
- Weed, Quarantine Weed, Naturalised
- 76, 86, 88, 220
- wH, cultivated.

Vallisneria aethiopica Fenzl
Hydrocharitaceae
- eelgrass, tape grass
- Weed
- 87, 88, 121
- wpH, cultivated.

Vallisneria americana Michx.
Hydrocharitaceae
Vallisneria asiatica Michx. (see)
- eelgrass, flumine Mississippi, tape grass, water celery, wild celery, American eelgrass, ribbon weed
- Weed, Naturalised
- 7, 23, 86, 87, 88, 98, 197, 203, 218
- wpH, cultivated, herbal.

Vallisneria asiatica Michx.
Hydrocharitaceae
= *Vallisneria americana* Michx.
- Weed
- 286
- wH, cultivated.

Vallisneria gigantea Graebn.
Hydrocharitaceae
- ribbon weed, eelweed, giant vallis, jungle val, eelgrass, sekishoumo
- Weed, Quarantine Weed, Naturalised, Native Weed, Environmental Weed
- 15, 87, 88, 246, 269, 287
- wpH, cultivated. Origin: Australia.

Vallisneria spiralis L.
Hydrocharitaceae
- eelweed, ribbon weed, tape grass, Italian vallisneria, corkscrew val, straight vallis, coiled vallisneria
- Weed, Naturalised, Environmental Weed, Casual Alien
- 70, 87, 88, 126, 217, 246, 272, 275, 280, 287, 297
- wpH, cultivated, herbal.

Vallota speciosa (L.f.) T.Durand & Schinz
Liliaceae/Amaryllidaceae
Amaryllis purpurea Ait., *Crinum speciosum* L.f., *Vallota purpurea* Herb.
- Casual Alien
- 280
- cultivated, herbal.

Vanda teres (Roxb.) Lindl. × *hookeriana* Rchb.f. cv. 'Miss Joachim'
Orchidaceae
- Miss Joachim
- Introduced
- 230
- cultivated.

Vandellia anagallis (Burm.) Yamaz.
Scrophulariaceae
- Weed
- 87, 88, 204

Vandellia anagallis Yamaz. var. *verbenaefolia* Yamaz.
Scrophulariaceae
- Weed
- 263
- aH.

Vandellia angustifolia Benth.
Scrophulariaceae
- Weed, Quarantine Weed
- 76, 87, 88, 203, 220, 263
- aH.

Vandellia cordifolia (Colsm.) G.Don
Scrophulariaceae
- Weed
- 235

Vandellia crustacea (L.) Benth.
Scrophulariaceae
= *Lindernia crustacea* (L.) F.Muell.
- urikusa
- Weed, Quarantine Weed
- 39, 76, 87, 88, 203, 220, 263
- aH, toxic.

Vandellia hirta (Cham. & Schltdl.) Yamaz.
Scrophulariaceae
- Weed
- 235

Vandellia pedunculata Benth.
Scrophulariaceae
- Weed, Quarantine Weed
- 76, 87, 88, 203, 220

Vandellia setulosa (Maxim.) Yamaz.
Scrophulariaceae
- Weed
- 87, 88

Vangueria madagascariensis J.F.Gmel.
Rubiaceae
- voa vanga, Indies tamarind
- Naturalised, Cultivation Escape
- 101, 261
- cultivated. Origin: tropical Africa, Madagascar.

Vanilla inodora Schiede
Orchidaceae
= *Vanilla mexicana* Mill.
- Weed
- 179
- herbal.

Vanilla mexicana Mill.
Orchidaceae
Vanilla inodora Schiede (see)
- Mexican vanilla
- Weed
- 179
- herbal.

Vanilla planifolia Jacks.
Orchidaceae
- vanilla
- Introduced, Cultivation Escape
- 230, 261
- H, cultivated, herbal.

Vanilla pompona Schiede
Orchidaceae
- West Indian vanilla
- Naturalised
- 101
- cultivated, herbal.

Vassobia breviflora (Sendtn.) Hunz.
Solanaceae
- falsa coerana
- Weed
- 255
- S. Origin: Brazil.

Vatica diospyoides Symington
Dipterocarpaceae
- Quarantine Weed
- 220

Vaucheria A.DC. spp.
Vaucheriaceae
- vaucheria
- Weed
- 88
- algae.

Velezia rigida L.
Caryophyllaceae
- velezia
- Naturalised, Introduced
- 34, 101
- aH.

Vella annua L.
Brassicaceae
= *Carrichtera annua* (L.) DC.
- Weed
- 87, 88

Vellereophyton dealbatum (Thunb.) Hilliard & Burtt
Asteraceae
- white cudweed
- Weed, Naturalised, Environmental Weed
- 7, 9, 72, 86, 88, 98, 176, 198, 203, 280
- aH. Origin: South Africa.

Venidium fastuosum (Jacq.) Stapf
Asteraceae
Arctotis fastuosa Jacq.
- cape daisy, monarch of the veld
- Weed, Naturalised
- 34, 101
- aH, herbal.

Ventenata dubia (Leers) Coss. & Durieu
Poaceae
Avena dubia Leers, *Avena fertilis* All., *Avena tenuis* Moench, *Avena triaristata* Vill., *Holcus triflorus* Poll., *Ventenata avenacea* Koeler
- vententata, ventenatagrass, North Africa grass, wiregrass, hairgrass
- Weed, Naturalised, Environmental Weed
- 79, 101, 161, 182, 212, 272
- pG, herbal. Origin: central and southern Europe, Asia, Africa.

Veratrum album L.
Liliaceae/Melanthiaceae
Helleborus albus Gueldenst., *Melanthium album* Thunb.
- white false hellebore, European false hellebore, white hellebore, false hellebore, elleboro bianco
- Weed
- 39, 70, 87, 88, 126, 161, 243, 272
- pH, cultivated, herbal, toxic.

Veratrum californicum Dur.
Liliaceae/Melanthiaceae

♦ western false hellebore, California false hellebore, cornlily, skunk cabbage
♦ Weed
♦ 39, 87, 88, 161, 189, 212, 218
♦ pH, cultivated, herbal, toxic.

Veratrum fimbriatum Gray
Liliaceae/Melanthiaceae
♦ fringed cornlily, fringed false hellebore
♦ Weed
♦ 161
♦ pH, toxic.

Veratrum insolitum Jeps.
Liliaceae/Melanthiaceae
♦ Siskiyou false hellebore, Del Norte false hellebore
♦ Weed
♦ 161
♦ pH, herbal, toxic.

Veratrum lobelianum Bernh.
Liliaceae/Melanthiaceae
Veratrum album L. ssp. *virescens* Gaudin, *Veratrum album* ssp. *lobelianum* (Bernh.) Arcang.
♦ Weed, Quarantine Weed
♦ 23, 76, 87, 88, 203, 220

Veratrum nigrum L.
Liliaceae/Melanthiaceae
Melanthium nigrum Thunb.
♦ false hellebore, black hellebore
♦ Weed
♦ 39, 272
♦ pH, cultivated, herbal, toxic.

Veratrum viride Aiton
Liliaceae/Melanthiaceae
♦ white hellebore, green hellebore, Indian poke, green false hellebore, American white hellebore, white false hellebore
♦ Weed
♦ 8, 39, 87, 88, 161, 218, 247, 292
♦ pH, cultivated, herbal, toxic.

Verbascum L. spp.
Scrophulariaceae
♦ mullein
♦ Weed, Naturalised, Environmental Weed
♦ 39, 181, 198, 272, 289
♦ herbal, toxic.

Verbascum barnadesii Vahl
Scrophulariaceae
= *Celsia bernadesii* G.Don. (NoR)
♦ Weed
♦ 88, 94
♦ Origin: Portugal, Spain.

Verbascum blattaria L.
Scrophulariaceae
♦ moth mullein, kesätulikukka
♦ Weed, Naturalised, Garden Escape, Environmental Weed, Casual Alien
♦ 23, 34, 39, 40, 42, 72, 80, 86, 87, 88, 94, 98, 101, 136, 146, 159, 161, 167, 176, 179, 195, 198, 203, 210, 212, 218, 269, 272, 280, 287, 294
♦ bH, cultivated, herbal, toxic. Origin: Europe.

Verbascum blattaria L. fo. *erubescens* Brügger
Scrophulariaceae
♦ Naturalised
♦ 287

Verbascum chaixii Vill.
Scrophulariaceae
♦ arotulikukka, nettle leaved mullein
♦ Casual Alien
♦ 42
♦ cultivated, herbal.

Verbascum cheiranthifolium Boiss.
Scrophulariaceae
♦ kaitatulikukka
♦ Casual Alien
♦ 42

Verbascum chinense (L.) Sant.
Scrophulariaceae
♦ Weed
♦ 66

Verbascum creticum (L.) Cav.
Scrophulariaceae
♦ Cretan mullein, mullein
♦ Weed, Naturalised, Environmental Weed, Casual Alien
♦ 7, 40, 72, 86, 88, 98, 176, 198, 203, 280
♦ b, cultivated. Origin: Mediterranean.

Verbascum densiflorum Bertol.
Scrophulariaceae
Verbascum cuspidatum Schrad., *Verbascum macratherum* Hal., *Verbascum thapsiforme* Schrad. (see), *Verbascum velenovskyi* Horak
♦ akantulikukka, denseflower mullein, wool mullein
♦ Weed, Naturalised, Casual Alien
♦ 40, 42, 70, 101, 241, 243, 272, 300
♦ bH, cultivated, herbal.

Verbascum × *kerneri* Fritsch
Scrophulariaceae
= *Verbascum phlomoides* L. × *Verbascum thaspus* L.
♦ Naturalised
♦ 101

Verbascum litigiosum Samp.
Scrophulariaceae
♦ Weed
♦ 70

Verbascum longifolium Ten.
Scrophulariaceae
♦ long leaved mullein
♦ Weed
♦ 272
♦ cultivated, herbal.

Verbascum lychnitis L.
Scrophulariaceae
♦ white mullein, käentulikukka
♦ Weed, Naturalised, Casual Alien
♦ 42, 87, 88, 101, 218, 272
♦ bH, cultivated, herbal.

Verbascum nigrum L.
Scrophulariaceae
♦ dark mullein, tummatulikukka, black mullein
♦ Weed, Naturalised
♦ 70, 87, 88, 94, 101, 272
♦ b/pH, cultivated, herbal.

Verbascum orientale Bieb.
Scrophulariaceae
♦ Weed
♦ 39, 87, 88
♦ toxic.

Verbascum ovalifolium Donn ex Sims
Scrophulariaceae
♦ soikkotulikukka
♦ Casual Alien
♦ 42

Verbascum phlomoides L.
Scrophulariaceae
Verbascum australe Schrad., *Verbascum bulgaricum* Velen., *Verbascum grandiflorum* Mill., *Verbascum italicum* Moric., *Verbascum nemorosum* Schrad., *Verbascum rugulosum* Willd., *Verbascum tomentosum* Lam., *Verbascum viminale* Guss.
♦ clasping mullein, orange mullein, rohtotulikukka, woolly mullein
♦ Weed, Naturalised, Casual Alien
♦ 39, 40, 42, 87, 88, 94, 101, 218, 272
♦ b/pH, cultivated, herbal, toxic.

Verbascum phoeniceum L.
Scrophulariaceae
♦ purppuratulikukka, purple mullein
♦ Weed, Naturalised, Cultivation Escape
♦ 42, 101, 272
♦ bH, cultivated, herbal.

Verbascum × *pterocaulon* Franch.
Scrophulariaceae
= *Verbascum blattaria* L. × *Verbascum thaspus* L.
♦ Naturalised
♦ 101

Verbascum pulverulentum Vill.
Scrophulariaceae
Verbascum floccosum Waldst. & Kit.
♦ hoary mullein
♦ Weed, Casual Alien
♦ 39, 272, 280
♦ bH, cultivated, herbal, toxic.

Verbascum pyramidatum Bieb.
Scrophulariaceae
♦ kartiotulikukka
♦ Casual Alien
♦ 42
♦ cultivated, herbal.

Verbascum schimperianum Boiss.
Scrophulariaceae
♦ Weed
♦ 221

Verbascum sinaiticum Benth.
Scrophulariaceae
♦ Weed
♦ 221
♦ herbal.

Verbascum sinuatum L.
Scrophulariaceae
♦ wavyleaf mullein, mullein
♦ Weed, Naturalised, Casual Alien
♦ 39, 40, 70, 86, 98, 101, 203, 221, 272
♦ bH, cultivated, herbal, toxic. Origin: southern Europe.

Verbascum speciosum Schrad.
Scrophulariaceae
♦ showy mullein
♦ Weed, Naturalised
♦ 101, 272
♦ cultivated, herbal.

Verbascum × spurium K.Koch
Scrophulariaceae
= *Verbascum lychnitis* L. × *Verbascum thaspus* L.
♦ Naturalised
♦ 101

Verbascum thapsiforme Schrad.
Scrophulariaceae
= *Verbascum densiflorum* Bertol.
♦ wool mullein
♦ Weed
♦ 23, 87, 88
♦ herbal.

Verbascum thapsus L.
Scrophulariaceae
Verbascum schraderi Mey.
♦ woolly mullein, flannel plant, common mullein, big taper, velvet dock, velvetplant, flannel leaved mullein, ukontulikukka, Jacob's staff, great mullein, Aaron's rod, blanket weed, candlewick, flannel leaf, shepherd's club
♦ Weed, Noxious Weed, Naturalised, Introduced, Garden Escape, Environmental Weed, Cultivation Escape
♦ 1, 4, 7, 15, 20, 23, 24, 34, 35, 39, 49, 52, 70, 72, 78, 80, 86, 87, 88, 97, 98, 101, 102, 104, 116, 129, 134, 136, 146, 147, 151, 159, 161, 165, 174, 176, 180, 181, 195, 198, 203, 207, 210, 211, 212, 218, 222, 229, 231, 233, 237, 241, 243, 249, 269, 272, 280, 286, 287, 290, 294, 295, 297, 299, 300
♦ bH, cultivated, herbal, toxic. Origin: Eurasia.

Verbascum thapsus L. fo. candicans House
Scrophulariaceae
♦ Naturalised
♦ 287

Verbascum virgatum Stokes
Scrophulariaceae
Verbascum blattarioides Lam.
♦ twiggy mullein, wand mullein, slender mullein, purplestamen mullein, mullein, virgate mullein, wand mullein, candlestick, moth mullein, Aaron's rod, green mullein
♦ Weed, Naturalised, Garden Escape, Environmental Weed
♦ 7, 15, 34, 72, 80, 86, 87, 88, 98, 101, 121, 134, 161, 167, 176, 179, 180, 198, 199, 203, 237, 241, 243, 251, 269, 280, 287, 290, 295, 300
♦ bH, cultivated, herbal. Origin: Eurasia.

Verbena aristigera S.Moore
Verbenaceae
♦ Naturalised
♦ 86
♦ cultivated. Origin: South America.

Verbena bipinnatifida Nutt.
Verbenaceae
= *Glandularia bipinnatifida* (Nutt.) Nutt. var. *bipinnatifida* (NoR) [see *Verbena ciliata* Benth.]
♦ prairie verbena
♦ Weed, Naturalised
♦ 98, 203
♦ cultivated, herbal.

Verbena bonariensis L.
Verbenaceae
♦ tall vervain, Argentine vervain, cluster flowered vervain, verbena, purpletop vervain, wild verbena, purpletop, purpletop verbena, jättiverbena
♦ Weed, Naturalised, Garden Escape, Environmental Weed, Cultivation Escape
♦ 7, 15, 24, 34, 39, 42, 51, 55, 72, 86, 87, 88, 98, 101, 121, 134, 158, 161, 165, 179, 180, 198, 203, 218, 237, 243, 255, 261, 269, 276, 280, 286, 287, 289, 295
♦ a/bH, cultivated, herbal, toxic. Origin: South America.

Verbena bonariensis L. var. bonariensis
Verbenaceae
♦ purpletop vervain
♦ Naturalised
♦ 101

Verbena bonariensis L. var. conglomerata Briq.
Verbenaceae
♦ purpletop vervain
♦ Naturalised
♦ 101

Verbena bracteata Lag. & Rodr.
Verbenaceae
♦ large bracted vervain, prostrate vervain, bracted vervain, bigbract verbena
♦ Weed, Naturalised, Native Weed
♦ 21, 34, 49, 80, 87, 88, 161, 174, 210, 212, 218, 243, 287
♦ a/pH, herbal. Origin: North America.

Verbena brasiliensis Vell.
Verbenaceae
♦ Brazilian vervain
♦ Weed, Naturalised, Environmental Weed
♦ 80, 98, 101, 179, 203, 257, 280, 286, 287
♦ herbal.

Verbena brasiliensis Vell. var. brasiliensis
Verbenaceae
♦ Naturalised
♦ 241

Verbena caracasana Kunth
Verbenaceae
♦ Naturalised
♦ 86
♦ cultivated. Origin: South America.

Verbena carolina L.
Verbenaceae
♦ Carolina vervain
♦ Weed
♦ 157, 199

♦ pH, arid.

Verbena ciliata Benth.
Verbenaceae
= *Glandularia bipinnatifida* (Nutt.) Nutt. var. *bipinnatifida* (NoR) [see *Verbena bipinnatifida* Nutt.]
♦ Weed, Quarantine Weed
♦ 76, 87, 88, 203, 220
♦ herbal.

Verbena corymbosa Ruiz & Pav.
Verbenaceae
♦ Weed
♦ 87, 88

Verbena dissecta Willd. ex Spreng.
Verbenaceae
= *Glandularia dissecta* (Willd. ex Schauer) Schnack & Covas
♦ Weed
♦ 87, 88

Verbena gracilescens (Cham.) Herter
Verbenaceae
♦ Weed
♦ 87, 88, 237, 295

Verbena hastata L.
Verbenaceae
♦ fake vervain, wild hyssop, wild verbena, ironweed, swamp verbena, blue vervain, American blue vervain
♦ Weed
♦ 23, 34, 49, 87, 88, 136, 161, 210, 212, 218
♦ pH, cultivated, herbal.

Verbena hispida Ruiz & Pav.
Verbenaceae
= *Glandularia hispida* Ruiz & Pav.
♦ hairy vervain
♦ Weed, Naturalised
♦ 86, 87, 88, 98, 203
♦ Origin: South America.

Verbena × hybrida hort. ex Groenl. & Rümpler
Verbenaceae
= *Verbena incisa* Hook. × *Verbena peruviana* (L.) Britt. × *Verbena phlogiflora* Cham. × *Verbena teucroides* Gillies & Hook. [possible parents]
♦ tarhaverbena, garden vervain, verbena
♦ Cultivation Escape
♦ 42
♦ aH, cultivated.

Verbena incompta P.W.Michael
Verbenaceae
♦ purpletop
♦ Naturalised
♦ 86, 176, 198

Verbena intermedia Gill. & Hook.
Verbenaceae
♦ verbena
♦ Weed
♦ 87, 88, 236, 237, 295

Verbena lasiostachys Link
Verbenaceae
♦ western vervain, common verbena
♦ Weed
♦ 161
♦ pH, herbal.

Verbena ligustrina Lag.
Verbenaceae
Junellia ligustrina (Lag.) Mold.
♦ Introduced
♦ 228
♦ arid.

Verbena litoralis Kunth
Verbenaceae
♦ seashore vervain, common verbena, shore vervain
♦ Weed, Naturalised
♦ 34, 55, 86, 87, 88, 98, 101, 134, 179, 203, 218, 237, 255, 257, 269, 280, 286, 287, 295
♦ b/pH, arid, herbal. Origin: tropical America.

Verbena menthaefolia Benth.
Verbenaceae
♦ Weed
♦ 199

Verbena montevidensis Spreng.
Verbenaceae
♦ Uruguayan vervain
♦ Naturalised
♦ 101

Verbena officinalis L.
Verbenaceae
♦ European vervain, common verbena, pigeon's grass, holy herb, vervain, nang dong laang, herb of the cross, wild verbena, herba sacra, herba veneris
♦ Weed, Naturalised, Introduced, Garden Escape, Environmental Weed, Casual Alien
♦ 8, 15, 23, 38, 39, 42, 44, 51, 55, 70, 86, 87, 88, 93, 94, 98, 101, 121, 158, 165, 176, 185, 191, 203, 218, 235, 238, 240, 243, 248, 253, 269, 272, 275, 280, 286, 297, 300
♦ a/pH, cultivated, herbal, toxic. Origin: Eurasia.

Verbena officinalis L. var. officinalis
Verbenaceae
♦ herb of the cross
♦ Naturalised
♦ 101

Verbena officinalis L. var. prostrata Gren. & Godr.
Verbenaceae
♦ prostrate verbena
♦ Naturalised
♦ 101

Verbena peruviana (L.) Britton
Verbenaceae
= *Glandularia peruviana* (L.) Small
♦ Weed
♦ 87, 88
♦ cultivated.

Verbena quadrangularis Vell.
Verbenaceae
♦ Naturalised
♦ 86
♦ Origin: South America.

Verbena recta H.B.K.
Verbenaceae
♦ Weed
♦ 199

Verbena rigida Spreng.
Verbenaceae
♦ stiff verbena, tuber vervain, roadside vervain, tuberous vervain
♦ Weed, Naturalised
♦ 7, 39, 55, 86, 87, 88, 98, 101, 161, 198, 203, 249, 269, 280, 287, 300
♦ cultivated, herbal, toxic. Origin: South America.

Verbena stricta Vent.
Verbenaceae
♦ hoary vervain, woolly verbena, tall vervain
♦ Weed, Naturalised, Native Weed
♦ 23, 87, 88, 161, 174, 210, 218, 287
♦ pH, cultivated, herbal. Origin: North America.

Verbena supina L.
Verbenaceae
♦ trailing verbena
♦ Weed, Naturalised
♦ 86, 87, 88, 98, 101, 185, 198, 203, 221, 269
♦ herbal. Origin: Mediterranean.

Verbena tenera Spreng.
Verbenaceae
= *Glandularia tenera* (Spreng.) Cabrera
♦ Weed
♦ 87, 88
♦ cultivated.

Verbena tenuisecta Briq.
Verbenaceae
= *Glandularia aristigera* (S.Moore) Tronc. (NoR)
♦ Mayne's pest, South American mock vervain, fine leaved verbena, moss verbena, tuber vervain, wild verbena
♦ Weed, Quarantine Weed, Naturalised, Introduced, Cultivation Escape
♦ 34, 39, 51, 55, 76, 87, 88, 93, 98, 121, 158, 161, 191, 203, 205, 261, 269, 287
♦ a/pH, cultivated, herbal, toxic. Origin: South America.

Verbena tridens Lag.
Verbenaceae
= *Junellia tridens* (Kuntze) Moldenke
♦ mata negra
♦ Weed
♦ 295

Verbena urticifolia L.
Verbenaceae
♦ white vervain, nokkosrautayrtti, nettle leaved vervain
♦ Weed, Casual Alien
♦ 23, 42, 87, 88, 161, 207, 210, 218
♦ pH, cultivated, herbal. Origin: eastern North America.

Verbena venosa Gillies & Hook.
Verbenaceae
♦ veined verbena
♦ Weed
♦ 39, 51, 121
♦ pH, herbal, toxic. Origin: South America.

Verbesina alata L.
Asteraceae
♦ capitaneja
♦ Weed
♦ 14, 87, 88
♦ herbal.

Verbesina alternifolia (L.) Britt. ex Kearney
Asteraceae
♦ wingstem, wingstem actinomeris
♦ Weed, Naturalised
♦ 161, 287
♦ cultivated, herbal.

Verbesina asteroides L.
Asteraceae
♦ Quarantine Weed
♦ 220

Verbesina caracasana Rob. & Greenm.
Asteraceae
♦ Weed
♦ 87, 88

Verbesina encelioides (Cav.) Benth. & Hook.f. ex A.Gray
Asteraceae
♦ crownbeard, golden crownbeard, wild sunflower, girasolillo, American dogweed, gold weed, South African daisy, yellowtop, cowpen daisy
♦ Weed, Quarantine Weed, Noxious Weed, Naturalised, Introduced, Environmental Weed
♦ 7, 39, 51, 76, 86, 87, 88, 93, 98, 147, 158, 161, 179, 198, 203, 228, 236, 237, 243, 261, 269, 270, 295, 300
♦ aH, arid, cultivated, herbal, toxic. Origin: Americas.

Verbesina encelioides (Cav.) Benth. & Hook.f. ex A.Gray ssp. encelioides
Asteraceae
♦ butter daisy, golden crownbeard, wild sunflower
♦ Weed
♦ 121
♦ aH. Origin: North America.

Verbesina encelioides (Cav.) Benth. & Hook.f. ex A.Gray ssp. exauriculata (Cav.) Gray (Rob. & Greenm.) Coleman
Asteraceae
Verbesina encelioides (Cav.) Benth. & Hook.f. ex A.Gray var. *exauriculata* Rob. & Greenm. (see)
♦ crownbeard, golden crownbeard
♦ Weed
♦ 34
♦ aH.

Verbesina encelioides (Cav.) Benth. & Hook.f. ex A.Gray var. exauriculata Rob. & Greenm.
Asteraceae
= *Verbesina encelioides* (Cav.) Benth. & Hook.f. ex A.Gray ssp. *exauriculata* (Rob. & Greenm.) Coleman
♦ crownbeard
♦ Weed
♦ 218

Verbesina fraseri Hemsl.
Asteraceae
♦ Weed
♦ 157
♦ pC.

Verbesina occidentalis (L.) Walter
Asteraceae
♦ yellow crownbeard, small yellow crownbeard
♦ Weed, Naturalised
♦ 161, 287
♦ cultivated, herbal.

Verbesina persicifolia DC.
Asteraceae
♦ Weed, Quarantine Weed
♦ 76, 87, 88, 203, 220

Verbesina subcordata DC.
Asteraceae
♦ Weed
♦ 87, 88, 237, 295

Vernicia fordii (Hemsl.) Airy Shaw
Euphorbiaceae
Aleurites fordii Hemsl. (see)
♦ tung oil tree
♦ Weed, Naturalised, Cultivation Escape
♦ 86, 98, 101, 161, 203, 261
♦ cultivated, toxic. Origin: China.

Vernonia adoensis Sch.Bip. ex Walp.
Asteraceae
Vernonia shirensis Oliv. & Hiern
♦ ironweed
♦ Native Weed
♦ 121
♦ pH. Origin: southern Africa.

Vernonia altissima Nutt.
Asteraceae
= *Vernonia gigantea* (Walter) Trel. ex Branner & Coville
♦ tall ironweed, ironweed
♦ Weed
♦ 87, 88, 161, 210, 218
♦ cultivated, herbal.

Vernonia ambigua Kotschy & Peyr.
Asteraceae
♦ Weed
♦ 87, 88

Vernonia anthelmintica (L.) Willd.
Asteraceae
Centratherum anthelminticum (L.) Kuntze
♦ ironweed
♦ Introduced
♦ 39, 228
♦ arid, cultivated, herbal, toxic.

Vernonia baccharoides Kunth
Asteraceae
♦ ocuera
♦ Weed
♦ 87, 88, 153

Vernonia baldwinii Torr.
Asteraceae
♦ western ironweed, Baldwin's ironweed
♦ Weed, Native Weed
♦ 23, 87, 88, 161, 174, 210, 218, 266
♦ pH, cultivated, herbal. Origin: North America.

Vernonia brasiliensis Less.
Asteraceae
♦ Weed
♦ 87, 88

Vernonia cainarahiensis Hieron.
Asteraceae
♦ Weed
♦ 87, 88

Vernonia camporum M.E.Jones
Asteraceae
♦ Weed
♦ 87, 88

Vernonia chamaedrys Less.
Asteraceae
Vernonanthura chamaedrys (Less.) H.Rob.
♦ vassoura branca, assapeixe branco
♦ Weed
♦ 255, 295
♦ Origin: Brazil.

Vernonia cinerea (L.) Less
Asteraceae
= *Cyanthillium cinereum* (L.) H.Rob.
♦ vernonia, little ironweed, ironweed, suea saam khaa, yaa sam wan, rabo de buey, yerba socialista
♦ Weed, Naturalised, Introduced, Environmental Weed, Casual Alien
♦ 6, 7, 13, 66, 86, 87, 88, 170, 179, 206, 209, 217, 218, 230, 238, 239, 243, 261, 262, 273, 276, 280, 286
♦ aH, arid, cultivated, herbal. Origin: tropical, subtropical Asia and Africa.

Vernonia cognata Less.
Asteraceae
Vernonia propinqua Hieron.
♦ assapeixe roxo
♦ Weed
♦ 255
♦ Origin: Brazil.

Vernonia colorata (Willd.) Drake
Asteraceae
Baccharis senegalensis Pers., *Eupatorium coloratum* Willd., *Gymnanthemum coloratum* (Willd.) H.Rob. & B.Kahn, *Vernonia senegalensis* (Pers.) Less.
♦ mjonso mlulunguja
♦ Weed
♦ 87, 88
♦ arid, herbal. Origin: Madagascar, southern Africa.

Vernonia divaricata Sw.
Asteraceae
♦ Weed
♦ 87, 88

Vernonia fasciculata Michx.
Asteraceae
♦ prairie ironweed, ironweed, smooth ironweed
♦ Weed
♦ 161
♦ cultivated, herbal.

Vernonia fastigiata Oliv. & Hiern
Asteraceae
Vernonia schinzii O.Hoffm.
♦ langbeenbossie
♦ Native Weed
♦ 121
♦ pH, arid. Origin: southern Africa.

Vernonia ferruginea Less.
Asteraceae
Vernonanthura ferruginea (Less.) H.Rob.

♦ assapeixe, assapeixe de santana
♦ Weed
♦ 255
♦ S/T. Origin: Brazil.

Vernonia flexuosa Sims
Asteraceae
♦ Weed
♦ 87, 88

Vernonia galamensis (Cass.) Less.
Asteraceae
Centrapalus galamensis Cass., *Conyza pauciflora* Willd., *Vernonia afromontana* R.E.Fr., *Vernonia pauciflora* (Willd.) Less. *nom. illeg.* (see), *Vernonia petitiana* A.Rich., *Vernonia senegalensis* Desf.
♦ ironweed
♦ Weed, Introduced
♦ 88, 228
♦ arid.

Vernonia gigantea (Walter) Trel. ex Branner & Coville
Asteraceae
Vernonia altissima Nutt. (see)
♦ tall ironweed, giant ironweed
♦ Weed
♦ 207
♦ herbal. Origin: North America.

Vernonia glabrata Less.
Asteraceae
Vernonia ensifolia Mart.
♦ assapeixe, assapeixe roxo
♦ Weed
♦ 255
♦ Origin: Brazil.

Vernonia incana Less.
Asteraceae
♦ matacampo
♦ Weed
♦ 237, 295

Vernonia intermedia DC.
Asteraceae
♦ Weed
♦ 87, 88

Vernonia kotschyana Sch.Bip.
Asteraceae
♦ Weed
♦ 87, 88

Vernonia lasiopus O.Hoffm.
Asteraceae
♦ Weed
♦ 87, 88

Vernonia menthaefolia (Poepp. ex Spreng.) Less.
Asteraceae
♦ Weed
♦ 14

Vernonia nigritiana Oliv. & Hiern
Asteraceae
♦ Weed
♦ 39, 87, 88
♦ herbal, toxic.

Vernonia noveboracensis (L.) Willd.
Asteraceae
♦ New York ironweed
♦ Weed
♦ 23, 88
♦ cultivated, herbal, toxic.

Vernonia nudiflora Less.
Asteraceae
Vernonia angustifolia Don ex Hook. &
Arn.
♦ alecrim do campo, falso alecrim
♦ Weed
♦ 33, 255
♦ Origin: South America.

Vernonia pallens Sch.Bip.
Asteraceae
♦ Weed
♦ 87, 88

Vernonia parviflora Reinw. in Blume
Asteraceae
♦ Weed
♦ 235

Vernonia patens Kunth
Asteraceae
♦ Weed
♦ 87, 88
♦ S/T, herbal.

Vernonia patula (Dry.) Merr.
Asteraceae
♦ Weed, Naturalised
♦ 13, 23, 87, 88, 235, 287
♦ herbal.

**Vernonia pauciflora (Willd.) Less. *nom.
illeg.***
Asteraceae
= *Vernonia galamensis* (Cass.) Less.
♦ Weed
♦ 87, 88

Vernonia perrottetii Sch.Bip. ex Walp.
Asteraceae
♦ Weed
♦ 88

Vernonia platensis (Spreng.) Less.
Asteraceae
Conyza platensis Spreng.
♦ assapeixe, orelha de mula
♦ Weed
♦ 255
♦ Origin: South America.

Vernonia polyanthes (Spreng.) Less.
Asteraceae
Eupatorium polyanthes Spreng.
♦ assapeixe, chamarrita
♦ Weed, Quarantine Weed
♦ 33, 76, 87, 88, 203, 220, 255
♦ Origin: Brazil.

Vernonia poskeana Vatke & Hildebr.
Asteraceae
Vernonia elegantissima Hutch. & Dalziel,
Vernonia samfyana G.V.Pope
♦ Weed
♦ 87, 88
♦ arid.

**Vernonia poskeana Vatke & Hildebr. var.
poskeana**
Asteraceae
♦ Native Weed
♦ 121
♦ aH. Origin: southern Africa.

Vernonia scabra Pers.
Asteraceae
♦ Weed

♦ 87, 88
♦ herbal.

Vernonia scorpioides (Lam.) Pers.
Asteraceae
Cacalia scorpioides (Lam.) Kuntze,
Cacalia tournefortioides (H.B.K.) Kuntze,
Conyza scorpioides Lam., *Lepidaploa
scorpioides* (Lam.) Cass., *Staehelina
solidaginoides* Willd. ex Less., *Vernonia
flavescens* Less., *Vernonia lanuginosa*
Gardner, *Vernonia longeracemosa*
C.Mart. ex DC., *Vernonia saepia* Ekman,
Vernonia subrepanda Pers., *Vernonia
tournefortioides* Kunth
♦ enxuga, capichingui de bicho
♦ Weed
♦ 87, 88, 255
♦ herbal. Origin: Brazil.

Vernonia sericea L. Rich.
Asteraceae
= *Lepidaploa sericea* (L.C.Rich.) H.Rob.
(NoR)
♦ longshoot
♦ Weed
♦ 87, 88

Vernonia staehinoides Harv.
Asteraceae
♦ Native Weed
♦ 121
♦ pH. Origin: southern Africa.

Vernonia stipulacea Klatt
Asteraceae
Vernonia ampla O.Hoffm.
♦ blue bitter tea, poison tree vernonia
♦ Native Weed
♦ 121
♦ S, toxic. Origin: southern Africa.

Vernonia tweediana Bak.
Asteraceae
Vernonia tweediana aff. Bak. (see)
♦ Weed
♦ 237

Vernonia tweediana aff. Bak.
Asteraceae
Vernonia tweediana Bak. (see)
♦ matacampo
♦ Weed
♦ 295

Vernonia undulata Oliv. & Hiern
Asteraceae
♦ Weed
♦ 87, 88
♦ pH.

Vernonia westiniana Less.
Asteraceae
Vernonia hebeclada DC.
♦ assapeixe, chamaritta
♦ Weed
♦ 255
♦ Origin: South America.

Veronica L. spp.
Scrophulariaceae
♦ speedwell
♦ Weed, Naturalised
♦ 198, 221, 237, 243, 272
♦ H, herbal.

Veronica acinifolia L.
Scrophulariaceae
♦ veronica acinifoglia
♦ Weed
♦ 88, 94, 253, 272
♦ herbal.

Veronica agrestis L.
Scrophulariaceae
♦ field speedwell, green field
speedwell, peltotädyke, procumbent
speedwell
♦ Weed, Naturalised, Introduced
♦ 23, 44, 70, 87, 88, 94, 101, 121, 161,
174, 218, 243, 253, 272, 280
♦ aH, cultivated, herbal. Origin:
Eurasia.

Veronica americana Schwein. ex Benth.
Scrophulariaceae
♦ American speedwell, brooklime,
American brooklime
♦ Weed, Naturalised
♦ 23, 87, 88, 280, 286
♦ pH, aqua, cultivated, herbal.

Veronica anagallis-aquatica L.
Scrophulariaceae
Veronica catenata Pennell (see)
♦ veronica, water speedwell, pink
water speedwell, greater water
speedwell, blue water speedwell, long
leaved water speedwell
♦ Weed, Naturalised, Native Weed,
Environmental Weed
♦ 15, 23, 42, 51, 72, 80, 86, 87, 88, 98,
121, 159, 161, 165, 180, 185, 198, 200,
203, 221, 241, 269, 272, 275, 280, 286,
287, 295, 297, 300
♦ waH, cultivated, herbal. Origin:
Eurasia, Africa.

Veronica anagalloides Guss.
Scrophulariaceae
♦ etelänojatädyke
♦ Weed, Naturalised, Casual Alien
♦ 42, 87, 88, 94, 287
♦ cultivated, herbal.

Veronica aquatica Bernh.
Scrophulariaceae
♦ Naturalised
♦ 287

Veronica arvensis L.
Scrophulariaceae
♦ corn speedwell, wall speedwell,
speedwell, rock speedwell, common
speedwell, ketotädyke
♦ Weed, Naturalised, Introduced,
Environmental Weed
♦ 7, 9, 15, 23, 34, 44, 70, 80, 86, 87, 88,
94, 98, 101, 134, 161, 174, 176, 198, 203,
204, 211, 217, 218, 236, 237, 241, 243,
249, 253, 256, 263, 269, 271, 272, 280,
286, 287, 295, 300
♦ aH, cultivated, herbal. Origin:
Europe.

Veronica austriaca L.
Scrophulariaceae
Veronica dentata F.W.Schmidt
♦ itävallantädyke, broadleaf speedwell
♦ Weed, Naturalised, Casual Alien
♦ 42, 101, 272
♦ cultivated.

Veronica austriaca L. ssp. *teucrium* (L.) D.A.Webb
Scrophulariaceae
= *Veronica teucrium* L. (NoR)
♦ broadleaf speedwell, germander leaf speedwell
♦ Naturalised
♦ 101
♦ cultivated, herbal.

Veronica beccabunga L.
Scrophulariaceae
♦ European brooklime, European speedwell, ojatädyke, brooklime
♦ Weed, Naturalised
♦ 23, 39, 70, 80, 87, 88, 101, 133, 197, 224, 241, 272, 287, 300
♦ pH, aqua, cultivated, herbal, toxic.

Veronica biloba L.
Scrophulariaceae
♦ bi-lobed speedwell, twolobe speedwell
♦ Weed, Naturalised
♦ 80, 101, 161, 212, 243, 248
♦ herbal.

Veronica campylopoda Boiss.
Scrophulariaceae
♦ mutkatädyke
♦ Naturalised, Casual Alien
♦ 42, 101

Veronica catenata Pennell
Scrophulariaceae
= *Veronica anagallis-aquatica* L.
♦ pink water speedwell, chain speedwell, vesitädyke
♦ Weed, Naturalised, Environmental Weed
♦ 15, 72, 86, 88, 98, 198, 203, 280
♦ wpH, promoted, herbal. Origin: northern hermisphere.

Veronica chamaedrys L.
Scrophulariaceae
♦ bird's eye speedwell, germander speedwell, nurmitädyke
♦ Weed, Naturalised
♦ 23, 70, 80, 87, 88, 101, 161, 218, 241, 243, 272, 280, 287, 300
♦ pH, cultivated, herbal.

Veronica chamaepithyoides Lam.
Scrophulariaceae
= *Veronica digitata* Vahl (NoR)
♦ Weed
♦ 88, 94

Veronica chamaepitys Griseb.
Scrophulariaceae
♦ Weed
♦ 87, 88

Veronica cymbalaria Bodard
Scrophulariaceae
♦ white speedwell, glandular speedwell, pale speedwell
♦ Weed, Naturalised, Casual Alien
♦ 40, 87, 88, 94, 101, 115, 253, 272
♦ cultivated.

Veronica didyma Ten.
Scrophulariaceae
= *Veronica polita* Fr.
♦ Weed

♦ 87, 88, 235, 243, 248, 275, 297
♦ aH, promoted.

Veronica didyma Ten. var. *lilacina* (Hara) Yamaz.
Scrophulariaceae
♦ Weed
♦ 286

Veronica dillenii Crantz
Scrophulariaceae
Veronica campestris Schmalh., *Veronica succulenta* All.
♦ hoikkatädyke, Dillenius's speedwell
♦ Weed, Naturalised, Casual Alien
♦ 42, 101, 243, 272

Veronica filiformis Sm.
Scrophulariaceae
♦ filiform speedwell, slender speedwell, creeping speedwell, creeping veronica, whetzel weed, kaukasiantädyke
♦ Weed, Naturalised, Introduced
♦ 23, 34, 40, 42, 44, 70, 80, 87, 88, 94, 101, 161, 211, 218, 280
♦ pH, cultivated, herbal. Origin: Caucasus.

Veronica gentianoides Vahl
Scrophulariaceae
♦ gentian blue speedwell, katkerotädyke
♦ Weed, Cultivation Escape
♦ 42, 80
♦ cultivated, herbal.

Veronica glauca Sibth. & Sm.
Scrophulariaceae
Veronica peloponnesiaca Boiss. & Orph., *Veronica chaubardii* Boiss. & Reut.
♦ Weed
♦ 88, 94

Veronica grandis Fisch. ex Spreng.
Scrophulariaceae
♦ heartleaf speedwell
♦ Naturalised
♦ 101

Veronica hederaefolia L.
Scrophulariaceae
= *Veronica hederifolia* L.
♦ Weed, Naturalised
♦ 286, 287

Veronica hederifolia L.
Scrophulariaceae
Veronica hederaefolia L. (see)
♦ ivy leaved speedwell, ivyleaf speedwell
♦ Weed, Naturalised, Introduced
♦ 23, 24, 34, 42, 44, 70, 80, 86, 87, 88, 94, 98, 101, 118, 161, 176, 198, 203, 218, 243, 253, 263, 272, 280
♦ aH, cultivated, herbal. Origin: Europe, west Asia, North Africa, Madeira and Canary Islands.

Veronica javanica Bl.
Scrophulariaceae
♦ Weed
♦ 87, 88, 235

Veronica longifolia L.
Scrophulariaceae
Pseudolysimachion longifolium (L.) Opiz., *Veronica maritima* L.

♦ long leaved speedwell, garden speedwell, rantatädyke
♦ Weed, Naturalised, Introduced, Garden Escape, Cultivation Escape
♦ 23, 40, 80, 88, 101, 272, 297
♦ pH, cultivated, herbal.

Veronica montana L.
Scrophulariaceae
♦ wood speedwell, vuoritädyke
♦ Casual Alien
♦ 42
♦ cultivated.

Veronica myriantha Tosh.Tanaka
Scrophulariaceae
= *Veronica anagallis* L. × *Veronica undulata* Wall. ex Roxb.
♦ Naturalised
♦ 287

Veronica officinalis L.
Scrophulariaceae
♦ common speedwell, heath speedwell, speedwell, common gypsyweed
♦ Weed, Naturalised, Casual Alien
♦ 8, 23, 39, 70, 80, 86, 87, 88, 161, 195, 198, 218, 241, 272, 280, 300
♦ pH, cultivated, herbal, toxic. Origin: Europe, west Asia, Azores and Madeira Islands.

Veronica opaca Fr.
Scrophulariaceae
♦ dark speedwell, broadsepal speedwell, himmeätädyke
♦ Weed, Naturalised
♦ 44, 70, 87, 88, 94, 253, 272, 287
♦ Origin: Europe.

Veronica peregrina L.
Scrophulariaceae
Veronica romana L.
♦ purslane speedwell, neckweed, wandering speedwell, American speedwell
♦ Weed, Naturalised, Casual Alien
♦ 23, 40, 42, 44, 86, 87, 88, 94, 98, 161, 176, 198, 203, 210, 212, 218, 249, 253, 263, 271, 286
♦ aH, cultivated, herbal. Origin: Americas.

Veronica peregrina L. ssp. *peregrina*
Scrophulariaceae
♦ neckweed, cudzia pravá
♦ Naturalised
♦ 241

Veronica peregrina L. ssp. *xalapensis* (Kunth) Pennell
Scrophulariaceae
Veronica peregrina L. var. *xalapensis* (Kunth) Pennell (see)
♦ purslane speedwell, neckweed, hairy purslane, speedwell
♦ Weed, Naturalised
♦ 180, 237, 243, 295, 300
♦ aH.

Veronica peregrina L. var. *xalapensis* (Kunth) Pennell
Scrophulariaceae
= *Veronica peregrina* L. ssp. *xalapensis* (Kunth) Pennell

◆ kemushikusa
◆ Weed
◆ 286

Veronica persica Poir.
Scrophulariaceae
Veronica buxbaumii Ten., *Veronica byzantina* (Sibth. & Sm.) Britton, Stern & Pogg., *Veronica diffusa* Raf., *Veronica meskhetica* Kem.-Nath., *Veronica precox* Raf., *Veronica rotundifolia* Sessé & Moç., *Veronica tournefortii* C.C.Gmel. (see)
◆ Persian speedwell, bird's eye speedwell, winter speedwell, common field speedwell, scrambling speedwell, Buxbaum's speedwell, Byzantine speedwell, veronica comune, creeping speedwell
◆ Weed, Naturalised, Introduced
◆ 7, 15, 23, 34, 40, 42, 44, 51, 70, 80, 86, 87, 88, 94, 98, 101, 118, 121, 134, 158, 161, 165, 176, 180, 198, 203, 204, 207, 212, 215, 217, 218, 237, 241, 243, 253, 255, 256, 263, 269, 271, 272, 275, 280, 286, 287, 295, 297, 300
◆ aH, arid, cultivated, herbal. Origin: Eurasia.

Veronica phyllostachya L.
Scrophulariaceae
◆ Weed, Naturalised
◆ 98, 203

Veronica plebeia R.Br.
Scrophulariaceae
◆ trailing speedwell
◆ Naturalised
◆ 101, 134, 280
◆ cultivated.

Veronica polita Fr.
Scrophulariaceae
Veronica didyma Ten. (see)
◆ field speedwell, wayside speedwell, kiiltotädyke, grey speedwell
◆ Weed, Naturalised
◆ 44, 80, 87, 88, 94, 101, 199, 218, 243, 253, 272, 280
◆ aH, cultivated, herbal.

Veronica praecox All.
Scrophulariaceae
◆ Breckland speedwell
◆ Weed
◆ 70, 88, 94, 253, 272

Veronica prostrata L.
Scrophulariaceae
◆ mätästädyke
◆ Weed, Cultivation Escape
◆ 42, 272
◆ cultivated, herbal.

Veronica scardica Griseb.
Scrophulariaceae
◆ kaakonojatädyke
◆ Casual Alien
◆ 42

Veronica scutellata L.
Scrophulariaceae
◆ marsh speedwell, skullcap speedwell, luhtatädyke, véronique à écus, veronica delle paludi
◆ Weed, Naturalised, Environmental Weed

◆ 23, 54, 86, 88, 176, 272, 280, 300
◆ pH, aqua, cultivated, herbal. Origin: North America, Europe, Asia.

Veronica serpyllifolia L.
Scrophulariaceae
◆ thyme leaved speedwell, thymeleaf speedwell, turf speedwell, orvontädyke
◆ Weed, Naturalised
◆ 15, 23, 44, 70, 80, 86, 87, 88, 101, 161, 176, 210, 218, 241, 243, 253, 272, 280, 287, 300
◆ pH, cultivated, herbal. Origin: Europe, Africa, Asia.

Veronica serpyllifolia L. ssp. humifusa (Dickson) Syme
Scrophulariaceae
Veronica borealis Kirsch., *Veronica humifusa* Dicks., *Veronica tenella* Land.
◆ bright blue speedwell
◆ Weed
◆ 34
◆ pH.

Veronica serpyllifolia L. ssp. serpyllifolia
Scrophulariaceae
◆ thymeleaf speedwell
◆ Naturalised
◆ 101
◆ pH.

Veronica spicata L.
Scrophulariaceae
Pseudolysimachium spicatum (L.) Opiz, *Pseudolysimachium spicatum* (L.) Opiz ssp. *spicatum, Pseudolysimachium spicatum* ssp. *hybrida* (L.) Holub., *Veronica hybrida* L.
◆ spiked speedwell, tähkätädyke
◆ Weed, Naturalised
◆ 23, 88, 101, 272
◆ a/pH, cultivated, herbal.

Veronica spuria auct. non L.
Scrophulariaceae
= *Pseudolysimachion spurium* (L.) Rauschert (NoR)
◆ bastard speedwell, speedwell
◆ Naturalised, Garden Escape
◆ 101
◆ pH, cultivated, herbal.

Veronica sublobata M.Fisch.
Scrophulariaceae
◆ Weed
◆ 88, 94

Veronica tournefortii C.C.Gmel.
Scrophulariaceae
= *Veronica persica* Poir.
◆ Weed
◆ 87, 88

Veronica triloba (Opiz) Opiz
Scrophulariaceae
Veronica hederaefolia L. ssp. *triloba* (Opiz) Hayek
◆ Weed
◆ 88, 94

Veronica triphyllos L.
Scrophulariaceae
◆ fingered speedwell, tahmatädyke
◆ Weed, Naturalised, Introduced, Casual Alien

◆ 23, 34, 42, 44, 70, 80, 87, 88, 94, 101, 243, 253, 272, 280, 287
◆ aH, herbal.

Veronica undulata Wall. ex Roxb.
Scrophulariaceae
◆ undulate speedwell, water speedwell
◆ Weed, Naturalised
◆ 87, 88, 101, 235, 274, 275, 286
◆ pH, promoted.

Veronica verna L.
Scrophulariaceae
◆ spring speedwell, kevättädyke, vernal speedwell
◆ Weed, Naturalised
◆ 70, 80, 88, 94, 101, 253, 272, 280
◆ herbal.

Veronica virginica L.
Scrophulariaceae
= *Veronicastrum virginicum* (L.) Farw.
◆ Weed
◆ 23, 39, 88
◆ cultivated, herbal, toxic.

Veronicastrum sibiricum (L.) Pennell
Scrophulariaceae
◆ Siberian veronicastrum
◆ Weed
◆ 297
◆ cultivated, herbal.

Veronicastrum virginicum (L.) Farw.
Scrophulariaceae
Leptandra virginica Nutt., *Veronica virginica* L. (see)
◆ Culver's physic, Culver's root, Beaumont's root, pink Culver's root
◆ Weed
◆ 8, 39, 161
◆ pH, cultivated, herbal, toxic.

Verticordia monadelpha Turcz.
Myrtaceae
◆ woolly feather flower
◆ Weed, Naturalised
◆ 7, 86
◆ cultivated. Origin: Australia.

Vesicularia dubyana Fleisch.
Hypnaceae
◆ Java moss
◆ Weed, Quarantine Weed
◆ 76, 88
◆ algae, aqua, cultivated.

Vestia foetida (Ruiz & Pav.) Hoffmanns.
Solanaceae
◆ huevil
◆ Casual Alien
◆ 280
◆ cultivated.

Vetiveria zizanioides (L.) Nash ex Small
Poaceae
Andropogon squarrosus Hook.f. *non* L.f., *Phalaris zizanioides* L.
◆ vertiver grass, khuskhus vetiver, baúl del pobre, khus khus, pacholi
◆ Weed, Naturalised, Introduced, Cultivation Escape
◆ 87, 88, 101, 117, 228, 261, 287
◆ G, arid, cultivated, herbal. Origin: Old World.

***Viburnum acerifolium* L.**
Adoxaceae/Caprifoliaceae/
Viburnaceae
♦ mapleleaf viburnum, dockmackie
♦ Weed
♦ 87, 88, 218
♦ cultivated, herbal.

***Viburnum alnifolium* Marsh.**
Adoxaceae/Caprifoliaceae/
Viburnaceae
♦ hobblebush viburnum
♦ Weed
♦ 87, 88, 218
♦ cultivated, herbal.

***Viburnum buddleifolium* C.Wright**
Adoxaceae/Caprifoliaceae/
Viburnaceae
♦ buddlejaleaf viburnum
♦ Naturalised
♦ 101
♦ cultivated.

***Viburnum × burkwoodii* Burkwood &
Skipw.**
Adoxaceae/Caprifoliaceae/
Viburnaceae
= *Viburnum carlesii* Hemsl. × *Viburnum
utile* Hemsl.
♦ burkwood viburnum
♦ Weed
♦ 80
♦ cultivated.

***Viburnum carlesii* (Weston) Rehder**
Adoxaceae/Caprifoliaceae/
Viburnaceae
♦ Korean spice viburnum
♦ Weed
♦ 80
♦ cultivated, herbal.

***Viburnum dentatum* L.**
Adoxaceae/Caprifoliaceae/
Viburnaceae
♦ arrowwood viburnum, arrowwood,
southern arrowwood
♦ Weed
♦ 87, 88, 218
♦ S, cultivated, herbal.

***Viburnum dilatatum* Thunb. ex Murray**
Adoxaceae/Caprifoliaceae/
Viburnaceae
♦ arrowwood, linden viburnum,
linden arrowwood, viburnum
♦ Weed, Naturalised
♦ 80, 101, 195
♦ S, cultivated, herbal.

***Viburnum lantana* L.**
Adoxaceae/Caprifoliaceae/
Viburnaceae
♦ wayfaring tree, twistwood, villaheisi
♦ Weed, Naturalised, Introduced,
Cultivation Escape
♦ 4, 39, 40, 42, 80, 88, 101, 195, 222
♦ S, cultivated, herbal, toxic.

***Viburnum lentago* L.**
Adoxaceae/Caprifoliaceae/
Viburnaceae
♦ sweet viburnum, nannyberry, sheep
berry, wild raisin
♦ Weed

♦ 87, 88, 218
♦ T, cultivated, herbal.

***Viburnum opulus* L.**
Adoxaceae/Caprifoliaceae/
Viburnaceae
Opulus vulgaris Borkh., *Viburnum
lobatum* Lam.
♦ guelder rose, cranberrybush,
viburnum, European highbush
cranberry, European cranberrybush,
highbush cranberry, crampbark,
snowball tree, cranberry tree, pimbina
♦ Weed, Naturalised, Native Weed,
Environmental Weed, Casual Alien
♦ 4, 8, 15, 39, 80, 88, 101, 151, 222, 280
♦ S, cultivated, herbal, toxic.

***Viburnum opulus* L. cv. 'Roseum'**
Adoxaceae/Caprifoliaceae/
Viburnaceae
♦ crampbark, snowball tree
♦ Casual Alien
♦ 15, 280
♦ S, cultivated, herbal, toxic.

***Viburnum opulus* L. var. *americanum*
Aiton**
Adoxaceae/Caprifoliaceae/
Viburnaceae
♦ European cranberrybush, American
cranberrybush, American cranberry
viburnum
♦ Weed
♦ 195

Viburnum opulus* L. var. *opulus
Adoxaceae/Caprifoliaceae/
Viburnaceae
♦ guelder rose, European
cranberrybush, cranberrybush
viburnum
♦ Weed, Naturalised, Garden Escape
♦ 101, 142, 195
♦ cultivated.

***Viburnum plicatum* Thunb.**
Adoxaceae/Caprifoliaceae/
Viburnaceae
♦ Japanese snowball
♦ Weed, Naturalised
♦ 101, 133, 195, 280
♦ S, cultivated, herbal.

***Viburnum prunifolium* L.**
Adoxaceae/Caprifoliaceae/
Viburnaceae
♦ blackhaw, stagberry, viburno
♦ Weed
♦ 39, 87, 88, 218
♦ T, cultivated, herbal, toxic.

***Viburnum rafinesquianum* Schult.**
Adoxaceae/Caprifoliaceae/
Viburnaceae
♦ Rafinesque viburnum, downy
arrowwood
♦ Weed
♦ 87, 88, 218
♦ cultivated, herbal.

***Viburnum × rhytidophylloides* Sur.**
Adoxaceae/Caprifoliaceae/
Viburnaceae
= *Viburnum lantana* L. × *Viburnum
rhytidophyllum* Hemsl.

♦ Naturalised
♦ 101

***Viburnum rhytidophyllum* Hemsl.**
Adoxaceae/Caprifoliaceae/
Viburnaceae
♦ leatherleaf arrowwood, leatherleaf
viburnum
♦ Naturalised
♦ 101
♦ cultivated, herbal.

***Viburnum rufidulum* Raf.**
Adoxaceae/Caprifoliaceae/
Viburnaceae
♦ rusty blackhaw, southern blackhaw
♦ Weed
♦ 87, 88, 218
♦ S, cultivated, herbal.

***Viburnum setigerum* Hance**
Adoxaceae/Caprifoliaceae/
Viburnaceae
♦ tea viburnum
♦ Naturalised
♦ 101
♦ S, cultivated.

***Viburnum sieboldii* Miq.**
Adoxaceae/Caprifoliaceae/
Viburnaceae
♦ Siebold's arrowwood, Siebold
viburnum
♦ Weed, Naturalised
♦ 80, 101, 133, 195
♦ S, cultivated, herbal.

***Viburnum suspensum* Lindl.**
Adoxaceae/Caprifoliaceae/
Viburnaceae
♦ viburnum
♦ Weed, Naturalised
♦ 98, 203
♦ S, cultivated.

***Viburnum tinus* L.**
Adoxaceae/Caprifoliaceae/
Viburnaceae
♦ laurustinus
♦ Weed, Naturalised, Garden Escape,
Environmental Weed
♦ 15, 39, 40, 72, 80, 86, 88, 98, 101, 198,
203, 215, 280, 290
♦ S, cultivated, herbal, toxic. Origin:
southern Europe.

***Vicia* L. spp.**
Fabaceae/Papilionaceae
♦ vetch, vervain
♦ Weed, Naturalised
♦ 23, 88, 154, 198, 236, 243, 272
♦ H, herbal, toxic.

***Vicia acutifolia* Elliott**
Fabaceae/Papilionaceae
♦ fourleaf vetch, sand vetch
♦ Weed
♦ 161, 249

***Vicia amoena* Fisch.**
Fabaceae/Papilionaceae
♦ tsurujibakama
♦ Weed
♦ 275, 297
♦ pH, promoted.

Vicia angustifolia (L.) Reichard
Fabaceae/Papilionaceae
= *Vicia sativa* L. ssp. *nigra* (L.) Ehrh.
♦ narrow leaved vetch, narrowleaf vetch
♦ Weed
♦ 39, 44, 88, 210, 218, 243, 263, 275, 286, 292, 295, 297
♦ aH, cultivated, herbal, toxic.

Vicia angustifolia L. fo. subtriflora Nageli & Thell.
Fabaceae/Papilionaceae
♦ Naturalised
♦ 287

Vicia angustifolia L. var. minor Ohwi
Fabaceae/Papilionaceae
♦ Weed, Naturalised
♦ 286, 287

Vicia articulata Hornem.
Fabaceae/Papilionaceae
Vicia monanthos (L.) Desf. *nom. illeg.* (see)
♦ oneflower tare, oneflower vetch, yksikukkavirvilä
♦ Weed, Naturalised, Casual Alien
♦ 42, 70, 88, 94, 101, 241, 272, 300
♦ aH, cultivated.

Vicia atropurpurea Desf.
Fabaceae/Papilionaceae
= *Vicia benghalensis* L.
♦ Weed
♦ 87, 88

Vicia benghalensis L.
Fabaceae/Papilionaceae
Vicia atropurpurea Desf. (see)
♦ narrow leaved purple vetch, purple vetch, wild purple vetch, reddish tufted vetch
♦ Weed, Naturalised, Introduced
♦ 7, 34, 86, 87, 88, 94, 98, 101, 121, 158, 203, 228, 241, 250, 253, 300
♦ a/bC, arid, cultivated, herbal. Origin: Eurasia.

Vicia bithynica L.
Fabaceae/Papilionaceae
Lathyrus bithynicus L., *Lathyrus tumidus* Willd., *Lathyrus turgidus* L., *Vicia serrata* Jacq.
♦ Bithynian vetch
♦ Weed
♦ 87, 88, 94, 253, 272
♦ herbal.

Vicia bungei Ohwi
Fabaceae/Papilionaceae
♦ Weed
♦ 275, 297
♦ pH.

Vicia calcarata Desf.
Fabaceae/Papilionaceae
= *Vicia monantha* Retz.
♦ Weed
♦ 87, 88

Vicia cassubica L.
Fabaceae/Papilionaceae
Ervilia cassubica Schur, *Ervum cassubicum* Peterm., *Vicia militans* Crantz
♦ pommerinvirna
♦ Weed
♦ 272
♦ herbal.

Vicia cinerea M.Bieb.
Fabaceae/Papilionaceae
♦ Weed
♦ 221

Vicia cracca (Velen.) P.Davis
Fabaceae/Papilionaceae
Vicia graccha Lepech., *Vicia heteropus* Freyn, *Vicia scheuchzeri* Brüg., *Vicia semicincta* Greene, *Vicia versicolor* Salisb.
♦ vetch, tufted vetch, bird vetch, cow vetch
♦ Weed, Naturalised, Introduced
♦ 34, 44, 52, 70, 80, 86, 87, 88, 94, 98, 101, 146, 161, 195, 203, 211, 218, 243, 272, 275, 280, 286, 292, 297
♦ pC, cultivated, herbal. Origin: Europe, Middle East, Caucasia.

Vicia cracca L. ssp. cracca
Fabaceae/Papilionaceae
♦ bird vetch
♦ Naturalised
♦ 101, 280

Vicia cracca L. ssp. tenuifolia (Roth) Gaudin
Fabaceae/Papilionaceae
Vicia tenuifolia Roth (see)
♦ cow vetch
♦ Naturalised
♦ 101

Vicia dasycarpa Ten.
Fabaceae/Papilionaceae
Cracca dasycarpa Alef., *Vicia plenigera* Form.
♦ hairyfruit vetch
♦ Weed
♦ 80
♦ cultivated, herbal.

Vicia dasycarpa Ten. var. glaberscens (Koch) Beck
Fabaceae/Papilionaceae
♦ Weed, Naturalised
♦ 286, 287

Vicia difformis Pourr.
Fabaceae/Papilionaceae
♦ intermediate periwinkle
♦ Weed
♦ 70

Vicia disperma DC.
Fabaceae/Papilionaceae
♦ European vetch, French tiny vetch, veccia a due semi
♦ Weed, Naturalised, Introduced
♦ 34, 86, 88, 94, 98, 101, 198, 203, 280
♦ aH, herbal. Origin: southern and north-west Europe.

Vicia dumetorum L.
Fabaceae/Papilionaceae
Abacosa dumetorum Alef., *Cracca dumetorum* (L.) Opiz, *Vicia desertorum* Link, *Vicia dumicola* Dulac, *Vicia patula* Moench, *Vicia variegata* Gilib.
♦ pensaikkovirna

♦ Weed, Casual Alien
♦ 42, 272
♦ cultivated, herbal.

Vicia ervilia (L.) Willd.
Fabaceae/Papilionaceae
Ervilia sativa Link, *Ervum ervilia* L., *Ervum plicatum* Moench, *Rhynchium plicatum* (Moench) Dulac
♦ blister vetch, bitter vetch, linssivirvilä
♦ Weed, Introduced, Casual Alien
♦ 39, 42, 87, 88, 94, 228, 272
♦ aH, arid, cultivated, herbal, toxic.

Vicia faba Pers.
Fabaceae/Papilionaceae
Faba bona Medik., *Faba minor* Roxb., *Faba vulgaris* Moench, *Vicia esculenta* Salisb.
♦ horse bean, faba bean, broad bean, English bean, Windsor bean, härkäpapu
♦ Weed, Naturalised, Introduced, Cultivation Escape
♦ 15, 34, 39, 42, 80, 87, 88, 101, 161, 247, 280
♦ aC, cultivated, herbal, toxic.

Vicia gracilis Loisel. nom. illeg.
Fabaceae/Papilionaceae
= *Vicia parviflora* Cav.
♦ Weed
♦ 87, 88

Vicia graminea Sm.
Fabaceae/Papilionaceae
♦ Weed
♦ 87, 88
♦ arid.

Vicia grandiflora Scop.
Fabaceae/Papilionaceae
♦ large flowered vetch
♦ Weed, Naturalised
♦ 80, 88, 94, 101, 272, 287
♦ cultivated, herbal.

Vicia hirsuta (L.) Gray
Fabaceae/Papilionaceae
Ervum hirsutum L., *Ervilia vulgaris* Godr., *Vicia coreana* Lév., *Vicia mitchelli* Raf., *Vicia taquetii* Lév.
♦ tiny vetch, vetch, hairy vetch, peltovirvilä, hairy tare, tiny purple vetch
♦ Weed, Naturalised, Introduced, Environmental Weed
♦ 7, 15, 24, 34, 39, 44, 70, 72, 80, 86, 87, 88, 94, 98, 101, 121, 134, 165, 176, 198, 203, 204, 235, 241, 243, 253, 263, 269, 272, 280, 286, 300
♦ aH, cultivated, herbal, toxic. Origin: Eurasia.

Vicia hybrida L.
Fabaceae/Papilionaceae
Hypechusa hybrida Ale., *Vicia linnaei* Rouy
♦ hairy yellow vetch
♦ Weed, Naturalised, Casual Alien
♦ 40, 87, 88, 101, 253, 272
♦ herbal.

Vicia lathyroides L.
Fabaceae/Papilionaceae
Vicia minima (Riv.) Lam., *Vicia praecox*
Jacq., *Wiggersia minima* Alef.
♦ spring vetch, pea vetch,
nätkelmävirna
♦ Weed, Naturalised
♦ 86, 87, 88, 98, 101, 203, 272, 280
♦ aC, herbal.

Vicia laxiflora Brot. nom. illeg.
Fabaceae/Papilionaceae
= *Vicia parviflora* Cav.
♦ slender vetch
♦ Naturalised
♦ 101

Vicia linearifolia Hook & Arn.
Fabaceae/Papilionaceae
♦ vicia
♦ Weed
♦ 295

Vicia lutea L.
Fabaceae/Papilionaceae
Vicia ciliata Schur, *Vicia flavida* Schur,
Vicia hirsutissima Ten., *Vicia laevigata*
Sm., *Vicia pauciflora* Form., *Vicia
tridentata* Gater.
♦ yellow vetch, smooth yellow vetch,
keltavirna
♦ Weed, Naturalised, Introduced,
Casual Alien
♦ 34, 42, 70, 87, 88, 94, 101, 221, 250,
253, 272, 280, 287
♦ aC, cultivated, herbal.

Vicia melanops Sibth. & Sm.
Fabaceae/Papilionaceae
♦ mustatäplävirna
♦ Weed, Casual Alien
♦ 42, 272

Vicia monantha Retz.
Fabaceae/Papilionaceae
Vicia calcarata Desf. (see)
♦ Syrian vetch, barn vetch
♦ Weed, Naturalised
♦ 7, 55, 87, 88, 98, 101, 185, 198, 203,
221, 243
♦ pH, promoted.

Vicia monantha Retz. ssp. monantha
Fabaceae/Papilionaceae
♦ spurred vetch, oneflower vetch
♦ Weed, Naturalised, Environmental
Weed
♦ 86, 93, 198, 205, 228
♦ Origin: Europe, Middle East.

**Vicia monantha Retz. ssp. triflora (Ten.)
B.L.Burtt & P.Lewis**
Fabaceae/Papilionaceae
♦ threeflower vetch, spurred vetch
♦ Weed, Naturalised
♦ 86, 93, 198, 205
♦ Origin: Europe, Middle East.

Vicia monanthos (L.) Desf. nom. illeg.
Fabaceae/Papilionaceae
= *Vicia articulata* Hornem.
♦ Weed
♦ 87, 88

Vicia nana Vogel
Fabaceae/Papilionaceae

♦ Weed
♦ 295

Vicia narbonensis L.
Fabaceae/Papilionaceae
Bona narbonensis Medik., *Faba
narbonensis* (L.) Schur, *Vicia platycarpos*
Roth
♦ purple broad vetch, French vetch,
narbonne vetch, broad leaved vetch
♦ Weed, Quarantine Weed,
Naturalised, Casual Alien
♦ 40, 42, 70, 76, 87, 88, 94, 101, 115, 203,
221, 243, 253, 272, 280
♦ aH, arid, cultivated. Origin: Eurasia.

Vicia onobrychioides L.
Fabaceae/Papilionaceae
♦ Weed
♦ 87, 88, 272
♦ arid.

Vicia orobus DC.
Fabaceae/Papilionaceae
Ervilia orobus Schur, *Ervum orobus*
Kitt., *Orobus aristatus* Lapeyr., *Orobus
silvaticus* L., *Vicilla orobus* Schur
♦ nummivirna, wood bitter vetch,
upright vetch
♦ Casual Alien
♦ 42

Vicia pannonica Crantz
Fabaceae/Papilionaceae
Hypechusa pannonica (DC.) Alef.,
Vicioides hirsuta Moench
♦ unkarinvirna, Hungarian vetch
♦ Weed, Naturalised, Introduced,
Casual Alien
♦ 34, 40, 42, 88, 94, 101, 253
♦ aC.

Vicia pannonica Cran. ssp. pannonica
Fabaceae/Papilionaceae
Vicia lineata Bieb.
♦ Hungarian vetch
♦ Weed
♦ 272

**Vicia pannonica Cran. ssp. striata (Bieb.)
Nyman**
Fabaceae/Papilionaceae
Vicia purpurescens Desf., *Vicia striata*
Bieb.
♦ Weed
♦ 272

Vicia parviflora Cav.
Fabaceae/Papilionaceae
Vicia gracilis Loisel. *nom. illeg.* (see),
Vicia laxiflora Brot. *nom. illeg.* (see),
Vicia tenuissima auct. (see)
♦ slender tare
♦ Weed
♦ 253

Vicia peregrina L.
Fabaceae/Papilionaceae
♦ yellow vetch, wandering vetch,
muukalaisvirna
♦ Weed, Naturalised, Casual Alien
♦ 42, 44, 87, 88, 94, 101, 243, 253, 272
♦ cultivated, herbal.

Vicia pisiformis L.
Fabaceae/Papilionaceae
Ervilia pisiformis Schur, *Ervum pisiforme*

Peterm., *Vicia ochroleuca* Gilib.
♦ hernevirna
♦ Casual Alien
♦ 42
♦ pH, cultivated.

Vicia pubescens (DC.) Link
Fabaceae/Papilionaceae
♦ Weed
♦ 253

Vicia sativa L.
Fabaceae/Papilionaceae
Vicia alba Medik., *Vicia bacla* Moench,
Vicia canadensis Zucc., *Vicia leucosperma*
Moench, *Vicia nemoralis* Steud., *Vicia
torulosa* Jord., *Vicia vulgaris* Uspen.
♦ spring vetch, narrowleaf vetch,
common vetch, rehuvirna, garden
vetch, tare
♦ Weed, Naturalised, Casual Alien
♦ 7, 9, 15, 39, 42, 44, 80, 87, 88, 94, 98,
101, 121, 136, 158, 161, 165, 176, 185,
198, 199, 203, 211, 212, 218, 221, 235,
237, 241, 243, 248, 253, 275, 280, 287,
292, 295, 297, 300
♦ a/bC, cultivated, herbal, toxic.
Origin: Mediterranean, west Asia.

**Vicia sativa L. ssp. cordata (Wulfen ex
Hoppe) Asch. & Graebn.**
Fabaceae/Papilionaceae
♦ common vetch
♦ Naturalised, Environmental Weed
♦ 86, 198
♦ Origin: Europe, Middle East.

Vicia sativa L. ssp. nigra (L.) Ehrh.
Fabaceae/Papilionaceae
Vicia abyssinica Alef., *Vicia angustifolia*
L. (see), *Vicia bobartii* Koch, *Vicia
cuneata* Guss., *Vicia debilis* Perez-
Lara, *Vicia heterophylla* C.Presl, *Vicia
lanciformis* Lange, *Vicia pilosa* M.Bieb.,
Vicia sativa L. var. *abyssinica* (Alef.)
Bak., *Vicia sativa* L. ssp. *angustifolia* (L.)
Gaudin, *Vicia sativa* L. var. *angustifolia*
(L.) Ser. (see), *Vicia sativa* L. ssp.
segetalis (Thuill.) Celak. (see), *Vicia
segetalis* Thuill.
♦ garden vetch, common vetch,
narrow leaved vetch
♦ Weed, Naturalised, Introduced,
Environmental Weed
♦ 34, 70, 72, 86, 101, 134, 176, 198, 211,
249, 250, 272
♦ aH, arid, cultivated. Origin: Europe,
Middle East.

Vicia sativa L. ssp. sativa
Fabaceae/Papilionaceae
Vicia communis Rouy, *Vicia sativa* L. ssp.
notata Asch. & Graebn.
♦ common vetch, pubescent common
vetch, garden vetch, spring vetch
♦ Weed, Naturalised, Environmental
Weed
♦ 9, 34, 40, 70, 72, 86, 101, 198, 269, 272
♦ aH, arid, cultivated. Origin: Europe.

**Vicia sativa L. ssp. segetalis (Thuill.)
Celak.**
Fabaceae/Papilionaceae
= *Vicia sativa* ssp. *nigra* (L.) Ehrh.

♦ Naturalised
♦ 40

Vicia sativa L. var. angustifolia (L.) Ser.
Fabaceae/Papilionaceae
= *Vicia sativa* ssp. *nigra* (L.) Ehrh.
♦ Weed
♦ 243

Vicia sepium L.
Fabaceae/Papilionaceae
Atossa sepium Alef., *Faba sepium* Bernh.,
Vicia rotundifolia Gilib., *Vicioides sepium*
Moench
♦ bush vetch, aitovirna
♦ Weed, Naturalised
♦ 44, 70, 80, 87, 88, 94, 101, 243, 272, 287
♦ pC, cultivated, herbal.

Vicia sepium L. var. montana W.D.J.Koch
Fabaceae/Papilionaceae
♦ bush vetch
♦ Naturalised
♦ 101

Vicia sepium L. var. sepium
Fabaceae/Papilionaceae
♦ bush vetch
♦ Naturalised
♦ 101

Vicia sibthorpii Boiss.
Fabaceae/Papilionaceae
♦ Weed
♦ 87, 88

Vicia tenuifolia Roth
Fabaceae/Papilionaceae
= *Vicia cracca* L. ssp. *tenuifolia* (Roth)
Gaudin
♦ fine leaved vetch, kaitahiirenvirna
♦ Weed, Casual Alien
♦ 42, 70, 87, 88, 272
♦ pH, promoted.

Vicia tenuissima auct.
Fabaceae/Papilionaceae
= *Vicia parviflora* Cav.
♦ hentovirvilä, slender tare
♦ Weed, Naturalised, Casual Alien
♦ 42, 70, 88, 94, 241, 300
♦ herbal.

Vicia tetrasperma (L.) Schreb.
Fabaceae/Papilionaceae
Ervilia tetrasperma (L.) Opiz, *Ervum
tetraspermum* L., *Viccia gemella* Crantz,
Vicia pusilla Mühlbg. ex Willd
♦ smooth tare, fourseed vetch, four
seeded vetch, slender vetch, lentil tare,
lentil vetch
♦ Weed, Naturalised, Introduced,
Environmental Weed
♦ 15, 34, 44, 70, 72, 80, 86, 87, 88, 94,
98, 101, 134, 161, 176, 198, 203, 211, 218,
235, 241, 243, 253, 269, 272, 280, 286,
292, 300
♦ aH, cultivated, herbal. Origin:
Europe, Middle East, North Africa.

Vicia unijuga A.Braun
Fabaceae/Papilionaceae
♦ siperianvirna
♦ Casual Alien
♦ 42
♦ pH, cultivated, herbal.

Vicia villosa Roth s.str.
Fabaceae/Papilionaceae
Ervum villosum (Roth) Traut., *Vicia
godroni* Rouy, *Vicia reuteriana* Boiss.
♦ hairy vetch, fodder vetch, winter
vetch, woollypod vetch, large Russian
vetch, ruisvirna
♦ Weed, Naturalised
♦ 15, 39, 39, 40, 44, 70, 80, 87, 88, 94, 98,
101, 121, 136, 146, 161, 176, 198, 199,
203, 211, 212, 218, 241, 243, 253, 280,
286, 287, 292, 295, 300
♦ aC, cultivated, herbal, toxic. Origin:
Eurasia.

**Vicia villosa Roth ssp. eriocarpa
(Hausskn.) P.W.Ball**
Fabaceae/Papilionaceae
♦ hairy vetch
♦ Naturalised
♦ 86, 176, 198
♦ Origin: Europe, Middle East.

**Vicia villosa Roth ssp. pseudocracca
(Bertol.) Ball**
Fabaceae/Papilionaceae
♦ winter vetch
♦ Naturalised
♦ 86, 101

Vicia villosa Roth ssp. varia (Host) Corb.
Fabaceae/Papilionaceae
Vicia dasycarpa auct., *Vicia glabrescens*
(W.Koch) Heimerl, *Vicia polyphylla*
Desf., *Vicia varia* Host, *Vicia villosa* Roth
var. *glabrescens* W.Koch, *Vicia villosa*
Roth ssp. *dasycarpa* (Ten.) Cav.
♦ winter vetch, woollypod vetch, hairy
vetch, winter vetch
♦ Weed, Naturalised, Introduced,
Casual Alien
♦ 34, 40, 86, 101, 272, 280
♦ aH.

Vicia villosa Roth ssp. villosa
Fabaceae/Papilionaceae
♦ hairy vetch, winter vetch
♦ Weed, Naturalised, Introduced
♦ 34, 86, 101, 198, 228, 272
♦ aH. Origin: Europe, Middle East.

Vicoa auriculata auct. non Cass.
Asteraceae
= *Pulicaria arabica* (L.) Cass.
♦ Weed
♦ 87, 88

Vicoa indica (L.) DC.
Asteraceae
♦ Weed
♦ 66, 87, 88
♦ herbal.

Vigna aconitifolia (Jacq.) Maréchal
Fabaceae/Papilionaceae
Dolichos dissectus Lam., *Phaseolus
aconitifolius* Jacq. (see), *Phaseolus
palmatus* Forssk.
♦ moth bean
♦ Introduced
♦ 228
♦ arid, cultivated, herbal.

**Vigna adenantha (G.Mey.) Maréchal,
Mascherpa & Stainier**

Fabaceae/Papilionaceae
Phaseolus adenanthus G.F.W.Mey. (see)
♦ moth bean, adzuki bean, wild pea
♦ Weed, Naturalised
♦ 86, 179, 206, 243
♦ arid, herbal. Origin: tropical
America.

Vigna angularis (Willd.) Ohwi & Ohashi
Fabaceae/Papilionaceae
♦ adzuki bean
♦ Naturalised
♦ 101
♦ cultivated, herbal.

**Vigna angularis (Willd.) Ohwi &
H.Ohashi var. nipponensis (Ohwi) Ohwi
& H.Ohashi**
Fabaceae/Papilionaceae
♦ Weed
♦ 286

Vigna angustifoliolata Verdc.
Fabaceae/Papilionaceae
Vigna stenophylla (Harv.) Burtt Davy
♦ wild sweetpea
♦ Native Weed
♦ 121
♦ pC. Origin: southern Africa.

Vigna hosei (Craig) Bak.
Fabaceae/Papilionaceae
♦ Sarawak bean
♦ Weed, Naturalised, Introduced
♦ 86, 101, 179, 230, 261, 287
♦ C, cultivated. Origin: Borneo, Java,
possibly Taiwan.

Vigna juruana (Harms) Verdc.
Fabaceae/Papilionaceae
♦ tropical cowpea
♦ Weed, Naturalised
♦ 101, 261

Vigna lanceolata Benth.
Fabaceae/Papilionaceae
♦ maloga bean
♦ Weed
♦ 55
♦ arid, cultivated.

Vigna luteola (Jacq.) Benth.
Fabaceae/Papilionaceae
Phaseolus luteolus (Jacq.) Gagnep.,
Vigna brachystachys Benth., *Vigna
bukobensis* Harms, *Vigna bukombensis*
Harms, *Vigna glabra* Savi, *Vigna nigerica*
A.Chev., *Vigna nilotica* (Del.) Hook.f.,
Vigna repens (L.) Kuntze (see)
♦ hairypod cowpea, kuanga,
dalrymple vigna
♦ Weed, Native Weed
♦ 87, 88, 121, 237, 295
♦ pC, arid, cultivated, herbal. Origin:
southern Africa.

Vigna marina (Burm.f.) Merr.
Fabaceae/Papilionaceae
♦ notched cowpea, dune bean,
tehsiluhte
♦ Weed
♦ 6, 87, 88
♦ C, arid, cultivated, herbal. Origin:
Africa, Asia, Australia.

Vigna mungo **(L.) Hepper**
Fabaceae/Papilionaceae
♦ black gram, urdi black bean, urd
♦ Naturalised
♦ 86, 101, 287
♦ cultivated, herbal.

Vigna oblongifolia **A.Rich. var.**
oblongifolia
Fabaceae/Papilionaceae
♦ Native Weed
♦ 121
♦ C. Origin: Africa.

Vigna parkeri **Bak.**
Fabaceae/Papilionaceae
Vigna gracilis auct. *non* (Guill. & Perr.)
Hook., *Dolichos maranguënsis* Taub.,
Vigna maranguënsis (Taub.) Harms.
♦ creeping vigna
♦ Quarantine Weed, Introduced
♦ 76, 81
♦ pH, cultivated. Origin: Africa.

Vigna peduncularis **(Kunth) Fawc. &**
Rendle var. *pusilla* **(Hassl.) Maréchal,**
Mascherpa & Stainier
Fabaceae/Papilionaceae
♦ Introduced
♦ 38

Vigna radiata **(L.) R.Wilczek**
Fabaceae/Papilionaceae
Phaseolus aureus Roxb. (see), *Phaseolus*
mungo auct. (*non* L.) Hepper, *Phaseolus*
radiatus L.
♦ mung bean
♦ Weed, Naturalised, Introduced,
Garden Escape
♦ 7, 98, 203, 228
♦ aC, arid, cultivated, herbal.

Vigna radiata **(L.) R.Wilczek var.** *radiata*
Fabaceae/Papilionaceae
♦ mung bean
♦ Naturalised, Environmental Weed
♦ 86
♦ aC, cultivated. Origin: Asia.

Vigna radiata **(L.) R.Wilczek var.**
sublobata **(Roxb.) Verdc.**
Fabaceae/Papilionaceae
Phaseolus sublobatus Roxb., *Phaseolus*
trinervius Wight & Arn., *Vigna*
opisotricho A.Rich., *Vigna sublobata*
(Roxb.) Babu & Sharma
♦ mung bean
♦ Introduced
♦ 228
♦ arid. Origin: Madagascar.

Vigna repens **(L.) Kuntze**
Fabaceae/Papilionaceae
= *Vigna luteola* (Jacq.) Benth.
♦ Weed
♦ 87, 88
♦ arid.

Vigna speciosa **(Kunth) Verdc.**
Fabaceae/Papilionaceae
♦ wondering cowpea
♦ Weed, Naturalised
♦ 101, 179

Vigna subterranea **(L.) Verdc.**
Fabaceae/Papilionaceae

Glycine subterranea L., *Voandzeia*
subterranea (L.) Thouars ex DC. (see)
♦ Congo goober, Madagascar peanut,
baffin pea, bambara bean, bambara
groundnut, earth pea, epi roui, juga
bean, juijiya, njuga bean, njuga mawe,
nlubu, nyimo, nzama, okpa otuanya,
voandzou
♦ Introduced
♦ 228
♦ arid.

Vigna trichocarpa **(C.Wright) A.Delgado**
Fabaceae/Papilionaceae
♦ Introduced
♦ 38

Vigna trilobata **(L.) Verdc.**
Fabaceae/Papilionaceae
Dolichos trilobatus L., *Phaseolus*
trilobatus (L.) Schreb., *Phaseolus trilobus*
auct. (see)
♦ Naturalised, Introduced,
Environmental Weed
♦ 7, 86, 228
♦ arid, cultivated, herbal. Origin: Asia.

Vigna umbellata **(Thunb.) Ohwi &**
Ohashi
Fabaceae/Papilionaceae
Azukia umbellata (Thunb.) Ohwi,
Dolichos umbellatus Thunb., *Phaseolus*
calcaratus Roxb., *Vigna calcarata* (Roxb.)
Kurz
♦ Japanese rice bean, climbing
mountain bean, mambi bean, oriental
bean, red bean, rice bean
♦ Naturalised, Introduced
♦ 86, 228
♦ arid, cultivated, herbal. Origin: Asia.

Vigna unguiculata **(L.) Walp.**
Fabaceae/Papilionaceae
Vigna sesquipedalis (L.) Fruw.
♦ cowpea, green snake bean,
Jerusalem pea, yard long bean,
snake bean, catjang, blackeye bean,
asparagus bean, frijol, lentejas
♦ Weed, Naturalised, Introduced,
Casual Alien
♦ 88, 98, 101, 161, 179, 203, 228, 255,
261
♦ aH, arid, cultivated, herbal. Origin:
tropical Africa.

Vigna unguiculata **(L.) Walp. ssp.**
cylindrica **(L.) Verdc.**
Fabaceae/Papilionaceae
♦ catjang, Jerusalem pea
♦ Naturalised
♦ 86
♦ cultivated. Origin: tropical Africa.

Vigna unguiculata **(L.) Walp. ssp.**
dekindtiana **(Harms) Verdc.**
Fabaceae/Papilionaceae
♦ black eyed pea
♦ Naturalised
♦ 86
♦ pH. Origin: tropical Africa.

Vigna unguiculata **(L.) Walp. ssp.**
unguiculata
Fabaceae/Papilionaceae
Dolichos melanophthalmus DC.,

Dolichos unguiculatus L., *Phaseolus*
sphaerospermus L., *Phaseolus*
unguiculatus (L.) Piper, *Vigna sinensis*
(L.) Savi ex Hassk.
♦ southern pea
♦ Naturalised
♦ 86
♦ cultivated. Origin: tropical Africa.

Vigna vexillata **(L.) A.Rich.**
Fabaceae/Papilionaceae
Vigia phaseoloides Bak., *Vigna*
angustifolia (Schumach. & Thonn.)
Hook.f., *Vigna capensis* (Thunb.) Burtt
Davy, *Vigna dolichoneura* Harms, *Vigna*
senegalensis A.Chev.
♦ masiva, mgcenga, murudji, musivha,
nyemba, obhombo, ulubombo,
umcwasibe, wild sweetpea, wilde
ertjie, wild cowpea, aka sasage, zombi
pea
♦ Weed, Naturalised, Native Weed,
Introduced
♦ 14, 98, 121, 157, 203, 228
♦ pC, arid, cultivated, herbal. Origin:
Africa, Asia.

Viguiera anchusaefolia **(DC.) Bak.**
Asteraceae
♦ catay
♦ Weed
♦ 87, 88, 237, 295

Viguiera annua **(Jones) Blake**
Asteraceae
= *Heliomeris longifolia* (Robins. &
Greenm.) Cockerell var. *annua*
(M.E.Jones) Yates (NoR)
♦ annual goldeneye
♦ Weed
♦ 39, 87, 88, 161, 218
♦ arid, herbal, toxic.

Viguiera dentata **(Cav.) Spreng.**
Asteraceae
♦ toothleaf goldeneye
♦ Weed, Quarantine Weed
♦ 14, 76, 87, 88, 199, 203, 220
♦ herbal.

Viguiera major **L.**
Asteraceae
♦ Weed
♦ 87, 88

Viguiera minor **L.**
Asteraceae
♦ Weed
♦ 87, 88

Viguiera stenoloba **Blake**
Asteraceae
♦ shrubby goldeneye, resin bush,
skeletonleaf goldeneye
♦ Weed
♦ 87, 88, 218
♦ herbal.

Villebrunea frutescens **(Thunb.) Bl.**
Urticaceae
= *Oreocnide frutescens* (Thunb.) Miq.
ssp. *frutescens* (NoR)
♦ Weed
♦ 87, 88

Vinca difformis Pourr.
Apocynaceae
♦ intermediate periwinkle
♦ Naturalised
♦ 40
♦ cultivated, herbal.

Vinca herbacea Waldst. & Kit.
Apocynaceae
♦ herbaceous periwinkle
♦ Weed, Naturalised
♦ 87, 88, 101
♦ cultivated, herbal.

Vinca major L.
Apocynaceae
Pervinca major Scop.
♦ bigleaf periwinkle, periwinkle, greater periwinkle, vinca, band plant, blue buttons, blue perwinkle
♦ Weed, Noxious Weed, Naturalised, Garden Escape, Environmental Weed, Cultivation Escape
♦ 7, 15, 20, 34, 35, 39, 40, 45, 72, 73, 78, 80, 86, 88, 97, 98, 101, 102, 116, 121, 134, 137, 151, 152, 165, 176, 198, 203, 218, 225, 231, 241, 246, 269, 272, 280, 286, 287, 289, 295, 296, 300
♦ pH, cultivated, herbal, toxic. Origin: Eurasia.

Vinca major L. cv. 'Variegata'
Apocynaceae
♦ Naturalised
♦ 280
♦ pH, cultivated, herbal, toxic.

Vinca minor L.
Apocynaceae
Pervinca minor Scop., *Vinca acutiflora* Bertol. ex W.Koch., *Vinca ellipticifolia* Stokes, *Vinca humilis* Salisb., *Vinca intermedia* Tausch.
♦ periwinkle, pikkutalvio, greater periwinkle, lesser periwinkle, common periwinkle, vinca
♦ Weed, Noxious Weed, Naturalised, Introduced, Garden Escape, Environmental Weed, Cultivation Escape
♦ 4, 39, 40, 42, 45, 80, 88, 101, 102, 104, 116, 121, 133, 137, 142, 146, 151, 154, 161, 195, 199, 218, 222, 251, 280
♦ pH, cultivated, herbal, toxic. Origin: Eurasia.

Vincetoxicum hirundinaria Medik.
Asclepiadaceae/Apocynaceae
= *Cynanchum vincetoxicum* (L.) Pers.
♦ common vincetoxicum, swallowwort, käärmeenpistoyrtti
♦ Weed
♦ 70, 215, 272
♦ pH, cultivated, toxic.

Vincetoxicum nigrum (L.) Moench
Asclepiadaceae/Apocynaceae
= *Cynanchum louiseae* Kartesz & Gandhi
♦ black swallowwort, dog strangling vine, Louis's swallowwort, climbing milkweed
♦ Weed, Quarantine Weed
♦ 39, 76, 88, 133, 195, 220, 224, 243
♦ pC, cultivated, herbal, toxic.

Vincetoxicum rossicum (Kleo.) Barb.
Asclepiadaceae/Apocynaceae
= *Cynanchum rossicum* (Kleo.) Barb.
♦ swallowwort, dog strangling vine
♦ Weed
♦ 133, 195, 224
♦ cultivated.

Viola L. spp.
Violaceae
♦ violet, pansy
♦ Weed
♦ 23, 39, 88, 243, 249, 272
♦ H, cultivated, herbal, toxic.

Viola ambigua Waldst. & Kit.
Violaceae
♦ Weed
♦ 272

Viola arvensis Murray
Violaceae
Viola agrestis Jord., *Viola arvatica* Jord., *Viola deseglisei* Jord., *Viola lloydii* Jord., *Viola pentela* Jord., *Viola ruralis* Jord., *Viola timbali* Jord., *Viola tricolor* L. ssp. *arvensis* (Murray) Gaudin
♦ field violet, wild pansy, field pansy, hearts ease, European field pansy, pelto orvokki, viola dei campi, corn speedwell, violeta silvestre
♦ Weed, Naturalised, Cultivation Escape
♦ 15, 34, 44, 70, 80, 86, 87, 88, 94, 98, 101, 118, 161, 165, 176, 198, 203, 207, 211, 212, 218, 236, 243, 249, 252, 263, 271, 272, 280, 291, 295
♦ aH, cultivated, herbal. Origin: Europe, Mediterranean.

Viola arvensis Murray ssp. *arvensis*
Violaceae
♦ Naturalised
♦ 241, 300

Viola betonicifolia Sm. var. *albescens* F.Maek. & Hashim.
Violaceae
♦ Weed
♦ 286

Viola biflora L.
Violaceae
♦ yellow wood violet, lapinorvokki, Arctic yellow violet, twoflower violet
♦ Weed
♦ 272
♦ pH, cultivated, herbal.

Viola cornuta L.
Violaceae
♦ sarviorvokki, horned violet, horned pansy
♦ Naturalised, Cultivation Escape
♦ 40, 42
♦ pH, cultivated, herbal.

Viola cucullata Ait.
Violaceae
♦ blue marsh violet
♦ Naturalised
♦ 287
♦ herbal.

Viola dacica Borbás
Violaceae

♦ fialka dácka
♦ Weed
♦ 272

Viola grypoceras A.Gray
Violaceae
♦ tachitsubosumire
♦ Weed
♦ 88, 204, 286
♦ pH, promoted. Origin: east Asia.

Viola hederacea Labill.
Violaceae
♦ native violet, Australian violet
♦ Weed, Naturalised, Garden Escape
♦ 87, 88, 280
♦ cultivated, herbal. Origin: Australia.

Viola hirta L.
Violaceae
♦ karvaorvokki, hairy violet
♦ Cultivation Escape
♦ 42
♦ cultivated, herbal.

Viola inconspicus Blume ssp. *nagasakiensis* (W.Becker) Wang & Huang
Violaceae
♦ arrowleaf violet
♦ Weed
♦ 273

Viola japonica Langsd.
Violaceae
♦ kosumire
♦ Weed
♦ 87, 88, 275, 286, 297
♦ pH, cultivated, herbal. Origin: east Asia.

Viola kitaibeliana Roem. & Schult.
Violaceae
Viola minima Presl, *Viola valesiaca* Thomas
♦ Johnny jumpup violet, dwarf pansy
♦ Weed
♦ 44, 87, 88, 243, 272

Viola kitaibeliana Roem. & Schult. var. *rafinesquii* (Greene) Fern.
Violaceae
♦ Johnny jumpup violet
♦ Weed
♦ 218

Viola lanceolata L.
Violaceae
♦ bog white violet, Amerikanorvokki
♦ Weed, Casual Alien
♦ 42, 80
♦ herbal.

Viola lutea Huds.
Violaceae
♦ keltaorvokki, mountain violet, mountain pansy
♦ Cultivation Escape
♦ 42
♦ cultivated, herbal.

Viola mandshurica W.Becker
Violaceae
♦ sumire
♦ Weed
♦ 87, 88, 204, 286
♦ herbal.

Viola minor (Makino) Makino
Violaceae
♦ Weed
♦ 286

Viola odorata L.
Violaceae
Viola hortensis Schur, *Viola sarmentosa* M.Bieb.
♦ sweet violet, violet, English violet, florist's violet, garden violet, sweet viola, violet, violet tea, common violet
♦ Weed, Naturalised, Garden Escape, Environmental Weed, Cultivation Escape
♦ 7, 15, 34, 39, 42, 72, 80, 86, 88, 98, 101, 121, 161, 198, 203, 241, 261, 272, 280, 287, 295, 296, 300
♦ pH, cultivated, herbal, toxic. Origin: Eurasia.

Viola ovato-oblonga (Miq.) Makino
Violaceae
♦ Weed
♦ 87, 88, 286

Viola palustris L.
Violaceae
♦ bog violet, marsh violet
♦ Weed
♦ 272
♦ pH, cultivated, herbal.

Viola papilionacea Pursh p.p.
Violaceae
= *Viola sororia* Willd.
♦ common blue violet, meadow violet, hooded blue violet
♦ Weed
♦ 161, 211
♦ cultivated, herbal.

Viola patrinii Ging.
Violaceae
♦ China violet
♦ Naturalised
♦ 101
♦ herbal.

Viola pedata L.
Violaceae
♦ bird's foot violet
♦ Weed
♦ 39, 161
♦ pH, cultivated, herbal, toxic.

Viola prionantha Bunge
Violaceae
♦ Weed
♦ 275, 297
♦ pH, promoted, herbal.

Viola rafinesquii Greene
Violaceae
= *Viola bicolor* Pursh (NoR)
♦ field pansy, Johnny jumpup
♦ Weed
♦ 161, 249

Viola reichenbachiana Jord.
Violaceae
Viola silvatica Fr.
♦ pale wood violet, early dog violet, pyökkiorvokki
♦ Weed

♦ 272
♦ cultivated, herbal.

Viola rhodopeia Beck.
Violaceae
♦ Weed
♦ 272

Viola riviniana Rchb.
Violaceae
♦ common dog violet, common violet, wood violet
♦ Weed, Naturalised, Garden Escape, Environmental Weed
♦ 70, 72, 86, 88, 98, 198, 203, 280
♦ pH, cultivated, herbal. Origin: Europe, North Africa.

Viola rupestris Schm.
Violaceae
Viola allionii Pio, *Viola arenaria* DC.
♦ Teesdale violet, hietaorvokki
♦ Weed
♦ 272
♦ cultivated.

Viola sororia Willd.
Violaceae
Viola papilionacea Pursh *p.p.* (see)
♦ woolly blue violet, common blue violet
♦ Weed, Naturalised
♦ 161, 287
♦ cultivated, herbal.

Viola tricolor L.
Violaceae
♦ wild violet, wild pansy, tricolor pansy, Johnny jumpup, hearts ease, pansy violet, European wild pansy, field pansy, pansy
♦ Weed, Naturalised, Cultivation Escape, Garden Escape
♦ 8, 15, 24, 34, 39, 44, 70, 87, 88, 94, 101, 121, 161, 218, 241, 243, 253, 280, 286, 287, 300
♦ a/pH, cultivated, herbal, toxic. Origin: Eurasia.

Viola tricolor L. ssp. tricolor
Violaceae
♦ wild pansy
♦ Weed
♦ 243, 272
♦ cultivated.

Viola triloba Schwein.
Violaceae
= *Viola palmata* L.
♦ Naturalised
♦ 287
♦ herbal.

Viola verecunda A.Gray
Violaceae
♦ tsubosumire
♦ Weed
♦ 87, 88, 204, 286
♦ pH, cultivated. Origin: east Asia.

Viola × wittrockiana Gams.
Violaceae
= *Viola tricolor* L. × *Viola lutea* Huds. × *Viola altaica* Ker Gawl. [most probable combination]
♦ tarhaorvokki, garden pansy, pansy

♦ Naturalised, Cultivation Escape
♦ 40, 42, 280
♦ pH, cultivated. Origin: horticultural hybrid.

Viola yedoensis Mak.
Violaceae
♦ Weed
♦ 275, 286, 297
♦ pH, promoted. Origin: Japan.

Virgilia capensis (L.) Poir.
Fabaceae/Papilionaceae
= *Virgilia oroboides* (Berg.) Salter
♦ Naturalised
♦ 198
♦ cultivated.

Virgilia oroboides (Berg.) Salter
Fabaceae/Papilionaceae
Virgilia capensis Lam. (see)
♦ blossom tree, snowdrop tree
♦ Weed, Native Weed, Casual Alien
♦ 121, 280
♦ T, cultivated. Origin: southern Africa.

Viscum album L.
Viscaceae
Viscum austriacum Wiesb., *Viscum laxum* Boiss., *Viscum stellatum* D.Don
♦ European mistletoe, vischio quercino
♦ Weed, Noxious Weed, Introduced, Casual Alien
♦ 34, 35, 39, 70, 87, 88, 161, 229, 272, 280
♦ S, parasitic, promoted, herbal, toxic.

Viscum cruciatum Sieb.
Viscaceae
♦ oriental mistletoe
♦ Weed
♦ 88, 115
♦ parasitic, herbal.

Vismia cayennensis (Jacq.) Pers.
Clusiaceae/Hypericaceae
♦ Weed
♦ 87, 88
♦ S/T.

Vismia guianensis (Aubl.) Choisy
Clusiaceae/Hypericaceae
Hypericum guianense Aubl.
♦ lacre
♦ Weed
♦ 255
♦ Origin: Brazil.

Vitex agnus-castus L.
Lamiaceae/Verbenaceae
♦ chastetree, chasteberry, monk's pepper, angus castus, lilac chastetree
♦ Weed, Naturalised, Introduced, Environmental Weed
♦ 80, 87, 88, 101, 151, 179, 261, 272
♦ S, cultivated, herbal. Origin: Mediterranean.

Vitex agnus-castus L. var. agnus-castus
Lamiaceae/Verbenaceae
♦ lilac chastetree
♦ Naturalised
♦ 101

Vitex agnus-castus L. var. caerulea Rehd.
Lamiaceae/Verbenaceae

- lilac chastetree
- Naturalised
- 101

Vitex cannabifolia Sieb. & Zucc.
Lamiaceae/Verbenaceae
- Naturalised
- 287
- herbal.

Vitex doniana Sweet
Lamiaceae/Verbenaceae
- black plum, lufulu
- Quarantine Weed
- 220
- cultivated, herbal.

Vitex ferruginea Schumach. & Thonn.
Lamiaceae/Verbenaceae
- kabulampako
- Quarantine Weed
- 220

Vitex glabrata R.Br.
Lamiaceae/Verbenaceae
- smooth chastetree
- Weed, Naturalised
- 22, 101, 179
- T, cultivated, herbal.

Vitex negundo L.
Lamiaceae/Verbenaceae
Vitex paniculata Lam.
- negundo chastetree, chastetree, vitex
- Naturalised, Introduced
- 101, 228
- S, arid, cultivated, herbal. Origin: Madagascar.

Vitex negundo L. var. heterophylla (Franch.) Rehd.
Lamiaceae/Verbenaceae
- negundo chastetree
- Naturalised
- 101
- cultivated, herbal.

Vitex negundo L. var. intermedia (P'ei) Moldenke
Lamiaceae/Verbenaceae
- negundo chastetree
- Naturalised
- 101

Vitex negundo L. var. negundo
Lamiaceae/Verbenaceae
- negundo chastetree
- Naturalised
- 101

Vitex parviflora Juss.
Lamiaceae/Verbenaceae
- small leaved vitex, lagundi, molave, smallflower chastetree
- Weed, Naturalised, Introduced
- 3, 101, 261
- Origin: south-east Asia.

Vitex quinata (Lour.) F.Williams
Lamiaceae/Verbenaceae
- Quarantine Weed
- 220

Vitex rotundifolia L.f.
Lamiaceae/Verbenaceae
- roundleaf chastetree, vitex
- Naturalised

- 101
- cultivated, herbal.

Vitex trifolia L.
Lamiaceae/Verbenaceae
- simpleleaf chastetree, namulega, beach vitex, three leaved chastetree
- Weed, Naturalised
- 6, 22, 80, 86, 87, 88, 101
- T, cultivated, herbal. Origin: Australia.

Vitex trifolia L. var. bicolor (Willd.) Moldenke
Lamiaceae/Verbenaceae
- simpleleaf chastetree, namulega
- Naturalised
- 101
- herbal.

Vitex trifolia L. var. subtrisecta (Kuntze) Moldenke
Lamiaceae/Verbenaceae
- simpleleaf chastetree
- Naturalised
- 101

Vitex trifolia L. var. trifolia
Lamiaceae/Verbenaceae
- simpleleaf chastetree
- Weed, Naturalised
- 101, 179

Vitex trifolia L. var. variegata Moldenke
Lamiaceae/Verbenaceae
- variegated chastetree
- Weed, Naturalised
- 101, 179

Vitis L. spp.
Vitaceae
- wild grape, grape, grapevine
- Weed, Naturalised
- 161, 198, 211
- herbal.

Vitis aestivalis Michx.
Vitaceae
- summer grape
- Weed
- 87, 88, 218
- pC, cultivated, herbal.

Vitis californica Benth.
Vitaceae
- grape, California wild grape
- Weed
- 116
- S, cultivated, herbal.

Vitis candicans Engelm. ex Gray
Vitaceae
= *Vitis mustangensis* Buckl. (NoR)
- mustang grape
- Weed
- 87, 88, 218

Vitis ficifolia Bunge var. lobata (Regel) Nakai
Vitaceae
- ebizuru
- Weed
- 286

Vitis hastata Miq.
Vitaceae
- Weed

- 87, 88
- herbal.

Vitis rotundifolia Michx.
Vitaceae
- muscadine grape, muscadine
- Weed
- 87, 88, 218
- pC, cultivated, herbal.

Vitis rupestris Scheele
Vitaceae
- sand grape
- Weed
- 87, 88, 218
- pC, arid, promoted, herbal.

Vitis tiliaefolia Roem. & Schult.
Vitaceae
- Weed
- 14, 87, 88
- herbal.

Vitis trifolia L.
Vitaceae
- Weed
- 87, 88
- herbal.

Vitis vinifera L.
Vitaceae
- cultivated grape, domestic grape, wine grape, parra, uva, vite, vindruva, viiniköynnös, weintraube, vine
- Weed, Naturalised, Cultivation Escape, Casual Alien
- 15, 34, 40, 42, 80, 101, 116, 261, 280
- C, cultivated, herbal.

Vitis vulpina L.
Vitaceae
Vitis riparia Michx., *Vitis illex* Bail.
- riverbank grape, frost grape, winter grape
- Weed
- 87, 88, 218
- pC, arid, cultivated, herbal.

Vittadinia australis A.Rich.
Asteraceae
- white fuzzweed
- Weed
- 165

Vittadinia cuneata DC.
Asteraceae
- Naturalised
- 280
- cultivated.

Vittadinia dissecta (Benth.) N.T.Burb.
Asteraceae
- Casual Alien
- 280
- cultivated.

Vittadinia gracilis (Hook.f.) N.T.Burb.
Asteraceae
- Naturalised
- 280
- cultivated.

Vittadinia muelleri N.T.Burb.
Asteraceae
- Casual Alien
- 280
- cultivated.

Vittadinia triloba (Gaudich.) DC.
Asteraceae
♦ fuzzweed
♦ Weed, Naturalised
♦ 87, 88, 287
♦ arid, cultivated.

Voandzeia subterranea (L.) Thouars ex DC.
Fabaceae/Papilionaceae
= *Vigna subterranea* (L.) Verdc.
♦ maapapu, bambara groundnut, Madagascan groundnut, bambara
♦ Casual Alien
♦ 42
♦ cultivated.

Vogelia apiculata (Fisch. Mey. & Avé.) Vierh.
Brassicaceae
= *Neslia paniculata* (L.) Desv. ssp. *thracica* (Velen.) Bornm.
♦ Weed
♦ 87, 88

Vogelia paniculata (L.) Hornem.
Brassicaceae
= *Neslia paniculata* (L.) Desv.
♦ Weed
♦ 87, 88

Volutarella divaricata Benth. & Hook.f.
Asteraceae
♦ Weed
♦ 87, 88

Volutarella ramosa (Roxb.) Santapau
Asteraceae
♦ Weed
♦ 87, 88

Vossia cuspidata (Roxb.) W.Griff.
Poaceae
♦ hippo grass, um soof reed, kaswegenda
♦ Weed, Noxious Weed, Native Weed
♦ 87, 88, 121, 229
♦ wpG, cultivated.

Vulpia C.Gmel. spp.
Poaceae
♦ fescue
♦ Weed, Naturalised
♦ 198, 243, 272
♦ G.

Vulpia alopecuros (Schousb.) Dumort.
Poaceae
♦ ketunhäntänata
♦ Casual Alien
♦ 42
♦ G.

Vulpia bromoides (L.) Gray
Poaceae
Bromus dertonensis All., *Festuca bromoides* L. (see), *Festuca sciuroides* Roth., *Vulpia dertonensis* (All.) Gola (see)
♦ squirreltail fescue, barren fescue, silvergrass, vulpia hairgrass, brome fescue, oravanhäntänätä
♦ Weed, Naturalised, Introduced, Environmental Weed, Casual Alien
♦ 7, 9, 15, 34, 38, 42, 68, 72, 86, 87, 88, 91, 98, 101, 121, 134, 158, 161, 176, 178, 198, 203, 228, 237, 241, 243, 261, 269, 272, 280, 287, 290, 295, 300
♦ aG, arid, cultivated, herbal. Origin: Eurasia.

Vulpia ciliata (Pers.) Link
Poaceae
Festuca barbata Gaud., *Festuca ciliata* Danth., *Vulpia danthonii* Volk.
♦ fringed fescue, bearded fescue
♦ Weed, Naturalised, Environmental Weed
♦ 72, 86, 88, 98, 101, 198, 203, 272
♦ aG, cultivated, herbal. Origin: Mediterranean.

Vulpia dertonensis (All.) Gola
Poaceae
= *Vulpia bromoides* (L.) Gray
♦ Weed
♦ 87, 88
♦ G, herbal.

Vulpia fasciculata (Forssk.) Fritsch
Poaceae
♦ dune fescue, vulpie à une seule glume
♦ Weed, Naturalised, Environmental Weed
♦ 7, 72, 86, 88, 98, 176, 198, 203
♦ aG, cultivated. Origin: western Europe, Mediterranean.

Vulpia geniculata (L.) Link
Poaceae
Festuca geniculata L.
♦ polvitähkänata
♦ Weed, Casual Alien
♦ 42, 87, 88, 253
♦ G.

Vulpia hybrida (Brot.) Pau
Poaceae
♦ Weed
♦ 87, 88
♦ G.

Vulpia megalura (Nutt.) Rybd.
Poaceae
= *Vulpia myuros* (L.) C.Gmel.
♦ foxtail fescue
♦ Weed, Naturalised
♦ 87, 88, 176, 269
♦ aG.

Vulpia membranacea (L.) Link
Poaceae
Festuca membranacea (L.) Dumort., *Festuca uniglumis* Sol., *Vulpia uniglumis* (Sol.) Dumort.
♦ dune fescue, oneglume fescue
♦ Naturalised, Environmental Weed
♦ 7, 86
♦ G.

Vulpia muralis (Kunth) Nees
Poaceae
♦ wall fescue
♦ Weed, Naturalised, Environmental Weed, Casual Alien
♦ 40, 86, 98, 198, 203
♦ G, cultivated. Origin: southern Europe.

Vulpia myuros (L.) C.Gmel.
Poaceae

Festuca linearis Gilib., *Festuca megalura* Nutt. (see), *Festuca myuros* L. (see), *Vulpia megalura* (Nutt.) Rybd. (see), *Vulpia pseudomyuros* Reichen., *Vulpia myuros* (L.) C.Gmel. var. *hirsuta* (Hack.) Asch. & Graebn.(see)
♦ rat's tail fescue, sixweeks grass, silvergrass, rat's tail fescue
♦ Weed, Naturalised, Introduced, Casual Alien
♦ 7, 9, 15, 38, 42, 51, 80, 87, 88, 91, 93, 98, 101, 121, 146, 158, 161, 176, 178, 198, 203, 205, 212, 228, 241, 243, 253, 269, 272, 280, 286, 287
♦ aG, arid, cultivated, herbal. Origin: Eurasia.

Vulpia myuros (L.) C.Gmel. fo. megalura (Nutt.) Stace & R.Cotton
Poaceae
♦ foxtail fesuce, rat's tail fescue
♦ Naturalised, Environmental Weed
♦ 72, 86, 134, 198
♦ G.

Vulpia myuros (L.) C.Gmel. fo. myuros
Poaceae
♦ rat's tail fescue
♦ Naturalised, Environmental Weed
♦ 72, 86, 198
♦ G.

Vulpia myuros (L.) C.Gmel. var. hirsuta (Hack.) Asch. & Graebn.
Poaceae
= *Vulpia myuros* (L.) C.Gmel.
♦ foxtail fescue, rat's tail fescue
♦ Weed, Introduced
♦ 38, 161
♦ aG.

Vulpia myuros (L.) C.Gmel. var. megalura (Nutt.) Auquier
Poaceae
♦ Naturalised
♦ 280, 287
♦ G.

Vulpia myuros (L.) C.Gmel. var. myuros
Poaceae
♦ false foxtail fescue, rat's tail fescue
♦ Naturalised
♦ 280, 300
♦ aG.

Vulpia octoflora (Walter) Rydb.
Poaceae
Festuca octoflora Walter, *Festuca tenella* Willd.
♦ sixweeks fescue, sixweeks grass, pullout grass
♦ Naturalised, Native Weed, Introduced
♦ 161, 174, 228, 286, 287
♦ aG, arid, herbal. Origin: North America.

W

Wachendorfia paniculata L.
Haemodoraceae
♦ rooikanol
♦ Weed, Naturalised
♦ 7, 86, 98, 203
♦ cultivated. Origin: South Africa.

Wachendorfia parviflora W.F.Barker
Haemodoraceae
♦ Quarantine Weed
♦ 220

Wachendorfia thyrsiflora Burm.
Haemodoraceae
♦ rooikanol
♦ Naturalised
♦ 86, 280
♦ cultivated. Origin: South Africa.

Wahlenbergia capensis (L.) A.DC.
Campanulaceae
♦ Weed, Naturalised, Environmental Weed
♦ 7, 86, 98, 203
♦ aH, cultivated. Origin: South Africa.

Wahlenbergia etbaica Vatke
Campanulaceae
♦ Weed
♦ 221

Wahlenbergia gracilenta Lothian
Campanulaceae
♦ Naturalised
♦ 280

Wahlenbergia gracilis (G.Forst.) A.DC.
Campanulaceae
= *Wahlenbergia marginata* (Thunb.) A.DC.
♦ Weed
♦ 87, 88
♦ cultivated, herbal. Origin: Australia.

Wahlenbergia linarioides (Lam.) DC.
Campanulaceae
Breweria linifolia Spreng., *Campanula linarioides* Lam.
♦ tuffy bells, linhito, falso linho
♦ Weed, Naturalised
♦ 101, 255, 295
♦ Origin: South America.

Wahlenbergia marginata (Thunb.) A.DC.
Campanulaceae
Campanula gracilis G.Forst., *Campanula marginata* Thunb., *Cervicina gracilis* (G.Forst.) Britten, *Wahlenbergia bivalvis* Merr., *Wahlenbergia gracilis* (G.Forst.) A.DC. (see)
♦ southern rockbell
♦ Weed, Naturalised
♦ 101, 179, 235, 273, 286
♦ herbal.

Wahlenbergia procumbens (Thunb.) A.DC.
Campanulaceae
♦ wild violet
♦ Native Weed
♦ 121
♦ pH, cultivated. Origin: southern Africa.

Wahlenbergia stellarioides Cham. & Schlechtd.
Campanulaceae
♦ Native Weed
♦ 121
♦ pH. Origin: southern Africa.

Wahlenbergia stricta (R.Br.) Sweet
Campanulaceae
♦ tall bluebell
♦ Casual Alien
♦ 280
♦ cultivated.

Wahlenbergia stricta (R.Br.) Sweet ssp. stricta
Campanulaceae
♦ Casual Alien
♦ 280

Wahlenbergia undulata (L.f.) A.DC.
Campanulaceae
Campanula undulata L.f., *Wahlenbergia bojeri* A.DC., *Wahlenbergia caledonica* Sond., *Wahlenbergia caledonica* Sond. var. *cyanea* (Engl. & Gilg) Brehmer, *Wahlenbergia cyanea* Engl. & Gilg, *Wahlenbergia dinteri* Brehmer, *Wahlenbergia engleri* Brehmer, *Wahlenbergia oatesii* Rolfe, *Wahlenbergia scoparia* Brehmer
♦ Native Weed
♦ 121
♦ aH, arid, cultivated, herbal. Origin: Madagascar.

Wahlenbergia vernicosa J.A.Pett.
Campanulaceae
♦ Casual Alien
♦ 280

Waldsteinia fragarioides (Michx.) Tratt.
Rosaceae
♦ Appalachian barren strawberry, barren strawberry
♦ Quarantine Weed
♦ 220
♦ cultivated, herbal.

Waldsteinia ternata (Stephan) Fritsch
Rosaceae
♦ rönsyansikka
♦ Cultivation Escape
♦ 42
♦ cultivated.

Wallichia disticha T.Anderson
Arecaceae
♦ Introduced
♦ 230
♦ T, cultivated.

Waltheria americana L.
Sterculiaceae
= *Waltheria indica* L.
♦ Florida waltheria
♦ Weed

♦ 12, 23, 88, 218
♦ herbal. Origin: Australia.

Waltheria douradinha St.-Hil.
Sterculiaceae
♦ doradinha
♦ Weed
♦ 255
♦ pH, herbal. Origin: Brazil.

Waltheria indica L.
Sterculiaceae
Waltheria americana L. (see), *Waltheria detonsa* A.Gray
♦ uhaloa, malva branca
♦ Weed, Naturalised, Introduced
♦ 87, 88, 121, 157, 158, 221, 228, 255, 300
♦ pS, arid, cultivated, herbal. Origin: possibly tropical America.

Waltheria ovata Cav.
Sterculiaceae
♦ Weed
♦ 87, 88

Waltheria tomentosa (J.R. & G.Forst.) H.St.John
Sterculiaceae
♦ falsa guanxuma
♦ Weed
♦ 255
♦ S. Origin: South America.

Washingtonia H.Wendl. spp.
Arecaceae
♦ fan palm, Washington palm
♦ Weed, Cultivation Escape
♦ 116, 122
♦ cultivated, herbal.

Washingtonia filifera (Linden ex André) H.Wendl.
Arecaceae
Neowashingtonia filamentosa Sudw., *Neowashingtonia filifera* Sudw., *Pritchardia filifera* L.Linden
♦ cotton palm, California fan palm, Washington palm, cabbage palmetto, palma de castilla, fan palm, American cotton palm
♦ Weed, Naturalised, Introduced, Garden Escape, Environmental Weed
♦ 7, 54, 72, 86, 88, 93, 205, 228
♦ T, arid, cultivated, herbal. Origin: south-western North America.

Washingtonia robusta H.Wendl.
Arecaceae
Neowashingtonia robusta (H.Wendl.) A.Heller, *Neowashingtonia sonorae* (S.Watson) Rose, *Washingtonia filifera* (L.Linden) H.Wendl. var. *sonora* (S.Watson) M.E.Jones, *Washingtonia gracilis* Parish, *Washingtonia sonorae* S.Watson
♦ Mexican washingtonia, Mexican fan palm, desert palm, palma colorada, skyduster, Mexican Washington palm, Washington fan palm
♦ Weed, Naturalised, Introduced, Garden Escape, Environmental Weed
♦ 54, 86, 88, 101, 179, 228
♦ arid, cultivated. Origin: south-eastern North America, Mexico.

Watsonia Mill. spp.
Iridaceae
- bugle lily, watsonia
- Naturalised, Environmental Weed
- 15, 86, 198

Watsonia aletroides (Burm.f.) Ker Gawl.
Iridaceae
- watsonia, bugle lily
- Weed, Naturalised, Environmental Weed
- 7, 54, 86, 88, 98, 198, 203
- cultivated. Origin: South Africa.

Watsonia ardernei Sander
Iridaceae
- Naturalised
- 280

Watsonia borbonica (Pourr.) Goldblatt
Iridaceae
- rosy watsonia, cape bugle lily
- Naturalised, Garden Escape, Environmental Weed
- 7, 86, 101, 198, 290
- cultivated, herbal. Origin: South Africa.

Watsonia borbonica (Pourr.) Goldblatt ssp. *ardernei* (Sander) Goldblatt
Iridaceae
- bugle lily
- Naturalised
- 40
- cultivated.

Watsonia bulbillifera J.W.Mathews & L.Bolus
Iridaceae
- = *Watsonia meriana* (L.) Mill. var. *bulbillifera* (J.W.Mathews & L.Bolus) D.A.Cooke
- wild watsonia, bulbil watsonia, bugle lily, Merian's bugle lily, watsonia
- Weed, Noxious Weed, Naturalised, Introduced, Garden Escape, Environmental Weed
- 9, 34, 86, 88, 98, 116, 147, 165, 203, 225, 246, 269, 280
- pH, cultivated, toxic. Origin: South Africa.

Watsonia leipoldtii L.Bolus
Iridaceae
- Weed, Naturalised
- 98, 203

Watsonia marginata (L.f.) Ker Gawl.
Iridaceae
- bordered watsonia, fairy watsonia, fragrant bugle lily, watsonia
- Weed, Naturalised, Garden Escape, Environmental Weed
- 7, 9, 72, 86, 88, 98, 101, 198, 203, 280
- pH, cultivated. Origin: South Africa.

Watsonia meriana (L.) Mill.
Iridaceae
- bulbil watsonia, bulbil bugle lily, rooikanol, watsonia
- Weed, Naturalised, Garden Escape, Environmental Weed
- 7, 9, 15, 72, 86, 88, 98, 101, 171, 176, 198, 203, 280
- pH, cultivated.

Watsonia meriana (L.) Mill. var. *bulbillifera* (J.W.Mathews & L.Bolus) D.A.Cooke
Iridaceae
- *Watsonia bulbillifera* J.F.Matthews & L.Bolus (see)
- bulbil watsonia, watsonia, wild watsonia, bugle lily, Merian's bugle lily
- Naturalised, Environmental Weed
- 176, 289, 296
- pH, cultivated. Origin: South Africa.

Watsonia pyramidata (Andrews) Klatt.
Iridaceae
- Weed, Naturalised
- 98, 203
- cultivated.

Watsonia versfeldii J.F.Mathews & L.Bolus
Iridaceae
- watsonia
- Weed, Naturalised, Garden Escape, Environmental Weed
- 7, 72, 86, 88, 98, 176, 198, 203, 290
- pH, cultivated. Origin: South Africa.

Watsonia versfeldii J.W.Mathews & L.Bolus var. *alba* J.W.Mathews & L.Bolus
Iridaceae
- Naturalised
- 176
- Origin: South Africa.

Watsonia wordsworthiana J.Mathews & L.Bolus
Iridaceae
- Weed, Naturalised
- 98, 203
- cultivated.

Weberbauerocereus Britton & Rose spp.
Cactaceae
- *Rauhocereus* Backeb. spp. (see)
- Weed, Quarantine Weed
- 76, 88, 203, 220

Weberocereus Britton & Rose spp.
Cactaceae
- *Werckleocereus* Britton & Rose spp. (see)
- Weed, Quarantine Weed
- 76, 88, 203, 220

Wedelia asperrima (Decne.) Benth.
Asteraceae
- Weed
- 39, 87, 88
- toxic.

Wedelia biflora (L.) DC.
Asteraceae
- *Melanthera biflora* (L.) Wild. *Verbesina biflora* L., *Wollastonia biflora* (L.) DC. (see)
- wedelia, beach sunflower
- Weed
- 13, 87, 88, 276
- S, herbal. Origin: tropical Asia.

Wedelia chinensis (Osbeck) Merr.
Asteraceae
- *Wedelia calendulacea* Less.
- Chinese wedelia
- Weed
- 87, 88, 235, 274
- herbal.

Wedelia glauca (Ortega) O.Hoffm. ex Hicken
Asteraceae
- = *Pascalia glauca* Ortega
- wedelia, sunchillo yuyo sapo, pascalia weed, pascalia
- Weed, Quarantine Weed, Naturalised, Casual Alien
- 76, 86, 87, 88, 98, 121, 198, 203, 236, 237, 270, 280, 295, 295
- pH, toxic. Origin: South America.

Wedelia gracilis Rich.
Asteraceae
- = *Sphagneticola gracilis* (L.C.Rich.) Pruski (NoR)
- Weed
- 87, 88

Wedelia lundii DC.
Asteraceae
- *Seruneum lundii* Kuntze
- Naturalised
- 287

Wedelia montana (Bl.) Boerl.
Asteraceae
- Weed
- 13

Wedelia padulosa DC.
Asteraceae
- Weed
- 87, 88

Wedelia parviceps Blake
Asteraceae
- Weed
- 157
- S.

Wedelia spilanthoides F.Muell.
Asteraceae
- Weed
- 276
- pH.

Wedelia trilobata (L.) Hitchc.
Asteraceae
- = *Sphagneticola trilobata* (L.C.Rich.) Pruski
- wedelia, Singapore daisy, dihpw ongohng, ngesil ra ngebard, rosrangrang, atiat, ate, creeping oxeye, manzanilla, margarita del pasto
- Weed, Naturalised, Introduced, Garden Escape, Environmental Weed, Cultivation Escape
- 3, 6, 54, 80, 86, 87, 88, 93, 107, 112, 122, 155, 161, 179, 201, 206, 230, 243, 249, 261, 268, 281, 287
- pH, cultivated, herbal.

Weigela floribunda (Siebold & Zucc.) K.Koch
Diervillaceae/Caprifoliaceae
- crimson weigela, Japanese wisteria
- Naturalised
- 101
- S, cultivated.

Weigela florida (Bunge) A.DC.
Diervillaceae/Caprifoliaceae
- oldfashioned weigela, weigela
- Naturalised, Cultivation Escape, Casual Alien

♦ 40, 101, 280
♦ S, cultivated, herbal.

Weigela hortensis (Siebold & Zucc.) K.Koch
Diervillaceae/Caprifoliaceae
♦ taniutsugi
♦ Weed
♦ 88, 204
♦ S, cultivated. Origin: east Asia.

Werckleocereus Britton & Rose spp.
Cactaceae
= *Weberocereus* Britton & Rose spp.
♦ Weed, Quarantine Weed
♦ 76, 88, 220

Westringia rosmariniformis Sm.
Lamiaceae
♦ Victoria rosemary
♦ Casual Alien
♦ 280
♦ cultivated.

Wiedemannia orientalis Fisch. & Mey.
Lamiaceae
♦ turkinpeippi
♦ Weed, Casual Alien
♦ 42, 87, 88, 243

Wigandia caracasana H.B.K.
Hydrophyllaceae
♦ wigandia
♦ Weed, Naturalised, Casual Alien
♦ 39, 86, 87, 88, 98, 121, 203, 280
♦ pS, cultivated, herbal, toxic. Origin: South America.

Wigandia urens (Ruiz & Pav.) Kunth
Hydrophyllaceae
♦ Caracus wigandia
♦ Weed, Naturalised
♦ 7, 39, 86, 101, 199
♦ cultivated, toxic. Origin: Central and South America.

Wigandia urens (Ruiz & Pav.) Kunth var. caracasana (Kunth) D.Gibson
Hydrophyllaceae
♦ Caracus wigandia
♦ Naturalised
♦ 101
♦ Origin: Mexico to Colombia.

Wikstroemia Endl. spp.
Thymelaeaceae
♦ false ohelo
♦ Weed
♦ 18, 88

Wikstroemia chamaedaphine Meissn.
Thymelaeaceae
♦ Weed
♦ 275

Wikstroemia gampi (Sieb. & Zucc.) Maxim.
Thymelaeaceae
♦ Weed
♦ 87, 88

Wikstroemia indica (L.) C.A.Mey.
Thymelaeaceae
Wikstroemia viridiflora Meissn.
♦ Weed
♦ 22, 39, 87, 88
♦ S, cultivated, herbal, toxic. Origin: Asia, Australia.

Winteria Ritt. spp.
Cactaceae
= *Cleistocactus* Lem. spp.
♦ Weed, Quarantine Weed
♦ 76, 88, 203

Winterocereus Backeb. spp.
Cactaceae
= *Cleistocactus* Lem. spp.
♦ Weed, Quarantine Weed
♦ 76, 88, 203

Wislizenia refracta Engelm.
Capparaceae/Cleomaceae/Oxystylidaceae
♦ jackass clover, California stinkweed, bee plant, spectacle fruit
♦ Weed
♦ 39, 87, 88, 161, 180, 218
♦ aH, herbal, toxic.

Wissadula amplissima (L.) R.E.Fr.
Malvaceae
♦ big yellow velvetleaf
♦ Weed
♦ 88

Wissadula excelsior (Cav.) Presl
Malvaceae
♦ Weed
♦ 87, 88

Wissadula subpeltata (Kuntze) R.E.Fr.
Malvaceae
♦ malva estrela
♦ Weed
♦ 255
♦ S. Origin: South America.

Wisteria Nutt. spp.
Fabaceae/Papilionaceae
♦ wisteria
♦ Weed
♦ 116, 154, 243, 247
♦ herbal, toxic.

Wisteria brachybotrys Sieb. & Zucc.
Fabaceae/Papilionaceae
♦ Weed
♦ 286
♦ cultivated.

Wisteria floribunda (Willd.) DC.
Fabaceae/Papilionaceae
♦ wisteria, Japanese wisteria
♦ Weed, Naturalised, Environmental Weed
♦ 39, 80, 88, 101, 102, 129, 151, 161, 189, 195, 286
♦ pC, cultivated, herbal, toxic.

Wisteria × formosa Rehd.
Fabaceae/Papilionaceae
= *Wisteria floribunda* (Willd.) DC. × *Wisteria sinensis* (Sims) DC.
♦ Naturalised
♦ 101

Wisteria frutescens Torr. & A.Gray
Fabaceae/Papilionaceae
Glycine frutescens L.
♦ wisteria, American wisteria
♦ Weed
♦ 80
♦ pC, cultivated, herbal.

Wisteria sinensis (Sims) DC.
Fabaceae/Papilionaceae

Glycine floribunda Willd., *Glycine sinensis* Sims, *Kraunhia floribunda* Taub., *Milletia chinensis* (DC.) Benth., *Wisteria chinensis hort.*, *Wisteria consequana* Loud.
♦ Chinese wisteria, wisteria
♦ Weed, Naturalised, Environmental Weed
♦ 3, 15, 39, 77, 80, 86, 88, 101, 102, 112, 129, 151, 161, 189, 280
♦ pC, cultivated, herbal, toxic. Origin: China.

Withania coagulans (Stocks) Dunal
Solanaceae
Puneeria coagulans Stocks
♦ paneer bandh
♦ Weed
♦ 87, 88
♦ arid, herbal.

Withania obtusifolia V.Tackh.
Solanaceae
= *Withania somnifera* (L.) Dunal ssp. *obtusifolia* (Tackh.) S.Abedin, M.A.Al-Yahya., S.A.Chaudhary & J.S.Mossa (NoR)
♦ warma
♦ Weed
♦ 221
♦ arid.

Withania somnifera (L.) Dunal
Solanaceae
♦ geneesblaren, poisonous gooseberry, ubab, aksan, asgandh, ashwahandha, aswagandha, cheparusiot, chepepterekiat, duffhro, emotoe, fuqqueish, girbah, gizara, gizewa, hidigaga, hindib, idi, idigaga, kabarra, khasraqul, kipkogai, labotwit, leekurun, lesayet, lopotwo, merjan, mgeda, morgan, mpwa, mtemua shimba, murambae, nhulapori, ofuyaendwa, olasaiyet, ouartinni, senn el far, serran, simm el ferakh, tatdra, tchintueumbuo, umuire, wintercherry, xharkhardii, xoxoriko
♦ Weed, Naturalised, Native Weed
♦ 39, 50, 51, 86, 87, 88, 98, 121, 185, 203, 221
♦ pS, arid, cultivated, herbal, toxic. Origin: Africa, Asia.

Wolffia Horkel ex Schleid. spp.
Lemnaceae
♦ duckweed, watermeal
♦ Weed, Quarantine Weed
♦ 76, 88, 220
♦ wH.

Wolffia arrhiza (L.) Wimm.
Lemnaceae
Bruniera vivipara Franch., *Grantia globosa* Griff., *Horkelia arrhiza* (L.) Druce, *Lemna arrhiza* L., *Lemna globosa* Roxb., *Wolffia michelii* Schleid.
♦ spotless watermeal, watermeal, rootless duckweed, least duckweed, dwarf duckweed, duckweed
♦ Weed, Naturalised, Native Weed
♦ 87, 88, 121, 272, 286, 287
♦ wpH, cultivated, herbal. Origin: Madagascar.

**Wolffia australiana (Benth.) Hartog &
Plas**
  Lemnaceae
  ♦ Environmental Weed
  ♦ 246
  ♦ wpH. Origin: Australia.

Wolffia brasiliensis Wedd.
  Lemnaceae
  ♦ Brazilian watermeal, South
  American watermeal
  ♦ Weed
  ♦ 255
  ♦ wpH. Origin: South America.

Wolffia columbiana Karst.
  Lemnaceae
  ♦ watermeal, Columbian watermeal
  ♦ Weed
  ♦ 87, 88, 218
  ♦ wpH, herbal.

Wolffia hyalina (Del.) Hegelm.
  Lemnaceae
  ♦ Weed
  ♦ 221
  ♦ wpH.

Wolffiella oblonga (Phil.) Hegelm.
  Lemnaceae
  ♦ lentilla de agua, saber bogmat
  ♦ Weed
  ♦ 295
  ♦ wpH.

Wollastonia biflora (L.) DC.
  Asteraceae
  = *Wedelia biflora* (L.) DC.
  ♦ ateate
  ♦ Introduced
  ♦ 230
  ♦ herbal.

Woodwardia areolata (L.) Moore
  Blechnaceae
  ♦ netvein chainfern, netted chainfern
  ♦ Weed
  ♦ 87, 88, 218
  ♦ cultivated, herbal.

Woodwardia radicans (L.) Sm.
  Blechnaceae
  ♦ rooting chainfern, chainfern
  ♦ Weed, Naturalised
  ♦ 40, 101, 179
  ♦ H, cultivated. Origin: southern
  Europe.

Wrangelia bicuspidata Børgesen
  Ceramiaceae
  ♦ red alga
  ♦ Weed
  ♦ 197
  ♦ algae.

Wulffia stenoglossa (Cass.) DC.
  Asteraceae
  ♦ cambará açu, cravo do campo
  ♦ Weed
  ♦ 255
  ♦ Origin: South America.

Wyethia amplexicaulis Nutt.
  Asteraceae
  Espeletia amplexicaulis Nutt.
  ♦ mule's ears, mule's ears wyethia
  ♦ Weed

  ♦ 87, 88, 161, 218
  ♦ pH, promoted, herbal.

Wyethia helenoides (DC.) Nutt.
  Asteraceae
  ♦ gray mule ears
  ♦ Weed
  ♦ 161, 180

X

Xanthium L. spp.
  Asteraceae
  ♦ cocklebur
  ♦ Weed, Quarantine Weed,
  Naturalised
  ♦ 23, 39, 76, 88, 191, 198, 221, 236
  ♦ toxic.

Xanthium abyssinicum Wallr.
  Asteraceae
  ♦ Weed
  ♦ 240
  ♦ aH.

Xanthium ambrosioides Hook. & Arn.
  Asteraceae
  ♦ cocklebur
  ♦ Weed, Noxious Weed, Naturalised
  ♦ 86, 87, 88, 237, 295
  ♦ Origin: South America.

Xanthium argenteum Widder
  Asteraceae
  ♦ Weed
  ♦ 87, 88
  ♦ Origin: Chile.

Xanthium brasilicum Vell.
  Asteraceae
  ♦ Weed
  ♦ 87, 88, 221, 242
  ♦ arid.

Xanthium californicum E.Greene
  Asteraceae
  = *Xanthium strumarium* L. var.
  canadense (Mill.) Torr. & Gray
  ♦ Californian burr, Noogoora burr
  ♦ Weed, Noxious Weed, Naturalised
  ♦ 86, 87, 88, 98, 169, 203
  ♦ aH.

Xanthium catharticum H.B.K.
  Asteraceae
  = *Xanthium spinosum* L.
  ♦ Naturalised
  ♦ 241, 300
  ♦ herbal.

Xanthium cavanillesii Schouw
  Asteraceae
  = *Xanthium strumarium* L. var.
  canadense (Mill.) Torr. & Gray
  ♦ Noogoora burr, abrojo grande, South
  American burr
  ♦ Weed, Quarantine Weed, Noxious
  Weed, Naturalised
  ♦ 62, 76, 86, 87, 88, 203, 236, 237, 269,
  270, 295, 300
  ♦ Origin: South America.

Xanthium chinense Mill.
Asteraceae
= *Xanthium strumarium* L. var.
canadense (Mill.) Torr. & Gray
♦ oriental cocklebur
♦ Weed
♦ 87, 88, 218
♦ herbal.

Xanthium echinatum Murray
Asteraceae
= *Xanthium strumarium* L. var.
canadense (Mill.) Torr. & Gray
♦ beach cocklebur
♦ Weed
♦ 87, 88, 218
♦ herbal.

Xanthium italicum Moretti
Asteraceae
= *Xanthium strumarium* L. var.
canadense (Mill.) Torr. & Gray
♦ Italian cockleburr
♦ Weed, Quarantine Weed, Noxious
Weed, Naturalised
♦ 62, 76, 86, 87, 88, 118, 203, 218, 269,
286, 287
♦ herbal. Origin: Americas.

Xanthium macrocarpum DC.
Asteraceae
= *Xanthium strumarium* (Mill.) Torr. &
A.Gray
♦ Weed
♦ 87, 88

Xanthium occidentale Bertol.
Asteraceae
= *Xanthium strumarium* L. var.
canadense (Mill.) Torr. & Gray
♦ Noogoora burr, beach cocklebur,
burrweed, clotbur, cocklebur,
European cocklebur, Italian cocklebur,
rough cocklebur, sheepbur
♦ Weed, Quarantine Weed, Noxious
Weed, Naturalised, Environmental
Weed
♦ 20, 62, 76, 86, 87, 88, 93, 147, 169, 177,
203, 232, 261, 269, 286, 287, 290
♦ aH. Origin: Americas.

Xanthium orientale L.
Asteraceae
= *Xanthium strumarium* L. var.
canadense (Mill.) Torr. & Gray
♦ Californian burr, beach cocklebur,
burrweed, clotbur, cocklebur,
European cocklebur, Italian cocklebur,
rough cocklebur, sheepbur
♦ Weed, Quarantine Weed, Noxious
Weed, Naturalised
♦ 62, 76, 86, 87, 88, 147, 203, 269
♦ herbal. Origin: Americas.

Xanthium pensylvanicum Wallr.
Asteraceae
= *Xanthium strumarium* L. var.
canadense (Mill.) Torr. & Gray
♦ common cocklebur
♦ Weed, Quarantine Weed
♦ 76, 87, 88, 203, 218, 220

Xanthium pungens Wallr.
Asteraceae
= *Xanthium strumarium* (Mill.) Torr. &

A.Gray
♦ Noogoora burr
♦ Weed, Noxious Weed
♦ 6, 39, 50, 55, 87, 88, 169, 276
♦ aH, arid, toxic.

Xanthium saccharatum Wallr.
Asteraceae
= *Xanthium strumarium* L. var.
canadense (Mill.) Torr. & Gray
♦ abrojo
♦ Weed
♦ 87, 88, 237, 295
♦ herbal.

Xanthium sibiricum Patrin ex Widder
Asteraceae
♦ Weed
♦ 88, 275, 297
♦ aH, promoted, herbal.

Xanthium speciosum Kearney
Asteraceae
= *Xanthium strumarium* L. var.
canadense (Mill.) Torr. & Gray
♦ showy cocklebur
♦ Weed
♦ 87, 88, 218

Xanthium spinosum L.
Asteraceae
Acanthoxanthium spinosum (L.) Fourr.,
Xanthium catharticum Kunth (see)
♦ spiny cocklebur, piikkisappiruoho,
boetebos, Bathurst burr, burrweed,
prickly burrweed, spiny clotbur,
dagger cocklebur, dagger weed, thorny
burrweed
♦ Weed, Quarantine Weed, Noxious
Weed, Naturalised, Introduced,
Environmental Weed, Casual Alien
♦ 1, 34, 39, 40, 42, 44, 45, 50, 51, 55, 62,
63, 70, 72, 76, 78, 80, 86, 87, 88, 93, 94,
95, 98, 101, 102, 115, 116, 121, 146, 147,
158, 161, 165, 169, 171, 176, 177, 185,
186, 198, 203, 205, 212, 218, 228, 229,
236, 237, 240, 241, 243, 246, 253, 256,
269, 270, 272, 278, 280, 283, 287, 295,
300
♦ aH, arid, cultivated, herbal, toxic.
Origin: tropical America.

Xanthium strumarium (Mill.) Torr. &
A.Gray
Asteraceae
Xanthium macrocarpum DC. (see),
Xanthium natalense Widder, *Xanthium*
pungens Wallr. (see), *Xanthium vulgare*
Hill
♦ Noogoora burr, Bathurst burr,
burrweed, clotbur, cocklebur,
sheepbur, ditchbur, heartleaf
cocklebur, common cocklebur,
rough cocklebur, buttonbur,
ditchbur, sea burdock, hedgehog
burrweed, sheepbur, kankerroos,
karheasappiruoho, abrojillo, kra chap
♦ Weed, Quarantine Weed, Noxious
Weed, Naturalised, Native Weed,
Introduced, Environmental Weed,
Casual Alien
♦ 8, 14, 34, 39, 40, 44, 51, 63, 66, 68, 70,
80, 87, 88, 93, 94, 95, 98, 102, 121, 133,

136, 136, 158, 161, 169, 171, 174, 180,
185, 186, 186, 195, 198, 203, 207, 210,
211, 212, 228, 229, 236, 239, 243, 246,
253, 255, 258, 263, 278, 280, 283, 286,
293, 299
♦ a/bS, arid, cultivated, herbal, toxic.
Origin: Mediterranean.

Xanthium strumarium L. ssp. italicum
(Moretti) D.Löve
Asteraceae
= *Xanthium strumarium* L. var.
canadense (Mill.) Torr. & Gray
♦ Italian cocklebur
♦ Weed, Naturalised, Introduced
♦ 228, 241, 272

Xanthium strumarium L. ssp. strumarium
Asteraceae
♦ cocklebur
♦ Weed
♦ 115, 243, 272

Xanthium strumarium L. var. canadense
(Mill.) Torr. & Gray
Asteraceae
Xanthium acerosum Greene, *Xanthium*
americanum Walt., *Xanthium*
californicum E.Greene (see), *Xanthium*
californicum Greene var. *rotundifolium*
Widder, *Xanthium calvum* Millsp. &
Sherff, *Xanthium campestre* Greene,
Xanthium canadense P.Mill., *Xanthium*
cavanillesii Schouw (see), *Xanthium*
cenchroides Millsp. & Sherff, *Xanthium*
chasei Fern., *Xanthium chinense* P.Mill.
(see), *Xanthium commune* Britt.,
Xanthium curvescens Millsp. & Sherff,
Xanthium cylindraceum Millsp. &
Sherff, *Xanthium echinatum* Murr. (see),
Xanthium echinellum Greene, *Xanthium*
glanduliferum Greene, *Xanthium*
globosum Shull, *Xanthium inflexum*
Mack. & Bush, *Xanthium italicum*
Moretti (see), *Xanthium macrocarpum*
DC. var. *glabratum* DC., *Xanthium*
macounii Britt., *Xanthium oligacanthum*
Piper, *Xanthium occidentale* Bertol. (see),
Xanthium orientale L. (see), *Xanthium*
oviforme Wallr., *Xanthium pensylvanicum*
Wallr. (see), *Xanthium saccharatum*
Wallr. (see), *Xanthium speciosum*
Kearney (see), *Xanthium strumarium* L.
var. *glabratum* (DC.) Cronquist (see),
Xanthium strumarium L. ssp. *italicum*
(Moretti) D.Löve (see), *Xanthium*
strumarium L. var. *oviforme* (Wallr.)
M.E.Peck, *Xanthium strumarium* L.
var. *pensylvanicum* (Wallr.) M.E.Peck,
Xanthium strumarium L. var. *wootonii*
(Cockerell) M.E.Peck, *Xanthium*
strumarium L. var. *wootonii* (Cockerell)
W.C.Martin & C.R.Hutchins, *Xanthium*
varians Greene, *Xanthium wootonii*
Cockerell
♦ Canada cocklebur, cocklebur,
common cocklebur
♦ Weed, Noxious Weed
♦ 49, 167, 229
♦ toxic.

Xanthium strumarium L. var. glabratum (DC.) Cronquist
Asteraceae
= *Xanthium strumarium* L. var. *canadense* (Mill.) Torr. & Gray
♦ cocklebur, rough cocklebur
♦ Noxious Weed
♦ 229

Xanthium strumarium L. var. japonica (Widder) Hara
Asteraceae
♦ Weed
♦ 235

Xanthium strumarium L. var. strumarium
Asteraceae
♦ cocklebur, rough cocklebur
♦ Noxious Weed
♦ 229

Xanthorrhoea preissii Endl.
Xanthorrhoeaceae
♦ western black boy
♦ Weed
♦ 87, 88
♦ cultivated.

Xanthosoma atrovirens K.Koch & Bouché
Araceae
♦ yautia amarilla, yautia vinola
♦ Naturalised, Cultivation Escape
♦ 101, 261
♦ cultivated, herbal. Origin: northern South America.

Xanthosoma caracu K.Koch & Bouché
Araceae
♦ yautia horqueta, yautia manola
♦ Naturalised, Cultivation Escape
♦ 101, 261
♦ cultivated. Origin: South America.

Xanthosoma cf. sagittifolium (L.) Schott
Araceae
♦ Weed
♦ 206

Xanthosoma helleborifolium (Jacq.) Schott
Araceae
♦ belembe silvestre
♦ Weed, Naturalised, Cultivation Escape
♦ 87, 88, 101, 261
♦ cultivated, herbal. Origin: Lesser Antilles, Central and South America.

Xanthosoma roseum Schott
Araceae
♦ rosy malanga, elephant's ear
♦ Quarantine Weed, Naturalised
♦ 101, 220

Xanthosoma sagittifolium (L.) Schott
Araceae
♦ melanga, elephant's ear, arrowleaf elephant's ear, sawawh wai
♦ Weed, Naturalised, Introduced, Environmental Weed, Cultivation Escape, Casual Alien
♦ 28, 88, 112, 134, 179, 230, 243, 261, 280
♦ H, cultivated, herbal. Origin: South America.

Xanthosoma undipes (K.Koch) K.Koch
Araceae
♦ tall elephant's ear, yautia palma
♦ Naturalised, Casual Alien
♦ 101, 261

Xanthosoma violaceum Schott
Araceae
♦ purplestem taro, blue taro
♦ Sleeper Weed, Naturalised, Environmental Weed, Cultivation Escape
♦ 86, 101, 155, 261
♦ cultivated, herbal.

Xeranthemum annuum L.
Asteraceae
Xeranthemum inodorum Moench, *Xeranthemum radiatum* Lam.
♦ pink everlasting, immortelle
♦ Weed
♦ 272
♦ aH, cultivated, herbal.

Xeranthemum cylindraceum Sibth. & Sm.
Asteraceae
Xeranthemum foetidum Moench.
♦ perpetuini piccoli
♦ Weed
♦ 70, 272
♦ cultivated, herbal.

Xerophyllum tenax (Pursh) Nutt.
Liliaceae/Melanthiaceae/Xerophyllaceae
Xerophyllum douglasii S.Watson, *Helonias tenax* Pursh
♦ beargrass, Indian basket grass, common beargrass, elk grass
♦ Weed
♦ 87, 88, 218
♦ pH, cultivated, herbal.

Ximenia americana L.
Olacaceae
Ximenia exarmata F.Muell., *Ximenia spinosa* Salisb.
♦ tallow nut, beach plum, false sandalwood, hog plum, lusantu, tallow wood
♦ Introduced
♦ 39, 228
♦ arid, cultivated, herbal, toxic.

Ximenia caffra Sond.
Olacaceae
♦ Weed
♦ 87, 88
♦ herbal.

Xiphidium caeruleum Aubl.
Haemodoraceae
♦ cola de paloma, mano poderosa
♦ Cultivation Escape
♦ 261
♦ H, cultivated, herbal.

Xylorhiza glabriuscula Nutt.
Asteraceae
♦ woody aster, smooth woodyaster
♦ Weed
♦ 161, 212
♦ herbal, toxic.

Xyris complanata R.Br.
Xyridaceae

♦ Hawai'i yellow eyed grass
♦ Weed, Naturalised
♦ 87, 88, 101
♦ herbal.

Xyris indica L.
Xyridaceae
= *Xyris torta* Sm. (NoR) [see *Xyris indica* L. var. *indica*]
♦ yellow eyed grass, kratin naa
♦ Weed
♦ 87, 88, 191, 204, 209, 239
♦ herbal.

Xyris indica L. var. indica
Xyridaceae
= *Xyris torta* Sm. (NoR) [see *Xyris indica* L.]
♦ Weed
♦ 170

Xyris jupicai Rich.
Xyridaceae
♦ Richard's yellow eyed grass
♦ Weed
♦ 179
♦ herbal.

Xyris melanocephala Miq.
Xyridaceae
♦ Weed
♦ 87, 88

Y

Youngia denticulata (Houtt.) Kitam.
Asteraceae
Prenanthes denticulata Houtt.
♦ yakushisou
♦ Weed
♦ 263, 286
♦ a/bH, promoted. Origin: east Asia.

Youngia heterophylla (Hemsl.) Babc. & Stebbins
Asteraceae
♦ Weed
♦ 275
♦ pH.

Youngia japonica (L.) DC.
Asteraceae
Crepis japonica (L.) Benth. (see)
♦ Asiatic hawksbeard, oriental false hawksbeard, oriental hawkbeard, native hawksbeard, onitabirako
♦ Weed, Naturalised
♦ 68, 87, 88, 93, 101, 121, 161, 179, 191, 204, 218, 235, 243, 249, 261, 263, 273, 274, 275, 276, 286, 297
♦ aH, promoted, herbal. Origin: Eurasia.

Yucca aloifolia L.
Agavaceae
♦ coastal yucca, dwarf yucca, Spanish bayonet, dagger plant, aloe yucca, aguja de adán, mata de huevos
♦ Weed, Sleeper Weed, Naturalised, Garden Escape, Environmental Weed, Cultivation Escape
♦ 7, 73, 86, 88, 98, 121, 155, 161, 179, 203, 261
♦ pS, cultivated, herbal, toxic. Origin: south-eastern North America.

Yucca campestris McKelvey
Agavaceae
♦ plains yucca
♦ Quarantine Weed
♦ 220
♦ cultivated.

Yucca elata Engelm.
Agavaceae
Yucca radiosa (Engelm.) Trel., *Yucca verdiensis* McKelvey, *Yucca angustifolia* Pursh var. *radiosa* Englem., *Yucca angustifolia* Pursh var. *elata* Engelm., *Yucca utahensis* McKelvey
♦ soaptree yucca, palmlilja jukka, soapweed, Utah yucca, verdi yucca, soaptree, pamilla, pamella, Spanish bayonet, datil, amole

♦ Weed
♦ 87, 88, 218
♦ S, cultivated, herbal.

Yucca filamentosa L.
Agavaceae
Yucca concava Haw.
♦ yucca, Adam's needle, spoonleaf yucca, palmlilja, hapsijukka
♦ Naturalised
♦ 7, 8, 8, 39, 86
♦ pS, cultivated, herbal, toxic. Origin: south-eastern North America.

Yucca glauca Nutt.
Agavaceae
♦ small soapweed, Great Plains yucca, soapweed yucca, beargrass, yucca, Spanish bayonet, soapweed
♦ Weed, Native Weed, Introduced
♦ 8, 39, 87, 88, 161, 174, 212, 218, 228
♦ pS, arid, cultivated, herbal, toxic. Origin: North America.

Yucca gloriosa L.
Agavaceae
Yucca ellacombei hort. ex Bak.
♦ Adam's needle, moundlily yucca, Spanish dagger, bayoneta Española
♦ Weed, Naturalised, Environmental Weed, Cultivation Escape
♦ 15, 86, 246, 261, 280
♦ S, cultivated, herbal. Origin: south-eastern North America.

Yucca guatemalensis Bak.
Agavaceae
Yucca elephantipes Regel.
♦ bluestem yucca, izote, palmito, bayoneta
♦ Introduced, Cultivation Escape
♦ 228, 261
♦ arid, cultivated. Origin: Mexico and Central America.

Yucca intermedia McKelvey
Agavaceae
= *Yucca baileyi* Woot. & Standl. var. *intermedia* (McKelvey) Reveal (NoR)
♦ Quarantine Weed
♦ 220
♦ cultivated.

Yucca navajoa J.M.Webber
Agavaceae
= *Yucca baileyi* Woot. & Standl. var. *navajoa* (J.M.Webber) J.M.Webber (NoR)
♦ Quarantine Weed
♦ 220
♦ cultivated.

Yucca L. spp.
Agavaceae
♦ yucca
♦ Weed
♦ 23, 88, 279
♦ herbal.

Yucca torreyi Schaf.
Agavaceae
♦ Torrey's yucca
♦ Weed
♦ 87, 88, 218
♦ cultivated, herbal.

Yungasocereus Ritt. spp.
Cactaceae
= *Samaipaticereus* Cárdenas spp.
♦ Weed, Quarantine Weed
♦ 76, 88, 203

Z

Zaleya decandra (L.) Burm.f.
Aizoaceae
♦ Weed
♦ 221
♦ cultivated.

Zaleya galericulata (Melville) H.Eichler
Aizoaceae
Trianthema galericulata Melville (see)
♦ hogweed
♦ Naturalised, Native Weed
♦ 198, 269
♦ pH, arid. Origin: Australia.

**Zaleya galericulata (Melville) H.Eichler
ssp. *australis* (Melville) S.W.L.Jacobs**
Aizoaceae
Trianthema australis Melville (see)
♦ Naturalised
♦ 86
♦ Origin: Australia.

Zaleya pentandra (L.) Jeffr.
Aizoaceae
Trianthema pentandra L. (see)
♦ African purslane, zaleya
♦ Weed
♦ 88, 121, 158, 242
♦ S, arid, herbal. Origin: obscure.

Zaluzianskya divaricata (Thunb.) Walp.
Scrophulariaceae
♦ spreading night phlox, creeping
night phlox, zed weed
♦ Weed, Naturalised, Environmental
Weed
♦ 7, 72, 86, 88, 98, 198, 203
♦ aH. Origin: South Africa.

Zamia furfuracea L.f.
Zamiaceae
♦ cardboard cycad, Jamaican sago tree,
cardboard palm
♦ Weed
♦ 179
♦ cultivated, herbal.

Zamia loddigesii Miq.
Zamiaceae
♦ camotillo
♦ Weed
♦ 87, 88
♦ herbal.

Zannichellia palustris L.
Potamogetonaceae/Zannichelliaceae
♦ horned pondweed
♦ Weed
♦ 23, 87, 88, 159, 200, 218, 221, 253, 272,
295, 297
♦ pH, aqua, promoted, herbal.

Zantedeschia aethiopica (L.) Spreng.
Araceae
Calla aethiopica L., *Richardia africana*
Kunth
♦ arum lily, calla lily, Egyptian lily,
florist's calla, garden calla, Jack in the
pulpit, lily of the Nile, pig lily, trumpet
lily, white arum lily
♦ Weed, Quarantine Weed, Noxious
Weed, Naturalised, Native Weed,
Introduced, Garden Escape,
Environmental Weed, Cultivation
Escape
♦ 7, 9, 15, 34, 39, 40, 62, 72, 76, 80, 86,
88, 98, 101, 116, 121, 132, 134, 147, 154,
165, 176, 189, 198, 203, 225, 246, 247,
261, 269, 280, 289, 296, 300
♦ pH, aqua, cultivated, herbal, toxic.
Origin: southern Africa.

Zantedeschia albomaculata (Hook.) Baill.
Araceae
Zantedeschia melanoleuca (Hook.f.) Engl.
♦ spotted calla lily
♦ Naturalised
♦ 101
♦ cultivated, herbal.

Zanthoxylum ailanthoides Sieb. & Zucc.
Rutaceae
♦ Weed
♦ 286
♦ T, promoted, herbal. Origin: south
China, Japan.

Zanthoxylum americanum Mill.
Rutaceae
Thylax fraxineum (Willd.) Raf.,
Zanthoxylum fraxineum Willd.,
Zanthoxylum fraxinifolium Marshall,
Zanthoxylum mite Willd., *Zanthoxylum
parvum* Shinners, *Zanthoxylum
ramiflorum* Michx.
♦ prickly ash, common prickly ash,
northern prickly ash
♦ Weed
♦ 87, 88, 161, 218
♦ S/T, cultivated, herbal, toxic.

Zanthoxylum clava-herculis L.
Rutaceae
♦ Hercules's club, southern prickly ash
♦ Weed
♦ 87, 88, 218
♦ S/T, cultivated, herbal.

Zanthoxylum fagara (L.) Sarg.
Rutaceae
Fagara affinis (Kunth) Schult. &
Schult.f., *Fagara culantrillo* (Kunth)
Krug & Urb., *Fagara culantrillo* (Kunth)
Schult. & Schult.f., *Fagara culantrillo*
var. *continentalis* Krug & Urb., *Fagara
culantrillo* var. *insularis* Krug & Urb.,
Fagara fagara (L.) Kuntze, *Fagara fagara*
(L.) Small, *Fagara hyemalis* (A.St.-
Hil.) Engl., *Fagara inermis* Willd. ex
Schult. & Schult.f., *Fagara lentiscifolia*
Humb. & Bonpl. ex Willd., *Fagara
nigrescens* R.E.Fr., *Fagara nigrescens*
Urb. & Ekman, *Fagara peckoltiana*
(Engl.) Engl., *Fagara peruviana* Willd.
ex Schult. & Schult.f., *Fagara pterota*

L., *Fagara pterota* var. *guaranitica*
Chodat & Hassl., *Fagaras fagara* (L.)
Kuntze, *Pterota fagara* (L.) Crantz,
Schinus fagara L., *Zanthoxylum
affine* Kunth, *Zanthoxylum aguilarii*
Standl. & Steyerm., *Zanthoxylum
atoyacanum* Lundell, *Zanthoxylum
atratum* Alain, *Zanthoxylum culantrillo*
Kunth, *Zanthoxylum culantrillo* var.
paniculatum Engl., *Zanthoxylum friesii*
P.G.Waterman, *Zanthoxylum hyemale*
A.St.-Hil., *Zanthoxylum insulare* Rose,
Zanthoxylum lentiscifolium (Humb.
& Bonpl. ex Willd.) Andersson,
Zanthoxylum marginatum Sessé &
Moç., *Zanthoxylum nicaraguense*
Standl. & L.O.Williams, *Zanthoxylum
nigrescens* J.Jiménez Alm., *Zanthoxylum
peckoltianum* Engl., *Zanthoxylum
praecox* A.St.-Hil., *Zanthoxylum pterota*
(L.) Kunth, *Zanthoxylum pterota* (L.)
St.-Lag., *Zanthoxylum pterota* var.
guaraniticum (Chodat & Hassl.)
P.G.Waterman, *Zanthoxylum sonorense*
Lundell
♦ colima, lime prickly ash, wild lime
♦ Weed, Quarantine Weed, Introduced
♦ 14, 76, 87, 88, 203, 218, 220, 228
♦ S/T, arid, cultivated, herbal.

Zanthoxylum martinicense (Lam.) DC.
Rutaceae
♦ white prickly ash
♦ Weed
♦ 87, 88
♦ herbal.

Zanthoxylum schinifolium Sieb. & Zucc.
Rutaceae
♦ Weed
♦ 286
♦ S, promoted, herbal. Origin: China,
Japan, Korea.

Zapoteca tetragona (Willd.) H.M.Hern.
Fabaceae/Mimosaceae
♦ Quarantine Weed
♦ 220

Zea luxurians (Durieu & Asch.) Bird
Poaceae
♦ teosinte
♦ Weed, Cultivation Escape
♦ 32
♦ G, cultivated.

Zea mays L.
Poaceae
Mays vulgaris Ser., *Mays zea* Gaertn.,
Thalysia mays Kuntze, *Zea alba* Mill.,
Zea americana Mill., *Zea curagua* Mol.,
Zea segetalis Salisb., *Zea vulgaris* Mill.
♦ Arab wheat, corn, corn of Mecca,
Indian corn, maize, mealie, Turkish
grain, sweet corn, maíz guacho,
teosinte
♦ Weed, Naturalised, Introduced,
Casual Alien
♦ 7, 39, 40, 42, 86, 98, 101, 121, 179, 198,
203, 230, 236, 241, 256, 261, 280
♦ aG, cultivated, herbal, toxic. Origin:
Americas.

Zea mays L. ssp. *mays*
Poaceae
- corn, maize
- Naturalised
- 101
- G.

Zea mays L. ssp. *mexicana* (Schrad.) Iltis
Poaceae
- = *Zea mexicana* (Schrad.) Kuntze
- teosinte
- Weed
- 179
- G.

Zea mays L. ssp. *parviglumis* Iltis & Doebley
Poaceae
- corn, maize
- Weed, Naturalised
- 101, 179
- G.

Zea mexicana (Schrad.) Kuntze
Poaceae
Zea mays L. ssp. *mexicana* (Schrad.) Iltis (see)
- Mexican teosinte
- Naturalised
- 86, 101
- G, cultivated. Origin: Mexico.

Zea perennis (A.S.Hitchc.) Reeves & Manglesd.
Poaceae
Euchlaena perennis Hitchc.
- perennial teosinte
- Naturalised
- 101
- G.

Zebrina pendula Schnizl.
Commelinaceae
= *Tradescantia zebrina* hort. ex Bosse
- zebrina, wandering zebrina, wandering Jew, inchplant, striped wandering creeper
- Weed, Naturalised, Native Weed, Environmental Weed, Cultivation Escape
- 3, 73, 80, 87, 88, 98, 121, 122, 126, 155, 157, 191, 201, 203, 287
- pH, cultivated, herbal, toxic. Origin: South America.

Zehneria indica (Lour.) Keraudren
Cucurbitaceae
Melothria indica Lour.
- Indian zehneria
- Weed
- 297

Zehneria scabra (L.f.) Sond.
Cucurbitaceae
- Native Weed
- 121
- pC, toxic. Origin: southern Africa.

Zehneria thwaitesii (Schweinf.) C.Jeffrey
Cucurbitaceae
- Weed
- 87, 88

Zehntnerella Britt. & Rose spp.
Cactaceae

= *Facheiroa* Britt. & Rose spp.
- Weed, Quarantine Weed
- 76, 88, 203, 220

Zelkova serrata (Thunb.) Makino
Ulmaceae
Abelicea hirta C.K.Schneid., *Corchorus hirtus non* L., *Corchorus serratus* Thunb., *Planera acuminata* Lindl., *Planera japonica* Miq., *Ulmus keaki* Sieb., *Zelkova acuminata* Planch., *Zelkova hirta* C.K.Schneid., *Zelkova keaki* Maxim., *Zelkova serrata* var. *tarokoensis* (Hayata) Li, *Zelkova tarokoensis* Hayata
- Japanese zelkova
- Naturalised
- 101
- T, cultivated, herbal.

Zephyranthes andersonii Nichols
Liliaceae/Amaryllidaceae
- Weed
- 87, 88

Zephyranthes atamasca (L.) Herb.
Liliaceae/Amaryllidaceae
Zephyranthes atamasco (L.) Herb. (see)
- atamasco lily
- Weed
- 161
- herbal, toxic.

Zephyranthes atamasco (L.) Herb.
Liliaceae/Amaryllidaceae
= *Zephyranthes atamasca* (L.) Herb.
- atamasco lily, rain lily, zephyr lily, fairy lily
- Weed
- 39, 87, 88, 218, 247
- pH, cultivated, herbal, toxic.

Zephyranthes candida (Lindl.) Herb.
Liliaceae/Amaryllidaceae
- autumn zephyr lily, zephyr lily
- Weed, Naturalised, Casual Alien
- 86, 98, 101, 203, 280, 287
- cultivated, herbal. Origin: Argentina, Paraguay.

Zephyranthes citrina Bak.
Liliaceae/Amaryllidaceae
- citron zephyr lily
- Weed, Naturalised, Garden Escape
- 101, 179, 261
- cultivated.

Zephyranthes drummondii D.Don
Liliaceae/Amaryllidaceae
= *Cooperia pedunculata* Herbert
- rain lily
- Naturalised
- 86
- cultivated. Origin: Texas, Louisiana, Mexico.

Zephyranthes eggersiana Urb.
Liliaceae/Amaryllidaceae
- Weed
- 87, 88

Zephyranthes grandiflora Lindl.
Liliaceae/Amaryllidaceae
Atamosco carinata (Herbert) P.Wilson
- pink storm lily, rosepink zephyr lily,

adelfa, duende rosado, rose fairy lily
- Weed, Sleeper Weed, Naturalised, Garden Escape, Environmental Weed, Cultivation Escape
- 86, 98, 155, 203, 261, 286, 287
- cultivated. Origin: Mexico.

Zephyranthes rosea (Spreng.) Lindl.
Liliaceae/Amaryllidaceae
Atamosco rosea (Lindl.) Greene
- Cuban zephyr lily, pileep, duende rojo, red atamosco lily
- Weed, Naturalised, Introduced, Garden Escape
- 87, 88, 101, 230, 261
- cultivated, herbal. Origin: Cuba.

Zeuxine strateumatica (L.) Schltdl.
Orchidaceae
- soldier's orchid
- Weed, Naturalised
- 101, 179, 261, 273
- herbal.

Zexmenia brachylepis (Griseb.) Cabrera
Asteraceae
- Weed
- 237

Zexmenia pinetorum Standl. & Steyerm.
Asteraceae
- Weed
- 157
- pH.

Zigadenus densus (Desr.) Fern.
Liliaceae/Melanthiaceae
- black snakeroot, Osceola's plume
- Weed
- 39, 161
- toxic.

Zigadenus elegans Pursh
Liliaceae/Melanthiaceae
- mountain deathcamas, white camas
- Weed, Noxious Weed
- 36, 39, 87, 88, 161, 218
- cultivated, herbal, toxic.

Zigadenus fremontii (Torr.) S.Wats.
Liliaceae/Melanthiaceae
- chaparral deathcamas, star zygadene, starlily, Fremont's deathcamas
- Weed
- 39, 87, 88, 161, 218
- pH, cultivated, herbal, toxic.

Zigadenus gramineus Rydb.
Liliaceae/Melanthiaceae
= *Zigadenus venenosus* S.Wats. var. *gramineus* (Rydb.) Walsh ex M.E.Peck (NoR)
- grassy deathcamas, deathcamas
- Weed, Noxious Weed
- 36, 87, 88, 161, 218
- herbal, toxic.

Zigadenus nuttallii (Gray) S.Wats.
Liliaceae/Melanthiaceae
- poison camas, Nuttall's deathcamas
- Weed
- 39, 161
- cultivated, herbal, toxic.

Zigadenus paniculatus (Nutt.) S.Wats.
Liliaceae/Melanthiaceae
♦ foothill deathcamas, meadow deathcamas, sandcorn
♦ Weed, Native Weed
♦ 39, 87, 88, 212, 218, 264
♦ pH, herbal, toxic. Origin: south-west North America.

Zigadenus Michx. spp.
Liliaceae/Melanthiaceae
♦ deathcamas, black snakeroot
♦ Weed, Quarantine Weed
♦ 76, 88, 154, 220, 247
♦ toxic.

Zigadenus venenosus S.Wats.
Liliaceae/Melanthiaceae
♦ meadow deathcamas, deathcamas, foothill deathcamas, deathcamas lily
♦ Weed, Noxious Weed, Native Weed
♦ 39, 88, 154, 161, 174, 212, 218, 219, 264, 299
♦ cultivated, herbal, toxic. Origin: North America.

Zilla spinosa (Turra) Prantl
Brassicaceae
Zilla myagroides Forssk.
♦ Weed
♦ 221
♦ arid, cultivated.

Zingeria trichopoda (Boiss.) Smirn.
Poaceae
♦ Weed
♦ 88
♦ G, cultivated.

Zingiber cassumunar Roxb.
Zingiberaceae
= *Zingiber purpureum* Roscoe
♦ jengibre colorado
♦ Cultivation Escape
♦ 261
♦ cultivated, herbal. Origin: east Asia, Malaysia.

Zingiber mioga (Thunb.) Roscoe
Zingiberaceae
♦ Mioga ginger
♦ Weed, Naturalised
♦ 286, 287
♦ herbal.

Zingiber officinale Roscoe
Zingiberaceae
♦ garden ginger, sinser, fiu, zenzero, jengibre
♦ Weed, Naturalised, Introduced, Cultivation Escape
♦ 39, 86, 98, 101, 203, 230, 261
♦ H, cultivated, herbal, toxic. Origin: east Asia.

Zingiber purpureum Roscoe
Zingiberaceae
Zingiber cassumunar Roxb. (see)
♦ Cassumunar ginger
♦ Naturalised
♦ 101

Zingiber Mill. spp.
Zingiberaceae
♦ ginger
♦ Weed

♦ 121
♦ pH. Origin: Eurasia.

Zingiber zerumbet (L.) Sm.
Zingiberaceae
♦ bitter ginger, shampoo ginger, wild ginger, avapui, ong en pele, jengibre amargo
♦ Weed, Naturalised, Cultivation Escape
♦ 86, 98, 101, 203, 261
♦ H, cultivated, herbal. Origin: obscure.

Zinnia angustifolia Kunth
Asteraceae
Zinnia linearis Benth.
♦ narrowlead zinnia, star zinnia
♦ Naturalised
♦ 101
♦ aH, cultivated, herbal.

Zinnia elegans Jacq.
Asteraceae
Zinnia violacea Cav. (see)
♦ wild zinnia, elegant zinnia, zinnia
♦ Naturalised, Garden Escape, Environmental Weed, Cultivation Escape
♦ 86, 261, 287
♦ aH, cultivated, herbal. Origin: Mexico.

Zinnia multiflora L.
Asteraceae
= *Zinnia peruviana* (L.) L.
♦ redstar zinnia
♦ Weed
♦ 51, 87, 88
♦ herbal.

Zinnia oligantha I.M.Johnst.
Asteraceae
♦ Weed
♦ 23, 88

Zinnia pauciflora L.
Asteraceae
= *Zinnia peruviana* (L.) L.
♦ Weed
♦ 87, 88

Zinnia peruviana (L.) L.
Asteraceae
Zinnia multiflora L. (see), *Zinnia pauciflora* L. (see)
♦ Peruvian zinnia, red spiderzinnia, kaffir daisy, redstar zinnia, wild zinnia
♦ Weed, Naturalised
♦ 86, 87, 88, 98, 101, 121, 158, 199, 203, 240, 256, 269
♦ aH, cultivated, herbal. Origin: South America.

Zinnia violacea Cav.
Asteraceae
= *Zinnia elegans* Jacq.
♦ elegant zinnia
♦ Weed, Naturalised
♦ 101, 179

Zizania aquatica L.
Poaceae
♦ annual wild rice, intiaaniriisi, Canadian wild rice, wild rice
♦ Weed, Quarantine Weed, Cultivation Escape

♦ 23, 42, 87, 88, 197, 218, 258
♦ G, aqua, cultivated, herbal.

Zizania latifolia (Griseb.) Turcz. ex Stapf
Poaceae
Hydropyrum latifolium Griseb., *Limnochloa caduciflora* Turcz. ex Trin., *Zizania aquatica* L. var. *latifolia* (Griseb.) Kom., *Zizania caduciflora* (Turcz. ex Trin.) Hand.-Mazz., *Zizania dahurica* Turcz. ex Steud., *Zizania mezii* Prod.
♦ Manchurian wild rice, perennial rice, water rice, Manchurian ricegrass, wild rice, Indian rice
♦ Weed, Quarantine Weed, Naturalised, Environmental Weed
♦ 88, 101, 181, 204, 208, 225, 246, 280, 286, 290, 297
♦ pG, aqua, cultivated, herbal.

Zizania palustris L.
Poaceae
♦ northern wild rice, wild rice
♦ Environmental Weed
♦ 246, 280
♦ G, herbal.

Zizaniopsis bonariensis (Bal. & Poitr.) Speg.
Poaceae
♦ espadaña
♦ Weed
♦ 87, 88, 237, 295
♦ G.

Zizaniopsis miliacea (Michx.) Döll & Asch.
Poaceae
♦ giant cutgrass
♦ Weed, Quarantine Weed
♦ 87, 88, 218, 258
♦ G, herbal.

Ziziphora capitata L.
Lamiaceae
♦ Weed
♦ 272

Ziziphora tenuior L.
Lamiaceae
♦ Weed
♦ 243, 248
♦ herbal.

Ziziphus jujuba (L.) Lam. nom. illeg. non Mill.
Rhamnaceae
= *Ziziphus mauritiana* Lam.
♦ jujube, Chinese jujube, common jujube, jujubier common, Chinese date, Chinesische dattel, brustbeerbaum, açofeifeira, azufaifo, jujube, Chinese apple, giuggiolo
♦ Quarantine Weed
♦ 76, 287
♦ T, arid, cultivated, herbal. Origin: east Asia.

Ziziphus jujuba Mill.
Rhamnaceae
Rhamnus zizyphus L., *Ziziphus sativa* Gaertn. *nom. illeg*, *Ziziphus spinosa* (Bunge) Hu ex F.H.Chen *nom. illeg.*, *Ziziphus vulgaris* Lam. *nom. illeg.*, *Ziziphus vulgaris* var. *spinosa* Bunge, *Ziziphus zizyphus* (L.) Meikle (see)

- jujube, jujuba
- Naturalised
- 76, 287
- T, cultivated, herbal.

Ziziphus mauritiana Lam.
Rhamnaceae
Paliurus mairei H.Lév., *Rhamnus jujuba* L., *Ziziphus jujuba* (L.) Gaertn. *nom. illeg.*, *Ziziphus mairei* (H.Lév.) K.Browicz & Lauener, *Ziziphus jujuba* (L.) Lam. *non* Mill. (see)
- chinee apple, Indian jujube, Chinese date, jujube
- Weed, Quarantine Weed, Noxious Weed, Naturalised, Garden Escape, Environmental Weed, Cultivation Escape
- 62, 76, 86, 87, 88, 93, 98, 101, 147, 151, 179, 191, 203, 261
- arid, cultivated, herbal. Origin: India.

Ziziphus mistol Griseb.
Rhamnaceae
- mistol
- Weed
- 295
- T, cultivated, herbal.

Ziziphus mucronata Willd.
Rhamnaceae
Ziziphus madecassus H.Perrier, *Zizyphus mucronatus* Willd.
- buffalo thorn, shiny leaf, cape thorn
- Weed, Sleeper Weed, Quarantine Weed, Naturalised, Environmental Weed
- 10, 22, 76, 86, 87, 88, 155, 203
- T, arid, cultivated, herbal. Origin: Madagascar.

Ziziphus mucronata Willd. ssp. *mucronata*
Rhamnaceae
- buffalo thorn, shiny leaf, wait a bit
- Native Weed
- 121
- S/T. Origin: southern Africa.

Ziziphus nummularia (Burm.f.) Wight & Arn.
Rhamnaceae
Rhamnus nummularia Burm.f., *Ziziphus rotundifolia* Lam.
- jujube, ber
- Weed, Quarantine Weed
- 76, 87, 88, 203, 220
- arid, herbal.

Ziziphus rotundifolia Lam.
Rhamnaceae
= *Ziziphus nummularia* (Burm.f.) Wight & Arn.
- Weed
- 87, 88

Ziziphus rugosa Lam.
Rhamnaceae
Ziziphus glabra Roxb.
- Weed
- 87, 88
- herbal.

Ziziphus spina-christi (L.) Desf.
Rhamnaceae
Ziziphus sphaerocarpa Tul., *Rhamnus spina-christi* L.
- Christ's thorn, nabbag, kurna, jujube
- Weed, Quarantine Weed, Noxious Weed, Naturalised, Introduced
- 76, 86, 88, 203, 220, 221, 228
- arid, cultivated, herbal. Origin: Asia.

Ziziphus Mill. spp.
Rhamnaceae
- jujube, Christ's thorn
- Weed, Quarantine Weed
- 76, 191
- herbal.

Ziziphus zeyheriana Sond.
Rhamnaceae
Ziziphus helvola Sond.
- buffalo thorn
- Weed
- 10

Ziziphus zizyphus (L.) Meikle
Rhamnaceae
= *Ziziphus jujuba* Mill.
- common jujube
- Naturalised
- 101
- herbal.

Zornia capensis Pers.
Fabaceae/Papilionaceae
- Native Weed
- 121
- pH. Origin: southern Africa.

Zornia curvata Mohlenbr.
Fabaceae/Papilionaceae
- Naturalised
- 257

Zornia diphylla auct. non Pers.
Fabaceae/Papilionaceae
= *Zornia gemella* (Willd.) Vogel (NoR)
- Weed, Introduced
- 38, 87, 88
- cultivated, herbal.

Zornia gibbosa Span.
Fabaceae/Papilionaceae
- Weed
- 66
- herbal. Origin: Australia.

Zornia gracilis DC.
Fabaceae/Papilionaceae
- Weed
- 295

Zornia latifolia Sm.
Fabaceae/Papilionaceae
- urinana
- Weed
- 87, 88, 255
- pH, cultivated, herbal. Origin: Brazil.

Zornia linearis E.Mey.
Fabaceae/Papilionaceae
- Native Weed
- 121
- pH. Origin: southern Africa.

Zornia ovata Vogel
Fabaceae/Papilionaceae
Zornia diphylla (L.) Pers. var. *latifolia* DC.
- zornia
- Weed
- 295

Zornia piurensis Mohlenbr.
Fabaceae/Papilionaceae
- Naturalised
- 257

Zornia reticulata Sm.
Fabaceae/Papilionaceae
- neto hoja zarzabacoa de dos hojas
- Weed
- 14
- herbal.

Zornia trachycarpa Vogel
Fabaceae/Papilionaceae
Zornia diphylla (L.) Pers. var. *trachycarpa* (Vogel) Benth.
- Weed
- 295

Zosima absinthifolia (Vent.) Link
Apiaceae
- Weed
- 221

Zostera japonica Asch. & Graebn.
Zosteraceae
Zostera nana Roth (see)
- dwarf eelgrass
- Weed, Naturalised
- 101, 197
- aqua.

Zostera marina L.
Zosteraceae
- eelgrass, common grass wrack, seawrack, meriajokas
- Weed
- 23, 88, 272
- pH, aqua, promoted, herbal.

Zostera nana Roth
Zosteraceae
= *Zostera japonica* Asch. & Graebn.
- Weed
- 221

Zosterella dubia (Jacq.) Small
Pontederiaceae
= *Heteranthera dubia* (Jacq.) MacMill.
- water stargrass
- Weed, Quarantine Weed
- 76, 88, 220
- aqua, cultivated, herbal.

Zoysia cultivars Willd.
Poaceae
- Weed
- 78, 116
- G.

Zoysia japonica Steud.
Poaceae
Osterdamia japonica (Steud.) Hitchc.
- Korean lawngrass, couch, Japanese lawngrass, Korean templegrass
- Weed, Naturalised
- 87, 88, 101, 179, 191, 204, 286
- pG, cultivated.

Zoysia matrella (L.) Merr.
Poaceae
Zoysia pungens Willd.
♦ Manila grass, siglap grass, Korean grass, Japanese carpet, Manila templegrass
♦ Weed, Naturalised
♦ 87, 88, 91, 101, 179
♦ a/pG, cultivated, herbal. Origin: Asia, Australia.

Zoysia matrella (L.) Merr. var. *matrella*
Poaceae
♦ Manila grass, Manila templegrass, yerba de Manila
♦ Naturalised, Cultivation Escape
♦ 101, 261
♦ G, cultivated. Origin: Asia.

Zoysia matrella (L.) Merr. var. *pacifica* P.C.Goudswaard
Poaceae
♦ Manila grass
♦ Naturalised, Introduced
♦ 101, 230
♦ G.

Zoysia matrella (L.) Merr. var. *tenuifolia* (Willd. ex Thiele) Sasaki
Poaceae
♦ Mascarene grass, zoisia, Manila templegrass
♦ Cultivation Escape
♦ 261
♦ G, cultivated. Origin: Asia.

Zoysia tenuifolia Willd. ex Thiele
Poaceae
♦ Mascarene grass
♦ Weed, Naturalised
♦ 80, 87, 88, 98, 101, 179, 203
♦ G, cultivated, herbal.

Zygophyllum album L.f.
Zygophyllaceae
♦ Weed
♦ 221

Zygophyllum apiculatum F.Muell.
Zygophyllaceae
♦ Weed
♦ 87, 88
♦ cultivated.

Zygophyllum coccineum L.
Zygophyllaceae
♦ Weed
♦ 39, 221
♦ toxic.

Zygophyllum fabago L.
Zygophyllaceae
♦ Syrian beancaper, beancaper
♦ Weed, Noxious Weed, Naturalised, Introduced
♦ 1, 34, 35, 39, 49, 80, 87, 88, 94, 101, 146, 156, 161, 218, 219, 229, 272, 275
♦ pH, cultivated, herbal, toxic.

Zygophyllum morgsana L.
Zygophyllaceae
♦ Native Weed
♦ 121
♦ pS, cultivated. Origin: southern Africa.

Zygophyllum simplex L.
Zygophyllaceae
♦ Weed, Native Weed
♦ 121, 221
♦ H, toxic. Origin: Africa, Middle East.

Zygophyllum L. spp.
Zygophyllaceae
♦ beancaper
♦ Weed
♦ 88, 221

INDEX

INDEX OF COMMON AND ALTERNATE NAMES

Index

Index